Heteroptera of Economic Importance

COVER PHOTOGRAPHS

Front Cover:
(clockwise, starting at bottom right)

Photos 1–5: Adult of *Loxa deducta* Walker (Pentatomidae) on fruits (berries) of privet, *Ligustrum lucidum* Ait. (Oleaceae).
Photo 6: Adults of the southern green stink bug, *Nezara viridula* (L.) (Pentatomidae), on star bristle, *Acanthospermum hispidum* DC (Compositae).

Back Cover:

Adult of *Piezodorus guildinii* (Westwood) (Pentatomidae) on hairy indigo, *Indigofera hirsuta* L. (Leguminosae). (All photographs by Antônio R. Panizzi.)

Heteroptera of Economic Importance

Edited by
Carl W. Schaefer
Antônio Ricardo Panizzi

CRC Press
Boca Raton London New York Washington, D.C.

Library of Congress Cataloging-in-Publication Data

Heteroptera of economic importance / edited by Carl W. Schaefer and Antônio Ricardo Panizzi.
    p. cm.
    Includes bibliographical references and index.
    ISBN 0-8493-0695-7 (alk. paper)
    1. Hemiptera. 2. Insect pests. 3. Beneficial insects. I. Schaefer, Carl W. (Carl Walter)
II. Panizzi, Antônio Ricardo.

OL521 .H48 2000
595.7′54165—dc21   00-036080
                                                                                         CIP

This book contains information obtained from authentic and highly regarded sources. Reprinted material is quoted with permission, and sources are indicated. A wide variety of references is listed. Reasonable efforts have been made to publish reliable data and information, but the authors and the publisher cannot assume responsibility for the validity of all materials or for the consequences of their use.

Neither this book nor any part may be reproduced or transmitted in any form or by any means, electronic or mechanical, including photocopying, microfilming, and recording, or by any information storage or retrieval system, without prior permission in writing from the publisher.

All rights reserved. Authorization to photocopy items for internal or personal use, or the personal or internal use of specific clients, may be granted by CRC Press LLC, provided that $.50 per page photocopied is paid directly to Copyright Clearance Center, 222 Rosewood Drive, Danvers, MA 01923 USA. The fee code for users of the Transactional Reporting Service is ISBN 0-8493-0695-7/00/$0.00+$.50. The fee is subject to change without notice. For organizations that have been granted a photocopy license by the CCC, a separate system of payment has been arranged.

The consent of CRC Press LLC does not extend to copying for general distribution, for promotion, for creating new works, or for resale. Specific permission must be obtained in writing from CRC Press LLC for such copying.

Direct all inquiries to CRC Press LLC, 2000 N.W. Corporate Blvd., Boca Raton, Florida 33431.

Trademark Notice: Product or corporate names may be trademarks or registered trademarks, and are used only for identification and explanation, without intent to infringe.

© 2000 by CRC Press LLC

No claim to original U.S. Government works
International Standard Book Number 0-8493-0695-7
Library of Congress Card Number 00-036080
Printed in the United States of America 1 2 3 4 5 6 7 8 9 0
Printed on acid-free paper

Preface

This book began as an all-morning conversation between its two editors, during the International Congress of Entomology, in Beijing, China, in 1992. The great importance of heteropterans as pests of crops, as vectors of a major disease (Chagas' disease), and as biological control agents of great but often untested importance persuaded us that a book on the subject was needed. Such a book, detailed and thorough, would help agriculturists in all parts of the world, and would we hoped lead to further work on the agricultural biologies and ecologies of these insects. By bringing this information together, the current and past status of pest species, or their current and past value as biological control agents, could be assessed. By so doing, we believe, it may even be possible to predict where and by what species other pests may emerge (for example, African relatives of a lace bug pest on crop X in Asia may one day become pests on crop X in Africa; see Schaefer 1998).

We thought first to put together a symposium for the next Congress. But the organizers turned down our proposal ("too narrow"), which solved a problem for us: there were too many groups to be covered in one symposium, and we were at a loss to decide who should be invited for the book, but *not* for the symposium. In any event, the book emerges un-Congressionally heralded. We do not feel it is the less for that.

Each chapter is written by one or a few authorities on a taxonomic group of heteropterans. We felt that such authorities would have the best knowledge of the general literature, and therefore the quickest access to the economic literature on their groups. One result of this reliance on the groups' authorities has been a remarkable completeness in the coverage of the literature — there are some 4500 references cited in this book.

Indeed, our most important instruction to each author was to be complete, to cover the relevant literature as completely as possible, and to interpret "relevant" as broadly as possible. Our aim has been to provide the reader with a starting point, a *point d'appui* from which to launch further studies on heteropterans of economic importance. We are quite and regretfully aware that completeness is impossible, that past literature (some of it important) has been overlooked, and that more important literature has appeared while this book has been in press. Each of the editors has added material to the chapters of others (and to our own), as we have come upon it.

The authors agreed with our desire for completeness. As a result, these chapters contain thorough and detailed accounts of the biologies of the economically important heteropterans, accounts often more detailed than occur elsewhere. We believe, therefore, that the book sets forth as complete a series of heteropteran biologies as can be found anywhere. Moreover, the bringing of this information together in one place makes possible for the first time various comparisons and analyses.

However, in many cases, life-history and biological data are scant. This reflects the state of the literature on these insects and not the authors' knowledge of that literature. It seems indeed remarkable that so little is known about so many pests. If control of these pests is ever to be permanently effected, their biologies must be studied; ignoring this simple truth will not allow one to avoid its consequences.

We note that some chapters are less detailed than others, although each certainly covers its subject thoroughly. Very recent books on Reduviidae and Miridae, by Ambrose (1999) and Wheeler (2001), respectively, and another, less recent one on Cimicidae (Usinger 1966), make detailed chapters on these families less necessary. Nevertheless, each chapter stands alone as a compendium of what is known of its group; each chapter will by itself bring readers up-to-date on its subject, and introduce them to the literature of that subject. That is the purpose of this book, and it is a purpose useful as much to students of heteropteran systematics, biology, and ecology, as it is to those more directly concerned with the economic significance of heteropterans.

The organization of each chapter is straightforward and, for the most part, uniform. However, we have not placed authors in organizational straitjackets, and we have allowed each some leeway. Each chapter is complete in itself, but the interested reader can move easily from one to another, confident that the format of each will be familiar. We have also tried to aid such movement by cross-referencing relevant chapters and sections.

The book itself has two divisions, each opening with a general chapter. The section on heteropterans as pests opens with a chapter on how these insects adversely affect plants, and the first chapter in the section on beneficial heteropterans considers how in fact predaceous heteropterans feed. However, we point out that not all the heteropterans in the first part of this book are harmful, nor are all of those in the second part useful. Some actual or potential biological control agents occur within families most of whose members harm plants, and a few predaceous groups contain members that pose a minor threat to fisheries. Like many of our friends, there is a little bit of bad in the best of them, and a little bit of good in the worst. This incomplete dichotomy is seen in the two Miridae chapters, where sometimes the same species is described as both harmful and useful; again, among the stilt bugs are some insects causing damage and others of some value in biological control.

The Tingidae exemplifies another problem we faced in this book — the decision whether a species is an "important species" or a "less important species" is often a subjective one. If a species does major damage only occasionally, is it important or less important? If a species does small damage often, is it important or less important? If a species does considerable damage in only one small part of the world, is it important or less important? When does "occasionally" merge into "often"? How small is "one small part of the world"? If a certain Indian farmer's brinjal crop is destroyed for several years running, the damage is certainly important to him and his family (see **Chapter 4, Section 3.18**). But if the brinjal lace bug is serious only in this farmer's state, do the editors of this book consider the bug an important pest? We editors can only state that we have tried in each case to follow the assessment of each chapter author(s), and in some cases to make our own best (if imperfect) judgment. Readers may disagree with these decisions, which, however, are less important than that all pests — major and minor — are (we believe) included.

Two final words of explanation. Readers should be aware that the names used herein are for the most part those of the authors reviewed. Yet applied entomologists, dealing as they are with immediate problems, may not take the time needed for accurate identification of their pests. As a result, pests may be misidentified in the literature, and these misidentifications perpetuated in the secondary literature (like this book). For example, the cotton stainers *Dysdercus koenigii* (F.) and *D. cingulatus* (F.) are distinct species whose distributions overlap only narrowly, in northeastern India and Bangladesh (Freeman 1947, Kapur and Vazirani 1956). Yet in the considerable literature on these bugs, there appears to have been little if any attempt to distinguish between them. The two names are used almost indiscriminately, and not only by applied entomologists — but by physiologists, endocrinologists, and others for whom correct systematics is more a burden than a useful predictor.

Where possible, we have corrected scientific names and identifications. Nevertheless, errors perforce occur. We only plead that we have done our best and that we ourselves have introduced new errors very rarely.

We have tried to give the authors of all scientific names. We believe we have succeeded with heteropterans, but have been less successful with other arthropods, and least successful with plants. We have also tried to use accepted scientific names, with the same declining degrees of success. We apologize for any nomenclatorial errors, while acknowledging that some are inevitable. With respect to the authors of plant names, we point out that our bible (Howes 1974) eschews authors completely. Finally, we spell out all authors' names except Linnaeus (L.) and Fabricius (F. or Fabr.), and those names which, abbreviated in our sources, we did not recognize.

Carl W. Schaefer
Storrs, Connecticut, U.S.A.

Antônio R. Panizzi
Londrina, Paraná, Brazil

REFERENCES CITED

Ambrose, D. 1999. Assassin Bugs. Science Publishers, Enfield, New Hampshire, U.S.A. 337 pp.
Freeman, P. 1947. A revision of the genus *Dysdercus* Boisduval (Hemiptera, Pyrrhocoridae), excluding the American species. Trans. R. Entomol. Soc. London 98: 373–424.
Howes, F. N. 1974. A Dictionary of Useful and Everyday Plants and Their Common Names. Cambridge University Press, Cambridge, U.K. 290 pp.
Kapur, A. P., and T. G. Vazirani. 1956. The identity and geographical distribution of the species of the genus *Dysdercus* Boisduval (Hemiptera: Pyrrhocoridae). Rec. Indian Mus. 54: 159–175.
Schaefer, C. W. 1998. Phylogeny, systematics, and practical entomology: the Heteroptera (Hemiptera). An. Soc. Entomol. Brasil 27: 499–511.
Usinger, R. L. 1966. Monograph of Cimicidae (Hemiptera-Heteroptera). Thomas Say Foundation. Entomological Society of America, College Park, Maryland, U.S.A. 585 pp.
Wheeler, A. G., Jr. 2001. Biology of the Plant Bugs: Pests, Predators, Opportunists. Cornell University Press, Ithaca, New York, U.S.A.

The Editors

Carl W. Schaefer, Ph.D., has taught at the University of Connecticut since 1966. He earned his B.S. from Oberlin College in 1956 and continued his studies at the University of California at Los Angeles from 1957 to 1958. He earned his Ph.D. in entomology from the University of Connecticut in 1964. He taught at Brooklyn College of the City University of New York, then returned to the University of Connecticut.

Dr. Schaefer has held several offices, including president, of the Connecticut Entomological Society, and president of the Eastern Branch and Chairman of Section A (Systematics and Morphology), both of the Entomological Society of America. Since 1997 he has been a Counselor of this Society's Entomological Foundation.

From 1973 to 1998 Dr. Schaefer was an editor of the *Annals of the Entomological Society of America*; in 1999, a special issue of the *Annals* honored him. He also serves on the editorial or advisory boards of the *European Journal of Entomology*, the *Annalen des Naturhistorischen Museums in Wien*, the *Anais da Sociedade Entomológica do Brasil*, and *The Journal of Entomological Research Society* (Ankara). He has presented papers and organized symposia on Heteroptera (and occasionally Hemiptera) at annual meetings of the Entomological Society of America for the past 30 years, and at several International Congresses of Entomology, and meetings of several Congressos Brasileiros de Entomologia. He is the co-convenor of the Systematics and Phylogeny Session of the 2000 International Congress of Entomology, and will participate at that Congress in two symposia and present two posters.

From 1973 to 1998, Dr. Schaefer compiled and disseminated *The Heteropterists' Newsletter*, which provided several hundred heteropterists with information on their colleagues' work and professional interests and needs. In 1997, at the invitation of the Brazilian Government, he consulted on heteropteran pests of rangeland and soybean. Dr. Schaefer has published more than 170 papers on various aspects of heteropteran systematics, morphology, phylogeny, biology, and ecology. He has edited or co-edited four books on such topics as hemipteran phylogeny, North American insect systematics, and the morphology of insect feeding. His interests are in the comparative morphology and biology, and their bearing on the phylogeny, of Heteroptera.

Antônio R. Panizzi, Ph.D., has been a research entomologist at the National Soybean Research Center of Embrapa, Londrina, Paraná, Brazil, since 1975. He earned his B.Sc. in agronomy in 1972 from the University of Passo Fundo, RS and his M.Sc. in entomology in 1975 from the Federal University of Paraná, both in Brazil, and his Ph.D. in entomology in 1985 from the University of Florida.

Dr. Panizzi is a member of several scientific societies, including the Entomological Society of Brazil, the Brazilian Society of Zoology, the Entomological Society of America, the Florida Entomological Society, and the Georgia Entomological Society. In 1992 he was awarded the Alexandre Rodrigues Ferreira Prize by the Brazilian Society of Zoology. He was editor-in-chief of the *Anais da Sociedade Entomológica do Brasil* from April 1993 to December 1998. He was an elected member of the Advisor Committee for Agronomy (Entomology) at the National Research Council (CNPq) of Brazil from 1997 to 1999. He currently serves the Entomological Society of Brazil as associate editor and president of the Society's Council.

Dr. Panizzi has participated in 53 national (Brazil) and 15 international scientific congresses and symposia. He was an invited speaker at the World Soybean Research Conference III at Ames (Iowa) in 1984, Conference IV in Buenos Aires in 1989, and Conference V in Chiang Mai (Thailand) in 1994; the XII International Plant Protection Congress in Rio de Janeiro in 1991; the XIX International Congress of Entomology in Beijing, China in 1992; Mercosoja 99 in Rosario, Argentina in 1999; the III Workshop on Integrated Sunn Pest Control, Ankara, Turkey in 1999; and the I Workshop on Agroecology, UNICAMP, Campinas, SP in 1999. He was an invited scientist at the National Institute of Agro-Environmental Sciences, Tsukuba, Japan in 1991. He is a scheduled invited speaker at the XXIII Brazilian Congress of Zoology, Cuiabá, MT, in 2000, and will participate in three symposia at the 2000 International Congress of Entomology.

Dr. Panizzi has co-edited *Insect Nutritional Ecology and Its Implications for Pest Management* [in Portuguese]. He has published extensively on Hemiptera (Heteroptera), including an *Annual Review* article on wild host plants of Pentatomidae. He teaches a course on Insect Nutritional Ecology at the State University in Londrina and at the Federal University of Paraná in Curitiba, both in Paraná state. His current research interests focus on the interactions of heteropterans (mostly Pentatomidae and Alydidae) with their wild and cultivated host plants, and the management of pest species on soybean.

Contributors

Imtiaz Ahmad
Department of Zoology
Karachi University
Karachi 7527
Pakistan
biosal@disicom.net.pk

Dunston P. Ambrose
Entomology Research Unit
Department of Zoology
St. Xavier's College
Palayankottai 627002
India
stxavier@md2.vsnl.net.in

Nils Møller Andersen
Zoologisk Museum
Universitetsparken 15
DK-2100 Copenhagen
Denmark

Miriam Becker
Departmento de Zoologia
Universidade Federal do
  Rio Grande do Sul
Av. Paulo Gama 110
Porto Alegre
RS 90040-060
Brazil

S. Kristine Braman
Department of Entomology
University of Georgia
Georgia Station
Griffin, GA 30223-1797
U.S.A.
kbraman@gaes.griffin.peachnet.edu

Allen Carson Cohen
Biological Control
  and Mass Rearing Unit
USDA, ARS, MSA
P.O. Box 5367
Mississippi State, MS 39762-5367
U.S.A.
acohen@bcmrru.msstate.edu

Patricia de Azambuja
Fundação Oswaldo Cruz
FIOCRUZ
Caixa Postal 926
Rio de Janeiro
RJ21045-900
Brazil
azambuja@gene.dbbm.fiocruz.br

Patrick De Clercq
Faculty of Agriculture
  & Applied Biological Sciences
University of Gent
Coupure Links 653
B-9000 Gent
Belgium
patrick.declercq@rug.ac.be

João C. P. Dias
Centro de Pesquisas René Rachou
Av. Augusto Lima 1715
Belo Horizonte
MB 30190-002
Brazil
jcpdias@netra.cpqrr.fiocruz.br

Eloi S. Garcia
Fundação Oswaldo Cruz
FIOCRUZ
Caixa Postal 926
Rio de Janeiro
RJ21045-900
Brazil
egarcia@gene.dbbm.fiocruz.br

Kari Heliövaara
Department of Applied Zoology
University of Helsinki
P.O. Box 27
Helsinki FIN-00014
Finland
kari.heliovaara@helsinki.fi

Thomas J. Henry
Systematic Entomology Laboratory
Plant Sciences Institute
c/o National Museum of Natural History
MRC-168
Washington, D.C. 20560
U.S.A.
thenry@sel.barc.usda.gov

Lionel Hill
Department of Primary Industry,
  Water and Environment
P.O. Box 303
Devonport
Tasmania 7310
Australia
lionel.hill@dpiwe.tas.gov.au

Koji Hori
Ozora-cho 5-12-2
Obihiro
Hokkaido 080-0838
Japan

David G. James
Irrigated Agriculture Research
  and Extension Center
Washington State University
24106 N. Bunn Road
Prosser, WA 99350
U.S.A.
djames@tricity.wsu.edu

M. Javahery
Macdonald College
21,111 Lakeshore Drive
P.O. Box 140
Ste. Anne de Bellevue
Québec H9X 3V9
Canada

Steven L. Keffer
Department of Biology
MSC 7801
James Madison University
Harrisonburg, VA 22807
U.S.A.
slkeffer@jmu.edu

Jill Kotulski
Department of Ecology
  and Evolutionary Biology
University of Connecticut, U-43
Storrs, CT 06269-3043
U.S.A.

John D. Lattin
Department of Entomology
Oregon State University
Corvallis, OR 97331-2907
U.S.A.
lattinj@bcc.orst.edu

Jerzy A. Lis
Department of Zoology
University of Opole
Oleska 22
PL-45-052 Opole
Poland

J. E. McPherson
Department of Zoology
Southern Illinois University
Carbondale, IL 62901
U.S.A.
mcpherson@zoology.siu.edu

Robert M. McPherson
University of Georgia
Coastal Plain Experiment Station
P.O. Box 748
Tifton, GA 31793
U.S.A.
pherson@tifton.cpes.peachnet.edu

Paula Levin Mitchell
Department of Biology
Winthrop University
Rock Hill, SC 29733
U.S.A.
mitchellp@winthrop.edu

Narisu
Department of Plant, Soil and Insect Sciences
University of Wyoming
Laramie, WY 82071
U.S.A.
narisu@uwyo.edu

Yosihiro Natuhara
Landscape Architecture and Conservation
Osaka Prefecture University
Graduate School of Agriculture
1-1 Gakuen-cho
Sakai 599-8531
Japan
natuhara@envi.osakafu-u.ac.jp

John W. Neal, Jr.
USDA, ARS
National Arboretum
10300 Baltimore Avenue
Building 010A
Beltsville, MD 20705-2350
U.S.A.

Antônio R. Panizzi
Laboratorio de Entomologia
Embrapa Soja
Caixa Postal 231
Londrína PR 86001-970
Brazil
panizzi@cnpso.embrapa.br

Miroslav Papáček
Department of Biology
Pedagogical Faculty
University of South Bohemia
Jeronymova 10
CS-371 15 České Budìjovice
Czech Republic
papacek@pf.jcu.cz

Carl W. Schaefer
Department of Ecology
 and Evolutionary Biology
University of Connecticut, U-43
Storrs, CT 06269-3043
U.S.A.
schaefer@uconnvm.uconn.edu

Robert W. Sites
Department of Entomology
1-87 Agriculture Building
University of Missouri
Columbia, MO 65211
U.S.A.
bugs@showme.missouri.edu

John R. Spence
Department of Entomology
University of Alberta
Edmonton
Alberta T6G 2E3
Canada

Merrill H. Sweet II
Department of Biology
Texas A&M University
College Station, TX 77843
U.S.A.
msweet@bio.tamu.edu

P. Venkatesan
Department of Zoology
Loyola College
Madras 600034
India

A. G. Wheeler, Jr.
Department of Entomology
114 Long Hall
Clemson University
Clemson, SC 29634-0365
U.S.A.
awhlr@clemson.edu

Acknowledgments

The help of many people made this book possible, and its editing endurable. We are grateful to our editor at CRC Press; John Sulzycki coaxed us toward deadlines, reminded us gently of them, sighed graciously when we missed them, and in general treated us far more kindly than we deserved and than, we fear, we ourselves treated the forgiving authors of this book. We are equally grateful to Christine Andreasen, our production editor at CRC Press, for very much the same reasons: an eye for detail that might be frightening in other contexts, and a patience that surpasseth all understanding.

Viviane R. Chocorosqui, Shirlei R. Cardoso, and Ana P. M. Mourão (Embrapa Soja, Paraná, Brazil) were indispensible in helping to edit various chapters, as was Michele Touchet (USDA, Washington, D.C.) in tracking down references obscure to the point of disappearance. G. J. Anderson and F. G. Coe (University of Connecticut) provided information on the economic importance of several plants mentioned in this book, and L. Mehrhoff (University of Connecticut) searched for the authors of many plant species.

To Carol Blow (University of Connecticut) we are particularly grateful. She patiently word-processed and word-reprocessed, chapter after chapter, until we editors finally got it right.

Antônio R. Panizzi thanks the National Research Council (CNPq, Brasília, Brazil) and Embrapa Soja, for financial support of the trips to the U.S. (May 1995, October 1998, and August 1999) to work on this long project with Carl W. Schaefer. The Research Foundation of the University of Connecticut helped support research for several of these chapters.

Above all, we thank the contributors to this book. Without them, of course, there would be no book. But also without them, we would not have learned how patient and forbearing good authors (that is, authors confident of their expertise) can be. The authors of this book accepted, without demur or complaint, our whining and nagging and wheedling and demanding. We asked for changes and we got them; we asked for rewrites and we got them; we asked for updated references and we got them. And so we thank the authors, on our own behalf and on behalf of the readers of this book.

Part of the Introduction was presented by Carl W. Schaefer in 1997, at the 16th Congresso Brasileiro de Entomologia (Salvador, Bahia) and at the Entomological Society of America meetings (Nashville, Tennessee). Carl W. Schaefer is grateful to the organizers and program chairmen of both societies for their invitations to speak, to the University of Connecticut for travel aid, and to the second editor (Antônio R. Panizzi) for hospitality. We are both grateful to John W. Neal, Jr., who read critically an earlier version of this Introduction, and much improved it.

Contents

INTRODUCTION

Chapter 1
Economic Importance of Heteroptera: A General View 3
Carl W. Schaefer and Antônio R. Panizzi

HARMFUL BUGS

Chapter 2
Possible Causes of Disease Symptoms Resulting from the Feeding
of Phytophagous Heteroptera..11
Koji Hori

Chapter 3
Plant Bugs (Miridae) as Plant Pests 37
A. G. Wheeler, Jr.

Chapter 4
Lace Bugs (Tingidae)..85
John W. Neal, Jr. and Carl W. Schaefer

Chapter 5
Palm Bugs (Thaumastocoridae)...139
Lionel Hill and Carl W. Schaefer

Chapter 6
Seed and Chinch Bugs (Lygaeoidea) 143
Merrill H. Sweet II

Chapter 7
Ash-Gray Leaf Bugs (Piesmatidae)..265
Narisu

Chapter 8
Cotton Stainers and Their Relatives (Pyrrhocoroidea: Pyrhocoridae and Largidae) 271
Carl W. Schaefer and Imtiaz Ahmad

Chapter 9
Scentless Plant Bugs (Rhopalidae)..309
Carl W. Schaefer and Jill Kotulski

Chapter 10
Broad-Headed Bugs (Alydidae)...321
Antônio R. Panizzi, Carl W. Schaefer, and Yosihiro Natuhara

Chapter 11
Leaf-Footed Bugs (Coreidae) 337
Paula Levin Mitchell

Chapter 12
Burrower Bugs (Cydnidae) 405
Jerzy A. Lis, Miriam Becker, and Carl W. Schaefer

Chapter 13
Stink Bugs (Pentatomidae) 421
**Antônio R. Panizzi, J. E. McPherson, David G. James, M. Javahery,
and Robert M. McPherson**

Chapter 14
Shield Bugs (Scutelleridae) 475
M. Javahery, Carl W. Schaefer, and John D. Lattin

Chapter 15
Several Small Pentatomoid Families (Cyrtocoridae, Dinidoridae, Eurostylidae,
Plataspidae, and Tessaratomidae) 505
Carl W. Schaefer, Antônio R. Panizzi, and David G. James

Chapter 16
Flat Bugs (Aradidae) 513
Kari Heliövaara

Chapter 17
Bed Bugs (Cimicidae) 519
Carl W. Schaefer

Chapter 18
Triatominae (Reduviidae) 539
Eloi S. Garcia, Patricia de Azambuja, and João C. P. Dias

Chapter 19
Adventitious Biters — "Nuisance" Bugs 553
Carl W. Schaefer

USEFUL BUGS

Chapter 20
How Carnivorous Bugs Feed 563
Allen Carson Cohen

Chapter 21
Creeping Water Bugs (Naucoridae) 571
Robert W. Sites

Chapter 22
Giant Water Bugs (Belostomatidae) 577
P. Venkatesan

Chapter 23
Waterscorpions (Nepidae) 583
Steven L. Keffer

Chapter 24
Small Aquatic Bugs (Nepomorpha) with Slight or Underestimated Economic Importance 591
Miroslav Papaček

Chapter 25
Semiaquatic Bugs (Gerromorpha) 601
John R. Spence and Nils Møller Andersen

Chapter 26
Minute Pirate Bugs (Anthocoridae)..607
John D. Lattin

Chapter 27
Damsel Bugs (Nabidae)...639
S. Kristine Braman

Chapter 28
Predacious Plant Bugs (Miridae)...657
A. G. Wheeler, Jr.

Chapter 29
Assassin Bugs (Reduviidae excluding Triatominae)...695
Dunston P. Ambrose

Chapter 30
Economic Importance of Predation by Big-Eyed Bugs (Geocoridae) 713
Merrill H. Sweet II

Chapter 31
Stilt Bugs (Berytidae) 725
Thomas J. Henry

Chapter 32
Predaceous Stinkbugs (Pentatomidae: Asopinae)..737
Patrick De Clercq

Insect Index..791

Plant Index...813

SECTION I

Introduction

CHAPTER 1

Economic Importance of Heteroptera: A General View

Carl W. Schaefer and Antônio R. Panizzi

The Heteroptera, or "true bugs," is a suborder of the insect order Hemiptera. The status of the other suborder, "Homoptera," has recently been questioned, because not all its members seem to have had a common ancestor. However, no one questions that the Heteroptera had a common ancestor. The close important similarities between Heteroptera and the other hemipterans argues very strongly that all are closely related; that is the reason we (and most other students of Heteroptera) treat all Hemiptera as a single order.

The suborder Heteroptera contains eight infraorders (Enicocephalomorpha, Dipsocoromorpha, Leptopodomorpha, Gerromorpha, Nepomorpha, Cimicomorpha, Pentatomomorpha, and Aradomorpha); of these, the last five contain species of economic importance and, of these, only Cimicomorpha and Pentatomomorpha contain many such species. Altogether, there may be some 37,000 described species of Heteroptera, and perhaps 25,000 species remaining to be described (C. W. Schaefer, unpublished). The common name for these insects is true bugs.

This common name, true bugs, has ancient etymological roots, and heteropterans are the only insects *correctly* called "bugs." The word is derived from the Anglo-Saxon *bugge* meaning a "wraith" or "specter" (compare modern English *bugbear, bugaboo,* perhaps even to *bug*). Almost certainly, a *bugge* was blamed for the mysterious itching welts people discovered on their bodies after a night's sleep in a room infested with *Cimex lectularius* L., an insect named by Linnaeus "the bug of the couch," or bed bug. Bed bugs slip at night from their hiding places to take the blood of sleeping people, and then return (still at night) to the cracks and crevices and crannies in the sleeper's room, bed, and bedclothes. The resulting welts on the awakening people seem to have had no cause — except for a malevolent secret attack by a wraith, a specter, a *bugge*. When the culprit was finally identified, the word *bugge* was applied to it and, by extension, to all its relatives — to all heteropterans.

The Heteroptera, like the other members of the Hemiptera, are characterized by elongate mouthparts designed for, and very efficient at, piercing other organisms (plant or animal) and sucking up their fluids. The bugs inject digestive enzymes through the mouthparts into the prey's body, and then suck up the digested and liquefied tissue. The mouthparts of hemipterans contain two channels, one through which is pumped salivary fluid into the organism (for external digestion, to disrupt tissues and cells, etc.), and the other through which the fluids are sucked back into the insect.

Like many other hemipterans, true bugs are for the most part diurnal, and their compound eyes are well developed. Their wings (two pairs, as is true of most insects) differ from one another; the front wings are partly leathery and partly membranous (hence, "Heteroptera, different wings"), and the hind wings are wholly membranous. In this they differ from most homopterans, whose front wings are completely leathery ("Homoptera"). True bugs differ also in possessing a shield-shaped modification of the thoracic dorsum, the scutellum ("little shield"), lying just between the front of the wings. And they differ most importantly in having scent glands, on the abdomen of both nymphs and adults and on the thorax of adults; some (often all) of the adults' abdominal glands may be nonfunctional. As noted below, these glands' secretions serve different functions in different species.

Immature heteropterans (*nymphs* in the United States, *larvae* in some other places) look like their adults, because they have incomplete metamorphosis; they lack wings and genitalia, however, and some body parts are disproportionately large.

All nonheteropteran hemipterans are herbivorous. Only some heteropterans take other food — many prey upon other arthropods, and a few (like the bed bug) feed on the blood of warm-blooded vertebrates. Except for a few aquatic bugs, no heteropteran feeds on cold-blooded vertebrates. Heteropterans are also the only marine (or quasi-marine) hemipterans; certain water striders live on the ocean's surface, far from land.

A far more detailed account of the Heteroptera, and of the individual families in the suborder, may be found in Schuh and Slater (1995); a general account is in Schaefer (2000). Other suggested references are at the end of this introductory chapter.

Heteroptera may be the most abundant of insect groups with incomplete metamorphosis, and it is also probably the most diverse of these groups. The suborder contains general and specific feeders on plants and on animals; feeding on vertebrate blood has evolved at least three times. Some bugs are specialized for feeding on ants, others on millipedes, and others on fungi under the bark of trees (and feeding on fungus seems to have evolved more than once); yet others live in spiders' webs and steal the spiders' prey. No other hemimetabolous order, much less a suborder, has so diverse an array of feeding habits.

In terms of diversity and of numbers of species, the suborder Heteroptera is evolutionarily very successful. One reason might appear to be that Heteroptera contains the family Miridae, whose nearly 10,000 species compose almost one third of the suborder's total. Yet this explanation begs the question of why this family is so successful.

But even without the Miridae, the great array of feeding types and preferences remains. Miridae contains both general and specific plant feeders, and general and specific feeders on other arthropods. But so do other groups of heteropterans; thus, removal of Miridae reduces the number of heteropteran species, but not their variety. Even without the Miridae, the Heteroptera would be a highly successful group.

There are eight major divisions of Heteroptera. Five of these, and part of a sixth, feed on animal fluids. Only two divisions and part of the sixth feed on plants. Yet these plant-feeding species are about 60% of the roughly 37,000 species of Heteroptera. It appears that Heteroptera is successful because of its herbivorous members, and became successful once some of its members turned to herbivory (the least advanced divisions contain relatively few species and are all predaceous). These herbivorous members are abundant, successful, and widespread across the world; it was inevitable that some of them would become pests of crops. It was inevitable as well that some of the 40% that are predaceous would feed on those and other pests and be actual or potential agents of biological control. In what follows, we shall consider first hemipterans, and then heteropterans, as feeders on plants.

In what ways are heteropterans — true bugs — special or unique as pests of crops? How does Hemiptera differ from other insect orders? How does Heteroptera differ from other hemipterans?

Members of many insect groups suck the internal fluids of other organisms — that is, pierce those other organisms and draw out their fluids (sap, hemolymph, blood, etc.). But it seems significant that, with one small exception, all these piercing–sucking groups feed upon animals — groups like some dipterans, most neuropterans, a few coleopterans, some hymenopterans. Very few such insects pierce and suck plants — some nematoceran dipterans, thrips, probably a few others.

Lepidopterans, of course, and many dipterans suck or lap fluids from the surfaces (or invaginated surfaces — flowers) of plants.

It is possible that the early hemipteran, before the division into the heteropteran and homopteran lines, sucked fluids from dead and dying plants and animals — that is, was a scavenger living on the ground (see Schaefer 1981, for some development of this idea). The heteropteran line may have come to specialize on piercing and sucking the fluids of other arthropods, and thus become more or less aggressive. The homopteran line fed similarly, but upon plants, for which a more passive, even sessile, way of life was more suitable. Early heteropterans probably had some competition from other arthropod groups also taking the internal fluids of arthropods, and this competition fostered an ability to move both evasively and aggressively. Homopterans, on the other hand, had considerably less competition (except perhaps from mites, for cell contents), because there were fewer groups taking plant fluids. [*Note:* In this paragraph and in what follows, by "homopteran," we mean merely "nonheteropteran," because Homoptera as a taxon is perhaps phylogenetically invalid, not being a monophyletic group; Heteroptera is monophyletic.]

This difference may help explain why Heteroptera became a major group only when several evolutionary lines within it turned from feeding on other arthropods to feeding on plants.

Many other insects feed on plants. But nearly all of these bite and chew the plants. Plants have evolved many defenses against biters and chewers, including defenses against vertebrate biters and chewers. But these defenses are *external*, on the surface of the plant: thickened cuticle or wax layers; thorns, spines, hairs, setae; scales; etc. Not all of these are exclusively defensive, but some are and many of the others have defensive roles as well as other ones. Other defenses are internal (e.g., tannins), but are effective primarily when ingested with chewed materials. These defenses do indeed deter biters and chewers, but deter hemipterans little. Hemipterans pierce the plant surface with their stylets to get the food. Moreover, the food they get is liquid and consists of dissolved nutrients or nutrients digested *in situ*; it is therefore easier to digest than the chewed surface materials of plants. The hemipteran thus bypasses many of the plant's defenses.

This way of feeding also tends to protect hemipterans against pesticides, many of which are stomach poisons that lie on the surface of the plant and can be bypassed by the hemipteran's stylets. Contact or systemic pesticides are more effective against hemipterans.

Thus, hemipterans — both homopterans and heteropterans — are pests because they suck the fluids of plants that humans want. They are serious pests (that is to say, they are evolutionarily successful pests) because they can penetrate many of the defenses of plants, because they lack much competition, and as a consequence they are abundant both in numbers and in variety. And as a consequence of this abundance, hemipterans are evolutionarily and ecologically "available" to attack a new crop when humans provide it (consider the situation with soybean in Brazil; Panizzi 1997a–d; Schaefer 1998).

How do homopterans and heteropterans differ? Do any of these differences cause homopterans and heteropterans to differ as pests?

A difference of major importance is how members of these two groups actually feed. Both penetrate plants to get their fluids, but not in the same way; nor do they get the fluids from the same parts of the plant. Many heteropterans (especially infraorder Pentatomomorpha) prefer the reproductive parts of plants: flowers, ovules, ovaries, ripening and ripened seeds. Others (e.g., many Miridae and Tingidae: infraorder Cimicomorpha) damage nonreproductive parts of the plant, causing the plant to mobilize nitrogen to make repairs (see **Chapter 2**). Homopterans feed either within plant cells or within the phloem or xylem vessels, in either case (especially in the latter) taking less nitrogen than do those heteropterans feeding on reproductive structures.

Because heteropterans prefer reproductive parts of the plant, they compete with humans, who also want seeds, fruits, and nuts, and for the same reason — these are the parts rich in nitrogen. Heteropterans thus compete more directly than do homopterans, which for the most part prefer — and with their water-recycling systems are adapted for feeding on — plant circulatory systems, or plant cells.

Heteropterans thus injure plants directly, by destroying or reducing the fecundity and fertility of reproductive structures desired also by humans, or by forcing the plant into undertaking energetically costly repairs.

However, homopterans do serious indirect damage in two ways. First, because many of them do not leave the plant (scale insects, aphids) and are often protected by ants, they may build large populations and seriously weaken the plant. Second, unlike heteropterans, homopterans are vectors of many plant disease organisms.

This difference (in vectoring) is directly correlated with the type of feeding of the two groups. Heteropterans thrust their stylets into plant tissues, between cells. Homopterans thrust their stylets either into plant cells (scale insects, aphids, other sternorrhynchans) or into plants' circulatory systems (leafhoppers, other auchenorrhychans). There the homopteran takes up fluids. This is an ideal system for the transmission of microorganisms living either in plant cells or in plant circulatory systems. On the other hand, heteropterans' stylets are less likely to encounter microorganisms, are less likely to take them up with the fluids, and are therefore only rarely vectors (see Nault 1997).

One may extend this argument to explain why more sternorrhynchous homopterans (aphids, etc.) transmit viruses than do auchenorrhynchous homopterans. The former penetrate plant cells; the latter do not. Aphids are also better vectors because, lacking chemosensory structures in their labia or stylets, they must actually take up fluids to test their suitability as food; acceptable fluids are retained and unacceptable ones regurgitated back into the plant (Nault 1997) — sometimes with viruses. Heteropterans, like auchenorrhychans, have these external sensory structures, and therefore do not regurgitate fluids.

Another difference between heteropterans and homopterans is the presence in the former of scent glands, both in the immature and the adult stages. These glands serve different functions, but defense is an important one, and perhaps the original function. Some of the secretions of the glands repel ants and other predators. Homopterans do not have scent glands, and their presence in heteropterans has (we believe) several consequences.

In general, heteropterans are larger than homopterans. Advantages of large size include greater dispersal ability, greater sexual reproductive potential, and the lack of appeal to small predators. Disadvantages include being more conspicuous to larger predators (especially visually orienting ones, like vertebrates). This disadvantage is lessened by the repellent effect of the scent. And being conspicuous has been turned to an advantage by many heteropterans, which in various ways have become unpalatable and warningly colored (aposematic), a phenomenon very rare in Homoptera. (Being bad-tasting may lessen the insect's need for scent, see Schaefer 1972.)

Large size requires a better food source and more food than does small size. Heteropterans' food is richer in nitrogen than that of many (most?) homopterans. Being relatively protected by scent permits heteropterans to build up large populations when food is abundant, and if that food is also desired by humans, these heteropterans are pests. Thus, scent glands are in part responsible for the economic importance of Heteroptera.

An important function of heteropteran scent is repelling ants (Carayon 1971), a universal predator. Hardly any heteropterans are attended by ants, but many homopterans have complex relationships with them, in a sense having coopted the enemy. One consequence, as mentioned above, is the buildup of large homopteran populations, guarded by ants, and this often leads to the buildup of honeydew. These large populations can weaken plants severely, and this accumulation of honeydew can foster growth of fungi and bacteria, which may in turn harm the plant.

Another consequence of the possession of scent glands is that heteropterans need not devote evolutionary resources to developing other types of defenses. Homopterans, on the other hand, have evolved numerous defenses — the many different waxy secretions of sternorrhynchans and the various hopping and jumping devices of auchenorrhychans.

Yet another consequence of heteropterans' larger size, and of auchenorrhychans' larger size than sternorrhynchans, is that the larger insects tend to be more mobile. Larger and more easily seen, those that are not warningly colored are more likely to move from their host plants, and more likely to find new host plants, or to find acceptable new kinds of host plants. Dispersal by the smaller homopterans is less effective; often it is passive, the aphid or scale being carried by the wind. But these groups have a rapid generation time (many generations in a short period), and they often have complex life cycles, including parthenogenesis. As a result of these features of the reproductive biology of sternorrhynchans, evolutionary change can occur rapidly, even though these insects may be less likely than their larger relatives to find new host plants. Thus, in the one case, the finding of new hosts is encouraged by the rapid discovery of them. In the other, the finding of new hosts is encouraged by the insects' ability to adapt rapidly to them once found.

In summary, both homopterans and heteropterans are distinctive in the way they feed — piercing plants or animals and sucking fluids from them. By doing so, these insects avoid many of the protections that plants have evolved, and serendipitously they also avoid some of the pesticides that humans have invented. Both heteropterans and homopterans may build up in great numbers and cause damage, but it is the latter whose large populations more often and in more places cause problems.

Homopterans and heteropterans differ, chiefly in size and in how they feed (i.e., how and exactly what they pierce and draw fluids from). Heteropterans take nitrogen-rich plant fluids, from wounds or from reproductive structures. Homopterans take cell contents or nitrogen-poor fluids from xylem or phloem. Each causes damage, heteropterans more directly than homopterans. However, it is likely that the indirect damage of homopterans is the more severe, on certain crops, because that damage is the transmission of plant diseases (mostly viral).

Finally, a few groups of heteropterans (but no homopterans) have become adapted for feeding on vertebrate blood. In a sense, those vertebrates and those plants that host heteropterans are not unlike one another. The vertebrates (bats for the most part, and humans and nesting rodents) live in groups and return at night to their homes; in this sense, they are semisessile, like plants. Both provide a predictable source of food for their heteropterans, and a source of food that is not killed by being fed upon. And thus both plants and these vertebrates differ from the arthropod prey of most predaceous heteropterans.

Many heteropterans are predaceous. One of us has suggested that predation was the original heteropteran way of life (see Schaefer 1981; for a counterview, see Sweet 1979). Many groups are entirely predaceous; many families, and many genera, contain some predaceous species and some herbivorous ones; and some species are both. Some groups are clearly primitively predaceous (their ancestors were predaceous) and others are secondarily predaceous (their ancestors were herbivorous) (see Schaefer 1997). Whatever the origin of their predaceous habit, or however "imperfect" that habit may be today, many of these heteropterans are excellent biological control agents, and many more (e.g., Reduviidae, see **Chapter 29**) have this potential. However, very recent study has shown that these insects feed differently than had been thought, and the application and use of them as biological control agents should be studied anew (see **Chapter 20**).

Heteropterans, then, or true bugs, are unique in several ways of concern to agricultural, medical, and veterinary entomologists. The more we know about these insects, the better able we shall be to control the harmful ones and to exploit the useful ones. The more we know also, the better able we shall be to live in a world compatible with our needs and yet sustainable. Remarkably little is known about the basic biology and ecology of most heteropteran pests (see Panizzi 1997a) and the same is true of other pests. We hope that this book will demonstrate the value of basic studies of biology, studies necessary if humans are to stop hobbling from day to day in their agriculture, and begin to take long-term strides.

SUGGESTED REFERENCES

Below (not starred) are some general references on Heteroptera. Despite the parochial titles of some, all give good information on Heteroptera in general and the subgroups of Heteroptera in particular. (Starred references are cited in this chapter.)

Blatchley, W. S. 1926. Heteroptera or True Bugs of Eastern North America, with Especial Reference to the Faunas of Indiana and Florida. Nature Publ. Co., Indianapolis, Indiana, U.S.A. 1116 pp.

Butler, E. A. 1923. Biology of the British Hemiptera-Heteroptera. H. F. & G. Witherby, London, U.K. 682 pp.

*****Carayon, J. 1971.** Notes et documents sur l'appareil odorant métathoracique des Hémiptères. Ann. Soc. Entomol. France (N.S.) 7: 737–770.

Dolling, W. R. 1991. The Hemiptera. Oxford University Press, Oxford, U.K. 274 pp.

*****Nault, L. R. 1977.** Arthropod transmission of plant viruses: a new synthesis. Ann. Entomol. Soc. Amer. 90: 521–541.

*****Panizzi, A. R. 1997a.** Wild hosts of pentatomids: ecological significance and role in their pest status on crops. Annu. Rev. Entomol. 42: 99–122.

*****1997b.** Entomofauna changes with soybean expansion in Brazil. Proc. World Soybean Res. Conf. 5: 166–169.

*****1997c.** Dynamics in the insect fauna adaptation to soybean in the tropics. Trends Entomol. 1: 71–88.

*****1997d.** A expansão da agricultura no centro-oeste brasileiro e os surtos recentes do percevejo-castanho. Rev. Plantio Direto Julho/Agosto 1997: 27–28.

*****Schaefer, C. W. 1972.** Degree of metathoracic scent gland development in the trichophorous Heteroptera (Hemiptera). Ann. Entomol. Soc. Amer. 65: 810–821.

*****1981.** The land bugs (Hemiptera: Heteroptera) and their adaptive zones. Pp. 67–83 *in* N. Ueshima [ed.], Phylogeny and Higher Classification of Heteroptera. Rostria No. 33, Suppl. 99 pp.

*****1997.** The origin of secondary carnivory from herbivory in Heteroptera (Hemiptera). Pp. 229–239 *in* A. Raman [ed.], Ecology and Evolution of Plant-Feeding Insects in Natural and Man-Made Environments. International Scientific Publ., New Delhi, India. 245 pp.

*****1998.** Phylogeny, systematics, and practical entomology: the Heteroptera (Hemiptera). An. Soc. Entomol. Brasil 27: 499–511.

2000. The Heteroptera, or true bugs. *In* S. Levin, et al. [eds.], Encyclopedia of Biodiversity. Academic Press, San Diego, California, U.S.A.

Schuh, R. T., and J. A. Slater. 1995. True Bugs of the World (Hemiptera: Heteroptera). Cornell University Press, Ithaca, New York, U.S.A. 337 pp.

Slater, J. A., and R. M. Baranowski. 1978. How to Know the True Bugs (Hemiptera–Heteroptera). Wm. C. Brown, Dubuque, Iowa, U.S.A. 256 pp.

Stonedahl, G. M., and W. R. Dolling. 1991. Heteroptera identification: a reference guide, with special reference on economic groups. J. Nat. Hist. 25: 1027–1066.

*****Sweet, M. H. 1979.** On the original feeding habits of the Hemiptera (Insecta). Ann. Entomol. Soc. Amer. 72: 575–579.

Weber, H. 1930. Biologie der Hemipteren: Eine Naturgeschichte der Schnabelkerfe. Verlag Julius Springer, Berlin, Germany. 543 pp.

SECTION II

Harmful Bugs

CHAPTER 2

Possible Causes of Disease Symptoms Resulting from the Feeding of Phytophagous Heteroptera

Koji Hori

1. INTRODUCTION

Symptoms of disease that develop on host plants under the influence of heteropterans have stimulated many entomologists to conduct investigations into the causes of deformation. In relation to plant-sucking insects, the following categories of phytotoxemias have been proposed by Carter (1962), based on external symptoms:

1. Local lesion at the insects feeding point,
2. Local lesions with development of secondary symptoms,
3. Tissue malformation, and
4. Symptoms indicating translocation of the causal entity.

These last may be (a) limited or (b) systemic. In relation to the genus *Lygus* (Miridae), on the other hand, Strong (1970) divided the damage into five types:

1. Abscission of fruiting body,
2. Catfacing or deformation of young fruits,
3. Necrosis,
4. Production of embryoless or shriveled seeds, and
5. Reduction of vegetative growth.

Tingey and Pillemer (1977) later modified this grouping and proposed the following five general types:

1. Localized wilting and tissue necrosis,
2. Abscission of fruiting forms,
3. Morphological deformation of fruit and seed,
4. Altered vegetative growth, and
5. Tissue malformations.

As mentioned above, classifications of damage are limited to that caused by a few species or that caused generally by plant-sucking insects; there is no classification for members of the Heteroptera as a group.

With respect to insect-induced galls, four main hypotheses on the stimulus for gall formation had been suggested by the end of the 19th century (Carter 1962). Norris (1979) stated that the malformations (including insect-induced galls) apparently involve imbalances in auxins in meristematic tissues and that these imbalances may result directly from (1) withdrawal of such chemicals from tissues along with the ingested liquefied contents of the disrupted cells; (2) injection of auxins into meristematic cells during feeding; and/or (3) injection of auxin inhibitors during feeding. The symptoms formed on host plants by the feeding of heteropterans may be attributed to some of the factors mentioned above and listed by Carter (1962). However, heteropterans cause many types of damage other than tissue malformations similar to insect-induced galls, and thus the causes of the damage must be discussed more broadly. Plant-sucking insects insert their stylets into the tissues of host plants and inject toxic saliva there, which is deposited or spreads through the surrounding tissues to affect their physiological condition (Smith 1920; Painter 1928; Leach and Smee 1933; Johnson 1936; Carter 1939, 1962; Kretovich 1944; Baker et al. 1946; Jeppson and MacLeod 1946; Addicott and Romney 1950; Flemion et al. 1951, 1952; Nuorteva and Reinius 1953; Flemion et al. 1954; Nuorteva 1954; Flemion 1955; Kloft 1960; Rastogi 1962; Lodos 1967; Miles 1969; Gopalan and Subramaniam 1977; Tingey and Pillemer 1977; Eggermann and Bongers 1980; Muimba-Kankolongo et al. 1987). Therefore, this chapter will describe those components of heteropteran saliva that may contribute to the damage of the host plant. The symptoms of the damage may vary with the different feeding methods and the different feeding sites of the bugs and, moreover, among the different organs of the injured plants (e.g., leaf, stem, root, fruit, seed). Therefore, the chapter must discuss the feeding methods and sites of heteropterans.

Physiological and biochemical conditions in host plant tissues may change because of the infestation of the bugs, inducing various symptoms. Such changes, especially in phenol metabolism (wound reaction that induces hypersensitive reaction in plant) and in hormonal balance resulting from the infestation of the bugs, may affect the appearance of the damage.

Therefore, these problems are discussed below.

2. FEEDING METHODS AND SITES

A heteropteran feeds in one of four ways:

1. Stylet-sheath feeding,
2. Lacerate-and-flush feeding,
3. Macerate-and-flush feeding, or
4. Osmotic pump feeding.

Most pentatomomorphans are of the first type, and seed-feeding pentatomomorphans (Lygaeidae and Pyrrhocoridae) and some pentatomids are of the second type; Cimicomorpha (Miridae) (feeding on fruits, growing tissues, and meristems) are of the third type; and coreids of the last type (Miles 1972, Miles and Taylor 1994). Some pentatomomorphans can employ two types of feeding, according to circumstances (Miles 1969); the pentatomid *Palomena angulosa* Motschulsky employs the stylet-sheath feeding method when feeding from the phloem of the host plant, but employs lacerate-and-flush feeding when feeding on the fruit (Hori et al. 1984). The pentatomid *Eurydema rugosum* Motschulsky uses the former type as young nymphs (mainly second instars) feeding from the phloem, but older nymphs or adults feeding on the mesophyll tissues use the latter type (Hori 1968b).

In the lacerate-and-flush feeding type, the bugs vigorously thrust their stylets back and forth until a number of cells has been lacerated. In the macerate-and-flush feeding type, the bugs inject salivary pectinase into the plant tissue to macerate the cells. The bugs of both types then flush out the contents of the cells with a more or less copious flow of saliva; as a result of such feeding, several cells are broken down. In osmotic pump feeding, the bugs secrete salivary sucrase into the plant tissue to increase the osmotic concentration of intercellular fluids and suck up the fluids, which contain sugars and amino acids exuded from the cells; as a result of this feeding, cells surrounding the stylets become empty. In the stylet-sheath feeding type, the bugs gently insert their stylets into the feeding site (mainly phloem), destroying only a few cells, and form a stylet sheath around the stylet track. Therefore, such feeding causes a minimum of immediate mechanical damage to the tissue of the host plant (Miles 1969, Miles and Taylor 1994).

The symptoms of damage vary with the site where the bugs feed. For example, only simple lesions occur at the feeding points when the bugs feed on mesophyll tissue; but, in many cases, tissue malformations occur when they feed on young growing tissues (e.g., young fruits) or meristematic tissues. Many *Lygus* bugs typically feed on growing and meristematic tissues — *L. hesperus* Knight (Strong 1970), *L. rugulipennis* Poppius = *L. disponsi* Linnavuori (Hori 1971a), *Lygus* spp. (Tingey and Pillemer 1977, Summerfield et al. 1982), *L. lineolaris* (Palisot de Beauvois) (Sapio et al. 1982).

Handley and Pollard (1993) have described changes in *L. lineolaris* feeding on strawberry. In early fruit development stages, the bugs repeatedly penetrate achenes (the site where IAA (iodo-acetic acid) is synthesized), causing buttoning malformation. In later stages of fruit development, the feeding sites of the bugs change to receptacle tissue, although usually near an achene, resulting in more localized damage, including creases and indentations.

3. COMPONENTS IN THE SALIVARY GLAND

Many investigators have detected various substances in the salivary glands of heteropterans in the salivary fluids in some cases and in extracts of the salivary glands (including tissues) in other cases. These substances are presumed to affect or disturb the physiological conditions of plants. Discussed below is whether or not the components contained in the salivary glands of heteropterans are fundamentally necessary to the feeding strategies of phytophagous Heteroptera and whether or not the substances are contained in the watery saliva.

Amylase has been detected in the salivary glands of Pentatomidae (18 species), Coreidae (9), Lygaeidae (10), Dinidoridae (1), Pyrrhocoridae (4), Miridae (25), Acanthosomatidae (4), Aradidae (1), Cydnidae (1), Largidae (1), Scutelleridae (1), Berytidae (1), and Tingidae (1) (Table 1). However, some of these reports may be based on methodologically questionable tests; for example, reported results for salivary amylase are worthless if the tests relied on the starch–iodine reaction without an oxidizing agent such as H_2O_2 (Taylor and Miles 1994). Moreover, when dealing with whole glands, there is a problem of contamination with hemolymph enzymes or even of microbial contamination during enzyme extraction. Therefore, some of the results above should perhaps be reexamined (P. W. Miles, personal communication). Nevertheless, amylase is a fundamental enzyme in the salivary glands of most phytophagous heteropterans, as many reliable results have shown. Several authors have confirmed the presence of the enzyme in the watery saliva of various heteropterans — e.g., in *Oncopeltus fasciatus* (Dallas) (Lygaeidae) (Miles 1967a), in *L. rugulipennis* = *L. disponsi* (Hori 1971b), in *L. rugulipennis, L. pratensis* (L.), *L. gemellatus* (Herrich-Schaeffer), and *L. punctatus* (Zetterstedt) (Varis et al. 1983). Possibly, many other heteropterans as well as these bugs have amylase in the watery saliva.

Here, the term α-glucosidase is based on a classical classification presented by Dadd (1970). α-Glucosidase has been detected in the salivary glands of Pentatomidae (2 species), Coreidae (6), Lygaeidae (3), Pyrrhocoridae (4), Miridae (10), and Scutelleridae (1) (Table 1). However, α-glucosidases in the salivary glands of these heteropterans (except Coreidae) were very weak in activity or were detected by histochemical methods. Therefore, these may be endoenzymes contained in the tissue of the salivary gland or may be unable to function in the plant tissue when injected, even though they are present in the watery saliva.

Table 1 Major Enzymes in the Salivary Glands of Phytophagous Heteroptera

Family Species	Enzyme	Ref.
Pentatomidae		
Aelia acuminata L.	Amylase, protease	Nuorteva (1954)
	Protease	Sossedov et al. (1969)
A. sibirica Reuter	Protease	Sossedov et al. (1969)
Agonoscelis rutilia F.	Phenoloxidase	Miles (1964)
Arma custos (F.)	Amylase	Hori (1969)
Carbula humerigera (Uhler)	Amylase	Hori (1969)
Carpocoris purpureipennis DeGeer	Amylase	Hori (1972)
Dolycoris baccarum (L.)	Amylase	Nuorteva (1954), Hori (1969)
	Protease	Nuorteva (1954), Nuorteva and Laurema (1961a)
Eumecopus australasiae (Donovan)	Phenoloxidase	Miles (1964a)
E. punctiventris Stål	Phenoloxidase	Miles and Sloviak (1970)
Eurydema rugosum Motschulsky	Amylase	Hori (1968a)
Eysarcoris aeneus (Scopoli)	Amylase	Hori (1972)
E. guttiger (Thunberg)	Amylase	Hori (1972)
Eysarcoris lewisi Distant	Amylase	Hori (1969)
E. ventralis (Westwood)	Amylase	Hori (1972)
Graphosoma lineatum (L.)	Amylase	Duspiva (1952)
G. rubrolineatum (Westwood)	Amylase	Hori (1969)
Nezara antennata Scott	Amylase	Hori (1972)
N. viridula (L.)	Phenoloxidase	Miles (1964a)
Oechalia schellembergi Guérin	Phenoloxidase	Miles (1964a)
Palomena angulosa Motschulsky	Amylase	Hori (1972)
	α-Glucosidase	Hori (1975a)
Pentatoma japonica (Distant)	Amylase	Hori (1972)
P. rufipes (L.)	Amylase	Hori (1972)
	α-Glucosidase	Baptist (1941)
Plautia crossata stali Scott	Amylase	Hori (1972)
Coreoidea		
Amblypelta sp.	Glucosidase	Miles (1987)
Clavigralla gibbosa Spinola	Amylase, Glucosidase, Protease	Mathur and Thakar (1969)
Cletus rusticus Stål	Amylase	Hori (1972)
C. signatus Walker	Amylase, α-glucosidase	Mall and Chattoraj (1968)
Coreus marginatus orientalis (Kiritschenko)	Amylase	Hori (1972)
	α-Glucosidase	Hori (1975a)
Corizus parumpunctatus Schilling	Amylase	Baptist (1941)
Leptocorisa varicornis F.	Amylase	Saxena (1954a,b)
	α-Glucosidase	Khan and Khan (1977)
Leptoglossus occidentalis Heideman	Pectinmethylesterase	Campbell and Shea (1990)
Mictis profana (F.)	α-Glucosidase	Taylor and Miles (1994)
	Phenoloxidase	Miles (1964a)
	Catecholoxidase	Taylor and Miles (1994)
Paraplesius unicolor Scott	Amylase	Hori (1972)
Rhopalus maculatus (Fieber)	Amylase	Hori (1969)
Stictopleurus punctatonervosus (Goeze)	Amylase	Hori (1969)
Lygaeidae		
Chilacis typhae Perr	Amylase	Baptist (1941)
Cymus glandicolor Hahn	Amylase	Hori (1972)

Table 1 (continued)   Major Enzymes in the Salivary Glands of Phytophagous Heteroptera

Family Species	Enzyme	Ref.
Elasmolomus sordidus (F.)	Phenoloxidase	Miles (1964a)
Gastrodes errugineus L.	α-Glucosidase	Baptist (1941)
Graptopeltus angustatus Montandon	Amylase	Hori (1972)
Heterogaster urtica F.	Amylase	Hori (1972)
Lygaeus equestris (L.)	Amylase	Hori (1972)
Nysius expressus Distant	Amylase	Hori (1972)
Oncopeltus fasciatus (Dallas)	Amylase	Bronskill et al. (1958), Feir and Beck (1961), Miles (1967a), Bongers (1970)
	α-Glucosidase	Bronskill et al. (1958), Feir and Beck (1961), Bongers (1970), Bongers and Muermann (1973)
	Protease	Bronskill et al. (1958), Feir and Beck (1961)
	Phenoloxidase	Miles (1960, 1967a), Miles and Helliwell (1961)
Oxycarenus hyalinipennis (Costa)	Amylase, α-glucosidase, protease	Saxena and Bhatnagar (1958)
Pachygrontha antennata (Uhler)	Amylase	Hori (1972)
Rhyparochromus sp.	Amylase	Hori (1969)
Dinidoridae		
Coridius janus (F.)	Amylase	Rastogi (1962)
	Protease	Rastogi and Datta Gupta (1962), Naqvi et al. (1973)
Pyrrhocoridae		
Dysdercus cingulatus F.	α-Glucosidase	Khan and Khan (1977)
D. fasciatus Signoret	Amylase	Ford (1962), Khan (1962a)
	α-Glucosidase	Ford (1962), Khan and Ford (1967)
D. howardi DeGeer	α-Glucosidase	Baptist (1941)
D. koenigii F.	Amylase	Janda and Munzarova (1980)
	Protease	Saxena (1955)
D. sidae Montz	Phenoloxidase	Miles (1964a)
D. singulatus F.	Amylase	Quayum (1966)
Odontopus nigricornis Stål	Amylase, protease	Rastogi and Datta Gupta (1961)
Pyrrhocoris apterus L.	α-Glucosidase	Baptist (1941)
	Pectinase	Courtois et al. (1968)
Miridae		
Adelphocoris lineolatus Goeze	Amylase, protease	Nuorteva (1954)
A. seticornis (F.)	Pectinase	Laurema and Nuorteva (1961)
A. suturalis (Jakovlev)	Amylase	Hori (1970a)
	α-Glucosidase	Hori (1975a)
	Pectinase	
A. triannulatus (Stål)	Amylase	Hori (1970a)
A. variabilis (Uhler)	Amylase	Hori (1972)
Bryocoropsis laticollis Schumacher	Amylase, α-glucosidase	Goodchild (1952)
	Protease	Kumar (1970)
Capsus ater L.	Amylase, protease	Nuorteva (1954)
	Pectinase	Laurema and Nuorteva (1961)
Creontiades dilutus (Stål)	Pectinase	Hori and Miles (1993)

continued

Table 1 (continued) Major Enzymes in the Salivary Glands of Phytophagous Heteroptera

Family Species	Enzyme	Ref.
C. modestum Distant	Phenoloxidase	Miles (1964a)
Deraeocoris punctulatus (Fallén)	Amylase	Hori (1972)
Distantiella theobroma (Distant)	Amylase, α-glucosidase	Goodchild (1952)
Eurystylus coelestialium (Kirkaldy)	Amylase	Hori (1972)
Helopeltis bergrothi Reuter	Amylase, α-glucosidase	Goodchild (1952)
H. clavifer (Walker)	Pectinase	Miles (1987)
H. corbisieri Schmitz	Protease	Kumar (1970)
Lyglineolaris sp. Palisot de Beauvois	Pectinase	Adams and McAllan (1958)
Lygocoris locorum (Meyer-Dur)	Amylase	Hori (1972)
L. nigritulus (Linnavuori)	Amylase	Hori (1972)
Lygus gemellatus (Herrich-Schaeffer)	Amylase, protease, pectinase	Varis et al. (1983)
L. hesperus Knight	Pectinase	Strong and Kruitwagen (1968)
L. pratensis (L.)	Amylase	Baptist (1941), Varis et al. (1983)
	α-Glucosidase	Baptist (1941)
	Protease	Varis et al. (1983)
	Pectinase	Laurema and Nuorteva (1961), Varis et al. (1983)
L. punctatus (Zetterstedt)	Amylase, protease, pectinase	Varis et al. (1983)
L. rugulipennis Poppius (=L. disponsi Linnavuori)	Amylase	Nuorteva (1954), Hori (1970a), Varis et al. (1983), Laurema et al. (1985)
	α-Glucosidase	Hori (1975a), Laurema et al. (1985)
	Protease	Nuorteva (1954), Hori (1970b), Varis et al. (1983), Laurema et al. (1985)
	Pectinase	Hori (1974a), Varis et al. (1983), Laurema et al. (1985)
	Phenoloxidase	Hori (1974a), Laurema et al. (1985)
L. saundersi Reuter	Amylase	Hori (1970a)
	α-Glucosidase, pectinase	Hori (1975a)
Miris dolabratus L.	Amylase	Nuorteva (1954)
	Protease	Nuorteva (1954; 1956a)
	Pectinase	Laurema and Nuorteva (1961)
Moissonia importunitas (Distant)	Amylase, protease, pectinase, phenoloxidase	Gopalan (1976)
Notostira erratica L.	Protease	Nuorteva (1954)
Onomaua lautus (Uhler)	Amylase	Hori (1972)
Orthocephalus funestus Jakovlev	Amylase	Hori (1972)
	α-Glucosidase, pectinase	Hori (1975a
Orthops sachalinus (Carvalho)	Amylase	Hori (1972)
Poeciloscytus unifasciatus (F.)	Pectinase	Laurema and Nuorteva (1961)
Pseudatomoscelis seriatus (Reuter)	Pectinase	Martin et al. (1988a)
Sahlbergella singularis Haglund	Amylase, α-glucosidase	Goodchild (1952)
Stenodema calcaratum Fallén	Amylase	Nuorteva (1954), Hori (1970a)
	Protease	Nuorteva (1954)
	Pectinase	Laurema and Nuorteva (1961)
Stenotus binotatus (F.)	Amylase	Hori (1972)
	α-Glucosidase, protease, pectinase	Takanona and Hori (1974)
Acanthosomatidae		
Elasmostethus humeralis Jakovlev	Amylase	Hori (1969)
Elasmucha dorsalis (Jakovlev)	Amylase	Hori (1972)
E. putoni Scott	Amylase	Hori (1972)
E. signoreti Scott	Amylase	Hori (1969)

Table 1 (continued) Major Enzymes in the Salivary Glands of Phytophagous Heteroptera

Family Species	Enzyme	Ref.
Aradidae		
Aradus orientalis Bergroth	Amylase	Hori (1969)
Cydnidae		
Macroscytus japonensis Scott	Amylase	Hori (1972)
Largidae		
Physopelta gutta (Burmeister)	Amylase	Hori (1972)
Scutelleridae		
Chrysocoris stollii Wolf	Amylase	Choudhuri et al. (1966)
	Protease	Sossedov et al. (1969)
Eurygaster integriceps Puton	Amylase	Kretovich et al. (1943), Vilkova (1968)
	α-Glucosidase	Vilkova (1968)
	Protease	Kretovich et al. (1943), Vilkova (1968)
Tectocoris lineola F.	Phenoloxidase	Miles (1964a)
Berytidae		
Metancanthus elegans L.	Amylase	Baptist (1941)
Tingidae		
Monanthia cardui L.	Amylase	Baptist (1941)
All the phytophagous bugs	Phenoloxidase	Miles (1968, 1969)

Taylor and Miles (1994) recently confirmed that α-glucosidase — maltose-hydrolyzing and sucrose-hydrolyzing enzyme(s) — was contained in the watery saliva of *Mictis profana* (F.) (Coreidae). Hori (1975a) detected strong activity of α-glucosidase in the salivary gland homogenate of *Coreus marginatus orientalis* (Kiritschenko) (Coreidae). Therefore, as Miles and Taylor (1994) insisted, a sucrase (α-glucosidase) with especially strong activity is contained only in the watery saliva of coreids and seems to be characteristic of this family.

Protease has been found in the salivary glands of Pentatomidae (4 species), Coreidae (1), Lygaeidae (2), Dinidoridae (1), Pyrrhocoridae (2), Miridae (13), and Scutelleridae (1) (Table 1). As mentioned above, some of these results may have to be reexamined. However, protease is a major component of the salivary gland at least in more phytophagous heteropterans, although the activity varies under different circumstances; in *L. rugulipennis* the enzyme is present in nymphs but not in adults, and in *Capsus ater* L. (Miridae) it is contained only in the salivary glands of bugs collected from oats (Nuorteva 1954). In *L. rugulipennis* (= *L. disponsi*), the enzyme's activity shows a seasonal change and differs even between one and another gland of the pair (Hori 1970b). Varis et al. (1983) confirmed that protease is contained in the watery saliva of *L. rugulipennis, L. gemellatus, L. pratensis,* and *L. punctatus*. As well as these species, many other heteropterans may possess protease in the watery saliva.

Pectinase has been detected in the salivary glands of Miridae (18 species), Coreidae (1), Lygaeidae (1), and Pyrrhocoridae (1) (Table 1). Miles (P. W. Miles, personal communication) believes that each family of Heteroptera has its own characteristics, and that, when reported results for a particular species fall outside a broad categorization, it is necessary to look at the work with particular care. For instance, a report (Campbell and Shea 1990) claiming a pectinase in a coreid is based on reactions with ruthenium red, which can give misleading results with nonenzymatic proteins and certain amino acids. Therefore, pectinase is probably an important and fundamental enzyme only in the family Miridae. Varis et al. (1983) confirmed that this enzyme is contained in the watery saliva of *L. pratensis, L. gemellatus,* and *L. punctatus.* Hori (1974a) found pectinase in the watery contents of the posterior lobe of the salivary gland of *L. rugulipennis.* Hori and Miles (1993) collected watery saliva which *Creontiades dilutus* (Stål) (Miridae) discharged from the tip of the stylets without any stimulant and found that it had strong pectinase. Thus, this enzyme may be a major component of the watery saliva of Miridae.

Phenoloxidase has been recorded in the salivary glands of Pentatomidae (5 species), Coreidae (1), Lygaeidae (2), Pyrrhocoridae (1), Miridae (3), and Scutelleridae (1) (Table 1). According to Miles (1968, 1969), polyphenoloxidase seems to be a universal component of the watery saliva of phytophagous Hemiptera.

Some other enzymes (polypeptidase, β-glucosidase, α-galactosidase, β-galactosidase, cellulase, trehalase, lipase/esterase, carboxylesterhydrolase, acid- or alkaline-phosphatase, phosphorylase) have also been found in the salivary glands of Pentatomidae (3 species), Coreidae (4), Lygaeidae (4), Dinidoridae (1), Pyrrhocoridae (3), Miridae (10), and Scutelleridae (1) (Table 2). In these heteropterans, the enzymes had only weak activity, and the results should be reexamined. In any case, these enzymes may be only minor components of the watery saliva or be endoenzymes present only in the glands.

Although the kinds of amino acids in the salivary glands (Table 3) vary among species and with the physiological conditions of the bugs, they are undoubtedly a component of the watery saliva in Heteroptera. This may be supported by the following statement and experimental confirmation: Nuorteva and Laurema (1961b) reported that amino acids in the salivary glands of *Aelia acuminata* (L.), *Dolycoris baccarum* (L.) (Pentatomidae), and *Capsus ater* (Miridae) came via hemolymph from the diet. Also, Miles (1967b) confirmed the transport in *O. fasciatus* (Lygaeidae) of amino acids from hemolymph to saliva. Moreover, Taylor and Miles (1994) detected amino acids in the watery saliva of *M. profana* (Coreidae).

Substances related to plant growth have been detected in the salivary glands of many heteropterans (Fisher et al. 1946; Nuorteva 1956b, 1958; Hori 1974b, 1975b, 1976; Hori and Miles 1977; Burden et al. 1989). In many cases, however, these substances have not been identified and it has not been explained whether they are present in the watery saliva or in the tissues of the salivary gland.

Nuorteva (1956b) found IAA in the salivary glands of *Stenodema calcaratum* (Fallén) (Miridae) and concluded that it was transferred via hemolymph from the diet. However, Strong (1970) and Hori (1974b, 1976) never detected IAA in the salivary glands of Miridae, Pentatomidae, and Coreidae and believed that, in general, the salivary glands of heteropterans do not contain an active amount of IAA. Moreover, Miles and Hori (1977), using a sucrose solution with ^{14}C-labeled IAA as the diet, confirmed that IAA was not transferred from the diet to the salivary gland of *C. dilutus* (Miridae) in a significant amount. This may be, in part, because of the ability of heteropterans to metabolize IAA in their gut and to excrete it conjugated with sugars or proteins (Hori and Endo 1977; Hori 1979, 1981; Hori et

Table 2   Other Enzymes in the Salivary Glands of Phytophagous Heteroptera

Family Species	Enzyme	Ref.
Pentatomidae		
Eurydema rugosum Motschulsky	Trehalase	Hori (1975a)
Palomena angulosa Motschulsky	α-Galactosidase, trehalase	Hori (1975a)
Coreoidea		
Cletus signatus Walker	Esterase	Mall and Chattoraj (1968)
Coreus marginatus Kiritschenko	β-Glucosidase; α-Galactosidase	Hori (1975a)
Leptocorisa varicornis F.	β-Glucosidase	Saxena (1954b)
Mictis profana (F.)	Phosphatase in gland extract	Taylor and Miles (1994)

Table 2 (continued) Other Enzymes in the Salivary Glands of Phytophagous Heteroptera

Family Species	Enzyme	Ref.
Lygaeidae		
Chilacis typhae Perr	Lipase	Baptist (1941)
Lygaeus sp.	Alkaline phosphatase	Kumar et al. (1980)
Oncopeltus fasciatus (Dallas)	Trehalase	Bongers (1970)
	Carboxylesterhydrolases	Bongers and Muermann (1973)
	Acid phosphatase	Beel and Feir (1977)
	Lipase	Bronskill et al. (1958), Feir and Beck (1961), Miles (1967a)
	Esterase	Salkeld (1960), Miles (1967a)
Oxycarenus hyalinipennis (Costa)	Esterase	Saxena and Bhatnagar (1958)
Dinidoridae		
Coridius janus F.	Acid phosphatase, alkaline phosphatase	Naqvi et al. (1973)
	Esterase	Rastogi (1962)
Pyrrhocoridae		
Dysdercus fasciatus Signoret	Lipase/esterase	Ford (1962), Khan (1974)
	β-Glucosidase	Ford (1962), Khan (1974)
	Peptidase	Ford (1962), Khan (1962b)
D. koenigii F.	Acid phosphatase, alkaline phosphatase	Kumar et al. (1978)
	Trehalase	Janda and Munzarova (1982)
Pyrrhocoris apterus L.	Phosphatase, phosphorylase	Kloft (1960)
Miridae		
Bryocoropsis laticollis Schumacher	Lipase/esterase	Goodchild (1952)
Distantiella theobroma (Distant)	Lipase/esterase	Goodchild (1952)
Helopeltis bergrothi Reuter	Lipase/esterase	Goodchild (1952)
Lygus gemellatus (Herrich-Schaeffer)	Phosphatase	Varis et al. (1983)
L. pratensis (L.)	Phosphatase	Varis et al. (1983)
L. punctatus (Zetterstedt)	Phosphatase	Varis et al. (1983)
L. rugulipennis Poppius (= Lygus disponsi Linnavuori)	Acid phosphatase	Hori (1974a), Laurema et al. (1985)
	Trehalase	Hori (1975a), Laurema et al. (1985)
	Phosphatase	Varis et al. (1983)
Orthocephalus funestus Jakovlev	β-Galactosidase	Hori (1975a)
Ragmus importunitus Distant	Cellulase	Gopalan (1976)
Sahlbergella singularis Haglund	Lipase/esterase	Goodchild (1952)
Scutelleridae		
Chrysocoris stollii Wolf	Polypeptidase	Choudhuri et al. (1966)
Eurygaster integriceps Puton	Lipase	Vilkova (1968)

Table 3 Amino Acids in the Salivary Glands of Phytophagous Heteroptera

Family Species	Contents	Ref.
Pentatomidae		
Aelia acuminata (L.)	15 in AL, 7 in PL	Nuorteva and Laurema (1961b)
Dolycoris baccarum (L.)		
Eumecopus australasiae (Donovan)		Miles (1964b)
Eurydema rugosum Motschulsky	16 kinds; 170 µg/50 bugs	Hori (1975c)
Coreidae		
Mictis profana (F.)	1.8 µg/µl leucine eq.	Taylor and Miles (1994)
Miridae		
Capsus ater (L.)	15 kinds; 207 µg/50 bugs	Nuorteva and Laurema (1961b), Hori (1975c)
Lygus gemellatus (Herrich-Schaeffer)	25 kinds; 1943 ng/SG	—
L. pratensis (L.)	25 kinds; 2148 ng/SG	Laurema and Varis (1991)
L. punctatus (Zetterstedt)	21 kinds; 913 ng/SG	—
L. rugulipennis Poppius	26 kinds; 2282 ng/SG	Laurema and Varis (1991)
Aradidae		
Aradus cinnamomeus Panzer		Heliovaara and Laurema (1988)

Note: AL = anterior lobe; PL = posterior lobe; SG = salivary gland.

al. 1979). Therefore, it may be safely said that IAA is not a component, or is a very minor component, of the saliva of heteropterans; but this does not mean that IAA never contributes to the production of damage. There is evidence in a report by Miles and Lloyd (1967) that *Elasmolomus sordidus* (F.) (Pentatomidae) can synthesize IAA while injecting saliva into plant tissue.

4. PHYSIOLOGICAL AND BIOCHEMICAL EFFECTS OF INJURY TO HOST PLANT

Heteropterans cause various physiological and biochemical changes in the tissues of host plants. However, there have been fewer studies on the changes caused by heteropterans than by homopterans, thrips, and mites. Kloft (1960) reported that known effects of free amino acids in the watery saliva of aphids when injected artificially into plants include streaming of cytoplasm, increased cell permeability, increased respiration rate, increased transpiration, and decreased photosynthesis. Such effects have been also found by a vast number of investigators in plants fed on by aphids, thrips, and mites.

Wheat kernels injured by the following heteropterans have been reported to have elevated levels of amylase and protease: *Eurygaster integriceps* Puton (Scutelleridae) and *Aelia* sp. (Pentatomidae) (Kretovich et al. 1943, Kretovich 1944, Atanassova and Popova 1968, Pokrovskaya et al. 1971), *L. rugulipennis* (Miridae) (Nuorteva 1954, Rautapaa 1969), *Miris dolabratus* L. (Miridae), *D. baccarum* L. (Pentatomidae) (Nuorteva 1954), *Nysius huttoni* (White) (Lygaeidae) (Cressey 1987, Every et al. 1990), and *Stenotus binotatus* F. (Miridae) (Every et al. 1992). The increase in enzyme activity in the injured grains has been considered the result of the injection of salivary enzymes by the bugs (Laurema et al. 1985). Apart from wheat kernels, the tissues of *Acacia iteaphylla* attacked by *Mictis profana* (Coreidae) show more sucrose hydrolyzing activity than do unattacked tissues (Miles and Taylor 1994).

A rise in phenoloxidase activities and an increase in the amount of phenol compounds in the host plant tissues injured by heteropterans have been proved and confirmed: in *Leptobyrsa rhododendri* Horváth (Tingidae) (Johnson 1936), in *Lygus rugulipennis* on sugar beet plant (Hori 1973a), in *Eurydema rugosum* (Pentatomidae) (Hori 1974c), in high infestation levels of *Moissonia* (= *Ragmus*) *importunitas* (Distant) (Miridae) (Gopalan and Subramaniam 1977), in *L. rugulipennis* on Chinese cabbage (Hori and Atalay 1980), in *Cyrtopeltis tenuis* Reuter (Miridae) (Raman and Sanjayan 1984, Raman et al. 1984), and in *L. rugulipennis* on pumpkin fruit (Hori et al. 1987). However, the pattern of change in activity after the infestation period differs with the species of bug and with the species of attacked plant. For example, in sugar beet leaf injured by *L. rugulipennis*, phenoloxidase activity increases rapidly from the first to the third day after the injury and is maintained at a high level until the 18th day (Hori 1973a); in pumpkin fruit injured by this bug, it rises just after infestation but soon drops and remains at the control level from the third day on (Hori et al. 1987); and in Chinese cabbage leaf injured by this bug, it shows a marked increase on the first day after infestation but drops rapidly toward a moderately high level on the third day and on the whole maintains this level until the 21st day (Hori and Atalay 1980).

In addition to these, several physiological and biochemical changes in host plants after infestation of heteropterans have been reported: for example, a rise in acid phosphatase activity (Hori 1973a, Raman and Sanjayan 1984, Hori et al. 1987); an increase in tannins (Raman and Sanjayan 1984); increases in respiratory rate, transpiration rate, respiratory quotient, catalase activity, and ascorbic acid oxidase activity, and a reduction in cytochrome oxidase activity (Gopalan and Subramaniam 1977); an increase in total amino acid content (Hori 1973b, 1975d; Gopalan 1975; Miles and Taylor 1994); a decrease in chlorophyll concentration (Hori 1974c, Hansen and Nowak 1985); a reduction in rate of photosynthesis (Brewer et al. 1979, Johnson and Knapp 1996); an increase in ethylene concentration (Martin et al. 1988a); a rise in IAA-oxidase activity (Hori et al. 1987); and a rise in sucrose-hydrolyzing activity (Miles and Taylor 1994). These physiological and biochemical changes in plant tissues may contribute to the production of damage and result in the various symptoms observed.

5. POSSIBLE CAUSES OF DAMAGE

5.1 Mechanical Damage

Most specific symptoms produced by the feeding of heteropterans can be explained by the chemical destruction of cell components, or by hormonal imbalance in plants as the result of the infestation and so on, in addition to any direct effects of mechanical destruction of cells around the inserted stylets. However, the overall damage to the plant is caused primarily by mechanical destruction, including the withdrawal of cell sap; and some authors stress that such mechanical effects are more important than other effects: in the chinch bug (Painter 1928), in *L. simonyi* Reuter (Miridae) on cotton (Taylor 1945), in *L. lineolaris* (Flemion 1955, 1958), in *L. rugulipennis* (Somermaa 1961), in *Moissonia* (= *Ragmus*) *importunitas* (Miridae) on sunn-hemp plant (Gopalan and Subramaniam 1977), in *Palomena angulosa* (Pentatomidae) on potato plant (Hori et al. 1984), and in *Bathycoelia thalassina* (Herrich-Schaeffer) (Pentatomidae) (Owusu-Manu 1990). Varis (1972) reproduced the appearance of symptoms formed on the sugar beet plant by *L. rugulipennis*, by artificially injuring the growing point of the plant. Mechanical destruction is especially important when meristematic tissues at the plant growing tip are attacked — in *Lygus* spp. on alfalfa (Jeppson and MacLeod 1946), in *L. rugulipennis* on *Pinus sylvestris* seedlings (Holopainen 1986), and on pine seedlings (Poteri et al. 1987).

Apart from the meristematic tissue, Hori (1968b) showed that the white circular spots formed on leaves of crucifers by the feeding of *E. rugosum* (Pentatomidae) were caused by the mechanical destruction of mesophyll cells by the stylets; the maximal radius of the feeding spot is equal to the maximal thrusting distance of the stylets (the length of the first two rostral segments). For the production of the white spots, of course, the withdrawal by the bug of cell contents, including chlorophyll, may be needed in addition to the mechanical destruction of cells. According to Hori (1974c), the amount of chlorophyll in the leaf tissue decreases as the result of the injury.

There is also a modified mechanical damage; Neal (1993) affirmed that coreid saliva is not intrinsically toxic. Instead, he linked the death of squash shoot tips caused by *Anasa tristis* (DeGeer) (Coreidae) with an interruption of xylem flow by a solidifying salivary secretion.

5.2 Chemical Damage

Allen (1947, 1951) stated that toxiniferous insects might cause growth changes in the plant by injecting salivary protease, which in turn might release active auxin from its protein component. Goodchild (1952), investigating the damage of cacao by the attack of the cacao capsid bugs (Miridae), *Sahlbergella singularis* Haglund, *Distantiella theobroma* (Distant), *Bryocoropsis laticollis* Schumacher, and *Helopeltis bergrothi* Reuter, found that the damage could not have been caused merely mechanically, and the affected cells are killed by the saliva and eventually shrivel, leaving a cavity. Therefore, it is probable that the watery saliva containing enzymes is injected into host plant tissues during the feeding of heteropterans. However, a portion of the saliva injected must be deposited in the tissues or spread to the tissues surrounding the feeding point without being sucked back by the bug to disturb the tissues physiology and biochemistry. In fact, the spreading of saliva in the tissues has been confirmed by Nuorteva and Reinius (1953), who found that ^{14}C-labeled saliva of the wheat bugs, *L. rugulipennis* (Miridae) and *Dolycoris baccarum* L. (Pentatomidae), injected into the host plant tissues during feeding, spread from the feeding point to the surrounding tissues; and by Miles (1959), who found that *O. fasciatus* (Lygaeidae) left deposits of watery saliva which seemed capable of extraintestinal digestion of starch (in the plant tissue). Such spreading of saliva is also supported by the following statements and an experimental confirmation: Tingey and Pillemer (1977) stated that salivary amylases and proteases appear to spread to the tissues and aid external digestion of insoluble organic nutrients there. Laurema et al. (1985) reported similarly for *L. rugulipennis*. Such diffusion of saliva from the feeding point is closely related to the feeding method of heteropterans. In lacerate-and-flush feeding and macerate-and-flush feeding, living cells surrounding the initial lesion will be subjected to the influence of any saliva left behind. On the other hand, in stylet-sheath feeding, cells surrounding the stylet sheath are to some extent protected by it from salivary action (Miles 1969, Miles and Taylor 1994). Therefore, in the first two types, the saliva and its enzymes are important for the production of damage (Laurema and Varis 1991). Apart from the spread of saliva, Miles (1987) found that a mirid, *Helopeltis clavifer* (Walker), and a coreid, *Amblypelta* sp., feeding on a variety of plant tissues, evacuated the contents of cells up to 3.5 mm from the farthest penetration of their mouthparts; that is, they can remove the cell contents without mechanical damage, suggesting either general maceration of tissues or the unloading of cytoplasmic content by osmotic pump feeding.

It is known that wheat grains injured by heteropterans are poor in quality; the following reports indicate this may be because of salivary enzymes injected into the kernels during feeding:

1. Kretovich (1944) stated that the destruction of gluten in infested wheat was caused by active proteolytic enzymes injected into the kernel by the wheat bug.
2. Nuorteva (1953) reported that the great degradation of flour quality caused by the wheat bugs may depend on the salivary protease.
3. Nuorteva (1954) concluded that an increase in maltose in kernels injured by wheat bugs is caused by salivary amylase injected during feeding.
4. Rautapaa (1969) ascribed the decrease of starch in wheat grains infested with *L. rugulipennis* to the action of amylase in the saliva.
5. Vilkova (1968) stated that the poorer quality of flour from grains injured by *Eurygaster integriceps* (Scutelleridae) was caused both by disaggregation of the gluten and hydrolysis of carbohydrates and fat in the flour. Both of these effects could be caused by amylase, α-glucosidase, protease, and lipase in the saliva of this bug.

6. Slepyan et al. (1972) found that the prick cavity in the endosperm of the kernels injured by
   E. integriceps was filled with the products of starch and gluten disintegration. This phenomenon
   also suggests the action of amylase and protease in the saliva injected.

Recently, in New Zealand, the damage to wheat caused by the wheat bugs, *N. huttoni* (White) (Lygaeidae) and *Stenotus binotatus* F. (Miridae), has been studied. Wheat grains infested by the bugs contain insect salivary protease and show degeneration in quality, possibly caused by the protease (Cressey and McStay 1984, Cressey et al. 1987, Every et al. 1990). Every et al. (1992) reported that the bugs deposited salivary enzymes into immature wheat grains while feeding, that these enzymes survived in the harvested wheat and destroyed gluten structure in dough, resulting in bread of poor quality, and that the considerable destruction of starch in the endosperm of kernels injured by *S. binotatus* may be caused by amylase in its saliva.

Apart from the wheat bugs injuring wheat grains, there have been some reports on the effects of salivary enzymes on host plants:

1. Dearman (1960) inferred that surface discoloration of tomato fruit might be caused by enzymes in the saliva of *Nezara viridula* (L.).
2. Brewer et al. (1979) found that some cytoplasmic organelles actually disappeared after attack by *Labops hesperius* (Miridae) and ascribed this phenomenon, in part, to an enzyme that dissolved them.
3. Sapio et al. (1982) reported that a split stem lesion on young populus trees caused by *Lygus lineolaris* was due to the chemical stimuli of salivary enzymes in addition to simple mechanical stimuli.
4. Summerfield et al. (1982) stated that the mouthparts of *Lygus* sp. attacking lentils caused localized mechanical damage to several cells, which was then compounded by the consequences of salivary secretion (enzymes) and/or removal of cell contents.

All these data show clearly that the salivary digestive enzymes of heteropterans play an important role in damaging plants: (1) degradation in the quality of wheat grains; (2) disturbance of the physiology and biochemistry of plant tissues surrounding the feeding puncture; and (3) direct formation of localized necrosis.

Recently, Miles and Taylor (1994) insisted that salivary sucrase, which operates the "osmotic pump," seems to be characteristic of coreids and is therefore more likely *a priori* to be the key factor associated with the characteristic lesions they produce.

Pectinase is an enzyme common in the Miridae and related closely with their macerate-and-flush feeding. Therefore, the enzyme may contribute to the appearance of the symptoms on attacked plants. Smith (1920) placed the salivary glands of *L. pubulinus* L. and *Plesiocoris rugicollis* (Fallén) on a potato slice and found that the tissue surrounding them was killed. Painter (1930) confirmed that the typical split lesions on cotton plant were caused by the injections of macerated cotton fleahopper, *Pseudatomoscelis seriatus* (Reuter). Pectinases are known to cause the disruption of the middle lamella of plant cell walls (Batemann and Miller 1966) and thus may be useful for softening and killing cells surrounding the injured area. Therefore, the death of the tissue on the potato slice and the lesions on cotton plant were probably caused by pectinases in the bugs (and presumably in the saliva). Strong and Kruitwagen (1968) found pectinase in the salivary gland of *L. hesperus* and considered that the bugs destructive capabilities are due primarily to the production of pectic enzymes and that the toxin produced by *Lygus* bugs and responsible for phytotoxic symptoms is a powerful pectic glycosidase(s). Strong (1970) concluded that the injury caused by *L. hesperus* is due principally to the enzymatic digestion of plant tissue by pectinase secreted during feeding.

The damage on sugar beets caused by the feeding of *L. rugulipennis* is accentuated by application of pectinase (Varis 1972). At least part of the damage caused by the cotton fleahopper is enzymatic in nature and could be caused partly or entirely by pectinase (Martin et al. 1988a). Moreover, this enzyme elicits ethylene production of cotton plants when injected into the tissue (Martin et al. 1988b, Burden et al. 1989). The injection of saliva containing pectinase into developing ovules and seeds of lucerne probably accounts for the etiology of damage caused by *Creontiades dilutus* (Stål) (Hori and Miles 1993). Also, Hori and Kishino (1992) inferred that pectinase injected into the flower tissues of lucerne with saliva during the feeding of *Adelphocoris suturalis* Jakovlev might be related to the withering (occurring within 3 days after the infestation) and abscission of flowers. *Lygus lineolaris* injures strawberry plants by causing the malformation of localized lesions on strawberry fruit. In early stages of fruit development, the achenes are destroyed mechanically by the stylets and chemically by the salivary pectinase, so the production of IAA and the translocation of IAA to the receptacle stop. As a result, the enlargement cells in the receptacle stops and the malformation occurs (Handley and Pollard 1993).

Apart from Miridae, there are two reports that, in Coreidae, pectinase may participate in the production of damage to plants. Epicarplesion of pistachio fruit caused by *Leptoglossus clypealis* Heidemann is suspected as being the result of an enzyme(s) injected into the fruit during feeding, as typical symptoms of the lesion are produced by applying commercial pectinase (Bolkan et al. 1984). Puncture wounds in cone scales caused by *L. occidentalis* Heidemann result from pectinmethyl-esterase activity in the saliva (Campbell and Shea 1990). However, the former authors could not detect pectinase in the salivary gland and the latter authors used ruthenium red to detect pectinase, which can give misleading results with nonenzymatic proteins and certain amino acids (P. W. Miles, personal communication). Therefore, these authors' presumption that the symptoms formed on plants by *Leptoglossus* spp. are caused by salivary pectinase should be reconsidered. I think that, as proved for other coreids (Miles and Taylor 1994), sucrase but not pectinase may participate in the production of the damage by these two coreids.

Hewett (1977) stated in his review that high levels of amino acids could be toxic in plants by disrupting fundamental enzyme systems. As mentioned above, the salivary glands of phytophagous heteropterans, as well as those of cecidogenic homopterans, contain many different amino acids. However, it is difficult to believe that insects' minute quantities of saliva can have any possible effect on plants, because the plants themselves contain so much more of the same amino compounds than is found in the salivary glands.

On the other hand, the amino acid pattern in sugar beet leaf is altered by an infestation of *Lygus rugulipennis* (= *L. disponsi*) (Hori 1973b). This alteration might disturb the metabolism of the plant tissue, which then could result in leaf malformation (Hori 1973b, 1975c).

5.3 Hormonal Mechanism

As mentioned above, auxin (or IAA) is not a main or an active component of heteropteran saliva, although it sometimes occurs in some species. Therefore, auxins from heteropterans usually cannot damage plants. There are a few exceptions. Nuorteva (1956b) confirmed that IAA is transferred from diet to saliva in *Stenodema calcaratum* Fallén (Miridae), and he assumed that the IAA may change the physiological state of host plants when injected. Scott (1970) found that plant growth from seed on which certain mirids had fed surpassed growth of plants from undamaged seed, and supposed that this might be due to a plant auxin injected into the seeds by the mirids, *L. hesperus* and *L. elisus*, during feeding. Recently, Martin et al. (1988a) found IAA and ACC (1-amino-cyclopropane-1-carboxylate) in the homogenate of *P. seriatus* (Miridae), and suggested that these two compounds directly affect ethylene production in plants (see below).

Hormonal balances in plants necessary for normal growth may be put out of order mechanically and/or chemically by the feeding of heteropterans. Miles (1989) stated that the components of saliva may release or spare the plant's own IAA. Strong (1970) reported that if feeding destroyed specific hormone-producing sites, the resultant damage was in accord with the known physiological responses of plants to hormonal manipulation. Hewett (1977) stated in his review that pest-induced changes in delicately balanced hormonal levels might be directly responsible for the observed reductions in photosynthesis and that any agent which can upset this balance may be responsible for abnormal growth.

Hormonal imbalance in plants may be also caused by auxin inhibitors and/or IAA synergists injected with the saliva while heteropterans feed. In fact, they may produce auxin inhibitor during feeding, because abscission caused by the feeding of the bugs is prevented by the application of α-naphthalene acetic acid (Carter 1962). Also, Allen (1947) stated that *Lygus* bugs injected a toxic auxin inhibitor into the plant tissues. Nuorteva (1956b) found plant growth-inhibiting substance(s), perhaps auxin inhibitor(s), in the salivary glands of *Miris dolabratus* (Miridae) and *D. baccarum* (Pentatomidae), and suggested that these inhibitors might disturb the physiological condition of the host plants. Moreover, the hypothesis that the malformations formed on plants by the feeding of heteropterans are partially due to the injection of auxin inhibitor and/or IAA synergist into the plants may also apply to the following statements:

1. The hypertrophy that occurs at the edges of the lesions of cocoa capsids (Miridae) must be due either to a gradient in concentration of substrates that cause necrosis at higher concentrations and increase growth at lower, or to differential rates of diffusion of necrosis-causing metabolites and growth-promoting ones (Miles 1972).
2. In the pumpkin fruit attacked by *L. rugulipennis* (Miridae), an IAA-oxidase inhibiting substance (IOIS) may penetrate the surrounding tissues from the feeding points with the saliva injected into those tissues, and may inhibit IAA-oxidase activity there. The quick decrease of IAA-oxidase within 3 days after infestation might be due to the spread of IOIS from the injured part. This, in turn, could stimulate the rise of IAA activity on the third and sixth day, when the cells surrounding the injured point swell and acid phosphatase activity necessary for energy metabolism in relation to cell hypertrophy rises (Hori et al. 1987).

According to Strong (1970) and Tingey and Pillemer (1977), increased rates in abscission of fruiting structures following feeding of *Lygus* bugs might be explained by destruction of meristematic tissues (IAA-producing site) in these organs, as auxin plays a major role in the prevention of abscission. Recently, the mechanism of abscission of cotton flower buds by the infestation of *P. seriatus* (Miridae) has been studied.

The cotton fleahopper induces stress ethylene production by detached, vegetative shot tips of young cotton plants, and the ethylene production rates reach abscission-inducing levels within 24 h and remain well in excess of threshold levels for 3 days (Grisham et al. 1987). The inducer (perhaps pectinase) seems to be in the saliva; salivary extracts are strong elicitors of stress ethylene production when injected into excised cotton shoot tips and the production also is induced by a solution of commercial pectinase (Martin et al. 1988a). On the other hand, IAA and ACC in the saliva can also contribute to abscission. When injected into cotton shoot tips at concentrations occurring in the bugs, IAA and ACC induced ethylene production to levels higher than in controls (Burden et al. 1989). But the IAA and ACC they detected was in extracts of whole insects. Therefore, it still needs to be shown that these auxins are contained in the watery saliva.

5.4 Plant and Insect Phenol–Phenoloxidase Systems in Relation to Damage

The phenol–phenoloxidase system (PPO system) of plants reacts to the feeding of heteropterans. Tissue damage from insect feeding initiates a series of wound-response reactions in the plant tissues, and in some circumstances (e.g., a hypersensitive reaction) phenols in the damaged and surrounding cells are oxidized to highly toxic quinones, which in plants is a nonspecific mechanism to resist invading organisms. The PPO system of the insect could participate in this process. These quinones may directly kill plant cells or they may transform tryptophan, by PPO-catalyzed oxidative deamination, to toxicogenic amounts of IAA (Miles 1969, Levin 1971, Norris 1979, Taylor and Miles 1994). Therefore, various deformations of plants could occur in this process. Then, the further oxidation of the quinones — a means by which some invading organisms have overcome this resistance — occurs to produce polymers and phenol–protein conjugates, which are nontoxic (brown melanin-like substances) (Miles 1969, Miles and Oertli 1993). Peng and Miles (1988a,b) confirmed experimentally that catechol oxidase contained in the saliva of an aphid, *Macrosiphum rosae* (L.), caused oxidative polymerization of catechin, which is an allelochemical in the buds of the rose. Because heteropterans have phenoloxidases in their saliva, they as well as aphids could cause oxidative polymerization of phenolics toxic to insects. Here, crops poor in quality (e.g., fruit with browned tissues) are produced. However, the situation is not so simple as mentioned above, because the stylet-sheath feeding method and the PPO system of insects participate in such a process in plants. The sheath could serve to reduce diffusion into the surrounding tissues of potentially necrosis-inducing phenols and quinones formed in the cells injured during formation of the stylet track. If the sheath is not formed, the phenols and quinones diffuse into the surrounding tissue (Miles 1990). In the former case, visible damage produced on plants is slight, whereas in the latter case it is severe. Because allelochemical phenolics (or quinones) are general biocides, their suppression by the insects (that is, by the PPO system) would also tend to preserve the plant tissues from hypersensitive responses; it would follow also that necrosis in mesophyll and cortical parenchyma could result from failure of the insect to repress completely the response of tissues to feeding. For the same reason, salivary components that in moderate quantities are innocuous or neutralize plant reactions might be able to cause toxic reactions if produced in excess (Miles 1990). These statements should be strengthened by the following rationale: Any rationale for such an interaction must presumably depend on the balance between oxidizing and reducing forces that occurs at the precise interface between plant and insect saliva. The plant response can be effective only if the plant can control any oxidation that occurs. In a well-balanced oxidation–reduction system, insect-deterrent phenolics will increase but necrosis-forming by-products will not be produced; under slightly more oxidizing conditions, hypersensitive necrosis may occur (perhaps due to quinones); under very rapidly oxidizing conditions, however, only local precipitation of polymeric phenolics and/or phenol–protein conjugates will occur. In the presence of peroxidase, moreover, even these may have no color and may simply condense on the walls of mechanically damaged cells or on the surface of any sheath material present. As a result of such rapid oxidation, the plant will not suffer necrosis nor will any insect deterrents accumulate. Even this analysis of the interaction of insect and plant oxidases ignores the possibility of oxidative reactions in regions beyond the immediate insect–plant interface, due either to diffusing insect saliva or, and perhaps more importantly, to diffusion of the plants own wound-reaction-inducing factors. What happens, therefore, will depend on the balance of phenolase and peroxidase within and between the insect saliva and the particular plant tissues that are being attacked, and also on any substances released into surrounding tissues (P. W. Miles, personal communication).

In relation to the malformation of sugar beet leaf formed by the infestation of *L. rugulipennis*, Hori (1973a, 1974b, 1975d) hypothesized that the causes of the damage are, in the first place, the particular feeding site (intrafascicular cambium in petiole and meristematic tissue in leaf vein) and the mechanical destruction of the tissue; second, the further destruction of the tissue by pectinase in the saliva; third, the factors that continuously keep polyphenoloxidase and peroxidase activities in the injured and surrounding tissues at high levels; and fourth, the long-lasting presence of quinones at a constant level in the injured and neighboring tissues. Presumably, some components of saliva injected into host plant tissue during feeding of the bug must spread from the point of injection into the surrounding tissues. Thus, it may be assumed that, in the injured part, quinones are produced constantly for a long time to destroy IAA or other auxins and inhibit growth, while in the surrounding tissues the substances promoting the activity of IAA or the substances inhibiting the activity IAA oxidase (IAA synergist) spread to promote growth. This imbalance of growth may possibly produce various types of malformations on the sugar beet leaf. However, several other experiments must be performed to strengthen this hypothesis, as pointed out by Tingey and Pillemer (1977). The role of quinones to form complex malformations on plants may be introduced by the following two reports:

1. The damage occurring in the leaf tissue of cabbage by the attack of *Eurydema rugosum* (Pentatomidae) is simple, but necrosis and malformation by the hypersensitive reaction do not result, because only low levels of quinones are produced in the injured tissues (Hori 1974c).
2. Not enough quinones are produced in the tissue of Chinese cabbage injured by the bug *L. rugulipennis* to account for the apparent chlorotic spots and/or malformation on the leaf (Hori and Atalay 1980).

6. CONCLUSION

Mentioned were possible causes of the damage to host plants caused by the feeding of phytophagous heteropterans; the factors were divided into four categories. However, a symptom of damage cannot be produced by only one factor but is produced by a complex combination of various factors, although the degree of importance of each factor differs with different types of damage. Figure 1 shows the possible relationships among the factors that induce damage.

Mechanical injury (rupture of cells by the stylets) is the first step in the incidence of the damage of host plant. Surely, a main cause of damage is mechanical destruction of cells by the stylets, as in the case of *E. rugosum* feeding on cabbage leaf (Hori 1968b). Even in this case, however, enzymes in the saliva may contribute to the formation of the white spots. Moreover, as pointed out by Varis (1972), the mechanical damage in the growing point of sugar beet caused by *L. rugulipennis* is accentuated by pectinase. It should be stressed that the reactivity of the particular plant tissues is a major factor in the effects of mechanical damage; some tissues of plants are very sensitive but most barely react at all (e.g., pricking meristematic tissues in the veins of some leaves can cause local swelling, yet the withdrawal of cell contents in mature tissues does no more than cause silver spots because of the replacement of cell contents with air). Chemical injury (degradation of cells by enzymes) is the step after mechanical injury. Of five factors related to the second step, three (phenoloxidase, auxins, and auxin synergists in the saliva) may not themselves cause tissue destruction but go directly to the third step. Although other factors (pectinase and digestive enzymes in the saliva) also go to the third step in the formation of the symptoms, they may also cause damage at this step. Pectinase macerates plant cells surrounding the stylet track and could cause localized necrosis in tissues. Digestive enzymes hydrolyze starch and protein in the plant tissue and could alter vegetative growth. Moreover, a sucrase in the saliva of coreids can function to unload cytoplasmic content from the cells surrounding the feeding point and cause localized necrosis.

The third step includes hormonal and physiological imbalances in the plant caused by the heteropterans, and the operation of plant PPO system triggered by the insect attack. As shown in Figure 1, some factors in the first and second steps can trigger the third step, followed by complex physiological and biochemical processes that cause tissue and fruit deformations and abscission of fruits and other reproductive structures.

The possible relationships shown in Figure 1 are a hypothesis. Much work is needed to confirm the various relations among contributing factors, and thus to explain the biochemical processes in detail.

7. ACKNOWLEDGMENT

The author expresses sincere thanks to P. W. Miles in Adelaide, South Australia, for his valuable comments and suggestions for improving the manuscript. Without his help, this review could not have been completed.

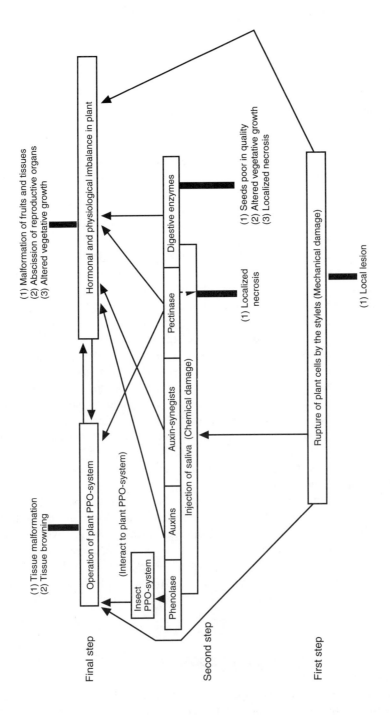

Figure 1 Possible relationship among various factors that induce the damage and resultant symptoms from the feeding of phytophagous Heteroptera.

8. REFERENCES CITED

Adams, J. B., and J. V. McAllan. 1958. Pectinase in certain insects. Canad. J. Zool. 36: 305–308.
Addicott, F. T., and V. E. Romney. 1950. Anatomical effects of Lygus injury to guayule. Bot. Gaz. 112: 133–134.
Allen, T. C. 1947. Suppression of insect damage by means of plant hormones. J. Econ. Entomol. 40: 814–817.
1951. Deformities caused by insects. Pp. 411–415 in F. Skoog [ed.], Plant Growth Substances. Univ. Wisconsin Press, Madison, Wisconsin, U.S.A. 476 pp.
Atanassova, E., and V. Popova. 1968. On some alterations in the composition of wheat grain damaged by wheat bugs (*Eurygaster* and *Aelia*). Plant Sci. 5: 81–87.
Baker, K. F., W. C. Snyder, and A. H. Holland. 1946. Lygus bug injury of lima bean in California. Phytopathology 36: 494–503.
Baptist, B. A. 1941. The morphology and physiology of the salivary glands of Hemiptera-Heteroptera. Q. J. Microsc. Sci. 83: 91–139.
Batemann, D. F., and R. L. Miller. 1966. Pectic enzymes in tissue degradation. Annu. Rev. Phytopathol. 4: 119–146.
Beel, C., and D. Feir. 1977. Effect of juvenile hormone on acid phosphatase activity in six tissues of the milkweed bug. J. Insect Physiol. 23: 761–763.
Bolkan, H. A., J. M. Ogawa, R. E. Rice, R. M. Bostock, and J. C. Crane. 1984. Leaffooted bug (Hemiptera: Coreidae) and epicarp lesion of pistachio fruits. J. Econ. Entomol. 77: 1163–1165.
Bongers, J. 1970. Die Carbohydrasen und Esterasen in Speicheldrusen und Mitteldarm von *Oncopeltus fasciatus* Dall. (Heteroptera: Lygaeidae). Z. Vergl. Physiol. 70: 382–400.
Bongers, J., and H.-M. Muermann. 1973. Einfluss von Nahrungsmangel und Kohlenhydratlosungen auf die Sekretion der Verdauungsenzyme und die Lebensdauer von *Oncopeltus fasciatus*. J. Insect Physiol. 19: 1555–1568.
Brewer, P. S., W. F. Campbell, and B. A. Haws. 1979. How black grass bugs operate. Utah Sci. 22: 21–24.
Bronskill, J. F., E. H. Salkeld, and W. G. Friend. 1958. Anatomy, histology, and secretions of salivary glands of the large milkweed bug, *Oncopeltus fasciatus* (Dallas) (Hemiptera: Lygaeidae). Canad. J. Zool. 36: 961–968.
Burden, B. J., P. W. Morgan, and W. L. Sterling. 1989. Indole-acetic acid and the ethylene precursor, ACC, in the cotton fleahopper (Hemiptera: Miridae) and their role in cotton abscission. Ann. Entomol. Soc. Amer. 82: 476–480.
Campbell, B. C., and P. J. Shea. 1990. A simple staining technique for assessing feeding damage by *Leptoglossus occidentalis* Heidemann (Hemiptera: Coreidae) on cones. Canad. Entomol. 122: 963–968.
Carter, W. 1939. Injuries to plants caused by insect toxins. Bot. Rev. 5: 273–326.
1962. Insects in Relation to Plant Disease. Inter-science Publisher, London, U.K. 705 pp.
Choudhuri, D. K., K. K. Das, and G. De. 1966. The pH and enzymes of the digestive system of *Chrysocoris stolli* Wolf. (Pentatomodae: Heteroptera). Proc. Natl. Acad. Sci. India, Sect. (B): Biol. Sci. 36: 281–284.
Courtois, J. E., F. Percheron, and M. J. Foglietti. 1968. Isolement étude du mode d'action d'une polygalacturonase d'insecte (*Pyrrhocoris apterus*). C.R. Hebd. Seanc. Acad. Sci. (Paris) 266D: 164–166.
Cressey, P. J. 1987. Wheat-bug damage in New Zealand wheats: some properties of a glutenin hydrolysing enzyme in bug-damaged wheat. J. Sci. Food Agric. 41: 159–165.
Cressey, P. J., and C. L. McStay. 1984. Wheat bug damage in New Zealand wheats. Development of a simple SDS-sedimentation test for bug damage. J. Sci. Food Agric. 38: 357–366.
Cressey, P. J., J. A. K. Farrell, and M. W. Stufkens. 1987. Identification of an insect species responsible for bug damage in New Zealand wheat. New Zealand J. Agric. Res. 30: 209–212.
Dadd, R. H. 1970. Digestion in insects. Pp. 117–145 in M. Florkin and B. T. Scheer [eds.], Chemical Zoology, Vol. 5. Academic Press, London, U.K. 460 pp.
Dearman, A. F. 1960. Salivary enzymes of *Nezara viridula* (L.) with reference to the tomato substrate. M.S. thesis, Louisiana State University, Baton Rouge, Louisiana, U.S.A.
Duspiva, F. 1952. Der Kohlenhydratumsatz im Verdauungstrakt der Rhynchoten: ein Beitrag zum Problem der stofflichen Wechselbeziehungen zwischen saugenden Insekten und ihren Wirstpflanzen. Mitt. Biol. Zent. Anst. Berl. 75: 82–89.

Eggermann, W., and J. Bongers. 1980. Die Bedeutung des wasserigen Speichelsekrets für die Nahrungsaufnahme von *Oncopeltus fasciatus* und *Dysdercus fasciatus*. Entomol. Exp. Appl. 27: 169–178.

Every, D., J. A. K. Farrell, and M. W. Stufkens. 1990. Wheat-bug damage in New Zealand wheats: the feeding mechanism of *Nysius huttoni* and its effect on the morphology and physiological development of wheat. J. Sci. Food Agric. 50: 297–309.

1992. Bug damage in New Zealand wheat grain: the role of various heteropterous insects. New Zealand J. Crop Hortic. Sci. 20: 305–312.

Feir, D., and S. D. Beck. 1961. Salivary secretions of *Oncopeltus fasciatus* (Dallas) (Hemiptera: Lygaeidae). Ann. Entomol. Soc. Amer. 54: 316.

Fisher, E. H., A. J. Riker, and T. C. Allen. 1946. Bud, blossom and pod drop of canning string beans reduced by plant hormones. Phytopathology 36: 504–523.

Flemion, F. 1955. Penetration and destruction of plant tissues during feeding by *Lygus lineolaris*. Pp. 1003–1007 *in* Proceedings, 14th Int. Hortic. Congr., The Hague, the Netherlands.

1958. Penetration and destruction of plant tissues during feeding by *Lygus lineolaris* P. de B. Pp. 475–478 *in* Proceedings, 10th Int. Congr. Entomol., Montreal, Canada.

Flemion, F., M. C. Ledbetter, and E. S. Kelley. 1954. Penetration and damage of plant tissues during feeding by the tarnished plant bug (*Lygus lineolaris*). Contrib. Boyce Thompson Inst. 17: 347–357.

Flemion, F., L. P. Miller, and R. M. Weed. 1951a. An estimation of the quantity of oral secretion deposited by *Lygus* when feeding on bean tissue. Contrib. Boyce Thompson Inst. 16: 420–433.

Flemion, F., R. M. Weed, and L. P. Miller. 1951b. Deposition of P32 into host tissue through the oral secretions of *Lygus oblineatus*. Contrib. Boyce Thompson Inst. 16: 285–294.

Ford, J. B. 1962. Studies on the digestive processes of *Dysdercus fasciatus* Sign. Ann. Appl. Biol. 50: 355.

Goodchild, A. J. P. 1952. A study of the digestive system of the West African cacao capsid bugs (Hemiptera: Miridae). Proc. Zool. Soc. London 122: 543–572.

Gopalan, M. 1975. Studies on feeding behaviour and salivary secretions of *Ragmus importunitas* Distant (Hemiptera: Miridae) and its influence on the physiology of sunn-hemp, *Crotalaria juncea* L. Ph.D. dissertation, Tamil Nadu Agricultural University, Coimbatore, India.

1976. Studies on salivary enzymes of *Ragmus importunitas* Distant (Hemiptera: Miridae). Curr. Sci. 45: 188–189.

Gopalan, M., and T. R. Subramaniam. 1977. Effect of infestation of *Ragmus importunitas* Distant (Hemiptera: Miridae) on respiration, transpiration, moisture content and oxidative enzymes activity in sunn-hemp plants (*Crotalaria juncea* L.). Curr. Sci. 47: 131–134.

Grisham, M. P., W. L. Sterling, R. D. Powell, and P. W. Morgan. 1987. Characterization of the induction of stress ethylene synthesis in cotton caused by the cotton fleahopper (Hemiptera: Miridae). Ann. Entomol. Soc. Amer. 80: 411–416.

Handley, D. T., and J. E. Pollard. 1993. Microscopic examination of tarnished plant bug (Heteroptera: Miridae) feeding damage to strawberry. J. Econ. Entomol. 86: 505–510.

Hansen, J. D., and R. S. Nowak. 1985. Evaluating damage by grass bug feeding on crested wheatgrass. Southwest. Entomol. 10: 89–94.

Heliovaara, K., and S. Laurema. 1988. Interaction of *Aradus cinnamomeus* (Heteroptera, Aradidae) with Pinus sylvestris: the role of free amino acids. Scand. J. For. Res. 3: 515–525.

Hewett, E. W. 1977. Some effects of infestation on plants: a physiological viewpoint. New Zealand Entomol. 6: 235–243.

Holopainen, J. K. 1986. Damage caused by *Lygus rugulipennis* Popp. (Heteroptera, Miridae), to *Pinus sylvestris* L. seedlings. Scand. J. For. Res. 1: 343–349.

Hori, K. 1968a. Some properties and developmental changes in occurrence of the salivary amylase of the cabbage bug, *Eurydema rugosum* Motschulsky (Hemiptera: Pentatomidae). Appl. Entomol. Zool. 3: 198–202.

1968b. Feeding behavior of the cabbage bug, *Eurydema rugosum* Motschulsky (Hemiptera: Pentatomidae) on the cruciferous plants. Appl. Entomol. Zool. 3: 26–36.

1969. Some properties of salivary amylases of *Adelphocoris suturalis* (Miridae), *Dolycoris baccarum* L. (Pentatomidae), and several other heteropteran species. Entomol. Exp. Appl. 12: 454–466.

1970a. Some properties of amylase in the salivary gland of *Lygus disponsi* (Hemiptera). J. Insect Physiol. 16: 373–386.

1970b. Some variation in the activities of salivary amylase and protease of *Lygus disponsi* Linnavuori (Hemiptera: Miridae). Appl. Entomol. Zool. 5: 51–61.
1971a. Studies on the feeding habits of *Lygus disponsi* Linnavuori (Hemiptera: Miridae) and the injury to its host plants. I. Histological observations of the injury. Appl. Entomol. Zool. 6: 84–90.
1971b. Physiological conditions in the midgut in relation to starch digestion and the salivary amylase of the bug *Lygus disponsi*. J. Insect Physiol. 17: 1153–1167.
1972. Comparative study of a property of salivary amylase among various heteropterous insects. Comp. Biochem. Physiol. 42B: 501–508.
1973a. Studies on the feeding habits of *Lygus disponsi* Linnavuori (Hemiptera: Miridae) and the injury to its host plant. III. Phenolic compounds, acid phosphatase and oxidative enzymes in the injured tissue of sugar beet leaf. Appl. Entomol. Zool. 8: 103–112.
1973b. Studies on the feeding habits of *Lygus disponsi* Linnavuori (Hemiptera: Miridae) and the injury to its host plant. IV. Amino acids and sugars in the injured tissue of sugar beet leaf. Appl. Entomol. Zool. 8: 138–142.
1974a. Enzymes in the salivary gland of *Lygus disponsi* Linnavuori (Hemiptera, Miridae). Res. Bull. Obihiro Univ. 8: 461–464.
1974b. Plant growth-promoting factor in the salivary gland of the bug, *Lygus disponsi*. J. Insect Physiol. 20: 1623–1627.
1974c. Chloropyll, phenol compounds, acid phosphatase and oxidative enzymes in the leaf tissue of cabbage injured by the cabbage bug, *Eurydema rugosum* Motschulsky (Hemiptera: Pentatomidae). Appl. Entomol. Zool. 9: 1–10.
1975a. Digestive carbohydrases in the salivary gland and midgut of several phytophagous bugs. Comp. Biochem. Physiol. 50B: 145–151.
1975b. Pectinase and plant growth-promoting factors in the salivary glands of the larva of the bug, *Lygus disponsi*. J. Insect Physiol. 21: 1271–1274.
1975c. Studies on the feeding habits of *Lygus disponsi* Linnavuori (Hemiptera: Miridae) and the injury to its host plant. VI. Amino acids in the artificially injured tissue of sugar beet leaf and in the cabbage leaf following attack by the cabbage bug *Eurydema rugosum* Motschulsky (Hemiptera: Pentatomidae). Appl. Entomol. Zool. 10: 203–207.
1975d. Plant growth-regulating factor, substances reacting with Salkovski reagent and phenoloxidase activities in vein tissue injured by *Lygus disponsi* Linnavuori (Hemiptera: Miridae) and surrounding mesophyll tissues of sugar beet leaf. Appl. Entomol. Zool. 10: 130–135.
1976. Plant growth-regulating factor in the salivary gland of several heteropterous insects. Comp. Biochem. Physiol. 53B: 435–438.
1979. Metabolism of ingested indole-3-acetic acid in the gut of various heteropterous insects. Appl. Entomol. Zool. 14: 149–158.
1981. Comparison of metabolic features of indole-3-acetic acid in the gut among five species of heteropterous insects. Appl. Entomol. Zool. 16: 503–504.
Hori, K., and R. Atalay. 1980. Biochemical changes in the tissue of Chinese cabbage injured by the bug *Lygus disponsi*. Appl. Entomol. Zool. 15: 234–241.
Hori, K., and M. Endo. 1977. Metabolism of ingested auxins in the bug *Lygus disponsi*: conversion of indole-3-acetic acid and gibberellin. J. Insect Physiol. 23: 1075–1080.
Hori, K., and M. Kishino. 1992. Feeding of *Adelphocoris suturalis* Jakovlev (Heteroptera: Miridae) on alfalfa plants and damage caused by the feeding. Res. Bull. Obihiro Univ. 17: 357–365.
Hori, K., and P. W. Miles. 1977. Multiple plant growth-promoting factors in the salivary glands of plant bugs. Marcellia 39: 399–400.
1993. The etiology of damage to lucerne by the green mirid, *Creontiades dilutus* (Stål). Austral. J. Exp. Agric. 33: 327–331.
Hori, K., Y. Kondo, and K. Kuramochi. 1984. Feeding site of *Palomena angulosa* Motschulsky (Hemiptera: Pentatomidae) on potato plants and injury caused by the feeding. Appl. Entomol. Zool. 19: 476–482.
Hori, K., D. R. Singh, and A. Sugitani. 1979. Metabolism of injested indole compounds in the gut of three species of Heteroptera. Comp. Biochem. Physiol. 64C: 217–222.
Hori, K., H. Torikura, and M. Kumagai. 1987. Histological and biochemical changes in the tissue of pumpkin fruit injured by *Lygus disponsi* Linnavuori (Hemiptera: Miridae). Appl. Entomol. Zool. 22: 259–265.

Janda, V., Jr., and M. Munzarova. 1980. Activity of digestive amylase and invertase in relation to development and reproduction of *Dysdercus koenigii* (Heteroptera). Acta Entomol. Bohemoslov. 77: 209–215.

1982. Trehalase activity in the tissue of *Dysdercus koenigii* (Heteroptera) during development and reproduction. Acta Entomol. Bohemoslov. 79: 1–6.

Jeppson, L. R., and G. F. MacLeod. 1946. Lygus bug injury and its effect on the growth of alfalfa. Hilgardia 17: 165–188.

Johnson, C. G. 1936. The biology of *Leptobyrsa rhododendri* Horváth (Hemiptera, Tingidae), the rhododendron lacebug. Ann. Appl. Biol. 24: 342–355.

Johnson, S. R., and A. K. Knapp. 1996. Impact of *Ischnodemus falicus* (Hemiptera: Lygaeidae) on photosynthesis and production of *Spartina pectinata* Wetlands. Environ. Entomol. 25: 1122–1127.

Khan, K. A., and M. A. Khan. 1977. Chlorides as activators and inhibitors of salivary invertase in *Periplaneta americana* Linn. (Dictyoptera: Blattidae), Dysdercus cingulatus Fabr. (Heteroptera: Pyrrhocoridae) and *Leptocorisa varicornis* Fabr. (Heteroptera: Lygaeidae [sic]). J. Entomol. Res. 1: 136–139.

Khan, M. A. 1962a. The secretary site of invertase and amylase in the digestive organs of *Dysdercus fasciatus* Dallas. Indian J. Entomol. 24: 142–143.

1962b. The distribution of dipeptidase activity in the digestive system of *Locust migratoria* L. and *Dysdercus fasciatus* Dallas. Comp. Biochem. Physiol. 6: 169–170.

Khan, M. R. 1974. Histochemical localization of the digestive enzymes in the alimentary canal of *Dysdercus fasciatus* Sign. (Heteroptera: Pyrrhocoridae). Pakistan J. Zool. 6: 13–16.

Khan, M. R., and J. B. Ford. 1967. The distribution and localization of digestive enzymes in the alimentary canal and salivary glands of the cotton stainer, *Dysdercus fasciatus*. J. Insect Physiol. 13: 1619–1628.

Kloft, W. 1960. Wechselwirkungen zwischen pflanzensaugenden Insekten und den von ihnen besogenen Pflanzengeweben. Teil I. Z. Angew. Entomol. 45: 337–381.

Kretovich, V. L. 1944. Biochemistry of the damage to grain by the wheat-bug. Cereal Chem. 21: 1–21.

Kretovich, V. L., A. A. Bundel, and K. V. Pshenova. 1943. Mechanism of wheat injury by *Eurygaster integriceps*. C.R. Acad. Sci. U.R.S.S. 39: 31–33.

Kumar, D., A. Ray, and P. S. Ramamurty. 1978. Histophysiology of the salivary glands of the red cotton bug, *Dysdercus koenigii* (Pyrrhocoridae-Heteroptera) — histological, histochemical, autoradiographic and electronmicroscopic studies. Z. Mikrosk. Anat. Forsch. 92: 147–170.

1980. Studies on the salivary glands of *Lygaeus* sp. (Lygaeidae — Heteroptera) — histological, histochemical, autoradiographic and electron microscopic investigations. Z. Mikrosk. Anat. Forsch. 94: 669–695.

Kumar, R. 1970. Occurrence of proteases in the salivary glands of cocoa capsids (Heteroptera: Miridae). J. New York Entomol. Soc. 78: 198–200.

Laurema, S., and P. Nuorteva. 1961. On the occurrence of pectin polygalacturonase in the salivary glands of Heteroptera and Homoptera Auchenorrhyncha. Ann. Entomol. Fenn. 27: 89–93.

Laurema, S., and A.-L. Varis. 1991. Salivary amino acids in *Lygus* species (Heteroptera: Miridae). Insect Biochem. 21: 759–765.

Laurema, S., A.-L. Varis, and H. Miettinen. 1985. Studies on enzymes in the salivary glands of *Lygus rugulipennis* (Hemiptera, Miridae). Insect Biochem. 15: 211–224.

Leach, R., and C. Smee. 1933. Gnarled stem canker of tea caused by the capsid bug (*Helopeltis bergrothi* Reut.). Ann. Appl. Biol. 20: 691–706.

Levin, D. A. 1971. Plant phenolics: an ecological perspective. Amer. Nat. 105: 157–181.

Lodos, N. 1967. Contribution to the biology of and damage caused by the cocoa coreid, *Pseudotheraptus devastans* Dist. (Hemiptera-Coreidae). Ghana J. Sci. 7: 87–102.

Mall, S. B., and A. D. Chattoraj. 1968. Hydrogen-ion concentration and the digestive enzymes in *Cletus signatus* Walker (Heteroptera: Coreidae). Indian J. Entomol. 30: 154–162.

Martin, W. R., Jr., P. W. Morgan, W. L. Sterling, and R. W. Meola. 1988a. Stimulation of ethylene production in cotton by salivary enzymes of the cotton fleahopper (Heteroptera: Miridae). Environ. Entomol. 17: 930–935.

Martin, W. R., Jr., P. W. Morgan, W. L. Sterling, and C. M. Kenerley. 1988b. Cotton fleahopper and associated microorganisms as components in the production of stress ethylene by cotton. Plant Physiol. 87: 280–285.

Mathur, L. M. L., and A. V. Thakar. 1969. Digestive physiology of Tur pod bug, *Clavigralla gibbosa* Spin. (Coreidae: Heteroptera). Indian J. Entomol. 31: 251–257.

Miles, P. W. 1959. The salivary secretions of a plant-sucking bug, *Oncopeltus fasciatus* (Dall.) (Heteroptera: Lygaeidae). J. Insect Physiol. 3: 243–255.

1960. The salivary secretions of a plant sucking bug, *Oncopeltus fasciatus* (Dall.) (Heteroptera: Lygaeidae). III. Origins in the salivary glands. J. Insect Physiol. 4: 271–283.

1964a. Studies on the salivary physiology of plant bugs: oxidase activity in the salivary apparatus and saliva. J. Insect Physiol. 10: 121–129.

1964b. Studies on the salivary physiology of plant bugs: the chemistry of formation of the sheath material. J. Insect Physiol. 10: 147–160.

1967a. The physiological division of labour in the salivary glands of *Oncopeltus fasciatus* (Dall.) (Heteroptera: Lygaeidae). Austral. J. Biol. Sci. 20: 785–797.

1967b. Studies on the salivary physiology of plant bugs: transport from haemolymph to saliva. J. Insect Physiol. 13: 1787–1801.

1968. Insect secretion in plants. Annu. Rev. Phytopathol. 6: 137–164.

1969. Interaction of plant phenols and salivary phenolases in the relationship between plants and Hemiptera. Entomol. Exp. Appl. 12: 736–744.

1972. The saliva of Hemiptera. Adv. Insect Physiol. 9: 183–255.

1987. Plant-sucking bugs can remove the content of cells without mechanical damage. Experientia 43: 937–939.

1989. The responses of plants to the feeding of Aphidoidea: principles. Pp. 1–21 *in* A. K. Minks and P. Harrewijn [eds.], Aphids. Their Biology, Natural Enemies and Control, Vol. C. Elsevier Science Publishers, Amsterdam, the Netherlands.

1990. Aphid salivary secretion and their involvement in plant toxicoses. Pp. 131–147 *in* R. K. Campbell and R. D. Eikenburry [eds.], Aphid–Plant Genotype Interactions. Elsevier Science Publishers, New York, New York, U.S.A. 378 pp.

Miles, P. W., and A. Helliwell. 1961. Oxidase activity in the saliva of a plant-bug. Nature 192: 374–375.

Miles, P. W., and K. Hori. 1977. Fate of ingested α-indolyl acetic acid in Creontiades dilutus. J. Insect Physiol. 23: 221–226.

Miles, P. W., and J. Lloyd. 1967. Synthesis of a plant hormone by the salivary apparatus of plant sucking Hemiptera. Nature 25: 801–802.

Miles, P. W., and J. J. Oertli. 1993. The significance of antioxidants in the aphid-plant interaction: the redox hypothesis. Entomol. Exp. Appl. 67: 275–283.

Miles, P. W., and D. Sloviak. 1970. Transport of whole protein molecules from blood to saliva of a plant-bug. Experientia 26: 611.

Miles, P. W., and G. S. Taylor. 1994. "Osmotic pump" feeding by coreids. Entomol. Exp. Appl. 73: 163–173.

Muimba-Kankolongo, A., E. R. Terry, and M. O. Adeniji. 1987. A disease-like injury on cassava caused by *Pseudotheraptus devastans* Dist. (Heteroptera: Coreidae). Trop. Pest Manage. 33: 35–38.

Naqvi, S. N. H., I. Ahmad, F. Yasmeen, and S. A. Khan. 1973. Digestive physiology with reference to enzymes in the alimentary canal of *Coridius janus* Fabr. (Heteroptera: Dinidoridae). Folia Biol. 21: 209–221.

Neal, J. J. 1993. Xylem transport interruption by *Anasa tristis* feeding causes Cucurbita pepo to wilt. Entomol. Exp. Appl. 96: 195–200.

Norris, D. M. 1979. How insects induce disease. Pp. 239–255 *in* J. G. Horsfall and E. B. Cowling [eds.], Plant Disease, Vol. IV. Academic Press, New York, New York, U.S.A. 466 pp.

Nuorteva, P. 1953. Die Bedeutung mechanischer Schädigung des Weizenkorns durch Wanzen für das Korn und für die Backfähigkeit des Mehles. Ann. Entomol. Fenn. 19: 29–33.

1954. Studies on the salivary enzymes of some bugs injuring wheat kernels. Ann. Entomol. Fenn. 20: 102–124.

1956a. Developmental changes in the occurrence of salivary proteases in *Miris dolabratus* L. (Hem., Miridae). Ann. Entomol. Fenn. 22: 117–119.

1956b. Studies on the effect of the salivary secretions of some Heteroptera and Homoptera on plant growth. Ann. Entomol. Fenn. 22: 108–117.

1958. Die Rolle der Speichelsekrete im Wechselverhaltnis zwischen Tier und Nahrungspflanze bei Homopteren und Heteropteren. Entomol. Exp. Appl. 1: 41–49.

Nuorteva, P., and S. Laurema. 1961a. Observations on the activity of salivary proteases and amylases in *Dolycoris baccarum* (L.) (Het., Pentatomidae). Ann. Entomol. Fenn. 27: 93–97.

1961b. The effect of diet on the amino acids in the haemolymph and salivary glands of Heteroptera. Ann. Entomol. Fenn. 27: 57–65.

Nuorteva, P., and L. Reinius. 1953. Incorporation and spread of C14-labeled oral secretions of wheat bugs in wheat kernels. Ann. Entomol. Fenn. 19: 95–104.

Owusu-Manu, E. 1990. Feeding behaviour and the damage caused by *Bathycoelia thalassina* (Herrich-Schaeffer) (Hemiptera: Pentatomidae). Café Cacao Thé 34: 97–104.

Painter, R. H. 1928. Notes on the injury to plant cells by chinch bug feeding. Ann. Entomol. Soc. Amer. 21: 232–242.

1930. A study of the cotton flea-hopper *Psallus seriatus* Reut., with special reference to its effect on cotton plant tissues. J. Agric. Res. 40: 485–516.

Peng, Z., and P. W. Miles. 1988a. Studies on the salivary physiology of plant bugs: function of the catechol oxidase of the rose aphid. J. Insect Physiol. 34: 1027–1033.

1988b. Acceptability of catechin and its oxidative condensation products to the rose ahid, *Macrosiphum rosae*. Entomol. Exp. Appl. 47: 255–265.

Pokrovskaya, N. F., G. I. Morozova, and N. M. Vinogradova. 1971. Proteins of *Eurygaster integriceps* Put. damaged wheat grain. Appl. Biochem. Microbiol. 7: 121–127 [in Russian with English summary].

Poteri, M., R. Heikkila, and L. Yuan-Yi. 1987. Development of the growth disturbance caused by *Lygus rugulipennis* in one-year-old pine seedlings. Folia For. 695: 1–14.

Quayum, M. A. 1966. Studies on the salivary glands of *Dysdercus cingulatus* Fabr. (Hemiptera). Pakistan J. Sci. Res. 18: 111–117.

Raman, K., and K. P. Sanjayan. 1984. Histology and histopathology of the feeding lesions by *Cyrtopeltis tenuis* Reut. (Hemiptera: Miridae) on *Lycopersicon esculentum* Mill. (Solanaceae). Proc. Indian Acad. Sci. 93: 543–547.

Raman, K., K. P. Sanjayan, and G. Suresh. 1984. Impact of feeding injury of *Cyrtopeltis tenuis* Reut. (Hemiptera: Miridae) on some biochemical changes in *Lycopersicon esculentum* Mill. (Solanaceae). Curr. Sci. 53: 1092–1093.

Rastogi, S. C. 1962. On the salivary enzymes of some phytophagous and predaceous Heteroptera. Sci. Cult. 28: 479–480.

Rastogi, S. C., and A. K. Datta Gupta. 1961. Studies on the physiology of digestion in the alimentary canal and salivary glands of *Odontopus nigricornis* Stål (Heteroptera: Pyrrhocoridae). Indian J. Entomol. 23: 106–115.

1962. The pH and the digestive enzymes of the red pumpkin bug, *Coridius janus* (Fabr.) (Heteroptera: Dinidoridae). Proc. Zool. Soc. 15: 57–64.

Rautapaa, J. 1969. Effect of *Lygus rugulipennis* Popp. (Hem., Capsidae) on the yield and quality of wheat. Ann. Entomol. Fenn. 35: 168–175.

Salkeld, E. H. 1960. A histochemical studies on localization and distribution of esterases in the salivary glands of the large milkweed bug, *Oncopeltus fasciatus* (Dallas) (Hemiptera: Lygaeidae). Canad. J. Zool. 37: 113–115.

Sapio, F. J., L. F. Wilson, and M. E. Ostry. 1982. A split-stem lesion on young hybrid *Populus* trees caused by the tarnished plant bug, *Lygus lineolaris* (Hemiptera [Heteroptera]: Miridae). Great Lakes Entomol. 15: 237–246.

Saxena, K. N. 1954a. Physiology of the alimentary canal of *Leptocorisa varicornis* Fabr. (Hemiptera: Coreidae). J. Zool. Soc. India 6: 111–122.

1954b. Physiology of digestion in *Leptocorisa varicornis* Fabr. (Hemiptera: Coreidae). Curr. Sci. 23: 132.

1955. Studies on the passage of food, hydrogenion concentration and enzymes in the gut and salivary glands of *Dysdercus koenigii* Fabr. (Pyrrhocoridae: Heteroptera). J. Zool. Soc. India 7: 145–154.

Saxena, K. N., and P. Bhatnagar. 1958. Physiological adaptations of dusky cotton bug, *Oxycarenus hyalinipennis* (Costa) (Heteroptera: Lygaeidae) to its host plant cotton. I. Digestive enzymes in relation to tissue preferences. Proc. Natl. Inst. Sci. India 24B: 245–257.

Scott, D. R. 1970. Feeding of *Lygus* bugs (Hemiptera: Miridae) on developing carrot and bean seed: increased growth and yields of plants grown from that seed. Ann. Entomol. Soc. Amer. 63: 1604–1608.

Slepyan, E. I., N. A. Vilkova, and I. D. Shapiro. 1972. Pathological changes in the structure of developing soft wheat grains infested with *Eurygaster integriceps* Put., depending on the peculiarities of its trophical regime in the process of nutrition. Izvest. Akad. Nauk. SSSR 2: 258–262 [in Russian with English summary].

Smith, K. M. 1920. Investigation of the nature and cause of the damage to plant tissue resulting from the feeding of capsid bugs. Ann. Appl. Biol. 7: 40–55.

Somermaa, K. 1961. Untersuchungen über die "Bollnaser Krankheit." III. Studien über die "Trübe Feldwanze" *Lygus rugulipennis*. Stat. Växtskyddsanstalt Medd. (Stockholm) 12 (86): 79–93.

Sossedov, N. I., N. N. Gurova, and Z. B. Drozdova. 1969. Comparative proteolytic activity of *Eurygaster integriceps* (Hemiptera: Pentatomidae) and Aelia (Pentatomidae) in infested wheat. Prikl. Biokhim. Mikrobiol. 5: 318–323 [in Russian with English summary].

Strong, F. E. 1970. Physiology of injury caused by *Lygus hesperus*. J. Econ. Entomol. 63: 808–814.

Strong, F. E., and E. C. Kruitwagen. 1968. Polygalacturonase in the salivary apparatus of *Lygus hesperus* (Hemiptera). J. Insect Physiol. 14: 1113–1119.

Summerfield, R. J., F. J. Muehlbauer, and R. W. Short. 1982. *Lygus* bugs and seed quality in lentils (*Lens culinaris* Medik.). USDA, ARS, Agric. Rev. Man. ARM-W-29: 1–43.

Takanona, T., and K. Hori. 1974. Digestive enzymes in the salivary gland and midgut of the bug *Stenotus binotatus*. Comp. Biochem. Physiol. 47A: 521–528.

Taylor, G. S., and P. W. Miles. 1994. Composition and variability of the saliva of coreids in relation to phytoxicoses and other aspects of the salivary physiology of phytophagous Heteroptera. Entomol. Exp. Appl. 73: 265–277.

Taylor, T. H. C. 1945. *Lygus simonyi* Reut., as a cotton pest in Uganda. Bull. Entomol. Res. 36: 121–148.

Tingey, W. M., and E. A. Pillemer. 1977. *Lygus* bugs: crop resistance and physiological nature of feeding injury. Bull. Entomol. Soc. Amer. 23: 277–287.

Varis, A.-L. 1972. The biology of *Lygus rugulipennis* Popp. (Het., Miridae) and the damage caused by this species to sugar beet. Ann. Agric. Fenn. 11: 1–56.

Varis, A.-L., S. Laurema, and H. Miettinen. 1983. Variation of enzyme activities in the salivary glands of *Lygus rugulipennis* (Hemiptera, Miridae). Ann. Entomol. Fenn. 49: 1–10.

Vilkova, N. A. 1968. On the physiology of nutrition in *Eurygaster integriceps* Put. (Heteroptera, Scutelleridae). Entomol. Rev. 47: 431–436.

CHAPTER 3

Plant Bugs (Miridae) as Plant Pests

A. G. Wheeler, Jr.

1. INTRODUCTION

The approximately 10,000 species of mirids, or plant bugs, represent almost a third of all described heteropterans. Members of this diverse family, which has persisted at least since the Jurassic, occur in all major zoogeographic regions.

Capsid bugs, a common name appearing in the older literature, is based on Capsidae, a former family name. The German common name *Blindwanzen* refers to the absence of ocelli (except in the small subfamily Isometopinae). *Blomstertaeger*, the Danish common name for mirids, refers to the bugs' frequent association with flowers.

Adults often are gray, brown, black, or yellowish green and are difficult to detect on the bark, leaves, and inflorescences of their hosts. The bright coloration of some species is considered aposematic. Ranging from 1.5 to about 15 mm long, adults are elongate or ovoid with the hemelytra subparallel. The body surface can be dull, polished, smooth, or punctate. Most species are pubescent, the setae or hairs being simple, silky, flattened, or scalelike (Henry and Wheeler 1988). These delicate, fragile bugs have a cuneus at the apex of the corium and usually one or two closed cells in the wing membrane (except in short-winged forms). Male terminalia are asymmetrical, the left paramere usually more strongly developed than the right (Cassis and Gross 1995, Schuh and Slater 1995). Wing polymorphism and myrmecomorphy (resemblance to ants) are common in this family (McIver and Stonedahl 1993).

Mirid higher classification is unsettled. In the recent world catalog (Schuh 1995) and catalog of the Palearctic region (Kerzhner and Josifov 1999) eight subfamilies are recognized: Bryocorinae, Cylapinae, Deraeocorinae, Isometopinae, Mirinae, Orthotylinae, Phylinae, and Psallopinae. Validity of the last-named subfamily, however, is problematic. An additional subfamily, the Palaucorinae, is sometimes recognized (Gorczyca 1997).

Plant bugs are among the most abundant insects on herbs, shrubs, and trees. Shrub- and tree-associated species tend to be univoltine and to overwinter in the egg stage. Most mirids are plant feeders, but perhaps one half of the species feed as scavengers or facultative predators. Some species are almost exclusively predacious. Mirids that are important as naturally occurring predators of crop pests or have been introduced and used successfully in biological control are treated in **Chapter 28**.

A few phytophagous Miridae are monophagous in a strict sense — that is, they develop on only a single host species (Wheeler 2001). In contrast, some polyphagous lygus bugs, such as *Lygus hesperus* (Scott 1977) and *L. lineolaris* (Young 1986) in North America and *L. rugulipennis* (Hori and Hanada 1970, Holopainen 1989, Holopainen and Varis 1991) in Eurasia, feed on more than 100 host plants. Most plant bugs, however, are oligophagous, occurring on members of one plant family or on plants belonging to a few closely related families.

Phytophagous mirids often exploit nutrient-rich mesophyll, apical meristems, or pollen. They feed on liquefied cell contents rather than imbibe plant sap, as is often misstated in the literature. The flexible stylets, combined with the use of polygalacturonase and other enzymes and a lacerate- or macerate-flush feeding mode, allow plant and animal tissues to be whipped into a slurry or soupy mixture that can be imbibed (see **Chapter 20**). Extraoral digestion, involving solid-to-liquid feeding, integrates morphological, behavioral, and biochemical adaptations that result in the efficient use of high-nutrient resources (Cohen 1998, Wheeler 2001).

The Miridae show greater trophic plasticity than do auchenorrhynchans and sternorrhynchans (Homoptera) and other heteropterans. Few other insectan families contain species that (1) are major crop pests; (2) are potentially useful in the biological control of weeds; (3) are key predators in certain agroecosystems; (4) have been used successfully in the biological control of crop pests; and (5) are detrimental predators that can impair the usefulness of herbivores released for the biocontrol of weeds and reduce the effectiveness of certain parasitoids (Wheeler 2001).

Trophic switching (Cohen 1996) results in frequent cannibalism, omnivory, and intraguild predation, as well as confusion about the status of a species on a particular crop. For example, certain plant bugs on apple or cotton have been categorized as both pests and important natural enemies. The bugs' status — injurious or beneficial — on a crop can depend on locality, season, host cultivar and stage of growth, availablity of prey or alternative food sources such as nectar and pollen, and pesticide applications (Wheeler 2001). Misconceptions or uncertainty about mirid feeding habits have led to needless applications of insecticides against species that are important predators of crop pests.

This chapter reviews the plant-feeding Miridae of major economic importance and some of the more significant minor pests; all species mentioned belong to the subfamilies Bryocorinae, Mirinae, Orthotylinae, and Phylinae. In the case of certain species, synonyms are listed in parentheses after the valid name; for complete synonymies, readers are referred to the Carvalho (1957–1960) catalog and its revision by Schuh (1995), as well as the catalog of Palearctic Miridae (Kerzhner and Josifov 1999). For discussion of additional pest species in the family, as well as additional information and references on mirid biology, readers should consult the world review by Wheeler (2001). This work provides not only more biological details on economically important phytophagous mirids but also data on the feeding habits of and injury symptoms (including color photographs) caused by noneconomic plant bugs. The coverage of the book includes mirids in relation to plant diseases, pollination, pollen feeding, and mycophagy; reviews oviposition injury resulting from the insertion of eggs in plant tissues; and discusses species that "bite" humans.

2. GENERAL STATEMENT OF ECONOMIC IMPORTANCE

Entomologists and growers can easily underestimate the importance of mirids as plant pests. Small populations of these bugs sometimes inflict substantial injury, which can be inconspicuous initially (some host plants remain symptomless) and be recognized only when plant growth intensifies the symptoms (Putschkov 1966, Becker 1974). Adults readily disperse to inflorescences of nonhost plants (i.e., those on which development does not occur), and those of some univoltine species die by the time injury is detected, complicating diagnosis of the problem. Accurate determination of the causal agent can be further complicated by symptoms that are similar to those caused by plant pathogens. Symptoms of mirid feeding also can be confused with those caused by chewing insects, drought, hail, high temperatures, insufficient pollination, mechanical injury, nutrient deficiency or toxicity, and pollution (Wheeler 2001).

Feeding symptoms in the Miridae probably are more diverse than in any other heteropteran family. Symptoms range from local lesions at feeding sites to systemic effects such as growth and differentiation disorders. One of the more familiar effects of plant bug feeding is chlorosis. Leaves fed on by bryocorines (mainly Eccritotarsini), certain mirines (but generally not lygus bugs), and orthotylines such as the Halticini typically become chlorotic. Mirid-induced chlorosis superficially resembles that caused by tingids, spider mites (Acari: Tetranychidae), and some mesophyll-feeding leafhoppers (Auchenorrhyncha: Cicadellidae) and is often accompanied by dark spots of excrement on lower leaf surfaces.

Other common symptoms of mirid feeding are blasting of new growth; foliar crinkling, crumpling, lesions, shot holing, and tattering; stem lesions and cankers; dimpling, scabbing, distortion, and abscission of fruits such as apple (*Malus domestica* Borkh.), guava (*Psidium guajava* L.), mango (*Mangifera indica* L.), and pear (*Pyrus communis* L.); and bushiness, multiple leadering, and witches'-brooming. More subtle effects of feeding by plant bugs, symptoms of which might be easily misdiagnosed, are wilted leaves, stem dieback, thickened leaves and stems, undeveloped flowers, and lagging growth rates (Wheeler 2001). Mirids are among the important causes of silvertop of grasses, a condition characterized by silvery white heads that appear mature but have mainly sterile flowers and lack viable seeds (e.g., Peterson and Vea 1969, 1971; Gagné et al. 1984; Wheeler 2001). The feeding of lygus bugs on seeds of carrot, *Daucus carota* L. subsp. *sativus* (Hoffm.), as well as on other Apiaceae leads to embryolessness. Although the embryo is destroyed, the endosperm remains normal in appearance so that worthless seeds cannot be detected by external appearance, size, or weight (Flemion and Hendrickson 1949). Carrot seeds fed on by *L. hesperus* (see **Section 3.8**), but which still are viable, germinate more slowly and the shoots grow more slowly than those from undamaged seeds. Later, however, plants from damaged seeds show accelerated growth, a phenomenon referred to as phytostimulation (Scott 1970, 1976, 1983). In addition, wheat (*Triticum aestivum* L.) grains of inferior baking quality can result from the feeding of certain mirids and other heteropterans (e.g., Morrison 1938, Meredith 1970).

At times, mirids injure nearly all major world crops. They are key pests of cotton (*Gossypium hirsutum* L.), and billions of dollars have been spent on plant bugs as cotton pests when costs of control are added to those associated with research. As the most important insect pests of cocoa (*Theobroma cacao* L.) in West Africa, mirids can cause annual losses in yield of 20 to 30% (Entwistle 1985). In some cases, recurrent attacks by plant bugs prevent cocoa and certain other tropical trees from attaining their normal size and shape (Crowe 1977).

Mirids also are major pests of apple, cashew (*Anacardium occidentale* L.), guava, mango, peach (*Prunus persica* (L.) Batsch), pear, strawberry (*Fragaria* × *ananassa* Duchesne), and other fruit and nut crops. As direct pests of fruit crops, mirids can render the fruit unmarketable. Their feeding on flowers can result in a failure to set fruit or abscission of young fruits, which is not necessarily of economic importance. Mirid-induced lesions also can allow secondary fungi to invade and to intensify fruit injury. Other crops that can sustain severe losses from mirids are alfalfa (*Medicago sativa* L.), sorghum (*Sorghum bicolor* (L.) Moench), and tea (*Camellia sinensis* (L.) Kuntze). Mirids also impair the aesthetic appearance of herbs, shrubs, and trees planted for their ornamental value (Wheeler 2001).

3. THE IMPORTANT SPECIES

Bryocorinae

Containing important pests of tropical crops, the subfamily Bryocorinae represents a diverse group that probably is not monophyletic. Schuh (1995) recognized the tribes Bryocorini, Dicyphini, and Eccritotarsini. Species of the small, fern-feeding tribe Bryocorini are not considered here. Most of the important bryocorine plant pests belong to the Dicyphini, a tribe in which Schuh (1995) recognized the subtribes Dicyphina, Monaloniina, and Odoniellina. The Dicyphina include species that feed on glandular-hairy plants such as tobacco (*Nicotiana tabacum* L.) and tomato (*Lycopersicon esculentum* Mill.) (Solanaceae); its members also show strong predatory tendencies (see **Chapter 28**). Species of the Monaloniina and Odoniellina, considered strictly phytophagous, typically cause lesions and cankers on tropical crops such as cashew, cocoa, and tea. Members of the phytophagous tribe Eccritotarsini often cause foliar chlorosis.

3.1   *Distantiella theobroma* (Distant)

Distantiella theobroma (and certain other bryocorines) is consistently devastating to cocoa production in West Africa. In the 1920s, cocoa mirids were responsible for the temporary abandonment of cocoa by West African farmers. As low-density pests, these bugs inflict damage disproportionate to the number of bugs present (Conway 1969, Southwood 1973, Wheeler 2001). Although at least 35 bryocorine species can be pests of cocoa in the Old and New Worlds, *D. theobroma* (and *Sahlbergella singularis*; **Section 3.6**) are usually the ones referred to as cocoa mirids (or capsids); both species are endemic to West Africa. Because cocoa is indigenous to the Neotropics, all Old World bryocorines associated with cocoa have adapted to this introduced plant mainly during the 20th century (Southwood 1960, Leston 1970).

Cocoa mirids are characterized by their complex interactions with the tropical forest environment. They especially need to be considered in relation to the ant mosaic of cocoa trees (e.g., Leston 1971, 1973; Majer 1972) and to two diseases, swollen shoot virus (Thresh 1960) and black pod (Wood and Lass 1985). Literature pertaining to cocoa mirids relative to the amount of light entering the canopy, the injury they cause in relation to host growth and physiology, and the buildup and movement of their populations is voluminous and often contradictory. The necessarily superficial treatment here of the literature on bionomics and control of *D. theobroma* (and *S. singularis*) should be supplemented by the thorough treatment in Entwistle's (1972) chapter on cocoa mirids in his *Pests of Cocoa*. Other useful references include Johnson (1962), Entwistle and Youdeowei (1964), Cochrane and Entwistle's (1964) bibliography of cocoa mirids, Lavabre's (1977) *Les Mirides du Cacaoyer*, and Wheeler's (2001) *Biology of the Plant Bugs*. The main types of injury — blast, staghead, and mirid pockets — are detailed by Williams (1953), Johnson (1962), and Leston (1970).

Largely because of the difficulty of culturing *D. theobroma* in the laboratory — adults are particularly difficult to maintain — bionomical information is incomplete (Lavabre 1969, Entwistle 1972). Mating is facilitated by a female sex pheromone (King 1973, Kumar and Ansari 1974). Youdeowei (1973) obtained life-history data from individuals confined in the field to chupons (upright stem or shoots) of mature cocoa trees. The most eggs laid by a female was 73. A mean fecundity of 100, with maximal egg production of 276, was obtained based on the bug's placement on cut shoots of the bombaceous tree *Ceiba pentandra* (L.) Gaertn. (Entwistle 1972). The incubation period is 12 to 16 days, with nymphal development requiring about 41 days (Youdeowei 1973). Cotterell (1926) reported a similar incubation period but a shorter nymphal period (average of 21 to 22 days). Although data on longevity are sketchy, adults probably live a month or less (Entwistle 1972).

Like many bryocorines associated with cocoa, *D. theobroma* feeds on superficial parenchyma of both vegetative and reproductive structures. Populations that feed on vegetative tissues can develop independently of those on pods, or the bugs can transfer to shoots after pods are harvested (Crowdy 1947, Gibbs et al. 1968). Shoots are fed on more than pods, although access to pods can improve fecundity (Houillier 1964a). This photophobic bug, which shows a tendency to aggregate, oviposits and feeds mostly at night. During the day, individuals are found mainly in protected sites on host trees (Entwistle 1972).

Elliptical, water-soaked areas up to 200 mm long begin to appear within 2 to 5 min after a bug begins to feed. The lesions blacken within a few hours and are more apparent on green stems and pods than on hardened stems (Entwistle 1972). A fifth instar can make 20 stem lesions during one night's feeding. Late instars and adults make larger and greater numbers of lesions than do early instars (Entwistle 1972, Kumar and Ansari 1974, Youdeowei 1977). Water tension of stems affects the size and form of lesions, which decrease in size as osmotic pressure increases (Cross 1971). Only a few lesions can cause death of an apical shoot (Leston 1970). Lesions dry after several weeks, dead tissue drops out, and the bark splits, leaving roughened areas on stems (Entwistle 1972).

Crop loss results more from attacks on shoots and branches than on pods. Pod feeding usually does not decrease yields once the cherelle (young fruit) stage is reached (Toxopeus and Gerard 1968, Marchart 1972). Feeding on young stems often causes wilting above the point of attack. Below the lesions, dieback does not occur unless pathogenic fungi, primarily *Calonectria rigidiuscula* ([Berk. & Br.] Sacc.), invade wound tissue. *Calonectria rigidiuscula*, which also enters other types of wounds, sometimes infects as many as 80 to 95% of primary, mirid-induced lesions; infections eventually invade the xylem, phloem, and medullary rays. This weak parasite can lie dormant for several years under callus tissue, spread to healthy tissues, and cause further dieback. The roughened bark that typifies *Calonectria* infections usually persists for several years (Entwistle 1972, Thorold 1975).

Mirid-induced lesions tend to be concentrated on the peduncular ends of pods (Houillier 1964b, Youdeowei 1977). Because mirid feeding mostly affects parenchymatous husk tissues outside the sclerotic layer, older pods sustain only slight damage. In some cases, the husk can crack, allowing phytopathogens to enter (Williams 1954, Entwistle 1972).

Like many other mirid pests of tropical crops, *D. theobroma* declines in numbers during the dry season (Entwistle 1972, Wheeler 2001). Its decline in density involves interactions among several factors, including water stress in cocoa tissues, desiccation of nymphs under conditions of lower humidity, and lack of protected sites due to low numbers of pods (Williams 1954, Entwistle 1972).

Synthetic organic insecticides such as HCH (= BHC) provided good control of *D. theobroma* during the 1960s, even though resistance to this chemical was reported in some Ghanaian populations. With heavy use of HCH placing intense selective pressure on the bugs, insecticide resistance increased and spread through West Africa in the 1970s (Johnson 1962, Collingwood 1977b, Entwistle 1985). Alternating insecticides of different classes (Bruneau de Miré 1985) and using synthetic pyrethroids (Idowu 1988) help alleviate the problem. The use of pyrethroids is limited by their high cost and the need for multiple applications each season (Owusu-Manu 1985). Although ants provide substantial natural control of cocoa mirids (e.g., Leston 1973, 1978), the practicality of relying on ant predation in an integrated control program is questionable (e.g., Squire 1947, Marchart 1971; see also Collingwood 1977a and Entwistle 1985).

3.2 *Helopeltis antonii* Signoret

A polyphagous bryocorine of the Oriental region, *H. antonii* is a pest of numerous tropical crops, feeding on both vegetative and reproductive parts of host plants. Although this bug can be a pest of cocoa and tea (Stonedahl 1991), it is particularly important on cashew and guava (Wheeler 2001). Some of the economic literature pertaining to *H. antonii* in South India and Sri Lanka may refer to the similar *H. bradyi* Waterhouse. Records of damage by *H. antonii* to various crops in Java and the Philippines definitely refer to other species of *Helopeltis* (Stonedahl 1991).

Adults of *H. antonii*, like other species of the genus, possess a scutellar process or spine; in nymphs, this process becomes apparent in the second instar. Eggs have two unequal respiratory filaments or horns arising from the anterior end.

Mating and oviposition behavior in *H. antonii* were described by Devasahayam (1988). Mean fecundity under laboratory conditions can be 40 to 50 (Puttarudriah 1952, Jeevaratnam and Rajapakse 1981a), but Devasahayam (1985) found that fecundity varies seasonally from 13 to 82 eggs. The life cycle can be completed in 16 to 17 (Pillai and Abraham 1975) to about 22 to 25 days (Jeevaratnam and Rajapakse 1981a); nymphal development on cashew in the laboratory required 17 to 28 days (Hari Babu et al. 1983). According to Jeevaratnam and Rajapakse (1981a), fecundity is greater and nymphal development more rapid when nymphs feed on tender shoots of cashew rather than on its fruits. Nymphs develop more rapidly on cocoa or guava (22 to 23 days) than on tea (Jeevaratnam and Rajapakse 1981a). Adults can live up to 3 months in the field (Puttarudriah 1952).

Nymphs and adults produce necrotic lesions on shoots of allspice *Pimenta dioica* (L.) Merr., black pepper *Piper nigrum* L. (Devasahayam et al. 1986), and neem *Azadirachta indica* A. Juss. in India (Ramakrishna Ayyar 1940). The bugs also can deform the leaves of black pepper in Malaysia (Blacklock 1954). As a fruit pest in India, *H. antonii* causes a scabbing of apples and a pitting of grape (*Vitis vinifera* L.); affected grapes can rot and drop prematurely (Puttarudriah and Appanna 1955). On guava it produces blisters and warts on the fruit, the injury resembling, and often attributed to, a fungal pathogen. Young fruit can drop prematurely, and the market value of harvested fruit is reduced (Puttarudriah 1952, Nair 1975, Sudhakar 1975).

On cashew, this plant bug might attain its greatest economic importance. A major cashew pest in Sri Lanka (Jeevaratnam and Rajapakse 1981a,b), *H. antonii* is the principal pest insect of this crop in southern India (e.g., Pillai and Abraham 1975, Hari Babu et al. 1983). Its feeding causes necrotic cankers that result in the drying of terminal shoots. Lesions, about 9 mm long and 2 mm wide, form on stems, the symptoms intensifying under higher temperatures (31°C) (Sathiamma 1977, Ambika and Abraham 1979). Nagaraja et al. (1994) described the biochemical changes occurring in leaves and young shoots fed on by *H. antonii*.

This bug's necrotic lesions on inflorescences and nuts of cashew allow secondary fungi to invade and cause inflorescence blight and shedding of fruits (Abraham 1958, Nambiar et al. 1973, Pillai and Abraham 1975, Thankamma Pillai and Pillai 1975, Ambika and Abraham 1979). With the coalescing of lesions, inflorescences can appear scorched (Nambiar et al. 1973). A fifth instar can cause 100 or more lesions on fruits in 24 hours (Devasahayam and Radhakrishnan Nair 1986). Because of their prolonged flowering period, cashew trees established by vegetative development are more susceptible to damage. Losses in yield of at least 30% can occur (Nair 1975, Pillai et al. 1976, Devasahayam and Radhakrishnan Nair 1986). A compensatory increase in numbers of fruits set occurs when fruitlets are severely damaged (Subbaiah 1983).

Populations of *H. antonii* on cashew are active mainly from October to May, increasing after the monsoon season and peaking in January when host trees are in full bloom (Pillai and Abraham 1975). To minimize losses in yield, three insecticide applications may be needed: at emergence of new flushes, at panicle development, and at fruit set (Pillai 1987). The use of synthetic organic insecticides, however, adversely affects parasitism and predation of *H. antonii* (Jeevaratnam and Rajapakse 1981a). Effective management of this pest should include host-resistant cultivars, manipulation of alternative host plants of the bug, and conservation of its natural enemies (Devasahayam and Radhakrishnan Nair 1986).

3.3 *Helopeltis schoutedeni* Reuter

This African species of warm, moist climates is an important, although sporadic and patchily distributed pest of cotton, colonizing fields when cotyledons become available (McDermid 1956). Many early reports of damage to cotton by *H. bergrothi* Reuter (e.g., Steyaert 1946) should be referred to *H. schoutedeni* (Carayon and Delattre 1948, McDermid 1956, Pearson 1958, Ripper and George 1965).

Females lay 30 to several hundred eggs, which hatch in about 2 weeks. The nymphal period lasts about 3 weeks. Adults might live 6 to 10 weeks. Fecundity and longevity are influenced by this polyphagous bug's host plants, temperature, and other factors; data obtained by different workers, therefore, vary widely (Pearson 1958, Ripper and George 1965, Cadou 1993).

Nymphs feed mainly on young cotton plants and adults mostly on bolls (Schmutterer 1969); seeds and woody tissues are avoided (Pearson 1958). This bug's feeding on leaves can result in relatively unimportant crumpling or puckering, whereas the lesions it produces on stems distort and even kill its hosts. Stem lesions, which resemble those associated with bacterial blight of cotton, can be misdiagnosed as representing a disease problem (Ripper and George 1965, Wheeler 2001). Boll feeding by *H. schoutedeni* results in a severely pitted surface; bolls sometimes split prematurely, especially after secondary pathogens invade. The mirid's first annual population peak tends to correspond to the main period of cotton's vegetative growth (mid-August) and the second to the period when young bolls grow most rapidly (September to October) (McDermid 1956, Pearson 1958). Prevalence of *H. schoutedeni* in cotton is affected by the proximity of alternative host plants (Ripper and George 1965).

On tea, the bug's water-soaked lesions on stems are marked by sunken areas. When the bug feeds on midribs, diamond-shaped lesions are produced. Foliage can become shot-holed when dead tissue drops from old lesions. Severely cankered stems become girdled, and the gall wood that forms around cankers gives stems a gnarled appearance (Leach and Smee 1933, Lever 1949, Benjamin 1968).

Chemical control on tea often is needed to prevent severe loss of crop; insecticides are used only after bug densities and economic thresholds are considered (Rattan 1992). The use of more tolerant tea cultivars (Rattan 1992) and annatto as a trap crop (Peregrine 1991) might minimize damage from this pest.

Helopeltis schoutedeni is one of several African species of the genus that feed on cashew. It occasionally causes the nuts to shrivel and abscise (Martin et al. 1997). It also causes foliar puckering on mango; shot holing of avocado (*Persea americana* Mill.) and quinine (*Cinchona* spp.) leaves (Gerin 1956); and stem and fruit cankers on avocado, guava, and mango (Leach 1935, Harris 1937, Gerin 1956).

3.4 *Helopeltis theivora* Waterhouse (= *Helopeltis theobromae* Miller)

The so-called tea mosquito bug, a key pest of tropical crops such as cocoa, guava, and tea, occurs mostly in Southeast Asia, Indonesia, Malaysia, India, and Sri Lanka (Stonedahl 1991). Its feeding habits and the resulting damage are generally similar to those of *H. antonii* (**Section 3.2**) and *H. schoutedeni* (**Section 3.3**). Adults vary considerably in size and coloration (Stonedahl 1991).

Das (1984) noted that in India eggs hatch in 5 to 7 days during June to August but require 20 to 27 days to hatch in January. Nymphs can develop in 9 to 10 days in June and 23 to 29 days in December to January. Adults live about a month when confined on cocoa in the laboratory; females lay 50 to 63 eggs (Lim and Khoo 1990).

Injury to tea was first noticed about 1865 (Watt and Mann 1903). Within 20 years of Peal's (1873) prediction that *H. theivora* would become a major pest of Indian tea culture, it began to cause "enormous" damage (Bamber 1893). The bugs, whose populations are highest during June to July and August to September, cause tea leaves to become shot holed and curled. Deposition of eggs in tender stems can contribute to the damage. Before the 1950s, crop losses of 100% were possible (Rao 1970; Muraleedharan 1987, 1992). The use of DDT provided satisfactory control, but once its use was curtailed or discontinued, *H. theivora* regained pest status on tea (Banerjee 1983, Das 1984).

Helopeltis theivora is a key pest of cocoa in West Malaysia (e.g., Khoo 1987, Khoo et al. 1991); the synonymic name *H. theobromae* is used in much of the Malaysian economic literature (Stonedahl 1991). It feeds on new flush shoots, cherelles, and pods, the seeds remaining unaffected (Miller 1941). Injury to young pods contributes to "cherelle wilt" (Khoo et al. 1991). Lesions made by *Helopeltis* species tend to be distributed over the entire pod surface rather than concentrated on the dark side of pods, as is the case with *D. theobroma* (**Section 3.1**) and *S. singularis* (**Section 3.6**) (Entwistle 1972). The availability of cocoa pods, which increases fecundity and longevity in *H. theivora*, is crucial to the bug's reproductive success. Precipitation also affects the bug's population dynamics, its densities usually decreasing with declining rainfall, even when pods are present (Tan 1974a).

Damage to cocoa from chronic infestations of *H. theivora* can include a witches'-brooming of stems, the symptoms resembling those caused by a fungal pathogen (Khoo 1987). Small pods fed on by this pest can wilt, and the weight of large pods is reduced. Damaged pods can be colonized by secondary invaders such as the coffee bean weevil, *Araecerus fasciculatus* (De Geer). Substantial losses in yield occur when the application of insecticides such as gamma-HCH (lindane) is withheld (Tan 1974b, Chung and Wood 1989).

Ants such as the black cocoa ant, *Dolichoderus thoracicus* (Smith), and the weaver ant, *Oecophylla smaragdina* (F.), prey on *H. theivora*. The interrelationships among ants, *Helopeltis*, and mealybugs were described by van der Goot (1917); investigations on this system were summarized by Giesberger (1983). The use of weaver ants to minimize damage to cocoa by *H. theivora* in Malaysia was discussed by Way and Khoo (1989, 1992).

In Malaysia, damage to guava by this important pest is sometimes attributed to fungi. Nymphs and adults cause lesions and cankers that can kill young terminals. The lesions produced on fruit, which sometimes result in cracking of the skin, can lower the market value or render fruit unsalable (Lim and Khoo 1990). Ants have relatively little effect on *H. theivora* populations on guava. Because trees with dense canopies are more susceptible to damage by this plant bug, the correct pruning of guava might minimize damage (Lim and Khoo 1990).

Helopeltis theivora sometimes severely damages the tender leaves of Indian long pepper (*Piper longum* L.) (Abraham 1991). References to this bug as an occasional pest of the camphor tree *Cinnamomum camphora* (L.) J. Presl and of quinine are given by Stonedahl (1991).

3.5 *Monalonion* Species

The diverse mirid fauna that typifies cocoa culture in the Old World, where cocoa is an introduced crop, does not develop on this plant in the New World, where it is indigenous to tropical rain forests (e.g., Toxopeus 1985). All cocoa mirids in the New World (about seven species) belong to the genus *Monalonion* (Entwistle 1972). These predominantly pod-feeding mirids can be considered ecological homologues of Old World *Helopeltis* species (**Sections 3.2 to 3.4 and 4.2**; Schmutterer 1977). Since the first reports of damage in the late 19th and early 20th centuries, *Monalonion* species have been recorded as cocoa pests in Brazil, Colombia, Costa Rica, Ecuador, Panama, Peru, and Venezuela (Entwistle 1965, 1972).

Life-history data for *Monalonion* species are scant (Entwistle 1972, Abreu 1977). Nymphs and adults feed on young shoots and twigs but seem to prefer developing fruits and pods. Green pods are marked by pustules and dark, warty spots; young fruit can wilt and die, and larger pods become deformed. Complete loss of crop can occur (Wille 1944, 1952).

After *M. dissimulatum* Distant was detected in Venezuelan cocoa estates in 1933, bean production during the following 15 years fell from 7.0 to 2.7 tons/ha (Entwistle 1972). This pest, known as the mosquito bug or "mosquilla," was once considered the most important cocoa pest in Peru. It thrives under conditions of high humidity, high temperature, and heavy shade (Wille 1944, 1952).

Injury similar to that inflicted by *M. dissimulatum* has been described in Central or South America for *M. annulipes* Signoret, *M. bahiense* Costa Lima, *M. bondari* Costa Lima, *M. schaefferi* Stål, and other species of the genus (Entwistle 1972, Abreu 1977). Abreu (1977) included additional biological information on *Monalonion* species and discussed their host plants, population fluctuations, host resistance, natural enemies, and chemical control.

3.6 *Sahlbergella singularis* Haglund

A more widespread cocoa pest than *D. theobroma* (**Section 3.1**), *S. singularis* has damaged the crop in West Africa since the early 20th century. The adaptation of this African plant bug to cocoa appears to have been facilitated by the proximity of cocoa estates to wild and cultivated hosts such as *Cola* species (Entwistle 1972, Youdeowei 1973).

Fecundity in this species is considered slightly greater than in *D. theobroma* (Johnson 1962). A mean incubation period of about 13 days in Youdeowei's (1973) study compares with 16 days recorded by Cotterell (1926). The nymphal period typically lasts 21 to 23 days (Cotterell 1926, Youdeowei 1973). In contrast to *D. theobroma*, a sex pheromone is unknown in *S. singularis*. Both *S. singularis* and *D. theobroma* generally oviposit and feed on the same tissues and cause similar damage. The former bug, however, is more common on older cocoa trees than is *D. theobroma* and feeds less often on chupons and more on fans (Entwistle 1972). Access to cocoa pods increased fecundity less in *S. singularis* than in *D. theobroma* (Houillier 1964a). Although *S. singularis* might not respond so quickly to insecticide pressures as *D. theobroma*, it also developed a resistance to commonly used chlorinated hydrocarbons in the 1960s (Entwistle 1972). The references listed under *D. theobroma* (**Section 3.1**) also provide for *S. singularis* detailed information on its ecology, damage, natural enemies, and control.

Mirinae

Largest of the mirid subfamilies, the Mirinae contain plant pests such as lygus bugs, as well as the mainly predacious genus *Phytocoris* (see **Chapter 28**). Many of the most familiar mirid plant pests belong to the Mirini, largest of this subfamily's six tribes. *Irbisia* and *Tropidosteptes* species typically cause foliar chlorosis, and some species of *Lygocoris* produce a shotholing and tattering of foliage (Wheeler 2001). The fourlined plant bug, *Poecilocapsus lineatus* (F.), is characterized by its rapid production of water-soaked lesions on leaves of herbs and shrubs (Wheeler and Miller 1981, Cohen and Wheeler 1998). Most Mirini, however, feed on meristems, new foliage, flower buds, flowers, and developing seeds of their hosts. All species of Mirinae discussed in this chapter as major or minor pests belong to the Mirini, except *Hyalopeplus pellucidus* Kirkaldy, which belongs to the small tribe Hyalopeplini. Trophic habits of members of the other mirine tribes — Herdoniini, Mecistoscelidini, Restheniini, and Stenodemini — are discussed by Wheeler (2001).

3.7 *Calocoris angustatus* Lethierry

The sorghum earhead bug, although polyphagous, prefers sorghum over other grain crops or grasses (Hiremath 1986). First reported to damage sorghum in India during the late 19th century, *C. angustatus* remains a key pest of this crop in the southern states of Andhra Pradesh, Karnataka, and Tamil Nadu (Jotwani 1983, Hiremath 1986, Gahukar 1991). The use of high-yielding sorghum varieties might allow this plant bug to become a problem farther north in India (Teetes et al. 1979).

Ballard (1916) conducted the first biological studies, noting that the life cycle can be completed in 15 to 17 days, which is 2 to 3 days longer than reported by Cherian et al. (1941). In more detailed studies by Hiremath and Viraktamath (1992), the incubation period was found to require 5.6 to 9.7 days in the laboratory (temperature not stated) and 6.3 to 11.0 days under field conditions; the nymphal period lasted 10 to 17 days in the field. Fecundity varied from about 58 in May to 267 eggs in October. Longevity ranged from 13 to 32 days (May) to 16 to 30 days (October). Hiremath and Viraktamath's (1992) paper contains descriptions of the immature stages and additional biological information, such as mating and oviposition behavior.

Adults in nearby crops are attracted to sorghum in the preflowering stage, which is preferred for oviposition (Hiremath and Thontadarya 1984a, Natarajan et al. 1988). Nymphs feed on the panicles (heads) as soon as they become available, causing grains to become chaffy, shriveled, and distorted. Grain set is inhibited and yields are decreased, the damage generally more severe on cultivars with compact rather than loose panicles (Cherian et al. 1941, Wheeler 2001). Grain quality is affected both physically and biochemically (Hiremath et al. 1983). Bug densities are higher during the milky stage of crop development (Hiremath and Thontadarya 1984b, Natarajan et al. 1989). Depending on the cultivar, economic injury levels range from 0.2 to 1.4 bugs per panicle

(Sharma and Lopez 1989; see also Natarajan and Sundara Babu 1988). Early-season buildup of populations subjects late-maturing sorghums to greater damage from *C. angustatus* (Leuschner et al. 1985). The effects that environmental factors (humidity, temperature, rainfall, sunshine) have on the bug's population dynamics are discussed by Sharma (1985), Natarajan et al. (1988), and Sharma and Lopez (1990).

Hiremath (1989) reported natural enemies of *C. angustatus* found during a 3-year survey. The screening of germplasm lines to develop sorghum cultivars resistant to the earhead bug was reported by Sharma and Lopez (1992) and summarized by Wheeler (2001).

3.8   *Lygus hesperus* Knight

The western tarnished plant bug ranges from southern British Columbia (Canada) to northern Mexico and east from Montana to West Texas (U.S.A) (Schwartz and Foottit 1998). This polyphagous species, known from more than 100 host plants, occurs most often on composites (Scott 1977, Domek and Scott 1985). Like other pestiferous species of the genus, *L. hesperus* is multivoltine, feeds mainly on reproductive plant parts, and during the growing season tracks a succession of flowering hosts, both wild and cultivated. Yet injury to crop seedlings and the vegetative stages of more mature plants can occur (Wheeler 2001).

Among the numerous studies on biology of *L. hesperus* (see bibliographies of Scott 1980, Graham et al. 1984) are those of life history under laboratory conditions (e.g., Shull 1933, Leigh 1963); influence of temperature on egg development (Champlain and Butler 1967) and on fecundity and longevity (Strong and Sheldahl 1970); oviposition-site preferences (Alvarado-Rodriguez et al. 1986); and effects of host genotype (five bean cultivars) on fecundity, longevity, intrinsic rate of increase, and other life-history parameters (Alvarado-Rodriguez et al. 1987). The effects on fecundity and longevity of host switching — transferring teneral adults reared on one host to a second plant species — were investigated by Al-Munshi et al. (1982). Other important studies on *L. hesperus* include Beards and Strong (1966) on the relationship between photoperiod and diapause; Strong et al. (1970) on reproductive biology, including mating behavior and use of a sex pheromone; and Strong (1970) on feeding behavior and physiology of host injury. Tingey and Pillemer (1977) reviewed crop resistance to lygus bugs and the physiological nature of the injuries they cause.

Field, forage, and vegetable crops (Wheeler 2001), as well as fruit (e.g., Barnett et al. 1976, Pickel and Bethell 1990) and nut (Rice et al. 1988) crops, and ornamental trees (Koehler 1963) and conifer seedlings (Schowalter et al. 1986, Schowalter 1987), can sustain injury from the western tarnished plant bug. The following abbreviated account of its pest status, particularly on alfalfa and cotton, should be supplemented by referring to additional information and references in Wheeler (2001), which includes discussion of the controversy surrounding the effects of mirids on cotton yields, the significance of damage thresholds, and the accuracy of sampling techniques. For information on *L. hesperus* as a cotton pest, Leigh and Matthews (1994) provided a useful summary of information; chapters in *Cotton Insects and Mites: Characterization and Management*, edited by King et al. (1996), should also be consulted for additional information. Scott (1983) discussed the relationships between *L. hesperus* and carrot, including phytostimulation, in which seeds fed on, but not destroyed, produce larger, more vigorous plants.

The western tarnished plant bug is often considered a key pest of cotton in California's San Joaquin Valley (U.S.A.) (e.g., Stern 1969, 1973; Leigh et al. 1988). Effects of its feeding are most critical during the beginning of the third week of squaring to the end of the sixth week (University of California 1984). Some researchers, however, feel its threat to cotton production has been exaggerated (e.g., Gutierrez 1995), which has led to unnecessary use of insecticides. Early-season spraying for lygus bugs often kills natural enemies of boll- and budworms, which then can reach outbreak levels later in the season (Wheeler 2001). A prophylactic use of insecticides against lygus bugs and other cotton pests resulted in the western tarnished plant bug's development of resistance to chlorinated hydrocarbons, carbamates, and organophosphates (e.g., Leigh et al. 1977).

Feeding by *L. hesperus* on cotton seedlings can retard growth, and that on terminals of later vegetative stages leads to loss of apical dominance, resulting in excessive branching and witches'-brooming and tall spindly plants with swollen nodes and shortened internodes (Wene and Sheets 1964, Hanny and Cleveland 1976, Hanny et al. 1977). Adults, however, are usually not attracted to cotton until squares (flower buds) form. Crops such as alfalfa, grain sorghum, safflower (*Carthamus tinctorius* L.), and sugar beet (*Beta vulgaris* L.) serve as sources of infestations. The bugs' migration into cotton is affected by the proximity and extent of alternative host crops, their time of harvest, composition and density of nearby weeds, and rainfall (Wheeler 2001).

When *L. hesperus* feeds on "pinhead" squares (about 2 mm long), the buds often drop prematurely. The effects of square abscission on cotton yield is difficult to determine because physiological stresses and phytopathogens also affect fruit set (e.g., Guinn 1982, Mauney and Henneberry 1984, Johnson et al. 1996). On squares 3 mm or more in diameter, the bugs feed mostly on developing pollen sacs. The squares, although normal in appearance, show necrotic anthers when the calyx and corolla are removed (Pack and Tugwell 1976, Tugwell et al. 1976, Mauney and Henneberry 1979). Boll feeding also occurs, particularly by adults. Symptoms on young bolls appear as dull, dark spots 1 to 2 mm in diameter (University of California 1984, Wilson et al. 1984). Internal evidence of feeding consists of dark necrotic areas and callouslike growths on the inner boll wall. Mature bolls can show a staining of lint that lowers the grade (University of California 1984, Wheeler 2001).

The development of cottons resistant to lygus bugs has received considerable attention. Wheeler (2001) reviewed the morphological traits of frego bract, glandlessness, and nectarilessness in relation to lygus bug populations.

As a pest of alfalfa seed crops, *L. hesperus* causes a blasting of flower buds and their premature abscission. The bugs feed mostly on basal flower parts but also on ovaries, causing bare rachises or "stripped racemes." Blossoms sometimes are so extensively injured that honeybees restrict their visits to alfalfa fields (e.g., Linsley and MacSwain 1947). Seeds that are fed on show small holes or depressions and shrivel; such seeds fail to germinate (e.g., Sorenson 1932, Carlson 1940, Wheeler 2001). The abscission of pods is related to the bugs' destruction of auxin-producing apical meristems and the creation of hormonal imbalances; auxin is chiefly responsible for preventing premature abscission (Strong 1970, Tingey and Pillemer 1977).

Adults often migrate into alfalfa seed fields when alfalfa hay and other crops are harvested (Stitt 1949, Gupta et al. 1980). *Lygus hesperus* prefers alfalfa to cotton (Cave and Gutierrez 1983). To protect cotton from lygus bugs, alfalfa can be interplanted as a trap crop that attracts adults when alfalfa is cut or safflower dries up. Strip cutting and border harvesting of alfalfa can protect cotton by retaining the bugs in uncut portions of alfalfa fields (Stern and Mueller 1968; Stern 1969; van den Bosch and Stern 1969; Sevacherian and Stern 1974, 1975).

As a key pest of strawberry in California, *L. hesperus* causes fruit distortion and catfacing (Allen and Gaede 1963, Maas 1984, Riggs 1990). Economic damage can result from an infestation level of 0.5 nymphs or adults per plant (Zalom et al. 1990). Readers should consult Flint (1990) and Zalom et al. (1990, 1993) for information on sampling methods and nonchemical means of managing lygus bug populations in California strawberries.

3.9 *Lygus lineolaris* (Palisot de Beauvois)

The tarnished plant bug might have the widest distribution of any New World mirid. Recorded in the United States from all 48 contiguous states and Alaska, it ranges transcontinentally across Canada, south to Guatemala, and occurs in Bermuda (Schwartz and Foottit 1998).

Young (1986) recorded *L. lineolaris* from more than 300 hosts. This plant bug tracks a succession of flowering plants, its populations often peaking at full bloom, remaining stable until seeds mature, and eventually collapsing (Woodside 1947). Common herbaceous hosts are annual fleabane *Erigeron annuus* (L.) Pers., horseweed *Conyza canadensis* (L.) Cronquist, and pigweed *Amaranthus retroflexus* L. (Woodside 1947, Fleischer and Gaylor 1987). Regional host lists of this bug are available for the Mississippi Delta (Snodgrass et al. 1984b) and the Canadian Prairie Provinces (Schwartz and Foottit 1992). Because so many of its hosts are crop plants, the applied entomological literature contains hundreds of papers on the tarnished plant bug as an agricultural pest. Crosby and Leonard (1914) reviewed the early economic literature. References to *L. lineolaris* in Graham et al. (1984) and Wheeler (2001) should be used to supplement the following account.

Adults mate when they are at least 4 days old (Bariola 1969). Following a preoviposition period of about 8 days (Stewart and Gaylor 1994a), eggs are laid in tender plant parts (Painter 1927, Fleischer and Gaylor 1988). Because oviposition preferences can change with time, short test periods can lead to erroneous conclusions regarding oviposition behavior (Gerber 1995). The incubation period is about 7 to 8 days (Stewart and Gaylor 1993). Temperature effects on incubation and nymphal development were reported by Ridgway and Gyrisco (1960). Host plant effects on fecundity were studied by Curtis and McCoy (1964), Khattat and Stewart (1977), and Stewart and Gaylor (1994b). Longevity in the laboratory varies with the food source and can be as long as 45 to 50 days (Curtis and McCoy 1964). Fleischer and Gaylor (1988) studied the influence of different hosts on the bug's population dynamics. Information on seasonality and number of generations is available in Gerber and Wise (1995). All nymphal stages were described and illustrated by Schwartz and Foottit (1992).

The tarnished plant bug not only is a generalist feeder, but also is considered a visual (Prokopy and Owens 1978) and a temperature generalist (Fleischer and Gaylor 1988). This mirid combines the attributes of a good colonizer (r strategist) and a good competitor (K strategist) (e.g., Croft and Hull 1983).

Perhaps a greater range of injury symptoms have been reported for the tarnished plant bug than for any other mirid. Symptoms include localized yellowing from feeding on leaf veins and chlorosis that precedes foliar wilting; death of growing tips; foliar shot holing, cupping, and puckering; split-stem lesions; and retarded growth, bushiness, witches'-brooming, shortened internodes, excessive branching, imperfect flowers, embryoless seeds, and blemishing and catfacing of fruit (Wheeler 2001). Despite the attention given this species by entomologists, the split-stem lesions it causes on hybrid poplar (*Populus* spp.) (Sapio et al. 1982, Wilson and Moore 1985) and its multiple leadering of conifer seedlings and transplants (South 1991) have been recognized only since the 1980s. Both types of injury initially were attributed to abiotic factors or to other biotic agents.

Known as an apple pest since the 1860s (Crosby and Leonard 1914), the tarnished plant bug causes flower bud abscission, which is of little economic consequence because of the small number of blossoms needed to set an apple crop. More significant is fruit malformation that results in the culling or downgrading of fruit. Papers dealing with *L. lineolaris* on apple are listed by Parker and Hauschild (1975) and discussed by Wheeler (2001).

On peach, overwintered and first-generation adults feed on terminal buds, causing loss of apical dominance and stunting; this injury is often referred to as "stop back" (Back and Price 1912). The bug's fruit feeding is even more damaging to peach. Young fruit can drop prematurely, and injured fruits that develop show catfacing: scarred, sunken areas that become corky and hard and lack fuzz (Howitt 1993). Such injury is undistinguishable from that caused by stink bugs (Pentatomidae). Catfacing involves the asymmetrical destruction of auxin-producing areas on fruits 8 to 20 mm in diameter (Strong 1970).

The tarnished plant bug can affect yields of lima bean (*Phaseolus lunatus* L.) (McEwen and Hervey 1960), green pepper (*Capsicum annuum* L.) (Huber and Burbutis 1967), and tomato (Davis et al. 1963). It is an important pest of oilseed rape and canola (*Brassica* spp.) (Butts and Lamb 1990a,b; Turnock et al. 1995; Leferink and Gerber 1997) and grain amaranth (*Amaranthus cruentus* L.) (Clark et al. 1995, Olson and Wilson 1990, Wilson and Olson 1992). An extensive literature, much of it discussed by Wheeler (2001), treats this plant bug as a pest of cotton. Feeding symptoms are similar to those of *L. hesperus* (**Section 3.8**). The bug causes the shedding of young squares, and its feeding on terminal buds leads to excessive branching or "crazy cotton" (e.g., Tugwell et al. 1976). The tarnished plant bug is considered an important or key cotton pest in the midsouth United States (e.g., Craig et al. 1997), but its threat to cotton in the midsouth and southeast United States might be exaggerated (Bacheler et al. 1990). Like populations of *L. hesperus*, those of *L. lineolaris* show tolerance or resistance to all major classes of insecticides (Wheeler 2001). An increased use of transgenic cottons — so-called Bt cotton — might increase injury from this species (e.g., Hardee and Bryan 1997, Mensah and Khan 1997).

Natural enemies attack all stages of the tarnished plant bug; Clancy and Pierce (1966) listed egg, nymphal, and adult parasitoids (see also King et al. 1996). Biological control efforts against *L. lineolaris* are reviewed by Hedlund and Graham (1987); the development and use of resistant cultivars of cotton are discussed by Wheeler (2001).

Orthotylinae

This cytologically and morphologically diverse subfamily contains relatively few crop pests, as well as a larger proportion of facultative predators than does the Mirinae. Predacious orthotylines important in classical and natural biological control are treated in **Chapter 28**. Schuh (1995) recognized three orthotyline tribes: Halticini, Nichomachini (whose habits are unknown), and Orthotylini. Halticines generally cause foliar chlorosis on their hosts. Some phytophagous members of the Orthotylini also cause chlorosis, but they usually do not induce shot-holing or foliar tattering, lesions and cankers, or abnormal growth patterns (Wheeler 2001). No Orthotylinae are categorized here as "important pests," but three species (see **Sections 4.16 through 4.18**) are treated later as minor crop pests.

Phylinae

Most members of this large subfamily probably are omnivores. Even pest species such as *Pseudatomoscelis seriata* (see **Section 3.10**) and *Spanagonicus albofasciatus* (Reuter) (not treated here) are facultative predators (Wheeler 2001). Schuh (1995) recognized five tribes: Auricillocorini, a small Oriental group; the predominantly Old World Hallodapini, antlike, often ground-inhabiting bugs; Leucophoropterini, which contain important predators, mainly in the Old World (see **Chapter 28**); Phylini, largest of this subfamily's tribes; and the widely distributed, generally predacious Pilophorini. All species of Phylinae discussed below as major or minor pests belong to the nominate tribe Phylini.

3.10 *Pseudatomoscelis seriatus* (Reuter) (= *Psallus seriatus* [Reuter])

The cotton fleahopper, widespread in the southern United States, ranges as far north as Saskatchewan and Minnesota in the west and New Jersey in the east. *Psallus seriatus* is the name used in the early economic literature; the spelling of the specific name — seriatus in *Pseudatomoscelis*—is a recent emendation (Wheeler 2001).

Reinhard (1926) described the immature stages, illustrated the nymphs, and provided life-history data. He reported a preoviposition period of 3 to 4 days, an incubation period of 6 to 12 days, and a nymphal development period of 7.5 to 19.0 days; development is largely temperature dependent. Eddy (1928) found that the preoviposition period ranged from 3 to 8 days, the incubation period 7 to 11 days, and nymphal development 7.5 to 31.5 (average = 16.3) days. Under laboratory conditions, maximum longevity is about 30 days for males and females (Reinhard 1926, Eddy 1928). Eddy (1928) recorded average fecundity of about 24 eggs.

This polyphagous bug develops mainly on reproductive structures of weeds belonging to numerous plant families. Some of the more commonly used hosts are species of *Croton, Monarda,*

Oenothera, and *Solanum* (e.g., Reinhard 1926, 1928; Fletcher 1940; Hixson 1941; Brett et al. 1946; Almand et al. 1976; Holtzer and Sterling 1980). When weed hosts mature or desiccate, *P. seriatus* adults often colonize cotton fields. The cotton fleahopper's massive emigration from weed hosts makes this otherwise innocuous bug an important pest (Sterling et al. 1989b). Although widespread outbreaks have occurred in the southern states — in 1926, for example (Isley 1927) — the cotton fleahopper is considered a significant pest only in Arkansas, Louisiana, Mississippi, and the eastern three quarters of Texas. Elsewhere in the Cotton Belt, it is merely a sporadic, minor pest (Sterling and Dean 1977).

Eggs of *P. seriatus* are laid mostly on hosts that are beginning to flower rather than on plants in a vegetative condition or that have finished blooming (e.g., Holtzer and Sterling 1980). Hosts near plants that harbor overwintered eggs are infested early in the season. Cotton fleahoppers then track a succession of flowering hosts and usually do not disperse from weed hosts into cotton until squares have formed, or at least do not develop large populations until squares are available (Almand et al. 1976, Snodgrass et al. 1984a). Irrigation and fertilization of cotton can promote greater fleahopper densities (Adkisson 1957), and populations sometimes increase after periods of rain. Once the bugs invade fields, they move from maturing to more succulent cotton plants (Glick 1983).

Although Howard (1898) gave circumstantial evidence for injury to cotton by *P. seriatus*, applied entomologists did not initiate studies of this plant bug until the 1920s (Hunter 1924, 1926; Reinhard 1926; Eddy 1927). Early work suggested that a recently appearing "mysterious disorder" of cotton was caused by a phytopathogen transmitted by the fleahopper (Hunter 1924). Subsequent research showed that typical hopper injury — shedding of small squares, stem lesions and swellings, and abnormal growth such as excessive branching — results from the bug's direct feeding and its salivary secretions (e.g., Painter 1930, King and Cook 1932). Insecticides applied to suppress fleahopper densities can trigger outbreaks of bud- and bollworms when their natural enemies are killed (e.g., Adkisson 1973, Adkisson et al. 1982).

Assessments of the cotton fleahopper's pest status and effects on yields should consider the ability of cotton to compensate for removal of squares (e.g., Sterling 1984; Stewart and Sterling 1988, 1989; Sterling et al. 1989b). For a discussion of square abscission in relation to ethylene production, concentration of indole-acetic acid (IAA) in the plant, and the contamination of *P. seriatus* by pathogenic and nonpathogenic fungi and bacteria, readers should consult the summary of research provided by Wheeler (2001). That reference also discusses this pest's natural enemies (see also Sterling et al. 1989a), the modeling of the impact of native predators and other natural enemies on fleahopper abundance (see also Breene et al. 1989, Sterling et al. 1989b), and development of cottons resistant to *P. seriatus* (see also Walker et al. 1974, Ring et al. 1993).

4. THE LESS IMPORTANT SPECIES

Bryocorinae

4.1   *Engytatus modestus* (Distant) (= *Cyrtopeltis modestus* Distant)

This widely distributed New World species is considered adventive in Hawaii, where it was detected in 1924. The tomato bug (or "tomato girdler") became increasingly important to Hawaiian tomato production in the late 1930s and early 1940s. This pest was nearly eliminated from commercial tomatoes once DDT became available for control in 1947 (Tanada and Holdaway 1954). The name *Cyrtopeltis modestus* is used in most of the economic literature on this species; *Engytatus*, formerly considered a subgenus of *Cyrtopeltis*, is now generally accorded generic status (Cassis 1984, Schuh 1995).

Nymphs and adults feed on flower buds of their hosts, retarding flower production. Infested tomato plants often become bushy and have swollen nodes and shortened internodes. The tomato bug also is a characteristic vascular feeder on stems; its punctures partially or completely encircle a stem or petiole. The resulting feeding scars or rings become visible within 24 hours, turn reddish brown, and become raised. Leaves above the feeding rings can wilt. Wound parenchyma that replaces elements of the vascular bundle often becomes so brittle that injured tomato stems break. Tomato bugs generally are more numerous on plants of high-nitrogen content (Tanada and Holdaway 1954).

Tanada and Holdaway's (1954) paper contains further information on the habits of *E. modestus*, including details of its feeding on vascular tissues of tomato, as well as chemical and histological analyses of the stem lesions it causes. In addition to being injurious to tomato, tobacco, and other plants outside Hawaii (Tanada and Holdaway 1954), this mirid is a facultative predator (see **Chapter 28**).

4.2 *Helopeltis* species

In addition to the species of *Helopeltis* discussed as major pests — *H. antonii*, *H. schoutedeni*, and *H. theivora* (**Sections 3.2 to 3.4**) — other species of the genus cause similar injury symptoms on tropical crops. The more important among these additional species are *H. anacardii* Miller in Africa and *H. bradyi* Waterhouse and *H. clavifer* (Walker) in the Oriental region.

At an average temperature of 23.6°C, the incubation period for *H. anacardii* caged on cashew shoots was 15 days and the nymphal period about 21 days. Nymphs and adults feed on cashew leaves, shoots, and fruits in Tanzania, retarding tree growth and causing undetermined losses in yield (Swaine 1959). De Silva (1961) reared *H. bradyi* (as *H. ceylonensis* De Silva) at 23.9°C, reporting an incubation period of 13 days and a nymphal period of 14.5 days when the bug was reared on pieces of cocoa. Damage by *H. bradyi* to cocoa pods in Sri Lanka was described by Fernando and Manickavasagar (1956) and De Silva (1961). *Helopeltis clavifer* is a well-studied pest of cocoa and tea in Papua New Guinea. Information on this pest includes a method for its continuous rearing in the laboratory (Smith 1973), evidence that females produce a sex pheromone (Smith 1977), description of the immature stages (Smith 1979), a list of this bug's cultivated and wild hosts in Papua New Guinea and discussion of its injury to plants (Smith 1978), relationship between overhead shade and the damage it causes on cocoa (Smith 1985), discussion of population fluctuations (Smith 1972), and assessment of its damage to tea production (Smith et al. 1985).

4.3 *Nesidiocoris tenuis* (Reuter) (= *Cyrtopeltis tenuis* Reuter)

Indigenous to the Old World, *N. tenuis* has been transported widely in commerce and apparently has colonized several island groups by aerial dispersal. Its distribution is essentially Paleotropical (Wheeler and Henry 1992). As is the case in *E. modestus*, this species has often been placed in the genus *Cyrtopeltis*.

El-Dessouki et al. (1976) reviewed previous studies of this plant bug as a pest of tomato, tobacco, and other crops. Solanaceous plants are the most common hosts. In biological studies in Egypt, El-Dessouki et al. (1976) found that eight annual generations can be completed in the laboratory: two in winter (17.4°C) and six during summer and autumn (28.2°C). The incubation period ranged from about 10 to 19 days and the nymphal period from about 16 to 23 days. Under laboratory conditions, fecundity ranged from 3 to 8 days in winter and 4 to 11 days in summer; longevity was 6 to 15 days in winter and 3 to 13 days in summer. Somewhat similar figures for longevity and nymphal development were obtained by Raman and Sanjayan (1984a) in laboratory studies (temperature and other conditions not stated) in India.

Nymphs and adults feed on all aerial portions of tomato, especially growing points and young leaves, and frequently return to previous feeding sites. The bugs cause brownish lesions or rings on stems, petioles, and leaf veins, as well as foliar crinkling. Affected plant parts often become brittle and drop prematurely. Infested parts showed slight increases in total protein content and slight decreases in total saccharides and disaccharides (El-Dessouki et al. 1976).

Wound responses in tomato are characterized by an increase in phenolic compounds and oxidative enzymes, which can lead to decreased growth-promoting activity (Raman and Sanjayan 1984a, Raman et al. 1984). Studies on *N. tenuis* in India include data on host relationships and population dynamics (Raman and Sanjayan 1984a), histology and histopathology of feeding lesions (Raman and Sanjayan 1984b), and the biochemical changes this plant bug induces in host tissues (Raman et al. 1984).

Like many dicyphine Bryocorinae, *N. tenuis* shows strong predacious tendencies. Its value as a predator of crop pests is noted in **Chapter 28**.

4.4 *Platyngomiriodes apiformis* Ghauri

Known only from the Malaysian state of Sabah, the "bee bug" feeds on branches, chupons, and pods of cocoa. The injury it causes on pods, which is considered more serious than that produced on cocoa by *H. clavifer* (**Section 4.2**), can lower yields (Pang 1981, Khoo et al. 1991).

Pang (1981) found that females in the laboratory and in field cages (environmental conditions not stated) laid an average of 71.6 eggs (range 13 to 172), the eggs hatching in 14 to 15 days. Nymphal development required about 21 days, and adults lived 17 to 39 days when caged on pods in the field and 6 to 21 days in the laboratory. The gregarious nymphs and adults feed mostly at the base of pods, creating sunken lesions 1 to 7 mm in diameter that can be invaded by secondary fungi. Khoo et al. (1991) provided excellent color photographs of injury to cocoa pods. Infestations generally are localized, the bugs preferring dense, shaded pods on older trees; the bugs are most abundant during April to May and September to October (Pang and Syed 1972, Pang 1981, Khoo et al. 1991).

4.5 *Pycnoderes quadrimaculatus* Guérin-Méneville

The bean capsid, found in the West Indies, Mexico, Central America, South America, and the southern United States, has become established in Hawaii through commerce. The informal common names "black squash mirid" (Maldonado 1969), "small squash bug" (Martorell 1976), and "humped-back melon bug" (Morrill 1925) have appeared in the economic literature.

In hot, dry regions of Hawaii this bryocorine has injured cucurbit crops such as squash (*Cucurbita* spp.), as well as sweet potato, *Ipomoea batatas* (L.) Lam., and white mustard, *Sinapsis alba* L. Nymphs and adults cause chlorosis or mottling of the upper leaf surfaces of their hosts and deposit black excrement on the lower surfaces (Holdaway and Look 1942). Affected foliage can wither and die, and fruit can drop prematurely. Bean capsids have damaged squash in Puerto Rico (Cotton 1918, Wolcott 1933); cucumbers (*Cucumis sativus* L.) in Mexico (Morrill 1925, 1927); christophine or chayote, *Sechium edule* (Jacq.) Sw., in Central America (King and Saunders 1984); and cucumbers (Cockerell 1899), gourds (*Cucumis* spp.) (Lebert 1935), and cantaloupe (*C. melo* var. *cantalupensis* Naudin) (Johnston 1940) in the southwestern United States.

Mirinae

4.6 *Adelphocoris lineolatus* (Goeze)

The alfalfa plant bug is native to Eurasia, including Great Britain, most of continental Europe, China, Japan, and Korea. The first North American record was Nova Scotia, with Iowa the first U.S. record. In North America this adventive species is known as far south as North Carolina and as far west as Alberta and Montana (Wheeler and Henry 1992).

Kullenberg (1944) reported the host plants, seasonality, and habits of *A. lineolatus* in Sweden. Hughes (1943) found that in Minnesota the incubation period ranged from 11 to 18 days, depending on temperature. The mean developmental period for nymphs was 18 to 28 days. Populations in

North America are bivoltine except in more northern parts of Canada, where they are univoltine (Craig 1963).

Considered an alfalfa (lucerne) pest in Eurasia (see Wheeler 2001), *A. lineolatus* is a pest of alfalfa seed crops in North America. Its feeding causes a blasting of buds, flower abscission, poor bloom, and shriveled seeds (Hughes 1943). Injury to alfalfa grown for forage — yellowing or reddening of leaves and plant stunting — resembles hopperburn caused by the potato leafhopper, *Empoasca fabae* (Harris) (Radcliffe and Barnes 1970). The alfalfa plant bug also injures reproductive parts of birdsfoot trefoil (*Lotus corniculatus* L.) (Wipfli et al. 1989, 1990). Additional information on bionomics is available in Hughes (1943), Craig (1963), and Soroka and Murrell (1993); Wheeler and Henry (1992) and Wheeler (2001) provide additional references.

4.7   *Closterotomus norwegicus* (Gmelin) (= *Calocoris norvegicus* [Gmelin])

Native to Europe, *C. norwegicus* is adventive in North America, where it likely has been introduced separately to the East and West Coasts. Early records from eastern North America date from the 1880s. This species should also be considered nonindigenous in New Zealand (Wheeler and Henry 1992). Until recently, it has been placed in *Calocoris*, but the former subgenus *Closterotomus* has been accorded generic status. The specific name, usually spelled *norvegicus* (an incorrect subsequent spelling), should be spelled with a "w" (Cassis and Gross 1995, Rosenzweig 1997, Kerzhner and Josifov 1999).

European populations of *C. norwegicus* (e.g., Kullenberg 1944, Lupo 1946, Southwood and Scudder 1956) are univoltine. Although this plant bug develops on grasses and other herbaceous hosts, its eggs are laid mainly on woody plants (Southwood and Leston 1959). The oviposition behavior was described by Smith (1925) and Massee and Steer (1928). Kullenberg (1944), Southwood and Scudder (1956), and Ehanno (1987) provided bionomic data for this mirid in Europe. Seasonality and host plants in California were reported by Purcell and Welter (1990). Kullenberg (1942) described and illustrated the egg, and Southwood and Scudder (1956) briefly described the nymphal stages and illustrated the fifth instar.

This plant bug, known as the potato mirid (or capsid) in Great Britain, is an occasional pest of numerous crops, the adults often dispersing and injuring woody plants when herbaceous hosts used for nymphal development mature or dry (e.g., Rice et al. 1988, Purcell and Welter 1990, Wheeler 2001). Injury ranges from a twisting and withering of asparagus (*Asparagus officinalis* L.) spears (Findlay 1975, Watson and Townsend 1981, Townsend and Watson 1982), multiple crowning of sugar beet (Bovien and Wagn 1951), shriveling of seeds and proliferation of secondary stems of legumes such as alfalfa, lotus (*L. pedunculatus* Cav.), and white clover (*Trifolium repens* L.) (Macfarlane et al. 1981, Wightman and Macfarlane 1982, Wightman and Whitford 1982, Clifford et al. 1983, Chapman 1984, Schroeder et al. 1998) to stunting of the growth of peach trees (Cravedi and Carli 1988) and epicarp lesion of pistachio (*Pistacia vera* L.) (Rice et al. 1985; Uyemoto et al. 1986; Purcell and Welter 1990, 1991). Other crops that can sustain minor injury are carrot, chrysanthemum (*Chrysanthemum* spp.), flax (*Linum usitatissimum* L.), potato, the small grains barley (*Hordeum vulgare* L.) and wheat, and strawberry (Wheeler and Henry 1992, Wheeler 2001).

4.8   *Creontiades dilutus* (Stål) (= *Megacoelum modestum* Distant)

The green mirid, endemic to Australia, has appeared in some of the economic literature under the synonym *Megacoelum modestum* Distant (Cassis and Gross 1995, Wheeler 2001). A mean incubation period of 16.0 days at 19°C and 5.6 days at 31°C and nymphal development of 26.3 days at 19°C and 9.3 days at 31°C was determined by Foley and Pyke (1985).

Chinajariyawong (1988) reported that *C. dilutus* causes wilted tips of cotton seedlings. Before reproductive parts of cotton become available, it feeds in early season on vegetative buds (Sadras and Fitt 1997). Its feeding on larger plants also leads to early-season loss of squares. Cotton compensates for some of this loss, but the bug's boll feeding can delay crop maturity by about a week (Chinajariyawong 1988, Mensah and Khan 1997, Murray and Lloyd 1997). Densities of the green mirid tend to be reduced on nectariless cultivars (Adjei-Maafo and Wilson 1983). Problems from this pest might intensify as insecticide use decreases with an increased planting of transgenic cottons (Fitt 1994, Mensah and Khan 1997). The use of an alfalfa interplanting or trap crops might help in managing populations of the green mirid in cotton (Mensah and Khan 1997).

Other types of injury caused by *C. dilutus* include deformed fruit of apple and gummosis of peach. Dark spots can also appear on peach fruit, with corky areas beneath (Pescott 1940, Victorian Plant Res. Inst. 1971).

The green mirid also is recorded as a pest of bean (*Phaseolus* spp.), carrot, grape, parsnip (*Pastinaca sativa* L.), passion fruit (*Passiflora edulis* Sims), and potato (*Solanum tuberosum* L.) (Hely et al. 1982, Foley and Pyke 1985). When caged on sunflower (*Helianthus* sp.) seedlings for more than 24 hours, five individuals can kill the plants; seedlings fed on for less than 24 hours eventually show shortened internodes and missing or deformed leaves (Miles and Hori 1977). Hori and Miles (1993) discussed the etiology of its injury to alfalfa. Nymphs feed on ovules of alfalfa and cannot complete their development on vegetative parts of the plants. Developing seeds show a yellowing or browning, and pods can wither (Miles and Hori 1977, Hori and Miles 1993).

4.9 *Creontiades pallidus* (Rambur)

Mainly a species of the Mediterranean region, northern and central Africa, and the Middle East, *C. pallidus* also is known from Brazil (Schuh 1995). This plant bug is sometimes called the boll (or bud) shedder bug (Wheeler 2001).

According to Soyer (1942), the incubation period is about 9 days and the nymphal development period 12 or 13 days under field conditions in the Democratic Republic of the Congo. In the laboratory, Ratnadass et al. (1994) obtained similar results when *C. pallidus* was reared on sorghum at 25°C: an incubation period of 5 to 7 days and total nymphal development of 8 to 14 days. Soyer (1942) described the immature stages and illustrated the second and fifth instars. Descriptions of all life stages and a summary of bionomics were provided by Cadou (1993).

As the common name suggests, this mirid is most important as a cotton pest. It can invade cotton when cereal crops ripen (Goodman 1955). Durra, a grain sorghum sometimes planted as a windbreak near cotton, can serve as a reservoir for infestations of *C. pallidus* in cotton (Ripper and George 1965). Nymphs are more injurious to cotton than the adults; both feed on squares and young bolls, which are usually pierced near or at a resin gland. Such punctures are marked by black spots where glands exude dark fluid. A sign of the bug's presence is bright yellow or brownish-yellow excrement. Its feeding causes the premature shedding of squares and young bolls. Older bolls that develop can split prematurely, fail to produce sound lint, and be colonized by secondary fungi (Willcocks 1922, Kirkpatrick 1923, Soyer 1942, Pearson 1958, Buyckx 1962, Ripper and George 1965).

Because cotton compensates for an early-season loss of squares, *C. pallidus* is usually considered an insignificant or minor pest (e.g., Goodman 1955, Schmutterer 1969). Stam's (1987) studies in Syria indicated that substantial yield loss can occur and that control is desirable when a level of seven individuals/50 sweeps is reached in June or July. Ripper and George (1965) stated that severe infestations of *C. pallidus* in cotton can delay crop maturity.

Among this bug's other host plants is sorghum. *Creontiades pallidus* is often encountered among the complex of panicle-feeding mirids on grain sorghum in Africa (e.g., Willcocks 1925, Gahukar et al. 1989, Steck et al. 1989, Sharma and Lopez 1990, Ratnadass et al. 1994) and sometimes feeds on the pistillate inflorescences of corn (*Zea mays* L.) (Willcocks 1925). It can be reared on the staminate inflorescences (Soyer 1942) but is not considered economically important on corn. Indigenous host plants of *C. pallidus* in Mali were reported by Ratnadass et al. (1997).

4.10 Eurystylus oldi Poppius (= Eurystylus immaculatus Odhiambo)

Restricted to the Ethiopian Region, *E. oldi* is widespread in the Sahel of Africa. In much of the economic literature on this sorghum pest, the synonym *E. immaculatus* has been used; references to *E. marginatus* Odhiambo and *E. rufocunealis* Poppius likely are misidentifications of *E. oldi* (Stonedahl 1995).

The preoviposition period is 2 to 3 days, the incubation period 4 to 7 days (23 to 27°C), and the nymphal development period 6 to 11 days (Ratnadass et al. 1994, 1995a). Only preliminary data on fecundity and longevity, based on a few laboratory observations, are available (Ratnadass et al. 1995a).

Eurystylus oldi is the most damaging species of the mirid complex that infests sorghum panicles in Africa. Its pest status might have been enhanced by an increased use of early-flowering cultivars with compact panicles and soft grains (e.g., Gahukar et al. 1989, Steck et al. 1989). Stonedahl (1995) reviewed field studies in which *E. oldi* is reported as the dominant panicle-feeding bug on sorghum.

Injury to sorghum grains involves direct feeding as well as oviposition (Steck et al. 1989, Ratnadass et al. 1995a). The bug's feeding removes nutrients and water, causing grains to shrivel, and its salivary secretions break down endosperm structure, altering the texture. The deposition of eggs into host tissues can cause grains to deteriorate, presumably from fungal infections (Steck et al. 1989). *Eurystylus oldi* reduces grain mass and quality as well as total yields (e.g., Steck et al. 1989, Ratnadass et al. 1995a). Efforts to assess damage and to screen cultivars resistant to *E. oldi* were reported by Sharma et al. (1992), Doumbia et al. (1995), and Ratnadass et al. (1995b). By reducing the mirid's densities on castor (*Ricinus communis* L.), an alternative host, sorghum might be protected from invasion and injury by this pest (Ratnadass et al. 1997).

4.11 Hyalopeplus pellucidus Kirkaldy

Once considered adventive in Hawaii, *H. pellucidus*, the transparentwinged plant bug, is actually endemic to the islands (Asquith 1997). Mau and Nishijima (1989) determined that the incubation period is about 7 days (26.7°C) and that nymphs develop in about 14 days.

This polyphagous mirid is economically important only on guava. Nymphs feed almost exclusively on flower buds, becoming common on guava trees when buds are numerous. The bugs cannot be reared on vegetative parts of guava. Nymphs injure the developing anthers, which are important in auxin production, and cause bud abscission. Nymphs (except first instars) are more injurious than are the adults. Continuation of management practices that result in sequential fruiting of guava and, therefore, a continuous food supply for the bugs, might intensify problems from this pest (Mau and Nishijima 1989).

4.12 Lygocoris pabulinus (L.)

This Holarctic species, known as the common green mirid (or capsid) in Europe, is widespread in Eurasia; references to its occurrence in the Oriental region (e.g., Yasunaga 1991) pertain to *L. viridanus* (Motschulsky) (Wheeler and Henry 1992). In North America, this mirine is transcontinental in Canada and ranges throughout much of the United States, including Alaska (Yasunaga 1991, Wheeler and Henry 1992). The Beringian distribution of *L. pabulinus* suggests that it is Holarctic in North America, but it also might have been accidentally introduced with nursery stock (Wheeler and Henry 1992). References to descriptions and illustrations of the immature stages are given by Wheeler and Henry (1992); additional references to the immature stages include Rostrup and Thomsen (1923) and Austin (1929, 1931a).

At 20°C the incubation period is about 16 days and the nymphal developmental period about 20 days (Mols 1990). Females mate after about 7 days, typically with males older than 8 days. As many as 200 eggs can be produced, with fecundity averaging about 100. Under laboratory conditions

(20°C, RH 65%, L:D 18:6 or 17:7), adults lived about 30 days (Blommers et al. 1997). Various other aspects of its reproductive behavior were reported by Blommers et al. (1988) and Groot et al. (1996, 1998).

Lygocoris pabulinus has a bivoltine life cycle in temperate Europe but is univoltine in Sweden (Austin 1931a, Kullenberg 1944, Southwood and Leston 1959, Blommers et al. 1997). This polyphagous plant bug (Austin 1931a) alternates between woody and herbaceous hosts, its eggs overwintering in shoots of woody plants such as fruit trees. Termination of egg diapause is discussed by Wightman (1969). Eggs hatch in spring, and early instars feed on developing fruits; the mid to late instars mostly move to herbaceous plants of the ground cover (e.g., Wightman 1968, Blommers et al. 1997). Nymphs apparently do not complete their development on the nutritionally inadequate resources of most woody plants (Blommers et al. 1997), although the life cycle can be completed on woody hosts (e.g., Hill 1952). Females from the spring generation, under long-day conditions, oviposit on herbaceous (summer) hosts. Second-generation females, responding to short-day conditions, leave their summer hosts to lay overwintering eggs on woody plants (Blommers et al. 1997). The paper by Blommers et al. (1997) should be consulted for additional information on bionomics of this mirid in Europe.

The life history of North American populations of *L. pabulinus* is little known. *Impatiens* spp. are the major hosts in eastern North America, but the bug seems more polyphagous in the West. Eastern and western populations of this species might not be conspecific (Wheeler and Henry 1992).

Injury by *L. pabulinus* to herbaceous weeds and field crops includes nearly transparent, pinholelike spots on leaves (e.g., Rostrup and Thomsen 1923, Petherbridge and Thorpe 1928, Austin 1931a), although the foliage of some plants remains asymptomatic (Smith 1931). Leaves of sugar beet can become puckered, followed by apical yellowing (Bovien and Wagn 1951). Similar yellowing or foliar crinkling occurs on artichoke (*Cynara scolymus* L.) (Smith 1920), bean (Carpenter 1920), and potato (Wightman 1974). Other symptoms associated with feeding by the common green mirid are foliar shot holing and tattering (e.g., Austin 1930, 1931a; Massee 1937; Hill 1952) and growth disorders such as excessive branching and stunting on currant, gooseberry (*Ribes* spp.), raspberry (*Rubus* spp.), and other woody plants (Austin 1931a; see also Wheeler 2001). Injury to developing fruits of apple and pear — malformation with corky, stony, or warty patches (e.g., Austin 1931a, Abraham 1935, Taksdal 1970) — was reviewed by Wheeler (2001). References dealing with monitoring *L. pabulinus* populations and other aspects of pest management were given by Blommers et al. (1997).

4.13 *Lygocoris rugicollis* (Fallén) (= *Plesiocoris rugicollis* [Fallén])

Like the preceding species, *L. rugicollis* shows essentially a circumboreal distribution (Scudder 1997) and is, therefore, considered a Holarctic mirid. The Palearctic distribution includes Great Britain, most of continental Europe, and northern Africa; this mirine ranges east in the former U.S.S.R. to Chukotka, Kamchatka, Magadan Province, and Sakhalin Island (Wheeler and Henry 1992). North American records include Alaska and the Canadian provinces Alberta, Manitoba, and Yukon in the west and Newfoundland in the east (Wheeler and Henry 1992, Scudder 1997). The name *Plesiocoris rugicollis* has been used in the economic literature, but *Plesiocoris* is now considered a synonym of *Lygocoris* (Schuh 1995).

Petherbridge and Husain (1918), Rostrup and Thomsen (1923), Austin (1931a), and Jacobi (1932) described the nymphal stages; the papers by Rostrup and Thomsen (1923) and by Jacobi (1932) also contain illustrations of all stages. Austin (1929) described the egg and oviposition sites.

The history of *L. rugicollis* as a plant pest in Europe parallels that of the apple red bugs *Heterocordylus malinus* (**Section 4.17**) and *Lygidea mendax* Reuter, whose rise and fall as pests in North America was described by Schaefer (1974). Like the previously obscure North American red bugs, *L. rugicollis* switched from native hosts — in this case, willow (*Salix* spp.) and alder (*Alnus* spp.) — to colonize and damage apples in the first quarter of the 20th century (Fryer 1929,

Austin 1931a, Wheeler 2001). Known as the apple capsid, it remained an important pest until the introduction of DDT in 1946 (Southwood and Leston 1959). Before synthetic organic insecticides were used in apple orchards, this pest was feared more than any other by British orchardists (Massee 1937). Since the elimination of the apple capsid from British orchards, it has not reinfested commercial orchards (Dicker 1967). The apple capsid also was once an important pest of currant, gooseberry, raspberry, and strawberry (e.g., Austin 1931a).

From about 1914 to the 1930s, European entomologists studied the bionomics of *L. rugicollis*. These workers observed that overwintering eggs hatch soon after budbreak on apple and that early instars feed on new foliage, causing characteristic brown or purple spots. Heavily damaged leaves become frayed or tattered (e.g., Austin 1931a). Nymphs, mainly late instars, also feed on developing fruit less than about 2.5 cm in diameter, which can result in abscission and distorted, unsalable fruit with corky scars. Adults generally feed little on fruit but often severely injure terminal shoots (e.g., Petherbridge and Husain 1918, Smith 1920, Rostrup and Thomsen 1923, Fryer 1929, Austin 1931a, Abraham 1935). A single generation is produced on apple, but populations might be bivoltine on wild hosts (Southwood and Leston 1959).

4.14   *Lygus rugulipennis* Poppius (= *Lygus disponsi* Linnavuori, *Lygus perplexus* Stanger)

The European tarnished plant bug is common and widespread in the Palearctic region, ranging from Great Britain and Spain east to Far East Russia and Japan (Schwartz and Foottit 1998). It was recognized as a Holarctic species when Schwartz and Foottit (1998) synonymized the Nearctic *L. perplexus* Stanger under *L. rugulipennis* (see also Leston 1952). In North America, *L. rugulipennis* is known from Alaska to northern California and southern Colorado (Schwartz and Foottit 1998).

Taxonomic confusion has surrounded the identity of *L. rugulipennis* in Eurasia. Many of the early records of *L. pratensis* (L.) as a crop pest refer to *L. rugulipennis*; the latter species is the more important pest in Britain and continental Europe (Southwood 1956, Varis 1972). Many of the early ecological studies in Britain, published under *L. pratensis* (e.g., Fox-Wilson 1925, 1938; Austin 1931b, 1932), likely refer to *L. rugulipennis* (Southwood 1956, Stewart 1969). *Lygus pratensis* apparently is rare in Britain (Woodroffe 1966). In the extensive literature on *L. rugulipennis* in Japan, based on ecological and physiological studies by K. Hori, the name *L. disponsi* Linnavuori is used; the latter name is now considered a synonym of *L. rugulipennis* (Schuh 1995, Kerzhner and Josifov 1999).

Life-history data such as the incubation period, nymphal development, preoviposition, fecundity, and longevity were provided by Boness (1963), Bech (1969), Stewart (1969), Hori and Hanada (1970), Varis (1972), and Hori and Kuramochi (1984). Total nymphal development varies among generations in Japan (Hori and Hanada 1970); the nymphal period and fecundity also vary among generations, as well as with temperature and other laboratory conditions, and are affected by the host plant (e.g., Boness 1963, Varis 1972, Hori and Kuramochi 1984). All immature stages of *L. rugulipennis* were described by Hori and Hanada (1970), the egg was described and illustrated by Kullenberg (1942), and nymphs were described by Boness (1963).

Bivoltine in most of Europe, *L. rugulipennis* is univoltine in Finland, Scotland, and Sweden (Kullenberg 1944, Stewart 1969, Varis 1972), but is bi- or trivoltine in Japan (Hori and Hanada 1970). The host plants of this polyphagous bug, which include more than 400 (357 determined to species or subspecies) in 57 families, were listed and discussed by Holopainen and Varis (1991). Host preferences were analyzed by Holopainen (1989).

Among the numerous biological and physiological studies on the European tarnished plant bug are those on feeding habits and host injury (Hori 1971a,b; Varis 1972), salivary enzymes and salivary constituents (e.g., Hori 1974a,b, 1975a,b; Laurema et al. 1985; Laurema and Varis 1991), flight activity (Johnson and Southwood 1949, Šedivý and Honěk 1983), and light and color reactions (Bech 1965). Although *L. rugulipennis* is not an important crop pest in Britain (Southwood and

Leston 1959, Stewart 1969), it injures various crops in continental Europe and Japan. Symptoms of its feeding include foliar crinkling, shot holing, tattering, and yellowing; decreased weight of roots, multiple crowns, multiple leaders, and retarded growth; and fruit deformities (e.g., Varis 1972, Tingey and Pillemer 1977, Wheeler 2001). *Lygus rugulipennis* can be a pest of field and forage crops such as alfalfa, clover (*Trifolium* spp.), rape (*Brassica napus* L.), sugar beet, and sunflower; wheat and other small grains; conifer seedlings and floral crops; cabbage (*B. oleracea* L.), potato, and other vegetables; and peach, pear, strawberry, and other fruits (e.g., Fox-Wilson 1938, Southwood and Leston 1959, Holopainen and Varis 1991, Varis 1997, Wheeler 2001). This plant bug has been studied extensively in relation to pine seedlings (Holopainen 1986, Holopainen and Rikala 1990), sugar beet (Varis 1972), and wheat (Nuorteva and Reinius 1953, Rautapää 1969, Varis 1991).

4.15 *Taylorilygus vosseleri* (Poppius) (= *Lygus vosseleri* Poppius)

This polyphagous mirine of the Ethiopian region is a cotton pest in more humid equatorial areas of Africa (Pearson 1958, Ripper and George 1965). In much of the applied literature the names *Lygus vosseleri* or *L. simonyi* (a misidentification) have been used (Pearson 1958).

The incubation period in *T. vosseleri* is 7 to 13 days, and nymphs can develop in about 11 days or require more than 40 days, depending on environmental conditions (Geering 1953, da Silva Barbosa 1959). Fecundity can be as high as 150 or 300 eggs (Ripper and George 1965). In the laboratory, longevity averaged about 43 days (Geering 1953), although under laboratory conditions adults often do not live more than about 10 days (da Silva Barbosa 1959). The egg and nymphal stages were briefly described by Cadou (1993).

Of the crops on which *T. vosseleri* develops, cotton usually sustains the most serious injury. *Taylorilygus vosseleri* can produce three generations on cotton (Ripper and George 1965). Nymphs feed on young vegetative tissues (stems and mature leaves are not injured), resulting in ragged, shredded, and shot-holed leaves (illustrated by Coaker 1957), as well as on squares and young bolls (Hancock 1935, Taylor 1945, Pearson 1958). Bolls become spotted, pitted, and can split prematurely and be invaded by secondary plant pathogens (Hancock 1935, Coaker 1957). The premature loss of squares causes normally dormant vegetative buds to sprout, which induces secondary growth and other abnormal growth patterns (Taylor 1945). Severity of injury depends partly on the numbers of bugs that disperse into cotton from alternative hosts, both the crop plants blackgram (*Vigna mungo* [L.] Heppner) and sorghum and wild plants such as *pseudarthria* (*Pseudarthria hookeri* Wight & Arn.) (McKinlay and Geering 1957, Stride 1968). Affected cotton plants often are abnormally tall, straggling, have shortened internodes, and show a cylindrical rather than a pyramidal form. Such injury resembles that caused by bacterial blight. Similar to the controversy surrounding the effects of lygus bugs on cotton yields in the United States, there is disagreement concerning actual losses caused by *T. vosseleri* in Africa (e.g., McKinley and Geering 1957, Pearson 1958). Despite the ability of cotton to compensate for the loss of squares, prolonged and severe infestations by *T. vosseleri* can reduce yields, especially in late plantings (Coaker 1957, McKinlay and Geering 1957). For additional information on this plant bug and references to its injury symptoms and economic importance on cotton, as well as on natural enemies, host resistance, and transmission of phytopathogens, readers should refer to reviews by Pearson (1958), Ripper and George (1965), and Cadou (1993).

Orthotylinae

4.16 *Halticus bractatus* (Say) (= *Halticus citri* [Ashmead])

The New World garden fleahopper, widespread in the United States, is also indigenous to portions of southern Canada, Mexico, and Central and South America (Henry and Wheeler 1988).

This species of the tribe Halticini is considered adventive in Hawaii (Davis et al. 1985). The synonym *H. citri* (Ashmead) has been used in some of the early applied literature (e.g., Beyer 1921).

Beyer (1921) reported an incubation period of 6 to 11 days and nymphal development of 10 to 18 days. The life-history data of Cagle and Jackson (1947), which show even wider variation in both life-history parameters, emphasize the effects of temperatures and generation on these measurements. Fecundity can reach 200 eggs, and longevity of both sexes can be nearly 100 days, but both vary widely depending on the generation. Beyer (1921) described and illustrated the immature stages as well as the sexually dimorphic adults. Males are fully winged (macropterous), but nearly all females are brachypterous, resembling (and sometimes confused with) flea beetles.

This polyphagous bug first attracted the attention of economic entomologists in the late 19th century (Webster 1897, Beyer 1921). The garden fleahopper has injured numerous cereal, flower garden, forage, greenhouse, and truck garden crops, but has seldom been a severe pest since the early 20th century. With hatching of overwintered eggs in spring, the nymphs feed on foliage of young plants, causing chlorosis on upper leaf surfaces resembling that caused by spider mites (Tetranychidae) (Beyer 1921). The chlorosis has been described as "white, snowflake like spots" (Day and Saunders 1990). Injury can assume an overall bleaching of infested leaves, followed by shriveling and death of foliage and retardation of plant growth. Severe infestations on alfalfa and clover can cause plant death (e.g., Beyer 1921). *Halticus bractatus* once destroyed several thousand acres of tomatoes on Mexico's western coast (Morrill 1925).

This multivoltine species can produce five or six generations in South Carolina (Beyer 1921) and five in Virginia (U.S.A.) (Cagle and Jackson 1947). Three generations apparently are produced on alfalfa in New Jersey (Day and Saunders 1990). Day and Saunders (1990) provided data on the garden fleahopper's abundance on alfalfa, noting it usually does not cause noticeable injury in established stands, and discussed parasitism of nymphs by a euphorine braconid. Day (1991) described the incidence of sexual dimorphism in alfalfa fields and the underrepresentation of adult males and nymphs in sweep-net samples.

4.17 *Heterocordylus malinus* Slingerland

Authorship of this species, traditionally attributed to Reuter, should be credited to Slingerland. M.V. Slingerland, an economic entomologist at Cornell University, unintentionally validated the name in 1909 in the published proceedings of the Western New York Horticultural Society. The formal taxonomic description by the Finnish hemipterist O. M. Reuter did not appear until later that year (Wheeler 1983).

Known as the apple red bug, dark apple red bug, or true apple red bug (Crosby 1911, Hodgkiss and Frost 1921), *H. malinus* bears no approved common name. Apple red bug is now the Entomological Society of America's common name for *Lygidea mendax* Reuter (Bosik 1997), a frequently co-occurring mirid. Like *Lygocoris rugicollis* (see **Section 4.13**) in British apple orchards, *H. malinus* has nearly been eliminated from commercial orchards (Schaefer 1974); now only rarely encountered in North American orchards, this plant bug is suppressed by modern management practices (Schaefer 1974, Boivin and Stewart 1983, Wheeler 1983).

This Nearctic member of the tribe Orthotylini ranges from Ontario and Québec (Canada) to Louisiana and Mississippi and west to Colorado (U.S.A) (Henry and Wheeler 1988). Its damage to apple was most severe in New England, New York, and Pennsylvania (U.S.A.) (Schaefer 1974). In the late 19th century, *H. malinus* appears to have shifted from native rosaceous hosts, such as crabapple (*Malus* spp.) (e.g., Cushman 1916), to colonize apple, an introduced crop plant (Chapman and Lienk 1971). Horticultural practices, such as fertilization and pruning of apple, might have resulted in a plant more succulent or suitable than this bug's native hosts (Schaefer 1974). Slingerland's (1909) report of *H. malinus* as an apple pest, based on observations dating from 1896, was followed by Crosby's (1911) notes on life history. In addition to describing the immature stages, Crosby (1911) stated that nymphal development in an insectary required about 35 days (temperature unspecified). Despite the research devoted to this pest in the early 19th century, detailed bionomic information is lacking. Knight (1915) described the oviposition behavior.

Overwintered eggs hatch at or near vegetative budbreak on apple. Early instars feed on the unfolding leaves, causing tiny red spots (Knight 1918, 1922; Hodgkiss and Frost 1921). Because eggs of *H. malinus* hatch about 7 to 10 days earlier than those of *L. mendax* — that is, before fruit has formed — nymphs feed sparingly or not at all on developing fruit. In New York (U.S.A.), *H. malinus* usually was not associated with the fruit scarring and malformation that characterized damage by *L. mendax* (Knight 1918, 1922). In other eastern states, the latter bug also was considered a more important apple pest than *H. malinus* (Schaefer 1974).

By the 1930s, *H. malinus* had been reduced to the status of only a minor or sporadic pest. This bug's "sudden rise, brief glory, and swift fall" were detailed by Schaefer (1974).

4.18 *Labops hesperius* Uhler

Known unofficially as a black grass bug (this common name also is applied to species of *Irbisia* [Mirinae]), this native North American plant bug ranges from British Columbia, Ontario, and perhaps Québec (Canada) south to California and New Mexico (U.S.A.); northeastern U.S. records likely pertain to *L. hirtus* Knight (Slater 1954, Larochelle and Larivière 1979, Kelton 1980, Larochelle 1984, Henry and Wheeler 1988). Because of the effects that this halticine mirid has on western range grasses, including reduced forage production (as well as costs of control and revegetation), it has been studied extensively (e.g., Haws 1978, Hansen et al. 1985b). Lattin et al. (1995) provided an annotated bibliography of *Labops* (and *Irbisia*) species. Outbreaks of this once innocuous plant bug began as early as the mid-1940s (Denning 1948) and appear to have been triggered by the revegetation of native rangelands with virtual monocultures of exotic grasses such as wheat grasses (*Agropyron, Thinopyrum* spp.) (Lattin et al. 1994, 1995; Wheeler 2001).

The univoltinism and egg diapause of *L. hesperius* hamper laboratory rearing and partly explain the lack of detailed descriptions of its immature stages or data on most of its life-history attributes. A preoviposition period of about 2 weeks is known (Todd and Kamm 1974), as is fecundity of about 50 eggs (Haws and Bohart 1986). Nymphal development in the field requires about 4 weeks (Higgins et al. 1977). Biological aspects that have received attention include the effects of temperature and photoperiod on egg diapause (Fuxa and Kamm 1976a), and wing polymorphism and its effects on the bug's dispersal ability (Fuxa and Kamm 1976b).

Overwintered eggs hatch in spring after snowmelt. In Oregon, hatching begins around late March (Todd and Kamm 1974), the nymphs causing foliar chlorosis characteristic of halticine mirids. Late instars cause substantially more damage than do the early instars (Ling et al. 1985). *Labops hesperius* often is most abundant at elevations between 1800 and 2800 m (Haws 1978, Haws and Bohart 1986). Heavy infestations are characterized by large, nearly circular, white or yellow patches within otherwise healthy areas of grass (Bohning and Currier 1967). Injury at the cellular level involves destruction of chloroplasts and the disappearance of cytoplasmic organelles such as endoplasmic reticulum and mitochondria (Brewer et al. 1979). Injury can also include reduction in seed number, seed weight, and percentage germination (Malechek et al. 1977, Hewitt 1980). Plant vigor can be impaired with reductions in crown carbohydrate reserves, leaf length, and seed-head height (Ansley and McKell 1982). Relatively low densities (156 bugs/m^2) may not reduce forage yields (Malechek et al. 1977), and insecticidal control usually is unwarranted (e.g., Vallentine 1989).

Crested wheat grass, *Agropyron cristatum* (L.) Gaertn., and intermediate wheat grass, *Thinopyrum intermedium* (Host) Barkw. & D. R. Dewey, are common hosts of *L. hesperius*. Some grass species and cultivars tolerate its feeding, and such less susceptible host grasses could be used to reseed western rangelands (Hewitt 1980; Hansen et al. 1985a,b). A key to reducing damage from this pest might also lie in diversifying rangelands through reseeding, emphasizing species diversity of native grasses (Spangler and MacMahon 1990, Lattin et al. 1994).

Phylinae

4.19 *Campylomma verbasci* (Meyer-Dür) (= *Campylomma nicolasi* Puton & Reuter)

This plant bug, native to central and southern Europe, has been known in Britain only since the 1930s (Thomas 1938) and might not be native to the British Isles (Wheeler and Henry 1992). In North America, where it clearly is adventive, *C. verbasci* might have been introduced separately with nursery stock to both the Northeast and the Pacific Northwest (Wheeler and Henry 1992). Movement of infested fruit trees from the East could have contributed to its westward spread (Smith 1991). Leonard (1915) described and illustrated the immature stages, and Collyer (1953) briefly described the nymphal stages and illustrated the egg and fifth instar.

An omnivore (see **Chapter 28** for discussion of its predatory tendencies), *C. verbasci* usually has been considered noneconomic in Britain and continental Europe (Collyer 1953, Southwood and Leston 1959, Niemczyk 1978). During the 1990s, however, it began to damage apple fruit in the Netherlands, and similar injury has occurred in Belgium and Bulgaria (Stigter 1996). The "mullein bug," as it is sometimes known in North America, is often a sporadic or an intermittent pest of apple. Studies of such pests have recently been emphasized by pest-management specialists, and Thistlewood and Smith (1996) suggested that more about the mullein bug's bionomics is known in North America than for any other small bug of similar feeding habits. The following biological account of *C. verbasci* should be supplemented by reference to Wheeler and Henry (1992) and especially the review of its ecology, sampling methods, and management on apple by Thistlewood and Smith (1996); both works give references to *C. verbasci* that are not mentioned here. Information on this plant bug as a minor pest of cotton has appeared under the name *C. nicolasi* Puton & Reuter (e.g., Cowland 1934, Ripper and George 1965; see also Wheeler 2001).

Fecundity of the mullein bug averaged about 37 eggs at 25°C, and laboratory-reared females lived about 17 days (Smith and Borden 1991). Niemczyk (1978) reported a 25-day longevity for males and females at 23°C; at that temperature, mean fecundity was only about 10 eggs. Eggs hatch in a mean of 7.2 days at 27°C or 13.0 days at 20°C (Smith and Borden 1991). The time required for overwintering eggs to hatch at different temperatures was determined by Judd and McBrien (1994), who evaluated simulation models for predicting the entire hatching period. Nymphal development required 16 days at 23°C (Niemczyk 1978), 21 days at 22°C (Smith and Borden 1991), and 23 days at 21°C (McMullen and Jong 1970). The early instars generally do not develop without access to mites, psyllids, or other arthropod prey (e.g., McMullen and Jong 1970, Niemczyk 1978, Smith 1991). Pollen, however, can substitute for an animal diet (Bartlett 1996).

Campylomma verbasci is bivoltine in Europe (Southwood and Leston 1959, Niemczyk 1978) and has two, three, or occasionally four generations in Canada (Thistlewood and Smith 1996). Eggs overwinter in the bark of apple and pear trees, mostly in lenticels of current-season shoots. Hatching occurs near full bloom of apple and pear (Thistlewood et al. 1990), but takes place later in southwestern Québec than in British Columbia, Nova Scotia, or New York (Boivin and Stewart 1982, Thistlewood et al. 1990). Nymphs live mostly within flower clusters or webbing made by lepidopteran larvae (Thistlewood and McMullen 1989). Most first-generation adults, which appear during June in Canada, disperse to herbaceous plants such as common mullein (*Verbascum thapsus* L.). In Canada, second-generation populations are lower on apple than in the first generation (the opposite pattern prevails in England; Collyer 1953). Adults return to fruit trees during August and September, and overwintering eggs are deposited in September and October (Thistlewood and Smith 1996).

Nymphal damage can be severe on susceptible apple varieties such as 'Golden Delicious' (Stigter 1996, Thistlewood and Smith 1996). Pickett (1938) reported that 90% of the fruit was rendered unmarketable in a Nova Scotian orchard. Injury appears at nymphal feeding sites as small, dark warts on young fruit, followed by the development of corky scars when the bugs' punctures are numerous. Fruit also can become malformed or misshapen (e.g., MacLellan 1979, Judd and

McBrien 1994). Pesticide treatments sometimes have been prescribed when *C. verbasci* is merely detected during the most critical stage of fruit development (Thistlewood et al. 1989) — that is, from the pink stage to 2 weeks after petal fall.

During the 1980s and 1990s, *C. verbasci* occurred consistently in Canadian apple orchards in at least low densities, but its damage has been sporadic, thus challenging pest managers. The availability of alternative foods such as mites and aphids, as well as weather and other factors, might affect the degree of fruit damage as much as do actual pest densities (e.g., Thistlewood et al. 1990, Reding and Beers 1996, Thistlewood and Smith 1996). Damage in orchards often follows the collapse of mite or aphid populations (Lord 1971, Niemczyk 1978).

Management practices in orchards include the elimination or reduction of alternative hosts such as mullein and the potential use of pheromone-based disruption of mating (Judd et al. 1995; McBrien et al. 1996, 1997; Thistlewood and Smith 1996). Pheromone-based monitoring can also allow nymphal densities to be predicted the following spring (McBrien et al. 1994). Certain insecticides applied at the pink and bloom stages of apple development can substantially reduce fruit damage (Reding and Beers 1996).

4.20 *Moissonia importunitas* (Distant) (= *Ragmus importunitas* Distant)

A plant bug of Paleotropical (Kerzhner and Josifov 1999) or Indo-Pacific distribution (Schuh 1984), *M. importunitas* is a pest of sunn hemp (*Crotalaria juncea* L.), which is grown in India for fiber and widely cultivated in the tropics as a green manure. van der Goot (1927) briefly described the immature stages of this phyline in reporting it as a pest of *Crotalaria* species in Indonesia. More detailed descriptions of the egg and nymphal stages, with illustrations, were provided by Gopalan and Basheer (1966).

Gopalan and Basheer's (1966) study of *M. importunitas* in India included a description of mating and oviposition behavior, as well as data on fecundity (38 to 82 eggs), longevity (about 21 days for males; 32 days, females), incubation period (7 to 8 days), and nymphal development ($\bar{x}$ = 18 days) (environmental conditions were not stated). Initial feeding symptoms on sunn hemp foliage are yellow spots, which often coalesce to form larger lesions; leaves may become distorted, desiccate, and drop prematurely. Injury can also result in stunting of plants and reduced seed set (Gopalan and Basheer 1966, Gopalan 1976b). Entire fields sometimes appear whitened from foliar chlorosis (Reddy 1958), and plant death can occur when infestations are severe (Agarwal and Gupta 1983). *Moissonia importunitas* is considered a minor pest of sunn hemp in Pakistan (Shabbir and Choudhry 1984).

Gopalan (1976a) determined the salivary enzymes of *M. importunitas*, including amylase, cellulase, polyphenol oxidase, and protease. The effects of this mirid's feeding on respiration, transpiration, moisture content, and oxidative enzyme activity of sunn hemp were reported by Gopalan and Subramaniam (1978).

5. PROGNOSIS

Some plant bugs might become more important as pests when little-used crops are planted more extensively. An increased planting of oilseed rape crops in western Canada, for example, was accompanied by damage from lygus bugs. Mirid problems similar to those associated with grain sorghums might appear with an increased use of other high-yielding tropical crops that lack resistance to certain insects.

Yet the pest status of the Miridae might be enhanced as much by more accurate diagnoses of their feeding symptoms (see Wheeler 2001) than through the development of new problems related to crop monocultures or the reappearance of former pests due to changing agricultural practices. Symptoms of mirid feeding such as stunting and lagging growth rates often go unrecognized by growers, pest-management consultants, and extension personnel. Other symptoms caused by mirids — lesions and cankers, shot holing, and witches'-brooming — are so similar to those induced by some phytopathogens that they are easily misdiagnosed. Not until the 1980s and 1990s was the otherwise well-studied tarnished plant bug demonstrated to cause split-stem lesions of hybrid poplar and to cause multiple-leadering of conifer seedlings. Similar examples of overlooked damage can be anticipated in this diverse family.

Heteropterans are mostly unimportant vectors of plant pathogens, particularly viruses (reviewed by Wheeler [2001]). Because the role of mirids as vectors of phytopathogens has received scant attention from researchers, additional examples of mirid-disseminated phytopathogens probably will be found. The overall perception of plant bugs as ineffective in disseminating bacterial and fungal pathogens, however, is unlikely to change. Moreover, the vectors of economically important plant viruses are well known, and mirids generally are poor vectors of plant viruses. Consequently, the possibility of discovering significant new vector species in the Miridae seems unlikely (Wheeler 2001).

6. CONCLUSIONS

Although several thousand papers have been published on plant bugs of economic importance, many aspects of mirid sensory and digestive physiology and their phytopathogenicity remain unknown or the information is equivocal. Life-history attributes, such as fecundity, longevity, and duration of the immature stages are lacking for many injurious species, or, when data are available in the literature, the researchers did not mention the rearing conditions, failed to maintain a controlled environment, or did not specify stage of the host plant or the plant parts (e.g., fruit) available to the bugs.

One can expect substantial advances in the knowledge of how mirids orient to host plants and how hosts are discriminated from nonhosts, as well as how secondary plant chemicals — attractants and stimulants, deterrents and inhibitors — are perceived. Researchers who monitor hemipteran feeding behavior electronically will begin to include the Miridae in their studies (so far only some sketchy work has been done; Wheeler 2001). In addition, the extent and significance of vascular tissue feeding by plant bugs will be clarified. Research also will elucidate the nature of plant injury, including the relationship between the bugs' salivary secretions and plant responses, and the role of auxins, inhibitors of IAA-oxidase, and amino acids. Information similar to that obtained by K. Hori on *Lygus rugulipennis* (see **Chapter 2**) thus will become available for additional pest mirids. Researchers also will begin to explore the question of how plant bugs might modify host physiology to create more nutrient-rich feeding sites.

Those who study mirids will provide detailed information on the trophic behavior of additional species and the bugs' associated feeding symptoms. Studies similar to those by Handley and Pollard (1993) on mirid injury to strawberry and Varis (1972) on injury to sugar beet would be especially useful. Sex pheromones undoubtedly will be discovered in additional mirid species and the attractant compounds identified and mimicked for more species of agricultural importance. Pest-management practices will include the use of pheromone-based disruption of mating and might involve the use of cations to inhibit polygalacturonase in the bugs' saliva. Readers interested in further comments on the future direction of both basic and applied research on mirids should consult the final chapter in Wheeler's (2001) *Biology of the Plant Bugs*.

7. ACKNOWLEDGMENTS

The author thanks the editors, Carl Schaefer and Antônio Panizzi, for the invitation to write this chapter; Tammy Morton (Clemson University) for generous assistance with manuscript preparation; Thomas Henry (Systematic Entomology Laboratory, USDA, c/o Smithsonian Institution, Washington, D.C.) for suggestions that improved the chapter; and Craig Stoops (Navy Disease Vector Ecology & Control Center, Bangor, Washington) for his encouragement during the lengthy period of manuscript preparation.

8. REFERENCES CITED

Abraham, C. C. 1991. Occurrence of *Helopeltis theivora* Waterhouse (Miridae: Hemiptera) as a pest of Indian long pepper *Piper longum* Linn. Entomon 16: 245–246.

Abraham, E. V. 1958. Pests of cashew *(Anacardium occidentale)* in South India. Indian J. Agric. Sci. 28: 531–543.

Abraham, R. 1935. Wanzen (Heteroptera) an Obstbäumen. (III. Mitteilung). Die anatomische Untersuchung geschädigter Früchte. Z. Pflanzenkr. 45: 463–474.

Abreu, J. M. 1977. Mirideos neotropicais associados ao cacaueiro. Pp. 85–106 *in* E. M. Lavabre [ed.], Les mirides du cacaoyer. Institut Français du Café et du Cacao, Paris, France. 366 pp.

Adjei-Maafo, I. K., and L. T. Wilson. 1983. Factors affecting the relative abundance of arthropods on nectaried and nectariless cotton. Environ. Entomol. 12: 349–352.

Adkisson, P. L. 1957. Influence of irrigation and fertilizer on populations of three species of mirids attacking cotton. FAO Plant Prot. Bull. 6: 33–36.

1973. The integrated control of the insect pests of cotton. Proc. Tall Timbers Conf. Ecol. Anim. Control Habitat Manage. 4: 175–188.

Adkisson, P. L., G. A. Niles, J. K. Walker, L. S. Bird, and H. B. Scott. 1982. Controlling cotton's insect pests: a new system. Science (Washington, D.C.) 216: 19–22.

Agarwal, R. A., and G. P. Gupta. 1983. Insect pests of fibre crops. Pp. 147–164 *in* P. D. Srivastva et al. [eds.], Agricultural Entomology. Vol. II. All India Scientific Writers' Society, New Delhi, India.

Allen, W. W., and S. E. Gaede. 1963. The relationship of lygus bugs and thrips to fruit deformity in strawberries. J. Econ. Entomol. 56: 823–825.

Almand, L. K., W. L. Sterling, and C. L. Green. 1976. Seasonal abundance and dispersal of the cotton fleahopper as related to host plant phenology. Texas Agric. Exp. Stn. B-1170: 1–15.

Al-Munshi, D. M., D. R. Scott, and H. W. Smith. 1982. Some host plant effects on *Lygus hesperus* (Hemiptera: Miridae). J. Econ. Entomol. 75: 813–815.

Alvarado-Rodriguez, B., T. F. Leigh, and K. W. Foster. 1986. Oviposition site preference of *Lygus hesperus* (Hemiptera: Miridae) on common bean in relation to bean age and genotype. J. Econ. Entomol. 79: 1069–1072.

Alvarado-Rodriguez, B., T. F. Leigh, K. W. Foster, and S. S. Duffey. 1987. Life tables for *Lygus hesperus* (Heteroptera: Miridae) on susceptible and resistant common bean cultivars. Environ. Entomol. 16: 45–49.

Ambika, B., and C. C. Abraham. 1979. Bio-ecology of *Helopeltis antonii* Sign. (Miridae: Hemiptera) infesting cashew trees. Entomon 4: 335–342.

Ansley, R. J., and C. M. McKell. 1982. Crested wheatgrass vigor as affected by black grass bug and cattle grazing. J. Range Manage. 35: 586–590.

Asquith, A. 1997. Hawaiian Miridae (Hemiptera: Heteroptera): the evolution of bugs and thought. Pac. Sci. 51: 356–365.

Austin, M. D. 1929. Observations on the eggs of the apple capsid (*Plesiocoris rugicollis* Fall.) and the common green capsid (*Lygus pabulinus* Linn.). J. South-East. Agric. Coll. Wye, Kent 26: 136–144.

1930. Capsid damage to the fruits of red currant. Gard. Chron. (London) 88: 94.

1931a. A contribution to the biology of the apple capsid (*Plesiocoris rugicollis* Fall.) and the common green capsid (*Lygus pabulinus* Linn.). J. South-East. Agric. Coll. Wye, Kent 28: 153–168.

1931b. Observations on the hibernation and spring oviposition of *Lygus pratensis* Linn. Entomol. Mon. Mag. 67: 149–152.

1932. A preliminary note on the tarnished plant bug (*Lygus pratensis* Linn.). J. R. Hortic. Soc. 57: 312–320.

Bacheler, J. S., J. R. Bradley, Jr., and C. S. Eckel. 1990. Plant bugs in North Carolina: dilemma or delusion? Pp. 203–205 *in* Proc. Beltwide Cotton Prod. Res. Conf., Jan. 9–14, 1990, Las Vegas, Nevada, National Cotton Council, Memphis, Tennessee, U.S.A.

Back, E. A., and W. J. Price, Jr. 1912. Stop-back of peach. J. Econ. Entomol. 5: 329–334.

Ballard, E. 1916. *Calocoris angustatus*, Leth. Bull. Agric. Res. Inst. (Pusa). 58: 1–8.

Bamber, M. K. 1893. A Text Book on the Chemistry and Agriculture of Tea, Including the Growth and Manufacture. Law-Publishing, Calcutta, India. 258 pp.

Banerjee, B. 1983. Pests of tea. Pp. 261–276 *in* P. D. Srivastva et al. [eds.], Agricultural Entomology. Vol. II. All India Scientific Writers' Society, New Delhi. 438 pp.

Bariola, L. A. 1969. The biology of the tarnished plant bug, *Lygus lineolaris* (Beauvois), and its nature of damage and control on cotton. Ph.D. dissertation, Texas A&M University, College Station, Texas, U.S.A. 102 pp.

Barnett, W. W., B. E. Bearden, A. Berlowitz, C. S. Davis, J. L. Joos, and G. W. Morehead. 1976. True bugs cause severe pear damage. California Agric. 30(10): 20–23.

Bartlett, D. 1996. Feeding and egg laying behaviour in *Campylomma verbasci* Meyer (Hemiptera: Miridae). M.S. thesis, Simon Fraser University, Burnaby, British Columbia, Canada. 100 pp.

Beards, G. W., and F. E. Strong. 1966. Photoperiod in relation to diapause in *Lygus hesperus* Knight. Hilgardia 37(10): 345–362.

Bech, R. 1965. Licht- und Farbreaktionen der *Lygus*-Arten. Biol. Zentralbl. 84: 635–640.

1969. Untersuchungen zur Systematik, Biologie und Ökologie wirtschaftlich wichtiger *Lygus*-Arten (Hemiptera: Miridae). Beitr. Entomol. 19: 63–103.

Becker, P. 1974. Pests of ornamental plants. Minist. Agric. Fish. Food Bull. 97. H.M.S. Office, London, U.K. 175 pp.

Benjamin, D. M. 1968. Economically important insects and mites on tea in East Africa. East Afric. Agric. For. J. 34: 1–16.

Beyer, A. H. 1921. Garden flea-hopper in alfalfa and its control. U.S. Dept. Agric. Bur. Entomol. Bull. 964: 1–27.

Blacklock, J. S. 1954. A short study of pepper culture with special reference to Sarawac. Trop. Agric. 31: 40–56.

Blommers, L. [H.M.], V. Bus, E. de Jongh, and G. Lentjes. 1988. Attraction of males by virgin females of the green capsid bug *Lygocoris pabulinus* (Heteroptera: Miridae). Entomol. Ber. (Amsterdam) 48: 175–179.

Blommers, L. H. M., F. W. N. M. Vaal, and H. H. M. Helsen. 1997. Life history, seasonal adaptations and monitoring of common green capsid *Lygocoris pabulinus* (L.) Hem., Miridae). J. Appl. Entomol. 121: 389–393.

Bohning, J. W., and W. F. Currier. 1967. Does your range have wheatgrass bugs? J. Range Manage. 20: 265–267.

Boivin, G., and R. K. Stewart. 1982. Identification and evaluation of damage to McIntosh apples by phytophagous mirids (Hemiptera: Miridae) in southwestern Quebec. Can. Entomol. 114: 1037–1045.

1983. Sampling technique and seasonal development of phytophagous mirids (Hemiptera: Miridae) on apple in southwestern Quebec. Ann. Entomol. Soc. Amer. 76: 359–364.

Boness, M. 1963. Biologisch-ökologische Untersuchungen an *Exolygus* Wagner (Heteroptera, Miridae) (ein Beitrag zur Agraökologie). Z. Wiss. Zool. 168: 376–420.

Bosik, J. J. 1997. Common names of insects & related organisms 1997. Entomological Society of America, Lanham, Maryland, U.S.A. 232 pp.

Bovien, P., and O. Wagn. 1951. Skadedyr på landbrugsplanter. Månedsovers. Plantesyg. No. 319: 67–77.

Breene, R. G., A. W. Hartstack, W. L. Sterling, and M. Nyffeler. 1989. Natural control of the cotton fleahopper, *Pseudatomoscelis seriatus* (Reuter) (Hemiptera, Miridae), in Texas. J. Appl. Entomol. 108: 298–305.

Brett, C. H., R. R. Walton, and E. E. Ivy. 1946. The cotton flea hopper, *Psallus seriatus* (Reut.), in Oklahoma. Oklahoma Agric. Exp. Stn. Tech. Bull. T-24: 1–31.

Brewer, P. S., W. F. Campbell, and B. A. Haws. 1979. How black grass bugs operate. Utah Sci. 40: 21–23.

Bruneau de Miré, P. 1985. Enquête sur la tolérance des mirides du cacaoyer aux insecticides au Cameroun. Café Cacao Thé 29: 183–196.

Butts, R. A., and R. J. Lamb. 1990a. Injury to oilseed rape caused by mirid bugs *(Lygus)* (Heteroptera: Miridae) and its effect on seed production. Ann. Appl. Biol. 117: 253–266.

1990b. Comparison of oilseed *Brassica* crops with high or low levels of glucosinolates and alfalfa as hosts for three species of *Lygus* (Hemiptera: Heteroptera: Miridae). J. Econ. Entomol. 83: 2258–2262.

Buyckx, E. J. E. [ed.]. 1962. Précis des maladies et des insectes nuisibles rencontrés sur les plantes cultivées au Congo, au Rwanda et au Burundi. Publ. Inst. Nat. Etude Agron. Congo, Brussels, Belgium. Hors Sér. 708 pp.

Cadou, J. 1993. Les Miridae du cotonnier en Afrique et a Madagascar. CIRAD-CA, Paris, France. 74 pp.

Cagle, L. R., and H. W. Jackson. 1947. Life history of the garden fleahopper. Virginia Agric. Exp. Stn. Tech. Bull. 107: 1–27.

Carayon, J., and R. Delattre. 1948. Les *Helopeltis* (Hem. Heteroptera) nuisibles de Côte d' Ivoire. Rev. Pathol. Veg. Entomol. Agric. Fr. 27: 185–194.

Carlson, J. W. 1940. Lygus bug damage to alfalfa in relation to seed production. J. Agric. Res. (Washington, D.C.) 61: 791–815.

Carpenter, G. H. 1920. Injurious insects and other animals observed in Ireland during the years 1916, 1917, and 1918. Econ. Proc. R. Dublin Soc. 2: 259–272.

Carvalho, J. C. M. 1957–1960. Catalogue of the Miridae of the World. Arq. Mus. Nac. Rio J. Part I. Cylapinae, Deraeocorinae, Bryocorinae 44(1): 1–158 (1957); Part II. Phylinae 45(2): 1–216 (1958); Part III. Orthotylinae 47(3): 1–161 (1958); Part IV. Mirinae 48(4): 1–384 (1959); Part V. Bibliography and General Index 51(5): 1–194 (1960).

Cassis, G. 1984. A systematic study of the subfamily Dicyphinae (Heteroptera: Miridae). Ph.D. dissertation, Oregon State University, Corvallis, Oregon, U.S.A. 389 pp.

Cassis, G., and G. F. Gross. 1995. Hemiptera: Heteroptera (Coleorrhyncha to Cimicomorpha) *in* W. W. K. Houston and G. V. Maynard [eds.], Zoological Catalogue of Australia. Vol. 27.3A. CSIRO Australia, Melbourne, Australia. 506 pp.

Cave, R. D., and A. P. Gutierrez. 1983. *Lygus hesperus* field life table studies in cotton and alfalfa (Heteroptera: Miridae). Canad. Entomol. 115: 649–654.

Champlain, R. A., and G. D. Butler, Jr. 1967. Temperature effects on development of the egg and nymphal stages of *Lygus hesperus* (Hemiptera: Miridae). Ann. Entomol. Soc. Amer. 60: 519–521.

Chapman, P. J., and S. E. Lienk. 1971. Tortricid fauna of apple in New York (Lepidoptera: Tortricidae); including an account of apples' occurrence in the State, especially as a naturalized plant. Special Publication, New York State Agric. Exp. Stn., Geneva, New York, U.S.A. 122 pp.

Chapman, R. B. 1984. Seed crop pests. Pp. 143–152 *in* R. R. Scott [ed.], New Zealand Pest and Beneficial Insects. Lincoln University College of Agriculture, Canterbury, New Zealand. 373 pp.

Cherian, M. C., M. S. Kylasam, and P. S. Krishnamurti. 1941. Further studies on *Calocoris angustatus* Leth. Madras Agric. J. 29: 66–69.

Chinajariyawong, A. 1988. The sap-sucking bugs attacking cotton: biological aspects and economic damage. Ph.D. dissertation, University of Queensland, Brisbane, Australia. 141 pp.

Chung, G. F., and B. J. Wood. 1989. Chemical control of *Helopeltis theobromae* Mill. and crop loss assessment in cocoa. J. Plant Prot. Trop. 6: 35–48.

Clancy, D. W., and H. D. Pierce. 1966. Natural enemies of lygus bugs. J. Econ. Entomol. 59: 853–858.

Clark, K. M., W. C. Bailey, and R. L. Myers. 1995. Alfalfa as a companion crop for control of *Lygus lineolaris* (Hemiptera: Miridae) in amaranth. J. Kansas Entomol. Soc. 68: 143–148.

Clifford, P. T. P., J. A. Wightman, and D. N. J. Whitford. 1983. Mirids in 'Grasslands Maku' lotus seed crops: friends or foes? Proc. New Zealand Grassland Assoc. 44: 42–46.

Coaker, T. H. 1957. Studies of crop loss following insect attack on cotton in East Africa. II. — Further experiments in Uganda. Bull. Entomol. Res. 48: 851–866.

Cochrane, T. W., and P. F. Entwistle. 1964. Preliminary world bibliography of mirids (= capsids) and other Heteroptera associated with cocoa (*Theobroma cacao* L.). Pp. 123–131 *in* Proc. Conf. Mirids and Other Pests Cocoa, Ibadan, Nigeria, 1964. West African Cocoa Research Institute, Ibadan, Nigeria.

Cockerell, T. D. A. 1899. Some insect pests of Salt River Valley and the remedies for them. Arizona Agric. Exp. Stn. Bull. 32: 273–295.

Cohen, A. C. 1996. Plant feeding by predatory Heteroptera: evolutionary and adaptational aspects of trophic switching. Pp. 1–17 *in* O. Alomar and R. N. Wiedenmann [eds.], Zoophytophagous Heteroptera: Implications for Life History and Integrated Pest Management. Thomas Say Publ. Entomol.: Proceedings Entomological Society of America, Lanham, Maryland, U.S.A. 202 pp.

1998. Solid-to-liquid feeding: the inside(s) story of extra-oral digestion in predaceous Arthropoda. Amer. Entomol. 44: 103–116.

Cohen, A. C., and A. G. Wheeler, Jr. 1998. Role of saliva in the destructive fourlined plant bug (Hemiptera: Miridae: Mirinae). Ann. Entomol. Soc. Amer. 91: 94–100.

Collingwood, C. A. 1977a. Biological control and relations with other insects. Pp. 237–255 *in* E. M. Lavabre [ed.], Les Mirides du Cacaoyer. Institute Français du Café et du Cacao (I.F.C.C.), Paris, France. 366 pp.

1977b. Insecticide resistance in West Africa. Pp. 279–284 *in* E. M. Lavabre [ed.], Les Mirides du Cacaoyer. Institut Français du Café et du Cacao (I.F.C.C.), Paris, France. 366 pp.

Collyer, E. 1953. Biology of some predatory insects and mites associated with the fruit tree red spider mite (*Metatetranychus ulmi* (Koch)) in south-eastern England II. Some important predators of the mite. J. Hortic. Sci. 28: 85–97.

Conway, G. R. 1969. Pests follow the chemicals in the cocoa of Malaysia. Nat. Hist. 78(2): 46–51.

Cotterell, G. S. 1926. A preliminary study of the life-history and habits of *Sahlbergella singularis* Hagl. and *Sahlbergella theobroma* Dist., attacking cocoa on the Gold Coast, with suggested control measures. Dept. Agric. Gold Coast Bull. 3: 1–26.

Cotton, R. T. 1918. Insects attacking vegetables in Porto Rico. J. Dept. Agric. Porto Rico 2: 265–317.

Cowland, J. W. 1934. Gezira Entomological Section, G.A.R.S. Final report on experimental work, 1932–33. Gezira Agric. Res. Serv. Sudan Gov. Rep. 1933: 107–125.

Craig, C., R. G. Luttrell, S. D. Stewart, and G. L. Snodgrass. 1997. Host plant preferences of tarnished plant bug: a foundation for trap crops in cotton. Pp. 1176–1181 *in* Proc. Beltwide Cotton Conf., Jan. 6–10, 1997, New Orleans, Louisiana, National Cotton Council, Memphis, Tennessee, U.S.A.

Craig, C. H. 1963. The alfalfa plant bug, *Adelphocoris lineolatus* (Goeze) in northern Saskatchewan. Canad. Entomol. 95: 6–13.

Cravedi, P., and G. Carli. 1988. Mirides nuisibles au pecher *in* H. Audemard [ed.], IOBC/WPRS Working Group "Integrated Protection in Fruit Orchards," sub-group "Peach Orchards." Proc. Workshop held at Valence (France), 31 Aug. to 2 Sept. 1988. Int. Organ. Biol. Control/West. Palearctic Reg. Sect. 11(7): 22–23.

Croft, B. A., and L. A. Hull. 1983. The orchard as an ecosystem. Pp. 19–42 *in* B. A. Croft and S. C. Hoyt [eds.], Integrated Management of Insect Pests of Pome and Stone Fruits. Wiley, New York, New York, U.S.A. 454 pp.

Crosby, C. R. 1911. Notes on the life-history of two species of Capsidae. Canad. Entomol. 43: 17–20.

Crosby, C. R., and M. D. Leonard. 1914. The tarnished plant-bug. Cornell Univ. Agric. Exp. Stn. Bull. 346: 463–526.

Cross, D. J. 1971. Water stress in cocoa and its effects on *Distantiella theobroma* in the laboratory. Pp. 252–256 *in* Proc. III Int. Cocoa Res. Conf., Accra, Ghana, 23–29 Nov. 1969. Cocoa Research Institute, Tafo, Ghana.

Crowdy, S. H. 1947. Observations on the pathogenicity of *Calonectria rigidiuscula* (Berk. & Br.) Sacc. on *Theobroma cacao* L. Ann. Appl. Biol. 34: 45–59.

Crowe, T. J. 1977. *Helopeltis* spp. Pp. 289–292 *in* J. Kranz, H. Schmutterer, and W. Koch [eds.], Diseases, Pests and Weeds in Tropical Crops. Parey, Berlin, Germany. 666 pp.

Curtis, C. E., and C. E. McCoy. 1964. Some host-plant preferences shown by *Lygus lineolaris* (Hemiptera: Miridae) in the laboratory. Ann. Entomol. Soc. Amer. 57: 511–513.

Cushman, R.A. 1916. The native food-plants of the apple red-bugs. Proc. Entomol. Soc. Washington 18: 196.

Das, S. C. 1984. Resurgence of tea mosquito bug, *Helopeltis theivora* Waterh., a serious pest of tea. Two Bud 31(2): 36–39.

da Silva Barbosa, A. J. 1959. The capsid complex of cotton in Moçambique. South Afric. J. Sci. 55: 147–153.

Davis, A. C., F. L. McEwen, and R. W. Robinson. 1963. Preliminary studies on the effect of lygus bugs on the set and yield of tomatoes. J. Econ. Entomol. 56: 532–533.

Davis, C. J., S. Matayoshi, and E. R. Yoshioka. 1985. *Halticus bractatus* Say. Proc. Hawaiian Entomol. Soc. 25: 7.

Day, W. H. 1991. The peculiar sex ratio and dimorphism of the garden fleahopper, *Halticus bractatus* (Hemiptera: Miridae). Entomol. News 102: 113–117.

Day, W. H., and L. B. Saunders. 1990. Abundance of the garden fleahopper (Hemiptera: Miridae) on alfalfa and parasitism by *Leiophron uniformis* (Gahan) (Hymenoptera: Braconidae). J. Econ. Entomol. 83: 101–106.

Denning, D. G. 1948. The crested wheat bug. Wyoming Agric. Exp. Stn. Circ. 33: 1–2.

De Silva, M. D. 1961. Biology of *Helopeltis ceylonensis* De Silva (Heteroptera- Miridae), a major pest of cacao in Ceylon. Trop. Agric. (Colombo) 117: 149–156.

Devasahayam, S. 1985. Seasonal biology of tea mosquito bug *Helopeltis antonii* Signoret (Heteroptera: Miridae) — a pest of cashew. J. Plant. Crops 13: 145–147.

1988. Mating and oviposition behaviour of tea mosquito bug *Helopeltis antonii* Signoret (Heteroptera: Miridae). J. Bombay Nat. Hist. Soc. 85: 212–214.

Devasahayam, S., and C. P. Radhakrishnan Nair. 1986. The tea mosquito bug *Helopeltis antonii* Signoret on cashew in India. J. Plant. Crops 14: 1–10.

Devasahayam, S., K. M. Abdulla Koya, and T. Prem Kumar. 1986. Infestation of tea mosquito bug *Helopeltis antonii* Signoret (Heteroptera: Miridae) on black pepper and allspice in Kerala. Entomon 11: 239–241.

Dicker, G. H. L. 1967. Integrated control of apple pests. Proc. 4th Br. Insectic. Fungic. Conf., Brighton (U.K.) 1: 1–7.

Domek, J. M., and D. R. Scott. 1985. Species of the genus *Lygus* Hahn and their host plants in the Lewiston-Moscow area of Idaho (Hemiptera: Miridae). Entomography 3: 75–105.

Doumbia, Y. O., K. Conare, and G. L. Teetes. 1995. A simple method to assess damage and screen sorghums for resistance to *Eurystylus marginatus*. Pp. 183–189 *in* K. F. Nwanze and O. Youm [eds.], Panicle Insect Pests of Sorghum and Pearl Millet: Proceedings of an International Consultative Workshop, 4–7 Oct. 1993, ICRISAT Sahelian Center, Miamey, Niger. International Crops Research Institute for the Semi-Arid Tropics, Patancheru, A.P., India.

Eddy, C. O. 1927. The cotton flea hopper. South Carolina Agric. Exp. Stn. Bull. 235: 1–21.

1928. Cotton flea hopper studies of 1927 and 1928. South Carolina Agric. Exp. Stn. Bull. 251: 1–18.

Ehanno, B. 1987. Les hétéroptères mirides de France. Tome II-A: inventaire et syntheses ecologiques. Inventaire Faune Flore 40: 1–647.

El-Dessouki, S. A., A. H. El-Kifl, and H. A. Helal. 1976. Life cycle, host plants and symptoms of damage of the tomato bug, *Nesidiocoris tenuis* Reut. (Hemiptera: Miridae), in Egypt. Z. Pflanzenkr. Pflanzenschutz 83: 204–220.

Entwistle, P. F. 1965. Cocoa mirids. Part 1. A world review of biology and ecology. Cocoa Grow. Bull. 5: 16–20.

1972. Pests of cocoa. Longman, London, U.K. 779 pp.

1985. Insects and cocoa. Pp. 366–443 *in* G. A. R. Wood and R. A. Lass. [eds.], Cocoa, fourth edition. Longman, London, U.K. 620 pp.

Entwistle, P. F., and A. Youdeowei. 1964. A preliminary world review of cacao mirids. Pp. 71–79 *in* Proc. Conf. Mirids and Other Pests Cacao, Ibadan, Nigeria, 1964. West African Cocoa Research Institute.

Fernando, H. E., and P. Manickavasagar. 1956. Economic damage and control of the cacao capsid, *Helopeltis* sp. (fam. Capsidae, ord. Hemiptera) in Ceylon. Trop. Agric. (Colombo) 112: 25–36.

Findlay, R. M. 1975. Pest of autumn-harvested asparagus. New Zealand J. Agric. 130: 56–57.

Fitt, G. P. 1994. Cotton pest management: part 3. An Australian perspective. Annu. Rev. Entomol. 39: 543–562.

Fleischer, S. J., and M. J. Gaylor. 1987. Seasonal abundance of *Lygus lineolaris* (Heteroptera: Miridae) and selected predators in early season uncultivated hosts: implications for managing movement into cotton. Environ. Entomol. 16: 379–389.

1988. *Lygus lineolaris* (Heteroptera: Miridae) population dynamics: nymphal development, life tables, and Leslie matrices on selected weeds and cotton. Environ. Entomol. 17: 246–253.

Flemion, F., and E. T. Hendrickson. 1949. Further studies on the occurrence of embryoless seeds and immature embryos in the Umbelliferae. Contrib. Boyce Thompson Inst. 15: 291–297.

Fletcher, R. K. 1940. Certain host plants of the cotton flea hopper. J. Econ. Entomol. 33: 456–459.

Flint, M. L. 1990. Pests of the garden and small farm: a grower's guide to using less pesticide. Univ. California Div. Agric. Nat. Resour. Publ. 3332. Oakland, California, U.S.A. 276 pp.

Foley, D. H., and B. A. Pyke. 1985. Developmental time of *Creontiades dilutus* (Stål) (Hemiptera: Miridae) in relation to temperature. J. Austral. Entomol. Soc. 24: 125–127.

Fox-Wilson, G. 1925. The egg of the tarnished plant bug, *L. pratensis* Linn. Entomol. Mon. Mag. 61: 19–20.

1938. II. The tarnished plant bug or bishop fly, *Lygus pratensis* L. — précis of present knowledge. J. R. Hortic. Soc. 63: 392–395.

Fryer, J. C. F. 1929. The capsid pests of fruit trees in England. Trans. 4th Int. Congr. Entomol., Ithaca, New York, U.S.A. 2: 229–236.

Fuxa, J. R., and J. A. Kamm. 1976a. Effects of temperature and photoperiod on the egg diapause of *Labops hesperius* Uhler. Environ. Entomol. 5: 505–507.

1976b. Dispersal of *Labops hesperius* on rangeland. Ann. Entomol. Soc. Amer. 69: 891–893.

Gagné, S., C. Richard, and C. Gagnon. 1984. La coulure des graminées: état des connaissances. Phytoprotection 65: 45–52.

Gahukar, R. T. 1991. Recent developments in sorghum entomology research. Agric. Zool. Rev. 4: 23–65.

Gahukar, R. T., Y. O. Doumbia, and S. M. Bonzi. 1989. *Eurystylus marginatus* Odh., a new pest of sorghum in the Sahel. Trop. Pest Manage. 35: 212–213.

Geering, Q. A. 1953. Studies of *Lygus vosseleri* Popp. (Heteroptera, Miridae). East and Central Africa. I. A method for breeding continuous supplies in the laboratory. Bull. Entomol. Res. 44: 351–362.

Gerber, G. H. 1995. Fecundity of *Lygus lineolaris* (Heteroptera: Miridae). Canad. Entomol. 127: 263–264.

Gerber, G. H., and I. L. Wise. 1995. Seasonal occurrence and number of generations of *Lygus lineolaris* and *L. borealis* (Heteroptera: Miridae) in southern Manitoba. Canad. Entomol. 127: 543–559.

Gerin, L. 1956. Les *Helopeltis* (Hemipt. Miridae), nuisible aux Quinqinas du Cameroun Français. J. Agric. Trop. Bot. Appl. 3: 512–540.

Gibbs, D. G., A. D. Pickett, and D. Leston. 1968. Seasonal population changes in cocoa capsids (Hemiptera, Miridae) in Ghana. Bull. Entomol. Res. 58: 279–293.

Giesberger, G. [1983]. Biological control of the *Helopeltis* pest of cocoa in Java. Pp. 91–180 *in* H. Toxopeus and P. C. Wessel [eds.], Cocoa research in Indonesia 1900–1950. Vol. 2. American Cocoa Research Institute, Washington, D.C.; International Office of Cocoa & Chocolate, Brussels, Belgium. 293 pp.

Glick, P. A. 1983. The influence of cultural practices on arthropod populations in cotton. U.S. Dept. Agric. Agric. Res. Serv. Agric. Rev. Man. ARM-S-32: 1–52.

Goodman, A. 1955. Observations on the status of certain insect pests of cotton at Tokar, Sudan. Emp. Cotton Grow. Rev. 32: 194–203.

Goot, P. van der. 1917. Dezwarte cacao-mier, *Dolichoderus bituberculatus*, Mayr, en haar beteekenis voor de cacao-cultuur op Java. Meded. Proefstn. Miden Java 25: 1–142.

1927. Het *Crotalaria*-wantsje. Korte Meded. Inst. Plantenziekt. (Buitenzorg) No. 6: 1–17.

Gopalan, M. 1976a. Studies on salivary enzymes of *Ragmus importunitas* Distant (Hemiptera: Miridae). Curr. Sci. (Bangalore) 45: 188–189.

1976b. Effect of infestation of *Ragmus importunitas* Distant (Hemiptera: Miridae) on the growth and yield of sannhemp [sic]. Indian J. Agric. Sci. 46: 588–591.

Gopalan, M., and M. Basheer. 1966. Studies on the biology of *Ragnus importunitas* D. (Miridae-Hemiptera) on sunnhemp. Madras Agric. J. 53: 22–23.

Gopalan, M., and T. R. Subramaniam. 1978. Effect of infestation of *Ragmus importunitas* Distant (Hemiptera: Miridae) on respiration, transpiration, moisture content and oxidative enzymes activity in sunnhemp plants (*Crotalaria juncea* L.). Curr. Sci. (Bangalore) 47: 131–134.

Gorczyca, J. 1997. Revision of the *Vannius*-complex and its subfamily placement (Hemiptera: Heteroptera: Miridae). Genus 8: 517–553.

Graham, H. M., A. A. Negm, and L. R. Ertle. 1984. Worldwide literature of the *Lygus* complex (Hemiptera: Miridae), 1900–1980. U.S. Dept. Agric. Bibliogr. Lit. Agric. No. 30: 1–205.

Groot, A. T., A. Schuurman, G. J. R. Judd, L. H. M. Blommers, and J. H. Visser. 1996. Sexual behaviour of the green capsid bug *Lygocoris pabulinus* L. (Miridae): an introduction. Proc. Sect. Exp. Appl. Entomol. Netherlands Entomol. Soc. (N.E.V., Amsterdam) 7: 249–252.

Groot, A. T., E. van der Wal, A. Schuurman, J. H. Visser, L. H. M. Blommers, and T. A. van Beek. 1998. Copulation behaviour of *Lygocoris pabulinus* under laboratory conditions. Entomol. Exp. Appl. 88: 219–228.

Guinn, G. 1982. Causes of square and boll shedding in cotton. U.S. Dept. Agric. Tech. Bull. 1672: 1–22.

Gupta, R. K., G. Tamaki, and C. A. Johansen. 1980. Lygus bug damage, predator-prey interaction, and pest management implications in alfalfa grown for seed. Washington State Univ. Coll. Agric. Res. Cent. Tech. Bull. 0092: 1–18.

Gutierrez, A. P. 1995. Integrated pest management in cotton. Pp. 280–310 *in* D. Dent [ed.], Integrated Pest Management. Chapman & Hall, London, U.K. 356 pp.

Hancock, G. L. R. 1935. Notes on *Lygus simonyi*, Reut. (Capsidae), a cotton pest in Uganda. Bull. Entomol. Res. 26: 429–438.

Handley, D. T., and J. E. Pollard. 1993. Microscopic examination of tarnished plant bug (Heteroptera: Miridae) feeding damage to strawberry. J. Econ. Entomol. 86: 505–510.

Hanny, B. W., and T. C. Cleveland. 1976. Effects of tarnished plant bug (*Lygus lineolaris*, Palisot de Beauvois), feeding on presquaring cotton (*Gossypium hirsutum* L.). P. 59 *in* Proc. Beltwide Cotton Prod. Res. Conf., Jan. 5–7, 1976, Las Vegas, Nevada (abstr.). National Cotton Council, Memphis, Tennessee, U.S.A.

Hanny, B. W., T. C. Cleveland, and W. R. Meredith, Jr. 1977. Effects of tarnished plant bug, *(Lygus lineolaris)*, infestation on presquaring cotton *(Gossypium hirsutum)*. Environ. Entomol. 6: 460–462.

Hansen, J. D., K. H. Asay, and D. C. Nielson. 1985a. Screening range grasses for resistance to black grass bugs *Labops hesperius* and *Irbisia pacifica* (Hemiptera: Miridae). J. Range Manage. 38: 254–257.

1985b. Feeding preference of a black grass bug, *Labops hesperius* (Hemiptera: Miridae), for 16 range grasses. J. Kansas Entomol. Soc. 58: 356–359.

Hardee, D. D., and W. W. Bryan. 1997. Influence of *Bacillus thuringiensis*-transgenic and nectariless cotton on insect populations with emphasis on the tarnished plant bug (Heteroptera: Miridae). J. Econ. Entomol. 90: 663–668.

Hari Babu, R. S., S. Rath, and C. B. S. Rajput. 1983. Insect pests of cashew in India and their control. Pesticides (Bombay) 17(4): 8–16.

Harris, W. V. 1937. *Helopeltis* bug. East Afric. Agric. J. 2: 387–390.

Haws, B. A. [compiler]. 1978. Economic impacts of *Labops hesperius* on the production of high quality range grasses. Final Rep. Utah Agric. Exp. Stn. to Four Corners Reg. Comm. Utah State University, Logan, Utah, U.S.A. 269 pp.

Haws, B. A., and G. E. Bohart. 1986. Black grass bugs *(Labops hesperius)* Uhler (Hemiptera: Miridae) and other insects in relation to crested wheatgrass. Pp. 123–145 *in* K. L. Johnson [ed.], Crested Wheatgrass: Its Values, Problems and Myths. Symposium proceedings. Utah State University, Logan, Utah, U.S.A.

Hedlund, R. C., and H. M. Graham. 1987. Economic importance and biological control of *Lygus* and *Adelphocoris* in North America. U.S. Dept. Agric. Agric. Res. Serv. ARS-64: 1–95.

Hely, P. C., G. Pasfield, and J. G. Gellatley. 1982. Insect Pests of Fruit and Vegetables in NSW. Inkata, Melbourne, Australia. 312 pp.

Henry, T. J., and A. G. Wheeler, Jr. 1988. Family Miridae Hahn, 1833 (= Capsidae Burmeister). The plant bugs. Pp. 251–507 *in* T. J. Henry and R. C. Froeschner [eds.], Catalog of the Heteroptera, or True Bugs, of Canada and the Continental United States. E. J. Brill, Leiden, the Netherlands. 958 pp.

Hewitt, G. B. 1980. Tolerance of ten species of *Agropyron* to feeding by *Labops hesperius*. J. Econ. Entomol. 73: 779–782.

Higgins, K. M., J. E. Bowns, and B. A. Haws. 1977. The black grass bug (*Labops hesperius* Uhler): its effect on several native and introduced grasses. J. Range Manage. 30: 380–384.

Hill, A. R. 1952. Observations on *Lygus pabulinus* (L.), a pest of raspberries in Scotland. East Malling (U.K.) Res. Stn. Rep. 1951: 181–182.

Hiremath, I. G. 1986. Host preference of sorghum earhead bug, *Calocoris angustatus* Lethierry (Hemiptera: Miridae). Entomon 11: 121–125.

1989. Survey of sorghum earhead bug and its natural enemies in Karnataka. J. Biol. Control 3: 13–16.

Hiremath, I. G., and T. S. Thontadarya. 1984a. Stage and part of sorghum earhead preferred for oviposition by sorghum earhead bug. Curr. Res. (Bangalore) 13: 37–38.

1984b. Seasonal incidence of the sorghum earhead bug, *Calocoris angustatus* Lethierry (Hemiptera: Miridae). Insect Sci. Appl. 6: 469–474.

Hiremath, I. G., and C. A. Viraktamath. 1992. Biology of the sorghum earhead bug, *Calocoris angustatus* (Hemiptera: Miridae) with descriptions of various stages. Insect Sci. Appl. 13: 447–457.

Hiremath, I. G., T. S. Thontadarya, and A. S. Nalini. 1983. Histochemical changes in sorghum grain due to feeding by sorghum earhead bug, *Calocoris angustatus* (Hemiptera: Miridae). Curr. Res. (Bangalore) 12: 15–16.

Hixson, E. 1941. The host relation of the cotton flea hopper. Iowa State Coll. J. Sci. 16: 66–68.

Hodgkiss, H. E., and S. W. Frost. 1921. The apple red bugs and their control. Pennsylvania Agric. Exp. Stn. Ext. Circ. 88: 1–8.

Holdaway, F. G., and W. C. Look. 1942. Insects of the garden bean in Hawaii. Proc. Hawaiian Entomol. Soc. 11: 249–260.

Holopainen, J. K. 1986. Damage caused by *Lygus rugulipennis* Popp. (Heteroptera, Miridae), to *Pinus sylvestris* L. seedlings. Scand. J. For. Res. 1: 343–349.

1989. Host plant preference of the tarnished plant bug *Lygus rugulipennis* Popp. (Het., Miridae). J. Appl. Entomol. 107: 78–82.

Holopainen, J. K., and R. Rikala. 1990. Abundance and control of *Lygus rugulipennis* (Heteroptera: Miridae) on Scots pine (*Pinus sylvestris* L.) nursery stock. New For. 4: 13–25.

Holopainen, J. K., and A.-L. Varis. 1991. Host plants of the European tarnished plant bug *Lygus rugulipennis* Poppius (Het., Miridae). J. Appl. Entomol. 111: 484–498.

Holtzer, T. O., and W. L. Sterling. 1980. Ovipositional preference of the cotton fleahopper, *Pseudatomoscelis seriatus*, and distribution of eggs among host plant species. Environ. Entomol. 9: 236–240.

Hori, K. 1971a. Studies on the feeding habits of *Lygus disponsi* Linnavuori (Hemiptera: Miridae) and the injury to its host plants I. Histological observations of the injury. Appl. Entomol. Zool. 6: 84–90.

1971b. Studies on the feeding habits of *Lygus disponsi* Linnavuori (Hemiptera: Miridae) and the injury to its host plant II. Frequency, duration and quantity of the feeding. Appl. Entomol. Zool. 6: 119–125.

1974a. Enzymes in the salivary gland of *Lygus disponsi* Linnavuori (Hemiptera: Miridae). Res. Bull. Obihiro Univ. 8: 173–260.

1974b. Plant growth-promoting factor in the salivary gland of the bug, *Lygus disponsi*. J. Insect Physiol. 20: 1623–1627.

1975a. Digestive carbohydrases in the salivary gland and midgut of several phytophagous bugs. Comp. Biochem. Physiol. 50B: 145–151.

1975b. Amino acids in the salivary glands of the bugs, *Lygus disponsi* and *Eurydema rugosum*. Insect Biochem. 5: 165–169.

Hori, K., and T. Hanada. 1970. Biology of *Lygus disponsi* Linnavuori (Hemiptera, Miridae) in Obihiro. Res. Bull. Obihiro Univ. (2) 6: 304–317.

Hori, K., and K. Kuramochi. 1984. Effects of food plants of the first generation nymph on the growth and reproduction of *Lygus disponsi* Linnavuori (Hemiptera, Miridae). Res. Bull. Obihiro Univ. 14: 89–93.

Hori, K., and P. W. Miles. 1993. The etiology of damage to lucerne by the green mirid, *Creontiades dilutus* (Stål). Austral. J. Exp. Agric. 33: 327–331.

Houillier, M. 1964a. Régime alimentaire et disponibilité de ponte des miridés dissimulés du cacaoyer. Rev. Pathol. Vég. Entomol. Agric. 43: 195–200.

1964b. Étude expérimentale de la répartition des piqures de miridés (*Sahlbergella singularis* Hagl. et *Distantiella theobromae* [sic] Distant) sur cabosse de cacaoyer. Rev. Pathol. Vég. Entomol. Agric. 43: 201–208.

Howard, L. O. 1898. Notes from correspondence. The so-called "Cotton Flea." P. 101 *in* Some miscellaneous results of the work of the Division of Entomology. III. U.S. Dept. Agric. Div. Entomol. Bull. 18 (n.s.).

Howitt, A. J. 1993. Common tree fruit pests. North Central Reg. Ext. Publ. 63. Michigan State University, East Lansing, Michigan, U.S.A. 252 pp.

Huber, R. T., and P. P. Burbutis. 1967. Some effects of the tarnished plant bug on sweet peppers. J. Econ. Entomol. 60: 1332–1334.

Hughes, J. H. 1943. The alfalfa plant bug *Adelphocoris lineolatus* (Goeze) and other Miridae (Hemiptera) in relation to alfalfa-seed production in Minnesota. Minnesota Agric. Exp. Stn. Tech. Bull. 161: 1–80.

Hunter, W. D. 1924. The so-called cotton flea. J. Econ. Entomol. 17: 604.

1926. The cotton hopper, or so-called "cotton flea." U.S. Dept. Agric. Dept. Circ. 361: 1–15.

Idowu, O. L. 1988. Comparative toxicities of new insecticides to the cocoa mirid, *Sahlbergella singularis* in Nigeria. Pp. 531–534 *in* Proc. 10th Int. Cocoa Res. Conf., Santo Domingo, Dominican Republic, 17–23 May 1987. Cocoa Producers' Alliance, Lagos, Nigeria.

Isely, D. 1927. The cotton hopper and associated leaf-bugs attacking cotton. Univ. Arkansas Coll. Agric. Ext. Circ. 231: 1–8.

Jacobi, E. F. 1932. De verschillen tusschen de larven van *Lygus pabulinus* en *Plesiocoris rugicollis*. Tijdschr. Plantenziekten 38: 213–219.

Jeevaratnam, K., and R. H. S. Rajapakse. 1981a. Biology of *Helopeltis antonii* Sign. (Heteroptera: Miridae) in Sri Lanka. Entomon 6: 247–251.

1981b. Studies on the chemical control of the mirid bug, *Helopeltis antonii* Sign., in the cashew. Insect Sci. Appl. 1: 399–402.

Johnson, C. G. 1962. Capsids: a review of current knowledge. Pp. 316–331 *in* J. B. Wills [ed.], Agriculture and Land Use in Ghana. Oxford University Press, London, U.K. 504 pp.

Johnson, C. G., and T. R. E. Southwood. 1949. Seasonal records in 1947 and 1948 of flying Hemiptera-Heteroptera, particularly *Lygus pratensis* L., caught in nets 50 ft. to 3,000 ft. above the ground. Proc. R. Entomol. Soc. London (A) 24: 128–130.

Johnson, D. R., C. D. Klein, H. B. Myers, and L. D. Page. 1996. Pre-bloom square loss, causes and diagnosis. Proc. Beltwide Cotton Conf., Jan. 9–12, 1996, Nashville, Tenn. 1: 103–105. National Cotton Council, Memphis, Tennessee, U.S.A.

Johnston, H. G. 1940. A melon bug (*Pycnoderes quadrimaculatus* Guer.). U.S. Dept. Agric. Insect Pest Surv. Bull. 20: 503.

Jotwani, M. G. 1983. Insect pests of sorghum, maize and pearl millet. Pp. 109–125 *in* P. D. Srivastva et al. [eds.], Agricultural Entomology, Vol. II. All India Scientific Writers' Society, New Delhi, India. 438 pp.

Judd, G. J. R., and H. L. McBrien. 1994. Modeling temperature-dependent development and hatch of overwintered eggs of *Campylomma verbasci* (Heteroptera: Miridae). Environ. Entomol. 23: 1224–1234.

Judd, G. J. R., H. L. McBrien, and J. H. Borden. 1995. Modification of responses by *Campylomma verbasci* (Heteroptera: Miridae) to pheromone blends in atmospheres permeated with synthetic sex pheromone or individual components. J. Chem. Ecol. 21: 1991–2002.

Kelton, L. A. 1980. The insects and arachnids of Canada. Part 8. The plant bugs of the Prairie Provinces. Heteroptera: Miridae. Agric. Canada Publ. 1703: 1–408.

Kerzhner, I. M., and M. Josifov. 1999. Cimicomorpha II: Miridae *in* B. Aukema and C. Rieger [eds.], Catalogue of the Heteroptera of the Palaearctic Region, Vol. 3. Netherlands Entomological Society, Amsterdam, the Netherlands. 577 pp.

Khattat, A. R., and R. K. Stewart. 1977. Development and survival of *Lygus lineolaris* exposed to different laboratory rearing conditions. Ann. Entomol. Soc. Amer. 70: 274–278.

Khoo, K. C. 1987. The Cocoa Mirid in Peninsular Malaysia and its Management. Planter (Kuala Lumpur) 63: 516–520.

Khoo, K. C., P. A. C. Ooi, and C. T. Ho. 1991. Crop Pests and Their Management in Malaysia. Tropical Press, Kuala Lumpur, Malaysia. 242 pp.

King, A. B. S. 1973. Studies of sex attraction in the cocoa capsid, *Distantiella theobroma* (Heteroptera: Miridae). Entomol. Exp. Appl. 16: 243–254.

King, A. B. S., and J. L. Saunders. 1984. The Invertebrate Pests of Annual Food Crops in Central America: A Guide to Their Recognition and Control. Overseas Development Administration, London, U.K. 166 pp.

King, E. G., J. R. Phillips, and R. J. Coleman [eds.]. 1996. Cotton Insects and Mites: Characterization and Management. Cotton Foundation, Memphis, Tennessee, U.S.A. 1008 pp.

King, W. V., and W. S. Cook. 1932. Feeding punctures of mirids and other plant-sucking insects and their effect on cotton. U.S. Dept. Agric. Tech. Bull. 296: 1–11.

Kirkpatrick, T. W. 1923. Preliminary notes on two minor pests of the Egyptian cotton crop (*Creontiades pallidus*, Ramb., and *Nezara viridula*, L.). Minist. Agric. Egyptian Tech. Sci. Serv. Bull. 33: 1–15.

Knight, H. H. 1915. Observations on the oviposition of certain capsids. J. Econ. Entomol. 8: 293–298.

1918. An investigation of the scarring of fruit caused by apple redbugs. Cornell Univ. Agric. Exp. Stn. Bull. 396: 187–208.

1922. Studies on insects affecting the fruit of the apple with particular reference to the characteristics of the resulting scars. Cornell Univ. Agric. Exp. Stn. Bull. 410: 447–498.

Koehler, C. S. 1963. *Lygus hesperus* as an economic insect on *Magnolia* nursery stock. J. Econ. Entomol. 56: 421–422.

Kullenberg, B. 1942. Die Eier der schwedischen Capsiden (Rhynchota) I. Ark. Zool. 33A(15): 1–16.

1944. Studien über die Biologie der Capsiden. Zool. Bidr. Uppsala. 23: 1–522.

Kumar, R., and A. K. Ansari. 1974. Biology, immature stages and rearing of cocoa-capsids (Miridae: Heteroptera). Zool. J. Linn. Soc. 54: 1–29.

Larochelle, A. 1984. Les punaises terrestres (Heteropteres: Geocorises) du Québec. Fabreries Suppl. 3: 1–513.

Larochelle, A., and M.-C. Larivière. 1979. Le genre *Labops* Burmeister au Québec, Canada (Heteroptera: Miridae): répartition géographique, habitat et biologie. Bull. Invent. Insect. Québec 1: 61–67.

Lattin, J. D., A. Christie, and M. D. Schwartz. 1994. The impact of nonindigenous crested wheatgrasses on native black grass bugs in North America: a case for ecosystem management. Nat. Areas J. 14: 136–138.

1995. Native black grass bugs *(Irbisia-Labops)* on introduced wheatgrasses: commentary and annotated bibliography (Hemiptera: Heteroptera: Miridae). Proc. Entomol. Soc. Washington 97: 90–111.

Laurema, S., and A.-L. Varis. 1991. Salivary amino acids in *Lygus* species (Heteroptera: Miridae). Insect Biochem. 21: 759–765.

Laurema, S., A.-L. Varis, and H. Miettinen. 1985. Studies on enzymes in the salivary glands of *Lygus rugulipennis* (Hemiptera, Miridae). Insect Biochem. 15: 211–224.

Lavabre, E. M. 1969. Recent progress in breeding certain pests of African cocoa in the laboratory. FAO Plant Prot. Bull. 17: 132–135.

Lavabre, E. M. [ed.]. 1977. Les mirides du cacaoyer. Institut Français du Café et du Cacao, Paris, France. 366 pp.

Leach, R. 1935. Insect injury simulating fungal attack on plants. A stem canker, an angular spot, a fruit scab and a fruit rot of mangoes caused by *Helopeltis bergrothi* Reut. (Capsidae). Ann. Appl. Biol. 22: 525–537.

Leach, R., and C. Smee. 1933. Gnarled stem canker of tea caused by the capsid bug (*Helopeltis bergrothi* Reut.). Ann. Appl. Biol. 20: 691–706.

Lebert, C. D. 1935. A plant bug (*Pycnoderes quadrimaculatus* Guer.). U.S. Dept. Agric. Insect Pest Surv. Bull. 15: 376.

Leferink, J. H. M., and G. H. Gerber. 1997. Development of adult and nymphal populations of *Lygus lineolaris* (Palisot de Beauvois), *L. elisus* Van Duzee, and *L. borealis* (Kelton) (Heteroptera: Miridae) in relation to seeding date and stage of plant development on canola (Brassicaceae) in southern Manitoba. Canad. Entomol. 129: 777–787.

Leigh, T. F. 1963. Life history of *Lygus hesperus* (Hemiptera: Miridae) in the laboratory. Ann. Entomol. Soc. Amer. 56: 865–867.

Leigh, T. F., and G. A. Matthews. 1994. *Lygus* (Hemiptera: Miridae) and other Hemiptera. Pp. 367–379 *in* G. A. Matthews and J. P. Tunstall [eds.], Insect Pests of Cotton. CAB International, Wallingford, U.K. 593 pp.

Leigh, T. F., C. E. Jackson, P. F. Wynholds, and J. A. Cota. 1977. Toxicity of selected insecticides applied topically to *Lygus hesperus*. J. Econ. Entomol. 70: 42–44.

Leigh, T. F., T. A. Kerby, and P. F. Wynholds. 1988. Cotton square damage by the plant bug, *Lygus hesperus* (Hemiptera: Heteroptera: Miridae), and abscission rates. J. Econ. Entomol. 81: 1328–1337.

Leonard, M. D. 1915. Further experiments in the control of the tarnished plant-bug. J. Econ. Entomol. 8: 361–367.

Leston, D. 1952. On certain subgenera of *Lygus* Hahn 1833 (Hem., Miridae), with a review of the British species. Entomol. Gaz. 3: 213–230.

1970. Entomology of the cocoa farm. Annu. Rev. Entomol. 15: 273–294.

1971. Ants, capsids and swollen shoot in Ghana: interactions and the implications for pest control. Pp. 205–221 *in* Proc. III Int. Cocoa Res. Conf., Accra, Ghana, 23–29 Nov. 1969. Cocoa Res. Institute, Tafo, Ghana.

1973. The ant-mosaic-tropical tree crops and the limiting of pests and diseases. PANS (Pest Artic. News Summ.) 19: 311–341.

1978. A Neotropical ant mosaic. Ann. Entomol. Soc. Amer. 71: 649–653.

Leuschner, K., S. L. Taneja, and H. C. Sharma. 1985. The role of host-plant resistance in pest management in sorghum in India. Insect Sci. Appl. 6: 453–460.

Lever, R. J. A. W. 1949. The tea mosquito bugs (*Helopeltis* spp.) in the Cameron Highlands. Malayan Agric. J. 32: 91–109.

Lim, T. K., and K. C. Khoo. 1990. Guava in Malaysia: Production, Pests and Diseases. Tropical Press, Kuala Lumpur, Malaysia. 260 pp.

Ling, Y. H., W. F. Campbell, B. A. Haws, and K. H. Asay. 1985. Scanning electron microscope (SEM) studies of morphology of range grasses in relation to feeding by *Labops hesperius*. Crop Sci. 25: 327–332.

Linsley, E. G., and J. W. MacSwain. 1947. Factors influencing the effectiveness of insect pollinators of alfalfa in California. J. Econ. Entomol. 40: 349–357.

Lord, F. T. 1971. Laboratory tests to compare the predatory value of six mirid species in each stage of development against the winter eggs of the European red mite, *Panonychus ulmi* (Acari: Tetranychidae). Canad. Entomol. 103: 1663–1669.

Lupo, V. 1946. Invasion of *Calocoris norvegicus* (Gml.) in the communes of the Vesuvius region. Int. Bull. Plant Prot. 20: 105M–108M.

Maas, J. L. [ed.]. 1984. Compendium of Strawberry Diseases. American Phytopathological Society, St. Paul, Minnesota, U.S.A. 138 pp.

Macfarlane, R. P., J. A. Wightman, R. P. Griffin, and D. N. J. Whitford. 1981. Hemiptera and other insects on South Island lucerne and lotus seed crops 1980–81. Proc. 34th New Zealand Weed Pest Control Conf. Pp. 39–42. New Zealand Weed and Pest Control Society, Palmerston North, New Zealand.

MacLellan, C. R. 1979. Pest damage and insect fauna of Nova Scotia apple orchards. Canad. Entomol. 111: 985–1004.

Majer, J. D. 1972. The ant mosaic in Ghana cocoa farms. Bull. Entomol. Res. 62: 151–160.

Maldonado Capriles, J. 1969. The Miridae of Puerto Rico (Insecta, Hemiptera). Puerto Rico Agric. Exp. Stn. Tech. Pap. 45: 1–133.

Malechek, J. C., A. M. Gray, and B. A. Haws. 1977. Yield and nutritional quality of intermediate wheatgrass infested by black grass bugs at low population densities. J. Range Manage. 30: 128–131.

Marchart, H. 1971. Ants, capsids and swollen shoot: a reply to Leston. Pp. 235–236 *in* Proc. III Int. Cocoa Res. Conf., Accra, Ghana, 23–29 Nov. 1969. Cocoa Research Institute, Tafo, Ghana.

1972. Effect of capsid attack on pod development. Cocoa Res. Inst. (Ghana) Rep. 1969–70: 99.

Martin, P. J. et al. 1997. Cashew nut production in Tanzania: constraints and progress through integrated crop management. Crop Prot. 16: 5–14.

Martorell, L. F. 1976. Annotated food plant catalog of the insects of Puerto Rico. Agricultural Experiment Station, University of Puerto Rico, Rio Piedras, Puerto Rico, U.S.A. 303 pp.

Massee, A. M. 1937. The Pests of Fruits and Hops. Crosby Lockwood, London, U.K. 294 pp.

Massee, A. M., and W. Steer. 1928. Capsid bugs. Gard. Chron. (London) (3) 84: 154–155.

Mau, R. F. L., and K. Nishijima. 1989. Development of the transparentwinged plant bug, *Hyalopeplus pellucidus* (Stål), a pest of cultivated guava in Hawaii. Proc. Hawaiian Entomol. Soc. 29: 139–147.

Mauney, J. R., and T. J. Henneberry. 1979. Identification of damage symptoms and patterns of feeding of plant bugs in cotton. J. Econ. Entomol. 72: 496–501.

1984. Causes of square abscission in cotton in Arizona. Crop Sci. 24: 1027–1030.

McBrien, H. L., G. J. R. Judd, and J. H. Borden. 1994. *Campylomma verbasci* (Heteroptera: Miridae): pheromone-based seasonal flight patterns and prediction of nymphal densities in apple orchards. J. Econ. Entomol. 87: 1224–1229.

1996. Potential for pheromone-based mating disruption of the mullein bug, *Campylomma verbasci* (Meyer) (Heteroptera: Miridae). Canad. Entomol. 128: 1057–1064.

1997. Population suppression of *Campylomma verbasci* (Heteroptera: Miridae) by atmospheric permeation with synthetic sex pheromone. J. Econ. Entomol. 90: 801–808.

McDermid, E. M. 1956. The insect pests of cotton in Zande district of Equatoria Province, Sudan 2. The insect pests, their status and control. Emp. Cotton Grow. Rev. 33: 44–66.

McEwen, F. L., and G. E. R. Hervey. 1960. The effect of lygus bug control on the yield of lima beans. J. Econ. Entomol. 53: 513–516.

McIver, J. D., and G. Stonedahl. 1993. Myrmecomorphy: morphological and behavioral mimicry of ants. Annu. Rev. Entomol. 38: 351–379.

McKinlay, K. S., and Q. A. Geering. 1957. Studies of crop loss following insect attack on cotton in East Africa. I. — Experiments in Uganda and Tanganyika. Bull. Entomol. Res. 48: 833–849.

McMullen, R. D., and C. Jong. 1967. New records and discussion of predators of the pear psylla, *Psylla pyricola* Forster, in British Columbia. J. Entomol. Soc. British Columbia 64: 35–40.

1970. The biology and influence of pesticides on *Campylomma verbasci* (Heteroptera: Miridae). Canad. Entomol. 102: 1390–1394.

Mensah, R. K., and M. Khan. 1997. Use of *Medicago sativa* (L.) interplantings/trap crops in the management of the green mirid, *Creontiades dilutus* (Stål) in commercial cotton in Australia. Int. J. Pest Manage. 43: 197–202.

Meredith, P. 1970. "Bug" damage in wheat. New Zealand Wheat Rev. No. 11, 1968–70: 49–53.

Miles, P. W., and K. Hori. 1977. Fate of ingested β-indolyl acetic acid in *Creontiades dilutus*. J. Insect Physiol. 23: 221–226.
Miller, N. C. E. 1941. Insects associated with cocoa *(Theobroma cacao)* in Malaya. Bull. Entomol. Res. 32: 1–16.
Mols, P. J. M. 1990. Forecasting orchard pests for adequate timing of control measures. Proc. Exp. Appl. Entomol. Neth. Entomol. Soc. (N.E.V., Amst.) 1: 75–81.
Morrill, A. W. 1925. Commercial entomology on the West Coast of Mexico. J. Econ. Entomol. 18: 707–716.
——— 1927. A plant bug (*Pycnoderes incurvatus* [sic] Dist.). U.S. Dept. Agric. Insect Pest Surv. Bull. 7: 111.
Morrison, L. 1938. Surveys of the insect pests of wheat crops in Canterbury and North Otago during the summers of 1936–37 and 1937–38. New Zealand J. Sci. Technol. 20(A): 142–155.
Muraleedharan, N. 1987. Entomological research in tea in southern India. J. Coffee Res. 17: 80–83.
——— 1992. Pest control in Asia. Pp. 375–412 *in* K. C. Willson and M. N. Clifford [eds.], Tea: Cultivation and Consumption. Chapman & Hall, London, U.K. 769 pp.
Murray, D. A. H., and R. J. Lloyd. 1997. The effect of spinosad (Tracer) on arthropod pest and beneficial populations in Australian cotton. Pp. 1087–1091 *in* Proc. Beltwide Cotton Conf., Jan. 6–10, 1997. National Cotton Council, Memphis, Tennessee, U.S.A.
Nagaraja, K. V., P. S. Bhavanishankara Gowda, V. V. Krishna Kurup, and J. N. John. 1994. Biochemical changes in cashew in relation to infestation by tea mosquito bug. Plant Physiol. Biochem. (New Delhi) 21: 91–97.
Nair, M. R. G. K. 1975. Insects and Mites of Crops in India. Indian Council of Agricultural Research, New Delhi, India. 404 pp.
Nambiar, K. K. N., Y. R. Sarma, and G. B. Pillai. 1973. Inflorescence blight of cashew (*Anacardium occidentale* L.). J. Plant. Crops 1: 44–46.
Natajaran, N., and P. C. Sundara Babu. 1988. Economic injury level for sorghum earhead bug, *Calocoris angustatus* Lethierry in southern India. Insect Sci. Appl. 9: 395–398.
Natarajan, N., P. C. Sundara Babu, and S. Chelliah. 1988. Influence of weather factors on sorghum earhead bug *Calocoris angustatus* Lethierry incidence. Trop. Pest Manage. 34: 413–420.
——— 1989. Seasonal occurrence of sorghum earhead bug, *Calocoris angustatus* Lethierry (Hemiptera: Miridae) in southern India. Trop. Pest Manage. 35: 70–77.
Niemczyk, E. 1978. *Campylomma verbasci* Mey-Dur (Heteroptera, Miridae) as a predator of aphids and mites in apple orchards. Pol. Pismo Entomol. 48: 221–235.
Nuorteva, P., and L. Reinius. 1953. Incorporation and spread of C^{14}-labeled oral secretions of wheat bugs in wheat kernels. Ann. Entomol. Fenn. 19: 95–104.
Olson, D. L., and R. L. Wilson. 1990. Tarnished plant bug (Hemiptera: Miridae): effect on seed weight of grain amaranth. J. Econ. Entomol. 83: 2443–2447.
Owusu-Manu, E. 1985. The evaluation of the synthetic pyrethroids for the control of *Distantiella theobroma* Dist. (Hemiptera, Miridae) in Ghana. Pp. 535–538 *in* Proc. 9th Int. Cocoa Res. Conf., Lomé, Togo, 12 Feb. 1984. Cocoa Producers' Alliance, Lagos, Nigeria.
Pack, T. M., and P. Tugwell. 1976. Clouded and tarnished plant bugs on cotton: a comparison of injury symptoms and damage on fruit parts. Arkansas Agric. Exp. Stn. Rep. Ser. 226: 1–17.
Painter, R. H. 1927. Notes on the oviposition habits of the tarnished plant bug, *Lygus pratensis* Linn, with a list of host plants. 57th Annu. Rep. Entomol. Soc. Ontario 1926: 44–46.
——— 1930. A study of the cotton flea hopper, *Psallus seriatus* Reut., with especial reference to its effect on cotton plant tissues. J. Agric. Res. (Washington, D.C.) 40: 485–516.
Pang, T. C. 1981. The present status of cocoa bee bug, *Platyngomiriodes apiformis* Ghauri, in Sabah and its life cycle study. Pp. 83–90 *in* Int. Symp. Problems of Insect Pest Management in Developing Countries. Trop. Agric. Res. Ser. 14. Trop. Agric. Res. Cent. Min. Agric., For., Fish., Yatabe, Tsukuba, Ibaraki, Japan.
Pang, T. C., and R. A. Syed. 1972. Some important pests of cocoa in Sabah. Pp. 45–49 *in* R. L. Wastie and D. A. Earp [eds.], Cocoa and coconuts in Malaysia. Proc. Conf. [on cocoa and coconuts], Kuala Lumpur, 25–27 Nov. 1971. Incorporated Society of Planters, Kuala Lumpur, Malaysia.
Parker, B. L., and K. I. Hauschild. 1975. A bibliography of the tarnished plant bug, *Lygus lineolaris* (Hemiptera: Miridae), on apple. Bull. Entomol. Soc. Amer. 21: 119–121.
Peal, S. E. 1873. The tea-bug of Assam. J. Agric. Hortic. Soc. India 4(1): 126–132.

Pearson, E. O. 1958. The Insect Pests of Cotton in Tropical Africa. Empire Cotton Growing Corp. Commonwealth Institute of Entomology, London, U.K. 355 pp.

Peregrine, W. T. H. 1991. Anatto — a possible trap crop to assist control of the mosquito bug (*Helopeltis schoutedeni* Reut.) in tea and other crops. Trop. Pest Manage. 37: 429–430.

Pescott, R. T. M. 1940. A capsid plant bug attacking stone fruits. J. Austral. Inst. Agric. Sci. 6: 101–102.

Peterson, A. G., and E. V. Vea. 1969. Silvertop, the elusive mystery. Minnesota Sci. 25(2): 12–14.

1971. Silvertop of bluegrass in Minnesota. J. Econ. Entomol. 64: 247–252.

Petherbridge, F. R., and M. A. Husain. 1918. A study of the capsid bugs found on apple trees. Ann. Appl. Biol. 4: 179–205.

Petherbridge, F. R., and W. H. Thorpe. 1928. The common green capsid bug *(Lygus pabulinus)*. Ann. Appl. Biol. 15: 446–472.

Pickel, C., and R. S. Bethell. 1990. Insects and mites. Pp. 1–31 *in* Apple Pest Management Guidelines (UCPMG Publ. 12). Univ. California Div. Nat. Resour. Publ. 3339.

Pickett, A. D. 1938. The mullein leaf bug — *Campylomma verbasci*, Meyer, as a pest of apple in Nova Scotia. 69th Annu. Rep. Entomol. Soc. Ontario. 1938: 105–106.

Pillai, G. B. 1987. Integrated pest management in plantation crops. J. Coffee Res. 17: 150–153.

Pillai, G. B., and V. A. Abraham. 1975. Tea mosquito — a serious menace to cashew. Indian Cashew J. 10(1): 5, 7.

Pillai, G. B., O. P. Dubey, and V. Singh. 1976. Pests of cashew and their control in India — a review of current status. J. Plant. Crops 4: 37–50.

Prokopy, R. J., and E. D. Owens. 1978. Visual generalist ["with"] visual specialist phytophagous insects: host selection behaviour and application to management. Entomol. Exp. Appl. 24: 609–620.

Purcell, M., and S. C. Welter. 1990. Seasonal phenology and biology of *Calocoris norvegicus* (Hemiptera: Miridae) in pistachios and associated host plants. J. Econ. Entomol. 83: 1841–1846.

1991. Effect of *Calocoris norvegicus* (Hemiptera: Miridae) on pistachio yields. J. Econ. Entomol. 84: 114–119.

Putschkov, V. G. 1966. The Main Bugs — Plant Bugs — as Pests of Agricultural Crops. Naukova Dumka, Kiev, U.S.S.R. 171 pp. [in Russian].

Puttarudriah [Puttarudraiah], M. 1952. Blister disease "Kajji" of guava fruits *(Psidium guava)*. Mysore Agric. J. 28: 8–13.

Puttarudriah, M., and M. Appanna. 1955. Two new hosts of *Helopeltis antonii* Signoret in Mysore. Indian J. Entomol. 17: 391–392.

Radcliffe, E. B., and D. K. Barnes. 1970. Alfalfa plant bug injury and evidence of plant resistance in alfalfa. J. Econ. Entomol. 63: 1995–1996.

Ramakrishna Ayyar, T. V. 1940. Handbook of Economic Entomology for South India. Government Press, Madras, India. 528 pp.

Raman, K., and K. P. Sanjayan. 1984a. Host plant relationships and population dynamics of the mirid, *Cyrtopeltis tenuis* Reut. (Hemiptera: Miridae). Proc. Indian Natl. Sci. Acad. B50: 355–361.

1984b. Histology and histopathology of the feeding lesions by *Cyrtopeltis tenuis* Reut. (Hemiptera: Miridae) on *Lycopersicon esculentum* Mill. (Solanaceae). Proc. Indian Acad. Sci. 93: 543–547.

Raman, K., K. P. Sanjayan, and G. Suresh. 1984. Impact of feeding injury of *Cyrtopeltis tenuis* Reut. (Hemiptera: Miridae) on some biochemical changes in *Lycopersicon esculentum* Mill (Solanaceae). Curr. Sci. (Bangalore) 53: 1092–1093.

Rao, G. N. 1970. Tea pests in southern India and their control. PANS (Pest Artic. News Summ.) 16: 667–672.

Ratnadass, A., B. Cissé, and K. Mallé. 1994. Notes on the biology and immature stages of West African sorghum head bugs *Eurystylus immaculatus* and *Creontiades pallidus* (Heteroptera: Miridae). Bull. Entomol. Res. 84: 383–388.

Ratnadass, A., Y. O. Doumbia, and O. Ajayi. 1995a. Bioecology of sorghum head bug *Eurystylus immaculatus*, and crop losses in West Africa. Pp. 91–102 *in* K. F. Nwanze and O. Youm [eds.], Panicle Insect Pests of Sorghum and Pearl Millet: Proceedings of an International Consultative Workshop, 4–7 Oct. 1993, ICRISAT Sahelian Center, Niamey, Niger. International Crops Research Institute for the Semi-Arid Tropics, Patancheru, A.P., India.

Ratnadass, A., O. Ajayi, G. Fliedel, and K. V. Ramaiah. 1995b. Host-plant resistance in sorghum to *Eurystylus immaculatus* in West Africa. Pp. 191–199 *in* K. F. Nwanze and O. Youm [eds.], Panicle Insect Pests of Sorghum and Pearl Millet: Proceedings of an International Consultative Workshop, 4–7 Oct. 1993, ICRISAT Sahelian Center, Niamey, Niger. International Crops Research Institute for the Semi-Arid Tropics, Patancheru, A.P., India.

Ratnadass, A., B. Cissé, D. Diarra, and M. L. Sangaré. 1997. Indigenous host plants of sorghum head-bugs (Heteroptera: Miridae) in Mali. Afric. Entomol. 5: 158–160.

Rattan, P. S. 1992. Pest and disease control in Africa. Pp. 331–352 *in* K. C. Willson and M. N. Clifford [eds.], Tea: Cultivation to Consumption. Chapman & Hall, London, U.K. 769 pp.

Rautapää, J. 1969. Effect of *Lygus rugulipennis* Popp. (Hem., Capsidae) on the yield and quality of wheat. Ann. Entomol. Fenn. 35: 168–175.

Reddy, D. B. 1958. Sannhemp [sic] and its insect fauna. Proc. 10th Int. Congr. Entomol. 3: 439–440.

Reding, M. E., and E. H. Beers. 1996. Influence of prey availability on survival of *Campylomma verbasci* (Hemiptera: Miridae) and factors influencing efficacy of chemical control on apples. Pp. 141–154 *in* O. Alomar and R. Wiedenmann [eds.], Zoophytophagous Heteroptera: Implications for Life History and Integrated Pest Management. Thomas Say Publ. Entomol.: Proceedings Entomological Society of America, Lanham, Maryland, U.S.A. 202 pp.

Reid, D. G. 1974. New records of *Hexamermis* (Nematoda: Mermithidae) parasitizing three species of *Slaterocoris* (Hemiptera: Miridae). Canad. Entomol. 106: 239.

Reinhard, H. J. 1926. The cotton flea hopper. Texas Agric. Exp. Stn. Bull. 339: 1–39.

1928. Hibernation of the cotton flea hopper. Texas Agric. Exp. Stn. Bull. 377: 1–26.

Rice, R. E., J. K. Uyemoto, J. M. Ogawa, and W. M. Pemberton. 1985. New findings on pistachio problems. California Agric. 39(1–2): 15–18.

Rice, R. E., W. J. Bentley, and R. H. Beede. 1988. Insect and mite pests of pistachios in California. Univ. California Div. Agric. Nat. Resour. Publ. 21452. 26 pp.

Ridgway, R. L., and G. G. Gyrisco. 1960. Effect of temperature on the rate of development of *Lygus lineolaris* (Hemiptera: Miridae). Ann. Entomol. Soc. Amer. 53: 691–694.

Riggs, D. I. 1990. Greenhouse studies of the effect of lygus bug feeding on 'Tristar' strawberry. Adv. Strawberry Prod. 9: 40–43.

Ring, D. R., J. H. Benedict, M. L. Walmsley, and M. F. Treacy. 1993. Cotton yield response to cotton fleahopper (Hemiptera: Miridae) infestations on the Lower Gulf Coast of Texas. J. Econ. Entomol. 86: 1811–1819.

Ripper, W. E., and L. George. 1965. Cotton Pests of the Sudan: Their Habits and Control. Blackwell, Oxford, U.K. 363 pp.

Rosenzweig, V. Ye. 1997. Revised classification of the *Calocoris* complex and related genera (Heteroptera: Miridae). Zoosyst. Ross. 6: 139–169.

Rostrup, S., and M. Thomsen. 1923. Bekaempelse af Taeger paa Aebletraeer samt Bidrag til disse Taegers Biologi. Tidsskr. Planteavl 29: 39–461.

Sadras, V. O., and G. P. Fitt. 1997. Apical dominance-variability among cotton genotypes and its association with resistance to insect herbivory. Environ. Exp. Bot. 38: 145–153.

Sapio, F. J., L. F. Wilson, and M. E. Ostry. 1982. A split-stem lesion on hybrid *Populus* trees caused by the tarnished plant bug, *Lygus lineolaris* (Hemiptera [Heteroptera]: Miridae. Great Lakes Entomol. 15: 237–246.

Sathiamma, B. 1977. Nature and extent of damage by *Helopeltis antonii* S., the tea mosquito on cashew. J. Plant. Crops 5: 58–62.

Schaefer, C. W. 1974. Rise and fall of the apple redbugs. Mem. Connecticut Entomol. Soc. 1974: 101–116.

Schmutterer, H. 1969. Pests of Crops in Northeast and Central Africa with Particular Reference to the Sudan. Fischer, Stuttgart, Germany. 296 pp.

1977. Other injurious Heteroptera. Pp. 300–301 *in* J. Kranz, H. Schmutterer, and W. Koch [eds.], Diseases, Pests and Weeds in Tropical Crops. Parey, Berlin, Germany. 666 pp.

Schowalter, T. D. 1987. Abundance and distribution of *Lygus hesperus* (Heteroptera: Miridae) in two conifer nurseries in western Oregon. Environ. Entomol. 16: 687–690.

Schowalter, T. D., D. L. Overhulser, A. Kanaskie, J. D. Stein, and J. Sexton. 1986. *Lygus hesperus* as an agent of apical bud abortion in Douglas-fir nurseries in western Oregon. New For. 1: 5–15.

Schroeder, N. C., R. B. Chapman, and P. T. P. Clifford. 1998. Effect of potato mirid *(Calocoris norvegicus)* on white clover seed production in small cages. New Zealand J. Agric. Res. 41: 111–116.

Schuh, R. T. 1984. Revision of the Phylinae (Hemiptera, Miridae) of the Indo-Pacific. Bull. Amer. Mus. Nat. Hist. 177: 1–476.

1995. Plant Bugs of the World (Insecta: Heteroptera: Miridae): Systematic Catalog, Distributions, Host List, and Bibliography. New York Entomological Society, New York, New York, U.S.A. 1329 pp.

Schuh, R. T., and J. A. Slater. 1995. True Bugs of the World (Hemiptera: Heteroptera): Classification and Natural History. Cornell University Press, Ithaca, New York, U.S.A. 336 pp.

Schwartz, M. D., and R. G. Foottit. 1992. Lygus bugs on the Prairies: biology, systematics, and distribution. Agric. Canad. Res. Branch Tech. Bull. 1992–4E: 1–44.

1998. Revision of the Nearctic Species of the Genus *Lygus* Hahn, with a Review of the Palaearctic Species (Heteroptera: Miridae). Mem. Entomol. Int. Vol. 10. Associated Publishers, Gainesville, Florida. 428 pp.

Scott, D. R. 1970. Feeding of *Lygus* bugs (Hemiptera: Miridae) on developing carrot and bean seed: increased growth and yields of plants grown from that seed. Ann. Entomol. Soc. Amer. 63: 1604–1608.

1976. Phytostimulation by lygus bugs feeding on developing seeds. Pp. 17–18 *in* D. R. Scott and L. E. O'Keeffe [eds.], Lygus Bug: Host Plant Interactions. University Press of Idaho, Moscow, Idaho, U.S.A. 38 pp.

1977. An annotated listing of host plants of *Lygus hesperus* Knight. Bull. Entomol. Soc. Amer. 23(1): 19–22.

1980. A bibliography of *Lygus* Hahn (Hemiptera: Miridae). Idaho Agric. Exp. Stn. Misc. Ser. 58: 1–71.

1983. *Lygus hesperus* Knight (Hemiptera: Miridae) and *Daucus carota* L. (Umbelliflorae: Umbelliferae): an example of relationships between a polyphagous insect and one of its plant hosts. Environ. Entomol. 12: 6–9.

Scudder, G. G. E. 1997. True bugs (Heteroptera) of the Yukon. Pp. 241–336 *in* H. V. Danks and J. A. Downes [eds.], Insects of the Yukon. Monogr. Ser. 2. Biological Survey of Canada (Terrestrial Arthropods), Ottawa, Ontario, Canada. 1034 pp.

Šedivý, J., and A. Honěk. 1983. Flight of *Lygus rugulipennis* Popp. (Heteroptera, Miridae) to a light trap. Z. Pflanzenkr. Pflanzenschutz 90: 238–243.

Sevacherian, V., and V. M. Stern. 1974. Host plant preferences of lygus bugs in alfalfa-interplanted cotton fields. Environ. Entomol. 3: 761–766.

1975. Movements of lygus bugs between alfalfa and cotton. Environ. Entomol. 4: 163–165.

Shabbir, S. G., and U. D. Choudhry. 1984. Pests of important fibre crops in Pakistan. Pp. 93–162 *in* M. K. Ahmed [ed.], Insect Pests of Important Crops in Pakistan. Department of Plant Protection, Karachi, Pakistan.

Sharma, H. C. 1985. Screening for host-plant resistance to mirid head bugs in sorghum. Pp. 317–335 *in* Proc. Int. Sorghum Entomol. Workshop, 15–21 July 1984, Texas A&M Univ., College Station. ICRISAT, Patancheru, A.P., India.

Sharma, H. C., and V. F. Lopez. 1989. Assessment of avoidable losses and economic injury levels for the sorghum head bug, *Calocoris angustatus* Leth. (Hemiptera: Miridae) in India. Crop Prot. 8: 429–435.

1990. Biology and population dynamics of sorghum head bugs (Hemiptera: Miridae). Crop Prot. 9: 164–173.

1992. Genotypic resistance in sorghum to head bug, *Calocoris angustatus* Lethiery [sic]. Euphytica 58: 193–200.

Sharma, H. C., Y. O. Doumbia, and N. Y. Diorisso. 1992. A headcage technique to screen sorghum for resistance to mirid head bug, *Eurystylus immaculatus* Odh. in West Africa. Insect Sci. Appl. 13: 417–427.

Shull, W. E. 1933. An investigation of the *Lygus* species which are pests of beans (Hemiptera, Miridae). Univ. Idaho Agric. Exp. Stn. Res. Bull. 11: 1–42.

Slater, J. A. 1954. Notes on the genus *Labops*, Burmeister in North America with the descriptions of three new species (Hemiptera: Miridae). Bull. Brooklyn Entomol. Soc. 49: 57–65, 89–94.

Slingerland, M. V. 1909. A red bug on apple. P. 91 *in* Proceedings, 54th Annual Meeting of the Western New York Horticultural Society, Jan. 27–28, Rochester, New York. Democrat and Chronicle Press, Rochester, New York, U.S.A.

Smith, E. S. C. 1972. Population fluctuations of the cocoa mirid in the Northern District of Papua New Guinea. Abstr. 14 Int. Congr. Entomol., Canberra, 22–30 Aug. 1972. P. 330. Australian Academy of Science, Canberra; Australian Entomological Society, Brisbane, Australia.

1973. A laboratory rearing method for the cacao mirid *Helopeltis clavifer* Walker (Hemiptera: Miridae). Papua New Guinea Agric. J. 24: 52–53.

1977. Presence of a sex attractant pheromone in *Helopeltis clavifer* (Walker) (Heteroptera: Miridae). J. Austral. Entomol. Soc. 16: 113–116.

1978. Host and distribution records of *Helopeltis clavifer* (Walker) (Heteroptera: Miridae) in Papua New Guinea. Papua New Guinea Agric. J. 29: 1–4.

1979. Descriptions of the immature and adult stages of the cocoa mirid *Helopeltis clavifer* (Heteroptera: Miridae). Pac. Insects 20: 354–361.

1985. A review of relationships between shade types and cocoa pest and disease problems in Papua New Guinea. Papua New Guinea J. Agric. For. Fish. 33: 79–88.

Smith, E. S. C., B. M. Thistleton, and J. R. Pippet. 1985. Assessment of damage and control of *Helopeltis clavifer* (Heteroptera: Miridae) on tea in Papua New Guinea. Papua New Guinea J. Agric. For. Fish. 33: 123–131.

Smith, K. M. 1920. The injurious apple capsid (*Plesiocoris rugicollis*, Fall.). J. Minist. Agric. (London) 27: 379–381.

1925. Note on the egg-laying of *Calocoris bipunctatus* Fab. Entomol. Mon. Mag. 61: 91–92.

1931. A Textbook of Agricultural Entomology. University Press, Cambridge, U.K. 285 pp.

Smith, R. F. 1991. The mullein bug, *Campylomma verbasci*. Pp. 199–214 *in* K. Williams [ed.], New Directions in Tree Fruit Pest Management. Good Fruit Grower, Yakima, Washington, U.S.A. 214 pp.

Smith, R. F., and J. H. Borden. 1991. Fecundity and development of the mullein bug, *Campylomma verbasci* (Meyer) (Heteroptera: Miridae). Canad. Entomol. 123: 595–600.

Snodgrass, G. L., W. P. Scott, and J. W. Smith. 1984a. A survey of the host plants and seasonal distribution of the cotton fleahopper (Hemiptera: Miridae) in the delta of Arkansas, Louisiana, and Mississippi. J. Georgia Entomol. Soc. 19: 34–41.

1984b. An annotated list of the host plants of *Lygus lineolaris* (Hemiptera: Miridae) in the Arkansas, Louisiana, and Mississippi Delta. J. Georgia Entomol. Soc. 19: 93–101.

Sorenson, C. J. 1932. Insects in relation to alfalfa-seed production. Utah Agric. Exp. Stn. Circ. 98: 1–28.

Soroka, J. J., and D. C. Murrell. 1993. The effects of alfalfa plant bug (Hemiptera: Miridae) feeding late in the season on alfalfa seed yield in northern Saskatchewan. Canad. Entomol. 125: 815–824.

South, D. B. 1991. *Lygus* bugs: a worldwide problem in conifer nurseries. Pp. 215–222 *in* J. R. Sutherland and S. G. Glover [eds.], Proc. First Meet. IUFRO Working Party S2.07-09 (Diseases and Insects in Forest Nurseries). For. Canad. Pacific Yukon Reg. Inf. Rep. BC-X-331.

Southwood, T. R. E. 1956. The nomenclature and life cycle of the European Tarnished Plant Bug, *Lygus rugulipennis* Poppius (Hem., Miridae). Bull. Entomol. Res. 46: 845–848.

1960. The flight activity of Heteroptera. Trans. R. Entomol. Soc. Lond. 112: 173–220.

1973. The insect/plant relationship — an evolutionary perspective. Pp. 3–30 *in* H.F. van Emden [ed.], Insect/Plant Relationships. Wiley, New York, New York, U.S.A. 215 pp.

Southwood, T. R. E., and D. Leston. 1959. Land and Water Bugs of the British Isles. Warne, London, U.K. 436 pp.

Southwood, T. R. E., and G. G. E. Scudder. 1956. The immature stages of the Hemiptera-Heteroptera associated with the stinging nettle (*Urtica dioica* L.). Entomol. Mon. Mag. 92: 313–325.

Soyer, D. 1942. Miride du cotonnier, *Creontiades pallidulus* Ramb. Capsidae (Miridae). Publ. Inst. Natl. Etude Agron. Congo Belge Ser. Sci. No. 29: 1–15.

Spangler, S. M., and J. A. MacMahon. 1990. Arthropod faunas of monocultures and polycultures in reseeded rangelands. Environ. Entomol. 19: 244–250.

Squire, F. A. 1947. On the economic importance of the Capsidae in the Guinean region. Rev. Entomol. (Rio de Janeiro) 18: 219–247.

Stam, P. A. 1987. *Creontiades pallidulus* (Rambur) (Miridae, Hemiptera), a pest on cotton along the Euphrates River and its effect on yield and control action threshold in the Syrian Arab Republic. Trop. Pest Manage. 33: 273–276.

Steck, G. J., G. L. Teetes, and S. D. Maiga. 1989. Species composition and injury to sorghum by panicle feeding bugs in Niger. Insect Sci. Appl. 10: 199–217.

Sterling, W. L. 1984. Action and inaction levels in pest management. Texas Agric. Exp. Stn. B-1480: 1–20.

Sterling, W. L., and D. A. Dean. 1977. A bibliography of the cotton fleahopper *Pseudatomoscelis seriatus* (Reuter). Texas Agric. Exp. Stn. MP-1342: 1–28.

Sterling, W. L., K. M. El-Zik, and L. T. Wilson. 1989a. Biological control of pest populations. Pp. 155–189 *in* R. E. Frisbie, K. M. El-Zik, and L. T. Wilson [eds.], Integrated Pest Management Systems and Cotton Production. Wiley, New York, New York, U.S.A. 437 pp.

Sterling, W. L., L. T. Wilson, A. P. Gutierrez, D. R. Rummel, J. R. Phillips, N. D. Stone, and J. H. Benedict. 1989b. Strategies and tactics for managing insects and mites. Pp. 267–305 *in* R. E. Frisbie, K. M. El-Zik, and L. T. Wilson [eds.], Integrated Pest Management Systems and Cotton Production. Wiley, New York, New York, U.S.A. 437 pp.

Stern, V. M. 1969. Interplanting alfalfa in cotton to control lygus bugs and other insect pests. Proc. Tall Timbers Conf. Ecol. Anim. Control Habitat Manage. 1: 55–69.

1973. Economic thresholds. Annu. Rev. Entomol. 18: 259–280.

Stern, V. M., and A. Mueller. 1968. Techniques of marking insects with micronized fluorescent dust with especial emphasis on marking millions of *Lygus hesperus* for dispersal studies. J. Econ. Entomol. 61: 1232–1237.

Stewart, R. K. 1969. The biology of *Lygus rugulipennis* Poppius (Hemiptera: Miridae) in Scotland. Trans. R. Entomol. Soc. Lond. 120: 437–457.

Stewart, S. D., and M. J. Gaylor. 1993. Age-grading eggs of the tarnished plant bug (Heteroptera: Miridae). J. Entomol. Sci. 28: 263–266.

1994a. Effects of age, sex, and reproductive status on flight by the tarnished plant bug (Heteroptera: Miridae). Environ. Entomol. 23: 80–84.

1994b. Effects of host switching on oviposition by the tarnished plant bug (Heteroptera: Miridae). J. Entomol. Sci. 29: 231–238.

Stewart, S. D., and W. L. Sterling. 1988. Dynamics and impact of cotton fruit abscission and survival. Environ. Entomol. 17: 629–635.

1989. Causes and temporal patterns of cotton fruit abscission. J. Econ. Entomol. 82: 954–959.

Steyaert, R. L. 1946. Plant protection in the Belgian Congo. Sci. Mon. (Washington, D.C.) 63: 268–280.

Stigter, H. 1996. *Campylomma verbasci*, a new pest on apple in the Netherlands. Pp. 140–144 *in* F. Polesny, W. Müller, and R. W. Olszak [eds.], Proc. Int. Conf. on Integrated Fruit Prod., 28 Aug.–2 Sept. 1995, Cedzyna, Poland. Bull. IOBC WPRS (Avignon, France) 19(4).

Stitt, L. L. 1949. Host-plant sources of *Lygus* spp. infesting the alfalfa seed crop in southern Arizona and southeastern California. J. Econ. Entomol. 42: 93–99.

Stonedahl, G. M. 1991. The Oriental species of *Helopeltis* (Heteroptera: Miridae): a review of economic literature and guide to identification. Bull. Entomol. Res. 81: 465–490.

1995. Taxonomy of African *Eurystylus* (Heteroptera: Miridae), with a review of their status as pests of sorghum. Bull. Entomol. Res. 85: 135–156.

Stride, G. O. 1968. On the biology and ecology of *Lygus vosseleri* (Heteroptera: Miridae) with special reference to its hostplant relationships. J. Entomol. Soc. South. Africa. 31: 17–55.

Strong, F. E. 1970. Physiology of injury caused by *Lygus hesperus*. J. Econ. Entomol. 63: 808–814.

Strong, F. E., and J. A. Sheldahl. 1970. The influence of temperature on longevity and fecundity in the bug *Lygus hesperus* (Hemiptera: Miridae). Ann. Entomol. Soc. Amer. 63: 1509–1515.

Strong, F. E., J. A. Sheldahl, P. R. Hughes, and E. M. K. Hussein. 1970. Reproductive biology of *Lygus hesperus* Knight. Hilgardia 40(4): 105–147.

Subbaiah, C. C. 1983. Fruiting and abscission patterns in cashew. J. Agric. Sci. (Cambridge) 100: 423–427.

Sudhakar, M. A. 1975. Role of *Helopeltis antonii* Signoret (Hemiptera: Miridae), in causing scab on guava fruits, its biology and control. Mysore J. Agric. Sci. 9: 205–206.

Swaine, G. 1959. A preliminary note on *Helopeltis* spp. damaging cashew in Tanganyika Territory. Bull. Entomol. Res. 50: 171–181.

Taksdal, G. 1970. Hagetege og stein i paere. Gartneryrket (Oslo) 60: 458–463.

Tan, G. S. 1974a. *Helopeltis theivora theobromae* on cocoa in Malaysia. I. Biology and population fluctuations. Malays. Agric. Res. 3: 127–132.

1974b. *Helopeltis theivora theobromae* on cocoa in Malaysia. II. Damage and control. Malays. Agric. Res. 3: 204–212.

Tanada, Y., and F. G. Holdaway. 1954. Feeding habits of the tomato bug, *Cyrtopeltis (Engytatus) modestus* (Distant), with special reference to the feeding lesion on tomato. Hawaii Agric. Exp. Stn. Tech. Bull. 24: 1–40.

Taylor, T. H. C. 1945. *Lygus simonyi*, Reut., as a cotton pest in Uganda. Bull. Entomol. Res. 36: 121–148.

Teetes, G. L., W. R. Young, and M. G. Jotwani. 1979. Insect pests of sorghum. Pp. 17–40 *in* Elements of integrated control of sorghum pests. FAO Plant Prod. Prot. Pap. 19.

Thankamma Pillai, P. K., and G. B. Pillai. 1975. Note on the shedding of immature fruits in cashew. Indian J. Agric. Sci. 45: 233–234.

Thistlewood, H. M. A., and R. D. McMullen. 1989. Distribution of *Campylomma verbasci* (Heteroptera: Miridae) nymphs on apple and an assessment of two methods of sampling. J. Econ. Entomol. 82: 510–515.

Thistlewood, H. M. A., and R. F. Smith. 1996. Management of the mullein bug, *Campylomma verbasci* (Heteroptera: Miridae), in pome fruit orchards of Canada. Pp. 119–140 *in* O. Alomar and R. N. Wiedenmann [eds.], Zoophytophagous Heteroptera: Implications for Life History and Integrated Pest Management. Thomas Say Publ. Entomol.: Proceedings Entomological Society of America, Lanham, Maryland, U.S.A. 202 pp.

Thistlewood, H. M. A., J. H. Borden, and R. D. McMullen. 1990. Seasonal abundance of the mullein bug, *Campylomma verbasci* (Meyer) (Heteroptera: Miridae), on apple and mullein in the Okanagan Valley. Canad. Entomol. 122: 1045–1058.

Thistlewood, H. M. A., R. D. McMullen, and J. H. Borden. 1989. Damage and economic injury levels of the mullein bug, *Campylomma verbasci* (Meyer) (Heteroptera: Miridae), on apple in the Okanagan Valley. Canad. Entomol. 121: 1–9.

Thomas, D. C. 1938. Report on the Hemiptera-Heteroptera taken in the light trap at Rothamsted Experimental Station, during the four years 1933–1936. Proc. Entomol. Soc. London (A) 13: 19–24.

Thorold, C. A. 1975. Diseases of Cocoa. Clarendon, Oxford, U.K. 423 pp.

Thresh, J. M. 1960. Capsids as a factor influencing the effect of swollen-shoot disease on cacao in Nigeria. Empire J. Exp. Agric. 28: 193–200.

Tingey, W. M., and E. A. Pillemer. 1977. Lygus bugs: crop resistance and physiological nature of feeding. Bull. Entomol. Soc. Amer. 23: 277–287.

Todd, J. G., and J. A. Kamm. 1974. Biology and impact of a grass bug *Labops hesperius* Uhler in Oregon rangeland. J. Range Manage. 27: 453–458.

Townsend, R. J., and R. N. Watson. 1982. Biology of potato mirid and Australian crop mirid on asparagus. Pp. 332–337 *in* M. J. Hartley [ed.], Proc. 35th New Zealand Weed and Pest Control Conf., Aug. 9–12, 1982. New Zealand Weed and Pest Control Society, Palmerston North, New Zealand.

Toxopeus, H. 1985. Botany, types and populations. Pp. 11–37 *in* G. A. R. Wood and R. A. Lass [eds.], Cocoa, fourth edition. Longman, New York, New York, U.S.A. 620 pp.

Toxopeus, H., and B. M. Gerard. 1968. A note on mirid damage to mature cacao pods. Nigerian Entomol. Mag. 1: 59–60.

Tugwell, P., S. C. Young, Jr., B. A. Dumas, and J. R. Phillips. 1976. Plant bugs in cotton: importance of infestation time, types of cotton injury, and significance of wild hosts near cotton. Arkansas Agric. Exp. Stn. Rep. Ser. 227: 1–24.

Turnock, W. J., G. H. Gerber, B. H. Timlick, and R. J. Lamb. 1995. Losses of canola seeds from feeding by *Lygus* species [Heteroptera: Miridae] in Manitoba. Canad. J. Plant Sci. 75: 731–736.

University of California. 1984. Integrated pest management for cotton in the western region of the United States. Univ. California Div. Agric. Nat. Resour. Publ. 3305: 1–144.

Uyemoto, J. K., J. M. Ogawa, R. E. Rice, H. R. Teranishi, R. M. Bostock, and W. M. Pemberton. 1986. Role of several true bugs (Hemiptera) on incidence and seasonal development of pistachio fruit epicarp lesion disorder. J. Econ. Entomol. 79: 395–399.

Vallentine, J. F. 1989. Range Development and Improvements, third edition. Academic Press, San Diego, California, U.S.A. 524 pp.

van den Bosch, R., and V. M. Stern. 1969. The effect of harvesting practices on insect populations in alfalfa. Proc. Tall Timbers Conf. Ecol. Anim. Control Habitat Manage. 1: 47–54.

Varis, A.-L. 1972. The biology of *Lygus rugulipennis* Popp. (Het., Miridae) and the damage caused by this species to sugar beet. Ann. Agric. Fenn. 11: 1–56.

1991. Effect of *Lygus* (Heteroptera: Miridae) feeding on wheat grains. J. Econ. Entomol. 84: 1037–1040.

1997. Seasonal occurrence of *Lygus* bugs on field crops in Finland. Agric. Food Sci. Finland 6: 409–413.

Victorian Plant Research Institute. 1971. Plant-feeding bugs in Victoria: a guide to identification and control. J. Agric. (Victoria) 69: 160–165.

Walker, J. K., G. A. Niles, J. R. Gannaway, J. V. Robinson, C. B. Cowan, and M. J. Lukefahr. 1974. Cotton fleahopper damage to cotton genotypes. J. Econ. Entomol. 67: 537–542.

Watson, R. N., and R. J. Townsend. 1981. Invertebrate pests on asparagus in Waikato. Pp. 70–75 *in* Proc. 34th New Zealand Weed Pest Control Conf. New Zealand Weed & Pest Control Society, Palmerston North, New Zealand.

Watt, G., and H. H. Mann. 1903. The Pests and Blights of the Tea Plant, second edition. Office of the Superintendent, Government Printing, Calcutta, India. 429 pp.

Way, M. J., and K. C. Khoo. 1989. Relationships between *Helopeltis theobromae* damage and ants with special reference to Malaysian cocoa smallholdings. J. Plant Prot. Trop. 6: 1–11.

1992. Role of ants in pest management. Annu. Rev. Entomol. 37: 479–503.

Webster, F. M. 1897. *Halticus bractatus* Say. Entomol. News 8: 209–210.

Wene, G. P., and L. W. Sheets. 1964. Lygus bug injury to presquaring cotton. Arizona Agric. Exp. Stn. Tech. Bull. 166: 1–25.

Wheeler, A. G., Jr. 1983. Outbreaks of the apple red bug: difficulties in identifying a new pest and emergence of a mirid specialist. J. Washington Acad. Sci. 73: 60–64.

2001. Biology of the Plant Bugs (Hemiptera: Miridae): Pests, Predators, Opportunists. Cornell University Press, Ithaca, New York, U.S.A.

Wheeler, A. G., Jr., and T. J. Henry. 1992. A Synthesis of the Holarctic Miridae (Heteroptera): Distribution, Biology, and Origin, with Emphasis on North America. Thomas Say Found. Monogr. Vol. 15. Entomological Society of America, Lanham, Maryland, U.S.A. 282 pp.

Wheeler, A. G., Jr., and G. L. Miller. 1981. Fourlined plant bug (Hemiptera: Miridae), a reappraisal: life history, host plants, and plant response to feeding. Great Lakes Entomol. 14: 23–35.

Wightman, J. A. 1968. A study of oviposition site, mortality and migration in the first (overwintering) generation of *Lygocoris pabulinus* (Fallen) (Heteroptera: Miridae) on blackcurrant shoots (1966–7). Entomologist 101: 269–275.

1969. Termination of egg diapause in *Lygocoris pabulinus* (Heteroptera: Miridae). Long Ashton (U.K.) Res. Stn. Rep. 1968: 154–156.

1974. Heteroptera beaten from potato haulms at Long Ashton Research Station. Entomol. Mon. Mag. 109: 132–139.

Wightman, J. A., and R. P. Macfarlane. 1982. The integrated control of pests of legume seed crops: 2. Summation and strategy of the 1980–81 season. Pp. 377–384 *in* Proc. 3rd Australasian Conf. Grassland Invert. Ecol., Adelaide, 30 Nov.–4 Dec. 1981. S. A. Government Printer, Adelaide, Australia.

Wightman, J. A., and D. N. J. Whitford. 1982. Integrated control of pests of legume seed crops. 1. Insecticides for mirid and aphid control. New Zealand J. Exp. Agric. 10: 209–215.

Willcocks, F. C. 1922. A survey of the more important economic insects and mites of Egypt. Sultanic Agric. Soc. Bull. 1: 1–482.

1925. The Insect and Related Pests of Egypt. Vol. II. Insects and Mites Feeding on Gramineous Crops and Products in the Field, Granary, and Mill. Sultanic Agricultural Society, Cairo, Egypt. 418 pp.

Wille, J. E. 1944. Insect pests of cacao in Peru. Trop. Agric. 21: 143.

1952. Entomologia Agricola del Peru. second edition revised and amplified. Ministerio de Agricultura, Lima, Peru. 543 pp.

Williams, G. 1953. Field observations on the cacao mirids, *Sahlbergella singularis* Hagl. and *Distantiella theobroma* (Dist.), in the Gold Coast. Part I. Mirid damage. Bull. Entomol. Res. 44: 101–119.

1954. Field observations on the cacao mirids, *Sahlbergella singularis* Hagl. and *Distantiella theobroma* (Dist.), in the Gold Coast. Bull. Entomol. Res. 45: 723–744.

Wilson, L. F., and L. M. Moore. 1985. Vulnerability of hybrid *Populus* nursery stock to injury by the tarnished plant bug, *Lygus lineolaris* (Hemiptera: Miridae). Great Lakes Entomol. 18: 19–23.

Wilson, L. T., T. F. Leigh, D. Gonzalez, and C. Foristiere. 1984. Distribution of *Lygus hesperus* (Knight) (Miridae: Hemiptera) on cotton. J. Econ. Entomol. 77: 1313–1319.

Wilson, R. L., and D. L. Olson. 1992. Tarnished plant bug, *Lygus lineolaris* (Palisot de Beauvois) (Hemiptera: Miridae): effect on yield of grain amaranth, *Amaranthus cruentus* L., in field cages. J. Kansas Entomol. Soc. 65: 450–452.

Wipfli, M. S., J. L. Wedberg, D. B. Hogg, and T. D. Syverud. 1989. Insect pests associated with birdsfoot trefoil, *Lotus corniculatus*, in Wisconsin. Great Lakes Entomol. 22: 25–33.

Wipfli, M. S., J. L. Wedberg, and D. B. Hogg. 1990. Damage potentials of three plant bug (Hemiptera: Heteroptera: Miridae) species to birdsfoot trefoil grown for seed in Wisconsin. J. Econ. Entomol. 83: 580–584.
Wolcott, G. N. 1933. An Economic Entomology of the West Indies. Entomological Society of Puerto Rico, San Juan, Puerto Rico, U.S.A. 688 pp.
Wood, G. A. R., and R. A. Lass. 1985. Cocoa, fourth edition. Longman, London, U.K. 620 pp.
Woodroffe, G. E. 1966. The *Lygus pratensis* complex (Hem., Miridae) in Britain. Entomologist 99: 201–206.
Woodside, A. M. 1947. Weed hosts of bugs which cause cat-facing of peaches in Virginia. J. Econ. Entomol. 40: 231–233.
Yasunaga, T. 1991. A revision of the plant bug, genus *Lygocoris* Reuter from Japan, Part I (Heteroptera, Miridae, *Lygus*-complex). Jpn. J. Entomol. 59: 435–448.
Youdeowei, A. 1973. The life cycles of the cocoa mirids *Sahlbergella singularis* Hagl. and *Distantiella theobroma* Dist. in Nigeria. J. Nat. Hist. 7: 217–223.
1977. Behaviour and activity. Pp. 223–236 *in* E. M. Lavabre [ed.], Les Mirides du Cacaoyer. Institut Français du Café et du Cacao, Paris, France. 366 pp.
Young, O. P. 1986. Host plants of the tarnished plant bug, *Lygus lineolaris* (Heteroptera: Miridae). Ann. Entomol. Soc. Amer. 79: 747–762.
Zalom, F. G., C. Pickel, and N. C. Welch. 1990. Recent trends in strawberry arthropod management for coastal areas of the western United States. Pp. 239–259 *in* N. J. Bostanian, L. T. Wilson, and T. J. Dennehy [eds.], Monitoring and Integrated Management of Arthropod Pests of Small Fruit Crops. Intercept, Andover, U.K. 301 pp.
Zalom, F. G., C. Pickel, D. B. Walsh, and N. C. Welch. 1993. Sampling for *Lygus hesperus* (Hemiptera: Miridae) in strawberries. J. Econ. Entomol. 86: 1191–1195.

CHAPTER 4

Lace Bugs (Tingidae)

John W. Neal, Jr. and Carl W. Schaefer

1. INTRODUCTION

Lace bugs, or Tingidae, are plant feeders that live mostly on the abaxial (lower) surface of the leaves of their hosts (Drake and Ruhoff 1965, Buntin et al. 1996) and stylet-feed to remove plant sap from palisade parenchyma tissue (Ishihara and Kawai 1981). The common name, "lace bug," comes from the thin outgrowths of the pronotum and the texture of the forewings, which appear lacy. These bugs are called *Netzwanzen* or *Gitterwanzen* in German, *netwantsen* or *netwerkwantsen* in Dutch, and *chinches de encaja* in Spanish. The vernacular names in French include that of the host plant, such as *le tigre du poirier, le tigre du cerisier,* and *le tigre du cafféier* (Drake and Ruhoff 1965). The family is widely distributed throughout the tropical and temperate zones of all continents and on most oceanic islands (Drake and Ruhoff 1965). Tingidae contains approximately 250 genera and 2000 described species (Stonedahl et al. 1992). Drake and Davis (1960) provided an excellent discussion of adult morphology, phylogeny, and higher classification. Péricart (1982) summarized the systematic position, morphology, and biology of the family. Development usually requires five instars; however, four have been reported for *Stephanitis rhododendri* Horvath, *Tanybryrsa cumberi* Drake (Štys and Davidová-Vilimová 1989), and *Oncochila simplex* (Herrich-Schaeffer) (Percora et al. 1992).

The proper spelling of the familial name was controversial among hemipterists for over a century. Other spellings of Tingidae (Tingididae and Tingitidae) were criticized by Parshley (1922a,b) and a decision by the International Commission on Zoological Nomenclature upheld the use of Tingidae (Hemming 1943).

Delicate, among other colorful terms, is often used to describe the ethereal appearance of the slow-flying adult, which is apparently defenseless. However, a wide variety of adaptive mechanisms protects the immatures. In addition to maternal care, some nymphs have alarm pheromones and their exoskeleton often bristles with setae that are needlelike or deliquescent. Other setae on the antennae as well as the body may secrete droplets with novel compounds with unusual biological activity. These various methods of protection are probably very common in Tingidae, and other defense systems may yet be discovered. Research on these defensive mechanisms may help explain why multiple aggregations of large numbers of immatures of the same species on single hosts are so successful and are relatively free of predation and parasitism.

Drake and Ruhoff (1965) observed that "all species are rather highly specialized in their food habits." Exceptions they identified were the polyphagous *Corythucha gossypii* (Fabricius), *C. marmorata* (Uhler), *S. typica* (Distant), and *Monosteria unicostata* (Mulsant and Rey). Livingstone

(1977) and Cobben (1978) concurred that most species are largely monophagous. However, in the three decades since Drake and Ruhoff's (1965) *World Catalog*, there has been an increase in reports on the economic impact of lace bugs and an increase in the identification of host plants; this increased knowledge now suggests that many species may be oligophagous (Tomokuni 1983; Wheeler 1977, 1981, 1987, 1989; Qi et al. 1991). Some species of Tingidae (e.g., *Copium* spp.) are among the only heteropterans which induce galling in their hosts (e.g., Monod and Carayon 1958).

Voltinism, the number of annual generations, remains either undefined or in a state of uncertainty for many species because of the lack of seasonal studies. Lace bugs are generally considered either uni- or bivoltine (Drake and Ruhoff 1965). Dolling (1991) agreed and stated that of the two dozen British species of lace bugs, "probably all" have one generation a year. However, multiple generations have been determined for species in many genera such as *Corythucha* (Neal and Douglass 1990), *Monosteira* (Maniglia 1983), *Oncochila* (Pecora et al. 1992), and *Stephanitis* (Dunbar 1974, Horn et al. 1980, Neal and Douglass 1988, Wang 1988, Braman et al. 1992). Although not recognized as such, multivoltinism has been reported in other genera, e.g., *Tingis* (Hall and Sosa 1994) and *Teleonemia* (Livingstone et al. 1981). Two mechanisms that regulate the number of generations in Tingidae are photoperiodic induction of diapause, reported in *Corythucha* (Neal et al. 1992) and perhaps present in genera that overwinter as adults. Also, eggs of *S. pyrioides* (Scott) have been found to be noncleidoic, that is, dependent on imbibing water from the host for the completion of embryonic development. Thus, during development, the absorption of moisture results in an increase in egg size and weight (J. W. Neal, personal observation). As a possible consequence of the moisture requirement, seasonal fluctuations in available moisture to the host could also affect egg development and voltinism.

The eggs of lace bugs are inserted either into the spongy mesophyll (e.g., *Stephanitis*) or set and held in place vertically or slanted on a pedicel on the abaxial leaf surface (e.g., *Corythucha*). The operculum of an inserted egg often is adjacent to a midvein or lateral vein and is artfully concealed with shellac-like frass, whose color varies by species. Eggs of lace bugs inserted in leaves can be dispersed through commercial trade and carried through quarantine. However, lace bug eggs, concealed or exposed, are not free of parasitoids (Yacoob and Livingstone 1983, Livingstone and Yacoob 1986).

2. GENERAL STATEMENT OF ECONOMIC IMPORTANCE

Feeding by immatures and adults results in stippling of leaves, and females' feeding causes substantially more injury than that of males or nymphs (Buntin et al. 1996). Pollard (1959), studying feeding on eggplant (*Solanum melongena* L.) by *Urentius hystricellus* (Richter), found that penetration of the epidermis was primarily intracellular, but could also be intercellular or stomatal. The feeding of *Stephanitis pyrioides* on azalea (*Rhododendron* sp.) and of *S. typica* on coconut (*Cocos nucifera* L.) begins with stomatal penetration (Ishiwara and Kawai 1981, Mathen et al. 1988), and Johnson (1937) earlier had suggested this occurred occasionally in *S. rhododendri* Horváth feeding on *Rhododendron*; see **Section 3.17**, *S. typica* (Distant), for more on stylet penetration of the phloem.

Stephanitis pyrioides, and doubtless other tingids, reduces leaf photosynthesis by damaging palisade parenchyma, thus restricting gas exchange through stomata and reducing chlorophyll content; as a result, chlorosis also occurs (Buntin et al. 1996).

As obligate plant feeders, lace bugs would seem ideally suited to become both economic pests and disease vectors. However, the potential for each does not always readily develop. Often when a lace bug is reported as a pest, it is a short-term outbreak with some economic damage noted. That lace bugs may be vectors of disease will always remain highly questionable, because of the aggregational feeding behavior of both nymphs and adults, the tendency of several species to remain on the same plant, and the fact the stylets do not enter vascular tissue as do stylets of other vectoring hemipterans; further, as weak fliers, adult dispersal is limited.

The economic importance of lace bugs will continue to increase as adventive species emigrate (e.g., as concealed eggs), and as minor crops gain importance and value and cultivation expands to serve a burgeoning world population. Deciding on the economic impact of a particular lace bug is often idiosyncratic, reflecting the view of the individual rather than economic consequences. For example, Essig (1958) stated lace bugs "normally infest native plants, though occasionally they attack cultivated crops and ornamental shrubs and trees, but are seldom present in sufficient numbers to warrant control measures." On the other hand, Johnson and Lyon (1988) wrote, "In terms of the injury they do to ornamental trees and shrubs, the Tingidae, or lace bugs, are the most important family in the order Hemiptera." However, of the 29 species of lace bugs identified by Johnson and Lyon (1988) as feeding on evergreen and deciduous trees and shrubs in North America, only a few ever achieve recognized pest status. Similarly, in Georgia (U.S.A.), 29 species of lace bugs are known, yet of these only 8 are considered economically important (Beshear et al. 1976). In a recent review of major insect pests of ornamental trees and shrubs, only the sycamore (*Corythucha ciliata*), the azalea *(S. pyrioides)*, and the rhododendron *(S. rhododendri)* lace bugs were considered major pests (Schultz and Shetlar 1994). Slater and Baranowski (1978) note that "… some species are destructive," identify *C. ciliata* (Say), *Gargaphia solani* (Heidemann), the immigrants *S. rhododendri* (in Europe), *S. pyrioides,* and *S. takeyai* Drake and Maa (in North America) as pests, and mention many more species only as abundant. Froeschner (1988) stated: "Most native species ordinarily are of little economic importance except under unusual circumstances," and cited *G. solani, C. cydoniae* (Fitch), and three species of *Stephanitis* as examples of pest species. Drake and Ruhoff (1965) listed 25 "useful plants upon which lace bugs breed" (their Appendix 1). Péricart (1983) identified several economically important species including the two most important in Europe, *S. pyri* (F.) and *M. unicostata*.

Stonedahl et al. (1992) in an identification guide to world tingid pests took a broader view and provided a key to 36 serious and widespread species; for each species, they listed the host plants and general distribution. Štusák (1977) noted that knowledge of tingids in the Oriental region is very poor, and provided significant references to the key pest species. Nayer et al. (1976) stated that in India, a few species are economically important and identified 11 as pests. In Agra, India there are about eight genera and ten species of economically important lace bugs (Livingstone 1959).

Two levels of economic pest status became apparent when preparing this review: that certain species of lace bugs frequently cause economic damage to their hosts and that other species may only occasionally attain pest status. Some species may experience periodic outbreaks, whereas others may be very serious in one locality and/or at only one time. Sometimes papers on a particular species are clustered over a short time, lending a false appearance of economic importance or concern. For example, after introduction of the rhododendron lace bug, *S. rhododendri*, apparently from North America into Europe, Europeans at first considered it a major threat, but it then inexplicably subsided (Dolling 1991, Judd and Rotherham 1992). A similar flush of papers followed the introduction of the sycamore lace bug, *C. ciliata*, into Europe. It dispersed rapidly and currently is a significant pest throughout Europe of the sycamore, *Platanus occidentalis* L. and its hybrid, the London plane tree.

Some species not listed here have very rarely become pests briefly and, of course, may do so in the future. For example, the elm lace bug, *C. ulmi* Osborn and Drake, of the northern United States which feeds on *Ulmus* spp. (Drake and Ruhoff 1965), caused some damage on elm in Connecticut in 1925 (Zappe 1926).

The mobile forms of lace bugs are highly susceptible to topical spray applications. Sprays must be directed to the undersurface (abaxial) of the leaf, and a second application is usually required 2 or 3 weeks later, to kill nymphs that hatched postspray, especially those hatched from eggs inserted into leaf tissue. However, because pesticides used and applications both change rapidly, methods of control with pesticides are not included in this chapter. Biological control possibilities are included.

Finally, a piesmatid, *Piesma quadratum* Fieber, was listed by Francki et al. (1981, p. 459) as a "lacebug vector" of beet leaf curl virus and Eisbein (1976) was cited as the source; yet Eisbein (1976) did not mention the family of *P. quadratum* (see **Chapter 7**).

3. THE IMPORTANT SPECIES

3.1 *Cochlochila bullita* (Stål) (= *Monanthia globulifera* Distant, *Monanthia bullita* Stål)

3.1.1 Distribution, Life History, and Biology

In addition to the plants mentioned below, this species attacks *Coleus* sp., *Ocimum sanctum* L., *Rosmarinus officinalis* L. (rosemary), *Salvia officinalis* L. (sage), and *Carthamus tinctorius* L. (safflower) (Stonedahl et al. 1992), and has been called the ocimum tingid (Livingstone and Yacoob 1987b).

There was an outbreak of *Cochlochila bullita* in India in 1950, on *O. kilmandscharicum* Guerke (grown for camphor in Kanpur), and the bug has been a common pest of *O. sanctum* (tulsi plant, shrubby basil) in India; and in southern India this lace bug is an important pest of *Coeus parviflorus* (Chinese potato) (Mohanasundaram and Rao 1973, Palaniswami and Pillai 1983).

This lace bug occurs in the Old World tropics. Iyengar (1924) described and illustrated the immatures and adult, and noted glandular hairs on the dorsum of each instar, these most abundant laterally on the abdomen. On *O. kilmandscharicum,* eggs are partially inserted obliquely into stems or shoots, and laid more often in groups than singly (Sharga 1953); however, on *Mentha* sp. (mint), they were laid near the margin of the leaf (Samuel 1939) and, on *O. basilicum* L. (sweet basil), they were laid singly or in clusters on leaves and young branches (Tigvattnanont 1989). Development time of the immatures is influenced by season. In Thailand, on *O. basilicum* in the laboratory, eggs hatched in 6.4 days, the five stadia took 8.9 days, male adults lived 58 days and females 44 days; females laid an average of 254 eggs each (Tigvattnanont 1989). The species is probably multivoltine and nondiapausing, because mating occurs throughout the year (but the bug is more abundant in India from March to June) (Sharga 1953).

3.1.2 Damage

Heavy feeding causes curling and drying of leaf tips, leaf dehiscence, and lowered production of inflorescence (Iyengar 1924, Mohanasundaram and Rao 1973, Palaniswami and Pillai 1983). On *Ocimum* spp. and *Mentha* spp., leaves turned yellow and drop (*Ocimum* spp.) or to turn yellow, then black, and drop (*Mentha* spp.) (Sharga 1953).

3.1.3 Control

Sharga (1953) reported that larvae of the coccinellids *Brumus saturalis* F., *Chilmenes sexmaculatus* F., and *Coccinella septempunctata* L. feed on nymphs of *Cochlochila bullita* on *Monanthia* and that predation by coccinellid adults and larvae is common on *Ocimum.* Livingstone and Yacoob (1987c) reported the mymarid *Parallelaptera polyphaga* from the eggs of *C. bullita*.

3.2 *Corythaica cyathicollis* (Costa) (= *Corythaica monacha* Stål 1873, *Corythaica planaris* Drake & Bruner, *Corythaica passiflorae* Monte)

3.2.1 Distribution, Life History, and Biology

This Neotropical species is a pest on many solanaceous crops, including eggplant (aubergine), potato, tobacco, and tomato; and also on *Brassica* sp., *Passiflora* sp., and *Ricinus* sp. (Drake and Ruhoff 1965, Stonedahl et al. 1992). In Puerto Rico, on eggplant, Cotton (1917) reported *C. cyathicollis* fed on the "so-called wild eggplant, *Solanum torvum*" [sic, not in Bailey 1976], a very abundant weed that serves as the alternate host for *Corythaica cyathicollis* during the intervals between crops. *Corythaica cyathicollis* females were highly fecund and reproduced continuously. Eggs were inserted flush into the leaf, leaving little except the operculum exposed. Nymphs fed in large aggregations, on both surfaces of the leaf of wild eggplant. Cotton (1917) included detailed descriptions, and measurements of the egg and five instars, and general comments on the adult stage, but no illustrations. Monte (1943, 1945) also described and measured the immatures (including egg) and adult, and illustrated the latter, and gave stadial lengths (in Brazil).

3.2.2   *Damage*

Heavy outbreaks of this species on eggplant in Puerto Rico completely denuded the plants (Cotton 1917). In Brazil, *C. cyathicollis* is a pest on tomato, potato, and giló (*S. gilo* Raddi) (Monte 1943, 1945), and more recently on cubiu (*S. sessiliflorum* Dunal) (Couturier 1988). Uncontrolled feeding damaged eggplant and giló more than potato, whose tubers were both smaller than normal and fewer. In all, 70% of the tomatoes were destroyed by feeding and consequent disease (Monte 1943, 1945).

3.2.3   *Control*

In Puerto Rico, the coccinelids *Megilla innonata* Vauls. and *Cycloneda sanguinea* L., and the reduviid *Zelus longipes* (L.), feed on this lace bug (Cotton 1917). In Brazil, the mymarid, *Anaphes tingitiphagus* Soares, was considered a good control (Monte 1943, 1945). Water extracts of crushed neem seeds (*Azadirachta indica* A. Juss) were highly effective in controlling *Corythaica cyathicollis* on eggplant in the Dominican Republic (Dreyer 1991).

3.3   *Corythauma ayyari* (Drake)

3.3.1   *Distribution, Life History, and Biology*

This lace bug occurs in India, Sri Lanka, Malaysia, and Singapore, where it attacks *Jasminum pubescens, Lantana* sp., and "chamela leaves" (Drake and Ruhoff 1965, Stonedahl et al. 1992); to this list Stonedahl et al. (1992) add *Athaea* (to which the marshmallow, *A. officinalis* L., belongs), although this genus occurs in the Palearctic (Willis and Airy Shaw 1973). It is a particular pest of jasmine in southern India (Nayer et al. 1976), where it was studied by Dorge (1971) and Nair and Nair (1974), on *J. sambac* (L.).

These authors differed on the lengths of eggs and of instars and, when their figures of fifth instars are compared, two species seem to be involved, not one. Nair and Nair (1974) reported the egg to be light yellow and to require 9 to 11 days incubation; the five stadia took an average of 15.6 days, at 22 to 30.7°C. Adults emerged at night. Mating occurred 1 day postemergence, followed by preoviposition (2.7 days), oviposition (8.2 days), and postoviposition (1.7 days) periods. Adults are short-lived: males 10 days, females 12.3 days. In Madras state, *C. ayyari* occurred occasionally in March and October and caused little injury to *J. sambac* (David 1958); however, in central Maharashtra, it was recorded as one of the most destructive pests of *Jasminum* spp. grown commercially (Dorge 1971).

3.3.2   *Damage*

Corythauma ayyari causes leaf curl on *Jasminum* spp. (Drake and Ruhoff 1965). More generally, feeding results in yellow spots, and heavy infestations cause leaves to become yellow and to dry. Slowed growth and flower production may result from severe infestations (Nair and Nair 1974).

3.3.3 Control

Livingstone and Yacoob (1987b) described a new trichogrammatid, *Lathromeromyia corythaumaii* Livingston & Yacoob, from the eggs of this species.

3.4 Corythucha (= Corythuca) ciliata (Say)

3.4.1 Distribution, Life History, and Biology

Once known as the buttonwood tingis (Morrill 1903), and later named the sycamore lace bug (Wade 1917), *C. ciliata* occurs in North America east of the Rocky Mountains (Bailey 1951, Drake and Ruhoff 1965). In the United States it is an annual, chronic pest of the popular ornamental tree, *Platanus occidentalis* L., the sycamore, and of its hybrid, the London plane tree.

Corythucha ciliata achieved international notoriety in 1964 (e.g., Groupe de Travail O.I.L.B./S.R.O.P "Lutte Integree contre *Corythuca ciliata*" 1986, 107 pp.) when it was first identified in Italy (Servadei 1966, Bin 1968). More than 20 years after its discovery in Europe, Maceljski (1986) reviewed the status and spread of *C. ciliata* in Austria, France, Hungary, Italy, Spain, Switzerland, and Yugoslavia. Heiss (1995) summarized much of the research on *C. ciliata* in Europe, and Stehlik's (1997) exhaustive bibliography documented its dispersal through Europe. *Corythucha ciliata* has also been reported in Chile (Prado 1990).

Corythucha ciliata has been described and measured by Wade (1917) (adult and instars), and Morrill (1903) has described and illustrated the instars. The major component of abdominal setal exudates was isolated and identified from late instars by Lusby et al. (1987).

Studies on the bionomics and biology of such a well-recognized pest oddly have been piecemeal; there is a major contribution by Wade (1917) and subsequent lesser studies summarized in part by Bailey (1951). Wade (1917) and Barber and Weiss (1922) reported eggs are deposited singly or in groups adjacent to veins on the abaxial leaf surface, the exposed egg perpendicular to the surface. The egg may be concealed in the dense pubescence (Heidemann 1911). In Georgia (U.S.A.), eggs were more abundant on the south and west sides of trees, but first instars were equally dense on all sides (Horn et al. 1983a).

Development is temperature dependent: eggs incubated 7 to 28 days (Wade 1917; Maceljeski and Balarin 1974; Horn et al. 1983b,c); in Italy, Santini and Crovetti (1986) reported egg development at six temperatures between 18 and 24°C. First stadia lasted 2 to 10 days (Horn et al. 1983c). In New Jersey (U.S.A.), all five stadia took about 3 weeks, and development from egg to adult about 5 weeks (Barber and Weiss 1922).

The adult overwinters under the loose, rougher bark of the host tree or may be found in cracks and crevices of fences and buildings near sycamore trees; emergent, overwintered adults feed for about 10 days before oviposition (Wade 1917). Barber and Weiss (1922) suggested *C. ciliata* is bivoltine; however, three generations were reported in Chile (Prado 1990) and Italy (Di Battisti 1983).

Horn et al. (1983a,b,c) developed a sampling method for estimating egg and first instar densities in Georgia, as well as the factors that affect the eggs and first instars. First instars tended to aggregate and fed on leaves that had contained the eggs from which they had hatched (Horn et al. 1983a). Egg mortality gradually increased during the summer, and about 50% of the eggs deposited during the 2-year study did not hatch (Horn et al. 1983c). Normal leaf moisture content and moderate temperatures (16.7 to 32.2°C) did not affect hatching (Horn et al. 1983b).

3.4.2 Damage

On sycamore, *C. ciliata* may occur in such numbers that the leaves whiten and drop prematurely (Slater and Baranowski 1978). This lace bug has been implicated in Spain in the transmission of the fungal pathogen *Ceratocystis fimbriata* var. *platani* (Ell. & Halst.) (Gil and Mansilla 1981).

However, Rogers et al. (1975, 1982), in Italy, did not find the pathogen on or in (gut) the bodies of *Corythucha ciliata* diapausing under the bark of plane trees. They did find a strong association between saprophytes of plane trees and the surface mycoflora of *C. ciliata*.

3.4.3   Control

Several studies have listed natural enemies of this lace bug, in its native region (Nearctic) and in Europe. However, none has helped achieve good control, nor have several pathogenic fungi.

Osborn and Drake reported a "red mite" parasitizing the adult, and Wade 1917 listed several predators, but said none was important. Horn et al. (1983b) reported parasitization of eggs by the mymarid *Erythmelus* sp. in September, and noted that nearly half of eggs on sycamore did not hatch in a 2-year study. In the laboratory, two predaceous thrips did not feed on *C. ciliata* eggs, but *Chrysopa rufilabris* Burmeister, *Deraeocoris nebulosus* (Uhler), and *Orius insidiosus* (Say) did (Horn et al. 1983b). Horn et al. (1983c) found high mortality in the first stadium, and in general young instars appear to be more susceptible to predation than older instars and adults.

In Yugoslavia, Maceljski and Balarin (1977) determined in a laboratory study that *Nabis pseudoferus* Remane, *Rhynocoris iracundus* (Poda), and *Himacerus mirmicoides* (O. Costa) readily feed on *Corythucha ciliata*. Balarin and Polenec (1984) identified 23 species of spiders in ten families from the plane tree. In feeding trials in the laboratory, the spiders *Chiracanthium mildei* L. Koch and *Theridion lunatum* (Oliv.) were voracious feeders on *Corythucha ciliata*. Tavella and Arzone (1987) identified 21 potential predator and 7 parasitoid species from the plane tree. Although the predators all fed on *C. ciliata*, their collective failure to achieve suppression was attributed to an asynchrony between their biological cycles and that of *C. ciliata*.

In Italy, seven fungal species, some suspected as pathogens, were isolated from overwintering adults of *C. ciliata* found dead under tree bark (Ozino Marletto and Menardo 1984). Arzone and Ozinio Marletto (1984) and Ozino Marletto and Arzone (1985) reported the effect of different temperatures and humidities on pathogenic fungal activity against *C. ciliata* adults by the deuteromycetes *Beauveria bassiana* (Bals.) Vuill., *Verticillium lecanii* (Zimm.) Viégas, and *Paecilomyces farinosus* (Helm ex S. F. Gray) Brown and Smith, isolated from *C. ciliata*. High rates of infertility occurred from the three pathogens and appeared inversely proportional to the increase in humidity.

3.5   *Corythucha (= Corythuca) cydoniae* (Fitch) (= *Tingis cydoniae* [Fitch]), (= *Corythuca arcuata* [nec Say])

3.5.1   Distribution, Life History, and Biology

This is the hawthorn lace bug, originally called the quince tingis by Fitch (1861). Its nomenclatorial history was worked out by Bailey (1951). *Corythucha cydoniae* occurs throughout North America (Drake and Ruhoff 1965, Slater and Baranowski 1978), where it is one of the most common lace bugs (although not in northern New England [U.S.A.]; Bailey 1951).

Seasonally, it occurs from May to September in the Washington, D.C. (U.S.A.) area (McAtee 1923), from late April to late September in south and central Indiana (U.S.A.) (Blatchley 1926), and from mid-May to late August in Massachusetts (U.S.A.) (Bailey 1951). It is abundant on hawthorn (*Crataegus* spp.), quince (*Cydonia oblonga* Mill.), cotoneaster (*Cotoneaster dammeri* Schneid.), *Pyrus* sp., and allied plants (Slater and Baranowski 1978), but it has adapted to more than 35 other woody rosaceous plants, as well as plants in Rubiaceae and Fagaceae (Wheeler 1981).

Although polyphagous on many woody species used in landscapes, the hawthorn lace bug in recent times is seldom reported as a pest. In a 1984 survey of urban/suburban landscapes, none of the 20 most common plants was a host of *Corythucha cydoniae* (Holmes and Davidson 1984). Yet *C. cydoniae* is recognized as the most destructive U.S. insect affecting *Pyracantha* sp. in Alabama (Manuel and Williams 1980), in Georgia (Beshear et al. 1976, Roberts et al. 1988), and in Florida

(Mead 1972). Resistance to *C. cydoniae* was identified in trials of selected *Cotoneaster* (*C. lactea* W. W. Smith) and *Pyracantha* spp. (Schultz and Parnell 1982; Schultz 1983a,b, 1985; Schultz and Coffelt 1987). When offered choices, the hawthorn bug preferred *Sorbus aria* to *Crataegus phaeopyrum* Medic., *P. crenato-serrata* Rehd., and *C. cotoneaster dammeri* (Schultz 1983b).

Life-history studies are few, and the immature stages have not been described or illustrated, except the egg: although described as smooth, whitish, glistening, and semitransparent (Comstock 1978, Bailey 1951), observations (J. W. Neal, personal observation) indicate the egg under the fecal deposit is black. The fifth instar can be identified with the key of Horn et al. (1979), and several drawings and photographs of the adult are available, beginning with Fitch (1861); the best is in Mead (1972).

Studies on firethorn, *P. coccinea* Roem. (Neal and Douglass 1990) and on cotoneaster, *Cotoneaster dammeri* (Braman and Pendley 1993), both found the third stadium to require the least time and the fifth the most, regardless of temperature. Although diapause is not reported for Tingidae, Neal and Douglass (1990) suspected its involvement in significant changes in the preoviposition response of the hawthorn lace bug to photoperiod. Later, Neal et al. (1992) showed the critical photoperiod for the induction of reproductive diapause in Maryland (U.S.A.) is between 13 and 14 hours. Elsewhere, population differences occur and, in Georgia, the critical photoperiod is between 12 and 13 hours (Braman and Pendley 1993). The adult is the ultimate sensitive stage (Neal et al. 1992).

Corythucha cydoniae was reported as univoltine by Bailey (1951), although the graphs in his paper suggest otherwise. Drake and Ruhoff (1965) believe that more temperate tingid species are univoltine or bivoltine.

Adults overwinter under loose bark on trees or under debris on the ground (Comstock 1879).

The instars have pronounced abdominal tubercles that support one or more secretory setae, and small denticles or spines form a row across each segment. The setal secretions comprise three major components, all acylcyclohexane-diones (Lusby et al. 1989).

3.5.2 Damage

Like that of most tingids, damage caused by the hawthorn lace bug is the discoloration of the leaves of the host. Because here the hosts are ornamental trees, discoloration is unwelcome. Rarely do these bugs reach sufficient numbers to cause permanent damage.

3.5.3 Control

Except for the discussion in Williams and Bedwell (1980), there apparently has been little account of biological control of the hawthorn lace bug.

3.6 Corythucha gossypii (F.)

3.6.1 Distribution, Life History, and Biology

This insect has been called the bean lace bug and the cotton lace bug; it is a serious pest of both crops, and feeds on other malvaceous and leguminous plants, as well as on species of *Annona, Citrus, Mangifera, Solanum,* and other crops from the southern United States to northern South America; it is particularly troublesome in the Caribbean islands and Central America (Stonedahl et al. 1992).

Leonard and Mills (1931) studied *C. gossypii* on lima bean in Puerto Rico; they provided an extensive summary of the distribution by country and the identification of a wide variety of host plants (similar to the host list in Drake and Ruhoff 1965) and stated the bug's economic importance in Puerto Rico was mainly from feeding on lima bean and the anona tree, *Annona diversifolia*.

Corythucha gossypii on the island feeds on various host plants and reproduces throughout the year (apparently multivoltine and nondiapausing). The egg and five instars are described, but the only illustration is of an egg.

The most thorough study of this species appears to be that of López M. et al. (1982), conducted in the laboratory on sunflower (of which this lace bug is a pest in Colombia). All stages are described and illustrated; eggs are laid on the undersurface of the leaves, along the midrib and major veins, embedded completely or partly at an angle; a female lays an average of 8.3 groups of eggs, each with 7.4 eggs, which measure 0.4 × 0.2 mm. Incubation takes 11.8 days and the nymphal stage 16.4 days; adults live about 25 days (López M. et al. 1982).

3.6.2 Damage

In Colombia *C. gossypii* may feed on the oil palm *Elaeis guineesis* Jacquin (Froeschner 1976). Mead (1989) reported that in Florida (U.S.A.) *C. gossypii* is abundant and has the widest range of hosts of any Florida tingid, including many ornamentals; the preferred hosts in Florida are castor bean, *Ricinus communis*; orchid trees, *Bauhinia* spp.; hibiscus, *Hibiscus* spp.; and Jamaica dogwood, *Piscidia piscipula*. Peña and Van Waddill (1988) reported *C. gossypii* on cassava, *Manihot esculenta*, in south Florida. The size and quality of anona beans were reduced, or their formation almost entirely prevented in Puerto Rico (Leonard and Mills 1931).

3.6.3 Control

Leonard and Mills (1931) indicate the exposed egg operculum would permit attack by a parasitoid, but reported none. Egg masses of the reduviid bug, *Zelus nugax* Stål, were observed on the anona trees and nymphs were observed feeding on immature *C. gossypii*.

3.7   *Corythucha* (= *Corythuca*) *marmorata* (Uhler)

3.7.1   Distribution, Life History, and Biology

This, the chrysanthemum lace bug, is widespread in the United States and southern Canada (Slater and Baranowski 1978). It is distinct from other *Corythucha* species by the whitish opaque ground color of the membranous structures and by their mottled appearance. The "marmorate" or marbled pattern is variable and more intensely displayed in males. Eggs are ovate and somewhat fusiform, 0.5 × 0.25 mm (Felt 1904). Abbott (1935) later redescribed the eggs from oak, *Quercus* sp. However, Bailey (1951) commented that Abbott's report apparently is in error; Bailey stated: "His [Abbott's] paper reveals an unfamiliarity with the literature on tingids, and, although possible, it seems unlikely that *C. marmorata* would use oak for a host. The eggs were more likely those of *C. arcuata* (Say) that normally lives on oak. Later in the season he found eggs in ragweed, which is a natural host plant." Bailey (1951) further summarized others' descriptions of the egg, including its placement and location. Felt (1904) measured, described, and figured features of the five instars. He also included Uhler's original description of the adult.

Corythucha marmorata may be found on either the upper or the lower leaf surface depending on the intensity of the sunlight and the temperature (Bailey 1951). Bailey (1951) reported that early in the season the lace bugs on chrysanthemum were mostly on the upper side of the leaves, but when collected from goldenrods and asters during midsummer they are usually on the lower surface. The number of generations was not reported. In Missouri, adults overwinter in grass clumps and in nearby ground debris (Froeschner 1944) and similar sites (Bailey 1951). Interesting host- and nonhost-related behavior by emergent adults in April was described by Bailey (1951). The significance of host patch edges to the colonization and development of *C. marmorata* on goldenrod, *Solidago altissima* L., was reported by Cappuccino and Root (1992).

3.7.2. Damage

Very abundant, *C. marmorata* occurs on various composites in fields, chiefly goldenrod (*Solidago*), and asters (*Aster*), and may severely injure chrysanthemums; it has also been reported from species of *Ambrosia, Helianthus, Rudbeckia,* and *Echinops* (Drake and Ruhoff 1965); and it may cause damage in greenhouses (Osborn and Drake 1916).

3.7.3 Control

No natural enemies have been recorded.

3.8 *Corythucha morrilli* Osborn & Drake (= *Corythucha decens* [Stål])

3.8.1 Distribution, Life History, and Biology

This, the Morrill lace bug, occurs mostly on asteraceous plants; its range is western North America south into Central America and the Caribbean, and it has been introduced into Hawaii (U.S.A.) (Drake and Ruhoff 1965). Little biology or taxonomy was reported prior to 1977 other than the original adult description. Tilden's (1950) early biological note in California (U.S.A.) on feeding, oviposition, and hatching was too general to be of value; however, the host was noted as coyote brush or chaparral broom, *Baccharis pilularis* de Candolle.

Three laboratory studies on different hosts have provided considerable insight into its biology. Studies by Rogers (1977) showed *C. morrilli* at 26.7°C preferred to feed and oviposit on the upper surface of sunflower, *Helianthus annuus* L., leaves; about two thirds of the eggs are positioned adjacent to the upper midrib. Head capsule widths of instars 1 to 5 ranged from 0.14 to 0.39 mm. The mean adult longevity of males was 69 days and of females 95 days. Females laid an average of 177 eggs. Development from egg to adult required about 30 days. In studies by Silverman and Goeden (1979), *C. morrilli*, at 27°C, fed gregariously on the abaxial surface of leaves of *Ambrosia dumosa*. They measured the three dimensions of the egg, and showed that its exposed end was capped with feces; 62% of the eggs were laid on the adaxial leaf surface (see Rogers 1977). Body lengths of five instars ranged from 0.32 to 1.77 mm. Nymphs fed in aggregations, but adults were less gregarious. During mating, the male was oriented approximately perpendicular to the female, her hemelytra and abdomen laterally between the male's hemelytra and abdomen. Males were polygamous and females polyandrous. Silverman and Goeden (1979) found *C. morrilli* is multivoltine (therefore nondiapausing) in southern California, although the number of generations was not reported. Stone and Watterson (1985) studied the development of the immature stages and oviposition rate of *C. morrilli* at seven temperatures on guayule. Development of nymphs was limited at the lowest temperature, 17.8°C, and no eggs or nymphs developed at temperature above 34°C. Adults usually fed singly, and nymphs preferred to feed in aggregations on the abaxial surface. Females inserted eggs singly or in clusters and seemed not to prefer the upper midrib area, as has been reported on previous hosts. The exposed cap was usually covered with feces.

3.8.2 Damage

The Morrill lace bug caused considerable damage to guayule (*Parthenium argentatum* Gray) in field plots and greenhouses in Texas (Stone and Watterson 1985); guayule is a perennial composite which potentially may yield a form of rubber (up to 22% dry weight) (Stone and Fries 1986).

3.8.3 Control

In the field in California, Silverman and Goeden (1979) observed a lygaeid, *Geocoris* sp., and a nabid, *Nabis* sp., feeding on both nymphs and adults. No parasites were detected.

3.9 Habrochila ghesquierei Schouteden

3.9.1 Distribution, Life History, and Biology

This species was described in 1953 as a common pest of coffee in several districts of Zaire by Schouteden (1953), who distinguished it from *H. placida* Horváth. The current known distribution includes Zaire and eastern Africa (Hill 1983a). Anonymous (1961) reported that eggs are embedded on the abaxial leaf surface and in the soft tissue of the terminal; many eggs embedded near a growing point can result in distortion of terminal growth; the operculum projects slightly from the surface and is visible. Hatch occurs in 22 to 32 days. The five instars and adults fed in aggregations only on the abaxial surface, and development was completed in 16 to 36 days. The female had a preoviposition period of 8 days.

3.9.2 Damage

Habrochila ghesquierei has been a sporadically severe pest of *arabica* coffee [*Note: H. placida* appears to prefer *robusta* coffee (D. S. Hill 1975)] over most of Kenya, and outbreaks followed indiscriminate use of DDT (Anonymous 1961; see also D. S. Hill 1975, 1983; Le Pelley 1969).

3.9.3 Control

This lace bug was the target of pesticide evaluations on coffee east of Zaire (Decelle 1955). A predaceous mirid (*Stethoconus* sp.) was reported by Anonymous (1961) and Hill (1975), and *S. distanti* Schouteden was reported by Le Pelley (1969).

3.10. Haedus vicarius (Drake)

3.10.1 Distribution, Life History, and Biology

Haedus vicarius is a serious pest of Congo jute, *Urena lobata* L., in Thailand (Tigvattnanont 1991). The bug also occurs in China, India, Indonesia, and the Philippine Republic (Drake and Ruhoff 1965), and Vietnam (Štusák 1984). Štusák (1984), studying the bug's biology on *Malus sylvestris* Mill. (crab apple), reported that the elongated eggs (0.44 to 0.17 mm) were deposited singly in the abaxial surface of the leaf; three different angles of insertion were described; most eggs were only partly inserted; and the multibranched trichomes of the leaf held the eggs upright. Štusák (1984) described and illustrated the egg. The five instars were 0.46, 0.60, 0.89, 1.25, and 1.62 mm long, and 0.23, 0.27, 0.38, 0.51, and 0.72 mm wide. The instars remained on the abaxial surface of the leaf. Štusák's (1984) descriptions of the instars are detailed, and he gives a key to them. The third to fifth instars have unusually large multibranched setae, and Štusák (1984) discusses their possible functions. Tigvattnanont (1991), studying the bug on Congo jute in Thailand, found much the same, except that occasionally eggs were laid in groups of two or three; about one fourth of the egg was inserted into the leaf. Each female laid an average of 179 eggs; incubation averaged 9 days, the five stadia required about 14 days, development from egg to adult took about 23 days; and males lived an average of 23 days and females lived much longer (data not given) (Tigvattnanont 1991).

3.10.2 Damage

Crab apple (*M. sylvestris*) was heavily damaged by feeding of *H. vicarilus* nymphs (Štusák 1984). As mentioned, the bug is also a serious pest of Congo jute, *U. lobata*.

3.10.3 Control

No natural enemies are recorded.

3.11 Leptodictya tabida (Herrich-Schaeffer) (= Monanthia tabida Herrich-Schaeffer)

3.11.1 Distribution, Life History, and Biology

This lace bug is a pest on monocotyledonous crops. It occurs from the southern United States into northern South America (Drake and Ruhoff 1965), and in Hawaii (Heu and Funasaki 1985). Heidemann (1913) reviewed earlier work, and reported the adult is 3.8 mm long and 1.6 mm across the widest part of the hemelytra; he included drawings of the fifth instar and adult, and emphasized the lateral expansion of the pronotum and abdomen, which are armed with stout irregularly shaped spines.

Chang's (1986) report is thus far the most comprehensive study of the bug's biology (done in Hawaii). The egg was inserted into parenchyma cells on the abaxial surface of the leaf, parallel with the leaf veins, its tip exposed and covered with clear frass. Eggs were laid in groups of 5 to 20 and hatched in 7 to 10 days. The five instars developed to adult in about 15 days. The first stadium required 3 days, the subsequent three stadia required about 2 days each, and the fifth 4 to 6 days. The length of development was affected by temperature and quantity of available food. The nymphs were described as flat, off-white, with erect spinelike projections along the sides and dorsum of the abdomen. The number of generations in Hawaii was not determined. However, during the summer, a single generation (egg to adult) was completed in 20 to 30 days. The adult was 3.5 mm long, flat, light straw to brown in color, and was distinctive with five long, erect spines projecting from the head and with a straight lateral expansion of the pronotum. Adults flew only in the early morning and late afternoon. In Hawaii, *L. tabida* was found in generally small populations on corn, guinea grass (*Panicum maximum* Jacq.), johnsongrass (*Sorghum halepense* L. Persoon), and barnyard grass (*Echinochloa crusgalli*) (a weed).

3.11.2 Damage

High populations of *L. tibida* reduced plant vigor, caused leaves to sensesce prematurely, and reduced the total area of photosynthesis. Infestations were more severe in calm areas than in windy areas of Maui (Hawaii, U.S.A.) (Chang 1986). Cane fields younger than 6 months seem to suffer more than older cane. On mainland United States, sugarcane production resumed in the lower Rio Grande Valley (U.S.A.) in the early 1970s following an end to production in the 1920s (Cowley and Sund 1973).

Meagher et al. (1991) reported *L. tabida* on sugarcane and recognized that corn, guinea grass, and johnsongrass were the probable alternative hosts that sustained *L. tabida* during the hiatus of sugarcane production in lower Texas. In Florida (U.S.A.), Hall and Sosa (1994) identified two annual population peaks of *L. tabida* on sugarcane, one during late spring/early summer, and one during fall. *Leptodictya tabida* was considered a pest of economic importance on sugarcane in Venezuela, where it was common and widely distributed; all stages lived in colonies on the abaxial surface of the cane foliage and, when abundant, caused a pronounced scorching of the leaves (Box and Guagliumi 1954).

3.11.3 Control

An attempt to introduce into Florida the mymarid parasitoid *Erythmelus* sp. from *L. tabida* eggs in Venezuela, was unsuccessful (Nguyen and Hall 1991, Hall and Sosa 1994).

3.12 Monosteira (= Monostira, Monanthia) unicostata (Mulsant and Rey)

3.12.1 Distribution, Life History, and Biology

Monosteira unicostata is a common pest of fruit trees (*Amygdalus, Cydonia, Malus, Prunus, Pyrus*), and also occurs on *Alnus, Crataegus, Juglans, Populus,* and *Salix* in the southern Palearctic region east to Turkestan (Stonedahl et al. 1992) and Iran (Péricart 1983).

There is some disagreement among the several reports on the bug's biology, perhaps because of development under different conditions and on different hosts. Bremond (1938), who referred to this species as the "false tiger" of fruit trees, reported the length of the egg and five instars (in Morocco) as 0.2, 0.4, 0.53, 0.7, 10.95, and 1.3 mm; Vidal (1939), also working in North Africa, reported the egg as 0.3 to 0.7 mm long (an oddly wide range), and the instars' lengths as about the same as Bremond (1938). The adult was 2.0 to 2.5 mm long (Vidal 1939), a length similar to those measured by Gomez-Menor Ortega (1950) (2.0 to 2.1 mm) and Péricart (1983) (2.2 to 2.8 mm). However, the drawings of instars by Péricart (1983) differ from those by Gomez-Menor Ortega (1954, 1955) (in Spain), especially those of the third and fourth instars. Other figures of the species are in Bremond (1938: adult, instars 1, 2, and 5, with an emphasis on the glandular setae), Gomez-Menor Ortega (1950, 1954, 1955: adult and all five instars), and Péricart (1983: egg, instars 2 to 5, and adult). Redescriptions of all stages occur in these papers as well.

In North Africa there are four generations (Bremond 1938, Vidal 1939) and in Italy there are three (Moleas 1985, 1987); the adult overwinters and appears in early April (North Africa) or later (Italy). Eggs are laid about 10 days later (Bremond 1938). The egg is inserted with only the operculum exposed in the abaxial surface of the leaf adjacent to the midrib, in a series of 5 to 15 (Bremond 1938). The fact that only females appear to overwinter raises the question of parthenogenesis (Bremond 1938), which remains to be confirmed. In Italy, on almond, the first generation, small numerically, developed when the fruit reached full size; then, feeding by large numbers of the second generation, present during July and August, caused leaf abscission. The third generation overlapped with the second in late July to early September, when the fruit ripened. The optimal temperature for development of nymphs was 30°C. Of the three generations the last two are much more damaging than the first (Moleas 1985, 1987).

Nymphs feed in aggregations and, when disturbed, disperse and regroup (Bremond 1938).

3.12.2 Damage

Monosteira unicostata was considered by Péricart (1983) as one of the two most important lace bugs in the western Palearctic; the other is *Stephanitis pyri* (F.); see **Section 3.13**). The most serious consequence of this bug's feeding is the defoliation that occurs in mid- to late summer, which results in smaller fruit which drops prematurely; poplar trees (*Populus* sp.) may be completely defoliated (Bremond 1938). In Apulia, Italy the bug caused premature fall of almond leaves when populations became high during midsummer (Moleas 1985, 1987).

3.12.3 Control

No predators or parasites have been reported.

3.13 Stephanitis pyri (F.)

3.13.1 Distribution, Life History, and Biology

Fabricius described this species twice: in 1775 (in *Acanthia*) and again in 1803 (in *Tingis*). Subsequent authors included it in *Tingis* until as recently as 1927, despite that Stål placed it in *Stephanitis* in 1873 (see nomenclatorial history in Drake and Ruhoff 1965). The insect has drawn much attention because it is highly polyphagous and widespread throughout Europe (including Russia), northern Africa, and the eastern Mediterranean to Armenia, Iraq, and Iran. Its hosts are primarily trees and shrubs of many unrelated genera: *Amygdalus, Castanea, Chaenomeles, Cormus, Cotoneaster, Crataegus, Cydonia, Juglans, Malus, Ligustrum, Populus, Prunus, Pyrus, Quercus, Ribes, Rosa, Sorbus, Robinia, Tinia, Ulmus,* and *Vaccinium* (Drake and Ruhoff 1965). It "is a very important pest of fruit trees in Turkey" (Önder and Lodos 1983).

The biology and ecology of *Stephanitis pyri* have been ably summarized by Péricart (1983), who also figured the egg, fifth instar, and adult; Gomez-Menor Ortega (1954, 1955) figured all five instars and described the adult; and Štusák (1959) described, figured, and measured the fifth instar. In France, the insect has two generations a year (Gregorio 1981; see also important references therein); but elsewhere it may have one to four (Péricart 1983); the adult overwinters. Gregorio (1981) and Péricart (1983) review the literature on this species.

3.13.2 Damage

Péricart (1983) lists this as one of the two most important tingids in the west Palearctic; *Monosteira unicostata* is the other; see **Section 3.12**. Its immatures often cause serious foliar damage to *Crataegus, Malus, Prunus,* and *Pyrus* (Péricart 1983), and it readily attacks pear and other orchard fruit trees.

3.13.3 Control

Host resistance in both apple and pear varieties has been demonstrated (Mohammad and Al-Mallah 1989a,b). Although abundant and found on many hosts, *S. pyri* has few significant parasites or predators.

The most important predator appears to be the host-specific mirid, *Stethoconus pyri* (Mella) (most frequently cited as *Stethoconus cyrtopeltis*; Kerzhner 1970). Henry et al. (1986) and Önder et al. (1986) give short reviews of *S. pyri* and, together, they provide an overview of the species. Unfortunately, this mirid is not abundant in the range of *Stephanitis pyri*. Generalist predators reported attacking *S. pyri* include Anthocoridae (*Anthocoris, Orius*), Chrysopidae (*Chrysopa*), and a thrips, *Cyrptothrips omnivorous* (Péricart 1983). The mymarid egg parasitoid, *Parallelaptera panis* Enock, was recovered from *S. pyri* in Russia (Goncharenko and Fursov 1988). Barnes (1930) repeated an unusual earlier report (Kieffer 1907) of a cecidomyid, *Endopsylla endogena* (Kieffer), emerging from an immature *S. pyri* in Portugal.

3.14 Stephanitis pyrioides (Scott)

3.14.1 Distribution, Life History, and Biology

The azalea lace bug has become of increasing importance as azaleas have become increasingly important as ornamentals. The insect occurs in China, Japan, Korea, and Taiwan (Maa 1957, Takeya 1963, Shen et al. 1985), and has been introduced into Argentina, Australia, Germany, Morocco, the Netherlands, and the eastern United States (Drake and Ruhoff 1965). Its origin is thought to be Japan, from where it was first described (Scott 1874). Its first U.S. appearance was in Washington, D.C., in 1910 (McAtee 1923), and it continues to spread through the eastern United States (Mead 1967, Torres-Miller 1989) and to become a major pest of azalea (Weiss 1916a,b; Weiss and Headlee 1918).

Stephanitis pyrioides was described from a single specimen by Scott (1874), who compared it with *S. pyri* F. It was later figured by Dickerson and Weiss (1917), Weiss and Headlee (1918), and Shen et al. (1985); the adult was redescribed and figured by Gomez-Menor Ortega (1954). The adult male is 2.8 to 3.3 mm long, and the female 2.9 to 3.3 mm (Shen et al. 1985). The immatures of *S. pyrioides* have also been described, figured, and measured by several authors: Dickerson and Weiss (1917: egg and five instars), Maa (1957: fifth instar), Lee (1969), and Shen et al. (1985) (both: instars); Shen et al. (1985) distinguished the instars by the number of eye facets.

Like most lace bugs, *S. pyrioides* is a weak flier. Dispersal within and among azaleas occurs usually as a consequence of local food exhaustion due to overcrowding. Courtship, if present, has not been described. When *in copulo*, the male is at approximately a 90° angle to the female. Oviposition of a single egg required 2 to 3 min (Dickerson and Weiss 1917). Barber and Weiss (1922), Dickerson and Weiss (1917), Weiss and Headlee (1918), and Bailey (1951) reported three broods in southern New Jersey and New England (U.S.A.). Four generations were reported in Maryland (Neal and Douglass 1988) and Georgia (Braman et al. 1992), and three in Japan (Nakasuga 1994). One of us (J. W. Neal, unpublished) has determined that the egg of *S. pyrioides* is noncleidoic and as a consequence hydrates significantly during embryonic development between days 2 and 8. Plant stress due to drought may affect voltinism. Adults, relatively cold-hardy, were observed in December, 1982 in Maryland (U.S.A.) (Neal 1985) and survived three consecutive winter conditions in central North Carolina (U.S.A.) (Nalepa and Baker 1994). Adult longevity under optimal rearing conditions at constant temperatures was exceptionally long (Neal and Douglass 1988). They found single females lived longer than females paired with males. At 26.1°C, paired females lived a mean of 73 days and the males lived for a mean of 104 days. Longevity of both sexes was increased at 20.6°C. Development time and thermal requirements for a nymphs at constant temperatures were studied in the United States by Neal and Douglass (1988) and Braman et al. (1992). They reported instar development across temperature was nonlinear, and that the third stadium required the least time for development (half the time required for the fifth instar). However, in Japan, development of overwintered eggs collected from host plants in February and March and incubated at constant temperatures was linear across temperature (Nakasuga 1994). Degree-day (°D) accumulations for generation development were determined by Braman et al. (1992) and Nakasuga (1994). Early instars feed in tight aggregations that dissipate with progressive instar development; fifth instars do not aggregate, probably as a result of a competitive search for a diminishing food supply from a single leaf rather than because of dissipation or change in the basic aggregation behavior (J. W. Neal, unpublished). The setal secretions of *S. pyriodes* are phenolic acetogenins (Oliver et al. 1985) and have high biological activity in bioassays unrelated to lace bug defense (Jurenka et al. 1989, Mason et al. 1991, Neal et al. 1995). However, it is likely that a major function of the droplets is related to defense against predators.

3.14.2 Damage

As azaleas (*Rhododendron* spp.) become more and more used as ornamental shrubs, this lace bug becomes more and more important as a pest. Other ericaceous hosts include *Kalmia latifolia* and *Pieris ovalifolia* (Drake and Ruhoff 1965), and *Lyonia neziki* Kakai et Hara ("nejiki") (Rhodoraceae) was listed as a host by Takeya (1963).

Rhododendron is a very large flowering plant genus in Asia (Cullen and Chamberlain 1978) and it is of interest that *S. pyrioides* has adapted primarily to azalea and is not found on rhododendrons. Although *S. pyriodes* is considered relatively common on azalea in Japan, the range has not naturally extended north to Hokkaido (Tomokuni 1987).

Location of the azalea shrub in relation to available sunlight has a pronounced effect on the amount of feeding damage that occurs. Lace bugs were reported by White (1933) to be "more severe on plants growing in the sun and [lace bugs] are of little importance in shaded areas." Azaleas in full sun all day were more than twice as likely to be infested as azaleas in the shade (Raupp 1984). When developing baseline data for an urban IPM pilot program, the azalea and *S. pyrioides* were often the most numerous shrub and pest requiring treatment (Hellman et al. 1982, Holmes and Davidson 1984, Raupp and Noland 1984). Collectively, developing nymphs contribute to foliar damage; individually, adult females caused substantially more leaf-feeding injury than males (Buntin et al. 1996). Ishihara and Kawai (1981) found that, when feeding, the stylets of *S. pyrioides* were adventitiously inserted through the stomata into the palisade parenchyma tissue, and, correspondingly, Buntin et al. (1996) found that feeding by *S. pyrioides* resulted in an increase in leaf stomatal resistance.

3.14.3 Control

Nymphs of *S. pyrioides* are highly susceptible to pesticides and any treatment that derails the first generation will be the most effective. Foliar sprays of several plant growth regulators, which may have affected nutritional quality, were found either to accelerate or to slow development of nymphs (Coffelt and Schultz 1988). Evidence for resistance has been reported in both evergreen azaleas (Schultz 1993) and deciduous ones (Braman and Pendley 1992, Wang et al. 1998). Balsdon et al. (1995) and Wang et al. (1999) examined the potential role of epicuticular lipids from azaleas on host plant acceptance.

In general, tingids appear to have very few specific predators and parasitoids. The absence of biocontrol agents of lace bugs was evident in a study of seven species in Missouri (U.S.A.) among three genera (Sheeley and Yonke 1977). However, *Anagrus takeyanus* Gordh, a mymarid egg parasitoid originally described from *S. takeyai* (Gordh and Dunbar 1977), was recovered from eggs of *S. pyrioides* (Balsdon et al. 1993). The biology of eggs of *S. pyrioides* was reported by Balsdon et al. (1995, 1996). The Japanese mirid *Stethoconus japonicus* Schumacher, first identified in the Western Hemisphere in Maryland as an obligate predator of *Stephanitis pyrioides* (Henry et al. 1986), was determined to be highly aggressive (Neal et al. 1991). The asynchronous seasonal hatch of the *Stethoconus japonicus* egg is regulated by plant phenology, and egg development is water dependent (Neal and Haldemann 1992). Although an effective predator, *S. japonicus* only occurs naturally among high populations of *Stephanitis pyrioides*.

3.15   *Stephanitis rhododendri* Horváth (= *Leptobyrsa explanata* Heidemann)

3.15.1   Distribution, Life History, and Biology

The rhododendron lace bug, *S. rhodendri*, was briefly described by Horváth (1905) from rhododendrons, *Rhododendron* sp., in the Netherlands. *Stephanitis rhododendri* was later described by Heidemann (1908) as *L. explanata* from several specimens from various locations that included the earliest known adult collected in 1887 in Pennsylvania. The taxonomic history of the species was reviewed by Johnson (1936). Hosts reported by Heidemann (1908) were *Kalmia latifolia* L. and *R. maximum* L.; another host is *Pieris* sp. (Drake and Ruhoff 1965); all are Ericaceae. In North America, *S. rhododendri* occurs from New England to Florida and west at least to Ohio, and has been reported in Oregon and Washington. Dickerson (1917) reviewed the early history of *S. rhododendri* in the United States.

Stephanitis rhododendri was apparently detected in England in 1901 prior to Heidemann's 1908 description (Wilson 1933) and finally confirmed in England in 1910 (Distant 1910). It is generally recognized that *S. rhododendri* was introduced from North America and spread rapidly through Europe (see review by Péricart [1983]). *Stephanitis rhododendri* was reported in Germany in 1915 (Steyer 1915), and later recognized in France, Switzerland, Scotland, Denmark, Sweden, Finland, and Belgium; the distribution now includes South Africa and Canada (Drake and Ruhoff 1965).

The origin of *S. rhododendri* is not without some controversy. Johnson (1936) discussed three hypotheses and concluded that the species was probably indigenous to North America. Another point, complementary to the first, has received little attention. This question is the correctness of the placement of *rhododendri* in *Stephanitis* instead of *Leptobyrsa*. There are 59 species described in *Stephanitis* (Drake and Ruhoff 1965), and most are distributed in Asia (including India). Only four species, *S. blatchleyi* Drake (probably extinct in North America; see Drake 1925), *S. mitrata* (Stål) (Brazil), *S. olyrae* Drake and Hambleton (Brazil), and *S. parana* Drake and Hambleton (Brazil) are from the Americas. All west Palearctic species and east Asiatic species of *Stephanitis* have five instars (Štys and Davidová-Vilimová 1989). The research of one of the present authors (J. W. Neal, unpublished) with *S. rhododendri* has confirmed four instars and that hybridization did not occur in reciprocal crosses with either *S. pyrioides* or *S. takeyai*. When studied for voltinism, *Stephanitis* species had three or four generations, whereas *S. rhododendri* has only one or two (Bruel 1947: one; Crosby and Hadley 1915, Dickerson 1917: one to two; Johnson 1936: one). These fundamental differences add another dimension to the original controversy of origin. Neal suggests that it will be determined that *S. rhododendri* does not belong in *Stephanitis*, that it may not be indigenous to North America, and that rhododendron may be an alternate host.

Stephanitis rhododendri is unusual in having only four instars (instead of the usual five). Theobald (1913) first reported this number, although Heidemann (1908) vaguely described four. The egg, four instars, and adult were described and illustrated by Johnson (1936), and the adult and last instar by Heidemann (1908).

[*Note:* Drake and Ruhoff (1965) list as a reference "H. W. Ascot 1930." This in fact refers to the letter to the editor by an "H. W.," who was writing from "Ascot" (= Ascot Heath, Berkshire, U.K.). Thus the author is anonymous, and we so list it in the references.]

3.15.2 Damage

Although this lace bug's potential for doing damage on flowering rhododendrons is often reported, actual damage varies. For example, it was "sighted" in England in 1960 (Allen 1962), but later was considered to have "disappeared" (Dolling 1991, Judd and Rotherham 1992). Jones (1993) commented on these and other reports of the bug's disappearance, described his unintentional "rediscovery" of *S. rhododendri*, and included a review of *S. rhododendri* in England. Jones (1993) did not mention it, but Johnson (1936) made it clear that in England "neither parasites nor predators of immature or adult stages" ... "are known." Therefore, in the absence of natural enemies, the failure of *S. rhododendri* to establish itself firmly raises the question of its country of origin and preferred host.

Johnson's (1937) study of *S. rhododendri* feeding and the histology of the feeding lesions was one of the most thorough studies of its kind to that time. When feeding, the stylet of adults and nymphs usually rupture an epidermal cell, and nymphs occasionally penetrate a stomatal pore (Johnson 1937). This description differs from that reported for *S. pyrioides* by Ishihara and Kawai (1981) and Buntin et al. (1996), who reported stomates were adventitious portals of entry. The difference may be due in part to improved instrumentation applied to these later studies.

3.15.3 Control

No parasites or predators have been recorded.

3.16 *Stephanitis takeyai* (Drake and Maa) (= *Tingis globulifera* Matsumura, *Stephanitis globulifera* (Matsumara))

3.16.1 Distribution, Life History, and Biology

The andromeda lace bug was first described, in Japan, by Matsumura (1905), and transferred to *Stephanitis* by Horváth (1912); as *globulifera* was preoccupied, Drake and Maa (1955) renamed it. Its distribution is similar to that of *S. pyrioides* (**Section 3.14**; Takeya 1963, Tomokuni 1987). It occurs also in the eastern United States, where it has been recorded from *Andromeda* sp., *Aperula* sp., *Cinnamomum* sp., *Lindera* sp., *Lynoia* sp., *Pieris* sp., and *Salix* sp. (Drake and Ruhoff 1965). Bailey (1974) collected *Stephanitis takeyai* from the rhododendron, *R. calendulaceum*, when the branches of *P. japonica* and *R. calendulaceum* were contiguous. However, despite these many hosts, *S. takeyai* prefers Japanese andromeda, *P. andromedae* (Schread 1968).

Schread (1953) gives a general description of this lace bug's biology in Connecticut (U.S.A.). He indicates four to five generations a year; overwintering eggs were scattered on the leaf, whereas summer eggs were deposited next to the midrib. However, Neal (1988) found that on evergreen azalea, *Rhododendron* sp., eggs are laid on the midrib. In New England (U.S.A.) overwintered eggs were not considered cold tolerant (Schread 1968, Dunbar 1974). Dunbar (1974) also found four generations, and described and measured the egg and five instars; discussed the differences among this species and *S. rhododendri* and *S. pyrioides*; and studied development times of the immature stages. Kawakami (1983) measured fifth instars of Japanese specimens, and Bailey (1974) the lengths and widths of New England specimens.

Major components of the exocrines from the setae on instars were identified by Oliver et al. (1990).

A unidirectional asymmetric sexual hybrid female resulted from reciprocal crosses between *S. takeyai* and *S. pyrioides* (Neal and Oliver 1991).

3.16.2 Damage

The andromeda lace bug can be a serious pest of the ornamental, Japanese andromeda, whose leaves may become blotched and yellowed; leaves may fall off, leading to dying off of twigs and small branches (Schread 1968).

3.16.3 Control

Mymarid parasitoids recovered from overwintering eggs by Schread (1968) were thought to be *Anaphes* sp. Dunbar (1974) collected an *Anagrus* sp. later described as *A. takeyanus* Gordh, by Gordh and Dunbar (1977), who also suggested that Schread's *Anaphes* sp. was probably *Anagrus* sp. This mymarid species is unusual because it is thelytokous and all other species of *Anagrus* in North America are arrhenotokous. Because egg parasitoids are not common on other species of *Stephanitis* in America, Gordh and Dunbar (1977) postulated that the new species was introduced with its host. This idea was strengthened by Tsukada (1992), when *A. takeyanus* was identified from Japan. The predaceous mirid, *Stethoconus japonicus* Schumacher, has been found with *Stephanitis takeyai* in Japan (Yasunaga et al. 1997).

3.17 *Stephanitis typica* (Distant) (= *Stephanitis typicus* Distant, *Cadamustus typicus* Distant)

3.17.1 Distribution, Life History, and Biology

Commonly known as the banana or plantain lacewing bug, this is a serious pest of crops in the Orient, where its range extends from Pakistan to Japan and east to Papua New Guinea (Drake and

Ruhoff 1965) and includes Thailand (Mohanasundaram and Boonyonk 1973). It is distinctive in having been reported as a vector of a plant disease.

Hoffman (1935) measured the egg (0.8 to 0.9 mm long) and described fourth and fifth instars; Livingstone and Yacoob (1987a) analyzed the egg morphometrically. Lee (1969) described and provided excellent figures of all immature stages; Stusák (1977) illustrated the fifth instar and its spiniform processes. The most thorough account of the life history was Mathen's (1960): development from egg to adult took about 25 days; at 12 to 33°C eggs hatched in 12 days and the five instars developed in 3, 2, 2, 2, and 4 days, respectively; when mating, the male and the female were at right angles. Tigvattnanont (1990a) also studied the bug's bionomics (on *Languas galaeata*) and listed some 30 host plants in Thailand.

3.17.2 Damage

Stephanitis typica has been reported as a pest on banana (*Musa*), coconut palm (*Cocos nucifera*), and various members of the Zingiberaceae — *Alpinia* spp., turmeric (= haldi), *Curcuma longa* L.; cardamom, *Elettaria cardamomum* and *Hedychium* spp.; *Phaeomeria* sp.; ginger, *Zingiber officinale*; also *Annona muricata, Artocarpus integrifolia*, and *Cinnamomum camphora* (Stonedahl et al. 1992).

The species was first described in 1903 under the genus *Cadamustus* on cardamom (*Hedychium* sp.) from Ceylon by Distant (1902, 1903). Hoffman (1935) and Mathen (1960) provide important early historical overviews. In India, the preferred hosts are turmeric (Thangavel et al. 1977) and cardamom (Butani 1985), both important cash crops used as spices and medicinally. *Stephanitis typica* was also considered a minor pest of coconut foliage, but it may be more serious as a vector of coconut root wilt disease.

First suggested by Nagaraj and Menon (1956), the role of this lace bug as a vector of coconut root wilt disease was confirmed by Shanta and Menon (1960) and Shanta et al. (1960).

Root wilt disease of coconuts greatly reduces the productivity of coconut palms. Mathen (1973) estimated an annual loss of more than 150 million rupees, and later Sunil (1992) estimated the disease caused the loss of 1 billion coconuts; with a 11.3% loss of oil per coconut, the total annual loss exceeded U.S.$100 million. The etiology of the disease was baffling for many years because the belief persisted that the causative agent was a soil-borne fungus, a theory developed after root (wilt) disease became prominent following floods in 1882 in Kerala. Pathogenicity trials with isolated fungal species failed to reproduce the symptoms of the disease.

Nagaraji and Menon (1956), Shanta and Menon (1960), and Shanta et al. (1960) were the first to implicate *S. typica* as the vector; and studies of abundance of *S. typica* and corresponding disease damage to coconut palm continued to implicate *S. typica* (Mathen 1982, 1985; Mathen et al. 1983). Curiously, for approximately 25 years, although bacteria and viruses were highly suspect, neither organism could be isolated or consistently observed in diseased plant tissues (review in Solomon 1991). Finally, Solomon et al. (1983), in ultrastructural studies of diseased vascular tissues, found a consistent association of a mycoplasma-like organism (MLO) in phloem elements in apical meristem, petioles of young leaves, and root tips of root (wilt) diseased palms. These findings were supported by two subsequent studies. The first described the detection of MLOs by Mathen et al. (1987) in salivary glands of feral *S. typica* 18 to 23 days after a 5-day confined exposure of *S. typica* to diseased palm; the positive identification was tempered by the absence of MLOs in *S. typica* incubated for fewer than 18 days. In the second study, Mathen et al. (1988) determined that, during feeding, the stylet tip of *S. typica* may terminate in the phloem, allowing acquisition of the phloem-borne MLOs. Joseph et al. (1972) reported earlier that, although a single *S. typica* could transmit the disease (thought to be a virus) to an indicator host (the cowpea *Vigna sinensis* (L.)), transmission was better effected by ten lace bugs. Transmission was later confirmed, when Mathen et al. (1990) exposed 2-year-old seedlings to >1000 adult *S. typica* with 5 days acquisition feeding. Indeed, a single lace bug has a low vector potential on palm (Shanta et al. 1964). Because disease-free coconut is uncertain in these regions of India, the oil palm, *Elaeis guineensis* Jacq., is recommended as an alternative that *S. typica* does not attack (Joseph and Shanta 1968).

One of the few detailed studies of lace bug feeding damage is that of Mathen et al. (1983), of *S. typica* feeding on coconut leaves. The stippled area caused by a single puncture is 0.11 mm^2, and one adult can cause 80 stipples in a day; in the laboratory, a single lace bug removes the chlorophyll from about 9 mm of leaves in a day. Stomatal penetration was not found, in this study or a later one (Mathen et al. 1988; but see *S. rhododendri*, **Section 3.15.2**). The end result of intracellular feeding may be penetration of the stylet tip into the phloem (Mathen et al. 1988). This, of course, would facilitate entry of the mycoplasma-like organisms.

3.17.3 Control

As early as 1935, Hoffman wrote of a plant bug attacking these lace bugs; this was almost certainly the mirid *Stethoconus praefectus* (Distant), whose description, life history, and predation on *Stephanitis typica* have since been given by Mathen and Kurian (1972). Henry et al. (1986) reviews *Stethoconus* species more generally.

There is some evidence of resistance to this lace bug: Mohanasundaram (1987) found 6 of 73 banana cultivars to be somewhat resistant.

3.18 Urentius hystricellus (Richter) (= Urentius aegyptiacus Bergevin, Urentius echinus Distant, Urentius sentis Distant)

3.18.1 Distribution, Life History, and Biology

This lace bug is a pest of *Solanum* species in Africa, the Middle East, India, Sri Lanka, and Malaysia (Stonedahl et al. 1992); it seems to prefer brinjal (aubergine, eggplant) (*S. melongena* L.), and is called the brinjal tingid. It has been reported on this plant in Thailand (Mohanasundaram and Boonyonk 1973) and Nigeria (Rodrigues 1977), and on *S. indicum* L. and *S. xanthocarpum* Schrad. & Wendl. (Tigvattnanont 1990b), tomato-fruited eggplant (*S. integrifolium* Poir.) (Brempong-Yeboah and Okoampah 1989), and even cotton (*Gossypium hirsutum* L.) (Rasool et al. 1986).

Although identified in 1869 (Drake and Ruhoff 1965), *U. hystricellus* received little attention until 1955 when, in India, Patel and Kulkarny (1955) provided the first detailed study on the bionomics of *U. hystricellus*, and included a valuable historical review. They described and illustrated the egg and instars, including the various tubercles of the latter; the egg measures 0.44 to 0.17 mm, and the instars are 0.6, 0.8, 1, 1.4, and 2.1 mm long, respectively. The egg (at 80 to 102°F) takes 5 to 7 days to incubate, and the stadia are 1.9, 1.3, 1.2, 1.5, and 2.7 (total: 8.1) days long. The adult too was described and measured; the female is shorter and wider than the male, and a view of her venter was given.

Oviposition occurs on the adaxial leaf surface (Thontadarya and Channa Basavanna 1959, Bhandari and Sohi 1962) in the field in India. Yet in Thailand, in the laboratory, Tigvattnanont (1990b) reported eggs were laid singly or in groups mainly on the abaxial surface, and development (egg to adult) required about 12 days.

In North Gujarat (India) *U. histricellus* was active during February through May (Patel and Kulkarny 1955); farther north (Punjab) all stages occurred from May into September (Singh and Mann 1986); and in Thailand March through June (Tigvattnanont 1990b). In the Punjab, adults overwinter November to March, and there were nine generations; populations peaked in June and August (Singh and Mann 1986). In all these studies, the lace bug preferred brinjal (aubergine, eggplant) to other hosts.

3.18.2 Damage

Brinjal is the chief crop attacked, although other *Solanum* species of economic significance may also be fed upon (see above).

Feeding occurs on either surface of the leaf; penetration is mainly intracellular, but may be intercellular or stomatal; and the stylets terminate usually in a palisade cell, sometimes a mesophyll cell, or rarely in the vascular system (Pollard 1959). Changes in carbohydrate, protein, and nucleic acid levels in brinjal as a result of feeding by the brinjal lace bug were reported by Jamal et al. (1979).

3.18.3 Control

Tigvattnanont (1990b) reported several predators.

4. THE LESS IMPORTANT SPECIES

[*Note:* Several of the species in this section may at certain times and at certain places do considerable crop damage. However, because these times or these places are few, these species are placed in this section.]

4.1, 4.2 *Abdastartus atrus* (Motschulsky) and *Abdastartus sacchari* Drake

These are both minor pests of sugarcane: the former in India and Sri Lanka (Stonedahl et al. 1992) and the latter in Java, Sumatra, and Taiwan (Drake 1930, Box 1953). *Abdastartus atrus* feeds on sugarcane 5 to 7 months old, and damages the leaves (Easwaramoorthy et al. 1982). No enemies of either species are known.

4.3 *Acanthochila armigera* (Stål)

This species has been reported on *Nicotiana, Ouratea,* and *Pisonia* in the southern United States south to Argentina (Drake and Ruhoff 1965, Stonedahl et al. 1992).

4.4 *Agramma atricapillum* (Spinola)

Agramma atricapillum has been reported on Cyperaceae (*Bolboschoenus, Carex, Schoenoplectus, Scirpus*), Juncaceae (*Juncus*), and rarely on Typhaceae (*Typha*) in the southern Palearctic region to Turkestan (Stonedahl et al. 1992). Štusák's (1972) host list includes Junacaceae (*J. actus* L., *J. maritimus* Lam., and *J. subnodulosus* Schrank), Cyperaceae (*Holoschoenus vulgaris* Link., *Carus* sp.), and Typhaceae (*Typha latifolia* L).

4.5 *Aconchus urbanus* (Horváth)

Aconchus urbanus is a common pest on papaya and *Urochloa* sp. in the Ethiopian and Oriental regions (Stonedahl et al. 1992). The fifth instar was described and illustrated by Štusák (1977), who noted the similarity of nymphal characters to those of *Stephanitis*. He also reported that *A. urbanus* apparently develops on the grass *Opismenus oppositus*.

4.6 *Compseuta ornatella* (Stål)

Compseuta ornatella occasionally reaches pest status on *Abutilon* sp., which is cultivated as an ornamental, and *Cordia* sp. in the Ethiopian region (Stonedahl et al. 1992). Rodrigues (1979b) described and illustrated the fifth instar, and stated the distribution was Kenya, Rwanda, South Africa, Spanish Guinea, Sudan, Tanzania, Uganda, Zaire, Zambia; and Rodrigues (1982) added Gambia.

4.7 Cysteochila ablusa Drake

Cysteochila ablusa is indigenous to India, and the host is *Bauhinia variagata* L. (Drake and Ruhoff 1965). Sandhu and Sohi (1979) reported that *Bauhinia* spp., commonly known as kachnar, is grown as an ornamental tree in lawns, public parks, and along roadsides. An outbreak of *C. ablusa* severely infested *B. blakeana, B. purpurea,* and *B. variagata* during July to August 1976. The population increase coincided with the sprouting of new shoots and leaves during the rainy season. The nymphs and adults were mostly confined to the terminal shoots, feeding on the abaxial leaf surface, buds, and flowers. The infested shoots dried, and the leaves fell.

4.8 Cysteochila endeca Drake

Cysteochila endeca develops primarily on tamarind, *Tamarindus indicus*, a fruit-bearing tropical evergreen, but was also reported from *Gardenia* sp. and potato, *Solanum tuberosum* L., in the Ethopian region (Stonedahl et al. 1992).

4.9 Diconocoris hewetti (Distant) (= Elasmognathus hewetti Distant, Diplogomphus hewitti [sic] [Distant])

The pepper tingid is a pest of black pepper (*Piper nigrum*) in Borneo and Sumatra (Drake and Ruhoff 1965), where it feeds on the flowers and perhaps the young fruit (van der Vecht 1935). It is also a major pest of pepper in Indonesia (Soetopo and Iskandar 1986). In Sarawak, *D. hewetti* caused about a 30% reduction in yield, by feeding on prefruit inflorescences; pepper was the second most important export crop (Rothschild 1968).

Eggs are usually laid in the flowering spikes with only the operculum exposed, and require at least 10 days to develop. The five stadia take 4, 3, 3, 4, and 5 days., and the period egg to adult is about 30 days. The instars do not move far from the oviposition site. Nymphs and adults fed only on inflorescences and earlier reports of *D. hewetti* feeding on fruits and leaves were not confirmed (Rothschild 1968). van der Vecht (1935) figured two late instars and provided dorsal and lateral views of the adult. Rothschild (1968) described and illustrated the egg, instars, and adult; the egg measured 0.22 × 0.75 mm, and the instars were 0.8, 1.1, 1.4, 2.1, and 3.4 mm long, respectively.

Gumbek (1986) developed four sampling methods to assessing pepper flower spike damage caused by *D. hewetti,* and, from the spatial patterns of nymphs, Karmawati (1989) developed a method for forecasting populations.

No parasites were recovered from any stage (Rothschild 1968).

4.10 Dictyla echii (Schrank)

Dictyla echii was frequently reported in the genus *Monanthia* prior to Drake and Ruhoff's (1965) catalog. *Dictyla echii* feeds on various Boraginaceae, including *Cynoglossum, Echium, Pulmonaria,* and *Symphytum* in the Palearctic region (Stonedahl et al. 1992). These plant genera include many cultivated ornamental species. After its initial collection in Pennsylvania in 1959, *D. echii* was recorded in North America from Maryland, Virginia, West Virginia, New York, and Ohio (U.S.A.), and Ontario (Canada) (Wheeler and Hoebeke 1985). In Pennsylvania, Wheeler and Hoebeke (1985) studied the seasonal history and habits on viper's bugloss or blueweed, *E. vulgare* L., and reported (with a thorough literature review) the following: eggs are generally inserted with only the operculum exposed in the midrib of the abaxial leaf surface; however, a few eggs were observed in the soft tissue of both surfaces. The exposed operculum is concealed by frass. There are five instars. *Dictyla echii* is bivoltine, and the adult overwinters in the soil near rosettes of *E. vulgare;* emergent adults were collected during mid-April and early May from Virginia pine, *Pinus virginiana* Mill. *Dictyla echii* has some potential for biological control on the noxious weed *E. plantagineum* L. (see **Section 5**). Control: The anthocorid *Orius niger* (Wolff) fed on eggs and first instar (Vayssieres 1983), and in Pennsylvania the reduviid *Sinea diadema* (F.) fed on a fifth instar (Wheeler and Hoebeke 1985).

4.11 *Dictyla monotropidia* (Stål) (= *Monanthia, Monotropida*)

Dictyla monotropidia has been frequently reported on species of *Cordia* (Drake and Ruhoff 1965) as well as cotton in Argentina (Fenton 1934), and has been intercepted on orchids from Mexico at the port of entry in Texas (U.S.A.) (Swezey 1945). *Cordia* contains several South American species that yield valuable timber, and other species are ornamental trees and shrubs. The biology has not been reported but its economic consequences have been recently described in Costa Rica, where nymphs and adults defoliate laurel, *C. alliodora*, during the dry season as well as during short dry periods (Hilje et al. 1991). Losses have been in the millions of dollars in nurseries and newly established plantings, from damage and the cost of control. During 1990 to 1991, *D. monotropidia* infested approximately 200 ha of *C. alliodora* in a plantation in the Provincia de Limón and U.S.$10,300 was expended for chemical control.

4.12 *Dictyla nassata* (Puton) (= *Monanthia nassata*)

Dictyla nassata occurs on Boraginaceae, including *Cynoglossum, Echium,* and *Onosma* in the southern Palearctic region and India. Štusák and Štys (1959) described the immatures. However, Rodrigues (1979a) noted the specimens he collected differed significantly from the descriptions of Štusák and Štys (1959). Specifically, Rodrigues (1979a) stated, "Our specimens are darker, smaller, and with shorter spinous processes on the body margins, head and dorsal body middle line." To complicate matters further, Vayssieres (1983) reported that *D. echii* and *D. nassata* frequently occur together on *E. plantagineum* and, further, Štusák and Štys (1959) were of the opinion that *D. nassata* seems to be a polytypic species that should be split into several subspecies. Therefore, until the species question is resolved, readers are referred to Štusák and Štys (1959) and Vayssieres (1983) for biology of immature stages.

4.13 *Dulinius conchatus* Distant

Dulinius conchatus is a pest on *Morinda* sp. and *Paederia foetida* in the Oriental region (Stonedahl et al. 1992), and especially a common and important pest on *M. tinctoria* in south India (Nayer et al. 1976).

4.14 *Dulinius unicolor* (Signoret) (= *Galeatus involutus* Drake)

Dulinius unicolor was first observed as a pest of coffee in Madagascar by Frappa (1931), who reported that feeding often kills bushes from defoliation. Drake and Ruhoff listed the only host plant as "musenoso" (1960), and later (1965) as "musenosa," from Kenya (Drake and Ruhoff 1965), but overlooked a second host when they cited Carayon's (1960) study of *Stethoconus frappini* Carayon, a mirid predator of *D. unicolor* on "caféier," the coffee tree. This Malagasy lace bug is recognized as a serious pest of coffee, and in Kenya is called "tigre du caféier" (LePelley 1969).

Decazy's (1975) detailed study of *D. unicolor* was conducted on the eastern coast of Madagascar. Meteorological factors had a major influence on the biology of *D. unicolor*, the number of nymphs being greatly reduced by rain. The embryonic and nymphal development were twice as long at 19°C as at 26°C and the minimal development temperature was approximately 10°C. There were two generations annually and the increase in the number of *S. frappai* predators increased proportionately to the number of lace bugs. The legume cover plant, *Flemingia* sp., had a deleterious effect on lace bug populations, and repellency was confirmed in olfactometry tests. Higher levels of infestations occurred in sunny plots than did in plots shaded by a species of *Crotalaria*. *Stethoconus frappai* Carayon is an aggressive mirid predator of *D. unicolor* (Carayon 1960).

4.15 *Eteonius sigillatus* Drake and Poor

This tingid occurs in India (Drake and Ruhoff 1965). Natural stands of wild olive, *Olea cuspidata* Wall (syn. *O. ferruginea* Royal), occur in the western Himalayas at 500 to 2000 m. In 1985, a heavy infestation of lace bugs on wild olive was identified by the CAB International Institute of Entomology as *E.* sp.? *sigilatus* [sic] subspecies *novum* (Gupta and Thakur 1991). This outbreak was of concern because improved cultivars of *O. europaea* L. had recently been introduced, and these lace bugs had not hitherto been reported on olive in India (or elsewhere). Gupta and Thakur (1991), studying the biology on wild olive, described and measured both sexes; males are slightly smaller.

Eteonius sigillatus feeds and develops on the abaxial leaf surface, to which the sticky excrement and exuviate of nymphs adheres. A heavily infested tree appears whitish-yellowish from a distance. The eggs were completely inserted into the leaf tissue and only the operculum was visible. *Eteonius sigilatus* was present on wild olive (although it fed on all varieties) throughout the year and overwintered as adults on the cultivated olive from the last week of November to late February. Oviposition began in March and continued until May. Males moved about more frequently on the host than females and began dying by the first week of April. Nymphs formed feeding aggregations, and the five stadia were 21 days long in March and 13 to 16 days from April to August 1987. The number of generations was not reported (Gupta and Thakur 1991).

4.16 *Galeatus scrophicus* Saunders (= *Cadmilos retiarius* Distant, *Galeatus helianthi* Önder and Lodos)

This lace bug has been reported from various composites, and is an occasional pest of *Chrysanthemum* and sunflower (*Helianthus* sp.), and less often of wormwood (*Artemisia* sp.), *Carthamus* sp., *Echinops* sp., *Launea* sp., *Pulicaria* sp., and *Senecio* sp. in southern Europe, Africa, west and central Asia, and India (Stonedahl et al. 1992); to this list Drake and Ruhoff (1965) add *Astragalus* sp. (Leguminosae) and *Plantago* sp. (Plantaginaceae) and Verma et al. (1974) add gaillardia (*Gaillardia* sp.), daisy (*Bellis perennis*), marigold (*Tagetes* sp.), and prickly poppy (*Argemone mexicana*). In India, on *H. annuus* L. and six other composites, eggs were deposited on the adaxial surface of the leaf, except on chrysanthemum where they were deposited on the abaxial surface. Eggs were 0.48 × 0.22 mm and consisted of three distinct regions: the operculum, the neck, and the body. The three regions were both illustrated and discussed in detail, and the five instars were also illustrated and described. Development times were not reported. No parental care was observed. Adults mated in both the morning and evening but never at midday. *Galeatus scrophicus* overwintered as an adult and began to appear in the middle of March. They spread to other species of Asteraceae in mid-May as a result of a rapid population increase. The bug was most apparent during July and August, and disappeared by September. By the end of July, most adults and nymphs fed on flower buds and stems of succulent herbs such as marigold, *Launea*, and *Vernonia* (Livingstone 1962). Also in India and on sunflower, Verma et al. (1974) studied the development in a greenhouse and reported that egg incubation averaged 6.9 days, and nymphal development was completed in 10.1 days; only 56% of the nymphs became adult. Štusák (1959) described, illustrated, and measured the fifth instar. Voltinism has not been reported. Singh et al. (1991) reported a serious outbreak of *G. scrophicus* in 1988 on sunflower grown for oilseed in June to August in Punjab, India. Livingstone (1962) reported an egg parasitoid, *Trichogramma* sp., and a parasitoid mite, *Leptus* sp., attached to both adults and nymphs.

Galeatus scrophicus (as *G. helianthi*, subsequently synonymized by Önder and Lodos 1980), was noted as a sunflower pest in southeast Turkey; males were 2.9 and females 3.1 mm long; dorsal and lateral views of the male are given, and a photograph of the damage (Önder and Lodos 1978). Feeding (by adults and nymphs) occurs on the underside of leaves, from June to September (Önder and Lodos 1983).

4.17 *Gargaphia lunulata* (Mayr)

This species feeds on many useful South American plants in several families: Euphorbiaceae (*Euphorbia* sp., *Manihot* sp., *Ricinus* sp.), Leguminosae (*Cassia* sp., *Glycine max* L., *Phaseolus* spp.), and Malvaceae (*Gossypium* sp., *Hibiscus* sp., and *Urena* sp.) (Stonedahl et al. 1992). Drake and Ruhoff (1965) also list passionflower (*Passiflora* sp.). Oddly, this lace bug is not listed in the available literature.

4.18 *Gargaphia sanchezi* Froeschner

Froeschner (1972) described this lace bug attacking beans (*Phaseolus vulgaris* L.) in Colombia, where Schoonhoven et al. (1975) studied its biology: adults occur on both surfaces of the leaf; females oviposit on the abaxial surface, where the five instars develop. At 25°C and 70% relative humidity, the first stadium took 10.4 days, and the fifth twice as long as any of the others. Females lived an average of 24.6 days and males 23.7. Because the life cycle of *G. sanchezi* takes about 20 days, and the bean crop matures in 80 days, bug populations may remain low in seasonal cropping. However, this lace bug can become a major pest on year-round plantings (Schoonhoven et al. 1975).

4.19 *Gargaphia solani* Heidemann

The eggplant lace bug is common on various *Solanum* spp., such as potato, eggplant (= brinjal, aubergine), tomato, and occurs also on other plants (*Althara* sp., *Cassia* sp., *Gossypium* sp., *Salvia* spp.) in Canada, the United States, and Mexico (Drake and Ruhoff 1965). It was first noticed, as a pest of eggplant, near Norfolk, Virginia (U.S.A.), in 1913 (Fink 1915), and was described in 1914 by Heidemann (1914). Fink (1915) realized that whenever eggplant was grown on a large scale, *Gargaphia solani* became a problem, and his study on the lace bug's biology and bionomics has become a lace bug classic. The adult is one of the largest U.S. species and differs from the others in its more prominent hood. Fink (1915) described and measured the five instars, and figured the fifth instar and adult. He also noted that nymphs bear spiny processes, "the function of which is not yet well understood." Eggs are deposited on the abaxial leaf surface in circular masses, attached at a slight angle and covered with frass. The nymphal stage requires about 10 days, and from egg to adult takes about 20. Adults mated in November, which suggests seven to eight generations a year; six occur on eggplant and the last on horse nettle, *Solanum carolinense* L. Froeschner (1944) found adults throughout the year in Missouri (U.S.A.); they hibernated in clumps of grass and sometimes under bark or mullein (*Verbascum thapsus* L.) leaves (Fink 1915).

Gargaphia solani is of interest because it was here that maternal care in Tingidae was first found (Fink 1915). Later, Kearns and Yamamoto (1981) showed that aggregated nymphs become alarmed when another nymph is crushed. The early (Fink 1915) observations of maternal care in an ostensibly nonsocial insect later generated considerable interest in insect altruistic behavior (Tallamy and Denno 1981a,b, 1982; Tallamy 1982, 1984, 1985, 1986; Tallamy and Dingle 1986; Tallamy and Horton 1990). The *raisons d'être* of nymphal spines and maternal care were later worked out in a study of the chemical defenses of *G. solani* and *Corythucha cydoniae* (Fitch) with different ecological survival strategies (Aldrich et al. 1991).

Several generalist predators feed on *G. solani* (Fink 1915); adults and larvae of the coccinellids *Hippodamia convergens* Guérin-Méneville and *Megilla maculata* DeGeer turn the lace bugs on their backs before feeding. Three species of spider also feed on this lace bug, as do the heteropterans *Podisus maculiventris* Say (Pentatomidae) and *Orius* (as *Triphleps*) *insidiosus* (Say) (Anthocoridae; on nymphs). Bailey (1951) summarized much of Fink's (1915) work.

4.20 Gargaphia tiliae (Walsh)

Large populations of *G. tiliae* caused severe browning in forests of basswood (*Tilia* sp.) throughout central and eastern Canada (Hall et al. 1998).

4.21 Gargaphia torresi Costa Lima

This polyphagous South American lace bug is a common pest primarily on species in Asteraceae (*Parthenium* sp., *Xanthium* sp.), Leguminosae (*Canavalia, Phaseolus*), Malvaceae (*Gossypium* sp., *Hibiscus* sp., *Sida* sp.), and on sweet potato (*Ipomoea* sp.), *Triumfetta* sp., and *Zea* sp. (Stonedahl et al. 1992). In the Paraguayan chaco it is an occasional pest of cotton, on the under surface of whose leaves it feeds. It was once so abundant that an entire field appeared yellow, from the many chlorotic leaves (Nickel 1958).

Nevertheless, *Gargaphia torresi* does not appear to be a major pest. It has also been reported in northeast Brazil on *Vigna unguiculata* ("caupi") (Moraes and Ramalho 1980) and, on beans, it has caused serious decline in production (Silva and Barbosa 1986).

The following predators of *G. torresi* were reported by Silva and Barbosa (1986): an anthocorid, *Xylocoris* sp.; a thrips, *Franklinothrips* sp.; and a trombiculid mite, *Bochartia* sp.

4.22 Habrochila laeta Drake

This species was described from India (Drake 1954), and no hosts were recorded by Drake and Ruhoff (1965). However, in 1976, Nayer et al. listed it as one of the few lace bug pests in India. It was considered the most important pest of barleria (*Barleria cristata* L.), a shrub grown in south Indian home gardens for its flowers (David and Rangarajan 1966). Asari (1972), studying the bug's biology in the laboratory (at 29°C), found that eggs are laid slanted into mesophyll on either side of midrib on the adaxial surface of the leaf; the egg and five instars were figured but not described. The five instars measured (lengths) 0.6, 0.8, 1.1, 1.4, and 1.8 mm and (widths) 0.2, 0.3, 0.5, 0.6, and 0.9 mm; lengths of antennae, rostrum, and legs were also given. Early instars aggregate. After a preoviposition period of 30 to 36 hours, the female oviposits for about 7 days laying an average of 40 eggs. Data on seasonal population fluctuations were provided, but voltinism was ill defined. *Habrochila laeta* occurred throughout the year on *B. cristata*, but populations shrank during monsoons and winter; peaks were from April to July (Asari 1972). Mohanasundaram and Basheer (1963) reported the impact of weather conditions on seasonal populations, and stated that most adults occur on the adaxial leaf surface, whereas the instars were on the abaxial surface.

4.23 Habrochila placida Horváth

Although described from cacao, *H. placida*, like *H. ghesquierei* Schouteden (**Section 3.9**), is a pest of coffee in Africa (Harris 1932; Mayné and Ghesquière 1934; LePelley 1969; Hill 1975, 1983), particularly of the *robusta, arabica*, and *stenophylla* varieties (Hargreaves 1940). Adults and nymphs feed in groups on the abaxial surface of leaves (Harris 1932, LePelley 1969), in the shade (Harris 1932). The feeding gives the leaves a silvery appearance (Harris 1932), and the foliage of heavily infested trees may appear yellowish at a distance (Hargreaves 1940). Although usually a minor pest, *H. placida* at times defoliates trees over small areas (Hargreaves 1940, Foucart 1954; also see LePelley 1969, a translation of Foucart 1954), which is especially serious when the trees are fruiting (Hargreaves 1940). The nymphs secrete "globules of liquid during feeding" from projections of the body (Harris 1932).

Hargreaves (1940) illustrated the egg, fifth instar, and adult, and reported that egg to adult takes 38 days, a time close to Foucart's (1954) 36 to 38 days. Foucart also reported egg, 13 to 16 days; and four stadia, 12 to 25 days. The discrepancy between Hargreaves' (1940) five instars and Foucart's (1954) four remains unresolved (five is more likely, as far more common in Tingidae and Heteroptera). The preoviposition period is 3 to 7 days, and the oviposition period is 2 to 3 weeks (Foucart 1954).

In Uganda, Hargreaves (1940) reported a mirid feeding eagerly on nymphs and adults of *H. placida*; this mirid was later identified as a *Stethoconus* sp. (LePelley 1969).

4.24 *Leptobyrsa decora* Drake

This lace bug has been recorded from teak (*Tectona grandis* L.) (Mishra and Sen Sarma 1986). It is of greater importance as a possible control agent: see **Section 5.9**.

4.25 *Leptopharsa gibbicarina* Froeschner

This species was described from Colombian specimens submitted for identification (Froeschner 1976). Accompanying correspondence stated that the lace bugs were moving from the foliage of *Pestalozzia* sp. (Cucurbitaceae) into plantations of African or oil palms, and causing losses. Attempts to control the bug with insecticides confirmed its economic importance (Reyes et al. 1988).

Genty et al. (1975) found a close association between leaf spot causing widespread damage in Colombia to oil palm (*Elaeis guineensis* Jacquin) and a lace bug later described as *L. gibbicarina*. Leaf spot, caused by the fungus *Pestatoliopsis* sp., was responsible for a bunch reduction of as much as 18 to 20 tons/ha (Hartley 1977) and as much as 40% crop loss (Genty et al. 1983). Reduction of leaf spot followed insecticidal control of the lace bug (Genty et al. 1975).

In Honduras, a complex of fungi including *Pestalotiopsis* spp., *Oxydothis* sp., and *Mycosphaerella* sp. was reported associated with a leaf spot of the oil palm, and aerial sprays of the fungicide Benlate did not reduce disease levels (Vessey 1981). Field observations (Vessey 1981) showed that fungal leaf-spot lesions on oil palm leaves started at the insect feeding wounds, and control of the disease by applications of insecticides reduced the number of sites of infection.

In Colombia, Ordonez-Giraldo (1993) evaluated isolates of the entomopathogenic fungus *Sporothrix insectorum* from *L. heveae* Drake and Poor in rubber plantations from Brazil, against *L. gibbicarina* on African oil palm. *Sporothrix insectorum* affected 73% of the lace bugs in a laboratory test and 47% of the lace bugs in a field test and was considered economically competitive with conventional insecticides.

4.26 *Leptopharsa heveae* Drake and Poor

This lace bug was described from Brazil on *Hevea brasiliensis*, "the best and most important source of natural rubber, much cultivated in Old World tropics; seeds used as food by aborigines in the northwestern part of the Amazon Valley; and the plant sometimes grown as an ornamental" (Bailey 1976).

Celestino-Filho and Magalhães (1986) found the naturally occurring entomopathogenic *S. insectorum* (Hoog and Evans) infesting up to 98% of *L. heveae* in a rubber plantation in Brazil. *Sporothrix insectorum* was isolated from *L. heveae* and in field applications provided control (Junqueira and Magalhães 1987, Junqueira et al. 1988). Lara and Tanzini (1997) identified clones of *H. brasiliensis* resistant to the feeding and oviposition of *L. heveae*.

4.27 Leptoypha minor McAtee

The ash lace bug occurs in California (Drake and Ruhoff 1965) and Arizona (U.S.A.) (Usinger 1946), where it feeds on the leaves of *Ceanothus* sp., *Fraxinus oregona* Nutt., *F. velutina* Torrey (Arizona ash), *Populus candicans*, as well as an "olive tree next to infested ash" (Drake and Ruhoff 1965).

Its biology was studied on ash in California by Usinger (1946) who reported that eggs first appeared in April, partially embedded on the abaxial leaf surface beside the veins; a distinctive prominent, oval, disklike cap projected from the leaf surface. Incubation in August was 14 days. The five stadia required 5, 7, 3, 3, and 6 days. The fifth instar was described in great narrative detail that included antennal segment ratios and color; head widths and body lengths of instars were given, as well as excellent drawings of the egg, five instars, and adult. Development from egg to adult required 38 days. Nymphs developed in aggregations on the abaxial surface. *Leptoypha minor* has four, possibly five, generations per year, and these by late summer collectively increase to an injurious population. Populations subsided in October and adults hibernated in and around trees. Usinger (1946) added Arizona to the distribution and stated, "I have never known *Leptoypha* to kill a tree, but at times nearly every leaf is covered with the bugs. At such times the entire foliage appears to be discolored."

4.28 Leptoypha mutica (Say)

This species occurs from Québec (Canada) to North Dakota (U.S.A.) and south to Florida and Texas (Blatchley 1926). It has been reported from several ornamental trees (Drake and Ruhoff 1965), especially as a seasonally abundant pest on the fringetree, or old-man's beard, *Chionanthus virginicus* L. (Dickerson and Weiss 1916). A heavy infestation can injure nearly every leaf; moderate injury is characterized by mottled leaves, and serious injury by uniformly yellow-brown and withered leaves (Mead 1975); in a nursery, the bugs could injure almost every leaf (Dickerson and Weiss 1916). Many eggs are laid as clusters in the leaf midrib of *C. virginicus*, on the abaxial surface; some are inverted adjacent to the midrib and covered with brownish frass; the egg is flask-shaped and only the operculum is exposed. The five instars are 0.6, 0.6, 0.9, 1.3, and 1.4 mm long and 0.3, 0.4, 0.6, 0.9, and 1.2 mm wide (Dickerson and Weiss 1916, who also described and figured the egg, instars, and adult). Nymphs were on the abaxial leaf surface of plants in the sun, but on the adaxial surface in the shade and where foliage was dense (Dickerson and Weiss 1916). There appear to be two generations (in Florida), and the adult overwinters.

4.29 Metasalis populi (Takeya) (= Hegesidemus harbus Drake; see Golub 1988)

Metasalis populi (as *H. harbus*) was reported a serious pest on clones of a poplar tree, *Populus americanus* (Dode), in China (Liang and Zhao 1987). These authors reported that eggs were laid on the abaxial leaf surface adjacent to the veins and were 0.43 mm long and 0.16 mm wide. Incubation during the four generations studied ranged from 9.4 to 11.7 days. Only four instars were reported, with the following measurements: lengths 0.52 to 0.53, 0.90 to 0.92, 1.38 to 1.39, and 2.17 to 2.18; and widths were 0.20 to 0.21, 0.45 to 0.46, 0.83 to 0.85, and 1.14 to 1.16 mm. The nymphal stage took 15.5 to 24 days. Adults spent 3 to 5 hours *in copulo*. Summer generation adults lived 23 to 28 days, and many adults overwintered in the seams of split tree bark; mortality was about 45%. Overwintered adults emerged in May and first-generation nymphs were present by the month's end. Feeding damage by adults and nymphs was described. Predators included *Leis axyridis* (Pallas), *Chrysopa septempunctata* Wesmal, *C. sinica* Tjeder, *Arma chinensis* Fallou, *Allotrobium* sp., and *Hierodula patellifera* (Serville).

4.30 *Monosteira edeia* Drake and Livingstone (= *Monosteira zizyphora* Ghauri)

This is a pest on jujube, or Chinese date (*Zizyphus jujuba* Miller), in northern India (Nayer et al. 1976), Pakistan, and the Mascarene Islands (Stonedahl et al. 1992). Described anew by Ghauri (1965), it was synonymized by Péricart (1981). Bhalla and Mann (1988) described damage and crop loss to *Z. jujuba*.

4.31 *Monosteira lobulifera* Reuter

This lace bug occurs in the Middle East (Drake and Ruhoff 1965) and, in Turkey, is an important pest of fruit trees, on the undersurface of the leaves the adults and nymphs feed (Önder and Lodos 1983, who provide a photograph of the damage).

4.32 *Phaenotropis cleopatra* (Horváth)

This is a common pest of *Oryza* and *Panicum* (Gramineae) and of *Tephrosia* (Leguminosae) in the circum-Mediterranean area east into the Indian subcontinent (Stonedahl et al. 1992). It is also a pest in southern Indian on indigo (*Indigofera* sp.) and *Tephrosia* (Nayer et al. 1976).

4.33 *Phatnoma marmorata* Champion

This lace bug has been reported a pest of cocoa and pineapple in Central America, Trinidad, Ecuador, and Brazil (Drake and Hambleton 1944, Drake and Ruhoff 1965, Stonedahl et al. 1992).

4.34 *Phatnoma varians* Drake

This northern South American (Drake and Ruhoff 1965) lace bug has been intercepted at a California port of entry on *Cattleya schroederae*, from Colombia (Swezey 1945); *Cattleya* is a genus of magnificent orchids.

4.35 *Plerochila australis* (Distant) (= *Teleonemia australis* Distant)

The olive bug (Pinhey 1946) was reported a pest of olive, *Olea europaea*, in the Ethiopian region and Mauritius (D. S. Hill 1983, Stonedahl et al. 1992). Pinhey (1946) observed the biology of *P. australis* on olive in Zimbabwe and reported that eggs were laid in clusters on the abaxial leaf surface; nymphs were green and gregarious; the adult was small with clubbed antennae; adults and nymphs were present May to December, adults most numerous June to November. No hosts were found other than wild olive. Pinhey (1946) stated that feeding by the adult and nymph may "… cause leaf fall, fruit withering, and undermining of the tree's vitality. There may be a considerable reduction in fruit formation. Young trees may have their growth checked, and consequently produce stunted trees."

A small black and white mirid found feeding on the lace bug's nymphs (Pinhey 1946) was probably a species of *Stethoconus*. Other predators were a lace bug larva (*Chrysopa* sp.) and a reduviid (*Rhynocoris tropicus* Herrich-Schaeffer) (Pinhey 1946).

4.36 *Pterochila horvathi* (Schouteden)

Pterochila horvathi has been reported a pest of olive, *O. europea*, and of jasmine, *Jasminum* sp., in the Ethiopean region and the Mascarene Islands (Drake and Mamet 1956, Stonedahl et al. 1992).

4.37 *Pseudacysta perseae* (Heidemann) (= *Acysta perseae* [Heidemann])

The avocado lace bug (Moznette 1922) was described (as *Acysta perseae*) by Heidemann (1908) from avocado (*Persea americana* P. Miller); earlier names of hosts are invalid. The bug occurs in the southern United States, Mexico, Puerto Rico, and the Dominican Republic (Drake and Ruhoff 1965, Abud 1991, Mead and Peña 1991). Heidemann (1908) includes drawings of the fifth instar and adult, and reports that the egg is laid vertically on the abaxial leaf surface and covered by fecal material. A late (fifth?) instar was described; and the adult was 2×0.8 mm and characterized as "this neat little tingid" (Heidemann 1908). Meade and Peña (1991) give detailed accounts of the lace bug's hosts, bionomics, and distribution in Florida.

Pseudacrysta perseae afflicts avocado in the winter (southern United States) and destroys leaf cells (Moznette 1922); in Puerto Rico it causes chlorosis and browning, especially along the midrib (Medina-Gaud et al. 1991). The thrips *Franklinothrips vespiformis* is an important predator in the Dominican Republic (Abud 1991).

4.38 *Stephanitis chinensis* Drake

This important pest of tea in southwestern China lays its eggs on the abaxial leaf surface during the winter; the eggs measure 0.3×12 mm and hatch in 30 to 34 days; there are four to five instars. After a preoviposition period of 8 to 12 days, the female lays an average of 110 eggs over a 16 to 39 day period; 70% of the eggs hatch. Females live 25 to 78 days and males 18 to 38 days; there are two or three generations a year, the number being influenced by elevation (800 to 1200 m). Some heteropteran predators were found: *Stethoconus japonicus* (Miridae), *Geocoris* sp. (Geocoridae), and *Anthocoris nemorum* (L.) (Anthocoridae) (Wang 1988, who also gives figures of egg, fifth instar, and adult).

4.39 *Stephanitis laudata* Drake and Poor

Stephanitis laudata causes severe damage to the upper canopy of the camphor tree, *Cinnamomum* sp., in China and Taiwan (Li et al. 1990). Li et al. (1990) report that the egg is 0.4×0.2 mm; the five instars are 0.6, 0.8, 0.9, 1.2, and 1.7 mm long and 0.3, 0.4, 0.5, 0.6, and 0.9 mm wide; nymphal development is seasonally (temperature) dependent and takes 14 to 25 days. Adult females are 3.4×1.8 mm and males 3×1.5 mm. Males live 8 to 26 and females 8 to 19 days. The egg overwinters in the leaf on the tree and hatch in April. There were five generations, the third (August and September) required the shortest development time. At a constant temperature of 25°C, the egg developed in 20.5 days and at 28°C in only 16 days. Beneficial insects yet to be identified include a predaceous mirid and a hymenopterous egg parasitoid. Pesticide application to nymphs of the first generation in May was highly efficacious (Li et al. 1990).

4.40 *Teleonemia nigrina* Champion

This United States to Central Mexican species feeds on several plants; those of commercial importance are sugar beets (*Beta vulgaris*) and snapdragon (*Antirrhinum majus*) (Drake and Ruhoff 1965). Hixson (1942) reported *T. nigrina* nearly destroying snapdragons in a commercial Oklahoma greenhouse, and also nearly killing a stand of the ornamental plant *Verbena* sp.

4.41 *Tingis beesoni* Drake

This lace bug attacks the fast-growing hardwood tree *Gmelina arborea* Roxb., an important timber tree, in Burma and India (Beeson 1941), where adults and nymphs feed in aggregations at the leaves' bases and on soft shoots. The leaves become spotted or brown-patched, and discoloration may extend several inches on both sides of the leaf blade. Leaves wither, and young plants may be completely defoliated; shoots become dry and black and covered with a sooty mold (Beeson 1941). In Burma, attacks may be so severe in nurseries and plantations that young trees die (Zeya 1983). In Kerala (India), *T. beesoni* did the most damage to *G. arborea* of 34 insect species associated with the tree; in a 10-ha plantation 67% of the year-old trees were infested, and 21% suffered total defoliation; *T. beesoni* attacked only saplings (Mathew 1986, Nair and Mathew 1988).

Eggs are inserted almost flush in the cortical tissue of the young shoot, aligned in a straight row and often covered with fecal material; eggs are 0.4 × 0.8 mm. Incubation ranged from 2 to 6 days, and the five stadia took 0.8, 1.1, 1.4, 2.1, and 2.4 days (these times varied with the temperature); the egg and the instars were described and the adult illustrated. Of the seven generations, six occurred in the summer and the seventh overwintered (as an egg inserted under bark) (Mathur 1955, 1979).

4.42   *Tingis buddleiae* Drake

This lace bug attacks *Buddle asiatica* Lour. (butterfly bush), *B. madagascarensis* Lam., and *Vitex trifolia* L., in China, India, and the Philippines (Drake and Ruhoff 1965, Livingstone 1968). Damage may be severe when populations are high; leaves and buds crinkle, dry, and then wither (Livingstone 1968). Eggs (12 to 50) are laid scattered along leaf veins and on the tips, on the abaxial surface; incubation is 10 to 12 days; the first and fifth (last) stadia take 3 days and the others 2 each (Nayer et al. 1976). The functional anatomy of the egg was studied by Livingstone (1967), and the insect's behavior and bionomics by Livingstone (1968). Štusák (1959) described all stages and figured the fifth instar, showing the long, robust, and ramified spinelike processes.

Nayer et al. (1976) reported the predators *Appolodotus* sp. (Miridae), *Chrysopa* sp., and *Coccinella* sp.

4.43   *Urentius euonymus* Distant

This, the hollyhock lace bug, is a pest of various Malvaceae (including *Sida cordifolia* and *Abutilon* spp.) in North and east Africa, the Middle East, and the Indian subcontinent; it is also a pest on *Cajanus cajan* (L.) Huth (not *C. indicus*, as reported by Bailey 1976) (Stonedahl et al. 1992). In northern India it occurs on *A. indicum, S. cordifolia,* and *Chrozophora rottleri* (Nayer et al. 1976).

Livingstone (1959) reported *U. euonymus* appeared on the perennial shrub *A. indicum* by the end of March, where the eggs were thought to overwinter. By mid-April, *Abutilon* was overcolonized and *U. euonymus* began to migrate to hollyhocks. By the end of May the leaves of infested hollyhocks showed pale-yellowish patches from feeding. By the first week of June, hollyhocks were no longer suitable and migration continued to local malvaceaous plants. In the fall, *U. euonymus* returned to *Abutilon*. Livingstone (1959) reported an unusual oviposition behavior: eggs were inserted in the leaf from the upper surface and usually were deposited along the margin. No eggs were observed in the midrib or lateral veins. The egg was smooth, and transparent, 0.19 × 0.39 mm.

The five instars were described and illustrated, and measured (lengths) 0.5, 0.7, 0.8, 1.3, and 1.7 mm, and (widths) 0.2, 0.4, 0.5, 0.8, and 0.9 mm (Livingstone 1959).

4.44   *Vatiga illudens* (Drake) (= *Leptopharsa illudens* Drake)

This is an important pest of cassava, *Manihot esculenta* Crantz, in the Caribbean, Guyana, and Brazil (Stonedahl et al. 1992) and in Colombia, Ecuador, and Venezuela (Froeschner 1993). Ohter hosts are *M. dulcis* and *M. utilissima* (Drake and Ruhoff 1965). (These species have since been synonymized under *M. esculenta*; Froeschner 1993.) Several *M. esculenta* cultivars (Cacau vermelhõ, Branca de Santa Catarina, Guaxupe, Pirassununga, and Sertaneja) are resistant to *V. illudens* (Cosenza 1982).

4.45 Vatiga manihotae (Drake) (= *Leptopharsa manihotae* Drake, *Tigava sesoris* Drake and Hambleton, *Vatiga viscosana* Drake and Hambleton)

This is the cassava lace bug. It occurs throughout South and Central America and in parts of the Caribbean (Drake and Ruhoff 1965, Froeschner 1993). Prolonged dry spells are favorable for the buildup of large populations, especially during the first 3 months of plant growth (Bellotti and van Schoonhoven 1978). There are five instars, which feed on the central part of the leaves. In the laboratory, adults (about 3 mm long) live about 50 days. Females lay an average of 61 eggs, which require about 8 days to hatch; the five stadia take about 16.5 days (Bellotti and van Schoonhoven 1978).

Vatiga manihotae is a pest of cassava in several countries, especially in South America (Bellotti and van Schoonhoven 1978; Melo 1990, and references therein), where in high numbers it can kill leaves. However, yield losses are not known, and the bugs' harmful effects seem to be in conjunction with those of other pests (Melo 1990).

Borrero and Bellotti (1983) reported biological control potential of a reduviid predator, *Zelus nugax* Stål.

[*Note:* The biology, control, and damage caused by this species and *V. illudens* (above) are briefly reviewed by Bellotti et al. (1999).]

5. BIOLOGICAL CONTROL

Because of their host specificity and, occasionally, their rapid buildup of populations, some tingids have from time to time been considered (and occasionally imported) as agents of biological control. One of the most vigorous attempts involved several species in control of the perennial woody shrub *Lantana camara* L. (reviewed by Harley and Kassulke 1971). None of these attempts has been highly successful.

5.1 Corythaica sp. nr. cyathicollis (Costa)

Tropical soda apple (*Solanum viarum* Dunal) is neotropical and an invasive pasture weed in many parts of the world, and recently (1990s) invaded the southern United States (Medal et al. 1996). A survey in Brazil and Paraguay of insects feeding on this plant reported *C.* sp. nr. *cyanthicollis* [sic] common on this plant, where it formed small aggregations on the undersides of the leaves (Medal et al. 1996).

5.2 Corythucha distincta Osborn and Drake

Called "the distinct lace bug" by Essig (1958), *C. ciliata* occurs in western North America, from British Columbia south and east into California, Colorado, and South Dakota (Drake and Ruhoff 1965). It may feed on several vegetable crops, but has not been reported as a pest (Essig 1958). Among its wild hosts are *Lathyrus nuttallii*, and the thistles *Carduus lanceolatus, Cnicus* sp. (Drake and Ruhoff 1965), and the Canada thistle *Cirsium arvense* (L.) (Story et al. 1985). On the last, *Corythucha ciliata* was the most common hemipteran (Story et al. 1985).

Thistles are fed upon from early May through the growing season, and heavily infested thistles fail to produce flower buds. However, in the laboratory, although the bug produces localized lesions and necrosis, plant growth was little affected. Control of the thistle was not achieved (Story et al. 1985).

5.3, 5.4 *Dictyla echii* (Schrank) and *Dictyla nassata* (Puton)

Wapshere (1981) reported both *D. echii* and *D. nassata* on *Echium plantagineum* L. (Boraginaceae) (Paterson's curse). This is an invasive weed of pastures in Europe into northern Africa, and occurs also in Australia. Both *Dictyla* species were found beneath the rosette leaves in spring, and in summer nymphs and adults became abundant on the flower shoot; abortion often resulted. Because of this damage Wapshere (1981) predicted these lace bugs "should be effective in reducing the weed's flower production in Australia."

Vayssieres (1983) studied the biology of *D. echii* in southwestern France and southern Portugal. Adults overwinter in the soil or debris near *Echium* spp. rosettes, and feed when the weather reached 15 to 20°C. Females deposit eggs (0.5 × 0.15 mm) later in the spring, on *Echium* spp. leaves perpendicular to the surface, in stems and midribs; the exposed egg is covered with a brown viscous material which hardens [frass?]. Eggs hatch in 6 days at 15 to 25°C, and development from egg to adult at 23.5°C requires 25 to 30 days; development rates at 8, 16, and 19°C were also reported. Adults are short, only 3.1 to 3.6 mm. The fifth instar and the adult were illustrated. Štusák (1957) provided measurements of the adult body, antenna, tibia, and tarsus as well as a general description of morphological features. The illustration of the caudal end of the fifth instar by Vayssieres (1983) had a single short process, whereas the first instar illustrated by Štusák (1957) had two short processes.

Vayssieres (1983) reported that no parasitoids have been reared from *D. echii* but the anthocorid predator, *Orius niger* Wolff, has been observed in southern France feeding on eggs and first instars.

Dictyla nassata was considered for use in biological control of *E. plantagineum* in Australia based on studies in the western Mediterranean region by Wapshere (1981) and in France and Portugal by Vayssieres (1983). The control potential of *D. nassata* was studied in Uzbekistan by Khaidarova (1975) on the toxic weed *Trichodesma incanum* (Bge.) DC, but its effectiveness has not been established.

5.5   *Dictyonata strichnocera* Fieber

This species has been recorded from *Larix decidua, Pinus silvestris, Sarothamus scoparius, Spartium scoparium,* and *Ulex* sp. (Drake and Ruhoff 1965). It also occurs on *U. europaeus* L. (gorse, furze-bush, whin) (Butler 1922, 1923), a common spiny evergreen weed shrub sometimes cultivated as a cover for game; but, introduced into New Zealand some 160 years ago as a cheap means of fencing, gorse was declared a noxious plant there in 1900 and some U.S.$22 million of pasture production was lost in 1985 (R. L. Hill 1986). Of 94 species of arthropods that attack gorse in Europe, *D. strichnocera* was among the four with the greatest potential for damaging gorse in New Zealand (R. L. Hill 1983).

Butler (1923) described the egg and several instars, illustrated the former, and indicated adults are present (in England) from June to October. Most adults matured in July and eggs laid in August hatched the following spring. Štusák (1957) included *D. strichnocera* in his key to Czech fifth-instar lace bugs, and later (1961) illustrated the operculum of the egg.

5.6   *Galeatus maculatur* (Herrich-Schaeffer)

Adults and nymphs of this lace bug were "common" in Hungary on *Hieracium pilosella* L., a Palearctic plant that has become a weed in dry pasturelands in North America, Japan, and New Zealand (Sárospataki 1999).

5.7, 5.8 *Gargaphia arizonica* Drake and Carvalho and *Gargaphia opacula* Uhler

The host plant of *G. arizonica* is silver nightshade, *Solanum elaeagnifolium* Cav., indigenous to Mexico and the southwestern United States (Goeden 1971), and now a weed in Australia, where it competes with crops and is toxic to livestock (Wapshere 1988). *Gargaphia opacula* also feeds on this weed, but has a broader range (Wapshere 1988). Both species were studied in Mexico to evaluate their potential for controlling the weed in Australia. Although Goeden (1971) discounted their effectiveness, Wapshere (1988) found they could completely destroy younger plants.

5.9 *Leptobyrsa decora* Drake

This lace bug (see also **Section 4.24**) feeds on the weed bitter or sour orange (*Citrus aurantium* L.) in northern South America (Drake and Ruhoff 1965) and on *Lantana* sp. in Peru and Trinidad (Harley and Kassulke 1971). On *Lantana* sp., the eggs are laid partly inserted next to the mid- and lateral veins of the abaxial surface of the leaf and are covered with frass. The rate of development depends upon temperature; in the summer (21 to 33°C) egg to adult took 31 days; at slightly lower temperatures (16 to 31°C), egg to adult took 44 days. Females care for their young up to late stadia, and during this period do not lay more eggs. If deprived of their young, females then oviposit again. Nymphs feed in aggregations on the abaxial surface of mature and young leaves, and adults feed singly or in smaller groups. The fed-upon leaves soon become chlorotic, then silvery, and may fall off. Scanning electron micrographs of egg, fifth instar, adult, and feeding damage have been published (Harley and Kassulke 1971).

In host preference studies (Harley and Kassulke 1971), *Leptobyrsa decora* fed and oviposited on 60 species of plant (including *Citrus*), but first instars died on all except *Lantana camara* and caused heavy damage. *Leptobyrsa decora* has since been released, and become established, in Australia and Hawaii to control *Lantana camara* (Harley and Kassulke 1971).

In India, *Leptobyrsa decora* was raised in quarantine by Misra (1985) for possible biological control. Misra (1985) provided drawings of a late (fourth?) instar and the dorsum and venter of the adult male. Mishra and Sen Sarma (1986) reared this lace bug on *Lantana camara*: the egg is 0.47 × 0.19 mm; oviposition lasts about 1 week, and eggs hatch in 10 to 15 days; nymphal development takes 17 to 18 days, and egg to adult requires 30 to 40 days; the female guards her eggs.

However, because this bug also attacks teak (*Tectona grandis* L.), Mishra and Sen Sarma recommend that the bug not be introduced into India.

5.10 *Oncochila simplex* (Herrich-Schaeffer) (= *Monanthia simplex* Herrich-Schaeffer)

Widely distributed throughout eastern Europe into Siberia, this lace bug feeds on various species of *Euphorbia* (Drake and Ruhoff 1965, Pecora et al. 1992). Leafy spurge (*Euphorbia esula* L.) was introduced into the United States from Eurasia early in this century, and has since become a serious problem in several northern and western states, where it may cause an annual loss of about U.S.$10.5 million (Pecora et al. 1992). *Oncochila simplex* was studied as a possible control agent by Pecora et al. (1992), who found (at 20°C, light:dark, 16:8) that the curved sac-shaped egg is laid under the mesophyll in stems; females produce an average of 175 eggs over 36 days; only four instars were reported and the four stadia took 33 to 340 days; adults lived 50 to 70 days. *Oncochila simplex* was multivoltine in the laboratory (11 generations) but, in northern Italy in the wild, there were 5 generations annually. Adults and nymphs both overwinter, 4 to 5 cm below the ground at the bases of the host plants. In no-choice tests of 26 plants, neonate nymphs developed on several species of *Euphorbia*, as well as on lettuce and corn, but in the field lettuce and corn did not support development. Pecora et al. (1992) concluded "that *O. simplex* should be introduced as a biological control agent against leafy spurge in the U.S."

5.11 *Teleonemia elata* Drake

Like other members of its genus, this lace bug feeds on *Lantana* spp. and has therefore been considered in the biocontrol of this weed.

Teleonemia elata was one of several South American tingids collected from *Lantana* spp. (in Brazil) during a survey by Harley and Kassulke (1971). Eggs were inserted singly, into stems, petioles, midribs, and the main veins of leaves; oviposition sites were inconspicuous, and the egg's operculum was flush with the surface. Development in the summer (at 21 to 33°C) took preoviposition 4, egg 20, nymphs 18, and egg to adult 42 days; at 16 to 31°C, the number of days required were preoviposition 7, egg 30, nymph 24, and egg to adult 61. Adults lived about 7 to 8 weeks. Adults usually fed on flowers, but nymphs fed principally on the adaxial leaf surface. Adult feeding caused the wilting and often the death of apical portions of stems, and feeding by nymphs caused the death and abscission of foliage. Figures include SEMs of the adult, fifth instar, and operculum, and a photograph of leaves wilted from feeding (Harley and Kassulke 1971). *Teleonemia elata* completed its life cycle only on *L. montevidensis* in feeding trials of 60 species for host suitability. Harley and Kassulke (1973) stated *T. elata* is more host specific on *Lantana* than *T. scrupulosa* (**Section 5.14**). *Teleonemia elata* was liberated for control of *L. camara* in Australia and was also imported from Australia into Hawaii and Zambia (Harley and Kassulke 1971), but there was no evidence in Australia of establishment of *T. elata* following release of insects in 1969 from Brazil (Harley 1971).

5.12 *Teleonemia harleyi* Froeschner

This lace bug was collected on Trinidad in a search for enemies of yellow sage, *L. camara* L. (Harley and Kassulke 1971, 1973), and described by Froeschner (1970). Harley and Kassulke (1973) reported that at 15.5 to 36.6°C, eggs were inserted singly into flower stalks, petioles, or next to veins on either leaf surface; incubation took 9 to 10 days and the nymphal stage 16 days. Both nymphs and adults preferred to feed on flowers or, if these were unavailable, on leaf buds; without flowers, reproduction was minimal. The preoviposition period was 7 to 9 days, and adults lived about 100 days, with 50% survival at about 58 days. Harley and Kassulke (1973) provide SEMs of the operculum of the egg, the fifth instar, and the adult. In cage trials, *T. harleyi* attacked with equal effectiveness all *Lantana* spp. naturalized in Australia; all flowers fed on were destroyed, and no seed was set. In 1972, *T. harleyi* was approved for release in Australia (Harley and Kassulke 1973).

5.13 *Teleonemia prolixa* Stål

This lace bug occurs throughout the neotropics and feeds on *Acacia riparia*, *Cinchona* sp. (quinine), and *Lantana camara*; however, Harley and Kassulke (1975), studying host preferences of this lace bug, concluded that only *L. camara* is a true host. The bug was released for biological control in Australia in 1974 (Harley and Kassulke 1975).

Mikania micrantha is a sprawling neotropical vine that was inadvertently introduced into southeast Asia, where it became a serious weed of such plantation crops as tea, teak, rubber, oil palm, and coconut. Cock's (1982) survey of enemies of *M. micrantha* identified a complex of tingid species near *T. prolixa* as one of nine major enemies. However, Cock's study (1982) indicated that *T. prolixa* caused no obvious damage to the plant. Cock (1982) recommended a taxonomic revision of the species group, if not the genus, should be undertaken before proceeding further with *T. prolixa*.

5.14 *Teleonemia scrupulosa* Stål (= *Teleonemia lantanae* Distant, *Teleonemia vanduzeei* Drake)

The lantana lace bug occurs from the southern United States south into South America (Habeck and Mead 1975). Because of its potential for controlling lantana, the bug has been introduced widely into the Old World tropics.

In laboratory studies, egg to adult took 17 days (Simmonds 1929) or 3 to 4 weeks; eggs hatched in 7 to 8 days and the five stadia required 15 to 18 days (Fyfe 1937). Eggs are laid in the mid- or lateral veins of the abaxial surface of a young terminal leaf; the operculum is covered with excrement; three overlapping generations may occur on a plant (Fiji) (Fyfe 1937), or 10 to 11 generations (Dehra Dun, India; Roonwal 1952). The egg is 0.5 × 0.2 mm, and the lengths of the five instars are 0.5, 0.8, 1.2, 1.7, and 2.5 mm; the adult is small (3.6 × 1.2 mm) and elongate (Roonwal 1952). In India (Dehra Dun), eggs laid in December to January overwintered and hatched in March (Roonwal 1952).

Illustrations of the fourth instar, egg, and egg *in situ* in the midvein are given by Simmonds (1929), of all instars by Roonwal (1952), and of the adult by Blatchley (1926), Simmonds, (1929), Roonwal (1952), and Habeck and Mead (1975); Drake (1918) describes the adult.

Yellow sage, *L. camara*, is a long-lived bushy perennial native to Mexico sometimes used as an ornamental. Its foliage is poisonous to livestock, and the plant is now considered one of the worst weeds of the world (Holm 1969, Cilliers and Neser 1991). The lantana lace bug has now been introduced to 15 countries for control of *L. camara*, from Hawaii (1902) through Zambia (1969) (Harley and Kassulke 1971).

Teleonemia scrupulosa was imported into Australia in 1935 to 1936 and became established throughout the lantana-infested area, but its value as a control agent was limited by a preference for white- and red-flowering varieties, and by its severe reduction during unusually wet or cold weather (Harley and Kassulke 1971). In Micronesia, *T. scrupulosa* was one of the five most important species of 13 exotic insect species introduced (Denton et al. 1991). Muniappan (1988, 1989) reviewed the biological control status of *T. scrupulosa* on lantana in Guam and Yap (Micronesia).

The lantana lace bug feeds primarily on the flowers, causing them to shrivel. Leaves also are lost, and flower and seed set much reduced (Cilliers 1987b).

Lantana camara was first introduced into India in 1809 and planted in the Calcutta Botanical Gardens (Beeson and Chatterjee 1940); its spread unrestricted, it soon became a serious pest (Roonwal 1952). In 1916 the Government of India studied the possibility of native insects' controlling *L. camara*, to no avail (Beeson and Chatterjee 1940). Several years later the biology and host preference of *T. scrupulosa* was extensively studied in quarantine on 13 tree species by Khan (1945) at the Forest Research Institute, Dehra Dun, for possible control of *L. camara*. Khan's (1945) assessment was that it "... feeds greedily on teak flowers" and would be a "... grave risk to seed production." Consequently, the research was ended and Khan (1945) reported "... all our living stock of the insect destroyed." However, the accidental finding of a single *T. scrupulosa* in 1951 resulted in a survey that concluded *T. scrupulosa* is established in India (Roonwal 1952). A further attempt occurred in 1969 to 1970, when *T. scrupulosa* was released in small numbers in Hyderabad (India) and rapidly radiated (Sivaramakrishnan 1976).

As Khan (1945) found that the lantana lace bug could live for several weeks on other Verbenaceae, Ramesh and Mukherjee (1992) evaluated the host potential of 31 plants of economic importance, in 15 families. *Teleonemia scrupulosa* nymphs survived only on *L. camara*, and Ramesh and Mukherjee (1992) concluded that the bug does not threaten other crops when used to control *L. camara*.

This echoed an earlier statement of Gupta and Pawar (1984), that "*T. scrupulosa* ... only suppresses growth of *Lantana*," and Livingstone et al. (1980) concluded that earlier accounts of the lantana lace bug's attacking teak were unfounded.

Teleonemia scrupulosa has been introduced into South Africa four times from 1968 to 1982 to control *L. camara* (Cilliers 1983). Cilliers (1987a,b) studied the bug for 3 years, and reported on its biology, seasonal history, and the impact of its feeding on *L. camara*. Cilliers and Nesser (1991) reviewed the biological control potential of the lace bug, and concluded that, in conjunction with two chrysomelids, *T. scrupulosa* exerted some degree of control.

Reported as predators on *T. scrupulosa* in Fiji were an unidentified mirid and the seed-feeding lygaeid *Germalus pacificus* Kirkaldy, both of which fed on all stages in the laboratory; other enemies included *Coccinella transversalis* (Coccinellidae) and larvae of *Hemerobius* sp. (Hemerobiidae). A fungus, *Hirsutella* sp., also attacked adults and fifth instars (Simmonds 1929, Fyfe 1935).

5.15 *Tingis ampliata* (Herrich-Schaeffer) (= *Monanthia ampliata* Herrich-Schaeffer)

The hosts of this Nearctic lace bug include the thistles *Carduus* sp. and *Cirsium* sp., and *Verbascus* sp. (Drake and Ruhoff 1965). Southwood and Scudder (1956) associated this species with creeping thistle (Canada thistle), *C. arvense* (L.) Scop., which had been thought the preferred host plant (Peschken 1977). However, the bug will develop fertile eggs while feeding on two economically important plants, globe artichoke (*Cynara scolymus* L.) and safflower (*Carthamus tinctorius* L.); nymphs will develop to adults on the artichoke, and Peschken (1977) concluded the bug should not be released in Canada. Peschken's study (1977) included a thorough and valuable review of the published work on this lace bug's biology and life history (see also Southwood and Scudder 1956; Eguagie 1972a,b, 1973, 1974a,b).

6. PROGNOSIS

It is clear from the foregoing that many lace bugs may sometimes become pests, but that far fewer are serious pests all the time. Lace bugs are generally quite host specific, and many are multivoltine; it follows then that species of lace bugs, harmless so far, may become pests, at least for a time and at least in one or two localities. A species of lace bug now feeding on the wild relative of economically important species may move to that species and become a pest, if only for a while (see Schaefer 1998, for a general discussion); phylogenetic work is therefore needed. In addition, of course, populations of a species once but no longer a pest may again resurge and cause damage.

Further work on the biology and, especially, the ecology, of lace bugs is needed, both on species that feed on valuable plants and on species that feed on wild relatives of economically important plants. Knowledge of these bugs' life histories and economical requirements will help one predict when and if lace bug species may become damaging.

For all these reasons, further knowledge will help predict when and if species that feed on harmful plants, or the relatives of harmful plants, may have value in biological control.

Finally, lace bugs are not generally considered vectors of plant pathogens. However, being quite host specific and feeding as they do on leaves and stems, two species (*Corythucha ciliata* and *Leptopharsa gibbicarina*; see **Sections 3.4 and 4.25**) have been suspected of vectoring pathogens of diseases found also on host plants' leaves and stems. The evidence for this association is circumstantial, and more work is needed.

7. CONCLUSIONS

Only occasionally does a lace bug become a long-term serious economic problem. Because most of them feed on leaves, their damage is less serious than if they fed on reproductive parts; larger populations of lace bugs are needed to effect serious damage. Being small and often multivoltine, such populations can, of course, build up rapidly, and this is the reason so many lace bug species are pests only occasionally.

Genera with the greatest number of economically important species (e.g., *Corythucha, Stephanitis, Teleonemia*) are merely very large genera and would be expected because of their size alone to have more species than smaller genera. There does not seem to be a biological or ecological propensity for certain genera to be of greater economic significance than others; this propensity lies at the species level.

Except for species of the mirid genus *Stethoconus*, there are no predators specifically attacking lace bugs. This is not unusual; most predators are generalists. Indeed, there may be more predators and parasites of lace bugs than have been reported. Lace bugs are small, feed mostly on the lower surface of leaves, and their eggs are deposited there, often inserted into tissue and covered with frass. Predators and (especially egg) parasites would be hard to see, and many have doubtless gone unreported. The fact that parental care has arisen in one genus (*Gargaphia;* various Tallamy references) suggests a need for protection from parasites, and therefore the presence of them.

The preference of lace bugs for the underside of leaves also makes them less susceptible to insecticide sprays, and their many generations a year increases the likelihood of their becoming resistant to insecticides.

8. REFERENCES CITED

Abbott, C. E. 1935. Notes on the oviposition and hatching of *Corythucha marmorata* Uhler. Bull. Brooklyn Entomol. Soc. 30: 13.
Abud A., A. J. 1991. Presencia de la chinche encaje del aguacate *Pseudocysia perseae* (Heidemann) (Hemiptera: Tingidae) en la Republica Dominicana. Primera Jornada Científica de Protección Vegetal. Univ. Autonoma de Santo Domingo, Dominican Republic.
Aldrich, J. R., J. W. Neal, Jr., J. E. Oliver, and W. R. Lusby. 1991. Chemistry vis-a-vis maternalism in lacebugs (Heteroptera: Tingidae): alarm pheromones and exudate defense in *Corythucha* and *Gargaphia* species. J. Chem. Ecol. 17: 2301–2322.
Allen, A. A. 1962. *Placotettix taeniatifrons* Kbm. (Hem., Cicadellidae) in S.E. London (N.W. Kent). Entomol. Mon. Mag. 98: 47.
Anonymous ("H. W."). **1930.** Rhododendron bug (*Leptobyrsa rhododendri*). Gard. Chron., Sept. 20. p. 238.
1961. An Atlas of Coffee Pests and Diseases. Coffee Board of Kenya, Nairobi, Kenya. 146 pp.
Arzone, A., and O. I. O. Marletto. 1984. Patogenicita di tre deuteromiceto nei confronti di *Corythucha ciliata* Say (Heteroptera Tingidae). Redia 67: 195–203.
Asari, K. P. 1972. Bionomics and immature stages of the Barleria lacebug *Habrochila laeta* Drake (Heteroptera: Tingidae). J. Bombay Nat. Hist. Soc. 72: 97–106.
Bailey, L. H. 1978. Hortus Third: A Concise Dictionary of Plants Cultivated in the United States and Canada, third edition. Cornell Univ. Macmillan Publ. Co. New York, New York, U.S.A. 1290 pp.
Bailey, N. S. 1950. An Asiatic tingid new to North America (Heteroptera). Psyche 57: 143–145.
1951. The Tingoidea of New England and their biology. Entomol. Amer. 31: 1–140.
1974. Additional notes on *Stephanitis takeyai* in New England (Heteroptera: Tingidae). Psyche 81: 534–538.
Balarin, I., and A. Polenec. 1984. Spiders, natural enemies of the sycamore lace bug. Zastita Bilja 35: 127–134.
Balsdon, J. A., S. K. Braman, and K. E. Espelie. 1996. Biology and ecology of *Anagrus takeyanus* (Hymenoptera: Mymaridae), an egg parasitoid of the azalea lace bug (Heteroptera: Tingidae). Environ. Entomol. 25: 383–389.
Balsdon, J. A., K. E. Espelie, and S. K. Braman. 1995. Epicuticular lipids from azalea (*Rhododendron* spp.) and their potential role in host plant acceptance by azalea lace bug, *Stephanitis pyrioides* (Heteroptera: Tingidae). Biochem. Syst. Ecol. 23: 477–485.
Balsdon, J. A., S. K. Braman, A. F. Pendley, and K. E. Espelie. 1993. Potential for integration of chemical and natural enemy suppression of azalea lace bug (Heteroptera: Tingidae). J. Environ. Hortic. 11: 153–156.
Barber, H. G., and H. B. Weiss. 1922. The lace bugs of New Jersey. New Jersey Dept. Agric. Circ. No. 54.
Barnes, H. F. 1930. Gall midges (Cecidomyidae) as enemies of the Tingidae, Psyllidae, Aleyrodidae and Coccidae. Bull. Entomol. Res. 21: 319–329.
Beeson, C. F. C. 1941. The Ecology and Control of the Forest Insects of India and the Neighbouring Countries. Vasant Press, Dehra Dun, India. 1007 pp.
Beeson, C. F., and N. C. Chatterjee. 1940. Possibilities of control of lantana (*Lantana aculeata* Linn.) by indigenous insect pests. Indian For. Rec. (1939) 6: 41–84.
Bellotti, A., and A. van Schoonhoven. 1978. Cassava pests and their control. Cassava Information Center, Series 09EC-2. Centro Internacional de Agricultural Tropical.
Bellotti, A. C., L. Smith, and S. L. Lapointe. 1999. Recent advances in cassava pest management. Annu. Rev. Entomol. 44: 343–370.

Beshear, R. J., H. H. Tippins, and J. O. Howell. 1976. The lace bugs (Tingidae) of Georgia. Georgia Agric. Exp. Stn. Bull 188. 29 pp.

Bhalla, J. S., and J. S. Mann. 1988. A new record of tingid bug, *Monosteira edeia* Drake and Livingstone (Tingidae: Heteroptera) on *Ziziphus* sp. at Ludhiana (PB) India. Punjab Hortic. J. 28: 1–2.

Bhandari, K. G., and G. S. Sohi. 1962. Bionomics and control of *Urentius sentis* Dist. (Hemiptera-Hetroptera: Tingidae). Punjab Hortic. J. 2: 44–61.

Bin, F. 1968. La diffusione della *Corythucha ciliata* Say, Tingidae neartico del Platano, nel Nord Italia. Boll. Zool. Agrar. Bachic., Ser., II. 9: 123–131.

Blatchley, W. S. 1926. Heteroptera or True Bugs of Eastern North America, with Especial Reference to the Faunas of Indiana and Florida. Nature Publ. Co., Indianapolis, Indiana, U.S.A. 1116 pp.

Borrero, H. M., and A. C. Bellotti. 1983. Estudio biologico de el chinche de encaje *Vatiga manihotae* (Hemiptera: Tingidae) y de uno de sus enimigos naturales *Zelus nugax* Stal [sic] (Hemiptera: Reduviidae). Yuca: Control Integrado de Plagas. PNUD/CIAT.

Box, H. E. 1953. List of Sugar-Cane Insects. Commonwealth Inst. Entomology, London, U.K. 101 pp.

Box, H. E., and P. Guagliumi. 1954. The insect affecting sugarcane in Venezuela. Proc. Intl. Soc. Sugarcane Tech. Br. West Indies.

Braman, S. K., and A. F. Pendley. 1992. Evidence for resistance of deciduous azaleas to azalea lace bug. J. Environ. Hortic. 10: 40–43.

1993. Temperature, photoperiod, and aggregation effects on development, diapause, reproduction, and survival in *Corythucha cydoniae* (Heteroptera: Tingidae). J. Entomol. Sci. 28: 417–426.

Braman, S. K., A. F. Pendley, B. Sparks, and W. G. Hudson. 1992. Thermal requirements for development, population trends, and parasitism of azalea lace bug (Heteroptera: Tingidae). J. Econ. Entomol. 85: 870–877.

Bremond, P. 1938. Le Faux-tigre des arbres fruitiers (*Monostira unicostata* Muls.) au Maroc. Rev. Pathol. Vég. Entomol. Agric. France 25: 294–307.

Brempong-Yeboah, C. Y., and N. D. Okoampah. 1989. A field experiment on the effects of some insecticides on the pests of garden egg (*Solanum integrifolium*) at Legon. Appl. Entomol. Zool. 24: 343–348.

Bruel van den, W. E. 1947. A propos du comportement en belgique de *Stephanitis rhododendri* Horv. (Tingitidae). Bull. Ann. Soc. Entomol. Belg. 83: 191–197.

Buntin, G. D., S. K. Braman, D. A. Gilbertz, and D. V. Phillips. 1996. Chlorosis, photosynthesis, and transpiration of azalea leaves after azalea lace bug (Heteroptera: Tingidae) feeding injury. J. Econ. Entomol. 89: 990–995.

Butani, D. K. 1985. Spices and pest problems: 10 — turmeric. Indian Agric. Res. Inst. Pestic. 19: 22–25.

Butler, E. A. 1922. A contribution towards the life-history of *Dictyonota strichnocera* Fieb. Entomol. Mon. Mag. 58: 179–182.

1923. A Biology of the British Hemiptera-Heteroptera. H. F. & G. Witherby, London, U.K. 682 pp.

Cappuccino, N., and R. B. Root. 1992. The significance of host patch edges to the colonization and development of *Corythucha marmorata* (Hemiptera: ingidae). Ecol. Entomol. 17: 109–113.

Carayon, J. 1960. *Stethoconus frappai* n. sp., miridé prédateur du tingidé du caféier, *Dulinius unicolor* (Sign.), à Madagascar. J. Agric. Trop. Bot. Appl. 7: 110–120.

Celestino-Filho, P., and F.E.L. Magalhães. 1986. Ocorrencia de fungo *Sporothrix insectorum* (Hoog and Evans) parasitando a mosca de renda *Leptopharsa hevae* (Drake and Poor) en seringal de cultivo. EMBRAPA-C.P.S.D., Pesquisa en andamento, No. 42, Manaus, A.M, Brazil.

Chang, V.C.S. 1986. The sugarcane lacebug: a new insect pest in Hawaii. Hawaiian Sugar Tech. 44th Annual Conf. Rep. A27–29.

Cheng, C. H. 1967. An observation on ecology of *Stephanitis typica* Distant (Hemiptera, Tingidae) on banana. J. Taiwan Agric. Res. 16: 54–69.

Cilliers, C. J. 1983. The weed, *Lantana camara* L., and the insect natural enemies imported for its biological control into South Africa. J. Entomol. Soc. South. Africa 46: 131–138.

1987a. Notes on the biology of the established insect natural enemies of *Lantana camara* L. (Verbenaceae) and their seasonal history in South Africa. J. Entomol. Soc. South. Africa 50: 1–13.

1987b. The evaluation of three insect natural enemies for the biological control of the weed *Lantana camara* L. J. Entomol. Soc. South. Africa 50: 15–34.

Cilliers, C. J., and S. Neser. 1991. Biological control of *Lantana camara* (Verbenaceae) in South Africa. Agric. Ecosyst. Environ. 37: 57–75.

Cobben, R. H. 1978. Evolutionary Trends in Heteroptera. Part II. Mouthpart-Structures and Feeding Strategies. H. Veenman & B.V. Zonen, Wageningen, the Netherlands. 407 pp.

Cock, M. J. W. 1982. Potential biological control agents for *Mikania micrantha* HBK from the Neotropical Region. Trop. Pest Manage. 28: 242–254.

Coffelt, M. A., and P. B. Schultz. 1988. Influence of plant growth regulators on the development of the azalea lace bug (Hemiptera: Tingidae). J. Econ. Entomol. 81: 290–292.

Comstock, J. H. 1879. The hawthorn tingis (*Corythucha arcuata*, Say, var.). U.S. Dept. Agric. Rep. Entomol. 1878–1883. 1: 221–222.

Cosenza, G. W. 1982. Cassava varieties resistant to lace bugs. Ann. Tech. Rept. EMBRAPA Agric. Res. Center. Cerrados 1980–1981, No. 6.

Cotton, R. T. 1917. The eggplant lace-bug in Porto Rico. J. Dept. Agric. Porto Rico. 1: 170–173.

Couturier, G. 1988. Alguns insetos do cubiu (*Solanum sessiliflorum* Dunal var. *sessiliflorum* Dunal, Solanaceae) na região de manaus-am. Acta Amazonica 18: 93–103.

Cowley, W. R., and K. A. Sund. 1973. Cane variety program for Lower Rio Grande Valley. Sugar J. 35: 25–27.

Crosby, C. R., and C. H. Hadley. 1915. The rhododendron lace-bug, *Leptobyrsa explanata* Heidemann. J. Econ. Entomol. 8: 409–414.

Cullen, J., and D. F. Chamberlain. 1978. A preliminary synopsis of the genus Rhododendron. Notes R. Bot. Gard. Edinburgh 36: 105–127.

Curri, G. A., and R. V. Fyfe. 1939. The lantana bug in Australia. J. Counc. Sci. Ind. Res. 12: 259–263.

David, B. V., and A. V. Rangarajan. 1966. Insects affecting *Baleria* spp. in Coimbatore. Indian J. Hortic. 23: 191–195.

David, S. K. 1958. Insects affecting jasmine in the Madras State. Madras Agric. J. 45: 146–150.

De Battisti, R. 1983. The plane and one of its injurious pests: *Corythucha ciliata* Say. L'Albero, l'Uomo, la Citta [edited by Giulini, P.], pp. 49–51.

Decazy, B. 1975. Contribution to the biological and ecological study of the "coffee tree tiger" *Dulinius unicolor* Sign. in Madagascar. Café Cacao Thé 19: 19–34.

Decelle, J. 1955. Quels sont les insecticides à utiliser contre deux ennemis due caféier d'Arabie, *Antestiopsis lineaticollis* et *Harochila ghesquierei*? Bull. Inst. Natl. Étude Agron. Congo Belg. 4: 67–75.

Denton, G. R. W., R. Muniappan, and M. Marutani. 1991. The distribution and biological control of *Lantana camara* in Micronesia. Micronesica (Suppl.) 3: 71–81.

Dickerson, E. L. 1917. Notes on *Leptobyrsa rhododendri* Horv. J. New York Entomol. Soc. 25: 105–112.

Dickerson, E. L., and H. B. Weiss. 1902. Contributions to a knowledge of the Rhynchota. Ann. Soc. Entomol. Belg. 47: 43–65.

1916. Notes on *Leptoypha mutica* Say (Hemip.). Entomol. News 27: 308–310.

1917. The azalea lace bug, *Stephanitis pyrioides* Scott (Tingitidae, Hemiptera). Entomol. News 28: 101–105.

Distant, W. L. 1902. The Fauna of British India Including Ceylon and Burna, Rhynchota, Vol. II. Taylor and Francis, London, U.K. 503 pp.

1903. Contributions to a knowledge of the Rhynchota. Ann. Soc. Entomol. Belg. 47: 43–65.

1910. An introduced pest to rhododendrons. Zoology 14: 395–396.

Dolling, W. R. 1991. The Hemiptera. Oxford. Natural History Museum, Oxford University Press, London, U.K. 274 pp.

Dorge, S. K. 1971. A note on *Corythauma ayyari* Drake (Tingidae) as a pest of *Jasminum* sp. in Maharashtra State. Sci. Cult. 37: 156–157.

Drake, C. J. 1918. The North American species of *Teleonemia* occurring north of Mexico. Ohio J. Sci. 18: 323–332.

1925. Concerning some Tingitidae from the Gulf states (Heteroptera). Florida Entomol. 9: 36–39.

1930. A new sugar-cane tingitid from Java and Sumatra (Hemiptera) Pan-Pac. Entomol. 7: 15–16.

1954. A miscellany of new Tingidae (Hemiptera). Proc. Biol. Soc. Washington 67: 1–15.

1966. A new species of lacebug from China (Hemiptera: Tingidae). Proc. Biol. Soc. Washington 79: 135–138.

Drake, C. J., and N. T. Davis. 1960. The morphology, phylogeny, and higher classification of the family Tingidae, including the description of a new genus and species of the subfamily Vianaidinae (Hemiptera: Heteroptera). Entomol. Amer. 39: 1–100.

Drake, C. J., and D. M. Frick. 1939. Synonymy and distribution of the lantana lace bug (Hemiptera: Tingidae). Proc. Hawaiian Entomol. Soc. 10: 199–202.

Drake, C. J., and E. J. Hambleton. 1944. Concerning Neotropical Tingitidae (Hemiptera). J. Washington Acad. Sci. 34: 120–129.
Drake, C. J., and D. Livingstone. 1964. Two new species of lacebugs from India (Hemiptera: Tingidae). Great Basin Nat. 24: 27–30.
Drake, C. J., and T. Maa. 1953. Chinese and other Oriental Tingoidea (Hemiptera). Q. J. Taiwan Mus. 6: 87–101.
1955. Chinese and other Oriental Tingoidea (Hemiptera). III. Q. J. Taiwan Mus. 8: 1–11.
Drake, C. J., and J. R. Mamet. 1956. Mauritian Tingidae (Hemiptera). Mauritius Inst. Bull. 3: 300–302.
Drake, C. J., and F. A. Ruhoff. 1960a. Tingidae: new genera, species, homonyms, and synonyms (Hemiptera). Great Basin Nat. 20: 29–38.
1960b. A necessary correction. Great Basin Nat. 20: 80.
1965. Lacebugs of the World, a Catalog (Hemiptera: Tingidae). U.S. Nat. Mus. Bull. 243., Smithsonian Institution, Washington, D.C., U.S.A. 634 pp.
Dreyer, M., C. 1991. Neem: a promising natural insecticide for small scale vegetable producers in the Dominican Republic. Rencontres-Caraibes-en-lutte-biologique. 491–500.
Dunbar, D. M. 1974. Bionomics of the andromeda lace-bug, *Stephanitis takeyai*. Mem. Connecticut Entomol. Soc. 277–289.
Easwaramoorthy, S., G. Santhalakshmi, and H. David. 1982. Record of a new tingid on sugarcane. Entomon 7: 121–122.
Eguagie, W. E. 1972a. Effects of temperature and humidity on the development and hatching of eggs of the thistle lace bug, *Tingis ampliata* (Heteroptera: Tingidae). Entomol. Exp. Appl. 15: 183–189.
1972b. Overwintering of the lacebug *Tingis ampliata* (Heteroptera) in Britain. Oikos 23: 63–68.
1973. Mating and oviposition of *Tingis ampliata* (Heteroptera: Tingidae). Entomol. Scal. 4: 283–292.
1974a. Cold hardiness of *Tingis ampliata* (Heteroptera). Entomol. Exp. Appl. 17: 204–214.
1974b. An analysis of movement of adult *Tingis ampliata* (Heteroptera) in a natural habitat. J. Anim. Ecol. 43: 521–535.
Eisbein, K. 1976. Investigations for electron microscopic demonstration of *Beta* virus 3 in *Beta vulgaris* L. and in *Piesma quadratum* Fieb. Arch. Phytopathol. Pflanzenschutz 12: 299–313.
Essig, E. O. 1958. Insects and Mites of Western North America. Macmillan, New York, New York, U.S.A. 1050 pp.
Fabricius, I. C. 1803. Systema Rhyngotorum Secundum Ordines, Genera, Species, Adiectis Synonymis, Locis, Observationibus, Descriptionibus. Carolum Reichard, Brunsvigae, Denmark. 314 pp.
Felt, E. P. 1904. Injurious insects, chrysanthemum lace bug *Corythuca marmorata* Uhler. New York State Mus. Bull. 76. Entomology 21 (1903).
Fenton, F. A. 1934. Tingitoidea affecting cotton. Canad. Entomol. 66: 198–199.
Fink, D. E. 1915. The eggplant lace bug. Bull. U.S.D.A. 239: 1–7.
Fitch, A. 1861. The quince tingis. Coun. Gent. 25: 114.
Foucart, G. 1954. Un nouvel ennemi du caféier d'Arabie au Kivu *Habrochila placida* (note preliminaire). Bull. Inst. Natl. Étude Agron. Congo Belg. 3: 51–62.
Francki, R. I. B., E. W. Kitajima, and D. Peters. 1981. Rhabdoviruses. Pp. 455–489 *in* E. Kurstak [ed.], Handbook of Plant Virus Infections and Comparative Diagnosis. Elsevier/North-Holland Biomedical Press, Amsterdam, the Netherlands.
Frappa, A. C. 1931. Notes sur deux nouveaux hémiptères nuisibles au caféier, à Madagascar. Rev. Pathol. Veg. Entomol. Agric. France 18: 212–214.
Froeschner, R. C. 1944. Contributions to a synopsis of the Hemiptera of Missouri, Pt. III. Am. Midland Nat. 31: 638–683.
1970. *Teleonemia harleyi*, a new species of lantana-feeding lace bug from Trinidad, W. I. Proc. Entomol. Soc. Washington 72: 470–472.
1972. A new species of *Gargaphia* lace bug from beans in Colombia. Proc. Entomol. Soc. Washington 74: 59–60.
1976. Description of a new species of lace bug attacking the oil palm in Colombia (Hemiptera: Tingidae). Proc. Entomol. Soc. Washington 78: 104–104.
1988. Family Tingidae Laporte, 1833. Pp. 708–733 *in* T. J. Henry and R. C. Froeschner [eds.], Heteroptera of Canada and the Continental United States. E. J. Brill, New York, New York, U.S.A.

1993. The neotropical lace bugs of the genus *Vatiga* (Heteroptera: Tingidae), pests of Cassava. Proc. Entomol. Soc. Washington 93: 457–462.

Fyfe, R. V. 1935. The lantana bug in Fiji. Agric. J. (Dept. Agric. Fiji). 8: 35–36.

1937. The lantana bug, *Teleonemia lantanae* Distant. J. Counc. Sci. Indian Res. 10: 181–186.

Gardner, J. C. M. 1944. A note on the imported Lantana bug (*Teleonemia scrupulosa* Stal.). Indian For. 70: 139–140.

Genty, P., M. A. Garzon, and R. Garcia. 1983. Damage and control of *Leptopharas pestalotiopsis* complex in oil palm. Oléagineux 38: 291–199.

Genty, P., J. G. Lopez, and D. Mariau. 1975. Pathogenic fungi damage induced by attacks of *Gargaphis* in Columbia. Oléagineux 30: 199–204.

Ghauri, M. S. K. 1965. New Tingidae from Oriental region. Ann. Mag. Nat. Hist. 13(8): 357–366.

Gil S., M. C., and J. P. Mansilla. 1981. Description de una nueva plaga del *Platanus* spp. en España. Commun. INIA, Ser. Prot. Veg. (Madris, Spain); No. 15: 11 pp.

Goeden, R. D. 1971. Insect ecology of silverleaf nightshade. Weed Sci. 19: 45–51.

Golub, V. B. 1988. Tingidae. *In* P.A. Lehr [ed.], Keys to the Insects of the Far East of the USSR, Vol. 2. Kauka, Leningrad [in Russian].

Gomez-Menor Ortega, J. 1950. La "Chincheta" del Almendro (*Mostire unicostata* Mulsant). Bol. Patol. Veg. Entomol. Agríc. 17: 97–109.

1954. Tingidos que viven sobre el peral. Bol. Patol. Veg. Entomol. Agríc. 20: 369–392.

1955. Hemípteros que atacan a los árboles y arbustos frutales. Bol. Patol. Veg. Entomol. Agríc. 21: 209–282.

Goncharenko, E. G., and V. N. Fursov. 1988. *Parallelaptera panis* (Hymenoptera: Mymaridae) — a parasitoid of the pear bug in Moldavia. Vestn. Zool. 6: 59–61.

Gordh, G., and D. M. Dunbar. 1977. A new *Anagrus* important in the biological control of *Stephanitis takeyai* and a key to the North American species. Florida Entomol. 60: 85–95.

Gregorio, De. R. 1981. *Stephanitis pyri*, cycle biologique en forêt de Bouconne. Bull. Soc. Entomol. France 86: 227–235.

Gumbek, M. 1986. Sampling methods for assessing pepper flower spike damage caused by tingid bug, *Diconocoris hewitti* Dist. Pp. 73–74 *in* Ann. Rept. Res. Branch Dept. Agric. 198.

Gupta, M., and A. D. Pawar. 1984. Role of *Teleonemia scrupulosa* Stal [sic] in controlling *Lantana*. Indian J. Weed Sci. 16: 221–226.

Gupta, P. R., and J. R. Thakur. 1991. New lace bug, *Eteoneus* sp.? *sigilatus* Drake and Poor, pest of olive in India. FAO Plant Prot. Bull. 39: 184–186.

Habeck, D. H., and F. W. Mead. 1975. Lantana lace bug, *Teleonemia scrupulosa* Stål. Florida Dept. Agric. Consumer Serv. Entomol. Circ. No. 156.

Hall, D. G. 1991. Sugarcane lace bug *Leptodictya tabida*, an insect pest new to Florida. Florida Entomol. 74: 148–149.

Hall, D. G., and O. Sosa, Jr. 1994. Populations levels of *Leptodictya tabida* (Hemiptera: Tingidae) in Florida sugarcane. Florida Entomol. 77: 91–99.

Hall, J. P., D. M. Stone, D. S. O'Brien, K. E. Pardy, W. J. Sutton, and G. C. Curry. 1998. Forest insect and disease conditions in Canada 1995. Forest Insect and Disease Survey. Natural Resources Canada, Canadian Forest Service, Ottawa, Canada.

Hargreaves, H. 1940. Coffee-pests. Pp. 348–351 *in* J. D. Tothill [ed.], Agriculture in Uganda. Oxford University Press., London, U.K.

Harley, K. L. S. 1971. Biological control of *Lantana*. PANS 17: 433–437.

Harley, K. L. S., and R. C. Kassulke. 1971. Tingidae for biological control of *Lantana camara* (Verbenaceae). Entomophaga 16: 389–410.

1973. The suitability of *Teleonemia harleyi* for biological control of *Lantana camara* in Australia. Entomophaga 18: 343–347.

1975. The suitability of *Teleonemia prolixa* (Stål) for biological control of *Lantana camara* in Australia. J. Austral. Entomol. Soc. 14: 225–228.

Harris, H. M. 1942. On the date of publication of Laporte's Essai. Pan-Pac. Entomol. 18: 161–162.

Harris, W. V. 1932. Report assistant entomologist. Rep. Dept. Agric. Tanganyika, 1931, pp. 87–93.

Harsh, N. S. K., Jamaluddin, and C. K. Tiwari. 1992. Top dying and mortality in provenance trial plantations of *Gemelina arborea*. J. Trop. For. 8: 55–61.

Hartley, W. W. S. 1977. The Oil Palm, second edition. Longman, London, U.K. 806 pp.

Heidemann, O. 1908. Two new species of North American Tingidae. Proc. Entomol. Soc. Washington 10: 103–109.

1911. Some remarks on the eggs of North American species of Hemiptera-Heteroptera. Proc. Entomol. Soc. Washington 10: 103–108.

1913. The sugar-cane tingid from Mexico. J. Econ. Entomol. 6: 249–251.

1914. A new species of North American Tingitidae. Proc. Entomol. Soc. Washington 16: 136–137.

Heiss, E. 1978. On the Heteroptera fauna of Nordtirol (Insecta: Heteroptera) part VII: Tingidae. Ber. Nat. Med. Ver. Innsbruck 65: 73–84.

1995. Die amerikanische Platanennetzwanze *Corythucha ciliata* — eine Adventivart im Vormarsch auf Europa (Heteroptera, Tingidae). Cat. Landesmus. (n.F.) 84: 143–148.

Hellman, J. L., J. A. Davidson, and J. Holmes. 1982. Urban ornamental and turfgrass integrated pest management in Maryland. Pp. 31–38 *in* Niemczyk, H. E., and B. G. Joyner [eds.], Advances in Turfgrass Entomology. Hammer Graphics, Inc., Picqua, Ohio, U.S.A.

Hemming, F. 1943. Opinions Rendered by the International Commission of Zoological Nomenclature. Vol. 2, Section A. Opinion 143, pp. 81–88.

Henry, T. J., J. W. Neal, Jr., and K. M. Gott. 1986. *Stethoconus japonicus* (Heteroptera: Miridae): a predator of *Stephanitis* lace bugs newly discovered in the United States, promising in the biocontrol of azalea lace-bug (Heteroptera: Tingidae). Proc. Entomol. Soc. Washington 88: 722–730.

Heu, R. A., and G. Y. Funasaki. 1985. New state records. Hawaii Pest Rep. 5: 1–4.

Hilje, L., L. Quiros, and F. Scorza Reggio. 1991. The present status of forest pests in Costa Rica. Manejo Integrado de Plagas No. 20–21, pp. 18–21.

Hill, D. S. 1975. Agricultural Insect Pests of the Tropics and Their Control. Cambridge University Press, Cambridge, U.K. 516 pp.

1983. Agricultural Insect Pests of the Tropics and Their Control, second edition. Cambridge University Press, Cambridge, U.K. 746 pp.

Hill, R. L. 1983. Prospects for the biological control of gorse. 1983. Proc. 36th New Zealand Weed and Pest Control Conf. 36: 56–58.

1986. Biological control of gorse: implications for the natural environment and for primary production. Entomology Division Rep. 6, DSIR, Private Bag, Christchurch, New Zealand. 130 pp.

Hixson, E. 1942. A new pest of snapdragon and verbena. J. Econ. Entomol. 35: 605–606.

Hoffman, R. L. 1953. A second case of lacebug bite (Hemiptera: Tingidae). Entomol. News 64: 176.

Hoffman, W. E. 1935. Observations on a hesperid leaf-roller and a lacebug. Two pests of banana in Kwangtung. Lingnan Sci. J. 14: 639–649.

Holm, L. 1969. Weeds: problems in developing countries. Weed Sci. 17: 113–118.

Holmes, J. J., and J. A. Davidson. 1984. Integrated pest management for arborists: implementation of pilot program. J. Arboricult. 10: 65–70.

Horn, K. F., M. H. Farrier, and C. G. Wright. 1979. Identification of the fifth nymphal stage of ten species of *Corythucha*. J. Georgia Entomol. Soc. 14: 131–136.

1980. Bionomics of *Corythucha associata* Osborne & Drake. J. Georgia Entomol. Soc. 15: 318–327.

Horn, K. F., M. H. Farrier, C. G. Wright, and L. A. Nelson. 1983a. A sampling method for estimating egg and first-instar densities of the sycamore lace bug, *Corythucha ciliata* (Say). J. Georgia Entomol. Soc. 18: 37–49.

Horn, K. F., C. G. Wright, and M. H. Farrier. 1983b. Some mortality factors affecting eggs of the sycamore lace bug, *Corythucha ciliata* (Say) (Hemiptera: Tingidae). Ann. Entomol. Soc. Amer. 76: 262–265.

Horn, K. F., M. H. Farrier, and C. G. Wright. 1983c. Estimating egg and first-instar mortalities of the sycamore lace bug, *Corythucha ciliata* (Say). J. Georgia Entomol. Soc. 18: 27–37.

Horváth, G. 1905. Tingitidae novae nel minus cognitae e regione palaearctica. Ann. Mus. Nat. Hungary 3: 556–572.

1912. Species generis Tingitidarum *Stephanitis*. Ann. Mus. Nat. Hungary 10: 319–339.

Howes, F. N. 1974. A Dictionary of Useful and Everyday Plants and Their Common Names. Cambridge University Press, Cambridge, U.K. 290 pp.

Ishihara, R., and S. Kawai. 1981. Feeding habits of the azalea lace bug *Stephanitis pyrioides* (Hemiptera: Tingidae). Jpn. J. Appl. Entomol. Zool. 25: 200–202.

Iyengar, M. O. T. 1924. The life-history of a tingid bug, *Monanthia globulifera*. Pp. 296–299 *in* [T. B. Fletcher, ed.], Report Proc. Fifth Entomol. Meeting, Pusa, India, 1923.

Jamal, A., A. Ahmad, A. Inam, and N. Adhami. 1979. Changes in carbohydrate, protein and nucleic acid levels of *Solanum melongena* Linn. eggplant leaves under pathogenic effect of *Urentius sentis* D. Indian J. Entomol. 41: 185–187.

Johnson, C. G. 1936. The biology of *Leptobyrsa rhododendri* Horváth (Hemiptera: Tingitidae), the rhododendron lace bug. I. Introduction, bionomics and life history. Ann. Appl. Biol. 23: 342–368.

1937. The biology of *Leptobyrsa rhododendri* Horváth (Hemiptera, Tingitidae), the rhododendron lacebug. II. Feeding habits and the histology of the feeding lesions produced in *Rhododendron* leaves. Ann. Appl. Biol. 24: 342–355.

Johnson, W. T., and H. H. Lyon. 1988. Insects that Feed on Trees and Shrubs, second edition. Comstock Publ. Assoc., Cornell University Press, Ithaca, New York., U.S.A. 556 pp.

Jones, R. A. 1993. The rhododendron lacebug, *Stephanitis rhododendri* Horváth, rediscovered in south-east London. British J. Entomol. Nat. Hist. 6: 139–140.

Joseph, T., and P. Shanta. 1968. Oil palm, *Elaeis guineensis* Jacq. a new host for *Stephanitis typicus* Dist. Curr. Sci. 37: 619.

Joseph, T., P. Shanta, and S. B. Lal. 1972. Role of *Stephanitis typicus* Distant in the spread of coconut root (wilt) pathogen. Indian J. Agric. Sci. 42: 414–417.

Judd, S., and I. D. Rotherham. 1992. The phytophagous insect fauna of *Rhododendron ponticum* L. in Britain. Entomologist 111: 134–150.

Junqueira, N., and F. E. L. Magalhaes. 1987. Isolamento e cultivo do fundo *Sporothrix insectorum* (Hoog and Evans) a ser utilizado para o controle de mosca de renda da seringueira. Comunicado técnico, EMBRAPA-C.N.P.S.D., No. 56, Manaus, A.M., Brazil. 129 pp.

Junqueira, N., M. V. B. Garcia, P. Celestino Filho, and L. A. C. Moraes. 1988. Controle biologico da mosca-de-renda (*Leptopharsa heveae*) em seringais de cultivo no estado do Amazonas. *In* Memorias Primer Simposio Nacional de Controle Biologico de Pragas e Vetores. Siconbiol, Rio de Janeiro, R.J. Brazil November 1988. 159 pp.

Jurenka, R. A., J. W. Neal, Jr., R. W. Howard, J. E. Oliver, and G. J. Blomquist. 1989. *In vitro* inhibition of prostaglandin H synthase by compounds from the exocrine secretion of lace bugs. Comp. Biochem. Physiol. 93C: 253–255.

Karmawati, E. 1989. Nymphal spatial pattern and sequential sampling plant for lace bug of pepper blossom in Bangka. J. Indian Crops Res. 1: 11–16.

Kawakami, Y. 1983. Sexual dimorphism in the nymphal stages of some Japanese Tingidae. Jpn. J. Appl. Entomol. Zool. 27: 197–202.

Kearns, R. S., and R. T. Yamamoto. 1981. Maternal behavior and alarm response in the eggplant lace bug, *Gargaphia solani* Heidemann (Tingidae: Heteroptera). Psyche 88: 215–230.

Kerzhner, I. M. 1970. New and little known mirid bugs (Heteroptera, Miridae) from the USSR and Mongolia. Entomol. Rev. 49: 392–399.

1913. Genera Insectorum, Diptera Fam. Cecidomyiidae. Endopsylla 221–222.

Khaidarova, Z. M. 1975. Phytophagous insects of the weed *Trichodesma incanum* (BGE.)D. C. in Uzbekistan. Entomol. Obozr. 54: 780–786.

Khan, A. H. 1945. On the lantana bug (*Teleonemia scrupulosa* Stål). Indian J. Entomol. 6: 149–161.

Kieffer, J. J. 1907. Eine neue endoparasitische Cecidomyidae. Z. Syst. Hymenopterol. Dipterol., pp. 129–130.

Knowlton, G. F. 1958. Tingidae are biters. Bull. Brooklyn Entomol. Soc. 53: 73.

Laporte, F. L. N. de C. 1833. Essai d'une classification systématique de l'ordre des Hémiptères (Hémiptères Hétéroptères, Latr.). Mag. Zool. 2: 1–88.

Lara, F. M., and M. R. Tanzini. 1997. Nonpreference of the lace bug *Leptopharsa heveae* Drake & Poor (Heteroptera: Tingidae) for rubber tree colones. An. Soc. Entomol. Brasil 26: 429–434.

Lee, C. E. 1969. Morphological and phylogenic studies on the larvae and male genitalia of the East Asiatic Tingidae (Heteroptera). J. Fac. Agric. Kyushu Univ. 15: 137–158, 203–204, 220–224.

Leonard, M. D., and A. S. Mills. 1931. Observations on the bean lace bug in Porto [sic] Rico. J. Dept. Agric. P. R. 15: 309–323.

Le Pelley, R. H. 1969. Pests of Coffee. Harlow, Longmans, London, U.K. 590 pp.

Li, Y. G., S. L. Chen, L. Li, X. Z. Liao, and L. Z. Yang. 1990. Bionomics and population dynamics of *Stephanitis laudata*. Insect Knowledge 27: 104–107.

Liang, C., and L. Zhao. 1987. Bionomics and control of *Hegesidemus harbus* Drake. Sci. Silve Sin. 23: 376–382.

Livingstone, D. 1959. On the bionomics and immature stages of *Urentius euonymus* Dist. (Heteroptera: Tingidae), a sap sucker on hollyhocks and other garden plants. Proc. First All-India Congr. Zool. Part 2: 510–519.

1962. On the biology and immature stages of *Cadmilos retiarius* Dist., a sap sucker on Compositae (Heteroptera: Tingidae). Agra Univ. J. Res. 11: 47–61.

1967. On the functional anatomy of the egg of *Tingis buddleiae* Drake (Heteroptera: Tingidae). J. Zool. Soc. India 19: 111–119.

1968. On the morphology and biology of *Tingis buddleiae* Drake (Heteroptera: Tingidae). Part I Bionomics. Agra Univ. J. Res. 17: 1–16.

1977. Host-specificity in Tingidae (Heteroptera) in relation to plants, parasites and predators. Pp. 23–28 *in* T. N. Ananthakrishnan [ed.], Insects and Host-Specificity. Macmillian, India.

1978. On the body outgrowths and the phenomenon of "sweating" in the nymphal instars of Tingidae (Hemiptera: Heteroptera). J. Nat. Hist. 12: 377–384.

Livingstone, D., and M. H. S. Yacoob. 1986. Natural enemies and biologies of the egg parasitoids of Tingidae of Southern India. Uttar Pradesh J. Zool. 6: 1–21.

1987a. Biosystematics of Tingidae on the basis of the biology and micromorphology of their eggs. Proc. Indian Acad. Sci. (Anim. Sci.) 96: 587–611.

1987b. A new species of *Lathromeromyia* of the subgenus *Lathromeromina* (Hymenoptera: Trichogrammatidae) from the eggs of *Corythauma ayyari* (Heteroptera: Tingidae). J. Bombay Nat. Hist. Soc. 84: 395–398.

1987c. A new species of *Parallelaptera* (Hymenoptera: Mymaridae) an egg parasitoid of Tingidae from southern India. J. Bombay Nat. Hist. Soc. 84: 628–631.

Livingstone, D., S. J. Bai, and M. H. S. Yacoob. 1980. The imported weed *Latana aculeata* L. (Verbenaceae) and the natural history of its imported sap sucker *Teleonemia scrupulosa* Stal (Heteroptera: Tingidae) in India. J. Indian Acad. Wood Sci. 11: 28–34.

1981. Population dynamics of the lantana bug *Teleonemia scrupulosa* Stål (Tingidae). Uttar Pradesh J. Zool. 1: 87–90.

López M., A. J., B. Villa M., and A. Madrigal C. 1982. Ciclo de vida de la chinche de encaje *Corythucha gossypii* (F.) (Hemiptera: Tingidae) en girasol (*Helianthus annuus* L.). Rev. Colomb. Entomol. 8: 19–27.

Lusby, W. R., J. E. Oliver, J. W. Neal, Jr., and R. R. Heath. 1987. Isolation and identification of the major component of setal exudate from *Corythucha ciliata*. J. Nat. Prod. 50: 1126–1130.

1989. Acylcyclohexanediones from setal exudate of hawthorn lace bug nymph *Corythucha cydoniae* (Hemiptera: Tingidae). J. Chem. Ecol. 15: 2369–2378.

Maa, T. 1957. Nymphal stages of certain Oriental Tingidae (Hemiptera) Q. J. Taiwan Mus. 10: 117–133.

Maceljski, M. 1986. Current status of *Corythucha ciliata* in Europe. Bull. OEPP/EPP 16: 621–624

Maceljski, M., and I. Balarin. 1974. Untersuchungen über einen neuen amerikanischen Schädling in Europa, die Platenen-Netzwanze, *Corythuca ciliata* (Say). Anz. Schädlingsk. Pflanz. Umweltschutz 47: 165–170.

1977. Beitrag zur Kenntnis natürlicher Feinde der Platanen-Netzwanze (*Corythucha ciliata* (Say), Tingidae, Heteroptera). Anz. Schädlingsk. Pflanz. Umweltschutz 50: 135–138.

Maniglia, G. 1983. Biological observations on *Monosteira unicostata* Muls. et Rey (Rhynchota-Tingidae) in Sicily. Phytophaga 1: 27–40.

Manuel, K., and M. Wiliams. 1980. Lace bug injury to ornamental trees and shrubs in Alabama. High. Agric. Res. Alabama Agric. Exp. Stn. 27: 10.

Mason, J. R., J. W. Neal, Jr., J. E. Oliver, and W. R. Lusby. 1991. Bird-repellent properties of secretions from nymphs of the azalea lace bug. Ecol. Appl. 1: 226–230.

Mathen, K. 1960. Observations on *Stephanitis typicus* Distant, a pest of coconut palm. 1. Description and life history. Indian Coconut J. 14: 8–27.

1973. The lace-winged enemy of coconut. Coconut Bull. 4: 2–4.

1982. Comparative abundance of the tingid vector *Stephanitis typica* (Distant) in root (wilt) disease affected and healthy coconut palms in Kerala. Proc. All India Symp. (Trivandrum, Kerala State, India): 29–33.

1985. Lace bug abundance in root (wilt) disease affected coconut palms — cause or effect? J. Plant. Crops 13: 56–59.

Mathen, K., and C. Kurian. 1972. Description, life-history and habits of *Stethoconus praefectus* (Distant) (Heteroptera: Miridae), predacious on *Stephanitis typicus* Distant (Heteroptera: Tingidae), a pest of coconut palm. Indian J. Agric. Sci. 42: 255–262.

Mathen, K., M. P. Govindankutty, and C. Mathew. 1983. Nature and extent of damage to coconut as a result of feeding by *Stephanitis typicus* (Distant) (Heteroptera: Tingidae). Placrosym II 1979: 466–473.

Mathen, K., J. J. Solomon, P. Rajan, and L. Geetha. 1987. Electron microscopic evidence on the role of *Stephanitis typica* (Distant) as vector of coconut root (wilt) disease. Curr. Sci. 56: 1239–1241.

Mathen, K., C. P. Radhakrishnan Nair, M. Gunasekharan, M. P. Govindankutty, and J. J. Solomon. 1988. Stylet course of lace bug *Stephanitis typica* (Distant) in coconut leaf. Proc. Indian Acad. Sci. 97: 539–544.

Mathen, K., P. Rajan, C. P. R. Nair, M. Sasikala, M. Gunasekharan, M. P. Govindankutty, and J. J. Solomon. 1990. Transmission of root (wilt) disease to coconut seedlings through *Stephanitis typica* (Distant) (Heteroptera: Tingidae). Trop. Agric. 67: 69–73.

Mathew, G. 1986. Insects associated with forest plantations of *Gmelina arborea* Roxb. in Kerala, India. Indian J. For. 9: 308–311.

Mathur, R. N. 1955. Immature stages of *Tingis beesoni* Drake (Heteroptera: Tingitidae). Entomologist 88: 248–251.

1979. Biology of *Tingis (Caenotingis) beesoni* Drake (Heteroptera: Tingitidae). Indian For. Bull. No. 276.

Matsumura, S. 1905. Thousand Insects of Japan. 2: 1–213.

Mayné, R., and J. Ghesquière. 1934. Hémiptères nuisibles aux végétaux du Congo belge. Ann. Gembloux 40: 3–41.

McAtee, W. L. 1923. Tingitoidea of the vicinity of Washington, D.C. (Heteroptera). Proc. Entomol. Soc. Washington 25: 143–151.

Mead, F. W. 1967. *Stephanitis* lace bugs of the United States. Florida Dept. Agric. Div. Plant Ind. Entomol. Circ. 62: 1–2.

1972. The hawthorn lace bug, *Corythucha cydoniae* (Fitch), in Florida Hemiptera: Tingidae. Florida Dept. Agric. Consumer Ser., Entomol. Circ. No. 127.

1975. The fringetree lace bug, *Leptopha mutica* (Say) (Hemiptera: Tingidae). Florida Dept. Agric. Consumer Serv. Entomol. Circ. No. 161.

1989. Cotton lace bug, *Corythucha gossypii*, in Florida (Hemiptera: Tingidae). Florida Dept. Agric. Consumer Serv. Entomol. Circ. No. 324.

Mead, F. W., and J. E. Peña. 1991. Avocado lace bug, *Pseudacysta perseae* Florida Dept. Agric. Consumer Serv. Entomol. Circ. No. 346.

Meagher, R. L., Jr., S. W. Wilson, R. S. Pfannenstiel, and G. Breene. 1991. Documentation of two potential insect pests of south Texas sugarcane. Southwest. Entomol. 16: 365–366.

Medal, J. C., R. Charudattan, J. J. Mullahey, and R. A. Pitelli. 1996. An exploratory insect survey of tropical soda apple in Brazil and Paraguay. Florida Entomol. 79: 70–73.

Medina-Gaud, S., A. E. Segarra-Carmona, and R. A. Franqui. 1991. The avocado lacewing bug, *Pseudacysta perseae* (Heidemann) (Hemiptera: Tingidae) in Puerto Rico. J. Agric. Univ. Puerto Rico 75: 185–188.

Melksham, J. A. 1984. Colonial oviposition and maternal care in two strains of *Leptobyrsa decora* Drake (Hemiptera: Tingidae) J. Austral. Entomol. Soc. 23: 205–210.

Melo, Q. M. S. 1990. Arthropod pests associated with cassava in Brazil. Pp. 132–138 *in* S. K. Hahn and F. E. Caveness [eds.], Integrated Pest Management for Tropical Root and Tuber Crops. International Institute of Tropical Agriculture, Ibadan, Nigeria.

Mishra, S. C., and P. K. Sen Sarma. 1986. Host specificity test and a note on life history of *Leptobyrsa decora* Drake (Hemiptera: Tingidae) on teak. Bull. Entomol. (New Delhi) 27: 81–86.

Misra, R. M. 1985. A note of *Leptobyrsa decora* Drake (Hemiptera: Tingitidae). A bio-control agent of *Lantana camara* (Verbenaceae). Indian For. 111: 641–644.

Mohammad, M. A., and N. M. Al-Mallah. 1989a. Effect of host plant on the lace bug *Stephanitis pyri* (F.) (Heteroptera: Tingidae). Mesopotamia J. Agric. 21: 271–281.

1989b. The influence of temperatures and host food on increase rate of the lace bug, *Stephanitis pyri* (F.) (Hemiptera, Tingidae). Mesopotamia J. Agric. 21: 28.

Mohanasundaram, M. 1987. Varietal susceptibility of bananas to the lace wing bug, *Stephanitis typicus* Distant (Tirgidae [sic]: Hemiptera). J. Madras Agric. 4: 465–468.

Mohanasundaram, M., and M. Basheer. 1963. Population studies of *Habrochila laeta* Drake (Tingidae-Hemiptera). Madras Agric. J. 50: 104.

Mohanasundaram, M., and C. Boonyonk. 1973. Preliminary list of the tingid fauna of Thailand. J. Madras Agric. 60: 644–645.

Mohanasundaram, M., and P. V. S. Rao. 1973. A note on *Cochlochila bullita* Horváth (Tingidae: Heteroptera) as a part of *Coleus parviflorus*, a tuber crop in Tamilnadu. Indian J. Entomol. 35: 346.

Moleas, T. 1985. Ecology and behavior of *Monosteira unicostata* Muls. and Rey in Apulia. Italian Nat. Congr. Entomol.: 437–444.

1987. Ethology, ecology and control of *Monosteira unicostata* Muls. et Rey on almond in Apulia. Difesa Delle Piante 10: 469–484.

Monod, T., and J. Carayon. 1958. Observations sur les *Copium* (Hemipt. Tingidae) et leur action cécidogène sur les fleurs de *Teucrium* (Labiées). Arch. Zool. Exp. Gen. 95: 1–31.

Monte, O. 1943. Notas sobre um percevejo, praga de várias solanáceas cultivadas. O Biologico 9: 113–120.

1945. Cultura do tomateiro especialmente as pragas e doenças e seu tratamento. Bibl. Agric. Pop. Bras., pp. 1–88.

Moraes, G. J. de, and F. S. Ramalho. 1980. Alguns insetos associados a *Vigna unguiculata* Walp. no Nordeste. Petrolina, EMBRAPA-CPATSA (EMBRAPA-CPATSA. Bol. pesq., 1). 10 pp.

Morrill, A. W. 1903. Notes on the immature stages of some tingitids of the genus *Corythuca*. Psyche 10: 127–134.

Moznette, G. F. 1922. The avocado, its insect enemies and how to combat them. U.S. Dept. Agric. Farmers' Bull. 1261. 31 pp.

Muniappan, R. 1988. Biological control of the weed, *Lantana camara* in Guam. J. Plant Prot. Trop. 5: 99–101.

1989. Biological control of *Lantana camara* L. in Yap. Proc. Hawaiian Entomol. Soc. 29: 195–196.

Nagaraj, A. N., and K. P. V. Menon. 1956. Note on the etiology of the wilt (root) disease of coconut palms in Travancore-Cochin. Indian Coconut J. 9: 161–165.

Nair, C. P. R., and M. R. G. K. Nair. 1974. Studies on the biology of the lace-wing *Corythauma ayyari* Drake a pest of jasmine. Agric. Res. J. Kerala 12: 172–173.

Nair, K. S. S., and G. Mathew. 1988. Biology and control of insect pests of fast-growing hardwood species. Kerala For. Res. Inst. Res. Rep. 51: 8 pp.

Nakasuga, T. 1994. Forecasting methods for the occurrence of the azalea lace bug *Stephanitis pyrioides* in Nagasaki prefecture. Proc. Assoc. Plant Prot. Kyushu 40: 137–142.

Nalepa, C. P., and J. R. Baker. 1994. Winter oviposition of the azalea lace bug (Hemiptera: Tingidae) in North Carolina. J. Entomol. Sci. 29: 482–489.

Nayer, K. K., T. N. Ananthakrishman, and B. V. David. 1976. General and Applied Entomology. Tata McGraw-Hill Publ. Co. Ltd., New Delhi, India. 589 pp.

Neal, J. W., Jr. 1985. Pest-free azaleas can be a reality. J. Azalea Soc. Am. 7: 25–29.

1988. Unusual ovipositional behavior on evergreen azalea by the Andromeda lace bug *Stephanitis takeyai* (Drake and Maa) (Heteroptera: Tingidae). Proc. Entomol. Soc. Washington 90: 520–554.

Neal, J. W., Jr., and L. W. Douglass. 1988. Development, oviposition, rate, longevity, and voltinism of *Stephanitis pryrioides* (Heteroptera: Tingidae), an adventive pest of azalea, at three temperatures. Environ. Entomol. 17: 827–831.

1990. Seasonal dynamics and the effect of temperature in *Corythucha cydoniae* (Heteroptera: Tingidae), Environ. Entomol. 19: 1299–1304.

Neal, J. W., Jr., and R. L. Haldemann. 1992. Regulation of seasonal egg hatch by plant phenology in *Stethoconus japonicus* (Heteroptera: Miridae), a specialist predator of *Stephanitis pyrioides* (Heteroptera: Tingidae). Environ. Entomol. 21: 793–798.

Neal, J. W., Jr., and J. E. Oliver. 1991. A unidirectional asymetric sexual hybrid in sympatric *Stephanitis* lace bugs (Hemiptera: Tingidae). Ann. Entomol. Soc. Amer. 84: 481–487.

Neal, J. W., Jr., R. H. Haldemann, and T. J. Henry. 1991. Biological control potential of a Japanese plant bug *Stethoconus japonicus* (Heteroptera: Miridae), an adventive predator of the azalea lace bug (Heteroptera Annals Tingidae). Ann. Entomol. Soc. Amer. 84: 287–293.

Neal, J. W., Jr., J. E. Oliver, and R. H. Fetterer. 1995. *In vitro* antimicrobial and nematocidal activity of acetogenins identified from exocrine secretions of *Stephanitis* and *Corythucha* lace bug nymphs (Heteroptera: Tingidae). Ann. Entomol Soc. Amer. 88: 496–501.

Neal, J. W., Jr., M. J. Tauber, and C. A. Tauber. 1992. Photoperiodic induction of reproductive diapause in *Corythucha cydoniae* (Heteroptera: Tingidae). Environ. Entomol. 21: 14–1418.

Nguyen, R., and D. G. Hall. 1991. The sugarcane lace bug, *Leptodictya tabida* (Herrich-Schaeffer) (Hemiptera: Tingidae). Florida Dept. Agric. Consumer Serv. Entomol. Circ. No. 348.

Nickel, J. L. 1958. Agricultural insects of the Paraguayan Chaco. J. Econ. Entomol. 51: 633–637.

Oliver, J. E., and W. R. Lusby. 1988. Synthesis of 2-acyl-3,6-dihydroxy-2-cyclohexen-1-ones. Tetra 44: 1591–1596.

Oliver, J. E., W. R. Lusby, and J. W. Neal, Jr. 1990. Exocrine secretion of the Andromeda lace bug *Stephanitis takeyai* (Hemiptera: Tingidae). J. Chem. Ecol. 16: 2243–2252.

Oliver, J., J. W. Neal, Jr., and W. R. Lusby. 1987. Phenolic acetogenins secreted by rhododendron lace bug, *Stephanitis rhododendri* Horvath (Hemiptera: Tingidae). J. Chem. Ecol. 13: 763–769.

Oliver, J. E., J. W. Neal, Jr., W. R. Lusby, J. R. Aldrich, and J. P. Kochansky. 1985. Novel components from secretory hairs of azalea lace bug *Stephanitis pyrioides* (Hemiptera: Tingidae). J. Chem. Ecol. 11: 1223–1228.

Önder, F., and N. Lodos. 1978. A new species of *Galeatus* Curtis (Heteroptera: Tingidae) from Turkey. Türk. Bit. Kor. Derg. 1: 23–27.

1980. A short note about *Galeatus helianthi* Önder and Lodos (Heteroptera: Tingidae). Türk. Bit Kor. Derg. 4: 231–232.

1983. Preliminary list of Tingidae with notes on distribution and importance of species in Turkey. Ege Univ. Ziraat Fak. Yayinl. No. 449: 1–51.

Önder, F., A. Onucar, and O. Ulu. 1986. Taxonomic status of *Stethoconous pyri* (Mella) (Het., Miridae) and some notes about its biology. Türk. Bit. Kor. Derg. 10: 149–153.

Ordonez-Giraldo, A. I. 1993. *Sporothrix insectorum*: méthode biologique de contrôl de la punaise *Leptopharsa gibbicarina* dans les cultures due palmier à huile en mérique Latine (Hemiptera, Tingidae). Bull. Soc. Entomol. France 98: 77–85.

Osborn, H., and C. J. Drake. 1916. The Tingitoidea of Ohio. Ohio Biol. Surv. Bull. No. 8.

1917. Notes on American Tingidae with descriptions of new species. Ohio J. Sci. 17: 295–307.

Ozino Marletto, O. I., and A. Arzone. 1985. Role of temperature and humidity in the action of pathogenic deuteromycetes on *Corythucha ciliata* (Say) (Rhynchota: Tingidae). Difesa Piante 8: 321–327.

Ozino Marletto, O. I., and R. Menardo. 1984. Micromycetes isolated from *Corythucha* ciliata Say. Boll. Lab. Entomol. Agric. *Filippo Silvestri* 41: 183–187.

Palaniswami, M. S., and K. S. Pillai. 1983. Biology of *Cochlochila bullita* S., a pest on Chinese potato. J. Root Crops 9: 59–62.

Parshley, H. M. 1922a. On the formation of family names like Tingidae. Science 56: 754–755.

1922b. Tingitidae or Tingidae. Science 56: 449.

Patel, R. C., and H. L. Kulkarny. 1955. Bionomics of *Urentius echinus* Dist. (Hemiptera-Heteroptera: Tingidae) an important pest of brinjal (*Solanum melongena* L.) in North Gujarat. J. Bombay Nat. Hist. Soc. 53: 86–96.

Pecora, P., A. Rizza, and M. Stazi. 1992. Biology and host specificity of *Oncochila simplex* (Hem.: Tingidae), a candidate for the biological control of leafy spurge *Euphorbia esula* L. "Complex." Entomophaga 37: 79–89.

Peña, J. E., and Van Waddill. 1982. Pests of cassava in south Florida. Florida Entomol. 65: 143–149.

Péricart, J. 1981. Révision systématique des Tingidae ouest-Paléarctiques. 7. Contribution a l'étude du genre *Monosteira costa* (Hemiptera). Ann. Soc. Entomol. France 17: 221–240.

1982. Les Hémiptères Tingidae: position systematique, morphologie, biologie et importance économique. Bull. Soc. France Entomol. 4: 4–6.

1983. Faune de France. 69. Hémiptères Tingidae Euro-méditerranéens. Paris. Fédération Française des Société de Sciences Naturelles, Paris, France. 618 pp.

Peschken, D. P. 1977. Host specificity of *Tingis ampliata* (Tingidae: Heteroptera): a candidate for the biological control of Canada thistle (*Cirsium arvense*). Canad. Entomol. 109: 669–674.

Pinhey, E. C. G. 1946. The olive bug. Rhodesia Agric. J. 43: 8–10.

Pollard, D. G. 1959. Feeding habits of the lace-bug *Urentius aegyptiacus* Bergevin (Hemiptera: Tingidae). Ann. Appl. Biol. 47: 778–782.

Prado, C. E. 1990. Presence in Chile of *Corythucha ciliata* (Say) (Hemiptera: Heteroptera: Tingidae). Rev. Chilena Entomol. 18: 53–55.

Qi, B., Nonnaizab, and C. W. Schaefer. 1991. The food plants of the Tingidae of Inner Mongolia, China. Phytophaga (Madras) 3: 109–120.
Ramesh, P., and D. Mukherjee. 1992. Host specificity of the lantana bug, *Teleonemia scrupulosa* Stal (Tingidae: Hemiptera) and its possible inclusion in the management of lantana weed (*Lantana camara* var. aculiata Linn.) in Himachal Pradesh. Indian J. Entomol. 54: 272–274.
Rasool, G., N. Ahmad, and N. A. Malik. 1986. The brinjal lace wing bug (*Urentius sentis* Distant) as a pest of cotton and its chemical control. J. Agric. Res. 321–323.
Raupp, M. J., 1984. Effects of exposure to sun on the frequency of attack by the azalea lacebug. J. Amer. Rhod. Soc. 38: 189–190.
Raupp, M. J., and R. M. Noland. 1984. Implementing landscape plant management programs in institutional and residential settings. J. Arboricult. 10: 161–169.
Reyes, A. R., M. A. Cruz, and P. Genty. 1988. The root absorption technique for controlling oil palm pests. Oléagineux 43: 363–370.
Robacker, C. D., and S. K. Braman. 1997. Field evaluation of azalea species and cultivars for resistance to azalea lace bug and cranberry rootworm. HortScience 32: 482.
Roberts J. R., Jr., R. D. Oetting, and W. L Corley. 1988. Insect pests on cotoneaster cultivars. Amer. Nurs. 167: 153.
Rodrigues, P. D. 1977. Four new species of African lacebugs (Hemiptera, Tingidae). Ann. Entomol. Fenn. 43: 75–80.
1979a. Notes on the fifth instar nymphs of *Dictyla nassata* (Puton) (Heteroptera: Tingidae). Arq. Mus. Bocage 6 (Suppl. 37): 1–4.
1979b. African Tingidae, V: nymphs of *Ammianus alaticollis* (Stål), *Dictyla flavipes* (Signoret) and *Compseuta ornatella* (Stål) (Heteroptera). Arq. Mus. Bocage 6: 481–488.
1982. Tingidae (Hemiptera, Heteroptera) from Gambia and Senegal. Entomol. Scand. 13: 77–80.
Rogers, C. E. 1977. Laboratory biology of a lace bug on sunflower. Ann. Entomol. Soc. Amer. 70: 144–145.
Rogers, J., R. Locci, and P. Vescovo. 1982. Contribution to tree pathology. III. On the association between *Corythucha ciliata* and saprophytic fungi in plane trees. Riv. Patol. Veg. 18: 149–155.
Rogers, L. J., and R. Locci. 1975. Contribution to tree pathology. II. On *Aureobasidium pullulans* in plane trees. Riv. Patol. Veg. Ser. IV. 11: 59–62.
Roonwal, J. L. 1952. The natural establishment and dispersal of an imported insect in India — the lantan bug *Teleonemia scrupulosa* Stål (= *lantanae* Distant) (Hemiptera, Tingidae), with a description of its egg, nymphs and adult. J. Zool. Soc. India 4: 1–16.
Rothschild, G. H. L. 1968. Notes on *Diconocoris hewetti* (Dist.) (Hemiptera: Tingidae) a pest of pepper in Sarawak (Malaysian Borneo). Bull. Entomol. Res. 58: 107–118.
Sailer, R. I. 1945. The bite of a lacebug, *Corythucha cydoniae* (Fitch). J. Kansas Entomol. Soc. 18: 81.
Samuel, C. K. 1939. Oviposition of the tingid *Monanthia globulifera* Wlk. Indian J. Entomol. 1: 89–99.
Sandhu, G. S., and A. S. Sohi. 1979. Chemical control of Kachnar tingid bug, *Cysteochila ablusa* Drake, a serious pest of *Bauhina* spp. with high-volume of ultra-low-volume spraying. Indian J. Entomol. 41: 160–163.
Santini, L., and A. Crovetti. 1986. Preliminary data on eggs development of the sycamore lace bug *Corythucha ciliata* (Say) (Rynchota Tingidae) at six different constant temperatures. Frustula Entomol. 7–8: 639–646.
Sarospataki, M. 1999. Phytophagous insects associated with *Hieracium pilosella* (Asteraceae) in Hungary, central Europe. Environ. Entomol. 28: 1–8.
Schaefer, C. W. 1998. Phylogeny, systematics, and practical entomology: the Heteroptera (Hemiptera). An. Soc. Entomol. Brasil 27: 499–511.
Schoonhoven, A. van, F. Burbano, and R. Arenas. 1975. Notes on the biology of the lace bug *Gargaphia sanchezi* (Hemiptera, Tingidae) a pest of beans (*Phaseolus vulgaris*). Turrialba 25: 327.
Schouteden, H. 1953. *Habrochila ghesquièrei* n. sp. parasite du caféier (Hem. Tingididae). Rev. Zool. Bot. Afr. 48: 104–105.
Schread, J. C. 1953. Andromeda lace bug. Bull. Connecticut Agric. Exp. Stn. 568: 3–13.
1968. Control of lace bugs on broadleaf evergreens. Bull. Connecticut Agric. Exp. Stn. 684: 3–7.
Schultz, P. B. 1983a. Feeding preference of hawthorn lace bug. Virginia J. Sci. 34: 108.
1983b. Evaluation of hawthorn lace bug (Hemiptera: Tingidae) feeding preference on *Cotoneaster* and *Pyracantha*. Environ. Entomol. 12: 1808–1810.

1985. Evaluation of selected *Cotoneaster* spp. for resistance to hawthorn lace bug. J. Environ. Hortic. 3: 156–157.

1993. Host plant acceptance of azalea lace bug (Heteroptera: Tingidae) for selected azalea cultivars. J. Entomol. Sci. 28: 230–235.

Schultz, P. B., and M. A. Coffelt. 1987. Oviposition and nymphal survival of the hawthorn lace bug (Hemiptera: Tingidae) on selected species of *Cotoneaster* (Rosaceae). Environ. Entomol. 16: 365–367.

Schultz, P. B., and L. A. Parnell. 1982. Preference of hawthorn lace bug to *Pyracantha* and *Cottoneaster* cultivars common to Virginia nurseries. Proc. South. Nurs. Assoc. Res. Conf. Annu. Rep. pp. 114–116.

Schultz, P. B., and D. J. Shetlar. 1994. Major insect pests of ornamental trees and shrubs. Pp. 237–247 *in* Leslie, A.R. [ed.], Handbook of IPM for Turf and Ornamentals. Lewis Publishers/CRC Press, Boca Raton, Florida, U.S.A. 660 pp.

Scott, J. 1874. On a collection of Hemiptera Heteroptera from Japan. Descriptions of various new genera and species. Ann. Mag. Nat. Hist. 4(14): 426–452.

Servadei, A. 1966. Un tingidae neartico comparso in Italia (*Corythucha ciliata* Say). Boll. Soc. Entomol. Ital. 96: 94–96.

Shanta, P., and K. P. V. Menon. 1960. Cowpea (*Vigna sinensis* Endl.), and indicator plant for the coconut wilt virus. Virology 12: 309–310.

Shanta, P., T. Joseph, and S. B. Lal. 1964. Transmission of root (wilt) disease of coconuts. Indian Coconut J. 18: 25–28.

Shanta, P., K. P. V. Menon, and K. P. Pillai. 1960. Aetiology of the wilt (root) disease: investigation on its virological nature. Indian Coconut J. 13: 56–66.

Sharga, U. S. 1953. Bionomics of *Monanthia globulifera* Walk. (Hemipera-Heteroptera: Tingidae). J. Bombay Nat. Hist. Soc. 51: 885–889.

Sheeley, R. D., and T. R. Yonke. 1977. Biological notes on seven species of Missouri tingids (Hemiptera: Tingidae). J. Kansas Entomol. Soc. 50: 342–356.

Shen, H. W., W. J. Wu, and P. S. Yang. 1985. The biology of the azalea lacebug, *Stephanitis pyrioides* (Scott) I. The morphology of the azalea lacebug, *Stephanitis pyrioides* (Scott). Mem. Coll. Agric. Nat. Taiwan Univ. 25: 143–154 [in Chinese, English summary].

Silva, C. C. A. D., and S. M. D. L. Barbosa. 1986. An outbreak of *Gargaphia torresi* in bean crop in Alagoas state. Pesq. Agropec. Bras. 21: 1003–1004.

Silverman, J., and R. D. Goeden. 1979. Life history of the lacebug, *Corythucha morrilli* Osborn and Drake, on the ragweed, *Ambrosia Dumosa* (Gray) Payne, in southern California (Hemiptera-Heteroptera: Tingidae). Pan-Pac. Entomol. 55: 305–308.

Simmonds, H. W. 1929. The life history of *Teleonemia lantanae*. Agric. J. Dept. Agric. Fiji 2: 36–39.

Singh, G., G. S. Grewal, and J. S. Bhalia. 1991. New record of tingid bug, *Galeatus scrophicus* (Saunders) as a serious pest of sunflower in Punjab. J. Insect Sci. 4: 93.

Singh, L., and H. S. Mann. 1986. Seasonal activity and population buildup of tingid bug, *Urentius hystricellus* Richter (Hemiptera: Tingidae) in the Punjab. Indian J. Ecol. 13: 301–306.

Sivaramakrishnan, V. R. 1976. Occurrence of lantana lace bug *Teleonemia scrupulosa* Stal (Hemiptera: Tingidae) in South India. Indian For. 102: 620–621.

Slater, J. A., and R. M. Baranowski. 1978. How to Know the True Bugs. Wm. C. Brown, Dubuque, Iowa, U.S.A. 256 pp.

Soetopo, D., and M. Iskandar. 1986. The efficacy of Mipcin WP and Bassa 50 EC against two major insect pests of pepper. Edisi Khusus Penelitian Tandman Reinpah dan Obat 11: 75–78.

Solomon, J. J. 1991. Recent advances in research on root (wilt) disease of coconut. J. Plant Crops 19: 153–162.

Solomon, J. J., Govindankutty, M. P., and F. Nienhaus. 1983. Association of mycoplasma-like organisms with the coconut root (wilt) disease in India. J. Plant Dis. Prot. 90: 295–297.

Southwood, T. R. E., and G. G. E. Scudder. 1956. The bionomics and immature stages of the thistle lace bugs (*Tingis ampliata* H.-S. and *T. cardui* L.) (Hem., Tingidae). Trans. Soc. Br. Entomol. 12: 92–112.

Stehlik, J. L. 1997. *Corythucha ciliata* (Say), a pest of plant trees, now also in the Czech Republic (Tingidae, Het.). Acta Mus. Moraviae Sci. Nat. 81: 299–306.

Steyer, --. [sic]. 1915. *Stephanitis rhododendri* Horvath (Hemip.) in Deutschland. Z. Angew. Entomol. 2: 434–435.

Stone, J. D., and J. N. Fries. 1986. Insect fauna of cultivated guayule, *Parthenium argentatum* Gray (Campanulatae: Compositae). J. Kansas Entomol. Soc. 59: 49–58.

Stone, J. D., and G. P. Watterson. 1985. Effects of temperature on the survival and development of the Morrill lace bug (Heteroptera: Tingidae) on Guayule. Environ. Entomol. 14: 329–331.

Stonedahl, G. M., W. R. Dolling, and G. J. duHeaume. 1992. Identification guide to common tingid pests of the world (Heteroptera: Tingidae). Trop. Pest Manage. 38: 438–449.

Story, J. M., H. DeSmet-Moens, and W. L. Morrill. 1985. Phytophagous insects associated with Canada thistle, *Cirsium arvense* (L.) Scop., in southern Montana. J. Kansas Entomol. Soc. 58: 472–478.

Štusák, J. M. 1957. A contribution to the knowledge of some last nymphal instars of the Czechoslovakian lace bugs (Hemiptera-Heteroptera, Tingidae). Čas. Česk. Spol. Entomol. 54: 132–141.

1959. Contributions to the knowledge of new or little known last nymphal instars of some tingid-bugs (Hemiptera-Heteroptera, Tingidae). Acta Soc. Mus. Nat. Prague 33: 363–376.

1961. Dritter Beitrag zur Kenntnis der Eier der Tingiden (Heteroptera, Tingidae). Acta Soc. Entomol. Čechoslov. 58: 71–88.

1972. Nymphs and host plants of *Agramma atricapillum* (Spinola), *Tingis auriculata* (Costa) and *Hyalochiton komaroffii* (Jakovelev) (Heteroptera, Tingidae). Acta Entomol. Bohemoslov. 69: 101–109.

1977. On nymphs of some Javanese *Tingidae* (Heteroptera). Acta Entomol. Bohemoslov. 74: 23–34.

1984. The biology and immature stages of *Haedus vicarius* (Heteroptera, Tingidae). Acta Entomol. Bohemoslov. 81: 246–258.

Štusák, J. M., and P. Štys. 1959. Investigations on the taxonomy and morphology of imagines and nymphs of some species of the genus *Monanthia* Le Peletier et Serville, 1825 (Hemiptera-Heteroptera: Tingidae). Acta Univ. Carol. Biol. 177–205.

Štys, P., and J. Davidová-Vilimová. 1989. Unusual numbers of instars in Heteroptera: a review. Acta Entomol. Bohemoslov. 86: 1–32.

Sunil, K. P. 1992. Coconut root wilt management: focus turns to in-vitro culture, 1992. Indian Coconut J. 22: 10.

Swezey, O. 1945. Insects associated with orchids. Proc. Hawaiian Entomol. Soc. 12: 343–403.

Takeya, C. 1953. Notes on the Tingidae of Shikoku, Japan (Hemiptera). Trans. Shikoku Entomol. Soc. 3: 167–176.

1963. Taxonomic revision of the Tingidae of Japan, Korea, the Ryukyus and Formosa. Part. 2. (Hemiptera). Mushi 37: 27–52.

Tallamy, D. W. 1982. Age specific maternal defense in *Gargaphia solani* (Hemiptera: Tingidae). Behav. Ecol. Sociobiol. 11: 7–11.

1984. Insect parental care. BioScience 34: 20–24.

1985. "Egg dumping" in lace bugs (*Gargaphia solani*, Hemiptera: Tingidae). Behav. Ecol. Sociobiol. 17: 357–362.

1986. Age specificity of "egg dumping" in *Gargaphia solani* (Hemiptera: Tingidae). Anim. Behav. 34: 599–603.

Tallamy, D. W., and R. F. Denno. 1981a. Alternative life history patterns in risky environments: an example from lace bugs (Hemiptera: Tingidae). Pp. 129–147 *in* R. F. Denno and H. Dingle [eds.], Insect Life History Patterns: Geographic and Habitat Variation. Springer-Verlag, New York, New York, U.S.A.

1981b. Maternal care in *Gargaphia solani* (Hemiptera: Tingidae). Anim. Behav. 29: 771–778.

1982. Life history trade-offs in *Gargaphia solani* (Hemiptera: Tingidae): the cost of reproduction. Ecology 63: 616–620.

Tallamy, D. W., and H. Dingle. 1986. Genetic variation in the maternal defensive behavior of the lace bug *Gargaphia solani*. Pp. 135–143 *in* M. D. Huettle [ed.], Evolutionary Genetics of Invertebrate Behavior: Progress and Prospects. Plenum Press, New York, New York, U.S.A. 335 pp.

Tallamy, D. W., and L. A. Horton. 1990. Costs and benefits of the egg-dumping alternative in *Gargaphia* lace bugs (Hemiptera: Tingidae). Anim. Behav. 39: 352–359.

Tavella, L., and A. Arzone. 1987. Investigations on the natural enemies of *Corythuca ciliata* (Say) (Rhynchota Heteroptera). Redia 70: 443–457.

Thangavel, P., C. V. Sivakumar, and A. A. Kareem. 1977. Control of lace-wing bug *Stephanitis typicus* Distant (Tingidae: Homoptera) in turmeric. Soc. Indian Hortic. 25: 170–172.

Theobald, F. V. 1913. The rhododendron bug (*Stephanitis rhododendri* Horvath). J. Southeastern Agric. Coll. Wye, Kent (England) 23: 297–302.

Thontadarya, T. S., and G. P. Channa Basavanna. 1959. Mode of egg laying in Tingidae (Hemiptera). Nature 184: 289–290.

Tigvattnanont, S. 1989. Studies on the bionomics and local distribution of some lace bugs in Thailand: I. *Monanthia globulifera* (Hemiptera: Tingidae). Khon Kaset (Khon Kaet Agric. J.) 17: 333–344.

1990a. Studies on the bionomics and local distribution of some lace bugs in Thailand. II. *Stephanitis typicus* (Hemiptera: Tingidae). Khon Kaset (Khon Kaet Agric. J.) 18: 200–212.

1990b. Studies on the bionomics and local distribution of some lace bugs in Thailand. III. *Urentius echinus* Distant (Hemptera: Tingidae). Khon Kaet Agric. J. 18: 251–260.

1991. Studies on the bionomics and local distribution of some lace bugs in Thailand. V. *Haedus vicarius* (Drake) (Hemiptera: Tingidae). Kasetsart J. Nat. Sci. 25: 153–161.

Tilden, J. W. 1950. Biological notes on *Corythucha morrilli* O. and D. (Hemiptera: Tingidae). Entomol. News 61: 135–137.

Tomokuni, M. 1983. Notes on the Japanese species of *Acalypta* (Hemiptera, Tingidae). Mem. Nat. Sci. Mus. Tokyo 16: 145–154.

1987. The Tingidae of Hokkaido, Japan (Insecta, Heteroptera). Mem. Nat. Sci. Mus. Tokyo 20: 115–121.

Torres-Miller, L. 1989. New records of lace bugs from West Virginia (Hemiptera: Tingidae). Insecta Mundi 3: 10.

Tsukada, M. 1992. A new record of *Anagrus takeyanus* Gordh et Dunbar (Hymenoptera, Mymaridae) from Japan. Jpn. J. Entomol. 60: 136.

Usinger, R. L. 1946. Biology and control of the ash lace-bug, *Leptoypha minor*. J. Econ. Entomol. 39: 286–289.

Vayssieres, J. F. 1983. Life histories and host specificities of the echium bugs *Dictyla echii* and *Dictyla nassata* (Hem.: Tingidae). Entomophaga 28: 135–144.

Vecht, J. van der 1935. Aanteekeningen over de pepernetwants (*Eleasmognathus hewitti* Dist.). "Landbouw." Tijdsch. Vereeniging Land Bouwcounsulten N-I. 10: 484–493.

Verma, N. E., D. S. Gupta, and H. V. Singh. 1974. Biology of tingid bug *Galeatus scrophicus* Saund (Hemiptera: Tingidae) on sunflower. J. Res. Haryana Agric. Univ. 4: 13–17.

Vessey, J. C. 1981. Control of leaf spot on oil palm in Honduras with insecticides. Oléagineux 36: 229–232.

Vidal, J. P. 1939. Le faux tigre du poirier (*Monostira unicostata* Mls. Hem. Heter.). Bull. Soc. Hist. Nat. Afr. Nord 30: 27–32.

Wade, O. 1917. The sycamore lace-bug. Oklahoma Exp. St. Bull. 116.

Wang, C. Y. 1988. The bionomics of *Stoephanitis* [sic] (Norba) *chinensis* Drake and its control. Insect Knowledge 25: 399–341.

Wang, Y., C. D. Robacker, and S. K. Braman. 1998. Identification of resistance to azalea lace bug among deciduous azalea taxa. J. Amer. Hortic. Sci. 123: 592–597.

Wang, Y., S. K. Braman, C. D. Robacker, and J. C. Latimer. 1999. Composition and variability of epicuticular lipids of azaleas and their relationship to azalea lace bug resistance. J. Amer. Hortic. Sci. 124: 239–244.

Wapshere, A. J. 1981. The biological control of Paterson's curse, *Echium plantagineum*: northern hemisphere studies. Pp. 599–602 *in* E. S. Del Fosse [ed.], Proc. V. Int. Symp. Biol. Control Weeds Brisbane. 1980. C.S.I.R.O. Melbourne, Australia.

1988. Prospects for the biological control of silver-leaf nightshade, *Solanum elaeagnifolium*, in Australia. Austral. J. Agric. Res. 39: 187–197.

Weiss, H. B. 1916a. Hemiptera. Entomol. News 27: 189–190.

1916b. Foreign pests recently established in New Jersey. J. Econ. Entomol. 9: 212–216.

Weiss, H. B., and T. J. Headlee. 1918. Some new insect enemies of greenhouse and ornamental plants in New Jersey. The rhododendron lace bug *Leptobyrsa rhododendri* Horv. Insect Enem. Greenhouse Orn. Plants, Circ. 100: 5–9.

Westwood, J. O. 1840. An Introduction to the Modern Classification of Insects. Vol. 2. Longman, Orme, Brown, Green, and Longmans, London, U.K. 587 pp.

Wheeler, A. G., Jr. 1977. Spicebush and sassafras as new North American hosts of andromeda lace bug, *Stephanitis takeyai* (Hemiptera: Tingidae). Proc. Entomol. Soc. Washington 79: 168–171.

1981. Hawthorn lace bug (Hemiptera: Tingidae), first record of injury to roses, with a review of host plants. Great Lakes Entomol. 14: 37–43.

1987. Hedge bindweed, *Calystegia sepium* (Covolvulaceae), an adventitious host of the chrysanthemum lace bug, *Corythucha marmorata* (Heteroptera: Tingidae). Proc. Entomol. Soc. Washington 89: 200.

1989. Late lilac, *Syringa villosa*: new host of the lace bug *Leptoypha mutica* (Heteroptera: Tingidae). Great Lakes Entomol. 22: 35–38.

Wheeler, A. G., Jr., and E. R. Hoebeke. 1985. *Dictyla echii*: seasonal history and North American records of an immigrant lace bug (Hemiptera: Tingidae). J. New York Entomol. Soc. 93: 1057–1063.
White, R. P. 1933. The insects and diseases of Rhododendron and azalea. J. Econ. Entomol. 26: 631–640.
Williams, M. L., and N. J. Bedwell. 1980. Insecticides for controlling hawthorn lace bug on *pyracantha*. Proc. South. Nurs. Assoc. Res. Conf. 96–97.
Willis, J. C., and H. K. Airy Shaw. 1973. A Dictionary of the Flowering Plants. Cambridge University Press, Cambridge, U.K. 1245 pp.
Wilson, G. F. 1933. Insect pests of hardy rhododendrons: their detection and control. New Flora Sylva 6: 36–44.
Wolfenbarger, D. O. 1963. Insect pests of the avocado and their control. Bull. Univ. Florida, Agric. Exp. Stn. Bull. 605A: 52 pp.
Yacoob, M., and D. Livingstone. 1983. Resource potentials of the egg parasitoids of Tingidae. Proc. Symp. Insect Ecol. Res. Man. 247–252.
Yasunaga, T., M. Taki, and Y. Nakatani. 1997. Species of the genus *Stethoconus* of Japan (Heteroptera, Miridae): predaceous deraeocorine plant bugs associated with lace bug (Tingidae). Appl. Entomol. Zool. 32: 261–264.
Zappe, M. P. 1926. Miscellaneous insect notes. Bull. Connecticut Agric. Exp. Stn. 275: 321–327.
Zeya, A. 1983. Further observations on *Tingis beesoni* (Hemiptera: Tingidae) incident on Yemane (*Gemelina arborea*). Socialist Republic of the Union of Burma, Ministry of Agriculture and Forests, Forest Research Institute, Leaflet No. 12. 9 pp.

CHAPTER 5

Palm Bugs (Thaumastocoridae)

Lionel Hill and Carl W. Schaefer

1. INTRODUCTION

The family Thaumastocoridae comprises two subfamilies, Thaumastocorinae in Australia and India and Xylastodorinae in South America and Cuba. The Cuban *Xylastodorus* species has been introduced into Florida (U.S.A.). Six genera and 19 species (Cassis et al. 1999) are included in these two subfamilies. Six of these live in the New World, one in India, and the remainder in Australia. In addition, Poinar and Santiago-Blay (1997) described an extinct xylastodorine genus from Dominican amber 20 to 40 million years old. They characterize the family as a Gondwanan relict with one species introduced by humans to Florida. Drake and Slater (1957) first revised the family and established one third of the species. Schaefer (1969) provided morphological and phylogenetic notes on the family. Family summaries are provided by Cassis and Gross (1995) and Schuh and Slater (1995). The phylogenetic relationships of the family are still being resolved (Schaefer 1969, Slater and Brailovsky 1983).

In 1957 only two hosts were known. Cassis et al. (1999) tabulate 24 host plants for 15 species, including several new records for Australian genera. Baranowski (1958b) and Hill (1988) provide the major biological studies.

The family name means "wonderful bug," but no common group name exists reflecting their cryptic habits and rare pest status. Schuh and Slater (1995) suggested the name "palm bugs" for the Xylastodorinae and this has been used by several authors. Body length in most genera is 2 to 3 mm but reaches 4.6 mm in *Discocoris imperialis* Slater and Schuh. *Baclozygum brachypterum* Slater is the shortest at 1.7 mm.

Despite their strong grip on the substrate, many species have been collected by beating. At least three species, representing both subfamilies (Baranowski 1958b; Slater 1973; L. Hill, unpublished data), can run sideways to retreat to the opposite leaf surface or to a crevice.

Adults of all species are flat to extremely flat. In general, the palm-feeding Xylastodorinae are more flattened and cryptic in feeding locations than the Thaumastodorinae, which are exposed during feeding on foliage of several plant families. The former are often translucent, whereas the Thaumastocorinae use red, brown, black, and cream cryptic coloration. The nymphs of *Thaumastocoris petilus* Drake and Slater are not flattened. Nymphs can vary in color from pale to dark within a genus.

All species have highly asymmetrical male genitalia. Both left- and right-handed terminal asymmetry occurs in males of *X. luteolus* Barber. The ovipositor is not sclerotized. The eggs are oblong, flattened, and relatively large. The eyes are large and, especially in Thaumastocorinae, protuberant.

The characteristic mandibular plates are often as long and broad as the clypeus. Their form can be sexually dimorphic. The labium, usually short, is long in *B. brachypterum* (Slater 1973). It feeds on *Xanthorrhoaea* (Cassis et al. 1999), which is a monocotyledonous mat-rush (Xanthorrhoeaceae) with hard narrow linear leaves all rising from the base. *Onymocoris* also has a long labium (Drake and Slater 1957). It feeds on *Banksia* and *Dryandra* of the Proteaceae (Cassis et al. 1999).

2. GENERAL STATEMENT OF ECONOMIC IMPORTANCE

Thaumastocoridae are rarely of economic importance. Two species, representing both subfamilies, have reached pest populations on ornamental plants growing away from their native habitats. These outbreaks have occurred at long intervals. Factors leading to outbreaks are unknown.

Thaumastocoridae suck sap from the surface cells of foliage. The appearance of this damage resembles that of thrips or some tingids, namely, a spotty silvering of foliage where air replaces sap in surface cells. In extreme cases foliage dies and becomes brown.

Discocoris with five species and the sole *Xylastodorus* species all feed on palms such as *Phytlephas*, *Socratea*, *Butia*, *Euterpe*, and *Roystonea*; but only *Xylastodorus* has reached pest populations. *Discocoris* lives around the inflorescence or infructescence (Schuh 1975, Slater and Brailovsky 1983, Slater and Schuh 1990). Xylastocorines may have co-evolved with New World palms (Couturier et al. 1998). Among the 12 Australian Thaumastocorinae feeding on Xanthorrhoeaceae, Myrtaceae, Proteaceae, Mimosaceae, Elaeocarpaceae, and Cunionaceae only *Baclozygum depressum* has been recorded at pest populations (Hill 1988, Cassis et al. 1999).

Thaumastocoridae are known from sea level in temperate and tropical regions to 1800 m altitude in Venezuela (Slater and Brailovsky, 1983).

3. THE IMPORTANT SPECIES

3.1 *Xylocoris luteolus* Barber

3.1.1 *Distribution, Life History, and Biology*

The royal palm bug, *X. luteolus*, occurs in Cuba and southern Florida (U.S.A.). It was probably introduced to Florida with its only known host, a royal palm *Roystonea regia*. The bug has not been found on three other royal palms that occur in Florida, including the native *R. elata*. Adults, 2.2 mm long, feed and mate in folded emerging leaflets. Males are slightly more common than females. Males exhibit right- or left-handed genital asymmetry suited to this cryptic microhabitat. Females have a soft ovipositor and lay about one egg per day in the flexible membranous scales that cover the undersurface of the leaflet midrib. The pale tan egg with a white cap is 0.5 mm long, oval in cross section, and hatches in 8 to 9 days. The five pale nymphal stages take 5 to 8 days each. Development from hatching to adult takes 23 to 37 days.

3.1.2 *Damage*

The bugs do their greatest damage by feeding on the folded pinnae of the fronds that have most recently broken away from the spike but, according to Baranowski (1958a), do not feed on younger compacted fronds. Damage first appears as small yellow spots caused by withdrawal of sap. These eventually coalesce and after extensive feeding entire fronds may turn brown. From a distance, this gives the palm a stratified appearance of fresh green new fronds, recently dead young fronds, older unaffected green fronds, and old, dead fronds. Moznette (1921) and Baranowski (1958a) illustrate the damage photographically. Smaller palm trees are not so badly infested as large trees. Palms in ornamental habitats appear susceptible perhaps because of disruption of biological control or stressed physiology.

3.1.3 Control

Satisfactory control with nicotine sulfate or chlorinated hydrocarbon insecticides was achieved in outbreaks in 1921 and 1957, respectively, provided a high-volume spray incorporating a spreader for penetration was applied to the terminal growth and repeated 2 weeks later to kill newly hatched nymphs from the impervious eggs (Moznette 1921, Baranowski 1958a). The carbamate aphicide pirimicarb with its fumigant and selective action warrants investigation for future control efforts against this sporadic pest. The mode of feeding and cryptic microhabitat of these bugs suggests that techniques extrapolated from thrips control may succeed.

3.2. *Baclozygum depressum* Bergroth

3.2.1 Distribution, Life History, and Biology

Baclozygum depressum occurs in temperate coastal Australia from west to east and up the east coast as far as southern Queensland with a disjunct population in Cape York; however, one of the present authors, L. Hill, doubts the conspecificity of Queensland specimens. Host plants are the myrtaceous trees *Eucalyptus globulus*, *E. viminalis*, *E. pulchella*, *E. trachyphloia*, and *Agonis flexuosa* (Cassis et al. 1999). Adults, nymphs, and eggs occur on foliage, stems, and seed capsules. Eggs are 0.65 mm long, black, oblong, and flattened. Females lay about 1.25 eggs per day on leaf tissue scarred by Lepidoptera and Coleoptera as well as in seed capsule crevices and petiole axils; undamaged leaf surface is not used. Although eggs are placed singly by females, clusters of 100 or more eggs can accumulate in favored sites such as between leaves scarred and webbed by Lepidoptera. One female can lay up to 80 eggs. Nymphs and adults feed predominantly on adult leaves which hang vertically and differ in form, orientation, and cuticle defense chemistry from juvenile leaves. Feeding occurs mostly on the sunny sides of leaves. It is not known whether feeding is nocturnal. At 20°C egg incubation requires 11 to 14 days; the five stadia require 7 to 8, 3 to 5, 4 to 6, 4 to 7, and 7 to 10 days; and adults can survive an additional 50 days including a preoviposition period of 7 to 15 days. Survival of early instars at 30°C is low (Hill 1988).

3.2.2 Damage

Feeding marks are pale epidermal spots several times larger than the stomata. These eventually coalesce patchily over the entire leaf area (except veins) to occupy perhaps half of the leaf area. Leaf death has not been observed. Bugs have been recorded in pest numbers only in Tasmania on ornamental trees (parkland, garden, and road verge), where they showed a marked preference for a dwarf cultivar of *E. globulus* over *E. viminalis* and *E. pulchella*. The dwarf *E. globulus* were growing in northern Tasmania away from their natural habitat of southern Tasmanian forests. Only one outbreak has been recorded. Hill (1988) illustrated the damage.

3.2.3 Control

Control has not been attempted. One parasitoid, an undetermined mymarid wasp, has been recorded from the eggs of *B. depressum* (Hill 1988).

4. PROGNOSIS

Species of *Discocoris* might become pests if their palm hosts were grown in ornamental environments. *Agonis flexuosa* and several *Eucalyptus* species are grown ornamentally in Australia outside their natural ranges. They are hosts for *B. depressum*. Cloned hybrids of *E. globulus* with other *Eucalyptus* species recently established in plantation forests may be at risk from *B. depressum*. Quarantine should be considered when trading *Eucalyptus* seed because the eggs of *B. depressum* can occur on seed capsules. *Banksia* and *Dryandra* shrubs grown ornamentally may be at risk from *Onymocoris* species. Similarly, *Thaumastocoris* species could threaten *Acacia*, *Eucalyptus*, *Elaeocoarpus*, and *Melaleuca* species, many of which are used ornamentally or commercially inside and outside Australia.

5. CONCLUSIONS

The Thaumastocoridae are not typical pest species. Their host ranges, fecundity, and perhaps mobility are generally those of specialist species rather than opportunists and they are likely to remain relatively innocuous except under special circumstances.

6. REFERENCES CITED

Baranowski, R. M. 1958a. The royal palm bug. Principes 2: 72–73.

1958b. Notes on the biology of the royal palm bug, *Xylastodoris luteolus* Barber (Hemiptera, Thaumastocoridae). Ann. Entomol. Soc. Amer. 51: 547–551.

Cassis, G., and G. F. Gross. 1995. Zoological Catalogue of Australia. Vol. 27.3A. Hemiptera: Heteroptera (Coleorrhyncha to Cimicomorpha). CSIRO Publishing, Melbourne, Australia. 506 pp.

Cassis, G., R. T. Schuh, and H. Brailovsky. 1999. A review of *Onymocoris* (Heteroptera: Thaumastocoridae), with a new species, and notes on hosts and distributions of other thaumastocorid species. Acta Soc. Zool. Bohem. 63: 19–36.

Couturier, G., F. Kahn, and M. S. Padilha de Oliveira. 1998. New evidence on the coevolution between bugs (Hemiptera: Thaumastocoridae: Xylastocorinae) and the New World palms. Ann. Soc. Entommol. France (n.s.) 34: 99–101.

Drake, C. J., and J. A. Slater. 1957. The phylogeny and systematics of the family Thaumastocoridae (Hemiptera: Heteroptera). Ann. Entomol. Soc. Amer. 50: 353–370.

Hill, L. 1988. The identity and biology of *Baclozygum depressum* Bergroth (Hemiptera: Thaumastocoridae). J. Austral. Entomol. Soc. 27: 37–42.

Moznette, G. F. 1921. Notes on the royal palm bug. Q. Bull. Florida State Plant Board 6: 10–15.

Poinar, G. O., and J. A. Santiago-Blay. 1997. *Paleodoris lattini* gen.n., sp.n., a fossil palm bug (Hemiptera: Thaumastocoridae, Xylastodorinae) in Dominican amber, with habits discernible by comparative functional morphology. Entomol. Scand. 28: 307–310.

Schaefer, C. W. 1969. Morphological and phylogenetic notes on the Thaumastocoridae (Hemiptera-Heteroptera). J. Kansas Entomol. Soc. 42: 251–256.

Schuh, R. T. 1975. Wing asymmetry in the thaumastocorid *Discocoris drakei* (Hemiptera). Rev. Peruana Entomol. 18: 12–13.

Schuh, R. T., and Slater, J. A. 1995. True Bugs of the World (Hemiptera: Heteroptera): Classification and Natural History. Cornell University Press, Ithaca, New York, U.S.A. 336 pp.

Slater, J. A. 1973. A contribution to the biology and taxonomy of Australian Thaumastocoridae with the description of a new species (Hemiptera: Heteroptera). J. Austral. Entomol. Soc. 12: 151–156.

Slater, J. A., and H. Brailovsky. 1983. The systematic position of the family Thaumastocoridae with the description of a new species of *Discocoris* from Venezuela (Hemiptera: Heteroptera). Proc. Entomol. Soc. Washington 85: 560–563.

Slater, J. A., and R. T. Schuh. 1995. A remarkably large new species of *Discocoris* from Colombia (Heteroptera: Thaumastocoridae). J. New York Entomol. Soc. 98: 402–405.

CHAPTER 6

Seed and Chinch Bugs (Lygaeoidea)

Merrill H. Sweet II

1. INTRODUCTION

The family Lygaeidae has recently been subdivided into 11 families, which are included with 5 other families in the Lygaeoidea (Henry 1997). As Henry noted, this author has long worked on this same subject, and agrees (Sweet 1992) with Henry that the long-established family concept of the Lygaeidae is not cladistically monophyletic. However, this arrangement, based on other data, differs in some taxa from the proposal of Henry. Unfortunately this work has not yet been published as Sweet had hoped to add molecular data to the analysis to substantiate such a seemingly radical proposal. As noted by Slater and O'Donnell (1995), the problem is that Lygaeidae *sensu lato* comprise a paraphyletic "stem" group. If treated as a family clade, to avoid paraphyly, the monophyletic taxon should include as well the Malcidae, Colobathristidae, Berytidae, Piesmatidae, and the superfamilies Pyrrhocoroidea and (probably) Coreoidea (Coreidae, Alydidae, Rhopalidae). All but the first two are treated elsewhere in this book. The Malcidae and the Colobathristidae are treated in this account: the former was originally included in the *Catalogue of the Lygaeidae* (Slater 1964a); and the latter was formerly included in the Lygaeidae by Stål (1874). This chapter will follow the arrangement of Henry, except that the Orsillinae (Orsillidae) and the Ischnorhynchinae (Ischnorhynchidae) are considered families distinct from the Lygaeinae (Lygaeidae *sensu stricto*) based on a large suite of synapomorphies (M. H. Sweet, unpublished).

In the traditional sense of the family, Slater and O'Donnell (1995) listed 4045 species and 651 genera in 15 subfamilies, an increase of 1297 species (47%) since 1964, when Slater published his monumental catalog of the Lygaeidae. This large increase indicates not only the intensity of work in the past 35 years, but also shows how incomplete the taxonomic knowledge must be because, as Slater and O'Donnell (1995) observed, most of the increases have occurred in taxa on which specialists happened to have worked. In addition, excellent studies have been carried out on natural populations of noneconomic species of the brightly colored aposematic Lygaeidae (*sensu stricto*). This work includes not only the extensive physiological work on *Oncopeltus fasciatus* Dallas, the large milkweed bug much used as a laboratory animal (see Feir 1974, and the bibliography of *Oncopeltus fasciatus* in Slater and O'Donnell 1995), but also the numerous experimental field studies on the brightly colored species of *Oncopeltus, Lygaeus, Neacoryphus, Spilostethus,* and *Horvathiolus* (see references in Slater and O'Donnell 1995). As Slater and O'Donnell stress, "When one reviews this body of work in perspective and relates it to the continuing physiological, molecular and biochemical studies, utilizing chiefly *Oncopeltus fasciatus* as a laboratory animal, it becomes apparent that in the Lygaeidae we have available for synthesis a body of information found in only a very few other insect groups." With the division of the family, this comment remains valid for the superfamily, and it should be stressed that this taxon, treated as a family or subfamily, includes many species of economic importance as well, enhancing the significance of the taxon Lygaeoidea.

The subfamilies recognized by Slater and O'Donnell (1995), with the number of known species in parentheses, are Lygaeinae (640), Orsillinae (255), Ischnorhynchinae (77), Cyminae (92), Blissinae (435), Artheneinae (18), Geocorinae (229), Psamminae (3), Henestarinae (20), Bledionotinae (26), Oxycareninae (147), Pachygronthinae (78), Heterogastrinae (97), Henicocorinae (1), and Rhyparochrominae (1927). Of these, Henry (1997) has transferred the Henicocorinae to the Idiostolidae and the Psamminae to the Piesmatidae.

2. GENERAL STATEMENT OF ECONOMIC IMPORTANCE

The economic species of the Lygaeoidea are concentrated in the families Blissidae, Orsillidae, Oxycarenidae, and Geocoridae; there are a few species of economic significance in the other families, especially in the large family Rhyparochromidae. It is not easy to decide whether a species is of major or minor importance because economic species that are restricted to one geographic area often would seem to be of minor importance to people living in other areas, but are very important to people where the insect occurs. Moreover, in the more "developed" parts of the world there may be an army of research workers publishing on an economic species and producing a large literature, but the insect may in the larger reality be but a minor nuisance such as some of the chinch bugs that infest expensive lawns and golf courses. On the other hand, a really serious pest that threatens the food supply in an "underdeveloped" poor area may have a meager literature and would thus seem of minor importance. Nevertheless, it is necessary to use the amount of literature to evaluate the subjective question of importance.

Although most of the Lygaeoidea are primarily seed feeders, so that the common name for the family Lygaeidae (*sensu lato*) is the seed bugs (Sweet 1960), the Blissidae (the chinch bugs), which include many economic species, are sap suckers that do not feed on seeds, with the single possible exception of *Ischnodemus brinki* Slater of South Africa (Slater 1976a). The mention of the well-studied *Blissus leucopterus* Say as feeding on seeds by Slansky and Panizzi (1987) is certainly incorrect. The Malcidae and the Colobathristidae, which were once included in the Lygaeidae, but are certainly in the Lygaeoidea, are also sap feeders that evidently do not feed on seeds and include species of economic importance (Štys 1963, 1966a,b, 1967; Sweet and Schaefer 1985). Similarly, the Piesmatidae, which were placed in a separate superfamily (Štys and Kerzhner 1975), but are actually in the Lygaeoidea (Henry 1997), are sap feeders and include species of economic importance (see Schaefer 1981; Heiss and Péricart 1983; and **Chapter 12** for reviews). The Berytidae, another member of the Lygaeoidea, are being treated as predators in this book (**Chapter 31**). This may be somewhat misleading, as the great majority of berytid species are phytophagous, some even pests (Schaefer and Wheeler 1982, Péricart 1984). Sweet stressed (1979) that plants are fundamentally low in protein, so consequently a great deal of secondary carnivory has evolved to take advantage of this excellent source of protein. I argued that from this need have evolved many ancient and recent lineages of the Heteroptera that became carnivorous, including some lineages in the Pentatomomorpha. The Geocoridae are a prime example of this in the Lygaeoidea, as are the Asopinae in the Pentatomidae (Schaefer 1996). It is therefore to be anticipated that phytophagous heteropteran species may also be on occasion carnivorous or cannibalistic, as is *Spilostethus pandurus* (see discussion of this species, **Section 3.8**).

The family Geocoridae, the big-eyed bugs, are predaceous, and these will be treated separately (**Chapter 30** on predaceous Lygaeoidea), although the geocorids will also feed on seeds and foliage of plants (Sweet 1960, Tamaki and Weeks 1972).

The Blissidae are probably the most economically important family of Lygaeoidea, and they will be emphasized in this review. Blissids are specialized for feeding on monocotyledonous plants, an ancient relationship that probably represents a Cretaceous radiation (Slater 1976a). Slater (1976a) gives in tabular form the known host plants of the blissid species and records of economic injury. Many Blissidae are notorious pests that attack graminaceous plants (Poaceae). These insects thus endanger the plants that produce not only the majority of the world's food but also other economic products. Economic grasses that have been attacked by blissids include grain crops: wheats, corn (maize), rice, sorghum, barley, rye, oats, and the many millets; grasses for sugar: sugarcane, corn, and sweet sorghum; grasses for alcoholic beverages, many of the above are fermented and distilled; grasses for pasture and hay forage for livestock and dairy production, many species of native and introduced grasses are used; several species of cool-season and warm-season grasses used for ground cover turf as in lawns, athletic fields, cemeteries, airfields, and golf courses; grasses used for sand and soil binders, and for ornamental plants; some grasses used for basketry, upholstery, ropes, and hats; grasses for roof thatching, brooms, beads, perfumes and herbs; and the bamboos that are well known for their extensive use in construction, poles, paper, for ornamental plants, as well as the young shoots for food (Stefferud 1948, Core 1955, Bailey 1949, Uphof 1968, Bailey and Bailey 1976, Brown 1979, Simpson and Conner-Ogorzaly 1995, Ohrnberger 1999). Given this great economic importance of the Gramineae (Poaceae), and that the potential for the development of new pest species is very high in the Blissidae (Slater 1967), blissids should be very closely monitored for potential new pest outbreaks and for accidental introductions. Slater (1967) is frequently cited as stating that species of the tropical grass tribes (actually subfamilies) Panicoideae and Eragrostoideae are the most extensively utilized hosts by blissids, and the temperate Festucoideae are the least utilized. However, as Slater himself noted, host ranges vary considerably, and *Blissus*, for example, attacks members of seven tribes, including the supposed little-utilized Festucoideae.

Even though the Blissidae have a worldwide distribution, in all zoogeographic regions, the status of the species as pests varies considerably from area to area, which would make accidental introductions a constant threat. Australia, for example, despite an ancient blissid fauna, has no blissid species recorded as pests, whereas Africa has a number of genera that include pest species (Slater 1976a). Also, although eastern North America and the eastern Palearctic and Oriental areas have several serious blissid pests, Europe of the western Palearctic appears to have no blissids of significant economic importance.

Interestingly, blissid species, even within the same genus, may differ considerably in host specificity, some being restricted to one host, and others being relatively general feeders on susceptible grasses (Slater 1976a). Moreover, a host-specific blissid species may exist in large populations, often with seemingly little visible effect upon the host plant, which is termed tolerance, while another host-specific blissid species may occur in very low populations on its host plant, which may reflect host resistance, despite host specificity. In general, the blissid species with the widest host ranges are the most likely to be economic pests and severely damage the plant as seen in *B. leucopterus* and *Macchiademus diplopterus* (Distant) (Leonard 1966, Slater 1976a).

The family Oxycarenidae, which feeds on both seeds and sap, chiefly of malvaceous plants, are of major economic importance because several species of *Oxycarenus,* the cottonseed bugs, feed on the seeds of cotton plants, causing the cotton to be stained (Fletcher 1914, Kirkpatrick 1923, Samy 1969, Anneke and Moran 1982). The widespread *O. hyalinipennis* (Costa), the dusky cottonseed bug, is the most studied species.

The Orsillidae, which include many economic species, especially of *Nysius* (the false chinch bugs), feed on seeds but also feed on vascular tissues, and much of the recorded damage occurs when large populations migrate from wild hosts to crop plants, especially during times of water stress (Ashlock 1967). Although the species are of no or little economic importance on the Hawaiian Islands (U.S.A.), this family has radiated there extensively, producing about 90 species and eight genera, and forms an excellent example of insular evolution (Usinger 1942, Usinger and Ashlock 1959, Ashlock 1967).

Most of the economic species of Orsillidae are contained in *Nysius*, a large worldwide genus that includes over 100 described species (Ashlock 1967). To quote Ashlock, "Virtually no place in the world where orsillines are found is lacking a species of *Nysius*. Members of the genus have been found on all the continents of the world except Antarctica, and in most island groups except southeastern Polynesia. Since the genus contains several species that occasionally are of economic importance, there has been more than casual interest in the group." However, again to quote Ashlock, "but since most Orsillinae are small, not very attractive, and taxonomically difficult, interest in the group — with few outstanding exceptions — has not been high." It is unfortunate, however understandable, that economic workers have generally focused on harmful outbreaks of *Nysius* bugs, not on the natural habits of the species or their taxonomy. Because of the apparent difficulty in identification of *Nysius*, a distressingly large number of reports in the economic literature, even in supposedly well-known North America, merely refer to "*Nysius* sp."

The Lygaeidae (*sensu stricto*), already mentioned with respect to physiological and ecological studies, have a few species, especially in the genus *Spilostethus,* that are recorded as pests on agricultural crops (Hoffman 1932, Bhattacharjee 1959, Slater 1964c). Like *O. fasciatus, Spilostethus pandurus* is rapidly becoming a laboratory animal in Europe and India, as well as being the most important economic pest of the Lygaeidae *sensu stricto.*

In the Rhyparochromidae, by far the largest family of the Lygaeoidea, there is a scattering of records of economic species, especially on peanuts, rice, and strawberries (Sweet 1960, Slater and O'Donnell 1995). The most unusual rhyparochromids that should be mentioned here because of their potential medical significance are the members of the Old World tropical tribe Cleradini, which are blood suckers on vertebrates (Sweet 1967; Wickramashinghe 1969; Wilton-Smith 1978; Malipatil 1983; Harrington 1988, 1990) similar to the largely New World triatomine Reduviidae (Ryckman and Blankenship 1984). As they are recorded to bite humans, cleradines therefore have the potential to become economically important as vectors of Chagas' disease (Illingworth 1917, Costa Lima 1940). As shown by the experimental work in Brazil by Lent (1939) and Ferreira and Deane (1939), the introduced cleradine species *Clerada apicicornis* Signoret can transmit *Trypanosoma* (= *Schizotrypanum*) *cruzi*, the causative agent of Chagas' disease. Interestingly, *C. apicicornis* is also a major predator on triatomines (Simmonds 1971), from which it probably secondarily obtains blood. The potential transmission of Chagas' disease may similarly be true of another rhyparochromid, the Neotropical udeocorine species *Bathycles* (= *Barnsleya*) *amarali* Corrêa, which is also hematophagous (Corrêa 1956).

The Malcidae, which were treated by Slater (1964a) as a subfamily of the Lygaeoidea and raised to family level by Štys (1967), are of major economic importance because members of the genus *Chauliops* attack leguminous crops in several tropical areas of the Old World, especially in the Orient (see review by Sweet and Schaefer 1985). *Malcus* itself is a minor pest of mulberries (Moraceae) in the Orient.

The Colobathristidae are long-legged slender insects with a tendency to ant mimicry that are found in the Neotropical and the Oriental–Australian regions. They all feed on grasses, and species of *Phaenacantha* are pests on sugarcane in Indonesia and Australia. Like the similar-appearing Berytidae, the Colobathristidae have had a checkered systematic history. Stål (1865) transferred *Colobathristes* Burmeister from the Coreidae to the Lygaeidae and established a separate subfamily for it. Bergroth (1910) elevated the taxon to family rank based on its coreidlike antennal insertion high on the head above a line between the base of the rostrum and the center of the eye. In addition, the valvulae of the ovipositor are platelike as in the Coreidae, not ensiform as in nearly all Lygaeoidea. However, based on their trichobothrial arrangement on sterna 5 and 6 of one trichobothrium anterior and two posterior to the spiracle, and other characters (Štys 1966b, Schaefer 1975, Sweet 1981), the Colobathristidae are cladistically clearly Lygaeoidea (lygaeid line) rather than Coreoidea (coreid line of Štys 1966a) and so are treated here under the Lygaeoidea.

An organizational problem is created in the Lygaeoidea, in that one is dealing with several families, each with species of greater and lesser importance. For clarity, the family designation of each species will be indicated beside the species name. For brevity, some of the economic species of minor importance will be mentioned in the context of the better-known related species in the taxon, usually a genus, as the biology, damage, and control often are very similar.

3. THE IMPORTANT SPECIES OF ECONOMIC IMPORTANCE

3.1 *Blissus leucopterus leucopterus* Say (Blissidae)

3.1.1 *Distribution, Life History, and Biology*

This North American species, the original chinch bug, is perhaps the most notorious of the pest species of the Lygaeoidea. It is now called the common chinch bug, because other species are also called chinch bugs. As a consequence of the economic importance of the chinch bugs, the Lygaeidae were known as the chinch bug family, an inappropriate name because chinch bug usually refers to only the three economically important species of the genus *Blissus* of North America, and most lygaeoids are seed feeders, not sap feeders like the chinch bugs (Sweet 1960, Werner 1982). The chinch bug family name is, however, appropriate for the taxon Blissidae. The name chinch bug is clearly derived from the Spanish *chinche* for bed bug, perhaps in reference to the buggy odor and small size of the insects (Fitch 1856; Webster 1896a, 1898). Such a vast economic literature exists on the chinch bug that it can only be summarized here; it should be noted that many papers are not listed in the reviews of the literature (Forbes 1890, Metcalf and Flint 1939, Metcalf et al. 1962, Slater 1960, Spike et al. 1994, Slater and O'Donnell 1995).

Chinch bugs of the genus *Blissus* are 3 to 4 mm long and 1 mm wide with a black body and black and white wings, the black color forming sort of an X. This coloration is especially distinctive in *B. leucopterus leucopterus* as the populations are almost completely macropterous, that is, long-winged. The brachypters, which are more abundant in most other species, are darker and the coloration pattern not so conspicuous, as the wings are short. The females are a little larger, generally over 4 mm long; the males smaller, generally under 4 mm. Aside from size, the adult sexes can be readily distinguished. The female abdomen is wider with a slit on the underside of the apex of the abdomen (segments 5 to 7) to accommodate the ensiform ovipositor that folds into the abdomen like a jackknife. The male abdomen is slenderer and has a rounded apex with an invaginated genital capsule that contains the phallus. This distinction is true of nearly all lygaeoid bugs.

Because of its economic importance, Slater and China (1961) proposed that the type species of *Blissus* be shifted to *B. leucopterus* because *B. leucopterus* is not congeneric with Palearctic *B. hirtulus* Burmeister 1835, the original type species for the genus *Blissus*. Unfortunately, the International Commission of Zoological Nomenclature (1964) rejected this proposal and instead placed *Blissus* on the official list of generic group names in zoology with *hirtulus* remaining as the type species. Slater (1979) and Slater and O'Donnell (1995) have rejected this decision, and continue to associate the name *Blissus* with *leucopterus*, and remove *hirtulus* to the genus *Geoblissus* Hidaka 1959, the type species of which, *G. rotundatus* Hidaka 1959, was synonymized with *hirtulus* itself by Slater, Ashlock, and Wilcox (1969). If *Blissus* is defined in the sense of *hirtulus,* as sanctioned by the Commission, then as Slater (1979) noted, the next available name for the Western Hemisphere chinch bugs of the *leucopterus* group would be *Neoblissus* Bergroth, which was synonymized with *Blissus* (in the sense of *leucopterus*) by Slater (1979). *Neoblissus,* however, is unusual in that it is based on a myrmecophilous species (*N. parasitaster* Bergroth) from Argentina and Brazil, which was found feeding on grasses in nests of fire ants (*Solenopsis saevissima* Buren), and was originally erected in distinction from *Blissus* (*sensu lato*) (Drake 1951).

There are approximately 25 described species in the genus *Blissus* defined as a New World taxon, 15 in the continental United States and Canada. Unfortunately, there is no key to the species, except to those of eastern North America (Leonard 1968a), a group of species Leonard called the *leucopterus* complex. The taxonomic chaos of *Blissus* at the time of Slater's catalog (1964) is shown by the distribution of the then included species in eight other genera (Slater and O'Donnell 1994).

Leonard (1966, 1968a) should be consulted for his excellent analysis of the literature, and for his contributions to the systematics, cytogenetics, zoogeography, ecology, and biology of the "*leucopterus* complex" of the genus *Blissus* in eastern North America. This complex includes several closely related, similar species (Leonard 1968a). The taxonomy of the complex was very confused and specialists had disagreed on the status of the taxa (Leonard 1966, 1968a) and as a consequence the literature must be treated with caution. For example, even some of the references and life-cycle information given in Spike et al. (1994) that were specifically attributed to *B. l. leucopterus* in the southern part of its range actually refer to *B. insularis* and, in the northern part of its range, to *B. l. hirtus*.

Fortunately, Leonard (1966, 1968a) thoroughly revised the classification of the "*leucopterus* complex" of *Blissus* in eastern North America. He recognized nine species and two subspecies, three of which are important pests, *B. l. leucopterus*, *B. l. hirtus* Montandon, and *B. insularis* Barber. He synonymized with *B. l. leucopterus* the variety names erected by Fitch (1856) and Riley (1875). He demonstrated the specific distinctiveness of the taxa *B. arenarius* Barber, and *B. insularis* Barber, of the *B. leucopterus* complex that were treated as varieties by Barber (1918), and he erected a new subspecies, *B. arenarius maritima*. Leonard treated *B. leucopterus* itself as two subspecies, *B. l. leucopterus* and *B. l. hirtus*. He restricted *B. l. leucopterus* to the east central United States from Virginia south to Georgia, and west to South Dakota and central Texas roughly west to the 101 parallel and between the 42nd to 37th parallels in the north and along the 31st parallel in the south. *Blissus leucopterus hirtus* extends from Virginia north to the maritime provinces of Canada and west to Ontario and Minnesota, and overlaps with *B. l. leucopterus* across Virginia, West Virginia, Ohio, Michigan, Wisconsin, and Minnesota. (See discussion below of *B. l. hirtus*.) Below the 31st parallel along the Gulf Coast and Florida the species found is apt to be the economic species *B. insularis*, which overlaps with *B. leucopterus* (Slater and Baranowski 1990). In the western United States some of the records attributed to *B. leucopterus* are definitely *B. mixtus* Barber (Drake 1951, Leonard 1966).

An important implication of Leonard's work is that the records of *B. leucopterus* outside of the eastern United States are very probably incorrect, or might possibly represent introductions of *B. leucopterus* into other areas, which is much less likely for the tropical records because of the temperate climate life-cycle adaptations of *B. leucopterus*. More likely, given the close similarity in appearance of the species of the *B. leucopterus* complex to one another, and with the presence of the *B. leucopterus* complex throughout the Western Hemisphere south to Argentina, these extralimital records are probably misidentifications or undescribed species. Clearly, a taxonomic revision of the genus *Blissus* (= *Neoblissus*) is very badly needed. As Slater (1979) observes, there is no key to the species of *Blissus* (nor of *Geoblissus*) aside from the species of eastern North America (Leonard 1968). Moreover, the pest status of many of these records shows the important potential of the *B. leucopterus* complex to produce new pest populations as follows. In Sonora, Mexico, "near *B. leucopterus*" is recorded as injuring maize (Carrillo et al. 1966), but it is interesting, given the virulent attacks on corn in the U.S. Midwest that there are so few records of chinch bugs on corn in Mexico, the probable place of origin of *Zea mays* (Doebley and Iltis 1980, Simpson and Conner-Ogorzaly 1995). In Tabasco, Mexico, "*B. leucopterus*" is a pest on introduced forage grasses, para grass, *Panicum (= Brachiaria) purpurascens* Raddi, and pangola grass, *Digitaria decumbens* Stent (Garcia 1974). In Colombia (one of *six Blissus* species!), "*B. leucopterus*" is a pest on rice, sorghum, liendre puerco (Japanese millet) *Echinochloa colonum* (L.) Link, and other forage grasses (Posado 1976, Vásquez and Sánchez 1991). In Brazil "*B. leucopterus*" is a pest on millo (millet) and the forage grasses *Brachiaria, Hemarthria, Setaria,* and *Cynodon* (Hadler Pupo 1976, Reis et al. 1976, Souza et al. 1995); and in Latin America in general "*B. leucopterus*" is listed as a pest on rice (Cheany and Jennings 1975, CIAT 1982).

Some of the species of *Blissus* are host specific on wild grasses, and often, even when present in large numbers, do not apparently injure or kill the plant, such as *B. iowensis* Andre on *Andropogon* (= *Schizachyrium*) *scoparium* (Michx.) Nash in Iowa (U.S.A.) (Decker and Andre 1937), *B. brevisculus* Barber on the same host in the northeastern United States (Leonard 1966, Wheeler 1989a), and *B. arenarius arenarius* Barber on American beach grass (*Ammophila breviligulata* Fern.) (Leonard 1966), which shows strong tolerance of the plant for the bugs. Often in fieldwork, species of *Blissus* are found existing on a natural host plant in very small numbers, which perhaps indicates strong host resistance. In contrast, other species, such as *B. leucopterus*, which feeds on many introduced hosts, both crops and weedy grasses (Headlee and McColloch 1913, Hayes and Johnson 1925, Burks 1934, Ahmad et al. 1984), and *B. insularis* Barber, which feeds on the introduced lawn grass *Stenotaphrum secundatum* (Walt.) Kuntze, are severely injurious (Leonard 1966). Slater (1976a) discusses the pest-feeding patterns among blissids in general, and *Blissus* exemplifies the full range of such feeding patterns:

1. The population may build up into large numbers and migrate to other, often unrelated, hosts and injure them, but cannot breed on them.
2. After the primary host dies, dries up, or is harvested, the bugs may migrate to a secondary host and may temporarily breed on it in the absence of the normal host, but return to the normal host when it is available.
3. In favorable times the bugs may migrate to a secondary host and actually breed on it, but survival is low and the colony does not persist.

A fourth pattern that is characteristic of some chinch bugs is to attack and injure introduced species that evidently lack tolerance or resistance "antibiosis" to the chinch bugs, while the insect could be relatively scarce on its native host. As Slater and Wilcox (1973) stressed, this capacity to adapt to new susceptible host plants makes blissid species a continual and potential danger to graminaceous crop plants and forage plants especially those that are introduced from other areas.

There is a real question about the native or original host plant(s) of *B. leucopterus*. Leonard (1966) and Slater (1967) noted that the prevailing pattern is for *Blissus* species to be host specific or have a narrow natural host range. However, each of the species listed by Leonard (1966) as hosts of *B. leucopterus* is either an introduced crop grass or an adventive weed grass not native to the United States. Hayes and Johnson (1925) observed that the native prairie grasses in Kansas were rarely attacked or damaged, in contrast to the susceptible introduced grasses, including the crop species. Ahmad et al. (1984b) evaluated noncrop grasses in Nebraska as hosts and similarly found that adventive species goosegrass *Eleusine indica* (L.) Gaertn., yellow foxtail (= yellow bristle grass) *Setaria glauca* (L.) Beauv. (= *S. lutescens* [Weigel] F. T. Hubb in Leonard 1966 and Slater 1976), and crabgrass *Digitaria sanguinalis* (L.) Scop. were preferred host species. Other weedy introduced grasses recorded as being attacked by *B. leucopterus* are quack grass *Agropyron repens*. L., orchard grass *Dactylis glomerata* L., barnyard grass *Echinochloa crusgalli* (L.) Beauv. (Flint et al. 1935), and green bristle grass *Setaria viridis* (L.) Beauv. (Leonard 1966). Ahmad et al. (1984b) noted in their studies, that, while the chinch bug would in the laboratory feed on the native species Indian grass *Sorghastrum nutans* (L.) Nash and switchgrass *Panicum virgatum* L., it did not do so in the field. Furthermore, they found that the lawn grasses Bermuda grass *Cynodon dactylon* (L.) Pers. and bluegrass *Poa pratensis* L., which were listed as host plants by Leonard, were antibiotic nonhosts in their studies. They attributed this difference either to other chinch bug species such as *B. insularis* or *B. l. hirtus* feeding on these grasses or to cultivar differences in the grasses used. This antibiotic effect on *B. l. leucopterus* was also true of St. Augustine grass *Stenotaphrum secundatum* (Walt.) Kuntze, the host grass of *B. insularis* Barb., indicating that the two species in the zone of overlap in the southern United States are ecologically segregated by host utilization. They also noted that crabgrass, *Digitaria sanguinalis* (L.) Scop., which often grows in lawns, might be heavily infested with *B. l. leucopterus,* while the bluegrass would be unaffected, and that one might erroneously believe in this situation that the chinch bugs were feeding on the bluegrass. Lynch et al. (1987) observed that chinch bugs in Georgia initially attacked goosegrass and, after killing it, migrated to Bermuda grass and damaged it, but as Slater (1976a) noted, this would be a temporary nonpersistent utilization of an unfavorable host plant. This is probably the explanation for the statement (Reinert et al. 1995), "Large numbers of these chinch bugs often migrate from maturing grain fields into turf areas of bermudagrass and Kentucky bluegrass and cause damage overnight, even though these grasses are not their preferred hosts." Lynch et al. (1987) suggested that perhaps *B. l. leucopterus* is in the process of adapting to a new host. Slater (1976a) believes that the few records of *B. leucopterus* on sugarcane actually pertain to *B. insularis,* but these may actually refer to records of other *Blissus* species from the Antilles, not of *B. insularis* (Leonard 1968a,b).

Webster (1895), in evaluating the hypothesis that the chinch bug migrated west from North Carolina (U.S.A.) where it was first recorded as a pest in 1783, gave evidence that the insect appeared very soon on corn in the same year that the sod was first broken in Illinois and so he believed that it must have already been present on native grasses, an assessment also made by Headlee and McColloch (1913) in Kansas. Interestingly, this early literature ignores that many Indians of eastern North America were agrarian and grew maize (corn) out to the eastern part of the Great Plains (Lowie 1954, Murdock 1967, Hughes 1983), and one would anticipate that chinch bugs had exploited maize before the advent of the European settlers. Webster also hypothesized that the chinch bug had originated in coastal Mexico and spread north into the United States; but this he based on specimens of *Blissus* reported in the literature, which specimens probably actually represented several undescribed species of *Blissus,* and on the presence of what he thought was *B. leucopterus* on beach grasses; however, this was actually another species, *B. arenarius* Barber. Some records of *B. leucopterus* (as the buffalo grass chinch bug) from the High Plains on buffalo grass, *Buchloë dactyloides* (Nutt.) Engelm. (Baxendale 1992), and from native short grass plains (Hewitt et al. 1974, Kumar et al. 1976, Watts et al. 1989) should be reexamined to identify the insects. These more western records may apply to *B. occiduus* Barber or *B. planarius* Barber, or perhaps an undescribed species (Reinert et al. 1995). The best candidate for the natural host plant of *B. l. leucopterus* may be the Carolina foxtail *Alopecurus carolinianus* Walt., a native species mentioned by Shelford and Flint (1943) as a host of the chinch bug. Gould (1975) described this as "a widespread native species found in moist, often-disturbed soils along ditches, streams, woods borders, and pasture lands." Another native grass listed by Shelford and Flint (1943) as a possible native host of *B. leucopterus* is tickle grass, *Agrostis hyemalis* (Walt.) B.S.P., a native meadow grass widespread throughout the range of *B. leucopterus* and "frequent in pastures, on road banks and ditch banks and open woods on sandy soils" (Gould 1975, Steyermark 1977). This species also appears to be a successional plant not common in the climax prairie communities. If these two species are indeed the original host species, then *B. leucopterus* may have been originally adapted to seek out these species, which only grew in disturbed habitats within the original prairie, and the strong migratory behavior and very low percentage of brachyptery (0.01% of *B. l. leucopterus*) (Leonard 1966) may have originally been an adaptation to seek out this scarce natural host after overwintering in native nonhost bunchgrasses in the prairies. This migratory behavior may be an important factor in the ability of this species to locate and exploit susceptible hosts such as introduced crop and weed grasses. Nevertheless, the presence of brachyptery (Barber 1937) may indicate a former time when *B. leucopterus* was tied to a susceptible host in a permanent habitat.

Although *Z. mays* is probably not native to the United States, and the probable ancestral species and its subspecific varieties are in southern Mexico and Guatemala (Doebley and Iltis 1980, Iltis and Doebley 1980, Doebley 1990), maize is certainly an ancient introduction into central and eastern North America and it had probably been attacked by chinch bugs for a long time before the advent of the Europeans.

There are many accounts of the bionomics of the chinch bug dating back to Fitch (1856), Walsh (1861), Riley (1870, 1875), Forbes (1890), Webster (1907), and Headlee and McColloch (1913), but the most thorough study is that of Luginbill (1922), who made many observations on the bionomics and behavior of chinch bugs. *Blissus leucopterus leucopterus* hibernates and diapauses as adults, and normally migrates to shelter in nonhost bunch-forming native grasses as little bluestem *Schizachyrium scoparium* (Michx.) Nash, big bluestem *Andropogon gerardii* Vitman, tall dropseed *Sporobolus asper* (Michx.) Kunth, and in the southern United States, false redtop *Triplasis purpurea* (Walt.) Chapm. and broomsedge bluestem *A. virginicus* L. (Headlee and McColloch 1913, Shelford 1932, Shelford and Flint 1943, Leonard 1966, Lamp and Hotzer 1980, Negrón and Riley 1991), which had led some workers erroneously to consider these species to be the wild host plants. The statement in Reinert et al. (1995) that *B. l. leucopterus* was a pest on many bunchgrasses probably stems from these records. Survival of the chinch bug is highest in such protected sites in nonhost clumps, especially under a snow cover in drier years, while mortality was greatest in wet, cold years at sites without the insulation of the clumps (Headlee and McColloch 1913, Decker and Andre 1937).

From these hibernating sites the chinch bugs migrate in the spring normally by flying during sunny days to attack small grains, like barley and wheat. During this migration the bugs were often reported to fill the air completely, forecasting a bad pest year (Riley 1870, 1875; Headlee and McColloch 1913; Luginbill 1922; Flint and Larrimer 1926). Later in the season, often after destroying the small grain crops, or after harvest, the bugs of the first new or spring generation often migrate by crawling to corn and sorghum that may be damaged considerably, reflecting the population buildup through two generations (Forbes 1890, Headlee and McColloch 1913, Flint et al. 1935, Snelling 1936, Wilde and Morgan 1978, Stuart et al. 1985). In the autumn, a third migration occurs made by the second, or summer, generation, usually by flight, to hibernation quarters in grass clumps (Riley 1870, 1875, Luginbill 1922, Flint and Larrimer 1926, Flint et al. 1935). Price and Waldbauer (1975) observed that the migration strategy of the chinch bug in seeking refuge in protected sites (grass clumps) in uncultivated areas and the return migration out to seek host plants maximizes the colonization of susceptible crop islands as, one might add, it probably once did for its original native, probably successional hosts.

There are evidently two generations a year throughout its range (Riley 1875, Headlee and McColloch 1913, Flint and Larrimer 1926, Leonard 1966), although Swenk (1925), Bigger (1935), Snelling (1936), and Meija-Ford (1997) thought there were probably three generations in the southern part of its range. Shelford (1932) was able to raise a third and even a partial fourth generation in the laboratory. However, this and other such laboratory records of a third generation may be an artifact of laboratory conditions, because Luginbill (1922), in his careful study, found that in the field there were only two generations of *B. l. leucopterus* at the latitude of South Carolina (U.S.A.) near the southern limit of the range of the species, the same as in the northern part of its range (Headlee and McColloch 1913, Drake et al. 1934). Leonard showed that under field conditions the second-generation fall adults entered into diapause and migrated to seek shelter for overwintering. The southern limit of the species may then reflect the need for cold to complete diapause under field conditions. Leonard (1966) documented the occurrence of reproductive diapause in a *B. l. leucopterus* strain from Oklahoma (U.S.A.), and that lengthening photoperiod appears to be involved in the natural termination of diapause. However, under laboratory conditions continuous high temperatures (29.5°C) can artificially break diapause in the absence of light. Leonard (1966) noted that because egg laying in *B. leucopterus* occurs over a long time, females of the overwintered generation may still be laying eggs when the first or summer generation is becoming adult and also laying eggs; later, when some bugs are maturing into diapausing fall adults, females of the summer generation would still be present, laying eggs (Luginbill 1922), leading to the mistaken belief that there is a partial third generation. Severin (1923) observed that under the short summer conditions in South Dakota (U.S.A.), many of the second-generation nymphs were unable to complete nymphal development and entered as nymphs into hibernating quarters but perished in the winter, as only the adults survived. Leonard (1966) found that in Connecticut (U.S.A.), the nymphs were similarly unable to survive the winter. Many workers such as Decker and Andre

(1937) had noted without comment that only the adults overwintered. It then seems probable that the shortness of the season to complete the second generation may thus impose the northern limit to the distribution of *B. l. leucopterus*.

The immature stages of the chinch bug have been informally described and illustrated by many authors (Fitch 1856; Riley 1870, 1875; Thomas 1879; Chambliss 1895; Severin 1923; Drake et al. 1934), but the most careful descriptions were made by Luginbill (1922), which were cited by Leonard (1968). The illustrations of the immature forms in Luginbill by E. H. Hart have been frequently reproduced in general texts and economic literature, but sometimes these figures have been erroneously attributed to Webster. There are, as usual in the Heteroptera, five instars. The following brief descriptions are from Luginbill (1922) and Severin (1923). The eggs are small, 0.9 mm long by 0.3 mm wide, elongate, slightly curved, with three to five, usually four small chorionic processes at the flattened cephalic end. The egg when laid is whitish with a smooth, shiny, somewhat iridescent chorion. The egg turns yellow and then red as it develops and the nymph within is readily visible before hatching. The first instar has a brown head and thorax. The abdomen is yellow at the base, light red in the middle, and black at the apex. The antennae are dusky and the legs pale. As it goes through its molts, the head and thoracic sclerites become black, the abdomen becomes redder and darker, the yellow basal abdominal segments become white, and the wing pads increase in size and length. In the fifth instar the head and thorax are black and the abdomen is a very dark red except for the pale basal two segments visible between the wing pads. On the dorsal midline of intersegmental sutures 4 to 5 and 5 to 6 are dark spots surrounding the scent glands. The five instars are easily discriminated as in Hemiptera in general, which makes such insects convenient to use in life-cycle studies. In the first instar, the head is about as wide as the thorax; in the second, the head is distinctly narrower than the thorax; in the third, the wing pads appear; in the fourth, the wing pads extend to the first abdominal segment; in the fifth, the wing pads extend to the second or third abdominal segment, covering most of the first two abdominal terga. The body lengths are first, 0.9 mm; second, 1.29 mm; third, 1.65 mm; fourth, 2.11 mm, and fifth, 2.97 mm.

In laboratory studies egg production per female ranges from 39 to 1091, much of the variation depending on the host cultivar used as food source, individual variation, and laboratory conditions (Janes 1935, Leonard 1966, Stuart et al. 1985, Wilde et al. 1986). Janes found that egg laying in *B. l. leucopterus* was influenced by temperature: the average number of eggs per female was 532 at 24.5°C ($n = 12$ females); 598 at 29.5°C ($n = 11$ females); 502 at 34.5°C ($n = 11$ females). Eggs are laid at a rate of 1 or 2 to 15 a day. The females died directly after the oviposition period (Luginbill 1922, Janes 1935, Leonard 1966). Janes found *B. leucopterus* unusual in that the males lived longer on average than females, as females are usually the longer lived in many insects. Luginbill observed that 8 to 9% of the eggs in his study were infertile. Leonard (1966) found that if unmated, the female laid some infertile eggs after a more extended preoviposition period, which may represent the unfertilized eggs of Luginbill. The eggs are laid preferentially in the leaf sheaths, but some are laid in the soil and at the base of the grass host plants (Luginbill 1922, Janes 1935).

Headlee and McColloch (1913) found that egg development time varied with temperature, and measured 17.3 days at 73°F, and 11.5 days at 78°F. Luginbill (1922) found the eggs to take from 31 days in cool spring conditions to 9 days in the hot summer to hatch. Shelford (1932) found egg development to be 16.5 days at 23°C.

Headlee and McColloch (1913) measured the development time from egg to adult to be 46 days in the summer. Decker and Andre (1938) in their excellent study on the univoltine species *B. iowensis* Andre found that *B. l. leucopterus* grew approximately twice as fast from eclosion to adult as *B. iowensis*: at 24.5°C, 40.0 compared to 74.8 days; at 31.5°C, 21.2 compared to 45.7 days. They interpreted these results as a consequence of *B. iowensis* being univoltine, and *B. leucopterus* bivoltine. Interestingly, *B. iowensis*, which in the field is host specific on little bluestem, *Andropogon* (= *Schizachyrium*) *scoparium*, grew as rapidly on succulent corn (*Z. mays*) as on its normal host, indicating the danger of using only laboratory data to assess host specificity. Burks (1934) reared *B. leucopterus* on 13 different grass species and found that the development time from egg to adult varied from 33 to 37 days in the summer. Shelford (1932) found the rate of development from the egg to the adult to vary from an average of 59.5 days at 22°C to 31.4 days at 31°C. He determined the optimal temperature for nymphal development to be between 29°C and 33°C. He further found that while high humidity was favorable, actual water was often fatal to the insects, especially the early instars. High temperatures are also fatal to the chinch bug and the insects succumb quickly to temperatures over 40°C (Forbes 1883, Headlee and McColloch 1913, Guthrie and Decker 1954). Mejia-Ford (1997) notes that the careful bionomic study on growth rates of the chinch bug by Vásquez and Sánchez (1991) in Colombia very probably refers to an undescribed *Blissus* species, not *B. leucopterus*.

Leonard (1966) observed that there was a definite courtship behavior in which the antennae vibrate in both males and females and the male on mounting the female flails her head with his antennae. Both he and Luginbill (1922) observed that the insects mate end to end, frequently at intervals of 5 to 8 days and may remain *en copula* for a long time, 2 to 12 hours, the female dragging the male behind her. Luginbill noted that the insects feign death when disturbed, and that in the cool of the day the insects are sluggish in contrast with their rapid movements at midday. He noted a strong social tendency even of first instars to huddle closely together at rest and during feeding. This social behavior may enhance the repellent effect of the scent gland secretions.

An interesting observation: Spike et al. (1991) showed that the fatty acid compositions of the triacylglycerols and phospholipids from *B. l. leucopterus* and *B. iowensis* were similar to the characteristic dipteran pattern, rather than to the pattern found in all other insects examined.

3.1.2 Damage

Enormous damage was done by *B. l. leucopterus* in the United States, probably greater than any other insect save the boll weevil and grasshoppers (Flint 1935, Metcalf, et al. 1962) and the literature gives many estimates of the immense damage in dollars, which should be adjusted to current value. Shimer in Packard (1870) estimated the damage to be over $100 million in 1864. This insect has caused severe damage to spring and winter wheat, spelt, maize (Indian corn), field corn, broom corn, millet, pearl millet, rye, barley, sorghum, grain sorghum, rice, and Sudan grass. Oats seems to be the least affected, barley the most (Riley 1875, Headlee and McColloch 1913, Drake et al. 1934, Flint et al. 1935, Packard and Benton 1937, Peairs 1941, Leonard 1966, Atkins et al. 1969). The earliest records of *B. l. leucopterus* cited it as being so destructive on small grains in Virginia and North Carolina (U.S.A.) that the wheat crop was destroyed in 1783 (Fitch 1856), which contributed to the abandonment of wheat cultivation in Virginia and North Carolina (Riley 1870; Webster 1896a,b, 1898; Simpson and Conner-Ogorzaly 1995). However, in more recent years wheat has been rarely damaged by chinch bugs, and, at present, damage lies more in the migration of chinch bugs from wheat to corn and sorghum (Wilde and Morgan 1978, Stuart et al. 1985).

The damage occurs through feeding on the phloem and xylem, often near the base of the plant where the grass meristem occurs. The feeding can cause wilting, chlorosis, stunting, and death in susceptible plants through the clogging of vascular transport systems (Painter 1928, 1951; Negrón and Riley 1990; Spike et al. 1991). Wilde and Morgan (1978) found that 30 adult chinch bugs were sufficient to kill sorghum plants 7 to 12.5 cm in height. The plants could survive lesser numbers of bugs and after insecticide control of the infestation, but the growth yield was reduced (Ahmad 1982, Ahmad et al. 1984a, Negrón and Riley 1985). As in similar feeding by leafhoppers, salivary sheaths were deposited by chinch bugs along feeding tracks within the tissues (Backus et al. 1988), showing characteristic feeding on phloem tissue bundles (Backus 1988). This is probably the salivary secretions that Painter (1928) observed to clog the vascular transport systems. Mejia-Ford (1997) demonstrated that, although chinch bugs do not feed directly on seeds of grains, the withdrawal of resources through sap feeding causes reduced seed yield and reduced seed development. For experimental work the insects can be readily reared in the laboratory, either on sorghum stalks in cartons (Parker and Randolph 1972) or on whole plants in pots in greenhouses (Shelford

1932, Leonard 1966). Shelford observed that in part some of the large variations in the literature, including his own work, were probably the result of inadequate laboratory conditions. Another important variable he noted was the "vigor" of the insect strain.

From the earliest records on, it has been noted that the damage was periodic, the greatest in dry years, and the least in wet years (Fitch 1856, Riley 1875, Thomas 1879, Shelford 1932, Flint 1935, Shelford and Flint 1943, Watts et al. 1989). Moreover, heavy rains can directly kill the insects, especially the young nymphs, by drowning them and beating them into the mud (mudding), so that a severe infestation may disappear magically after a hard rain (Riley 1875, Headlee and McColloch 1913, Swenk 1925, Drake et al. 1934). Swenk (1925) and Shelford and Flint (1943) give good discussions of the historical outbreaks of the chinch bug. Thomas (1879) proposed that a succession of dry years and high temperatures led to severe outbreaks. However, Shelford (1932) and Shelford and Flint (1943), by carefully analyzing meteorological records and recorded damage, noted that this hypothesis appears to be basically true, but other important factors were involved in the outbreaks. Shelford (1932), through experimental work, demonstrated that the vigor of a chinch bug strain in the field was very important in causing damage, which helped to explain why greater damage would occur in some areas but not other areas with evidently similar weather conditions. Shelford and Flint (1943) showed that there was a distinct correlation of damage with high sunspot numbers and ultraviolet radiation, which is in turn related to higher solar output. The study by Souza et al. (1995) in Brazil on supposed *B. leucopterus* is probably on an undescribed species of *Blissus*, but it is interesting to note that they found its most serious damage was during periods of highest precipitation, the opposite pattern of the North American *B. leucopterus*.

Although the chinch bug has not been recently the severe pest on corn and small grains it once was, it reappeared in the 1970s as a serious pest of grain milo sorghum (Wood and Starks 1972, Wilde and Morgan 1978), and caused losses in Kansas estimated at $19 million in 1989 (Spike et al. 1994). On corn, which is not so good a host as sorghum, the chinch bug has reappeared as a pest in recent years (Negrón and Riley 1985, 1990) in Louisiana (U.S.A.). Spike et al. (1994) hypothesized that the lessening of the importance of chinch bug on corn in the midwestern United States was probably related to reductions in acreage of its overwintering sites in wild grasses and reductions in acreage of its small grain hosts. Although it was earlier considered only a minor rice pest (Ingram 1927, Douglas and Ingram 1942), *B. l. leucopterus* is progressively becoming a more serious rice pest in Texas (U.S.A.) (Texas Agricultural Extension Service 1997, Mejia-Ford 1997).

3.1.3 Control

Riley (1875) and Webster (1909) listed several bird species that consume chinch bugs, in particular the bobwhite, *Colinus virginicus* (L.). In Iowa (U.S.A.) 20 species of birds, especially quail, devoured chinch bugs (Drake et al. 1934). Howard (1887) noted that the reduviid bug *Pselliopus cinctus* (Fabr.) was an important predator of chinch bugs. Shimer (in Riley 1875) observed that the lady bird beetle *Coccinella munda* Say and the lacewing *Chrysopa plorabuna* Fitch feeds on chinch bugs in areas of high bug density. Ants, as important ground insects, were cited as probable controls (Riley 1875; Forbes 1883, 1916), and Leonard (1966) noted that in his laboratory cultures the presence of the ant *Leptothorax curvispinosus ambiguus* Emery apparently caused the decline of chinch bugs. Dahms and Kagan (1938) found that the beetle *Collops quadrimaculatus* F., abundant in the field with chinch bugs, ate their eggs.

Two insect parasites are known from *B. l. leucopterus*. More important is a scelionid egg parasite, *Eumicrosoma benefica* Gahan (McColloch 1913; McColloch and Yuasa 1914, 1915), which parasitized from 30 to 50% of the eggs (Packard and Benton 1937, Wright and Danielson 1992). A tachinid fly *Phorantha occidentis* Walker was reared from chinch bugs in South Carolina (U.S.A.) (Luginbill 1922). This fly is also known from *Nysius angustatus* Uhl., *N. "ericae"*(= *raphanus* Howard), and *Miris (= Leptopterna) dolabratus* L. (Arnaud 1978). Arnaud noted that the taxonomic status of *Phorantha occidentis* is uncertain. *Mermis* nematodes (hair snakes) have been reared from the common chinch bug (Webster 1909, Drake et al. 1934), but apparently no effort has been made to study or utilize this parasite.

Much effort has been made to utilize fungal diseases to control chinch bugs after it was realized that they could devastate chinch bug populations (Shimer 1867). Steinhaus (1949) listed six fungi species as parasitic on *B. l. leucopterus,* two being significant: the gray fungus *Empusa aphidis* Hoff. and the white muscardine fungus *Beauveria bassiana* (Balsomo) Vuillemin (= *Sporotrichum globuliferus*). The latter species, which consists of six strains (Macleod 1954), is the first pathogen applied to control a pest insect (Snow 1893, Forbes 1895). However, it was shown that the spores were abundantly present in nature (Billings and Glenn 1911), and the disease was effective only in wet years when it often caused dramatic reductions in chinch bug populations (Headlee and McColloch 1913); in dry years the disease was ineffective, so interest in using it for control has dwindled (Coppel and Mertens 1977). Krueger et al. (1991, 1992) demonstrated that clumps of little bluestem *Schizachyrium (= Andropogon) scoparium* served as a stable perennial reservoir for *B. bassiana,* which promotes the infection of overwintering chinch bugs. The practice of burning the bunchgrasses thus reduces the chance of infection of the chinch bugs by the fungus. Contrary to common belief, the infection of chinch bugs by *Beauveria* was similar in dry and wet years, but the effect of dry years was to inhibit the propagation of the fungus through reduced conidia development (Ramoska 1983, Krueger et al. 1992). The effect of the fungus was much less on chinch bugs feeding on sorghum and corn than on those feeding on other plants, which suggests a fungal inhibitor in corn and sorghum (Ramoska and Todd 1985).

Glasgow (1914) has described bacterial symbionts in gastric mycetomes from this species. These microbes are probably mutualistic (Steinhaus et al. 1956) and may play a role in host plant utilization; this seems deserving of further study, given the varying patterns of host specificity and damage by *Blissus* species.

Because of the specificity of blissids to grasses, the rotation of crops is a common and effective strategy to control chinch bugs when the alternate crop is not a grass but a dicot, such as soybeans. It also helps to grow only one grass crop each year such as summer corn or spring wheat because this would deprive one generation of chinch bugs of food (Riley 1875, Flint et al. 1935, Wilde et al. 1986a). However, rotating sequentially two grass species, such as wheat with maize (corn), sorghum, or rice, would set up the potential for the development of severe population outbreaks (Luckman and Metcalf 1975, Weber and Parada 1994).

Because chinch bugs overwinter in clumps of native grasses, an early control method was to burn native bunchgrasses near agricultural fields, especially small grains; but this method tends to eliminate natural enemies, including diseases, and should be used only during severe infestations that have successfully overwintered (Riley 1875, Flint 1919, Flint and Larrimer 1926, Flint et al. 1935).

Because chinch bugs of the first or spring generation crawl to new hosts after the first, usually a small grain crop, have either died or been harvested, barrier techniques using creosote ridge lines and paper strips or dust lines were widely used to prevent this invasion from one field to another (Riley 1875, Flint and Larrimer 1926, Harris and Decker 1934, Metcalf and Flint 1939, Metcalf et al. 1962). The dust line technique is enhanced by the susceptibility of *B. L. leucopterus* to hot conditions especially if the humidity is high (Guthrie and Decker 1954). The effectiveness of creosote was due to the excitorepellency of 4,6-dinitro-*o*-cresol, which causes "acute disturbances of the sensory nervous system and uncoordinated behavior, causing the insect to escape the treated area" (Metcalf et al. 1962, Metcalf and Metcalf 1975). The chinch bug similarly invades rice from spring wheat in Arkansas and Louisiana (U.S.A.), before the rice is tall enough to be submerged. Once submerged, rice is protected from the chinch bugs (Ingram 1927, Isely and Horsfall 1931, Douglas and Ingram 1942, Meija-Ford 1997).

Much effort has been made to develop strains of crop grasses that are resistant to the chinch bug (Packard and Benton 1937, Painter 1951): corn (Flint 1921, Holbert et al. 1937, Dahms and

Sieglinger 1954, Davis et al. 1996); wheat (Jones 1937, Stuart et al. 1985, Wilde et al. 1986b); barley (Dahms and Johnson 1954); sorghum (Snelling et al. 1937; Dahms and Fenton 1940; Wilde and Morgan 1978; Wilde 1979; Smith et al. 1981; Ahmad et al. 1984a; Stuart et al. 1985, Mize and Wilde 1986a,b,c; Meehan and Wilde 1989; Wilde and Cox 1990; Shah et al. 1994); pearl millet (Starks et al. 1982, Merckle et al. 1983). Dahms et al. (1936) discovered that cultivars of corn that were more susceptible to chinch bug attack were higher in nitrogen (thus more nutritious) than the more resistant cultivars. The same is true of sorghum (Dahms and Fenton 1940). Wilde et al. (1986) showed that low levels of antibiosis in wheat cultivars could be measured through reduced egg production. Schalk and Ratcliffe (1977) emphasized that the use of host plant resistance has saved millions of dollars by not using insecticides in the control of chinch bugs.

Before the advent of modern insecticides, the chief direct control was barriers of creosote-impregnated paper and shallow ditches in which calcium cyanide or dinitro orthocresol dust was often placed as an insecticide (Flint and Balduf 1924, Flint and Larrimer 1926, Harris and Decker 1934, Flint et al. 1935, Packard and Benton 1937); the chemicals, however, were very toxic to corn as well as to animals and humans (Richardson et al. 1937). For a while after its invention 10% DDT dust was heavily used, along with rotenone, nicotine, and sadadilla dusts (Packard, cited in Stefferud 1948). The most effective of the modern insecticides for the control of *B. l. leucopterus* on wheat is an organic phosphate phorate used as granules (Polivka and Irons 1966, Peters 1983). Polivka and Irons (1966) found phorate granules applied at 1 lb actual/acre very effective against chinch bugs invading corn from wheat because the insects move into crop fields on the ground over a long time period, which requires a long-residual-effect insecticide like phorate. However, phorate should be used very cautiously because it is extremely poisonous (Hill 1975, Thomson 1998). Most other insecticides lose their effectiveness in a few days. On sorghum at planting time the carbamates carbofuran and carbosulfan sprays and granules were effective and also relatively persistent (Wilde and Morgan 1978; Mize et al. 1980; Kindler and Staples 1982; Kindler et al. 1982; Peters 1982, 1983; Wilde et al. 1986; Bauernfeind 1987; David et al. 1991), but a major problem is the long period of repeated invasions of chinch bugs into sorghum, which requires repeated applications of less persistent insecticides, especially as the insects are highly mobile so that chemical contact may be uncertain in the field (Brooks and Higgins 1985, Bauernfeind 1987). Control was better in a rotation of soybeans and sorghum when carbofuran was used in alternate years (Wilde et al. 1986a). However, these carbamates are very toxic; carbosulfan is banned for use in the United States, and the granule preparation of carbofuran is greatly restricted in the United States (Thomson 1998). The use of imidicloprid seed treatment proved better than using soil carbofuran in protecting seeding sorghum (Wilde 1997) and is less toxic (Thomson 1998). Mejia-Ford (1997) showed that seed treatment with fipronil and imidicloprid at 0.356 and 3.74 g (Al)/seed provided good control against adult chinch bugs on rice. Acephate, chlorpyrifos, methyl parathion, lambda cyhalothrein carbaryl, and filpronil applied early postemergence all provided excellent control of adult chinch bugs on rice. However, the herbicide propanil greatly increases the mortality of young rice to chinch bug attack.

3.2 *Blissus leucopterus hirtus* Montandon (Blissidae)

3.2.1 Distribution, Life History, and Biology

In addition to the work of Leonard (1966, 1968a), Tashiro (1967) gives an excellent review of the biology and economic importance of *B. l. hirtus*, the hairy chinch bug. Reinert et al. (1995) is also a useful publication on *B. l. hirtus*.

Leonard (1966, 1968a) formally reduced *B. hirtus* Montandon to subspecies rank, not variety as was stated by Slater and O'Donnell 1994, although others have used *hirtus* as a "variety" name (Barber 1918, Torre-Bueno 1946). *Blissus leucopterus hirtus* has distinctly golden-yellow long setae and a dark abdomen, whereas *leucopterus* has shorter silvery setae and the abdomen is lighter in color (Leonard 1968a). Both subspecies can be distinguished from *B. insularis,* the third economic species, which is found in St. Augustine lawns, in having a labium that does not exceed the middle of the metasternum, whereas in *B. insularis* the labium nearly reaches the abdomen. In addition, the hairy chinch bugs are generally more robust and a little shorter (3.0 to 3.6 mm) than *B. l. leucopterus* or *B. insularis* (Leonard 1966, Tashiro 1987). Leonard (1968a) could also distinguish the fourth instars of these three taxa. However, as Leonard notes, because of the great individual variation within each taxon, not all specimens can be placed readily in either subspecies or species and good separation requires an adequate population sample, a generalization also true of the other species of the *B. leucopterus* complex. In his work testing reproductive isolation, Leonard found that although *insularis* × *leucopterus* would cross and lay fertile eggs, the nymphs did not develop, indicating that the two are reproductively isolated and thus distinct species, not subspecies. On the other hand, *B. l. leucopterus* from Oklahoma (U.S.A.) would readily mate with *B. l. hirtus* of Connecticut (U.S.A.), lay eggs, and some nymphs developed into adults, which led Leonard, based on an evident lack of reproductive isolation in laboratory work, to treat these two taxa as subspecies.

However, these two subspecies are biologically very distinctive: *leucopterus* of the central and southeastern United States is nearly entirely macropterous, only 0.01% brachyptery (Leonard 1966) and is a pest on small grains, corn, sorghum, and rice, whereas *hirtus* in the northeastern United States and eastern Canada is pterygopolymorphic, the brachypterous form predominant (63.7%), and is a pest on lawn grasses (Leonard 1966). The absence of *B. l. hirtus* on corn and small grains is surprising, given the cultivation of corn since precolonial times in northeastern United States and the early records in the late 1700s indicating that *B. l. leucopterus* was destructive on small grains in Virginia and North Carolina (U.S.A.). Moreover, *B. l. leucopterus* is not recorded as a pest breeding on lawn grasses in the midwestern United States. Ahman et al. (1984) showed that Kentucky bluegrass (*Poa pratensis* L.) was not a host as stated by Leonard (1966), but an antibiotic nonhost to *B. l. leucopterus.* These differences in host plants, in wing polymorphism, and in their parapatric to narrowly sympatric distributions indicate that *B. l. hirtus* may be more of a "semispecies" than a subspecies of *B. leucopterus,* and it could well be accorded species recognition. This possibility Leonard (1966) himself recognized, and he advised further study in the zone of apparent parapatry in the east and the zone of overlap in the west through Michigan, Wisconsin, and Minnesota (U.S.A.) to resolve this question. The recognition that these are different taxa evidently genetically isolated in nature is important in dealing with the control of these pests.

As with *B. l. leucopterus,* the original host of *B. l. hirtus* is unclear. Each of the grass species listed by Leonard and Tashiro (1987) as injured by *B. l. hirtus* is an introduced species. The cool-season grasses attacked include creeping red fescue *Festuca rubra* L., tall fescue *F. arundinacea* Schreb., perennial ryegrass *Lolium perenne* L., Italian or annual ryegrass *L. multiflorum* L., creeping bent grass *Agrostis palustris* (Huds.) Farw. (= *A. stolonifera* L.), colonial bent grass *A. tenuis* Sibth., velvet bent grass *A. canina* L., and Kentucky bluegrass *Poa pratensis* L., annual bluegrass *P. annua* L., and rough bluegrass *P. trivialis* L. Among the warm season grasses, *B. l. hirtus* attacks only zoysiagrass, *Zoysia japonicum* Steud, *Z. matrella* (L.) Merr., and *Z. tenufolia* Willd. exTrin. Crabgrass, *Digitaria sanguinalis* (L.), a pest in lawns, is not fed on by *B. l. hirtus* (Leonard 1966), although this species is attacked by *B. l. leucopterus* in the Midwest (U.S.A.) in lawns. A meadow grass, the naturalized (Fernald 1950) timothy *Phleum pratense* L., which is much used for forage, is the only host evidently fed on by both *B. l. hirtus* and by *B. l. leucopterus* (Leonard 1966). This grass is incidentally the earliest record of damage by the hairy chinch bug (Lintner 1883).

The one species that may be a possible native host species is red fescue, which is a Holarctic species, although the lawn cultivars attacked by *B. l. hirtus* are all of European origin. There are also several other species of *Festuca* native in North America that might be hosts (Gould 1975). Leonard (1966) notes that the lawn bent grasses (*Agrostis* spp.) that are native to North Africa and Asia Minor are especially favored hosts (Tashiro 1987). There are also several North American species of creeping *Agrostis* (Gould 1975) that may serve as native hosts. This subject is clearly one worthy of further study. However, care must be exerted in determining the natural host plants because in the laboratory, but not in the field, the hairy chinch bug could be reared year-round on young corn stalks (*Zea mays*) (Baker et al. 1981b).

The question of the natural host plant is important in assessing the significance of the relatively high percentage of the flightless brachypterous form (67%, Leonard 1966) in *B. l. hirtus*. Leonard discusses wing polymorphism in the *B. leucopterus* complex in relation to migration and temporary environments (Southwood 1960, 1961, 1962; Sweet 1964) and concludes that there is a good correlation between the degree of brachyptery and habitat permanence as Sweet found with New England Rhyparochromidae. The area inhabited by *B. l. hirtus* is at present a mosaic of wooded areas, open fields, pastures, and turfgrass areas. What was the habitat of the hairy chinch bug before the advent of the European colonists who created this habitat mosaic? This is a question well worth investigating in the light of Sweet's work showing that rhyparochromid species of open areas that are largely brachypterous occurred naturally in special habitat islands that undergo ecological succession very slowly. Tashiro (1987) suggests that the high brachyptery percentage may be the result of lawn type environments that are permanent enough to have selected for the brachypterous form, but Sweet believes that the brachypterous strategy is much more ancient as shown by the predominance of brachypters in the earliest records of *B. l. hirtus*.

The hairy chinch bug overwinters as an adult in plant debris in meadows, along the margins of meadows in woodlands, and in tufts of meadow grasses. When populations are high, they may hibernate around foundations of houses, even in houses (Maxwell and McLeod, 1936, Leonard 1966). In contrast to *B. l. leucopterus*, *B. l. hirtus* has not been recorded to hibernate in large numbers in clump-forming grasses such as the common little bluestem (*Schizachyrium scoparium* (Michx.) Nash, which is common in northeastern old fields. Leonard (1966) found the winter mortality to be high (34%) in a population at Mansfield, Connecticut (U.S.A.). Similarly, in New Jersey (U.S.A.) during 1974 and 1975 the winter mortality in the populations studied was 68 and 48%, respectively, which was considered to be the major source of population control in the populations of chinch bugs (Mailloux and Streu 1981). The overwintered adults leave their hibernation sites when a threshold temperature of 7°C (45°F) is reached (Mailloux and Streu 1981). This temperature was also determined by Liu and McEwen (1979) to be the threshold temperature for ovary development and the production of eggs. For about 2 weeks in the spring then, the adults search for host plants, mate, and begin to lay eggs (Reinert et al. 1995). As in *B. l. leucopterus* diapause can be broken by 10 to 14 days of continuous exposure to a temperature of 29.5°C (85°F) in the absence of light (Leonard 1966, Baker et al. 1981b). As the populations are mostly brachypterous, dispersal occurs mainly by foot, so the overwintering sites would be expected to be relatively near the host plants. Some macropters (long-winged adults) were found by Leonard (1966) on sheets on clotheslines, which indicates that flight dispersal probably occurs during the day. Mailloux and Streu (1981) captured a small number of macropters in flight traps throughout the active season from May to October.

Unlike *B. l. leucopterus*, which is quickly killed by exposure to hot ground over 40°C (Forbes 1883, Headlee and McColloch 1913, Guthrie and Decker 1954), *B. l. hirtus* tolerates well very hot ground surfaces much over 40°C in the sunny areas where it frequently damages lawns (D. E. Leonard 1966, personal observations). [*Note:* The reference in Tashiro (1987) to the contrary refers to *B. l. leucopterus*, not *B. l. hirtus* (Guthrie and Decker 1954).] Behaviorally, in collecting the three pest *Blissus* taxa, it was noticed that *B. l. hirtus* ensconces itself much closer to the soil substrate than does *B. l. leucopterus* or *B. insularis*, evidently as a consequence of its better tolerance of hot conditions. This tolerance of hot conditions in an insect of a more northern distribution may be a clue that the original habitat of *B. l. hirtus* was in patches of grass, probably a *Festuca* sp. or a creeping species of *Eragrostis* (Fernald 1950), in scattered, open, hot, semipermanent habitats undergoing slow ecological succession, where the brachypterous form was strongly selected for, but the macropter is relatively abundant to allow adequate dispersal within the ecological mosaic of largely forested northeastern North America. This would essentially preadapt *B. l. hirtus* for exploiting lawn grasses in the anthropogenic mosaic of turf habitats.

Blissus l. hirtus differs from *B. l. leucopterus* in that in the northern part of its range in northern New York and Michigan (U.S.A.), and in Ontario (Canada), it has one generation a year while in the southern part of its range in southern New York and New Jersey to Ohio (U.S.A.) it has, like *B. l. leucopterus,* two generations a year (Maxwell and McLeod 1936, Liu and McEwen 1979, Sears et al. 1980, Mailloux and Streu 1981, Niemczyk 1982, Tashiro 1987, Niemczyk et al. 1992, Reinert et al. 1995). Niemczyk (1982) advised that the overwintering adults, despite their small populations, should be controlled early in spring to prevent the multiplication into high populations in late summer by the second generation.

Tashiro has excellent photographs of the adults (male and female), and eggs. Leonard (1966, 1968a) did not describe the egg, and the description in Tashiro is of the egg of *B. l. leucopterus* from Chambliss (1895) and Luginbill (1922), which however, is probably similar to the egg of *B. l. hirtus.* Tashiro (1987) did illustrate with photos the early (young) and late (mature) eggs of the hairy chinch bug. Meiosis and early embryology were studied by Choban and Gupta (1972) who measured the egg as averaging 264.1 μm wide by 831.5 μm long.

The descriptions of the nymphs in Tashiro are similarly of *B. l. leucopterus* from Luginbill. Because the species of *Blissus* of eastern North America have similar nymphs, Leonard (1966) gave a general overall description, followed by specific measurements and other details of each species. His is the first description of *B. l. hirtus.* The lengths of the nymphs, which are given in parallel with those of *B. l. leucopterus* (Luginbill 1922), are a little larger: first, 0.99; second, 1.38; third, 1.82; fourth, 2.06; fifth, 3.23 mm. Mailloux and Streu (1981) found that the head capsule widths show the least overlap among instars. In coloration, the nymphs of *B. l. hirtus* become distinctly orange in the third and fourth instars as compared with reddish in *B. l. leucopterus* (Luginbill 1922, Niemczyk 1981, Tashiro 1987). Tashiro (Plate 10, Photographs A and B) illustrates the orange first and second instars, but the fifth instar is unfortunately that of a *Nysius* sp. (false chinch bug). Leonard (1968) discriminates the fourth instars of *B. l. hirtus* as strongly rufescent on the underside of the head and thorax and femora and tibia as compared with having a red tint on the underside and dusky femora and tibiae in *B. l. leucopterus.*

As in *B. l. leucopterus,* the eggs are laid in the grass leaf sheaths, on the stolons, and in soil next to the plant. In the laboratory the insects would lay eggs in cheesecloth rolls (Parker and Randolph 1972, Baker 1981b). Estimates of the number of eggs range from 1.5 to 20 eggs a day, total eggs from 54 to 170 eggs, and ovipositing females can live as long as 102 days (Baker et al. 1981b, Kennedy 1981). Mailloux and Streu (1981) surprisingly gave a seemingly low fecundity of only 15.6 eggs per female in the spring generation and 6.9 per female in the summer generation. Fecundity here was apparently calculated from the total adult females in the population that were produced by adults of the previous generation after mortality had taken its toll, hence the apparent low fecundity. They found the threshold temperature for egg development in the bivoltine New Jersey (U.S.A.) population to be 14.6°C (59°F) and the egg development time to vary with temperature: 21°C, 18.5 days; 22°C, 14 days; 24°C, 12 days; 30°C, 8 days; and 35°C, 6 days. In terms of degree-days, the first-generation egg hatch was completed near 115 degree-days (in early June), and the second-generation egg hatch was completed at 850 degree-days (after mid-August). The peak abundance of adult bugs occurred at 1159 degree-days (during mid-October). Not determined were conditions initiating diapause that allows univoltine and bivoltine conditions to exist across the range of *B. l. hirtus.*

Baker et al. (1981b) found the laboratory-rearing method of Parker and Randolph (1972) for *B. l. leucopterus* to be inadequate because of mold growth, and modified the technique by using 1 to 2% sodium hypochlorite solution as a sterilant for eggs and corn; but nymphal mortality was still high (~70%). The average development times in days in the laboratory for the stadia at 26°C and 16-hour photoperiod were first, 12.3; second, 5.4; third, 5.2; fourth, 4.9; fifth, 7.1; total 35.5 days, which approximates the 4 to 6 weeks in the field in Maryland (Baker et al. 1981b). Baker et al. (1981b) found the preoviposition period in the female to be 10.8 days and the longevity of laboratory females to be 73 days compared with 102 days for field adults. He noted that the total number of eggs per female, 118 (laboratory) and 170 (field), was considerably less than that reported for *B. l. leucopterus* (532 eggs/female) (Janes 1935), and attributed this to host plant differences or inherent differences among the *Blissus* species (subspecies).

For field studies of hairy chinch bug populations there are several useful sampling techniques (Leonard 1966, Tashiro 1987). The most common one is to drive a metal cylinder, such as a gallon (3.8-l) can with the ends cut off, into the ground and to fill it part way with water. Within 10 min the chinch bugs float to the surface and can be counted (Streu and Vasvarly 1966, Mailloux and Streu 1979, Sears et al. 1980, Reinert et al. 1995). Schread (1970) laid a white sheet over a flooded area of lawn and counted the insects that crawled out of the thatch and on to the sheet. The next technique, which is advantageous for obtaining eggs as well, uses water pressure to flush out the grass thatch, and then to centrifuge the residue to obtain the immature stages. Mailloux and Streu (1979) showed a remarkable 90 to 100% recovery of planted eggs using this technique. They therefore could sample egg populations in the field, and found a maximal egg population of 193 eggs/930 cm^2 in a Kentucky bluegrass turf. Using this method, Mailloux and Streu found that populations of hairy chinch bugs were overdispersed in lawns and showed the most aggregation in early nymphal stages. The technique used by Niemczyk (1982) is to place soil plugs 10.8 cm diameter by 7.6 cm deep in a Berlese funnel grass side down and allow the heat of a lamp to drive the insects down and into a jar of 70% alcohol. A simple but destructive method that works well for wild populations is to dig up a clump of grass and beat it apart over a white plastic sheet and use an aspirator to collect the insects released. When used carefully, this has the advantage of isolating the host grasses, as grasses often grow intermingled together.

3.2.2 Damage

In northeastern North America the hairy chinch bug is a serious pest of lawns, golf courses, cemeteries, wherever a short turf is desired (Schread 1950, 1960, 1963; Tashiro 1987; Reinert et al. 1995). However, it was first recognized as a pest on the introduced meadow grass timothy *Phleum pratense* L. (Lintner 1883). Howard (1887) first observed the hairy chinch bug damaging lawns in Brooklyn, New York (U.S.A.). Although long recorded in the area (Drake 1951, Leonard 1966), *B. l. hirtus* as a pest has been progressively moving northward in northeastern North America, evidently following the general climatic warming trend seen in many other animals and plants. Tashiro (1987) noted that in the early literature the hairy chinch bug was a severe pest in lawns in southern New York, New Jersey, and Connecticut (U.S.A.), but beginning about 1970 the insect has become a pest in northern and western New York. Similarly, beginning in 1971, the insect began damaging lawns in Guelph, Ontario (Canada), and is now an established serious pest (Maxwell and McLeod 1936; Leonard 1966; Goble 1971, 1972; Sears 1978; Sears et al. 1980; Mailloux and Streu 1980, 1981, 1982).

In Connecticut, Schread (1950, 1960, 1963, 1970) considered the hairy chinch bug to be the most serious pest of lawn grasses. When cool-season turfgrasses are kept at normal lawn depth, the grass is most susceptible because there is a thick mat of stolons (the thatch) where the bugs can conceal themselves and feed on the grasses (Mailloux and Streu 1979, 1981, 1982; Davis and Smitley 1990). The use of fungicides as mancozeb promotes the development of the thatch, which in turn promotes the development of the chinch bug (Davis and Smitley 1990). When the grass is kept very short as in putting greens of golf courses, such a mat does not build up and the grass becomes relatively immune from damage (Maxwell and McLeod 1936, Baker et al. 1981b). Tashiro (1987) notes that the hairy chinch bug feeds on the basal stems and the crown of the grasses hidden in the thatch. Because the bugs tend to aggregate when feeding, the first sign of damage is localized yellow areas that turn brown when the grass is killed (Reinert et al. 1995). These localized areas coalesce to create large patches of dead or dying grass as illustrated (Plate 9, Photograph A, in Tashiro 1987). The most favorable physical environments for the bugs are sunny, hot, dry areas with thatch where the moisture deficiency leaves the grass less tolerant to injury (Maxwell and

McLeod 1936; Niemczyk 1980, 1981). Under these conditions entire lawns can be killed in less than one season (Leonard 1966). Often this damage does not become visible until after the rains return and the dead turfgrass fails to respond to the rain (Maxwell and McLeod 1936, Polivka 1963, Leonard 1966, Streu and Vasvarly 1966, Schread 1970). The hairy chinch bug has become a much more serious pest because of the development of resistance to the insecticides chlordane and dursban, and because the insecticides have changed the invertebrate fauna in the thatch (Streu and Cruz 1972, Streu 1973, Baker et al. 1981a).

3.2.3 Control

Leonard (1966) listed *Geocoris bullatus* (Say) and *G. uliginosus* (Say) as potential predators of *B. l. hirtus* because they were often found together in the same habitat. This is true. *Geocoris* adults have often been seen feeding on nymphs of *B. l. hirtus,* and Tashiro (1987) illustrates (Plate 10, Photograph F) a big-eyed bug (*G. bullatus*) feeding on a third instar. Mailloux and Streu (1981) found in lawns in New Jersey (U.S.A.), a carabid, *Amara* sp., that ate eggs of *B. l. hirtus,* but judged *Amara* populations to be too low to be an important egg predator. Cruz (1972) found a mite, *Pergamasus crassipes* (L.), that fed on first and second instars of hairy chinch bugs (among many other small soil invertebrates), and he considered it and *G. bullatus* to be the major predators. Mailloux and Streu (1981) confirmed these as predators and further showed in laboratory trials that staphylinids *Philonthus varius* Gyllenhal and *Hyponygrus* sp., the nabid *Nabis ferus* L., linyphiid spiders *Erigona blaea* Grosvy and Bishop, *Grammonota inornata* Emerton fed on hairy chinch bugs nymphs. However, Mailloux and Streu, through a statistical analysis, determined that the population densities of each predator were too low, even in the aggregate of all predators, to affect chinch bug populations. Although Leonard (1966) found no parasites in hairy chinch bug populations in Connecticut (U.S.A.), Mailloux and Streu (1981) found *Eumicrosoma benefica* Gahan (Hymenoptera: Scelionidae) in hairy chinch bug eggs in New Jersey (U.S.A.), but noted that the parasite's life cycle was not synchronized with the host eggs and considered the effect on the parasite to be negligible on chinch bug populations in their 1974 study. They did not rear any parasites from the adults or nymphs. As to vertebrate predators being effective against *B. l. leucopterus,* Hudson (1914) stated that the (hairy) chinch bug in Ontario "enjoys almost practical immunity from attack by birds."

Leonard (1966) observed a fungus that he presumed was the white muscardine fungus *Beauveria bassiana* (Balsamo) Vuillemin attacking hairy chinch bugs in the field and the laboratory. He noted that as in *B. l. leucopterus,* the effect of the fungus was evident, for in wet seasons the hairy chinch bug is not a problem, and it is abundant only during prolonged periods of hot, dry weather unfavorable for sporulation of the fungus. Mailloux and Streu (1981) measured the incidence of the fungus in New Jersey (U.S.A.) as varying from a minimum of 20% to a maximum of 80 to 90% during a 2-year study. Liu and McEwen (1977) described a new microsporidian protozoan pathogen, *Nosema blissi,* from Guelph, Ontario (Canada), that resides within the Malpighian tubules. This organism may have a potential for biological control if pathogenic.

With the development of insecticide resistance (Streu and Vasvary 1966, Streu and Cruz 1972, Streu 1973), it became imperative to seek other means of control, especially as predator control seems ineffective (Mailloux and Streu 1981, Ratcliffe 1982); and studies were initiated on host plant resistance or tolerance. Some strains showed greater resistance or tolerance to the hairy chinch bug. Kentucky bluegrass Baron and Newport cultivars were more tolerant than the susceptible Adelphi cultivar. In perennial ryegrasses, Score, Pennfine, and Manhattan, showed the least infestation when compared with 12 other cultivars. In fine-leaved fescue, 76 G1–332 and FL-1 were least infested while Jamestown was the most infested (Baker and Ratcliffe 1978, Baker et al. 1981a, Ratcliffe 1982). A fungal endophyte, *Acremonium lolii* Latch, Christensen and Samuels (Clavicipitaceae: Belansiae), conferred antibiotic insect resistance in perennial ryegrass, *Lolium perenne* (L.) (Prestidge et al. 1982, Barker et al. 1984), through the production of clavine ergot alkaloid neurotoxins, which were earlier implicated in livestock poisoning known as the staggers disorder (Gallagher et al. 1982). This led to research that demonstrated that the fungal alkaloids inhibited feeding by hairy chinch bugs on the cultivar Repell that has a high endophyte content (Saha et al. 1987, Mathias et al. 1990). However, Carriere et al. (1998) attempted to protect Kentucky bluegrass cultivar "Merit" by planting it with an endophyte-protected (with *Neotyphodium*) cultivar Prelude II of perennial ryegrass, but found that the highly mobile third instars avoided Prelude II and fed on Merit despite the grasses being mixed together. This relationship may explain the results of a study by Davis and Smitley (1990), who found that chinch bug populations increased following the application of the fungicide mancozeb, which in turn led to heavier thatch development favorable to the chinch bug.

As Tashiro (1987) noted, despite the best efforts to avoid insecticides, i.e., integrated pest management (IPM), the first line of defense for controlling turf-infesting insects continues to be insecticidal control. When there is a sudden and unexpected heavy infestation of a detrimental insect, there is no alternative but to depend on a recommended insecticide. However, as Reinert (1980) emphasized, the turfgrass industry has relied almost solely on insecticides. This reliance on insecticides has led to the progressive development of resistance (Reinert 1982b, Tashiro 1987). Chlordane, one the most commonly used insecticides, has become ineffective against the hairy chinch bug in Connecticut (U.S.A.) (Schread 1963), in New Jersey (U.S.A.) (Streu and Vasvary 1966, Streu 1969) and in Ontario (Canada) (Sears et al. 1980). Many of the earlier effective insecticides, such as DDT, phorate (Polivka 1963), malathion, parathion, dieldrin, aldrin, and heptachlor, have proved to be too poisonous or persistent in the environment, and were discontinued by the U.S. Environmental Protection Agency for home use against lawn insects (Niemczyk 1980). Sears et al. (1980) compared effectiveness of available insecticides and found the best control was obtained with chlorpyrifos (dursban), carbaryl, and diazinon, in that order. Chlorpyrifos binds the most tightly with the thatch (Niemczyk 1982), which made it especially effective against the turf-living chinch bugs that feed on the stolons of the grasses. To help minimize the use of insecticides, Liu and McEwen (1979) developed a program of using temperature accumulations and sequential sampling to pinpoint the best time to spray to control the hairy chinch bug in Guelph, Ontario (Canada), where the species is univoltine. This was when the first generation is mostly third instars, the eggs of the overwintered females have already hatched, and the first generation adults have all died. This was approximately in the period 750 to 950 degree-days above the 7°C threshold. The population threshold they suggested was 30 nymphs/35 in.^{2} (225 cm^2). Spraying at this critical time yielded best control with minimal use of insecticides.

3.3   *Blissus insularis* Barber (Blissidae)

3.3.1   Distribution, Life History, and Biology

Blissus insularis is the third member of the *leucopterus* complex that is a major economic pest. It attacks St. Augustine grass lawns, and in parallel with its host grass, it ranges through the southern United States from the Carolinas to Texas north to approximately the 31st parallel, and it may extend into northeast Mexico (Leonard 1966). Although originally described as a variety, Leonard (1966) showed that despite the chromosome complement being similar to that of *B. leucopterus* and *B. hirtus*, *B. insularis* was genetically isolated from the other taxa of the *B. leucopterus* complex and gave it species rank. Originally known as the lawn chinch bug, it was renamed the southern chinch bug when it was given species rank (Stringfellow 1969b). Although similar, it is distinguished morphologically from *B. leucopterus* by being shorter (3.05 mm) and narrower (0.69 mm), compared with 3.57 by 1.02 mm for *leucopterus*; by having a silvery-gray pubescence on the anterior half of the pronotum; by having a labium that exceeds the midline of metasternum; and by having veins in corium that are white, not yellowish.

Blissus insularis appears to be related to an Antillean complex of *Blissus*. Indeed, chinch bugs identified as *B. insularis* are recorded from Mexico, Antigua, Bahamas, Cuba, Dominica, Grenada,

Jamaica, Martinique, Panama, Virgin Islands, St. Domingo, St. Vincent, and Trinidad (Slater 1964c). This insular pattern should reward careful study such as was done by Leonard on the eastern U.S. populations. Also interesting is the report by Reinert et al. (1995) that damaging populations of *B. insularis* occur in California (U.S.A.) and Mexico, as Leonard did not find *B. insularis* in these areas, although he thought it might occur in northeastern Mexico, adjacent to the populations of *B. insularis* in Texas. This needs study soon to ascertain whether these are new introductions of *B. insularis* through St. Augustine grass cultivation, as the grass is planted in lawns by plugs or squares which would encourage the accidental transport and introduction of *B. insularis*. It is a common sight in the southern United States to see large truckloads of St. Augustine grass being transported to make lawns around newly built homes, and so provide new habitats for *B. insularis*. Leonard (1968a) excluded from the original *B. insularis* type series the chinch bug populations from the Antilles that Barber (1918) had originally included in *B. insularis*, hence the name *insularis*. Later, Leonard (1968b) described three species, *B. slateri* from Puerto Rico (U.S.A.), *B. planus* from Grenada, West Indies, and *B. antillus* from Puerto Rico, the latter a pest in lawns of *Polytria amaura*, a grass from Java. Slater (personal communication) doubts the specific identity of these taxa, a question that can only be resolved by further revisional studies and experimental work such as was carried out by Leonard (1966, 1968a). *Blissus slateri* Leonard has been reported from Mato Grosso do Sul of Brazil (Valério et al. 1998), a location far from the type locality of Puerto Rico. This is very probably not actually *B. slateri*, and is probably an undescribed species, but it does indicate the potential of *Blissus* to become a pest, as was emphasized by Schaefer (1998b).

Blissus insularis is pterygopolymorphic but is unusual in that the macropterous form predominates, only 28% brachyptery being present. Leonard noted that in hammocks in southern Florida (U.S.A.) the populations were entirely macropterous and suggested that in this environment there was strong selection against the brachypter. However, given the huge areas devoted to St. Augustine grass lawns, its main host plant in the southeast, it would seem that dispersal by foot would be adequate, leading to selection for the brachypter instead. This question should be amenable to experimental studies. It is possible that the southern chinch bug was once largely brachypterous on its original host plant before the introduction of St. Augustine grass provided a new opportunity that favored the macropterous form.

Kerr (1956) noted that *B. insularis* and *B. leucopterus* not only differed morphologically but biologically in that *B. insularis* is not a pest on corn and small grains, but, instead, feeds on only a few other grasses, the most famous of which is St. Augustine grass, *Stenotaphrum secundatum* (Walt.) Kuntze, which is the most important warm-season lawn grass in the southeastern United States. The grass is, however, not native to the United States (Gould 1975), Mexico (Beetle 1983), or the Antilles (Liogier and Martorell 1982), but is evidently an ornamental cultivar derived from the Old World (Africa) where the greatest diversity occurs, both within the species and in the genus (Reinert et al. 1986). The only other grass recorded as being an important host is torpedo grass (*Panicum repens* L.), a minor pasture grass in Florida (U.S.A.) (Kerr and Kuitert 1955, Kelsheimer and Kerr 1957, Kerr 1966). However, this grass, which has a pantropical coastal distribution, was very probably introduced from the Old World tropics (Uphof 1968, Gould 1975). Other hosts recorded are Pangola grass *Digitaria decumbens* Stent. (Kerr and Kuitert 1955), centipede grass *Eremochla ophiuroides* (Monroe) Hack, Bermuda grass *Cynodon dactylon* (L.) Pers. (Kelsheimer and Kerr 1955, Kerr 1966), bahiagrass *Paspalum notatum* Flugge., and Zoysiagrass (*Zoysia* spp.) (Kerr 1966). However, as Kerr (1966) stressed, development was very slow on these other grasses and feeding on them was incidental to large population buildups on St. Augustine grass. Slater (1990) mentioned *Oryza sativa* L. and *Panicum maximum* Jacq. as being hosts, but this is based on work by Wolcott (1936) in Puerto Rico and St. Croix, American Virgin Islands, and may refer to one of the two species described by Leonard (1968b) from Puerto Rico that were originally included as paratypes in the type series of *B. insularis* by Barber (1918). Probably based on these older records is the statement by Reinert et al. (1995) that damaging populations of *B. insularis* occur throughout the Caribbean archipelago. Similarly questionable may be Slater's (1976) recording *B. insularis* from the Lesser Antilles as breeding on

Panicum bartowense Scribn. & Merr. and occurring on carpet grass, *Axonopus compressus* (Swartz) Beauv., the latter a widespread weedy Neotropical grass sometimes used in lawns. Slater (1976) considered the records of *B. leucopterus* on sugarcane to pertain to *B. insularis* but this based on the records from Wolcott (1941) of *B. leucopterus* (= *B. insularis sensu* Barber) on sugarcane, but Leonard (19668a, 1968b) had determined that these records were not *B. insularis*. Sugarcane is grown in Louisiana and Texas (U.S.A.). No definite records of *Blissus* on sugarcane in the United States can be found. Leonard (1968a) reported adult specimens of *B. insularis* from rice and collected it on "native grasses" in hammocks in Florida (U.S.A.) but these are evidently not breeding records. Unfortunately these "native grasses" were not identified. Leonard (1968a) was able to rear *B. insularis* in the laboratory on corn seedlings and Sudan grass (*Sorghum vulgare* var. *sudanense* Hitchc.), although it does not attack these species in the field. Nevertheless, as with the other economic species of *Blissus* discussed here, all these grass species are introduced grasses, so the original native host of *B. insularis* evidently still remains unknown.

In southern Florida (U.S.A.), *B. insularis* shows no diapause, and breeds continuously with six to ten generations a year; but the adults nevertheless predominate during the winter. In northern Florida and along the Gulf Coast to Texas (U.S.A.) there are three to four generations and only adults are found in the winter (Kelsheimer and Kerr 1957, Eden and Self 1960, Leonard 1966, Kerr 1966, Reinert and Kerr 1973, Tashiro 1987). This needs further study as a possible case of ongoing evolution of a life-cycle adaptation to a temperate climate if *B. insularis*, indeed, has an Antillean origin.

The southern chinch bug differs from *B. leucopterus* in not having a distinct pattern of a spring migratory flight, and flight is evidently of minor importance. Instead, most dispersal occurs by walking and streams of bugs can been seen moving from heavily infested areas, where they may travel several hundred feet in a half hour (Reinert and Kerr 1973). The adults tend to aggregate and prefer sandy soils to muck soils (Kerr 1966, Crocker and Simpson 1981).

The egg appears not be described but is said to be identical to that of *B. l. leucopterus* and turns from tan to reddish as it develops, except that, when parasitized, it remains tan and then turns black when the parasite pupates (Reinert 1972, Tashiro 1987). The nymphal stages were described by Leonard (1968) and are basically similar to those of *B. leucopterus* (Tashiro 1987). The lengths of the instars are first, 0.83; second, 1.44; third, 1.92; fourth, 2.11, and fifth, 3.05 mm. The legs, head, and pronotum are yellowish in the early instars, become darker ochraceous in the fourth, finally castaneous in the fifth instar. In *B. leucopterus* (*sensu lato*) the head and thorax becomes castaneous to piceous in the fourth instar.

Reproductive activity begins as soon as warm conditions begin which in southern Florida (U.S.A.) leads to a large surge of first instar in February and in late March to early April in northern Florida. In sunny open areas in St. Augustine lawns, aggregations of 500 to 1000 nymphs/0.1 m^2 is common and as many as 2300/0.1 m^2 have been counted (Reinert and Kerr 1973). In the laboratory the females mate repeatedly, often for a long as 2 hours, and are often seen in the field walking about end to end, the female pulling the male along. As in *B. leucopterus,* courtship involves a touching of antennae. Eden and Self (1960) stated that the females begin to lay eggs 7 to 10 days after mating.

The eggs are inserted into protected places, especially the crevices of the grass nodes and at the junction of blades and stems (Slater and Baranowski 1990). The adults live for about 70 days and females deposit a few eggs a day over several weeks for a total of 100 to 300 eggs (Burton and Hutchins 1958, Leonard 1966). According to Eden and Self (1960) in Mobile, Alabama (U.S.A.) area the eggs hatched in 14 days, while in summer in Florida the eggs hatch in 7 to 10 days (Kelsheimer and Kerr 1957). Total development from egg to adult varies with locality and probably ambient temperature: 49 to 59 days in Alabama (Eden and Self 1960); 30 to 45 days in Mississippi (U.S.A.) (Burton and Hutchins 1958); 35 days in south Florida (Kelsheimer and Kerr 1957). Kerr (1966) did the most thorough laboratory study. He found that a generation is completed in 93 days at 70°F (21°C), in 35 days at 83°F (28°C); egg development, 24.6 and 8.8

days; first instar, 18.9 and 6.5; second instar 13.1 and 4.0; third instar, 10.6 and 4.1; fourth instar, 10.6 and 4.1; fifth instar, 14.8 and 6.7 days. He mentioned a few sixth and seventh instars, and also only four instars but these must either be misinterpretations or extremely unusual data needing further study given the near-universality of five instars in the Hemiptera, and the Lygaeoidea in particular (Sweet 1964). The preoviposition periods were 14.2 and 5.6 days. At 83°F (28°C), the female longevity was 70.4, the male, 42.1 days; the number of eggs laid, 289 per female, at a rate of about 4.5 per day. He found that virgin females would lay a few eggs that did not hatch.

3.3.2   Damage

Although St. Augustine grass is not native to North America, it is an old introduction and the earliest record of damage by the southern chinch bugs dates to the turn of the 19th century in Florida (U.S.A.) (Kerr 1957). The grass is well known for its vulnerability to the southern chinch bug, which has a distribution similar to St. Augustine grass (Kerr 1966). With the increased planting of St. Augustine grass, the insect has spread along the Gulf Coast through Louisiana and Texas (U.S.A.) (Oliver and Komblas 1968). It first damaged lawns in Texas in the 1950s and the populations and damage have progressively increased (Hamman 1969, Tashiro 1987).

The southern chinch feeds at the nodes and basal parts of the plants, but not on the leaves. The first instars are vigorous and often push into narrow space where the grass leaf leaves the node and feed protected from predators and rain for a week or two (Kerr 1966). This feeding by all stages causes the plant to become dwarfed, to yellow, and eventually to die. The grass is the most susceptible under dry conditions when only 25 to 30 insects/0.1 m^2 can cause severe damage. Both nymphs and adults are destructive and tend to aggregate rather than disperse, causing scattered spots to develop, which allows ample time to apply control before the damage spreads to the entire lawn (French 1964, Kerr 1966, Stringfellow 1969b, Oliver and Komblas 1981, Tashiro 1987). Heavily fertilized grass with lush growth suffers the most from the insect. In South Florida (U.S.A.) where the grass grows continuously as there is little winter frost, the thatch grows very thick and produces an ideal habitat for the bug. Although moderate moisture favors the bug, excessive moisture suppresses it (Kerr 1966). With more than 368,700 ha (911,000 acres) of St. Augustine grass (37% of the total) in Florida alone, the money spent on the control of the southern chinch bug in lawns exceeds that for all other insects in Florida with the possible exception of the citrus rust mite. In Texas (U.S.A.) about 56% of the lawns are composed of St. Augustine grass with a correspondingly large requirement for chinch bug control (Carter and Duble 1976). In 1980, more than $25 million annually went to control this pest, and with as many as six insecticidal applications a year to a given lawn, it is a lucrative business (Kerr 1966; Stringfellow 1967, 1968, 1969b; Strobel 1971; McGregor 1976; Reinert 1978a; Toshiba 1987).

3.3.3   Control

Reinert (1978b) made a thorough experimental study of the predators present in lawns that may feed on the southern chinch bug. The eggs and all stages of the chinch bugs were fed on by *Geocoris uliginosus* (Say) and *G. bullatus* (Say); the eggs and early instars by the anthocorid bugs *Xylocoris vicarius* (Reuter) and *Lasiochilus pallidulus* Reuter; and the nymphs and adults were fed on by the earwig *Labidura riparia* Pallas, the ant *Solenopsis geminata* (F.), and a lycosid spider, *Lycosa* sp. Reinert noted that the big-eyed bugs hold promise as they were abundant in the lawns, with abundances as high as 17/0.1 m, and in the laboratory big-eyed bugs fed on 9.6 nymphs a day. The earwigs, although not numerous, search over a large area and one earwig was observed to eat 50 adult chinch bugs a day. However, Kerr (1966) concluded from his work on lawn insects in Florida (U.S.A.) that predators do not appear to help greatly in suppressing chinch bug numbers with the exception of large populations of fire ants (*Solenopsis* spp.), which are hardly desirable in lawns.

He also observed that in a decade of work with *B. insularis* being brought in the laboratory, no insect parasites have emerged from the bugs.

Reinert (1972a) discovered that the scelionid egg parasite, well known from *B. leucopterus*, *Eumicrosoma benefica* Gahan, also parasitizes eggs of the southern chinch bug. He found the wasp to have an average abundance of $34.5/0.1$ m^2 in lawns as compared to chinch bug population of $90/0.1$ m^2 and he concluded that the wasp had been in south Florida undetected for a long time and could be an important biological control on the southern chinch bug.

Kerr (1958) tried unsuccessfully to control *B. insularis* in field tests using both *Beauveria globulifera* and *Metarrhizium anisphilae* (Metch.) Sorokin. Kerr (1966) inferred the natural presence of the pathogen through the effect that moist rainy conditions had in reducing the chinch bug populations. Reinert (1978b) found the fungus to be pathogenic on all life-history stages, but it produced effective killing infections only under moist conditions and when chinch bug populations were high.

For cultural controls, first, it is important to keep thatch depth at a minimum, because the thatch provides shelter for chinch bugs and will bind with the insecticide, making control less effective. For optimal turfgrass health no more than 35 to 40% of the leaves should be cut at a time during mowing and the lawn should be mowed no less often than once a week. It also helps to keep the thatch aerated and moist, but not overwatered, as this will inhibit grass breakdown. Water sufficiently to wet the ground to 6 in. deep. Because grass is more susceptible to chinch bug damage when heavily fertilized, reduce fertilization to no more than 3 to 4 lb/1000 ft^2; 2 lb in the shade.

Beginning with Reinert's work (1972b), many cultivars of St. Augustine grass have been screened for host resistance to the southern chinch bug. Floratam, a polyploid cultivar has been the most successful, and it was released for use in 1973 jointly by the University of Florida and Texas A&M University experiment stations (Horn et al. 1973). Moreover, Floratam is also resistant to the St. Augustine decline virus. However, Floratam is a coarse grass cultivar, while the cultivars with better lawn characteristics are the more susceptible to both the chinch bugs and the virus. However, Floratam is not as cold hardy as other cultivars (Reinert and Dudeck 1974; Reinert 1978a, 1982a,b; Burton et al. 1979; Reinert et al. 1980; Reinert and Niemczyk 1982; Crocker et al. 1982, 1989; Merchant and Crocker 1998; R. L. Crocker, personal communication). However, as can be expected, some Florida populations of *B. insularis* have overcome the Floratam strain (Busey and Center 1987; Busey 1990a,b, 1995) but the Texas populations of *B. insularis* as yet have not overcome the Floratam cultivar (Crocker et al. 1989; R. L. Crocker, personal communication). Busey and Zaenker (1992) showed that the resistance of Floratam stems from an inherited lack of feeding responses by the chinch bugs. Reinert et al. (1986) were able to find some African polyploid strains of *Stenotaphrum secundatum* and *S. dimidiatum* (L.) Brongn. that showed higher resistance than Floratam to *B. insularis,* but these African St. Augustine grasses were unacceptable in their growth and agronomic characteristics, so future breeding research is needed, which is difficult because St. Augustine grass is propagated by asexual clones. However, aesthetics has won out and there is relatively little interest now in Floratam and plant breeders have largely discontinued this research (R. L. Crocker, personal communication). It would seem that this should be an excellent opportunity for genetic engineering by gene transfers.

In Florida (U.S.A.) the southern chinch bug, with its continuous breeding and the near monoculture of St. Augustine lawn habitat available, coupled with an extensive use of insecticides by the large lawn-spray industry, has repeatedly evolved resistance to insecticides, beginning with DDT in 1957 (Kerr 1958). As Reinert (1982b) noted, in Florida an excellent new insecticide for the southern chinch bug may be totally useless within a few years. He charts the development of resistance in Florida: by 1977, the southern chinch bug has showed resistance to chlordane, parathion, diazinon, and chlorpyrifos. Other newer organophosphate insecticides have shown early resistance developing (Reinert and Niemczyk 1982, Reinert and Portier 1983). Outside of Florida, these insecticides still provide adequate control and are at present being recommended (R. L. Crocker, personal communication), but it will be only a matter of time until resistance appears.

The insecticides chlorpyrifos, acephate, and lambda-cyhalothrin used at recommended rates killed nymphs and adults but not the eggs, so resistance was through egg survival. Only chlorpyrifos could kill the egg, but only after topical application in the laboratory, not in the field where the eggs are well protected by oviposition in crevices (Nagata and Cherry 1999). At present in Florida the carbamate propoxur and the synthetic pyrethroids fenvaleratre and permethrin (Spectra-cide®) provide good control for resistant populations (Reinert 1982c). Where resistance is not yet a problem as in Texas (U.S.A.), good control is provided by diazinon (except to golf courses and sod farms) (2 to 3 oz/1000 ft^2); chlorpyrifos (Dursban®) (11/2 oz/1000 ft^2), acephate (Orthene®) (1.2 to 2.4 oz/1000 ft^2) (Merchant and Crocker 1998; R. L. Crocker, personal communication). For commercial professional control, bendiocarb, bifenthrin, carbaryl, cyfluthrin (not on golf courses and sod farms), ethoprop (for golf courses), fonofos, imidacloprid, isazophos, isofenphos, lambdo-cyhalothrin, and permethrin are all available. It is wise to alternate the different insecticides to lessen the development of resistance (Crocker and Merchant 1998, Pinkston 1998, Short 1998). Tashiro (1987) emphasized that such chemicals are nonetheless poisons that should be used as little as possible. When the natural enemy predator–parasite complex was present, and with proper cultural controls, the southern chinch bug did not spread so rapidly and the populations often crashed in August; conversely, insecticide-treated lawns continued to have severe chinch bug problems. If the chinch bug populations were carefully monitored and lawns were treated with insecticides only when the economic thresholds of slightly more than 20 bugs/0.1 m^2 was exceeded, insecticide applications could be reduced by 90% (Reinert 1978b, 1982d). But this may require more forebearance than most homeowners can muster.

3.4   *Cavelerius saccharivorus* (Okajima) (Blissidae)

3.4.1   Distribution, Life History, and Biology

Slater and Miyamoto (1963) and Slater (1979) placed the economically important species once placed in *Blissus* and *Ischnodemus* (*saccharivorus*) and *Macropes* (*excavatus*) along with seven other species from Southeast Asia in the genus *Cavelerius*. *Cavelerius saccharivorus*, known in economic literature as the Oriental chinch bug, is a pest on sugarcane (*Saccharum officinarum* L.). Two other species, *C. excavatus* (Distant) and *C. sweeti* Slater and Miyamoto, are also recorded as pests on sugarcane in India (Slater 1964a). *Cavelerius saccharivorus* was described from Taiwan (Okajima 1922) but it had been earlier recorded on sugarcane in 1914 from the southern Japanese islands, the Ryukyus, in Oshima and Okinawa, into which it was certainly introduced with sugarcane, probably from Taiwan (Murai 1975, Azuma 1977, Fujisaki 1985). Zheng and Zou (1981) in a manual of identification of Chinese insects key out *C. saccharivorus* and *C. excavatus*. Feng and Liu (1993) reported *C. saccharivorus* to be a major pest on sugarcane in Jiangxi Province, China, a subtropical location about 300 miles inland from the coast of Taiwan Strait. This leaves open the question of the origin of the Oriental chinch bug. The other seven species of *Cavelerius* are known from Burma, Nepal, Thailand, Malaya, and East India. These species may perhaps also feed on sugarcane or its relatives. Ashlock (Slater et al. 1969) collected *C. illustris* Distant (the type species) and *C. minor* Slater and Miyamoto in Thailand probably on different grass hosts, but unfortunately the grasses were not identified. The genus *Saccharum* contains about a dozen species of tall grasses in the Old World tropics (Bailey 1949), but sugarcane itself is a cultigen that is believed to be originally derived from New Guinea (Simpson and Conner-Ogorzaly 1995) and so sugarcane itself is evidently not native to the areas where it is attacked by *Cavelerius* bugs. It is therefore interesting that the blissid *Dentisblissus venosus* (Breddin) attacks *Saccharum* spp. (including sugarcane and Pit Pit) in New Guinea (Slater 1968). Actually, sugarcane is attacked by many insects. Takara and Azuma (1968) listed 165 insects as infesting sugarcane, but considered *C. saccharivorus* to be the third most serious pest in the Ryukyu Islands. *Cavelerius saccharivorus* is also reported to injure millet (*Panicum mileaceum* L.) and bamboo (*Arundinaria cobinii*) in Japan (Shiraki 1952), but these records need to be confirmed, as neither are mentioned in Hasegawa (1980) or Zheng (1994). However, Yamada et al. (1984) stated that they could rear *C. saccharivorus* successfully in the laboratory on maize, Kentucky bluegrass, and sorghum, but the best results were obtained with sugarcane leaves, on which the insects could be reared for two generations. Murai (1975) reported *C. saccharivorus* macropters to migrate to *Miscanthus sinensis* Anders grassland adjacent to sugarcane fields in Okinawa, and to utilize *Miscanthus* as an alternate host in September to November as eggs and nymphs were found on it. However, Oshiro (1979) reexamined the *Miscanthus* populations and found no clear evidence that *Cavelerius* could successfully breed on *Miscanthus,* and considered these totally macropterous populations to be the result of dispersal from sugarcane fields.

The Oriental chinch bug is an elongate black insect 7 to 8 mm long with contrasting pale wings. The nymphs are striking in coloration, dark, with a conspicuous white lateral spot on each side of the body. The earlier instars are paler with the spots not as conspicuous (Murai 1975). The species is pterygopolymorphic with three forms: macropterous (long-winged, extending to segment 6 or 7), brachypterous (short-winged, extending to segment 4 or 5), and micropterus (very reduced wings, extending to segment 2 or 3). In the long-winged form the pale membrane has a large irregular central brown patch (Chu 1995).

Its wing polymorphism and economic importance has created much interest in the biology of this species. Earlier in Taiwan (then Formosa), and later in Okinawa, the Oriental chinch bug has been the focus of a considerable amount of basic work. Matsumoto (1935) described its mating behavior, its overwintering as an egg, and its life cycle as having three generations a year in Taiwan. While Fujisaki (1993c) agreed that there were three generations a year in Okinawa, and Feng and Liu (1993) reported the same in Jiangxi, China. Murai (1975) stated that in Okinawa there were actually two, and perhaps a partial third generation, both of which could lay diapause eggs in the autumn.

Maki (1937) described the egg and nymphs of *C. saccharivorus* and the scelionid egg parasite *Eumicrosoma blissae* (Maki). Sadoyama (1998) described the complex oviposition behavior of the parasitic wasp, which helped to ensure that only one wasp egg is laid in the *Cavelerius* egg. Sadoyama (1999) developed for this research a simple method for acquiring the eggs in the laboratory using a glass tube and laboratory film with narrow gaps provided to stimulate oviposition. He noted that this oviposition behavior into gaps effectively protected the eggs from ant predation in the field. Takano and Yanagihara (1939) summarized knowledge about natural enemies in Taiwan. Yamada and Yagi (1984) were able to induce metamorphosis of a precocious adult from a fourth instar nymph by using precocene II derived from *Ageratum houstonianum* Mill.

Considerable research has been done on the ecological significance and development of wing polymorphism (Murai 1975, 1977, 1979; Azuma 1977; Oshiro 1979, 1981; Fujisaki 1985, 1986a,b, 1989a,b, 1992, 1993a,b; Chu 1995). Under low population density the brachypterous form was more fecund than the macropterous form, but under high densities the macropters were more fecund, as they were larger and longer lived. The macropterous adult takes longer to become sexually mature than the brachypter, probably to permit more time for dispersal. Murai (1979) showed that the macropter was promoted to disperse in high densities because of a lack of suitable oviposition sites. The appearance of the macropters for dispersal was promoted by high population densities, larger body size, high temperatures, and long photoperiods, the brachypter by the converse conditions. Oshiro (1981) showed temperature and photoperiod to be the more important in affecting wing length. Under high population densities, body size is positively related with macroptery, but the micropter appears as well, producing a bimodal population (Fujisaki 1986b). The fifth instar alone responds to crowding to increase macroptery (Fujisaki 1989b). Based on these experimental data, wing polymorphism then appears to be polygenic, although strongly affected by environmental conditions, and the male reacts differently from the female to maximize the fitness of each sex with reference to wing polymorphism (Fujisaki 1993a,b). Fujisaki (1985, 1993a) proposed that pterygopolymorphy was a mixed strategy that allows the insect on one hand to be sedentary enough as a flightless brachypter to exploit the relatively tolerant sugarcane in ratooned fields, but on the other hand as a macropter to disperse to new fields of sugarcane. About half of the females were mated before the dispersal flight. He noted that the extremely brachypterous (micropterous) males, since they matured earlier than the macropters, frequently mated with the macropterous females before dispersal, permitting genetic dispersal of the brachypterous condition, which thus confers male fitness. It should be noted here that this pterygopolymorphic strategy must be a preadaptation for this is an artificial habitat with an introduced plant and an introduced insect species. What was the role of wing polymorphism on the original host plant of *C. saccharivorus*? Fujita (1977) observed that the macropter was relatively abundant as compared with the blissids, *Dimorphopterus pallipes* Distant and *D. japonicus* Hidaka, that exploit more permanent wild hosts, and he attributed this to the dependence of *C. saccharivorus* on cultivated sugarcane. Nevertheless, as Fujisaki (1986) cautioned, the sugarcane habitat in the ratooning cultivation system is not so temporary as it might seem, as it provides food for the insect population over several years. The life cycle of *C. saccharivorus* thus does fit well the cultivation pattern of sugarcane with a mixture of old ratoon fields and new fields.

The insects lay their eggs in masses on the margin of the inside of the leaf sheaths. In Japan oviposition of the first generation begins in early June and ends in late July. In the fall from late October to late January the females lay diapause eggs. The eggs of macropters showed a more intense diapause (Suzuki et al. 1979, Hokyo et al. 1983, Fujisaki 1993a,c). Dispersal of the macropters occurs in June, when the immature macropters climb to the tops of the sugarcane, indicating that dispersal normally occurs before ovarian maturation. Brachypters of the same age remain under the leaf sheaths and are already reproductive (Fujisaki 1985) before the later-maturing macropters. The following bionomic data comes from Fujisaki (1986a) who reared the insects at 25°C and 16:18 L:D photoperiod. The percent brachyptery from a low-density population was 93%, from a high-density population, 56%. The fecundity of brachypters from low-density population was 173 eggs; from high-density population, 106 eggs; and of macropters from high-density population, 142 eggs. The eggs were laid in 7 to 9 clutches over a maximum of 80 to 100 days of adult longevity. The premating period for brachypters was 10 days; for macropters 20 days. The preoviposition period for brachypters was 14 days, and for macropters 33 days, showing the delay to promote dispersal. The teneral period was 7 days. There appears to be no data on the length of time of the individual stadia.

Slater and Miyamoto (1963) described the alimentary organs and the male and female reproductive organs of *C. saccharivorus*. The testes are unusual among the Blissidae in having only four follicles. They noted the close similarity of the male phallus to that of *C. excavatus* as compared with the phalli of other blissids.

3.4.2 Damage

As Fujisaki (1986a) emphasized, sugarcane is relatively tolerant of severe infestations of this insect. Abundances of the insect reach 100 per stalk of sugarcane without evidently damaging the plant, although growth was inhibited. He notes that this is in contrast to the attacks of the brown planthopper (*Nilaparvata lugens* Stål), which causes hopperburn, severely injuring the host plant (Kisimoto 1965). This tolerance promotes the survival of the bug populations on the host plant as reservoirs to invade new fields.

3.4.3 Control

Important biological control of the Oriental chinch bug is provided by the egg parasitoid *Eumicrosoma blissae,* which has exceeded 50% parasitism rate in the field (Azuma 1977, Sadoyama 1999). Although the behavior of oviposition into the eggs was analyzed in detail, the control provided by the parasite has not been discussed, nor has any work on predators controlling populations been reported.

Sukuki (1990) in a report on varieties of plants registered with the National Agricultural Research Center of Japan listed one cultivar, Ni6, as being resistant to *C. saccharivorus*.

Hara et al. (1971) reported that aerial applications of mixture of fenitrothion (MEP) (2%) with malathion (1%) at 20 kg/ha and diazinon (3%) at 30 kg/ha gave good control of *C. saccharivorus*.

The extensive use of chemical control is shown indirectly by the following: Nagamine et al. (1975) hypothesized that the extensive use of organochlorinated insecticides against *C. saccharivorus* had caused the ecological release of the cicada *Mogannia iwaskii* Mats., causing the emergence of a new pest. Ito (1976) similarly believes that the widespread application of organochlorinated insecticides in sugarcane has, through the suppression of natural enemies, actually encouraged the spread of such introduced pest species as *C. saccharivorus* as well as the emergence of secondary indigenous pests.

3.5 *Cavelerius excavatus* (Distant) and *Cavelerius sweeti* Slater and Miyamoto

3.5.1 Distribution, Life History, and Biology

These species together are known as the black bug of sugarcane, and are serious pests in India and Pakistan. Interestingly, the black bug was not recorded as a pest of sugarcane until 1921 by Fletcher, despite Fletcher's (1914) having recorded 18 other insect pests of sugarcane in India at that time. Distant (1904) also made no special mention of it, in contrast to stating that *Dimorphopterus gibbus* (F.) was a serious pest of sugarcane in India. *Cavelerius* (as *Macropes*) was first reported as a serious pest on sugarcane in 1930 (Rahman and Nath 1939). This may indicate a relatively recent host transfer and adaptation to sugarcane.

There is an important question concerning the identity of the black bug of sugarcane, which is the reason *C. excavatus* and *C. sweeti* are listed together here. The specimen localities of *C. excavatus* given in Slater and Miyamoto (1963) are from far southern India in Tamil Nadu, although Distant (1903) originally described the species (lectotype) from Assam, at Shillong (now a state of Meghalaya) where the black bug is not a pest (Rahman and Nath 1939). Also important is that Distant described *C. excavatus* as brachypterous, with the membrane extending only to the fourth abdominal segment, and he used this as a major key character (Distant 1904). In both *C. sweeti* and *C. excavatus sensu* Slater and Miyamoto the hemelytra extend to the sixth segment, and the populations appear to be macropterous — there is no mention of a brachypterous form as discussed in *C. saccharivorus*. *Cavelerius sweeti* Slater and Miyamoto is recognized as the species that is the serious pest on sugarcane in western India (Punjab, Haryana, and Uttar Pradesh) (Chopra 1970, 1989) and Pakistan (Safti 1986). The western India and Pakistan records of *C.* (as *Macropes*) *excavatus* listed in Slater (1964) therefore very probably refer to *C. sweeti*, the type specimens of which are from Khadro, Sind, Pakistan (attacking sugarcane), and Dehra Dun, Rajpur (in northwest Uttar Pradesh), India (Slater and Miyamoto 1963). At present, it seems that all economic records of *C. excavatus* on sugarcane, including the major biological studies by Rahman and Nath (1939), Rahman et al. (1947), Atwal and Singh (1971), and Sardana and Mukunthan (1991), may refer to *C. sweeti* with the possible exception of the more eastern records from Pusa, Bihar State, India, where the black bug is scarcely a pest (Fletcher 1914, Rahman and Nath 1939, Kumar et al. 1987) and in the Nadia district of West Bengal (Tewari 1980). Fletcher (1919), who first listed *C.* (as *Macropes*) *excavatus* as injurious to sugarcane, stated that it was found throughout India, and feeds on rice, maize, Job's tears, and various grasses, as well as sugarcane. This is a suspiciously broad host range, given that there are several species of *Macropes* and *Cavelerius* in India (Slater and O'Donnell 1995). Moreover, Rahman and Nath (1939) stated that in the Punjab the insect is not known from any other host. Perhaps workers in India and Pakistan can verify the species of *Cavelerius* present on sugarcane, especially in other sugarcane growing areas.

The two species may be readily distinguished by bright yellowish femora in *C. excavatus* and dark brown femora in *C. sweeti* as well by other characters used in the key by Slater and Miyamoto

(1963). In his key to the nymphs of the Blissinae (Blissidae) of the world, Slater (1979) only had the Japanese *C. saccharivorus* to study and did not include the Indian species.

Also best noted here is that *Cavelerius* (as *Macropes*) *tinctus* (Distant 1904) was recorded as a pest on sugarcane in Pakistan (Marwat and Khan 1987). This identification should be confirmed because *C. tinctus* was described from Tenasserim by Distant (1904), but was erroneously placed in India by Slater (1964, 1979). Tenasserim is actually in far lower (southern) Burma (Myanmar), equivalent to the modern province of Tanintharyi. In any case, *C. tinctus*, which is not in the key by Slater and Miyamoto (1963), differs from the above in having all three basal antennal segments bright orange-yellow (Slater 1979). The Pakistan specimens may perhaps be a misidentification of *C. sweeti*.

Rahman and Nath (1939) made a major bionomic study of the black bug, which was damaging sugarcane in Punjab, India. They described and figured the egg and the five instars. The egg is cigar-shaped with five chorionic processes. It is creamy white when laid and becomes vinaceous-brown later on. The egg is 0.32 by 1.15 mm. The eggs are often laid in parallel rows strung together at their width (Garg and Chaudhary 1979). The first instar has yellow-ochraceous head and thorax; pale first to third abdominal segments; and fourth to seventh orange tinged with red. From the illustrations, the later instars of *C. sweeti* (as *M. excavatus*) are similar to *C. saccharivorus* in having conspicuous large lateral white spots on the abdomen, which is otherwise dark reddish-brown. The head and thorax are dark brown, suffused with reddish. The lengths of the instars are first, 1.35; second, 2.8; third, 3.7; fourth, 4.6; fifth 6.5 mm; the adult is 7 mm with a large black medial area on the membrane.

Rahman and Nath (1939) determined that there were three generations a year, two in the summer on the plants and one in the winter, the latter represented by adults on the ground laying eggs. However, they record that the adults are very long lived, and the female live for 118 days and the male for 112 days in the summer, and during the winter for 242 and 212 days, respectively. The winter eggs are laid 2 to 3 in. deep in the soil where the adults feed on the base of the plants. The summer generations are passed on the plant, including the winter eggs that hatch in March to April and climb up into the plant to feed on the leaves of ratooned sugarcane. The summer eggs were laid in clusters of 14 to 67 on the inner side of the sheathing bases of the side leaves. Copulation occurred in all seasons, the duration of copulation being for about 2 hours in the summer and 74 hours in the winter.

Atwal and Singh (1971) followed this work with a careful bionomic study in the laboratory, and measured the effect of temperature on growth at 20, 25, and 30°C: egg development was respectively 33.1, 17.2, and 12 days; the egg viability was 57, 65, and 71%. Stadia under 90% relative humidity, the most favorable humidity, were first, 12.8, 10.3, 8.5 days; second, 7.9, 6.8, 5.4 days; third, 11.5, 7.8, 5.3 days; fourth, 13.8, 7.4, 4.9 days; fifth, 14.8, 8.0, 7.6 days. At 35°C, the egg and first to third instars did not develop. Nymphal survival was greatest at 90% relative humidity. The preoviposition periods were, respectively, 30.3, 17.1, 13.9 days; the oviposition periods were 49.2, 36.6, 22.6 days. The fecundity was 94, 148, and 166 eggs per female. However, Rahman and Nath (1939) recorded one female as laying 478 eggs. Lower humidities (40%) significantly reduced the rates of development. Under field conditions of varying temperatures (Rahman and Nath 1939), the eggs hatched in 9 to 17 days from May to September; in 85 to 159 days from October to January; from 23 to 25 days from February to March. Garg and Chaundry (1979) observed the same phenomenon, 15 to 20 days incubation of egg in the summer, 100 to 150 days in the winter. This very slow development in the dry winter months indicates a diapause condition similar to that of *C. saccharivorus*. However, Rahman and Nath (1939) and Atwal and Singh (1971) both stated that the adults hibernated in the soil, although laying eggs in the soil during this period. If so, this would an unusual case of two stages — egg and adult — hibernating (or estivating). Actually, as the adults were laying eggs, this may not be a true diapause, although winter adult longevity was much greater and the time to complete development for the winter generation was 195 to 202 days, while summer generations took 35 to 50 days. Rahman et al.

(1947) observed the black bug populations on the plants increased from spring to summer to a peak in August and declined into the winter. This pattern, including the optimal growth and fecundity at 30°C and 90% relative humidity, corresponds, as Atwal and Singh (1971) noted, to an adaptation to the rainy monsoon period. Interestingly, this is the reverse of the North American chinch bugs (*Blissus* spp.) discussed earlier, which are favored by hot dry conditions. Sardana and Mukunthan (1991) demonstrated statistically that during the rainy season on ratoon sugarcane the black bug population was at its maximum in colonies in the middle or whorl level of the plant. Knowing this habit, Brar and Avasthy (1988) developed an efficient sampling technique for estimating the abundance of *Cavelerius* in the sugarcane fields. Bains and Dhaliwal (1983) observed that outbreak populations were highly discontinuous and attributed this to the number of adults that succeeded in overwintering. They also noted that the years with the more severe outbreaks had moderate temperatures and lower levels of rainfall in the dry season.

3.5.2 Damage

With the practice of ratooning of sugarcane, *C. sweeti* has become a major pest in subtropical India as the insects survive under the central leaf whorl, under the sheathing base of leaves, in crop residues, and in the winter in the soil at the base of the plant. The main sources of the outbreaks are the overwintered eggs that hatch simultaneously with the return of moderate humid conditions of the monsoon. The continuous infestation causes the leaves to turn pale yellow with brown-rust colored eye-shaped patches, and growth is greatly retarded. When very severe, the leaves become riddled with holes as a result of several bugs puncturing them at the same spot. Garg and Chaudhary (1979) described the damage as causing the sugar cane leaves to look pale and sick with reddish brown spots, presenting a scorched appearance visible in the distance. Rahman and Nath (1939) stated that the bugs secrete a foul-smelling fluid that evidently promotes decay. Populations can rise to as much as 1000 bugs per cane shoot. The yield of the crop declines in terms of millable canes and sugar (gur) extracted (Rahman and Nath 1939, Atwal 1965, Aggarwal and Tiwari 1967, Tewari 1980, Maheshwari and Malik 1988, Sardana and Mukunthan 1991, Jaipal 1996).

3.5.3 Control

Jaipal (1996) compared mechanical methods of control with chemical methods. Mechanical control consisted of removing the trash and shaving the stubble before and after the sprouting of the canes. The insecticide used was endosulfan. He found no statistical difference between the two methods of control in the amount of millable stalks and yield of sugar. However, shaving should be done before sprouting to avoid damaging the new canes. A drawback of chemical control is the potential destruction of natural enemies. Jaipal believes that the practice of trash burning *in situ*, as had been recommended (Rahman and Nath 1939, Gupta and Avasthy 1954, Khan and Sharfuddin 1966, Garg and Chaudhary 1979, Mrig and Chaudhary 1994a), should be avoided for this also would be detrimental to natural enemies. However, Rahman and Nath (1939) found that this burning treatment of ratooned fields completely destroyed the pest, and the yield was similar to that from plots not burned.

Rahman and Nath (1939) discovered two parasitic wasp species that attack the eggs of *C. excavatus* (= *sweeti*). These were described by Nixon (1938) as a new genus and species, *Nardo cumaeus* and *N. phaeax* (Scelionidae). The genus is interesting in that the bodies of the wasps are extremely flattened, which undoubtedly facilitates their moving in tight crevices and gaining access to the eggs of *C. sweeti*. Rahman and Nath (1939) noted that the parasites attacked the eggs heavily, especially the summer eggs, and reported parasitism rates varying from 27 to 84%. It is surprising that subsequent workers appear to have ignored this potential for biological control.

Varma et al. (1988) discovered the fungus *Beauveria bassiana* attacking a weevil *Phytoscaphus* sp., which is the first record of the fungus in India. They found this fungal strain to be pathogenic to *C. sweeti* and other sugarcane pests. Varma and Tandan (1996) demonstrated that the fungus is highly pathogenic and proposed that it be used for biologic control. In laboratory trials it killed 99% of the adults and 95% of the nymphs of *C. sweeti* although the insect was resistant to other fungi. Whereas the fungus in North America was too dependent on weather conditions to be reliably effective (see *Blissus leucopterus*), it would seem that the moist monsoon summer climate, which is favorable to the Oriental chinch bug, should be favorable to the fungus as well, affording good biological control.

Similar to the later work with *C. saccharivorus,* Rahman and Nath (1939), and Rahman et al. (1947) showed that cultivars with broad, loosely attached leaf sheaths harbored much higher insect populations than the cultivars with tight sheaths. They also found that the insects migrated from the sheaths to feed all over the plant in the morning and sought cover under the sheaths as the day progressed. Jaipal (1991) evaluated 23 cultivars of sugarcane, and found most were susceptible, but one, CoS 767, was the least susceptible and showed no apparent signs of damage. Such host resistance would be another important method of control to explore.

However, most research has been on control using insecticides. The earliest used in India was a fish oil–nicotine sulfate emulsion (Rahman and Nath 1939, Maheshwari and Malik 1988). Modern synthetic organic insecticides that have been recommended to give good control in India are as follows: endrin (0.02%), the most effective (Gupta and Avasthy 1954), has been banned, but phosphamidon 0.02%, metasystox (0.05%), and endosulfan (0.05%) in that order were nearly as effective as endrin, and much less hazardous to the operator (Sandhu and Madan 1972); malathion, sevimol (carbayl + molasses); endosulfan and phenthoate were effective and inexpensive (Madan et al. 1978); also to replace endrin, endosulfan and quinalphos were effective (Tewari 1980); endosulphan at 0.3 kg/ha, which can be easily sprayed by hand with a sprinkling can or a foot sprayer was effective and economical (Duhra and Sandhu 1983a, b); lindane and endosulfan at 1 kg/ha (Sandhu and Duhra 1984); 0.05% endosulfan, 0.2% BHC (HCHL), 0.05% fenitrothion, and 0.05% phenthoate (Chaudhary et al. 1984); endosulfan 35% at 1 l/ha (Saini 1985); granulated endosulfan (thiodan 5G) at 12 kg/ha (Varma and Mehrotra 1985); endosulfan and oftanol (isofenphos) at 200 ml/ha; quinalphos and cypermethrin at 200 ml/ha; chlorpyrifos and fenvalerate at 100 g/ha (Maheshwari and Malik 1988); of these latter, the most effective was fenvalerate dust (Singh 1990); with cultivar Co 1148, phenthate 50 EC at 750 ml/ha and lindane at 1250 ml/ha were most effective and economical (Singh et al. 1991); endosulfan at 0.3 kg/ha was most effective (Duhra and Singh 1992); endosulfan 1.4 l/ha, fenitrothion 1.0 l/ha, and dichlorvas 400 ml/ha were similarly effective, with dichlorvas being the most economical (Madan et al. 1993); lindane appeared to give the best overall control of various pests as well as the Oriental chinch bug (Mrig and Chaudhary 1994b). In Pakistan, Marwar and Khan (1987) recommended insecticide treatment with granules of disulton 10G and carbofuran 3G for *C. tinctus* (probably *C. sweeti*) and other sucking pests on sugarcane.

3.6 *Macchiademus diplopterus* (Distant) (Blissidae)

3.6.1 Distribution, Life History, and Biology

This species, known as the grain stink bug or stinkbesies (Smit 1964), is a serious pest of small grains in South Africa similar to the chinch bug of North America (Herring 1973). It was originally described in the genus *Blissus,* under which it is listed in most of the older literature. It is a small black and white insect, similar to *B. leucopterus*, but more slender and longer, about 4 to 5 mm long. Slater (1967) first moved this species to *Atrademus* and, later, Slater and Wilcox (1973) erected the genus *Macchiademus* to receive this and four other closely related South African species of the Cape region of South Africa, three of which they described as new at this time. Based on the nymphs, there are several other as-yet-undescribed species. Only *M. diplopterus* is of economic importance.

Like the chinch bugs of North America, this is a native species that was able to host transfer to introduced crop grasses. In the drier grain-growing area of the western Cape Province, South Africa, around Touws River, Sutherland, Citrusdal, Porterville, Worcester, Graafwater, Prince Albert, Picketsberg, and Oudtshoorn, its attacks on wheat, barley, and oats have often made their cultivation uneconomical (Lounsbury 1918, Smit 1964, Sim 1965, Matthee 1974, Annecke and Moran 1982). Interestingly, it does not attack sorghum, maize, or sugarcane; nor does any other member of the rich South African blissid fauna attack these three important economic grasses (Slater and Wilcox 1973a, Carnegie 1974, Matthee 1974, Annecke and Moran 1982), but as Slater and Wilcox predict, it is perhaps only a matter of time before such a transfer may occur, as had evidently happened with *M. diplopterus* itself, transferring to small grains. The native known hosts are *Ehrharta longiflora* J. E. Smith, *E. erecta* Lam., *E.* near *calycina* Sm., and *Pentaschistus thunbergii* Stapf. Myburgh and Kriegler (1967) found it, including its immature stages, on several weedy introduced grasses: *Hordeum murinum* L., *Avena fatua* L., *Bromus catharticus* Vahl, *Bromus diandrus* Roth, *Poa annua* L., and *Lolium multiflorum* Lam (= *L. perenne* L.). Slater and Wilcox noted that most of the collecting records of *M. diplopterus* were in lowland, relatively dry disturbed areas where the grasses had straggling growth patterns and did not form dense clumps. In contrast, the other species of *Macchiademus* were in highland permanent locations, with largely brachypterous populations that are restricted to particular host plants. *Macchiademus diplopterus,* although mostly macropterous, has a small percentage of brachypters, which are similar to the brachypters of the other species, the wings attaining the third or fourth abdominal tergum. Slater and Wilcox observed that this is similar to *Blissus leucopterus* in North America and proposed that the predominance of macroptery in both cases is the consequence of strong selection for migration to and from new host plants (human-introduced cultigens and weeds) to hibernation/estivation quarters, and the presence of the brachypter indicates a former restriction to a more permanent habitat prior to the introduction of these new host plants. Wheat has been cultivated in the Cape region since the middle of the 17th century (Sim 1965), which may be sufficient time for such selection to occur. Goliomee (1959) observed that the insects fly actively at temperatures above 21°C in tremendous numbers in late spring to seek sheltering places to survive the dry summer, which is when fruits start to ripen. The insects hide under loose bark of trees, in cavities of fruit, under shrubs, in clumps of veld grasses, in stubble, or even under thatch roofs. This has given rise to the many records since 1937 of the insects being intercepted on apricots, peaches, apples, and pears, as well as in wheat and grass seeds from South Africa, so that Slater (1964c) and Herring (1973) stated that the insect was a probable fruit feeder, at least as an adult. But later Slater and Wilcox (1973) realized that this was an accidental consequence of the insect's migratory shelter-seeking behavior. Anneke and Moran (1982) noted that the insects may travel several kilometers and may be found congregating in large numbers in the branches and under the bark of shrubs and trees such as *Eucalyptus,* deciduous fruit trees, vines, fig trees, and pomegranates. The insects are quite cold hardy, which leads to their surviving being transported under refrigeration (31 to 34°F) for 6 to 10 weeks on fruit and were thus intercepted alive on apricots and peaches at U.K. and U.S. quarantine stations (Myburgh and Kriegler 1967). This is important, because it indicates that the insect has the dangerous potential to invade new areas and survive temperatures well below those experienced in South Africa (Herring 1973).

The insect emerges in the spring in June and July in warmer areas to the end of August or even into December in colder areas (Matthee 1974). Cool, wet weather will delay the commencement of the migration to the wheat fields (Sim 1965). The following life-history information comes from Sim (1965) and Matthee (1974). There is one generation a year. The insects mate on the host plant and lay eggs in rows on the inside of the grass leaf-sheaths. The eggs are laid in masses of between 14 and 25. A total of 50 to 150 eggs are laid per female. The newly laid eggs are white to light yellow, gradually changing color to orange just before hatching. The duration of the egg stage is approximately 20 days to 1 month. There are five stadia, taking about 6 weeks to reach the adult stage, during which time the adults of the original population die. The adults of the new generation fly back to their oversummering quarters largely during November. Slater and Wilcox (1973) described the fifth, fourth, and third instars, and the egg. The egg is elliptical with four chorionic processes, and laid in offset rows, glued together. The length and width are not given. The nymphs have an abdomen banded in white and red, the anterior abdominal segments more white, the posterior more red, with a dark chocolate brown head, thorax, and wing pads. The lengths of the instars are fifth, 3.28 mm; fourth, 2.02 mm; third, 1.66 mm.

Oliver et al. (1996) determined that the scent gland secretions in both males and females consisted mostly of tridecane, and some (*E*)-2-octenal, and (*E*)-1-hexenel present and concluded that these were defensive secretions, not pheromones.

3.6.2   Damage

The sap is sucked out of the stems of the grasses and this impairs growth, resulting in stunted appearance and drying of leaves (Sim 1965, Matthee 1974). The plants often die before producing ears of grain. The infestations are, like those of the North American chinch bug, most severe in the drier areas and in drier, warmer years. The ears are also said to be attacked and the developing grains sucked out (Matthee 1974). This seed-feeding behavior needs to be verified, as only the Cape of South Africa species *Ischnodemus brinki* Slater among the Blissidae is known to be a seed feeder. *Ischnodemus brinki* feeds in seed heads of *Wachendorfia paniculata* Burm. and *Dilatris viscosa* L. (Haemodoraceae) (Slater and Wilcox 1973a).

3.6.3   Control

For cultural control, Matthee (1974) recommends burning the harboring sites, especially near the grain fields. For chemical control he recommends spraying harboring sites that cannot be burned, before egg laying commences, normally before the beginning of August in early spring. Dieldrin is recommended by both Matthee (1974) and Smit (1964). Smit recommends using an 18% emulsion at the rate 1/2 pint/acre, mixed with 25 gal of water. However, because of its great toxicity and persistence, dieldrin is no longer registered for use in South Africa (Matthee 1974) and not listed in Thomson (1998). Sim (1965) also recommends 1/2 pint parathion, 40 to 50% emulsion/morgen (= 2.1 acres in Afrikaans), or 1 pint telodrin 15%/morgen, or dusting with BHC at either 7.5 lb of 40% BHC/morgen or 20 lb of 5% HCL/morgen. However, Annecke and Moran (1982) cautioned that insecticidal control is risky and the size of the crop may not justify the expense of control.

3.7   *Dimorphopterus gibbus* (F.) (Blissidae)

3.7.1   Distribution, Life History, and Biology

Dimorphopterus is a large Old World genus that is evidently the ecological equivalent of the New World *Blissus* (Hamid and Slater 1980) and most of the species are listed in the literature under *Blissus*. Of about 35 known species of *Dimorphopterus*, 16 occur in the Ethiopian region. Hamid and Slater consider the economic potential of this genus to be very high. The most widespread species is *D. gibbus* that extends from West Africa to the Oriental and Australian regions. Although Distant (1904) reported it damaging sugarcane in India from Bombay and Cawnpore, it had evidently been confused with the black bug, *Cavelerius excavatus* (Distant) (probably = *C. sweeti* Slater and Miyamoto) of sugarcane until Chaudhary and Khanna (1964) recognized it as a separate species, and differentiated it on the basis of its smaller size of 4 to 5 mm as compared with 6 to 7 mm for the black bug, the lack of black spot on the margin of the forewing and the lack of spines on the slender femora. Slater (1974, 1976a) mentioned briefly that this species was found on sugarcane in India. Given that sugarcane is a cultigen originally from New Guinea

(Simpson and Conner-Ogorzaly 1995), it is interesting that *D. gibbus* was recorded in West Africa, on the thatching grass *Hyparrhenia involucrata* Stapf. (Hamid and Slater 1978). There are six other species of *Dimorphopterus* in India (Slater 1974), but *D. gibbus* can be easily recognized in having a glabrous, not hairy pronotum and being a little larger, the others being under 4.0 mm in length. The biologies of the other species are unknown and *D. gibbus* in India is known only from sugarcane. The species is pterygopolymorphic and Distant (1904) illustrated the brachypter, mistaking it for the immature form. Slater noted that in the Nigerian population the brachypterous wings are transparent, unlike the opaque wings of the Oriental specimens.

Mathur and Prakash (1989a) made a careful study of the bionomics of this species in the laboratory and in the field over the year from March to December. The insects live low on the host plant under the leaf sheaths during the day but climb at night to feed on the upper tender shoots. The population peaked in the field in May and June, but declined greatly with the heavy rains of the monsoon in late June. However, in the laboratory the hot humid conditions favored the species, so the decline was probably due to drowning at the base of the plants. In the field, mating occurred hidden in the sheaths, and in the laboratory the insects mated frequently end to end, the females with several males, the males with several females, the number of matings averaged 33 times per individual, and the length of copulation averaged 11 hours, the total time in copulation averaged 242 hours. The preoviposition period after mating averaged 4.2 days. The females laid the eggs in the field inside tightly fitted leaf sheaths, but will lay eggs at random if no proper oviposition site is available. The eggs are normally thrust deeply into cut edges of leaf sheaths in the laboratory. The eggs are normally laid in groups of two to nine arranged in linear rows. The average number of eggs laid per day was 5, but varied from 1 to 31 eggs. The total number of eggs laid averaged 285, laid over an average of 56 days. There was no relation between the age of the female and the number of eggs laid except toward the end of life. The postmating period (between last mating and death) was 2.6 days for males, 9.5 for females. They describe and illustrate the immature stages. The eggs are creamy white when laid, changing to dirty white and finally pinkish. The eggs average 0.31 by 1.31 mm. They are elongated with a bluntly rounded posterior end, and the anterior end bears two to five, usually four, micropylar processes, the variation in number occurring in the same female and the eggs of the same day. The first instars are creamy white except for the slightly darker prothorax and mesothorax. In succeeding instars the coloration changes to a pinkish orange color. The incubation period ranged from 7 to 11 days, averaging 8.5 days. The nymphs after hatching search first for a concealed abode. Each instar feeds 15 to 20 hours after ecdysis. The first instar is creamy white, with a slightly darker head and thorax. The later instars become uniformly pinkish orange color with a brown head and thorax. The instar lengths are first, 0.99 to 1.32 mm; second, 1.26 to 1.96 mm; third, 1.87 to 2.6 mm; fourth, 2.64 to 3.58 mm; and fifth, 3.25 to 4.8 mm. The average stadia for the nymphs are first, 5.6 days; second, 3.7 days; third, 3.0 days; fourth, 4.1 days; fifth, 5.5 days, for a total of 25 days in April, 22 days in June, all at average temperature of 33°C and relative humidity of 48%. The longevity after becoming reproductive is an average of 44 days for the males and 53 days for females of the first generation (April to June), and for the second generation (May to August) 52 days for males, 71 days for females. There are two generations a year, second-generation adults going into diapause through the cooler, dry/winter season. Both Mathur and Parkash (1989a) and Chaudhary and Khanna (1964) noted that the macropterous form was more abundant during the warmer months the brachypter during the cooler months.

3.7.2 *Damage*

The damage done by this insect is not clear from the literature, but in view of the statements that this is a severe pest confused with the black bug, and that the damage is similar, it may be the actual cause of the recorded damage by black bugs of sugarcane (see *Cavelerius excavatus/C. sweeti*).

3.7.3 Control

Mather and Prakash (1989b) observed a great variation in the population sizes of this pest on sugarcane. They studied the susceptibility of sugarcane varieties to the sugarcane lygaeid bug. In the laboratory they found no significant differences in nymphal development rates on the different varieties. They concluded as did Rahman and Nath (1939) for *C. sweeti* Slater and Miyamoto (as *Macropes excavatus* Distant), that the morphological characteristic responsible for this difference in varieties was the sheathing leaf base. The varieties with a small, tight leaf sheath were much more resistant than varieties with a loose broad sheath.

By using a ratoon crop of sugarcane in a test of ten insecticides, the best control was obtained with quinalphos at 0.2 kg/ha (93.7% mortality) or chlorpyrifos at 0.2 kg/ha (93.6% mortality). The poorest control was with Dipel (*Bacillus thuringiensis*) at 0.2 kg/ha (28% mortality) (Pandey et al. 1977).

3.8 *Spilostethus pandurus* (Scopoli) (Lygaeidae *sensu stricto*)

3.8.1 Distribution, Life History, and Biology

As noted in the introduction to this chapter, the Lygaeidae *sensu stricto* here are limited to the Lygaeinae of Slater and O'Donnell (1995). As also noted, physiologists and ecologists have extensively studied members of this family, and this species is of considerable past and potential economic importance. The genus *Spilostethus* is a moderately large Old World genus, with 24 described species, of which 18 are Ethiopian. One, *S. pandurus*, is extremely widespread, found throughout the Old World tropics and subtropics from South Africa to southern Europe to China. A number of varieties has been proposed, some of which may be worthy of subspecies designation; but the existing keys emphasize coloration, not geography. Moreover, as is usually true of any common, brightly colored insect that occurs in Europe, the synonymy is complex, and involves the species name itself. *Cimex militaris* Fabricius 1775 (type locality the "Orient") is the type species of *Spilostethus* as fixed by Slater (1964c), but it is treated by Slater as the junior synonym of *C. pandurus* Scopoli 1763 (type locality Italy). However, some authors consider *militaris* a separate species, others as a subspecies of *S. pandurus* (see Slater 1964c). *Lygaeus elegans* Wolff 1802 was described from India where *S. pandurus* is the most important, economically, but following Stål (1874), the name has been applied primarily as a subspecific name to African populations (Slater 1964c, Slater and Sperry 1973). Two insular names have been proposed, *S. pandurus tetricus* Horváth for the Canary Islands population and *S. p. asiaticus* Kolenati for the Madagascan population (the latter name being an unfortunate locality error for the Caucasus Mountains). In Europe this insect is known as the pandar, and Wachmann (1989) has in his handbook to the bugs fine photographs of the adult and the nymphs of what is presumably the European form, or subspecies, *S. p. pandurus*. A study of the variation in this widespread conspicuous insect should be very rewarding (Slater 1964c, Slater and Sperry 1973) and it is badly needed, because as well as being a pest on crops this species is becoming extensively used for physiological studies (Varma and Prakash 1980; Singh and Sharma 1984; Ibanez et al. 1993a,b; Garcera and Tamarelle 1995). Even more perplexing is the status of *S. macilentus* (Stål), which was described from Africa, and for which Slater (1964c) had fixed as lectotype a specimen from South Africa; and later, Slater and Sperry (1973) noted that *S. macilentus* is accurately known only from South Africa and Malawi. However, Chopra (1972) stated that the populations known as *S. pandurus* in India were actually *S. macilentus*. Therefore, on this authority much of the literature, especially the economic and physiological studies in India, had reported this species as *S. macilentus*. Thangavelu (1981) reassessed the evidence and stated, based on the distributional evidence and long usage, that the name of the economically important milkweed bug of India is *S. pandurus*. With this decision Mukhopadhyay (1983) evidently concurs, so the records of

S. macilentus in India are included here under *S. pandurus* until more evidence is forthcoming, such as in a badly needed review of the highly variable widespread species *S. pandurus* and its relatives. Similarly, the references to *L. militaris* or *S. militaris* in the physiological literature in India very probably refer to *S. pandurus*.

Like many other Lygaeidae *sensu stricto*, *S. pandurus* feeds preferentially on members of the Asclepidaceae (see summary of asclepiad species attacked in Slater and Sperry 1973), from which the insect can sequester cardiac glucoside poisons (Veneuw et al. 1972). This capacity to sequester poisons is important from the economic perspective because relatively few predators, especially birds and lizards, feed on the brightly colored *S. pandurus* (Abushama and Ahmed 1976, Mukhopadhyay 1983). Abushama and Ahmed showed experimentally that the secretions of the scent glands also incorporated cardiac glucosides from the plants and that the secretions were repugnant to scorpions, birds, and cats. Aldrich et al. (1997) similarly demonstrated the use of the plant's noxious chemicals in lygaeid defensive secretions. Ecologically, the species of *Spilostethus* especially in Africa, evidently form sympatric aposematic species complexes, and Marshall (1902) proposed that *S. p. elegans* forms in South Africa a Mullerian mimicry association with *S. rivularis* (Germar), *S. crudelis* (F.), and *Graptostethus servus* (F.), and a Bastesian mimicry association with *Reduvius* sp. To this list could be added, in South Africa, the similarly bright red and black *S. lemniscatus* (Stål), *S. taeniatus* (Stål), *S. furculus* (H.S.) and *S. macilentus* Stål. Also interesting are the frequent reports of this species in natural habitats, especially in Europe, feeding and breeding on members of the Compositae (Boselli 1932, Gomez-Menor 1955, Slater and Sperry 1973), so that the Compositae may serve as alternate wild hosts and also as sources of poisonous plant compounds. Boldt (1989) summarized the large number of poisonous compounds found in species of the composite *Baccharis* on which feeds the lygaeid *Ochrimnus mimulus* (Stål) (Palmer 1986). This makes understandable the records of *S. pandurus* feeding on *Vernonia cinerea*, a weed in India. However, the genus contains a large number of medicinal plants and many contain vernonin, a bitter poisonous glucoside (Uphof 1968), so the attraction of *S. pandurus* may be for its allelochemic compounds. Noteworthy in this context are records of *S. pandurus* feeding on poisonous plants other than the Asclepidaceae, such as the apocynaceous oleander *Nerium oleander* L. (Lindberg 1932, Ramade 1960, Thangavelu 1978d), a plant much used in horticulture for landscaping and as an ornamental (Bailey 1949), and on *Datura stramonium* L. (Lepelley 1960), a medicinal plant introduced from the New World and commonly called jimsonweed (Simpson and Conner-Ogorzaly 1995) although other poisonous species of *Datura* occur in the Old World, notably *D. mentel* L. of India (Uphof 1968). Perhaps then, this insect may sequester toxins other than the cardiac glucosides of its asclepiaceaous hosts. Mukopadhyay (1983) observed that the adults preferentially fed on seeds of the asclepiad *Calotropis* while the nymphs, especially the earlier instars, fed more on leaves and stems (which is where the toxins are available). The small nymphs often aggregated together in feeding, representing, she observed, not only a feeding response, but presenting a mass of red that collectively helped protect the nymphs from vertebrate predation. Gamberale and Tullberg (1996) tested this hypothesis and found that aggregating prey of *S. pandurus* were a more effective aposematic signal to naive chicks (*Gallus gallus* L.) than single prey. Mukhopadhyay and Saha (1992a, 1992b, 1993) addressed a paradox: while the preference by *S. pandurus* for *Calotropis* was strongest, the actual production growth was less as compared with other food plants despite higher consumption, and they showed that the metabolic respiration was higher feeding on *Calotropis* than on other seeds. They proposed that this was because of the large diversion of energy that detoxifies the allelochemicals present in *Calotropis*. This, then, would be the cost the insect pays to sequester cardiac glucosides, the trade-off being a defense mechanism against predators. The polyphagy of the *S. pandurus* discussed below can then in part be seen as the acquisition of metabolically inexpensive food, especially after the defense compounds are already in place. Although little is said in the literature, *S. pandurus* is entirely macropterous and evidently disperses readily to find new hosts to exploit from its permanent base on *Calotropis*, a point emphasized by Sanjayan (1993) with reference to the sympatric, ecologically similar *S. hospes*. In this context it is of considerable theoretical interest with respect to the origin of pterygopolymorphism that Solbreck and Anderson (1989) discovered in a laboratory culture of *S. pandurus,* derived from a colony on *Nerium oleander* L. in Cyprus, a mutant short-winged form that is based on a recessive gene.

Although *S. pandurus* does prefer to feed and breed on members of the Asclepidaceae, it is highly polyphagous and is reported from 15 or 16 families of plants, and is sometimes a severe, if secondary, pest on many crop plants (Bhattacherjee 1959, Slater 1964c, Gentry 1965, Slater and Sperry 1973, Thangavelu 1979, Mukhopadhyay 1983). However, as Slater and Sperry (1973) caution, many of the records in the literature may be merely sitting records or sporadic records of adults temporarily feeding on plants. Nevertheless, there are many economic records of *S. pandurus* damaging crops. Gingelly, til, or sesame (*Sesamum indicum* L.) (Pedaliaceae) in India was severely damaged by adults and nymphs (Thangvelu 1979, Ahuja 1989). Thangvelu noted that the damage to sesame was especially severe when the crop was grown near its primary asclepidaceous host plants madar or akk, *Calotropis gigantea* (L.) R. Br. and *C. procera* (Ait.) R. Br. The feeding of *Spilostethus pandurus* and other Lygaeidae (*sensu stricto*) (*S. hospes* (F.), and *Graptostethus servus* (F.), probably reduces the viable seed crop of madar, but promoting such feeding was not recommended as a control of this weed because of the tendency of *S. pandurus* to invade various crops (Mukhopadhyay 1983).

Spilostethis pandurus attacks durra millet or jowar, *Sorghum bicolor* (L.) Moench (= *Sorghum vulgare* Pers.) var. *durra* Hubbard and Rehd. (Poaceae), (Brain 1929; Vishakantaiah and Gowda 1973; Prabhakar et al. 1981, 1986), so much so that Theobald (1906) in India called *S. pandurus* the "dura plant bug." Other grains attacked are ragi, or finger millet *Eleusine coracana* Gaertn. and pearl millet *Pennisetum americanum* Auth. and *P. typhoideum* Rich., both probably synonyms of *P. glaucum* (L.) R. Br. (Vishakantaiah and Gowda 1973, Sandhu et al. 1974, Kapoor et al. 1982). Cotton (*Gossypium* spp.) is commonly attacked (Maxwell-Lefroy 1909; Fletcher 1914, 1919; Ritchie 1925; Harris 1936; Lepelley 1960; Butani 1971). Other crops listed by Bhattacherjee (1959) as being damaged in India are roselle *Hibiscus sabdariffa* L., grown for edible calyx and fiber; bhindi or okra *Hibiscus (= Abelmoschus) esculentus* L., gram or garbanzo *Cicer arietinum* L., urid or black gram *Phaseolus mungo* L., broad bean *Vicia faba* L., arahar, red gram, or pigeon pea *Cajanus cajan* Millsp., peanuts *Arachis hypogea* L. (Varma and Shukla 1976), country beans *Cyanopsis tetragonoloba* (L.) Taubert (Thangavelu 1979), the source of guar gum (Uphof 1968), lucerne or alfalfa *Medicago sativa* L., Egyptian clover *Trifolium alexandrinum* L., brinjal or eggplant *Solanum melongena* L., tomato *Lycopersicum esculentum* Mill., tobacco (*Nicotiana* spp.), sugarcane *Saccharum officinale* L., tori or vegetable-sponge *Luffa acutangula* Roxb., which was a preferred host (Chopra and Yadav 1974), Lolichi *Litchi chinensis* Sonn. (= *Nephelium litchi* Cambess), jamon or Java plum *Eugenia jambolana* Lam., fig (*Ficus* sp.), apricot *Prunus armeniaca* L., peach *P. persica* Batsch, mango *Mangifera indica* L. (Ahmad 1946), guava *Psidium guajava* L., grape *Vitis vinifera* L., and citrus fruit (Isaac 1946); in Italy, hazelnuts *Corylus avellana* L., pista, or pistachio *Pistacia vera* L., trees were damaged (Romeo 1929); in Ethiopia, Chinese artichokes *Stachys sieboldii* Miq. were damaged (Gentry 1965); and in the Congo, sweet potato *Ipomoea batatas* Poir. was attacked (Mayné and Ghesquière 1934). Prasad and Kalavati (1987) found it feeding in clusters on used and discarded soapnuts, probably the Chinese soapnut tree *Sapindus mukorossi* Gaertn. The nuts are much used in India and China as a substitute for soap and for beads (Uphof 1968). Gentry (1965) records *Spilostethis pandurus* on many of the above crop plants in Libya, Lebanon, Egypt, Sudan, Pakistan, Iraq, Iran, Ethiopia, Somalia, East Africa, as well as India.

Although a weed in wetter parts of the Indian subcontinent, in desert areas *Calotropis procera* is valuable and grown on plantations (Parihar 1981). It is used for fiber, which is of great strength, and the floss of the seeds is used for stuffing mattresses, quilts, etc. (Uphof 1968).

Although many workers place emphasis on whether or not the insect breeds on the plant (Slater 1964c, Slater and Sperry 1973), for economic damage even sporadic feeding can be important. Ershad and Barkhordary (1974a,b) demonstrated experimentally that the phytopathogenic fungus

Nematospora coryli Peglion that causes yeast-spot disease was probably transmitted to at least 12 species of fruit and nut trees mechanically by feeding by *Spilostethus* and several pentatomid bugs, as their mouthparts were shown experimentally to have sufficient spores to initiate infections of the fungus on fruits and nuts.

S. pandurus can be reared on and prefers seeds of *Calotropis,* but it can be reared very successfully on hulled sunflower seeds (Kugelberg 1973, Slater and Sperry 1973, Varma and Skukla 1976, Adell et al. 1988). However, there is very little mention of *S. pandurus* on commercial sunflowers aside from a listing in Annecke and Moran (1982) in South Africa. Similarly, Malipatil (1979) could easily raise *S. hospes* on sunflower seeds in Australia. However, in neither India nor Australia did *Spilostethus* bugs feed on sunflower seeds in the field despite their ready proximity and availability in sunflower fields next to *Spilostethus* colonies on *Calotropis* (Mukhopadhyay 1983). In work with feeding in rhyparochromid bugs (Sweet 1960, 1964), the key was to shell or peel the sunflower seeds, whereupon the various species of bugs would usually recognize the seeds as food and proceed to feed readily. This was the case with a great many lygaeoid species that were reared in the United States, South Africa, Mexico, South America, and Australia (M. H. Sweet, personal observation). Nevertheless, often growth was not optimal as compared with that on the normal host seed. However, with *S. pandurus*, El Sherif and El-Shazly (1993) demonstrated that under laboratory conditions almond, sesame, sunflower, peanuts, nuts, and pumpkin seeds were all fed on, but sunflower gave the fastest development, the best survivorship, and the greatest fecundity, and sesame, which is attacked, gave the least. El-Shazley (1993) measured carefully the metabolic productivity of the bug on sunflower seeds to demonstrate the value of this diet. On the other hand, Mukhopadhyay and Saha (1992a,b, 1994) in comparing four host seeds (*Calotropis*, eggplant or aubergine, sesame, and sunflower) in both *S. hospes* and *S. pandurus* found the lowest fecundity and hatchability for both species on sunflower seeds, which they noted had the highest phenolic content of the food seeds. Gupta and Gandhi (1996) found *Calotropis* to be the most preferred host plants, but feeding occurred on other foods in this order of preference: red gram > okra > green gram (*Vigna radiata*) > gram > cotton > sorghum. Chopra and Yadav (1974) found the crop preferences to be tori (*Luffa*) > lucerne > lobia (*V. unguiculata*) > tomato > eggplant (brinjal) > cotton > sugarcane > bhindi (okra).

The immature forms have been described by Bhattacherjee (1959) in India, by Hamid and Ahmen (1972) in Pakistan, and by Slater and Sperry (1973) in South Africa. The egg is broadly oval 0.65 by 1.2 mm, with 10 chorionic processes, and pearly white when laid, becoming purple just before hatching. The nymph is bright red after hatching, and in later instars the wing pads become dark brown and shining. Dark markings develop on the thorax and abdomen, but these colors differ in the descriptions of Bhattacherjee and of Slater and Sperry. In any study of the geographic variation, the nymphs should be included. The nymphal lengths from Bhattacherjee are first, 1.66 mm; second, 2.2 mm; third, 3.25 mm; fourth, 5.5 mm; fifth, 10 mm. The stadial lengths in the summer were egg, 4.0; first, 3.2; second, 3.4; third, 4.4; fourth, 5.7; fifth, 7.2 days. Kugelberg (1973) got similar development rates in a colony raised on sunflower seeds at 25°C, 18-hour photoperiod: egg, 9 days; first, 6.1 days; second, 4.6 days; third, 4.8 days; fourth, 5.7 days; and fifth, 10.4 days. The stadia measured by Verma and Shula (1976) at 26°C are similar: first, 6.2 days; second, 4.0 days; third, 5.4 days; fourth, 4.4 days; fifth, 7.8 days. The adults are about 12 mm long, bright red with a transverse black band on the corium and a milky white membrane as illustrated by Wachmann (1989).

Bhattacherjee (1959) reported 50 to 60 eggs laid per female with as many as 90 laid in one or more clusters underneath fallen leaves or flowers. The eggs were laid loose and not cemented together. Thangavelu (1979) recorded 45 to 90 eggs with an exceptional case of 130 eggs. Mukopadhyay (1983) found much higher oviposition of 150 average, 75 to 232 range, the difference she attributed to there being an adequate supply of *Calotropis* seeds because, in her work, diets without *Calotropis* seeds showed greatly reduced oviposition. Verma and Shukla (1976) similarly had an average of 130 and up to 221 eggs on a *Calotropis* diet, and found it difficult to raise the insects on other foods.

On the other hand, Kugelberg (1973) using shelled sunflower seeds in the laboratory at 25°C, found females to lay up to 10 egg batches, each batch containing 50 to 60 eggs. The preoviposition period of the adults was 5 to 7 days. The insects mated for several hours. At temperatures lower than 60°F there was no copulation or oviposition (Thangavelu 1979). In the winter in northern India the insects disappeared from the host plants and hibernated in leaf litter especially at higher elevations (Bhattacherjee 1959, Chopra and Yadav 1974, Thangavelu 1979). Thangavelu dissected the resting individuals and found degenerated ovaries and fat body, which indicated a diapause condition. The average longevity of the adults in captivity was 24 to 32 days in the summer and a little longer in the winter, 28 to 48 days (Bhattacherjee 1959). Varma and Shukla (1976) similarly found longevity to be 22 to 31 days. These results seem a little short, as Kugelberg (1973) found the adults to live for about 3 months and some individuals survived for 7 months. Perhaps these differences reflect different geographic populations or ecotypes. In south India under warmer conditions there are six to seven overlapping generations (Thangavelu 1979). Chopra and Yadav (1974) found five generations in North India and insects reached their peak abundance in the crop fields in mid-August. In Giza, Egypt, El Shazly (1996) also measured six overlapping generations.

The main predation described by Bhattacherjee and Mukhopadhyay was cannibalism of late instars on earlier instars or on weak individuals. Mukhopadhyay (1983) wondered if this might be a mechanism of population control because there appeared to be so little natural predation occurring. Thangavelu (1978b, 1979) agreed with the apparent paucity of predation but he did observe that on gingelly (sesame) the common crow *Corvus splendens* L. fed on the nymphs, as did spiders. In India there appears to be no record of any parasites of either the eggs or the adults in the economic literature. However, Ramamurty (1970) in his excellent study of the reproductive organs of "*Lygaeus* sp." mentioned that a tachinid fly often parasitized the adults and nymphs of the population that he had collected on *Calotropis gigantea*. Based on the location of the collection site on the campus at Banares Hindu University, Varanasi, India, and the described year-round abundance of the bugs, it is highly probable that the insects the Commonweath Institute of Entomology identified for him as "*Lygaeus* sp." were *S. pandurus;* the uncertainty of the identification is understandable because of the great variability and confused taxonomy of this species. In Italy, Anderson (1991) studied the parasitism of *S. pandurus* by *Leucostoma crassa*. He showed that the tachinid fly parasitizes several lygaeids, *L. creticus* Lucus, *L. equestris* (L.), and *Tropidothorax leucopterus* (Goeze). He showed that the fly varies in size according to the host size, and also in fecundity, the smaller flies laying fewer eggs. Muller and Fritsche (1993), working with the same parasite, showed that the adult *Spilostethus* reacted to the presence of the parasite differently and the males suffered more attacks. Anderson (1991) also worked in the laboratory with an egg parasitoid, *Telenomus* sp. Anderson and Solbreck (1992) noted that the *S. pandurus* nymphs directly after they hatched attacked the unhatched sibling eggs, fertile, unfertile, and parasitized, showing a preference for the unfertilized eggs while the parasite preferred to attack the fertilized eggs. But the parasite develops more slowly than the unparasitized egg, leaving it, too, open to attack by the newly hatched nymphs. Thus, the title of their paper: "Who eats whom?"

Archibald (1911) described in *S. pandurus* protozoan gut infections by *Herpetomonas* (= *Leptomonas*) *lygaei* Archibald and by *Blastocrithia* (= *Crithidia*) *familiaris* Gibbs (Tieszen et al. 1985, 1986, 1989), protozoans that probably also enter into the host plants, perhaps *Calotropis*, similar to what was proposed by Holmes (1925) and was shown by McGhee and Hanson (1962, 1964) with *L. oncopelti* Noguchi and Tilden in *Oncopeltus fasciatus* Dallas and *Asclepias syriaca* L. In India *L. bakeri* Prasad and Kalavati (1987) were described from the mid- and hindgut of *S. militaris* (= *S. pandurus*), coexisting with *L. lygaei*.

3.8.2 Damage

Bhattacherjee (1959) placed broad bean and jamon branches in muslin bags and showed that the bugs had a devastating effect on the number of fruit developing. He noted that the bugs suck the sap from the flowers, fruits, the epidermis of tender branches, shoots, and leaves. They can pierce the seed endocarp even after it becomes completely lignified. Thangavelu (1979) described that the bugs fly into gingelly (sesame) fields just prior to flowering. Their severe attacks wither completely the plant, the seeds develop poorly, and the pods dry and wither prematurely. Vishakantaiah and Gowda (1973) noted that the nymphs and adults sucked the sap from tender spikelets of the millet ragi (*Eleusine coracana*) and caused chaffy grains to form.

3.8.3   Control

Bhattacherjee (1959) recommends the removal and destruction of the reservoir species, like milkweeds (*Calotropis*) and Vernonia, and the use of a long-lasting contact insecticide to protect the crop plant. Yaday and Chopra (1973) compared 12 insecticides with DDT as the standard, and found the most effective against *S. macilentus* (= *S. pandurus*) was methyl-parathion (26.4 times as toxic), followed by methyl-demeton (17.7 times), phosphamidon (16.9 times), dimethoate (13.1 times), malathion (9.8 times), carbaryl (8.1 times), endosulfan (6.7 times), chlordane (6.6 times), BHS (5.3 times), endrin (5.2 times), and toxaphene (3.1 times). Sandhu et al. (1974) found that on pearl millet (*Pennisetum typhoides*) in India, a low-volume-concentrate spray of fenitrothion at 300 g/ha and an ultralow-volume spray of monocrotophos at 75 g/ha gave effective and persistent treatments.

3.9   *Nysius niger* Baker 1906 (Orsillidae)

3.9.1   *Distribution, Life History, and Biology*

In contrast with the extensive synonymy in the African and European fauna discussed earlier, *N. ericae* of North America, the original false chinch bug, had been distinguished from *N. ericae* Schilling of the Palaearctic as a separate species, *N. niger* Baker (Ashlock 1977). In an apparent oversight, Ashlock's action was not recorded in the *Catalogue of the Lygaeidae of the World* (1960–1994) (Slater and O'Donnell 1995), nor in "Lygaeidae of Florida" (Slater and Baranowski 1990) in which works they refer to *N. ericae* in the old sense. However in Slater and Baranowski's (1978) "How to Know the Hemiptera," this species is referred to as *N. niger*. The true *N. ericae* is judged here, based on the literature, to be a lesser pest in the Palearctic region and is treated under minor pests.

It is, however, very difficult to be sure of the identity of the *Nysius* species in North America in the older economic literature, as Barber (1947) discussed very well. Consequently, much of the economic literature referring to *N. ericae* (= *niger*), actually refers to *N. raphanus* Howard, *N. angustatus* Uhler, *N. tenellus* Barber, or *N.* (= *Xyonysius*) *californicus* (Stål), or to other *Nysius* species, as Barber distinguished 14 species in North America. These species look alike superficially, all being small insects, about 3 to 4 mm long, gray-brown mottled with darker gray spots, and with glassy wings, but the common economic species of North America can be distinguished as follows (Barber 1947). If the bucculae under the head are very short and there is a lateral wing strigil, it is *Xyonysius* (4 spp.) (see below). If the bucculae are high anteriorly, and taper posteriorly, this is *N. tenellus* of the western United States, which overlaps with *N. raphanus* in the southwestern United States and extends through Mexico and into the West Indies and, strangely, into far southern Florida (Slater and Baranowski 1990). If the bucculae are low throughout, the pronotum as short as head, and membrane very long, extending much beyond the apex of the corium and the abdomen, it is *N. raphanus*, the southern false chinch bug of the southern United States. If the pronotum is longer than the head and the membrane is relatively short, extending only as long as the membrane before the apex of the corium and only a little beyond the apex of the abdomen, it is *N. niger*, the northern false chinch bug of the northern United States and southern Canada. As *N. raphanus* extends as far north as South Dakota (Hantsbarger 1957), it overlaps considerably with *N. niger*, which is listed by Ashlock and A. Slater (1988) as far south as Texas, Missouri, and Virginia, in the United States. Because these species have such large overlapping distributions (see Slater 1964c and Ashlock and A. Slater 1988), care must be taken in identifying the species and reference to Barber's and Ashlock's papers is strongly advised.

Slater (1964c) tried valiantly to sort out the North American economic literature by species, but in most cases they were educated guesses because of the overlapping distributions of these species and the lack of taxonomic understanding until after Barber (1947) and Ashlock (1977). Similarly Ashlock and A. Slater (1988) in the catalog to the North American Hemiptera show in their records an almost complete overlap with the most northern records of *N. ericae* in *N. niger*, and the most southern records in *N. raphanus,* and they emphasized that, as Slater (1964c) said, the literature records for these two species are irrecoverably confused. Therefore, the older references cited here must be evaluated carefully. Slater and Baranowski concluded from their fieldwork that *N. niger* does not occur in Florida (U.S.A.), although *N. scutellatus* (Dallas) of the West Indies does, and because it resembles *N. ericae* it may be the source of some of the *N. ericae* records in Florida in Ashlock and Slater (1988). *Nysius scutellatus* is easily distinguished as it has high rather than the low bucculae of *N. niger*. However, it helps to sort the relevant literature to note that *N. niger* was originally described from the state of Washington (U.S.A.) by Baker (1906) as a variety of *N. angustatus* Uhl., a variety that was synonymized with *N. ericae* by Barber (1947) and so became the name of the North American "*N. ericae*." Critical here as discussed later under that species, *N. raphanus* was described from Kansas (U.S.A.).

The nymphs of *Nysius* species are remarkably similar in appearance, with brown and white striped dorsal head and thorax, and the abdomen a mottled reddish-brown and white. As Essig (1929) wrote, "the coloration matches the dry weeds and soil to a remarkable degree." Tashiro (1987) shows an excellent photograph of a fifth instar, which is probably of this species as it comes from a lawn situation in New York (U.S.A.), but it is labeled erroneously as *Blissus leucopterus hirtus*. Wilson (1910) figured the adult and fifth instar from New Jersey (U.S.A.), which is also probably this species. Burgess and Weegar (1986), whose goal was to rear the insects in the laboratory to study yeast transmission, found that egg development took 6 to 12 days (average, 8), and the first adults emerged within 16 to 19 days (average, 21). The adults laid eggs until they were 7 weeks old. They stated that the species was multivoltine and mated and laid eggs year-round. However, they kept the insects at 16-hour photoperiod and at 23°C, which probably overcame the natural life cycle in Saskatchewan, Canada, in which declining short photoperiod probably initiated the production of diapause eggs. Further, they stored the eggs for months at 5°C, evidently for convenience of hatching the nymphs as needed, but this could have provided sufficient cold to complete diapause development in the eggs.

Nysius niger, the northern false chinch bug, is a serious pest of cruciferous crops. Its preference for cruciferous plants was shown experimentally by its attraction to mustard oils, which was most effective at the time of blooming of the European cultivated cruciferous plants canola (rape seed or colza) *Brassica napa* L., the Indian mustard *Brassica juncea* (L.) Czern et Coss., and the flixweed *Descurainia sophia* (L.) Webb, a weed introduced from Europe, where, however, its seeds have been used to make a pungent mustard (Uphof 1968). As these plants are all European introductions, the original or natural hosts of *N. niger* are probably to be sought among the native Cruciferae. Burgess et al. (1983) discovered that *N. niger* alone among the 18 available insects was evidently responsible for the transmission of a serious yeast pathogen (*Nematospora coryli* Peglion) to mustard in Saskatchewan, Canada. But where did the spores come from, as the fungus did not survive on the host plant vascular tissues of either the mustard or flixweed, or in the soil, or in the insect? This led Burgess and McKenzie (1991) to study the biology of *Nysius niger* and they discovered, contrary to the available economic literature, that *N. niger* overwinters in diapause as an egg. The eggs, too, tested negative for the yeast. However, they discovered that the young nymphs did not feed on the sprouting young plants but on the overwintered seeds of flixweed, which contained the pathogen, and this feeding continued through to the adult stage, after which the adults emigrated to the flowering flixweed and mustard, infecting both with the seed pathogen *Nematospora* acquired from the flixweed seeds. This makes *Nysius niger* a far more serious pest than would be realized merely from damage to the vegetative growth, because the economic value of these cruciferous crops, mustard and canola, is from the seeds, which Burgess and McKenzie (1991) have demonstrated are infected with the seed pathogen *Nematospora* by *Nysius niger*.

It should be mentioned here that in 1958 when the author was first working on the biology of lygaeoid bugs, before specializing on the rhyparochromid bugs (Sweet 1964), he observed *N. niger* (= *ericae*, then) at Storrs, Connecticut (U.S.A.), feeding vigorously in the autumn on pepper-grass seeds (*Lepidium sativum* L.), mating, and laying in the soil eggs that did not hatch in the laboratory, which extends the range of the egg diapause to Connecticut, and makes it probable that overwintering as an egg and feeding on Cruciferae is characteristic of *N. niger*. In this context, it is interesting that Böcher (1972, 1975, 1976) demonstrated that in Greenland *N. groenlandicus* (Zett.), which Schmitz and Péricart (1993) had synonymized with the Palearctic *N. ericae* (once the name of the false chinch bug of North America), also overwinters as an egg. However, although generally oligophagous, *N. groelandicus* feeds preferentially on seeds of Compositae rather than on the seeds of Cruciferae. Certainly, in evaluating the species of *Nysius,* it should prove very important to understand the ecology and biology of the species as well as its morphology.

However, *N. niger* will attack a variety of crop plants, evidently after its wild cruciferous hosts are dehydrated or depleted (Beirne 1972). This would mean that *Nematospora coryli* would have an infection avenue to other fruits and seeds than to mustard and canola seeds, as demonstrated by the work of Ershad and Barkhordary (1974a,b) with *Spilostethus pandurus*. This dangerous possibility should be monitored carefully, as *Nematospora coryli* apparently has been only recently introduced into North America from Eurasia.

3.9.2 *Damage*

Essig (1929) expressed well the economic importance of *Nysius ericae* (? = *niger;* exactly which species he was referring to is perhaps immaterial here). "This very interesting and often destructive species normally breeds and feeds in the native grasslands [of western North America, probably California] where it multiplies in countless numbers. In the late summer and fall of the year ... and again the spring ... the wingless nymphs swarm from the dry grasslands into the adjoining cultivated areas and infest all green and growing plants and soon do great damage .... Alfalfa, clover, grains, grasses; and other forage crops, truck crops, sugar beets, cotton, grapevines, fruit trees and young fruit, berries, native flowers, and weeds are commonly infested and often destroyed." Knowlton (1935) found it to damage beets severely in Utah (U.S.A.). Smith (1944) lists it as the false chinch bug damaging broccoli, kale, cabbage, and lettuce in the western United States. Knowlton and Wood (1943) noted that in Utah, "The insect breeds in large numbers on roadsides and deserted farms among Russian-thistle, atriplexes, mustards and grasses. When the hordes of false chinch bugs swarm ... on to farms and gardens, they frequently damage sugarbeets, alfalfa, cabbage, clover, potatoes, turnips, radishes, grasses and many other garden and forage crops." This scenario probably applies to nearly all the other economic species of *Nysius,* but the insect species appears to prefer a basic group of plants. Peairs (1941), in summarizing the literature, probably is referring to *N. niger* when he indicated that it was a pest especially of the cruciferous crops, attacking beets, cabbage, and is one of the more important pests of sugarbeets. He stated that the nymphs feed almost entirely on noncultivated plants and only the adults commonly become injurious in gardens. The general pattern in the literature is often that the adult stage of *Nysius* is more polyphagous and flies to injure a greater variety of plants, although, as Essig indicated, nymphs, too, can migrate and do harm when conditions demand. In British Columbia (Canada), Parshley (1919) recorded this species as appearing in great numbers, swarming over and inside houses, and complaints were even made that they bit children.

3.9.3 Control

As to natural controls, and one can reasonably assume by the distribution that this applies to *N. niger*, Arnaud (1978) list a tachinid parasite *Hyalomyia (= Hylalomya) aldrichii* Townsend from *N. "ericae"* in Manitoba (Canada). This parasitic fly also occurs in *N. raphanus* (Milliken and Wadley 1923).

With the great numbers of false chinch bugs reported in the field, it can be anticipated that vertebrates would be important predators. Knowlton and Wood (1943) examined the stomach contents of 2525 birds that ate *Nysius*, representing 46 species, and found 4597 specimens of *N. "ericae."* The heaviest predation was by the American pipit *Anthus spinoletta* and the Western chipping sparrow *Spizella passerina*.

Because this insect breeds initially in natural or weedy habitats, its control has been difficult. Essig (1929) recommended burning over or cultivating the adjoining grasslands, which may ameliorate invasions by the nymphs, but not the adults that can fly. Even devices that would mechanically remove the bugs from the surrounding fields have been proposed, but, as Peairs (1941) noted, it is probably cheaper to combat the insects directly on the occasions that they invade crops. In the older literature (Milliken 1916, Peairs 1941, Smith 1944), it was recommended to use nicotine and soap sprays or dusting with 25% calcium cyanide. For "modern" chemical controls, see *N. raphanus* and *N. vinitor* for the full arsenal of insecticides used against *Nysius* and which would very likely control *N. niger* if the crop is sprayed at the time that the bugs migrate from the ground to the crops.

3.10 *Nysius raphanus* Howard (Orsillidae)

3.10.1 Distribution, Life History, and Biology

As discussed under *N. niger*, the records in the economic literature of this species are hopelessly ensnared with *N. niger* and also *N. tenellus*. Indeed, Slater (1964c) placed some literature records under both species. However, one can be reasonably sure that the economic records of *Nysius* in the southeastern United States refer to *N. raphanus*, and most of the economic records in the midwestern and southwestern United States probably also refer to *N. raphanus* (M. H. Sweet, personal observation). Because *N. raphanus* was originally described from Kansas (Howard 1872), many of the papers referring to *N. ericae* actually refer to *N. raphanus*. From his description of the species, Blatchley's (1926) discussion of *N. ericae* clearly refers to *N. raphanus*. Also, the older records of *N. strigosus* Uhler and of *N. minutus* Uhler (or as var. *minutus*) as in Smith (1942) also undoubtedly refer to *N. raphanus* based on Barber's (1947) synonymy of these names. Nevertheless, to be certain, this chapter will only refer to the more recent records with the exception of the work of Milliken (1916, 1918) and Milliken and Wadley (1922, 1923) in Kansas (U.S.A.) because the excellent illustration in Milliken (1918) shows plainly, based on the length of the membrane, that they were working with *N. raphanus*.

Like *N. niger*, *N. raphanus* has a definite association with cruciferous plants. Howard (1872) in describing this species called it the radish bug, and the specific name is from the generic name for the radish. However, there is also a strong association with the Compositae as shown by the following. Slater and Baranowski (1990) reported breeding populations in Florida (U.S.A.) on the cudweed *Gnaphalium purpureum* L. Palmer (1987) reported it on *Baccharis neglecta* Britton in Texas (U.S.A.) and Boldt (1989) listed it also from *B. halimifolia* L. and *B. sarothroides* (Gray). This association with the Compositae helps to explain the tremendous numbers of *N. raphanus* as reported by Latham (1975) near the summit of Capulin National Monument in Mexico in April and May. In west Texas (U.S.A.) in May on the cool higher elevations, as on Mt. Emory, remarkable aggregations of *N. raphanus* have been frequently observed. Later on in the summer rainy season, in lowland desert areas below Mt. Emory, numerous *N. raphanus* on many species of roadside and desert Compositae were often found (M. H. Sweet, personal observations). It is also noteworthy that *N. tenellus* Barber in the western United States and Mexico also shows a close affinity to the Compositae, and in Barber's (1947) type-series list the large majority of the host collecting records are from composites. Similarly, in central and eastern Texas *N. raphanus* has frequently been collected from seed heads of composites (M. H. Sweet, personal observation). Watts et al. (1989) observed that *N. raphanus* builds up to very large densities on the western ranges in New Mexico and Utah (U.S.A.), but there was no evidence of the insect injuring the grasses. Hewitt et al. (1972) measured 7000 specimens of *N. raphanus*/m^2 in the buffalo grass–blue grama range in western Texas. With its abundance in the southwestern United States, it seems probable that *N. raphanus* is the *Nysius* sp. considered a potential pest on cultivated guayule (*Parthenium argentatum* Gray), a composite being studied as a potential source of latex for rubber (Stone and Fries 1986). Another explanation for the use of the Compositae was suggested by Böcher (1972) in his study of *N. groenlandicus* (Zett.) and also by Carrillo (1967), who observed an heavy infestation of *N. tenellus* Barber on the introduced composite ox-tongue *Picris echioides* L. They both observed that the eggs were commonly fastened to the pappus bristles of the fruit and realized that wind could blow the seeds with the eggs attached, providing a special means of phoretic dispersal. This may help explain the tremendous spread potential of *Nysius* and its ready colonization of distant islands, which had given rise to the remarkable radiation on the Hawaiian Islands (U.S.A.), as noted earlier.

However, as is generally the case with *Nysius*, *N. raphanus* is polyphagous and can damage many plants. Riley (1873) described this species as *N. destructor*, and gave it the common name of the false chinch bug. He recorded it in Missouri (U.S.A.) as injurious to potatoes, turnips, beets, radishes, cabbages, strawberries, and grapevines. It has even damaged such an unlikely host as a salt cedar tree, *Tamarix pentandra* Pall (= *ramossissima* Ledeb [Bailey and Bailey 1976]) (Watts et al. 1989). *Nysius raphanus* is undoubtedly the *Nysius* first recorded in the United States as a pest on cotton in Texas (Sanderson 1906), in California as *N. minutus* (Smith 1942), and in Arizona (Wene et al. 1965, Terry 1991). It attacked tobacco in Florida and Georgia (Tappan 1967, 1970) and lettuce, cabbage, and turnips in the lower Rio Grande Valley (Wene 1953, 1958). In Colorado *N. raphanus* destroyed stands of quinoa (*Chenopodium quinoa* Willd.) being grown for their seeds (Cranshaw et al. 1990). On grain sorghum in Oklahoma (Wood and Starks 1972) and Texas (Hall and Teetes 1981, 1982; Archer et al. 1990), the nymphs and adults first concentrated on the leaves, but, later on, they attacked the seed heads, and the populations reached several thousand per plant. They caused a drastic reduction in weight, number, and germinability of the sorghum seed. However, there is apparently no similar attack on the seeds of range and pasture grasses (Watts et al. 1989). In California, it was a major pest on walnuts (Haley and Baker 1982) and severely damaged vineyards there (Barnes 1970). Leigh (1961) and Barnes (1970) both recognized a very important factor that led to the ready genesis of *N. raphanus* as a pest. This was the invasion into the agricultural fields of the London rocket *Sisymbrium irio* L., a cruciferous plant that was introduced into California from Europe (Robbins et al. 1951), and on which *N. raphanus* feeds vigorously in early spring and produces a generation of nymphs. The bugs also fed other weedy mustards, as the shepherd's purse *Capsella bursa-pastoris* (L,) Medic., and peppergrass *Lepidium nitidium* Nutt. After these weeds die back or are "reduced" by cultivators, the bugs move in great numbers on to the agricultural plants, here cotton or grapevines.

Nysius raphanus clearly overwinters as an adult in California and Arizona (U.S.A.) as shown by the above description of the invasions of this pest. In Kansas, Byers (1973) observed in early April a mating aggregation of *N. raphanus* on a clay tile with no nymphs present, which indicated that the adults had survived the winter. Similarly, in Texas adult *N. raphanus* were often found from late fall to early spring overwintering on the ground especially under *Aster ericoides* L., which is an autumn and winter weed common in lawns (M. H. Sweet, personal observations). Leigh (1961) and Barnes (1970) observed in their work that the bugs completed a full generation in the spring on the mustards before moving to the crop plant. One can reasonably surmise that the species is multivoltine, the number of generations depending on the climate.

Milliken (1918) described the egg and illustrated the instars. The illustrated egg is of the macroparoid type (Putschkova 1956), elongate, and slightly curved, tapering to both ends and smooth, 0.4 by 1.5 mm. He described the eggs as being laid in soil, in composite flowers, in leaf clusters as of thyme-leaved spurge (*Chamaesyce serpylliforia* Pers.) and carpetweed (*Mollugo verticillata* L.), and between glumes of grasses. The first instar is 0.7 mm long. Watts et al. (1989) illustrated the adult and several instars with photographs. Milliken (1918) measured at ambient temperature of 80°F (27°C) egg development as 4 days and the stadia as first, 5 days; second, 3.8 days; third, 3.6 days; fourth, 4.5 days; and fifth, 3.8 days. The length of the life cycle was 28.4 days from egg to adult, and he calculated that there are five to seven generations a year, which is perhaps an exaggeration. His data on oviposition seems too low as he recorded only one egg a day or less. Yet he described a female laying as many as eight eggs at a time. Milliken also gave the adult life as only 9 to 19 days (average, 12), which also seems too low. He stated that the insect probably overwinters as an egg or an early nymph because the fall eggs hatched in November. It is more likely, given the above evidence of the adults overwintering, that the species is multivoltine with some individuals still laying eggs late in the year.

3.10.2 Damage

The remarks on general damage by *N. niger* probably apply with equal force to *N. raphanus*. Leigh (1961) illustrated the damage done by the bugs attacking the young seedlings of cotton, causing the plants to wilt and become malformed and covered with the feces of the bug. What makes the damage especially severe is the gregarious feeding behavior of the insects when as many as 50 or more bugs attack a single seedling. The insects shelter on the ground in the night and climb up into the plant to feed when the temperatures are high. Barnes (1970) observed that the feeding by the adults on grapevines has a less severe effect on the plant than the feeding by the nymphs. He noted that within hours after the nymphs begin to feed, the foliage of grapes becomes wilted, and within 2 to 3 days the vineyard appears as if killed by frost. He proposed that the saliva of the nymphs contains a toxin to which the grape is sensitive. Because vineyard holdings are large, efforts to control the pest are likely be too late and the damage is frequently irremediable.

3.10.3 Control

Milliken and Wadley (1923) reared from *N. "ericae."* (probably *N. raphanus)* in Kansas (U.S.A.) a tachinid fly *Phasia (= Phorantha) occidentis* Walker. Arnaud (1978) noted that the correct taxonomic placement of this tachinid is unknown, although Aldrich (1915), who identified the fly, stated that it is the most abundant of all Tachinidae he had collected. He observed that he had very commonly collected it in the flowers of composites — which is where *Nysius* is to be found — feeding on the seeds. *Phasia occidentis* has also been reared from *Blissus leucopterus* by Luginbill (1922) who noted a specimen was also reared from the introduced plant bug *Miris (= Leptopterna) dolabratus* L., which is unusual as otherwise only the tachinid genus *Alophorella* is known to parasitize Miridae in North America (Arnaud 1978). Although Aldrich (1915) had concluded from his fieldwork that the *P. occidentalis* was probably single brooded, Milliken and Wadley (1923) concluded from their work that the tachinid was multivoltine with the generation time being about 26 days, similar to that of the host *Nysius*. Because a parasite emerged from a newly molted isolated adult, they reasoned that parasitism began in the nymphal stage. This fast life cycle is supported by the larvae emerging and pupating from May to November. Next, the adult tachinids lived only 3 to 7 days even when sugar water was provided, and the pupa developed into an adult fly in 5 to 7 days in warm weather and 8 to 11 days in the cooler weather of May and September. Oddly, the parasitism affected almost exclusively the female bugs. The parasitized females laid no eggs and died soon after the maggot emerged. Because pupae were emerging as late as November, they believed that fly probably hibernated as a pupa. Milliken and Wadley (1923) observed that the incidence of parasitism was at best low (7.7%) and so concluded that the tachinid was only minor check on the populations of the false chinch bug. In southern California (U.S.A.) Clancy and Pierce (1966) reared from *N. raphanus* the tachinid *Hyalomyia aldrichii* Townsend, which is in the same subfamily and tribe as *P. occidentalis* (Arnaud 1978). *Hyalomyia aldrichii* also parasitized *Geocoris punctipes* Say, *G. bullatus* Say, and *G. uliginosus* Say.

Wadley later (1940) published on the earlier discovery in 1914 in Kansas of an egg parasite, *Telenomus ovivorus* (Ashmead), attacking the eggs of *N. "ericae"* (probably *N. raphanus*). He noted that in the spring the bug placed its eggs in the soil and the parasite was much less abundant then. In early June *Nysius* moved to Compositae seed heads, especially of *Gaillardia pulchella,* the firewheel, and laid eggs abundantly in the seed heads, at which time the parasite became much more abundant. Later *Nysius* moved to the grass *Eragrostis* on which the eggs, hidden in the tight glumes, suffered little parasitism, and the parasite numbers dropped. In the laboratory the parasites took 11 to 13 days to complete development. Wadley (1940) noted in his work that unfertilized female wasps produced only males.

Nysius raphanus and *N. tenellus* are attacked by a sphecid wasp *Solierella peckhami* (Ashmead), which hunts the nymphs to provision its nests. Some 10 to 15 late instars are stung and paralyzed to provision each cell. The wasp lays an egg on one of the nymphs. After the egg hatches, the wasp larva feeds on the nymphs and completes its life cycle. However, the chrysidid wasp *Pseudolopya taylori* searches for first or second instars of *Nysius* and lays a single egg in the tiny nymph. The larva of *Pseudolopya* hatches in the nymph but does not grow beyond the first stage, and causes the nymph evidently no harm and the nymph grows to the last instar. However, if the nymph is parasitized by *S. peckhami,* the *Pseudolopya* larva grows rapidly, consumes the nymph, emerges, and finds the *Solierella* egg, which is promptly destroyed, and the *Pseudolopya* larva proceeds to consume the *Nysius* nymphs. If the *Nysius* nymph is not paralyzed, the chrysidid larva perishes (Carrillo and Caltagirone 1970). Such a complex predator–prey–parasite situation gives testimony to a long history of large numbers of *Nysius* in nature to permit such a complex relationship to evolve.

The work of Knowlton and Wood (1943) on *N. "ericae"* described under *N. niger,* could well be this species. Milliken and Wadley (1922) studied the relationship of *G. pallens* var. *decoratus* Uhl. to *Nysius "ericae,"* probably *N. raphanus,* and showed that *Geocoris* was the primary predator on *Nysius;* to quote them: "The largest numbers of this predator have been found among swarms of [*Nysius*] ... and as the host is driven by the death of its food plants to change its breeding places at intervals through the summer, it is followed by its bloodthirsty enemy, large numbers of which soon congregate at each new location." Horner (1972) found the spider *Metaphidippus galathea* Walckenaer to feed on *N. "ericae"* (probably *N. raphanus*), and recommended that it be used for biological control.

As noted under *N. niger* the older literature on control is badly confused regarding what species was being discussed, but the paper by Milliken (1916), which is undoubtedly on *N. raphanus,* serves well to introduce the older methods. First, he recommends burning the natural wild food plants, and that a strong-blast gasoline torch is effective. The best insecticide for the bugs on cabbages or sugar beets is a spray of fish-oil soap or strong laundry soap at 1 lb to 5 gal of water. On turnips and radishes, it is better to use 1 part nicotine sulfate to 1000 parts water. He also recommended using a sticky trap and beating the insects off the crop plant into the trap.

Barnes (1970) recommended that the most cost-effective method of control was to destroy the crucifers by cultivation in the vineyards one month before the grapes begin to leaf out, and to extend the cultivation area about 200 ft beyond the stands of London rocket and the vineyard. Failing this, the infested area should be treated with an appropriate insecticide. Wene (1958) and Leigh (1961) found malathion to be the most toxic to *N. raphanus* but dieldrin, dipterex, and endrin were all effective, whereas sevin, DDT, and toxophene were ineffective. Daniels (1969) similarly found malathion applied at a rate of 0.75 lb/acre to wheat and parathion at 0.5 lb/acre sugar beets gave good control to *Nysius* sp., which in the panhandle of Texas (U.S.A.) was very probably *N. raphanus.*

3.11 *Nysius vinitor* Bergroth and *Nysius clevelandensis* Evans (Orsillidae)

3.11.1 *Distribution, Life History, and Biology*

These two Australian species are considered together here because they are of similar economic importance and they often occur together, so the literature is much confused; *N. vinitor*, the rutherglen bug, is more important than *N. clevelandensis*, the gray cluster bug. Both are economically very important because of their feeding on sunflowers, but these insects can invade and harm many other economically important plants as well (Smith 1927, Hassan 1977). Furthermore, several of the important studies have compared the two species. Woodward (1964) reviewed the distribution of the species, and noted that whereas both species occur throughout continental Australia, *N. clevelandensis* occurs more in warm, humid areas and extends north into New Guinea and does not extend into Tasmania, whereas *N. vinitor* occurs in drier areas and does not extend into New Guinea, but does extend into Tasmania, Australia, where it coexists with *N. turneri* Evans. Because of their broad sympatry, the records before Evans (1929, 1936) are uncertain regarding which species was being studied, and, furthermore, agriculturalists often do not distinguish them as they are similar in appearance as is the case with *Nysius* species in general. Most records do apply to *N. vinitor*, but as Evans (1929) noted, some important papers such as Smith (1927) in Queensland (Australia), actually refer to his new species *N. clevelandensis*. Evans (1936) similarly noted that the work by Nicholls (1932) in Tasmania actually pertains to another new species, *N. turneri*, an endemic of Tasmania. Smith (1927) studied carefully a species he called the rutherglen bug as understood in Queensland. While Smith was uncertain of its specific identity, many later workers cited his work as being on *N. vinitor*. The three Australian species are distinguished by Evans (1929, 1936) as follows: if the bucculae are short, reaching posteriorad only to antenniferous tubercles and the costal margin is glabrous and convex, it is *N. vinitor;* the other species have long bucculae reaching to the posterior margin of the head and the costal margins are sinuate and hairy. In *N. clevelandensis* the bucculae are high anteriorly and taper to the back of the head, whereas in *N. turneri* the bucculae are lightly convex and relatively high through most of their length.

These species are native pests that inhabit weedy vegetation often near agricultural areas. Smith (1927) recorded *N. clevelandensis* from introduced species of weedy Compositae, cudweed (*Gnaphalium purpureum* L.), sowthistle (*Sonchus oleraceus* L.), and flossflower (*Agaretum conyzoides* L.), the latter sometimes cultivated as an ornamental (Bailey 1949). Elshafie (1976) found *N. vinitor* abundant in all stages under the low prostate herbs pigweed *Portulaca oleracea* L., and corn sperry *Spergula arvensis* L. He observed that despite the large populations of the insects, the host plants appeared healthy. He attributed this to the insects feeding mainly on the numerous seeds of the plants; the plants also provided water and cover for the insects beneath their prostate stems. Attia and Elshafie (1974) used soil containing *P. oleracea* seeds in their successful rearing method for *N. vinitor*. McDonald and Smith (1988) also found the weedy low plants cape weed *Artotheca calendula* L., wireweed *Polygonum aviculare* L., and stone crop *Crassula* sp. (probably *Sedum acre* L., a widely introduced European weed) to be favorable host plants. Noteworthy here is that all these weedy hosts are introduced plants, as are the agricultural crops afflicted by these insects. It would seem that the advent of European agriculture, including the weeds, had benefited these insects very greatly.

Nysius vinitor and *N. clevelandensis* often become serious pests when they leave their weedy hosts en masse and invade agricultural crops (Smith 1927, Evans, 1936, Sloan 1942, Hassan 1977, Broadley and Ironside 1980, Emmett et al. 1992), partly because of their breeding on sunflower *Helianthus annuus* L., safflower, flaxseed (linseed), and rapeseed (canola), and partly because in times of dry weather, when plants are moisture-stressed, the bugs, as adults, attack a great variety of other plants: potatoes, tomatoes, cruciferous crops (cabbage, cauliflower, broccoli, Brussels sprout, turnip, and radish), beans, carrots, capsicums, beetroots, tobacco, grain sorghum, lucerne, strawberries, cotton, grapevines, young citrus, pawpaw (*Carica papaya* L.), stone fruits (cherry, plum, peach, apricot) and pome fruits (apple, pear), and even the native *Eucalyptus* (French 1918, Smith 1927, Nichols 1932, Evans 1936, Hassan 1977, Murray 1980, Malipatil 1996). Sometimes the plants damaged are ones that the insect cannot persist on, such as cotton (*Gossypium* sp.) (Chinajariyawong et al. 1989). The plants most damaged is the sunflower *Helianthus annuus* L. and the safflower *Carthamus tinctorius* L., on both of which the insects can breed successfully by feeding on the crop seeds (Broadley 1978, 1980). Smith (1927) discovered that the first and second instars of *N. clevelandensis* required the flower head of sowthistle to develop, and could not grow on the leaves, whereas the later instars could develop on the leaves.

Because the *Nysius* bugs build up into large populations, especially *N. vinitor*, and because of their strong dispersal behavior, they often become nuisances in homes; they are small enough to enter through screens, attack people, and cause painful bites and skin irritation reactions (Gellatley and Forrester 1985, Southcott 1988).

Because of its economic importance, much work has been done on *Nysius* in Australia, especially on its migratory behavior as this is the source of most of the damage to crops, when the insects leave weedy hosts when the hosts senesce or dehydrate (Kehat and Wyndham 1972a,b, 1973a, 1974a). Kehat and Wyndham showed that most migration measured near the fields was of immature females, but McDonald and Farrow (1988), who sampled the air column at both 2 m and at 100 to 300 m, found males and older fertilized females well represented. Furthermore, they gave good evidence for long-distance migrations of 200 to 300 km from subtropical locations to more southern temperate locations; the migratory behavior occurred mostly at night after dusk and was maximal with disturbed weather associated with prefrontal airflows. A sharp difference was found between *N. vinitor* and *N. clevelandensis*. *Nysius vinitor* was one of the most abundant insects in the upper air and comprised over 99% of the *Nysius* specimens. This substantiates the position of Kehat and Wyndham that *N. vinitor* is a strongly r-selected species that is well adapted to exploit temporary habitats such as weedy fields or agricultural crops. McDonald and Farrow noted that in the outback natural areas *N. vinitor* especially exploited patches of Compositae, which requires both local and long-distance dispersal, the latter more related to the seasonal changes in rainfall and temperature in Australia, leading to the movement southward in the spring from the drier interior areas to the moister, more southern areas. As *N. clevelandensis* is better adapted to more humid subtropical areas, the bug disperses largely by more local movements and is therefore much less prone to be an migrant pest invading crop fields. Woodward (1964) noted that although *N. clevelandensis* is much more abundant in the field in Queensland, *N. vinitor* is much more abundant in light traps in the same area, which probably shows this difference in migratory behavior. The last three instars are relatively agile and fast moving and can readily disperse a considerable distance on ground to new available host plants, as sunflower plants, and readily climb up into the plant to feed on sunflower seeds in the head of the plant and down again to reach another sunflower plant (Ramesh and Laughlin 1994a; Ramesh 1994a,b, 1995).

Smith (1927) made an early life-history study of what he called the rutherglen bug, as noted, actually *N. clevelandensis*. He illustrated and described the eggs and nymphal stages. The nymphs have the striped heads and mottled bodies characteristic of *Nysius*. The egg is creamy white, elongate oval (the macroparoid egg of Putschkova 1956), 0.025 mm wide by 1 mm long. The instar lengths are first, 0.75; second, 1.25; third, 1.5; fourth, 1.75; fifth, 2.25; adult male, 3.5; female 4.5 mm. The stages of *N. vinitor* are similar (Slater 1976b).

Smith used sowthistle rather than cudweed because it was easier to count the eggs. The insects laid their eggs not in the ground but on the plants. Cudweed was highly preferred for oviposition because the plant had a heavy coating of fine down. Sowthistle did not, so the eggs were laid conveniently in the seed heads. He found some eggs laid in the feathery awns of the introduced blady grass or lalung (*Imperata arundinacea* Cyrill.), a grass used in the Orient for weaving mats and hats (Uphof 1968). It is unknown from his paper whether this is a host record or merely a convenient oviposition site. He raised the bugs through two generations and described the mating and oviposition behavior, hatching, and molting. He found the total number of eggs per female to vary from 134 to 435 over an oviposition period of 13 to 25 days. The duration of egg incubation was 6 days, range 4 to 8 days; for stadia, first, 5 days, range 1 to 8; second, 3 to 4 days, range 2 to 9 d; third, 3 to 4 days, range 2 to 9 d; fourth, 3 to 4 days, range 1 to 6 d; fifth, 5 days, range 3 to 7 d; the average nymphal development was 20 to 21 days.

Attia (1982) compared the biology of *N. vinitor* with *N. clevelandensis* in a laboratory study using the rearing techniques of Smith (1927), Attia and Elshafie (1973), and Elshafie (1977) for *N. vinitor*. The insects were supplied weekly with crushed sunflower seeds, water, and a sprig of pigweed and, for *N. clevelandensis*, some flowers of the composite fleabane *Erigeron* sp., probably the introduced weed *E. annuus* (L.) Pers., and crushed sunflower seeds were supplied, because *N. clevelandensis* was shown in field studies to be more restricted to composites (Smith 1927). The two species were found to be similar in mating behavior and no effective hybridization occurred between them. The precopulatory period for both was about 3 days, and both species were not able to feed at low temperature (4.5°C) and starved to death in 26 to 27 days, but survived if allowed to feed at warmer temperatures for 4 hours during the day. The adult longevity at room temperature was statistically similar at different relative humidities and the maximal longevity varied from 24 to 44 days. However, the species were different in that *N. clevelandensis* had an increased egg productivity at higher relative humidities (180/female to 140/female at 90% relative humidity), while *N. vinitor* was higher at lower humidities (268/female to 149/female at 45% relative humidity). In development times, the species were similar, development varied from winter temperatures of 14 to 16°C to summer temperatures of 27 to 28°C as follows: mean total nymphal development from 38 to 42 days in the winter, 17 to 20 days in the summer; mean preoviposition period from 10 to 12 days to 3 to 4 days; oviposition periods from 35 to 40 to 14 to 15 days; egg incubation period from 12 to 13 to 5 to 6 days. Vattanakul and Rose (1982) determined the stadia of *N. clevelandensis* at 25°C to be egg, 7.0; first, 6.5; second, 4.5; third, 4.5; fourth, 4.5; fifth, 6.0 days; and the preoviposition period, 5.5 days. For each species this would permit seven to eight generations. This compares well with Kehat and Wyndam's (1972) field estimate of three to four generations in Adelaide, South Australia, recalling that this is a Mediterranean climate with a cool winter and a dry summer so that development cannot be continuous. An important factor in the field was the adults and nymphs sunning themselves, raising body temperature above ambient temperature and thus achieving higher growth rates (McDonald and Smith 1988). Because the number of immatures reduces considerably in the autumn, and the adults overwinter in southern Australia, Allsopp (1984) investigated the role of photoperiod on the life cycle of *N. vinitor*, comparing growth at 10:14 L:D long days to 14:10 L:D but could find no significant differences. He concluded that other factors, such as temperature and egg parasitism, resulted in the decline in nymphs in the autumn. However, he did not investigate the more natural effect of changing photoperiods in combination with changing temperature.

3.11.2 Damage

A list of the plants injured by *N. vinitor* and *N. clevelandensis* was given earlier. The damage is of two types: on seeds, which are highly nutritious and often support breeding, and on plant tissues, which are attacked especially in times of water stress, but the plant can be severely damaged by the saliva, which is evidently toxic to some plants. Both types of injury happen to sunflowers, safflowers, flaxseed, and rapeseed (Hassan 1977, Broadley 1978, Broadley and Rossiter 1980). On sunflowers the adult insects arrive in numbers at the time of budding and congregate on the upper parts of the plants to feed. Especially in times of drought stress the bugs can afflict severe damage causing the flower heads to collapse, wilt, and blacken (Hassan 1977, Broadley and Rossiter 1980). Forrester and Saini (1982) showed that in a comparison with irrigated and nonirrigated sunflower fields the damage and reduction in sunflower grain yield was much greater in moisture-stressed crops. At the completion of flowering a dramatic increase in the number of adults occurs and eggs are laid abundantly on the seed heads. The adult insects and nymphs feed directly on the seeds in the heads, which affects adversely the seed yield, oil content, quality, and germination (Broadley and Rossiter 1980, 1982; Broadley et al. 1986). Strawberries are often severely damaged, the flower buds and flowers destroyed, and the fruits deformed by the feeding on the fruits (Murray 1980). On potatoes, *N. clevelandensis* attacks the terminal shoots first, causing wilting and death, and eventually kills the whole plant. If the plant survives, the potato tubers are too small to be marketable (Smith 1927). On beets, *N. clevelandensis* attacks the leaf bases at the head of the root, causing the beet to be reduced in size proportional to the infestation (Smith 1927). Tomatoes are often attacked heavily, causing severe damage to the fruit and killing the terminals (Hassan 1977). Although *N. clevelandensis* does not breed on them, the stone and pome fruit are often heavily damaged and disfigured, greatly reducing their market value (Hassan 1977). On citrus fruit, *N. clevelandensis* damage on the fruit results from repeated puncturing of the young fruits, blemishing them and causing them to develop unevenly and produce a crop of small fruits, deficient in juice (Smith 1927). Similarly, grapevines are attacked in dry weather causing the grapes to shrivel and dry, and the shoots to wilt and die (Emmett et al. 1992). Unlike *N. niger* and *N. raphanus* in North America and *N. huttoni* in New Zealand, there is no close specificity with the Cruciferae, and the damage reported to cruciferous crops is only occasional (Hamilton 1980).

The only probable record of disease transmission is the discovery that DNA sequence analysis of phytoplasmas responsible for papaya (*Carica papaya* L.) dieback, yellow crinkle, and mosaic diseases were similar to sequences derived from the *N. vinitor* (White et al. 1997).

3.11.3 Control

Attia (1973) reared from both *N. vinitor* and *N. clevelandensis* the tachinid *Alophora (Mormonomyia) lepidofera* (Malloch). It is interesting, given that Crosskey (1973) cautioned Attia that these tachinids are poorly known and the identifications are provisional, that Malloch (1929) had described this species in the genus *Hyalomyia* (= *Hyalomya*), in which genus is placed *H. aldrichii* Townsend that parasitizes *N. raphanus* and other lygaeids in North America (Arnaud 1978) and in Europe *H. pusilla* (Meigen) has been reared from many small lygaeoids, including *N. cymoides* (Spinola), *N. jacobaea* (Schilling), and *N. lineatus* (Costa) (Dupuis 1963). Crosskey (1973) noted that the subgenus *Mormonomyia* is otherwise an Ethiopian taxon. It might be added here that the generic and tribal classifications of the tachinids of the subfamily Phasiinae appear quite different (Dupuis 1963, Crosskey 1973, Arnaud 1978), and these authors did not apparently cite one another. Chadwick and Nikitin (1985) reared a specimen of another tachinid of the Phasiinae, *Leucostoma simplex* (Fallen), from *N. vinitor*. This tachinid species has a remarkable host and geographic distribution, being reared in North America and Europe from nabids, rhopalids, and even the grasshopper *Melanoplus* (Dupuis 1963, Arnaud 1978). Attia was able to confirm the host–parasite relationship of *Alophora* by mating the flies and effecting parasitism of uninfected adult *N. vinitor* and *N. clevelandensis* and rearing an F_1 generation. The tachinid larvae emerged from *Nysius* in 10 days. The field parasitism rate was similar for the two species, the females of *N. vinitor* 11 to 62%, and for *N. clevelandensis*, 13 to 61%; the males no more than 5%, often none. The bias to females is similar to that reported for *N. raphanus* tachinid parasitism (Milliken and Wadley 1923). Loudon and Attia (1981) also reared *A. lepidofera* from *Oxycarenus luctuosus* (Montrouzier & Signoret) and described the larvae of the tachinid. They hypothesized, based on its form, that the larva enters into the host through the metathoracic spiracle. By dissecting the tachinid they found it to be oviparous and the eggs to develop externally, not within the female, which is a primitive characteristic of the subfamily Phasiinae (Dupuis 1963). The ovaries of the host are consumed so the parasitized female *Nysius* lays no eggs. The bug dies shortly after the maggot emerges.

Gellatley and Forrester (1985) note that the natural enemies of *N. vinitor* include the fungus *Beauveria bassiana* and a scelionid egg parasite that attacks the eggs in the autumn. Forrester (1980) reported the parasite to be *Telenomus* sp. As to predators, Sadana and Goel (1996) demonstrated experimentally that the spider *Oxyopes pandae* Tikader, which occurs in great abundance in sunflower fields, feeds voraciously on *N. vinitor.*

Attia (1974) found *N. vinitor* and *N. clevelandensis* react similarly to insects, so they can be discussed together. He evaluated 21 insecticides in the laboratory, and found Promecarb, mevinphos, parathion, nalid, methidathion, and diazinon to be the most effective in contact toxicity. He recommended that these be used instead of the widely used malathion (maldison), which was recommended by Leigh (1961) for *N. raphanus*. McDonald et al. (1986) found in field trials that acephate, cypermetrin, deltamethrin, endosulfan, fenvalerate, methamidophos, and permethrin were effective. One difficulty of control of insects on sunflowers is that honeybees are needed to pollinate the flowers. Insecticides can either harm the bees directly or the honey will become contaminated with the insecticide, which is often the case with carbaryl. It is thus imperative that the spraying on sunflowers or any other flowering crop be carried out well before or well after pollination (Broadley and Rossiter 1982), or that an insecticide be used at dusk (such as endosulfan, which breaks down in 5 hours, before daybreak the next day) that shows a low toxicity to bees as well (Broadley 1980). Endosulfan, methidathion, and maldison are the most recommended insecticides for use against *Nysius* with different concentrations recommended for the many crops damaged by the bugs (Broadley and Rossiter 1980, Hassan 1980, Gellatley and Forrester 1985). Broadley (1978) recommended against routine control, noting that sunflower plants can tolerate considerable damage and still produce economical crops, and he recommended control only when necessary. Allsop (1988) tried to establish a sequential sampling program to develop action thresholds for decisions about managing pest populations, i.e., when to spray for control, but noted that the present estimates were only guesses and more knowledge is needed for appropriate mathematical treatment. Furness (1976) suggested caution in using insecticides to control sporadic attacks of *N. vinitor* because secondary infestations of other pests such as long-tailed mealy bugs, *Pseudococcus longispinus* (Targioni-Tozzetti), often follow such chemical control.

3.12 *Nysius huttoni* Buchanan-White (Orsillidae)

3.12.1 Distribution, Life History, and Biology

This insect, known in the economic literature as the wheat bug, is the most widespread species of *Nysius* in New Zealand. The variability is such that Eyles (1960) was led to believe that *Brachynysius convexus* Usinger was one of the forms of *N. huttoni*. Eyles (1960) in his careful study of this remarkable variation distinguished three forms in both sexes based on wing development — macropterous, subbrachypterous, and brachypterous — and also distinguished three independent groups based on size. He demonstrated that these three size groups could be distinguished by size differences in the eggs and the instars, which also showed this trimodality. There was also considerable variation in coloration and pilosity. However, Eyles demonstrated by mating tests that the wing forms were of the same species, and he had partial evidence that the size groups also were of the same species. When working in New Zealand, this author, too, was impressed by the variation and noted that it seemed to have a population basis so that several species seemed present, not one (M. H. Sweet, personal observations). Later, Eyles and Ashlock (1969) discovered that *B. convexus* of Usinger was indeed a valid species, restricted, like their new species *N. liliputanus*, to a few localities in the Southern Alps of the South Island, New Zealand, in contrast to *N. huttoni*, which is found throughout the North and South Islands including Stewart Island and Chatham Islands. *Nysius huttoni* also shows a wide ecological distribution from seaside locations to high altitudes (1830 m). It is surprising then that Eyles found no geographic or elevational parameters in the diversity of *N. huttoni*. If this variation is indeed sympatric, it may be of considerable theoretical interest to determine how it came to be. This question gains greater interest because *N. huttoni* is an economically important pest species. In contrast to *Nysius,* Eyles (1990) found *Rhypodes,* another orsillid genus that is endemic to New Zealand, to consist of 28 species, which makes New Zealand second only to Hawaii (U.S.A.)

in the number of orsillid species present in a fauna. Only one of these, *R. anceps* (White), is of minor economic importance and has been implicated in causing runny dough (Morrison 1939).

As is usual with *Nysius*, *N. huttoni* is polyphagous and feeds on weedy plants from which it invades crop species. It has been recorded as feeding on and injuring linen flax, strawberry, lucerne, red clover, white clover, subterranean clover, Nassella tussock grass, rape, beets, wheat, and especially crucifers, both wild and cultivated (Myers 1921, 1926; Gurr 1957; Eyles 1965b; Schroeder and Clifford 1996). However, it has been definitely reared only from the crucifers shepherd's purse *Capsella bursa-pastoris* (L.) Medik. (Gurr 1957) and twin cress *Coronopus didymus* (L.) Sm. (Eyles 1963a). Gurr also found it to feed heavily on the crucifer Curnow's curse *Calandrinia caulescens* H.B.K. Nevertheless, Eyles (1965) has no doubt that *N. huttoni* feeds on the following largely weedy plants because there is a close association of all stages of the bug with them: sheep sorrel *Rumex acetosella* L., wireweed *Polygonum aviculare* L., suckling clover *Trifolium dubium* Sibth., sand spurrey *Spergularia rubra* (L.) J. & C. Presl., catch fly *Silene gallica* L., white clover *T. repens* L., Onehunga weed *Soliva sessilis* Ruiz & Pav., red clover *T. pratense* L., lucerne *Medicago sativa* L., and the scarlet pimpernel *Anagallis arvensis* L. Farrell and Stufkens (1993) found that in Canterbury, South Island, New Zealand, chickweed *Stellaria media* (L.) Vill. was the most abundant spring host plant and the fathen *Chenopodium album* L., as well as wheat was an important summer food source when the annual weeds dry up. It should be noted that nearly all of these host plants are introduced weedy plants. What were the natural hosts before the advent of the Europeans? The only native plant recorded as having been attacked is the common composite shrub tauhinu *Cassinia leptophyla* (Forster f.) R. Br., which is common on beaches and used in horticulture (Bailey and Bailey 1976, Poole and Adams 1990). It seems surprising that little feeding on other grasses was observed, given the pest status of *N. huttoni* on wheat. A possible record is Kelsey (1957) who studied the fauna of the tussock grassland community *Festuca novae-zelandiae* Cockayne, *Poa caespitosa* Forst, and *P. colensoi* Hook. f., and observed the presence of *N. huttoni* feeding on developing seed heads. Eyles listed some other plants as being shelter hosts, low plants under which the insects can overwinter, but apparently did not breed on. Syrett and Smith (1998) found *N. huttoni* (and *Rhypodes* spp.) numerous on four introduced weedy hawkweeds (*Hieracium* spp.), but concluded that these and other native species were insufficient to control these weeds and recommended the introduction of specialist species from Europe to control these weeds in the paddocks.

Eyles (1960) in his study of the variability of *N. huttoni* described in detail and figured the egg and the nymphal instars. The egg measures an average of 0.28 by 0.77 mm and is from water white to very pale yellowish orange. The nymphs are similar to other *Nysius* in being mottled brown and black with the distinctive longitudinal striping on the head. The mean lengths of the instars are first, 0.84 mm; second, 1.23 mm; third, 1.5 mm; fourth, 2.05 mm; fifth, 2.52 mm. In the adults, recalling the size diversity, the males vary from 3.86 to 2.38 mm, and the females from 4.34 to 2.47 mm.

Eyles (1965) described in detail the mating behavior, and noted that copulation varied from a few minutes to as long as 6.5 hours and occurred throughout the day. Gurr (1957) found the eggs to be laid in crevices in the soil rather than on the host plants as had been recorded for other species of *Nysius*. Gurr measured egg productivity to be 25 to 174 eggs per female, and the preoviposition period as 3 to 11 days. Farrell and Stufkens (1993) measured egg productivity under long days as 98.6 eggs per female with a maximum of 248 eggs. Eyles (1963b) made a remarkably detailed study of fecundity in greenhouse conditions in glass tubes over 2 years using as food sprigs of twin cress supplied each day during oviposition periods in which the eggs were laid in cotton wool. He found the total egg productivity to be maximal in spring and summer females that had continual access to males and minimal in fall females and females with limited or no access to males. He found that the eggs were laid in bursts of several days and with intervening breaks of several days. Unmated females laid eggs at the rate of a female fertilized only once. Interestingly, while a single copulation is sufficient to fertilize all eggs in the life of a female, the female lays a much higher number when there is continual access to males.

Eyles (1963) in his study of the development of *N. huttoni*, examined the ability of the different stages to survive environmental extremes. The eggs could survive 32 days at 3°C, and develop at 47°C if the relative humidity was high. He measured the mean incubation period of the eggs to be 9.5 days, in summer 8 days and in winter 26 days. He described in detail the process of hatching, which took 9 to 11 min and was similar to that described by Usinger (1942) for *N. coenosulus* Stål (= *N. kinbergi* Usinger). Most hatchings occurred between 9 A.M. and noon. At 2 hours after hatching the nymphs began to feed on stems and seed cases of twin cress. As he expressed the data in ranges over the year at ambient temperature, the stadia were first, 2 to 17; second, 3 to 15; third, 3 to 13; fourth, 3 to 14; fifth, 4 to 15 days; total nymphal development, 23 to 52 days. Gurr (1957) estimated that there were two generations at Nelson in the South Island, New Zealand. Eyles (1963) could rear four generations a year in the greenhouse in the North Island, New Zealand. Gurr (1952) stated that the adults begin to overwinter about the end of April in Nelson, and emerge again about the beginning of August. Eyles (1965) calculated that with the overlap in generations in the North Island there were three generations per year. He concluded the adults were not in a true diapause, but in a quiescent state as they would become reproductive when brought into warmer conditions and, similarly, fifth instars would molt in the fall under warmer conditions and become reproductive.

However, Farrell and Stufkens (1993) found that at Canterbury in the South Island there were only two generations. They noted that the second generation adults were sexually immature, and demonstrated that the insects under short-day regimes (12:12 L:D) as third and fourth instars went into diapause as adults, whereas under long day (16:8 L:D hours) the insects remained reproductive. The diapausing adults migrate to hibernacula in litter and other sheltered places and emerged in spring to feed on annual weeds. In years with dry early springs, the wheat bugs would migrate directly to wheat and damage the developing kernels.

3.12.2 Damage

Cruciferous crops (*Brassica oleracea* L.) and rape (*B. napus* L.), swede (*B. napobrassica* Mill), and turnips (*B. rapa* L.) are subject to attacks by *N. huttoni*. When fed on by *N. huttoni*, the brassicas developed cankerous growths on the stems, causing the leaf to wither and die and the stems to collapse from the damage (Gurr 1957, Eyles 1965a, Ferguson 1994). Fodder beet fields have also been attacked by *N. huttoni* (Hartley 1980). Myers (1921) found *N. huttoni* to be a severe pest on lucerne and red clover. Attacks by *N. huttoni* on white clover reduced seed fill and resultant seed quality (Schroeder and Clifford 1996).

The greatest attention has been to the damage caused by the attacks on wheat (*Triticum* spp.) in New Zealand, which was first discovered by Morrison (1938). In the Middle East pentatomids (sunn bugs) *Eurygaster integriceps* Put. and *Aelia rostrata* Boh. (Berliner 1931, Lorenz and Meredith 1988, Swallow and Every 1991) cause similar damage in which the enzymatic action of the saliva degrades the gluten giving rise to slimly runny dough that is unsuitable for baking. It was thought that a number of heteropterans in New Zealand, especially mirids, caused the damage (Morrison 1938). A careful study of the available heteropterous bugs feeding on wheat in New Zealand showed that all the bug species damaged the wheat kernel, but N. *huttoni* was the major cause of the sticky dough (Cressey et al. 1987; Every et al. 1989, 1992, 1998). The glutenin hydrolyzing serine proteinase enzyme of *N. huttoni* is remarkably potent and is effective across a broad pH and temperature range and shows little inhibition by most metal salts and EDTA (Cressey et al. 1987; Every 1992, 1993). Bug damage to flour is determined by counting the number of damaged kernels with characteristic puncture marks. A 3% content is sufficient to produce poor-quality bread.

3.12.3 Control

Because the wheat bug feeds on weedy plants of waste lots and roadsides, often in the vicinity of crop plants, it is notoriously difficult to control. Moreover, it is only in dry years that it migrates to crops and severely attacks wheat (Eyles 1965a, Farrell and Stufkens 1993). There is hope that wheat cultivars can be developed that provide resistance against the bug as there is a considerable difference among the cultivars in the damage inflicted upon the kernels (Every et al. 1998). However, Gurr (1957) considered the greatest danger and need for control to be on cruciferous crops, not wheat.

Possibly because of the zoogeographic isolation of New Zealand, there are no records of parasitism by insects of the eggs or adults of *N. huttoni*, and no records of pathogenic fungi. Because of the potential sensitivities of the native noneconomic insects, the importation of such biological control agents is not recommended (M. H. Sweet, personal observation). Similarly, there are no direct records of predators, expect that the introduced starling *Sturnus vulgaris* L. fed on *N. huttoni* throughout the year (Lobb and Wood 1971). Schroeder and Clifford (1996) list chrysopid larvae and adults and larvae of *Coccinella undecimpunctata* as associated potential predators.

Trought (1975) found that dicrotophos, trichloronate, chlopyrifos, and a mixture of omethoate and azinphos-ethyl gave equivalent control for *N. huttoni* on brassicas.

3.13 *Nysius plebeius* Distant (Orsillidae)

3.13.1 *Distribution, Life History, and Biology*

In most of the literature, this species name is emended to *plebejus*, but this is not Distant's original spelling of the name. This species was described from Japan (Slater 1964a) and later recorded in the economic literature from Korea and Taiwanese China. One record from Midway Island, if correct, probably represents an introduction (Suehiro 1959). It is evidently the only *Nysius* species of economic importance in the eastern Orient (China, Korea, and Japan). In Japan it is recorded as a pest of rice and strawberries (Hasegawa 1980). Kisimoto (1984) studied this species in the context of other rice stink bugs in Japan. It is the most injurious pest of florists' chrysanthemum (*Chrysanthemum morifolium* Ramat.) (Kim et al. 1994a,b) in Korea but, oddly, not in Japan where chrysanthemum is much cultivated. In Taiwan it also damaged grain sorghum (Hong 1986).

The species was dominant in natural habitats in the Korea Republic on Mt. Yeogi (Gjoh and Lee 1988) and Mt. Kyeryongsan (Choe and Jang 1992). Kim et al. (1994a) in their sampling around chrysanthemum fields found *N. plebeius* on 20 herbaceous host plants, mostly weeds. It was most abundant on chrysanthemum itself, fleabane (*Erigeron linifolius* L.), strawberry *(Fragaria chisoensis* Duchart), and Italian ryegrass (*Lolium multiflorum* Lam.). Hong (1986) in China found near the sorghum fields the insect feeding on 22 species of weeds, mostly cosmopolitan, belonging to 8 plant families; 12 species were composites and 4 Amaranthaceae and, interestingly, given that sorghum is a grass, none was listed on weedy grasses.

In Japan Kisimoto (1984) recorded two generations a year on rice with the adult overwintering. Kim et al. (1994b) in field and insectary studies in Korea found *N. plebeius* to have three generations a year with adult peaks in mid-April, mid-June, and late August. The insect appears in light traps in late August through mid-September, probably representing a migratory period, as the fall adults were nonreproductive. The insect overwinters as an adult (90%) or fifth instar (10%) and mating occurs actively in the following spring. The eggs are laid in masses on the surface of leave or stems of the host plants. The number of eggs laid seems low, per generation — 41.5, 37.2, and 35.4 eggs per female. The oviposition periods were 25.6, 23.4, and 21 days for each generation. The longevities for females were 32.6, 27.5, and 25.1 days, and less for males, 15.2, 12.3, and 12.1 days. The overwintered generation took 6 to 7 weeks to produce new adults in the spring, while the two summer generations took 5 to 6 weeks to complete development. The egg incubation periods were 14.2, 9.3, and 7.2 days, respectively, for each generation. Total nymphal development periods were 45.2, 32.1, and 31.2 days per generation. The sex ratio was approximately 1:1. The females were about 4.8 mm, and the males were about 4.5 mm in length. The nymphs are not described, but are illustrated in popular Japanese literature and show the characteristic color pattern of *Nysius* with striped heads and mottled brown and white abdomens.

3.13.2 Damage

Kim et al. (1994a,b) refer to heavy infestations and severe damage and illustrated the chrysanthemum leaves and flowers wilted and discolored by the attacks of the bugs. They found all 37 cultivars damaged, 9 especially so. Hong (1986) noted that the adult bugs invaded the sorghum heads from the weeds, fed on the kernels from the milky stage to the dough stage, and decreased in population as the grain matured. Infestations of 40 or more bugs resulted in yield reductions. The damaged grains rarely developed fully and were considerably smaller, softer, and lighter than the undamaged ones. Kisimoto (1984) found *N. plebeius* to be one of a complex of 15 bugs feeding on rice and causing pecky rice.

3.13.3 Control

Kim et al. (1994a) recommended using insecticides applied three to four times at ten intervals. They found the insecticides elsan (phenthoate), sumithion (fenitrothion), bassa (fenobucarb), and mopox (mercarbam), all effective, in that order, in controlling the insect. Kisimoto (1984) found among several insecticides that carbamate compounds, such as BPMC (*o*-sec-butylphenyl methylcarbamate) and MTMC (*m*-tolyl methylcarbamate), are effective against lygaeids on rice.

3.14 *Oxycarenus hyalinipennis* Costa and *Oxycarenus* spp. (Oxycarenidae)

3.14.1 Distribution, Life History, and Biology

In the family Oxycarenidae there are 24 small genera and one large genus, *Oxycarenus*, which contains about 55 species, of which at least 6 species are listed as pests, especially on malvaceous plants such as cotton and hibiscus. Most, 36, of the species of *Oxycarenus* are in Africa, so the genus is probably of Ethiopian origin (Horváth 1909, 1926; Samy 1969) although most of the genera are Palearctic. The systematics of the genus is complex, and the species are often difficult to tell apart, compounded by taxonomic errors, so the literature records are often badly confused (Samy 1969, Slater 1972c). The genus as a whole needs to be revised as the studies of Samy and of Slater were of the African and South African faunas, respectively, and the Palearctic and Oriental faunas, including three anomalous species that are listed from New Caledonia, which is also the type locality for the common Australian species *O. luctuosus* (Slater 1964a). However, most of the economic species appear have a similar biology, feeding on seeds, especially of cotton (*Gossypium* spp.) and other Malvaceae. Therefore, to avoid undue repetition, the most widespread species, *O. hyalinipennis*, the dusky cottonseed bug, will be treated, and the other economic species only mentioned, as they may differ from *O. hyalinipennis*. In addition, *O. hyalinipennis* has been extensively used in physiological studies so there is a relatively large literature on this species. *Oxycarenus hyalinipennis* ranges throughout Africa into southern Europe and through subtropical and tropical Asia to the Philippine Islands. It has been introduced into South America probably with cottonseed and is recorded as a pest on cotton in Brazil, Paraguay, Argentina, and Bolivia (see Slater 1964a). As it has recently been collected in the West Indies (North Caicos and Providenciales) (Slater and Baranowski 1994), *O. hyalinipennis* should therefore be considered an important potential pest on cotton in the North America, the only continent from which *Oxycarenus* is at present absent. Moreover, as far as is known, the native Oxycarenidae of North America feed on Compositae, not Malvales, so *O. hyalinipennis* will have a large ecological niche to fill and it should have interesting effects on the natural guild of insects feeding on the seeds of Malvaceae, many of which plants are exceedingly abundant weeds in agricultural fields and roadsides.

Oxycarenus hyalinipennis is also recorded as a pest on the malvaceous food plant okra *Hibiscus* (= *Abelmoschus*) *esculentus* L., especially in India and Pakistan, where *O. laetus* Kirby evidently displaces *O. hyalinipennis* on cotton. *Oxycarenus hyalinipennis* is also recorded from *Abutilon* spp. (Malvaceae), many species of which are ornamentals and used for fiber, and an especially fine fiber comes from the country mallow *A. indicum* L. (= *mauritianum* [Jacq.] Medic) (Samy 1969, Abu-Mensah and Kumar 1977). Kirkpatrick (1923), who studied this species on cotton in Egypt, observed that it fed almost entirely on members of the Malvaceae. He recorded it on cultivated kenaf, *Hibiscus cannabinus* L., which is an important source of fiber and oil, on the ornamentals the cotton rose *H. mutabilis* L., the flower of an hour *H. trionum* L., the hollyhock *Althea* (= *Alcea*) *rosea* (L.) Cav., the marsh mallow *Althea officinalis* L., *Pavonia hastata* Cav., and the weeds globemallow (*Sphaeralcea* sp.), *Sida* spp., especially *S. mollissima,* and *Malva* spp. The only nonherbaceous host was the exotic Australian tree kurrijong *Sterculia* (= *Brachychiton*) *diversifolia* Don (= *populneum* R. Br.) (Sterculiaceae). Kirkpatrick noted that this diversity of alternate host plants makes the control of *O. hyalinipennis* difficult. Moreover, different hosts will provide food at different times of the year, maintaining the *O. hyalinipennis* populations. Ewete (1984) found *O. hyalinipennis* in Nigeria attacking roselle, *Hibiscus sabdariffa* L., a shrub whose fruit is used for food and the stems for cordage fiber, and on Aramina, *Urena lobata* L., the source of a valuable fiber (Uphof 1968). Mayné and Ghesquière (1934) found *O. hyalinipennis* on *Abutilon cabrae*, *H. lancibracteatus,* and *H. surattensis,* as well as on cotton. They noted that bolls damaged by caterpillars of Lepidoptera allowed *O. hyalinipennis* access to the seeds earlier. They also found *O. hyalinipennis* on *Sterculia tragacantha* and *S. bequaerti.* Dimetry (1971) found *O. hyalinipennis* to feed, but more sparingly, on the weeds *Malva rotundifolia* L. and on the globemallows *Sphaeralcea umbellata* Don and *S. miniata* Don. In India it attacked a medical plant, kasturi blendi, *H. abelmoschus* L. (= *Abelmoschus moschatus* Medik.) (Rajashekhargouda et al. 1984). Ram and Chopra (1984) recorded *O. hyalinipennis* from over 57 plant species, but most of these were for shelter and a partial food supply. They reported *O. hyalinipennis* to destroy the fruits of the raisin bush *Grewia subinaequalis* (Tiliaceae). Most unusual is their reporting all stages on pigeon pea, *Cajanus cajan* (L.) Millsp., and on bajra or pearl millet, *Pennisetum glaucum* (L.) R. Br. If correct, these are the only records of an *Oxycarenus* feeding and breeding on a legume or a grass. It is predicted that these will prove to be sporadic attacks by populations migrating from malvaceous host plants. This may also be true of the record by Goyal (1974) of *O. hyalinipennis* feeding on sunflowers (M. H. Sweet, personal observations).

The other *Oxycarenus* species of economic importance attacking cotton in Africa are listed as follows, along with the other economic host plants attacked: *O. gossipinus* Distant, also on okra, kenaf, roselle and aramina; *O. fieberi* Stål, also on hemp, *Sida acuta* Berm., a plant used for its fiber, on cola, *Cola nitida* Scott and Endl. (Sterculiaceae), which is much used as a stimulant, on *Triumfetta rhomboidea* Jacq. (Tiliaceae), a woody plant used for its fiber, on roselle and Queensland hemp, *S. cordifolia* L. and *S. rhombifolia* L., which species are used as sources of fiber; *O. albidipennis* Stål, also on flowers of *Dombeya* sp. (Sterculiaceae), species of which are ornamental and used as sources of cordage fiber; *O. rufiventris* (Germar), also on bolls of okra, roselle, kenaf, *T. macrophylla* (Tiliaceae), a tree used for fiber, *H. fuscus* Garcke (= *H. gossypinus*), *H. meeusei* Exell, and on the ornamental plant rose-of-China *H. rosa-sinensis* L.; *O. dudgeoni* Distant, also on kapok, *Ceiba pentandra* Gaertn. (Bombacaceae); *O. multiformis* Samy, also on *Pavonia, Sida, Abutilon,* and *Hibiscus* spp; and *O. bokalae* Samy on cotton and *Hibiscus tiliaceus* L. (Theobald 1905; Fletcher 1914; Mansfield-Aders 1919; Mellor 1932; Mayné and Ghesquière 1934; Hargreaves 1937, 1948; Evans 1952; Odhiambo 1957; Slater 1964a,c; Gentry 1965; Samy 1969; Leston 1970; Slater 1972c; Abu-Mensah and Kumar 1977; Ewete and Osisanya 1985).

In the Oriental region, in addition to *O. hyalinipennis, O. laetus* Kirby, *O. gossypii* Horváth, *O. bicolor* Fieber, and *O. lugubris* Motschulsky all attack cotton. *Oxycarenus laetus* Kirby, a species that tends to replace *O. hyalinipennis* in India on cotton, is a common insect with many host plants (Distant 1904; Maxwell-Lefroy 1907, 1909), and as the following records attest it reproduces only on Malvales. Misra (1921) evidently named the insect the dusky cotton bug. Although ecologically similar to *O. hyalinipennis*, it does not show the biogeographic spread potential of *O. hyalinipennis*, as it is restricted to Iraq and India east to Burma, Malaya, and Indochina (Slater 1964a). In addition to cotton, *O. laetus* feeds on kenaf (or gogu or Deccan hemp), okra or bhindi, hollyhock, cotton-rose, the muskmallow *H. abelmoschus* L., the rose-of-China *H. rosa-sinensis* L., the India mallow or kadghi *Abutilon indicum* (L.) Sweet, the portia tree *Thespesia populnea* (L.) Soland, the pink dombeya or Persddrolpeer *Dombeya mastersii Hooker* (= *D. burgessiae* Gerr. Ex Harv) (Sterculiaceae), an ornamental shrub from South Africa (Palgrave 1977), parusha, *Grewia asiatica* L. (Tiliaceae), a tree with edible fruits (Uphof 1968), *Melochia corchorifolia* L. (Sterculiaceae), a plant whose leaves are consumed in India (Uphof 1968), and, important for pest insects, the weeds *Malva rotundifolia* L., *Malva* sp. *Malvastrum* sp., and *Malvascus* sp. (Mukhopadhyay et al. 1981; Thangavelu 1978a,b,c, 1983; Dhiman and Krishna 1989; Qureshi et al. 1992). [*Note:* Bailey and Bailey (1976) place *Dombeya* in the family Byttneriaceae, but most authorities like Palgrave (1977) place it in Sterculiaceae and consider the Byttneriaceae a synonym or subdivision of the Sterculiaceae (Cronquist 1968). In any case *Dombeya* is a member of the Malvales.]

With such a rich fauna of *Oxycarenus,* especially in Africa, and its potential economic importance, this should be a fine opportunity to study the ecology of these species carefully, and also to test the reality of some of these closely related species, many of which are sympatric. The value of such an approach is shown by the excellent work of Leston (1970) and Abu-Mensah and Kumar (1977) in Nigeria who worked on the ecological relationship of *O. hyalinipennis, O. fieberi,* and *O. dudgeoni.* Each species showed a preference for a host species in the natural habitat: *O. hyalinipennis* for country mallow, *O. fieberi* for hemp, and *O. dudgeoni* for kapok, but they were able to use other plants for food. There is considerable overlap in host plants, and many alternate weedy hosts are listed. The key position of cultivated cotton was shown by seed preference studies. Although cotton was not a native host, each *Oxycarenus* species showed a strong preference for cottonseeds. The authors recommended that the wild malvaceous host plants should be destroyed before cotton plantations were established in Nigeria. Ewete and Osisanya (1985) showed that in Nigeria a suite of host plants provides nearly continuous breeding for *O. gossipinus:* okra in February to May; cotton, kenaf, and roselle in June to December; and okra in September to December. *Oxycarenus dudgeoni* is unusual in that it is mainly arboreal and that it could exploit host plants of three families: Malvaceae, Sterculiaceae, and also the Bombaceae. Dimetry (1971) in Egypt showed that, of 12 species of Malvaceae, the seeds of cotton and kenaf were the most preferred and supported the best growth and the highest fecundities.

Samy (1969) believed that the Malvales (Malvaceae, Sterculiaceae, and Tiliaceae) were the only true hosts of the African *Oxycarenus.* The Bombaceae are also part of the Malvales (Cronquist 1968). Samy considered all the records other than Malvales to be sitting or resting records. Kirkpatrick discovered in Egypt that *O. hyalinipennis* cannot reproduce if there are no malvaceous seeds to feed on. Samy extended this idea to the Malvales in general, stating that *Oxycarenus* must have seeds of Malvales to reproduce.

However, this specialization to the Malvales is not true of all *Oxycarenus.* Species of the *Oxycarenus* subgenus *Pseudoxycarenus* Samy, as far as is known, feed on the seeds of *Protea* (Proteaceae) (Slater 1972). With the development of cut-flower industry in South Africa, *O. maculatus* is becoming considered a pest species to be controlled (Annecke and Moran 1982). In the *Oxycarenus* subgenus *Euoxycarenus, O. (E.) pallens* (Herrich-Schaeffer) feeds on the composites *Centaurea splendens* and *Centaurea* sp. (Putschkov 1956, Kerzhner and Yaczewski 1964) and *Cirsium segetum* Bunge (Zheng and Zou 1981). *Oxycarenus modestus* Fallén is most unusual in that it is arboreal, feeding on and injuring the cones of alder (*Alnus* sp.), and has a boreal distribution in the Palearctic realm (Gulde 1936, Putschkov 1956, Slater 1964a, Zheng and Zou 1981). Wachmann (1989) has an excellent photograph of the insect, which is called in German the *Spitznase* after its pointed head. However, this species may not be related to the true *Oxycarenus*, and Samy's generalization may hold for *O. hyalinipennis* and its allies. This shows the need for a world revision of *Oxycarenus* and related genera in the Oxycarenidae.

In Australia, where they are called coon bugs, the more tropical *O. arctatus* (Walker) and the more temperate *O. luctuosus* (Montrouzier) are also pests on cotton, and during dry years they will also invade and injure figs, cherries, raspberries, apricots, peaches, and plums (Froggatt 1901, Hill 1926, Hargreaves 1948, Pearson 1958, Gross 1959, Hassan 1977, Hely et al. 1982, Wilson et al. 1983).

Oxycarenus luctuosus is well known for forming swarming aggregations on malvaceous weedy hosts (Tillyard 1926). This remarkable behavior has been observed frequently in fieldwork in South Australia (M. H. Sweet, personal observations). This "swarming" is a common phenomenon in the Oxycarenidae. An uncommon species like the eastern North American *Crophius disconotus* Say that feeds on *Solidago canadensis* will suddenly occur in large numbers in a small area. The insects will aggregate in large masses on a few plants such as *Sida rhombifolia* L. and *Malva rotundifolia* L. but will be absent from the same host plant only a few feet away. One wonders whether this behavior is to facilitate feeding, promote reproduction, or is an antipredator strategy (or all of these). Most unusual is the carefully documented note by Woodward (1984) that *O. arctatus* can feed and develop in large numbers from egg to adult by feeding on cattle dung in which there were, moreover, no seeds. Woodward found the guts of the insects filled with a brownish fluid, but as the new adult females had no eggs developing, it is possible that original adults had first fed on the seeds of a malvaceous plant and then subsequently invaded cattle dung. Woodward suggested that this feeding was a strategy for surviving the extended drought conditions that often occur in Australia.

Observing that the seeds of the Australian *Gossypium sturtianum* Willis, a wild cotton, do not readily germinate, as is characteristic of many arid-land plants, Karban and Lowenberg (1992) put seeds of *G. sturtianum* with *O. luctuosus*. They discovered that seeds fed on by *O. luctuosus* had a much higher germination rate than the control seeds not fed on. This was probably due to scarification allowing water to enter the seed. They noted that excessive feeding will kill the seed, but moderate feeding, as an ecological trade-off, benefits *G. sturtianum,* allowing the plant to respond more quickly to the erratic rainfall of central Australia. There are many species of wild *Gossypium*, mostly in subtropical semiarid areas, but relatively little attention has been focused on these species, which is surprising as the behavior of *Oxycarenus* to native *Gossypium* may provide some important clues useful for controlling the pest species of *Oxycarenus*.

Kirkpatrick (1923) and Pearson (1958) made some interesting observations about the life cycle of *O. hyalinipennis* in Egypt that indicate how difficult it is to control this pest. Although *O. hyalinipennis* is found throughout the Tropics, where it can breed continuously if malvaceous seeds are available, in Temperate Zone areas it shows a facultative diapause. If the minimum and maximum temperatures drop below 15°C and 30°C, the bugs cease to feed on cotton and enter into sexual diapause, in which the ovaries remain undeveloped and copulation does not occur. The bugs will seek water but the excreta is colorless in contrast to the normal yellow-brown appearance. The bugs now fly away to resting places and tend to be more gregarious, clustering together in sheltered places, as on trees, in dried flower heads, under leaves, in pods of legumes, habits that lead to erroneous host determinations, as Samy (1969) observed. On warm days the bugs will often fly about seeking other shelters. During this overwintering period, the weight of the bugs decreases steadily and there is heavy mortality. In April when the new cotton plants are well grown, the bugs will colonize the plants, resting in the bracteoles of the buds and bolls. They may not reproduce until the bolls open in July or August. At this time breeding begins and the population grows rapidly, peaking in October. However, this will be modified if there is an earlier fruiting malvaceous plant and reproduction will commence much earlier. This shows the great importance of alternate hosts. Some, such as okra and kenaf and the weed *Sida mollissima,* produce seeds in spring or summer, which may allow another generation that may then invade cotton. The ornamental plant cotton-rose *Hibiscus mutabilis* L., on the other hand, produces seed in the winter so in warm years the cotton-rose may permit another generation. On the other hand, if a population breeds on *S. mollissima*, an early spring seed producer, it may enter diapause in the early summer if seeds are lacking, and this population cannot be induced to breed on cotton in August, for it is in diapause in preparation for the winter to come. This reproductive system gives *O. hyalinipennis* great life-cycle flexibility, making more understandable its worldwide distribution in many different climates.

There is some controversy about the feeding on cotton by *Oxycarenus*. Kirkpatrick (1923) noted that the insects were often sheltering on the cotton plant, not feeding on it, and so did no damage to the plant or to the lint directly. Instead, the insects preferentially fed on ripe seeds. Saxena and Bhatnagar (1958) stated that the insects fed preferentially on the cotton leaf, drawing food mostly from mesophyll and phloem, and only attacked the seed coat, not the kernel of the cottonseed. They used evidence from the distribution of sugar and enzymes to substantiate their conclusion, which was in direct contradiction to Kirkpatrick's fieldwork. However, Bhatnagar (1962) carefully reinvestigated the feeding behavior of *O. hyalinipennis* and demonstrated that the insects did specifically feed on the kernel of the seeds, and required seeds to reproduce and grow. She verified Kirkpatrick's observation that the bugs do not feed on the host plant unless the insects are highly desiccated, and that this suggested that this desiccated state had led to the erroneous results of Saxena and Bhatnagar (1958). This specificity for seeds was especially true of the nymphs. When feeding, the insects with their slender stylets were able to avoid the pigment gland cells with toxic gossypol (Khan and Agarwal 1981), both in green plant tissues and in the seeds. Bhatnagar (1962) found that the presence of sugar in solution did not affect the feeding or drinking activities, verifying that the feeding on liquids was for water, not sugar. Bhatnagar further showed that only by starving the bugs would they be induced to feed on non-Malvales seeds. Gandhi has continued this research on the feeding physiology, growth, behavior, uptake, and utilization of sugars and amino acids in *O. hyalinipennis,* and many of these studies were comparative studies with an auchenorrhynchan, *Oxyrhachis tarandus* Fabr. (Gandhi 1975, 1977, 1978, 1979a,b,c, 1980a,b, 1981a,b, 1988a,b, 1989; Joshi and Gandhi 1986).

However, the question can be fairly raised, given that *O. hyalinipennis* can feed on the plant for water with no ill effect, and the plant does indeed carry valuable sugars and amino acids in the sap, why then does not the bug feed on the plant when seeds are not yet formed or after the seeds have been shed, but instead goes into facultative diapause? True, the seeds are very rich food, but as the above research shows the leaves are nutritious, too, and indeed many other heteropterans do feed and reproduce on cotton (Hargreaves 1948), which should logically tide the bugs over to the next seed crop. Instead, as Kirkpatrick (1923) demonstrated, the bugs avoid feeding on the plant as much as possible, utilizing rainwater, dew, and extrafloral nectaries, and draw on the plant as a last resource, even when concealing themselves in the foliage of the cotton plant. The saliva of the bugs is evidently toxic, as shown by the damage to nonhosts. A working hypothesis is proposed: the bugs avoid feeding on the plant to avoid harming the hosts that will produce much more valuable food seeds at a later time. This is the strategy of a good parasite: do not kill the goose that lays the golden eggs.

Oxycarenus species are small, oval, somewhat depressed insects about 1.5 mm wide by 4 mm long, the males a little smaller than the females. The head is narrow, somewhat pointed, with small subglobose eyes. The general coloration of *O. hyalinipennis* is a black body and white forewing. Other species often show patterns of red on the pronotum and abdomen that suggest aposematic coloration. The forefemora of the species in this genus in both sexes are armed beneath with two to three to five spines, so it seems probable that the insects contest the seeds (Sweet 1964).

Several authors (Bedford 1923, Kirkpatrick 1923, Barbosa 1950, Dimetry 1971) have described the nymphs and eggs of *O. hyalinipennis*. The egg is 0.29 mm wide by 0.97 mm long, slender, subcylindrical, with 25 longitudinal ribs or corrugations. The anterior end is broadly rounded, and bears six chorionic processes; the posterior is distinctly pointed. Putschkova (1956) called this the oxycarenoid-type egg. During development, the eggs change from straw yellow to orange as seen through the transparent chorion. The nymphs are orange-red on hatching and later have a dark red abdomen (Kirkpatrick 1923) that has a greenish tint (Ewete 1984). In the aggregate when in a feeding swarm, the nymphs are very conspicuous. The investiture is of the peculiar glandular peglike hairs characteristic of the Oxycarenidae (Slater and Sweet 1970, Dolling 1991). There are two abdominal scent glands located between terga 4 to 5, and 5 to 6. The orifices are close together with no scent gland plates around them (Dimetry 1971).

Kirkpatrick (1923) measured the average instar lengths of *O. hyalinipennis* as first, 1.2; second, 1.53; third, 2.25; fourth, 3.28; fifth, 4.27 mm. The male average length is 3.82; the female, 4.41 mm (Samy 1969). The egg incubation period varied from 43 days at 14.5°C to 4 days at 35°C. At temperatures above 37.8°C the eggs did not hatch. The whole life cycle can be completed in 3 weeks but cooler, more shaded conditions will prolong development time. Dimetry (1971) measured development times at different relative humidities and temperatures in Egypt. For egg incubation and nymphal development relative humidity had no effect, but temperature was important. At 25°C, 30°C, and 35°C incubation rates were 6.5, 5.4, and 4.1 days; the total nymphal development periods were 25.9, 17.2, and 14 days, respectively. If diapause does not intervene, copulation can occur with 24 hours of becoming an adult. Kirkpatrick (1923) states that oviposition is completed within 1 to 4 days and only 25 to 40 eggs are laid. These numbers seem much too low, and probably indicate inadequate rearing conditions. In contrast to Kirkpatrick, Dimetry (1971) measured the fecundities of *O. hyalinipennis* at 25°C to be 90 eggs; at 30°C, 110 eggs; and at 35°C, 57 eggs. He found the greatest hatching percentage (97.5%) at 30°C. He found adult longevities to be 35, 29, and 20 days for males and 42, 36, and 23 days for females at the same temperatures. Hammad et al. (1972) in Egypt found copulation to last 1 to 4 hours in an end-to-end position, with the larger female leading. After 1 to 3 hours, the female may copulate again, but the male is often a different individual. The preoviposition period was 2 to 3 days. They noted that the eggs were laid either singly or in groups of two to four, and eggs were stuck to each other by a sticky material. Like Kirkpatrick, Hammad et al. (1972) found an average fecundity of only 20 eggs, which is much too low. They measured the adult life span to be 5 days, again much too low. On hatching, a spinelike egg burster on the embryonic head capsule ruptures the egg chorion transversely. Hatching takes 15 to 30 min and nymphs usually cluster together for about 30 min before dispersing to look for food. Kirkpatrick (1923) measured three to four generations a year depending on food and the advent of diapause. Hammad et al. (1972) measured seven generations a year in the laboratory, the whole life cycle having an average of 55 days.

Awan and Qureshi (1996) found the reproductive behavior of *O. laetus* to be similar to *O. hyalinipennis*. The egg incubation time was 4.5 days. The stadia were first, 6.2; second, 7.0, third, 7.2; fourth, 7.5; fifth, 8.2 days. Thangavelu (1978c) found the development time to be 31 to 39 days from egg to adult. Thangavelu (1978c) determined that of seven plant species, *Abutilon indicum* was the most preferred species and okra the least, with cotton intermediate in host preference. Raman (1987) compared the development rates on difference foods and found that *O. laetus* grew the most rapidly, had the highest food uptake, the highest fecundity, and the greatest longevity on cottonseed.

Malipatil (1979) reared *O. luctuosus* on *Sida rhombifolia* seeds and obtained fecundities of 86 to 259 eggs, at 1 to 9 eggs a day. The eggs were laid in rows with their sides touching one another and the substratum. The stadial lengths for *O. luctuosus* were egg, 5 to 9; first, 3 to 6; second, 3 to 8; third, 3 to 4; fourth, 5 to 9; fifth 5 to 11 days. The total development time in the summer was 31 to 36 days and in the winter 56 to 64 days. The males lived for 107 to 149 days and one egg-laying female lived for 81 days.

Slater (1972) reviewed the South African Oxycarenidae fauna and described the nymphs of several of the economic species. For *O. rufiventris,* he described all five instars. The fifth instar has a bright orange-red abdomen and reddish-brown head and thorax. The instar lengths are first, 1.07; second, 1.50; third, 1.60; fourth, 2.88; fifth, 3.16 mm. For *O. fieberi,* he described the fifth, fourth, and second instars. The nymphal coloration is similar to *O. rufiventris* but the legs are more uniformly red. The lengths of the instars are second, 1.5; fourth, 2.56; fifth, 3.20 mm. Woodward (1984) mentioned that the nymphs of *O. arctatus* are bright red in color.

Ewete (1984) described the immature stages of *O. gossipinus*. The egg is similar with a mean length of 0.88 mm. The nymphs are yellow on hatching and become progressively darker as they grow, the later instars with yellow to pinkish head and thorax and a distinctly brown, not greenish, abdomen. The mean instar lengths are first, 1.28; second, 1.83; third, 2.39; fourth, 3.21; fifth, 3.92 mm. The mean male length is 4.02 mm, and the female 4.29 mm. Ewete and Osisanya (1985) showed the nymphal development period at 25 to 33°C to vary with diet: on okra, 15.5; kenaf, 15.7; roselle, 15.8; and cotton, 18.1 days. The corresponding life spans were on okra, 59.2 male (41.1 female); kenaf, 57.5 (40.1); roselle, 57.5 (39.3); cotton, 43.5 (32.5) days. Ewete and Osisanya (1988) showed that the insects preferred higher (31.5 to 34°C) to lower temperatures (19 to 24°C).

The scent gland secretions of both the adults and the nymphs may be important in deterring potential parasites and predators. The scent gland physiology of the metathoracic gland has been studied by Olagbemiro and Staddon (1983) and Knight et al. (1984). They determined the scent gland components and found (Z,E)-alpha-farnesene to be the major component, among many others, and unusual in Heteroptera. The nymphal glands are unusual in having cuticular scent canals that bear the defensive secretions.

3.14.2 Damage

Oxycarenus hyalinipennis are sometimes called dusky cotton stainer bugs, as well as cotton seed bugs, because when the insects gather inside the boll to feed on the seeds, they may get entrapped and be crushed accidentally during the ginning process, sometimes staining the cotton (Hassan 1977, Ezueh 1978, Annecke and Moran 1982), but Pearson (1958) considers this actually to be a minor problem. In any case, the staining is much less than that caused by the much larger *Dysdercus* bugs (Pyrrhocoridae), which are more generally known as the cotton stainer bugs.

As discussed earlier, *Oxycarenus* does *not* damage the green cotton plants' foliage and stems and other hosts of the Malvales, but instead is a seed feeder. Because cottonseed is a valuable source of oil, the feeding by the bugs on the seeds is much more important. The large populations that can build up in the boll may reduce the oil content and the weight of the seed by 15%, and reduce the vitality of the seeds. The gregarious feeding of the bugs seriously promotes this damage, producing light seeds (Annecke and Moran 1982).

Kirkpatrick (1923) observed that the damage to cottonseed by *O. hyalinipennis* was greatly facilitated by the pink boll worm *Pectinophora gossypiella* (Saunders), which eats a hole into the boll, thus allowing the bugs easier and earlier access to the seeds. Similarly, Mukhopadhyay et al. (1981) observed that the caterpillars of the noctuid moth *Earias fabis* Stoll bored into the capsules of mushkdana, *Hibiscus abelmoschus* L., which allows *O. laetus* earlier access to the seeds.

In light of this specialization on the Malvales, the most puzzling records are of *Oxycarenus* damaging fruit, as was earlier mentioned for *O. luctuosus; O. exitiosus* Distant (= *O. annulipes* [Germar]) was originally taken in South Africa injuring peaches. Also, *O. amygdali* Distant was named after an association with peach trees and the bug was also collected on apple blossoms (Samy 1969). Samy discounted these records, recalling that in Egypt he had observed, as Kirkpatrick had done, *O. hyalinipennis* in large numbers in the opened cotton boll, but the cotton plant itself was evidently unharmed by the dense population of bugs. Indeed, Swart (1978) called these insects the fruit tree stink bug, *vrugteboomstinkbesie* in Afrikaans. He noted that *O. exitiosus* Distant (= *O. annulipes* Germar) and a number of closely related stink bug species attack apricot and peach fruit sporadically in swarms in the southern and western Cape of South Africa. He described the insect to insert its long, thin mouthparts deep into the flesh of rather developed apricot or peach fruit, and to suck out sap. Under severe attacks, the cells of the fruit flesh just underneath the skin are sucked dry to such an extent that a corky layer develops. Sucking marks are visible externally as pinprick spots. In Israel, Nakache and Klein (1992) stated that *O. hyalinipennis* injures dates, figs, avocados, and persimmons, which appears to be the same phenomenon. Even more remarkable is that Nakache and Klein (1993) found *O. hyalinipennis* attacking sunflowers and other crops. They describe the attacks as causing deformation, shrinking, and change of color of the seeds. Not clear here is whether the bugs could reproduce on sunflowers or whether as with the attacks on fruit this is a reaction to toxic saliva of

Oxycarenus. In any case, the damage was severe, and the numerous rotting heads of sunflowers in the fields showed the extensive damage, much worse than on the native malvaceous hosts. The record by Goyal (1974) of *O. hyalinipennis* attacking sunflowers may fall in this category.

In view of all the research showing specificity for Malvales and feeding only on the seeds, avoiding the green foliage and stems, why is this severe damage occurring? It is no wonder Samy (1969) doubted the veracity of the reports. This pattern, however, seems similar to reports of *Nysius* causing similarly severe damage on fruit trees, while on its natural weedy host plant damage seemed minimal. Perhaps the same phenomenon occurs here. It is proposed that the saliva of *Oxycarenus* may be toxic to fruit and leaves that the bugs fed on, albeit only briefly, and perhaps mainly for water (M. H. Sweet, personal comment). In contrast, the natural malvaceous hosts are tolerant of the saliva and are essentially unharmed. Further studies are badly needed to discover why the bugs attack and injure fruits and sunflower plants.

Because the insects fly to places to overwinter (Kirkpatrick 1923), the insects often enter homes to aggregate. The smell of the secretions is very strong, and with the insect's habit of flying to lights at night, it can become a severe household pest, especially in the winter when the insects aggregate (Nakache and Klein 1992).

3.14.3 Control

Abu-Mensah and Kumar (1977), after studying the ecology of *Oxycarenus* in Nigeria, concluded that the best control measure was to eliminate the herbaceous malvaceous hosts that serve as the reservoirs for *Oxycarenus* populations. They noted that the experimental feeding tests demonstrated that cottonseeds are very attractive to the *Oxycarenus*, more so than their wild host plant seeds. However, so abundant may these weedy or wild hosts be in many situations that their extermination may not be practical. Kirkpatrick (1923) recommended eliminating alternate host plants near and in the cotton fields and reducing as much as possible places for the bugs to rest. Especially important is to store seed cotton away from moisture, as the bugs need moisture to feed on the seed. Porcelli and Palmieri (1996) showed that in Italy the infestations of *O. hyalinipennis* on kenaf was the consequence of populations' immigrating from wild malvaceous hosts, so they noted that control, especially biological control, of this pest will be difficult in the presence of wild reservoir hosts.

Kirkpatrick (1923) found that strong rainstorms were very detrimental to populations of *O. hyalinipennis*, as a great many drown. Abu-Mensah and Kumar (1977) showed that in Nigeria the dry season was favorable to population development of all three *Oxycarenus* species, but the onset of the rainy season resulted in mass destruction of colonies, especially of *O. dudgeoni*. Moreover, they observed that seeds would sprout in wet weather and be unavailable for food. Termite activity also increased in wet weather, resulting in premature destruction of seeds.

Unfortunately, biological control using parasites and predators is evidently not practical for *Oxycarenus*. No egg parasites and very few tachinids, *Alophora lepidofera* (Mall.), have been reared from *O. luctuosus*, much fewer than from *N. vinitor* and *N. clevelandensis* (Loudon and Attia 1981). Perhaps the swarming behavior repels the tachinid fly. Kirkpatrick (1923) in his careful study of *O. hyalinipennis* in Egypt found no parasites. Abu-Mensah and Kumar (1977) in their extensive survey found no egg parasites and found tachinid parasites, Alophora (*Mormonomyia* sp.), only in O. *dudgeoni*. None was reared from *O. hyalinipennis* or *O. fieberi*. In *O. dudgeoni* the parasitism rate was very low, less than 1.0%. The most prominent predators on *Oxycarenus* were assassin bugs (Reduviidae), especially *Rhynocoris bicolor* (F.), *R. loratus* (Stål), *R. carmelita* (Stål), and *R. albopilosus* (Signoret). When starved, the reduviids could kill about 11 bugs in a day. They found a common lizard in the field, *Agama agama* L., not to feed on bugs and suggested that the defensive secretions repelled the lizard. Dimetry (1973) in Egypt noted that very few predators ate the bugs, but that the ant lion *Chrysopa vulgaris* preyed on the fifth instars of *O. hyalinipennis*. Malipatil (1979) found in laboratory studies in Queensland (Australia) that *Geocoris capricornutus* Kirkaldy and *G. lubra* Kirkaldy would feed on the nymphs of *O. luctuosus*, but the latter species was highly cannibalistic. The *Geocoris* would also feed on sunflower seeds as Sweet (1960) showed with other species. It remains to be seen if these *Geocoris* species might be useful in the control of *O. luctuosus*. The potential for host plant resistance was shown by the reductions in *O. laetus* populations as the density of gossypol glands increased (Khan and Agarwal 1981). It is surprising that the Indian workers had not found any parasitoid insects in their extensive work on *O. hyalinipennis* and *O. laetus*.

For chemical control, Hill (1975) recommends using BHC (lindane), to control the cottonseed bug *O. hyalinipennis* on okra and cotton. This is a toxic insecticide and care should be taken using it around bees and vertebrates. Swart (1978) noted that no insecticide is registered in South Africa for the control of the fruit tree stink bug *O. annulipes (= O. exitiosus)*. However, the bug can be readily controlled on apricots and peaches when it occurs by full-cover sprays of endosulfan 47.5% wp or 35% ec at 100 g/100 l of water. Materu et al. (1974) sprayed kenaf fields in Tanzania by airplane with 26% DDT and 6.6% monocrotophos which resulted in a 96% reduction in *O. hyalinipennis* populations. In Brazil, Santos et al. (1977) found that spraying cotton infested with *O. hyalinipennis* with 50% malathion 1.2 l/ha, and 50% dieldrin 0.72 kg/ha gave good yields of seed cotton in the following year. With *O. laetus* Dhingra and Verma (1984) found in laboratory testing of 15 insecticides that fenthion was the most toxic, followed by monocrotophos and parathion. Interestingly enough, but to be expected, Melou and Yana (1964), based on the field records, concluded that *O. hyalinipennis* is developing geographic races based on resistance to insecticides, reflecting the selection pressure of the use of different insecticides.

3.15 *Chauliops choprai* Sweet and Schaefer and *Chauliops* spp. (Malcidae)

3.15.1 *Distribution, Life History, and Biology*

As mentioned earlier, Štys (1967) separated the family Malcidae from the Lygaeidae and included in it the small subfamilies Malcinae Stål and the Chauliopinae Breddin. The move presaged the present division of the family. The Malcinae consist of one genus, a few species of which are of minor economic importance and which are discussed under this heading. The other subfamily, the Chauliopinae, consists of only two genera, the moderately large *Chauliops* of Africa and Asia and the monotypic *Neochauliops*, which is restricted to Africa (Štys 1963). The genus *Chauliops* has a subtropical to tropical distribution from Japan and China through Malaya and India to Nigeria, Central Africa. *Chauliops fallax* Scott, the type species, occurs in Japan and China, and is known in the Japanese literature as the small bean bug; *C. choprai* Sweet and Schaefer and *C. nigrescens* Distant occur in India, the former as the little bean bug, cowpea, or soybean bug, and the latter as the black bean bug; *C. lobatula* Breddin occurs in Ceylon; *C. bisontula* Banks in Malaya, Indonesia, and China; *C. horizontalis* Zheng in China; and *C. rutherfordi* Distant in East Africa where *Neochauliops laciniata* Bergroth also occurs. All the species are recorded as pests except *C. lobatula, C. horizontalis,* and *N. laciniata,* but this is negative knowledge as virtually nothing is known about these species aside from their distributions. Zheng (1981) noted that of the three species in China only *C. fallax* is of known economic importance and that *C. horizontalis* is known from an unidentified legume. In general, these insects are most damaging in the more temperate parts of their range, not the more tropical (Sweet and Schaefer 1985).

The Chauliopinae are unusual in that all the species, as far as is known, feed on a closely related complex of genera in the Phaesoleae, one of the ten or more tribes of the subfamily Papilionoidea, itself one of three subfamilies in the Leguminosae (Sweet and Schaefer 1985). See Sweet and Schaefer (1985, Table 2) for a list of the host plants of each species annotated by their economic significance. The species with the widest host range and the most damaging records is the Indian species *C. choprai,* which is known in the literature before 1985 (and after 1985!) as *C. fallax*. Summarizing from Sweet and Schaefer, the economic host plants include soybean *Glycine max* (L.) Merrill, kidney or French bean *Phaseolus vulgaris* L., golden gram or mung *Phaseolus aureus* Roxb., cowpea or black-eyed pea *Vigna (= Dolichos) unguiculata* (L.) (= *V. sinensis* Endl.), moong or green gram *V. radiata* (L.), urid or black gram *V. (=Phaseolus) mungo* (L.), mothbean *V. (=Phaseolus) acontifolia* (Jacq.), lablab or bonavist *Dolichos lablab* (L.), horse gram *D. biflorus* L. Lal (1974, 1981) found French beans to be the most preferred and black gram the least preferred for *C. choprai. Chauliops nigrescens* also attacks asparagus bean, *Psophocarpus tetragonolobus* (L.). *Chauliops fallax* is also known from the infamous kudzu *Pueraria lobata* (Willd.) (Tayutivutikul, and Yano 1989), which was imported into the United States as an ornamental ground cover, but has become a serious pest vine in the southeastern United States, literally burying entire trees. Sweet and Schaefer (1985) noted that kudzu, at least in Japan, does not seem to be a pest plant and perhaps the plant is kept in check by *Chauliops* as the infestations on kudzu seemed relatively heavy. Despite the economic impact of kudzu in the United States, they recommended against importing the little bean bug for the control of kudzu, because it is a potential pest on bean crops. Incidentally, the roots of kudzu are edible and are the source of Japanese arrowroot (Uphof 1968). In Africa, *C. rutherfordi* is recorded as feeding in Nigeria (Golding 1931, 1937) on the leaves of *Calopogonium mucunoides* Desv. and *Desmodium maurtianum* DC, legumes grown as green manure and ground cover crops throughout the Tropics (Uphof 1968). However, the records from cotton (Hargreaves 1948) are probably only sitting records because of the host specificity exhibited by the Chauliopinae for a restricted group of Leguminosae.

Chauliopines feed on leaves, that is, the mesophyll and sap of the plants, not on the seeds. The blissids, although plant feeders, are more on the stems where they feed on vascular tissues as discussed earlier. *Malcus* bugs similarly feed on leaves (Štys 1967, see *Malcus*). Among the other Pentatomomorpha, leaf mesophyll feeding is also seen in the lygaeoid families Piesmatidae (Drake and Davis 1958, Schaefer 1981, Heiss and Péricart 1983), Colobathristidae (Illingworth 1920, Tan and Johnson 1974), and the Berytidae (Readio 1923, Wheeler and Schaefer 1982, Péricart 1984). Sweet and Schaefer (1985) wondered if mesophyll feeding were a function of small size, but some colobathristids and berytids are relatively large.

Chauliopines are chunky little insects about 1.25 to 1.5 mm wide by 2.0 to 3.5 mm long, with unusual stylate eyes. For diagnoses and a key to the Asian species see Schaefer and Sweet (1985), who also describe the egg and nymphs of *C. fallax*. Rawat and Sahu (1973), Chopra and Rustagi (1982), and Singh et al. (1987) described the nymphs of *C.˙choprai* (as *C. fallax*). The nymphs are unusual in that they bear remarkably thick flat hairs with glandular apices. There are two scent glands on the abdomen between segments 3 to 5 and 5 to 6. The lengths of the nymphs are first, 0.80; second, 0.86; third, 1.25; fourth, 1.56; fifth, 2.06 mm. The egg is short and stout, 0.18 by 0.38 mm, with an unusual caplike pseudoperculum divided into four quarters, each with a chorionic process.

The eggs are laid glued by a thick, dark secretion on the underside of the leaf along the leaf veins and in plant hairs on stems and shoots. The insects live on the leaf surface, usually the undersurface, and when disturbed, the adults drop off quickly, falling to the ground. The nymphs cling tightly to the underside of the leaf and tend to aggregate together (Rawat and Saha 1973, Bhardwaj and Bhalla 1975, Sweet and Schaefer 1985). Singh et al. (1987) noted that in the field the insects mated for 90 to 140 min, but they could be induced to mate in the laboratory. Lal (1981) records the following bionomic data. There were 30 to 60 eggs per female, which appears to be a little low. However, for *C. nigrescens,* Sharma et al. (1993) found the females to average 45 eggs in an average of 15.5 days, which is comparable. The egg incubation period was 7 to 23 days depending on the temperature. Lal (1981) found the nymphs to require from 19 to 24 days to complete development. Each instar feeds, and the stadia were first, average 3.4 (range, 3 to 4); second, 3.8 (3 to 5); third, 4.5 (4 to 6); fourth, 4.5 (3 to 4); fifth, 5.9 (5 to 9) days. The preoviposition period was 2 to 5 days. Lal found the adults of the summer are surprisingly short lived; the females averaged 13 (7 to 17), the males 9 (5 to 12) days. Similarly, Bhardwaj and Bhalla (1975) found that for *C. nigrescens* the average life span for females was 10.2 days (range 7 to 13) and males, 7.3 (5 to 10) days. On the other hand, Rawat and Sahu (1973) and Singh et al. (1987) found the adults of *C. choprai* to live from 20 to 28 days. The female oviposition period was only 4 to 7 days. Lal (1981) gave evidence for three overlapping generations for *C. choprai* in Himachal Pradesh in North

India between July and October. Bhardwaj and Bhalla (1975) similarly found for *C. nigrescens* three generations in the summer. Population density peaked in August. In the winter, the bug overwintered at high altitudes, under the bark of kosh trees (*Alnus nitida* Gaertn.), where it survived cold conditions of –3 to –6°C with 50 to 150 cm of snow (Lal 1981).

3.15.2 Damage

When feeding on the leaves, the leaves become covered with tiny pale spots representing the removal of parenchyma tissues and chlorophyll. Singh et al. (1987) illustrates the leaf of French bean riddled with spots and notes that the leaves then gradually turned yellow, withered, and dropped from the plant. The feeding weakens the plants and causes great damage to crops, especially in India and Japan (see references in Sweet and Schaefer 1985). Damage is especially serious during the rainy season and, as Lal (1981) noted, "French bean crops can be so badly damaged by the pest that it did not give any return to the farmer and were just ploughed into the field." The damage of *C. nigrescens* is evidently different, as Sharma et al. (1993) report that on the same host plant, French bean, *C. nigrescens* causes characteristic round yellow spots on the leaves.

3.15.3 Control

Gyawali (1987) discovered in Nepal that intercropping rice with soybeans reduced the incidence of *C. fallax* (probably *C. choprai*) and other pests of soybeans by 42%. It is interesting that for a bug that lays its eggs exposed on the surface of the leaf, where the bugs feed, there appears to be no record of any parasitism of the insect or any predation recorded, despite relatively intensive research on *Chauliops*.

Kashyap et al. (1980) compared 15 insecticides and found the most toxic (presumably the most effective) to be phosphamidon, methyl, parathion, endosulphan, quinalphos, and fenitrothion, in that order. Lal (1981) tested seven insecticides and found phosphamidon at 0.05% the most effective and malathion at 0.1% the least effective.

3.16 *Elasmolomus pallens* Dallas and *Dieuches* spp. (Rhyparochromidae)

3.16.1 Distribution, Life History, and Biology

Unfortunately, there is a nomenclatorial difficulty with this species and with *Dieuches* spp. so that the literature records are confused (Eyles 1973); thus, the two are considered together here for accuracy. The species formerly known as *E. sordidus* (Fabricius, 1787) must change its name because the original name, *Cimex sordidus*, is a junior primary homonym of *C. sordidus* Goeze 1778, as well as of *C. sordidus* Thunberg, a coreid. The next available name is *Rhyparochromus pallens* Dallas, 1852. In much of the older literature this species was placed in the genera *Pachymerus*, *Dieuches*, and *Aphanus* as well as *Rhyparochromus* and as a subgenus, *Elasmolomus*, of these genera. Slater (1964c) treated the subgenus as a genus. At present, the genus consists of 17 species. However, as Slater (1964c) noted, *E. pallens*, the type species of *Elasmolomus*, is distinctly different from the other species in the genus, and appears more related to *Dieuches*, so future revisions may change the nomenclature considerably.

Elasmolomus pallens has the most widespread distribution of the included species and extends from the Cape Verde Islands through Africa and southeast Asia to Indonesia, China, and Japan (Slater 1964a). Probably this broad distribution is at least in part anthropogenic, as shown by its introduction into Brazil at Piracicaba, in the São Paulo region (Slater 1972a). An important factor in the distribution of the bug is the association of this species with *Arachis hypogaea* L., an unusual legume known as peanuts by Americans, or as they are called in the Old World, groundnuts. In the Spanish-speaking world, they are called cacahuetes, which is a name close to the original common name, as the peanut is a native of eastern Bolivia and Peru and was originally domesticated by the native Indians, spread by commerce and agriculture to the tropical Old World, and then to the West Indies and North America (Simpson and Conner-Ogorzaly 1995). Slater warned that the introduction of this species may create a serious pest on many crop plants as well as peanuts.

This may help explain the introduction of *D. armatipes* (Walker) into the New World in the Dominican Republic, Jamaica, and the Grand Cayman Islands (Henry and Froeschner 1993). As they noted, Walker attributed receiving the specimen from the New World through Captain Friend. All subsequent workers discredited the record, believing the specimen actually came from Africa. This new evidence makes it probable that this is an old introduction into the West Indies. Here it is proposed that it occurred through shipments of peanuts from Africa, where peanut cultivation was successful early on (M. H. Sweet, personal comment); indeed, the odd common name for peanuts — goobers — came with slaves from Africa (Simpson and Conner-Ogorzaly 1995). Southwood and Leston (1959) noted that *Elasmolomus* was frequently introduced into England during World War II in shipments of groundnuts from Africa. The huge success of peanuts is well deserved. Peanuts are highly nutritious and have many uses in agriculture and commerce. Moreover, they provide a good protein and oil source that grows well under tropical conditions and in relatively poor soils (Uphof 1968): this also promoted the dispersal of the insect.

Because the pod of the peanut grows into the ground, the pods must be dug up and allowed to dry in stooks and ricks to prevent the growth of fungi that produce the dangerous seed toxin aflatoxin. However, in many areas, especially in Africa, this leaves the peanuts exposed to feeding by ground bugs of the Rhyparochromidae. The most important of these bugs is evidently *E. pallens* in Africa (Roubaud 1916, Risbec 1941, Mackie 1944, Cotterell 1952, Howe 1952, Conway 1976) and in India (Fletcher 1914, Kasargode and Deshpande 1921, Nair 1975, Chopra and Singal 1982). Other bugs that infest the outside drying sheds in Africa are *Naphius* (= *Aphanus*) *apicalis* Dallas, *D. armipes* (Walker), *D. albostriatus* (Fabricius), and *D. patruelis* (Stål), which, however, had not been clearly differentiated from each other (Eyles 1973). Unfortunately, according to Eyles (1973) in the economic literature what had been called *D. armipes* is actually *D. albostriatus*, and what had been called *D. patruelis* is actually *D. armatipes* (Walker). *Dieuches patruelis* appears not to be a pest, but *D. consimilis* Distant and, *D. humilis* Reuter were also found feeding on peanuts in Africa (Eyles 1973). Often the *Dieuches* species are more important than *E. pallens* (Risbec 1941, Cancela da Fonseca 1956). *Dieuches* itself is a huge genus of 131 species, with 87 in Africa with eight closely related genera. No wonder the taxonomic confusion occurred. Because of this taxonomic confusion, and the similarity of their economic impact, these peanut-feeding ground bugs are considered together here with *Elasmolomus* rather than sorted out into a long series of minor species. Of course, the *Dieuches* species feed on a variety of host seeds, many of no economic importance, others of potential importance as discussed by Eyles (1973). For example, *D. annalatus* Sign. feeds on castor beans in Madagascar (Frappa 1931). *Dieuches triangulus* Eyles feeds on strawberries in Kenya. Several species are closely associated with coffee, and others are associated with cucurbit fruits (Eyles 1973).

In addition to the feeding records on peanuts, this complex of ground bugs also feeds on another oil seed, sesame or benniseed (*Sesame orientale* L.) (Fletcher 1914, Hoffman 1932, Nair 1975, Chopra and Singal 1982). Mukhopadhyay and Saha (1992a,b,c, 1995) compared feeding *Elasmolomus* on four different seeds — peanuts, sesame, wheat, and fig seeds of the peepal or bo tree *Ficus religiosa* L. They discovered that sesame was the most preferred seed, and it supported the highest productivity in proportion to feeding.

These are large rhyparochromids. *Elasmolomus* female is about 10 mm and the male is about 8.5 mm long. *Dieuches armatipes* by comparison is even larger; the male is 10.5 mm and the female is 11.5 mm long. This gives the insects considerable power and they commonly move seeds to hiding places. Maxwell-Lefroy (1909) remarks that *A. sordidus* (= *E. pallens*) in India "have been found to infest threshing floors and carry off the wheat grains to the margins of the floor and hide them. What nourishment they can extract from a dry wheat grain seems doubtful, but they carry off the grains so abundantly that the cultivators require to collect them again every morning."

Two workers have given accounts of the bionomics of *E. pallens*. Corby (1947) described its life history in Nigeria, including descriptions and figures of the egg and all the instars. Oddly, he figures six instars rather than the normal five. It appears he may have considered the shrunken newly hatched first instar and the full-grown first instar about to molt as two different instars, which is an easy error to make, recalling that the first and second instars are the most difficult instars to tell apart. Hoffman (1932) in Canton (China) also studied this species. He found the insect in China to feed on the common nightshade *Solanum nigrum* L., driving its mouthparts deep into the seed of the fruit. The nymphs have shining brown head and thorax, and the abdomen brown with lateral pale spots. His figure shows well the distinctive y-suture of the Rhyparochromini. The length of the nymphs are first, 3.03; second, 4.38; third, 5.0; fourth, 6.09; fifth, 8.70 mm. The first stadium was 5 to 6 days; the last, 6 to 7 days. There are several generations a year. Corby (1947) found the insects are active chiefly at night and that the life cycle takes about 2 months.

3.16.2 Damage

As Conway (1976) noted, many earlier workers believed that the insects were essentially harmless because the seeds seemed unmolested, unlike the visible damage of chewing beetles or small rodents. Corby (1947) described that at Yola, Nigeria, the peanuts were stored in mud and thatch huts, and here the populations of *E. pallens* reached such proportions that their movements when disturbed were readily audible, and accumulations of exuviae were as much as a half inch. By careful measurements he showed that in Gambia there were distinct reductions in the weight of peanuts, sometimes so great that the shrunken kernels rattled inside the pods. The farmers in Gambia realized this and they called the bugs "Mingna," meaning suckers, and they blamed the bugs for the light weight of the nuts. The bugs caused losses in oil content and germination as well as rises in free fatty acids (Gillier 1970). The damage caused the nuts to become bitter and rancid. Germinative ability of the peanuts was reduced by one half (Risbec 1941).

3.16.3 Control

Risbec (1941) reported a bethylid wasp (*Cephalonomia*) as parasitizing the eggs of *E. pallens*. He also reports the reduviid *Coranus pallidus* as predaceous on both nymphs and adults. Conway (1976) tested surface insecticidal dusting and layering of dust with the stacks of peanuts with 0.5% lindane and 2% malathion. He found that a surface application of lindane at 45 g/m was very effective. In contrast, he found malathion to be comparatively ineffective.

4. THE LESS IMPORTANT SPECIES

4.1 *Blissus canadensis* Leonard (Blissidae)

Other members of the *B. leucopterus* complex outside of eastern North America show a dangerous potential to develop quickly into a pest species by host-transferring to economically important introduced grasses. In Saskatchewan and Alberta, Canada, and in Montana (U.S.A.), *B. canadensis* was recorded as a pest species on barley and on crested wheatgrass *Agropyron cristatum* (L.) Gaertn., both introduced grasses (Leonard 1970). Leonard believes that the northern records of *B. occiduus* Barber as being a pest in Saskatchewan and Montana (Parker 1920; Slater 1964a, records from Canada) actually refer to *B. canadensis*. If so, then the biology of this species is quite different from the other species, as Parker found that there was only one generation a year that matured on native prairie grasses relatively early in the spring. His observation of mating in the summer, the death of the adults in the fall, and the early appearance of nymphs in the spring, suggest that this insect overwinters as an egg. This shows the advantage of an egg diapause in that the whole generation of eggs is ready to start exploiting a host plant as early as possible upon hatching in the spring. After becoming adult in early summer, the insect spends the remainder of the summer producing diapause eggs. This strategy requires a predictable habitat (Sweet 1964). Similarly, in adaptation to a permanent habitat, this insect is pterygopolymorphic, and roughly equal numbers of macropters and brachypters are present. Interestingly, Parker found it to damage oats, but it did not injure corn or wheat, a different feeding preference from *B. leucopterus leucopterus,* which damaged oats the least. Parker could readily rear the nymphs on barley, which would correspond to the records of *B. canadensis* damaging barley.

If *B. occiduus* is a species of Colorado and New Mexico (U.S.A.) (Barber 1918), it seems unlikely that this is the species injuring sugarcane in Baja California (Mexico) (Ferris 1920; Box 1953; Slater 1976a). It seems much more likely, given the ecological geography of Baja California, that this is another undescribed species. *Blissus occiduus* is listed in *Common Names of Insects* (Bosik 1997) as the western chinch bug. The economic records evidently refer to *B. canadensis* Leonard or to an undescribed species, not to *B. occiduus* itself.

4.2 *Blissus antillus* Leonard

Leonard (1968) described *B. antillus* from Puerto Rico (U.S.A.) as damaging lawns of *Polytria amaura* (Buese) Kuntze, a grass introduced from Java (Liogier and Martorell 1982) and, in addition, one specimen was collected on goosegrass, *Eleusine indica* (L.) Gaertn., a widespread weedy grass from South America (Gould 1975). As these specimens were once included in *B. insularis,* it should be observed that this is an entirely different host plant than that used by *B. insularis* in the United States. Moreover, Leonard described two species from the same island, *B. slateri* as well as *B. antillus,* so it does not seem to be merely a case of insular variation. Leonard is probably correct in recognizing these populations as species distinct from *B. insularis.* Further experimental work is needed to resolve this question, which is important given the potential for *Blissus* species to develop into pests.

4.3 *Blissus richardsoni* Drake

In Peru, *B. richardsoni* Drake was recorded as damaging maize (corn), *Zea mays* L., and sugarcane, *Saccharum officinarum* L. (Lamas 1945), neither of which are native to Peru, and as damaging rice, *Oryza sativa* L., in Cuba (Bruner et al. 1945, Slater 1976a). However, these records need to be confirmed because *B. richardsoni* was described from Buenos Aires (Argentina) (Drake 1940). One doubts that these host records actually refer to *B. richardsoni;* again, we are probably dealing with an undescribed species.

4.4 *Blissus pulchellus* Montandon

This author has a specimen of *B. pulchellus* Montandon that was collected from Tiquizate, Guatemala on Dec. 12, 1959 by F. A. Bianchi, and labeled as damaging golf greens. *B. pulchellus* was originally described from *Panicum* sp. in Honduras and para grass, *Panicum purpureum* Ruiz & Pav., in Panama (Drake 1951), which are forage grasses, not lawn grasses. Interestingly, Georgiou and Taylor (1977) listed *B. pulchellus* as an example of the evolution of pesticide resistance in insects. However, no other records of this species could be found.

4.5 *Dimorphopterus similis* (Slater) (Blissidae)

Dimorphopterus similis, which was originally described from South Africa (Slater 1964c), attacked upland rice in Nigeria and caused the complete failure of the crop (Ene and Hamid 1982).

4.6 Dimorphopterus cornutus novaeguieneae Ghauri (Blissidae)

This subspecies was reported in New Guinea to be a pest on dryland rice planting, and to attack a lawn of carpetgrass, *Axonopus compressus* (Swartz) Beauv., an introduced Neotropical weedy grass (Uphof 1968).

4.7 Dimorphopterus brachypterus (Rambur) (Blissidae)

This widespread pterygopolymorphic species ranges through Africa to southern Europe. Slater (1974) noted that it has been taken from millet in Senegal and in Khartoum (Sudan) on durra, the latter being a variety of sorghum (var. *durra* Hubbard and Rehd). Slater (1976) identifies the millet as *Panicum mileaceum* L. It has also been collected in South Africa on breeding host plants *Urochloa mosambicensis* (Hack.) Dandy and *Cynodon plectostachys* (K. Schum.). In southern Europe it breeds on Bermuda grass, *Cynodon dactylon* (L.) Pers. (Wagner and Slater 1965). The latter record is important because this is a lawn grass used extensively in the southern United States and the introduction of this species may produce another serious pest blissid on Bermuda grass as *B. insularis* is on St. Augustine grass.

4.8 Dimorphopterus hessei (Slater) (Blissidae)

This species is known from Rhodesia (Zimbabwe) and the Transvaal, South Africa. It was reported as attacking lawn grass in Gwelo, Zimbabwe (Slater 1974).

4.9 Dimorphopterus spinolae (Signoret) (Blissidae)

This species was recorded as a serious pest on common reed *Phragmites australis* (Cav.) Trin ex Steud. in China, causing great damage to a reed marsh at Lake Lu in Hunan province. The Ovid abstracting service, translating from Chinese, listed the reed as a sedge, but *Phragmites* is a grass. The pest overwinters as an adult in reed debris or 2 to 3 cm underground. There are two generations a year (Li 1982). Li states that effective control was obtained using Cartap (Padan) at 0.1 kg/m^2. Padan is a mixture of 1.5% methyl parathion, 3% HCH (BHC), and 1.5% Kitazin. Li recommends use of the pesticide be strictly controlled to protect natural enemies present. *Dimorphopterus spinolae* is also recorded as a pest on common reed in Jiangsu province where it was referred to as the sorghum chinch-bug (Hao et al. 1992). Hao et al. noted that scelionid wasps were important natural enemies. They recommended burning the plant residues and soaking the eggs of the pest in late development stages. A systemic insecticide such as omethoate produced 64 to 83% control. You et al. (1997) recorded that the eggs of *D. spinolae* in the reed marshes are eaten by the earwig *Proreus simulans* (Stål).

However, this insect, the type of the genus, is a widespread European insect that breeds on the grass *Calamagrostis epigeios* Roth. (Gulde 1921; Putschkov 1958a, 1958b). As *D. spinolae* is unrecorded from the common reed in Europe, this identification must be confirmed, although the Chinese species of *Dimorphopterus* were revised and included *D. spinolae* as one of 11 species (Cheng and Tsou 1982). In this context, Slater (1974) noted that *D. spinolae* is closely related to *D. blissoides* (Baerensprung), which is common in the eastern Palearctic on *Phragmites australis* (= *communis*) (Putschkova 1956, Putschkov 1958b). Putschkov (1958b) described the nymphs, and Putschkova (1956) the eggs of both of these species. Slater (1974), however, doubted that the species are distinct as he could separate them only on the yellow-brown femora of *D. blissoides* compared with the dark femora of *D. spinolae*. However, it seems likely that the species are distinct based on the host plant records. Interestingly, Slater and Wilcox (1973) found *D. zuluensis* (Slater) on *P. communis* (= *australis*) Trin. and *P. mauritanicus* Kunth. in South Africa. Zheng and Wang (1987) report *D. spinolae* from *Phyllostachys viridis* (Young) McClure, which is a giant bamboo of commercial value. However, this record is evidently the species *D. japonicus* (Hidaki) and is probably a sitting or hibernation record, not a host record (Zheng 1994).

4.10 *Dimorphopterus pallipes* (Distant) (Blissidae)

This species is listed as a pest on rice in the Fukuoka perfecture Japan (Yokoyama et al. 1972). The trichogrammatid parasite *Oligosita itoi* Yashiro was reared from the eggs of *D. pallipes* (Yashiro 1979).

4.11 *Geoblissus hirtulus* (Burmeister) (Blissidae)

Geoblissus includes seven Old World species formerly placed in *Blissus* with *G. hirtulus* the original type species for the genus *Blissus* (Slater and O'Donnell 1995). As discussed under *B. leucopterus,* Slater and China (1961) proposed, because of its great economic importance, that type species designation be transferred from *hirtulus to leucopterus*, but the zoological Commission ruled otherwise. *Geoblissus hirtulus* itself is a pest species on turfgrasses in Japan (Yoshida et al. 1979, Hotsukade et al. 1979) and it attacks sugarcane in Khartoum (Sudan) (Slater 1964b). The type locality of *G. hirtulus* is Alexandria, Egypt (Slater 1964b). Lindberg (1948) reports *G. hirtulus* from the grass *Aeluropus villosus* Trin. in Cyprus, which might be a natural host record. Members of the genus are distinctive in having heavily spined legs that they employ to dig themselves quickly into loose sand (Slater et al. 1969).

4.12 *Ischnodemus diplanche* Slater and Harrington (Blissidae)

This species, which was described from northern South Africa, Botswana, and Zimbabwe from several grasses, including *Diplanchne fusca* (L.) Beauv., was recorded under *I. congoensis* Slater (Slater and Harrington 1970) as "attacking ears of ropoko and rice" in Zimbabwe (Slater 1964b). The identity of ropoko could not be found, but it is probably a local name for a variety of sorghum.

4.13 *Ischnodemus falicus* (Say) (Blissidae)

Wheeler (1996) recorded this species as a minor pest of an ornamental grass, the cultivar "Aureomarginata," which is derived from the prairie cordgrass *Spartina pectinata* Link, a grass widespread in wetlands of the eastern North America (Gould 1975). The insect is host specific to this grass species as shown by its absence from nearby clumps of two other species of *Spartina* as well as other grass species (Harrington 1972, Wheeler 1996). The bug has been recorded as a minor household nuisance in swarming around houses (Farrier 1958). As determined by Harrington (1972), there is one generation a year and the adults overwinter. Both Harrington and Wheeler observed that the insect aggregates in large numbers of adults and nymphs on certain clumps of the host plant. Wheeler (1966) noted that on disturbance the aggregation together releases a strong buggy odor, and he speculated that the odor may afford the host some protection from ungulate herbivores, as was suggested by Slater and Wilcox (1973a) for other blissid species. Wheeler observed that large populations of the bugs caused foliar chlorosis and the blades became brown at the tips. Johnson and Knapp (1996) showed that the bugs in high populations reduced photosynthetic productivity and reduced aboveground biomass productivity of *S. pectinata*. However, low densities of *I. falicus* did not affect the plant, and the authors concluded that the insect has little long-term effect on the plant and the plant tolerates well occasional outbreaks of the insect.

4.14 *Ischnodemus fulvipes* (De Geer) (Blissidae)

This widespread Neotropical species is unusual in its host range. There are several records from the horticultural plant *Canna indica* L., which despite its specific epithet is a Neotropical species with a number of related species in the genus *Canna*. The genus forms the small family Cannaceae, which most authorities consider endemic to the Americas (Core 1955). Two species occur in the United States, *C. flaccida* Salisb. from Florida to South Carolina and in southern Texas and *C. glauca* in southern Texas (Correll and Correll 1975). Many *Canna* species are now widely introduced throughout the tropics through horticultural trade, and one species, *C. edulis* Kerr, has tubers that are eaten throughout the tropics. It is known as Queensland arrowroot, although it appears to have originated in the West Indies (Simpson and Conner-Ogorzaly 1995), where *I. fulvipes* also occurs, and is the only *Ischnodemus* recorded from the West Indies (Slater and Wilcox 1969). *Ischnodemus fulvipes* was also found breeding on *Thalia geniculata,* a horticultural plant of the family Maranthaceae, which is closely related to the Cannaceae and is also largely restricted to the Neotropics. One species, *T. dealbata* Fraser, occurs in the southern United States from Texas to South Carolina (Correll and Correll 1975). Slater (1976a) doubted the records of *I. fulvipes* from Texas, but with *Canna* and *Thalia* native to Texas the insect may occur there. The species is also recorded from bromeliads and bamboo (Slater and Wilcox 1969), which is probably the widest host range of any blissid, spanning four families of monocotyledenous plants. Baranowski (1979) described the immature stages of this species.

4.15   *Ischnodemus noctulus* Distant (Blissidae)

Like *I. fulvipes,* this species breeds on an unusual host. Distant (1904) wrote, citing Green the collector, that *I. noctulus* was "injurious to foliage of Mysore Cardamon plant. The irritation caused by the punctures of the insects causes the leaves to roll up longitudinally, under which cover the bugs live and breed." Uphof (1968) lists the Mysore cardamon plant as *Elettaria cardomomum* Maton, the true cardamom, a valuable spice of commerce. *Elettaria* is a member of the Zingiberaceae, the ginger family. Slater et al. (1969) list *I. noctulus* from *Alpinia,* another genus of the Zingiberaceae. *Alpinia* is a large genus, with some species that have highly ornamental flowers and others that are used medically (Bailey 1949, Core 1955, Uphof 1968). Interestingly, the Zingiberaceae with the Cannaceae and the Marantaceae discussed above under *I. fulvipes* are placed with the Musaceae in the order Zingiberales. However, Slater (1979) in his analysis of the genus *Ischnodemus* did not find these two species closely related. If indeed so, then the unusual similarity in host–plant relationship would be due merely to convergence.

4.16   *Ischnodemus perplexus* Slater and Harrington (Blissidae)

This is a species widespread in tropical Africa (Slater and Harrington 1970). Some specimens of the type series were taken on rice in Senegal (Slater 1976).

4.17   *Macropes obnubilus* (Distant) (Blissidae)

This species is known in Japan as the bamboo chinch bug (Hasegawa 1980). It has been collected in the Bonin Islands (Slater and Ahmad 1967) and in Guam and Vietnam (Slater et al. 1969), and as the synonym *M. hedini* Lindberg, from China (Slater and Wilcox 1973b). Ashlock collected this species breeding on a dwarf bamboo, *Arundinaria simoni* A. and C. Riviere, at Hikosan Kyushu, Japan, but it occurs on other species (Slater et al. 1969). Dr. S. Asahina (in Slater et al. 1969) stated that this species feeds upon a commercial species of bamboo, and the feeding punctures become surrounded by dark rings. Because of the decorative effect produced, such bamboo commands a higher price than bamboo not so marked. Slater et al. noted that this might be the only case of a member of the Blissidae being economically beneficial. Asahina also noted that the species makes sounds, but Slater et al. could find no stridulating apparatus. Perhaps the sounds are made by a tymbal apparatus. Yokoyama et al. (1972) reported this species as also feeding on rice in Japan.

Miyamoto and Kifune (1984) described a new species of Strepsiptera, *Blissoxenos esakii,* from this species. They erected a new subfamily, the Blissoxeninae of the family Corioxenidae, a primitive family that parasitizes, as far as is known, only members of the Heteroptera: Pentatomomorpha. Miyamoto and Kifune also present a useful key to the genera of the Corioxenidae. During the course of an anatomical study of the alimentary organs of the Heteroptera, Miyamoto (1961) collected many specimens of *Blissoxenos* from *M. obnubilus* and only a few from *Iphicrates spinicaput* Scott and *Dimorphopterus japonicus* (Hidaka). From its numbers, perhaps this strepsipteron is an important agent in controlling the populations of *M. obnubilus*.

4.18 Macropes harringtonae Slater Ashlock & Wilcox (Blissidae)

Zheng (1994) reports this species to be a pest on *Phyllostachys pubescens* Mazel. This tall bamboo species is much used for construction and the young shoots are consumed for food (Uphof 1968). Zheng and Wang (1987) found the bug overwintering as an adult on *Pleioblastus amarus* (Keng) Keng f., a small ornamental bamboo species (Bailey 1949).

4.19 Macropes maai Slater & Wilcox (Blissidae)

Zheng (1994) recorded this species from *Phyllostachys viridis* (not *virescens*) (Young) McClure, which is the most important commercial species of bamboo in China, and therefore this insect species has a high potential to become a pest. Other species of Blissidae that are listed by Slater (1976) from unidentified bamboos and may be of economic importance are *M. subauratus* Distant, *M. privus* Distant, *M. varipennis* (Walker), and *M. spinimanus* Motschulsky.

4.20 Macropes punctatus (Walker) (Blissidae)

Slater and Wilcox (1973b) recorded this species in India and Pakistan from bamboo and from *Saccharum bengalense* Retz., a tall grass species sometimes cultivated as an ornamental grass (Bailey and Bailey 1976). In India and Pakistan, the grass is much used as the source of munj, a valuable fiber that is used to make ropes, mats, and baskets. The species *S. munga* Roxb. and *S. arundinaceum* Retz. are also used for munj in India (Uphof 1968). Slater and Wilcox (1973b) described the fifth instar. The abdomen is brick red with yellow spots, and the head and thoracic sclerites are reddish brown.

4.21 Pirkimerus japonicus (Hidaki) (Blissidae)

Zheng (1994) records this species as a serious pest on *Phyllostachys pubescens* (not *virescens*) Mazel ex Lehaie, an important tall bamboo much used for construction and many other purposes in the Yangtze Valley (China) (Uphof 1968). The bug retards the growth of the bamboo stems, causing the bamboo stem to crack, lowering its value as a building material, and decreasing the production of the bamboo shoots, an important vegetable in China. Eventually, the insect causes the death of individual plants and the deterioration of the bamboo plantation. *Pirkimerus* is unusual in that it is the only blissid that lives inside the bamboo stalk rather than under the leaf sheaths. It enters into the stem through holes made by moth larvae and feeds on the bamboo from inside the stem (Zheng 1994), which may explain its small size and cylindrical shape. Slater and Ahmad (1965) record two other species, *P. davidi* in China and *P. esakii,* from Tokara Island (Japan), on bamboo.

4.22 Iphicrates weni Zheng (Blissidae)

Zheng (1994) reports this species to feed on *Sinocalamus* (= *Bambusa*) *beecheyanus* (Munroe) McClure, which is an important source of edible bamboo shoots in the Orient (Uphof 1968). He also records *I. spinicaput* (Scott) in China on *Indocalamus migoi* (Nakai) Keng f. Slater (1976a) reports *Iphicrates angulatus* Slater to breed on *Arundinaria cobonii* in New Guinea. Two other species of *Iphicrates* are recorded from New Guinea on "bamboo": *I. nigritus* Slater and *I. papuensis* Slater.

4.23 *Spilostethus hospes* (F.) (Lygaeidae *sensu stricto*)

The taxonomy of this species is complex with much synonymy and described forms and, like *S. pandurus*, is in need of a careful revision (Slater 1964c). *S. hospes*, like *S. pandurus*, is becoming much used in physiological studies (Groeters 1996, Sanjayan et al. 1996). Groeters (1996) demonstrated that a bacterium was transmitted in the eggs that destroyed males and promoted a female-biased sex ratio.

The biology of this species is similar to *S. pandurus* and the two species are often found together on the same asclepidaceous host in India, madar or akk, *Calotropis gigantea* (L.) R. Br. and *C. procera* (Ait.) R. Br. (Mukhopadhyay 1983). Interestingly, Hoffman makes no mention of *S. hospes* on asclepidaceous plants in China, but instead he noted it feeds heavily on the common nightshade *Solanum nigrum* L., which is a noxious weed. Hoffman recommends it as a biological control on the nightshade. However, it appears to be a myth that the berries of the nightshade are poisonous. They have been called wonderberries and made into pies (Uphof 1968, Bailey and Bailey 1971). *Spilosthethus hospes* has a more eastern distribution than *S. pandurus*, not extending into Africa, is much less abundant, and the records of it as a pest are much fewer (for which reason it is listed among the less important species), although some of the same species are attacked by *S. hospes* in India (Mukhopadhyay 1983): *Sorghum vulgare* Pers. (= *Sorghum bicolor* [L.] Moench.), eggplant (*Solanum melongena* L.), cotton, on green bolls (*Gossypium* spp.) (Hoffman 1932, 1934, in China; Gentry 1965, in Pakistan). *Spilostethus hospes* is a pest in India on the Malabar nut tree *Adhatoda vasica* Nees (Swamy and Rajagopal 1995a), a medicinal shrub also used for dyes and making beads, and a pest on the glory lily *Gloriosa superba* L. (Swamy and Rajagopal 1995b), a climbing lily used in medicine and as an ornamental plant, but which is very poisonous; it was prescribed in Burma to commit ritual suicide (Uphof 1968). In China Hoffman (1932) found it a pest on the Cape gooseberry *Solanum peruviana* L., and on Hung Pin Tsoi *Emilia sonchifolia* (L.) DC, whose leaves are used as food and as an herb tea (Uphof 1968); and on the aramina plant *Urena lobata* L., whose bark is used to make a fine fiber (Uphof 1968). Malipatil (1979) found *Spilosthethus hospes* in Australia breeding not only on *Asclepias*, but also on the introduced *Solanum sodomaeus* L., the apple of Sodom or yellow popolo, which is sometimes grown as an ornamental and a medicinal plant (Uphof 1968, Bailey and Bailey 1976). Sanjayan (1993) observed that *Spilosthethus hospes* effectively utilizes the seeds of eggplant (aubergine), soushumber, *S. torvum* Swartz, which is used as a vegetable in the tropics (Uphof 1968), tomatoes, *Lycopersicon esculentum* Mill, and *Euphorbia hirta* L., a medicinal plant, because it uses the perennial asclepiad *Calotropis* and the composite *Vernonia* as out-of-season hosts, from which to attack economically important crops. The same statement could be made for *S. pandurus*.

As described by Hoffman (1932, 1934) in China and Malipatil (1979) in Australia, the eggs and nymphs of *S. hospes* are similar to *S. pandurus*. The nymphs are bright red and the wing pads become a shining dark brown in the fifth instar. The stadia as measured by Hoffman were first, 3 to 5 days; second, 3 to 6 days; third, 3 to 6 days; fourth, 4 to 6 days; fifth, 6 to 8 days. Mukhopadhyay (1983) found that *S. hospes* laid a lower number of eggs (150 vs. 385), but had greater hatching success (69 to 62%). *Spilostethus hospes* also showed a more rapid rate of development (Mukhopadhyay 1983) on milkweed seeds. Malipatil (1979) reared the insect in Australia and found similar development rates, but the fecundity was much greater, varying from 168 to 419 eggs, and the longevity was greater, from 127 to 272 days for females compared with 15 to 32 days for Mukhopadhyay. Ecologically, Mukhopadhyay found that *S. hospes* was more abundant when madar had a discontinuous distribution; *S. pandurus* more abundant in continuous thickets of madar. Thangavelu (1980) found *S. hospes* to be more abundant than *S. pandurus* in the cooler months of the year and that *S. hospes* had a narrower preferred host range than *S. pandurus* (Thangavelu 1978d), which helps to explain its lesser pest status. Two species of gut flagellates, *Leptomonas inhospes* and *L. indica* Prasad and Kalavati (1987) are known from *S. hospes*.

4.24   *Spilostethus rivularis* (Germar) (Lygaeidae *senso stricto*)

This African species is recorded on cotton and nymphs and adults from *Sorghum vulgare* (= *S. bicolor*) in Nigeria (Golding 1931), on cotton in Portuguese East Africa (= Mozambique) (Hargreaves 1948); from *Hibiscus*, cotton, and sweet potato in the Congo (Mayné and Ghesquière 1934). Slater and Sperry (1973) report taking large numbers of this species in Pretoria, Transvaal, South Africa from a composite, *Lopholaena coriifolia* Phillips and Sm., a common woody shrub in the area.

Slater and Sperry (1973) described the egg and all instars of this species. The nymphs are a conspicuous yellow-orange to orange marked with brown.

4.25   *Spilostethus furculus* (Herrich-Schaeffer) (Lygaeidae *sensu stricto*)

There are several records of this African species injuring crops: on okra (*H. esculentus* L.) and tobacco (*Nicotiana tabacum* L.) in Sierra Leone (Hargreaves 1937); on cotton (*Gossypium* sp.) in Tanganyika (Tanzania) (Ritchie 1925); in the Congo with other Malvaceae (Mayné and Ghesquière 1934); and in Nigeria on sunberry (*Physalis minima* L.), a solanaceous plant with edible fruits grown throughout the tropics. The latter record is most interesting because Slater and Sperry (1973) reported this species not to breed on Asclepidaceae and were puzzled about the diversity of food plants, having taken nymphs on seed heads of *Albuca setosa* Jacq. (Liliaceae) and *Cotyledon mollissima* Schonl. (Crassulaceae). Members of both genera are found in cultivation as ornamentals (Bailey and Bailey 1976), and the latter genus is very poisonous. Several weedy *Solanum* species, *S. elaeagnifolium* Cav. and *S. mauritanum* Scop., had invaded South Africa, and Olckers and Hulley (1989, 1991) and Olckers et al. (1995) are investigating the potential for *S. furcula* to be used for biological control on both introduced and indigenous weedy species of *Solanum*, as the insect feeds on the seeds of *Solanum*. However, caution should be used, as several species of *Solanum* are agricultural crop plants and the insect is not host specific in its feeding behavior. Slater and Sperry (1973) described the egg and the five instars. The nymphs, similar to the adults, are bright orange-red marked with black; however, even within South Africa the coloration of the nymphs varied considerably.

4.26   *Spilostethus longulus* Dallas (Lygaeidae *sensu stricto*)

This species is known from North Africa south to Nigeria and Ethiopia and east to Iran and India. It is listed as damaging cotton in Nigeria (Golding 1937) and Ethiopia (Hargreaves 1948). Gentry (1965) lists it as injuring cotton and beans in Libya, Egypt, and the Sudan.

4.27   *Graptostethus servus* F. (Lygaeidae *sensu stricto*)

Members of this genus are brightly colored red and black similar to *Lygaeus* and *Spilostethus*. Of 33 species listed in the genus *Graptostethus*, only *G. servus* appears to be of economic importance, although evidently minor. This species has the widest distribution of any Lygaeidae (*senso stricto*), ranging throughout the Old World tropics and subtropics from South Africa to the southern Palearctic through Asia to Australia and Hawaii (U.S.A.). Like *Spilostethus pandurus* this species is highly variable and needs careful revision within the species and with reference to its congeners. A. Slater (1978, 1985) cautioned that some of the species related to *G. servus* are difficult to identify, and some synonymies and economic records may be incorrect. Five subspecies have been described, but all the economic records evidently pertain only to the nominate species (Slater 1964c).

As Slater and Sperry (1973) noted, the economic host records are puzzling, as are the noneconomic host records. In China, Hoffman (1934) completely described the life history of a species he listed as near *Graptostethus servus* that was feeding on *Ipomoea caraica* (L.) Sweet, a convolvulaceous species that is used more for fiber than for food but is also used in horticulture as an ornamental (Bailey 1949, Uphof 1968). Hoffman (1934) does list *G. servus* itself on *I. purpurea* (L.) Lam., the common morning glory or clockflower of horticulture (Bailey 1949). Fletcher (1914, 1919) reported it on sweet potato, *I. batatas* (L.) Lam., in India. But in Africa, it is not recorded on Convolvulaceae or on any Asclepidaceae. In his fieldwork in South Africa, Slater (in Slater and Sperry 1973) found it copulating on *Salsola kali*. var. *tenuifolia* Tausch. (Chenopodiaceae), the infamous noxious weed from Eurasia well known as the Russian thistle, although there are about 13 native species of *Salsola* in South Africa. In India, Mukhopadhyay (1983) specifically searched for *G. servus* on madar (*Calotropis gigantea* and *C. procera*) (Asclepidaceae), in her study of *Spilostethus pandurus* and *S. hospes*, but reported none, noting as Maxwell-Lefroy (1909) stated, there were no accurate observations of its natural food.

As Slater and Sperry (1973) summarized, *G. servus* is recorded as injuring cotton in Africa, India, the Orient, and New Guinea although they doubted that any of the records indicated actual breeding on cotton (Fletcher 1919, Mayné and Ghesquière 1934, Harris 1936, Chatterjee 1937, Hargreaves 1948, Lepelley 1960). Golding (1947) records *G. servus* adults and, nymphs feeding in Nigeria on the seeds of the jute plants pat or koshta (*Corchorus capsularis* L. and *C. olitorius* L.) (Tiliaceae), two Indian species that are cultivated widely for the fibrous bark used in making burlap, twine, gunnysacks, carpets, etc., and the young foliage used for food in Africa, India, and the Orient (Bailey 1949, Uphof 1968). Fletcher (1919) also records it from jute in India. The Tiliaceae are placed in the order Malvales with the Malvaceae, the family that includes cotton. However, there is a great diversity of records from other plants showing a dispersal to a variety of alternate hosts: on redgram, probably *Phaseolus aureus* Roxb. or *P. mungo* L.; sugarcane (Fletcher 1919, Chatterjee 1937); on *Plectronia* (= *Canthium*) *didymum* Bedd., the source of Ceylon boxwood; *Zizyphus* sp., probably *Z. jujuba* Mill, a fruit much consumed in India; on sunflower (Chatterjee 1937); in Italy on the pistachio nut *Pistacia vera* L. (Romeo 1929); in China (Hoffman 1934) on *Gardenia augusta* (L.) (synonym of *G. florida* L.), an ornamental plant from which perfume is extracted (Uphof 1968); from *Sorghum vulgare* Pers. (synonym of *S. bicolor* [L.] Moench.) in Nigeria (Golding 1931) and India (Prabhakar et al. 1981); in India (Dhamdhere and Rawat 1969) adults and nymphs on sunnjute or sannhemp (*Crotalaria juncea* L.), a leguminous plant much used in commerce for its tough fiber; on soybeans in Thailand (Gyawali 1987); in Sierra Leone on jack bean (*Canavalia ensiformis* DC) (Hargreaves 1937), a legume from the West Indies whose pods are used for food, although the fresh seeds are poisonous, and which is also used as an ornamental plant (Bailey 1949, Uphof 1968); in India on Malabar nut tree (*Adhatoda vasica* Nees) (Swamy and Rajagopal 1995a), a medicinal shrub also used for dyes and making beads; and in India on the glory lily *Gloriosa superba* L. (Swamy and Rajagopal 1995b).

With the above extensive literature on the economic importance of this species, it is astonishing to find it listed by its synonym *Graptostethus manillensis* (Stål) (A. Slater 1978) as the wood rose bug. This is fairly obscure as Uphof (1968) has no record of this plant and Bailey (1949) says *Ipomoea tuberosa* L. is sometimes known as the wooden rose. In Mabberley (1981) under the convolvulaceous genus *Operculina* (treated by Bailey as a synonym of *Ipomoea*) there is a reference to "wooden roses" as the swollen capsules used for decorations. Neither the plant name nor the insect name is correct.

Slater and Sperry (1973) described the egg and first instar. The egg is stoutly elliptical with seven chorionic processes; the nymph is brownish in color. Dhamdhere and Rawat (1969) briefly described all stages. The egg stage took 3 to 7 days, and the nymphal stage 15 to 17 days, the adult male lived 3 to 4 days, the female 6 to 11 days, for a total life cycle of 17 to 25 days, which seems rather short. The females laid 136 eggs in 9 days and 46 in 4 days. The females mated several times, and laid the eggs singly on the seeds of sannhemp, on which the nymphs and adults fed.

4.28 Oncopeltus famelicus (F.) (Lygaeidae sensu stricto)

This striking red and black species, as summarized by Slater and Sperry (1973) from work in South Africa, feeds primarily on milkweeds (*Asclepias* spp.). Vickermann (1962) found it to feed on an asclepiadaceous vine *Pergularia extensa* N. E. Brown and to transmit *Phytomonas elmassiani* (Migone) to the plant. Marshall (1902) proposed that *O. famelicus* formed a Mullerian complex with cantharid beetles on milkweeds. However, this species has been reported on a surprising variety of economic plants in Africa. Golding (1927) in Nigeria reports adults and nymphs from detanve, *Pouzolzia guineensis* Benth. in Hook, an urticaceous plant used for food; from *Mucuna urens* Medic., a legume probably used for a dye; from *Triumfetta rhomboidea* Jacq., a tiliaceous woody plant used for fiber; and on cotton (*Gossypium hirsutum* L.). Mayné and Ghesquière (1934) reported it in the Congo from sweet potato (*Ipomoea batatas* Poir.), from candillo (*Urena lobata* L.), a shrub used for fiber, and as abundant on the passion fruit *Passiflora edulis* Sims and also on cotton. Brain (1929) reports it on sugarcane in Zululand (South Africa). Lepelley (1960) reported it on the fig *Ficus exasperata* Vahl. in Uganda.

Slater and Sperry (1973) described the first, third, fourth, and fifth instars. The nymphal coloration was a variegated red, brown, yellow, and white, evidently aposematic like the adults.

4.29 Caenocoris nerii (Germar) (Lygaeidae sensu stricto)

This large handsome black species, conspicuously marked with red (Wachmann 1989), is found around the Mediterranean into India and feeds almost exclusively on oleander (*Nerium oleander* L.), an apocynaceous shrub, but it also feeds on Asclepidaceae (Slater and Sperry 1973), and in India on the asclepidaceous weed madar (*Calotropis gigantea*) (Thangavelu 1980). Another species of Apocynaceae, *Nerium indicum* Mill., which is also very poisonous and is used in horticulture (Bailey and Bailey 1976), occurs in India and southeast Asia (Uphof 1968) and is probably the species referred to as *Nerium* sp. as a host of *Caenocoris nerii* by Thangavelu (1978d). Slater (1964c), based on its disjunct distribution in the Mediterranean area and in South Africa, proposed that *C. nerii* was probably introduced into South Africa on oleander. Oleander is a favorite horticultural shrub widely grown in warmer areas for its beautiful red, pink, and white flowers and evergreen foliage and also for its ability to live in dry areas and relatively saline soils; however, it is highly poisonous to humans because of its toxic levels of cardiac glycogens (Bailey 1949, Uphof 1968, Ellefson et al. 1992). The potential for *C. nerii* to become an introduced pest on oleander should therefore be considered very high.

4.30 Lygaeus equestris L. (Lygaeidae sensu stricto)

This brightly colored Palaearctic species, known as the "Ritterwanzen" in German (Wachmann 1989), is the type species of the genus *Lygaeus* to which belongs the North American little milkweed bug *L. kalmii* Stål, which has been used in ecological and physiological studies, but to a lesser extent than the large milkweed bug *Oncopeltus fasciatus* Dallas (Slater 1964a, A. Slater 1985). Although feeding on the asclepiad swallowswort *Cyanchium vincetoxicum* B. Bs., itself a medicinal plant (Uphof 1968), the host range of *L. equestris* is larger but it is generally not thought of as an economic insect. However, Popov (1973) found it injuring a number of medicinal plants: *A. absintium* L., *Artemisia maritima* L., *Digitalis amandiana* Sampaio, *D. chinensis* Hook et Arn., *Pyrethrum cinerariaefolium* Trev., *Lysimachia vulgaris* L., *Echinops sphaerocephalus* L., *Rheum caspicum* Pall, *Centaurea sibirica* L. One wonders if the insect can sequester toxins from these plants other than the cardiac glucosides of Asclepidaceae.

There is a large literature on this species in Europe, but as this species is of minor importance, and most records noneconomic, the reader is referred to the compilation by Slater (1963a).

4.31 Nysius ericae (Schilling) (Orsillidae)

This species, once thought to have a Holarctic distribution, is now restricted to the Palearctic (Ashlock 1977), although Schmitz (1976) extended *N. ericae* into Africa through his synonymy of *N. binotatus* Germar of South Africa and *N. albidus* Dallas of West Africa, an action, as stated earlier, that is doubtful; Slater (1964c) is correct that a worldwide revision of the genus is badly needed (M. H. Sweet, personal comments).

In contrast to the Nearctic records of *N. niger* and *N. raphanus* (both formerly *N. ericae*) on crucifers, the Palearctic records make no report on crucifers. However, the identity of the insect is uncertain because many of the articles in the European literature cite the voluminous research on the false chinch bug in North America even when acknowledging that *N. ericae* does not occur in North America. Dolling (1991) observes that *N. thymi* is a coastal species in Britain, while *N. ericae* is a species of open, bare ground habitats where it feeds on many plants, especially composites, and where the ground may be found teeming with thousands of the bugs. Similar to other species of *Nysius*, when there is a lack of water, the bugs migrate from weedy plants to cultivated plants, the severity of attack varying with the buildup of the insect in the spring on weedy host plants and the degree of drought in late spring and early summer, which makes this insect especially dangerous in areas with a Mediterranean climate. Called the "chinche gris," or gray bug in Spain, *N. ericae* especially attacks and damages vineyards; to a lesser extent, but sometimes seriously, it damages pear, prune, peach, loquat, and citrus trees (Gomez-Menor 1955; Rivero and Marí 1983). Similarly, in Italy it attacks vineyards, causing the leaves and grapes to wilt rapidly, and the vine shoots show hypertrophic reactions (Pennacchio and Marullo 1986). In the area of northeast Africa and southwest Asia, it is listed as polyphagous, attacking the listed species with the addition of tobacco (Gentry 1965).

In India, where it is called the dusky bug, *N. ericae* was observed feeding in large numbers on the milky grains of pearl millet *Pennisetum typhoides* (Burm. f.) L. C. Rich. (Deol 1985). Pal (1975) found *N. ericae* to be pest on lucerne (alfalfa) (*Medicago sativa* L.). It is also a major pest in India on the annual wormwood (*Artemisia annua* L.), a plant cultivated as a medical herb, for artemisinin, which is a potent drug in the control of choroquine-resistant malaria. The insect was most numerous at the time that the *Artemisia* plants came into flower and set their seeds (Mehta et al. 1996).

For control to prevent invasions from weedy hosts it is important to eliminate the weeds around cultivated plants, especially in vineyards (Rivero and Marí 1983, Pennacchio and Marullo 1986). Once an invasion has occurred, it is important to use a fast-acting insecticide such as lindane, malathion, diazinon, and endosulfan as the damage occurs quickly (Rivero and Marí 1983). Deol (1985) tested eight insecticides and found carbaryl, monocrotophos, chlopyriphos, phosalone, and dimethoate to be effective in controlling the bug on pearl millet.

4.32 Nysius cymoides Spinola (Orsillidae)

In Israel, mass infestations of *N. cymoides* have occurred on cabbage and cauliflower sown for seeds, on wild mustard, and on grapevines. It was also collected on clover and alfalfa fields in the summer (Rivnay 1962). Rivnay stated that the bugs preferentially attacked the cruciferous inflorescences and pods when the seeds are in the "milk " stage. In the Negev, it attacked the introduced river red gum *Eucalyptus amaldulensis* Dehnh. and the flat-topped yate *E. occidentalis* Endl. (Gilat and Louis 1974). In Egypt it is part of a complex of pests attacking leguminous plants (Harakly and Assem 1978). In Italy, Parenzan (1985) reported *N. cymoides* to attack the newly introduced buxaceous plant, jojoba *Simmondsia chinensis* (Link.) Kellog (= *Simmondsia californica* Nutt.), a plant valued for its excellent oil and used as a horticultural substitute for the hedge shrub *Buxus* in arid areas (Bailey and Bailey 1976). The bug appeared to have built up large populations in the plantation on the weed *Portulaca oleracea* L. and then had migrated to jojoba.

Rivnay (1962) observed that the nymphs were only present in the spring and that the adults when they matured in the laboratory and adults in the field in summer did not reproduce, so he concluded that this insect goes into an adult diapause in the dry summer and becomes reproductive in the spring. The number of generations was unknown, but he observed that the insects could go through their life cycle in a few weeks. It seems probable that this species is univoltine in contrast to other species of *Nysius* discussed here. This would be a useful adaptation to the Mediterranean climate of the area. Nevertheless, Rivnay reported that a garden looked as though several anthills were disturbed as the plants were totally covered with the pest. One year, millions of nymphs had developed in the wild mustard fields and, when the short-lived mustard plants died back, the nymphs migrated en masse to adjoining vineyards, destroying them.

For control Rivnay (1962) recommended the systemic insecticides metaisosystox and phosphamidon on the grapevines, and using strips of land around the vineyards dusted with DDT or dieldrin "to ward off the oncoming march of bugs."

4.33 *Nysius graminicolus* Kolenati (Orsillidae)

This species, which is the type species of Stål's subgenus *Macroparius*, has a distribution similar to that of *N. cymoides* but appears to be a lesser pest. It is recorded in Italy as damaging grain sorghum, as well as rape, sunflower, seed beets, *Salvia*, and various Gramineae, to which it had migrated after its wild hosts had dried up (Bin and Colazza 1986). *Nysius graminicolus* damaged the seedpods and seeds of kenaf, causing a large number of damaged or aborted seeds (Parenzan et al. 1994). Because of the records from grasses, this is probably the *Nysius* listed as a Sunn pest with the pentatomids *Aelia* and *Eurygaster* in Anatolia, Turkey (Kinaci et al. 1997). It attacked several medicinal herbs in Bulgaria: *herba absinthi* (*Artemisia maritima* L.), chamomile (*Matricaria chamomilla* L.), and brown knapweed (*Centaurea jacea* L.) (Popov 1973).

4.34 *Nysius thymi* (Wolff) (Orsillidae)

This Holarctic species, the type species of *Nysius*, is known from dry habitats where it feeds on thyme (*Thymus*), wormwood (*Artemisia*), and heather (*Calluna*) (Southwood and Leston 1959, Wachmann 1989). There is a large European literature (Slater 1964a) on this species, but very few economic records. Studzinski and Malachowska (1973) found it feeding heavily on both wild and cultivated crucifers in Poland.

4.35 *Nysius senecionis* (Schilling) (Orsillidae)

Gentry (1965) reports this species as other *Nysius* to invade from weed species to damage vineyards, especially newly planted vines. The insect in the Ukraine overwinters as an egg, and the eggs are laid in cracks in the ground (Putschkova 1956).

4.36 *Nysius binotatus* (Germar) (Orsillidae)

This South African species, which was made a subspecies of the Palearctic *N. senecionis* by Linnavuori (1989), was reported by Evans (1936) as destroying vegetable crops and garden flowers, and as attacking peaches. Annecke and Moran (1982) reported *N. binotatus* as sporadically numerous on turnips. It has been reported to attack sunflower seed heads (Annecke and Moran 1982). Nixon (1946) in Uganda reported *N. binotatus* from the composites *Erigeron* and *Vernonia* and found it parasitized by the wasp *Euphorus carcinus* Nixon. Slater (1964c) notes that this species is less abundant than *N. natalensis* and some of these records may well be referable to that species.

4.37 *Nysius natalensis* Evans (Orsillidae)

Slater (1964c) lists this species as the most abundant of the four species of *Nysius* recorded from South Africa. This species is referred to in the South African economic literature as the false chinch bug. Matthee (1974) and Annecke and Moran (1982) list it as an occasional pest on wheat. They reported heavy infestations in which the insects suck sap from the seed husks, and as the infestations are heaviest on the margins of the fields the insects may be invading from the surrounding veld (grasslands). The insect also attacks onions, leeks, garlic, alfalfa, and sunflowers. They observed that little is known about the biology of this species. Matthee (1974) recommended using demeton-*S*-methyl to control the insect.

4.38 *Nysius albidus* Dallas (Orsillidae)

This species of West Africa was observed in Nigeria on cowpea *Vigna unguiculata*, on which it destroyed the seeds in the pods (Lukefahr 1981).

4.39 *Nysius stali* Evans (Orsillidae)

In Nigeria this species was the most numerous pest on sunflowers during flowering and head formation (Misari 1990).

4.40 *Nysius inconspicuus* Distant (Orsillidae)

This species severely infested sesame (gingelly) in the Indian state of Tamil Nadu (Thangavelu 1978a) and also caused serious damage to sunflower heads in the same area (Logiswaran et al. 1982). Joseph (1959) found *N. inconspicuus* severely infesting tobacco capsules. In Pakistan, they reported the pest laid 12 eggs/female, which seems very low. At 30.5°C the incubation period was 5 days and the nymphal development period was 20 days. Thangavelu (1978a) found that the species bred on the weedy plants *Aerva tomentosa, Amaranthus bengalensis, A. viridis, Celosia argentea, Ageratum conyzoides, Euphorbia pilulifera,* and *Mollugo* sp. Of these weeds, *Aerva* was the preferred food plant. On this reservoir the bugs can persist between the harvests of sesame or sunflowers, making this a potentially difficult pest to control. Joseph (1959) found that in India three species of reduviids (*Rhynocoris lapidicola* Samuel and Joseph, *R. nysiiphagus* Samuel and Joseph, and *Coranus* sp.) fed on *N. inconspicuus* both in the field and in laboratory tests. He found the reduviids approach the prey from behind and pierce it from behind in the abdomen, paralyzing the prey. The young nymphs of the reduviids could feed on the smaller instars of *Nysius* but not the older ones.

4.41 *Nysius minor* Distant (Orsillidae)

Goel (1983) found this species to be a pest on sunflowers in Uttar Pradesh (India) on crops in the winter, summer, and monsoon periods.

4.42 *Nysius ceylonicus* Mots. (Orsillidae)

Lal (1989) reported *N. ceylonicus* to cause damage to lettuce, asparagus, cowpeas, sunflowers, cosmos, sweet william in Himchal Pradesh, India. He found the insects to hibernate as adults under the bark of the forest tree *Alnus nitidula* between 45 to 150 cm from the ground, and to survive temperatures as low as –3 to –6°C.

4.43 Nysius turneri Evans (Orsillidae)

Nysius turneri, the Invermay bug, is a minor pest in Tasmania, Australia, where it coexists with *N. vinitor* (Evans 1936, 1938). Evans stated that this species is found in moister environments than *N. vinitor*, but when the vegetation dries up this bug like *N. vinitor* attacks various vegetable crops and fruit trees and is an occasional serious pest. Evans stated that the species studied by Nicholls (1932) was *N. turneri*. Nicholls estimated from fieldwork that there were five generations a year of *N. vinitor* in Tasmania. This seems exaggerated, given the cool climate of Tasmania.

4.44 Nysius caledoniae Distant and other Hawaiian Nysius (Orsillidae)

Nysius caledoniae, which was introduced into Hawaii and Oahu (U.S.A.), has become very abundant in the lowlands, breeding on a variety of weeds, chiefly introduced composites (*Emilia sonchifolia*, *Erigeron canadensis*, *Sonchus oleraceus*, *Pluchea indica*, and *P. odorata*). It evidently moved from weeds to commercial plantings of vanda orchids and severely injured them (Beardsley 1965, 1979; Yoshioka 1965). Usinger (1949) observed that *N. caledoniae* has a very great spread potential and was readily dispersed by humans especially during World War II when he recorded it as introduced into Guam (Usinger 1946). Herring (1974) warned that there is great danger that *N. caledoniae* would be introduced into the southern United States and become very injurious in commercial flower-growing areas. It is listed in Bosik (1997) as the Caledonia seed bug.

In addition in Hawaii, *N. kinbergi* Usinger (listed as *N. coenosulus* Stål) also attacked vanda orchids (Beardsley 1977, 1979). A grower stated that he lost 80% of his crop of orchids. This is the species whose immature forms were described by Usinger (1942) under *N. coenosulus*. It breeds primarily on the weed *Erigeron canadensis* (Beardsley 1977) and lays its eggs in the composite heads (Usinger 1942). Ashlock (1967) observed that when feeding on its normal composite host plant, little or no damage occurred, but on the orchid the damage was severe, indicating that the saliva of the bug was very toxic to the plant. *Nysius coenosulus* (under the name *N. nigriscutellus* Usinger) is an occasional pest of beets and potatoes (Holdaway 1944), which it invades from its wild introduced hosts *Amaranthus* sp., *Chenopodium* sp., and *Portulaca oleracea* L. (Beardsley 1977). *Nysius nemorivagus* White is listed as damaging Chinese cabbage and potatoes (Holdaway 1944). In addition, an undetermined, evidently introduced species near the Australian *N. vinitor* [*Note:* Ashlock had cautioned Beardsley that there were several species under the name vinitor] was discovered in the lowlands of Oahu feeding on the weeds *A. spinosus*, *A. retroflexus*, *Euphorbia hirta*, and *P. oleracea*. It, too, was found damaging vanda orchid flowers (Beardsley 1977). This confusion of names by the leading specialists about these Hawaiian species shows the difficulty of the taxonomy of *Nysius*!

4.45 Nysius spp. (Orsillidae)

There are many references in the literature to *Nysius* sp., which is very unfortunate, as one does not know what bug is being studied and usually no mention is made of voucher specimens. This attitude is shown by Bosik (1997), who refers to *Nysius* spp. as seed bugs when this should be the name for the family (now superfamily) Lygaeidae (Sweet 1960). This shows how important systematic research is and why it should be supported even by those who are chiefly interested in controlling the insect regardless of its name. Some papers contain important biological information, some of potential use in controlling the insect, but not knowing the insect, it is useless to review this literature, which includes studies in both the so-called developed and the undeveloped worlds. One recent reference, however, may be mentioned here. In Kuwait a *Nysius* sp. was reported in swarms alighting on the bare hands, legs, or other exposed areas of the human body and inflicting painful bites and sucking blood (Al-Houty 1990). What is this insect? Is it an already described species, perhaps an economic one?

4.46 *Xyonysius californicus* (Stål) (Orsillidae)

This species resembles a large *Nysius* but has a distinctive forewing margin strigil (Ashlock and Lattin 1963). Although it is not a pest of cotton in California (U.S.A.), it has been reported in Mexico as damaging the fruiting forms, the squares, and bolls of cotton (Rude 1950). Stone and Fries (1986) found *X. californicus* to be very numerous in Texas (U.S.A.) on guayule (*Parthenium argentatum* Gray) and is probably a potentially serious pest. Guayule is being cultivated as an alternative source of rubber. Slater and Baranowski (1990) mention that this bug is common on both wild and cultivated plants.

4.47 *Xyonysius major* Berg (Orsillidae)

Schaefer (1998a) recorded this species as being a pest on sunflowers in Londrina, in the state of Paraná, Brazil. Schaefer observed that this species from Brazil is not *X. californicus* and raised *X. major* from synonymy with *X. californicus*. He noted that this is probably the *Nysius* reported as damaging sunflowers in Mato Grosso state (Boiça et al. 1984) and elsewhere in Brazil (Zucchi et al. 1993). Rodriguez and Eberhard (1994) described the mating behavior of *Xyonysius*.

4.48 *Rhypodes (= Hudsona) anceps* (White) (Orsillidae)

Of the many species of *Rhypodes* on New Zealand (Eyles 1990), only *R. anceps* (White) is of minor economic importance in New Zealand. It feeds on grasses and occasionally attacks wheat. It has been implicated in causing runny dough (Morrison 1938), but it has a much more minor role than *N. huttoni* (see *N. huttoni*). The species is unusual in that it is entirely brachypterous and, compared with other *Rhypodes,* it has conical, laterally pointed posterior pronotal corners, and a linear abdomen.

4.49 *Orsillus depressus* Dallas (Orsillidae)

This unusual flattened orsillid is a pest of conifers in Europe and feeds on the seeds in the cones of cypress (*Cupressus* spp.), pine (*Pinus* spp.), juniper (*Juniperus communis* L. and *J. excelsa* Bieberstein), and the Lawson or Port Orford cypress, *Chamaecyparis lawsoniana* (A. Murr.) Parl., the latter a common ornamental species introduced from western North America (Stichel 1957, Slater 1964a, Dupuis 1965, Aukema 1988). It is spreading north from southern Europe (Aukema 1988) and has recently been introduced into Britain on Lawson cypress (Hawkins 1989). *Orsillus maculatus* Fieb. has a similar range and habits as *O. depressus* and lives in cypress cones, arbor vitae or white cedar (*Thuja* spp.), and junipers (*Juniperus* spp.), feeding on the seeds (Putschkov 1958, Kerzhner and Yaczewski 1964). *Orsillus maculatus* lacks the dark medial line on the pronotum found in *O. depressus*, and the labium is very long, attaining the end of the abdomen. It seems surprising that *Orsillus* is not in North America and the genus should be considered as having a high potential for accidental introductions.

4.50 *Belonochilus numenius* (Say) (Orsillidae)

Heidemann (1902, 1911) discovered this insect to oviposit on, hide inside of, and feed on the seed balls of sycamore or American plane tree (*Platanus occidentalis* L.). He described the eggs and nymphs of this insect. Ashlock (1967) found that the insect in all stages feeds on the seeds of sycamore and he could rear the insects from egg to adult. *Belonochilus numenius* is unusual among the Orsillidae of North America in being an arboreal species. Wheeler (1984) listed this species as a pest on the London plane tree, *Platanus* × *acerfolia* (Ait.) Willd., a tree much used as an ornamental in cities, as it is not only beautiful with its multicolored bark, but it also withstands air pollution well. However, the natural habitat of the native sycamore is along riversides. Wheeler described the insect's life cycle, immature stages, and feeding habits. He found the species at least in the more southern parts of its range to be multivoline, but to survive the winter as eggs in the seed balls. The bug extends south to Florida (U.S.A.) and Mexico (Slater 1964a, Slater and Baranowski 1990). It is common in central Texas (U.S.A.), where the nymphs feed on the overwintered seed balls in the spring after hatching from the eggs that have overwintered in the seed balls on the tree (M. H. Sweet, personal observations). The extremely long labium of this insect is clearly an adaptation for feeding on the sycamore seed balls. A good common name for this insect would be the sycamore seed bug. It is also found on the western plane trees or sycamores *P. racemosa* Nutt. and *P. wrightii* S. Watts (Van Duzee 1914), and as *B. numenius* is known from Mexico by the synonym *B. mexicanus* Distant it is probably on the Mexican sycamore *P. mexicana* Moric.

As relatively few lygaeoids are pests in dicotyledonous trees, it should be noted that in Europe some species of the genus *Arocatus* (Lygaeidae *sensu stricto*) are adapted to arboreal life. *Arocatus melanocephalus* feeds on elms (*Ulmus* spp.), *A. roeseli* Schilling on alder (*Alnus* spp.), and *A. longiceps* Stål on maples (*Acer* spp.) (Kerzhner and Yaczewski 1964). The genus *Ficus* also has a number of lygaeoids especially the Heterogastridae and Rhyparochromidae feeding on figs, an activity that may be of economic significance (Slater 1972b).

4.51   *Kleidocerys resedae* (Panzer) (Ischnorhychidae)

Wheeler (1975, 1976) named this insect the birch catkin bug. He listed it as a nuisance pest species because it sometimes invaded homes in the autumn in large numbers (Wheeler 1975). It has evidently a Holarctic distribution, although the other species in the genus especially in the Palearctic have much more restricted distributions (Slater 1964a). However, *K. geminates* (Say) was described from Indiana and Kansas (U.S.A.), and, following Barber (1953), it is given subspecific status in both Slater (1964a) and A. Slater and Ashlock (1988), However, the distributions of these subspecies very broadly overlap in northern North America, *K. resedae resedae* to the north, and *K. r. geminatus* to the south, so their systematic status should be reviewed, especially as *K. resedae* was originally described from Germany.

The species was recorded from 43 species of plants of 14 families (Wheeler 1976a,b), and many of these records are of breeding populations. In Germany, it is known as the "Birkenwanzen" because it lives on birch and alder trees (*Betula* and *Alnus* spp.), feeding and breeding on the dry seed catkins (Jordan 1933, Wachmann 1989). The insects feed and reproduce on the overswintered dry cattails (*Typha* spp.) (Claassen 1921). On cultivated hybrid spirea (*Spirea vanhouttei* Zabel) Blatchley (1926) found hundreds of adults and nymphs. Slater and Baranowski (1990) found *K. resedae* to breed on morning glory and blackberries (*Rubus allegheniensis* Porter ex Bailey). With such a disparate pattern of host plants Slater and Baranowski suggested, "this species merits careful investigation to see if there might be a complex of sibling species involved." This possibility is enhanced by the genus having a wing–abdomen stridulatory apparatus, and the different English species including *K. resedae* have distinctive calls (Leston 1957). Heinrich in a letter to Slater and Baranowski discovered a population in Maine in which hundreds of adults stridulated in unison when disturbed. Melber et al. (1980) in Germany in a study of the social behavior of the acanthostomatid *Elasmucha grisea* (L.) found *K. resedae* eating the eggs of this arboreal stink bug, which shows again the facultative predation so common in phytophagous Heteroptera. In Europe, Kinzelbach (1970) described a new Strepsiptera *Loania canadensis* from *K. resedae*. It became the basis for the new subfamily Loaniinae of the Callipharixenidae.

4.52   *Malcus japonicus* Ishihara and Hasegawa (Malcidae)

The genus *Malcus* is the only genus in the subfamily Malcinae, the other subfamily being the Chauliopinae, which contains many serious pests of legumes as discussed earlier. There are 19 species and very little is known about the species (Štys 1967). *Malcus japonicus* is known in Japan as the mulberry bug and feeds on the leaves of *Morus bombycis* Koidz. The insect is listed as a pest species in Japan (Hasegawa 1980). The genus is evidently not specific to *Morus,* as in India *Malcus scutellatus* Distant (= *M. flavidipes* Stål) were found breeding on cucurbits in the Coimbatore District of South India (Mohanasundaram 1972).

4.53 *Phaenacantha australiae* Kirk (Colobathristidae)

As discussed earlier, this small interesting family properly belongs in the Lygaeoidea, and was originally included in Lygaeidae by Stål. Štys (1966b) used this species to study the morphology of the Colobathristidae. He argued that the family was in the malcid not the coreoid or largid lineages. In the context of the Lygaeoidea, it is then interesting that the colobathristids like the malcids are parenchyma and plant sap feeders and do not feed on seeds. Unfortunately, nothing is known about the biology of the primitive Oriental genus *Dyakiella* (Štys 1966b). The species are highly elongated insects, with slender antennae and legs and all species, where known, feed on monocots; the Old World species are all known from grasses to which the slender form probably is procryptic. It is predicted that *Dyakiella* will prove to be a grass feeder (M. H. Sweet, personal comment). There are several species of *Phaenacantha* that are reported as minor pests, damaging sugarcane. Illingworth (1920) studied in tropical northeastern Queensland *P. australiae,* which is appropriately known as the linear bug. He determined that the insect fed on the leaves and stems of various grasses and became occasionally a pest of sugarcane, a plant that is not native to Australia. Myatt (1973) noted that an outbreak of *P. australiae* on young sugarcane in 1972 was a consequence of unusual weather conditions in which populations built up in the surrounding grasslands and later dry conditions caused the bugs to migrate in numbers into sugarcane fields. Normally, he noted, the natural enemies in the grasslands keep this species in check. Illingworth (1920) figured and described the immature forms, and Malipatil and Kumar (1975) added to his descriptions, describing the egg and the nymphs in more detail and illustrating the fifth instar. The egg is spindle-shaped with the ventral side flattened for adhesion to the plant. There are 7 to 10 apical chorionic processes in a circle on the widened anterior end. The width of the egg is 0.43 mm, the length, 1.58 mm. There are three abdominal scent glands in the nymphs between segments 3 to 4, 4 to 5, and 5 to 6. The nymphs are pale, slender, and delicate. The lengths of the instars are first, 1.4; second, 3.4; third, 4.7; fourth, 6.0; fifth, 7.1 mm.

Tan and Johnson (1974) reported that *P. saccharicida* Karsch in Malaya caused the yellowing and subsequent necrosis of cane leaves caused by the bugs' feeding on the partially opened spindle leaves in both planted and ratooned cane. A heavy infestation not only caused a reduction in active photosynthetic tissue, but also caused a growth restriction in the plant. In 1970, this species was the principal pest in the sugar plantations in Malaysia, and the infestation, while worst on the margins next to uncultivated areas, occurred throughout the plantation. Tan and Johnson (1974) noted that during wet weather the insects were attacked by a white fungus that rapidly reduced the number of the bugs. They also found a natural predator in the fields, *Sycanus leucomesus* Walker, a reduviid, feeding on adults and nymphs of *P. saccharicida,* and suggested that the reduviids could be a good agent of biological control. As the colobathristid feeds directly on the stems and leaves of the sugarcane, they suggested that there is a possibility for viral transmission from wild grasses to sugarcane that would explain the yellowing and necrosis of the sugarcane tissues in reaction to the feeding by the bugs. They reported that good control of *P. saccharicida* was obtained with 0.07% endosulfan (Thiodan). Lim and Pan (1976) reported that *P. saccharicida* and *P. bicolor* (Distant) also attack maize and other graminaceous plants in Malaysia.

4.54 *Phaenacantha viridipennis* Horváth (Colobathristidae)

Zheng (1994) in his review of the heteropterans on bamboo, listed *P. viridipennis* and *P. bicolor* on bamboo, the former on *Dinochloa utilis* McClure, a species that is used for paper pulp in China.

Zheng was uncertain whether this species is specific for bamboo, noting that colobathristids commonly live on tall grasses as *Miscanthus sinensis* Andersson and *Saccharum arundinaceum* Retz. However, he considered a shift to bamboos to be entirely possible.

4.55 *Peruda brasiliana* Carvalho and Costa (Colobathristidae)

Couturier et al. (1996) in a study of the pests of the peach palm or pijuayo, *Bactris gasipaes* H. B. K. (Palmae), in the Peruvian Amazonia, found this colobathristid to be a minor pest on the peach palm. They determined experimentally that the bug fed on foliage of the peach palm and that the injury to the plant by the bug was greater in the presence of an eriophytid mite, *Retracrus*.

4.56 *Oxycarenus lavaterae* (F.) (Oxycarenidae)

This species of *Oxycarenus* is treated separately because it is not a treated as a pest on cotton in most of the literature, although Gentry (1965) lists it on cotton in Tunisia, Morocco, and Algeria. A good illustration of the species is in Wachmann (1989). It is a dark black species, 5 to 6 mm long, with reddish-brown corium and pale membrane. It is a Palearctic species that is a pest in southern Europe on small-leaved linden, *Tilia cordata* Mill. (= *parvifolia* Ehrh.), an ornamental tree much used along roads and in parks. The insect can develop dense, large populations on the host trees (Perini and Tamanini 1961, Ciampolini and Trematerra 1986–1987, Velimirovic et al. 1992, Gyorgy 1997). These immense population masses are often called swarming and the insect will sometimes densely aggregate on only a few plants. This is a quite different usage of swarming than in the well-known nuptial lekking swarms as in mayflies, caddisflies, enicocephalid Heteroptera (Wygodzinsky and Schmidt 1991), and some nematocerous Diptera, or in honeybee swarming. The mass of bugs often causes alarm, especially when they enter houses (Eritja et al. 1997, Goula et al. 1999).

Velimirovic et al. (1992) describe the natural history of *O. lavaterae* in Montenegro. The bug overwinters in colonies on the bark of the linden tree in crevices, cracks, and under the bark. In the spring the bugs mate and lay eggs in the cracks of bark on the trunk and branches of the linden tree. In dense colonies 10 to 25 cm long, the bugs, adults and nymphs, feed directly on the trunk and branches of the linden tree. On the tree in the colonies the bugs maintain a head-up position. Through the summer the insects move into the higher branches, but do not feed on the leaves. The larvae are dark reddish color and resemble the adults as they get older. Velimirovic et al. (1992) estimated that there are three to four overlapping generations a year. Heavy infestations of the tree cause fading of the leaves and early leaf loss. Although Perini and Tamanini (1961) reported the insects similarly in masses on linden in Italy, Ciampolini and Trematerra (1986–1987) report the insects to be on the trees to overwinter, and to develop on herbaceous plants such as *Malva sylvestris* and other Malvaceae, including *Hibiscus* spp. in nurseries. They report two generations a year, the second generation the one that overwinters. The adults damage stone fruits in midsummer. For control Ciampolini and Trematerra recommend clearing the wild Malvaceae from around orchards and overwintering areas.

Velimirovic et al. (1992) found phospomidone to provide adequate control on the insect, but had to use a forklift to reach the higher branches. In Spain, Eritja et al. (1997) tested a number of insecticides and found Trilomethrin at 22.7 mg/l yields an LT-50 of 30.6 min and was considered the most appropriate product.

4.57 *Leptodemus minutus* Jakovlev (Oxycarenidae)

Bergevin (1923) reported this species to suck human blood. Slater (1972) suspected that *L. minutus* was merely the piercing of skin either in an exploratory manner or for moisture as do a number of other phytophagous hemipterons. He found nothing in the behavior of a South African

Leptodemus (*L. irroratus* Slater) to suggest anything other than the normal seed-feeding habits of an oxycarenid. It fed on the seeds of a composite, *Matricaria* sp. However, Selim et al. (1990) reported skin lesions on people in Kuwait that they attributed to this species. The species *L. minutus* is of minor economic importance and was found feeding on olive, orange, and other plants in Libya (Gentry 1965). Gentry also recorded *Metopoplax ditomoides* (Costa) infesting the seedpods of the ornamental plant hollyhock *Althea rosea* (L.) Cav.

4.58 *Togo hemipterus* Scott (Rhyparochromidae)

Rice (*Oryza sativa* L.) is a very important crop in the Orient, and the Heteroptera "stink bugs" are serious pests that damage rice kernels, causing pecky rice (Tachikawa et al. 1976). About 90 species of stinkbugs invade the rice paddies, and half of them are considered pests causing yield reduction or pecky rice. Of these, about 15 species are serious pests in Japan (Kisimoto 1983, Tomokuni 1993). These stinkbugs include eight Pentatomidae, one Coreidae (*Cletus punctiger* Dallas), two Miridae, one Rhopalidae (*Aeschynteles maculatus* Fieber), one Alydidae *Leptocorisa chinensis* F., and two Lygaeoidea. One, *Nysius plebeius* Distant (not *plebejus* of the authors), was already discussed under the Orsillidae; the other is *Togo hemipterus,* a member of the Myodochini. Even though rice paddies are a temporary environment *T. hemipterus* is pterygopolymorphic. It is a mottled brown insect about 7 mm long (see illustration in Tomokuni 1993). The species and *Togo* itself, with two other scarce species of no known economic importance, *T. praetor* Bergroth and *T. victor* Bergroth, appear to be endemic to Japan (Slater 1964a). Other rhyparochromids that are sometimes pests in rice fields are *Pachybrachius luridus* Hahn, *P. flavipes* (Motschulsky), *Stigmatonotum sparsum* Lindberg, *S. rufipes* (Motschulsky), *Paromius exguus* (Distant), *P. gracilis* (= *P. pallidus* Montrouzier), *P. piratoides* (Costa), *P. jejunus* (Distant), *Pachybrachius pacificus* (Stål), *P. rusticus* (Scott), *P. lateralis* (Scott), *Elasmolomus pallens* Dallas, *Graptopeltus amurensis* (Lindberg), *G. albomaculatus* (Scott), and G. *angustatus* (Montandon). In addition, two Cymidae, *Cymus aurescens* Distant and *C. basicornis* Motshulsky, are listed (Tachikawa et al. 1976, Hasegawa 1980, Tomokuni 1993). This is a remarkable assemblage of lygaeoids attacking rice and may reflect that rice originated in Southeast Asia (Simpson and Conner-Ogorzaly 1995) with a correspondingly long history of host adaptation.

Togo hemipterus lays one to several eggs in the glumes of the seed spikelets of rice. The species is multivoltine and has two or more generations a year. Because *Togo* is largely brachypterous, it tends to be found along the margins and on rice hills of the rice paddies (Kisimoto 1983). When collecting in Japan, *Togo* was found in grassy fields away from rice, so it is probably a general feeder on grasses that invades rice fields (M. H. Sweet, personal observation).

When the danger of pecky rice is high, Kisimoto (1983) recommends using such carbamate compounds as BPMC (*o*-sec-butyl-phenyl-methylcarbamate) and MTMC (*m*-tolyl-methylcarbamate) that are effective against *Togo* and other lygaeoids.

4.59 *Horridipamara nietneri* Dohrn (Rhyparochromidae)

In India, the adults and nymphs of *H. nietneri* attack rice during the milky and soft-dough stages of the rice fields producing pecky rice (Gupta et al. 1990). The bug populations vary from three to five bugs per plant in the dry season and from five to nine bugs in the wet season.

4.60 *Paromius longulus* (Dallas) (Rhyparochromidae)

In Louisiana and Texas (U.S.A.), *P. longulus* has been recognized as a pest in rice fields causing pecky rice similar to the damage made by the rice stink bug, *Oebalus* (= *Solubea*) *pugnax* (Fabr.) (Douglas and Ingram 1942). It is an interesting contrast with the fauna of Japan in which there is a large contingent of rhyparochromids feeding on rice and the small lygaeoid fauna in

North America that invades rice fields. *Paromius longulus* is certainly an opportunistic species, as it has been reported to injure strawberries (Watson 1931, Fletcher 1939, Watson and Tissot 1942) and to feed on a variety of plants, as discussed in Sweet (1960). In fieldwork in the southern United States this species was found to be a general feeder on grasses, to which its elongate form is clearly procryptic, but it will feed on other seeds as opportunity arises (M. H. Sweet, personal observation).

4.61 *Cymoninus notabilis* (Distant) (Cymidae)

In southern Brazil, *Cymoninus* is recorded as feeding on rice panicles and may cause damage to the rice fields (Albuquerque 1990). This is surprising, because nearly all of the Cymidae feed on members of the Cyperaceae, not Gramineae. This may make it more relevant that two Cymidae, *Cymus aurescens* Distant and *C. basicornis* Motshulsky, are listed from rice in Japan (Tachikawa et al. 1976).

4.62 *Myodocha serripes* (Olivier) (Rhyparochromidae)

To the general entomologist this species makes a lasting impression because of its elongate neck and heavily spinous legs that give the impression that it is predaceous. The species is oligophagous on seeds, among them the achenes of the strawberry plant *Fragaria* spp. Strawberries are a favorite food not only for *Myodocha* (Bryson 1939, Neiswander 1944) but also for *Pachybrachius* (now *Pseudopachybrachius) vinctus (*Say) (Quaintance 1897, Watson 1931, Watson and Tissot 1942) and for *Pachybrachius* (now *Neopamera*) *bilobata* (Say). Both of these myodochine rhyparochromids are relatively polyphagous and feed on variety of seeds, *P. vinctus* more on grass seeds, and *P. bilobata* more on dicotyledenous seeds. These are all common species in the southern United States that live in relatively temporary environments, are entirely macropteros, disperse readily, often flying to lights in the summer, are bivoltine to multivoltine, and are closely related to species complexes of Neotropical rhyparochromids; in short, these are opportunistic *r*-selected species (Sweet 1964) that can be expected to be common in agroecosystems, often eating weed seeds as well. *Pachybrachius vinctus* reported on rice in Brazil (Albuquerque 1990) is then to be expected. Sweet (1964) described the ecology and life cycle of this species in New England (U.S.A.). Larivière and Larochelle (1991) extended its distribution in Canada and discussed its bionomics.

4.63 *Euander lacertosis* (Erichson) (Rhyparochromidae)

This is a member of the Southern Hemisphere tribe Udeocorini that dominates Australia. In this fauna this species is unusually opportunistic and occurs in a variety of habitats, temporary or permanent, and feeds on a great variety of seeds (Slater 1976b). In Australia it is known as the strawberry bug for very large populations can develop in cultivated strawberry fields. When working in Tasmania, it was possible to collect some of these insects and study their feeding habitats. As with the North American strawberry bugs, the primary focus of their feeding was on the achenes on the outside of the fruits; very little feeding occurred on the strawberry fruit itself. The owner of the strawberry farm acknowledged that the number of bugs was disconcerting, but because he could see no direct harm being done to the strawberries, he preferred not to spray the strawberries with insecticides as he wished not to contaminate the strawberries. However, he asked that the bugs be collected when his customers were not around, so as not to call attention to the astonishing numbers of bugs coexisting with his strawberries (M. H. Sweet, personal observations). Because Tasmania has a moist climate, the bugs probably did not need to feed on the strawberries for water. Several other lygaeoids were in the strawberries, but in much lower abundance. Worldwide, bugs are found associated with strawberries; the strawberry achenes must be a rich food. In southern

Sweden, for example, four lygaeoids are reported from commercial strawberry fields (Gertsson 1979, 1980). It is predicted that wherever strawberries are grown, the local lygaeoids will be found feeding on them, but the damage will be minimal, because of the feeding on the superficial achenes (M. H. Sweet, personal comment).

4.64 Conifer Cone Rhyparochromidae

Already discussed were the cone bugs of the genus *Orsillus* (see Orsillidae), which are restricted to the Palearctic region. An interesting assemblage of rhyparochromids, mainly of the tribe Drymini, feed on the seeds of coniferous trees, and as this affects the available seed stocks, these have been treated as pest insects (Yates 1986). On the ground, feeding on fallen seeds in eastern North America, are *Eremocoris ferus* (Say) and *E. borealis* (Dallas) (Sweet 1964). When there is a good seed year the populations can become very large, and both species are recorded as invading homes and being nuisance pests (Wheeler 1989). In the coniferous forests of western North America there is an interesting fauna of 12 species of *Eremocoris* from Canada to the highlands of Mexico. As far as is known from fieldwork and laboratory studies in western North America, these species all feed on the fallen seeds of conifers (M. H. Sweet, personal observations). Many of the Palaearctic *Eremocoris* probably also feed on conifer seeds. Indeed, Linnaeus named the type of genus *Eremocoris, Cimex abietis; abietis* is derived from the old Latin name for spruce and fir, and *Abies* is now the generic name for fir.

The next set includes the species that live on trees and are often found living inside the cones, and are therefore flattened insects. These are mostly species of the genus *Gastrodes*, some of which occur in the Palearctic and the others in the Nearctic (Usinger 1938). *Gastrodes abietum* Berg. is found in the cones of common spruce, *Picea excelsa* Link (= *P. abies* [L.] Karst). *Gastrodes grossipes* DeGeer lives in pine cones, usually Scots pine (*Pinus sylvestris* L.) (Putschkov 1956, Kerzhner and Yaczewski 1964). In addition, there is one species of *Eremocoris, E. depressus* Barber, which is flattened for living inside pine cones (Slater and Baranowski 1990, Slater et al. 1993). Wheeler (1996) collected the insect from four pine species, *Pinus palustris* Mill., *P. pungens* Lamb., *P. rigida* Mill. and *P. taeda* L. An unusual Myodochini, *Slaterobius quadristriatus* Barber, although not flattened, feeds up in the tree on the seeds of the jack pine *P. banksiana* Lamb. (Slater et al. 1993).

5. SPECIES OF MINOR BENEFIT

There is a tendency to divide the world into the beneficial and the malevolent. From the ecological perspective bugs are part of the wonderful tapestry we call nature, always interesting, never boring. Because most lygaeoid bugs are feeders on small seeds, especially the many species that live in great numbers in temporary habitats, these insects help to destroy many seeds that we would call weeds. In this larger picture of population regulation of natural populations of plants, the Lygaeoidea are perhaps not a minor but a major benefit to humans as well as to nature. Palmer (1986) considered introducing *Ochrimnus mimulus* Stål into Australia from Texas (U.S.A.) to control the pernicious woody composite *Baccharis* spp. because *O. mimulus* appeared virtually host specific with a life cycle adapted to *Baccharis*, but he found a few individuals on other plants and decided the danger of a host transfer in Australia would be a great risk. This seems a correct assessment (M. H. Sweet, personal comment).

On the other hand (Bennett 1966, 1975) investigated, using rhyparochromid bugs for the control of several introduced floating aquatic plants, three *Lipostemmata* species (*L. humeralis* Berg, *L. major* Ashlock, and *L. scutellatus* Ashlock) on the water fern *Salvinia* and a *Valtissius* sp. on the araceous plant water lettuce (*Pistia stratiotes* L.).

Already discussed under *Blissus* is that while some valuable grasses are attacked, chinch bugs also attack some of the weedy grasses in fields and in lawns, effecting some weed control. Also noted is that the feeding punctures of *Macropes obnubilus* (Distant) enhance the commercial value of bamboo in Japan.

6. PROGNOSIS

There is little doubt that as our numbers increase and we dominate more and more the living space of our planet, and convert more and more land from natural communities to the unnatural communities we call agroecosystems, we are providing new opportunities for our friendly and not-so-friendly competitors, the insects, to exploit such systems. Some lygaeoid taxa such as the Blissidae and the Orsillidae are ecological bombs, waiting to explode with outbreaks of new pests as the opportunity may arise. We need to be aware of this potential problem.

7. CONCLUSIONS

Very few of the lygaeoids have been studied in sufficient detail to produce the ecological knowledge necessary to control the pest insects to levels below the economic threshold, or even to determine what that threshold might be. It is a paradox that knowledge of the charismatic true Lygaeidae (*Oncopeltus* and its allies) has far outstripped knowledge about the economic taxa. Despite all the fine words about IPM, there is a distressing tendency to reach for the spray can before calmly considering alternative strategies for managing insect populations. It is perhaps easier to develop new insecticides (Casida and Quistad 1998) than to avoid using them. Just because there are insects at the same dinner table does not necessarily mean that they constitute a dangerous pest population. It takes knowledge to determine this, and knowledge is usually not adequate to prevent the use of pesticides.

The wretched state of taxonomic knowledge about the economic species is impressive. Nearly all the economic genera — *Blissus, Cavelerius, Nysius, Oxycarenus, Spilostethus,* and *Geocoris*, as well — are large, complex sets of taxa, for which understanding of the species is rudimentary. Without good systematic understanding, ecological understanding is bound to falter, and be unable to provide the tools with which to manage or live in harmony with these friendly competitors.

8. ACKNOWLEDGMENTS

First, the author wants to thank all those who suggested references, sent references, and in general encouraged this work. Those wonderful folks at the library, who so patiently searched for many nearly inaccessible articles, are thanked a thousand times over. Finally, thanks to the patient editors for having faith that the chapters would get done.

9. REFERENCES CITED

Abu-Mensah, K., and R. Kumar. 1977. Ecology of *Oxycarenus* species (Heteroptera: Lygaeidae) in southern Ghana. Biol. J. Linn. Soc. 9: 349–377.

Abushama, F. T., and A. A. Ahmed. 1976. Food-plant preference and defence mechanism in the lygaeid bug *Spilostethus pandurus* (Scop.). Z. Angew. Entomol. 80: 206–213.

Adell, J. C., A. Moya, and L. M. Botella. 1988. Larval arrest in the development of the lygaeid bug *Spilostethus pandurus* under competition for food and space. Evol. Biol. 2: 251–260.

Aggarwal, R. A., and C. B. Tiwari. 1967. Black bug a summer enemy of sugar cane and its control. Prog. Farming (India) 3: 5.

Ahmad, T. 1946. Short notes and exhibits. Indian J. Entomol. 7: 240–241.
Ahmad, T. R. 1982. Comparative biology and economics of the chinch bug, *Blissus leucopterus leucopterus* (Say) (Hemiptera: Lygaeidae) on wild and cultivated grasses. Diss. Abstr. Int. (B) 43: 37.
Ahmad, T. R., S. D. Kindler, and K. P. Pruess. 1984a. Recovery of 2 sorghum varieties from sublethal infestations of chinch bug *Blissus leucopterus leucopterus* (Say) (Hemiptera: Lygaeidae) J. Econ. Entomol. 77: 151–152.
Ahmad, T. R., K. P. Pruess, and S. D. Kindler. 1984b. Non-crop grasses for the chinch bug *Blissus leucopterus leucopterus* (Hemiptera: Lygaeidae). J. Kansas Entomol. Soc. 57: 17–20.
Ahuja, D. B. 1989. Seasonal occurrence of insect pests in *Sesamum indicum* Linn. J. Entomol. Res. 13: 116–120.
Albuquerque, G. S. 1990. Ocorrencia de *Cymoninus notabilis* (Distant) e *Pseudopachybrachius vincta* (Say) (Hemiptera: Lygaeidae) na cultura do arroz no Rio Grande do Sul. An. Soc. Entomol. Brasil 20: 217–218.
Aldrich, J. M. 1915. Results of twenty-five years collecting in the Tachinidae, with notes on some common species. Ann. Entomol. Soc. Amer. 8: 79–84.
Aldrich, J. R., W. S. Leal, R. Nishida, A. P. Khrimian, C. J. Lee, and Ê. Sakuratai. 1997. Semiochemistry of aposematic seed bugs. Entomol. Exp. Appl. 84: 127–135.
Al-Houty, W. 1990. *Nysius* (Hem., Lygaeidae) sucking human blood in Kuwait. Entomol. Mon. Mag. 126: 95–96.
Allsopp, P. G. 1984. The effect of day length on the rate of development, longevity and fecundity of the Rutherglen bug *Nysius visitor* Bergroth (Hemiptera Lygaeidae), Pp. 27–34. *in* P. Bailey and D. Swinger [eds.], Proceedings, Fourth Austr. Appl. Entomol. Res. Conf., Adelaide, 24–28 Sept. 1984. Pest Control: Recent Advances and Future Prospects. South Australia Government Printer, Adelaide, Australia.
1988. Spatial distribution and sequential sampling of *Nysius* spp. (Hemiptera: Lygaeidae) on sunflowers. Austral. J. Exp. Agric. 28: 279–282.
Amuh, I. K. A. 1972. Survey of cotton pests in southern Ghana. Ghana J. Sci. 12: 29–40.
Ananthakrishnan, T. N., K. Raman, and K. P. Sanjayan. 1982. Comparative growth rate, fecundity and behavioral diversity of the dusky cotton bug, *Oxycarenus hyalinipennis* Costa (Hemiptera; Lygaeidae) on certain malvaceous host plants. Proc. Indian Nat. Sci. Acad. 48: 577–584.
Anderson, D. B. 1991. Seed bugs in trophic webs; interactions with resources, competitors and enemies. III. Host choice and its consequences in the parasitic fly *Leucostoma crassa*. Vaxtskyddsrapporter. Avhandlingar. 22: 1–15.
Anderson, D. B., and C. Solbreck. 1991. Seed bugs in trophic webs; interactions with resources, competitors and enemies. IV. Who eats whom? Complex interactions between an egg parasitoid and its cannibalistic host. Vaxtskyddsrapporter. Avhandlingar. 22: 16–22.
1992. Who eats whom? Complex interactions between an egg parasitoid and its cannibalistic host. Oikos 63: 459–464.
Annecke, D. P., and V. G. Moran. 1982. Insects and Mites of Cultivated Plants in South Africa. Butterworths, Durban, South Africa.
Archer, T. L., J. C. ves Losada, and E. D. Bynum, Jr. 1990. Influence of planting date on abundance of panicle-feeding insects associated with sorghum. J. Agric. Entomol. 7: 233–239.
Arnaud, P. H., Jr. 1978. A host-parasite catalog of North American Tachinidae (Diptera). U.S. Dept. Agric. Misc. Publ. No. 1319. 860 pp.
1967. A generic classification of the Orsillinae of the World (Hemiptera-Heteroptera: Lygaeidae. Univ. California Publ. Entomol. 48: 1–82.
Ashlock, P. D. 1967. New records and name changes of North American Lygaeidae (Hemiptera: Heteroptera: Lygaeidae). Proc. Entomol. Soc. Washington 79: 575–582.
Ashlock, P. D., and J. D. Lattin. 1963. Stridulatory mechanisms in the Lygaeidae, with a new American genus of Orsillinae (Hemiptera: Lygaeidae). Ann. Entomol. Soc. Amer. 56: 693–703.
Ashlock, P. D., and A. Slater. 1988. Family Lygaeidae. The seed bugs and chinch bugs. Pp. 167–245 *in* T. J. Henry and R. C. Froeschner [eds.] Catalog of the Heteroptera, or True Bugs, of Canada and the Continental United States. E. J. Brill, New York, New York, U.S.A.
Atkins, I. M., M. E. McDaniel, and J. H. Gardenhire. 1969. Growing oats in Texas. Texas Agric. Exp. Stn. Bull. 1091: 3–28.

Attia, F. I. 1973. *Alophora lepidofera* a native parasite of the Rutherglen bug, *Nysius visitor* and the grey cluster bug, *Nysius clevelandensis* (Hemiptera: Lygaeidae) in Australia. J. Austral. Entomol. Soc. 12: 353–354.
1974. Laboratory evaluation of insecticides against *Nysius vinitor* Bergroth and *Nysius clevelandensis* Evans (Hemiptera: Lygaeidae). J. Austral. Entomol. Soc. 13: 161–164.
1982. Comparative studies of the biology of *Nysius visitor* Bergroth and *N. clevelandensis* Evans (Hemiptera: Lygaeidae). Gen. Appl. Entomol. 14: 15–20.
Attia, F. I., and M. Elshafie. 1974. A technique for culturing rutherglen bug *Nysius vinitor* Bergroth (Hemiptera: Lygaeidae). J. Entomol. Soc. Australia 8: 37.
Atwal, A. S., 1965. Pests of crops and their control. Punjab Agric. Univ. Ludhiana: 8–9.
Atwal, A. S. and S. Singh. 1971 (1972). Influence of different levels of temperature and relative humidity on the speed of development, survival and fecundity of *Macropes excavatus* Distant (Hemiptera: Lygaeidae). Indian J. Entomol. 33: 166–171.
Aukema, B. 1988. *Orsillus depressus* nieuw voor Nederland en Belgi (Heteroptera: Lygaeidae). Entomol. Ber. (Amsterdam) 48: 181–183.
Awan, M. S., and M. B. Qureshi. 1996. Study of biological parameters of dusky cotton bug, *Oxycarenus loetus* [sic] Kirby (Lygaeidae: Heteroptera) in Sindh. Proc. Pakistan Congr. Zool. 16: 265–268.
Azuma, S. 1977. Biological studies on the sugar cane insect pests in Okinawa, with a special reference to the change of their composition and infestation in relation to the interaction of new commercial sugar cane varieties. Bull. Coll. Agric. Univ. Ryukyus 24: 1–158 [in Japanese, with English summary].
Backus, E. A. 1988. Sensory systems and behaviors which mediate hemipteran plant-feeding: a taxonomic overview. J. Insect Physiol. 34: 151–165.
Backus, E. A., W. B. Hunter, and C. N. Arne. 1988. Technique for staining leafhopper salivary sheaths and eggs within unsectioned plant tissue. J. Econ. Emtomol. 81: 1819–1823.
Bailey, L. H. 1949. Manual of Cultivated Plants Most Commonly Grown in the Continental United States and Canada. Macmillan, New York, New York, U.S.A. 1116 pp.
Bailey, L. H., and E. Z. Bailey. 1976. Hortus Third. A Concise Dictionary of Plants Cultivated in the United States and Canada. Macmillan, New York, New York, U.S.A.
Bains, S. S., and Z. S. Dhaliwal. 1983. Outbreak of the sugarcane black bug *Cavelerius excavatus* (Hemiptera Heteroptera Lygaeidae) on ratoon crop of sugarcane in the Punjab India. Indian J. Ecol. 10: 161–163.
Baker, C. F. 1906. Notes on the *Nysius* and *Ortholomus* of America. Invertebr. Pac. 1: 133–140.
Baker, P. B., and R. H. Ratcliffe. 1977 (1978). Evaluation of bluegrasses for tolerance to *Blissus leucopterus hirtus* (Hemiptera: Lygaeidae). J. New York Entomol. Soc. 85: 165–166.
Baker, P. B., R. H. Ratcliffe, and A. L. Steinhauer. 1981a. Tolerance to hairy chinch bug feeding in Kentucky blue grass. Environ. Entomol. 10: 153–157.
1981b. Laboratory rearing of the hairy chinch bug. Environ. Entomol. 10: 226–229.
Baranowski, R. M. 1979. Notes on the biology of *Ischnodemus oblongus* and *Ischnodemus fulvipes* with descriptions of the immature stages (Hemiptera: Heteroptera: Lygaeidae). Ann. Entomol. Soc. Amer. 72: 655–658.
Barber, G. W. 1937. Number of short-winged chinch bugs produced under laboratory conditions. J. Econ. Entomol. 30: 802–804.
Barber, H. G. 1918. A new species of *Leptoglossus*: a new *Blissus* & varieties. Bull. Brooklyn Entomol. Soc. 13: 35–39
1947. Revision of the genus *Nysius*. Washington Acad. Sci. 37: 354–366.
1953. A revision of the genus *Kleidocerys* Stephens in the United States (Heteroptera: Lygaeidae). Proc. Entomol. Soc. Washington 55: 273–283.
Barbosa, A. J. S. 1950. Estudo da biologia dos percevejos da semente do algodão e dos prejuízos causados pelos mesmos. Rev. Fac. Cien. Univ. Lisboa 1: 117–132.
Barker, G. M., R. P. Pottinger, P. J. Addison, and R. A. Prestidge. 1984. Effect of *Lolium* endophyte fungus infections and behavior of adult Argentine stem weevil. New Zealand J. Agric. Res. 27: 271–277.
Barnes, M. M. 1970. Genesis of a pest: *Nysius raphanus* (Hemiptera: Lygaeidae) and *Sisymbrium irio* in vineyards. J. Econ. Entomol. 63: 1462–1463.
Bauernfeind, R. J. 1987. Residual effectiveness of insecticides for the control of chinch bugs (Heteroptera: Lygaeidae) in sorghum. J. Kansas Entomol. Soc. 60: 336–339.

Baxendale, F. P. 1992. Insects and related pests of turfgrass. Pp. 16–35 *in* F. P. Baxendale and R. E. Gaussoin [eds.]. Integrated Management Guide for Nebraska Turfgrass. University of Nebraska Publ. EC 92–1557-S.

Beardsley, J. W., Jr. 1965. *Nysius caledoniae* Distant. *In* Notes and exhibitions. Proc. Hawaiian Entomol. Soc. 19: 14.

1977. The *Nysius* seed bugs of Haleakala National Park, Maui (Hemiptera: Lygaeidae: Orsillinae). Proc. Hawaiian Entomol. Soc. 22: 443–450.

1979. Notes on two *Nysius* species accidentally introduced into Hawaii (Hemiptera: Lygaeidae: Orsillinae). Proc. Hawaiian Entomol. Soc. 23: 51–54.

Bedford, H. W. 1923. The pests of cotton in the Anglo-Egyptian Sudan. Bull. Wellcome Trop. Res. Lab. Entomol. Sect. 19: 1–38.

Beetle, A. A. 1983. Las Gramíneas de México. Tomo I. Secretaria de Agricultura y Recursos Hidraulicos, Cotecoca, Mexico.

Beirne, B. P. 1972. Pest insects of annual crop plants in Canada. IV. Hemiptera-Homoptera. Mem. Entomol. Soc. Canada 85: 1–37.

Bennett, F. D. 1966. Investigations on the insects attacking the aquatic ferns. *Salvinia* spp. in Trinidad and Northern South America. Proc. South. Weed Conf. 19: 497–500.

1975. Insect and plant pathogens for the control of *Salvinia* and *Pistia*, Pp. 28–35 *in* Proceedings: Symposium on Water Quality Management through Biological Control. Dept. Environ. Eng. Sci. Univ. Florida. 164 pp.

Bergevin, E. 1923. A propos de quelques nouveaux hemipteres piqueurs. Bull. Soc. Hist. Nat. Afr. N. 14: 27–28.

Berliner, E. 1931. 'Leimkleber Weizen' ist 'Wanzenweizen'! Mühenlab 1: 25–26.

Bhardwaj, S. P., and O. P. Bhalla. 1975 (1977). Laboratory observations on the biology of the bean bug *Chauliops nigrescens* (Hemiptera: Lygaeidae). Indian J. Entomol. 37: 408–410.

Bhat, A. A. 1988. *Chauliops* sp. (Hemiptera: Lygaeidae) as a serious pest of beans. *Phaseolus vulgaris* in Kashmir. Geobios New Rep. 7: 84–85.

Bhatnagar, P. L. 1962. Certain factors influencing the establishment of *Oxycarenus hyalinipennis* Costa (Heteroptera: Lygaeidae) on various plants. Part 1. Ingestion of food. Indian J. Entomol. 25: 48–62.

Bhattacherjee, N. S. 1959. Studies on *Lygaeus pandurus* Scopoli (Heteroptera: Lygaeidae). 1. Bionomics, descriptions of the various stages, biology and control. Indian J. Entomol. 21: 259–272.

Billings, F. H., and P. A. Glenn. 1911. Results of the artificial use of the white-fungus disease in Kansas. U.S. Dept. Agric. Bur. Entomol. Bull. 107. 58 pp.

Blatchley, W. S. 1926. Heteroptera or True Bugs of Eastern North America, with Especial Reference to the Faunas of Indiana and Florida. Nature Publ. Co., Indianapolis, Indiana, U.S.A. 1116 pp.

Böcher, J. 1972. Feeding biology of *Nysius groenlandicus* (Zett.) (Heteroptera: Lygaeidae) in Greenland, with a note on oviposition in relatioin to food-source and dispersal of the species. Medd. Groenl. 191: 1–41.

1975. Notes on the reproductive biology and egg-diapause in *Nysius groenlandicus* (Zett.) (Heteroptera: Lygaeidae). Vidensk. Medd. Dan. Naturhist. Foren. København 138: 21–38.

1976. Population studies on *Nysius groenlandicus* (Zett.) (Heteroptera: Lygaeidae) in Greenland with particular reference to climatic factors, especially the snow cover. Vidensk. Medd. Dan. Naturhist. Foren. København 139: 61–89.

Boiça, A. L., Jr., A. C. Bolonhezi, and J. P. Neto. 1984. Levantamento de insetos-praga e seus inimigos naturais em girassol (*Helianthus annuus* L.), cultivado em primeira e segunda época, no município de Selvíria-MS. An. Soc. Entomol. Brasil 13: 189–196.

Boldt, P. E. 1989. *Baccharis* (Asteraceae) a review of its taxonomy, phytochemistry, ecology, economic status, natural enemies, and the potential for its biological control in the United States. Texas Agric. Exp. Stn. Misc. Pub. 1674. College Station, Texas, U.S.A. 32 pp.

Boselli, F. B. 1932. Studio biologico degli emitteri che attaccano le noccinole in Sicilia. Boll. Lab. Zool. Portici 26: 142–309.

Bosik, J. J. (chairman) 1997. Common Names of Insects and Related Organisms. Committee on Common Names of Insects. Entomological Society of America, Lanham, Maryland, U.S.A. 232 pp.

Box, H. E. 1953. List of Sugar-Cane Insects. Commonwealth Institute of Entomology, London, U.K.

Brain, C. K. 1929. Insect Pests and Their Control in South Africa. Nasionale Pers Beperk, Cape Town, South Africa. 468 pp.

Brar, R. S., and P. N. Avasthy. 1988. A new sampling technique for estimating population of sugarcane black bug. Indian J. Ecol. 15: 109–110.

Broadley, R. H. 1978. Insect pests of sunflower. Queensland Agric. J. 104: 4, 307–314.

1980. Pesticides, bees, and sunflowers. Queensland Agric. J. 106: 3, 255–258.

Broadley, R. H., and D. A. Ironside. 1980. Insect pests of sunflower. Part 1. Queensland Agric. J. 106: 5, xxv–xxviii.

Broadley, R. H., and P. D. Rossiter, 1980. Rutherglen bug control in sunflowers. Queensland Agric. J. 102: 408–410.

1982. Incidence of *Nysius* spp. (Hemiptera: Lygaeidae) in south Queensland sunflowers. Gen. Appl. Entomol. 14: 69–71

Broadley, R. H., B. W. Simpson, and C. H. S. Beavis. 1986. Damage by *Nysius* spp. (Hemiptera: Lygaeidae) in non-stressed sunflower *Helianthus annuus* L. crops. Gen. Appl. Entomol. 18: 17–24.

Broodryk, S. W., and G. A. Matthews 1994. *Dysdercus* and other Heteroptera. Pp. 267–284 *in* G. A. Matthews and J. P. Turnstall [eds.], Insect Pests on Cotton. CAB International, Oxon, U.K.

Brooks, H. L., and R. A. Higgins. 1985. Sorghum insect management for 1984. Kansas State University MF-742. 8 pp.

Brown, L. 1979. Grasses: An Identification Guide. Houghton Mifflin, Boston, Massachusetts, U.S.A.

Bryson, H. R. 1939. U.S.D.A. Bur. Entomol. Plant Q. Insect Pest Surv. Bull. 19: 153.

Bruner, S. C., L. C. Scaramuzza, and A. R. Ortero. 1945. Catalogo de los insectos que atacan alas plantas economicas de Cuba. Bol. Estac. Agron. Cuba N. 63. 246 pp.

Burgess, L., and D. L. McKenzie. 1991. Role of the insect *Nysius niger* (Hemiptera: Lygaeidae), and flixweed, *Descurainia sophia*, in infection of Saskatchewan mustard crops with a yeast, *Nematospora sinecauda*. Plant Dis. Surv. 7: 37–41.

Burgess, L., and H. H. Weegar. 1986. A method for rearing *Nysius ericae* the false chinch bug (Hemiptera: Lygaeidae). Canad. Entomol. 1998: 1059–1062.

Burgess, L., J. Dueck, and D. L. McKenzie. 1983. Insect vectors of the yeast *Nematospora coryli* in mustard *Brassica junea* crops in southern Saskatchewan, Canada. Canad. Entomol. 115: 25–30.

Burks, B. D. 1934. Food plants of the chinch bug. J. Econ. Entomol. 28: 1100–1101.

Burton, B. D., J. A. Reinert, and R. W. Toler. 1979. Effects of the southern chinch bug (*Blissus insularis*) the St. Augustine decline strain of Panicum mosaic virus on seventeen accessions and two cultivars of St. Augustine grass. Phytopathology 69: 525–526.

1983. Combined resistance in St. Augustine grass *Stenotaphrum secundatum* to the southern chinch bug *Blissus insularis* (Hemiptera: Lygaeidae) and the St. Augustine decline strain of *Panicum* mosaic virus. Plant. Dis. 67: 171–172.

Burton, M., and R. E. Hutchins. 1958. Insect damage of lawns checked. Mississippi Farm Res. 21: 1–7.

Busey, P. 1990a. Inheritance of host adaptation in the southern chinch bug (Hemiptera: Lygaeidae). Ann. Entomol. Soc. Amer. 83: 563–567.

1990b. Polyploid *Stenotaphrum* D. L. germplasm resistance to the polyploid damaging population southern chinch bug (Hemiptera: Lygaeidae). Crop Sci. 30: 588–593.

Busey, P., and P. J. Center. 1987. Southern chinch bug (Hemiptera: Lygaeidae) overcomes resistance in St. Augustine grass. J. Econ. Entomol. 80: 608–611.

Busey, P., and E. I. Zaenker. 1992. Resistance bioassay from southern chinch bug (Heteroptera: Lygaeidae) excreta. J. Econ. Emtomol. 85: 2032–2038.

Butani, D. K. 1971 (1972). Some new insects associated with cotton (*Gossypium hirsutum*) in northern hirsutum-arboreum region. Indian J. Entomol. 33: 227–228.

Byers, G. W. 1973. A mating aggregation of *Nysius raphanus* (Hemiptera: Lygaeidae). J. Kansas Entomol. Soc. 46: 281–282.

Cancela da Fonseca, J. P. 1956. Aspectos fitossanitários do amendoim (mancarra) armazenado na Guiné Portuguesa. II. Montes e benténs no campo. III. Armazéna de entidades comerciais. Garcia Orta 6: 605–621.

Carnegie, A. J. M. 1974. Insects of sugarcane in South Africa. Entomol. Mem. Depart. Agric. Tech. Serv. Republ. South Africa 39: 1–14.

Carriere, Y., A. Bouchard, S. Bourassa, and J. Brodeur. 1998. Effect of endophyte incidence in perennial ryegrass on distribution, host-choice, and performance of the hairy chinch bag (Hemiptera: Lygaeidae). J. Econ. Entomol. 91: 324–328.

Carrillo, S. J. L. 1967. A mechanism for egg dispersal in *Nysius tenellus* Barber (Hemiptera: Lygaeidae). Pan-Pac. Entomol. 43: 80.

Carrillo, S. J. L., and L. E. Caltagirone. 1970. Observations on the biology of *Solierella peckhami*, *S. blasidelli* (Sphecidae) and two species of Chrysididae (Hymenoptera). Ann. Entomol. Soc. Amer. 63: 672–681.

Carrillo, S. J. L., A. Ortega C., and W. M. Gibson. 1966. Lista de insectos en la coleccion entomologica del Instituto Nacional de Investigaciones Agricolas. Instituto Nacional de Investigaciones Agricolas, S. A. G. Mexico. Foll. Misc. No. 14. 133 pp.

Carter, R. R., and R. L. Duble. 1976. Variety evaluations in St. Augustine grass for resistance to the southern lawn chinch bug, PR-3374. Pp. 16–17 *in* Turfgrass Research in Texas, Consolidated PR-3364-3376. Texas Agricultural Experiment Station, College Station, Texas.

Casida, J. E., and G. B. Quistad. 1998. Golden age of insecticide research: past, present, or future? Annu. Rev. Entomol. 43: 1–16.

Chadwick, C. E., and M. I. Nikitin. 1985. Records of parasitism by members of the family Tachinidae (Diptera: Tachinidae). Austral. Zool. 21: 587–598.

Chambliss, C. E. 1895. The chinch bug, *Blissus leucopterus* (Say) (Hemiptera: Lygaeidae). Tennessee Agric. Exp. Stn. Bull. 8: 41–55.

Chatterjee, N. C. 1937. Entomological investigations on the spike disease of sandal. Lygaeidae (Hemipt.) Indian Forest Rec. (n.s.) 3: 105–121.

Chaudhary, J. P., S. R. Yadav, and J. N. Gupta 1984. Chemical control of sugarcane black bug, *Cavelarius excavatus* (Dist.) (Hemiptera; Lygaeidae). Indian J. Entomol. 46: 442–446.

Cheany, R. L., and P. R. Jennings. 1975. Field problems of rice in Latin America. Series GE-15. Centro Internacional de Agricultura Tropical (CIAT), Cali, Colombia.

Chen, Y. 1984. Bamboo and Its Cultivation. Chinese Forestry Press, Beijing, China [in Chinese, *fide* Zheng 1994].

Cheng, L. I., and H. G Tsou. 1982. A preliminary study on Chinese *Dimorphopterus* Stål (Hemiptera: Lygaeidae: Blissinae). Act. Entomol. Sin. 25: 423–430 [in Chinese with English summary].

Chinajariyawong, A., G. H. Walter, and V. E. Harris. 1989. Pest status of *Nysius clevelandensis* Evans and *Nysius visitor* Bergroth (Hemiptera: Lygaeidae) on cotton. J. Austral. Entomol. Soc. 28: 287–290.

Choban, R. G. and A. P. Gupta. 1972. Meiosis and early embryology of *Blissus leucopterus hirtus* Montandon (Heteroptera: Lygaeidae). Int. J. Insect Morph. Embryol. 1: 301–314.

Choe, K. R., and C. Jang. 1992. Communities analysis of superfamily Lygaeoidea (Hemiptera) in Mt. Kyeryongsan. Korean J. Entomol. 22: 175–185 [in Korean, with English abstract].

Chopra, N. P. 1970. Sugarcane black bug — a nomenclatorial correction. Indian J. Entomol. 32: 394–395.

1971 (1972). Milkweed bug — a nomenclatorial correction. Indian J. Entomol. 33: 359–360.

1989. Sugarcane black bugs (Hemiptera: Lygaeidae: Blissinae). Agric. Sci. Dig. 9: 45–46.

Chopra, N. P., and R. S. Kashyap. 1973. Relative toxicity of different insecticides against sugarcane black bug. Pesticides 1973: 21–22.

Chopra, N. P., and K. B. Rustagi. 1982. The subfamily Chauliopinae of India and Sri Lanka (Hemiptera: Malcidae). Orient. Insects 16: 19–28.

Chopra, N. P., and S. K. Singal. 1982. Studies on the genus *Elasmolomus* Stål (Hemiptera: Lygaeidae: Rhyparochrominae) of India. Bull. Entomol. 23: 16–23.

Chopra, N. P., and S. R. Yadav. 1974. Observations on seasonal history and food preference of *Spilostethus macilentus* Stal (Hemiptera: Lygaeidae). Indian J. Entomol. 36: 361–362.

Chu, Y. I. 1995. Wing polymorphism in insects and its strategy for survival — specially referred to the polymorphism of *Cavelerius saccharivorus* (Okajima) (Lygaeidae: Heteroptera). Plant Prot. Bull. (Taichung) 37: 1–14.

Ciampolini, H., and P. Trematerra. 1986–1987. Biological studies on *Oxycarenus lavaterae* (F.) (Rhynchota Heteroptera Lygaeidae). Boll. Zool. Agric. Bachic. 19: 187–197 [in Italian].

CIAT (Centro Internacional de Agricultura Tropical). 1982. Descripción y daño de los insectos que atacan al arroz en América Latina. Centro Internacional de Agricultura Tropical, Cali, Colombia.

Claassen, P. W. 1921. *Typha* insects: their ecological relationships. Cornell Univ. Agric. Exp. Stn. Mem. 47: 463–509.

Clancy, C. A., and H. D. Pierce. 1966. Natural enemies of some *Lygus* bugs. J. Econ. Entomol. 59: 853–858.

Conway, J. A. 1976. The significance of *Elasmolomus sordidus* (Hemiptera: Lygaeidae) attacking harvested ground nuts in the Gambia. Trop. Sci. 18: 187–190.

Coppel, H. C., and J. W. Mertins. 1977. Biological Insect Pest Suppression. Springer-Verlag, Berlin, Germany. 314 pp.

Corby, H. D. L. 1947. *Aphanus* (Hem. Lygaeidae) in stored groundnuts. Bull. Entomol. Res. 37: 609–617.

Core, E. L. 1955. Plant Taxonomy. Prentice-Hall, Englewood Cliffs, New Jersey, U.S.A.

Corrêa, R. R. 1956. Barnsleyia amarali g. n. e sp. n de cleradino hematófago (Hemiptera-Heteroptera: Lygaeidae). Rev. Brasil. Malarial. Doenças Trop. 8: 117–125.

Correll, D. S., and H. B. Correll. 1975. Aquatic and Wetland Plants of the Southwestern United States. Stanford University Press, Stanford, California, U.S.A.

Costa Lima, A. da. 1940. Insectos do Brasil II. Hemipteros: pp. 104–106. Escola Nacional de Ins. Rio de Janeiro, Brazil.

Cotterell, G. S. 1952. The insects associated export produce in southern Nigeria. Bull. Entomol. Res. 43: 145–152.

Couturier, G., E. Tanchiva, H. Inga, J. Vásquez, and R. R. Riva. 1996. Notas sobrae los artrópodos que viven en el pijuayo (*Bactris gasipaes* H. B. K.: Palmae) en la Amazonía peruana. Rev. Peruana Entomol. 39: 135–142.

Cranshaw, W. S., B. C. Kondratieff, and T. R. Qian. 1990. Insects associated with quinoa, *Chenopodium quinoa*, in Colorado. J. Kansas Entomol. Soc. 63: 195–199.

Cressey, P. J. 1987. Wheat-bug damage in New Zealand wheats: some properties of a glutenin hydrolysing enzyme in bug-damaged wheat. J. Sci. Food Agric. 41: 159–165.

Cressey, P. J., J. A. K. Farrell, and M. W. Stufkens. 1987. Identification of an insect species causing bug damage in New Zealand wheats. New Zealand J. Agric. Res. 30: 209–212.

Crocker, R. L., and C. L. Simpson. 1981. Pesticide screening test for the southern chinch bug *Blissus insularis* (Hemiptera: Lygaeidae). J. Econ. Entomol. 74: 730–731.

Crocker, R. L., R. W. Toler, and C. L. Simpson. 1982. Bioassay of St. Augustine grass lines for resistance to southern chinch bug (Hemiptera: Lygaeidae) and to St. Augustine decline virus. J. Econ. Entomol. 75: 515–516.

Crocker, R. L., R. W. Toler, J. B. Beard, M. C. Engelke, and J. S. Kubica-Breier. 1989. St. Augustine grass antibiosis to southern chinch bug (Hemiptera: Lygaeidae) and to St. Augustine decline strain of *Panicum* mosaic virus. J. Econ. Entomol. 82: 1729–1732.

Cronquist, A. 1968. The Evolution and Classification of Flowering Plants. Houghton Mifflin Company, Boston, Massachusetts, U.S.A. 396 pp.

Crosskey, R. W. 1973. A conspectus of the Tachinidae (Diptera) of Australia, including keys to the supraspecific taxa and taxonomic and host catalogues. Bull. Br. Mus. Nat. Hist. (Entomol.) Suppl. 21: 1–221.

Dahms, R. G., and F. A. Fenton. 1940. The effect of fertilizers on chinch bug (Hemiptera: Lygaeidae) resistance in sorghum. J. Econ. Entomol. 33: 688–692.

Dahms, R. G., and T. H. Johnson. 1954. Reaction of barley varieties to chinch bug (Hemiptera: Lygaeidae). Oklahoma Agric. Exp. Stn. Manuscript Rep. Abstr. 10. 4 pp.

Dahms, R. G., and J. B. Sielinger. 1954. Reaction of sorghum varieties to the chinch bug (Hemiptera: Lygaeidae). J. Econ. Entomol. 47: 536–537.

Daniels, N. E. 1969. Chemical control of a species of the false chinch bug group on wheat and sugar beets. Texas Agric. Exp. Stn. Prog. Rpt. 2667. 2 pp.

David, B. C., T. J. Riley, and M. S. Smith. 1991. Chinch bug control in sorghum with liquid and granular soil insecticides, 1989. Ins. Acar. Tests 16: 216.

Davis, F. M., W. P. Williams, and J. Van den Berg. 1996. Screening maize for resistance to chinch bug (Heteroptera: Lygaeidae) under greenhouse conditions. J. Econ. Entomol. 89: 1318–1324.

Davis, M. G. K., and D. R. Smitley. 1990. Association of thatch with populations of hairy chinch bug (Hemiptera: Lygaeidae) in turf. J. Econ. Entomol. 83: 2370–2374.

Decker, G. C., and F. Andre. 1937. Winter mortality of chinch bugs (Hemiptera-Lygaeidae) in Iowa. J. Econ. Entomol. 30: 928–934.

1938. Biological notes on *Blissus iowensis* Andre. (Hemiptera-Lygaeidae). Ann. Entomol. Soc. Amer. 31: 457–466.

Deol, G. S. 1985. New record of *Nysius ericae* (Schilling) (Lygaeidae: Hemiptera) on pearl millet and efficacy of some insecticides for its control. Indian J. Plant Prot. 13: 51–52.

Dhamdhere, S. V., and R. R. Rawat. 1969. Preliminary studies on the biology of *Graptostethus* sp. (Lygaeidae, Hemiptera) feeding on stubbles of sannhemp. J. Bombay Nat. Hist. Soc. 66: 217–221.

Dhiman, S. C., and K. S. Krishna. 1989. Occurrence of *Oxycarenus laetus* Kirby in the western districts of U.P. (India) and records of its new host plants. Bull. Entomol. 30: 126–129.

Dhingra, S., and S. Verma. 1984. Relative toxicity of insecticides to the adults of *Oxycarenus laetus* Kiraby, a pest of okra. J. Entomol. Res. 8: 111–113.

Dimetry, N. 1971. Studies on the host preference of the cotton seed bug, *Oxycarenus hyalinipennis* (Costa) (Hemiptera: Lygaeidae). Z. Angew. Entomol. 68: 63–67.

1974. Contributions to the biology of the cotton seed bug, *Oxycarenus hyalinipennis* (Costa). Bull. Soc. Entomol. Egypt 57: 193–199.

Distant, W. L. 1901. Rhynchotal notes. XI. Heteroptera: Fam. Lygaeidae. Ann. Mag. Nat. Hist 7 (8): 464–486.

1904. Fauna of British India, including Ceylon & Burma. Order Rhynchota. Vol. II. (Heteroptera): 1–506. Taylor and Francis, London, U.K.

Doebley, J. F. 1990. Molecular evidence of gene flow among *Zea* species. Bioscience 40: 443–448.

Doebley, J. F., and H. H. Iltis. 1980. Taxonomy of *Zea* (Gramineae). I. A subgeneric classification with key to taxa. Amer. J. Bot. 67: 982–993.

Dolling, W. R. 1991. The Hemiptera. Oxford University Press, Oxford, U.K. 274 pp.

Douglas, W. A., and J. W. Ingram. 1942. Rice-field insects. U.S. Dept. Agric. Circ. No. 632. 32 pp.

Drake, C. J. 1940. Dos nuevas especies del genero *Blissus* Klug. (Hemiptera: Lygaeidae) de la Argentina. Notas Mus. La Plata 5: 223–226.

1951. New American chinch bugs (Hemiptera: Lygaeidae). J. Washington Acad. Sci. 41: 319–323.

Drake, C. J., and N. T. Davis. 1958. The morphology and systematics of the Piesmatidae (Hemiptera), with keys to the world genera and American species. Ann. Entomol. Soc. Amer. 51: 567–581.

Drake, C. J., G. C. Decker, and A. D. Worthington. 1934. The chinch bug in Iowa. Iowa Agric. Exp. Circ. 199. 16 pp.

Duhra, M. S., and J. S. Sandhu. 1983a. Evaluation of insecticide application methods for the control of sugarcane black bug. Maharashtra Sugar 8: 47–48.

1983b (1985). Evaluation of insecticide application methods for the control of sugarcane black bug. Pp. 31–35 in Proceedings, 47th Annual Convention of the Sugar Technologists' Association of India. 1983. Sugar Technologists' Association of India, Kanpur, India.

Duhra, M. S., and D. P. Singh. 1992. Chemical control of sugarcane black bug? *Cavelerius excavatus*. Coop. Sugar 23: 391–393.

Dunbar, D. M. 1971. Big-eyed bugs in Connecticut lawns. Connecticut Agric. Exp. Stn. Circ. No. 244. 6 pp.

Dupuis, C. 1963. Essai monographique sur les Phasiinae (Diptères Tachinaires parasites d'Hétéroptères). Mém. Mus. Natl. Hist. Nat. Sér. A, Zool. 26: 1–146.

1965 (1966). Étude de l'oligophagie de trois punaises des Genévriers et revue des plantes-hôtes des Hétéroptères voisins. Cah. Nat. Bull. Natl. Par. (n.s.) 21: 105–122.

Eden, W. C., and R. L. Self. 1960. Controlling chinch bugs on St. Augustine grass lawns. Auburn Agric. Exp. Stn. Prog. Rpt. 79. 3 pp.

Ellefson, C. L., T. L. Stephens, and D. Welsh. 1992. Xeriscape Gardening: Water Conservation for the American Landscape. Macmillan, New York, New York, U.S.A.

Elshafie, M. 1976. *Nysius vinitor* Berg. (Hemiptera: Lygaeidae) infesting pig weed, *Portulaca oleracea* L. J. Entomol. Soc. Australia 9: 54.

1977. Cultural methods for *Nysius vinitor* Bergroth (Hemiptera: Lygaeidae). Circ. Entomol. Soc. Australia No. 280: 44.

El Shazly, M. M. 1993. Quantitative evaluation of food intake and assimilation by *Spilostethus pandurus* (Scopoli) (Heteroptera: lygaeidae). Bull. Entomol. Soc. Egypt 71: 109–117.

1995. Effect of temperature on development and population growth of *Spilostethus pandurus* (Scopoli) (Hemiptera: Lygaeidae) in Giza, Egypt. Insect Sci. Appl. 16: 17–25.

El Sherif, H. A., and M. M. El-Shazly. 1993. Effect of different foods on certain biological aspects of *Spilostethus pandurus* (Scopoli) (Heteroptera, Lygaeidae). Bull. Entomol. Soc. Egypt 71: 61–73.

Emmett, R. W., G. A Buchanan, and P. A. Magarey. 1992. Grapevine diseases and pest management. Austral. New Zealand Wine Ind. J. 7: 149–171.

Ene, J. C., and A. Hamid. 1982. *Dimorphopterus similis* (Hemiptera: Lygaeidae: Blissinae) a new rice pest from Nigeria. Entomol. Mon. Mag. 118: 7–8.

Eritja, R., C. Aranda, M. Goula, and M. Espinosa. 1997. Laboratory tests of pyrethroid and organophosphate insecticides on *Oxycarenus lavaterae* (Heteroptera: Lygaeidae). J. Econ. Entomol. 90: 1508–1513.

Ershad, D., and M. Barkhordary. 1974a. Host range and vectors of *Nematospora coryli* Peglion in Kerman of Iran. Iranian J. Plant Pathol. 10: 34–39; 85–91 [in Persian, English].

1974b. A new method for isolation of *Nematospora coryli* Peglion from vector bugs. Entomol. Phytopathol. Appl. 37: 22–23; 4–5 [in Persian, English].

Esaki, T. 1926. Verzeichnis der Hemiptera-Heteroptera der Insel Formosa. Ann. Hist. Nat. Mus. Hungary 24: 136–189.

Essig, E. O. 1929. Insects of Western North America. Macmillan, New York, New York, U.S.A. 1035 pp.

Evans, J. W. 1929. A new species of *Nysius* (Hem. Lygaeidae) from Australia. Bull. Entomol. Res. 19: 351–354.

1936. A new species of *Nysius* from Tasmania, and notes on the economic importance of the genus. Bull. Entomol. Res. 27: 673–676.

1938. The Invermay bug. Tasmanian J. Agric. 9: 196–198.

Every, D. 1992. Relationship of bread baking quality to levels of visible wheat-bug damage and insect proteinase activity in wheat. J. Cereal Sci. 16: 183–193.

1993. Purification and characterization of a glutenin hydrolysing proteinase from wheat damaged by the New Zealand wheat bug, *Nysius huttoni*. J. Cereal Sci. 18: 239–250.

Every, D., J. A. Farrell, and M. W. Stufkens, 1989. Effect of *Nysius huttoni* on the protein and baking properties of two New Zealand wheat cultivars. New Zealand J. Crop Hortic. Sci. 17: 55–60.

1990. Wheat-bug damage in New Zealand wheats: the feeding mechanism of *Nysius huttoni* (Hemiptera: Lygaeidae) and its effect on the morphological and physiological development of wheat. J. Sci. Food Agric. 50: 297–309.

1992. Bug damage in New Zealand wheat grain: the role of various heteropterous insects. New Zealand J. Crop Hortic. Sci. 20: 305–312.

Every, D., Farrell, J. A., M. W. Stufkens, and A. R. Wallace. 1998. Wheat cultivar susceptibility to grain damage by the New Zealand wheat bug, *Nysius huttoni*, and cultivar susceptibility to the effects of bug proteinase on baking quality. J. Cereal Sci. 27: 37–46.

Ewete, F. K. 1984. The immature stages of *Oxycarenus gossypinus* [sic] (Heteroptera: Lygaeidae). Rev. Zool. Afr. 98: 596–605.

Ewete, F. K., and E. O. Osisanya. 1985. Effect of various diets seeds on development of the cotton seed bug *Oxycarenus gossypinus* [sic] Distant (Heteroptera: Lygaeidae). Insect Sci. Appl. 6: 543–546.

1988. The reactions of *Oxycarenus gossypinus* [sic] Distant (Heteroptera: Lygaeidae) to temperature and relative humidity. Insect Sci. Appl. 9: 31–36.

Eyles, A. C. 1960. Variation in the adult and immature stages of *Nysius huttoni* White (Heteroptera: Lygaeidae) with a note on the validity of the genus *Brachynysius* Usinger. Trans. R. Entomol. Soc. London 112: 53–72.

1963a. Incubation period and nymphal development in *Nysius huttoni* White (Heteroptera: Lygaeidae: Orsillinae). New Zealand J. Sci. 6: 446–461.

1963b. Fecundity and oviposition rhythms in *Nysius huttoni* White (Heteroptera: Lygaeidae). New Zealand J. Sci. 6: 186–207.

1965a. Damage to cultivated Cruciferae by *Nysius huttoni* White (Het. Lyg.). New Zealand J. Agric. Res. 8: 363–366.

1965b. Notes on the ecology of *Nysius huttoni* White (Hemiptera: Lygaeidae). New Zealand J. Sci. 8: 494–502.

1973. Monograph of *Dieuches* Dohrn (Heteroptera: Lygaeidae). Otago Daily Times Ltd., Dunedin, New Zealand. 465 pp.

1974. Terrestrial bugs. New Zealand's Nat. Heritage 3: (34) 953–956.

1990. A review and revision of *Rhypodes* Stål (Hemiptera: Lygaeidae). New Zealand J. Zool. 17: 347–418.

Eyles, A. C. and P. D. Ashlock. 1969. The genus *Nysius* in New Zealand (Heteroptera: Lygaeidae). New Zealand J. Sci. 12: 713–727.

Ezueh, M. L. 1978. Introductory Economic Entomology for Nigeria. Ethiope Publishing Co., Benin City, Nigeria.

Farrell, J. A., and M. W. Stufkens. 1993. Phenology, diapause, and overwintering of the wheat bug, *Nysius huttoni* (Hemiptera: Lygaeidae) in Canterbury, New Zealand. New Zealand J. Crop Hortic. Sci. 21: 123–131.

Farrier, M. H. 1958. Summary of insect conditions 1957: North Carolina. U.S. Dept. Agric. Coop. Econ. Insect Rep. 8: 240–245.
Feir, D. 1974. *Oncopeltus fasciatus:* a research animal. Annu. Rev. Entomol. 19: 81–91.
Feng, G. H., and G. H. Liu. 1993. A study on the spatial distribution pattern and sampling method of *Cavelerius saccharivorus* (Okajima). Entomol. Knowledge 30: 150–153 [in Chinese with English abstract].
Ferguson, C. M. 1994. *Nysius huttoni* White (Heteroptera: Lygaeidae) a pest of direct drilled brassicas. Pp. 196–197 *in* Proceedings, 47th New Zealand Plant Protection Conference, Waitangi Hotel, New Zealand, 9–11 Aug., 1994. New Zealand Plant Protection Society, Rotorua, New Zealand.
Fermanian, T. W., M. C. Shurtleff, R. Randell, H. T. Wilkinson, and P. L. Nixon, 1997. Controlling Turfgrass Pests, second edition. Prentice-Hall, Upper Saddle River, New Jersey, U.S.A. 655 pp.
Fernald, M. L. 1950. Gray's Manual of Botany, eighth edition. American Book Company, New York, New York, U.S.A.
Ferris, G. F. 1920. Insects of economic importance in the Cape region of Lower California, Mexico. J. Econ. Entomol. 13: 463–467.
Fitch, A. 1856. Chinch bug. Trans. New York State Agric. Soc. 15: 277–297.
Fletcher, R. K. 1939. U.S.D.A. Bur. Entomol. Plant Quar. Insect Pest Surv. Bull. 19: 85.
Fletcher, T. B. 1914. Some South Indian insects and other animals of importance considered especially from an economic point of view. Government Press, Madras, India. xxii + 546 pp.
1919. Reports and proceedings, 3rd Entomological Meeting, Pusa, India.
Flint, W. P. 1919. Burn the chinch bug. Univ. Illinois Exten. Circ. No. 28. Urbana, Illinois, U.S.A. 7 pp.
1935. The chinch bug. J. Econ. Entomol. 28: 333–341.
Flint, W. P., and W. V. Balduf. 1924. Calcium cyanide for chinch-bug control. Univ. Illinois Agric. Exp. Stn. Bull. 249: 71–84.
Flint, W. P., and W. H. Larrimer. 1926. The chinch bug and how to fight it. U.S. Dept. Agric. Farmers' Bull. No. 1498. 16 pp.
Flint, W. P., G. H. Dugan, and J. H. Bigger. 1935. Fighting the chinch bug on Illinois farms. Univ. Illinois Coll. Agric. Circ. 431. 16 pp.
Forbes, S. A. 1883. Studies on the chinch bug (*Blissus leucopterus* Say). Illinois State Entomol. Rep. No. 16: 57 pp.
1890. Contribution to an economic bibliography of the chinch bug. 1785–1888. 16th Rep. Nox. Ben. Ins. State Illinois: 71–127.
1895. On the contagious disease in the chinch bug. (*Blissus leucopterus,* Say)). 19th Rep. Nox. Ben. Ins. State Illinois: 16–176.
Forbes, S. A. 1916. The chinch-bug outbreak of 1910–1915. 29th Rep. Nox. Benefic. Ins. State Illinois: 71–127.
Forrester, N. W. 1980. Insect pests of sunflowers. Agric. Gaz. New South Wales 91: 34–37.
Forrester, N. W., and H. Saini. 1982. Effect of moisture stress on damage to sunflowers by Rutherglen bug (*Nysius vinitor*). Pp. 130–132 *in* Proceedings, 10th International Sunflower Conference. Australian Sunflower Association, Toowoomba, Australia.
Frappa, C. 1931. Notes biologiques sur quelques insectes nouveaux ou peu connus et nuisibles aux plantes cultivées à Madagascar. Bull. Soc. Entomol. France 186–192.
French, C., Jr., 1918. The rutherglen bug (*Nysius vinitor*). A destructive pest to potatoes, tomatoes, peaches, etc. J. Dept. Agric. Victoria 16: 199–200.
French, J. C. 1964. Chinch bugs. Univ. Georgia Coll. Agric. Leafl. No. 20: 2 pp.
Froggatt, W. W. 1901. Notes on Australia Hemiptera. Agric. Gaz. New South Wales 12: 1592–1599.
Fujisaki, K. 1985. Ecological significance of the wing polymorphism of the oriental chinch bug *Cavelerius saccharivorus* Okajima (Hemiptera: Lygaeidae). Res. Popul. Ecol. 17: 125–136.
1986a. Reproductive properties of the Oriental chinch bug *Cavelerius saccharivorus* Okajima (Hemiptera: Lygaeidae) in relation to its wing polymorphism. Res. Popul. Ecol. 28: 42–52.
1986b. Genetic variation of density responses in relation to wing polymorphism in the Oriental chinch bug *Cavelerius saccharivorus* Okajima (Hemiptera: Lygaeidae). Res. Popul. Ecol. 28: 219–230.
1989a. Morphometric traits of adults of the oriental chinch bug *Cavelerius saccharivorus* Okajima (Hemiptera: Lygaeidae) in relation to its wing polymorphism. Jpn. J. Appl. Entomol. Zool. 24: 20–28.
1989b. Wing form determination and sensitivity of stages to environmental factors in the Oriental chinch bug, *Cavelerius saccharivorus* Okajima (Hemiptera: Lygaeidae). Jpn. J. Appl. Entomol. Zool. 24: 287–294.

1992. A male fitness advantage to wing reduction in the Oriental chinch bug, *Cavelerius saccharivorus* Okajima (Hemiptera: Lygaeidae). Res. Popul. Ecol. 34: 173–183.

1993a. Genetic correlation of wing polymorphism between females and males in the Oriental chinch bug, *Cavelerius saccharivorus* Okajima (Heteroptera: Lygaeidae). Res. Popul. Ecol. 35: 317–324.

1993b. Wing reduction in the autumn generation of the Oriental chinch bug *Cavelerius saccharivorus* Okajima (Heteroptera: Lygaeidae). Jpn. Appl. Entomol. Zool. 28: 112–115.

1993c. Reproduction and egg diapause of the Oriental chinch bug, *Cavelerius saccharivorus* Okajima (Heteroptera: Lygaeidae), in the subtropical winter season in relation to its wing polymorphism. Res. Popul. Ecol. 35: 171–181.

Fujita, K. 1977. Wing form composition in the field population of the two species of lygaeid bugs, *Dimorphopterus pallipes* and *D. japonicus*, and its relation to environmental conditions. Jpn. J. Ecol. 27: 263–267 [in Japanese with English summary].

Furness, G. O. 1976. The dispersal, age-structure and natural enemies of the long-tailed mealybug, *Pseudococcus longispinus* (Targioni-Tozzetti), in relation to sampling and control. Austral. J. Zool. 24: 237–247.

Gallagher, R. T., A. C. Campbell, A. D. Hawks, P. T. Holland, D. A. McGaveston, and E. A. Pansier. 1982. Ryegrass staggers: the presence of Lolitrem neurotoxins in perennial ryegrass seed. New Zealand Vet. J. 30: 183–184.

Gamberale, G., and B. S. Tullberg. 1996. Evidence for a more effective signal in aggregated aposematic prey. Anim. Behav. 52: 597–601.

Gandhi, J. R. 1975. Nutritive value, absorbability, and assimilation of food ingested by *Oxycarenus hyalinipennis* (Hemiptera: Lygaeidae). Indian J. Entomol. 37: 415–417.

1977. Water relations in heteropteran *Oxycarenus hyalinipennis* Costa and homopteran *Oxyrhachis tarandus* Fabr. Indian J. Entomol. 39: 65–69.

1978. Utilization of plant food in heteropteran *Oxycarenus hyalinipennis* Costa and homopteran *Oxyrhachis tarandus* Fabr. Indian J. Entomol. 39: 107–113.

1979a. Utilization of dietary constituents in heteropteran *Oxycarenus hyalinipennis* and homopteran *Oxyrhachis tarandus*. Indian J. Entomol. 41: 13–17.

1979b. Acceptance of different extracts and purified chemicals by adults of *Oxycarenus hyalinipennis* (Hemiptera: Lygaeidae) and *Dysdercus fasciatus* (Hemiptera: Pyrrhocoridae). Indian J. Entomol. 41: 155–159.

1979c. Growth and longevity of *Oxycarenus hyalinipennis* (Hemiptera: Lygaeidae) on different seeds. Indian J. Entomol. 41: 184.

1980a. Quantitative intake of different sugars by *Oxycarenus hyalinipennis* Costa. Indian J. Entomol. 44: 303–304.

1980b. Fate of fructose in heteropteran *Oxycarenus hyalinipennis* and homopteran *Oxyrhachis tarandus*. Indian J. Entomol. 42: 723–727.

1981a. Factors governing utilization of food in hompteran *Oxyrhachis tarandus* Fabr. and heteropteran *Oxycarenus hyalinipennis* Costa II. Rate of conduction, digestion and absorption of sucrose. Indian J. Entomol. 43: 17–28.

1981b. Factors governing the utilization of food in homopteran *Oxyrhachis tarandus* Fabr. and heteropteran *Oxycarenus hyalinipennis* Costa. III. Concentration of invertase enzyme digestion. Indian J. Entomol. 43: 35–38.

1981c. Factors governing utilization of food in homopteran *Oxyrhachis tarandus* Fabr. and heteropteran *Oxycarenus hyalinipennis* Costa. IV. Metabolism of absorbed sugars. Indian J. Entomol. 43: 49–58.

1988a. Orientation and feeding activity of heteropteran *Oxycarenus hyalinipennis* Costa on different seeds. Indian J. Entomol. 50: 383–385.

1988b. Fate of some amino acids in heteropteran *Oxycarenus hyalinipennis* Costa. Indian J. Entomol. 50: 390–395.

1989 (1990). Degree of ingestion of water by heteropteran *Oxycarenus hyalinipennis* Costa and homopteran *Oxyrhachis tarandus* Fabr. Indian J. Entomol. 51: 472–473.

Garcera, M. D., and M. Tamarelle. 1995. Gastrin-cholecystokinin-like immunoreactivity in the central nervous system of the milkweed bug, *Spilostethus pandurus*. Ultrastructural aspects of the reactive A cells of the brain. Biol. Cell. 84: 205–213.

Garcia, M. B. 1974. Ecological study of the entomofauna in para grass *Panicum purpurascens* and pangola grass *Digitaria decumbens* in la Chontalpa, Tabasco, Mexico. Folia Entomol. Mex. 29: 68–70.

Garg, D. O., and J. P. Chaudhary. 1979. Insect pests of sugarcane in Punjab and their control. I. Sucking pests. Indian Sugar 28: 697–701.

Gellatley, J. G., and N. W. Forrester. 1985. Rutherglen bug. Agfacts. No. AE. 41. 4 pp.

Gentry, J. W. 1965. Crop insects of Northeast Africa–Southwest Asia. Agriculture Handbook No. 273. U.S. Dept. Agric., Washington, D.C., U.S.A.

Georgiou, G. P., and C. E. Taylor. 1977. Pesticide resistance as an evolutionary phenomenon. Proc. 15th Int. Congr. Entomol., Washington, D.C., U.S.A.

Gertsson, C.-A. 1979. Bugs in strawberry fields. Vaxtskyddsnotiser 43: 81–86.

1980. The occurrence of bugs (Hemiptera Heteroptera) in strawberry fields in southern Sweden. Entomol. Tidskr. 101: 71–74.

Ghauri, M. S. K. 1982. A new subspecies of *Dimorphopterus cornutus* Slater (Hemiptera: Lygaeidae) from Papua New Guinea on rice and carpet grass. Bull. Entomol. Res. 72: 133–137.

Gilat, B. M., and B. M. Louis. 1974. Biennial Report 1971/1973. Negev Research Institute for Silviculture and Applied Ecology, Tel Aviv, Israel. 13 pp.

Giliomee, J. H. 1959. Grain stink-bug can be controlled effectively. Farming South Africa 35: 47–48.

Gillier, P. 1970. Influence d'attaques d'*Aphanus sordidus* sur la quality des graines d'arachide. Oléagineux 25: 465–466.

Gjoh, H. G., and J. O. Lee. 1988. Species of bugs from Mt. Yeogi in Suwon and their seasonal prevalence. Res. Rep. Rural Dev. Admin., Crop Protect. 30: 1–5 [in Korean, with English abstract].

Glasgow, H. 1914. The gastric caeca & caecal bacteria of the Heteroptera. Biol. Bull. Wood's Hole 26: 101–171.

Goble, H. W. 1971. Insects of the season 1971 related to fruit, vegetables, field crops, and ornamentals. Proc. Entomol. Soc. Ontario 102: 5–6.

1972. Insects of the season 1972 related to fruit, vegetables, field crops, and ornamentals. Proc. Entomol. Soc. Ontario 103: 1–2.

Goel, S. C. 1983. Insect succession and community organization in relation to sunflower. Pp. 147–152 *in* Insect Ecology and Resource Management. Sanatan Dharm College, Muzaffarnagar, India.

Goh, H. G., Y. H. Kim, Y. I. Lee, and K. M. Choi. 1988. Species and seasonal fluctuation of rice ear injurious bugs and pecky rice. Res. Rep. Rural Dev. Admin. 30: 47–52.

Golding, F. D. 1927. Notes on the food plants & habits of some southern Nigerian insects. Bull. Entomol. Res. 18: 95–96.

1931. Further notes on the food plants of Nigerian insects. Bull. Entomol. Res. 22: 221–223.

1937. Further notes on the food plants of Nigerian insects. IV. Bull. Entomol. Res. 28: 5–9.

1947. Further notes on the food plants of Nigerian insects. VI. Bull. Entomol. Res. 38: 75–80.

Gomez-Menor, J. 1955. Hemípteros que atacan a los arboles y arbustos frutales. Minist. Agric. Inst. Nac. Inv. Agron. (Estac. Fitopath. Agric. Madrid. Ser. Fitopath.) No. 286: 1–74.

Goula, M., M. Espinosa, R. Eritja, and C. Aranda, 1999. *Oxycarenus lavaterae* (Fabricius, 1787) in Cornella de Llobregat (Barcelona, Spain) (Heteroptera, Lygaeidae). Bull. Soc. Entomol. France 104: 1, 39–43 [in Spanish].

Gould, F. W. 1975. The Grasses of Texas. Texas A&M University Press, College Station, Texas, U.S.A.

Goyal, N. P. 1974. Heavy population of dusky cotton bug on sunflower. Indian Bee J. 36: 21.

Gracen, V. E., Jr., and W. B. Guthrie. 1986. Host plant resistance for insect control in some important crop plants. Crit. Rev. Plant Sci. 4: 277–291.

Groeters, F. R. 1996. Maternally inherited sex ratio distortion as a result of a male-killing agent in *Spilostethus hospes* (Hemiptera: Lygaeidae). Heredity 77: 201–208.

Gross, G. F. 1959. On the Australian species of coon bugs (*Oxycarenus* Fieber, Heteroptera–Lygaeidae). Rec. Austral. S. Austral. Mus. 13: 359–368.

Gulde, J. 1921. Die Wanzen (Hemiptera-Heteroptera) der Umgebung von Frankfurt a. M. und des Mainzer Beckens. Abh. Senckenb. Naturforsch. Ges. (Frankfurt) 37: 329–503.

1936. Die Wanzen Mitteleuropas. Hemiptera Heteroptera Mitteleuropas. Frankfurt a. M. 5 (1): 1–104.

Gupta, B. D., and P. N. Avasthy. 1954. Some recommendations for the control of sugarcane pests in India. Indian Sugar 4: 387–405.

Gupta, K. K., and J. R. Gandhi. 1996. Foraging strategies and feeding preference of *Lygaeus militaris*. Indian J. Entomol. 58: 4, 337–341.

Gupta, S. P., A. Prakash, A. Choudbury, J. Rao, and A. Gupta. 1990. New records of bugs infesting paddy fields in Orissa. J. Insect Sci. 3: 185.

Gurr, L. 1957. Observations on the distribution, life history, and economic importance of *Nysius huttoni* (Lygaeidae: Hemiptera). New Zealand J. Sci. Tech. 38 A: 710–714

Guthrie, F. E., and G. C. Decker. 1954. The effect of humidity and other factors on the upper thermal death point of the chinch bug (Hemiptera: Lygaeidae). J. Econ. Entomol. 47: 882–887.

Gyawali, B. K. 1987. Influence of rice on lygaeid and mirid bugs of soyabean. Q. Newsl. Asia Pac. Plant Prot. Comm., FAO, Thailand. 30: 27–30.

Gyorgy, B. 1997. New pests of ornamental trees of streets squares, parks in Hungary. Med. Fac. Land. Toegepaste Biol. Wetens. Univ. Gent 62: 321–329 [in German].

Hadler Pupo, N. I. 1976. Pastagens e forrageras; pragas doencas, plantas invasoras e tóxicas, controles. Campinas, Brasil. 180 pp. [in Portuguese].

Haley, M. J., and L. Baker. 1982. Integrated Pest Management for Walnuts. University of California, Berkeley, California, U.S.A.

Hall, D. G., IV, and G. L. Teetes. 1981. Alternate host plants of sorghum panicle feeding bugs in southeast central Texas. Southwest Entomol. 6: 220–228.

1982. Yield loss-density relationships of four species of panicle-feeding bugs in sorghum. Environ. Entomol. 11: 738–741.

Hamid, A. 1985. Blissinae as potential pests of graminaceous crops in Nigeria and West Africa: a review. Nigerian J. Sci. Technol. 3: 99–107.

Hamid, A., and A. S. Ahmed. 1972. Immature stages of *Spilostethus pandurus militaris* Fabricus [sic]. (Heteroptera: Lygaeidae) Pakistan J. Sci. Ind. Res. 15: 181–183.

Hamid, A., and J. A. Slater. 1978 (1980). The Blissinae of West Africa (Hemiptera: Lygaeidae). Bull. Inst. Fond. Afr. Noire. Ser A. 40: 852–892.

Hamilton, J. T. 1980. What's eating your cabbages. Agric. Gaz. New South Wales 91: 35–36.

Hammad, S. M., N. Armanius, and A. A. El-Deeb. 1972. Some biological aspects of *Oxycarenus hyalinipennis* Costa (Hemiptera: Lygaeidae). Bull. Soc. Entomol. Egypt. 56: 33–38.

Hamman, P. J. 1969. Control of southern chinch bug, *Blissus insularis*, in Brazos County, Texas. Pp. 15–17 *in* H. T. Streu and R. T. Bangs [eds.], Proceedings, Scott's Turfgrass Res. Conf. 1. 89 pp.

Hantsbarger, W. M. 1957. *Nysius* of South Dakota (Lygaeidae-Hemiptera). J. Kansas Entomol. Soc. 30: 156–159.

Hao, K. S., C. L. Hu, L. Wan, and Z. Q. Wu. 1992. The bionomics and control of the sorghum chinch-bug (*Dimorphopterus spinolae*). Plant Prot. 18: 20–21 [in Chinese with English abstract].

Hara, K., S. Fukamachi, and M. Osikawa. 1971. Control of Oriental chinch bug, *Cavelerius saccharivorus* (Okajima), by aerial application of fine granules. Proc. Assoc. Plant Prot. Kyushu. 17: 135–136 [in Japanese, with English abstract].

Harakly, F. A., and M. A. H Assem. 1978. Ecological studies on the truly pests of leguminous plants in Egypt. II. Piercing and sucking pests. Pp. 237–242 *in* Proceedings, Fourth Conference of Pest Control, Sept. 30 to Oct. 3, 1978. (Part I). Academy of Scientific Research and Technology and National Research Centre, Cairo, Egypt.

Hargreaves, E. 1937. Some insects and their food plants in Sierra Leone. Bull. Entomol. Res. 28: 505–520.

1948. Lists of recorded cotton insects of the World. Commonwealth Inst. Entomol., London, U.K. 50 pp.

Harrington, J. 1972. Notes on the biology of *Ischnodemus* species of American north of Mexico Hemiptera: Lygaeidae: Blissinae). Univ. Connecticut Occus. Pap. (Biol. Sci. Ser.) 2: 47–56.

1988. Comments on the blood-feeding tribe Cleradini (Hemiptera: Lygaeidae: Rhyparochrominae) and description of a new genus and new species with the legs modified for grasping. Ann. Entomol. Soc. Amer. 81: 578–580.

1990. Detecting evidence of hematophagy in dry museum specimens of *Clerada apicicornis* (Hemiptera: Lygaeidae: Rhyparochrominae). Ann. Entomol Soc. Amer. 83: 545–548.

Harris, H. M., and G. C. Decker. 1934. Paper barriers for chinch bug control. J. Econ. Entomol. 27: 854–857.

Harris, W. V. 1936. Annotated list of insects injurious to cotton in Tanganyika. Bull. Entomol. Res. 27: 523–527.

Hartley, M. J. 1980. A preliminary examination of pests of fodder beet in Canterbury. Pp. 211–214 *in* Proceedings of the Thirty-third New Zealand Weed and Pest Control Society Inc., Palmerston North, New Zealand.

Hasegawa, H. [ed.]. 1980. Major Insect and Other Pests of Economic Plants of Japan. Jpn. Plant Protection Association, Tokyo, Japan.

Hassan, E. 1977. Major Insect and Mite Pests of Australian Crops. Ento Press, Gatton, Queensland, Australia. 238 pp.

1980. Control of Insect and Mite Pests of Australian Crops. Ento Press, Gatton, Queensland, Australia. 298 pp.

Hatsukade, M., K. Yamashita, and M. Yoshida 1979. Studies on hemipterous insects injuring turfgrass II. Selection of several insecticides for control of *Geoblissus hirtulus* Burmeister. J. Jpn. Turfgrass Res. Assoc. 8: 137–143.

Hawkins, P. D. 1989. *Orsillus depressus* Dallas (Hem., Lygaeidae) an arboreal ground bug new to Britain. Entomol. Mon. Mag. 125: 241–242.

Hayes, W. P., and C. O. Johnson. 1925. The reaction of certain grasses to chinch-bug attack. J. Agric. Rec. 6: 575.

Headlee, T. J., and J. W. McColloch. 1913. The chinch bug. Kansas State Agric. Exp. Stn. Bull. 19: 287–353.

Heidemann, O. 1902. Notes on *Belonochilus numenius* Say. Proc. Entomol. Soc. Washington 5: 11–12.

1911. Some remarks on the eggs of North American species of Hemiptera-Heteroptera. Proc. Entomol. Soc. Washington 13: 128–140.

Heiss, E., and J. Péricart. 1983. Revision of Palaearctic Piesmatidae (Heteroptera). Mitt. Munch. Entomol. Ges. 73: 61–171.

Hely, P. C., G. Pasfield, and J. G. Gellatley. 1982. Insect Pests of Fruit and Vegetables in N.S.W. Inkata Press, Melbourne, Australia.

Henry, T. J. 1997. Phylogenetic analysis of family groups with the infraorder Pentatomomorpha (Hemiptera: Heteroptera), with emphasis on the Lygaeoidea. Ann. Entomol. Soc. Amer. 90: 275–301.

Henry, T. J., and R. C. Froeschner. 1993. *Dieuches armatipes* (Walker) (Heteroptera: Lygaeidae) newly discovered in the Western Hemisphere. Proc. Entomol. Soc. Washington 95: 449–452.

Herring, J. L. 1973. Insects not known to occur in the United States. South African grain bug. U.S. Dept. Agric. Coop. Econ. Inst. Rep. 23: 733–734.

1974. Insects not known to occur in the United States. New Caledonian orchid pest (*Nysius caledoniae* Distant). U.S. Dept. Agric. Coop. Econ. Inst. Rep. 24: 590–591.

Hewitt, G. B, E. W. Huddleston, R. J. Vavigne, D. N. Ueckert, and J. Gordon Watts. 1974. Rangeland Entomology. Range Science Series No. 2. Society for Range Management, Denver, Colorado, U.S.A. 127 pp.

Hill, D. S. 1975. Agricultural Pests of the Tropics and Their Control. Cambridge University Press, Cambridge, U.K.

Hill, G. F. 1926. Insects affecting cotton in Australia. Proc. Pan Pac. Sci. Congr. Australia 1923. 1: 406–408.

Hoberlandt, L. 1960. Spedizione Italianial Karakorum ed al Hindu-Kush (1954–1955). Hemiptera-Heteroptera. Atti Mus. Civ. Stor. Nat. Trieste 22: 55–65.

Hoffman, W. E. 1932. The economic status of lygaeids and notes on the life-history of *Lygaeus hospes* Fab. and *Aphanus sordidus* Fabr. Lingnan Sci. J. 11: 119–135.

Hoffman, W. E. 1934. The life history of a species of *Graptostethus* (Hemiptera, Lygaeidae). Lingnan Sci. J. 13: 171–176.

Hokyo, N., H. Suzuki, and M. Murai. 1983. Egg diapause in the Oriental chinch bug. *Cavelerius saccharivorus* Okajima (Heteroptera: Lygaeidae) 1. Incidence and intensity. Jpn. J. Appl. Entomol. Zool. 18: 382–391.

Holdaway, F. G. 1944. Insects of vegetable crops in Hawaii today. Proc. Hawaiian Entomol. Soc. 12: 59–80.

Holmes, F. O. 1925. The relation of *Herpetomonas elmassiani* (Migone) to its plant and insect hosts. Biol. Bull 49: 323–327.

Hong, S.-C. 1986. Damage to sorghum by a lygaeid bug, *Nysius plebejus* [sic] Distant (Hemiptera: Lygaeidae). China J. Entomol. 6: 119–125.

Horn, G. C. 1962. Chinch bugs and fertilizer, is there a relationship? Florida Turfgrass Assoc. Bull. 9: 3, 5.

Horn, G. C., A. E. Dudeck, and R. W. Toler. 1973. Floratam St. Augustine grass. Univ. Florida Agric. Exp. Stn. Circ. s-224: 13 pp.

Horner, N. V. 1972. *Metaphidippus galathea* as a possible biological control agent. J. Kansas. Entomol. Soc. 45: 324–327.

Horváth, G. 1909. Les relations entre les faunes Hémipterologiques de l'Europe et de l'Amerique du Nord. Proc. 7th Int. Cong. Zool. Boston, Massachusetts, U.S.A. pp. 1–12.

1926. Sur les *Oxycarenus* nuisibles aux cotonniers, avec la description d'une espéce nouvelle (Hem. Lygaeidae). Bull. Soc. Entomol. France 1926: 135–136.

Horváth, Z., and G. Bujaki. 1991. *Lygaeus equestris* L. a pest of sunflower. Novenyvedelem 27: 508–512.
Howard, L. O. 1887. The chinch bug (*Blissus leucopterus*) Say, Order Hemiptera; family Lygaeidae. Rep. Entomol. USDA: 51–58.
Howard, W. R. 1872. The radish bug — a new insect (*Nysius raphanus*, n. sp.). Canad. Entomol. 4: 219–220.
Howe, R. W. 1952. Entomological problems of food storage in northern Nigeria. Bull. Entomol. Res. 43: 111–144.
Hughes, J. D. 1983. American Indian Ecology. Texas Western Press, University of Texas, El Paso, Texas, U.S.A.
Ibanez, P., M. D. Garcera, E. Alcacer, and R. Martinez. 1993a. Juvenile hormone effects on *Spilostethus pandurus* (Hemiptera: Heteroptera) vitellogenesis. Rev. Esp. Fisiol. 49: 87–92 [in Spanish].
Ibanez, P., R. Garcera, M. D. Alcacer, E. Conill, and F. Martinez. 1993b. Effect of starvation on haemolymph vitellogenins and ovary uptake in *Spilostethus pandurus*. Comp. Biochem. Physiol. B: Comp. Biochem. 104: 531–536.
Illingworth, J. F. 1917. *Clerada apicicornis* sucking blood (Hemip.). Proc. Hawaiian Entomol. Soc. 3: 277–279.
1920. Cane grub investigation. Queensland Agric. J. 14: 148–152.
Iltis, H. H., and J. F. Doebley. 1980. Taxonomy of *Zea* (Gramineae). II. Subspecific categories in the *Zea* complex and a generic synopsis. Amer. J. Bot. 67: 994–1004.
Ingram, J. W. 1927. Insects injurious to the rice crop. U.S. Dept. Agric. Bull. No. 1543: 16 pp.
International Commission Zoological Nomenclature. 1964. Opinion 705, *Blissus* Burmeister, 1835 (Insecta, Hemiptera): added to the official list of generic names. Bull. Zool. Nomen. 21: 198–201.
Isely, D., and W. B. Horsfall. 1931. The chinch bug as a rice pest. J. Kansas Entomol. Soc. 4: 70–73.
Issac, P. V. 1946. Report of the entomologists. Sci. Rep. Indian Agric. Rec. Inst., 1944–45: 73–79.
Ito, Y. 1976. Status of insect pests of sugarcane in the southwestern islands of Japan. Jpn. Agric. Res. Q. 10: 63–69.
Jaipal, S. 1991. Infestation and severity of damage by black bug *Cavelerius sweeti* to shoots of nineteen sugarcane accessions and cultivars. Ann. Appl. Biol. 118 (Suppl.): 120–121.
1996. Evaluation of mechanical and chemical methods for control of black bug (*Cavelerius sweeti*) in sugarcane ratoon crop. Ann. Appl. Biol. 128 (Suppl.): 4–5.
Janes, M. J. 1935. Oviposition studies on the chinch bug *Blissus leucopterus* Say (Hemiptera: Lygaeidae). Ann. Entomol. Soc. Amer. 25: 109–142.
Johnson, S. R., and A. K. Knapp. 1996. Impact of *Ischnodemus falicus* (Hemiptera: Lygaeidae) on photosynthesis and production of *Spartina pectinata* wetlands. Environ. Entomol. 25: 1122–1127.
Jones, E. T. 1937. Differential resistance to chinch bug attack in certain strains of wheat. Trans. Kansas Acad. Sci. 40: 135–142.
Jordan, K. H. C. 1933. Beiträge zur Biologie heimischer Wanzen (Heterop.). Stettin. Entomol. Zeitg. 94: 212–236.
Joseph, M. T. 1959. Biology, bionomics and economic importance of some reduviids collected from Delhi. Indian J. Entomol. 21: 48–58.
Joshi, S., and J. R. Gandhi. 1986. Aggregation, feeding and growth of *Oxycarenus hyalinipennis* Costa (Hemiptera: Lygaeidae) on different seeds. Ann. Entomol. 4: 15–19.
Kakakhel, S. A., and M. Amjad. 1997. Biology of *Nysius inconspicuus* Distant and its economic impact on sunflower (*Helianthus annuus* L.). Helia 20: 9–14.
Kapoor, K. N., S. V. Dhamdhere, O. P. Singh, and U. S. Misra. 1982. Population dynamics of insect pests of pearl millet in northern Madhya Pradesh. Indian J. Plant Prot. 9: 69–73.
Karban, R., and G. Lowenberg. 1992. Feeding by seed bugs and weevils enhances germination of wild *Gossypium* species. Oecologia 92: 196–200.
Kasargode, R. S., and V. G. Deshpande. 1921. The groundnut bug and its control. Bull. Dept. Agric. Bombay No. 105: 20 pp.
Kashyap, N. P., and R. L. Adlakha. 1971 (1972). New records of insect pests of soybean crop. Indian J. Entomol. 33: 467–468.
Kashyap, N. P., S. F. Hameed, D. N. Vaidya, and G. S. Dogra. 1980. Relative toxicity of some insecticides to soybean bug *Chauliops fallax* Scott. Indian J. Entomol. 42: 263–265.
Kehat, M., and M. Wyndam. 1972a. The effect of food and water on development, longevity and fecundity in the Rutherglen bug, *Nysius vinitor* (Hemiptera: Lygaeidae). Austral. J. Zool. 20: 119–130.

1972b. The influence of temperature on development, longevity, and fecundity in the Rutherglen bug, *Nysius vinitor* (Hemiptera: Lygaeidae). Austral. J. Zool. 20: 67–78.
1973a. Flight activity and displacement in the Rutherglen bug *Nysius vinitor* (Hemiptera: Lygaeidae). Austral. J. Zool. 21: 413–426.
1973b. The relation between food age, and flight in the Rutherglen bug, *Nysius vinitor* (Hemiptera: Lygaeidae). Austral. J. Zool. 21: 427–434.
1974a. Differences in flight behavior of male and female *Nysius vinitor* Bergroth (Hemiptera: Lygaeidae). J. Austral. Entomol. Soc. 13: 27–29.
1974b. The effect of temperature and relative humidity extremes on the survival of the Rutherglen bug *Nysius vinitor* (Hemiptera: Lygaeidae). J. Austral. Entomol. Soc. 13: 81–84.
Kelsey, J. M. 1957. Insects attacking tussock. New Zealand J. Sci. Technol. 10: 638–642.
Kelsheimer, E. G., and S. H. Kerr. 1957. Insects and other pests of lawns and turf. Florida Agric. Exp. Stn. Cir. 5-96: 2–5.
Kennedy, M. K. 1981. Chinch bugs: biology and control. Amer. Lawn Appl. July/Aug.: 12–15.
Kerr, S. H. 1958. Tests on chinch bugs, and the current status of controls. Proc. Florida Hortic. Soc. 71: 400–403.
1962. Lawn insect studies 1962. Proc. Univ. Florida Turf Manager. Conf. 10: 201–208.
1966. Biology of the chinch bug *Blissus insularis* (Hemiptera: Lygaeidae). Florida Entomol. 49: 9–18.
Kerr, S. H., and L. C. Kuitert. 1955. Biology and control of insect and arachnid pests of turf grasses. Florida Agric. Exp. Stn. Ann. Rep. 101–102.
Kerzhner, I. M., and T. L. Yaczewski. 1964. Hemiptera. Pp. 851–1118 *in* G. Ya. Bienko [ed.], Keys to the Insects of the European USSR. Vol. 1. Apterygota, Palaeoptera, Hemimetabola. Acad. Sci. USSR. Zoological Institute. Keys to the fauna of the USSR. No. 84. 1214 pp. [in Russian; translated in 1967 by J. Salkind, Israel Program for Scientific Translations, Jerusalem, Israel].
Khan, M. R., and I. Sharfuddin. 1966. Pest control by non-insecticidal methods. Tech. Bull. No. 2. Bureau Agricultural Information of Government of Pakistan. 9 pp.
Khan, Z. R., and R. A. Agarwal. 1981 (1983). Relationship between gossypol glands and incidence of some important pests on different genotypes of cotton. J. Entomol. Res. 5: 169–172.
Kim, J.-B., T.-S. Kim, D.-S. Kang, W.-K. Shin, and Y.-S. Lee. 1994a. Investigation of pentatomid species of chrysanthemum of host plants of *Nysius plebejus* [sic] Distant (Hemiptera: Lygaeidae) and its control. Korean J. Appl. Entomol. 33: 1–5 [in Korean with English abstract].
Kim, J.-B., D.-S. Kang, T.-S. Kim, W.-K. Shin, and Y.-S. Lee. 1994b. Studies on the life history of *Nysius plebejus* [sic] Distant (Hemiptera: Lygaeidae) an insect pest of chrysanthemum. Korean J. Appl. Entomol. 33: 56–59 [in Korean with English abstract].
Kinaci, E., G. Kinaci, A. F. Yildirim, and A. Atli. 1997. Sunn pest problems in Central Anatolia and the role of wheat varieties in integrated control. Wheat: prospects for global improvement. Pp. 121–125 *in* Proceedings, 5th International Wheat Conference, Ankara, Turkey, 10–14 June 1996. Kluwer Academic Publishers, Dordrecht, the Netherlands.
Kindler, S. D., and R. Staples. 1982. Foliar applications of insecticides for control of chinch bugs on grain sorghum, 1981. Ins. Acar. Tests 7: 176.
Kindler, S. D., K. P. Pruess, and S. M. Spooner. 1982. Granular application of phorate granules for control of chinch bugs on winter wheat. Ins. Acar. Tests 7: 198.
Kinzelbach, R. K. 1970. *Loania canadensis* n. gen., n. sp. und die Untergliederung der Callipharixenidae. Senckenburgiana Biol. 51: 99–107.
Kirkpatrick, T. W. 1923. The Egyptian cotton seed bug (*Oxycarenus hyaliniipennis* Costa) (Hemiptera: Lygaeidae). Bull. Minist. Agric. Egypt, Tech. Sci. Serv. 35: 1–107.
Kisimoto, R. 1965. Studies on the polymorphism and its role in the population growth of the brown planthopper, *Nilaparvata lugens* Stal. Bull Shikoku. Agric. Exp. Stn. 13: 1–107 [in Japanese with English summary].
1984. Damage caused by rice stink bugs and their control. Japan Pesticide Information No. 43: 9–13.
Knight, D. W., M. Rossiter, and B. W. Staddon. 1984. Z e-alpha farnesene major component of secretion from metathoracic scent gland of cottonseed bug *Oxycarenus hyalinipennis* (Hemiptera: Lygaeidae). J. Chem. Ecol. 10: 641–650.
Knowlton, G. F., and S. L. Wood. 1943. Utah bird predators of the false chinch bug. J. Econ. Entomol. 36: 332–333.

Kobayashi, T. 1981. Insect pests of soybean in Japan. Misc. Publ. Tohoku Natl. Agric. Stn. 2: 1–39.

Kobayashi, T., and L. Nugaliyadde. 1988. The damage of rice particles caused by insect pests in Sri Lanka. Tropical Agriculture Research Center. Reprinted from JARQ 2: 314–322.

Kortier Davis, M. G., and D. R. Smitley. 1990a. Association of thatch with populations of hairy chinch bug (Hemiptera: Lygaeidae) in turf. J. Econ. Entomol. 83: 2370–2374.

1990b. Relationship of hairy chinch bug (Hemiptera: Lygaeidae) presence and abundance to parameters of the turf environment. J. Econ. Entomol. 83: 2375–2379.

Krueger, S. R., J. R. Nichols, and W. A. Ramoska. 1991. Infection of chinch bug *Blissus leucopterus leucopterus* (Hemiptera: Lygaeidae) adults from *Beauveria bassiana* (Deuteromycotina Hyphomycetes) conidia in soil under controlled temperature and moisture conditions. J. Invertebr. Pathol. 58: 19–26.

1992. Habitat distribution and infection rates of the fungal pathogen, *Beauveria bassiana* (Balsamo) Vuillemin in endemic populations of the chinch bug, *Blissus leucopterus leucopterus* (Say) (Hemiptera: Lygaeidae) in Kansas. J. Kansas Entomol. Soc. 65: 115–124.

Kugelberg, O. 1973. Notes on the rearing of *Spiostethus pandurus* (Heteroptera, Lygaeidae) on sunflower seeds. Entomol. Exp. Appl. 16: 552–553.

Kumar, K., S. C. Gupta, U. K. Mishra, G. P. Dwivedi, and N. N. Sharma. 1987. Sugarcane pests in Bihar: retrospect and prospect — a review. Agric. Rev. 8: 59–66.

Kumar, R., R. J. Lavigne, J. E. Lloyd, and R. E. Pfadt. 1976. Insects of the Central Plains Experiment Range, Pawnee National Grassland. Science Monograph 32. Agric. Exp. Stn. Univ. Wyoming, Laramie, Wyoming, U.S.A. 74 pp.

Kundu, G. C., and J. K. Sharma. 1974 (1976). New insect pests of sorghum ears in Rajasthan. Ann. Arid Zone 13: 273–274.

Labrador, S. J. R. 1972. *Blissus* sp. related to *Blissus insularis* (Hemiptera: Lygaeidae) a pangola grass pest in Zulia State. Folia Entomol. Mexico 23–24: 53–55.

Lal, O. P. 1974. (1975). Occurrence of *Chauliops fallax* Scott (Hemiptera: Lygaeidae) on French Bean and Horse Gram in Himachal Pradesh. Indian J. Entomol. 36: 67–68.

1981. A contribution to the knowledge of ecology, biology, host range and control of the lygaeid bug *Chauliops fallax* Scott (Hemiptera: Lygaeidae) a pest of pulse crops in Kulu Valley. Riv. Agric. Subtrop. Trop. 75: 381–403.

1992. Evaluation of certain insecticides against the pest complex of French bean seed crop under field conditions. J. Entomol. Res. 16: 57–61.

Lal, O. P. 1989. Hibernation of the bug, *Nysius ceylonicus* Mots. (Hemiptera: Lygaeidae) in Kulu valley, Himachal Pradesh (India). J. Entomol. Res. 13: 91–92.

Lal, O. P., and K. Sengupta. 1977. Laboratory observations on the biology of the bean bug, *Chauliops nigrescens* Distant (Hemiptera: Lygaeidae). Indian J. Entomol. 37: 408–413.

Lamas, C. J. M. 1945. Observaciones sobre insectos del algodo nero en chira, Piura Pativilea, Sryse y tluaura. Inf. Estac. Exp. Agric. La Molina N. 59. 75 pp.

Lamp, W. O., and T. O. Holtzer. 1980. Distribution of overwintering chinch bugs, *Blissus leucopterus leucopterus* (Hemiptera: Lygaeidae). J. Kansas Entomol. Soc. 53: 320–324.

Lanham, U. N. 1975. A mountain-top swarm of the hemipteran *Nysius raphanus* (Hemiptera: Lygaeidae) in New Mexico, with notes on other insects. Pan-Pac. Insects 51: 166–167.

Larivière, M. C., and A. Larochelle. 1991. Notes on the distribution and bionomics of *Myodocha serripes* (Heteroptera: Lygaeidae). Entomol. News 102: 31–32.

Lavigne, R. J., R. Kumar, J. Leetham, and V. Keith. 1972. Population densities and biomass of above ground arthropods under various grazing and environmental stress treatments on the Pawnee Site, 1971 US IBP Grassland Biome Tech. Rep. 204.

Leigh, T. F. 1961. Insecticide susceptibility of *Nysius raphanus* (Hemiptera: Lygaeidae), a pest of cotton. J. Econ. Entomol. 54: 120–122.

Leonard, D. E. 1966. Biosystematics of the *"leucopterus* complex*"* of the genus *Blissus* (Heteroptera: Lygaeidae). Bull. Connecticut Agric. Exp. Stn. 677: 47 pp.

1968a. A revision of the genus *Blissus* (Heteroptera: Lygaeidae) in eastern North America. Ann. Entomol. Soc. Amer. 61: 239–250.

1968b. Three new species of *Blissus* from the Antilles (Heteroptera: Lygaeidae). Proc. Entomol. Soc. Washington 70: 150–153.

1970. A new North American species of *Blissus* (Heteroptera: Lygaeidae). Canad. Entomol. 102: 1531–1533.
Lepelley, R. H. 1960. Agricultural Insects of East Africa. East Africa High Commission. Nairobi, Kenya. 307 pp.
Leston, D. 1957. The stridulatory mechanisms in terrestrial species of Hemiptera Heteroptera. Proc. Zool. Soc. London 128: 369–386.
1970. The identity and pest potential of *Oxycarenus spp.* (Hem. Lygaeidae) in Ghana. Bull. Entomol. Res. 60: 285–289.
Li, H.-K. 1982. *Dimorphopterus spinolae* a serious reed insect pest Hemiptera Heteroptera Lygaeidae. Act. Entomol. Sin. 25: 24–30.
Lim, G. T., and Y. C. Pan. 1976. Colobathristid bugs. Entomol. Newsl. South. Africa 3: 2.
Lindberg, H. 1932. Inventa entomologica itineris Hispanici & Maroccani, quod a 1926 fecerunt Harald & Hakan Lindberg. XIII. Hemiptera Heteroptera (excl. Caspidae & Hydrobiotica). Comment. Biol. 3: 19: 1–53.
Linnavuori, R. E. 1989. Heteroptera of Yemen and South Yemen. Act. Entomol. Fenn. 54: 1–40.
Lintner, J. S. 1883. The chinch bug in New York. Science 2: 540.
Liogier, H. A., and L. F. Martorell. 1982. Flora of Puerto Rico and Adjacent Islands: A Systematic Synopsis. Editorial de la Universidad de Puerto Rico, Rio Piedras, Puerto Rico, U.S.A.
Liu, H. J., and F. L. McEwen. 1977. *Nosema blissi* sp.n. (Microsporida: Nosematidae) a pathogen of the chinch bug, *Blissus leucopterus hirtus* (Hemiptera: Lygaeidae). J. Invertebr. Pathol. 29: 141–146.
1979. The use of temperature accumulations and sequential sampling in predicting damaging populations of *Blissus leucopterus hirtus* (Hemiptera: Lygaeidae). Environ. Entomol. 8: 512–515.
Lobb, W. R., and J. Wood. 1971. Insects in the food supply of starlings in mid-Canterbury. New Zealand Entomol. 5: 17–24.
Lorenz, K., and P. Meredith. 1988. Insect-damaged wheat: history of the problem, effects on baking quality, remedies. Lebensm. Wiss. Technol. 21: 183–187
Loudon, B. J., and F. I. Attia. 1981. The immature stages of *Alophora lepidofera* a native parasite of Lygaeidae Hemiptera in Australia. Austral. Entomol. Mag. 7: 61–67.
Lounsbury, P. P. 1918. Division of Entomology. Annual Report 1917–18. Union of S.A. Dept. Agric. Rep. pp. 87–107.
Lowie, R. H. 1954. Indians of the Plains. McGraw-Hill, New York, New York, U.S.A. 258 pp.
Luckman, W. H., and R. L. Metcalf. 1975. The pest management concept. Pp. 3–35 *in* R. L. Metcalf and W. H. Luckman [eds.], Introduction to Insect Pest Management. John Wiley & Sons, New York, New York, U.S.A.
Luginbill, P. 1922. Bionomics of the chinch bug. U.S. Dept. Agric. Bur. Entomol. Bull. 1016: 14 pp.
Lukefahr, M. J. 1981. Occurrence of injurious insects in Nigeria and Brazil. Tropical Grain Legume Bulletin No. 23. 29 pp.
Lynch, R. E., S. Some, I. Dicko, H. D. Wells, and W. G. Monson. 1987. Chinch bug damage to Bermuda grass. J. Entomol. Sci. 22: 153–158.
Mabberley, D. J. 1981. The Plant-Book. A Portable Dictionary of the Higher Plants. Cambridge University Press, Cambridge, U.K. 706 pp.
Mackie, J. R. 1944. Annual Report for the Agricultural Department (Nigeria) for the year 1943. 34 pp.
MacLeod, D. M. 1954. Investigations on the genera *Beauveria* Vuill. and *Tritrachium* Dimber. Canad. J. Bot. 32: 818–890.
Madan, Y. P., B. S. Chillar, and D. Singh. 1993. Evaluation of insecticides for the control of sugarcane black bug, *Cavelerius sweeti* Dist. [sic] Agric. Sci. Dig. 13: 137–140.
Madan, Y. P., J. N. Gupta, S. C. Bhardwaj, and R. A. Singh. 1978. Evaluation of some insecticides against sugarcane black bug, *Macropes excavatus* Distt. [sic] (Hemiptera Lygaeidae). Indian Sugar Crops J. 5: 26–27.
Maheshwari, B. K., and S. S. Malik. 1988. Some new insecticides to control black bug in sugarcane. Pesticides 22: 23–24.
Mailloux, G., and H. T. Streu. 1979. A sampling technique for estimating hairy chinch bug (*Blissus leucopterus hirtus* Montandon, Hemiptera: Lygaeidae) populations and other arthropods from turf grass. Ann. Soc. Entomol. Québec 24: 139–143.
1981. Population biology of the hairy chinch bug (*Blissus leucopterus hirtus* Montandon: Hemiptera: Lygaeidae). Ann. Soc. Entomol. Québec 26: 51–90.

1982. Spatial distribution pattern of hairy chinch bug (*Blissus leucopterus hirtus* Montandon: Hemiptera: Lygaeidae) populations in turf grass. Ann. Soc. Entomol. Québec 27: 111–131.

Malipatil, M. B. 1979. The biology of some Lygaeidae (Hemiptera: Heteroptera) of south-east Queensland. Austral. J. Zool. 27: 231–249.

Malipatil, M. B. 1983. Revision of World Cleradini (Heteroptera: Lygaeidae) with a cladistic analysis of relationships within the tribe. Austral. J. Zool. 31: 205–225.

Malipatil, M. B., and R. Kumar. 1975. Biology and immature stages of some Queensland Pentatomomorpha (Hemiptera: Heteroptera). J. Austral. Entomol. Soc. 14: 113–128.

Mansfield-Aders, W. 1919. Insects injurious to economic crops in the Zanzibar protectorate. Bull. Entomol. Res. 10: 145–155.

Marshall, G. A. K. 1902. Five years observations and experiments (1896–1901) on the bionomics of South African insects, chiefly directed to the investigation of mimicry and warning colours. Trans. Entomol. Soc. London 1902: 287–584.

Marwat, G. N. K., and A. A. Khan. 1987. Using granules against sucking insect pest of sugarcane. Pakistan J. Agric. Res. 8: 57–60.

Mathias, J. K., R. H. Ratcliffe, and J. L. Hellman. 1990. Asssociation of an endophytic fungus in perennial ryegrass and resistance to the hairy chinch bug (Hemiptera: Lygaeidae). J. Econ. Entomol. 83: 1640–1646.

Mathur, R. B., and O. Prakash. 1989a. Biology of sugarcane lygaeid bug, *Blissus gibbus* F. (Lygaeidae: Hemiptera). Bull. Entomol. 30: 164–171.

1989b. Relative susceptibility of certain promising varieties of sugarcane to sugarcane lygaeid bug, *Blissus gibbus* F. (Lygaeidae: Hemiptera). Bull. Entomol. 30: 234–235.

Matsumoto, F. 1935. On the life history of *Ischnodemus saccharivorus* Okajima injurious to sugar cane in Formosa (Lygaeidae, Heteroptera). 1. Trans. Nat. Hist. Soc. Formosa 25: 314–319 [in Japanese].

Matthee, J. J. 1974. Pests of graminaceous crops in South Africa. Entomol. Mem. Dep. Agric. Tech. Serv. Republ. South Africa 40: 1–23.

Maxwell, K. E., and G. F. McLeod. 1936. Experimental studies on the hairy chinch bug. J. Econ. Entomol. 29: 339–343.

Maxwell-Lefroy, H. 1907. The more important insects injurious to Indian agriculture. Mem. Dept. Agric. India I. 113–252.

1909. Indian Insect Life. Govt. India Publication, Agricultural Research Institute of Pusa, India. 786 pp.

Mayné, R., and J. Ghesquière. 1934. Hémiptères nuisibles aux végétaux du Congo Belge. Ann. Gembloux 40: 3–41.

McColloch, J. W. 1913. A parasite of the chinch bug egg. Canad. Entomol. 45: 342–343; J. Econ. Entomol. 6: 425–426.

McColloch, J. W., and N. Yuasa. 1914. A parasite of the chinch bug egg. J. Econ. Entomol. 14: 219–227.

1915. A new parasite of the chinch bug egg. Entomol. News 26: 147–149.

McDonald, G., and R. A. Farrow. 1988. Migration and dispersal of the Rutherglen bug, *Nysius vinitor* Bergroth (Hemiptera: Lygaeidae), in eastern Australia. Bull. Entomol. Res. 78: 493–509.

McDonald, G., and A. M. Smith. 1988. Phenological development and seasonal distribution of the rutherglen bug *Nysius vinitor* Bergroth (Hemiptera: Lygaeidae) on various hosts in Victoria, South-eastern Australia. Bull. Entomol. Res. 78: 673–682.

McDonald, G., R. H. Broadley, A. M. Smith, and M. D. Blackburn. 1986. Evaluation of insecticides for *Nysius vinitor* Bergroth. Gen. Appl. Entomol. 18: 11–16.

McEwen, G. L. 1973. Insects of the season in Ontario 1973 related to fruit, vegetables, field crops and ornamentals. Proc. Entomol. Soc. Ontario 104: 1–2.

McGhee, R. B., and W. L. Hanson. 1962. Growth and reproduction of *Leptomonas oncopelti* in the milkweed bug, *Oncopeltus fasciatus*. J. Protozool. 9: 488–493.

1964. Comparison of the life cycle of *Leptomonas oncopelti* and *Phytomonas elmassiani*. J. Protozool. 11: 555–562.

McGregor, R. A. 1976. Florida turfgrass survey, 1974. Florida Crop and Livestock Rep. Serv. 33 pp.

Meehan, J., and G. Wilde. 1989. Screening for sorghum line and hybrid resistance to chinch bug (Hemiptera: Lygaeidae) in the greenhouse and growth chamber. J. Econ. Entomol. 82: 616–620.

Mehta, S. S., D. Singh, J. Singh, J. Tripathi, and S. Kumar. 1996. Arthropods associated with *Artemisia annua* in north Indian plains. J. Med. Arom. Plant Sci. 18: 26–33.

Mejia-Ford, O. 1997. Studies of chinch bug, *Blissus leucopterus leucopterus* (Say) (Hemiptera: Lygaeidae) in rice, *Oryza sativa* L. — an integrated pest management approach. Dissertation, Texas A&M University, College Station, Texas, U.S.A. 129 pp.

Melber, A., L. Hoelscher, and G. H. Schmidt. 1980. Further studies on the social behavior and its ecological significance in *Elasmucha grisea* (Hem.-Het.: Acanthosomatidae). Zool. Anz. 205: 27–38.

Mellor, J. E. M. 1932. Notes from Zanzibar, Tanganyika, Kenya, Uganda and the Sudan: August to December 1928. Entomol. Mon. Mag. 68: 234–252.

Melou, J. P., and A. Yana. 1964. Sur l'existence de races geographiques-distinctes par leur resistence aux insecticides chez *Oxycarenus hyalinipennis* (Hemiptera: Lygaeidae). Methodes de controle de l'efficacite des insecticides. Doc. Tech. Inst. Nat. Rech. Agron. Tunise 4: 13–19.

Merchant, M. E., and R. L. Crocker. 1998. Chinch bugs in St. Augustine lawns. L-1766. Texas Agricultural Extension Service, College Station, Texas.

Merckle, O. G., K. J. Starks, and A. J. Casady. 1983. Registration of pearl millet germplasm lines with chinch bug resistance. Crop Sci. 23: 601.

Metcalf, C. L., and W. P. Flint. 1939. Destructive and Useful Insects, Their Habits and Control. McGraw-Hill, New York, New York, U.S.A. 908 pp.

Metcalf, C. L., W. P. Flint, and R. L. Metcalf. 1962. Destructive and Useful Insects, fourth edition. McGraw-Hill, New York, New York, U.S.A. 1087 pp.

Metcalf, R. L., and R. A. Metcalf. 1975. Attractants, repellents, and genetic control in pest management. Pp. 275–306 in R. L. Metcalf and W. H. Luckman [eds.], Introduction to Insect Pest Management. John Wiley & Sons, New York, New York, U.S.A.

Milliken, F. B. 1916. The false chinch bug and measures for controlling it. Farmers' Bull. USDA 76: 1–4.

1918. *Nysius ericae*, the false chinch bug. J. Agric. Res. 13: 571–578.

Milliken, F. B., and F. M. Wadley. 1922. *Geocoris pallens* Stål var. *decoratus* Uhl., a predaceous enemy of the false chinch bug. Bull. Brooklyn Entomol. Soc. 17: 143–146.

1923. *Phasia (Phorantha) occidentis*, Walker, an internal parasite of the false chinch bug. Bull. Brooklyn Entomol. Soc. 18: 28–31.

Misari, S. M. 1990. Pest complex of sunflower (*Helianthus annuus* L.) in parts of Nigerian savanna. Savanna 11: 1–11.

Miyamoto, S. 1961. Comparative morphology of alimentary organs of Heteroptera, with the phylogenetic consideration. Sieboldia 2: 197–259.

Miyamoto, S., and T. Kifune. 1984. Descriptions of a new genus and two new species of Strepsiptera parasitic on Japanese Heteroptera. Kontyu 52: 137–149.

Mize, T. W., and G. Wilde. 1986a. New resistant germplasm to the chinch bug in grain sorghum: contribution of tolerance and antixenosis as resistance mechanisms. J. Econ. Entomol. 79: 42–45.

1986b. New grain sorghum sources of antibiosis to the chinch bug (Heteroptera: Lygaeidae). J. Econ. Entomol. 79: 176–180.

1986c. Reproduction of the chinch bug (Heteroptera: Lygaeidae) on new resistance sources in grain sorghum. J. Econ. Entomol. 79: 664–667.

Mize, T., W. G. Wilde, and M. T. Smith. 1980. Chemical control of chinch bug (Heteroptera: Lygaeidae) and greenbug (Homoptera: Aphididae) on seedling s*orghum* with seed, soil, and foliar treatments. J. Econ. Entomol. 73: 544–547.

Mohanasundaram, M. 1972. On the new host records for some south Indian crop pests. Indian J. Entomol. 34: 259–261.

Morgan, D. L., G. W. Frankie, and M. J. Gaylor. 1978. Potential for developing insect resistant plant materials for use in urban environments. Pp. 267–294 in G. W. Frankie and C. S. Koehler [eds.], Perspectives in Urban Entomology. Academic Press, New York, New York, U.S.A. 417 pp.

Morrison, L. 1938 (1939). Surveys of the insect pests of wheat crops in Canterbury and North Otago during the summers of 1936–1937 and 1937–1938. New Zealand J. Sci. Technol. A20: 142–155.

Mrig, K. K., and J. P. Chaudhary. 1994a. Effect of date of trash burning on populations of black bug, *Cavelerius sweeti* Slater and Miyamoto in sugarcane ratoon. Coop. Sugar 26: 111–112.

1994b. Comparative efficacy of soil insecticides on incidence of shoot, root borers and black bug in sugarcane ratoons. Indian Sugar 43: 923–929.

Mukhopadhyay, A. 1983. Some aspects of ecology of lygaeid bugs (Heteroptera: Insecta) of the host plant madar and effect of the latter on the fecundity of *Spilostethus hospes* (Fabr.). Pp. 91–101 *in* Proceedings, Symp. Host. Environ. Zool. Surv., India.

Mukhopadhyay, A., and B. Saha. 1992a. Performance of the polyphagous pest, *Spilostethus pandurus* (Hemiptera: Lygaeidae) in relation to the nutritional status of four host-seeds. Phytophaga 4: 111–120.

1992b. Nutritional values of four host seeds and their relationship with adult and nymphal performance of seedbug (*Elasmolomus sordidus*) (Hemiptera: Lygaeidae). Indian J. Agric. Sci. 62: 42–46.

1992c. Nutritional value of four host seeds and their relationship with adult and nymphal performance of seed bug *Elasmolomus sordidus* (Heteroptera Lygaeidae). Indian J. Agric. Sci. 62: 834–837.

1993. A mass budget of milkweed bug, *Spilostethus pandurus* (Scopoli) on four host seeds. Ann. Entomol. 11: 19–23.

1994. On the life-performance of *Spilostethus hospes* (Fabr.) (Heteroptera: Lygaeidae) in response to the nutritional quality of four host seeds. J. Bengal Nat. Hist. Soc. 13: 5–14.

1995. Nutrient requirement of seed bug (*Elasmolomus sordidus*) (Heteroptera: Lygaeidae) on four host seeds. Indian J. Agric. Sci. 65: 154–155.

Mukhopadhyay, A., T. N. Ananthakrisnan, and K. Thangavelu. 1981. Record of new host plants of *Oxycarenus laetus* Kirby (Hemiptera: Lygaeidae) with notes on its ecology from eastern India. Bull. Zool. Surv. India 4: 117–118.

Muller, C. B., and F. Fritsche. 1993. Unequal costs of parasitism for male and female lygaeid bugs. J. Insect Behav. 6: 265–269.

Murai, M. 1975. Population studies of *Cavelerius saccharivorus* Okajima (Hemiptera: Lygaeidae) a few findings on population interchange. Res. Popul. Ecol. (Japan) 17: 51–63.

1977. Population studies of *Cavelerius saccharivorus* Okajima (Hemiptera: Lygaeidae) adult dispersal in relation to the density. Res. Popul. Ecol. (Japan) 21: 153–163.

1979. Dispersal and reproductive properties of sugarcane bugs *Cavelerius saccharivorus* (Hemiptera: Lygaeidae). Res. Popul. Ecol. (Japan) 21: 153–163.

Murdock, G. P. 1967. Ethnographic Atlas. University of Pittsburgh Press, Pittsburgh, Pennsylvania, U.S.A. 128 pp.

Murray, D. A. H. 1980. Strawberry pests. Queensland Agric. J. 106: 248–254.

Myatt, O. W. D. 1973. Linear bug damage in the Herbert River district. Cane Growers Q. Bull. 36: 134–135.

Myburgh, A. C., and P. J. Kreigler. 1967. The grain stink-bug, *Blissus diplopterus* Dist., as a pest of export fruit, with special reference to its cold-hardiness. J. Entomol. Soc. South. Africa 29: 90–95.

Myers, J. G. 1921. Insect pests of lucerne and clover. New Zealand J. Agric. 23: 156–162.

1926. Biological notes on New Zealand Heteroptera. Trans. New Zealand Inst. 56: 449–411.

Nagamine, M., R. Teruya, and Y. A. Ito. 1975. Life table of *Mogannia iwasakii* (Homoptera: Cicadiidae) in sugarcane field of Okinawa. Res. Popul. Ecol. 17: 39–50.

Nagata, R. T., and R. H. Cherry. 1999. Survival of different life stages of the southern chinch bug (Hemiptera: Lygaeidae) following insecticidal applications. J. Entomol. Sci. 34: 126–131.

Nair, M. R. G. K. 1975. Insects and mites of crops in India. Pp. 80–90. Indian Council of Agricultural Research, New Delhi, India.

Nakache, Y., and M. Klein. 1992. The cotton seed bug, *Ocycarenus* [sic] *hyalinipennis*, attacked various crops crops in Israel in 1991. Hassadeh 72: 773–775 [in Hebrew with English summary].

1993. The cotton seed bug, *Oxycarenus hyalinipennis*, attacks sunflowers in the Beisan Valley. Hassadeh 73: 853–856 [in Hebrew with English summary].

Negrón, J. F., and T. J. Riley. 1985. Effect of chinch bug (Hemiptera: Lygaeidae) feeding in seedling field corn. J. Econ. Entomol. 78: 1370–1372.

1990. Long-term effects of chinch bug (Hemiptera: Lygaeidae) feeding on corn. J. Econ. Entomol. 83: 618–620.

1991. Seasonal migration and overwintering of the chinch bug (Hemiptera: Lygaeidae) in Louisiana. J. Econ. Entomol. 84: 1681–1685.

Neiswander, R. B. 1944. Insect pests of strawberries in Ohio. Ohio Agric. Exp. Stn. Bull. No. 651. 37 pp.

Newman, L. J. 1928. The rutherglen bug (*Nysius vinitor*), Order: Hemiptera. Family: Lygaeidae. J. Dept. Agric. W. Australia 5: 322–324.

Nicholson, K. 1988. Insect detection: a seed bug *Eremocoris depressus* Barber (Hemiptera: Lygaeidae). Tribology 27: 7.

Nicholls, H. M. 1932. The rutherglen bug. Tasmanian J. Agric. 3: 51–53.

Niemczyk, H. D. 1980. Insects and their control. Lawn Care Ind., Mar.: 34–46.
1981. Destructive Turf Insects. HDN Books, Wooster, Ohio, U.S.A. 48 pp.
1982. Chinch bug and bluegrass billbug control with spring application of chlorpyrifos. Pp. 85–89 *in* H. D. Niemczyk and B. G. Joyner [eds.], Advances in Turfgrass Entomology. Hammer Graphics, Piqua, Ohio, U.S.A. 149 pp.
Niemczyk, H. D., and B. G. Joyner [eds.]. 1982. Advances in Turfgrass Entomology. Hammer Graphics, Piqua, Ohio, U.S.A. 149 pp.
Niemczyk, H. D., R. A. J. Taylor, M. P. Tolley, and K. T. Power. 1992. Physiological time-driven model for predicting first generation of the hairy chinch bug (Hemiptera: Lygaeidae) on turfgrass in Ohio. J. Econ. Entomol. 85: 821–829.
Nixon, G. E. J. 1938. XXXI. — Three new Telenominae. Ann. Mag. Nat. Hist. 2: (1) 279–288.
1946. Euphorine parasistes of capsid & lygaeid bugs in Uganda (Hymenoptera, Braconidae). Bull. Entomol. Res. 37: 113–129.
Odhiambo, T. R. 1957. The bionomics of *Oxycarenus* species (Hemiptera, Lygaeidae), and their status as cotton pests in Uganda. J. Entomol. Soc. South. Africa 20: 235–249.
Ohrnberger, D. 1999. The Bamboos of the World. Elsevier, Amsterdam. 585 pp.
Okajima. 1922. Concerning a serious new pest on the sugar cane. Nogakukaiho (Tokyo) 236: 363–371 [in Japanese].
Olagbemiro, T. O., and E. W. Staddon. 1983. Isoprenoids from metathoracic scent gland of cotton seed bug *Oxycarenus hyalinipennis* (Hemiptera: Lygaeidae). J. Chem. Ecol. 9: 1397–1412.
Olckers, T., and P. E. Hulley. 1989. Insect herbivore diversity on the exotic weed *Solanum mauritianum* Scop. and three other *Solanum* species in the eastern Cape Province. J. Entomol. Soc. South. Africa 52: 81–93.
1991. Seed damage by three local Heteroptera species on the exotic weed *Solanum elaeagnifolium* (Solanaceae). J. Entomol. Soc. South. Africa 54: 269–270.
Olckers, T., P. E. Hulley, and M. P. Hill. 1995. Insect herbivores associated with indigenous species of *Solanum* (Solanaceae) in the Transvaal, South Africa, and in Namibia. Afric. Entomol. 3: 49–57.
Oliver, A. D., and K. M. Komblas. 1968. Controlling chinch bugs in St. Augustine grass. Louisiana Agric. 11: 3–16.
Oliver, J. E., A. J. Reinecke, and S. A. Reinecke. 1996. Verdedigingsekresies van die graanstinkluis *Macchiademus diplopterus* (Heteroptera: Lygaeidae). S.-Afr. Tydskr. Natuurwet. Tegnol. 15 [in Afrikaans with English abstract].
Oshiro, Y. 1979. Population dynamics of the Oriental chinch bug *Cavelerius saccharivorus* (Hemiptera: Lygaeidae) in sugar cane fields. 1. Occurrence of macropterous form and its migration to the spring and summer sugar cane fields. Kontyu 47: 606–615.
1981. Studies on the population dynamics of the Oriental chinch bug, *Cavelerius saccharivorus* (Hemiptera: Lygaeidae) in sugar cane field. Part 2. Effects of temperature, day length and population density on the appearance of the macropterous adult. Kontyu 49: 385–389.
Packard, A. S., Jr., 1870. Treatise on Injurious and Beneficial Insects to Crops. Reprinted by Logos Press, New Delhi, India. 702 pp.
Packard, C. M., and C. Benton. 1937. How to fight the chinch bug. U.S. Dept. Agric. Farm. Bull. 1780. 21 pp.
Painter, R. H. 1928. Notes on the injury to plant cells by chinch bug (Hemiptera: Lygaeidae) feeding. Ann. Entomol. Soc. Amer. 21: 232–242.
1951. Insect Resistance in Crop Plants. University Press of Kansas, Lawrence, Kansas, U.S.A.
Painter, R. H., H. R. Bryson, and D. A. Wilbur. 1954. Insects and mites that attack wheat in Kansas. Bull. Kansas State Coll. Agric. Exp. Stn. 367: 1–47.
Pal, S. K. 1975. Pest records of desert grasses and legumes. Entomol. Newsl. 5: 40.
Palgrave, K. C. 1977. Trees of Southern Africa. C. Struik, Cape Town, South Africa.
Palmer, W. A. 1986. Host specificity of *Ochrimnus mimulus* (Stål) (Hemiptera: Lygaeidae) with notes on its phenology. Proc. Entomol. Soc. Washington 88: 451–454.
Pandey, B. N., B. P. Singh, R. Dayal, and R. Sanehi. 1977. Efficacy of quinalphos and chlorpyriphos emulsions in controlling black bug of sugarcane. Indian Sugar Crops J. 4: 89–91.
Parenzan, P. 1985. Damage to jojoba (*Simmondsia chinensis*) by *Nysius (Macroparius) cymoides* Spin. (Rhynchota–Heteroptera–Lygaeidae) in Apulia. Entomologica 20: 99–108 [in Italian].

Parenzan, P., F. Porcelli, and A. Sinacori. 1994. *Nysius graminicola* (Klt. 1858) e *Oxycarenus hyalinipennis* (Costa 1838) (Rhynchota, Lygaeidae) su kenaf in Basilicata. Inf. Fitopatol. 44: 62–64 [in Italian with English abstract].

Parihar, D. R. 1981 (1983). Some ecological observations of insect pests of aak (*Calotropis procera*) and their significance in Rajasthan desert. Indian J. For. 4: 191–195.

Parker, F. W., and N. M. Randolph. 1972. Mass rearing of the chinch bug (Hemiptera: Lygaeidae) in the laboratory. J. Econ. Entomol. 65: 894–895.

Parker, J. R. 1920. The chinch bug in Montana. J. Econ. Entomol. 13: 318–321.

Parshley, H. M. 1919. On some Hemiptera from western Canada. Occ. Pap. Mus. Zool. Michigan 7: 35.

Peairs, L. M. 1941. Insect Pests of Farm, Garden and Orchard, fourth edition. John Wiley & Sons, New York, New York, U.S.A. 549 pp.

Pearson, E. O. 1958. The Insect Pests of Cotton in Tropical Africa. CIE, London, U.K. 355 pp.

Pearson, J. F., and S. L. Goldson. 1980. A preliminary examination of pests of fodder beet in Canterbury. Pp. 211–214 *in* M. J. Hartley [ed.], Proceedings of the 33rd New Zealand Weed and Pest Control Conference. Willow Park Motor Hotel, Tauranga, Aug. 12–14, 1980. New Zealand Weed and Pest Control Society, Inc., Palmerston North, New Zealand.

Pennacchio, F., and R. Marullo. 1986. Danni da *Nysius ericae* sulla vite in Campania. Inf. Fitopatol. 36: 23–25 [in Italian with English summary].

Péricart, J. 1984. Hémiptères Berytidae Euro-Méditerranéens. Faune de France. France et régions limitrophes, Vol. 70. Féderation française des Sociétés de Sciences Naturelles, Paris, France.

Perini, T., and L. Tamanini. 1961. Osservazioni sulla coomparsa in massa dell'*Oxycarenus lavaterae* (F.). Studi. Trentini Sci. Nat. 38: 57–66.

Peters, L. L. 1982. Susceptibility of chinch bugs to selected insecticides — laboratory study (Hemiptera: Lygaeidae). J. Kansas Entomol. Soc. 55: 317–322.

1983. Chinch bug (Heteroptera: Lygaeidae) control with insecticides on wheat, field corn, and grain sorghum, 1981. J. Econ. Entomol. 76: 178–181.

Pinkston, K. 1998. 1999 OSU Extension Agent's handbook of insect, plant disease and weed control. Oklahoma State University Publ. E-832. Oklahoma Cooperative Extension Service, Stillwater, Oklahoma.

Pivnick, K. A., D. W. Reed, J. G. Millar, and E. W. Underhill. 1991. Attraction of northern false chinch bug *Nysius niger* (Heteroptera: Lygaeidae) to mustard oil. J. Chem. Ecol. 17: 931–941.

Polivka, J. B. 1963. Control of the hairy chinch bug, *Blissus leucopterus hirtus* Mont. (Hemiptera: Lygaeidae) in Ohio. Ohio Agric. Exp. Stn. Res. Circ. 122: 8 pp.

Polivka, J. B., and F. Irons. 1966. Experimental control of chinch bugs (Hemiptera: Lygaeidae) on corn. J. Econ. Entomol. 59: 759.

Poole, L., and N. Adams. 1990. Trees and Shrubs of New Zealand. Revised Edition. DSIR Publishing, Wellington, New Zealand. 256 pp.

Popov, P. 1973. Insect pests on medicinal plants in Bulgaria. I. Hemiptera. Rastenievud. Nauki 10: 157–164 [in Bulgarian with English summary].

Porcelli, F., and F. A. Palmieri. 1996. *Oxycarenus hyalinipennis* (Costa 1838) (Rhynchota, Lygaeidae) on kenaf in Basilicata. Entomologica 30: 197–205 [in Italian].

Posado Ochoa, L., I. Zenner de Polania, I. S. de Arevalo, A. Saldarria-Ga V., F. Garcia R., and R. Cardenas, M. 1976. Lista insectos dañinos y otras plagas en Colombia. Bogotá, Instituto Colombiano Agropecuario, Programa de Entomología. Bol. Tecn. No. 144: 484 pp.

Prabhakar, B., P. K. Rao, and B. H. K. M. Rao. 1981. Note on hemipterous species complex on sorghum at Hyderabad. Indian J. Agric. Sci. 51: 818–819.

1986. Studies on the seasonal prevalence of certain Hemiptera occurring on sorghum. Entomon 11: 95–99.

Prestidge, R. A., R. P. Pottiner, and G. M. Barker. 1982. An association of *Lolium* endophyte with ryegrass resistance to Argentine stem weevil. Pp. 119–122 *in* Proceedings, 35th New Zealand Weed Pest Control Conference. New Zealand Weed and Pest Control Society, Palmerston North, New Zealand.

Price, P. W., and G. P. Waldbauer. 1975. Ecological aspects of pest management. Pp. 37–73 *in* R. L. Metcalf and W. H. Luckmann [eds.], Introduction to Pest Management. John Wiley & Sons, New York, New York, U.S.A.

Pupedis, R. J., C. W. Schaefer, and R. P. Duarte. 1985. Postembryonic changes in some sensory structures of the Tingidae (Hemiptera: Heteroptera). J. Kansas Entomol. Soc. 58: 277-289.

Putschkov, V. G. 1956. Basic trophic groups of phytophagous hemipterous insects and changes in the character of their feeding during the process of development. Zool. Zh. 35: 32–44 [in Russian].
1958a. Larvae of Hemiptera-Heteroptera. I. Lygaeidae. Rev. Entomol. U.S.S.R. 37: 392–213 [translated from Russian].
1958b. Larvae of the subfamily Blissinae (Heteroptera, Heteroptera) in the fauna of the European part of the USSR. Akad. Nauk Ukrain. RsR. N. 11: 1272–1274 [translated from Russian].
Putschkova, L. V. 1956. The eggs of Hemiptera-Heteroptera. II. Lygaeidae. Rev. Entomol. 35: 262–284 [translated from Russian].
Quaintance, A. L. 1897. Strawberry insects. Florida Agric. Exp. Stn. Bull. 42: 551–600.
Qureshi, M. B., M. S. Awan, and A. M. Dharejo. 1992. Population dynamics, dispersal pattern and host records of dusky cotton bug, *Oxycarenus laetus* Kirby (Lygaeidae: Heteroptera) in Sindh. Pakistan J. Zool. 4: 223–225.
Rahman, K. A., and P. S. Cheema. 1947. Field-studies on sugarcane black bug (*Macropes excavatus* Dist.) in the Punjab. Indian J. Agric. Sci. 17: 291–295.
Rahman, K. A., and R. Nath. 1939. The black bug of sugar cane. *Macropes excavatus* Dist (Lygaeidae: Heteroptera). Indian J. Entomol. 1: 25–34.
Rajashekhargouda, R. M., C. Devaiah, and S. Yelshetty. 1984. New record of insect pests infesting kasturi bhendi, *Hibiscus abelomoschus* Linnaeus, a medicinal plant. J. Bombay Nat. Hist. Soc. 81: 212–213.
Ram, P., and N. P. Chopra. 1984. Host plant relationships of *Oxycarenus hyalinipennis* (Costa) (Hemiptera: Lygaeidae: Oxycareninae). Bull. Entomol. 25: 111–116.
Ramamurty, P. S. 1970. A histological study of the internal organs of reproduction in *Lygaeus* sp. (Lygaeidae-Insecta). Indian J. Zool. 11: 67–78.
Raman, K. 1987. Nutritional value of malvaceous seeds and related life-table analysis in terms of feeding and reproductive indices in the dusky cotton bug, *Oxycarenus laetus* Kirby (Hemiptera: Lygaeidae). Proc. Anim. Sci. – Indian. Acad. Sci. 96: 195–206.
Ramesh, P., and R. Laughlin. 1984. Dispersal of nymphs of *Nysius vinitor* Berg. and its importance in the management of sunflower crops. Pp. 15–20 *in* P. T., Bailey and D. E. Swincer [eds.], Proceedings, 4th Austral. Appl. Entomol. Res. Conf. Adelaide. S. Austral. Dept. Agric.
Ramesh, P. 1994a. Movement of *Nysius vinitor* Berg. (Hemiptera–Lygaeidae) nymphs on the soil surface. Bull. Entomol. 35: 40–47.
1994b. Dispersal capacity of the nymphs of Rutherglen bug, *Nysius vinitor* Berg (Hemiptera: Lygaeidae). Bull. Entomol. 35: 48–52.
1995 (1996). Monitoring up and down movements of the Rutherglen bug *Nysius vinitor* Berg. (Hemiptera: Lygaeidae) on the stems of sunflower (*Helianthus annuus* L.) plants. Indian J. Entomol. 57: 240–244.
Ramoska, W. A. 1983. The influence of relative humidity on *Beauveria bassiana* infectivity and replication in the chinch bug *Blissus leucopterus* (Hemiptera: Lygaeidae). J. Invertebr. Pathol. 43: 389–394.
Ramoska, W. A., and T. Todd. 1985. Variation in efficacy and viability of *Beauveria bassiana* in the chinch bug (Hemiptera: Lygaeidae) as a result of feeding activity on selected host plants. Environ. Entomol. 14: 146–148.
Rao, B. S. 1963. Pests of leguminous covers in Malaya and their control. Plant. Bull. Rubber Res. Inst. 68: 172–186.
Ratcliffe, R. H. 1982. Evaluation of cool-season turfgrasses for resistance to the hairy chinch bug. Pp. 13–18 *in* H. D. Niemczyk and B. J. Joyner [eds.], Advances in Turfgrass Entomology. Hammer Graphics, Piqua, Ohio.
Rawat, R. R., and H. R. Sahu. 1973 (1974). A study on the bionomics of *Chauliops fallax* Scott (Heteroptera: Lygaeidae) at Sehore (Madhya Pradesh). J. Bombay Nat. Hist. Soc. 70: 475–479.
Readio, P. A. 1923. The history of *Jalysus spinosus* (Say) (Neididae, Heteroptera). Canad. Entomol. 55: 230–236.
Reinert, J. A. 1972a. New distribution and host record for the parasitoid *Eumicrosoma benefica*. Florida Entomol. 55: 143–144.
1972b. Turf-grass insect research. Proc. Florida Turfgrass Manage. Conf. 20: 79–84.
1978a. Antibiosis to the southern chinch bug by St. Augustine grass accessions. J. Econ. Entomol. 71: 21–24.
1978b. Natural enemy complex of the southern chinch bug in Florida. Ann. Entomol. Soc. Amer. 71: 728–731.

1982a. A review of host resistance in turfgrasses to insects and acarines with emphasis on the southern chinch bug. Pp. 3–12 *in* H. D. Niemczyk and B. G. Joyner [eds.], Advances in Turf Grass Entomology. Hammer Graphics, Piqua, Ohio. 149 pp.
1982b. Insecticide resistance in epigeal insect pests of turfgrass. I. A review. Pp. 71–76 *in* H. D. Niemczyk and B. G. Joyner [eds.], Advances in Turf Grass Entomology. Hammer Graphics, Piqua, Ohio. 149 pp.
1982c. Carbamate and synthetic pyrethroid insecticides for control of organophosphate-resistant southern chinch bugs (Heteroptera: Lygaeidae) J. Econ. Entomol. 75: 716–718.
1982d. Southern chinch bug resistance to insecticides: a method for quick diagnosis of chlopyrifos (OP) resistance and alternate controls. Florida Turfgrass Proc. 30: 64–78.
Reinert, J. A., and A. E. Dudeck. 1974. Southern chinch bug resistance in St. Augustine grass. J. Econ. Entomol. 67: 275–277.
Reinert, J. A., P. R. Heller, and R. L. Crocker. 1995. Pest information. chinch bugs. Pp. 38–42 *in* R. L. Brandenburg and M. G. Villani [eds.] Handbook of Turfgrass Insect Pests. Entomological Society of America, Lanham, Maryland, U.S.A. 140 pp.
Reinert, J. A., B. D. Bruton, and R. W. Toler. 1980. Resistance of St. Augustine grass to southern chinch bug and St. Augustine decline strain of *Panicum* mosaic virus. J. Econ. Entomol. 73: 602–604.
Reinert, J. A., P. Busey, and F. G. Bilz. 1986. Old World St. Augustine grasses resistant to the southern chinch bug. J. Econ. Entomol. 79: 1073–1075.
Reinert, J. A., and K. M. Portier. 1983. Distribution and characterization of organophosphate resistant southern chinch bugs in Florida. J. Econ. Entomol. 76: 1187–1190.
Reinert, J. A., and S. H. Kerr. 1973. Bionomics and control of lawn chinch bugs. Bull. Entomol. Soc. Amer. 19: 91–92.
Reinert, J. A., and H. D. Niemczyk. 1982. Insecticide resistance in epigeal insect pests of turfgrass. II. Southern chinch bug resistance to organophosphates in Florida. Pp. 77–80 *in* H. D. Niemczyk and B. G. Joyner [eds.], Advances in Turf Grass Entomology. Hammer Graphics, Piqua, Ohio. 149 pp.
Reis, P. R., A. Costa, Jr., and L. Lobato. 1976. *Blissus leucopterus*, nova praga de Gramineas introduzida no estado de Minas Gerais. An. Soc. Entomol. Brasil 5: 241–242.
Remade, F. 1960. Contribution a l'étude des Rhynchotes Hetéroptères terrestres de Provence. Ann. Soc. Entomol. France 129: 201–222.
Richardson, C. H., C. C. Deonier, and W. A. Simanton. 1937. The toxicity of certain insecticides to the chinch bug. J. Agric. Res. 54: 59–78.
Richie, A. H. 1925. Entomological report 1924–1925. Tanganyika Terr. Rept. Dept. Agric. 1924–1925: 25: 41–44.
Riley, C. V. 1870. The chinch bug — *Micropterus leucopterus* Say. Pp. 15–37 *in* Second Annual Report on the Noxious and Beneficial Insects of the State of Missouri, etc. 2nd Annu. Rep. State Board Agric. for 1869.
1873. Eighth Annual Report on the Noxious, Beneficial and Other Insects of the State of Missouri, etc., 98th Annu. Rep. State Board Agric. for 1872. 160 + 8 pp.
1875. The chinch bug — *Micropterus leucopterus*, Say. Pp. 19–71 *in* Seventh Annual Report on Noxious and Beneficial Insects of the State of Missouri, etc., 7th Ann. Rept. St. Board Agric. for 1874.
Risbec, J. 1941. Les insectes de l'arachide. Trans. Lab. Entomol. Sect. Soudan Rech. Agron. 21 pp. [Abstract in Rev. Appl. Entomol. (A) 36: 44].
Rivero, J. M. del, and F. G. Marí. 1983. El hemíptero heteróptero chinche gris, *Nysius ericae* (Schill.), como plaga. Bol. Serv. Def. Plagas Insp. Fitopatol. 9: 3–13 [in Spanish with English abstract].
Rivnay, E. 1962. Field Crops in the Near East. W. Junk, The Hague, the Netherlands.
Robbins, W. W., M. K. Bellue, and W. S. Bali. 1951. Weeds of California. California Dept. Agric. Exp. Stn. Bull. 547 pp.
Rodriguez, R. L., and W. G. Eberhard. 1994. Male courtship before and during copulation in two species of *Xyonysius* bugs (Hemiptera, Lygaeidae). J. Kansas Entomol. Soc. 67: 37–45.
Romeo, A. 1929. Contributo alla conscenza del pitstaccio (nota preliminaire). 8 pp. [Abstract in Rev. Appl. Entomol. (A). 33: 51].
Roubaud, E. 1916. Les insectes et la dégénérescence des arachides au Sénégal. Annu. Mém. Com. Etud. Hist. Sci. Afr. Occid. Françoise 76 pp. [Abstract in Rev. Appl. Entomol. (A) 5: 338–339].
Rude, C. S. 1950. Cotton injured by *Nysius californicus*. J. Econ. Entomol. 43: 548–549.

Ryckman, R. E., and C. M. Blankenship. 1984. The Triatominae and Triatominae-borne trypanosomes of North and Central America and the West Indies: a bibliography with index. Bull. Soc. Vector Ecol. 9: 112–430.

Sadana, G. L., and N. Goel. 1996. Predation on some insects of oilseed crops by spider *Oxyopes pandae* Tikader (Oxopidae: Arachnidae). J. Insect Sci. 9: 71–72.

Sadoyama, Y. 1998. Oviposition behavior of *Eumicrosoma blissae* (Maki) (Hymenoptera, Scelionidae), an egg parasitoid of the Oriental chinch bug, *Cavelerius saccharivorus* Okajima (Heteroptera: Lygaeidae). Appl. Entomol. Zool. 33: 207–213.

1999. A simple method for collecting eggs of the Oriental chinch bug, *Cavelerius saccharivorus* Okajima (Heteroptera: Lygaeidae). Appl. Entomol. Zool. 34: 155–159.

Safti, M. 1986. Outbreak of pests and diseases. Pakistan Q. Newsl., Asia Pac. Plant Prot. Comm., FAO, Thailand 29: 25.

Saha, D. C., J. M. Johnson-Cicalese, P. M. Halisky, M. I. Van Heemstra, and C. R. Funk. 1987. Occurrence and significance of endophytic fungi in the fine fescues. Plant Dis. 70: 1021–1024.

Saini, S. S. 1985. Outbreak of sugarcane black bug, *Cavelerius excavates* Distant (Hemiptera: Lygaeidae) during summer 1977 in Punjab. Plant. Prot. Bull. 37: 39–40.

Sales, F. M., M. de F. B. Goncalves, O. F. G. Martins, and C. Mendes 1979. Insetos e outros artrópodes de importancia âgricola, em perimetros irrigados e de sequeiro do Estado do Piauí. Fitossanidade 3: 12–18.

Samy, O. 1969. A revision of the African species of *Oxycarenus* (Hemiptera: Lygaeidae). Trans. R. Entomol. Soc. London 121: 79–165.

Sanderson, E. D. 1906. Report on miscellaneous cotton insects in Texas. U.S. Dept. Agric. Bur. Entomol. Bull. 57: 29–31.

Sandhu, J. S., and M. S. Duhra. 1984. Control of sugarcane black bug. Indian Sugar Crops J. 10: 7–8.

Sandhu, J. S., and Y. P. Madan. 1972. Studies on the chemical control of the sugarcane lygaeid bug (*Macropes excavatus*). J. Res. India 9: 1: 64–67.

Sandhu, G. S., B. Singh, and J. S. Bhalla. 1974. Notes on the chemical control of milkweed bug (Hemiptera: Lygaeidae) on pearl millet with different low volume concentrate insecticides in Punjab India. Indian J. Agric. Sci. 44: 588–599.

Sanjayan, K. P. 1993. Patterns in the population drift and seed resource utilization of *Spilostethus hospes* (Fab.). Entomon 18: 159–168.

Sanjayan, K. P., T. Ravikumar, and S. Albert. 1996. Changes in the haemocyte profile of *Spilostethus hospes* (Fab.) (Heteroptera: Lygaeidae) in relation to eclosion, sex and mating. J. Biosci. (Bangalore) 21: 781–788.

San Martin, P. R. 1966. Notas sobre "*Neoblissus parasitaster*," Bergroth, 1903 (Hemiptera: Lygaeidae). Rev. Bras. Biol. 26: 247–251.

Santos, J. H. R. dos, F. V. Vieira, and J. F. Alves. 1977. The influence of application of insecticides at the time *Oxycarenus hyalinipennis* (Costa, 1847) is present on the yield of 'moco' cotton in the following season's crop. Fitossanidade 2: 11–13 [in Portuguese].

Sardana, H. R., and N. Mukunthan. 1991. Ecological studies relevant to the management of black bug, *Cavelerius excavates* Slater and Mugomoto [sic] (Hemiptera: Lygaeidae) on sugarcane. Indian J. Entomol. 53: 608–618.

Saxena, K. N., and P. L. Bhatnagar. 1958. Physiological adaptations of dusky cotton bug, *Oxycarenus hyalinipennis* (Costa) (Heteroptera; Lygaeidae) to its host plant, cotton. Proc. Nat. Inst. Sci. India 24: 245–257.

Schaefer, C. W. 1975. Heteropteran trichobothria (Hemiptera: Heteroptera). Int. J. Insect Morphol. Embryol. 4: 193–264.

1981. Improved cladistic analysis of the Piesmatidae and consideration of the known host plants. Ann. Entomol. Soc. Amer. 74: 536–539.

1997. The origin of secondary carnivory from herbivory in the Heteroptera. Pp. 229–239 *in* A. Raman [ed.], Ecology and Evolution of Plant-feeding Insects in Natural and Man-made Environments. International Scientific Publications, New Delhi, India.

1998a. The taxonomic status of *Xyonysius major* (Berg) (Hemiptera: Lygaeidae), an occasional pest of sunflower in Brazil. An. Soc. Entomol. Brasil 27: 55–58.

1998b. Phylogeny, systematics, and practical entomology: the Heteroptera (Hemiptera). An. Soc. Entomol. Brasil 27: 449–511.
Schalk, J. M., and R. H. Ratcliffe. 1977. Evaluation of the United States Department of Agriculture program on alternative methods of insect control: host plant resistance to insects. FAO Plant Prot. Bull. 25: 141–146.
Scheel, C. A., S. D. Beck, and J. T. Medler. 1957. Nutrition of plant-sucking Hemiptera. Science 125: 444–445
Schmitz, G. 1976. La faune terrestre de L'ile de Sainte-Hélène. Troisieme Partie. 2. Insectes (suite et fin). 20. Heteroptera. 5. Fam. Lygaeidae. Ann. Mus. R. Afr. Cent., Zool., Ser. 8. Sci. Zool. 215: 391–410.
Schmitz, G., and J. Péricart. 1993. Contribution à une mise en ordre de la nomenclature du genre *Nysius* Dallas (*sensu lato*) pour la région Paléarctique (Hemiptera, Lygaeidae). Nouv. Rev. Entomol. 10: 173–186.
Schread, J. C. 1950. Chinch bug control in lawns. Connecticut Agric. Exp. Stn. Circ. 168. 6 pp.
1960. Insect pests of Connecticut lawns. Connecticut Agric. Exp. Stn. Circ. 212. 10 pp.
1963. The chinch bug and its control. Connecticut Agric. Exp. Stn. Circ. 223. 4 pp.
1970. Chinch bug control. Connecticut Agric. Exp. Stn. Circ. 233: 6 pp.
Schroeder, N., and P. Clifford. 1996. The incidence of insect pests and their arthropod predators in 24 Canterbury white clover seed crops. Pp. 29–33 *in* Joint Symposium, White Clover: New Zealand's Competitive Edge. Lincoln University, New Zealand, 21–22 November, 1995. Agronomy Society of New Zealand, Christchurch, New Zealand.
Sears, M. K. 1978. Hairy chinch bugs in lawns. Ontario (Canada) Ministry Agric. Food Factsheet AGDEX 626: 2 pp.
Sears, M. K., F. L. Mcewen, G. Ritcey, and R. R. McGraw. 1980. Evaluation of insecticides for the control of the hairy chinch bug *Blissus leucopterus hirtus* (Hemiptera: Lygaeidae) on Ontario, Canada Lawns. Proc. Entomol. Soc. Ontario 111: 13–20.
Selim, M. M., R. Dvorak, T. Khalifa, I. Al-Awadi, A. Al-Humaidi, and M. Al-Faris. 1990. Insect bite lesions in Kuwait possibly due to *Leptodemus minutus*. Int. J. Dermatol. 29: 507–510.
Severin, H. C. 1923. The chinch bug. South Dakota Agric. Mech. Arts. Bull. 202: 562–576.
Shah, G. S., G. E. Wilde, and P. J. Bramel-Cox. 1994. Resistance to chinch bug *Blissus leucopterus leucopterus* (Say). (Heteroptera: Lygaeidae) among greenbug resistant grain sorghum sources. Sarhad J. Agric. 10: 419–424
Sharma, K. C., U. Chauhan, A. K. Verma, and A. K. Sood. 1993. Biology of the black bean bug, *Chauliops nigrescens* Distant (Hemiptera: Lygaeidae) infesting French bean, *Phaseolus vulgaris* (L.). J. Entomol. Res. (New Delhi) 17: 305–307.
Sharma, P. L., and O. P. Balla. 1964. Survey study of insect pests of economic importance in Himachal Pradesh. Indian J. Entomol. 26: 318–331.
Shelford, V. E. 1932. Experimental and observational study of the chinch bug (Hemiptera: Lygaeidae) in relation to climate and weather. Illinois Nat. Hist. Surv. 19: 478–547.
Shelford, V. E., and W. P. Flint. 1943. Populations of the chinch bug (Hemiptera: Lygaeidae) in the upper Mississippi Valley from 1823–1940. Ecology 24: 435–456.
Shimer, H. 1867. Notes on *Micropus (Lygaeus) leucopterus,* Say. (the chinch bug). With an account of the great epidemic disease of 1865 among insects. Proc. Acad. Nat. Sci. Philadelphia 19: 75–80.
Shiraki, T. 1952. Catalogue of injurious insects in Japan (exclusive of animal parasites). Prelim. Stud. Econ. Sci. Sect. Nat. Resources Div. G. H. Q. Allied Powers, Tokyo 3: 71: 1–162.
Short, D. 1998. Turf insect management suggestions (commercial and non-commercial). Cooperative Extension Service, University of Florida, Gainesville, Florida.
Sim, J. T. R. 1965. Wheat production in South Africa. Dep. Agric. Tech. Serv., South Africa Bull. 377: 1–75.
Simmonds, F. J. 1971. Report on investigations into natural enemies of *Triatoma* for the World Health Organization. Commonwealth Institute of Biological Control, Trinidad.
Simpson, B. B., and M. C. Conner-Ogorzaly. 1995. Economic Botany: Plants in Our World, second edition, McGraw-Hill, New York, New York, U.S.A.
Singh, A. 1990. Control of black bug, *Cavelerius sweeti* Snm. [sic] in sugarcane. Pp. 151–155 *in* Proceedings, 52nd Annual Convention of the Sugar Technologists' Association of India, 1989. Sugar Technologists' Association of India, Kanpur, India.
Singh, J. P., and S. D. Rai. 1967. Record of seed bug *Aphanus* in Tarai area. Allahabad Farmer 41: (3) 143–144.

Singh, K. J., and O. P. Singh. 1993. Efficacy of biopesticides on eggs and adults of cowpea bug (*Chauliops fallax*) infesting soybean (*Glycine max*). Indian J. Agric. Sci. 63: 756–758.

Singh, M., R. A. Singh, L. K. Vishwakarma, and V. D. Verma. 1991. Effective and economical doses of some recommended insecticides for the control of black bug (*Macropes excavatus* Dist.) in U.P. Bharatiya Sugar 16: 33–34.

Singh, O. P., K. J. Singh, and K. K. Nema. 1987a. Bionomics and host plants of *Chauliops fallax* Scott (Hemiptera: Lygaeidae), a potential pest of soybeans in Madhya Pradesh, India. Trop. Grain. Leg. Bull. N. 34: 16–20.

Singh, P. P., and M. C. Sharma. 1984. Hydrogen-ion concentration and distribution of digestive enzymes in alimentary canal and salivary glands of *Spilostethus macilentus* Stål (Lygaeidae: Heteroptera). Indian J. Entomol. 46: 12–17.

Singh, O. P., K. K. Nema, and S. N. Verma. 1987b. New insect pests to soybean in Madhya Pradesh, India. FAO Plant Prot. Bull. 35: 100–102.

Singh, S. R. 1990. Insect Pests of Tropical Food Legumes. John Wiley & Sons, New York, New York, U.S.A.

Slansky, F., Jr., and A. R. Panizzi. 1987. Nutritional ecology of seed-sucking insects. Pp. 283–320 *in* F. Slansky, Jr. and J. G. Rodriguez [eds.], Nutritional Ecology of Insects, Mites, Spiders and Related Invertebrates. John Wiley & Sons, New York, New York, U.S.A. 1016 pp.

Slater, A. 1978. Taxonomic notes on Lygaeinae from Australia and neighboring areas (Heteroptera: Lygaeidae). Ann. Entomol. Soc. Amer. 71: 854–858.

1985. A taxonomic revision of the Lygaeinae of Australia (Heteroptera: Lygaeinae of Australia (Heteroptera: Lygaeidae). Univ. Kansas Sci. Bull. 52: 301–481.

Slater, J. A. 1964a. A Catalogue of the Lygaeidae of the World. 2 vols. University of Connecticut, Storrs, Connecticut, U.S.A. 1668 pp.

1964b. Ethiopian Lygaeidae II. The subfamily Blissinae in tropical Africa (Hemiptera: Lygaeidae). Rev. Zool. Bot. Africa 70: 305–355.

1964c. Results of the Lund University Expedition in 1950-1951. Chapter II. Hemiptera (Heteroptera): Lygaeidae. South Afric. Anim. Life 10: 15–228.

1967. Insectes Hétéroptères Lygaeidae Blissinae. Faune Madagascar 25: 1–53.

1968. A contribution to the systematics of Oriental and Australian Blissinae (Hemiptera: Lygaeidae). Pac. Insects 10: 275–294.

1972a. The occurrence of *Elasmolomus sordidus* (F.), a potential pest of peanuts in Brazil (Hemiptera: Lygaeidae). O Biológico 38: 394–397.

1972b. Lygaeid bugs as seed predators of figs. Biotropica 4: 145–151.

1972c. The Oxycareninae of South Africa (Hemiptera: Lygaeidae). Univ. Connecticut Occ. Pap. (Biol. Sci. Ser.) 2: 59–105.

1974. The genus *Dimorphopterus* (Hemiptera: Lygaeidae: Blissinae). Trans. R. Entomol. Soc. London 126: 57–89.

1976a. Monocots and chinch bugs: a study of host plant relationships in the lygaeid subfamily Blissinae (Hemiptera: Lygaeidae). Biotropica 8: 143–165.

1976b. The immature stages of Lygaeidae (Hemiptera: Heteroptera) of southwest Australia. J. Austral. Entomol. Soc. 15: 101–126.

1979. The systematics, phylogeny and zoogeography of the Blissinae of the World (Hemiptera, Lygaeidae) Bull. Amer. Mus. Nat. Hist. 165: 1–180.

1994. The occurrence of *Oxycarenus hyalinipennis* (Hemiptera: Lygaeidae) in the West Indies and New Lygaeidae records for the Turks and Caicos Islands of Providenciales and North Caicos. Florida Entomol. 77: 495–497.

Slater, J. A., and I. Ahmad. 1965. A revision of the genus *Pirkimerus* Distant (Hemiptera: Lygaeidae, Blissinae). Trans. R. Entomol. Soc. London 117: 313–328.

1967. A new species of *Macropes* from Afghanistan with synonymic notes on other species of the genus (Hemiptera, Lygaeidae). Acta Entomol. Mus. Nat. Pragae 37: 255–259.

Slater, J. A., and R. M. Baranowski. 1978. How to Know the True Bugs. The Pictured Key Nature Series. Wm. C. Brown Co., Dubuque, Iowa, U.S.A.

1990. Lygaeidae of Florida (Hemiptera: Heteroptera). Vol 14. Arthropods of Florida and neighboring land areas. Florida Dept. Agric. Consumer Serv., Gainesville, Florida, U.S.A. i–xv + 211 pp.

1994. The occurrence of *Oxycarenus hyalinipennis* (Costa) (Hemiptera: Lygaeidae) in the West Indies and new Lygaeidae records for the Turks and Caicos Islands of Providenciales and North Caicos. Florida Entomol. 77: 495–497.

Slater, J. A., and W. E. China. 1961. *Blissus* Burmeister, 1835 (Insecta, Hemiptera): proposed designation of a type species under the plenary powers. Z. N. (S.) 1471. Bull. Zool. Nomen. 18: 346–348.

Slater, J. A., and B. J. Harrington. 1970. A revision of the genus *Ischnodemus* Fieber in the Ethiopian Region (Hemiptera: Lygaeidae, Blissinae). Ann. Transvaal Mus. 26: 211–275.

Slater, J. A., and S. Miyamoto. 1963. A revision of the sugar cane bugs of the genus *Cavelerius* (Lygaeidae: Blissinae). Mushi 37: 139–154.

Slater, J. A., and J. E. O'Donnell. 1995. A Catalogue of the Lygaeidae of the World (1960–1994). New York Entomological Society, New York, New York, U.S.A.

Slater, J. A., and B. Sperry. 1973. The biology and distribution of the South African Lygaeinae, with descriptions of new species (Hemiptera: Lygaeidae). Ann. Transvaal Mus. 28: 117–201.

Slater, J. A., and M. H. Sweet. 1970. Two new species of ant-mimetic Lygaeidae from South Africa. J. Kansas Entomol. Soc. 43: 221–237.

Slater, J. A. and D. B. Wilcox. 1969. A revision of the genus *Ischnodemus* in the Neotropical Region (Hemiptera: Lygaeidae: Blissinae). Misc. Publ. Entomol. Soc. Amer. 6: 197–238.

1973a. The chinch bugs or Blissinae of South Africa (Hemiptera: Lygaeidae). Mem. Entomol. Soc. South Africa 12: 1–135.

1973b. A revision of the genus *Macropes* (Hemiptera: Lygaeidae: Blissinae). Pac. Insects 15: 213–258.

Slater, J. A., P. D. Ashlock, and D. B. Wilcox. 1969. The Blissinae of Thailand and Indochina. Pac. Insects 11: 671–733.

Slater, J. A., M. H. Sweet, and H. Brailovsky. 1993. Two new species of *Slaterobius* Harrington with comments on the ecology and distribution of the genus (Hemiptera: Lygaeidae). Proc. Entomol. Soc. Washington 95: 590–602.

Sloan, W. J. S. 1942. The control of tomato pests. Queensland Agric. J. 56: 277–294.

Smit, B. 1964. Insects in Southern Africa: How to Control Them. Oxford University Press, Thibault House, Cape Town, South Africa.

Smith, E. C. 1855. The chinch bug. Cultivator 3 (3): 237–238.

Smith, G. L. 1942. California cotton insects. California Agric. Exp. Stn. Bull. 57: 29–31.

1944. Insects affecting vegetable seed crops in the Western States. J. Econ. Entomol. 37: 363–370.

Smith, J. H. 1927. Life history notes on the rutherglen bug. Queensland Agric. J. 27: 285–302.

Smith, M. T., G. Wilde, and T. Mize. 1981. Chinch bug damage and effects of host plant and photoperiod. Environ. Entomol. 10: 122–124.

Snelling, R. O. 1936. Third generation and method of migration of chinch bug in southwestern Oklahoma. J. Econ. Entomol. 2: 797–803.

Snelling, R. O., R. H. Painter, J. H. Parks, and W. M. Osborn. 1937. Reaction of the sorghums to the chinch bug. U.S. Dept. Agric. Tech. Bull. 585: 56 pp.

Snow, F. H. 1893. Contagious diseases of the chinch bug. Second Annual Report of the Director of Kansas State University for the year 1892. 56 pp.

Solbreck, C., and D. B. Anderson. 1989. Wing reduction; its control and consequences in a lygaeid bug, *Spilostethus pandurus*. Hereditas 111: 1–6.

Southcott, R. V. 1988. Some harmful Australian insects. Med. J. Australia 149: 656–662.

Southwood, T. R. E. 1960. The flight activity of Heteroptera. Trans. R. Entomol. Soc. London 112: 173–220.

1961. A hormonal theory of the mechanism of wing polymorphism in Heteroptera. Proc. R. Entomol. Soc. London (A) 36: 63–66.

1962. Migration of terrestrial arthropods in relation to habitat. Biol. Rev. 37: 171–214.

1969. Population studies of insects attacking sugar cane. Pp. 237–459 *in* J. R. Williams, J. R. Metcalfe, R. W. Mungomery, and R. Mathes [eds.], Pests of Sugar Cane. International Society of Sugar Cane Technologists, Elsevier, Amsterdam, the Netherlands.

Southwood, T. R. E., and D. Leston. 1959. Land and Water Bugs of the British Isles. Warne, London, U.K. 436 pp.

Souza, S. O. D., J. Santana, F. J. Lisboa, and A. Shimoya. 1995. Performance of six tropical grasses to the incidence of *Blissus leucopterus* (Say, 1832) (Hemiptera: Lygaeidae) in the presence and absence of fertilization. Ciencia Pratica 19: 369–374.

Spike, B. P., R. J. Wright, S. Danielson, and D. W. Stanley-Samuelson. 1991. The fatty acid compositions of phospholipids and triacylglycerols from two chinch bug species *Blissus leucopterus leucopterus* and *Blissus iowensis* (Hemiptera: Lygaeidae) are similar in the characteristic dipteran pattern. Comp. Biochem. Physiol. 998: 799–802.

Spike, B. P., G. E. Wilde, T. W. Mize, R. J. Wright, and S. D. Danielson. 1994. Bibliography of the chinch bug, *Blissus leucopterus leucopterus* (Hemiptera: Lygaeidae) since 1888. J. Kansas Entomol. Soc. 67: 116–125.

Staddon, B. W., and T. O. Olagbemire. 1984. Composition and novel pattern of emission of defensive scent oils in the larva of the cotton seed bug *Oxycarenus hyalinipennis* (Hemiptera: Lygaeidae). Experientia 40: 114–116.

Stål, C. 1874. Enumeratio Hemipterorum. K. Svenska Vetensk. Akad. Handl. Forh., Stockholm.

Starks, K. J., A. J. Casady, O. G. Meekle, and D. Boozaya-Angoon. 1982. Chinch bug resistance in Pearl millet, *Pennisetum americanum*. J. Econ. Entomol. 75: 337–339.

Stefferud, A. 1948. Grass. Yearbook of Agriculture. U.S. Department of Agriculture, Washington, D.C., U.S.A.

Steinhaus, E. A. 1949. Principles of Insect Pathology. McGraw-Hill, New York, New York, U.S.A. 757 pp.

Steinhaus, E. A., M. M. Batey, and C. L. Boerke 1956. Bacterial symbioses from the caeca of certain Heteroptera. Hilgardia 24: 495–518.

Streu, H. T., and L. M. Vasvary. 1966. Pesticide activity and growth response effects in turfgrass. Bull. New Jersey Agric. Exp. Stn. Bull. 17–21.

Steyermark, J. A. 1977. Flora of Missouri. Iowa State University Press, Ames, Iowa, U.S.A.

Stichel, W. 1957. Illustrierte Bestimmungstabellen der Wanzen. — II. Europa 4 (3): 65–96.

Stone, J. D., and J. N. Fries. 1986. Insect fauna of cultivated guayule, *Parthenium argentatum* Gray (Companulatae: Compositae). J. Kansas Entomol. Soc. 59: 1, 49–58.

Streu, H. T. 1973. The turfgrass ecosystem: impact of pesticides. Bull. Entomol. Soc. Amer. 19: 89–91.

Streu, H. T., and C. Cruz. 1972. Control of the hairy chinch bug in turf grasses in the northeast with Dursban insecticide. Down Earth 28: 1–4.

Stringfellow, T. L. 1967. Studies on turfgrass insect control in south Florida. Proc. Florida State Hortic. Soc. 80: 486–491.

1968. Studies on turfgrass insect control in south Florida. Proc. Florida State Hortic. Soc. 81: 447–454.

1969a. Developments in Florida turfgrass insect control, 1969. Proc. Florida Turfgrass Manage. Conf. 17: 94–100.

1969b. Turfgrass insect research in Florida. Pp. 19–33 *in* H. T. Streu and R. T. Bangs [eds.], Proceedings of Scott's Turfgrass Research Conference. O. M. Scott & Sons, Marysville, Ohio, U.S.A. 89 pp.

Strobel, J. 1971. Turfgrass Proc. Florida Turfgrass Manage. Conf. 19: 19–28.

Stuart, J., G. Wilde, and J. H. Hatchett. 1985. Chinch bug (Heteroptera: Lygaeidae) reproduction, development and feeding preference on various wheat cultivars and genetic sources. Environ. Entomol. 14: 539–543.

Studzinski, A., and D. Malachowska. 1973. Hemipterous bugs (Heteroptera) occurring on wild cruciferous plants (Cruciferae) in Poland in 1970. Rocz. Nauk Rolni., E. 3: 79–100 [in Polish, with English abstract].

Štys, P. 1963. Notes on the taxonomy, distribution and evolution of the Chauliopinae (Lygaeidae: Heteroptera). Acta Univ. Carolinae Biol. 2: 209-216.

1966a. Revision of the genus *Dayakiella* Horv. and notes on its systemical position (Heteroptera: Colobathristidae). Acta Entomol. Bohemoslov. 63: 27-39.

1966b. Morphology of the wings, abdomen and genitalia of *Phaenacantha australiae* Kirk. (Heteroptera, Colobathristidae) and notes on the phylogeny of the family. Acta Entomol. Bohemoslov. 63: 266–280.

1967. Monograph of Malcinae, with reconsideration of morphology and phylogeny of related groups (Heteroptera, Malcidae). Acta Entomol. Mus. Nat. Pragae 37: 351–516.

Štys, P., and I. Kerzhner. 1975. The rank and nomenclature of higher taxa in recent Heteroptera. Acta Entomol. Bohemoslov. 72: 65–79.

Suehiro, A. 1959. Insects and other arthropods from Midway Atoll. Proc. Hawaiian Entomol. Soc. 17: 289–298.

Suzuki, M. 1990. New summer crop varieties registered by the Ministry of Agriculture, Forestry and Fisheries in 1990: paddy rice, sweet potato, soybean and sugarcane. Jpn. J. Breed. 40: 537–547 [in Japanese, with English abstract].

Suzuki, H., M. Murai, and N. Hokyo. 1979. Temperature dependence in oviposition of overwintering adults of *Cavelerius saccharivorus* Okajima (Heteroptera: Lygaeidae). Appl. Entomol. Zool. 14: 126–127.

Swallow, W. H., and P. J. Cressey. 1987. Historical overview of wheat-bug damage in New Zealand wheats. New Zealand J. Agric. Res. 30: 341–344.

Swallow, W. H., and D. Every. 1991. Insect enzyme damage to wheat. Cereal Foods World 36: 505–508.

Swamy, B. C. H., and D. Rajagopal. 1995a. Insect pests of vasaka — a medicinal plant. Univ. Agric. Sci. Bangalore Curr. Res. 24: 7, 129.

1995b. Insect pests of *Gloriosa superba* Linn. — an Indian medicinal plant. Indian J. For. 18: 158–160.

Swart, P. L. 1978. Less important insect pests of stone fruits. Decid. Fruit Grow. 28: 238–242, 244–247.

Sweet, M. H. 1960. The seed bugs: a contribution to the feeding habits of the Lygaeidae (Hemiptera: Heteroptera). Ann. Entomol. Soc. Amer. 53(3): 317–321.

1964. The biology and ecology of the Rhyparochrominae of New England. (Heteroptera: Lygaeidae). Pts. I and II. Entomol. Amer. 43: 1–124, 44: 1–201.

1967. The tribal classification of the Ryparochrominae (Heteroptera: Lygaeidae). Ann. Entomol. Soc. Amer. 60: 208–226.

1979. On the original feeding habits of the Hemiptera (Insecta). Ann. Entomol. Soc. Amer. 72: 575–579.

1981. External morphology of the pregenital abdomen and its evolutionary significance in the order Hemiptera. Rostria 33: 41–51.

1992. The paraphyly of the Lygaeidae. Int. Congr. Entomol. Beijing, China. 1992 (Abstr.).

Sweet, M. H., and C. W. Schaefer. 1985. Systematic status and biology of *Chauliops fallax* Scott, with a discussion of the phylogenetic relationships of the Chauliopinae (Hemiptera: Malcidae). Ann. Entomol. Soc. Amer. 76: 526–536.

Swenk, M. H. 1925. The chinch bug and its control. Nebraska Agric. Exp. Stn. Circ. 28. 34 pp.

Syrett, P., and L. A. Smith. 1998. The insect fauna of four weedy *Hieracium* (Asteraceae) species in New Zealand. New Zealand J. Zool. 25: 73–83.

Tachikawa, S., T. Kobayashi, and H. Hasegawa. 1976. Identification of some heteropterans producing pecky rice. Shokubutsu Bōeki 30: 149–154 [in Japanese].

Takano, H., and M. Yanagihara. 1939. Investigation on sugar cane pests and beneficial insects and animals. Taiwan Gov. Sugar Exp. Stn. Bull. No. 2.

Takara, T., and S. Azuma. 1969. Important insect pests affecting sugarcane and problems of their control in the Ryukyu Islands. Pp. 1424–1432 *in* Proceedings, International Society of Sugarcane Technologists. Thirteenth Congress, Taiwan, March 2–17, 1968. Elsevier Publishing Company, Amsterdam, the Netherlands.

Tamaki, G., and R. E. Weeks. 1972. Biology and ecology of two predators, *Geocoris pallens* Stål and *G. bullatus* (Say). USDA Tech. Bull. 1446. 46 pp.

Tamanini, L. 1961. Ricerche zoologiche sul Massiccio del Pollino. XXX. Emitteri Eterotteri. Ann. Ist. Mus. Zool. Univ. 13: 1–128.

Tan, S. W., and C. A. Johnson. 1974. Yellowing of sugarcane leaves caused by *Phaenacantha saccharicida* Karsch (Hemiptera: Lygaeidae [sic]). Pp. 453–456 *in* Proceedings, 15th Congress of the International Society Sugar Cane Technology. Vol. 1. Durban, South Africa.

Tappan, W. B. 1967. Chemical control and physical autecology of insects attacking cigar-wrapper tobacco. Florida Agric. Exp. Stn. Annu. Rep. 1967: 156.

1970. *Nysius raphanus* attacking tobacco in Florida and Georgia. J. Econ. Entomol. 63: 658–600.

Tashiro, H. 1987. Turfgrass Insects of the United States and Canada. Cornell University Press, Ithaca, New York, U.S.A. 391 pp.

Tayutivutikul, J., and K. Yano. 1989. Biology of insects associated with Kudzu plant 1. *Chauliops fallax*. Jpn. J. Entomol. 57: 831–842.

Terry, L. I. 1991. Pest and predator populations following early-season cotton insect control in Arizona. Southwest. Entomol. 16: 51–62.

Tewari, R. K. 1980. Field evaluation of some modern insecticides against sugarcane black bug (*Cavelerius excavatus* Dist.) (Hemiptera: Lygaeidae). Pestology 4: 21–22.

Texas Agricultural Extension Service. 1997. Rice production guidelines. Publ. No. D-1253. Texas Agricultural Extension Service, College Station, Texas, U.S.A.

Thangavelu, K. 1978a. First record of host plants and additional distribution of *Nysius inconspicuous* Distant (Hemiptera: Lygaeidae). Curr. Sci. 47: 249.

1978b. Population dynamics of the dusky cotton bug *Oxycarenus laetus* Kirby (Hemiptera: Lygaeidae) in relation to climatic variation. Proc. Indian Acad. Sci. Sect. B 87: 387–396.

1978c. Some aspects of host specificity in the Indian dusky cotton bug, *Oxycarenus laetus* Kirby (Hemiptera: Lygaeidae). J. Nat. Hist. 12: 481–486.

1978d. Some notes on the ecology of three milkweed bugs in India (Heteroptera: Lygaeidae). J. Nat. Hist. 12: 641–647.

1979. The pest status and biology of *Spilostethus pandurus* (Hemiptera: Lygaeidae). Entomon 4: 137–142.

1980. Influence of climate on the population of three milkweed bugs in south India (Heteroptera: Lygaeidae). Proc. Indian Acad. Sci. Anim. Sci. 89: 579–586.

1981. Present systematic status of the Indian milkweed bug (Heteroptera: Lygaeidae). Orient. Insects 15: 93–95.

1983. Investigation on the pest status of the dusky cotton bug (*Oxycarenus laetus* Kirby) (Lygaeaidae: Heteroptera) on cotton. Turrialba 33: 94–97.

Theobald, F. V. 1906. Notes on African cotton insects. Entomologist 39: 27–30.

Thomas, C. 1879. The chinch bug. U.S. Dept. Interior Entomol. Comm. Bull. 5: 1–44.

Thomson, W. T. 1998. Agricultural Chemicals. Book 1. Insecticides, Acaricides and Ovicides. 14th revision. Thomson Publications, Fresno, California, U.S.A.

Tieszen, K. L., D. H. Molyneux, and S. K. Abdel-Hafez. 1983. Investigation of the pest status of the dusky cotton bug *Oxycarenus laetus* Kirby (Hemiptera: Lygaeidae) on cotton. Turrialba 33: 94–97.

1985. Ultrastructure of cyst formation in *Blastocrithidia familiaris* in *Lygaeus pandurus* (Hemiptera: Heteroptera: Lygaeidae). Z. Parasitenk. 71: 179–188.

1986. Host-parasite relationships of *Blastocrithidia familiaris* in *Lygaeus pandurus* (Hemiptera: Heteroptera: Lygaeidae). Parasitology 92: 1–12.

1989. Host-parasite relationships and cysts of *Leptomonas*. Parasitology 98: 395–400.

Tillyard, P. 1926. The Insects of Australia and New Zealand. Angus and Robertsson, Sidney, Australia.

Tomokuni, M. 1993. A Field Guide to Japanese Bugs: Terrestrial Heteropterans. Zenkoku Noson Kyoiku Kyokai Publ. Co. Tokyo, Japan. 380 pp.

Torre-Bueno, J. R. 1946. A synopsis of the Hemiptera-Heteroptera of America north of Mexico. III. Family XI. Lygaeidae. Entomol. Amer. 26: 1–141.

Tremewan, W. G. 1986. *Spilostethus pandurus* (Scopoli) (Hemiptera: Lygaeidae) feeding on *Zygaena (Agumenia) johannae* le Cerf (Lepidoptera: Zygaenidae). Gazette 37: 74.

Trought, T. E. T. 1975. A comparison of chemicals for the control of *Nysius* on brassicas. Pp. 226–229 *in* Proceedings of the Twenty-eighth New Zealand Weed and Pest Control Conference, Angus Inn, Hastings, Aug. 5–7, 1975, New Zealand Weed and Pest Control Society, Inc., Hamilton, New Zealand.

Uphof, J. C. Th. 1968. Dictionary of Economic Plants. J. Cramer, Stecher-Hafner Service Agency, New York, New York, U.S.A.

Usinger, R. L. 1938. A review of the genus *Gastrodes* (Lygaeidae, Hemiptera). Proc. California Acad. Sci. (Ser. 4) 23: 298–301.

1942. The genus *Nysius* and its allies on the Hawaiian Islands (Hemiptera, Lygaeidae, Orsillini). Bull. Bishop Mus. 173: 1–167.

1946. Insects of Guam. II. Hemiptera-Heteroptera of Guam. Bull. Bishop Mus. 189: 1–237.

1949. War-time dispersal of Pacific Island *Nysius* (Hemiptera: Lygaeidae). Proc. Hawaiian Entomol. Soc. 13: (3) 447.

Usinger, R. L., and P. D. Ashlock. 1959. Revision of the Metrargini (Hemiptera, Lygaeidae). Proc. Hawaiian Entomol. Soc. 17: 93–116.

Valério, J. R., J. M. Viera, and C. S. Valle. 1998. Ocorrência de *Blissus slateri* (Heteroptera, Blissidae) em pastagem no Mato Grosso do Sul. Resumos XVII Congr. Bras. Entomol., 205 pp.

Van Duzee, E. P. 1914. A preliminary list of the Hemiptera of San Diego County, California. Trans. San Diego Soc. Nat. Hist. 2: 1–57.

Varma, A., and A. K. Mehrotra. 1985. Evaluation of different granular insecticides in the control of black bug, *Cavelerius excavatus* Dist. Indian Sugar Crops J. 11: 55–56.

Varma, A., and G. S. Shukla. 1976 (1981). Rearing technique of *Lygaeus militaris* Fabricius (Lygaeidae: Hemiptera). Indian J. Zool. 17: 187–190.

Varma, A., and S. K. Srivastava. 1977. Seasonal occurrence and feeding behavior of *Lygaeus militaris* (Hemiptera: Lygaeidae). Geobios 4 (6): 254–256.

Varma, A. B., and K. Tandan. 1996. Pathogenicity of three entomogenous fungi against insect pests of sugarcane. J. Biol. Control 10: 87–91.

Varma, A., B. K. Tandan, and K. Singh. 1988. First record of Beauveria bassiana (Balsamo) Vuillemin, an entomogenous fungus from the sugarcane defoliator *Phytoscaphus* sp. (Coleoptera: Curculionidae) from India. Curr. Sci. 57: 396.

Varma, B. R., and M. M. Prakash. 1980. Variations in sterol and sterol esters during development of *Lygaeus militaris* (Hemiptera: Lygaeidae). Sci. Cult. 46: 434–435.

Vásquez J., M. N., and G. Sánchez G. 1991. Biologia habitos y huespedes de la chinche de las raices *Blissus leucopterus* (Say) (Hemiptera: Lygaeidae). Rev. Colombiana Entomol. 17: 8–15.

Vattanakul, J., and H. A. Rose. 1982. The culture of the grey cluster bug, *Nysius clevelandensis* Evans (Hemiptera: Lygaeidae) under laboratory conditions. Austral. Entomol. Mag. 8: 71–72.

Velimirovic, V., Z. Djurovic, and M. Raicevic. 1992. Bug *Oxycarenus lavaterae* Fabricius (Hemiptera: Lygaeidae) new pest on lindens in southern part of Montenegro. Zast. Bilja 43: 69–72.

Veneuw, J., T. Reichstein, and M. Rothschild. 1972. Heart poisons in the lygaeid bugs *Caenocoris nervii* and *Spilostethus pandurus*. Insect Biochem. 1: 273–384.

Vickerman, K. 1962. Observation of the life cycle of *Phytomonas elmassiani* (Migone) in East Africa. J. Protozool. 9: 26–33.

Vishakantaiah, M., and B. L. B. Gowda. 1973. Two new hosts of *Lygaeus pandurus* Scopoli (Hemiptera: Lygaeidae) in Mysore. Madras Agric. J. 60: 340.

Wachmann, E. 1989. Wanzen Beobachten — Kennen Lernen. Newmann-Neudamm, Melsungen, Germany.

Wadley, F. M. 1940. *Telonomus ovivorus* (Ashmead), an egg parasite of the false chinch bug. J. Kansas Entomol. Soc. 13: 6–7.

Wagner, E., and J. A. Slater. 1965. Zur Systematik der Blissinae Stål in der Paläarktis (Hemiptera-Heteroptera: Lygaeidae). Entomol. Ber. (Amsterdam), 24: 66–76.

Wang, Z. F., and L. Y. Zheng. 1985. Preliminary reports on four heteropterous pests of conifers from Zhejiang Province. J. Zhejiang For. Coll. 2: 64–66.

Watson, J. R. 1931. Florida truck and garden insects. Florida Agric. Exp. Stn. Bull. 232: 1–112.

Watson, J. R., and A. N. Tissot. 1942. Insects and other pests of Florida vegetables. Florida Agric. Exp. Stn. Bull. 370: 1–118.

Watts, J. G., G. B. Hewitt, E. W. Huddleston, H. Grant Kinzer, R. J. Vavigne, and D. Ueckert. 1989. Rangeland Entomology, second edition. Range Science Series No. 1. 388 pp.

Weber, G., and O. Parada. 1994. Development of an integrated pest management for rice in Latin America. Pp. 733–748 *in* E.A. Heinrichs [ed.], Biology and Management of Rice Insects. John Wiley & Sons, New York, New York, U.S.A.

Webster, F. M. 1896a. The chinch bug *Blissus leucopterus* Say. Ohio Agric. Exp. Stn. Bull. 69. 23 pp.

1896b. The probable origin and diffusion of *Blissus leucopterus* and *Murgantia histrionica*. J. Cincinnati Soc. Nat. Hist. 18: 141–155.

1898. The chinch bug: its probable origin and diffusion, its habits and development, natural checks, and remedial and preventive measures, with mention of the habits of an allied European species. U.S. Dept. Agric. Div. Entomol. Bull. 15: 82 pp.

1909. The chinch bug (*Blissus leucopterus* Say). U.S. Dept. Agric. Bur. Entomol. Circ. 113: 1–27.

Wen, D. Y., S. S. Zhao, and X. Y. Deng. 1984. A list of bamboo pests from Guilin, Liu-shou, Nan-ning of China. J. Central-South For. Coll. 4: 2: 177–183.

Wene, G. P. 1953. The false chinch bug. Proc. Rio Grande Valley Hortic. Inst. 7: 75–76.

1958. Control of *Nysius raphanus* Howard (Hemiptera: Lygaeidae) attacking vegetables. J. Econ. Entomol. 51: 250–251.

Wene, G. P., L. A. Carruth, A. D. Telford, and L. Hopkins. 1965. Descriptions and habits of Arizona cotton insects. Arizona Agric. Exp. Stn. Bull. A-23: 61 pp.

Werner, G. F. [chairman]. 1982. Common names of insects and related organisms. Entomol. Soc. Amer. 132 pp.

Wheeler, A. G., Jr. 1975. Birch catkin bug — a nuisance insect. Pennsylvania Pest Control Q. Spring: 4.

1976. Life history of *Kleidocerys resedae* (Hemiptera: Lygaeidae) on European white birch and ericaceous shrubs. Ann. Entomol. Soc. Amer. 69: 459–463.

1982. Bedbugs and other bugs. Pp. 319–351 *in* A. Mallis and F. Franzak [eds.], Handbook of Pest Control, sixth edition. Franzak & Foster, Cleveland, Ohio.

1984. Seasonal history, habits, and immature stages of *Belonochilus numenius* (Hemiptera: Lygaeidae). Proc. Entomol. Soc. Washington 86: 790–796.

1996. *Ischnodemus falicus* (Heteroptera: Lygaeidae): first record from ornamental grasses, and seasonality on prairie cordgrass in Pennsylvania. Proc. Entomol. Soc. Washington 98: 195–198.

1989a. *Blissus breviusculus:* new distribution records of a little-known chinch bug (Hemiptera: Lygaeidae). J. New York Entomol. Soc. 97: 265–270.

1989b. *Eremocoris borealis* and *E. ferus* as household pests in Pennsylvania and Connecticut. Entomol. News 100: 165–168.

Wheeler, A. G., Jr., and C. W. Schaefer. 1982. Review of stilt bug (Hemiptera: Berytidae) host plants. Ann. Entomol. Soc. Amer. 75: 498–506.

White, D. T., S. J. Billington, K. B. Walsh, and P. T. Scott. 1997. DNA sequence analysis supports the association of phytoplasmas with papaya (*Carica papaya*) dieback, yellow crinkle and mosaic. Austral. Plant Pathol. 26: 28–36.

Wickramashinghe, M. B. 1969. Developmental stages of *Edulica ornata* (Heteroptera: Lygaeidae) associated with the Ceylon palm squirrel *Funambulus palmatum.* Spolia Zeylan. 31: 395–397.

Wilde, G. E. 1979. Chinch bug (Heteroptera: Lygaeidae) resistance in grain sorghum. Proc. 34th Annu. Corn and Sorghum Industry Res. Conf. Amer. Seed Trade Assoc. Pub. 34: 188–192.

1997. Effect of imidicloprid seed treatment and planting time applications of insecticides on chinch bug (Hemiptera, Lygaeidae) and resulting yields of sorghum. J. Agric. Entomol. 14: 385–391.

Wilde, G. E., and P. Cox. 1990. Recent studies of chinch bug resistance in sorghum. Proc. 45th Annu. Corn and Sorghum Industry Res. Conf. Amer. Seed Trade Assoc. Pub. 45: 121–130.

Wilde, G. E., and T. W. Mize. 1979. Using foliar and planting time insecticides to control chinch bugs in grain sorghum. Kansas Agric. Exp. Stn. Keeping Up with Research, 42.

Wilde, G., and J. Morgan. 1978. Chinch Bug (Hemiptera, Lygaeidae) on sorghum: chemical control, economic injury levels, plant resistance. J. Econ. Entomol. 71: 908–910.

Wilde, G., T. Mize, J. Stuart, J. Whitworth, and R. Kinsinger. 1984. Comparison of planting-time applications of granular or liquid insecticides and liquid fertilizer plus insecticide combinations for control of chinch bugs (Hemiptera: Lygaeidae) and greenbugs (Homoptera: Aphididae) on seedling sorghum. J. Econ. Entomol. 77: 706–708.

Wilde, G., O. Russ, and T. W. Mize. 1986a. Tillage, cropping, and insecticide use practice; effects on efficacy of planting time treatments for controlling greenbug (Homoptera: Aphididae) and chinch bug (Heteroptera: Lygaeidae) in seedling sorghum. J. Econ. Entomol. 79: 1364–1365.

Wilde, G., J. Stuart, and J. H. Hatchett. 1986b. Chinch bug (Heteroptera: Lygaeidae) reproduction on selected small grains and genetic sources. J. Kansas Entomol. Soc. 59: 550–551.

Wilson, C. G., and G. J. Flanagan. 1990. The phytophagous insect fauna of the introduced shrubs *Sida acuta* Burm. F. and *Sida cordifolia* L. in the Northern Territory, Australia. Austral. Entomol. Mag. 17: 7–15.

Wilson, L. T., P. M. Room, and A. S. Bourne. 1983. Dispersion of arthropods, flower buds and fruit in cotton fields: effects of population density and season on the fit of probability distributions. J. Austral. Entomol. Soc. 22: 129–134.

Wilton-Smith, F. D. 1978. Two hematophagous *Clerada* spp. (Heteroptera: Lygaeidae) inhabiting the nest of the ringtail possum *Pseudocheirus peregrinus.* J. Austral. Entomol. Soc. 17: 1–4.

Wolcott, G. N. 1936. Insectae Borinquenses. A revised annotated check-list of the insects of Puerto Rico. With a host plant index by J. Otero. J. Agric. Univ. Puerto Rico 20: 1–627.

1950 (1951). The insects of Puerto Rico. J. Agric. Univ. Puerto Rico 1948. 32: 1–975.

Wood, E. A., Jr., and K. J. Starks. 1972. Damage to sorghum by a lygaeid bug, *Nysius raphanus* (Heteroptera: Lygaeidae). J. Econ. Entomol. 65: 1507–1508.

Woodward, T. E. 1964. Preliminary note on the distribution of *Nysius visitor* Bergroth and *Nysius clevelandensis* Evans (Hemiptera: Lygaeidae). J. Entomol. Soc. Queensland 3: 85.

1984. The plant-sucking bug *Oxycarenus arctatus* (Hemiptera: Lygaeidae) developing in cattle dung in drought and post-drought conditions. J. Austral. Entomol. Soc. 23: 147–148.

Wright, R. J., and S. D. Danielson. 1992. First report of the chinch bug (Heteroptera: Lygaeidae) egg parasitoid *Eumicrosoma beneficum* Gahan (Hymenoptera: Scelionidae) in Nebraska. J. Kansas Entomol. Soc. 65: 346–348.

Wrightman, J. A., K. M. Dick, G. V. Ranga Rao, T. G. Shanower, and C. G. Gold. 1990. Pests of groundnut in the semi-arid tropics. Pp. 243–322 *in* S. R. Singh [ed.], Insect Pests of Tropical Food Legumes. John Wiley & Sons, New York, New York, U.S.A.

Wygodzinsky, P., and K. Schmidt. 1991. Revision of the New World Enicocephalomorpha (Heteroptera). Bull. Amer. Mus. Nat. Hist. 200: 1–265.

Xu, J. 1989. The spatial distribution pattern and sampling methods for the nymph of Oriental chinch bug (Hemiptera: Lygaeidae) in the seedling period of sugarcane. J. Fujian Agric. Coll. 18: 88–93.

Yadav, S. R., and N. P. Chopra. 1973. Effectiveness of various insecticides against *Spilostethus macilentus* Stal (Hemiptera: Lygaeidae). HAU J. Res. Hissar 3: 187–189.

Yamada, S., and S. Yagi. 1984. Induction of precocious metamorphosis in the Oriental chinch bug *Cavelerius saccharivorus* (Hemiptera: Lygaeidae) by precocene II. Jpn. J. Appl. Entomol. Zool. 19: 521–523.

Yamada, S., S. Yagi, K. Fujisaki, and N. Hokyo. 1984. An easy and successive rearing of the Oriental chinch bug *Cavelerius saccharivorus* (Hemiptera: Lygaeidae). Jpn. J. Appl. Entomol. Zool. 28: 25–29.

Yashiro, N. 1979. Studies on the Japanese species of *Oligosita* (Hymenoptera: Trichogrammatidae). Trans. Shikoku Entomol. Soc. 14: 195–203.

Yates, H. D. 1986. Checklist of insect and mite species attacking cones and seeds of world conifers. J. Entomol. Sci. 21: 142–168.

Yokoyama, S., T. Takasaki, and N. Fujiyoshi. 1972. Studies on the forecasting of Heteroptera feeding on the rice plant. I. Species and distribution in Fukuoka prefecture in 1971. Proc. Assoc. Plant Prot. Shikoku (Kyushu). 18: 51–53 [in Japanese].

Yoshida, M., T. Hayakawa, and M. Hatsukade. 1979. Studies on hemipterous insects injuring the turfgrass I. Bionomics and damage to turfgrass by the *Geoblissus hirtulus* Burmeister (Hemiptera: Lygaeidae). J. Jpn. Turfgrass Res. Assoc. 8: 129–135.

Yoshioka, E. 1965. *Nysius caledoniae* Distant. *In* Notes and exhibitions. Proc. Hawaiian Entomol. Soc. 19: 19.

Youdeowei, A. 1973. Some Hemiptera-Heteroptera associated with cacao farms in Nigeria. Turrialba 23: 162–171.

Young, V. L. 1960. Preliminary studies on *Chauliops fallax* Scott (Hemiptera: Lygaeidae) in western part of Hunan Province. Acta Entomol. Sin. 10: 67–71.

Young, W. R. 1970. Sorghum insects. *In* J. S. Wall and W. M. Ross [eds.], Sorghum Production and Utilization. Avi Publishing Co., Westport, Connecticut, U.S.A. 702 pp.

Zhang, S. M. 1985. Economic Insect Fauna of China. Fasc. 31 — Hemiptera. Zhongguo jingi kunchong zhi. 31: 242 pp.

Zheng, L.-Y. 1987. Lygaeidae. Pp. 185–188 *in* F. S. Huang and L.Y. Zheng [eds.], Forest Insects of Yunnan. Yunnan Science and Technology Press, Kumming, China.

1994. Heteropteran insects feeding on bamboos in China. Ann. Entomol. Soc. Amer. 87: 91–96.

Zheng, L. Y., and Z. F. Wang. 1987. Notes of some blissine bugs (Hemiptera: Lygaeidae) on bamboos from China. J. Bamboo Res. 6: 65–69 [in Chinese, with English summary].

Zheng, L. Y., and H. G. Zou. 1981. Lygaeidae. Pp. 1–214 *in* Hsiao, T. Y [ed.], A Handbook for the Identification of the Chinese Hemiptera-Heteroptera. Science Press, Peking, China [in Chinese].

Zucchi, R. A., S. S. Neto, and O. Nakono. 1993. Guia de identificação de pragas agrícolas. FEALQ, Piracicaba, São Paulo, Brazil. 139 pp.

CHAPTER 7

Ash-Gray Leaf Bugs (Piesmatidae)

Narisu

1. INTRODUCTION

Six genera in Piesmatidae, *Thaicoris, Miespa, Mcateella, Afropiesma, Parapiesma*, and *Piesma*, are known (Schaefer 1981, Schuh and Slater 1995, Heiss and Péricart 1997); the largest genus, *Piesma*, comprises half the world fauna. Piesmatids are usually small (1.5 to 4.0 mm long), mostly brownish or greenish gray (hence their common name), and often have ornate processes and decoration dorsally. Most live on well-exposed sandy soils, steppes, and sublittoral areas. Some piesmatids are important vectors of virus diseases of some economically important plants.

2. GENERAL STATEMENT OF ECONOMIC IMPORTANCE

Piesmatids are phytophagous and usually feed on the leaves, stems, and flowers of Chenopodiaceae, Caryophyllaceae, and Leguminosae (Schaefer 1981, 1983). *Piesma quadratum* (Fieber) and *P. cinereum* (Say) are important vectors of viral diseases of sugar beets. Other species, like *P. capitatum* (Wolff) and *P. maculatum* (Laporte de Castelnau), have also been reported from sugar beets (Heiss and Péricart 1983). However, feeding by these species on sugar beet does not transmit viral disease and causes little damage. The great majority of piesmatids is of no economic importance.

3. THE IMPORTANT SPECIES

3.1 *Piesma quadratum* (Fieber)

3.1.1 *Distribution, Life History, and Biology*

Adults are macropterous to brachypterous; oval; uniformly greenish-yellow to dark gray, mottled with brown. Length = 2.15 to 3.40 mm; width across hemelytra = 1.1 to 1.5 mm (Heiss and Péricart 1983).

According to Heiss and Péricart (1983) and the present author's own observations, eclosion begins after 2 to 3 weeks, nymphal development takes 4 to 6 weeks, and there are four to six instars independent of sex. All instars are green to greenish gray. Most studies of *P. quadratum* have been done in Germany and Poland, where it is the essential vector of a viral disease, called beet Kräuselkrankheit virus, a beet leaf curl disease of sugar beets (*Beta vulgaris* L.) (Wille 1929; Völk and Krzal 1957; Proeseler 1964, 1966a,b, 1978a,b,c, 1980; Schmutterer 1980). Although *P. quadratum* occurs throughout Europe, the disease does not occur in them (Dunning and Byford 1982). The author has observed the disease in northern China, where it is probably more widespread than stated. The following life-history statement is mostly based on work done in Germany and Poland.

Piesma quadratum generally has two generations a year. Adults leave the hibernating place in late April. Mating is preceded by a period of stridulation by males which causes aggregations, followed by slower calls (Leston 1957, Haskell 1958, Southwood and Leston 1959). Oviposition takes place in May, sometimes extended into August (Thomas 1956). Each female deposits 100 to 160 eggs in 1 to 3 days, with some interruptions. Eggs are deposited on the stems or leaves of the host plants, or even directly on soil or stones; they are glued frequently near and parallel to veins of the leaves and deposited singly, or sometimes in very small groups. The most precocious adults are developed in July and start a second generation. They lay eggs, producing summer nymphs, which complete development before the beginning of fall. A brief migration to hibernating places takes place from August to October. Hibernation occurs at a maximal depth of 5 to 15 cm beneath litter at the base of the host plants, preferably on dry, sandy, and sun-exposed places like slopes and roadsides.

The adult *P. quadratum* is resistant to starvation; it may survive without feeding for 2 to 3 weeks at 20°C and 4 to 6 weeks at 10 to 12°C. The lack of nourishment causes a progressive lethargy ending in quiescence with retraction of the appendages. The same effect, which does not differ from the condition during hibernation may be caused by excessive heat or cold; activity resumes after some minutes to some hours when conditions are again favorable (Heiss and Péricart 1983).

Piesma quadratum often feed in large groups, different instars and adults together on plants of *Chenopodium, Beta, Atriplex, Salsola*, and *Halimione* (Schaefer 1981, 1983).

3.1.2 Damage

The beet Kräuselkrankheit virus is a bacilliform particle measuring about 225 × 80 nm. It has an electron-dense core with a central channel and an enveloping membrane with small protrusions (Bugbee et al. 1986). The development of the virus within the insects appears not to have been studied.

The virus is transmitted through the saliva. Adults and nymphs of *P. quadratum* can infect plants 21 to 28 days after acquisition from feeding on diseased plants. The virus is not transmitted to the eggs by females (Heiss and Péricart 1983). The insect may acquire the virus in 30 min and may inoculate plants in feeding periods of about the same length after a latent period of 7 to 35 days. After a single feeding, the insect remains infective throughout life. The virus must feed directly on the tissues of the insect and *P. quadratum* is the essential host for the survival of the virus (Bugbee et al. 1986).

The disease is first visible when whitish spots appear on the leaves where the insects have fed. Leaves curl inward, forming a structure like a lettuce head. Later, leaves fade as if already aged. Young plants are affected by vein clearing, vein deformation, and inhibition of growth. The result is an important loss of sugar beet yield, which may reach 75% in the most serious cases (Heiss and Péricart 1983).

Eisbein (1976) and Francki et al. (1981) also mentioned *P. quadratum* as a vector of beet leaf curl virus, but the latter refer to the bug as a lace bug (Tingidae).

In Germany, beet rosette virus, or beet latent rosette disease, which can be transmitted to sugar beets by *P. quadratum*, has been reported, so far only in greenhouses (Nienhaus and Schmutterer 1976, Schmutterer 1976). The disease differs from both leaf curl and savoy (see below). It infects young plants, which do not produce green leaves beyond those of the first rosette; the plants die after some weeks. The microorganism, which can be transmitted by both adults and nymphs of *P. quadratum*, is a rickettsialike organism (150 × 700 to 1000 nm) (Bugbee et al. 1986). Under field conditions, sugar beets can be infected without considerable consequences. *Piesma quadratum* becomes a vector after 10 to 30 days of acquiring the pathogen, and remains infective for life (Heiss and Péricart 1983).

3.1.3 Control

Both chemical and cultural protection of sugar beets against these diseases has been used in the past. The cultural approach was the application of baiting stripes — several rows of sugar beets planted before the bugs emerged from hibernation and, once the bugs had oviposited, plowed under. The use of chemical pesticides such as parathion after 1954 was more successful, but the disease persisted. Other methods of control, such as the release of males sterilized by irradiation or the use of juvenile hormones blocking embryonic and nymphal development, have been investigated (Lefevre 1976, Weiss 1976–1977). However, the fight against this pest has proved difficult because wild Chenopodiaceae serve as reservoirs. The bugs' potential hibernacula and the considerable range of ovipositional period also protect *P. quadratum*.

3.2 Piesma cinereum (Say)

3.2.1 Distribution, Life History, and Biology

Adults are macropterous; oval; cream tinged with green to fuscous, occasionally with dark markings. Length = 2.8 to 3.3 mm; width = 1.3 to 1.4 mm (Bailey 1951).

Life history and development (which are similar to those of *P. quadratum*) have been studied more in the United States, where savoy disease occurs, than elsewhere. Eggs are deposited singly and fastened lightly to the leaf tissue and sides of veins of host plants. Nymphs are green or greenish with white markings and live on the lower leaf surfaces and flower heads, but the adults occur on any portion of the plant. There are five instars and reproduction is continuous throughout June to late August; in North America, there appear to be two generations per year (Barber 1924, Harry and Lott 1924, Bailey 1951).

In North America, wild host plants of *P. cinereum* are *Amaranthus, Chenopodium, Atriplex, Salsola, Hyptis, Scirpus*, and *Vitis* (Bailey 1951, Coons et al. 1958, Drake and Davis 1958, Schaefer 1981). *Piesma cinereum* also transmits a viral disease, beet savoy virus, or savoy, to sugar beet (Coons et al. 1937, 1950, 1958).

3.2.2 Damage

Affected sugar beet (*B. vulgaris* L.) shows dwarfed, down-curled, savoyed leaves, particularly the innermost leaves. Preliminary symptoms are clearing and then thickening of veinlets, giving the dorsal leaf surface a netted appearance. Roots of affected plants show, in late stages, phloem necrosis and flesh discoloration, simulating curly-top effects. Affected roots test 4.21 percentage points lower in sucrose content than healthy plants (Coons et al. 1937, 1950, 1958; Gaskill and Schneider 1966).

The virus is transmitted by adults of *P. cinereum*. It overwinters in affected plants and in the adult insects. The incubation period in sugar beet is 3 to 4 weeks. Attempts at transfer by means of juice, the aphids *Myzus persicae* (Sulzer) and *Aphis rumicis* L., and leafhoppers (including *Eutettix tenellus* Baker) have been unsuccessful. Apparently, this disease differs from beet leaf curl disease and beet latent rosette disease. The disease occurs in the midwestern United States, west to Colorado and Wyoming. Incidence ranges from a trace to 5%. The disease will probably remain a relatively minor factor affecting crop production (Coons et al. 1950, 1958).

Piesma cinereum has also been reported to damage the leaves and flowers of grapevines (Weiss and Lott 1924, Schaefer 1981), but it is of little economic importance.

3.2.3 Control

Control of the disease has been similar to that of *P. quadratum*. It is usually effective to eliminate wild hosts of *P. cinereum* near beet fields. Coons et al. (1958) stated that U.S. 215 × 216 variety of sugar beet, a leaf-spot-resistant variety, and U.S. 104 × 400, were both resistant to savoy disease in the western United States.

4. THE LESS IMPORTANT SPECIES

4.1, 4.2 *Piesma capitatum* and *Piesma maculatum*

Piesma capitatum and *P. maculatum* have been noxious to sugar beet in Russia and Finland, respectively (Vasilev 1911, Brunner 1954, Varis 1973, Heiss and Péricart 1983). They attack the very young sprouts, and in some cases influence normal growth, but clearly are of little economic importance; they do not transmit diseases. Cultural control, like that for *P. quadratum*, has been used. Treatment with insecticides such as dimethoates and parathion has also proven effective.

5. PROGNOSIS

Most information comes from studies of only a few common species. Because of their small size, piesmatids have been little studied, and their potential harm for the sugar beet industry is probably more serious than recognized. Even though leaf curl is not so serious as before in some European countries, it has caused severe damage in Germany and Poland, and might spread to other areas. In the United States, beet savoy has been a problem and the real damage done by the disease is probably more serious than realized.

6. CONCLUSIONS

Further research on the biology and behavior of these piesmatid vectors and the viruses will help renew control methods. Although piesmatids are of secondary importance compared to other families of Heteroptera, the species mentioned may be harmful because of their ability to transmit viral diseases, and also because they are not successfully controlled by management techniques. Clearly, more research focusing on the transmitting process is needed and a better understanding of the interaction of insect–virus may prove useful to mitigate the effects on plants such as sugar beets.

7. REFERENCES CITED

Bailey, N. S. 1951. The Tingoidea of New England and their biology. Entomol. Amer. 31: 1–140.
Barber, G. W. 1924. Notes on *Piesma cinerea* Say. Psyche 31: 229–232.
Brunner, Y. N. 1954. Specific composition and development of insect complexes, noxious to sugar beets in Khazakhastan and Middle Asia. Zool. Zh. 33: 1236–1244 [in Russian, English summary].

Bugbee, W. M., J. E. Duffus, B. B. Fischer, F. J. Hills, L. D. Leach, E. G. Ruppel, C. L. Schneider, E. E. Schweizer, A. E. Steele, A. Ulrich, E. D. Whitney, and Y. M. Yun. 1986. Biotic diseases and disorders. Pp. 3–39 *in* E. D. Whitney and J. E. Duffus [eds.], Compendium of Beet Diseases and Insects. APS Press, St. Paul, Minnesota, U.S.A.

Coons, G. C., J. E. Kotila, and D. Stewart. 1937. Savoy, virus disease of beet transmitted by *Piesma cinerea*. Phytopathology 27: 125.

1950. Savoy, virus disease of beet transmitted by *Piesma cinerea*. Proc. Amer. Soc. Sugar Beet. 6: 500–501.

Coons, G. H., D. Stewart, H. W. Bockstahler, and C. L. Schneider. 1958. Incidence of savoy in relation to the variety of sugar beets and to the proximity of wintering habitat of the vector, *Piesma cinerea*. Plant Dis. Rep. 42: 502–511.

Drake, C. J., and N. T. Davis. 1958. The morphology and systematics of the Piesmatidae and American species. Ann. Entomol. Soc. Amer. 51: 567–581.

Dunning, A., and W. Byford. 1982. Diseases. Pp. 51–98 *in* Pests, Diseases and Disorders of the Sugarbeet. Deleplanque & Cie, Paris, France.

Eisbein, J. 1976. Investigations for electron microscope demonstration of *Beta* virus 3 in *Beta vulgaris* L. and in *Piesma quadratum* Fieb. Arch. Pytopathol. Pflanzenschutz 12: 299–313.

Francki, R. I. B., E. W. Kitajima, and D. Peters. 1981. Rhabdoviruses. Pp. 455–489 *in* E. Kurstak [ed.], Handbook of Plant Virus Infections and Comparative Diagnosis. Elsevier/North-Holland Biomedical Press, Amsterdam, the Netherlands.

Gaskill, J. O., and C. L. Schneider. 1966. Savoy and yellow vein diseases of sugarbeet in the Great Plains in 1963–64–65. Plant Dis. Rep. 50: 457–459.

Harry, B. W., and R. B. Lott. 1924. Notes on *Piesma cinerea* Say in New Jersey (Hemiptera). Psyche 31: 233–235.

Haskell, P. T. 1958. Stridulation and its analysis in certain Geocorisae (Hemiptera Heteroptera). Proc. Zool. Soc. London 129: 351–358.

Heiss, E., and J. Péricart. 1983. Revision of Palaearctic Piesmatidae (Heteroptera). Mitt. Münch. Entomol. Ges. 73: 61–171.

1997. Revised taxonomic status of some Old World Peismatidae (Heteroptera). Z. Arbeitsgem. Österr. Entomol. 49: 119–120.

Lefevre, M. 1976. Der Einfluss von Juvenoiden mit unterschiedlicher Wirkungsweise auf die Embryonal- und Postembryonalentwicklung der Rübenblattwanze *Piesma quadratum* Fieb. Z. Angew. Entomol. 82: 187–192.

Leston, D. 1957. The stridulatory mechanisms in terrestrial species of Hemiptera Heteroptera. Proc. Zool. Soc. London 128: 369–386.

Nienhaus, F., and H. Schmutterer. 1976. Rickettsialike organisms in latent rosette (witches' broom) diseased sugar beet (Beta vulgaris) and spinach (Spinacia oleracea) plants and in the vector *Piesma quadratum* Fieb. Z. Pflanzenkr. Krankh. Pflanzenschutz 83: 641–646.

Proeseler, G. 1964. Der Nachweis der Vermehrung des Rübenkräuselkrankheits-Virus in *Piesma quadrata* (Fieb.) mit Hilfe der Injektionstechnik. Naturwissenschaften 51: 150–151.

1966a. Beziehung zwischen der Rübenblattwanze *Piesma quadratum* Fieb. und dem Rübenkräuselvirus. I. Virusübertragungsversuche und Zucht des Vektors. Phytopathol. Z. 56: 191–211.

1966b. Physiologische und histologische Untersuchungen an virusfreien und virustragenden *Piesma quadratum* Fieb. Biol. Zentralbl. 58: 211–229.

1978a. Entwicklungsdauer und Anzahl der Larvenstadien von *Piesma quadratum* (Fieb.) dem Vektor des Rübenkräuselvirus. Nachrichtenbl. Pflanzenschutz 32: 6–7.

1978b. Larven von *Piesma quadratum* (Fieb.) als Vektoren des Rübenkräuselvirus. Nachrichtenbl. Pflanzenschutz 32: 254–256.

1978c. Die Akquisitions- und Zirkulationszeit des Rübenkräuselvirus. Arch. Phytopathol. Pflanzenschutz 14: 95–98.

1980. Piesmids. Pp. 97–113 *in* K. F. Harris and K. Maramorosch [eds.], Vectors of Plant Pathogens. Academic Press, New York, New York.

Schaefer, C. W. 1981. Improved cladistic analysis of the Piesmatidae and consideration of known host plants. Ann. Entomol. Soc. Amer. 74: 536–539.

1983. Host plants and morphology of the Piesmatidae and Podopinae (Hemiptera: Heteroptera): further notes. Ann. Entomol. Soc. Amer. 76: 134–137.

Schmutterer, H. 1976. Die Rübenblattwanze *Piesma quadratum* Fieb. als Vektor eines neuen Pathogens bei Zucker- und Futterrüben. Z. Pflanzenkr. Pflanzenschutz 83: 606–610.

1980. Übertragung von zwei verschiedenen Herkunften des Rübenkräuselvirus durch Larven von *Piesma quadratum* Fieber. Z. Pflanzenkr. Pflanzenschutz 87: 145–149.

Schuh, R. T., and J. A. Slater. 1995. True Bugs of the World (Hemiptera: Heteroptera): Classification and Natural History. Cornell University Press, Ithaca, New York, U.S.A. 336 pp.

Southwood, T. R. E., and D. Leston. 1959. Land and Water Bugs of the British Isles. F. Warne, London, U.K. 436 pp.

Thomas, D. C. 1956. Notes on the biology of some Hemiptera Heteroptera. V. — Piesmatidae and Tingidae. Entomologist 89: 13–14.

Varis, A. L. 1973. *Piesma maculatum* Lap. (Het.: Piesmidae) as a pest on sugar beet in Finland. Ann. Agric. Fenn. 12: 105–112.

Vasilev, E. M. 1911. Zametka o povrezhdenii sakharnoy sveklovitsy klopom *Piesma capitata* Wolff i o merakh borby s nim. Vest N. Sakharn. Promyshl. 5: 140–142.

Völk, G., and H. Krzal. 1957. Übertragungsversuche mit *Piesma quadratum* Fieber, dem Vektor der Kräuselkrankheit der Zucker- und Futterrübe. Nachrichtenbl. Dtsch. Pflanzenschutzdienst 9: 17–28.

Weiss, H. B., and R. B. Lott. 1924. Notes on *Piesma cinerea* Say in New Jersey (Hemiptera). Psyche 31: 233–235.

Weiss, M. 1976–1977. Untersuchungen zur Anwendbarkeit von Autzidmethoden auf *Piesma quadratus* Fieb. (Het.: Piesmidae) 1. Dosis-Effekt-Kurven und Beeinflussung der Lebensdauer von Männchen bei Behandlung geschlechtsreifer Imagines, mit Gammastrahlen. Z. Angew. Entomol. 82: 241–246.

Wille, J. 1929. Die Rübenblattwanze *Piesma quadrata* Fieb. Monogr. Pflanzenschutz 21: 1–114.

CHAPTER 8

Cotton Stainers and Their Relatives (Pyrrhocoroidea: Pyrhocoridae and Largidae)

Carl W. Schaefer and Imtiaz Ahmad

1. INTRODUCTION

Members of the superfamily Pyrrhocoroidea are of moderate size (usually 5 to 25 mm long) and are often brightly and contrastingly colored red, yellow, or white, and black. The bugs are characterized by having (usually) two incomplete cells basally in the membrane of the hemelytron, abdominal sterna 2 through 7 fused, and (always) no ocelli. The phylogenetic relationships and the higher classification of the two families, Largidae (largid bugs) and Pyrrhocoridae (cotton stainers), are being studied by the present authors.

The superfamily is represented in all zoogeographic regions, but is particularly abundant in the tropics and subtropics, with some exceptions, including the strange and unique occurrence of the European *Pyrrhocoris apterus* L. in New Jersey (U.S.A.) (Barber 1911). The superfamily's only members of serious economic importance are the cotton stainers, species of the worldwide genus *Dysdercus*.

The only world catalogs of the two families (treated as subfamilies of Pyrrhocoridae *sensu lato*) are by Bergroth (1913) and Hussey (1929). Hussey recognized 14 genera and 99 species in Largidae (= Euryphthalminae) and 29 genera and 262 species in Pyrrhocoridae. We estimate that in the 70 years since, these numbers have increased to about 65 genera and 400 species, total. The lists and systematic work of Blöte (1931) and Schmidt (1931, 1932) are very useful and consider both families.

The most important genus, and the largest, is *Dysdercus* (Pyrrhocoridae), with 75 species (Hussey 1929); it is also the only genus that occurs in both the Old and the New Worlds. These are the cotton stainers, so named because both their excreta and the disease organisms they admit to the cotton stain it. There is no comprehensive study of world Pyrrhocoroidea or of *Dysdercus*. However, Beccari and Gerini (1970) catalog the species of *Dysdercus* and list host plants; and the New World *Dysdercus* species have been revised by Costa Lima et al. (1962), and especially by Van Doesburg (1968), who also lists food plants, discusses biology and phylogenetic relationships, and illustrates all species. Based on Van Doesburg's work, Zrzavý and Nedvěd (1997) have worked out the phylogeny of the New World *Dysdercus*. Henry (1988b) cataloged the eight species found along the southern border of the United States.

There is no similar study of Old World *Dysdercus*. Freeman (1947) revised those Old World species represented in the British Museum (Natural History); thus limited, he did not see all species, nor types of some species in the Museum; the revision of Costa Lima et al. (1962) was similarly limited, to specimens in Brazilian museums. Kapur and Vazirani (1956) discussed in detail the eight species of *Dysdercus* then known from the Indian subcontinent; and Schouteden (1912) and Couilloud (1989) list the 12 species of importance to cotton in sub-Saharan Africa (including Madagascar), and provide a general account of their biology (Couillard unfortunately does not distinguish among the several species; he treats them collectively). Freeman (1947), Kapur and Vazirani (1956), Costa Lima et al. (1962), and Van Doesburg (1968) provide good keys to the species they consider, as does Villiers (1947) for the West African *Dysdercus*.

Largidae are distinguished from Pyrrhocoridae by their lanceolate ovipositors (platelike in Pyrrhocoridae), and consequent divided seventh sterna, and by differences in the metathoracic scent gland apparatus and the aedeagus.

The family has not been revised, although a revision is under way by the present authors. Of the two largest genera (Hussey 1929), one (*Largus*) is New World and the other (*Physopelta*) is Old World. Papers by Van Doesburg (1966) and Brailovsky (especially 1981, 1991, 1993, 1997) will help identify New World specimens. The U.S. species are cataloged by Henry (1988a).

A few studies have considered the position of Pyrrhocoroidea within the Heteroptera (e.g., Schaefer 1964, Kumar 1968), and Henry (1997) has placed that position on a solid basis: sister group of Coreoidea (*sensu stricto*), these taken together sister group of Lygaeoidea (*sensu lato*). Zrzavý (1990) considers the evolution of warning coloration within this large complex, and Zrzavý and Nedvěd (1999) the evolution of mimicry complexes in New World *Dysdercus*, Ahmad and his students have looked at various aspects of comparative morphology and phylogeny within the superfamily (e.g., Ahmad et al. 1988, Mohammad and Ahmad 1991); this latter work is continuing (I. Ahmad and C. W. Schaefer, personal observation).

The morphologies of many Pakistani Pyrrhocoroidea have been described by Ahmad and his students; these are not cited here but will be discussed fully in forthcoming revisions. Other studies have emphasized the scent gland systems: the nymphal abdominal glands (Stein 1966a,b, 1967, 1969) (*P. apterus* and *D. intermedius* Distant); and the metathoracic gland of these species (Schumacher 1971a,b,c; Schumacher and Stein 1971). Farine et al. (1992) extracted some 40 compounds from the posterior-most abdominal gland of *P. apterus* nymphs, and 35 (mostly the same ones) from the metathoracic glands of adults; 13 of these compounds had not been found before in Heteroptera; "the biological role of all the identified chemicals is almost unknown in *P. apterus*" (p. 1673). Altogether, a remarkable total of 63 compounds, in 11 chemical categories, have been isolated from the metathoracic (adult) and the posterior-most abdominal (nymph) glands of three pyrrhocorid species (Farine et al. 1993). The structure of epidermal glands found posteriously in females ("uradénies") was studied by Thouvenin (1965) and Chevaillier (1965), the latter in *D. fasciatus* Signoret.

Of interest too have been the distribution and fine structure of sensilla on the labium (Peregrine 1972, *D. fasciatus*; Gaffal 1981, *D. intermedius*), antenna (Gaffal 1976, *D. intermedius*), or both (Gaffal 1979, *D. intermedius*; Rani and Madhavendra 1995, *Odontopus nigricornis* (Stål)). Zrzavý (1995) analyzed postembryonic development of abdominal color pattern in *P. apterus* and *D. cingulatus* (F.), and Sláma (1964a) worked out the influence and effect of juvenile hormone on epidermal cells of *P. apterus* during metamorphosis.

Ludwig (1926) described and illustrated in detail the copulatory structures of *P. apterus*, and how they function during mating. The external genitalia (both sexes) of three neotropical *Dysdercus* species were compared by Jurberg et al. (1982), and the structure and development of the ovaries and associated structure have been studied in various pyrrhocorids by Pluot (1970, 1973), and Robert (1975, includes both sexes; 1976; 1979). Allatectomy inhibits egg maturation in *D. koenigii* (F.) (Tiwari and Srivastava 1979), as it does in general.

The chromosome complement of Pyrrhocoroidea is variable. In Pyrrhocoridae, there are 9 to 16 autosome pairs depending on the species, plus XO or X_1X_2; some variation occurs within *Dysdercus* itself (Toledo Piza 1947, Ueshima 1979, Papeschi et al. 1998). Largidae have fewer autosomes, 4 to 5 autosomal pairs in Larginae and 4 to 7 pairs in Physopeltinae; largids too have an XO system, and they also have an m chromosome ("microchromosome"), which may ally this family with Lygaeoidea (but see Schaefer 1993, p. 114). Manna (1957), Ueshima (1979), and Ahmad (1991) review the literature, and Ahmad (1991) discusses it (see also Wilson 1909). The sex chromosome mechanism in some physopeltines is discussed in Ray-Chaudhuri and Manna (1955), Manna and Deb-Mallick (1981), and Manna et al. (1985): the results suggest (to the present authors) that the sex chromosome mechanism here is not so different from that in Pyrrhocoridae.

A bacterium has been isolated from the gut of *P. apterus* that apparently aids in anaerobic digestion (Haas and König 1988); in 1994 the first inducible antibacterial peptides from a hemimetabolan were isolated from the same species (Cociancich et al. 1994).

Other accounts of morphology, physiology, etc. may be found in the individual species accounts.

Most pyrrhocoroids live on the ground or on low plants; some live on higher bushes and trees. Those on the ground occur in the upper layer of the soil, under dead leaves, and in crevices. Here at least one species, *P. apterus*, can find water in the driest of habitats (Toms 1983). Ahmad and Schaefer (1987) discuss the habitats of pyrrhocoroids, and suggest that the ground is the more primitive habitat (which is generally the case in the Heteroptera; C. W. Schaefer, personal observation).

Copulation takes place end to end and may continue for several days. Males may guard their mates (Carroll and Loye 1990), and oviposition may continue for 2 to 6 days (e.g., *D. koenigii*; Ahmad and Mohammed 1983). Eggs are laid singly or in clusters in sand, soil, or on plants, and are sensitive to changes in temperature and humidity. Usually 80 to 90% of all eggs hatch. The first instars of *D. obscuratus* Distant are said to feed underground, on food (*Sida* sp.) provided by their mother (Barber 1925). Like those of nearly all heteropterans, pyrrhocoroid nymphs have five instars. Nymphs of pyrrhocorids are bright red, sharply contrasting with largid nymphs, which are dark. [*Note:* The biologies of only a very few pyrrhocoroids have been studied, so it is unclear how widely the facts above may be generalized.]

Many New World largids (Larginae: Araphini) mimic ants, as do two Old World genera of Pyrrhocoridae (*Courtesius* Distant and *Myrmoplasta* Gerstaecker).

Only one species is sexually dimorphic. The male *Macrocheraia grandis* (Gray) has a much more elongated abdomen than the female; because the wings of both sexes are of the same length, the male appears brachypterous (see **Section 4.19**). Of the species so far studied, only *D. fasciatus* differs between the sexes in its scent composition (Farine et al. 1993).

Ahmad and Schaefer (1987) summarize the known host plants of the two families. The few *Largus* records suggest these species feed are polyphagous; *Physopelta* spp. prefer *Mallotus* spp. (Euphorbiaceae); and *Macrocheraia grandis*, the only species of Physopeltinae: Lohitini, is an occasional pest on several crops (see **Section 4.19**).

Non-*Dysdercus* Pyrrhocoridae feed widely, but prefer members of the Malvalves, a preference narrowed in the many species of *Dysdercus* to Malvaceae; however, *Dysdercus* spp. feed on many nonmalvaceous plants, and these then allow cotton stainers to survive between cotton crops (Ahmad and Schaefer 1987).

Pyrrhocoroids feed for the most part on reproductive parts of plants. Because seeds are often unavailable throughout the pyrrhocoroid's life, migration among plants may occur (see Sands 1917, Van Doesburg 1968, Janzen 1972; some discussion in Ahmad and Schaefer 1987). The relationship among food supply, reproduction, and migration has been well worked out by many authors (Dingle and Arora 1973, Davis 1975, Derr et al. 1981), as has the relationship among these and flight-muscle histolysis and its hormonal control by Davis (1975). Mukhopadhyay (1993) discusses some aspects of seed feeding in *D. koenigii*. The mouthparts and mode of feeding of *D. koenigii* and *D. fasciatus* are described in detail by Saxena (1963) and Rathore (1961), respectively.

There are several reports of nonplant feeding by pyrrhocoroids (Ahmad and Schaefer 1987, Steinbauer 1996); these instances may reflect a need for highly concentrated protein. More interesting are the reports of carnivory; these (including cannibalism, on eggs, nymphs, or dead conspecifics) are scattered throughout the literature (Ahmad and Schaefer 1987, Schaefer and

Ahmad 1987). At least two genera, *Antilochus* and *Raxa* (Pyrrhocoridae), are facultatively predaceous, and may be (or become) obligatorily predaceous (see Schaefer 1999).

Most pyrrhocorids are aposematic and therefore presumably distasteful to predators, such as vertebrates. But some, despite their warning colors, are eaten by birds and amphibians. In Britain it is recorded that amphibians and birds prey on *Pyrrhocoris apterus* and a mermithid worm parasitizes this fire bug (Ullrich 1953, Southwood and Leston 1959). Birds play a minor role in the control of some *Dysdercus* spp. in Trinidad; Sands (1917) mentioned *Tyrannus rostratus* (Tyrannidae) as an enemy of *D. delauneyi* on St. Vincent Island; stomachs of these birds contained many cotton stainers; Van Doesburg (1968) listed several birds reported to prey on *Dysdercus* to a limited extent in the neotropics. Myers (1927) recorded that the fish *Gambusia punctata* Poey eats various stages of *D. andreae* and *D. mimulus* Hussey fallen from *Sida* overhanging the water. A large Cuban tree frog, *Hyla septentriomalis* Boulenger, in captivity ate numbers of *D. andreae* at night.

Important natural enemies are found among the Arthropoda. Tachinid flies insert their eggs into the body of second instars (Wille 1952). The full-grown maggot then leaves the victim and pupates in the soil. *Acaulona peruviana* Townsend and *Paraphorantha peruviana* Townsend were found on *Dysdercus* spp. in Peru (Townsend 1913, Wille et al. 1958). *Dysdercus howardi* Ballou and *D. howardi* var. *minor* Ballou are occasionally parasitized by a species of *Trichopoda* (Ulrich 1916). *Megilla maculata* var. (Coleoptera) has been observed eating a young larva of *D. andreae* (L.) (Ballou 1906).

A small mite and a small arachnid externally parasitize stainers (Myers 1927). Ectoparasitic Acari from *D. mendesi* Blöte and *D. ruficollis* (L.) have been found at Campinas state, Brazil (Mendes 1938). *Treatia dysderci* Evans, a parasitic mite, was described from *D. howardi* Ballou from Trinidad (Evans 1963). Ballou (1916) noted that eggs of *D. delauneyi* may be attacked by a thrips species.

The analysis of these data lead to the conclusion that with one possible exception none of these predators and parasites is sufficiently abundant or host specific to become valuable in controlling cotton stainers, although an othopheidemid mite kills its host *Dysdercus* when in heavy infestation (Shahi and Krishna 1981b).

The reduviid *Rhynocoris albopunctatus* Stål feeds on several pests in Ugandan cotton fields, including "*Dysdercus* sp." (Nyiira 1970). Members of one reduviid genus are closely associated with Pyrrhocoridae. Several species of the African *Phonoctonus* are the only predators whose preferred food appears to be pyrrhocoroids, although Stride (1956) considered other prey. So specialized are these predators that they form a mimetic assemblage with their prey, various *Dysdercus* spp. and other pyrrhocorids (Schouteden 1916, Stride 1956). This phenomenon is sometimes termed "aggressive mimicry" (Schouteden 1916, Bourdouxhe and Jolivet 1981), a wildly anthropocentric interpretation that assumes first that pyrrhocoroids see reduviids and second that they see reduviids as humans do. Neither assumption is in the least likely (see Schaefer and Ahmad 1987); indeed, Fuseini and Kumar (1975b) found no relationship between mimicry and predation in *Dysdercus* and associated reduviids (*Phonoctonus*). More probably, reduviids looking like pyrrhocorids share whatever protection the latter gain from their aposematicism (Schaefer 1999). The mimicked pyrrhocorids (mostly Old World *Dysdercus* spp.) live in large aggregations with smaller numbers of their mimicking predators. Aggregations of organisms similarly aposematic may warn off predators from a greater distance than would a single such organism (e.g., as may be true of the bug *Parastrachia japonensis* (Scott); Tachikawa and Schaefer 1985; see also Gamberale and Tullberg 1998, and references therein).

There are far fewer *Phonoctonus* than pyrrhocorids in the same assemblage (35:1) (Schouteden 1916, Fuseini and Kumar 1975b), but even under these conditions with a rough calculation of the prey consumed and total life span of the predator and the prey, and also considering the fecundity of both predator and prey, it is estimated that the *Phonoctonus* would diminish a *Dysdercus* population by somewhat more than half (35/20) (Schaefer and Ahmad 1987). After that time, the full complement of *Phonoctonus* will reproduce, as will the remainder of the *Dysdercus*. It is likely that in a short time good control of *Dysdercus* could be accomplished. We suggested that the efficacy of *Phonoctonus* as a control agent of *Dysdercus* be tested through a rearing-and-release programs. Release could be of the eggs and hatchings early in the first stadium (1 to 7 days), when they do not feed (Stride 1956). The successful biological control of the cotton stainers on a safe and permanent basis of *Phonoctonus* spp. seems possible (see Schaefer and Ahmad 1987).

Further accounts of biology are found below, under the accounts of individual species.

Much of the biological, ecological, and especially the physiological, biochemical, and endocrinological work on pyrhhocoroids has been done with *Pyrrhocoris apterus* (L.) and with several species of *Dysdercus*. Work with *P. apterus* has been admirably reviewed by Socha (1993) and will not be reviewed again here. Unfortunately, no such review exists for *Dysdercus*, whose species are important cotton pests. Much of the work has been done in the warmer cotton-growing areas of the world and published in occasionally obscure journals. We do not claim that what follows in this chapter is an exhaustive, or even an adequate, survey of this literature. The reader should realize also that information here is organized by species, but that what has been learned about one *Dysdercus* species may well apply to others.

2. GENERAL STATEMENT OF ECONOMIC IMPORTANCE

The most important — indeed, the only important — economic impact pyrrhocoroids have is the devastating effect on cotton by *Dysdercus* species. Because this damage is the same regardless of where or by what *Dysdercus* species it is caused, we describe it here. This account is based on that by Whitfield (1933), who does not make clear if it is based on one or all four of the species he considers (*D. superstitiosus* (F.), *D. fasciatus* Signoret, *D. nigrofasciatus* Stål, and *D. cardinalis* Gerstaecker).

Dysdercus damages cotton in two ways, both serious: direct and indirect, Whitfield's (1933) primary and secondary. Direct damage varies with age of the boll. In young bolls, the developing seed may be destroyed, and the boll may dry up; an older attacked boll may also dry up. Although the excreta of the instars is colorless, that of adults is yellowish and stains the cotton lint. As a result, fibers are rendered useless, and seeds are emptied of their oil.

Indirect damage is more common. "Both fungi and bacteria enter the bolls through the stylet punctures, giving rise to a variety of diseases almost all of which completely destroy the bolls. The symptoms of such attack vary from a slimy wet rot ... to a dry rot which completely eats out the interior of the boll" (Whitfield 1933, p. 302).

Of these indirect effects, the most serious is internal boll disease, which is caused by *Nematospora gossypii* and related fungi; these are probably transmitted from contaminated boll to uncontaminated boll mechanically, on the mouthparts of the bug (Frazer 1944). The ecology of the bug–fungus association, and the early history of its study, are recounted by Frazer (1944), who also provides in detail the way in which the bug becomes infected and the way in which it passes the fungus on to the plant. In the New World, boll disease is caused by fungi and bacteria, including *Xanthomonas malvacearum* var. *campestris* and *Colleotrichum gossypii* (Moreira et al. 1994).

There are also passing references to "cotton stainers" or "*Dysdercus* sp.(p.)" on several crops; that these insects are not identified suggests they are of no great importance: "cotton stainer," on cotton in the United States (Coad 1929); "*Dysdercus* sp.," on *Hibiscus cannabinus* L., *Abelmoschus esculentus* L., and *Urena lobata* L. in Togo (Poutouli 1992); and "*Dysdercus* sp.," on the cowpea *Vigna unguiculata* (LK.) Walp., in Benin (Dreyer and Baumgärtner 1995).

[*Chauvinistic note:* Several species of *Dysdercus* occur in the United States (Mead 1966, Henry 1988), and early in this century were recorded from cotton fields; but cotton stainers no longer are a problem here, perhaps adequately controlled by methods used to control other cotton pests. In addition, the destruction of cotton waste has removed overwintering sites and, in Florida at least, cotton is a less important crop than it was (Mead 1966).]

A further statement is necessary: Much of what is known about insect metamorphosis and its control, especially in insects with incomplete metamorphosis, has been learned from work on *P. apterus*, the only pyrrhocorid found widely in Europe. This work is summarized in Socha's thorough review (1993); and it includes the exciting work on "paper factor" (see Sláma and Williams 1965; 1966a,b; Williams and Sláma 1966; also Carlisle and Ellis 1967), which led to the discovery, exploitation, and synthesis of juvenile hormone analogues (see Sláma 1962; 1964b,c,d, for early work), now so widely used in control (Carayon and Thouvenin 1966). For its contributions both to general knowledge of insect development, metamorphosis, physiology, endocrinology, and biochemistry; and for its less direct contributions to arthropod pest control, the firebug, *P. apterus*, is surely of economic importance.

3. THE IMPORTANT SPECIES

3.1 *Dysdercus cingulatus* (F.) (= *Dysdercus koenigii* (F.))

3.1.1 Distribution, Life History, and Biology

This red cotton bug is 12 to 18 mm long, mostly reddish with a white collar and black corial spots. According to Freeman (1947) and Kapur and Vazirani (1956), this species occurs in Sumatra, Borneo, the Philippines, Australia (type locality), the Indo-Chinese peninsula, and Sri Lanka; in the Indian subcontinent it is found only in Bangladesh and northeastern India. Much of the literature on this species in fact refers to *D. koenigii* (F.). The two species differ: *D. cingulatus* slightly larger, femora with some or much black; *D. koenigii* femora completely red (Freeman 1947, Kapur and Vazirani 1956). *Dysdercus cingulatus* was abundant in cotton fields in Papua New Guinea (Ballard 1927).

In this chapter we at first tried to distinguish between the two species on the basis of the studied specimens' localities, but these are rarely presented. Therefore, we have been obliged to treat the species as the authors of the papers identified them, knowing sadly that often we, like those authors, are wrong. However, the two species are closely related, and doubtless what has been written about the one is true for the other.

Despite the large literature on this species, we have found nothing on its basic biology or life cycle.

Mehta (1930) showed that in Punjab Province (now Pakistan), temperature and humidity were the most important factors in controlling *D. cingulatus* populations. Bugs are few in May to August, when humidity is low and temperatures are high. As temperatures drop, populations increase, to drop again as the weather turns cold. Similarly, in Bihar (India), nymphal survival of *D. koenigii* increased with increasing humidity (Pathak and Sinha 1993).

Aggregations of the insect occur, of course, because they feed on a monoculture, cotton. But aggregating is also stimulated by olfaction and thigmotactism (Farine and Lobreau 1984). Aggregation of *P. apterus* nymphs stimulates development (Schmuck 1994), and it may in *D. cingulatus* as well.

A few aspects of the morphology of *D. cingulatus* have been considered, and of course morphological studies on other *Dysdercus* species for the most part apply as well to *D. cingulatus*. The structure and working of the mouthparts were well and thoroughly worked out by Rathore (1961), who also described how feeding is accomplished. Farine (1987, 1988) described the scent (exocrine) glands of nymphal and adult *D. cingulatus*, and for the first time (1988) showed accessory glands in the male. The secretions of some of these glands were later analyzed (Farine et al. 1992).

The embryology of *D. cingulatus* has also been published (Tandon 1969).

Dysdercus cingulatus has been the basis of many studies (especially Indian) of physiology and structure. The hemocytes have been described and shown to resemble those of *Rhodnius prolixus* Stål (Zaidi and Khan 1974a) and to change qualitatively with age and reproductive cycle (Zaidi and Khan 1975a) and with metamorphosis (Zaidi and Khan 1975b); hemolymph cholesterol has been measured (Zaidi and Khan 1972), and total cholesterol also varies with the insect's age and reproductive and metamorphosic status (Zaidi and Khan 1974b). Varying also with age and reproductive status are whole-body proteins (Sifat and Khan 1974), and the protein content of thoracic (flight) muscles decreases with starvation (Muraleedharan and Prabhu 1977). This latter phenomenon is correlated with the degeneration of the indirect flight muscles and the incorporation of their proteins into the ovaries; the control of this is described by Nair and Prabhu (1985a,b,c). Doubtless also correlated with muscle degeneration is the decrease in midgut protease and invertase activities following protein starvation (Muraleedharan and Prabhu 1978, 1979a), a decrease perhaps mediated by the median neurosecretory cells of the brain (Muraleedharan and Prabhu 1979b). The influence of hormones on feeding in *D. cingulatus* was studied by Muralheedran and Prabhu (1981). The molting hormone was analyzed by Aldrich et al. (1982).

3.1.2 Damage

Like other *Dysdercus* species, this one is a vector of internal boll disease. The best accounts of the transmission of the causative fungi are in Whitfield (1933) and Frazer (1944); see **Section 2**.

This species has caused considerable damage to ripening seed heads of wheat (*Triticum aestivum* Lamk.) in Rajasthan (India), during February to March. One variety of wheat, N.P. 718, was resistant, but a local variety suffered considerable damage. Cotton had earlier been grown in these wheat fields.

3.1.3 Control

Three types of control have been applied to *D. cingulatus*; none has been especially successful, and one remains experimental. Chemical control uses pesticides, what might be termed "natural chemical control" uses natural compounds from other organisms, and biological control uses parasites and predators.

The chemicals so far tested are sterilants. "Apholate" sterilized females exposed for 18 hours to 0.7 mg apholate per square inch; normal females mated to males similarly exposed laid far fewer eggs than normal (Mustafa and Naidu 1964). The related compound triphenytin acetate is less effective, and not recommended (Ansari and Khan 1973). The chitin inhibitors diflubenzuron at concentrations above 100 parts per million applied topically lowered egg hatch to 0 to 30% of normal (Chockalingam and Noorjahan 1984). When applied to surfaces touched by the bugs, diflubenzuron more greatly affected females than males, and did not affect viability or transfer of sperm (Gupta et al. 1994). However, *D. cingulatus* may become resistant to diflubenzuron (Gupta et al. 1993). The related compound penfluron caused 42 to 100% sterility in females treated topically with dosages from 0.001 to 0.1% (Ahmad 1980, Satyanarayana et al. 1985). Tepa also sterilizes, in doses of 0.05% applied topically to females (Ansari 1973). Treated adults mated as readily as untreated (Sukumar and Naidu 1980) and affected older adults more than younger ones (Sukumar 1980).

"Natural" chemicals include the defensive secretion of the adult tessaratomid (Hemiptera) *Tessaratoma javanica* (Thunberg), which acts like a juvenile hormone mimic by disrupting the fifth-instar–adult molt (Ashok et al. 1978). In addition, several hormone-mimicking extracts of plants have been tested: plumbagin (from *Plumbago* spp.) (Joshi et al. 1988; see also **Section 3.2.3**); extract from *Polyscias guilfoylei* Bailey (Rajendran and Gopalan 1982); from *Lantana camara, Anthocephalus cadamba, Tectona grandis, Calophyllum* sp., and *Phyllanthus emblica* (Prabhu and

John 1975); and from *Nepeta hindostana* (Roth) (Labiatae) (Srivastava and Neraliya 1995). These all adversely affected growth and/or metamorphosis.

Infected by injection of a water suspension of *Aspergillus flavus* conidia, late-instar and adult *D. cingulatus* greatly increased oxygen consumption, lost weight (partly from water loss), and died (Prabhakar et al. 1992). In the laboratory, *Bacillus thuringiensis* was "quite virulent against *D. cingulatus*" (Kaushik and Chopra 1969), as was the fungus *Paecilomyces farinosus* (Dickson ex Fries) (Kuruvilla and Jacob 1980).

D. P. Ambrose and his students have been studying the use of Reduviidae in biological control (see **Chapter 29**); they are among the first to do so, and they are the first to do so comprehensively. Among other pest species, they have worked with *D. cingulatus* but, because of the concerns we express in **Section 3.1.1** above, we believe this species may in fact be *D. koenigii*. Nevertheless, they report the insect as *D. cingulatus*, and we so report it here. This work is (so far) with five species of Reduviidae, and involves studies of prey preference and prey suitability, and mass rearing (Ambrose and Sahayaraj 1993; Ambrose and Claver 1995; Sahayaraj and Ambrose 1993, 1994, 1995–1996, 1996; Ambrose and Claver 1999).

3.2 *Dysdercus koenigii* (F.) (= *Dysdercus cingulatus*)

3.2.1 Distribution, Life History, and Biology

As noted above (**Section 3.1.1**), this red cotton bug has been confused with *D. cingulatus*, and some of the literature referring to the one may in fact refer to the other. It occurs throughout India, Sri Lanka, Pakistan, and Burma (Kapur and Vazirani 1956), and Afghanistan (Stehlík and Kerzhner 1999). The bug is 11 to 15.5 mm long and, except for the white collar and a pair of black corial spots, the dorsum is uniformly reddish (Freeman 1947, Kapur and Vazirani 1956).

Like *D. cingulatus*, *D. koenigii* is a pest of cotton. More has been published on its biology than on that of *D. cingulatus*. Mating occurs 2 to 3 days after molting and continues 3 to 5 days; oviposition may then continue for 2 to 6 days. A single female (in southern Pakistan) may lay 10 to 50 eggs, scattered on the substrate (Ahmad and Mohammad 1983); or a single female (in Maharashtra State, India) may lay 27 to 144 eggs, in the soil litter (Kamble 1971). Eggs are 0.8×1.2 mm, and the five instar lengths are 1.6–2.5, 2.1–3.4, 4.1–5.7, 5.5–7.8, and 9.4–10.1 mm; in each case, the smaller number is of the Indian specimens, the larger of the Pakistani. The egg hatches in 5.4 (Pakistan) or 6.2 (India) days, and the stadial lengths are 4.4 (3.4), 4.2 (3 to 4), 5.3 (3.5), 4.0 (6 to 8), and 4.6 (12 to 16) days (India in parentheses). It is not clear if these differences reflect different environmental conditions, or incorrect species identification(s). Chatterjee and Raychoudhuri (1960) give figures for instar and stadial lengths of Calcutta *D. koenigii* similar (but not identical) to those of Kamble (1971). The immature stages are briefly described in Chatterjee and Raychoudhuri (1960), Kamble (1971), and Ahmad and Mohammad (1983), and are illustrated in the last; the stages are keyed in Ahmad (1985), which has a color photograph of the eggs. In the field (Maharashtra State) the bugs reached a population peak in December and January, on cotton (Kamble 1971).

Dysdercus koenigii could be reared on okra fruits or opened cotton bolls (Kamble 1971) or on *Withenia somnifera* plants (Ahmad and Mohammad 1983). They will feed on a variety of other malvaceous plants (Kamble 1971); Singh and Ram (1987) found that of five malvaceous plants, deccan hemp (*H. cannabinus*) was the best for nymphal survival and fertility, cotton being second. When fed boiled seed of okra, the fecundity, fertility, life span, and survival rate of *D. koenigii* all decreased (Chakraborti 1998), suggesting the importance of water-soluble nutrients. Like other *Dysdercus* species, *D. koenigii* can maintain itself on wild malvaceous plants, such as *Abutilon* sp. in Maharashtra State (India) (Wadnerkar et al. 1979). Although somatic tissues may sustain the bugs, fruits or seeds of the plants — cotton, or other Malvaceae — are required for successful reproduction (Shahi and Krishna 1981a).

Of three species of *Dysdercus* — *koenigii, ruficollis* (L.) (as *fulvoniger* [DeGeer]), and *voelkeri* — *D. koenigii* was the only one to prefer an aqueous gossypol solution to water alone (Schoonhoven and Derksen-Koppers 1973), which suggests (to the present authors) a greater prediliction for cotton by *D. koenigii* than by the other species. [*Note:* It is unclear whether by *D.* "*voelkeri*" these authors refer to *D. transversalis* Blöte or to *D. superstitiosus* (F.): see Beccari and Gerini (1970).] The activities of some digestive enzymes were studied by Janda and Munzarová (1980), and the mode of feeding by Saxena (1963).

The anatomy and histology of the *D. koenigii* ovary was described by Deshpande and Srivastava (1981), and of the internal genitalia of both sexes by Khanna (1979), who describes for the first time in Heteroptera a pair of mesadene glands, each opening into a vas deferens. Reproduction itself is described in several papers: The endocrine control of attraction and courtship, by Sharma et al. (1975) and Hebbalkar and Sharma (1982); the diminished fucundity of older males and females, and the loss of response to females by antennectomized males, by Shahi and Krishna (1979a); the greater fecundity and fertility of females mated for longer periods of time, by Shahi and Krishna (1979b). Hebbalkar and Sharma (1991) showed a cyclical consumption of oxygen in female (but not male) *D. koenigii*, and consumption was reduced in females (but not males) treated with precocene II.

Finally, the histophysiology of the salivary glands was studied by Kumar et al. (1978), the protein metabolism of these glands by Kumar (1982), the hematocytes of the adult by Sharma et al. (1998), and the cytogenetics of the species by Ahmad (1991).

Nymphal survival of *D. koenigii* is highest at high humidities, in Bihar (India) (Pathak and Sinha 1993); similarly, in Punjab (Pakistan), population increase of *D. cingulatus* (**Section 3.1**) is promoted at higher humidities (Mehta 1930). The discovery of an orange color mutant may help in studies of the studies of the population genetics of the species (Gujar and Rao 1993).

3.2.2   Damage

Like the other *Dysdercus* species, *D. koenigii* damages cotton by feeding on it, excreting into it, and especially by introducing fungi into it. In Pakistan, cotton is a major export crop, cultivated (in 1980) on some 4 million acres, and earning some U.S.$420 million annually (Ahmad 1985). *Dysdercus koenigii* does not appear to be a major pest of cotton everywhere in the Indo-Pakistan subcontinent, but it is an important one in Pakistan (Ahmad and Mohammad 1983, Ahmad 1985) and India (Kamble 1971, Wadnerkar et al. 1979).

Dysdercus koenigii may also be a pest on other plants of economic importance: okra, egg plant, and hollyhock (Kamble 1971, Wadnerkar et al. 1979). Bhindi (*Abelmoschus esculentus*: Malvaceae) is also attacked, in Bihar (India) (Pathak and Sinha 1993). It has also been a minor pest of early varieties of the legumes red gram, pigeon pea (*Cajanus cajan* [L.] Mill sp.) (Singh and Singh 1978a,b,c) and peanut (groundnut) (*Arachis hypogaea* L.) (Jayanthi et al. 1993a,b,c). both near New Delhi (India). And Chandel et al. (1984) write without comment that "the red cotton bug, *Dysdercus koenigii* Fabr. is the most damaging pest of solanacious [sic] vegetables."

3.2.3   Control

In his life table of *D. koenigii*, Ahmad (1983) showed that the egg and early instars are probably the most chemical-sensitive stages, against which the least amount of pesticides might be used. Among compounds tested (Ahmad 1985) were dimilin (LD_{50} = 3.0 µg/nymph) and methoprene (LD_{50}=10.0 µg/nymph). Methoprene and two other juvenile hormone analogs (hydroprene, kinoprene) inhibited mating behavior of cotton stainers when applied to them as nymphs (Hebbalkar and Sharma 1980). Several new juvenile hormone analogs show considerable promise in controlling this cotton stainer (and presumably others) (Tikku et al. 1999). Two substituted thiourea compounds sterilized females in a chamber treated with the compounds (Sandhu et al. 1984); tepa, applied to third instars, also sterilized adults but did not affect reproductive activity (Sehgal et al. 1980, and references therein). The growth regulator hydroprene also sterilized both sexes (Revathy et al. 1979), as did several substituted pyrimidine-2-thiols (Kaur et al. 1993).

Several natural plant products have been tested for sterilizing or growth-inhibiting effects against *D. koenigii*. Stem, leaf, root, and reproductive-structure extracts of 21 plants were applied to cotton stainer eggs, and the ID_{50} (presumably, inhibition of development) calculated. Some ID_{50} values were as low as 4 to 5 µg (*Cassia siamea, Desmodium polycarpum, Mimusops elengi, Vetiveria zizaniodes*) (Suryakalla et al. 1995), and these plants (we believe) deserve a closer look. Similarly, extracts of *Azadirachta indica* A. Juss. (azadirachtin) (Koul 1984); *A. indica, Calophyllum inophyllum* L., and *Swietenia macrophylla* King (Agrawal 1993); *Callistemon lanceolatus* (Australian bottle brush) (Katiyar and Srivastava 1982, Katiyar et al. 1984); and eucalyptus oil (R. K. Srivastava and Krishna 1990, 1992; R. K. Srivastava et al. 1995; S. K. Srivastava and Krishna 1992) all affect reproduction and/or development adversely. *Artemisia annua* oil applied topically had an LD_{50} of 0.48 µg/l/fifth instar; development was delayed as well (Rao et al. 1999). Plumbagin was toxic to eggs at concentrations of 0.0044 to 0.0066% (Rao and Gujar 1995), and volatiles from *Allium sativum* bulbs significantly reduced egg hatch (Gurusubramanian and Krishna 1996). Aristolochic acid (from *Aristolochia bracteata* Retz.), given to late instars and early adults in their drinking water, significantly reduced egg hatch at doses of 0.001 to 0.002% (Saxena et al. 1979). The Australian bottle brush extract stimulates oocyte production when applied to adult females (Katiyar and Srivastava 1982). An ether extract of the aquatic alga *Chara zeylanica* Klein ex Wild induced significant sterility in *Dysdercus koenigii* (Sarkar et al. 1995). Of 12 lamiaceous plants tested, 9 deterred oviposition by *D. koenigii* to some extent (Sharma et al. 1981).

The defensive secretion of adult *Tessaratoma javanica* (Thunberg) mimics juvenile hormone by disrupting the final molt of *D. koenigii* (Ashok et al. 1978).

Of several chemicals tested, mercuric chloride proved to be an effective repellent of *D. koenigii*, and to remain effective for up to 70 days; malonic acid and 4-phenyl phenol were also effective for about 2 months (Ahmad and Khan 1980).

X-irradiation caused histopathological effects and both infertility and infecundity in females (lower doses) and males (Srivastava and Deshpande 1983a,b; Srivastava et al. 1985). The present authors suspect the biology of *D. koenigii* renders it an unlikely candidate for sterile-male control.

We have found only few reports of biological control. Chatterjee and Raychoudhuri (1960) record "a spider, and a reduviid bug" as preying on *D. koenigii*, and "a helminth" as occasionally parasitizing females. Both Shahi and Krishna (1981b) and Banerjee and Datta (1980) report a mite, *Hemipteroseius indicus* (Krantz and Khot), reducing populations of this red cotton bug. [*Note:* Although dated later, Shahi and Krishna's claim to be the first to report this association is doubtless correct; the journal in which Banerjee and Datta published always appeared several years after the cover date.]

The frequently predaceous pyrrhocorid *Antilochus coquebertti* F. (see review in Schaefer 1999) will feed on *D. koenigii* (Kamble 1974). [*Note:* Based on the brief description in this paper, I. Ahmad believes this species actually to be *A. russus* Stål.]

[*Note:* The excellent work by D. P. Ambrose and his students on the use of reduviids for biological control of *D. cingulatus* (see **Section 3.1.3**) may in fact refer to *D. koenigii*, if the distributional restrictions outlined above (**Section 3.1.1**) are correct. See also **Chapter 29**.]

3.3 *Dysdercus fasciatus* Signoret

3.3.1 *Distribution, Life History, and Biology*

This red cotton bug's basic color is red-reddish yellow, with a white collar and some black spotting; the black corial and thoracic fasciae are distinctive; it is 10.5 to 10.5 mm long (Freeman 1947, who also figures the paramere and a lateral body view). *Dysdercus fasciatus* occurs throughout sub-Saharan Africa, including Madagascar (Hussey 1929, Freeman 1947, Beccari and Gerini 1970). Cowland and Ruttledge (1927) and Whitfield (1933) note that *D. fasciatus* occurs in cooler wetter parts of Sudan than do other *Dysdercus* species. [*Note:* Janzen (1972) writes that *D. fasciatus* is abundant in Costa Rica, and that it "is very likely a recent introduction to the New World tropics from Africa (H. Dingle, personal communication)" (p. 351).]

Interest in this species appears to have begun with the introduction of cotton by the British into southern Sudan and South Africa. It is highly unfortunate that most accounts of the biology of African *Dysdercus* do not distinguish among the species! From Cowland and Ruttledge (1927) through Whitfield (1933) to Fuseini and Kumar (1975a,b), three to five species, not always the same ones, are lumped together and treated as "*Dysdercus*." What follows is gleaned as best we can from these papers, but we realize (as these authors did not) that the biologies and ecologies of all African species are *not* the same. Occasionally, an author singles out a particular species.

Whitfield (1933) writes that the baobab, *Adansonia digitata*, is the natural host of *D. fasciatus*, in southern Sudan, from which it moved to cotton when that crop was introduced in 1925 (Cowland and Rutledge 1927). Mating occurs 2 to 3 days after emerging as adults, and after a brief courtship. A female lays 7 to 10 batches of eggs, each with 56 to 106 eggs (Fuseini and Kumar 1975a). Eggs are laid just under the soil surface and hatch in 7 days, if temperature and humidity are favorable (Whitfield 1933), or 4.3 days in southern Ghana (Fuseini and Kumar 1975a). Stadial durations are (Sudan): 2, 6, 5, 5, 7 days (Whitfield 1933), or (Ghana): 4, 4, 4, 6, 8 days (Fuseini and Kumar 1975a). Eggs are 1.43 × 0.96 mm, and the five instars are 1.9, 3.3, 6.0, 7.3, and 11.4 mm long (Fuseini and Kumar 1975a, who also figure the instars). Fecundity, fertility, and longevity did not differ between mated and unmated males (Odhiambo 1968), although virgins and females that mated only once lived longer than females kept with males (Hodjat 1969).

The immatures are "highly gregarious," presumably responding to the same stimuli as does *D. intermedius* Distant (**Section 3.4**); first instars much prefer to cluster on wet surfaces than on dry (Fuseini and Kumar 1975a). However, crowding affected the insects adversely, adults being smaller and the sex ratio becoming male biased, and females laying fewer eggs (Hodjat 1969). Crowded *Pyrrhocoris apterus* nymphs developed faster than uncrowded (Schmuck 1994), as may also those of *D. intermedius*.

In southern Sudan, breeding occurs throughout the year, but decreases in the dry season (Cowland and Rutledge 1927, Whitfield 1933), when adults may shelter in crevices in the baobab trees (Whitfield 1933); in Ghana, it is the wet season that causes population decrease, more so in the forest than in the savannah regions (Fuseini and Kumar 1975b). [*Note:* It is not clear that *D. fasciatus* occurs in both; this uncertainty is one sad consequence of the lumping the accounts of different species into one.] However, Golding (1928) writes that in Nigeria, *D. fasciatus* is common in more arid savannah country.

Temperature directly affects growth and reproduction. By testing several temperatures, Clarke and Sardesai (1959) found 28°C the best: here the rate of increase was the greatest, in part because (presumably) both longevity and growth were greatest; fecundity is directly correlated with body size (Clarke and Sardesai 1959). In a study of the allometry of growth, Blackith et al. (1963) showed a discontinuity in pattern in the fifth instar–adult transition (it is suggested that the same is true of other heteropterans; C. W. Schaefer, unpublished analyses). There is no diapause (Fuseini and Kumar 1975a) or "true resting period" (Whitfield 1933).

Dysdercus fasciatus, like other cotton stainers, feeds widely on malvalian plants. As noted, Cowland and Ruttledge (1927) and Whitfield (1933) consider baobab (tebeldi) tree the natural host. Other hosts are given in Fuseini and Kumar (1975b), and they include the sterculiaceous plant *Sterculia foetida* L. However, Tengecho (1994) did not find *D. fasciatus* on baobab (*Adansonia digitata*) in Kenya, but this is compromised by the author's confusion over whether *D. fasciatus* feeds on *S. rhynchcarpa*: Tengecho's text says it does not, but the reference is to "Fig. 1," which presents no data on native hosts of *D. fasciatus*. However, a later study (Chemengich and Khaembra 1998) from the same laboratory, although not mentioning Tengecho (1994), confirmed that

D. fasciatus does poorly on *A. digitata*, as well as on the malvalian plant *Abutilon mauritianum* (Jacq. Medic.); *Gossypium hirsutum, Ceiba pentrandra* Gaertn., and *S. rhynchocarpa* (K. Schum.) were suitable in both nymphal survival and growth. Tengecho (1994) also lists the hosts of this cotton stainer given in the literature, as do Ahmad and Schaefer (1987). In general, *D. fasciatus*, like other cotton stainers, has an abundance of wild host plants, from which it can move to cotton (or other malvalien crop plants) when these become available.

Dysdercus fasciatus is less sensitive to low concentrations of cotton seed-kernel extract than is the lygaeid cotton pest *Oxycarenus hyalinipennis* Costa (Gandhi 1979), which suggests the lygaeid is more closely tied to cotton than is *D. fasciatus*. Nevertheless, this cotton stainer requires some level of gossypol, a chemical found only in cotton seed, in its diet; none, or too much, gossypol adversely affects development, fecundity, and fertility of the bugs (Almeida 1980a,b,c).

The structure, development, and some ideas of function of the epidermis of *D. fasciatus* were examined by Lawrence and Staddon (1975), and the internal and external structure of the head and mouthparts by Khan (1972). Berridge (1965a,b,c) studied in impressive detail the physiology of this insect's excretion.

3.3.2 Damage

Dysdercus fasciatus does considerably less damage to noncotton crops than do its Indian congeners. But its damage to cotton is considerable; indeed, the general account above (**Section 2**) is taken from Whitfield's (1933) account of Sudanese *Dysdercus*.

3.3.3 Control

Early workers (Cowland and Ruttledge 1927, Whitfield 1933) emphasized cultural control measures in Sudan, such as using blow lamps to burn out residual populations sheltering in baobab trees (successful in low-population years) (Cowland and Rutledge 1927), and destroying postharvest cotton debris as well as nearby baobab trees (Whitfield 1933). Chemengich and Khaembra (1998) suggest planting unsuitable malvalian plants (see above) as trap crops.

Whitfield (1933) also recommends spraying, but does not say with what. The juvenile hormone analogue tepa (topical; 10 and 15 µg/third or early fifth instar) caused some reduction in percent surviving adults and some reduction in fecundity (Almeida 1978).

The eggs of *D. fasciatus* show increasing resistance with age to the insecticides allethrin and the triethanolamine salt of 3:5 dinitro-ortho-cresol, but decreasing resistance with age to HETP (Sakeld and Potter 1953). Some resistance by fifth instars to carbaryl and (to a lesser extent) lindane and methidathion occurred in Kenyan *D. fasciatus*; carbaryl resistance, but not lindane resistance, was increased after six generations of laboratory selection (Nyamasyo and Karel 1982).

Juvenile hormone and several mimics interrupted or disrupted embryonic development of *D. fasciatus* (Wall 1974).

Dysdercus fasciatus has the usual number of occasional arthropod predators, but in the Sudan and Ghana none of these is effective (Whitfield 1933, Fuseini and Kumar 1975b). In more southern Africa, reduviids (Hemiptera) of the genus *Phonoctonus* appear to specialize on *Dysdercus* spp. as prey (Evans 1962, Fuseini and Kumar 1975b), and indeed members of these genera mimic one another (Stride 1956, Fuseini and Kumar 1975b). (One of the ineffective Sudanese predators is *Phonoctonus lutescens* [Whitfield 1933] which also occurs in Ghana [Fuseini and Kumar 1975b]). These reduviids may provide effective control; see Schaefer and Ahmad (1987).

Fuseini and Kumar (1975b) did not find any egg parasites. However, the tachinids (Diptera) *Bogosiella pomeroyi* Villen. (Sudan: Whitfield 1933) and *Alophora nasalis* (Bezzi) emerged from dead adults, which had been parasitized as second instars; percent parasitism was low (Ghana: Fuseini and Kumar 1975b).

3.4 *Dysdercus intermedius* Distant

3.4.1 Distribution, Life History, and Biology

Dysdercus intermedius is a large (14.5 to 22.5 mm) species, red or orange, and yellow-gray dorsally, with a white collar and narrow black corial markings (Freeman 1947); it occurs throughout southern Africa (Freeman 1947, Beccari and Gerini 1970).

As seems to be the case with so many other *Dysdercus* species, very little seems to have been done on the biology, ecology, and life history of *D. intermedius*. However, it has been a favorite subject for other studies, especially by Youdeowei on aggregation.

In a series of papers, Youdeowei (1966, 1967a,b, 1968, 1969) showed that *D. intermedius* are attracted to plants, remain on cotton in response to specific chemical signals (perhaps linalool, Everton et al. 1979), and aggregate with conspecifics again on the basis of vision Youdeowei (1966, 1969); a moist substrate tended to cause aggregations to break up (Youdeowei 1967a), as did temperatures above 30°C (Youdeowei 1968); high relative humidity increases the tendency to aggregate, as do coolness and shade (Melber 1979a). In this preference for lesser light, *D. intermedius* (and *D. cardinalis*) apparently differ from *D. fasciatus* and *D. nigrofasciatus*. The latter two migrate the most during the full moon, and the former two during the new moon (Robertson 1977). On the other hand, *D. intermedius* nymphs foraged and fed in the day, not at night (Schmidt et al. 1992), although at least part of that time may have been crepuscular.

Crowding caused synchrony of molting, mortality especially during the second stadium, and a quickening of ovarian development in virgins (Youdeowei 1967b). In the laboratory, the flight muscles histolyze after ecdysis and females do not fly, a phenomenon apparently associated with being well fed (Edwards 1969), a condition which, presumably, in the field makes flight to seek food unnecessary.

Egg to adult took 27.7 days, in the laboratory in Tanzania. Oviposition began 6 to 13 days after eclosion; females laid four clutches (some laid up to seven), each containing 95 to 126 eggs (Kasule 1985). This bug seems to prefer trees; the only wild hosts Kasule (1985) found were *S. africana* (Lour) Fiori and *S. quinqueloba* (Gorcke) K. Schum. Two other species, *D. nigrofasciatus* and *D. cardinalis*, were found on more plants (Kasule 1985). Frazer (1944) gives a thorough and detailed description of the process of feeding.

Everton et al. (1979) illustrate and describe the metathoracic scent gland, and Youdeowei and Calam (1969) describe and illustrate this gland, as well as the abdominal glands of the nymphs. The mouthparts are described and illustrated by Frazer (1944), these and the head by MacGill (1947), and the salivary pump and brain by Popham (1962a,b, respectively). Bentz and Kallenborn (1995) describe the fine structure of the gastric ceca, and Dittman and Biczkowski (1995) the induction of yolk formation.

The control of cuticular pattern formation in the abdomen of *D. intermedius* has been studied by Nübler-Jung (1974, 1977, 1979).

3.4.2 Damage

See **Section 2**.

3.4.3 Control

See Control Sections of other *Dysdercus* species.

3.5 *Dysdercus superstitiosus* (F.) (= *Dysdercus völkeri* Schmidt, *Dysdercus voelkeri* Schmidt)

3.5.1 Distribution, Life History, and Biology

The name *D. völkeri* was synonymized with *D. superstitiosus* by Freeman (1947), but some of the work since has been done under the former name. The species occurs throughout sub-Saharan Africa, including Madagascar (Freeman 1947, Beccari and Gerini 1970). This is a relatively large species, 12 to 17.5 mm long; the head and prothorax are reddish (except for the white pronotal collar and a black fascia), and the corium is yellow-gray with a pair of black spots (Freeman 1947). Two forms occur in Nigeria, a banded one appearing earlier and remaining longer than a spotted one (Golding 1928, who believes the two forms do not interbreed).

In the Sudan, *D. superstitiosus* is scarce in March and April. In May they mate and lay eggs, on the surface of the ground or in small holes dug 10 to 20 mm deep (Whitfield 1933, Fuseini and Kumar 1975b); in Ghana, females laid six to nine batches, each with an average of 81.9 eggs; an egg is 1.0 × 1.5 mm (Fuseini and Kumar 1975b). In Ghana the incubation period was 4.5 days, and the five stadia were 3, 4, 3 to 4, 5 to 7, and 6 to 7 days (Fuseini and Kumar 1975b). In Sudan the five stadia were 3, 4, 5, 6, and 8 days long. The bugs feed on various wild malvalian plants. As the rains end in October, the bugs move to cotton, and then to *S. cinerea*, and feed on both in November and December; in December and January, as the dry season arrives, the bugs remain on *S. cinerea*, and move also to wild Malvales. The early rains of May stimulate the cycle to begin anew (Whitfield 1933). Cowland and Ruttledge (1927) note that in their area of the Sudan, the bugs' chief wild host is tebeldi, or baobab, *Adansonia digitata* (see **Section 3.3.1**). Fuseini and Kumar (1975b) describe and illustrate the instars.

Dysdercus superstitiosus is an important pest of cotton, as well as of okra, *Abelmoschus* (= *Hibiscus*) *esculentus* Moench; like other *Dysdercus*, it feeds widely on seeds of various malvalian plants. In southern Nigeria, the highest fecundity was recorded on silk cotton tree (*Bombax buonopozense* P. Beauv.) (309 eggs) and okra (307 eggs); cotton was next (249), then roselle (*H. sabdariffa* L.) (164 eggs); no eggs were laid by bugs fed kenaf (*H. cannabinus* L.), flame of the forest (*Delonix regia* Boj. ex Hook), aramina fiber (*Urena lobata* L.), or green gram (*Vigna* [= *Phaseolus*] *aureus*). Total nymphal durations varied little among the plants, except that development was incomplete with green gram and flame of the forest; percent emergence (to adult) ranged from 57 to 75%, except 38% on kenaf. Males lived up to 50% longer than females, and females lived longest on okra (17 days). Overall, okra seems to have been the best plant tested, and cotton intermediate (Egwuatu 1980). There is some evidence that *D. superstitiosus* can survive on nonmalvalian plants: Geering and Coaker (1960) raised this cotton stainer successfully on *Sorghum vulgare*, *Vigna unguiculata* (but see *V. aureus*, above), and the graminaceous *Pennisetum typhoides* and *Zea mays*; however, cotton promotes higher fecundity and longer life. In Ghana, on a different series of food plants (all Malvales), *Ceiba pentandra* was preferred over three other species, but comparative tests of fecundity and longevity were not made (Fuseini and Kumar 1975a).

Because the bugs require ripe seeds, they move from host to host, as the seeds of each become available (Edmunds 1978); this pattern appears to characterize *Dysdercus* species in general (see Van Doesburg 1968, Janzen 1972). However, in *D. superstitiosus* at least (perhaps others?), migration is more complex. There seems to be an innate (genetic) basis for migration, which is further influenced broadly by wind patterns in West Africa and more narrowly by temperature and humidity in Côte d'Ivoire (where this work was done) (Duviard 1972, 1973, 1977). A consequence of migration or dispersal is that crowding is reduced. Crowding hastens development, but increases nymphal mortality and egg production. Decreasing crowding thus increases fecundity and longevity, and thus the damage done to cotton (Egwuatu and Opara 1985).

3.5.2 Damage

In addition to being an important pest of cotton (and Pomeroy 1924 illustrates this damage), *D. superstitiosus* can at times damage other malvalian crops (Youdeowei 1973; and see **Section 3.5.1**), and Geering and Coaker (1960) are concerned it may be a pest of sorghum as well. In Nigeria it is a minor pest of soybean, whose leaves it attacks (Ezueh and Dina 1980).

Urena lobata (Malvaceae) (aramina fiber) is a fiber crop grown for some years in Sierra Leone as a substitute for jute (Harris 1981). *Dysdercus superstitiosus* does considerable damage to the seeds, inducing up to 70% nonavailability; the abundance of wild alternative hosts makes control of this cotton stainer difficult (Harris and Bindi 1983), as is the case also with other species on other crops.

3.5.3 Control

See **Section 3.3.3**. Ochou et al. (1998) recommend that although three quarters of the cotton farmers in Côte d'Ivoire are illiterate, they can and should be trained to recognize cotton pests (most can already) and diagnose the best means of combating them. Several African *Dysdercus* species, including *D. superstitiosus,* show some resistance to carbaryl, and slight resistance to lindane and methidathion (Nyamasyo and Karel 1982).

Species of *Phonoctonus* (Hemiptera: Reduviidae) are important predators of *Dysdercus* species in Africa, including *D. superstitiosus* (Stride 1956, Galichet 1956), and their possible use in biological control has been discussed (Schaefer and Ahmad 1987). The tachinid (Diptera) parasite *Epineura helva* Wied has been found in Côte d'Ivoire (Galichet 1956).

3.6 *Dysdercus nigrofasciatus* Stål (Pyrrhocoridae)

3.6.1 Distribution, Life History, and Biology

This is a moderately large (11.5 to 18 mm) (although Villiers 1947 gives 10 to 16 mm) cotton stainer, whose head varies from yellowish to nearly black, and which has a white collar and much of whose dorsal surface is yellowish gray; the corium has a pair of dark elongate spots (Freeman 1947). *Dysdercus nigrofasciatus* occurs throughout central and southern Africa (Freeman 1947, Beccari and Gerini 1970). Beccari and Gerini (1970) report a record from "Trinidad," but this species is not mentioned by Van Doesburg (1968) and the record is doubtless based on a misidentification.

There appears to be little information on *D. nigrofasciatus* specifically. It is lumped with other African *Dysdercus* species in many accounts (Schouteden 1912, Cowland and Ruttledge 1927, Whitfield 1933); indeed, Whitfield (1933) writes that "the life-cycle of this species is identical with that of *D. superstitiosus* [in the Sudan]," except it has not "been found on *Sterculia*" (p. 302). The account of Ullyett (1930) also claims to be general, but it is based upon *D. nigrofasciatus*, as is the following.

As do other *Dysdercus* species, the female *D. nigrofasciatus* oviposits small batches in shallow depressions in the soil. These depressions are near the host plant's base, about 1 to 1.5 cm deep, and once covered up are difficult to discern; the female uses her front legs in making them, and takes about an hour, usually at night. Incubation takes 7 to 8 days in warm weather and 9 to 12 in cool (in South Africa); about 80% hatch. The first stadium occurs under the soil and lasts 3 to 4 (warm) or up to 12 (cool) days; the other four stadia are 4 to 7 (warm) and 9 to 14 (cool), 5 to 7 and up to 13, 6 to 8 and up to 16, and 11 to 15 and up to 26 days. Second instars are gregarious and feed on leaves and bracts (of cotton); third instars begin to disperse and to feed on the green boll, as do subsequent instars. Males (in the laboratory) lived 39 to 86 (average 66) days, and females 39 to 84 (58) days. Mating occurs 2 to 6 days after the final molt, and feed while mating. Females laid 481 to 1354 (average nearly 900) eggs in 5 to 12 batches (average 8); food supply, not temperature, influences the number. No diapause was found, although after four generations in June eggs did not hatch, suggesting either cold (Ullyett) or a diapause (C. W. Schaefer, personal observation). Populations can increase rapidly, as the large numbers of eggs indicate. The insects feed on wild malvalian plants as well as cotton, and "they have also been recorded as feeding on citrus fruits (Ullyett 1930, p. 8), another example of a *Dysdercus* turning to Rutaceae.

Ullyett (1930) writes of two forms, a "small pale form" of winter and a larger darker one of spring. It is likely that these forms are those accorded varietal rank by other authors (see synonymies in Freeman 1947). The larger form migrates in February or March (in South Africa) to cotton, and gives rise to the smaller, which remains for the rest of the season (Ullyett 1930); the migration itself may be influenced by the moon, larger numbers of *D. nigrofasciatus* having been found (in Tanzania) near full moon, not new moon (Robertson 1977).

Dysdercus nigrofasciatus feeds on other malvalian plants, as do most (all?) *Dysdercus* species; Ullyett (1930) lists some of these plants, as do Beccari and Gerini (1970) and Ahmad and Schaefer (1987). In tests with seeds of cotton, kapok (*C. petandra*), and okra (*H. esculentus*), this cotton stainer survived on all three, but reproduced better on cotton seeds, as did *D. superstitiosus* in a similar study (Geering and Coaker 1960) (see **Section 3.5.1**).

3.6.2 Damage

For damage done to cotton, see **Section 2**. Ullyett (1930) mentions that *D. nigrofasciatus* has "been recorded as feeding on citrus fruits"; and, like all Old World *Dysdercus* species, this one too feeds on such malvalian crops as okra (bhindi) and kapok; it also attacks pigeon pea (*Cajanus cajan*; Lateef and Reed 1990).

3.6.3 Control

Ullyett (1930) recommends removing wild malvalian plants, the wild hosts of this cotton stainer. An alternative might be to *introduce* such plants, as traps (Schouteden 1912 appears to suggest this). Other control measures are as outlined in other sections above.

Resistance to three insecticides applied topically to fifth instars and expressed in LC_{50} values were carbaryl (337 mg/l), lindane (294), and methidathion (110) (Nyamasyo and Karel 1982).

4. THE LESS IMPORTANT SPECIES

4.1 *Dysdercus sidae* Montrouzier (Pyrrhocoridae)

This small (8 to 14.5 mm) species occurs throughout Australasia (Freeman 1947, Beccari and Gerini 1970; the latter also list host plants) and in Australia is a pest of cotton (Ballard and Holdaway 1926), not so much because of direct damage as because of internal boll rot (Ballard and Evans 1928). The cotton stainer is mostly red, marked with some (or only a little) black; a black spot behind each eye is distinctive (Freeman 1947).

What follows is from work of Ballard and Evans (1928) in Queensland (Australia). Adults arrive in the cotton fields at times determined in a given year not by the stage of cotton, but probably by weather and especially the condition of wild host plants. Hibernation apparently does not occur. Eggs are 0.9 × 1.3 mm; first instars are 1.65 mm long; second, 3; third, 4.5; fourth, 6.9; fifth, about 9 mm. Eggs and instars are described, as is the adult. First instars do not feed; the others feed on seeds. All instars are gregarious, especially before molting. Adults begin to mate 3 to 4 days after eclosion; however, if food (especially cotton seed) is unavailable, "sex consciousness does not seem to be aroused" (p. 409). Eggs are laid as much as an inch below the ground, or under debris. Ten females each laid an average of 6.6 batches, each batch with an average of 75 eggs (13 to 151); each female laid in her lifetime an average of 496 eggs. Adults appear to require unripe (green) seeds for reproduction; immatures up to the fifth stadia can survive well on ripe seeds alone. On cotton, the bugs prefer the seeds at the base of the bolls, where the bugs may cluster. (Ballard and Evans, 1928, describe feeding and food choice in fine detail.) The bugs are also cannibalistic, even when plant food is available; oddly, younger (smaller) nymphs are more likely to attack older than the reverse.

Rates of development (stadial lengths) varied with temperature, slowest at 60°F, speeding up from 65 to 80°F, and leveling off above 80°F. Above 90°F the insects sought shelter; 90 to 100°F was fatal after 3 to 4 days.

Dysdercus sidae was considered by Ballard and Holdaway (1926) as less important on cotton than the scutellerid *Tectocoris diophthalmus* (Thunberg) (see **Chapter 14, Section 4.3**), because the latter species spends more time feeding on more stages of the cotton plant.

Ballard and Evans (1928) dismiss destruction of wild hosts, and use of insecticides, because the former is impossible in vast country like Queensland, and the latter is (or was) impracticable because "by the time *Dysdercus* comes into the cotton no horse-drawn machine could get into it" (p. 431). Doubtless things have changed since. The authors do suggest trapping fifth instars by spreading such aggregating and molting sites as slabs of bark, and smearing these with adhesive. Lures, too, they recommend, but note that none is available.

An unidentified reduviid, some mites, and a mantid prey on *D. sidae* in Queensland (Ballard and Evans 1928), and two tachinid parasites (which emerge from the anus of the adult bug) were collected (Ballard and Evans 1928) and described by Curran (1927).

4.2 *Dysdercus cardinalis* Gerstaecker (Pyrrhocoridae)

This cotton stainer is found in eastern Africa (Freeman 1947, Beccari and Gerini 1970, who also list host plants) and extends into the cotton-growing regions of the Sudan, where it is limited by long and severe dry seasons (Whitfield 1933). Dorsally it is red and yellow-gray, and fairly large (12.5 to 19 mm) (Freeman 1947). This appears to be a less serious pest of African cotton than other species of *Dysdercus*. In Machakos District, Kenya, it was found more on wild hosts than on cotton, in contrast to *D. fasciatus* **(Section 3.3)** (Tengecho 1994). Of five malvalian plants tested, also in Kenya, baobab (*Adansonia digitata*) and *Abutilon mauritianum* were unsuitable for nymphal growth and survival, and *Gossypium hirsutum*, *Ceiba pentrandra* and *Sterculia rhynchocarpa* were suitable (Chemengich and Khaemba 1998, who suggested the first two plants be planted as trap crops). It is also a pest of pigeon pea (*Cajanus cajan*) (Lateef and Reed 1990). The effects of aggregating were studied by Melber (1979a,b): all rates of development were increased and the precopulation and preoviposition periods were decreased by aggregation (positive effects), but adult longevity, egg size, fertility, and fecundity were all decreased (negative effects); these all doubtless work together to maximize survival and continuance in ecologically good times and bad.

The juvenile hormone analogue ethylfarnesoate dihydrochloride, applied to fifth instars, caused 50% inhibition of adult characters at 0.08 µg when applied topically, and caused 50% inhibition when applied as an aerosol (1%) for 1.3 s (Bransby-Williams 1971). Resistance to carbaryl applied topically (expressed as an LC_{50}) was 147 mg/l (Nyamasyo and Karel 1982).

4.3 *Dysdercus melanoderes* Karsch (Pyrrhocoridae)

This large species (15 to 22 mm) is found in Nigeria and nearby west African countries (Golding 1928, Freeman 1947, Beccari and Gerini 1970). Beccari and Gerini (1970) and Fuseini and Kumar (1975a) list host plants, and the former list the several species of *Phonoctonus* (Reduviidae) that prey on this cotton stainer. *Dysdercus melanoderes* is dorsally yellowish; the upturned thoracic margins and the very pale membrane separate it from other species (Freeman 1947). It feeds on fallen cotton seeds in dense aggregations, and attacks also okra and "kenaf," but not cocoa, in Nigeria (Youdeowei 1973), where it occurs with *D. superstitiosus* in cotton fields (Golding 1928; see **Section 3.5**). The immatures are figured and described by Fuseini and Kumar (1975b) (Ghana): egg, 0.98 × 1.47 mm; first instar, 2.23; second, 3.85; third, 5.53; fourth, 8.81; fifth, 13.55 mm. Five females laid four to eight batches of eggs each, a batch averaging 76 eggs; each female laid *in toto* 332 to 695 eggs. The incubation period averaged 4.5 days, and the five stadial altogether 32 days. The nymphs aggregate, and greatly prefer a damp substrate to a dry (Fuseini and Kumar 1975b).

In Ghana, the preferred wild host of this and several other *Dysdercus* species is *Ceiba pentandra* (L.) Gaertn., tested against *S. foetida* L., *Bombax sessilis* (Benth.) Bakh., and *Hibiscus micranthus* L. (Fuseini and Kumar 1975a). *Dysdercus melanoderes* prefers the forest to the savannah (Golding 1928); in the forest it is preyed upon by the reduviid *P. subimpictus* Stål, which it closely resembles; few other predators or parasites (and no egg parasites) were found (Fuseini and Kumar 1975a). Singh et al. (1978) list it as a minor pest of grain legumes.

4.4., 4.5 *Dysdercus maurus* Distant (= *Dysdercus howardi minor* Ballou, *Dysdercus ruficollis* F. (*nec* L.), *Dysdercus pallidus* Blöte), and *Dysdercus ruficollis* (L.) (Pyrrhocoridae)

These are Neotropical cotton stainers. They are treated here together because it is not clear in some papers which species was studied: "*Dysdercus ruficollis*" may refer to either, and even when authors provide the describer's name (F. or L.), the lack of any reference to Van Doesburg's (1968) paper renders the identification suspect, for it was Van Doesburg (1968) who straightened out the taxonomy. The two species have roughly similar distributions, in Brazil and north-coastal South America (Van Doesburg 1968), but these distributions may merely reflect collecting bias. Both are small (7.5 to 14.5 mm), and reddish to black (see Van Doesburg 1968, Schaefer 1998a). The two are best separated by their males' parameres: singly hooked in *D. ruficollis*, doubly hooked in *D. maurus* (Van Doesburg 1968, Jurberg et al. 1982; the latter authors also describe other aspects of genitalic and somatic morphology). Almeida et al. (1986) list wild hosts of both (?) these species, and show geographic variation in host preferences in Brazil. Fifth instars and adults fed more on seeds than flowers of *Sida carpinifolia* L., whereas earlier instars showed no preference (Azevedo-Ramos et al. 1991), perhaps because seeds provide something necessary for reproduction (see **Section 4.1**). In Rio de Janeiro state (Brazil), high temperatures and an increase in the number of rainy days in 1 month were correlated with a decrease in populations of this bug in the next month; the bugs were established on *Sidasthrum micranthum* (St. Hil.) (Malvaceae) (Xerez et al. 1984).

Townsend (1913) calls *D. ruficollis* (but almost certainly *D. maurus*; see Van Doesburg 1968, p. 138) the "common cotton-stainer of the Peruvian coast region" (p. 92), and Wille (1952) considered it the most damaging pest in Peru (but see **Section 4.9**); this species (as *D. howardi minor*) and *D. f. fulvoniger* (De Geer) (as *D. howardi* Ballou) were also the worst cotton pests in Trinidad (Urich 1916). *Dysdercus maurus* is also a pest in Paraguay, where it attacks green bolls and aborts the seeds; other hosts in the Paraguayan Chaco include the kapok tree (*Chorista speciosa*) (Nickel 1958, as *D. pallidus*). It also may feed on *Citrus* spp., although the bugs here may have been migrating (Moizant and Terán B. 1970); however, other species of *Dysdercus* feed on citrus, and this record may presage a problem.

A dust composed of 3% gamma isomer of BHC, 5% DDT, and 40% sulfur reduced populations of *D. maurus* in Paraguay, but only temporarily (Nickel 1958). A juvenile hormone analogue, applied topically to fifth instars, disrupted normal development (Jurberg et al. 1983). Townsend (1913) reared from his Peruvian specimens and described a new species of Tachinidae (Diptera). And in Trinidad Urich (1916) recommended hand-picking, and destruction of wild host plants (but see **Section 4.1**).

4.6 *Dysdercus obscuratus* Distant (Pyrrhocoridae)

Dysdercus obscuratus and its several subspecies are distributed from southern Mexico into northern South America (Van Doesburg 1968). Yet Barber (1925) records it, in some numbers, from the lower Rio Grande Valley of Texas (U.S.A.), and across the river in northern Mexico. Van Doesburg cites Barber's paper, but does not include Barber's record; this suggests to us that Van Doesburg was uncertain of the identification, which, however, was made by a respected heteropterist, W. L. McAtee; the question remains open. The following is from Barber (1925).

This cotton stainer was originally found on wild plants (especially *Sida carpinifolia*, but also two *Ambrosia* species), and later also on many cotton plants in many locations. The egg and the first four instars are illustrated and described. Durations: egg, 7.9 days; first stadium, 6.3 (spent entirely underground); second, 8.4; third, 9.4; fourth, 4 to 5; fifth, 18.2 days; adults lived 49.8 (females) and 48.3 (males) days. Adults and later instars are gregarious, several hundred congregating on a plant. Mating occurs after 2 days, eggs are placed under the ground, and a dead leaf or seed capsule is provided for food; egg counts were difficult, but two females together laid 107 eggs.

4.7 *Dysdercus mimus* (Say) (= *Dysdercus albidiventris* Stål) (Pyrrhocoridae)

Dysdercus mimus ranges from the southernmost tip of Texas (U.S.A.) through northern South America (Van Doesburg 1968, map 37); it also occurs in Arizona (Morrill 1918, as *D. albidiventris*), where in 1916 it destroyed on some farms up to 90% of the cotton bolls (Morrill gives a photograph of damage). Almeida et al. (1986) list host plants, and show geographic differences in host-plant preferences.

4.8 *Dysdercus mimulus* Hussey (Pyrrhocoridae)

This species occurs from southern Florida and Texas (U.S.A.) south through Central America (Van Doesburg 1968, map 151). In 1938 it was recorded as a pest of cotton, although not a major one (Cassidy and Barber 1938) and, a year later, "it was not considered of much economic importance" (Cassidy and Barber 1940, p. 2). Chemicals used to control other insects, especially other heteropterans, perhaps controlled this insect as well.

4.9 *Dysdercus fulvoniger* (De Geer) (= *Dysdercus howardi* Ballou) (Pyrrhocoridae)

This cotton stainer occurs in South and Central America, and many of the Caribbean islands (Van Doesburg 1968); in Trinidad it together with *D. maurus* (see **Section 4.4**) are the "worst pest[s] of cotton on [the] island" (Urich 1916, p. 18). Host plants are listed by Urich (1916), Almeida et al. (1986), and Van Doesburg (1968). Several immatures were captured by sphecid wasps in Trinidad (Vesey-Fitzgerald 1956).

4.10 *Dysdercus peruvianus* Guérin Méneville (Pyrrhocoridae)

"This pyrrhocorid is considered to be the most serious pest of cotton in Peru" (Anonymous 1961; but see **Section 4. 4**), where in 1957–1958 it caused some U.S.$10 million in damage and destroyed up to 70% of the fiber. In early spring, the insects feed in the Andean foothills on wild malvalian plants and, in summer, move down to cotton fields, where adult females lay batches of about 50 eggs just beneath the soil surface. Eggs hatch in 5 to 7 days and the first two instars congregate, whereas the last three disperse; adults eclose 30 to 40 days after hatching, and move from field to field (Anonymous 1961). In the laboratory in Brazil, nymphs did not develop at 32°C, but did at temperatures from 18 to 30°C. Optimal (briefest) development occurred at 30°C (25 days), but nymphal mortality was higher than at lower temperatures. The lower threshold temperature for egg–adult development was 12.86°C, and the thermal requirements were 419.66 degree-days (Milano et al. 1999). The bugs feed on a variety of hosts, including citrus (Anonymous 1961, Van Doesburg 1968, Almeida et al. 1986).

Once the preferred insecticide, BHC is being replaced by organophosphates, because resistance to it has developed; moreover, the movement of large populations from foothills into cotton fields often swamps pesticide use (Wille et al. 1958).

4.11 *Dysdercus suturellus* (Herrich-Schaeffer) (Pyrrhocoridae)

This relatively small (10.0 to 12.5 mm long) bug occurs from the southeastern United States (Henry 1988b) into the Bahamas and Jamaica; *D. s. capitatus* Distant occurs in Central America (Van Doesburg 1968). The nominal subspecies is red (head, much of pronotum, scutellum), blackish to gray (rest of pronotum, corium), and dirty white (pronotal collar, borders of corium) (Van Doesburg 1968).

Many years ago, this insect was reported from cotton fields in Florida, where it fed also on wild Malvaceae, as well as on guava (*Psidium* sp.) leaves and several wild Solanaceae (Morrill 1910); some of these records are dubious, although Morrill (1910) reports large numbers breeding on Spanish cocklebur (*U. lobata*). Earlier, Howard (1897) stated the bug has moved from cotton, and "has largely transferred its attention to the orange fruit"; but, by 1910, Morrill could write that it is the "most destructive cotton pest in Florida." Howard (1897) further suggests a "rich orange-yellow dye" can be made from *D. suturellus* which, with a suitable mordant, can dye wool and silk.

Roselle (*H. sabdariffa* L.) is grown in the Old World tropics, and sometimes in the American tropics; the fruits are prepared in various ways for eating, have medicinal value, and the seeds are good feed for chickens. A minor pest, presumably only in the New World, is *D. suturellus* (Morton 1987).

Four females laid an average of 116 eggs each, loosely in the sand in the laboratory (Morrill 1910). A parasitic trypanosomatid flagellate has been isolated from the gut of *D. suturellus* (Wallace 1977), but seems to do the bug no harm.

4.12 *Antilochus coquebertii* (F.) (Pyrrhocoridae)

This pyrrhocorid is best known as a predator of other pyrrhocorids, especially *Dysdercus* (although this apparent preference may simply reflect the large numbers and populations of *Dysdercus* species; see brief reviews in Schaefer and Ahmad 1987, and Schaefer 1999). However, *A. coquebertii* also feeds on plants, and at times may become a pest of cotton and bhindi (Singh and Tomar 1977) wherever in its range these crops are grown (the plains of India, Burma, Sri Lanka, Malaysia, and the former Indochina) (Hussey 1929, Chopra and Zamal 1988). The immatures have been described and figured by Quayum and Nahar (1980), and the adult morphology by Quayum and Nahar (1980) and Chopra and Zamal (1988), the latter more accurately.

Of nine insecticides tested against *A. coquebertii*, the carbamate Baygon was the most effective (100% mortality within 12 hours after topical treatment with 0.06% in acetone) (Singh and Tomar 1977).

4.13 *Dindymus versicolor* (Herrich-Schaeffer) (Pyrrhocoridae)

"*Dindymus versicolor* is regarded as a minor pest of soft fruit orchards, market gardens and home gardens in southeastern Australia and Tasmania" (Stahle 1981, p. 375). Stahle goes on to show that the damage locally may become more than minor: significant losses of sunflower and apple crops as well as the introduction into peaches of fungal infections via the bugs' feeding punctures.

As is suggested by the number of crops attacked, *D. versicolor* feeds on a wide variety of plants (many not native to Australia), encompassing 20 families, and also on the carcass of various insects and even the occasional vertebrate. This eclecticism allows the bugs to move from plant to plant as the season progresses. Yet the bug prefers sunflower (not native), "minced beef" (!), and the dead bodies of its conspecifics (Stahle 1981). In the field, its wide food range, and presumably the buildup of populations that the wide range permits, make it a potential pest on many crops, much like other polyphagous bugs such as *Nezara viridula* (L.) (**Chapter 13, Section 3.17**) and *Leptoglossus australis* (F.) (**Chapter 11, Section 3.6.1**).

The five instars are figured and described by Stahle (1979); the first instar is 1.9 mm long; second, 3.1; third, 5.0; fourth, 7.3; fifth, 8.7. The post-first instars of *D. versicolor* may be distinguished from those of the occasionally sympatric *Dysdercus sidae* (see **Section 4.1**) by the presence in the latter of a pale line running the length of the thoracic dorsum, and (in fifth instars) by the presence in *D. sidae* of a red spot above and below the eye (Stahle 1979).

4.14 *Dindymus flavipes* (Signoret) (Pyrrhocoridae)

This is an occasional pest of cotton in the Côte d'Ivoire, although because it may at times be confused with *Dysdercus* (species not given) (Sauvaut 1949), it is unclear how often damage can truly be attributed to *D. flavipes*.

4.15 *Melamphaus faber* (F.) (Pyrrhocoridae)

Large numbers of this bug at times attack the fruits of the trees *Hydnocarpus anthelmintaca* Pieree and *H. wightiana* Blume, in Malaysia. The oil of these trees is (or was) used to treat leprosy. The insects feed by inserting their mouthparts full-length through the hard covering of the fruit, to reach the nut inside. It is possible the punctures allow entry of drosophilid dipterans and *Carpophilus* sp. beetles (Nitidulidae), but apparently fungi are not introduced (Miller 1932)

Copulation lasts 1 to 2 days and eggs are laid in the soil, in batches of 110 to 180 eggs. Miller (1932) described and illustrated the immatures: the egg is 1 × 1.8 mm, and the five instars are 2.5, 3.5, 7, 12, and 16 mm long. Adults are black dorsally and red ventrally and large: male 21 to 22 mm, female 24.5 to 26.5 mm (Miller 1932).

4.16 *Scantius volucris* (Gerstäcker) (Pyrrhocoridae)

This is described as a "potential pest of cotton in the ... Punjab" (Pakistan), from which it has been occasionally collected; it also feeds on the weed *Withenia somnifera* L. (Ahmad and Mohammad 1983). The immatures are illustrated, described, and keyed; the five instars are 1.7, 2.6, 4.0, 5.5, and 6.4 mm long (Ahmad and Mohamadd 1983).

4.17 *Odontopus confusus* Distant (Pyrrhocoridae)

This African species may carry to several species of *Sterculia* the fungus (*Nematospora gossypii*) which causes internal boll disease in cotton (Frazer 1944). Several of these *Sterculia* species are of some economic significance and, more important, they thus serve as reservoirs for the fungus.

4.18 *Largus* spp. (Largidae)

Largus is the largest genus of the exclusively New World subfamily Larginae. It ranges from the southern United States well into central South America; the subfamily is being revised by the authors (C. W. Schaefer and I. Ahmad, in preparation), and the North American species of *Largus* by Schaefer and Z. W. Kalinowski (in preparation); until this revision is complete, the status of several names of North American *Largus* remains in doubt. Little is known of *Largus* species biology. The most complete account is Booth's (1990), of *L. californicus* (Van Duzee), studied in southern California (U.S.A.). It is likely that the data outlined here for this differ little for other species. After mating (described in Booth 1992), eggs are laid in the ground in clusters, 12 to 286 eggs/cluster (average: 133). Hatching occurs in 14 days (average); each stadium lasts 1 to 21 days, but mortality can be very high; adults live 2 to 7 months; that no adults were seen in March or April suggests there is one generation a year. First instars aggregate on the empty eggshells, cannibalizing the unhatched ones, and then disperse to plants, where they again aggregate; in the field, also, tight clusters of mixed instars occurred. Fifth instars are differently colored than adults. *Largus californicus* seems to prefer *Lupinus arboreus*, but a remarkably large array of plants is also fed upon (in 19 dicot families), as well as dead insects and vertebrate feces. Parasites include a tachinid dipteran, a fungus, and the scelionid egg parasite *Gryon largi* (Ashmead) (see also Masner 1983). Among the predators were a lynx spider, and a reduviid (*Apiomerus* sp.); feeding trials with two species of lizard indicated that *Largus californicus* is distasteful. *Largus californicus* is not economically significant. However, other species of the genus are occasionally minor pests. *Largus cinctus* (Herrich-Schaeffer) occurs on physic nut (*Jatropha curacas*; for the importance of this plant, see **Chapter 9, Section 3.7**). *Largus succinctus* Herrich-Schaeffer was found on cotton in the southern United States (especially Texas) early in the twentieth century, but seems no longer to be thought a problem (Sanderson 1905, Morrill 1910, Pierce 1917). And *Largus* sp., *L. humilis* (Drury), and *L. rufipennis* (Laporte), all been found on cotton in Brazil (Silva et al. 1968). *Largus humilis* (Brazil) has also been found on *Citrus* sp. (Silva et al. 1968) and *L. succinctus* (Texas) on peaches and tomatoes (Sanderson 1905, Morrill 1910).

In the Peruvian Amazon region, *Eugenia uniflora* (Myrtaceae) (pitanga, or Surinam cherry) is widely cultivated as a hedge plant; its edible fruit is used in jellies, etc., and its crunched leaves are supposed to repel insects. *Largus balteatus* Stål feeds on its leaves and fruits, but are not serious pests (Couturier et al. 1996).

4.19 *Macrocheraia grandis* Gray (= *Lohita grandis* [Gray], *Macroceroea grandis* [Gray]) (Largidae)

The giant red bug is known almost universally as *Lohita grandis*; but see Stehlik and Kerzhner (1999). It is a pest on several crops, especially trees, in northeastern India (Joshi and Khan 1990), and its range extends to the north (Hussey 1929); a subspecies, *M. grandis sumatrana* (Distant), occurs in Malaysia (Hussey 1929). It may become locally abundant on cotton (*Gossypium herbuceum*), bhindi or okra (*Abelmoschus* [= *Hibiscus*] *esculentus*), and the trees gutel (*Trewia nudiflora*) and semul (*Bombax ceiba*), on whose seeds, fruits, shoots, and foliage it feeds, retarding growth and inhibiting seed germination (Joshi and Khan 1990, Dhiman and Dhiman 1986).

Macrocheraia grandis is deep red, with some black markings. The bug is notable for its extreme sexual dimorphism: males are 39 to 54 mm long and females 29 to 33 mm (Dhiman and Chatterjee 1980). This dimorphism begins with the last (fifth) instar (Joshi and Khan 1990); the difference is exaggerated by the fact the wings in each sex are of the same length, reaching the end of the female's abdomen but barely reaching the middle of the male's. The male's antenna is 63 mm long and the female's 36 mm (Dhiman and Chatterjee 1980), which renders the male's antenna three times longer than his head and the female's only 33% longer than hers (Ahmad and Abbas 1985). Further comparative measurements are given by Dhiman and Chatterjee (1980). The genus and species are redescribed by Ahmad and Abbas (1985), who also describe and illustrate the genitalia and the metathoracic scent gland apparatus; Dhiman (1983) described the soft anatomy of these glands.

Mating takes 60 to 90 min, and a female lays 10 to 21 eggs in one or two batches on the soil or in ground litter. The eggs (2.9 × 1.5 mm) hatch (25 to 100% viability) in 11 (28 to 35.5°C) or

33 (12.5 to 29°C) days. The five instars are 4.3, 7.3, 11.4, 15.6, 30.7 (male) and 24.2 (female) mm long. In Assam, there are two generations a year. The five stadia take about 64 to 90 days (winter, 12.5 to 31°C) and 64 to 79 days (summer, 25 to 32.5°C). Adults live 7 to 38 days, depending upon weather and food (Joshi and Khan 1990).

Dhiman and Dhiman (1986; in Uttar Pradesh, India) believe *Trewia nudiflora* is the preferred host, and list nine other plants on which it feeds, in addition to the economically important ones above; they also list nine plants on which it shelters. Cotton and bhindi were not fed upon in their study area (Dhiman and Dhiman 1986).

Like certain lygaeids (Sweet 1964), late instars and adults of *M. grandis* may carry seeds with their forelegs, and all stages aggregate, the younger nymphs often sheltered beneath the older; up to 16 individuals may feed on a single seed. If seeds become scarce, the bugs may feed on younger nymphs, and occasionally on bird excreta (Dhiman and Dhiman 1990). Food is located with setae on the fourth antennal segment (illustrated in Dhiman and Dhiman 1990).

The most effective insecticides were carbaryl and chloropyriphos, 39 and 35 times more toxic than DDT when applied in an emulsion to foliage of *T. nudiflora*, which was then presented to the bugs (Joshi and Khan 1990).

5. CONCLUSIONS

Certainly the most economically important group in Pyrrhocoroidea is *Dysdercus*, perhaps because it is the largest genus in a superfamily so many of whose members feed on relatives of cotton, and on cotton itself. Of the roughly 60 species of *Dysdercus*, the most serious cotton pests belong to the group of 25 (or so) in the Old World tropics; the 35 (or so) New World species are much less serious. This may reflect a more recent establishment of cotton in the Neotropics, or a descent of the New World species from one or a few Old World species for which cotton is not a favored food plant. Whatever the reason, *Dysdercus* spp. are more important cotton pests in Africa and Asia than they are in the New World, but this should not encourage complacency in the latter (see **Section 6, Prognosis**).

Other pyrrhocoroids are minor pests on cotton (although these may become serious locally for short periods) and a few other crops. And some species may actually prey on pest *Dysdercus*. But the only real importance of the family is as stainers of cotton, and as mechanical vectors of boll rot disease pathogens; that is, the only really important members of the superfamily are species of *Dysdercus*.

6. PROGNOSIS

One reading through the several species accounts in this chapter will be struck by the amount of work being done (especially in India) in two important areas: the use of Reduviidae for the biocontrol of these pests (see also **Chapter 29**) and the treatment of these bugs by a wide array of natural plant products which have ovicidal effects or which adversely affect development or fecundity (see also Enslee and Riddiford 1977, Kumar and Thakur 1989). We strongly urge that similar work be done in the cotton-growing regions of Africa and, indeed, those of all parts of the world.

It is also important to monitor the occurrence and the potential of New World *Dysdercus* species, which so far have done little damage (see Schaefer 1998b). As the world warms, *Dysdercus* may appear in the United States more frequently, where it could become at least a minor pest. And as cotton increases as a crop in Brazil (and elsewhere in Latin America), so does the possibility of damage by local cotton stainers (see also Schaefer 1998b).

Lacking are basic studies on the biologies, ecologies, and life histories of many of these cotton stainers. Some work on African species is marred nearly to the point of being useless by the lumping of several species together and treating them as one. The value of some of the Indian work is vitiated by poor taxonomy. Our recommendations therefore are (1) further detailed studies of the biologies of individual and accurately identified pest *Dysdercus* and of nonpest *Dysdercus* species; (2) analytical studies on the biological control potential of parasites and predators, especially of reduviids, along the lines of D. P. Ambrose's work (**Section 3.1.3, Chapter 29**) and of Schaefer and Ahmad (1987).

These insects often feed on nonmalvalian plants and, of course, on wild relatives of cotton. The use of some of these plants as trap crops might be explored (see el Heneidy and Sekematte 1996).

Finally, *D. fasciatus* appears to have been introduced into, and to have become established in, Costa Rica (elsewhere?) (Janzen 1972). Little comment has been made on this interesting fact, although Janzen does suggest in passing that this bug may replace human-eliminated seed-feeding vertebrates as competition for malvalian seeds needed by autochthonous pyrrhocorids. *Dysdercus fasciatus* is a serious pest of cotton in Africa (**Section 3.3**) and it must be monitored in cotton-growing regions of the neotropics.

7. REFERENCES CITED

Agrawal, I. L. 1993. Ovicidal activity of four plant extracts on *Dysdercus koenigii* Fabr. (Heteroptera: Pyrrhocoridae). Indian J. Entomol. 55: 440–443.

Ahmad, I., 1983. Age-specific life table of red cotton stainer *Dysdercus koenigii* (Fabricius) (Hemiptera: Pyrrhocoridae). Chemosphere 12: 1541–1544.

1985. Biology, population dynamics and some control strategies of red cotton stainer, *Dysdercus koenigii* (Fabr.) (Hemiptera: Pyrrhocoridae). Proc. Entomol. Soc. Karachi 13: 136–144.

1991. Cytological studies on cotton stainer *Dysdercus koenigii* (F.) (Hemiptera: Pyrrhocoridae) and their bearing on classification. Pakistani J. Entomol. (Karachi) 6: 15–22.

Ahmad, I., and N. Abbas. 1985. Redescription of *Lohita grandis* (Gray) (Hemiptera: Pyrrhocoroidea: Largidae) from Bangladesh with reference to its relationships. Proc. Entomol. Soc. Karachi 14–15: 13–20.

Ahmad, I., and F. A. Mohammad. 1983. Biology and immature systematics of red cotton stainer *Dysdercus koenigii* (Fabr.) (Hemiptera: Pyrrhocoridae) with a note on their relationships. Bull. Zool. 1: 1–9.

Ahmad, I., and C. W. Schaefer. 1987. Food plants and feeding biology of the Pyrrhocoroidea (Hemiptera). Phytophaga 1: 75–92.

Ahmad, I., F. A. Mohammad, and M. Afzal. 1988. Comparative morphology of alimentary organs of some pyrrhocoroids (Hemiptera: Trichophora). Pakistan J. Sci. Ind. Res. 31: 38–42.

Ahmad, M. 1980. Cytological effects of tepa on the reproductive organs of *Dysdercus cingulatus* Fabr. J. Entomol. Res. 4: 171–186.

Ahmad, M., and N. H. Khan. 1980. Chemical repellants for *Dysdercus koenigii* (Fabr.). Indian J. Entomol. 42: 820–821.

Aldrich, J. R., T. J. Kelly, and C. W. Woods. 1982. Larval molting hormone of trichophoran Hemiptera-Heteroptera: makisterone A, not 20-hydroxyecdysone. J. Insect Physiol. 28: 857–861.

Almeida, A. A. de 1978. Influence of tepa on the development of nymphal stages of the cotton stainer *Dysdercus fasciatus* Signoret. Rev. Bras. Biol. 38: 813–817.

1980a. Influence of the glandless and glanded cottonseeds on development, fecundity fertility of *Dysdercus fasciatus* Signoret (Hemiptera, Pyrrhocoridae). Rev. Bras. Biol. 40: 475–483.

1980b. Food consumption and comparative preference of *Dysdercus fasciatus* Signoret for glandless and glanded cottonseeds (Hemiptera, Pyrrhocoridae). Rev. Bras. Biol. 40: 485–489.

1980c. Uptake and excretion of gossypol in *Dysdercus fasciatus* Signoret. Rev. Bras. Biol. 40: 659–662.

Almeida, J. R. de, S. B. Almeida, and R. de Xerez. 1986. Variação geográfica na dieta das espécies brasileiras de percevejos "manchadores de algodão" (Hemiptera, Pyrrhocoridae, *Dysdercus* spp.). Rev. Bras. Biol. 46: 329–337.

Almeida, J. R. de., Y. Mizuguchi, R. de Xerez, and G. M. da Solva. 1981. Parasitose em percevejos manchadores de algodão, *Dysdercus* spp. (Hemiptera, Pyrrhocoridae). Rev. Brasil. Entomol. 25: 55–60.

Ambrose, D. P., and M. A. Claver. 1995. Food requirement of *Rhynocoris kumarii* Ambrose & Livingstone (Heteroptera, Reduviidae). J. Biol. Control 9: 47–50.

1999. Suppression of cotton leafworm *Spodoptera litura*, flower beetle, *Mylabria pustulata*, and red cotton bug, *Dysdercus cingulatus* by *Rhynocoris marginatus* (Fabr.) (Het., Reduviidae) in cotton field cages. J. Appl. Entomol. 123: 225–229.

Ambrose, D. P., and K. Sahayaraj. 1993. Predatory potential and stage preference of a reduviid predator *Allaeocranum quadrisignatus* Reuter on *Dysdercus cingulatus* Fabricius. J. Biol. Control 7: 12–14.

Anonymous. 1961. Insects not known to occur in the United States, Peruvian cotton stainer (*Dysdercus peruvianus* Guerin). U.S. Dept. Agric. Co-op. Econ. Insect. Rep. 11: 19–20.

Ansari, H. J., and M. A. Khan. 1973. Effect of triphenyltin acetate on the fecundity and fertility of *Dysdercus cingulatus* F. (Heteroptera: Pyrrhocoridae). Curr. Sci. 42: 280–281.

Ansari, M. A. 1973. Biological effects of tepa on adults of *Dysdercus cingulatus* Fabr. Indian J. Entomol. 35: 289–292.

Ashok, S., B. K. Rao, S. S. Thakur, P. Judson, N. S. Raj, and G. M. Ram. 1978. Defensive secretion of the soap-nut bug, *Tessaratoma javanica* (Thunberg) (Heteroptera, Pentatomidae) as a juvenile hormone mimic. Zool. Jahrb. Physiol. 82: 383–394.

Azevedo-Ramos, C., P. R. S. Moutinho, and P. Guimarães. 1991. Food exploitation of reproductive structures of *Sida carpinifolia* (Malvaceae) by *Dysdercus ruficollis* L. 1764 (Hemiptera, Pyrrhocoridae). Rev. Bras. Entomol. 35: 761–765.

Ballard, B. E., and F. G. Holdaway. 1926. The life history of *Tectocoris lineola* F. and its connection with internal boll rot in Queensland. Bull. Entomol. Res. 16: 329–346.

Ballard, E. 1927. Some insects associated with cotton in Papua and the mandated territory of New Guinea. Bull. Entomol. Res. 17: 295–300.

Ballard, E., and M. G. Evans. 1928. *Dysdercus sidae*, Montr., in Queensland. Bull. Entomol. Res. 18: 405–432.

Ballou, H. A. 1906. Cotton stainers. West Indian Bull. 7: 64–85 [not seen, *fide* Van Doesburg 1968].

1916. Insect pests. Part. I. Pp. 3–16 *in* Report on the prevalence of some pests and diseases in the West Indies during 1915. West Indian Bull. 16: 3–16 [not seen, *fide* Van Doesburg 1968].

Banerjee, P., and S. Datta. 1980. Biological control of red cotton bug, *Dysdercus koenigii* Fabricius by mite, *Hemipteroseius indicus* (Krantz and Khot). Indian J. Entomol. 42: 265–267.

Barber, H. G. 1911. *Pyrrhocoris apterus* Linn. in the United States. J. New York Entomol. Soc. 19: 111–112.

Barber, T. C. 1925. Preliminary observations on an insect of the cotton stainer group new to the United States. J. Agric. Res. 31: 1137–1147.

Beccari, F., and V. Gerini. 1970. Catalogo delle specie appartenenti al genere *Dysdercus* Boisduval (Rhynchota, Pyrrhocoridae). Riv. Agric. Subtrop. Trop. 64: 1–67.

Bentz, C., and H. G. Kallenborn. 1995. Feinstruktur der Mitteldarm-Caeca der Baumwollwanze, *Dysdercus intermedius* (Heteroptera, Pyrrhocoridae). Entomol. Generalis 20: 27–36.

Bergroth, E. 1913. Supplementum catalogi heteropterorum Bruxellensis. II. Coreidae, Pyrrhocoridae, Colobathristidae, Neididae. Mem. Soc. Entomol. Belgique 22: 126–183.

Berridge, M. J. 1965a. The physiology of excretion in the cotton stainer, *Dysdercus fasciatus* Signoret. I. Anatomy, water excretion and osmoregulation. J. Exp. Biol. 43: 511–521.

1965b. The physiology of excretion in the cotton stainer, *Dysdercus fasciatus* Signoret. II. Inorganic excretion and ionic regulation. J. Exp. Biol. 43: 523–533.

1965c. The physiology of excretion in the cotton stainer, *Dysdercus fasciatus* Signoret. III. Nitrogen excretion and excretory metabolism. J. Exp. Biol. 43: 535–552.

Blackith, R. E., R. G. Davies, and E. A. Moy. 1963. A biometric analysis of development in *Dysdercus fasciatus* Sign. (Hemiptera: Pyrrhocoridae). Growth 27: 317–334.

Blöte, H. C. 1931. Catalogue of the Pyrrhocoridae in the 'sRijks Museum van Natuurlijke Historie. Zool. Meded. (Leiden) 14: 97–136.

Booth, C. L. 1990. Biology of *Largus californicus* (Hemiptera: Largidae). Southwest Entomol. 35: 15–22.

1992. Ontogenetic color change and mating cues in *Largus californicus* (Hemiptera: Largidae). Ann. Entomol. Soc. Amer. 85: 351–354.

Bourdouxhe, L., and P. Jolivet. 1981. Nouvelles observations sur le complexe mimetique de Mesoplatys cincta Olivier (Coleoptere Chrysomelidae) au Senegal. Bull. Mens. Soc. Linn. Lyon 50: 46–48.

Brailovsky, H. 1981. Araphe H. S. descripción de nuevas especies (Hemiptera: Heteroptera: Largidae). Folia Entomol. Mex. 47: 81–109.

1991. Four new species of the neotropical genus *Theraneis* Spinola (Hemiptera: Heteroptera: Largidae). J. New York Entomol. Soc. 99: 630–636.
1993. El género *Astemma* con descripción de nuevas especies (Hemiptera: Heteroptera: Largidae). An. Inst. Biol. Univ. Autón. México (Zool.) 64: 39–47.
1997. An analysis of the genus *Stenomacra* Stal [sic] with description of four new species, and some taxonomic rearrangements (Hemiptera: Heteroptera: Largidae). J. New York Entomol. Soc. 105: 1–14.
Bransby-Williams, W. R. 1971. Juvenile hormone activity of ethylfarnesoate dihydrochloride with the cotton stainer *Dysdercus cardinalis* Gerst. Bull. Entomol. Res. 61: 41–47.
Carayon, J., and M. Thouvenin. 1966. Emploi d'une substance mimétique de l'hormone juvénile pour la lutte contre les *Dysdercus*, hémiptères nuisibles au connonier. Acad. Agric. France 1966: 340–346.
Carlisle, D. B., and P. E. Ellis. 1967. Abnormalities of growth and metamorphosis in some pyrrhocorid bugs: the paper factor. Bull. Entomol. Res. 57: 405–417.
Carroll, S. P., and J. E. Loye. 1990. Male-biased sex ratios, female promiscuity, and copulatory mate guarding in an aggregating tropical bug, *Dysdercus* bimaculatus. J. Insect Behav. 3: 33–48.
Cassidy, T. P., and T. C. Barber. 1938. Hemipterous cotton insects of Arizona and their economic importance and control. U.S. Dept. Agric., Bur. Entomol. Plant Quarantine. E-239. 14 pp.
1940. Investigations in control of hemipterous cotton insects in Arizona by the use of insecticides. U.S. Dept. Agric., Bur. Entomol. Plant Quarantine. E-506. 19 pp.
Chakraborti, S. 1998. Effects of starvation and differential feeding on boiled and normal seeds of lady's finger on the developmental biology of *Dysdercus koenigii* Fab. (Heteroptera: Pyrrhocoridae). J. Entomol. Res. 22: 17–21.
Chandel, B. A., U.K. Pandoy, and A. K. Singh. 1984. Insecticidal evaluation of some plant products against red cotton bug, *Dysdercus koenigii* Fabr. Indian J. Entomol. 46: 187–191.
Chatterjee, N. B., and D. N. Raychoudhuri. 1960. Bionomics of *Dysdercus koenigii* (Fabr.). Indian Agric. 4: 104–112.
Chemengich, B. T., and B. M. Khaemba. 1998. Comparative studies on biological performance of the cotton stainers *Dysdercus cardinalis* Gerst [sic] and *Dysdercus fasciatus* Signoret bred on different hosts. Insect Sci. Appl. 18: 25–29.
Chevaillier, P. 1965. Étude des phénomènes sécrétoires dans les uradénies des femelles de *Dysdercus fasciatus* Sign. (Hemiptera, Pyrrhocoridae). Ann. Entomol. Soc. France (n.s.) 1: 989–994.
Chockalingam, S., and A. Noorjahan. 1984. The ovicidal effect of difflubenzuron on hemipteran bugs, *Dysdercus cigulatus* and *Chrysocoris purpureus*. Curr. Sci. 53: 1112–1113.
Chopra, N. P., and M. D. S. Zamal. 1988. Identity of *Antilochus coquebertii* (Fabricius) (Hemiptera: Pyrrhocoridae). Bull. Entomol. 29: 100–103.
Clarke, K. U., and J. B. Sardesai. 1959. An analysis of the effects of temperature upon the growth and reproduction of *Dysdercus fasciatus* Sign. (Hemiptera, Pyrrhocoridae). I. The intrinsic rate of increase. Bull. Entomol. Res. 49: 387–405.
Coad, B. R. 1929. Cotton insect problems in the United States. Trans. IV Int. Congr. Entomol.: 241–247.
Cociancich, S., A. Dupont, G. Gehy, R. Lanot, F. Holder, C. Hetru, J. A. Hoffmann, and P. Bulet. 1994. Novel inducible antibacterial peptides from a hemipteran insect, the sap-sucking bug *Pyrrhocoris apterus*. Biochem. J. 300: 567–575.
Costa Lima, A. M. da, N. Guitton, and O. V. Ferreira. 1962. Sobre as espécies americanas do gênero *Dysdercus* Boisduval (Hemiptera, Pyrrhocoridae, Pyrrhocorinae). Mem. Inst. Oswaldo Cruz 60: 21–58.
Couilloud, R. 1989. Hétéroptères déprédateurs du cotonnier en Afrique et à Madagascar (Pyrrhocoridae, Pentatomidae, Coreidae, Alydidae, Rhopalidae, Lygaeidae). Coton Fibres Trop. 44: 3–185.
Couturier, G., L. Quiñones R., I. González R., R. Riva R., and F. Young R. 1996. Los insectos plaga de las Myrtaceae frutales en Pucallpa, Amazonia peruana. Rev. Peruana Entomol. 39: 125–130.
Cowland, J. W., and W. Ruttledge. 1927. Notes on cotton-stainers (*Dysdercus*) in the Sudan. Bull. Entomol. Res. 18: 159–163.
Curran, C. H. 1927. Three new Tachinidae attacking injurious insects in Queensland. Bull. Entomol. Res. 18: 165–167.
Davis, N. T. 1975. Hormonal control of flight muscle histolysis in *Dysdercus fulvoniger*. Ann. Entomol. Soc. Amer. 68: 710–714.
Derr, J. A., B. Alden, and H. Dingle. 1981. Insect life histories in relation to migration, body size, and host plant array: a comparative study of *Dysdercus*. J. Anim. Ecol. 50: 181–193.

Deshpande, D. J., and K. P. Srivastava. 1981. Histological and histochemical studies on the ovaries of the red cotton bug, *Dysdercus koenigii* (Heteroptera: Pyrrhocoridae). Z. Mikroskop. Anat. Forsch. 95: 1035–1047.

Dhiman, S. C. 1983. Metasternal scent glands of *Lohita grandis* Gray. Geobios New Rep. 2: 19–22.

Dhiman, S. C., and V. C. Chatterjee. 1980. Sexual dimorphism in *Lohita grandis* Gray (Heteroptera-Pyrrhocoreidae [sic]. J. Bombay Nat. Hist. Soc. 77: 529.

Dhiman, S. C., and S. D. Dhiman. 1986. Host plants and seasonal occurrence of *Lohita grandis* Gray (Heteroptera-Pyrrhocoroidea-Largidae). Indian For. 112: 272–274.

Dhiman, S. D., and S. C. Dhiman. 1990. Feeding behaviour of *Lohita grandis* Gray (Heteroptera: Largidae). Uttar Pradesh J. Zool. 10: 140–142.

Dingle, H., and G. Arora. 1973. Experimental studies of migration in bugs of the genus *Dysdercus*. Oecologia 12: 119–140.

Dittmann, F., and M. Biczkowski. 1995. Induction of yolk formation in hemipteran previtellogenic oocytes (*Dysdercus intermedius*). Invertebr. Reprod. Dev. 28: 63–70.

Dreyer, H., and J. Baumgärtner. 1995. The influence of post-flowering pests on cowpea seed yield with particular reference to damage by Heteroptera in southern Benin. Agric. Ecosyst. Environ. 53: 137–149.

Duviard, D. 1972. Les vols migratoires de *Dysdercus voelkeri* Schmidt (Hemiptera, Pyrrhocoridae) en Côte d'Ivoire. I. Le rythme endogène fondamental. Coton Fibres Trop. 27: 379–388.

1973. Les vols migratoires de *Dysdercus voelkeri* Schmidt (Hemiptera: Pyrrhocoridae) en Côte d'Ivoire. II. Les rythmes exogènes. Coton Fibres Trop. 28: 239–252.

1977. Migrations of *Dysdercus* spp. (Hemiptera: Pyrrhocoridae) related to movements of the Inter-Tropical Convergence Zone in West Africa. Bull. Entomol. Res. 67: 185–204.

Edmunds, M. 1978. Contrasting methods of survival in two sympatric cotton stainer bugs (Hem., Pyrrhocoridae) in Ghana during food shortage. Entomol. Mon. Mag. 114: 241–244.

Edwards, F. J. 1969. Environmental control of flight muscle histolysis in the bug *Dysdercus intermedius*. J. Insect Physiol. 15: 2013–2020.

Egwuatu, R. I. 1980. The role of food plants in the development and survival of *Dysdercus superstitiosus* F. (Hemiptera, Pyrrhocoridae). Rev. Zool. Afr. 94: 780–790.

Egwuatu, R. I., and G. O. Opara. 1985. Some effects of rearing density on the developmental biology, fecundity, and longevity of Dysdercus voelkeri (F.) Schmidt (Heteroptera, Pyrrhocoridae). Beitr. Trop. Landwirtsch. Veterinärmed. 23: 315–324.

Enslee, E. C., and L. M. Riddiford. 1977. Morphological effects of juvenile hormone mimics on embryonic development in the bug, *Pyrrhocoris apterus*. Wilhelm Roux Arch. 181: 163–181.

Evans, D. E. 1962. The food requirements of *Phonoctonus nigrofasciatus* Stål (Hemiptera, Reduviidae). Entomol. Exp. Appl. 5: 33–39.

Evans, G. O. 1963. Observations on the classification of the family Otopheidomenidae (Acari: Mesostigmata) with descriptions of two new species. Ann. Mag. Nat. Hist. (13) 5: 609–620 [not seen, *fide* Van Doesburg 1968].

Everton, I. J., D. W. Knight, and B. W. Staddon. 1979. Linalool from the metathoracic scent gland of the cotton stainer *Dysdercus intermedius* Distant (Heteroptera: Pyrrhocoridae). Comp. Biochem. Physiol. 63B: 157–161.

Ezueh, M. I., and S. O. Dina. 1980. Pest problems of soybeans and control in Nigeria. Pp. 275–283 *in* [F. T. Corbin, ed.], World Soybean Research Conference II. Proceedings. Westview Press, Boulder, Colorado, U.S.A.

Farine, J.-P. 1987. The exocrine glands of *Dysdercus cingulatus*: morphology and function of nymphal glands. J. Morphol. 194: 195–207.

1988. The exocrine glands of *Dysdercus cingulatus* F. (Heteroptera: Pyrrhocoridae): morphology and function of adults' glands. Ann. Soc. Entomol. France (n.s.) 24: 241–256.

Farine, J.-P., and J. P. Lobreau. 1984. La grégarisme chez *Dysdercus cingulatus* Fabr. (Heteroptera, Pyrrhocoridae): nouvelle méthode d'interpretation statistique. Insect. Soc. 31: 277–290.

Farine, J.-P., O. Bonnard, R. Brossut, and J.-L. le Quere. 1992. Chemistry of pheromonal and defensive secretions in the nymphs and the adults of *Dysdercus cingulatus* Fabr. (Heteroptera: Pyrrhocoridae). J. Chem. Ecol. 18: 65–76.

Farine, J.-P., X. Everaerts, R. Brossut, and J.-L. le Quere. 1993. Defensive secretions of nymphs and adults of five species of Pyrrhocoridae (Insecta: Heteroptera). Biochem. Syst. Ecol. 21: 363–371.

Frazer, H. L. 1944. Observations on the method of transmission of internal boll disease of cotton by the cotton stainer bug. Ann. Appl. Biol. 31: 271–290.

Freeman, P. 1947. A revision of the genus *Dysdercus* Boisduval (Hemiptera: Pyrrhocoridae), excluding the American species. Trans. R. Entomol. Soc. 98: 373–424.

Fuseini, B. A., and R. Kumar. 1975a. Ecology of cotton stainers (Heteroptera: Pyrrhocoridae) in southern Ghana. Biol. J. Linn. Soc. 7: 113–146.

1975b. Biology and immature stages of cotton stainers (Heteroptera: Pyrrhocoridae) found in Ghana. Biol. J. Linn. Soc. 7: 83–111.

Gaffal, K. P. 1976. Die Feinstruktur der Sinnes- und Hüllzellen in den antennalen Schmecksensillen von *Dysdercus intermedius* Dist. (Phyrrhocoridae [sic], Heteroptera). Protoplasma 88: 101–115.

1979. An ultrastructural study of the tips of four classical bimodal sensilla with one mechanosensitive and several chemosensitive receptor cells. Zoomorphologie 92: 273–291.

1981. Terminal sensilla on the labium of *Dysdercus intermedius* Distant (Heteroptera: Pyrrhocoridae). Int. J. Insect Morphol. Embryol. 10: 1–6.

Galichet, P. F. 1956. Quelques facteurs de réduction naturelle dans une population de *Dysdercus superstitiosus* Fab. (Hemiptera, Pyrrhocoridae). Rev. Pathol. Veg. Entomol. Agric. France 35: 27–49.

Gamberale, G., and B. S. Tullberg. 1998. Aposematism and gregariousness: the combined effect of group size and coloration on signal repellence. Proc. R. Soc. London (B) 265: 889–894.

Gandhi, J. R. 1979. Acceptance of different extracts and purified chemicals by adults of *Oxycarenus hyalinipennis* Costa and *Dysdercus fasciatus* Dallas. Indian J. Entomol. 41: 155–159.

Geering, Q. A., and T. H. Coaker. 1960. The effects of different plant foods on the fecundity, fertility and development of a cotton stainer. Empire Cotton Grow. Corp. Res. Mem. No. 39: 61–76.

Golding, F. D. 1928. Notes on the bionomics of cotton stainers (*Dysdercus*) in Nigeria. Bull. Entomol. Res. 18: 319–334.

Grimm, C., and E. Führer. 1998. Population dynamics of true bugs (Heteroptera) in physic nut (*Jatropha curcas*) plantations in Nicaragua. J. Appl. Entomol. 122: 515–521.

Gujar, G. T., and R. V. S. Rao. 1993. Biology and genetics of orange mutant of the cotton stainer, *Dysdercus koenigii*. Insect Sci. Appl. 14: 545–549.

Gupta, M., R. Singh, and P. K. Gupta. 1993. Development of resistance to diflubenzuron in *Dysdercus cingulatus* Fabr. Indian J. Entomol. 55: 393–395.

1994. Effect of diflubenzuron on the fecundity and fertility of *Dysdercus cingulatus*. Indian J. Entomol. 56: 431–432.

Gurusubramanian, G., and S. S. Krishna. 1996. The effects of exposing eggs of four cotton pests to volatiles of *Allium sativum* (Liliaceae). Bull. Entomol. Res. 86: 29–31.

Haas, F., and H. T. König. 1988. *Coriobacterium glomerans* gen. nov., sp. nov. from the intestinal tract of the red soldier bug. Int. J. Syst. Bacteriol. 38: 382–384.

Harris, P. J. C. 1981. Seed viability, dormancy, and field emergence of *Urena lobata* L. in Sierra Leone. Trop. Agric. (Trinidad) 58: 205–213.

1983. Effect of cotton stainers (*Dysdercus* spp.) on *Urena lobata* seed quality. Proc. 10th Int. Cong. Plant Prot. 1: 107.

Harris, P. J. C., and F. Bindi. 1983. *Dysdercus* spp. as pests of *Urena lobata* in Sierra Leone. Trop. Pest Manage. 29: 1–6.

Hebbalkar, D. S., and R. N. Sharma. 1980. Repression of mating behaviour by nymphal exposure to sublethal doses of juvenile hormone analogues in *Dysdercus koenigii* (Fabricius) (Hemiptera: Pyrrhocoridae). Curr. Sci. 49: 457–459.

1982. An experimental analysis of mating behavior of the bug *Dysdercus koenigii* F. (Hemiptera: Pyrrhocoridae). Indian J. Exp. Biol. 20: 399–405.

1991. Effects of age, sex, oogenetic cycles, body weight, juvenoids and precocene II on the oxygen consumption of the red cotton bug *Dysdercus koenigii* F. Indian J. Exp. Biol. 29: 962–966.

el Heneidy, A. H., and M. B. Sekematte. 1996. Contribution of trap crops in the integrated control of insect pests of cotton in Uganda. Ann. Agric. Sci. Moshtohor 34: 1229–1246.

Henry, T. J. 1988a. Family Largidae Amyot and Sereville, 1843. Pp. 159–165 *in* T. J. Henry and R. C. Froeschner [eds.], Catalog of the Heteroptera, or True Bugs, of Canada and the Continental United States. E. J. Brill, Leiden, the Netherlands. 958 pp.

1988b. Family Pyrrhocoridae Fieber, 1860. The cotton stainers. Pp. 613–615 *in* T. J. Henry and R. C. Froeschner [eds.], Catalog of the Heteroptera, or True Bugs, of Canada and the Continental United States. E. J. Brill, Leiden, the Netherlands. 958 pp.

1997. Phylogenetic analysis of family groups within the infraorder Pentatomomorpha (Hemiptera: Heteroptera), with emphasis on the Lygaeoidea. Ann. Entomol. Soc. Amer. 90: 275–301.

Hodjat, S. H. 1969. The effects of crowding on the survival, rate of development, size, colour and fecundity of *Dysdercus fasciatus* Sign. (Hem., Pyrrhocoridae) in the laboratory. Bull. Entomol. Res. 58: 487–504.

Howard, L. O. 1897. Insects affecting the cotton plant. U.S. Dept. Agric. Farmers' Bull. 47: 31 pp.

Hussey, R. F. 1929. General Catalogue of the Hemiptera. Fasc. III. Pyrrhocoridae. Smith College, Northampton, Massachusetts, U.S.A. 144 pp.

Janda, V., Jr., and M. Munzarová. 1980. Activity of digestive amylase and invertase in relation to development and reproduction of *Dysdercus koenigii* (Heteroptera). Acta Entomol. Bohemoslov. 77: 209–215.

Janzen, D. H. 1972. Escape in space by *Sterculia apetala* seeds from the bug *Dysdercus fasciatus* in a Costa Rican deciduous forest. Ecology 53: 350–361.

Jayanthe, M., K. M. Singh, and R. N. Singh. 1993a. Succession of insect pests on high-yielding variety MH 4 of groundnut under Delhi conditions. Indian J. Entomol. 55: 24–29.

1993b. Pest complex of a high yielding groundnut variety MH4 under Delhi conditions. Indian J. Entomol. 55: 30–33.

1993c. Population build up of insect pests on MH 4 variety of groundnut influenced by abiotic factors. Indian J. Entomol. 55: 109–123.

Joshi, K. C., and H. R. Khan. 1990. Biology and control of the giant red bug *Lohita grandis* Gray (Hemiptera: Pyrrhocoridae: Largidae). Indian For. 116: 312–319.

Joshi, N. K., K. M. Lathika, A. Banerji, and M. S. Chadha. 1988. Effects of plumbagin on growth and development of red cotton bug, *Dysdercus congulatus* Fab. Proc. Indian Nat. Sci. Acad. B54: 43–46.

Jurberg, J., E. F. Rangel, and T. C. M. Gonçalves. 1982. Estudo morfológico comparativo da genitália de três espécies do gênero *Dysdercus* Guérin Méneville 1831 — (Hemiptera, Pyrrhocoridae). Rev. Bras. Biol. 47: 387–407.

Jurberg, J., T. C. M. Gonçalves, S. B. Almeida, and J. S. Almeida. 1983. Alteracão morfológicas provocadas pela aplicacão de um análogo de hormônio juvenil em *Dysdercus ruficollis* (Linnaeus, 1764) (Hemiptera, Pyrrhocoridae, Pyrrhocorinae). Mem. Inst. Oswaldo Cruz 78: 61–65.

Kamble, S. T. 1971. Bionomics of *Dysdercus koenigii* Fabr. (Hemiptera: Pyrrhocoridae). J. New York Entomol. Soc. 79: 154–157.

1974. Notes on a predator of *Dysdercus koenigii* (Hemiptera: Pyrrhocoridae). Curr. Sci. 43: 159.

Kapur, A. P., and T. G. Vazirani. 1956. The identity and geographical distribution of the Indian species of the genus *Dysdercus* (Boisduval (Hemiptera: Pyrrhocoridae). Rec. Indian Mus. 54: 159–175.

Kasule, F. K. 1985. A comparison of the reproductive strategies of three species of *Dysdercus* from Africa. Oecologia 65: 260–265.

Katiyar, R. L., and K. P. Srivastava. 1982. An essential oil from the Australian bottlebrush, *Callistemon lanceolatus* (Myrtaceae) with juvenoid properties against the red cotton bug, *Dysdercus koenigii* Fabr. (Heteroptera: Pyrrhocoridae). Entomon 7: 463–468.

Katiyar, R. L., K. P. Srivastava, and V. K. Awasthi. 1984. Effect of a plant juvenoid on the vitellogenin synthesis in the bug, *Dysdercus koenigii* Fabr. Curr. Sci. 53: 826–828.

Kaur, H., R. Sandhu, and S. S. Dhillon. 1993. Substituted pyrimidine-2-thiols: a newly discovered group of antifertility agents against red-cotton bug. Indian J. Entomol. 55: 396–403.

Kaushik, S. N., and N. P. Chopra. 1969. Susceptibility of *Dysdercus cincgulatus* to *Bacillus thuringiensis*. Indian J. Entomol. 31: 285–286.

Khan, M. R. 1972. The anatomy of the head-capsule and mouthparts of *Dysdercus fasciatus* Sign. (Pyrrhocoridae, Hemiptera). J. Nat. Hist. 6: 289–310.

Khanna, S. 1979. The reproductive system of *Dysdercus koenigii* (F.) (Hem., Pyrrhocoridae). Entomol. Mon. Mag. 114: 45–55.

Koul, P. 1984. Azadirachtin-I: interaction with the development of red cotton bugs. Entomol. Exp. Appl. 36: 85–88.

Kumar, B. H., and S. S. Thakur. 1989. Effect of certain non-edible seed oils on growth regulation in Dysdercus similis [sic] (F). J. Anim. Morphol. Physiol. 36: 209–218.

Kumar, D. 1982. Autoradiographic studies of the protein metabolism in the salivary glands of the red cotton bug *Dysdercus koenigii* (Pyrrhocoridae: Hemiptera). Entomon 7: 173–179.

Kumar, D., A. Ray, and P. S. Ramamurty. 1978. Histo-physiology of the salivary glands of the red cotton bug, *Dysdercus koenigii* (Pyrrhocoridae-Heteroptera) — histological, histochemical, autoradiographic and electron-microscopic studies. Z. Mikrosp. Anat. Forsch. 92: 147–170.

Kumar, R. 1968. Aspects of the morphology and relationships of the superfamilies Lygaeoidea, Piesmatoidea and Pyrrhocoroidea (Hemiptera: Heteroptera). Entomol. Mon. Mag. 103: 251–261.

Kuruvilla, S., and A. Jacob. 1980. Pathogenicity of the entomogenous fungus *Paecilomyces farinosus* (Dickson ex Fries) to several insect pests. Entomon 5: 175–176.

Lateef, S. S., and W. Reed. 1990. Insect pests on pigeon pea. Pp. 193–242 *in* S. R. Singh [ed.], Insect Pests of Tropical Food Legumes. John Wiley & Sons, Chichester, U.K. 451 pp.

Lawrence, P. A., and B. W. Staddon. 1975. Peculiarities of the epidermal gland system of the cotton stainer *Dysdercus fasciatus* Signoret (Heteroptera: Pyrrhocoridae). J. Entomol. (A) 49: 121–136.

Ludwig, W. 1926. Untersuchungen über den Copulationsapparat der Baumwanzen. Z. Morphol. Oekol. Tiere 5: 291–380.

MacGill, E. I. 1947. The anatomy of the head and mouth-parts of *Dysdercus intermedius* Dist. Proc. Zool. Soc. London 117: 115–128.

Manna, G. K. 1957. Sex mechanism in *Dysdercus*. Curr. Sci. 26: 187.

Manna, G. K., and S. Deb-Mallick. 1981. Sex chromatin elimination in the polymorphic male pyrrhocorid bug, *Iphita limbata*. Entomon 6: 287–296.

Manna, G. K., N. Ueshima, S. K. Dey, and S. Deb-Mallick. 1985. Marked sex chromosome variations between Indian and Japanese species of *Physopelta* (Largidae, Heteroptera). Cytologia 50: 621–630.

Masner, L. 1983. A revision of *Gryon* Haliday in North America (Hymenoptera: Proctotrupoidea: Scelionidae). Canad. Entomol. 115: 123–174.

Mead, F. W. 1966. Cotton stainers, *Dysdercus* spp., in Florida (Hemiptera: Pyrrhocoridae). Florida Dept. Agric. Entomol. Circ. 48: 2 pp.

Mehta, D. R. 1930. Observations on the influence of temperature and humidity on the bionomics of *Dysdercus cingulatus* Fabr. Bull. Entomol. Res. 21: 547–562.

Melber, A. 1979a. Influence of abiotic factors and physiological conditions on the formation of aggregations in cotton-bug (*Dysdercus* spp., Heteroptera). Entomol. Exp. Appl. 25: 196–202.

1979b. Einfluss der Populationsdichte und des Gruppenlebens auf die Bionomie von *Dysdercus cardinalis* Gerst. (Heteroptera, Pyrrhocoridae). Z. Angew. Entomol. 88: 144–158.

Mendes, L. O. T. 1938. Lista dos inimigos naturais de *Dysdercus* spp., observados no Estado de São Paulo. Rev. Entomol. (Rio de Janeiro) 9: 215–217 [not seen, *fide* Van Doesburg 1968, Almeida et al. 1981].

Milano, P., F. L. Cônsoli, N. G. Zério, and J. R. P. Parra. 1999. Exigências térmicas de *Dysdercus peruvianus* Guérin-Méneville (Heteroptera: Pyrrhocoridae), o percevejo manchador do algodoeiro. An. Soc. Entomol. Brasil 28: 233–238.

Miller, N. C. E. 1932. Observations on *Melamphaus faber*, F. (Hem., Pyrrhocoridae) and descriptions of early stages. Bull. Entomol. Res. 23: 195–201.

Mohammad, F. A., and I. Ahmad. 1991. Comparative morphology of male and female reproductive organs of some pyrrhocorids (Hemiptera: Trichophora) of Pakistan and its bearing on classification. Pakistani J. Sci. Ind. Res. 34: 459–462.

Moizant, R. C., and J. Terán B. 1970. *Dysdercus maurus* Distant (Hemiptera, Pyrrhocoridae) sobre *Citrus* spp. Agron. Trop. (Maracay) 20: 267–269.

Moreira, P. H. R., J. J. Soares, A. C. Busoli, V. R. da Cruz, M. H. L. Pimentel, and G. J. B. Pelinson. 1994. Causas do apodrecimento de maçãs do algodoeiro. Pesq. Agropec. Bras. 29: 1503–1507.

Morrill, A. W. 1910. Plant-bugs injurious to cotton bolls. U.S. Dept. Agric. Bur. Entomol. Bull. 86: 1–110.

1918. Insect pests of interest to Arizona cotton growers. Univ. Arizona Coll. Agric. Agric. Exp. Stn. Bull. 87: 173–205.

Morton, J. 1987. Roselle, *Hibiscus sabdariffa* L. Pp. 281–296 *in* J. F. Morton [ed.], Fruits of Warm Climates. Julia F. Morton, Miami, Florida, U.S.A. 505 pp.

Mukhopadhyay, A. 1993. Chemical ecology of some seed-feeding Hemiptera. Pp. 180–195 *in* T. N. Anathakrishnan and A. Raman [eds.], Chemical Ecology of Phytophagous Insects. Oxford and IBH Publ. Co., New Delhi, India.

Muraleedharan, D., and V. K. K. Prabhu. 1977. Protein content of the thoracic muscles of the red cotton bug, *Dysdercus cingulatus* Fabr. (Pyrrhocoridae: Heteroptera). Entomon 2: 255–257.
1978. Food intake and midgut protease activity in the red cotton bug, *Dysdercus cingulatus* Fabr. (Heteroptera: Pyrrhocoridae). Entomon 3: 11–17.
1979a. Effect of quality of food on midgut invertase activity in the red cotton bug *Dysdercus cingulatus* Fabr. Indian J. Exp. Biol. 17: 1262–1263.
1979b. Role of the median neurosecretory cells in secretion of protease and invertase in the red cotton bug, *Dysdercus cingulatus*. J. Insect Physiol. 25: 237–240.
1981. Hormonal influence on feeding and digestion in a plant bug, *Dysdercus cingulatus*, and a caterpillar, *Hyblaea puera*. Physiol. Entomol. 6: 183–189.
Mustafa, M., and M. B. Naidu. 1964. Chemical sterilization of *Dysdercus cingulatus* F. (red cotton bug). Indian J. Exp. Biol. 2: 55–56.
Myers, J. G. 1927. Ethological observations on some Pyrrhocoridae of Cuba. Ann. Entomol. Soc. Amer. 20: 279–300.
Nair, C. R. M., and V. K. K. Prabhu. 1985a. The role [sic] of feeding, mating and ovariectomy on degeneration of indirect flight muscles of *Dysdercus cingulatus* (Heteroptera: Pyrrhocoridae). J. Insect Physiol. 31: 35–39.
1985b. The role of endocrines in flight-muscle degeneration in *Dysdercus cingulatus* (Heteroptera: Pyrrhocoridae). J. Insect Physiol. 31: 223–227.
1985c. Entry of proteins from degenerating flight muscles into oöcytes in *Dysdercus cingulatus* (Heteroptera: Pyrrhocoridae). J. Insect Physiol. 31: 383–388.
Nickel, J. L. 1958. Agricultural insects of the Paraguayan Chaco. J. Econ. Entomol. 51: 633–637.
Njau, M. A., and J. R. Mainoya. 1984. Effects of feeding on host plant seeds on the development and fecundity of the cotton stainer *Dysdercus nigrofasciatus* Stal [sic] (Pyrrhocoridae: Hemiptera). Kenya J. Sci. Technol. (B) 5: 3–10.
Nübler-Jung, K. 1974. Cell migration during pattern reconstitution in the insect integument (*Dysdercus intermedius* Dist., Heteroptera). Nature 248: 610–611.
1977. Pattern stability in the insect segment I. Pattern reconstitution by intercalary regeneration and cell sorting in *Dysdercus intermedius* Dist. Wilhelm Roux Arch. 183: 17–40.
1979. Pattern stability in the insect segment II. The intersegmental region. Wilhelm Roux Arch. 186: 211–233.
Nyamasyo, G. H. N., and A. K. Karel. 1982. Studies on insecticide resistance in cotton stainers, *Dysdercus* spp. (Hemiptera: Pyrrhocoridae), in Kenya. Bull. Entomol. Res. 72: 461–465.
Nyiira, Z. M. 1970. The biology and behavior of *Rhinocoris albopunctatus* (Hemiptera: Reduviidae). Ann. Entomol. Soc. Amer. 63: 1224–1227.
Ochou, G. O., G. A. Matthews, and J. D. Mumford. 1998. Farmers' knowledge and perception of cotton insect pest problems in Côte d'Ivoire. Int. J. Pest Manage. 44: 5–9.
Odhiambo, T. R. 1968. The effects of mating on egg production in the cotton stainer, *Dysdercus fasciatus*. Entomol. Exp. Appl. 11: 379–388.
Papeschi, A. G., M. J. Bressa, L. M. Mola, and M. L. Larramendy. 1998. Sex chromosome determining system variation in *Dysdercus* (Pyrrhocoridae, Heteroptera). Cytogenet. Cell Genet. 81: 107 (Suppl. S).
Pathak, B., and D. P. Sinha. 1993. Effect of relative humidity on the nymphs of *Dysdercus koenigii* (Hemiptera: Pyrrhocoridae) in relation to survival, instar duration and growth. Environ. Ecol. 11: 71–73.
Peregrine, D. J. 1972. Fine structure of sensilla basiconica on the labium of the cotton stainer, *Dysdercus fasciatus* (Signoret) (Heteroptera: Pyrrhocoridae). Int. J. Insect Morphol. Embryol. 1: 241–151.
Pierce, W. D. 1917. How insects affect the cotton plant and means of combatting them. U.S. Dept. Agric. Farmers' Bull. 890. 27 pp.
Pluot, D. 1970. La spermathèque et les voies génitales femelles des pyrrhocorides (Hemiptera). Ann. Soc. Entomol. France 6 (n.s.): 777–807.
1973. Effets d'une discontinuité de l'oviducte sur l'ovaire et les pédicelles correspondants chez les *Dysdercus* (Hem. Pyrrhocoridae). Ann. Soc. Entomol. France (n.s.) 9: 813–839.
Pomeroy, A. W. J. 1924. Further observations on *Dysdercus superstitiosus* F., and other insects affecting cotton in southern Africa. Bull. Entomol. Res. 15: 173–176.
Popham, E. J. 1962a. On the salivary pump of *Dysdercus intermedius* Dist. (Hemiptera Heteroptera) and other bugs. Proc. Zool. Soc. London 139: 489–493.

1962b. Prognathism and brain structure in *Forficula auricularis* L. (Dermaptera) and *Dysdercus intermedius* Dist. (Hemiptera Heteroptera). Zool. Anz. 168: 35–43.

Poutouli, W. 1992. Plantes-hôtes secondaires des hètèroptéres recenses sur coton, mais, niebe au Togo. Med. Fac. Landbouw. Univ. Gent 57: 627–636.

Prabhakar, J. D., N. R. Jagtap, S. S. Y. Hakin, M. R. Encily, and S. B. Kahemkalyani. 1992. Study of the physiology of the host parasite relationships of *Dysdercus cingulatus* and *Aspergillus flavus*. Indian J. Entomol. 54: 80–83.

Prabhu, V. K. K., and M. John. 1975. Ovarian development in juvenilised adult *Dysdercus cingulatus* affected by some plant extracts. Entomol. Exp. Appl. 18: 87–95.

Quayum, M. A., and G. Nahar. 1980. External morphology of the pyrhhocorid bug, *Antilochus coqueberti* [sic] (Fabr.), a predator of *Dysdercus koenigii* (Fabr.) Bangladesh J. Zool. 8: 119–126.

Rajendran, B., and M. Gopalan. 1982. Note on the biochemical changes induced by *Polyscias guilfoylei* Bailey (Araliaceae) in *Dysdercus cingulatus* Fabricius (Hemiptera: Pyrrhocoridae). Indian J. Agric. Sci. 52: 46–48.

Rani, P. U., and S. S. Madhavendra. 1995. Morphology and distribution of antennal sense organs and diversity of mouthpart structures in *Odontopus nigricornis* (Stall [sic]) and *Nezara viridula* L. (Hemiptera). Int. J. Insect Morphol. Embryol. 24: 119–132.

Rao, P. J., K. M. Kumar, S. Singh, and B. Subrahmanyam. 1999. Effect of *Artemsia annua* oil on development and reproduction of *Dysdercus koenigii* F. (Hem., Pyrrhocoridae). J. Appl. Entomol. 123: 315–318.

Rao, R. V. S., and G. T. Gujar. 1995. Toxicity of plumbagin and juglone to the eggs of the cotton stainer, *Dysdercus koenigii*. Entomol. Exp. Appl. 77: 189–192.

Rathore, Y. K. 1961. Studies on the mouth-parts and feeding mechanism in *Dysdercus cingulatus* Fabr. (Pyrrhocoridae: Heteroptera). Indian J. Entomol. 23: 163–185.

Ray-Chaudhuri, S. P., and G. K. Manna. 1955. Evidence of a multiple sex-chromosome mechanism in a pyrrhocorid bug, *Physoelta schlanbuschi* Fabr. Proc. Zool. Soc. (Bengal) 8: 65–77.

Revathy, D., B. K. Rao, S. S. Thakur, and N. S. Raj. 1979. Effectivity of an insect growth regulator, hydoprene, on the cotton stainer *Dysdercus koenigii* F. J. Exp. Biol. (India) 17: 945–947.

Robert, A. 1975. Étude du cycle biologique et de l'influence de quelques facteurs de l'environment sur la reproduction en laboratoire d'un hétéroptère Pyrrhocoridae, *Roscius elongatus* Stål. Ann. Univ. Brazzaville (C) 11: 173–184.

1976. Étude comparée de l'evolution des ovarioles chez deuz espèces sympatriques d'hétéroptères, Pyrrhocoridae: *Roscius elongatus* et *Roscius brazzavilliensis;* moment d'apparition de la méïose. C. R. Acad. Sci. Paris (D) 282: 565–568.

1979. Les premiers stades de l'ovogenese et les variations de la neurosecretion cerebrale chez deux especes sympatriques *Roscius elongatus*, Stål et *R. brazzavilliensis*, Robert (Heteroptera: Pyrrhocoridae). Int. J. Insect Morphol. Embryol. 8: 11–31.

Robertson, I. A. D. 1977. Records of insects taken at light traps in Tanzania. IV — Seasonal changes in catches and effect of the lunar cycle on insects of the genus *Dysdercus* (Heteroptera: Pyrrhocoridae). Ministry of Overseas Dev., Centre Overseas Pest Res. [London, U.K.] Misc. Rep. 30. 8 pp.

Sahayaraj, K., and D. P. Ambrose. 1993. Biology and predatory potential of *Coranus nodulosus* Ambrose and Sahayaraj on *Dysdercus cingulatus* Fabr. and *Oxycarenus hyalipennis* Costa (Heteroptera: Reduviidae). Hexapoda 5: 17–23.

1994. Prey influence on laboratory mass rearing of *Neohaematorrhophus therasii* Ambrose and Livingstone a potential biocontrol agent Insecta: Heteroptera: Reduviidae. Bio-Sci. Res. Bull. 10: 35–40.

1995–1996. Short term functional response and stage preference of the reduviid predator *Ectomocoris tibialis* Distant on the cotton stainer *Dysdercus cingulatus* Fabr. Z. Angew. Entomol. 81: 219–225.

1996. Biocontrol potential of the reduviid predator *Neohaematorrhopus therasii* Ambrose & Livingstone (Heteroptera: Reduviidae). J. Adv. Zool. 17: 49–53.

Sakeld, E. H., and C. Potter. 1953. The effect of the age and stage of development of insect eggs on their resistance to insecticides. Bull. Entomol. Res. 44: 527–580.

Sanderson, E. D. 1905. Miscellaneous cotton insects in Texas. U.S. Dept. Agric. Farmers' Bull. 223. 23 pp.

Sandhu, R., D. J. Kaur, and H. Kaur. 1984. Biological effects of two substituted thiourea compounds in *Dysdercus koenigii* (Fabr.), Heteroptera. Zool. Orient. 1: 43–44.

Sands, W. R. 1917. Observations on the cotton stainer in St. Vincent. West Indian Bull. (Trinidad) 16: 235–255 [not seen, *fide* Almeida et al. 1981].
Sarkar, G. M., R. D. Banerjee, M. L. Chatterjee, S. Dutta, and A. Ghosh. 1995. Antimicrobial and insecticidal activity of an aquatic alga, *Chara zeylanica* Klein ex Willd. Int. J. Environ. Stud. (A: Environ. Stud.) 48: 29–39.
Satyanarayana, K., C. Keena, and K. Sukumar. 1985. Penfluron induced sterility and ovarian inhibition in adults of *Dysdercus cingulatus* F. Int. Pest Control 27: 47–48.
Sauvaut, A. 1949. Sur *Dindymus flavipes* (Pyrrhocoridae), insecte rare en Côte d'Ivoire. Bull. Soc. Hist. Nat. Toulouse 84: 154–156.
Saxena, B. P., O. Koul, K. Tikku, V. K. Atal, O. P. Suri, and K. A. Suri. 1979. Aristolochic acid — an insect chemosterilant from *Aristolochia bracteata* Retz. Indian J. Exp. Biol. 17: 354–360.
Saxena, K. N. 1963. Mode of ingestion in a heteropteran insect *Dysdercus koenigii* (F.) (Pyrrhocoridae). J. Insect Physiol. 9: 47–71.
Schaefer, C. W. 1964. The morphology and higher classification of the Coreoidea (Hemiptera-Heteroptera): Parts I and II. Ann. Entomol. Soc. Amer. 57: 679–684.
1993. The Pentatomomorpha (Hemiptera: Heteroptera): an annotated outline of its systematic history. Eur. J. Entomol. 90: 105–122.
1998a. Notes on *Dysdercus* from Brazil (Hemiptera: Pyrrhocoridae). An. Soc. Entomol. Brasil 27: 485–488.
1998b. Phylogeny, systematics, and practical entomology: the Heteroptera (Hemiptera). An. Soc. Entomol. Brasil 27: 499–511.
1999. Review of *Raxa* (Hemiptera: Pyrrhocoridae). Ann. Entomol. Soc. Amer. 92: 14–19.
Schaefer, C. W., and I. Ahmad. 1987. Parasites and predators of Pyrrhocoroidea (Hemiptera), and possible control of cotton stainers by *Phonoctonus* spp. (Hemiptera: Reduviidae). Entomophaga 31: 269–275.
Schmidt, E. 1931. Zur Kenntnis der Familie Pyrrhocoridae Fieber. (Hemiptera-Heteroptera). Teil I. Stettiner Entomol. Z. 92: 1–51.
1932. Zur Kenntnis der Familie Pyrrhocoridae Fieber. (Hemiptera-Heteroptera). Teil II. Wiener Entomol. Z. 49: 236–281.
Schmidt, G. H., A. Scheer, and A. Melber. 1992. Untersuchungen zur Tagesrhythmik im Lauf- und Nahrungssuchverhalten bei Baumwollwanzen (*Dysdercus intermedius* Dist.) (Insecta: Heteroptera). Zool. Jahrb. Physiol. 96: 481–509.
Schmuck, R. 1994. Adaptive value of aggregation behavior in the fire bug *Pyrrhocoris apterus* (Heteroptera, Pyrrhocoridae). Entomol. General. 19: 143–156 [original not seen, *fide* abstract].
Schoonhoven, L. M., and I. Derksen-Koppers. 1973. Effects of secondary plant substances on drinking behaviour in some Heteroptera. Entomol. Exp. Appl. 16: 141–145.
Schouteden, H. 1912. Les hémiptères parasites des cotonniers en Afrique. Rev. Zool. Africaine 1: 297–321.
1916. Cas de mimetismé chez les Hémiptères Africains. Rev. Zool. Afric. 4: 251–258.
Schumacher, R. 1971a. Zur funktionellen Morphologie der imaginalen Duftdrüsen zweier Landwanzen. III. Mitteilung: die Drüsenzelle des imaginalen Duftdrüsenkomplexes der Feuerwanze *Pyrrhocoris apterus* L. (Geocorisae, Fam.: Pyrrhocoridae). Z. Wiss. Zool. (Leipzig) 183: 71–82.
1971b. Zur funktionellen Morphologie der imaginalen Duftdrüsen zweier Landwanzen. IV. Mitteilung: das ableitende Kanalsystem und das Reservoir des imaginalen Duftdrüsenklompexes der Feuerwanze *Pyrrhocoris apterus* L. (Geocorisae, Fam.: Pyrrhocoridae). Z. Wiss. Zool. (Leipzig) 183: 83–96.
1971c. Zur funktionelle Morphologie der imaginalen Duftdrüsen zweier Landwanzen. II. Mitteilung: das Reservoir und das "Nierenförmige Organ" des imaginalen Duftdrüsenkomplexes der Baumwollwanze *Dysdercus intermedius* Dist. Z. Wiss. Zool. (Leipzig) 182: 411–426.
Schumacher, R., and G. Stein. 1971. Zur funktionelle Morphologie der imaginalen Duftdrüsen zweier Landwanzen. I. Mitteilung: Drüsenzellen und ableitendes Kanalsystem des imaginalen Duftdrüsenkomplexes der Baumwollwanze *Dysdercus intermedius* Dist. Z. Wiss. Zool. (Leipzig) 182: 395–410.
Sehgal, S. S., S. C. Maheshwari, and K. D. Chaudhary. 1980. Sterilization of red cotton bug with tepa: effects on longevity, duration of copulation, number of matings, and sexual competitiveness. Canad. J. Zool. 58: 2015–2017.
Shahi, K. P., and S. S. Krishna. 1979a. Reproductive programming in *Dysdercus koenigii* (Fabr.) (Heteroptera, Pyrrhocoridae) in relation to age of sexes and antennectomy in males. Nat. Acad. Sci. Let. (India) 2: 115–116.

1979b. Oviposition and fertility of eggs in *Dysdercus koenigii* (Fabr.) (Heteroptera: Pyrrhocoridae). Mitt. Zool. Mus. Berlin 55: 291–296.
1981a. Influence of adult nutrition or some environmental factors on sexual activity and subsequent reproductive programming in the red cotton bug, *Dysdercus koenigii*. J. Adv. Zool. (India) 2: 25–31.
1981b. A new host of *Hemipteroseius indicus*. Experientia 37: 1072.
Sharma, P. R., K. Tikku, and B. P. Saxena. 1998. A light and electron microscopic study of the haemocytes of adult red cotton bug *Dysdercus koenigii*. Biologia 53: 759–764.
Sharma, R. N., V. Joshi, G. Zadu, and A. S. Bhosale. 1981. Oviposition deterrence activity in some Lamiaceae plants against some insect pests. Z. Naturforsch. 36C: 122–125.
Sharma, U., S. L. Sahni, and D. P. Sinha. 1975. Endocrine control of mating instinct in *Dysdercus koenigii* (Hemiptera: Pyrrhocoridae). Experientia 31: 995–996.
Sifat, S., and M. A. Khan. 1974. Changes in total proteins of the whole body and ovaries of Dysdercus cingulatus Fabr. [sic] (Hemiptera: Pyrrhocoridae) in relation to age and reproduction. J. Anim. Morphol. Physiol. 21: 64–70.
Silva, A. G. d'A., C. R. Gonçalves, D. M. Galvão, A. J. L. Gonçalves, J. Gomes, M. N. Silva, and L. Simoni. 1968. Quarto Catálogo dos Insetos que Vivem nas Plantas do Brasil — Seus Parasitos e Predadores. Parte II, Vol. 1. Ministr. Agric., Rio de Janeiro, Brazil.
Singh, D. R., and S. S. Tomar. 1977. Comparative toxicity list of nine insecticides for laboratory control of Antilochus conqueberti [sic] Fabr. Dtsch. Entomol. Z. (n. F.) 2: 181–185.
Singh, H., and B. Ram. 1987. Effect of different host plants on the development of red cotton bug, *Dysdercus koenigii* Fabr. Indian J. Entomol. 49: 345–350.
Singh, R. M., and K. N. Singh. 1978a. Succession of insect-pests in early varieties of red gram, *Cajanus cajan* (L.) Millsp. Indian J. Entomol. 40: 1–6.
1978b. Incidence of insect-pests in early varieties of red gram, *Cajanus cajan* (L.) Millsp. Indian J. Entomol. 40: 229–240.
1978c. Influence of intercropping on succession and population build up of insect-pests in early variety of red gram, *Cajanus cajan* (L.) Millsp. Indian J. Entomol. 40: 361–375.
Singh, S. R., H. F. van Emden, and T. A. Taylor. 1978. Check list of insect and mite pests of grain legumes. Pp. 399–417 *in* S. R. Singh, H. F. van Emden, and T. A. Taylor [eds.], Pests of Grain Legumes. Academic Press, London, U.K. 454 pp.
Sláma, K. 1962. The juvenile hormone like effect of fatty acids, fatty alcohols, and some other compounds in insect metamorphosis. Acta Soc. Entomol. Čechoslov. 59: 323–340.
1964a. Die Einwirkung des Juvenilhormons auf die Epidermiszellen der Flügelanlagen bei künstlich beschleunigter und verzögerter Metamorphose von *Pyrrhocoris apterus* L. Zool. Jahrb. Physiol. 70: 427–454.
1964b. Hormonal control of respiratory metabolism during growth, reproduction, and diapause in female adults of *Pyrrhocoris apterus* L. (Hemiptera). J. Insect Physiol. 10: 283–303.
1964c. Hormonal control of haemolymph protein concentration in the adults of *Pyrrhocoris apterus* L. (Hemiptera). J. Insect Physiol. 10: 773–782.
1964d. Hormonal control of respiratory metabilism during growth, reproduction, and diapause in male adults of *Pyrrhocoris apterus* L. (Hemiptera). Biol. Bull. 127: 499–510.
Sláma, K., and C. M. Williams. 1965. Juvenile hormone activity for the bug *Pyrrhocoris apterus*. Fed. Proc. (Fed. Amer. Soc. Exp. Biol.) 54: 411–414.
1966a. "Paper factor" as an inhibitor of the embryonic development of the European bug, *Pyrrhocoris apterus*. Nature 210: 329–330.
1966b. The juvenile hormone. V. The sensitivity of the bug, *Pyrrhocoris apterus*, to a hormonally active factor in American aper-pulp. Biol. Bull. 130: 235–246.
Socha, R. 1993. *Pyrrhocoris apterus* (Heteroptera) — an experimental model species: a review. Eur. J. Entomol. 90: 241–286.
Southwood, T. R. E., and D. Leston. 1959. Land and Water Bugs of the British Isles. Frederick Warne & Co., London, U.K. 436 pp.
Srivastava, K. P., and D. J. Deshpande. 1983a. Effect of X-irradiation on fecundity, fertility & longevity of the red cotton bug *Dysdercus koenigii* Fabr. Indian J. Exp. Biol. 21: 604–606.
1983b. Histopathological effects of X-irradiation on the ovaries of the red cotton bug, *Dysdercus koenigii* Babr. (Heteroptera: Pyrrhocoridae). Entomon 8: 277–286.

Srivastava, K. P., D. J. Deshpande, and R. L. Katiyar. 1985. X-irradiation induced histochemical changes in the ovaries of *Dysdercus koenigii* Fabr. Entomon 10: 29–33.

Srivastava, R. K., and S. S. Krishna. 1990. Temperature- and exposure duration-related eucalyptus oil odour effects on egg hatchability and subsequent postembryonic development in *Dysdercus koenigii* (F.) (Heteroptera: Pyrrhocoridae). J. Adv. Zool. [India] 11: 103–105.

1992. Effect of exposure of eggs or first instar nymphs of *Dysdercus koenigii* (F.) (Heteroptera: Pyrrhocoridae) to eucalyptus oil odour on the insect's postembryonic development and/or reproductive potential. Phytophaga 4: 47–52.

Srivastava, R. K., G. Gurusubramanian, and S. S. Krishna. 1995. Postembryonic development and reproduction in *Dysdercus koenigii* (F.) (Heteroptera, Pyrrhocoridae) on exposure to eucalyptus oil volatiles. Biol. Agric. Hortic. 12: 81–88.

Srivastava, R. P., and R. S. Gupta. 1971. *Dysdercus cingulatus* Fabr., a new pest of wheat — a possible case of host cross-over. Indian J. Entomol. 33: 354.

Srivastava, S. K., and S. S. Krishna. 1992. Eucalyptus oil odour treatment effects on biochemistry of some tissues of female nymphs/adults of *Dysdercus koenigii*. Insect Sci. Appl. 13: 145–149.

Srivastava, U. S., and S. Neraliya. 1995. An insect growth regulatory factor from the plant *Napeta hindostana* (Roth) Haines (Labiatae). Natl. Acad. Sci. Let. (India) 18: 185–188.

Stahle, P. P. 1979. The immature stages of the harlequin bug, *Dindymus versicolor* (Herrich-Schaeffer) (Hemiptera: Pyrrhocoridae). J. Austral. Entomol. Soc. 18: 271–276.

1981. Food preference in the harlequin bug *Dindymus versicolor* (Herrich-Schaeffer) (Hemiptera: Pyrrhocoridae), a minor pest of fruit in south eastern Australia. Austral. J. Ecol. 6: 375–382.

Stehlik, J., and I. M. Kerzhner. 1999. On taxonomy and distribution of some Palaearctic and Oriental Largidae and Pyrrhocoridae (Heteroptera). Zoosyst. Rossica 8: 121–128.

Stein, G. 1966a. Über den Feinbau der Duftdrüsen von Feuerwanzen (*Pyrrhocoris apterus* L., Geocorisae). I. Mitteilung. Zur funktionellen Morphologie der Drüsenzelle. Z. Zellforsch. 74: 271–290.

1966b. Über den Feinbau der Duftdrüsen von Feuerwanzen (*Pyrrhocoris apterus* L., Geocorisae). II. Mitteilung. Das ableitende Kanalsystem und die nichtdrüsigen Anteile. Z. Zellforsch. 75: 501–516.

1967. Über den Feinbau der Duftdrüsen von Feuerwanzen (*Pyrrhocoris apterus* L., Geocorisae). Die 2. larvale Abdominaldrüse. Z. Zellforsch. 79: 49–63.

1969. Über den Feinbau der Duftdrüsen von Heteropteren. Die hintere larvale Abdominaldrüse der Baumwollwanze *Dysdercus intermedius* Dist. (Insecta, Heteroptera). Z. Morphol. Tiere 65: 374–391.

Steinbauer, M. J. 1996. Notes on extra-phytophagous food sources of *Gelonus tasmanicus* (Le Guilolou) (Hemiptera: Coreidae) and *Dindymus versicolor* (Herrich-Schaeffer) (Hemiptera: Pyrrhocoridae). Austral. Entomol. 23: 121–124.

Stride, G. O. 1956. On the mimetic association between certain species of *Phonoctonus* (Hemiptera, Reduviidae) and the Pyrrhocoridae. J. Entomol. Soc. South. Africa 19: 12–28.

Sukumar, K. 1980. Tepa induced sterility of *Dysdercus cingulatus* F. in relation to age of insect. Int. Pest Control 1980: 2 pp.

Sukumar, K., and M. B. Naidu. 1980. Sexual behavior of *Dysdercus cingulatus* F. after treatment with tepa, a chemosterilant. Indian J. Exp. Biol. 18: 652–654.

Suryakala, S., S. Thakur, and B. K. Rao. 1995. Ovicidal activity of plant extracts on *Spodoptera litura* and *Dysdercus koenigii*. Indian J. Entomol. 57: 192–197.

Sweet, M. H. 1964. The biology and ecology of the Rhyparochrominae of New England (Heteroptera: Lygaeidae). Parts I and II. Entomol. Amer. 43: 1–124, 44: 1–201.

Tachikawa, S., and C. W. Schaefer. 1985. Biology of *Parastrachia japonensis* (Hemiptera: ?-idae). Ann. Entomol. Soc. Amer. 78: 387–397.

Tandon, N. 1969. Embryology of the red cotton bug, *Dysdercus cingulatus* (Fabricius). I. Early embryonic development. Proc. Zool. Soc. Calcutta 22: 139–149.

Tengecho, B. 1994. Distribution and occurrence of some cotton stainers (Heteroptera: Insecta) on different host plants at Masongaleni Machakos District, Kenya. Insect Sci. Appl. 15: 49–54.

Thouvenin, M. 1965. Étude préliminaire des "uradénies" chez certains hétéroptères pentatomomorphes. Ann. Entomol. Soc. France (n.s.) 1: 973–987.

Tikku, K., S. C. Taneja, S. Koul, K. L. Dhar, and B. P. Saxena. 1999. Comparative effects of five new synthetic juvenile hormone analogues against the red cotton bug *Dysdercus koenigii* F. Curr. Sci. 76: 89–90.

Tiwari, R. K., and K. P. Srivastava. 1979. Corpus allatum in relation to egg maturation in the red cotton bug, *Dysdercus koenigii* (Heteroptera). Acta Entomol. Bohemoslov. 76: 175–180.

Toledo Piza, Jr., S. de. 1947. Chromosômios de Dysdercus (Hemiptera-Pyrrhocoridae). An. Esc. Super. Agric. "Luiz de Queiroz" Univ. São Paulo 4: 209–216.

Toms, S. V. 1983. The consumption of soil moisture condensate by the bugs *Stenocephalus marginatus* and *Pyrrhocoris apterus*. Zool. Zh. 10: 1587–1588 [in Russian, English summary].

Townsend, C. H. T. 1913. Muscoid parasites of the cotton-stainer and other lygaeids. Psyche 20: 21–94.

Ueshima, N. 1979. Insecta 6. Hemiptera II: Heteroptera. Pp. 1–117 *in* B. John, H. Bauer, H. Kayano, A. Levin, and M. White [eds.], Animal Cytogenetics. Gebrüder Borntraeger, Berlin, Germany.

Ullrich, V. W. 1953. Beobachtung zur Frage der Schutztracht der Feuerwanzen *Pyrrhocoris apterus* L. Beitr. Entomol. 3: 406–411.

Ullyett, G. C. 1930. The life-history, bionomics and control of cotton stainers (Dysdercus spp.) in South Africa. Sci. Bull. Union South Africa (Pretoria) 94: 1–9.

Urich, F. W. 1916. Insects infecting the cotton plant in Trinidad. Bull. Dept. Agric. Trinidad and Tobago 15: 18–19.

Van Doesburg, Jr., P. H. 1966. Heteroptera of Suriname I. Largidae and Pyrrhocoridae. Stud. Fauna Suriname other Guyanas 9: 1–60.

1968. A revision of the New World species of *Dysdercus* Guérin Méneville (Heteroptera, Pyrrhocoridae). Zool. Verh. 97: 1–215.

Vesey-Fitzgerald, D. F. 1956. Notes on Sphecidae (Hym.) and their prey from Trinidad and British Guiana. Entomol. Mon. Mag. 92: 286–287.

Villiers, A. 1947. Étude systématique des *Dysdercus* de l'Ouest Africain. C. R. Sem. Agric. Yangambi Commun. 49: 685–690.

Wadnerkar, D. W., B. B. Gaikawad, and U. T. Thombre. 1979. New record of an alternate host plant of red cotton bug, *Dysdercus koenigii* (Fabr.). Indian J. Entomol. 41: 185.

Wall, C. 1974. Disruption of embryonic development by juvenile hormone and its mimics in *Dysdercus fasciatus* Sign. (Hemiptera, Pyrrhocoridae). Bull. Entomol. Res. 64: 421–433.

Wallace, F. G. 1977. *Leptomonas seymouri* sp. n. from the cotton stainer *Dysdercus suturellus*. J. Protozool. 24: 483–484.

Whitfield, F. G. S. 1933. The bionomics and control of *Dysdercus* (Hemiptera) in the Sudan. Bull. Entomol. Res. 24: 301–313.

Wille, J. E. 1952. Entomología Agricola del Peru. Ministerio de Agricultura, Lima, Peru. 543 pp.

Wille, J. E., O. Beingolea G., and J. E. González B. 1958. Insectos e insecticidas en la campaña algodonera 1957–58. Informe no. 106, Estacion Experimental Agricola de "La Molina," Ministerio de Agricultura, Lima, Peru. 25 pp.

Williams, C. M., and K. Sláma. 1966. The juvenile hormone VI. Effects of the "paper factor" on the growth and metamorphosis of the bug, *Pyrrhocoris apterus*. Biol. Bull. 130: 247–253.

Wilson, E. B. 1909. Studies on chromosomes IV. The "accessory" chromosome in Syromastes and Pyrrochoris [sic] with a comparative review of the types of sexual differences of the chromosome groups. J. Exp. Zool. 6: 69–102.

Xerez, R. de, J. R. Almeida, and L. Gonçalves. 1984. Flutuação na densidade de uma população de *Dysdercus maurus* em Itaguaí, Estado do Rio de Janeiro. Arq. Univ. Fed. Rur. Rio de Janeiro (Itaguaí) 7: 111–116.

Youdeowei, A. 1966. Laboratory studies on the aggregation of feeding *Dysdercus intermedius* Distant (Heteroptera: Pyrrhocoridae). Proc. R. Entomol. Soc. London (A) 41: 45–50.

1967a. The reactions of *Dysdercus intermedius* (Heteroptera, Pyrrhocoridae) to moisture, with special reference to aggregation. Entomol. Exp. Appl. 10: 194–210.

1967b. Observations on some effects of population density on *Dysdercus intermedius* Distant (Heteroptera: Pyrrhocoridae). Bull. Entomol. Soc. Nigeria 1: 18–26.

1968. The behaviour of a cotton stainer *Dysdercus intermedius* (Heteroptera, Pyrrhocoridae) in a temperature gradient and the effect of temperature on aggregation. Entomol. Exp. Appl. 11: 68–80.

1969. The behaviour of a cotton stainer, *Dysdercus intermedius*, Distant (Heteroptera: Pyrrhocoridae) towards and its significance for aggregation. Anim. Behav. 17: 232–237.

1973. Some Hemiptera-Heteroptera associated with cacao farms in Nigeria. Turrialba 23: 162–171.

Youdeowei, A., and D. H. Calam. 1969. The morphology of the scent glands of *Dysdercus intermedius* Distan [sic] (Hemiptera: Pyrrhocoridae) and a preliminary analysis of the scent gland secretions of the fifth instar larva. Proc. R. Entomol. Soc. London (A) 44: 38–44.

Zaidi, Z. S., and M. A. Khan. 1972. Cholesterol in the haemolymph of *Dysdercus cingulatus* Fabr. (Hemiptera: Pyrrhocoridae). Curr. Sci. 41: 787.

1974a. Free haemocytes in the adult red cotton bug, *Dysdercus cingulatus* Fabr. (Hemiptera: Pyrrhocoridae). Indian J. Entomol. 36: 302–307.

1974b. Cholesterol concentration in the haemolymph, fat body and gonads of the red cotton bug, *Dysdercus cingulatus* Fabr. (Hemiptera: Pyrrhocoridae) in relation to metamorphosis, age and reproduction. Indian J. Entomol. 36: 11–16.

1975a. Inverse relationship between plasmatocytes and adipohaemocytes of *Dysdercus cingulatus* Fabr. (Hemiptera: Pyrrhocoridae) related to age and reproductive cycles. Curr. Sci. 44: 346–347.

1975b. Changes in the total and differential haemocyte counts of *Dysdercus cingulatus* Fabr. (Hemiptera: Pyrrhocoridae) related to metamorphosis and reproduction. J. Anim. Morphol. Physiol. 22: 11–119.

Zrzavý, J. 1990. Evolution of the aposematic colour pattern in some Coreoidea s. lat. (Heteroptera): a point of view. Acta Entomol. Bohemoslov. 87: 470–474.

1995. Morphological organization of abdominal colour patterns in Pyrrhocoridae bugs (Hemiptera-Heteroptera: Pentatomomorpha). J. Zool. Syst. Ecol. Res. 33: 3–8.

Zrzavý, J., and O. Nedvěd. 1997. Phylogeny of the New World *Dysdercus* (Insecta: Hemiptera: Pyrrhocoridae) and evolution of their colour patterns. Cladistics 13: 109–123.

1999. Evolution of mimicry in the New World *Dysdercus* (Hemiptera: Pyrrhocoridae). J. Evol. Biol. 12: 956–969.

CHAPTER 9

Scentless Plant Bugs (Rhopalidae)

Carl W. Schaefer and Jill Kotulski

1. INTRODUCTION

The Rhopalidae is a family of 20 genera and about 200 species (Göllner-Scheiding 1983), distributed worldwide and phylogenetically related to the Coreidae and its relatives (Schaefer 1964). The common name, "scentless plant bugs," refers to the reduction of the adults' metathoracic scent glands; these glands are functional, although externally they are difficult or impossible to discern (Aldrich et al. 1990; see also Davidová-Vilímová et al. in press). The name also refers to the fact the bugs are all phytophagous, the members of one subfamily (Rhopalinae) feeding on many different plants, and those of the other subfamily (Serinethinae) preferring members of the Sapindaceae (Schaefer and Chopra 1982, Schaefer and Mitchell 1983).

Of the two subfamilies, the Rhopalinae is the larger, and appears to be centered in the Palearctic; the Serinethinae is mostly Tropical (Schaefer 1992). The classification and phylogeny of tribes and subfamilies have been considered by Chopra (1967), Schaefer and Chopra (1982), Putschkov (1986), and Li and Zheng (1994). Schaefer and Chopra (1982) recognized the two subfamilies mentioned above; however, as they carefully noted, Serinethinae is paraphyletic, a fact that recognizes the considerable anagenetic evolution of this group, but a fact that nevertheless has deeply upset the fiercer cladists. Li and Zheng (1994) propose that only tribes be recognized between family and genus. This seems a reasonable compromise, although, again, the considerable anagenetic advance of the clade containing *Boisea, Leptocoris,* and *Jadera* (= Serinethinae *sensu* Schaefer and Chopra) becomes obscured. The history of rhopalid higher classification is reviewed briefly in Davidová-Vilímová et al. (in press).

Most genera have been revised (see titles of the following papers), by Chopra (1968, 1973), Gross (1960), and especially by Göllner-Scheiding (1975, 1976, 1977, 1978a,b, 1979, 1980a,b, 1984, 1988). The New World genera of Rhopalinae were probably derived from Old World emigrants (Schaefer 1993).

Rhopalids are phytophagous, feeding (like many Coreoidea) on green or ripe seeds. Of the two subfamilies, the smaller, Serinethinae, is largely restricted to the order Sapindales, especially the family Sapindaceae (Schaefer and Chopra 1982, Schaefer and Mitchell 1983). The feeding biology and ecology of several *Jadera* species have been worked out very nicely by Carroll and Loye (1987) and Carroll and Boyd (1992), and the intrapopulation evolutionary significance of different responses to different host plants have been analyzed, in *J. haematoloma* (Herrich-Schaeffer), by Carroll et al. (1997).

There are also occasional references to less usual food: one serinethine fed on bird droppings (see Adler and Wheeler 1984), and *B. rubrolineata* (Barber) was once found preying on caterpillars (Horn 1973).

The larger subfamily, Rhopalinae, feeds more widely. However, it does not feed on the more primitive plant subclasses (few heteropterans do), or upon grasses (except some Chorosomini). Some rhopaline groups prefer Malvaceae, some prefer Labiatae, and others prefer composites (Schaefer and Chopra 1982, Schaefer and Mitchell 1983).

The biology of only a few species has been worked out, and then not in detail. Some scattered data are available also for other species. In general, however, the biology of Rhopalidae is not at all well known. Readio (1928), Hambleton (1909), and Yonke and Medler (1967) describe the biology of a few North American *Corizus* species; Wheeler and Hoebeke (1988) present some notes on *Rhopalus tigrinus* Schilling in the United States (into which it recently emigrated), and Wheeler and Henry (1984) some notes on *Arhyssus hirtus* (Torre-Bueno). The laboratory life history, and descriptions of immatures, of *A. lateralis* (Say) are given by Packewitz and McPherson (1983), and a brief account of the biology of *Harmostes reflexulus* (Say) by Yonke and Medler (1967). The ecology and feeding preferences of *Myrmus miriformis* Fallén, and its interactions with several Miridae, were described by Gibson (1976). All the preceding are Rhopalinae. In addition to the fine work of Carroll and his colleagues (see accounts below) on several economically important Serinethinae, Wolda and Tanaka have considered the seasonal ecology of several *Jadera* spp. (Wolda and Tanaka 1987, Tanaka et al. 1987, Tanaka and Wolda 1988, Wolda 1989).

There are very few records of predators and parasites. Among the former are yellow jackets, *Vespula pensylvanica* (Saussure), preying on *Ithamar hawaiiensis* (Kirkaldy) in Hawaii (U.S.A.) (Gambino et al. 1987, Gambino 1992; the rhopalid cited as belonging to "Homoptera" in one case, and to "Alydidae" in both); a lynx spider, *Peucetia veridans* (Hentz), on *Harmostes reflexulus* in Florida (U.S.A.) (Randall 1982); and a single *H. serratus* (F.) in a nest of the wasp *Bicyrtes spinosa* (F.) (Sanchez and Genaro 1992). Parasites include the tachinid (Diptera) *Leucostoma acirostre* Reinhard on adult *H. reflexulus* (Wisconsin, U.S.A.) (Yonke and Walker 1970); an undetermined nematode on an adult male *Stictopleurus punctiventris* (Dallas), identified as *Corizus crassicornis* (L.), but see Göllner-Scheiding (1983, p. 140), also in Wisconsin (Yonke and Medler 1967). Mites, *Leptus* sp., have been collected from *Agrophopus* sp. in India (Dhiman and Singh 1988a).

In addition, see individual accounts below of economically important species.

2. GENERAL STATEMENT OF ECONOMIC IMPORTANCE

Scentless plant bugs are not of great importance. Some serinethines become annoying in the northeastern United States when they come to houses for warmth in late fall. Other rhopalids occasionally are or may become minor pests of cotton, or such ornamentals as soapberry and goldenrain trees. One species, *Niesthrea louisianica* (Sailer), is beneficial, helping to control the invasive weed velvetleaf.

3. SPECIES

3.1 *Boisea trivittata* (Say) (= *Leptocoris trivittata* Say)

3.1.1 *Distribution, Life History, and Biology*

In 1880, the boxelder bug, *B. trivittata*, was known from Mexico and the western United States east to Kansas and Missouri (Howard 1909). Since then, the scentless plant bug has moved east with its host plant, the boxelder tree (*Acer negundo* L.), also known as ash-leaf maple or three-leaf maple (Slater and Schaefer 1963). Adult boxelder bugs are elongated-oval, about 12 mm long, dark brown to black, with conscious red markings dorsally (Savos 1968) including three red longitudinal lines on the prothorax (Wollerman 1965). This species is very close to *B. rubrolineata* (Barber) (see **Section 3.2**).

Adult males live 1 to 22 days, and females 1 to 20 days. The boxelder bug is bivoltine, depending on the geographic location (Wollerman 1965, Yoder and Robinson 1990). The female will lay her eggs in leaf litter, tree bark, in houses, and on host plants (Yoder and Robinson 1990).

The most complete account of the insect's biology is by Tinker (1952), who worked in Minnesota (U.S.A.). Adults hibernate, often near buildings, emerge in spring, and in a few weeks migrate; during this time they are found on or near the ground, not on boxelder trees; from midsummer until September, the numbers on boxelder trees increase, and then decrease until mid-October. Movement has then begun to the places of hibernation. The timing of this movement corresponds with that of fruit development of the female trees, whose ovules increase 1 to 10 mm from mid-June to late August. Feeding is closely tied to temperature, the amount of feeding by fifth instars and adults quadrupling with each 5°C increase, in temperature (from 15 to 25°C) (calculated from Tinker's Fig. 2).

An egg mass averaged 11 eggs, and a female in the laboratory laid an average of 230 eggs in her lifetime (Tinker 1952).

Woolley's studies of internal anatomy (1951, and especially 1949) of the boxelder bug, are among the most complete for any heteropteran.

3.1.2   Damage

The boxelder bug becomes most noticeable in the autumn, when adults move to the warmth of houses, often in very large numbers (Howard 1909; Knowlton 1935, 1953; Wollerman 1965; Yoder and Robinson 1990). The bugs may cluster on the sides of houses, or infiltrate inside, causing some annoyance but posing no threat to health. This problem is exacerbated by the fact that the host plant, boxelder, is often planted near homes as an ornamental.

Boisea trivittata feeds on the seeds, and thus occurs only on the female trees of the monoecious *A. negundo* (Wollerman 1965). However, it has not been reported to occur in such high numbers as to endanger the tree or its reproduction. It will feed also on ash, maple, plum, cherry, peaches, pears, and grapes (Savos 1968). The damage done to host plants consists of scarring fruit and killing seeds and leaves.

3.1.3   Control

The boxelder bug may be controlled with such pesticides and agents as chlordane and kerosene, and also by the removal of unwanted boxelder trees (Howard 1909, Knowlton 1953). These agents should be applied to the bottom of the trunk where the bugs aggregate. When boxelder bugs enter homes as the days get colder, commercial insect sprays should be used (Knowlton 1935), or the bugs can be vacuumed up (Savos 1968). A commercial strain of the fungus *Beauveria bassiana* appears effective in control (Reinert et al. 1999).

3.2   *Boisea rubrolineata* (Barber) (= *Leptocoris rubrolineata* Barber)

3.2.1   Distribution, Life History, and Biology

This species, the western boxelder bug, was described by Barber (1956) for what had been treated as the West Coast (U.S.A.) boxelder bug. The two species are very close, *B. rubrolineata* differing consistently only in having red corial veins; although the ranges of the two species overlap in the U.S. southwest, there is little evidence of hybridization (Schaefer 1975).

The biology of this species is less well known than is that of *B. trivittata*; it feeds on *A. macrophyllum* Pursch (Schowalter 1986); and its biology is probably much like that of its eastern relative. There is a single report of this bug feeding on oakworm pupae (Horn 1973).

3.2.2 Damage and control

Like *B. trivittata*, *B. rubrolineata* overwinters in aggregations (Schowalter 1986), and when these occur on or in homes, the bug becomes a pest. It has also caused serious if occasional damage to almonds and pistachios in California (Michailides et al. 1988). [*Note:* the species here was identified as *B. trivittata*, but undoubtedly was *B. rubrolineata*.]

3.3   *Leptocoris augur* (F.) (= *Serinetha augur* F.)

This species occurs in southeast Asia (Göllner-Scheiding 1983). In India it is a pest of the macassor oil tree — lac tree, kusum tree: *Schleichera oleosa* (Lour) (Sapindaceae) (Dhawan and Gandhi 1989) — to whose seeds it causes considerable damage (Dhiman and Singh 1988a). It feeds also on the fruit of this plant, and occasionally on the leaves (Malhotra 1958, who provides a photograph of the bugs on the fruit). Although called a cotton pest, and fed "bolls of cotton," by Basker et al. (1994), Malhotra (1958) writes that this bug is not a cotton pest. [*Note:* The Indian literature, following Distant (1904), continues to call this species "*Serinetha*" *augur*.]

In the laboratory, 2% of seeds fed on by *L. augur* germinated; 50% of the controls germinated (Malhotra 1958). Newton (1984) writes that the bugs have difficulty reaching the seeds in ripe fallen fruit, and (at least during the hot season, in Madhya Pradesh state), rely on langurs to open the fruits; the bugs (mostly nymphs) then feed on the seeds. When the rains come, the bugs (now mostly adults) disperse to, and feed on, various grasses and herbs (Newton 1984) or, in Bihar, on *Lantana* (Malhotra 1958). Newton (1984) suggests this association with langurs allows the nymphs (perhaps too small easily to reach the seeds of mature fruits, C. W. Schaefer, personal observation) to survive the hot dry season.

The bugs are often heavily aggregated; Newton (1984) counted more than 500 in one group. An aggregation phermone, together with the attractive scent of the kusum fruit, bring the bugs together (Dhawan and Gandhi 1989).

Eggs are laid in batches or singly, up to 16 in a batch; a female may lay as many as 200 eggs in her lifetime (of about 100 days). An egg is 1.8 × 1.2 mm and, like those of many rhopalids, attached by a short pedicel. Eggs hatch in 6 to 9 (summer) or 16 to 19 days (winter; in Bihar). The nymphal stage lasts 31 to 44 days; first instars may not feed (those that appeared to feed, we suspect sought water, C. W. Schaefer and J. Kotulski, personal observation) (Malhotra 1958). Dhiman (1983) describes and illustrates the egg.

Leptocoris augur has two parasites, *Hexameris* sp. (Nematoda: Mermithidae) (Dhiman 1984) and *Leptus* sp. (Acarina: Eruthraeidae) (Dhiman and Singh 1988b). Males are less parasitized than females by the mite, but more parasitized than females by the nematode. Males infected by *Hexamermis* sp. have enlarged abdomens, an inability to fly, reduction in hind wings, reduced testes and less developed vasa deferentia, complete absorption of fat bodies by the parasite, and sometimes sterility (Dhiman 1984, Dhiman and Singh 1988b).

No insecticides against *L. augur* are known, but penfluron caused sterility in the closely related *B. coimbatorensis* (Gross) (Satyanarayana and Sukumar 1985).

3.4   *Jadera haematoloma* (Herrich-Schaeffer)

The soapberry bug, *J. haematoloma*, is a Neotropical bug that feeds primarily on the mature and nearly mature seeds of sapindaceous plants (Carroll and Boyd 1992); these include the soapberry tree and the goldenrain tree (Ribeiro 1989). The young nymphs feed upon low-growing weeds and adults are sometimes found on bolls of cotton, where they do minor damage (Sanderson 1905). In large numbers, the bugs can damage ornamentals and, like the boxelder bug, they may cluster on houses as the weather cools (Reinert et al. 1999). Nymphs and adults of *J. haematoloma* are toxic and distasteful, especially when aggregated (Ribeiro 1989).

A commercial preparation of the fungus *Beauveria bassinana* provided good control in laboratory tests (Reinert et al. 1999).

3.5 Jadera sanguinolenta (F.)

This species was found on cotton in the Paraguayan Chaco, where it may inflict some damage, especially to green bolls (Nickel 1958). It also occurs on rice and coffee in São Paulo (Brazil) (Silva et al. 1968).

3.6 Niesthrea louisianica (Sailer)

3.6.1 Distribution, Life History, and Biology

Niesthrea louisianica is a largely beneficial insect. It ranges from New York to Florida, through the Mississippi Valley to Iowa, and west to Arizona (U.S.A.); it prefers warmer areas (Wheeler 1977). It occurs on malvaceous plants, including velvetleaf (*Abutilon theophrasti* Medikus), cotton (*Gossypium* sp.), okra (*Hibiscus esculentus* L.), and rose of Sharon (*H. syriacus* L.), feeding on the fruits and seeds (Wheeler 1977).

Aspects of the life history of *N. louisianica* have been studied on rose of Sharon (Wheeler 1977) and several other plants, including velvetleaf (Jones et al. 1985). The post-egg development of the bug parallels that of its host, and populations decline in late fall when seeds become old and scarce; adults overwinter in ground litter (North Carolina, U.S.A.), mate in late April to May, and oviposit on the lower surfaces of leaves; first instars emerge in June and apparently do not feed; second and third instars and later instars then feed on flowers and early seed. There are several overlapping generations and females mate more than once; mating may take 12 to 40 hours; the egg is stalked (as are those of many — all? — rhopalids), and eggs are laid in clusters of 3 to 35 (average, 14); eggs of later generations are laid within flowers or on fruits; four females laid an average of 738 eggs each over their lifetimes. Eggs hatch in 4 to 8 days, and the nymphal stage is 3 to 4 weeks long; longevity of adults averaged 61.5 days (females) and 49.5 days (males) (Wheeler 1977).

Based on developmental time and mortality, *N. louisianica* does best on fruit and seed of velvetleaf and fruit of rose of Sharon, and worst on cotton and okra (Jones et al. 1985).

3.6.2 Damage

Velvetleaf is an increasingly serious problem in corn (*Zea mays* L.), soybean (*Glycine max* L.), cotton (*Gossypium hirsutum* L.), and sorghum (*Sorghum bicolor* L.) fields (Patterson et al. 1987), where it overgrows and strangles the other crops. In addition, the trichome exudate of velvetleaf disturbs and even traps parasitoids of some insect pests (Gruenhagen and Perring 1999). *N. louisianica* attacks the reproductive parts of the female plant (Patterson et al. 1987). Seeds attacked by the insect are paler, smaller, and shriveled or with sunken areas (Kremer and Spencer 1989b). In controlled chambers *Niesthrea louisianica* damaged reproductive structures of velvetleaf and reduced viable seed production by up to 99% (Patterson et al. 1987).

Wounding and penetration of the velvetleaf seed coat by this bug may promote microbial infection by allowing entry of seed-associated microorganisms (Kremer and Spencer 1989a,b). The insect feeds only on immature seeds because they are softer and easier to penetrate. This action could favor both the insect and microorganism. The microorganism rides on *N. louisianica* and "mops up" the leaking seed juice (Kremer 1988). Only about 1 to 5% of the seeds attacked by the pair are able to germinate (Kremer 1988). This scentless plant bug may save U.S.$383 million in velvetleaf damage each year in the United States (Spencer 1984, 1987), and an inundative release of the bug in five U.S. states provided good control (Spencer 1988). However, in China, velvetleaf is used for bast fiber, cordage, cloth, fishing nets, and sometimes for food (Spencer 1984).

3.6.3 Control

An egg parasite, *Telenomus* sp. (Scelionidae), has been found (Wheeler 1977, Jones et al. 1985), as well as a tachinid parasite of the adult (Jones et al. 1985). Spiders will feed on the instars and adults (Wheeler 1977, Jones et al. 1985).

3.7 Niesthrea sidae (F.)

Jatropha curcas is a plant with many uses in the northern Neotropics. It is a purgative; its oil is used in the making of soap, candles, paint, and in the wool industry; its kernels are roasted and eaten; and when infested by certain lac insects their xudates are valued as varnishes; the ashes are a substitute for salt. The plant is important in the local economy of Mesoamerica and the Caribbean.

Niesthrea sidae was reported, first as *N.* sp. (Grimm and Maes 1997) and then as *N. sidae* (Grimm and Führer 1998), on fruit of *J. curcas* in Nicaragua. It was not so serious a pest as the scutellerid *Pachycoris klugii* Burmeister (see **Chapter 14**). This rhopalid occurs from the southern United States into South America; however, because there is much variation from region to region, more than one species may be involved (Chopra 1973), and the Nicaraguan pest on *J. curcas* may not be *N. sidae*.

Niesthrea sidae has also been recorded from São Paulo state (Brazil), on *Abutilon* sp., cotton, and rice; here, too, it seems not to be a major pest. Its eggs were parasitized by *Telenomus* sp. (Scelionidae) (Silva et al. 1968).

3.8 Liorhyssus hyalinus (F.)

3.8.1 Distribution and biology

The hyaline grass bug feeds on sorghum (Hall and Teetes 1981) and pistachio fruit (Michailides et al. 1987) in North America. The nymphs damage fruit and seeds. This bug moves onto these crops from nearby wild hosts, mostly grasses (Hall and Teetes 1981, Michailides et al. 1987).

Liorhyssus hyalinus occurs worldwide (Göllner-Scheiding 1983).

3.8.2 Damage

It is common on cotton in Egypt and occasionally common on tobacco in Afghanistan; in neighboring countries it feeds also on cereals and legumes (Anonymous 1965). In California (U.S.A.) it and several other heteropterans are responsible for "early season epicarp lesion" in April and May of pistachio, the damaging and darkening of tissue and subsequent abortion of the young fruit (Michailides et al. 1987).

3.8.3 Control

To control these insects the host trees must be sprayed with insecticides. Spraying twice is most efficient; using carbaryl insecticide and a permethrin spray in combination seems to wipe out populations of hylaline grass bugs (Michailides et al. 1987). A tachinid (Diptera) parasite was recorded from *L. hyalinus* in Arizona (U.S.A.) (Butler et al. 1982).

3.9 *Aufeius impressicollis* Stål

This rhopalid has been collected on several economically important crops, but Wheeler (1984), in the only account of its biology, shows that species of Amaranthaceae are probably the true hosts. To the extent that some of these plants are pests, this bug is economically important.

It differs from other rhopalines in its laterally dilated abdomen and broadly exposed connexivum. Females lay several dozen eggs over a week; eggs hatch in about 8 days and adults emerge in about 3 weeks; all stages feed only on flower parts (Wheeler 1984).

3.10 *Arhyssus parvicornis* (Signoret)

This species is found "only rarely" on cotton in the Paraguayan Chaco (Nickel 1958).

3.11 *Leptocoris amicta* (Germar) (= *Serinetha amicta* [Germar])

Leptocoris amicta has been observed on coffee in sub-Saharan Africa (Schouteden 1912, Villiers 1952). It almost certainly is not, nor will it become, a pest on coffee. Schouteden (1912) illustrates the adult (twice).

3.12 *Leptocoris hexophthalma* (Thunberg)

This bug occurs in southern and eastern Africa; it is pale red to reddish-brown, and 10 to 11.5 mm long (Prins 1983, who illustrates two nymphs and the adult). In South Africa it sometimes aggregates in large numbers under the ornamentals bergvygie (*Drosanthemum* sp.) and skaapbossie (*Justicia orchiodes*), on whose foliage the nymphs do considerable damage (Prins 1983).

4. CONCLUSIONS

Few members of the Rhopalidae are of economic concern. A few (especially *Leptocoris augur* and *Liorhyssus hyalinus*) may occasionally attack certain crops. The two North American *Boisea* species may from time to time become nuisances as they seek to overwinter in homes.

The biologies of very few rhopalids have been worked out, and members of the Rhopalinae are rather small and undistinguished. It is possible, therefore, that undetected rhopalids cause some damage that has gone as yet unremarked. More study of their biologies would make it easier to predict what damage may be caused, and what damage might be caused, by Rhopalidae.

5. REFERENCES CITED

Adler, P. H., and A. G. Wheeler, Jr. 1984. Extra-phytophagous food sources of Hemiptera-Heteroptera: bird droppings, dung, and carrion. J. Kansas Entomol. Soc. 57: 21–27.

Aldrich, J. R., S. P. Carroll, J. E. Oliver, W. R. Lusby, A. A. Rudmann, and R. M. Waters. 1990. Exocrine secretions of scentless plant bugs: *Jadera, Boisea* and *Niesthrea* species (Hemiptera: Heteroptera: Rhopalidae). Biochem. Syst. Ecol. 18: 369–376.

Anonymous. 1965. Rhopalidae (= Coreidae). P. 12 *in* Anonymous [ed.], Crop Insects of Northeast Africa-Southwest Asia. Agriculture Handbook 273. U.S. Dept. of Agriculture.

Barber, H. G. 1956. A new species of Leptocoris (Coreidae: Leptocorini). Pan-Pac. Entomol. 32: 9–11.

Basker, P., L. S. Ranganathan, and P. Padmanabhan. 1994. The regulatory mechanism of the male accessory reproductive gland (ARG) of *Serinetha augur* (Fabr [sic: no period]) (Heteroptera: Coreidae) — a cotton pest. Insect Sci. Appl. 15: 351–360.

Butler, G. D., Jr., T. J. Henneberry, F. G. Werner, and J. M. Gillespie. 1982. Seasonal distribution, hosts, and identification of parasites of cotton insects. U.S. Dept. Agriculture, Agric. Res. Serv., Agric. Rev. Manuals ARM-W-27. 54 pp.

Carroll, S. P., and C. Boyd. 1992. Host race radiation in the soapberry bug: natural history with the history. Evolution 46: 1053–1069.

Carroll, S. P., and J. E. Loye. 1987. Specialization of *Jadera* species (Hemiptera: Rhopalidae) on seeds on the Sapindaceae (Sapindales), and coevolutionary responses of defense and attack. Ann. Entomol. Soc. Amer. 80: 373–378.

Carroll, S. P., H. Dingle, and S. P. Klassen. 1997. Genetic differentiation of fitness-associated traits among rapidly evolving populations of the soapberry bug. Evolution 51: 1182–1188.

Chopra, N. P. 1967. The higher classification of the family. Rhopalidae (Hemiptera). Trans. R. Entomol. Soc. London 119: 363–399.

1968. A revision of the genus *Arhyssus* Stål. Ann. Entomol. Soc. Amer. 61: 629–655.

1973. A revision of the genus *Niesthrea* Spinola (Rhopalidae: Hemiptera). J. Nat. Hist. 7: 441–459.

Davidová-Vilímová, J., M. Nejedlá, and C. W. Schaefer. Dorso-abdominal scent glands and metahoracic evaporatoria in adults of central European Rhopalidae (Hemiptera), with a discussion of phylogeny and higher systematics. Europ. J. Entomol. (in press).

Dhawan, N., and J. R. Gandhi. 1989. Aggregation and feeding behaviour of *Serinetha augur* Fabr. (Heteroptera: Coreidae). Indian J. Entomol. 51: 19–23.

Dhiman, S. C. 1983. Studies on the oviposition and egg structure of *Serinetha augur* Fabr. (Heteroptera: Coreidae). Uttar Pradesh J. Zool. 3: 166–169.

1984. A new record of *Hexamermis* sp. (Nematoda-Mermithidae) parasitizing *Serinetha augur* (Fabr.) (Heteroptera-Coreidae). Curr. Sci. 53: 1310.

Dhiman, S. C., and R. K. Singh. 1988a. New record of host plants and mite parasite of *Agraphopus* sp. (Coreoidea: Rhopalidae). Uttar Pradesh J. Zool. 8: 103–104.

1988b. Studies on the parasitism of *Leptocoris augur* (Fabr.) (Heteroptera: Rhopalidae) by *Leptus* sp. of mite (Acarina: Eruthraeidae). Uttar Pradesh J. Zool. 8: 180–185.

Distant, W. L. 1904. The Fauna of British India, Including Ceylon and Burma. Rhynchota. Vol. 1. (Heteroptera). Taylor and Francis, London, U.K. 438 pp.

Gambino, P. 1992. Yellowjacket (*Vespula pensylvanica*) predation at Hawaii volcanoes and Haleakala National Parks: identity of prey items. Proc. Hawaiian Entomol. Soc. 31: 157–164.

Gambino, P., A. C. Medeiros, and L. L. Loope. 1987. Introduced vespids *Paravespula pensylvanica* prey on Maui's endemic arthropod fauna. J. Trop. Ecol. 3: 169–170.

Gibson, C. W. D. 1976. The importance of food plants for the distribution and abundance of some Stenodemini (Heteroptera: Miridae) of limestone grassland. Oecologia 25: 55–76.

Göllner-Scheiding, U. 1975. Revision der Gattung Stictopleurus Stål, 1872 (Heteroptera, Rhopalidae). Dtsch. Entomol. Z. (n. F.) 22: 1–60.

1976. Revision der Gattung Liorhyssus Stål, 1870 (Heteroptera, Rhopalidae). Dtsch. Entomol. Z. (n. F.) 23: 181–206.

1977. Revision der Gattungen Agrophopus Stål, 1872, und Leptoceraea Jakovlev, 1873 (Heteroptera, Rhopalidae). Dtsch. Entomol. Z. (n. F.) 24: 223–249.

1978a. Revision der Gattung *Harmostes* Burm., 1835 (Heteroptera, Rhopalidae) und einige Bemerkungen zu den Rhopalinae. Mitt. Zool. Mus. Berlin 54: 257–311.

1978b. Bemerkungen zur Gattung *Rhopalus* Schilling einschliesslich *Brachycarenus* Fieber (Heteroptera, Rhopalidae). Mitt. Zool. Mus. Berlin 54: 313–331.

1979. Die Gattung Jadera Stål, 1862 (Heteroptera, Rhopalidae). Dtsch. Entomol. Z. (n. F.) 26: 47–75.

1980a. Revision der afrikanischen Arten sowie Bemerkungen zu weiteren Arten der Gattung Leptocoris Hahn, 1833, und Boisea Kirkaldy, 1910. Dtsch. Entomol. Z. (n. F.) 27: 103–148.

1980b. Einige Bemerkungen zu den Gattung *Corizus* Fallén, 1814, und *Xenogenus* Berg, 1883 (Heteroptera, Rhopalidae). Mitt. Zool. Mus. Berlin 56: 111–121.

1983. General-Katalog der Familie Rhopalidae (Heteroptera). Mitt. Zool. Mus. Berlin 59: 37–189.

1984. Ergänzungen zu den Gattungen *Liorhyssus* Stål, 1970, *Niesthrea* Spinola, 1837, und *Rhopalus* Schilling, 1827 (Heteroptera, Rhopalidae). Mitt. Zool. Mus. Berlin 60: 115–121.

1988. Die Gattung Peliochrous Stål, 1873 (Heteroptera: Rhopalidae). Dtsch. Entomol. Z. (n. F.) 35: 293–297.

Grimm, C., and E. Führer. 1998. Population dynamics of true bugs (Heteroptera) in physic nut (*Jatropha curcas*) plantations in Nicaragua. J. Appl. Entomol. 122: 515–521.

Grimm, C., and J.-M. Maes. 1997. Insectos asociados al cultivo de tempate (*Jatropha curcas*) en el Pacifico de Nicaragua. III. Coreoidea (Heteroptera). Rev. Nicaraguense Entomol. 42: 15–34.

Gross, G. F. 1960. A revision of the genus Leptocoris Hahn (Heteroptera: Coreidae: Rhopalinae) from the Indo-Pacific and Australian Regions. Rec. South. Austral. Mus. 13: 403–451.

Gruenhagen, N. M., and T. M. Perring. 1999. Velvetleaf: a plant with adverse impacts on insect natural enemies. Environ. Entomol. 28: 884–889.

Hall, D. G., IV, and G. L. Teetes. 1981. Alternate host plants of sorghum panicle-feeding bugs in southeast central Texas. Southwest. Entomol. 6: 220–228.

Hambleton, J. C. 1909. Life history of *Corizus lateralis* Say. Ann. Entomol. Soc. Amer. 2: 272–276.

Horn, D. J. 1973. *Leptocoris rubrolineatus*, an occasional predator of the California oakworm, *Phyrganida californica* (Hemiptera, Rhopalidae; Lepidoptera: Dioptidae). Pan-Pac. Entomol. 49: 196.

Howard, L. O. 1909. The boxelder plant-bug. U.S. Dept. Agriculture Circ. (2nd series). No. 28: 1–3.

1935. The boxelder bug. Utah Agric. Exp. Stn. Leafl. 56: 2 pp.

Jones, W. A., Jr., H. E. Walker, P. C. Quimby, and J. D. Ouzts. 1985. Biology of *Niesthrea louisianica* (Hemiptera: Rhopalidae) on selected plants, and its potential for biocontrol of velvetleaf, *Abutilon theophrasti* (Malvaceae). Ann. Entomol. Soc. Amer. 78: 326–330.

Knowlton, G. F. 1935. The boxelder bug. Utah Agric. Exp. Stn. Leafl. 56: 2 pp.

1953. Controlling the boxelder bug. Utah State Agric. Coll. Ext. Circ. 162: 2 pp.

Kremer, R. J. 1988. Fungus sides with bug against weed. Agric. Res. (USDA) 36 (7): 6.

Kremer, R. J., and N. R. Spencer. 1989a. Interaction of insects, fungi, and burial on velvetleaf (*Abutilon theophrasti*) seed viability. Weed Technol. 3: 322–328.

1989b. Impact of a seed-feeding insect and microorganisms on velvetleaf (*Abutilon theophrasti*) seed viability. Weed Sci. 37: 211–216.

Li, X., and L. Zheng. 1994. A preliminary study on the phylogeny of Rhopalidae (Hemiptera: Coreoidea). Acta Zootax. Sin. 19: 78–89 [in Chinese, English summary].

Malhotra, C. P. 1958. Bionomics of *Serinetha augur* Fabr. and its association with *Dysdercus cingulatus* Fabr., the red cotton bug. Indian For. 94: 669–671.

McNally, A. G. 1957. Notes on the boxelder bug: *Leptocoris trivittatus* (Say). Annu. Rep. Entomol. Soc. Ontario (1956) 87: 80.

Michailides, T. J., R. E. Rice, and J. M. Ogawa. 1987. Succession and significance of several hemipterans attacking a pistachio orchard. J. Econ. Entomol. 80: 398–406.

Michailides, T. J., J. M. Ogawa, R. E. Rice, and R. D. Sanders. 1988. Association of the boxelder bug (Hemiptera: Rhopalidae) with epicarp lesion of pistachio fruits. J. Econ. Entomol. 81: 1148–1151

Newton, P. N. 1984. A feeding association between a heteropteran bug and langurs. J. Bombay Nat. Hist. Soc. 81: 180–181.

Nickel, J. L. 1958. Agricultural insects of the Paraguayan chaco. J. Econ. Entomol. 81: 633–637.

Packewitz, S. M., and J. E. McPherson. 1983. Life history and laboratory rearing of *Arhyssus lateralis* (Hemiptera: Rhopalidae) with descriptions of immature stages. Ann. Entomol. Soc. Amer. 76: 477–482.

Patterson, D. T., R. D. Coffin, and N. R. Spencer. 1987. Effects of temperature on damage to velvet leaf (*Abutilion theophrasti*) by scentless plant bug (*Niesthrea louisianica*). Weed Sci. 35: 324–327.

Prins, A. J. 1983. Morphological and biological notes on some South African arthropods associated with decaying organic matter. Part 1. Chilopoda, Diplopoda, Arachnida, Crustacea, and Insecta. Ann. South African Mus. 92: 53–112.

Putschkov, V. G. 1986. True bugs of the family Rhopalidae of the fauna of the USSR. Keys to the Fauna of the USSR. Zool. Inst. Acad. Sci. USSR. Vol. 146. 132 pp. [in Russian].

Randall, J. B. 1982. Prey records of the green lynx spider, *Peucetia viridans* (Hentz) (Aranceae, Oxyopidae). J. Arachnol. 10: 19–22.

Readio, P. A. 1928. Studies on the biology of the genus *Corizus* (Coreidae, Hemiptera). Ann. Entomol. Soc. Amer. 21: 189–201.

Reinert, J. A., T. A. Knauf, S. J. Maranz, and M. Bishr. 1999. Effect of *Beauveria bassiana* fungus on the boxelder and red shouldered bugs (Hemiptera: Rhopalidae). Florida Entomol. 82: 469–474.

Ribeiro, S. T. 1989. Group effects and aposematism in *Jadera haematoloma* (Hemiptera: Rhopalidae). Ann. Entomol. Soc. Amer. 82: 466–475.

Sanchez, C. S., and J. A. Genaro. 1992. Observaciones sobre la conducta de nidificación en esfécidos de Cuba (Hymenoptera). *Bicyrtes spinosa* (Fabr.). Poeyana 411:8 pp.

Sanderson, E. D. 1905. Miscellaneous cotton insects in Texas. U.S. Dept. Agric. Farmers' Bull. No. 223: 1–23.

Satyanarayana, K., and K. Sukumar. 1985. Sterility and retardation of growth by penfluron in soapnut bug *Leptocoris coimbatorensis* (Gross) (Hemiptera, Coreidae). Z. Angew. Entomol. 100: 367–372.

Savos, M. G. 1968. Boxelder bugs and their control. Co-op. Extens. Serv., Coll. Agric., Univ. Connecticut 68–61: 2 pp.

Schaefer, C. W. 1964. The morphology and higher classification of the Coreoidea (Hemiptera: Heteroptera). Parts I and II. Ann. Entomol. Soc. Amer. 57: 670–684.

1975. A re-assessment of North American *Leptocoris* (Hemiptera-Heteroptera: Rhopalidae). Ann. Entomol. Soc. Amer. 68: 537–541.

1992. The Rhopalinae (Hemiptera: Rhopalidae) and the Palearctic. Proc. 4th Eur. Congr. Entomol. (Gödöllö), pp. 652–654.

1993. Origins of the New World Rhopalinae (Hemiptera: Rhopalidae). Ann. Entomol. Soc. Amer. 86: 127–133.

Schaefer, C. W., and N. P. Chopra. 1982. A cladistic analysis of the Rhopalidae, with a list of food plants. Ann. Entomol. Soc. Amer. 75: 224–233.

Schaefer, C. W., and P. L. Mitchell. 1983. The food plants of the Coreoidea (Hemiptera: Heteroptera). Ann. Entomol. Soc. Amer. 76: 591–615.

Schouteden, H. 1912. Les Hémiptères parasites du caféier en Agrique. Rev. Zool. Afr. 2: 19–34.

Schowalter, T. D. 1986. Overwintering aggregation of *Boisea rubrolineata* (Heteroptera: Rhopalidae) in western Oregon. Environ. Entomol. 15: 1055–1056.

Silva, A. G. d'A., C. R. Conçalves, J. Gomes, M. N. Silva, and L. Simoni. 1968. Quarto Catálogo dos Insetos que Vivem nas Plantas do Brasil — Seus Parasitos e Predadores. Parte II, Vol. 1. Minist. Agric., Rio de Janeiro, Brazil. 622 pp.

Slater, J. A., and C. W. Schaefer. 1963. *Leptocoris trivittatus* (Say) and *Coriomeris humilis* Uhl. in New England (Hemiptera: Coreidae). Bull. Brooklyn Entomol. Soc. 57: 114–117.

Spencer, N. R. 1984. Velvet leaf, *Abutilon theophrasti* (Malvaceae), history and economic impact in the United States. Econ. Bot. 38: 407–416.

1987. *Niesthrea louisianica* (Hemiptera: Rhopalidae) reduces velvet leaf, *Abutilon theophrasti* (Malvaceae), seed production and viability. Environ. Entomol. 16: 963–966.

1988. Inundative biological control of velvetleaf, *Abutilon theophrasti* (Malvaceae) with *Niesthrea louisianica* (Hem.: Rhopalidae). Entomophaga 33: 421–429.

Tanaka, S. and H. Wolda. 1988. Oviposition behavior and diel rhythms of flight and reproduction in two species of tropical seed bugs. Proc. K. Ned. Akad. Wet. (C) 91: 165–174.

Tanaka, S., H. Wolda, and D. L. Denlinger. 1987. Seasonality and its physiological regulation in three neotropical insect taxa from Barro Colorado Island, Panama. Insect Sci. Appl. 8: 507–514.

Tinker, M. E. 1952. The seasonal behavior and ecology of the boxelder bug *Leptocoris trivittatus in Minnesota* [sic: italics]. Ecology 33: 407–414.

Villiers, A. 1952. Hémiptères de l'Afrique Noire (Punaises et Cigales). Initiations Africaines IX. Institut Français d'Afrique Noire, Dakar, Senegal. 256 pp.

Walker, A. J., Jr., H. E. Walker, P. C. Quimby, and J. D. Ouzts. 1985. Biology of *Niesthrea louisianica* (Hemiptera: Rhopalidea) on selected plants, and its potential for bio-control of velvet leaf, *Abutilon theophrasti* (Malvaceae). Ann. Entomol. Soc. Amer. 78: 326–330.

Wheeler, A. G., Jr. 1977. Life history of *Niesthrea louisianica* (Hemiptera: Rhopalidae) on rose of Sharon in North Carolina. Ann. Entomol. Soc. Amer. 70: 631–634.

1984. *Aufeius impressicollis* (Hemiptera: Rhopalidae) easternmost U.S.-record, host plant relationships, and laboratory rearing. J. New York Entomol. Soc. 92: 174–178.

Wheeler, A. G., Jr., and T. J. Henry. 1984. Host plants, distribution, and description of fifth-instar nymphs of two little-known Heteroptera, *Arhyssus hirtus* (Rhopalidae) and *Esperanza texana* (Alydidae). Florida Entomol. 67: 521–529.

Wheeler, A. G., Jr., and R. E. Hoebeke. 1988. Biology and seasonal history of *Rhopalus (Brachycarenus) tigrinus*, with descriptions of immature stages (Heteroptera: Rhopalidae). J. New York Entomol. Soc. 96: 381–389.

Wolda, H. 1989. Energy requirements of tropical insects during an adverse season. Nutr. Ecol. Ins. Env., pp. 1–9.

Wolda, H., and S. Tanaka. 1987. Dormancy and aggregation in a tropical insect *Jadera obscura* (Hemiptera: Rhopalidae). Proc. K. Ned. Akad. Wet. (C) 90: 351–366.

Wollerman, E. H. 1965. The boxelder bug. U.S. Dept. Agric. Forest Pest Leaf. 95. 6 pp.

Woolley, T. A. 1949. Studies on the internal anatomy of the boxelder bug. *Leptocoris trivittatus* (Say). Ann. Entomol. Soc. Amer. 42: 203–226.

1951. The circulatory sysstem of the boxelder bug *Leptocoris trivittatus* (Say). Amer. Midl. Nat. 46: 634–639.

Yoder, K. M., and W. H. Robinson. 1990. Seasonal abundance and habitats of the boxelder bug, *Boisea trivittata* (Say), in an urban environment. Proc. Entomol. Soc. Washington 92: 802–807.

Yonke, T. R., and J. T. Medler. 1967. Observations on some Rhopalidae (Hemiptera). Proc. North. Cent. Branch Entomol. Soc. Amer. 22: 74–75.

Yonke, T. R., and D. L. Walker. 1970. Field history, parasites, and biology of *Harmostes reflexulus* (Say) (Hemiptera: Rhopalidae). J. Kansas Entomol. Soc. 43: 444–450.

CHAPTER **10**

Broad-Headed Bugs (Alydidae)

Antônio R. Panizzi, Carl W. Schaefer, and Yosihiro Natuhara

1. INTRODUCTION

The Alydidae, or broad-headed bugs, is a relatively small herbivorous family. In North America these bugs are usually found in foliage and flowers of various plants along roadsides and woods (Froeschner 1988).

The family is divided into two subfamilies, Alydinae and Micrelytrinae, the latter with two tribes, Micrelytrini and Leptocorisini (Schaefer 1965, 1999; Li and Zheng 1993); Ahmad (1965) raised the two tribes to subfamily rank, but without explanation. Schaefer (1999) extracted characteristics from this literature and determined their derived states, in an attempt to establish better the higher classification of the Alydidae. The Leptocorisini have been revised by Ahmad (1965) and the Alydinae (to genus) by Schaffner (1964); Siwi and Van Doesburg (1984) reviewed the *Leptocorisa* of Indonesia, and Linnavuori (1987) the African species of the family.

The Alydinae feed almost exclusively on legumes (Schaefer 1972, 1980; Schaefer and Mitchell 1983). The Leptocorisini feed on grasses (Schaefer 1972). Thus, these two groups may become serious pests on leguminous and graminaceous crops, especially on soybean and rice, respectively. The food plants of micrelytrines are not known, and none is of economic significance.

There are several references to alydines feeding on carrion and vertebrate fecal matter (Schaefer 1980, Adler and Wheeler 1984); the bugs are probably attracted to those rich sources of semiliquid nitrogen, but it is unclear why alydines should be attracted more (apparently) than other bugs. Perhaps they require higher concentrations of nitrogen than do other bugs, hence their association with legumes, nitrogen-rich plants.

The nymphs of all alydines (as far as is known) mimic ants in form and behavior (e.g., Oliveira 1985), as do the adults of many Micrelytrini (Schaefer 1972, 1999). The chromosome number is $10A + 2m + X0$ or (in some *Leptocorisa*) $7A + 2m + X0$ (Ueshima 1979, Sands 1982a); Sands (1982b) suggests a mechanism by which changes in chromosome numbers may have occurred.

This chapter will cover the species that interact with plants of economic importance, in particular species in *Leptocorisa* (Leptocorisini), and *Megalotomus*, *Neomegalotomus*, and *Riptortus* (all Alydinae).

2. GENERAL STATEMENT OF ECONOMIC IMPORTANCE

Broad-headed bugs are of little economic importance in most of the Holarctic. However, some species are harmful to major crops in other parts of the world: the tropics (South America and Africa) and parts of Asia. Pest species are *Leptocorisa* spp., which are important pests of rice in southern and southeastern Asia (Srivastava and Saxena 1967); *Neomegalotomus parvus* (Westwood), widely scattered in South America, particularly in Brazil, and detrimental to common beans (Chandler 1984, 1989) and a potential pest of soybean (Panizzi 1988); and finally, *Riptortus* spp., minor pests of pigeon pea (*Cajanus cajan*) (Shanower et al. 1999), but major pests of soybean in Asia and Africa (see references in Jackai et al. 1990).

Two species may be of some economic benefit (**Section 5**).

3. THE IMPORTANT SPECIES

3.1 *Leptocorisa acuta* (Thunberg) (= *Leptocorisa varicornis* F.)

3.1.1 Distribution, Life History, and Biology

The rice bug, *L. acuta* (Thunberg) (formerly known as *L. varicornis* F.) (see Ahmad 1965), is an important pest of rice throughout Asia. It is very common in India, Indonesia, Sarawak, and New Guinea, and is found in the Philippines, Samoa, New Caledonia, and Fiji; but it has not been recorded from Java and Sri Lanka (Dresner 1958, Matthysse 1957, Ahmad 1965, Pathak 1968). Along with *L. oratorius* (F.) and *L. chinensis* Dallas, it forms a group of pests of paddy crops, but it can be easily separated from them by the parameres (bifurcated at apex) and by the triangular projection of the 7th abdominal sternum in females (Ahmad 1965). Adults are slender and cylindrical, pale brown, and 15.0 to 16.0 mm long. Females may lay 25 to 87 eggs during their life time (Haque 1980). Newly emerged nymphs are pale yellowish-green, and 2 mm long; second, third, and fourth instars are similar in color, and 6, 11, and 14 mm long. Fifth instars are pale brown dorsally and white green below, 17 mm long. The nymphal stage takes 20 to 31 days (Rai 1981). The list of host plants is similar to that described for *L. oratorius* (which see), rice being the preferred host. It has also been recorded on nongraminaceous alternate hosts such as nutmeg, mango, guava, rubber, and tea (Abraham and Mony 1977). Based on specific differences in immature stages, oviposition sites, and hatching patterns, Cobblah and Hollander (1992) provided information useful for quick identification of these two species, which may occur simultaneously in rice fields.

Leptocorisa acuta may also occur in some numbers on Akh (Madar, *Calotropis procera*), a plant of some medicinal importance in India (Verma et al. 1978). However, we suspect it not likely to become an important pest of this asclepiadaceous plant.

3.1.2 Damage and Control

The damage done to rice, and control measures, are similar to those described for *L. oratorius* (**Section 3.2**). *Leptocorisa acuta* is reported to transmit the sheath rot disease, *Sarocladium oryzae* (Sawada) W. Gam & D. Hawksw, to rice in India (Lakshmanan et al. 1992). Dresner (1958) estimated that *L. acuta* may reduce national rice yield in Indonesia by 10%, and in many areas 100% loss of yield is noted. He also recommended the use of a poison bait using a crustacean (*Potamon* sp.) dipped in 0.2% dieldrin suspension, tied near the head of a flowering rice plant. *Leptocorisa acuta* will be attracted to this bait and become intoxicated. In the Philippines, the most effective insecticides against this and other rice pests were endrin and EPN (Matthysse 1957). In India, fenitrothion at 1125 ml/ha proved useful (Krishnamurthy et al. 1977); BHC (10%) and carbaryl (5%) were ineffective, fenthion, phosphamidon, monocrotophos, and dichlorvos (all at 0.1%) were highly effective (Velusamy et al. 1977).

3.2 Leptocorisa oratorius (F.) (= Leptocorisa acuta (Thunberg), Leptocorisa maculiventris Dallas)

3.2.1 Distribution, Life History, and Biology

This rice bug (formerly known as *L. maculiventris* Dallas and sometimes confused with *L. acuta;* see Ahmad 1965), is a major pest of rice, widely distributed in southern and southeastern Asia and Australia (Ahmad 1965, Pathak 1968). It appears to be a minor pest of rice in northern India (where it is known as the gundhi bug) (Kushwaha 1988). The list of host plants includes mostly Gramineae, cultivated or not, such as grain sorghum, grasses (*Brachiaria, Cenchrus, Digitaria, Echinochloa,* and *Panicum*), the euphorb *Euphorbia geniculata* Ortega (Li 1985), and legumes such as *Vigna* sp.

The *L. oratorius* adult has a robust body, 18.0 to 18.5 mm long, pale brown, very similar to *L. acuta* and *L. chinensis* Dallas, but can easily be separated by a series of ventrolateral black dots on the abdomen (Ahmad 1965).

The biology of *L. oratorius* has been studied by several authors. Adults, especially females, are attracted to green and blue (Dhiman and Chatterjee 1980). Li (1985) reported that eggs are laid in rows on the leaf blade of rice near the tip, and occasionally on the stem or the panicle. Incubation period is 5 to 8 days; nymphal developmental time is 15 to 32 days; mean number of eggs/female is 40.5. Seven overlapping generations a year were observed in Papua New Guinea. *Leptocorisa oratorius* prefers rice to wheat for oviposition (Morrill and Arida 1991). Other common weed grasses are as preferred for oviposition over rice, but nymphs have greater survivorship and a shorter developmental time on rice (Morrill et al. 1990). In Maharashtra State, India, adults overwintered on nonrice plants, and moved to rice in February; one alternative host plant was kulthi (*Dolichos biflorus*), a crop legume to which some damage was caused (Dalvi et al. 1985).

Dhiman (1981) described the tentorium, a structure rarely studied in Heteroptera.

3.2.2 Damage and Control

Damage by *L. oratorius* to rice is severe. Nymphs and adults suck the sap and attack the grains, particularly during the milky stage, leading to empty or "pecked" grains. The resulting damage varies with the time of attack. For example, weight loss in the panicles is greater when grains suffer injury during the milky stage, than when attacks occur during the "dough" stage, and feeding during flowering causes failure of panicles (Li 1985). Heavy infestations can result in 80% (Maharashtra State, India) (Dalvi et al. 1985) or total (Malaysia) loss of the crop (Lim 1971). Ito et al. (1992) estimated that *L. oratorius* may damage 6.4 to 7.7 rice grains/day/adult, when released in caged rice plants with the panicle at the flowering stage.

Leptocorisa oratorius is efficiently controlled by insecticides, such as monocrotophos (Li 1985). Integrated pest management programs for seed bugs on rice are based on the number of nymphs or adults per sample unit; the economic threshold level is reached when two to three insects per hill are found during the ripening stage from milk to hard dough. Because this bug is highly mobile, a more sophisticated threshold is based on the average counts over a number of days to give "rice–bug days" (Way et al. 1991, and references therein). Also, weed control may help mitigate the impact of this pest on rice, because weeds support development of nymphs, and make fields more attractive to adults (Morrill et al. 1990). Indeed, the control of weeds, especially the *Euchinochloa crus-galli* complex, is the more effective and practical measure to reduce rice damage caused by ear-sucking bugs in West Malaysia (Ito et al. 1992). Other measures to manage

L. oratorius include selecting uniformly flowering varieties, planting one to two varieties per area, and implementing natural enemies (Li 1985).

Natural enemies of *L. oratorius* include egg parasitoids, such as *Gryon flavipes* Ashmead, which can destroy 40 to 90% of the eggs in Sarawak (Rothschild 1970a,b); and *G. nixoni* Masner and *Ooencyrtus erionotae* Ferriere, which may parasitize approximately 15 and 4% of the eggs, respectively, in Papua New Guinea (Li 1985). The rate of egg parasitism is influenced by egg age and plant species hosting the eggs (Morrill and Almazon 1990). Burdeos and Gabriel (1995) found the entomopathogenic fungus *Metarhizium anisopliae* (Metsch.) Sorok. effective (in the laboratory) against this insect. The orthopteran *Conocephalus longipennis* (DeHaan) (Tettigoniidae) may prey on this rice bug's eggs in Malaysia (Ito et al. 1995).

3.3 *Neomegalotomus parvus* (Westwood)

3.3.1 Distribution, Life History, and Biology

Neotropical members of the genus *Megalotomus* have proved not to be congeneric with *Megalotomus* species from the rest of the world. Therefore, the new genus *Neomegalotomus* Schaffner & Schaefer has been erected to include the Neotropical species formerly included in *Megalotomus* (Schaffner and Schaefer 1998, Schaefer and Panizzi 1998).

The Neotropical *N. parvus* (Westwood) is common in Brazil. Other species, such as *N. rufipes* (Westwood) and *N. pallescens* (Stål), are mentioned in the Brazilian literature (Silva et al. 1968, Kobayashi and Cosenza 1987), but it is almost certain that only one species is involved, for which *N. parvus* is the preferred name (J. C. Schaffner, personal communication).

In Brazil, *N. parvus* was first reported by Lima (1919) in Rio de Janeiro, on *Crotalaria* sp. Other hosts include common bean, soybean, cotton, tomato, lupin, pigeon pea, and lablab (Panizzi 1988, Chandler 1989, Santos and Panizzi 1998a).

Nymphs mimic ants and are darkish. Adults are light brown (males), with a whitish band along both sides of the thorax. Females are entirely darkish brown. Body length is ~10 mm (Lima 1919, Paradela F° et al. 1972).

Little has been published on the biology of this species. Panizzi (1988) and Santos and Panizzi (1998a) studied nymphal and adult biology of *N. parvus* on selected legumes. Nymphs showed 100% mortality on vegetative soybean, and on immature or mature pods of several indigo legume species; over 80% mortality on lupin or green bean seed; and ~80% mortality on soybean pods (mature or immature). Best performance (i.e., approximately 10 to 25% nymph mortality) was observed on mature seeds of pigeon pea, lablab, and soybean, and on pods (mature or immature) of pigeon pea. Female fecundity may vary from 12 on lupin pod, to 118 eggs/female on mature pigeon pea pods. Egg hatchability was also variable, from ~4% on lupin to ~81% on mature pigeon pea seeds. These laboratory results partially explain the great abundance of adults on maturing pigeon pea, lablab, and soybean plants often observed in the field (Ventura and Panizzi 1997). On soybean, female *N. parvus* prefer to oviposit on the upper third of the plant, mainly on the lower surface of leaves, and on the region close to the midrib (Panizzi et al. 1996).

3.3.2 Damage and Control

The damage by *N. parvus* to economically important crops has not been properly evaluated. Paradela F° et al. (1972) refer to *N. parvus* as a pest of common bean in São Paulo state, particularly because of its ability to transmit the yeast spot disease caused by *Nematospora coryli* Peglion. In Minas Gerais state, it can damage seeds and be responsible for seedling mortality of common bean (Chandler 1984, 1989). It is also extremely abundant in soybean (Panizzi 1988), particularly during the time of seed maturation. This may explain why no conspicuous damage to this crop has been observed in soybean fields, despite the bug's great abundance. Apparently there is an asynchrony between the vulnerable stages of seed development, i.e., pod-filling stage, and *N. parvus* populations, and an escape in time of the soybean plants occurs. Therefore, there is usually no need to control this insect on soybean. However, studies conducted in the greenhouse indicated that if soybean plants are artificially infested during the pod-filling stage, a reduction in seed vigor and viability will occur (Santos and Panizzi 1998b). On common bean they seem to inflict greater damage, and control might be necessary. *Neomegalotomus parvus* is easily killed by conventional insecticides used to control other hemipteran pests of soybean. Not much is known about natural enemies of *N. parvus*. Santos and Panizzi (1997) refer to three species of tachinids, *Hyalomyia* sp., *Hyalomyodes* sp., and *Trichopoda* sp., attacking adults in northern Paraná state, Brazil.

3.4   Riptortus clavatus (Thunberg)

3.4.1   Distribution, Life History, and Biology

The bean bug is perhaps the most studied species in *Riportus*. It is a major pest of grain legumes in Africa and Asia, particularly of soybean (Ezueh and Dina 1979, Talekar 1987). In Japan, where it is also an important pest of soybean (Kobayashi 1976, 1981) and kidney bean (*Phaseolus vulgaris* L.) (Ishihara 1950), many aspects of its biology have been investigated, such as adult diapause, mating behavior, sound production, and chemical communication (Numata 1985; Numata et al. 1986, 1989; Leal et al. 1995; Toda et al. 1995; Sakurai 1996a,b, 1998; Morita and Numata 1997). Kono (1988) studied the effects of crowding on laboratory populations of this species, and reports on the rearing of *R. clavatus* in the laboratory have appeared (Kamano 1980, Noda and Kamano 1983, Kikuchi and Kobayashi 1986). Adults are cylindrical, dark brown, with characteristic whitish or yellow lines on the venter (Kobayashi 1978, Singh and van Emden 1979).

Riptortus clavatus feeds on seeds of about 30 plant species of the families Leguminosae, Gramineae, Convolvulaceae, Rosaceae, and Pedaliaceae. It produces two to three generations a year, and adults of the third generation appear from early October to early November and overwinter in sunny tussocks of grass (Kobayashi 1981). Adults lay eggs beginning in mid-May (Kobayashi 1978, Natuhara 1985). The egg is described by Ishihara (1950), who also describes and keys the instars.

3.4.2   Damage and Control

The damage of *R. clavatus* to soybean is similar to that of other seed-sucking insects. Pods injured earlier during the pod-elongation and pod-filling stages become flattened, with sterile seeds, and usually wither and fall (Kobayashi 1978, 1981). It is suspected that *R. clavatus* persists more intensely in pod feeding than do other seed-suckers, such as the southern green stink bug, *Nezara viridula* (L.) (Suzuki et al. 1991), but this needs confirmation.

Control measures against *R. clavatus* on soybean, and other *Riptortus* species on other grain legumes, include use of insecticides such as endosulfan, phosphamidon, monocrotophos, and fenitrothion at about 500 g/ha, sprayed two times at 20-day intervals, starting 10 days after the beginning of flowering (Taylor 1978, Sawada 1988). Diflubenzuron, a benzoylphenyl urea insecticide that blocks the biosynthesis of chitin, reduces adult longevity and fecundity of females (Kim et al. 1992). These types of insecticides are highly desirable because they are species specific, nonpersistent, and effective against insects resistant to other groups of insecticides, such as the broad-spectrum organophosphorus insecticides. However, the so-called insect growth regulators, which include diflubenzuron and pyriproxyfen, may also harm predaceous hemipterans such as *Podisus maculiventris* (Say) (De Clercq et al. 1995), and, therefore, they should be used cautiously in integrated pest management programs. The identification of pheromones of *R. clavatus* (Leal and Kadosawa 1992) may open new perspectives for managing this pest species; one component of an aggregation pheromone attracts the egg parasitoid

Ooencyrtus nezarae (Hymenoptera: Encyrtidae) (Mizutani et al. 1997). This parasitoid is quite effective, parasitizing more than 70% of the eggs on soybean leaves (in Fukuoka, Japan), although only 50% of eggs on stems and pods (Takasu et al. 1998). Ishimoto and Kitamura (1993) reported inhibitory effects of a mungbean, *Vigna radiata* (L.), genotype on nymphal growth of *R. clavatus*, which also opens up the possibility of pest management through host plant resistance. Cysteine proteinase inhibitors from rice seeds, oryzacystatons I and II, retarded growth of *R. clavatus* at concentrations of 0.3 to 0.5% (weight/weight), and killed at 1% (Kuroda et al. 1996). Natural enemies of *R. clavatus* include the egg parasitoids *Gryon japonicum* (Ashmead), *Gryon* sp., *O. nezarae* Ishii, *Ooencyrtus* sp., and *Anastatus japonicus* Ashmead, on soybean and other legumes in Japan (Noda 1989).

3.5 Riptortus dentipes (F.) (= *Riptortus tenuicornis* [Dallas])

3.5.1 Distribution, Life History, and Biology

Riptortus dentipes is a major pest of cowpea in Africa, probably among the most destructive of all pests of developing pods of cowpea plants in western Nigeria and Benin (Aina 1975, Akingbohungbe 1977, Dreyer et al. 1994, Dreyer and Baumgärtner 1995). It is also regarded as a secondary pest of cowpea in Asia (Singh et al. 1990), and of cowpea and soybean in Africa (Jackai et al. 1990, Ntonifor et al. 1996). Abate and Ampofo (1996) in their review of pests of beans (*Phaseolus vulgaris*) refer to *R. dentipes* as a pest of this crop in Africa. It has also been reported to damage other leguminous plants and cocoa (Lodos 1969).

The biology of *R. dentipes*, with descriptions of its developmental stages, was studied by Aina (1975) and Akingbohungbe (1977) in Nigeria. Adults are brown, about 20.0 mm long, cylindrical, with a characteristic whitish or yellow line on both sides of the venter. Adult longevity is variable, unmated males living for ~28 days, and unmated females 16 days. Mated adults lived ~8 days. A fertilized adult female laid on average 40 eggs. Eggs are dark brown, flattened dorsally, usually deposited on the brown peduncle or at the base of pods. Eggs are laid in batches of 4 to 16, and hatching occurs after ~6 days. First instars are 3.5 mm long, and dark brown to black. Second instars are 4.9 mm long, similar to first instars in color. Later instars are 6.8, 8.5, and 13.6 mm long, respectively, generally dark brown to black.

In laboratory tests, *R. dentipes* preferred cowpea to soybean, even when nymphs had been raised on the latter; however, in the field, when podding of the two crops was synchronized, the bugs showed no preference. It seems possible, therefore, that this bug could become a more serious pest on soybean (Ntonifor et al. 1996).

3.5.2 Damage and Control

The damage caused by *R. dentipes*, and its control, are as described for *R. clavatus* (above). In Côte d'Ivoire, it is a more serious pest on soybean than *Nezara viridula* (Pentatomidae), causing 72% pod damage and significantly reducing yield and quality (Acle and Rolim 1994). Studies conducted in Africa (Alghali 1992, 1993; Ogunwolu 1992) add dimethoate as an efficient insecticide to control this pest, sprayed once at flowering and once at podding. Also, the intercropping of cowpea with sorghum reduces the damage by this bug on cowpea One difficulty in controlling this pest is reinvasion from adjacent areas. Natural enemies include egg parasitoids, observed in screen cages over cowpea plants in the field in Nigeria (Singh et al. 1990). Writing generally about cowpea pests in Nigeria, Egwuatu and Taylor (1977) comment that "the real answer ... for pest control on pigeon pea ... should be based on a fundamental understanding of the crop/pest complex relationship" (p. 275). There is concern that this bug will shift to soybean in Africa (Ntonifor et al. 1996); its congeners, *R. clavatus* and *R. linearis*, are already soybean pests in Africa and Asia (**Sections 3.4 and 3.6**).

3.6 Riptortus linearis (F.)

3.6.1 Distribution, Life History, and Biology

This is another member of the *Riptortus* spp. complex that is a pest of grain legumes; considered a major pest of soybean in Africa and Asia, it may feed on some Solanaceae and Convolvulaceae as well (Gangrade 1974, Ezueh and Dina 1979, Talekar 1987, Talekar et al. 1995).

Adults are ferrugineous to castaneous, and 12 to 14 mm long (Schaffner 1964). The histology of the internal genitalia, and some notes on the development of the male system, are given by Ramamurty (1969), and Singh and Sharma (1989) studied the structure of the foregut of fed and starved bugs. Talekar et al. (1995) studied nymph development, oviposition, and damage to soybean by *R. linearis*. Egg incubation and the nymphal stage took 12.7 and 35.5 days, and 4.1 and 11.5 days, respectively, at 20 and 35°C. Females lay eggs singly, mostly on leaves and only on soybean plants having pods. The seed size and germination rate decreased substantially as the time of infestation of adults on soybean plants increased. *Riptortus linearis* also caused extensive damage to soybean (30 to 40% of the plants injured) during August–early October in India (Kashyap and Adlakha 1971), where it is also a pest of pigeon pea and chickpea (Srivastava 1964, Davies and Lateef 1975). In the Philippines it is a soybean pest mostly during May planting (Litsinger et al. 1978).

3.6.2 Damage and Control

Damage done and control measures are similar to those reported for *R. clavatus* (**Section 3.4.2**).

4. THE LESS IMPORTANT SPECIES

4.1 Alydus pilosulus Herrich-Schaeffer

This broad-headed bug occurs throughout the United States (Blatchley 1926, Froeschner 1988). It is a slender bug, grayish to light brown, with white lateral margins on the pronotum, and with prominent humeri. The male is 9.5 to 10.9 mm long and the female 10.5 to 11.2 mm (Wilkinson and Daugherty 1967).

It has a preoviposition period of 6.6 days and fecundity of 46 eggs/female. Eggs are dark brown, smooth, shining, and flattened on one side. Nymphal developmental takes ~27 days and nymphal mortality is ~85%. Adult males live about 16.5 days, and females about 23.3 days. It completes two generations a year in southern Wisconsin (Wilkinson and Daugherty 1967, Yonke and Medler 1968).

Alydus pilosulus feeds on several legumes (Schaefer 1980). It was first recorded as a pest of lima bean and cowpea early this century (Chittenden 1902). Later, there were several reports of damage to legumes, soybean being a preferred host (Underhill 1943, Blickenstaff and Huggans 1962). Since then, it has been considered a potential pest of soybean, because of its feeding damage and because it may transmit yeast-spot disease (Wilkinson and Daugherty 1967), but *A. pilosulus* has thus far remained of secondary importance as an economic pest species.

4.2 Megalotomus quinquespinosus Say

This is a northern Nearctic species, occurring from Québec (Canada), New England, and North Carolina (U.S.A.) west across North America (Blatchley 1926). Like other alydines, it feeds mostly on legumes (Yonke and Medler 1965, Tugwell et al. 1973); feeding on carrion has also been reported (Bromley 1937).

The biology of *M. quinquespinosus* was studied by Yonke and Medler (1965) in the field, and by rearing it on mature seeds of red clover and soybean in the laboratory. Eggs were laid singly, immediate incubation period is ~13 days, and delayed incubation period may last ~81 days. Nymphs required about 5 days in each of the first three stadia. The fourth stadium took 6.5 days, and the 5th took 9.8 days. The abdomens of nymphs greatly expanded after feeding. Adults are 14 to 16 mm long, and light brown. In the field, *M. quinquespinosus* completes two generations a year in southern Wisconsin, these overlapping from the end of June through August. Yonke and Medler (1965) reported it as of no economic importance. We have found no reports of it as a pest species. However, it will feed on soybean, and if it should extend its range to the soybean production areas of the United States, it may become a pest of this crop.

4.3   *Riptortus pedestris* (F.)

This alydine is generally a secondary pest of cowpea, mung bean, soybean, common bean, and pigeon pea in India (Srivastava 1964, Bhardwaj and Bhalla 1976, Visalakshi et al. 1976, Garg 1992).

The adult is dark brown with two black bands ventrally on the abdomen, and about 15 mm long. Adult longevity is ~45 days, and mating takes place at night. Females lay ~115 eggs during an oviposition period of about 30 days; eggs are laid singly, usually on pods. The egg is brownish-black and hemispherical. Nymphs are elongated, early instars dark brown, and late instars brownish-black (Visalakshi et al. 1976).

Nymphs and adults suck sap from leaves of mung bean; whitish spots appear, which later turn yellowish, and the leaf dies (Garg 1992). Like the other species of *Riptortus*, however, *R. pedestris* feeds mostly on seeds, causing variable damage to these legume crops (see *R. clavatus* for damage description). Usually no insecticide is needed to control this insect, because of its secondary importance; but the heavy damage on cowpea, as illustrated by Visalakshi et al. (1976), may justify control measures.

4.4   *Riptortus serripes* (F.)

Riptortus serripes (F.) is a secondary pest of soybean in Australia (Turner 1978). Adults are dark, with a pale yellow band along the side of the body, 14 to 18 mm long (Shepard et al. 1983). Brier and Rogers (1991) infested caged soybean plants with the southern green stink bug, *N. viridula*, and *R. serripes*. They found that *R. serripes* was less detrimental to soybean than *N. viridula*, causing 64 to 72% of the seed yield loss caused by the latter. Usually, there is no need to control *R. serripes*, but insecticides applied against *N. viridula*, such as endosulfan, will also eliminate *R. serripes* (Turner 1978).

4.5   *Leptocorisa* spp. and *Stenocoris* spp.

Species of *Leptocorisa* other than *L. acuta* and *L. oratorius*, considered above, and *L. solomonensis* Ahmad, pests of rice, do not feed on rice, even when it is available: In Papua New Guinea, *L. palawanensis* Ahmad and *L. discoidalis* Walker feed on native grasses; the latter indeed cannot develop on rice, nor will its females lay eggs when provided only rice (Sands 1977). For other species, that do infest rice, native grasses provide an alternate food source (Corbett 1930, Sands 1977), one less attractive than rice to *L. oratorius* and *L. acuta* but equally attractive to *L. solomonensis* (Sands 1977). Therefore, the presence of native grasses in the vicinity of rice fields is important to the nutritional ecology of these bugs, as they move from grasses to rice, and vice versa.

Stenocoris tipuloides (DeGeer) was listed as "observed by author" in the southern U.S. rice-growing areas by Bowling (1967). It and *S. filiformis* (F.) were earlier recorded as "of no economic importance" by Hussey (1951), who wrote also that "both species ... occur extensively in Florida."

Leptocorisa sp. was recently found to infest soybean and rubber plant, *Hevea brasiliensis*, in Brazil, but this seems to be very occasional (A. M. Faria, personal communication to A. R. Panizzi, 1998).

4.6 Other Minor Pest Species

There are several species of alydids considered minor pests of crops. Among them, *R. attricornis* Stål is reported to damage soybean in the Philippines (Rodrigo 1947), and *Hyalymenus tarsatus* (F.) to damage macadamia (Garcia M. and Carballo V. 1995) and tempate, *Jatropha curcas* L., Euphorbiaceae, in Central America (Grimm and Maes 1997).

5. SPECIES OF MINOR BENEFIT

5.1 *Zulubius acaciaphagus* Schaffner

Zulubius comprises two species, *Z. maculatus* (Thunberg) and *Z. acaciaphagus* Schaffner, and is restricted to sub-Saharan Africa (Schaffner 1987).

Zulubius acaciaphagus is somewhat linear and flattened, about 12.5 mm long, general coloration is reddish brown to brown, occasionally fuscous, and is found in South Africa, Swaziland, and Tanzania (Schaffner 1987). It feeds on seeds of *Acacia cyclops* A. Cunn. ex G. Don, a small evergreen tree introduced into South Africa from Australia for dune stabilization and shelterbelts (Shaughnessy 1980). This plant has become an invasive weed with a very aggressive expansion (MacDonald 1985). *Zulubius acaciaphagus* destroyed up to 84% of the *A. cyclops* seed crop (Holmes and Rebelo 1988), and may be an important agent controlling this weed invader (Holmes et al. 1987). In Australia, *A. cyclops* is attacked by many seed-feeding insects, including nine hemipterans (van den Berg 1980), and this may explain the low seed banks found there. The bugs make minute feeding holes in the testae, causing seeds to imbibe moisture and either germinate or rot (Neser 1984). Therefore, the economic importance of the bug relies on its potential as a biological control agent of *A. cyclops* in South Africa.

5.2 *Hyalymenus tarsatus* (F.)

This species has been found feeding on the fruits of *Sapium sebiferum* (L.) Roxb., a major invader from Asia of tall grass prairie and hardwood forest in Louisiana and Texas (Johnson and Allain 1998). Although not so abundant on this plant as the coreid *Leptoglossus zonatus* (L.) (Johnson and Allain 1998), and although *S. sebiferum* is certainly not the most important host of *H. tarsatus* (C. W. Schaefer, personal comment), this bug may help control this plant.

6. PROGNOSIS

So far, most alydid species are secondary pests. However, species of *Leptocorisa*, *Megalotomus*, *Neomegalotomus*, and *Riptortus*, which have been reported as causing substantial damage to different grain crops (leguminous and nonleguminous plants), may become of greater concern as these crops expand in tropical areas. For example, *N. parvus* was not even cited in the Brazilian literature as a pest of soybean 15 to 20 years ago. Today, with the expansion of this crop toward the central-western and northeastern areas of the country, this bug is becoming more and more abundant. This fact, which is leading to an increase in numbers of *N. parvus*, may result in greater damage to soybean and other crops, such as common bean, as well. Therefore, monitoring of populations of this and other species, on different agroecosystems, is strongly recommended to predict possible outbreaks of these bugs.

7. CONCLUSIONS

Although some aspects of the biologies of some Alydidae are known, and some species, such as *R. clavatus*, have been extensively studied, some basic features of the biology of alydids nevertheless remain unclear. For example, their feeding habits (do alydids feed regularly on carrion or fecal matter? do first instars take food?), their dispersal behavior (high mobility with common reinvasion of field crops after insecticide treatment), and their chemical ecology (alarm and sex pheromones) are, among others, basic research issues that remain poorly understood and await further investigation.

In the area of management of alydid pest species, a similar situation is observed. Research is needed to evaluate the extent of damage these insects may be causing to economically important plants throughout the world. Except for the paper by Brier and Rogers (1991), measuring damage by *R. serripes* to soybean, most reports refer to injury made to grains or other plant structures in a general fashion. Little information was found addressing such issues as the amount of damage vs. insect numbers, or interactions of the damage with different phenological stages of plant development. Moreover, the importance of alydids in transmitting plant diseases has been little studied.

In conclusion, future research on both basic and applied aspects of the biology of alydids should be conducted. This will yield valuable information on the bugs' biology and ecology, and on the extent of the bugs' damage and need for management.

8. ACKNOWLEDGMENTS

Antônio R. Panizzi thanks the Conselho Nacional de Desenvolvimento Científico e Tecnológico (CNPq) and the Centro Nacional de Pesquisa de Soja of Embrapa, of the federal government of Brazil, for providing financial support for a stay at the University of Connecticut to work on this chapter with Carl W. Schaefer.

9. REFERENCES CITED

Abate, T., and J. K. O. Ampofo. 1996. Insect pests of beans in Africa: their ecology and management. Annu. Rev. Entomol. 41: 45–73.

Abraham, C. C., and K. S. R. Mony. 1977. Occurrence of *Leptocorisa acuta* Fabr. (Coreidae, Hemiptera) as a pest of nutmeg trees. J. Bombay Nat. Hist. Soc. 74: 553.

Acle, D., and R. B. Rolim. 1994. Infestation of soybean with two species of pod-sucking bugs. Insect Sci. Appl. 15: 337–341.

Adler, P. H., and A. G. Wheeler, Jr. 1984. Extra-phytophagous food sources of Hemiptera-Heteroptera: bird droppings, dung, and carrion. J. Kansas Entomol. Soc. 57: 21–27.

Ahmad, I. 1965. The Leptocorisinae (Heteroptera: Alydidae) of the world. Bull. British Mus. Nat. Hist., Entomol. Suppl. 5: 1–156.

Aina, J. O. 1975. The life history of *Riptortus dentipes* F. (Alydidae, Heteroptera). A pest of growing cowpea pods. J. Nat. Hist. 9: 589–596.

Akingbohungbe, A. E. 1977. Notes on the identify and biology of the *Riptortus* spp. (Heteroptera: Alydidae) associated with cowpeas in Nigeria. J. Nat. Hist. 11: 477–483.

Alghali, A. M. 1992. Insecticide application schedules to reduce grain yield losses caused by insects of cowpea in Nigeria. Insect Sci. Appl. 13: 725–730.

1993. Intercropping as a component in insect pest management for grain cowpea, *Vigna unguiculata* Walp production in Nigeria. Insect Sci. Appl. 14: 49–54.

Bhardwaj, S. P., and O. P. Bhalla. 1976. Record of insect pests of soybean in Himalachal Pradesh. Indian J. Entomol. 38: 286–289.

Blatchley, W. S. 1926. Heteroptera or True Bugs of Eastern North America, with Especial Reference to the Faunas of Indiana and Florida. Nature Publ. Co., Indianapolis, Indiana, U.S.A. 1116 pp.

Blickenstaff, C. C., and J. L. Huggans. 1962. Soybean insects and related arthropods in Missouri. Missouri Agric. Exp. Stn. Bull. 803: 1–51.

Bowling, C. C. 1967. Insect pests of rice in the United States. Pp. 551–580 *in* [no editor], The Major Insect Pests of the Rice Plant. Int. Rice Research Inst., Johns Hopkins Press, Baltimore, Maryland, U.S.A. 729 pp.

Brier, H. B., and D. J. Rogers. 1991. Susceptibility of soybeans to damage by *Nezara viridula* (L.) (Hemiptera: Pentatomidae) and *Riptortus serripes* (F.) (Hemiptera: Alydidae) during three stages of pod development. J. Austral. Entomol. Soc. 30: 123–128.

Bromley, S. W. 1937. Food habits of alydine bugs (Hemiptera: Coreidae). Bull. Brooklyn Entomol. Soc. 32: 159.

Burdeos, A. T., and B. P. Gabriel. 1995. Virulence of different *Metarhizum anisopliae* (Metsch.) Sorokin isolates against the rice bug, *Leptocorisa oratorius* Fabr. (Hemiptera: Alydidae). Philippine Entomol. 9: 467–478.

Chandler, L. 1984. Crop life table studies of the pests of beans (*Phaseolus vulgaris* L.) at Goiânia, Goiás. Rev. Ceres 31: 284–298.

1989. The broad-headed bug, *Megalotomus parvus* (Westwood), (Hemiptera: Alydidae), a dry season pest of beans in Brazil. Annu. Rep. Bean Improv. Coop. 32: 84–85.

Chittenden, F. H. 1902. Some insects injurious to vegetable crops. U.S. Dept. Agric., Div. Entomol. Bull. 33 (n.s.): 106 [not seen, *fide* Wilkinson and Daugherty 1967].

Cobblah, M. A., and J. D. Hollander. 1992. Specific differences in immature stages, oviposition sites and hatching patterns in two rice pests, *Leptocorisa oratorius* (Fabricius) and *L. acuta* (Thunberg) (Heteroptera: Alydidae). Insect Sci. Appl. 13: 1–6.

Corbett, G. H. 1930. The bionomics and control of *Leptocorisa acuta* Thunb. with notes on other *Leptocorisa* spp. in Malaya. Sci. Ser. Dept. Agric. Straits Settl. Fed. Malay. 4: 1–40.

Dalvi, C. S., R. B. Dumbre, and V. G. Khanvilkar. 1985. Off season biology of rice earhead bug. J. Maharashtra Agric. Univ. 10: 234–235.

Davies, J. C., and S. S. Lateef. 1975. Insect pests of pigeon pea and chickpea in India and prospects for control. Pp. 319–331 *in* Proc. Int. Workshop on Grain Legumes, ICRISAT, Hyderabad.

De Clercq, P., A. De Cock, L. Tirry, E. Vinuela, and D. Degheele. 1995. Toxicity of diflubenzuron and pyriproxyfen to the predatory bug *Podisus maculiventris*. Entomol. Exp. Appl. 74: 17–22.

Dhiman, S. C. 1981. Tentorium in *Leptocorisa varicornis* Fabr. (Heteroptera — Coreidae). Folia Morphol. 29: 336–338.

Dhiman, S. C., and V. C. Chatterjee. 1980. Phototropism in *Leptocorisa varicornis* Fabr. Geobios 7: 329–333.

Dresner, E. 1958. A poison bait for the control of *Leptocorixa* [sic] *acuta* (Thunb.). J. Econ. Entomol. 51: 405.

Dreyer, H., and J. Baumgärtner. 1995. The influence of post-flowering pests on cowpea seed yield with particular reference to damage by Heteroptera in southern Benin. Agric. Ecosyst. Environ. 53: 137–149.

Dreyer, H., J. Baumgärtner, and M. Tamò. 1994. Seed-damaging field pests of cowpea (*Vigna unguiculata* L. Walp.) in Benin: ocurrence and pest status. Int. J. Pest Manage. 40: 252–260.

Egwuatu, R. I., and T. A. Taylor. 1997. Insect species associated with *Cajanus cajan* (L.) Druce (pigeon pea) in Nigeria and their monthly abundance. East Afric. Agric. For. J. 42: 271–275.

Ezueh, M. I., and S. O. Dina. 1979. Pest problems of soybeans and control in Nigeria. Pp. 275–283 *in* F. T. Corbin [ed.], Proceedings World Soybean Research Conference II, Westview Press, Boulder, Colorado, U.S.A.

Froeschner, R. C. 1988. Family Alydidae Amyot and Serville, 1843. Pp. 4–11 *in* T. J. Henry and R. C. Froeschner [eds.], Catalog of the Heteroptera, or True Bugs, of Canada and the Continental United States. E. J. Brill, London, U.K. 658 pp.

Gangrade, G. A. 1974. Insects of soybean. Jawaharlal Nehru Krishi Vishwa Vidyalaya, Jabalpur, India, Tech. Bull. 24, 88 pp.

Garcia M., G., and M. Carballo V. 1995. Evaluacion de aislados de *Beauvaria bassiana* y *Metarhizium anisopliae* en el control de *Hyalymenus tarsatus* (Hemiptera: Alydidae) en macadamia. Manejo Int. Plagas, Costa Rica 36: 7–11.

Garg, D. K. 1992. Occurrence of *Riptortus pedestris* (Fabricius) (Coreidae: Heteroptera) as a major pest of blackgram in Kumaon Hills of Uttar Pradesh. Indian J. Entomol. 54: 367–368.

Grimm, C., and F. Guharay. 1998. Control of leaf-footed bug *Leptoglossus zonatus* and shield-backed bug *Pachycoris klugii* with entomopathogenic fungi. Biocontrol Sci. Technol. 8: 365–376.

Grimm, C., and J. M. Maes. 1997. Insectos asociados al cultivo de tempate (*Jatropha curcas*) enel Pacifico de Nicaragua. III. Coreoidea (Heteroptera). Rev. Nicaraguense Entomol. 42: 15–34.

Haque, G. 1980. Some observations on the copulation, ovulation and ovipositional behaviour of *Leptocorisa varicornis* Fabr. (Coreidae: Hemiptera) [sic]. Allahabad Farmer 51: 93–95.

Holmes, P. M., and A. G. Rebelo. 1988. The occurrence of seed-feeding *Zulubius acaciaphagus* (Hemiptera, Alydidae) and its effects on *Acacia cyclops* seed germination and seed banks in South Africa. South Afric. J. Bot. 54: 319–324.

Holmes, P. M., G. B. Dennill, and E. J. Moll. 1987. Effects of feeding by native alydid insects on the seed viability of an alien invasive weed, *Acacia cyclops*. South Afric. J. Sci. 83: 580–581.

Hussey, R. F. 1951. Leptocorixa filiformis in the United States (Hemiptera: Coreidae). Florida Entomol. 33: 150–154.

Ishihara, T. 1950. The developmental stages of some bugs injurious to the kidney bean (Hemiptera). Trans. Shikoku Entomol. Soc. 1: 17–31.

Ishimoto, M., and K. Kitamura. 1993. Inhibitory effects of adzuki bean weevil-resistant mungbean seeds on growth of the bean bug. Jpn. J. Breed. 43: 75–80.

Ito, K., H. N. Kin, and C. P. Min. 1995. *Conocephalsu longipennis* (DeHaan) (Orthoptera, Tettigoniidae) — a suspected egg-predator of the rice bug in the Muda area, West Malaysia. Appl. Entomol. Zool. 30: 599–601.

Ito, K., H. Sugiyama, and N. S. Nik Mohd. Noor. 1992. Damages in rice plants caused by ear-sucking bugs in the Muda Area, West Malaysia. JARQ 26: 67–74.

Jackai, L. E. N., A. R. Panizzi, G. G. Kundu, and K. P. Srivastava. 1990. Insect pests of soybean in the tropics. Pp. 91–156 *in* S. R. Singh [ed.], Insect Pests of Tropical Food Legumes. John Wiley & Sons, Chichester, U.K. 451 pp.

Johnson, S. R., and L. K. Allain. 1998. Observations on insect use of Chinese tallow [*Sapium sebiferum* (L.) Roxb.] in Louisiana and Texas. Castanea 63: 188–189.

Kamano, S. 1980. Artificial diet for rearing bean bug, *Riptortus clavatus* Thunberg. Jpn. J. Appl. Entomol. Zool. 24: 184–188.

Kashyap, N. P., and R. L. Adlakha. 1971. New records of insect pests of soybean crop. Indian J. Entomol. 33: 467–468.

Kikuchi, A., and T. Kobayashi. 1986. A simple rearing method of *Piezodorus hybneri* Gmelin and *Riptortus clavatus* Thunberg (Hemiptera: Pentatomidae, Alydidae) supplying dried seeds. Bull. Nat. Agric. Res. Cent., Tsukuba, No. 6: 1–42.

Kim, G. A., Y. J. Ahn, and K. Y. Cho. 1992. Effects of diflubenzuron on longevity and reproduction of *Riptortus clavatus* (Hemiptera: Alydidae). J. Econ. Entomol. 85: 664–668.

Kobayashi, T. 1976. Insect pests of soybean in Japan and their control. PANS 22: 336–349.

1978. Pests of grain legumes including soybeans and their control in Japan. Pp. 59–65 *in* S. R. Singh, H. F. van Emden, and T. A. Taylor [eds.], Pests of Grain Legumes: Ecology and Control. Academic Press, London, U.K. 454 pp.

1981. Insect pests of soybeans in Japan. Misc. Publ. Tohoku Natl. Agric. Exp. Stn. 2: 1–39.

Kobayashi, T., and G. W. Cosenza. 1987. Studies on the integrated control of soybean stink bugs in the Cerrados. Jpn. J. Agric. Res. Q. 20: 229–236.

Kono, S. 1988. Effect of population density on bionomics and demographic parameters of bean bug, *Riptortus clavatus* Thunberg (Heteroptera: Alydidae). Jpn. J. Appl. Entomol. Zool. 32: 277–282 [in Japanese, English summary].

Krishnamurthy, P., S. Venkatachalam, and S. Syamprakasam. 1977. A note on the chemical control of earhead bug *Leptocorisa acuta* (Thumb [sic]) Coreidae: Hemiptera on paddy. Pesticides (India) 11: 38–39.

Kuroda, M., M. Ishimoto, K. Suzuki, H. Kondo, K. Abe, K. Kitamura, and S. Arai. 1996. Oryzacystatins exhibit growth-inhibitory and lethal effects on different species of bean insect pests, *Callosobruchus chinensis* (Coleoptera) and *Riptortus clavatus* (Hemiptera). Biosci. Biotechnol. Biochem. 60: 209–212.

Kushwaha, K. S. 1988. Insect pest complex of rice in Haryana. Indian J. Entomol. 50: 127–130.

Lakshmanan, P., S. M. Kumar, and R. Velusamy. 1992. Role of earheadbug (*Leptocorisa acuta*) feeding on sheath rot disease caused by *Sarocladium oryzae* in *Oryza sativa* in India. Phytoparasitica 20: 107–112.

Leal, W. S., and T. Kadosawa. 1992. (E)-2-Hexenyl hexanoate, the alarm pheromone of the bean bug *Riptortus clavatus* (Heteroptera: Alydidae). Biosci. Biotechnol. Biochem. 56: 1004–1005.

Leal, W. S., H. Higuchi, N. Mizutani, H. Nakamori, T. Kadosawa, and M. Ono. 1995. Multifunctional communication in *Riptortus clavatus* (Heteroptera: Alydidae): conspecific nymphs and egg parasitoid *Ooencyrtus nezarae* use the same adult attractant pheromone as chemical cue. J. Chem. Ecol. 21: 973–985.

Li, C. S. 1985. Biological and ecological studies of the rice bug, *Leptocorisa oratorius* (F.) (Hemiptera: Alydidae) and its control in Papua New Guinea. Mushi 50: 1–12.

Li, X., and L. Zheng. 1993. Preliminary study of the phylogeny of Alydidae (Hemiptera: Coreoidea). Acta Zootax. Sin. 18: 330–343.

Lim, G. S. 1971. The rice bug *Leptocorisa* and its control in West Malaysia. Malaysia Min. Agric. Lands Tech. Leafl. No. 6.

Lima, A. C. 1919. Nota sobre o mimetismo da nympha do *Alydus* (*Megalotomus*) *pallescens* com formiga e considerações relativas a espécie *Galeottus formicarius* (Hemiptera-Coreidae). Arch. Esc. Sup. Agric. Niterói 4: 5–8.

Linnavuori, R. E. 1987. Alydidae, Stenocephalidae and Rhopalidae of West and Central Africa. Acta Entomol. Fenn. 49: 1–36.

Litsinger, J. A., C. B. Quirino, M. D. Lumaban, and J. P Bandong. 1978. The grain legume pest complex of rice-based cropping systems at three locations in the Philippines. Pp. 309–320 *in* S. R. Singh, H. F. van Emden, and T. A. Taylor [eds.], Pests of Grain Legumes: Ecology and Control. Academic Press, London, U.K. 454 pp.

Lodos, N. 1969. Minor pests and other insects associated with *Theobroma cacao* L. in Ghana. Ghana J. Agric. Sci. 2: 61–72.

MacDonald, I. A. W., M. L. Jarman, and P. M. Beeston. 1985. Management of invasive alien plants in the fynbos biome. South Africa Nat. Sci. Prog. Rep. 111, Pretoria.

Matthysse, J. G. 1957. Research on insecticidal control of Philippine crop pests. J. Econ. Entomol. 50: 517–518.

Mizutani, N., T. Wada, H. Higuchi, M. Ono, and W. S. Leal. 1997. A component of a synthetic aggregation pheromone of *Riptortus clavatus* (Thunberg) (Heteroptera: Alydidae), that attracts an egg parasitoid, *Ooencyrtus nezarae* Ishii (Hymenoptera: Encyrtidae). Appl. Entomol. Zool. 32: 504–507.

Morita, A., and H. Numata. 1997. Distribution of photoperiodic receptors in the compound eyes of the bean bug, *Riptortus clavatus*. J. Comp. Physiol. A 180: 181–185.

Morrill, W. L., and L. P. Almazon. 1990. Effect of host plant species and age of rice bug (Hemiptera: Alydidae) eggs on parasitism by *Gryon nixoni* (Hymenoptera: Scelionidae). J. Entomol. Sci. 25: 450–452.

Morrill, W. L., and G. S. Arida. 1991. Oviposition preference and survival of selected rice insects on wheat. J. Econ. Entomol. 84: 656–658.

Morrill, W. L., N. Pen-Elec, and L. P. Almazon. 1990. Effects of weeds on fecundity and survival of *Leptocorisa oratorius* (Hemiptera: Alydidae). Environ. Entomol. 19: 1469–1472.

Natuhara, Y. 1985. Migration and oviposition in the bean bug, *Riptortus clavatus* Thunberg (Heteroptera). Plant Prot. 39: 153–156.

Neser, S. 1984. Natural enemies of *Acacia* species in Australia. Pp. 124–125 *in* B. Dell [ed.], Proceedings IV International Conference on Mediterranean Ecosystems, University of Western Australia, Nedlands.

Noda, T. 1989. Seasonal occurrence of egg parasitoids of *Riptortus clavatus* (Thunberg) (Heteroptera: Alydidae) on several leguminous plants. Jpn. J. Appl. Entomol. Zool. 33: 257–259.

Noda, T., and S. Kamano. 1983. Effects of vitamins and amino acids on the nymphal development of the bean bug, *Riptortus clavatus* Thunberg. Jpn. J. Appl. Entomol. Zool. 27: 295–299.

Ntonifor, N. N., L.E. N. Jackai, and F. K. Ewete. 1996. Influence of host plant abundance and insect diet on the host selection behavior of *Maruca testulalis* Geyer (Lepidoptera: Pyralidae) and *Riptortus dentipes* Fab. (Hemiptera: Alydidae). Agric. Ecosyst. Environ. 60: 71–78.

Numata, H. 1985. Photoperiodic control of adult diapause in the bean bug, *Riptortus clavatus*. Mem. Fac. Sci., Kyoto Univ. 10: 29–48.

Numata, H., N. Matsui, and T. Hidaka. 1986. Mating behavior of the bean bug, *Riptortus clavatus* Thunberg (Heteroptera: Coreidae): behavioral sequence and the role of olfaction. Appl. Entomol. Zool. 21: 119–125.

Numata, H., M. Kon, H. Fujii, and T. Hidaka. 1989. Sound production in the bean bug, *Riptortus clavatus* Thunberg (Heteroptera: Alydidae). Appl. Entomol. Zool. 24: 169–173.

Ogunwolu, E. O. 1992. Field infestation and damage to soybean and cowpea by pod sucking bugs in Benue State, Nigeria. Insect Sci. Appl. 13: 801–805.

Oliveira, P. S. 1985. On the mimetic association between nymphs of *Hyalymenus* spp. (Hemiptera: Alydidae) and ants. Zool. J. Linn. Soc. 83: 371–384.

Panizzi, A. R. 1988. Biology of *Megalotomus parvus* (Hemiptera: Alydidae) on selected leguminous food plants. Insect Sci. Appl. 9: 279–285.

Panizzi, A. R., E. Hirose, and E. D. M. Oliveira. 1996. Egg allocation by *Megalotomus parvus* (Westwood) (Heteroptera: Alydidae) on soybean. An. Soc. Entomol. Brasil 25: 537–543.

Paradela Fº, O., C. J. Rossetto, and A. S. Pompeu. 1972. *Megalotomus parvus* Westwood (Hemiptera: Alydidae), vector de *Nematospora coryli* Peglion em feijoeiro. Bragantia 31: 5–10.

Pathak, M. D. 1968. Ecology of common insect pests of rice. Annu. Rev. Entomol. 13: 257–294.

Rai, P. S. 1981. Life cycle of rice earhead *Leptocorisa acuta* (Thunberg) Coreidae: Hemiptera. J. Maharastra Agric. Univ. 6: 252–253.

Ramamurty, P. S. 1969. Histological studies of the internal genitalia of *Riptortus linearis* Fabr. with a preliminary account of the development of the male efferent system (family Coreidae-Heteroptera, Insecta). Indian J. Zoot. 10: 13–26.

Rodrigo, P. A. 1947. Soybean culture in the Philippines. Philippine J. Agric. 13: 1–22.

Rothschild, G. H. L. 1970a. Observation on the ecology of the rice-ear bug *Leptocorisa oratorius* (F.) (Hemiptera: Alydidae) in Sarawak (Malaysian Borneo). J. Appl. Ecol. 7: 147–167.

1970b. *Gryon flavipes* (Ashmead) [Hymenoptera, Scelionidae], an egg-parasite of the rice earbug *Leptocorisa oratorius* (Fabricius) [Hem. Alydidae]. Entomophaga 15: 15–20.

Sakurai, T. 1996a. Multiple mating and its effect on female reproductive output in the bean bug *Riptortus clavatus* (Heteroptera: Alydidae). Ann. Entomol. Soc. Amer. 89: 481–485.

1996b. Effects of male cohabitation on female reproduction in the bean bug, *Riptortus clavatus* Thunberg (Heteroptera: Alydidae). Appl. Entomol. Zool. 31: 313–316.

1998. Receptivity of female remating and sperm number in the sperm storage organ in the bean bug, *Riptortus clavatus* (Heteroptera: Alydidae). Res. Popul. Ecol. 40: 167–172.

Sands, D. P. A. 1977. The biology and ecology of *Leptocorisa* (Hemiptera: Alydidae) in Papua New Guinea. Dept. Primary Ind., Port Moresby, Res. Bull. 18: 1–104.

Sands, V. E. 1982a. Cytological studies of the Coreidae and Alydidae (Hemiptera: Heteroptera). I. Male meiosis in Malaysian species. Caryologia 35: 291–305.

1982b. Cytological studies of the Coreidae and Alydidae (Hemiptera: Heteroptera). II. Karyological changes exemplified by Malaysian genera. Caryologia 35: 335–345.

Santos, C. H., and A. R. Panizzi. 1997. Tachinid parasites of adult *Megalotomus parvus* West. (Hemiptera: Alydidae). An. Soc. Entomol. Brasil 26: 577–578.

1998a. Nymphal and adult performance of *Neomegalotomus parvus* (Hemiptera: Alydidae) on wild and cultivated legumes. Ann. Entomol. Soc. Amer. 91: 445–451.

1998b. Danos qualitativos causados por *Neomegalotomus parvus* (Westwood) em sementes de soja. An. Soc. Entomol. Brasil 27: 387–393.

Sawada, M. 1988. Insect pests of late cultivated soyabean and their control. Bull. Chiba Ken Agric. Exp. Stn. 29: 159–172.

Schaefer, C. W. 1965. The morphology and higher classification of the Coreoidea (Hemiptera: Heteroptera). Part III. The families Rhopalidae, Alydidae, and Coreidae. Misc. Publ., Entomol. Soc. Amer. 5: 1–76.

1972. Clades and grades in the Alydidae. J. Kansas Entomol. Soc. 45: 135–141.

1980. The host plants of the Alydinae, with a note on heterotypic feeding aggregations (Hemiptera: Coreoidea: Alydidae). J. Kansas Entomol. Soc. 53: 115–122.

1999. The higher classification of the Alydidae (Hemiptera: Heteroptera). Proc. Entomol. Soc. Washington 101: 94–98.

Schaefer, C. W., and P. L. Mitchell. 1983. Food plants of the Coreoidea (Hemiptera: Heteroptera). Ann. Entomol. Soc. Amer. 76: 591–615.

Schaefer, C. W., and A. R. Panizzi. 1998. The correct name of *"Megalotomus"* pests of soybean (Hemiptera: Alydidae). An. Soc. Entomol. Brasil 27: 669–670.

Schaffner, J. C. 1964. A taxonomic revision of certain genera of the tribe Alydini (Heteroptera: Coreidae). Ph.D. dissertation, Iowa State University, Ames, Iowa, U.S.A.

1987. The genus *Zulubius* Bergroth (Heteroptera: Alydidae). J. Entomol. Soc. South. Africa 50: 313–322.

Schaffner, J. C., and C. W. Schaefer. 1998. *Neomegalotomus* new genus (Hemiptera: Alydidae: Alydinae). Ann. Entomol. Soc. Amer. 91: 395–396.

Shanower, T. G., J. Romeis, and E. M. Minja. 1999. Insect pests of pigeonpea and their management. Annu. Rev. Entomol. 44: 77–96.

Shaughnessy, G. L. 1980. Historical ecology of alien woody plants in the vicinity of Cape Town, South Africa. Ph.D. dissertation, University of Cape Town, Cape Town, Rep. South Africa [not seen, *fide* Holmes and Rebelo 1988].

Shepard, M., R. J. Lawn, and M. A. Schneider. 1983. Insects on grain legumes in northern Australia. A survey of potential pests and their enemies. University of Queensland Press, St. Lucia, Australia. 81 pp.

Silva, A. G. d'A., C. R. Gonçalves, D. M. Galvão, A. J. L. Gonçalves, J. Gomes, M. N. Silva, and L. Simoni. 1968. Quarto catálogo dos insetos que vivem nas plantas do Brasil — seus parasitas e predadores. Parte II, Vol 1, Min. Agric., Rio de Janeiro, Brazil. 622 pp.

Singh, P. P., and M. C. Sharma. 1989. Anatomy and histology of the alimentary canal of *Riptortus linearis* Fabr. (Heteroptera: Coreidae) under fed and starved condition. Pt. I Foregut. Bull. Entomol. 30: 209–215.

Singh, S. R., and H. F. van Emden. 1979. Insect pests of grain legumes. Annu. Rev. Entomol. 24: 255–278.

Singh, S. R., L. E. N. Jackai, J. H. R. dos Santos, and C. B. Adalla. 1990. Insect pests of cowpea. Pp. 43–89 *in* S. R. Singh [ed.], Insect Pests of Tropical Food Legumes. John Wiley & Sons, Chichester, U.K. 451 pp.

Siwi, S. S., and P. H. Van Doesburg. 1984. *Leptocorisa* Latreille in Indonesia (Heteroptera, Coreidae, Alydinae). Zool. Med. 58: 117–129.

Srivastava, A. S., and H. P. Saxena. 1967. Rice bug *Leptocorisa varicornis* Fabricius and allied species. Pp. 525–546 *in* [no editor], Symposium on the Major Insect Pests of the Rice Plant. Johns Hopkins Press, Baltimore, Maryland, U.S.A. 729 pp.

Srivastava, B. K. 1964. Pests of pulse crops. Pp. 83–91 *in* N. C. Pant [ed.], Entomology in India. Entomol. Soc. India, New Delhi, India.

Suzuki, N., N. Hokyo, and K. Kiritani. 1991. Analysis of injury timing and compensatory reaction of soybean to feeding of the southern green stink bug and the bean bug. Appl. Entomol. Zool. 26: 279–287.

Takasu, K., Y. Hirose, and M. Takago. 1998. Occasional interspecific competition and within-plant microhabitat preference in egg parasitoids of the bean bug, *Riptortus clavatus* (Hemiptera: Alydidae) in soybean. Appl. Entomol. Zool. 33: 391–399.

Talekar, N. S. 1987. Insects damaging soybean in Asia. Pp. 25–45 *in* S. R. Singh, K. O. Rachie, and K. E. Dashiell [eds.], Soybeans for the Tropics: Research, Production and Utilization. John Wiley & Sons, Chichester, U.K. 230 pp.

Talekar, N. S., L.-Y. Huang, H.-H. Chou, and J.-J. Ku. 1995. Oviposition, feeding and developmental characteristics of *Riptortus linearis* (Hemiptera: Alydidae), a pest of soybean. Zool. Stud. 34: 111–116.

Taylor, W. E. 1978. Recent trends in grain legume pest research in Sierra Leone. Pp. 93–98 *in* S. R. Singh, H. F. van Emden, and T. A. Taylor [eds.], Pests of Grain Legumes: Ecology and Control. Academic Press, London, U.K. 454 pp.

Toda, S., K. Fujisaki, and F. Nakasuji. 1995. The influence of egg size on development of the bean bug, *Riptortus clavatus* Thunberg (Heteroptera, Coreidae). Appl. Entomol. Zool. 30: 485–487.

Tugwell, P., E. P. Rouse, and R. G. Thomson. 1973. Insects in soybeans and a weed host (*Desmodium* sp.). Agric. Exp. Stn. Rep. Ser. 214, Univ. Arkansas, U.S.A.

Turner, J. W. 1978. Pests of grain legumes and their control in Australia. Pp. 73–81 *in* S. R. Singh, H. F. van Emden, and T. A. Taylor [eds.], Pests of Grain Legumes: Ecology and Control. Academic Press, London, U.K. 454 pp.

Ueshima, N. 1979. Hemiptera II: Heteroptera. Pp. 1–117 *in* B. John [ed.], Animal Cytogenetics. 3. Insecta 6. Gebrüder Borntraeger, Berlin, Germany. 117 pp.

Underhill, G. W. 1943. Two pests of legumes: *Alydus eurinus* Say and *A. pilosulus* Herrich-Schaeffer. J. Econ. Entomol. 36: 289–293.

van den Berg, M. A. 1980. Natural enemies of *Acacia cyclops* A. Cunn. ex. G. Don. and *Acacia saligna* (Labill.) Wendl. in western Australia. III. Hemiptera. Phytophylactica 12: 223–226.

Velusamy, R., I. P. Janaki, and S. Jayaraj. 1977. Control of rice gundhi bug *Leptocorisa acuta* (Thunberg) (Coreidae, Hemiptera). Madras Agric. J. 64: 274–275.

Ventura, M. U., and A. R. Panizzi. 1997. *Megalotomus parvus* West. (Hemiptera: Alydidae): Inseto adequado para experimentação e didática entomológica. An. Soc. Entomol. Brasil 26: 579–581.

Verma, A., S. K. Srivastava, and T. B. Sinha. 1978. Seasonal occurrence in the population of different pests of *Calotropis procera*. Indian J. Entomol. 40: 204–210.

Visalakshi, A., A. Jacob, and M. R. G. K. Nair. 1976. Biology of *Riptortus pedestris* F. (Coreidae: Hemiptera), a pest of cowpea. Entomon 1: 139–142.

Way, M. O., A. A. Grigarick, J. A. Litsinger, F. Palis, and P. Pingali. 1991. Economic thresholds and injury levels for insect pests of rice. Pp. 67–105 *in* E. A. Heinrichs and T. A. Miller [eds.], Rice Insects: Management Strategies. Springer-Verlag, New York, New York, U.S.A. 347 pp.

Wilkinson, J. D., and D. M. Daugherty. 1967. Biology of the broadheaded bug *Alydus pilosulus* (Hemiptera: Alydidae). Ann. Entomol. Soc. Amer. 60: 1018–1021.

Yonke, T. R., and J. T. Medler. 1965. Biology of *Megalotomus quinquespinosus* (Hemiptera: Alydidae). Ann. Entomol. Soc. Amer. 58: 222–224.

1968. Biologies of three species of *Alydus* in Wisconsin. Ann. Entomol. Soc. Amer. 61: 526–531.

CHAPTER 11

Leaf-Footed Bugs (Coreidae)

Paula Levin Mitchell

1. INTRODUCTION

Bugs in the family Coreidae are commonly known as squash bugs or leaf-footed bugs, although neither term is truly appropriate as a family name. "Squash bug" refers specifically to the New World genus *Anasa*, species of which are notorious pests of cucurbitaceous crops. Worldwide, however, legumes are more likely than cucurbits to be damaged by coreids. "Leaf-footed bug" applies only to those species characterized by leaf-like dilations of the hind tibiae. Often the third antennal segment shows a similar foliation, especially in nymphs. In several species the hind femora in one or both sexes is thickened and adorned with large spines. Hogue (1993) recently coined the term "big-legged bugs" for the family, because of the latter characteristic.

As a group, coreids may be distinguished from most heteropteran families (but not Alydidae, Rhopalidae, or Stenocephalidae) by the numerous veins in the membrane of the hemelytra. Overall body length ranges from 7 to 45 mm, and the head is small in comparison with the body (Schuh and Slater 1995). Color is variable; temperate species are mainly brown or gray as adults, but nymphs may be brightly (perhaps warningly) colored. Adults of several tropical species are quite colorful. The family is distributed worldwide, although the number of species decreases markedly from tropical to temperate regions.

Coreids are primarily but not exclusively phytophagous. Several instances of coprophagy and carrion feeding have been reported (Adler and Wheeler 1984, Belthoff and Ritchison 1991, Steinbauer 1996a). Many coreids are gregarious, especially as nymphs, although territorial defense by males has been observed in five species (Mitchell 1980a; Fujisaki 1981; Miyatake 1993, 1995, 1997; Eberhard 1998). All species in which male combat has been documented are characterized by sexually dimorphic, enlarged hind femora, a trait common in several coreid tribes; thus, territorial defense may be relatively widespread in the family.

Like other heteropterans, coreids have well-developed scent glands ventrally on the metathorax of adults but dorsally on the abdomen in nymphs. These glands release an odor that serves both to deter predators and as an alarm pheromone, causing aggregations to scatter. Composition of the gland secretion is variable, but may include acids, aldehydes, alcohols, and acetate or butyrate esters of these alcohols (Aldrich et al. 1978, Aldrich 1988). In addition, males of some coreid species release species-specific volatiles, which probably serve as attractant pheromones (Aldrich et al. 1993).

The number of generations per year varies, depending on species, geographic location, and host plant specificity. In temperate zones, adults are the primary overwintering stage, although overwintering eggs have been reported (Jones 1993). Eggs may be ovoid (most species), hemicylindrical (*Phthia, Narnia, Leptoglossus, Mygdonia, Anoplocnemis, Mictis, Elasmopoda*), or spheroid (*Acanthocephala*), and are laid singly or in clusters, on or near the host plant. Developing nymphs usually pass through five stadia. As in pentatomids, it appears that first instars of some species require only a source of water (Mitchell 1980b, Ito 1984a, Jackai 1989, Steinbauer et al. 1998). From second instar through adult, coreids may feed preferentially in vascular tissue (e.g., *Cloresmus, Notobitus*), mesophyll cells (*Chelinidea*), or various reproductive parts (e.g., *Leptoglossus*). Host plant specificity ranges from extreme polphagy to restricted feeding on a single genus (Schaefer and Mitchell 1983).

2. GENERAL STATEMENT OF ECONOMIC IMPORTANCE

Relatively few coreid species are considered to be of economic importance. Of the approximately 2000 species worldwide (Dolling 1991), 82 have been mentioned in the literature as potentially or actually damaging. Similarly, only 6 (7%) of the 87 coreid species listed by Froeschner (1988) for the United States and Canada have attained even minor pest status. Of the 13 species of coreids associated with agricultural crops in Honduras, only 3 are described as potential, minor, or actual pests (Passoa 1983).

Grain legumes, rice, cassava, cucurbits, tomatoes and other garden vegetables, and various fruit and nut trees are among the crops attacked by coreids worldwide. Coreids are also a concern in forestry; *Eucalyptus* plantations in Australia and conifers (pines and Douglas fir) in North America are damaged by various coreid species. Economic losses on some crops may be substantial. Uncontrolled, high infestations of the pod-sucking bug complex (coreids and alydids) on legumes can lead to complete crop failure in West Africa. On the other hand, seemingly low population densities of pest coreids on coconut (e.g., *Pseudotheraptus wayi* Brown, *Amblypelta* spp.) can cause extensive economic losses.

Two very different types of damage associated with coreid feeding are reflected in the common names of different species complexes; e.g., "tip-wilters" vs. "pod-sucking bugs." Bugs that typically feed in or near the vascular tissue cause wilting of the shoot apical to the point of attack; heavy infestation may cause the entire plant to be stunted or wilted. An osmotic pump feeding mechanism has been demonstrated in some coreid species (see *Amblypelta*, **Section 3.1.1.2**), in which cell contents distant from the point of stylet entry are emptied and a lesion forms on the stem (Miles 1987, Miles and Taylor 1994). Species feeding on reproductive parts cause direct injury to fruits and developing seeds. In this case, the type and extent of damage often depend on the developmental stage of the fruit. For example, pecans attacked in the liquid endosperm stage by *L. phyllopus* L. develop a condition known as black pit and are eventually aborted, whereas damage later in development causes kernel discoloration but not fruit drop (Yates et al. 1991). Some coreids appear to be obligate fruit or pod feeders, whereas others switch from tender shoots to reproductive tissue as the season progresses. A few species are described as exclusively sap-feeding (e.g., *Amorbus*); Kumar (1966) considers this characteristic to be primitive.

No coreids have been specifically implicated as virus vectors. However, penetration of the stylets into plant tissue facilitates entry of various other pathogenic microorganisms such as bacteria or fungi. Coreid feeding lesions have been associated with transmission of a fungal disease of cassava (candlestick disease) (Boher et al. 1983), and several coreids have been implicated in transmission of the fungus *Nematospora coryli* Peglion. Recent studies suggest that transmission of trypanosomatids (*Phytomonas, Herpetomonas*) is associated with feeding by coreids. A trypanosome parasitic (but not necessarily pathogenic) within the lactiferous tubes of *Euphorbia pinea* L. is transmitted by the stenocephalid *Dicranocephalus agilis* (Scopoli) (Dollet et al. 1982, Dollet 1984). Similarly, *Phthia picta* Drury has been shown to vector a trypanosome pathogen of tomato (Fiorini et al. 1993), and a trypanosomatid from the digestive tract and salivary glands of

L. zonatus (Dallas) can be transmitted to corn under laboratory conditions (Jankevicius et al. 1993). A survey of Brazilian Heteroptera showed *Phytomonas* spp. to be present in the salivary glands or digestive tract of at least five coreid species (Sbravate et al. 1989). In South America, *P. staheli* is strongly associated with sudden wilt disease of oil palms (although a causal relationship has not yet been established) (Thurston 1998). As oil palm production in South America and the Caribbean increases, the ability of coreids to vector such protozoan pathogens may become economically important.

A few coreid genera have received attention as potential biological control agents (see **Section 5**). Species of *Chelinidea*, specialists on the pads of prickly pear (*Opuntia* spp.), have been released in areas where this cactus has been introduced or spread into former rangeland (Hamlin 1924a, Goeden et al. 1967). Similarly, *Pomponatius typicus* Distant has been considered as a possible weed control agent for the introduced tree *Melaleuca quinquuenervia* (Cav.) S. T. Blake in Florida (U.S.A.) (Balciunas and Burrows 1993), and *Mozena obtusa* Uhler is a potentially important native control agent of mesquite (*Prosopis glandulosa* Torrey) (Ueckert 1973).

3. THE IMPORTANT SPECIES

3.1 *Amblypelta* Stål

The genus *Amblypelta* includes a number of economically important species in the South Pacific. Coconut, rubber, cocoa, guava, papaya, macadamia, and cassava are among the crops attacked. In all, 15 species have been described from the South Pacific and southeastern Indian Oceans (Ghauri 1984), many of which are restricted in distribution to a single island (Brown 1958a). A key and distribution map for the Australian species are provided by Donaldson (1983), and for all species by Brown (1958a) and Ghauri (1984). Extensive studies by Brown (1958a,b, 1959a,b), Lever (1981), and Phillips (1940) examined the role of these insects in immature nutfall of coconut in the Solomon Islands, and the biological control provided by ants. Many workers on this genus and related dasynines have noted that the damaging effects of these bugs are disproportionate both to the size of the feeding puncture and the population density (Brown 1958b); in other words, densities less than one bug per plant can cause extensive crop losses, and a nearly invisible external puncture gives rise long afterward to internal lesions in the fruit. Damage by species in the related genera *Pseudotheraptus* and *Paradasynus* is strikingly similar to that of *Amblypelta* species (see discussion below, **Sections 3.7 and 4.10**). Thus, for many years the pest status of *Amblypelta* and related genera remained unrecognized in coconut despite intensive field studies (Lever 1981).

In addition to the species discussed in detail below, four others feed on cacao: *A. bukharii* Ghauri (Papua New Guinea), *A. danishi* Ghauri (Irian Jaya, West New Guinea) (Ghauri 1984), *A. ardleyi* Brown (Papua New Guinea), and *A. madangana* Brown and Ghauri (Papua New Guinea) (Szent-Ivany 1963, Lever 1981). *Amblypelta costalis* Van Duzee feeds on cassava, cacao, coconut, and rubber, and *A. cristobalensis* on coconut and cassava. Lever (1981), Brown (1958b), and Phillips (1941) list host plants of other species. A coconut pest in Kerala, India provisionally identified as *Amblypelta* sp. (Kurian et al. 1972), is *P. rostratus* Distant (see **Section 4.10**).

3.1.1 *Amblypelta lutescens* Distant

3.1.1.1 Distribution, Life History, and Biology — Known in Australia as banana-spotting bugs, these insects are damaging to a variety of tree fruits and nuts in Queensland. Two subspecies have been described, *A. l. lutescens* from Australia (Queensland, the Northern Territory, and Western Australia) and *A. l. papuensis* from Papua New Guinea. The Australian form is described as pale green in life (Brown 1958a), although pinned specimens appear pale brown, with pale to chestnut brown hemelytra. Body length ranges from 11 to 15 mm, and the antennae are slightly shorter than the total body length. The Papua New Guinea form, known as the tip-wilt bug, is darker, with longer antennae (equal to or slightly greater than total body length). Differences in the shape of the male pygophore have also been noted (Brown 1958a).

3.1.1.2 Damage — In Papua New Guinea, *A. l. papuensis* damages mango, sweet potato, pawpaw, cacao pods, and stems of papaya and cassava; it causes wilting of new growth on rubber and immature nutfall of coconuts (Szent-Ivany and Catley 1960a, Lamb 1974). In Australia, Donaldson (1983) notes that confusion in the older literature between *A. l. lutescens* and the closely related *A. nitida* makes correct assignment of early host plant records difficult. Crops attacked include avocado, banana, cashew, custard apple, guava, lychee, macadamia, and pawpaw (Donaldson 1983, Smith 1985). Both nymphs and adults are damaging, either feeding directly on reproductive parts or ingesting the contents of parenchma cells in growing stem tissue. Generally, it had been thought that these insects fed by physically piercing cell membranes to withdraw the contents. However, Miles (1987), working with an undetermined species of *Amblypelta* from Papua New Guinea, showed that the bugs can actually empty the contents of parenchyma cells located as far as 3.5 mm from the tips of their stylets — in other words, without physically penetrating the cell. Cells emptied in this manner subsequently collapse, and a "feeding lesion" forms, which extends beyond the plant tissue actually penetrated by the stylets. The bugs were observed to cause feeding lesions on stems of cassava and sweet potato; whether this mode of feeding is also used in fruit and nut feeding is unknown. Further studies of this feeding method in other coreids (Miles and Taylor 1994) elucidated the process: a salivary enzyme, sucrase, alters the osmotic pressure, causing the intercellular spaces to fill with cell contents leaked from surrounding parenchyma and phloem.

Damage by *A. l. lutescens* is considered to be the major factor inducing fruit abcission of green lychees in some areas of northern Queensland. Up to 98% of fallen fruit was found to be bug damaged. Dissection of lychee fruit showed internal, elongate-oval lesions on developing seeds, although no indications of damage were visible externally (Waite 1990). In contrast, damaged avocados develop "cracks and craters" (Ryan 1994); on pawpaw, dark sunken areas on the fruit, swelling of the trunk, or dark lesions on new growth are signs of *A. l. lutescens* infestation. Pawpaw trees adjacent to rain forest are significantly more damaged than those in open areas, suggesting that a clear border around plantings, or a border of nonhost vegetation, could provide some level of control (Ryan 1994). Distribution of *A. l. papuensis* in rubber plantations was described as "fairly even" (Szent-Ivany and Catley 1960a), whereas Ryan (1994) found a distinctly clumped distribution of *A. l. lutescens* damage in pawpaw plantings.

3.1.1.3 Control — Chemical control of *A. l. lutescens* in Australia is obtained using endosulfan (Fay and Huwer 1993). Three egg parasites (*Anastatus* sp., *Ooencyrtus* sp., *Gryon* sp.) have been reared from eggs laid in noncrop environments (Fay and Huwer 1993); a similar complex of egg parasites was reared from eggs of *A. l. papuensis* (Szent-Ivany and Catley 1960a), and is known from other tropical dasynine species as well (see listing in Fay and Huwer 1993). A native ant species, *Oecophylla smaragdina* (F.), reduced populations of *A. l. lutescens* and other pest insects in cashew orchards in northern Australia (Peng et al. 1995).

Chemical analysis of pheromones has shown that *A. l. lutescens* and *A. nitida* males produce species-specific volatile blends which may operate as attractants, in addition to the metathoracic and dorsal abdominal defensive secretions produced by both sexes (Aldrich et al. 1993). The major component of the blend produced by *A. l. lutescens* males is an unusual enantiomer of nerolidol, which is thought to be used as a kairomone by the parasitic tachinid, *Trichopoda pennipes* (F.), in seeking its North American coreid host, *Leptoglossus phyllopus* (L.) (Aldrich et al. 1993). The authors suggest that perhaps this parasite could be useful in a control program for *A. l. lutescens*.

3.1.2 Amblypelta nitida (Stål)

3.1.2.1 Distribution, Life History, and Biology — Although *Amblypelta* species overlap in Australia, *A. nitida* has a more southerly distribution, extending from Queensland to New South Wales. The species are similar in appearance but may be separated on the basis of male and female genitalia or length ratio of the antennal segments; Donaldson (1983) provides the most recent key. Male-specific volatiles of *A. nitida* are chemically quite distinct from those of *A. l. lutescens* (Aldrich et al. 1993). Eggs are oval and are laid singly. Development time from egg to adult is estimated at 34 to 38 days, and three generations are estimated to occur in southern Queensland (Brimblecombe 1948; this research refers to the bug as *A. lutescens*, but a subsequent publication, Brimblecombe 1962, clarifies the species). Detailed descriptions of the instars are given by Brimblecombe (1962).

3.1.2.2 Damage — *Amblypelta nitida* is known as the fruit-spotting bug, and has been observed to feed on macadamia, pecan, avocado, peach, and passionfruit (Donaldson 1983). Damage to macadamia nuts may be severe, particularly at the soft-shelled stage (Ironside 1984). Feeding on small nuts results in premature nutfall, but there are no external signs of damage on the husk. Wilting of shoots also occurs. Photographs of nut damage are given by Brimblecombe (1948). In lychee, which is damaged by both species of *Amblypelta*, 93.5% of green fruit drop was attributed to *A. nitida* in south Queensland. Losses were estimated equivalent to 50 to 60 kg of mature fruit per tree in 1986 and 1987. The early-maturing Tai So variety was more susceptible to fruit drop than later-maturing lychee varieties (Waite 1990).

3.1.2.3 Control — Deltamethrin and endosulfan are recommended for control, although the former is very damaging to the natural enemy complex on macadamia (Ironside 1984).

3.1.3 Amblypelta manihotis (Blöte) (= Dasynus manihotis Blöte)

This species is restricted to Java, where it causes sporadic damage to cassava. Illustrations of adults and typical stem injury are given by Phillips (1941). Drawings of the egg, first instar, and a later instar are given by Leefmans (1935), who was unable to rear the insects in captivity on cassava. Damage occurs when the insects feed on the upper stems, causing leaf drop and withering of the shoots. Feeding lesions appear as irregular brown sunken areas on the stem. *Acacia villosa* and several species of *Albizia* Durazini (Leguminosae) are listed as wild host plants (Phillips 1941). Egg parasites reared from *A. manihotis* include *Hadronotus* (= *Gryon*) *homoeoceri* Nixon, *Ooencyrtus malayensis* Ferrèire, and a species of *Anastatus* (Phillips 1941).

3.1.4 Amblypelta theobromae Brown

Restricted in distribution to Papua New Guinea, this species is considered mainly a pest of cacao, although individuals have been collected on annatto, cassava, rubber, coconut, and cashew nut, and damage reported to the last two crops (Brown 1958a, Lever 1982, Smith 1984). Ghauri (1984) gives the distribution as east and central Papua New Guinea, whereas Smith (1984) places the species "along the north coast." This species differs from the three above in having a pronounced pronotal collar. Eggs are ovoid; nymphs may be yellow, orange, brown, or green, depending on stadium and color morph. Detailed descriptions of egg and nymphs are given by Smith (1984).

Damage to cacao pods appears as round, brown scars, and can be distinguished from mirid damage in that the lesions are larger and found all over the fruit surface. Malformed fruit and fungal infection may follow (Brown 1958b). Damage to branch tips has also been observed (Szent-Ivany 1963). Distribution of *A. theobromae* was found to be fairly even across cocoa plantations, densities being as high as four adults per pod. Nymphs and copulating pairs also are found on pods (Szent-Ivany 1961). Populations of less than one insect per tree were found to be damaging; however, infestations of the ant *Oecophylla smaragdina* (F.) protected cacao trees from damage by *A. theobromae* (Brown 1958b).

3.1.5 Amblypelta cocophaga China

3.1.5.1 Distribution, Life History, and Biology — *Amblypelta cocophaga* is a serious pest of cocoa and coconut in the Solomon Islands; two subspecies, *A. c. cocophaga* and *A. c. malaitensis*, have been described. Like *A. theobromae*, this species has a pronotal collar. Descriptions of adults are given by China (1934) and Brown (1958a), and drawings of the nymphs by Phillips (1940). Time from oviposition to adult emergence is 36 to 41 days, and the preoviposition period is 15 to 19 days. Female lifetime fecundity is over 100 eggs. The main oviposition site is stated by Phillips (1940) to be the central, unopened leaf of the palm; Brown (1959b) reports eggs evenly distributed on fronds of different ages including this terminal spike, but more rarely on the spadices. Eggs are laid singly, in protected areas on the underside of the leaflets, and second instars move from there to the spadix. Bugs feed preferentially on the male flowers when the spadix first opens, shifting subsequently to female flowers and small nuts. Most feeding occurs at the base of the nut. Even a single bug is enough to destroy all the nuts of a spadix (Phillips 1940). *Amblypelta cocophaga* is polyphagous, and is found on various plantation undergrowth shrubs and on plants in the bush (Phillips 1940). It appears to have a preference for pioneer species in disturbed forest (Bigger 1985), making it a likely candidate for pest status on newly introduced plantation crops. In addition to coconut and cacao, this species has been recorded damaging shoots of papaya, kapok, cassava, and *Eucalyptus* (Phillips 1940, Bigger 1985).

3.1.5.2 Damage — Coconuts damaged by *A. cocophaga* show dark, elongate furrows beneath the calyx, also visible internally as brown stains (Phillips 1940). On cacao pods, damage is similar to that of *A. theobromae* — the brown external lesions are circular, and may lead to malformed and shriveled fruit but not to fruit drop. Terminal shoots may also be fed upon by *A. c. malaitensis*, causing dieback of the tips. On cassava, the damage resembles that of *A. manihotis*, in that feeding on petioles and stems causes wilting of the leaves (Brown 1958b).

Recent attempts to develop *Eucalyptus deglupta* plantations in the Solomon Islands were threatened by *A. cocophaga*, and ultimately abandoned because of insect and weed problems. Adults and nymphs feed on young shoots, producing a black lesion and eventual wilting. Bugs feed preferentially on the apical shoot, which leads to a profusion of side shoots and branching. In cases of heavy infestation by nymphs, the crown of the tree becomes bushy and malformed. Experiments comparing infestations in cleared and uncleared (strip) plantings showed no significant differences in proportion of infested trees after 1 year, although the initial rate of infestation was higher in the cleared plots because these trees grew faster. Overall, between 37% (cleared plots) and 41% (uncleared plots) of trees were so damaged by *A. cocophaga* as to be useless for timber (Bigger 1985).

3.1.5.3 Control — The indigenous ant *O. smaragdina* (F.) and the nonnative invader *Anoplolepis longipes* (Jerdon) protect coconut trees from *A. cocophaga*. However, colonies of other ant species, *Pheidole megacephala* (F.), *Iridomyrmex myrmecodiae* Emery, drive *O. smaragdina* from the coconut plantations (Phillips 1940), leaving the palms susceptible to nutfall. Yet *I. myrmecodiae* does provide some protection against *A. cristobalensis* Brown on San Cristobal (Solomon Islands) (Brown 1959a). The eupelmid *Anastatus axiagasti* Ferrière parasitized an average of 33% of *Amblypelta cocophaga* eggs on Guadalcanal. Release of the tachinid fly *Trichopoda pennipes* (F.) was unsuccessful (Brown 1959b).

3.2 *Anasa* Amyot and Serville

This large, primarily Neotropical, genus has been revised by Brailovsky (1985). Many of the 63 described species feed on cucurbits, but not all necessarily cause economic damage. For example, *A. repetita* Heidemann appears to be restricted to wild cucumber (*Sicyos* spp.) (Blatchley 1926) and has been reported on cultivated cucurbits only once (Parshley 1918). In contrast, *A. armigera* (Say) feeds on both native (*Sicyos angulatus* L.) and cultivated cucumber (Parshley 1918), and is reported from squash and cantaloupe as well (see below). Of the eight species reported to be of economic importance, the most famous is unquestionably the squash bug, *A. tristis* De Geer, which has given its common name to the family as a whole. In addition to the four species discussed below, *A. guttifera* Berg is a pest of various cucurbits in Argentina, and *A. litigiosa* Stål, *A. maculipes* Stål, and *A. uhleri* Stål attack squash or other cultivated cucurbits in Mexico (Brailovsky 1985).

3.2.1 *Anasa tristis* De Geer

3.2.1.1 *Distribution, Life History, and Biology* — Commonly known as the squash bug, this species occurs from Québec and Ontario (Canada) to Brazil (Froeschner 1988). The adult is a variable, mottled shade of brown dorsally, with a black hemelytral membrane and a yellow connexivum alternating with dark blotches. The venter and the legs are mottled yellow. Hind legs are slender and unarmed in both sexes. Antennae are uniformly dark, or yellow marked with dark dots. At the base of each antenna is a small tubercle (Blatchley 1926). Body length ranges from 13 to 18 mm; females are significantly heavier than males. Beard (1940) describes the reproductive system of both males and females.

Eggs are oval, 1.55 mm in length, white or yellow when first laid and turning eventually to a dark golden bronze. The antennae and legs of nymphs are covered with hairs, which become shorter and less noticeable as development progresses. Instars are described in detail and illustrated by Chittenden (1899).

Adults emerge from overwintering sites, mate, and oviposit on seedling cucurbits. Egg production by females is affected by temperature, ranging from 13.2 eggs/female/day at 30°C to 3.9 at 20°C (Woodson and Fargo 1991). Average mass size was 21 eggs in a laboratory study (Bonjour et al. 1993), and ranged from 17.4 to 19.5 in the field (Bonjour et al. 1990). Nymphal development times have been estimated from cohort rearing data (Fargo 1986) as 4, 2.75, 5.5, 2.75, 4.38, and 4.91 days for egg and first through fifth stadia, respectively, at 30°C and 16:8 (L:D) hours, for a total of 24.29 days. Eggs fail to hatch at a temperature of 15.6°C or above 33.3°C. Stadia and degree-day requirements have been determined for insects reared on potted summer squash; development data for 12 constant temperatures between 20 and 40°C were used to devise models relating developmental rate to temperature (Fargo and Bonjour 1988). Similar data have been collected for insects reared on pumpkin seedlings (*Cucurbita moschata* Duchesne) to develop a predictive model of squash bug development in Illinois (Fielding and Ruesink 1988). The effect of temperature on egg development was found to be linear, and the degree-day model provided accurate predictions. The relationship between temperature and nymphal development was nonlinear, and the model underestimated both field development rates and laboratory development rates under variable temperature regimes.

In the United States, one generation is completed annually in Connecticut (Beard 1940), one and a partial second generation in central Illinois (Fielding and Ruesink 1988), 1.5 generations in Kansas (Nechols 1987), and two to three in Oklahoma (Fargo et al. 1988), although a subsequent study in Oklahoma reported fewer than two generations completed during the growing season (Palumbo et al. 1993). Reproductive diapause, characteristic of overwintering adults, may be induced by photoperiods less than 14:10 (L:D) hours, and maintained by short day lengths (Nechols 1988). Seasonal history has been documented in detail for Kansas (Nechols 1987) and Oklahoma (Fargo et al. 1988) populations.

On summer squash seedlings with fewer than 10 leaves, adults are more likely to be found on the ground than on plants. Egg masses are placed mainly on leaves (94.5%), although cotyledons, petioles, or stems may also serve infrequently as oviposition sites. Of the eggs placed on leaves during the vegetative period of plant growth, 92.2% were located on the leaf underside, and 90% within the lower half of the canopy (Palumbo et al. 1991a). Oviposition site is not affected by host plant species (Bonjour et al. 1990). All life stages exhibit aggregated patterns of dispersion; nymphs are more strongly aggregated than adults, and small nymphs (first and second instars) more so than larger ones (Palumbo et al. 1991b).

3.2.1.2 Damage — The bug is considered a pest of cultivated crops in the genus *Cucurbita*, particularly summer squash and pumpkin, throughout most of the United States and in Mexico. Damage is particularly severe in the midwest and southwest (Nechols 1987, Palumbo et al. 1993). In Honduras, this species is known as the "chinche de ayote," and is considered a potential pest on pumpkin and squash. Feeding damage by adults or nymphs on leaves initially appears as a brown spot. Gregarious feeding will cause the leaf to mottle and then turn entirely brown, referred to by Beard (1940) as the "burning effect." Feeding by adults on seedling squash reduces yield in two ways: direct seedling mortality and delayed growth (Woodson and Fargo 1992). Seedling size at infestation significantly affects yield, and total crop loss may occur at infestation levels of two adults per plant at the two-leaf stage. In laboratory studies, production of male flowers was not affected by squash bug density, but vegetative growth rate and ovulate flower production were reduced (Woodson and Fargo 1991). Later in the season, heavy infestation during harvest can reduce yield by >50%, from leaf wilting and necrosis, direct damage to fruit, and plant mortality (Palumbo et al. 1993).

Squash bugs have definite oviposition preferences among squash cultivars and cucurbit species, which reflect the suitability of these hosts for nymphal survivorship and adult reproduction (Bonjour and Fargo 1989; Bonjour et al. 1990, 1993). In Oklahoma, crookneck and yellow straightneck "Hyrific," *Cucurbita pepo* L., were preferred for oviposition over zucchini, acorn, spaghetti squash, *C. pepo*, and butternut squash, *C. moschata* (Duchesne) Poiret. The former two cultivars had more egg masses per leaf, and a higher percentage of leaves with eggs; but total egg masses per plant did not differ among the six squashes tested. Pumpkin and squash, *C. pepo*, were preferred over muskmelon, *C. melo* L., cucumber, *Cucumis sativus* L., and watermelon, *Citrullus lanatus* (Thunberg), having significantly more total eggs per plant, and a higher percentage of leaves with eggs (Bonjour et al. 1990). Survivorship from egg to adult was 70% on pumpkin, 49.02% on yellow straightneck squash, and considerably lower on watermelon (14.44%). On cucumber, only 0.32% of nymphs reached adulthood, and none survived on muskmelon. Nymphal development was significantly delayed on watermelon. Males reared on squash were significantly heavier than those reared on watermelon; female weights followed a similar trend, but differences were not significant (Bonjour and Fargo 1989). Daily fecundity on squash was higher than that on watermelon and muskmelon, but did not differ significantly from pumpkin. Because of differences in longevity, females on pumpkin had significantly higher lifetime fecundity than those provided with squash. Males lived longer on pumpkin, squash, and watermelon than on cucumber and muskmelon (Bonjour et al. 1993). Somewhat similar results were found for a Kansas population; survival to adulthood was significantly lower on *Cucurbita moschata* and *C. maxima* Duchesne (cv. Green Striped Cushaw pumpkin) than on *C. pepo*, although development time (egg to adult), oviposition period, and longevity were similar on the three cultivars. Survivorship on butternut increased in bugs of the second generation, suggesting that selection was operating (Vogt and Nechols 1993a).

Histological aspects of feeding damage have been examined by several authors (Beard 1940, Bonjour et al. 1991, Neal 1993). The stylet path is intercellular, with an external stylet sheath and a more diffuse deposition of salivary material within the plant tissue (Neal 1993). Feeding on squash leaves produces localized injury to epidermal and palisade cells, and more extensive, generalized injury to the spongy mesophyll (Beard 1940). On stems, the squash bug ingests both xylem contents and cell cytoplasm, and most stylet insertions penetrate to the inner phloem. Wilting of seedling plants results from blockage of all xylem elements, effectively girdling the stem, and no evidence of a salivary toxin has been found (Neal 1993). Preliminary electronic monitoring studies using first instars feeding on leaves indicated three distinct waveforms in addition to insertion and pullout spikes. One waveform was associated with stylet insertion into vascular bundles. Probing duration per probe was shorter on muskmelon than on squash, pumpkin, cucumber, or buffalo gourd (*C. foetidissima* Humboldt, Bonpland, & Kunth) (Bonjour et al. 1991).

3.2.1.3 Control — In the past, carbaryl provided limited control for the home garden, and parathion was recommended for commercial use (Little 1972). Recent studies show both carbaryl and disulfoton to be ineffective against adults (Palumbo and Fargo 1989a,b). Early instars are highly susceptible to insecticides (Criswell 1987, cited in Palumbo et al. 1993), and weekly sprays with cypermethrin provided good control (Palumbo et al. 1993). However, such a spray schedule is neither practical nor cost-effective. Equivalent yields were obtained using an action threshold of one egg mass per plant. The initial spray should be followed by a maintenance treatment timed precisely at 98 degree-days above 15.6°C. This second application is aimed at nymphs emerging from egg masses unhatched at the time of the first spray, as cypermethrin has no ovicidal activity. Application of insecticides must begin prior to harvest, before populations reach damaging levels (Palumbo et al. 1993).

Most registered insecticides currently provide poor control for adults and large nymphs (Zavala 1991, cited in Olson et al. 1996). Because squash bugs aggregate and oviposit on the undersurface of leaves, adequate spray coverage is difficult to obtain on large plants, and tractor-drawn spray equipment cannot be used (Fargo and Bonjour 1988). Therefore, late-season control is difficult to achieve using insecticides. Furthermore, growers are concerned that foliar sprays at flowering could harm pollinators (Palumbo et al. 1991c). Alternative management tactics are needed (Palumbo et al. 1991c, Olson et al. 1996). Recommended cultural controls include cleanup of crop residues and other areas where adults may overwinter. In small gardens, handpicking of adults and eggs may be effective when populations are low (Little 1972). Various types of mulch (aluminum, black plastic, white plastic) were found to increase, rather than decrease, squash bug populations compared with bare ground; the mulch provided a protected aggregation site for adults and nymphs. The increased attractiveness of mulched plots to squash bug may negate the advantages of these production systems (Cartwright et al. 1990).

Planting dates may be manipulated to manage infestations of overwintering bugs in spring crops (Palumbo et al. 1991c). Adults preferentially colonize larger, many-leaved squash plants, and the number of squash bugs per leaf is correlated with the number of leaves per plant. Therefore, eggs and nymphs reach significantly higher densities on early-planted squash, suggesting that trap cropping could be an effective management strategy (Palumbo et al. 1991c). Sequential sampling plans based on Taylor's power law have been developed for adults, egg masses, and nymphs on summer squash, but do not always provide estimates at the levels of precision needed (Palumbo et al. 1991b).

Predators of squash bug nymphs include the spined soldier bug, lacewings, nabids, and spiders, and adults have been dissected from the digestive tracts of various birds; but none of these predators is considered to have a significant effect on squash bug populations (Beard 1940). The parasitic nematode *Neoaplectana carpocapsae* (Weiser) may be used to introduce a bacterium, *Achromobacter nematophilus*, that kills squash bugs; but host infection by the nematode was low in initial field trials (Wu 1988).

Adults are parasitized by the tachinid *Trichopoda pennipes* (F.) (Beard 1940, Arnaud 1978). Ashmead (1887) first reported an encyrtid, *Ooencyrtus anasae* Ashmead (as *Encyrtus anasae* n. sp.); a eupelmid, *Anastatus reduvii* Howard (as *Eupelmus reduvii* Howard); and a scelionid, *Gryon* (= *Hadronotus*) *anasae* Ashmead (as *Telonomus anasae* n. sp.) parasitizing squash bug eggs in Florida, parasitism by the last wasp reaching 30%. The distribution of *G. anasae* extends from

South Carolina to Panama (Masner 1983), although *Anasa tristis* is not the sole host throughout this range. Beard (1940) reported no egg parasitism in Connecticut. Two other scelionid wasps, *G. carinatifrons* (Ashmead) (as *H. carinatifrons* Ashmead) and *G. pennsylvanicum* (Ashmead) (as *H. ajax* (Girault)) have been reared from the eggs of *A. tristis* (Girault 1913, 1920; Schell 1943; Masner 1983). Oviposition behavior of *G. carinatifrons* (as *Hadronotus*) is described by Girault (1913); mating, oviposition, and larval development of *G. pennsylvanicum* (as *H. ajax*) are described by Schell (1943), who reported field parasitism ranging from 23 to 42%.

Several species of *Ooencyrtus* and *G. pennsylvanicum* have been intensively studied as potential biological controls of the squash bug in the midwestern United States (Tracy and Nechols 1987, 1988; Nechols et al. 1989; Vogt and Nechols 1991, 1993a,b). For all parasitoid species, parasitization rates declined with increasing host age. Older eggs were also less suitable for wasp larval development. Compared with three species of *Ooencyrtus*, *G. pennsylvanicum* parasitized significantly more host eggs, had the highest fecundity, net reproductive rate, and intrinsic rate of increase, and was judged to have the greatest potential for augmentative release (Nechols et al. 1989).

Mated *G. pennsylvanicum* females show a pronounced diel activity pattern, oviposition occurring more frequently in the morning hours (Vogt and Nechols 1991). Females of *G. pennsylvanicum* have never been observed to host-feed (Vogt and Nechols 1991). Exudates from squash leaf trichomes, which produce glucose, galactose, and protein, served as adequate sources of nutrition for adult females (Olson and Nechols 1995). Availability of host eggs did not affect daily egg deposition or duration of oviposition, but females repeatedly deprived of host eggs for intervals of 3 days had lower lifetime fecundity than females supplied daily with an egg mass. Resorption of oocytes was not induced by host deprivation for 3-day intervals, but longer periods of deprivation (5 to 9 days) resulted in a high percentage of abnormal oocytes (Vogt and Nechols 1993b).

Tritrophic interactions have been noted: eggs from squash bugs reared on a susceptible summer squash cultivar (*C. pepo*, Early Prolific Straightneck) produced wasps with greater longevity (expressed as a longer postreproductive period) than their counterparts on a resistant winter squash (*C. moschata*, Waltham butternut) (Vogt and Nechols 1993a,b). However, no other attributes of the life history were affected by resistant cultivars; larval development, survivorship, sex ratio, oviposition period, fecundity, and intrinsic rate of increase did not differ among wasps reared from host eggs derived from populations on *C. pepo*, *C. moschata*, and *C. maxima* (Vogt and Nechols 1993a). Therefore, plant resistance and biological control using *G. pennsylvanicum* would appear to be compatible management strategies. A preliminary field evaluation indicated that augmentative release alone was not more economically feasible than chemical control using esfenvalerate. Although parasitoid releases reduced squash bug densities fourfold on pumpkin, populations were higher throughout the season in augmentative release plots than in insecticide-treated plots, and yields were lower. Furthermore, projected costs of rearing the parasitoid were high, so that net income would decrease considerably (by U.S.$5400/ha) if a farmer were to substitute biological for chemical control. However, combining augmentative release with use of a resistant cultivar may have potential for squash bug management, because the two strategies combined appear to have a more than additive effect in suppressing bug populations (Olson et al. 1996).

3.2.2 *Anasa armigera* Say

Although of considerably lesser economic importance than the squash bug, this species has been collected from asparagus, squash, cantaloupe, melon, and cucumber, damage being particularly severe in autumn on the latter crop (Chittenden 1899, Brailovsky 1985). The insect is reported from Ontario (Canada) to Florida (U.S.A.) (Froeschner 1988), but economic damage is reported mainly from the southern United States. Adults are dull yellow, with numerous dark punctures. The membrane of the hemelytra is brown, and the connexivum is black alternating with yellow. Length is 13 to 17 mm. This species may be separated from *A. tristis* by the large spine at the base of each antenna, and from *A. scorbutica* (F.) by the black dots on the basal segment of the antennae.

The adults, four of the five instars, and the egg are described and illustrated by Chittenden (1899). *Trichopoda pennipes* (F.) is reported to parasitize adults (Arnaud 1978).

3.2.3   Anasa scorbutica (F.)

The adult, 12 to 14 mm long, is pale yellow, with a prominent spine at the base of each antenna. The membrane of the hemelytra is dark brown. It may be separated from the similar *A. armigera* by the absence of black spots on the yellow antennae, and from *A. tristis* by the spines on the head. Distribution ranges from Oklahoma, Texas, and Florida (U.S.A.) southward through Mexico and into Central and South America and the West Indies. It has also been reported from the Galápagos Islands (Froeschner 1988). Although there is no record of this species being injurious to cucurbits in the United States, it is recorded in Honduras on *C. pepo* (L.) in the flowering stage and on maize in the grain-filling stage, and is categorized as a rare pest (Passoa 1983). In Cuba it is reported from loofah (*Luffa aegyptiaca* Miller) and squash (Barber and Bruner 1947), in Costa Rica on *C. pepo*, in Nicaragua on physic nut (*Jatropha curacas* L.), and in Surinam on *Lagenaria leucantha* (Brailovsky 1985, Grimm 1996). Brailovsky (1985) summarizes the known distribution and habitats of *A. scorbutica*: it is common in tropical and subtropical regions; attacks plantings of squash, tomato, loofah, and gourd; and is found on cotton and in sugarcane plantations.

3.2.4   Anasa andresii (Guérin-Méneville)

This species resembles *A. tristis* in that the bases of the antennae do not bear long spines, and the femora are unarmed. It may be distinguished from *A. tristis* by the orange or yellow coloration of the fourth antennal segment, contrasting with the darker preceding segments. Body color of the adult apparently varies, being described by different authors as "dull greenish-yellow with black punctures" (Blatchley 1926), "yellowish brown ... with black tubercles" (Jones 1916), and "dull yellowish brown or greenish brown with numerous black punctures over entire dorsal surface" (Baranowski and Slater 1986). Males are 13.4 mm (12.5 to 14) and females are 15.7 mm (15 to 16.5) long. Eggs are oval, reddish brown, and 1.45 mm long. They are usually laid in batches (mean = 15 eggs, range 2 to 50 per cluster) on the undersurface of leaves, but also may be placed on other plant surfaces (Jones 1916). Detailed descriptions of instars are given by Jones (1916).

Distribution is from the southern United States south through Mexico to Colombia and the West Indies (Blatchley 1926). The species is a pest of squash in Louisiana (Jones 1916) and Cuba (Barber and Bruner 1947); in Mexico it has been observed on the fruit of *C. ficifolia* Bouché and on bromeliads (Brailovsky 1985). Damage on squash is similar to that of *A. tristis*; nymphs and adults cluster and feed on the undersurface of leaves, causing the leaves to wilt and die. Fruit is also fed upon, although this occurs mainly when the vines are already dead (Jones 1916). Brailovsky (1985) reports predation by a reduviid, *Apiomerus* sp.

3.3   Anoplocnemis Stål

These large, impressive bugs are found primarily in the warmer regions of the Old World. The sexual dimorphism of the hind legs is pronounced; femora of males are more enlarged, curved, and spined than those of females (Villiers 1952). Eggs are hemicylindrical and laid in chains. Three species have been reported as crop pests: *A. curvipes* and *A. tristator* in Africa, and *A. phasiana* in India. A list of species and geographic distributions is given by O'Shea (1980).

3.3.1   Anoplocnemis curvipes F.

3.3.1.1   Distribution, Life History, and Biology — This species is found commonly throughout sub-Saharan Africa. Adults are entirely black or gray except for the last antennal segment which is bright yellow-orange. Adult measurements are given by Aina (1975a) as males, "about 38 mm," females, "about 30 mm," but by Villiers (1952) as 22 to 26 mm. Scaled drawings (Hartwig and de Lange 1978) are consistent with the latter estimate. The hind femora of males are strongly incrassate and curved, with an enormous preapical spine; those of females are thin and relatively straight, with small spines. Posterolateral angles of the pronotum are pointed and curved forward (Villiers 1952). Geographic variation in color, markings, and proportions is reported in South African populations (Hartwig and de Lange 1978). Egg chains are laid on stems, on leaf petioles, or on the abaxial surface of leaves, and range from 5 to 36 per batch. In cowpea, eggs are not laid directly on the plants; rather, oviposition occurs on leguminous trees and other wild hosts near cultivated fields (Singh and Jackai 1985). First (and often second) instars are gregarious (Aina 1975a). Parental care was observed in South African populations: females remained near their eggs and early instars, and a male was often observed feeding near such an aggregation (Hartwig and de Lange 1978). Drawings, measurements, and descriptions of the egg, nymphs, and nymphal development are given by Aina (1975a); detailed adult drawings are provided by Hartwig and de Lange (1978). Egg to adult emergence is completed on average in 32 days (four instars) or 33 days (five instars) on cowpea under laboratory conditions (23 to 29°C), the preoviposition period ranging from 8 to 17 days (Aina 1975a). Two to four generations per year are completed in South Africa (Hartwig and de Lange 1978).

3.3.1.2 Damage — Adults and nymphs feed on tender stems and shoots of a wide variety of host plants, causing them to wither. Damage to cotton, orange, mango, fig, eggplant, and occasionally cassava is reported by Villiers (1952); Agunloye (1986) reports this species attacking guava foliage. Golding (1935, 1937, 1948), in Nigeria, observed adults and nymphs on shoots of *Citrus paradisi* (grapefruit) and *C. aurantium* L., and nymphs on *Canavalia ensiformis*. In Ghana, pods and young leaves of *Phaseolus aureus* are damaged (Forsyth 1966). Damage to new growth and petioles causes severe necrosis in peanut grown in central Africa; necrotic lesions in rubber become entry points for a fungus, *Lasiodiplodia theobromae* (Pat.) Grif. et Maubl., resulting in dieback (Buyckx 1962). Over 100 hosts are known in South Africa, where the bug is considered a serious pest of nursery plants and several cultivated crops. Developing seed or shoots are damaged. Bugs feed 5 to 10 cm from the shoot tip, causing withering and death of the apical shoot (Hartwig and de Lange 1978). In Somalia and Sudan, the species is considered only a minor, although polyphagous, pest, the principal damage being to tender shoots (USDA 1965). On cotton, *A. curvipes* feeds on bolls, stems, and shoots, but primarily causes the tips of the main stem and side branches to collapse and die; however, the bugs are described as "never abundant" on cotton in tropical Africa and are considered with rare exceptions to be only a minor pest on this crop (Pearson 1958).

Anoplocnemis curvipes is perhaps best known, and certainly most damaging, as a member of the pod-sucking bug complex that attacks various legumes including cowpea in sub-Saharan Africa (see *Clavigralla*, below), often causing complete crop failure. Feeding by *A. curvipes* produces brown, slightly raised lesions on the pod, and causes abortion of punctured seeds. Pathogen entry into pods is facilitated by these feeding lesions (Aina 1975a). Pod age at infestation determines damage; shriveling of pods and damage to seeds were more severe in pods <1 week old (Khaemba and Khamala 1981).

The importance of *A. curvipes* in relation to the other pod-sucking bugs varies geographically. At Mokwa, Nigeria, in the southern Guinea savanna, *A. curvipes* comprised less than 1% of the "pod-sucking bugs" (PSB) in visual counts (Suh et al. 1986). Farther south in Nigeria, at Ile-Ife, *A. curvipes* populations peaked in November and December, but bugs were present in April, August, and September plantings (Akingbohungbe 1982). In Ghana, the species is classified as a minor pest of cowpea; both adults and nymphs are reported to feed on the pods (van Halteren 1971).

3.3.1.3 Control — Chemical treatment was not successful in controlling *A. curvipes* infestations in nurseries, because of a continual influx of adults from weed hosts. Eliminating breeding sites through early season weed control is recommended (Hartwig and de Lange 1978). On mango in South Africa, no insecticides are registered for control of *A. curvipes*, and hand-collecting is the recommended control tactic (Villiers 1994). Endosulfan, fenitrothion, and dimethoate can be used for control in cowpea (Singh and Jackai 1985). In cowpea–maize intercrops, oviposition occurs on maize rather than on cowpea. High nymphal mortality subsequently occurs when the maize dries up (Ochieng, unpublished, cited in Singh et al. 1978a). Eight cowpea cultivars (out of >4000 screened) showed resistance to damage by this species and *Riptortus dentipes* (F.). Pod color, peduncle length, and external pod morphology were among the characters affecting susceptibility. Dark-colored pods with short peduncles were more resistant, and damage was greater to pods with uneven or ridged walls, because seeds were more accessible (Khaemba 1985). The ant *Pheidole megacephala* F. attacks young nymphs, but *Anoplolepis custodiens* (Smith) does not. Parasitism by the tachinids *Paraclara magnifica* (Bezzi) and *Hermya confusa* Curran reaches 13% in Nigeria and 14.5% in Tanzania (Matteson 1980).

3.3.2   *Anoplocnemis phasiana* F.

3.3.2.1   Distribution, Life History, and Biology — This species is found in China, the Philippines, India, and islands of the Indian Ocean. Adults, 22 to 28 mm in length, resemble *A. curvipes*, in that the sexual dimorphism of the hind femora is pronounced. Humeral angles are rounded, rather than spined. The description by Hingston (1929) is delightful enough to justify a direct quote:

> It is a stout insect, black throughout except for some yellow near the tips of its antennae, and on its back an orange patch which is ordinarily hidden by the closed wings. It possesses a peculiar geometrical appearance, as though built on some cubist plan. Its thorax is a sloping equilateral triangle; its abdomen is isosceles in shape. Its wings, owing to the way in which they overlap, make a pattern of geometrical figures on its back. But look at its hind legs. They are very extraordinary. Those of the female show nothing in particular, but in the male they are distended into clubs and furnished with a strong tooth. On the whole we have an insect of grotesque appearance, repulsive by reason of its black colour and the extravagant conformation of its parts.

Eggs are reddish-brown, hemicylindrical, and laid cryptically in chains along stems (Hingston 1929, Hoffmann 1933). Development time (egg to adult) ranges from 53 to 63 days (Maheswariah and Puttarudriah 1956). The first instar is described as spiderlike and gregarious (Hingston 1929). The tendency to aggregate diminishes with each molt, and third through fifth instars are likely to be solitary. When disturbed, nymphs drop to the ground. Early instars have dilated front tibiae, just as in *A. curvipes*. Hingston (1929) suggests that mimicry of small, dead leaves (often produced at the shoot tip as a result of bug damage) is the advantage provided by this unusual appearance. Drawings of the adult, egg chain, and several instars are given by Hingston (1929). Detailed descriptions of all stages are given by Hoffmann (1933) and Maheswariah and Puttarudriah (1956).

3.3.2.2   Damage. This species is reported as a pest of mung bean in India, feeding on the leaves and pods (Nair 1986, cited in Sehgal and Ujagir 1988). Major host plants in central India are reported to be *Cassia occidentalis* L. and *Aegle marmelos* (L.) Correa de Serra (Hingston 1929); in southwestern India, severe damage to grapevine and brinjal (eggplant) has also been noted (Maheswariah and Puttarudriah 1956). In Malaya, it is reported to feed on rubber, passionflower, lime, groundnut, eggplant, and *Luffa acutangula* (L.) Roxburgh, inflicting minor damage to shoots and "leaf-stems" (Miller 1941). Shoots and petioles of cacao are also occasionally damaged (Entwhistle 1972). In China, cultivated food plants include lima bean, Chinese long bean (*Vigna sesquipedalis* L.), pigeon pea, groundnut, bamboo, and guava, and the species is described as a major, and potentially serious, pest (Hoffmann 1933). Feeding on young shoots of tea and dadap, *Erythrina subumbrans* (Hasskarl) Merrill, was noted by Green (1905) in Ceylon.

3.3.2.3 Control — Recommended control measures earlier in the century involved hand-picking and brushing bugs into a container of water and oil or kerosene (Green 1905, Hoffmann 1933, Miller 1941).

3.3.3 Anoplocnemis tristator Germar

Found in west and central Africa, this species is smaller (15 to 19 mm) than *A. curvipes*, with a less-pronounced sexual dimorphism in hind femoral size and shape, and the pronotal angles more rounded. Adults are brownish-red or brownish-yellow, with short yellow pubescence. Crop damage is similar to that of *A. curvipes*. These bugs are especially prevalent on cotton in central Africa (Villiers 1952). In Nigeria, Golding (1927, 1937) reports adults on shoots of *Lonchocarpus cyanescens* and *Indigofera* sp., and on the leaf midrib of *Mangifera indica* L. and *Paullinia pinnata* L. (as *A. testator* F., adults and nymphs). Forsyth (1966) recorded adults in Ghana feeding on leaves of *Manihot esculenta* Crantz.

3.4 Clavigralla Spinola

The taxonomic status of African and Oriental species of *Clavigralla* has been clarified by Dolling (1978, 1979). The genus name *Acanthomia* Stål (*Acanthomyia* or *Acanthomya*), extensively used in the earlier literature, has been synonymized (Dolling 1979). However, much confusion exists in the early economic literature. References to the legume pest *Acanthomia horrida* Germar, for example, may refer to *Clavigralla elongata* Signoret (in eastern Africa), or to *C. shadabi* Dolling (in western Africa), whereas *C. horrida* (Germar) is restricted in distribution to South Africa, and is not recorded as a serious pest of any crop. Similarly, references to *A. brevirostris* Stål in northeast Africa and India probably represent *C. scutellaris* (Westwood), but elsewhere in Africa this name probably refers to the very similar *C. tomentosicollis* Stål (Dolling 1979). Particularly in cases of geographic overlap, correct assignment of host plant and economic damage records is not always possible. Furthermore, complexes of pod-sucking coreids and alydids, e.g., *C. shadabi*, *C. tomentosicollis*, *Anoplocnemis curvipes* (F.), *Riptortus dentipes* (F.), and *Mirperus jaculus* (Thunberg), are often studied collectively, without distinguishing species; assessment of damage by specific coreid species in such cases is impossible.

3.4.1 Clavigralla tomentosicollis Stål

3.4.1.1 Distribution, Life History, and Biology — Found throughout sub-Saharan Africa, this species has been intensively studied as part of a destructive complex of bugs attacking grain legumes. This PSB complex consists of various combinations of coreid and alydid species, depending on geographic location, and may include *C. shadabi*, *C. elongata*, *A. curvipes*, *R. dentipes*, and *M. jaculus*.

Like most species of *Clavigralla*, *C. tomentosicollis* is pubescent and spiny, with tubercles on the pronotal disk, and spines on the posterolateral angles of the pronotum and on the femora. Body color is brown, with silvery and brown pubescence. Males are 8.3 to 9.7 mm long and females 9.3 to 11.5 mm (Dolling 1979). Photographs and detailed descriptions of adults, eggs, and nymphs are given by Materu (1972). The egg micropyles are described by Taylor and Omoniyi (1970).

Eggs are hemispherical, creamy-white when laid, and gradually turning brown. In Tanzania, they are laid in clusters of 5 to 40 on pods or leaf surfaces (Materu 1972); the leaf underside is preferred for oviposition (Olatunde et al. 1991a). In Nigeria, larger egg batch sizes are reported (mean = 20, range 2 to 99) and total lifetime fecundity averaged over 200 eggs (Egwuatu and Taylor 1977d). Nymphs are gregarious, and large aggregations may form on pods, resulting in exceedingly high densities on single plants (134 bugs per plant) (Materu 1971). Pods and peduncles may be completely covered by nymphal aggregations during an outbreak; when disturbed, the nymphs fall to the ground (Cobblah 1991). Development time from oviposition to adult emergence averaged 15.7 days on pigeon pea under field conditions, and 18.2 days in an insectary on detached pods (Egwuatu and Taylor 1977d). Laboratory studies using cowpea showed slightly longer development times: 16.5 days on fresh seeds, 17.1 days on dry seeds, and 17.2 days on fresh pods excluding the egg incubation period (Jackai 1989). The effects of humidity and temperature on rate of development, mortality, and behavior have been reported by Egwuatu (1978, 1980), Egwuatu and Taylor (1977e), and Jackai and Inang (1992); a method for mass-rearing on dry seed is described by Jackai (1989). The preoviposition period is 6 to 9 days (Egwuatu and Taylor 1977d). Adults less than 6 days old are weak flyers; thus, dispersal of adults to new host plants would not begin until a week after emergence. Continuous tethered flight longer than 1 hour was recorded in the laboratory (Egwuatu 1982).

In Benin, field counts and mark-release-recapture studies showed that colonization by immigrating adults begins soon after the appearance of cowpea pods. The immigration rate is high (approximately 0.6 adults/day/2 m of row) from 45 to 55 days after sowing, and decreases thereafter. Oviposition occurs when pods are fully elongated and seeds are beginning to mature, and newly emerged adults appear at pod harvest. Mobility of the adults, coupled with the common practice of staggered plantings, necessitates regional, rather than local, management strategies (Dreyer and Baumgärtner 1997).

Field studies in Nigeria provide an estimate of five to nine generations per year on pigeon pea, the insect being present all year. However, population density varies seasonally, higher populations being associated with reproductive stages of the crop, and with the dry season of the humid tropics, characterized by high temperatures, and low humidity and rainfall (Egwuatu and Taylor 1983). The long oviposition period and life span (13 weeks) allow adults to colonize plantings of both pigeon pea and cowpea (Egwuatu and Taylor 1977d).

3.4.1.2 Damage — This species is considered to be the most damaging of the PSB complex on pigeon pea (Materu 1971, Egwuatu and Taylor 1977d), and along with *R. dentatus*, the most important of the PSB on cowpea in Nigeria and Kenya (Jackai and Daoust 1986). Sampling based on visual counts showed *C. tomentosicollis* to compose 99% of the PSB population on mature cowpeas in the savanna region of Nigeria (Suh et al. 1986). In Ghana, *Clavigralla* spp. were accorded only minor pest status by van Halteren (1971), but Cobblah (1991) recorded severe reduction in yields attributed to *C. tomentosicollis* and *C. shadabi*. In Benin, *C. tomentosicollis* was the most damaging of the PSB complex in the majority of fields surveyed, and the only one of the PSB complex to produce appreciable nymphal populations in cowpea fields (Dreyer et al. 1994, Dreyer and Baumgärtner 1995).

Effects of feeding by *C. tomentosicollis* and related species are similar on pigeon pea and cowpea: pods abcise or dry and shrivel prematurely. When pods are retained on the plant, the seeds do not develop completely, but may abort, shrivel, or wrinkle (Jackai et al. 1989, Gethi and Khaemba 1991). Detailed investigation of feeding showed that seed damage correlated closely with number of pod punctures, and that females were more damaging than males (Jackai 1984). On bean, *Phaseolus vulgaris* L., damage by *C. tomentosicollis* and other PSBs appears as wrinkling of the seed coat and browning or shriveling of the seeds (Abate and Ampofo 1996).

3.4.1.3 Control — In sub-Saharan Africa, pigeon pea and hyacinth bean are grown mainly as garden crops. Cowpea, grown traditionally in an intercrop system with sorghum or maize, is gradually shifting in some areas to a monocrop. Although current control tactics for cowpea rely on chemical insecticides (cypermethrin + dimethoate), these are not readily available to all farmers. Therefore, economic and environmental concerns have redirected research efforts toward plant resistance (Olatunde et al. 1991a), locally obtainable plant extracts (Jackai et al. 1992), intercropping systems, and natural enemies (Matteson 1982). Although cowpea can compensate to some extent for insect damage during flowering, PSB damage can lead to total loss of the crop. High levels of infestation or extended periods of infestation reduce the ability of the plant to compensate. In Benin, PSBs were the most important pest group in 80% of fields, seed damage exceeding 50% in some locations (Dreyer et al. 1994). Losses from these bugs mask the damage potential of other insects because of competition (Dreyer and Baumgärtner 1995). In Nigeria, studies of damage at different stages and bug population densities yielded action threshold levels on cowpea as follows: for the flowering stage, two *C. tomentosicollis* (fourth instar) per ten plants, and for the pod stage, four *C. tomentosicollis* per ten plants. (Jackai et al. 1989). Minimal effective insecticide application (cypermethrin + dimethoate) was found to be two sprays at 10-day intervals, beginning at the flower-bud stage (Amatobi 1995). Bifenthrin is slightly less effective than other pyrethroids in controlling *Clavigralla* spp. (Amatobi 1994). Efficacy of various organochlorine, organophosphate, and carbamate compounds on PSB are given by Taylor (1968).

An aqueous extract of neem seeds applied to cowpea plants and detached pods caused 100% mortality of nymphs and adults; dry powder applied to the plants was less effective. Neem is used as a malaria suppressant in Nigeria, and therefore may be considered by farmers to be an acceptable control means (Jackai et al. 1992). Varieties of cowpeas were screened for resistance to bug damage using dry seeds (rather than pods or fresh seeds), and results were compared with those of field and greenhouse trials (Jackai 1990). Two wild cowpea relatives showed antibiosis; nymphal mortality was significantly higher on these lines, and seed damage was reduced in field tests. Additional screening tests identified several cultivars with low seed damage ratings (Olatunde et al. 1991b), and found evidence of feeding preferences but not ovipositional preferences among cowpea cultivars (Olatunde et al. 1991a). These cultivars exibiting antixenosis also possessed variable levels of antibiosis (Olatunde and Odebiyi 1991b), which appeared to be related to differences in crude protein content (Olatunde and Odebiyi 1991a). Presence of trichomes did not contribute to antibiosis (Jackai and Oghiakhe 1989). In Ghana, one variety with a particularly tough seed coat was found to be moderately resistant to PSB feeding (Agyen-Sampong 1978). Despite active research programs at the International Institute of Tropical Agriculture and the Institute for Agricultural Research in Nigeria, no cultivars with resistance to *Clavigralla* spp. are at present available to farmers, nor are there yet any effective cultural control methods (Amatobi 1995).

In Tanzania, the egg parasite *Hadronotus* (= *Gryon*) *gnidas* Nixon was reared from eggs collected on several crops. Percent parasitism was 5.6% on bean, 11% on pigeon pea, and 23.8% on *Dolichos lablab* L. (Materu 1971). The same species parasitizes *C. tomentosicollis* eggs in Nigeria (Taylor 1975; Egwuatu and Taylor 1979, 1983). The biology of this scelionid has been extensively studied (Taylor 1975; Egwuatu and Taylor 1977a,b,c); it is highly host-specific and will not attack *C. shadabi* or other coreoids. Parasitization can reach 76% in pigeon pea, but averages 38%; although *G. gnidas* has been implicated in the periodic fluctuations of *C. tomentosicollis* populations, it does not provide sufficient natural control to prevent economic loss (Egwuatu and Taylor 1979, 1983). Two other egg parasites (*Anastatus* sp., *Ooencyrtus* sp.) have been reared from *C. tomentosicollis* eggs on pigeon pea, but the incidence of parasitism was <0.3% (Egwuatu and Taylor 1979). On cowpea in northern Nigeria, *G. gnidas* was the major egg parasite, but the encyrtids *Ooencyrtus patriciae* Subba Rao and *O. kuvanae* (Howard) were also common. Although total egg parasitism reached peaks of 51 and 62%, bug population levels nonetheless exceeded the economic injury level in both years of the study (Matteson 1981). Hymenopteran egg parasitoids reported from *Clavigralla* spp. are summarized by Shanower et al. (1999). Low levels of adult parasitism by the tachinid *Alophora nasalis* Bezzi were recorded in Nigeria (0.2%) and Tanzania (0.9%) (Matteson 1980). Various general predators are listed by Egwuatu and Taylor (1983), but none is considered to be important in regulating bug populations. One study of intercropping indicated higher initial PSB numbers in maize–cowpea than in sorghum–cowpea cropping systems. Bug numbers appeared to increase on intercropped cowpea compared with a monocrop system in one study site in Nigeria, but decreased in another (Matteson 1982). Jackai et al. (1985) and Jackai and Daoust (1986) summarize other studies of cowpea cropping systems, showing mixed results. Although intercropping alone is not sufficient, Alghali (1993) suggests that

C. tomentosicollis populations may be effectively managed by combining sorghum–cowpea intercropping with reduced insecticide usage.

In east Africa, the species is considered along with *C. elongata* to be a pest of beans (*Phaseolus* spp.), pigeon pea, cowpea, and hyacinth bean (*D. lablab*), and much economic research focuses on the two species together, or as part of a complex including alydids and pentatomids. The PSB complex is considered to be among the most important economic pests on grain legumes in Tanzania (Kayumbo 1978). Details of economic damage and control methods for this complex are given under *C. elongata* below.

3.4.2 Clavigralla shadabi Dolling

In the economic literature of west Africa, this species has usually been referred to as *Acanthomia horrida* Germar; however, references to populations of *A. horrida* in east Africa are probably *C. elongata* Signoret (see below). Body length of males ranges from 7.9 to 9.8 mm, and that of females 8.5 to 10.5 (Dolling 1979). Adults are dark brown, with light gray wings; both the pronotal and abdominal spines are pronounced (Aina 1972).

The oval, white egg is laid singly on leaves and petioles of cowpea. Preoviposition period ranged from 6 to 9 days in mated females, and average lifetime fecundity averaged 284 eggs. Drawings of first, third, and fifth instars are provided by Aina (1972). Total nymphal development (egg to adult emergence) on cowpea averaged 22.4 days at 23 to 29°C (Aina 1972).

On cowpea in Nigeria, this species can complete two generations before harvest. Populations of the bug then remain on ratooning stands. Large populations especially occur in the drier season. No alternate host was noted (Aina 1972). In another study, populations of *C. shadabi* were found to peak in late May to June, in contrast to those of *C. tomentosicollis*, which reached highest numbers in November and December. Cowpea planted in June and July was least infested by PSB (Akingbohungbe 1982). A reduviid, *Rhynocoris bicolor* F., preyed on *C. shadabi*, but was found only at low density in the field (Aina 1972). In Ghana, the reduviid *Harpactor segmentarius* Germar fed on nymphs (Cobblah 1991).

3.4.3 Clavigralla elongata Signoret

3.4.3.1 *Distribution, Life History, and Biology* —
This species is found in central, eastern, and southern Africa, but is replaced by *C. shadabi* in the west African mainland. The two species overlap only in Zaire. It is often referred to in the economic literature of east Africa as *Acanthomia horrida*. *Clavigralla elongata* can be distinguished from *C. tomentosicollis* and *C. gibbosa* by the uniform pubescence of the pronotum; in the latter two species the anterior pronotal pubescence differs in color and texture from that of the posterior. It is very similar in appearance and habits to *C. shadabi*, but can be distinguished from the latter species by the relative size of the spines on the pronotal disk. In *C. elongata*, the two anterior pairs of spines are roughly equal in size; in *C. shadabi*, the front pair is noticeably larger. Females are 8.3 to 10.8 mm long; males are smaller (8.3 to 9.6 mm) (Dolling 1979). The white, oval eggs are laid singly (Materu 1972). Photographs and detailed descriptions of the adults, eggs, and nymphs are given by Materu (1972).

3.4.3.2 *Damage* —
In Tanzania, *C. elongata* and *C. tomentosicollis* are pests of beans, pigeon pea, cowpea, and hyacinth bean (Materu 1972, Kayumbo 1977). Beans (*P. vulgaris* L.) grown for seed are also damaged; however, nymphs were not observed on this crop. Adults moved into bean fields from other hosts at flowering, and fed on developing pods. Seed damage ranged from 0.6 to 30.8%, and several insecticides were effective (Swaine 1969). In Kenya, *C. elongata* and *C. tomentosicollis* adults infest pigeon pea, hyacinth bean, and gram, *Vigna radiata* (L.) R. Wilczek and *V. mungo* (L.) Hepper, in addition to cowpea (Khamala 1978).

Damage by adult *C. elongata* and *C. tomentosicollis* was compared on caged beans, using densities higher than those found in the field. Bug feeding by both species reduced all parameters measured: seed weight, seed number, and number of "clean" (i.e., high-quality) seeds. Visible damage included dimpled and wrinkled seed coats, and brown or shriveled seeds. Field experiments with pigeon pea showed that loss was proportional to bug density, ranging from 10.5 to 44.4% damaged seed; bug-damaged seeds also had lowered germination. The fungus *Nematospora coryli* Peglion was associated only with bug-damaged beans and peas, never with "clean" seeds (Materu 1970). The two species were differently distributed in the field: *C. elongata* was concentrated in areas of dense growth, particularly in beans, whereas *C. tomentosicollis* densities were highest on pigeon pea. Both species moved from small, peasants' fields into larger plantings of these crops after flowering, and completed two generations (Materu 1971).

3.4.3.3 Control — For cowpea, small farmers in Tanzania do not generally use any control techniques other than intercropping (Kayumbo 1977). Intercropping with maize in Kenya increased densities of these PSBs (Gethi and Khaemba 1991), as it did with the PSB complex in Nigeria (Matteson 1982). Overall bug density in the Kenya study was higher during the first planting (i.e., during the wet season). Endosulfan applied only to cowpea in the mixed cropping system did not provide control, because bugs moved to maize (Gethi and Khaemba 1991). Kayumbo (1977) compared direct counts, removal sampling, and mark-release-recapture techniques for the PSB complex on cowpea in Tanzania, and found visual count estimates of population density to be consistently lower than those from the other two methods. However, reliability of direct counts improved during the morning hours (0700 to 1030 hours).

Matteson (1980, 1981) reported egg parasitism by *O. patriciae* and several scelionid species, and adult parasitism by the tachinid *Alophora nasalis* Bezzi, in Tanzania.

3.4.4 Clavigralla gibbosa (Spinola)

3.4.4.1 Distribution, Life History, and Biology — Known as the tur pod bug, *C. gibbosa* is considered a major pest of pigeon pea (tur) and other legumes on the Indian subcontinent (Prasad and Chand 1990). A second species, *C. scutellaris* (Westwood), also occurs on pigeon pea, but until recently the two species were not differentiated in field studies (Shanower et al. 1999). Four detailed studies of the biology of *C. gibbosa* on pigeon pea have been conducted in India (Bindra 1965, Nawale and Jadhav 1978, Choudhary and Dhamdhere 1981a, Kashyap and Mehta 1992). Bindra (1965) summarizes the early literature. Anatomical and histological studies of the alimentary canal are reported by Kurup (1964); descriptions of the internal morphology of the reproductive organs are given by Ahmad (1979). Adults are reddish-brown and pubescent; the angles of the pronotum are spined. Males and females are similar but the female is somewhat larger (9.2 to 9.8 mm) with a more rounded abdomen. Body length of males ranges 8.1 to 8.9 mm (Dolling 1978). Estimates of life history parameters vary with rearing conditions, geographic region, temperature, and time of year. Nawale and Jadhav (1978) report average life spans varying from 25.4 to 31.8 days for males and 28.0 to 34.8 days for females. Maximal longevity of >150 days has been recorded (Bindra and Singh 1971). Multiple mating is common, but a single mating is sufficient to fertilize all eggs. Estimates of mean lifetime fecundity range 60 to 202 eggs per female. Adults first appear in the field around the time of pod formation (October or November), and the bugs remain active on pigeon pea until late May or until harvest. Generations overlap, and estimates of generations per year range from six (Bindra and Singh 1971) to eight (Choudhary and Dhamdhere 1981a). Whether the adults move to another leguminous host or estivate during June to October is not known. Eggs are laid in clusters of 2 to 62 (Shanower et al. 1999), mainly on pods, but also on leaves or buds of the host plant. They are white when laid, changing to dark brown. Nymphs are gregarious. Drawings of the egg, nymph, and adult are provided by Bindra and Singh (1971); Choudhary (1969) describes all instars. Depending on time of year, locality, and temperature, the egg incubation period may range from 5 to 18 days. Similarly, average nymphal development rate is reported as 15 days in March, but >20 days in December (Nawale and Jadhav 1978). A variable preoviposition period is also reported by most authors.

3.4.4.2 Damage — On pigeon pea, adults and nymphs mainly damage green (unripe) pods, but have also been observed to feed on stems, leaves, and flower buds. Feeding on the pods produces pale yellow blotches. Heavy infestation causes the pods to shrivel, and the seeds within remain small (Bindra and Singh 1971). Economically damaging populations have been reported in Madhya Pradesh, Maharashtra, Uttar Pradesh, and Himachal Pradesh (India) (Nawale and Jadhav 1978, Srivastava and Singh 1979, Choudhary and Dhamdhere 1981a, Kashyap and Mehta 1992). In Uttar Pradesh, population densities were considerably higher in rabi (spring) pigeon pea than in the kharif (fall) cropping season (Singh and Singh 1991). In Hyderabad, farther south, *C. gibbosa* is reported to feed on pigeon pea, but does not occur in large enough populations to necessitate control measures (Davies and Lateef 1978). In the laboratory, adults survived on detached pods of various legumes, but first instars could be reared to adulthood only on pigeon pea (Bindra 1965). On mung bean, high populations of *C. gibbosa* and *Riptortus* sp. cause severe damage to inflorescences, flower stalks, and tender pods in Uttar Pradesh. Premature flower drop and drying of pods of the summer crop are attributed to this bug complex (Sehgal and Ujagir 1988). Hyacinth bean and cowpea are also reported as food plants of this species (Hoffman 1933).

3.4.4.3 Control — Traditional control methods include shaking infested plants over oil, or hand-collecting (Bindra and Singh 1971). The scelionid egg parasite, *Hadronotus* (= *Gryon*) *antestiae* Dodd, has been recorded; estimates of parasitism were as high as 56% (Bindra 1965). Egg clusters of *C. gibbosa* and *C. scutellaris* are also heavily parasitized by *G. clavigrallae* Mineo (Shanower et al. 1999). Up to 20% parasitism of nymphs and adults by a predatory erythraeid mite, *Bochartia* sp., has been observed (Rawat et al. 1969). *Clavigralla gibbosa* showed significant ovipositional preferences among varieties of pigeon pea in small-plot field tests, but no resistance to adult or nymphal infestation was demonstrated (Choudhary and Dhamdhere 1981b). Studies of intercropping with other legumes, sorghum, or millet have yielded variable results: in New Delhi, *C. gibbosa* colonization was delayed by 8 weeks and populations were reduced, but no significant effect of intercropping was found in Orissa, farther to the southeast (Singh and Singh 1978, Patnaik et al. 1989).

Clavigralla gibbosa can be controlled with monocrotophos or endosulfan, but synthetic pyrethroids were not found to be so effective (Singh 1985). Eggs of *C. gibbosa* are highly susceptible to several carbamate insecticides (Singh et al. 1982).

3.5 *Cletus* Stål

Although *C. punctiger* Dallas has received the most attention in recent years as a rice pest in East Asia, other *Cletus* spp. are also of economic importance. *Cletus notatus* Thunberg is mentioned as a pest of cowpea in Nigeria (references in Aina 1972), and *C. bipunctatus* (Westwood) is an economic pest in Pakistan (Ahmad et al. 1983). Soybean pods are attacked by *C. trigonus* (Thunberg) in Taiwan (Wang 1980). *Cletus signatus* Walker is a major pest of rice in Manipur, India (10 to 40% infestation, reaching 2 to 3 bugs/10 panicles) and is also reported from groundnut (Singh and Singh 1987, CABI 1997). This species harbors symbiotic bacteria, which are transmitted transovarially. In laboratory culture, these bacteria were able to degrade DDT, parathion, and carbaryl (Singh 1974, Singh and Pant 1974).

3.5.1 *Cletus punctiger* Dallas

3.5.1.1 Distribution, Life History, and Biology — This species is distributed throughout north, central, and southern Japan, but is not found in the mountains (Kisimoto 1983). *Cletus punctiger* is polyphagous, feeding on various grasses and only dispersing to rice when ears of preferred wild hosts are unavailable (Ito 1982). Wang (1980) reports *C. punctiger* on soybean pods in Taiwan. Other host plant families include Polygonaceae and Cyperaceae (Ito 1978, Kisimoto 1983). Both adults and nymphs cause injury to the rice head. Movement of bugs into late-heading fields is rare; thus, early-heading fields are more likely to be damaged. This difference is attributed partly to a greater availability of alternate hosts later in the season (Ito 1982), and partly to a decrease in flight ability over time (Ito 1980). Both tethered flight and flight mill studies showed that the proportion of bugs flying continuously for >1 min. decreased from June through August, then increased in October, prior to the migration to overwintering sites (Ito 1980). The preferred overwintering site for *C. punctiger* is dry, sunny locations on leaves (Ito 1978).

In central Japan *Cletus punctiger* is thought to complete one generation annually, reproduction occurring in July and August (Ito 1980). Elsewhere in Japan, one to three generations are estimated, depending on location (Ito 1978, Kisimoto 1983). Eggs are laid singly. Nymphs are light colored with dark antennae; adults are brown, slender, with spines on the humeral angles of the pronotum. Egg development requires 114 degree-days, and no eggs survive temperatures below 13.0°C. For nymphs fed rice ears these values are 356.6 degree-days, 12.6°C for females, and 347.0 degree-days, 13.1°C for males (Ito 1978). The life span of overwintering adults was 240 to 360 days when provided with hulled rice and grass ears. Preoviposition period was 216 degree-days when females were fed ears of rice (Hiroshima Prefectural Agricultural Experiment Station 1977, cited in Ito 1978). Critical daylength required to induce diapause is 13.5 to 14.0 hours; individuals past the fourth stadium are susceptible (Shimizu and Maru 1976, cited in Ito 1978).

Starvation-longevity in this species has been extensively studied, to determine how the insects survive the periods when suitable host plants are scarce (Ito 1984a, 1985, 1986). In June and July, spring host plants are mostly senescent or plowed under, but rice plants are not yet reproductive. Vegetative (seedling) hosts are thought to provide inadequate nutrition for *C. punctiger* (Ito 1984a). Rice heads, however, are an adequate food for nymphal development (Ito 1977, cited in Ito 1978). Adults emerging from overwintering hibernation in April were shown experimentally to survive in large numbers until early June, when provided only with a water source (plant seedlings). However, such adults would not live long enough to colonize early-heading rice. The adults that successfully colonize rice are the ones that locate a spring host plant, feed until late May or June, and accumulate stores of lipids to survive until rice or other summer hosts become reproductive (Ito 1984a). Starvation-longevity is directly related to lipid reserves, whose accumulation is affected by diapause (Ito 1986). Lipid concentration reaches a peak in posthibernating adults in May and June, and again in prehibernating adults in Septembe to November; it is lowest in reproductive adults (Ito 1985).

3.5.1.2   Damage — The condition known as "pecky rice" is caused by a number of heteropteran insects; in Japan, this complex includes *C. punctiger*. When rice in the milk stage is damaged by pentatomids, a fungus may enter the grain through the feeding puncture. During milling, these damaged rice grains do not break, and the quality and marketing grade of the rice is therefore lowered. This condition is referred to as "peckiness" (Hill 1975). In Japan, rice grains showing any symptoms of stink bug feeding, including black or brown discolorations or raised blemishes, are classified as "pecky rice." At least 14 other heteropteran species, including pentatomid, mirid, lygaeid, rhopalid, and alydid bugs, have been implicated in addition to *C. punctiger* (Kisimoto 1983). An adult *C. punctiger* can damage 1.5 grains of milk-stage rice per day (Ito 1984b).

3.5.1.3   Control — Light traps, which are effective for some species that cause pecky rice, are not effective for *C. punctiger*, although these bugs are strong fliers (Ito 1978). Weed management near paddy fields is recommended; organophosphorus insecticides (fenitrothion, fenthion, trichlorfon, phenthoate) can provide control when necessary (Kisimoto 1983). Egg parasitism is reported to reach 60%; eight species of parasitic wasps are reported (Hiroshima Prefectural Agricultural Experiment Station 1978, cited in Ito 1978). Both *C. punctiger* and *C. rusticus* eggs are acceptable hosts for *G. japonicum* (Ashmead) (Hymenoptera: Scelionidae) (Noda 1990). Thresholds based on sweep netting have been established for *C. punctiger* (Shimizu and Maru 1978, cited in Ito 1978). For "second grade rice," no more than 15 to 30 bugs per 60 sweeps can be tolerated during the period from heading to the milk stage.

It should be noted that much valuable information on *C. punctiger* is published in Japanese, without English abstracts. This discussion, therefore, relies heavily on descriptive summaries of such research cited in English publications by Japanese authors; other information was simply not obtainable.

3.5.2   Cletus rusticus Stål

This species overlaps in distribution with *C. punctiger*, but is restricted to polygonaceaous hosts and does not damage rice. Interspecific matings occur in the field, however, and viable (albeit infertile) first-generation progeny can be obtained from laboratory crosses. The hybrid and *C. punctiger* are equally damaging to rice in the milk stage. Photographs of both species and the hybrid are provided in Ito (1984b). Adult *C. punctiger* may be distinguished from *C. rusticus* by the smaller body size, lighter color, and absence of a black stripe on the first antennal segment of the former species.

3.5.3   Cletus bipunctatus (Westwood)

Distributed throughout Pakistan, this species feeds on amaranth (*Amaranthus viridis* L.) and purslane (*Portulaca oleracea* L.). Eggs are white, oblong, 1.1 mm long, and are laid singly on plant surfaces. Egg incubation requires 3 to 7 days, and total nymphal development ranges 15 to 25 days (Ahmad et al. 1983). Detailed descriptions and measurements of each instar are also given by these authors, with a key to distinguish *C. bipunctatus* nymphs from those of *Cletomorpha hastata* F. Chromosomal studies are reported by Ahmad (1979).

3.6   Leptoglossus Guérin-Méneville

This is a large genus with more than 40 species (Brailovsky and Sánchez 1983), all but one restricted to the Western Hemisphere. The revision by Osuna (1984) divided *Leptoglossus sensu lato* into several genera: *Veneza* gen. nov., *Fabrictilis* gen. nov., *Theognis* Stål, *Stalifera* gen. nov., and *Leptoglossus sensu stricto*; however, this arrangement was not accepted by Froeschner (1988) or Henry and Froeschner (1992); see Packauskas and Schaefer (in press).

All species are characterized by flattened, dilated, hind tibiae, ranging in shape from lanceolate to broadly scalloped. The head is elongate and narrow, extending forward beyond the bases of the antennae. The humeral angles of the pronotum are often expanded, and the hemelytra in many species bear a white or pale yellow stripe, zigzag, or series of dots. Femora are spined beneath, and in some species males have thicker hind femora than females (Allen 1969). Detailed drawings of the Mexican species are given by Brailovsky and Sánchez (1983). These bugs differ from most other genera in the tribe Anisoscelini in that the eggs are hemicylindrical (like those of *Phthia*, see **Section 4.11**) rather than ovoid or dome-shaped (Osuna 1984). Feeding habits are variable, ranging from extreme polyphagy, e.g., *L. phyllopus* (L.), *L. gonagra* (F.), to specialization on a single genus, e.g., *L. ashmeadi* Heidemann and *L. brevirostris* Barber on *Phoradendron*, *L. fulvicornis* (Westwood) on *Magnolia* (Wheeler and Miller 1990). In addition to the ten economically damaging species discussed below, *L. chilensis* (Spinola) has been recorded from peaches, nectarines, plums, grapes, figs, and grapefruit (Allen 1969); *L. impictus* (Stål) from tomato and potato (Bosq 1937); and *L. cinctus* (Herrich-Schaeffer) from cucurbits and tomato (CABI 1997). *Leptoglossus dilaticollis* Guérin has been observed on cacao (Monte 1937, cited in Costa Lima 1940). Fruit fall in a cultivated palm, *Bactris gasipaes* Kunth, is caused by *L. lonchoides* Allen; this species is reported from Brazilian and Peruvian Amazonia (Couturier et al. 1991, 1996b). An infestation of *L. concolor* (Walker) was observed in a Florida orange grove (Mead 1974, cited in Baranowski and Slater 1986), and this species has also been reported on guava (CABI 1997). *Leptoglossus balteatus* (L.) feeds on several crops including tomatoes, oranges, cowpeas, guava, and *Luffa* spp. in Cuba, although the latter two are the only reported breeding hosts (Barber and Bruner 1947).

3.6.1 Leptoglossus australis (F.) (= Leptoglossus membranaceous F., Leptoglossus bidentatus Montrouzier)

3.6.1.1 Distribution, Life History, and Biology — This is the only *Leptoglossus* found outside the Western Hemisphere, and is widely distributed from Africa to India, Australia, Southeast Asia, Taiwan, Japan, and the South Pacific islands (Allen 1969). Several authors have noted the similarity between *L. australis* and the Neotropical *L. gonagra* (F); Baranowski and Slater (1986) consider the Eastern Hemisphere bugs to represent a geographically separated population of *L. gonagra*, and Carver et al. (1991) use *gonagra* for the sole Australian species. Allen (1969), although noting the close resemblance between *gonagra* and *australis*, maintains their taxonomic status as separate species. He proposes *gonagra* as the ancestral stock for *australis*, the original dispersal to the Eastern Hemisphere having been through Africa. Froeschner (1988) reports only New World distribution records for *L. gonagra*, clearly maintaining the two species as distinct.

Adults are black or dark brown, 18 to 24 mm long, with a distinctive reddish-orange or yellow curved band on the anterior of the pronotum. The venter is spotted with orange-red. Males have enlarged hind femora. Other characteristics such as tibial dilations and humeral angles are extremely variable over the geographic range (Fernando 1957, Allen 1969, Miyatake 1993).

Eggs are pale brown, 1.6 mm long, and laid in batches of up to 32 (Visalakshi et al. 1980). The preferred oviposition site on cucurbit vines is the tendrils. The early instars are reddish, and the distinctive foliated hind tibiae appear in the third instar. In all instars the abdomen is spined laterally. Fourth instars develop yellow and black markings, and the fifth instar is dark brown to black (Fernando 1957, Visalakshi et al. 1980). All instars are illustrated by Visalakshi et al. (1980), and described by Pagden (1928). Duration of the egg stage is given as 6 to 7 days in India and 8 to 10 days in Papua New Guinea; nymphal development required 52 days in western India (Visalakshi et al. 1980, Szent-Ivany and Catley 1960b).

Nymphs of all stages are gregarious (Szent-Ivany and Catley 1960b). In Japan, adults overwinter in large aggregations on the withered leaves of the host plant *Luffa aegyptiaca* Miller (= *L. cylindrica* [L.]), and on several nonhost species. On Okinawa, these aggregations begin to disperse in early March (Miyatake 1992). On Ishigaki Island, waves of immigrating adults moved into cultivated fields during June to July, presumably from wild melon or other cucurbits. Cages containing males attracted other individuals, suggesting the presence of a male pheromone (Yasuda 1990, Yasuda and Tsurumachi 1994). However, males show aggressive behavior toward one another, using the enlarged hind femora in combat on *L. aegyptiaca* fruits (Miyatake 1993).

The seasonal history and ecology of *Leptoglossus australis* have been studied in detail in subtropical Japan (Yasuda 1990). The developmental threshold was 15.4°C for eggs and 17.3°C for nymphs. The effective accumulative temperature in day-degrees was estimated at 81.1 for eggs and 450.6 for nymphs. Four generations per year are estimated to occur in this region. The wild melon, *Bryonopsis laciniosa*, serves as the first breeding host of the year; adults immigrate to bitter gourd and other cultivated cucurbits and complete two generations. A final generation develops on the wild host during October to November and these adults, the overwintering population, may damage citrus.

3.6.1.2 Damage — This species is highly polyphagous. It is reported as a common, widely occurring, but minor pest of cucurbits and a wide variety of alternative hosts (Hill 1975). In India, the bugs damage citrus fruits, pomegranate, and snake gourd (*Trichosanthes cucumerina* L.). On snake gourd, the vine stems and fruits are both subject to feeding damage (Visalakshi et al. 1980, and references therein). Damage to pomegranate is caused primarily by adults, although all stages are found on the crop. Feeding on ripening pomegranates causes the fruit to rot and drop prematurely (Jadhav et al. 1976). In Nigeria, feeding records include *Citrus paradisi* Macfadyen and *C. sinensis* (L.), *Passiflora foetida* (L.), *Cucumis sativus* L., and *T. cucumerina* L; nymphs were observed only on the last three hosts (Golding 1937). Similarly, in Sudan, Rhodesia, and southern Africa, damage is reported to citrus fruits and cucurbits (USDA 1965). Crops attacked by adults and nymphs in Japan are bitter gourd, cucumber, and loofah (*Luffa aegyptiaca* Miller); adults damage citrus and may be found occasionally on tomato and guava (Yasuda 1990, and references therein). In Ceylon, damage is particularly severe to orange, tangerine, and cucurbits, but other recorded hosts include the tree tomato *Cyphomandra crassicaulis* (Ortega) Kuntze, passion fruit, peach, plum, the cape gooseberry *Physalis peruviana* L., bean, pea, vegetable marrow, and cabbage (Green 1912, Fernando 1957). The bugs are not regular pests, however; only in occasional years does an infestation reach epidemic levels (Fernando 1957). Adults damage oranges in Malaya, and all stages are injurious to cucurbits, cotton, and bean (Pagden 1928). Host records in Papua New Guinea encompass 26 plant species from 15 families, but economic damage is restricted to the families Cucurbitaceae, Passifloraceae, and Rutaceae. *Passiflora edulis* Sims, *P. quadrangularis* L., and the wild *P. foetida* (L.) are heavily attacked along with *Cucumis melo* (L.), *C. sativus* (L.), and mandarin orange (*Citrus reticulata* Blanco) (Szent-Ivany and Catley 1960b). Injury to the latter crop results in premature fruit drop (Szent-Ivany and Stevens 1966). One incidence of damage to eucalypts has been reported (Szent-Ivany and Catley 1960b). Other crops on which these bugs have been observed are cabbage, cacao, cassava, coffee, cowpea, groundnut, mango, millet, oil palm, pigeon pea, rice, rubber, sweet potato, and yam (Szent-Ivany and Catley 1960b, Hill 1975, Singh et al. 1978b, Schaefer and Mitchell 1983).

Although many host plants have been documented for this species, clear feeding preferences exist. Bitter gourd (*Momordica charantia* L.) was significantly preferred over seven other fruit tested in the laboratory (Jadhav et al. 1979). Cotton and banana were the least preferred among the fruit tested; and preference for snake gourd, smooth gourd (*L. aegyptiaca* Miller), ridged gourd (*L. acutangula* Roxburgh), sweet orange (*C. sinensis* (L.) Osbeck), and pomegranate (*Punica granatum* L.) was intermediate. These results are consistent with field observations in Ceylon showing bitter gourd to be considerably preferred over snake gourd (Fernando 1957).

Feeding on bitter gourd or other vines results in wilting of the plant beyond the point of attack. Nymphs feed primarily on vegetative parts of the vine and on dry, exposed seeds; adults feed on developing and mature fruit, penetrating to the seeds. Damaged stems turn yellowish-brown and may die if an infestation is severe. Feeding on leaves causes browning and withering. Fruit damage is visible as a superficial color change from green to orange-red, collapsed pulp, and empty seeds. Light infestations result in patchy orange-red discoloration, which reduces market value. Mature fruit may split following heavy feeding damage. *Luffa* fruit exhibit similar empty seeds and collapsed pulp, followed by shriveling, necrosis, and premature shedding. Oranges and tangerines also drop prematurely when attacked, and exhibit collapsed pulp and patches of surface discoloration (Fernando 1957).

3.6.1.3   Control — Early recommendations for control include hand-picking, capturing on trays smeared with tar, or kerosene emulsions (Green 1912); DDT and malathion have also been used (Szent-Ivany and Catley 1960b). Parathion and BHC are effective, although control measures are not often needed (Hill 1975). Because of inadequate coverage in vine crops like cucurbits, liquid spray formulations are not always effective; dusting with gamma-BHC was found to provide the more reliable control. Susceptibility of nymphs to various organochlorine insecticides did not precisely match that of adults; for example, endrin was highly effective against nymphs but caused low adult mortality. Reinfestation from both eggs in the field and adults from wild hosts necessitated repeated dusting of the crop (Fernando 1957).

Two reduviid species have been observed preying on nymphs in Ceylon (Szent-Ivany and Catley 1960b). Mantids (*Hierodulam patellifea*) and spiders (*Neoscona theisi*) prey on nymphs in Japan, and reduviids have been observed puncturing eggs. Interspecific competition with the melon fly, *Dacus cucurbitae* Coquillett, may also reduce nymphal populations in bitter gourd, because fly eggs hatch faster, and the larvae cause fruit to drop before bug eggs hatch (Yasuda 1990). (This, unfortunately, is of little practical value to the squash grower.) A major source of egg mortality is parasitism, primarily by the scelionid wasp *Gryon pennsylvanicum* (Ashmead), but occasionally by *Anastatus japonicus* (Eupelmidae). The scelionid female is attracted to caged *Leptoglossus australis* males, indicating that a bug pheromone is being used by the wasp in searching for host eggs. Ant predation also accounts for some loss of eggs (Yasuda and Tsurumachi 1995).

3.6.2 *Leptoglossus clypealis* Heidemann

3.6.2.1 Distribution, Life History, and Biology — The spine extending forward from the front of the head readily distinguishes this species, which has been reported as a pest of almonds, plums, and pistachio in the southwestern United States (Heidemann 1910, Anonymous 1985, Rice et al. 1985). Adults are tan to light brown, with a pale brown transparent membrane. The band on the corium is a broad zigzag, whitish-yellow to orange-yellow. Dilations of the hind tibiae are lanceolate, as in *L. corculus* and *L. occidentalis* (see below). Adults range in length from 16 to 20 mm (Heidemann 1910). The species is distributed from Iowa and Kansas west to California and south into Mexico (Allen 1969, Brailovsky and Sánchez 1983). Chemistry of the alarm pheromones is given by Aldrich and Yonke (1975).

In central Texas, the breeding host is juniper (*Juniperus* spp.) (Mitchell 1980b), and in Missouri, nymphs are found on sumac, *Rhus aromatica* Aiton (Froeschner 1942). Adults have also been collected from arborvitae (*Thuja* sp.) in Texas (J. C. Schaffner, Texas A&M University, personal communication). Nymphs, adults, and mating pairs have been collected from three species of juniper in Arizona; mating pairs were also found on manzanita, *Arctostaphylos pungens* Kunth (Ericaceae), and a few nymphs and adults on *Dasylirion wheeleri* S. Watson (Agavaceae) (Jones 1993). In Texas, adults overwinter and are active on juniper from April until early November. Eggs, nymphs, or gravid females were found in May to June and again in September to October, suggesting two generations per year (Mitchell 1980b). Nymphs can be reared in the laboratory on sunflower seeds, green beans, and juniper cuttings in water. At 26°C and 16:8 (L:D) hours, the duration of nymphal development from hatch to adult was 31.5 (±3.1) and 34.0 (±3.0) days for males and females, respectively (Mitchell 1980b).

3.6.2.2 Damage — On pistachio in California, adults first appear in late July, followed by nymphs in mid-August; both occur continuously for the remainder of the growing season (Michailides et al. 1987). The bugs are found in all major pistachio-growing areas of California (Bolkan et al. 1984). Feeding on pistachio (*Pistacia vera* L.) causes epicarp lesion, a condition in which the outer hull darkens and hull tissues collapse, leading to abortion of immature fruit. The fruit shows a large, sunken discoloration with dried resinous material. Necrotic spots on the surface or internal browning of the mesocarp and endocarp may also be visible. Occurrence of epicarp lesion ranged from 4 to 49% in infested orchards in California in 1984. When fruit are attacked in later stages of development, the feeding on partially hardened endocarp tissues produces kernel necrosis and distortion of the nutmeat. The bugs are capable of penetrating the shells until harvest. Up to 16% of nuts have been found to be damaged in this manner. Bugs feed preferentially at the stem end and along sutures, so much of the damage is localized at the fruit base (Bolkan et al. 1984, Michailides et al. 1987, Michailides 1989). *Leptoglossus clypealis* is not the only culprit; *L. occidentalis* Heidemann and several species of pentatomids have also been implicated in kernel necrosis, and mirids can cause epicarp lesion (Rice et al. 1985).

Several hemipterans, including *L. clypealis*, are suspected to vector the pistachio disease stigmatomycosis, which causes affected kernels to become rancid and slimy. Fungi, particularly *Nematospora* spp., are associated with this disease. In laboratory and field-cage studies, *N. coryli* Peglion was transmitted by *L. clypealis*, producing symptoms of stigmatomycosis (Michailides and Morgan 1990, 1991).

Damage to almonds includes gumming, black-spotted hulls, and shriveled and collapsed nutmeats. Infestations in almond orchards are not a widespread problem in California but may be locally severe, and appear to be associated with dry spring weather (Anonymous 1985).

3.6.2.3   Control — Carbaryl and permethrin sprays were ineffective in controlling epicarp lesion in pistachio, because of continuous invasion over time by leaf-footed bugs and other hemipteran species (Michailides et al. 1987). The parasitic wasp *G. pennsylvanicum* (Ashmead) (Scelionidae) is reported from eggs of an unidentified coreid on almond (Masner 1983); it is possible that this host was *L. clypealis*.

3.6.3   *Leptoglossus conspersus* Stål

Allen (1969) reports this species only from Colombia and Brazil, but the range was extended to Mexico by Brailovsky (1976). Brailovsky and Sánchez (1983) include Venezuela in the range as well, and it seems likely that additional collections will eventually resolve this apparently discontinuous distribution. Adults are reddish-brown with an irregular whitish-yellow stripe on the corium. The pronotum bears two large yellow-orange spots. The tibial foliations are scalloped on the outside but lanceolate on the inner side. The species is very similar in appearance to *L. zonatus*, but may be separated by the larger pronotal spots in *L. conspersus* and the coloration of the antennal segments, bicolored in *L. zonatus* but unicolorous in *L. conspersus* (Allen 1969, Brailovsky and Sánchez 1983). In Brazil, this species is reported to feed on tomato fruits (Monte 1934, 1937; cited in Costa Lima 1940), but is not considered among the more important species of *Leptoglossus* (Costa Lima 1940). In Colombia, *L. conspersus* is listed as a pest of passion fruit, *Passiflora edulis* var. *flavicarpa* Sims and *P. quadrangularis* L. The bugs are reported to feed by extracting sap, and may cause flowers and young fruit to drop. On *P. quadrangularis*, deformed fruit may also result from bug damage (Anonymous 1975).

3.6.4   *Leptoglossus corculus* (Say)

3.6.4.1   Distribution, Life History, and Biology — Known as the leaf-footed pine seedbug, this species is considered to to be one of the most destructive insects attacking pine seed orchards in the southeastern United States. These orchards, which produce not lumber but improved seeds for reforestation efforts, may suffer up to 50% seed loss from sucking insect damage (DeBarr 1967). Adults are light or dark brown, with small black spots on the pronotum and venter. The pronotum is rounded. The stripe on the corium is a narrow, pale yellow zigzag, restricted to the veins, and may be absent entirely. The expanded hind tibia are lanceolate, the outer dilation the longest, extending along 85 to 95% of the total length of the tibia. Length is 18 to 19 mm. This species is restricted to the eastern half of the United States, from New York to Florida and west to Texas and Wisconsin; it is not reported from Mexico (Allen 1969, Brailovsky 1976, Katovich and Kulman 1987). Records from farther west are probably *L. occidentalis* (see below) (Allen 1969).

The hemicylindrical eggs, 1.75 mm long, have a large dorsal operculum and are laid end to end in a linear arrangement of 6 to 20 eggs (mean = 10) on pine needles. Eggs are cream colored, turning dark reddish-brown over time (DeBarr 1967). Detailed descriptions of adult and nymphs, with a key to the instars and information on duration of each stadium, are provided by Ebel et al. (1981). At 24°C, 16:8 (L:D) hours, development required 12.43 days for the egg and 43.17 days from hatch to adult molt. Adults overwinter, and are continuously present from late March until

November in Georgia. Reproductively active females occur from mid-April through late August, and nymphs are evident from May until the end of the growing season in late October. Two full generations and a partial third are estimated (Ebel 1977).

3.6.4.2 Damage — Pine is the only known breeding host. Younger nymphs feed on needles and on ovules in first-year female strobili ("conelets"), whereas third through fifth instars and adults damage seeds on more mature cones. First instars aggregate and feed on needles near the egg mass. Second instars feed on conelets, destroying individual ovules and inducing conelet abortion. Caging second instars on shortleaf pine (*Pinus echinata* Miller) conelets resulted in 100% abortion. Conelets protected from nymphs by exclusion cages, or exposed to adult bugs, did not abort. Results were similar for loblolly pine, *P. taeda* L. (DeBarr and Ebel 1974). The nymphs penetrate nucellar tissue with their stylets, feeding on cytoplasm (DeBarr and Kormanik 1975). Adults may feed on developing male flowers, shoots, or buds, but the primary feeding site is second-year cones. Stylets of adults and older nymphs penetrate the cone scales and seed wings to the endosperm of the seed, causing this layer to collapse. Depending on the duration of feeding, the seed may appear hollow, or the endosperm may be collapsed, shrunken, or missing entirely, and the seed coat may be discolored in the vicinity of the penetration site (DeBarr 1970). Often, no external damage is evident although the endosperm is entirely destroyed and the seed is nonviable. Even minor necrosis of the endosperm, produced by brief feeding, may lower the chance of germination, and pathogen invasion becomes more likely once the seed coat is punctured by the stylets. Radiography can be used to detect otherwise inconspicuous damage to mature seeds (DeBarr 1970). Most southeastern species of *Pinus* are subject to feeding by these bugs, except jack pine (*P. banksiana* Lamb) (Ebel 1977) and possibly sand pine (Hedlin et al. 1981). In the southeastern United States, populations occur regularly in high numbers on shortleaf pine (*P. echinata*), spruce pine (*P. glabra* Walter), Virginia pine (*P. virginiana* Miller), and the exotic Japanese red pine (*P. densiflora* Siebold and Zuccarini), and in moderate numbers on slash pine (*P. elliottii* Engelmann), longleaf pine (*P. palustris* Miller), pitch pine (*P. rigida* Miller), loblolly pine (*P. taeda* L.), and eastern white pine (*P. strobus* L.) (Ebel 1977). One *L. corculus* adult was observed to feed on cones of Fraser fir, *Abies fraseri* (Pursh) Poiret in North Carolina (Bradley et al. 1981), but no breeding records on a conifer other than pine are recorded. In Wisconsin, *L. corculus* adults and nymphs have been collected from red pine, *P. resinosa* Aiton, with damage to seed evident from radiographic analysis (Katovich and Kulman 1987).

3.6.4.3 Control — Intensive research has been conducted on chemical and biological control in the 30 years since the economic importance of this species was first recognized (DeBarr 1967). Laboratory screening of insecticides applied topically has shown that second instars are highly susceptible to many commercially available insecticides. Aldicarb, aminocarb, azinphosmethyl, carbofuran, dicrotophos, monocrotophos, propoxur, and several experimental pyrethroid compounds were shown to have LD_{90} values <1 µg/g of body weight. Natural pyrethrins were comparatively nontoxic, although sublethal doses resulted in developmental loss of the hind legs in nymphs (DeBarr and Nord 1978). Tests on adults showed similarly high levels of toxicity: decamethrin, carbofuran, aminocarb, fenvalerate, and bioethanomethrin were more toxic than the standard, azinphosmethyl (LD_{90} = 1.76 µg/g for females) (Nord and DeBarr 1983). Deltamethrin, azinphosmethyl, and fenvalerate have high residual contact toxicity against adults on loblolly pine foliage; propoxur, malathion, and chlorpyrifos had low residual toxicity but were effective as fumigants (Nord 1990). Persistence of insecticides under field conditions varied. Effectiveness of fenvalerate against *L. corculus* adults remained high even after 6 weeks, but that of azinphosmethyl (applied as a wettable powder) diminished within 1 to 2 weeks (Nord and DeBarr 1992). Field tests of insecticides applied for control of coneworms confirmed that azinphosmethyl, carbofuran, and dicrotophos were effective against *L. corculus* (DeBarr and Nord 1978). Both low- and high-volume sprays of fenvalerate, azinphosmethyl, and permethrin at monthly intervals are effective, as are high-volume applications of phosmet (Nord et al. 1984, 1985). Carbofuran was the first insecticide registered for control of this species in southern pine seed orchards (DeBarr and Nord 1978).

Biological control of *L. corculus* has also been intensively studied as an alternative to chemical spraying. One adult parasite, the tachinid *Trichopoda pennipes* (F.), and one predator, the wheelbug *Arilus cristatus* (L.) (Hemiptera: Reduviidae), have been identified (DeBarr 1969, cited in Fedde 1982). Eggs are attacked by three hymenopteran parasites: a scelionid, *G. pennsylvanicum* (Ashmead) (Masner 1983); a eupelmid, *Anastatus* sp. (G. Fedde, personal communication, cited in Fedde 1982); and an encyrtid, *Ooencyrtus trinidadensis* Crawford (Yoshimoto 1977, as *O. leptoglossi* n. sp.). The biology of the encyrtid was intensively studied by Fedde (1982), and its potential as a biological control agent was assessed. The parasite produces multiple progeny per host egg, and the generation time is short; thus, if orchard conditions favored parasite survival, this species could provide effective control. Eggs of two other *Leptoglossus* species, as well as various pentatomids and lepidopterans, were found suitable for laboratory rearing of this parasite, or possibly as alternate field hosts. Optimal rearing conditions were determined to be 24 to 30°C, 76% RH.

3.6.5 Leptoglossus gonagra F. (= Fabrictilis gonagra (F.))

3.6.5.1 Distribution, Life History, and Biology — Closely allied and morphologically very similar to *L. australis*, this species is distributed widely in the Neotropics, from Mexico to Argentina. In the United States, it has been collected from Florida, Louisiana, Texas, and Missouri. It may be distinguished from *L. australis* by the longer outer tibial dilations, occupying 85 to 90% of the hind tibia, and the yellow spots on the metathoracic scent gland openings (Allen 1969). Commonly known in Florida as the citron bug, this species is considered only a minor pest of citrus groves in Florida (Bullock 1974), but a major pest on several crops in South America.

Eggs are dark brown, just under 1.4 mm in length, and laid in rows of 12 to 14. In laboratory and field, eggs are laid on varied substrates, and not always on the host plant. Early instars are bright red with black appendages, and highly gregarious. The last two instars are dull brown with the flattened hind tibiae well developed (Leonard 1931, Amaral and Storti 1981). Egg incubation required 8.21 ± 0.03 days at ambient temperatures ranging from 25.76 to 29.25°C and 65% RH. Nymphs reared individually on pumpkin had an average development time of 54.42 days, but mortality was high, particularly in the second stadium. Adult longevity was 36.91 ± 4.61 and 37.15 ± 4.67 days for males and females, respectively, with a preoviposition period of 20.09 ± 2.95 days. Average clutch size was 13.12 ± 0.66 (maximum 40 eggs), and average lifetime fecundity was 61.54 ± 14.87 eggs (maximum 169) (Amaral and Storti 1981).

3.6.5.2 Damage — In Brazil, *L. gonagra* feeds on Brazilian guava (araçá), guava, oranges, passion fruit, pomegranate, pumpkin, and vegetable marrows; nymphs are found mainly on *Momordica charantia* L. and *Solanum americanum* Miller (Costa Lima 1940, and references therein; Amaral and Storti 1981). Bosq (1940a) reports this species (as "*Leptoglossus gonager*") to be abundant on tobacco, and also attacking citrus, cultivated cucurbits, and passionflower. In Puerto Rico, ripening oranges and grapefruits are damaged but *M. charantia* is reported to be the breeding host (Leonard 1931). *Luffa aegyptiaca* Miller (= *L. cylindrica* [L.]) supports large breeding populations in Cuba, but adults also feed on guava and oranges (Barber and Bruner 1947). In Florida, populations develop on citron melons or in old watermelon fields; when these plants dry up, the bugs migrate to cultivated tangerines and oranges (Thompson 1934, Mead 1971b). Other crops on which *L. gonagra* has been reported include corn, squash, tobacco, physic nut, and mango (Allen 1969, Baranowski and Slater 1986, Grimm and Führer 1998). Studies in Brazil have shown *L. gonagra* (as *Fabrictilis gonagra*) collected on citrus to carry *Phytomonas* spp. (a trypanosomatid) in the salivary glands and digestive tract. *Phytomonas* spp. are plant parasites thought to be transmitted by insect saliva (Sbravate et al. 1989). Nymphs and adults transmit the fungal pathogen *Nematospora coryli* Peglion to citrus fruit (Dammer and Ravelo 1990).

Stylet insertion and damage to grapefruit and oranges by *L. gonagra* and *L. phyllopus* (L.) are described in detail by Albrigo and Bullock (1977). Some feeding sites, mainly in parenchyma tissue between the oil glands, developed secondary fungal infections; punctures through the oil glands did not, but release of oil did result in necrotic spots on the rind. Alkali treatment of bug-damaged grapefruit (for commercial peeling and processing) resulted in reddish-brown spots forming on the surfaces of the peeled segments where bug feeding had occurred. Fruit drop caused by bug damage has been noted by other authors, but was not documented in this study (Albrigo and Bullock 1977, and references therein). Damage to ripening oranges is initially visible on the surface as small, pale, watery spots. Beneath the rind, this discoloration is more pronounced, and a region of drier pulp cells extends from the spot toward the center of the fruit. Damaged pulp acquires a bitter flavor, and multiple damage spots infrequently caused fruit drop (Leonard 1931). Calza et al. (1964) also noted fruit drop associated with feeding damage to citrus.

3.6.5.3 Control — To prevent infestations, Florida growers are advised to time their mowing activity in the citrus groves to destroy known host plants (e.g., citron) before nymphs complete development (Albrigo and Bullock 1977). Pyrethrum-oil contact sprays have been used sucessfully in the past, but removal of the ground cover on which nymphs developed was strongly recommended as a control tactic (Leonard 1931). Azinphosmethyl, carbofuran, parathion, and malathion provided equivalent control when bugs were caged on treated foliage in the field (Bullock 1974); propoxur (Baygon) and mevinphos (Phosdrin) were effective in laboratory tests (Calza et al. 1964). Specimens of a scelionid wasp, *G. carinatifrons* (Ashmead), are known from *L. gonagra* eggs in Florida (Masner 1983), but no recent research on biological control has been done. Turkeys, maintained in citrus groves, were used at one time as a control measure (Watson 1940, cited in Bullock 1974).

3.6.6 Leptoglossus occidentalis Heidemann (= Theognis occidentalis [Heidemann])

3.6.6.1 Distribution, Life History, and Biology — Originally restricted in distribution to the western United States, Canada, and Mexico, this economically important conifer pest has recently moved eastward in the United States, and is now reported in the Midwest (as *Theognis occidentalis*), North Atlantic, and New England states, and Ontario (Canada) (Schaffner 1967, Allen 1969, Katovich and Kulman 1987, McPherson et al. 1990, Marshall 1991, Gall 1992). The eastward spread has been mapped by Gall (1992), who notes that the first specimen from Pennsylvania occurred in a corn shipment from the Midwest, and the first Connecticut specimen was found indoors on a Douglas fir Christmas tree. Thus, although the bugs are strong fliers and much of the eastward movement may be attributed to direct dispersal following commercial pine plantings and landscaping, transcontinental shipping has probably played a role as well (Gall 1992).

The species is similar in appearance to *L. corculus*, from which it may be distinguished by the somewhat shorter outer tibial dilation, occupying 66 to 72% of the tibia and equal in length to the inner dilation (Allen 1969). Gall (1992) has suggested that the dorsal abdominal coloration is also useful as a distinguishing character; that of *L. occidentalis* is patterned distinctively in orange and black whereas the tergites of *L. corculus* are uniformly dark or have small cream or white patches.

One generation occurs annually in California; three in Mexico (Hedlin et al. 1981, Cibrián-Tovar et al. 1986). Adults overwinter in large clusters initiated by a male aggregation pheromone (Blatt and Borden 1996a), and feed in spring on male flowers, causing deformation and reducing pollen production. Egg masses are laid in clusters linearly along needles of the host tree. In the laboratory, females oviposit on average 12 eggs per day; average lifetime fecundity is 73 eggs per female. The egg, 2 mm in length, is light brown when first laid but turns reddish-brown as the embryo develops (Koerber 1963, Hedlin et al. 1981, Cibrián-Tovar et al. 1986). Adults and nymphs aggregate on cones and tips of branches; distribution is patchy and preferences for particular clones within a host plant species are evident in seed orchards (Blatt and Borden 1996b).

3.6.6.2   Damage   — Bugs feed by piercing the cone scales and penetrating the developing seeds. A detailed description of feeding activity is given by Koerber (1963). If the seed is damaged before the seed coat has hardened, the seed appears empty and flattened. If the seed is damaged when mature, the tissues become spongy and shrunken, but the seed does not collapse (Hedlin et al. 1981). The female gametophyte is partially to completely disorganized (Krugman and Koerber 1969). In addition to cones, feeding has also been observed on terminal shoots (Schaffner 1967), and on exposed seed in fully opened cones (Cibrián-Tovar et al. 1986).

Host plants include *Pseudotsuga menziesii* (Mirbel) Franco (Douglas fir), *Pinus ponderosa* Lawson, *P. contorta* Douglas, *P. banksiana* Lambert, and *P. attenuata* Lemm. in California (Koerber 1963, Krugman and Koerber 1969); *P. sylvestris* L. in Iowa (Schaffner 1967); *P. resinosa* Aiton in Minnesota and Wisconsin (Katovich and Kulman 1987); *P. nigra* J. F. Arnold in Illinois; *P. strobus* L. and *Picea glauca* (Moench) Voss (white spruce) in Michigan (McPherson et al. 1990). In New York, nymphs and adults have been collected from *Tsuga canadensis* (L.) Carrière, and adults from *P. mugo* Turra (Gall 1992). In Mexico, 15 species of *Pinus* are reported as hosts (Cibrián-Tovar et al. 1986). Laboratory studies have shown that whole or broken seeds of nearly all pines and firs of commercial importance in California are suitable for nymphal development; the one exception was *Chamaecyparia lawsoniana* (A. Murray) Parlatore (Port-Orford cedar) (Koerber 1963). Adults have been collected on almond trees in California, and one batch of nymphs developed successfully when caged on pistachio (Uyemoto et al. 1986), indicating that the apparently strong reliance on Pinaceae as a food source is not absolute. Adults are frequently found overwintering in houses (Marshall 1991), constituting a nuisance for homeowners in addition to the damage this species causes in seed orchards.

Seed losses up to 41% in Douglas fir, and 26% in western white pine, have been attributed to this species (Hedlin et al. 1981). In Mexico, it is considered one of the most important pine pests, causing seed crop losses of 30% in *P. cembroides* (Cibrián-Tovar et al. 1986). However, in British Columbia, populations on lodgepole pine and Douglas fir do not reach threshold levels, and management is rarely necessary (Blatt and Borden 1996b).

Experiments using inclusion and exclusion cages on western white pine (*P. monticola* Douglas ex D. Don) showed that early feeding on second-year cones reduced total seed yield per cone fivefold over controls, and caused 40% conelet abortion. Later feeding on second-year cones did not lead to abortion, but significantly increased the number of partially filled seed and seed with poorly defined endosperm, as visible in radiographs (Connelly and Schowalter 1991). Similar caging studies on Douglas fir also showed that the effect of bug feeding varies significantly with time of year. Although no conelet abortion was observed, the number of aborted seed was increased 50% by early and midseason feeding by first through third instars, and the number of partially filled seed increased with late-season damage by fourth and fifth instars (Schowalter and Sexton 1990). Over a 3-year period, cones exposed in July and August showed the highest incidence of partially filled seed, characteristic of *L. occidentalis* damage (Schowalter 1994). Feeding by nymphs during early cone development caused a significant increase in the number of seeds fused to the scale, possibly caused by bug stylets rupturing the seed coat early in its development. These fused seeds represent a further impact of *L. occidentalis* on seed loss, likely to be undetected in standard assessments of damage (Bates et al. 1997). Exclusion cage studies on ponderosa pine in Nebraska showed that damage by *L. occidentalis* can cause up to 41% seed loss, including damage to first- and second-year ovules, removal of seed contents, and possibly conelet abortion (Pasek and Dix 1988). Saliva of *L. occidentalis* contains polygalacturonases and pectinmethylesterases. Staining for the latter enzyme in the saliva of *L. occidentalis* can locate feeding punctures in the cone scales and estimate damage to Douglas fir, sugar pine, and western white pine (Campbell and Shea 1990). Histological studies of internal damage to seeds from *L. occidentalis* feeding are described by Krugman and Koerber (1969).

Damage to pistachio by *L. occidentalis* has also been reported, similar to that caused by *L. clypealis* (see **Section 3.6.2.2**) (Rice et al. 1985). Caged in the field on fruit clusters,

L. occidentalis adults induced epicarp lesions on pistachio fruit and damaged the endocarp and cotyledons; however, mortality in the mesh cages was high. These bugs are also considered an urban nuisance in the northern United States and Canada, as they enter buildings in autumn to overwinter and can be found in large aggregations reaching >2000 individuals (Gall 1992, Blatt 1994, Hall 1996).

3.6.6.3   Control — Systemic insecticides (acephate, dimethoate, carbofuran) implanted into Douglas fir were tested against various seed and cone insects (Stein et al. 1993), but low sample sizes and infestations did not allow any conclusions to be drawn regarding *L. occidentalis*. Sprays of malathion, dimethoate, and permethrin, and application of diatomaceous earth caused mortality in laboratory tests; dimethoate and permethrin were also effective in field tests with residual effects evident for 2 and >3 weeks after spraying, respectively (Summers and Ruth 1987).

3.6.7   *Leptoglossus oppositus* (Say)

3.6.7.1   Distribution, Life History, and Biology — This species is most common in the northern and eastern United States, although the range extends from New York to Florida on the East Coast, to Iowa and Wisconsin in the Midwest, Arizona and Texas in the Southwest, and into central Mexico (Allen 1969, Brailovsky 1976). Adults are brown, with an incomplete stripe across the corium that appears in most specimens as two small dots or dashes. Hind tibial dilations are deeply scalloped, and, unlike those of other related species (*L. zonatus*, *L. stigma*), the side margins of the pronotum are not serrated. Mean length of males from a Texas population was 17.15 ± 0.84 mm, that of females was 18.48 ± 0.82 mm (Mitchell 1980b).

Early instars are bright orange, darkening to brown in the final instar. Detailed descriptions of nymphs, and drawings of all instars, are given by Chittenden (1902). Aggregations of up to 30 bugs, consisting of adults and various instars, have been found on catalpa pods. At 26°C and 16:8 (L:D) hours, the duration of nymphal development from hatch to adult was 40.0 ($\pm$ 4.4) and 41.38 ($\pm$ 4.6) days for males and females, respectively (Chittenden 1925).

The composition of the scent fluid has been described for both nymphs and adults (Aldrich and Yonke 1975). Trapping experiments have shown that males, but not females, also produce cyclic sesquiterpenes, in addition to benzyl alcohol secreted by the ventral abdominal gland (Aldrich et al. 1979, Aldrich 1988).

3.6.7.2   Damage — Chittenden (1899) described this species as "an enemy to the growth of cucurbits, although not one of prime importance," and noted damage to melon, squash, and cucumber. Large numbers of *L. oppositus* have been reported from peach fruit and cowpea in Georgia, although *L. phyllopus* was more commonly found (Bissell 1929). Adults of both species fly to pecan orchards when preferred hosts are destroyed or senesce, and feeding causes two distinct types of damage, black pit and kernel spot. On eastern white pine, *L. oppositus* nymphs and adults have been collected along with the more common pine species, *L. corculus* (Bradley et al. 1981). Mitchell (1980b) summarized literature records and unpublished observations of feeding and breeding hosts, adult collection records, and economic damage comprising 18 plant families; at least 11 of these represent breeding records.

3.6.7.3   Control — The only recommended control measure in years past was to prevent the growth of alternate host plants in the vicinity of the pecan orchard, and thereby prevent infestations. One entomologist noted: "If at any time they migrate to an orchard, there is nothing that can be done to stop their damage" (Hill 1938). Adults are parasitized by the tachinid *Trichopoda pennipes* (F.) (Arnaud 1978). Intensity of parasitism is greater in males, suggesting that the flies are using a male-specific pheromone to locate hosts (Aldrich 1988).

3.6.8 Leptoglossus phyllopus (L.)

3.6.8.1 Distribution, Life History, and Biology

These leaf-footed bugs resemble *L. oppositus*, but are somewhat smaller with a broad, straight white stripe across the corium. The range extends from New York to California (U.S.A.), throughout Mexico and into Central America (Allen 1969). Average length of males from a Texas population was 15.74 ± 0.90 mm; that of females was 16.66 ± 1.01 mm (Mitchell 1980b). Eggs, 1.7 mm long (Menzies 1941), are laid in linear clusters of 4 to 42 eggs ($X = 18.9 \pm 8.6$) (Mitchell 1980b). Detailed descriptions of the instars are given by Menzies (1941). Development time from hatch to adult at 26°C and 16:8 (L:D) hours was 35.18 ± 4.0 for males, 37.56 ± 5.3 for females (Mitchell 1980b). Bugs can be reared on green bean and sunflower seed (Mitchell 1980b), green bean and peanut, or on a meridic diet (Brewer and Jones 1985).

3.6.8.2 Damage

Feeding behavior is described by Girault (1906). Histological studies of stylet penetration showed the probes to reach developing seeds within green cherry tomatoes, and vascular bundles of plant stems (Mitchell 1980b). However, as *L. phyllopus* cannot develop on vegetative tissue alone, the vascular bundles were presumed to be mainly a source of moisture (Mitchell 1980b). This species, although restricted to breeding on plant reproductive parts, is highly polyphagous with respect to plant species. Literature records and unpublished observations of feeding and breeding hosts, adult collection records, and economic damage comprising 30 plant families are summarized by Mitchell (1980b). In Florida (U.S.A.), *L. phyllopus* is a minor pest of several crops and an occasional major pest of citrus (Mead 1971a). In Mexico, *L. phyllopus* occurs on maize, pomegranate, guava, soybean, sorghum, linseed, and turnips (Brailovsky and Sánchez 1983). As part of a complex of heteropterans attacking tomato in South Carolina (U.S.A.), *L. phyllopus* can cause considerable crop loss (Muckenfuss et al. 1994). In Texas (U.S.A.), peaches may be disfigured by leaf-footed bug feeding punctures, and such damage is referred to as "cat-facing." Along with *L. oppositus* and several pentatomids, adult *L. phyllopus* migrate to pecan orchards and feed on developing nuts, causing black pit and kernel spot (Hill 1938). Black pit is a darkening of the inside of the nut, with little external evidence of injury, and leads to shedding of immature nuts. It results from hemipteran feeding during the water stage. Kernel spot is caused by insect punctures at later stages of kernel development (Adair 1932). Damage to pecans from the hemipteran complex can be assessed by staining the outside of the pecan shells with a red fluorescent dye. Stylet sheaths can also be observed internally, connecting the inside of the shell to the seed coat. Hemipteran damage to pecan results in losses in yield and kernel quality (Yates et al. 1991).

Cage studies on South Carolina cowpeas during early bloom showed *L. phyllopus* at a density of three adults per plant causes 22% pod damage, 63% seed abortion, and 54% loss in total seed. The same density at late bloom caused 22% loss in total seed yield. The bugs did not affect pod abscission or vine weight (Schalk and Fery 1982). Seeds from infested plants were smaller and often discolored (Fery and Schalk 1981). Damage to grain sorghum in Texas has also been demonstrated; *L. phyllopus* fed on the stem, branches, and glumes of the panicles, but primarily damaged the grain. Early grain development was the most susceptible stage; the damage threshold was calculated at six *L. phyllopus* adults per panicle from the milk stage until maturity, but 16 bugs from the soft or hard dough stage (Hall and Teetes 1980, 1982). Seed weight and seed germination were significantly reduced. A complex of leaf-footed bugs and pentatomids caused losses in late season (soft dough stage) sorghum in Georgia. Feeding caused scarring and shrinking of the kernels, and the fecal matter deposited on the heads encouraged the growth of microorganisms (Wiseman and McMillian 1971). In Texas, *L. phyllopus* has been recorded in moderate numbers on sunflower, *Helianthus annuus* L., and aggregations can kill individual plants (Adams and Gaines 1950). Although these authors classified *L. phyllopus* as a stem-feeding species, Mitchell (1980b) observed aggregations primarily on the developing seed heads of sunflower.

Damage to grapefruit is similar to that of *L. gonagra* (see above; Albrigo and Bullock 1977); maxillary stylets penetrate to the juice vesicle area within the fruit, and some punctures result in secondary infection by fungi. On oranges, buds and tender shoots may be damaged at bloom, and feeding on ripening fruit may cause fruit drop and premature color break (Hubbard 1885, Mead 1971a).

3.6.8.3 Control — Recommended cultural controls include cleanup to eliminate hibernation sites, removal of alternate host plants, and shaking aggregations into kerosene containers when the weather is cool (Mead 1971a, Little 1972). Dusting or spraying with carbaryl is effective (Little 1972). Adult bugs are parasitized by the tachinids *Trichopoda pennipes* (F.) and *T. plumipes* (F.). (Arnaud 1978). Parasites reared from eggs of *L. phyllopus* include *G. pennsylvanicum* (Masner 1983), *G. carinatifrons* (Mitchell et al. 1999), *Anastatus semiflavidus* (Gahan), and an undescribed species of *Anastatus* (Mitchell and Mitchell 1986). Integrated pest-management (IPM) programs for tomato have been designed to enhance egg parasitoid survival. Such programs were shown to produce marketable crops not significantly different in yield from calendar-based spraying with esfenvalerate. Percent damage from sucking bugs was significantly higher in IPM plots for some harvests, but fruit from all treatments was considered marketable (Muckenfuss et al. 1994).

3.6.9 Leptoglossus stigma (Herbst) (= Veneza stigma [Herbst])

3.6.9.1 Distribution, Life History, and Biology — This species is difficult to distinguish from *L. concolor*; both have prominent, dark, pronotal calli, a wide zigzag cream or yellow line across the hemelytra, and dilations of the hind tibiae phylliform on the outer side but lanceolate on the inner side. Occasional individuals of both species have a narrow marking on the corium, or none at all. The two species are somewhat geographically separated; *L. stigma* occurs east of the Andes in South America, in Surinam, Ecuador, Brazil, and Paraguay, while *L. concolor* is found farther north, in Central America, Mexico, Greater Antilles, and Florida (U.S.A.) (Allen 1969). However, Brailovsky (1976) reports *L. stigma* from Mexico, using Allen's (1969) diagnostic characters of the male genitalia. Published reports of *L. stigma* from the U.S. mainland and the Greater Antilles probably represent *L. concolor* (Allen 1969, Baranowski and Slater 1986).

Laboratory rearing of *L. stigma* (as *Veneza stigma*) has been accomplished using a diet of guava leaves and pieces of guava, pumpkin, *Cucurbita pepo* L., and chayote, *Sechium edule* (Jacquin) Swartz) fruit (Amaral and Cajueiro 1977). At ambient temperatures ranging from 20.5 to 29.8°C, and 60 to 78.4% RH, the duration of the egg stage was 13.10 ± 0.13 days. Nymphal development required 77.07 days; nymphs were separated 1 day after hatch and reared individually. Eggs, 1.4 mm in length, are laid in a line, and such egg masses have been observed along the midvein of the leaf (Mariconi and Soubihe 1961, cited in Amaral and Cajueiro 1977). Mass size ranges 4 to 61 eggs. Newly hatched nymphs aggregate on the eggshells, and both first and second instars are very active and gregarious. Third through fifth instars are slower and were more dispersed throughout the rearing container (Amaral and Cajueiro 1977).

3.6.9.2 Damage — In Brazil, *L. stigma* is an agricultural pest (Amaral and Cajueiro 1977). The adult bugs are often found on Myrtaceae, especially guava (*Psidium guajava* L.) and Brazilian guava (*P. araca*). Both buds and fruit are fed upon, resulting in bud drop. Mature fruit remain on the plant, but black patches and hard spots on guava are attributed to feeding damage. The bugs also feed on oranges, both green and ripe, occasionally causing decay when punctures in the fruit facilitate entry of *Penicillium* spores. Other crops on which this species is reported to occur are cashew, pumpkin, carambola (starfruit, *Averrhoa carambola* L.), pomegranate, tangerine, and mango (Costa Lima 1940; Amaral and Cajueiro 1977, and references therein). Adults collected from carambola were positive for trypanosomatids, thought to be *Phytomonas* spp. (Sbravate et al. 1989). In Mexico, adults and nymphs have been collected from fruit of guava (Brailovsky and

Sánchez 1983). References in the West Indian literature to *L. stigma* on guava, cashew, lychee, and annatto (*Bixa orellana* L.) probably refer to *L. concolor* (Allen 1969).

3.6.10   *Leptoglossus zonatus* (Dallas)

3.6.10.1   Distribution, Life History, and Biology — This species ranges from the southwestern United States through Mexico and Central America south to Brazil. The adults are a variable shade of brown, usually with a wide zigzag band across the corium, and two large whitish-yellow spots on the pronotum. Dilations of the hind tibiae are externally phylliform with deep scallops, but thinner and lanceolate on the inner side (Allen 1969).

Eggs, 1.47 mm long, are bright green when first laid but turn brown during incubation (Fernandes and Grazia 1992). On pomegranate, egg chains are placed on small stems and leaf midribs (Jones 1993). Detailed descriptions of nymphs are given by Fernandes and Grazia (1992). First instars form aggregations, arranged in concentric circles with their heads pointing outward. Second instars are also gregarious, and dispersion begins in the third stadium (Fernandes and Grazia 1992). Nymphal aggregations respond to alarm pheromones of both adults and nymphs by dispersing (Leal et al. 1994). Chemical composition of the alarm pheromones in *L. zonatus* was determined by Leal et al. (1994); descriptions of the chromosomes and meiotic events are given by Packauskas (1990).

Under laboratory-rearing conditions of 12:12 (L:D) hours, 28°C, and 78% RH, the egg incubation period was 9.6 days. Nymphal development required 28.7 days on corn grains and 31.6 days on sorghum. Adult females laid 5.5 egg masses each with an average of 15.2 eggs, placed in the characteristic linear arrangement. Longevity was 71 days for females, 54.3 days for males when grain sorghum was provided as the food source (Matrangolo and Waquil 1994). Nymphs can also be successfully reared on soybean and green bean, but mortality is higher, development time is longer, and body weight of adults is lower than on corn; furthermore, females reared on soybean lay few or no eggs (Panizzi 1989). A meridic diet originally developed for *Lygus* sp. (Miridae) can be used to rear *L. zonatus* with survivorship >60%. At the optimal rearing temperature, 25°C, fecundity was 347.7 ± 76 eggs per female, the preoviposition period was 22 days, and eggs hatched in 10.2 days. Data are given for oviposition and developmental parameters at temperatures from 20 to 35°C (Jackson et al. 1995).

3.6.10.2   Damage — *Leptoglossus zonatus* is a highly polyphagous species. In Mexico, it is reported to damage cotton, limes, oranges, guava, melons, avocado, sorghum, tomato, cucurbits, eggplant, and pomegranate; breeding populations have been observed on *Actinocheita filicina* (D. C. Barkely) (Anacardiaceae) and *Schizocarpum reflexum* Rose (Convovulaceae) (Brailovsky and Sánchez 1983, and references therein). In east Texas (U.S.A.), nymphs and adults have been observed feeding in large numbers on the fruits of Chinese tallow, *Sapium sebiferum* (L.) Roxburgh (Johnson and Allain 1998). In Honduras, the common name is "chinche patona" (large-legged bug); it is considered a rare pest, unlikely to be found in densities high enough to cause significant economic damage. However, it has been collected from beans, pepper fruits, flowering maize, sunflower, guava, citrus, ripe sorghum, tomato, and cucumber (Passoa 1983). In plantations of physic nut (*Jatropha curcas* L.) in Nicaragua, this bug is the second most common heteropteran; it feeds on flowers and developing fruit and is capable of completing the entire life cycle on this host (Grimm and Maes 1997a,b; Grimm and Führer 1998). Damage includes aborted fruit, malformed seeds, reduced seed weight, and lowered oil content (Grimm and Maes 1997b). Additional host plants in Nicaragua are given by Grimm and Führer (1998).

In Brazil, both adults and nymphs are found on pomegranate; adults copulate while feeding on the pomegranate fruit and bugs feed on fruit of all sizes. Damage to pomegranate is visible as depressed brown spots with a darker central region; the spots grow larger and darker over time. The damaged tissue does not grow along with the healthy tissue, leading to cracks and fissures. A large number of lesions cause complete splitting and consequent decomposition of the fruit (Raga et al. 1995). Other crops attacked in Brazil include oranges, guava, passion fruit, citrus, corn, soybean, sorghum, and tomato (Kubo and Batista Filho 1992; Raga et al. 1995, and references therein). On sorghum, nymphs and adults are found mainly on the panicles; on corn ears adults feed with their stylets piercing the husk, whereas nymphs are found mainly on the leaves (Matrangolo and Waquil 1994).

In Arizona (U.S.A.), heart rot of pomegranates is said to be transmitted by this species (Burgess and Hawkins 1945). Large nymphal aggregations may be found on pomegranate fruit (Jackson et al. 1995). Another breeding host is *Chilopsis linearis* (Cavanilles) Sweet (desert willow), which, like pomegranate, is planted in urban areas as an ornamental (Jones 1993). A report of this species as a minor pest of sorghum in Kansas (U.S.A.) (Hayes 1922) may refer to *L. phyllopus*, as Allen (1969) does not include Kansas in the range of *L. zonatus*, whereas *L. phyllopus* is a known pest of this crop in Texas (Hall and Teetes 1980, 1982).

Individuals collected from corn in Brazil harbored trypanosomatids in the salivary glands and gut, and this species is therefore considered a potential vector of these plant flagelloses (Sbravate et al. 1989). Seeds, but not stems and leaves, of corn harbored the protozoans. Pure cultures of the trypanosomatids in bug salivary glands and digestive tract were used successfully to infect corn plants in the laboratory. The strain was identified as a new species, *Herpetomonas macgheei* (Jankevicius et al. 1993).

3.6.10.3 Control — Parasitic wasps from the following genera have been reared from *L. zonatus* eggs in Arizona: *Gryon*, *Ooencyrtus*, *Anastatus*, and *Neorileya* (Jones, unpublished, cited in Jones 1993); the latter, a eurytomid, is not known to attack eggs of any other *Leptoglossus* species. Parasitoids reared from *L. zonatus* in Brazil include *Gryon* sp. and a tachinid, *Trichopoda* sp. (Souza and Amaral Filho 1998). Two entomopathogenic fungi, *Beauveria bassiana* (Bals.) Vuil. and *Metarhizium anisopliae* (Metsch.) Sorok., provided satisfactory control of *L. zonatus* in a physic nut plantation in Nicaragua (Grimm and Guharay 1998).

3.7   *Pseudotheraptus* Brown

Similar in general appearance and feeding habits to *Amblypelta*, these bugs are restricted in distribution to tropical Africa. They can be distinguished from *Amblypelta* by the longer antennae of the latter, as well as differences in male genitalia, wing venation, and scent gland structure (Brown 1955). Of the three species placed in the genus by Brown (1955), two are considered economically damaging. It should be noted that early reports prior to 1955 refer to these as *Theraptus* spp.; in fact, the generic name *Pseudotheraptus* was selected intentionally to minimize confusion (Brown 1955). Brown (1955) provides a key to the genus and a description of the species.

3.7.1   *Pseudotheraptus wayi* Brown

3.7.1.1   Distribution, Life History, and Biology — Initially recognized as the cause of gummosis and premature nutfall of coconut in East Africa (Way 1951), this insect is now considered a pest of macadamia nut, avocado, guava, cashew nut, cacao, and mangoes as well (Warui 1983; Dennill and Erasmus 1992; van der Meulen 1992, and references therein). The original range (given by Brown 1955, as coastal East Africa, including Zanzibar and other islands off the coast) has expanded in the last two decades to include South Africa (van der Meulen 1992, and references therein). Some research reports from West Africa (e.g., Julia and Mariau 1978) refer to *P. wayi* as a new pest of coconut in the Côte d'Ivoire, but later papers (Mariau et al. 1981, Douaho 1984) identify the species in question as *P. devastans*. Adults are brownish-red, with black membrane and hind wings. The veins of the hemelytra and the lateral pronotal carinae are pale. Length of males ranges from 12 to 13 mm, and of females from 14 to 15.5 mm (Brown 1955). Nymphs are red-brown, brown, or green-brown (Wheatley 1961). Vanderplank (1958) provides photographs of third and fourth instars.

Adults breed continuously in tropical East Africa (Way 1953a); seven generations per year are estimated in Kenya (Warui 1983). Eggs are laid singly at a rate of 2 to 3 per day, for a total lifetime fecundity of over 100 eggs per female (Warui 1983). In the field, eggs are found primarily in the crown, on the spadix or on nearby fibers, but may also be found on the leaves and trunk (Oswald 1990). Nymphs are also generally found on the spadices. Field observations are difficult because of the height of the palm trees, coupled with the behavior of adults and nymphs. Adults are agile and take flight when disturbed; nymphs hide in crevices and debris in the crown during periods of bright sunlight (Vanderplank 1960a). Thus, nymphal activity varies with time of day, making unbiased sampling difficult. Distribution of nymphs among trees was significantly nonrandom (Yeo and Foster 1958), although these results may have been affected by sampling problems (Vanderplank 1960a). Laboratory studies have shown the total nymphal development time on cashew nut to be 35 days (egg through final molt); preoviposition period is 12 to 13 days at 27 to 30°C (Warui 1983). On coconut, Tait (1954) reports 31 days for nymphal development at 28° C, and 39.5 days at 24°C; Way (1953a) gives 41 days at 24.6°C.

3.7.1.2 Damage — Damage to coconut is similar to that caused by *Amblypelta cocophaga* in the Solomon Islands, and has been extensively described by several workers (Tait 1954; Vanderplank 1958, 1959a; Way 1953a; Wheatley 1961). In trees free of insect damage, abscission of a large percentage of nutlets occurs naturally shortly after fertilization, and no further nutfall occurs (Vanderplank 1958). Feeding by *P. wayi* third to fifth instars or adults through the calyx produces an elongate necrotic lesion which causes nutlets less than 3 months old to drop from the tree. Flower buds and flowers will also abscise following bug damage. The proportion of damaged nutlets may be used as a comparative indicator of *P. wayi* populations, when direct counts of nymphs are impractical (Vanderplank 1958). Because a single puncture can result in abscission, even low populations of *P. wayi* can cause economic losses on coconut (Way 1953a); up to 98% reduction in yield has been reported). During the course of nymphal development, a single individual damages an average of 42.4 flowers (equivalent to the average monthly flower production); the palm cannot compensate for this damage by producing more flowers per spadix (Oswald 1989). Older nuts (aged 3 to 6 months), also penetrated through the calyx, do not abscise but instead become distorted and scarred. A gummy material is secreted by the plant into the damaged area, and hardens on the outside of the coconut. Copra yield from such nuts is reduced. Yield from full-size nuts (older than 6 months) is not reduced by *P. wayi* feeding. Photographs of damaged nutlets are provided by Vanderplank (1958); Way (1953a) illustrates gummosis and damage to older nuts.

In guava and mango, feeding also produces lesions on the immature fruit, and may lead to malformed fruit or abortion of smaller fruit. The round, external, brown depressions on guava are often mistaken by growers for hail damage. Internally, lesions develop into woody knobs, rendering the fruit unsuitable for market although still edible (van der Meulen 1992). Damage resulting in loss of 12.2 to 44.7% of the crop was observed in South Africa; this estimate does not include aborted fruit. Feeding by *P. wayi* produces lesions on the fruit surface of avocados; these external marks cause rejection of the fruit for export (Dennill and Erasmus 1992). Among the complex of avocado pests, coconut bugs are ranked first in importance, causing 4.7% culling of harvested fruit. Again, these figures do not take into consideration possible fruit loss before harvest. Severity of infestation and economic loss is predicted to increase over time, as the avocado industry expands in South Africa (Dennill and Erasmus 1992). Feeding by *P. wayi* on young shoots of cashew increases infection by pathogens, and results in dieback of the shoots. Developing nuts or whole inflorescences may also abort if damaged (Warui 1983).

3.7.1.3 Control — Early efforts at control focused on the observation that coconut trees infested with the red tree ant, or weaver ant, *Oecophylla longinoda* (Latreille), were undamaged by *P. wayi*

(Way 1951, 1953a,b; Vanderplank 1960a,b), as was also found by workers in the Solomon Islands (see *Amblypelta*, **Section 3.1.5.3**). When *O. longinoda* were experimentally removed by spraying lower trunks and leaves with dieldrin, subsequent nut production dropped to zero (Way 1953b). Colonies of this predatory ant occupy and defend the crowns of coconut palms, and compete with several other ant species. In East African coconut plantations, *O. longinoda* is often displaced by *Anoplolepis longipes*, *A. custodiens* (F. Smith), or *Pheidole* spp. (Way 1953b, Vanderplank 1959a). Thus, research was directed at increasing the percentage of trees inhabited by the preferred ant species. The outcome of ant competition depended to some extent on environmental factors; for example, *Oecophylla* outcompete *Pheidole* under wet conditions (Vanderplank 1960b). However, coconut palm height, leafiness, and variety appear to have no effect on *O. longinoda* distribution (Vanderplank 1960b). Interplanting of cashew nut or citrus trees encourages *O. longinoda* activity, as does increased ground cover (weeds) (Oswald 1989, Wheatley 1961). AMDRO™ (hydramethylnon) may be used to eliminate colonies of competing ant species. If 60 to 70% of palms are infested with *O. longinoda*, control of *P. wayi* is equivalent to that obtained with threshold insecticide treatment (Oswald 1989).

Neither *A. longipes* nor *A. custodiens* was considered by early workers to be of any use in controlling populations of coconut bug in Africa (Way 1953b, Vanderplank 1959a); in fact, these species were viewed quite negatively because they destroyed *O. longinoda* colonies. However, recent studies of *A. custodiens* in Tanzania (Löhr 1992) show a significant inverse relationship between ant activity and nutlet damage, although control by *O. longinoda* is more effective.

Chemical control of *P. wayi* in coconut has not been overly successful in the past. Hand-dusting with γ BHC (lindane) controlled the pest but this eliminated *O. longinoda* and induced an unusually high population of diaspidid scale (Way 1953a). Vanderplank (1959b) attempted to develop a resinous formulation of DDT and other chlorinated hydrocarbons that would kill *P. wayi* and *Anoplolepis* sp. without interfering with *O. longinoda* populations. Hand-spraying, which at that time had to be done at the crown of the tree by individual climbers, was understandably not popular among small farmers (Wheatley 1961); climbing palms inhabited by *O. longinoda* is "an uncomfortable experience and is not willingly undertaken" (Yeo and Foster 1958). The current recommended threshold for spraying is 0.14 bugs per palm in Zanzibar; endosulfan is recommended over dimethoate (Oswald 1989). As of 1990, no insecticide was registered for control of this insect on guava in South Africa (Vermeulen et al. 1990, cited in van der Meulen 1992).

Although the major focus of IPM programs is still on *O. longinoda*, other potential biological control organisms have been investigated. An unidentified strepsipteran and two egg parasite species were observed by Tait (1954) in Zanzibar. More recent studies also identified two egg parasites, *Anastatus* sp. (Eupelmidae) and *Ooencyrtus albicrus* (Prinsloo) (Encyrtidae). The latter parasite shows promise as a control agent through mass releases, although natural parasitism did not maintain *P. wayi* populations below threshold levels (Oswald 1990).

3.7.2   *Pseudotheraptus devastans* Distant

3.7.2.1   Distribution, Life History, and Biology — Almost identical in general appearance to *P. wayi*, this bug is a pest of coconut and several other crops in West and Central Africa. The two species may be distinguished by characteristics of the male genitalia, or by lengths of the antennal segments: in *P. devastans*, the first segment is shorter than segments 2 and 4, which are subequal; in *P. wayi*, the first and second segments are subequal, and each is slightly longer than segment 4. The species do not overlap geographically (Brown 1955). Eggs are oval, and usually laid separately or in loose clusters; in cassava, they are laid on the undersurface of leaves (Boher et al. 1983). Nymphs are elongate-oval, reddish-brown with dark brown head and thorax (see Lodos 1967, for detailed descriptions). Early instars have extremely long antennae with flattened antennal segments (Julia and Mariau 1978). Total development time (egg to final molt) averaged 39 days in an outdoor insectary in Ghana; bugs were fed cacao pods, mango fruit, or guava fruit. Preoviposition period was 12.5 days. Nymphs and adults are present year-round in Ghana (Lodos 1967).

3.7.2.2 Damage — Feeding by adults or older nymphs initially produces round, dark lesions on cocoa pods. These lesions expand into necrotic scars and eventually cause deformed pods. Small pods (2 to 5 cm) cease development entirely. Saprophytic fungi may invade the lesions. Deformation is apparently caused by development of hardened tissue internally in the area of the feeding scar. Damage to guava, cashew nut, and fruit of *Sterculia* sp. is similar (Lodos 1967).

On coconut, *P. devastans* attacks nuts at all stages, from recently fertilized to fully developed; however, nuts 3 to 6 months old are most frequently damaged. Symptoms resemble those caused by *P. wayi*: an elongate brown blemish at the edge of the calyx develops into a crevice filled with gummy material. The wounds often appear as rings around the fruit. Nut development is affected, and copra yield is reduced (Mariau 1969). Only minor crop losses were reported in West Africa (Côte d'Ivoire and Dahomey, Benin) until 1972, when the insect became a more serious problem in the Côte d'Ivoire (Julia and Mariau 1978, Mariau et al. 1981).

On cassava, bug-feeding lesions become infected with a fungus, *Colletotrichum gloeosporioides* Penz., which then invades the cortical parenchyma and vascular tissue. In the absence of bug damage, the fungus remains in an inactive form on the stem surface. Fungal infection causes leaf drop and withering of the shoots; the condition is known as "candlestick disease" (Boher et al. 1983). The symptoms of this disease were initially attributed directly to *P. devastans* feeding and the necrotic action of the saliva (Dubois and Mostade 1973, cited in Boher et al. 1983).

3.7.2.3 Control — Egg parasitism by a scelionid, and egg predation by the ant *Pheidole megacephala* F. have also been observed (Lodos 1967). However, the ant *Oecophylla longinoda* is considered the only efficient biological control agent (Julia 1978). Carbaryl, used against mirid pests of cocoa, was found by Lodos (1967) to increase coreid pest populations in Ghana. In contrast, Julia (1978) showed carbaryl to be the most effective for repeated treatments of eight insecticides tested against *P. devastans* in the Côte d'Ivoire. Propoxur is the most recent recommendation for treatment of individual trees lacking ant colonies (Douaho 1984). Economic thresholds based on nut damage are given by Julia and Mariau (1978); no chemical control is deemed necessary if 60 to 70% of trees examined have *O. longinoda* colonies (Julia 1978). Older plantings usually have 80 to 90% of trees infested by *O. longinoda*, which ensures adequate control. Younger trees may not have established colonies, and an integrated control program involves encouraging ant infestation by maintaining contact between forest regrowth and coconut palms; by allowing palm leaves to touch the ground, serving as bridges for colonizing ants; and by physical transfer of nests (Douaho 1984). When protection by *Oecophylla* is insufficient, the critical threshold is 30 bugs (adults plus nymphs) per hectare (Mariau et al. 1981).

4. THE LESS IMPORTANT SPECIES

4.1 *Acanthocoris* Amyot et Serville

Several species of *Acanthocoris* damage crops in south and east Asia, primarily Solanaceae and Convolvulaceae. However, this genus is perhaps best known for Fujisaki's (1975, 1980, 1981) extensive behavioral analyses of the winter cherry bug, *A. sordidus* Thunberg, including studies of nymphal aggregation and dispersal, spatial distribution patterns of adults, and harem defense polygyny. In addition to the Asian species discussed below, *A. obscuricornis* Distant feeds in Kenya on *Solanum indicum* L., a plant used in African folk medicine (Prestwich 1976).

4.1.1 *Acanthocoris scabrator* (F.)

This species is found in southern China, Indonesia, Myanmar, Malaysia, India, and Sri Lanka. Adults are gray-brown, with yellowish-brown hairs, and are 13 to 14.5 mm long. The connexivum is prominent, and the posterior femora of males are somewhat incrassate (Hoffman 1927, CABI 1997). Descriptions and illustrations of the egg, five instars, and the adult are given by Hoffman (1927) and Koshy et al. (1977), along with descriptions of mating and oviposition. Eggs are laid in clusters of 18 to 20 on the underside of leaves. Nymphs feed gregariously on stems and leaves of solanaceous plants and on the vines of *Ipomoea* spp. Adults may feed on larger stems, terminal branches, and leaves of various solanaceous plants, and on vines of *Ipomoea*, but on mango they also attack unripe fruits (Hoffman 1933, Koshy et al. 1977).

Cape gooseberry (*Physalis peruviana* L.), cayenne pepper (*Capsicum annuum* L.), eggplant (*S. melongena* L.), and squash (*Cucurbita maxima* Duchesne) are attacked by this species in southern China; damage to cape gooseberry and pepper is particularly severe. Other hosts are *S. nigrum* L., morning glory (*Ipomoea* sp.), *Cestrum nocturnum* L., and *S. torvum* Swartz (Hoffman 1927). In Kerala, India, damage to mango (*Mangifera indica* L.) has been noted. Feeding by adults on fruits allows invasion by secondary pathogens, eventually leading to rotting and premature fruit drop (Koshy et al. 1977).

Cultural control, involving removal of alternate host plants from orchards and fields, is strongly recommended (Hoffman 1927, Koshy et al. 1977), along with hand-picking or brushing into tins of kerosene. Observed natural enemies include spiders and a chalcidoid egg parasite (Hoffman 1927).

4.1.2   *Acanthocoris sordidus* Thunberg

The winter cherry bug feeds on red pepper (*Capsicum frutescens* L.), eggplant, tomato, sweet potato, and mulberry (*Morus alba* L.) in addition to winter cherry (*Physalis alkekengi* L.). In general, it is considered a pest of cultivated Solanaceae and Convolvulaceae (Fujisaki 1980, Lu and Lee 1987, CABI 1997). In Japan, mating and oviposition occur during June and July, and males defend mating territories. Females aggregate on stems, in the middle or lower portion of the host plant. Egg masses are laid on the lower leaf surface, and early instars are highly gregarious. First instars remain by the egg mass, and move to the stem after the first molt. Nymphs respond to predatory coccinellid beetles by dispersing and then reforming aggregations (Fujisaki 1975, 1980, 1981). One scelionid parasite, *Trissolcus* sp., has been recorded from eggs in Taiwan (Lu and Lee 1987). In laboratory tests, eggs of *A. sordidus* were unsuitable for development of *Gryon japonicum* (Ashmead); females drummed, drilled, and marked eggs, but no progeny emerged (Noda 1990).

4.2   *Amorbus* Dallas

About 15 to 20 species of *Amorbus* are known in Australia (Carver et al. 1991, Steinbauer 1995); all feed on eucalypts. Some species are currently minor pests of plantation-grown *Eucalyptus*, and are likely to increase in importance along with the growth of plantation forestry in Australia (Steinbauer 1995). The Tasmanian species *A. obscuricornis* (Westwood) has been intensively studied (Steinbauer and Clarke 1995, Steinbauer and Davies 1995, Steinbauer 1997, Steinbauer et al. 1997).

4.2.1   *Amorbus obscuricornis* (Westwood)

Adults of *A. obscuricornis* are brown, about 15 mm in length, and sexually dimorphic; males have thicker hind femora than females. Both adults and nymphs produce defensive secretions, although the actual compounds differ (Steinbauer and Davies 1995). Eggs are green initially, changing to purple as development progresses. Eyespots become visible through the chorion early in development (Steinbauer 1997). The first instar is green and black, and later instars are orange and black (Elliott and deLittle 1985, Steinbauer 1995). Photographs of *A. obscuricornis*, with typical damage, are provided by Elliott and deLittle (1985) and Steinbauer et al. (1997); drawings of the egg, five instars, and adult of *A. alternatus* may be found in Carver et al. (1991).

Amorbus obscuricornis completes one generation per year, overwintering as adults (Steinbauer 1997). Fat content increases in autumn; metabolism of these reserves, measured as starvation-longevity, is affected by temperature and relative humidity (Steinbauer 1995). Developmental data for eggs and nymphs at a range of temperatures are presented by Steinbauer (1997); 147 degree-days were required for egg development, and the developmental threshold temperature was estimated to be 11.8°C. For nymphal development, these values were 509 degree-days and 12.0°C.

Feeding by nymphs and adults in spring and summer on *Eucalyptus* trees causes wilting of the terminal shoots. Shoots 2 mm in diameter on young, growing trees are preferred feeding sites (Elliott and deLittle 1985). *Amorbus obscuricornis* is found in both wet and dry sclerophyll forest, and all three groups of eucalypts (gum, peppermint, ash) are attacked (Steinbauer 1995). Heavy feeding on saplings can result in stunting and coppicing, similar to the effects of manual pruning (Elliott and deLittle 1985, Steinbauer 1995, Steinbauer et al. 1997). Bug infestations are unpredictable, reaching high numbers in some years but not others; thus, insecticides are not routinely used to control bug populations (Steinbauer 1995). Manna is exuded from bug feeding sites, and is attractive to ants; however, the ants do not prey upon the adult bugs (Steinbauer 1996b). An encyrtid parasitoid, *Xenoencyrtus hemipterus* (Girault), was reported attacking eggs of *A. obscuricornis* colonies in a greenhouse, but no parasitoids were obtained from eggs placed in the field, suggesting that this coreid is not the preferred host. Other natural enemies include erythraeid mites, spiders, and tachinid flies (Steinbauer and Clarke 1995).

4.3 *Chelinidea* Uhler

The status of cactus bugs in the genus *Chelinidea* depends on whether one views their host plant, *Opuntia*, as a weed or an economically valuable crop. Where *Opuntia* has spread thickly into rangeland, from accidental introduction or overgrazing, the cactus is considered a weed and the bug, therefore, a beneficial (see **Section 5** for discussion of *Chelinidea* as a biological control agent). However, prickly pear is used by ranchers in the southwestern states as emergency forage for cattle during drought, and fruits and spineless pads are also sold as food in the produce section of grocery stores. Thus, from a different perspective, *Chelinidea* spp. may be considered minor rangeland and crop pests.

All species of *Chelinidea* feed in the mesophyll of pads (joints), leaving a distinctive pale spot at the point of stylet insertion. Such spots may coalesce if numerous, eventually leading to withering of the pad. Fruit may be attacked occasionally. Both adults and nymphs are damaging. Mann (1969) notes that feeding on young joints and fruit impedes growth, and is more destructive than damage to the older joints. Fungi are more likely to invade pads with *Chelinidea* feeding damage. Weakened joints are also more susceptible to frost, and heavily infested plants may collapse (Hunter et al. 1912). Fire-damaged cactus are preferred over undamaged plants as feeding and mating locations (Sickerman and Wangberg 1983).

Herring (1980) provides a key to species.

4.3.1 *Chelinidea vittiger* Uhler

This species occurs from southern Canada to northern Mexico wherever *Opuntia* is found, and is the most widely distributed species in the United States (Herring 1980). It varies in color from pale to yellow with black stripes; body length is 10 to 13 mm. Eggs are laid in batches, attached to spines of the cactus plant. Nymphs are polymorphic, the abdomen either reddish or dark green, and all stages (particularly the early instars) are gregarious. In southern California (U.S.A.), the species is univoltine (De Vol and Goeden 1973), but elsewhere two or more generations occur annually, and extensive overlap occurs between generations. Adults overwinter, sheltering under pads or debris at the base of the plant and emerging on warm winter days. At 31°C, total nymphal development time on *O. polyacantha* Haworth was 69.9 days from hatch to adulthood, and the preoviposition period was 5.86 weeks. Adult life span varied with temperature, but was longest (>300 days) at 24°C (Carroll and Wangberg 1981). At 27°C, egg incubation averaged 13 days, and nymphal development times varied with host plant, from 72.9 to 85.9 days (De Vol and Goeden 1973). On *O. vaseyi* (Coulter) Britton and Rose, the usual host, mean lengths of first through fifth stadia were 9, 8, 14, 17, and 25 days. Spiders (*Phidippus* spp.) prey on cactus bugs; and a tachinid fly, *Trichopoda pennipes* (F.), parasitizes adults (Hamlin 1924b, Arnaud 1978). Two egg parasites of *C. vittiger* are known: *Ooencyrtus johnsoni* Howard (Encyrtidae) (De Vol and Goeden 1973) and *Gryon chelinideae* Masner (Scelionidae). *Gryon pennsylvanicum* (Ashmead) is also reported from eggs of *Chelinidea* sp. (Masner 1983). Up to 80% egg parasitism has been reported (Mann 1969).

4.3.2   Chelinidea tabulata (Burmeister)

This species has a more southern distribution, from the southwestern United States through Mexico and Central America to Venezuela (Herring 1980). The biology of *C. tabulata* is similar to that of *C. vittiger*, except the nymphs are light green, and are not markedly gregarious.

4.4   Crinocerus sanctus (F.)

This medium-sized, robust bug may be recognized by the unique dorsal color pattern, "a light brown St. Andrew's cross on a darker brown background" (O'Shea 1980). The posterior femora are curved, incrassate in males, and covered with tubercles or spines. Distribution includes Argentina and Brazil (O'Shea 1980), and the species is a pest on a wide variety of crops. Photographs of adults, nymphs, and eggs are given by Amaral Filho (1986). Three females reared from field-collected fifth instars laid 80 ± 46.19 eggs (8.0 ± 0.66 per cluster) over an average life span of 215 ± 36 days. At 22°C and a photoperiod of 12:12 (L:D) hours, development time was 12.27 ± 0.06 days for eggs and 58.41 ± 1.25 days for nymphs (5.13 ± 0.12, 14.22 ± 0.49, 12.31 ± 0.44, 13.55 ± 0.43, and 18.10 ± 0.62 for first through fifth stadia, respectively) (Amaral Filho 1986). At 27°C, 6 to 11 days for eggs and 56 days for nymphal development (5.38 ± 0.36, 14.78 ± 3.68, 7.75 ± 2.06, 14.33 ± 10.48, and 14.0 ± 8.02 for first through fifth stadia, respectively) was reported by Soglia et al. (1998).

Reported host plants include cotton, mimosa, guava, and citrus (Amaral Filho 1986, and references therein), but cultivated legumes (pigeon pea, various beans, cowpea) are the most heavily damaged. Feeding on leaves and pods of *Canavalia* spp. is reported in Argentina and Brazil (Bosq 1937, Costa Lima 1940).

Crinocerus sanctus is considered one of the two most important species of pod bug on cowpea in Brazil. It is ranked as the principal pest in the northern part of Brazil (Pará, Amazonas, and Acre) and either principal or sporadic in the northeastern states. Stems, new foliage, and green pods are attacked (Daoust et al. 1985). *Crinocerus sanctus* is also a pest of acerola, or Barbados cherry (*Malpighia puncifolia* L.), feeding on shoots and green fruit and causing wilting and deformation. Infestations of up to 56% have been reported in acerola plantings (Soglia et al. 1998). Control recommendations in cowpea are to apply dimethoate, diazinon, or methyl parathion at initiation of fruiting (Daoust et al. 1985).

4.5   Dasynus piperis China

The "pepper-bug" is distributed throughout Malaysia and Indonesia wherever pepper (*Piper nigrum* L.) is cultivated. Adults are green, with brown hemelytra, measuring 11.5 to 13 mm (China 1928, Miller 1936). The egg and all instars are described and illustrated by van der Vecht (1933) and Miller (1936). Clusters of two to ten eggs are laid on the leaves or (less frequently) the fruit of cultivated pepper; adults and nymphs feed on the fruit (van der Vecht 1933, Miller 1936). Nymphs and adults have also been observed on *Citrus* sp. (Miller 1936).

The species is accorded only minor pest status; only on the island of Bangka (Indonesia) has it been considered an important pest. Second instars prefer fruits at an early stage, but subsequent instars and adults feed on starch contained in fruit $4\frac{1}{2}$ to 9 months old (half-ripened to ripe). Damage appears as a discoloration, and the fruit contents turn brownish-black and decay. Damaged fruit fall prematurely, and an entire spike will fall even if only some fruits are attacked (van der Vecht 1933, Miller 1936). Cultivation techniques affect bug density on pepper plantations, and younger plantings tend to be more heavily infested (van der Vecht 1933).

Hand-picking is recommended in the older literature (Miller 1936). Several species of egg parasites have been reared, including *Ooencyrtus malayensis* Ferrière (Hymenoptera:Encyrtidae), *Anastatus dasyni* Ferrèire (Hymenoptera: Eupelmidae), and *Gryon* (= *Hadronotus*) spp. (van der Vecht 1933, Miller 1936, CABI 1997).

4.6   *Gelonus tasmanicus* (Le Guillou)

This species, along with *Amorbus obscuricornis* (see *Amorbus*, above), is classified as a minor pest on plantation-grown eucalypts in Australia (Elliott and deLittle 1985). However, recent research (Steinbauer et al. 1997) has shown that the latter species causes more severe damage and poses a considerably greater threat than *G. tasmanicus*. Adults of *G. tasmanicus* are about 13 mm long, brown and black, with white markings on the antennae and legs (Elliott and deLittle 1985, Steinbauer 1995). Eggs are opaque (no visible eyespots), pale blue at first but turn brown shortly after oviposition (Steinbauer 1997). Nymphs are brownish black throughout development (Elliott and deLittle 1985, Steinbauer 1996a). Both nymphs and adults produce defensive odors; the scent of nymphs is predominantly 2-hexenal, whereas that of adults is composed mainly of *n*-butyl butanoate (Steinbauer and Davies 1995).

The species is univoltine, overwintering as an adult (Steinbauer 1997). The seasonal phenology is similar to that of *A. obscuricornis*, although *G. tasmanicus* has a narrower host range and appears to tolerate lower temperatures (Steinbauer 1995). In Tasmania, only eucalypts in the gum and ash groups are used as host plants (Steinbauer 1995). The minimal developmental temperature for eggs was 10.8°C; 136 degree-days were required for hatch. Nymphs could not be reared in captivity in sufficient numbers to calculate developmental parameters (Steinbauer 1997). Reported natural enemies are the same as for *A. obscuricornis* (see **Section 4.2.1**) (Steinbauer and Clarke 1995).

4.7   *Homoeocerus* Burmeister

Agunloye (1986) notes that "the literature on the genus *Homoeocerus* in relation to horticultural crops is scanty." In the economic literature in general, the notation "*Homoeocerus* spp." occurs with frustrating frequency. With the exception of *H. pallens* F. (see **Section 4.7.1**), which has been thoroughly studied, most references are brief notes or lists of species feeding on plantation crops such as cocoa or coffee. Thus, Forsyth (1966) in Ghana records *Homoeocerus* spp. puncturing cacao (*Theobroma cacao* L.) pods, fruit of avocado (*Persea americana* Miller) and Surinam cherry (*Eugenia uniflora* L.), and young shoots of rubber, *Hevea brasiliensis* (A. Jussieu) Mueller. *Homoeocerus* spp. are reported as proportionately less abundant than *Pseudotheraptus devastans* on cassava, *Manihot esculenta* Crantz (Boher et al. 1983). *Homoeocerus ignotus* Schouteden is found on leguminous trees, coffee, and cacao in West Africa; it pierces young cacao pods but the damage is of little importance (Alibert 1951). *Homoeocerus puncticornis* (Burmeister) is included in a checklist of insect pests of grain legumes (Singh et al. 1978b).

In Malaya, *H. serrifer* Westwood is found on sweet potato and green gram, *Vigna radiata* (L.) R. Wilczek, and is classified as a minor pest of the latter (Hoffman 1933, Ooi 1973). Two species, *H. laevilineus* Stål and *H. inornatus* Stål, damage the timber tree *Dalbergia sissoo* Roxburgh ex de Candolle in India (Hoffman 1933). Two species, *H. walkeri* Kirby and *H. striicornis* Scott, breed on *Albizia chinensis* (Osbeck) Merrill and *A. lebbeck* (L.) Bentham in China, although only the former coreid is accorded pest status. New growth, terminal branches, and seedlings are damaged (Hoffman 1933). Descriptions of all stages and some biological information for these two species are provided by Hoffman (1933).

Some information is available on natural enemies of various species. In southern India, *H. prominulus* (Dallas) feeds on *Prosopis cineraria* (L.) Druce, a local timber tree with edible fruits. The oval eggs are deposited in a line on stems or leaves. Two hymenopteran parasites, *Trissolcus* sp. (Scelionidae) and *Anastatus* sp. (Eupelmidae), have been reared from the eggs, but the scelionid is more common (Chandrasekar 1988). Eggs of an Indonesian crop pest, *H. marginellus* Herrich-Schaeffer, are a suitable host for the encyrtid *Ooencyrtus malayensis* Ferrière (van der Vecht 1933). In Japan, *O. nezarae* Ishii attacks eggs of *H. unipunctatus* Thunberg, which feeds on kudzu vine (*Pueraria lobata* Willdenow); parasitism by *Anastatus gastropache* Ashmead and an unidentified *Gryon* species was also recorded (Takasu and Hirose 1986).

4.7.1 Homoeocerus pallens F.

This African species has been intensively studied as a pest of guava (Agunloye 1986), but is reported from a variety of cultivated plants in Ghana, Nigeria, and the Congo. The dark green or brown adults oviposit on the leaf surface or stems of guava; in laboratory breeding studies, females laid two to four eggs per day. At 28°C, egg development required 6 ± 2.5 days, and nymphal development 23 days. In Nigeria, *H. pallens* occurs on guava primarily in the southern rain forest region, and is less common in the drier savanna. Insects were present on guava during all months except November and December, reaching a maximum of 14 per tree in September. High population densities from June through September coincided with the guava fruiting season and with periods of high relative humidity (Agunloye 1986). The species is also listed as a pest of grain legumes (Singh et al. 1978b), and has been observed on citrus fruit, cassava leaves, tobacco leaves, cotton, *Albizia* sp., and cacao pods (Golding 1927, Forsyth 1966). Actual damage to cacao appears to vary geographically; *H. pallens* is an occasional pest of cacao in the Congo Republic, and feeds on cherelles in Nigeria (Entwistle 1972, Youdeowei 1973). However, in Ghana this species is attracted to leguminous shade trees such as *Gliricidia sepium* (Jacquin) Walpers in cocoa plantations. It has not been observed to feed on cacao, nor can it be reared on this plant in the laboratory (Lodos 1967).

Damage appears as black spots on the guava fruit surface; swelling and cracking at the puncture site develop after several days, and infestation by fungi may follow. The swollen areas are hard internally. Damaged mature fruits remain on the tree, but immature fruits begin to drop 11 days after infestation. Nymphs are more damaging than adults, in that initial signs of damage on immature fruit appear significantly faster (Agunloye 1986).

4.8 Mictis Leach

In addition to the well-known crusader bug, *M. profana* (F.) (see **Section 4.8.1**), several other species in this Old World genus are of economic importance. In Malaya, *M. tenebrosa* (F). damages young shoots of lime, and this species was also noted to damage terminal branches of forest trees in China (Hoffmann 1933). Damage to mango in Brunei is attributed to *M. longicornis* Westwood (CABI 1997). Kumar (1966) collected *M. caja* Stål on tomato, bean, and apple-of-Peru (*Nicandra physalodes* Gaertn.) in Australia.

4.8.1 Mictis profana (F.)

Adults of this species are notable for the pale yellow cross on the hemelytra, which gave rise to the common name "crusader bug." The species occurs in Fiji, Indonesia, Papua New Guinea, and throughout the Australian mainland, where it is a minor pest of citrus and *Acacia*. Males have larger, more curved hind femora than females, spined tibiae, and distinct pheromone glands on the abdomen. Eggs are hemicylindrical, laid in chains averaging 12 eggs per mass. Descriptions of the instars are given by Kumar (1966). Development time (egg to final molt) averages 38.66 (± 3.82) days, followed by a long preoviposition period. Adults can live for 3.5 months. Durations of stadia, adult longevity, and fecundity data are given by Flanagan (1994). The species is polyphagous, and has been recorded to feed on a variety of native and introduced plants, and cultivated crops, including grape, pawpaw, cowpea, tomato, beans, and green gram (Hoffmann 1933; Kumar 1966; Carver et al. 1991; Flanagan 1994, and references therein). Damage is to young shoots and new growth. Older literature recommends hand-picking for control on citrus. However, *M. profana* is also of interest from a biological control standpoint, as it feeds in Australia on the introduced weed *Mimosa pigra* L. (Flanagan 1994).

4.9   *Notobitus meleagris* (F.)

Although four species of *Notobitus* are reported to feed on bamboo in China (Zheng 1994), only *N. meleagris* is considered to be of potential economic importance (Hill 1983). Adults of this species are brown and 20 to 24 mm long. Hill (1983) describes the legs as "normal"; presumably he was referring to a female, as Miyatake (1995) notes a sexual dimorphism and enlarged, spined hind femora in males. Nymphs and adults inhabit the space between the leaf sheath and the stem internode, but may also be found in the open on branches (Hill 1983, Zheng 1994). Miyatake (1994, and references therein) reports 14 bamboo species as host plants, including the commercially important *Phyllostachys viridis* (Young) McClure (Zheng 1994). Distribution includes southern China, India, several islands in the Malay archipelago, Taiwan, and, more recently, Okinawa and other Japanese islands (Hoffman 1933; Miyatake 1994, 1995, and references therein). Feeding occurs on stems and shoots (Hill 1983). On Okinawa, adults overwinter in aggregations, and breed on *Bambusa oldhamii* Munro in spring and summer. Mating and feeding occur on both lateral shoots (those that develop from bamboo stems) and bamboo shoots (those that will become new stems). However, in southern China, matings were restricted to bamboo shoots (Zhen-Yao 1989, cited in Miyatake 1994). Males defend territorial mating aggregations; the number of females in such an aggregation is related to the diameter of the shoot and the size of the defending male (Miyatake 1995). Eggs are placed mostly on the backs of leaves (Miyatake 1994).

Damage is of two types, necrotic stem lesions and death of shoots. The stem lesions enlarge, often longitudinally, which could weaken the bamboo for construction purposes (Hill 1983). Bamboo shoots are also used as a vegetable; thus, shoot death caused by bug feeding represents another potential economic problem (Miyatake 1994). No species of *Notobitus* is reported to reach damaging population densities in southern China (Zheng 1994), but Hoffman (1933) describes a *Notobitus* sp., possibly *N. meleagris*, as a "pest of considerable importance" in Kwangtung Province, China. Hill (1983) classifies this species as "more of academic interest than directly agricultural." If the spread of *N. meleagris* in Japan continues northward, it could pose an economic threat to bamboo shoot cultivation (Miyatake 1994). No control measures have been recommended; Hill (1983) notes that because only ornamental plants were damaged, control was not considered worthwhile.

4.10   *Paradasynus* China

An undetermined coreid, provisionally identified as *Amblypelta* sp. (see **Section 3.1**), but now thought to be *P. rostratus* Distant, was reported in 1972 to attack coconut in Kerala, India (Kurian et al. 1972, 1976). Damage to tender cashew nuts and guava by *Paradasynus* sp. has also been noted. The species on cashew is identified only as "*Paradasynus* sp. (*rostratus* Distant?)" (Nair and Remany 1964), whereas the guava species is noted to be *P. rostratus* and clearly identified with the earlier "coconut bug of Kerala" (Beevi et al. 1989).

4.10.1 *Paradasynus rostratus* Distant

Adults are brown, with a white ring on the terminal antennal segment (Beevi et al. 1989). Length measurements have been variously reported as males, 15 to 17 mm, females, 21 to 22 mm (Beevi et al. 1989); and males, 15 to 16 mm, females, 16 to 17 mm (Kurian et al. 1979). Eggs are elongate-oval, yellow or orange-yellow when laid but turning dark red during development. They are laid in clusters on fruit, leaves, or branches of guava, leaves of cashew, and on or near the coconut inflorescence. Nymphs are reddish-brown, and the third segment of the antenna is foliated in all instars. Detailed descriptions of all stages are provided by Kurian et al. (1979). First and second instars are gregarious on guava; third instars through adult may be found individually on fruit. Nymphal development requires 25 to 34 days ($X = 29.5$) (Kurian et al. 1979). All instars, egg, and adult are figured in Nair and Remany (1964). However, the illustration in Kurian et al. (1976) of "*Paradasynus* sp." bears little resemblance to the *Paradasynus* sp. pictured in Nair and Remany (1964); clearly there is some confusion here.

Damage to coconut is similar to that caused by *Amblypelta cocophaga* and *Pseudotheraptus wayi* (**Sections 3.1.5.2 and 3.7.1.2**); feeding produces necrotic lesions, furrows, and gummosis on the immature nuts, and eventual shedding (Kurian et al. 1972, 1979). Low populations, even single individuals, can cause extensive nut loss (Kurian et al. 1976). Feeding on guava results in discoloration of the pericarp and malformed fruits, whereas immature (tender) cashew nuts shrivel and turn black. Damage to rubber, tapioca, and tamarind has also been reported (Kurian et al. 1976, 1979), the losses in tamarind approaching 65%. Carbaryl has been used successfully for control in coconut plantings (Kurian et al. 1979). Scelionid and eupelmid egg parasites have been reared from eggs on cashew nut (Nair and Remany 1964).

4.11 *Phthia picta* (Drury) (= *Dallacoris pictus* [Drury])

At least six species of *Phthia* are known from the Neotropics (Blatchley 1926, Froeschner 1981), although only *P. picta* is reported to be of economic importance. Osuna (1981) described the new genus *Dallacoris* for this species, which was accepted by Henry and Froeschner (1992). However, *P. picta* continues to be used in the economic literature, and will therefore be retained here. The species occurs from the southern United States and the Antilles throughout the Neotropics to Argentina (Osuna 1981). Adults of *P. picta* are dark brown or black dorsally, and black or dull red with black dots ventrally. Coloration, particularly of the pronotum, varies, ranging from solid black to black marked with vivid yellow, orange, ochre, or reddish spots and bands. The lateral margins of the pronotum are toothed. The femora are long and slender, armed with small spines, and the tibia are simple (Blatchley 1926, Serantes 1973, Osuna 1981). Length is given as 14 to 16 mm for Florida (U.S.A.) specimens (Blatchley 1926), 17.4 mm for Argentine specimens (Serantes 1973). Eggs are hemicylindrical, 1.43 mm in length, and are laid in chains. Early instars are bright orange with black markings; fifth instars are darker, brown or black with orange markings, and covered with fine pubescence. Detailed descriptions of each instar are given by Serantes (1973), with drawings of the egg, first and fifth instars, and adult.

In Argentina, female lifetime fecundity on tomato was 35 to 45 eggs, laid in several batches of 10 to 20. However, an occasional female laid 70 eggs, and lifetime fecundity of >100 eggs was observed on rare occasions. From eclosion to adult molt required 30 to 35 days. Adults lived 33 to 35 days (maximum of 70), and were found in all months of the year. All instars were found between October and May, whereas in June only fifth instars were encountered. Three to five generations per year are estimated in the Buenos Aires, Argentina area (Serantes 1973).

In Brazil, egg incubation ranges from 8 to 12 days at different times of year. Nymphal development time on tomato required 40 to 88 days (mean = 54.87) in spring to summer, when nymphs were reared in aggregations, but development was extended, with considerably lower survivorship, when nymphs were reared in isolation at other times of year. Survivorship of nymphs in isolation was improved with the addition of pumpkin leaves and fruit, although survivorship on pumpkin alone was poor (Amaral 1981b).

Courtship, precopulatory behavior, and copulation are described and illustrated by Amaral (1981a). Mated females laid an average of 66 eggs (range 12 to 242) over a period of 24.6 days (range 1 to 80). Average number of clutches was 3.65 ± 0.57, with 18.10 ± 0.84 eggs per mass. The preoviposition period averaged 24.69 ± 1.88 days. Longevity of mated adults was significantly different from that of unmated individuals; however, males and females did not differ (Amaral 1980).

First and second instars are normally gregarious and active; the last three instars tend to be dispersed and move more slowly. On pumpkin, nymphs may be found on the underside of leaves, sucking from the veins (Amaral 1981a).

These bugs are highly polyphagous; nymphs and adults are found in Buenos Aires on plants of several families, including tomato (*Lycopersicon esculentum* Miller), various *Solanum* spp. (including eggplant), and *Cucurbita maxima* Duchesne (turban squash). An extensive list of Argentine host plants is provided by Serantes (1973), who notes that damage is most severe on tomato. Fruit develop abnormally when attacked, and may decay. The common name of the bug in Argentina is, in fact, "chinche de tomate," although it is reported by various authors from potato, eggplant, pimiento, tobacco, various squashes, melon, peas, beans, rice, pomegranate, passion fruit, cotton, white clover, and sunflower, among others (Serantes 1973). In Honduras, *P. picta* is listed as occurring on maize, rice, potato, tomato, sweet potato, and eggplant, but is not accorded even minor pest status on these crops (Passoa 1983). Similarly in Cuba, occurrence on tomato, squash, and cowpea is infrequent, and this species is not considered a pest (Barber and Bruner 1947). In Brazil, adults and nymphs attack tomato and pumpkin (*C. pepo* L.) in summer (Amaral 1981a,b).

In addition to direct damage, feeding punctures in tomato fruit permit the entry of fungal pathogens and leave the fruit susceptible to attack by other insects (references in Amaral 1980). *Phthia picta* has also recently been implicated in the transmission of a trypanosomatid parasite of tomatoes, *Phytomonas serpens* (Gibbs) (Jankevicius et al. 1989, Silva and Roitman 1990, Fiorini et al. 1993). This trypanosomatid was isolated from field-collected tomato fruit in southeastern Brazil, but was never found in stems or leaves. Fruit with evidence of bug feeding damage were more likely to harbor *Phytomonas serpens*, and field-collected adults and nymphs of *Phthia picta* carried this flagellate in the digestive tract. Cultures isolated from field-collected tomatoes could be used to infect both laboratory-reared fruit and nymphs of *P. picta* from a laboratory colony. Transmission from bug to plant and plant to bug was obtained. Infected insects had promastigote-type flagellates in salivary glands, digestive tract, and feces (Jankevicius et al. 1989). A defined diet has recently been developed on which both fruit- and bug-derived strains can be cultured (Silva and Roitman 1990). Light and electron microscopic examination of *Phytomonas* sp. strains from field-collected tomatoes and field-collected *Phthia picta* showed the isolates to be essentially identical. Both strains were shown to agglutinate with the hemolymph of *P. picta*, which is reminiscent of the relationship between the trypanosome causing Chagas' disease and one of its triatomine hosts, *Rhodnius prolixus* Stål (Fiorini et al. 1993). Although *Phytomonas serpens* is referred to as a parasite, Jankevicius et al. (1989) had no evidence that this flagellate was actually pathogenic to tomato; however, some *Phytomonas* species are known to cause severe epidemics in other crops. Ripe fruit infected with a *Phytomonas* sp. transmitted by *Phthia picta* were described as showing "light colour spots on the surface and internal decay" (Fiorini et al. 1993); the former probably results from direct bug damage, but the internal symptoms may not. A similar, possibly identical, trypanosomatid was reported from tomato and the pentatomid *Nezara viridula* (L.) in South Africa as *Leptomonas serpens* (Gibbs 1957, cited in Jankenvicius et al. 1989), but it was not grown in laboratory culture. Trypanosomatids have been cultured from a variety of other Brazilian coreids in addition to *P. picta* (Sbravate et al.

1989). Thus, the geographic and taxonomic extent of this potential vectoring problem is probably not yet fully understood.

4.12 *Spartocera* Laporte (= *Corecoris* Hahn)

Species in this genus are primarily Neotropical, and are easily recognized by the widely expanded and rounded abdomen and the much narrower hemelytra. In addition to the species discussed in detail below, *S. brevicornis* (Stål) can be found on cultivated solanaceous plants in Argentina, although its host plant is "yerba mora" (*Solanum nigrum* L.). This species is also reported as injurious to tobacco, *Nicotiana tabacum* L., in Brazil (Bosq 1937, and references therein). Damage to sweet potato, *Ipomoea batatas* (L.), is caused by nymphs and adults of *Spartocera batatas* (F). in Puerto Rico and Jamaica (Wolcott 1933); this species also occurs in Brazil (Amaral Filho and Vieira 1978), and is parasitized in both localities by the tachinid *Trichopoda pennipes* (F.). Other species, such as *S. lativentris* (Stål), are restricted to wild solanaceous hosts and are at present of no economic importance; however, ecological studies of such related species are exceedingly valuable (Becker and Prato 1982).

4.12.1 *Spartocera dentiventris* (Berg)

This Brazilian species, known as the ash gray tobacco bug (*percevejo cinzento do fumo*), has been intensely studied in recent years as a serious pest of tobacco. Adults measure 20 mm long by 10 mm wide, and have teeth posteriorly on the connexivum. Descriptions and illustrations of the egg and the five instars are given by Caldas et al. (1998). Eggs are deposited in clusters, and the preferred oviposition site is the abaxial leaf surface. Durations of the first through fifth stadia on tobacco were found to be 5.4 ± 0.08, 9.7 ± 0.19, 4.5 ± 0.08, 5.3 ± 0.08, and 9.0 ± 0.13 days; overall nymphal mortality is 35.29% (Redaelli et al. 1998a,b). Following a preoviposition period of 11.8 days, females laid an average of 355.9 eggs over 48.61 days. Longevity of adult males was 62.6 days, that of females was 66.1 days (Redaelli et al. 1998a).

Feeding by *S. dentiventris* on tobacco causes wilting and rolling of new leaves (Costa 1941, cited in Caldas et al. 1998). Several hymenopteran parasites attack the eggs, including *Gryon gallardoi* (Brethes) (Scelionidae) and *Neorileya ashmeadi* Crawford (Eurytomidae) (Santos et al. 1998). These same species are reported to parasitize eggs of *S. lativentris* (Becker and Prato 1982).

4.12.2 *Spartocera diffusa* (Say)

The "Florida potato plant bug" occurs in the southern United States, from North Carolina to New Mexico, and damages potatoes in Florida. The brownish yellow adults are 17 to 20 mm long. (Blatchley 1926, Chittenden 1927). Descriptions of first, second, and fifth instars and the egg are provided by Chittenden (1927). Eggs are oval and laid on leaves in batches of 8 to 10. Feeding on tomato fruit, canna leaves, grape leaves, and *Solanum nigrum* is reported in addition to cultivated potato. Hills of potato inhabited by *S. diffusa* are described as "non-productive" (Chittenden 1927).

4.12.3 *Spartocera fusca* (Thunberg)

Distributed from the southern United States to Argentina (Brailovsky and Sánchez 1983), this large (20 to 24 mm), yellow-orange bug is found on weeds and in gardens, and on the edges of fields and orange groves (Blatchley 1926). It feeds primarily on solanaceous hosts (*Solanum nigrum*, tomato, and *Capsicum annuum* L.), but is also reported from cotton, sweet potato, and beans (Bosq 1937; Costa Lima 1940; Brailovsky and Sánchez 1983; CABI 1997, and references therein).

4.13 OTHER MINOR PEST SPECIES

Coreid species for which minimal biological or economic information could be found are summarized in Table 1. Information is limited to the crops listed as damaged, the plant parts punctured (if known), and the locality. References that provide further information are indicated.

5. BENEFICIAL SPECIES

Only one coreid genus has received much attention as a beneficial insect in weed control. *Chelinidea*, a specialist on the pads of prickly pear (*Opuntia* spp.), has been used as a biological control agent in areas where this cactus has been introduced or spread into former rangeland. *Chelinidea vittiger* Uhler, *C. tabulata* (Burmeister), and *C. canyona* Hamlin were imported into Australia (Hamlin 1924a), and *C. vittiger* was released on Santa Cruz Island, California (U.S.A.), for prickly pear control (Goeden et al. 1967). However, the failure of these bugs to disperse much beyond the initial point of introduction rendered them less effective than other control agents of *Opuntia*. In Hawaii, both *C. vittiger* and *C. tabulata* were examined as possible biological control agents for the tree cactus, *Opuntia megacantha*. When nymphs were caged on pineapple plants, scarring of the pineapple leaves was observed, and these coreids were therefore judged unsuitable for release (Fullaway 1954).

Mozena obtusa Uhler is a potentially important native control agent of mesquite (*Prosopis glandulosa* Torrey) in areas of the United States where this invasive weed has become a problem in rangeland (Ueckert 1973), and it is also being considered as a biological control agent in Australia (Parsons and Cuthbertson 1992). Pod feeding by these coreids significantly reduces pod moisture, pod weight, seed weight, seed germination, and sprout weight, while increasing the percentage of aborted immature pods. However, as with the cactus-feeding *Chelinidea* spp. (see **Section 4.3**), *M. obtusa* would be viewed as detrimental rather than beneficial by ranchers who use mesquite pods as alternate sources of food for cattle during times of drought.

Recently, another coreid species, *Pomponatius typicus* Distant, has been considered as a possible weed control agent. *Melaleuca quinquenervia* (Cav.) S.T. Blake is a tree native to Australia but an introduced weed in Florida (U.S.A.). The host range of *P. typicus* is limited to *Melaleuca* and *Callistemon* spp. (Burrows and Balciunas 1998). Growth of saplings in Australia was significantly limited by tip-wilting caused by *P. typicus* feeding, suggesting that this species may have potential in a biological control program for introduced *M. quinquenervia* populations in southern Florida (Balciunas and Burrows 1993; also see Brailovsky and Monteith 1996, for food plant records and taxonomy of *Pomponatius* spp.).

6. PROGNOSIS

Only 4% of coreid species worldwide are even mentioned in the economic literature and, of those species, even fewer are considered serious pests. However, losses attributed to these few major pests, particularly *Clavigralla* spp. in sub-Saharan Africa, may be devastating. Heavily damaged crops, such as pigeon pea, are traditionally grown in gardens by small farmers, and insecticides are neither environmentally nor economically desirable. No single approach to control, whether chemical, cultural, or biological, has shown consistent success. Promising strategies may be those that combine several types of control; for example, intercropping alone has shown mixed results for control of pod-sucking bugs, but intercropping in combination with reduced insecticide applications may be a practical approach (Alghali 1993). Similarly, biological control of squash bug on pumpkin is not cost effective, but could become economically viable in combination with the use of resistant varieties (Olson et al. 1996).

Table 1 Summary of Coreid Species of Minor Economic Importance Not Covered in the Text

Species	Plant	Plant part	Locality	Source
Athaumastus haematicus (Stål)	Cotton, potato, *Phaseolus vulgaris*, eggplant, sunflower		Argentina	Bosq 1937, 1940b
	Oranges		Brazil	Costa Lima 1940
Aulacosternum nigrorubrum Dallas	*Hibiscus rosa-sinensis* L.	Buds	Australia	Kumar 1966,* Gough and Hamacek 1989
Brachylybus inflexus Blöte	Cacao	Pods, stalks	Papua New Guinea	Szent-Ivany 1963
B. variegatus Le Guillou	Taro (*Colocasia esculenta* (L.) Schott)	Leaves and petioles	Fiji, Pacific Islands	Mitchell and Maddison 1983
Camptischium clavipes (F.) (as *Acanthocerus clavipes* F.)	Castor bean, guaco		Brazil	Monte 1937, cited in Costa Lima 1940
Cletomorpha lancigera F.	Cacao	Leaves, cherelles	Nigeria	Golding 1941,1948; Forsyth 1966; Youdeowei 1973
Cletomorpha fuscescens Walker	Cowpea	Pods	Ghana	van Halteren 1971
	Cowpea	Pods	Benin	Dreyer and Baumgärtner 1995
	Cowpea	Pods, shoots	Nigeria	Aina 1975b
	Amaranthus spp.		Uganda	Le Pelley 1959
C. unifasciata Blöte	Cowpea	Pods, shoots	Nigeria	Aina 1975b
	Cacao	Leaves, new flush	Ghana	Alibert 1951,* Forsyth 1966, Lodos 1967
Carlisis wahlbergi Stål	Gardenia, azaleas	Twigs	South Africa	Hartwig and de Lange 1978*
Cloresmus modestus Distant	Bamboo	New culms	China	Hoffman 1933*
C. montanus Hsiao	Bamboo	Culm sheath	China	Zheng 1994
Dasynus piperis China	Pepper	Fruits	Indonesia	van der Vecht 1933*
Dersagrena flaviventris (Berg) (as *Athaumastus flaviventris* (Berg))	Cotton		Argentina	Bosq 1937

Species	Plant	Part	Country	Reference
Diactor bilineatus F.	Passiflora quadrangularis L.		Brazil	Costa Lima 1940
Elasmopoda valga (L.)	Ornamentals	Shoots	South Africa	Hartwig and de Lange 1978
Holhymenia histrio F.	Passiflora spp.	Shoots, flowers	Brazil	Costa Lima 1940, Boiça et al. 1998
Hydara tenuicornis Westwood	Cacao		Nigeria	Youdeowei 1973
Hypselonotus intermedius Distant	Jatropha curcas L.	Flowers	Nicaragua	Grimm and Führer 1998
Hypselonotus lanceolatus var. gentilis Horváth	Eugenia stipitata McVaugh (arazá)	Fruit	Peru	Couturier et al. 1996a
Hypselonotus lineatus detersus Horvath	Jatropha curcas L.	Flowers	Nicaragua	Grimm and Führer 1998
Lybindus dichrous Stål	Pineapple	Fruit	Brazil	Mariconi 1953*
Machtima cruciger (F.)	Roses, dahlias		Brazil	Costa Lima 1940
Mozena obtusa Uhler	Mesquite	Pods	USA	Ueckert 1973*
Omanocoris versicolor (Herrich-Schaeffer)	Khejri (Prosopis spicigera)	Shoots, leaves	India	Chandra and Chandra 1985
	String bean, cotton		Iraq	Linnavuori 1993
Plectrocnemia oblongipes F.	Mango	Petioles	Nigeria	Golding 1937
	Legumes in cacao farms		Nigeria	Youdeowei 1973
Physomerus grossipes F.	Sweet potato, Phaseolus vulgaris L., cowpea	Stems	Malaya	Miller 1929*
Priocnemicoris flaviceps Guérin	Cacao	Young trees	Papua New Guinea	Szent-Ivany 1961
Thlastocoris laetus Mayr	Pineapple	Fruit	Brazil, Peru, Surinam, Guyana, Venezuela	Couturier et al. 1993

Sources that provide some biological or economic information are indicated by an asterisk (*).

Another area of concern is the current trend toward increased monocrop plantings of legumes (e.g., cowpea in sub-Saharan Africa, soybean in Brazil). Some species may shift from wild to cultivated legumes, or move from potential to actual pest status, as has been noted for alydids such as *Neomegalotomus parvus* (Westwood) in Brazil (Santos and Panizzi 1998). Finally, the ability of coreids to transmit trypanosomatids (see *Phthia picta*, **Section 4.11**), coupled with the surprising number of coreids surveyed that have been shown to harbor these potential plant pathogens (*Holhymenia histrio* F., *Hypselonotus* spp., *Leptoglossus* spp.) (Sbravate 1989), certainly warrants further investigation.

7. CONCLUSIONS

Several species of Coreidae may cause extensive damage and economic losses, particularly among legume and cucurbit crops and tree fruit. Severe damage may be inflicted when populations increase to uncontrolled high levels (e.g., *Clavigralla*), or populations may remain low but a single feeding puncture may be sufficient to induce abcission or loss of marketability; in such cases (e.g., *Amblypelta*, *Pseudotheraptus*), a few individuals may be capable of great damage.

Current control strategies rely heavily on insecticides. Biological control using ants has proved successful in some cases, and research on scelionid and encyrtid egg parasites appears promising. Future research should focus on combined control strategies, and should investigate the compatibility of plant resistance, biorational insecticides, biological control, and cultural controls.

Basic biological research is still needed for several species of economic importance, and for potential pests in areas where legume cultivation is expanding. Finally, continued studies are needed on the mechanisms of coreid feeding, and the different types of damage inflicted by "tip-wilters" and fruit- or seed-feeding species. An improved understanding of stylet action and feeding behavior is important not only in regard to physical damage to plant tissue but also to transmission of plant pathogens such as fungi and trypanosomatids.

8. ACKNOWLEDGMENTS

The author is grateful to many colleagues who offered access to databases and references, including T. J. Henry, A. R. Panizzi, L. H. Rolston, C. W. Schaefer, B. M. Shepard, A. G. Wheeler, Jr., and the late Dennis Leston. Invaluable help with translations was provided by Sarah Ralston (Portuguese), Guillermo Castillo (Spanish), and Ed Haynes (German and Hindi). John Schmidt patiently (and uncomplainingly) helped me search for botanical names and describers, and Martin Steinbauer read and edited portions of the manuscript. Finally, the author thanks the interlibrary loan faculty and staff at Dacus Library — Fern Hieb, Sarah McIntyre, Ann Thomas, and David Weeks — without whom this project could not possibly have been completed.

9. REFERENCES CITED

Abate, T., and J. K. O. Ampofo. 1996. Insect pests of beans in Africa: their ecology and management. Annu. Rev. Entomol. 41: 45–73.
Adair, H. S. 1932. Black pit of the pecan and some insects causing it. U.S. Dept. Agric. Circ. No. 234. 14 pp.
Adams, A. L., and J. C. Gaines. 1950. Sunflower insect control. J. Econ. Entomol. 43: 181–184.
Adler, P. H., and A. G. Wheeler, Jr. 1984. Extra-phytophagous food sources of Hemiptera-Heteroptera: bird droppings, dung, and carrion. J. Kansas Entomol. Soc. 57: 21–27.
Agunloye, O. 1986. *Homoecerus pallens* F. (Hemiptera; Coreidae), a pest of a guava (*Psidium guajava*) in Nigeria. Crop Res. 25: 81–88.

Agyen-Sampong, M. 1978. Pests of cowpea and their control in Ghana. Pp. 85–92 *in* S. R. Singh, H. F. van Emden, and T. A. Taylor [eds.], Pests of Grain Legumes: Ecology and Control. Academic Press, London, U.K. 454 pp.

Ahmad, I. 1979. Systematics and Biology of Pentatomomorphous Superfamilies Coreoidea and Pentatomoidea of Pakistan: Final Report. University of Karachi, Karachi, Pakistan. 625 pp.

Ahmad, I., N. Abbas, and M. U. Shadab. 1983. Biology and immature systematics of two gonocerines, *Cletus bipunctatus* (Westwood) and *Cletomorpha hastata* Fabr. (Coreidae: Coreinae), with reference to their relationship. Biologia 29: 55–68.

Aina, J. O. 1972. The biology of *Acanthomia horrida* Germ. (Heteroptera: Coreidae), a pest of fresh cowpea pods. Bull. Entomol. Soc. Nigeria 3: 85–90.

1975a. The life-history of *Anoplocnemis curvipes* F. (Coreinae, Coreidae, Heteroptera), a pest of fresh cowpea pods. J. Nat. Hist. 9: 685–692.

1975b. The distribution of coreids infesting cowpea pods in southwestern Nigeria. Nigerian J. Entomol. 1: 119–123.

Akingbohungbe, A. E. 1982. Seasonal variation in cowpea crop performance at Ile-Ife, Nigeria, and the relationship to insect damage. Insect Sci. Appl. 3: 287–296.

Albrigo, L. G., and R. C. Bullock. 1977. Injury to citrus fruits by leaffooted and citron plant bugs. Proc. Florida State Hortic. Soc. 90: 63–67.

Aldrich, J. R. 1988. Chemical ecology of the Heteroptera. Annu. Rev. Entomol. 33: 211–238.

Aldrich, J. R., and T. R. Yonke. 1975. Natural products of abdominal and metathoracic scent glands of coreoid bugs (Coreidae). Ann. Entomol. Soc. Amer. 68: 955–960.

Aldrich, J. R., M. S. Blum, and H. M. Fales. 1979. Species-specific natural products of adult male leaf-footed bugs (Hemiptera: Heteroptera). J. Chem. Ecol. 5: 53–62.

Aldrich, J. R., M. S. Blum, A. Hefatz, H. M. Fales, H. A. Lloyd, and P. Roller. 1978. Proteins in a nonvenomous defensive secretion: biosynthetic significance. Science 201: 452–454.

Aldrich, J. R., G. K. Waite, C. Moore, J. A. Payne, W. R. Lusby, and J. P. Kochansky. 1993. Male-specific volatiles from Nearctic and Australasian true bugs (Heteroptera: Coreidae and Alydidae). J. Chem. Ecol. 19: 2767–2781.

Alghali, A. M. 1993. Intercropping as a component in insect pest management for grain cowpea, *Vigna unguiculata* Walp production in Nigeria. Insect Sci. Appl. 14: 49–54.

Alibert, H. 1951. Les insectes vivant sur les cacaoyers en Afrique occidentale. Mem. Inst. Fr. Afr. Noire 15: 1–171.

Allen, R. C. 1969. A revision of the genus *Leptoglossus* Guerin (Hemiptera: Coreidae). Entomol. Amer. 45: 35–140.

Amaral, B. F. do Filho. 1980. Observações biológicas em adultos acasa lados e não acasalados de um coreídeío praga — *Phthia picta* (Drury, 1770). An. Soc. Entomol. Brasil 9: 75–79.

1981a. Aspectos compartamentais de *Phthia picta* (Drury, 1770) em condições de laboratório (Hemiptera, Coreidae). Rev. Bras. Biol. 41: 441–446.

1981b. Efeito de dietas naturais e de fatores ambientais na biologia de *Phthia picta* (Drury, 1770) sob condições de laboratorio (Hemiptera, Coreidae). Rev. Bras. Biol. 41: 845–853.

1986. Observações sobre o ciclo biológico de *Crinocerus sanctus* (Fabricius 1775) (Hemiptera: Coreidae) sob condições de laboratório. An. Soc. Entomol. Brasil 15: 5–18.

Amaral, B. F. do Filho, and I. V. M. Cajueiro. 1977. Observações sobre o ciclo biológico de *Veneza stigma* (Herbst, 1794) Osuna, 1975 (Hemiptera, Coreidae) em laboratório. An. Soc. Entomol. Brasil 6: 164–172.

Amaral, B. F. do Filho, and A. Storti. 1981. Laboratory studies on the biology of *Leptoglossus gonagra* (Fabricius, 1775), (Coreidae, Hemiptera) = Estudos biologicos sobre *Leptoglossus gonagra* (Fabricius, 1775), (Coreidae, Hemiptera) em laboratório. Saad Publications. Translation Division, Karachi, Pakistan. 11 leaves [translated from An. Soc. Entomol. Brasil 5: 130–137, 1976].

Amaral, B. F. do Filho, and A. O. S. Vieira. 1978. Ocorrencia de *Trichopoda pennipes* (Fabricius, 1794) (Diptera, Tachinidae) parasitando *Corecoris batatas* (Fabricius, 1798) (Hemiptera Coreidae), no Brasil. An. Soc. Entomol. Brasil 7: 67–68.

Amatobi, C. I. 1994. Field evaluation of some insecticides for the control of insect pests of cowpea (*Vigna unguiculata* (L.) Walp.) in the Sudan savanna of Nigeria. Int. J. Pest Manage. 40: 13–17.

1995. Insecticide application for economic production of cowpea grains in the northern Sudan savanna of Nigeria. Int. J. Pest Manage. 41: 14–18.

Anonymous. 1975. Plagas de los cultivados de badea, curuba, maracuya, papayo, y vid en el occidente Antioqueño. Sec. Agric. Fomento. Medellín, Colombia. 8 pp.

1985. Attack of the leaffooted bug. Almond Facts 50: 12–13.

Arnaud, P. H. 1978. A host-parasite catalog of North American Tachinidae (Diptera). U.S. Dept. Agric. Misc. Publ. No. 1319. USDA, Washington, D.C., U.S.A. 860 pp.

Ashmead, W. H. 1887. Report on insects injurious to garden crops in Florida. U.S. Dept. Agric. Bur. Entomol. Bull. 14: 9–28.

Balciunas, J. K., and D. W. Burrows. 1993. The rapid suppression of the growth of *Melaleuca quinquenervia* saplings in Australia by insects. J. Aquatic Plant Manage. 31: 265–270.

Baranowski, R. M., and J. A. Slater. 1986. Coreidae of Florida (Hemiptera, Heteroptera). Arthropods of Florida and Neighboring Land Areas. Vol. 12, no. 630. Contribution / Florida. Bureau of Entomology, Florida Dept. Agric. Consumer Serv., Division of Plant Industry, Gainesville, Florida, U.S.A. 82 pp.

Barber, H. G., and S. C. Bruner. 1947. The Coreidae of Cuba and the Isle of Pines with the description of a new species (Hemiptera-Heteroptera). Mem. Soc. Cubana Hist. Nat. 19: 77–88.

Bates, S. L., J. H. Borden, A. R. Kermode, C. G. Lait, and R. G. Bennett. 1997. New impact of the western conifer seed bug, *Leptoglossus occidentalis*, on developing Douglas-fir seeds. Seed and Seedling Extension Topics, British Columbia Minist. For. 10: 24–26.

Beard, R. L. 1940. The biology of *Anasa tristis* DeGeer with particular reference to the tachinid parasite *Trichopoda pennipes* Fabr. Connecticut Agric. Exp. Stn. Bull. 440: 597–679.

Becker, M., and M. D. Prato. 1982. Natality and natural mortality of *Spartocera lativentris* Stål, 1870 (Heteroptera: Coreidae) in the egg stage. An. Soc. Entomol. Brasil 11: 269–281.

Beevi, S. N., A. Vilsalakshi, K. K. R. Nair, K. S. Remamony, and N. M. Das. 1989. Guava as a potential host of *Paradasynus rostratus* Dist. (Coreidae), the coreid bug of coconut in Kerala. Entomon 14: 363–364.

Belthoff, J. R., and G. Ritchison. 1991. *Acanthocephala terminalis* (Coreidae) associated with eastern screech-owl droppings. J. Kansas Entomol. Soc. 64: 458–459.

Bigger, M. 1985. The effect of attack by *Amblypelta cocophaga* China (Hemiptera: Coreidae) on growth of *Eucalyptus deglupta* in the Solomon Islands. Bull. Entomol. Res. 75: 595–608.

Bindra, O. S. 1965. Biology and bionomics of *Clavigralla gibbosa* Spinola, the pod bug of pigeon-pea. Indian J. Agric. Sci. 35: 322–334.

Bindra, O. S, and H. Singh. 1971. Tur pod bug *Clavigralla gibbosa* Spinola (Coreidae: Hemiptera). Pesticides 5: 3–4, 32.

Bissell, T. L. 1929. Notes on *Leptoglossus phyllopus* L. and *L. oppositus* Say. J. Econ. Entomol. 22: 597–598.

Blatchley, W. S. 1926. Heteroptera or True Bugs of Eastern North America, with Especial Reference to the Faunas of Indiana and Florida. Nature Publ. Co., Indianapolis, Indiana, U.S.A. 1116 pp.

Blatt, S. E. 1994. An unusually large aggregation of the western conifer seed bug, *Leptoglossus occidentalis* (Hemiptera: Coreidae), in a man-made structure. J. Entomol. Soc. British Columbia 91: 71

Blatt, S. E., and J. H. Borden. 1996a. Evidence for a male-produced aggregation pheromone in the western conifer seed bug, *Leptoglossus occidentalis* Heidemann (Hemiptera: Coreidae). Canad. Entomol. 128: 777–778.

1996b. Distribution and impact of *Leptoglossus occidentalis* Heidemann (Hemiptera: Coreidae) in seed orchards in British Columbia. Canada. Entomol. 128: 1065–1076.

Boher, B., J. F. Daniel, G. Fabres, and G. Bani. 1983. Action de *Pseudotheraptus devastans* (Distant) (Het., Coreidae) et de *Colletotrichum gloeosporioides* Penz. dans le développement de chancres et la chute des feuilles chez le manioc (*Manihot esculenta* Crantz). Agron. Sci. Prod. Veg. Environ. 3: 989–993.

Boiça, A. L. Jr., E. L. L. Baldin, and J. C. Oliveira. 1998. Determinação da preferência alimentar de *Holhymenia histrio* F. (Hemiptera: Coreidae) por frutos e botões florais de genótipos de maracujazeiro. P. 32 in Proceedings, 17th Congresso Brasileiro de Entomologia, 9–14 August 1998, Rio de Janeiro, Brazil.

Bolkan, H. A., J. M. Ogawa, R. E. Rice, R. M. Bostock, and J. C. Crane. 1984. Leaffooted bug (Hemiptera: Coreidae) and epicarp lesion of pistachio fruits. J. Econ. Entomol. 77: 1163–1165.

Bonjour, E. L., and W. S. Fargo. 1989. Host effects on the survival and development of *Anasa tristis* (Heteroptera: Coreidae). Environ. Entomol. 18: 1083–1085.

Bonjour, E. L., W. S. Fargo, and P. E. Rensner. 1990. Ovipositional preference of squash bugs (Heteroptera: Coreidae) among cucurbits in Oklahoma. J. Econ. Entomol. 83: 943–947.

Bonjour, E. L., W. S. Fargo, J. A. Webster, P. E. Richardson, and G. H. Brusewitz. 1991. Probing behavior comparisons of squash bugs (Heteroptera: Coreidae) on cucurbit hosts. Environ. Entomol. 20: 143–149.

Bonjour, E. L., W. S. Fargo, A. A. Al-Obaidi, and M. E. Payton. 1993. Host effects on reproduction and adult longevity of squash bugs (Heteroptera: Coreidae). Environ. Entomol. 22: 1344–1348.

Bosq, J. M. 1937. Lista preliminar de los Hemipteros (Heteropteros), especialmente relacionados con la agricultura nacional. Rev. Soc. Entomol. Argentina 9: 111–134.

1940a. Apuntes sôbre insectos que pueden ser de interes para la agricultura Argentina. Rev. Chilena Hist. Nat. Pura Appl. 43: 49–51.

1940b. Lista preliminar de los Hemipteros (Heteropteros), especialmente relacionados con la agricultura nacional (continuatión). Rev. Soc. Entomol. Argentina 10: 399–417.

Bradley, E. L., B. H. Ebel, and K. O. Summerville. 1981. *Leptoglossus* spp. observed on eastern white pine and Fraser fir cones. U.S. Dept. Agric. Forest Service Research Note SE-310: 3 pp.

Brailovsky, H. 1976. Contribución al estudio de los Hemiptera-Heteroptera de México: VIII. Una nueva especie de *Leptoglossus* Guerin (Coreidae-Coreinae) y datos sobre distributión de las especies Mexicanas del género. An. Inst. Biol. Univ. Nal. Autón. México 47, Ser. Zool. (2): 35–42.

1985. Revision del genero *Anasa* Amyot-Serville (Hemiptera — Heteroptera — Coreidae — Coreinae, Coreini). Monogr. Inst. Biol. Univ. Nal. Autón. México, No. 2: 1–266.

Brailovsky, H., and G. B. Monteith. 1996. A new species of *Pomponatius* Distant from Australia (Hemiptera: Heteroptera: Coreidae: Acanthocorini). Mem. Queensland Mus. 39: 205–210.

Brailovsky, H. C., and C. Sánchez. 1983. Hemiptera-Heteroptera de México XXIX. Revisión de la familia Coreidae Leach. Parte 4. Tribu Anisoscelidini Amyot-Serville. An. Inst. Biol. Univ. Nal. Autón. México 53 (1982), Ser. Zool. (1): 219–275.

Brailovsky, H., C. W. Schaefer, E. Barrera, and R. J. Packauskas. 1994 (published 1995). A revision of the genus *Thasus* (Hemiptera: Coreidae: Coreinae: Nematopodini). J. New York Entomol. Soc. 102: 318–343.

Brewer, F. D., and W. A. Jones, Jr. 1985. Comparison of meridic and natural diets on the biology of *Nezara viridula* (Heteroptera: Pentatomidae) and eight other phytophagous Heteroptera. Ann. Entomol. Soc. Amer. 78: 620–625.

Brimblecombe, A. R. 1948. Fruit-spotting bug as a pest of the Macadamia or Queensland nut. Queensland Agric. J. 67: 206–211.

1962. Descriptions of the life cycle stages of the fruit spotting bug *Amblypelta nitida* (Stål) (Hemiptera: Coreidae). J. Entomol. Soc. Queensland 1: 16–20.

Brown, E. S. 1955. *Pseudotheraptus wayi*, a new genus and species of coreid (Hemiptera) injurious to coconuts in East Africa. Bull. Entomol. Res. 46: 221–240.

1958a. Revision of the genus *Amblypelta* Stål (Hemiptera: Coreidae). Bull. Entomol. Res. 49: 509–541.

1958b. Injury to cacao by *Amblypelta* Stål (Hemiptera: Coreidae) with a summary of food plants of species of this genus. Bull. Entomol. Res. 49: 543–554.

1959a. Immature nutfall of coconuts in the Solomon Islands. I. Distribution of nutfall in relation to that of *Amblypelta* and of certain species of ants. Bull. Entomol. Res. 50: 97–133.

1959b. Immature nutfall of coconuts in the Solomon Islands. III. Notes on the life history and biology of *Amblypelta*. Bull. Entomol. Res. 50: 559–566.

Bullock, R. C. 1974. Toxicity of selected insecticides to the citron bug, *Leptoglossus gonagra*, a minor pest of Florida citrus. Proc. Florida State Hortic. Soc. 87: 17–19.

Burgess, P. S., and R. S. Hawkins. 1945. Fifty-fifth Annual Report, Arizona Agric. Exp. Stn. 54 pp.

Burrows, D. W., and J. K. Balciunas. 1998. Biology and host range of *Pomponatius typicus* Distant (Heteroptera: Coreidae), a potential biological control agent for the paperbark tree, *Melaleuca quinquenervia*, in southern Florida. Austral. J. Entomol. 37: 168–173.

Buyckx, E. J. E. 1962. Précis des maladies et des insectes nuisibles rencontrés sur les plantes cultivées au Congo, au Rwanda et au Burundi. Publications de L'Institut National pour l'Étude agronomique du Congo (INEAC). Hors Ser. 1962, Brussels.

CAB International. 1997. Crop Protection Compendium: Module 1. South-East Asia and Pacific. CAB International, Wallingford, Oxon, U.K.

Caldas, B.-H. C., L. R. Redaelli, and L. M. G. Diefenbach. 1998. Descrição dos estágios imaturos de *Corecoris dentiventris* Berg (Hemiptera: Coreidae). An. Soc. Entomol. Brasil 27: 405–412.

Calza, R., A. Orlando, V. Rossetti, and J. T. Nakadaira. 1964. O percevejo *Leptoglossus gonagra* (Fabr. 1775) em citros, no estado de São Paulo. O Biológico 30: 188–197.

Campbell, B. C., and P. J. Shea. 1990. A simple staining technique for assessing feeding damage by *Leptoglossus occidentalis* Heidemann (Hemiptera: Coreidae) on cones. Canad. Entomol. 122: 963–968.

Carroll, S. C., and J. K. Wangberg. 1981. Temperature effects on nymphal development and reproduction of the cactus bug *Chelinidea vittiger* Uhler (Hemiptera: Coreidae). J. Kansas Entomol. Soc. 54: 804–810.

Cartwright, B., J. C. Palumbo, and W. S. Fargo. 1990. Influence of crop mulches and row covers on the population dynamics of the squash bug (Heteroptera: Coreidae) on summer squash. J. Econ. Entomol. 83: 1988–1993.

Carver, M., G. F. Gross, and T. E. Woodward. 1991. Hemiptera. Pp. 429–509 *in* The Insects of Australia, ed. CSIRO, second edition. Cornell University Press, Ithaca, New York, U.S.A. 1137 pp.

Chandra, H., and S. Chandra. 1985. *Omanocoris versicolor* (Coreidae: Heteroptera), a new pest on khejri (*Prosopis spicigera*). Plant Prot. Bull. 37: 39.

Chandrasekar, S. S. 1988. Bioecological and host relationship studies in some coreids (Insecta: Heteroptera: Coreidae) from southern India. Ph.D. thesis, University of Madras, Chennai, India. 150 pp.

Chiang, H. S., and L. E. N. Jackai. 1988. Tough pod wall: a factor involved in cowpea resistance to pod sucking bugs. Insect Sci. Appl. 9: 389–393.

China, W. E. 1928. A new species of *Dasynus*, Burm., injurious to pepper in Java (Heteroptera Coreidae). Bull. Entomol. Res. 19: 253–254.

1934. A new species of Coreidae (Heteroptera) injurious to coconut in the Solomon Islands. Bull. Entomol. Res. 25: 187–189.

Chittenden, F. H. 1899. Some insects injurious to garden and orchard crops. U.S. Dept. Agric. Div. Entomol. Bull. No. 19 (n.s.). 99 pp.

1902. Some insects injurious to vegetable crops. U.S. Dept. Agric. Div. Entomol. Bull. No. 33 (n.s.). 117 pp.

1925. Note on the behavior of *Leptoglossus oppositus* Say. Bull. Brooklyn Entomol. Soc. 20: 148–149.

1927. The Florida potato plant-bug. State Plant Board Florida Q. Bull. 11: 115–118.

Choudhary, M. J. 1969. Bionomics of *Clavigralla gibbosa* Spin. (Coreidae: Hemiptera) a pest of *Cajanus cajan*. Labdev J. Sci. Technol. 7 (Part B): 200–207.

Choudhary, M. R., and S. V. Dhamdhere 1981a. Biology of *Clavigralla gibbosa* Spinola (Heteroptera: Coreidae), a pest of *Cajanus cajan*. Food Farming Agric. 14: 67–69.

1981b. Varietal preference of *Clavigralla gibbosa* Spinola to pigeon pea. Food Farming Agric. 14: 85–86.

Cibrián-Tovar, D., B. H. Ebel, H. O. Yates, and J. T. Méndez-Montiel. 1986. Cone and seed insects of the Mexican conifers. USDA Forest Service, Southeastern Forest Experiment Station, Asheville, North Carolina. 110 pp. [in English and Spanish].

Cobblah, M. A. 1991. Some pod-sucking bugs of cowpea, *Vigna unguiculata* (L.), in Ghana. Discovery Innovation 3: 77–79.

Connelly, A. E., and T. D. Schowalter. 1991. Seed losses to feeding by *Leptoglossus occidentalis* (Heteroptera: Coreidae) during two periods of second-year cone development in western white pine. J. Econ. Entomol. 84: 215–217.

Costa Lima, A. da. 1940. Insetos do Brasil. Vol. 2. Hemípteros. Escola Nacional de Agronomia. Série didática no. 3, Rio de Janeiro, Brazil. 351 pp.

Couturier, G., H. Brailovsky, and R. Zucchi. 1993. *Thlastocoris laetus* Mayr, 1866 (Hemiptera: Coreidae: Acanthocerini), a new pineapple pest. Sci. Agric. 50: 517–520.

Couturier, G., C. R. Clement, and P. Viana Filho. 1991. *Leptoglossus lonchoides* Allen (Heteroptera: Coreidae), causante de la caída de los frutos de *Bactris gaipaes* (Palmae) en la Amazonia central. Turrialba 41: 293–298.

Couturier, G., L. Quiñones, I. González R., R. Riva, and F. Young. 1996a. Los insectos plaga de las Myrtaceae frutales en Pucallpa, Amazonía Peruana. Rev. Peruana Entomol. 39: 125–130.

Couturier, G., E. Tanchiva, H. Inga, J. Vásquez, and R. Riva. 1996b. Notas sobre los artrópodos que viven en el pijuayo (*Bactris gasipaes* H. B. K.: Palmae) en la Amazonía Peruana. Rev. Peruana Entomol. 39: 135–142.

Cuda, J. P., and C. J. DeLoach. 1998. Biology of *Mozena obtusa* (Hemiptera: Coreidae), a candidate for the bilogical control of mesquite, *Prosopis* spp. (Fabaceae). Biol. Control 13: 101–110.

Dammer, K. H., and H. G. Ravelo. 1990. Infection of *Leptoglossus gonagra* (Fabr.) with *Nematospora coryli* Peglion and *Ashbya gossypii* (Ashby & Nowell) Guillermond in a citrus plantation in the Republic of Cuba. Arch. Phytopathol. Pflanzenschutz 26: 71–78 [CABI Abstracts].

Daoust, R. A., D. W. Roberts, and B. P. DasNeves. 1985. Distribution, biology and control of cowpea pests in Latin America. Pp. 249–264 *in* S. R. Singh and K. O. Rachie [eds.], Cowpea Research, Production and Utilization. John Wiley & Sons, Chichester, U.K. 460 pp.

Davies, J. C., and S. S. Lateef. 1978. Recent trends in grain legume pest research in India. Pp. 25–31 *in* S. R. Singh, H. F. van Emden, and T. A. Taylor [eds.], Pests of Grain Legumes: Ecology and Control. Academic Press, London, U.K. 454 pp.

DeBarr, G. L. 1967. Two new sucking insect pests of seed in southern pine seed orchards. U.S. Forest Service Research Note SE-78: 3 pp.

1970. Characteristics and radiographic detection of seed bug damage to slash pine seed. Florida Entomol. 53: 109–117.

DeBarr, G. L., and B. H. Ebel. 1974. Conelet abortion and seed damage of shortleaf and loblolly pines by a seedbug, *Leptoglossus corculus*. For. Sci. 20: 165–170.

DeBarr, G. L., and P. P. Kormanik. 1975. Anatomical basis for conelet abortion on *Pinus echinata* following feeding by *Leptoglossus corculus* (Hemiptera: Coreidae). Canad. Entomol. 107: 81–86.

DeBarr, G. L., and J. C. Nord. 1978. Contact toxicity of 34 insecticides to second-stage nymphs of *Leptoglossus corculus* (Hemiptera: Coreidae). Canad. Entomol. 110: 901–906.

Dennill, G. B., and M. J. Erasmus. 1992. The insect pests of avocado fruits: increasing pest complex and changing pest status. J. Entomol. Soc. South. Africa 55: 51–57.

De Vol, J. E., and R. D. Goeden. 1973. Biology of *Chelinidea vittiger* with notes on its host-plant relationships and value in biological weed control. Environ. Entomol. 2: 231–240.

Dollet, M. 1984. Plant diseases caused by flagellate protozoa (*Phytomonas*). Annu. Rev. Phytopathol. 22: 115–132.

Dollet, M., D. Cambrony, and D. Gargani. 1982. Culture axenique in vitro de *Phytomonas* sp. (Trypanosomatidae) d'euphorbe, transmis par *Stenocephalus agilis* Scop. (Coreidae). C. R. Seances Acad. Sci. Ser. III Sci. Vie 295: 547–550.

Dolling, W. R. 1978. A revision of the Oriental pod bugs of the tribe Clavigrallini (Hemiptera: Coreidae). Bull. Brit. Mus. Nat. Hist. Entomol. 36: 281–321.

1979. A revision of the African pod bugs of the tribe Clavigrallini (Hemiptera: Coreidae) with a checklist of the world species. Bull. Brit. Mus. Nat. Hist. Entomol. Ser. 39: 1–84.

1991. The Hemiptera. Oxford University Press, New York, New York, U.S.A. 274 pp.

Donaldson, J. F. 1983. The Australian species of *Amblypelta* Stål (Hemiptera: Coreidae). J. Austral. Entomol. Soc. 22: 47–52.

Douaho, A. 1984. Les ravageurs et maladies du palmier à huile et du cocotier. Lutte biologique contre *Pseudotheraptus* et espèces voisines. Oléagineux 39: 257–260.

Dreyer, H., and J. Baumgärtner. 1995. The influence of post-flowering pests on cowpea seed yield with particular reference to damage by Heteroptera in southern Benin. Agric. Ecosyst. Environ. 53: 137–149.

1997. Adult movement and dynamics of *Clavigralla tomentosicollis* (Heteroptera: Coreidae) populations in cowpea fields of Benin, West Africa. J. Econ. Entomol. 90: 421–426.

Dreyer, H., J. Baumgärtner, and M. Tamo. 1994. Seed-damaging field pests of cowpea (*Vigna unguiculata* L. Walp.) in Benin: occurrence and pest status. Int. J. Pest Manage. 40: 252–260.

Ebel, B. H. 1977. *Leptoglossus corculus* (Say) — a study of its biology, life states, and laboratory colonization. Ph.D. dissertation, University of Georgia, Athens, Georgia, U.S.A. 101 pp.

Ebel, B. H., H. O. Lund, and H. O. Yates III. 1981. Nymphal instars of *Leptoglossus corculus* (Say) with comparisons to the adult (Hemiptera: Coreidae). J. Georgia Entomol. Soc. 16: 156–171.

Eberhard, W. G. 1998. Sexual behavior of *Acanthocephala declivis guatemalana* (Hemiptera: Coreidae) and the allometric scaling of their modified hind legs. Ann. Entomol. Soc. Amer. 91: 863–871.

Egwuatu, R. I. 1978 (published 1981). The effects of relative humidity on the development of eggs and nymphs of *Acanthomia tomentosicollis* Stål (Hemiptera, Coreidae). East Afric. Agric. For. J. 44: 93–101.

1980. Adaptation of *Acanthomia tomentosicollis* Stål (Hemiptera, Coreidae) a major pest of pigeonpeas to its environment: the behaviour in a temperature gradient and the influence of humidity on thermal death point and water loss. Z. Angew. Entomol. 90: 347–354.
1982. Laboratory studies on free and tethered flight in *Clavigralla* (= *Acanthomia*) *tomentosicollis* Stål (Hemiptera, Coreidae). Z. Angew. Entomol. 93: 147–151.
Egwuatu, R. I., and T. A. Taylor. 1977a. The immature stages of *Gryon gnidus* Nixon (Hymenoptera), a scelionid egg-parasite of *Acanthomia tomentosicollis* Stål (Hemiptera: Coreidae) in Nigeria. Rev. Zool. Afr. 91: 1042–1046.
1977b. Bionomics of *Gryon gnidus* Nixon (Hymenoptera: Scelionidae), an egg parasite of *Acanthomia tomentosicollis* Stål (Hemiptera: Coreidae) in Nigeria. Appl. Entomol. Zool. 12: 76–78.
1977c. Development of *Gryon gnidus* (Nixon) (Hymenoptera: Scelionidae) in eggs of *Acanthomia tomentosicollis* (Stål) (Hemiptera: Coreidae) killed either by gamma irradiation or by freezing. Bull. Entomol. Res. 67: 31–33.
1977d. Studies on the biology of *Acanthomia* tomentosicollis (Stål) (Hemiptera: Coreidae) in the field and insectary. Bull. Entomol. Res. 67: 249–257.
1977e. The effects of constant and fluctuating temperatures on the development of *Acanthomia tomentosicollis* Stål (Hemiptera, Coreidae). J. Nat. Hist. 11: 601–608.
1979. Field parasitisation of the eggs of *Acanthomia tomentosicollis* Stål (Hemiptera: Coreidae) at Ibadan, Nigeria. Nigerian J. Entomol. 3: 133–144.
1983. Studies on the seasonal changes in the population of *Acanthomia tomentosicollis* Stål (Hemiptera: Coreidae) on Cajanus cajan (Linnaeus) Druce (pigeon pea) in Ibadan, Nigeria. Nigerian J. Entomol. 4: 20–30.
Elliott, H. J., and D. W. deLittle. 1985. Insect Pests of Trees and Timber in Tasmania. Forestry Commission, Hobart, Tasmania, Australia. 90 pp.
Entwistle, P. F. 1972. Pests of Cocoa. Longman Group Limited, London, U.K. 779 pp.
Fargo, W. S. 1986. Estimation of life stage development times based on cohort data. Southwest. Entomol. 11: 89–94.
Fargo, W. S., and E. L. Bonjour. 1988. Development rate of the squash bug, *Anasa tristis* (Heteroptera: Coreidae), at constant temperatures. Environ. Entomol. 17: 926–929.
Fargo, W. S., P. E. Rensner, E. L. Bonjour, and T. L. Wagner. 1988. Population dynamics in the squash bug (Heteroptera: Coreidae)–squash plant (Cucurbitales: Cucurbitaceae) system in Oklahoma. J. Econ. Entomol. 81: 1073–1079.
Fay, H. A. C., and R. K. Huwer. 1993. Egg parasitoids collected from *Amblypelta lutescens lutescens* (Distant) (Hemiptera: Coreidae) in North Queensland. J. Austral. Entomol. Soc. 32: 365–367.
Fedde, V. H. 1982. Bionomics of the Caribbean bug egg parasite, *Ooencyrtus trinidadensis* Crawford (Hymenoptera: Encyrtidae). Ph.D. dissertation, North Carolina State University, Raleigh, North Carolina, U.S.A. 119 pp.
Fernandes, J. A. M., and J. Grazia. 1992. Estudo dos estágios imaturos de *Leptoglossus zonatus* (Dallas, 1852) (Heteroptera-Coreidae). An. Entomol. Soc. Brasil 21: 179–188.
Fernando, H. E. 1957. The biology and control of *Leptoglossus membranaceus* Fabricius (Fam. Coreidae Ord. Hemiptera). Trop. Agric. 113: 107–118.
Fery, R. L., and J. M. Schalk. 1981. An evaluation of damage to southernpeas (*Vigna unguiculata* (L.) Walp.) by leaffooted and southern green stink bugs. Hortscience 16: 286.
Fielding, D. J., and W. G. Ruesink. 1988. Prediction of egg and nymphal developmental times of the squash bug (Hemiptera: Coreidae) in the field. J. Econ. Entomol. 81: 1377–1382.
Fiorini, J. E., P. M. Silva de F., R. P. Brazil, M. Attias, M. J. G. Esteves, and J. Angluster. 1993. Axenic cultivation of a pathogenic *Phytomonas* species isolated from tomato fruit, and from its phytophagic insect vector, *Phthia picta* (Hemiptera: Coreidae). Cytobios 75: 163–170.
Flanagan, G. J. 1994. The Australian distribution of *Mictis profana* (F.) (Hemiptera: Coreidae) and its life cycle on *Mimosa pigra* L. J. Austral. Entomol. Soc. 33: 111–114.
Forsyth, J. 1966. Agricultural Insects of Ghana. Ghana Universities Press, Accra, Ghana. 163 pp.
Froeschner, R. C. 1942. Contributions to a synopsis of the Hemiptera of Missouri. Part II. Coreidae, Aradidae, Neididae. Amer. Midl. Nat. 27: 591–609.
1981. Heteroptera or true bugs of Ecuador: a partial catalog. Smithsonian Contrib. Zool. No. 322. 147 pp.

1988. Family Coreidae Leach, 1815. Pp. 69–92 *in* T. J. Henry and R. C. Froeschner [eds.], Catalog of the Heteroptera, or True Bugs, of Canada and the Continental United States. E. J. Brill, Leiden, the Netherlands. 958 pp.

Fujisaki, K. 1975. Breakup and reformation of colony in the first-instar larvae of the winter cherry bug, *Acanthocoris sordidus* Thunberg (Hemiptera: Coreidae), in relation to the defence against their enemies. Res. Pop. Ecol. 16: 252–264.

Fujisaki, K. 1980. Studies on the mating system of the winter cherry bug, *Acanthocoris sordidus* Thunberg (Heteroptera: Coreidae) I. Spatio-temporal distribution patterns of adults. Res. Pop. Ecol. 21: 317–331.

Fujisaki, K. 1981. Studies on the mating system of the winter cherry bug, *Acanthoris sordidus* Thunberg (Heteroptera: Coreidae) II. Harem defence polygyny. Res. Pop. Ecol. 23: 262–279.

Fullaway, D. 1954. Biological control of cactus in Hawaii. J. Econ. Entomol. 47: 696–700.

Gall, W. K. 1992. Further eastern range extension and host records for *Leptoglossus occidentalis* (Heteroptera: Coreidae): well-documented dispersal of a household nuisance. Great Lakes Entomol. 25: 159–171.

Gethi, M., and B. M. Khaemba. 1991. Damage by pod-sucking bugs on cowpea when intercropped with maize. Trop. Pest Manage. 37: 236–239.

Ghauri, M. S. K. 1984. Two new species of *Amblypelta* Stål attacking cacao in Papua New Guinea and Irian Jaya (West New Guinea), with a key to its species (Heteroptera, Coreidae). Reichenbachia 22: 51–64.

Girault, A. A. 1906. The method of feeding in *Leptoglossus*. Entomol. News 17: 382–383.

1913. Fragments of North American insects III. Entomol. News 24: 53–63.

1920. New serphidoid, cynipoid, and chalcidoid Hymenoptera. Proc. U.S. Natl. Mus. 58: 177–216.

Goeden, R. D., C. A. Fleshner, and D. W. Ricker. 1967. Biological control of prickly pear cacti on Santa Cruz Island, California. Hilgardia 38: 579–606.

Golding, F. D. 1927. Notes on the food-plants and habits of some southern Nigerian insects. Bull. Entomol. Res. 18: 95–99.

1935. Further notes on the food-plants of Nigerian insects III. Bull. Entomol. Res. 26: 263–265.

1937. Further notes on the food-plants of Nigerian insects IV. Bull. Entomol. Res. 28: 5–9.

1941. Capsid pests of cacao in Nigeria. Bull. Entomol. Res. 32: 83–89.

1948. Further notes on the food-plants of Nigerian insects VI. Bull. Entomol. Res. 38: 75–80.

Gough, N., and E. L. Hamacek. 1989. Insect induced bud fall in cultivated hibiscus and aspects of the biology of *Macroura concolor* (Macleay) (Coleoptera: Nitidulidae). J. Austral. Entomol. Soc. 28: 267–277.

Green, E. E. 1905. Entomological notes — July–August. Trop. Agric. 31: 408–411.

1912. The paddle-legged bug (*Leptoglossus membranaceous*). Trop. Agric. 38: 529–530.

Grimm, C. 1996. Cuantificacion de daños por insectos en los frutos del tempate, *Jatropha curcas*, (Euphorbiaceae) a traves de una tabla de vida. Manejo Integrado de Plagas (Costa Rica) No. 42: 23–30.

Grimm, C., and E. Führer. 1998. Population dynamics of true bugs (Heteroptera) in physic nut (*Jatropha curcas*) plantations in Nicaragua. J. Appl. Entomol. 122: 515–521.

Grimm, C., and F. Guharay. 1998. Control of leaf-footed bug, *Leptoglossus zonatus* and shield-backed bug *Pachycoris klugii* with entomopathogenic fungi. Biocontrol Sci. Technol. 8: 365–376.

Grimm, C. and J.-M. Maes. 1997a. Insectos associados al cultivo de tempate (*Jatropha curcas*) en el Pacifico de Nicaragua. III. Coreoidea (Heteroptera). Rev. Nicaraguense Entomol. 42: 15–34.

1997b. Arthropod fauna associated with *Jatropha curcas* in Nicaragua: a synopsis of species, their biology and pest status. Pp. 31–39 *in* G. M. Gübitz, M. Mittelbach, and M. Trabi [eds.], Biofuels and Industrial Products from *Jatropha curcas*. Dbv-Verlag, Graz, Austria.

Hall, D. G., and G. L. Teetes. 1980. Damage to sorghum seed by four common bugs. Texas Agric. Exp. Stn. Tech. Bull. No. PR-3647. 8 pp.

1982. Damage to grain sorghum by southern green stink bug, conchuela, and leaffooted bug. J. Econ. Entomol. 75: 620–625.

Hall, J. P. 1996. Forest insect and disease conditions in Canada 1994. Forest Insect and Disease Survey, Canadian Forest Service, Ottawa, Canada.

Hamlin, J. C. 1924a. Biological control of prickly-pear in Australia: contributing efforts in North America. J. Econ. Entomol. 17: 447–460.

1924b. A review of the genus *Chelinidea* (Hemiptera-Heteroptera) with biological data. Ann. Entomol. Soc. Amer. 17: 193–208.

Hartwig, E. K, and H. C. de Lange. 1978. Observations on three coreids, *Carlisis wahlbergi* Stål, *Anoplocnemis curvipes* (Fabr.) and *Elasmopoda valga* (L.). Tech. Commun. Dept. Agric. Tech. Serv. Pretoria. No. 141: 12 pp.

Hayes, W. P. 1922. Observations on insects attacking sorghums. J. Econ. Entomol. 15: 349–356.

Hedlin, A. F., H. O. Yates III, D. C. Tovar, B. H. Ebel, T. W. Koerber, and E. P. Merkel. 1981. Cone and seed insects of North American conifers. 2nd print. Canad. For. Serv, U.S. Dept. Agric. For. Serv., and Sec. Agric. Recur. Hidraul. México. 122 pp.

Heidemann, O. 1910. New species of *Leptoglossus* from North America (Hemiptera: Coreidae). Proc. Entomol. Soc. Washington 12: 191–197.

Henry, T. J., and R. C. Froeschner. 1992. Corrections and additions to the "Catalog of the Heteroptera, or True Bugs, of Canada and the Continental United States." Proc. Entomol. Soc. Washington 94: 263–272.

Herring, J. L. 1980. A review of the cactus bugs of the genus *Chelinidea* with the description of a new species [*Chelinidea staffilesi*] (Hemiptera: Coreidae). Proc. Entomol. Soc. Washington 82: 237–251.

Hill, D. 1975. Agricultural Insect Pests of the Tropics and Their Control. Cambridge University Press, Cambridge, U.K. 516 pp.

1983. Agricultural Insect Pests of the Tropics and Their Control, second edition. Cambridge University Press, Cambridge, U.K. 746 pp.

Hill, S. O. 1938. Important pecan insects of northern Florida. Florida Entomol. 21: 9–13.

Hingston, R. W. G. 1929. A study in insect protection. J. Bombay Nat. Hist. Soc. 33: 341–346.

Hoffman, W. E. 1927. Notes on a squash bug of economic importance. Lingnan Sci. J. 5: 281–292.

1933. Life history notes on some Kwangtung, China, coreids (Hemiptera, Coreidae). Lingnan Sci. J. 12: 97–127.

Hogue, C. L. 1993. Latin American Insects and Entomology. University of California Press, Berkeley, California, U.S.A. 536 pp.

Hubbard, H. G. 1885. Insects affecting the orange. U.S. Dept. Agric. Div. Entomol. Ser. 2326, Vol. 40. 227 pp.

Hunter, W. D., F. C. Pratt, and J. D. Mitchell. 1912. The principal cactus insects of the United States. U.S. Dept. Agric. Bur. Entomol. Bull. 113: 1–17.

Ironside, D. A. 1984. Insecticidal control of fruitspotting bug, *Amblypelta nitida* Stål (Hemiptera: Coreidae) and macadamia nutborer, *Cryptophlebia ombrodelta* (Lower) (Lepidoptera: Tortricidae). Queensland J. Agric. Anim. Sci. 41: 101–107.

Ito, K. 1978. Ecology of the stink bugs causing pecky rice. Rev. Plant. Prot. Res. 11: 62–78.

1980. Seasonal change of flight ability of *Cletus punctiger* (Heteroptera: Coreidae). Appl. Entomol. Zool. 15: 36–44.

1982. Immigration of *Cletus punctiger* Dallas (Heteroptera: Coreidae) into paddy field in relation to food plant preference in the adult. Jpn. J. Appl. Entomol. Zool. 26: 300–304.

1984a. The effect of feeding on the subsequent starvation longevity in post-hibernating *Cletus punctiger* (Heteroptera: Coreidae). Appl. Entomol. Zool. 19: 461–467.

1984b. Interspecific hybridization between *Cletus punctiger* Dallas and *C. rusticus* Stål (Heteroptera: Coreidae). Appl. Entomol. Zool. 19: 142–150.

1985. Seasonal changes of lipid content in adult *Cletus punctiger* (Heteroptera: Coreidae). Appl. Entomol. Zool. 20: 350–351.

1986. Starvation longevity and lipid accumulation in non-diapausing and diapasuing adults of the coreid bug *Cletus punctiger*. Entomol. Exp. Appl. 40: 281–284.

Jackai, L. E. N. 1984. Studies on the feeding behaviour of *Clavigralla tomentosicollis* (Stål) (Hemiptera, Coreidae) and their potential use in bioassays for host plant resistance. Z. Angew. Entomol. 98: 344–350.

1989. A laboratory procedure for rearing the cowpea coreid, *Clavigralla tomentosicollis* Stål (Hemiptera), using dry cowpea seeds. Bull. Entomol. Res. 79: 275–281.

1990. Screening of cowpeas for resistance to *Clavigralla tomentosicollis* Stål (Hemiptera: Coreidae). J. Econ. Entomol. 83: 300–305.

Jackai, L. E. N., and R. A. Daoust. 1986. Insect pests of cowpeas. Annu. Rev. Entomol. 31: 95–119.

Jackai, L. E. N., and E. E. Inang. 1992. Developmental profiles of two cowpea pests on resistant and susceptible *Vigna* genotypes under constant temperatures. J. Appl. Entomol. 113: 217–227.

Jackai, L. E. N., and S. Oghiakhe. 1989. Pod wall trichomes and resistance of two wild cowpea, *Vigna vexillata*, accessions to *Maruca testulalis* (Geyer) (Lepidoptera: Pyralidae) and *Clavigralla tomentosicollis* Stål (Hemiptera: Coreidae). Bull. Entomol. Res. 79: 599–605.

Jackai, L. E. N., P. K. Atropo, and J. A. Odebiyi. 1989. Use of the response of two growth stages of cowpea to different population densities of the coreid bug, *Clavigralla tomentosicollis* (Stål) to determine action threshold levels. Crop Prot. 8: 422–428.

Jackai, L. E. N., E. E. Inang, and P. Nwobi. 1992. The potential for controlling post-flowering pest of cowpea, *Vigna unguiculata* Walp. using neem, *Azadirachta indica* A. Juss. Trop. Pest Manage. 38: 56–60.

Jackai, L. E. N., S. R. Singh, A. K. Raheja, and F. Wiedijk. 1985. Recent trends in the control of cowpea pests in Africa. Pp. 233–243 *in* S. R. Singh and K. O. Rachie [eds.], Cowpea Research, Production, and Utilization. John Wiley & Sons, Chichester, U.K.

Jackson, C. G., M. S. Tveten, and P. J. Figuli. 1995. Development, longevity and fecundity of *Leptoglossus zonatus* on a meridic diet. Southwest. Entomol. 20: 43–48.

Jadhav, L. D., D. S. Ajri, M. V. Kadam, and S. K. Dorge. 1976. Leaf-footed plant bug on pomegranate in Maharashtra. Entomol. Newslet. 6: 57.

Jadhav, L. D., M. V. Kadam, D. S. Ajri, and R. N. Pokharkar. 1979. Host preference of leaf-footed plant bug, *Leptglossus membranaceus* Fabricius. Indian J. Entomol. 41(3): 299–300.

Jankevicius, J. V., S. I. Jankevicius, M. Campaner, I. Conchon, L. A. Maeda, M. M. G. Teixeira, E. Freymuller, and E. P. Camargo. 1989. Life cycle and culturing of *Phytomonas serpens* (Gibbs), a trypanosomatid parasite of tomatoes. J. Protozool. 36: 265–271.

Jankevicius, S. I., M. L. de Almeida, J. V. Jankevicius, M. Cavazzana, Jr., M. Attias, and W. de Souza. 1993. Axenic cultivation of trypanosomatids found in corn (*Zea mays*) and in phytophagous hemipterans (*Leptoglossus zonatus* Coreidae) and their experimental transmission. J. Eukaryot. Microbiol. 40: 576–581.

Johnson, S. R., and L. K. Allain. 1998. Observations on insect use of Chinese tallow (*Sapium sebiferum* (L.) Roxb.) in Louisiana. Castanea 63: 188–189.

Jones, T. H. 1916. Notes on *Anasa andresii* Guér., an enemy of cucurbits. J. Econ. Entomol. 9: 431–434.

Jones, W. A. 1993. New host and habitat associations for some Arizona Pentatomoidea and Coreidae. Southwest. Entomol. Suppl. No. 16. 29 pp.

Julia, J. F. 1978. La punaise du cocotier: *Pseudotheraptus* sp. en Côte-d'Ivoire II. Méthode de lutte intégrée en Côte-d'Ivoire. Oléagineux 33: 113–118.

Julia, J. F., and D. Mariau. 1978. La punaise du cocotier: *Pseudotheraptus* sp. en Côte-d'Ivoire I. Études préalables à la mise en point d'une méthode de lutte intégrée. Oléagineux 33: 65–75.

Kashyap, N. P., and P. K. Mehta. 1992. Biology of *Clavigralla gibbosa* Spinola, a pest of pigeonpea in Himachal Pradesh. J. Insect Sci. 5: 209–210.

Katovich, S. A., and H. M. Kulman. 1987. *Leptoglossus corculus* and *Leptoglossus occidentalis* (Hemiptera: Coreidae) attacking red pine, *Pinus resinosa*, cones in Wisconsin and Minnesota. Great Lakes Entomol. 20: 119–120.

Kayumbo, H. Y. 1977. Insect pest populations in mixed crop ecosystems. Trop. Legume Grain Bull. 8: 24–27.

1978. Pests of cowpea and their control in Tanzania. Pp. 123–126 *in* S. R. Singh, H. F. van Emden, and T. A. Taylor [eds.], Pests of Grain Legumes: Ecology and Control. Academic Press, London, U.K. 454 pp.

Khaemba, B. M. 1985. Sources of resistance to the common cowpea pod sucking bugs *Riptortus dentipes* (Fabricius) and *Anoplocnemis curvipes* (Fabricius) in cowpea, *Vigna unguiculata* (L.) Walp. Turrialba 35: 209–213.

Khaemba, B. M., and C. P. M. Khamala. 1981. Relation of pod age to the expression of resistance in cowpea *Vigna unguiculata* (L.) to common pod sucking bugs *Riptortus dentipes* (F.) and *Anoplocnemis curvipes* (F.) (Hemiptera: Coreidae) Nigeria. Kenya J. Sci. Technol. Ser. B. Biol. Sci. 2: 47–52.

Khamala, C. P. M. 1978. Pests of grain legumes and their control in Kenya. Pp. 127–134 *in* S. R. Singh, H. F. van Emden, and T. A. Taylor [eds.], Pests of Grain Legumes: Ecology and Control. Academic Press, London, U.K. 454 pp.

Kisimoto, R. 1983. Damage caused by rice stink bugs and their control. Jpn. Pestic. Inf. No. 43: 9–13.

Koerber, T. W. 1963. *Leptoglossus occidentalis* (Hemiptera: Coreidae), a newly discovered pest of coniferous seed. Ann. Entomol. Soc. Amer. 56: 229–234.

Koshy, G., A. Visalakshy, and M. R. G. K. Nair. 1977. Biology of *Acanthocoris scabrator* Fabr., a pest of mango. Entomon 2: 145–147.

Krugman, S. L., and T. W. Koerber. 1969. Effect of cone feeding by *Leptoglossus occidentalis* on ponderosa pine seed development. For. Sci. 15: 104–111.

Kubo, R. K., and A. Batista Filho. 1992. Ocorrência e danos provocados por *Leptoglossus zonatus* (Dallas, 1852) (Hemiptera: Coreidae) An. Soc. Entomol. Brasil 21: 467–470.

Kumar, R. 1966. Studies on the biology, immature stages, and relative growth of some Australian bugs of the superfamily Coreoidea (Hemiptera: Homoptera). Austral. J. Zool. 14: 895–991.

Kurian, C., G. B. Pillay, V. A. Abraham, and K. Mathen. 1972. Record of a coreid bug (nut crinkler) as a new pest of coconut in India. Curr. Sci. 41: 37.

Kurian, C., V. A. Abraham, and A. Koya. 1976. A new enemy of coconut in India [*Paradasynus*, Coreidae]. Indian Farming 26: 11–12.

1979. Studies on *Paradasynus rostratus* Dist. (Heteroptera: Coreidae) a pest of coconut. Pp. 484–503 *in* Placrosym II. Second Symposium on Plantation Crops, Indian Society for Plantation Crops, Ootamund, India. 555 pp.

Kurup, N. G. 1964. A study of the digestive tract of four Indian gymnocerate Hemiptera. Ann. Entomol. Soc. Amer. 5: 200–209.

Lamb, K. P. 1974. Economic Entomology in the Tropics. Academic Press, London, U.K. 195 pp.

Leal, W. S., A. R. Panizzi, and C. C. Niva. 1994. Alarm pheromone system of leaf-footed bug *Leptoglossus zonatus* (Heteroptera: Coreidae). J. Chem. Ecol. 20: 1209–1216.

Leefmans, S. 1935. Biological notes on *Dasynus manihotis* Blöte. Zoöl. Meded. 18: 237–240.

Leonard, M. D. 1931. *Leptoglossus gonagra* injuring citrus in Porto Rico. J. Econ. Entomol. 24: 765–767.

Le Pelley, R. H. 1959. Agricultural Insects of East Africa. East African High Commission, Nairobi, Kenya. 307 pp.

Lever, R. J. A. W. 1981. New locality and host plant records of *Amblypelta* spp. (Hemiptera, Coreidae) with notes on early studies. Proc. Trans. Brit. Entomol. Nat. Hist. Soc. 14: 112–114.

1982. *Amblypelta* spp. (Hem. Coreidae), new Australian records. Proc. Trans. Brit. Entomol. Nat. Hist. Soc. 15: 88.

Linnavuori, R. E. 1993. Hemiptera of Iraq. II. Cydnidae, Thaumastellidae, Pentatomidae, Stenocephalidae, Coreidae, Alydidae, Rhopalidae, and Pyrrhocoridae. Entomol. Fenn. 4: 37–56.

Little, V.A. 1972. General and Applied Entomology, third edition. Harper & Row, New York, New York, U.S.A. 527 pp.

Lodos, N. 1967. Contribution to the biology of and damage caused by the cocoa coreid, *Pseudotheraptus devastans* Dist. (Hemiptera: Coreidae). Ghana J. Sci. 7: 87–102.

Löhr, B. 1992. The pugnacious ant, *Anoplolepis custodiens* (Hymenoptera: Formicidae), and its beneficial effect on coconut production in Tanzania. Bull. Entomol. Res. 82: 213–218.

Lu, F. M., and H. S. Lee. 1987. Seasonal occurrence of the insect pests on eggplants. Plant Prot. Bull., Taiwan. 29: 61–70.

Maheswariah, B. M., and M. Puttarudriah. 1956. Some observations on the life-history and habits of *Anoplocnemis phasiana* Fabricius. Mysore Agric. J. 31: 248–255.

Mann, J. 1969. Cactus feeding insects and mites. U.S. Natl. Mus. Bull. No. 256: 158 pp.

Mariau, D. 1969. *Pseudotheraptus*: un nouveau ravageur du cocotier en Afrique occidentale. Oléagineux 24: 21–25.

Mariau, D., R. Desmier de Chenon, J.-F. Julia, and R. Philippe. 1981. Les ravageurs du palmier à huile et du cocotier en Afrique occidentale. Oléagineux (No. speciale) 36: 206–207.

Mariconi, F. A. M. 1953. O "percevejo do abacaxi" (*Lybindus dichrous* Stål, 1859). O Biológico 19: 155–162.

Marshall, S. A. 1991. A new Ontario record of a seed eating bug (Hemiptera: Coreidae) and other examples of the role of regional insect collections in tracking changes to Ontario's fauna. Proc. Entomol. Soc. Ontario 122: 109–111.

Masner, L. 1983. A revision of *Gryon* Haliday in North America (Hymenoptera: Proctotrupoidea: Scelionidae). Canad. Entomol. 115: 123–174.

Materu, M. E. A. 1970. Damage caused by *Acanthomia tomentosicollis* Stål and *Acanthomia horrida* Germ. (Hemiptera, Coreidae). East Afric. Agric. For. J. 35: 429–435.

1971. Population dynamics of *Acanthomia* spp. (Hemiptera, Coreidae) on beans and pigeon peas in the Arusha area of Tanzania. East Afric. Agric. For. J. 36: 361–383.

1972. Morphology of adults and description of the young stages of *Acanthomia tomentosicolis* Stål and *Acanthomia horrida* Germ. (Hemiptera, Coreidae). J. Nat. Hist. 6: 427–450.

Matrangolo, W. J. R., and J. M. Waquil. 1994. Biologia de *Leptoglossus zonatus* (Dallas) (Hemiptera: Coreidae) alimentados com milho e sorgo. An. Soc. Entomol. Brasil 23: 419–423.

Matteson, P. C. 1980. Rearings of Tachinidae (Dipt.) from pod-sucking Hemiptera from cowpea plots in Nigeria and Tanzania. Entomol. Mon. Mag. 116: 66.

1981. Egg parasitoids of hemipteran pests of cowpea in Nigeria and Tanzania, with special reference to *Ooencyrtus patriciae* Subba Rao (Hymenoptera: Encyrtidae) attacking *Clavigralla tomentosicollis* Stål (Hemiptera: Coreidae) Bull. Entomol. Res. 71: 547–554.

1982. The effects of intercropping with cereal and minimal permethrin application on insect pests of cowpeas and their natural enemies in Nigeria. Trop. Pest. Manage. 28: 373–380.

McPherson, J. E., R. J. Packauskas, S. J. Taylor, and M. F. O'Brien. 1990. Eastern range extension of *Leptoglossus occidentalis* with a key to *Leptoglossus* species of America north of Mexico (Heteroptera: Coreidae). Great Lakes Entomol. 23: 99–104.

Mead, F. W. 1971a. Leaffooted bug, *Leptoglossus phyllopus* (Linnaeus) (Hemiptera: Coreidae). Florida Dept. Agric. Consumer Serv., Div. Plant Industry, Entomol. Circ. No. 107. 2 pp.

1971b. Annotated key to leaffooted bugs, *Leptoglossus* spp., in Florida (Hemiptera: Coreidae). Florida Dept. Agric. Consumer. Serv., Div. Plant Industry, Entomol. Circ. No. 113. 4 pp.

Menzies, G. C. 1941. A study of the leaf-footed plant bug *Leptoglossus phyllopus* Linné. M.S. thesis, Texas A&M University, College Station, Texas, U.S.A. 30 pp.

Michailides, T. J. 1989. The "Achilles heel" of pistachio fruit. California Agric. 43(5): 10–11.

Michailides, T. J., and D. P. Morgan. 1990. Etiology and transmission of stigmatomycosis disease of pistachio in California. Pp. 88–95 *in* California Pistachio Industry [ed.], Annu. Rep. Crop Year 1989–90. Fresno, California, U.S.A. 136 pp.

1991. New findings on the stigmatomycosis disease of pistachio in California. Pp. 106–110 *in* California Pistachio Industry [ed.], Annu. Rep. Crop Year 1990–91. Fresno, California, U.S.A.

Michailides, T. J., R. E. Rice, and J. M. Ogawa. 1987. Succession and significance of several hemipterans attacking a pistchio orchard. J. Econ. Entomol. 80: 398–406.

Miles, P. W. 1987. Plant sucking bugs can remove the contents of cells without mechanical damage. Experientia 43: 937–939.

Miles, P. W., and G. S. Taylor. 1994. Osmotic pump feeding by coreids. Entomol. Exp. Appl. 73: 163–173.

Miller, N. C. E. 1929. *Physomerus grossipes* F. (Coreidae, Hemiptera-Heteroptera). A pest of Convolvulaceae and Leguminosae. Malay. Agric. J. 17: 403–420.

1936. VIII. *Dasynus piperis* China. (Heteroptera — Coreidae). A minor pest of pepper (*Piper nigrum* L.). J. Fed. Malay States Mus. 18: 109–116.

1941. Insects associated with cocoa in Malaya. Bull. Entomol. Res. 32: 1–16.

Mitchell, P. L. 1980a. Combat and territorial defense of *Acanthocephala femorata* (Hemiptera: Coreidae). Ann. Entomol. Soc. Amer. 73: 404–408.

1980b. Host plant utilization by leaf-footed bugs: an investigation of generalist feeding strategy. Ph.D. dissertation, University of Texas, Austin, Texas, U.S.A. 226 pp.

Mitchell, P. L., and F. L. Mitchell. 1986. Parasitism and predation of leaffooted bug (Hemiptera: Heteroptera: Coreidae) eggs. Ann. Entomol. Soc. Amer. 79: 854–860.

Mitchell, P. L., E. S. Paysen, A. E. Muckenfuss, M. Schaffer, and B. M. Shepard. 2000. Natural mortality of leaffooted bug (Hemiptera: Heteroptera: Coreidae) eggs in cowpea. J. Agric. Urban Entomol. 16: in press.

Mitchell, W. C. and P. A. Maddison. 1983. Pests of taro. Pp. 181–235 *in* J. Wang [ed.], Taro — A Review of *Colocasia esculenta* and Its Potentials. University of Hawaii Press, Honolulu, Hawaii, U.S.A.

Miyatake, T. 1992. Overwintering aggregations of the leaf-footed plant bug, *Leptoglossus australis* Fabricius (Heteroptera: Coreidae) in Okinawa Island. Proc. Assoc. Pl. Prot. Kyushu 38: 118–121.

1993. Male-male aggressive behavior is changed by body size difference in the leaf-footed plant bug, *Leptoglossus australis* Fabricius (Heteroptera, Coreidae). J. Ethol. 11: 63–65.

1994. Seasonal abundance of the bamboo bug, *Notobitus meleagris* Fabricius (Heteroptera: Coreidae) in Okinawa Island. Appl. Entomol. Zool. 29: 601–603.

1995. Territorial mating aggregation in the bamboo bug, *Notobitus meleagris*, Fabricius (Heteroptera: Coreidae). J. Ethol. 13: 185–189.

1997. Functional morphology of the hind legs as weapons for male contests in *Leptoglossus australis* (Heteroptera: Coreidae). J. Insect Behav. 10: 727–735.

Muckenfuss, A., A. Keinath, and M. Shepard. 1994. Refining tomato integrated pest management. Pp. 104–110 *in* R. Dufault [ed.], 1992/94 Clemson University Vegetable Report. Vol. 6. Clemson University, Clemson, South Carolina, U.S.A. 148 pp.

Nair, M. R. G. K., and K. S. Remany. 1964. *Paradasynus* sp. (Hemiptera: Coreidae) as a pest of cashewnut in Kerala. Indian J. Entomol. 26: 461–462.

Nawale, R. N., and L. D. Jadhav. 1978. Bionomics of tur pod bug *Clavigralla gibbosa* Spinola (Coreidae: Hemiptera). J. Maharashtra Agric. Univ. Pune 3: 275–276.

Neal, J. J. 1993. Xylem transport interruption by *Anasa tristis* feeding causes *Cucurbita pepo* to wilt. Entomol. Exp. Appl. 69: 195–200.

Nechols, J. R. 1987. Voltinism, seasonal reproduction, and diapause in the squash bug (Heteroptera: Coreidae) in Kansas. Environ. Entomol. 16: 269–273.

1988. Photoperiodic responses of the squash bug (Heteroptera: Coreidae): diapause induction and maintenance. Environ. Entomol. 17: 427–431.

Nechols, J. R., J. L. Tracy, and E. A. Vogt. 1989. Comparative ecological studies of indigenous egg parasitoids (Hymenoptera: Scelionidae; Encyrtidae) of the squash bug, *Anasa tristis* (Hemiptera: Coreidae). J. Kansas Entomol. Soc. 62: 177–188.

Noda, T. 1990. Laboratory host range test for the parasitic wasp, *Gryon japonicum* (Ashmead) (Hymenoptera: Scelionidae). Jpn. J. Appl. Entomol. Zool. 34: 249–252.

Nord, J. C. 1990. Toxicities of insecticide residues on loblolly pine foliage to leaffooted pine seed bug adults (Heteroptera: Coreidae). J. Entomol. Sci. 25: 3–9.

Nord, J. C., and G. L. DeBarr. 1983. Contact toxicities of 12 insecticides to *Leptoglossus corculus* adults (Hemiptera: Coreidae). Canad. Entomol. 115: 211–214.

1992. Persistence of insecticides in a loblolly pine seed orchard for control of the leaffooted pine seed bug, *Leptoglossus corculus* (Say) (Hemiptera: Coreidae). Canad. Entomol. 124: 617–629.

Nord, J. C., G. L. Debarr, N. A. Overgaard, W. W. Neel, R. S. Cameron, and J. F. Godbee. 1984. High-volume applications of azinphosmethyl, fenvalerate, permethrin, and phosmet for control of coneworms (Lepidoptera: Pyralidae) and seed bugs (Hemiptera: Coreidae and Pentatomidae) in southern pine seed orchards. J. Econ. Entomol. 77: 1589–1595.

Nord, J. C., G. L. DeBarr, L. R. Barber, J. C. Weatherby, and N. A. Overgaard. 1985. Low-volume applications of azinphosmethyl, fenvalerate, and permethrin for control of coneworms (Lepidoptera: Pyralidae) and seed bugs (Hemiptera: Coreidae and Pentatomidae) in southern pine seed orchards. J. Econ. Entomol. 78: 445–450.

Olatunde, G. O., and J. A. Odebiyi. 1991a. The relationship between total sugar, crude protein and tannic acid contents of cowpea, *Vigna unguiculata* L. Walp. and varietal resistance to *Clavigralla tomentosicollis* Stål. (Hemiptera: Coreidae). Trop. Pest. Manage. 37: 393–396.

1991b. Some aspects of antibiosis in cowpeas resistant to *Clavigralla tomentosicollis* Stål. (Hemiptera: Coreidae) in Nigeria. Trop. Pest Manage. 37: 273–276.

Olatunde, G. O., J. A. Odebiyi, and L. E. N. Jackai. 1991a. Cowpea antixenosis to the pod sucking bug, *Clavigralla tomentosicollis* Stål. (Hemiptera: Coreidae). Insect Sci. Appl. 12: 449–454.

Olatunde, G. O., J. A. Odebiyi, H. S. Chiang, and L. E. N. Jackai. 1991b. Identification of sources of resistance in cowpea, *Vigna unguiculata* L. Walp. to *Clavigralla tomentosicollis* Stål. (Hemiptera: Coreidae). Insect Sci. Appl. 12: 455–461.

Olson, D. L., and J. R. Nechols. 1995. Effects of squash leaf trichome exudates and honey on adult feeding, survival, and fecundity of the squash bug (Heteroptera: Coreidae) egg parasitoid *Gryon pennsylvanicum* (Hymenoptera: Scelionidae). Environ. Entomol. 24: 454–458.

Olson, D. L., J. R. Nechols, and B. W. Schurle. 1996. Comparative evaluation of population effect and economic potential of biological suppression tactics versus chemical control for squash bug (Heteroptera: Coreidae) management on pumpkins. J. Econ. Entomol. 89: 631–639.

Ooi, A. C. P. 1973. Some insect pests of green gram, *Phaseolus aureus*. Malay. Agric. J. 49: 131–141.

O'Shea, R. 1980. A generic revision of the Acanthocerini (Hemiptera: Coreidae: Coreinae). Stud. Neotrop. Fauna Environ. 15: 57–80.

Osuna, E. 1981. Revision generica de la tribu Leptoscelidini (Hemiptera–Heteroptera, Coreidae). Universidad Central de Venezuela, Instituto de Zoologia Agricola Publication, Maracay, Venezuela. 113 pp.

1984. Monografia de la tribu Anisoscelidini (Hemiptera, Heteroptera, Coriedae) I. Revisión genérica. Bol. Entomol. Venez. (n.s.) 3: 77–148.

Oswald, S. 1989. Untersuchungen zur integrierten Bekampfung der Kokoswanze *Pseudotheraptus wayi* Brown (Heteroptera: Coreidae) auf Sansibar. Giessen: (s.n.) 190 pp.

1990. Possibilities for the use of *Ooencyrtus albicrus* (Prinsloo) (Hym., Encyrtidae) in an integrated pest management approach against the coconut bug *Pseudotheraptus wayi* Brown (Hem., Coreidae) in Zanzibar. Z. Angew. Entomol. 110: 198–202.

Packauskas, R. J. 1990. Cytology of *Leptoglossus zonatus* (Hemiptera: Coreidae). Entomol. News 101: 203–206.

Packauskas, R. J., and C. W. Schaefer. 2000. Clarification of some taxomonic problems in Anisoscelini and Leptoscelini (Hemiptera: Coreidae: Coreinae). Proc. Entomol. Soc. Washington 102: in press.

Pagden, H.T. 1928. *Leptoglossus membranaceus* F. a pest of Cucurbitaceae. Malay. Agric. J. 16(12): 387–403. [Biol. Abstr. 1930: 28841].

Palumbo, J. C., and W. S. Fargo. 1989a. Systemic insecticides for control of summer squash, 1987. Insectic. Acar. Tests 14: 160.

1989b. Control on summer squash, 1988. Insectic. Acar. Tests. 14: 161.

Palumbo, J. C., W. S. Fargo, and E. L. Bonjour. 1991a. Within-plant distribution of squash bug (Heteroptera: Coreidae) adults and egg masses in vegetative stage summer squash. Environ. Entomol. 20: 391–395.

1991b. Spatial dispersion patterns and sequential sampling plans for squash bugs (Heteroptera: Coreidae) in summer squash. J. Econ. Entomol. 84: 1796–1801.

1991c. Colonization and seasonal abundance of squash bugs (Heteroptera: Coreidae) on summer squash with varied planting dates in Oklahoma. J. Econ. Entomol. 84: 224–229.

Palumbo, J. C., W. S. Fargo, R. C. Berberet, E. L. Bonjour, and G. W. Cuperus. 1993. Timing insecticide applications for squash bug management: impact on squash bug abundance and summer squash yields. Southwest. Entomol. 18: 101–111.

Panizzi, A. R. 1989. Desempenho de ninfas e adultos de *Leptoglossus zonatus* (Dallas, 1852) (Hemiptera: Coreidae) em diferentes alimentos. An. Soc. Entomol. Brasil 18: 375–389.

Parshley, H. M. 1918. Three species of *Anasa* injurious in the North (Hemiptera, Coreidae). J. Econ. Entomol. 11: 471–472.

Parsons, W. T., and E. G. Cuthbertson. 1992. Noxious Weeds of Australia. Inkata Press, Melbourne, Australia. 692 pp.

Pasek, J. E., and M. E. Dix. 1988. Insect damage to conelets, second-year cones, and seeds of ponderosa pine in southeastern Nebraska. J. Econ. Entomol. 81: 1681–1690.

Passoa, S. 1983. Lista de los insectos asociados con los granos basicos y otros cultivos selectos en Honduras. CEIBA (Escuela Agricola Panamericana) 25: 1–97.

Patnaik, N. C., A. N. Dash, and B. K. Misra. 1989. Effect of intercropping on the incidence of pigeonpea pests in Orissa, India. Int. Pigeonpea Newsl. 9: 24–25.

Pearson, E. O. 1958. The Insect Pests of Cotton in Tropical Africa. Commonwealth Institute of Entomology, London, U.K. 355 pp.

Peng, R. K., K. Christian, and K. Gibb. 1995. The effect of the green ant, *Oecophylla* (Hymenoptera: Formicidae), on insect pests of cashew trees in Australia. Bull. Entomol. Res. 85: 279–284.

Phillips, J. S. 1940. Immature nutfall of coconuts in the Solomon Islands. Bull. Entomol. Res. 31: 295–316.

1941. A search for parasites of dasynine bugs in the Netherlands Indies. Trans. R. Entomol. Soc. London 95: 119–144.

Prasad, D., and P. Chand. 1990. Pests of redgram and their management. Recent Researches in Ecology, Environment and Pollution No. 5. Paper presented at the 3rd Conference of the Mendelian Society of India, Birsa Agricultural University, Ranchi, India, 4–7 April 1989.

Prestwich, G. D. 1976. Composition of the scents of eight East African hemipterans. Nymph-adult chemical polymorphism in coreids. Ann. Entomol. Soc. Amer. 69: 812–814.

Raga, A., C. de T. Piza, Jr., and M. F. de Souza Filho. 1995. Ocorrência e danos de *Leptoglossus zonatus* (Dallas) (Heteroptera: Coreidae) em romã, *Punica granatum* L., em Campinas, São Paulo. An. Soc. Entomol. Brasil 24: 183–185.

Rawat, R. R., K. N. Kapoor, U. S. Misra, and S. V. Dhamdere. 1969. A record of predatory mite, *Bochartia* sp. (Erythraeidae: Acarina) on *Clavigralla gibbosa* Spinola. J. Bombay Nat. Hist. Soc. 66: 403–404.

Redaelli, L. R., B. C. Caldas, and L. M. G. Diefenbach. 1998a. Parâmetros reprodutivos de *Corecoris dentiventris* Berg, 1884 (Hemiptera, Coreidae) em cultura de fumo (*Nicotiana tabacum*). P. 618 *in* Proceedings, 17th Congresso Brasileiro de Entomologia, 9–14 August 1998, Rio de Janeiro, Brazil.

1998b. Biologia de *Corecoris dentiventris* Berg, 1884 (Hemiptera, Coreidae) em cultura de fumo (*Nicotiana tabacum*). P. 944 *in* Proceedings, 17th Congresso Brasileiro de Entomologia, 9–14 August 1998, Rio de Janeiro, Brazil.

Rice, R. R., J. K. Uyemoto, J. M. Ogawa, and W. M. Pemberton. 1985. New findings on pistachio problems. California Agric. 39: 15–18.

Ryan, M. A. 1994. Damage to papaw trees by the banana-spotting bug, *Amblypelta lutescens lutescens* (Distant) (Hemiptera: Coreidae), in North Queensland. Int. J. Pest Manage. 40: 280–282.

Santos, C. H., and A. R. Panizzi. 1998. Nymphal and adult performance of *Neomegalotomus parvus* (Hemiptera: Alydidae) on wild and cultivated legumes. Ann. Entomol. Soc. Amer. 91: 445–451.

Santos, R. S. S., L. R. Redaelli, L. M. G. Diefenbach, and J. C. Salazar. 1998. Parasitóides associados à fase de ovo de *Corecoris dentiventris* Berg, 1884 (Hemiptera, Coreidae), P. 250 *in* Proceedings, 17th Congresso Brasileiro de Entomologia, 9–14 August 1998, Rio de Janeiro, Brazil.

Sbravate, C., M. Campaner, L. E. A. Camargo, I. Conchon, M. M. G. Teixeira, and E. P. Camargo. 1989. Culture and generic identification of trypanosomatids of phytophagous Hemiptera in Brazil. J. Protozool. 36: 543–547.

Schaefer, C. W., and P. L. Mitchell. 1983. Food plants of the Coreoidea (Hemiptera: Heteroptera). Ann. Entomol. Soc. Amer. 76: 591–615.

Schaffner, J. C. 1967. The occurrence of *Theognis occidentalis* in the midwestern United States. J. Kansas Entomol. Soc. 40: 141–142.

Schalk, J. M., and R. L. Fery. 1982. Southern green stink bug and leaffooted bug: effect on cowpea production. J. Econ. Entomol. 75: 72–75.

Schell, S. C. 1943. The biology of *Hadronotus ajax* Girault (Hymenoptera–Scelionidae), a parasite in the eggs of squash-bug (*Anasa tristis* DeGeer). Ann. Entomol. Soc. Amer. 36: 625–635.

Schowalter, T. D. 1994. Cone and seed insect phenology in a Douglas-fir seed orchard during three years in western Oregon. J. Econ. Entomol. 87: 758–765.

Schowalter, T. D., and J. M. Sexton. 1990. Effect of *Leptoglossus occidentalis* (Heteroptera: Coreidae) on seed development of Douglas-fir at different times during the growing season in western Oregon. J. Econ. Entomol. 83: 1485–1486.

Schuh, R. T., and J. A. Slater. 1995. True Bugs of the World (Hemiptera: Heteroptera): Classification and Natural History. Cornell University Press, Ithaca, New York, New York, U.S.A. 336 pp.

Sehgal, V. K., and R. Ugajir. 1988. Insect pests and pest management of mungbean in India. Pp. 315–328 *in* Mungbean: Proceedings of the Second International Symposium, Asian Vegetable Research and Development Center, Shanhua, Taiwan. 730 pp.

Serantes, G. H. E. 1973. Biologia de *Phthia picta* (Drury) (Hemiptera: Coreidae). Fitotec. Lat. Amer. 9: 1–9.

Shanower, T. G., J. Romeis, and E. M. Minja. 1999. Insect pests of pigeonpea and their management. Annu. Rev. Entomol. 44: 77–96.

Sickerman, S. L., and J. K. Wangberg. 1983. Behavioral responses of the cactus bug, *Chelinidea vittiger* Uhler, to fire damaged host plants. Southwest. Entomol. 8: 263–267.

Silva, J. B. T., and I. Roitman. 1990. Growth of *Phytomonas serpens* in a defined medium; nutritional requirements. J. Protozool. 37: 521–523.

Singh, G. 1974. Endosymbiotic microorganisms in *Cletus signatus* Walker (Coreidae: Heteroptera). Experientia 30: 1406–1407.

Singh, G., and N. C. Pant. 1974. Degradation of insecticides by the cultured symbiotes of *Cletus signatus* Walker (Coreidae: Heteroptera). Curr. Sci. 43: 624–625.

Singh, H. K., and H. N. Singh. 1991. Some major pest incidence on certain late cultivars of pigeon pea, *Cajanus cajan* L., during half pod formation stage. Indian J. Entomol. 53: 298–303.

Singh, M. P., and M. I. Singh. 1987. First recorded incidence of rice bugs in Manipur, India. Int. Rice Res. Newsl. 12: 31.

Singh, O. P., U. S. Misrah, and S. I. Ali. 1982. Susceptibility of insect eggs to carbamates. Indian J. Plant Prot. 10: 79–80.

Singh, R. N., and K. M. Singh. 1978. Influence of intercropping on succession and population build up of insect pests in early variety of red gram, *Cajanus cajan* (L.) Millsp. Indian J. Entomol. 40: 361–375.

Singh, S. R. 1985. Insects damaging cowpeas in Asia. Pp. 245–248 *in* S. R. Singh and K. O. Rachie [eds.], Cowpea Research, Production, and Utilization. John Wiley & Sons, Chichester, U.K.

Singh, S. R., and L. E. N. Jackai. 1985. Insect pests of cowpea in Africa: their life cycle, economic importance and potential for control. Pp. 217–231 *in* S. R. Singh and K. O. Rachie [eds.], Cowpea Research, Production and Utilization. John Wiley & Sons, Chichester, U.K. 460 pp.

Singh, S. R., H. F. van Emden, and T. A. Taylor. 1978a. The potential for the development of integrated pest management systems in cowpeas. Pp. 329–335 *in* S. R. Singh, H. F. van Emden, and T. A. Taylor [eds.], Pests of Grain Legumes: Ecology and Control. Academic Press, London, U.K. 454 pp.

1978b. Checklist of insect and mite pests of grain legumes. Pp. 399–417 *in* S. R. Singh, H. F. van Emden and T. A. Taylor [eds.], Pests of Grain Legumes: Ecology and Control. Academic Press, London, U.K. 454 pp.

Smith, E. S. C. 1984. Studies on *Amblypelta theobromae* Brown (Heteroptera: Coreidae) in Papua New Guinea. I. Descriptions of the immature and adult stages. Bull. Entomol. Res. 74: 541–547.

1985. New host records of *Amblypelta lutescens lutescens* (Distant) (Hemiptera: Coreidae) in north-western Australia. Austral. Entomol. Mag. 12: 55–56.

Soglia, M. C. M., W. M. S. Sá, and A. S. Nascimento. 1998. Aspectos bioecológicos do percevejo *Crinocerus sanctus* (Fabr., 1775) (Heteroptera: Coreidae), praga da aceroleira (*Malpighia puncifolia*). P. 621 *in* Proceedings, 17th Congresso Brasileiro de Entomologia, 9–14 August 1998, Rio de Janeiro, Brazil.

Souza, C. E. P., and B. F. Amaral Filho. 1998. Registro de parasitóides em *Leptoglossus zonatus* Dallas, 1892 (Heteroptera: Coreidae). P. 518 *in* Proceedings, 17th Congresso Brasileiro de Entomologia, 9–14 August 1998, Rio de Janeiro, Brazil.

Srivastava, K. M., and R. P. Singh. 1979. Upsurge of *Clavigralla gibbosa* Spinola on red gram in Uttar Pradesh. Indian J. Entomol. 41: 395–396.

Stein, J. D., R. E. Sandquist, T. W. Koerber, and C. L. Frank. 1993. Response of Douglas-fir cone and seed insects to implants of systemic insecticides in a northern California forest and a southern Oregon seed orchard. J. Econ. Entomol. 86: 465–469.

Steinbauer, M. J. 1995. The biogeography and host plant utilisation of eucalypt feeding Coreidae (Hemiptera: Heteroptera). Ph.D. thesis, University of Tasmania, Hobart, Tasmania, Australia. 326 pp.

1996a. Notes on extra-phytophagous food sources of *Gelonus tasmanicus* (Le Guillou) (Hemiptera: Coreidae) and *Dindymus versicolor* (Herrich-Schäffer) (Hemiptera: Pyrrocoridae). Austral. Entomol. 23: 121–124.

1996b. A note on manna feeding by ants. J. Nat. Hist. 30: 1185–1192.

1997. Seasonal phenology and developmental biology of *Amorbus obscuricornis* (Westwood) and *Gelonus tasmanicus* (Le Guillou) (Hemiptera: Coreidae). Austral. J. Zool. 45: 49–63.

Steinbauer, M. J., and A. R. Clarke. 1995. *Xenoencyrtus hemipterus* (Girault) (Hymenoptera: Encyrtidae), an egg parasitoid of Coreidae (Hemiptera) in Tasmania. J. Austral. Entomol. Soc . 34: 63–64.

Steinbauer, M. J., and N. W. Davies. 1995. Defensive secretions of *Amorbus obscuricornis* (Westwood), *A. rubiginosus* (Guérin-Méneville) and *Gelonus tasmanicus* (Le Guillou) (Hemiptera: Coreidae). J. Austral. Entomol. Soc. 34: 75–78.

Steinbauer, M. J., A. R. Clarke, and S. C. Paterson. 1998. Changes in eucalypt architecture and the foraging behaviour and development of *Amorbus obscuricornis* (Hemiptera: Coreidae). Bull. Entomol. Res. 88: 641–651.

Steinbauer, M. J., G. S. Taylor, and J. L. Madden. 1997. Comparison of damage to *Eucalyptus* caused by *Amorbus obscuricornis* and *Gelonus tasmanicus*. Entomol. Exp. Appl. 82: 175–180.

Suh, J. B., L. E. N. Jackai, and W. N. O. Hammond. 1986. Observations on pod-sucking bug populations on cowpea at Mokwa, Nigeria. Trop. Grain Legume Bull. 33: 17–19.

Summers, D., and D. S. Ruth. 1987. Effect of diatomaceous earth, malathion, dimethoate and permethrin on *Leptoglossus occidentalis* (Hemiptera: Coreidae): a pest of conifer seed. J. Entomol. Soc. British Columbia 84: 33–38.

Swaine, G. 1969. Studies on the biology and control of pests of seed beans (*Phaseolus vulgaris*) in North Tanzania. Bull. Entomol. Res. 59: 323–338.

Szent-Ivany, J. J. H. 1961. Insect pests of *Theobroma cacao* in the territory of Papua and New Guinea. Papua New Guinea Agric. J. 13: 127–147.

1963. Further records of insect pests of *Theobroma cacao* in the territory of Papua and New Guinea. Papua New Guinea Agric. J. 16: 37–43.

Szent-Ivany, J. J. H., and A. Catley. 1960a. Notes on the distribution and economic importance of the Papuan tip-wilt bug, *Amblypelta lutescens papuensis* Brown (Heteroptera: Coreidae). Papua New Guinea Agric. J. 13: 59–65.
1960b. Observations on the biology of the black leaf-footed bug *Leptoglossus australis* (F.) (Heteroptera, Coreidae) in the territory of Papua and New Guinea. Papua New Guinea Agric. J. 13: 70–75.
Szent-Ivany, J. J. H., and R. M. Stevens. 1966. Insects associated with *Coffea arabica* and some other crops in the territory of Papua New Guinea. Papua New Guinea Agric. J. 18: 101–119.
Tait, E. M. 1954. Some notes on the life history and habits of *Theraptus* sp. Bull. Entomol. Res. 45: 429–432.
Takasu, K., and Y. Hirose. 1986. Kudzu-vine community as a breeding site of *Ooencyrtus nezarae* Ishii (Hymenoptera: Encyrtidae), an egg parasitoid of bugs attacking soybean. Jpn. J. Appl. Entomol. Zool. 30: 302–304.
Taylor, T. A. 1968. The effects of insecticide application on insect damage and the performance of cowpea in Southern Nigeria. Nigerian Agric. J. 5: 29–37.
1975. *Gryon gnidus*, a seclionid egg-parasite of *Acanthomia tomentosicollis* (Hemiptera: Coreidae) in Nigeria. Entomophaga 20: 129–134.
Taylor, T. A., and O. Omoniyi. 1970. Variation in number of micropyles in eggs of *Acanthomia tomentosicollis* (Stål) (Hemiptera: Coreidae) Nigeria. Entomol. Mag. 2: 70–73.
Thompson, W. L. 1934. Food habits of *Leptoglossus gonagra*. Florida Entomol. 18: 46.
Thurston, H. D. 1998. Tropical Plant Diseases, second edition. APS Press, St. Paul, Minnesota, U.S.A. 200 pp.
Tracy, J. L., and J. R. Nechols. 1987. Comparisons between the squash bug egg parasitoids *Ooencyrtus anasae* and *O.* sp. (Hymenoptera: Encyrtidae): development, survival, and sex ratio in relation to temperature. Environ. Entomol. 16: 1324–1329.
1988. Comparison of thermal responses, reproductive biologies, and population growth potentials of the squash bug egg parasitoids *Ooencyrtus anasae* and *O.* sp. (Hymenoptera: Encyrtidae). Environ. Entomol. 17: 636–643.
Ueckert, D. N. 1973. Effects of leaf-footed bugs on mesquite reproduction. J. Range Manage. 26: 227–229.
[USDA] U.S. Department of Agriculture. 1965. Crop Insects of Northeast Africa-Southwest Asia. Agric. Handbook No. 273. USDA, Beltsville, Maryland, U.S.A.
Uyemoto, J. K. J. M. Ogawa, R. E. Rice, H. R. Teranishi, R. M. Bostock, and W. M. Pemberton. 1986. Role of several true bugs (Hemiptera) on incidence and seasonal development of pistachio fruit epicarp lesion disorder. J. Econ. Entomol. 79: 395–399.
van der Meulen, T. 1992. Assessment of damage caused by the coconut bug *Pseudotheraptus wayi* (Brown) (Hemiptera: Coreidae) on guavas. Fruits 47: 317–320.
Vanderplank, F. L. 1958. Studies on the coconut pest *Pseudotheraptus wayi* Brown (Coreidae) in Zanzibar I. Method of assessing the damage caused by the insect. Bull. Entomol. Res. 49: 559–584.
1959a. Studies on the coconut pest *Pseudotheraptus wayi* Brown (Coreidae) in Zanzibar II. Some data on the yields of coconuts in relation to damage caused by the insect. Bull. Entomol. Res. 50: 135–149.
1959b. Studies on the coconut pest *Pseudotheraptus wayi* Brown (Coreidae) in Zanzibar III. A selective residual formulation and its effects [sic] the ecology of the insect. Bull. Entomol. Res. 50: 151–164.
1960a. The availability of the coconut bug *Pseudotheraptus wayi* Brown. Bull. Entomol. Res. 51: 57–60.
1960b. The bionomics and ecology of the red tree ant *Oecophylla* sp. and its relationship to the coconut bug *Pseudotheraptus wayi* Brown (Coreidae). J. Anim. Ecol. 29: 15–33.
van der Vecht, J. 1933. De Groote Peperwants of Semoenjoeng (*Dasynus piperis* China). Proefschr. Rijksuniv., Leiden, the Netherlands. 101 pp.
van Halteren, P. 1971. Insect pests of cowpea, *Vigna unguiculata* (L.) Walp., in the Accra plains. Ghana J. Agric. Sci. 4: 121–123.
Villiers, A. 1952. Hémiptères de l'Afrique noire (punaises et cigales). Initiations Africaines. IX. Institut Français d'Afrique Noire, Dakar, Senegal. 256 pp.
de Villiers, E. A. 1994. Control tip wilters on young mango trees. Inligtingsbulletin — Instituut vir Tropiese en Subtropiese Gewasse. 264: 9–10 [CAB Abstr.].
Visalakshi, A., S. N. Beevi, T. Premkumar, and M. R. G. K. Nair. 1980. Biology of *Leptoglossus australis* (Fabr.) (Coreidae: Hemipera) a pest of snake gourd. Entomon 5: 77–79.
Vogt, E. A., and J. R. Nechols 1991. Diel activity patterns of the squash bug egg parasitoid *Gryon pennsylvanicum* (Hymenoptera: Scelionidae). Ann. Entomol. Soc. Amer. 84: 303–308.

1993a. Responses of the squash bug (Hemiptera: Coreidae) and its egg parasitoid, *Gryon pennsylvanicum* (Hymenoptera: Scelionidae) to three *Cucurbita* cultivars. Environ. Entomol. 22: 238–245.

1993b. The influence of host deprivation and host source on the reproductive biology and longevity of the squash bug egg parasitoid *Gryon pennsylvanicum* (Ashmead) (Hymenoptera: Scelionidae). Biol. Control 3: 148–154.

Waite, G. K. 1990. *Amblypelta* spp. (Hemiptera: Coreidae) and green fruit drop in lychees. Trop. Pest Manage. 36: 353–355.

Wang, C. L. 1980. Soybean insects occurring at podding stage in Taichung. J. Agric. Res. China 29: 283–286.

Warui, C. M. 1983. A laboratory study of the growth and fecundity of *Pseudotheraptus wayi* Brown (Hemiptera: Coreidae): a cashewnut tree pest in Kenya. Kenya J. Sci. Technol. Ser. B. 4: 67–70.

Way, M. J. 1951. An insect pest of coconuts and its relationship to certain ant species. Nature 168: 302.

1953a. Studies on *Theraptus* sp. (Coreidae): the cause of the gumming disease of coconuts in East Africa. Bull. Entomol. Res. 44: 657–667.

1953b. The relationship between certain ant species with particular reference to biological control of the coreid, *Theraptus* sp. Bull. Entomol. Res. 44: 669–691.

Wheatley, P. E. 1961. The insect pests of agriculture in the coast province of Kenya. IV. Coconut. East Afric. Agric. For. J. 27: 33–35.

Wheeler, A. G., Jr., and G. L. Miller. 1990. *Leptoglossus fulvicornis* (Heteroptera: Coreidae), a specialist on *Magnolia* fruits: seasonal history, habits, and descriptions of immature stages. Ann. Entomol. Soc. Amer. 83: 753–765.

Wiseman, B. R., and W. W. McMillian. 1971. Damage to sorghum in south Georgia by Hemiptera. J. Georgia Entomol. Soc. 6: 237–242.

Wolcott, G. N. 1933. An Economic Entomology of the West Indies. The Entomological Society of Puerto Rico. San Juan, Puerto Rico, U.S.A. 688 pp.

Woodson, W. D., and W. S. Fargo. 1991. Interactions of temperature and squash bug density (Hemiptera: Coreidae) on growth of seedling squash. J. Econ. Entomol. 84: 886–890.

1992. Interactions of plant size and squash bug (Hemiptera: Coreidae) density on growth of seedling squash. J. Econ. Entomol. 85: 2365–2369.

Wu, H.-J. 1988. Biocontrol of squash bug with *Neoaplectana carpocapsae* (Weiser). Bull. Inst. Zool., Acad. Sin. 27: 195–203.

Yasuda, K. 1990. Ecology of the leaf footed bug, *Leptoglossus australis* Fabricius (Heteroptera: Coreidae), in the sub-tropical region of Japan. Trop. Agric. Res. Ser. No. 23: 229–238.

Yasuda, K., and M. Tsurumachi. 1994. Immigration of leaf-footed plant bug, *Leptoglossus australis* (Fabricius) (Heteroptera: Coreidae), into cucurbit fields on Ishigi Island, Japan, and role of male pheromone. Jpn. J. Appl. Entomol. Zool. 38: 161–167.

1995. Influence of male adults of the leaf-footed plant bug *Leptoglossus australis* (Fabricius) (Heteroptera: Coreidae), on host-searching of the egg parasitoid, *Gryon pennsylvanicum* (Ashmead) (Hymenoptera: Scelionidae). Appl. Entomol. Zool. 30: 139–144.

Yates, I. E., W. L. Tedders, and D. Sparks. 1991. Diagnostic evidence of damage on pecan shells by stink bugs and coreid bugs. J. Amer. Soc. Hortic. Sci. 116: 42–46.

Yeo, D., and R. Foster. 1958. Preliminary note on a method for the direct estimation of populations of *Pseudotheraptus wayi* Brown on coconut palms. Bull. Entomol. Res. 49: 585–590.

Yoshimoto, C. M. 1977. A new species *Ooencyrtus leptoglossi* (Hymenoptera: Chalcidoidea, Encyrtidae) reared from eggs of *Leptoglossus corculus* (Hemiptera: Coreidae). Canad. Entomol. 109: 1009–1012.

Youdeowei, A. 1973. Some Hemiptera-Heteroptera associated with cacao farms in Nigeria. Turrialba 23: 162–171.

Zheng, L.-Y. 1994. Heteropteran insects (Hemiptera) feeding on bamboos in China. Ann. Entomol. Soc. Amer. 87: 91–96.

CHAPTER 12

Burrower Bugs (Cydnidae)

Jerzy A. Lis, Miriam Becker, and Carl W. Schaefer

1. INTRODUCTION

The family Cydnidae *sensu lato* comprises eight subfamilies (Amnestinae, Corimelaeninae, Cydninae, Garsauriinae, Parastrachiinae, Scaptocorinae, Sehirinae, and Thyreocorinae). About 120 genera and 750 species are included in these eight subfamilies (J. A. Lis, unpublished). The New World species have been revised by Froeschner (1960), the Oriental ones by Lis (1994), and the four species found in New Zealand have been described and their biologies recorded by Larivière (1995). The cydnids of the Iberian Peninsula have been redescribed and keyed by de la Fuente (1972).

Cydnids are known popularly as "burrower bugs" (or "burrowing bugs"); most of these bugs are black or brown, usually 2 to 20 mm long, live deep in the soil, and are root-feeders. However, some are not fossorial and may be seed-feeders (Schaefer 1988). Although the family is regarded as quite primitive in the superfamily Pentatomoidea, its species possess some derived characters (broad and flattened head, often armed with rows of strong setae or spines, broadened and heavily spinose anterior tibiae, tarsi reduced or even absent in a few taxa, for example), which make them very well adapted for digging in the ground.

Cydnids are of interest because several show varying degrees of maternal care of eggs and early instars (Tachikawa and Schaefer 1985; Filippi-Tsukamoto 1995; Kight 1995, 1996, 1997), a subsocial behavior found only sporadically in other heteropteran groups (see Tallamy and Schaefer 1997).

2. GENERAL STATEMENT OF ECONOMIC IMPORTANCE

Burrowing bugs have generally been considered of little economic importance. Nevertheless, in some years some species may become abundant and cause damage to cultivated crops. Up to now 27 species in three subfamilies (Scaptocorinae, Cydninae, and Sehirinae) have been reported as pests, about half of them (mostly scaptocorines) as serious pests. They mostly attack field crops and cereals, sometimes also fruits and legumes. Because the damage is often to the roots, the damage can be mistaken for such problems as soil deficiency. Thus, the presence of burrower bugs may be overlooked. Some species can be detected by their pungent odor when the soil is dug. Also, for root crops, the insect damage may not reduce the yield but can greatly reduce the commercial value of the crop. In these cases, the damage is assessed only after harvest.

Even for species of economic importance, the original native hosts are not known. Practically nothing is known of the native plants and habitats, for field records are taken only when damage is caused to plants of economic importance. Moreover, misidentifications are common, which render reported distributions still less reliable.

An unidentified cydnid was reported feeding on the corms of taro (Mitchell and Maddison 1983).

In field crops, burrower bugs are reported to occur in patches. No explanation thus far has been offered for this aggregated distribution in the soil. Thus, field tests with insecticides often fail to give reliable results because the bug distribution pattern in the soil is far from random.

In the Neotropics, the main problems with cydnids occur in Brazil. In the rest of the Neotropical region damage by burrower bugs is generally not important, except perhaps in Colombia. Schuh and Slater (1995) attach no economic importance to cydnids as pests of cultivated crops in America north of Mexico.

3. THE IMPORTANT SPECIES

3.1 *Scaptocoris castanea* Perty

3.1.1 *Distribution, Life History, and Biology*

Scaptocoris castanea Perty is a "catch-all name," applied whenever damage by a brown burrower bug has been detected, particularly in Brazil. Thus far there are no records of *S. castanea* for countries other than Brazil and Argentina. Illustrations of the species in such papers as Brisolla et al. (1985) or Puzzi and Andrade (1957) and Brewer (1972) do not help identify the pest. On the contrary, they provide the false impression that, based on these representations, the species can be characterized. This is not true. As pointed out by Schaefer (1995), "... cydnids are small, drab, and unassuming bugs which look remarkably alike." Identification to species level is based on characters that must be evaluated in detail under a binocular microscope (Froeschner 1960; Becker 1967, 1996). Therefore, references under this name do not necessarily refer to the actual *S. castanea*.

The species attacks the roots of maize, cotton, rice, groundnuts, lucerne, sugarcane, potato, peas, tomatoes, pimentos, and *capsicum* (Costa Lima 1940; Hayward 1943; Andrade and Puzzi 1951, 1953; Froeschner 1960; Becker 1967; Brewer 1972). It is one of the principal pests of groundnuts (Bastos Cruz et al. 1962, Brewer 1972). Adults and nymphs feed also in banana plantations and in areas planted with beans and maize (Brisolla et al. 1985).

Despite the economic importance of this species, consistent investigations on its biology and ecology have not been carried out. What follows concerns Brazil unless otherwise stated and consists mainly of occasional or unsystematic field observations. Also, some incorrect statements have been repeated without dispute and have attained the status of "truth" — for example, the occurrence of cyclic population fluctuations, night migration, and the production of subterranean tunnels. Despite the nearly 80 years since the first record of *S. castanea* in an agricultural crop (Moreira 1923), very little has been added.

The first record of *S. castanea* was on rice in Minas Gerais state (Moreira 1923). In 1940, Costa Lima reported it also on cotton and black beans, as well as rice, again in Minas Gerais, and for the first time in São Paulo state. In 1951, *S. castanea* is mentioned as a new pest species of sugarcane (Andrade and Puzzi 1951); and, in 1953, it was reported as attacking sugarcane over a much larger area of São Paulo state (Andrade and Puzzi 1953). According to Puzzi and Andrade (1957), 5 years later *S. castanea* attained pest status in São Paulo not only on sugarcane but also on cotton, rice, maize, and black beans. The authors suggest that this species is a pastureland dweller only recently infesting crops of economic importance. In newly established coffee crops and *Eucalyptus* plantations, damage by the burrower bug was rather severe if the land had previously been covered by pasturelands with high densities of the bug. On the other hand, in mature coffee stands the bugs preferred the roots of weeds. This paper (Puzzi and Andrade 1957) contains the first account of the biology and ecology of this species. The authors call attention to the aggregated nature of the infestation and to the difficulty of studying the bugs, especially when they dig as much as 1.5 mm deep into the soil.

In the Gallo et al. (1970) and Zucchi et al. (1993) textbooks on entomology, *S. castanea* is referred to from cotton, peanuts, rice, sugarcane, maize, and sorghum. In 1985, Brisolla et al. mentioned the occurrence of *S. castanea* patchily distributed in banana crops by the seaside in São Paulo. This is the first record for banana crops; the crop was established in a site where maize and black beans used to be cultivated.

In 1993 a severe attack on pasturelands was attributed to *S. castanea* in Minas Gerais and São Paulo states as reported by M. C. Mendes et al. (unpublished) during the 14th Brazilian Congress of Entomology. In 1995, J. S. Correia et al. (unpublished) reported (15th Brazilian Congress of Entomology) an attack in pasturelands in Bahia state; the infestation was said to be in patches and more intense in sandy soils. *Brachiaria decumbens*, *B. brisantha*, and *B. humidicola* were the grasses severely attacked, and an average of as many as 312 bugs/m^2 was reported. [Voucher specimens should be examined to ascertain if the damage was really caused by *S. castanea* or by *Atarsocoris brachiariae* Becker or even by *S. carvalhoi* Becker, already reported on tobacco in Bahia state by Lavigne (1959).]

Recently (16th Brazilian Congress of Entomology), Z. Ramiro (1997, unpublished) recorded *S. castanea* in soybean crops in São Paulo state, since 1992–1993. More recently, this species has been registered as a devastating pest of soybean in central Brazil (A. R. Panizzi, personal communication).

In Argentina *S. castanea* thus far has not attained pest status. Records occur for potatoes, tomatoes, and pimentos by Hayward (1943) under the name of *S. terginus* Schiødte. Brewer (1972) reports the same species to occur in alfalfa in Cordoba, as much as 70 cm deep in the soil.

The odor released by this bug has interesting effects on other organisms. It repels ants and inhibits development of 15 kinds of soil-inhabiting fungi, including that which causes fusarium "wilt" of banana trees (Froeschner and Chapman 1963; see also **Sections 4.1 and 4.3**).

3.1.2 Damage

It damages roots in its nymphal and adult stages (Puzzi and Andrade 1957, Bastos Cruz et al. 1962), and harms rooting and germination (Andrade and Puzzi 1953).

3.1.3 Control

Dust of chlordane, aldrin, lindane or toxaphene, and BHC (all mixed with a fertilizer) applied to the open furrows before the planting of sets gives the best results (Andrade and Puzzi 1953, Puzzi and Andrade 1957, Bastos Cruz et al. 1962). The species should also be dealt with by flooding wherever possible (Moreira 1923). More recently, application of organophosphate granular pesticides in the soil in the row near the seeds, or seed treatment with carbamate pesticides, has been used in Brazil (J. Libério, personal communication). In addition, deep plowing has been recommended as a cultural practice to control *S. castanea*, by disturbing nymphs and/or adults and exposing them to the action of sunlight and predators (J. Libério, personal communication).

3.2 *Stibaropus formosanus* (Esaki)

3.2.1 Distribution, Life History, and Biology

This Oriental species injures the roots of wheat, maize, millet, and sorghum, and is the main pest of sugarcane (Takano and Yanagihara 1939, Esaki and Ishihara 1951, Box 1953). Its life cycle lasts more than 2 years; nymphs and adults overwinter in the soil; it is oligophagous, preferring Graminae (Anonymous 1977). It can survive prolonged starvation and its infestation usually occurs in fields of drained loamy-sand and, to a lesser extent, of light alkali soil (Anonymous 1977).

3.2.2 Damage

See damage under *Scaptocoris castanea* (**Section 3.1.2**).

3.2.3 Control

Integrated control consists of changing the cropping system by rotating wheat with cotton or sesame, sweet potato, and so forth, increasing irrigation and manuring, and the rational application of insecticides (CS_2) (Anonymous 1977).

3.3 *Stibaropus indonesicus* Lis (= *Stibaropus molginus*, not of Schiødte: records from Java)

3.3.1 Distribution, Life History, and Biology

This species was misidentified and reported from Java (in many papers) under the name of *Stibaropus molginus* (Schiødte). The species has only recently been shown to be a new one (Lis 1991). This Oriental species has been found feeding on young sugarcane roots (Wilbrink 1912, Kalshoven 1950, Box 1953). All developmental stages occur together throughout the year; total development is estimated to last about 6 months; and the bugs can survive for several weeks without feeding (Wilbrink 1912). The eggs are white, oval, 1.75 × 1.00 mm, and are laid scattered in the ground. Egg production obviously is small and continues over a long period; nevertheless, the bugs produce many offspring, and up to 55 specimens were found at the roots of a single plant (Wilbrink 1912).

3.3.2 Damage

It damages roots, harming rooting and germination (Wilbrink 1912). Manifestation as a pest species is limited to fields, where sugarcane is cultivated without rotation with other crops on loose and not too wet soils.

3.3.3 Control

CS_2 was successfully used as a control agent in the past, either by putting 2 cm³ in a small hole in the ground, or, if the pest was not yet widespread, by pouring 5 to 10 cm³ in the furrows between the plants. Collecting bugs — often two petrol cans per field daily — was not a successful control method. Crop rotation obviously prevents outbreaks of this species (Wilbrink 1912, Kalshoven 1950).

3.4 *Atarsocoris brachiariae* Becker

3.4.1 Distribution, Life History, and Biology

This burrower bug is thus far known only from Brazil. This species was only recently named (Becker 1996), and many former reports of plant root damage attributed to *Scaptocoris castanea* may in fact refer to *A. brachiariae*. Thus, the actual distribution of *A. brachiariae* may be much larger than thus far reported. It is also unknown whether the area where "pasture land bug" (the common name proposed by Becker 1996) causes damage is actually in expansion or, conversely, whether the new pasturelands are actually invading areas where the bug used to live in a patchwork of favorable and unfavorable areas. This remains to be investigated, because nearly nothing is known of cydnids from central Brazil, their habitats, and native host plants.

Atarsocoris brachiariae can be easily distinguished from *S. castanea* by the absence of tarsi (Becker 1996). In addition, the body size (smaller), the color (lighter, amber yellow), the clypeus (expanded toward the apex, distal margin truncate), the middle tibia (setae evenly distributed on the dorsal surface), and the hemelytra (usually shorter) distinguish the pastureland bug from the brown bug.

3.4.2 Damage

This species causes serious damage to extensive areas of pastureland in central Brazil. For example, in a single district of Mato Grosso state (Don Aquino) 10,000 ha of cultivated grasses were rapidly killed by the bug. Serious damage has also been reported to vast areas of cultivated pastures in São Paulo (Ramiro et al. 1989, Costa and Forti 1993), Mato Grosso (J. C. Mendes, unpublished), and Bahia (J. S. Correia et al., unpublished) states.

3.4.3 Control

See control measures under *S. castanea* (**Section 3.1**).

3.5 Byrsinus varians (F.) (= Aethus laticollis orientalis Ghauri)

3.5.1 Distribution, Life History, and Biology

This is a serious pest of roots of Bajra, *Pennisetum typhoides* (Burm.), a millet crop cultivated in semiarid areas in India (Ghauri 1975). It is found also on the mesocotyl of young seedlings of wheat (*Triticum*) (Ghauri 1975, Sandhu and Deol 1976) and sorghum (Lis 1994).

3.5.2 Damage

The bugs cause lesions to roots and eventually may kill host plants.

3.5.3 Control

No control measures were found in the literature against this pest. However, those control measures reported for other cydnid species may be effective against this species (see details under *S. castanea*, **Section 3.1**).

3.6 Cyrtomenus (Cyrtomenus) mirabilis (Perty)

3.6.1 Distribution, Life History, and Biology

This species occurs in Brazil, Peru, Paraguay, and Argentina (Froeschner 1960). In the guide to agricultural pests of Brazil (Zucchi et al. 1993), the geographic distribution of *C. mirabilis* is incorrectly said to be from the southern United States to Argentina. According to Froeschner's revision of the family for the Western Hemisphere (1960), Peru is the northernmost area occupied by this species. Also, according to Froeschner's catalog (1988), the *Cyrtomenus* species occurring in the United States are *C. (C.) ciliatus* (Palisot de Beauvois) and *C. (C.) crassus* Walker. The presence of a *C. (Syllobus) emarginatus* Stål population in Florida needs verification (Froeschner

1960, 1988). According to Becker and Galileo (1982), *C. (C.) mirabilis* Perty and *C. (C.) bergi* Froeschner may be the same species, and thus the latter would be a junior synonym of *C. mirabilis*. These species are morphologically very close.

In Brazil, the "root black bug" causes losses to peanuts and pasturelands (Bastos Cruz et al. 1962, Calcagnolo and Tella 1965, Zucchi et al. 1993). This species has been recorded from São Paulo state sucking the roots and pods of the peanut plant. According to Calcagnolo and Tella (1965), it occurs especially in crops following the use of the land as pasture and the distribution of the insects is influenced by the type of soil and the humidity.

3.6.2 Damage

No detailed descriptions were found in the literature regarding the damage done to peanuts and pasturelands. However, this should be similar to what have been described to former cydnid species (see damage under *S. castanea*, **Section 3.1**).

3.6.3 Control

Best results were obtained by mixing a 2.5% aldrin or heptachlor dust lightly into furrows before sowing at the rate of 1 oz/5 yd (Calcagnolo and Tella 1965). This gave over 90% mortality of the adults and nymphs, and did not affect plant emergence or development (Calcagnolo and Tella 1965). A dust of 10% carbaryl (Sevin) applied at the same rate gave a significant although smaller reduction in infestation; disulfoton (Disyston) was less effective and phytotoxic to the young plants, but promising results were given by 1% gamma BHC (lindane) and 10% toxaphene (Calcagnolo and Tella 1965). Insecticide treatments for control are listed also by Bastos Cruz et al. (1962).

3.7 *Cyrtomenus bergi* Froeschner

3.7.1 Distribution, Life History, and Biology

It has been suggested that this species is a synonym of *C. mirabilis* (Becker and Galileo 1982; see **Section 3.6.1**). It is found from Mexico into Argentina (Froeschner 1960), and is commonly collected in light traps (Cividanes et al. 1981).

The species lives on roots of maize, onion, sorghum, peanut, and sugarcane (Cividanes et al. 1981, Garcia and Belloti 1982, Bellotti and Garcia 1983, Shenk and Saunders 1984). It also attacks the roots of cassava (Centro Internacional de Agricultura Tropical 1981, 1984; Gold et al. 1991).

In Colombia this species is known as the "subterranean chinch bug" or "small pox bug" and has been observed since about 1980 in several important cassava-growing regions. Garcia and Bellotti (1982) and Bellotti and Garcia (1983) report biological information obtained under laboratory conditions (average temperature 23°C and relative humidity 65%, range not given). The incubation period lasted 13.6 (11 to 18) days; the total nymphal period lasted 111.2 (91 to 134) days; the preoviposition period lasted 10.5 (8 to 11) days. An average of 250 (200 to 280) eggs was deposited per female, and on average a female oviposited 6 eggs per week. Total adult longevity was more than 250 days. Riis and Esbjerg (1998a,b) studied the effects on *C. bergi* of different soil and air humidities during different seasons.

The biology and pest status of this bug has been reviewed by Bellotti et al. (1999).

3.7.2 Damage

In plowed plots *C. bergi* reduced the emergence of maize seedlings by nearly 50% (Carballo 1982, *fide* Shenk and Sounders 1984). Damage to maize caused by *C. bergi*, among other insects, was significantly less with no-tillage than in plowed systems. In Colombia, *C. bergi* is present in the soil throughout the cassava crop cycle and root damage is initiated during the first or second month (Bellotti et al. 1988).

Nymphs and adults introduce the stylet into the parenchyma of the cassava root. Soil microorganisms may invade the cassava root through the wounded area, causing the "small pox" look. Root damage cannot be detected until roots are harvested and peeled. Thus, 20 to 30% of roots damaged will often result in complete rejection by traders. Therefore, even a light infestation of *C. bergi* in a cassava field can result in severe losses.

3.7.3 Control

In humid regions of Costa Rica, no-tillage systems result in less damage to maize than plowed systems, probably because they encourage development of more predators and parasites. (Carballo 1982, *fide* Shenk and Saunders 1984). In Colombia, on cassava, Bellotti et al. (1988) suggest intercropping with *Crotalaria* or similar crops, because these repel *Cyrtomenus bergi*, which can otherwise complete development on cassava and therefore infest the next year's crop.

More recently, Caicedo V. and Bellotti (1994) have shown that the nematode *Steinernema carpocapsae* Weiser can control *C. bergi* under laboratory conditions. Adults are the most susceptible to parasitization, and first and second instars the least. Another nematode, *Heterorhabditis bacteriophora* Poinar, is effective against fifth instars of this cydnid and, indeed, may be more effective overall than *S. carpocapsae* (Fernanda Barberenc and Bellotti 1998).

Other methods of control were proposed by Centro International de Agricultura Tropical (1984).

3.8 *Microporus nigrita* (F.) (= *Aethus nigrita* (F.))

3.8.1 Distribution, Life History, and Biology

This is a Palearctic burrower bug found throughout most of Europe, the former Soviet Union, and parts of Asia. When it occurs in large numbers it injures crop plants, e.g., lupine, potatoes, and rye, and sometimes also the seedlings of apricot and other trees (*Quercus, Acer*) (Schumacher 1916; Reclaire 1936; Petrucha 1949; Medvedev et al. 1952; Otten 1956; Putschkov 1961, 1972). It is generally found in sandy areas (dunes, fields), and may occur up to 15 cm deep at the roots of weeds like *Artemisia campestris* L., *Achillea, Calluna*, and grasses, particularly *Corynephorus canescens* (L.) Beauv. (Otten 1956, Hertzel 1983). Adults overwinter about 5 cm deep in loose sand and become active on warm days in early spring (Schumacher 1916). Mating occurs during April and May (Hertzel 1983).

Microporus nigrita has also been introduced into the United States, probably in soil ballast (Hoebeke and Wheeler 1984). These authors provide illustrations for the identification of *M. nigrita*, and summarize the present distribution in the United States and the biological information available. Thus far, this sporadic pest of the Old World has not attained pest status in the United States.

3.8.2 Damage

The development of crop plants stops when six to ten individuals are found in 1 m^2 (Petrucha 1949, Putschkov 1961).

3.8.3 Control

Hexachlorane applied at the beginning of spring gives the best results (Petrucha 1949, Putschkov 1961).

3.9 *Pangaeus bilineatus* (Say)

3.9.1 *Distribution, Life History, and Biology*

This species occurs throughout the United States (except the Pacific Northwest), and in Mexico, Guatemala, and Bermuda (Froeschner 1960, 1988); it may also occur in Brazil (Cole 1988), although Froeschner's (1981) synonymy of the Ecuadoran *P. bilinaetus* with *P. aethiops* (F.) casts doubt on the authenticity of Brazilian records. *Pangaeus bilineatus* is a highly variable species (Froeschner 1960). In Texas peanut (*Arachis hypogaea* L.) fields, adult *P. bilineatus* overwinter 15 to 18 cm deep in the soil until very early spring; about 50% of the population survives the winter. At least 85% of the population is in diapause through late February. Eggs are laid singly in the soil, near roots and pods (in peanut fields), by immigrating females (Cole 1988). At times, adults in large numbers have been attracted to the eugenol or geranol used in Japanese beetle traps (Froeschner 1960).

Early in the season this species has been found with *Tominotus communis* (Uhler) (see **Section 4.8**) in large numbers in litter near peanut fields. The onset of immigration was synchronized with pod maturation. Adults were obtained in soil samples only when some maturing pods were present on the plants. Immigration of adults of both sexes into the peanut fields continued for about 45 days. At the same time, adults were found near peanut fields, in the soil under wild grasses and weeds. After the initial infestation, all developmental stages were found on peanuts. Eggs were laid singly in the soil, near roots and pods. Smith and Pitts (1974) conclude that *P. bilineatus* could become a pest of peanuts throughout the United States.

Gould (1931) also describes the biology of the species, in Virginia spinach fields. His account taken with that of Smith and Pitts (1974) suggests that *P. bilineatus* is not strictly an underground dweller. He describes "enormous numbers of these burrower-bugs scrambling about over the loose soil of the spinach beds." The insects were seldom seen in bright sunlight and "ran" rapidly from one clod to another, or "dug into the soil for concealment if no lumps were present." A decided preference was shown for hiding in the cracked earth caused by the germination of the seed. Gould (1931) concluded that the bugs immigrated into the fields from the neighboring woods. He also mentioned that farm laborers said the bugs, as they migrated from woods to fields, were "marching across the roads and paths like army worms."

3.9.2 *Damage*

Pangaeus bilineatus has been reported doing serious damage to cotton seedlings and pepper seedbeds, and damages peanut, spinach, vegetable crops, and strawberry (Gould 1931, Otten 1956, Smith and Pitts 1974, Froeschner 1988). Nymphs and adults of both sexes pierce the pods of peanuts and feed on the kernels inside and, in Texas, may be major pests of this crop, reducing nut quality and gross income by as much as 20%; it may also be a vector of certain plant viral diseases (Smith and Pitts 1974).

The insects attacked spinach plants shortly after the cotyledons were out of the seed coat (Gould 1931, Smith and Pitts 1974). The sucking of the stem resulted in a withered and dying plant which could not push through the ground. Up to 30 bugs were present on each inch of row. Also, a cooperative behavior was reported: 4 to 5 bugs, although sometimes as many as 20, carried off the seed coat of the attacked seed, and buried it.

3.9.3 *Control*

The best result is obtained by treatments of G formulations of diazinon and Dyfonate (*O*-ethyl *s*-phenylethylphosphone-dithioate) applied directly over the row in a 33-cm band, with a tractor-mounted G applicator. Application should be made when the adult bugs begin invading the peanut field. Within 48 hours postapplication, a 2-acre-in. irrigation must be applied to incorporate the insecticide (Smith and Pitts 1974).

4. THE LESS IMPORTANT SPECIES

4.1 *Scaptocoris talpa* Champion

This species is known from Mexico, Guatemala, Honduras, and Panama. It is a large burrowing bug (more than 8.5 mm long). Its eggs are laid near the bases of plants (Bianchi 1935, Otten 1956).

Scaptocoris talpa attacks sugarcane (Champion 1900, Bianchi 1935) and banana (Timonin 1958). However, there is some evidence that the bug protects the banana plant from the wilt disease caused by *Fusarium oxysporum f. cubense*. The bug's volatile scent has a fungicidal and fungistatic action and is not phytotoxic to the banana plant (see also **Sections 3.1 and 4.3**). It also protects the banana roots from nematode parasites (Timonin 1958, 1961a,b).

4.2 *Scaptocoris carvalhoi* Becker

Thus far, this species has been recorded only from Bahia state (northeast Brazil). The original description was based on specimens collected by Lavigne (1959) and reported as *S. castanea* damaging tobacco and blackbeans in Antas, Bahia. According to Lavigne (1959), where beans and tobacco are grown together, the insect definitely prefers the latter. The bugs were found 20 to 30 cm deep in the soil, feeding on the deeper roots. Unfortunately, in later years no more specimens from northeast Brazil have been provided for identification. *Scaptocoris carvalhoi* is rather similar to *S. castanea* and the damage caused to crops in this region might have been wrongly attributed to the latter.

4.3 *Scaptocoris divergens* Froeschner

This species occurs in Cuba, Central America, Trinidad, Venezuela, and Colombia. Thus far, no economic importance has been attached to this species. However, it should be mentioned in the context of IPM programs for its potential use as a coadjutant in plant protection.

Roth (1961) described the scent glands of adults and nymphs of *S. divergens*. In Costa Rica the insects were found in sandy soils in patches where banana plants were healthy although in areas attacked by *Fusarium* wilt disease. Roth's findings confirm those of Timonin (1961a,b) for *S. talpa* and also point out that the volatile secretion of the scent glands repels *Pheidole* sp. ants (see also **Sections 3.1 and 4.1**).

Willis and Roth (1962) studied the relationship of *S. divergens* to soil and moisture. In the laboratory the insects burrow into light soils more easily than into heavy soils and do not burrow into very compact or very humid soils. The authors report that when removed from the soil the bugs lose water and soon die. Even so, there is evidence that they fly actively: they are attracted to artificial light (Martorell 1939, in Venezuela; M. Becker, unpublished, in Colombia).

This species can be rather easily separated from its congeners by the divergently expanded clypeus, which surpasses the margins of the head (paraclypei) (see Froeschner 1960, Becker 1967).

4.4 *Stibaropus tabulatus* (Schiødte)

This species sucks the roots of tobacco in India (Ayyar 1930, Otten 1956, Lis 1994), and up to 30 bugs may be found attached to the roots of a single plant (Ayyar 1930). Like those of other cydnids, its distribution is patchy, occurring in some numbers in certain areas and absent from others; it prefers loose loamy soil, especially that with alluvial deposits from earlier flooding (Ayyar

1930). Bugs were not found associated with other plants (Ayyar 1930), which suggests some degree of host specificity. Other aspects of its biology (Ayyar 1930): It forms small chambers in the soil, in which mating pairs are sometimes found; eggs are laid near the roots of tobacco plants, 7 to 15 cm deep; eggs are 0.8 × 1.7 mm; the egg stage lasts 4 to 15 days; the first instar is 2.0 mm long and whitish (subsequent instars died). Crude oil emulsion and tobacco extract (1 lb oil/6 gal water soaked with 3 lb of tobacco decoction) applied at the base of 80 plants killed 95% of the bugs.

4.5, 4.6 *Stibaropus pseudominor* Lis, and *S. molginus* (Schiødte)

These species were reported as damaging roots of sugarcane and palm trees (Box 1953, Otten 1956, Lis 1994).

4.7 *Cyrtomenus ciliatus* (Palisot de Beauvois)

Nymphs and adults were reported injuring chufa, or edible sedge-root (*Cyperus esculentus* L.) (Tissot 1939). Its feeding also causes yellow to dark brown spots on the kernel of peanuts (Smith and Pitts 1974).

4.8 *Tominotus communis* (Uhler)

Adults and nymphs of the species cause peanut kernels to be bumpy, discolored, and badly flavored (Smith and Pitts 1974); see also **Section 3.9**.

4.9 *Aethus philippinensis* Dallas

This species causes sporadic crop damage to germinating corn (Kalshoven 1950, under the name of *A. indicus*). The species damages roots of rice, *Glycine*, and clover (*Trifolium*) (Kobayashi 1964, 1974). It is reported also as a sugarcane pest from Taiwan and Hawaii (Takano and Yanagihara 1939, Box 1953). Adults overwinter and lay eggs on or in the ground in April to June; new adults appear in July or August (Kobayashi 1964). The biology is summarized in Lis (1994).

4.10, 4.11 *Lactistes minutus* Lis, and *Macroscytus transversus* (Burmeister)

These species were reported as sugarcane pests (the first under the name of *L. rastellus*) (Box 1953).

4.12 *Canthophorus melanopterus* (Herrich-Schaeffer)

Suspected of doing harm to olives, this species is found often in clusters under the bark of trees (olive, grapevine) (Priesner and Alfieri 1953). The hibernated adults usually lay eggs on or in the ground surface in May, and a new generation appears in June or July (Kobayashi 1964). The female protects the egg mass, and the larvae are gregarious (Kobayashi 1959, 1964).

4.13 *Sehirus luctuosus* Mulsant et Rey

When occurring in large numbers, this species can injure potatoes and beet roots (Sorauer 1932, Putschkov 1961); it sometimes damages raspberries, bilberries, and fruit trees (Schumacher 1916; Reclaire 1936; Otten 1956; Putschkov 1961, 1972).

4.14 *Tritomegas bicolor* (L.)

In large numbers this species can injure fruit trees and shrubs (grapes) (Stellwaag 1928; Otten 1956; Putschkov 1961, 1972). It was observed sporadically to damage strawberries, and plants of Fabaceae and Rosacae (Gulde 1921, Cohrs and Kleindienst 1934, Otten 1956, Hertzel 1983).

4.15 *Tritomegas sexmaculatus* (Rambur)

Adults may be pests of fruit trees, and nymphs may injure carrots and other crops (Sorauer 1932; Otten 1956; Putschkov 1961, 1972).

4.16 *Crocistethus waltli* (Fieber)

This species has been recorded as harmful to young seedlings and fruit of grapes (Stellwaag 1928, Otten 1956).

4.17 *Fromundus pygmaeus* (Dallas) (= *Geotomus pygmaeus* [Dallas])

This species has at least once attacked a human (in fact, a heteropterist) (Miller 1931, 1971; see **Chapter 19**).

4.18 *Adrisa* spp.

Three species of *Adrisa* live in the debris beneath several *Acacia* spp. trees, and feed on *Acacia* seeds (Van den Berg 1982).

5. PROGNOSIS

Only nine species of Cydnidae are consistently serious pests of plants. The remaining species are injurious only occasionally, and especially when many individuals occur in a small area. Nevertheless, other species of the family (especially of the subfamily Scaptocorinae) may become serious pests.

6. CONCLUSIONS

Future research should concentrate on the biology and life cycle of certain species of Cydnidae found naturally in or near areas under cultivation or about to be cultivated. Such species may become a problem for the cultivated crops. Studies on the environmental conditions that might stimulate the transformation of a harmless species into a pest should also be made.

7. ACKNOWLEDGMENTS

The authors are very grateful to Drs. A. C. Bellotti and L. Riis (Cassava Project, CIAT, Cali, Colombia) for providing reprints and unpublished information on *Cyrtomenus bergi*.

8. REFERENCES CITED

Andrade, A. C., and D. Puzzi. 1951. Resultados preliminares de experiências para controlar o percevejo castanho em cana-de-açúcar. O Biológico 17: 44–49.

1953. Experiências com inseticidas orgânicos para controlar o "percevejo castanho" (*Scaptocoris castaneus*) em cana-de-açúcar. O Biológico 19: 187–189.

Anonymous. 1977. The biology and integrated control of the root bug *Stibaropus formosanus* (Ish. et Yan.). Acta Entomol. Sin. 20: 276–278.

Ayyar, P. N. K. 1930. A note on *Stibaropus tabulatus* Schiöt. (Hem., Pent.), a new pest of tobacco in south India. Bull. Entomol. Res. 21: 29–31.

Bastos Cruz, B. P., M. B. Figueiredo, and E. Almeida. 1962. Principais doenças e pragas do amendoim [sic] no Estado de São Paulo. O Biológico 28: 189–195.

Becker, M. 1967. Estudos sobre a subfamília Scaptocorinae na região neotropical (Hemiptera: Cydnidae). Arq. Zool. (São Paulo) 15: 291–325.

1996. Uma nova espécie de percevejo-castanho (Heteroptera: Cydnidae: Scaptocorinae) praga de pastagens do centro-oeste do Brasil. An. Soc. Entomol. Brasil 25: 95–102.

Becker, M., and M. H. M. Galileo. 1982. A genitália de macho em cinco gêneros neotropicais da subfamília Cydninae (Heteroptera: Cydnidae). Rev. Bras. Biol. 42: 21–30.

Bellotti, A. C., and C. Garcia. 1983. The subterranean chinch bug, a new pest of cassava. Cassava Newsletter 7: 10–11.

Bellotti, A. C., L. Smith, and S. L. Lapointe. 1999. Recent advances in cassava pest management. Annu. Rev. Entomol. 44: 343–370.

Bellotti, A. C., O. Vargas H., B. Arias, O. Castano, and C. Garcia. 1988. *Cyrtomenus bergi* Froeschner, a new pest of cassava: biology, ecology and control. VIIth Symposium of the International Society for Tropical Root Crops, Gosier (Guadeloupe), 1–6 July 1985, Ed. INRA, Paris, France. 551–561.

Bianchi, F. A. 1935. Two interesting pests of sugarcane in Guatemala *Podischnus agenor* Burmeister and *Scaptocoris talpa* Champion. Hawaiian Plant Rec. 39: 191–197.

Box, H. E. 1953. List of Sugar-Cane Insects. Commonwealth Institute of Entomology, London, U.K. 101 pp.

Brewer, M. 1972. *Scaptocoris castaneus* Perty, chinche dañina a raíces de alfalfa (Hemiptera-Cydnidae). IDIA 194: 27–28.

Brisolla, A. D., E. L. Furtado, M. C. F. Cardim, and O. S. Kawamoto. 1985. Ocorrência do percevejo castanho — *Scaptocoris castaneus* Perty, 1830 — em bananal na região litorânea do Estado de São Paulo. O Biológico 51: 135–137.

Caicedo V., A. M., and A. C. Bellotti. 1994. Evaluacion del potencial del nematodo entomógeno *Steinernema carpocapsae* (Rhabditida: Steinernematidae) para el control de *Cyrtomenus bergi* Froeschner (Hemiptera: Cydnidae) en condiciones de laboratorio. Rev. Colomb. Entomol. 20: 241–246.

Calcagnolo, G., and R. de Tella. 1965. Resultados dos experimentos de combate ao *Cyrtomenus mirabilis* [sic] Perty, 1834 — percevejo preto da raiz do amendoinzeiro. O Biológico 31: 31–31.

Cassidy, T. P. 1939. U.S. Dept. Agric., Bur. Entomol. Plant Quar., Ins. Pest Surv. Bull. 19: 322.

Centro International de Agricultura Tropical. 1981. Entomology. Pp. 1–93 *in* Cassava Program Annual Report for 1980. CIAT 1981, Cali, Colombia.

1984. Entomology. Pp. 71–95, 325–369 *in* Cassava Program Annual Report for 1982 and 1983. CIAT 1984, Cali, Colombia.

Champion, G. C. 1900. A species of *Scaptocoris* Perty, found at roots of sugarcane. Entomol. Mon. Mag. 36: 255–256.

Cividanes, F. J., S. Silveira Neto, and P. S. M. Botelho. 1981. Flutuação populacional de cidnídeos coletados em regiões canavieiras de São Paulo. Científica 9: 241–247.

Cohrs, C., and C. Kleidienst. 1934. Hemiptera-Heteroptera (Wanzen) Zentralsachsens. Ber. Naturwiss. Ges. Chemnitz 1934: 143–181.

Cole, C. L. 1988. Stratification and survival of diapausing burrowing bugs. Southwest. Entomol. 13: 243–246.

Costa, C., and L. C. Forti. 1993. Ocorrência de *Scaptocoris castanea*, Perty 1830 em pastagens cultivadas no Brasil. Pesq. Agropec. Bras. 28: 977–979.

Costa Lima, A. M. da. 1940. Insetos do Brasil, 2, Hemípteros. Escola Nacional de Agronomia, Série Didática No. 3: 351 pp.

Dolling, W. R. 1981. A rationalized classification of the burrower bugs (Cydnidae). Syst. Entomol. 6: 61–67.

Esaki, T., and T. Ishihara. 1951. Hemiptera of Shansi, North China. II. Pentatomoidea. Mushi 22: 22–44.
Fernanda Barberena, M., and A. C. Bellotti. 1998. Parasitismo de dos razas nematodo *Heterorhabditis bacteriophora* sobre la chinche *Cyrtomenus bergi* (Hemiptera: Cydnidae) en laboratorie. Rev. Colomb. Entomol. 24: 7–11.
Filippi-Tsukamoto, L., S. Nomakuchi, K. Kuki, and S. Tojo. 1995. Adaptiveness of parental care in *Parastrachia japonensis* (Hemiptera: Cydnidae). Ann. Entomol. Soc. Amer. 73: 387–397.
Froeschner, R. C. 1960. Cydnidae of the Western Hemisphere. Proc. U.S. Natl. Mus. 111: 337–680.
1981. Heteroptera or true bugs of Ecuador: a partial catalog. Smithsonian Contrib. Zool. 322: 1–147.
1988. Family Cydnidae Billberg, 1820, burrowing bugs. Pp. 119–129 *in* T. J. Henry and R. C. Froeschner [eds.], Catalog of the Heteroptera, or True Bugs, of Canada and the Continental United States. E. J. Brill, Leiden, the Netherlands. 958 pp.
Froeschner, R. C., and Q. L. Chapman. 1963. A South American cydnid, *Scaptocoris castaneus* Perty, established in the United States. Entomol. News 74: 95–98.
Froeschner, R. C., and J. W. E. Steiner. 1983. Second record of South American burrowing bug, *Scaptocoris castaneus* Perty (Hemiptera: Cydnidae) in the United States. Entomol. News 94: 176.
de la Fuente, J. A. 1972. Revisión de los pentatómidos ibéricos. Familia Cydnidae Billberg, 1820. Bol. R. Soc. Esp. Hist. Nat. (Biol.) 70: 33–78.
Gallo, D., O. Nakano, F. M. Wiendl, S. Silveira Neto, and R. P. L. Carvalho. 1970. Manual de Entomologia — Pragas das Plantas e Seu Controle. Editora Agronômica Ceres, São Paulo, Brazil. 858 pp.
Garcia, G., and A. C. Bellotti. 1982. Estudio preliminar de la biologia y morfologia de Cyrtomenus bergi [sic] Froeschner, nueva plaga de la yuca. Rev. Colomb. Entomol. 6: 55–61.
Ghauri, M. S. K. 1975. On a new subspecies of *Aethus laticollis* Wagner (Hemiptera: Heteroptera: Cydnidae) as a serious pest of *Pennisetum typhoides* (Burm.) in India. Bombay Nat. Hist. Soc. 72: 226–227.
Gidayatov, D. A. 1982. True-Bugs of the Pentatomomorpha Group of Azarbaidjan. Institute of Zoology, Academy of Sciences of SSR Azarbaidjan, Baku. 160 pp. [in Russian].
Gold, C. S., O. Vargas, and J. Wightman. 1991. Intercropping effects on insect pests of cassava in Colombia and of groundnut in India. Pp. 329–332 *in* G. K. Veeresh, D. Rajagopal, and C. A. Viraktamath [eds.]. Advances in Management and Conservation of Soil Fauna. Oxford University Press, New Delhi, India. 925 pp.
Gould, G. E. 1931. *Pangaeus uhleri*, a pest of spinach. J. Econ. Entomol. 24: 484–486.
Gulde, J. 1921. Die Wanzen (Hemiptera-Heteroptera) der Umgebung Frankfurt a. M. und des Mainzer Beckens. Arch. Senckenb. Naturf. Ges. 37: 239–503.
Hayward, K. J. 1943. Memoria anual del año 1942. Rev. Ind. Agric. Tucuman 33: 66–84.
Hertzel, G. 1983. Beitrage zur Insektenfauna der DDR: Heteroptera — Plataspidae und Cydnidae (Insecta). Faun. Abh. 10: 111–124.
Hoebeke, E. R., and A. G. Wheeler, Jr. 1984. *Aethus nigritus* (F.), a palearctic burrower bug established in eastern North America (Hemiptera-Heteroptera: Cydnidae). Proc. Entomol. Soc. Washington 86: 738–744.
Kalshoven, L. G. E. 1950. De Plagen van de Cultuurgewassen in Indonesie. Vol. I. N. V. Uitgeverij W. Van Hoewe. 'S-Gravenhage-Bandoeng, the Netherlands. 512 p.
Kight, S. L. 1995. Do maternal burrower bugs, *Sehirus cinctus* Palisot (Heteroptera: Cydnidae), use spatial and chemical cues for egg discrimination? Canad. J. Zool. 73: 815–817.
1996. Concaveation and maintenance of maternal behavior in a burrower bug (*Sehirus cinctus*): a comparative perspective. J. Comp. Physiol. 110: 69–76.
1997. Factors influencing maternal behaviour in a borrower bug, *Sehirus cinctus* (Heteroptera: Cydnidae). Anim. Behav. 53: 105–112.
Kobayashi, T. 1959. Biological notes on *Sehirus niveimarginatus* (Scott) and *Urostylis westwoodi* Scott. Shin Konchû 12: 8–10 [in Japanese].
1964. Developmental stages of *Geotomus pygmaeus* (Dallas) and *Sehirus niveimarginatus* (Scott) (Cydnidae). Kontyu 32: 21–27.
1974. A note on *Aethus indicus* (Westwood) and *Geotomus pygmaeus* (Dallas) in Ishigahi Island. Rostria 23: 123 [in Japanese].
Larivière, M.-C. 1995. Cydnidae, Acanthosomatidae, and Pentatomidae (Insecta: Heteroptera): systematics, geographical distribution, and bioecology. Fauna of New Zealand. No. 35. Menaaki Whenua Press, Lincoln, New Zealand. 111 pp.

Lavigne, G. L. 1959. Nova praga da cultura do fumo na Bahia: *Scaptocoris castaneus* Perty, 1830. Bol. Inst. Biol. Bahia 4: 27–33.

Lis, J. A. 1991. Studies on Oriental Cydnidae. IV. New species, new synonyms and new records (Heteroptera). Ann. Upper Silesian Mus. Entomol. 2: 165–190.

1994. A Revision of Oriental Burrower Bugs (Heteroptera: Cydnidae). Upper Silesian Museum, Bytom, Poland. 349 pp.

Martorell, L. F. 1939. Insects observed in the State of Aragua, Venezuela, South America. J. Agric. Univ. Puerto Rico 23: 177–264.

Medvedev, S. I., A. G. Treml, and D. S. Szapiro. 1952. Fauna vreditelei agrolesomeliorativnych pitomnikov v lesostepnoi i stepnoi zonach Ukrainy. Zastch. Lesonas.Vred. Bolezh. 1952: 47–60.

Miller, N. C. E. 1931. *Geotomus pygmaeus* Dallas (Heteroptera-Cydnidae) attempting to suck human blood. Entomologist 64: 214.

1971. The Biology of the Heteroptera, second edition. E. W. Classey, Hampton, Middlesex, U.K. 206 pp.

Mitchell, W. C., and P. A. Maddison. 1983. Pests of taro. Pp. 180–235 *in* J. Wang [ed.], Taro — A Review of *Colocasia esculenta* and Its Potentials. University of Hawaii Press, Honolulu, Hawaii, U.S.A.

Moreira, C. 1923. Insectos nocivos aos arrozaes e seu controle. Almanac. Agric. Bras. 1923: 193–194.

Otten, E. 1956. Heteroptera, Wanzen, Halbflüger. Pp. 1–149 *in* H. Blunck [ed.], Tierische Schädlinge an Nutzpflanzen, 2. Teil, 3. Lieferung: Heteroptera, Homoptera I. Teil. P. Parey, Berlin, Germany. 399 pp.

Petrucha, O. I. 1949. Schkidniki bobovych roslin, zachrdi borobt'bi z nimi. Vid-vo KDU. 1949: 28–39.

Priesner, H., and A. Alfieri. 1953. A review of the Hemiptera-Heteroptera known to us from Egypt. Bull. Soc. Fouad. I. Entomol. Egypt 37: 1–119.

Putschkov, V. G. 1961. Shield-bugs — Pentatomoidea. Fauna Ukrainy 21(1): 1–338. Academy of Sciences of Ukrainian SSR, Kiev, Ukraine [in Ukranian].

1972. Hemiptera (Heteroptera) — true-bugs. Nasekomyje i klestschi vrediteli selskohozjastvennyh kultur 1: 222–262 [in Russian].

Puzzi, D., and A. C. Andrade. 1957. O "percevejo castanho" — *Scaptocoris castaneus* (Perty) — no Estado de São Paulo. O Biológico 23: 157–163.

Ramiro, A. Z., J. B. M. Araujo, and L. A. Rodrigues. 1989. Ocorrência do "percevejo castanho," *Scaptocoris castanea* Perty, 1830, em pastagens da Dira de Marília, SP. O Biológico 55: 13–14.

Reclaire, A. 1936. 2e vervolg op de naamlijst der in Nederland en het omliggend gebied waargenomen wantsen (Hemiptera-Heteroptera). Entomol. Ber. 210: 243–260.

Riis, L., and P. Esbjerg. 1998a. Movement, distribution, and survival of *Cyrtomenus bergi* (Hemiptera: Cydnidae) within the soil profile in experimentally simulated horizontal and vertical soil water gradients. Environ. Entomol. 27: 1175–1181.

1998b. Season and soil moisture effect on movement, survival, and distribution of *Cyrtomenus bergi* (Hemiptera: Cydnidae) within the soil profile. Environ. Entomol. 27: 1182–1189.

Roth, L. M. 1961. A study of the odoriferous glands of *Scaptocoris divergens* (Hemiptera: Cydnidae). Ann. Entomol. Soc. Amer. 54: 900–911.

Sandhu, G. S., and G. S. Deol. 1976. New records of pest on wheat. Indian J. Entomol. 37: 85–86.

Schaefer, C. W. 1988. The food plants of some "primitive" Pentatomoidea (Hemiptera: Heteroptera). Phytophaga 2: 19–45.

1995. Book review: a revision of Oriental Burrower Bugs (Heteroptera: Cydnidae). Ann. Entomol. Soc. Amer. 88: 595.

Schuh, R. T., and J. A. Slater. 1995. True Bugs of the World (Hemiptera: Heteroptera) Classification and Natural History. Cornell University Press, Ithaca, New York, U.S.A. 336 pp.

Schumacher, F. 1916. Die faunistischen und biologischen Verhältnisse der einheimischen Cydniden. Dtsch. Entomol. Z. 1916: 210–213.

Shenk, M., and J. L. Saunders. 1984. Vegetation management systems and insect responses in the humid tropics of Costa Rica. Trop. Pest Manage. 30: 186–193.

Smith, J. W., Jr., and J. T. Pitts. 1974. Pest status of *Pangeus bilineatus* attacking peanuts in Texas. J. Econ. Entomol. 67: 111–113.

Sorauer, P. 1932. Handbuch der Pflanzenkrankheiten, 5: 420–505. P. Parey, Berlin, Germany.

Stellwaag, F. 1928. Die Weinbauinsekten der Kulturländer. P. Parey, Berlin, Germany. 884 pp.

Tachikawa, S., and C. W. Schaefer. 1985. Biology of *Parastrachia japonensis* (Hemiptera: Pentatomoidea: ?-idae). Ann. Entomol. Soc. Amer. 73: 387–397.

Takano, S., and M. Yanagihara. 1939. [Researches on injurious and beneficial animals of sugarcane.] Taiwan Sugar. Exp. Stn., Extra Rep. 2: 1–311 [in Chinese].

Tallamy, D. W., and C. W. Schaefer. 1997. Maternal care in the Hemiptera: ancestry, alternatives, and current adaptive value. Pp. 94–115 *in* J. C. Choe and B. J. Crespi [eds.], The Evolution of Social Behavior in Insects and Arachnids. Cambridge University Press, Cambridge, U.K. 541 pp.

Timonin, M. 1958. *Scaptocoris talpa* on roots of banana and other plants in Honduras. Plant Prot. Bull. 6: 74–75.

1961a. The interaction of plant, pathogen, and *Scaptocoris talpa* Champion. Canad. J. Bot. 39: 695–703.

1961b. Effect of volatile constituents of *Scaptocoris talpa* Champion on the growth of soil fungi. Plant Soil 14: 323–334.

Tissot, A. N. 1939. [Note]. U.S. Dept. Agric., Bur. Entomol. Plant Quarantine, Ins. Pest Surv. Bull. 19: 455.

Van den Berg, M. A. 1982. Hemiptera attacking *Acacia dealbata* Link., *Acacia decurrens* Willd., *Acacia longifolia* (Andr.) Willd., *Acacia mearnsii* de Wild, and *Acacia melanoxylon* R. Br. in Australia. Phytophylactica 14: 47–50.

Wilbrink, G. 1912. Die kedirische Wortelwants. Meded. Proefs. Java-Suikerindustrie 22: 1111–1123.

Willis, E. R., and L. M. Roth. 1962. Soil and moisture relations of *Scaptocoris divergens* Froeschner (Hemiptera: Cydnidae). Ann. Entomol. Soc. Amer. 55: 21–33.

Zucchi, R. A., S. Silveira Neto, and O. Nakano. 1993. Guia de Identificação de Pragas Agrícolas. FEALQ, Piracicaba, Brazil. 139 pp.

CHAPTER 13

Stink Bugs (Pentatomidae)

Antônio R. Panizzi, J. E. McPherson, David G. James, M. Javahery,
and Robert M. McPherson

1. INTRODUCTION

The Pentatomidae is one of the largest families within the Heteroptera. Of the estimated 36,096 described species of Heteroptera, 4,123 species belong to the family Pentatomidae. This family is ranked the third largest together with the Lygaeidae; it is surpassed by the Reduviidae (second largest family) and the Miridae, which is by far the largest family of Heteroptera (C. W. Schaefer, personal communication).

Within this family, there are eight subfamilies: Asopinae, Cyrtocorinae, Discocephalinae, Edessinae, Pentatominae, Phyllocephalinae, Podopinae, and Serbaninae (Schuh and Slater 1995). The subfamily Pentatominae is the largest of the subfamilies, and its members are plant feeders. Phytophagous stink bugs, in general, are characterized by being round or ovoid, with five-segmented antennae, three-segmented tarsi, and a scutellum that is short, usually narrowed posteriorly, and more or less triangular. They are called stink bugs because they produce a disagreeable odor by means of scent glands that open in the region of the metacoxae. Some nymphs have scent glands located on the dorsum of the abdomen (Borror et al. 1989).

Stink bugs feed by inserting their stylets into the food source to suck up nutrients. By doing this, they cause injury to plant tissues, resulting in plant wilt and, in many cases, abortion of fruits and seeds. During the feeding process, they also may transmit plant pathogens, which increases their damage potential. Because some feed on several plant species of economic importance, they are regarded as major pests.

2. GENERAL STATEMENT OF ECONOMIC IMPORTANCE

The phytophagous pentatomids of economic importance, which will be covered in this chapter, belong to the subfamilies Edessinae, with some pest species in the genus *Edessa*, and Pentatominae, which contains the majority of species that are pests of crops (Schuh and Slater 1995).

The economic importance of these insects varies greatly from species to species, and within a species, depending on the plant attacked. Edessines feed on Solanaceae and are pests of tomato and potato in South America. Among the Pentatominae, the pest species injure a wide range of plants, from vegetables to trees. Among them, the generalist southern green stink bug, *Nezara viridula* (L.), worldwide; *Piezodorus* spp. in South America, Africa/Orient; *Euschistus* spp. in the New World; and *Eurydema* spp. in the Orient attack grain, legumes, and vegetables. Members of the genera *Oebalus, Mormidea*, and *Aelia* are major pests of Gramineae, particularly on rice and wheat. Pests of trees include the spined citrus bug, *Biprorulus bibax* Breddin, on *Citrus* spp. in Australia; *Lincus* spp. on coconut in South America; *Bathycoelia thalassina* (Herrich-Schaeffer) on cocoa in Africa; and *Plautia* spp. on orchard plants in the Orient. The wide ranges of bug species and host plants fed upon, with frequently detrimental effects to plant production, make this, perhaps, the most economically important group of insects among the Heteroptera. As an example, total losses from Heteroptera during 1985, including stink bugs, were estimated at $3.5 million for the pecan industry in Georgia alone (Douce and Suber 1986). In addition, pentatomids may disturb humans by invading houses in large numbers for overwintering (Watanabe et al. 1994a,b,c,d).

This chapter will cover the important and less important species and will provide information on their distribution, biology, and damage, and control measures to mitigate their impact to plant production.

3. THE IMPORTANT SPECIES

3.1 *Acrosternum hilare* (Say)

3.1.1 *Distribution, Life History, and Biology*

This species is widely distributed in North America, from Quebec and New England west to the Pacific Coast and south to Florida, Texas, and California (U.S.A.) (Torre-Bueno 1939, McPherson 1982). It is polyphagous, feeding on over 30 plant species such as apple, pear, and buckthorn (*Rhamnus* spp.) in its northern distribution (Whitmarsh 1917, Schoene and Underhill 1933, Underhill 1934, Esselbaugh 1948, Sailer 1953, McPherson 1982, Javahery 1990). In its southern distribution (below 40° latitude), soybean and green beans are the crops attacked most frequently, although damage also is caused to other legumes (Sailer 1953, Miner 1966).

Adults are green and 13 to 19 mm long. The eggs (Whitmarsh 1917, Underhill 1934, Esselbaugh 1946, Miner 1966, Deitz et al. 1976) and nymphal instars (Whitmarsh 1917, Underhill 1934, DeCoursey and Esselbaugh 1962, Miner 1966, Deitz et al. 1976) have been described. The stadia for the first to fifth instars average 7, 8.9, 7.9, 8.9, and 12.8 days, respectively (Miner 1966).

Acrosternum hilare is a semimigratory, late-season pentatomid that overwinters as adults and completes one generation per year in the northern part of its range. Nymphal development is completed during August and new adults appear from mid-September to early October (Miner 1966, McPherson 1982, Javahery 1990). However, it completes two generations per year under favorable conditions such as those encountered in southern Illinois and the southeastern and Gulf states (Sailer 1953, Jones and Sullivan 1982, McPherson 1982, McPherson and Tecic 1997). It is more abundant in early maturing soybean varieties planted earlier in the growing season than later (McPherson et al. 1988). Its populations are not influenced by row width in soybeans (McPherson et al. 1988) or by strip intercropping and no-tillage in corn/soybean production systems (Tonhasca and Stinner 1991).

3.1.2 *Damage*

The damage done to soybean is similar to that described for *N. viridula*. *Acrosternum hilare* also damages peaches, causing catfacing. On cotton, *A. hilare* can reduce harvestable cotton, lower seed germination, and increase the amount of immature fibers (Barbour et al. 1990).

3.1.3 Control

See control measures under *N. viridula* (**Section 3.17.3**). Natural enemies include egg parasitoids such as *Telenomus podisi* Ashmead (McPherson 1982, Javahery 1990).

3.2 *Aelia furcula* Fieb.

3.2.1 Distribution, Life History, and Biology

This species has been reported from western Iran, eastern Turkey, Syria, and the southwestern part of the independent states of the former U.S.S.R. (Alexandrov and Mirzayan 1949, China and Lodos 1959, Wagner 1960, Putschkov 1961, Brown 1962). High densities of *A. furcula* may occur from time to time in certain areas such as Central, Fars, and Hamadan Provinces, Iran (Brown 1962; Javahery 1972, 1978). The Palearctic species occurring in the Near East, southwest Asia, and the Mediterranean basin commonly are known as "Sunn pests" or "Soun pests" and feed on small grains such as wheat and barley.

The life cycle of *A. furcula* takes 2.5 months on wheat, barley, and other Gramineae. Nymphs and adults feed actively during spring and early summer. Later, the new adults migrate to high altitudes in the mountains and for approximately 9 months overwinter within or under bushes such as *Astragalus* spp. and dead leaves where high mortality may occur (M. Javahery, unpublished data). *Aelia furcula* completes one generation per year in all areas of its distribution (Putschkov 1961, Brown 1962; M. Javahery, unpublished data).

Eggs are cup-shaped, creamy-colored, and 0.8 to 0.9 mm long. They are laid in tight clusters of approximately 12 and are attached by their sides in two rows on wheat and barley heads, mostly on the awn (Javahery 1994). Incubation takes 10 to 20 days under field conditions. First instars cluster on or around the eggshells, later dispersing toward the heads. Nymphal developmental time is variable (40 to 60 days), depending on food availability and weather conditions.

3.2.2 Damage

During spring, overwintered adults attack leaves, shoots, and stems of newly formed heads of wheat and barley. Damaged leaves, central shoots, or heads change to a brownish or creamy color, and, when damaged at an early growth stage, the shoots may die before ear formation. Later, the new generation of nymphs and young adults feed on the "milky" stage of developing grain leading to a reduction in baking quality (Paulian and Popov 1980, Javahery 1995). One adult per square meter is estimated to cause a loss in yield of 40 to 340 kg/ha (3 to 12%). The damage caused by the bugs varies according to the types of wheat under cultivation; the so-called "hard" wheats are attacked 1.5 times more than "soft" wheats and, therefore, have lower economic thresholds (Paulian and Popov 1980).

3.2.3 Control

Two chemical treatments usually are applied against *A. furcula* when population densities reach the economic threshold. The first treatment is targeted to control overwintered adults at population density of $1/m^2$ during the tillering stage of wheat and barley. The second treatment is targeted against nymphs (3 to $5/m^2$) at the beginning of grain filling (Javahery 1995). Several insecticides have been used to control *A. furcula* such as the organophosphorus compounds Lebaycid 50 EC and Sumithion 50 EC (Javahery 1972, 1978, 1993). Cultural control involving resistant varieties (Javahery 1993, 1996) and two-stage harvesting (i.e., cutting plants and allowing them to cure before threshing, which will reduce maturation time and damage by *A. furcula*; Javahery 1995, 1996) are other control strategies against this pest. Natural enemies include adult (Tachinidae) and egg (Hymenoptera) parasitoids such as *Trissolcus rufiventris* Mayr (M. Javahery, unpublished data).

3.3 *Aelia germari* Kuster

3.3.1 Distribution, Life History, and Biology

This is a west Mediterranean species that attacks wheat crops in Algeria, Morocco, and north central Spain (Wagner 1960, Voegelé 1969, Bensebbane 1981). It also has been reported from Turkey (Puton 1892, Puton and Noualhier 1895), Syria (Stichel 1961), and Israel (China and Lodos 1959). During spring, overwintered adults feed for several days on wild graminaceous plants, particularly *Hordeum murinum* L., colonizing wheat fields thereafter. Eggs are laid in two parallel rows; the incubation period lasts 10 to 15 days. The nymphal stadia are 45 to 50 days, with adults of the second generation appearing in late July to early August. After harvesting, adults of this second generation migrate to higher altitudes for aestivation and subsequent hibernation. This species constitutes approximately 80% of the pentatomids in the field in Algeria and Morocco (Voegelé 1968, Bensebbane 1981).

3.3.2 Damage

Damage done to cereals is similar to that described for *A. furcula* (see **Section 3.2.2**). Also, the damage by this bug to wheat grains causes loss in baking quality and reduces the developmental time of the rice weevil, *Sitophilus oryzae* (L.), when compared with weevils fed on undamaged grains (Fourar and Fleurat-Lessard 1997).

3.3.3 Control

Many insecticides are used to control *A. germari*, Fenthion being one of the more common. Both conventional and ultralow-volume (ULV) sprays have been used to control this pest in wheat fields and in overwintering sites (Bensebbane 1981). Natural enemies include the egg parasitoid *T. grandis* Thomson, which may kill 90% of the eggs (Laraichi and Voegelé 1975).

3.4 *Aelia melanota* Fieber

3.4.1 Distribution, Life History, and Biology

This species occurs mainly in north central and northwestern Iran and in the southwestern part of the independent states of the former U.S.S.R. It also is reported from southeastern Turkey and northwestern Afghanistan (China and Lodos 1959, Putschkov 1961, Brown 1962). It is a dominant species in northern Iran where both nymphs and adults attack wheat, barley, and other Gramineae (Brown 1962).

Aelia melanota is closely related to *A. rostrata* Boheman and *A. obtusa* Fieber. *Aelia melanota* is longer and broader and somewhat paler than the other two species (Brown 1962). Eggs are laid in two rows on stalks or ears of food plants, with the first instars emerging 10 days later. Nymphal development requires 45 days under field conditions. Similar to other species of *Aelia* inhabiting Iran, *A. melanota* completes one generation per year and is migratory, moving to higher altitudes to overwinter and returning to lower altitudes during the spring to colonize cereal fields (Brown 1962; M. Javahery, unpublished data).

3.4.2 Damage

The damage done to cereals is similar to that described for *A. furcula* (see **Section 3.2.2**).

3.4.3 Control

Control measures are similar to those described for *A. furcula* (see **Section 3.2.3**). Natural enemies include tachinid flies and the egg parasitoids *T. rufiventrice* Mayr and *T. grandis* Thomson, which parasitize up to 50% of the eggs (M. Javahery, unpublished data).

3.5 *Aelia rostrata* Boheman

3.5.1 Distribution, Life History, and Biology

This species occurs mostly in the Near Eastern countries such as central and southeastern Turkey, the southwestern part of the newly independent states of the former U.S.S.R., and northeastern Iran. It attacks wheat, barley, rye, oats, and a number of other gramineaceous plants, wheat being the preferred host (Putschkov 1961, Brown 1962).

Adults are elongate, shield-shaped, and 10.5 to 11.5 mm long. New adults are pale, becoming creamy gray after reproductive dormancy in late autumn and winter. Eggs are laid in two rows, usually on stalks and ears of their food plants, with each cluster averaging 12 eggs (Putschkov 1961). The incubation period lasts 7 to 15 days under field conditions. First instars are creamy-dark and approximately 1.1 mm long; later instars have several gray or brownish lines. Nymphal development takes 35 to 60 days. *Aelia rostrata* completes one generation per year (Putschkov 1961). Adults are migratory, overwintering on mountains in *Quercus* scrub and migrating back to lower altitudes during spring to colonize cultivated cereals (Putschkov 1961, Brown 1965).

3.5.2 Damage

Damage to cereals is similar to that described for other species of *Aelia* (see *A. furcula*, **Section 3.2.2**).

3.5.3 Control

Insecticides are used to control this pest (see *A. furcula*, **Section 3.2.3**). Natural enemies include tachinid flies that parasitize adults and scelionid wasps (e.g., *T. rufiventrice* Mayr and *T. grandis* Thomson) that parasitize eggs (M. Javahery, unpublished data).

3.6 *Arvelius albopunctatus* (De Geer)

3.6.1 Distribution, Life History, and Biology

The tomato stink bug, *A. albopunctatus*, is a Neotropical pentatomid, which also occurs in the southwestern United States, Mexico, and the West Indies (Froeschner 1988). In Brazil, it is reported as an important pest of plants of the family Solanaceae including tomato, potato, and several wild species. It also feeds on sweet potato, green beans, sunflower, pepper, eggplant, and okra (Silva et al. 1968, Grazia 1977). It feeds on soybean in Argentina (Rizzo 1976) and in other countries of the Americas (see review by Panizzi and Slansky 1985a).

Adults are 11 to 12 mm long with dark spines on each side of the pronotum. Overall coloration is light yellow to green with dark spots on the pronotum and scutellum and white spots on the hemelytra (Rizzo 1968). Eggs are light yellow, barrel-shaped, and deposited in clusters of 7 to 65 (Grazia et al. 1984). First instars are oval, 1.8 to 2.0 mm long, light yellow with two gray lines on the thorax. Second instars disperse and feed on fruit, leaves, and stems. They are oval, and 2.0 to 2.9 mm long. The third, fourth, and fifth instars are oval shaped, 3.4 to 5.0, 5.2 to 7.2, and 8.3 to 10.6 mm long, respectively.

Data on the biology of *A. albopunctatus* are relatively scarce, despite the importance of this stink bug as a pest of several plants of economic importance.

3.6.2 Damage

Arvelius albopunctatus causes wilting of potato leaves by its feeding (Dias et al. 1985). Although it is called the tomato stink bug, the present authors could not find a description of its damage to this plant.

3.6.3 Control

Arvelius albopunctatus is controlled by organophosphate insecticides. It is parasitized by a tachinid fly, *Trichopodopsis pennipes* (F.) (Silva et al. 1968).

3.7 *Biprorulus bibax* Breddin

3.7.1 Distribution, Life History, and Biology

The spined citrus bug, *B. bibax*, is an important pest of citrus grown in irrigated, inland areas of southeastern Australia. Prior to the mid-1980s, it was confined to southern Queensland and northern and coastal New South Wales and rarely was considered to be of economic concern. However, during 1985 to 1990, *B. bibax* rapidly extended its range to include the major citrus-growing areas of southern New South Wales, Victoria, and South Australia. Large, damaging populations developed in these areas and substantial fruit losses were incurred, particularly on the favored host varieties, lemon and mandarin. *Biprorulus bibax* is confined to Australia but clearly would pose a serious threat in other citrus-producing areas of the world should it be introduced accidentally.

Adults are 15 to 22 mm long with a prominent black, sharp spine on each side of the pronotum. Spine tip to spine tip measures 12 to 18 mm, with females generally larger than males (James 1989). Overall coloration is lemon green, tending to paler green on the ventral surface. Older individuals tend to be considerably paler. Adults are extremely difficult to see on their native host, desert lime, *Eremocitrus glauca* Lindl., which is precisely the same shade of green and has spines similar to those of the bugs. Males and females generally occur in a 1:1 ratio (James 1989).

Eggs are laid in clusters of 6 to 36, usually in two parallel rows. The mean number of eggs per cluster ranges from 5 to 15 during summer (James 1989, 1990a), but larger clusters occur during spring (10 to 30) (James 1990a). The egg is 1 mm across, almost globular and pearly white when first laid. Red, yellow, and black markings develop as the embryo matures. The first instar is almost oval, 1.5 to 2 mm long, with a short rostrum. First instars do not feed, although some individuals have been observed penetrating unhatched eggs with their stylets. They cluster atop the eggshells until the first molt. The second instar is 4 to 5 mm long and has the general shape of the first instar. The rostrum becomes conspicuous, and the stylets invariably are held half-exerted and erect in front of the head. Second instars disperse and feed on fruit, stems, and leaves. The third to fifth instars are 4 to 6, 9.5 to 11, and 13 to 16 mm long, respectively.

This species overwinters as nonreproductive adults, with reproduction beginning in early spring (late September to early October in southern Australia and approximately a month earlier in Queensland). In the south, the first generation of nymphs matures in December (midsummer), followed by a second generation during late December to early February. Adults produced by the third generation enter reproductive dormancy in late March and April (autumn) (James 1989, 1990a). Adults occur in all months with the greatest numbers found during winter (James 1990a). Females begin ovipositing 2 to 3 weeks after maturity and can lay up to 250 eggs in their lifetime (D. G. James, unpublished data). Oviposition occurs from October to April with peaks in November (late spring) and February to March (late summer to early autumn). Nymphs are found in all months except August and September (late winter) and are most common from December to April (midsummer to midautumn). *Biprorulus bibax* is capable of developing large populations in commercial citrus orchards. James (1989) estimated adult populations of up to 11,670 individuals in a 1-ha lemon grove in southern New South Wales. Adults are long-lived with most living for at least 6 months; many of those that overwinter live for 12 months or more. One female marked in December 1987 subsequently was recaptured in July 1989 in its second winter of dormancy and probably lived for 2 years (James 1989).

Development of immature stages occurs at constant temperatures from 20 to 32.5°C with a lower developmental threshold of approximately 14°C and an upper threshold of between 35 and 37.5°C (James 1990c). An estimated 455.4 degree-days is required for complete development, and survivorship is highest between 25 to 30°C. During late March and April (midautumn), adult *B. bibax* enter reproductive dormancy in response to decreasing photoperiod and remain nonreproductive until spring (James 1989, 1990a,b, 1991). This population comprises both young adults that have not yet reproduced and individuals that formerly were gravid. Reproductive adults and developing nymphs held under a 12-hour photoperiod in the laboratory become nonreproductive dormant adults. Females appear to pass the winter in a photoperiodically maintained reproductive diapause (James 1991).

Overwintering of *B. bibax* in southern Australia is based on physiological and behavioral strategies that appear to maximize fitness and survival. Once bugs enter reproductive dormancy, they move from summer breeding sites (usually lemon groves) to adjacent overwintering sites, usually in orange, mandarin, or grapefruit trees. In a study on overwintering *B. bibax*, James (1990a) found 80% of bugs congregated on the first row of oranges adjacent to lemons. Few bugs were found in the second row and virtually none in the third row. Bug distribution among trees in the first row usually is uneven with some trees containing up to 500 bugs in two or three tightly packed clusters, whereas adjacent trees have none. Winter aggregations of nonreproductive bugs also occur on the native host *E. glauca* in southwestern Queensland (James 1993a). Interestingly, *B. bibax* adults in this region and in far western New South Wales also undergo a summer reproductive dormancy coinciding with a period when fruit often is not available on *E. glauca* due to very hot and dry conditions (James 1992b). Although most adults overwinter in aggregations, a few individuals occur singly on nonlemon or lemon citrus. Clustering bugs in southwestern New South Wales were significantly larger and contained twice as much lipid as those found alone on the same (orange) hosts (James 1990a). Clearly, these bugs were better equipped physiologically to survive the winter, and the behavioral strategy of clustering might further improve survival. It is likely that the shoulder spines of *B. bibax* evolved primarily as an aid to crypsis and not as an antipredator device. However, a large number of adult *B. bibax* clustered on a branch presents a formidable prickly mass, which may be a more effective predator deterrent than that presented by a single bug. Overwintering bugs remain inactive and in the inner canopy during cool to cold conditions. However, during milder (>15°C), sunny periods, they move to the outer canopy and bask in the sunshine. Little feeding appears to occur during winter. Overwintering colonies break up during early spring (September) with bugs usually moving back into lemon groves.

Biprorulus bibax adults in southern Australia show substantial variation in weight and energy reserves associated with the reproductive condition. During spring and summer, individuals are reproductive and have low wet and dry weights and correspondingly low lipid reserves (James et al. 1990). During autumn, reproduction ceases and body weights and lipid reserves increase. Lipid accumulation begins after induction of reproductive diapause and before bugs leave lemons. Nonreproductive bugs overwintering on oranges, mandarins, or grapefruit contain substantial amounts of lipid (mean, 50 to 120 mg/bug) that are maintained until spring; in contrast, bugs that remain on lemon during winter contain low reserves of lipid, similar to those found in summer reproductive bugs (James 1990b, James et al. 1990). Lean bugs overwintering on lemons may represent a portion of the population that is unable to enter reproductive diapause. Bugs collected from autumn populations on lemon (newly entered into reproductive diapause) and confined to either oranges or lemons as overwintering hosts show a sex difference in efficiency of lipid storage. Males show no difference in lipid storage when held on either host but females contain significantly greater amounts of lipid when held on oranges (James 1994b). This suggests that *B. bibax* females can only maintain high levels of lipid if they move to nonlemon hosts, whereas males do not have the same necessity to leave lemon groves during autumn. The accumulation and storage of energy reserves to enable survival during seasonal adversity appear to be fundamental to the success of *B. bibax* as a citrus pest. Lean dry weights of autumn/spring bugs are greater than those of summer individuals. Size of bugs is consistent throughout the year, and it is possible that an additional energy source such as glycogen or protein also may be stored by nonreproductive *B. bibax*. Water content of bugs follows a seasonal trend that is consistent with the hypothesis that young individuals maintain a better water balance than old or reproductively dormant individuals. The most highly synchronized emergence of new adults occurs in January (James 1990a) and is reflected by a steep increase in water content (James et al. 1990). As bugs age and enter reproductive dormancy, water content declines.

The major native host of *B. bibax* is desert lime or native kumquat, *E. glauca*, although populations also have been found on the finger lime, *Citrus australasica*, which is confined to coastal scrub areas of southern Queensland (Summerville 1931). *Biprorulus bibax* feeds on all varieties of commercial citrus although lemons and mandarins are clearly favored as summer hosts (Summerville 1931, Hely et al. 1982). However, since invading the citrus regions of southern Australia, many instances of reproducing populations of *B. bibax* damaging oranges have occurred (D. G. James, unpublished data). Citrus host variety significantly affects oviposition, fecundity, and longevity of *B. bibax* females (James 1992a). Immature and mature oranges and lemons and immature mandarins produce good survival and oviposition, whereas grapefruit, calomondin, kumquat, and trifoliata do not. Different strains of *B. bibax* from native and commercial hosts demonstrate considerable differences in performance on different citrus host varieties. Some strains are well adapted to reproducing on oranges, whereas others are not. *Biprorulus bibax* can survive, but not reproduce, on a noncitrus diet. Adult bugs have been kept alive in the laboratory for many months on various potted houseplants (D. G. James, unpublished data). This ability to utilize the sap of noncitrus hosts clearly is an important trait enabling prolonged survival when citrus hosts are scarce. In the native habitat of *B. bibax* (Western Queensland and Western New South Wales), host tree (*E. glauca*) patches invariably are scattered and separated by long distances.

3.7.2 Damage

Biprorulus bibax primarily feeds on and damages immature lemons, mandarins, and oranges with a preference for the first two citrus varieties. However, fruit at any stage can be attacked and damaged. Bugs also will feed on stems, leaves, and flowers, but these suffer no apparent damage. The first sign of damage to immature fruit often is gum extruding through the puncture that the bug has made. This is followed by premature coloring of the fruit and fruit drop. Internally, fruits show drying, staining, and gumming of segments. Immature mandarins fall soon after attack. Mature or semimature lemons and mandarins tend to hang on the tree for a long period after bug attack but show the same drying, staining, and gumming of segments. Lemons often are harvested with considerable internal damage yet appear undamaged externally. Mature oranges rarely show any damage internally. Most damage to lemons and mandarins occurs during late summer to autumn although, where large overwintering populations occur, damage can be severe in spring.

3.7.3 Control

An integrated management program for *B. bibax* recently has been developed and now is used widely in the Australian citrus industry (James 1994a). Management is based on conservation of natural enemies, physical or chemical treatment of overwintering populations, and strategic low-rate applications of a selective insecticide. Estimated adult populations of 10 to 35,000/1.5 ha and 70 to 90% fruit damage in untreated crops were reduced to <500/1.5 ha and <5%, respectively, following implementation of the management program in orchards in southern New South Wales (James 1994a). Prior to the development of this integrated management program, a serious impact of *B. bibax* on citrus production was disruption to the citrus ecosystem by application of broad-spectrum insecticides for its control. Many citrus orchards in southern Australia had not been exposed to insecticides for 15 to 20 years before the arrival of *B. bibax*, and broad-spectrum treatments for this pest threatened the biological control of other citrus pests such as scale insects.

Biprorulus bibax is subject to attack by a large number of parasitoids and predators. Hymenopterous egg parasitoids are the single most effective group of natural enemies affecting this bug. At least 13 species of egg parasitoids have been recorded from *B. bibax*: *Trissolcus oenone* (Dodd) (= *T. biproruli* Girault), *T. flaviscapus* Dodd, *T. latisulcus* (Crawford), *T. mitsukurii* (Ashmead), *T. ogyges* (Dodd), *T. oeneus* (Dodd), *Psix glabiscrobis* (Girault) (Scelionidae), *Anastatus biproruli* (Girault) (Eupelmidae), *Acroclisoides tectacorisi* Girault and Dodd (Pteromalidae), *Centrodora darwini* (Girault) (Aphelinidae), *Ooencyrtus* sp. n., *Ectopiognatha major* Perkins, and *Xenoencyrtus hemipterus* (Girault) (Encyrtidae) (Summerville 1931, James 1993b, Johnson 1991). Spring and early summer parasitism by hymenopterous parasitoids is effective in inhibiting *B. bibax* population development and preventing fruit damage (James 1990d, 1994a). The abundance of egg parasitoids of *B. bibax* in southern Australia appears to be increasing and may be part of a natural strengthening of the natural enemy complex associated with this bug following its invasion into southern citrus-growing areas. A number of releases of *T. oenone* were made in some citrus areas soon after the arrival of *B. bibax* (James 1994a). The reduction in effectiveness of parasitism during late summer (January to February) is a fundamental problem in biological control of *B. bibax* and usually requires chemical intervention to prevent economic damage. The efficacy of egg parasitoids may be reduced in the hot, dry conditions that prevail during late summer in southern Australia. It also is possible that some of these parasitoids may have a summer reproductive diapause that ensures synchronization with *B. bibax* in the ancestral habitat (James 1992b). Although southern populations of *B. bibax* do not appear to have retained summer reproductive diapause, it is possible that the parasitoids have.

Many generalist predators (e.g., mantids, predatory pentatomids, spiders, ants, lacewing larvae, assassin bugs, and mirid bugs) feed on all stages of *B. bibax*, and this complex appears to be highly important in regulation of the bug's populations. One general predator, in particular, the assassin bug, *Pristhesancus plagipennis* Walker, is considered to play an important role in regulation of *B. bibax* populations in Queensland (James 1992c).

Laboratory bioassays with endosulfan against *B. bibax* show high toxicity of this compound even at very low rates (James 1993c). Orchard applications of endosulfan at 8 to 10 ml/100 l control *B. bibax* and prevent resurgences of damaging populations for 11 to 12 weeks. In addition, these rates have low toxicity to a number of beneficial insects including the egg parasitoids of *B. bibax* (James 1993c). The use of endosulfan against *B. bibax* is recommended only for treatment of late summer populations that are not being regulated by parasitoids, and for winter populations.

Control of overwintering *B. bibax* is an important part of integrated management of this pest. Most problems with *B. bibax* arise following spring dispersal of large overwintering populations. Low spring populations enhance the efficacy of biological control and ensure that populations remain below damaging levels for a longer period of time. The well-defined reproductive dormancy and aggregation behavior of adult *B. bibax* enable effective targeting of large populations with minimal cost, effort, or disruption to the ecosystem. Considerable numbers of *B. bibax* can be collected by hand (up to 500 per tree), or an endosulfan spray can be applied to the rows of trees that harbor bugs. In most instances, 80% or more of the overwintering population is found in the first row of nonlemon citrus adjacent to lemons (James 1990b). A higher rate of endosulfan (30 ml/100 l) is recommended for overwintering *B. bibax* as individuals often are inactive. It also is important to spray during mild, sunny weather in early or late winter when *B. bibax* individuals are more likely to be near the outside of the tree canopy.

Improvements in the management of *B. bibax* by reducing dependence on endosulfan are being investigated currently. The main area of research centers on development of the aggregation pheromone and/or alarm pheromones as management tools (James et al. 1994).

3.8 Dolycoris baccarum L.

3.8.1 Distribution, Life History, and Biology

The berry bug, *D. baccarum*, a pentatomid of western Mediterranean origin (Polivanova 1957), occurs throughout the Palearctic region. It is an established pest in the former U.S.S.R., Scandinavia, Germany, and most Mediterranean countries and can cause significant damage when its populations are high. Although known mainly as a cereal pest, it is polyphagous, attacking a variety of crops including sunflower, tobacco, cherry, bean, potato, and artichoke.

The biology of *D. baccarum* has been studied in northern Germany (Tischler 1938, 1939), the former U.S.S.R. (Kamenkova 1958), and southern Norway (Conradi-Larsen and Somme 1973). In northern locations (Scandinavia, U.S.S.R.), *D. baccarum* has one generation per year, whereas in more southerly regions two generations may occur. Under laboratory conditions, development takes 28 to 33 days at 29°C (Perepelitza 1969), and the first four nymphal stadia last 20 days at 23 to 27°C (Tanskii 1971). At 21°C, development from egg to adult takes 48 to 52 days (Conradi-Larsen and Somme 1973). Fecundity is influenced strongly by diet of the females. It is greater on sunflower than tobacco, but the most favorable diet appears to be a mixed one (e.g., tobacco, sunflower, henban, and weed species) (Kamenkova 1958). Overwintered adults begin ovipositing during May/June in northern regions. In Cyprus, mating of overwintered bugs occurs by the end of March and oviposition begins in April (Krambias 1987). In Israel, oviposition begins in March (Yathom 1980). Egg clusters average 23 to 25 eggs. In Scandinavia, nymphs occur from early June until the end of August, and the first adults appear at the end of July (Conradi-Larsen and Somme 1973). In Cyprus, *D. baccarum* overwinters as nonreproductive adults at low altitudes under stones or in thorny bushes and estivates during June to November in large aggregations at higher elevations (Krambias 1987). Adults may live for up to 1 year in Cyprus, although the feeding period is limited to 2.5 months, 15 days of which are the most critical for crop damage (Yathom 1980). Prior to estivation, *D. baccarum* may cause serious damage to crops. Babrakzai and Hodek (1987) demonstrated that reproductive dormancy in a population of *D. baccarum* in the Czech Republic was a case of facultative reproductive diapause, regulated primarily by photoperiod. In contrast, the results of Conradi-Larsen and Somme (1973) suggest that *D. baccarum* in Scandinavia undergoes an obligatory diapause.

3.8.2 Damage

Dolycoris baccarum attacks cereal grains in the milk stage. On sunflower, flowering heads and stems are attacked with the stems extruding sap and sometimes breaking at the affected site. Up to 50 bugs may occur on a single flower head, feeding on the developing seeds. Infested plants produce degenerated, empty seeds (Yathom 1980). Leaves that are attacked develop light-colored spots at the punctured loci, wilt, and desiccate. *Dolycoris baccarum* is highly polyphagous and many plants have been recorded as hosts.

3.8.3 Control

Current control measures of *D. baccarum* include the use of insecticides such as diazinon or fenitrothion. In the past, hexachlorocyclohexane was applied against estivating *D. baccarum* in

Cyprus (Krambias 1987). The use of smoke generators (DDVP/gas oil) in areas covered with plastic has given good control. A further improvement has been achieved by placing a substrate on the soil surface, which encourages the bugs to aggregate. The bugs assemble and remain below the substrate instead of stones. In this way, it is not necessary to turn numerous stones to expose the bugs to insecticide applications. Natural enemies, particularly egg parasitoids, appear to be important in population regulation of *D. baccarum*. Kamenkova (1958) reported parasitism of 72.5 to 98.3% in the summer egg generation but did not provide information on identity of the parasitoids. In Cyprus, Morris (1929) recorded the tachinid fly *Gymnosoma rotundatum* L. as an important parasitoid of *D. baccarum* with approximately 50% of hibernating bugs affected. In Morioka, Japan, Honda (1985) found 15 to 33% of adults of *D. baccarum* parasitized by the tachinid fly *Cylindromyia brassicaria* (F.). Krambias (1987) reported that the eggs of *D. baccarum* are "often infested by *Telenomus* sp." (Scelionidae).

3.9 *Edessa meditabunda* (F.)

3.9.1 Distribution, Life History, and Biology

This Neotropical pentatomid is a pest of many species of Solanaceae, including tomato and potato, and of Leguminosae, including peas, soybean, and alfalfa. It also feeds on cotton, eggplant, tobacco, sunflower, papaya, and grapes (Silva et al. 1968, Rizzo 1971, Lopes et al. 1974). Among the hosts, soybean is perhaps the most important. In certain areas of the southernmost Brazilian state of Rio Grande do Sul, *E. meditabunda* is the second most abundant pentatomid species on soybeans, particularly at the beginning of the crop reproductive period (Galileo et al. 1977). It also is abundant in the western areas of soybean cultivation in Brazil (A. R. Panizzi, unpublished data). In Argentina, it is considered an important pest of this crop (Rizzo 1972).

Rizzo (1971) described the morphology of the developmental stages. Adults are oval and green with dark brown hemelytra; males and females are 11.8 to 12.2 and 12 to 13 mm long, respectively. Eggs are light green, oval, 1.5 mm long, and laid in parallel rows. Egg clusters contain 12 to 14 eggs firmly stuck together. The first to fifth instars are 2.0 to 2.2, 3.0, 4.6 to 4.9, 7.0 to 9.0, and 9.5 to 10.5 mm long, respectively.

Nymphal development usually lasts 50 to 65 days (Rizzo 1971), but this is strongly dependent on the food ingested. For example, total developmental time (from the second stadium to adult) may take from 47 days on soybean leaves to 78 days on sunflower fruit; usually, nymphal mortality is high under laboratory conditions (Panizzi and Machado-Neto 1992). Two to three generations per year occur in Argentina (Rizzo 1971).

3.9.2 Damage

Edessa meditabunda prefers to feed on vegetative tissues. Intense feeding activity on potato plants may cause plant death (Rizzo 1971). This bug also causes cotton squares to rot by transmitting the yeast *Nematospora gossypii* (Mendes 1956). *Edessa meditabunda* feeds on soybean stems, which become dark at the feeding sites (Galileo and Heinrichs 1979, Panizzi and Machado-Neto 1992). Its preference for stems is unique among the species of stink bugs that colonize soybean worldwide (Kogan and Turnipseed 1987) and may represent its use of an unfilled niche. This preference for vegetative plant tissue somewhat decreases its effect on seed yield, as reported for soybean (Costa and Link 1977).

3.9.3 Control

There is little information available on control measures for this pest species. On soybean, chemical insecticides such as monocrotophos and endosulfan are recommended against stink bugs, including *E. meditabunda* (see control under *N. viridula*, **Section 3.17.3**). In Argentina, two natural enemies are reported: *Bycyrtes discisa* (Tachenberg) (Hymenoptera: Sphecidae) and *Trissolcus caridei* Bréthes (Hymenoptera: Scelionidae) (De Santis and Esquivel 1966); in Uruguay, a tachinid fly *Xenopyxis edessae* (Townsend) is mentioned (Silva et al. 1968); in Brazil, *Telenomus edessae* Bréthes was found to parasitize 7% of eggs collected in soybean fields (Corrêa-Ferreira 1986); and in southern Brazil, *Trissolcus urichi* (Crawford) was reported from *E. meditabunda* eggs (Corrêa-Ferreira and Moscardi 1995).

3.10, 3.11 *Eurydema rugosum* Motschulsky and *Eurydema pulchrum* Westwood

3.10.1, 3.11.1 Distribution, Life History, and Biology

These cabbage stink bugs are important pests of crucifers in the Orient, particularly in Japan. Several species in this genus also are pests of crucifers in the former U.S.S.R., central Europe, and central and east Asia (e.g., Rusanova 1926, Petropavlovskaya 1955, Linnavuori 1993). *Eurydema pulchrum* also has been recorded from India (Pajni and Sidhu 1982) but does not appear to be a pest in this country.

Eurydema rugosum females lay clusters of approximately 12 eggs, which hatch in 6 to 7 days at 25°C (Kiritani and Kimura 1966). The first instars form a dense aggregation on or around the eggshells. Second instars leave the eggshells but remain aggregated. Some aggregative behavior persists throughout nymphal development, which lasts 26 to 27 days at 25°C (Kiritani and Kimura 1966). *Eurydema rugosum* and *E. pulchrum* overwinter as adults in reproductive diapause, which is induced by short-day photoperiod and/or host plant status (Numata and Yamamoto 1990). In spring, temperatures above 13.2°C are required to initiate reproduction, and egg laying usually occurs from mid-April onwards (Ikeda-Kikue and Numata 1992). Host plant status does not influence reproductive activity in spring as it does in autumn. In central Japan, *E. rugosum* produces three generations annually, whereas *E. pulchrum* has two. Populations of *E. rugosum* generally reach a peak in early June, whereas those of *E. pulchrum* peak in mid-October; however, there is much variation from year to year (Morimoto et al. 1991). In dense stands of crucifers, adult and nymphal *E. rugosum* tend to aggregate on the upper parts of the plants; however, if plants are sparse, bugs generally are distributed evenly from the lower to upper parts of the plants (Okamoto and Kuramochi 1984). This probably is a phototactic response similar to the basking behavior described for *N. viridula* (Waite 1980).

3.10.2, 3.11.2 Damage

Bugs in this genus are pests of all crucifers, feeding on leaves, seeds, flowers, and stems, causing shriveling and desiccation of affected parts. Some preference is shown for feeding on flowers and young pods.

3.10.3, 3.11.3 Control

Little information is available on the control of these pests. Insecticides appear to be the primary method of control. Natural enemies of *Eurydema* spp. do not appear to have been studied.

3.12 *Euschistus heros* (F.)

3.12.1 Distribution, Life History, and Biology

This is a Neotropical pentatomid occurring in South America and possibly in Panama (Rolston 1974). Most reports refer to its distribution in Brazil where it is a major component of the pentatomid–pest complex on soybean. Williams et al. (1973) probably were the first to record

E. heros on soybean in São Paulo. Initially considered a secondary pest, *E. heros* more recently has increased in numbers and may inflict severe damage to soybean. It is the predominant species on this crop in some regions of São Paulo (Calcagnolo et al. 1977, Rodini and Grazia 1979) and Mato Grosso do Sul (Degaspari and Gomez 1979). It seems to be adapted to warmer regions (Cividanes 1992), being more abundant from north of Paraná to the central and western regions of Brazil. *Euschistus heros* may occur in low numbers in the southernmost state, Rio Grande do Sul (Link 1979). Corrêa et al. (1977) recorded it in several states, with greatest abundance in Goiás.

Besides soybean, this insect feeds on several other plants in Brazil, including species of Leguminosae, Solanaceae, Brassicaceae, and Compositae (Link 1979, Ferreira and Panizzi 1982, Link and Grazia 1987, Malaguido and Panizzi 1998, Panizzi and Oliveira 1998). Even though it is polyphagous and, therefore, able to colonize alternative hosts during the mild winter of northern Paraná state, *E. heros* also overwinters underneath dead leaves in this area (Panizzi and Niva 1994). It accumulates lipids during overwintering, which allows it to survive during nonfeeding periods and to break dormancy with the start of favorable conditions (Panizzi and Hirose 1995, Panizzi and Vivan 1997). During summer, there are numerous weeds associated with the soybean crop, and this bug apparently is adapted to exploit some of these weeds as a nutritional resource including the euphorb *Euphorbia heterophylla* L. (Meneguim et al. 1989) and the star bristle, *Acanthospermum hispidum* DC; this latter plant already has been studied with regard to its role in the phenology of *E. heros* (Panizzi and Rossi 1991a).

Data on the biology of *E. heros* are relatively scarce due to this species' minor importance previously as a pest of economic crops. However, with the bug's increasing incidence on soybean, its biology on this host and on alternate hosts such *Euphorbia heterophylla* has been studied (Villas Bôas and Panizzi 1980, Pinto and Panizzi 1994). Depending on the type of food ingested and whether or not a switch in food from nymph to adult has occurred, reproductive performance of females will vary. In general, females will lay 6 to 11 egg clusters (although it can be as high as 34) during their life spans, with a total of 61 to 99 eggs/female (a high of 287 eggs/female has been reported).

Grazia et al. (1980) described the diagnostic characteristics of the various stages of *E. heros*. The first to fifth instars are 1.47 to 1.72, 1.81 to 2.55, 3.36 to 3.98, 4.84 to 5.95, and 6.87 to 12.08 mm long, respectively. Summer adults have longer pronotal spines and usually are dark brown compared with overwintering adults, which have shorter pronotal spines and usually are light brown (Panizzi and Niva 1994, Mourão 1999).

3.12.2 Damage

Euschistus heros feeds on pods of soybeans. The extent of the damage is related to the stage of seed development. During early pod and seed development, feeding generally results in pod-abscission or abortion of young seeds; during pod-fill, it results in shriveled and deformed seeds; during seed maturation, it results in little deformation of the seed. Feeding activity on soybean will delay leaf maturation, cause foliar retention, and cause development of abnormal leaflets and pods close to the main stem. Transmission of microorganisms also can occur (see detailed description of damage to soybean under *N. viridula*, **Section 3.17.2**). This bug's damage to sunflower seeds has been studied (Malaguido and Panizzi 1998). Yield (kg) and weight of 1,000 seeds (g) were reduced significantly with infestations of 8 bugs/plant from R3 (internode immediately below the reproductive bud continues to lengthen, lifting the inflorescence head above the surrounding leaves in excess of 2 cm) to harvest.

3.12.3 Control

Control measures against *E. heros* mainly involve the use of chemical insecticides such as monocrotophos and endosulfan. In addition, this bug is managed efficiently in southern Brazil by the release of the egg parasitoid *Trissolcus basalis* (Wollaston) (Corrêa-Ferreira 1993). Additional egg parasitoids of *E. heros* are *Telenomus mormideae* Lima, *Telenomus* sp., *Trissolcus scuticarinatus* Lima, *Ooencyrtus* sp., and *Neorileya* sp. (Corrêa-Ferreira 1986). Recently, Corrêa-Ferreira et al. (1998) reported adult *E. heros* parasitized by the encyrtid *Hexacladia smithi* Ashmead in northern Paraná state. In this same area, Panizzi and Oliveira (1999) found the tachinids *Trichopoda giacomellii* (Blanchard) (= *Eutrichopodopsis nitens* Blanchard) and *Hyalomyodes* sp. attacking *Euschistus heros*; parasitism reached 37% but did not occur from July to October, when the bugs hid beneath dead leaves on the ground. An effective pest management program to control this and other species of heteropteran pests of soybeans in Brazil is discussed (see details under *N. viridula*, **Section 3.17.3**).

3.13 Euschistus servus (Say)

3.13.1 Distribution, Life History, and Biology

The brown stink bug, *E. servus*, is the most economically important of the *Euschistus* species occurring in America north of Mexico. According to McPherson (1982) it is divided into two subspecies, *E. s. servus* (Say) and *E. s. euschistoides* (Vollenhoven). The first subspecies is distributed in the southeastern United States west through Louisiana, Texas, New Mexico, and Arizona into California. The second subspecies occurs across the northern part of the continent (see references in McPherson 1982). This stink bug completes two generations per year and overwinters as adults (McPherson and Mohlenbrock 1976). The eggs (Esselbaugh 1946, Deitz et al. 1976, Munyaneza and McPherson 1994) and first to fifth instars (DeCoursey and Esselbaugh 1962, Munyaneza and McPherson 1994) have been described. It has been reared in the laboratory, and its mating behavior has been studied (Youther and McPherson 1975).

3.13.2 Damage

Euschistus servus causes yield and quality losses to several crops including soybean (Daugherty et al. 1964, McPherson et al. 1979), corn (Townsend and Sedlacek 1986; Sedlacek and Townsend 1988; Apriyanto et al. 1989a,b), cotton (Barbour et al. 1988), alfalfa (Russell 1952), sorghum (Wiseman and McMillian 1971, Hall and Teetes 1981), fruit (Chandler 1950, Rings 1957, Phillips and Howell 1980), pecan (Dutcher and Todd 1983, Yates et al. 1991), and tobacco (Reich 1991). Its feeding on pecan causes fruit drop (Dutcher and Todd 1983).

3.13.3 Control

See control measures under *N. viridula* (**Section 3.17.3**). Natural enemies include the encyrtid parasitoid *Hexacladia smithi* Ashmead; the egg parasitoids *Telenomus podisi* Ashmead, *Telenomus ashmeadi* Morrill, *Trissolcus basalis* (Wollaston), and *T. euschisti* (Ashmead); and several species of tachinid parasitoids of adults (see references in McPherson 1982).

3.14 Lincus lobuliger Breddin

3.14.1 Distribution, Life History, and Biology

Among the 33 species of *Lincus* described (Rolston 1983a, 1989), *L. lobuliger* is one of the most important pests of palm trees. *Lincus* spp. are Neotropical and vectors of diseases of cultivated palms in South America caused by flagellates (*Phytomonas* sp.).

Lincus lobuliger is restricted to the northeast areas of Brazil (Rolston 1983a). It is a darkish pentatomid, approximately 9.4 to 12.0 mm long. Recently, Moura and Resende (1995) described its feeding behavior on coconut, *Cocos nucifera* L. Adults feed on the axils of the leaf petioles, sucking the sap of the epidermis of young leaves near the trunk. They also feed on African oil palm, *Elaeis guineensis* Jacq. Little is known of this bug's biology. Llosa et al. (1990) studied the ecology of two other species, *L. spurcus* Rolston and *L. malevolus* Rolston, on native palm species of the Peruvian Amazonia. They found that the bugs hide between the spines and the brown fibers of the leaf sheaths. As a result, it is difficult to see and capture them; however, they are detected easily because of their characteristic odor. Adults are attracted by the aromatic volatile compounds of the inflorescences, new palms being colonized as a result. Adults are photophobic and probably disperse by night flight.

3.14.2 Damage

The damage by *L. lobuliger* to coconut trees in northeast Brazil is severe. The major damage resulting from its feeding is the transmission of the flagellate protozoan *Phytomonas* sp., which causes the palm leaves to yellow and the progressive death of the inflorescences, then fruit fall, and finally plant death (Resende et al. 1986, Resende and Bezerra 1990). *Lincus* spp. are reported to cause "sudden death" in oil palms in Peru by transmitting *Phytomonas* sp. If the bugs are not controlled, 50 to 90% of the plants can be destroyed (Liceras and Hidalgo 1987). Dolling (1984) reported *Lincus* spp. transmitting *P. staheli* McGhee & McGhee to cultivated palms in South America; this flagellate causes the disease 'marchitez' in oil palms in Ecuador and hartrot (or heartrot) in coconut palms in French Guiana. In Surinam, *Lincus* spp. cause hartrot in coconut palms (Asgarali and Ramkalup 1985).

3.14.3 Control

Chemical control seems to be the most effective control measure against *L. lobuliger*. In Brazil, Moura and Resende (1995) reported that monocrotophos applied via root was 100% effective until 15 days, when toxicity started to decrease. After 25 days, insects were no longer killed. Positive results also were obtained by Reyes et al. (1988) using monocrotophos against a tingid, *Leptopharsa gibbicarina* Froeschner, on palm oil trees in Colombia. Other pests of coconut also are controlled effectively with monocrotophos via root, such as lepidopterans in Indonesia (Ginting and Chenon 1987).

Little is known about the natural enemies of *L. lobuliger*. Moura and Resende (1995) referred to a complex of egg parasitoids in northeast Brazil, but no data are available on the species involved or their impact on *L. lobuliger* populations. Rasplus et al. (1990) mentioned *Hexacladia linci* Rasplus (Hymenoptera: Encyrtidae) parasitizing nymphs and adults of *L. malevolus* Rolston in Peru.

3.15 *Mormidea quinqueluteum* (Lichtenstein)

3.15.1 Distribution, Life History, and Biology

This Neotropical pentatomid occurs in Argentina, Brazil, Paraguay, and Uruguay. Adults are fuscous to black dorsally, light castaneous ventrally, and 7.3 to 8.9 mm long (Rolston 1978). Martins et al. (1986) described the immature stages of *M. quinqueluteum*. The first to fifth instars are 1.38, 1.65, 2.72, 4.10, and 5.74 mm long, respectively. Lima (1940) refers to its occurrence in Brazil on several species of cultivated Gramineae. It is a major pest of rice in southern Brazil (Cruz and Corseuil 1970) and has been reported on soybean (Costa and Link 1974). Little is known of its biology.

3.15.2 Damage

The damage to rice by *M. quinqueluteum* is similar to that described for other major pests of this crop (see more details under discussion of *Oebalus poecilus*, **Section 3.18.2**).

3.15.3 Control

Mormidea quinqueluteum is controlled on rice efficiently by chemical insecticides (see further details under *O. poecilus*, **Section 3.18.3**).

3.16 Mormidea notulifera Stål

3.16.1 Distribution, Life History, and Biology

This stink bug occurs in Argentina, Brazil, Peru, and Uruguay; adults are light brown to fuscous dorsally and 7.0 to 8.5 mm long (Rolston 1978). It is a pest of rice but also feeds on oat, wheat, and lupin (Silva et al. 1968, Link and Grazia 1987). Weber et al. (1988) described the immature stages of *M. notulifera*. The first to fifth instars are 1.28, 1.72, 2.61, 3.78, and 4.86 mm long, respectively. Few data were found related to its biology. Weber et al. (1988) recorded the developmental time from egg to adult as 23 to 30 days.

3.16.2 Damage

Mormidea notulifera is one of the main pests of rice in southern Brazil. Damage to this crop is similar to that described for *O. poecilus* (see **Section 3.18.2**).

3.16.3 Control

Mormidea notulifera is controlled effectively, as are the other pentatomid pests of rice in southern Brazil, with chemical insecticides (see details under *O. poecilus*, **Section 3.18.3**).

3.17 Nezara viridula (L.)

3.17.1 Distribution, Life History, and Biology

The southern green stink bug, *Nezara viridula*, has a worldwide distribution, occurring throughout the tropical and subtropical regions of Europe, Asia, Africa, and the Americas (Lethierry and Severin 1893). A distribution map published by the Commonwealth Institute of Entomology was reproduced by DeWitt and Godfrey (1972) and updated later (Todd and Herzog 1980). To this distribution should be added California in the United States and some areas in South America, wherein the bug is expanding its distribution, particularly in Paraguay, south Argentina, and the central west toward the northeast of Brazil. The expansion in South America is the result of increased acreage for soybean production (Panizzi and Slansky 1985a; Hoffmann et al. 1987; A. R. Panizzi, unpublished data).

Nezara viridula is highly polyphagous, attacking both monocots and dicots. Kiritani et al. (1965) mentioned that as many as 145 species within 32 plant families have been recorded as hosts. Since 1965, many more plants, including cultivated and uncultivated species, have been recorded as hosts worldwide (see references in Todd 1989, Panizzi 1997). However, this stink bug shows a preference for leguminous plants (Todd and Herzog 1980, Panizzi 1997, and references therein).

There are several reports on sequences of host plants used by *N. viridula* during the season. For example, in the United States, adults leave overwintering shelters during spring and begin feeding and ovipositing in clover, small grains, early spring vegetables, corn, tobacco, and weed hosts, where they complete the first generation. A second generation is completed on leguminous weeds, vegetables and row crops, cruciferous plants, and okra, which are typical midsummer hosts. Third-generation adults move into soybean where they complete the fourth and fifth generations (Todd 1989, see also references therein). Another example of host plant sequence for *N. viridula* was reported in Paraná state, southern Brazil (Panizzi 1997). The bug concentrates on soybean during the summer, which is very abundant at this time, but also is found on common bean, *Phaseolus vulgaris* L.; two or three generations are completed on these crops. During fall, adults move to wild hosts, which include the star bristle, *Acanthospermum hispidum*, and the castor bean, *Ricinus communis* L., where they feed but do not reproduce. They also move to wild legumes such as the beggar weed, *Desmodium tortuosum* (Swartz) DC, and *Crotalaria* spp., where they feed and reproduce, with the resulting offspring completing a fourth generation. During late fall and early winter, *N. viridula* completes a fifth generation on radish, *Raphanus raphanistrum* L., and mustards, *Brassica* spp. During winter, it may feed but will not reproduce on wheat, *Triticum aestivum* L. During spring, a sixth generation is completed on Siberian motherwort, *Leonurus sibiricus* L. Many other cases of host plant sequences for this bug have been studied elsewhere, indicating the importance of knowing these sequences to manage this pest species better. The literature on wild hosts and the ecological significance and role they play in the pest status of this stink bug and other pentatomids on crops was reviewed recently (Panizzi 1997).

Several laboratory studies have examined the rearing procedures for, and development of, *N. viridula* (e.g., Harris and Todd 1980a, 1981; Brewer and Jones 1985; Jones 1985; Panizzi and Rossini 1987; Panizzi and Meneguim 1989; Panizzi and Saraiva 1993; Panizzi and Hirose 1995). Nymphs feeding on developing pods appear to survive better on seeds that are more mature than on those at the beginning or in mid-development (Panizzi and Alves 1993). Both pods and seeds of legumes generally provide a suitable food source for nymphs, whereas survival is poor on leaves and stems (Panizzi and Slansky 1991). Among the many food plants, a privet species, *Ligustrum lucidum* Ait. (which was misidentified as *L. japonicum* Thunb.), recently was found to be a suitable food plant for rearing *N. viridula* in the laboratory, using immature fruits (berries) (Panizzi et al. 1996). On this plant, *N. viridula* shows greater mating and ovipositional activity, resulting in higher fecundity when compared with soybean, one of its preferred hosts (Panizzi and Mourão 1999). This plant, which is common in southern areas of Brazil, hosts over a dozen species of pentatomids; these species feed on this plant and most are able to reproduce on it (A. R. Panizzi, unpublished data).

Attraction of *N. viridula* males to females first was demonstrated by Mitchell and Mau (1971). Pavis and Malosse (1986) isolated compounds emitted by males that elicit long-range behavioral sequences in females leading them to the vicinity of males (Borges et al. 1987). The components of the male-produced pheromone were identified (Aldrich et al. 1987). This pheromone appears to be a sex pheromone that attracts females only (Brézot et al. 1993), although Harris and Todd (1980b) demonstrated long-range field attraction of both sexes and fifth instars to males. Subsequent short-range courtship is believed to be mediated by sound communication through substrate-borne vibrations (Čokl et al. 1972, 1978; Ryan and Walter 1992). Males begin producing a pheromone at approximately the same time as they become sexually mature, usually 7 to 10 days after adult ecdysis. Females become responsive to the male pheromone 4 days after adult ecdysis. Reproductively dormant males do not produce the pheromone and nonreproductive females are unresponsive to the pheromone (Brennan et al. 1977).

The behavior of the nymphs, including gregariousness, feeding, and defensive behavior; and of the adults, including dispersion and aggregation, mating, defense, and diapause has been reviewed (Todd and Herzog 1980, Todd 1989). Adults are green and 12 mm long. Descriptions and illustrations of the eggs, nymphs, and adults have been published by many authors (e.g., Jones 1918, Drake 1920, Kobayashi 1959, Rizzo 1968, Todd and Herzog 1980). Nymphs in the early instars are strongly gregarious, but the gregarious habit disappears in the fourth instar (Panizzi et al. 1980). During the summer, the developmental time from egg to adult is approximately 35 days.

3.17.2 Damage

The more recent reviews on the damage caused by *N. viridula* include DeWitt and Godfrey (1972), Todd and Herzog (1980), Panizzi and Slansky (1985a), and Todd (1989). Significant reductions in yield, quality, and germination can result from feeding by this pest (e.g., Jensen and Newsom 1972; Nilakhe et al. 1981a,b; Schalk and Fery 1982; Negron and Riley 1987; Lye et al. 1988a). The oil and protein content of soybean seeds apparently is unaffected by moderate to heavy levels of feeding (Thomas et al. 1974), but the chemical composition of the fatty acids can change (Todd et al. 1973). Oil and protein can affect nymphal developmental time and adult weight (Calhoun et al. 1988). Also, soybean seeds damaged by *N. viridula* are more susceptible to attack by stored product pests (Todd and Womack 1973); damaged seeds can have an increased incidence of pathogenic organisms (Ragsdale et al. 1979, Payne and Wells 1984, Russin et al. 1988, Reilly and Tedders 1989). The feeding punctures on seeds cause minute darkish spots, and chalky-appearing air spaces generally are produced when the cell contents are withdrawn; later, dark discoloration may surround the punctures and the inner membrane of the seed coat may be abnormally fused to the cotyledons (Miner 1966).

Nezara viridula apparently feeds preferentially on soybean seeds in the upper half of the plant until high infestations develop, and damage-free seeds compensate for damaged seeds with increases in 100-seed weight (Russin et al. 1987). Feeding damage potential is comparable for adults and fifth instars; less damage is caused by third and fourth instars (McPherson et al. 1979). The damage caused by adults also is variable with age and physiological condition of the bugs. For example, the number of feeding punctures made by adults increased from day 1 to day 4 and leveled off thereafter; adults previously deprived of food for 24 hours made the same number of punctures as those fed continuously, but starved adults fed longer than adults not starved (Panizzi 1995). Also, *N. viridula* feed more frequently on seeds closest to the pedicel compared with other seeds (Panizzi et al. 1995).

The effects of *N. viridula* feeding on corn are more severe on younger plants, and yield reductions are attributed to total ear loss rather than reduction in kernel weight (Negron and Riley 1987). On tomato, the bug's feeding is associated mostly with reduced quality (Lye et al. 1988b); the bugs prefer mature green tomatoes to mature red tomatoes if the fruits are of equal size (Lye and Story 1988).

Nezara viridula also attacks nut crops. It is one of the most damaging pests of macadamia nuts (Mitchell et al. 1965, Jones and Caprio 1992), causing nut abortion in small nuts and quality damage in larger nuts in the trees or on the ground (Jones and Caprio 1994). When caged on pecan clusters before shell-hardening in Georgia, *N. viridula* caused 34% fruit drop from mid- to late June and 63% fruit drop from late June to mid-July (Dutcher and Todd 1983). Its feeding resulted in black pit and kernel spot, with a mean volume of kernel spot feeding sites of 5.7 mm^3; the average ingestion of adults feeding on pecan kernels is 14 calories/feeding site compared 6.5 calories/feeding site on soybean seeds (Dutcher and Todd 1983).

In recent years, *N. viridula* has been reported as a pest of flue-cured tobacco (Reich 1991). Both adults and nymphs extract plant fluids from younger tender growth, causing the leaves to wilt and flop over due to loss of turgor pressure. Generally, the plants recover from this injury, even at maintained population densities of five adults per plant (R. M. McPherson, unpublished data), although some leaves dry up similarly to sunscald.

3.17.3 Control

Several tactics have been studied and implemented to manage *N. viridula* on many crops. The use of chemical control still is the major tool used by growers, but many other tactics within the context of cultural practices and biological control have been investigated and used. For chemical control, chlorinated hydrocarbon, organophosphate, and organochlorophosphate insecticides, such as methyl parathion, monocrotophos, trichlorphon, dimethoate, and endosulfan have been recommended (see references in Jackai et al. 1990). In Brazil, some of these insecticides are recommended to control *N. viridula* on soybean using reduced dosages mixed with sodium chloride, which increase their efficacy (Corso 1990, Sosa Gómez et al. 1993, Corso and Gazzoni 1998).

Sodium chloride was found to affect the behavior of *N. viridula* by increasing the time spent on food-touching (Niva and Panizzi 1996), which, in turn, increased the bug's contact with the insecticides. Therefore, these chemical treatments (i.e., reduced dosages mixed with sodium chloride) have been adopted widely by soybean growers. Also, the use of a natural insecticide, such as an extract of neem seed, decreases the scars on pecan nuts caused by the feeding of *N. viridula* (Seymour et al. 1995).

Trap cropping has been utilized to suppress *N. viridula* and other pentatomid species on soybean (Newsom and Herzog 1977, Panizzi 1980, McPherson and Newsom 1984, Kobayashi and Cosenza 1987, Todd and Schumann 1988). It takes advantage of the fact that stink bugs colonize soybeans during the pod-set and pod-filling stages of plant development. Thus, early-maturing or early-planted soybeans are highly attractive to stink bugs and can be used to exploit this behavioral response by attracting large populations of the bugs into small areas containing these trap crops (usually less than 10% of the total area). The concentration of ovipositing females and the subsequent nymphal population can be controlled effectively with insecticides before the next generation of adults disperses to the surrounding fields. The key to success of this cultural practice is to control nymphs before they become adults and disperse to adjacent fields. If insecticidal controls are not timed properly, then this management strategy is useless and, in fact, actually intensifies the stink bug problem in the main crop. Inoculative releases of the egg parasitoid *Trissolcus basalis* (Wollaston) (Scelionidae) in early-maturing soybean, used as a trap crop, caused a reduction and delay in the population peak of this stink bug (Corrêa-Ferreira and Moscardi 1996).

Some progress has been made in the development of host plant resistance to feeding by *N. viridula* on some crops such as soybean (see references in Panizzi and Slansky 1985a and Jackai et al. 1990), and pecan (Dutcher and Todd 1983). But, so far, cultivars with variable levels of resistance have been used little, mainly because the potential production of these cultivars is usually lower than with other commercial cultivars.

Nezara viridula is attacked by numerous natural enemies including parasitoids, predators, and entomopathogens (see references in Panizzi and Slansky 1985a and Todd 1989). As many as 57 parasitoid species are recorded for *N. viridula* (Jones 1988). The most important of these are the egg parasitoid *Trissolcus basalis* and, in the New World, the adult parasitoids *Trichopoda pennipes* (F.) in North America (Corrêa-Ferreira 1986, Jones 1988) and *T. giacomelli* (Blanchard) (= *Eutrichopodopsis nitens* Blanchard) (Corrêa-Ferreira 1984). *Trissolcus basalis*, two additional species of egg parasitoids, and two species of tachinids were introduced in Hawaii to control *N. viridula*; *T. basalis* and one of the tachinid species became established (Davis 1967). In Brazil, there is a current program that supplies soybean growers with egg clusters of *N. viridula* parasitized by *T. basalis*. These egg clusters are sent to growers through the mail attached to tags. Each tag contains approximately 5,000 eggs. These tags are tied to soybean plants (one tag per hectare), and adult parasitoids emerging from these eggs can maintain populations of *N. viridula* below the economic threshold levels (Corrêa-Ferreira 1993).

Several arthropod predators feed on *N. viridula* (references in Todd 1989). There is a complex of nonspecific predators, and the fire ant, *Solenopsis invicta* Buren, is mentioned as an effective predator of eggs and early instars in the United States (Krispyn and Todd 1982, Stam et al. 1987). There is an interesting report of substantial kill of *N. viridula* by birds in Argentina (Beltzer et al. 1988).

3.18 *Oebalus poecilus* (Dallas)

3.18.1 *Distribution, Life History, and Biology*

The small rice stink bug, *O. poecilus* [also known as *Solubea poecila* (Dallas) (see Sailer 1944 and 1957), is, perhaps, the most important pest of rice in South America (Lima 1940, Sailer

1944, Grist and Lever 1969, van Halteren 1972). Although showing preference for Gramineae such as rice, barley, oat, corn, and wheat, it also is associated with soybean, cotton, and guava (Silva et al. 1968, Lopes et al. 1974). Albuquerque (1989) listed 42 species of host plants for this pest in the state of Rio Grande do Sul, Brazil, and (1990) noted its increasing numbers on sorghum in this state (Albuquerque 1990). Adults are ferrugineous to dark castaneous dorsally, with yellowish spots; the venter usually is darker in females than males. Males and females are 6.9 to 8.3 and 7.4 to 9.5 mm long, respectively (Sailer 1944). The small rice stink bug has distinctive morphs that distinguish nondiapausing and diapausing adults. The adult key characters are the pronotal shape and color of the body. In the nondiapausing summer morph, the lateral angles of the pronotum are spinose and the predominant dorsal coloration is dark brown, almost black; in the overwintering diapausing morph, the lateral angles are rounded and the predominant dorsal coloration is light brown (Albuquerque 1989). Despite the importance of *O. poecilus* as a pest, little has been published on its biology. Silva (1992a) reported that the incubation period at 27°C was 4.1 days, and egg hatchability was approximately 97%. The egg color changed due to the embryonic development, which allowed the age of the egg to be estimated. *Oebalus poecilus* was reared successfully in the laboratory using panicles of the weed *Polygonum punctatum* Elliot.

3.18.2 Damage

Rosseto et al. (1972) described the damage by *O. poecilus* to rice. Nymphs and adults feed on the developing grains. The nature and extent of the damage depend on the stage of grain development at the time of attack. Florets fed upon during early endosperm formation (milk stage) result in either empty glumes or severely atrophied kernels. Feeding during later stages of endosperm development (dough and hard stages) results in a chalky discoloration around the feeding site. Rice with this damage, known as "pecky rice," is weakened structurally and often breaks under mechanical stress during milling. Furthermore, the pecky rice that escapes breakage is of inferior quality as a finished product because of the discoloration. Other types of damage such as withering of young plants, reduction of tillers, and production of empty grains due to feeding activities of other species of stink bugs on rice have been described (Kisimoto 1983).

3.18.3 Control

Although *O. poecilus* is a major pest, no effective method has been developed to manage this stink bug in rice fields. The use of chemical insecticides has failed partly because the bugs constantly move into and among rice fields and because the insecticides used are of low persistence. Nevertheless, insecticides such as methyl parathion (1.5% as dust), 20 kg/ha.; malathion (4% as dust), 20 kg/ha; monocrotophos (240 to 270 grams of active ingredient/ha); phosphamidon (1.5% as dust) 20 kg/ha; and endosulphan, carbaryl, and trichlorphom have been recommended against this pest (Rai 1971; Corseuil and Cruz 1972; Costa and Link 1988, 1992c).

Albuquerque (1993) proposed a pest management strategy for *O. poecilus* in southern Brazil based on planting time. This method consisted of delaying and restricting the time of rice planting. Planting the crop in the first half of December results in an asynchrony between the pest and the crop because the developing grain is available in March when the bugs reaching the adult stage are diapausing morphs that migrate to hibernating sites rather than remaining in the crop. The result is that damage is reduced considerably.

Natural enemies of *O. poecilus* include the egg parasitoid *Telenomus mormideae* Lima, and the nymph and adult parasitoid *Beskia cornuta* Brauer & Bergenstan (Diptera: Tachinidae) (Silva et al. 1968). In Rio Grande do Sul, over 30% of *O. poecilus* eggs can be killed by *T. mormideae* in paddy rice fields (Silva 1992b).

3.19 Oebalus pugnax (F.)

3.19.1 Distribution, Life History, and Biology

The rice stink bug, *O. pugnax*, is found throughout the continental United States east of the Rocky Mountains as far north as New York, southern Minnesota, and southern Michigan; it is also known from the West Indies and the northern Gulf Coast of Mexico (Sailer 1944, McPherson 1982). It is brownish yellow and has a slender body, and is 8.0 to 12.0 mm long. The eggs (Douglas and Ingram 1942, Esselbaugh 1946, Odglen and Warren 1962) and various instars (Douglas and Ingram 1942, DeCoursey and Esselbaugh 1962, Odglen and Warren 1962) have been described. It is a serious economic pest of rice in the southern United States (Swanson and Newsom 1962, Jones and Cherry 1986, Harper et al. 1993). In southern Florida, adults are more common than nymphs in the first rice crop, whereas nymphs are more common in the ratoon crop (Foster et al. 1989). It feeds on wheat, oat, barley, rye, grain sorghum, corn, and forage grasses (Hall and Teetes 1981, Harper et al. 1993). Nymphal survival is generally higher on rice and grain sorghum than on vasey grass, *Paspalum urvillei* Steudel, one of its common wild hosts (Naresh and Smith 1983). *Oebalus pugnax* feed on the stems, branches, and glumes of grain sorghum but prefer seeds (Hall and Teetes 1982a).

3.19.2 Damage

Adults and nymphs feeding on rice kernels cause conspicuous areas of discoloration, resulting from the introduction of fungi and other microbes during feeding (Daugherty and Foster 1966, Marchetti and Petersen 1984, Lee et al. 1993). Kernels affected in this way are called pecky rice and are reduced in acceptability and value (Ito 1978, Marchetti and Petersen 1984, Lee et al. 1993). Speckback, a site-specific lesion on kernels of milled rice, is not associated with stink bug feeding (Cogburn and Way 1991).

3.19.3 Control

Oebalus pugnax populations need to be monitored from panicle exsertion to head maturation because a significant amount of peck can occur before the milk and the grain maturation stages (Harper et al. 1993). Severe losses in rice yields occur when infestation levels average 230 stink bug adults and/or nymphs per 1,000 panicles. Numerous rice lines have been evaluated for resistance to *O. pugnax* feeding and some possess moderate resistance (Nilakhe 1976, Robinson et al. 1981). Pest management programs have been developed for managing *O. pugnax* populations on rice (Drees 1983, Harper et al. 1994).

Natural enemies of *O. pugnax* include the egg parasitoids *Ooencyrtus anasae* (Ashmead) and *Telenomus podisi* Ashmead and the tachinid flies *Gymnoclytia immaculata* (Macquart), *Beskia aelops* (Walker), *Euthera tentatrix* Loew, and *Cylindromyia euchenor* (Walker). Predators, including the red-winged blackbird, are referred to in the literature (references in McPherson 1982).

3.20 Oebalus ypsilongriseus (De Geer)

3.20.1 Distribution, Life History, and Biology

This stink bug, along with *O. poecilus* (Dallas), is considered an important pest of rice in South America, particularly in Brazil. *O. ypsilongriseus* and *O. ornatus* (Sailer) were reported from seven Latin America countries and showed the widest geographic range among the stink bugs collected in Colombia and 12 selected rice-producing countries of Latin America (Pantoja et al. 1995). Recently, it has been recorded feeding on rice in Florida (U.S.A.) (Cherry et al. 1998). In Brazil, it is considered a key rice pest from the southernmost state Rio Grande do Sul (Link et al. 1989) up to the central areas of Goiás State (Ferreira 1980) and toward the far northern state of Pará (Silva and Magalhães 1981). Pantoja (1990) stated that it is an important pest of rice in Colombia. It also is associated with cotton, barley, oat, and wheat (Silva et al. 1968, Lopes et al. 1974).

Adults are dark yellowish, with yellow marks on the scutellum; males and females are 8.0 to 8.75 and 8.5 to 10.0 mm long, respectively (Sailer 1944). Little has been published on this bug's biology. The eggs (including embryonic development) (Del Vecchio and Grazia 1992a) and first to fifth instars (Del Vecchio and Grazia 1993a) have been described. Del Vecchio and Grazia (1992b) reported that the leaves of *Polygonum punctatum* Elliot were a preferred oviposition site, the number of eggs/cluster was approximately 17, and the number of eggs/female was less for hibernating morphs than for nonhibernating morphs. Del Vecchio and Grazia (1993b) also determined the duration (in days) of the incubation period and first to fifth stadia and the percent mortality associated with each stage.

3.20.2 Damage

The damage of *O. ypsilongriseus* to rice is similar to that described for *O. poecilus* (see **Section 3.18.2**).

3.20.3 Control

As with *O. poecilus*, the impact of *O. ypsilongriseus* to rice is mitigated with chemical insecticides (see **Section 3.18.3**). Natural enemies include the egg parasitoid *Telenomus mormideae* Lima (Silva et al. 1968). However, no studies are available that evaluate the effect of *T. mormideae* on the pest population.

3.21 *Palomena angulosa* Motschulsky

3.21.1 Distribution, Life History, and Biology

This polyphagous pentatomid occasionally is an important pest of various crops in Japan and other countries in Southeast Asia.

Adults overwinter and produce one generation annually. *Palomena angulosa* has a long oviposition period (June to early August) and is considered to have poor ability for oviposition site selection. Nymphs hatching in late summer or developing on unsuitable host plants are smaller, and the resulting adults may have difficulty in accumulating sufficient energy reserves for overwintering and reproducing the following spring. However, decreasing photoperiods appear to play a role in synchronizing and hastening nymphal development late in the season (Hori 1986, 1987). Eggs often are laid in poor locations (for example, the walls of houses, firewood, withered trees, tree trunks) and nymphs are forced to "forage" for food, often moving from one plant to another. When there are no fruiting plants in spring and early summer, the bugs have to survive on other plant parts. Young nymphs feeding on alternate food are smaller but partly recover in later stages when they find optimal food (Hori and Kuramochi 1986).

3.21.2 Damage

Palomena angulosa attacks a number of field crops including potato, rapeseed, alfalfa, and beans. Hori et al. (1984) showed that *P. angulosa* removed sap from leaf veins and petioles of potato plants more often than from stems. The bugs often feed at growing points, and the plants begin to wither 1 to 7 days after attack. Entire plants usually wither when attack is severe. Fruiting bodies of plants usually are attacked preferentially and may be necessary for optimum nymphal development.

3.21.3 Control

Little information is available although insecticides are probably the major control option. No information is available on natural enemies.

3.22 Piezodorus guildinii (Westwood)

3.22.1 Distribution, Life History, and Biology

The small green stink bug, *P. guildinii*, is a Neotropical pentatomid found from the southern United States to Argentina. It first was described from the island of St. Vincent (Stoner 1922) and frequently has been reported from Central and South America (see references in Panizzi and Slansky 1985a). It is a major pest of soybean in South America. In Brazil, this stink bug was seldom found on soybean until the early 1970s. Subsequently, it has become more common, ranging from Rio Grande do Sul (32° S latitude) to Piauí (5° S latitude). With the expansion of soybean production to the central, west, and northeast regions of the country, *P. guildinii* is, perhaps, the most important pest of soybean in Brazil.

The list of food plants includes some economically important plants in addition to soybean, mostly legumes such as common bean, pea, and alfalfa; it occasionally is reported on sunflower, cotton, and guava, but it is not believed to be a serious pest of these crops (Panizzi and Slansky 1985a). Native host plants include species of indigo legumes, *Indigofera* spp., in the southern United States (Panizzi and Slansky 1985b), Colombia (Hallman 1979), and Brazil (Panizzi 1992). It also feeds on legumes of the genera *Sesbania* and *Crotalaria* (Panizzi 1985, 1987; Panizzi and Slansky 1985b).

The diagnostic characteristics of the various stages, including adults, have been described and illustrated (Fraga and Ochoa 1972, Galileo et al. 1977, Grazia et al. 1980). Adults are light green to yellowish, with a red band at the base of the scutellum, particularly on females. Eggs are blackish and laid in two parallel rows. The first to fifth instars are 1.30, 2.25, 2.58, 4.60, and 7.87 mm long, respectively.

The biology of *P. guildinii* has been studied in detail, including nymphal developmental time and survivorship, adult longevity and reproduction, and nymphal and adult dispersal on soybean and on alternate host plants (e.g., Fraga and Ochoa 1972; Panizzi and Smith 1977; Panizzi et al. 1980; Costa and Link 1982; Panizzi and Slansky 1985c; Panizzi 1987, 1992). These parameters and others will vary depending on the food source and if the food for nymphs and adults has been switched. For example, the number of egg clusters/female can vary from approximately 3 on soybean to 37 in some indigo species. Total number of eggs/female can vary approximately from 28 on soybean to 200 on *I. hirsuta* L. and 500 on *I. truxillensis* H.B.K. Adults live approximately 50 days on soybean (mean of female and male) and 90 days on some of the indigo legumes. Despite its relatively low performance on soybean compared with native legumes, *P. guildinii* is highly detrimental to soybean, as noted previously. Food-switching from nymph to adult may have a positive, negative, or no effect on the performance of the nymphs or adults, depending on the suitability of the foods involved. For example, females fed on *Sesbania aculeata* Pers. (= *S. bispinosa* [Jacq.] W. F. Wight.) pods during the nymphal and adult stages will lay twice as many eggs as those fed on soybean as nymphs and on *S. aculeata* as adults.

3.22.2 Damage

Piezodorus guildinii, as with most of the phytophagous pentatomids associated with soybean, feeds primarily on pods. The nature and extent of the damage to this crop is similar to that reported to *N. viridula* (see **Section 3.17.2**). *Piezodorus guildinii* is associated with several microorganisms present on damaged soybean seeds (Panizzi et al. 1979). It usually is the first species to appear in soybean fields during flowering or even earlier, similarly to *Piezodorus* sp. as reported by Ezueh and Dina (1980) for Nigeria. Apparently, *P. guildinii* is adapted to feeding on flowering plants better than other pentatomid species. However, it does have to feed on reproductive structures to thrive.

3.22.3 Control

Chemical insecticides, such as monocrotophos and endosulfan, are the major weapons used to control *P. guildinii*. This bug is managed on soybean by means of an integrated pest management program used widely in Brazil (see details under *N. viridula*, **Section 3.17.3**). Natural enemies include several species of egg parasitoids such as *Telenomus mormideae* Lima, *Telenomus* sp., *Trissolcus basalis* (Wollaston), *T. scuticarinatus* Lima, *Ooencyrtus submetallicus* (Howard), and *Ooencyrtus* sp. (Corrêa-Ferreira 1986); *T. mormideae* Lima is the most important. Fifth instars of *P. guildinii* are attacked by nymphs and adults of the pentatomid *Tynacantha marginata* Dallas (Panizzi and Smith 1976a). Adults are reported to be a substantial part of the diet of birds in Argentina (Beltzer et al. 1988).

3.23 Piezodorus hybneri Gmelin

3.23.1 Distribution, Life History, and Biology

This pentatomid has been reported as a pest of soybean in Thailand, Korea, Taiwan, and Japan. It also is a pest of this crop in some parts of India under the synonym of *P. rubrofasciatus* F. (Singh et al. 1989). *P. hybneri* also occurs in Australia and is known as the redbanded shield bug. However, it only is a minor pest that occasionally attacks soybean in Queensland (Shepard et al. 1983).

Piezodorus hybneri adults overwinter and first appear in soybean fields in late June to early July. Little is known about the ecology of this stink bug prior to its appearance in soybean each year. It is likely that the offspring from overwintered adults develop on noncrop hosts, with adults from this generation migrating into soybean fields as the crop reaches podding stage (Higuchi 1992). Immigrant females usually are gravid and mated, and oviposition usually begins in early July and continues until mid-August.

Higuchi (1994) found 55% of egg clusters were laid on pods. The number of eggs per cluster ranged from 2 to 47 with an average of approximately 20, and each female laid approximately 10 clusters (Higuchi and Mizutani 1993). A new generation of adults appeared in late July and August. Mark-recapture experiments indicated adult *P. hybneri* actively moved in and out of soybean crops; residence times averaged 3 to 5 days (Higuchi 1992). The sex ratio of *P. hybneri* in soybean is usually 1:1. Reproductive adults are fairly long-lived in the laboratory (50 to 130 days) (Kadosawa and Santa 1981), whereas nonreproductive (overwintering) adults probably live 8 to 9 months.

3.23.2 Damage

Soybean is the crop most frequently attacked by *P. hybneri* although some other crops such as beans, alfalfa, and lentils may be damaged. In India, damage to linseed, potato, tomato, chili, and brinjal has been reported (Joseph 1953). Pods and seeds of soybean are attacked preferentially, causing pod shedding or poor ripening (see detailed description of damage to soybean under *N. viridula*, **Section 3.17.2**). Kawamoto et al. (1987) showed *P. hybneri* has a high feeding frequency on soybean and spends approximately 50% of each day feeding.

3.23.3 Control

The abundance and efficiency of scelionid egg parasitoids (primarily *Telenomus triptus* Nixon) in regulating populations of *P. hybneri* suggest there are good prospects for the development of an integrated pest management strategy for this pentatomid incorporating biological control and egg parasitoid–friendly insecticides. Current control methods are based on insecticides, but the high mobility of *P. hybneri* reduces their efficacy. Singh et al. (1989) found quinalphos was the most effective insecticide against *P. hybneri*. High rates of egg parasitism by scelionid wasps have been recorded for *P. hybneri* in soybean crops. *Telenomus triptus* was reported to be the dominant parasitoid by Higuchi (1993). *Ooencyrtus nezarae* Ishii also occurred but was less important. Parasitism by *T. triptus* generally is high during the entire oviposition period of *P. hybneri* on soybean. Higuchi (1993) recorded parasitism rates of 60 to 95%, resulting in very low numbers of nymphs. Adult females of *T. triptus* enter soybean fields after the first *P. hybneri* immigrants lay eggs; however, their arrival usually is so rapid that only the first egg clusters escape parasitism.

3.24, 3.25 *Plautia stali* Scott and *Plautia affinis* Dallas

3.24.1, 3.25.1 Distribution, Life History, and Biology

The brown-winged green bug, *Plautia stali*, is a serious pest in parts of southeastern Asia including Japan, Taiwan, and South China. In Japan, it is one of the most serious pests of fruit trees. In 1967, it was discovered for the first time in Hawaii (Mau and Mitchell 1978). It damages mulberry, peach, plum, persimmon, cherry, grape, guava, strawberry, pomegranate, bean, and pea (Shiga and Moriya 1984, Shiga 1991). *Plautia affinis*, a related species, is a minor pest of fruits and vegetables in eastern Australia (McDonald 1971). It feeds on raspberry, *Rubus idaeus* L., in southeastern Queensland (Coombs and Khan 1998) and on soybean in northern Australia (Shepard et al. 1983).

Plautia stali is a small green pentatomid with brown wings. There are two or three generations per year in Japan with overwintering occurring as adults in reproductive diapause (Kotaki and Yagi 1987). Overwintering adults are reddish-brown and revert to green when diapause is terminated. Diapause is induced by short-day photoperiods and mediated by juvenile hormone (Kotaki and Yagi 1989). Critical daylength for ovarian development is between 13.5 and 14.0 hours (Fukuda and Fujiie 1988). The lower threshold for development was estimated to be 12.7°C, and 430 degree-days are required for development from egg to adult. First instars are gregarious, inactive, and do not feed. Some gregariousness also occurs in second instars, but later instars disperse. The females are polyandrous and males are polygamous (Mau and Mitchell 1978). Females copulate up to twice a week and remain *in copula* for a few minutes to 5 days. The frequency of female mating suggests that the amount of sperm transferred during a single mating is insufficient for the entire ovipositional period. Females begin ovipositing as early as 8 days after molting and as soon as 1 day after mating. Adults readily migrate to trees bearing fruit and are powerful fliers (Moriya 1987). Males appear to produce an aggregation pheromone similar to that reported for *Biprorulus bibax* Breddin (James et al. 1994). Caged males strongly attract males and females of the same species, peaking at dusk or just before complete darkness (Moriya and Shiga 1984). The chemical basis of this attraction has not been reported. A synthetic aggregation pheromone for *P. stali* would be a useful monitoring tool for this pest.

3.24.2, 3.25.2 Damage

Plautia stali attacks a wide range of tree fruits and some vegetables. Fruit mostly is attacked when ripe or near ripening, which causes blemishing and often internal damage. There also are many noncrop hosts. *Plautia affinis*, when feeding directly on developing or mature raspberry fruit, causes damage (Coombs and Khan 1998).

3.24.3, 3.25.3 Control

Insecticides appear to be the primary weapon for control of *P. stali*. A range of natural enemies attack *P. stali*, the most important of which probably are egg parasitoids. *Trissolcus plautiae*

(Watanabe) is the predominant parasitoid in southwestern Japan and attacks eggs of *P. stali* through all stages of their development (Ohno 1987). The tachinid *Gymnosoma rotundatum* L. has been reported as a parasitoid of adult *P. stali* (Yamada and Miyahara 1979). The occurrence of significant natural enemies and the likely presence of a male-produced aggregation pheromone offer good prospects for development of an integrated pest management strategy for this pest. *Plautia affinis* eggs are attacked by the egg parasitoids *T. basalis* (Wollaston), *T. oenone* (Dodd), and *Telenomus* sp. Late instars and adults are attacked by the braconid *Aridelus* sp. (Euphorinae), a tachinid fly, the predatory pentatomid *Amyotea hamatus* Walker, the reduviid *Pristhesancus plagipennis* Walker, and flower spiders (Thomisidae) (Coombs and Khan 1998).

3.26 Thyanta perditor (F.)

3.26.1 Distribution, Life History, and Biology

This stink bug, although inhabiting much of northern South America (Callan 1948), seems to be more typical of the West Indies and Mexico (Van Duzee 1904). In the United States it has been reported from Florida, Texas, and Arizona (Rider and Chapin 1992). Fennah (1935) reported *T. perditor* as feeding on soybean in Trinidad, and Waldbauer (1977) reported it as an important pest of soybean in Colombia. Rosseto et al. (1978) found this insect infesting soybean, sorghum, and sesame in Brazil, and Gomez (1980) reported it from rice, soybean, and wheat in the state of Mato Grosso do Sul. Amaral Fº. et al. (1992) found it infesting rice in São Paulo and studied its biology under different temperature regimes.

Brief notes on this insect's biology on wheat were published by Perez et al. (1980). They recorded the maximum egg cluster size as 24, the incubation period as 3 to 7 days, egg fertility as 75%, a nymphal period of 65 days, and adult longevity of over 65 days. Panizzi and Herzog (1984) studied its biology on wheat, soybean, and on a native host, *Bidens pilosa*. They found that wheat and, to a larger extent, soybean were inadequate food but that *B. pilosa* was suitable. These results suggest that *T. perditor* found in soybean or wheat fields may be related to the presence of *B. pilosa* rather than soybean or wheat, although feeding may occur on these latter two species. Indeed, seed yield and quality of wheat are affected by *T. perditor* (Ferreira and Silveira 1991).

Adults are green with a red band on the pronotum. Grazia et al. (1982c) described and illustrated the five nymphal instars. The first to fifth instars are 1.47 to 2.01, 1.84 to 2.62, 2.36 to 3.87, 4.48 to 6.40, and 6.21 to 8.83 mm long, respectively.

3.26.2 Damage

Prior to its first appearance in significant numbers during the late 1970s in Brazil, *T. perditor* was not considered a pest species, and no published information was found in the literature related to its economic importance. Today, it is considered a pest of some Gramineae, although it only is a minor pest of soybean, at least in Brazil.

Thyanta perditor feeds on seeds and gives sorghum heads a "wilted-wrinkled" appearance (Busoli et al. 1984). It also causes grain failure of wheat with a consequent reduction in seed yield (Ferreira and Silveira 1991).

3.26.3 Control

Chemical control of *T. perditor* on wheat is recommended when populations reach one adult per five heads (Ferreira and Silveira 1991). Natural enemies include the tachinid fly *Trichopoda giacomellii* Blanchard (= *Eutrichopodopsis nitens* Blanchard), which can parasitize up to 50% of field-collected adults, and the egg parasitoid *Trissolcus scuticarinatus* (Lima), which can parasitize 53% of field-collected egg clusters in Paraná state (Panizzi and Herzog 1984). Other egg parasitoids include *T. basalis* (Wollaston), *T. scuticarinatus* Lima, and *Neorileya* sp. (Corrêa-Ferreira 1986).

3.27   *Tibraca limbativentris* Stål

3.27.1   Distribution, Life History, and Biology

The stem rice stink bug, *T. limbativentris*, a Neotropical pentatomid, is a major pest of paddy rice in southern Brazil, Argentina, and Uruguay. It also attacks soybean, tomato, wheat, and native Gramineae (Lopes et al. 1974, Rizzo 1976). Adults are light brown dorsally, dark brown ventrally, and 13.0 mm long. The developmental stages have been described (Rizzo 1976). First and second instars are 1.5 and 2.3 mm long, respectively, reaching 9.5 mm in the last instar.

Little is known about this stink bug's biology. In Argentina, reproduction starts during September and October when adult females lay eggs on leaves and stems of rice plants. Nymphal development lasts 35 to 42 days. During summer months, adult longevity reaches 30 days. Costa and Link (1992a) studied adult dispersion on rice paddy fields in southern Brazil.

3.27.2   Damage

The impact of feeding by *T. limbativentris* on paddy rice plants was evaluated by Costa and Link (1992b). The degree of damage depends on plant growth stage. During the reproductive stage, each insect may damage 3.2 stems during a 15-day period. Grain loss varies from approximately 59 kg/ha during the vegetative stages to 65 kg/ha during the reproductive stages. Because of feeding, seeds will become broken and opaque, decreasing their quality. Stems will have dark spots at the feeding sites.

3.27.3   Control

Chemical insecticides such as trichlorphon (500 to 1,000 g active ingredient/ha), endosulphan (420 to 525), malathion (1,000), methomyl (450), methylparathion (600), and monocrotophos (300) usually are recommended to control *T. limbativentris* in paddy rice fields in Brazil (Corseuil and Cruz 1971, Costa and Link 1992c). Natural enemies include a predatory fly *Efferia* sp. (Asiliidae) in Argentina (Rizzo 1976) and the egg parasitoid *Ooencyrtus fasciatus* Mercet in Brazil (Silva et al. 1968). Recently, Boff and Boff (1995) reported a parasitic fungus, *Cordyceps nutans* Pat., colonizing *T. limbativentris* in Santa Catarina State, Brazil. Such parasitism mostly was observed on overwintering insects that remained on weeds (Gramineae) and when no chemicals (fungicides) were sprayed. The entomopathogenic fungi *Metarhizium anisopliae* (Metsch.) Sorok and *Beauveria bassiana* (Bals.) Vuill. were reported as potential agents to control *T. limbativentris* in paddy rice fields in southern Brazil (Martins and Lima 1994).

4.   THE LESS IMPORTANT SPECIES

4.1   *Acrosternum* spp.

There are several species of *Acrosternum* in the Eastern and Western Hemispheres, the largest number occurring in the Americas (Rolston 1983b). Among them, the Neotropical *A. armigera* (Stål) commonly is noted as a secondary pest of soybean in Brazil (e.g., Costa and Link 1974, Lopes et al. 1974, Panizzi and Smith 1976b). However, Del Vecchio et al. (1988) believes that these references to *A. armigera* are, in fact, dealing with another species, *A. bellum*, described by Rolston (1983b). This last species also has been recovered on cabbage, okra, pea, spinach, common bean, tobacco, and wheat in several areas of the southernmost state of Rio Grande do Sul (Link and Grazia 1987). These two species are similar, the predominant color being green (Rolston 1983b).

Another species, *A. impicticorne* (Stål), has been recorded on soybean in different areas of Brazil and also is considered a secondary pest. It also feeds on cotton (Silva et al. 1968). Grazia et al. (1982a) described the nymphs, which are similar to those of *Nezara viridula*. Adults are green with a black spot present in each basal angle of the scutellum.

Acrosternum marginatum (Palisot de Beauvois) feeds on soybean in North and South America (Panizzi and Slansky 1985a). It also feeds on physic nut, *Jatropha curcas* L. (Euphorbiaceae), in Nicaragua (Grimm and Führer 1998). Hallman et al. (1992) studied its biology on common beans in Colombia. *Acrosternum marginatum* is considered the most common species of *Acrosternum* in Central America; it is found from the southwestern United States to Venezuela and Ecuador and throughout the Caribbean from Florida to Guadeloupe (Rolston 1983b). *Acrosternum acuta* Dallas feeds on pods and seeds of soybean in Africa (Turnipseed and Kogan 1976, Jackai and Singh 1987) and is a pest of pigeon pea in Kenya (Khamala et al. 1978).

These *Acrosternum* species are polyphagous and, in general, secondary pests. No published data were found related to their control. In Brazil, where they commonly are found on soybean, insecticides used to eliminate the major stink bug pest species also will control them because the bugs occur concurrently during the reproductive period. Egg parasitoids, such as *Telenomus mormideae* Lima, *Trissolcus basalis* (Wollaston), *T. scuticarinatus* Lima, and *Neorileya* sp., parasitize as much as 67% of the eggs of *Acrosternum* spp. in soybean fields in Paraná state (Corrêa-Ferreira 1986).

4.2 *Aelia acuminata* L.

This species has the widest distribution among the *Aelia* pests of wheat. It occurs in the western Palearctic region and throughout Europe from 55° north latitude southward to 30° in northern Africa. It is found as far west as Ireland, as far east as central Russia, and southward to Kazakhstan, Afghanistan, and Iran (Southwood and Leston 1959, Wagner 1960, Putschkov 1961, Brown 1962, Javahery 1967, Voegelé 1969). Adults are pale in color with fine black spots and are 7.5 to 9.5 mm long. The eggs are pale with fine hairs and 0.7 to 0.8 mm long. Developmental time from egg to adult takes 45 to 79 days under field conditions and 29 to 31 days at 28°C in the laboratory. Fecundity averages 76 eggs per female in the field, and 30 and 291 eggs/female at 20 and 28.5°C, respectively, in the laboratory (Tischler 1938, Javahery 1967). *Aelia acuminata* feeds on a wide range of graminaceous plants, particularly *Festuca, Poa, Agrostis, Dactylis, Lolium,* and *Bromus* spp. In the absence of these preferred food plants, it will feed on wheat in rain-fed and irrigated cultivation (Tischler 1937, 1938; Putschkov 1961; Brown 1962; Javahery 1967; Voegelé 1969). Natural enemies of *A. acuminata* include tachinid parasitoids of adults and scelionid parasitoids, such as *Trissolcus waloffae* Javahery and *T. rufiventris* Mayr, of eggs (Delucchi 1961; Javahery 1968, 1995).

4.3 *Aelia americana* Dallas

This species occurs in North America from British Columbia south to Arizona and east to Manitoba, Michigan, and Illinois (McPherson 1982). Similar to the Palearctic species in this genus, *A. americana* is associated with wild and cultivated Gramineae, these last including wheat, barley, and rye, and is reported to cause some damage to them in Kansas (McPherson 1982). Adults are creamy colored with four longitudinal blackish stripes on the dorsum and are 7.0 to 9.0 mm long. Eggs are dirty white, 1 mm long, and usually laid in two rows on the stems and leaves of grasses in clusters of approximately 12 eggs. The incubation period lasts 10 days, and nymphal development takes 35 to 45 days at 23 to 25°C in the laboratory (Javahery 1994). Usually, there is no need to control this insect on cereals.

4.4 Aelia virgata Klug

This species occurs from southwestern Iran to north of the Caspian Sea (in the independent states of southwestern Russia) and in the eastern Mediterranean region, particularly in Turkey, Syria, Lebanon, Israel, and northern Iraq (Alexandrov and Mirzayan 1949, China and Lodos 1959, Brown 1962). *A. virgata* is distinguished easily from other closely related species by the presence of four longitudinal black stripes on its dorsum. It usually is found with *Eurygaster integriceps* Puton (Scutelleridae) in overwintering sites at high altitudes (Brown 1962, Javahery 1972). This stink bug occurs in low numbers and is unlikely to cause significant damage to cultivated cereals (Brown 1962, Javahery 1995).

4.5 Agnoscelis rutila (F.)

The horehound bug, *A. rutila*, is usually found on the weed horehound, *Marrubium vulgare* L., and, occasionally, on other labiate plants and on grain legumes in Australia (Gross 1976, Shepard et al. 1983). It sometimes swarms on foliage and blossoms of fruit trees and ornamental plants but rarely causes injury (Fletcher 1985). In New Guinea, it has been found on citrus. It is distributed widely in southeast Australia, New Guinea, and surrounding islands. The adult is bright orange with black markings and approximately 12 mm long. Javan specimens are much yellower than those from other areas. This species overwinters in the adult stage on horehound plants. Eggs often are parasitized by *Trissolcus basalis* (Wollaston), the major parasitoid of *Nezara viridula* (L.) in Australia. It thus serves as an important "reservoir" for this parasitoid, maintaining populations in the absence of *N. viridula* and enabling rapid parasitism of the latter species when it appears.

4.6 Antestiopsis spp.

These bugs, known as "antestia bugs," are reported as pests of coffee in many African countries. The main pest species are *Antestiopsis orbitalis* (Westwood) with the subspecies *bechuana* (Kirkaldy) and *ghesquierei* Carayon, *A. intricata* (Ghesq. & Carayon), *A. facetoides* Greathead, *A. crypta* Greathead, and *A. falsa* Schouteden. All of these species formerly were included in the genus *Antestia*, which now includes the following species reported to breed on coffee: *A. usambarica* Schouteden, *A. cincticollis* (Schaum), and *A. trivialis* Stål (Greathead 1966a). *Antestiopsis clymeneis* (Kirkaldy) may reach pest status in Madagascar (Greathead 1968). These species are variable in color and size. *Antestiopsis* adults are brownish with white markings and yellow to red spots and 6.7 to 8.4 mm long (Greathead 1966a); *Antestia* adults are brownish without spots and 13 to 15 mm long. The longevity of *Antestiopsis orbitalis bechuana* is approximately 80 days, fecundity is 9.7 egg clusters/female with 9.8 eggs/cluster, and viability of eggs is approximately 70% (Abasa 1973). Food plants are, in general, restricted to the family Rubiaceae with coffee, *Coffea arabica* L. and *C. robusta* Lind., being preferred host plants, particularly the former. *Antestia cincticollis* is reported on cotton in Africa (Couilloud 1989). On coffee, bugs feed on the berries, causing small berries to drop and big berries to become soft or rotten. High populations of bugs will damage growing points of coffee trees, which will yield lower production and lower quality beans (McNutt 1979).

Recommendations of sampling procedures were proposed for *Antestiopsis orbitalis* attacking coffee plantations in Burundi (Cilas et al. 1998). Chemical control of antestia bugs is the most effective control measure. In Uganda, fenitrothion is recommended when populations of the bugs reach one to two adults per tree (McNutt 1975, 1979). Natural enemies include several species of egg parasitoids and nymphal and adult parasitoids and predators, with the tachinid *Bogosia rubens* Villeneuve being effective in controlling adult populations of *Antestiopsis* spp. in East Africa (Greathead 1966b, Greathead and Bigger 1967).

4.7   *Aspavia armigera* (F.)

This is an occasional pest of cowpea in Africa, and it may occur in Asia (Singh et al. 1990). Adults are dark brown with a large scutellum and three white or orange angular spots. Several species of *Aspavia* are known to occur in Africa, but *A. armigera* is the most common; its eggs frequently are found on cowpea (Singh and Jackai 1985). The adults and nymphs feed on cowpea pods. Signs of damage are similar to those for other pod-sucking bugs. It also is a pest of soybean (Ezueh and Dina 1980), rice (Jackai et al. 1990), and cotton (Couilloud 1989). Chemical control is similar to that recommended for other stink bugs, such as *N. viridula* (see **Section 3.17.3**).

4.8   *Axiagastus cambelli* Distant

The coconut spathe bug, *A. cambelli*, occurs throught out the Papua New Guinea region but large populations only occur on the Lihir group of islands. The low populations observed on the Gazelle Peninsula, East New Britain, and elsewhere suggest that natural enemies may be more dominant in these regions. Brief notes on the biology of *A. cambelli* were given by Tothill (1929) and Lever (1933). Eggs are laid on coconut leaflets, leaves, the undersides of spadix stipules, and trunks. Although the coconut palm, *Cocos nucifera*, is its principal host, *A. cambelli* also has been recorded from betel nut palm, *Areca* sp. (Lever 1933). Eggs hatch in 6 to 8 days, and nymphal development takes approximately 45 days. A survey of the natural enemies of *Axiagastus cambelli* occurring on the Gazelle Peninsula revealed the presence of three egg parasitoids, and one nymph and one adult parasitoid. *Trissolcus painei* Ferr. was the dominant egg parasitoid followed by *Anastatus* sp. On the Lihir group of islands, only two egg parasitoids, *T. painei* and *Anastatus* sp., were recorded (Baloch 1973).

4.9   *Bagrada cruciferarum* Kirkaldy

The painted bug, *Bagrada cruciferarum*, is a pest of oilseeds and vegetables in India. In the midhill regions of Himachal Pradesh, it is a serious pest of cauliflower during April to May at siliqua-formation stage (Dhaliwal and Goma 1979). Verma et al. (1993) studied the bioecology of *B. cruciferarum*. The pest appeared on cauliflower during March, and peak populations occurred in May during the siliqua-formation stage. Adults are black with yellow and pink spots; males and females are 5.29 and 7.12 mm long, respectively. Eggs are oval, dirty white, and 0.876 mm long. First to fourth instars are 1.12, 1.39, 1.50, and 2.45 mm long, respectively. Fifth instars are similar to adults and 5.29 mm long. Adult longevity is 23 to 27 days, and fecundity is 36 to 173 eggs/female. This insect can be reared easily in the laboratory on Indian mustard, *Brassica juncea* (L.) (Batra and Sarup 1962). Little information was found regarding its control. Chemical insecticides probably are the main control measure against this pest (Vora et al. 1985).

4.10   *Bathycoelia thalassina* (Herrich-Schaeffer)

This stink bug is a pest of cocoa in most of the cocoa-growing countries in West Africa (Wood 1970, Owusu-Manu 1971). Its flight activity, reproduction in association with cocoa plants, and dispersal were studied in Ghana (Owusu-Manu 1977). Adults were collected in greater numbers during September and February; egg laying occurred mostly during February. Adults fly from one tree to another, but in open fields they may fly long distances on a straight course. Host plants include cash crops such as kola, *Cola acuminata* (Beauv.) Schott & Endl.; mangoes, *Mangifera indica* L.; and *Citrus* spp. (Owusu-Manu 1978). The main damage to cocoa is premature ripening of the fruits (Lodos 1967). Adults have long stylets (approximately 2.2 cm) with which they penetrate the pod cortex to suck liquid from the developing cocoa beans; this results in malformed or atrophied beans, brown, instead of pink, and dry, lacking the sugary mucilage that covers normally developed beans (Gerard 1965). Owusu-Manu (1990) gave details on this bug's feeding habits including the duration, frequency, and amount of food ingested; and the damage to cocoa. Chemical control has not been successful (Owusu-Manu 1973), and alternative methods based on cultural practices and biological control agents should be implemented to control this pest effectively.

4.11   *Carpocoris fuscispinus* Boheman

This species extends from central and southern Iran northward to the southern independent states of the former U.S.S.R. and is considered a minor pest of cereals (Alexandrov and Mirzayan 1949; Putschkov 1961; Javahery 1972, 1993). Overwintered adults feed on wheat during the filling period, and on wild and graminaceous plants, in central and southern Iran (Putschkov 1961; M. Javahery, unpublished data). Adults are brownish-green and 12 to 15 mm long. Eggs are barrel-shaped, brownish, and laid in clusters of 14 eggs in four rows attached to kernels of food plants. The incubation period takes 14 to 15 days in the field, and 14 to 15 days at 20°C and 8 to 9 days at 27°C; nymphal development takes approximately 45 days (Putschkov 1961, Javahery 1995). *Carpocoris fuscispinus* migrates to higher altitudes during early summer and returns to the breeding areas during the following spring. Overwintered adults appear in wheat fields during April to May and feed on green shoots, leaves, and stems. *Carpocoris fuscispinus* is not considered an important pest (Javahery 1995).

4.12   *Chlorochroa ligata* (Say)

Chlorochroa ligata, the conchuela, is a member of a genus that has had a confused taxonomic history (Buxton et al. 1983, Thomas 1983). Several species have been listed from numerous cultivated crops and wild hosts (Buxton et al. 1983). *Chlorochroa sayi* (Stål) is mentioned frequently in the older literature, but there is doubt whether this actually was the species under investigation. Thus, it is difficult to attribute the results of these older studies to *C. sayi* with any certainty. *Chlorochroa ligata* occurs over much of the western United States and adjacent Canada (Buxton et al. 1983). The eggs and first to fifth instars have been described (e.g., Morrill 1910). This stink bug feeds on numerous cultivated crops, including sorghum. Hall and Teetes (1982b) found that nonseed feeding on sorghum generally increased as the density of the bugs increased and as seed development progressed. They also found that feeding on the panicles during the milk, soft-dough, and hard-dough stages reduced yield and germination of seeds, but that the effects of feeding were less severe as the grain matured.

4.13   *Cuspicona simplex* Walker

Gross (1975, 1976) reported this species as associated with solanaceous plants and sometimes as a pest of potatoes and tomatoes. *Cuspicona simplex* is common in southern Australia and readily found during the summer months. It also occurs in New Zealand, having been introduced accidentally from Australia. Adults are bright green and overwinter at the base of plants or in grass or loose soil. Two or three generations occur annually in New South Wales (McDonald and Grigg 1980). Eggs are laid in clusters of 14. No information was found on control measures against this pest.

4.14   *Dichelops furcatus* (F.)

This Neotropical pentatomid occurs in Argentina, Bolivia, Brazil, Paraguay, and Uruguay (Grazia 1978). Adults are brownish-green dorsally and green ventrally; males and females are 10.62 and 11.73 mm long, respectively. The first to fifth instars are 1.67, 2.38, 4.59, 6.93, and 9.42 mm long, respectively (Grazia et al. 1982b). Host plants include mostly legumes such as soybean, alfalfa, and beans; the bugs also may feed on tobacco and strawberry (Lopes et al. 1974, Rizzo 1976). Few data are available on the biology of *D. furcatus*. This probably is because this species rarely is found in great numbers.

Dichelops furcatus is reported as a secondary pest of soybean in several southern states of Brazil (Costa and Link 1974, Panizzi and Smith 1976b, Galileo et al. 1977). The damage caused to soybean and control measures against this bug are similar to those described for *N. viridula* (see **Sections 3.17.2 and 3.17.3**).

4.15   *Dichelops melacanthus* (Dallas)

This pentatomid occurs in several countries of South America and is similar in appearance to *D. furcatus* and *D. phoenix* Grazia. These three species are separated by the morphology of the genitalia (Grazia 1978). *Dichelops melacanthus* occurs on soybean in Brazil (Corso 1984, Corrêa-Ferreira 1986). It is possible that misidentifications involving *D. furcatus* and *D. melacanthus* have occurred. Recently, Ávila and Panizzi (1995) reported an unusually massive attack of *D. melacanthus* on seedling corn in the western state of Mato Grosso do Sul and in the more southern state of Paraná. Apparently, this pentatomid is associated more with Leguminosae than Gramineae; further research is needed to determine whether or not its association with corn is fortuitous. Natural enemies include the egg parasitoids *Telenomus mormideae* Lima and *Trissolcus basalis* (Wollaston), which may parasitize up to 60% of the eggs of this stink bug in soybean fields in Paraná state (Corrêa-Ferreira 1986).

4.16   *Dictyotus caenosus* (Westwood)

The brown shield bug, *D. caenosus*, is common in southern Australia, New Zealand, and New Caledonia. It sometimes is found on ornamentals and vegetables and field crops such as soybean and sunflower (Fletcher 1985). It often is found in association with *Nezara viridula* on soybean but apparently does little damage in its own right. The adult is evenly dull brown, and approximately 9 mm long. Adults overwinter at the bases of plants or under leaves or debris. This species often is most common in coastal areas and frequently is found in colonies (Gross 1975).

4.17   *Dolycoris penicillatus* Horvath

This species is found from southern Iran to the north of the Caspian Sea, occasionally inflicting damage to wheat; it reproduces on wild Gramineae (Javahery 1995; M. Javahery, unpublished data). Adults are brownish and 9 to 12 mm long. Overwintered adults appear mostly during spring on rain-fed wheat cultivations, feed on stems and leaves, and oviposit on leaves and ears. Eggs are pale, and 1 mm long; the incubation period lasts 10 to 15 days. Nymphs feed on sap and milky grains and only one generation is completed per year (Javahery 1995; M. Javahery, unpublished data). Occasionally, this stink bug must be controlled with insecticides such as fenthion and fenitrothion (M. Javahery, unpublished data).

4.18   *Edessa rufomarginata* (De Geer)

This Neotropical pentatomid is a secondary pest of a wide variety of plants, mostly those in the family Solanaceae. Major economic host plants include tobacco, potato, eggplant, soybean, sunflower, rice, and corn (Silva et al. 1968, Rizzo and Saini 1987). Tobacco plants, when heavily infested, wilt and may die (Monte 1939). Descriptions of the developmental stages and aspects of this bug's biology were studied by Rizzo and Saini (1987) in Argentina and by Fortes and Grazia (1990) in Brazil. Adults are oval, green dorsally, and orange-brown ventrally; males and females are 18.5 to 19 and 19.5 to 20.0 mm long, respectively. Eggs are light green and 1.8 mm long. The first to fifth instars are 3.0 to 3.3, 5.0 to 5.5, 7.5 to 8.0, 8.6 to 9.2, and 12.0 to 15.0 mm long, respectively. Egg clusters contain 12 to 14 eggs and are laid in two parallel rows. Nymphal development takes 52 to 89 days (mean, 65 days). In Argentina, *E. rufomarginata* completes one generation per year (Rizzo and Saini 1987), but in Brazil it probably has several generations. Eggs are parasitized by the scelionids *Telenomus edessae* Bréthes, *T. schrottkyi* Bréthes, and *Dissolcus paraguayensis* Bréthes. Adults are attacked by the tachinid flies *Neobrachelia* sp. and *Xenopyxis edessae* (Townsend) (Silva et al. 1968).

4.19 *Euschistus conspersus* Uhler

Euschistus conspersus occurs in western North America and, occasionally, can cause damage to crops. It has been associated particularly with damage to alfalfa, sorghum, cotton, and sugar beets. The bugs prefer to feed on alfalfa seeds rather than on surrounding crops until alfalfa matures or is harvested; then, they disperse to surrounding fields of cotton and sorghum. Higher populations are present on sorghum than on cotton when the bugs are dispersing from alfalfa in mid- to late summer; however, chemical control is necessary on both crops to prevent stink bug–induced crop injury (Toscano and Stern 1976a). *Euschistus conspersus* causes reductions in cotton seed weight, germination, and lint yields (Toscano and Stern 1976b). It is the most common stink bug on tomatoes in California (Hoffmann et al. 1987). The bug's feeding causes light-colored blemished areas on the fruit, which can lower quality at harvest. It also has been reported to transmit yeasts, *Nematospora* spp., that may cause tomato decay (Hoffmann et al. 1987). Stink bug management in tomatoes grown in California is discussed by Zalom et al. (1997).

4.20 *Euschistus tristigmus* (Say)

The dusky stink bug, *E. tristigmus*, is one of the least economically important members of the genus in North America. Of the two subspecies reported in the literature (i.e., *tristigmus* and *luridus*), *tristigmus* seems to be the most damaging. This subspecies ranges across much of North America south of about latitude 41° north (McPherson 1982). It feeds on soybean (Jones and Sullivan 1979), to which it transmits the yeast-spot disease (Daugherty 1967). It also feeds on cotton, apple, sweet corn, and several uncultivated plants (references in McPherson 1982). The eggs (Esselbaugh 1946, Javahery 1994) and first to fifth instars (DeCoursey and Esselbaugh 1962) have been described. *Euschistus tristigmus* has two generations per year in southern Illinois (U.S.A.) (McPherson 1975) but only one generation in southeastern Canada (M. Javahery, unpublished data). Natural enemies include the egg parasitoids *Trissolcus euschisti* (Ashmead) and *Telenomus podisi* Ashmead, and the tachinid fly *Euthera tentatrix* Loew (McPherson and Mohlenbrock 1976; M. Javahery, unpublished data).

4.21 *Euschistus variolarius* (Palisot de Beauvois)

The onespotted stink bug, *E. variolarius*, is distributed widely in North America south to the West Indies (Froeschner 1988). It is a minor pest, feeding on a wide variety of plants from vegetables and graminaceous plants to fruit trees (McPherson 1982). Its feeding can affect growth and yield of corn (Apriyanto et al. 1989b, Townsend and Sedlacek 1986, Sedlacek and Townsend 1988) and cause scarring of pears (Wilks 1964) and other fruits. No-till planted corn in wheat stubble appears to favor *E. variolarius* damage to corn (Annan and Bergman 1988). Corn plants injured by *E. servus* and *E. variolarius* produce tillers and have delayed silking (Apriyanto et al. 1989b). Adults are yellowish brown and 11.0 to 15.0 mm long. The eggs (Parish 1934, Esselbaugh 1946, Munyaneza and McPherson 1994) and first to fifth instars (Parish 1934, DeCoursey and Esselbaugh 1962, Munyaneza and McPherson 1994) have been described. The incubation period takes 10 to 15 days in the field and 9 days in the laboratory at 25°C; nymphal development lasts 35 to 50 days in the field and 32 to 40 days in the laboratory (23 to 25°C) (Esselbaugh 1948, McPherson 1982). Adults appear in July, feed on fruits or buds of food plants, and overwinter under leaf litter in bushes or in woodlands. This species completes one generation per year above latitude 40° north (Blatchley 1926, Rolston and Kendrick 1961, McPherson 1982). Natural enemies include the egg parasitoid *Telenomus podisi* Ashmead (Yeargan 1979) and the tachinid fly *Trichopoda pennipes* (F.) (Parish 1934).

4.22, 4.23 *Halyomorpha mista* Uhler and *Halyomorpha marmorea* F.

The brown marmorated stink bug, *H. mista*, *Plautia stali* Scott, and *Nezara antennata* Scott have become horticultural pests of some importance in Japan. *Halyomorpha mista* overwinters as adults and has a large host range including plants of economic importance such as pea, cucumber, and kidney beans (Kawada and Kitamura 1983a). Little is known about its life history. This bug is gregarious during overwintering periods. Its fecundity and oviposition period increase with the number of matings (Kawada and Kitamura 1983b). Another species, *H. marmorea*, has been reported as a severe pest of arecanut in India (Vidyasagar and Bhat 1986).

4.24, 4.25 *Loxa flavicollis* (Drury) and *Loxa deducta* (Walker)

These two stink bugs are secondary pests of a few crops, particularly in South America. *Loxa flavicollis* occurs on olive and tomato in Brazil (Silva et al. 1968). In Argentina, Rizzo (1976) referred to its occurrence in low numbers on cultivated plants but did not list the plant species. It also feeds on soybean in North and South America (Panizzi and Slansky 1985a). Adults are green and 18.0 to 20.0 mm long.

Loxa deducta feeds on soybean in South America (Panizzi and Slansky 1985a) and on *Citrus* sp. (D. Link, personal comunication). In general, it occurs in low numbers. Panizzi and Rossi (1991b) studied its nymphal and adult biology on leucaena, *Leucaena leucocephala* (Lam.), which is used as cattle food, and on soybean. Nymphal mortality in the laboratory was high (>50%); nymphal development (excluding the first instar) lasted approximately 36 to 57 days. Females laid two to five egg clusters during their life spans, with 27 to 66 eggs/female. In another study, however, Panizzi et al. (1998) found *Loxa deducta* in relatively high numbers on privet, *Ligustrum lucidum* Ait. (Oleaceae), on which they feed and reproduce. Laboratory studies indicated a much better performance on privet fruits compared with data in the literature when bugs were raised on other hosts (see references in Panizzi 1997). As with the former species, *Loxa deducta* is green. No data on damage to soybean or to any cultivated plant were found in the literature related to this species. Natural enemies include an egg parasite not yet identified (A. R. Panizzi, unpublished data).

4.26 *Murgantia histrionica* (Hahn)

The harlequin bug, *M. histrionica*, an exotic pest from Central America, was first recorded in the United States in 1864 (Walsh 1866). It has a variable number of generations, from three to four in North Carolina (Brett and Sullivan 1974) and from two to three in Virginia (Ludwig and Kok 1998a). It has been reported to feed on over 50 species of plants. Crucifers, including cabbage, collards, broccoli, Brussels sprouts, kale, mustard, turnip, and cauliflower are a few of the economically important crops attacked by this pest (Walker and Anderson 1933, White and Brannon 1933). Streams and Pimentel (1963) reported that the average life spans for males and females were 68.2 and 82.0 days, respectively. The eggs and various instars have been described (see references in McPherson 1982). Overwintering adults emerge in spring, mate after 6 to 9 days, and lay eggs 2 to 8 days later. Canerday (1965) reported that the first to fifth stadia on cabbage at 25°C were 4 .4, 9.5, 7.6, 9.9, and 14.8 days, respectively. Adults and nymphs feed on the leaves and stems of plants, causing blotching at the feeding site. Small plants attacked by the harlequin bug wilt, turn brown, and eventually die. Larger plants are able to withstand attack, but plant growth may be stunted

(White and Brannon 1933). Mustard and rape may be used as trap crops to prevent populations from reaching the main broccoli crop (Ludwig and Kok 1998b). Natural enemies of the harlequin bug include the egg parasitoids *Ooencyrtus johnsoni* (Howard) and *Trissolcus murgantiae* Ashmead (Ludwig and Kok 1998a). Swallows feed on the adults (Walker and Anderson 1933).

4.27   *Nezara antennata* Scott

The Oriental green stink bug, *N. antennata*, is distributed widely in Japan except for Hokkaido. It is a polyphagous pest, attacking approximately 80 plant species in 25 families. It has two or three generations per year with adults overwintering. The following spring, females begin ovipositing when the maximum temperature reaches 14 to 15°C, and adults of the first generation appear from late June to the middle of September. Oviposition is accelerated by short photoperiods (Noda 1984). *N. antennata* and *N. viridula* are sympatric in southern Japan, and the latter species generally is more important as a pest.

4.28   *Piezodorus lituratus* (F.)

This pentatomid occurs in most countries of the Palearctic region (Southwood and Leston 1959, Putschkov 1961, Javahery 1967). Host plants include gorse, *Ulex europaeus*; broom, *Sarothamus scoparius*; the legumes *Lupinus angustifolium*, *L. albus*, *Trifolium medium*, *Melilotus* sp., and *Medicago* sp.; and the woody Papilionaceae of the tribe Genisteae (Southwood and Leston 1959, Putschkov 1961, Javahery 1967).

Adults are reddish-brown in late summer and in early fall but become bright green during spring. Males and females are 9.53 and 11.26 mm long, respectively. The first to fifth instars are 2.3, 2.9, 3.9, 8.3, and 9.3 mm long, respectively. Eggs are barrel-shaped, blackish with two circles of white or creamish bands, and 1.1 mm long (Javahery 1967). The embryonic development and hatching mechanisms have been studied (Javahery 1967). Nymphal development on broom was 73.2 days. Females oviposited on buds, pods, twigs, and leaves in clusters of 14 to 21 eggs. Fecundity varied from 42 to 148 eggs per female on gorse and on broom, respectively. Longevity on broom for males and females was 59 and 100 days, respectively (Javahery 1967). Newly emerged adults migrated to forest or woodlands during sunny warm days and estivated for 2 to 4 months on trees in temperate climate or at higher altitude on mountains in subtropical climate. In the early fall after rain and cool periods, they migrated again or moved down to the ground and hibernated under grasses and leaves. *Piezodorus lituratus* completes one generation per year in its northern distribution with an obligatory diapause (Putschkov 1961, Javahery 1967).

Natural enemies include the scelionid egg parasitoids *Trissolcus semistriatus* Ness, *T. grandis* Thomson, *Telenomus truncatus* Mayr, and *T. sokolovi* Mayr, which can parasitize 65% of eggs during May to June and up to 98% of eggs at the end of summer (Javahery 1967). Tachinid flies are reported as parasitoids of adult *P. lituratus* (Putschkov 1961).

4.29   *Plautia affinis* Dallas

This green stink bug is common in southeastern Australia and occasionally is numerous on and harmful to ripening stone fruit, mulberry, grape, and vegetables (Fletcher 1985). It is closely related to *P. stali* Scott and is similar in size and color. As with *Dyctyotus caenosus* (Westwood), this bug also is commonly found in association with *Nezara viridula*, which is always the more important pest.

4.30   *Proxys punctulatus* (Palisot de Beauvois)

This is a pentatomid found in North, Central (Blatchley 1926), and South America (Silva et al. 1968). Ashmead (1895) reported it on cotton, and it feeds on soybean in the United States (Panizzi and Slansky 1985a). In Brazil, Silva et al. (1968) recorded it from white cherry. Vangeison and McPherson (1975) described the life history, immature stages, and laboratory rearing of this species. Another species of *Proxys*, *P. albopunctulatus* (Palisot de Beauvois), occurs in the Neotropical region and is a secondary pest of velvetbean (Silva et al. 1968, Froeschner 1988). *Proxys* sp. is rarely found feeding on soybean in Brazil (A. R. Panizzi, unpublished data). Therefore, these species are considered of minor economic importance, requiring no control measures.

4.31 *Scotinophara coarctata* (F.)

The Malayan black bug, *S. coarctata*, is a pest of rice in several Asian countries, including Malaysia and the Philippines (Corbett and Yusope 1972, Barrion et al. 1982). The bugs are attracted to rice fields where they concentrate and quickly reach high populations (Ito et al. 1993). Field infestations occur at any plant growth stage. Insect feeding on young plants causes plant discoloration with desiccation of leaves, and plants may die under heavy attack. Attacks on older plants may cause dead panicles. Newly emerged panicles had the number of filled grains reduced with one or two bugs feeding during 4 days (Morrill et al. 1995). The injury level of three bugs per hill has been established (Heinrichs et al. 1987), and insecticides commonly are applied to avoid bug outbreaks.

4.32 *Thyanta custator accerra* McAtee

The redshouldered stink bug, *T. custator accerra*, is a minor pest of crops in North America, having been recorded from numerous cultivated and uncultivated plants. Among the cultivated plants, corn, sorghum, wheat, soybean, alfalfa, bean, and peach are mentioned in the literature (references in McPherson 1982). As with other members of the genus, it frequently is observed on graminaceous plants. This species overwinters as adults and has two generations per year. Natural enemies include the egg parasitoid *Telenomus podisi* Ashmead and the tachinid flies *Cylindromyia fumipennis* (Bigot), *Gymnoclytia occidua* (Walker), and *Euclytia flava* (Townsend) (Oetting and Yonke 1971).

4.33 *Thyanta pallidovirens* (Stål)

This pentatomid is distributed through the far western states of the United States and into British Columbia (Canada) (Schotzko and O'Keeffe 1990a). It is a pest of nut crops in California causing epicarp lesions, nut abortion, and kernel necrosis (Rice et al. 1985, 1988; Michailides et al. 1987). It also damages various legumes, tomatoes, and brassica crops (Schotzko and O'Keeffe 1990b, Zalom et al. 1997). It overwinters on the ground in sheltered places in and around orchards (Rice et al. 1988). In the laboratory, it takes approximately 25 days to develop from egg to adult under long-day conditions, and adults can live for over 2 months (Schotzko and O'Keeffe 1990a). The reproductive behavior of *T. pallidovirens* has been investigated (e.g., Schotzko and O'Keeffe 1990a, Wang and Millar 1997).

4.34 *Vitellus* spp.

A number of species in this genus occasionally are found on citrus in Australia, particularly *V. warreni* Gross and McDonald, *V. antenna* Breddin, and *V. rufolineus* (Walker). These bugs are similar to each other, both bright green to olive green with dark brown or carmine pronotal spines (Gross and McDonald 1994). They look like smaller versions of the major citrus pest *Biprorulus bibax* Breddin (spined citrus bug). *Vitellus warreni* often is found on citrus in the Murrumbidgee Irrigation Area of southern New South Wales and will feed on oranges in the laboratory. However, *Vitellus* spp. are never present in sufficient numbers to cause economic damage.

5. PROGNOSIS

Many species of phytophagous pentatomids are important economic pests. Despite the vast amount of information regarding pest species and control measures used, their damage potential to crop production remains high. In general, there is a lack of holistic studies of the interactions of these bugs with their wild and cultivated host plants and with their overwintering sites. Particularly in the control area, more information is needed regarding the time of invasion of overwintering populations into crops and the movement of bugs between wild and cultivated plants. This knowledge will increase the efficacy and efficiency of insecticide use and the potential manipulation of natural enemies inhabiting these systems.

6. CONCLUSIONS

Managing economically damaging populations of stink bugs is a complex and diverse activity. It requires good knowledge of the cropping systems involved, the economics of control practices and potential damage losses, the environmental impact of various management practices, the ecological and biological responses associated with specific plant/pest interactions, and the impact of cultural practices on current and subsequent stink bug pest populations. Most, if not all, of the stink bug pests reviewed in this chapter are attracted to the reproductive stages of the host plants they infest. Until these hosts reach the desired reproductive stage, the bugs often feed and reproduce on numerous wild and cultivated plants (see review by Panizzi 1997). Then they move into the major planting of row crops, vegetables, fruit, and nut-producing trees where extensive economic losses can occur because of pesticide control costs and quality and yield reductions.

Many crops escape major stink bug injury or sustain little damage because of various biotic and/or abiotic agents that keep bug populations low directly (i.e., parasitoids, predators, entomopathogens) or indirectly (e.g., temperature extremes, humidity, food and habitat availability). Crop phenology also may not be synchronized with the stink bug's life cycle, preventing significant damage. However, if an economically damaging infestation of stink bugs develops, several management options usually are available.

One option is that the producer/processor/consumer can tolerate the stink bug problem and accept the crop quality and yield reductions that occur. Another is to tolerate the infestation up to an economic injury level, then treat with an insecticide to keep crop losses to a minimum.

Other practices can be applied to avoid stink bug infestations in the primary crop. These include planting trap crops, planting early- or late-maturing varieties that escape stink bug migrations, and planting varieties that are resistant to stink bug feeding, all of which can help reduce or avoid stink bug damage. Finally, enhancing parasitoid populations by innoculative/inundative releases, two-stage harvesting in cereals, destroying alternate plant hosts in or around cultivated fields, avoiding the planting of certain ground cover crops in orchards, and spraying alternate crops with insecticides before the bugs move to the primary crop also can be effective stink bug management practices. All these practices help minimize the use of pesticides, and thereby help protect the environment.

7. ACKNOWLEDGMENTS

A.R. Panizzi thanks the Conselho Nacional de Desenvolvimento Científico e Tecnológico (CNPq) and the Centro Nacional de Pesquisa de Soja of Embrapa of the federal government of Brazil for providing financial support for a stay at the University of Connecticut to work on this chapter.

8. REFERENCES CITED

Abasa, R. O. 1973. Oviposition, fertility, and longevity and their relation to copulation in *Antestiopsis lineaticollis* (Heteroptera: Miridae). Entomol. Exp. Appl. 16: 178–184.

Albuquerque, G. S. 1989. Ecologia de populações, biologia e estratégias da história de vida de *Oebalus poecilus* (Dallas, 1851) (Hemiptera: Pentatomidae). M.Sc. thesis, Universidade Federal do Rio Grande do Sul, Porto Alegre, RS, Brazil. 309 pp.

1990. Primeiro registro de ocorrência de *Oebalus poecilus* (Dallas, 1851) (Hemiptera: Pentatomidae) na cultura do sorgo (*Sorghum bicolor* (L.) Moench). An. Soc. Entomol. Brasil 19: 219–220.

1993. Planting time as a tactic to manage the small rice stink bug, *Oebalus poecilus* (Hemiptera, Pentatomidae), in Rio Grande do Sul, Brazil. Crop Prot. 12: 627–630.

Aldrich, J. R., J. E. Oliver, W. R. Lusby, J. P. Kochansky, and J. A. Lockwood. 1987. Pheromone strains of the cosmopolitan pest, *Nezara viridula* (Heteroptera: Pentatomidae). J. Exp. Zool. 244: 171–175.

Alexandrov, N. V., and H. Mirzayan. 1949. Les punaises des céréales appartenant au genre *Aelia*. Entomol. Phytopathol. Appl. 9: 27–32.

Amaral Fº., B. F., C. C. Lima, C. M. R. Silva, and F. L. Consoli. 1992. Influência da temperatura no estágio de ovo e adulto de *Thyanta perditor* (Fabricius, 1794) (Heteroptera, Pentatomidae). An. Soc. Entomol. Brasil 21: 15–20.

Annan, I. B., and M. K. Bergman. 1988. Effects of the onespotted stink bug (Hemiptera: Pentatomidae) on growth and yield of corn. J. Econ. Entomol. 81: 649–653.

Apriyanto, D., J. D. Sedlacek, and L. H. Townsend. 1989a. Feeding activity of *Euschistus servus* and *E. variolarius* (Heteroptera: Pentatomidae) and damage to an early growth stage of corn. J. Kansas Entomol. Soc. 62: 392–399.

Apriyanto, D., L. H. Townsend, and J. D. Sedlacek. 1989b. Yield reduction from feeding by *Euschistus servus* and *E. variolarius* (Heteroptera: Pentatomidae) on stage V2 field corn. J. Econ. Entomol. 82: 445–448.

Asgarali, J., and P. Ramkalup. 1985. Study of *Lincus* sp. (Pentatomidae) as the possible vector of hartrot in coconut. Surinam Agric. 33: 56–61.

Ashmead, W. H. 1895. Notes on cotton insects found in Mississippi. Insect Life 7: 320–326.

Ávila, C. J., and A. R. Panizzi. 1995. Occurrence and damage by *Dichelops* (*Neodichelops*) *melacanthus* (Dallas) (Heteroptera: Pentatomidae) in corn. An. Soc. Entomol. Brasil 24: 193–194.

Babrakzai, Z. H., and I. Hodek. 1987. Diapause induction and termination in a population of *Dolycoris baccarum* (Heteroptera: Pentatomoidea) from central Bohemia. Vestn. Cesk. Spol. Zool. 51: 85–88.

Baloch, G. M. 1973. Natural enemies of *Axiagastus cambelli* Distant (Hemiptera: Pentatomidae) on the Gazelle Peninsula, New Britain. Papua New Guinea Agric. J. 24: 41–45.

Barbour, K. S., J. R. Bradley, Jr., and J. S. Bacheler. 1988. Phytophagous stink bugs in North Carolina cotton: an evaluation of damage potential. Beltwide Cotton Prod. Res. Conf. Proc.: 280–282.

1990. Reduction in yield and quality of cotton damaged by green stink bug (Hemiptera: Pentatomidae). J. Econ. Entomol. 83: 842–845.

Barrion, A. T., O. Machida, and J. A. Litsinger. 1982. The Malayan black bug, *Scotinophara coarctata* (F.) (Hemiptera: Pentatomidae): a new rice pest in the Philippines. Int. Rice Res. Newsl. 7: 6–7.

Batra, H. N., and S. Sarup. 1962. Technique of mass-rearing of the painted bug, *Bagrada cruciferarum* Kirk. (Heteroptera: Pentatomidae). Indian Oilseeds J. 6: 135–145.

Beltzer, A. H., M. R. Salusso, and E. H. Bucher. 1988. Alimentacion del nacunda (*Podager nacunda*) en Parana (Entre Rios). Hornero 13: 47–52.

Bensebbane, C. G. 1981. Les punaises des blés en Algèrie. Bull. OEPP 11: 33–38.

Blatchley, W. S. 1926. Heteroptera or True Bugs of Eastern North America, with Especial Reference to the Faunas of Indiana and Florida. Nature Publ. Co., Indianapolis, Indiana, U.S.A. 1116 pp.

Boff, M. I. C., and P. Boff. 1995. Ocorrência de *Cordyceps nutans* Pat. em percevejo do colmo do arroz, *Tibraca limbativentris* Stål (Heteroptera: Pentatomidae). An. Soc. Entomol. Brasil 24: 177–178.

Borges, M., P. C. Jepson, and P. E. Howse. 1987. Long range mate location and close range courtship behaviour of the green stink bug, *Nezara viridula* and its mediation by sex pheromones. Entomol. Exp. Appl. 44: 205–212.

Borror, D. J., C. A. Triplehorn, and N. F. Johnson. 1989. An Introduction to the Study of Insects. Saunders College Publ., Philadelphia, Pennsylvania, U.S.A. 875 pp.

Brennan, B. M., F. Chang, and W. C. Mitchell. 1977. Physiological effects on sex pheromone communication in the southern green stink bug, *Nezara viridula*. Environ. Entomol. 6: 167–173.

Brett, C. H., and M. J. Sullivan. 1974. The use of resistant varieties and other cultural practices for control of insects on crucifers in North Carolina. North Carolina Agric. Exp. Stn. Bull. 449: 1–31.

Brewer, F. D., and W. A. Jones, Jr. 1985. Comparison of meridic and natural diets on the biology of *Nezara viridula* (Heteroptera: Pentatomidae) and eight other phytophagous Heteroptera. Ann. Entomol. Soc. Amer. 78: 620–625.

Brézot, P., C. Malosse, and M. Renou. 1993. Étude del'atractivité de la phéromone des mâles chez *Nezara viridula* L. (Heteroptera: Pentatomidae). C. R. Acad. Sci. Paris, Sci. Vie 316: 671–675.

Brown, E. S. 1962. Notes on the systematics and distribution of some species of *Aelia* Fabr. (Hemiptera, Pentatomidae) in the middle East, with special reference to the *rostrata* group. Ann. Mag. Nat. Hist. 5: 129–145.

1965. Notes on the migration and direction of flight of *Eurygaster* and *Aelia* species (Hemiptera, Pentatomoidea) and their possible bearing on invasions of cereal crops. J. Anim. Ecol. 34: 93–107.

Busoli, A. C., F. M. Lara, J. Grazia, and O. A. Fernandes. 1984. Ocorrência de *Thyanta perditor* (Fabricius, 1794) (Heteroptera, Pentatomidae) danificando sorgo em Jaboticabal, São Paulo, Brasil. An. Soc. Entomol. Brasil 13: 179–181.

Buxton, G. M., D. B. Thomas, and R. C. Froeschner. 1983. Revision of the species of the *sayi*-group of *Chlorochoa* Stal (Hemiptera: Pentatomidae). California Dept. Food Agric., Occas. Pap. Entomol. 29: 1–25.

Calcagnolo, G., A. A. Massariol, and D. A. Oliveira. 1977. Estudo da eficiência de inseticidas no combate de percevejos pentatomídeos em soja. O Biológico 43: 97–102.

Calhoun, D. S., J. E. Funderburk, and I. D. Teare. 1988. Soybean seed crude protein and oil levels in relation to weight, developmental time, and survival of southern green stink bug (Hemiptera: Pentatomidae). Environ. Entomol. 17: 727–729.

Callan, E. McC. 1948. The Pentatomidae, Cydnidae and Scutelleridae of Trinidad, B.W.I. Proc. R. Entomol. Soc. London (B)17: 115–124.

Canerday, T. D. 1965. On the biology of the harlequin bug, *Murgantia histrionica*. Ann. Entomol. Soc. Amer. 58: 931–932.

Chandler, S. C. 1950. Peach insects of Illinois and their control. Illinois Nat. Hist. Surv. Circ. 43: 1–63.

Cherry, R., D. Jones, and C. Deren. 1998. Establishment of a new stink bug pest, *Oebalus ypsilongriseus* (Hemiptera: Pentatomidae) in Florida rice. Florida Entomol. 81: 216–220.

China, W. E., and N. Lodos. 1959. A study of the taxonomic characters of some species of *Aelia* F. (Hemiptera, Pentatomidae). Ann. Mag. Nat. Hist. 2: 577–602.

Cilas, C., B. Bouyjou, and B. Decazy. 1998. Frequency and distribution of *Antestiopsis orbitalis* Westwood (Hem., Pentatomidae) in coffee plantations in Burundi: implications for sampling techniques. J. Appl. Entomol. 122: 601–606.

Cividanes, F. J. 1992. Determinação das exigências térmicas de *Nezara viridula* (L., 1758), *Piezodorus guildinii* (West., 1837) e *Euschistus heros* (Fabr., 1798) (Heteroptera: Pentatomidae) visando ao seu zoneamento ecológico. Ph.D. thesis, University of São Paulo, Piracicaba, Brazil.

Cogburn, R. R., and M. O. Way. 1991. Relationship of insect damage and other factors to the incidence of speckback, a site-specific lesion on kernels of milled rice. J. Econ. Entomol. 84: 987–995.

Čokl, A., M. Gogala, and M. Jez. 1972. The analysis of the acoustic signals of the bug *Nezara viridula* (L.). Biol. Vestn. 20: 47–53.

Čokl, A., M. Gogala, and A. Blazevic. 1978. Principles of sound recognition in three pentatomid bug species (Heteroptera). Biol. Vestn. 26: 81–94.

Conradi-Larsen, E., and L. Somme 1973. Notes on the biology of *Dolycoris baccarum* L. (Het., Pentatomidae). Nor. Entomol. Tidsskr. 20: 245–247.

Coombs, M., and S. A. Khan. 1998. Population levels and natural enemies of *Plautia affinis* Dallas (Hemiptera: Pentatomidae) on raspberry, *Rubus idaeus* L., in south-eastern Queensland. Aust. J. Entomol. 37: 125–129.

Corbett, G. H., and M. Yusope. 1972. *Scotinophara coarctata* (F.) (the black bug of padi). Malays. Agric. J. 12: 91–107.

Corrêa-Ferreira, B. S. 1984. Incidência do parasitóide *Eutrichopodopsis nitens* Blanchard, 1966 em populações do percevejo verde *Nezara viridula* (Linnaeus, 1758). An. Soc. Entomol. Brasil 13: 321–330.

1986. Ocorrência natural do complexo de parasitóides de ovos de percevejos da soja no Paraná. An. Soc. Entomol. Brasil 15: 189–199.

1993. Utilização do parasitóide de ovos *Trissolcus basalis* (Wollaston) no controle de percevejos da soja. CNPSo/EMBRAPA, Londrina, Circ. Tec. 11: 1–40.

Corrêa-Ferreira, B. S., and F. Moscardi. 1995. Seasonal occurrence and host spectrum of egg parasitoids associated with soybean stink bugs. Biol. Control 5: 196–202.

1996. Biological control of soybean stink bugs by inoculative releases of *Trissolcus basalis*. Entomol. Exp. Appl. 79: 1–7.

Corrêa, B. S., A. R. Panizzi, G. G. Newman, and S. G. Turnipseed. 1977. Distribuição geográfica e abundância estacional dos principais insetos-pragas da soja e seus predadores. An. Soc. Entomol. Brasil 6: 40–50.

Corrêa-Ferreira, B. S., M. C. Nunes, and L. D. Uguccioni. 1998. Ocorrência do parasitóide *Hexacladia smithii* Ashmead em adultos de *Euschistus heros* (F.) no Brasil. An. Soc. Entomol. Brasil 27: 495–498.

Corseuil, E., and F. Z. Cruz. 1971. Ação de alguns inseticidas em pulverização sobre *Tibraca limbativentris*. Agron. Sulriogr. 7: 47–52.

1972. Ensaio com inseticidas em pó sobre *Oebalus poecilus*. Pesq. Agropec. Bras. 7: 173–175.

Corso, I. C. 1984. Constatação do agente causal da mancha-de-levedura em percevejos que atacam a soja no Paraná. An. III Semin. Nac. Pesq. Soja, Campinas, SP, pp. 152–157.

1990. Uso de sal de cozinha na redução da dose de inseticida para controle de percevejos da soja. Comunic. Téc., Embrapa Soja, 45: 1–7.

Corso, I. C., and D. L. Gazzoni. 1998. Sodium chloride: an insecticide enhancer for controlling pentatomids on soybeans. Pesq. Agropec. Bras. 33: 1563–1571.

Costa, E. C., and D. Link. 1974. Incidência de percevejos em soja. Rev. Cent. Cien. Rur. 4: 397–400.

1977. Danos causados por algumas espécies de Pentatomidae em duas variedades de soja. Rev. Cent. Cien. Rur. 7: 199–206.

1982. Dispersão de adultos de *Piezodorus guildinii* e *Nezara viridula* (Hemiptera: Pentatomidae) em soja. Rev. Cent. Cien. Rur. 12: 51–57.

1988. Eficiência de alguns inseticidas, no controle do percevejo *Oebalus poecilus* (Dallas, 1851), na cultura do arroz. An. Reun. Cult. Arroz Irrig. 17: 240–243.

1992a. Dispersão de *Tibraca limbativentris* Stål 1860 (Hemiptera, Pentatomidae) em arroz irrigado. An. Soc. Entomol. Brasil 21: 197–202.

1992b. Avaliação de danos de *Tibraca limbativentris* Stal, 1860 (Hemiptera, Pentatomidae) em arroz irrigado. An. Soc. Entomol. Brasil 21: 187–195.

1992c. Avaliação de inseticidas no controle de percevejos na cultura do arroz irrigado. Lavoura Arrozeira 45: 21–23.

Couilloud, R. 1989. Hétéroptères de predateurs du cotonnier en Afrique et a Madagascar (Pyrrhocoridae, Pentatomidae, Coreidae, Alydidae, Rhopalidae, Lygaeidae). Cotton Fibres Trop. 44: 185–227.

Cruz, F. Z., and E. Corseuil. 1970. Notas sobre o percevejo grande do arroz. Lavoura Arrozeira 23: 53–56.

Daugherty, D. M. 1967. Pentatomidae as vectors of yeast-spot disease of soybeans. J. Econ. Entomol. 60: 147–152.

Daugherty, D. M., and J. E. Foster. 1966. Organism of yeast-spot disease isolated from rice damaged by rice stink bug. J. Econ. Entomol. 59: 1282–1283.

Daugherty, D. M., M. H. Neustadt, C. W. Gehrke, L. E. Cavanah, L. F. Williams, and D. E. Green. 1964. An evaluation of damage to soybeans by brown and green stink bugs. J. Econ. Entomol. 57: 719–722.

Davis, C. J. 1967. Progress in the biological control of the southern green stink bug, *Nezara viridula* variety *smaragdula* (Fabricius) in Hawaii (Heteroptera: Pentatomidae). Mushi 39: 9–16.

DeCoursey, R. M., and C. O. Esselbaugh. 1962. Descriptions of the nymphal stages of some North American Pentatomidae (Hemiptera-Heteroptera). Ann. Entomol. Soc. Amer. 55: 323–342.

Degaspari, N., and S. A. Gomez. 1979. Distribuição geográfica e abundância estacional de insetos pragas da soja e seus inimigos naturais. III Reun. Pesq. Soja, Região Central, pp. 182–185.

Deitz, L. L., J. W. Van Duyn, J. R. Bradley, Jr., R. L. Rabb, W. M. Brooks, and R. E. Stinner. 1976. A guide to the identification and biology of soybean arthropods in North Carolina. North Carolina Agric. Exp. Stn. Tech. Bull. 238: 1–264.

Delucchi, V. L. 1961. Le complexe des *Asolcus* Nakagawa (*Microphanurus* Kieffer) (Hymenoptera, Proctotrupoidea), parasites oophages des punaises des céréals au Moroc et an Moyan-orient. Cah. Rech. Agron. 14: 41–67.

Del Vecchio, M. C., and J. Grazia. 1992a. Estudo dos imaturos *Oebalus ypsilongriseus* (De Geer, 1773): I — Descrição do ovo e desenvolvimento embrionário (Heteroptera: Pentatomidae). An. Soc. Entomol. Brasil 21: 375–382.

1992b. Obtenção de posturas de *Oebalus ypsilongriseus* (De Geer, 1773) em laboratório (Heteroptera: Pentatomidae). An. Soc. Entomol. Brasil 21: 367–373.

1993a. Estudo dos imaturos de *Oebalus ypsilongriseus* (De Geer, 1773): II — Descrição das ninfas (Heteroptera: Pentatomidae). An. Soc. Entomol. Brasil 22: 109–120.

1993b. Estudo dos imaturos de *Oebalus ypsilongriseus* (De Geer, 1773): III — Duração e mortalidade dos estágios de ovo e ninfa (Heteroptera: Pentatomidae). An. Soc. Entomol. Brasil 22: 121–129.

Del Vecchio, M. C., J. Grazia, and R. Hildebrand. 1988. Estudo dos imaturos de pentatomídeos (Heteroptera) que vivem sobre soja (*Glycine max* (L.) Merrill): V — *Acrosternum bellum* Rolston, 1983 com a descrição da genitália da fêmea. An. Soc. Entomol. Brasil 17: 467 482.

De Santis, L., and L. Esquivel. 1966. Tercera lista de himenopteros parasitos y predadores de los insectos de la Republica Argentina. Rev. Mus. La Plata 9: 47–215.

DeWitt, N. B., and G. L. Godfrey. 1972. The literature of arthropods associated with soybeans. II. A bibliography of the southern green stink bug *Nezara viridula* (Linneaus) [sic] (Hemiptera: Pentatomidae). Illinois Nat. Hist. Surv. Biol. Notes 78: 1–23.

Dhaliwal, H. S., and B. D. Goma. 1979. Seasonal abundance of various pests on cauliflower seed crop lower hills at Solan. Indian J. Ecol. 6: 101–109.

Dias, J. A. C. S., A. S. Costa, V. A. Yuki, and N. P. Granja. 1985. Murcha transitória do ponteiro da batata associada a alimentação do percevejo-do-tomateiro. An. Soc. Entomol. Brasil 14: 335–340.

Dolling, W. R. 1984. Pentatomid bugs (Hemiptera) that transmit a flagellate disease of cultivated palms in South America. Bull. Entomol. Res. 74: 473–476.

Douce, G. K., and E. F. Suber [eds.]. 1986. Summary of losses from insect damage and costs of control in Georgia, 1985. Univ. Georgia Spec. Publ. 55: 27.

Douglas, W. A., and J. W. Ingram. 1942. Rice-field insects. U.S. Dept. Agric. Circ. 632: 1–32.

Drake, C. J. 1920. The southern green stink-bug in Florida. Florida State Plant Board Q. Bull. 4: 41–94.

Drees, B. M. 1983. Rice insect management. Texas Agric. Ext. Serv. Publ. B-1445: 1–12.

Dutcher, J. D., and J. W. Todd. 1983. Hemipteran kernel damage of pecan. Pp. 1–11 *in* J. A. Payne [ed.], Pecan Pest Management — Are We There? Misc. Publ. Entomol. Soc. Amer. 13: 1–140.

Esselbaugh, C. O. 1946. A study of the eggs of the Pentatomidae (Hemiptera). Ann. Entomol. Soc. Amer. 39: 667–691.

Esselbaugh, C. O. 1948. Notes on the bionomics of some midwestern Pentatomidae. Entomol. Amer. 28: 1–73.

Ezueh, M. I., and S. O. Dina 1980. Pest problems of soybeans and control in Nigeria. Pp. 275–283 *in* F. T. Corbin [ed.], World Soybean Research Conference II, Westview Press, Boulder, Colorado, U.S.A. 897 pp.

Fennah, R. G. 1935. A preliminary list of the Pentatomidae of Trinidad, B.W.I. Trop. Agric. 12: 192–194.

Ferreira, E. 1980. Efeitos da integração de meios de controle sobre os insetos do arroz de sequeiro. Ph.D. dissertation, Universidade de São Paulo, Piracicaba, Brazil. 129 pp.

Ferreira, E., and P. M. Silveira. 1991. Dano de *Thyanta perditor* (Hemiptera: Pentatomidae) em trigo (*Triticum aestivum* L.). An. Soc. Entomol. Brasil 20: 165–171.

Ferreira, B. S. C., and A. R. Panizzi. 1982. Percevejos-pragas da soja no norte do Paraná: abundância em relação a fenologia da planta e hospedeiros intermediários. An. II Semin. Nac. Pesq. Soja, vol. II, pp. 140–151.

Fletcher, M. 1985. Plant bugs. New South Wales Agric. Agfact AE. 38: 1–7.

Fortes, N. D. F., and J. Grazia. 1990. Estudo dos estágios imaturos de *Edessa rufomarginata* (De Geer, 1773) (Heteroptera–Pentatomidae). An. Soc. Entomol. Brasil 19: 191–200.

Foster, R. E., R. H. Cherry, and D. B. Jones. 1989. Spatial distribution of the rice stink bug (Heteroptera: Pentatomidae) in Florida rice. J. Econ. Entomol. 82: 507–509.

Fourar, R., and F. Fleurat-Lessard. 1997. Effects of damage by wheat bug, *Aelia germari* [Hemiptera: Pentatomidae], on grain quality and on reproductive performance of the rice weevil, *Sitophilus oryzae* [Coleoptera: Curculionidae], on harvested grain. Phytoprotection 78: 105–116.

Fraga, C. P., and L. H. Ochoa. 1972. Aspectos morfologicos y bioecologicos de *Piezodorus guildinii* (West.) (Hemiptera, Pent.). IDIA-Supl. 28: 103–117.

Froeschner, R. C. 1988. Family Pentatomidae Leach, 1815. The stink bugs. Pp. 544–597 *in* T. J. Henry and R. C. Froeschner [eds.], Catalog of the Heteroptera, or True Bugs, of Canada and the Continental United States. E. J. Brill, New York, New York, U.S.A. 958 pp.

Fukuda, H., and A. Fujiie. 1988. Life cycle of *Plautia stali* Scott. Bull. Chiba Exp. Stn. 29: 173–180.

Galileo, M. H. M., and E. A. Heinrichs. 1979. Danos causados à soja em diferentes níveis e épocas de infestação durante o crescimento. Pesq. Agropec. Bras. 14: 279–282.

Galileo, M. H. M., H. A. O. Gastal, and J. Grazia. 1977. Levantamento populacional de Pentatomidae (Hemiptera) em cultura de soja (*Glycine max* (L.) Merr.) no município de Guaíba, Rio Grande do Sul. Rev. Bras. Biol. 37: 111–120.

Gerard, B. M. 1965. *Bathycoelia thalassina* (Herrich-Schaeffer), (Hemiptera: Pentatomidae); a pest of *Theobroma cacao* L. Nature 207: 881.

Ginting, C. M., and D. Chenon. 1987. Utilization de la technique d'absorption racionaire d'insecticides sistemiques pour une protection à long term des cocotiers et autres cultures industrialles. Oléagineux 42: 63–70.

Gomez, S. A. 1980. Informações preliminares sobre os danos causados ao trigo pelo percevejo *Thyanta perditor* (F.) (Hemiptera: Pentatomidae). Pesqui. Andam., UEPAE-Dourados, No. 3.

Grazia, J. 1977. Revisão dos pentatomídeos citados no "Quarto Catálogo dos Insetos que Vivem nas Plantas do Brasil" (Hemiptera-Pentatomidae-Pentatomini). Dusenia 10: 161–174.

1978. Revisão do genero *Dichelops* Spinola, 1837 (Heteroptera, Pentatomidae, Pentatomini). Iheringia 53: 3–119.

Grazia, J., M. C. Del Vecchio, F. M. P. Balestieri, and Z. A. Ramiro. 1980. Estudo das ninfas de pentatomídeos (Heteroptera) que vivem sobre soja (*Glycine max* (L.) Merrill): I — *Euschistus heros* (Fabricius, 1798) e *Piezodorus guildinii* (Westwood, 1837). An. Soc. Entomol. Brasil 9: 39–51.

Grazia, J., M. C. Del Vecchio, and R. Hildebrand. 1982a. Estudo das ninfas de pentatomídeos (Heteroptera) que vivem sobre soja [*Glycine max* (L.) Merrill]: IV — *Acrosternum impicticorne* (Stal, 1872). An. Soc. Entomol. Brasil 11: 261–268.

Grazia, J., M. C. Del Vecchio, C. T. Teradaira, and Z. A. Ramiro. 1982b. Estudo das ninfas de pentatomídeos (Heteroptera) que vivem sobre soja (*Glycine max* (L.) Merrill): II — *Dichelops* (*Neodichelops*) *furcatus* (Fabricius, 1775). An. II Semin. Nac. Pesq. Soja, Brasilia, Distrito Federal, Brazil. pp. 92–103.

Grazia, J., M. C. Del Vecchio, R. Hildebrand, and Z. A. Ramiro. 1982c. Estudo das ninfas de pentatomídeos (Heteroptera) que vivem sobre soja (*Glycine max* (L.) Merrill): III — *Thyanta perditor* (Fabricius, 1794). An. Soc. Entomol. Brasil 11: 139–146.

Grazia, J., R. Hildebrand, and A. Mohr. 1984. Estudo das ninfas de *Arvelius albopunctatus* (De Geer, 1773) (Heteroptera, Pentatomidae). An. Soc. Entomol. Brasil 13: 141–150.

Greathead, D. J. 1966a. A taxonomic study of the species of *Antestiopsis* (Hemiptera: Pentatomidae) associated with *Coffea arabica* in Africa. Bull. Entomol. Res. 56: 515–554.

1966b. The parasites of *Antestiopsis* spp. (Hem. Pentatomidae) in East Africa, and a discussion of the possibilities of biological control. Commonw. Inst. Biol. Control Tech. Bull. 7: 113–137.

1968. On the taxonomy of *Antestiopsis* spp. (Hem., Pentatomidae) of Madagascar, with notes on their biology. Bull. Entomol. Res. 59: 307–315.

Greathead, D. J., and M. Bigger. 1967. Notes on the biology of *Bogosia rubens* Villeneuve (Dipt., Tachinidae) a parasite of *Antestiopsis* spp. (Hem., Pentatomidae) and its introduction to Mt. Kilimanjaro, Tanzania Commonw. Inst. Biol. Control Tech. Bull. 9: 1–8.

Grimm, C., and E. Führer. 1998. Population dynamics of true bugs (Heteroptera) in physic nut (*Jatropha curcas*) plantations in Nicaragua. J. Appl. Entomol. 122: 515–521.

Grist, D. H., and R. J. A. W. Lever. 1969. Pests of Rice. Longmans, London, U.K. 520 pp.

Gross, G. F. 1975. Plant-Feeding and Other Bugs (Hemiptera) of South Australia: Heteroptera-Part I. S. Austral. Govt. Printer, Adelaide, Australia. 250 pp.

1976. Plant-Feeding and Other Bugs (Hemiptera) of South Australia: Heteroptera. Part II. S. Austral. Govt. Printer, Adelaide, Australia. 501 pp.

Gross, G. F., and F. J. D. McDonald. 1994. Revision of *Vitellus* Stål and *Avicenna* Distant in Australia (Hemiptera; Pentatomidae). J. Austral. Entomol. Soc. 33: 265–274.

Hall, IV, D. G., and G. L. Teetes. 1981. Alternate host plants of sorghum panicle-feeding bugs in southeast central Texas. Southwest. Entomol. 6: 220–228.
1982a. Damage by rice stink bug to grain sorghum. J. Econ. Entomol. 75: 440–445.
1982b. Damage to grain sorghum by southern green stink bug, conchuela, and leaffooted bug. J. Econ. Entomol. 75: 620–625.
Hallman, G. 1979. Importancia de algumas relaciones naturales plantas-artropodos en la agricultura de la zona calida del Tolima central. Rev. Colomb. Entomol. 5: 19–26.
Hallman, G. J., C. G. Morales, and M. C. Duque. 1992. Biology of *Acrosternum marginatum* (Heteroptera: Pentatomidae) on common beans. Florida Entomol. 75: 190–196.
Harper, J. K., M. O. Way, B. M. Drees, M. E. Rister, and J. W. Mjelde. 1993. Damage function analysis for the rice stink bug (Hemiptera: Pentatomidae). J. Econ. Entomol. 86: 1250–1258.
Harper, J. K., J. W. Mjelde, M. E. Rister, M. O. Way, and B. M. Drees. 1994. Developing flexible economic thresholds for pest management using dynamic programming. J. Agric. Appl. Econ. 26: 134–147.
Harris, V. E., and J. W. Todd. 1980a. Duration of immature stages of the southern green stink bug, *Nezara viridula* (L.), with a comparative review of previous studies. J. Georgia Entomol. Soc. 15: 114–124.
Harris, V. E., and J. W. Todd. 1980b. Male-mediated aggregation of male, female and 5th instar southern green stink bug and concomitant attraction of a tachinid parasite, *Trichopoda pennipes*. Entomol. Exp. Appl. 27: 117–126.
1981. Rearing the southern green stink bug, *Nezara viridula*, with relevant aspects of its biology. J. Georgia Entomol. Soc. 16: 203–210.
Heinrichs, E. A., I. T. Domingo, and E. H. Castillo. 1987. Resistance and yield responses of rice cultivars to the black bug, *Scotinophara coarctata* (F.) (Hemiptera: Pentatomidae). J. Plant Prot. Trop. 4: 55–64.
Hely, P. C., G. Pasfield, and J. G. Gellatley. 1982. Insect Pests of Fruit and Vegetables in NSW. Inkata Press, Melbourne, Australia. 312 pp.
Higuchi, H. 1992. Population prevalence of occurrence and spatial distribution pattern of *Piezodorus hybneri* adults (Heteroptera: Pentatomidae) on soybeans. Appl. Entomol. Zool. 27: 363–369.
1993. Seasonal prevalence of egg parasitoid attacking *Piezodorus hybneri* (Heteroptera: Pentatomidae) on soybeans. Appl. Entomol. Zool. 28: 347–352.
1994. Seasonal prevalence and mortality factors of eggs of *Piezodorus guildinii* Gmelin (Heteroptera: Pentatomidae) in a soybean field. Jpn. J. Appl. Entomol. Zool. 38: 17–21.
Higuchi, H., and N. Mizutani. 1993. Ovarian development state and oviposition of adult females of *Piezodorus hybneri* (Heteroptera: Pentatomidae) collected in soybean field. Jpn. J. Appl. Entomol. Zool. 37: 5–9.
Hoffmann, M. P., L. T. Wilson, and F. G. Zalom. 1987. Control of stink bugs in tomatoes. California Agric. 41 (5–6): 4–6.
Honda, K. 1985. A note on *Cylindromyia brassicaria* (Fabr.) (Diptera: Tachinidae), a parasite of the sloe bug, *Dolycoris baccarum* Linnaeus (Hemiptera: Pentatomidae). Jpn. J. Appl. Entomol. Zool. 29: 78–80.
Hori, K. 1986. Effects of photoperiod on nymphal growth of *Palomena angulosa* Motschulsky (Hemiptera: Pentatomidae). Appl. Entomol. Zool. 21: 597–605.
1987. Effects of stationary and changing photoperiods on nymphal growth of *Palomena angulosa* Motschulsky (Hemiptera: Pentatomidae). Appl. Entomol. Zool. 22: 528–532.
Hori, K., and K. Kuramochi. 1986. Effects of temporal foods in early nymphal stage on later growth of *Palomena angulosa* Motschulsky and *Eurydema rugosum* Motschulsky (Hemiptera: Pentatomidae). Appl. Entomol. Zool. 21: 39–46.
Hori, K., Y. Kondo, and K. Kuramochi. 1984. Feeding site of *Palomena angulosa* Motschulsky (Hemiptera: Pentatomidae) on potato plants and injury caused by the feeding. Appl. Entomol. Zool. 19: 476–482.
Ikeda-Kikue, K., and H. Numata. 1992. Effects of diet, photoperiod and temperature on the post diapause reproduction in the cabbage bug, *Eurydema rugosa*. Entomol. Exp. Appl. 64: 31–36.
Ito, K. 1978. Ecology of the stink bugs causing pecky rice. Rev. Plant Prot. Res. 11: 62–78.
Ito, K., H. Sugiyama, N. M. Salleh, and C. P. Min. 1993. Effects of lunar phase on light trap catches of the Malayan black bug, *Scotinophara coarctata* (Heteroptera: Pentatomidae). Bull. Entomol. Res. 83: 59–66.
Jackai, L. E. N., and S. R. Singh. 1987. Entomological research on soybeans in Africa. Pp. 17–24 *in* S. R. Singh, K. O. Rachie, and K. E. Dashiell [eds.], Soybean for the Tropics: Research, Production and Utilization. John Wiley & Sons, Chichester, U.K. 230 pp.

Jackai, L. E. N., A. R. Panizzi, G. G. Kundu, and K. P. Srivastava. 1990. Insect pests of soybean in the tropics. Pp. 91–156 *in* S. R. Singh [ed.], Insect Pests of Tropical Food Legumes. John Wiley & Sons, Chichester, U.K. 451 pp.

James, D. G. 1989. Population biology of *Biprorulus bibax* Breddin (Hemiptera: Pentatomidae) in a southern New South Wales citrus orchard. J. Austral. Entomol. Soc. 28: 279–286.

1990a. Seasonality and population development of *Biprorulus bibax* Breddin (Hemiptera: Pentatomidae) in south-western New South Wales. Gen. Appl. Entomol. 22: 61–66.

1990b. Energy reserves, reproductive status and population biology of overwintering *Biprorulus bibax* (Hemiptera: Pentatomidae) in Southern New South Wales. Austral. J. Zool. 38: 415–422.

1990c. Development and survivorship of *Biprorulus bibax* (Hemiptera: Pentatomidae) under a range of constant temperatures. Environ. Entomol. 19: 874–877.

1990d. Incidence of egg parasitism of *Biprorulus bibax* Breddin (Hemiptera: Pentatomidae) in southern New South Wales and northern Victoria. Gen. Appl. Entomol. 22: 55–60.

1991. Maintenance and termination of reproductive dormancy in *Biprorulus bibax* (Hemiptera: Pentatomidae). Entomol. Exp. Appl. 60: 1–5.

1992a. Effect of citrus host variety on oviposition, fecundity and longevity in *Biprorulus bibax* Breddin (Heteroptera). Acta Entomol. Bohemoslov. 89: 65–67.

1992b. Summer reproductive dormancy in *Biprorulus bibax* (Hemiptera: Pentatomidae). Austral. Entomol. Mag. 19: 65–68.

1992c. Effect of temperature on development and survivorship of *Pristhesancus plagipennis* (Hemiptera: Reduviidae). Entomophaga 37: 259–264.

1993a. Apparent overwintering of *Biprorulus bibax* (Hemiptera: Pentatomidae) on *Eremocitrus glauca* Rutaceae. Austral. Entomol. 20: 129–132.

1993b. New egg parasitoids for *Biprorulus bibax* (Breddin) (Hemiptera: Pentatomidae). J. Austral. Entomol. Soc. 32: 67–68.

1993c. Toxicity and use of endosulfan against spined citrus bug, *Biprorulus bibax*, and some of its egg parasitoids. Plant Prot. Q. 8: 54–56.

1994a. The development of suppression tactics for *Biprorulus bibax* (Heteroptera: Pentatomidae) as part of an integrated pest management program in citrus in inland south-eastern Australia. Bull. Entomol. Res. 84: 31–38.

1994b. Effect of citrus host variety on lipid reserves in overwintering *Biprorulus bibax* (Breddin) (Heteroptera). Victorian Entomol. 24: 42–45.

James, D. G., R. J. Faulder, and G. N. Warren. 1990. Phenology of reproductive status, weight and lipid reserves of *Biprorulus bibax* (Hemiptera: Pentatomidae). Environ. Entomol. 19: 1710–1715.

James, D. G., K. Mori, J. R. Aldrich, and J. E. Oliver. 1994. Flight-mediated attraction of *Biprorulus bibax* Breddin (Pentatomidae: Hemiptera) to natural and synthetic aggregation pheromone. J. Chem. Ecol. 20: 71–80.

Javahery, M. 1967. The biology of some Pentatomoidea and their egg parasites. Ph.D. dissertation, University of London, London, U.K. 475 pp.

1968. The egg parasite complex of British Pentatomoidea (Hemiptera): taxonomy of Telenominae (Hymenoptera: Scelionidae). Trans. R. Entomol. Soc. London 120: 417–436.

1972. Annual technical report on the cereal sunn pests. Plant Pests Res. Inst., Tehran, Iran. 72 pp.

1978. Management of economically important Pentatomoidea in Iran. Antenna 2: 74.

1990. Biology and ecological adaptation of the green stink bug (Hemiptera: Pentatomidae) in Quebec and Ontario. Ann. Entomol. Soc. Amer. 83: 201–206.

1993. Sunn pests survey and their control. Proc. Sunn Pests Conf., Univ. Tehran, Iran. pp. 43–48.

1994. Development of eggs in some true bugs (Hemiptera–Heteroptera). Part I. Pentatomoidea. Canad. Entomol. 126: 401–433.

1995. A technical review of sunn pests (Heteroptera: Pentatomoidea) with special reference to *Eurygaster integriceps* Puton. FAO, Regional. Office for the Near East, Cairo, Egypt. 80 pp.

1996. Sunn pest of wheat and barley in the I. R. of Iran: chemical and cultural methods of control. Pp. 61–74 *in* R. H. Miller and G. J. Morse [eds.], Sunn pests and their control in the Near East. FAO Plant Prot. No. 138.

Jensen, R. L., and L. D. Newsom. 1972. Effect of stink bug-damaged soybean seeds on germination, emergence, and yield. J. Econ. Entomol. 65: 261–264.

Johnson, N. F. 1991. Revision of Australasian *Trissolcus* species (Hymenoptera: Scelionidae). Invertebr. Taxon. 5: 211–239.

Jones, D. B., and R. H. Cherry. 1986. Species composition and seasonal abundance of stink bugs (Heteroptera: Pentatomidae) in southern Florida rice. J. Econ. Entomol. 79: 1226–1229.

Jones, T. H. 1918. The southern green plant-bug. U.S. Dept. Agric. Bull. 689: 1–27.

Jones, V. P., and L. C. Caprio. 1992. Damage estimates and population trends of insects attacking seven macadamia cultivars in Hawaii. J. Econ. Entomol. 85: 1884–1890.

1994. Southern green stink bug (Hemiptera: Pentatomidae) feeding on Hawaiian macadamia nuts: the relative importance of damage occurring in the canopy and on the ground. J. Econ. Entomol. 87: 431–435.

Jones, W. A., Jr. 1985. *Nezara viridula*. Pp. 339–343 *in* P. Singh and R. F. Moore [eds.], Handbook of Insect Rearing. Vol. 1, Elsevier, New York, New York, U.S.A. 488 pp.

1988. World review of the parasitoids of the southern green stink bug, *Nezara viridula* (L.) (Heteroptera: Pentatomidae). Ann. Entomol. Soc. Amer. 81: 262–273.

Jones, W. A., Jr., and M. J. Sullivan. 1979. Soybean resistance to the southern green stink bug, *Nezara viridula*. J. Econ. Entomol. 72: 628–632.

1982. Role of host plants in population dynamics of stink bug pests of soybeans in South Carolina. Environ. Entomol. 11: 867–875.

Joseph, T. 1953. On the biology, bionomics, seasonal incidence and control of *Piezodorus rubrofasciatus* Fabr., a pest of linseed and lucerne at Delhi. Indian J. Entomol. 15: 33–37.

Kadosawa, T., and H. Santa. 1981. Growth and reproduction of soybean pod bugs (Heteroptera) on seeds of legumes. Bull. Chigoku Natl. Agric. Exp. Stn. E19: 75–97.

Kamenkova, K. V. 1958. Biology and ecology of berry bug, *Dolycoris baccarum* L. An auxiliary host for the stink bug eater in Krasnodar territory. Entomol. Rev. 37: 478–500.

Kawada, H., and C. Kitamura. 1983a. Bionomics of the brown marmorated stink bug, *Halyomorpha mista*. Jpn. J. Appl. Entomol. Zool. 27: 304–306.

1983b. The reproductive behavior of the brown marmorated stink bug, *Halyomorpha mista* Uhler (Heteroptera: Pentatomidae). I. Observation of mating behavior and multiple copulation. Appl. Entomol. Zool. 18: 234–242.

Kawamoto, H., N. Ohkubo, and K. Kiritani. 1987. Modeling of soybean pod feeding behavior of stink bugs. Appl. Entomol. Zool. 22: 482–492.

Khamala, C. P. M., L. M. Oketch, and J. B. Okeyo-Owuor. 1978. Insect species associated with *Cajanus cajan*. Kenya Entomol. Newsl. 8: 3–5.

Kiritani, K., and K. Kimura. 1966. A study on the nymphal aggregation of the cabbage stink bug *Eurydema rugosum* Motschulsky (Hemiptera: Pentatomidae). Appl. Entomol. Zool. 1: 21–28.

Kiritani, K., N. Hokyo, K. Kimura, and F. Nakasuji. 1965. Imaginal dispersal of the southern green stink bug, *Nezara viridula* L., in relation to feeding and oviposition. Jpn. J. Appl. Entomol. Zool. 9: 291–297.

Kisimoto, R. 1983. Damage caused by rice stink bugs and their control. Jpn. Pestic. Inf. 43: 9–13.

Kobayashi, T. 1959. The developmental stages of some species of the Japanese Pentatomidae (Hemiptera). VII. Developmental stages of *Nezara* and its allied genera. Jpn. J. Appl. Entomol. Zool. 3: 221–231.

Kobayashi, T., and G. W. Cosenza. 1987. Integrated control of soybean stink bugs in the Cerrados. Jpn. Agric. Res. Q. 20: 229–236.

Kogan, M., and S. G. Turnipseed. 1987. Ecology and management of soybean arthropods. Annu. Rev. Entomol. 32: 507–538.

Kotaki, T., and S. Yagi. 1987. Relationship between diapause development and coloration change in brown-winged green bug, *Plautia stali* Scott (Heteroptera: Pentatomidae). Jpn. J. Appl. Entomol. Zool. 31: 285–290.

1989. Hormonal control of adult diapause in the brown-winged green bug, *Plautia stali* Scott (Heteroptera: Pentatomidae). Appl. Entomol. Zool. 24: 42–51.

Krambias, A. 1987. Host plant, seasonal migration and control of the berry bug, *Dolycoris baccarum* L. in Cyprus. FAO Plant Prot. Bull. 35: 25–26.

Krispyn, J. W., and J. W. Todd. 1982. The red imported fire ant as a predator of the southern green stink bug on soybean in Georgia. J. Georgia Entomol. Soc. 17: 19–26.

Laraichi, M., and J. Voegelé. 1975. Lutte biologique au Maroc contre la punaise des blés *Aelia germari* Valeur comparative des deux parasites oophages: *Asolcus grandis* et *Ooencyrtus fecundus* [Hym. Scelionidae et Encyrtidae]. Ann. Soc. Entomol. France 11: 783–789.

Lee, F. N., N. P. Tugwell, S. J. Fannah, and G. J. Weidemann. 1993. Role of fungi vectored by rice stink bug (Heteroptera: Pentatomidae) in discoloration of rice kernels. J. Econ. Entomol. 86: 549–556.
Lethierry, L., and G. Severin. 1893. Catalogue Général des Hémiptères. Tomo I. Hétéroptères Pentatomidae. F. Hayez, Académie Royale de Belgique, Bruxelles, Belgium. 286 pp.
Lever, R. J. A. W. 1933. Notes on two hemipterous pests of the coconut in the British Solomon Islands. Agric. Gaz. British Solomon Islands. 1: 2–6.
Liceras, L., and J. L. Hidalgo. 1987. *Lincus* sp. (Hem.: Pentatomidae), agente vector de la "marchitez subita" de la palma aceitera en el Peru. Rev. Peruana Entomol. 30: 103–104.
Lima, A. M. C. 1940. Insetos do Brasil. Hemipteros. Tomo II. Esc. Nac. Agron., Rio de Janeiro, Brazil. 351 pp.
Link, D. 1979. Percevejos do gênero *Euschistus* sobre soja no Rio Grande do Sul (Hemiptera: Pentatomidae). Rev. Cent. Cien. Rur. 9: 361–364.
Link, D., and J. Grazia. 1987. Pentatomídeos da região central do Rio Grande do Sul. An. Soc. Entomol. Brasil 16: 116–129.
Link, D., E. C. Costa, and M. F. S. Tarrago. 1989. Ocorrência de percevejos pentatomídeos em lavouras de arroz na região central do Rio Grande do Sul. An. Reun. Cult. Arroz Irrig. 18: 346–353.
Linnavuori. R. 1993. Hemiptera of Iraq. 2. Cydnidae, Thaumastellidae, Pentatomidae, Stenocephalidae, Coreidae, Alydidae, Rhopalidae and Pyrrhocoridae. Entomol. Fenn. 4: 37–56.
Llosa, J. J., G. Couturier, and F. Kahn. 1990. Notes on the ecology of *Lincus spurcus* and *L. malevolus* (Heteroptera: Pentatomidae: Discocephalinae) on Palmae in forests of Peruvian Amazonia. Ann. Soc. Entomol. France 26: 249–254.
Lodos, N. 1967. Studies on *Bathycoelia thalassina* (H.-S.) (Hemiptera: Pentatomidae) the cause of premature ripening of cocoa pods in Ghana. Bull. Entomol. Res. 57: 289–299.
Lopes, O. J., D. Link, and I. V. Basso. 1974. Pentatomídeos de Santa Maria — lista preliminar de plantas hospedeiras. Rev. Cent. Cien. Rur. 4: 317–322.
Ludwig, S. W., and L. T. Kok. 1998a. Phenology and parasitism of harlequin bugs, *Murgantia histrionica* (Hahn) (Hemiptera: Pentatomidae), in southwest Virginia. J. Entomol. Sci. 33: 33–39.
1998b. Evaluation of traps crops to manage harlequin bugs, *Murgantia histrionica* (Hahn) (Hemiptera: Pentatomidae) on broccoli. Crop Prot. 17: 123–128.
Lye, B.-H., and R. N. Story. 1988. Feeding preference of the southern green stink bug (Hemiptera: Pentatomidae) on tomato fruit. J. Econ. Entomol. 81: 522–526.
Lye, B.-H., R. N. Story, and V. L. Wright. 1988a. Damage threshold of the southern green stink bug, *Nezara viridula* (Hemiptera: Pentatomidae) on fresh market tomatoes. J. Entomol. Sci. 23: 366–373.
1988b. Southern green stink bug (Hemiptera: Pentatomidae) damage to fresh market tomatoes. J. Econ. Entomol. 81: 189–194.
Malaguido, A. B., and A. R. Panizzi. 1998. Danos de *Euschistus heros* (Fabr.) (Hemiptera: Pentatomidae) em aquênios de girassol. An. Soc. Entomol. Brasil 27: 535–542.
Marchetti, M. A., and H. D. Petersen. 1984. The role of *Bipolaris oryzae* in floral abortion and kernel discoloration in rice. Plant Dis. 68: 288–291.
Martins, J. F. S., and M. G. A. Lima. 1994. Fungos entomopatogênicos no controle do percevejo do colmo do arroz *Tibraca limbativentris* Stal: virulência de isolados de *Metharizium anisopliae* (Metsch.) Sorok. e *Beauveria bassiana* (Bals.) Vuill. An. Soc. Entomol. Brasil 23: 39–44.
Martins, J. F. S., M. C. Del Vecchio, and J. Grazia. 1986. Estudo dos imaturos de pentatomídeos (Heteroptera) que vivem sobre arroz (Oryza sativa L.): I — *Mormidea quinqueluteum* (Lichtenstein, 1796). An. Soc. Entomol. Brasil 15: 349–359.
Mau, R. F. L., and W. C. Mitchell. 1978. Development and reproduction of the Oriental stink bug, *Plautia stali* (Hemiptera: Pentatomidae). Ann. Entomol. Soc. Amer. 71: 756–757.
McDonald, F. J. D. 1971. Life cycle of the green stink bug, *Plautia affinis* Dallas (Hemiptera: Pentatomidae). J. Austral. Entomol. Soc. 10: 271–275.
McDonald, F. J. D., and J. Grigg. 1980. The life cycle of *Cuspicona simplex* Walker and *Monteithiella humeralis* (Walker) (Hemiptera: Pentatomidae). Gen. Appl. Entomol. 12: 61–71.
McNutt, D. N. 1975. Pests of coffee in Uganda, their status and control. PANS 21: 9–18.
1979. Control of *Antestiopsis* spp. on coffee in Uganda. PANS 25: 5–15.
McPherson, J. E. 1975. Life history of *Euschistus tristigmus tristigmus* (Hemiptera: Pentatomidae) with information on adult seasonal dimorphism. Ann. Entomol. Soc. Amer. 68: 333–334.

1982. The Pentatomoidea (Hemiptera) of Northeastern North America with Emphasis on the Fauna of Illinois. Southern Illinois University Press, Carbondale, Illinois, U.S.A. 240 pp.

McPherson, J. E., and R. H. Mohlenbrock. 1976. A list of the Scutelleroidea of the La Rue-Pine Hills Ecological Area with notes on biology. Great Lakes Entomol. 9: 125–169.

McPherson, J. E., and D. L. Tecic. 1997. Notes on the life histories of *Acrosternum hilare* and *Cosmopepla bimaculata* (Heteroptera: Pentatomidae) in southern Illinois. Great Lakes Entomol. 30: 79–84.

McPherson, R. M., and L. D. Newsom. 1984. Trap crops for control of stink bugs in soybean. J. Georgia Entomol. Soc. 19: 470–480.

McPherson, R. M., L. D. Newsom, and B. F. Farthing. 1979. Evaluation of four stink bug species from three genera affecting soybean yield and quality in Louisiana. J. Econ. Entomol. 72: 188–194.

McPherson, R. M., G. W. Zehnder, and J. C. Smith. 1988. Influence of cultivar, planting date, and row width on abundance of green cloverworms (Lepidoptera: Noctuidae) and green stink bugs (Heteroptera: Pentatomidae) in soybean. J. Entomol. Sci. 23: 305–313.

Mendes, L. 1956. Podridão interna dos capulhos do algodoeiro obtida por meio de insetos. Bragantia 15: 9–11.

Meneguim, A. M., M. C. Rossini, and A. R. Panizzi. 1989. Desempenho de ninfas e adultos de *Euschistus heros* (F.) (Hemiptera: Pentatomidae) em frutos verdes de amendoim-bravo *Euphorbia heterophylla* (Euphorbiaceae) e em sementes e vagens de soja. Res. Congr. Bras. Entomol. 12, Vol. 1, p. 43.

Michailides, T. J., R. E. Rice, and J. M. Ogawa. 1987. Succession and significance of several hemipterans attacking a pistachio orchard. J. Econ. Entomol. 80: 398–406.

Miner, F. D. 1966. Biology and control of stink bugs on soybeans. Arkansas Agric. Exp. Stn. Bull. 708: 1–40.

Mitchell, W. C., and R. F. L. Mau. 1971. Response of the southern green stink bug and its parasite, *Trichopoda pennipes*, to male stink bug pheromones. J. Econ. Entomol. 64: 856–859.

Mitchell, W. C., R. M. Warner, and E. T. Fukunaga. 1965. Southern green stink bug, *Nezara viridula* (L.), injury to macadamia nut. Proc. Hawaiian Entomol. Soc. 19: 103–109.

Monte, O. 1939. Hemipteros fitofagos. O Campo 4: 51–63.

Morimoto, N., M. Fujino, N. Tanahashi, and H. Kishino. 1991. Coexistence of the two closely related species of cabbage stinkbug, *Eurydema rugosum* and *E. pulchrum* (Heteroptera: Pentatomidae), in the field in central Japan. 1. Distribution, life cycle and host plant preferences of the two species. Appl. Entomol. Zool. 26: 435–442.

Moriya, S. 1987. Automatic data acquisition systems for study of the flight ability of brown-winged green bug, *Plautia stali* Scott (Heteroptera: Pentatomidae). Appl. Entomol. Zool. 22: 19–24.

Moriya, S., and M. Shiga. 1984. Attraction of the brown-winged green bug, *Plautia stali* Scott (Heteroptera: Pentatomidae) for males and females of the same species. Appl. Entomol. Zool. 19: 317–322.

Morrill, A. W. 1910. Plant-bugs injurious to cotton bolls. U.S. Dept. Agric. Bur. Entomol. Bull. 86: 1–110.

Morrill, W. L., B. M. Shepard, G. S. Rida, and M. Parducho. 1995. Damage by the Malayan black bug (Heteroptera: Pentatomidae) in rice. J. Econ. Entomol. 88: 1466–1468.

Morris, H. M. 1929. A note on "Vromousa" (*Dolicoris baccarum* L.). Cyprus Agric. J. 24: 149–150.

Moura, J. I. L., and M. L. V. Resende. 1995. Eficiência de monocrotophos aplicado via raiz no controle de *Lincus lobuliger* Bred. em coqueiro. An. Soc. Entomol. Brasil 24: 1–6.

Mourão, A. P. M. 1999. Influência do fotoperíodo na indução da diapausa do percevejo-marrom *Euschistus heros* (Fabr.) (Hemiptera: Pentatomidae). M.Sc. thesis, Universidade Estadual de Londrina, Londrina, Paraná, Brazil. 76 pp.

Munyaneza, J., and J. E. McPherson. 1994. Comparative study of life histories, laboratory rearing, and immature stages of *Euschistus servus* and *Euschistus variolarius* (Hemiptera: Pentatomidae). Great Lakes Entomol. 26: 263–274.

Naresh, J. S., and C. M. Smith. 1983. Development and survival of rice stink bugs (Hemiptera: Pentatomidae) reared on different host plants at four temperatures. Environ. Entomol. 12: 1496–1499.

Negron, J. F., and T. J. Riley. 1987. Southern green stink bug, *Nezara viridula* (Heteroptera: Pentatomidae), feeding in corn. J. Econ. Entomol. 80: 666–669.

Newsom, L. D., and D. C. Herzog. 1977. Trap crops for control of soybean pests. Louisiana Agric. 20 (3): 14–15.

Nilakhe, S. S. 1976. Rice lines screened for resistance to the rice stink bug. J. Econ. Entomol. 69: 703–705.

Nilakhe, S. S., R. B. Chalfant, and S. V. Singh. 1981a. Evaluation of southern green stink bug damage to cowpeas. J. Econ. Entomol. 74: 589–592.

1981b. Field damage to lima beans by different stages of southern green stink bug. J. Georgia Entomol. Soc. 16: 392–396.

Niva, C. C., and A. R. Panizzi. 1996. Efeitos do cloreto de sódio no comportamento de *Nezara viridula* (L.) (Heteroptera: Pentatomidae) em vagem de soja. An. Soc. Entomol. Brasil 25: 251–257.

Noda, T. 1984. Short day photoperiod accelerates the oviposition in the oriental stink bug, *Nezara antennata* Scott (Heteroptera: Pentatomidae). Appl. Entomol. Zool. 19: 119–120.

Numata, H., and K. Yamamoto. 1990. Feeding on seeds induces diapause in the cabbage bug, *Eurydema rugosum*. Entomologia 57: 281–284.

Odglen, G. E., and L. O. Warren. 1962. The rice stink bug, *Oebalus pugnax* F. [sic] in Arkansas. Arkansas Agric. Exp. Stn. Rep. Ser. 107: 1–23.

Oetting, R. D., and T. R. Yonke. 1971. Biology of some Missouri stink bugs. J. Kansas Entomol. Soc. 44: 446–459.

Ohno, K. 1987. Effect of host age on parasitism by *Trissolcus plautiae* (Watanabe) (Hymenoptera: Scelionidae), an egg parasitoid of *Plautia stali* Scott (Heteroptera: Pentatomidae). Appl. Entomol. Zool. 22: 646–648.

Okamoto, K., and K. Kuramochi. 1984. Distribution, feeding site and development of the cabbage bug *Eurydema rugosum* Motschulsky (Hemiptera: Pentatomidae) on cruciferous plants. Appl. Entomol. Zool. 19: 273–279.

Owusu-Manu, E. 1971. *Bathycoelia thalassina*. Another serious pest of cocoa in Ghana. C.M.B. Newsl. 47: 12–14.

1973. *Bathycoelia* field spraying trials. Rep. Cocoa Res. Inst. Ghana 1971/72: 138–140.

1977. Flight activity and dispersal of *Bathycoelia thalassina* (Herrich-Schaeffer) Hemiptera, Pentatomidae. Ghana J. Agric. Sci. 10: 23–26.

1978. Host plants of *Bathycoelia thalassina* (H.-S.) (Hem., Pentatomidae) in Ghana. Entomol. Mon. Mag. 114: 201–202.

1990. Feeding behaviour and the damage caused by *Bathycoelia thalassina* (Herrich-Schaeffer) (Hemiptera: Pentatomidae). Café Cacao Thé 34: 97–104.

Pajni, H. R., and C. S. Sidhu. 1982. A report on the pentatomid fauna of Chandigarh and surrounding areas (Heteroptera: Hemiptera). Res. Bull. Sci., Parts III–IV, 33: 177–181.

Panizzi, A. R. 1980. Uso de cultivar armadilha no controle de percevejos em soja. Trigo e Soja 47: 11–14.

1985. *Sesbania aculeata*: nova planta hospedeira de *Piezodorus guildinii* no Paraná. Pesq. Agropec. Bras. 20: 1237–1238.

1987. Impacto de leguminosas na biologia de ninfas e efeito da troca de alimento no desempenho de adultos de *Piezodorus guildinii* (Hemiptera: Pentatomidae). Rev. Bras. Biol. 47: 585–591.

1992. Performance of *Piezodorus guildinii* on four species of *Indigofera* legumes. Entomol. Exp. Appl. 63: 221–228.

1995. Feeding frequency, duration and preference of the southern green stink bug (Heteroptera: Pentatomidae) as affected by stage of development, age, and physiological condition. An. Soc. Entomol. Brasil 24: 437–444.

1997. Wild hosts of pentatomids: ecological significance and role in their pest status on crops. Annu. Rev. Entomol. 42: 99–122.

Panizzi, A. R., and R. M. L. Alves. 1993. Performance of nymphs and adults of the southern green stink bug (Heteroptera: Pentatomidae) exposed to soybean pods at different phenological stages of development. J. Econ. Entomol. 86: 1088–1093.

Panizzi, A. R., and D. C. Herzog. 1984. Biology of *Thyanta perditor*. Ann. Entomol. Soc. Amer. 77: 646–650.

Panizzi, A. R., and E. Hirose. 1995. Seasonal body weight, lipid content, and impact of starvation and water stress on adult survivorship and longevity of *Nezara viridula* and *Euschistus heros*. Entomol. Exp. Appl. 76: 247–253.

Panizzi, A. R., and E. Machado-Neto. 1992. Development of nymphs and feeding habits of nymphal and adult *Edessa meditabunda* (Heteroptera: Pentatomidae) on soybean and sunflower. Ann. Entomol. Soc. Amer. 85: 477–481.

Panizzi, A. R., and A. M. Meneguim. 1989. Performance of nymphal and adult *Nezara viridula* on selected alternate host plants. Entomol. Exp. Appl. 50: 215–223.

Panizzi, A. R., and A. P. M. Mourão. 1999. Mating, ovipositional rhythm and fecundity of *Nezara viridula* (L.) (Heteroptera: Pentatomidae) fed on privet, *Ligustrum lucidum* Thunb., and on soybean, *Glycine max* (L.) Merrill fruits. An. Soc. Entomol. Brasil 28: 35–40.

Panizzi, A. R., and C. C. Niva. 1994. Overwintering strategy of the brown stink bug in northern Paraná. Pesq. Agropec. Bras. 29: 509–511.

Panizzi, A. R., and E. D. M. Oliveira. 1998. Performance and seasonal abundance of the Neotropical brown stink bug, *Euschistus heros* nymphs and adults on a novel food plant (pigeonpea) and soybean. Entomol. Exp. Appl. 88: 169–175.

1999. Seasonal occurrence of tachinid parasitism on stink bugs with different overwintering strategies. An. Soc. Entomol. Brasil 28: 169–172.

Panizzi, A. R., and C. E. Rossi. 1991a. The role of *Acanthospermum hispidum* in the phenology of *Euschistus heros* and of *Nezara viridula*. Entomol. Exp. Appl. 59: 67–74.

1991b. Efeito da vagem e da semente de *Leucaena* e da vagem de soja no desenvolvimento deninfas e adultos de *Loxa deducta* (Hemiptera: Pentatomidae). Rev. Bras. Biol. 51: 607–613.

Panizzi, A. R., and M. C. Rossini. 1987. Impacto de várias leguminosas na biologia de ninfas de *Nezara viridula* (Hemiptera: Pentatomidae). Rev. Bras. Biol. 47: 507–512.

Panizzi, A. R., and S. I. Saraiva. 1993. Performance of nymphal and adult southern green stink bug on an overwintering host and impact of nymph to adult food-switch. Entomol. Exp. Appl. 68: 109–115.

Panizzi, A. R., and F. Slansky, Jr. 1985a. Review of phytophagous pentatomids (Hemiptera: Pentatomidae) associated with soybean in the Americas. Florida Entomol. 68: 184–214.

1985b. New host plant records for the stink bug *Piezodorus guildinii* in Florida (Hemiptera: Pentatomidae). Florida Entomol. 68: 215–216.

1985c. Legume host impact on performance of adult *Piezodorus guildinii* (Westwood) (Hemiptera: Pentatomidae). Environ. Entomol. 14: 237–242.

1991. Suitability of selected legumes and the effect of nymphal and adult nutrition in the southern green stink bug (Hemiptera: Heteroptera: Pentatomidae). J. Econ. Entomol. 84: 103–113.

Panizzi, A. R., and J. G. Smith. 1976a. Observações sobre inimigos naturais de *Piezodorus guildinii* (Westwood, 1837) (Hemiptera, Pentatomidae) em soja. An. Soc. Entomol. Brasil 5: 11–17.

1976b. Ocorrência de Pentatomidae em soja no Paraná durante 1973/74. O Biológico 42: 173–176.

1977. Biology of *Piezodorus guildinii*: oviposition, development time, adult sex ratio, and longevity. Ann. Entomol. Soc. Amer. 70: 35–39.

Panizzi, A. R., and L. M. Vivan. 1997. Seasonal abundance of the Neotropical brown stink bug, *Euschistus heros* in overwintering sites and the breaking of dormancy. Entomol. Exp. Appl. 82: 213–217.

Panizzi, A. R., J. G. Smith, L. A. G. Pereira, and J. Yamashita. 1979. Efeitos dos danos de *Piezodorus guildinii* (Westwood, 1837) no rendimento e qualidade da soja. An. I Semin. Nac. Pesq. Soja, 2: 59–78.

Panizzi, A. R., M. H. M. Galileo, H. A. O. Gastal, J. F. F. Toledo, and C. H. Wild. 1980. Dispersal of *Nezara viridula* and *Piezodorus guildinii* nymphs in soybeans. Environ. Entomol. 9: 293–297.

Panizzi, A. R., C. C. Niva, and E. Hirose. 1995. Feeding preference by stink bugs (Heteroptera: Pentatomidae) for seeds within soybean pods. J. Entomol. Sci. 30: 333–341.

Panizzi, A. R., L. M. Vivan, B. S. Corrêa-Ferreira, and L. A. Foerster. 1996. Performance of southern green stink bug (Heteroptera: Pentatomidae) nymphs and adults on a novel food plant (Japanese privet) and other hosts. Ann. Entomol. Soc. Amer. 89: 822–827.

Panizzi, A. R., A. P. M. Mourão, and E. D. M. Oliveira. 1998. Nymph and adult biology and seasonal abundance of *Loxa deducta* (Walker) on privet, *Ligustrum lucidum*. An. Soc. Entomol. Brasil 27: 199–206.

Pantoja, A. 1990. Lista preliminar de plagas del arroz in Colombia. Arroz Amer. 11: 1–9.

Pantoja, A., E. Daza, C. Garcia, O. I. Mejía, and D. A. Rider. 1995. Relative abundance of stink bugs (Hemiptera: Pentatomidae) in Southwestern Colombia rice fields. J. Entomol. Sci. 30: 463–467.

Parish, H. E. 1934. Biology of *Euschistus variolarius* P. de B. (Family Pentatomidae; Order Hemiptera). Ann. Entomol. Soc. Amer. 27: 50–54.

Paulian, F., and C. Popov. 1980. Wheat Documenta. Pp. 69–74. Ciba-Geigy Ltd., Basel, Switzerland.

Pavis, C., and C. Malosse. 1986. Mise en evidence d'un attractif sexuel produit par les males de *Nezara viridula* (L.) (Heteroptera: Pentatomidae). C. R. Acad. Sci. Ser. III 303: 273–276.

Payne, J. A., and J. M. Wells. 1984. Toxic penicillia isolated from lesions of kernel-spotted pecans. Environ. Entomol. 13: 1609–1612.

Perepelitza, L. V. 1969. The breeding of *Dolycoris baccarum* in the laboratory. Zool. J. 48: 757–759.

Perez, C. A., J. L. Souza Fº., and O. Nakano. 1980. Observações sobre a biologia e hábito do percevejo *Thyanta perditor* (F.) (Hemiptera-Pentatomidae) em planta de trigo. Solo 72: 61–62.

Petropavlovskaya, M. B. 1955. The biology of the pentatomid pests of vegetable crops. Tr. Altaisk. Skh. Inst. 2: 146–156.

Phillips, K. A., and J. O. Howell. 1980. Biological studies on two *Euchistus* [sic] species in central Georgia apple orchards. J. Georgia Entomol. Soc. 15: 349–355.

Pinto, S. B., and A. R. Panizzi. 1994. Performance of nymphal and adult *Euschistus heros* (F.) on milkweed and on soybean and effect of food switch on adult survivorship, reproduction and weight gain. An. Soc. Entomol. Brasil 23: 549–555.

Polivanova, E. N. 1957. Factors affecting the abundance of cereal bugs (Pentatomidae) in the southern grain-growing regions of the European part of the Soviet Union. Dokl. Akad. Nauk SSSR. 112: 538–541.

Puton, A. 1892. Hémiptères nouveaux ou peu connues et notes diverses (IV. Hémiptères d'Akbes, region de l'Amunus (Syrie septentrionale) recoltes par M. Delagrange). Rev. Entomol. 11: 34–36.

Puton, A., and M. Noualhier. 1895. Supplement a la liste des Hémiptères d'Akbes. Rev. Entomol. 14: 170–177.

Putschkov, V. G. 1961. Fauna of Ukraine. Shieldbugs. Vol. 21, Acad. Sci., UK, SSR, Kiev, Ukraine, USSR.

Ragsdale, D. W., A. D. Larson, and L. D. Newsom. 1979. Microorganisms associated with feeding and from various organs of *Nezara viridula*. J. Econ. Entomol. 72: 725–727.

Rai, B. K. 1971. Laboratory and field testing of insecticides against paddy bug, *Oebalus poecila* (Dallas) and a technique for low-volume drift spraying of paddy, for its control. Int. Rice Commun. Newsl. 20: 8–17

Rasplus, J. Y., D. Pluot-Sigwalt, J. F. Llosa, and G. Couturier. 1990. *Hexacladia linci*, n.sp. (Hymenoptera: Encyrtidae) endoparasite de *Lincus malevolus* Rolston (Heteroptera: Pentatomidae) au Perou. Ann. Soc. Entomol. France 26: 255–263.

Reich, R. C. [ed.]. 1991. Flue-Cured Tobacco Field Manual, third edition. R. J. Reynolds Tobacco Co., Winston-Salem, North Carolina, U.S.A. 86 pp.

Reilly, C. C., and W. L. Tedders. 1989. Bacterial involvement in late season drop and kernel discolor of pecan. Proc. Annu. Conv. Southeast. Pecan Growers Assoc. 82: 195–199.

Resende, M. L. V., and J. L. Bezerra. 1990. Transmissão da murcha de *Phytomonas* a coqueiros por *Lincus lobuliger* (Hemiptera: Pentatomidae). Summa Phytopat. 16: 27.

Resende, M. L. V., R. E. L. Borges, J. L. Bezerra, and D. P. Oliveira. 1986. Transmissão da murcha de *Phytomonas* a coqueiros e dendezeiros por *Lincus lobuliger* Bred., 1908 (Hemiptera: Pentatomidae). Rev. Theobroma 16: 149–154.

Reyes, R., M. A. Cruz, and P. Genty. 1988. La absorcion radicular en el control de plagas en palma africana. Palmas 2: 19–23.

Rice, R. E., J. K. Uyemoto, J. M. Ogawa, and W. M. Pemberton. 1985. New findings on pistachio problems. California Agric. 39: 15–18.

Rice, R. E., W. J. Bentley, and R. H. Beade. 1988. Insect and mite pests of pistachios in California. Univ. California Div. Agric. Nat. Res. Coop. Ext. Publ. 21452: 1–26.

Rider, D. A., and J. A. Chapin. 1992. Revision of the genus *Thyanta* Stål, 1862 (Heteroptera: Pentatomidae). II. North America, Central America, and the West Indies. J. New York Entomol. Soc. 100: 42–98.

Rings, R. W. 1957. Types and seasonal incidence of stink bug injury to peaches. J. Econ. Entomol. 50: 599–604.

Rizzo, H. F. E. 1968. Aspectos morfológicos y biológicos de *Nezara viridula* (L.) (Hemiptera, Pentatomidae). Agron. Tropical 18: 249–274.

1971. Aspectos morfologicos y biologicos de *Edessa meditabunda* (F.) (Hemiptera, Pentatomidae). Rev. Peruana Entomol. 14: 272–281.

1972. Enemigos animales del cultivo de la soja. Rev. Inst. Bol. Cer. No. 285l: 6 pp.

1976. Hemípteros de Interés Agrícola. Editorial Hemisferio Sur, Buenos Aires, Argentina. 69 pp.

Rizzo, H. F., and E. D. Saini. 1987. Aspectos morfologicos y biologicos de *Edessa rufomarginata* (De Geer) (Hemiptera, Pentatomidae). Rev. Fac. Agron. 8: 51–63.

Robinson, J. F., C. M. Smith, G. B. Trahan, and M. Hollay. 1981. Evaluation of 32 uniform rice nursery lines for rice stink bug resistance. Louisiana Agric. Exp. Stn., Rice Exp. Stn. Annu. Prog. Rep. 73: 278–285.

Rodini, E. S. O., and J. Grazia. 1979. Abundância de algumas espécies de insetos (Coleoptera e Hemiptera) em soja [*Glycine max* (L.) Merrill] no município de Aguaí, SP. An. I Semin. Nac. Pesq. Soja, 2: 17–22.

Rolston, L. H. 1974. Revision of the genus *Euschistus* in middle America (Hemiptera, Pentatomidae, Pentatomini). Entomol. Amer. 48: 1–102.

1978. A revision of the genus *Mormidea* (Hemiptera: Pentatomidae). J. New York Entomol. Soc. 86: 161–219.

1983a. A revision of the genus *Lincus* Stål (Hemiptera: Pentatomidae: Discocephalinae: Ochlerini). J. New York Entomol. Soc. 91: 1–47.

1983b. A revision of the genus *Acrosternum* Fieber, subgenus *Chinavia* Orian, in the Western Hemisphere (Hemiptera: Pentatomidae). J. New York Entomol. Soc. 91: 97–176.

1989. Three new species of *Lincus* (Hemiptera: Pentatomidae) from palms. J. New York Entomol. Soc. 97: 271–276.

Rolston, L. H., and R. L. Kendrick. 1961. Biology of the brown stink bug, *Euschistus servus* Say. J. Kansas Entomol. Soc. 34: 151–157.

Rosseto, C. J., S. Silveira Neto, D. Link, J. Grazia-Vieira, E. Amante, D. M. Souza, N. V. Banzatto, and A. M. Oliveira. 1972. Pragas do arroz. Reun. Com. Arroz Amer., No. 2, Pelotas, Brazil. pp. 149–238.

Rosseto, C. J., J. Grazia, and A. Savy F°. 1978. Ocorrência de *Thyanta perditor* (Fabricius, 1974) como praga no Estado de São Paulo (Hemiptera: Pentatomidae). V Congr. Bras. Entomol., Bahia, Brazil. [np].

Rusanova, V. N. 1926. On the biology and coloration of members of the genus *Eurydema* Lap. Zashch. Rast. Vred. 3: 378–383.

Russell, E. E. 1952. Stink bugs on seed alfalfa in southern Arizona. U.S. Dept. Agric. Circ. 903: 1–19.

Russin, J. S., M. B. Layton, D. B. Orr, and D. J. Boethel. 1987. Within-plant distribution of, and partial compensation for, stink bug (Heteroptera: Pentatomidae) damage to soybean seeds. J. Econ. Entomol. 80: 215–220.

Russin, J. S., D. B. Orr, M. B. Layton, and D. J. Boethel. 1988. Incidence of microorganisms in soybean seeds damaged by stink bug feeding. Phytopathology 78: 306–310.

Ryan, M. A., and G. H. Walter. 1992. Sound communication in *Nezara viridula* (L.) (Heteroptera: Pentatomidae): further evidence that signal transmission is substrate-borne. Experientia 48: 1112–1115.

Sailer, R. I. 1944. The genus *Solubea* (Heteroptera: Pentatomidae). Proc. Entomol. Soc. Washington 46: 105–127.

1953. A note on the bionomics of the green stink bug (Hemiptera: Pentatomidae). J. Kansas Entomol. Soc. 26: 70–71.

1957. *Solubea* Bergroth, 1891, a synonym of *Oebalus* Stål, 1862, and a note concerning the distribution of *O. ornatus* (Sailer) (Hemiptera: Pentatomidae). Proc. Entomol. Soc. Washington 59: 41–42.

Schalk, J. M., and R. L. Fery. 1982. Southern green stink bug and leaffooted bug: effect on cowpea production. J. Econ. Entomol. 75: 72–75.

Schoene, W. J., and G. W. Underhill. 1933. Economic status of the green stinkbug with reference to the sucession of its wild hosts. J. Agric. Res. 46: 863–866.

Schotzko, D. J., and L. E. O'Keeffe. 1990a. Ovipositional rhythms of *Thyanta pallidovirens* (Hemiptera: Pentatomidae). Environ. Entomol. 19: 630–634.

1990b. Effect of pea and lentil development on reproduction and longevity of *Thyanta pallidovirens* (Stål) (Hemiptera: Heteroptera: Pentatomidae). J. Econ. Entomol. 83: 1333–1337.

Schuh, R. T., and J. A. Slater. 1995. True Bugs of the World (Hemiptera: Heteroptera). Classification and Natural History. Cornell University Press, Ithaca, New York, U.S.A. 336 pp.

Sedlacek, J. D., and L. H. Townsend. 1988. Impact of *Euschistus servus* and *E. variolarius* (Heteroptera: Pentatomidae) feeding on early growth stages of corn. J. Econ. Entomol. 81: 840–844.

Seymour, J., G. Bowman, and M. Crouch. 1995. Effects of neem seed extract on feeding frequency of *Nezara viridula* L. (Hemiptera: Pentatomidae) on pecan nuts. J. Austral. Entomol. Soc. 34: 221–223.

Shepard, M., R. J. Lawn, and M. A. Schneider. 1983. Insects on Grain Legumes in Northern Australia. A Survey of Potential Pests and Their Enemies. Univ. Queensland Press, St. Lucia, Australia. 81 pp.

Shiga, M. 1991. Dispersal, food resource utilization, and life history strategy of the brown-winged green bug, *Plautia stali*. Proc. Int. Seminar on Migration and Dispersal Agric. Insects, Tsukuba, Japan, pp. 81–90.

Shiga, M., and S. Moriya. 1984. Utilization of food plants by *Plautia stali* Scott (Hemiptera, Heteroptera, Pentatomidae), an experimental approach. Bull. Fruit Tree Res. Stn., Series A, 11: 107–121.

Silva, A. B., and B. P. Magalhães. 1981. Insetos nocivos a cultura do arroz no Estado do Pará. EMBRAPA/CPATU, Circ. Tec. 22: 1–14.

Silva, A. G. D.'A., C. R. Gonçalves, D. M. Galvão, A. J. L. Gonçalves, J. Gomes, M. N. Silva, and L. Simoni. 1968. Quarto Catálogo dos Insetos que Vivem nas Plantas do Brasil — Seus Parasitas e Predadores. Parte II, Vol. I. Min. Agric., Rio de Janeiro, Brazil. 622 pp.

Silva, C. P. 1992a. Aspectos biológicos básicos de *Oebalus poecilus* (Dallas, 1851) (Heteroptera: Pentatomidae) por ataque de parasitóides de ovos na cultura de arroz. An. Soc. Entomol. Brasil 21: 225–231.

1992b. Mortalidade de *Oebalus poecilus* (Dallas, 1851) (Heteroptera: Pentatomidae) por ataque de parasitóides de ovos na cultura de arroz. An. Soc. Entomol. Brasil 21: 289–296.

Singh, O. P., K. J. Singh, and R. D. Thakur. 1989. Studies on the bionomics and chemical control of stink bug, *Piezodorus rubrofasciatus* Fabricius, a new pest of soybean in Madhya Pradesh. Indian J. Plant Prot. 17: 81–83.

Singh, S. R., and L. E. N. Jackai. 1985. Insect pests of cowpeas in Africa: their life cycle, economic importance, and potential for control. Pp. 217–231 *in* S. R. Singh and K. O. Rachie [eds.], Cowpea Research, Production and Utilization. John Wiley & Sons, Chichester, U.K. 448 pp.

Singh, S. R., L. E. N. Jackai, J. H. R. Santos, and C. B. Adalla. 1990. Insect pests of cowpea. Pp. 43–89 *in* S. R. Singh [ed.], Insect Pests of Tropical Food Legumes. John Wiley & Sons, Chichester, U.K. 451 pp.

Sosa Gómez, D. R., C. Y. Takachi, and F. Moscardi. 1993. Determinação de sinergismo e susceptibilidade diferencial de *Nezara viridula* (L.) e *Euschistus heros* (F.) (Hemiptera: Pentatomidae) à inseticidas em mistura com cloreto de sódio. An. Soc. Entomol. Brasil 22: 569–576.

Southwood, T. R. E., and D. Leston. 1959. Land and Water Bugs of the British Isles. Wayside Woodl. Ser. Vol. 9, London, U.K. 436 pp.

Stam, P. A., L. D. Newsom, and E. N. Lambremont. 1987. Predation and food as factors affecting survival of *Nezara viridula* (L.) (Hemiptera: Pentatomidae) in a soybean ecosystem. Environ. Entomol. 16: 1211–1216.

Stichel, W. 1961. Illustrierte Bestimmungs-Tabellen der Wanzen. II. Europa 4, 18: 548–556.

Stoner, D. 1922. Report on the Scutelleroidea collected by the Barbados-Antigua expedition from the University of Iowa in 1918. Univ. Iowa Studies Nat. Hist. 10: 3–17. Dept. Agric. Stock, Div. Entomol. Plant Pathol. Bull. No. 8.

Streams, F. A., and D. Pimentel. 1963. Biology of the harlequin bug, *Murgantia histrionica*. J. Econ. Entomol. 56: 108–109.

Summerville, W. A. T. 1931. The larger horner citrus bug. Queensland Dept. Agric. Stock, Div. Entomol. Plant Pathol., Brisbane, Bull. 8.

Swanson, M. C., and L. D. Newsom. 1962. Effect of infestation by the rice stink bug, *Oebalus pugnax*, on yield and quality in rice. J. Econ. Entomol. 55: 877–879.

Tanskii, V. I. 1971. A comparative study of food specialization in oligophagous and polyphagous pentatomid pests of wheat in northern Kazakhstan. Zool. Zh. 50: 1335–1340.

Thomas, D. B., Jr. 1983. Taxonomic status of the genera *Chlorochoa* Stål, *Rhytidilomia* [sic] Stål, *Liodermion* Kirkaldy, and *Pitedia* Reuter, and their included species (Hemiptera: Pentatomidae). Ann. Entomol. Soc. Amer. 76: 215–224.

Thomas, G. D., C. M Ignoffo, C. E. Morgan, and W. A. Dickerson. 1974. Southern green stink bug: influence on yield and quality of soybeans. J. Econ. Entomol. 67: 501–503.

Tischler, W. 1937. Untersuchungen über Wanzen an Getreide. Arb. Physiol. Angew. Entomol. Berlin. 4: 193–231.

Tischler, W. 1938. Zur Ökologie der wichtigsten in Deutschland an Getreide schadlichen Pentatomiden I. Z. Morph. Ökol. Tiere 34: 317–366.

Tischler, W. 1939. Zur Ökologie der wichtigsten in Deutschland an Getreide schadlichen Pentatomiden II. Z. Morph. Ökol. Tiere 35: 252–287.

Todd, J. W. 1989. Ecology and behavior of *Nezara viridula*. Annu. Rev. Entomol. 34: 273–292.

Todd, J. W., and D. C. Herzog. 1980. Sampling phytophagous Pentatomidae on soybean. Pp. 438–478 *in* M. Kogan and D. C. Herzog [eds.], Sampling Methods in Soybean Entomology. Springer-Verlag, New York, New York, U.S.A. 587 pp.

Todd, J. W., and F. W. Schumann. 1988. Combination of insecticide applications with trap crops of early maturing soybean and southern peas for population management of *Nezara viridula* in soybean (Hemiptera: Pentatomidae). J. Entomol. Sci. 23: 192–199.

Todd, J. W., and H. Womack. 1973. Secondary infestations of cigarette beetle in soybean seed damaged by southern green stink bug. Environ. Entomol. 2: 720.

Todd, J. W., M. D. Jellum, and D. B. Leuck. 1973. Effects of southern green stink bug damage on fatty acid composition of soybean oil. Environ. Entomol. 2: 685–689.

Tonhasca, A., Jr., and B. R. Stinner. 1991. Effects of strip intercropping and no-tillage on some pests and beneficial invertebrates of corn in Ohio. Environ. Entomol. 20: 1251–1258.

Torre-Bueno, J. R. 1939. A synopsis of the Hemiptera-Heteroptera of America north of Mexico. Part I. Families Scutelleridae, Cydnidae, Pentatomidae, Aradidae, Dysodiidae and Termitaphididae. Entomol. Amer. 19: 141–304.

Toscano, N. C., and V. M. Stern. 1976a. Dispersal of *Euschistus conspersus* from alfalfa grown for seed to adjacent crops. J. Econ. Entomol. 69: 96–98.

1976b. Cotton yield and quality loss caused by various levels of stink bug infestations. J. Econ. Entomol. 69: 53–56.

Tothill, J. D. 1929. A reconnaissance survey of agricultural conditions in the British Solomon Islands Protectorate. F'scap. Fol. Suva, Fiji, pp. 1–17.

Townsend, L. H., and J. D. Sedlacek. 1986. Damage to corn caused by *Euschistus servus*, *E. variolarius*, and *Acrosternum hilare* (Heteroptera: Pentatomidae) under greenhouse conditions. J. Econ. Entomol. 79: 1254–1258.

Turnipseed, S. G., and M. Kogan. 1976. Soybean entomology. Annu. Rev. Entomol. 21: 247–282.

Underhill, G. W. 1934. The green stinkbug. Virginia Agric. Exp. Stn. Bull. 294: 1–26.

Van Duzee, E. P. 1904. Annotated list of the Pentatomidae recorded from America north of Mexico, with descriptions of some new species. Trans. Amer. Entomol. Soc. 30: 1–80.

Vangeison, K. W., and J. E. McPherson. 1975. Life history and laboratory rearing of *Proxys punctulatus* (Hemiptera: Pentatomidae) with descriptions of immature stages. Ann. Entomol. Soc. Amer. 68: 25–30.

van Halteren, P. 1972. Some aspects of the biology of the paddy bug, *Oebalus poecilus* (Dallas), in Surinam. Surinaamse Landbouw 2: 23–33.

Verma, A. K., S. K. Patyal, O. P. Bhalla, and K. C. Sharma. 1993. Bioecology of painted bug (*Bagrada cruciferarum*) (Hemiptera: Pentatomidae) on seed crop of cauliflower (*Brassica oleracea* var. *botrytis* subvar. *cauliflora*). Indian J. Agric. Sci. 63: 676–678.

Vidyasagar, S. P. V., and K. S. Bhat. 1986. A pentatomid bug causes tender nut drop in arecanut. Curr. Sci. 55: 1096–1097.

Villas Bôas, G. L., and A. R. Panizzi. 1980. Biologia de *Euschistus heros* (Fabricius, 1798) em soja [*Glycine max* (L.) Merrill]. An. Soc. Entomol. Brasil 9: 105–113.

Voegelé, J. 1968. Importance des processus superieurs des genitalia males pour la systématique des *Aelia* palearctiques (Hem. Pentatomidae). Ann. Soc. Entomol. France 4: 185–195.

1969. Les *Aelia* du Maroc. El-Awamia 30: 1–136.

Vora, V. J. R., R. K. Bharodia, and M. N. Kapadia. 1985. Pests of oilseed crops and their control: rape and mustard. Pesticides 19: 38–40.

Wagner, E. 1960. Die paläarktischen Arten der Gattung *Aelia* Fabricius 1803 (Hem. Het. Pentatomidae). Sond. Z. Angew. Entomol. 47: 149–195.

Waite, G. 1980. The basking behavior of *Nezara viridula* L. (Pentatomidae: Hemiptera) on soybeans and its implication in control. J. Austral. Entomol. Soc. 19: 157–159.

Waldbauer, G. P. 1977. Damage to soybean seeds by South American stink bugs. An. Soc. Entomol. Brasil 6: 223–229.

Walker, H. G., and L. D. Anderson. 1933. Report on the control of the harlequin bug, *Murgantia histrionica* Hahn, with notes on the severity of an outbreak of this insect in 1932. J. Econ. Entomol. 26: 129–135.

Walsh, B. J. 1866. The Texas cabbage bug. Pract. Entomol. 1: 110.

Wang, Q., and J. G. Millar. 1997. Reproductive behavior of *Thyanta pallidovirens* (Heteroptera: Pentatomidae). Ann. Entomol. Soc. Amer. 90: 380–388.

Watanabe, M., R. Arakawa, Y. Shinagawa, T. Kohama, and Y. Kosuge. 1994a. Application of insecticide aerosols to window-frames for the house-invading stink bugs control. Sci. J. Pestol. 9: 17–21.

Watanabe, M., R. Arakawa, Y. Shinagawa, and T. Okazawa. 1994b. Overwintering flight of brown-marmorated stink bug, *Halyomorpha mista* to the buildings. Hyg. Zool. 45: 25–31.

Watanabe, M., R. Arakawa, Y. Shinagawa, and T. Okazawa. 1994c. Anti-invading methods against the brown marmorated stink bug, *Halyomorpha mista*, in houses. Hyg. Zool. 45: 311–317.

Watanabe, M., T. Inaoka, R. Arakawa, Y. Shinagawa, T. Kohama, and Y. Kosuge. 1994d. Susceptibility of four species of house-invading stink bugs to several insecticides. Hyg. Zool. 45: 239–244.

Weber, M. A., M. C. Del Vecchio, and J. Grazia. 1988. Estudo dos imaturos de pentatomideos (Heteroptera) que vivem sobre arroz (*Oryza sativa* L.): II — *Mormidea notulifera* Stål, 1860. An. Soc. Entomol. Brasil 17: 161–173.

White, W. H., and L. W. Brannon. 1933. The harlequin bug and its control. U.S. Dept. Agric. Farmers' Bull. 1712: 1–10.

Whitmarsh, R. D. 1917. The green soldier bug, *Nezara hilaris* Say. Order Hemiptera. Family, Pentatomidae. A recent enemy in northern Ohio peach orchards. Ohio Agric. Exp. Stn. Bull. 310: 519–552.

Wilks, J. M. 1964. The spined stink bug: cause of cottony spot in pear in British Columbia. Canad. Entomol. 96: 1198–1201.

Williams, R. N., J. R. Panaia, F. Moscardi, W. Sichmann, G. E. Allen, G. L. Greene, and D. H. C. Lasca. 1973. Principais pragas da soja no estado de São Paulo: reconhecimento, métodos de levantamento e melhor época de controle. Secr. Agric., CATI, pp. 1–18.

Wiseman, B. R., and W. W. McMillian. 1971. Damage to sorghum in south Georgia by Hemiptera. J. Georgia Entomol. Soc. 6: 237–242.

Wood, G. A. R. 1970. *Bathycoelia thalassina*, a potentially major pest in West Africa. Cocoa Grow. Bull. 14: 32–34.

Yamada, K., and M. Miyahara. 1979. Studies on the ecology and control of some stink bugs infesting fruit. II. *Gymnosoma rotundatum* as a natural enemy to the brown-winged green bug. Bull. Fukuoka Hortic. Exp. Stn. 17: 54–62.

Yates, I. E., W. L. Tedders, and S. Sparks. 1991. Diagnostic evidence of damage on pecan shells by stink bugs and coreid bugs. J. Amer. Soc. Hortic. Sci. 116: 42–46.

Yathom, S. 1980. An outbreak of *Dolycoris baccarum* L. (Heteroptera: Pentatomidae) on sunflower in Israel. Israel J. Entomol. 14: 25–28.

Yeargan, K. V. 1979. Parasitism and predation of stink bug eggs in soybean and alfalfa fields. Environ. Entomol. 8: 715–719.

Youther, M. L., and J. E. McPherson. 1975. A study of fecundity, fertility, and hatch in *Euschistus servus* (Hemiptera: Pentatomidae) with notes on precopulatory and copulatory behavior. Trans. Illinois State Acad. Sci. 68: 321–338.

Zalom, F. G., J. M. Smilanick, and L. E. Ehler. 1997. Fruit damage by stink bugs (Hemiptera: Pentatomidae) in bush-type tomatoes. J. Econ. Entomol. 90: 1300–1306.

CHAPTER 14

Shield Bugs (Scutelleridae)

M. Javahery, Carl W. Schaefer, and John D. Lattin

1. INTRODUCTION

The Scutelleridae is a family of heteropterans related to Pentatomidae and other Pentatomoidea. There are about 80 genera and 450 to 500 species that occur worldwide, especially in the tropics and subtropics (Kirkaldy 1909, Lattin 1964, Schuh and Slater 1995). Although all species are phytophagous, only relatively few have been reported as pests. However, one group of these pests, *Eurygaster* spp., or Sunn pests (less frequently Senn, Shüne, or soun pests), devastate wheat crops in the Middle East and Near East (see below). In addition, several species may become pests locally, and, because many feed on developing seeds, their damage may be difficult to detect.

The name, "shield bug," refers to the broad scutellum, which in these bugs covers (or in a few species nearly covers) the entire abdomen. Some species are brilliantly colored, and sometimes vividly iridescent: "Some scutellerids are among the most spectacularly colored of all Heteroptera" (Schuh and Slater 1995). In size, species range from about 5 to 20 mm.

There has been some question whether Scutelleridae should be treated as a pentatomid subfamily (as for many years it was), or as a family; recent (i.e., since about the 1970s) authors treat it as a family (see McDonald and Cassis 1984, Schuh and Slater 1995). The tribal and subfamilial categories remain unclear; in a series of papers Leston (1952a,b,c,d, 1953, 1954) analyzed the problem but, in the end, arrived at no firm conclusion. The question remains both open and vexed. (See also de la Fuente 1973, pp. 250, 255; Afzal et al. 1982.)

There is no catalog of Scutelleridae, except for Kirkaldy (1909). Several regional or local lists exist. Both outdated and unsurpassed, two works by Schouteden are indispensable to the student of Scutelleridae: his account (with keys, species lists, and color plates) of the genera of the family (1904, 1906) and his monograph (also with keys and color plates) of the African species (1903). Hoffmann (1932, 1935, 1948) has cataloged the Scutelleridae from China and surrounding countries, and Yang (1934) revised the Chinese species; Ahmad et al. (1979) lists and keys the scutellerids of Pakistan (and Bangladesh); McDonald and Cassis (1984) revised the Australian members of the family as did de la Fuente (1973) for the Iberian (Europe) species (creating two new tribes in the process); Alayo D. (1967) lists and discusses the Cuban Scutelleridae; and Lodos et al. (1998) catalog the Scutelleridae of Turkey, with brief biological notes. In addition, Froeschner (1988) catalogs the scutellerids of North America, Lattin (1964) in an unpublished dissertation revises them, and Eger and Lattin (1995) consider some nomenclatorial problems.

Although often large and brightly colored, scutellerid genera have been surprisingly little studied. Several have been revised, e.g., the Australian *Theseus* (Baehr 1989), the Neotropical *Polytes* (Eger 1990), the Palearctic *Odontoscelis* (Kis 1979, Göllner-Scheiding 1986), the tropico-Asian *Cantao* (McDonald 1988), the Indo-Pakistani species of *Poecilocoris* (Ahmad et al. 1980, Ahmad and Kamaluddin 1982), and the Australasian *Calliphara* (Lyal 1979), to list a few of the more recent revisions.

Various aspects of scutellerid morphology are presented briefly in the papers above; purely morphological work is largely restricted to the genitalia of a few species — male genitalia (McDonald 1961, 1963c; Singh 1971; Agarwal and Baijal 1984), female genitalia (McDonald 1963a, Singh 1968a), both sexes' genitalia (Baijal and Agarwal 1980; Vodjdani 1954, 1961a) — and accounts of more general morphology (Ahmad and Mushtaq 1977, Moizuddin and Ahmad 1980, Ahmad and Moizuddin 1980); of the Malpighian tubules (Singh 1965), the midgut epithelium (Singh 1968), and the nervous system (Singh 1969). Chorionic patterns of the eggs of *Eurygaster* spp. have been studied by Putschkova (1959), Grigorov (1988), and Javahery (1994), and yolk formation by Verma and Basiston (1974).

The eggs and instars of several economically unimportant species are described by Kobayashi (1954, 1956, 1967), Harris and Andre (1934), McDonald (1960, 1963b), and Reid and Barton (1989). Grigorov (1988) gives electron micrographs of the chorions of three species of *Eurygaster*.

The secretions from the metathoracic and abdominal scent glands of several scutellerid species, adult and nymph, have been analyzed (Ubik et al. 1975; Smith 1978a,b; Kumari et al. 1984; Gough et al. 1985, 1986; Janaiah et al. 1988); in addition, secretions of the androcone glands (Carayon 1984) of the male's abdominal venter have been studied in *Tectocoris purpureus* (Thunberg) (Knight et al. 1985). In some of the former preparations, there may have been some contamination of the latter secretions, judging from the methods of preparation (e.g., Ubik et al. 1975). The first example in Insecta of a branched carbon-chain spiroketal has been found in the abdominal glands of adult *Cantao parentum* (White) (Moore et al. 1994). It is possible that synthetic scent-gland secretions may prove useful in control of scutellerid pests. Indeed, these secretions both have antifungal activity (Surender et al. 1987a) and cause cytological aberrations in plant cells (Surender et al. 1987b).

The few species of Scutelleridae whose karyotypes have been studied all have 12 (10 + XY) chromosomes (Manna and Deb-Mallick 1981; Mittal and Lelamma 1981; and, especially, Ueshima 1979), except *Chrysocoris stollii* (Wolff) (12 + XY) (Singh and Singh 1966).

Scutellerids as a family feed widely; no group of genera appears to specialize on a particular group of plants (C. W. Schaefer, unpublished compilation from the literature; see also Leston 1973); the preference of *Eurygaster* spp. for grasses and their relatives has predisposed them to become pests on grain crops. The family varies also in its preference for plant parts, some species feeding on seeds, others on somatic tissues, and others on fruits (Harris and Andre 1934, Leston 1973, McDonald 1960, Bérenger and Lupoli 1991). Not surprisingly, the worst pests are those that feed on reproductive structures. One species, *Scutiphora pedicellata* (Kirby), apparently is a pollinator (Webb 1989).

Two species have been reared in the laboratory (Walt and McPherson 1972, Reid and Barton 1989).

Hussey (1934) discusses parental care in several scutellerids. And Londt and Reavell (1982) record *Solenosthedium* (as *Solenostethium*) *liligerum* (Thunberg) "sucking goat dung."

2. GENERAL STATEMENT OF ECONOMIC IMPORTANCE

As mentioned, except for the Sunn pests, few scutellerids are of major economic importance. Sunn pests (*Eurygaster* spp., especially *E. integriceps* Puton), however, do enough damage to warrant considering the entire family economically significant (Javahery 1995). [*Note:* However, Lodos (1961) comments that the Turkish word *Süne* refers only to *E. integriceps* Puton.]

3. THE IMPORTANT SPECIES

3.1 *Eurygaster* species

- *Eurygaster* species have a broad body and a length-to-width ratio of 1.56. The scutellum is curved and extends to the end of the abdomen. Currently, 18 species are known. These include 13 species and one fossil from the Palearctic (Puton 1881; Vodjdani 1954; Southwood and Leston 1959; Putschkov 1961, 1965; Javahery 1995), and four living species from the Nearctic (Bliven 1956, 1962; Vodjdani 1961a; Lattin 1964; McPherson 1982; Froeschner 1988). Of the Palearctic species, only three, *E. austriaca* Schranck, *E. integriceps* Puton, and *E. maura* L., cause damage to wheat and barley in the Near East, Middle East, and in southwestern Asian countries (Brown and Eralp 1962; Brown 1963; Popov 1972; Paulian and Popov 1980; Gaffour-Bensebbane 1981; Lodos 1981; Stamenković 1990; Mohaghegh et al. 1991; Javahery 1995, 1996). The principal scutellerid Sunn pest species are in the genus *Eurygaster* and are discussed in this chapter.

In spring, the overwintered adult bugs in the Palearctic attack the leaves, shoots, and stems as well as the newly formed ears of the branch, causing the plants to die before ear formation. Later, the new generation of nymphs (second to fifth instars) and young adults feed on the milky stage or mature grain, which reduces baking quality (Putschkov 1961, 1965; Martin et al. 1969; Paulian and Popov 1980; Javahery 1995). Sunn pests are a major factor in the production of wheat and to a lesser extent of barley. During outbreaks, infestations may cause 100% crop loss. About 8 million ha of wheat in the Middle East and Near East, plus areas in the southwestern republics of the former U.S.S.R., are infested annually by Sunn pests. In the Middle East and Near East about 2 million ha are treated annually with insecticides at an estimated cost of about U.S.$42 million (Paulian and Popov 1980; Javahery 1992, 1995, 1996; Skaf 1996).

Chemical control has not been satisfactory as evidenced by increasing Sunn pest populations and periodic outbreaks over the past five decades. Sunn pests are adapted to a wide range of habitats and exhibit variation in life cycle and behavior over the range of their distribution (28 to 55° N latitude) (Javahery 1995). These bugs are so important that they have engaged the attention of several international organizations, such as the Food and Agricultural Organization, in organizing and coordinating research and training programs (Brown 1962a; Javahery 1967, 1978a, 1992, 1995; Martin et al. 1969; FAO 1993; Miller and More 1996).

3.1.1 *Eurygaster integriceps* Puton

3.1.1.1 *Distribution, Life History, and Biology*

Eurygaster integriceps is the most important species, occurring between 28° and 55° latitude, and 10° west to 60° east longitude. This Sunn or Senn pest is a serious pest of wheat and, to a lesser extent, barley, in Bulgaria, Romania, Greece, Turkey, southwest Asia, and particularly in the Middle East (Syria, Lebanon, Jordan, Iraq, Iran), Afghanistan, and several of the independent countries in the southwest of the former U.S.S.R., where wheat and barley are the main crops and the principal source of food (Alexandrov 1947–49; Wagner 1951; Vodjdani 1954; Putschkov 1961; Brown and Eralp 1962; Yüksel 1968; Martin et al. 1969; Javahery 1978a,b, 1995; Paulian and Popov 1980; Popov 1982a; Lodos and Önder 1983; Lodos and Kavut 1991). This species has never been reported from western Europe, North Africa, Arabia, or the southern states of the Persian Gulf. Critchley (1998) provides a comprehensive review of the literature (some 230 references).

Eurygaster integriceps is 10 to 12 mm long, and varies from gray to creamy brown, to reddish or black; most, however, are gray or clay-colored (Javahery 1995). Detailed descriptions of eggs, nymphs, and adults are given by Puton (1881), Makhotin (1947), Wagner (1951), Vodjdani (1954), Putschkov (1961), Lodos (1961), Javahery (1994, 1995), and structure of scent gland by Daroogheh (1990). *Eurygaster integriceps* is morphologically similar to *E. maura* but can be distinguished by the genitalia (male: Vodjdani 1954; Putschkov 1961; Brown and Eralp 1962; Javahery 1967, 1995; Mohaghegh et al. 1991; female: Southwood and Leston 1959; Brown and Eralp 1962; Gaffour-Bensebbane 1994).

Egg characteristics are useful in distinguishing *Eurygaster* species (see Grigorov 1988). The eggs of *E. integriceps* are green, shiny, spherical, about 1 mm in diameter, and possess 16 to 23 micropylar processes (Putschkova 1959; Javahery 1994, 1995). *Eurygaster integriceps* is univoltine, and has an obligate diapause throughout its range; its life cycle is divided into an active and a passive period (see Vodjdani 1954, Javahery 1995). The active period begins in early spring when, at about 18°C, the insects leave their hibernation sites and migrate to the fields. This period persists for 2.5 to 3 months during which mating, oviposition, and development of immature stages take place, on wheat, barley, or other graminaceous plants. The time spent in each stage varies considerably and depends on various factors, particularly weather, in the development of the egg and first instar. Both food and weather are critical for the second instar to the mature adult. This insect is adapted to a wide range of climatic conditions (between 28° and 55° latitude). Both adults and nymphs are active at high temperature and feed intensively on stems, leaves, and particularly on ears of their hosts; attacked plants die. Toward the end of this active period, the young adults migrate to higher altitudes in the mountains, where they rest at lower temperatures (Putschkov 1961; Brown 1965; Javahery 1978b, 1995).

The passive period is from midsummer to early the following spring, a period of about 9 months known as the estivation–hibernation or the "overwintering" period. During the hot and dry months of late summer and autumn, *E. integriceps* estivates within or under plants at high altitudes; and, before the cool weather sets in, it migrates to lower altitudes for hibernation. This occurs mostly on southern slopes under bushes and dead leaves for the rest of the autumn and the winter months. During the estivation period the bugs are partially mobile, whereas during hibernation (cold weather) they are almost completely inactive (Brown 1962a, Martin and Javahery 1968, Javahery 1995). The insects use their stored reserves (fat) during estivation and hibernation (Fedotov 1947–1960, Brown 1962a). In early spring, *E. integriceps* leave their hibernation sites and migrate to the fields, where they repeat their annual cycle.

Diapause is characterized by an arrest in reproductive development. During this period, the fat body and weight of the insect gradually decrease (Popov 1979), reaching a minimum after the insects descend to the fields in spring. Attempts to break diapause before early winter have not been successful (Javahery 1967, 1995; Martin et al. 1969; Daroogheh 1993b; Zarnegar 1995).

The new generation of *E. integriceps* becomes sexually mature after termination of diapause in early winter. Overwintered adults appear in wheat and barley fields from February to April each year and feed during the warmer hours of the day (temperatures above 10 to 12°C). About 1 to 4 weeks later, mating and oviposition commence (temperatures above 20°C). This species copulates five to ten times during its life, but females produce nearly the same number of eggs from either single or multiple mating (Martin et al. 1969; Popov 1982b; Javahery 1967, 1995, 1996). The sex ratio normally approaches 1:1, but this may shift toward females under conditions such as high rainfall over several seasons (Brown 1962a, Martin et al. 1969, Popov 1982b, Javahery 1995).

Many data are available on the fecundity of *E. integriceps* (Makhotin 1947; Alexandrov 1947–49; Putschkov 1961; Javahery 1967, 1995, 1996; Martin et al. 1969; Paulian and Popov 1980; Abdollahi 1988). Until 1958, fecundity was studied on green shoots of cereals. However, during 1958 to 1960, culture of Sunn pests was established on dry wheat grain supplemented with water on wet cotton (Zomorrodi 1959, Voegelé 1960, Javahery 1967). Fecundity is related mostly to food, temperature, and humidity (Javahery 1995, 1996).

Both fecundity and longevity in *E. integriceps* are much higher on grain supplemented with water (at about 27 to 28°C) than on vegetative tissue. Fecundity is slightly higher on grain and green shoots of wheat than on barley. This is interesting because under natural conditions these insects mostly attack wheat even when barley fields are close by. The average fecundity per female of *E. integriceps* on wheat grain with water is 190 to 200 eggs, and on barley grain and water about 180 to 185. Fecundity, however, decreases to half when the insects are bred on green shoots of wheat or barley at the same temperature, and also at low temperatures from 20 to 22°C.

Fecundity of *E. integriceps* bred on either wheat or barley grain supplemented with water sharply decreases under crowded conditions. Variability in the number of eggs per batch is also more common under crowded conditions: usually 14 eggs per batch are produced but batches of 10, 13, or 15 are produced under crowded conditions (Martin et al. 1969; Javahery 1967, 1993, 1995, 1996).

Eurygaster integriceps oviposits in two to three rows in batches of 14 eggs attached together on leaves, but rarely on stems or on soil. Oviposition behavior is not the same in other Sunn pest species (Makhotin 1947; Martin et al. 1969; Javahery 1994, 1995).

Incubation takes 10 to 20 days in the field. Hatching occurs during the day and first instars aggregate on or around the eggshells, where they remain up to about the middle of the second stadium, living on the reserves remaining in the eggshell. During the last days of the second stadium, the nymphs disperse mainly toward the ears. The duration of the five stadia varies (45 to 60 days) with weather conditions and availability of food (Makhotin 1947, Paulian and Popov 1980, Javahery 1995). Under laboratory conditions at 27 to 28°C, the five stadia take 4, 4.5, 4, 6, and 10 to 11 days, respectively (Javahery 1995).

The young of the new generation appear from late May to the middle or end of July. The instars can be identified on the basis of size and wing development. It is quite usual to observe second to fifth instars feeding simultaneously on ears, although they will also feed on leaves and stems. Intensive feeding takes place from the third stadium on, and particularly in the fifth and by the young adult (Makhotin 1947, Yüksel 1968, Martin et al. 1969, Paulian and Popov 1980, Javahery 1995).

Adults of the new generation appear in June and July and accumulate a large store of reserves. The weight of the female increases from 110 mg to 145 to 155 mg. Accumulated fat accounts for 36 to 42% of body weight in females, and 32 to 38% in males. When food supplies are abundant and the weather is favorable, the insect can accumulate the necessary reserves within approximately 10 days. However, during rainy periods and under cooler temperatures, this process may take up to 3 weeks. In a late-developing population, however, feeding may continue on the ears of harvested crops lying in the fields until sufficient reserves are accumulated to permit migration. Under favorable conditions the female *E. integriceps* can produce between 150 and 200 progeny on wheat or barley with one or more matings (Fedotov 1947–1960; Martin et al. 1969; Paulian and Popov 1980; Daroogheh 1994; Javahery 1973a, 1995, 1996).

In spring and early summer, *E. integriceps* breed and feed in wheat and barley fields at relatively lower altitudes (1000 to 1200 m). In late summer the new adults usually leave the breeding areas when the temperature is 25 to 43°C and migrate about 10 to 50 km to higher altitudes (2000 to 2500 m) where temperatures are cooler (15 to 30°C). At these altitudes they estivate within or under bushes from July to October before moving (in short flights or by walking) to lower altitudes (1700 to 1900 m) in November where they pass the rest of the winter in a state of low metabolic activity. During the overwintering period about 25% of the fat reserves are consumed and mortality is usually between 10 and 30%. However, mortality may increase up to 50% if there is very cold weather just before migration to the fields. These areas where the pests spend the winter are known as "overwintering sites" and the fields to which they migrate are referred to as "invasion areas" (Fedotov 1947–1960; Brown 1962a, 1965; Paulian and Popov 1980; Daroogheh 1993b; Javahery 1995).

The up-and-down movement of this scutellerid pest species from the breeding areas to the mountains and back to the fields is recognized as an obligatory migratory behavior that may be described as a kind of safety valve associated with a change in the insect's requirements. *Eurygaster integriceps* is migratory in most areas of its distribution in the Near East and Middle East, including the Commonwealth of Independent States of the former U.S.S.R. However, in some localities, such as the central region of Iraq, in the Diyarbakir area of eastern Turkey, and in the Orzouieh plain of south central Iran, little or no migration apparently takes place (Brown 1965, Yüksel 1968, Martin el al. 1969, Javahery 1995). It has been observed that some individuals may remain in the mountains as a settled population instead of moving down to the plains in spring (Arnoldi 1955, Javahery 1995). This nonmigratory portion of the population is advantageous to the species in that it safeguards the species against unfavorable conditions elsewhere and ensures its continued existence. It also preserves part of the population from being exposed to the selective pressure of pesticides in treated fields, thereby delaying or minimizing development of insecticide resistance (Brown 1965; Javahery 1995, 1996; Popov et al. 1996).

This cereal bug is found on several wild graminaceous plants such as *Agrostis*, *Bromous*, *Dactylis*, *Festuca*, *Lolium*, and *Poa*. In the absence of these, however, the insect can feed and develop on cultivated cereal plants, especially both dry-farmed and irrigated wheat. Gerini (1968) lists more than 15 species, of eight to nine families, of nongrain host plants. *Eurygaster integriceps* has become adapted to over 15 different wheat varieties, in addition to a number of wild graminaceous species such as *Heteranthelium pliliferum* (Banks and Soland.) Hochst; *H. pliliferum* grows in the mountainous regions of the Middle East at altitudes of about 2000 m (Arnoldi 1955; Brown 1962a; Javahery 1972, 1995, 1996; Radjabi and Termeh 1988).

A long-term study of population development of *E. integriceps* indicated that changes in population density of this insect and the occurrence of outbreaks are determined by the availability of food plants, by climatic conditions (especially temperature and rain), and by parasitism of both eggs and adults (Javahery 1973a, 1995).

3.1.1.2 Damage

The losses of feed crops in all countries where *E. integriceps* occurs is so great, that, with the exception of Turkey, there is a deficit between production and consumption. This shieldbug is adapted to a wide range of habitats and exhibits variation in life cycle and behavior over the range of its distribution (28 to 55° latitude) (Brown 1962a; Paulian and Popov 1980; Javahery 1995, 1996).

Most damage to wheat seedlings is to the seed-bud region, which damages seed set and germination, and the seed itself becomes more water absorbent (Grigorov 1989; see also **Section 3.1.2.2**) The grain itself of *E. integriceps*–damaged wheat has more water-soluble nitrogen (because of a decrease in glutenin and gladin), and this nitrogen has higher enzymatic activity than occurs in normal grain (Pokrovskaya et al. 1971).

3.1.1.3 Forecasting Population Densities and Outbreaks

Fluctuations in population density from one year to another are related to the physiological condition of the individuals and the biotic and abiotic ecological factors. It is possible from a study of the physiological conditions of the internal organs to forecast the degree of abundance of *E. integriceps* during the forthcoming invasions of wheat fields. Survival of individuals depends on the quantity of food reserves (size of fat body) from the time of migration to resting sites (summer, autumn, winter) to the moment they return to the fields the following year in the spring (Fedotov 1947–1960, Brown 1962a, Popov et al. 1996).

Monthly estimation of fat content by chemical analysis or dissection to assess the physiological condition of the internal organs should be carried out during both estivation and hibernation. Changes in population density can also be observed by monitoring populations over consecutive years in cultured fields and overwintering areas (Brown 1962a, Martin et al. 1969, Javahery 1995). The use of insecticides, and local movements after the main migration, may affect population densities; but in the overwintering areas the population densities are more constant (Brown 1962a, Javahery 1995, Popov et al. 1996). Jovanié and Stamenković (1978) have found this to be true also of *E. maura* (L.) and *E. austriaca* Schrank.

This method was first used by Russian workers to forecast the population of *E. integriceps* (Fedotov 1947–1960). The estimates of population density in the overwintering areas appear to be more reliable than the data taken from fields. A long-term study of *E. integriceps* in two localities has provided valuable information on forecasting population densities and periodical outbreaks of this very important wheat and barley pest (Martin and Javahery 1968; Martin et al. 1969; Javahery 1995, 1996).

3.1.1.4 Control Methods

3.1.1.4.1 Chemical Control — A wide range of chemical compounds has been used against Sunn pest (Javahery 1978a, 1995, 1996). Today, however, preference is given to organophosphorus compounds such as Lebaycid (= Baytex or Fenthion) and Sumithion (= Folithion) 50 EC. Long-term use of insecticides has led to development of slight resistance in fourth and fifth instars of Sunn pests. However, several new chemical compounds are at present under investigation (Djavadzadeh and Esmaili 1991, Haghshenas 1993, Javahery 1995).

The decision to treat and the timing of treatments should be carefully considered to reduce costs and the number of applications, as well as to obtain more effective results. These decisions will also lead to a more rational use of insecticides and ultimately to conservation of natural enemies (Javahery 1972, 1973b, 1978b, 1995). Ideally, one treatment should be applied against the overwintered adults before oviposition and, if necessary, another against the second instars. Overwintered *E. integriceps*, however, appear in the fields over a 1 to 4 week period, and treatment is only effective for about 10 days. When the population density reaches the economic threshold, two treatments are often applied: the first at the population density of the overwintered adults above one individual/m^2 which is usually toward the end of the tillering stage in wheat and barley; the second at three to five nymphs/m^2 (second and a few third instars), which is about the time of flowering to initiation of grain filling. The most effective time to apply insecticides against any Sunn pest is during grain filling. This will reduce both the population of early instars and the late overwintered migratory adults (Javahery 1978a, 1978b, 1995, 1996).

The second treatment is usually delayed until most nymphs (50 to 60%) are third, fourth, and even fifth instars, which coincides with the "milky stage" of winter wheat. Failure to control nymphal development at this point can lead to considerable loss in yield as a result of the intensive feeding of the fourth and fifth instars and the adults of the new generation (Javahery 1978a, 1995).

It must be noted that because first instars and early second instars do not feed, any treatment within this period of development is premature, unnecessary, and has little effect on population reduction (Javahery 1978a, 1995).

Wheat and barley are planted both on the plains and in mountainous areas in the Near East, Middle East, and southwestern Asia. About two thirds of these cereals are cultivated in a dry-farming system at the higher altitudes where rainfall is fairly sufficient. On the plains, however, cereal crops are cultivated in irrigated fields, quite often in small blocks, within cash crops, and often under high-power transmission lines or surrounded by trees. Such fields are usually not treated and serve as reservoirs for pests.

At present, systems of varying capacities are used for both ground and aerial application and are also determined by field size and location (FAO 1993; Javahery 1972, 1973b, 1978b, 1995, 1996).

Each insecticide used against *E. integriceps* has had an effective usefulness of about 10 to 15 years. Pesticide use started during World War II with the introduction of dinitro-ortho-creasol. This compound was sprayed on adults of *E. integriceps* as they aggregated on and under their host plants (*Artemisia* spp., *Astragalus* spp., *Acantholimon* spp., and *Acanthophyllum* spp.) during estivation and hibernation on the Alborz mountains in Iran (Garah-Aghatch in the east, and Alamout to the west of Tehran) (Javahery 1978a, 1995, 1996).

In 1952, DDT gave good control against *Eurygaster* spp. in the field until 1960, when nymphs of *E. integriceps* became resistant. In 1959, Dipterex, an organophosphate, was highly effective on adults and nymphs of this species and was recommended for its control. Both insecticides were commonly used until 1965, and were replaced by Lebaycid (= Baytex or Fenthion) in 1970. Sumithion (= Folithion) was subsequently introduced in 1973. *E. integriceps* has become resistant to both these compounds, since 1975 to Lebaycid, and since 1987 to Sumithion (Javadzadeh-Siahkalrudi 1992). Pesticide applications against this pest have increased dramatically during the past 20 years, from about 50,000 ha treated in 1968 to 1970 in both Iran and Turkey, to about 1 million and 600,000 ha, respectively, in each year between 1991 and 1998. In most regions (Afghanistan, Iraq, Iran, Jordan, Syria, and Turkey) more than 2 million ha are treated annually at an estimated cost of U.S.$42 million (Javahery 1995, 1996).

Other insecticides have been tested and used to control *E. integriceps* during the past 30 years. At present, most affected countries in the Near East, Middle East, and southwestern Asia depend on insecticides for Sunn pest control (Javahery 1972, 1973b, 1978a, 1993, 1995; FAO 1993).

The estimated cost of chemical control varies between U.S.$15.00 to $60.00/ha in the affected countries. Without control in epidemic years, damage may reach 90 to 100% (Javahery 1978b, 1995; FAO 1993).

Chemical control of *E. integriceps* is costly, given the low yields of both wheat and barley. However, this method is applied because it is paid by the governments to guarantee the food security of their nations. The cost of insecticides has also increased sharply during the past two decades and this species has become resistant to chemical compounds. New compounds will be even more costly (Javahery 1995, 1996).

Although chemical treatments increase the yield of infested wheat and barley, a long-term study indicates that chemical control has more and greater disadvantages than benefits. Therefore, it is apparent that other control methods, such as cultural and biological ones, should be applied (Yüksel 1968; Javahery 1978a, 1978b, 1979, 1995, 1996; FAO 1993). Other control methods, such as behavioral, genetic, and entomopathogenic agents with high pathogenicity, should be studied (Daroogheh 1993b, Rezabeigi 1995, Zarnegar 1995, Najafi Mirak 1998). For four decades it was thought that use of organophosphates would end future outbreaks of Sunn pests. This has been shown to be incorrect. Outbreaks of *E. integriceps* have occurred periodically, with population densities of overwintered adults estimated at 2 to 20 individuals/m^2 and nymphs at 10 to 300/m^2. The extensive use of pesticides has also resulted in a significant decrease in natural enemies of Sunn pests, in particular the highly effective scelionid egg parasitoids. Other adverse effects on the environment and on human health have also become evident, in addition to which chemicals have become increasingly costly in recent years (FAO 1993; Javahery 1995, 1996).

All insecticides used to control *E. integriceps* are highly toxic to its natural enemies, particularly to hymenopterous egg parasitoids; pesticides are generally more toxic to the natural control agents than to their hosts. Surveys carried out in Iran and Turkey indicate that the number of egg parasitoids has significantly decreased in areas where chemical treatments were applied against *E. integriceps* from 1987 to 1995. A comparative study carried out in fields and orchards in Waramin, around Karaj (near Tehran) and in Isfahan, Iran, indicates many more of these useful egg parasitoids in nontreated areas.

The relative toxicity of insecticides to the two principal groups of *E. integriceps* parasitoids (the egg parasitoids and the dipterous parasitoids) should be investigated. Attempts should be made to apply insecticides that are more selective and less toxic to these natural enemies (Javahery 1995).

3.1.1.4.2 Biological Control — Several groups of parasitoids are important in reducing population of *E. integriceps*. Tachinid flies (subfamily Phasiinae) attack the adults and occasionally fifth instars; and hymenopterans, mainly Scelionidae (subfamily Telenominae) and certain Encyrtidae, parasitize the eggs. Egg parasitoids play a significant role in control of this cereal pest (Vassiliev 1913; Alexandrov 1947–1949; Talhouk 1961; Brown 1962b; Javahery 1967, 1968, 1995; Safavi 1968; Martin and Javahery 1968; Martin et al. 1969); Gerini (1968) lists several predators and parasites.

The tachinids are of less importance than the hymenopterous parasitoids in the degree of parasitism. The taxonomic status of several species needs to be clarified. The following species parasitize adult Sunn pests: *Phasia subcoleopterata* L., *P. oblonga* F., *Ectophasia crassipennis* F., *Ectophasia rubra* Girschner, *Heliozeta helluo* F., *Elomyia lateralis* Meigen, *Clytiomyia helvola* Meigen, *Gymnosoma desertorum* Rohd., *Helomyia lateralis* Meigen, *Cylindromyia brassicaria* F. (*dominant species) (Putschkov 1961, Amir-Maafi et al. 1991, Javahery 1995). These flies overwinter as third-stage larvae in the bodies of their hosts. Shortly after the return of *E. integriceps* to the cereal fields, the mature parasitoid larvae leave their hosts through the end of the abdomen and pupate in the soil (Amir-Maafi el al. 1991).

In almost all species, the main emergence of adult flies is in the spring after migration of the parasitized hosts to the fields, but free puparia and adults occasionally may be found in the hibernation areas in autumn and spring. The adult flies appear in the field during April and May and lay their eggs, one to six per host, on the dorsal surface — on the pronotum, the connexivum of the abdomen, the compound eyes, or on either side of the scutellum, but never on its dorsal surface (Alexandrov 1947–1949, Putschkov 1961, Brown 1962b, Amir-Maafi et al. 1991). After 2 to 4 weeks, the eggs hatch and the first instars enter the host body. Development of the immature stages varies among the different species and ranges from 26 to 46 days. When the parasitoid larva leaves its host, the Sunn pest dies. The degree of parasitism by dipterous parasitoids changes from year to year and in different cereal fields. However, average parasitism is 7 to 73% (Brown 1962b, Amir-Maafi et al. 1991, Javahery 1995).

Most hymenopterous egg parasitoids belong to the genera *Telenomus* Haliday and *Trissolcus* (= *Asolcus* Nakagawa, and *Microphanurus* Kieffer). Meier (1940) gives a key to, and brief descriptions of, the important egg parasites in the former U.S.S.R. These wasps are the most important agents of biological control of pentatomoids attacking wheat and barley in the Near East, Middle East, southwestern Asia, and Mediterranean basin (e.g., Turkey, Greece, Italy, Morocco, and Spain). Although much useful work has been done on the systematics of these minute wasps (Mayr 1879, 1903; Kieffer 1926; Nixon 1939; Masner 1958; Delucchi 1961; Javahery 1967, 1968; Kozlov 1968) there is at present some confusion among several closely related species (Javahery 1968, Kozlov and Kononova 1983). The following scelionids have been reared from eggs of *E. integriceps* in the regions mentioned above: *Trissolcus grandis* Thomson, *T. semistriatus* Ness, *T. vassilievi* Mayr, *T. basalis* Wollaston, *T. nigribasalis* Voegelé, *T. rufiventris* Mayr, *T. simoni* Mayr, *T. waloffae* Javahery, *Telenomus* chlorops Thomson, *T. sokolovi* Mayr, *T. truncatus* Mayr. The most effective and widespread of these egg parasitoids appears to be *Trissolcus grandis*.

Much information is available on the systematics, biology, ecology, oviposition behavior, reproductive potential, parasitism, and mass breeding of these useful egg parasitoids (Vassiliev 1913; Makhotin 1947; Delucchi 1961; Wilson 1961; Voegelé 1965; Javahery 1967, 1968, 1995, 1996; Safavi 1968; Martin and Javahery 1968; Martin et al. 1969; Popov 1980; Asgary 1995; Iranipour 1996; Shahrokhi 1997). Ophagous parasitoids are by far the most important naturally occurring control agents in Sunn pest areas. About 15 species of egg parasitoids are reported to attack Sunn pests, with different degrees of effectiveness. Parasitism fluctuates 10 to 98%. Information on egg parasitoids of Sunn pests has increased during the past five decades, particularly between 1955 and 1970; most studies, however, were focused on systematics, egg parasitoids/host relationships, and their distributions, particularly in Iran, Morocco, Romania, Syria, Turkey, the United Kingdom, and the Commonwealth of Independent States (Alexandrov 1947–49, Voegelé 1966, Martin and Javahery 1968, Safavi 1968, Martin et al. 1969, Javahery 1995).

Mass rearing and release have been pursued only in Iran, from 1953 to 1963. The egg parasitoids were bred in cardboard boxes (25 × 30 × 25 cm) on eggs of *E. integriceps* and released at a rate of 15,000 wasps/ha during the oviposition period of Sunn pests in May and early June in wheat fields in Isfahan and Waramin. As many as 210 million egg parasitoids, consisting of *Trissolcus grandis*, *T. semistriatus*, and *T. vassilievi*, were released in 1959 (Vassiliev 1913; Makhotin 1947; Alexandrov 1947–1949; Javahery 1967, 1968, 1995; Safavi 1968; Martin et al. 1969; Shahrokhi 1997).

Among other parasitoids and predators of Sunn pests, only tachinid flies have a relatively significant impact in some areas, but the degree of parasitism is seldom sufficient to exert any major control on populations of *E. integriceps*. Studies on taxonomy, life history, and parasitism have been made in several countries, such as Iran, Turkey, and the Commonwealth of Independent States (Alexandrov 1947–1949, Dupuis 1948, Brown 1962b, Amir-Maafi et al. 1991).

Parasitism varies from 7 to 73% depending on the fly species and environmental conditions. Generally higher rates of tachinid parasitism occur in areas with more than 200 mm of precipitation. In north central Iran, parasitism in overwintered bugs before oviposition was 7.56 and 11.58%, respectively, in 1987 and 1988 in Fashand; 5.12 and 10.25%, respectively, in 1987 and 1988 in Dorvan. Later in the spring during oviposition of the hosts, parasitism in the Fashand area reached a maximum of 66.92 and 62.59%, respectively, in 1989 and 1990, and 52.89 and 64.72%, respectively, in the Said-Abad region (Amir-Maafi 1991). In wheat fields around Bingol in eastern Turkey, a level of 73% parasitism was reported in adults of *E. austriaca* (Brown 1962b, Yüksel 1968).

Using serological methods, Kuperstein (1979) showed that five sympatric species of Carabidae (Coleoptera) at a combined density of nearly $5/m^2$, consumed about three *E. integriceps*/m^2 as late instars; it is believed this work was carried out in Russia, although the author does not say so.

3.1.1.4.3 Potential for Behavioral, Entomopathogenic Fungi, and Genetic Control —

All species of scutellerid Sunn pests have one generation per year. Reproduction normally takes place by sexually mature overwintered males and females after migration to the cereal fields, as well as on wild graminaceous plants in the hibernation sites.

A monumental amount of work has been done on factors regulating population in *E. integriceps* (Fedotov 1947–60; Putschkov 1961; Brown 1962b; Javahery 1995, 1996; Popov et al. 1996) and controlling this cereal pest, by both biological and chemical control (Martin and Javahery 1968; Martin et al. 1969; Javahery 1972, 1978a, 1995, 1996; Rosca et al. 1996; Voegelé 1966).

However, long-term studies of population density and controlling *E. integriceps* by means of mass breeding and release of scelionid parasitoids as well as application of several chlorinated hydrocarbons, e.g., DDT; carbamates, e.g., Sevin; and particularly organophosphates, e.g., Lebaycid and Sumithion, indicate that these control methods are not satisfactory, and that serious outbreaks of this noxious cereal pest still occur (FAO 1993; Javahery 1995, 1996). Thus, other controlling methods that can be applied against Sunn pests need to be investigated and used.

The following new programs are currently under study:

Entomopathogenic fungi — Sunn pest mortality caused by entomopathogenic fungi has been observed in their overwintering sites in northern Iran, eastern Turkey, Romania, Russia, and the Commonwealth of Independent States of the former U.S.S.R. (Putschkov 1961, Brown 1962a, Javahery 1995, Rosca et al. 1996). Mass mortality of *E. integriceps* caused by *Beauveria bassiana* Vuill. has been noticed under bushes of *Artemisia heba-alba* and *Astragalus* spp. at altitudes of 1700 to 2200 m after migration of overwintered adults to the nearby wheat fields in the Alborz mountains (Ghrah-Aghach and Alamout) north of Iran (M. Javahery, unpublished data). However, very little is known about the effect and pathogenicity of entomopathogenic fungi in depressing the population of the Sunn pests. A research program has started to investigate this at the entomology research laboratory of the University of Vermont with collaboration of the International Center for Agricultural Research in the Dry Areas (ICARDA) in Aleppo, Syria. During the past 2 years exploratory activities to collect entomopathgenic fungi from overwintering *E. integriceps* has been conducted in southern Turkey, southeastern Syria, Russia, and Uzbekistan (Parker et al., personal communication). Over 150 isolates of *B. bassiana*, *Verticillium lecanii*, *V. psalliotae*, *Paecilomyces farinosus*, *P. liliacinus*, and *Aphanocladium album* have been collected. In 1999, 25 of the isolates, selected for their high pathogenicity, were assayed directly against Sunn pest at ICARDA in Syria, and many caused high mortality in Sunn pest 5 to 7 days. Further pathogenicity trials are needed and planned to facilitate selection of the most promising ones for further development as biological control agents (Parker et al., personal communication).

Sex pheromone — Chemical attraction in a number of bugs has been studied recently (e.g., Aldrich 1996). Sexual communication of individuals calling a mate has been recently investigated in *E. integriceps*. Ždárek and Kontev (1975) showed females of *E. integriceps* were stimulated to mate by a male–produced pheromone. Ubik et al. (1977) and Vrkoč et al. (1977) reported the identification of vanillin and ethyl acrylate from extracts of male volatiles captured on filter papers within glass containers. Later, the same group (Kontev et al. 1978) reported the identification of sex pheromone of the Sunn pest (*E. integriceps*) from similarly trapped volatiles. Staddon et al. (1994) and Abdollahi (1995) reported that males of this scutellerid attract nearby females with vanillin and ethyl acrylate. They identified the major component in male sex pheromone as a homosesquiterpenoid. Gries et al. (unpublished) are further investigating isolation and identification of sex pheromones of this Sunn pest (personal communication). At this stage, it would be extremely valuable to conduct a replicated field experiment that proves that caged virgin male *E. integriceps* do indeed attract females. Once demonstrated, it will be possible to identify the signal that mediates attraction.

Genetic control — Evolutionary shifts in the diets of herbivorous insects require genetic variation in the behavioral or physiological traits that determine host use (Jaenike 1990). Reviewing the genetic factors affecting expression and stability of resistance, Gallun and Khush (1980) state that major genes generally impart high levels of resistance. But for every major gene found in the host there is a corresponding gene for virulence in the insect. Thus, there is a greater chance of an insect biotype being selected where resistant varieties are grown widely.

Developing wheat resistance is perhaps the easiest, most economical, and most effective way of controlling Sunn pests with no environmental hazards. It is also cheaper to develop wheat resistant to Sunn pests than new insecticides. Another advantage of using Sunn pest resistance varieties, for cereal grown in the Near East, Middle East, and southwestern Asian countries, is that no skill in Sunn pest control or cash investment is required of the grower.

Resistant varieties of wheat and barley have been used for the past three decades, and have been successful against such other pests of wheat and barley as the Hessian fly, *Mayetiola* (= *Phytophaga*) *destructor* (Say). Preliminary observations on the response of *E. integriceps* to four varieties of wheat (Roushan, Omid, Tabassi, Mexican) have indicated a variable preference, which suggests that some varieties may be less acceptable than others (M. Javahery, unpublished data).

More recent results from experimental field studies in Iran have indicated the following tolerance levels of wheat varieties to *E. integriceps*: (1) high-level tolerance, Sardari and Bezostaya; (2) medium-level tolerance, Roshan, Tabassi, and Omid; and (3) susceptible, Ghods, Karaj No. 1. The barley varieties Ariwat and Valfajr showed medium tolerance (Talai et al. 1991). Mirkarimi (1992) indicates that Sadari cultivar has high tolerance to *E. integriceps*. Rezabeigi (1995) indicated that wheat varieties show less resistance to overwintering adults, nymphs, and new adults. He further stated that wheat varieties of Navid and falat were resistant and Sardari, Bistoon, and Sabalan were susceptible to Kermanshah populations of the insect, whereas Gaphgaz, Rashid, and Azadi were resistant and Bistoon, Sabalan, Blich2, Omid, Ghods, and Zardak were susceptible to the insect populations of Waramin. Gaphgaz, Rashid, and Navid were most resistant and Arvand, Sabalan, Bistoon, and Sardary were susceptible for both Kermanshah and Waranin populations of *E. integriceps*. Najafi Mirak (1998) in a genetic study of resistance to *E. integriceps* in wheat, stated that Falat and Bezostaya X Golestan were the best cultivar and hybrid for increasing resistance to ear damage by overwintered Sunn pest. For grain injury, Golestan X Gafgas cross was the best hybrid for increasing resistance to grain injury by nymphs of this Sunn pest.

Genetic manipulation of resistance in controlling *E. integriceps* is under investigation and early results are promising. It should be realized, however, that developing wheat resistance to Sunn pests is a complex research program and requires cooperation among the disciplines of entomology, mycology, agronomy, and physiology (see also **Section 3.1.3.2**).

3.1.1.4.4 Cultural Control — Two-stage mechanized harvesting is a common harvesting practice for wheat, used in Canada since 1934 to save crops from the first frost. The method consists of first cutting the crop and laying it in windrows to cure, and then gathering it for threshing. This method reduces the time required for the crops to mature. In the Commonwealth of Independent States, two-stage mechanized harvesting became widespread in 1957 to prevent shriveling and seeding of grain when harvesting was prolonged. At present, a region can be harvested by this method in 5 to 7 days, which prevents continued feeding and development of new generation bugs. As a result, *E. integriceps* may not be able to complete nymphal development or accumulate the necessary reserves for migration to overwintering sites, and for survival during estivation and hibernation. In the Commonwealth of Independent States two-stage harvesting has proved to be a very effective and economic method of controlling this pest. It is used in irrigated areas, where wheat ripens at the same time as the Sunn pests mature. This method is at present undergoing field trials in Iran (Javahery 1995), where Esmaili and Dehghan (1991) found it effective in a study in wheat fields of Karaj, west of Tehran.

The ripening of wheat at about the same time as Sunn pests mature contributes to an increase in their population in the following season, because the intensively feeding fifth instars and new adults have prolonged access to a highly nutritive food source, leading to a maximal accumulation of reserves and greater reproductive potential in the spring. During this period of intensive feeding the potential for grain damage is at its greatest (Martin et al. 1969, Popov et al. 1996).

Late sowing, early grazing, and irrigation delay the ripening of wheat and ensure a longer feeding period for Sunn pest development. On the other hand, early planting (about 15 days earlier) leads to early ripening. As a consequence, earlier harvesting of ripe wheat protects (within 5 to 10 days of cutting and laying in windrows) the crop from attack by fifth instars and, particularly, new adults. In addition, the process also leads to a better quality of grain because of less shriveling (Donskoff 1996, Skaf 1996).

Ripe, unharvested fields of wheat and barley are often left standing for 2 to 3 weeks, because of a shortage of labor and of combine harvesters. When this occurs, losses are caused not only by Sunn pest feeding but also by seeding of the grain (Javahery 1995).

3.1.2 Eurygaster maura (L.)

3.1.2.1 Distribution, Life History, and Biology

Eurygaster maura causes damage to wheat and barley in northern Iran and central Turkey. It is common southeast of the Caspian Sea (Gonbad, Gorgan, Mazendran) and in the eastern part of Azerbaijan around Ardebil and the Moghan steppe. In these regions *E. maura* is often collected with *E. integriceps* (Javahery 1995).

Eurygaster maura completely replaces *E. integriceps* in the northern and central parts of Turkey (Modarres Awal 1992). It occurs on wheat and graminaceous plants in the southwestern newly independent countries of the former U.S.S.R. It continues east into western Pakistan (Ahmad and Moizuddin 1978), and to continental Europe and southern England, and south to Spain, Morocco, and Algeria (Tischler 1939, Vojdani 1954, Southwood and Leston 1959, Putschkov 1961, Brown 1962a, Paulian and Popov 1980, Gaffour-Bensebbane 1981, Javahery 1995).

This species is closely related to *E. integriceps*. It is somewhat broader, about 9 to 12 mm long, and brownish–gray. The male genitalia are the best character for distinguishing *E. maura* from *E. integriceps* (Putschkov 1961, Paulian and Popov 1980, Mohaghegh et al. 1991, Javahery 1995). Its immature stages were described and figured by Ahmad and Moizuddin (1978).

The life history of *E. maura* is similar to that of *E. integriceps*. In spring, the overwintered adults attack the leaves, shoots, and stems of newly formed ears; the central shoots die. Later, the new generation of nymphs (second to fifth instars) and young adults feed on the milky stage or mature grain. This leads to a reduction in the baking quality of the wheat (Putschkov 1961; M. Javahery, unpublished data).

Eurygaster maura is a semimigratory, monovoltine insect, having obligate diapause throughout the range of its geographic distribution. The overwintered adults emerge from hibernation in late March and early April and migrate to graminaceous plants, especially cereals. However, in Yugoslavia they may also feed on as many as 57 plant species representing 17 families; many, but not all, are graminaceous (Stamenković 1977; see also **Section 3.1.1.1**). Oviposition starts 3 to 4 days after mating. The eggs are laid in batches of 14, usually in two rows on ears or on the underside of cereal leaves. The eggs are green, shiny, spherical, about 1 mm in diameter, and have 18 to 23 micropylar processes. Females copulate three to five times and lay five to eight batches of fertilized eggs under field conditions. Incubation takes 10 to 15 days and nymphal development 50 to 65 days. The adults of the new generation appear in July. They feed for about 2 weeks on milky or hard grains of cereals before migrating to their estivation sites (at cooler temperatures) in the mountains or nearby forests. *Eurygaster maura* overwinters as an adult under bushes or dead leaves (Southwood and Leston 1959, Putschkov 1961, Brown 1962a, Mohaghegh et al. 1991, Javahery 1995).

Eurygaster maura is found with *E. integriceps* in wheat fields and at higher altitudes in the estivation and hibernation sites in the southeast of the Caspian Sea (Gorgan), Iran. In central Turkey it occurs together with *Aelia rostrata* Boheman (see **Chapter 13, Section 3.5**). It breeds and feeds in wheat and barley fields at relatively low altitudes (100 to 1200 m). In late summer it usually leaves the breeding areas when the temperature is 25 to 35°C and migrates about 5 to 15 km to higher altitudes (500 to 2100 m) in the mountains or the forests where temperatures are cooler (16 to 25°C). At these altitudes it estivates within or under bushes from July to October before moving to lower altitudes in November, where it passes the rest of the winter in a state of low metabolic activity. Little is known about the mortality of this species during its overwintering period (Tischler 1939, Putschkov 1961, Brown 1962a, Mohaghegh et al. 1991, Modarres Awal 1992, Javahery 1995).

Eurygaster maura is a northern species adapted to lower temperatures and high humidity such as exists around the Caspian Sea, Black Sea, and Europe. In these regions it is commonly found on wild graminaceous plants such as *Agrostis, Bromous, Dactylis, Festuca, Lolium*, and *Poa* species. However, in the absence of these, this scutellerid can feed and develop on cultivated cereals, particularly wheat. Population development of this insect is determined by both the availability of food, and the response to abiotic and biotic factors in the breeding areas and overwintering sites (Modarres Awal 1992, Mohaghegh et al. 1991).

3.1.2.2 Damage

The general and the biochemical damage are similar to that caused by *E. integriceps* (**Section 3.1.1.2**). Damage to the plant by *E. maura* was described by Stamenković (1976): the primary shoot of the central leaf may wither from the insect puncture; later, because the bugs attack the lower part of the internodes, breakage of the plant may occur. Spikes may be damaged while still ensheathed; after heading, the spike above the puncture withers. Individual spikelets may be destroyed. Spikes may also be attacked at later stages (premilk, milk, wax, full maturity), damaging the kernels (indicated by yellow spots with black centers).

3.1.2.3 Control

Several tachinids and hymenopterans reduce populations of *E. maura*. *Phasia crassipennis* F. attacks adults and *Trissolcus semistriatus* Ness, *T. rufiventris* Mayr, *T. simoni* Mayr, and *T. vassilievi* Mayr parasitize the eggs (Putschkov 1961). *Trissolcus basalis* Wollaston, *T. grandis* Thomson, and

T. rufiventris are important parasitoids in reducing populations of *E. maura* in the southeast of the Caspian Sea, particularly in the Tchmestan area (Javahery 1995).

Several insecticides have been used against this scutellerid at high population densities. At present, the organophosphorus compounds Lebaycid (= Baytex or Fenthion) and Sumithion (= Folithion) 50 EC are used when necessary.

3.1.3 Eurygaster austriaca (Schranck)

3.1.3.1 Distribution, Life History, and Biology

This species occurs from Turkey and the newly independent southwestern countries of the former U.S.S.R. westward to continental Europe and southern England, and south to the Mediterranean basin in Algeria and Morocco (Wagner 1951, Vodjdani 1954, Southwood and Leston 1959, Voegelé 1960, Putschkov 1961, Brown 1962a, Gaffour-Bensebbane 1981, Mohaghegh et al. 1991, Modarres Awal 1992). This species is quite distinctive; it is 11 to 14 mm long and brownish.

Eurygaster austriaca occurs on wild graminaceous plants and cereals. It appears on them in late March and April and, after feeding for several days (on the green parts), it commences to mate and oviposit. Oviposition is in batches of 14 eggs on the underside of a grass or cereal leaf; the female lays five to seven batches. Embryonic and nymphal development take about 2 and 6 weeks, respectively. The adults of the new generation appear in July. They fly to their estivation sites in nearby woodlands or forests after feeding for about 2 weeks on grains of wheat or wild graminaceous plants. Adults overwinter under bushes or dead leaves in the woodlands or forests close to the breeding fields. There is one generation per year throughout the range (Putschkov 1961).

3.1.3.2 Control

Among its natural enemies, tachinids flies and scelionid egg parasitoids reduce the population of this pest; the latter are more important in regulating *E. austriaca* (Putschkov 1961).

The two organophosphorus compounds Lebaycid (= Baytex or Fenthion) and Sumithion (= Folithion) 50 EC are used at high population densities.

Studies of wheat resistance to this bug have been under way in Yugoslavia (Stamenković 1985, 1993). In field trials, cultivars of wheat have shown a range of resistance (measured as percentage of kernels damaged) from 1.3 to 5.1; over 9 years, this range has remained fairly constant. [*Note:* This suggests the degree of resistance is genetic.]

3.2 Tetyra bipunctata (Herrich-Schaeffer)

3.2.1 Distribution, Life History, and Biology

The shieldbacked pine seedbug occurs in the eastern United States and southern Canada, west into Michigan and south to the Gulf of Mexico (McPherson 1982, Froeschner 1988, Larivière and Larochelle 1988). The bug is associated with pine, upon whose seeds and twigs it feeds, occasionally causing damage to commercial pine orchards. Several other species of *Tetyra* occur in southern North America (Froeschner 1988), and in Central and South America, but none of these has been recorded as a pest.

This seedbug is 12 to 17 mm long, and brownish-yellow mottled with black. Its biology has been studied in Wisconsin (U.S.A.) (Gilbert et al. 1967), in the southern United States (Ebel et al. 1975), and in the laboratory in Texas (Cameron 1981); the following is drawn from these accounts. There is one generation a year, and adults overwinter, usually in debris on the ground.

Eggs are laid in late summer (Texas: Cameron 1981) or early summer (Wisconsin: Gilbert et al. 1967), the difference probably reflecting the lengths of the growing seasons in these parts of the United States. The ovoid to spherical eggs are laid in groups of 14 (so stated by Cameron 1981, so inferred from Gilbert et al. 1967), in two rows, on pine needles. They hatch in about 9 days. First instars are gregarious and do not feed (Gilbert et al. 1967); the older instars and adults feed on seed within the cones (Ebel et al. 1975). The first four stadia took 6 to 7 days and the last about 12, in Wisconsin; the five instars are about 2.1, 3.9, 5.8, 9, 11 mm long (Gilbert et al. 1967).

3.2.2 Damage

Damage is primarily to nearly mature seed (Ebel et al. 1975), on various *Pinus* species (Gilbert et al. 1967; Larivière and Larochelle 1988; "its hosts include all the major pine species," Ebel et al. 1975). On loblolly pine (*P. taeda* L.) plantations in Louisiana (U.S.A.), *T. bipunctata* and the coreid *Leptoglossus corculus* (Say) (see **Chapter 11, Section 3.6.4**) caused up to 50% seed loss; damage by *T. bipunctata* (especially its fifth instar) was less than that by *L. corculus* up to seed-coat formation, but equaled it thereafter (Williams and Goyer 1980).

3.2.3 Control

McPherson (1982) reports from the literature that *T. bipunctata* is parasitized by a eupelmid wasp, *Anastatus reduvii* (Howard), and is used for nest-provisioning by the nyssonid wasp *Bicyrtes quadrifasciata* (Say). Cameron (1981) discusses in some detail sampling and various control methods for *T. bipunctata* as part of a complex of southern pine-seed orchards.

3.3 *Pachycoris klugii* Burmeister

3.3.1 Distribution, Life History, and Biology

Physic nut, *Jatropha curcas* L., is used widely throughout Central America, northern South America, and the Caribbean medicinally (as a purgative); its oil is used in many products from paint to lubricating oil; its seed kernels are roasted and eaten; the ashes of its burned roots are used as salt; and, when certain scale insects are allowed to infest it, their exudate is used for varnishes of high quality (e.g., for guitars). The seeds contain 25 to 40% oil. It is an important crop in this region, although it appears not yet to have achieved the status of a major crop (F. Coe, University of Connecticut, personal communication).

Several heteropterans attack the fruits of this plant; the two most important are the coreid *Leptoglossus zonatus* (Dallas) (see **Chapter 11, Section 3.6.10**) and *Pachycoris klugii* (Grimm and Guharay 1998). The latter occurs in Mexico and Central America (Grimm and Somarriba 1998). The bug is bluish, with large reddish or red-orange spots, which may obscure the blue (Grimm and Maes 1997).

In Nicaragua (and doubtless elsewhere) this bug is multivoltine; the bugs estivate in the dry season, in dry litter. Adults move to trees when the rains begin; females are apparently already mated, for they lay eggs at once, on the underside of physic nut leaves; a female laid an average of 2.4 egg masses, each with 30 eggs. In the laboratory, 95% of the eggs hatched. Eggs took 7 to 8 days to hatch, and the nymphal period lasted 32.5 days. Instars 1 to 5 were (lengths): 3, 5, 7, 9, and 14 mm (Grimm and Somarriba 1998). Populations are directly correlated with the abundance of fruit (Grimm and Führer 1998).

No host was found other than *J. curcas*; eggs were sometimes laid on other plants, but the hatchlings either died or moved to *J. curcas*. This bug is therefore considered a more serious pest than *L. zonatus*, which has other hosts (Grimm and Führer 1998). The instars are gregarious, and the female protects the eggs with scent-gland secretions (Grimm and Somarriba 1998).

3.3.2. Damage

Adults and nymphs (not including first instars?) feed on ripe or unripe fruit, in whose absence adults will attack somatic tissues (Grimm and Somarriba 1998). Heteropterans collectively (including *L. zonatus*, and especially including *P. klugii*) in one instance reduced seed yield by 18.5% (Grimm and Führer 1998).

3.3.3 Control

Pachycoris klugii has several parasites and predators. The parasite *Telenomus pachycoris* Costa Lima (Scelionidae) is the most effective, destroying up to 100% of all eggs, despite the female's attempts to protect them with scent-gland secretions (Grimm and Somarriba 1998). *Procheiloneuris* sp. (Encyrtidae) also parasitizes the eggs, and several reduviids and pentatomids (Heteroptera), as well as ants and spiders, occasionally prey upon *Pachycoris klugii* (Grimm and Maes 1997, Grimm and Somarriba 1998).

Because the physic nut fruits are synchronous, Grimm and Führer (1998) suggest harvesting them as soon as they ripen, as a form of cultural control.

In controlled tests, Grimm and Guharay (1998) found several entomopathogenic fungi effective against *P. klugii* (and *L. zonatus*). In particular, *Metarhizium anisopliae* (Metsch.) Sorok. increased fruit yield by 28%, and caused 65% mortality.

4. THE LESS IMPORTANT SPECIES

4.1   *Coleotichus blackburniae* White

Two species of *Coleotichus* may feed on *Acacia* species. One of these, *C. blackburniae*, occurs on the native koa, *A. koa* Gray, on all islands of Hawaii (U.S.A.) (Stein 1983; see also **Section 5**).

The metathoracic scent gland secretion of the related Fijian *Coleotichus* (as "*Coleostichus*") *sordidus* Walker has been analyzed (Smith 1978a).

4.2   *Calidea* spp.

Several species of *Calidea* are occasional pests of cotton in tropical Africa (Schouteden 1912, Kaufmann 1966, Couilloud 1989). The insects may become abundant at the end of the wet season, when the cotton fruits upon which they feed are maturing (Couilloud 1989). The life history of *C. dregii* Germar has been partly worked out in the laboratory, on its usual host, *Jatropha podagrica* Hooker, by Kaufmann (1966). Females start to lay eggs about 10 days after mating, and continue to lay at 5-day intervals; each lays 150 to 200 eggs, in batches that average 30 eggs each. The eggs are almost spherical. A female does not stand guard over the eggs. Eggs hatch in 4 to 5 days, and first instars remain on the eggs, probably ingesting symbionts. Second instars begin to feed on the host plant, and early instars are gregarious. The five stadia lasted (averages) 3.5, 9, 11, 18, and 23 days, respectively, and the instars' body lengths increased from 2.5 (first) to 9 (fifth) mm; nymphal mortality was 65%. Adults are diurnal and, like nymphs, feed on the reproductive parts (especially the ovaries) of the plant. Mating follows a brief courtship. In Ghana, *C. duodicimpunctata* (Signoret) feeds on *J. podagrica* and *J. gossypiifolia* L. (Leston 1973).

4.3 *Tectocoris diophthalmus* (Thunberg) (= *Tectocoris lineola* F.)

The cotton harlequin bug is an occasional pest in parts of Australia (Ballard and Holdaway 1926, Wilson et al. 1983), where its natural hosts are probably wild malvaceous plants (Knight et al. 1985). The abundance of color varieties has caused taxonomic confusion (Gross 1975, McDonald and Cassis 1984). In general, the bug is light to dark orange with patches of iridescent blue or green (Ballard and Holdaway 1926). Based on the structure of both male and female genitalia, McDonald (1961, 1963a) suggests *Tectocoris* be raised to subtribal level (in Scutellerini); because Scutelleridae is now a family, this action would create the tribe Tectocorini in Scutellerinae (s.s.).

The bug's immatures have been described by Ballard and Holdaway (1926) and McDonald (1963b), and the latter keys the instars of this and other Queensland (Australia) scutellerids. Instars lengths are 2.2, 4.4, 6, 8.9, and about 13 mm (first to fifth instars); the egg is 1.9 mm long (Ballard and Holdaway 1926). The life history, based on bugs on caged cotton plants, is as follows (Ballard and Holdaway 1926). [*Note:* These authors are vague about some details.] High temperatures and humidity stimulate mating; a total of 45 egg masses contained (each) an average of 110 eggs. Hatching occurs 16 to 22 days after oviposition. Stadial lengths (days): first: 8; second: 11 to 13 (p. 335) or 12 to 13 (p. 336); third: no data; fourth: 9 to 11; fifth: no data.

The metathoracic scent gland secretion has been analyzed by Smith (1978b), as has the secretion of the males' androcone glands (Knight et al. 1985).

Damage to cotton is less often direct than indirect: boll rot organisms are introduced as the insect probes (Ballard and Holdaway 1926, who provide photographs of normal cotton seeds and seeds attacked by a fungus thus introduced; also Wilson et al. 1983); see also *Dysdercus* sp. (**Chapter 8**).

In the past, chemicals used to control more important cotton pests also controlled the cotton harlequin bug (Page 1970, not seen, *fide* Wilson et al. 1983), but as control measures change, the pest status of this bug may change as well (Wilson et al. 1983). Natural enemies (in Queensland, Australia) include several egg parasites, a spider, and several birds (Ballard and Holdaway 1926).

Tectocoris biophthalmus also causes some bud fall in flowering hibiscus (*Hibiscus rosa-sinensis*), also in Queensland (Gough and Hamacek 1989).

4.4 *Scutellera* spp.

Scutellera perplexa Stoll (= *S. nobilis* Distant, *S. nobilis* [F.]), a greenish to purple bug with some spotting and striping of black, occurs in northern India, Pakistan, Burma, and Siam (Atkinson 1887–1889). Adults and nymphs of this species feed on ripe fruits of khata mitha (country gooseberry), *Phyllanthus distichus* Muell., in India. The fruits are used in pickles and tarts, and when fed upon by *S. perplexa* are rendered useless. Carbaryl, malathione, lindane, and BHC provided control, with toxicities in that (descending) order (Kavadia et al. 1971).

In Pakistan, *S. fasciata* Panzer is a pest of plum, *Zizyphus numularis* (Burm.); its egg and instars have been described (Ahmad and Moizuddin 1978).

4.5 *Diolcus irroratus* (F.)

This Neotropical species feeds on *Croton discolor* Willd., which it prefers to the sympatric and closely related *C. linearis* Jacq. The latter plant is of medicinal and insecticidal importance in the Caribbean, and similar value may be found in *C. discolor* (Loc and Elliott 1996). If so, or if *D. irroratus* comes to feed on *C. linearis*, the bug could become a pest.

4.6 Lampromicra senator (F.)

This species occurs occasionally on cherry in southern Queensland (Australia) (McDonald 1963a). The following is from McDonald's account. Adults overwinter in crevices on the ground, mate after emergence, and oviposit from early October until January (summer in Queensland). One female lays 20 to 25 eggs, in batches of 5 to 16 eggs each. Hatching occurs in 7 to 15 days (depending on temperature); the five instars' lengths in mm (and the stadial durations in days) are: first: 1.8 (7); second: 2.7 (9); third: 4 (7); fourth: 5.3 (8); and fifth: 7.7 (7). Adults and nymphs feed on the fruit, especially of the (apparently preferred) host, the euphorb *Breynia oblongifolia* J. Muell.

4.7 Chrysocoris spp.

Two species of this large brilliantly colored genus have been recorded as minor pests. *Chrysocoris stollii* (Wolff) feeds on the flowers of *Calotropis procera* (Asclepiadaceae), a commercial medicinal plant known in India as Akh or Madar (Verma et al. 1978). Several aspects of *Chrysocoris stolli* morphology have been studied: nervous system (Singh 1969), internal male genitalia and their development (Singh 1971), the external genitalia of both sexes (Baijal and Agarwal 1980), and the origin and composition of the egg's yolk (Verma and Basiston 1974). The general external morphology of *C. stockerus* (L.) has also been described and compared with that of several other Pakistani scutellerids.

Crysocoris purpureus Westwood, an iridescent greenish scutellerid with dark spots, 15 to 17 mm long, at times attacks poplar *Populus deltoides* Bartr. ex Marsh, an increasingly important timber tree in India (Roychoudhury et al. 1994). In Madhya Pradesh (India), the bug is most abundant from April to September, and goes through its life cycle in 6 to 7 weeks. This bug also "breeds freely" on *Jatropha curcas* and may "be injurious in gardens." It damages young nursery poplars by removing sap from the shoots, which wither (Roychoudhury et al. 1994). *Crysocoris purpureus* has also been found on nursery seedlings and plants of *Acacia auriculiformis*, in Madhya Pradesh (India), causing 10 to 15% damage by taking sap from the leaves (Meshram et al. 1992).

The secretions of its abdominal (Janaiah et al. 1988) and metathoracic (Kumari et al. 1984) scent glands have been analyzed; the former disturbed mitosis in onion root tips (Surender et al. 1987b), and the latter has antifungal activity (Surender et al. 1987a).

The chitin-synthesis inhibitor, diflubenzuron, prevents hatching of *C. purpureus* eggs when applied topically to the eggs themselves or to the gravid female (Chockalingam and Nooerjahan 1984).

4.8 Sphaerocoris testudogrisea (De Geer)

This species has been recorded in large numbers on cacao leaves (Nigeria) in the dry season (Youdeowei 1973). It is not clear if the bug does damage.

The secretion of the adult male's first abdominal accent gland (a possible sex pheromone) of the related species *S. annulus* (F.) (West Africa) has been analyzed (Gough et al. 1986).

4.9 Pachycoris fabricii (L.)

This scutellerid is an occasional pest of guava in Brazil (A. R. Panizzi, personal communication, 1997). See also *Pachycoris klugii* (**Section 3.3**).

4.10 Nearctic Eurygaster

Environmental conditions, particularly temperature, are not favorable for *Eurygaster* north of 40° north latitude in the Nearctic. Therefore, population density will be low in this region (Lattin

1964; McPherson 1982; M. Javahery, unpublished data). In central California (U.S.A.), however, *E. amerinda* Bliven and *E. minidoka* Bliven occur in open woodlands in the foothills area and feed principally on wild oats. Under certain conditions, however, they shift to other host plants (*Amsinckia intermedia* F. & M., *Ranunculus callifornia* Benth.) (Vojdani 1961b, Lattin 1964).

Comparing the biologies and life cycles of these two Californian *Eurygaster* species with two of the several *Eurygaster* pest species of the Palearctic such as *E. integriceps* and *E. maura* indicates that *E. amerinda* at least can disperse and even migrate to a more favorable environment when feeding is disrupted near shelter sites. Wild oats, the principal host of *Eurygaster* spp. in central California, is among the new species of grasses introduced and established in California less than 200 years ago, and which now cover 50 to 90% of California grassland (Vojdani 1961b). The adaptation to this plant and not to other grasses is significant. Perhaps because wild oat is softer and richer in nutrients than the other grasses, and its grain is larger, it could provide more starch and protein (Vojdani 1961b). However, the present situation of *Eurygaster* south of latitude 40°, and particularly in areas such as central California, may not always exist. Any one of several possible modifications of present agricultural practices or land use could cause an escape from the relatively low general equilibrium position which now prevails below an economic threshold, to a higher injurious average population density (Vojdani 1961b; M. Javahery, unpublished data).

Observations by the authors in Nearctic and Palearctic regions also support the above assumptions and it would not be surprising to find that *Eurygaster* species become pests of wheat and barley in favorable southern states (U.S.A.) when agriculture encroaches into its habitat.

5. SPECIES OF MINOR BENEFIT

Two species of *Coleotichus* may feed on *Acacia* species. Two wattle species, *A. cyclops* A. Cunn. et G. Don and *A. saligna* (Labill.) Wendl., are becoming pests in South Africa. In western Australia, nymphs and adults of the yellow to yellow brown, 14 to 19 mm long (Gross 1975) *C. costatus* (F.) feed on the developing and mature seeds of these trees (van den Berg 1980; see also **Section 4.1**).

6. PROGNOSIS

It is probable that the usual control measures used against more serious pests will control the occasional scutellerid pest. Control against Sunn pests, however, will take considerably more knowledge and (especially) international cooperation and study than now occurs (see recommendations in Javahery 1995).

Four species of *Eurygaster* occur in Canada and the United States, where, like their Old World congeners, they feed mostly on grasses (Vojdani 1961a, Lattin 1964). A few specimens have been collected from wheat in western North America, but the bugs have never occurred in large numbers. Yet continual efforts in wheat breeding might result in varieties that could be fed upon by native North America *Eurygaster* species. Wheat has long been resistant to the Hessian fly, but little seems to be known about resistance to *Eurygaster* species in North America. The developing seeds of wheat are the main part of the plant damaged, and the same is true for grass-feeding native North American species (Vodjdani 1961b). Thus, the potential exists in North America.

Finally, an anonymous U.S. Department of Agriculture pamphlet warns that "Sunn pest," *E. integriceps*, may appear in the United States (Anonymous 1963).

7. CONCLUSIONS

With one major exception, scutellerids are not important pests. In several tropical regions some species do damage from time to time, but steady damage is rare, major damage is rarer, and steady serious damage does not occur — except for the complex of scutellerids (*Eurygaster* spp.) and pentatomids (see **Chapter 13, Sections 3.2 to 3.5**) known as Sunn pests. These pentatomoids ravage wheat in a broad swath from southeastern Europe into the Asian republics of the former U.S.S.R. (see Javahery 1995); they alone grant the family Scutelleridae major pest status.

8. ACKNOWLEDGMENTS

The first author thanks the Lyman Entomological Museum and Research Laboratory of McGill University, the Food and Agricultural Organization of the United Nations, and the Plant Protection Department of Tehran University for their cooperation in research on Sunn pests.

9. REFERENCES CITED

Abdollahi, G. A. 1988. Influences of the food plants on reproduction of cereal Sunn pest *Eurygaster integriceps* Put.(Het. Scutelleridae) in Karaj. M.Sc. thesis, Tehran University, Tehran, Iran.

1995. Morphological and physiological approaches to the control of Heteroptera of economic importance in Iran. Ph.D. thesis. University of Wales, Cardiff, U.K.

Afzal, M., S. S. Shaukat, and I. Ahmad. 1982. Taxometric study of the family Scutelleridae Leach (Hemiptera: Pentatomoidea) from Pakistan. Orient. Insects 16: 373–383.

Agarwal, S. B., and H. N. Baijal. 1984. Morphology of male genitalia of Scutellerinae and Pentatominae (Heteroptera) with remarks on inter-relationship within the family Pentatomidae. J. Entomol. Res. 8: 52–60.

Ahmad, I., and S. Kamaluddin. 1982. A revision of the genus *Poecilocoris* Dallas (Pentatomoidea: Scutelleridae) from IndoPakistan subcontinent with description of three new species. Orient. Insects 16: 259–295.

Ahmad, I., and M. Moizuddin. 1978. Eggs and nymphal systematics of two species of shield bugs (Pentatomoidea: Scutelleridae) of Pakistan, with reference to phylogeny. Pakistan J. Zool. 10: 95–102.

1980. Aspects of internal anatomy of *Alpocoris lixoides* Germar (Pentatomidae: Scutelleridae: Odontoscelinae) of Pakistan with phylogenetic considerations. Proc. 1st Pakistan Congr. Zool. (B): 195–201.

Ahmad, I., and S. Mushtaq. 1977. External morphology of *Chrysocoris stockerus* (Linn.) (Pentatomoidea: Scutelleridae) with comparative accounts on fourteen related species from Pakistan, Azad Kashmir and Bangladesh and their bearing on classification. Pakistan J. Sci. Ind. Res. 20: 18–40.

Ahmad, I., S. Kamaluddin, and M. Moizuddin. 1980. New status of Parapoecilocoris Schouteden 1904 (Pentatomoidea: Scutelleridae: Scutellerinae) with redescription of type species *P. interruptus* (Westwood). Proc. 1st Pakistan Congr. Zool. (B): 203–207.

Ahmad, I., M. Moizuddin, and A. A. Khan. 1979. Generic and supergeneric keys with reference to check list of lower pentatomoid fauna of Pakistan (Heteroptera: Pentatomomorpha) with notes on their distribution and food plants. Suppl. Entomol. Soc. Karachi, Pakistan 4, Part 4: 50 pp.

Alayo, D. P. 1967. Catalogo de la fauna cubana. XIX. Hemipteros de Cuba — IV. Familia Scutelleridae. Trab. Divulgacion Mus. "Felipe Poey" Acad. Cienc. Cuba No. 44: 13 pp.

Aldrich, J. R. 1996. Sex pheromones in Homoptera and Heteroptera. Pp. 199–233 *in* C. W. Schaefer [ed.], Studies on Hemipteran Phylogeny. Thomas Say Publications in Entomology, Entomological Society of America, Lanham, Maryland, U.S.A. 244 pp.

Alexandrov, N. 1947–1949. *Eurygaster integriceps* Put. et ses parasites à Varamine. Entomol. Phytopath. Appl. 5: 11–14, 29–41; 6–7; 8–17, 28–47; 8: 13–20,16–52 [in Persian with French summary].

Amir-Maafi, M. 1991. An investigation for identifying and efficiency of parasitoid flies of cereal Sunn pest (*Eurygaster integriceps* Put.) in Karaj, Iran. M.Sc. thesis, Tehran University, Tehran, Iran [in Persian with English summary].

Amir-Maafi, M., A. Kharrazi–Pakdel, and M. Esmaili. 1991. Identification and biology of parasitoid flies of Sunn pests (*Eurygaster integriceps* Put.) in Karaj, Iran. P. 12 *in* Proc. Tenth Plant Prot. Congr. Iran. Kerman, Iran.

Anonymous. 1963. The Senn pest. U.S. Dept. Agric. PA-582. 4 pp.

Arnoldi, K. V. 1955. Hibernation of *Eurygaster integriceps* in the mountains of Kuban in the light of investigations in 1949–53. Pp. 171–237 *in* D. M. Fedotov [ed.], *Eurygaster integriceps* 3 [in Russian].

Asgari, S. 1995. A study of possibility of mass rearing of Sunn bug egg parasitoids on its alternative host, *Graphosoma lineatum* L. (Het. Pentatomidae). M.Sc. thesis, Tehran University, Tehran, Iran [in Persian, with English summary].

Atkinson, E. T. 1887–1889. Indian Hemiptera. Family Pentatomidae. Privately publ., Calcutta, India. 205, 184, 109 pp.

Baehr, M. 1989. Review of the Australian shield bug genus *Theseus* Stål. Spixiana 11: 243–258.

Baijal, H. N., and S. B. Agarwal. 1980. Morphology of the genitalia of *Chrysocoris stollii* Wolff. (Heteroptera: Scutelleridae). J. Entomol. Res. 4: 57–62.

Ballard, E., and F. G. Holdaway. 1926. The life-history of *Tectocoris lineola* F. and its connection with internal boll rots in Queensland. Bull. Entomol. Res. 16: 329–346.

Bérenger, J. M., and R. Lupoli. 1991. Notes sur un Scutelleridae peu commun en France: *Psacasta tuberculata* F. 1781. Entomologiste 47: 229–234.

Bliven, B. P. 1956. New Hemiptera from the western states with previously described species and new synonymy in the Psyllidae. Privately published by author, Eureka, California, U.S.A. 27 pp.

1962. Contributions to a knowledge of the Scutelleridae I: on the identity of *Eurygaster minidoka*. Occident. Entomol.1: 66–67.

Brown, E. S. 1962a. Researches on the ecology and biology of *Eurygaster integriceps* Put. (Hemiptera, Scutelleridae) in Middle East countries, with special reference to the overwintering period. Bull. Entomol. Res. 53: 445–514.

1962b. Notes on parasites of Pentatomidae and Scutelleridae (Hemiptera–Heteroptera) in Middle East countries, with observations on biological control. Bull. Entomol. Res. 53: 241–256.

1963. Report on research on the soun pest (*Eurygaster integriceps* Put.) and other wheat pentatomids in Middle East countries, 1958–1961. Dept. Tech. Coop. Tech. Assist. Progr., CENTO Misc. 4: 1–43.

1965. Notes on the migration and direction of flight of *Eurygaster* and *Aelia* species (Hemiptera, Pentatomoidea) and their possible bearing on invasions of cereal crops. J. Anim. Ecol. 34: 93–107.

Brown, E. S., and M. Eralp. 1962. The distribution of the species of *Eurygaster* Lap. (Hem., Scutelleridae) in Middle East countries. Ann. Mag. Nat. Hist. (13) 5: 65–81.

Cameron, R. S. 1981. Toward insect pest management in southern pine seed orchards with emphasis on the biology of *Tetyra bipunctata* (Hem., Pentatomidae) and the pheromone of *Dioryctria clarioralis* (Lep., Pyralidae). Texas For. Serv. Publ. 126: 149 pp.

Carayon, J. 1984. Les androconies de certains Hémiptères Scutelleridae. Ann. Soc. Entomol. France (n.s.) 20: 113–134.

Chockalingam, S., and A. Noorjahan. 1984. The ovicidal effect of diflubenzuron on hemipteran bugs, *Dysdercus cingulatus* and *Chrysocoris purpureus*. Curr. Sci. 53: 1112–1113.

Couilloud, R. 1989. Hétéroptères déprédateurs du cotonnier en Afrique et à Madagascar (Pyrrhocoridae, Pentatomidae, Coreidae, Alydidae, Rhopalidae, Lygaeidae). Coton Fibres Trop. 44: 185–227.

Critchley, B.R. 1998. Literature review of sunn pest *Eurygaster integriceps* Put. (Hemiptera: Scutelleridae). Crop Prot. 17: 271–287.

Daroogheh, H. 1990. Structure of the scent gland system of the Sunn pest *Eurygaster integriceps* Put. Appl. Entomol. Phytopathol. 57: 13–19.

1993a. A report on male sternal pheromone glands in *Eurygaster integriceps*. J. Entomol. Soc. Iran 12, 13: 5–6.

1993b. Effect of juvenile hormone on diapausing adult of Sunn bug *Eurygaster integriceps*. J. Agri. Sci. Technol. 1: 3–6.

1994. The effect of juvenile hormone mimic on metamorphosis of Sunn pest. J. Entomol. Soc. Iran 1: 19–26.

DeBarr, G.L. 1970. Characteristics and radiographic detection of Tetyra and Leptoglossus seed bug damage to slash pine seed. Florida Entomol. 53: 109–117.

Delucchi, V. 1961. Le complexe des *Asolcus* Nakagawa (*Microphanurus* Keiffer) (Hym., Proctotrupoidea) parasites oöphages des punaises des céréales au Maroc et au Moyen-Orient. Cah. Rech. Agron. 14: 41–67.

Djavadzadeh, M., and M. Esmaili 1991. An investigation on susceptibility of different populations of Sunn pest (*Eurygaster integriceps* put.) to fenitrothion. Proc. Tenth. Plant Prot. Congr. Iran. Kerman, Iran. 9 [in Persian, with English abstract].

Donskoff, M. 1996. Prospects for international cooperation on Sunn pest research and control. Pp. 17–21 *in* R. H. Miller and J. G. Morse [eds.], Sunn Pests and Their Control in the Near East. FAO Publ. No. 138. Rome, Italy. 165 pp.

Dupuis, C. 1948. Notes á propos des *Eurygaster* (Hemipt. Pentatomoidea, fam. Scutelleridae). Systematique-biologique-parasites. Entomologiste 4: 202–205.

Ebel, B. H., T. H. Flavell, L. E. Drake, H. O. Yates III, and G. L. DeBarr. 1975. Southern pine seed and cone insects. General Service Report/SE-8, U.S. Dept. Agric. For. Serv. Southeast. For. Exp. Stn., Asheville, North Carolina, U.S.A. 40 pp.

Eger, J. E., Jr. 1990. Revision of the genus *Polytes* Stål (Heteroptera: Scutelleridae). Ann. Entomol. Soc. Amer. 83: 115–141.

Eger, J. E., Jr., and J. D. Lattin. 1995. Generic placement and synonymy of some New World Scutelleridae (Hemiptera: Heteroptera) in the British Museum (Natural History). J. New York Entomol. Soc. 103: 412–420.

Esmaili, M., and A. Dehghan. 1991. Study on two-stage harvesting of wheat in relation to *Eurygaster integriceps* Put. Proc. Sunn Pests Conf. Tehran. Pp. 96–115 [in Persian].

FAO. 1993. Sunn pest problem and its control in the Near East region. FAO Publ. Aleppo, Syria. 28 pp.

Fedotov, D. M. [ed.]. 1947–60. The Noxious Pentatomid *Eurygaster integriceps* Put. Vols. I–IV. (Vol. I, 272 pp., 1947; Vol. II, 271 pp., 1947; Vol. III, 278 pp., 1955; Vol. IV, 239 pp., 1960. Moscow. Akad. Nauk, U.S.S.R. [in Russian].

Froeschner, R. C. 1988. Family Scutelleridae Leach, 1815. The shield bugs. Pp. 684–693 *in* T. J. Henry and R. C. Froeschner [eds.], Catalog of the Heteroptera, or True Bugs, of Canada and the Continental United States. E. J. Brill, Leiden, the Netherlands. 958 pp.

de la Fuente, J. A. 1973. Revisión de los pentatómidos ibéricos. Familia Scutelleridae Leach, 1815, con adiciones y corecciones a un trabajo anterior. Bol. R. Soc. Esp. Hist. Nat. (Biol.) 71: 235–270.

Gaffour-Bensebbane, C. 1981. Les punaises des blés en Algérie. Bull. OEPP 11(2): 33–38.

1994. Les variations morphologiques de l'appareil génital ectodermique des femelles de Scutelleridae (Heteroptera, Pentatomoidea) 1. Les pièces internes. Nouv. Rev. Entomol. (n.s.) 11: 267–281.

Gallun, R. L., and G. D. Khush. 1980. Genetic factors affecting expression and stability of resistance. Pp. 63–85 *in* F. G. Maxwell and P. R. Jennings [eds.], Breeding Plants Resistant to Insects. John Wiley, New York, New York, U.S.A.

Gallun, R. L., K. J. Starks, and W. D. Guthrie. 1975. Plant resistance to insects attacking cereals. Annu. Rev. Entomol. 20: 337–357.

Gerini, V. 1968. Contributo alla conoscenza dell *Eurygaster integriceps* Puton. Riv. Agric. Subtrop. Trop. 62: 386–399.

Gilbert, B. L., S. J. Barras, and D. M. Norris. 1967. Bionomics of *Tetyra bipunctata* (Hemiptera: Pentatomidae: Scutellerinae) as associated with *Pinus banksiana* in Wisconsin. Ann. Entomol. Soc. Amer. 60: 698–701.

Göllner-Scheiding, U. 1986. Revision der Gattung *Odontoscelis* Laporte de Castelnau, 1832 (Heteroptera: Scutelleridae). Dtsch. Entomol. Z. (n.F.) 33: 95–127.

Gough, N., and E. L. Hamacek. 1989. Insect induced bud fall in cultivated hibiscus and aspects of the biology of *Macroura concolor* (Macleay) (Coleoptera: Nitidulidae). J. Austral. Entomol. Soc. 28: 267–277.

Gough, A. G. E., J. G. C. Hamilton, D. E. Games, and B. W. Staddon. 1985. Multichemical defense of plant bug *Hotea gambiae* (Westwood) (Heteroptera: Scutelleridae). Sesquiterpenoids from abdominal gland in larvae. J. Chem. Ecol. 11: 343–352.

Gough, A. G. E., D. E. Games, B. W. Staddon, D. W. Knight, and T. O. Olagbemiro. 1986. C9 aliphatic aldehydes: possible sex pheromone from male tropical west African shield bug, *Sphaerocoris annulus*. Z. Naturforsch. 41c: 1073–1076.

Grigorov, P. 1988. Electron-microscopic study of the egg's horion [sic] in species of the genus *Eurygaster* Lap. (Heteroptera: Scutelleridae). Plant Sci. (Sofia) 25: 94–99 [in Bulgarian, with English summary].

1989. Effect of damages caused by *Eurygaster integriceps* on wheat seeding properties. Plant Sci. (Sofia, Bulgaria) 26: 23–29 [in Bulgarian, with English summary].

Grimm, C., and E. Führer. 1998. Population dynamics of true bugs (Heteroptera) in physic nut (*Jatropha curcas*) plantations in Nicaragua. J. Appl. Entomol. 122: 515–521.
Grimm, C., and F. Guharay. 1998. Control of leaf-footed bug *Leptoglossus zonatus* and shield-backed bug *Pachycoris klugii* with entomopathogenic fungi. Biocontrol Sci. Technol. 8: 365–376.
Grimm, C., and Maes. 1997. Insectos asociados al cultivo de tempate (*Jatropha curcas* L.) (Euphorbiaceae) en el Pacifico de Nicaragua. I. Scutelleridae (Heteroptera). Rev. Nicaraguense Entomol. 39: 13–26.
Grimm, C., and A. Somarriba. 1998. Lifecycle and rearing of the shield–backed bug *Pachycoris klugii* in Nicaragua (Heteroptera: Scutelleridae). Entomol. Gen. 22: 211–221.
Gross, G. F. 1975. Plant-Feeding and Other Bugs (Hemiptera) of South Australia. Heteroptera — Part I. South Australian Government [no city], Australia. 250 pp.
Haghshenas, A. R. 1993. The effect of five insecticides compounds with granule and emulsion formulation on Sunn pest *Eurygaster integriceps* Put. M.Sc. thesis, Tehran University, Tehran, Iran [in Persian, with English summary].
Harris, H. M., and F. Andre. 1934. Notes on the biology of *Acantholoma denticulata* Stål [sic] (Hemiptera: Scutelleridae). Ann. Entomol. Soc. Amer. 27: 5–15.
Hoffmann, W. E. 1932. A list of Pentatomidae, Plataspidae, and Coreidae (order Hemiptera) of China, Korea, and Indo-China. J. Pan-Pac. Res. Inst. 7: 6–11.
1935. An abridged catalogue of certain Scutelleroidea (Plataspidae, Scutelleridae, and Pentatomidae) of China, Chosen, Indo-China, and Taiwan. Lingnan Univ. Sci. Bull. 7: 1–294.
1948. First supplement to catalogue of Scutelleroidea. Lingnan Sci. J. 22: 1–41.
Hussey, R. F. 1934. Observations on *Pachycoris torridus* (Scopy), with remarks on parental care in other Hemiptera. Bull. Brooklyn Entomol. Soc. 29: 135–145.
Iranipour, S. 1996. A study on population fluctuation of the egg parasitoids of *Eurygaster integriceps* Put. (Heteroptera: Scutelleridae) in Karaj, Kamalabad, and Fashand. M.Sc. thesis, Tehran University, Tehran, Iran [in Persian, with English summary].
Jaenike, J. 1990. Host specialization in phytophagous insects. Annu. Rev. Ecol. Syst. 21: 243–273.
Janaiah, C., S. J. L. Kumari, P. S. Rao, and F. Surender. 1988. Chemical analysis of secretion from the abdominal scent glands of *Chrysocoris purpureus* (Heteroptera: Pentatomidae). Proc. Indian Acad. Sci. (Anim. Sci.) 97: 111–115.
Javadzadeh-Siahkalrudi, M. 1992. A study of Sunn pest (*Eurygaster integriceps*) in respect to their susceptibility to fenitrothion. M.Sc. thesis, Tehran University, Tehran, Iran [in Persian, with English summary].
Javahery, M. 1967. The biology of some Pentatomoidea and their egg parasites. Ph.D. thesis, University of London, London, U.K. 475 pp.
1968. The egg parasite complex of British Pentatomoidea (Hemiptera): taxonomy of Telenominae (Hymenoptera: Scelionidae). Trans. R. Entomol. Soc. London. 120: 417–436.
1972. Annual Technical Report on Sunn Pests to the Plant Pests Research Institute. Tehran, Iran. 72 pp. [in Persian].
1973a. Population dynamics of *Aelia* and *Eurygaster* spp. in some regions in Iran. CENTO Publication, pp. 20–21. Tehran, Iran.
1973b. Chemical controlling applications that are used against *Aelia* and *Eurygaster* in Iran. CENTO Publication, pp. 55–56. Tehran, Iran.
1978a. Management of economically important Pentatomoidea in Iran. Antenna (Bull. Entomol. Soc. London) 2: 74.
1978b. Importance and methods of control of Sunn pest in Iran. Bull. 27. Plant Protection Organization, Tehran, Iran [in Persian].
1979. Les punaises nuisibles aux céréales en Iran et les méthodes de prevention de dégâts. Pp. 18–21 *in* Bull. IOBC/WPRS.
1992. Economic importance and application of IPM for controlling Sunn pests. Proc. Sunn Pests Conf. Tehran. Pp. 43–48 [in Persian].
1993. Sunn pests (Pentatomoidea) of cereal crops in Iran: chemical and cultural method of control. FAO/ICARDA. Expert Consultation on Sunn Pest. FAO Pub. Aleppo, Syria. 16 pp.
1994. Development of eggs in some true bugs (Hemiptera:Heteroptera). Part I. Pentatomoidea. Canad. Entomol. 126: 401–433.

1995. A technical review of Sunn pests (Heteroptera: Pentatomoidea) with special reference to *Eurygaster integriceps* Puton. FAO/RNE Publ. Priv. 80 pp.

1996. Sunn pest of wheat and barley in the I. R. of Iran; chemical and cultural methods of control. Pp. 61–74 *in* R. H. Miller and J. G. Morse [eds.], Sunn Pest and Their Control in the Near East. FAO Pub. No. 138. Rome, Italy. 165 pp.

Jovanić, M., and S. Stamenković. 1978. Prognose des Getreidewanzenauftretens (*Eurygaster austriaca* und *E. maura*) in nordöstlichen Jugoslawien. Pp. 173–178 *in* [T. Wetzel, ed.], II. Symposium mit Beiteiligung sozialistischer Länder. Martin-Luther-Univ., Halle-Wittenberg, [East] Germany.

Kaufmann, T. 1966. Note on the life history and morphology of *Calidea dregii* (Hemiptera: Pentatomidae: Scutellerini) in Ghana, West Africa. Ann. Entomol. Soc. Amer. 59: 654–659.

Kavadia, V. S., S. K. Verma, P. C. Jain, and H. C. L. Gupta. 1971. Relative toxicity of some insecticides to *Scutellera nobilis* Fabr. [sic] (Hemiptera: Pentatomidae). Indian J. Entomol. 33: 372–373.

Kieffer, J. J. 1926. Scelionidae (Hymenoptera, Proctotrupidae). *In* Das Tierreich, Berlin Universität, Leipzig, Germany.

Kirkaldy, G. W. 1909. Catalogue of the Hemiptera (Heteroptera) with biological and anatomical references, list of food plants and parasites, etc. Vol. 1. Cimicidae. Felix L. Dames, Berlin, Germany. 392 pp.

Kis, B. 1979. Beiträge zur Kenntnis der Gattung *Odontoscelis* Lap. (Heteroptera: Scutelleridae). Trav. Mus. Hist. Nat. Grigore Antipa (Bucharest, Romania) 20: 203–209.

Knight, D. W., B. W. Staddon, and M. J. Thorne. 1985. Presumed sex pheromone from androconical glands of male cotton harlequin bug, *Tectocoris purpureus* (Heteroptera: Scutelleridae) identified as 3,5-dihydroxy-4-pyrone. Z. Naturforsch. 40c: 851–853.

Kobayashi, T. 1954. The developmental stages of some species of the Japanese Pentatomoidea (Hemiptera). III. Trans. Shikoku Entomol. Soc. 4: 63–68.

1956. The developmental stages of some species of the Japanese Pentatomoidea (Hemiptera). V. Trans. Shikoku Entomol. Soc. 4: 120–130.

1958. The developmental stages of some species of the Japanese Pentatomoidea (Hemiptera). VI. Trans. Shikoku Entomol. Soc. 5: 121–132.

1967. Developmental stages of *Poecilocoris* and its allied genera of Japan (Hemiptera: Scutelleridae) (The developmental stages of some species of the Japanese Pentatomoidea, XVII). Trans. Shikoku Entomol. Soc. 9: 86–94.

Kontev, C., J. Zdárek, J. Vrkoc, and K. Ubik. 1978. Sex pheromone of the Senn pest (*Eurygaster integriceps*) Put. P. 161 *in* Schadenerreger in der Industriemassigen Getreideproduktion, Martin-Luther-Universität, Halle-Wittenberg, [East] Germany.

Kozlov, M. A. 1968. *Telenomus* (Hym., Sce., Telenominae) Kavkaza jajcejedy vrednoj cerepaski (*E. integriceps*) I drugih klopov. Tr. Vses. Entomol. Ova, 52: 188–223 [in Russian].

Kozlov, M. A., and S. V. Kononova. 1983. Telenominae of the fauna of the U.S.S.R. Determinations of the fauna of the U.S.S.R. Vol. 1. 136. Zoological Institute, Academy of Sciences, Moscow, U.S.S.R. 336 pp. [in Russian].

Kumari, S. J. L., C. Janaiah, and N. Chari. 1984. Structure and volatile constituents of metathoracic scent glands of a pentatomid bug, *Chrysocoris purpureus* (Westw.). Uttar Pradesh J. Zool. 4: 156–161.

Kuperstein, M. L. 1979. Estimating carabid effectiveness in reducing the Sunn pest, *Eurygaster integriceps* Puton (Heteroptera: Scutelleridae) in the U.S.S.R. Misc. Publ. Entomol. Soc. Amer. 11: 79–84.

Larivière, M.-C., and A. Larochelle. 1988. Further data on the establishment of *Tetyra bipunctata* (Herrich-Schaeffer) (Heteroptera: Scutelleridae) in Quebec. J. New York Entomol. Soc. 96: 281–283.

Lattin, J. D. 1964. The Scutellerinae of America north of Mexico (Hemiptera: Heteroptera: Pentatomidae). Ph.D. dissertation, University of California, Berkeley, California, U.S.A. 350 pp.

Leston, D. 1952a. Notes on the Ethiopean Pentatomidae II: a structure of unknown function in the Sphaerocorini Stål (Hem. Het.). Entomologist 85: 179–180.

1952b. Notes on the Ethiopian Pentatomoidea (Hemiptera): V. On the specimens collected by Mr. A. L. Capener, mainly in Natal. Ann. Mag. Nat. Hist. (12)5: 512–520.

1952c. Notes on the Ethiopean Pentatomoidea (Hemiptera): VI. Some insects in the Hope Department, Oxford. Ann. Mag. Nat. Hist. (Ser. 12) 5: 892–904.

1952d. Notes on the Ethiopean Pentatomoidea (Hemiptera): VIII. Scutelleridae Leach of Angola, with remarks upon the male genitalia and classification of the subfamily. Publ. Cult. Ca. Diamante Angola 16: 9–26.

1953. The Entomologist label list of British shield-bugs (Hemiptera: Pentatomoidea). Entomologist 86: 83; Suppl. to Entomologist 86: xvi–xvii. [*Note:* Eurygastrinae here raised to subfamily.]

1954. Wing venation and male genitalia of *Tessaratoma* Berthold, with remarks on Tessaratominae Stål (Hemiptera: Pentatomidae). Proc. R. Entomol. Soc. London (A) 29: 9–16.

1973. The natural history of some West African insects. Entomol. Mon. Mag. 108: 110–122.

Loc, L., and N. B. Elliott. 1996. Host plant selection by the shield bug *Diolcus irroratus* (F.) (Hemiptera: Scutelleridae). Pp. 83–86 *in* N. B. Elliott, D. C. Edwards, and P. J. Godfrey [eds.], Proc. 6th Symp. Nat. Hist. Bahamas. Bahamian Field Station, San Salvador, Bahamas.

Lodos, N. 1961. Türkiye, Irak, Iran ve suriye'de süne (*Eurygaster integriceps* Put.) problemi üzerinde incelemeler. Ege Üniv. Ziraat Fak. Yayinl. 51: 1–115 [in Turkish, with English summary].

1981. Pentatomoid pests of wheat in Turkey. EPPO Bull. 11: 9–12.

Lodos, N., and H. Kavut. 1991. Süne (*Eurygaster integriceps* Put. — Heteroptera, Scutelleridae) 'nin Türkiye 'de yayisili ile ilgili yeni bilgiler. Türk. Entomol. Derg. 15: 107–112 [in Turkish, with English summary].

Lodos, N., and F. Önder. 1983. Süne (*Eurygaster integriceps* Put.) 'nín Türkíye'de yayilisi üzerinde düsünceler. Bitki Koruma Bültení 23: 53–60 [in Turkish, with English summary].

Lodos, N., F. Önder, E. Pehlivan, R. Atalaya, E. Erkin, Y. Karsavuran, S. Tezcan, and S. Aksoy. 1998. Faunistic studies on Pentatomoidea (Plataspidae, Acanthosomatidae, Cydnidae, Scutelleridae, Pentatomoidae) of western Black Sea, central Anatolia and Mediterranean regions of Turkey. Ege Üníversítesí, Bornova–Izmír, Turkey. 75 pp.

Londt, J. G. H., and P. E. Reavell. 1982. Records of coprophagy in pentatomids. J. Entomol. Soc. South Africa 45: 275.

Lyal, C. H. C. 1979. A review of the genus *Calliphara* Germar, 1839 (Hemiptera: Scutelleridae). Zool. Med. 54: 149–181.

Makhotin, A. A. 1947. Materials on development of the noxious "little turtle," *E. integriceps* Put. Pp. 19–48 *in* D. M. Fedotov [ed.], *Eurygaster integriceps*, Vol II. Acad. Sci. U.S.S.R. Moscow, U.S.S.R. [in Russian].

Manna, G. K., and S. Deb-Mallick, 1981. Meiotic chromosome constitution in forty-one species of Heteroptera. Chrom. Inf. Ser. 31: 9–11.

Martin, H. E., and M. Javahery. 1968. Note sur la punaise des céréales *Eurygaster integriceps* Put. et ses parasites du genre *Asolcus* en Iran. Proc. XIII Int. Cong. Entomol. Moscow. Vol. 2. pp. 166–167.

Martin, H. E., M. Javahery, and G. Radjabi. 1969. Note sur la punaise des céréales *E. integriceps* Put. et de ses parasites du genre *Asolcus* en Iran. Entomol. Phytopathol. Appl. 28: 38–46.

Masner, L., 1958. Some problems of the taxonomy of the subfamily Telenominae (Hym., Scelionidae). Pp. 375–381 *in* 1. Int. Conf. Insect. Path. Biol. Control, Prague, Czechoslovakia.

Mayr, G., 1879. Über die Schlupfwespengattung *Telenomus*. Verh. Zool. Bot. Ges. Wien 29: 697–714.

1903. Hymenopterologische Miszellen. Verh. Zool. Bot. Ges. Wien 53: 399.

McDonald, F. J. D. 1960. Studies on the life-history and biology of *Choerocoris paganus* (Fabricius) (Heteroptera: Pentatomidae: Scutellerinae). Univ. Queensland Paper (Entomology) 1(9): 135–147.

1961. A comparative study of the male genitalia of Queensland Scutellerinae Leach (Hemiptera: Pentatomidae). Univ. Queensland Pap. (Entomol.) 1(12): 173–186.

1963a. A comparative study of the female genitalia of Queensland Scutellerinae Leach (Hemiptera: Pentatomidae). Univ. Queensland Pap. (Entomol.) 1(15): 229–236.

1963b. Nymphal systematics and life histories of some Queensland Scutellerinae Leach (Hemiptera: Pentatomidae). Univ. Queensland Pap. (Entomol.) 1: 277–295.

1963c. Morphology of the male genitalia of five Queensland Scutellerinae (Hemiptera: Pentatomidae). J. Entomol. Soc. Queensland 2: 24–30.

1988. A revision of *Cantao* Amyot and Serville (Hemiptera: Scutelleridae). Orient. Insects 22: 287–299.

McDonald, F. J. D., and G. Cassis. 1984. Revision of the Australian Scutelleridae Leach (Hemiptera). Austral. J. Zool. 32: 537–572.

McPherson, J. E. 1982. The Pentatomoidea (Heteroptera) of Northeastern North America with Emphasis on the Fauna of Illinois. Southern Illinois University Press, Carbondale, Illinois, U.S.A. 240 pp.

Meier, N. F. 1940. Parasites cultured in USSR in 1938–1939 on the eggs of corn-bug (*Eurygaster integriceps* OSCH). Vestn. Zash. Rast. 3: 77–82 [English translation prepared and published by Indian Natl. Sci. Doc. Centre, Delhi].

Meshram, B. C., S. C. Pathak, and Jamaluddin. 1992. A new report of *Chrysocoris purpureus* Westw. (Hemiptera: Scutelleridae) as a pest on *Acacia auriculiformis*. Indian For. 118–169.

Miller, R. H., and J. G. More [eds.]. 1996. Sunn pests and their control in the Near East. FAO publ. No. 138. Rome, Italy. 165 pp.

Mirkarimi, A. 1992. Study of resistance in several wheat varieties against *Eurygaster integriceps* Put. in Waramin. Proc. Sunn Pest Conf., Tehran University, Tehran, Iran. Pp. 116–134 [in Persian].

Mittal, O. P., and J. Lelamma. 1981. Chromosome number and sex-mechanism in twenty-eight species of Indian pentatomoid bugs. Chrom. Inf. Ser. 30: 6–7.

Modarres Awal, M. 1992. Biology, ecology of *Eurygaster integriceps* Put. and its outbreak in Trakia (European part of Turkey). Proc. Sunn Pest Conf., Tehran University, Tehran, Iran. Pp. 51–70 [in Persian].

Mohaghegh, J., M. Esmaili, and E. Bagheri-zonouz. 1991. A systematic review of genus *Eurygaster* Laporte (Heteroptera: Scutelleridae). Proc. Tenth Plant Prot. Congr. Iran. Kerman, Iran. P. 66 [in Persian].

Moizuddin, M., and I. Ahmad. 1980. Some aspects of morphology of two species of shield bugs (Pentatomoidea: Scutelleridae) of Pakistan with reference to phylogeny. Sind Univ. Res. J. (Sci. Ser.) 12–13: 17–30.

Moore, C. J., A. Hubener, and Y. Q. Tu. 1994. A new spiroketal type from the insect kingdom. J. Org. Chem. 59: 6136–6138.

Najafi Mirak, T. 1998. Genetic study of resistance to Sunn pest (*Eurygaster integriceps* Put.) in wheat. M.Sc. thesis, Tehran University, Tehran, Iran [in Persian, with English abstract].

Nixon, G. E. J. 1939. Parasites of hemipterous grain-pests in Europe (Hymenoptera; Proctotrupoidea). Arb. Morph. Taxon. Entomol. Berlin-Dahlem 6: 129–136.

Paulian. F., and C. Popov. 1980. Wheat documenta. Pp. 69–74 *in* CIBA-GEIGY Ltd., Basel, Switzerland.

Pokrovskaya, N. E., G. I. Morozova, and N. M. Vinogradova. 1971. Proteins of *Eurygaster integriceps* Put. damaged wheat grain. Appl. Biochem. Microbiol. 7: 121–127 [in Russian, with English summary].

Popov, C. 1972. Cercetări privind aria de răspindire ş intensitatea atacului la *Eurygaster* Lap. in Romănia. An. I.C.C.P.T. 38: 77–90 [in Romanian, with English summary].

1979. Cercetări privind reducerea corpului gras la ploşniţa cerealelor (*Eurygaster integriceps* Put.) pe timpul diapauzei. An. I.C.C.P.T. 45: 363–370 [in Romanian, with English summary].

1980. Activitatea paraziţilor cofagi în perioada de ponta a ploşniţelor cerealelor (*Eurygaster* sp.). An. I.C.C.P.T. 46: 347–353 [in Romanian, with English summary].

1982a. Plo-ni-a asiatica a cerelelor *Eurygaster integriceps* Put., important daunător al grîului in Romănia. An. I.C.C.P.T. 50: 379–390 [in Romanian, with English summary].

1982b. Cercetări privind influenţa raportului dintre sexe asupra capaciţătii de reproducere la *Eurygaster integriceps* Put. An. I.C.C.P.T. 49: 217–223 [in Romanian, with English summary].

Popov, C., A. Barbulescu, and I. Vonica. 1996. Population dynamics and management of Sunn pest in Romania. Pp. 47–59 *in* R. H. Miller and J. G. Morse [eds.], Sunn Pests and Their Control in the Near East. FAO Publ. No. 138. Rome, Italy. 165 pp.

Puton, A. 1881. Synopsis des Hémiptères Hétéroptères de France, 4e partie. Remiremont, France. 129 pp.

Putschkov, V. G. 1961. Fauna of Ukraine. Shieldbugs, Vol. 21, Ac. Sci. U.K. SSR. Kiev, Ukraine, U.S.S.R. 338 pp. [in Ukrainian].

1965. Shieldbugs of Central Asia. Acad. Sci. Kinghiz SSR. 329 pp.

Putschkova, L. V. 1959. The eggs of true bugs (Hemiptera: Heteroptera). V. Pentatomoidea. Rev. Entomol. URSS 40: 131–143 [in Russian].

Radjabi, G., and F. Termeh. 1988. Reproduction of *Eurygaster integriceps* Put. and *Aelia furcula* in their hibernation sites and its connection with these bugs extension in Iran. Appl. Entomol. Phytopathol. 55: 3–32 [in Persian, with English summary].

Reid, J. L., and H. E. Barton. 1989. Laboratory rearing of *Chelysomidea guttata* (Hemiptera: Scutelleridae) with descriptions of immature stages. Ann. Entomol. Soc. Amer. 82: 737–740.

Rezabeigi, M. 1995. Morphological and biochemical aspects of 25 wheat cultivars in relation to resistance to cereal Sunn pest (*Eurygaster* inegriceps Put.). M.Sc. thesis, Tehran University, Tehran, Iran [in Persian, with English summary].

Rosca, I., C. Popov, A. Barbulescu, I. Vonica, and K. Fabritius. 1996. The role of natural parasitoids in limiting the level of Sunn pest populations. Pp. 35–45 in R. H. Miller and J. G. Morse [eds.], Sunn Pests and Their Control in the Near East. FAO Publ. No. 138. Rome, Italy. 165 pp.

Roychoudhury, N., K. C. Joshi, and P. S. Rawat. 1994. A new record of *Chrysocoris purpureus* Westwood (Heteroptera: Scutelleridae) on poplar, *Populus deltoides* Bartr. ex Marsh. Indian For. 120: 1126–1128.

Safavi, M. 1968. Étude biologique et écologique des hyménoptères parasites des oeufs des punaises des céréal. Entomophaga 13: 381–495.

Schouteden, H. 1903a. Faune entomologique de l'Afrique tropicale. Rhynchota aethiopica. I. Scutellerinae et Graphosomatinae. Tome 1 (Fasc. 1). Ann Mus. Congo (Zool., Ser. III), Brussels, Belgium. 8 unnumbered, 131 pp.

1903b. Genera Insectorum. Heteroptera. Fam. Pentatomidae Subfam. Scutellerinae. M. P. Wytsman, Brussels, Belgium. 98 pp.

1906. Genera insectorum. Heteroptera. Fam. Pentatomidae Subfam. Scutellerinae. M. P. Wytsman, Brussels, Belgium. Pp. 99–100.

1912. Les hémiptères parasites des cotonniers en Afrique. Rev. Zool. Afric. 1: 297–321.

Schuh, R. T., and J. A. Slater. 1995. True Bugs of the World (Hemiptera: Heteroptera). Cornell University Press, Ithaca, New York, U.S.A. 336 pp.

Shahrokhi, S. 1997. A study on mass rearing of *Trissolcus grandis* on *Graphosoma lineatum* eggs and the quality control for biological control of Sunn Pest, *Eurygaster integriceps* (Hem. Scutelleridae). M.Sc. thesis, Tehran University, Tehran, Iran [in Persian, with English summary].

Singh, D. N., and S. D. Singh. 1966. Meiotic studies in the bug, *Chrysocoris stollii* Wolff. Naturwissenschaften 53: 91.

Singh, M. P. 1965. Structure and postembryonic development of the Malpighian tubules of *Chrysocoris stollii* Wolff (Heteroptera: Pentatomidae). Entomologist 1965: 268–279.

1968a. Female reproductive organs and their development in *Chrysocoris stollii* Wolff (Heteroptera: Pentatomidae). Bull. Entomol. (India) 9: 25–35.

1968b. Histological changes in the midgut epithelium during metamorphosis in *Chrysocoris stollii* Wolff (Heteroptera, Pentatomidae). Mikroskopie 23: 162–165.

1969. Nervous system of *Chrysocoris stolli* Wolff. (Heteroptera: Pentatomidae). Bull. Entomol. Res. 10: 60–67.

1971. Development of male reproductive organs of *Chrysocoris stolli*. J. Kansas Entomol. Soc. 44: 433–440.

Skaf, R. 1996. Sunn pest problems in the Near East. Pp. 9–16 *in* R. H. Miller and J. G. Morse [eds.], Sunn Pests and Their Control in the Near East. FAO Publ. No. 138. Rome, Italy. 165 pp.

Smith, R. M. 1978a. The defensive secretion of *Coleostichus* [sic] *sordidus* (Heteroptera: Scutelleridae) from Fiji. New Zealand J. Sci. 21: 121–122.

1978b. The defensive secretion of the bugs *Lamprophara bifasciata*, *Adrisa numensis*, and *Tectocoris diophthalmus* from Fiji. New Zealand J. Zool. 5: 821–822.

Southwood, T. R. E., and D. Leston. 1959. Land and Water Bugs of the British Isles. Frederick Warne & Co. Ltd., London, U.K. 436 pp.

Staddon, B. W., A. Abdollahi, J. Parry, and M. Rossiter. 1994. Major component in male sex pheromone of cereal pest *Eurygaster integriceps* Puton identification as a homosesquiterpenoid. J. Chem. Ecol. 20: 2721–2731.

Stamenković, S. 1976. Vidovi oštećenja od žitnih stenica. Zaš. Bilja 27: 335–348 [in Serbian, with English summary].

1977. Dopunska ishrana žitnih stenica iz roda *Eurygaster* Lap. na mestima prezimljavanja u vojvodini. Zaš. Bilja 28: 419–427 [in Serbian, with English summary].

1985. Evaluating winter wheat and barley for resistance to major insect pests in Yugoslavia. Pp. 174–177 *in* [no editor], Schaderreger in der Getreideproduktion. Martin-Luther-Univ., Halle-Wittenberg, Germany.

1990. The Suni bugs on small grains in Yugoslavia. Scopolia, Suppl. 1: 113–116.

1993. Proučavanje otpornosti ozime pšenice prema žitnoj stenici (*Eurygaster austriaca* Schrk., Pentatomidae, Heteroptera). Zaš. Bilja 44: 31–37 [in Serbian, with English summary].

Stein, J. D. 1983. Insects infesting *Acacia koa* (Legumosae [sic]) and *Metrosideros polymorpha* (Myrtaceae) in Hawaii: an annotated list. Proc. Hawaiian Entomol. Soc. 24: 305–316.

Surender, P., C. Janaiah, V. K. Reddy, and S. M. Reddy. 1987a. Antifungal activity of secretions of scent glands from heteropteran bugs. Indian J. Exp. Biol. 25: 233–234.

Surender, P., T. Mogili, D. Thirupathi, C. Janaiah, and Vidyati. 1987b. Effect of scent components on somatic cells of *Allium sativum* L. Curr. Sci. 56: 964–967.

Talai, R. J. 1989. Influences of the food plants on reproduction of cereal Sunn pest *Eurygaster integriceps* Put. (Het. Scutelleridae) in the field. M.Sc. thesis, Tehran University, Tehran, Iran [in Persian, with English summary].

Talai, R., M. Esmaili, C. Abd-mishani, P. Azmayesh-fard, and E. Bagheri-zenouz. 1991. Evaluation of baking quality of wheat in relation to Sunn pest (*Eurygaster integriceps* Put.) damage. Proc. Tenth Plant Prot. Congr. Iran. Kerman, Iran. P. 47.

Talhouk, A. S., 1961. Biological control of Sunn Pest through its egg–parasite, *Asolcus* (*Microphanurus*) *semistriatus* (NEES). Bull. Fac. Agric. Sci. Amer. Univ. Beirut. No. 12, 38 pp.

Tischler, W., 1939. Zur Ökologie der Wichtigsten in Deutschland an Getreide schädlichen Pentatomiden. II. Z. Morph. Ökol.Tiere 35: 251–287.

Ubik, D., J. Vrkoč, J. Ždárek, and C. Kontev. 1975. Vanillin: possible sex pheromone of an insect. Naturwissenschaften 62: 348.

Ueshima, N. 1979. Hemiptera II: Heteroptera. B. John [ed.], Animal Cytogenetics. Vol. 3: Insecta 6. Gebrüder Borntraeger, Berlin, Germany. 117 pp.

Van den Berg, M. A. 1980. Natural enemies of *Acacia cyclops* A. Cunn. ex G. Don and *Acacia saligna* (Labill.) Wendl. in western Australia. III. Hemiptera. Phytophylactica 12: 223–226.

Vassiliev, J. V. 1913. *Eurygaster integriceps* Put. and new methods of fighting it by the aid of parasites. S. Petersbourg, 81 pp. [in Russian].

Verma, A., S. Kumar, and T. B. Sinha. 1978. Seasonal occurrence in the populations of different pests of *Calotrophis procera*. Indian J. Entomol. 40: 204–210.

Verma, G. P., and A. K. Basiston. 1974. Cytochemical studies on the origin and composition of yolk in *Chrysocoris stollii* (Heteroptera) 1. Cytologia 39: 619–631.

Vojdani, S. 1954. Contribution à l'étude des punaises des céréales et en particulier d'*E. integriceps*. Ann. Epiphyt. 5: 105–160.

1961a. The Nearctic species of the genus *Eurygaster* (Hemiptera: Pentatomidae: Scutellerinae). Pan-Pac. Entomol. 37: 97–107.

1961b. Bio-ecology of some *Eurygaster* species in central California (Pentatomidae-Scutellerinae). Ann. Entomol. Soc. Amer. 54: 567–578.

Voegelé, J. 1960. Inventaire des espèces de punaises des genres *Aelia* et *Eurygster* existant au Maroc, basé, sur l'étude du squelette génital. Cah. Ech. Agron.10: 5–25.

1965. Nouvelle méthode d'étude systematique des espèces du genre *Asolcus* cas d'*Asolcus rungsi*. Al-Awamia (Rabat) 14: 95–113.

1966. Review of biological control of Sunn pest. Pp. 23–33 *in* R. H. Miller and J. G. Morse [eds.], Sunn Pests and Their Control in the Near East. FAO Publ. No. 138. Rome, Italy. 165 pp.

Vrkoč, J., K. Ubik, J. Ždárek, and C. Kontev. 1977. Ethyl acrylate and vanillin as components of the male sex pheromone complex in *Eurygaster integriceps*. Acta Entomol. Bohemoslov. 74: 205.

Wagner, E. 1951. Über die Variationen bei *Eurygaster* Arten Hemipt. Heter. Scut. Comment. Biol. 12(2): 43.

Walt, J. F., and J. E. McPherson. 1972. Laboratory rearing of *Stethaulax marmoratus* (Hemiptera: Scutelleridae). Ann. Entomol. Soc. Amer. 65: 1242–1243.

Webb, G. A. 1989. Insects as potential pollinators of *Micromyrtus ciliata* (Sm.) Druce, Myrtaceae. Victorian Nat. (Australia) 106: 148–151.

Williams, V. G., and R. A. Goyer. 1980. Comparison of damage by each life stage of *Leptoglossus corculus* and *Tetyra bipunctata* to loblolly seed pines. J. Econ. Entomol. 73: 497–501.

Wilson, F. 1961. Adult reproductive behaviour in *Asolcus basalis* (*Hymenoptera*: *Scelionidae*). Austral. J. Zool. 5: 737–751.

Wilson, L. T., D. R. Booth, and R. Morton. 1983. The behavioural activity and vertical distribution of the cotton harlequin bug *Tectocoris diophthalmus* (Thunberg) (Heteroptera: Scutelleridae) on cotton plants in a glasshouse. J. Austral. Entomol. Soc. 22: 311–317.

Yang, W.-I. 1934. Notes on the Chinese Scutelleridae. Bull. Fan Mem. Inst. Biol. (Peiping) 5: 237–284.

Youdeowei, A. 1973. Some Hemiptera-Heteroptera associated with cacao farms in Nigeria. Turrialba 23: 162–171.

Yüksel, M. 1968. Güney ve güneydogu anadulu'da Süne *Eurygaster integriceps* Put. un vayilisi, biyolojisi, ekoloji epidemiyolojisk ve arari Üzerinde arastirmalar. Ankara, Turkey. 260 pp. [in Turkish].

Zarnegar, A. 1995. The effect of juvenile hormone mimics on diapause and reproduction of Sunn pest *Eurygaster integriceps* Put. (Het. Scutelleridae). M.Sc. thesis, Tehran University, Tehran, Iran [in Persian with English summary].

Ždárek, J., and C. Kontev. 1975. Some ethological aspects of reproduction in *Eurygaster integriceps*. Acta Entomol. Bohemoslov. 72: 239.

Zomorrodi, A., 1959. La lutte biologique contre la punaise du blé *Eurygaster integriceps* Put. par *Microphanurus semistriatus* Nees., en Iran. Rev. Pathol. Veg. Entomol. Agric. France 38: 167–174.

CHAPTER 15

Several Small Pentatomoid Families (Cyrtocoridae, Dinidoridae, Eurostylidae, Plataspidae, and Tessaratomidae)

Carl W. Schaefer, Antônio R. Panizzi, and David G. James

1. INTRODUCTION

These small families contain a few species of economic importance.

2. CYRTOCORIDAE

Cyrtocoris gibbus (F.) feeds on the branches of bracatinga (*Mimosa scrabella* Benth.) (Leguminosae) (Costa Lima 1940), a tree of southern Brazil planted as an ornamental; its wood is used for furniture, and to make charcoal (Lorenzi 1992). The family Cyrtocoridae, perhaps related to Cydnidae, has been revised, with illustrations and keys, by Packauskas and Schaefer (1998).

3. DINIDORIDAE

This small Old World tropical family has been revised recently by Durai (1987) and cataloged by Rolston et al. 1996). Earlier, the species from China and surrounding areas were cataloged by Hoffmann (1935, 1948), and the world species by Kirkaldy (1909). The most important species economically are members of *Aspongopus* (formerly *Coridius*), including *A. janus* (F.), the red pumpkin bug (see below). Several papers consider some aspects of this bug's anatomy and physiology (exact subject matter in titles of these papers: Rastogi 1961; 1962a,b; 1965; Rastogi and Datta Gupta 1962; Kaushik et al. 1977, 1979); the metathoracic scent of an *Aspongopus* sp. (from Kenya) was analyzed by Prestwich (1976).

The red pumpkin bug (*A. janus*) is, as its name suggests, a pest of cucurbit crops in India (Chatterjee 1934, Mishra and Sharma 1989), although it also attacks akh, or madar (*Calotropis procera*), a wild herb (Asclepiadaceae), in Uttar Pradesh, India (Verma et al. 1978). In the former Indo-China, Sri Lanka, and Malaysia, *A. fuscus* (Westwood) is also a cucurbit pest, and feeds too on eggplant (*Solanum melongena* L.) (Solanaceae) and lablab (*Dolichos lablab* L.), castor oil bean (*Ricinus communis* L.), and lima bean (*Phaseolus lunatus* L., these last two in Canton, China (all Leguminosae) (Hoffman 1932). Adults and nymphs of *A. brunneus* (Thunberg) have caused 20 to

30% damage to nursery seedlings of karanj (*Pongamia planata*, Leguminosae) in Madhya Pradesh, India (Meshram et al. 1990). In Ghana, *A. remipes* Stål occasionally infests cocoa trees, although it prefers cucurbits (Lodos 1969). Lefroy (1909) reports that in Assam, *A. nepalensis* (Westwood) is ground up with rice and eaten.

Dinidorids seem to prefer cucurbits. However, the food plants of only a few dinidorid species are known, and this preference may be restricted to *Coridius* and its relatives (Schaefer and Ahmad 1987). Other members of the family may feed more widely, or may prefer plants in the orders Theales and Violales (to which Cucurbitaceae belongs), both of the suborder Dilleniidae (Schaefer 1987).

Studies on the biology of these insects is fragmentary and scattered. What follows is a composite of bits and tatters of data from several species. Adults are large and robust; *A. janus* is a deep orange and some 2.5 cm long. Eggs are barrel-shaped and laid end to end in a single row of 12 to 14 eggs (*A. remipes*, Lodos 1969), or 3 to 30 eggs, on the underside of the host plant's midrib (*A. fuscus*, Hoffmann 1932). Nymphs occur on the ground [*Note:* These may have been overwintering, C. W. Schaefer, personal comment] (*A. brunneus* and *A. obscurus*, Meshram et al. 1990), and adults and nymphs feed on stems [not reproductive parts] (*A. janus*, Mishra and Sharma 1989; *Aspongopus* sp., in Malaysia, Hoffmann 1932; *A. brunneus*, Meshram et al. 1990). Damage is enhanced because nymphs and adults are gregarious (*A. fuscus*, Hoffmann 1932). In southern China, adults occur from early spring to late fall; eggs have been found from mid-May through late August (*A. fuscus*, Hoffmann 1932); in Uttar Pradesh (India), population increases occurred in late summer (August) through fall (into October), increases apparently stimulated by moderately high temperatures and relative humidity and decreases by sudden rises in temperature and decreases in relative humidity (*A. janus*, Verma et al. 1978).

"Hymenopterous egg parasites appeared late in July... ," in Canton (Hoffmann 1932).

4. UROSTYLIDAE

This small family contains two subfamilies, Urostylinae (25 to 30 species) and Saileriolinae (about five species). Saileriolinae was erected by China and Slater (1956) and discussed further by Schaefer and Ashlock (1970); it may be sufficiently distinct to warrant full family rank (Zheng and Schaefer, unpublished). It is of no economic importance.

The systematics of Urostylinae have been most thoroughly worked out by Yang (1938, 1939), Ren (1984a, 1986a,b), and Ahmad et al. (1992); Ahmad et al. (1992) key the species of the Indo-Pakistan region and analyze all genera cladistically. The egg and five instars of several Japanese species have been described and illustrated by Kobayashi (1953, 1965).

Urostylines attack some trees of value as ornamentals or fruit producers; Prof. Ren Shu-zhi (personal communication, February 1999) lists the following: *Urostylis yangi* Maa (oaks, *Quercus* spp.; *Castanea* spp.), *U. annulicornis* Scott (*Quercus* spp.), *Urochela luteovaria* Distant (*Prunus* sp., *Pyrus* sp., *Malus* sp.), *U. quadrinotata* Reuter (elm, *Ulmus* sp.). Yang (1936) records an outbreak in the Lushan mountains, Kiukiang, China, of *Urochela distincta* Distant, a reddish-black insect, somewhat flat, with "a very stinking odour" (p. 57). A large swarm of them flew in such numbers as to obscure the sky and, falling into a reservoir, rendered the water undrinkable. They entered houses, creating great annoyance. In nature, the adults cluster among the rocks above 1500 m, and do not occur lower down. When the weather warms, they fly forth in large numbers, and then become pests. They apparently cause no crop damage, but are annoying enough for all that. Recommended long-term control consisted of filling the rock crevices with clay, and replacing the shrub fed upon (after warm weather), *Diervilsa japonica* var. *sinica* Rehd., with pine trees. In the short term, the best control was using gunpowder placed in a bamboo gun (directions here provided), or bomb, placed in the rock crevices. This works.

5. PLATASPIDAE (= COPTOSOMATIDAE)

Members of this relatively small (about 150 species) family are characterized by their small size (1.5 to 15 mm) and globose shape. The scutellum completely covers the abdomen, and is usually continuous with the pronotum, rendering the bugs semicircular in side view. Most plataspids are uniformly brown or black, but a few have various amounts of patterning. The family is almost completely Old World tropical, and most species are Oriental (Kirkaldy 1909). The Chinese species were revised and keyed by Yang (1934) and cataloged by Hoffmann (1935, 1948), and the systematics and evolutionary relationships of the western Palearctic species were studied by Davidová-Vilimová and Štys (1980); this paper also keys the species and illustrates morphology. Ahmad and Moizuddin (1992) redescribe, key, and illustrate the Indo-Pakistani species, and describe the new genus *Neobozius*. The taxonomic papers of Miller (1955), Ghauri (1965), and Jessop (1983) also illustrate various morphological features (especially metathoracic scent gland region, wing venation, and external genitalia), as well as describe new taxa. The metathoracic scent apparatus of *Coptosoma cribraria* (F.) is discussed by Ahmad and Moizuddin (1975a), and the spermathecae of several Korean species by Kim and Lee (1993). The internal anatomy of two *Coptosoma* species was described and illustrated by Ahmad and Moizuddin (1975b) and Moizuddin and Ahmad (1975). Ren (1984b, 1985) describes the fine structure of the chorions of several species; and Carayon (1949) describes the arrangement of the eggs and their oviposition.

Plataspidae is one of a few heteropteran groups that feed mostly on legumes (Schaefer 1988). Several therefore have occasionally become pests on leguminous crops, especially various beans, in the Old World tropics. *Coptosoma cribraria* is a pest of *Dolichos lablab* in Pakistan (Ahmad and Moizuddin 1975a) and in India (Chatterjee 1934); *C. nubila* Germar and *Brachyplatys testudonigra* De Geer feed on *Cajanus indicus* and *Gliricidia* sp., respectively (Golding 1931); and *Coptosoma punctissimum* Montandon feeds on soybean and kidney bean in Japan (Ishihara 1950), and on many crop legumes, and also on sugarcane (*Saccharum officinarum* L., *Corchorus capsularis* L.), sweet potato (*Ipomoea batatas* Lam.), and rice (*Oryza sativa* L.) throughout the Orient (these records may apply to *Coptosoma cribraria*) (Hoffmann 1932). *Coptosoma ostensum* Distant attacks the tree *Butia frondosa* (F.R.I.) in Coimbatore (India); "trees are covered with millions of this bug and their nymphs every year in March-April. Attacked leaves turn brown" and drop (Chatterjee 1934, p. 4)

Several species of *Coptosoma* (*C. cribraria, siamicum, ostensum, fuscomaculatum* Distant, and *variegatum* Montandon) all attack the sandal tree (*Santalum album* L.), in India; these bugs do not appear to transmit the virus causing spike disease (Chatterjee 1934).

Enarsiella spp. (Hymenoptera: Aphelinidae) may parasitize plataspids in China (Huang and Polaszek 1996).

6. TESSARATOMIDAE

These are large (greater than 15 mm) bugs, ovate or elongate-ovate, similar in appearance to pentatomids, with a characteristic small triangular head; the pronotum is large and the triangular scutellum does not cover the corium (Schuh and Slater 1995). Members of this family are Old World tropical; species of the genus *Piezosternum* Amyot and Serville occur in the Neotropical region (Froeschner 1988).

A systematic catalog has been published (Rolston et al. 1993), following the earlier catalogs of the fauna of China and surrounding regions (Hoffmann 1935, 1948). Yang (1935) redescribed and keyed the 15 species then known from China, and illustrated the genitalia. Sinclair (1989) analyzed the genera cladistically and made some systematic changes, in an unpublished dissertation; and Kumar (1974) keyed the genera of the subfamily Nataticolinae, whose relationships he had earlier (1969) discussed. Respiratory structures and their development are considered by Devi et al. (1974, 1978). A list of food plants of ten genera (13 species) indicates that the family feeds widely but shows a slight preference for plants of the orders Rosales and Sapindales (Schaefer and Ahmad 1987). Guarding of the eggs and early instars occurs in some species (Tachikawa 1991, Gogala et al. 1998).

The only species known to be of economic importance is the bronze orange bug, *Musgraveia sulciventris* Stål. This tesseratomid is native to Australia, and is host-specific on citrus, both indigenous and introduced. The first specimens were collected in 1863 from Queensland and it was first recorded as a pest of cultivated citrus in 1892 (Oliff 1892). It is now common on citrus trees in coastal Queensland and north-coastal New South Wales. The bronze orange bug can be a severe pest of backyard and home garden citrus, causing heavy fall of young fruits (Hely et al. 1982). However, it is rarely a problem in commercial citriculture. The bugs are notable for their foul odor and their corrosive secretions, which burn foliage. The chemical composition of scent glands of tessaratomids (e.g., *Tessaratoma javanica* Thunberg) has been studied in detail (Janaiah et al. 1979a,b), and considered a mimic of the juvenile hormone (Ashok et al. 1978).

The bronze orange bug is univoltine with a nymphal diapause. The eggs are laid on the undersides of leaves in batches, usually consisting of 14 eggs, arranged in four rows. Fecundity is poor: only 12 to 32 eggs per female and an average hatchability of 85%, and thus the number of offspring per adult is low (Cant 1994). The eggs are glossy, green, and spherical, about 3 mm in diameter, and hatch in 8 to 14 days. The light green first instars remain near the eggshells and do not feed. After a few days they molt to the second stage, in which they overwinter. The second instars are about 6 mm long, pale green or yellow, and almost transparent. They are thin, flat, and this is often called the "tissue paper" stage; they remain inactive on the undersides of leaves until spring. It is very difficult to see them but a "stink bug odor" emanates when they are disturbed. The empty eggshells remain on the tree for a long time and are fairly obvious. When tree growth begins in spring, the second instars become active, feeding on young succulent leaves. The third instar is reached after about 3 to 4 weeks of feeding, and the third, fourth, and fifth stadia last about 3 weeks each. The third instar is green and may be margined with black. The fourth instar is green, pink, or salmon-colored and has wing buds. The fifth instar, which appears in late spring is green-gray, salmon, or pink and has large wing buds. It feeds voraciously and is very aggressive in discharging a defensive secretion which can burn tender skin. The adult stage is reached in early summer and adults are present on trees throughout summer. Most die by midautumn. Adults are 25 mm long, stoutly built, and bronze-colored at first, becoming nearly black as they age. Groups of about six bugs often feed together on shoot growth and fruit stalks. When young growth is scarce, the bugs may not feed until new growth appears. Mating and egg laying usually follow resumption of feeding. During summer, adults sometimes cluster on the trees in masses which may be as big as a football. If disturbed, the bugs fly around like angry bees. The adults are very active and, like fifth instars, readily discharge a defensive secretion which can stain and burn human skin. The adults can squirt this secretion for up to 60 cm, often directed toward human or avian eyes. Adult *M. sulciventris* are strong fliers and can migrate long distances between citrus trees. Before establishment of the citrus industry in Australia, the range of *M. sulciventris* was limited by the distribution of one of its indigenous hosts, *Microcitrus australasica*.

The damage by *Musgraveia sulciventris* is generally restricted to spring and early summer when they feed on young shoot growth and the stalk of flowers and young fruits. The shoots die and the flowers and fruits fall. As the supply of sappy growth is destroyed, the bugs feed on the stalks of older leaves, which then fall. Further damage can result from the caustic defensive secretion produced by the bugs; brown spots of dead tissue appear on affected areas, sometimes causing leaf fall. Trees that have suffered a severe attack take on an unthrifty appearance and carry a sparse crop. When bugs are removed, trees recover well but an out-of-season cropping habit may follow. Summer-feeding adults do not appear to cause significant damage.

Musgraveia sulciventris rarely requires control in commercial orchards. In home gardens insecticide sprays are used or the bugs are simply removed by hand. High temperatures, particularly in association with low humidity, can cause significant mortalities. Natural enemies include birds and the assassin bug, *Pristhesancus plagipennis* Walker (Reduviidae), but egg parasitoids are uncommon. An Argentine tachinid parasitoid, *Trichopoda giacomellii* (Blanchard), released to control the pentatomine *Nezara viridula* (L.), was not attracted to *M. sulciventris* (Sands and Coombs 1999).

Here *Asiarcha nigridorsis* (Stål) is included with some trepidation, for the authors of the only article found on it evidently have confused this "green bug" with the southern green bug, *N. viridula*, deduced from the authors' bibliography of papers on *N. viridula*. These authors (Andal et al. 1991) report *A. nigridorsis* a pest in Nilgiris (India) on the leaves of *Elaegnus latifolia* (whose fruits are eaten) and on peach (*Prunus persica*); the nymphs do more damage than the adults.

The bug is dark above, its wings' membranes coppery green, and is 29 mm long (Distant 1904) or 20 to 22 mm long (Andal et al. 1991). Mating occurs 3 days after eclosion and lasts 60 to 75 min; males mate with many females. Oviposition occurs in the early morning; the laying of the 12 to 14 eggs on a leaf midrib takes about 25 min. Males live 41 to 63 days and females 75 to 92 days. The egg is 1.8 × 2.0 mm. The five stadia last (summer/winter) 9 to 12/14 to 20, 10 to 12/16 to 23, 6 to 9/16 to 30, 19 to 27/13 to 15, 7 to 11/13 to 22 days, respectively.

7. PROGNOSIS AND CONCLUSIONS

With a few exceptions, it is unlikely that any members of these small pentatomoid families will become serious pests; most will do local and sporadic damage, sometimes serious at the time, but not lasting and not widespread. One exception is the bronze orange bug, *M. sulviventris* (Tessaratomidae), which it is believed will continue to damage citrus crops in Australia. and may also become a problem in commercial citrus as pesticide use declines. Similarly, the red pumpkin bug, *Aspongopus janus* (Dinidoridae) and its relatives will probably be pests — serious at times — of cucurbit crops in India.

Methods used to control more serious pests of these various crops will in general control these minor pentatomoid pests as well.

8. ACKNOWLEDGMENTS

The authors are very grateful to Prof. Ren Su-zhi, for providing host plants of several urostylid and plataspid species.

9. REFERENCES CITED

Amad, I., and M. Moizuddin. 1975a. Scent apparatus morphology of bean plataspid *Coptosoma cribrarium* (Fabricius) (Pentatomoidea: Plataspidae) with reference to phylogeny. Pakistan J. Zool. 7: 45–49.

1975b. Some aspects of internal anatomy of *Coptosoma cribrarium* (Fabr.) (Pentatomoidea: Plataspidae) with reference to phylogeny. Folia Biol. 23: 53–61.

1992. Plataspidae Dallas (Hemiptera: Pentatomoidea) from Pakistan and Bangladesh with keys including Indian taxa. Annot. Zool. Bot. 208: 32 pp.

Ahmad, I., M. Moizuddin, and S. Kamaluddin. 1992. A review of cladistics of Urostylidae Dallas (Hemiptera: Pentatomoidea) with keys to taxa of Indian subregion. Philippine J. Sci. 121: 263–297.

Andal, S., C. Sarasu, and A. S. Begum. 1991. A field study of the green bug *Asiarcha nigridorsis* (Hemiptera: Tessaratomidae). Indian Zool. 15: 157–158.

Ashok, B. S., B. K. Rao, S. S. Bhakur, P. Judson, N. S. Raj, and G. M. Ram. 1978. Defensive secretion of the soap-nut bug, *Tessaratoma javanica* (Thunberg) (Heteroptera, Pentatomidae) as a juvenile hormone mimic. Zool. Jahrb. Physiol. 82: 383–394.

Cant, R. 1994. The mysteries of the bronze orange bug. Circ. Entomol. Soc. New South Wales, Inc., pp. 47–48.
Carayon, J. 1949. L'oothèque d'Hémiptères Plataspidés de l'Afrique tropicale. Bull. Soc. Entomol. France 5: 66–69.
Chatterjee, N. C. 1934. Entomological investigations on the spike disease of sandal (24) Pentatomidae (Hemipt.). Indian For. Rec. 20: 1–31.
China, W. E., and J. A. Slater. 1956. A new subfamily of Urostylidae from Borneo (Hemiptera: Heteroptera). Pac. Sci. 10: 410–414.
Costa Lima, A. M. 1940. Insetos do Brasil. 2° tomo, capitulo XXII. Hemipteros. Escola Naçional de Agronomia, Rio de Janeiro, Brazil. 351 pp.
Davidová-Vilimová, J., and P. Štys. 1980. Taxonomy and Phylogeny of West Palaearctic Plataspidae (Heteroptera). Československá Akad. Věd, Prague, Czechoslovakia. 155 pp.
Devi, S., K. Rao, S. S. Thakur, and S. Devi. 1974. A study on the morphology of spiracles and structural changes during development of a pentatomid bug — *Tessaratoma javanica* (Thunberg). J. Anim. Morphol. Physiol. 25: 27–31.
1978. Morphological studies on the tracheal system of a pentatomid bug, *Tessaratoma javanica* (Thunberg). J. Anim. Morphol. Physiol. 25: 32–39.
Distant, W. L. 1904. The Fauna of British India including Ceylon and Burma. Rhynchota. — Vol. I (Heteroptera). Taylor and Francis, London, U.K. 438 pp.
Durai, P. S. S. 1987. A revision of the Dinidoridae of the world (Heteroptera: Pentatomoidea). Orient. Insects 21: 163–360.
Froeschner, R. C. 1988. Family Tessaratomidae Stål, 1864. The tessaratomids. Pp. 694–695 *in* T. J. Henry and R. C. Froeschner [eds.], Catalog of the Heteroptera, or True Bugs, of Canada and the Continental United States. E. J. Brill, Leiden, the Netherlands. 958 pp.
Ghauri, M. S. K. 1965. New Plataspidae from the Ethiopean Region (Hemiptera). Proc. R. Entomol. Soc. London (B) 34: 127–131.
Gogala, M., H.-S. Yong, and C. Brühl. 1998. Maternal care in *Pygoplatys* bugs (Heteroptera: Tessaratomidae). Eur. J. Entomol. 95: 311–315.
Golding, F. D. 1931. Further notes on the food-plants of Nigerian insects. Bull. Entomol. Res. 22: 221–223.
Hely, P. C., G. Pasfield, and J. G. Gellatley. 1982. Insect Pests of Fruit and Vegetables in NSW. Inkata Press, Melbourne, Australia. 312 pp.
Hoffmann, W. E. 1932. Notes on the bionomics of some Oriental Pentatomidae (Hemiptera). Arch. Zool. Ital. 16: 1010–1027.
1935. An abridged catalogue of certain Scutelleroidea (Plataspidae, Scutelleridae, and Pentatomidae) of China, Chosen, Indo-China, and Taiwan. Lingnan Univ. Sci. Bull. 7: 1–294.
1948. First supplement to catalogue of Scutelleroidea. Lingnan Sci. J. 22: 1–41.
Huang, J., and A. Polaszek. 1996. The species of *Encarsiella* (Hymenoptera: Aphelinidae) from China. J. Nat. Hist. 30: 1649–1659.
Ishihara, T. 1950. The developmental stages of some bugs injurious to the kidney bean (Hemiptera). Trans. Shikoku Entomol. Soc. 1: 17–31.
Janaiah, C., N. Chari, and P. V. Reddy. 1979a. Glycogen and total proteins in flight muscles and lateral scent glands of *Tessaratoma javanica* Thunberg (Pentatomidae: Hemiptera). Entomon 4: 65–66.
Janaiah, C., P. S. Rao, N. Chari, and P. V. Reddy. 1979b. Chemical composition of the scent glands of adults and nymphs of the bug *Tessaratoma javanica* Thunberg. Indian J. Exp. Biol. 17: 1233–1235.
Jessop, L. 1983. A review of the genera of Plataspidae (Hemiptera) related to *Libyaspis*, with a revision of *Cantharodes*. J. Nat. Hist. 17: 31–62.
Kaushik, S. C., G. Kumar, and P. K. Gupta. 1977. Studies on the organs of circulation of the red pumpkin bug Coridus [sic] janus Fabr. (Heteroptera: Dinidorinae: Pentatomidae). J. Anim. Morphol. Physiol. 24: 244–250.
1979. Functional anatomy of the male reproductive organs of Coridus [sic] janus (Heteroptera-Dinidorinae, Pentatomidae) with special reference to the bulbous [sic] ejaculatorius. J. Anim. Morphol. Physiol. 26: 211–219.
Kim, H. R., and C. E. Lee. 1993. The spermathecae of the Plataspidae from Korea (Heteroptera, Hemiptera). Nat. Life (Korea) 23: 115–120.

Kirkaldy, G. W. 1909. Catalogue of the Heteroptera (Hemiptera) with Biological and Anatomical References, Lists of Food Plants and Parasites, etc., Prefaced by a Discussion on Nomenclature, and an Analytical Table of Families. Vol. I. Cimicidae. Felix L. Dames, Berlin, Germany. 392 pp.

Kobayashi, T. 1953. Developmental stages of some species of the Japanese Pentatomoidea, II. Sci. Rep. Matsuyama Agric. Coll. 11: 73–89.

1965. Developmental stages of *Urochela* and an allied genus of Japan (Hemiptera: Urostylidae). (The developmental stages of some species of the Japanese Pentatomoidea, XIII). Trans. Shikoku Entomol. Soc. 8: 94–104.

Kumar, R. 1969. Morphology and relationships of the Pentatomoidea (Heteroptera). III. Natalicolinae and some Tessaratomidae of uncertain position. Ann. Entomol. Soc. Amer. 62: 681–695.

1974. A key to the genera of Natalicolinae Horvath, with the description of a new species of Tessaratominae Stål and with new synonymy (Pentatomoidea: Heteroptera). J. Nat. Hist. 8: 675–679.

Lefroy, H. M. 1909. Indian Insect Life. Today and Tomorrow's Printers and Publishers, New Delhi, India. 786 pp.

Lodos, N. 1969. Minor pests and other insects associated with *Theobroma cacao* L. in Ghana. Ghana J. Agric. Sci. 2: 61–72.

Lorenzi, H. 1992. Árvores Brasileiras. Manual de Identificação de Plantas Arbóreas Nativas do Brasil. Editora Plantarum Ltda., Nova Odessa, São Paulo, Brazil. 352 pp.

Meshram, P. B., S. C. Pathak, and Jamaluddin. 1990. A new report of *Aspongopus brunneus* Thunberg (Hemiptera: Pentatomidae) as a pest on karanj (*Pongami pinata*). Indian For. 116: 926.

Miller, N. C. E. 1955. New genera and species of Plataspidae Dallas 1851 (Hemiptera–Heteroptera). Ann. Mag. Nat. Hist. (12) 8: 576–596.

Mishra, R. K., and R. N. Sharma. 1989. Host preference by adult *Aspongopus janus* F. Indian J. Entomol. 51: 480–481.

Moizuddin, M., and I. Ahmad. 1975. Some aspects of internal anatomy of *Coptosoma noualhieri* Mont (Pentatomoidea: Plataspidae) with reference to phylogeny. Rec. Zool. Surv. Pakistan 7: 101–110.

Oliff, A. S. 1892. Entomological notes. Bronzy orange bug (*Oncoscelis sulciventris* Stål). Agric. Gaz. New South Wales 3: 368–370.

Packauskas, R. J., and C. W. Schaefer. 1998. Revision of the Cyrtocoridae (Hemiptera: Pentatomoidea). Ann. Entomol. Soc. Amer. 91: 363–386.

Prestwich, G. D. 1976. Composition of the scents of eight East African hemipterans, nymph-adult chemical polymorphism in coreids. Ann. Entomol. Soc. Amer. 69: 812–814.

Rastogi, S. C. 1961. Salivary proteases and pH of the blood in the nymphal stages of *Coridius janus* Fabr. Sci. Cult. 27: 249–250.

1962a. The status of ileum in Heteroptera. Sci. Cult. 28: 31–32.

1962b. On the salivary enzymes of some phytophagous and predaceous Heteroptera. Sci. Cult. 28: 479–480.

1965. The food pump and associated structures in *Coridius janus* (Fabr.) (Heteroptera: Dinidoridae). Proc. R. Entomol. Soc. London (A) 40: 125–134.

Rastogi, S. C., and A. K. Datta Gupta. 1962. The pH and the digestive enzymes of the red pumpkin bug, *Coridius janus* (Fabr.) (Heteroptera: Dinidoridae). Proc. Zool. Soc. [India] 15: 57–64.

Ren Shu-zhi. 1984a. New species of Urostylidae from Hengduan Mountains in southwest China (Hemiptera: Heteroptera). Entomotaxonomia 6: 9–13 [in Chinese, with English summary].

1984b. Studies on the fine structure of egg-shells and the biology of *Megacopta* Hsiao et Jen from China (Hemiptera: Plataspidae). Entomotaxonomia 6: 327–332 [in Chinese, with English summary].

1985. Fine surface structure of eggs and classification of five species of Coptosoma Laporte (Hemiptera, Plataspidae). Anim. Mondo 2: 235–242 [in Chinese, with English summary].

1986a. New species and new record of Urostylidae from China (Hemiptera: Heteroptera). Entomotaxonomia 8: 133–138 [in Chinese, with English summary].

1986b. Notes on new species of *Urochela* Dallas and *Urostylis* Westwood from Yunnan (Heteroptera: Urostylidae). Zool. Res. [China] 7: 139–146 [in Chinese, with English summary].

Rolston, L. H., R. L. Aalbu, M. J. Murray, and D. A. Rider. 1993. A catalog of the Tessaratomidae of the world. Papua New Guinea J. Agric. For. Fish. 36: 36–108.

Rolston, L. H., D. A. Rider, M. J. Murray, and R. L. Aalbu. 1996. Catalog of the Dinidoridae of the world. Papua New Guinea J. Agric. For. Fish. 39: 22–101.

Sands, D. P. A., and M. T. Coombs. 1999. Evaluation of the Argentinian parasitoid, *Trichopoda giacomellii* (Diptera: Tachinidae), for biological control of *Nezara viridula* (Hemiptera: Pentatomidae) in Australia. Biol. Control 15: 19–24.

Schaefer, C. W. 1987. Food plants of the Dinidoridae (Hemiptera): a further note. Phytophaga 1: 161–162.

1988. The food plants of some "primitive" Pentatomoidea (Hemiptera: Heteroptera). Phytophaga 2: 19–45.

Schaefer, C. W., and I. Ahmad. 1987. The food plants of four pentatomoid families (Hemiptera: Acanthosomatidae, Tessaratomidae, Urostylidae, and Dinidoridae). Phytophaga 1: 21–34.

Schaefer, C. W., and P. D. Ashlock. 1970. A new genus and species of Saileriolinae (Hemiptera: Urostylidae). Pac. Insects 12: 629–639.

Schuh, R. T., and J. A. Slater. 1995. True Bugs of the World (Hemiptera:Heteroptera). Classification and Natural History. Cornell University Press, Ithaca, New York, U.S.A. 336 pp.

Sinclair, D. P. 1989. A cladistic, generic revision of the Oncomeridae Stål n. stat. and Tessaratomidae Schilling n. stat. (Hemiptera: Heteroptera: Pentatomoidea). Ph.D. dissertation, University of Sydney, Sydney, Australia. 324 pp.

Tachikawa, S. 1991. Studies on the subsociality of Japanese Heteroptera. Tokyo Agric. Press, Tokyo, Japan. 167 pp. [in Japanese].

Verma, A., S. K. Srivastava, and T. B. Sinha. 1978. Seasonal occurrence in the population of different pests of *Calotropis procera*. Indian J. Entomol. 40: 204–210.

Yang, We.-I. 1934. Revision of Chinese Plataspidae. Bull. Fan Mem. Inst. Biol. 5: 137–235.

1935. Notes on the Chinese Tessaratominae with description of an exotic species. Bull. Fan Mem. Inst. Biol. 6: 103–144.

1936. The outbreak of the *Urochela distincta* Distant in Lushan. Bull. Fan Mem. Inst. Biol. 7: 57–61.

1938. A new method for the classification of urostylid insects. Bull. Fan Mem. Inst. Biol. (Zool. Ser.) 8: 35–48.

1939. A revision of Chinese urostylid insects (Heteroptera). Bull. Fan. Mem. Inst. Biol. (Zool. Ser.) 9: 5–66.

CHAPTER **16**

Flat Bugs (Aradidae)

Kari Heliövaara

1. INTRODUCTION

The flat bugs of the family Aradidae constitute a well-defined family of the Aradoidea (Gyllensvärd 1964) and contain eight closely related subfamilies: Chinamyersiinae, Calisiinae, Aradinae, Isoderminae, Prosympiestinae, Aneurinae, Carventinae, and Mezirinae (Usinger and Matsuda 1959, Vásárhelyi 1987). The only economically significant species belongs to the genus *Aradus*, whose members are usually called the bark bugs. Most species of flat bugs are small, usually dark, black or brown, and, as their name suggests, they are dorsoventrally very flat. Usinger and Matsuda (1959) provide a systematic account of the family, and Kormilev and Froeschner (1987) a catalog.

The most important factor affecting the behavior of Aradidae is their mycetophagous habit. Their flattened bodies, short legs, and short, thick antennae enable them to enter under loose bark in the first years of tree decay when the mycelia of the higher fungi abound. In the anterior part of the head (clypeus) the bugs have coiled setae composed of mandibulary and maxillary stylets. In the resting position both stylets are coiled up to form a compact double spiral (Weber 1930), whereas when the bug is feeding the length of the uncoiled stylets may be as much as five to six times that of the insect.

Most aradids live under the bark of old, dead, or dying trees attacked by fungi. They usually have very specialized nutrient and microenvironmental requirements; some require or at least favor burned trees; others live in old, virgin forests. Species of this mycetophagous family are probably closely associated with particular fungi that occur under bark. The presence of a particular flat bug species may be linked to the stage of succession and deterioration of the log or tree that provides a suitable microhabitat for specific fungal development.

Being rather uncommon or at least only locally abundant insects, only a few specimens are known of a large number of species (Gyllensvärd 1964, Vásárhelyi 1976). Several species have suffered from modern silvicultural practices. For instance, there are 21 species of flat bugs in Sweden and 18 species in Finland, but in both countries as many as 11 species have been classified endangered (Ehnström et al. 1993).

The genus *Aradus* has a preponderantly Holarctic distribution (Heiss 1994). Although flat bugs exist even in countries with high degrees of agriculture and forestry, both the number of individuals and species seem to be highest in the large primary taiga forests of Russia and eastern Europe.

2. GENERAL STATEMENT OF ECONOMIC IMPORTANCE

The fungal feeding of aradids has been confirmed repeatedly, and is of no economic importance. However, the only economically important species, the pine bark bug *A. cinnamomeus* Panzer, pushes its stylets into phloem, cambium, and xylem tissues of living pine saplings, disturbing their growth. However, it is possible that even this species utilizes some fungi living on pine (Usinger and Matsuda 1959).

3. THE IMPORTANT SPECIES

3.1 *Aradus cinnamomeus* Panzer

3.1.1 Distribution, Life History, and Biology

The pine bark bug, *A. cinnamomeus* Panzer, is actually a complex of three species. Two species (*A. antennalis* Parshley, *A. kormilevi* Heiss) have a Nearctic distribution and one species, the true *A. cinnamomeus*, the pine bark bug, has a west Palearctic distribution (Heiss 1980). Only the true *A. cinnamomeus* is known to be a pest of pines.

Adult pine bark bugs are 3.5 to 4.5 mm long but only 0.75 mm thick. Adults are brown, but small nymphs are bright red. When disturbed, pine bark bugs secrete a characteristic scent of pear. *Aradus cinnamomeus* differs from other species of the family in displaying exceptional wing polymorphism. Most females (~97%) are brachypterous. The proportion of macropterous females is usually 3%, but it may increase as living conditions on the host tree worsen (Heliövaara 1984a). Males are "stenopterous"; i.e., their hemelytra almost reach the tip of the abdomen but are greatly narrowed in their apical three-quarters. Only the few macropterous females can fly. Migration takes place in August on warm, sunny days, just after the final molt (Heliövaara 1984a).

Aradids in general have one or two generations per year, but *A. cinnamomeus* has a 2-year life span in most parts of Europe; in northern Europe close to the tree limit, the life span is 3 years (Brammanis 1975, Heliövaara and Väisänen 1987). Eggs are laid in May, nymphs hatch in June, and reach the fourth instar during the first summer. After hibernation at the base of the tree or in surrounding litter (Brammanis 1975, Heliövaara 1982), the bugs become adult in July to August, after which they hibernate. It is not until the following spring that the bugs mate and begin egg-laying (Tropin 1949, Brammanis 1975, Heliövaara 1984b, Vásárhelyi and Böröcz 1987).

The 2-year life span of the pine bark bug has led to periodicity and a unique biogeographic pattern of distribution. In most of Europe, there seem to be two major alternate-year populations that live in different geographic areas (Heliövaara and Väisänen 1984, 1987, 1988). For instance, the bugs reproduce in even-numbered years in eastern Finland and eastern Sweden, and in odd-numbered years in the western parts of these countries. In each area, however, there always seems to be a very small proportion of bugs (fewer than 1/1000) that reproduce in off years (Heliövaara et al. 1994).

3.1.2. Damage

The pine bark bug is a pest of pines, especially the Scots pine (*Pinus sylvestris* L.). It has been occasionally recorded on several other pine species, such as *P. mugo* Turra, *P. nigra* Arnold, *P. strobus* L., and *P. contorta* Douglas ex Loudon (Brammanis 1975) and even on *Larix* species (Turček 1964). The species seems to favor areas characterized by poor soil and sparse undergrowth. In light, young, dry, upland pine forests hundreds or even thousands of bugs can be found on a single stem (Tropin 1949, Brammanis 1975).

Pine bark bugs live in crevices in the bark of the trunk. Both adults and nymphs suck sap with their highly specialized piercing mouthparts from the young tissues surrounding the cambium, thus disturbing the conduction of fluids in the tree. Presence of the bugs in large numbers causes noticeable growth retardation, yellowing of needles, and withering of the pine (Brammanis 1975, Heliövaara 1984b). The damaged tissue can be clearly seen as dark areas in cross sections of the trunk of an affected tree. Part of the damage caused by the bugs evidently results from the loss of sap and nutrients from the host tree but, because the attack symptoms often appear to be more prominent than the number of bugs would suggest, the involvement of other factors such as a salivary toxin has been suggested. Careful analysis of the free amino acids of the bugs has revealed that the concentration of glutamic acid is low in the hemolymph but high in the salivary glands. However, no distinctly phytotoxic compounds have been identified in these insects so far (Heliövaara and Laurema 1988).

In central and eastern Europe the pine bark bug has long been regarded as a harmful pest of pines (Sajó 1895, Krausse 1919, Strawinski 1925, Vásárhelyi 1983). It has caused serious damage, especially in pine stands planted in nitrogen-poor sandy soils in the Ukraine, Byelorussia, Latvia, and Poland (Stark 1933, Tropin 1949, Ozols 1960). The damage typically continues for several years in a given area without the appearance of any striking symptoms. The yield of the stand is gradually reduced or, in the worst case, the whole stand must be renewed. In Lithuania the area of forest chronically damaged by this pest has been estimated to be at least 1000 to 1300 ha (Valenta et al. 1980). The pine bark bug has also caused problems in the Netherlands (Doom 1976, Cobben 1987), Finland (Hokkanen et al. 1987), and Sweden, where several hundred hectares of young pine stands had to be felled because of this pest (Brammanis 1975). It also occurs as a minor pest in Norway (Pettersen 1975) and Spain (Muñoz and Soria 1986).

3.1.3 Control

Generally, 25% of the laid eggs are parasitized by a scelionid wasp, *Telenomus aradi* Kozlov (Heliövaara et al. 1982). Parasitoids of nymphs or adults are not known. Other invertebrate predators include *Raphidia* (Neuroptera) larvae, a bdellid mite *Bdella longicornis* (L.), and spiders such as *Philodromus fuscomarginatus* De Geer, *Clubiona subsultans* Thorell, *Drapetisca socialis* (Sundevall), *Micaria subopaca* Westring, *Salticus cingulatus* (Panzer), and *Moebilia penicillata* (Westring). However, spiders only rarely seem to eat pine bark bugs (Hokkanen et al. 1987).

Several chemical and biological methods including systemic insecticides and fungal diseases have been used with variable results in the control of *A. cinnamomeus*. In Russia, the insecticides have generally been used in winter, chemicals being spread on the hibernation places of the nymphs, which are more susceptible than the adults (Brammanis 1975, Valenta et al. 1980). The bug seems to be a tenacious species highly resistant to chemicals, and the results obtained with a wide range of chemicals have not been encouraging. Systemic insecticides have recently been used with better results. Both dimethoate and lindane treatments decreased the bug density, but the recovery of height growth of the treated trees was not clearly noticeable (Heliövaara et al. 1983). No real progress has yet been achieved in the biological control of the pine bark bug. Fungal diseases, especially *Beauveria* species, have increased the mortality of the bugs on a few occasions (Smirnov 1954, Turček 1964).

Forest fertilization has been tested as a control method against the pine bark bug, but the results are contradictory. Valenta et al. (1980) found that fertilizers increase the secretion of resin and change the level of carbohydrates, causing death of the bugs. Brammanis (1975) noticed a strong decrease in the number of bugs in fertilized areas and recommended forest fertilization for control. In a more detailed investigation, nitrogen fertilization increased both the height growth of the pines and the bug population (Heliövaara et al. 1983), which is a characteristic response in sap-sucking insects. However, the usage of fertilization could be justified in that the most susceptible phase of the stand is passed more rapidly.

Because no single effective method of controlling *A. cinnamomeus* is yet available, the use of several prophylactic silvicultural practices probably gives the most favorable results. As the bugs thrive best in warm and light conditions, high stocking densities should be used in pine stands in potential damage areas. The structure of a resistant pine stand should be even and dense without gaps, and heavy early thinnings should be avoided (Brammanis 1975, Heliövaara 1984b, Hokkanen et al. 1987). Mixed species stands (*Pinus*, *Betula*) should be favored when possible. The influence of birches has been partly attributed to the shading effect, and partly to the litter formed from birch leaves. This litter is very favorable for many pathogens, and the bugs overwintering in it are easily infected and destroyed.

4. PROGNOSIS

During recent decades the pine bark bug has increased in abundance. The increase in number and probably in the expansion of the bug's distribution is apparently a result of supplementary planting, favoring the Scots pine in the regeneration of forests, construction of highways, air pollution, and many other kinds of human activities (Heliövaara and Väisänen 1983, 1986). Although the bug is very abundant in suitable habitats, it has been controlled only locally in small areas. Recently adopted silvicultural practices favoring biodiversity probably decrease the probability of damage caused by the pine bark bug, but the potentially warming climate works in an opposite direction.

5. REFERENCES CITED

Brammanis, L. 1975. Die Kiefernrindenwanze, *Aradus cinnamomeus* Panz. (Hemiptera–Heteroptera). Ein Beitrag zur Kenntnis der Lebensweise und der forstlichen Bedeutung. Stud. For. Suec. 123: 1–81.

Cobben, R. H. 1987. *Aradus signaticornis* in Nederland, met opmerkingen ovar enkele andere met *Pinus* geassocieerde Hemiptera (Heteroptera: Aradidae; Homoptera: Cercopidae). Entomol. Ber. 47: 33–38.

Doom, D. 1976. Über Biologie, Populationsdichte und Feinde der Kiefernrindenwanze, *Aradus cinnamomeus*, an stark geschädigten Kiefern. Z. Pflanzenkr. Pflanzenschutz 83: 45–52.

Ehnström, B., U. Gärdenfors, and Å. Lindelöw. 1993. Rödlistade evertebrater i Sverige 1993 (Red data book on insects in Sweden 1993). Databanken för Hotade Arter, Uppsala, Sweden, 69 pp.

Gyllensvärd, N. 1964. A key to Swedish Aradidae (Hem. Het.) with figures of the male genitalia. Opusc. Entomol. 29: 110–116.

Heiss, E. 1980. Nomenklatorische Änderungen und Differenzierung von *Aradus crenatus* Say, 1831, und *Aradus cinnamomeus* Panzer, 1806, aus Europa und U.S.A. (Insecta: Heteroptera, Aradidae). Ber. Nat. Med. Ver. Innsbruck 67: 103–116.

1994. *Aradus pericarti*, a new species from South America (Heteroptera, Aradidae). J. New York Entomol. Soc. 102: 102–106.

Heliövaara, K. 1982. Overwintering sites of the pine bark-bug, *Aradus cinnamomeus* (Heteroptera, Aradidae). Ann. Entomol. Fenn. 48: 105–108.

1984a. Alary polymorphism and flight activity of *Aradus cinnamomeus* (Heteroptera, Aradidae). Ann. Entomol. Fenn. 50: 69–75.

1984b. Ecology of the pine bark bug, *Aradus cinnamomeus* (Heteroptera, Aradidae). A forest entomological approach. Univ. Helsinki, Dept. Agric. Forest Zool., Rep. 7: 1–38 + appendix.

Heliövaara, K., and S. Laurema. 1988. Interaction of *Aradus cinnamomeus* (Heteroptera, Aradidae) with *Pinus sylvestris*: the role of free amino acids. Scand. J. For. Res. 3: 515–525.

Heliövaara, K., and R. Väisänen. 1983. Environmental changes and the flat bugs (Heteroptera, Aradidae and Aneuridae). Distribution and abundance in Eastern Fennoscandia. Ann. Entomol. Fenn. 49: 103–109.

1984. The biogeographical mystery of the alternate-year populations of *Aradus cinnamomeus* (Heteroptera, Aradidae). J. Biogeogr. 11: 491–499.

1986. Industrial air pollution and the pine bark bug, *Aradus cinnamomeus* Panz. (Het., Aradidae). Z. Angew. Entomol. 101: 469–478.
1987. Geographic variation in the life-history of *Aradus cinnamomeus* and a breakdown mechanism of the reproductive isolation of allochronic bugs (Heteroptera, Aradidae). Ann. Zool. Fenn. 24: 1–17.
1988. Periodicity of *Aradus cinnamomeus* (Heteroptera, Aradidae) in northern Europe. Entomol. Tidskr. 109: 53–58.
Heliövaara, K., E. Terho, and E. Annila. 1983. Effect of nitrogen fertilization and insecticides on the population density of the pine bark bug, *Aradus cinnamomeus* (Heteroptera, Aradidae). Silva Fenn. 17: 351–357.
Heliövaara, K., E. Terho, and M. Koponen. 1982. Parasitism in the eggs of the pine bark-bug, *Aradus cinnamomeus* (Heteroptera, Aradidae). Ann. Entomol. Fenn. 48: 31–32.
Heliövaara, K., R. Väisänen, and C. Simon. 1994. Evolutionary ecology of periodical insects. Trends Ecol. Evol. 9: 475–480.
Hokkanen, T., K. Heliövaara, and R. Väisänen. 1987. Control of *Aradus cinnamomeus* (Heteroptera, Aradidae) with special reference to pine stand condition. Commun. Inst. For. Fenn. 142: 1–27.
Kormilev, N. A., and R. C. Froeschner. 1987. Flat bugs of the world. A synonymic list (Heteroptera: Aradidae). Entomography 5: 1–246.
Krausse, A. 1919. Über *Aradus cinnamomeus* Panz., die Kiefernrindenwanze. Z. Angew. Entomol. 5: 134–136.
Muñoz, M. C., and Y. Soria. 1986. *Aradus cinnamomeus* (Panzer) (Hem. Heteroptera) un factor de debilitación del *Pinus sylvestris* L. en el Sistema Central. Bol. Serv. Plagas 12: 163.
Ozols, G. E. 1960. Vrediteli sosnovyh kul'tur na primorskih djunah rizkogo zaliva. (Pests of pine stands on the coastal dunes of the Riga Bay). Zool. Zhurnal 39: 63–70 [in Russian, English summary].
Pettersen, H. 1975. Furubarktegen, *Aradus cinnamomeus* Panz. (Hemiptera, Heteroptera) — et lite kjent skadeinsekt på skog i Norge [The pine flatbug, *Aradus cinnamomeus* Panz. (Hemiptera, Heteroptera) — a little known pest in Norwegian forestry]. Tidsskr. Skogbr. 2: 259–269 [in Norwegian, English summary].
Sajó, K. 1895. Über Insektenfeinde von *Pinus sylvestris* und *P. austriaca*. Z. Pflanzenkr. 5: 129–134.
Smirnov, B. A. 1954. Microbiological methods in the control of the pine bark bug. Lesn. Hozjajstvo 12: 72 [in Russian].
Stark, V. N. 1933. Materialy k izukeniju klopov roda Aradus europejskojtajgi (Hemiptera, Aradidae). [Beitrag zur Kenntnis der Aradus-Arten der europäischen Taiga (Hemiptera, Aradidae)]. Entomol. Obozr. 25: 69–82 [in Russian, German summary].
Strawinski, K. 1925. Historja naturalna korowca sosnowego, *Aradus cinnamomeus* Pnz. (Hemiptera-Heteroptera). I. (Natural-History of *Aradus cinnamomeus* Pnz. I.). Rep. Inst. Forest Prot. Entomol. 2(1): 1–51 [in Polish, English summary].
Tropin, I. V. 1949. The Pine Bark Bug and Its Control. Goslesbumizdat, Moscow, Russia. 55 pp. [in Russian].
Turček, F. J. 1964. Beiträge zur Ökologie der Kiefernrindenwanze *Aradus cinnamomeus* Panz. (Heteroptera, Aradidae). Biologia (Bratislava) 19: 762–777.
Usinger, R. L., and R. Matsuda. 1959. Classification of the Aradidae (Hemiptera-Heteroptera). British Museum (Natural History), London, U.K. 410 pp.
Valenta, V. T., A. K. Zegas, and G. K. Kilikjavitsjus. 1980. The pine bark bug and the ways to protect young pine stands. Kaunas-Girionis, 12 pp. [in Russian].
Vásárhelyi, T. 1976. Notes on the genus *Aradus* Fabricius, 1803 (Heteroptera: Aradidae). Acta Zool. Acad. Sci. Hung. 22: 198–195.
1983. Egy elfelejtett kartevo, az *Aradus cinnamomeus* Panzer (Heteroptera: Aradidae) eletmodja Magyarorszagon. Summary: Bionomy of a forgotten pest, the pine bark-bug (*Aradus cinnamomeus* Panzer) in Hungary (Heteroptera: Aradidae). Allattani Közlemenyek 70: 91–97 [in Hungarian, English summary].
1987. On the relationships of the eight aradid subfamilies (Heteroptera). Acta Zool. Hung. 33: 263–267.
Vásárhelyi, T., and P. Böröcz. 1987. Studies on the life-history of *Aradus cinnamomeus* in Hungary (Heteroptera: Aradidae). Folia Entomol. Hung. 48: 233–239.
Weber, H. 1930. Biologie der Hemipteren. Biologische Studienbücher, XI. J. Springer, Berlin, Germany. 543 pp.

CHAPTER **17**

Bed Bugs (Cimicidae)

Carl W. Schaefer

1. INTRODUCTION

"The family Cimicidae includes the human bed bugs, the bat bugs, chicken bugs, swallow bugs, pigeon bugs, and others for which no common names have been proposed" (Usinger 1966a, p. 1), and contains an estimated 100 species. Ryckman et al. (1981) provide a checklist to and bibliography of the species of the New World. The definitive study of the family is Usinger's monograph (1966a), a classic not merely in the literature of Heteroptera, but of Insecta. Its 14 chapters cover everything from the etymology of *bug* through the medical significance of these insects, to their embryology, morphology, physiology, and systematics. There are also chapters on the strange reproductive system by Carayon, on reproductive physiology by Davis, and on cytogenetics by Ueshima and McKean. The book is so thorough and covers the literature so completely (580 references), and is moreover so available, that this chapter will treat this family lightly, and emphasize the more recent literature. A briefer and more recent review is by Forattini (1990).

All members of the family are temporary ectoparasites on a limited array of warm-blooded vertebrates: bats, people, and birds that roost together. The bugs live separately from their hosts, moving to them only to feed; other aspects of the bugs' lives are spent in cracks and crevices in the walls of caves or birdhouses or humans' rooms. All stages take vertebrate blood as their only food. The bugs may accumulate in large numbers and as a result the loss of blood may be serious; this is the chief, perhaps the only, damage done.

Cimicids are rather small, 3 to 13 mm, wingless even as adults, flattened, rounded to oval in general shape, with asymmetric male genitalia and an asymmetric abdominal "pouch" for the reception of sperm in the female (except in the most primitive genus). Cimicids range in color from pale brown to dark brown, some being a reddish brown. The family is distributed worldwide, mostly in the tropics and subtropics; several economically important species occur in temperate regions; and, of course, the bed bug itself occurs wherever there are humans. The ten species known from India are cataloged with their hosts by Bhat (1974), and the eight African genera are keyed by Coetzee and Segerman (1992). Schofield (1994) suggests the abundance of cimicids in Africa is correlated with — and may explain why — no *reduvii*ds in Africa have become hematophagous. Finally, Ueshima (1968) describes several new species, and gives updated keys to the species of *Cimex*, *Cacodmus*, *Stricticimex*, and *Leptocimex*.

The morphology, biology, and physics of feeding in bloodsucking insects, including cimicids, have been described in several papers. Such insects can ingest much blood in a short time, an ability made necessary by the infrequency of feeding opportunities. Bed bugs (*Cimex lectularius*

L.) can fully engorge in 2.5 (first instars) to nearly 5 min (adult male), taking up from 1.8 to 24 µg of blood per second, the blood meal weighing from 3.7 (first instars) to 4.9 (fifth instar) times the body weight of the unfed insect (Tawfik 1968). Two early but very useful reviews of blood feeding by arthropods (including bed bugs) are those by Hocking (1971) and Friend and Smith (1977). The mechanics of feeding, especially the means by which fluids containing large particles (blood cells) are moved through narrow tubing, is described by Dickerson and Lavoipierre (1959), Tawfik (1968), Daniel and Kingsolver (1983), Kingsolver and Daniel (1993), and Loudon and McCulloh (1999). Valenzuela et al. (1996a,b; 1998) have worked out in the bed bugs the biochemistry of blood-clotting inhibition in the host.

That molting and egg production are both triggered by — and require — a blood meal, is by now well known (e.g., references in Usinger 1966a; Davis 1964; 1965a,b; Rao and Davis 1969). The physiological and biochemical relationship between feeding and hormone release is reviewed by Adams (1999), who, however, considers *Rhodnius prolixus* Stål (Reduviidae: Triatominae) and *Aedes aegypti* (L.) (Diptera: Culicidae) and not Cimicidae; the principles remain much the same, however.

Reproduction in Cimicidae (and in certain related groups) is unique, to the point of being bizarre; it is thoroughly described by Carayon (1966a, 1969, 1977); its morphology is described by Davis (1956) and its control by Davis (1964; 1965a,b; 1966) and Rao and Davis (1969). *Briefly*: One clasper of the male slices into the female's abdomen and deposits sperm; in primitive cimicids the point of puncture and of deposition appear to be random, whereas in more advanced cimicids (including the bed bugs), a pad of tissue receives the sperm after the clasper has penetrated the overlying cuticle. Following this "traumatic insemination," the sperm mass, activated by the seminal fluid (Davis 1965a, Rao and Davis 1969), then moves to the base of the ovarioles, where the sperm is stored (Davis 1964). Variations of this process doubtless occur in other cimicid species. The biochemistry of sperm migration has been worked out by Davis (references above) and Rao and Davis (1969) for *C. lectularius* and some aspects for *C. hemipterus* (F.) by Ruknudin and Raghavan (1988).

Several methods and devices for rearing bed bugs on blood or blood components have been described (Davis 1956, Wattal and Kalra 1961, Gilbert 1964, Usinger 1966a, Hall et al. 1979, Ogston and Yanovski 1982, Bell and Schaefer 1966) [*Note:* the "blood" mentioned in the first paper is doubtless bed bug excreta]. Lee (1955) and Milward de Azevedo and de Oliveira (1980) describe a device for feeding bird cimicids.

2. GENERAL STATEMENT OF ECONOMIC IMPORTANCE

Cimicids have yet to be definitely associated with a disease pathogen; Burton (1963a) lists many pathogens that have survived up to nearly a year in bed bugs, but there is no proof of any transmission. One possible exception is the hepatitis B virus (see **Section 3.1.2.2**). Moore et al. (1991), reviewing invertebrate viruses, write, "[s]urprisingly, there are only two reports to suggest that bed bugs may transmit human pathogenic viruses" (p. 638): Kaeng Khoi bunyvirus in Thailand (see **Sections 4.2, 4.3**) and hepatitis B virus in Senegal (Moore et al. 1991); since then, there have been other reports of the latter virus in Africa (see **Section 3.1.2.2**).

Human immunodeficiency virus (HIV) was retained in *C. hemipterus* for a week after taking it up from virus highly concentrated in blood meals. Yet the virus did not replicate, was not detected in excreta, and was not transmitted (Webb et al. 1989). This bed bug is thus neither a biological nor a mechanical vector of HIV, and it appears that *C. lectularius* is not either (Jaenson 1985).

The chief economic importance of cimicids is the blood they draw from vertebrates of interest to humans (including humans themselves). Because they are secretive and usually not seen, populations of cimicids may build up, to the point that the resulting blood loss seriously weakens the hosts. Meat and egg production of domestic poultry, for example, may decrease. It is not impossible that humans — especially the very young and the very old — may be sufficiently weakened by loss of blood that they become susceptible to diseases that might ordinarily have had little effect.

In addition to the species considered below, people entering bat caves are occasionally bitten by cimicids not usually associated with humans (e.g., Overal and Wingate 1976); a few of these cases are mentioned in **Chapter 19**.

3. THE IMPORTANT SPECIES

3.1 *Cimex lectularius* L.

3.1.1 *Distribution, Life History, and Biology*

The bed bug is perhaps the most widely known insect. Usinger (1966a) recounts the depth in time and the breadth in geographic space of peoples' acquaintance with the bed bug; every language has a word for it, and references to it go back 2500 years at least (Usinger 1966a). Some names have forgotten political implications, such as "Norfolk-Howards" (see Sailer 1952). Indeed, the word *bug* itself refers to the bed bug, derived as it is from the Middle English *bugge*, originally meaning "wraith" or "ghost" (compare *bugbear, bogeyman*, etc.). Of course, this is how bed bugs are perceived — or, rather, not perceived: They slip onto the sleeping host, feed, leave, and are gone; awaking, the person is aware only of itching welts and, sometimes, a faint spicy odor (or, in England, odour). Clearly, a wraith, a *bugge*, has visited in the night.

The bed bug, *C. lectularius*, is one of very few insects associated only with humans, until a single discovery of a population in an Afghanistan bat cave, in 1965 (Usinger and Povolný 1966). This discovery reinforces the idea that the ancestors of bed bugs lived in bat caves in the Near East and there became associated with humans, perhaps as humans moved in time from natural shelters (such as caves) to villages to cities, in the Fertile Crescent. The evidence for this view, outlined in Usinger and Povolný (1966) and Usinger (1966a), is biological, morphological, cytogenetic, and even etymological. (Opposed to this view, I believe, is the improbable claim that until the 1980s the bed bug was unknown in Iraq; Abul-Hab et al. 1989.) In this regard, the discovery by Roer (1969, 1975) of *C. lectularius* in bat caves near human dwellings may be significant.

In general, *C. lectularius* occurs in temperate or subtropical regions, and is replaced in the tropics by *C. hemipterus*, but there is considerable overlap and some hybridization (see **Section 3.2.1**). *Cimex lectularius* may be increasing in numbers at the expense of *C. hemipterus* in Brazil, especially in the more temperate south, perhaps because of the influx of European immigrants (brief discussion in Nagem 1985). The reverse seems to be occurring in parts of South Africa: *C. hemipterus* may be replacing *C. lectularius* (see **Section 3.2.1**).

The bed bug will also hybridize in the laboratory with the pigeon bug, *C. columbarius* Jenyns, but fertility is low (Ueshima 1964); moreover, introgression does not occur (Usinger 1966a).

Despite the importance and ubiquity of *C. lectularius*, there are surprisingly few detailed studies of its morphology. The best general account is Ferris and Usinger (1957; paraphrased in Usinger 1966a). The internal and external genitalia have been described and discussed in marvelous detail by Davis (1956), the development of their asymmetry by Ludwig (1937), and the metathoracic scent gland system by Carayon (1966); other aspects of cimicid morphology have been described in other species (see subsequent sections). Ueshima (1966, 1967) considers chromosome structure and cytogenetics of the family.

Walpole (1987) describes in detail the legs and their sensory structures. The fine structure of the sensilla of the antenna was studied by Steinbrecht and Müller (1976), and the distribution of sensilla on the terminal antennal segments by Levinson et al. (1974a). The latter study was part of an investigation of the bed bug's alarm pheromone (Levinson et al. 1974a,b): This pheromone, produced when disturbed by both nymphs and adults (and therefore from different glands), causes other bugs to become agitated and, eventually, to disperse. Another scent, produced by adults only, induces aggregation (Levinson and Bar Ilan 1971). Bed bugs live in homes, usually hidden from sight. Heavy infestations or evidence thereof (often spots of excreta) may be clearly visible (King et al. 1989). To quote Groves (1967), who was writing about London:

> The bed bug is nocturnal in habit spending the daytime in cracks in walls, beneath skirting boards, behind wallpaper, and in corners of mattresses and beds, etc., in rooms used by man for sleeping quarters, coming out at night in search of a blood meal from its host. In spite of generally improved living conditions nowadays the bed bug is still present in slums and other unhygienic dwellings [*Note:* Others insist that bed bugs are less class conscious than this: for example, Negromonte et al. (1991) found no correlation between infestation and socioeconomic position.] throughout the London Area. The paucity of available records is however due, in the main, to the reluctance with which the presence of the bug is declared or admitted by the owners of such properties to the Public Health authorities.

The number of bugs is independent of the number of persons in the home (Newberry and Jansen 1986), and presumably increases with readily available and plentiful food, and the length of time without control.

That the bugs are hardy increases the difficulty of their control. Bed bug eggs survive a month at temperatures of 40 to 50°F, and 2 days at 28 to 32°F (10 to 15 days is fatal); newly hatched bugs can survive 28 to 32°F for up to 18 days, can endure such chilling, thawing, rechilling for shorter lengths of time, and can withstand 96°F for a week; high humidity is worse for them than high (or low) temperatures. (But others — e.g., Usinger 1966a, Marx 1955 — suggest humidity is only a minor factor in the bugs' lives.) Eggs can last up to a week buried in wet or dry sand (45 to 50°F) or submerged in water (60 to 63°F), and for 2 days in ice (Bacot 1914). After a single blood meal, the various stages of *C. lectularius* can survive without feeding at 18°C (and 37°C) as follows: first instar, 114 (17); second, 171 (30); third, 214 (35); fourth, 234 (37); fifth, 161 (33); adult female, 277 (32); adult male, 176 (29) days (Usinger 1966a). The debilitating effect of high temperatures may be the result of inactivation of rickettsial symbionts (Chang 1974).

The vast literature on bed bugs' biology and ecology is summarized by Usinger (1966a, Chapter 2; see also Johnson 1941); it will only be outlined here. Normal development occurs between about 14°C and about 36°C (see paragraph above). At 18°C the preovipositional period is about 8 days and at 33°C about 3 days. The various stages at 18°C (and at 33°C) last as follows: egg, 21 (4) days; first stadium, 19 (4); second, 18 (4); third, 17 (6); fourth, 19 (8); fifth, 26 (8).

Molting and oviposition must be preceded by a blood meal. Occasionally a few eggs develop without a meal (autogeny), the eggs having developed from reserves "left over" from the fifth stadium. Different stages take different amounts of blood; in general, instars take nearly four times their own weight at each feeding, males about 1.5 times, and females somewhat more than twice their own weight. Nymphs may breed a day after molting (Usinger 1966a) and gain from their feeding an increase of about 30% of their body weight (Johnson 1960).

Although humans are the usual hosts, bed bugs may feed on other hosts. Johnson (1937) found mouse, chicken, and humans, in that order, yielded the most eggs; Usinger (1966a) writes that rabbit-reared bed bugs laid the most eggs when fed on rabbit, the least on pigeon, and chicken and humans (in that order) were intermediate; nymph to adult durations were shortest for chicken-fed bugs and longest for those fed on humans. [*Note:* Left to others is speculation on the phylogenetic significance of these differences.] The initial attractant for the bugs is warmth and carbon dioxide (Marx 1955), both given off in quantities by warm-blooded vertebrates.

In a natural infestation studied in England, females mated about once a week and fed (under the best conditions) about every 5 to 6 days; they laid an average of three eggs a day (Mellanby 1939), but the average longevity under these conditions is not reported, although it is probably 4 to 6 weeks. In a somewhat similar study in Minas Gerais, Brazil, 40 females yielded an average of 60 eggs over a 30-day period at an average temperature of 22°C (Nagem et al. 1988).

3.1.2. Damage

The bed bug's chief damage is the blood removed from the host, and the irritation caused by its feeding. Unpleasant also is the smell of bed bugs (although to some the spicy scent is not displeasing!). Of continuing concern too is the possibility that bed bugs may be, or become, vectors (biological or mechanical) of pathogens.

3.1.2.1 *Damage: as a Blood Feeder* — Because bed bugs feed on human blood, the damage they do depends upon the condition of the host. A large infestation may cause serious harm over time to very young children, to the very old, or to the sick; the damage is simply the unacceptable loss of blood. In addition, the itching welts raised are annoying. Another harm is less easily described. In many places the presence of bed bugs (like that of cockroaches) is considered to reveal "poor personal hygiene and bad housekeeping" (Reid 1990, p. 48), or a "socially deprived human habitation" (Moore et al. 1991, p. 638). This social stigma may be severe, although in truth the bed bug is an equal-opportunity feeder and may be found in the best of homes.

Bed bugs may also attack other hosts, especially poultry. It occurs only occasionally in poultry houses, but in the large mechanized ones its removal may be difficult because of the large size of the facility, the many available hiding places for the bugs, and the ample and close-packed poultry. The economic impact of bed bugs on poultry facilities has not been estimated, but include allergic reactions of workers, lowered egg and meat production, fecal spots on poultry eggs, and increased feed consumption. Just as bugs may be carried by workers to their homes, bugs may be transferred to other poultry facilities in flats and trays. (The above is taken from a review by Axtell and Arends 1990.)

Cimex lectularius has also been found on small mammals in a German zoological garden (Petzsch 1953). It may occur in similar situations elsewhere, but this author knows of no other reports.

3.1.2.2. *Damage: as a Vector of Pathogens* — Several attempts over many years have failed to show conclusively that *C. lectularius* is a consistent vector, biological or mechanical, of any human or domestic animal pathogen. A possible — indeed, probable — exception is hepatitis B virus (see below).

One semipopular circular states, "... bed bugs can carry the causal agents for a number of diseases (anthrax, plague, yellow fever, and typhus), [but] they are not considered to be important in transmitting these diseases" (Reid 1990, p. 48) (this rather overstates bed bugs' importance; C. W. Schaefer, personal comment).

In an early review of bed bugs as vectors of pathogens, Hase (1938) quotes others' work indicating that several pathogens may survive brief periods in bed bugs: pathogens of plague, yellow fever, typhus, tularemia, and a few others. The bed bug has been implicated as a vector of none of these, however. Nevertheless, Hase notes that the length of time some pathogens remain viable in bed bugs may create two problems: infected bugs from other places may reach uninfected areas, as travel times decrease; and the building of large apartment complexes in large cities ("Riesenwohnblöcke in den Grosstädten") may provide islands for the spread of infected bed bugs. [*Note:* Hase lamented the difficulty of getting current literature, a problem doubtless exacerbated by the time and place he was working (1938, in Germany): "Leider ist die fremdländische, namentlich die überseeische Literatur z.Z. ausserordentlich schwer, oder gar nicht beschaffbar ..." (Hase 1938, p. 3, footnote).]

The absence of subsequent reports of bed bugs as such vectors makes Hase's concerns less significant. Yet the potential would seem to remain, however remote, for bed bugs to be mechanical vectors of certain pathogens.

In a comprehensive review of human pathogens found in *C. lectularius* and *C. hemipterus* (see **Section 3.2.2**), Burton (1963a) listed 25 pathogens found often or rarely in the common bed bug; these ranged from viruses (poliomyelitis, yellow fever, smallox) to filarial nematodes (see below). However, after reviewing some 93 studies (of both bed bug species), Burton (1963a) concluded that there was no evidence of either biological or mechanical transmission of a pathogen to humans by bed bugs.

Several stages of the filarial worms *Brugia malayi* (India) and *Wucheraria bancrofti* (India and British Guiana) have been found in both *C. lectularius* and *C. hemipterus*, but these bugs almost certainly do not transmit these nematodes (Burton 1962, 1963b).

Three trypanosomatid protozoans of the same genus as *Trypanosoma cruzi* (the pathogen of Chagas' disease: see **Chapter 18**) occur in bats and were cultured *in vitro* and *in vivo* (Paterson and Woo 1984). None of these cultures was infective to nonbat mammals, and the authors hoped the system might be used to study Chagas pathogens safely.

For some time it has been thought that blood-feeding arthropods might transmit hepatitis B virus (see Newkirk et al. 1975). Although the evidence that other arthropods do so is slight at best (Jaenson 1985), bed bugs are able in the laboratory to transmit the virus mechanically, and have been implicated in such transmission in the field in Africa (Jupp and McElligott 1979, Jupp et al. 1982). (The virus undergoes no qualitative or quantitative change in the bug, so biological transmission does not occur; Jupp and McElligott 1979.) Fed to the bugs via membrane, the virus has been detected in the excreta of bed bugs (Jupp et al. 1982, Silverman et al., 1998); it is passed transstadially (from one instar to the next) (Newkirk et al. 1975, Jupp and McElligott 1979) but not transovarially (Jupp and McElligott 1979).

It remains unclear, however, if and to what extent bed bugs are important in actually transmitting the disease. The evidence consists largely of hepatitis B virus in the bodies of bed bugs from infested dwellings. Yet when bed bugs were controlled in a 2-year experiment in the Gambia, the incidence of hepatitis B among children (the frequent victims) did not decrease (Mayans et al. 1994). The means of mechanical transmission (if indeed it occurs) is also somewhat obscure; perhaps viral particles from excreta or a crushed bug may be scratched into the skin (Jupp et al. 1982), and/or interrupted feeding (feeding begun on an infected person but completed on an uninfected) may be the means of transition (Jupp et al. 1982, Jupp and Williamson 1990).

Thus, laboratory tests indicate that "mechanical transmission ... is probably an important means of hepatitis B virus transmission among humans in south Africa" (Jupp and McElligott 1979, abstract), and fieldwork suggests that "[t]he major mode of transmission of hepatitis B in childhood remains unknown" (Mayans et al. 1994). Yet even these assertions are ambiguous. Jupp and McElligott do not repeat theirs in a subsequent paper (Jupp et al. 1982), and Mayans et al. (1994) may have been studying *C. hemipterus*; they mention the name of the "bed bug" only once, and do not make absolutely clear that *C. hemipterus* is the species they studied! Nevertheless, the two species are sufficiently close phylogenetically and biologically (Usinger 1966a) that what is medically true of one is probably true of the other.

3.1.3 Control

Because bed bugs hide, do so in places difficult to access, and do so in the living places of humans, many pesticides and many means of application are either ineffective or too dangerous. Or, as Usinger (1966a) puts it, "Control of bed bugs has tested man's ingenuity for centuries" (p. 43), and it continues to do so. Usinger (1966a, Chapter 5) reviews the history of bed bug control into the 1960s, including the use of DDT. Fumigation of homes, especially with pesticides with residual action, is most often recommended; DDT, however, is no longer recommended, because of its toxicity to humans and also because bed bugs are becoming resistant (e.g., Gaaboub 1971, Kapanadze and Sukhova 1981, Nagem and Williams 1992; the first also considers dieldrin resistance and the second considers resistance to chlorophos). Moreover, in the absence of strong evidence that bed bugs transmit diseases, the need for complete control is less. One consequence of DDT resistance is that homes sprayed with DDT to combat mosquitoes soon have an increase in bed bugs; the residents then complain, and DDT spraying sometimes stops (see Rafatjah 1971, Newberry et al. 1990).

Other early means of control included fumigation with burning sulfur or hydrocyanic acid, spraying beds with kerosene pyrethrum, "heating rooms or entire buildings to a temperature of 120–125°F for one hour" (Anonymous 1941) or 10 to 12 hours (McDaniel 1940), and spraying beds with kerosene or kerosene plus pyrethrum (Marlatt 1907, 1916; McDaniel 1940; Anonymous 1941; there are libraries bulging with other references on bed bug control). In the United States, the reduviid *Reduvius personatus* (L.) preys upon bed bugs often enough to have been called the "masked bed bug hunter" (Readio 1927), and Clausen (1950) writes that "as early as 1726" the European predaceous asopine pentatomid *Perillus bidens* "was recommended for the biological control of bed bugs" (p. 587); Usinger (1966a) lists other occasional predators on various stages of *C. lectularius*.

Systemic pesticides have also been tried: Usinger (1966a) reports that Lindquist fed DDT and pyrethrum to rabbits, and that bed bugs which then fed upon the rabbits died. Similarly, fifth-instar bed bugs feeding on rabbits fed dipterex suffered mortalities increasing with increased dosage of dipterex (Adkins and Arant 1957).

Other chemicals have been and are being tested: Precocene 2, an antijuvenile hormone compound, applied on filter paper in containers with mated females, inhibited egg development; applied similarly to instars, the compound caused precocious and teratological development (Feldlaufer et al. 1981). The oil of *Eucalyptus saligna* Sm. (a complex of several compounds) killed all bed bugs in a few minutes (Kambu et al. 1982). On the other hand, several juvenoids were not particularly effective (Radwan and Sehnal 1979). Fletcher and Axtell (1993) exposed bed bugs to filter paper impregnated with several insecticides. Of these, dichlorvos, pirimiphos methyl, and bendiocarb were the most effective; permethrin, malathion, lamba-cyhalothrin, carbaryl, and tetrachlorvinphos were less so, but were nevertheless toxic; and fenvalerate was not very toxic. Different surfaces were also tested. In general, the more absorbant surfaces (cloth, wood, cardboard) held the insecticide the longest (cotton cloth longer than cotton/polyester cloth), but in some cases metal was the best.

Bednets impregnated with permethrin controlled bed bugs in Papua New Guinea for several months, and may have worked longer, but reinfection probably occurred; lice were also controlled, and people traveled several miles to have their bednets retreated (Charlwood and Dagoro 1989).

As mentioned above, bed bugs in tropical areas are often controlled as a by-product of mosquito control against malaria. Newberry et al. (1984) provides evidence for this serendipitous result (using DDT). Lambda-cyhalothrin also is effective against mosquitoes and bed bugs. The residual efficacy of the insecticide was as little as 2 weeks up to 14 weeks (Lesueuer et al. 1993).

In England, a very heavily infested flat (apartment), recently vacated, was treated first with 20% pirimiphosmethyl and then, because the population of bed bugs was so high, again with 4% phenothrin and 2% tetramethrin (King et al. 1989).

As a general rule, the best control appears to be achieved with a pesticide that can penetrate into the crevices where the bed bugs spend their days, which remains there for some time (at least until eggs hatch, eggs being particularly resistant), and, of course, which is nontoxic to humans and to domestic animals. The pesticide must also be one to which bed bugs do not become resistant; and ideally it is one that also kills far more serious arthropods, such as mosquitoes.

In poultry houses, bed bugs are controlled as part of a strategy for all ectoparasites. However, because the bugs do not live on the birds and because it is not usually convenient to remove the birds, spraying must be nontoxic to the birds and must reach the small spaces in which bed bugs live (see Axtell and Arends 1990).

Some measure of control of *C. lectularius* may occur where small populations co-occur with *C. hemipterus* (F.). Cross-mating between the two renders female *C. lectularius* sterile or dead (see **Sections 3.2.1 and 3.2.3**).

3.2 Cimex hemipterus (F.) (= Cimex rotundatus [Signoret])

3.2.1 Distribution, Life History, and Biology

Cimex hemipterus, the tropical bed bug, is distributed throughout the world tropics, where it is frequently sympatric with the closely related common bed bug, *C. lectularius* (Usinger 1966a). This overlap is often more apparent than real, however, because where they co-occur in the same geographic region, *C. hemipterus* prefers warmer and *C. lectularius* cooler habitats. Thus, the two species in the same geographic area may in fact be separated by altitude; in India, for example, *C. hemipterus* "is widely distributed in the plains," whereas *C. lectularius* occurs "[i]n the hilly regions" (Wattal and Kalra 1961, p. 157).

The chief morphological difference between the two species is a greater width-to-length ratio of the *C. lectularius* pronotum; in addition, *C. lectularius* on average is somewhat longer than *C. hemipterus*. More important differences are karyotypical, at the population level: the chromosome complement of *C. hemipterus* is constant over its range (as far as it has been checked), at $14A/X_1X_2Y$, whereas that of *C. lectularius* varies widely from population to population (Usinger 1966a, Ueshima 1966).

The two species are sufficiently closely related that hybridization between them can occur in the laboratory (Usinger 1966a) and in nature (Walpole and Newberry 1988, in South Africa). This cross-mating may result in sterility (Usinger 1966a, Walpole and Newberry 1988), and female *C. lectularius* mated by *C. hemipterus* die (Omori 1941; N. T. Davis unpublished, *in* Usinger 1966a). Indeed, this fact has led to the suggestion such cross-mating may help control bed bugs (see **Section 3.2.3**).

This phenomenon — unrestrained cross-mating with deleterious results — has fueled controversy over speciation mechanisms (e.g., Ferguson 1990, 1992, Newberry and Brothers 1990, Coetzee et al. 1995); this controversy is beyond the scope of this book. However, one should note that male bed bugs will mate with *anything*, including other males and even bits of cork (C. W. Schaefer, unpublished); it is not the propensity to mate that is important here, but its consequences. When the male *C. lectularius* mates with the *C. hemipterus* female, the results are not harmful (Omori 1941). Because the fuel for the speciation controversy is so far solely African, two complications may intrude. First, the field evidence for the reduction of *C. lectularius* populations by *C. hemipterus* comes from dwellings already sprayed with DDT, as Coetzee et al. (1995) point out. Second, *C. hemipterus* may only recently (in evolutionary time) have arrived into South Africa, with the 19th century movement of Indians to that continent (Usinger and Povolný 1966). Finally, in two thorough comparisons of the two species, Hase (1930) and Omori (1941) showed that *C. lectularius* does less well (longevity, fecundity) than *C. hemipterus* under conditions of high temperature and high humidity.

Scattered aspects of the morphology of *C. hemipterus* have been published. The antennal sensilla of immature and adult bugs was discussed by Prakash et al. (1996) and sensory organs of the adult body as well as the structure of the central nervous system by Singh et al. (1996); the setae and setal patterns of the legs were described and compared in *C. hemipterus* and *C. lectularius* by Walpole (1987). Bhatnagar (1972) describes the spiracles of *C. hemipterus,* and Pinto (1927) has discovered a "stigmate respiratoire sur les tarses" in this insect and in *C. lectularius*. [*Note:* This appears to be an enlargement of the tracheal apex in the tarsus.] Description of internal anatomy include Malpighian tubules (Bahadur 1964), hemocytes (Sonawane and More 1993), and structure (and anomalies) where the female receives sperm (Costa Leite 1948).

The most detailed accounts of the tropical bed bug's biology are by Omori (1941, in Taiwan), Hase (1931, in Venezuela), and Wattal and Kalra (1961, in Delhi, India). In Venezuela, females laid about 2 to 3 eggs per day; and, in India, total eggs per female ranged from 78 to 125; in each case, once-mated females laid fewer eggs than multiply mated females. Again in India, eggs took 3 days to hatch, and stadia lasted 4 days (first), 3 (second), 3 (third), 4 (fourth), and 6 (fifth); adult females lived 41 days and males 198 days (averages) (Wattal and Kalra 1961); that males live longer than females appears to be generally true in Cimicidae. Wattal and Kalra (1961) describe and illustrate the egg and instars of *C. hemipterus*.

Females take more blood than males, and this is true of fifth instars as well (Yanovski and Ogston 1982).

Although the tropical bed bug prefers humans to other hosts (Wattal and Kalra 1961), it has also been recorded from bats in India, Indonesia, and Mexico (reviewed in Marinkelle 1967) and in Colombia from bats near human dwellings (Marinkell 1967).

3.2.2 Damage

The damage caused by the tropical bed bug is much the same as that caused by the common bed bug (see **Section 3.1.2**): chiefly irritation and sleeplessness to humans. It may also attack poultry (Rosen et al. 1987, in Israel).

Although bed bugs will take up various stages of the filarial (nematode) human parasites *Wucheraria bancrofti* (New and Old World tropics) and *Brugia malayi* (Old World tropics), they do not develop, or do not develop to the infective stage, in bed bugs, and therefore these bugs are not vectors (Wharton and Omar 1962; Burton 1962, 1963b; see also **Section 3.1.2**).

Again like *C. lectularius*, *C. hemipterus* has been implicated in the spread of hepatitis B virus. There is less evidence, however, that the latter bug is a vector than that the former is (see **Section 3.1.2**). Most of the evidence is by implication: that an antigen to the virus persists up to 6 weeks in the bugs and may explain the high rate of infection in the field (Ogston et al. 1979) and that the bugs transmitted a radio-tracer and thus *could* transmit the virus (Ogston 1981). However, *if* the paper by Mayans et al. (1994) refers to *C. hemipterus* (this is not clear), transmission of the virus by this bug may occur in Africa.

After exposure to a high concentration of HIV, the virus was detectable in the bugs for 8 days. However, the virus did not replicate in the bugs, could not be detected in their feces, and was not transmitted mechanically; thus, the bugs are neither biological nor mechanical vectors of HIV (Webb et al. 1989).

Finally, in his compendium of human disease organisms suspected of being transmitted by *C. hemipterus*, Burton (1963a) lists five species (not including HIV and *B. malayi*), ranging from filarial worms to the yellow fever virus. There is no good evidence that any of them has actually been transmitted by these bugs to humans (Burton 1963a).

3.2.3 Control

Where the tropical bed bug (*C. hemipterus*) and the common bed bug (*C. lectularius*) coexist, cross-mating occurs (Walpole and Newberry 1988, Walpole 1988, Newberry 1989); and, when cross-mating occurs, female *C. lectularius* may be killed (see **Section 3.2.1**). Over time, populations of the common bed bug may be reduced (Newberry and Mchunu 1989, in South Africa), especially if they are small (Newberry 1989). This, of course, merely replaces the common bed bug with the tropical bed bug. Moreover, in parts of Brazil, *C. lectularius* appears to be replacing *C. hemipterus* (see **Section 3.1.1**).

The latter can be controlled with contact insecticides used in malaria control (see **Section 3.1.3**); malathion, sprayed on the houses after removal of the poultry, gave good results (Rosen et al. 1987). Bisazir, to which bugs were exposed in petri dishes, achieved more than 50% sterility of both sexes at concentrations above 0.12% (Mishra 1980). Kumar et al. (1995) found no repellent activity with diethy-*m*-toluamide (DEFT), diethyl phenyl-acetamide (DEPA), or dimethylphthalate (DRIP), although as concentrations increased, feeding by the bugs decreased.

The occasionally predaceous pyralid moth *Pyralis pictalis* Curt. will feed on the eggs of *C. hemipterus* (Wattal and Kalra 1960).

3.3, 3.4 *Oeciacus vicarius* Horváth and *Oeciacus hirundinis* (Lamarck)

3.3.1, 3.4.1 Distribution, Life History, and Biology

These are the only members of *Oeciacus*; both feed on swallows and occasionally and accidentally on humans and are closely related to *Cimex*, from which they differ in smaller size, longer bristles, thicker third and fourth antennal segments, a few other minor characters, and, of course, in host (Usinger 1966a, who seems almost to apologize for retaining the genus as separate). The genus is Holarctic: *O. hirundinis* lives throughout western and central Europe south into North Africa; *O. vicarius* occurs mostly in the transmontane western United States (Usinger 1966a), although there are records from a scattering of more eastern states and Canadian Provinces — Iowa, Maine, New Hampshire, New York, Ontario, and "NS" [?Nova Scotia] (Froeschner 1988). The bug is not associated with the swallows in their South American wintering quarters (Usinger 1966a), in the laboratory (Usinger 1966a), and in nature (Walpole and Newberry 1988, in South Africa), nor have they been found in the southeastern United States part of the swallows' range (Usinger 1966a, Smith and Eads 1978, Froeschner 1988). *Oeciacus vicarius* feeds mostly on swallows (especially cliff swallows), but *O. hirundinis* prefers the house martin as well as swallows and many other birds (Usinger 1966a).

The most complete modern account of the biology of *O. vicarius* is that by Loye (1985b; in Oklahoma, U.S.A.) who here and elsewhere (see below) also discusses the bugs' association with their host. In April the bugs begin to assemble on the empty nests of the cliff swallows. When migrating birds return to these nests, a few weeks later, the bugs feed, disperse to other nests on the birds, and within 24 hours lay eggs. Young swallows hatch in mid-June, fledge in July, and then leave. Up to this time the bugs move over the cliff face during the day, and inside and just outside the birds' nests at night. These Oklahoma bugs seem (from Loye's 1985b data) to be most active in June to July, when the swallows are present; in Texas, however, George (1987) writes that they are the most active the week after the birds have departed. In July, they cluster in nests and crevices and become nonrandomly dispersed: in August, 3 of 11 nests contained no bugs, and 4 other nests contained 1300 to 6048 bugs each. These bugs overwinter, as third to fifth instars and adults; some of the overwintering females are inseminated and can lay fertile eggs the following spring. A blood meal is required for the production of eggs, and females lay an average of 16 (7 to 27) eggs. Eggs hatch after 3 to 5 days, and adults appear in 60 days.

For 9 months of the year, swallows are absent and the bugs have no food. Some may find alternative hosts nearby (bats, other birds), but most must subsist on internal food reserves. Mortality is high — up to 99% — and bugs do not survive more than a year without food (Loye 1985b). The alternative hosts vary in their efficacy (Loye 1985b). Adults fed in the laboratory on swallow nestlings, suckling mice, and baby chicks took the same amounts of blood and lived equally long; nymphs live longer on swallow blood. More males survived than females. Females laid fewer eggs on chicks than on swallows, however (Loye 1985b), but this seems less important than the equality of survival when there are no swallows.

In the laboratory, fed on mice, *O. vicarius* survived less time than wild bugs (adult median survival times: wild males, 97 days, laboratory males, 77 days; wild females, 81 days; laboratory females, 55 days), but laboratory females produced more eggs per week (average: 8.6 vs. 4.6) and began laying eggs earlier (Rush 1981). It remains unclear if the greater fertility of laboratory bugs compensates for their lessened survivorship; it is unclear also what the effect of an unnatural food (mouse) might be.

The European bug, *O. hirundinis* is, despite its specific epithet, more an ectoparasite of nestling house martins than of swallows (Møller et al. 1994). However, it has also been recorded from at least 16 additional bird as well as three mammalian host species (Országh et al. 1990). In Slovakia, the martin bug has two generations a year, with peaks of population in early spring and again in

December. Spring is when the martins return, and presumably the winter peak (comprised mostly of immatures) represents bugs hatched from the spring's population. Although the martins by then have left for the winter, other hosts (such as house sparrows) remain available, and have perhaps been sought out during the wanderings of the bugs mentioned by Országh et al. (1990). In Poland, the seasonality may be somewhat different. Data in Kaczmarek (1991, Table) suggest peak populations in house martins in July–September and again in November–December. These twin-peaked populations probably result from the ability of the European martin bug, *O. hirudinis*, to use more hosts than does the swallow bug, *O. vicarius*, and to sustain itself in the migratory absence of its preferred host.

3.3.2, 3.4.2 Damage

In the 1890s and 1900s, in the United States, swallow bugs were thought to be bed bugs, and "a virtual war of extermination was waged throughout the United States against the beautiful and useful barn swallow" (Sailer 1952, p. 2). Despite this error, *O. vicarius* does at times enter human habitations, presumably during the long bleak period when the swallows are away. This occurs when the swallows' nests are built near or on houses; the problem is exacerbated by the fact that in many places these swallows are protected by law and their nests cannot be destroyed (Eads et al. 1980).

The bugs may do more than annoy humans: Several viruses related to those causing western equine encephalitis ("Fort Morgan" virus) have been isolated from *O. vicarius*, which seems to act as a reservoir and vector to swallows, but (so far) not to humans (Hayes et al. 1977; Rush et al. 1980, 1981; Monath et al. 1980; Loye and Hopla 1983). More (47 to 71%) swallow bugs became infected when exposed to infected birds in the spring than when exposed in the summer (11 to 26%); this seasonal difference may be caused by seasonal differences in reproductive physiology (of birds and/or bugs) (Rush et al. 1981)

In addition, of course, cliff (and other) swallows are protected in many U.S. states and are everywhere prized for their appearance, and their voracious appetite for small and unpleasant dipterans. High densities of swallow bugs can disturb these birds, and very high densities can interfere with the birds' reproduction.

Brown and Brown (1986) show that the bugs reduce nestling survivorship by nearly half, and were more important ectoparasites than fleas; ticks (*Ornithodoros concanensis* Cooley and Kohls) were not compared, although their ability to wait 4 years without feeding (Loye and Hopla 1983) suggests they are important recolonizers of swallows arriving at long abandoned nests.

These effects on the bird hosts, and the effects of other ectoparasites, in turn affect colonization by colonial birds returning from overwintering (Loye and Carroll 1998). Many of these birds are protected, in some places may become endangered, and they may avoid choice nesting sites if these are heavily infested with populations of swallow bugs from the preceding year (see review and discussion by Loye and Carroll 1998).

The European *Oeciacus hirundinis* affects its hosts (chiefly the house martin) much as *O. vicarius* affects its, causing substantial loss of body mass (Møller et al. 1994). To the extent that these European hosts are desired birds, the European bug may be considered also of economic importance. In addition, there is at least one report (Grešíková et al. 1980) of *O. hirundinis* as a possible vector (in Slovakia) of paramyxovirus type 4 (infection rate 0.1% in adult bugs, 0.4% in second to fifth instars, thus indicating transstadial transmission).

3.3.3, 3.4.3 Control

Steinbrink (1982) writes that contact insecticides help control *O. hirundinis* on martins that nest in homes, but that removing the nests will only lead the birds to renest nearby.

4. SPECIES OF MINOR IMPORTANCE

4.1 *Leptocimex boueti* (Brumpt)

This West African cimicid parasitizes bats, but also may infest human homes in rural areas (Kettle, no date). It is distinguished by its thorax being not much wider than its head, by its greatly elongated third antennal segment (Harwood and James 1979), and by its long legs (Usinger 1966a); the last is presumably an adaptation for living in caves (Usinger 1966a).

4.2, 4.3 *Cimex insuetus* Ueshima and *Stricticimex parvus* Ueshima

These bat bugs "attack humans ferociously" when they enter Thai caves, and the local people believe "that bed bug bites cause an illness in persons entering bat caves for the first time" (p. 365), which they do to collect guano (Williams et al. 1976). Bats and bugs (and ticks) were tested for Kaeng Khoi virus; bugs were 0.4 to 1.1% infected, *C. insuetus* more than *S. parvus*. Whether the bugs are or may become biological or (more probably) mechanical vectors of the virus to humans, is not known (Williams et al. 1976).

4.4 *Haematosiphon inodorus* (Duges)

The Mexican chicken bug occurs in the western United States (Usinger 1966b) south into Mexico (Usinger 1966a); the Florida record (Hussey 1928) is actually *Ornithocoris pallidus* Usinger (Usinger 1966a) (see **Section 4.5**). It feeds on many different large birds (Grubb et al. 1986), including owls and chickens, and has been recovered from a California (U.S.A.) condor, *Gymngyps californianus* (Shaw), nest (Usinger 1947) and a bald eagle nest in Arizona (U.S.A.) (Grubb et al. 1986). It may be distinguished from other cimicids by its long beak, which reaches nearly to the end of the body (Usinger 1946). It differs also from other cimicids and most other heteropterans in having only four instars (Lee 1955).

The egg is 0.89 × 0.44 mm, and the four instars are 1.14, 1.45, 2.14, and 2.94 mm long. All instars are pale amber or amber. [*Note:* They become red directly after a meal, and darker as the meal is digested and becomes more concentrated. This is generally true of cimicid instars.] The immatures and the adult are described and illustrated by Lee (1955). The egg requires 5 days to hatch (average), and the four stadia last 8.5, 7.8, 7.4, and 8.3 days. The sex ratio was 57 male to 42 female. Starved bugs lived (from last meal to death) 5.1 (first instar), 6.1 (second), 8.3 (third), 14.9 (adult female), and 12.4 (adult male) days; fourth instars were not tested. Attempts to feed bugs on mammals (white rats, several bats) were unsuccessful (Lee 1955). Unlike the bed bug, the Mexican chicken bug is not negatively phototropic; indeed, it appears to be slightly positively phototropic (Lee 1954a). Females lay eggs only after a blood meal; a female lays an average of 4.4 eggs per day, for about 3 weeks; a few mated but unfed females laid some eggs (autogeny). Of the eggs laid, an average of 56.9 eventually yielded adults (Lee 1954b).

The Mexican chicken bug occasionally becomes a pest in its rather limited distribution, where it is not uncommon. At times, its populations build up greatly in poultry houses; for example, in southern New Mexico (U.S.A.) this bug may "swarm in great numbers, ... covering the poultry eggs with its excrement, which show as black specks" (Townsend 1893, p. 40). In poultry houses the bug is very difficult to control; the best control, Townsend suggests, is just to keep the poultry permanently out-of-doors! *Haematosiphon inodorus* may also occur in such numbers in the nests of large birds as to cause death: 1425 and 1778 bugs each from two barn owl nests (Lee 1954b) and 21,000 to 31,000 bugs in a bald eagle nest, where the deaths of two eaglets were caused indirectly or directly by these bugs' feeding (Grubb et al. 1986).

4.5, 4.6 *Ornithocoris pallidus* Usinger and *Ornithocoris toledoi* Pinto

Ornithocoris toledoi, the Brazilian chicken bug, differs from *O. pallidus* in being slightly larger and having the pronotum nearly twice as wide as the head (about 1.7 larger in *O. pallidus*). Both species occur at times in chicken houses in South America, and *O. pallidus* occurs also in Mississippi (Wilson et al. 1986), Florida, and Georgia (U.S.A.), to which states it may have been introduced from South America (Usinger 1966a). The *H. inodorus* record from Florida (Hussey 1928) actually refers to *O. pallidus* (Usinger 1966a).

The immature stages of *O. toledoi* have been described and illustrated by Milward de Azevedo and Jansen (1981), and the external morphology of the adults by Jurberg and Milward de Azevedo (1982).

Stadial lengths of *O. toledoi* are 7.5 (egg), 6.5 (first stadium), 6.5 (second), 7.9 (third), 7.9 (fourth), and 7.0 (fifth) days (Snipes et al. 1940, *in* Usinger 1966a).

4.7 *Cimex columbarius* Jenyns

The pigeon bug is so closely related to, and so closely resembles, the common bed bug, that the two have often been considered conspecific; the main difference is the ratio between head width and third antennal segment (1.45 in *C. lectularius*, 1.78 in *C. columbarius*). *Cimex columbarius* is European and, unlike *C. lectularius*, feeds on pigeons (and may become a pest on domestic pigeons) and at least one other bird, the pied flycatcher (*Muscicapa atricapilla* L.) in man-made birdhouses (Usinger 1966a).

Hybridization will occur in the laboratory between the two species, but fertility is low (Ueshima 1964). Johnson (1939) raised both on rabbit for several generations and found the antennal:head ratio to remain constant. For this reason, and because the third-antenna:head ratio also remains constant no matter on what host the bugs are raised (*C. lectularius* on pigeon, *C. columbarius* on rabbit), Usinger (1966a) concludes that the two are reproductively isolated species.

4.8 *Hesperocimex sonorensis* Ryckman

This is an ectoparasite of purple martins, *Progne subis* (L.), so far found only in the southwestern United States and adjoining Mexico (Ryckman 1958, Usinger 1966a). To the extent that the purple martin is a bird of interest to humans, this bug can be considered a pest. The bug is described by Ryckman (1958) and redescribed by Usinger (1966a). Its karyology and the results of some intrageneric hybridization experiments are given by Ryckman and Ueshima (1964). Fertility, fecundity, and measurements of immatures (including egg) are provided by Ryckman (1958).

5. CONCLUSIONS

In the absence of conclusive evidence that bed bugs serve as vectors of human disease pathogens (or, indeed, of pathogens of veterinary importance), the bugs' economic importance resides in the blood they take, and the collateral importance as indicators (often unjustly) of unclean and "primitive" living conditions. Blood loss may be serious when infestations are large and the human hosts are weak, whether from extreme youth or age, or from disease.

Cimicids are also of importance as ectoparasites of domestic birds, whose reproductive potential and whose productivity (meat, eggs) they may lessen. And some cimicids may harm birds that humans like for aesthetic reasons — martins, for example, and swallows.

6. PROGNOSIS

Research continues on the possibility that bed bugs (*C. lectularius, C. hemipterus*) may transmit hepatitis B virus. Doubtless other claims will arise that these bugs are vectors — mechanical if not biological — of other human pathogens. Given the history of the many earlier such claims, these seem unlikely to be substantiated. But the hepatitis B claim remains open.

Finally, the vast majority of cimicids feed on bats. If human beings were ever able to look on bats favorably, and without Transylvanian shudders, the economic significance of this family would skyrocket.

7. REFERENCES CITED

Abul-Hab, J., H. All-Baerzangi, S. Kassal, M. Fuad, and T. Al-Obaidi. 1989. The common bed bug *Cimex lectularius* L. (Hemiptera, Cimicidae) in Baghdad City, Iraq. J. Biol. Sci. Res. (Iraq) 20: 455–462.

Adams, T. S. 1999. Hematophagy and hormone release. Ann. Entomol. Soc. Amer. 92: 1–13.

Adkins, T. R., Jr., and F. S. Arant. 1957. Systemic effect of dipterex on the bed bug and Gulf Coast tick when administered to rabbits. J. Econ. Entomol. 50: 166–168.

Anonymous. 1941. Bed bugs. Insects in Relation to National Defense Circ. 10. 6 pp.

Axtell, R. C., and J. J. Arends. 1990. Ecology and management of arthropod pests of poultry. Annu. Rev. Entomol. 35: 101–126.

Bacot, A. W. 1914. The influence of temperature, submersion and burial on the survival of eggs and larvae of Cimex lectularius. Bull. Entomol. Res. 5: 111–117.

Bahadur, J. 1964. The Malpighian tubules of a blood sucking insect, *Cimex rotundatus* (Hemiptera). Sci. Cult. 30: 405–406.

Bell, W., and C. W. Schaefer. 1966. Longevity and egg production of female bed bugs, *Cimex lectularius*, fed various blood fractions and other substances. Ann. Entomol. Soc. Amer. 59: 53–56.

Bhat, H. R. 1974. A review of Indian Cimicidae (Hemiptera-Heteroptera). Orient. Insects 8: 545–550.

Bhatnagar, B. S. 1972. Spiracles in certain terrestrial Heteroptera (Hemiptera). Int. J. Insect Morphol. Embryol. 1: 207–217.

Brown, C. R., and M. B. Brown. 1986. Ectoparasitism as a cost of coloniality in cliff swallows (*Hirundo pyrrhonota*). Ecology 67: 1206–1218.

Burton, G. J. 1962. Observations on filarial larvae in bed bugs feeding on human carriers of *Wucheraria bancrofti* and *Brugia malayi* in south India. Amer. J. Trop. Med. Hyg. 11: 68–75.

1963a. Bedbugs in relation to transmission of human disease. Public Health (U.S. Dept. HEW) Rep. 78: 513–524.

1963b. Natural and experimental infection of bedbugs with *Wucheraria bancrofti* in British Guiana. Amer. J. Trop. Med. Hyg. 12: 541–547.

Carayon, J. 1966a. Traumatic insemination and the paragenital system. Pp. 81–166 *in* R. L. Usinger [ed.], Monograph of Cimicidae (Hemiptera-Heteroptera). Thomas Say Foundation, Vol. 7, Entomological Society of America, College Park, Maryland, U.S.A. 585 pp.

1966b. Metathoracic scent apparatus. Pp. 69–80 *in* R. L. Usinger [ed.], Monograph of Cimicidae (Hemiptera-Heteroptera). Thomas Say Foundation, Vol. 7, Entomological Society of America, College Park, Maryland, U.S.A. 585 pp.

1969. Comparaison du spermalege des *Cimex* (Hem. Cimicidae) avec les réactions inflammatoires de type "granulome." Ann. Zool. (Ecol. Anim.) 1: 73–82.

1977. Insémination extra-génitale traumatique. Pp. 349–390 *in* P. P. Grassé [ed.], Traité de Zoologie, tome 8, fasc. VA. Masson et Cie, Paris, France.

Chang, K. P. 1974. Effects of elevated temperature on the mycetome and symbiotes of the bed bug *Cimex lectularius* (Heteroptera). J. Invertebr. Pathol. 23: 333–340.

Charlwood, J. D., and H. Dagoro. 1989. Collateral effects of bednets impregnated with permethrin against bedbugs (Cimicidae) in Papua New Guinea. Trans. R. Soc. Trop. Med. Hyg. 83: 261.

Clausen, C. P. 1950. Entomophagous Insects, first edition. McGraw-Hill Book Co., New York, New York, U.S.A. 688 pp.

Coetzee, M., and J. Segerman. 1992. The description of a new genus and species of cimicid bug from South Africa (Heteroptera Cimicidae Cacodminae). Trop. Zool. 5: 229–235.

Coetzee, M., R. H. Hunt, and D. E. Walpole. 1995. Interpretation of mating between two bedbug taxa in a zone of sympatry in KwaZulu, South Africa. Pp. 175–190 *in* D. M. Lambert and H. G. Spencer [eds.], Speciation and the Recognition Concept: Theory and Application. Johns Hopkins University Press, Baltimore, Maryland, U.S.A.

Costa Leite, I. 1948. Cimex hemipterus [sic] (Hemiptera, Cimicidae). An. Acad. Bras. Cienc. 20: 273–276.

Daniel, T. L., and J. G. Kingsolver. 1983. Feeding strategy and the mechanics of blood sucking in insects. J. Theor. Biol. 105: 661–672.

Davis, N. T. 1956. The morphology and functional anatomy of the male and female reproductive systems of *Cimex lectularius* L. (Heteroptera, Cimicidae). Ann. Entomol. Soc. Amer. 49: 466–493.

1964. Studies of the reproductive physiology of Cimicidae (Hemiptera) — I. Fecundation and egg maturation. J. Insect Physiol. 10: 947–963.

1965a. Studies of the reproductive physiology of Cimicidae (Hemiptera) — II. Artificial insemination and the function of the seminal fluid. J. Insect Physiol. 11: 355–366.

1965b. Studies of the reproductive physiology of Cimicidae (Hemiptera) — II. The seminal stimulus. J. Insect Physiol. 11: 1199–1211.

1966. Reproductive physiology. Pp. 167–182 *in* R. L. Usinger [ed.], Monograph of Cimicidae (Hemiptera-Heteroptera). Thomas Say Foundation, Vol. 7, Entomological Society of America, College Park, Maryland, U.S.A. 585 pp.

Dickerson, G., and M. M. J. Lavoipierre. 1959. Studies on the methods of feeding of blood-sucking arthropods — II. The method of feeding adopted by the bed-bug (*Cimex lectularius*) when obtaining a blood-meal from the mammalian host. Ann. Trop. Med. Parasitol. 53: 347–357.

Eads, R. B., D. B. Francy, and G. C. Smith. 1980. The swallow bug, *Oeciacus vicarius* Horvath (Hemiptera: Cimicidae), a human household pest. Proc. Entomol. Soc. Washington 82: 81–85.

Feldlaufer, M. F., M. W. Eberle, and G. A. H. McClelland. 1981. Developmental and teratogenic effects of precocene 2 on the bed bug, *Cimex lectularius* L. Insect Sci. Appl. 1: 389–392.

Ferguson, J. W. H. 1990. Interspecific hybridization in bedbugs supports the Recognition Concept of species. South Afric. J. Sci. 86: 121–124.

1992. Bedbug hybridization and the Recognition Concept of species: the weakness of defining species in terms of isolated gene pools. South Afric. J. Sci. 88: 10–11.

Ferris, G. F., and R. L. Usinger. 1957. Notes and descriptions of Cimicidae. Microentomology 22: 1–37.

Fletcher, M. G., and R. A. Axtell. 1993. Susceptibility of the bedbug, *Cimex lectularius*, to selected insecticides and various treated surfaces. Med. Vet. Entomol. 7: 69–72.

Forattini, O. P. 1990. Os cimicideos e sua importância em saúde pública (Hemiptera-Heteroptera: Cimicidae. Rev. Saúde Públ. 24 (Suppl.): 1–37.

Friend, W. G., and J. J. B. Smith. 1977. Factors affecting feeding by bloodsucking insects. Annu. Rev. Entomol. 22: 309–331.

Froeschner, R. C. 1988. Family Cimicidae Latreille, 1802. The bed bugs. Pp. 64–68 *in* T. J. Henry and R. C. Froeschner [eds.], Catalog of the Heteroptera, or True Bugs, of Canada and the Continental United States. E. J. Brill, Leiden, the Netherlands. 988 pp.

Gaaboub, I. A. 1971. Present status of DDT and dieldrin resistance in the bed-bug, *Cimex lectularius*, in Alexandria District, United Arab Republic. Z. Angew. Entomol. 68: 176–180.

George, J. E. 1987. Field observations on the life cycle of *Ixodes baergi* and some seasonal and daily activity cycles of *Oeciacus vicarius* (Hemiptera: Cimicidae), *Argas cooleyi* (Acari: Argasidae), and *Ixodes baergi* (Acari: Ixodidae). J. Med. Entomol. 24: 683–688.

Gilbert, I. H. 1964. Laboratory rearing of cockroaches, bed-bugs, human lice and fleas. Bull. World Health Organ. 31: 561–563.

Grešíková, M., J. Nosek, F. Čiampor, M. Sekeyová, and R. Turek. 1980. Isolation of paramyxovirus type 4 from *Oeciacus hirundinis* bugs. Acta Virol. 24: 222–223.

Groves, E. W. 1967. Hemiptera-Heteroptera of the London area. Part IV. London Nat. 46: 82–104.

Grubb, T. G., W. L. Eakle, and B. N. Tuggle. 1986. *Haematosiphon inodorus* (Hemiptera: Cimicidae) in a nest of a bald eagle (*Haliaeetus leucocephalus*) in Arizona. J. Wild. Dis. 22: 125–127.

Hall, R. D., E. C. Turner, Jr., and W. B. Gross. 1979. A simple apparatus for providing blood diets at constant temperature and with different corticosterone levels to individual bed bug colonies (Hemiptera: Cimicidae). J. Med. Entomol. 16: 259–261.

Harwood, R. F., and M. T. James. 1979. Entomology in Human and Animal Health, seventh edition. Macmillan Publ. Co., New York, New York, U.S.A. 658 pp.

Hase, A. 1930. Weitere Versuche zur Kenntnis der Bettwanzen Cimex lectularius L. und Cimex rotundatus Sign. (Hex.-Rhynch.). Z. Parasitenk. (Abt. Z. Wiss. Biol.) 2(3): 368–418.

1931. Über Lebensbedingungen, Verhalten und Fruchtbkarkeit der tropischen Hauswanze Cimex rotundatus Sign. (Hex. rhynch.) in Venezuela. Z. Parasitenk. (Abt. Z. Wiss. Biol.) 3(4): 837–893.

1938. Zur hygienischen Bedeutung der parasitarën Haus- und Vogelwanzen sowie über Wanzenpopulationen und Wanzenkreuzungen. Z. Parasitenk. 10: 1–30.

Hayes, R. O., D. B. Francy, J. S. Lazuick, G. C. Smith, and E. P. J. Gilbbs. 1977. Role of the cliff swallow bug (*Oeciacus vicarius*) in the natural cycle of a western equine encephalitis-related alpha virus. J. Med. Entomol. 14: 257–262.

Hocking, B. 1971. Blood-sucking behavior of terrestrial arthropods. Annu. Rev. Entomol. 16: 1–26.

Hussey, R. L. 1928. The Mexican chicken bug in Florida. Florida Entomol. 12: 43–44.

Jaenson, T. G. T. 1985. Kan insekter sprida hepatit B-virus och AIDS-associerat retrovirus (LAV/HTLV-III)? Läkartidningen 82: 4133–4136.

Johnson, C. G. 1937. The relative values of man, mouse, and domestic fowl as experimental hosts for the bed-bug, *Cimex lectularius* L. Proc. Zool. Soc. London (A) 107: 107–126.

1939. Taxonomic characters, variability, and relative growth in *Cimex lectularius* L. and *C. columbarius* Jenyns (Heteropt. Cimicidae). Trans. R. Entomol. Soc. 89: 543–568.

1941. The ecology of the bed-bug, *Cimex lectularius*, in Britain. J. Hyg. 41: 345–461.

1960. The relation of weight of food ingested to increase in body-weight during growth in the bed-bug, *Cimex lectularius* (L.) (Hemiptera). Entomol. Exp. Appl. 3: 238–240.

Jupp, P. G., and S. E. McElligott. 1979. Transmission experiments with hepatitus B surface antigen and the common bedbug (*Cimex lectularius* L.). South Afric. Med. J. 56: 54–57.

Jupp, P. G., and C. Williamson. 1990. Detection of multiple blood-meals in experimentally fed bedbugs (Hemiptera: Cimicidae). J. Entomol. Soc. South. Africa 53: 137–140.

Jupp, P. G., S. E. McElligott, and G. Lecatsas. 1982. The mechanical transmission of hepatitis B virus by the common bedbug (*Cimex lectularius* L.) in South Africa. South Afric. Med. J. 63: 77–81.

Jurberg, J., and E. M. V. Milward de Azevedo. 1982. Contribuição para o estudo da morfologia de *Ornithocoris toledoi* Pinto, 1927 (Hemiptera, Cimicidae). Rev. Bras. Biol. 42: 255–262.

Kaczmarek, S. 1991. *Oeciacus hirundinis* z gniad jaskólek *Delichon urbica* i *Hirundo rustica*. Wiad. Parazytol. 37: 277–280.

Kambu, K., N. di Phanzu, C. Coune, J.-N. Wauters, and L. Angenot. 1982. Contribution a l'étude des propriétés insecticides et chimiques d'*Eucalyptus saligna* du Zaire. Plant. Med. Phytother. 16: 34–38.

Kapanadze, E. I., and M. N. Sukhova. 1981. [Changes in the sensitivity levels of DDT- and chlorophos-resistant populations of bed bugs (*Cimex lectularius* L.) under experimental conditions.] Med. Parasitol. Parazit. Bolezni. 50: 38–41 [English abstract only seen].

Kettle, D. S. no date. Medical and Veterinary Entomology. John Wiley & Sons, New York, New York, U.S.A. 658 pp.

King, F., I. Dick, and P. Evans. 1989. Bed bugs in Britain. Parasitol. Today 5: 100–102.

Kingsolver, J. G., and T. L. Daniel. 1993. Mechanics of fluid feeding in insects. Pp. 149–162 in C. W. Schaefer and R. A. B. Leschen [eds.], Functional Morphology of Insect Feeding. Thomas Say Publ. Entomol.: Entomological Society of America, Lanham, Maryland, U.S.A. 162 pp.

Kumar, S., S. Prakesh, and K. M. Rao. 1995. Comparative activity of 3 repellants against bedbugs *Cimex hemipterus* (Fabr.). Indian J. Med. Res. 102: 20–23.

Lee, R. D. 1954a. The absence of negative phototrophism in the Mexican chicken bug, Haematosiphon inodorus (Duges). Pan-Pac. Entomol. 30: 159–160.

1954b. Oviposition by the poultry bug. J. Econ. Entomol. 47: 224–226.

1955. The biology of the Mexican chicken bug, Haematosiphon inodorus (Duges). Pan-Pac. Entomol. 31: 47–61.

Lesueur, D., B. L. Sharp, C. Fraser, and S. M. Ngxongo. 1993. Assessment of the residual efficacy of lambda-cyhalothrin. 1. A laboratory study using *Anopheles arabiensis* and *Cimex lectularius* (Hemiptera, Cimicidae) on treated daub wall substrates from Natal, South Africa. J. Amer. Mosq. Control. Assoc. 9: 408-413.

Levinson, H. Z., and A. R. Bar Ilan. 1971. Assembling and alerting scents produced by the bedbug *Cimex lectularius*. Experientia 27: 102–103.

Levinson, H. Z., A. R. Levinson, B. Müller, and R. A. Steinbrecht. 1974a. Structure of sensilla, olfactory perception, and behaviour of the bedbug, *Cimex lectularius*, in response to its alarm pheromone. J. Insect Physiol. 20: 1231–1248.

Levinson, H. Z., A. R. Levinson, and U. Maschwitz. 1974b. Action and composition of the alarm pheromone of the bedbug *Cimex lectularius* L. Naturwissenschaften 61: 684–685.

Loudon, C., and K. McCulloh. 1999. Application of the Hagen-Poiseuille equation to fluid feeding through short tubes. Ann. Entomol. Soc. Amer. 92: 153–158.

Loye, J. E. 1985a. The life history and ecology of the cliff swallow bug, *Oeciacus vicarius* (Hemiptera: Cimicidae). Cah. ORSTOM (Entomol. Méd. Parasitol.) 23: 133–139.

1985b. Host-effects on feeding and survival of the polyphagous cliff swallow bug, *Oeciacus vicarius* (Hemiptera: Cimicidae). Bull. Soc. Vector Ecol. 10: 7–13.

Loye, J. E., and S. P. Carroll. 1998. Ectoparasite behavior and its effect on avian nest site selection. Ann. Entomol. Soc. Amer. 91: 159–163.

Loye, J. E., and C. E. Hopla. 1983. Ectoparasites and microorganisms associated with the cliff swallow in west-central Oklahoma. II. Life history patterns. Bull. Soc. Vector Ecol. 8: 79–84.

Ludwig, W. 1937. Über die Genese der Asymmetrieform bei Bettwanzen. Wilhelm Roux Arch. Entwicklungsmech. Org. 136: 294–312.

Marinkelle, C. J. 1967. *Cimex hemipterus* (Fabr.) from bats in Colombia, South America (Hemiptera: Cimicidae). Proc. Entomol. Soc. Washington 69: 179–180.

Marlatt, C. L. 1907. The bedbug. U.S. Dept. Agric. Circ. 47 (revised). 8 pp.

1916. The bedbug. U.S. Dept. Agric. Farmers' Bull. 754. 12 pp.

Marx, R. 1955. Über die Wirtsfindung und die Bedeutung des Artspezifischen Duftstoffes bei *Cimex lectularius* Linné. Z. Parasitenk. 17: 41–73.

Mayans, M. V., A. J. Hall, H. M. Inskip, S. W. Lindsay, J. Chotard, M. Mendy, and H. C. Whittle. 1994. Do bedbugs transmit hepatitis B? Lancet 343: 761–763.

McDaniel, E. L. 1940. Bedbugs. Michigan State Coll. Extens. Bull. 211: 2 pp.

Mellanby, K. 1939. The physiology and activity of bed-bug (*Cimex lectularius* L.) in a natural infestation. Parasitology 31: 200–211.

Milward de Azevedo, E. M. V., and A. M. Jansen. 1981. Morfologia externa dos estádios imaturos pós-embrionários de *Ornthicoris toledoi* Pinto, 1927 (Hemiptera, Cimicidae). Rev. Bras. Biol. 41: 111–115.

Milward de Azevedo, E. M. V., and J. L. de Oliveira. 1980. Um novo suporte e métodos para estudos em cimicideos parasitos de aves em laboratório (Hemiptera, Cimicidae). Rev. Bras. Entomol. 24: 53–57.

Mishra, R. K. 1980. Sterilization of bed-bug, *Cimex hemipterus* [sic] Fabr. with bisazir. Z. Angew. Entomol. 89: 247–249.

Møller, A. P., D. de Lope, J. Moreno, G. González, and J. J. Pérez. 1994. Ectoparasites and host energetics: house martin bugs and house martin nestlings. Oecologia 98: 263–268.

Monath, T. P., J. S. Lazuick, C. P. Cropp, W. A. Rush, C. H. Calisher, R. M. Kinney, D. W. Trent, G. E. Kemp, G. S. Bowen, and D. B. Francy. 1980. Recovery of tonate virus ("Bijou Bridge" strain), a member of the Venezuelan equine encephalomyelitis virus complex, from cliff swallow nest bugs (*Oeciacus vicarius*) and nestling birds in North America. Amer. J. Trop. Med. Hyg. 29: 969–983.

Moore, N. F., S. M. Eley, R. A. Gardner, and D. H. Molyneux. 1991. Viruses from bedbugs. Pp. 638–643 *in* J. R. Adams and J. R. Bonami [eds.], Atlas of Invertebrate Viruses. CRC Press, Boca Raton, Florida, U.S.A. 684 pp.

Nagem, R. L. 1985. Ocorrência de *Cimex lectularius* L., 1758 (Hemiptera, Cimicidae) em algumas habitações humanas de Belo Horizonte e municípios vizinhos. Rev. Bras. Entomol. 29: 217–220.

Nagem, R. L., and P. Williams. 1992. Susceptibility tests of the bed-bug *Cimex lectularius* L. (Hemiptera, Cimicidae) to DDT in Belo Horizonte, MG (Brazil). Rev. Saude Publ., São Paulo 26: 125–128.

Nagem, R. L., J. R. Botalho, and M. R. S. Negromonte. 1988. Bionomia de *Cimex lectularius* L., 1758 (Hemiptera, Cimicidae) em Belo Horizonte, MG. I. Observação e periodo de incubação dos ovos em laboratório. Rev. Bras. Entomol. 32: 323–329.

Negromonte, M. R. S., P. M. Linardi, and R. L. Nagem. 1991. Prevalência, intensidade e fluxo da infestação por *Cimex lectularius* L., 1758 (Hemiptera, Cimicidae) em uma comunidade de Belo Horizonte, MG. Rev. Bras. Entomol. 35: 715–720.

Newberry, K. 1989. The effects on domestic infestations of *Cimex lectularius* bedbugs of interspecific mating with *C. hemipterus*. Med. Vet. Entomol. 3: 407–414.

Newberry, K., and D. J. Brothers. 1990. Problems in the Recognition Concept of species: an example from the field. South Afric. J. Sci. 86: 4–6.

Newberry, K., and E. J. Jansen. 1986. The common bedbug *Cimex lectalarius* in African huts. Trans. R. Soc. Trop. Med. Hyg. 80: 653–658.

Newberry, K., and Z. M. Mchunu. 1989. Changes in the relative frequency of infestations of two sympatric species of bedbug in northern Natal and KwaZulu, South Afric. Trans. R. Soc. Trop. Med. Hyg. 83: 262–264.

Newberry, K., E. J. Jansen, and A. G. Quann. 1984. Bedbug infestation and intradomiciliary spraying of residual insecticide in KwaZulu, South Africa. South Afric. J. Sci. 80: 377.

Newberry, K., Z. M. Mchunu, and S. Q. Cebekhulu. 1990. The effect of bedbug control on malaria control operations. South Afric. J. Sci. 86: 211–212.

Newkirk, M. M., A. E. R. Downe, and J. B. Simon. 1975. Fate of ingested hepatitis B antigen in blood-sucking insects. Gastroenterology 69: 982–987.

Ogston, C. W. 1981. Transfer of radioactive tracer by the bedbug *Cimex hemipterus* (Hemiptera: Cimicidae): a model for mechanical transmission of hepatitis B virus. J. Med. Entomol. 18: 107–111.

Ogston, C. W., and A. D. Yanovski. 1982. An improved artificial feeder for bloodsucking insects. J. Med. Entomol. 9: 42–44.

Ogston, C. W., F. S. Wittenstein, W. T. London, and I. Millman. 1979. Persistence of hepatitis B surface antigen in the bedbug *Cimex hemipterus* (Fabr.). J. Infect. Dis. 140: 411–414.

Omori, N. 1939. Experimental studies on the cohabitation and crossing of two species of bedbugs, *Cimex lectularius* L. and *Cimex hemipterus* (F.), and on the effects of interchanging males of one species for the other, every alternate day, upon the fecundity and longevity of females of each species. Acta Jpn. Med. Trop. 1: 127–154.

1941. Comparative studies on the ecology and physiology of common and tropical bed bugs, with special reference to the reactions to temperature and moisture. Taiwan Igakkai J. Med. Assoc. Formosa 60: 555–729.

Országh, I., M. Krumpál, and D. Cyprich. 1990. Contribution to the knowledge of the martin bug Oeciacus hirundinis (Heteroptera, Cimicidae) in Czechoslovakia. Zbr. Slov. Nár. Múz. Prír. Vedy 36: 43–60.

Overal, W. L., and L. R. Wingate. 1976. The biology of the batbug Stricticimex antennatus (Hemiptera: Cimicidae) in South Africa. Ann. Natal. Mus. 22: 821–828.

Paterson, W. B., and P. T. Woo. 1984. The development of the culture and bloodstream forms of three *Trypanosoma* (*Schizotrypanum*) spp. (Protista: Zoomastigophorea) from bats in *Cimex lectularius* (Hemiptera: Cimicidae). Canad. J. Zool. 62: 1581–1587.

Petzsch, H. 1953. *Cimex lectularius* L. als Parasit verschiedener warmblütiger Zoo-Tiere, insbesondere von gehaltenen Kleinsäugetieren. Beitr. Entomol. 3: 404–405.

Pinto, C. 1927. De la présence d'un stigmate respiratoire sur les tarses du *Cimex hemipterus, C. lectularius, Pediculus humanus, Haematopinus eurysternus* et chez les larves de *Triatoma megista*. Bol. Biol. (São Paulo, Brazil) 8: 115–128.

Prakash, S., R. S. Chauhan, B. D. Parashar, S. Chandna, and K. M. Rao. 1996. Morphology and distribution of antennal sensilla in the postembryonic developmental stages of *Cimex hemipterus* Fabricius. Ital. J. Zool. 63: 131–133.

Radwan, W. A., and F. Sehnal. 1979. Preliminary tests of juvenoids on the bed bug, *Cimex lectularius* L. Bull. Entomol. Soc. Egypt 11: 77–80.

Rafatjah, H. 1971. The problem of resurgent bed-bug infestation in malaria eradication programmes. J. Trop. Med. Hyg. 74: 53–56.

Rao, H. V., and N. T. Davis. 1969. Sperm activation and migration in bed bugs. J. Insect Physiol. 15: 1815–1832.

Readio, P. A. 1927. Studies on the biology of the Reduviidae of America north of Mexico. Univ. Kansas Sci. Bull. 17: 5–291.

Reid, B. 1990. Don't let the bedbugs bite. Pest Control, June 1990: 48–50.

Roer, H. 1969. Über Vorkommen und Lebensweise von *Cimex lectularius* und *Cimex pipistrelli* (Heteroptera, Cimicidae) Fledermausquartieren. Bonner Zool. Beitr. 4: 355–359.

1975. Zur Übertragung von in Fledermauswanzen (Heteroptera, Cimicidae) durch ihre Wirte. Myotis 13: 62–64.

Rosen, S., A. Hadani, A. Gur Kavi, E. Berman, U. Bendheim, and A. Y. Hisham. 1987. The occurrence of the tropical bedbug (*Cimex hemipterus*, Fabricius) in poultry barns in Israel. Avian Pathol. 16: 339–342.

Ruknudin, A., and V. V. Raghavan. 1988. Initiation, maintenance and energy metabolism of sperm motility in the bed bug, *Cimex hemipterus*. J. Insect Physiol. 34: 137–142.

Rush, W. A. 1981. Colonization of the swallow bug in the laboratory. Ann. Entomol. Soc. Amer. 74: 556–559.

Rush, W. A., D. B. Francy, and R. E. Bailey. 1981. Seasonal changes in susceptibility of a population of swallow bug (Hemiptera: Cimicidae) to Fort Morgan virus. J. Med. Entomol. 18: 425–428.

Rush, W. A., D. B. Francy, G. C. Smith, and C. B. Cropp. 1980. Transmission of an arbovirus by a member of the family Cimicidae. Ann. Entomol. Soc. Amer. 73: 315–318.

Ryckman, R. E. 1958. Description and biology of Hesperocimex sonorensis, new species, an ectoparasite of the purple martin (Hemiptera: Cimicidae). Ann. Entomol. Soc. Amer. 51: 33–47.

Ryckman, R. E., and N. Ueshima. 1964. Biosystematics of the *Hesperocimex* complex (Hemiptera: Cimicidae) and avian hosts (Piciformes: Picidae; Passeriformes: Hirundinidae). Ann. Entomol. Soc. Amer. 57: 624–638.

Ryckman, R. E., D. G. Bentley, and E. F. Archbold. 1981. The Cimicidae of the Americas and oceanic islands, a checklist and bibliography. Bull. Soc. Vector Ecol. 6: 93–142.

Sailer, R. I. 1952. The bedbug, an old bedfellow that's still with us. Pest Control 1952: 4 pp.

Schofield, C. J. 1994. Triatominae Biology & Control. Eurocommunica Publ., Bognor Regis, U.K. 77 pp.

Silverman, A. L., J. A. Blow, Qu L. H., I. M. Ziron, E. D. Walker, and S. C. Gordon. 1998. Persistence of the hepatitis B viral genome in the common bedbug (Hemiptera: Cimicidae). Ann. Entomol. Soc. Amer. 51: 33–47.

Singh, R. N., K. Singh, S. Prakash, M. J. Mendki, and K. M. Rao. 1996. Sensory organs on the body parts of the bed-bug *Cimex hemipterus* Fabricius (Hemiptera: Cimicidae) and the anatomy of its central nervous system. Int. J. Insect Morphol. Embryol. 25: 183–204.

Smith, G. C., and R. B. Eads. 1978. Field observations on the cliff swallow, *Petrochelidon pyrrhonota* (Vieillot), and the swallow bug, *Oeciacus vicarius* Horváth. J. Washington Acad. Sci. 68: 23–26.

Snipes, T., J. C. M. Carvalho, and O. E. Tauber. 1940. Biological studies of *Ornithocoris toledoi* Pinto, the Brazilian chicken bedbug. Iowa State Coll. J. Sci. 15: 27–35 [not seen, *fide* Usinger 1966a].

Sonawane, Y. S., and N. K. More. 1993. The circulating hemocytes of the bed bug *Cimex rotundatus* (Sign.). J. Anim. Morphol. Physiol. 40: 79–86.

Steinbrecht, R. A., and B. Müller. 1976. Fine structure of the antennal receptors of the bed bug, *Cimex lectularius* L. Tissue Cell 8: 615–636.

Steinbrink, H. 1982. Schwalbenparasiten als Gesundheitsschädlinge. Angew. Parasitol. 23: 155–157.

Tawfik, M. S. 1968. Feeding mechanisms and the forces involved in some blood-sucking insects. Quaest. Entomol. 4: 92–111.

Townsend, C. H. T. 1893. Note on the cornuco, a hemipterous insects which infests poultry in southern New Mexico. Proc. Entomol. Soc. Washington 3: 40–42.

Ueshima, N. 1964. Experiments on reproductive isolation in *Cimex lectularius* and *Cimex columbarius*. Pan-Pac. Entomol. 40: 47–53.

1966. Cytology and cytogenetics. Pp. 183–245 *in* R. L. Usinger [ed.], Monograph of Cimicidae (Hemiptera-Heteroptera). Thomas Say Foundation, Vol. 7, Entomological Society of America, College Park, Maryland, U.S.A. 585 pp.

1967. Supernumerary chromosomes in the human bed bug, *Cimex lectularius* Linn. (Cimicidae: Hemiptera). Chromosoma (Berlin) 20: 311–331.

1968. New species and new records of Cimicidae with keys (Hemiptera). Pan-Pac. Entomol. 44: 264–279.

Usinger, R. L. 1946. Household bugs. Pest Control Sanitation 1(12): 19.

1947. Native hosts of the Mexican chicken bug, *Haematosiphon inodora* (Dugés). (Hemiptera, Cimicidae). Pan-Pac. Entomol. 23: 140.

1966a. Monograph of Cimicidae (Hemiptera-Heteroptera. Vol. 7. Thomas Say Foundation. Entomological Society of America, College Park, Maryland, U.S.A. 585 pp.

1966b. Distributional patterns of selected western North American insects: distributional patterns of Cimicidae in western North America. Bull. Entomol. Soc. Amer. 12: 112.

Usinger, R. L., and D. Povolný. 1966. The discovery of a possibly aboriginal population of the bed bug (Cimex lectularius [sic] Linnaeus, 1758). Acta Mus. Moraviae 51: 237–241.

Valenzuela, J. G., J. A. Guimaraes, and J. M. C. Ribeiro. 1996a. A novel inhibitor of factor X activation from the salivary glands of the bed bug *Cimex lectularius*. Exp. Parasitol. 83: 184–190.

Valenzuela, J. G., O. M. Chuffe, and J. M. C. Ribeori. 1996b. Apyrase and anti-platelet activities from the salivary glands of the bed bug *Cimex lectularius*. Insect Biochem. Mol. Biol. 21: 557–562.

Valenzuela, M. G., R. Charlab, M. Y. Galperini, and J. M. C. Ribeiro. 1998. Purification, cloning, and expression of an apyrase from the bed bug *Cimex lectularius*. J. Biol. Chem. 273: 30583–30590.

Walpole, D. E. 1987. External morphology of the legs of two species of bed bugs (Hemiptera: Cimicidae). J. Entomol. Soc. South. Africa 50: 193–201.

1988. Cross-mating studies between two species of bedbugs (Hemiptera: Cimicidae) with a description of a marker of interspecific mating. South Afric. J. Sci. 84: 215–216.

Walpole, D. E., and K. Newberry. 1988. A field study of mating between two species of bedbug in northern KwaZulu, South Africa. Med. Vet. Entomol. 2: 293–296.

Wattal, B. L., and N. L. Kalra. 1960. *Pyralis pictalis* Curt. (Pyralidae: Lepidoptera) larvae as predators of eggs of bed bug, *Cimex hemipterus* Fab. (Cimicidae: Hemiptera). Indian J. Malariol. 14: 77–79.

1961. New methods for the maintenance of a laboratory colony of bed-bug, *Cimex hemipterus* Fabricius, with observations on its biology. Indian J. Malariol. 15: 157–171.

Webb, P. A., C. M. Happ, G. O. Maupin, B. J. B. Johnson, C.-Y. Ou, and T. P. Monath. 1989. Potential for insect transmission of HIV: experimental exposure of *Cimex hemipterus* and *Toxorhynchites amboinensis* to human immunodeficiency virus. J. Infect. Dis. 160: 970–977.

Wharton, R. H., and A. H. bin Omar. 1962. Failure of *Wucheraria bancrofti* and *Brugia malayi* to develop in the tropical bedbug *Cimex hemipterus*. Ann. Trop. Med. Parasitol. 56: 188–190.

Williams, J. E., S. Imlarp, F. H. Top, Jr., D. C. Cavanaugh, and P. K. Russell. 1976. Kaeng Khoi virus from naturally infected bedbugs (Cimicidae) and immature free-tailed bats. Bull. World Health Organ. 53: 365–369.

Wilson, N., G. C. Smith, and D. L. Sykes. 1986. An additional record of *Ornithocoris pallidus* (Hemiptera: Cimicidae) for the United States. J. Med. Entomol. 23: 575.

Yanovski, A. D., and C. W. Ogston. 1982. Sex differences in size of the blood meal in the bed bug *Cimex hemipterus* (Hemiptera: Cimicidae). J. Med. Entomol. 19: 45–47.

CHAPTER 18

Triatominae (Reduviidae)

Eloi S. Garcia, Patricia de Azambuja, and João C. P. Dias

1. INTRODUCTION

The subfamily Triatominae of the family Reduviidae contains more than 110 species, of which several are vectors or potential vectors of Chagas' disease (American trypanosomiasis). Most species, and all known vectors, occur in the New World (Lent and Wygodzinsky 1979). One genus and several species occur in India, where the possibility of Chagas' disease exists, if only in theory (Schaefer 1998). Brief accounts of the biology and ecology of these Indian triatomines may be found in Ambrose (1999). Chagas' disease is of great importance in South and Central America (see below), and a few rare cases have been reported in the United States (Hays et al. 1961, Ryckman 1981). The trypanosome, the triatomine vector, and wild reservoirs all occur in the United States (see Burkholder et al. 1980).

Lent and Wygodzinky (1979) consider the phylogenetic relationships of these bugs; others (e.g., Brodie and Ryckman 1967) have studied relationships of groups of triatomines. Ryckman provides checklists of the species of South America (1986a), of the New World north of South America (1986b), of western North America (1976), and of the Old World (1981); he also lists the vertebrate hosts of North and Central American, and West Indian, triatomines (Ryckman 1986c).

Except for such rare reports as that of Kitselman and Grundman (1940) of the equine encephalitis virus having been isolated from *Triatoma sanguisuga* LeConte, and the retention of the plague bacillus in several species of triatomine (Ames et al. 1954), the importance of Triatomine is exclusively that they are vectors of *Trypanosoma cruzi* Chagas. The discoverer of Chagas' disease, the Brazilian Carlos Chagas, described many clinical, anatomical–pathological, and epidemiological elements as well as the flagellate protozoan *T. cruzi* as the etiologic agent of the disease, and identified the vector as an insect of the order Hemiptera (Chagas 1909). Therefore, since that time the medical importance of the reduviid subfamily Triatominae as a vector of disease has been recognized.

Because of the importance of Chagas' disease, the literature on it is vast. This is no less true of the literature on its vectors, the Triatominae. An excellent brief historical account of the disease is in Morel (1998), a longer account is in Bastien (1998), and much of the work in the triatomine vectors is summarized in Usinger et al. (1966), Zeledón and Rabinovich (1981), Schofield (1994), and Carcavallo et al. (1998a).

Triatomines are 20 to 40 mm long, brownish and often patterned with darker brown, and elongate. The subfamily has been revised by Lent and Wygodzinsky (1979).

From the epidemiological point of view, two main cycles of *T. cruzi* can be distinguished: the *sylvatic* and the *domestic* (Barretto 1979). The older sylvatic cycle involves the circulation of the parasite among wild species of triatomines and small mammals, in different bioecological situations of tropical and subtropical America. In such an interaction, the parasite also can be transmitted orally, when susceptible mammals ingest either infected vectors or reservoirs. Some triatomines are experimental vectors of Indonesian trypanosomes of monkeys (Weinman et al. 1978). Sylvatic triatomines do not impact the flora and fauna and generally do not colonize domestic dwellings. *Trypanosoma cruzi* also does not harm wild mammals, despite being very common among them. The sylvatic cycle does not affect humans, except as the activity of human beings disturbs the natural environment through deforestation and extensive agriculture (Forattini 1980). This human intervention has occurred throughout Latin America during the last 400 years, frequently producing an ecological imbalance that modifies fauna and flora and sometimes results in the introduction of human Chagas' disease (HCD) in colonized areas (Forattini 1980, Dias 1987, Schofield and Dias 1991).

The domestic cycle is historically much more recent, and involves human beings, domestic mammals, and a few species of triatomines that are able to colonize domestic habitats. There is evidence to show that scarce and rare foci of HCD were present among pre-Incaic people during the first millennium A.D.; nevertheless, the real dispersion of the disease occurred after Columbus, as a result of the social and ecological changes caused by the European conquest (Schofield and Dias 1991, Dias 1992). These harsh political and economic changes, occurring from the 17th through the 19th centuries, disturbed the natural balance and habitats of *T. cruzi*, its triatomine vectors, and its mammalian reservoirs, thus bringing the disease to more people. At the same time, these changes displaced and impoverished the local populations, rendering them more susceptible to the disease (Dias 1986, 1992; Schofield and Dias 1991).

2. GENERAL STATEMENT OF ECONOMIC IMPORTANCE

HCD has been traditionally an endemic rural disease, associated with poor houses invaded and colonized by infected vectors. Nevertheless, in the last few decades, migration of thousands of infected individuals to urban centers has brought the parasite to nonendemic centers. This has not only spread the disease, but provided *T. cruzi* with a new way of infecting: via blood transfusion (Dias 1992, Schmunis 1995). Other alternative routes of *T. cruzi* transmission to humans are congenital, by organ transplant, by laboratory accidents, and orally (when susceptible individuals eat uncooked meat infected with the parasite) (WHO 1991).

The medical and social impact of HCD is very high. About 16 to 18 million individuals are infected and more than 90 million are at risk of infection each year (WHO 1991). In the acute phase, generally affecting young children, mortality is estimated from 1 to 5% with a direct cost in Bolivia, for instance, reaching approximately U.S.$2.150 million/year (Bolivia 1994). In the chronic phase, the incidence of severe cardiopathy can reach 20 or 30%, while digestive problems affect around 10% of infected individuals in certain areas. According to a recent estimate by the World Bank, Chagas' disease is among the highest socially damaging of transmissible diseases in Latin America, being surpassed only by acute respiratory diseases, diarrheaic diseases, and AIDS (Schmunis 1995). This estimation was based on the difference in productivity of diseased individuals and healthy individuals, considering the lost working days due to this disease and to death caused by it. A single estimate for Bolivia shows for 420,000 persons with chronic symptoms and an average disability of 0.25, a shocking sum of 105,000 years of productive life lost each year; the same study verified an annual loss of U.S.$100,000/year in that country, for medical treatment (Bolivia 1994). In Brazil, the costs of lost labor of HCD may reach about U.S.$625 million/year, and the losses due to social security (premature retirement) were estimated to be U.S.$400,000 in only one Brazilian state (Dias 1992, Schmunis 1995).

3. THE IMPORTANT SPECIES OF TRIATOMINAE

Triatomines are common from the southern United States through all of Latin America to 50° south latitude, in Patagonia, thus in 17 countries in the Western Hemisphere (WHO 1991). Common names for triatomines are barbeiros, benchuca, vinchuca, kissing bugs, cone-nosed bugs, and chince de monte. The most important species are *Triatoma infestans* Klug, *T. dimidiata* Latreille, *T. barberi* Usinger, *T. brasiliensis* Neiva, *T. maculata* Erichson, *T. pseudomaculata* Correa & Espinola, *T. rubrovaria* Blanchard, *T. sordida* Stål, *Rhodnius prolixus* Stål, *R. ecuatoriensis* Lent & Leon, *R. pallescens* Barber, and *Panstrongylus megistus* Burmeister (see details in Lent and Wygodzinsky 1979).

4. BIOLOGY AND REARING OF TRIATOMINAE

Some triatomines are sylvatic and are found in the safety of burrows and nests of wild vertebrates (opossums, rodents, and birds), rocks (especially associated with small rodents), fallen timber, hollow trees, roots, palms, and bromeliads. Many species are found in peridomestic locations and in domestic animal houses, and others are domestic, occurring inside poor human habitations even when only minimal conditions of shelter and food are offered. Domiciliary triatomines are quite opportunistic in their host selection and feed well on humans (for review see Carcavallo et al. 1998a). They are poor fliers, and dispersal (often of their eggs) usually is related to the transportation of humans, their furniture, goods, or migratory animals. However, in some cases dispersive flight is associated with poor nutritional status of the bug and high ambient temperatures (Schofield et al. 1991, 1992; Lehane et al. 1992).

Rhodnius prolixus is an important triatomine and an example of the biology of the group. The operculated eggs are 2.5 mm long and bright red. The first instar is about the size of the egg. Each of the five instars feeds and then molts, increasing approximately 1.5 to 2.0 times its original size at each molt. All triatomines are obligatorily hematophagous; thus, all five instars require blood meals for their complete development, and the adult needs blood for reproduction. Adults lay eggs in small batches of 14 eggs each and about 20 batches are deposited during the lifetime. The life cycle in the laboratory requires 80 to 100 days from egg to adult. After laying eggs, the females usually feed again on blood. The life span is about 5 to 6 months, unusually long for an adult insect. Development depends upon temperature between 27 to 30°C. Most triatomines have morphological anomalies, do not molt or reproduce, or die, at temperatures above 35°C or below 22°C. Humidity requirements vary and depend on the triatomine species; 60 to 85% relative humidity is considered ideal for most species. Recently, several attempts have been made to find a reliable climate indicator of the distribution of triatomines, in particular to predict changes caused by regional and global climate modification. Microenvironmental conditions, mainly temperature and relative humidity, are strategic factors for triatomine habitats and behavior in nature (WHO 1991, Carcavallo et al. 1998c).

Sherlock et al. (1987) summarized the species found infected with *Trypanosoma cruzi* and other flagellates. Some trypanosomatids are pathogenic on triatomine species. For example, *Trypanosoma rangeli* in *R. prolixus* and *Blastocrithidia* sp. in *Triatoma infestans* may cause behavioral alterations and disturbances of organ systems affecting reproduction and development of these species (Schaub et al. 1992, Schaub 1994).

Two methods are available for large-scale rearing of these insects. A more conventional technique involves feeding insects on immobilized living hosts, such as rabbits (Ryckman and Ryckman 1967, Hill and Campbell 1975), sheep (Gardiner and Maddrell 1972), chickens (Correa 1954), or mice, rats, guinea pigs (Lent and Wygodzinsky 1979). In such cases, triatomines quickly, skillfully, and painlessly remove from their host, in only a few minutes, large volumes of blood (usually, 8 to 12 times their body weight). An alternative method involves the use of membrane feeding devices and blood with anticoagulants or defibrinated blood from vertebrates (Gardiner and Maddrell 1972, Pimley and Langley 1978, Garcia et al. 1984b). The blood meal is heated to 37 to 38°C to attract the insects to the membrane. The frequency of feeding depends on a combination of rearing temperature and humidity, as well as on the triatomine species. For large-scale rearing, all five instars are fed once every 3 weeks. This period allows all nymphs to digest their meal, to molt, and to be ready for the next feeding without excessive starvation. Adults are fed every 15 days, any eggs being removed immediately before feeding.

5. FEEDING AND DIGESTION

Hosts are located by a diversity of cues, including air temperature (Wigglesworth 1972), air currents, and chemicals (Lazzari 1994; review in Hocking 1971). On the host, the bug penetrates the skin with the mandibles in a series of rapid movements. Then the maxillae are thrust into the tissues in a back-and-forth motion and finally rest inside a blood vessel (Lavoipierre et al. 1959). Once the maxillae are in the meal, probing of the diet and salivation occurs and, if a feeding stimulant is present, pumping begins at a high rate (Friend and Smith 1971).

Bloodsucking is complex and made difficult by the presence of hemostatic mechanisms used by the vertebrate host to prevent blood loss (Ribeiro 1987). Triatomines cause small lesions during feeding (Ribeiro and Garcia 1981). Thus, the main hemostatic components are the formation of a platelet plug (in response to ADP and collagen) and vasoconstriction. The salivary secretion of *R. prolixus* contains anticoagulants (Hellmann and Hawkins 1964, 1965), apyrase (Ribeiro and Garcia 1980, Smith et al. 1980), platelet antiaggregating activity, antiserotonin and antihistamine activities, antithromboxane A2 (for review see Ribeiro 1987), and a vasodilator that releases nitric oxide in the blood (Ribeiro et al. 1993). Thus the antihemostatic action of saliva has a degree of redundancy both with regard to positioning of the maxillae inside the blood vessel and to maintaining feeding by assuring a continuous flow of blood.

The blood ingested by triatomines is stored in the dilated anterior midgut (also called crop and stomach). Because more than half the blood proteins are inside the erythrocytes, a few days after feeding most erythrocytes are hemolyzed by a lytic factor, rendering the hemoglobin accessible to digestion (Azambuja et al. 1983). A considerable diversity of digestive enzymes in triatomines has been described: midgut glycosidases and lysozyme (Ribeiro and Pereira 1984); and Garcia and Guimarães (1979), Terra et al. (1988), and Ferreira et al. (1988a,b) showed the presence of soluble β-acetylglucosaminidase, β-galactosidase, α-glucosidase, and α-mannosidase predominating in the anterior midgut lumen; soluble aminopeptidase, membrane-bound α-glucosidase, and membrane-bound α-mannosidase in the posterior midgut tissue; and aminopeptidase, cathepsin B-like and cathepsin D-like enzymes, carboxypeptidase-A, carboxypeptidase-B, di- and tripeptidases, amylase, β-glucosidase, and β-mannosidase in the posterior midgut. Digestion possibly starts in the lumen of the midgut and ends with the action of enzymes trapped between and associated with double extracellular membranes, also named perimicrovillar membranes, or on the surface of midgut cells (Terra et al. 1988).

6. DEVELOPMENT AND EGG LAYING

Wigglesworth (1934, 1940) found that there exists a critical period in *R. prolixus* before which the head is necessary for molting in each instar, and that the stretching of the abdomen that occurs when the insect takes a blood meal initiates the release of a brain factor and leads to ecdysis. This factor was called brain hormone (Wigglesworth 1934, 1940) and, recently, prothoracicotropic hormone (Gilbert et al. 1980). Recently, the relationship between blood feeding and endocrinology has been reviewed (Adams 1999).

Since the earliest studies of Wigglesworth (1934, 1936), it has become evident that the corpus allatum affects the intermolt period of triatomines. The prolongation of the intermolt period in allatectomized *R. prolixus* has been counteracted by juvenile hormone analogue replacement therapy (Dumser and Davey 1975) or by treatment with 20-hydroxyecdysone (Baehr 1975). Similarly, the delay in molting induced by allatectomy or by chemicals like precocenes and isopentenylphenol (two allatotoxins), was also reversed by juvenile hormone or ecdysone therapies (Azambuja et al. 1981).

Wigglesworth (1936) clearly indicated that the full expression of egg development depends on the corpus allatum and juvenile hormone production. This hormone acts both on the fat body to induce synthesis of vitellogenin (Coles 1965) and on the ovary by inducing large spaces between the follicle cells, which allow the passage of the vitellogenin into the oocyte and cause oocyte maturation (Davey 1980). The brain exerts both stimulatory and inhibitory effects on the corpus allatum (Baehr 1974, Mundall and Engelmann 1977, Davey 1987). *Rhodnius prolixus* ovulation and oviposition require a myotropic hormone that increases the contractility of the ovary muscles, the ovary inducing the expulsion of the egg from the ovary. The release of myotropic hormone is related to the mating and the presence of eggs in the ovary (Ruegg et al. 1981, 1982). Triatomines exhibit cyclical patterns of egg production, which are associated with the intake of a blood meal. Protein meals are sufficient to initiate oogenesis, and nonprotein diets do not initiate oogenesis in *R. prolixus* females (Garcia and Azambuja 1985).

7. *TRYPANOSOMA CRUZI* AND INSECT VECTOR INTERACTIONS

The main aspects of the interactions between *T. cruzi* and other trypanosomatids on their insect hosts were reviewed by Gonzalez et al. (1998a) and Azambuja et al. (1998), covering the vast literature. This chapter summarized the most important features of the life cycle and factors involved in the development of the flagellates in the triatomines. Other reviews are of little value, despite their titles, because they focus almost completely on dipterans (Balashov 1984, Dye 1992).

Trypanosoma cruzi comprises a pool of parasite populations circulating among humans, vectors, sylvatic reservoirs, and domestic animals. It displays quite distinct morphological and functional forms related to the vertebrate or invertebrate hosts (Garcia and Azambuja 1991). Bloodstream trypomastigotes occur sparsely in host blood and, if they are ingested by the insect vector, the first parasite differentiation is in the crop. The differentiation to epimastigote starts a few hours after parasite ingestion. The rate of trypomastigote transformation and development of the parasite in the vector vary and are related to the strains of the parasites and vector species (Garcia et al. 1984c).

The eventual differentiation of epimastigotes to metacyclic trypomastigotes (the infective form of the parasite in the insect vector) occurs in the whole gut but predominantly in the rectum. Zeledón et al. (1977) demonstrated that feeding and defecation behavior of triatomines differ among species and among different stages of the development of the same vector species. The time required for a full blood meal is related to the size of the insect. Schofield (1994) showed that postfeeding excretion times vary among species of triatomines and agrees with Zeledón et al. (1977) that only those species that excrete quickly before leaving the host are efficient vectors.

Metacyclic trypomastigotes are released in the feces and urine, and reach the mammalian host either through a skin wound or directly through the mucosa. The parasites invade a wide range of vertebrate host cells, and differentiate into nonflagellated amastigotes that pass through various division cycles as amastigotes inside the host cells. The amastigotes differentiate into trypomastigotes that are liberated on cell rupture to either initiate the next infection cycle with other cells or to complete the life cycle within the insect vector if ingested with the blood meal (Brener 1973, Zeledón 1997).

The development and metacyclogenesis of different *T. cruzi* strains and/or clones are related to the nature of the parasite and its susceptibility to a triatomine species (Garcia and Dvorak 1982,

Garcia et al. 1986b). The establishment of *T. cruzi* infection in the gut of the vector may be dependent on, and possibly regulated by, biochemical factors. Blood meals induce digestive proteases that are active in the midgut regions where the epimastigotes develop, but they do not affect the rate of *T. cruzi* growth (Garcia and Gilliam 1980). However, a crop lytic factor lysed many strains of trypanosomes, with different lysis kinetics (Azambuja et al. 1989). The facts that the Y strain of *T. cruzi*, which is very sensitive to the crop lytic factor, has a very low rate of development in *R. prolixus* (Garcia et al. 1984a) and that the clone Dm28c, which is more resistant to it, grows very well in the digestive tract of this vector indicate a selective advantage for the development of certain strains of *T. cruzi* over other strains in the vector (Garcia et al. 1989a,b).

Lectins occur in the midgut and crop of *R. prolixus*, and lectin receptors occur in epimastigotes but not in trypomastigotes of *T. cruzi*; thus, distinct lectins in the gut of bugs and their selective reaction with the parasite may represent a regulatory mechanism in interaction (Pereira et al. 1981).

Whole blood feeding is much more efficient than plasma feeding in supporting parasite survival and differentiation to metacyclic trypomastigotes. It is thus supposed that hemoglobin is important to the parasite life cycle in the vector, a hypothesis confirmed by Garcia et al. (1995). It is postulated that globin fragments released by proteolytic enzymes attacking hemoglobin modulate the dynamics of *T. cruzi* differentiation from epimastigotes into metacyclic trypomastigotes in the vector's midgut (Garcia et al. 1995). In the rectum, *T. cruzi* begins adhering by the flagellum, which expands its membrane, thus increasing the contact area with the cell substrate. Metacyclogenesis occurs following torsion and elongation of the epimastigote body (Zeledón et al. 1984). The metacyclic trypomastigotes accumulate in the rectum and are eliminated in the urine and feces together with the nontransformed epimastigotes (Pessoa and Martins 1977).

Recently, Gonzalez et al. (1998b, 1999) demonstrated that the head, possibly the neurosecretory system of the brain, is important for development of the perimicrovillar membranes and for maintaining the organization of the midgut cells; they observed a direct relationship between the development of the perimicrovillar membranes in the midgut and the growth of *T. cruzi*. Notwithstanding other regulatory considerations, observations and considerations strongly suggest that a brain factor, which stimulates ecdysteroid production of the prothoracic glands, is responsible, directly or indirectly, for the ultrastructural organization of the *R. prolixus* midgut, and that the perimicrovillar membranes are involved in the development of *T. cruzi* in the gut of this insect vector.

8. CONTROL OF CHAGAS' DISEASE

Control constitutes the most feasible perspective for the problem of Chagas' disease in the social and political reality of Latin America, particularly vector control (Schofield and Dias 1991, WHO 1991); regional approaches may be the most productive (see Schofield 1992). Blood banks are also very important at the present time and must be controlled. HCD can be controlled by direct and specific interventions, chiefly by eliminating domestic vectors and selecting blood donors. The most significant financial benefit of Chagas' disease control accrues from savings in consultation, care, and supporting treatment of chronic patients (Schofield and Dias 1991).

No effective treatment exists for Chagas' disease (Schmunis 1995). Chemoprophylaxis is a possible and effective approach in transfusional Chagas' disease, with the addition of gentian violet to stored blood. In organ transplantation (chiefly of kidneys and heart), prevention is preferably made by an adequate selection of donors discharging those who are seropositives.

8.1 Triatomine Control in Public Health

More than 80% of HCD is attributable to vector transmission, so control of domestic triatomines is considered the main target of existing control programs. A direct fight against sylvatic triatomines and wild reservoirs of *T. cruzi* is not feasible, except via a good environmental management plan to maintain them separated from domestic environments, as well as to detect and eliminate potential invasions of sylvatic vectors.

Vector control in Chagas' disease presupposes programs to be carried out in endemic regions, basically using insecticides and housing improvement. Both strategies generally result in a considerable reduction of domestic triatomine populations, as well as in a significant reduction in the incidence of HCD (Schofield and Dias 1991, WHO 1991). Chemical spraying is considered a short-term approach, whereas housing improvement is long term. These strategies are clearly complementary and must be encouraged in endemic areas, according to the local conditions and possibilities. When domestic triatomines are controlled, the transmission of HCD generally is interrupted. A very close correlation between house density of triatomines and natural infection of triatomines by *T. cruzi* seems to be the rule at the level of the domestic cycle of the parasite. Immediately after well-conducted insecticide spraying, acute cases disappear in endemic areas. The reduction of infected triatomines also results in the displacement of the epidemiological curve of HCD transmission from lower to higher age groups, i.e., minor morbidity and mortality (Dias 1991, 1992). When control activities are interrupted or not followed by surveillance, in endemic areas, the possibilities of housing reinfestation are very high (Dias 1958, Schofield and Dias 1991). Health education is the third fundamental tool in HCD control, chiefly to improve the consolidation of the program, in its surveillance phase (Dias 1987, Schofield and Dias 1991).

Biological control has shown little promise, although a number of nonhost-specific predators will feed on triatomines, and there are several hymenopteran egg parasites (de Santis et al. 1981, 1987). Perhaps the peridomiciliary habitats of the disease vectors are inaccessible to predators and parasites.

8.1.1   Chemical Fight against Chagas' Disease Vectors

The rational application of residual chemical insecticides in and around homes is the most practicable approach in the control of HCD vectors. After World War II, the great impact of DDT in malaria control excited health technicians and health program managers about the possibilities of vector control in other vector-borne diseases. Unfortunately, DDT was not effective against triatomines, in spite of some relative impact observed in certain geographic areas where both diseases coexisted. Since 1947, when Dias and Pellegrino carried out the first field trials of triatomine control using the organochlorine BHC in Brazil, there have been several Chagas' disease vector control trials and campaigns in different parts of Latin America (Dias 1958, 1986; Nocerino 1976; Schofield and Dias 1991). Until the 1980s, these campaigns relied chiefly on BHC and dieldrin (organochlorines), and sometimes also organophosphates (malathion) and carbamate (Propoxur) compounds were employed, generally with good results in short-term evaluations (Tonn 1988, Schofield and Dias 1991). The residual effect of these compounds normally lasted 2 or 3 months and triatomines commonly did not develop resistance against them. Even with an acceptable environmental impact of these compounds on public health, many human and ecological problems on the agricultural side were registered (especially with organochlorine compounds), and some of these products were prohibited in several countries. During the 1970s and 1980s, trials with synthetic pyrethroids proved to be very effective against Chagas' disease vectors, and lacked the toxicologic and ecological negative impact of traditional insecticides. The most active group of the new generation of pyrethroids is that derived from Permethrin, with its characteristic alpha-cyano substitution. These products have a strong "knock-down" effect against different species of triatomines, with a residual activity of 8 months or more indoors and a poorer activity in the peridomestic environment (residual action of 4 months or less) (Dias 1987, Schofield and Dias 1991).

Some alternatives are being tried in the chemical fight against triatomines, looking for better effectiveness (knock-down and residual effects). They are basically a smoke generator canister with insecticide properties (employing DDVP lindane and pyrethroids) and an insecticide paint (malathion or lindane in PVA matrices). The canister is very practical and has a good knock-down power, being able to kill triatomines even when they hide inside wall cracks. Unfortunately this tool does not work in "open" spaces like most peridomestic foci. The available paints have a very good residual effect (at least 2 or 3 years), but their initial cost is high and they also do not work at the peridomestic level.

8.1.2 Housing Improvement and House Management

This is the other strategy to avoid the human–vector contact. This strategy is much more permanent and provides a much better social impact than insecticides. The best thing to do is to combine insecticide with housing improvement, thus reaching immediate long-term results (Dias 1987, Schofield and Dias 1991, Bolivia 1994). In a recent trial in a hyperendemic region of Bolivia, the reinfestation rate of only houses sprayed was 23.5% after 6 months, whereas houses both improved and sprayed had 10.1% reinfestation (Bolivia 1994).

There are different approaches to house improvement, from the total replacement of miserable huts to small modifications of specific house components. Generally, certain triatomine species invade and colonize houses even when only minimal conditions of shelter and food are available. Some species prefer to install themselves in particular points of the house (*Triatoma infestans* in the cracks of the walls of bedrooms and some mudroofs, *R. prolixus* in the roof of palm-leaf covered houses, *T. dimidiata* in the mud floor, etc.) or of peridomestic ecotopes (*T. infestans* in rabbit and chicken houses, *T. sordida* in chicken and pigeon houses, *T. brasiliensis* and *T. rubrovaria* in stone piles, *T. dimidiata* in firewood piles, etc.).

Dirtiness, darkness, general disorganization, cracks in the walls, primitive roofs and floors, accumulation of debris, lack of hygiene, and the presence of vertebrates (chiefly dogs, cats, rodents, poultry, and marsupials) are considered the basic factors of triatomine colonization in domestic habitats (Barretto 1979). Poor rural huts offer an abundance of these conditions in all parts of Latin America characterized by illiterate population, temporary employment, lack of land ownership, hygiene, and self-protection, etc. (Dias 1986, Schofield and Dias 1991). This is the general frame of the chagasic areas, where generally people lack economic conditions to construct good and safe houses, or to maintain them. This was exemplified in Costa Rica in the 1970s, when the socioeconomic improvement of the country practically extinguished the mud floors and the wood-fire stoves in rural areas, thereby reducing enormously house infestation by *T. dimidiata* in endemic regions (Zeledón 1974).

Educational aspects are critical to maintain the house free of triatomines. For example, in an endemic area of Brazil where all the huts were recently replaced by brick houses in a hydroelectric project, after 2 years 85% of the houses had been enlarged by new rooms constructed of local mud or had chicken houses constructed of mud and other primitive materials (Brasil 1994).

9. FINAL REMARKS

The control of HCD is an attainable goal and depends, in Latin America, basically on a political decision of the governments of the affected countries. The control programs must make the fight against domestic triatomines and against the transmission of HCD by blood transfusion a priority. Most of the existing constraints concerning the implementation and/or the maintenance of the regional programs depend on political, financial, and administrative circumstances. Nevertheless, consolidation and success of these programs also will depend on community participation, at the surveillance level. Considering the vectors, the main strategies must be chemical control plus housing improvement, thus pursuing the best results for short and long term. The control of sylvatic triatomines and reservoirs is pointless, but the house must be protected from occasional invasions. The final cost–benefit relation is considered very positive in Chagas' disease control. Research needs concerning HCD control are chiefly related to the challenge of controlling triatomines at the peridomiciliary level and to community participation in the surveillance phase of the existing programs. Housing improvement is a very important tool in HCD control, but presupposes complementary activities such as education and community organization.

10. ACKNOWLEDGMENTS

This work was supported by the Conselho Nacional de Desenvolvimento Científico e Tecnológico (CNPq), Escola Brasil-Argentina de Biotecnologia (MCT), Programa de Apoio à Pesquisa Estratégica em Saúde (PAPES, FIOCRUZ), the ProBal project from CAPES/DAAD, and the Commission of European Communities (CEC) (ERB3504PL921057).

11. REFERENCES CITED

Adams, T. S. 1999. Hematophagy and hormone release. Ann. Entomol. Soc. Amer. 92: 1–13.
Ambrose, D. P. 1999. Assassin Bugs. Science Publishers, Enfield, New Hampshire, U.S.A. 337 pp.
Ames, C. T., S. F. Quan, and R. E. Ryckman. 1954. Triatominae in experimental transmission of plague. Amer. J. Trop. Med. Hyg. 3: 890–896.
Azambuja, P., E. S. Garcia, and J. M. C. Ribeiro 1981. Effects of ecdysone on the metamorphosis and ecdysis prevention induced by precocene II in *Rhodnius prolixus*. Gen. Comp. Endocrinol. 45: 100–104.
Azambuja, P., J. A. Guimarães, and E. S. Garcia. 1983. Haemolytic factor from the crop of *Rhodnius prolixus*: evidence and partial characterization. J. Insect Physiol. 29: 833–837.
Azambuja, P., C. B. Mello, L. N. D'Escoffier, and E. S. Garcia. 1989. *In vitro* cytotoxicity of *Rhodnius prolixus* hemolytic factor and mellitin towards different trypanosomatids. Brazilian J. Med. Biol. Res. 22: 597–599.
Azambuja, P., C. B. Mello, D. Feder, and E. S. Garcia. 1998. Influence of the triatomine cellular and humoral defense system on the development of trypanosomatids. Pp. 709–733 *in* R. U. Carcavallo, I. Galindez Girón, J. Jurberg, and H. Lents [eds.], Atlas of Chagas' Disease Vectors in the Americas. Editora FIOCRUZ, Rio de Janeiro, Brazil. 733 pp.
Baehr, J. C. 1974. Contributions a l'étude des variations naturalles et expérimentales de la protéinémie chez les femelles de *Rhodnius prolixus*. Gen. Comp. Endocrinol. 22: 146–153.
1975. Comparison des effects de l'ecdysone et de l'ecdysterone sur la qualité de la mue chez un Insect Hémiptére, Reduviidae: *Rhodnius prolixus* (Stal) [sic]. C. R. Acad. Sci. Paris 280: 1465–1468.
Baker, C. A., and G. A. Schaub. 1984. Scanning electron microscopic studies of *Trypanosoma cruzi* in the rectum of the vector *Triatoma infestans*. Z. Parasitenkd. 70: 459–469.
Balashov, Yu. S. 1984. Interaction between blood-sucking arthropods and their hosts, and its influence on vector potential. Annu. Rev. Entomol. 29: 137–156.
Barretto, A. 1979. Epidemiologia. Pp. 89–151 *in* Z. Brener and Z. Andrade [eds.], *Trypanosoma cruzi* e Doença de Chagas. Editora Guanabara Koegan, Rio de Janeiro, Brazil. 323 pp.
Bastien, J. W. 1998. The Kiss of Death: Chagas' Disease in the Americas. University of Utah Press, Salt Lake City, Utah, U.S.A. 301 pp.
Bolivia. 1994. Chagas en Bolivia. El trabajo del programa piloto de control del Chagas. SNS/CCH, Ministerio de Desarrollo Humano, La Paz, Bolivia, pp. 1–81.
Brasil. 1994. Controle da doença de Chagas. Ministério da Saúde, Fundação Nacional de Saúde, Brasília, Brazil. pp. 1–80.
Brener, Z. 1973. Biology of *Trypanosoma cruzi*. Annu. Rev. Microbiol. 27: 347–382.
Brodie, H. D., and R. E. Ryckman. 1967. Molecular taxonomy of Triatominae (Hemiptera: Reduviidae). J. Med. Entomol. 4: 497–517.
Burkholder, J. E., T. C. Allison, and V. P. Kelly. 1980. *Trypanosoma cruzi* (Chagas) (Protozoa: Kinetoplastida) in invertebrate, reservoir, and human hosts of the lower Rio Grande Valley of Texas. J. Parasitol. 66: 305–311.
Carcavallo, R. U., I. Galindez Girón, J. Jurberg, and H. Lent. 1998a. Atlas of Chagas' Disease Vectors in the Americas, Vol. I and II. Editora FIOCRUZ, Rio de Janeiro, Brazil. 733 pp.

Carcavallo, R. U., C. Galvão, D. S. Rocha, J. Jurberg, and S. I. Casas. **1998b**. Predicated effects of warming on Chagas' disease vectors and epidemiology. Entomol. Vectores 5: 137–138 (abstr.).

Carcavallo, R. U., M. E. Franca Rodrigues, R. Salvatella, S. I. Curto de Casas, I. A. Sherlock, C. Galvão, D. S. Rocha, I. Galindez Girón, M. A. Otero Arocha, A. Martinez, J. A de Rosa, D. M. Canale, T. H. Far, and J. M. S. Barata. **1998c**. Habitats and related fauna. Pp. 561–600 *in* R. U. Carcavallo, I. Galindez Girón, J. Jurberg, and H. Lent [eds.], Atlas of Chagas' Disease Vectors in the Americas. Editora FIOCRUZ, Rio de Janeiro, Brazil. 733 pp.

Chagas, C. **1909**. Nova tripanomíase humana. Estudos sobre a morfologia e o ciclo evolutivo do *Schizotrypanum cruzi*, n. gen., n. sp., agente etiológico de nova entidade mórbida do homem. Mem. Inst. Oswaldo Cruz 1: 159–218.

Coles, G. C. **1965**. Studies on the hormonal control of metabolism in *Rhodnius prolixus* Stal [sic]. I. The adult female. J. Insect Physiol. 11: 1325–1330.

Correa, R. R. **1954**. Alguns dados sobre a criação de Triatomíneos em laboratório. Folia Clin. Biol. 22: 51–56.

Davey, K. G. **1980**. The physiology of reproduction in *Rhodnius prolixus* and other insects: some questions. Pp. 325–344 *in* M. Locke and D. S. Smith [eds.], Insect Biology in the Future. Academic Press, New York, New York, U.S.A. 678 pp.

1987. Effect of the brain and corpus cardiacum on egg production in *Rhodnius prolixus*. Arch. Insect Biochem. Physiol. 4: 243–249.

de Santis, L., M. C. Coscarón, and M. S. Loiácono. **1987**. Nuevos aportes al conocimiento de los insectos entomofagos que destroyen a las vinchucas. Rev. Soc. Entomol. Argent. 44: 169–177.

de Santis, L., M. S. Loiácono de Silva, and M. C. Coscarón de Larramendy. **1981**. Lucha biologica contra las vinchucas (Hem. Reduviidae). El empleo de insectos entomofagos. Rev. Mus. La Plata (n.s.) (Zool.) 12: 239–260.

Dias, E. **1958**. Profilaxia da doença de Chagas. O Hospital 51: 285–289.

1986. Aspectos socio-culturales y económicos relativos al vector de la enfermedad de Chagas. Pp. 67–86 *in* R. Carvallo and T. Tonn [eds.], Factores Biológicos y Ecológicos en la Enfermedad de Chagas, Vol. II. PAHO, Buenos Aires, Argentina. 357 pp.

1987. Control of Chagas disease in Brazil. Parasitol. Today 3: 336–341.

1991. Chagas disease control in Brazil: which strategy after the attack phase? Ann. Soc. Belg. Med. Trop. 71: 75–86.

1992. Epidemiology. Pp. 49–80 *in* S. Wendel, Z. Brener, M. E. Camargo, and A. Rassi [eds.], Chagas Disease (American Trypanosomiasis): Its Impact on Transfusion and Clinical Medicine. ISBT, São Paulo, Brazil. 425 pp.

1994. Ecological aspects of the vectorial control of Chagas' disease in Brazil. Cad. Saúde Públ. 10: 352–358.

Dumser, J. B., and K. G. Davey **1975**. The *Rhodnius* testis: hormones, differentiation of the germ cells, and duration of the moulting cycle. Canad. J. Zool. 1673–1681.

Dye, C. **1992**. The analysis of parasite transmission by bloodsucking insects. Annu. Rev. Entomol. 16: 1–26.

Ferreira, C., E. S. Garcia, and W. R. Terra. **1988a**. Properties of midgut hydrolases from nymphs and adults of the hematophagous bug *Rhodnius prolixus*. Comp. Biochem. Physiol. 90B: 433–437.

Ferreira, C., A. F. Ribeiro, E. S. Garcia, and W. R. Terra. **1988b**. Digestive enzymes trapped between and associated with the double plasma membranes of *Rhodnius prolixus* posterior midgut cells. Insect Biochem. 18: 521–530.

Forattini, O. P. **1980**. Biogeografia, origem e distribuição da domiciliação de triatomíneos no Brasil. Rev. Saúde Públ. 14: 265–299.

Fraidenraich, D., C. Peña, E. Isola, E. Lammel, O. Coso, A. Díaz Añel, S. Pongor, H. N. Torres, and M. M. Flawiá. **1993**. Stimulation of *Trypanosoma cruzi* adenylyl cyclase by an α^D-globin fragment from Triatoma hindgut: effect of differentiation of epimastigote to trypomastigote forms. Proc. Natl. Acad. Sci. U.S.A. 90: 10140–10144.

Friend, W. G., and J. J. B. Smith. **1971**. Feeding in *Rhodnius prolixus*: mouthpart activity and salivation, and their correlation with changes of electric resistance. J. Insect Physiol. 17: 233–243.

Garcia, E. S., and P. Azambuja. **1985**. A protein diet initiates oogenesis in *Rhodnius prolixus*. Brazil. J. Med. Biol. Res. 18: 195–199.

1991. Development and interactions of *Trypanosoma cruzi* within the insect vector. Parasitol. Today 7: 240–244.

Garcia, E. S., and J. A. Dvorak. 1982. Growth and development of two *Trypanosoma cruzi* clones in the arthropod *Dipetalogaster maximus*. Amer. J. Trop. Med. Hyg. 31: 259–262.

Garcia, E. S., and F. C. Gilliam. 1980. *Trypanosoma cruzi* development is independent of protein digestion in the gut of *Rhodnius prolixus*. J. Parasitol. 66: 1052–1053.

Garcia, E. S., and J. A. Guimarães. 1979. Proteolytic enzymes in the *Rhodnius prolixus* midgut. Experientia 25: 305–306.

Garcia, E. S., P. Azambuja, and W. S. Bowers. 1984a. Comparison of structurally analogous allatoxins on the molting and morphogenesis of *Rhodnius prolixus*, and the reversal of ecdysial stasis by ecdysone. Arch. Insect Biochem. Physiol. 1: 367–373.

Garcia, E. S., P. Azambuja, and V. T. Contreras. 1984b. Large-scale rearing of *Rhodnius prolixus* and preparation of metacyclic trypomastigotes of *Trypanosoma cruzi*. Pp. 43–46 *in* C. Morel [ed.], Genes and Antigens of Parasites. A Laboratory Manual. Fundação Oswaldo Cruz, Rio de Janeiro, Brazil. 634 pp.

Garcia, E. S., E. Vieira, J. E. P. L. Gomes, and A. M. Gonçalves. 1984c. Molecular biology of the interaction *Trypanosoma cruzi*/invertebrate host. Mem. Inst. Oswaldo Cruz 79: 33–37.

Garcia, E. S., M. S. Gonzalez, and H. Rembold. 1989a. Curing effect of azadirachtin A in the triatomine host, *Rhodnius prolixus*, from its parasite, *Trypanosoma cruzi*. Pp. 263–269 *in* D. Borovsky and A. Spielman [eds.], Host-Regulated Developmental Mechanisms in Vector Arthropods. Florida University Press, Vero Beach, Florida, U.S.A. 287 pp.

Garcia, E. S., M. S. Gonzalez, P. Azambuja, and H. Rembold. 1989b. Chagas' disease and its insect vector. Effect of azadirachtin A on the interaction of a triatomine host (*Rhodnius prolixus*) and its parasite (*Trypanosoma cruzi*). Z. Naturforsch. 44c: 317–322.

Garcia, E. S., M. S. Gonzalez, P. Azambuja, F. E. Baralle, D. Fraidenaich, H. N. Torres, and M. M. Flawiá. 1995. Induction of *Trypanosoma cruzi* metacyclogenesis in the hematophagous insect vector by hemoglobin and peptides carrying globin sequences. Exp. Parasitol. 81: 255–261.

Gardiner, B. O. C., and S. H. P. Maddrell. 1972. Techniques for routine and large-scale rearing of *Rhodnius prolixus* (Hem., Reduviidae). Bull. Entomol. Res. 61: 505–516.

Gilbert, L. I., W. E. Bollenbacher, N. Agui, N. A. Granger, B. J. Sedlak, D. Gibbs, and C. M. Buys. 1980. The prothoracicotropic hormone. Amer. Zool. 21: 641–653.

Gonzalez, M. S., P. Azambuja, and E. S. Garcia. 1998a. The influence of triatomine hormonal regulation on the development of *Trypanosoma cruzi*. Pp. 665–707 *in* R. U. Carcavallo, I. Galindez Girón, J. Jurberg, and H. Lents [eds.], Atlas of Chagas' Disease Vectors in the Americas. Editora FIOCRUZ, Rio de Janeiro, Brazil. 733 pp.

Gonzalez, M. S., N. F. S. Nogueira, D. Feder, W. de Souza, P. Azambuja, and E. S. Garcia. 1998b. Role of the head in the ultrastructural midgut organization in *Rhodnius prolixus* larvae: evidence from head transplantation experiments and ecdysone therapy. J. Insect Physiol. 44: 553–560.

Gonzalez, M. S., N. F. S. Nogueira, C. B. Mello, G. A. Schaub, P. Azambuja, and E. S. Garcia. 1999. Influence of the brain on the midgut arrangement and *Trypanosoma cruzi* development in the vector, *Rhodnius prolixus*. Exp. Parasitol. (in press).

Hays, K. L., H. F. Turner, and P. F. Olsen. 1961. Chagas' disease in Alabama. Highlights Agric. Res. (Auburn Univ.) 8: 1.

Hellman, K., and R. I. Hawkins. 1964. Anticoagulant and fibrinolytic activities from *Rhodnius prolixus* Stål. Nature (London) 201: 1008–1009.

1965. Prolixin S and prolixin G: two anticoagulants from *Rhodnius prolixus* Stahl [sic]. Nature (London) 207: 265–268.

Hill, P., and J. A. Campbell. 1975. Improved rearing of symbiont-free and infected *Rhodnius prolixus* using rabbit-ear, warmers and simple microbiological isolators. Lab. Prac. 24: 746–747.

Hocking, B. 1971. Blood-sucking behavior of terrestrial arthropods. Annu. Rev. Entomol. 16: 1–26.

Kitselman, C. H., and A. W. Grundmann. 1940. Equine encephalomyelitis virus isolated from naturally infected *Triatoma sanguisuga* LeConte. Agric. Exp. Stn. Kansas State Coll. Agric. Appl. Sci. Tech. Bull. 50: 1–14.

Lavoipierre, M. M. J., G. Dickerson, and R. M. Gordon. 1959. Studies on the methods of feeding of blood sucking arthropods. I. The manner in which triatomine bugs obtain their blood meal, as observed in the tissues of the living rodent, with some remarks on the effects of the bite on human volunteers. Ann. Trop. Med. Parasitol. 53: 235–250.

Lazzari, C. R. 1994. Sensory ecology of *Triatominae*: modification of vectors' behaviour. Mem. Inst. Oswaldo Cruz 89: 22–23.

Lehane, M. J., P. K. McEwen, C. J. Whitaker, and C. J. Schofield. 1992. The role of temperature and nutritional status in flight initiation by *Triatoma infestans*. Acta Trop. 52: 27–38.

Lent, H., and M. Wygodzinsky. 1979. Revision of the Triatominae (Hemiptera, Reduviidae), and their significance as vectors of Chagas' disease. Bull. Amer. Mus. Nat. Hist. 163: 127–520.

Morel, C. M. 1998. Chagas disease from discovery to control — and beyond. Benjamin Osuntokun Memorial Lecture. WHO/RPS/ACHRS, Geneva, Switzerland. 30 pp.

Mundall, E., and F. Engelman. 1977. Endocrine control of vitellogenin synthesis and vitellogenesis in *Triatoma infestans*. J. Insect Physiol. 23: 825–826.

Nocerino, F. 1976. Susceptibilidad de *R. prolixus* y *T. maculata* a los insecticidas ern Venezuela. Bol. Dir. Malariol. Saneam. Ambiental 16: 276–283.

Pereira, M. E. A., A. F. B. Andrade, and J. M. C. Ribeiro. 1981. Lectins of distinct specificity in *Rhodnius prolixus* interact selectively with *Trypanosoma cruzi*. Science 211: 597–600.

Pessoa, S. B., and A. B. Martins. 1977. *Tryapnosoma cruzi* e moléstia de Chagas. Pp. 143–191 *in* Parasitologia Médica. Editora Guanabara Koegan, Rio de Janeiro, Brazil. 456 pp.

Pimley, R. W., and P. A. Langley. 1978. Rearing triatomine bugs in the absence of a live host and some effects of diet on reproduction of *Rhodnius prolixus* Stal [sic] (Hemiptera, Reduviidae). Bull. Entomol. Res. 68: 243–250.

Ribeiro, J. M. C. 1987. Role of saliva in blood-feeding by arthropods. Annu. Rev. Entomol. 32: 463–478.

Ribeiro, J. M. C., and E. S. Garcia. 1980. The salivary and crop apyrase activity of *Rhodnius prolixus*. J. Insect Physiol. 26: 303–307.

1981. The role of salivary glands in feeding in *Rhodnius prolixus*. J. Exp. Biol. 94: 219–230.

Ribeiro, J. M. C., and M. Pereira. 1984. Midgut glycosidases of *Rhodnius prolixus*. Insect Biochem. 14: 103–108.

Ribeiro, J. M. C., J. M. Hazzard, R. H. Nussenzveig, D. Champagne, and F. A Walker. 1993. Reversible binding of nitric oxide by a salivary nitrosylhemeprotein from the blood sucking bug, *Rhodnius prolixus*. Science 260: 539–541.

Ruegg, R. P., F. L. Kriger, K. G. Davey, and C. G. H. Steel. 1981. Ovarian ecdysone elicits release of a myotropic ovulation hormone in *Rhodnius* (Insect: Hemiptera). Int. J. Invertebr. Reprod. 3: 357–361.

Ruegg, R. P., I. Orchard, and K. G. Davey. 1982. 20-Hydroxyecdysone as a modulator of electrical activity in neurosecretory cells of *Rhodnius prolixus*. J. Insect Physiol. 13: 1629–1636.

Ryckman, R. E. 1981. The kissing bug problem in western North America. Bull. Soc. Vector Ecol. 6: 167–169.

1986a. The Triatominae of South America: a checklist with synonymy (Hemiptera: Reduviidae: Triatominae). Bull. Soc. Vector Ecol. 11: 199–208.

1986b. Names of the Triatominae of North and Central American and the West Indies: their histories, derivations and atymology (Hemiptera: Reduviidae: Triatominae). Bull. Soc. Vector Ecol. 11: 209–220.

1986c. The vertebrate hosts of the Triatominae of North and Central America and the West Indies (Hemiptera: Reduviidae: Triatominae). Bull. Soc. Vector Ecol. 11: 221–241.

Ryckman, R. E., and E. F. Archbold. 1981. The Triatominae and triatominae-borne trypanosomes of Asia, Africa, Australia and the East Indies. Bull. Soc. Vector Ecol. 6: 143–166.

Ryckman, R. E., and M. A. Casdin. 1976. The Triatominae of western North America, a checklist and bibliography. California Vector News 23: 35–52.

Ryckman, R. E., and A. E. Ryckman. 1967. Reduviid bugs. Pp. 183–200 *in* Insect Colonization and Mass Production. Academic Press, New York, New York, U.S.A. 387 pp.

Schaefer, C. W. 1998. Phylogeny, systematics, and practical entomology: the Heteroptera (Hemiptera). An. Soc. Entomol. Brasil 27: 499–511.

Schaub, G. A. 1994. Pathogenicity of trypanosomatids on insects. Parasitol. Today 10: 463–466.

Schaub, G. A., B. Rohr, and S. Wolf. 1992. Pathological effects of *Blastocrithidia triatominae* (Trypanosomatidae) on *Triatoma infestans* with different infection rates (Heteroptera, Reduviidae). Entomol. Gen. 17: 21–27.

Schmunis, G. A. 1995. Prólogo. *In* J. C. P. Dias and J. R. Moura [eds.], Clínica e Terapêutica da Doença de Chagas. Um Manual para o Clínico Geral. Fiocruz, Rio de Janeiro, Brazil.

Schofield, C. J. 1992. Eradication of *Triatoma infestans*: a new regional programme for southern Latin America. Ann. Soc. Belg. Méd. Trop. 72 (Suppl.): 69–70.

1994. Triatomine: Biology and Control. Eurocommunica Publication, Bognor Regis, U.K. 77 pp.
Schofield, C. J., and J. C. P. Dias. 1991. A cost–benefit analysis of Chagas disease control. Mem. Inst. Oswaldo Cruz 86: 285–295.
Schofield, C. J., M. J. Lehane, P. McEwan, S. S. Catalá, and D. E. Gorla. 1991. Dispersive flight by *Triatoma sordida*. Trans. R. Soc. Trop. Med. Hyg. 85: 676–678.
1992. Dispersive flight by *Triatoma infestans* under natural climatic conditions in Argentina. Med. Vet. Entomol. 6: 51–56.
Sherlock I. A., R. U. Carcavallo, and I. Galindez Giréon. 1987. List of natural and experimental flagellate infections in several triatomine species. Pp. 289–298 *in* R. U. Carcavallo, I. Galindez Girón, J. Jurberg, and H. Lents [eds.], Atlas of Chagas' Disease Vectors in the Americas. Editora FIOCRUZ, Rio de Janeiro, Brazil. 391 pp.
Smith, J. J. B., R. Cornish, and J. Wilkes. 1980. Properties of a calcium-dependent apyrase in the saliva of the blood-feeding bug, *Rhodnius prolixus*. Experientia 36: 898–900.
Terra, W. R., C. Ferreira, and E. S. Garcia. 1988. Origin, distribution, properties and functions of the major *Rhodnius prolixus* midgut hydrolases. Insect Biochem. 18: 423–534.
Tonn, R. B. 1988. Review of recent publications on the ecology, biology and control of vectors of Chagas' disease. Rev. Argentina Microbiol. 2: 4–24.
Usinger, R. L., P. Wygodzinsky, and R. E. Ryckman. 1966. Biosystematics of Triatominae. Annu. Rev. Entomol. 11: 309–330.
Weinman, D., R. C. Wallis, W. H. Cheong, and S. Mahadevan. 1978. Triatomines as experimental vectors of trypanosomes of Asian monkeys. Amer. J. Trop. Med. Hyg. 27: 232–237.
WHO, 1991. Control of Chagas disease. Geneva. WHO Tech. Series No. 811: 1–95.
Wigglesworth, V. B. 1934. The physiology of ecdysis in *Rhodnius prolixus* (Hemiptera). II. Factors controlling moulting and metamorphosis. Q. J. Microsc. Sci. 77: 191–222.
1936. The function of the *corpus allatum* in the growth and reproduction of *Rhodnius prolixus*. Q. J. Microsc. Sci. 79: 91–121.
1940. The determination of characters at metamorphosis in *Rhodnius prolixus* (Hemiptera). J. Exp. Biol. 17: 201–222.
1972. The Principles of Insect Physiology, seventh edition. Chapman & Hall, London, U.K. 827 pp.
Zeledón, R. 1974. Epidemiology, modes of transmission and reservoir hosts of Chagas' disease. Ciba Found. Symp. 20: 51–77.
1997. Infection of the insect host by *Trypanosoma cruzi*. Pp. 271–287 *in* R. U. Carcavallo, I. Galindez Girón, J. Jurberg, and H. Lents [eds.], Atlas of Chagas' disease vectors in the Americas. Editora FIOCRUZ, Rio de Janeiro, Brazil. 391 pp.
Zeledón, R., and J. E. Rabinovich. 1981. Chagas' disease: an ecological appraisal with special emphasis on its insect vectors. Annu. Rev. Entomol. 26: 101–133.
Zeledón, R., R. Alvarado, and L. F. Jirou. 1977. Observations on the feeding and defecation patterns of three triatomine species (Hemiptera: Reduviidae). Acta Trop. 34: 65–77.
Zeledón, R., R. Belanos, and M. Rojas. 1984. Scanning electron microscopy of the final phase of the life cycle of *Trypanosoma cruzi* in the insect vector. Acta Trop. 41: 39–43.

CHAPTER **19**

Adventitious Biters — "Nuisance" Bugs

Carl W. Schaefer

1. INTRODUCTION

Very few heteropterans attack humans: triatomine reduviids and bed bugs take vertebrate blood, and some willingly or preferentially take blood from humans (**Chapters 17 and 18**). Nearly all other heteropterans leave vertebrates alone, and no other heteropteran attacks people — except adventitiously. This chapter considers the many accounts of attacks on humans by bugs that feed on other organisms. These attacks cause no real harm, but they are at best annoying and at worst important nuisances.

Such feeding may be termed *accidental*, but this word suggests the bug thought it was doing something else, but became distracted or bemused. *Adventitious* is a better word, for the bug is biting the human on purpose, however unusual such biting may be. The usual purposes appear to be defensive, or the obtaining of water and/or solutes. The attraction often appears to be perspiration, although Waage (1979) cites several authors who believe unusual environmental conditions (e.g., strong light) or extreme hunger may also evoke such biting.

In a remarkably thorough review, Ryckman (1979) and Ryckman and Bentley (1979) write that "fifteen families of" Heteroptera "have been incriminated as having bitten or annoyed man" (abstract). Although their bibliography of some 300 references includes many triatomines and bed bugs, it also includes very many references to bugs as adventitious biters; these references are annotated. Bequaert (1926), too, lists several adventitious biters, and Myers (1929) lists a number of homopterans (these included also in Ryckman 1979; and see **Section 3**).

What follows will list and briefly discuss nonhematophagous heteropterans recorded as having bitten humans.

2. ADVENTITIOUS BITERS BY FAMILY

2.1 Anthocoridae

This family appears to contain most of these biters, although the many European reports of *Lyctocoris campestris* (F.) may inflate the total. In Britain, this bug may infest people's clothing and bedding (Southwood and Leston 1959), especially bedding stuffed with grasses or straw (Štys and Daniel 1957). People are often attacked in their sleep (Štys and Daniel 1957, Štys 1973) by this bug, which usually lives in birds' nests and beneath bark (Štys 1973, a paper that describes several cases, both previously published and not, in then Czechoslovakia). Štys (1973) recommends, as control, sterilizing hay before it is used as mattress stuffing, or fumigating the home with methyl bromide.

Of this species, China (1933) records Reuter's suspicion that, when the bug sucks blood from horses and cattle, it may "transmit the dreaded foot and mouth disease" (p. 517); this worry is unwarranted.

Several other anthocorid species are recorded as attacking people: several *Anthocoris* spp. in Europe (Péricart 1972) and *A. nemorum* (L.) in England (Verdcourt 1990, the attacks occurring over several years; Atkinson 1990; Smith 1990). *Blepharidopterus angulatus* (Fallén) (England) and *Orius insidiosus* (Say) (Illinois, U.S.A.) also have attacked people (Smith 1990, and Malloch 1916, respectively). These bites tend to occur in gardens, near fruit trees, in summer; the itching and small welts persist several days.

In Sudan Republic, *A. kingi* Brumpt and *Piezostethus afer* Reuter have bitten humans, the latter "causing irritation and swelling" (Lewis 1958, p. 45).

There are at least two accounts of anthocorids biting workers. *Xylocoris nigromarginatus* Carayon, an African species usually found in and around bark, persistently bit a South African miner 1200 m below ground. The bugs were abundant on wooden poles brought down as supports for the roof. Where there was light and refrigerated water, the bugs did not occur (Ledger et al. 1982). Earlier, Dolling (1977) reported a heavy infestation of *Dufouriellus ater* (Dufour), also a bark-inhabiting species, in a clothing factory in England; how they got there is not clear, but the steamy heat and humidity, and the small insects living in the debris, surely supported them. The annoyance was enough to cause concern, and the factory was fumigated successfully with a compound whose active ingredient was 16% gamma HCH. Dolling (1977) also reported *X. galactinus* (Fallén) biting workers in a biscuit factory, again causing concern.

Doubtless anthocorids also adventitiously attack other warm-blooded vertebrates, although the naked skin of humans, when heated and damp, probably emits more attractants. But nest-dwelling vertebrates, birds and mammals, must be attacked too. The present author has seen only one account, however: two specimens of *Orius insidiosus* attacking a white-footed mouse (*Peromyscus leucopus*), in Connecticut (U.S.A.); the bugs' mouthparts were embedded in the mammal's skin (Krinsky et al. 1980).

2.2 Cimicidae

Although cimicids are in fact hematophagous, most do not feed on human blood (see **Chapter 17**). Occasionally, however, some cimicids are adventitious feeders on humans. *Cimex pipistrelli* Jenyns, when experimentally fed "on the previously warmed skin of [the author's] arm," produced eggs, and a fifth instar molted (Southwood 1954, p. 35); this, however, appears to be less an example of adventitious feeding and more an example of a bug led unwitting into the path of temptation.

The Mexican chicken (or poultry) bug, *Haematosiphon inodorus* (Duges), feeds willingly on people, and poultry porters have complained; occasionally, it may infest humans' dwellings (Lee 1955). Similarly, *Cimexopsis nyctalis* List, whose usual host is the chimney swift, *Chaetura pelagica* (L.), has at least three times bitten people in Connecticut (U.S.A.) (Welch 1990). The bat cimicid, *Stricticimex antennatus* Ferris and Usinger, bit humans entering an abandoned mine in South Africa; the bite was painless, but later the area became reddened, swollen, and itched (Overal and Wingate 1976).

2.3 Miridae

Although the largest of the heteropteran families (more than 30% of the species), Miridae does not contain the most adventitious biters. Bequaert (1926) and Myers (1929) list seven species. Altogether in the lists of Bequaert (1926), Myers (1929), and Ryckman (1979), about ten species occur, several of them listed more than once. To these may be added *Lygocoris* (as *Lygus*) *communis novascotiensis* (Knight), which has "stung" people passing below apple trees; "if left undisturbed, it will sometimes pierce the skin ... and remain feeding until completely engorged with blood" (Brittain 1917, p. 17). An "early instar" of the tarnished plant bug, *Lygus lineolaris* Palisot de Beauvois, also "will pierce human skin and suck blood" (Khattat and Stewart 1977). J. D. Lattin (personal communication) describes the result of letting a mature nymph of *Deraecoris* sp. (on *Corylus*), feed on him (in Oregon, U.S.A.): "A small lump arose and stayed hard for about two weeks." And the azalea plant bug, *Rhinocapsus vanduzeei* Uhler, may bite workers on azaleas painfully, causing "a chigger-like welt" (Wheeler and Herring 1979, p. 12; also Wheeler 1979). *Irbisia solani* (Heidemann) bit someone briefly, "leaving a small red spot which soon disappeared" (Usinger 1934, p. 98).

Further accounts may be found in Wheeler (2001).

2.4   Tingidae

It is interesting that although there are several accounts of tingids' biting humans, none of the older reviews mentions this family (i.e., Bequaert 1926, Myers 1929, Usinger 1934, Ryckman 1979). *Corythucha cydoniae* (Fitch) bit several people in North Carolina (U.S.A.), once on a crowded bus; the biting consisted more of probing than of sustained feeding, and resulted in a red and itching area (Sailer 1945). Another *Corythucha* species, *C. ciliata* (Say), bit someone in Virginia (U.S.A.) with much the same results (Hoffman 1953). In Utah (U.S.A.), "thousands of *Corythucha morrilli* Osborne and Drake" were on rabbit brush, *Chrysothamnus nauseosus*, causing browning of the foliage. "It was hot and I was perspiring freely." ... "Hundreds of individuals were scattered over my clothing and skin," and "[t]he skin irritation ... was severe and their biting persisted for some time." Other attacks occurred elsewhere (Knowlton 1958; quotes from p. 73, not in this sequence). More recently (1981), in Connecticut (U.S.A.), *Corythucha* sp. bit several people gathered under a sycamore tree (R. J. Packauskas, personal communication); and (1999, also in Connecticut) *Stephanitis takeyai* Drake and Maa (identified by M. Wall), bit a woman on the cheek, causing it to swell (C. W. Schaefer, personal observation).

It should be remarked that the preponderance of *Corythucha* species in these accounts more probably reflects the ubiquity of these species on ornamental plants than it reflects a greater propensity of these species to bite people.

2.5   Reduviidae

There are several records of nontriatomine reduviids biting humans. Many of these bugs are large, and their approach — accidental or not — to people would be noticed (unlike the approach of the small tingids or tiny anthocorids); therefore, some of the bitings may have been defensive, the bug defending against the person's warding-off action.

Readio (1927) gives several records of reduviids biting people in the United States: these include *Melanolestes picipes* (Herrich-Schaeffer) (the "black kissing bug," or "black corsair"), which is a fairly frequent attacker [*Note:* "Kissing bug," a name of this and occasionally other species, derives from the bugs' propensity to bite people's lips; and "corsair" I suppose because of the way these bugs sometimes sail fiercely toward one, like a Barbary privateer]; *Rasahus biguttatus* (Say) ("two-spotted corsair"), and *R. thoracicus* Stål. Readio (1927) writes that this last species is in fact responsible for many of the "spider" bites reported in California. Less frequently biting are the wheel bug, *Arilus cristatus* (L.), *Apiomerus spissipes* (Say), and nymphs of *Acholla multispinosus* (De Geer). Also in the United States, *Arilus cristatus* again, immature *Zelus* sp., and *M. picipes* have bitten people in Illinois (U.S.A.) (J. E. McPherson, personal communication), whereas despite the thousands of *Sinea* and *Zelus* spp. handled by A. C. Cohen in the course of his research, none has bitten him (personal communication). In Sudan Republic *Pasira* (as *Eremovescia*) *lewisi* (Miller) and *P. basiptera* Stål have bitten people (Lewis 1958); and *Oncocephalus pacificus* Kirkaldy bit a woman on the arm in Hawaii (U.S.A.) (Gagne 1971).

In Japan, several reduviids have adventitiously bitten people: *Isyndus obscurus* (Dallas), *Cleptocoris* (as *Peirates*) *turpis* (Walker), and *Sirthenea flavipes* (Stål) (Yasunaga et al. 1993).

Ryckman (1979) lists references to other records, and Readio (1927) suggests that members of the subfamily Peiratinae (which includes *Rasahus* and *Melanolestes*) are rather more likely to bite humans than members of other subfamilies (except, of course, Triatominae).

2.6 Other Heteropteran Groups

Aquatic bugs — Like reduviids, these are relatively large predaceous bugs, and not reluctant to bite; some people are more attractive than others (see **Section 4**, below), and the bite can be very painful. Not surprisingly, the larger the bug, the more painful the bite. Giant water bugs (Belostomatidae) are among the largest of all heteropterans, and some have been called "toe-biters" because of the willingness with which they sample incautious waders. As reported in **Chapter 22**, giant water bugs inject a paralysant into the wound, which produces an intense burning sensation that may last for several hours. Aquatic heteropterans bite most probably in defense, for neither moisture nor salts are likely to stimulate them.

Nabidae — Gyotoku (1960) reports a bite by *Nabis kinbergii* Reuter.

Pentatomoidea — Members of the families Cydnidae and Pentatomidae have, rarely, been reported to bite people. The cydnid, *Fromundus* (as *Geotomus*; see Lis 1994) *pygmaeus* (Dallas) bit a man's thumb in Kuala Lumpur (Malaysia): "After exploring a small area of the surface of the skin it ... chose a spot and then plunged its stylets in ... apparently sucking for ... about ten minutes" (p. 214); a slight itching and a white spot remained for about a day; however, the following day the bug contained no blood (Miller 1931). Another report (Takai et al. 1975) is of mass occurrences of the cydnid *Aethus indicus* (Westwood) in homes in Japan; biting was not reported. Most Pentatomidae are plant feeders. Of these, *Melanaethus subglaber* (Walker) (as *M. parvulus* [Signoret]) "were really 'biters'" in Utah (U.S.A.) (Knowlton 1960, p. 53). In Sudan Republic, "a species of *Mecidea* bit a man" (Lewis 1958, p. 41). Asopine pentatomids are predators on other insects. It might be expected that more of this group would attack humans than of the other, plant-feeding groups of Pentatomidae. However, the reports are few (see **Chapter 32**). Indeed, asopines seem rather less inclined to bite humans than cydnids. Warren and Wallis (1971) write that fifth instars of the spined soldier bug, *Podisus maculiventris* (Say) were "enticed to attack a human finger" (p. 114), suggesting a certain reluctance on the part of the bugs. A. C. Cohen writes (personal communication) that in his years of working with asopines, he has not been bitten, and P. De Clercq (personal communication) has been bitten only twice in 10 years of handling asopines. No records of pentatomids biting are in Ryckman's (1979) comprehensive list.

Lygaeoidea — In Kuwait, in summer, large numbers of *Nysius* (not further identified) landed on and bit exposed arms and legs, causing some pain, and swelling that lasted several days; populations decreased in the autumn (al-Houty 1990). *Nysius* sp. was also reported biting a person in Sudan, as was a *Pamera* (as *Orthaea*) sp. (Lewis 1958). Also in Sudan, as well as in Egypt, in desert areas, *Leptodemus bicolor ventralis* Schmidt bit several people in several locations, causing irritation and fever (Lewis 1958). Two species of *Clerada* take human blood in houses in eastern Brazil (Ferreira and Deane 1938); *Clerada* belongs to a group of lygaeids that probably feeds on vertebrate blood. The genera mentioned above (except *Pamera*) are listed also by Bequaert (1926), and de Bergevin (1924, 1925) mentions both *Geocoris* and *Leptodemus*. It seems likely that many of these records are independent of one another. Many of these adventitious bitings occurred in desert areas of northern Africa. In the laboratory, *G. punctipes* Say will bite (A. C. Cohen, personal communication). Berytids may, very rarely, bite humans; *Pronotacantha annulata* Uhler has at least once inflicted a painful bite (Henry 1997, p. 80).

Pyrrhocoridae — In their lists, Ryckman (1979) and Ryckman and Bentley (1979) cite several authors' reports of *Dysdercus superstitiosus* (F.) biting people, especially during the dry season when food (and water?) was scarce.

Rhopalidae — Having learned that starved boxelder bugs would feed on a piece of beef, Usinger (1934) conducted a strange experiment. He fed blood serum [human?] to starved boxelder bugs, identified by him as *Leptocoris trivittatus* (Say) but, almost certainly, *Boisea rubrolineata* (Barber), a species which, like Usinger, occurs in California (U.S.A.) but which, in 1934, had not yet been described by Barber (1956). Some bugs had earlier been induced to bite a person's arm. Starved first instars survived up to 2 weeks, and a few molted to second instars; starved adults also fed, but Usinger does not say if they reproduced, nor how long they lived.

3. HOMOPTERA

Several of the cited papers also mention homopterans — mostly auchenorrhynchans — adventitiously biting humans (Myers 1929, Usinger 1934, Ryckman 1979, also McCrae 1974).

4. CONCLUSIONS

No adventitious biter has been shown to transmit any disease. The effects of such biting range from pinpricks to serious pain, depending for the most part on the size of the biter (and presumably the susceptibility to pain of the bitten). Such biting is likely to occur when the bugs are in large numbers and/or when the human is hot and perspiring; when the population is very high, and when the human is particularly damp, the chances of such biting seem to be increased. Consider, for example Knowlton's (1958) experience: When it was hot, he perspired; and the tingids had caused considerable damage to their host plant (see **Section 2.3**).

Just as some insects bite people on purpose, some people are more attractive to adventitiously biting bugs than are others. John Spence (a contributor to this book), offers the following:

> Once, soon after I arrived in Edmonton [Alberta, Canada], I went out to a lakeside cottage, at the behest of someone in my dean's office, for a bowl of wonderful homemade ice cream and a short swim on a hot summer's day. The question was, what was biting the daughter of the wealthy lady who turned the ice cream, every time the daughter jumped into the lake? It turned out to be notonectids, although for some strange reason they did not bite me. The wee lass was indeed covered with red spots from the accumulated bites and I watched in horror as the notonectids did literally attack her. I know not why …[, nor why t]hey still have not started to bite me.

Similarly, A. C. Cohen and his colleagues have not been bitten by the many reduviids and asopine pentatomids they have worked with over the years, and P. De Clercq has been bitten only twice in the 10 years he has been studying asopine biology and ecology. Many are called, but few are chosen.

Usinger's (1934) experiments with boxelder bugs were designed to test the idea that phytophagous bugs can move fairly easily to vertebrate blood; they often appear to be attracted to uric acid (although the boxelder bugs on beef were not). More important, "the same [chemical] elements are found in plants as in blood and often in very similar combinations although in very different proportions" (p. 99). As a result, "the jump from plant sucking to blood sucking may be quite within the range of possibility" (p. 99), a hypothesis repeated as recently as 1990 (al-Houty 1990).

The evolutionary significance of adventitious biting is obscure, if it exists at all. The fact that some lyctocorine anthocorids live in birds' nests and there feed at times on avian blood may be of some importance, as these anthocorids are related to Cimicidae (see discussion in Péricart 1972). This idea is broached by Weber (1930) and discussed by Waage (1979), but one suspects that sometimes a blood-filled lyctocorinae may have been feeding not on a bird but on an avian ectoparasite full of avian blood.

One may conclude that adventitious biting causes no great harm. Yet, as Hoffman (1953) wonders, "[t]here remains unanswered the question of what manner of spiritual or material satisfaction rewards the [insects] from their ventures into anthropophagism" (p. 176).

5. ACKNOWLEDGMENTS

The author is grateful to many of the contributors of this book, who sent references and personal accounts of these biters. In particular, I thank Jack Lattin, Yosihiro Natuhara, John Spence, Kris Braman, and R. J. Packauskas. I also thank Michael Wall, for identifying *Stephanitis takeyai*.

6. REFERENCES CITED

al-Houty, W. 1990. Nysius [sic] (Hem., Lygaeidae) sucking human blood in Kuwait. Entomol. Mon. Mag. 126: 95–96.
Atkinson, M. D. 1990. Reports of human biting by the flower bug, Anthocoris nemorum [sic] (L.) Hem. Anthocoridae). Entomol. Mon. Mag. 126: 96.
Barber, H. G. 1956. A new species of Leptocoris [sic] (Coreidae: Leptocorini). Pan-Pac. Entomol. 32: 9–11.
Bequaert, J. 1926. Medical and economic entomology. Pp. 160–257 *in* Medical Report of the Hamilton Rice Seventh Expedition to the Amazon. Contrib. Harvard Inst. Trop. Biol. Med. no. 4 [not seen, *fide* Myers 1929, Ryckman 1979].
de Bergevin, E. 1924. Nouvelles observations sur les hémiptères suceurs de sang humain. Bull. Soc. Hist. Nat. Afr. Nord 15: 259–262 [not seen, *fide* Ryckman 1979].
1925. Les hémiptères suceurs de sang. Arch. Inst. Pasteur Alger. 3: 28–41 [not seen, *fide* Ryckman 1979].
Brittain, W. H. 1917. The green apple bug in Nova Scotia. Proc. Nova Scotia Dept. Agric. Bull. 8: 1–63.
China, W. E. 1933. A new genus and species of Anthocoridae (Hemiptera) from New Zealand. Ann. Mag. Nat. Hist. (10) 11: 514–518.
Dolling, W. R. 1977. *Dufouriellus ater* (Dufour) (Hemiptera: Anthocoridae) biting industrial workers in Britain. Trans. R. Soc. Trop. Med. Hyg. 71: 355.
Ferreira, L. de C., and L. Deane. 1938. Encontro de um nová hematophago do homen com hábitos domiciliares. Brasil. Med. 52: 1137–1141 [not seen, *fide* Ryckman 1979].
Gagne, W. 1971. *Oncocephalus pacificus* Kirkaldy. Proc. Hawaiian Entomol. Soc. 21: 25.
Gyotoku, N. 1960. A case of bite by *Nabis (Tropiconabis) kinbergii* Reuter. Pulex 25: 98 [in Japanese; not seen, *fide* Y. Natuhara, personal communication].
Henry, T. J. 1997. Monograph of the stilt bugs, or Berytidae (Heteroptera), of the Western Hemisphere. Mem. Entomol. Soc. Washington 19: 1–149.
Hoffman, R. L. 1953. A second case of lacebug bite (Hemiptera: Tingidae). Entomol. News 64: 176.
Khattat, A. M., and R. K. Stewart. 1977. Development and survival of *Lygus lineolaris* exposed to different laboratory rearing conditions. Ann. Entomol. Soc. Amer. 70: 274–278.
Knowlton, G. F. 1958. Tingidae are biters. Bull. Brooklyn Entomol. Soc. 53: 73.
1960. An unusual flight of Hemiptera in southern Utah. Proc. Utah Acad. 37: 53–54.
Krinsky, W. L., A. B. Carey, and M. G. Carey. 1980. *Orius insidiosus* (Say) (Hemiptera: Anthocoridae) biting white-footed mice. Entomol. News 91: 31.
Ledger, J. A., J. B. Rossiter, and J. M. C. Ossthuizen. 1982. Anthocorid bug bites in a Transvaal goldmine. A case report. South Afric. Med. J. 62: 69–70.
Lee, R. D. 1955. The biology of the Mexican chicken bug, Haematosiphon inodorus [sic] (Duges). Pan-Pac. Entomol. 31: 47–61.
Lewis, D. J. 1958. Hemiptera of medical interest in the Sudan Republic. Proc. R. Entomol. Soc. (A) 33: 43–47.
Lis, J. A. 1994. A Revision of Oriental Burrower Bugs (Heteroptera: Cydnidae). Upper Silesian Museum, Bytom, Poland. 349 pp.
Malloch, J. R. 1916. Triphleps insidiosus [sic] (Say) sucking blood (Hem., Het.). Entomol. News 27: 200.

McCrae, A. W. R. 1974. An instance of man-biting by *Empoasca* sp. (Hem., Cicadellidae) in Britain. Entomol. Mon. Mag. 109: 238–239.

Miller, N. C. E. 1931. Geotomus pygmaeus [sic] Dallas (Heteroptera: Cydnidae) attempting to suck human blood. Entomologist 62: 214.

Myers, J. G. 1929. Facultative blood-sucking in phytophagous Hemiptera. Parasitology 21: 472–480.

Overal, W. L., and L. R. Wingate. 1976. The biology of the batbug Stricticimex antennatus [sic] (Hemiptera: Cimicidae) in South Africa. Ann. Natal Mus. 22: 821–828.

Péricart, J. 1972. Hémiptères Anthocoridae, Cimicidae, Microphysidae de l'Ouest-Paléarctique. Fauna de l'Europe et du Bassin Méditerranéen, Vol. 7. Masson et Cie, Paris, France. 402 pp.

Readio, P. A. 1927. Studies on the biology of the Reduviidae of America north of Mexico. Univ. Kansas Sci. Bull. 17: 5–291.

Ryckman, R. E. 1979. Host reactions to bug bites (Hemiptera, Homoptera): a literature review and annotated bibliography. Part I. California Vector News 26: 1–24.

Ryckman, R. E., and D. G. Bentley. 1979. Host reactions to bug bites (Hemiptera, Homoptera): a literature review and annotated bibliography. Part II. California Vector News 26: 25–49.

Sailer, R. I. 1945. The bite of a lacebug, *Corythucha cydoniae* (Fitch). J. Kansas Entomol. Soc. 18: [no pagination].

Smith, K. G. V. 1990. Hemiptera (Anthocoridae and Miridae) biting man. Entomol. Mon. Mag. 126: 96.

Southwood, T. R. E. 1954. The production of fertile eggs by *Cimex pipistrelli* Jenyns (Hem., Cimicidae) when fed on human blood. Entomol. Mon. Mag. 90: 35.

Southwood, T. R. E., and D. Leston. 1959. Land and Water Bugs of the British Isles. Frederick Warner & Co., Ltd., London, U.K. 436 pp.

Štys, P. 1973. Cases of facultative parasitism of Lyctocorinae (Heteroptera: Anthocoridae) on man in Czechoslovakia. Folia Parasitol. (Praha) 20: 103–104.

Štys, P., and M. Daniel. 1957. *Lyctocoris campestris* (F.) (Heteroptera, Anthocoridae) jako fakultativní ektoparasit …lov-ka. Acta Soc. Entomol. Čechosloveniae 54: 88–97 [in Czech, English summary].

Takai, R., T. Yamaguchi, and T. Kurihara. 1975. Mass occurrence of *Aethus indicus* (Hemiptera: Cydnidae) as a house-frequenting pest in the Amami Islands. Jpn. J. Sanit. Zool. 26: 61–63 [in Japanese; not seen, *fide* Y. Natuhara, personal communication].

Usinger, R. L. 1934. Blood sucking among phytophagous Hemiptera. Canadian Entomol. 66: 97–100.

Verdcourt, R. 1990. *Anthocoris nemorum* (L.) (Hem., Anthocoris nemorum (L.) (Hem., Anthocoridae) attacking man. Entomol. Mon. Mag. 126: 96.

Waage, J. K. 1979. The evolution of insect/vertebrate associations. Biol. J. Linn. Soc. 12: 187–224.

Warren, L. O., and G. Wallis. 1971. Biology of the spined soldier bug, *Podisus maculiventris* (Hemiptera: Pentatomidae). J. Georgia Entomol. Soc. 6: 109–116.

Weber, H. 1930. Biologie der Hemipteren, eine Naturgeschichte der Schnabelkerfe. Julius Springer, Berlin, Germany. 543 pp.

Welch, K. A. 1990. First distributional records of *Cimexopsis nyctalis* List (Hemiptera: Cimicidae) in Connecticut. Proc. Entomol. Soc. Washington 92: 811.

Wheeler, A. G., Jr. 1979. Azalea lace bug, *Rhinocapsus vanduzeei* Uhler. Regul. Hortic. 5: 19–20.

2001. Biology of the Plant Bugs: Pests, Predators, Opportunists. Cornell University Press, Ithaca, New York, New York, U.S.A.

Wheeler, A. G., Jr., and J. L. Herring. 1979. A potential pest of azaleas. Q. Bull. Amer. Rhododendron Soc. 33: 12–14.

Yasunaga, T., M. Takai, I. Yamashita, M. Kawamura, and T. Kawasawa. 1993. Field guide to Japanese bugs. Zenkoku Noson Kyoiku Kyoaki, Tokyo, Japan [in Japanese; not seen, *fide* Y. Natuhara, personal communication].

SECTION III

Useful Bugs

CHAPTER 20

How Carnivorous Bugs Feed

Allen Carson Cohen

1. INTRODUCTION

Most families in the hemipteran suborder Heteroptera contain at least some species of zoophages. If Henry and Froeschner's (1988) accounting of U.S. and Canadian Heteroptera reflects worldwide proportions, 65% of all families in this suborder are partially or entirely composed of carnivorous species. Although consideration of the original (carnivorous or phytophagous) trophic habits of this group makes for a fascinating discussion (see Cobben 1979; Sweet 1979; Schaefer 1993, 1997; Cohen 1996, 1998a,b), the purpose here is to describe the nature of the feeding method as it pertains to carnivory. No suggestions about the origin of predation or phytophagy in the Heteroptera will be made. Also avoided will be another related and equally fascinating area of discussion: that many heteropterans are well adapted for straddling both phytophagous and carnivorous feeding niches (Rastogi 1962, Slater and Carayon 1963, Ridgway and Jones 1968, Tamaki and Weeks 1972, Bryan et al. 1976, Wheeler 1976, Cobben 1978, Braimah et al. 1982). This explanation of the feeding method in predaceous Heteroptera should clarify some misunderstandings, misconceptions, and oversimplifications of a delightfully complex and efficient process.

2. NUTRITIONAL ECOLOGY OF PREDACEOUS HETEROPTERA

The most important point of this chapter is that heteropteran predators do not confine their ingestion to the body fluids of their prey, as is so often stated or implied. Instead, they use a solid-to-liquid feeding method by attacking the nutrient-rich solid or semisolid organs and tissues of their prey. Table 1 (Cohen and Patana 1984, 1985; Cohen 1998a, and unpublished data) shows the range of protein, fat, and cholesterol composition of whole larvae and hemolymph of a "typical" lepidopteran prey. These data demonstrate that if a predator confined itself to ingestion of hemolymph, it would ingest food that was between one third and one tenth as concentrated with protein and lipids and about one twentieth as concentrated with the essential nutrient cholesterol than it would in consuming the whole prey. A predaceous heteropteran, therefore, would have to consume between 3 and 20 times as much biomass of hemolymph as it would mixed prey biomass (solid tissues and hemolymph combined) to get a given amount of protein, fat, or cholesterol.

Table 1 Composition of Proteins, Fat, and Cholesterol in Hemolymph and Whole Body of a Larval Lepidopteran[a]

Component	Total Larva	Hemolymph	Tissues[b]
Water	70–80%	93–95%	65–75%
Protein	6–16%	0.5–2%	16–24%
Fat	5–13%	0.2–1.5%	10–22%
Cuticle	1.5–3%	—	—
Cholesterol	~1000 µg/g	~50 µg/g	~950 µg/g

[a] Ranges determined for *Trichoplusia ni*, *Heliothis virescens*, and *Spodoptera exigua* larvae (Cohen and Patana 1984, 1985; Cohen 1998a; A. C. Cohen, unpublished data).
[b] Hemolymph-free and cuticle-free.

Table 2 illustrates the consequences of choosing hemolymph over solid tissues. If a predator consumed only the hemolymph of cabbage looper larvae, *Trichoplusia ni* Hübner, that weighed 100 mg, it would get less than one third of the protein biomass available in solid tissues, one tenth of the fat, and one nineteenth of the cholesterol. If the same predator had a lifetime requirement of 0.5 mg of cholesterol, it would have to devour about 100 larvae if each weighed about 100 mg. But if the same individual were to consume solid tissues instead of hemolymph, it would require only about five or six such prey to satisfy its sterol requirements. Because sterols are an absolute requirement for most, if not all, insects (House 1974), these calculations closely approximate "real-world" values (also see Cohen 1984, and Cohen and Brummett 1997).

Table 2 Composition (in mg/100 mg larva) of Hemolymph and Hemolymph-Free Tissues in *Trichoplusia ni*

Component	Hemolymph	Hemolymph-Free Tissues
Water	47	36
Protein	2	6.5
Fat	0.5	5.0
Cholesterol	0.005	0.095

Source: Cohen, A. C., *Amer. Entomol.* 44: 103–116, 1998.

Hemolymph feeding, obviously, is not nutritionally profitable. The disparity between the nutritional rewards from hemolymph-restricted feeding vs. ingestion of solid material has tremendous ecological implications. For theoretical ecologists and their applied science counterparts, the biological control practitioners, it must have been easy to accept the idea of multiple killing by heteropteran predators that were putatively "skimming" hemolymph from each prey item. This hemolymph-skimming idea allowed ecologists to use Holling's (1966) functional response models with little concern for how predators were able to collapse handling time as they increased kill rate. And it is tempting to suppose that a predator that skims nutrient-poor, highly dilute liquids would be likely to kill as many prey as it could capture. Cohen and Tang (1997), however, questioned how accurately functional response models fit the real world of heteropteran predators. Heteropteran predators are adapted to feeding on large prey. The reduviids in the Cohen and Tang (1997) study required an hour or more per prey item to use their extra-oral feeding mechanism to extract nutrients. Similarly, O'Neil (1998) discussed how functional response kinetics determined in the laboratory do not apply to field situations. Moreover, Cohen and Byrne (1992) and Cohen (1998a) showed that several species of Heteroptera exhibited classical Type II functional responses when confronted with small prey (0.10 × the predators' body mass). With large prey (~0.75 × the predators' body mass), there was a flat line averaging slightly more than 1.0 large prey attacked during an 8-hour feeding period. In choice tests with several predator species (Cohen 1998a), predatory Heteroptera selected large prey over small prey almost exclusively.

3. GUT AND SALIVARY GLAND STRUCTURE IN PREDACEOUS HETEROPTERA

A second argument against restricted hemolymph feeding is that the structures of the digestive tracts of predaceous Heteroptera are not suited to exclusive liquid feeding. These digestive tracts are relatively straight tubes with none of the complexities of the tracts that characterize plant sap feeders such as aphids, coccids, cicadids, etc. (Goodchild 1966, Cohen 1998b). These mainly xylem and phloem feeders use filter chambers or similar structures to ingest a large amount of liquid that is processed rapidly; their hosts' water content is removed anteriorly and shunted into the hindgut for removal prior to its dilution in the hemocoel (Goodchild 1966).

The structure of the salivary gland complex in Heteroptera was explained by Nishijima and Sogawa (1963), and that of predators was discussed by Cohen (1998b). Briefly, all predaceous heteropterans have a pair of main glands, each consisting of an anterior and posterior lobe. The pentatomomorphans also have at least one lateral lobe. The posterior lobe secretes digestive enzymes (Miles 1972; Cohen 1993, 1998b). In heteropterans that secrete salivary flanges, the anterior lobe produces the flange material. In mirids, which do not secrete flanges, an anterior lobe is present, but the function is not known (Cohen 1998b). Another pair of structures present in all predaceous heteropterans are the accessory glands, which are lateral to the anterior midgut and appressed closely to it by a network of tracheoles. The accessory glands are critical in recycling water from the gut back into the main salivary glands, into the prey, and then back to the gut. Because the digestive enzymes must be carried into the prey in repeated intervals during feeding, water must be recycled throughout the feeding process.

4. VERTEBRATE BLOOD FEEDERS

The issues of food concentrations and recycling of water raise questions about hematophagous Heteroptera, triatomine reduviids, and cimicids. Hematophages from both families ingest a diet that is already in liquid form. However, their food — vertebrate blood — has a hematocrit (cellular component) of about 20 to 45% of the total volume of the blood meal, and >75% of the blood proteins are found in red blood cells (Lavoipierre et al. 1959, Kingsolver and Daniel 1993). Nutrient concentration of vertebrate blood is thus about 10 to 100 times higher than that of arthropod hemolymph. Furthermore, hematophagous arthropods undertake an extremely rapid diuresis during and after the blood meal characterized by an excretion of a copious and dilute urine (Wigglesworth 1972). Predaceous Heteroptera generally do not produce copious or dilute urine after feeding on prey (A. C. Cohen 1982, unpublished data).

5. THE FEEDING PROCESS IN PREDACEOUS HETEROPTERA

The feeding process follows these stereotypic steps. First, the prey is debilitated with a venom, mechanical action, or both. Next, a likely step is the removal (via the food canal of the maxillary stylets) of some of the prey's hemolymph. Hemolymph water is recycled from the gut, into the accessory salivary gland, and thence carried by the accessory duct to the more anterior main salivary glands. The water is deposited in the salivary gland lumen until enzymes can be secreted into the lumen (Goodchild 1966, Cohen 1998a). The watery saliva (Miles 1972) is injected, via the salivary duct of the maxillary stylets, into the prey, where it can effect liquefaction of a patch of prey solids. It is further diluted by watery saliva, reducing the viscosity of the slurry. The importance of viscosity reduction cannot be overemphasized in light of the fact that highly viscous liquids would be difficult to conduct through long (>1.0 mm) stylets with very narrow (<0.02 mm) diameter food canals (Mittler 1967, Cohen 1998b). The diluted slurry is removed from the prey, via the food canal, through the buccal cavity and into the gut. Once in the gut, water from the liquefied prey is again recycled via the accessory gland and through the entire cycle. This process is repeated until the prey is emptied of nearly all its digestible materials or until the predator has reached satiety. This highly incremental process cannot be bypassed or rushed beyond the kinetics capability of the enzymes or the water-exchange process (Cohen 1995, 1998b; Cohen and Tang 1997).

Many predaceous Heteroptera inject venom that rapidly debilitates prey (Edwards 1961, Cohen 1996). The distinction between venom and digestive enzymes is problematic, and it is not within the scope of the present discussion to argue this issue. It is evident, however, that several components of heteropteran salivary secretions have hydrolytic activities that also help them function as venom (e.g., proteinases, phospholipase A_2, and hyaluronidase). Baptist (1941) and Hori (1968) should be consulted for classical discussions of the salivary glands and their histological changes during the feeding process of heteropteran predators.

It is not clear whether predaceous heteropterans immediately inject salivary secretions into prey as they penetrate the cuticle or if they remove a small amount of hemolymph, allowing room for the saliva. Many heteropterans produce a salivary flange on the prey's cuticle (including all predators from Pentatomomorpha and many from Cimicomorpha such as Nabidae and Anthocoridae). Flange production takes place early in the feeding bout. The flange cements the labium to the prey, sealing the labium and stylets, and serves as a fulcrum for stylet movements within the prey (Cohen 1990, 1996, 1998b). Some predators, such as mirids, produce no salivary flange. These predators use their mandibular stylets to stabilize the labium and allow maxillary stylets to move within the prey's body to perform their digestive/ingestive functions (Cobben 1978).

Heteropteran stylets are highly modified mandibles and maxillae. For a more detailed treatment of the structure and function of mouthparts in predaceous and hematophagous Heteroptera, the following references should be consulted: Cobben (1978), Cohen (1990), Faucheux (1975), Friend and Smith (1971), Miles (1972). Mandibular stylets vary from one species to another. Dentition varies from a very regular sawtooth configuration, with dozens of equal-sized teeth in reduviids, to extremely barbed and recurved teeth that are used in a harpoonlike fashion by asopine pentatomids. Although the mandibular stylets are always paired in predaceous Heteroptera, they are not always symmetrical. For example, in the lygaeoid, *Geocoris punctipes* (Say), one mandibular stylet has rasplike dentition, whereas the other has a few sawtooth points on the tip. Cobben (1979) presented an array of photographs that capture some of the bizarre variations of mandibular stylets in the Heteroptera.

Maxillary stylets are among the most interesting and highly modified mouthparts in the class Insecta. They are less varied than mandibular stylets, although some (e.g., in Reduviidae) contain brushlike, filtering structures close to the tips. Others are unadorned. Throughout Heteroptera, except in the Corixidae, the maxillary stylets interlock over nearly their entire length. On their inner face, maxillary stylets have apposing grooves that form a food canal and a salivary canal. The stylets, as a bundle or fascicle, are held together, in part, by a groove in the labium (beak) and by the labrum. They are also held together by interlocking "lips" that extend nearly the entire length of the stylets.

The range of movements of the stylet bundle and their flexibility and alacrity are amazing. The stylet bundle is projected into the prey by protractor muscles in the predator's head. Because stylets lack muscles, all twisting and bending motions are achieved through appositional forces, differential pressures of one stylet against the others, or from leverage gained from pressure against the salivary flange and structures within the prey. As a result of these forces, stylets can move with split-second speed within the legs, antennae, and every recess in the prey. They whip around inside the prey, cutting, tearing, and rasping tissues, and performing rapid mechanical breakdown.

As they mechanically disrupt their prey, heteropteran predators also use their stylets to inject potent digestive enzymes that chemically macerate the prey's tissues. Digestive salivary enzymes in the Heteroptera include proteinases, lipases, phospholipase A_1, amylase, pectinase, invertase, hyaluronidase, and nucleases (Nuorteva 1958, Miles 1972, Cohen 1998a). All these enzymes, except for pectinase and invertase, are found in predaceous heteropterans (Cohen 1998a). Salivary proteinases in predaceous Heteroptera have been seldom characterized specifically, but most of those that have been belong to the category trypsin-like peptidase (Cohen 1998a). Trypsin-like peptidases are endoproteinases that attack sites where lysine or arginine residues are present within protein molecules. Chymotrypsin-like activity, however, has not been reported in salivary secretions of heteropterans (Cohen 1993). The other category of alkaline or serine proteinases is elastase-like enzyme, which until recently has received little attention in biochemical studies of arthropods (Christeller et al. 1990); it has not been reported in heteropteran saliva. Recently, however, an elastase-like enzyme was found in the saliva of the reduviid *Zelus renardii* Kolenati (A. C. Cohen and D. L. Brummett, unpublished data).

The salivary enzymes and stylets armed with teeth, barbs, or rasps equip heteropteran predators to perform the chemical and physical maceration of prey. The maceration process, which also takes place in many phytophagous Heteroptera, is suited to "solid-to-liquid feeding," a phrase that was coined to describe the action used by heteropteran and other predators that use extra-oral digestion (Cohen 1998a). The term, "solid-to-liquid feeding" is meant to distinguish this type of feeding from the simple ingestion of materials that were already liquids before feeding began. No predaceous heteropteran (in fact, no predaceous arthropod, in general) is known that feeds exclusively on materials that are already liquid. This contrasts with many homopterans and a few heteropterans, such as chinch bugs (Blissidae) that actually feed strictly on liquids (plant xylem and phloem saps). Many phytophagous heteropterans that are stated to be strict liquid feeders are actually solid-to-liquid feeders. For example, *Lygus* bugs (Miridae) are often said to feed on cell saps or fluids from vascular bundles, but they feed on mesophyll cells and their solid contents, especially from fruits. Examination of the insides of corpses of prey that *Lygus* bugs have eaten provides vivid testimony of the adeptness of the solid-feeding abilities of these insects.

6. SUMMARY

Predaceous heteropterans feed mainly on the solid contents of their prey. They insert their stylets into the prey, injecting venom and digestive enzymes, and mechanically disrupt prey with the dentition on the mandibular stylets. They use their maxillary stylets, which contain a food canal and a salivary canal, to inject salivary secretions that liquefy the prey through the process of solid-to-liquid feeding. The stylets, which are held together ventrally by the labium and dorsally by the labrum, are protracted, retracted, twisted, and bent by muscles in the head. Contrary to frequent assertions of strict liquid feeding, heteropteran predators feed mainly on the solid materials of their prey. By feeding mainly on solid food they tend to use fewer prey than they would if they were hemolymph (body fluid) feeders. Their extra-oral digestion predisposes them to feed on large prey (usually more than 10% and often 50 to 300% of their own body weight). Their feeding mechanism permits highly efficient selection of nutrients and utilization of prey materials (Cohen 1989), but it demands a substantial commitment of time and materials. As a result, heteropteran predators can be considered "major investors" (Cohen and Tang 1997), and therefore can be expected to attack a few large prey rather than many small prey as a general feeding strategy. The feeding system of heteropteran predators can also be regarded as one of the most sophisticated, highly specialized, and elaborate feeding apparati in the entire animal kingdom. Certainly, an understanding of this apparatus should not only enhance appreciation for its complexities but also set the stage for a deeper understanding of the ecological and evolutionary significance of the predaceous Heteroptera.

An important application of understanding the feeding mechanism of predaceous Heteroptera is in biological control. As pointed out previously (Cohen 1992, 1998a), misunderstanding of the feeding mechanism was a major impediment to the development of artificial diets for predators from this suborder. Assumptions that predaceous heteropterans were strictly "hemolymph" feeders led to efforts (including this author's own early work on artificial diets) to develop strictly liquid artificial diets. As was later learned, such diets were far too dilute to support nutritional needs of the predators, and liquids that were presented in large feeding packets diluted the digestive enzymes, disallowing the reingestion of these salivary enzymes that are required to complete digestion in the midgut. It was only after understanding the actual feeding process, that this author was able to simulate natural prey contents in artificial diets. A second aspect of a biological control application of knowledge of the feeding mechanism was discussed by Cohen and Tang (1997) and Cohen (1998a). The incorrect assumption that heteropteran predators "skim" the liquid components from their prey probably has led to expectations of exaggerated voracity of these very efficient insects. It is very important to temper expectations of the killing capacity of these predators, which play an important potential role in pest management. As argued previously (Cohen 1998a), biological control strategies may need to be revised in terms of releasing greater numbers of small heteropteran predators. The examples behind this discussion underscore the importance of developing and applying a thorough understanding of basic biology to practical pest management problems.

7. ACKNOWLEDGMENTS

The author thanks C. W. Schaefer and A. G. Wheeler, Jr. for their critical reviews of an earlier version of this chapter, and D. L. Brummett and J. L. Cohen for their excellent technical assistance.

8. REFERENCES CITED

Baptist, B. A. 1941. The morphology and physiology of the salivary glands of the Heteroptera–Hemiptera. Q. J. Microsc. Sci. 83: 91–139.
Braimah, S. A., L. A. Kelton, and R. K. Stewart. 1982. The predaceous and phytophagous plant bugs (Heteroptera: Miridae) found on apple trees in Quebec. Nat. Canad. (Rev. Ecol. Syst.). 109: 153–180.
Bryan, D. E., C. G. Jackson, R. L. Carranza, and E. G. Neemann. 1976. *Lygus hesperus*: production and development in the laboratory. J. Econ. Entomol. 69: 127–129.
Christeller, J. T., W. A. Laing, B. D. Shaw, and E. P. J. Burgess. 1990. Characterization and partial purification of digestive proteases of the black field cricket, *Teleogryllus commodus* (Walker): elastase is a major component. Insect Biochem. 20: 157–164.
Cobben, R. H. 1978. Evolutionary Trends in Heteroptera. Part II. Mouthpart-structures and Feeding Strategies. Mededelingen Landbouwhogeschool, Wageningen, the Netherlands. 407 pp.
1979. On the original feeding habits of the Hemiptera (Insecta): a reply to Merrill Sweet. Ann. Entomol. Soc. Amer. 72: 711–715.
Cohen, A. C. 1982. Water and temperature relations of two hemipteran members of a predator–prey complex. Environ. Entomol. 11: 715–719.
1984. Food consumption, food utilization and metabolic rates of *Geocoris punctipes* (Het.: Lygaeidae) fed *Heliothis virescens* (Lep.: Noctuidae) eggs. Entomophaga 29: 361–367.
1989. Ingestion and food consumption efficiency in a predaceous heteropteran. Ann. Entomol. Soc. Amer. 82: 495–499.
1990. Feeding adaptations of some predaceous Heteroptera. Ann. Entomol. Soc. Amer. 83: 1215–1223.
1992. Using a systematic approach to develop an artificial diet for predators. Pp. 71–91 *in* T. E. Anderson and N. C. Leppla [eds.], Advances in Insect Rearing for Research and Pest Management. Westview Press, Boulder, Colorado, U.S.A. 519 pp.
1993. Organization of digestion and preliminary characterization of salivary trypsin-like enzymes in a predaceous heteropteran, *Zelus renardii*. J. Insect Physiol. 39: 823–829.
1995. Extra-oral digestion in predatory Arthropoda. Annu. Rev. Entomol. 40: 85–103.
1996. Plant feeding by predatory Heteroptera: evolutionary and adaptational aspects of trophic switching. Pp. 1–17 *in* O. Alomar and R. N. Wiedenmann [eds.], Zoophagous Heteroptera: Implications for Life History and Integrated Pest Management. Thomas Say Publ. Entomol.: Entomological Society of America, Lanham, Maryland, U.S.A. 202 pp.

1998a. Solid-to-liquid feeding: the inside(s) story on extra-oral digestion in predaceous Arthropoda. Amer. Entomol. 44: 103–116.

1998b. Biochemical and morphological dynamics and predatory feeding habits in terrestrial Heteroptera. Pp. 21–32 *in* M. Coll and J. R. Ruberson [eds.], Predatory Heteroptera in Agroecosystems: Their Ecology and Use in Biocontrol. Thomas Say Publ. Entomol.: Entomological Society of America, Lanham, Maryland, U.S.A. 233 pp.

Cohen, A. C., and D. L. Brummett. 1997. The non-abundant nutrient (NAN) concept as a determinant of predator–prey fitness. Entomophaga 42: 85–91.

Cohen, A. C., and D. N. Byrne. 1992. *Geocoris punctipes* as a predator of *Bemisia tabaci*: a laboratory evaluation. Entomol. Exp. Appl. 64: 195–202.

Cohen, A. C., and J. W. Debolt. 1983. Rearing *Geocoris punctipes* (Say) on insect eggs. Southwest Entomol. 8: 61–64.

Cohen, A. C., and R. Patana. 1984. Ontogenetic and stress-related changes in hemolymph chemistry of beet armyworms. Comp. Biochem. Physiol. 71A: 193–198.

1985. Chemical composition of tobacco budworm eggs during development. Comp. Biochem. Physiol. 81B: 165–169.

Cohen, A. C., and R. Tang. 1997. Relative prey weight influences handling time and extracted biomass in predatory heteropterans. Environ. Entomol. 26: 559–565.

Edwards, J. S. 1961. The action and composition of the saliva of an assassin bug *Platymeris rhadamantus* Gaerst. (Hemiptera, Reduviidae). J. Exp. Biol. 38: 61–77.

Faucheux, M. M. 1975. Relations entre l'ultrastructure des stylets mandibulaires et maxillaires et la prise de nourriture chez les Insectes Hémiptères. C.R. Acad. Sci. Paris. 281: 41–44.

Friend, W. G., and J. J. B. Smith. 1971. Feeding in *Rhodnius prolixus*: mouthpart activity and salivation, and their correlation with changes in electrical resistance. J. Insect Physiol. 17: 223–243.

Goodchild, A. J. P. 1966. Evolution of the alimentary canal in the Hemiptera. Biol. Rev. 41: 97–140.

Henry, T. J., and R. C. Froeschner. 1988. Introduction. Pp. ix–xix *in* T. J. Henry, and R. C. Froeschner [eds.], Catalog of the Heteroptera, or True Bugs, of Canada and the Continental United States. E. J. Brill, New York, New York, U.S.A. 958 pp.

Holling, C. S. 1966. The functional response of invertebrate predators to prey density. Mem. Entomol. Soc. Canada 48: 1–86.

Hori, K. 1968. Histological and histochemical observations on the salivary gland of *Lygus disonsi* Linnavouri (Hemiptera, Miridae). Res. Bull. Obihiro Zootech. Univ. 5: 735–744.

House, H. L. 1974. Nutrition. Pp. 1–62 *in* M. Rockstein [ed.], The Physiology of Insecta. Vol. V. Academic Press, New York, New York, U.S.A. 648 pp.

Kingsolver, J. G., and T. L. Daniel. 1993. Mechanics of fluid feeding in insects. Pp. 141–161 *in* C. W. Schaefer and R. A. B. Leschem [eds.], Functional Morphology of Insect Feeding. Thomas Say Publ. Entomol.: Proceedings Entomological Society of America, Lanham, Maryland, U.S.A. 162 pp.

Lavoipierre, M. M. J., G. Dickerson, and R. M. Gordon. 1959. Studies on the methods of feeding of bloodsucking arthropods. Ann. Trop. Med. Parasitol. 53: 235–250.

Miles, P. W. 1972. The saliva of Hemiptera. Adv. Insect Physiol. 9: 183–256.

Mittler, T. E. 1967. Flow relationships for hemipterous stylets. Ann. Entomol. Soc. Amer. 60: 1112–1114.

Nishijima, Y., and K. Sogawa. 1963. Morphological studies on the salivary glands of Hemiptera I. Heteroptera. Res. Bull. Obihiro Zootech. Univ. 3: 512–521.

Nuorteva, P. 1958. On the occurrence of proteases and amylases in the salivary glands of *Cinarapiceae* (Panz.) (Hom.: Aphididae). Suom. Hyont. Aikak. 24: 89.

O'Neil, R. J. 1998. Functional response and search strategy of *Podisus maculiventris* (Heteroptera: Pentatomidae) attacking Colorado potato beetle (Coleoptera: Chrysomelidae). Environ. Entomol. 26: 1183–1190.

Rastogi, S. C. 1962. On the salivary enzymes of some phytophagous and predaceous heteropterans. Sci. Cult. 28: 479–480.

Ridgway, R. L., and S. L. Jones. 1968. Plant feeding by *Geocoris pallens* and *Nabis americoferus*. Ann. Entomol. Soc. Amer. 61: 232–233.

Schaefer, C. W. 1993. The Pentatomomorpha (Hemiptera: Heteroptera): an annotated outline of its systematic history. Eur. J. Entomol. 90: 105–122.

1997. The origin of secondary carnivory from herbivory in Heteroptera (Hemiptera). Pp. 229–239 *in* A. Raman [ed.], Ecology and Evolution of Plant-Feeding Insects in Natural and Man-made Environments. International Scientific Publications, New Delhi, India. 245 pp.

Slater, J. A., and J. Carayon. 1963. Ethiopian Lygaeidae. IV: A new predatory lygaeid from Africa with a discussion of its biology and morphology (Hemiptera: Heteroptera). Proc. R. Entomol. Soc. London 38: 1–11.

Sweet, M. H. 1979. On the original feeding habits of Hemiptera (Insecta). Ann. Entomol. Soc. Amer. 72: 575–579.

Tamaki, G., and R. E. Weeks. 1972. Biology and ecology of two predators, *Geocoris pallens* Stål and *G. bullatus* (Say). U.S. Dept. Agric. Tech. Bull. 1446: 46 pp.

Wheeler, A. G., Jr. 1976. *Lygus* bugs as facultative predators. Pp. 28–35 *in* D. R. Scott and L. E. O'Keeffe [eds.], *Lygus* Bug: Host–Plant Interactions. University Press of Idaho, Moscow, Idaho, U.S.A., iii + 38 pp.

Wigglesworth, V. B. 1972. Principles of Insect Physiology. Chapman & Hall, London, U.K. 827 pp.

CHAPTER 21

Creeping Water Bugs (Naucoridae)

Robert W. Sites

1. INTRODUCTION

The family Naucoridae (*sensu lato*), commonly known as creeping water bugs or saucer bugs, is represented worldwide by 413 described extant species in 39 genera and 7 subfamilies. Most species occur in the New and Old World tropics with moderate representation in temperate regions. These predaceous, aquatic insects are the predominant members of macroinvertebrate communities in tropical streams in which they are found (Stout 1982) and are keystone consumers (Sites and Willig 1991). Naucorids are ovate with retentorial prothoracic legs (Sites and Nichols 1990) and natatorial meso- and metathoracic legs, and attain various degrees of dorsoventral flattening. All life stages are passed underwater. Although naucorids probably greatly influence the biota of tropical streams, ecological parameters associated with these insects are poorly known (Stout 1981, Sites and Willig 1991). The family occupies a variety of both lotic and lentic mesohabitats, and species are often specific in microhabitat association. Several species are able to inhabit brackish or saline waters (Lindberg 1948, La Rivers 1951, Thorpe 1965, Nieser 1975). Respiration in naucorids differs among genera and life stages. Most adults employ compressible gas gill respiration, although plastron respiration is used by some (Thorpe and Crisp 1947a,b; Thorpe 1950; Parsons and Hewson 1974). All known nymphs use cutaneous respiration, including those of the plastron respiring adults (Thorpe and Crisp 1947a, Sites and Nichols 1993).

2. ECONOMIC IMPORTANCE

Despite their painful bites, these bugs must be considered beneficial because they often feed on dipteran larvae. Immatures of several families of noxious Diptera have been used as food in rearing experiments. Sites and Nichols (1990) reared *Ambrysus lunatus* Usinger on larvae of *Prosimulium* sp. (Simuliidae), Hungerford (1927) fed *Pelocoris femoratus* Palisot de Beauvois on mosquito (Culicidae) and midge (Chironomidae) larvae, and McPherson et al. (1987) reared *P. femoratus* on *Chaoborus* sp. (Chaoboridae) larvae. In feeding trials, Clarke and Baroudy (1990) showed that *Laccocoris limigenus* (Stål) prefers midge larvae to oligochaetes. Several papers cite *Ilyocoris cimicoides* (L.) as feeding on mosquito larvae and pupae (i.e., Eysell 1905, Hamlyn-Harris 1929), including *Anopheles* sp. (Federici 1920). Wladimirow and Smirnov (1932) found that *I. cimicoides* would consume eight mosquito larvae per day.

The possibility that naucorids serve as food for fish has not been examined, although the potential for rejection after being bitten in mouth tissues suggests that naucorids do not constitute a major portion of the diet of most fish, even in tropical areas where naucorids abound. Research on trophic interactions is needed.

The biological information available for Naucoridae is based on only a few Nearctic and Palearctic species. In the laboratory, temperature-dependent stadia have been determined for a few species; oviposition to adult was reported as 114 days for *Ambrysus lunatus* (Sites and Nichols 1990); 105 days for *A. mormon* (Usinger 1946); 95 days for *I. cimicoides* (Rawat 1939); and 77, 88 to 102, and 81 days for *P. femoratus* (Torre Bueno 1903, Hungerford 1927, McPherson et al. 1987, respectively). *Pelocoris binotulatus* (Stål) was calculated from field data to require 29 days for completion of the five instars, exclusive of egg incubation (López Ruf and Kehr 1994). Peak fecundity occurred generally in December in Argentina, and the adult sex ratio was skewed toward females (Estevez et al. 1989). The overall survival rate, based on instar-specific survivorship data, was about 15%, the greatest mortality occurring in the later instars (López Ruf and Kehr 1994).

The number of generations per year in Naucoridae appears to increase with milder seasonal extremes. The temperate species *I. cimicoides* (Rawat 1939) and *P. femoratus* (McPherson et al. 1987) are univoltine. *Pelocoris binotulatus* also appears to be univoltine, based on seasonal production of eggs (Estevez et al. 1989). In a transitional, southern temperate region with mild, stable winter conditions, *A. lunatus* (Sites and Nichols 1990) and *Cryphocricos hungerfordi* Usinger (Sites and Nichols 1993) were reported as bivoltine. Studies of voltinism of tropical species are lacking.

In extreme latitudes, adults overwinter, e.g., *I. cimicoides* (Rawat 1939) and *P. femoratus* (Torre Bueno 1903, McPherson et al. 1987); however, in milder climates (e.g., central Texas, U.S.A.), third to fifth instars and adults overwinter, e.g., *A. lunatus* (Sites and Nichols 1990) and *C. hungerfordi* (Sites and Nichols 1993). The only naucorid known to overwinter in hibernation is *P. femoratus* (Polhemus 1979), at the bottoms of ponds and pools in muck and detritus (Uhler 1884, Blatchley 1926, Bobb 1974, McPherson et al. 1987). An unusual case of overwintering was reported for an unspecified *Naucoris* (probably *I. cimicoides*; see Rawat 1939); adults occur on the ground not far from the water (Kramer 1935). Torre Bueno (1903) reported that overwintering adults of *I. cimicoides* survive until the following autumn.

Nymphal growth has been addressed in a variety of alpha-taxonomic and other reports focusing on descriptions of immature stages — *A. lunatus* (Sites and Nichols 1990), *A. mormon* (Usinger 1946), *C. hungerfordi* (Sites and Nichols 1993), *L. limigenus* (Clarke and Baroudy 1990), *P. femoratus* (McPherson et al. 1987), *P. poeyi* (Sites 1991). Patterns of nymphal growth were examined by Sites and Willig (1994a), who suggested that ontogenetic trajectories established with principal components analysis could be used in studies of systematic relationships. In other studies of nymphal growth, conspecific populations of *A. mormon* that develop in different thermal environments exhibit concomitant differences in allometries (Sites et al. 1996).

Adult sexual dimorphism is evident in both dorsal and ventral aspects of the terminal abdominal segments, which frequently are the only available diagnostic characters for species identification. Further, in morphometric studies, Sites and Willig (1994a,b) found that sexual dimorphism varies not only intergenerically, but also interspecifically, and, in each case, the dimorphism was attributable to females being consistently larger than males. Wing polymorphism occurs in various genera and involves the degree of forewing or hind wing development (Sites 1990). Several genera with brachypterous morphs are those that are known to use plastron respiration (i.e., *Aphelocheirus, Cryphocricos*).

Most naucorid species for which oviposition is known either adhere eggs to plants or rock substrata (Hinton 1981). Specific examples include *Ambrysus mormon* to pebbles (Usinger 1946); *Aphelocheirus aestivalis* (Larsén 1927) and *C. hungerfordi* to rocks (Sites and Nichols 1993); *L. limigenus* to hard substrata (Clarke and Baroudy 1990); and *Ambrysus lunatus* (Sites and Nichols 1990), *Naucoris maculatus* F. (Lebrun 1960), and *P. femoratus* (Torre Bueno 1903, Hungerford 1927, McPherson et al. 1987) to plants. In contrast, *I. cimicoides* inserts eggs into plant tissue underwater (Cobben 1968), such as into stems of *Ranunculus* or water peppermint (Rawat 1939). Cobben (1968), Hinton (1981), and Sites and Nichols (1999) provided summaries of naucorid egg information.

Dispersal among naucorids is poorly understood. Stream capture was recently suggested as the method of dispersal of *A. hungerfordi* Usinger from western Mexico to Texas across the Sierra Madre Occidental mountains (Sites and Bowles 1995). In other dispersal mechanisms, Larsén (1970) suggested the possibility of passive transport by waterfowl, and Usinger (1956) reported adults to fly diurnally, although rarely. Although Torre Bueno (1903) hypothesized that naucorids fly at night for dispersal, Usinger (1956) reported that naucorids do not fly to lights at night, although they are attracted to a light held near the water. Nonetheless, the only genus known to be attracted to lights is *Potamocoris* (La Rivers 1950, Polhemus and Polhemus 1982). It is probable that flight occurs because some species occur in habitats that can be reached only by flight (Polhemus 1979, Clarke and Baroudy 1990). Rietschel (1985) found live specimens of *N. maculatus* in the Mediterranean Sea in France during a full moon. Further, Stout (1982) hypothesized that the winged morph of *C. latus* Usinger is responsible for colonization of upstream reaches via flight. Other species crawl upstream in compensation for downstream displacement (Stout 1982). Views of dispersal for *I. cimicoides* ostensibly are contradictory: Rawat (1939) hypothesized that this species crawls from pond to pond; whereas Kirkaldy (1905) reported that it flies at night, and Larsén (1970) hypothesized flight based on the discovery of functional flight musculature and the sudden appearance in ponds and pools. Because different morphs occur with and without fully developed flight musculature and associated structure, the possibility exists that both mechanisms of dispersal are employed by this species.

Dispersion patterns have been reported for only one species: López Ruf and Kehr (1995) showed first instars of *P. binotulatus nigriculus* Berg to be contagiously distributed, whereas all other instars were randomly distributed. Males and females were distributed randomly in 64% of the samples. Distribution patterns were influenced by attributes associated with floating vegetation.

Mating behavior has been reported for a few species of Naucoridae, including *A. lunatus* (Sites and Nichols 1990), *Aphelocheirus aestivalis* (Larsén 1927), and *I. cimicoides* (Larsén 1938, Rawat 1939). Also, Kramer (1935) described the mating behavior of an unspecified species of *Naucoris* (probably *I. cimicoides*). In more comprehensive studies, Constantz (1974) and Brewer and Sites (1994) reported behavioral inventories for *A. occidentalis* La Rivers and *P. femoratus*, respectively, which included detailed accounts of mating behavior. In other studies of behavior, movement and feeding were reported for *C. latus* and *L. insularis* (Stout 1982), feeding was reported for *L. limigenus* (Clarke and Baroudy 1990), and a Type II functional response for density-dependent prey selection was reported for fifth instar and adult *I. cimicoides* (Venkatesan and Cloarec 1988).

Naucorid sound production was reported in 1727 by Frisch for *I. cimicoides*. Subsequently, the structure of the stridulatory mechanism was described (Hofeneder 1937, Polhemus 1994). More specifically, all *Limnocoris* species have a stridulitrum of sclerotized ridges on the lateral margins of at least abdominal sterna II–III and a plectrum on the distal ridge on the hind femur (Polhemus 1994).

Certain genera of naucorids flourish in tropical streams from which riparian trees have been removed. Species of *Aphelocheirus* can become particularly dense in these disturbed areas because increased sunlight allows increased algal growth, which in turn serves as food for naucorid prey (Polhemus and Polhemus 1988). The same phenomenon has been noted for several common Neotropical genera (e.g., *Ambrysus, Cryphocricos, Limnocoris*). The propensity for naucorid population densities to increase with increasing tropical deforestation could become beneficially important because these insects feed on larvae of black flies (Simuliidae), vectors of the filarial nematode that causes onchocerciasis (river blindness). As humans become increasingly environmentally invasive in the tropics, pesticide poisoning of nontarget aquatic insects will become serious. Gelbic et al. (1994) characterized five categories of molting damage caused by the juvenile hormone analogue methoprene S to *I. cimicoides*, and found the pesticide to be toxic to naucorids at concentrations ≥0.2 ppm/ml water.

3. SUMMARY

Naucorid ecology and life histories have been little studied, with the exception of only a few temperate species. Virtually no information exists for tropical species in this primarily tropical family. There appears to be no negative economic impact of naucorids and they are beneficial in their control of larvae of noxious Diptera, an avenue of research warranting further exploration.

4. REFERENCES CITED

Blatchley, W. S. 1926. Heteroptera or True Bugs of Eastern North America, with Especial Reference to the Faunas of Indiana and Florida. Nature Publ. Co., Indianapolis, Indiana, U.S.A. 1116 pp.
Bobb, M. L. 1974. The insects of Virginia: No. 7. The aquatic and semi-aquatic Hemiptera of Virginia. Virginia Polytech. Inst. State Univ. Res. Div. Bull. 87: 1–195.
Brewer, D. W., and R. W. Sites. 1994. Behavioral inventory of *Pelocoris femoratus* (Hemiptera: Naucoridae). J. Kansas Entomol. Soc. 67: 193–198.
Clarke, F., and E. Baroudy. 1990. Studies on *Laccocoris limigenus* (Stal.) [sic] (Hemiptera: Naucoridae) in Lake Naivasha, Kenya. Entomologist 109: 240–249.
Cobben, R. H. 1968. Evolutionary trends in Heteroptera. Part I. Eggs, architecture of the shell, gross embryology and eclosion. Centre for Agricultural Publishing and Documentation, Wageningen, the Netherlands. 475 pp.
Constantz, G. D. 1974. The mating behavior of a creeping water bug, *Ambrysus occidentalis* (Hemiptera: Naucoridae). Amer. Mid. Nat. 92: 234–239.
Estevez, A. L., M. L. López Ruf, and J. A. Schnack. 1989. Ciclo anual, fecundidad y proporcion de sexos de una poblacion de *Pelocoris (P.) binotulatus nigriculus* Berg (Hemiptera Limnocoridae). Limnobios 2: 729–732.
Eysell, A. 1905. Die Stechmücken. Pp. 78–80 *in* C. Mense [ed.], Handbuch der Tropenkrankheiten, Bd. 2, Barth, Leipzig, Germany.
Federici, E. 1920. Sulla lotta naturale control le larve di *Anopheles* per mezzo degli insetti acquatici. R. C. Accad. Lincei 29: 170–173, 219–222, 244–247.
Frisch, J. L. 1727. Beschreibung von allerley Insekten in Deutschland. Teil 6. Nicolai, Berlin, Germany. 10 + 34 pp.
Gelbic, I., M. Papácek, and J. Pokuta. 1994. The effects of methoprene S on the aquatic bug *Ilyocoris cimicoides* (Heteroptera, Naucoridae). Ecotoxicology 3: 89–93.
Hamlyn-Harris, R. 1929. The relative value of larval destructors and the part they play in mosquito control in Queensland. Proc. R. Soc. Queensland 41(3): 23–38.
Hinton, H. E. 1981. Biology of Insect Eggs. Vols. 1–3. Pergamon Press, Oxford, U.K. 1125 pp.
Hofeneder, K. 1937. Über das Stridulationsorgan von Naucoris cimicoides L. M. Festschrift zum 60 Geburtstage von Professor Dr. Embrik Strand 3: 355–360.
Hungerford, H. B. 1927. The life history of the creeping water bug, *Pelocoris carolinensis* Bueno (Naucoridae). Bull. Brooklyn Entomol. Soc. 22: 77–83.
Kirkaldy, G. W. 1905. A guide to the study of British waterbugs (aquatic Rhynchota). Entomologist 38: 173–178, pl. II.
Kramer, H. 1935. Beiträge zur Biologie von *Naucoris* mit besonderer Berücksichtigung der Atmung. Arch. Hydrobiol. 28: 523–554.
Larsén, O. 1927. Über die Entwicklung und Biologie von *Aphelocheirus aestivalis* Fabr. Entomol. Tidskr. 48: 181–206.
1938. Untersuchungen über den Geschlechts-Apparat der aquatilen Wanzen. Opusc. Entomol. Suppl. I, 388 pp.
1970. The flight organs of *Ilyocoris cimicoides* L. (Hem., Naucoridae). Entomol. Scand. 1: 227–235.
La Rivers, I. 1950. A new species of the genus *Potamocoris* from Honduras. Proc. Entomol. Soc. Washington 52: 301–304.
1951. A revision of the genus *Ambrysus* in the United States (Hemiptera: Naucoridae). Univ. California Publ. Entomol. 8: 277–338.

Lebrun, D. 1960. Recherches sur la biologie et l'éthologie de quelque Hétéroptères aquatiques. Ann. Soc. Entomol. France 129: 179–199.

Lindberg, H. 1948. Zur Kenntnis der Insektenfauna im Brackwasser des Baltischen Meeres. Soc. Sci. Fenn. Commun. Biol. 10: 1–206.

López Ruf, M. L., and A. I. Kehr. 1994. Estimacion y analisis de la frecuencia de ninfas por estadio, en una poblacion de *Pelocoris (P.) binotulatus nigriculus* Berg (Hemiptera–Limnocoridae). Rev. Bras. Biol. 54: 71–75.

1995. Dispersion espacial y temporal de una poblacion de *Pelocoris (P.) binotulatus nigriculus* Berg (Heteroptera, Limnocoridae). Rev. Bras. Biol. 55: 141–146.

McPherson, J. E., R. J. Packauskas, and P. P. Korch, III. 1987. Life history and laboratory rearing of *Pelocoris femoratus* (Hemiptera: Naucoridae), with descriptions of immature stages. Proc. Entomol. Soc. Washington 89: 288–295.

Nieser, N. 1975. The Water Bugs (Heteroptera: Nepomorpha) of the Guyana Region. Studies on the Fauna of Suriname and Other Guyanas No. 59. 310 pp.

Parsons, M. C., and R. J. Hewson. 1974. Plastral respiratory devices in adult *Cryphocricos* (Naucoridae: Heteroptera). Psyche 81: 510–527.

Polhemus, D. A., and J. T. Polhemus. 1988. The Aphelocheirinae of tropical Asia (Heteroptera: Naucoridae). Raffles Bull. Zool. 36: 167–300.

Polhemus, J. T. 1979. Family Naucoridae. Pp. 131–138 *in* A. S. Menke [ed.], The Semiaquatic and Aquatic Hemiptera of California (Heteroptera: Hemiptera). Bull. Calif. Insect Surv. 21.

1994. Stridulatory mechanisms in aquatic and semiaquatic Heteroptera. J. New York Entomol. Soc. 102: 270–274.

Polhemus, J. T., and D. A. Polhemus. 1982. Notes on Neotropical Naucoridae II. A new species of *Ambrysus* and review of the genus *Potamocoris* (Hemiptera). Pan-Pac. Entomol. 58: 326–329.

Rawat, B. L. 1939. On the habits, metamorphosis and reproductive organs of *Naucoris cimicoides* L. (Hemiptera-Heteroptera). Trans. R. Entomol. Soc. London 88: 119–138.

Rietschel, S. 1985. Naucoriden im Mittelmeer als Modellfall für das Vorkommen von Wasserwanzen in den Solnhofener Plattenkalken. Carolinea 42: 143–144.

Sites, R. W. 1990. Morphological variations in the hemelytra of *Cryphocricos hungerfordi* Usinger (Heteroptera: Naucoridae). Proc. Entomol. Soc. Washington 92: 111–114.

1991. Egg ultrastructure and descriptions of nymphs of *Pelocoris poeyi* (Guérin Méneville) (Hemiptera: Naucoridae). J. New York Entomol. Soc. 99: 622–629.

Sites, R. W., and D. E. Bowles. 1995. *Ambrysus hungerfordi* (Hemiptera: Naucoridae) occurrence in the United States. J. Kansas Entomol. Soc. 68: 476–478.

Sites, R. W., and B. J. Nichols. 1990. Life history and descriptions of immature stages of *Ambrysus lunatus lunatus* (Hemiptera: Naucoridae). Ann. Entomol. Soc. Amer. 83: 800–808.

1993. Voltinism, egg structure, and descriptions of immature stages of *Cryphocricos hungerfordi* (Hemiptera: Naucoridae). Ann. Entomol. Soc. Amer. 86: 80–90.

1999. Egg architecture of Naucoridae (Heteroptera): internal and external structure of the chorion and micropyle. Proc. Entomol. Soc. Washington 101: 1–25.

Sites, R. W., and M. R. Willig. 1991. Microhabitat associations of three sympatric species of Naucoridae (Insecta: Hemiptera). Environ. Entomol. 20: 127–134.

1994a. Efficacy of mensural characters in discriminating among species of Naucoridae (Insecta: Hemiptera): multivariate approaches and ontogenetic perspectives. Ann. Entomol. Soc. Amer. 87: 803–814.

1994b. Interspecific affinities in *Ambrysus* (Hemiptera: Naucoridae). Proc. Entomol. Soc. Washington 96: 527–532.

Sites, R. W., M. R. Willig, and R. S. Zack. 1996. Morphology, ontogeny, and adaptation in *Ambrysus mormon* (Hemiptera: Naucoridae): quantitative comparisons among populations in different thermal environments. Ann. Entomol. Soc. Amer. 89: 12–19

Stout, R. J. 1981. How abiotic factors affect the distribution of two species of tropical predaceous aquatic bugs (family: Naucoridae). Ecology 62: 1170–1178.

1982. Effects of a harsh environment on the life history patterns of two species of tropical aquatic Hemiptera (family: Naucoridae). Ecology 63: 75–83.

Thorpe, W. H. 1950. Plastron respiration in aquatic insects. Biol. Rev. 25: 344–390.

1965. The habitat of *Aphelocheirus aestivalis* (F.) (Hem.-Het., Aphelocheiridae). Entomol. Mon. Mag. 101: 251–253.

Thorpe, W. H., and D. J. Crisp. 1947a. Studies on plastron respiration. I. The biology of *Aphelocheirus* [Hemiptera, Aphelocheiridae (Naucoridae)] and the mechanisms of plastron retention. J. Exp. Biol. 24: 227–269.
1947b. Studies on plastron respiration. II. The respiratory efficiency of the plastron in *Aphelocheirus*. J. Exp. Biol. 24: 270–303.
Torre Bueno, J. R. de la. 1903. Brief notes toward the life history of *Pelocoris femorata* Pal. B. with a few remarks on habits. J. New York Entomol. Soc. 11: 166–173.
Uhler, P. R. 1884. Order VI-Hemiptera. Pp. 204–296 *in* J. S. Kingsley [ed.], The Standard Natural History. Vol. II. Crustacea and Insects. S. E. Casino and Company, Boston, Massachusetts, U.S.A. 555 pp.
Usinger, R. L. 1946. Notes and descriptions of *Ambrysus* Stål with an account of the life history of *Ambrysus mormon* Montd. (Hemiptera, Naucoridae). Univ. Kansas Sci. Bull. 31: 185–210.
1956. Aquatic Hemiptera. Pp. 182–228 *in* R. L. Usinger [ed.], Aquatic Insects of California with Keys to North American Genera and California Species. University of California Press, Berkeley, California, U.S.A. 508 pp.
Venkatesan, P., and A. Cloarec. 1988. Density dependent prey selection in *Ilyocoris* (Naucoridae). Aquat. Insects 10: 105–116.
Wladimirow, M., and E. Smirnov. 1932. Experimente an Wasserinsekten, die sich von Culicidenlarven ernähren. Zool. Anz. 99: 192–206.

CHAPTER 22

Giant Water Bugs (Belostomatidae)

P. Venkatesan

1. INTRODUCTION

Belostomatid bugs (giant water bugs) are medium-sized to very large (10 to 110 mm) brownish aquatic bugs. They are dorsoventrally flattened and have strong, thick, raptorial front legs and broad, flat middle and hind legs fringed with swimming hairs. The head is drawn out in front of the eyes as a triangular extension (Polhemus 1982). The visible rostrum is three segmented. The tarsi are two or three segmented. Ocelli are absent and the four-segmented antennae are usually concealed in pockets beneath the head. These bugs obtain air with a pair of retractable, straplike appendages at the abdominal apex. Some aspects of the structure and process of breathing are given by Miller (1961, 1977) and Parsons (1972a,b, 1973).

Although belostomatids are strong swimmers, they are sedentary hunters (Menke 1979). They wait on submerged vegetation or other support, their raptorial forelegs held in readiness to seize quickly any moving object that passes. The diet of belostomatids is varied and they probably will suck dry anything that they can subdue, also liquefying the soft tissues. DeCarlo (1959) showed that the paralyzing toxin injected by belostomatids is important in subduing the prey. A bite on a human hand by a belostomatid produces an immediate burning sensation which may last several hours.

An interesting feature of the biology of these bugs is that in several species the males care for the eggs, which are laid upon the males' backs (see Smith 1997, for a review and discussion of paternal care in giant water bugs).

Attraction to lights by most species has earned them the name "electric light bugs." Some have a migratory flight periodicity correlated with the lunar cycle, e.g., *Diplonychus rusticus* (F.) (= *indicus* Venkatesan and Rao) (Duviard 1974).

An early but useful account of the structure and physiology of two North American *Belostoma* species is given by Locy (1884).

2. ECONOMIC IMPORTANCE

Belostomatid bugs play an important role in freshwater ecosystems and knowledge of the bugs is important for the study of fish biology and for the proper management of hatcheries (Dimmock 1886, Wilson 1958).

In general, belostomatids are intermediate-stage predators in the food chain of their communities. Many have little importance to humans other than as natural food for fish. But the Old World

Limnogeton species are obligate feeders on freshwater snails that serve as intermediate hosts of human and veterinary schistosomiasis (Voelker 1966, 1968). Some New World *Belostoma* also feed on snails and are important in controlling gastropod populations (Cullen 1969, Kesler and Munns 1989). Juvenile *B. flumineum* Say fed at the highest rates on the largest cladoceran prey (Cooper 1983). *Belostoma* and *Lethocerus* can be nuisances in swimming pools, to which they may be attracted by lights. Some belostomatids are eaten by people of certain cultures, such as *Lethocerus* spp. in China (Hoffmann 1931). Adult *Lethocerus* spp. are the most ferocious predators, feeding voraciously on fishes as much as 20 cm long (Hoffmann 1931, Tawfik 1970): these bugs may at times be of concern to fisheries (Wilson 1958). They also feed on frogs and their tadpoles (Rankin 1935). The salivary secretion of *L. fakir* (Gistel) contains a potent mixture of hydrolytic enzymes (proteases) which liquefies the soft tissues of frogs (Rees and Offord 1969).

The importance of *Belostoma* spp. in biocontrol has been little investigated, except in control of mosquitoes (Bay 1974). Late instars of *Belostoma* feed effectively on larvae and pupae of *Aedes* mosquitoes (Consoli et al. 1989), but *B. flumineum* in California fed more readily on mosquito fish (Miura et al. 1984). Early instars of *Lethocerus* also feed on mosquito larvae (Cullen 1969), as do all instars of *Sphaerodema urinator* Dufour (Tawfik et al. 1978). There is also increasing evidence that *Diplonychus* spp. too may help control mosquitoes (Venkatesan et al. 1986). And as mentioned above, some giant water bugs may help control the intermediary hosts of medically important schistosomes.

3. SPECIES OF ECONOMIC IMPORTANCE

Some species of Belostomatidae are of recognized economic importance, and the role of additional forms is being investigated. They are voracious and potential predators on mosquito larvae, snails, fishes, and frogs.

3.1   *Diplonychus* Laporte spp.

3.1.1   Distribution, Life History, and Biology

This genus occurs in Africa, southern Asia, the East Indies, and Australia. Speciation has apparently been more profuse in Africa. The genus is characterized by strongly dilated front femora bearing two grooves for the reception of the tibia, and by two small, equal claws; the phallus is diagnostic. There are approximately 20 known species (Lauck and Menke 1961).

Diplonychus rusticus feeds on mosquito and chironomid larvae, tadpoles, snails, and fingerlings, and it shows some potential for mosquito control in southern India, where its biology, behavior, and life history have been well studied (Venkatesan 1981, Sankaralingam 1990, Venkatesan et al. 1993, Manickannu and Venkatesan 1994, Venkatesan and Arivoli 1995, Jayanthi and Venkatesan 1997).

That habitats with this belostomatid lack mosquitoes suggests the bugs' potential for biological control of mosquitoes (Venkatesan and Sivaraman 1984, Venkatesan 1985, Venkatesan et al. 1986) and tolerance to pesticides (Venkatesan et al. 1991).

3.1.2   Damage

Larger instars attack prey more successfully and have shorter handling times than smaller instars (Venkatesan and Sivaraman 1984), and males — with and without eggs — prefer large prey and kill more than they eat (Venkatesan and D'Sylva 1990). The bugs attack more prey in shallow than in deep water and kill more prey in habitats without vegetation; the predatory efficiency of this species reflects discontinuous predation (Venkatesan and Rao 1981). The instars grow maximally when *Culex* larvae are the prey (Venkatesan and Muthukrishnan 1987).

3.1.3 Control

Pesticides used to control mosquito larvae may inadvertently reach nontarget organisms. Aquatic insect species differ greatly in their susceptibility to pesticides or other toxicants. When exposed to deltramethrine compound, the male *D. rusticus* is more resistant to the pesticide than the female (Venkatesan et al. 1991).

3.2 *Lethocerus* Mayr spp.

3.2.1 Distribution, Life History, and Biology

Lethocerus has a worldwide distribution but reaches its greatest development in the Western Hemisphere, especially the Neotropical region.

This genus is characterized by the broad, thin, flat hind tibiae and tarsi, by the division of the abdominal sterna into median and parasternites, the projections on the fourth antennal segment, and the stout beak. The front tarsus is three segmented and terminates in one large claw (Nieser 1975). There are nearly 24 species in this genus (Menke 1961). Among several studies of the morphology of *Lethocerus* spp. is a series on *L. indicum* Lep. et Serv. by Bhargava (1967a,b, 1972a,b, 1976). The life histories of several New World *Lethocerus* have been studied by Hunderford (1925), Rankin (1935), and DeCarlo (1962). *Lethocerus* spp. show a lunar periodicity in their flight pattern; most fly at full moon, which may facilitate recognition of topography, especially water (Cullen 1969). In South America, heavy rains induce flight activity regardless of lunar periodicity (Nieser 1975).

3.2.2 Damage

Lethocerus spp. feed on a variety of organisms, including fish, tadpoles, and small frogs, and they may attack prey much larger than themselves. Reared in the field on tadpoles, *L. americanus* (Leidy) went from egg to adult in an average of 33.4 days (Rankin 1935). When handled, several *Lethocerus* species can squirt a foul-smelling fluid more than a meter.

3.2.3 Control

There are no specific protocols or recommendations for the control of belostomatids in fisheries, aquaria, or ponds. The resistance of these bugs to pesticides used to control immature mosquitoes should be tested (see **Section 3.1.3**). *Diplonychus rusticus* is less susceptible to the pyrethroid K-othrine than are several other aquatic hemipterans (Jebanesan and Angelo 1991).

4. PROGNOSIS

Further work is needed on the interactions between belostomatids and their prey, especially at the population level. This work will determine whether the harm done by these bugs outweighs their biocontrol value in any particular situation or habitat. Mass rearing of such potentially valuable species as *D. rusticus* should be undertaken, because so far the most work on it has already been done (see **Section 3.1**); therefore, it is the best species with which to test the effects of giant water bugs on mosquito populations.

Other aspects of mosquito–giant water bug interactions should be studied as well. For example, Carvalho-Pinto et al. (1995) have shown that a bacterium which helps control mosquito larvae persists in the dead bodies and the feces of the giant water bugs (*Belostoma micantulum* Stål) that have fed upon infected mosquitoes. This secondary value of these bugs and other predators in the biocontrol of mosquito larvae warrants further study.

5. CONCLUSIONS

Belostomatids are both useful and harmful, but neither their value nor their harm is well understood. It seems likely that further study will show that their value in controlling immature mosquitoes and the intermediate hosts of medically important schistosomes far overweighs the harm they do to fisheries and the vertebrate denisons of small ponds.

6. REFERENCES CITED

Bay, E. C. 1974. Predator–prey relationship among aquatic insects. Annu. Rev. Entomol. 19: 441–453.
Bhargava, S. 1967a. Morphology, musculature and innervation of the reproductive organs of *Lethocerus indicum* (Hemiptera: Belostomatidae). J. Kansas Entomol. Soc. 40: 109–124.
1967b. Studies on the morphology of *Lethocerus indicum* Lep. et Serv. (Hemiptera: Belostomatidae) Part VI. The histomorphology of the central nervous system. Bull. Entomol. [India] 8: 23–28.
1972a. Neurosecretory cells in the ventral ganglia of Lethocerus indicum [sic] (Heteroptera). Dtsch. Entomol. Z. (n.F.) 19: 65–72.
1972b. Changes in the neurosecretory system of female Lethocerus [sic] (Heteroptera). Dtsch. Entomol. Z. (n.F.) 19: 141–149.
1976. Morphology of Lethocerus indicum [sic] (Het. Belostomatidae). Dtsch. Entomol. Z. 23: 69–82.
Carvalho-Pinto, C. J., L. Rabinovitch, R. S. A. Alves, C. M. B. Silva, and R. A. G. B. Consoli. 1995. Fate of *Bacillus sphaericus* after ingestion by the predator *Belostoma micantulum* (Hemiptera: Belostomatidae). Mem. Inst. Oswaldo Cruz 90: 329–330.
Consoli, R. A. G. B., M. H. Pereira, A. L. Melo, and L. H. Pereira. 1989. *Belostoma micantulum* Stål 1858 (Hemiptera: Belostomatidae) as a predator of larvae and pupae of *Aedes fluviatilis* (Diptera: Culicidae) in laboratory conditions. Mem. Inst. Oswaldo Cruz 84: 577–578.
Cooper, S. D. 1983. Selective predation on cladocerans by common pond insects. Canad. J. Zool. 61: 879–886.
Cullen, M. J. 1969. The biology of giant water bugs in Trinidad. Proc. R. Entomol. Soc. London 44: 123–136.
DeCarlo, J. A. 1959. Hemipteros cryptocerata. Efectos de sus picaduras. Prim. J. Entomoepid. Argent. 715–719.
1962. Consideraciones sobre la biologia de *Lethocerus mazzani* DeCarlo. Psyche 23: 143–151.
Dimmock, G. 1886. Belostomatidae and some other fish destroying bugs. Annu. Rep. Fish and Game. Commonw. Massachusetts Publ. Doc. 25(D): 67–74.
Duviard, D. 1974. Flight activity of Belostomatidae in central Ivory Coast. Oecologia 15: 321–328.
Hoffmann, W. E. 1931. Studies on the bionomics of the water bug *Lethocerus indicus* (Hemiptera: Belostomatidae) in China. Verh. Intern. Verein. Theor. Angew. Limnol. 5: 661–667.
Hungerford, H. B. 1925. Notes on giant water bugs. Psyche 32: 88–91.
Irwin, M. E. 1962. Observations on hibernation in *Belostoma*. Pan-Pac. Entomol. 38: 3–162.
Jayanthi, M., and P. Venkatesan. 1997. Population dynamics of the water bug, *Diplonychus indicus* Venk. & Rao, a bioagent in mosquito breeding pond. J. Entomol. Res. 21: 117–123.
Jebanesan, A., and J. Angelo. 1991. Acute toxicity of K-othrine to the aquatic hemipterans, *Ranatra filiformis* (Fabr.), *Ranatra elongata* (Fabr.), *Anisops bouvieri* (Kirkaldy) and *Diplonychus indicus* Venk. & Rao. Pollut. Res. 10: 157–160.
Kesler, D. H., and W. R. Munns, Jr. 1989. Predation by *Belostoma flumineum* (Hemiptera): an important cause of mortality in freshwater snails. J. North Amer. Benthol. Soc. 8: 342–350.
Lauck, D. R., and A. S. Menke. 1961. The higher classification of the Belostomatidae (Hemiptera). Ann. Entomol. Soc. Amer. 54: 644–657.

Locy, W. A. 1884. Anatomy and physiology of the family Nepidae. Amer. Nat. 18: 250–255, 353–367.

Manickannu, R., and P. Venkatesan. 1994. Accessary sex glands in female *Diplonychus indicus* Venk. & Rao. J. Entomol. Res. 18: 255–260.

Menke, A. S. 1961. A review of the genus *Lethocerus* (Hemiptera: Belostomidae) in the Eastern Hemisphere with the description of a new species from Australia. Austral. J. Zool. 8: 285–288.

1979. Nepidae and Belostomatidae. Pp. 70–86 *in* A. S. Menke [ed.], The Semiaquatic Hemiptera of California (Heteroptera: Hemiptera). University of California Press, London, England. 166 pp.

Miller, P. L. 1961. Some features of the respiratory system of *Hydrocyrius columbiae* Spin. (Belostomatidae, Hemiptera). J. Insect Physiol. 6: 243–271.

1977. Motor responses to changes in the volume and pressure of the gas stores of a submerged water-bug, *Lethocerus cordofanus*. Physiol. Entomol. 2: 27–36.

Miura, T., R. M. Takahashi, and W. H. Wilder. 1984. Impact of the mosquitofish (*Gambusia affinis*) on a rice field ecosystem when used as a mosquito control agent. Mosq. News 44: 510–517.

Nieser, N. 1975. The water bugs (Heteroptera: Nepomorpha) of the Guyana Region. Stud. Fauna Suriname other Guyanas 59: 1–310, 24 plates.

Parsons, M. C. 1972a. Respiratory significance of the thoracic and abdominal morphology of *Belostoma* and *Ranatra* (Insecta, Heteroptera). Z. Morphol. Tiere 73: 163–194.

1972b. Morphology of the three anterior pairs of spiracles of *Belostoma* and *Ranatra* (aquatic Heteroptera: Belostomatidae, Nepidae). Canad. J. Zool. 50: 865–876.

1973. Morphology of the eighth abdominal spiracles of *Belostoma* and *Ranatra* (aquatic Heteroptera, Belostomatidae, Nepidae). J. Nat. Hist. 7: 255–265.

Polhemus, T. J. 1982. Marine Hemiptera of the Northern Territory including the first freshwater species of *Halobates eschscholtz* (Gerridae, Veliidae, Hermatobatidae and Corixidae). J. Austral. Entomol. Soc. 21: 5–11.

Rankin, K. P. 1935. Life history of *Lethocerus americanus* (Leidy) (Hemiptera: Belostomatidae). Univ. Kansas Sci. Bull. 15: 479–491.

Rees, A. R., and R. E. Offord. 1969. Studies on the protease and other enzymes from the venom of *Lethocerus cordofanus*. Nature 221: 675–677.

Sankaralingam, A. 1990. Dynamics of predation and reproduction in the aquatic insect *Diplonychus indicus* Venk. & Rao. Ph.D. dissertation, University of Madras, Madras, India.

Smith, R. L. 1997. Evolution of paternal care in the giant water bugs (Heteroptera: Belostomatidae). Pp. 116–149 *in* J. C. Choe and B. J. Crespi [eds.], Social Behavior in Insects and Arachnids. Cambridge University Press, Cambridge, U.K. 541 pp.

Tawfik, M. F. S. 1970. The life history of the giant water bug *Lethocerus niloticus* Stål. Bull. Soc. Entomol. Egypt 53: 299–310.

Tawfik, M. F. S., El-Sherif, and A. F. Lutfallah. 1978. The biology of *Spherodema urinator* Duf. (Hemiptera, Belostomatidae). Z. Angew. Entomol. 86: 266–273.

Venkatesan, P. 1981. Influence of temperature and salinity variations on an aquatic bug population in a tropical pond. Hydrobiology 79: 33–50.

1985. Biological control potential of water bugs in the antimosquito work. Proc. Seminar Entomophagous Insects (Calicut), India, pp. 78–83.

Venkatesan, P., and S. Arivoli. 1995. Food intake and energy expenditure pattern in *Diplonychus indicus*, J. Appl. Zool. Res. 6: 38–40.

Venkatesan, P., and S. Muthukrishnan. 1987. Impact of predation and feed utilization on reproduction of *Diplonychus indicus* and *Ranatra filiformis*. Proc. Indian Acad. Sci. (Anim. Sci.): 233–238.

Venkatesan, P., and T. K. Raghunatha Rao. 1981. Description of a new species and a key to Indian species of Belostomatidae. J. Bombay Nat. Hist. 77: 299–303.

Venkatesan, P., and S. Sivaraman. 1984. Changes in the functional response of instars of *Diplonychus indicus* Venk. & Rao (Hemiptera: Belostomatidae) in its predation of two species of mosquito larvae of varied size. Entomon 9: 191–196.

Venkatesan, P., and T. D'Sylva. 1990. Influence of prey size on choice by the water bug, *Diplonychus indicus* Venk. & Rao (Hemiptera: Belostomatidae). J. Entomol. Res. 14: 130–138.

Venkatesan, P., M. Bagyalakshmi, and T. D'Sylva. 1991. Effect of Decamethrin on bioagent in mosquito control. J. Ecobiol. 3: 261–266.

Venkatesan, P., K. Elumalai, and S. Durairaj. 1993. Respiratory efficiency of the water bug *Diplonychus indicus* under salinity and temperature stress. J. Ecobiol. 5: 241–245.

Venkatesan, P., G. C. Guillareme, and F. Halberg. 1986. Modeling prey-predator cycle using hemipteran predators of mosquito larvae for leading world-wide mosquito borne disease incidence. Chronobiology 13: 351–354.

Voelker, J. 1966. Wasserwanzen als obligatorische Schneckenfresser im Nildelta *(Limnogeton fieberi* Mayr) (Belostomatidae, Hemiptera). Z. Tropenmed. Parasitol. 17: 155–165.

1968. Untersuchungen zur Ernährung, Fortpflanzungsbiologie und Entwicklung von *Limnogeton fieberi* Mayr (Belostomatidae, Hemiptera) als Beitrag zur Kenntnis von natürlichen Feinden tropischer Süsswasserschnecken. Entomol. Mitt. Zool. Staatsinst. Zool. Mus. Hamburg 3: 201–224.

Wilson, C. A. 1958. Aquatic and semiaquatic Hemiptera of Mississippi. Tulane Stud. Zool. 6: 115–170.

CHAPTER 23

Waterscorpions (Nepidae)

Steven L. Keffer

1. INTRODUCTION

The family Nepidae is divided into two subfamilies, Nepinae and Ranatrinae, and five tribes: Nepini, Curictini, Goondnomdanepini, Austronepini, and Ranatrini (Lansbury 1974). Most of the family's approximately 225 species are found in 2 of the 14 genera: *Laccotrephes* with ~60 species and *Ranatra* with ~110 species.

Nepids derive their common name, waterscorpions, from their elongate, caudal respiratory siphon. This movable siphon consists of two filaments, derived from the eighth abdominal tergum, which, when pressed together, form an air channel. Thrust through the water surface from below, the siphon conducts air to a subelytral air store via the insect's tracheal system and spiracles on the dorsum of the first abdominal segment (Parsons 1972).

Waterscorpions exhibit plant mimicry: the broad, dorsoventrally flattened Nepini resemble leaves; Ranatrini, elongate and cylindrical, resemble sticks. The other three tribes are intermediate in shape between the leaflike and sticklike extremes.

Waterscorpions are usually found along the margins of still water. Nepini are found in the mud amid plant debris in very shallow water where their leaflike shape makes them difficult to detect. Ranatrini generally occur in deeper water, perched on overhanging and floating vegetation alongside a pond, lake, or stream where their sticklike shape renders them cryptic. The Curictini and Austronepini are found in tangled vegetation in water of varying depth. The Goondnomdanepini, found exclusively in flowing water on the undersurface of rocks, are the exception to the water margin and still water preferences of the rest of the family.

Waterscorpion eggs bear two or more respiratory horns at their anterior pole and are oviposited either in mud (Nepini, Curictini) or vegetation (Ranatrini). The oviposition preferences of Austronepini and Goondnomdanepini are not known.

Waterscorpions are ambush predators, remaining motionless until a passing prey can be captured with a swift strike of their raptorial prothoracic legs. Prey capture is facilitated by their acute vision and mechanoreceptors in their profemora (Cloarec 1976). Bailey (1985) studied the multiprey capture behavior of waterscorpions. Cloarec (1986) discovered that waterscorpions can strike simultaneously at two different prey, the forelegs acting independently.

An early but useful general account of the structure and physiology of *Nepa* sp. and *Ranatra* sp. (both North American) is given in Locy (1884). The anatomy of *N. cinerea* L. was given a lengthy treatment by Hamilton (1931). For a detailed account of the reproductive anatomy and behavior of *N. cinerea* and *R. linearis* L. see Larsén (1938).

2. GENERAL STATEMENT OF ECONOMIC IMPORTANCE

Nepids are generalist predators, and have been reported to feed on a great variety of aquatic organisms including fish, tadpoles, small frogs, aquatic insects and other invertebrates, and terrestrial insects that have fallen to the water surface. They are also noted for their cannibalism (Radinovsky 1964).

In the late 19th century, it was reported that *R. linearis* killed carp spawn (Anonymous 1878) and that draining and restocking of experimental ponds was necessary to control this waterscorpion. However, another communication in the same year reported that *R. linearis*, when accidentally introduced into an aquarium, failed to attack carp, instead preferring to predate on "small English fishes" (Ormerod 1878). Kirkaldy (1906) reported that *N. cinerea* predated on both fish eggs and small fish. Hoffmann (1930) noted that *R. chinensis* Mayr fed on aquatic insects that would serve as fish food but concluded that this waterscorpion species was not detrimental because it was rarely found in the same habitat as commercially important fish. Bisht and Das (1981) observed that five waterscorpions, *Laccotrephes griseus* (Guérin), *L. ruber* [sic] (no *Laccotrephes* species with this epithet exists), *R. varipes* Stål, *R. filiformis* F., and *R. elongata* F. destroyed fish larvae, fish fry, and the aquatic invertebrates that fish eat. In a laboratory experiment testing predation on fish and fish food by *L. ruber* and *R. filiformis* and six species of dytiscid beetle, *L. ruber* ranked second and fourth in predation upon fish fingerlings and fish fry, and second or third in predation upon aquatic invertebrates. *Ranatra filiformis*, on the other hand, ranked seventh and eighth in predation upon fish fingerlings and fry and seventh in predation upon fish food. Thus, the *Laccotrephes* species posed more of a threat to fish, and their food, than the *Ranatra* species.

On the beneficial side of the economic ledger, it has often been noted that waterscorpions consume mosquito larvae (Jenkins 1964) and thus have the potential to be used in mosquito control efforts. However, only a handful of studies has examined experimentally the predation of waterscorpions on mosquitoes. Here species will be discussed as economically "important" if laboratory and/or field studies have indicated a strong potential for mosquito control. Species for which the evidence for control is less compelling will be discussed as "less important."

3. THE IMPORTANT SPECIES

3.1 *Laccotrephes grossus* F. (= *Laccotrephes kohlii* Ferrari)

This nepine occurs throughout China, southeast Asia, and India. Hoffmann (1927) described the natural history, reproductive behavior, and life history of this species (as *L. kohlii*). The adults measured about 30 to 35 mm long and 7 to 9 mm wide at the posterior pronotal angle. Females oviposited in mud in early spring and again in summer. One female was observed to lay 99 eggs over 61 days. The whitish eggs had five to nine respiratory horns. Eggs eclosed in 12 to 13 days. Development of the five instars averaged 39.4 days. Average stadial lengths were as follows: first, 4.5 days ($n = 20$); second, 4.9 days ($n = 27$); third, 6.4 days ($n = 24$); fourth, 8.5 days ($n = 23$); fifth, 13.9 days ($n = 18$). Females ovipositing in spring did so again in summer. Immatures developing from the spring oviposition became adult in summer, overwintered as adults, and reproduced the following spring. Immatures resulting from the summer oviposition overwintered as immatures and reproduced the following summer. Hoffmann was unable to determine adult longevity.

Based on field and laboratory observations, Hoffmann concluded that *L. grossus* could be employed in mosquito-eradication efforts. He collected waterscorpions from permanent bodies of water and distributed them into the temporary aquatic environments caused by the spring rains. Ordinarily these temporary bodies of water would be colonized by mosquitoes whose first generations would escape predation. By seeding these habitats with aquatic bugs, Hoffmann was able to achieve control of these early mosquitoes. Although several insects were "successfully used in this way, *Laccotrephes* and *Sphaerodema* [a belostomatid waterbug] proved the most successful" (p. 11). Further, Hoffmann noted that the simplicity and low cost of this method, coupled with the local abundance of the waterscorpions, made this a particularly practical means of mosquito control. In conclusion, Hoffmann stated that *L. grossus* should "be considered a beneficial insect of considerable importance" (p. 11).

In 1933 Hoffmann reported on a field and laboratory study of another *Laccotrephes* species (no epithet given). Hoffmann was able to control mosquito larvae in the standing water in potted lotus plants by the introduction of this waterscorpion.

3.2   *Laccotrephes maculatus* F.

This species occurs in Java, India, Formosa, and Japan. Chockalingam and Somasundaram (1987) tested the predation efficiency of *L. maculatus* in the laboratory in three ways. First, using adults of four different ages, they found that predation peaked at 15 days and then declined. Second, presented with either a culture of mixed mosquito instars or pure cultures of first to fourth instars, *L. maculatus* adult predation was highest with the mixed culture. Finally, *L. maculatus* adult daily consumption of larvae increased as the density of mosquito first instars increased. Chockalingam and Somasundaram concluded that *L. maculatus* could be used in the biological control of early instar mosquitoes.

3.3   *Ranatra fusca* Palisot de Beauvois

This ranatrine species is found across the northern United States and southern Canada (Sites and Polhemus 1994). Packauskas and McPherson (1986) studied the life history of *R. fusca* (see their paper for references to previous studies of the biology of this species). Adults overwintered and oviposition commenced in early May and continued through late August. Eggs were laid out of the water in the mud bank ringing the study pond but in the laboratory females oviposited in vegetation. The cream-colored eggs had two respiratory horns. In the laboratory immatures were reared under constant temperature (~26.7°C) and photoperiod (16:8 / L:D). Eggs eclosed after an average of 11.3 days and total nymphal development was completed after an average of 35.7 days. Average stadial lengths were as follows: first, 6.1 days ($n = 154$); second, 5.7 days ($n = 139$); third, 6.1 days ($n = 133$); fourth, 6.8 days ($n = 129$); fifth, 10.9 days ($n = 121$). Field evidence suggested that this species is bivoltine.

There are numerous recordings in the literature of *R. fusca* predating on mosquitoes. Radinovsky (1964) observed an adult capturing two mosquito larvae, one per foreleg, while feeding on a third impaled on its beak. Howard et al. (1912) reported 98 mosquito larvae killed in 1 hour by three *R. fusca* nymphs.

Weissmann (1986) compared the predation of the mosquito fish, *Gambusia affinis*, with *R. fusca* on mosquitoes in several laboratory experiments. He found that *G. affinis* was a more effective predator than *R. fusca* in both dense and sparse vegetation. In addition, he discovered that *R. fusca* adults were more effective as predators than nymphs in sparse vegetation but that nymphs were slightly more effective in dense vegetation. Finally, his results demonstrated that the nymphs were ineffective predators in sparse vegetation.

Although *R. fusca* proved to be less efficient than *G. affinis* at mosquito predation, Weissmann argued that the waterscorpion had a place in mosquito control programs in Colorado (U.S.A.) because, unlike *G. affinis*, *R. fusca* could successfully overwinter. Thus, he concluded it might be better to stock permanent ponds with the winter-hardy *R. fusca* rather than restock every year with *G. affinis*.

3.4   *Ranatra parvipes vicina* Signoret

This species is found in tropical Africa and Egypt. Adults measure 23 to 29 mm (Poisson 1965). Tawfik et al. (1986) compared the predation by first instars and adults of this species on mosquito larvae with 13 other insect predators. In the laboratory, the adult waterscorpions were found to have the third highest predation rate, behind an adult dytiscid beetle and a dragonfly nymph. Darriet and Hougard (1993) studied the predation of all five instars and the adults on mosquitoes. Stadia were as follows: first, 6 days; second, 4 days; third, 4 days; fourth, 6 days; fifth, 11 days. Over the full course of nymphal development, female waterscorpions killed 1576 mosquito larvae with a predation rate of 99%. Male nymphs killed 1314 larvae at a rate of 83%. Adult females and males killed 95 and 90 mosquito larvae per day, respectively, over a 7- to 10-month period.

4. THE LESS IMPORTANT SPECIES

4.1 *Cercotmetus* Stål

The nine species of *Cercotmetus* are known from southeast Asia to New Guinea (Lansbury 1973). Little is known about the biology of these ranatrine bugs. Elongate and cylindrical, they differ from the related *Ranatra* in their short forelegs and respiratory siphons.

Williamson (1949) observed that an unidentified *Cercotmetus* sp. in Malaysia was able to keep an area of 5 to 6 yd around it clear of mosquito larvae and ate 20 larvae per day. Unlike *Ranatra* spp., which typically perch on vegetation, Williamson's *Cercotmetus* sp. was floating in the water. In the photographs accompanying Williamson's brief description, a *Cercotmetus* sp. adult is pictured with a larva impaled on its beak and two additional larvae captured, one by each foreleg. Williamson concluded by noting that this species' life history would need to be studied before it could be exploited for mosquito control.

Laird (1956, pp. 76–77) was so impressed by Williamson's claims for *Cercotmetus* that he wondered about the feasibility of introducing this species into permanent malarial ponds in Melanesia.

4.2 *Ranatra chinensis* Mayr

This species is known from China and Japan. Males are 75 to 101 mm long including the respiratory siphons; females are 90 to 105 mm (Lansbury 1972).

Hoffmann (1930) studied the life history of this species much as reported for two species of *Laccotrephes* (see **Section 3.1**). Eggs were oviposited in the bank around the pond, inches from the water. Average stadial lengths in days for the five instars were as follows: first, 5.93 ($n = 33$); second, 5.66 ($n = 53$); third, 5.59 ($n = 52$); fourth, 6.72 ($n = 47$); fifth, 12.26 ($n = 15$). In the laboratory, *R. chinensis* nymphs preferred mosquito larvae and pupae although the nymphs had trouble handling the pupae. Large mosquito larvae, however, were observed to kill *R. chinensis* first instars. Hoffmann concluded that *R. chinensis* was a beneficial insect that could be used in mosquito control efforts. However, he did not test the efficacy of this waterscorpion as a mosquito predator either in the field or in the laboratory.

4.3 *Ranatra elongata* F.

This species is known from India and Sri Lanka. Males measure 81 to 101 mm long including the respiratory siphons; females measure 93 to 105 mm (Lansbury 1972). Rao (1962) described the life history of this species. Breeding occurred in October and November and the female oviposited 10 to 20 eggs in the substratum (Rao 1976), 24 to 48 hours after copulation. The eggs were pale yellow and had two respiratory horns. Eclosion occurred approximately 1 week after oviposition. Stadial lengths, in days, were as follows: first, 5 to 6; second, 8 to 9; third, 7; fourth, 9; fifth, 9. Each instar demonstrated a definite food preference and the first instar's preference was mosquito larvae. No claims were made about the use of *R. elongata* for mosquito control.

4.4 Other Waterscorpion Species

Several other workers have studied nepid consumption of mosquito larvae. For example, Pruthi (1928) reported that *L. griseus* consumed 26 larvae daily and *R. longipes* Stål ate 34 larvae daily. Twinn (1931) exposed third and fourth instars of two species of the mosquito *Culex* to a *Ranatra* sp. and found that 24 of 30 larvae were destroyed in 9 days. Dubitskii (1985) reported on the predation of the five instars and adult of a *Ranatra* sp., and two other aquatic heteropterans, *Notonecta* sp. and *Naucoris* sp., upon the four instars of the mosquito *C. modestus*. The adult *Ranatra* sp. had the highest levels of predation of the three heteropterans. Ouda et al. (1986) studied predation of mosquito larvae by eight insect predators including nymphs of a *Nipa* [sic] sp. In the laboratory the waterscorpions consumed more mosquito larvae than predaceous diving beetle larvae and damselfly nymphs but they were outperformed by predaceous diving beetle adults, notonectid nymphs, and three species of dragonfly nymphs.

5. PROGNOSIS

Although it is well known that waterscorpions predate upon mosquito larvae, only four species have been carefully studied for their potential in mosquito control. Given the increasing interest in integrated pest management, now seems an opportune time for further laboratory and field testing of waterscorpions, and other nepomorphan predators of mosquitoes, e.g., Notonectidae and Belostomatidae, as agents of mosquito control.

6. CONCLUSIONS

Waterscorpions feed upon and, in turn, are fed upon by a variety of aquatic organisms. Thus, both Weissmann (1986) and Darriet and Hougard (1993) were careful to point out that although their studies demonstrated significant mosquito consumption by waterscorpions, their results could not be extrapolated to the field because their experiments were conducted in the absence of alternative prey for the waterscorpions and waterscorpion predators. Thus, further, more complex experiments are needed both to determine what the natural food of waterscorpions is (see Giller 1982), and to test the effects of alternative prey and predators on waterscorpion consumption of mosquitoes. Ultimately, field tests will be necessary to determine whether the introduction of waterscorpions into either temporary aquatic habitats, as advocated by Hoffmann (1927, 1933), or permanent habitats, as advocated by Weissmann (1986), can effectively control mosquitoes. Care must be taken in such studies to ensure that waterscorpions do not decimate important nontarget organisms, i.e., commercial or recreational fish and the invertebrate organisms that the fish feed on. In addition, it will be important to know to what extent commercial pesticides may harm waterscorpions (e.g., Mathavan and Jayakumar 1987).

If additional laboratory and field studies continue to indicate that waterscorpions can play a role in mosquito control then it will also be necessary to determine whether or not waterscorpions can be cultured in the numbers that are needed for commercial biocontrol efforts. The highly successful laboratory rearing of Packauskas and McPherson (1986) of *R. fusca* and McPherson and Packauskas (1987) of *Nepa apiculata* Uhler suggests that this should be possible.

7. REFERENCES CITED

Anonymous. 1878. *Ranatra linearis* attacking carp eggs. Entomologist 11: 95.
Bailey, P. C. E. 1985. "A prey in hand," multi-prey capture behavior in a sit-and-wait predator, *Ranatra dispar* (Heteroptera: Nepidae), the water stick insect. J. Ethol. 3: 105–112.

Bisht, R. S., and S. M. Das. 1981. Observation on aquatic insects as food of fishes and the predatory action of some aquatic insects on fish and fish food. J. Inland Fish. Soc. India 13: 80–86.

Chockalingam, S., and P. Somasundaram. 1987. Efficiency of mosquito predation by an aquatic insect, *Laccotrephes maculatus* (Hemiptera: Nepidae). Arch. Hydrobiol. Beih. Ergebn. Limnol. 28: 525–528.

Cloarec, A. 1976. Interactions between different receptors involved in prey capture in *Ranatra linearis*. Biol. Behav. 1: 251–266.

1986. Distance and size discrimination in a water stick insect, *Ranatra linearis* (Heteroptera). J. Exp. Biol. 120: 59–77.

Darriet, F., and J.-M. Hougard. 1993. Étude en laboratoire de la biologie et des capacités prédatrices de l'Hétéroptère aquatique *Ranatra parvipes vicina* (Signoret, 1880) á l'encontre des larves de moustiques. Rev. Hydrobiol. Trop. 26: 305–311.

Dubitskii, A. M. 1985. The biological regulation of Culicidae in the Soviet Union. Pp. 323–360 *in* M. Laird and J. W. Miles [eds.], Integrated Mosquito Control Methodologies, Vol. 2. Academic Press, London, U.K.

Giller, P. S. 1982. The natural diets of waterbugs (Hemiptera-Heteroptera); electrophoresis as a potential method of analysis. Ecol. Entomol. 7: 233–237.

Hamilton, M. A. 1931. The morphology of the water-scorpion, *Nepa cinerea* Linn. (Rhynchota, Heteroptera). Proc. Zool. Soc. Lond 3: 1067–1136.

Hoffmann, W. E. 1927. Biological notes on *Laccotrephes* (Hemiptera, Nepidae). Lingnan Agric. Rev. 4: 77–93.

1930. Notes on the life history of *Ranatra chinensis* Mayr. Proc. Nat. Hist. Soc. Fukien Christian Univ. 3: 31–37.

1933. The life history of a second species of *Laccotrephes* (Hemiptera, Nepidae). Lingnan Sci. J. 12: 245–256.

Howard, L. O., H. G. Dyar, and F. Knab. 1912. The mosquitoes of North and Central America and the West Indies, Carnegie Institution, Washington, D.C., U.S.A. Vol. 1: 164–172.

Jenkins, D. W. 1964. Pathogens, parasites and predators of medically important arthropods. Bull. World Health Organ. 30, suppl: 60–61.

Kirkaldy, G. W. 1906. List of the genera of the pagiopodous Hemiptera-Heteroptera, with their type species, from 1758 to 1904 (and also of the aquatic and semiaquatic Trochalopoda). Trans. Amer. Entomol. Soc. 32: 117–156.

Laird, M. 1956. Studies of mosquitoes and freshwater ecology in the south Pacific. R. Soc. New Zealand Bull. 6: 1–213.

Lansbury, I. 1972. A review of the Oriental species of *Ranatra* Fabricius (Hemiptera–Heteroptera: Nepidae). Trans. R. Entomol. Soc. London 124: 287–341.

1973. A review of the genus *Cercotmetus* Amyot & Serville, 1843 (Hemipera–Heteroptera: Nepidae). Tijdsch. Entomol. 116: 84–106.

1974. A new genus of Nepidae from Australia with a revised classification of the family (Hemiptera: Heteroptera). J. Austral. Entomol. Soc. 13: 219–227.

Larsén, O. 1938. Untersuchungen über den Geschlechtsapparat der aquatilen Wanzen. Opusc. Entomol. (Suppl.), 1: 1–388.

Locy, W. A. 1884. Anatomy and physiology of the family Nepidae. Amer. Nat. 18: 250–255, 353–367.

Mathavan, S., and E. Jayakumar. 1987. Long-term effects of pesticides (fenthion and temophos) on growth and fecundity of an aquatic bug *Laccotrephes griseus* (Guerin). Indian J. Exp. Biol. 25: 48–51.

McPherson, J. E., and R. J. Packauskas. 1987. Life history and laboratory rearing of *Nepa apiculata* (Heteroptera: Nepidae), with descriptions of immature stages. Ann. Entomol. Soc. Amer. 80: 680–685.

Ormerod, E. A. 1878. *Ranatra linearis*. Entomologist 11: 119–120.

Ouda, N. A., Abdul-Hussain, M. Al-Faisal, and H. H. Zayia. 1986. Laboratory observations on the efficiency of seven mosquito larvae predators. J. Biol. Sci. Res. 17: 245–252.

Packauskas, R. J., and J. E. McPherson. 1986. Life history and laboratory rearing of *Ranatra fusca* (Hemiptera: Nepidae) with descriptions of immature stages. Ann. Entomol. Soc. Amer. 79: 566–571.

Parsons, M. C. 1972. Respiratory significance of the thoracic and abdominal morphology of *Belostoma* and *Ranatra* (Insecta, Heteroptera). Z. Morphol. Tiere 73: 163–194.

Poisson, R. A. 1965. Catalogue des Hétéroptères Hydrocorises africano-malgaches de la famille des Nepidae (Latreille) 1802. Bull. l'I.F.A.N. 27: 229–269.

Pruthi, H. S. 1928. Some insect and other enemies of mosquito larvae. Indian J. Med. Res. 16: 153–157.

Radinovsky, S. 1964. Cannibal of the pond. Nat. Hist. 73 (Nov.): 16–25.
Rao, T. K. R. 1962. On the biology of *Ranatra elongata* Fabr. (Heteroptera: Nepidae) and *Sphaerodema annulatum* Fabr. (Heteroptera: Belostomatidae). Proc. R. Entomol. Soc. London (A) 37: 61–64.
1976. Bioecological studies on some aquatic Hemiptera-Nepidae. Entomon 1: 123–132.
Sites, R. W., and J. T. Polhemus. 1994. Nepidae of the United States and Canada. Ann. Entomol. Soc. Amer. 87: 27–42.
Tawfik, M. F. S., M. M. El-Hussieni, and H. A. Bakr. 1986. Ecological observations on aquatic insests [sic] attacking mosquitoes in Egypt. Bull. Soc. Entomol. Egypte 66: 117–126.
Twinn, C. R. 1931. Observations on some aquatic animal and plant enemies of mosquitoes. J. Canad. Entomol. 63: 51–61.
Weissmann, M. J. 1986. Biology of the waterscorpion *Ranatra fusca* Palisot de Beauvois in Colorado, with notes on mosquito control capability. M.S. thesis, University of Colorado, Boulder, Colorado, U.S.A.
Williamson, K. B. 1949. Naturalistic measures of anopheline control. Pp. 1360–1384 *in* Boyd, M. F. [ed.], Malariology, Vol. 2. Saunders, Philadelphia, Pennsylvania, U.S.A.

CHAPTER 24

Small Aquatic Bugs (Nepomorpha) with Slight or Underestimated Economic Importance

Miroslav Papáček

1. INTRODUCTION

The suborder Nepomorpha comprises 11 families, 2 of which are ripicolous, the others aquatic. Belostomatids, nepids, and also naucorids, treated in previous chapters, are of some economic importance. Most other families (Notonectidae, Corixidae, Pleidae, Helotrephidae, Aphelocheiridae, Gelastocoridae, Ochteridae) seem economically unimportant; however, there is the question whether the members of these families really lack economic importance or whether their roles in those food webs that are important to humans or in the conservation of nature are merely not known. Small predaceous or omnivorous aquatic bugs are often very abundant in various types of water bodies, and under certain conditions they may be as important as the large predaceous species. Biologically they are probably the most important adult insects in the littoral region of lakes and ponds (see Hutchinson 1993). The aim of this chapter is to show their real or anticipated importance.

2. GENERAL STATEMENT OF ECONOMIC IMPORTANCE

Various species of these families can have positive economic importance or can be accepted as pests in some cases. They have positive importance as biological control agents of mosquitoes (especially Notonectidae, Corixidae, Pleidae, and Helotrephidae) or other blood-sucking dipterans' larvae and pupae (all families mentioned), as food for some fish, some endangered species of amphibians, birds, and bats (Notonectidae, Corixidae), as feedstuffs for aquarium fish and poultry, or food for humans themselves, and as a relatively good indicator of water quality (Corixidae). Their negative importance lies in the fact they are predators of fish larvae, juvenile crayfish, and larvae of some endangered amphibians (especially Notonectidae, some Corixidae). The control of these bugs as pests is not developed.

It is not possible to define which species are economically the most important. Their influence is often cumulative in food webs. For this reason this chapter does not deal with the individual species but rather attempts to analyze the economic importance of the families.

3. THE RELATIVELY IMPORTANT FAMILIES

3.1 Notonectidae Leach

3.1.1 Distribution, Life History, and Biology

The backswimmers (boat flies, wherry men) occur worldwide in temperate as well as in tropical regions. Basic data on their life cycles, mating behavior, reproductive competence, and diapause were published, for example, by Hungerford (1919, 1933), Clark (1928), Streams and Newfield (1972), Vanderlin and Streams (1977), Lang and Markl (1981), Hilsenhoff (1984), and Papáček (1989) (*Notonecta* spp.), Leong (1962), Young (1970, 1978) (*Anisops* spp.), and Gittelman and Severance (1975) (*Buenoa* spp.). Embryonal and postembryonal development was studied by Ellis and Borden (1969) and Gittelman (1978), among others. The best-known model species from the point of view of life cycles and biology are probably the Old World species *N. glauca* L. (see, e.g., Hoppe 1912, Lehmann 1923, Weber 1930, Larsén 1938, Papáček and Soldán 1987), and the New World species *N. hoffmanni* Hungerford (see, e.g., McPherson 1965, Fox 1975). A more detailed review of ecology has been published recently by Savage (1989).

Macropterous backswimmers fly very well, and are excellent invaders in various types of water. They are often among the first successive stages in aquatic habitats. Sympatric, often closely related species, are vertically and horizontally stratified, and different instars of the same species may be similarly stratified (Gittelman 1976, 1977; Cook and Streams 1984). This stratification can also be influenced by both abiotic and biotic conditions (Streams and Shubeck 1982; Sih 1982; Cockrell 1984a,b; Streams 1986, 1992a,b; Bailey 1987).

Notonectids are very aggressive predators, attacking many pelagic and benthic invertebrates, including their own larvae, invertebrates that fall onto the water surface, and even small vertebrates (e.g., Murdoch et al. 1984, Giller 1986). They detect prey by visual and vibratory signals (e.g., Schwind 1980, Streams 1982). The backswimmers are, at least in ponds, size-limited predators (Cooper 1983, Scott and Murdoch 1983, Cronin and Travis 1986), their food niches changing with their developmental stages: the older the stage, the larger their prey (Fox 1975). The largest species of *Notonecta* attack prey about 2.5 to 3.5 cm long (Dahm 1972, Dubský 1991).

3.1.2 Positive Economic Importance

Notonectids are predators of mosquito and some blood-sucking Diptera (e.g., Ceratopogonidae) larvae and pupae. Their predatory strategy can be considered a "strategy of water surface cleaning." This behavioral phenomenon makes backswimmers possible control agents worldwide of mosquitoes (Laird 1947; Ellis and Borden 1970; Hoy and Reed 1970; Toth and Chew 1972; Gittelman 1974a; Hazelrigg 1974; Fox 1975; Zalom 1978; McDonald and Buchanan 1981; Giller and McNeil 1981; Aly and Mulla 1987; Miura and Takahashi 1987, 1988; Streams 1992a; Wattal et al. 1996; Neri-Barbosa et al. 1997). Mosquito density is also reduced because mosquitoes less frequently lay eggs in localities with notonectids (Chesson 1984, 1989; Blaustein et al. 1995). *Bacillus thuringiensis* var. *israelensis*, used to control mosquito larvae, appears to have little effect on *N. glauca* L., and the two together may provide effective control (Olejníček and Maryškova 1986).

Notonectids are a common part of the natural diet of some predaceous or insectivorous fishes (e.g., Macan 1977, Sih 1987, Cook and Streams 1984, Miura et al. 1984, Bendell 1986, Bendell and McNicol 1987) and anuran adults (Morin et al. 1988).

3.1.3 Damage

Backswimmers can be pests in pond culture of various small fishes, as well as of the larvae of fish species living in still freshwater habitats (Bueno 1905, Clark 1928, Gorai and Chaudhuri 1962,

Leong 1962). Berezina (1955, 1962) found that *N. glauca* in ponds of ~0.01 ha can kill 2.5 to 3.5 thousand fish larvae per day. A laboratory experiment showed that the one *N. glauca* adult killed an average of 2.6 fish larvae per day (Dahm 1972). Predation on small juvenile crayfish was noted as well (Dye and Jones 1975, Hirvonen 1992).

Notonectids are also predators on some endangered amphibians (Cronin and Travis 1986, Giller 1986, Dubský 1991).

3.2 Corixidae Leach

3.2.1 Distribution, Life History, and Biology

Water boatman are the largest nepomorphan group. They occur worldwide in various types of stable and temporary, continental and insular, fresh and saline waters. A relatively large literature deals with life cycles of corixids and with various aspects of their reproductive biology (see, e.g., Hungerford 1919, 1948; Larsén 1938; Wróblewski 1958; Young 1965; Kaiser 1966; Pajunen 1970; Jansson and Scudder 1974; Scudder 1976, 1987; Aiken 1982, 1985; Papáček and Bohoněk 1989; Papáček and Třiska 1992). The most comprehensive recent outlines or reviews of corixid biology were published by Scudder (1976), Jansson (1986), Savage (1989), and Hutchinson (1993).

Corixids have a high dispersal potential, which allows them to utilize various available habitats (Jansson 1986). They seem to be ecologically adaptable to a wide range of environmental conditions, but individual species show distinct preferences for water habitats of a certain water quality (Bosmans 1982; Savage 1994a,b). Most fly very well, and are typical insect invaders of water bodies, including newly developed ones.

Most corixids are omnivorous, with a wide spectrum of potential food. Many species are largely phytophagous (Puchkova 1969); many are carnivorous (e.g., Sutton 1951, Zwart 1965, Scudder 1966, Jansson 1978, Popham et al. 1984). Corixidae represent one of the important predators, especially in acidified or inland saline waters (Scudder 1976, Henrikson and Oscarson 1981). The composition of their diet changes with the developmental stage and with particular conditions of the habitat (e.g., Jansson 1973a,b). Corixids can feed on algal cells, filamentous blue-greens, diatoms, microscopic protozoans and rotifers, small invertebrates, fish eggs, detritus, etc. (e.g., Hungerford 1919, 1948; Walton 1943; Griffith 1945; Sutton 1951; Jaczewski 1961; Wróblewski 1963; Scudder 1969a,b; Sokolovskaya and Zhitneva 1973; Jansson and Scudder 1974; Reynolds 1975; Bakonyi 1976; Pajunen 1979; Reynolds and Scudder 1987a,b).

3.2.2 Positive Economic Importance

Various corixid species are incidental or obligatory predators on mosquito larvae in various regions of the world (Sailer and Lienk 1954; Sokolovskaya and Zhitneva 1973; Reynolds 1975; Henrikson and Oscarson 1981; Nyman et al. 1985; Reynolds and Scudder 1987a,b) and may be important control agents of mosquitoes.

Corixidae are prey for insectivorous or predaceous fish (Frost and Macan 1948; Fernando 1956; Macan 1965; Washino and Hokama 1967; Ahmed et al. 1970; Scudder 1976; Applegate and Kieckhefer 1977; Farley and Jounce 1977; Miura et al. 1979, 1984; Henrikson and Oscarson 1978, 1985; Oscarson 1987). They can even be of significant importance for the development of some young stages of fish, as their food in small water bodies (see Rask 1983). Corixids have also marginal importance as a part of the diet of some endangered or protected salamanders (Griffith 1945), anurans (Morin et al. 1988), water birds (Munro 1945; Eriksson 1976; M. Papáček, unpublished data). Flying corixids may fall prey to birds (Fernando 1959) or bats (Walton 1943).

The adults and eggs of some corixids are dried and used as food for aquarium fish, caged birds, and other birds bred by humans (Hutchinson 1993). They are also eaten by some human populations (Mexico, Egypt) (Hungerford 1948, Devey 1957, Hutchinson 1993).

Savage (1994a,b) and Sládeček and Sládečková (1994) summarize data documenting the relationship of corixids to saprobity (organic pollution) of waters and show that the presence or absence, and sometimes the numbers of individuals, can help estimate the saprobic index.

3.2.3 Damage

Corixids can be pests of fish culture. Some species are facultative predators of fish eggs and larvae. A secretion of their saliva induces irreversible changes in the blood of fish fry (Sokolovskaya and Zhitneva 1973).

4. THE LESS IMPORTANT FAMILIES

4.1 Pleidae Fieber and Helotrephidae Esaki and China

Pleidae (pygmy backswimmers) and Helotrephidae are phylogenetically and biologically closely related families. Pleids are distributed worldwide; most species live in tropical regions like most species of helotrephids. These predaceous small water bugs also invade very small temporary water bodies (lithotelms or ceramics water bowls, among others). They may be at least somewhat useful as control agents of mosquito larvae (Gittelman 1974b, Takahashi et al. 1979, Šafař 1991, Papáček 1993).

4.2 Aphelocheiridae Fieber

Aphelocheirids are benthic predaceous bugs with "plastron respiration," living in lakes, streams, and rivers in the Old World. They migrate seasonally and have a relatively wide ecological valence, but they prefer aerated water (e.g., Messner et al. 1983, Saettem 1986). Their economic importance is questionable. On the one hand they suppress populations of blood-sucking dipterans (larvae of Simuliidae); on the other hand they also suppress populations of predators (larvae of Trichoptera) that feed on simuliids too (M. Papáček, unpublished data).

4.3 Gelastocoridae Kirkaldy and Ochteridae Kirkaldy

The toad bugs and velvety shore bugs are riparian or amphibian predaceous insects. They are distributed mostly in tropical and subtropical regions. Their feeding niches are only poorly known (Hungerford 1922, Todd 1955, Poisson 1957), but our knowledge suggests that they help to suppress populations of some blood-sucking Diptera whose aquatic larvae breathe the air on the water surface, or live in small water bodies, in holes in the sand, or in rock pools, or have saprophagous larvae.

5. PROGNOSIS

The economic effects of these Heteroptera (especially notonectids and corixids) on human activity and health have not been calculated. Details are lacking on the energetic demands of the bug production, and the energetic needs of food relations that may have economic effects. Not known, as well, is the percentage of the biomass of fish or mosquito larvae that water bugs destroy under various conditions at the locality. Missing, therefore, are reliable quantitative data to determine economic damage or profit. Getting better knowledge of the biology of water bugs and learning the quantitative characteristics of their food relations are important tasks for future basic research. Because the assumed economic importance of these families appears slight, there is little pressure for applied research. Nevertheless, applied research can pursue both the need for monitoring the economically important levels of water bugs as pests of fish culture where they do damage, and the levels for active control of these pests. It is also necessary to test how insecticides, used now or in the future against mosquito larvae, affect mosquito predators including water bugs, and to look for selective insecticides that harm them less.

6. CONCLUSIONS

Although these insects (especially Notonectidae, Corixidae, Pleidae, and Helotrephidae) differ in their life strategies, they can achieve a similar spatial distribution in individual habitats and can have a similar function in food webs in some cases. The most important role of these families is as biological control agents of some blood-sucking Diptera (especially mosquitoes), and as a pest of fish culture.

7. REFERENCES CITED

Ahmed, W., R. K. Washino, and P. A. Gieke. 1970. Further biological and chemical studies on *Gambusia affinis* (Baird and Girard) in California. Proc. California Mosq. Control Assoc. 38: 95–97.

Aiken, R. B. 1982. Sound production and mating in waterboatman *Palmacorixa nana* (Heteroptera: Corixidae). Anim. Behav. 50: 54–61.

1985. Sound production by aquatic insects. Biol. Rev. 60: 163–211.

Aly, Ch., and M. S. Mulla. 1987. Effect of two microbial insecticides on aquatic predators of mosquitoes. Z. Angew. Entomol. 103: 113–118.

Applegate, R. L., and R. W. Kieckhefer. 1977. Ecology of Corixidae (water boatmen) in Lake Poinsett, South Dakota. Amer. Midl. Nat. 97: 198–208.

Bailey, P. C. 1987. Abundance and age-specific spatial and temporal distribution in two waterbug species, *Anisops danaei* (Notonectidae) and *Ranatra dispar* (Nepidae) in three farm dams in South Australia. Oikos 49: 83–90.

Bakonyi, G. 1976. Contribution to the knowledge of the feeding habitats of some water boatmen: *Sigara* spp. (Heteroptera: Corixidae). Folia Entomol. Hung. 31: 19–24.

Bendell, B. E. 1986. The effects of fish and pH on the distribution and abundance of backswimmers (Hemiptera: Notonectidae). Canad. J. Zool. 64: 2696–2699.

Bendell, B. E., and D. K. McNicol. 1987. Fish predation, lake acidity and the composition of aquatic insect assemblages. Hydrobiologia 150: 193–202.

Berezina, N. A. 1955. O pitanii nekotorykh vodnykh klopov kak konkurentov i vreditelei molodi ryb. [On the feeding of some waterbugs — concurrents and pest of fish fry.] T. Mosk. Tek. Inst. Rybn. Promsti. 7: 142–148 [in Russian].

1962. Borba s khishchnymi vodnymi nasekomymi [Control of predatory aquatic insects]. Rybovodstvo Rybolovstvo 5: 25–26 [in Russian].

Blaustein L., B. P. Kotler, and D. Ward. 1995. Direct and indirect effects of a predatory backswimmer (*Notonecta maculata*) on community structure of desert temporary pools. Ecol. Entomol. 20: 311–318.

Bosmans, R. 1982. Distribution and ecology of Belgian Corixidae (Hemiptera). Acad. Analecta 44: 23–60.

Bueno, J. R. de la Torre. 1905. The genus *Notonecta* in America north of Mexico. J. New York Entomol. Soc. 13: 143–167.

Chesson, J. 1984. Effect of notonectids (Hemiptera, Notonectidae) on mosquitoes (Diptera: Culicidae): Predation or selective oviposition? Environ. Entomol. 13: 531–538.

1989. The effect of alternative prey on the functional response of *Notonecta hoffmani*. Ecology 70: 1227–1235.

Clark, L. B. 1928. Seasonal distribution and life history of *Notonecta undulata* in the Winnipeg Region, Canad. Ecology 9: 383–403.

Cockrell, B. J. 1984a. Effect of water depth on choice of spatially separated prey by *Notonecta glauca* L. Oecologia (Berlin) 62: 256–261.

1984b. Effect of temperature and oxygenation on predator–prey overlap and prey choice of *Notonecta glauca*. J. Anim. Ecol. 53: 519–532.

Cook, V. L., and F. A. Streams. 1984. Fish predation on *Notonecta* (Hemiptera): relationship between prey risk and habitat utilization. Oecologia (Berlin) 64: 177–183.

Cooper, S. D. 1983. Selective predation on cladocerans by common pond insects. Canad. J. Zool. 61: 879–886.

Cronin, J. T., and J. Travis. 1986. Size-limited predation on larval *Rana areolata* (Anura: Ranidae) by two species of backswimmer (Insecta: Hemiptera: Notonectidae). Herpetologica 42: 171–174.

Dahm, E. 1972. Zur Biologie von *Notonecta glauca* (Insecta, Hemiptera) unter besonderer Berücksichtigung der fischereilichen Schadwirkung. Int. Rev. Ges. Hydrobiol. 57: 429–461.

Devey, E. S. 1957. Limnologic studies in Middle America. Trans. Connecticut Acad. Arts Sci. 39: 217–238.

Dubský, L. 1991. Predace larev obojživelníků (Amphibia, Anura) plošticemi rodu *Notonecta* (Heteroptera). [Predation on the amphibian larvae (Amphibia, Anura) by bugs of the genus *Notonecta* (Heteroptera)]. M.Sc. thesis, University of South Bohemia, Ped. Fac., České Budějovice Czech Republic. 72 pp. [in Czech].

Dye, L., and P. Jones. 1975. The influence of density and invertebrate predation on the survival of young-of-the-year *Orconectes virilis*. Freshwater Crayfish 2: 529–538.

Ellis, R. A., and J. H. Borden. 1969. Effects of temperature and other environmental factors on *Notonecta undulata* Say. Pan-Pac. Entomol. 45: 20–25.

1970. Predation by *Notonecta undulata* (Heteroptera: Notonectidae) on larvae of the yellow fever mosquito. Ann. Entomol. Soc. Amer. 63: 963–973.

Eriksson, M. O. G. 1976. Food and feeding habits of downy goldeneye *Bucephala clangula* (L.) ducklings. Ornis Scand. 7: 159–169.

Farley, D. G., and L. C. Younce. 1977. Effects of *Gambusia affinis* (Baird and Girard) on selected nontarget organisms in Fresno County rice fields. Proc. California Mosq. Vector Control Assoc. 45: 87–94.

Fernando, C. H. 1956. On the food of four common freshwater fish of Ceylon. Ceylon J. Sci. C7: 201–217.

1959. The colonization of small freshwater habitats by aquatic insects. 2. Hemiptera (the water-bugs). Ceylon J. Sci., Biol. Sci. 2: 5–32.

Fox, L. R. 1975. Some demographic consequences of food shortage for the predator, *Notonecta hoffmanni*. Ecology 56: 868–880.

Frost, W. E., and T. T. Macan. 1948. Corixidae (Hemiptera) as food of fish. J. Anim. Ecol. 17: 174–179.

Giller, P. S. 1986. The natural diet of the Notonectidae: field trials using electrophoresis. Ecol. Entomol. 11: 163–172.

Giller, P. S., and S. McNeil. 1981. Predation strategies, resource partitioning and habitat selection in *Notonecta* (Hemiptera/Heteroptera). J. Anim. Ecol. 50: 789–808.

Gittelman, S. H. 1974a. *Martarega hondurensis* and *Buenoa antigone* as predators of mosquito larvae in Costa Rica (Hemiptera: Notonectidae). Pan-Pac. Entomol. 50: 84–85.

1974b. The habitat preference and immature stages of *Neoplea striola* (Hemiptera, Pleidae). J. Kansas Entomol. Soc. 47: 491–503.

1976. Swimming ability of Notonectidae (Hemiptera). Psyche 83: 319–323.

1977. Leg segment proportions, predatory strategy and growth in backswimmers (Hemiptera: Pleidae, Notonectidae). J. Kansas Entomol. Soc. 50: 161–171.

1978. Optimum diet and body size in backswimmers (Heteroptera: Notonectidae, Pleidae). Ann. Entomol. Soc. Amer. 71: 737–747.

Gittelman, S. H., and P. Severance. 1975. The habitat preference and immature stages of *Buenoa confusa* and *B. margaritacea* (Hemiptera: Notonectidae). J. Kansas Entomol. Soc. 48: 507–518.

Gorai, A. K., and D. N. Chaudhuri. 1962. Food and feeding habits of *Anisops bouvieri* Kirk. (Heteroptera: Notonectidae). J. Asiat. Soc. 4(3–4): 135–139.

Griffith, M. E. 1945. The environment, life history, and structure of the water boatman, *Rhamphocorixa acuminata* (Uhler) (Hemiptera, Corixidae). Univ. Kansas Sci. Bull. 30: 241–365.

Hazelrigg, J. E. 1974. *Notonecta unifasciata* as predators of mosquito larvae in simulated field habitats. Proc. California Mosq. Control Assoc. 42: 60–65.

Henrikson, L., and H. Oscarson. 1978. Fish predation limiting abundance and distribution of *Glaenocorisa p. propinqua*. Oikos 31: 102–105.

1981. Corixids (Hemiptera-Heteroptera), the new top predators in acidified lakes. Verh. Int. Verein. Theor. Angew. Limnol. 21: 1616–1620.

1985. Waterbugs (Corixidae, Hemiptera-Heteroptera) in acidified lakes: habitat selection and adaptations. Ecol. Bull. 37: 232–238.

Hilsenhoff, W. L. 1984. Aquatic Hemiptera of Wisconsin. Great Lakes Entomol. 17: 29–50.

Hirvonen, H. 1992. Effects of backswimmer (*Notonecta*) predation on crayfish (*Pacifastacus*) young: autotomy and behavioural responses. Ann. Zool. Fenn. 29: 261–271.

Hoppe, J. 1912. Die Atmung von *Notonecta glauca* (Hemiptera, Heteroptera). Zool. Jahrb. Anat. Allg. Zool. Physiol. 31: 189–244.

Hoy, J. B., and D. E. Reed. 1970. Biological control of *Culex tarsalis* in a California rice field. Mosq. News 30: 222–230.

Hungerford, H. B. 1919. The biology of aquatic and semi-aquatic Hemiptera. Univ. Kansas Sci. Bull. 11: 1–341.

1922. The life history of the toad bug. Univ. Kansas Sci. Bull. 24: 145–171.

1933. The genus *Notonecta* of the world. Univ. Kansas Sci. Bull. 21: 5–195.

1948. The Corixidae of the Western Hemisphere (Hemiptera). Univ. Kansas Sci. Bull. 32: 1–827.

Hutchinson, G. E. 1993. A Treatise on Limnology IV. The Zoobenthos. John Wiley & Sons, New York, New York, U.S.A. 944 pp.

Jaczewski, T. 1961. Notes on the biology of Corixidae (Heteroptera). Pol. Pismo Entomol. 31: 295–300.

Jansson, A. 1973a. Stridulation and its significance in the genus *Cenocorixa* (Hemiptera, Corixidae). Behaviour 46: 1–36.

1973b. Diel periodicity of stridulating activity in the genus *Cenocorixa* (Hemiptera, Corixidae) Ann. Zool. Fenn. 10: 378–383.

1978. Viability of progeny in experimental crosses between geographically isolated populations of *Arctocorixa carinata* (C. Sahlberg) (Heteroptera, Corixidae). Ann. Zool. Fenn. 15: 77–83.

1986. The Corixidae (Heteroptera) of Europe and some adjacent regions. Acta Entomol. Fenn. 47: 1–94.

Jansson, A., and G. G. E. Scudder. 1974. The life cycle and sexual development of *Cenocorixa* species (Hemiptera, Corixidae) in the Pacific Northwest of North America. Freshwater Biol. 4: 73–92.

Kaiser, E. W. 1966. *Micronecta*-artene i Denmark (Hemiptera, Corixidae). [*Micronecta* species in Denmark]. Flora Fauna 72: 139–147 [in Danish, English summary].

Laird, M. 1947. Some natural enemies of mosquitoes in the vicinity of Palmalmal, New Britain. Trans. R. Soc. New Zealand 76: 453–476.

Lang, H. H., and H. Markl. 1981. Sex discrimination in the back swimmer *Notonecta glauca* upon contact with conspecific (Heteroptera: Notonectidae). Entomol. Generalis 7: 175–191.

Larsén, O. 1938. Untersuchungen über den Geschlechtsapparat der aquatilen Wanzen. Opusc. Entomol. Suppl. I: 1–388.

Lehmann, H. 1923. Biologische Beobachtungen an *Notonecta glauca*. Zool. Jahrb. Syst. 46: 121–159.

Leong, C. Y. 1962. The life-history of *Anisops breddini* Kirk. (Hemiptera, Notonectidae). Ann. Mag. Nat. Hist. 5, Ser. 13: 377–383.

Macan, T. T. 1965. Predation as a factor in the ecology of water bugs. J. Anim. Ecol. 34: 691–698.

1977. The fauna in the vegetation of a moorland fishpond as revealed by different methods of collecting. Hydrobiology 55: 3–15.

McDonald, G., and G. A. Buchanan. 1981. The mosquito and predatory insect fauna inhabiting fresh-water ponds, with particular reference to *Culex annulirostris* Skuse (Diptera: Culicidae). Austral. J. Ecol. 6: 21–27.

McPherson, J. E. 1965. Notes on the life history of *Notonecta hoffmanni* (Hemiptera: Notonectidae). Pan-Pac. Entomol. 41: 86–89.

Messner, B., I. Groth, and D. Taschenberger. 1983. Zum jahreszeitlichen Wanderverhalten der Grundwanze *Aphelocheirus aestivalis*. Zool. Jahrb. Syst. 110: 323–331.

Miura, T., and R. M. Takahashi. 1987. Augmentation of *Notonecta unifasciata* eggs for suppressing *Culex tarsalis* larval population densities in rice fields. Proc. Pap. California Mosq. Vector Control Assoc. 55: 45–49.

1988. The relationship between the absolute population density and sweep net samples of notonectids in California rice fields. Proc. Pap. California Mosq. Vector Control Assoc. 56: 164–169.

Miura, T., R. M. Takahashi, and R. J. Stewart. 1979. Habitat and food selection by the mosquitofish *Gambusia affinis*. Proc. California Mosq. Vector Control Assoc. 47: 46–50.

Miura, T., R. M. Takahashi, and W. H. Wilder. 1984. Impact of the mosquitofish (*Gambusia affinis*) on a rice field ecosystem when used as a mosquito control agent. Mosq. News 44: 510–516.

Morin, P. J., P. S. Lawler, and E. Johnson. 1988. Competition between aquatic insects and vertebrates: interaction strength and higher order interactions. Ecology 69: 1401–1409.

Munro, J. A. 1945. The birds of the Cariboo Parklands, British Columbia. Can. J. Res. D23: 17–103.

Murdoch, W. W., M. Scott, and P. Ebsworth. 1984. Effects of the general predator *Notonecta* (Hemiptera) upon a freshwater community. J. Anim. Ecol. 47: 581–592.

Neri-Barbosa, J. F., H. Quiroz-Martinez, M. L. Rodriguez-Tovar, L. O. Tejada, and M. H. Badii. 1997. Use of Bactimos® briquets (B.t.i. formulation) combined with the backswimmer *Notonecta irrorata* (Hemiptera: Notonectidae) for control of mosquito larvae. J. Amer. Mosq. Control Assoc. 13: 87–89.

Nyman, H. G., H. G. Oscarson, and J. A. E. Stenson. 1985. Impact of invertebrate predators on the zooplankton composition in acid forest lakes. Ecol. Bull. 37: 239–243.

Olejníček, J., and B. Maryškova. 1986. The influence of *Bacillus thuringiensis* var. *israelensis* on the mosquito predator *Notonecta glauca*. Folia Parasitol. 33: 279–280.

Oscarson, H. G. 1987. Habitats segregation in a water boatman (Corixidae) assemblage — the role of predation. Oikos 49: 133–140.

Pajunen, V. I. 1970. Adaptation of *Arctocorisa carinata* (Sahlb.) and *Callicorixa producta* (Reut.) populations to a rock pool environment. Proc. Adv. Study Inst. Dynamics Popul., Oosterbeek, 1970: 148–158.

1979. Competition between rock pool corixids. Ann. Zool. Fenn. 16: 138–143.

Papáček, M. 1989. Životiní cykly univoltinních vodních ploštic (Heteroptera, Nepomorpha) v Československu. [Life cycles of univoltine water bugs (Heteroptera, Nepomorpha) in Czechoslovakia]. Pr. e Slov. Ent. Spol. SAV, Bratislava 8: 45–52 [in Czech, English summary].

1993. K problematice morfologie a bionomie vodních ploštic nadčeledi Pleoidea a Notonectoidea (Heteroptera: Nepomorpha). [On the morphology and bionomy of water bugs of the superfamilies Pleoidea and Notonectoidea (Heteroptera: Nepomorpha)] Assoc. Prof. thesis, University of South Bohemia, České Budějovice, Czeck Republic. 232 pp. [in Czech].

Papáček, M., and K. Bohoněk. 1989. The life cycle and ovarian development of *Corixa punctata* (Heteroptera, Corixidae) in South Bohemia. Acta. Entomol. Bohemoslov. 86: 96–110.

Papáček, M., and T. Soldán. 1987. Development of the female internal reproductive system of *Notonecta glauca* (Heteroptera, Notonectidae) and the life cycle in South Bohemia. Acta Entomol. Bohemoslov. 84: 161–180.

Papáček, M., and K. Třiska. 1992. Lebenszyklus und Entwicklung der Ruderwanze *Cymatia coleoptrata* (Heteroptera, Corixidae) in Südböhmen (Mitteleuropa). Zool. Jahrb. Syst. 119: 425–435.

Poisson, R. 1957. Hétéropteres aquatiques. Faune France 61: 1–263.

Popham, E. J., M. T. Bryant, and A. A. Savage. 1984. The role of front legs of British corixid bugs in feeding and mating. J. Nat. Hist. 18: 445–464.

Putschkova, L. V. 1969. Troficheskyie svyazi greblyakov i vliyanie *S. striata* L. na vodnuyu rastitelnost. [On the trophic relationships of waterboatmans and influence of *Sigara striata* on the water plants (Corixidae).] Zool. Zhurn. 48: 1581–1583 [in Russian, English summary].

Rask, M. 1983. Differences in growth of perch (*Perca fluviatilis*) in two small forest lakes. Hydrobiologia 101: 139–144.

Reynolds, J. D. 1975. Feeding in corixids (Heteroptera) of small alkaline lakes in central B.C. Verh. Int. Verein. Limnol. 19: 3073–3078.

Reynolds, J. D., and G. G. E. Scudder. 1987a. Experimental evidence of the fundamental feeding niche in *Cenocorixa* (Hemiptera: Corixidae). Canad. J. Zool. 65: 967–973.

1987b. Serological evidence of realised feeding niche in *Cenocorixa* species (Hemiptera: Corixidae) in sympatry and allopatry. Canad. J. Zool. 65: 974–980.

Saettem, L. M. 1986. The life history of *Aphelocheirus aestivalis* Fabricius (Hemiptera) in Norway. Arch. Hydrobiol. 106: 245–250.

Šafař, P. 1991. pH jako jeden z abiotických faktorů vodního prostředí [pH as abiotic factor of water environment]. M.Sc. thesis, University of South Bohemia, Ped. Fac., České Budějovice, Czech Republic. 69 pp. [in Czech].

Sailer, R. I., and S. E. Lienk. 1954. Insect predators of mosquito larvae and pupae in Alaska. Mosq. News 14: 14–16.

Savage, A. A. 1989. Adults of the British aquatic Hemiptera Heteroptera. A key with ecological notes. Sci. Publ. Freshwater Biol. Assoc. 50: 1–173.

1994a. The distribution of Corixidae in relation to the water quality of British lakes: a monitoring model. Freshwater Forum 4: 32–61.

1994b. Corixidae and water quality. Freshwater Forum 4: 214–216.

Schwind, R. 1980. Geometrical optics of the *Notonecta* eye: adaptations to optical environment and way of life cycle. J. Comp. Physiol. 140: 59–68.

Scott, M. A., and W. W. Murdoch. 1983. Selective predation by the backswimmer, *Notonecta*. Limnol. Oceanogr. 28: 362–366.

Scudder, G. G. E. 1966. The immature stages of *Cenocorixa bifida* (Hung.) and *C. expleta* (Uhler) (Hemiptera: Corixidae). J. Entomol. Soc. British Columbia 63: 33–40.

1969a. The distribution of two species of *Cenocorixa* in inland saline lakes of British Columbia. J. Entomol. Soc. British Columbia 66: 32–41.

1969b. The fauna of saline lakes on the Fraser Plateau in British Columbia. Verh. Int. Verein. Limnol. 17: 430–439.

1976. Water boatmen of saline waters (Hemiptera: Corixidae). Pp. 263–289 *in* L. Cheng [ed.], Marine Insects. Elsevier North-Holland Publ. Co., New York, New York, U.S.A. 581 pp.

1987. Aquatic and semiaquatic Hemiptera of peatlands and marshes in Canada. Mem. Entomol. Soc. Canada 140: 65–98.

Sih, A. 1982. Foraging strategies and the avoidance of predation by an aquatic insect, *Notonecta hoffmanni*. Ecology 63: 786–798.

1987. Predators and prey lifestyles: an evolutionary and ecological overview. Pp. 203–224 *in* W. C. Kerfoot and A. Sih [eds.], Predation: Direct and Indirect Impacts on Aquatic Communities. University Press of New England, Hanover, New Hampshire, U.S.A. 563 pp.

Sládeček, V., and A. Sládečková. 1994. Corixidae as indicators of organic pollutions. Freshwater Forum 4: 211–213.

Sokolovskaya, N. P., and L. D. Zhitneva. 1973. O biologii klopov greblyakov (Heteroptera, Corixidae) vredyashchikh rybovodstvu v rostovskykh oblastyakh. [A contribution to the biology of Corixidae (Heteroptera), pests of fish culture in the Rostov district]. Zool. Zh. 52: 1330–1335 [in Russian, English summary].

Streams, F. A. 1982. Diel foraging and reproductive periodicity in *Notonecta undulata* Say (Heteroptera). Aquat. Insects 4: 111–119.

1986. Foraging behavior in a notonectid assemblage. Amer. Midl. Nat. 117: 353–361.

1992a. Age-dependent foraging depths of two species of *Notonecta* (Heteroptera: Notonectidae) breeding together in a small pond. Aqua. Insects 14: 183–191.

1992b. Intrageneric predation by *Notonecta* (Hemiptera: Notonectidae) in the laboratory and in nature. Ann. Entomol. Soc. Amer. 85: 265–273.

Streams, F. A., and S. Newfield. 1972. Spatial and temporal overlap among breeding populations of New England *Notonecta*. Univ. Connecticut Occas. Pap. (Biol. Sci. Ser.) 2: 139–157.

Streams, F. A., and T. P. Shubeck. 1982. Spatial structure and intraspecific interactions in *Notonecta* populations. Environ. Entomol. 11: 652–659.

Sutton, M. F. 1951. On the food, feeding mechanism and alimentary canal of Corixidae (Hemiptera, Heteroptera). Proc. Zool. Soc. London 121: 465–499.

Takahashi, R. M., R. J. Stewart, C. H. Schefer, and R. D. Sjogren. 1979. An assessment of *Plea striola* (Hemiptera: Pleidae) as a mosquito control agent in California. Mosq. News 39: 514–519.

Todd, E. L. 1955. A taxonomic revision of the family Gelastocoridae (Hemiptera). Univ. Kansas Sci. Bull. 37: 277–475.

Toth, R. S., and R. M. Chew. 1972. Development and energetics of *Notonecta undulata* during predation on *Culex tarsalis*. Ann. Entomol. Soc. Amer. 65: 1270–1279.

Vanderlin, R. L., and F. A. Streams. 1977. Photoperiodic control of reproductive diapause in *Notonecta undulata*. Environ. Entomol. 6: 258–262.

Walton, G. A. 1943. The water bugs (Rhynchota-Hemiptera) of North Somerset. Trans. Soc. Brit. Entomol. 8: 231–290.

Washino, R. K., and Y. Hokama. 1967. Preliminary report on the feeding pattern of two species of fish in a rice habitat. Proc. California Mosq. Vector Control Assoc. 35: 84–87.

Wattal, S., T. Adak, R. C. Dhiman, and V. P. Sharma. 1996. The biology and predatory potential of notonectid bug, *Enithares indica* (Fabr.) against mosquito larvae. Southeast Asian J. Trop. Med. Pub. Health 27: 633–636.
Weber, H. 1930. Biologie der Hemipteren. Julius Springer Verlag, Berlin, Germany. 543 pp.
Wróblewski, A. 1958. The Polish species of the genus *Micronecta* Kirk. (Heteroptera, Corixidae). Ann. Zool. Warszawa 17: 247–381.
1963. Notes on Micronectinae from USSR (Heteroptera, Corixidae). Bull. Acad. Pol. Sci. 10: 319–324.
Young, E. C. 1965. Teneral development in British Corixidae. Proc. R. Entomol. Soc. A40: 159–168.
1970. Seasonal changes in populations of Corixidae and Notonectidae (Hemiptera, Heteroptera) in New Zealand. Trans. R. Soc. New Zealand. 12: 113–133.
1978. Seasonal cycles of ovarian development in Corixidae and Notonectidae, aquatic Hemiptera-Heteroptera. New Zealand Entomol. 6: 361–362.
Zalom, F. G. 1978. Backswimmer prey selection with observations on cannibalism (Hemiptera: Notonectidae). Southwest Nat. 23: 617–622.
Zwart, K. W. R. 1965. On the influence of some food substances on survival of Corixidae (Heteroptera). Pp. 411–412 *in* Proceedings, XIIth International Congress of Entomology. London, U.K.

CHAPTER 25

Semiaquatic Bugs (Gerromorpha)

John R. Spence and Nils Møller Andersen

1. INTRODUCTION

The infraorder Gerromorpha comprises eight families (Mesoveliidae, Hebridae, Paraphrynoveliidae, Macroveliidae, Hydrometridae, Hermatobatidae, Veliidae, and Gerridae). About 150 genera and 1670 species are included in these eight families, of which the most species-rich are the Veliidae (720 species) and the Gerridae (620 species) (N. M. Andersen 1982a, and unpublished). For the last mentioned family, modern taxonomic revisions are available for the primarily Holarctic genera *Aquarius* (Andersen 1990), *Gerris* (Andersen 1993a), and *Limnoporus* (Andersen and Spence 1992), for the Old World genera *Limnogonus* and *Neogerris* (Andersen 1975), including the Oriental genera *Metrocoris* (Chen and Nieser 1993) and *Ventidius* (Chen and Zettel 1999). The Trepobatinae have recently been reviewed by Polhemus and Polhemus (1993, 1994, 1995, 1996) and the Australian gerrids by Andersen and Weir (1994, 1997, 1998).

These bugs are mainly predator-scavengers rather tightly adapted for life on water surfaces (Andersen 1982a, Spence and Andersen 1994), although several genera have secondarily invaded damp, terrestrial habitats (Andersen 1982b). Gerromorphans are relatively common inhabitants of both still and running freshwater systems throughout the world; several lineages carry the insect banner into the marine realm and are the only hexapods inhabiting the open ocean (Andersen and Polhemus 1976; Andersen 1989, 1991, 1992; Andersen and Weir 1994). Adaptations for life on water surfaces center on (1) elongation of legs and associated modification of thoracic structure and musculature for highly specialized locomotion (Andersen 1976, 1982a); (2) diverse life-history patterns including considerable variation in wing length, wing musculature, and diapause regulation (Vepsäläinen 1978, Köpfli et al. 1987, Muraji et al. 1989, Spence 1989, Harada 1991, Andersen 1993); and (3) relatively complex mating behaviors (Andersen 1994, Rowe et al. 1994, Spence and Andersen 1994, Rowe and Arnqvist 1996, Arnqvist 1997).

Most gerromorphans have five instars, although only four are known for *Mesovelia* and in some species of *Rhagovelia*; intraspecific variation in instar number (4 to 5) occurs in several species of *Microvelia* (Andersen 1982a). Most species from temperate areas spend the winter on land as adults in reproductive diapause; overwintering diapause in the egg stage is known from a few genera (see Spence and Andersen 1994). Many tropical species breed throughout the year but some exhibit ovarian diapause during the dry season (Selvanayagam and Rao 1986). The ocean-going species of *Halobates* appear to reproduce throughout the year.

Gerromorphans are attracting increasing attention from behaviorists and evolutionary biologists, especially those interested in the evolution and implications of reproductive and life-history adaptations (Spence and Andersen 1994). However, only a few possible direct links to economic importance have been suggested. These reports are mostly casual and provide grounds for further study rather than a firm basis for evaluating the economic significance of the infraoder. They fall into three categories as summarized below.

2. IMPORTANCE IN THE RICE CROP

Given their semiaquatic habits, rice is the only major agricultural system in which gerromorphans might be expected to play some sort of economically significant role. It appears that they are a mixed blessing. Although gerromorphans are not generally known to be phytophagous, Gupta et al. (1990) wrote that *L. nitidus* (Mayr) (given as *Gerris nitida* by Gupta et al., but see Andersen 1975) causes significant losses of rice in northeastern India. These insects were fairly common (4 to 9 bugs/m^2) in rain-fed lowland rice paddies and grain infestation rates of 1.9 to 4.2% were claimed. Damage was caused by both adults and nymphs sucking sap from the grains during the milky stage of their development; adults were also reported to suck sap from the outer side of the boot-sheath leaf. Such findings deserve further investigation as this is the only report known to the authors of phytophagy among gerromorphans. However, this same species and the congeneric *L. fossarum* (F.) have also been reported to be significant predators of the brown plant hopper, *Nilaparvata lugens* (Stål), one of the world's most important rice pests (Samal and Misra 1982, Selvanayagam and Rao 1986), although detailed studies of their significance have not yet been published.

A detailed assessment of the predatory impact of another gerromorphan, *Microvelia douglasi atrolineata* (Bergroth), on *N. lugens* has been presented by Nakasuji and Dyck (1984). They concluded that *M. d. atrolineata* is "one of the most important natural enemies of the brown planthopper." Several elements of planthopper and microveliid behavior were critical to their analysis. It is common for brown planthoppers to fall from rice plants onto the water surface of paddy fields. For example, over four paddy fields, they estimated that 12 to 43% of the brown planthopper nymphs fell to the water surface, the dropoff rate being generally higher for adults than for nymphs. Before these plant-feeding bugs can move back onto a rice plant, some are picked off by roving *M. d. atrolineata*. The percentage of successful attacks, over a range of predator–prey instar combinations, varied from 4.8 to 17.7%, but the microveliids generally performed better as predators when attacking in groups of up to eight individuals. Nakasuji and Dyck also observed the highest attack efficiency at a density of about 450 *Microvelia*/m^2, which is roughly the peak density observed in paddy fields. Because the Japanese subspecies, *M. d. douglasi* Scott is known to be highly susceptible to organic insecticides (Miyahara and Oho 1957), use of broad-spectrum insecticides probably diminishes the utility of microveliids as biological control agents in highly managed paddies.

3. NATURAL ENEMIES OF BITING FLIES

Several species of gerromorphans have been observed consuming mosquito eggs (Laird 1956, Gupta 1985) and larvae (Hungerford 1920, Frick 1949, Sprague 1956, Kurihara 1974, Gupta 1985, Selvanayagam and Rao 1986), but their actual significance as natural enemies of biting flies has not been established. Although it is possible that gerrids do catch incidental larvae and a few emerging adults, it seems more likely that mosquitoes, like other dipterans with aquatic stages (cf. Foster and Treherne 1986), are consumed after they are trapped in the surface tension and thus no longer threats to human welfare. Nonetheless, Miura and Takahashi (1988) showed that *M. pulchella* Westwood can survive and reproduce wholly on a diet of mosquitoes.

Nummelin (1988) studied the pest management potential of two common East African gerrids, *G. swakopenis* (Stål) and *L. cereiventris* (Signoret), on newly emerged adults of *Aedes aegypti* L., the principal vector of yellow fever. In his experiments, adult gerrids of both species, without access to other food, killed 5 to 20% of teneral mosquitoes before they could take flight in simple laboratory experiments run under rather high densities of water striders. Gerrid nymphs, however, were not particularly effective at catching the hatchling mosquitoes, which escaped easily by walking across the surface, even before they could take flight. Nummelin concluded that the potential of these two gerrids was "too low for use of water striders as a controlling agent of mosquitoes." Thus, it appears that although waterstriders provide some background mortality of mosquitoes, their potential as biological control agents for these pests is actually rather low.

4. INDICATORS OF ENVIRONMENTAL CONDITIONS

Gerromorphans are conspicuous components of undisturbed aquatic and wetland habitats worldwide. Although these bugs are perhaps not critical functional elements of such ecosystems, their presence and local diversity in anthropogenic landscapes may provide good indications of how human activities affect these systems in comparison with areas less dominated by human activity. For example, the large-bodied *Aquarius najas* DeGeer appears to be threatened in the Netherlands because of an overall reduction of suitable habitat (Higler 1967, Nieser and Wasscher 1986). It has also disappeared from several river systems in Denmark where it was abundant during the 19th century (Damgaard and Andersen 1996). Nieser and Wasscher (1986) suggest that the species depends on small streams that run through forested areas and that are not polluted by high quantities of organic waste associated with animal agriculture. Declines in *A. najas* populations could indicate that conservation attention is required for the larger suite of species associated with such habitats that were once more common in western and northern Europe.

Gerromorphan populations may also increase in response to anthropogenic effects. For example, Nieser and Wasscher (1986) attribute a conspicuous increase in the records of *A. paludum* F. to an increase in numbers of "recreative" ponds built and stocked for angling in the southern part of the Netherlands. In contrast, Danish populations of this species are at present found on only two to three forest ponds in Zealand, and all but one of these appears to be declining (Damgaard and Andersen 1996). With sufficient study of metapopulation dynamics (Hanski and Gilpin 1997), such dispersive gerromorphan species might have value as indicators of landscape level changes in wetland habitat.

Bendell and McNicol (1987) reported an increase in abundance of gerrids with increasing acidity in Ontario (Canada), and attributed this change to the drastic reduction in fish populations that accompanies acidification. However, Bendell (1988) suggested that the system is more complicated. In this survey of 53 lakes, there was no evidence of association between acidity (pH range 4.0 to 7.5) and the presence or apparent population size of five gerromorphan species. However, there was a significant tendency for populations of *Metrobates hesperius* Uhler and *Trepobates inermis* Esaki to be found on lakes with higher pH; *M. hesperius* was absent below a pH of 5.1. Bendell (1988) also reported that density of *Rheumatobates rileyi* Bergroth increased strongly with pH, but that this was not also significantly associated with changes in fish abundance. Thus, semiaquatic bugs may have potential value in economic monitoring, but more work is required to link changes in distribution and abundance to causative changes in environmental factors.

5. CONCLUSIONS

The semiaquatic bugs, including eight families and 150 genera, are mainly predators with little known direct economic importance. *Limnogonus nitidus* has been implicated as a potential pest of rice in the milky stage of development, but this single finding of phytophagy within the Gerromorpha requires corroboration. Clearly, this same species and several other gerromorphans appear to play more significant roles as predators of the brown planthopper and other rice pests in flooded paddies. Although it has also been suggested that several tropical gerrids are predators of mosquitoes, and that they might have a role in control of vectors of human disease, the few data available do not establish their importance in this regard. Gerromorphans appear to be vulnerable to certain types of pollution and their population sizes may reflect changes to the area and health of regional wetlands. Although further work is required, semiaquatic bugs appear to have potential as indicator taxa for monitoring quantitative and qualitative changes in regional wetlands.

6. REFERENCES CITED

Andersen, N. M. 1975. The *Limnogonus* and *Neogerris* of the Old World with character analysis and a reclassification of the Gerrinae (Hemiptera: Gerridae). Entomol. Scand. Suppl. 7: 1–96.

1976. A comparative study of locomotion on the water surface in semiaquatic bugs (Insecta, Hemiptera, Gerromorpha). Vidensk. Medd. Dan. Naturhis. Foren. 139: 337–396.

1982a. The Semi-aquatic Bugs (Hemiptera, Gerromorpha). Phylogeny, Adaptations, Biogeography and Classification. Scandinavian Science Press, Klampenborg, Denmark. 455 pp.

1982b. Semiterrestrial water striders of the genera *Eotrechus* Kirkaldy and *Chimarrhometra* Bianchi (Insecta, Hemiptera, Gerridae). Steenstrupia 9: 1–25.

1989. The coral bugs, genus *Halovelia* Bergoth (Hemiptera, Veliidae). II. Taxonomy of the *H. malaya*-group, cladistics, ecology, biology and biogeography. Entomol. Scand. 20: 179–227.

1990. Phylogeny and taxonomy of water striders, genus *Aquarius* Schellenberg (Insecta, Hemiptera, Gerridae), with a new species from Australia. Steeenstrupia 16: 37–81.

1991. Marine insects: genital morphology, phylogeny and evolution of sea skaters, genus *Halobates* (Hemiptera: Gerridae). Zool. J. Linn. Soc. 103: 21–60.

1992. A new genus of marine water striders (Hemiptera, Veliidae) with five new species from Malesia. Entomol. Scand. 22: 389–404.

1993a. Classification, phylogeny, and zoogeography of the pond skater genus *Gerris* Fabricius (Hemiptera: Gerridae). Canad. J. Zool. 71: 2473–2508.

1993b. The evolution of wing polymorphism in water striders (Gerridae): a phylogenetic approach. Oikos 67: 433–443.

1994. The evolution of sexual size dimorphism and mating systems in water striders (Hemiptera: Gerridae): a phylogenetic approach. Ecoscience 1: 208–214.

Andersen, N. M., and J. T. Polhemus. 1976. Water-striders (Hemiptera: Gerridae, Veliidae, etc.). Pp. 187–224 *in* L. Cheng [ed.]: Marine Insects. North-Holland Publishing Company, Amsterdam, the Netherlands. 581 pp.

Andersen, N. M., and J. R. Spence. 1992. Classification and phylogeny of the Holarctic water strider genus *Limnoporus* Stål (Hemiptera: Gerridae). Canad. J. Zool. 70: 753–785.

Andersen, N. M., and T. A. Weir. 1994. The sea skaters, genus *Halobates* Eschscholtz (Hemiptera, Gerridae), of Australia: taxonomy, phylogeny, and zoogeography. Invertebr. Taxon. 8: 861–909.

1997. The gerrine water striders of Australia (Hemiptera: Gerridae): taxonomy, distribution, and ecology. Invertebr. Taxon. 11: 203–299.

1998. Australian water striders of the subfamilies Trepobatinae and Rhagadotarsinae (Hemiptera: Gerridae). Invertebr. Taxon. 12: 509–544.

Arnqvist, G. 1997. The evolution of water strider mating systems: causes and consequences of sexual conflicts. Pp. 146–163 *in* J. C. Choe and B. J. Crespi [eds.], The Evolution of Mating Systems in Insects and Arachnids. Cambridge University Press, Cambridge, U.K.

Bendell, B. E. 1988. Lake acidity and the distribution and abundance of waterstriders near Sudbury, Ontario. Canad. J. Zool. 66: 2209–2211.

Bendell, B. E., and D. K. McNicol. 1987. Fish predation, lake acidity and the composition of aquatic insect assemblages. Hydrobiologia 150: 193–202.

Chen, P., and N. Nieser. 1993. A taxonomic revision of the Oriental water strider genus *Metrocoris* Mayr (Hemiptera, Gerridae). Steenstrupia 19: 1–82.

Chen, P., and H. Zettel. 1999. A taxonomic revision of the Oriental water strider genus *Ventidius* Distant (Hemiptera, Gerromorpha, Gerridae). Tijdschr. Entomol. 141: 137–208.
Damgaard, J., and Andersen, N. M. 1996. Distribution, phenology, and conservation status of the larger water striders in Denmark. Entomol. Medd. 64: 289–306.
Foster, W. A., and J. E. Treherne. 1986. The ecology and behaviour of a marine insect, *Halobates fijiensis* (Hemiptera: Gerridae). Zool. J. Linn. Soc. 86: 391.412.
Frick, K. E. 1949. The biology of *Microvelia capitata* Guérin, 1857, in the Panama Canal Zone and its role as a predator on Anopheline larvae. Ann. Entomol. Soc. Amer. 42: 77–100.
Gupta, S. P., A. Prakash, A. Choudhury, J. Rao, and A. Gupta. 1990. New records of bugs infesting paddy fields in Orissa. J. Insect Sci. 3: 185.
Gupta, Y. C. 1985. Bionomics of *Gerris adelaidis* Dohrn (Heteroptera: Gerridae). Bull. Entomol. 26: 129–139.
Hanski, I. A., and M. E. Gilpin [eds.]. 1997. Metapopulation Biology. Academic Press, San Diego, California, U.S.A. 512 pp.
Harada, T. 1991. Effects of photoperiod and temperature on phototaxis in a water strider, *Gerris paludum insularis* (Motschulsky). J. Insect Physiol. 37: 27–34.
Higler, L. W. G. 1967. Some notes on the distribution of the waterbug, *Gerris najas* (DeGeer, 1773) in the Netherlands (Hemiptera-Heteroptera). Beaufortia 14: 87–92.
Hungerford, H. B. 1920. The biology and ecology of aquatic and semiaquatic Hemiptera. Kansas Univ. Sci. Bull. 11: 1–328.
Köpfli, R., R. Hauser, and M. Zimmermann. 1987. Diapausedetermination bei Wasserläufern (Hemiptera, Gerridae). Rev. Suisse Zool. 94: 533–543.
Kurihara, T. 1974. Natural enemies of mosquito. Seikatsu to Kankyo 19(4): 21–26.
Laird, M. 1956. Studies of mosquitoes and freshwater ecology in the South Pacific. Bull. R. Soc. New Zealand 6: 1–213.
Miura, T., and R. Takahashi. 1988. Predation of *Microvelia pulchella* (Hemiptera: Veliidae) on mosquito larvae. J. Amer. Mosq. Control Assoc. 4: 91–93.
Miyahara, K., and N. Oho. 1957. On the resistance of *Micorvelia douglasi* Scott to insecticides. Proc. Assoc. Plant Prot. Kyushu 3: 62–64.
Muraji, M., T. Miura, and F. Nakasuji. 1989. Change in photoperiodic sensitivity during hibernation in a semi-aquatic bug, *Microvelia douglasi* (Heteroptera: Veliidae). Appl. Entomol. Zool. 24: 450–457.
Nakasuji, F., and V. A. Dyck. 1984. Evaluation of the role of *Microvelia douglasi atrolineata* (Bergoth) as predator of the brown planthopper *Nilaparvata lugens* Stål. Res. Popul. Ecol. 26: 134–149.
Nieser, N., and M. Wasscher. 1986. The status of the large waterstriders in the Netherlands (Heteroptera: Gerridae). Entomol. Ber. (Amsterdam) 46: 68–76.
Nummelin, M. 1988. Waterstriders (Het: Gerridae) as predators of hatchling mosquitoes. Bicovasc. Proc. 1: 121–125.
Polhemus, J. T., and D. A. Polhemus. 1993. The Trepobatinae (Heteroptera: Gerridae) of New Guinea and surrounding regions, with a review of the world fauna. Part 1. Tribe Metrobatini. Entomol. Scand. 24: 241–284.
1994. The Trepobatinae (Heteroptera: Gerridae) of New Guinea and surrounding regions, with a review of the world fauna. Part 2. Tribe Naboandelini. Entomol. Scand. 25: 333–359.
1995. The Trepobatinae (Heteroptera: Gerridae) of New Guinea and surrounding regions, with a review of the world fauna. Part 3. Tribe Trepobatini. Entomol. Scand. 26: 97–118.
1996. The Trepobatinae (Heteroptera: Gerridae) of New Guinea and surrounding regions, with a review of the world fauna. Part 4. Tribe Stenobatini. Entomol. Scand. 27: 279–346.
Rowe, L., and G. Arnqvist. 1996. Analysis of the causal components of assortative mating in water striders. Behav. Ecol. Sociobiol. 38: 279–286
Rowe, L., G. Arnqvist, A. Sih, and J. J. Krupa. 1994. Sexual conflict and the evolutionary ecology of mating patterns: water striders as a model system. Trends in Ecol. Evol. 9: 289–293.
Samal, P., and B. C. Misra. 1982. Biological notes on the waterstrider, *Limnogonus niditus* (Mayr), as a predator on the rice brown planthopper, *Nilaparvata lugens* (Stål). Ann. Entomol. Soc. Amer. 75: 12–13.
Selvanayagam, M., and T. K. R. Rao. 1986. Feeding behaviour in Gerridae (Hemiperta: Heteroptera). Pp. 155–164 *in* Proc. II Oriental Entomol. Symp., 21–24 February 1984, Association for Advancement of Entomology, Trivandrum, India.

Spence, J. R. 1989. The habitat templet and life history strategies of pondskaters (Heteroptera: Gerridae): reproductive potential, phenology and wing dimorphism. Canad. J. Zool. 67: 2432–2447.

Spence, J. R., and N. M. Andersen 1994. Biology of water striders: interactions between systematics and ecology. Annu. Rev. Ecol. Syst. 39: 101–128.

Sprague, I. B. 1956. The biology and morphology of *Hydrometra martini* Kirkaldy. Kansas Univ. Sci. Bull. 38: 579–693.

Vepsäläinen, K. 1978. Wing dimorphism and diapause in *Gerris*: determination and adaptive significance. Pp. 218–253 *in* H. Dingle [ed.], Evolution of Insect Migration and Diapause. Springer-Verlag, Berlin, Germany.

CHAPTER 26

Minute Pirate Bugs (Anthocoridae)

John D. Lattin

1. INTRODUCTION

The purpose of this chapter is to provide the reader with examples of different species of Anthocoridae that are known to occur in managed ecosystems. Only a small percentage of the described species is thus far known in this regard. Selected references to each species are included to provide further information. The interested reader should be able to identify which species may be useful in any particular situation. There are surely local species as yet poorly known that may have similar habits and be adapted to local conditions. There is much to learn about the role and position of these small predators in managed agroecosystems.

The family Anthocoridae contains between 500 to 600 species worldwide. These bugs are small in body size and number of species. Siemann et al. (1996) proposed that groups of small physical size might contain fewer species than larger-sized taxa. The numbers reflect the scattered nature of the literature, and the absence of a recent world catalog. Over 1000 citations have been reviewed during this current preparation and will serve as the basis for such a catalog. Schuh and Slater (1995) have given a superb treatise on the bug fauna of the world and their treatment of the three families they include from the former family Anthocoridae provided an excellent overview of these taxa (treated as three families: Anthocoridae, Lasiochilidae, and Lyctocoridae). Most species are known only from the original description and perhaps a few subsequent distribution notes. In contrast, the Palearctic region is quite well known from the publications of Southwood and Leston (1959), Péricart (1972), Zheng and Bu (1990), Hiura (1959, 1960, 1966, 1967), and the recent catalog by Péricart (1996). Kelton (1978) published on the fauna of Canada and Henry (1988) produced a catalog of Canada and continental United States. There are other publications dealing with different localities around the world, for example, Gross (1954, 1955, 1957) and Cassis and Gross (1995) for Australia.

Study of the fauna of the Indian subcontinent had its beginnings in the works of Distant (1906, 1910) and Poppius (1913). Subsequently, a long series of publications has added substantially to those early efforts; some of these publications include Narayanan and Chatterji (1952, 1953), Chatterji (1955), Ghauri (1964, 1972a), Rajasekhara and Chatterji (1970), Rajasekhara (1973), Muraleedharan and Ananthakrishnan (1971, 1974, 1978), Muraleedharan (1977a, 1977b, 1978), Basu (1982), Kumar and Ananthakrishnan (1984), and Naseer and Abdurahiman (1993). Péricart (1996) provided an amplified distribution of those species that extend into the Palearctic.

Only a few species of minute pirate bugs have received extensive treatment in the literature, chiefly but not entirely in the Northern Hemisphere. These small bugs are important components of the predatory fauna found in many agroecosystems (Carayon 1961, Lattin 1999). Increased interest in biological control has resulted in greater attention to the Anthocoridae and other predatory Heteroptera in these managed ecosystems. Although not all publications can be included here, for obvious reasons, an attempt is made to cover as many species as possible and to provide the reader with an introduction to the members of this family and their differing roles in agroecosystems around the world.

1.1 Systematics

There are between 500 and 600 species of Anthocoridae distributed over the world. Schuh and Slater (1995) placed this group of insects in the Cimicomorpha following Schuh and Štys (1991); some of the other common families included in this latter category are the Cimicidae, Miridae, Nabidae, Reduviidae, and Tingidae. Within the Cimicomorpha, the Anthocoridae are placed in the superfamily Cimicoidea where Schuh and Slater included the Anthocoridae, Lasiochilidae, Lyctocoridae, Cimicidae, Plokiophilidae, and Polyctenidae. The first three taxa are included in the single family, the Anthocoridae, by many authors (Reuter 1884; Péricart 1972, 1996; Cassis and Gross 1995; Lattin, 1999). Here a single family, Anthocoridae, with three well-defined subfamilies, the Anthocorinae, Lasiochilinae, and Lyctocorinae, is considered. There seems no doubt about the status of these three categories, only about their systematic level. Present knowledge is still rudimentary, as most taxa are known poorly. This is especially true of species from South America, much of Africa, and southeast Asia. A world catalog is now under way that will give us a more accurate picture of the world's fauna. The catalogs of Henry (1988), Cassis and Gross (1995), and Péricart (1996) set the stage for such a volume.

1.2 Food and Feeding

All anthocorids have sucking mouthparts and use them to take nourishment. Most known species are predatory as nymphs and adults, chiefly on other small arthropods. It must be stated that knowledge is limited to a select few species that have been studied in any detail, chiefly those known to be of economic importance. Their known prey include mites, aphids, psyllids, scales, woolly aphids, thrips, bark beetles, and the eggs of some Lepidoptera. Carayon (1961) recognized their potential value in managed landscapes and included these small insects among the other predatory true bugs found in agroecosystems. He did question their ability to be major predators under such conditions, but recognized that knowledge was limited at the time. He went on to add substantially to that knowledge base, in addition to elegant work on the bugs' functional morphology (see especially, Carayon 1972a). There is now a substantial literature on the role played by a few species of Anthocoridae, and selected examples of these works are included here. A more general treatment of predaceous Heteroptera, including some coverage of the Anthocoridae, was provided by the papers in Alomar and Wiedenmann (1996).

Their predatory habits have been known for some time; less is known about their plant feeding activities, but such feeding does occur. *Orius insidiosus* (Say) long had been known to include corn pollen in a diet that includes selected pests of corn, and Reid and Lampman (1989) demonstrated that the same bug was attracted to an extract of corn silk perhaps related to the attraction of pest species to the same volatiles (Cantelo and Jacobson 1979). Carayon and Remade (1962) documented the pollen-feeding habits of *O. pallidicornis* (Reuter) in Europe. Alborn et al. (1997) reported on the plant volatiles produced in response to oral secretions by the beet armyworm. The females of a parasitic wasp were attracted to these plant volatiles. Chu (1969) demonstrated that *Lyctocoris beneficus* (Hiura) and *Xylocoris galactinus* (Fieber) could survive on moldy grain. Bacheler and Baranowski (1975) detailed the plant-feeding habits of *Paratriphleps laeviusculus* Champion in Florida (U.S.A.). Salas-Aguilar and Ehler (1977) showed that *O. tristicolor* (White), too, could survive on plant food. Armer et al. (1998) has shown the relevance of plant feeding to the life of *O. insidiosus* in soybean fields. Cohen (1996) presented a fine overview of plant feeding by predatory Heteroptera, including a few Anthocoridae, and Naranjo and Gibson (1996) reviewed the effects of plant feeding on life history and population dynamics. Enough is now known to alert one to the fact that plants will provide clues to some of the predators associated with them. More will surely be discovered about this interesting area of chemical ecology.

1.3 Plant Structure

The structure or architecture of plants has a significant influence on the associated insect fauna. Lawton (1983) provided an excellent review and analysis of the various ways that insect diversity is influenced by the physical structure of plants. Although speaking chiefly of phytophagous insects, he also reflected on the diversity of elements of the predatory fauna on the plant feeders. Strong et al. (1984) provided a useful review of the deployment and diversity of insects on plants, which expanded the original work by Lawton. A realistic understanding of the structure of crop plants, for example, is useful in understanding the capabilities of predatory taxa such as the Anthocoridae.

Lawton (1983) described the forb–shrub–tree system and demonstrated that increased architectural complexity resulted in increased insect species richness and diversity. An examination of the anthocorid fauna associated with these broad categories of plant structure shows a similar trend. The chief occupants of the forb layer are members of the widely occurring genus *Orius*. Most anthocorids reported from such crop types belong to this genus and only rarely is another genus represented (e.g., *Xylocoris*). It is likely, of course, that other genera will be found in other parts of the world, but present knowledge is limited. In the more complex shrub plants, the diversity of the anthocorid fauna increases. Species of *Orius* may occur and other genera appear (e.g., *Anthocoris* and *Montandoniola*). Thus, the potential for other predatory taxa increases.

The greatest diversity of anthocorids occurs on trees, both deciduous and coniferous. It appears that conifers harbor a greater diversity of genera and species (Péricart 1972, Lattin and Stanton 1992). Deciduous trees support such genera as *Anthocoris* and *Temnostethus*, and coniferous trees accomodate *Acompocoris, Elatophilus, Melanocoris, Scoloposcelis, Tetraphleps*, and *Xylocoris*. It is possible that the deciduous nature of some trees influences the fauna in contrast to the long-lasting needles of conifers. A detailed study of Anthocoridae found on species of *Larix*, a deciduous group of conifers, would be instructive. Knowledge of tropical conditions is quite rudimentary, but a similar study of evergreen broad-leafed trees should provide similar comparisons. There is extensive literature on the anthocorid fauna of apple and pear trees in the North Temperate Zone. Curiously, only *Anthocoris* species have been associated with these tree crops and the broad-leafed tree dwellers of *Temnosthethus* have not (Anderson 1962a, Madsen et al. 1963, Péricart 1972, Drukker and Sabelius 1990.). A few species of the genus *Orius* have been reported on these trees besides *Anthocoris* (Mészáros 1984b, Kabicek and Hejzlar 1996). The works by Drukker, Sabelius, and associates in the Netherlands are of great interest here (Drukker and Sabelius 1990; Drukker et al. 1992, 1995; Scutareanu et al. 1994) as they unravel the chemical ecology of *Anthocoris* species. One wonders about the anthocorids found on apple and pear in temperate regions outside the Northern Hemisphere.

Artificial domiciles (greenhouses) represent a special case because they allow greater manipulation of the environment. Several anthocorids (i.e., species of *Anthocoris* and *Orius*) have proved useful in this scenario (e.g., Jacobson 1991, Gillespie and Quiring 1992, Chyzik et al. 1995, Cloutier et al. 1995, Kawai 1995, van Houten et al. 1995, van de Veire and Degheel 1995). Experimentally, these "habitats" suggest some interesting possibilities for studying other anthocorids.

1.4 Habitats

As mentioned previously, plant structure increases in complexity as one goes from grasses to forbs, shrubs, and trees (Lawton 1983). These small bugs occupy an amazing variety of habitats in a variety of natural and managed ecosystems. A few (e.g., *Lasiochilus* spp. and *Xylocoris* spp.) are found in the litter layers of forests and grasslands (Zimmerman 1948, Péricart 1972, Parsons et al. 1991). Here, they seem to feed on a variety of small arthropods, including Collembola and Psocoptera. They even occur in accumulations of berries that fall to the ground under bushes. Some of these berries are often brought into wood-rat nests (*Neotoma* spp.) as a food store, and are eaten by lepidopteran larvae which, in turn, are fed upon by the bugs (*Nidicola* spp.) (Peet 1973, 1979). It has been suggested that these feeding activities might have preceded the move into stored products (Peet 1979). Linsley (1944) provided a fine review of possible sources of arthropods now found in stored food habitats. Zimmerman (1948) documented the occurrence of several species of *Lasiochilus* in the leaf litter on the ground in Hawaii; other species of this genus occur in clusters of dead leaves in trees and shrubs (Usinger 1946, Herring 1967). Here the prey is probably psocids and other small arthropods. Lattin (2000) documented a species of *Cardiastethus* and one of *Lasiochilus* in dead leaf clusters of *Quercus* containing a variety of small arthropods.

Grasses show minimal architectural complexity and the associated fauna of Anthocoridae reflects this simplicity. Most known species are found around the bases of grasses, a habitat very similar to the leaf litter. Here, several species of *Xylocoris* are found (Péricart 1972, Carayon 1972b, Kerzhner and Elov 1976, Kelton 1978). This is another possible pathway toward stored food products. Several species of *Xylocoris* occur here and are important predators on stored grain pests (Tawfik and El-Husseini 1971b, Kerzhner and Elov 1976, Arbogast 1984, Dunkle and Ivie 1994).

1.4.1 Forb layer

The forb layer has only a moderate fauna of Anthocoridae, largely species of the genus *Orius*. Many annual crops fall into this category; it is no surprise, then, that species of *Orius* are found commonly on crops, almost worldwide (see below). An occasional specimen of *Anthocoris* will be taken in such cropping systems but more probably the bugs will be *Orius* spp. The diversity of habitats within the forb category in which *Orius* species are found probably reflects their adaptability.

1.4.2 Shrub layer

Plants of the shrub layer are more complex than the plants of the grass and forb layers and are accompanied by an increase in the diversity of the anthocorid fauna. Species of *Orius* are well represented here and feed on a wide variety of prey items, as do some species of *Anthocoris*. Although many plant species have this growth form, the number of agricultural crops of such structure are more limited: crops like tea, cotton, grapevines, coffee, hazel, or ornamentals such as *Bucus* and *Ficus*. Indeed, anthocorids have been reported from all of these (e.g,. Buchholz et al. 1994). Specific examples might include *A. butleri* Le Quesne on *Bucus* where it feeds on psyllids (Cobben and Arnoud 1969), and Carayon (1982) described a species of *Wollastoniella* from Java feeding on insects on tea. Waloff (1968) provided an extensive study of the dynamics of insects on broom, *Sarothamnus scoparius* (L.) Wimmer, in England. The common *Anthocoris* species found were *A. nemorum* (L.), *A. nemoralis* (F.), and *A. sarothamni* Douglas and Scott. Earlier, Anderson (1962a) studied six species of *Anthocoris* in England, including the three studied by Waloff. *Anthocoris sarothamni* has been reported from broom (Anderson 1962a), and Duso and Girolami (1983) found *O. vicinus* Ribaut and *O. majusculus* Reuter in vineyards in Italy. The present author has taken *O. tristicolor* and *A. antevolens* (White) from commercial filbert orchards in Oregon (U.S.A.), where the introduced filbert aphid, *Myzocallis coryli* (Goetze), a pest species, was a prey item. The introduced species, *O. minutus* (L.), was found in canefields (Lattin et al. 1989). Certainly most studies have been on cotton. Many papers deal with the predator fauna found in many parts of the world; some examples include van den Bosch and Hagen (1966), Schuster et al. (1976), and Wilson and Gutierrez (1980).

1.4.3   Trees

The greatest architectural complexity is found on trees, whether conifer or broadleaf, deciduous or evergreen. The long life of most trees may be a factor because it provides long duration of habitats (e.g., coarse-woody debris) and successional stages of different sizes and complexity. Curiously, the best information comes from coniferous trees, perhaps because of their large stands and economic importance. Now one sees vast plantations and replanted, rotational forests, often of single species. Reproductive forestry is increasing dramatically as purely extractive forestry declines. Efforts are being made to utilize biological control efforts against pest species because the vastness of these forests makes it difficult to use more common pest control strategies, especially pesticides that would have widespread ecological impact.

Coniferous trees support a variety of anthocorids, more particularly members of the Pinaceae than of the Cupressaceae. Most plantation efforts involve species of such genera as *Abies, Larix, Pinus, Pseudotsuga,* and *Tsuga* where these small predators feed on a variety of potential pests. Species of the bug genera *Acompocoris, Melanocoris,* and *Tetraphleps* are surface predators, largely on Homoptera such as aphids and woolly adelgids (Péricart 1972, Lattin and Stanton 1993b, Lattin 1999). Several species of *Tetraphleps* were introduced into North America against the balsam woolly aphid, an introduced pest, but apparently the bugs did not become established (Ghauri 1964, Mitchell and Wright 1967, Clark et al. 1971). Chacko (1973) reported on the impact of *T. raoi* Ghauri on *Pineus* sp. in India, and *T. raoi* was introduced into Kenya against a species of *Pineus* occurring on several pines grown in plantations and has become established and contributes significantly to the control of that homopteran pest (Aloo and Karanja 1986, Karanja and Aloo 1990). Various species of *Lyctocoris, Scoloposcelis,* and probably *Xylocoris* are known predators of bark beetles (Scolytidae) on a variety of conifers (e.g., Drake 1921; Carayon 1953; Péricart 1972; Moore 1972; Ohmart 1981; Schmitt and Goyer 1983a,b; Riley and Goyer 1986).

Some remarkable work has been done on several species of *Elatophilus* (Reuter). Many species of this genus are specialist predators on species of the scale genus *Matsucoccus,* found on North Temperate pine species (Mendel et al. 1991). These scales occur deep under the bark where they feed on the tree. Even more remarkable, the tiny predators are able to find their way to these scales and feed upon them (Lussier 1965, Mendel et al. 1991). Because the ranges of some *Elatophilus* species occur beyond the range of *Matsucoccus,* some other prey items must exist as Sands (1957), Cobben and Arnoud (1969), and Lattin and Stanton (1993a) pointed out. Lattin and Stanton (1993a) found that only one species of *Elatophilus* was positively associated with *Matsucoccus* in North America although other such relationships are likely (Lussier 1965). Excellent work has been done on the chemical ecology of *E. nigricornis* by Biliotti and Riom (1967) and *E. hebraicus* by Mendel et al. (1995a,b), and these works should be consulted by the interested reader.

Less is known about species of anthocorids on broad-leafed trees (deciduous and evergreen) than about conifer anthocorids, except for the genus *Anthocoris* where several species are well known in apple and pear agroecosystems in Europe and North America (e.g., Anderson 1962a, Fields et al. 1981, Nicoli et al. 1989, Scutareanu et al. 1996). As with the work on *Elatophilus,* some exceptional efforts are being made in the Netherlands on several species of *Anthocoris* found on pear (see for example Drukker and Sabelius 1990; Drukker et al. 1995; Scutareanu et al. 1996). There is much to learn about the roles of Anthocoridae on agroecosystems of the tropics, especially where evergreen broad-leafed crop trees are concerned, but the work cited above gives much to build on.

2. SPECIES AND GENERA OF ECONOMIC IMPORTANCE

2.1 Acompocoris Reuter

Mitchell (1962) reported *A. lepidus* (Van Duzee) feeding on the balsam woolly aphid in the mountains of western Oregon (U.S.A.). Rozhkov (1966) cited a species of *Acompocoris* feeding on aphids on Siberian larch. Péricart (1972) reported on several species of *Acompocoris* in Europe; one, *A. pygmaeus* (Fallén), lives predominantly on *Pinus sylvestris*, and another, *A. alpinus* Reuter, is found on *Abies, Picea*, and *Larix*; both anthocorid species feed on aphids, mainly *Cinara*.

2.2 Anthocoris Fallén

Anthocoris Fallén is a widespread taxon of about 70 species and subspecies whose distribution is chiefly, but not exclusively, in the Northern Hemisphere. Péricart (1972, 1996) reported 47 species and subspecies in the Palearctic; Jessop (1983) dealt with 10 species in the United Kingdom; and Henry (1988) reported 12 species in Canada and the United States. Much earlier, Champion (1900) reported 6 species from Mexico into Central America. A single species, *A. austropicus* Gross, was reported from Australia by Cassis and Gross (1995), known only from the type locality. Although a few species have been found in the forb layer (e.g., *A. nemorum* in potato; Turka 1987), most species are found in the shrub and tree layers (e.g., Valenti et al. 1996, Sechser and Engelhardt 1988). A representative group of species of *Anthocoris* is presented here to give the reader a cross section of their activities and capabilities in agricultural settings. A few have been found occasionally on conifer trees (Lattin and Stanton 1992), but species of *Anthocoris* are mostly found on shrubs and broad-leafed trees, including fruit trees.

There is a rich literature on a small number of species of *Anthocoris,* which includes those associated with a variety of shrub and tree crops. The nonagricultural species have been studied to provide additional background for other species of particular interest. Sands (1957) gave a base study of the eggs and nymphs of many anthocorids in the United Kingdom, including *Anthocoris* spp. Anderson (1962b) published an account of *A. antevolens* White in Canada and the United States. There is a series of papers on the widespread *A. confusus* Reuter in the Palearctic (e.g., Anderson 1962a; Hill 1968; Péricart 1972; Evans 1976b,c,d), and Procter (1946), Anderson and Kelton (1963), Kelton (1978), and Scudder (1986) reported on the same species introduced into North America. *Anthocoris gallarum-ulmi* (DeGeer) was treated by Anderson (1962a), Parker (1984), and these papers are useful even though their subjects are not associated with crops. Anderson (1962a) researched *A. minki* Dohrn in the United Kingdom. It, too, is widespread in the Palearctic (Péricart 1972, Al-Marouf 1990). MacPhee and Sanford (1954) and Kelton (1978) outlined the activities of *A. musculus* (Say) in eastern North America. *Anthocoris nemoralis* (F.) is widespread in the Palearctic; it, too, has been introduced into North America (Anderson 1962b, Péricart 1972) and has evoked a massive literature, as has *A. nemorum* (L.), also widely spread across the Old World (Hill 1957, Anderson 1962a, Evans 1976a). *Anthocoris sarothamni*, which occurs in Europe and North Africa on broom (Hill 1961, Anderson 1962a), has not yet been found in North America, but broom has been introduced widely so the bug may have been introduced accidentally, too. *Anthocoris tomentosus* Péricart (1972) occurs in western North America as far north as Alaska (U.S.A.) and the Yukon Territory (Canada). [*Note:* Much of the earlier literature referred to it as *A. melanocerus*.] Although *Salix* is its usual "host," it is now found commonly on fruit trees (Madsen 1961, Anderson 1962b, Tamaki and Olsen 1977, Kelton 1978, Horton et al. 1998). Thus, there is literature on many species of *Anthocoris* to provide the reader an extensive background on their habits and behavior. Péricart (1996) provided a catalog of the Palearctic species with great detail on distribution, and, earlier (1972), gave a fine review of western European species. Henry (1988) published a catalog of the species in Canada and the United States. The recent work by Drukker, Sabelius, Scutareanu, and their associates in the Netherlands on the chemical ecology of *Anthocoris* spp. in relation to their prey, especially the pear psylla, clearly demonstrates some extremely interesting and profitable lines of investigation. Aldrich (1998) included this work in his review of the chemical work on the predaceous Heteroptera.

2.2.1 Anthocoris antevolens White

Anthocoris antevolens White is a common species in western North America (Anderson 1962b, Kelton 1978, Henry 1988). Anderson (1962b) provided coverage of this species in British Columbia (Canada) and Oregon (U.S.A.), reporting on its biology and occurrence on fruit trees. *Salix* spp. seems to be a common, natural host. Madsen et al. (1963) reviewed its status as a control element of the pear psylla, *Psylla pyricola* Förster, in California (U.S.A.). Lattin and Stanton (1992) reported this species as an occasional visitor to lodgepole pine. The present author has seen specimens from western Oregon on broom. Curiously, although many introduced species of insects occur on this introduced plant, *A. sarothamni* has never been reported from North America.

2.2.2 Anthocoris confusus Reuter

Anthocoris confusus Reuter was first reported from North America by Procter (1946) from Maine (U.S.A.), and Anderson and Kelton (1963) reported it from Ontario (Canada) and Tennessee (U.S.A.). Specimens were collected on *Fagus grandifolia* Ehrhart, and hibernating on a peach tree. Subsequently, it has been recorded from British Columbia (Canada) (Scudder 1986).

2.2.3 Anthocoris musculus (Say)

Anthocoris musculus (Say) is the common species in eastern North America (Kelton 1978, Henry 1988), where *Salix* spp. are common hosts (Kelton 1978). There exists a modest literature on it. MacPhee and Sanford (1954) dealt with this species as a predator of mites and the eye-spotted bud moth (eggs and larvae) in Ontario (Canada). Lord (1965, 1972) included this species in his studies on the fauna of apple in Nova Scotia (Canada), as did MacLellan (1977). Alleyne and Morrison (1978) found this species to be a predator on the lettuce root aphid, *Pemphigus bursarius* (L.), in Québec (Canada). It was even reported from *Pemphigus* sp. galls on *Populus* in Florida (U.S.A.) (Florida Div. Plant Ind. 1993).

2.2.4 Anthocoris nemoralis (F.)

This species was recorded by Anderson and Kelton (1963) from Ontario (Canada) where a specimen had been collected on peach. Subsequently, the species was recorded from British Columbia (Canada) by McMullen and Jong (1967) after being introduced from Switzerland to combat the pear psylla (Clausen 1978b). Burts (1971), Brunner (1973), and Brunner and Burts (1975) dealt with the newly introduced *A. nemoralis* as a predator of the pear psylla in Washington (U.S.A.). Hagley and Simpson (1983) found it was a predator of the pear psylla, *Psylla pyricola*, in Ontario. Hagen and Dreistadt (1990) reported this species from California (U.S.A.), where it was found feeding on the psylla, *Acissa uncatoides* (Ferris and Klyver), occurring on *Acacia longifolia*. Dreistadt and Hagen (1994) published a more substantial paper on the biological control of this psyllid in California.

An extensive economic literature exists on *A. nemoralis* in Europe. Anderson (1962a) and Péricart (1972) provided basic information on the biology, ecology, and systematics of *A. nemoralis*, making later work more substantial. As with *A. nemorum*, most applied literature on *A. nemoralis* is based upon the activities of this predator on pear, where it is an important predator of the pear psylla. Drukker and Sabelius (1990), Drukker et al. (1992), and Scutareanu et al. (1994, 1996) published on the chemical ecology of the pear pest and its predators and on the population dynamics of the same pest and its anthocorid predators.

Fauvel et al. (1994) and Rieux et al. (1994) worked on the biological control of the pear psylla in France in systems utilizing *A. nemoralis*. Some difficulty was encountered with the system used to introduce the eggs of the predator but the anthocorid nymphs did flourish and brought the pear psylla under control. Berrada et al. (1996) showed differential responses of *A. nemoralis* to different insecticides, suggesting caution in the compounds used and their application. Forti et al. (1993) found that certain insecticides used to control the tortricid *Cydia pomenella* on pears resulted in a larger population of the pear psylla. Earlier, Nicoli et al. (1990) found that the use of *Bacillus thuringiensis* Berl. ssp. *kurstaki* provided the same control against tortricids as did chemical insecticides and actually enhanced the activity of *A. nemoralis*, a predator of the pear psylla.

A 10-year study on the pear psyllid in Switzerland (Staubli et al. 1992) provided sampling protocols, descriptions of the life stages of the pest, and economic thresholds of adults and nymphs. Chemical control was discussed and included the impact of these compounds on the pest, as well as on *A. nemoralis*. Solomon and Fitzgerald (1990) reported no dangerous impacts on *A. nemoralis* from the application of fenoxycarb to control the pear psylla in the United Kingdom. Care must be used in selecting appropriate compounds for use in biological control programs against this important pear pest.

As with *A. nemorum*, similar studies have been made on hawthorn psyllids (Novak and Achtziger 1995) and in apple ecosystems against mites. Toxicity to nymphs and adult *A. nemoralis* differed. Curiously, *A. nemoralis* has been introduced into North America, but *A. nemorum* has not been, although both are abundantly represented in the western European fauna.

2.2.5 Anthocoris nemorum (L.)

This was the first species of *Anthocoris* to be described, in 1758. Péricart (1972) provided a thorough review of it with a distribution map that shows a range extending from the British Isles to the former Soviet Far East (Péricart 1996). It is interesting that the male claspers of this species and the equally wide range of *A. limbatus* Fieber show great similarity to that of the North American *A. antevolens* White, *A. dimorphicus* Kelton and Anderson, and *A. musculus* (Say). Besides the fine review by Péricart (1972), there are basic papers on its biology and ecology by Hill (1957), Anderson (1962a), Waloff (1968), and Evans (1976a) which provide essential information for anyone studying *A. nemorum*.

From the standpoint of economic value, both *A. nemorum* and *A. nemoralis* have been studied intensively in the western Palearctic region as major predators on pear and apple. The reader is referred to a brilliant series of papers by the Dutch group (e.g., Drukker and Sabelius 1990, Scutareanu et al. 1994, Scutareanu et al. 1996). The efforts of this group have focused on the chemical ecology of *A. nemorum*, especially in its role as a predator on the pear psylla. The bug responds to chemicals emanating from the leaves fed upon by the psyllid. Artigues et al. (1996) described the dynamics of pest species on pear and their associated beneficial arthropods in Spain. The anthocorids appear later in the temporal sequence than do the predatory Miridae. Sechser and Engelhardt (1988) described the control in Switzerland of the pear psylla with an insect growth regulator that appeared to have minimal impact on both *A. nemorum* and *A. nemoralis*, allowing them to help keep the pest under control. Ledee (1995), working in France on apple, and Novak and Achtziger (1995), working in Germany on hawthorn, provided good information on this bug species' effect on the pest populations on their respective hosts. *Anthocoris nemorum* has been studied in cucumber crops grown in greenhouses in England by Jacobson (1991) where the bug was a predator on the introduced western flower thrips, *Frankliniella occidentalis*. After an initial surge by the thrips, their numbers dropped to ten per leaf or fewer. The bug also fed on pest aphids and mites.

2.2.6 *Anthocoris pilosus* (Jakovlev) (= *Anthocoris sibiricus* Jakovlev *nec* Reuter)

This was considered one of three major pests of fruit tree pests in Kazakhastan (Zlatanov 1990). Jonsson (1983) reported *A. pilosus* as one of four species from apple trees in southeastern Norway. Mészáros (1984b) found this species on apple trees in Hungary and Mészáros (1984a) in maize fields. Alaoglou (1994) reported *A. pilosus* from Turkey, where the bug was feeding on *Cryptomyzus ribis* (L.), a pest of red currant. The reader is referred to Péricart (1972, 1996), Elov (1976), and Elov and Kerzhner (1977) for a clarification of the differences between *A. pilosus* and *A. sibiricus* Reuter.

2.2.7 *Anthocoris sibiricus* Reuter

As currently recognized, this species occurs in east and west Siberia, northern China, and Mongolia (Péricart 1972, and especially Péricart 1996). Péricart (1972) considered *A. tomentosus* of North America a subspecies of *A. sibiricus* Reuter, but that was before the tentative conclusions were reached that *A. pilosus* (Jakolev) and *A. sibiricus* (Reuter) may be distinct species. Elov and Kerzhner (1977) provided important information on this taxonomic problem. *Anthocoris tomentosus* Péricart is now considered a distinct species by Elov (1976), Kelton (1978), and Henry (1988).

2.2.8 *Anthocoris tomentosus* Péricart

This is a North American species whose range is largely in the Pacific northwest of the United States (Kelton 1978, Henry 1988). It occurs at higher elevations in the southern part of its range. *Salix* spp. seems to be its favored native habitat but it has adapted to fruit trees in the Pacific northwest. Madsen (1961), Anderson (1962b), Fields and Beirne (1973), Tamaki and Olson (1977), Kelton (1978), and Horton et al. (1998) discussed its role as a predator in fruit trees, especially on the pear psylla, and Wilde (1965) reported its introduction into Ontario (Canada) for the same purpose.

2.3 *Blaptostethus* Fieber

Tawfik and El-Husseini (1971a) provided life history information on *B. pallescens* Poppius (as *B. piceus pallescens*) in Egypt. These authors considered this species to have been accidentally introduced into Egypt, as its normal distribution is in tropical and subtropical regions. In Egypt, this predaceous bug was found commonly on corn, attacking a variety of insects and mites. The life-history information was derived from laboratory rearings. Illustrations of the egg and first and fifth instars are given. Although a variety of potential prey was observed, the usual prey included the eggs and young larvae of *Chilo agamemnon* Bleszynski, *Ostrinia nubilalis* Huebner, *Spodoptera littoralis* (Boisduval), *Pyroderces simplex* Walsingham, and *Cryptoblabes gnidiella* Miller. The thrips *Bolacidothrips graminis* Priesner, the aphid *Aphis maidis* Fitch, as well as several species of red spider mites (*Tetranychus* spp.) also were noted as prey. The small larvae of *P. simplex* and *C. gnidiella* seemed to be favored over the other species in the field studies. Specific information on the abundance of *P. simplex* and its bug predator is presented in chart form, derived from several corn plantations in Giza. It is of interest that the bug was able to live for a short time feeding on plants only, although no reproduction occurred. Similar observations have been made on other species of anthocorids (e.g., *Orius insidiosus*, *O. pallidicornis*, *O. tristicolor*) (Carayon and Steffan 1959, Salas-Aguilar and Ehler 1977, Armer et al. 1998, Lattin 1999). Rajasekhara (1973) reported and described a species found on sugarcane in India.

2.4 *Dufouriellus* Kirkaldy

Dufouriellus ater (Dufour) is native to the western Palearctic Region and is widely distributed there (Péricart 1972). Péricart gives a fine discussion of this anthocorid, including appropriate illustrations and a distribution map. Later, Péricart (1996) reported it from China. Its normal habitat seems to be beneath the bark of broad-leafed and a few coniferous trees where it feeds on bark beetles (Scolytidae) and, probably, other small insects (Strawínski 1964). It has been distributed by commerce into North and South America, where it has been recovered from wood products and stored products (Arbogast 1984). (See U.S. Congress, Off. Tech. Assess. 1993, for a more general discussion on the impact of nonindigenous species on the United States.) In California (U.S.A.), it was collected beneath the bark of a dead tree in the company of another nonindigenous bug, *Patapius spinosus* (Rossi) (Leptopodidae). The present author collected specimens of this species under the bark of sycamore trees on the campus of the University of California at Davis, and it has been beaten from a juniper tree in eastern Oregon (U.S.A.). Arbogast (1984) reported it from similar habitats in Massachusetts and Texas (U.S.A.). This species is discussed here because it is sometimes found in stored foodstuffs (peanuts) in Georgia (U.S.A.) (Arbogast 1984), and was studied in Egypt as a potential biological control agent of stored product insects (Awadallah et al. 1984).

Arbogast (1984) provided detailed information on the occurrence of *D. ater* in stored peanuts, including data on the duration of the life stages in the laboratory where it was fed the paralyzed larvae of the Indian meal moth, *Plodia interpunctella* (Huebner). Further, he compared *D. ater* with several species of the anthocorid genus *Xylocoris* (*X. galactinus, X. hiurai,* and *X. sordidus*). He felt that *D. ater* was a useful biological control agent against stored grain pests. After this comparison, he stated that *X. flavipes* was an even better predator but that *D. ater* also was useful. Some difficulties were experienced in large-scale rearing of *D. ater*, and these were considered a drawback. Awadallah et al. (1984) studied this species in the laboratory in Egypt (based upon specimens from Arbogast) and provided further biological information on its life stages. They, too, considered it a potential biological control agent against stored grain insects.

There is an enigma about its occurrence in North America. It has been taken from beneath the bark of trees in widely scattered localities. It has been reported, as well, from stored food products at some localities. Kelton (1978) considered it as probably introduced. This is the only species of the genus and it only occurs in the fully winged state. There is no good information regarding why it should be found in stored food products. Perhaps it is naturally Holarctic but has been introduced in stored products as well.

2.5 *Elatophilus* Reuter

The genus *Elatophilus* Reuter contains about 20 species. It is Holarctic, distributed chiefly in the North Temperate region, where it occurs on various species of *Pinus*. Lussier (1965), Biliotti and Riom (1967), Riom et al. (1971), Delages (1978), and Mendel and associates (1991, 1995a,b) have demonstrated the relationships between some species of *Elatophilus* and some species of the scale genus *Matsucoccus* on pines. These efforts now include some elegant chemical ecology that links the males of *M. josephi* and its predator, *E. hebraicus* Péricart, to a synthetic female sex pheromone of the females of *M. josephi*. Jactel et al. (1996) presented a sampling scheme for *M. feytaudi* Ducasse and its relation to male scales collected by pheromone trap. Similar chemical connections are being discovered between anthocorid predators of pear psylla (e.g., Drukker et al. 1995). There is more to be learned about the prey of *Elatophilus* spp. besides *Matsucoccus*, as has been pointed out by Sands (1957), Cobben and Arnoud (1969), and Lattin and Stanton (1993a). Some species of *Elatophilus* have been reported to feed on aphids and some species of *Elatophilus* range far north of the known ranges of species of *Matsucoccus* (see above). Only a single North American species of *Elatophilus, E. inimica* Drake and Harris, has been positively associated with a species of *M. resinosae* (Lussier 1965). The prey of the other species of *Elatophilus* remains to be determined. Several species of *Matsucoccus* are serious pests of plantation-grown pines in both the Old and New World and these small predators are important elements in biological control efforts (Biliotti and Riom 1967; Mendel et al. 1991, 1995a,b).

2.6 *Lyctocoris* Hahn

This is a modest-sized genus with about 25 species (Chu 1969). Although it is chiefly found in the Northern Hemisphere, and especially in the North Temperate region, at least one species, *L. campestris* (F.), has been introduced accidentally into many parts of the world. These introductions are usually into stored food accumulations, although the insect is believed to be naturally subcorticular (Péricart 1972). Trematerra and Dioli (1993) reported this species from Italy in association with insects of stored *Triticus spelta* L. This record is of especial interest because *L. campestris* is a native of Europe and more commonly found under the bark of trees (Péricart 1972). Štys and Daniel (1957) provide an interesting account of the bug in central Europe and document its occurrences in domiciles, where it bit some of the occupants (and see **Chapter 19**). Hicks (1959, 1962, 1971) provided an extensive bibliography of this species as an inhabitant of bird nests. Parajulee and Phillips (1992) gave results of laboratory rearing and observations in the field on this important stored pest predator, and later (1995) studied the seasonal population numbers and the spatial patterns of *L. campestris* in shelled corn over a 2-year period. Three trapping methods were used for this bug, the most efficient being probe traps. Both sexes of *L. campestris* showed aggregated distribution within the shelled corn. More recently, Parajulee et al. (1995) and Parajulee and Phillips (1995) discussed the life history of nymphal stages and the survivorship of this species in stored grain habitats. Chu (1969) provided a detailed study of a related species, *L. benicus* Hiura, also in stored grain, along with *X. galactinus*. Besides its predatory activities, Chu found that both species could survive for some time on moldy grain; the absence of animal prey did not seem to influence the reproductive capabilities of the male, but did impact the egg-producing abilities of the female. This seems to be true also of several other anthocorids (e.g., *Orius tristicolor*). Although some species may have much lower reproductive ability, others merely stay alive without reproduction.

Where known, most species of *Lyctocoris* occur under the bark of trees where they feed on bark beetles (Scolytidae) and other subcorticular insects (Péricart 1972). The present author once took *L. campestris* from under the bark of a dead oak in western Oregon (U.S.A.), well removed from any stored product facility. The species also occurs in stacks of dead vegetation. *Lyctocoris elongatus* (Reuter) has been observed feeding on Scolytidae under the bark of pine trees in Mississippi (U.S.A.) (Moore 1972; Schmitt and Goyer 1983a,b) and Texas (U.S.A.) (Moser et al. 1971).

2.7 *Macrotracheliella* Champion

Macrotracheliella nigra was described by Parshley (1917) from Massachusetts (U.S.A.). Nothing was known of its habits, and little known of its distribution, for many years (Henry 1988). A related species, *M. laevis* Champion, described from Mexico in 1900, was found in galls of *Ficus* in Puerto Rico by Wolcott (1953). *Cardiastethus rugicollis* Champion also was reported from *Ficus* galls in Puerto Rico. These galls were the result of the thrips, *Gynaikothrips ficorum* Marchal. Much later, *Montandoniola moraguesi* and two unidentified anthocorids were introduced into southern California (U.S.A.) in an effort to control the thrips on *Ficus* in that area (Clausen, 1978a), but no establishment was reported. According to Paine (1992), *Macrotracheliella nigra* was later recovered feeding on the thrips in California. It was not known if this species might have been one of the unidentified anthocorids introduced earlier. Paine worked on thrips predation by *M. nigra* (along with a species of *Chrysopa*), studying its development and biology in that area. According to this author, adult thrips are preferred over nymphal thrips, but these predators may cause substantial mortality of thrips occurring in the galls on the *Ficus* plants. *Macrotracheliella nigra* is a nonindigenous species in southern California and has been moved far beyond its natural range. Both *M. nigra* and *M. laevis* are now known to feed on Cuban laurel thrips, the first species in California and the second in Puerto Rico. *Macrotracheliella nigra* joins *Montandoniola moraguesi* as a biological control agent against this thrips.

2.8 *Montandoniola* Puton

Montandoniola moraguesi (Puton) is a well-known biological control agent of the Cuban laurel thrips, *G. ficorum* (Marchal). This thrips species has become a serious pest of ornamental plants in many parts of the world (Mound et al. 1995), where these plants have been introduced. Curiously, the anthocorid was described originally from southern Europe by Puton (Carayon and Ramade 1962), even though its original range is now believed to be southeast Asia (Herring 1966a). Similarly, the so-called Cuban laurel thrips was given its common name because of its occurrence in Cuba, even though the thrips also came from southeast Asia (Mound et al. 1995). Thus, the initial discovery of the anthocorid in southern Europe and the naming of the thrips were both based upon accidentally introduced organisms. Tawfik and Nagui (1965) studied the life history of this anthocorid in Egypt. The anthocorid was purposefully introduced into Hawaii as a biological control in 1964 and quickly spread through several of the islands, where it became a competent control agent of the introduced Cuban laurel thrips. Two other anthocorids, *Orius insidiosus* (Say) and *O. tristicolor* (White), were introduced into the Hawaiian Islands for other purposes (Funasaki 1966) about the same time, but both introductions were regarded as failures. [*Note:* It now appears that at least *O. tristicolor* is established in the Hawaiian Islands (J. D. Lattin, unpublished), perhaps from this original introduction or a more recent one.] A closer study of the comparison between these deliberate introductions and their success/failure would certainly be instructive. More recently, Reimer (1988) has demonstrated that *M. moraguesi* now feeds as well on another thrips, *Liothrips urichi* Karny, itself brought into Hawaii to control a serious weed, *Clidemia hirta* (L.) David Don. This unanticipated result was considered biotic interference by Reimer. Elsewhere, *M. moraguesi* has been introduced accidentally into Europe and deliberately into California — in the latter instance, to control the Cuban laurel thrips, but without apparent success (Paine 1992).

2.9 *Oplobates* Reuter

Oplobates woodwardi Gross was found feeding on the scale insect, *Saccharicoccus sacchari* (Cockerell), a widespread pest of sugarcane in Australia and New Zealand (Carver et al. 1987).

2.10 *Orius* Wolff

Orius is a large (75 spp.) worldwide genus. Most known species are predaceous, but some include plants in their diet (*O. insidiosus*), and a few are almost entirely plant feeders, e.g., *O. pallidicornis* (Reuter). Carayon and Steffan (1959) provide a fine review of 13 species, and include detailed information on the phytophagous habits of *O. pallidocornis*. They reviewed the diverse predatory habits of other species, whose prey include Homoptera, Thysanoptera, Lepidoptera, Heteroptera, Diptera, Coleoptera, and Acarina. These are small predators whose prey are also small. Most *Orius* are found in the forb layer, and a few on shrubs and even fewer on trees. Space allows only a brief discussion of some of the known species, selected to give the reader an idea of their habits. Péricart (1972) covered the western Palearctic species and Herring (1966b) provided a review of *Orius* of the Western Hemisphere. References to species of *Orius* were long found under the earlier generic name *Triphleps*. Searches of earlier literature should utilize *Triphleps* and *Orius* to ensure complete coverage (Herring 1966b).

2.10.1 *Orius albidipennis* (Reuter)

Orius albidipennis was cited as a predator of pests on cotton in Pakistan by Salim et al. (1987). The prey included aphids, leafhoppers, aleyrodids, mites, thrips, and early stages of moths. The same species was cited by Zaki (1989) from Egypt as a predator in cotton and maize fields and it was reared on lepidopterous larvae and the larvae of the confused flour beetle.

2.10.2 Orius australis (China)

Orius australis was described from Australia (China 1926) where it was feeding on the eggs of the lepidopteran *Heliothis obsoleta* (Herrich-Schaeffer) in Queensland. Carayon and Steffan (1959) included this species in their review of the feeding habits of *Orius*. Cassis and Gross (1995) gave a thorough review of *O. australis* in their recent catalog.

2.10.3 Orius indicus (Reuter)

This species feeds on thrips in India, where Rajasekhara and Chatterji (1970) described its biology on *Taeniothrips nigricornis* (Schmutz).

2.10.4 Orius insidiosus (Say)

This anthocorid is widespread in eastern North America and extends south into South America (Herring 1966b). It is one of the most abundant species of Anthocoridae in eastern North America and has been long known for its predaceous habits on a number of small insects and arthropods (Marshall 1930, Barber 1936). Carayon and Steffan (1959), reviewing the range of predators taken by this anthocorid, indicated they had located 40 references on such prey as early as 1958. The massive literature that followed extended their findings. The reader is referred to the electronic databases available for complete coverage since 1970, for only selected references are cited here.

Barber (1936) reported that *O. insidiosus* can develop on the corn plant alone, but the females were infertile. Kiman and Yeargan (1985) provided detailed information on the results of rearing this bug on a combination of plant and animal foodstuffs. They compared the results with the plant-feeding habits of *Nabis* spp. and *Geocoris* spp. Richards and Schmidt (1996) provided information on the impact of additions to the diet of *O. insidiosus*; greatest productivity was achieved with a combination of animal and plant materials. Coll and Izraylevich (1997) discussed the consequences of *O. insidiosus*' feeding on plants and the impact it might have on prey populations, and developed a model based on the activities of this bug. The reader is referred to Armer et al. (1998) for current knowledge on the role of plant materials in development of *O. insidiosus*.

Orius insidiosus has been reported a predator on arthropods on a variety of crops, especially forbs, but also including shrub types and fruit trees (Coll and Ridgway 1995). Soybeans have received considerable attention as a habitat for this predator (e.g., Shepard et al. 1974; Raney and Yeargan 1977; Isenhour and Marston 1981; Isenhour and Yeargan 1981, 1982). Irwin and Shepard (1980) published an extensive review of sampling soybeans for predatory Heteroptera, and Bechinski and Pedigo (1982) provided an evaluation of various sampling procedures. Busoli et al. (1984) dealt with this predator occurring on sorghum in Brazil, and Dicke and Jarvis (1962) and Picket and Gilstrap (1986) considered the role of *O. insidiosus* as a predator in corn ecosystems. Martinez and Pienkowski (1982) provided useful information on this bug as a predator of the potato leafhopper, a pest of alfalfa in Virginia. Many papers have documented this bug as part of the predator complex on cotton (e.g., Iglinsky and Rainwater 1950, Schuster et al. 1976, and Reilly and Sterling 1983). *Orius insidiosus* was reported as a predator in apple orchard ecosystems by McCaffrey and Horsburgh (1986) and Welty (1995), although *O. insidiosus* is not usually thought to be an inhabitant of trees. *Orius insidiosus* has been introduced into Europe for biological control of the western flower thrips in greenhouses (van den Meiracker 1994). *Orius insidiosus* was introduced into Hawaii in 1951 in an effort to control the corn earworm where it was expected to feed on the eggs of the pest (Weber 1953). It did not become established, according to Oatman (1978).

The sycamore lace bug, *Corythucha ciliata* (Say) (Tingidae), has been introduced from North America into Europe, where it has become a pest (see **Chapter 4, Section 3.4**). The present author found it in numbers under the bark of the plane tree along the shores of Lake Balathon (Hungary). Horn et al. (1983) reported *O. insidious* feeding on the eggs of this tingid. In combination with several other predators, up to 25% of the eggs were destroyed in the laboratory. Perhaps it might be useful against this accidentally introduced tingid.

2.10.5 Orius laevigatus (Fieber)

This species is western Palearctic (including North Africa) (Péricart 1972). Carayon and Steffan (1959) cite several papers on its prey which includes mites, thrips, aphids, whiteflies, and noctuid moths. The biology and morphology of *O. laevigatus* were reviewed by Péricart (1972). Recently, Frescata and Mexia (1996) reported this bug as a predator of the western flower thrips on strawberries in Portugal, and Fiume (1996) found the same bug feeding on thrips and the aphid *Myzus persicae* (Sulzer) on sweet peppers in Italy. Details of the life history of this bug species were given by Alauzet et al. (1994).

2.10.6 Orius maxidentes Ghauri

Orius maxidentes is a predator of the sorghum earhead midge (Cecidomyiidae) in India. Thontadarya and Rao (1987) reared it on this prey and on the sorghum flower thrips. According to these authors, there were only four instars of the bug. [*Note:* The usual number is five.]

2.10.7 Orius minutus (L.)

This is a well-known species in the Palearctic region ranging from western Europe to China (Péricart 1972, 1996). It has been reported as a predator in caneberry crops and fruit trees, where it feeds on thrips, mites, psyllids, mirids, and cicadellids (Collyer 1953, Carayon and Steffan 1959, Péricart 1972). There are several citations on this species in the European literature (e.g., Mészáros 1984b). Tonks (1953) was the first to document its occurrence (as a nonindigenous species) in North America. Several other references to its occurrence in Canada and the United States followed. Lattin et al. (1989) published the present distribution of this accidental introduction, providing a map and drawings of the nymph and adult in addition to a summary of its feeding habits.

2.10.8 Orius niger (Wolff)

Orius niger is widely distributed in the Palearctic region where it is well known as a predator of aphids, thrips, and lepidopteran, pentatomid, and mite eggs (Carayon and Steffan 1959, Péricart 1972). Of special interest is the statement by Carayon and Steffan (1959) that it is a predator of the tingid, *Stephanitis pyri* (F.). Berest (1980) reported this predator feeding on aphids on cereals in the Ukraine. Rácz and Visnyovszky (1985) found it a predator in stands of maize in Hungary, as did Mészáros (1984a). Josifov (1962) reported it as a predator in fields of alfalfa in Bulgaria, and Mészáros (1984b) reported it from apple orchards.

2.10.9 Orius sauteri (Poppius)

Orius sauteri is another species reported to feed on thrips (*Thrips palmi* Karny) in China (Wang 1994). The thrips were damaging eggplant in China, as were spider mites and aphids, prey also acceptable to *O. sauteri*. Funao and Yoshiyasu (1995) found the same bug in Japan where it was reared using the aphid, *Aphis gossyii* Glover, and corn pollen — another example of a species of *Orius* able to use pollen in its diet. This bug also was a predator of aphids in Japanese potato fields (Nakata 1995).

2.10.10 Orius tantillus (Motschulsky)

Orius tantillus was considered a major predator on broadcast rice in Sri Lanka (Rajendram 1994).

2.10.11 Orius thripoborus (Hesse)

Orius thripoborus Hesse was cited by Dennill (1992) as a likely control of a thrips on avocado in the Transvaal of South Africa.

2.10.12 Orius tristicolor (White)

This species occurs throughout western North America, extending south into South America (Herring 1966a). Carayon and Steffan (1959) summarized the early literature on this species, whose prey includes mites, aphids, thrips, psyllids, and other small insects. Anderson (1962b) studied the life history of this species in British Columbia (Canada) and Oregon (U.S.A.). Askari and Stern (1972) presented detailed information on its life history using *Tetranychus pacificus* McGregor as the prey. They noted that whereas *O. tristicolor* might feed at times on a plant (lima bean), it could not survive without animal food. Salas-Aguilar and Ehler (1977) reported that although thrips were the preferred prey of this bug, a combination of thrips, green beans, and pollen was most satisfactory. Manuel and Mojica (1982) examined the interactions between *O. tristicolor* and *O. thyestes* Herring feeding on *T. urticae* Koch in greenhouses in Mexico. Their field studies showed that both bug species fed on a variety of prey, partially explaining why *O. tristicolor* is considered such a good predator. Tamaki and Olson (1977) felt that *O. tristicolor* was not a useful predator on aphids, but Letoureau and Altieri (1983) regarded this bug species as a predator of the western flower thrips, *Frankliniella occidentalis* (Nesbitt), a pest of considerable importance. The thrips-feeding habit resulted in the bugs' being introduced into Hawaii against the Cuban laurel thrips (Clausen 1978a), but apparently they did not become established. More recently, the bug has been introduced into Europe as a biological control agent against the western flower thrips (van den Meiracker 1994). Teerling et al. (1993) described its use of the alarm pheromone as a prey-finding kairomone by predators. More theoretically, Wilson and Gutierrez (1980) studied the within-plant distribution of a variety of predators, including *O. tristicolor*, found on cotton. They included information on predator efficiencies and sampling techniques. Ellington et al. (1984) reviewed various sampling techniques for organisms in cotton fields. There is a substantial literature on this species of *Orius*. The reader is referred to the many references to be found in the several electronic databases.

2.10.13 Orius vicinus Ribaut

This is another species that includes plant materials, chiefly pollen, in its diet. This species is well known for feeding on mites and aphids on apples (Péricart 1972). Fauvel (1971), in France, elaborated the role of pollen in the life cycle of the bug. This plant material seemed most commonly utilized in the spring. When confined to this foodstuff, the immature stages develop, but the resulting adults are smaller and structural alterations may occur. When arthropod prey is included in the diet, fecundity and adult life are normal. Fauvel et al. (1978) discussed the role of this species as a predator of the gall-forming mite, *Eriophyes fraxinicorus*, on *Fraxinus excelsior*. Heitmans et al. (1986) discussed in detail its role as a predator of both predaceous and phytophagous mites in an orchard in the Netherlands. Duso and Girolami (1983) found this species as well as *O. majusculus* (Reuter) to be predators of the mite *Panonychus ulmi* Koch in vineyards in Italy. More recently, Larivière and Wearing (1994) collected it as an introduction into New Zealand, where it has become a predator of orchard pests.

2.11 *Paratriphleps* Champion

Paratriphleps laeviusculus Champion is reported on cotton in Peru feeding on the eggs and early instars of *Heliothis virescens* (F.) (Cuevo et al. 1975). One should see the paper by Bacheler and Baranowski (1975) for remarks on this species by Herring, for there has been some confusion as to its identity. Oatman (1978) reported on this species' introduction into Texas (U.S.A.), where it failed to become established. A species of *Paratriphleps* was reported to feed on other Noctuidae occurring on beans, maize, and potato in Peru by Javier and Peralta (1976a). The same authors (1976b) reported *Paratriphleps* sp. feeding on Noctuidae on "choclo" maize crops, also in Peru. They reported that *P. laeviusculus* also fed upon the cotton plant itself, as well as the Noctuidae. Bacheler and Baranowski (1975) reported pollen feeding by this same species of bug in Florida.

2.12 *Scoloposcelis* Fieber

Scoloposcelis Fieber contains about 12 species, most of which are found in the Old and New World temperate regions (Péricart 1972, 1996; Kelton 1978; Henry 1988). Species of *Scoloposcelis* are predators of bark beetles and other invertebrates under the bark of conifers and deciduous trees (Péricart 1972, Kelton 1978). Schmitt and Goyer (1983a,b) described the activity of *S. flavicornis* Reuter (as *S. mississipensis* Drake and Harris) feeding on the southern pine beetle in southern United States. Lattin and Stanton (1992) provided information on the same bug occurring under the bark of lodgepole pine in the western United States.

2.13 *Tetraphleps* Fieber

Tetraphleps Fieber has about 14 described species throughout the Holarctic region (Kelton 1966; Péricart 1972, 1996; Henry 1988). Species of this genus are surface feeders on certain Homoptera, especially Aphididae and Adelgidae (Péricart 1972, Rao and Ghani 1972, Kelton 1978, Lattin and Stanton 1992). Their selective feeding habits have attracted the attention of workers in biological control, and several species have been introduced from Pakistan and India into Canada and the United States against the balsam woolly aphid (Mitchell and Wright 1967, Amman and Speers 1971, Clark et al. 1971). *Tetraphleps abdulghani* Ghauri and *T. raoi* Ghauri were the two species involved (Ghauri 1964), but apparently neither species introduced from 1961 to 1965 became established. Only *T. abdulghani* was reported feeding on *Adelges* on a species of *Abies* in Pakistan by Ghauri; *T. raoi* was reported feeding on a species of *Pineus* on pine in India (Ghauri 1964). Péricart (1972) reported a *T. bicuspis* (Herrich-Schaeffer) from larch, and sometimes from *Abies, Picea*, and *Pinus*. The prey was reported as the aphids *Cinara* spp. Péricart also found that *T. aterrimus* (J. Sahlberg) lived on *A. sibirica*, including Siberian fir grown in plantations in western Russia. No prey species were listed by him. The differences between the host trees and prey should be noted. Probably this fact, plus the difference in geographic locations, contributed to the unsuccessful attempts at introduction. Other anthocorids may be better adapted. Foottit and Mackauer (1963) reviewed the subspecies of the balsam woolly aphid, *Adelges piceae*, and provided a sound systematic base for the prey for future work. Kelton (1978) cited four species of *Tetraphleps* feeding on the same introduced pest (balsam woolly aphid); all these bug species are native to North America.

Clear evidence of the value of attention to host plant and prey in the native locations and the likelihood of successful establishment comes from the work of Aloo and Karanja (1986) and Karanja and Aloo (1990). *Tetraphleps raoi* was introduced into Kenya to control *Pineus pini*, a pest in pine plantations in that country (*Pinus occidentalis* Schwartz, *P. oocarpa* Schiede, and *P. radiata* D. Don). Originally, Ghauri (1964) had reported *T. raoi* from pine in northeastern India, where it was feeding on a species of *Pineus*, and Chacko (1973) provided biological observations on *T. raoi* feeding on *Pineus* spp. found on *Pinus insularis* Endlicher in India, indicating that it was the chief predator of the *Pineus*. Karanja and Aloo (1990) reported successful colonization and establishment of the predator in Kenya. A decline of *P. pini* was reported in some cases, reducing the damage to the trees, demonstrating that establishment is possible. The widespread plantations of *Pinus radiata* increase the possibility that exotic *Pineus* species will be introduced. This anthocorid, and others, are likely candidates for introduction as biological control agents against these pests.

2.14 *Wollastoniella* Reuter

Carayon (1982) described *W. testudo* Carayon from Java, where it was reported to be feeding on the mite *Brevipalpus phoenicus* (Geijabes) that was damaging tea in plantations. This tiny anthocorid (less than 1.5 mm) seemed adapted to the small size of the mite upon which it fed. This supports the idea of Carayon (1961) regarding the scaling of the body size of heteropteran predators feeding on different-sized prey in agroecosystems. Yasunaga and Miyamoto (1993) described the new species, *W. rotunda*, from Thailand where it was feeding on *Thrips palmi* Karny. Several years later, Yasunaga (1995) described another species of *Wollastoniella*, *W. parvicuneus*, also from Thailand, where it, too, fed on *T. palmi* found feeding on eggplant.

2.15 *Xylocoris* Dufour

This genus contains about 60 species and is widely distributed, especially in the Northern Hemisphere (Chu 1969; Schuh and Slater 1995; J. D. Lattin, unpublished). J. Carayon (1972b) provided a superb generic revision in which he treated the four subgenera. This effort provided the basis for further work over the ensuing years. *Xylocoris cursitans* (Fallén), the type of the genus, typifies the habits of species found beneath the bark of trees. Many others occur on the ground around the bases of grasses and other plants (Péricart 1972). Some species occur in both short- and long-winged forms. At least three major species of *Xylocoris* are found in stored food products, especially grain — *X. flavipes* (Reuter), *X. galactinus* (Fieber), and *X. sordidus* (Reuter) — although their original habitats were probably plant debris on the ground (e.g., *X. galactinus*). *Xylocoris flavipes* is so widespread today in stored grain that it is difficult to determine its original range (Péricart 1972). Originally described from Algeria, it is widely distributed around the world. Most publications on stored grain species have dealt with *X. flavipes*; only a few deal with the other species.

2.15.1 *Xylocoris flavipes* (Reuter)

Although described from Algeria, the native range of this species has been the subject of considerable discussion (Jay et al. 1968, Carayon 1972b, Péricart 1972, Cassis and Gross 1995). It occurs in North Africa and in parts of western Europe, South America, and the United States. Cassis and Gross report it from the Oriental region, including Sri Lanka and Fiji, and a wide distribution in Australia. The true native range is difficult to determine, in part because of the occurrence of other species of *Xylocoris*, but it appears to be the warmer parts of the world. As one of the dominant species of Anthocoridae found in stored grain habitats around the world, such occurrences complicate the identification of natural occurrence. Some of the species placed in synonymy with *X. flavipes* come from Argentina, India, and the Canary Islands (Cassis and Gross 1995).

Jay et al. (1968) provided basic biological information on this species in the United States and concluded it was a good predator against four major stored grain pests. Arbogast et al. (1971) described the life stages of *X. flavipes* and described its predaceous habits on stored-products prey. Awadallah and Tawfik (1972) described the life history of this species based on laboratory studies. LeCato and Davis (1973) indicated that the effectiveness of this predation seemed to be associated with the relative size of the prey. Over the next few years, information was gained on the impact of feeding and the influence of temperature and humidity on development (Awadallah and Tawfik 1972; Tawfik et al. 1983a,b). Later efforts examined the interactions between *X. flavipes* and *Trichogramma* egg parasites of storage insects (Brower and Press 1988), as well as *Bracon* parasites of the same insects (Kraszpulski and Davis 1988); and Press (1989) examined the compatibility of the bug with a species of Ichneumonidae parasitic on the almond moth. More recently, Phillips et al. (1995) described different terpenes produced by *X. flavipes* that proved toxic to *Tribolium castaneum* (Herbst) and *Oryzaephilus surinamensis* (L.), both important stored-product pests. Baker and Throne (1995) gave information on pesticide resistance of beneficial insects, including *X. flavipes*, in the southeast United States.

2.15.2 Xylocoris galactinus (Fieber)

Xylocoris galactinus was described from Europe, where it has a wide distribution (Péricart 1972, 1996). Its principal habitat seems to be plant litter and occasionally on the ground around the bases of grasses. Hall (1950) published a thorough study of its biology and habits in England, and this information was broadened by Péricart (1972). This anthocorid, always fully winged, has become an inhabitant of stored grain and has been distributed in many parts of the world (Chu 1969, Péricart 1972). This led Péricart (1972) to refer to the species as a "quasi-cosmopolite." It is a predator on a variety of insects found in the stored grain habitat. Chu (1969) provided an extensive study of this species in stored grain in Japan (along with *Lyctocoris beneficus*). His study must rank among the most detailed on any anthocorid. Besides feeding on a variety of pests, it can also survive on moldy grain. Tawfik and El-Husseini (1971a) published the life history of this species in stored grain in Egypt, and Afifi and Ibrahim (1991) studied the effect of prey on this same species. Finally, Dunkle and Ivie (1994) reported *X. galactinus* in piles of wheat outside in Montana (U.S.A.), a habitat that provides some clue how this species might have become associated with stored grain, an association that led to its movement to many parts of the world.

2.15.3 Xylocorus sordidus (Reuter)

Described from Brazil, this species is known to occur in South and Central America (Henry 1988); it also occurs in North America and the West Indies. Some of these records represent natural distribution and others are probably associated with stored products. Arbogast et al. (1983) published a base study of this species followed by another (Arbogast et al. 1985) on its life stages. Tawfik et al. (1986) and Awadallah et al. (1986) provided laboratory studies on this species in Egypt, examining the impact of relative humidity and temperature on development and the effect of prey on the same species. Thus far, this species has more limited occurrence in stored grain facilities, but may be moved via commerce to other localities.

3. CONCLUSIONS

This brief presentation provides an entrée to the applied work that has been done on the family Anthocoridae. Some basic information on this taxon is included, but more is found in Lattin (1999) and publications referenced there. Péricart (1972) provided a masterly review of the western Palearctic fauna, and his catalog (1996) expanded the scope and included an excellent bibliography of the region, including works from the former Soviet Union, China, and Japan. The many papers by J. Carayon and R. H. Cobben laid a fine foundation for the work. Basic biological work was done by Hill (1957), Sands (1957), Anderson (1962a,b), and Evans (1976a,b,c,d). Lawton (1983) provided the insight into the importance of plant structure and its influence on the local fauna.

Biological control has become a major emphasis in control efforts, as Carayon (1961) predicted and as is discussed by Kogan and Lattin (1999). Many of the citations in this chapter deal with the roles played by these tiny predators, extending even into the controlled environments of greenhouses. Aldrich (1998) provides an overview of the efforts in chemical ecology of bugs; for example, one should examine the work on *Anthocoris* spp. (e.g., Drukker and Sabelius 1990; Drukker et al. 1992, 1995; Scutareanu et al. 1994, 1996) and on *Elatophilus* spp. (e.g., Biliotti and Riom 1967; Mendel et al. 1991, 1995a,b). Much has been learned over the past years, but much lies ahead.

4. PROGNOSIS

Most reviews disclose shortcomings in the knowledge base. This chapter is no exception. Some identified needs include regional faunal studies of species found on crops; biologies of most species, especially field studies; expanded studies on eggs and nymphs; feeding studies, including the role of plants as a dietary supplement; continued studies on the chemical ecology of nymphs and adults as in *Anthocoris* and *Elatophilus*; regional faunal studies, especially of Africa, Central and South America, and southeast Asia; worldwide/regional generic revisions including those of *Anthocoris, Cardiastethus, Lasiochilus, Orius,* and *Xylocoris.*

5. DEDICATION

This chapter is dedicated to my friend and colleague, the late Dr. René H. Cobben of the Laboratorie voor Entomologie, Landbouwhogeschool (now Landbouwuniversiteit), in Wageningen, the Netherlands. It was my great fortune to have worked with René during 1965–66 and 1973–74 while on sabbatical leaves in Wageningen. Tragically, he died as a young man but not before he left us a rich legacy of elegant scientific work whose quality will long remain (e.g., Cobben 1968, 1978). Much of this chapter was written in Wageningen during July and August, 1997, while on a return visit with my family. His presence was strongly felt and deeply missed. *Wel te rusten René.*

6. ACKNOWLEDGMENTS

A project of this magnitude builds upon a vast legacy of prior efforts of over 150 years. In addition to the authors of the earlier works, many others contributed to this effort and include the following colleagues: N. H. Anderson, A. Asquith, J. Carayon, G. Cassis, A. Christie, R. H. Cobben, C. J. Drake, H. M. Harris, T. J. Henry, J. L. Herring, D. R. Horton, L. A. Kelton, I. M. Kerzhner, T. M. Lewis, L. Parks, J. Péricart, T. W. Phillips, V. Racz, C. W. Schaefer, R. T. Schuh, G. G. E. Scudder, J. A. Slater, Sir Richard Southwood, N. L. Stanton, P. Štys, D. Tilles, R. L. Usinger. Deepest thanks are due those individuals for their inspiration, assistance, and patience. All have contributed to my education. Regrettably, several are no longer alive (J. Carayon, R. H. Cobben, C. J. Drake, H. M. Harris, J. H. Herring, R. L. Usinger), but their many contributions remain.

7. REFERENCES CITED

Afifi, A. I., and A. M. A. Ibrahim. 1991. Effect of prey on various stages of the predator *Xylocoris galactinus* (Fieber) (Hemiptera: Anthocoridae). Bull. Fac. Agric. Univ. Cairo 42: 139–150.
Alaoglu, O. 1994. Parasitoid and predator fauna of *Cryptomyzus ribis* (L.) (Homoptera: Aphididae) a pest of red currents, *Ribes aureum* L. in Erzurum. Pp. 49–58 *in* Biyolojik Mucadele Kongresi Bildirileer, 25–28 Ocak 1994, Vol. III, Ege Universitesi Ziraat Fakultesi Bitki Koruma Bolumu, Izmir, Turkey [in Turkish].
Alauzet, C., D. Dargagnon, and J. C. Malausa. 1994. Bionomics of a polyphagous predator: *Orius laevigatus* (Het.:Anthocoridae). Entomophaga 39: 33–40.

Alborn, H. T., T. C. K. Turlinge, T. J. Jones, G. Stenhagen, J. H. Loughnin, and J. H. Tumlinson. 1997. An elicitor of plant volatiles from beet armyworm oral secretions. Science 276: 945–949.

Aldrich, J. R. 1998. Status of semiochemical research on predatory Heteroptera. Pp. 33–48 *in* M. Coll, and J. R. Ruberson [eds.], Predatory Heteroptera: Their Ecology and Use in Biological Control. Entomol. Soc. Amer., Lanham, Maryland, U.S.A. 233 pp.

Alleyne, E. H., and F. O. Morrison. 1978. The natural enemies of the lettuce root aphid, *Pemphigus bursarius* (L.) in Quebec, Canada. Ann. Entomol. Soc. Québec 22: 181–187.

Al-Marouf, I. N. 1990. Ecological studies on poplar leaf psyllid *Camarotoscena speciosa* Flor. (Homoptera: Psyllidae) in Mosul area. Arab J. Plant Prot. 8: 16–20.

Alomar, O., and R. N. Wiedenmann [eds.]. 1996. Zoophagous Heteroptera: Implications for Life History and Integrated Pest Management. Thomas Say Pub. Entomol.: Proceedings Entomological Society of America, Lanham, Maryland, U.S.A. 202 pp.

Aloo, T. C., and M. K. Karanja. 1986. The biological control of pine woolly aphid in Kenya using *Tetraphleps raoi* (Hem, Anthocoridae). Egerton Univ. Coll. (Kenya) Res. Pap. Ser. No. 10: 16 pp.

Amman, G. D., and C. P. Speers. 1971. Introduction and evaluation of predators from India and Pakistan for control of the balsam woolly aphid (Homoptera: Adelgidae) in North Carolina. Canad. Entomol. 103: 528–533.

Anderson, N. H. 1962a. Bionomics of six species of *Anthocoris* (Heteroptera: Anthocoridae) in England. Trans. R. Entomol. Soc. London 114: 67–95.

1962b. Anthocoridae of the Pacific Northwest with notes on distribution, life histories and habitats (Heteroptera). Canad. Entomol. 94: 1325–1334.

Anderson, N. H., and L. A. Kelton. 1963. A new species of *Anthocoris* from Canada, with distribution records for three other species (Heteroptera: Anthocoridae). Canad. Entomol. 95: 439–442.

Arbogast, R. T. 1984. Demography of the predaceous bug *Dufouriellus ater* (Hemiptera: Anthocoridae). Environ. Entomol. 13: 990–994.

Arbogast, R. T., M. Carthon, and J. R. Roberts, Jr. 1971. Developmental stages of *Xylocoris flavipes* (Hemiptera: Anthocoridae), a predator of stored-product insects. Ann. Entomol. Soc. Amer. 64: 1131–1134.

Arbogast, R. T., B. R. Flaherty, R. V. Byrd, and J. W. Press. 1983. Demography of the predaceous bug *Xylocoris sordidus* (Reuter). Amer. Midland Nat. 109: 398–405.

1985. Developmental stages of *Xylocoris sordidus* (Hemiptera: Anthocoridae). Entomol. News 96: 53–58.

Armer, C. A., R. N. Wiedenmann, and D. R. Bush. 1998. Plant feeding site selection on soybean by the facultatively phytophagous predator *Orius insidiosus*. Entomol. Exp. Appl. 86: 109–118.

Artigues, M., J. Avilla, A. M. Jauset, M. J. Sarasua, F. Polesny [ed.], W. Muller [ed.], and R. W. Olszak. 1996. Predators of *Cacopsylla pyri* in NE Spain. Heteroptera: Anthocoridae and Miridae. International Conference on Integrated Fruit Production, Cedzyna, Poland, 28 August–2 September 1995. Bull. OILB, SROP. 19: 231–235.

Askari, A., and V. M. Stern. 1972. Biology and feeding habits of *Orius tristicolor* (Hemiptera: Anthocoridae). Ann. Entomol. Soc. Amer. 65: 96–100.

Awadallah, K. T., and M. F. S. Tawfik. 1972. The biology of *Xylocoris* (= *Piezostethus*) *flavipes* (Reut.) (Hemiptera: Anthocoridae). Bull. Soc. Entomol. Egypte 56: 177–189.

Awadallah, K. T., M. F. S. Tawfik, N. Abou-Zeid, and M. M. El-Husseini. 1984. The life history of *Dufouriellus ater* (Duf.) (Hemiptera: Anthocoridae). Bull. Soc. Entomol. Egypte 63: 191–197.

Awadallah, K. T., M. F. S. Tawfik, M. M. El-Husseini, and A. M. A. Ibrahim. 1986. Effect of larval preys on the biocycle of *Xylocoris sordidus* (Reuter) (Anthocoridae, Hemiptera). Arch. Phytopathol. Pflanzenschutz 22: 237–242.

Bacheler, J. S., and R. M. Baranowski. 1975. *Paratriphleps laeviusculus*, a phytophagous anthocorid new to the United States (Hemiptera: Anthocoridae). Florida Entomol. 58: 157–163.

Baker, J. E., and J. E. Throne. 1995. Insecticide resistance in beneficial insects associated with stored grain in the southeastern United States. Resistant Pest Manage. 7: 17.

Barber, G. W. 1936. *Orius insidiosus* (Say), an important natural enemy of the corn ear worm. U.S. Dept. Agric. Tech. Bull. No. 504. 1–24.

Basu, R. C. 1982. A review on the taxonomical works done on the hemipterous group of insects during the period 1970–1980 in India. Proc. Zool. Soc. Calcutta 34: 55–77.

Bechinski, E. J., and L. P. Pedigo. 1982. Evaluation of methods for sampling predatory arthropods in soybeans. Environ. Entomol. 11: 756–761.

Berest, Z. L. 1980. Entomophages regulating the number of cereal aphids in the wheat fields of the Ukr. SSR. Right Bank Steppe Zone. Vestn. Zool. 5: 84–87 [in Russian].

Berrada, S., T. X Nguyen, and D. Fournier. 1996. Comparative toxicities of some insecticides to *Cacopsylla pyri* L. (Hom. Psyllidae) and one of its important biological control agents, *Anthocoris nemoralis* F. (Het., Anthocoridae). J. Appl. Entomol. 120: 181–185.

Biliotti, E., and J. Riom. 1967. Faune corticole du pin maritime: *Elatophilus nigricornis* (Hemiptera: Anthocoridae). Ann. Soc. Entomol. France (n.s.) 3: 1103–1108.

Brower, J. H., and J. W. Press. 1988. Interaction between the egg parasite *Trichogramma pretiosum* (Hymenoptera: Trichogrammatidae) and a predator, *Xylocoris flavipes* (Hemiptera: Anthocoridae) of the almond moth. *Cadra cautella* (Lepidoptera: Pyralidae). J. Entomol. Sci. 23: 342–349.

Brunner, J. F. 1973. Biology and behavior of *Anthocoris nemoralis* (Hemiptera: Anthocoridae) a predator of the pear psylla, *Psylla pyricola*. Proc. Washington State Entomol. Soc. 34: 353–354.

Brunner, J. F., and E. C. Burts. 1975. Searching behavior and growth rates of *Anthocoris nemoralis* (Hemiptera: Anthocoridae). Ann. Entomol. Soc. Amer. 68: 311–315.

Buchholz, U., S. Schmidt, and G. Schruft. 1994. The use of immunological technique in order to evaluate the predation on *Eupoecilis ambiguella* (Hbn.) (Lepidoptera: Cochylidae) in vineyards. Biochem. Syst. Ecol. 22: 671–677.

Burts, E. C. 1971. *Anthocoris nemoralis* — a new predator for control of pear psylla. Proc. Washington State Hortic. Assoc. 67: 110–112.

Busoli, A. G., F. M. Lara, S. Gravena, and E. B. Malheiros. 1984. Aspectos bioecologicos da mosca do sorgo (*Contarinia sorghicola* (Coquillet), 1898) (Diptera-Cecidomyiidae) e inimigos naturais, na região de Jaboticabal, SP. An. Soc. Entomol. Brasil 13; 167–176.

Cantelo, W. W., and M. Jacobson. 1979. Corn silk volatiles attracts many pest species of insects. J. Environ. Sci. Health A 14: 695–707.

Carayon, J. 1953. Observations sur *Scoloposcelis obscurella* (Zett.) (Hemipt. Anthocoridae) espèce nordique, nouvelle pour la France, où est une relicte glaciaire. Rev. Sci. Nat. Auvergne 19: 65–72.

1961. Quelques remarques sur les Hémiptéres: leur importance comme insectes auxiliaries et les possibilités de leur utilisation dans la lutte biologique. Entomophaga 6: 133–141.

1972a. Cautéres systématiques et classification des Anthocoridae (Hemipt.), Ann. Soc. Entomol. France (n.s.) 8: 309–49

1972b. Le genre *Xylocoris* subdivision et espècies nouvelles (Hem. Anthocoridae). Ann. Soc. Entomol. France (n.s.) 8: 579–606.

1982. *Wollastoniella testudo* n. sp., Hémiptère Anthocoridae prédateur d'un acarien nuisible au théier a Java. Bull. Zool. Mus. Univ. Amsterdam 8: 141–144.

Carayon, J., and F. Ramade. 1962. Note sur la présence en France et en Italie de *Montandoniola moraguese* (Puton) avec quelques observations sur cet Hétéroptère Anthocoridé. Bull. Soc. Entomol. France 67: 207–211.

Carayon, J., and J. R. Steffan. 1959. Observations sur le régime alimentaire des *Orius* et particuliérement, d'*Orius pallidicornis* (Reuter) (Heteroptera: Anthocoridae). Cah. Nat. Bull. Nat. Parisiens (n.s.) 15: 53–63.

Carver, M., F. A. Inkerman, and N. J. Ashbolt. 1987. *Anagyrus saccharicola* Timberlake (Hymenoptera: Encerytidae) and other biota associated with *Saccharicoccus sacchari* (Cockerell) (Homoptera: Pseudococcidae) in Australia. J. Austral. Entomol. Soc. 26: 367–368.

Cassis, G., and G. F. Gross. 1995. Hemiptera: Heteroptera (Coleorrhyncha to Cimicomorpha). Pp. 1–506 *in* W. W. A. Houston and G. V. Maynard [eds.], Zool. Cat. Australia. Vol. 27. 3A. CSIRO, Melbourne, Australia.

Chacko, M. J. 1973. Observations on some natural enemies of *Pineus* sp. (Hem., Adelgidae) at Shillong (Meghalaya), India, with special reference to *Tetraphleps raoi* Ghauri (Hem.: Anthocoridae). Tech. Bull. Commonw. Inst. Biol. Control No. 16: 41–46.

Champion. G. C. 1897–1901. Insecta: Rhynchota (Hemiptera–Heteroptera). Pp. 305–344 *in* F. O. Goodwin and O. Salvin [eds.], Biologia Centrali-Americana, Vol. II, 1900, London, U.K. 416 pp.

Chatterji, S. M. 1955. On the biology of *Triphleps sinui* Narayanan and Chatterji (Rhynchota: Anthocoridae) a predator of stored cereal pests in India. Proc. Zool. Soc. Bengal 8: 31–38.

China, W. E. 1926. A new species of *Triphleps* (Heteroptera, Anthocoridae) preying on the eggs of *Heliothis obsoleta*. H. S. in Queensland. Bull. Entomol. Res. 16: 361–362.

Chu, Y. I. 1969. On the bionomics of *Lyctocoris beneficus* (Hiura) amd *Xylocoris galactinus* (Fieber) (Anthocoridae, Heteroptera). J. Fac. Agric. Kyushu Univ. 15: 1–136.

Chyzik, R., M. Klein, and Y. Ben-Dov. 1995. Overwintering biology of the predatory bug *Orius albidipennis* (Hemiptera: Anthocoridae) in Israel. Biocontrol Sci. Tech. 5: 287–296.

Clark, R. C., D. O. Greenbank, D. G. Bryant, and J. W. E. Harris. 1971. *Adelges piceae* (Ratz) balsam woolly aphid (Homoptera: Adelgidae). Pp. 113–127 *in* Biological Control Programme Against Insects and Weeds in Canada. 1959–68. Commonw. Agric. Bur., Tech. Commun. No. 4. 266 pp.

Clausen, C. P. 1978a. Phlaeothripidae. Cuban laurel thrips. Pp. 18–19 *in* C. P. Clausen [ed.], Introduced Parasites and Predators of Arthropod Pests and Weeds: A World Review. U.S. Dept. Agric., Agric. Res. Ser., Agric. Handb. No. 480. 545 pp.

1978b. Psyllidae. P. 170 *in* C. P. Clausen [ed.], Introduced Parasites and Predators of Arthropod Pests and Weeds: A World Review, U.S. Dept. Agric., Agric. Res. Ser., Agric. Handb. No. 480. 545 pp.

Cloutier, C., D. Arodokoun, S. G. Johnson, and L. Gelinas. 1995. Thermal dependence of *Amblyseius cucumberis* (Acari: Phytoseliidae) and *Orius insidiosus* (Heteroptera: Anthocoridae) in greenhouses. Pp. 231–235 *in* NATO-ASI. Ser. A, Life Science. Plenum, New York, New York, U.S.A. 276 pp.

Cobben, R. H. 1968. Evolutionary Trends in Heteroptera. Part 1. Eggs, architecture of the shell, gross embryology, and eclosion. Centre for Agric. Publ. and Doc., Wageningen, the Netherlands. 475 pp.

1978. Evolutionary Trends in Heteroptera. Part 2. Mouthpart-structures and feeding strategies. Mededel. Landbouwhogeschool 78-5. H. Veenman and Zonen, Wageningen, the Netherlands. 407 pp.

Cobben, R. H., and B. Arnoud. 1969. Anthocoridae van *Viscum, Buxus*, en *Pinus* in Nederland (Heteroptera). Natuurhist. Maandbl. 47: 15–21.

Cohen, A. C. 1996. Plant feeding by predatory Heteroptera: evolutionary and adaptational aspects of trophic switching. Pp. 1–17 *in* O. Alomar and R. M. Wiedenmann [eds.], Zoophytophagous Heteroptera: Implications for Life History and Integrated Pest Management. Thomas Say Pub. Entomol.: Proceedings Entomological Society of America, Lanham, Maryland, U.S.A. 202 pp.

Coll, M., and S. Izraylevich. 1997. When predators also feed on plants: effects of competition and plant quality on omnivore-prey population dynamics. Ann. Entomol. Soc. Amer. 90: 155–161.

Coll, M., and R. L. Ridgway. 1995. Functional and numerical responses of *Orius insidiosus* (Heteroptera: Anthocoridae) to its prey in different vegetable crops. Ann. Entomol. Soc. Amer. 88: 732–738.

Collyer, E. 1953. Biology of some predatory insects and mites associated with the fruit tree red spider mite [*Metatetranychus ulmi* (Koch)] in southeastern England. II. Some important predators of the mite. J. Hortic. Sci. 28: 85–97.

Cuevo, C. M., P. D. Gjeda, and G. C. Korytkowski. 1975. Ciclo biologico, morfologia y compartamientos de *Paratriphleps laeviusculus* (Hemip. Anthocoridae). Rev. Peruana Entomol. 17: 32–39.

Delages, H. 1978. La lutte contre la cochenille devastatrice des pins maritimes. Phytoma (No. 279): 19–23.

Dennill, G. B. 1992. *Orius thripoborus* (Anthocoridae), a potential biocontrol agent of *Heliothrips haemorrhoidalis* and *Selenothrips rubrocinctus* (Thripidae) on avocado fruits in the eastern Transvaal. J. Entomol. Soc. South. Africa 55: 255–258.

Dicke, F. F., and J. L. Jarvis. 1962. The habits and seasonal abundance of *Orius insidiosus* (Say) (Hemiptera: Heteroptera: Anthocoridae) on corn. J. Kansas Entomol. Soc. 35: 339–344.

Distant, W. L. 1906. Rhynchota, Vol. III: 1–10 *in* C. T. Bingham [ed.], The Fauna of British India. Taylor and Francis, London, U.K.

1910. Rhynchota, Vol. V. Pp. 295–309 *in* C. T. Bingham [ed.], The Fauna of British India. Taylor and Francis, London, U.K.

Drake, C. J. 1921. A new beetle from the Adirondacks, notes on the work of *Xyloterinus politus* Say. Ohio J. Sci. 21: 201–205.

Dreistadt, S. H, and K. S. Hagen. 1994. Classical biological control of the *Acacia* psyllid, *Acissia uncatoides* (Homoptera: Psyllidae), and predator–prey-plant interactions in the San Francisco Bay Area. Biol. Control 4: 319–327.

Drukker, B., and M. W. Sabelius. 1990. Anthocorid bugs respond to odour emanating from *Psylla* infested pear trees. Proc. Sec. Exp. Appl. Entomol., Netherlands Entomol. Soc. No. 1: 88–89.

Drukker, B., P. Scutareanu, L. H. M. Blommers, and M. W. Sabelius. 1992. Olfactory response of migrating anthocorids to *Psylla*-infested pear trees in an orchard. Proc. Sec. Exp. Appl. Entomol. Netherlands Entomol. Soc. No. 3: 51–56.

Drukker, B., P. Scutareanu, and M. W. Sabelius. 1995. Do anthocorid predators respond to synomones from Psylla-infested pear trees under field conditions? Entomol. Exper. Appl. 77: 193–203.

Dunkle, F. V., and M. A. Ivie. 1994. *Xylocoris galactinus* (Fieber) (Hemiptera: Anthocoridae) newly discovered in Montana stored grain. Pan-Pac. Entomol. 70: 327–328.

Duso, C., and V. Girolami. 1983. Ruolo degli Antocoridi nel controllo del *Panonychus ulmi* Koch nei vigneti. Bull. Inst. Entomol. Univ. Degli Studi Bologna 38: 157–169 (1982).

Ellington, J., K. Kiser, G. Ferguson, and M. Cardenas. 1984. A comparison of sweepnet, absolute, and insect-vac sampling methods in cotton ecosystems. J. Econ. Entomol. 77: 599–605.

Elov, E. S. 1976. Bugs of the family Anthocoridae (Heteroptera) in Soviet Central Asia and Kazakhstan. Entomol. Rev. 55: 369–380 [in Russian; English translation, 55: 74–81].

Elov, E. S., and I. M. Kerzhner. 1977. Bugs of the families Anthocoridae, Cimicidae and Microphysidae (Heteroptera) of the Mongolian People's Republic. Nasekomye Mongolii 5: 203–220 [in Russian].

Evans, H. F. 1976a. The role of predator–prey size ratio in determining the efficiency of capture by *Anthocoris nemorum* and the escape reactions of its prey, *Acyrthosiphon pisum*. Ecol. Entomol. 1: 85–90.

1976b. The effect of prey density and host plant characteristics on oviposition and fertility in *Anthocoris confusus* (Reuter). Ecol. Entomol. 1: 157–161.

1976c. The searching behavior of *Anthocoris confusus* (Reuter) in relation to prey density and plant surface topography. Ecol. Entomol. 1: 163–169.

1976d. Mutual interference between predatory anthocorids. Ecol. Entomol. 1: 283–286.

Fabricius, J. C. 1794. Entomologia systematica emendata et aucta, secundum classes, ordines, genera, species, adjectis synonymis, locis, observationibus. C. G. Proft, Hafniae, Denmark 4: 1–472.

Fauvel, G. 1971. Influence de l'alimentation sur la biologie *d'Orius (Heterorius) vicinus* Ribaut (Heteroptera, Anthocoridae). Ann. Zool. Ecol. Anim. 8: 31–42.

Fauvel, G., A. Rambier, and D. Cotton. 1978. Activite predatrice et multiplication *d'Orius (Heterorius) vicinus* (Het. : Anthocoridae) dans les galles *d'Eriophyes fraxinivorus* (Acarina: Eriophyidae). Entomophaga 23: 261–270.

Fauvel, G., R. Rieux, F. Faivre d'Arcier, and A. Lyoussoufi. 1994. Essai de lutte biologique contre *Cacopsylla pyri* (L.) en verger de poirier par un apport experimental *d'Anthocoris nemoralis* F. au stade oeuf: I. methodologie. Bull. OILB-SROP 17: 81–85.

Fields, G. J., and B. P. Beirne. 1973. Ecology of anthocorid (Hemipt.: Anthocoridae) predators of the pear psylla (Homopt.: Psyllidae) in the Okanagan Valley, British Columbia. J. Entomol. Soc. British Columbia 70: 18–19.

Fields, G. J., R. W. Zwick, and H. R. Moffitt. 1981. A bibliography of *Psylla* (Homoptera: Psyllidae) on pear trees. U.S. Dept. Agric., Sci. Ed. Adm., Bibliog. Lit. Agric. No. 17. 13 pp.

Fiume, E. 1996. Comparative efficacy study of different strategies for control of aphids and thrips on sweet peppers (*Capsicum annum* L.) Boll. Lab. Entomol. Agrar. Filippo Silvesri 51: 37–49.

Florida Division of Plant Industry. 1993. Tri-ology — Technical Report 1993. 32(3): 1–3.

Foottit, R. G., and M. Mackauer. 1963. Subspecies of the balsam woolly aphid, *Adelges piceae* (Homoptera: Adelgidae) in North America. Ann. Entomol. Soc. Amer. 76: 299–304.

Forti, D., G. Angeli, and C. Ioriatti. 1993. Effects secondaires sur l' equilibre *Cacopsylla pyri — Anthocoris nemoralis* de deux I. C. I. utilises pour la lutte contre *Cydia pomonella* dans un verger de poirier du trentino (Italie). Meded. Fac. Landbouwwet. Univ. Gent. 58: 533–542.

Frescata, C., and A. Mexia. 1996. Biological control of thrips (Thysanoptera) by *Orius laevigatus* (Heteroptera: Anthocoridae) in organically grown strawberries. Biol. Agric. Hortic. 13: 141–148.

Funao, T., and Y. Yoshiyasu. 1995. Development and fecundity of *Orius sauteri* (Poppius) (Hemiptera: Anthocoridae) reared on *Aphis gossypii* Glover and corn pollen. Jpn. J. Appl. Entomol. Zool. 39: 84–85 [in Japanese].

Funasaki, G. Y. 1966. Studies on the life cycle and propagation technique of *Montandoniola moraguesi* (Puton) (Heteroptera: Anthocoridae). Proc. Hawaiian Entomol. Soc. 19: 209–211.

Ghauri, M. S. K. 1964. Notes on the Hemiptera from Pakistan and adjoining areas. Ann. Mag. Nat. Hist. (13) 7: 673–688.

1972a. Notes on the Hemiptera from Pakistan. J. Nat. Hist. 6: 279–288.

1972b. The identity of *Orius tantillus* (Motschulsky) and notes on other Oriental Anthocoridae (Hemiptera, Heteroptera). J. Nat. Hist. 6: 409–421.

Gillespie, D. R., and D. J. M. Quiring. 1992. Competition between *Orius tristicolor* (White) (Hemiptera: Anthocoridae) and *Amblyseius cucumeris* (Oudemans) (Acari: Phytoseiidae) feeding on *Frankliniella occidentalis* (Pergand) (Thysanoptera: Thripidae). Canad. Entomol. 124: 1123–1125.

Gross, G. C. 1954. A revision of the flower bugs (Heteroptera: Anthocoridae) of the Australian and adjacent Pacific regions. Part I. Rec. S. Austral. Mus. 11: 129–164.

1955. A revision of the flower bugs (Heteroptera: Anthocoridae) of the Australian and adjacent Pacific regions. Part II. Rec. S. Austral. Mus. 11: 409–422.

1957. A revision of the flower bugs (Heteroptera: Anthocoridae) of the Australian and adjacent Pacific regions. Part III. Rec. S. Austral. Mus. 13: 131–142.

Hagen, K. S., and S. H. Dreistadt. 1990. First California record for *Anthocoris nemoralis* (Fabr.) (Hemiptera: Anthocoridae), a predator important in the biological control of psyllids (Homoptera: Psyllidae). Pan-Pac. Entomol. 66: 323–324.

Hagley, E. A. C., and C. M. Simpson. 1983. Effect of insecticides on predators of the pear psylla, *Psylla pyricola* (Hemiptera: Psyllidae), in Ontario. Canad. Entomol. 115: 1409–1414.

Hall, D. W. 1950. Observations on the distribution, habits and life-history of the bug *Piezostethus galactinus* (Fieb.) (Hem-Anthocoridae). Entomol. Mon. Mag. 87: 45–52.

Heitmans, W. R. B., W. P. J. Overmeer, and L. P. S. van der Geest. 1986. The role of *Orius vicinus* Ribaut (Heteroptera: Anthocoridae) as a predator of phytophagous and predaceous mites in a Dutch orchard. J. Appl. Entomol. 102: 391–402.

Henry, T. J. 1988. Family Anthocoridae. Pp. 12–28 *in* T. J. Henry and R. C. Froeschner [eds.], Catalog of the Heteroptera, or True Bugs, of Canada and the Continental United States. E. J. Brill, Leiden, the Netherlands. 958 pp.

Herring, J. L. 1966a. The correct name for an anthocorid predator of the Cuban laurel thrips. Proc. Entomol. Soc. Washington 68: 93.

1966b. The genus *Orius* of the Western Hemisphere (Hemiptera: Anthocoridae). Ann. Entomol. Soc. Amer. 59: 1093–1109.

1967. Insects of Micronesia — Heteroptera: Anthocoridae. Bernice P. Bishop Mus. Insects of Micronesia 7: 391–414.

Hicks, E. A. (N. D. - 1959). Checklist and Bibliography on the Occurrence of Insects in Birds' Nests. Iowa State College Press, Ames, Iowa, U.S.A. 681 pp.

1962. Check-list and bibliography on the occurrence of insects in birds' nests. Suppl. I. Iowa State J. Sci. 36: 233–348.

1971. Check-list and bibliography on the occurrence of insects in birds' nests. Suppl. II. Iowa State J. Sci. 46: 123–338.

Hill, A. R. 1957. The biology of *Anthocoris nemorum* (L.) in Scotland. (Hemiptera: Anthocoridae) Trans. R. Entomol. Soc. London 109: 379–394.

1961. The biology of *Anthocoris sarothamni* Douglas and Scott in Scotland (Hemiptera: Anthocoridae). Trans. R. Entomol. Soc. London 113: 41–54.

1968. The bionomics and ecology of *Anthocoris confusus* Reuter in Scotland. Trans. Soc. British Entomol. 18: 29–48.

Hiura, I. 1959. Contributions to the knowledge of Anthocoridae from Japan and its adjacent territories (Hemiptera: Heteroptera) 1. Bull. Osaka Mus. Nat. Hist. 11: 1–10.

1960. Contributions to the knowledge of Anthocoridae from Japan and its adjacent territories (Hemiptera: Heteroptera). 2. Bull. Osaka Mus. Nat. Hist. 12: 43–55.

1966. Contribution to the knowledge of Anthocoridae from Japan and its adjacent territories (Hemiptera: Heteroptera). 3. Bull. Osaka Mus. Nat. Hist. 9: 29–37.

1967. Contribution to the knowledge of Anthocoridae from Japan and its adjacent territories (Hemiptera: Heteroptera). 4. Bull. Osaka Mus. Nat. Hist. 20: 61–63.

Horn, K. F, M. H. Farrier, and C. G. Wright. 1983. Some mortality factors affecting eggs of the Sycamore lace bug, *Corythucha ciliata* (Say) (Hemiptera: Tingidae). Ann. Entomol. Soc. Amer. 76: 262–265.

Horton, D. R., T. M. Lewis, T. Hinojosa, and D. A. Broers. 1998. Photoperiod and reproductive diapause in the predatory bugs *Anthocoris tomentosus, A. antevolens*, and *Deraeocoris brevis* (Heteroptera: Anthocoridae, Miridae) with information on overwintering sex ratio. Ann. Entomol. Soc. Amer. 91: 81–86.

Iglinsky, W., and C. F. Rainwater. 1950. Observations and life history notes on *Orius insidiosus* (Say), an important natural enemy of the red spider mite, *Spoganychus* spp., on cotton in Texas. J. Econ. Entomol. 43: 567–568.

Irwin, M. E, and M. Shepard. 1980. Sampling predaceous Hemiptera on soybean. Pp. 505–531 *in* M. Kogan and D. C. Herzog [eds.], Sampling Methods in Soybean Entomology. Springer-Verlag, New York, New York, U.S.A. 587 pp.

Isenhour, D. J., and N. L. Marston. 1981. Seasonal cycles of *Orius insidiosus* (Hemiptera: Anthocoridae) in Missouri soybeans. J. Kansas Entomol. Soc. 54: 129–142.

Isenhour, D. J., and K. V. Yeargan. 1981. Predation by *Orius insidiosus* on the soybean thrips, *Sericothrips variabilis:* effect of prey stage and density. Environ. Entomol. 10: 496–500.

1982. Oviposition sites of *Orius insidiosus* (Say) and *Nabis* spp. in soybean (Hemiptera: Anthocoridae and Nabidae). J. Kansas Entomol. Soc. 55: 65–72.

Jacobson, R. 1991. Integrated control of *Frankliniella occidentalis* in UK cucumber crops — use of *Anthocoris nemorum*. Meded. Fac. Landbouwwet. Rijksuniv. Gent 56: 2a, 235–240.

Jactel, H., N. Perthuisot, P. Menassieu, G. Raise, and C. Burban. 1996. A sampling design for within tree larval populations of the maritime pine scale, *Matsucoccus feytaudi* Duc. (Homoptera: Margarodidae), and the relationship between larval population estimates and male catch in pheromone traps. Canad. Entomol. 128: 1143–1156.

Javier, T. G., and S. T. Peralta. 1976a. Evaluacion cuentitativa del control biologio entres cultivos del Valle Montaro. Rev. Peruana Entomol. 18: 69–71.

1976b. Tendencia del control biologico en tres sistemes de cultivo de maiz "choclo. " Rev. Peruana Entomol. 18: 72–76.

Jay, E., R. Davis, and S. Brown. 1968. Studies on the predaceous habits of *Xylocoris flavipes* (Reuter). J. Georgia Entomol. Soc. 3: 126–130.

Jessop, L. 1983. The British species of *Anthocoris* (Hem. Anthocoridae). Entomol. Mon. Mag. 119: 221–223.

Jonsson, N. 1983. The bug fauna (Hem. Heteroptera) on apple trees in south-eastern Norway. Fauna Norv., Ser. B. 30: 9–13.

Josifov, M. 1962. Quantitative and qualitative study of the entomogauna of lucerne fields in the region of Sophia, with an attention to the suborder Heteroptera. Izv. Zool. Inst. Sofia, 11: 117–140 [in Russian].

Kabicek, J., and P. Hejzlar. 1996. Predation by *Orius majusculus* (Heteroptera: Anthocoridae) on the apple aphid *Aphis pomi* (Sternorrhyncha: Aphididae) on apple tree. Ochr. Rostlin 32: 57–63. [in Czech].

Karanja, M. K., and T. C. Aloo. 1990. The introduction and establishment of *Tetraphleps raoi* Ghauri as a control of woolly aphid in Kenya. Kenya For. Res. Inst., Tech. Note No. 12. 11 pp.

Kawai, A. 1995. Control of *Thrips palmi* Karny (Thysanoptera: Thripidae) by *Orius* spp. (Heteroptera: Anthocoridae) on greenhouse eggplant. Appl. Entomol. Zool. 30: 1–7.

Kelton, L. A. 1966. Synopsis of the genus *Tetraphleps* Fieber in North America (Hemiptera: Anthocoridae). Canad. Entomol. 98: 199–204.

1978. The insects and arachnids of Canada. Part 4. The Anthocoridae of Canada and Alaska (Heteroptera: Anthocoridae). Res. Branch, Canada Dept. Agric., Publ. No. 1639. 101 pp.

Kerzhner, I. M., and E. S. Elov. 1976. Bugs of the genus *Xylocoris* Duf. from the subgenus *Proxylocoris* Carayon (Heteroptera: Anthocoridae) of the fauna of the USSR and adjacent regions. Entomol. Obozr. 55: 364–368 [in Russian].

Kiman, Z. B., and K. V. Yeargan. 1985. Development and reproduction of the predator *Orius insidiosus* (Hemiptera: Anthocoridae) reared on diets of selected plant material and arthropod prey. Ann. Entomol. Soc. Amer. 78: 464–467.

Kogan, M., and J. D. Lattin. 1999. Agricultural systems as ecosystems. Pp. 1–33 *in* J. R. Ruberson [ed.], Handbook of Pest Management. Marcel Dekker, New York, New York, U.S.A.

Kraszpulski, P., and R. Davis. 1988. Interaction of a parasite, *Bracon hebetor* (Hymenoptera: Braconidae), and a predator *Xylocoris flavipes* (Hemiptera: Anthocoridae) with populations of *Tribolium castaneum* and *Plodia interpunctella*. Amer. Midl. Nat. 119: 71–76.

Kumar, N. S., and T. N. Ananthakrishnan. 1984. Predator–thrips interactions with reference to *Orius maxidentex* Ghauri and *Carayonocoris indicus* Muraleedharan (Anthocoridae: Heteroptera). Proc. Indian Nat. Sci. Acad. B50 No. 2: 139–145.

Larivière, M.-C., and C. H. Wearing. 1994. *Orius vicinus* (Ribaut) Heteroptera: Anthocoridae), a predator of orchard pests new to New Zealand. New Zealand Entomol. 17: 17–21.

Lattin, J. D. 1999. Bionomics of the Anthocoridae. Annu. Rev. Entomol. 44: 207–231.

2000. Dead-leaf clusters as habitats for adult *Calliodis temnostethoides* and *Cardiastethus luridellus* and other Anthocoridae (Hemiptera: Heteroptera). Great Lakes Entomol. 32: 33–38.

Lattin, J. D., and N. L. Stanton. 1992. A review of the species of Anthocoridae (Hemiptera: Heteroptera) found on *Pinus contorta*. J. New York Entomol. Soc. 100: 424–479.

1993a. Taxonomic and biological notes on North American species of *Elatophilus* Reuter (Hemiptera: Heteroptera: Anthocoridae). J. New York Entomol. Soc. 101: 88–94.

1993b. A review of the genus *Melanocoris* Champion with remarks on distributions and host–tree associations (Hemiptera: Heteroptera: Anthocoridae). J. New York Entomol. Soc. 101: 95–107.

Lattin, J. D., A. Asquith, and S. Booth. 1989. *Orius minutus* (Linnaeus) in North America (Hemiptera: Heteroptera: Anthocoridae). J. New York Entomol. Soc. 97: 409–416.

Lawton, J. H. 1983. Plant architecture and the diversity of phytophagous insects. Annu. Rev. Entomol. 28: 23–39.

LeCato, G. L., and R. Davis. 1973. Preferences of the predator *Xylocoris flavipes* (Hemiptera: Anthocoridae) for species and instars of stored-product insects. Florida Entomol. 56: 57–59.

Ledee, S. 1995. Inventaire des hétèropterès dans les vergers de pommers, du nord de la France: interêt des haies composites et de la strate herbacee. Meded. Fac. Landhouwkund. Toegepast Biol. Wet. Univ. Gent. 60: 793–797.

Letourneau, D. K., and M. A. Altieri. 1983. Abundance patterns of a predator, *Orius tristicolor* (Hemiptera: Anthocoridae), and its prey, *Frankiniella occidentalis* (Thysanoptera: Thripidae): habitat attraction in polycultures versus monocultures. Environ. Entomol. 12: 1464–1469.

Linnaeus, C. 1758. Systema Nature, editio decima, reformata. Laurentii Salvii, Holmiae, Sweden 1: 1–823.

Linsley, E. G. 1944. Natural sources, habitats and reservoirs of insects associated with stored food products. Hilgardia 16: 187–224.

Lord, F. T. 1965. Sampling predator populations on apple trees in Nova Scotia. Canad. Entomol. 97: 287–298.

1972. Comparisons of the abundance of the species composing the foliage inhabiting fauna of apple trees. Canad. Entomol. 104: 731–749.

Lussier, S. J. 1965. A study of *Elatophilus inimica* D. & H. (Hemiptera: Anthocoridae) and its role in the natural control of the red-pine scale *Matsucoccus resinosae* D. & G. (Homoptera: Margarodidae). M.S. thesis, University of Massachusetts, Amherst, Massachusetts, U.S.A. 42 pp.

MacLellan, C. R. 1977. Populations of some major pests and their natural enemies on young and semidwarf apple trees in Nova Scotia. Canada. Entomol. 109: 797–806.

MacPhee, A. N., and K. H. Sanford. 1954. The influence of spray programs on the fauna of apple orchards in Nova Scotia. VII. Effects on some beneficial arthropods. Canad. Entomol. 86: 128–135.

Madsen, H. F. 1961. Notes on *Anthocoris melanocerus* Reuter (Hemiptera: Anthocoridae) as a predator of the pear psylla in British Columbia. Canad. Entomol. 93: 660–662.

Madsen, H. F., P. H. Westigard, and R. L. Sission. 1963. Observations on the natural control of the pear psylla, *Psylla pyricola* Foster, in California. Canad. Entomol. 95: 837–846.

Manuel, F. A., and H. B. Mojica. 1982. Biologia, comportamiento y capacidad depredatora de *Orius tristicolor* (White) y *Orius thyestes* (Hemiptera: Anthocoridae) y su efecto sobre *Tetranychus urticae* Koch (Acarina: Tetranychidae) en cacahuate. Colegiode Postgraduado Escuela National de Agriculture, Chapingo, México. No. 49: 67–79.

Marshall, G. E. 1930. Some observations on *Orius (Triphleps) insidiosus* (Say). J. Kansas Entomol. Soc. 1: 29–32.

Martinez, D. G., and R. L. Pienkowski. 1982. Laboratory studies on insect predators of potato leafhopper eggs, nymphs, and adults. Environ. Entomol. 11: 361–362.

McCaffrey, J. P., and R. L. Horsburgh. 1986. Biology of *Orius insidiosus* (Heteroptera: Anthocoridae): a predator in Virginia apple orchards. Environ. Entomol. 15: 984–988.

McMullen, R. D., and C. Jong. 1967. New records and discussion of predators of the pear psylla, *Psylla pyricola* Foerster, in British Columbia. Entomol. Soc. British Columbia 64: 35–50.

Mendel, Z., E. Carmi, and H. Podoler. 1991. Relation between the genera *Matsucoccus* (Homoptera: Margarodidae) and *Elatophilus* (Hemiptera: Anthocoridae) and their significance. Ann. Entomol. Soc. Amer. 84: 502–507.

Mendel, Z., E. Carmi-Gera, H. Podoler, and F. Assael. 1995a. Reproductive behavior of the specialist predator *Elatophilus hebraicus* (Hemiptera: Anthocoridae). Ann. Entomol. Soc. Amer. 88: 856–861.

Mendel, Z., L. Zegelman, A. Hassner, F. Assael, M. Harel, S. Tam, and E. Dunkelblum. 1995b. Outdoor attractancy of males of *Matsucoccus josephi* (Homoptera: Matsucoccidae) and *Elatophilus hebraicus* (Hemiptera: Anthocoridae) to synthetic female sex pheromone of *Matsucoccus josephi*. J. Chem. Ecol. 21: 331–341.

Mészáros, Z. [ed.]. 1984a. Results of faunistical studies in Hungarian maize stands (Maize Ecosystem Research No. 16). Acta Phytopathol. Acad. Sci. Hungary 19: 65–90.

1984b. Results of faunistical and floristical studies in Hungarian apple orchards (Apple Ecosystem Research No. 26). Acta Phytopathol. Acad. Sci. Hungary 19: 91–176.

Mitchell, R. G. 1962. Balsam woolly aphid predators native to Oregon and Washington. Oregon State Univ., Agric. Exp. Stn., Tech. Bull. No. 62: 1–63.

Mitchell, R. G., and K. H. Wright. 1967. Foreign predator introductions for control of the balsam woolly aphid in the Pacific Northwest. J. Econ. Entomol. 60: 142–147.

Moore, G. E. 1972. Southern pine beetle mortality in North Carolina by parasites and predators. Environ. Entomol. 1: 58–65.

Moser, J. C., R. C. Thatcher, and L. S. Packard. 1971. Relative abundance of southern pine beetle associates in East Texas. Ann. Entomol. Soc. Amer. 54: 72–77.

Mound, L. A., C. L. Wang, and S. Okahma. 1995. Observations in Taiwain on the identity of the Cuban laurel thrips (Thysanoptera, Phlaeothripidae). J. New York Entomol. Soc. 103: 185–190.

Muraleedharan, N. 1977a. Some genera of Anthocorinae (Heteroptera: Anthocoridae) from south India. Entomon 2: 231–235.

1977b. A new genus of Anthocoridae (Heteroptera) from south India. Orient. Insects 11: 463–466.

1978. A new species of *Lasiochilus (indicus)* Reuter (Heteroptera: Anthocoridae) from India. Bull. Zool. Surv. India 1: 267–269.

Muraleedharan, N., and T. N. Ananthakrishnan. 1971. Bionomics of *Montandoniola moraguesi* (Puton) (Heteroptera: Anthocoridae), a predator on gall thrips. Bull. Entomol. 12: 4–10.

1974. New and little-known species of *Orius* Wolff from India (Hemiptera: Anthocoridae). Orient. Insects 8: 37–41.

1978. Bioecology of four species of Anthocoridae (Hemiptera: Insecta) predaceous on thrips with key to genera of anthocorids from India. Rec. Zool. Surv. India. Misc. Publ., Occas. Pap. No. 11. 32 pp.

Nakata, T. 1995. Population fluctuations of aphids and their natural enemies on potato in Hokkaido, Japan. Appl. Entomol. Zool. 30: 129–138.

Naranjo, S. E., and R. L. Gibson. 1996. Phytophagy in predaceous Heteroptera: effects on life history and population dynamics. Pp. 57–93 *in* O. Alomar and R. N. Wiedenmann [eds.], Zoophagous Heteroptera: Implications for Life History and Integrated Pest Management. Thomas Say Publ. Entomol.: Proceedings Entomological Society of America, Lanham, Maryland, U.S.A. 202 pp.

Narayanan, E. S., and S. M. Chatterji. 1952. A new species of *Triphleps* (Hemiptera, Heteroptera: Anthocoridae) predaceous on storage pests in India. Proc. Zool. Soc. Bengal 5: 163–166.

1953. On a new macropterous species of *Triphleps* (Hemiptera, Heteroptera: Anthocoridae) predaceous on storage pests in India. Proc. Zool. Soc. Bengal 6: 121–123.

Naseer, M., and U. C. Abdurahiman. 1993. Cannibalism in *Cardiastethus exiguus* Poppius (Hemiptera: Anthocoridae), a predator of the coconut caterpillar *Opisina arenolella* Walker (Lepidoptera: Xylorictidae). J. Adv. Zool. 14: 1–6.

Nicoli, G., L. Corazza, and R. Cornale. 1990. Lotta biologica contro i Lepidotteri Tortricidi ricamatori del pero con *Bacillus thuringiensis* Berl. ssp. *kurstaki*. Inf. Fitopatol. 40: 6, 55–62.

Nicoli, G., R. Cornale, L. Corazza, and I. Marzocchi. 1989. Attivita di *Anthocoris nemoralis* (F.) (Rhyn. Anthocoridae) nei confronti di *Psylla pyri* (L.) (Rhy. Psyllidae) in pereti a diversa gestione fitoiatrica. Boll. Inst. Entomol. Guido-Grandi Univ. Stud. Bologna 43: 171–186.

Novak, H., and R. Achtziger. 1995. Influence of heteropteren predators (Het., Anthocoridae, Miridae) on larval populations of hawthorne psyllids (Hom. Psyllidae). J. Appl. Entomol. 119: 479–486.

Oatman, E. R. 1978. Noctuidae. Pp. 205–210 *in* C. P. Clausen [ed.], Introduced Parasites and Predators of Arthropod Pests and Weeds: A World Review. U.S. Dept. Agric., Agric. Res. Ser., Agric. Handb. No. 480. 545 pp.

Ohmart, C. P. 1981. An annotated list of insects associated with *Pinus radiata* D. Don in California. Commonwealth Scientific and Industrial Research Organisation, Div. Forest Research, Div. Report No. 8. 50 pp. (Melbourne, Australia).

Paine, T. D. 1992. Cuban laurel thrips (Thysanoptera: Phlaeothripidae) biology in southern California: seasonal abundance, temperature, temperature-dependent development, leaf suitability, and predation. Ann. Entomol. Soc. Amer. 85: 164–172.

Parajulee, M. N., and T. W. Phillips. 1992. Laboratory rearing and field observations of *Lyctocoris campestris* (Heteroptera: Anthocoridae), a predator of stored-product insects. Ann. Entomol. Soc. Amer. 85: 736–743.

1995. Survivorship and cannibalism in *Lyctocoris campestris* (Hemiptera: Anthocoridae): effects of density, prey availability, and temperature. J. Entomol. Sci. 30: 1–8.

Parajulee, M. N, T. W. Phillips, J. E. Throne, and E. V. Nordheim. 1995. Life history of immature *Lyctocoris campestris* (Hemiptera: Anthocoridae): effects of constant temperatures and relative humidities. Environ. Entomol. 24: 4, 889–897.

Parker, N. J. B. 1984. Biology and bionomics in Scotland of *Anthocoris gallarum-ulmi*. Ecol. Entomol. 9: 55–67.

Parshley, H. M. 1917. A species of *Macrotrachiella* found in New England (Hemip., Anthocoridae). Entomol. News 28: 37–38.

Parsons, G. L., G. Cassis, A. R. Moldenke, J. D. Lattin, N. H. Anderson, J. C. Miller, P. Hammond, and T. D. Schowalter. 1991. Invertebrates of the H. J. Andrews Experimental Forest, Oregon. V: An annotated list of insects and other arthropods. U.S. Dept. Agric., For. Ser., Pacific Northwest Res. Stn., Gen. Tech. Rep. PNW-GTR. 168 pp.

Peet, W., Jr. 1973. Biological studies on *Nidicola marginata* (Hemiptera: Anthocoridae). Ann. Entomol. Soc. Amer. 66: 344–348.

1979. Description and biology of *Nidicola jaegeri*, n. sp., from southern California (Hemiptera: Anthocoridae). Ann. Entomol. Soc. Amer. 72: 430–437.

Péricart, J. 1971. Observations diverses et nouvelles synonymies concernant les Anthocoridae et Microphysidae Palearctiques (Heteroptera). Bull. Soc. Linn. Lyon 40: 93–114.

1972. Faune de l'Europe et du Bassin Méditerréen. No. 7. Hémiptères Anthocoridae, Cimicidae et Microphysidae de l'Ouest-Paléarctique. Masson et Cie Éditeurs, Paris, France. 402 pp.

1996. Family Anthocoridae Fieber. 1836. Pp. 108–140 *in* B. Aukema and C. Rieger [eds.], Catalogue of the Heteroptera of the Palaearctic Region, Vol. 2. Netherlands Entomological Society, Amsterdam, the Netherlands. 361 pp.

Phillips, T. W., M. N. Parajulee, and D. K. Weaver. 1995. Toxicity of terpenes secreted by the predator *Xylocoris flavipes* (Reuter) to *Tribolium castaneum* (Herbst) and *Oryzaephilus surinamensis* (L.) J. Stored Prod. Res. 31: 131–138.

Pickett, C. H., and F. E. Gilstrap. 1986. Natural enemies associated with spider mites (Acari: Tetranychidae) infesting corn in the High Plains region of Texas. J. Kansas Entomol. Soc. 59: 524–536.

Poppius, B. 1913. Zur Kenntnis der Miriden, Isometopiden, Anthocoriden, Nabiden und Schizopteriden Ceylons. Entomol. Tidskr. 34: 239–260.

Press, J. W. 1989. Compatibility of *Xylocoris flavipes* (Hemiptera: Anthocoridae) and *Venturia canescens* (Hymenoptera: Icheumonidae) for suppression of the almond moth, *Cadra cautella* (Lepidoptera: Pyralidae). J. Entomol. Sci. 24: 156–160.

Procter, W. 1946. Biological survey of the Mount Desert Region. Part VII. Being a revision of Parts I and VI with the addition of 1100 species. The insect fauna with reference to methods of capture, foot plants, the flora and other biological features. Wistar Inst. Anat. Biol., Philadelphia, Pennsylvania, U.S.A. 566 pp.

Rácz, V., and E. Visnyovszky. 1985. Changes in the abundance of aphidophagous Heteroptera and syrphids occurring in maize stands of different management types. Phytopathol. Acad. Sci. Hung. 20: 193–200.

Rajasekhara, K. 1973. A new species of *Blaptostethus* (Hemiptera: Anthocoridae) from Mysore, India. Ann. Entomol. Soc. Amer. 66: 86–87.

Rajasekhara, K., and S. Chatterji. 1970. Biology of *Orius indicus* (Hemiptera: Anthocoridae), a predator of *Taeniothrips nigricornis* (Thysanoptera). Ann. Entomol. Soc. Amer. 63: 364–367.

Rajendram, F. G. 1994. Population sampling of planthoppers, leafhoppers, and insect predators on broadcast rice, treated with carbofuran, in eastern Sri Lanka. Insect Sci. Appl. 15: 139–143.

Raney, H. G., and K. V. Yeargan. 1977. Seasonal abundance of common phytophagous and predaceous insects in Kentucky soybeans. Trans. Kentucky Acad. Sci. 38: 83–87.

Rao, V. P., and M. A. Ghani [eds.]. 1972. Studies on predators of *Adelges* spp. in the Himalayas. Commonw. Inst. Biol. Control, Trinidad. Misc. Publ. No. 3. 116 pp.

Reid, C. D., and R. L. Lampman. 1989. Olfactory responses of *Orius insidiosus* (Hemiptera: Anthocoridae) to volatiles of corn silks. J. Chem. Ecol. 15: 1109–1115.

Reilly, J. J., and W. L. Sterling. 1983. Dispersion patterns of the red imported fire ant (Hymenoptera: Formicidae), aphids, and some predaceous insects in east Texas cotton fields. Environ. Entomol. 12: 380–385.

Reimer, N. J. 1988. Predation on *Lirothrips urichi* Karny (Thysanoptera: Phlaeothripidae): a case of biotic interference. Environ. Entomol. 17: 132–134.

Reuter, O. M. 1884. Monographia Anthocoridarum Orbis Terrestris: 1–204. Helsingforisae (Finland). [Also published in Acta Soc. Sci. Fenn. 16 (1885): 555–758].

Richards, P. C., and J. M. Schmidt. 1996. The effects of selected dietary supplements on survival and reproduction of *Orius insidiosus* (Say) (Hemiptera: Anthocoridae). Canad. Entomol. 128: 171–176.

Rieux, R., G. Fauvel, F. Faivre d'Arcier, G. Fournage, and A. Lyoussoufi. 1994. Essai de lutte bioloque contre *Cacopsylla pyri* (L.) en verger de poinier par un apport experimental d'*Anthocoris nemoralis* F. au stade oeuf. II. Results et discussion. Bull. OILB, SROP, 17: 120–124.

Riley, M. A., and R. A. Goyer. 1986. Impact of beneficial insects on *Ips* spp. (Coleoptera: Scolytidae) bark beetles in felled loblolly and slash pines in Louisiana. Environ. Entomol. 15: 1220–1224.

Riom, J., B. Gerbinot, A. Roulbria, and A. P. Fabre. 1971. Elements de la bioecologia de *Matsucoccus feytaudi* Duc. (Coccoidea, Margarodidae) et de ses predateurs dans le sud-est et le sud-ouest de la France. Ann. Zool. Ecol. Anim. 3: 153–176.

Rozhkov, A. S. [ed.] 1966. Pests of Siberian larch. Academy of Sciences of the USSR, Siberian Department, East-Siberian Biological Institute. 393 pp. [in Russian; English translation, 1970].

Salas-Aguilar, J., and L. E. Ehler. 1977. Feeding habits of *Orius tristicolor*. Ann. Entomol. Soc. Amer. 70: 60–62.

Salim, M., S. A. Masud, and A. M. Khan. 1987. *Orius albidipennis* (Reut.) (Hemiptera: Anthocoridae) — a predator of cotton pests. Philipp. Entomol. 7: 37–42.

Sands, W. A. 1957. The immature stages of some British Anthocoridae (Hemiptera). Trans. R. Entomol. Soc. London 109: 295–310.

Schmitt, J. J., and R. A. Goyer. 1983a. Consumption rates and predatory habits of *Scoloposcelis mississippensis* and *Lyctocoris elongatus* (Hemiptera: Anthocoridae) on pine bark beetles (*Dendroctonus frontalis*). Environ. Entomol. 12: 363–367.

1983b. Laboratory development and description of immature stages of *Scoloposcelis mississippensis* Drake and Harris and *Lyctocoris elongatus* (Reuter) (Hemiptera: Anthocoridae) predators of southern pine bark beetles (Coleoptera: Scolytidae) (*Dendroctonus frontalis*). Ann. Entomol. Soc. Amer. 76: 868–872.

Schuh, R. T., and J. A. Slater. 1995. True Bugs of the World (Hemiptera: Heteroptera). Comstock Publ. Assoc., Ithaca, New York, U.S.A. 336 pp.

Schuh, R. T., and P. Štys. 1991. Phylogenetic analysis of cimicomorphan family relationships (Heteroptera). J. New York Entomol. Soc. 99: 298–350.

Schuster, M. F., D. G. Holder, E. T. Cherry, and F. G. Maxwell. 1976. Plant bugs and natural enemy insect populations on Frego bract and smoothleaf cottons. Mississippi State Univ., Agric. For. Exp. Stn., Tech. Bull. No. 75: 11 pp.

Scudder, G. G. E. 1986. Additional Heteroptera new to British Columbia, Canada. J. Entomol. Soc. British Columbia 83: 63–65.

Scutareanu, P., B. Drukker, and M. W. Sabelis. 1994. Local population dynamics of pear psylla and their anthocorid predators. Bull. OILB-SROP. 12: 18–22.

Scutareanu, P., B. Drukker, J. Bruin, M. A. Posthumus, and M. W. Sabelis. 1996. Leaf volatiles and polyphenols in pear trees infested by *Psylla pyricola*. Evidence of simultaneously induced responses. Chemoecol. 7: 1, 34–38.

Sechser, B., and M. Engelhardt. 1988. Strategy for the control of pear pests by the use of an insect growth regulatory as a key element. Mitt. Schweiz. Entomol. Ges. 61: 217–221.

Shepard, M., G. R. Carner, and S. G. Turnipseed. 1974. Seasonal abundance of predaceous arthropods in soybeans. Environ. Entomol. 3: 985–988.

Siemann, E., D. Tilman, and J. Haarstad. 1996. Insect species diversity, abundance and body size relationships. Nature 380: 704–706.

Solomon, M. G., and J. D. Fitzgerald. 1990. Fenoxycarb, a selective insecticide for inclusion in integrated pest management systems for pear in the U.K. J. Hortic. Sci. 65: 535–539.

Southwood, T. R. E., and D. Leston. 1959. Land and Water Bugs of the British Isles. Warne, London, U.K. 436 pp.

Staubli, A., M. Hachler, D. Pasquier, P. Antonin, and C. Minaz. 1992. Dix annees d'experiences et observations sur le psylle commun du poirier *Cacopsylla* (= *Psylla*) *pyri* L. en Suisse romande. Rev. Suisse Vitic. Arbor. Hortic. 2412: 89–104.

Strawínski, K. 1964. Zoophagism of terrestrial Hemiptera: Heteroptera occurring in Poland. Ekol. Pol., S. A. 12, 27: 429–452.

Strong, D. R., J. H. Lawton, and Sir Richard Southwood. 1984. Insects on Plants. Harvard University Press, Cambridge, Massachusetts, U.S.A. 313 pp.

Štys, P., and M. Daniel. 1957. *Lyctocoris campestris* (F.) (Heteroptera, Anthocoridae) jako fakultativni ektoparasit člověka. Acta Soc. Entomol. Čechosloveniae 54: 1–10.

Tamaki, G., and D. Olson. 1977. Feeding potential of predators of *Mysus persicae*. J. Entomol. Soc. British Columbia 74: 23–26.

Tawfik, M. F. S., and M. M. El-Husseini. 1971a. The life history of the *Blaptostethus piceus* Fieber, var. *pallescens* Poppius (Hemiptera: Anthocoridae). Bull. Soc. Entomol. Egypt 55: 239–252.

1971b. The life history of *Xylocoris* (=*Piezostethus*) *galactinus* (Fieber) (Hemiptera: Anthocoridae). Bull. Soc. Entomol. Egypt 55: 171–183.

Tawfik, M. F. S., and A. Nagui. 1965. The biology of *Montandoniola moraguesi* (Puton), a predator of *Gynaikothrips ficorum* Marchal, in Egypt. Bull. Soc. Entomol. Egypt 49: 181–200.

Tawfik, M. F. S., K. T. Awadallah, and N. A. Abou-Zeid. 1983a. Effect of temperature and relative humidity on the adult of *Xylocoris flavipes* (Reuter) (Anthocoridae: Hemiptera: Heteroptera). Ann. Agric. Sci. Moshtohor, Moshtohor: Zagazig Univ. 19: 427–434.

1983b. Effect of feeding on various preys on the biocycle of *Xylocoris flavipes* (Reuter) (Anthocoridae: Hemiptera: Heteroptera). Ann. Agric. Sci. Moshtohor, Moshtohor: Zagazig Univ. 19: 435–441.

Tawfik, M. F. S., M. M. El-Husseini, and A. M. A. Ibrahim. 1986. Effect of temperature and relative humidity on the biocycle of *Xylocoris sordidus* (Reuter) (Anthocoridae, Hemiptera). Arch. Phytopathol. Pflanzenschutz 22: 115–129.

Teerling, C. R., D. R. Gillespie, and J. H. Borden. 1993. Utilization of western flower thrips alarm pheromone as a prey-finding kairomone by predators. Canad. Entomol. 125: 413–437.

Thontadarya, T. S., and K. J. Rao. 1987. Biology of *Orius maxidentes* Ghauri (Hemiptera: Anthocoridae), a predator of the sorghum earhead midge, *Contarinia sorghicola* (Coquillet). Mysore J. Agric. Sci. 21: 21–31.

Tonks, N. V. 1953. Annotated list of insects and mites collected on brambles in the lower Frazer Valley, British Columbia, 1951. Proc. Entomol. Soc. British Columbia 49: 27–28.

Trematerra, P., and P. Dioli. 1993. *Lyctocoris campestris* (F.) (Heteroptera: Anthocoridae) in stores of *Triticum spelta* L. in central Italy. Boll. Zool. Agrar. Bachicolt. 25: 251–257.

Turka, I. 1987. Predatory bugs in the entomofauna of potato-natural enemies of virus vectors. Trudy-Latviiskoi Sel'skokhozy-aistvennoi Akademii No. 236: 73–78 [in Russian].

U.S. Congress, Office of Technology Assessment. 1993. Harmful non-indigenous species in the United States. OTA-F-565, U.S. Govt. Print. Office, Washington, D.C., U.S.A. 391 pp.

Usinger, R. L. 1946. Heteroptera of Guam. Insects Guam 2: 13–103.

Valenti, M. A., A. A. Berryman, and G. T. Ferrell. 1996. Anthocorids associated with a manzanita gall induced by the aphid *Tamalia coweni* (Cockerell) (Homoptera: Aphididae). Canad. Entomol. 128: 839–847.

van den Bosch, R., and K. Hagen. 1966. Predaceous and parasitic arthropods in California cotton fields. California Agric. Exp. Stn. Bull. 820. 32 pp.

van den Meiracker, R. A. F. 1994. Induction and termination of diapause in *Orius* predatory bugs. Entomol. Exp. Appl. 73: 127–137.

van de Veire, M., and D. Degheel. 1995. Comparative laboratory experiment with *Orius insidiosus* and *Orius albidipennis* (Het: Anthocoridae), two candidates for biological control in glasshouses. Entomophaga 40: 341–344.

van Houten, Y. J., P. C. J. van Rijn, L. K. Tanigoshi, P. van Stratum, and J. Bruin. 1995. Preselection of predatory mites to improve year-round biological control of western flower thrips in greenhouse crops. Entomol. Exp. Appl. 74: 225–234.

Waloff, N. 1968. Studies on the insect fauna on Scotch broom *Sarothamnus scoparius* (L.) Wimmer. Adv. Ecol. Res. 5: 87–208.

Wang, C. L. 1994. The predaceous capacity of two natural enemies of *Thrips palmi* Karny, *Campylomma chinensis* Schuh (Hemiptera: Miridae) and *Orius sauteri* (Poppius) (Hemiptera: Anthocoridae). Plant Prot. Bull. (Taichung) 36: 141–154 [in Chinese].

Weber, P. W. 1953. Recent liberations of beneficial insects in Hawaii — II. Proc. Hawaiian Entomol. Soc. 15: 127–130.

Welty, C. 1995. Survey of predators associated with European red mite (*Panonychus ulmii*, Acari: Tetranychidae) in Ohio apple orchards. Great Lakes Entomol. 28: 171–184.

Wilde, W. H. A. 1965. The pear psylla, *Psylla pyricola* Foerster in Ontario (Homoptera: Chermidae). Proc. Entomol. Soc. Ontario 95: 5–10.

Wilson, L. T., and A. P. Gutierrez. 1980. Within plant distribution of predators on cotton: comments on sampling and predator efficiencies. Hilgardia 48(2): 3–11.

Wolcott, G. N. 1953. Control of the Cuban laurel thrips *Gynaikothrips ficorum*. J. Agric. Univ. Puerto Rico 37: 234–240.

Yasunaga, T. 1995. A new species of the genus *Wollastoniella* Reuter (Heteroptera: Anthocoridae), predator of *Thrips palmi* (Thysanoptera) in eggplant gardens of Thailand. Appl. Entomol. Zool. 30: 203–205.

Yasunaga, T., and S. Miyamoto. 1993. Three anthocorid species (Heteroptera: Anthocoridae), predators of *Thrips palmi* (Thysanoptera) in eggplant gardens in Thailand. Appl. Entomol. Zool. 28: 227–232.

Zaki, F. N. 1989. Rearing of two predators, *Orius albidipennis* (Reut.) and *Orius laevigatus* (Fieber) (Hem. Anthocoridae) on some insect larvae. J. Appl. Entomol. 107: 107–109.

Zheng, L. Y., and W. J. Bu. 1990. A list of Anthocoridae from China. Contrib. Tianjin Nat. Hist. Mus. 7: 23–27 [in Chinese].

Zimmerman, E. C. 1948. Insects of Hawaii. Vol. 3, Heteroptera. Univ. Hawaii Press, Honolulu, Hawaii, U.S.A. 255 pp.

Zlatanov, B. V. 1990. Are the predatory bugs harmful? Zashch. Rast. Moskva 1990. No. 6: 43–44.

CHAPTER 27

Damsel Bugs (Nabidae)

S. Kristine Braman

1. INTRODUCTION

The Nabidae, or damsel bugs, contains 31 genera and approximately 380 species (Lattin 1989, and references therein). Nabids, generalist predators, feed on a wide variety of small arthropods including both pest and beneficial species (e.g., Jervis 1990). They also probe plants, but are unable to complete development in the absence of prey (Stoner 1972). Probing the plant for moisture appears to do little injury and may help sustain the predator in periods of prey scarcity, although *Nabis alternatus* Parshley carries the plant pathogenic yeast *Nematospora coryli* (Burgess et al. 1983). Nabidae have a wide geographic distribution and vary considerably in habitats exploited. Much of the current knowledge stems from their prevalence in managed agroecosystems, but they are known to occupy many terrestrial habitats including arboreal environments (e.g., Larivière 1992a). An examination of 17 species of Nabidae in Romania revealed 4 species only from crop habitats, 3 only in the mountains, and 10 from a variety of habitats (Rosca 1982).

Position within the Heteroptera and family classification were reviewed by Lattin (1989). Nabidae have been placed within the superfamily Cimicoidea in the group Cimicomorpha (Leston et al. 1954, Schuh 1986, Schuh and Štys 1991). Morphological adaptations and trends in evolution of life-forms of Cimicomorpha of Kara Kum are discussed by Kaplin (1993). There is no clear consensus concerning organization of the major subdivisions of the family. The Velocipedinae and Medocostinae, for example, have sometimes been considered separate families (China and Miller 1959, Schuh 1986). However, Carayon (1970), Kerzhner (1981), and Péricart (1987) retained these groups within the Nabidae, and recognized four subfamilies (Nabinae, Prostemminae, Velocipedinae, and Medocostinae). Kerzhner further divided the Nabinae into four tribes, Nabini, Arachnocorini, Carthasini, and Gorpini.

The subfamily Nabinae contains most of the North American species, and the genus *Nabis* includes the most numerous and better known members of the family. Aerial dispersal of *N. capsiformis* Germar contributes to its wide distribution (Kerzhner 1983), although other nabid species have sometimes been mistaken for *N. capsiformis* (Kerzhner et al. 1982, Woodward 1982).

Confusion can arise with inconsistent use of scientific names. *Nabis roseipennis* Reuter, for example, is synonymous with *N. punctipes, Corsicus roseipennis,* and *Reduviolus roseipennis*. *Nabis americoferus* (Carayon) was previously referred to as *N. ferus* (L.) (also *Cimex ferus, Miris ferus, M. vagans, C. ferus,* and *R. ferus*). Carayon (1961) named the species found in America *americoferus* to distinguish it from the dissimilar European specimens of *N. ferus. Nabis rufusculus*

Reuter differs from *N. kalmii* Reuter primarily in distribution and degree of macroptery and may be the same species (Harris 1928, Mitri 1960).

Characters used in identifying species include such external morphological features as size and shape of the male claspers, interocular distances, degree of fuscous markings, color and patterns on the body, as well as the internal genitalia of males and females. Adults of selected species of agronomic importance may be identified using keys found in Benedict and Cothran (1975a), Dietz et al. (1976), Hormchan et al. (1976), Irwin and Shepard (1980), and Frank and Slosser (1991). Elvin and Sloderbeck (1984) provided a key to the nymphs of selected species of nabids in the southeastern United States.

2. NABIDS AS PREDATORS

Although much of the world literature is taxonomic (e.g., Carayon 1961, 1970; Kerzhner 1968, 1981, 1983, 1992; Larivière 1992a,b; Péricart 1987; Remane 1964), a growing body of literature has identified the Nabidae as among the most frequently encountered predatory Heteroptera in North American agroecosystems (Yeargan 1998). Among the most common Nabidae in North American agroecosystems are *N. americoferus, N. alternatus* Parshley, *N. roseipennis,* and *N. capsiformis* (Henry and Lattin 1988). Additional species that are less well represented in faunistic surveys include *N. rufusculus, N. kalmii,* and *Hoplistoscelis deceptivus* (Harris). Nabidae, with Lygaeidae, Anthocoridae, Pentatomidae, Miridae, and Berytidae, vary widely in relative abundance among such dominant field crops as corn, soybean, wheat, hay, and tobacco. Heteropteran predators in soybean, including *Nabis* spp., comprise 40 to 89% of the total predatory insects (Irwin and Shepard 1980). Benedict and Cothran (1975b) determined that species of *Orius, Nabis,* and *Geocoris* were the most abundant heteropterans in alfalfa in California (U.S.A.). Knowledge of predator impact and population estimations are necessary in allowing successful integration of natural mortality factors from predation into economic injury level assessments (Irwin and Shepard 1980). Nabids alone or together with other predators have been evaluated for their effectiveness in laboratory, greenhouse, and field tests against arthropod pests of crops including cotton, alfalfa, and soybean.

Nabis americoferus attacked eggs and pink bollworm early instars in laboratory tests and destroyed an average of 10.9 eggs per potted cotton plant in the greenhouse (Orphanides et al. 1971). Under these experimental conditions, nabids were less efficient than several other predators tested. Spatial location, searching activity, and effectiveness in reducing numbers of first instar bollworm on cotton of *N. roseipennis* were examined in the greenhouse and laboratory by Donahoe and Pitre (1977). Adults were located primarily in the upper third of the plant on the undersurface of leaves. The greatest activity was during crepuscular hours with peaks at 7:00 to 8:00 P.M. and 7:00 to 8:00 A.M. First, third, and fifth instar nabids consumed 1.9, 7.9, and 38.2 larvae per day and feeding effectiveness decreased with an increase in number of predators or an increase in plant age or size. Acceptability of tobacco budworm larvae to *N. roseipennis* varied with age of predator and prey (Nadgauda and Pitre 1986). Fifth instar and adult nabids were able to feed on early to mid-fourth instars, but not on larger larvae. Third and fourth instars consumed larvae as old as the mid-third stadium, whereas first and second instars *N. roseipennis* could kill only first and second instars of the prey.

Field studies of predators of the bollworm in cotton demonstrated that piercing and sucking insect predators, including nabids, destroyed more eggs than chewing predators (Bell and Whitcomb 1964) and that *N. americoferus* reduced caged larval populations by 77.9% (100 nabids per cage) or 85% (200 nabids per cage) (van den Bosch et al. 1969). Field observations of 1605 bollworm second instars released in cotton revealed that nine individuals were preyed upon by nabids (Whitcomb 1967). When six predator species were tested in field cages, *N. americoferus* and *Geocoris pallens* Stål (Lygaeidae) showed the best potential as pink bollworm egg predators on cotton (Irwin et al. 1973). Efficiency of *N. americoferus*, *G. pallens*, and *Chrysopa carnea* Stephens (Neuroptera), alone and in combination, was evaluated in field cages on cotton with *Lygus hesperus* as prey (Leigh and Gonzales 1976). *Nabis americoferus* was effective in sleeve cages but not in large field cages in those experiments. Consumption of pink bollworm eggs in the field was highest for *Collops vittatus* (Say) (Coleoptera), *Hippodamia convergens* Guérin-Méneville, and *Chrysopa carnea*, followed by *Nabis* spp., *Sinea confusa* Caudell, *Geocoris* spp., and *Orius tristicolor* (White) (Henneberry and Clayton 1985). Hagler and Naranjo (1994) determined the frequency of heteropteran predation on sweetpotato whitefly and pink bollworm in Arizona cotton using ELISAs. *Orius tristicolor* and *G. punctipes* tested positive for *Bemisia* sp. eggs much more often than *N. alternatus*. Interspecies association among cotton arthropods was determined on 31 sampling dates over 6 years in nine fields in New Mexico (U.S.A.) (Ellington et al. 1997). *Nabis* spp. were found associated with other genera, followed in abundance by *Geocoris* and *Orius* and nine other numerically prominent arthropod taxa. Predators were associated with various primary consumers 163 times and with other predators 191 times, indicating that switching may readily occur and that predators feed on other predators.

Nabis roseipennis was the most frequently labeled predator in an autoradiographic study of predaceous arthropods in soybean (McCarty et al. 1980). Predators of nabids and big-eyed bugs included several species of spiders. Nabids were shown also to be cannibalistic on conspecific nymphs. More than 50% of *Geocoris* and *Nabis* collected in soybean in Blackville, South Carolina (U.S.A.), tested positive for velvetbean caterpillar in precipitin tests designed to evaluate predator–prey encounters (Gardner et al. 1981).

Nabids, chrysopids, and spiders were recorded as velvetbean caterpillar predators most frequently in Florida soybean when using the radiographic tracers, whereas ants and earwigs were the predators spotted most often during continuous observations (Buschman et al. 1977). Elvin et al. (1983) demonstrated that total mortality inflicted by predators on velvetbean caterpillar larvae in soybean can be the same for predator complexes of different species composition. At one study site, nabids consumed the greatest proportion of larvae, and at another site imported fire ants were the major consumers; yet at both sites similar total mortalities were recorded.

In a comparison of *N. americoferus* and *N. roseipennis* as predators of the green cloverworm, consumption rates ranged from one egg or fifth instar per day for predator first instars to 12 to 26 eggs or first instars for predator fifth instars and adults (Sloderbeck and Yeargan 1983). The two species were similar in their predatory capabilities, acceptability of prey varying with age of prey and predator, as noted by Nadgauda and Pitre (1986) for the tobacco budworm. Using potted soybean plants, mortality inflicted on eggs and on first and second instar green cloverworms was measured when *N. roseipennis* and *O. insidiosus* Say searched either alone or in combination (Clements and Yeargan 1997a,b). *Nabis roseipennis* always consumed far more prey than *O. insidiosus* when the predators were tested separately. Mortality inflicted by the two predators when present together as adults was found to be additive. Combined predation by the two species present in stages other than adult did not differ from that observed for *N. roseipennis* alone.

Field experiments demonstrated the ability of *N. roseipennis* to reduce populations of green cloverworm (Braman and Yeargan 1989). The seasonal abundance and within-plant distribution of *N. roseipennis* overlap with early instars of the green cloverworm, suggesting its practical potential as a predator. Further, *N. roseipennis* nymphs successfully reduced numbers of green cloverworm larvae in the field in the presence of alternative prey under minimally modified conditions. Starting densities of 20 to 25 third instar predators per 0.9 m row plot resulted in up to 50% reduction in subsequent green cloverworm populations.

Pedigo et al. 1972 concluded that predation by *O. insidiosus* and *Nabis* spp. was the primary cause of differential mortality between exposed and protected populations of green cloverworms in small plots in Iowa. In a life-table analysis of the green cloverworm, however, late-season epizootics of *Nomuraea rileyi* (Farlow) were deemed the major regulating factors, although parasites and predators were important mortality factors during years of endemic populations (Pedigo et al. 1983).

Nabis americoferus was one of three predators evaluated alone and in combination in a greenhouse study of predation on three insect pest species (Tamaki and Weeks 1972). Nabids alone or in some combination were more effective than all other treatments in reducing populations of noctuid larvae when these larvae were small. Coccinellids, however, were much more efficient aphid predators. Plant density and temperature played an important role in predator–prey interactions in field cages with *Myzus persicae* (Sulz), *N. alternatus*, and *G. bullatus* (Say) (Tamaki et al. 1981).

All stages of *Lygus* spp. were effectively preyed upon by *N. alternatus* (Tamaki et al. 1978). Females consumed more prey, especially when prey were large. Adult nabids consumed a daily average of 0.23 adults to 4.41 second instars. Analysis of partial life tables of western yellow-striped armyworm, *Spodoptera praefica* (Grote), in hay alfalfa in California (U.S.A.) revealed that most real mortality was inflicted by a complex of polyphagous predators including *N. americoferus* and *N. alternatus* (Bisabri-Ershadi et al. 1981).

Propp (1982) determined that although attack rates of *N. americoferus* using *L. hesperus* Knight or *S. exigua* (Hübner) as hosts varied with prey type and age and complexity of the experimental universe, the rates were highest and handling times were lowest for reproductively mature females. Prereproductive females had the lowest attack rates and highest handling times of all stages. With the exception of prereproductive females in a simple universe, the fit to Holling's (type 2) functional response model was good. Numerical response of aphid predators to varying prey densities in alfalfa was examined by Evans and Youssef (1992). Specialist coccinellids were strongly positively correlated with aphid densities, whereas the generalist nabids and lygaeid predators showed an absence of such a response.

Potato leafhopper eggs, nymphs, and adults were preyed upon under laboratory conditions by *N. americoferus* (Martinez and Pienkowski 1982) but *O. insidiosus* was the more efficient egg predator in those experiments. Adult nabids killed an average of 2.2 nymphs per day and 1.9 adult leafhoppers per day. Nabids may be important predators of leafhoppers in alfalfa; however, they are reported to prefer pea aphids and may switch to this prey, thus reducing their effectiveness as predators of the potato leafhopper (Flinn et al. 1985). Predation rates of nymphal and adult male and female *N. roseipennis* were determined using nymphal and adult potato leafhoppers as prey (Rensner et al. 1983). Female nabids preyed upon more leafhoppers than did male nabids. The number of leafhoppers killed per day during that study were two to four times higher than those reported for the smaller *N. americoferus* by Martinez and Pienkowski (1982). The Mexican bean beetle, a pest of soybean, was also preyed upon by *Nabis* spp. (Waddill and Shepard 1974); eggs and first, second, and third instars were accepted as prey by *N. roseipennis* in laboratory tests.

Field evaluations of the effectiveness of nabids against noctuid pests in soybean were conducted by Barry et al. (1974), Richman et al. (1980), and Reed et al. (1984). These studies were conducted in large field cages where often both the pest and the predator were introduced into the field cage and the resultant impact recorded. In all cases nabids had higher predation rates than other predators tested. High predation rates per predator combined with numerical abundance suggest the potential value of these predators to growers.

Mack et al. (1987) found that *Nabis* spp. and *Geocoris* spp. were the most abundant foliar-dwelling predators in peanut fields in Alabama (U.S.A.). Nabids are also components of the predator complex on tobacco (Semptner and Roberts 1978), but heteropteran predators are less common on that crop than on corn, soybean, or cotton (Roach 1980). Species of Chrysomelidae, including the Colorado potato beetle, have also been reported as prey of nabids (Burgess 1982, Culliney 1986, Lattin 1989). Ground-living predators, including Nabidae, helped slow the growth of apple aphid populations in apple orchards in central Washington (U.S.A.) (Carroll and Hoyt 1984). Anthocorids, mirids, and nabids composed the bulk of predaceous Heteroptera important in the management of pear psyllid, *Psylla pyri* (L.) in southeastern France (Severin et al. 1984). Anthocorids were the most abundant group. It was suggested that a mean of one predaceous bug collected per branch-beat sample could indicate adequate natural control of psyllids. The most numerous predatory insects found in the nests of the arctiid, *Hyphantria cunea* (Drury), belonged to the families Nabidae, Pentatomidae, Chrysopidae, and Coccinellidae. Predators caused 5 to 62% mortality of caterpillars, spiders giving the most effective control (Bel'skaya et al. 1985).

Nabis punctatus Costa was common in alfalfa, clover, sunflower, soybean, chamomile, kenaf, and various horticultural crops in central Italy (Czepak et al. 1994). Eggs were most frequently deposited in the petioles of upper and middle portions of plants. In sunflower fields, 12.1% of the nabid eggs were parasitized by the scelionid *Telenomus* sp. and the mymarid *Polynema* sp. In soybean, 6.2% of the egg masses of the pentatomid *Nezara viridula* (L.) were preyed upon by *Nabis punctatus*. In Turkey, predation on the larval stage of *Ostrinia nubilalis* Hübner ranged from 21 to 52% during 1989 and 1990 (Kayapinar and Kornosor 1993). Chiappini and Reguzzi (1994) further discuss the morphology, biology, and ecology of this species.

Nabis species were among the main predators responsible for reducing aphid numbers on hops, as determined using exclusion studies (Campbell and Cone 1994). Cantharids and nabids (*N. ferus* and *N. pseudoferus*) were important aphidophagous predators in winter wheat in Germany (Lobner and Hartwig 1994). In feeding trials, 10 to 15 aphids per day were consumed by nabids. Nabidae was the most common group of polyphagous predators of aphids on wheat in Romania (Banita et al. 1995), comprising 36% of the beneficial fauna and a 1:33 ratio with prey. The role of five nabid species, *N. limbatus, N. flavomarginata, N. ferus, N. punctatus,* and *N. brevis*, was significant in rye, where they constituted the major group of predators and hindered early-season aphid population increase (Krotova 1991). First, second, and third instars and adult *N. sinoferus* Hsiao consumed 19, 79, 141, and 189 aphids, respectively (Fan and Mu 1989). However, Rosenheim et al. (1993) determined that the addition of *Nabis* spp. or *Zelus renardii* to a guild of predators caused sufficient predation on the predator *Chrysopa carnea* to release *Aphis gossypii* from regulation by *C. carnea*.

Predation of lepidopteran pests in vineyards in southwestern Germany was detected by ELISA in 15.1% of nabids, *Aptus mirmicoides* (O. Costa), examined (von Buchholz et al. 1994). *Aptus mirmicoides* successfully preyed upon large larvae of *Eupocilia ambiguella* (Hbn) (von Buchholz and Schruft 1994), and other predators attacked different stages. *Nabis pseudoferus* was among the most common predators in the insect fauna of sugarbeet in Turkey (Yildirim and Ozbek 1992), and was one of four main predators of *Mamestra brassicae* (L.) eggs, although natural control was insufficient to keep the pest from exceeding the damage threshold (Injac and Krnjajic 1990). *Nabis alternatus* was the most important insect predator of black grass bugs *Labops hesperius* in Utah (U.S.A.) grasslands (Araya and Haws 1991).

The role of nabids in regulation of pest populations of importance to urban agriculture remains largely unknown. Reinert (1978) reported a species of *Pagasa* among the natural enemy complex of the southern chinch bug, an important turfgrass pest. Leddy (1996) examined the role of plant architectural complexity on predator abundance and azalea lace bug, *Stephanitis pyrioides* (Scott), mortality. Generalist predators, including Nabidae, were more abundant in complex habitats than in simple landscape habitats for the vast majority of dates sampled, and for the season total.

3. BIOLOGY, ECOLOGY, AND BEHAVIOR

Aspects of the biology and behavior of many nabid species were described briefly by Harris (1928). Mundinger (1922) described the life history of *N. roseipennis* and *N. rufusculus*. Kerzhner (1983, 1992) and Péricart (1987) offer summaries of the family Nabidae and provide extensive bibliographies, as does Lattin (1989). Nabids are all terrestrial and usually foliar inhabitants, although there are some ground-dwelling species and inhabitants of semiaquatic environments. Nabids insert their eggs into plant tissue with only the operculum remaining visible. The five instars typical of Heteroptera occur in most species studied, except *Nabicula limbata*, where four have been reported. Nabids in general probably overwinter as adults in reproductive diapause under debris left in the summer habitat. Some *Nabis* spp. may also overwinter as eggs (Harris 1928).

The chromosome numbers of many nabids have been characterized (Leston 1957; Mikolajski 1965, 1967). The degree of genetic differentiation among local populations of *N. alternatus, N. roseipennis,* and *N. americoferus* was determined by Grasela and Steiner (1993) in an electrophoretic study.

3.1 Developmental and Reproductive Biology

Developmental times and/or reproductive potential of several *Nabis* spp. have been determined under controlled laboratory conditions (Perkins and Watson 1972, Hormchan et al. 1976, Braman et al. 1984, Guppy 1986, Nadgauda and Pitre 1986, Braman and Yeargan 1988). *Nabis americoferus* develops more rapidly than *N. roseipennis* when fed the same diet under identical temperature conditions. Development of these two species required a minimum of 18.6 days at 33°C to a maximum of nearly 4 months near the developmental threshold (Braman et al. 1984). *Nabis americoferus* fed alfalfa blotch leaf miners at constant 23°C developed more rapidly than those reared on pea aphids, 25.8 vs. 28.8 days, respectively. Females collected from the field had a mean fecundity of 157 ± 11.2 eggs per female, 7 to 10 day preoviposition period, and an 18.6 ± 1.2 day oviposition period (Guppy 1986).

Nymphal developmental periods of *N. roseipennis* fed tobacco budworm larvae or tarnished plant bug nymphs ranged from 27.3 to 30.5 days (30 to 20°C) and 24.4 to 33.2 days on the respective diets. Mean fecundities at different temperatures and when fed either tobacco budworm larvae or plant bugs were 22 to 52 eggs per female or 28 to 66 eggs per female (Nadgauda and Pitre 1986). *Nabis capsiformis* and *N. alternatus* have average nymphal durations very similar to those of *N. americoferus* (Taylor 1949, Perkins and Watkins 1972, Hormchan et al 1976).

Nabis rufusculus completed development at temperatures ranging from 18 to 33°C (Braman and Yeargan 1988). Successful development did not occur at 15°C. Time required to complete development ranged from 82.8 to 27.1 days and was similar for males and females. An alternating temperature regime of 21 and 33°C (mean of 27°C) resulted in slower development than that which occurred at constant 27°C because the upper temperature of the alternating regime was detrimental. Development under an alternating regime of 12 and 24°C with a mean of 18°C, however, was accelerated over that which occurred at a constant 18°C. Developmental thresholds calculated for egg, nymphal, and total development of *N. rufusculus* were 13.1, 11.6, and 11.9°C, respectively. *Nabis rufusculus* develops at a rate similar to *N. roseipennis* and more slowly than *N. americoferus*. Total egg production was similar for all three species at an optimal temperature (21°C), but *N. americoferus* deposited eggs over a shorter period of time than did the other two species. *Nabis americoferus* has a minimal weight threshold for egg production (O'Neill 1992) and is more efficient at producing eggs than *N. roseipennis*.

Nabis stenoferus developed more rapidly with increasing temperature from 20 to 28°C, but a delay in development was seen at 30°C (Kitamura and Kondo 1995). Threshold temperatures for development of eggs and nymphs were 13.3 and 13.5°C, respectively. Thermal unit requirements were 321.1 degree-days from egg to adult. Survival was optimized at 24°C. Base temperature for development and thermal unit requirements for the eggs of *N. sinoferus* Hsiao were 19.2°C and 15.9 degree-days. The threshold for nymphal development was 18.3°C, and 121.5 degree-days were required for development (Fan and Mu 1989).

Consumption by adult *N. roseipennis* of *Pseudoplusia includens* larvae infected with nuclear polyhedrosis virus significantly reduced the predator's longevity and reduced total fecundity. Nymphal diet (infected vs. uninfected larvae) impacted nymphal growth rate and daily oviposition rates of subsequent adults (Ruberson et al. 1991). Nymphs that fed on virus-infected larvae excreted the virus for several days (Young and Yearian 1992). Velvetbean caterpillar larval mortality was similar whether or not the source of infection was a diseased caterpillar or a contaminated fifth instar nabid. *Nabis roseipennis* preferred virus-infected to healthy larvae (Young and Kring 1991) in petri dishes and on soybean, presumably because of a reduction in defensive responses of prey.

Richards and Harper (1978) found *N. alternatus* eggs most commonly inserted in the main alfalfa stem at stem widths of 1.5 to 1.7 mm. When stems were over 50 cm ong, however, most eggs were in branches and petioles, possibly because of the increased hardness or width of the main stem. The distribution of *N. roseipennis* eggs in greenhouse-grown soybean (Pfannenstiel and Yeargan 1998) closely approximated the distribution of eggs observed in the field (Isenhour and Yeargan 1982). Eggs deposited in soybean were generally in the petioles, and the lower third of the soybean plant was generally avoided.

The critical photoperiod for induction of reproductive diapause of *N. americoferus* and *N. roseipennis* at 24°C is between 13.5 and 14.0 hours light per day (Yeargan and Barney 1996). The nymphal stage is the most sensitive to diapause-inducing photoperiods. The adult stage must continue to experience a short photophase for diapause to continue, conditions that individuals experience in the field beginning in late summer.

3.2 Occurrence and Abundance, Dispersion, and Impact of Parasitoids

Estimates of seasonal population trends for nabids have often been reported at the generic level. Reports regarding number of generations per year for particular nabid species are scarce, and are usually based on studies in perennial plant ecosystems. Stoner et al. (1975) reported that *N. americoferus* and *N. alternatus* probably have five generations per year in alfalfa in Arizona (U.S.A.). The sex ratio of both species was approximately 1:1, with perhaps a slight majority of females. Guppy (1986) reported that *N. americoferus* produces two generations in eastern Ontario (Canada). Wheeler (1977) also noted the appearance of two generations in alfalfa in New York State (U.S.A.).

Studies of predator abundance necessarily concentrate on population events that occur within the growing season for that crop. *Nabis* spp. were the most abundant of the three predatory insect groups studied in cotton in Mississippi (U.S.A.) (Dinkins et al. 1970), *N. roseipennis* being more commonly collected than *N. alternatus* or *N. americoferus*. The three *Nabis* spp. were most abundant in June. Cotton and soybean fields were sampled in distinctly different ecological habitats in Mississippi by Pitre et al. (1978). Populations of beneficials, including nabids, reached highest levels in mid-June to mid-July and were appreciably higher in the Blackbelt, where fields were smaller and were surrounded by woodland and grassland vegetation. Damsel bugs collected included *N. deceptivus* Harris, *N. sordidus* Reuter, *N. capsiformis, N. alternatus,* and *N. roseipennis*. Among four specific crop habitats (soybean, corn, tomato, and tobacco), nabids were associated primarily with soybean; only small numbers were found in the other crops (Pfannenstiel and Yeargan 1998b).

Predaceous heteropterans are usually more abundant than all other insect predators in soybean and are primarily composed of the genera *Nabis, Orius,* and *Geocoris* (Irwin and Shepard 1980). Percent capture varies with location, time of season, and sampling method. *Nabis* spp. composed the largest group of predators in seasonal samples taken in South Carolina soybean (Shepard et al. 1974). Populations of nymphs occurred as a bimodal curve, indicating two generations; and populations peaked in September. In Brazilian soybean, *Nabis* spp. were the most common predators throughout Brazil and were most abundant in February (Correa et al. 1977). In the Backa region of Serbia *Lygus* and *Nabis* were the most numerous heteropterans in soybean, comprising 90% of the 2787 individuals collected (Keresi 1993). *Nabis feroides, N. ferus,* and *N. pseudoferus* were represented in the samples.

In a survey of sunflower fields in eastern South Dakota (U.S.A.), *O. insidiosus, Nabis* spp., *Chrysopa* spp., and four coccinellid species were 78% of the foliage-dwelling predators (Royer and Walgenbach 1991). Within the *Nabis* complex of central Iowa soybean, *N. americoferus* made up 55.9% of adults from sweep samples, *N. roseipennis* 23.3%, *N. alternatus* 8.6%, and *N. kalmii* 4.9%. Populations were particularly abundant during pod-fill (Bechinski and Pedigo 1981).

The seasonal abundance and phenology of three *Nabis* spp. and their parasitoids were studied in alfalfa and soybean in central Kentucky (U.S.A.) (Braman and Yeargan 1990). Although all three species were present in both crops, there were differences. *Nabis americoferus* was numerically dominant in alfalfa, where this species produced three generations per year. In soybean, where one generation is produced, *N. roseipennis* rather than *N. americoferus* or *N. rufusculus* produced the most nymphs for three consecutive years. Sweep sampling recovered fewer nymphs than suction or shake sampling and, thus, was less satisfactory in assessing population trends. Visual and suction sampling gave equal results for *N. capsiformis* in tomato in Brazil (Raga et al. 1990). Sampling methods were evaluated for *Nabis* spp. in alfalfa (Gyenge et al. 1997a) and a sequential sampling plan developed (Gyenge et al. 1997b). Iwao's (1968) regression procedure applied to shake samples in soybean (Iwao et al. 1981) showed that nymphal *Nabis* spp. populations consisted of groups of individuals that are slightly aggregated. Adult *Nabis* spp, however, existed as single individuals randomly distributed. A sequential count plan for use in a pest management survey was developed. Differences in within-plant distribution of nabid adults and nymphs of different species helped to explain why these predators are sampled with varying degrees of efficiency (Wilson and Gutierrez 1980, Braman and Yeargan 1990, Clements and Yeargan 1997b). In cotton, for example, adult nabids consistently occurred higher in the canopy than nymphs and were captured more frequently in sweep net samples (Wilson and Gutierrez 1980).

Seven species of *Nabicula* occur in Canada (Larivière 1994). Relationships between macroptery and geographic distribution, ecological assemblages, and keys to all taxa were presented in Larivière (1994). Distribution and biology of eight species of Irish Nabinae are discussed by Roth and O'Connor (1993).

Parasitoids reared from nabids that had been collected as adults or nymphs in alfalfa or soybean included the tachinid *Leucostoma simplex* (Fallén) and the braconid *Wesmaelia pendula* Foerster (Hendrick and Stern 1970, Stoner 1973). Parasitism by the mymarid egg parasitoid *Polynema boreum* Girault was much higher in alfalfa than in soybean (50 vs. 5%) (Braman and Yeargan 1990). *Nabis* spp. deposited the majority of their eggs in the apical 30 cm of alfalfa in Arizona (U.S.A.), 35% of which were parasitized by *P. boreum* (Graham and Jackson 1982). This parasitoid inserts its eggs through the plant tissue and into the host egg (Hendrick and Stern 1970). Development of the mymarid requires 20 to 23 days at 21°C. *Leucostoma simplex* and *W. pendula* attack late instars and adults. *Wesmalia pendula* emerges primarily from males, whereas *L. simplex* attacks either sex. In alfalfa in central California (U.S.A.), *P. boreum* attacks up to 70% of the *N. americoferus* eggs deposited in September, whereas *L. simplex* parasitizes a maximum of 12% of the adults in June and 12% of the nymphs in early August (Benedict and Cothran 1978). Peaks in parasitism by *P. boreum* in Arizona (U.S.A.) were 20 to 45% in lucerne grown for hay and 75% in lucerne grown for seed (Lakin et al. 1984). Mean progeny production was 31.2 when *N. alternatus* was the host and 23.6 when *N. americoferus* was parasitized.

3.3 Cultural Practices That Affect Nabid Distribution and Abundance

Nabis spp. were more abundant in narrow-row soybean (Sprenkel et al. 1979, Ferguson et al. 1984). Double-cropped fields apparently support fewer nabids than conventionally plowed or cultivated or drilled (narrow row) fields until after the beans have fully formed (McPherson et al. 1982, Ferguson et al. 1984). Buschman et al. (1984) reported that nabid populations were not so strongly affected by cultural practices as are lepidopteran larvae, yet the predators appeared to be slightly more abundant in later-planted and later-maturing cultivars. Although *Geocoris* and spider populations were higher in narrow-row soybean, the authors were unable to observe a clear response to row spacing in populations of nabids or coccinellids. Sprenkel et al. (1979), however, reported more immature nabids in early-planted, narrow-row treatments than in any other combination of planting date or row width. When soybean was planted in every combination of early and late planting and narrow- and wide-row spacing, none of the most common predators, *Nabis* spp., *Geocoris* spp., *O. insidiosus*, and spiders, was substantially affected by canopy closure (Anderson and Yeargan 1998). *Nabis* species were more abundant in soybean habitats with grasses and mixed weeds than in weed-free soybean or soybean with broadleaf weeds alone (Shelton and Edwards 1983). Slight increases in abundance of *N. americoferus* were observed in bean fields when adjacent hay fields were cut (Stoltz and McNeal 1982). Proximity to seminatural habitats influences time of colonization and relative production rates in agricultural habitats (Katz et al. 1989).

Annual and perennial legumes were evaluated as potential ground covers to supply nitrogen and increase beneficial arthropods in Oklahoma (U.S.A.) pecan orchards (Smith et al. 1994). Density of lady beetles correlated positively with that of aphids that colonized legumes each year. Other predators, including *Nabis* spp., were either not abundant or their densities did not correlate with those of the aphids. Alfalfa interplanted with cotton has been used to manage lygus bugs in cotton (Godfrey and Leigh 1994). It was concluded that to optimize *Lygus hesperus* management, while still producing high levels of nabids, strip-cut alfalfa should be cut on a 28-day cycle.

Nabis roseipennis and *N. americoferus* were more abundant in dicultures of corn and beans than in monocultures (Coll and Bottrell 1995). Mortality of Mexican bean beetle larvae was density independent in *Phaseolus vulgaris* L. monocultures and inversely density dependent in dicultures. Irrigation and soil fertility may influence nabid abundance in agricultural habitats. *Nabis alternatus* was more abundant in alfalfa under sprinkler than with flood irrigation (Schaber and Entz 1994). More of *N. alternatus* were also observed in cotton plots irrigated weekly than biweekly (Flint et al. 1994). Plant water stress, as measured by leaf water potential, was 13% greater at 13 days than at 7 days in plants irrigated biweekly. High-phosphorus application to soybean favored dense populations of nymphs of *Nabis* spp. and *Hoplistoscelis* spp. in Florida (U.S.A.) during the first year of a 2-year study (Funderburk et al. 1994), but high potassium and magnesium did not. It was concluded that increased outbreaks of pests from overfertilization of soybean were not the result of fertility effects on population densities of nabids, spiders, or big-eyed bugs.

Nabids have been the subjects of laboratory screening of insecticides (Lingren and Ridgway 1967, Hamilton and Kieckhefer 1969, Martinez and Pienkowski 1983, Chatenet and Zhu 1985), but most available data are from field tests conducted primarily in alfalfa, cotton, and soybean. Hamilton and Kieckhefer (1969) found that *N. americoferus* was less sensitive to malathion and parathion than was the English grain aphid and suggested the use of selected dosages of these insecticides in aphid control. Bidrin, methyl parathion, demetron, phosphamidon, and trichlorfon were all extremely toxic to *N. americoferus*, *G. punctipes*, and *O. insidiosus*, while displaying varying levels of toxicity to convergent lady beetles and green lacewings in laboratory studies (Lingren and Ridgway 1967). Six insecticides registered for use on alfalfa were determined to be less toxic to *N. americoferus* than to the potato leafhopper in alfalfa (Martinez and Pienkowski 1983). Tarnished plant bugs displayed an intermediate susceptibility and were suggested as an alternate prey because plant bugs might be useful in conserving beneficials when potato leafhoppers were reduced by insecticides.

Field tests in alfalfa have shown adverse effects of several insecticides including trichlorfon, parathion, DDT, DDT-toxaphene, carbaryl, mevinphos, methidathion, dimethoate, methomyl, and oxamyl (Stern et al. 1959, Stern 1963, Summers et al. 1975, Harper 1978). Some of these insecticides proved to be less disruptive than other chemicals, namely, heptachlor, *Bacillus thuringiensis*, demetron, malathion, and primicarb. In one case, entomophagous species became more abundant in fungicide-treated plots, probably because aphid populations increased in those plots (Radcliffe et al. 1976).

Secondary outbreaks of cabbage loopers or beet armyworms in cotton following insecticide applications (Falcon et al. 1968, Eveleens et al. 1973, Ehler et al. 1973, Stoltz and Stern 1979) were attributed to reductions in predator populations, including *Nabis* spp. The insecticide used most often in these experiments was dimethoate. In one case an increase in *N. americoferus* mortality from untreated control to aldicarb to dimethoate was reflected in proportional increases in seasonal larval populations of the cabbage looper (Stoltz and Stern 1979).

Nabis spp. were unaffected by applications of diflubenzuron (Keever et al. 1977, Deakle and Bradley 1982). When carbaryl, endrin, azinophosmethyl, malathion, and toxaphene were compared, toxaphene proved the most toxic to *Nabis* spp. and azinophosmethyl the least toxic (Laster and Brazzel 1968). Adkisson (1958) reported that neither heavy application of fertilizer nor weekly applications of endrin produced significant decreases in *Nabis* populations.

In soybean, Shepard et al (1977) suggested that a reduction in arthropod predators by foliar applications of insecticides was responsible for resurgence of lepidopteran pests; aldicarb-caused reduction in heteropteran predators, including *Nabis* spp., was similar to the effects of methyl parathion and resulted in as much as a sevenfold increase in corn earworm larvae compared with controls (Morrison et al. 1979). Although similar results were noted in one experiment by Lentz et al. (1983), no significant reductions in *Nabis* spp. were observed in three other experiments using eight nematocides including aldicarb. Failure to demonstrate consistent resurgence in late-season pest populations was also reported by Farlow and Pitre (1983).

Toxaphene applied as an herbicide for sicklepod control produced no measurable effects on *Nabis* spp. (Huckaba et al. 1983). However, populations of *N. roseipennis* decreased with applications of toxaphene to weeds in Florida soybean (Domiciano and Herzog 1990). Whitcomb and Baker (1974) reported that the effect of toxaphene and parathion was similar to that of parathion alone, and that some concentrations of methyl parathion, methomyl, and carbaryl that reduced or failed to reduce numbers of *Nabis* spp. and *Geocoris* spp. Endosulfan alone or with dimethoate was the most injurious to *Nabis* spp. and *Geocoris* spp. followed by trichlorfon among six pesticides examined for control of soybean pests in Brazil (Lorenzato and Corseuil 1982). *Orius* spp. and *Nabis* spp. were significantly reduced by most of the 11 organophosphate and synthetic pyrethroid insecticides examined in alfalfa by Rotrekl (1994). Low numbers of predators and parasitoids 2 weeks after insecticide applications appeared to be the result of both direct effect of pesticides and a reduction in aphid prey.

Lambda cyhalothrin and monocrotophos reduced populations of *N. capsiformis* for approximately 24 days on soybeans in Brazil (White et al. 1992), but it was concluded the effects on predators and parasitoids were transient and would not reduce the potential for natural enemies to control subsequent infestations of *Anticarsia gemmatalis* Hübner on soybeans. Toxicity of selected insecticides to the hemipteran predators *G. punctipes, N. capsiformis, N. roseipennis,* and *Podisus maculiventris* (Say) was evaluated by contact with foliar residues and indirectly through the consumption of soybean looper, *Pseudoplusia includens* (Walker), previously exposed to insecticides (Boyd and Boethel 1998a). Methyl parathion and permethrin generally were more toxic than newer insecticides after predators were exposed to treated foliage. Chlorfenapyr caused contact toxicity equal to permethrin and methyl parathion. Exposure to foliage treated with emamectin benzoate resulted in lower mortality as compared with chlorfenapyr. Foliage treated with *Bacillus thuringiensis* Berliner subsp. *kurstaki* had the lowest contact toxicity to hemipteran predators of all insecticides tested. Standard insecticides (i.e., methyl parathion and thiodicarb) caused low indirect toxicity to hemipteran predators after consumption of treated prey. Chlorfenapyr caused significantly greater indirect toxicity than emamectin benzoate, permethrin, and thiodicarb to adult *N. roseipennis*. Consumption of chlorfenapyr-treated prey also caused significantly greater mortality than imidacloprid, permethrin, spinosad, and thiodicarb to *G. punctipes* adults. Duration of chlorfenapyr residues was evident up to 72 hours after application (Boyd and Boethel 1998b).

Oien and Ragsdale (1993) examined the susceptibility of nontarget hosts to a microsporidian, *Nosema furnacalis*, a potential biocontrol agent for European corn borer; none of the nine predators tested, including *Nabis americoferus,* displayed any active infection. Of the 424 adult predators fed spores or infected larvae, only 2 tested weakly positive with ELISA.

4. FUTURE WORK AND CONCLUSIONS

Clearly the Nabidae play an important role in the natural suppression of pest populations in managed agroecosystems. Their polyphagous habit, sometimes considered a detriment to their usefulness in control programs, may also allow them to persist in times of low target pest density. Although the individual contribution of nabids to natural control is of interest, nabids generally occur as members of a guild of heteropteran predators. In this context additional research defining actual and potential impact of these predators would be particularly useful. Although intrinsic rates of increase and numerical responses of varying nabid species are not well known, it is generally understood that the capacity for population growth of most agricultural pest species far exceeds that of most nabid species, limiting their usefulness in outbreak situations. Mortality inflicted by heteropteran guilds, including nabids, on early growth stages of lepidopteran pests in particular, represents an important contribution to pest management in sustainable systems.

Fundamental knowledge of the nabid species of many areas of the world is still lacking, particularly in the tropics. There continues to be a particular need for biological information on tropical nabid species, as Lattin (1989) has indicated. The many papers published on aspects of the biology of nabid species worldwide indicate their importance as predators in a diversity of managed settings. Evidence suggests that nabids respond to habitat manipulation, providing direction for future research. What biotic and abiotic factors influence habitat selection by nabids and affect their capacity for population increase? Very little is known about the potential of nabids as predators in urban agricultural settings. Certainly, in urban situations, where human populations are concentrated, mandates to produce alternatives to traditional pesticide use provide impetus to examine the role of these and other candidate natural enemies in ornamental plant production and landscape maintenance.

5. REFERENCES CITED

Adkisson, P. L. 1958. The influence of fertilizer applications on populations of *Heliothis zea* (Boddie) and certain insect predators. J. Econ. Entomol. 51: 757–759.

Anderson, A. C., and K. V. Yeargan. 1998. Influence of soybean canopy closure on predator abundances and predation on *Helicoverpa zea* (Lepidoptera: Noctuidae) eggs. Environ. Entomol. 27: 1488–1495.

Araya, J. E., and B. A. Haws. 1991. Arthropod populations associated with a grassland infested by black grass bugs, *Labops hesperius* and *Irbisia brachycera* (Hemiptera: Miridae) in Utah, U.S.A. FAO Plant Prot. Bull. 39: 75–81.

Banita, E., D. Searpe, F. Vilau, M. Voicu, B. Kis, R. Serafim, M. Sterghiu, and E. Luca. 1995. Cercetari privind relatiile dintre populatiile de afide si speciile de pradatori polifagi la graul de Toamna din Campia olteniei. Prob. Prot. Plant. 23: 231–245.

Barry, R. M., J. H. Hatchett, and R. D. Jackson. 1974. Cage studies with the predators of the cabbage looper, *Trichoplusia ni* and corn earworm *Heliothis zea* in soybeans. J. Georgia Entomol. Soc. 9: 71–78.

Bechinski, E. J., and L. P. Pedigo. 1981. Ecology of predaceous arthropods in Iowa soybean agroecosystems. Environ. Entomol. 10: 771–778.

Bell, K. O., and W. H. Whitcomb. 1964. Field studies on egg predators of the pink bollworm, *Heliothis zea* (Boddie). Florida Entomol. 47: 171–180.

Bel'skaya, E. A., A. A. Sharov, and S. S. Izhevskii. 1985. Predators of the American white web-worm moth (*Hyphantria cunea*) in the south of the European part of the USSR. Zool. Zh. 64: 1384–1391.

Benedict, J. H., and W. R. Cothran. 1975a. Identification of the damsel bugs, *Nabis alternatus* Parshley and *N. americoferus* Carayon (Heteroptera: Nabidae). Pan-Pac. Entomol. 51: 170–171.

1975b. A faunistic survey of the Hemiptera-Heteroptera found in northern California hay alfalfa. Ann. Entomol. Soc. Amer. 68: 897–900.

1978. Parasitism of the common damsel bug *Nabis americoferus*, in central California hay alfalfa. Southwest. Entomol. 3: 37–42.

Bisabri-Ershadi, B., L. E. Ehler, and B. Ershadi. 1981. Natural biological control of the western yellow-striped armyworm, *Spodoptera praefica* (Grote), in hay alfalfa in northern California. Hilgardia 49: 1–23.

Boyd, M. L., and D. J. Boethel. 1998a. Susceptibility of predaceous Hemipteran species to selected insecticides on soybean in Louisiana. J. Econ. Entomol. 91: 401–409.

1998b. Residual toxicity of selected insecticides to Heteropteran predaceous species (Heteroptera: Lygaeidae, Nabidae, Pentatomidae) on soybean. Environ. Entomol. 27: 154–160.

Braman, S. K., P. E. Sloderbeck, and K. V. Yeargan. 1984. Effects of temperature on the development and survival of *Nabis americoferus* and *N. roseipennis* (Hemiptera: Nabidae). Ann. Entomol. Soc. Amer. 77: 592–596.

1989. Intraplant distribution of three *Nabis* species (Hemiptera: Nabidae), and impact of *N. roseipennis* on green cloverworm populations in soybean. Environ. Entomol. 18: 240–244.

1990. Phenology and abundance of *Nabis americoferus, N. roseipennis,* and *N. rufusculus* (Hemiptera: Nabidae) and their parasitoids in alfalfa and soybean. J. Econ. Entomol. 83: 823–830.

Braman, S. K., and K. V. Yeargan. 1988. Comparison of developmental and reproductive rates of *Nabis americoferus, N. roseipennis,* and *N. rufusculus* (Hemiptera: Nabidae). Ann. Entomol. Soc. Amer. 81: 923–930.

von Bucholz, U., and G. Schruft. 1994. Staatliches Arthropoden auf Bluten und Fruchten der Weinrebe (*Vitis vinifera* L.) als Antagonisten des einbindigen Traubenwicklers (*Eupocilia ambiguella* Hbn.) (Lep., Cochylidae). J. Appl. Entomol. 118: 31–37.

von Bucholz, U., S. Schmidt, and G. Schruft. 1994. The use of an immunological technique in order to evaluate the predation on *Eupoecilia ambiguella* (Hbn.) (Lepidoptera: Cochylidae) in vineyards. Biochem. Syst. Ecol. 22: 671–677.

Burgess, L. 1982. Predation on adults of the flea beetle *Phyllotrete cruciferae* by the western damsel bug, *Nabis alternatus* (Hemiptera: Nabidae). Canad. Entomol. 114: 763–764.

Burgess, L., J. Dueck, and D. L. McKenzie. 1983. Insect vectors of the yeast *Nematospora coryli* in mustard, *Brassica juncea*, crops in southern Saskatchewan. Canad. Entomol. 115: 25–30.

Buschman, L. L., H. N. Pitre, and H. F. Hodges. 1984. Soybean cultural practices: effects on populations of geocorids, nabids, and other soybean arthropods. Environ. Entomol. 13: 305–317.

Buschman, L. L. W. H. Whitcomb, R. E. Hemenway, D. L. Mays, N. Ru, N. C. Leppla, and B. J. Smittle. 1977. Predators of velvetbean caterpillar eggs in Florida soybeans. Environ. Entomol. 6: 403–407.

Campbell, C. A. M. and W. W. Cone. 1994. Influence of predators on population development of *Phorodon humuli* (Homoptera: Aphididae) on hops. Environ. Entomol. 23: 1391–1396.

Carayon, J. 1961. Valeur systématique des voies ectodermiques de l'appareil génital femelle chez les Hémiptères Nabidae. Bull. Mus. Natl. Hist. Nat. 33: 183–196.

1970. Étude des Alloeorhynchus d'Afrique Centrale avec quelques remarques sur la classification des Nabidae (Hemiptera). Ann. Soc. Entomol. France (n.s.) 6: 899–931.

Carroll, D. P., and S. C. Hoyt. 1984. Natural enemies and their effects on apple aphid, *Aphis pomi* DeGeer (Homoptera: Aphididae), colonies on young apple trees in central Washington. Environ. Entomol. 13: 469–481.

Chatenet, B., and Z. Y. Zhu. 1985. The safety of phosalone to beneficial arthropods. Chin. J. Biol. Control 1: 56.

Chiappini, E., and M. C. Reguzzi. 1994. Observations on *Nabis punctatus* Costa (Rhyncota, Nabidae). Redia 77: 79–99.

China, W. E., and N. C. E. Miller. 1959. Check list and keys to the families and subfamilies of the Hemiptera-Heteroptera. Bull. Brit. Mus. Nat. Hist. Entomol. 8: 1–45.

Clements, D. J., and K. V. Yeargan. 1997a. Comparison of *Orius insidiosus* (Hemiptera: Anthocoridae) and *Nabis roseipennis* (Hemiptera: Nabidae) as predators of the green cloverworm. Environ. Entomol. 26: 1482–1487.

1997b. Seasonal and intraplant distribution of *Orius insidiosus* (Hemiptera: Anthocoridae) on soybean and possible interactions with *Nabis roseipennis* (Hemiptera: Nabidae). J. Kansas Entomol. Soc. 70: 94–99.

Coll, M. and D. G. Bottrell. 1995. Predator–prey association in mono- and dicultures: effect of maize and bean vegetation. Agric. Ecosyst. Environ. 54: 115–125.

Correa, B. S., A. R. Panizzi, G. G. Newman, and S. G. Turnipseed. 1977. Distribuição geográfica e abundância estacional dos principais insetos-pragas da soja e seus predadores. An. Soc. Entomol. Brasil 6: 40–50.

Culliney, T. W. 1986. Predation on adult *Phyllotreta* flea beetles by *Podisus maculiventris* (Hemiptera: Pentatomidae) and *Nabicula americolimbata* (Hemiptera: Nabidae). Canad. Entomol. 188: 731–732.

Czepak, C., E. Conti, and F. Bin. 1994. Biological observations on *Nabis punctatus* Costa (Heteroptera: Nabidae). Inf. Fitopatol. 44: 55–60.

Deakle, J. P., and J. R. Bradley, Jr. 1982. Effects of early season applications of diflubenzuron and azinophosmethyl on population levels of certain arthropods in cotton fields. J. Georgia Entomol. Soc. 17: 200–204.

Dietz, L. L., J. W. Van Duyn, J. R. Bradley, Jr., R. L. Rabb, W. M. Rooks, and R. E. Stinner. 1976. A guide to the identification and biology of soybean arthropods in North Carolina. North Carolina Agric. Exp. Stn. Tech. Bull. 238: 1–264.

Dinkins, R. L., J. R. Brazzel, and C. A. Wilson. 1970. Seasonal incidence of major predaceous arthropods in Mississippi cotton fields. J. Econ. Entomol. 63: 814–817.

Domiciano, N. L., and D. C. Herzog. 1990. Population dynamics of pests and some of their predators in soybean field under influence of herbicidal applications of toxaphene. Pesq. Agropec. Bras. 25: 253–273.

Donahoe, M. C., and H. N. Pitre. 1977. *Reduviolus roseipennis* behavior and effectiveness in reducing numbers of *Heliothis zea* on cotton. Environ. Entomol. 6: 872–876.

Ehler, L. E., K. G. Eveleens, and R. van den Bosch. 1973. An evaluation of some natural enemies of cabbage looper on cotton in California. Environ. Entomol. 2: 1009–1015.

Ellington, J., M. Southward, and T. Carrillo. 1997. Associations among cotton arthropods. Environ. Entomol. 26: 1004–1008.

Elvin, M. K., and P. E. Sloderbeck. 1984. A key to the nymphs of selected species of Nabidae (Hemiptera) in the southeastern U.S.A. Florida Entomol. 67: 269–273.

Elvin, M. K., J. L. Stimac, and W. H. Whitcomb. 1983. Estimating rates of arthropod predation on velvetbean caterpillar larvae in soybeans. Florida Entomol. 66: 319–330.

Evans, E. W., and N. N. Youssef. 1992. Numerical responses of aphid predators to varying prey density among Utah alfalfa fields. J. Kansas Entomol. Soc. 65: 30–38.

Eveleens, K. G., R. Van den Boosch, and L. E. Ehler. 1973. Secondary outbreak induction of beet armyworm by experimental insecticide applications in cotton in California. Environ. Entomol. 2: 497–503.

Falcon, L. A., R. Van den Bosch, C. A. Ferris, L. K. Stromberg, L. K. Etzel, R. E. Stinner, and T. F. Leigh. 1968. A comparison of season long cotton pest control programs in California during 1966. J. Econ. Entomol. 61: 633–642.

Fan, B. H., and J. Y. Mu. 1989. A study on the ecological characteristics of *Nabis sinoferus* Hsiao and its controlling effects on aphids. Insect Knowledge 26: 79–81.

Farlow, R. A., and H. N. Pitre. 1983. Effects of selected of selected pesticide application routines on pest and beneficial arthropods on soybean in Mississippi. Environ. Entomol. 12: 552–557.

Ferguson, H. J., R. M. McPherson, and W. A. Allen. 1984. Effect of four soybean cropping systems on the abundance of foliage inhabiting insect predators. Environ. Entomol. 13: 1105–1112.

Flinn, P. W., A. A. Hower, and R. A. J. Taylor. 1985. Preference of *Reduviolus americoferus* (Hemiptera: Nabidae) for potato leafhopper nymphs and pea aphids. Canad. Entomol. 113: 365–369.

Flint, H. M., F. D. Wilson, W. D. Hendrix, J. Leggett, S. Naranjo, T. J. Henneberry, and J. W. Radin. 1994. The effect of plant water stress on beneficial and pest insects including the pink bollworm and the sweetpotato whitefly in two short-season cultivars of cotton. Southwest. Entomol. 19: 11–22.

Frank, W. A., and J. E. Slosser. 1991. An illustrated guide to the predaceous insects of the northern Texas Rolling Plains. Misc. Publ. Tex. Agric. Exp. Stn. No. MP 1718: 23 pp.

Funderburk, J. E., F. M. Rhoads, and I. D. Teare. 1994. Modifying soil nutrient level affects soybean insect predators. Agron. J. 86: 581–585.

Gardner, W. A., M. Shepard, and R. Noblet. 1981. Precipitin test for examining predator–prey interactions in soybean fields. Canad. Entomol. 113: 365–369.

Godfrey, L. D., and T. F. Leigh. 1994. Alfalfa harvest strategy effect on *Lygus* bug (Hemiptera: Miridae) and insect predator population density: implications for use as trap crop in cotton. Environ. Entomol. 23: 1106–1118.

Graham, H. M., and C. G. Jackson. 1982. Distribution of eggs and parasites of *Lygus* spp. (Hemiptera: Miridae), *Nabis* spp. (Hemiptera: Nabidae), and *Spissistilus festinus* (Say) (Homoptera: Membracidae) on plant stems. Ann. Entomol. Soc. Amer. 75: 56–60.

Grasela, J. J., and W. W. M. Steiner. 1993. Population genetic structure among populations of three predaceous nabid species: *Nabis alternatus* Parshley, *Nabis roseipennis* Reuter, and *Nabis americoferus* Carayon (Hemiptera: Nabidae). Biochem. Syst. Ecol. 21: 813–823.

Guppy, J. C. 1986. Bionomics of the damsel bug *Nabis americoferus* Carayon (Hemiptera: Nabidae) a predator of the alfalfa blotch leafminer (Diptera: Agromyzidae). Canad. Entomol. 118: 745–751.

Gyenge, J. E., J. D. Edelstein, and E. V. Trumper. 1997a. Comparacion de tecnicas de muestreo de artropodos depredadores en alfalfa y efecto de factores ambientales sobre sus estimaciones de abundancia. CEIBA 38: 13–18.

Gyenge, J. E., E. V. Trumper, and J. D. Edelstein. 1997b. Diseno de planes de muestreo de artropodos depredadores en alfalfa con niveles fijos de precision. CEIBA 38: 23–28.

Hagler, J. R., and S. E. Naranjo. 1994. Determining the frequency of heteropteran predation on sweetpotato whitefly and the pink bollworm using multiple ELISAs. Entomol. Exp. Appl. 72: 59–66.

Hamilton, E. W., and R. W. Kieckhefer. 1969. Toxicity of malathion and parathion to predators of the English grain aphid. J. Econ. Entomol. 62: 1190–1192.

Harper, A. M. 1978. Effect of insecticides on the pea aphid, *Acyrthosiphon pisum* (Homoptera: Aphidae) and associated fauna on alfalfa. Canad. Entomol. 110: 891–894.

Harris, H. M. 1928. A monographic study of the hemipterous family Nabidae as it occurs in North America. Entomol. Amer. 9: 1–97.

Hendrick, R. D., and V. M. Stern. 1970. Biological studies of three parasites of *Nabis americoferus* (Hemiptera: Nabidae) in southern California. Ann. Entomol. Soc. Amer. 63: 382–391.

Henneberry, T. J., and T. E. Clayton. 1985. Consumption of pink bollworm (Lepidoptera: Gelechidae) and tobacco budworm (Lepidoptera: Noctuidae) eggs by some predators commonly found in cotton fields. Environ. Entomol. 14: 416–419.

Henry, T. J., and J. D. Lattin. 1988. Family Nabidae. Pp. 500–512 *in* R. C. Froeschner and T. J. Henry [eds.], Catalog of the Heteroptera of Canada and the Continental United States. E. J. Brill, Leiden, the Netherlands. 958 pp.

Hormchan, P., L. W. Hepner, and M. F. Schuster. 1976. Predaceous damsel bugs: identification and distribution of the subfamily Nabinae in Mississippi. Mississippi Agric. Exp. Stn. Tech. Bull. 76: 1–4.

Huckaba, R. M., J. R. Bradley, and J. W. van Duyn. 1983. Effects of herbicidal applications of toxaphene on soybean thrips, certain predators, and corn earworm in soybean. J. Georgia Entomol. Soc. 18: 195–200.

Injac, M., and S. Krnjajic. 1990. The role of natural enemies on reduction of *Mamestra brassicae* L. population density in the region of Belgrade. Zast. Bilja 41: 111–124.

Irwin, M. E., and M. Shepard. 1980. Sampling predaceous Hemiptera on soybeans. Pp. 503–531 *in* M. Kogan and D. Herzog [eds.], Sampling Methods in Soybean Entomology. Springer-Verlag, New York, New York, U.S.A.

Irwin, M. E., B. W. Gill, and D. Gonzalez. 1973. Field cage studies of native egg predators of the pink bollworm in southern California cotton. J. Econ. Entomol. 67: 193–196.

Isenhour, D. J., and K. V. Yeargan. 1982. Oviposition sites of *Orius insidiosus* (Say) and *Nabis* species in soybean. J. Kansas Entomol. Soc. 55: 65–72.

Iwao, S., L. R. Taylor, E. J. Bechinski, and L. P. Pedigo. 1981. Population dispersion and development of sampling plans for *Orius insidiosus* and *Nabis* spp. in soybeans. Environ. Entomol. 10: 956–959.

Jervis, M. A. 1990. Predation of *Lissonota coracinus* (Gmelin) (Hymenoptera: Ichneumonidae) by *Dolichonabis limbatus* (Dahlbom) (Hemiptera: Nabidae). Entomol. Gaz. 41: 231–233.

Kaplin, V. G. 1993. Life form of bugs of the infraorder Cimicomorpha (Heteroptera) of Kara Kum. Entomol. Oboz. 2: 304–320.

Katz, E., P. Duelli, and P. Wiedemeier. 1989. Der Einfluss der Nachbarschaft naturnaher Biotope auf Phänologie und Produktion von entomophagen Arthropoden in Intensivkulturen. Mitt. Dtsch. Ges. Allg. Angew. Entomol. 7: 306–310.

Kayapinar, A., and S. Kornosor. 1993. *Ostrinia nubilalis* Hubner (Lep., Pyralidae) in larva donemleri uzerinde avci boceklerin etkisinin arastirilmasi. Türk. Entomol. Derg. 17: 69–76.

Keever, D. W., J. R. Bradley, and M. C. Ganyard. 1977. Effects of diflubenzuron (Dimilin) on selected beneficial arthropods in cotton fields. Environ. Entomol. 6: 732–736.

Keresi, T. 1993. Fauna Heteroptera no soji u Backoj. Zast. Bilja 44: 189–195.

Kerzhner, I. M. 1968. New and little known palearctic bugs of the family Nabidae (Heteroptera). Acta Entomol. Mus. Nat. Pragae 40: 517–525.

1981. Fauna SSSR. Nasekomye Khobotnye, t. 13, vyp. 2, Poluzhestttkokrylye Semejstva Nabidae. Akad. Nauk, Leningrad, U.S.S.R.

1983. Airborne *Nabis capsiformis* (Heteroptera: Nabidae) from the Atlantic, Indian and Pacific oceans. Int. J. Entomol. 25: 273–275.
1992. New and little known Nabidae from North America (Heteroptera). Zoosyst. Rossica. 1: 37–45.
Kerzhner, I. M., T. E. Woodward, and N. Strommer. 1982. *Nabis kingbergii* Reuter, the current name for *Tropiconabis nigrolineatus* (Distant), and its Australian distribution (Hemiptera: Nabidae). J. Austral. Entomol. Soc. 21: 306.
Kitamura, K., and H. Kondo. 1995. Influence of temperature and prey density on development, survival rate and predation of *Nabis stenoferus* (Hemiptera: Nabidae). Jpn. J. Appl. Entomol. Zool. 39: 261–263.
Kretzschmar, G. P. 1948. Soybean insects in Minnesota with special reference to sampling techniques. J. Econ. Entomol. 41: 586–591.
Krotova, I. G. 1991. Bugs of the family Nabidae (Hemiptera) predaceous on aphids in priob' forest steppe. Zool. Zh. 70: 59–68.
Lakin, K. R., C. G. Jackson, and H. M. Graham. 1984. Field and laboratory studies on parasitism of *Nabis* spp. by the wasp *Polynema boreum*. Southwest. Entomol. 9: 391–396.
Larivière, M.-C. 1992a. *Hoplistoscelis sordidus* (Heteroptera: Nabidae) in Canada. Great Lakes Entomol. 25: 47–50.
1992b. *Himacerus apterus* (Fabricius), a Eurasian Nabidae (Hemiptera) new to North America: diagnosis, geographical distribution, and bionomics. Canad. Entomol. 124: 725–728.
1994. Biodiversity of *Nabicula* Kirby species (Hemiptera: Nabidae) in Canada: faunistic review, bioecology, biogeography. Canad. Entomol. 126: 327–378.
Laster, M. L., and J. R. Brazzel. 1968. A comparison of predator populations in cotton under different control programs in Mississippi. J. Econ. Entomol. 61: 714–719.
Lattin, J. D. 1989. Bionomics of the Nabidae. Annu. Rev. Entomol. 34: 383–400.
Leddy, P. M. 1996. Factors influencing the distribution and abundance of azalea lace bug, *Stephanitis pyrioides*, in simple and complex landscape habitats. Ph.D. dissertation, University of Maryland, College Park, Maryland, U.S.A.
Leigh, T. F., and D. Gonzalez. 1976. Field cage evaluation of predators for control of *Lygus hesperus* Knight in cotton. Environ. Entomol. 5: 948–952.
Lentz, G. L., A. Y. Chambers, and R. M. Hayes. 1983. Effects of systemic insecticide-nematocides on mid-season pest and predator populations in soybean. J. Econ. Entomol. 76: 836–840.
Leston, D. 1957. Cytotaxonomy of Miridae and Nabidae (Hemiptera). Chromosoma 8: 609–616.
Leston, D., J. G. Pendergast, and T. R. E. Southwood. 1954. Classification of the terrestrial Heteroptera (Geocorisae). Nature 174: 91–92.
Lingren, P. D. and R. L. Ridgway. 1967. Toxicity of five insecticides to several insect predators. J. Econ. Entomol. 60: 1639–1641.
Lobner, U., and O. Hartwig. 1994. Soldier beetles (Col. Cantharidae) and nabid bugs (Het., Nabidae) occurrence and importance as aphidophagous predators in winter wheat fields in the surroundings of Halle/Saale (Sachsen Anhalt). Bull. OILB SROP. 17: 179–187.
Lorenzato, D., and E. Corseuil. 1982. Efeitos de diferentes meios de controle sobre as principais pragas de soja (*Glycine max* (L.) Merrill) e seus predadores. Agron. Sulriograndense 18: 61–84.
Mack, T. P., R. H. Walker, and G. Wehtje. 1987. Impact of sicklepod control on several insect pests and their arthropod natural enemies in Florunner peanuts. Crop Prot. 6: 185–190.
Martinez, D. G., and R. L. Pienkowski. 1982. Laboratory studies on insect predators of potato leafhopper eggs, nymphs, and adults. Environ. Entomol. 11: 361–362.
1983. Comparative toxicity of several insecticides to an insect predator, nonpest species, and a pest prey species. J. Econ. Entomol. 76: 933–935.
McCarty, M. T., M. Shepard, and S. G. Turnipseed. 1980. Identification of predaceous arthropods in soybeans by using autoradiography. Environ. Entomol. 9: 199–203.
McPherson, R. M., J. C. Smith, and W. A. Allen. 1982. Incidence of arthropod predators in different soybean cropping systems. Environ. Entomol. 11: 685–689.
Mikolajski, M. 1965. Chromosome numbers in *Nabis* Lt. (Heteroptera: Nabidae). Experientia 21: 445.
1967. Chromosome studies in the genus *Nabis* (Heteroptera, Nabidae). Zool. Pol. 17: 313–334.
Mitri, T. K. 1960. The taxonomic value of the external female genitalia in the genus *Nabis* Latreille (Hemiptera: Heteroptera: Nabidae). M.S. thesis, University of California, Davis, California, U.S.A.

Morrison, D. E., J. R. Bradley, and J. W. van Duyn. 1979. Populations of corn earworm and associated predators after applications of certain soil applied pesticides to soybeans. J. Econ. Entomol. 72: 97–100.

Mundinger, F. G. 1922. The life history of two species of Nabidae (Hemiptera: Heteroptera). New York State Coll. For. Publ. 16: 149–167.

Nadgauda, D., and H. N. Pitre. 1986. Effects of temperature on the feeding, development and longevity of *Nabis roseipennis* (Hemiptera: Nabidae) fed tobacco budworm (Lepidoptera: Noctuidae) larvae and tarnished plant bug (Hemiptera: Miridae) nymphs. Environ. Entomol. 15: 536–539.

Oien, C. T., and D. W. Ragsdale. 1993. Susceptibility of nontarget hosts to *Nosema furnacalis* (Microsporida: Nosematidae), a potential biological control agent of European corn borer, *Ostrinia nubilalis* (Lepidoptera: Pyralidae). Biol. Control 4: 323–328.

O'Neill, R. J. 1992. Body weight and reproductive status of two nabid species (Heteroptera: Nabidae) in Indiana. Environ. Entomol. 21: 191–196.

Orphanides, G. M., D. Gonzales, and B. R. Bartlett. 1971. Identification and evaluation of pink bollworm predators in southern California. J. Econ. Entomol. 64: 421–424.

Pedigo, L. P., E. J. Bechinski, and R. A. Higgins. 1983. Partial life tables of the green cloverworm (Lepidoptera: Noctuidae) in soybean and a hypothesis of population dynamics in Iowa. Environ. Entomol. 12: 186–195.

Pedigo, L. P., J. D. Stone, and G. L. Lentz. 1972. Survivorship of experimental cohorts of the green cloverworm on screenhouse and open field soybean. Environ. Entomol. 1: 180–186.

Péricart, J. 1987. Hémiptères Nabidae d'Europe Occidentale et du Maghreb. Faune France 71: 185 pp.

Perkins, P. V., and T. F. Watson. 1972. Biology of *Nabis alternatus* (Hemiptera: Nabidae). Ann. Entomol. Soc. Amer. 65: 54–57.

Pfannenstiel, R. S., and K. V. Yeargan. 1998a. Ovipositional preference and distribution of eggs in selected field and vegetable crops by *Nabis roseipennis* (Hemiptera: Nabidae). J. Entomol. Sci. 33: 82–89.

1998b. Association of predaceous Hemiptera with selected crop habitats. Environ. Entomol. 27: 232–239.

1998c. Partitioning two- and three-trophic-level effects of resistant plants on the predator, *Nabis roseipennis* (Hemiptera: Nabidae). Entomol. Exp. Appl. 88: 203–209.

Pitre, H. N., T. L. Hillhouse, M. C. Donahue, and H. C. Kinard. 1978. Beneficial arthropods on soybean and cotton in different ecosystems in Mississippi. Mississippi Agric. Exp. Stn. Tech. Bull. 90: 1–9.

Propp, G. D. 1982. Functional response of *Nabis americoferus* to two of its prey, *Spodoptera exigua* and *Lygus hesperus*. Environ. Entomol. 5: 1195–1207.

Radcliffe, E. B., R. W. Weires, R. E. Studer, and D. K. Barnes. 1976. Influence of cultivars and pesticides on pea aphid and associated arthropod taxa in a Minnesota alfalfa ecosystem. Environ. Entomol. 5: 1195–1207.

Raga, V., S. Gravena, S. A. de Bortoli, J. Arai, and G. N. Wassano. 1990. Amostragem de insetos e atividade de artrópodos predadores na cultura do tomateiro de crescimnento determinado. An. Soc. Entomol. Brasil 19: 253–271.

Reed, T., M. Shepard, and S. G. Turnipseed. 1984. Assessment of the impact of arthropod predators on noctuid larvae in cages in soybean fields. Environ. Entomol. 13: 954–961.

Reinert, J. A. 1978. Natural enemy complex of the southern chinch bug in Florida. Ann. Entomol. Soc. Amer. 71: 728–731.

Remane, R. 1964. Weitere Beiträge zur Kenntnis der Gattung *Nabis* Latr. (Hemiptera-Heteroptera: Nabidae). Zool. Beitr. (n.F.) 10: 253–314.

Rensner, P. E., W. O. Lamp, R. J. Barney, and E. J. Armbrust. 1983. Potato leafhopper *Empoasca fabae* (Homoptera: Cicadellidae) and their movement into spring planted alfalfa. J. Kansas Entomol. Soc. 56: 446–450.

Richards, L. A., and A. M. Harper. 1978. Oviposition by *Nabis alternatus* (Hemiptera: Nabidae) in alfalfa. Canad. Entomol. 110: 1359–1362.

Richman, D. B., R. C. Hemenway, Jr., and W. H. Whitcomb. 1980. Field cage evaluation of predators of the soybean looper *Pseudoplusia includens* (Lepidoptera: Noctuidae). Environ. Entomol. 9: 315–317.

Roach, S. H. 1980. Arthropod predators on cotton, corn, tobacco, and soybeans in South Carolina. J. Georgia Entomol. Soc. 15: 131–138.

Rosca, I. 1982. Raspindirea speciilor familiei Nabidae–Costa 1852, in Romania. An. Inst. Cerce. Cereale Plante Teh. Fundulea 49: 225–233.

Rosenheim, J. A., L. R. Wilhoit, and C. A. Armer. 1993. Influence of intraguild predation among generalist insect predators on the suppression of an herbivore population. Oecologia 96: 439–449.

Roth, S., and J. P. O'Connor. 1993. A review of the Irish Nabinae (Insecta: Hemiptera: Nabidae). Bull. Irish Biogeogr. Soc. 16: 6–17.

Rotrekl, J. 1994. Vliv insekticidu na uzitecnou entomofaunu vojtesky (*Medicago sativa* L.). Ochr. Rostlin 30: 67–77.

Royer, T. A., and D. D. Walgenbach. 1991. Predaceous arthropods of cultivated sunflower in eastern South Dakota. J. Kansas Entomol. Soc. 64: 12–116.

Ruberson, J. R., S. Y. Young, and T. J. Kring. 1991. Suitability of prey infected by nuclear polyhedrosis virus for development, survival, and reproduction of the predator *Nabis roseipennis* (Heteroptera: Nabidae). Environ. Entomol. 20: 1475–1479.

Schaber, B. D., and T. Entz. 1994. Comparison of insect abundance under sprinkler and flood irrigation systems in alfalfa (*Medicago sativa* (L.)) grown for seed. Proc. Entomol. Soc. Manitoba 49: 16–25.

Schuh, R. T. 1986. The influence of cladistics on heteropteran classification. Annu. Rev. Entomol. 31: 67–93.

Schuh, R. T., and P. Štys. 1991. Phylogenetic analysis of cimicomorphan family relationships (Heteroptera). J. New York Entomol. Soc. 99: 298–350.

Semptner, R. J., and J. E. Roberts. 1978. Occurrence of beneficial insects on flue-cured tobacco treated with soil insecticides. J. New York Entomol. Soc. 86: 321.

Severin, F., J. P. Bassino, M. Blanc, D. Bony, J. P. Gendrier, J. N. Reboulet, and M. Tisseur. 1984. Cultures fruitieres. Importance des héteroptères predateures des psylles du poirier dans le sud-est de la France. Phytoma 354: 33–35.

Shelton, M. D., and C. R. Edwards. 1983. Effects of weeds on the diversity and abundance of insects in soybeans. Environ. Entomol. 12: 296–298.

Shepard, M., G. R. Carner, and S. G. Turnipseed. 1974. Seasonal abundance of predaceous arthropods in soybeans. Environ. Entomol. 3: 985–988.

1977. Colonization and resurgence of insect pests of soybean in response to insecticides and field isolation. Environ. Entomol. 6: 501–506.

Sloderbeck, P. E., and K. V. Yeargan. 1983. Comparison of *Nabis americoferus* and *Nabis roseipennis* (Hemiptera: Nabidae) as predators of the green cloverworm (Lepidoptera: Noctuidae). Environ. Entomol. 12: 161–165.

Smith, M. W., R. D. Eikenbary, D. C. Arnold, B. S. Landgraf, G. G. Taylor, G. E. Barlow, B. L. Carroll, B. S. Cheary, N. R. Rice, and R. Knight. 1994. Screening cool season legume cover crops for pecan orchards. Amer. J. Alternative Agric. 3: 127–135.

Sprenkel, R. K., W. M. Brooks, J. W. VanDuyn, and L. L. Deitz. 1979. The effects of three cultural variables on the incidence of *Nomuraea rileyi*, phytophagous lepidoptera, and their predators on soybean. Environ. Entomol. 8: 334–339.

Stern, V. M. 1963. The effect of various insecticides on *Trichogramma semefumatum* and certain predators in southern California. J. Econ. Entomol. 56: 348–349.

Stern, V. M., R. Van den Bosch, and H. T. Reynolds. 1959. Effects of dylox and other insecticides on entomophagous insects attacking field crop pests in California. J. Econ. Entomol. 53: 67–72.

Stoltz, R. L., and C. D. McNeal, Jr. 1982. Assessment of insect emigration from alfalfa hay to bean fields. Environ. Entomol. 11: 578–580.

Stoltz, R. L., and V. M. Stern. 1979. Comparative insecticide induced mortality of *Nabis americoferus* in cotton. Environ. Entomol. 8: 48–50.

Stoner, A. 1972. Plant feeding by *Nabis*, a predaceous genus. Environ. Entomol. 1: 557–558.

1973. Incidence of *Wesmaelia pendula* (Hymenoptera: Braconidae), a parasitoid of male *Nabis* species in Arizona. Ann. Entomol. Soc. Amer. 66: 471–473.

Stoner, A., A. M. Metcalfe, and E. R. Weeks. 1975. Seasonal distribution, reproductive diapause, and parasitization of three *Nabis* species in southern Arizona. Environ. Entomol. 4: 211–214.

Summers, C. G., R. L. Coviello, and W. R. Cothran. 1975. The effect on selected entomophagous insects of insecticides applied for pea aphid control in alfalfa. Environ. Entomol. 4: 612–614.

Tamaki, G., and R. E. Weeks. 1972. Efficiency of three predators, *Geocoris bullatus, Nabis americoferus*, and *Coccinella transverssoguttata*, used alone or in combination against three insect prey species, *Myzus persicae, Ceramica picata*, and *Mamestra configurata*, in a greenhouse study. Environ. Entomol. 1: 258–263.

Tamaki, G., D. Olsen, and R. K. Gupta. 1978. Laboratory evaluation of *Geocoris bullatus* and *Nabis alternatus* as predators of *Lygus*. Entomol. Soc. British Columbia 75: 35–37.

Tamaki, G, M. A. Weiss, and G. E. Long. 1981. Evaluation of plant density and temperature in predator–prey interactions in field cages. Environ. Entomol. 10: 716–720.

Taylor, E. J. 1949. A life history study of *Nabis alternatus*. J. Econ. Entomol. 42: 991.

van den Bosch, R., T. F. Leigh, D. Gonzalez, and R. E. Stinner. 1969. Cage studies on predators of the bollworm in cotton. J. Econ. Entomol. 62: 1486–1489.

Waddill, V., and M. Shepard. 1974. Potential of *Geocoris punctipes* (Hemiptera: Lygaeidae) and *Nabis* spp. (Hemiptera: Nabidae) as predators of *Epilachna varivestis* (Coleoptera: Coccinellidae). Entomophaga 19: 421–426.

Wheeler, A. J., Jr. 1977. Studies on the arthropod fauna of alfalfa. VII. Predaceous insects. Canad. Entomol. 109: 423–427.

Whitcomb, W. H. 1967. Field studies of predators of the second instar bollworm, *Heliothis zea* (Boddie) (Lepidoptera: Noctuidae). J. Georgia Entomol. Soc. 2: 113–118.

Whitcomb, W. H. and R. Baker. 1974. Minimum rates of insecticide on soybeans: *Geocoris* and *Nabis* populations following treatment. Florida Entomol. 57: 114.

White, J. S., R. A. Brown, A. L. Bettencourt, and C. M. S. Soares. 1992. Lambda cyhalothrin: effects on natural pest control in Brazilian soyabeans. Proc. Brighton Crop Prot. Conf., Pests and Diseases. 8 pp.

Wilson, L. T., and A. P. Gutierrez. 1980. Within-plant distribution of predators on cotton: comments on sampling and predator efficiencies. Hilgardia 48: 3–11.

Woodward, T. E. 1982. The identity of the species commonly known in Australia as *Nabis capsiformis* Germar (Hemiptera: Nabidae). J. Austral. Entomol. Soc. 21: 143–146.

Yeargan, K. V. 1998. Predatory Heteroptera in North American agroecosystems: an overview. Pp. 7–19 *in* M. Coll and J. Ruberson [eds.], Predatory Heteroptera in Agroecosystems: Their Ecology and Use in Biological Control. Thomas Say Publ.: Entomological Society of America, Lanham, Maryland, U.S.A.

Yeargan, K. V., and W. E. Barney. 1996. Photoperiodic induction of reproductive diapause in the predators *Nabis americoferus* and *Nabis roseipennis* (Heteroptera: Nabidae). Ann. Entomol. Soc. Amer. 89: 70–74.

Yeargan, K. V., and W. R. Cothran. 1972. An escape barrier for improved suction sampling of *Pardosa ramulosa* and *Nabis* spp. populations in alfalfa. Environ Entomol. 3: 189–191.

Yildirim, E., and H. Ozbek. 1992. Erzurum seker fabrikasina bagli sekerpancari uretim alanlarindaki zararli ve yararli bocek turleri. Proc. Second Turkish National Congress of Entomol., pp. 621–635.

Young, S. Y., and T. J. Kring. 1991. Selection of healthy and nuclear polyhedrosis virus (NPV) infected *Anticarsia gemmatalis* (Lep.: Noctuidae) as prey by nymphal *Nabis roseipennis* (Hemiptera: Nabidae) in laboratory and on soybean. Entomophaga 36: 265–273.

Young, S. Y., and W. C. Yearian. 1992. Movement of nuclear polyhedrosis virus into velvetbean caterpillar (Lepidoptera: Noctuidae) larval populations on soybean by *Nabis roseipennis* (Heteroptera: Nabidae) nymphs. J. Entomol. Sci. 27: 126–134.

CHAPTER 28

Predacious Plant Bugs (Miridae)

A. G. Wheeler, Jr.

1. INTRODUCTION

Ideas regarding mirid feeding habits have fluctuated widely over the past 150 years (Wheeler 1976; see also Wheeler 2001). Some mid-19th century authors overemphasized predation, sometimes excluding mention of phytophagy in a family that contains important plant pests (see **Chapter 3**). Yet by the late 19th century, some workers were characterizing the Miridae as a strictly phytophagous group (e.g., Kirby 1892, Saunders 1892). Mirids continued to be underappreciated as predators even after several species had been used successfully in classical biological control in the 1920s and 1930s.

It now is evident that the Miridae contain a relatively high proportion of omnivores — that is, species that feed at more than one trophic level. Kullenberg (1944) considered 25% of the species ($n = 100$) he studied in Sweden to be at least facultatively predacious. A similar percentage of omnivorous species ($n = \sim200$) can be inferred for the British mirid fauna (Southwood and Leston 1959). Zoophagy might actually occur in at least a third of the 9800+ described species of Miridae. Among the principal subfamilies, mainly certain tribes and subtribes of the Bryocorinae (Bryocorini, Eccritotarsini, Monaloniina, and Odoniellina; *sensu* Schuh 1995) and the stenodemine Mirinae lack (or show slight) predacious tendencies (Wheeler 2001).

Even now, few entomologists realize that some mirids are specialized predators of lace bugs (see **Section 4.10**) or thrips (**Section 4.11**), and that certain others specialize on scale insects (**Section 4.12**). In some textbooks of entomology (e.g., Evans 1984), the characterization of mirids as mostly phytophagous, with only a few predatory species, ignores the prevalence of omnivory in the family. Similarly misleading is New's (1991) comment in *Insects as Predators* that at least some species are facultatively predacious. Moreover, a recent review of carnivory in phytophagous insects (Whitman et al. 1994) fails to depict mirids as frequent predators. Some field guides (e.g., Borror and White 1970) omit any mention of predation by mirids. On the other hand, an appropriate description of the Miridae, including their feeding habits, is that by Sweetman (1958): plant bugs are among the most common insects, the majority are phytophagous, many are destructive pests, but a large number are predacious. One of the few heteropterans and perhaps the only mirid ever featured on a postage stamp (Turkey) is *Deraeocoris rutilus* (Herrich-Schaeffer), shown preying on an aphid (see Hamel 1990: fig. 18).

Mirids might be overlooked as predatory because even obligate carnivores in the family are not morphologically specialized for prey capture as are bugs in certain other heteropteran families (e.g., Myers 1927b, Kiritshenko 1951). But members of the ancestrally predaceous family Miridae (e.g., Schuh 1976, Cobben 1978, Schuh and Štys 1991, Schuh and Slater 1995, Wheeler 2001) feed in a carnivore-like manner, some species possessing potent salivary enzymes and venoms (Cohen 1996). The family thus comprises species that feed on arthropod eggs, neonate larvae, or dying or weakened individuals, but also "true predators" (Miles 1972) whose saliva can rapidly immobilize prey. Tending also to hinder appreciation of plant bugs as predators is the inherent difficulty of quantifying their effects on prey populations (Greenstone and Morgan 1989, Hagler and Naranjo 1994, Ruberson and Coll 1998, Wheeler 2001). The impact of sucking predators is less easily evaluated than that of chewing predators.

Even though in this book separate chapters are devoted to mirids as plant pests (**Chapter 3**) and as predators, categorization of species as either phytophagous or zoophagous often is subjective (Wiedenmann and Wilson 1996, Wheeler 2001). No other heteropteran family contains so many chiefly predatory species that once were considered injurious to crops, or so many omnivores that can have either a pest or beneficial status on the same crop.

This chapter treats the most important naturally occurring predacious mirids and those used in biological control and integrated pest management (IPM), as well as selected less important predatory species. A recent review of heteropteran predators useful in biological control (Hagen et al. 1999) includes information on 18 species, with generic accounts for *Lygus* and *Phytocoris*. Species treated by Hagen et al. (1999) that are not included in the present chapter are *D. lutescens* (Schilling), *D. ruber* (L.), *Heterotoma merioptera* (Scopoli), and *Spanagonicus albofasciatus* (Reuter). Two of these species — *D. ruber* and *H. merioptera* — are discussed in Wheeler and Henry's (1992) review of North American mirid species common to the Old and New World. Information on additional predatory mirids and a more extensive literature review are provided by Wheeler (2001); also included in that reference book are a review of cannibalism in the Miridae and discussion of predation by mirids on beneficial arthropods such as phytoseiid mites, parasitic wasps, and coccinellids.

In the present chapter, distinctions between "major" and "minor" predators are arbitrary, being determined not only by the role a particular species plays in helping suppress prey populations, but also by the amount of research data available and by the economic importance of the bug's prey and associated host plants. For the purposes of this chapter, a mirid that attacks a key pest of major food or fiber crops was likely to be considered more important than one that feeds on pests of landscape or ornamental plants. The spelling "predacious," rather than "predaceous," follows Frank and McCoy (1989).

2. GENERAL STATEMENT OF ECONOMIC IMPORTANCE

Specialists in biological control have devoted more attention to parasitoids than to predators (e.g., Döbel and Denno 1994). Among predacious arthropods, researchers have tended to emphasize groups such as phytoseiids, chrysopids, or coccinellids to the neglect of heteropterans (e.g., Coll and Ruberson 1998b). The value of predacious Heteroptera in agroecosystems and their use in IPM, however, were the subject of two recent symposia (Alomar and Wiedenmann 1996, Coll and Ruberson 1998a). Contrasted with predatory heteropterans such as anthocorids, geocorid lygaeoids, and asopine pentatomids, mirids remain understudied for the role they play in natural and managed systems.

Yet some of the greatest successes of classical biocontrol involve mirids that feed mostly on eggs of delphacid planthoppers (see **Sections 3.4 and 3.6**). Plant bugs also contribute to the natural (passive) control of plant pests. Predation by mirids on arthropod eggs, neonate caterpillars, and early instars of other arthropods might represent the most important contributions of these natural enemies to pest suppression (Yeargan 1998). Even plant bugs that are key or otherwise significant crop pests can be considered beneficial under certain conditions. Examples of occasionally beneficial omnivores are lygus bugs (*Lygus* spp.; **Section 4.13**), the "mullein bug" *Campylomma verbasci* (**Section 4.20**), and the cotton fleahopper *Pseudatomoscelis seriatus* (**Section 4.23**) (see also **Chapter 3** for discussion of these bugs as plant pests).

An increasing interest in generalist predators (e.g., Murdoch et al. 1985, Ehler 1998) and the realization that omnivory helps these natural enemies persist during periods of prey scarcity (e.g., Döbel and Denno 1994, Alomar and Albajes 1996, Naranjo and Gibson 1996, Wheeler 2001) have stimulated research on predacious mirids. In Europe, the use of IPM in greenhouses (and polyhouses) allowed large numbers of naturally occurring dicyphine mirids to develop on protected crops (Alomar et al. 1990, Malausa and Trottin-Caudal 1996), which focused attention on these useful predators. Since the mid-1980s, researchers have evaluated the effects of several dicyphine species on populations of thrips and whiteflies that are key pests of greenhouse crops. Especially in the Mediterranean region, predatory mirids have been incorporated into management programs for field- and greenhouse-grown vegetables and greenhouse-grown floral crops (e.g., Alomar and Albajes 1996, Malausa and Trottin-Caudal 1996). Because the importation of Palearctic dicyphine mirids used successfully against pests of greenhouse crops in Europe is disallowed by the U.S. and Canadian governments (Wheeler 2001), North American researchers are assessing the potential for indigenous dicyphines to control pests such as whiteflies in greenhouses (McGregor et al. 1999). Also under way are studies to enhance the use of predatory mirids in row crops, in fruit and nut crops, and in landscape plantings and nurseries.

3. THE IMPORTANT SPECIES

Bryocorinae

For a brief overview of this subfamily and discussion of bryocorines as plant pests, readers should see **Chapter 3**. Bryocorine predators include several dicyphine species that have received recent attention as biocontrol agents of greenhouse pests.

3.1 *Dicyphus tamaninii* Wagner

This Old World dicyphine, a polyphagous predator, occurs in Europe (especially the Mediterranean region), North Africa (Tunisia), and the Near East (Israel) (Goula and Alomar 1994, Kerzhner and Josifov 1999). Occurring in outdoor tomato (*Lycopersicon esculentum* Mill.) and other crops (e.g., Gabarra et al. 1988, Riudavets and Castañé 1998), *D. tamaninii* is being used to manage greenhouse populations of pests such as the western flower thrips *Frankliniella occidentalis* (Pergande) and the greenhouse whitefly *Trialeurodes vaporariorum* (Westwood).

Nymphal development can be completed on a diet of thrips larvae (Riudavets et al. 1993). When fed second-instar larvae of *F. occidentalis* under experimental conditions (25°C, 75% RH, and 16:8 L:D), *D. tamaninii* had an incubation period of 12 days, a nymphal development period of about 19 days, fecundity of 4.5 eggs, and longevity of about 21 days. Nymphs can consume about four second-instar thrips larvae per day, whereas adults consume more than ten individuals per day (Riudavets and Castañé 1998). In cage trials, this mirid can maintain western flower thrips at low densities on cucumber (Gabarra et al. 1995); substantial predation also occurs when the thrips infest beans (Riudavets et al. 1993). The feeding behavior of *D. tamaninii* varies according to the prey's host plant (Gessé Solé 1992).

Different release rates of the predacious mirid on cucumber result in consistent reductions in densities of the western flower thrips, suggesting the bug's potential in inoculative biocontrol. A 3:10 predator:prey ratio (with initial infestation of 5 adult thrips/leaf) was able to keep thrips populations on cucumber below the economic injury level (Castañé et al. 1996).

Dicyphus tamaninii also has been studied as a whitefly predator. When provided second and third instars of the greenhouse whitefly, the bugs completed their nymphal development in about 22 days (22°C, 16:8 L:D) and fed on a mean of 57 prey during development (Salamero et al. 1987). Whitefly densities in unsprayed tomato plots are negatively correlated with densities of *D. tamaninii*. Predator:prey ratios of 1:5 can provide adequate whitefly control in outdoor tomatoes in Spain. Development of a decision chart allows tomato growers to spray only when needed to keep whitefly injury below threshold levels or to prevent fruit injury by this omnivorous plant bug (Alomar and Albajes 1996). Similar fruit injury by the plant bug does not occur on cucumber (Gabarra et al. 1995). The mirid offers considerable potential as a predator of the sweetpotato whitefly *Bemisia tabaci* (Gennadius), even when this pest co-occurs in greenhouses with the greenhouse whitefly (Barnadas et al. 1998).

Alvarado et al. (1997) evaluated *D. tamaninii* as a predator of the cotton aphid *Aphis gossypii* Glover on cucumbers, *Cucumis sativus* L., and the potato aphid *Macrosiphum euphorbiae* (Thomas) on tomato. Females showed a type II functional response when provided varying densities of the cotton aphid.

3.2 *Macrolophus melanotoma* (Costa) (= *Macrolophus caliginosus* Wagner)

A Eurasian species, *M. melanotoma* is particularly common in the Mediterranean region (Malausa and Trottin-Caudal 1996, Schelt et al. 1996); its distribution by country is given by Kerzhner and Josifov (1999). Although *M. caliginosus* Wagner was synonymized with *M. melanotoma* by Carapezza (1995), the former name continues to be used in the biocontrol literature. This dicyphine is not easily distinguished from *M. pygmaeus* (Rambur) (Goula and Alomar 1994, Kerzhner and Josifov 1999).

A polyphagous predator found mostly on plants of the Asteraceae and Solanaceae (Schelt et al. 1996), *M. melanotoma* is not impeded by glandular hairs of its host plants (Koppert 1993). It often occurs in unsprayed vegetable crops such as tomato (e.g., Riudavets et al. 1993, Goula and Alomar 1994, Riudavets and Castañé 1998). An important native host is the herbaceous composite *Inula viscosa* (L.) Ait. (Carayon 1986). The bugs apparently prefer the stationary instars of the greenhouse whitefly (Koppert 1993, Constant et al. 1994), but they will also feed on the sweetpotato whitefly (Barnadas et al. 1998), as well as on other greenhouse pests, such as mites, thrips, and aphids. The last-named prey are more readily attacked than are spider mites (Fauvel et al. 1987, Foglar et al. 1990). *Macrolophus melanotoma* females showed a type II functional response when provided varying densities of mites (Foglar et al. 1990). Nymphs of this plant bug can develop on *A. gossypii* on cucumber and on *Macrosiphum euphorbiae* on tomato, although the consumption rate of both prey species is lower than for *D. tamaninii* (**Section 3.1**) (Alvarado et al. 1997). *Macrolophus melanotoma* similarly consumes fewer thrips (Riudavets and Castañé 1998) and whiteflies (Riudavets et al. 1995) than does *D. tamaninii*.

Fauvel et al. (1987) determined fecundity when *M. melanotoma* fed on different prey. Egg production was highest on a diet of pyralid eggs and lowest when the bugs fed on spider mites. Fecundity was greater at 20°C than at 15, 25, 30 or 10°C (in decreasing order of mean fecundity). Readers are referred to the paper by Fauvel et al. (1987) for details of the influence of different temperatures and different prey on this mirid's fecundity, longevity, incubation period, and nymphal development. Additional data on life-history parameters of *M. melanotoma* are cited by Riudavets et al. (1995), Alvarado et al. (1997), and Riudavets and Castañé (1998). Constant et al. (1994) described embryonic development and analyzed biochemical characteristics of the egg, Ferran et al. (1996) described oviposition behavior, and Constant et al. (1996b) studied oviposition-site preferences in relation to hardness of host plant tissues.

The ability of *M. melanotoma* to develop on important pests such as the greenhouse whitefly, sweetpotato whitefly, and western flower thrips (Riudavets et al. 1993, 1995; Barnadas et al. 1998) increases its potential in biological control. This predator also is active at low temperatures (Carayon 1986). The bug is generally mass reared on eggs of the Mediterranean flour moth (*Ephestia kuehniella* Zeller), but an ability to culture the bugs on an artificial medium and to use an artificial substrate for oviposition (Grenier et al. 1989, Constant et al. 1996a) would further enhance its usefulness in biological control. This bug's omnivorous habits could facilitate its use in preventive introductions for whitefly suppression (Schelt et al. 1996). To be effective in controlling whiteflies in the greenhouse, *M. melanotoma* should be introduced when prey densities are low and may need to be used with the parasitoid *Encarsia formosa* Gahan (e.g., Malézieux et al. 1995, Schelt et al. 1996), but the mirid also can be used on early tomato crops when temperatures are too low for the parasitoid to exert effective whitefly control (Koppert 1993).

Macrolophus melanotoma, now commercially available (Barnadas et al. 1998), has been used successfully against the greenhouse whitefly on tomatoes, especially when it was released (four predators/plant) early in the whitefly cycle, followed by a later release at a similar rate (Malausa and Trottin-Caudal 1996). This dicyphine does not generally feed on tomato fruit, and plant injury has not been a problem in commercial greenhouses (Alomar et al. 1990, Koppert 1993, Malausa and Trottin-Caudal 1996). In the field, a release rate economically feasible for growers might be two bugs per square meter (Schelt et al. 1996). A recent decline in densities of *M. melanotoma* in tomato may have been the result of toxicity from insecticides used to control the western flower thrips on seedlings (Figuls et al. 1999).

Deraeocorinae

Species of the Deraeocorinae, a subfamily of worldwide distribution, are sometimes considered strict or obligate predators and can be specific to certain plant species (e.g., Knight 1941, Southwood and Leston 1959, Slater and Baranowski 1978, Dolling 1991, Kerzhner and Josifov 1999). Some plant feeding occurs in this subfamily, although its extent and importance require clarification. Certain mostly predacious species of *Deraeocoris* were once regarded as crop pests (see Wheeler 2001). The Deraeocorinae comprise six tribes (Schuh 1995), including Clivinematini, in which predation on scale insects is known; Deraeocorini, the largest tribe, members of which often are generalist predators; and Hyaliodini, composed of mite feeders as well as specialized predators of lace bugs. Although some species of *Deraeocoris* warrant consideration among the more important predacious Miridae, they are treated later in this chapter as less important predators (**Sections 4.5 to 4.8**).

Isometopinae

The largely tropical subfamily Isometopinae, owing to the possession of ocelli, has often been treated as a separate family. Carayon (1958) argued convincingly for the inclusion of these bugs in the Miridae. The Isometopinae are generally considered the most primitive of the mirid subfamilies (Schuh 1976, Schuh and Slater 1995). Biological data on isometopids are scant, but several species (see **Section 4.12**) are specialized predators of armored scales (Diaspididae).

Mirinae

The mostly predacious genus *Phytocoris* (**Section 4.14**), largest in the Miridae, belongs to the Mirinae (see **Chapter 3** for an overview of the subfamily and treatment of mirines as crop pests). Otherwise, this subfamily lacks many predators of economic importance. Omnivory, however, is common in several mirine genera, and even the injurious lygus bugs (**Section 4.13**) are facultative predators.

Orthotylinae

The diverse Orthotylinae (see **Chapter 3** for a brief summary of the subfamily and their role as plant pests) is characterized by numerous omnivores, especially in the largest tribe, Orthotylini.

Of the three species discussed here, one (*Cyrtorhinus fulvus*; **Section 3.4**) has been used successfully in classical biological control, and another (*C. lividipennis*; **Section 3.5**) is a major predator of rice planthoppers.

3.3 *Blepharidopterus angulatus* (Fallén)

Known as the black-kneed mirid (or capsid) in Britain, *B. angulatus* is a widespread Eurasian orthotyline (Kerzhner and Josifov 1999) that also occurs in North America. Its detection in the Nearctic region at Halifax, Nova Scotia, a Canadian port city from which numerous Palearctic insects have been reported (e.g., Hoebeke and Wheeler 1996), plus subsequent records from the Pacific Northwest, suggest multiple North American introductions and an adventive Nearctic status (Wheeler and Henry 1992).

Although *B. angulatus* has not been used in classical biological control, it is one of the best-studied predatory mirids. Much of the extensive biological literature has been summarized by Wheeler (2001); this reference should be consulted for additional information on *B. angulatus* as a predator of the European red mite *Panonychus ulmi* (Koch), leafhoppers, aphids, and the codling moth *Cydia pomonella* (L.). Huffaker et al. (1970) summarized British studies on this plant bug as a European red mite predator.

The egg was described and illustrated by Kullenberg (1942); Sanford (1964a) described the oviposition sites on apple. Collyer (1952) described and illustrated all immature stages, as well as the adult male and female. Collyer's (1952) paper also covers seasonality of this univoltine bug in Britain, duration of the nymphal stages, and consumption of European red mites by each nymphal instar and by the adult male and female. Kullenberg (1944), who referred only to this mirid's phytophagous habits, discussed the bionomics of Swedish populations (as *Aetorrhinus angulatus*). Ehanno (1987) provided information on seasonal history and host plants in France. Common hosts of *B. angulatus* are fruit trees such as apple (*Malus domestica* Borkh.), as well as alder (*Alnus*), birch (*Betula*), and other deciduous trees (Collyer 1952, Southwood and Leston 1959). Parasitism of *B. angulatus* was studied by Glen (1977a).

Blepharidopterus angulatus is a polyphagous predator that feeds on spider mites; auchenorrhynchans and sternorrhynchans ("homopterans") such as aphids, leafhoppers, and psyllids; and lepidopteran eggs and larvae. Murray and Solomon (1978) reported larvae of the European alder leafminer *Fenusa dohrnii* (Tischbein) as prey in England. In Nova Scotia, Canada, *B. angulatus* apparently is an important natural enemy of the white apple leafhopper *Typhlocyba pomaria* McAtee in the prey's second generation (MacPhee 1979). This mirid will cannibalize when its typical prey are scarce (Collyer 1965). Other natural enemies, such as phytoseiid mites and coccinellid larvae, also serve as alternative prey, and as do mummified aphids containing hymenopteran parasitoids (e.g., Collyer 1952, Krämer 1961, Glen 1973, Wheeler 2001).

Attention was focused on the black-kneed mirid as a mite predator in Britain after Austin and Massee (1947) noted that some insecticide trials (1943 to 1944) had to be discontinued because the mirid had so reduced European red mite populations. Elsie Collyer's investigations began in 1945 (Collyer 1952). Although the mirid is one of the mite's most important natural enemies, this predator often allows mite densities to fluctuate widely from year to year so that long-term suppression of the pest is not achieved (Collyer 1953c, 1964b). Predation is generally most effective on second-generation mites because overwintered eggs of the black-kneed mirid hatch 4 to 5 weeks later than those of its prey (Muir 1965b). The plant bug proved sensitive to pesticides used during the 1950s in full spray programs in British orchards (Collyer and Massee 1958); pesticide use thus interfered with effective mite control. In a series of papers, R. C. Muir continued to study predation by *B. angulatus* as a factor in stabilizing European red mite populations. Such work included the effects of pesticides on both mirid and mite densities (Muir 1965a,b), effect of temperature on diapause termination in overwintered eggs of *B. angulatus* (Muir 1966a), and recolonization of apple orchards following the bug's elimination by DDT applications (Muir 1966b). In addition, a capture–recapture technique was used in orchards to estimate the size and rates of change in mirid numbers (Muir 1958).

In other studies, researchers have looked at *B. angulatus* as a predator of codling moth eggs in Britain (e.g., Glen 1975a); predation intensity appears to be unrelated to egg density (Glen 1977b). Glen and Brain (1978) included the mirid in a predation model that might prove useful in managing codling moth infestations in commercial orchards. The use of alder windbreaks facilitates the bug's colonization of orchards (e.g., Solomon 1975). Gange and Llewellyn (1989) studied population dynamics of the mirid on alder (*Alnus* spp.) and determined the factors that affect its colonization of orchards from windbreaks. Predation on codling moth eggs (and neonate larvae) also occurs in Canada (MacLellan 1962, 1973), but apparently this mirid is a less important mortality factor than it is in Britain (Glen 1975a).

Glen (1975b) studied the searching behavior and prey-density requirements of *B. angulatus* as a predator of the aphid *Eucallipterus tiliae* (L.) and the leafhopper *Alnetoidea alneti* (Dahlbom) on linden (lime) trees (*Tilia* × *vulgaris* Hayne) in Britain. Included in this research was the construction of a model that allowed prediction of the prey densities needed for optimal development by each mirid instar, and determination of prey consumption relative to aphid or leafhopper densities. Glen (1973) assessed the plant bug's food requirements when feeding on lime aphids, noting that the fifth instar contributed 50% to total weight of prey consumed and that overall growth efficiency of females is significantly greater than that of males. Glen and Barlow (1980) investigated the population dynamics of *B. angulatus* when feeding on lime aphids.

3.4 *Cyrtorhinus fulvus* Knight

Endemic to the Pacific and Indo-Malay regions (Matsumoto and Nishida 1966), *C. fulvus* has effected long-term control of its prey, the taro delphacid *Tarophagus colocasiae* (Matsumura). This egg predator provides an outstanding example of the use of classical biological control.

Cyrtorhinus fulvus was introduced to Hawaii (with several hymenopteran parasitoids) from the Philippines in 1938, following outbreaks of the adventive taro delphacid ("leafhopper" in the early literature) in 1930. The edible roots of taro are an important food source for Hawaiians. The mirid soon increased in abundance and brought the delphacid under control (Fullaway 1940). In 1947, *C. fulvus* was introduced to Guam, where it also suppressed populations of the taro delphacid (Pemberton 1954). Nearly 50 years after its initial releases in Hawaii, this plant bug was said to keep the delphacid in check in the Pacific Islands unless the application of pesticides reduced predator effectiveness (Mitchell and Maddison 1983, Waterhouse and Norris 1987).

Substantial biological data on *C. fulvus* and results from experimental work on predator–prey relationships became available well after pest populations had been reduced to subeconomic levels. Matsumoto and Nishida (1966) determined that the incubation period is about 13 days and that nymphal development required about 15 days (temperature and humidity were not controlled). Either four or five instars occur in this specialized predator of *T. colocasiae* eggs. Ovipositing females are attracted by the presence of prey eggs. Adults of *C. fulvus* are stronger fliers than those of their prey; the mirid also shows a strong numerical response. Fluctuations in predator numbers are highly correlated with those of its fulgoroid prey, spatial distributions of predator and prey overlapping (Döbel and Denno 1994, Wheeler 2001). Populations of *C. fulvus* consist mostly of females when prey densities are low, but contain mostly males when delphacid numbers are high (Matsumoto and Nishida 1966). Egg cannibalism (Matsumoto 1964) helps this predator to persist during periods of prey scarcity.

Matsumoto and Nishida (1966) demonstrated that hand removal of *C. fulvus* from taro leads to increases in prey populations. As Döbel and Denno (1994) noted, the mirid's dispersal advantage over its prey compensates for the predator's reproductive disadvantage.

3.5 *Cyrtorhinus lividipennis* Reuter

Occurring in southern Asia and the Pacific Islands (Carvalho and Southwood 1955, Liquido and Nishida 1983), *C. lividipennis* has an Indo-Pacific distribution (*sensu* Schuh 1984). Liquido and Nishida (1983) noted that this predator generally occurs in humid, flooded rice fields but not in upland areas.

Cyrtorhinus lividipennis affects the population dynamics of co-occurring rice pests, such as the brown planthopper *Nilaparvata lugens* (Stål), the whitebacked planthopper *Sogatella furcifera* (Horváth), and the green leafhopper *Nephotettix virescens* (Distant). This predator also has been introduced to Hawaii to help suppress the corn delphacid *Peregrinus maidis* (Ashmead). As might be expected, such an important predator has been the subject of numerous studies. Döbel and Denno (1994) summarized the life-history attributes of the mirid and the brown planthopper, emphasizing factors that influence suppression and stability of prey populations. Both the brown planthopper and whitebacked planthopper are wing-dimorphic pests whose polymorphic life histories tend to promote escape from predators (see Döbel and Denno 1994). Readers should supplement the brief account that follows by referring to Döbel and Denno (1994), Hagen et al. (1999), and Wheeler's (2001) treatment of *C. lividipennis*.

Reyes and Gabriel (1975) described and illustrated the immature stages of *C. lividipennis*. The number of nymphal instars varies from three to five, such variation apparently being intrinsic to a particular population (Liquido and Nishida 1985c). For the three-instar, four-instar, and five-instar groups, mean nymphal development times are 14.6, 18.6, and 20.2 days, respectively (Napompeth 1973). Data on life-history attributes, such as incubation period, preoviposition, fecundity, longevity, and duration of the nymphal stages, were provided by Napompeth (1973), Reyes and Gabriel (1975), Pophaly et al. (1978), Misra (1980), Liquido and Nishida (1985a), and Chua and Mikil (1989). The effects of prey and stage on fecundity, longevity, development time, size, and food consumption were reported by Chua and Mikil (1989). Liquido and Nishida (1985b) described eclosion, molting, and oviposition and mating behavior. The numbers of delphacid prey consumed were determined by Reyes and Gabriel (1975) and Chua and Mikil (1989), and predation rates under laboratory conditions were measured by Sivapragasam and Asma (1985). Song and Heong (1997) assessed the mirid's functional response when fed brown planthopper eggs at six constant temperatures (20 to 35°C). For both brown planthopper and green leafhopper eggs, the functional response of *C. lividipennis* conforms to Holling's type II (Sivapragasam and Asma 1985, Heong et al. 1990, Song and Heong 1997). Although *C. lividipennis* feeds mostly on plant- and leafhopper eggs, it preys more on auchenorrhynchan nymphs than do other specialized egg predators in the family, such as *C. fulvus* (**Section 3.4**) and *Tytthus mundulus* (**Section 3.6**) (Rapusas et al. 1996; see also references in Wheeler 2001).

Döbel and Denno (1994) noted examples of both the success and failure of *C. lividipennis* to suppress populations of rice hoppers (see also Wheeler 2001). This predator's ability to respond numerically to increases in pest densities helps offset destabilizing effects of the mirid's functional response. The mirid can aggregate in areas of high prey density and increase its reproduction in response to increasing prey numbers (Döbel and Denno 1994; Wheeler 2001, and references therein). When predator–prey populations are temporally synchronized, *C. lividipennis* can track and suppress its planthopper prey. Data on the plant bug's effectiveness have accrued from predator-removal experiments and from studies that show a negative relationship between predator density and rate of increase in prey populations (Benrey and Lamp 1994, Döbel and Denno 1994). A late colonization of rice and, therefore, a delayed numerical response is mostly responsible for the failure of *C. lividipennis* to suppress rice planthopper populations (e.g., Dyck and Orlido 1977, Bentur and Kalode 1987, Manti 1989). An ability of the mirid to switch to alternative prey, to cannibalize, and to feed to a limited extent on rice enhances it persistence (O'Connor 1952; Hinckley 1963; Napompeth 1973; Sivapragasam and Asma 1985; Chua and Mikil 1989; van den Berg et al. 1988, 1992; Ooi and Shepard 1994).

The mirid also can migrate to distant areas (e.g., Kisimoto 1979; Ooi 1979; Riley et al. 1987, 1995; Reynolds and Wilson 1989; Rutter et al. 1998). In Japan, *C. lividipennis* cannot overwinter, but it migrates there annually from mainland China. The mirid's late colonization of Japanese rice fields, compared with its prey, results in asynchrony in predator–prey populations. *Cyrtorhinus lividipennis* thus plays a less important role in limiting brown planthopper densities in Japan than it does in tropical Asia or the Pacific region (Kuno and Hokyo 1970, Kiritani 1979, Benrey and Lamp 1994, Döbel and Denno 1994).

The use on rice of pyrethroids and certain other insecticides tends to uncouple *Cyrtorhinus*–prey interactions. In addition to insecticide-induced decreases in *C. lividipennis* populations (e.g., Stapley 1976, Dyck and Orlido 1977, Reissig et al. 1982a, Kalode 1983, Fabellar and Heinrichs 1984), a resurgence in hopper numbers may result from the stimulatory effect of an insecticide on prey fecundity or on plant growth (e.g., Chelliah and Heinrichs 1980, Heinrichs et al. 1982, Reissig et al. 1982b). Insecticide use in rice sometimes can be reduced by planting resistant cultivars, although resistant rices can be less attractive to the mirid than are prey-susceptible cultivars (Rapusas et al. 1996). Hare (1994) and Rapusas et al. (1996) stressed the need for long-term evaluations of compatibility between the use of resistant rice cultivars and the use of natural enemies such as *C. lividipennis*.

Cyrtorhinus lividipennis has been introduced from Guam to Hawaii to help control the adventive corn delphacid, which was not being effectively suppressed by another introduced predatory mirid, *Tytthus mundulus* (see **Section 3.6**). The former species has adapted to the relatively dry corn agroecosystem, which contrasts with the humid environments this plant bug occupies in its native region (Liquido and Nishida 1985c). The mirid responds numerically to increases in prey populations (Napompeth 1973). In some years, a lack of synchrony between predator and prey populations explains the mirid's failure to maintain corn delphacid numbers below threshold levels (Napompeth 1973; see also Döbel and Denno 1994, Wheeler 2001).

Phylinae

The Phylinae are characterized by a relatively high proportion of omnivorous species. This mirid subfamily not only contains important plant pests (see **Chapter 3**) and useful predators, but also several species that have been variously categorized as key pests or as key predators.

3.6 *Tytthus mundulus* (Breddin)

A member of the small tribe Leucophoropterini, *T. mundulus* occurs in southern Asia, the Pacific islands, and Australia, and has been introduced to Hawaii (U.S.A.) for biological control (Usinger 1939; Zimmerman 1948a,b; Liquido and Nishida 1983; Schuh 1984; Cassis and Gross 1995). The record from Japan apparently was based on misidentification (Kerzhner and Josifov 1999). *Tytthus mundulus* was released in Florida (U.S.A.) after the sugarcane delphacid was detected there in the early 1980s, but the mirid did not become established in sugarcane fields (Frank and McCoy 1993). Species of *Tytthus* were placed in the orthotyline genus *Cyrtorhinus* (see **Sections 3.4, 3.5**) before Carvalho and Southwood (1955) showed that two subfamilies were involved: Orthotylinae (*Cyrtorhinus* spp.) and Phylinae (*Tytthus* spp.).

The use of *T. mundulus* against the sugarcane delphacid *Perkinsiella saccharicida* Kirkaldy represents a successful episode in biological control, providing an example of long-term effective suppression of an exotic plant pest (Huffaker et al. 1971). For more information on how this mirid saved the Hawaiian sugarcane industry, readers should consult detailed summaries of this biocontrol project by Muir (1920), Imms (1926), Timberlake (1927), Williams (1931), Swezey (1936), Pemberton (1948), or Wheeler (2001).

Following the sugarcane delphacid's ("leafhopper" in early economic literature) detection in Hawaii in 1900, this adventive pest soon threatened the island sugarcane industry. A search for natural enemies began in areas where the delphacid is native. In 1920, Frederick Muir observed *T. mundulus* preying on the delphacid in Queensland, Australia. Even though he had experimented (1905 to 1906) with this plant bug as a predator of planthopper nymphs and adults in Fiji, he was then unaware of the bug's largely specialized egg-feeding habits and, therefore, did not appreciate its biocontrol potential. After an initial release of *T. mundulus* from Australia failed to establish the predator in sugarcane, additional material was shipped from Fiji later in 1920. Within 3 years, the mirid brought the sugarcane delphacid under control in Hawaii (e.g., Williams 1931, Swezey 1936).

Although it was known at the time that *T. mundulus* was able to complete about ten annual generations (contrasted to three or four for its prey) (Swezey 1936), only later (1950s) did researchers identify life-history attributes of the plant bug (and those of the delphacid) that facilitated pest suppression. It is now known that the mirid's effects on sugarcane delphacid populations are density dependent; that its shorter generation time, compared with its prey, helps offset the planthopper's greater fecundity; and that the mirid can track its prey reproductively (Verma 1955, Bull 1981, Döbel and Denno 1994). The sugarcane delphacid, unlike the even more fecund corn delphacid (Verma 1955), is not wing dimorphic; wing dimorphism promotes instability in predator–prey interactions and tends to favor pest outbreaks (see Döbel and Denno 1994).

Verma (1955) provided information on oviposition habits, preoviposition period, fecundity, longevity, and duration of the nymphal stages. Napompeth (1973) discovered variation in the number of nymphal instars of *T. mundulus* (mostly five, occasionally four).

Because *T. mundulus* will feed on auchenorrhynchans other than the sugarcane delphacid, researchers had hoped that this predator would also suppress Hawaiian populations of the corn delphacid. Nymphs are able to develop on eggs or nymphs of this pest (Verma 1955). Williams (1931), however, had pointed out that *T. mundulus* is not attracted to corn fields. Even though this predator had proved so successful in controlling the sugarcane delphacid, studies by Verma (1955) and Napompeth (1973) revealed aspects of the corn–delphacid–mirid relationship that preclude successful control of *Peregrinus maidis*, or at least make long-term pest suppression unlikely. Among these factors is the mirid's preference for ovipositing in different tissues and sites from those of its prey. The corn delphacid often oviposits in soft tissues of roots of young plants (it avoids hard, dry tissues of old hosts that are favored by the mirid), where its eggs are less vulnerable to mirid attack as a result of protection by ants. Because of the mirid's oviposition preferences, delphacid eggs laid in roots are actually unavailable to this predator. Corn plants also do not provide a suitable windbreak for *T. mundulus* (Verma 1955). Moreover, *T. mundulus* typically shows a delayed numerical response, migrating into corn later than its delphacid prey because hardened plant tissues are needed for oviposition. As noted above, the planthopper's wing dimorphism promotes escape from predators (see Döbel and Denno 1994, Wheeler 2001).

Tytthus mundulus also has been evaluated for possible release against other auchenorrhynchan pests, for example, a tropiduchid planthopper (*Numicia viridis* Muir) on sugarcane in southern Africa (Simmonds 1969). That the mirid can be cultured on alternative (i.e., nonhomopteran) prey in the laboratory (Stephens 1975, Takara and Nishida 1981) might allow mass rearing of this predator and enhance its further use in biological control (Wheeler 2001).

4. THE LESS IMPORTANT SPECIES

Bryocorinae

4.1 *Campyloneura virgula* (Herrich-Schaeffer)

This common Old World dicyphine (see Kerzhner and Josifov 1999, for distribution) is adventive in North America, where it is known from British Columbia (Canada) to California (U.S.A.) (Lattin and Stonedahl 1984). An illustration of the egg was given by Cobben (1968) and a color illustration and brief description of the fifth instar by Butler (1923).

Found on trees and shrubs such as alder, European filbert (*Corylus avellana* L.), hawthorn (*Crataegus* spp.), and oak (*Quercus* spp.) (Ehanno 1987), *C. virgula* is parthenogenetic (Carayon 1989, Wheeler and Henry 1992) and univoltine (Ehanno 1987), the eggs overwintering in Britain but adults overwintering in France (Ehanno 1987). Prey consist of small arthropods and their eggs, including spider mites, aphids, and psocids; chrysomelid larvae also are attacked (e.g., Schumacher 1917, Southwood and Leston 1959, Strawiński 1964, Lattin and Stonedahl 1984).

4.2 Dicyphus errans (Wolff)

A Eurasian species (see Kerzhner and Josifov 1999, for distribution), *D. errans*, like most dicyphines, is omnivorous. Like many other dicyphines, it lives on plants with glandular trichomes. Southwood (1986) described the pretarsal adaptations and walking behavior that allow the bugs to traverse plant surfaces without being entrapped by sticky exudates.

In reporting injury to zonal geranium (*Pelargonium* × *hortorum* L. H. Bailey) in German greenhouses, Schewket Bey (1930) noted that nymphal development time decreased when the diet included aphids (rather than only geranium leaves), and that this mirid could not develop without some animal food. In France, *D. errans* has been combined with hymenopteran aphid parasitoids to produce tomato crops in greenhouses without the need for insecticides (Lyon 1986; see also Malausa and Trottin-Caudal 1996). This predator occurs on greenhouse-grown tomatoes in Italy, where it preys not only on the green peach aphid *Myzus persicae* (Sulzer), but also on eggs and adults of the greenhouse whitefly. *Dicyphus errans* is considered an important predator in Italy; its occurrence on tomatoes in greenhouses tends to coincide with declines in aphid densities during periods when insecticides are not used (Petacchi and Rossi 1991, Quaglia et al. 1993).

4.3 Engytatus modestus (Distant) (= Cyrtopeltis modestus Distant)

This facultative predator was discussed in **Chapter 3** as a plant pest; that chapter's **Section 4.1** contains a summary of its distribution and reviews key papers on its bionomics. Tanada and Holdaway (1954) noted that *E. modestus* feeds on aphids, mealybugs, lepidopteran eggs and small larvae, and on other small insects. It preys on aphids on tomato in Hawaii (U.S.A.) (e.g., Holdaway and Look 1940), and, in Florida (U.S.A.), *E. modestus* so reduced infestations of the green peach aphid in an experimental tobacco seedbed that insecticides could no longer be screened for aphid control (Wilson 1948). Larvae of agromyzid leafminers are also attacked within their mines. Parrella and Bethke (1982) and Parrella et al. (1982) assessed the biocontrol potential of this plant bug as a natural enemy of a serpentine leafminer, *Liriomyza trifolii* (Burgess), on chrysanthemum (*Chrysanthemum* × *morifolium* Ramat.) and tomato. *Engytatus modestus* also preys on other Miridae, for example, the dicyphine *Tupiocoris notatus* (Distant) on tobacco (*Nicotiana tabacum* L.) (Thomas 1945). The latter mirid species, not treated in this chapter, is also a facultative predator.

4.4 Nesidiocoris tenuis (Reuter) (= Cyrtopeltis tenuis Reuter)

An omnivore treated as a plant pest in **Chapter 3**, *N. tenuis* is one of several Miridae considered injurious to a particular crop by some authors but a beneficial natural enemy on the same crop by other workers (e.g., Libutan and Bernardo 1995). The conflicting reports of this bug as either a harmful or useful species on tobacco were reviewed by Torreno and Magallona (1994).

Life-history data were presented by El-Dessouki et al. (1976), who focused on plant feeding by *N. tenuis*, and by Torreno and Magallona (1994), who emphasized predation. The latter authors reviewed the biological literature on this mirid, including papers not cited by El-Dessouki et al.

(1976). Torreno and Magallona's (1994) study contains information on mating, preoviposition period, fecundity, longevity, nymphal development, feeding sites and spatial distribution on tobacco, alternative host plants, and natural enemies. They noted that fecundity, longevity, and survival were enhanced by a diet of tobacco plus prey (noctuid larvae) contrasted with a tobacco-only diet. Similarly, Libutan and Bernardo (1995) reported that development is more rapid on a diet of noctuid eggs than on a plant (tomato) diet.

This facultative predator accepts a wide range of prey and will cannibalize and also feed on dead insects trapped on its sticky host plants (e.g., El-Dessouki et al. 1976, Torreno and Magallona 1994). Neonate noctuid larvae (<8 hours old) are preferred over 1- or 2-day-old larvae. Second instars are usually attacked only during or immediately after a molt (Torreno 1994). As in many other predacious mirids, zoophagous tendencies and prey consumption are greater in later instars (Torreno 1994, Libutan and Bernardo 1995). In the Philippines, *N. tenuis* is considered an important predator of the cutworm *Spodoptera litura* F. on tobacco (Torreno 1994).

Kajita (1978) reported that *N. tenuis* feeds on all stages of the greenhouse whitefly under laboratory conditions. The mirid occurs naturally in tomato fields of the Mediterranean region and sometimes invades and becomes established in greenhouses (Goula and Alomar 1994, Malausa and Trottin-Caudal 1996). Predation on whiteflies and on larvae of the agromyzid leafminer *Liriomyza trifolii* has been observed on gerbera (*Gerbera jamesonii* Bolus ex Hook. f.) and tomato in unheated plastic greenhouses (Nucifora and Calabretta 1986), and on greenhouse-grown zucchini (*Cucurbita pepo* L.) in Italy (Arzone et al. 1990). In surveys of potential predators of the western flower thrips in Spain, *N. tenuis* was rarely found in tomato fields; in the laboratory, nymphs readily consumed thrips larvae (Riudavets and Castañé 1998).

Deraeocorinae

4.5 *Deraeocoris brevis* (Uhler)

A multivoltine generalist predator indigenous to western North America, *D. brevis* ranges from Alaska, Yukon, and British Columbia to Manitoba south to California, Colorado, Arizona, and New Mexico (Henry and Wheeler 1988, Scudder 1997). This species, in contrast to host-plant-specific members of the genus, develops on numerous plants, and is especially common on sagebrush (*Artemisia*) in sagebrush communities (Scudder 1997). Westigard (1973) determined that nymphal development requires about 25 days (21°C) and reported the average daily consumption of pear psylla (*Cacopsylla pyricola* Foerster) eggs and nymphs by nymphs (I to V) and adults of *D. brevis*. An average of about 400 immature pear psylla are consumed during the predator's nymphal development. Adults overwinter in protected places, the populations often female biased (Westigard 1973, Horton et al. 1998). Two to four annual generations have been recorded (McMullen and Jong 1967, Westigard 1973). Horton et al. (1998) demonstrated that the critical photoperiod inducing reproductive diapause is between 16:8 and 15:9 hours (L:D). The preoviposition period of non-diapausing females is 10 to 11 days (Horton et al. 1998).

Most of the biological literature on *D. brevis* concerns its role as a natural enemy in tree fruit ecosystems of the Pacific Northwest (Horton et al. 1998). It is usually more common on pear than other predatory heteropterans, such as anthocorids (Riedl 1991); nymphs and adults often are the most numerous natural enemies in commercial pear orchards during late July and August (Booth and Riedl 1996). Although this predatory mirid sometimes occurs on pear in large numbers during its first generation, it typically is not an early-season colonist in orchards (Westigard 1973, Fye 1981, Gut et al. 1982). Suppression of pear psylla populations may not occur if cool, early-season conditions delay immigration of *D. brevis* adults and buildup of its populations (Westigard et al. 1979). The use of certain insecticides disrupts *D. brevis*–pear psylla interactions (e.g., Westigard 1973, Hagley and Simpson 1983, van de Baan and Croft 1990).

Deraeocoris brevis has also been studied as a predator of the apple aphid *Aphis pomi* De Geer (Carroll and Hoyt 1984, Haley and Hogue 1990) and the filbert aphid *Myzocallis coryli* (Goetze) (Messing and AliNiazee 1985, 1986) in the Pacific Northwest. This plant bug might reduce densities of the McDaniel spider mite *Tetranychus mcdanieli* McGregor on apple (Hoyt 1969) and is considered a potential enemy of the grape mealybug *Pseudococcus maritimus* (Ehrhorn) (Grasswitz and Burts 1995). Westigard (1973) observed *D. brevis* on scale insect-infested Ponderosa pine, *Pinus ponderosa* Douglas ex Lawson & C. Lawson. Additional prey include the European red mite, leafhopper nymphs, and neonate tortricid larvae. Occasional feeding on leaves and immature fruit may help sustain its populations in orchards during times when arthropod prey are at low densities (McMullen and Jong 1967). *Deraeocoris brevis* is the only native mirid available from suppliers of beneficial organisms in North America; it appears in Hunter's (1997) list of suppliers as an aphid, a whitefly, or a generalist predator.

4.6   *Deraeocoris nebulosus* (Uhler)

This widely distributed generalist predator occurs from New Brunswick and Nova Scotia (Canada) to Florida (U.S.A.) west only to Ontario in Canada but to California, Colorado, Arizona, and New Mexico in the United States (Henry and Wheeler 1988). McCaffrey and Horsburgh (1980) described the egg and the oviposition site (leaf midveins) on apple; Wheeler et al. (1975) described and illustrated all nymphal instars.

Long known as a predator, *D. nebulosus* was listed as a natural enemy of cankerworms (Geometridae) by Uhler (1876) and as a predator of soft scales or coccids by Murtfeldt (1894). Mary Murtfeldt also observed predation on an oak-inhabiting phylloxeran in 1888 (Knight 1921).

Wheeler et al. (1975) noted that *D. nebulosus* is trivoltine in Pennsylvania (U.S.A.), where it tracks a succession of prey, such as mites, aphids, scale insects, psyllids, whiteflies, and lace bugs on landscape trees and shrubs; determined nymphal development times (total time averages approximately 20 days at 20 to 21°C); and demonstrated that consumption of oak lace bug, *Corythucha arcuata* (Say), nymphs by adult mirids conforms to a type II functional response. About 108 lace bug nymphs, on average, are consumed during nymphal development. Snodgrass (1991) found sizable populations (>137,000/ha) in Mississippi (U.S.A.) cotton fields and suggested that this plant bug might be an important predator of cotton pests. In further work on the bug's potential as a natural enemy in cotton, Jones and Snodgrass (1998) reported predation on the silverleaf whitefly *Bemisia argentifolii* Bellows & Perring. Nymphs fed a whitefly diet developed in an average of about 33 days; mean fecundity was about 242 eggs.

Deraeocoris nebulosus is often common in apple orchards, where it preys on pests such as the codling moth (e.g., Stultz 1955) and tortricoid leafrollers (Parrella et al. 1981). The latter authors considered this mirid a major predator that warrants emphasis in IPM programs for apple pests. Occasionally, *D. nebulosus* preys on other beneficial insects in orchards, for example, the anthocorid *Orius insidiosus* (Say) (McCaffrey and Horsburgh 1986). This plant bug also may be an important predator of aphids, spittlebugs, and the pecan nut casebearer *Acrobasis nuxvorella* Neunzig in pecan, *Carya illinoensis* (Wangenh.) K. Koch, orchards (Mizell and Schiffhauer 1987, Tedders 1995, Ellington et al. 1998).

4.7   *Deraeocoris pallens* (Reuter)

An Old World species known primarily from the Middle East (Kerzhner and Josifov 1999), *D. pallens* is a generalist predator that feeds mainly on aphids and whiteflies. It is one of the more common predatory insects in cotton fields in Israel (Susman 1988). Under laboratory conditions (25 to 28°C), fecundity ranged from 23 to 268 eggs, the preoviposition period was 4 to 5 days, and longevity of females was 14 to 34 days. When this deraeocorine was reared on the sweetpotato whitefly, the incubation period averaged 6.4 days and nymphs developed in 11.1 days. The whitefly and the cotton aphid were preferred as prey over the citrus mealybug *Planococcus citri* (Risso) (Susman 1988).

This multivoltine mirid also is an important predator of cotton pests in Turkey. Ghavami et al. (1998) determined that nymphs and adults feed not only on aphids and whiteflies but also on spider mites and lepidopteran larvae. Consumption increases with increasing prey densities. In Iran, *D. pallens* is considered an important natural enemy of the alfalfa weevil *Hypera postica* (Gyllenhal) (Vojdani and Doftari 1963). Occasional intraguild predation occurs; Susman (1988), for example, noted that *Orius* sp. in cotton can serve as prey of this mirid.

4.8 *Deraeocoris signatus* (Distant)

Known from Australia, India, and Sri Lanka (Cassis and Gross 1995), *D. signatus* is one of several species of the genus that were once thought to be crop pests. Chinajariyawong and Harris (1987), however, demonstrated that the brown smudge bug does not injure cotton in Australia and is mostly predacious. Newly hatched nymphs provided only cotton squares or tips died within 3 days; nymphs developed to adults when fed cotton plus cotton aphids or aphids alone. Fecundity and longevity were enhanced by an aphid diet. Under laboratory conditions, nymphs also prey on European red mites and noctuid eggs and larvae. This predator also occurs in alfalfa, *Medicago sativa* L., in peanut, *Arachis hypogaea* L., in soybean, *Glycine max* (L.) Merr., and in sunflower, *Helianthus* spp. (Chinajariyawong and Harris 1987, and references therein).

4.9 *Hyaliodes harti* Knight

Hyaliodes harti, belonging to the deraeocorine tribe Hyaliodini, is widespread in North America. In Canada, it occurs from Nova Scotia to British Columbia; the U.S. range is Maine to Georgia west to Colorado and North Dakota (Henry and Wheeler 1988, Polhemus 1994). Although *H. harti* was imported from Nova Scotia to New Zealand by biological control specialists, the mirid did not become established (Walker et al. 1989). This species resembles the often co-occurring *H. vitripennis* (Say); Knight (1941) described *H. harti* as a distinct species. Some older references to the former species (e.g., Gilliatt 1935) refer to *H. harti*. *Hyaliodes vitripennis* is not discussed further here; the most detailed biological study of *H. vitripennis* is that by Horsburgh (1969).

Only sketchy bionomic data are available for *H. harti*. Sanford (1964a) described the typical oviposition site on apple; eggs usually are laid singly in thick wood near a bud on current-season growth, the operculum flush with the bark surface. This predacious mirid occurs on grape (*Vitis* spp.), raspberry (*Rubus* spp.), and fruit trees, as well as on maple (*Acer* spp.), oak, and other hardwood trees (e.g., Kelton 1983). In Nova Scotia, Gilliatt (1935) stated that overwintered eggs of this univoltine species hatch on apple in late June, that adults appear in about 4 weeks, and that adults are numerous only through early August.

Hyaliodes harti feeds mainly on mites, mite eggs, aphids, and psyllids (e.g., Braimah et al. 1982, Kelton 1983). Gilliatt (1935), Lord (1949), and Sanford and Herbert (1966) considered it an important predator of all stages of the European red mite on apple. The bugs suppress mite numbers in some years, but in others they allow densities to fluctuate widely (Sanford and Lord 1962). The apple rust mite *Aculus schlechtendali* (Nalepa), which is readily consumed in laboratory feeding trials, might serve as alternative prey in orchards (Herbert and Sanford 1969). Lord (1968) found that *H. harti* preferred apple trees or limbs that bear a high proportion of fruiting clusters. Many of the pesticides once used to control pests of apple proved toxic to *H. harti* (e.g., MacPhee and Sanford 1956, 1961). Croft (1990), however, reported that this plant bug was one of the most pesticide-tolerant species among the common natural enemies in his database.

Hyaliodes harti, perhaps best known as a mite predator on apple, also attacks other apple pests. It preys on eggs and larvae of the eyespotted bud moth *Spilonota ocellana* (Denis & Schiffermüller) in cages, and nymphs fed on the apple mealybug *Phenacoccus aceris* (Signoret) under experimental conditions (Chachoria 1967). In some years, *H. harti* substantially reduces codling moth populations in eastern Canada (Shteynberg 1962; MacLellan 1963, 1972).

4.10 *Stethoconus japonicus* Schumacher

Most heteropteran predators are generalists. Species of *Stethoconus* (Deraeocorinae: Hyaliodini), however, are obligate predators of lace bugs. Adults of *Stethoconus* species somewhat resemble those of their lace bug prey, particularly in their strongly projected scutellum and the reticular markings on the hemelytra or forewings (Yasunaga et al. 1997). Carayon (1960) and Henry et al. (1986) reviewed known *Stethoconus*-tingid associations (see also Wheeler 2001). In this small genus (nine spp.; Schuh 1995), substantial biological data are available only for *S. japonicus* and *S. praefectus* (Distant). The latter species, not treated in this chapter, was studied by Mathen and Kurian (1972).

Stethoconus japonicus, native to Asia, is known from China, Japan, Korea, and Russia (Far East) (Kerzhner and Josifov 1999). It is considered adventive in the New World (District of Columbia, Maryland, and New York, U.S.A.), probably having been accidentally introduced with shipments of azalea (*Rhododendron*) nursery stock (Wheeler and Henry 1992).

In first reporting *S. japonicus* from North America, Henry et al. (1986) redescribed and illustrated the adult. Neal et al. (1991) described the egg and oviposition sites on azalea, reported an average fecundity of 236 eggs, female longevity of 20 to 30 days in the laboratory (~26°C), an incubation period (nondiapausing eggs) of about 19 days, and a nymphal development period of about 9 days. When reared on their preferred prey, the azalea lace bug *Stephanitis pyrioides* (Scott), nymphs consumed about 18 prey nymphs.

Neal and Haldemann (1992) emphasized the significance of the mirid's change in oviposition sites with progression of the season. Eggs, which are laid predominantly in the midribs of new leaves during June, are deposited later in the season in current-year stems or leaf scars of year-old growth. This shift in oviposition sites correlates with declining leaf moisture, which the bugs apparently can perceive — that is, females respond behaviorally to changes in the water potential of their host plants (Neal et al. 1991, Neal and Haldemann 1992). Overwintered eggs of *Stethoconus japonicus* do not hatch in Maryland (U.S.A.) until late June, when second instars of second-generation lace bugs are present. Synchrony of the predator with its prey's second generation might be an adaptation that prevents overexploitation of its sole food source. Moreover, an asynchrony in the life cycles of *S. japonicus* and the azalea lace bug could enhance this predator's biocontrol potential or its incorporation into pest-management programs in urban and suburban landscapes; the mirid may escape toxic effects of insecticides applied to control first-generation lace bugs (Neal and Haldemann 1992).

Stethoconus japonicus not only is a natural enemy of the azalea lace bug on ornamental azaleas in Japan, but also is associated with another Asian lace bug pest, *Stephanitis takeyai* Drake & Maa, on *Pieris japonica* (Thunb.) D. Don ex G. Don. (Yasunaga et al. 1997). This Asian shrub, known as Japanese andromeda, is commonly used as a landscape plant in North America and is often seriously injured by *S. takeyai* (Dunbar 1974), a lace bug that has been accidentally introduced into North America (Bailey 1950). Perhaps *Stethoconus japonicus* will increase its U.S. distribution on azalea and also will begin to colonize Japanese andromeda infested with *Stephanitis takeyai*, the so-called andromeda lace bug.

4.11 *Termatophylidea* spp.

In the immediately preceding account of *Stethoconus japonicus* (**Section 4.10**), the existence of specialized predators in the Miridae was said to be uncommon. Species of *Termatophylidea*, a small Neotropical genus (7 spp.) in the small tribe Termatophylini (30 spp. worldwide; Cassis 1995), provide another example of specialized habits; at least several species appear to be obligate thrips predators. Callan (1975) commented that these bugs occupy a niche filled in other regions of the world by anthocorids.

Wheeler (2001) reviewed *Termatophylidea*–thrips associations, noting that three species — *T. maculata* Usinger, *T. opaca* Carvalho, and *T. pilosa* Reuter & Poppius — prey on the redbanded thrips *Selenothrips rubrocinctus* (Giard). Van Doesburg (1964) described the immature stages and illustrated the egg and the first, third, and fifth instars of *T. opaca*. In Suriname, this species occurred on the lower leaf surfaces of cocoa, *Theobroma cacao* L., mostly among colonies of the redbanded thrips. All adults ($n = \sim 30$) reared on the thrips in the laboratory were females (van Doesburg 1964), but parthenogenesis in *T. opaca* has not been confirmed. Callan (1975) considered parthenogenesis unlikely because males are known in five other species of the genus.

Nymphs and adults of *T. maculata* (as *T. "maculosa"*) were observed in Trinidad to prey on redbanded thrips larvae (adults were not attacked) on leaves of cocoa and cashew (*Anacardium occidentale* L.), and this plant bug was reared on thrips larvae (Callan 1943).

Although van Doesburg (1964) and Callan (1975) regarded *Termatophylidea* species as relatively unimportant predators of thrips, no experimental studies have been conducted. Wheeler (2001), noting at least a potential for *Termatophylidea* species to suppress populations of pestiferous thrips in greenhouses, acknowledged the possibility that these bugs might not be able to reproduce on novel prey such as the western flower thrips, to adapt to herbaceous host plants of thrips, or to search thrips-infested parts of novel hosts.

Isometopinae

4.12 *Corticoris signatus* (Heidemann)

In reviewing the scant information available on feeding habits of isometopines, Wheeler and Henry (1978a) described and illustrated the fifth instar of *C. signatus* and provided a key to fifth-instar isometopines in eastern North America (five spp.). They discussed *C. signatus* — and *Myiomma cixiiforme* (Uhler) — in Pennsylvania (U.S.A.) as a predator of armored scale insects (Diaspididae): obscure scale, *Melanaspis obscura* (Comstock), on pin oak, *Quercus palustris* Münchh., and walnut scale, *Quadraspidiotus juglansregiae* (Comstock) (host tree not specified). *Corticoris signatus* also has been associated with gloomy scale, *M. tenebricosa* (Comstock), on silver maple, *Acer saccharinum* L., in Alabama (Miller and Williams 1985).

Nymphs and adults of the bivoltine *C. signatus* feed on obscure scale by probing around and under the cover of a female before curving their stylets under a cover to prey on a female. Overwintered eggs, which are laid under old scale covers, hatch in Pennsylvania during April, adults are most numerous in early June, early instars of a second generation are present by mid-June, and adults can be found until early August (Wheeler and Henry 1978a).

Once considered rare insects (see Wheeler and Henry 1978a), isometopine mirids (also referred to as isometopids or jumping tree bugs) can be abundant on trunks and branches of hardwood trees infested by diaspidid scales. Isometopine species, such as *C. signatus*, that appear specialized for scale predation, might be among the most important natural control agents of armored scale insects infesting trees in landscape plantings and nurseries. Other isometopines attack scale insects that are pests of citrus trees (Hesse 1947; see also Wheeler 2001). No isometopine species has been the subject of detailed studies on its bionomics or assessments of its potential for suppressing scale populations.

Mirinae

4.13 *Lygus* species

Among Heteroptera, only the Miridae contain important or even key pests that also can be considered beneficial (see also **Sections 4.20, 4.23**). Some omnivorous berytids might be important predators in managed systems (**Chapter 31**) but they are not considered key plant pests — that is, ones causing consistently serious crop-production problems.

Lygus bugs are some of the more important pests in the family (see **Chapter 3, Sections 3.8, 3.9, 4.14**), yet their facultative predation on various pest arthropods deserves attention from researchers. The salivary constitutents of at least *L. hesperus* Knight contain amylase and pectinase as adaptations for phytophagy, but also phospholipase A and a venom, which "indicate an advanced state of adaptation to predation" (Cohen 1996).

Wheeler (1976) reviewed zoophagy in *Lygus*, noting that *L. hesperus*, *L. lineolaris* (Palisot de Beauvois), and *L. rugulipennis* Poppius are omnivorous. Prey include aphids and leafhoppers in addition to members of the Coleoptera, Diptera, Hymenoptera, and Lepidoptera. Lygus bugs are known to feed on pests such as the alfalfa weevil, the beet armyworm *Spodoptera exigua* (Hübner), the Colorado potato beetle *Leptinotarsa decemlineata* (Say), the pea aphid *Acyrthosiphon pisum* (Harris), and the potato leafhopper *Empoasca fabae* (Harris). Records of predation by *Lygus* species published after Wheeler's (1976) review were mentioned by Wheeler (2001), including feeding by *L. lineolaris* on natural enemies such as aphidiid (Wheeler et al. 1968) and braconid wasps (Culliney et al. 1986).

Bryan et al. (1976) demonstrated that in the case of *L. hesperus* addition of beet armyworm larvae to a plant diet is superior to a plant-only diet. Cleveland (1987) determined that *L. lineolaris* nymphs and adults feed on eggs and neonate larvae of the tobacco budworm *Heliothis virescens* (F.) and remarked that tarnished plant bugs warrant recognition as predators of cotton pests. Hagler and Naranjo (1994) used pest-specific immunodiagnosis (ELISA) to assess predation by anthocorid, geocorid, mirid (*Lygus*), nabid, and reduviid bugs on eggs of the sweetpotato whitefly and the pink bollworm *Pectinophora gossypiella* (Saunders) in Arizona (U.S.A.) cotton fields. Nearly 20% of field-collected *L. hesperus* nymphs ($n = 1342$) and more than 30% of adults ($n = 1709$) tested positive for bollworm egg antigen; about 28% of *L. hesperus* nymphs and 20% of adults were positive for whitefly predation. About 12.5% of nymphs and adults scored positive for eggs of both pests.

Wheeler (1976) concluded from literature review that predation by *Lygus* species occurs infrequently on larvae and adults of pests in holometabolous orders, but that egg predation might be important in some managed systems. Hagler and Naranjo (1994), based on their own work and that of others (e.g., Bryan et al. 1976, Cleveland 1987), stated that "perhaps we need to assess carefully the beneficial impact of *L. hesperus* in the cotton agroecosystem." The difficulties inherent in incorporating zoophytophagous heteropterans, such as lygus bugs, into IPM strategies were discussed by Wiedenmann and Wilson (1996).

4.14 *Phytocoris* species

Phytocoris Fallén, largest genus in the Miridae (~650 spp.; Schuh and Slater 1995), is considered predominantly zoophagous (e.g., Stonedahl 1988). Yet substantial biological data are available for few species; most of the information is limited to scattered records on feeding habits and seasonal history (reviewed by Stonedahl 1988). No member of the genus has been the subject of detailed studies similar to those on many of the other predatory mirids covered in this chapter.

Overwintering occurs as eggs; most species apparently are univoltine; and feeding on mites, aphids, and other arthropods is common. Many are bark inhabitants, nymphs and adults resting in a flattened position on branches and trunks of trees. The bugs' often dark coloration renders them cryptic on host trees and shrubs; some *Phytocoris* species are a mottled green (or greenish white) and dark, which resembles the color of lichens on their woody hosts (Knight 1941, Southwood and Leston 1959, Stonedahl 1988). The genus contains prey-restricted species associated with a single plant genus, as well as generalists that track prey on woody plants of several families (e.g., Wheeler and Henry 1977, Stonedahl 1988). A relatively small proportion of *Phytocoris* species lives on grasses or other herbaceous plants (Stonedahl 1988).

Seasonality, host plants, and feeding habits are known for several species that are common in Britain and continental Europe, including *P. dimidiatus* Kirschbaum, *P. intricatus* Flor, *P. populi* (L.), *P. tiliae* (F.), *P. ulmi* (L.), and *P. varipes* Boheman (e.g., Butler 1923, Kullenberg 1944, Collyer 1953b, Southwood and Leston 1959, Strawínski 1964, Ehanno 1987). Literature pertaining to common Old World species that have been accidentally introduced into North America — all species cited above except *P. intricatus* — has been summarized by Wheeler and Henry (1992).

Effects of *Phytocoris* species on population dynamics of pest arthropods are little known. Kelton (1983) listed 15 *Phytocoris* species from apple, pear, and other fruit crops in Canada. *Phytocoris canadensis* Van Duzee and *P. erectus* Knight (see Kelton 1983 and Stonedahl 1988, for nomenclatural clarifications) prey on the immature stages of the codling moth on apple in Nova Scotia; *P. conspurcatus* Knight feeds on the apple aphid in Quebec (LeRoux 1960). Under experimental conditions, Bouchard et al. (1988) studied the consumption of apple aphids by *P. canadensis*. In British apple orchards, *P. reuteri* Saunders, *P. tiliae*, and *P. ulmi* attack the European red mite (Collyer 1953b). *Phytocoris* species also are natural enemies of forest pests such as the Douglas-fir tussock moth *Orgyia pseudotsugata* (McDunnough), the larch casebearer *Coleophora laricella* (Hübner), and the larch sawfly *Pristiphora erichsonii* (Hartig) (Stonedahl 1988, and references therein). On juniper (*Juniperus* spp.), *P. breviusculus* Reuter preys on armored scale insects (*Carulaspis* spp.) and has been associated with populations of the spruce spider mite *Oligonychus ununguis* (Jacobi) on juniper (Wheeler and Henry 1977).

Orthotylinae

4.15 *Blepharidopterus provancheri* (Burque) (= *Diaphnocoris pellucida* [Uhler])

Blepharidopterus provancheri is a widely distributed plant bug known from Nova Scotia (Canada) to Tennessee and Missouri (U.S.A.) west to British Columbia (Canada) and California (U.S.A.) (Henry and Wheeler 1988). Much of the economic literature on this species appears under the synonymic name *pellucida* (in the genera *Diaphnidia* or *Diaphnocoris*) or as *Diaphnidia* or *Diaphnocoris provancheri* (see Schuh 1995, for nomenclatural clarification).

Found on various trees and shrubs (Knight 1941, Kelton 1980), *B. provancheri* occurs consistently on apple and other fruit crops (Kelton 1983). The oviposition sites on apple and the egg were described by Sanford (1964a); the fifth instar was briefly described by Akingbohungbe et al. (1973). Steiner (1938) summarized this bivoltine bug's seasonal history on apple. Gilliatt (1935) described its prey-searching behavior and provided data on consumption of European red mites. This plant bug was considered one of the more important European red mite predators in Nova Scotia (Gilliatt 1935).

This orthotyline feeds not only on injurious mites but also on beneficial species such as the predatory phytoseiid *Typhlodromus pyri* Scheuten (Herbert 1962). Other prey include aphids (e.g., Bouchard et al. 1988), leafhoppers (Stear 1925, Steiner 1938), mealybugs (Chachoria 1967), psyllids (McMullen and Jong 1967), and codling moth eggs and neonate larvae (MacLellan 1972, Kelton 1983). Cannibalism also occurs in *B. provancheri* (Steiner 1938).

4.16 *Ceratocapsus* species

A New World genus of more than 130 species, *Ceratocapsus* Reuter at times has been placed in the separate orthotyline tribe Ceratocapsini, but ceratocapsines are currently recognized as members of the large tribe Orthotylini (Schuh 1995). Biological data on *Ceratocapsus* species are scant, but the genus usually is considered predatory. Many of the North American species are restricted to particular host plant genera, such as *Quercus* (e.g., Slater and Baranowski 1978, Kelton 1983, Wheeler 1991). At least in temperate regions of North America, most species are univoltine (Wheeler 1991).

Kelton (1983) recorded six species as mite, aphid, or whitefly predators on Canadian fruit crops; specific prey records were not mentioned. One of the species Kelton (1983) considered predacious, *C. modestus* (Uhler), feeds on eggs of the grape phylloxera *Daktulosphaira vitifoliae* (Fitch). In Pennsylvania (U.S.A.), nymphs and adults of *C. modestus* were found within curled, phylloxeran-infested leaves of grape (*Vitis* spp.). On the abaxial leaf surface, a bug inserts its stylets into the opening of a phylloxeran gall to prey on eggs of this pest (Wheeler and Henry 1978b). Their paper also includes a description and illustration of the fifth instar.

The Neotropical *C. dispersus* Carvalho & Fontes and *C. mariliensis* Carvalho & Fontes prey on eggs and larvae of the cotton leafworm *Alabama argillacea* (Hübner) and the tobacco budworm in Brazil (Carvalho et al. 1983, Gravena and Pazetto 1987). Encalada and Viñas (1989) reared *C. dispersus* under insectary conditions on eggs of the pink bollworm and larvae of the cotton leafperforator *Bucculatrix thurberiella* Busck. During development, nymphs fed on an average of 9.6 pink bollworm eggs and 4.3 cotton leafperforator larvae; adults fed on 6.7 eggs and 5.5 larvae. An undetermined species of *Ceratocapsus* feeds on cotton leafworm eggs in Texas (U.S.A.) (Gravena and Sterling 1983); *C. punctulatus* Reuter preys on bollworm (= corn earworm), *Helicoverpa zea* (Boddie), eggs in Texas (Nuessly and Sterling 1994).

4.17 *Malacocoris chlorizans* (Panzer)

A Eurasian orthotyline widespread in Britain and continental Europe (Kullenberg 1944, Southwood and Leston 1959, Kerzhner and Josifov 1999), *M. chlorizans* was recently reported as an adventive species in North America (British Columbia, Canada, and Washington, U.S.A.) (Schwartz and Scudder 1998). The egg was described and illustrated by Kullenberg (1942). In contrast to the endophytic oviposition habits of nearly all other Miridae, *M. chlorizans* deposits its eggs mostly superficially or semiexposed on host plants (Collyer 1953b, Southwood 1956). The fifth instar was described by Collyer (1953b) (see also Foschi and Carlotti 1956).

Partially phytophagous (e.g., Kullenberg 1944, Collyer 1953b), this plant bug lives on various deciduous trees, particularly apple, elm (*Ulmus* spp.), European filbert, and willow (*Salix* spp.) (Kullenberg 1944, Southwood and Leston 1959, Schwartz and Scudder 1998). The seasonality of this bivoltine mirid on apple in Switzerland was studied by Geier and Baggiolini (1952) and in Italy by Foschi and Carlotti (1956). Known in Britain as the delicate apple capsid (or mirid) (Southwood and Leston 1959), *M. chlorizans* feeds on eggs of the European red mite and can be one of this pest's most important natural enemies (Geier and Baggiolini 1952, Niemczyk 1963; see also Fauvel 1999). Solomon (1982) noted that this bug is less able to survive at low prey densities than is another mirid (*Blepharidopterus angulatus*; see **Section 3.3**) that preys on European red mites. The delicate apple capsid also is adversely affected (and sometimes eliminated from commercial orchards) by certain pesticides used to control apple pests (e.g., Geier and Baggiolini 1952, Foschi and Carlotti 1956, Hesjedal 1986). This generalist predator also feeds on aphids, psyllids, and eggs and larvae of lepidopteran leafminers (Kullenberg 1944, Southwood and Leston 1959, Schwartz and Scudder 1998, Fauvel 1999).

Phylinae

4.18 *Atractotomus mali* (Meyer-Dür)

Atractotomus mali is a widespread Palearctic phyline (see Kerzhner and Josifov 1999, for distribution) that was first reported from North America (Nova Scotia, Canada) by Knight (1924). Lord (1949) recorded this species from Nova Scotian apple orchards under the misidentification *Criocoris saliens* (Reuter). North American populations, which are known from the Canadian maritime provinces to Virginia (U.S.A.), are considered adventive (Wheeler and Henry 1992). In reviewing the literature on this plant bug, Wheeler and Henry (1992) listed references that contain descriptions and illustrations of the immature stages (see also Finţescu 1914) and noted that Stonedahl (1990) treated *A. mali* as *incertae sedis* because *mali* is not congeneric with the type species of the genus. The proper generic placement of *mali* remains uncertain (Kerzhner and Josifov 1999).

This omnivorous mirid (Kullenberg 1944, Sanford 1964b), called the apple brown bug in Canada (e.g., MacPhee and MacLellan 1972), lives primarily on rosaceous trees and shrubs, being particularly common on apple and hawthorn (Southwood and Leston 1959, Ehanno 1987, Wheeler and Henry 1992). Its status on apple — injurious or beneficial — was discussed by Sanford (1964b) and Wheeler (2001). In Europe and Canada, this univoltine (Kullenberg 1944, Ehanno 1987) predator feeds on the European red mite, aphids (and their honeydew), psyllids, and larvae and pupae of lepidopterans (Finţescu 1914, Kullenberg 1944, Sanford 1964b, Strawiński 1964, Lord 1971, Kelton 1983). Harizanova (1989) reported that 290 to 360 apple aphids are consumed during nymphal development. Consumption rates for *A. mali* reared on the apple aphid and on the apple sucker were reported by Jonsson (1987). In addition to the apple sucker *Cacopsylla mali* (Schmidberger), psyllid prey include *Cacopsylla* spp. on hawthorn (Novak and Achtziger 1995). *Atractotomus mali* also is known to cannibalize (Collyer 1953b).

4.19 *Campylomma liebknechti* (Girault)

Campylomma liebknechti is indigenous to Australia, where it bears the common name apple dimpling bug (e.g., Chinjariyawong and Walter 1990). Before 1992, this phyline plant bug was known in the applied literature as *C. livida* Reuter. Chinajariyawong and Walter (1990) suggested that *C. livida* might represent a complex of cryptic or sibling species; in revising the Indo-Pacific Phylinae, Schuh (1984) had noted that species discrimination in *Campylomma* can depend on characters of the male genitalia and he questioned the validity of *C. livida* records from Australia because he had not been able to examine Australian material. Indeed, Malipatil (1992) determined that specimens of *"C. livida"* from Australia had been misidentified — true *C. lividum* (the correct spelling) is restricted to the Oriental region — and that the correct name of the Australian bug is *C. liebknechti*.

The apple dimpling bug is polyphagous, nymphs and adults having been recorded from numerous hosts in unrelated plant families (Chinajariyawong and Walter 1990, Malipatil 1992). Although this plant bug injures apple fruit (e.g., Lloyd 1969) and is a potential early-season pest of cotton (Bishop 1980, Chinajariyawong and Walter 1990), it also is a facultative predator of mites in orchards and lepidopteran eggs and neonate larvae in cotton (Readshaw 1971, 1975; Room 1979). This mirid also feeds on larvae of the light brown apple moth *Epiphyas postvittana* Walker (MacLellan 1973). Chinajariyawong and Walter (1990) demonstrated that a mixed diet (cotton tips or squares plus noctuid eggs) was needed for nymphal development and reproduction; a mixed diet also enhanced fecundity.

4.20 *Campylomma verbasci* (Meyer-Dür) (= *Campylomma nicolasi* Puton & Reuter)

This omnivorous phyline was discussed as an apple pest in **Chapter 3** (see **Section 4.19**). Native to Eurasia and accidentally introduced into North America (Wheeler and Henry 1992), *C. verbasci* is known as the mullein bug in Canada (e.g., Thistlewood and Smith 1996). Much of the economic literature treats this bug's injury to apple.

As a predator, *C. verbasci* feeds on the European red mite, and to a lesser extent, on aphids in European apple orchards (Collyer 1953a, Niemczyk 1978). In feeding preference tests, the bugs consumed about equal numbers of European red mites and twospotted spider mites, but few apple aphids (Arnoldi et al. 1992). Even so, it might be important in suppressing small colonies of apple aphids (Niemczyk and Pruska 1986). This bug's early instars appear to require animal food (Niemczyk 1978). Canadian prey records include several important pests: the apple aphid, codling moth, European red mite, pear psylla, and western flower thrips (McMullen and Jong 1967, 1970; Thistlewood and Smith 1996). Under experimental conditions, each *C. verbasci* nymph consumed an average of 631 pear psylla eggs during nymphal development (McMullen and Jong 1970). Thistlewood and Smith (1996), however, concluded that predation by *C. verbasci* does not balance the economic damage from its fruit feeding on apple. On pear, a crop in which the economic injury level is "at least 10 times that on apple," *C. verbasci* is an important predator of the pear psylla and, therefore, conservation of this plant bug might be warranted (Thistlewood and Smith 1996).

4.21 *Pilophorus perplexus* Douglas & Scott

This common Palearctic bug (see Kerzhner and Josifov 1999, for Eurasian distribution) is thought to have been accidentally introduced into North America, perhaps with shipments of fruit trees from Europe. In North America, *P. perplexus* is known from the eastern United States and Pacific Northwest, suggesting multiple introductions from the Old World (Wheeler and Henry 1992). Kullenberg (1942) described and illustrated the egg; Sanford (1964a) described this bug's oviposition sites on apple in Nova Scotia, Canada.

A univoltine mirid (Ehanno 1987), *P. perplexus* occurs on fruit trees, such as apple and pear, and on oaks and other deciduous trees (Kullenberg 1944, Strawínski 1964, Ehanno 1987). The bugs feed on leaves and young fruits of their hosts (Kullenberg 1944), and this species sometimes is considered predominantly phytophagous (Strawínski 1964). But this pilophorine mirid also is zoophagous. Prey include mites, aphids, psyllids, scale insects, and larvae of the codling moth and other lepidopterans (Kullenberg 1944, Southwood and Leston 1959, Kelton 1983, Wheeler and Henry 1992). In English apple orchards, *P. perplexus* feeds on the armored scale *Quadraspidiotus pyri* (Lichtenstein) (Solomon 1976), but apparently also on the oystershell scale *Lepidosaphes ulmi* (L.) (Easterbrook et al. 1985). Fulton (1918) observed *P. perplexus* (as *P. walshii* Uhler) in New York (U.S.A.) apple orchards concealed among leaves curled from aphid feeding and commented on the bug's potential as an aphid predator. In Pennsylvania (U.S.A.), Stinner (1975) reported its feeding on crawlers and females of the globose scale *Sphaerolecanium prunastri* (Fonscolombe).

In the 1920s, *P. perplexus* was found in France feeding on larvae of a pear midge, *Dasyneura pyri* (Bouché) (Myers 1927a). This predator was to have received further evaluation as a possible biocontrol agent of the midge in New Zealand, where this injurious cecidomyiid had been accidentally introduced. Before *P. perplexus* could be fully evaluated, it was discovered that this mirid already was present in New Zealand, apparently having been unintentionally introduced with shipments of hymenopteran parasitoids of the pear midge (Wheeler and Henry 1992, and references therein; see also Wheeler 2001, for further discussion).

4.22 *Psallus ambiguus* (Fallén)

Psallus ambiguus is a widely distributed Palearctic phyline (Kerzhner and Josifov 1999). The egg was described and illustrated by Kullenberg (1942), the egg and oviposition sites were figured by Abraham (1937), and the nymphs were briefly described by Petherbridge and Husain (1918) and Collyer (1953b); the last-named work also contains an illustration of the fifth instar. Morris (1965) described and illustrated all preimaginal stages.

Occurring on numerous tree and shrub species, this univoltine bug is consistently found on alder, apple, hawthorn, and willow (Kullenberg 1944, Collyer 1953b, Southwood and Leston 1959, Strawínski 1964, Morris 1965, Ehanno 1987). Injury by this omnivorous plant bug to apple and pear was reviewed by Wheeler (2001). Although Morris (1965) was able to rear two adults on apple flowers (nymphs soon died when fed apple leaves), some animal food seems necessary for nymphal development (Niemczyk 1968).

Prey of *P. ambiguus* are mites, aphids, leafhoppers, psyllids, and lepidopteran larvae (Kullenberg 1944, Collyer 1953b, Strawínski 1964, Morris 1965); arthropod eggs generally are not consumed (Morris 1965). Plant pests that serve as prey include the apple aphid, apple sucker, and European red mite (Niemczyk 1968). The last-named prey may represent only an alternative food source because *P. ambiguus* females reared on European red mites do not produce eggs (Niemczyk 1966). The most detailed biological studies are those in England by Morris (1965), who covered seasonality, termination of egg diapause, nymphal development, dispersal, feeding habits, and mortality, and by Niemczyk (1967, 1968) in Poland.

4.23 *Pseudatomoscelis seriatus* (Reuter)

The cotton fleahopper, an important pest in portions of the U.S. Cotton Belt, was discussed as a phytophage in **Chapter 3** (see **Section 3.10**). Its predatory tendencies were not discovered until the late 1970s.

McDaniel and Sterling (1979) reported that two *P. seriatus* adults fed on radioactively labeled tobacco budworm eggs in Texas (U.S.A.). Since then, nymphs and adults have been confirmed as predators of budworm and bollworm eggs (Agnew et al. 1982, McDaniel and Sterling 1982, Nuessly and Sterling 1994). Eggs of the cotton leafworm also are attacked (Gravena and Sterling 1983). Predation on lepidopteran eggs is highest on squares and terminals (Nuessly and Sterling 1994), with individual consumption potentially one budworm or bollworm egg per day (Sterling 1982). A reduction in percentage predation on tobacco budworm eggs in May, when cotton fleahopper densities were increasing, might be explained by changes in this bug's trophic habits — from mostly zoophagous during presquaring in cotton to mostly phytophagous once squares became available (McDaniel and Sterling 1982). Despite the pest status of *P. seriatus* in much of Texas, this plant bug might be beneficial in cotton after flowering ceases (Sterling 1982). Sterling (1982) also noted that the cotton fleahopper is important in the cotton agroecosystem because it provides prey for other predators.

4.24 *Sejanus albisignatus* (Knight)

Known only from Australia and New Zealand (Cassis and Gross 1995), *S. albisignatus* lives on apple and pear, as well as on other deciduous trees, where it apparently is bivoltine (Dumbleton 1938, 1964). Dumbleton (1938) illustrated and described the egg and noted oviposition sites on fruit trees.

Sejanus albisignatus can be reared on young shoots and leaves of apple and pear, but this mirid also is predacious. It feeds on all stages of the European red mite, other tetranychids (*Bryobia* spp.), aphids, psyllids, and codling moth eggs and neonate larvae (Collyer 1964a, 1976; Dumbleton 1964; Collyer and Geldermalsen 1975), and is a probable predator of the midge *Dasyneura pyri* on pear (Dumbleton 1938). Even though use of an integrated spray program on apple in New Zealand allowed *S. albisignatus* to attain abundance for several seasons, this bug (and other mite predators) did not suppress populations of the European red mite (Collyer 1976).

5. PROGNOSIS

Predacious mirids will continue to be discovered, prey records will accumulate in species already known to be predatory, and entomologists will provide life-history data on additional Miridae, both Old and New World species. Especially useful would be studies that quantify the effects of predacious mirids on pest densities. Following the lead of European entomologists, who have incorporated dicyphine mirids into IPM programs for greenhouse pests such as thrips and whiteflies, North American workers are determining whether native dicyphines (the Palearctic mirids used in biocontrol are prohibited from being imported into North America) might similarly suppress pest populations in greenhouses.

Despite the inherent unpredictability of zoophytophagous mirids, researchers can be expected to determine predator–prey ratios needed to assure predation and to avoid injury to the host plant. For outdoor tomato crops in Spain, pest managers already have been provided decision-making guidelines — that is, the need for pesticide intervention — pertaining to whitefly densities and those of the whitefly predator *Dicyphus tamaninii*. Even with this successful use of dicyphine mirids in commercial tomato fields, the use of predacious mirids in most other managed systems creates problems for applied entomologists. In apple orchards, for instance, fruit injury by omnivorous mirids can occur even when alternative prey are available (e.g., Thistlewood and Smith 1996).

Researchers, however, will become better able to cope with dynamic agroecosystems and the unpredictability of plant injury by omnivorous mirids, including species that are considered pests requiring consistent suppression by insecticides. At least in certain regions of a crop's production, or in portions of a crop cycle, it might eventually be possible to take advantage of predatory tendencies of lygus bugs in row crops such as cotton, or to exploit zoophagy by sporadic pests such as the mullein bug *Campylomma verbasci* in apple or pear orchards. One might also expect to see predacious mirids such as *Deraeocoris* and *Stethoconus* species used in landscape plantings or in nurseries to help control aphids, lace bugs, and other pest arthropods. Moreover, underappreciated mirids — for example, the scale-feeding isometopines — will warrant measures to conserve their populations in urban and suburban landscapes; perhaps these predatory mirids will even be manipulated to minimize damage from armored scale insects such as the obscure scale on oaks.

6. CONCLUSIONS

In the introduction to this chapter, predatory habits were estimated to occur in at least one third of the nearly 10,000 described species of Miridae. To emphasize further the extent of predation in this family, the same number of species (or generic) accounts — 30 — is provided for predatory mirids as was done in **Chapter 3** for mirids as plant feeders. Many more predaceous mirids could have been included, but prey records are available in Wheeler (2001) for additional species.

As is the case in other groups of heteropteran predators, large gaps exist in the knowledge of predatory mirids. Research on predacious mirids important in agroecosystems actually lags behind that on most other predacious heteropterans — anthocorids, geocorids (*Geocoris* spp.), nabids, and pentatomids (Asopinae). Future research needs for zoophytophagous Heteroptera, including mirids, were proposed by Wiedenmann and Wilson (1996). Predatory mirids in particular offer abundant opportunities for research; specific areas needing attention were highlighted by Wheeler (2001).

7. ACKNOWLEDGMENTS

In **Chapter 3**, gratitude was expressed to Carl Schaefer, Antônio Panizzi, Tammy Morton, Thomas Henry, and Craig Stoops; this expression of appreciation also extends to the present chapter.

8. REFERENCES CITED

Abraham, R. 1937. Beobachtungen über die Eiablage einiger Capsiden. Arb. Physiol. Angew. Entomol. Berlin 4: 321–324.

Agarwal, R. A., and G. P. Gupta. 1983. Insect pests of fibre crops. Pp. 147–164 *in* P. D. Srivastva et al. [eds.], Agricultural Entomology, Vol. II. All India Scientific Writers' Society, New Delhi, India.

Agnew, C. W., W. L. Sterling, and D. A. Dean. 1982. Influence of cotton nectar on red imported fire ants and other predators. Environ. Entomol. 11: 629–634.

Akingbohungbe, A. E., J. L. Libby, and R. D. Shenefelt. 1973. Nymphs of Wisconsin Miridae. Hemiptera: Heteroptera. Univ. Wisconsin-Madison Coll. Agric. Res. Div. R2561: 1–25.

Alomar, O., and R. Albajes. 1996. Greenhouse whitefly (Homoptera: Aleyrodidae) predation and tomato fruit injury by the zoophytophagous predator *Dicyphus tamaninii* (Heteroptera: Miridae). Pp. 155–177 *in* O. Alomar and R. N. Wiedenmann [eds.], Zoophytophagous Heteroptera: Implications for Life History and Integrated Pest Management. Thomas Say Publ. Entomol.: Proceedings Entomological Society of America, Lanham, Maryland, U.S.A. 202 pp.

Alomar, O., and R. N. Wiedenmann [eds.]. 1996. Zoophytophagous Heteroptera: Implications for Life History and Integrated Pest Management. Thomas Say Publ. Entomol.: Proceedings Entomological Society of America, Lanham, Maryland, U.S.A. 202 pp.

Alomar, O., C. Castañé, R. Gabarra, and R. Albajes. 1990. Mirid bugs — another strategy for IPM on Mediterranean vegetable crops? Proc. Working Group, "Integrated Control in Glasshouses," Copenhagen, Denmark, 5–8 June 1990. Int. Organ. Biol. Control/West. Palearctic Reg. Sect. Bull. 13(5): 6–9.

Alvarado, P., O. Baltà, and O. Alomar. 1997. Efficiency of four Heteroptera as predators of *Aphis gossypii* and *Macrosiphum euphorbiae* (Hom.: Aphididae). Entomophaga 42: 215–226.

Arnoldi, D., R. K. Stewart, and G. Boivin. 1992. Predatory mirids of the green apple aphid *Aphis pomi*, the two-spotted spider mite *Tetranychus urticae* and the European red mite *Panonychus ulmi* in apple orchards in Québec. Entomophaga 37: 283–292.

Arzone, A., A. Alma, and L. Tavella. 1990. Ruolo dei Miridi (Rhynchota Heteroptera) nella limitazione di *Trialeurodes vaporariorum* Westw. (Rhynchota Aleyrodidae): nota preliminare. Boll. Zool. Agrar. Bachic. 22: 43–52.

Austin, M. D., and A. M. Massee. 1947. Investigations on the control of the fruit tree red spider mite (*Metatetranychus ulmi* Koch) during the dormant season. J. Pomol. 23: 227–253.

Bailey, N. S. 1950. An Asiatic tingid new to North America (Heteroptera). Psyche 57: 143–145.

Barnadas, I., R. Gabarra, and R. Albajes. 1998. Predatory capacity of two mirid bugs preying on *Bemisia tabaci*. Entomol. Exp. Appl. 86: 215–219.

Benrey, B., and W. O. Lamp. 1994. Biological control in the management of planthopper populations. Pp. 519–550 *in* R. F. Denno and T. J. Perfect [eds.], Planthoppers: Their Ecology and Management. Chapman & Hall, New York, New York, U.S.A. 799 pp.

Bentur, J. S., and M. B. Kalode. 1987. Off-season survival of the predatory mirid bug, *Cyrtorhinus lividipennis* (Reuter). Curr. Sci. (Bangalore) 56: 956–957.

Bishop, A. L. 1980. The potential of *Campylomma livida* Reuter, and *Megacoelum modestum* Distant (Hemiptera: Miridae) to damage cotton in Queensland. Austral. J. Exp. Agric. Anim. Husb. 20: 229–233.

Booth, S. R., and H. Riedel. 1996. Diflubenzuron-based management of the pear pest complex in commercial orchards of the Hood River Valley in Oregon. J. Econ. Entomol. 89: 621–630.

Borror, D. J., and R. E. White. 1970. A Field Guide to the Insects of America North of Mexico. Houghton Mifflin, Boston, Massachusetts, U.S.A. 404 pp.

Bouchard, D., J. G. Pilon, and J. C. Tourneur. 1988. Voracity of mirid, syrphid, and cecidomyiid predators under laboratory conditions. Pp. 231–234 *in* E. Niemczyk and A. F. G. Dixon [eds.], Ecology and Effectiveness of Aphidophaga. Proc. Int. Symp., Teresin, Poland, Aug. 31–Sept. 5, 1987. SPB, The Hague, the Netherlands. 341 pp.

Braimah, S. A., L. A. Kelton, and R. K. Stewart. 1982. The predaceous and phytophagous plant bugs (Heteroptera: Miridae) found on apple trees in Quebec. Nat. Canada (Quebec) 109: 153–180.

Bryan, D. E., C. G. Jackson, R. L. Carranza, and E. G. Neemann. 1976. *Lygus hesperus*: production and development in the laboratory. J. Econ. Entomol. 69: 127–129.

Bull, R. M. 1981. Population studies on the sugar cane leafhopper (*Perkinsiella saccharicida* Kirk.) in the Bundaberg District. Proc. Austral. Soc. Sugar Cane Technol. 1981: 293–303.

Butler, E. A. 1923. A Biology of the British Hemiptera-Heteroptera. Witherby, London, U.K. 682 pp.

Callan, E. McC. 1943. Natural enemies of the cacao thrips. Bull. Entomol. Res. 34: 313–321.

1975. Miridae of the genus *Termatophylidea* [Hemiptera] as predators of cacao thrips. Entomophaga 20: 389–391.

Carapezza, A. 1995. The specific identities of *Macrolophus melanotoma* (A. Costa, 1853) and *Stenodema curticolle* (A. Costa, 1853) (Insecta Heteroptera, Miridae). Nat. Sicil. 19: 295–298.

Carayon, J. 1958. Études sur les Hémiptères Cimicoidea. Mém. Mus. Natl. Hist. Nat. Ser. A. Zool. 16: 141–172.
1960. *Stethoconus frappai* n. sp., miridé prédateur du tingidé du caféier, *Dulinius unicolor* (Sign.), a Madagascar. J. Agric. Trop. Bot. Appl. 7: 110–120.
1986. *Macrolophus caliginosus*, Hémiptère Miridae, à reproduction hivernale. Entomologiste (Paris) 42: 257–262.
1989. Parthénogénèse constante prouvée chez deux Hétéroptères: Le miride *Campyloneura virgula* et l'anthocoride *Calliodis maculipennis*. Ann. Soc. Entomol. France 25: 387–391.
Carroll, D. P., and S. C. Hoyt. 1984. Natural enemies and their effects on apple aphid, *Aphis pomi* DeGeer (Homoptera: Aphididae), colonies on young apple trees in central Washington. Environ. Entomol. 13: 469–481.
Carvalho, J. C. M., and T. R. E. Southwood. 1955. Revisão do complexo *Cyrtorhinus* Fieber–*Mecomma* Fieber (Heteroptera, Miridae). Bol. Mus. Para. Emilio Goeldi 11: 1–72.
Carvalho, J. C. M., A. V. Fontes, and T. J. Henry. 1983. Taxonomy of the South American species of *Ceratocapsus*, with descriptions of 45 new species (Hemiptera: Miridae). U.S. Dept. Agric. Tech. Bull. 1676: 1–58.
Cassis, G. 1995. A reclassification and phylogeny of the Termatophylini (Heteroptera: Miridae: Deraeocorinae), with a taxonomic revision of the Australian species, and a review of the tribal classification of the Deraeocorinae. Proc. Entomol. Soc. Washington 97: 258–330.
Cassis, G., and G. F. Gross. 1995. Hemiptera: Heteroptera (Coleorrhyncha to Cimicomorpha). *In* W. W. K. Houston and G. V. Maynard [eds.], Zoological Catalogue of Australia. Vol. 27.3A. CSIRO Australia, Melbourne, Australia. 506 pp.
Castañé, C., O. Alomar, and J. Riudavets. 1996. Management of western flower thrips on cucumber with *Dicyphus tamaninii* (Heteroptera: Miridae). Biol. Control 7: 114–120.
Chachoria, H. S. 1967. Mortality in apple mealybug, *Phenacoccus aceris* (Homoptera: Coccidae), populations in Nova Scotia. Canad. Entomol. 99: 728–730.
Chelliah, S., and E. A. Heinrichs. 1980. Factors affecting insecticide-induced resurgence of the brown planthopper, *Nilaparvata lugens* on rice. Environ. Entomol. 9: 773–777.
Chinajariyawong, A., and V. E. Harris. 1987. Inability of *Deraeocoris signatus* (Distant) (Hemiptera: Miridae) to survive and reproduce on cotton without prey. J. Austral. Entomol. Soc. 26: 37–40.
Chinajariyawong, A., and G. H. Walter. 1990. Feeding biology of *Campylomma livida* Reuter (Hemiptera: Miridae) on cotton, and some host plant records. J. Austral. Entomol. Soc. 29: 177–181.
Chua, T. H., and E. Mikil. 1989. Effects of prey number and stage on the biology of *Cyrtorhinus lividipennis* (Hemiptera: Miridae): a predator of *Nilaparvata lugens* (Homoptera: Delphacidae). Environ. Entomol. 18: 251–255.
Cleveland, T. C. 1987. Predation by tarnished plant bugs (Heteroptera: Miridae) of *Heliothis* (Heteroptera: Noctuidae) eggs and larvae. Environ. Entomol. 16: 37–40.
Cobben, R. H. 1968. Evolutionary trends in Heteroptera. Part I. Eggs, Architecture of the Shell, Gross Embryology and Eclosion. Centre for Agricultural Publishing & Documentation, Wageningen, the Netherlands. 475 pp.
1978. Evolutionary trends in Heteroptera. Part II. Mouthpart-structures and Feeding Strategies. Meded. Landbouwhogesch. Wageningen 78–5: 1–407.
Cohen, A. C. 1996. Plant feeding by predatory Heteroptera: evolutionary and adaptational aspects of trophic switching. Pp. 1–17 *in* O. Alomar and R. N. Wiedenmann [eds.], Zoophytophagous Heteroptera: Implications for Life History and Integrated Pest Management. Thomas Say Publ. Entomol.: Proceedings Entomological Society of America, Lanham, Maryland, U.S.A. 202 pp.
Coll, M., and J. R. Ruberson [eds.]. 1998a. Predatory Heteroptera: Their Ecology and Use in Biological Control. Thomas Say Publ. Entomol.: Proceedings Entomological Society of America, Lanham, Maryland, U.S.A. 233 pp.
1998b. Predatory Heteroptera: an important yet neglected group of natural enemies. Pp. 1–6 *in* M. Coll and J. R. Ruberson [eds.], Predatory Heteroptera: Their Ecology and Use in Biological Control. Thomas Say Publ. Entomol.: Proceedings Entomological Society of America, Lanham, Maryland, U.S.A. 233 pp.

Collyer, E. 1952. Biology of some predatory insects and mites associated with the fruit tree red spider mite (*Metatetranychus ulmi* (Koch)) in south-eastern England 1. The biology of *Blepharidopterus angulatus* (Fall.) (Hemiptera-Heteroptera, Miridae). J. Hortic. Sci. 27: 117–129.

1953a. Biology of some predatory insects and mites associated with the fruit tree red spider mite (*Metatetranychus ulmi* (Koch)) in south-eastern England II. Some important predators of the mite. J. Hortic. Sci. 28: 85–97.

1953b. Biology of some predatory insects and mites associated with the fruit tree red spider mite (*Metatetranychus ulmi* (Koch)) in south-eastern England III. Further predators of the mite. J. Hortic. Sci. 28: 98–113.

1953c. Biology of some predatory insects and mites associated with the fruit tree red spider mite (*Metatetranychus ulmi* (Koch)) in south-eastern England IV. The predator–mite relationship. J. Hortic. Sci. 28: 246–259.

1964a. Phytophagous mites and their predators in New Zealand orchards. New Zealand J. Agric. Res. 7: 551–568.

1964b. A summary of experiments to demonstrate the role of *Typhlodromus pyri* Scheut. in the control of *Panonychus ulmi* (Koch) in England. Pp. 363–371 *in* Proc. 1st Int. Congr. Acarology, Fort Collins (Colorado, U.S.A.), 2–7 Sept. 1963. F. Paillart, Abbeville, France. 439 pp.

1965. Cannibalism as a factor affecting mortality of *Blepharidopterus angulatus* (Fall.) (Heteroptera: Miridae). East Malling (U.K.) Res. Stn. Rep. 1964: 177–179.

1976. Integrated control of apple pests in New Zealand 6. Incidence of European red mite, *Panonychus ulmi* (Koch), and its predators. New Zealand J. Zool. 3: 39–50.

Collyer, E., and M. van Geldermalsen. 1975. Integrated control of apple pests in New Zealand 1. Outline of experiment and general results. New Zealand J. Zool. 2: 101–134.

Collyer, E., and A. M. Massee. 1958. Some predators of phytophagous mites, and their occurrence, in southeastern England. Proc. 10th Int. Congr. Entomol., Montreal (1956) 4: 623–626.

Constant, B., S. Grenier, and G. Bonnot. 1994. Analysis of some morphological and biochemical characteristics of the egg of the predaceous bug *Macrolophus caliginosus* (Het.: Miridae) during embryogenesis. Entomophaga 39: 189–198.

Constant, B., S. Grenier, and G. Bonnot. 1996a. Artificial substrate for egg laying and embryonic development by the predatory bug *Macrolophus caliginosus* (Heteroptera: Miridae). Biol. Control 7: 140–147.

Constant, B., S. Grenier, G. Febvay, and G. Bonnot. 1996b. Host plant hardness in oviposition of *Macrolophus caliginosus* (Hemiptera: Miridae). J. Econ. Entomol. 89: 1446–1452.

Croft, B. A. 1990. Arthropod Biological Control Agents and Pesticides. Wiley, New York, New York, U.S.A. 723 pp.

Culliney, T. W., D. Pimentel, O. S. Namuco, and B. A. Capwell. 1986. New observations of predation by plant bugs (Hemiptera: Miridae). Can. Entomol. 118: 729–730.

Döbel, H. G., and R. F. Denno. 1994. Predator–planthopper interactions. Pp. 325–399 *in* R. F. Denno and T. J. Perfect [eds.], Planthoppers: Their Ecology and Management. Chapman & Hall, New York, New York, U.S.A. 799 pp.

Doesburg, Jr., P. H. van. 1964. *Termatophylidea opaca* Carvalho, a predator of thrips (Hem.-Het.). Entomol. Ber. (Amsterdam) 24: 248–253.

Dolling, W. R. 1991. The Hemiptera. Oxford University Press, Oxford, U.K. 274 pp.

Dumbleton, L. J. 1938. Notes on a new mirid bug (*Idatiella albisignata* Knight). New Zealand J. Sci. Technol. 20(B): 58B–60B.

1964. Notes on insects. New Zealand Entomol. 3(3): 24–25.

Dunbar, D. M. 1974. Bionomics of the andromeda lacebug, *Stephanitis takeyai*. Mem. Connecticut Entomol. Soc. 1974: 277–289.

Dyck, V. A., and G. C. Orlido. 1977. Control of the brown planthopper *(Nilaparvata lugens)* by natural enemies and timely application of narrow-spectrum insecticides. Pp. 58–72 *in* Food and Fertilizer Technology for the Asian and Pacific Region [compil.]. Taipei, Taiwan. 258 pp.

Easterbrook, M. A., M. G. Solomon, J. E. Cranham, and E. F. Souter. 1985. Trials of an integrated pest management programme based on selective pesticides in English apple orchards. Crop Prot. 4: 215–230.

Ehanno, B. 1987. Les hétéroptères mirides de France. Tome II-A: Inventaire et syntheses ecologiques. Inventaire Faune Flore 40:1–647. Secretariat de la Faune et de la Flore, Paris, France.

Ehler, L. E. 1998. Invasion biology and biological control. Biol. Control 13: 127–133.

El-Dessouki, S. A., A. H. El-Kifl, and H. A. Helal. 1976. Life cycle, host plants and symptoms of damage of the tomato bug, *Nesidiocoris tenuis* Reut. (Hemiptera: Miridae), in Egypt. Z. Pflanzenkr. Pflanzenschutz 83: 204–220.

Ellington, J., D. Richman, S. Meeks, T. Carrillo, S. Liesner, and S. T. Ball. 1998. Biological control of insect pests in pecans. New Mexico State University, Las Cruces, New Mexico, U.S.A. 15 pp.

Encalada, E., and L. Viñas. 1989. *Ceratocapsus dispersus* (Hemiptera, Miridae) en Piura: biologia y capacidad predatora en insectario. Rev. Peruana Entomol. 32: 1–8.

Evans, H. E. 1984. Insect Biology: A Textbook of Entomology. Addison-Wesley, Reading, Massachusetts, U.S.A. 436 pp.

Fabellar, L. T., and E. A. Heinrichs. 1984. Toxicity of insecticides to predators of rice brown planthoppers, *Nilaparvata lugens* (Stål) (Homoptera: Delphacidae). Environ. Entomol. 13: 832–837.

Fauvel, G. 1999. Diversity of Heteroptera in agroecosystems: role of sustainability and bioindication. Agric. Ecosyst. Environ. 74: 275–303.

Fauvel, G., J. C. Malausa, and B. Kaspar. 1987. Etude en laboratoire des principales caractéristiques biologiques de *Macrolophus caliginosus* (Heteroptera, Miridae). Entomophaga 32: 529–543.

Ferran, A., A. Rortais, J. C. Malausa, J. Gambier, and M. Lambin. 1996. Ovipositional behaviour of *Macrolophus caliginosus* (Heteroptera: Miridae) on tobacco leaves. Bull. Entomol. Res. 86: 123–128.

Figuls, M., C. Castañé, and R. Gabarra. 1999. Residual toxicity of some insecticides on the predatory bugs *Dicyphus tamaninii* and *Macrolophus caliginosus*. BioControl 44: 89–98.

Finţescu, G. N. 1914. Contributions à la biologie de l'hémiptère "*Capsus Mali*" (Meyer) (syn. *Capsus magnicornis* Fallen), *Plytocoris* [sic] *magnicornis* (Macq), *Atractotomus mali* (Fieber), *Capsus plenicornis* [sic]. Bucarest Bull. Acad. Româna 3: 132–140.

Foglar, H., J. C. Malausa, and E. Wajnberg. 1990. The functional response and preference of *Macrolophus caliginosus* [Heteroptera: Miridae] for two of its prey: *Myzus persicae* and *Tetranychus urticae*. Entomophaga 35: 465–474.

Foschi, S., and G. Carlotti. 1956. *Malacocoris chlorizans* Pz. var. *smaragdina* Fieb. predatore del "ragno rosso." Redia 41: 105–111.

Frank, J. H., and E. D. McCoy. 1989. Introduction to attack and defense: behavioral ecology of parasites and parasitoids and their hosts. Behavioral ecology: From fabulous past to chaotic future. Florida Entomol. 72: 1–6.

1993. Introduction to the behavioral ecology of introduction: the introduction of insects into Florida. Florida Entomol. 76: 1–53.

Fullaway, D. T. 1940. An account of the reduction of the immigrant taro leaf-hopper *(Megamelus proserpina)* population to insignificant numbers by the introduction and establishment of the egg-sucking bug *Cyrtorhinus fulvus*. Pp. 345–346 *in* Proc. 6th Pac. Sci. Congr. Pac. Sci. Assoc., Univ. California, Berkeley, July 24 to Aug. 12, 1939. University of California Press, Berkeley, California, U.S.A.

Fulton, B. B. 1918. Observations on the life history and habits of *Pilophorus walshii* Uhler. Ann. Entomol. Soc. Amer. 11: 93–96.

Fye, R. E. 1981. An analysis of pear psylla populations, 1977–79. U.S. Dept. Agric. SEA ARM-W 24: 1–32.

Gabarra, R., C. Castañé, E. Bordas, and R. Albajes. 1988. *Dicyphus tamaninii* as a beneficial insect and pest in tomato crops in Catalonia, Spain. Entomophaga 33: 219–228.

Gabarra, R., C. Castañé, and R. Albajes. 1995. The mirid bug *Dicyphus tamaninii* as a greenhouse whitefly and western flower thrips predator on cucumber. Biocontrol Sci. Technol. 5: 475–488.

Gange, A. C., and M. Llewellyn. 1989. Factors affecting orchard colonisation by the black-kneed capsid (*Blepharidopterus angulatus* (Hemiptera: Miridae)) from alder windbreaks. Ann. Appl. Biol. 114: 221–230.

Geier, P., and M. Baggiolini. 1952. *Malacocoris chlorizans* Pz. (Hem. Het. Mirid.), prédateur des Acariens phytophages. Mitt. Schweiz. Entomol. Ges. 25: 257–259.

Gessé Solé, F. 1992. Comportamiento alimenticio de *Dicyphus tamaninii* Wagner (Heteroptera: Miridae). Bol. Sanid. Veg. Plagas 18: 685–691.

Ghavami, M. D., A. F. Özgür, and U. Kersting. 1998. Prey consumption by the predator *Deraeocoris pallens* Reuther [sic] (Hemiptera: Miridae) on six cotton pests. Z. Pflanzenkr. Pflanzenschutz 105: 526–531.

Gilliatt, F. C. 1935. Some predators of the European red mite, *Paratetranychus pilosus* C. & F., in Nova Scotia. Canad. J. Res. 13: 19–38.

Glen, D. M. 1973. The food requirements of *Blepharidopterus angulatus* (Heteroptera: Miridae) as a predator of the lime aphid, *Eucallipterus tiliae*. Entomol. Exp. Appl. 16: 255–267.

1975a. The effects of predators on the eggs of codling moth *Cydia pomonella*, in a cider-apple orchard in south-west England. Ann. Appl. Biol. 80: 115–119.

1975b. Searching behaviour and prey-density requirements of *Blepharidopterus angulatus* (Fall.) (Heteroptera: Miridae) as a predator of the lime aphid, *Eucallipterus tiliae* (L.), and leafhopper, *Alnetoidea alneti* (Dahlbom). J. Anim. Ecol. 44: 115–134.

1977a. Ecology of the parasites of a predatory bug, *Blepharidopterus angulatus* (Fall.). Ecol. Entomol. 2: 47–55.

1977b. Predation of codling moth eggs, *Cydia pomonella*, the predators responsible and their alternative prey. J. Appl. Ecol. 14: 445–456.

Glen, D. M., and N. D. Barlow. 1980. Interaction of a population of the black-kneed capsid, *Blepharidopterus angulatus*, and its prey, the lime aphid. Ecol. Entomol. 5: 335–344.

Glen, D. M., and P. Brain. 1978. A model of predation on codling moth eggs *(Cydia pomonella)*. J. Anim. Ecol. 47: 711–724.

Goula, M., and O. Alomar. 1994. Míridos (Heteroptera Miridae) de interés en el control integrado de plagas en el tomate. Guía para su identificación. Bol. Sanid. Veg. Plagas 20: 131–143.

Grasswitz, T. R., and E. C. Burts. 1995. Effect of native natural enemies on the population dynamics of the grape mealybug, *Pseudococcus maritimus* (Hom.: Pseudococcidae), in apple and pear orchards. Entomophaga 40: 105–117.

Gravena, S., and J. A. Pazetto. 1987. Predation and parasitism of cotton leafworm eggs, *Alabama argillacea* [Lep.: Noctuidae]. Entomophaga 32: 241–248.

Gravena, S., and W. L. Sterling. 1983. Natural predation on the cotton leafworm (Lepidoptera: Noctuidae). J. Econ. Entomol. 76: 779–784.

Greenstone, M. H., and C. E. Morgan. 1989. Predation on *Heliothis zea* (Lepidoptera: Noctuidae): An instar-specific ELISA assay for stomach analysis. Ann. Entomol. Soc. Amer. 82:45–49.

Grenier, S., J. Guillaud, B. Delobel, and G. Bonnot. 1989. Nutrition et élevage du prédateur polyphage *Macrolophus caliginosus* [Heteroptera, Miridae] sur milieux artificiels. Entomophaga 34: 77–86.

Gut, L. J., P. H. Westigard, C. Jochums, and W. J. Liss. 1982. Variation in pear psylla (*Psylla pyricola* Foerster) densities in southern Oregon orchards and its implications. Acta Hortic. (Wageningen) 124: 101–111.

Hagen, K. S., N. J. Mills, G. Gordh, and J. A. McMurtry. 1999. Pp. 383–503 *in* T. S. Bellows and T. W. Fisher [eds.], Handbook of Biological Control: Principles and Applications of Biological Control. Academic Press, San Diego, California, U.S.A. 1046 pp.

Hagler, J. R., and S. E. Naranjo. 1994. Determining the frequency of heteropteran predation on sweetpotato whitefly and pink bollworm using multiple ELISAs. Entomol. Exp. Appl. 72: 59–66.

Hagley, E. A. C., and C. M. Simpson. 1983. Effect of insecticides on predators of the pear psylla, *Psylla pyricola* (Hemiptera: Psyllidae), in Ontario. Canad. Entomol. 115: 1409–1414.

Haley, S., and E. J. Hogue. 1990. Ground cover influence on apple aphid, *Aphis pomi* DeGeer (Homoptera: Aphididae), and its predators in a young apple orchard. Crop Prot. 9: 225–230.

Hamel, D. R. 1990. Insects on stamps. Amer. Entomol. 36: 273–281.

Hare, J. D. 1994. Status and prospects for an integrated approach to the control of rice planthoppers. Pp. 615–632 *in* R. F. Denno and T. J. Perfect [eds.], Planthoppers: Their Ecology and Management. Chapman & Hall, New York, New York, U.S.A. 799 pp.

Harizanova, V. 1989. Biological peculiarities of the predatory apple bug *Atractotomus mali* (Heteroptera: Miridae). Rastenieved. Nauki 26(9): 98–102 [in Bulgarian, English summary].

Heinrichs, E. A., G. B. Aquino, S. Chelliah, S. L. Valencia, and W. H. Reissig. 1982. Resurgence of *Nilaparvata lugens* (Stål) populations as influenced by method and timing of insecticide applications in lowland rice. Environ. Entomol. 11: 78–84.

Henry, T. J., and A. G. Wheeler, Jr. 1988. Family Miridae Hahn, 1833 (= Capsidae Burmeister). The plant bugs. Pp. 251–507 *in* T. J. Henry and R. C. Froeschner [eds.], Catalog of the Heteroptera, or True Bugs, of Canada and the Continental United States. E. J. Brill, Leiden, the Netherlands. 958 pp.

Henry, T. J., J. W. Neal, Jr., and K. M. Gott. 1986. *Stethoconus japonicus* (Heteroptera: Miridae): a predator of *Stephanitis* lace bugs newly discovered in the United States, promising in the biocontrol of azalea lace bug (Heteroptera: Tingidae). Proc. Entomol. Soc. Washington 88: 722–730.

Heong, K. L., S. Bleih, and A. A. Lazaro. 1990. Predation of *Cyrtorhinus lividipennis* Reuter on eggs of the green leafhopper and brown planthopper in rice. Res. Popul. Ecol. (Kyoto) 32: 255–262.

Herbert, H. J. 1962. Overwintering females and the number of generations of *Typhlodromus (T.) pyri* Scheuten (Acarina: Phytoseiidae) in Nova Scotia. Canad. Entomol. 94: 233–242.

Herbert, H. J., and K. H. Sanford. 1969. The influence of spray programs on the fauna of apple orchards in Nova Scotia XIX. Apple rust mite, *Vasates schlechtendali*, a food source for predators. Canad. Entomol. 101: 62–67.

Hesjedal, K. 1986. Skadedyrmiddel i ulike konsentrasjonar på blad-og nebbteger i frukthagar. Forsk. Fors. Landbruket 37: 213–217.

Hesse, A. J. 1947. A remarkable new dimorphic isometopid and two other new species of Hemiptera predaceous upon the red scale of citrus. J. Entomol. Soc. South. Africa 10: 31–45.

Hinckley, A. D. 1963. Ecology and control of rice planthoppers in Fiji. Bull. Entomol. Res. 54: 467–481.

Hoebeke, E. R., and A. G. Wheeler, Jr. 1996. *Meligethes viridescens* (F.) (Coleoptera: Nitidulidae) in Maine, Nova Scotia, and Prince Edward Island: diagnosis, distribution, and bionomics of a Palearctic species new to North America. Proc. Entomol. Soc. Washington 98: 221–227.

Holdaway, [F. G.], and [W. C.] Look. 1940. Ecology of the tomato bug, *Cyrtopeltis varians*. Hawaiian Agric. Exp. Stn. Rep. 1939: 36–37.

Horsburgh, R. L. 1969. The predaceous mirid *Hyaliodes vitripennis* (Hemiptera) and its role in the control of *Panonychus ulmi* (Acarina: Tetranychidae). Ph.D. dissertation, Pennsylvania State University, University Park, Pennsylvania, U.S.A. 106 pp.

Horton, D. R., T. M. Lewis, T. Hinojosa, and D. A. Broers. 1998. Photoperiod and reproductive diapause in the predatory bugs *Anthocoris tomentosus, A. antevolens,* and *Deraeocoris brevis* (Heteroptera: Anthocoridae, Miridae) with information on overwintering sex ratios. Ann. Entomol. Soc. Amer. 91: 81–86.

Hoyt, S. C. 1969. Population studies of five mite species on apple in Washington. Pp. 117–133 *in* Proc. 2nd Int. Congr. Acarol., Sutton Bonington (England), 19–25 July 1967. Akadémiai Kiadó, Budapest, Hungary. 652 pp.

Huffaker, C. B., M. van de Vrie, and J. A. McMurtry. 1970. Ecology of tetranychid mites and their natural enemies: a review II. Tetranychid populations and their possible control by predators: an evaluation. Hilgardia 40(11): 391–458.

Huffaker, C. B., P. S. Messenger, and P. DeBach. 1971. The natural enemy component in natural control and the theory of biological control. Pp. 16–67 *in* C. B. Huffaker [ed.], Biological Control. Plenum Press, New York, New York, U.S.A. 511 pp.

Hunter, C. D. 1997. Suppliers of beneficial organisms in North America. Calif. Environ. Prot. Agency, Dep. Pestic. Regul. Environ. Monitoring Pest Manage. Branch. Sacramento, California, U.S.A. 32 pp.

Imms, A. D. 1926. The biological control of insect pests and injurious plants in the Hawaiian Islands. Ann. Appl. Biol. 13: 402–423.

Jones, W. A., and G. L. Snodgrass. 1998. Development and fecundity of *Deraeocoris nebulosus* (Heteroptera: Miridae) on *Bemisia argentifolii* (Homoptera: Aleyrodidae). Florida Entomol. 81: 345–350.

Jonsson, N. 1987. Nymphal development and food consumption of *Atractotomus mali* (Meyer-Dür) (Hemiptera: Miridae), reared on *Aphis pomi* (DeGeer) and *Psylla mali* Schmidberger. Fauna Norv. Ser. B 34: 22–28.

Kajita, H. 1978. The feeding behaviour of *Cyrtopeltis tenuis* Reuter on the greenhouse whitefly, *Trialeurodes vaporariorum* (Westwood). Rostria (Osaka) 29: 235–238.

Kalode, M. B. 1983. Leafhopper and planthopper pests of rice in India. Pp. 225–245 *in* W. J. Knight et al. [eds.], Proc. 1st Int. Workshop on Biotaxonomy, Classification and Biology of Leafhoppers and Planthoppers (Auchenorrhyncha) of Economic Importance, London, 4–7 October 1982. Commonwealth Institute of Entomology, London, U.K. 500 pp.

Kelton, L. A. 1980. The Insects and Arachnids of Canada. Part 8. The Plant Bugs of the Prairie Provinces. Heteroptera: Miridae. Agric. Canada Publ. 1703: 1–408.

1983. Plant Bugs on Fruit Crops in Canada. Heteroptera: Miridae. Res. Branch Agric. Canada Monogr. No. 24: 1–201.

Kerzhner, I. M., and M. Josifov. 1999. Cimicomorpha II: Miridae *in* B. Aukema and C. Rieger [eds.], Catalogue of the Heteroptera of the Palaearctic Region, Vol. 3. Netherlands Entomological Society, Amsterdam, the Netherlands. 577 pp.

Kirby, W. F. 1892. Elementary Text-book of Entomology, second edition revised and augmented. Swan Sonnenschein, London, U.K. 281 pp.
Kiritani, K. 1979. Pest management in rice. Annu. Rev. Entomol. 24: 279–312.
Kiritshenko, A. N. 1951. True bugs of the European USSR. Key and bibliography. Opred. Faune USSR 42: 1–423 [in Russian].
Kisimoto, R. 1979. Brown planthopper migration. Pp. 113–124 in Brown Planthopper: Threat to Rice Production in Asia. International Rice Research Institute, Los Banõs, the Philippines. 369 pp.
Knight, H. H. 1921. Monograph of the North American Species of *Deraeocoris* (Heteroptera, Miridae). 18th Rep. State Entomol. Minnesota pp. 76–210. Agricultural Experiment Station, St. Paul, Minnesota, U.S.A.
1924. *Atractotomus mali* (Meyer) found in Nova Scotia (Heteroptera, Miridae). Bull. Brooklyn Entomol. Soc. 19: 65.
1941. The plant bugs, or Miridae, of Illinois. Illinois Nat. Hist. Surv. Bull. 22: 1–234.
Koppert. 1993. A new predatory bug for *Bemisia* control in tomato? Koppert Bio-Journal (Berkel en Rodenrijs) No. 6: 4.
Krämer, P. 1961. Untersuchungen über den Einfluss einiger Arthropoden auf Raubmilben (Acari). Z. Angew. Zool. 48: 257–311.
Kuno, E., and N. Hokyo. 1970. Comparative analysis of the population dynamics of rice leafhoppers, *Nephotettix cincticeps* Uhler and *Nilaparvata lugens* Stål, with special reference to natural regulation of their numbers. Res. Popul. Ecol. (Kyoto) 12: 154–184.
Lattin, J. D., and G. M. Stonedahl. 1984. *Campyloneura virgula*, a predaceous Miridae not previously recorded from the United States (Hemiptera). Pan-Pac. Entomol. 60: 4–7.
LeRoux, E. J. 1960. Effects of "modified" and "commercial" spray programs on the fauna of apple orchards in Quebec. Ann. Soc. Entomol. Québec 6: 87–121.
Libutan, G. M., and E. N. Bernardo. 1995. The host preference of the capsid bug, *Cyrtopeltis tens* [sic] Reuter (Hemiptera: Miridae). Philippine Entomol. 9: 567–586.
Liquido, N. J., and T. Nishida. 1983. Geographical distribution of *Cyrtorhinus* and *Tytthus* (Heteroptera: Miridae), egg predators of cicadellid and delphacid pests. FAO Plant Prot. Bull. 31: 159–162.
1985a. Population parameters of *Cyrtorhinus lividipennis* (Heteroptera: Miridae) reared on eggs of natural and factitious prey. Proc. Hawaiian Entomol. Soc. 25: 87–93.
1985b. Observations on some aspects of the biology of *Cyrtorhinus lividipennis* Reuter (Heteroptera: Miridae). Proc. Hawaiian Entomol. Soc. 25: 95–101.
1985c. Variation in number of instars, longevity, and fecundity of *Cyrtorhinus lividipennis* Reuter (Hemiptera: Miridae). Ann. Entomol. Soc. Amer. 78: 459–463.
Lloyd, N. C. 1969. The apple dimpling bug, *Campylomma livida* Reut. (Hemiptera: Miridae). Agric. Gaz. New South Wales 80: 582–584.
Lord, F. T. 1949. The influence of spray programs on the fauna of apple orchards in Nova Scotia. III. Mites and their predators. Canad. Entomol. 81: 217–230.
1968. Influence of the proportion of fruiting to non-fruiting clusters on the distribution of insect predators on apple trees. Canad. Entomol. 100: 308–312.
1971. Laboratory tests to compare the predatory value of six mirid species in each stage of development against the winter eggs of the European red mite, *Panonychus ulmi* (Acari: Tetranychidae). Canad. Entomol. 103: 1663–1669.
Lyon, J. P. 1986. Use of aphidophagous and polyphagous beneficial insects for biological control of aphids in greenhouse. Pp. 471–474 in I. Hodek [ed.], Ecology of Aphidophaga. Proc. 2nd Symp. at Zvíkovské Podhradí, Sept. 2–8, 1984. Junk, Dordrecht, the Netherlands. 562 pp.
MacLellan, C. R. 1962. Mortality of codling moth eggs and young larvae in an integrated control orchard. Canad. Entomol. 94: 655–666.
1963. Predator populations and predation on the codling moth in an integrated control orchard — 1961. Mem. Entomol. Soc. Canada 32: 41–54.
1972. Codling moth populations under natural, integrated, and chemical control on apple in Nova Scotia (Lepidoptera: Olethreutidae). Canad. Entomol. 104: 1397–1404.
1973. Natural enemies of the light brown apple moth, *Epiphyas postvittana*, in the Australian Capital Territory. Canad. Entomol. 105: 681–700.

MacPhee, A. W. 1979. Observations on the white apple leafhopper, *Typhlocyba pomaria* (Hemiptera: Cicadellidae), and on the mirid predator *Blepharidopterus angulatus*, and measurements of their cold-hardiness. Canad. Entomol. 111: 487–490.

MacPhee, A. W., and C. R. MacLellan. 1972. Ecology of apple orchard fauna and development of integrated pest control in Nova Scotia. Proceedings Tall Timbers Conf. Ecol. Anim. Control Habitat Manage. 3: 197–208.

MacPhee, A. W., and K. H. Sanford. 1956. The influence of spray programs on the fauna of apple orchards in Nova Scotia. X. Supplement to VII. Effects on some beneficial arthropods. Canad. Entomol. 88: 631–634.

1961. The influence of spray programs on the fauna of apple orchards in Nova Scotia. XII. Second supplement to VII. Effects on beneficial arthropods. Canad. Entomol. 93: 671–673.

Malausa, J. C., and Y. Trottin-Caudal. 1996. Advances in the strategy of use of the predaceous bug *Macrolophus caliginosus* (Heteroptera: Miridae) in glasshouse crops. Pp. 178–189 *in* O. Alomar and R. N. Wiedenmann [eds.], Zoophytophagous Heteroptera: Implications for Life History and Integrated Pest Management. Thomas Say Publ. Entomol.: Proceedings Entomological Society of America, Lanham, Maryland, U.S.A. 202 pp.

Malézieux, S., C. Giradet, B. Navez, and J.-M. Cheyrias. 1995. Contre l'Aleurode des serres en cultures de tomates sous abris: utilisation et développement de *Macrolophus caliginosus* associé a *Encarsia formosa*. Phytoma 471: 29–32.

Malipatil, M. B. 1992. Revision of Australian *Campylomma* Reuter (Hemiptera: Miridae: Phylinae). J. Austral. Entomol. Soc. 31: 357–368.

Manti, I. 1989. The role of *Cyrtorhinus lividipennis* Reuter (Hemiptera, Miridae) as a major predator of the brown planthopper *Nilaparvata lugens* Stål (Homoptera: Delphacidae). Ph.D. dissertation, University of the Philippines, Los Baños, the Philippines. 126 pp.

Mathen, K., and C. Kurian. 1972. Description, life-history and habits of *Stethoconus praefectus* (Distant) (Heteroptera: Miridae) predaceous on *Stephanitis typicus* Distant (Heteroptera: Tingidae), a pest of coconut palm. Indian J. Agric. Sci. 42: 255–262.

Matsumoto, B. M. 1964. Predator–prey relationships between the predator, *Cyrtorhinus fulvus* Knight, and the prey, the taro leafhopper, *Tarophagus proserpina* (Kirkaldy). M.S. thesis, University of Hawaii, Honolulu, Hawaii, U.S.A. 70 pp.

Matsumoto, B. M., and T. Nishida. 1966. Predator–prey investigations on the taro leafhopper and its egg predator. Hawaii Agric. Exp. Stn. Tech. Bull. 64: 1–32.

McCaffrey, J. P., and R. L. Horsburgh. 1980. The egg and oviposition site of *Deraeocoris nebulosus* (Hemiptera: Miridae) on apple trees. Canad. Entomol. 112: 527–528.

1986. Biology of *Orius insidiosus* (Heteroptera: Anthocoridae): A predator in Virginia apple orchards. Environ. Entomol. 15: 984–988.

McDaniel, S. G., and W. L. Sterling. 1979. Predator determination and efficiency on *Heliothis virescens* eggs in cotton using $^{32}P^2$. Environ. Entomol. 8: 1083–1087.

1982. Predation of *Heliothis virescens* (F.) eggs on cotton in east Texas. Environ. Entomol. 11: 60–66.

McGregor, R. R., D. R. Gillespie, D. M. J. Quiring, and M. R. J. Foisy. 1999. Potential use of *Dicyphus hesperus* Knight (Heteroptera: Miridae) for biological control of pests of greenhouse tomatoes. Biol. Control 16: 104–110.

McMullen, R. D., and C. Jong. 1967. New records and discussion of predators of the pear psylla, *Psylla pyricola* Forster, in British Columbia. J. Entomol. Soc. British Columbia 64: 35–40.

1970. The biology and influence of pesticides on *Campylomma verbasci* (Heteroptera: Miridae). Canad. Entomol. 102: 1390–1394.

Messing, R. H., and M. T. AliNiazee. 1985. Natural enemies of *Myzocallis coryli* (Hom.: Aphididae) in Oregon hazelnut orchards. J. Entomol. Soc. British Columbia 82: 14–18.

1986. Impact of predaceous insects on filbert aphid, *Myzocallis coryli* (Homoptera: Aphididae). Environ. Entomol. 15: 1037–1041.

Miles, P. W. 1972. The saliva of Hemiptera. Adv. Insect Physiol. 9: 183–255.

Miller, G. L., and M. L. Williams. 1985. Notes on some little known scale insect predators recently collected in Alabama. J. Alabama Acad. Sci. 56: 81 (abstr.).

Misra, B. C. 1980. The Leaf and Planthoppers of Rice. Central Rice Research Institute, Cuttack, India. 182 pp.

Mitchell, W. C., and P. A. Maddison. 1983. Major taro pests in the Pacific Islands. P. 167 *in* Proc. Pac. Sci. Assoc. 15th Congr., Dunedin, New Zealand, 1–11 Feb. 1983. Vol. I. Royal Society of New Zealand, Wellington, New Zealand.

Mizell, R. F., III, and D. E. Schiffhauer. 1987. Seasonal abundance of the crapemyrtle aphid, *Sarucallis kahawaluokalani*, in relation to the pecan aphids, *Monellia caryella* and *Monelliopsis pecanis* and their common predators. Entomophaga 32: 511–520.

Morris, M. G. 1965. Some aspects of the biology of *Psallus ambiguus* (Fall.) (Heteroptera: Miridae) on apple trees in Kent. Entomologist 98: 14–31.

Muir, F. 1920. Report of entomological work in Australia, 1919–1920. Hawaiian Plant. Rec. 23: 125–130.

Muir, R. C. 1958. On the application of the capture–recapture method to an orchard population of *Blepharidopterus angulatus* (Fall.) (Hemiptera-Heteroptera, Miridae). East Malling (U.K.) Res. Stn. Rep. 1957: 140–147.

1965a. The effect of sprays on the fauna of apple trees. I. The influence of winter wash, captan, and limesulphur on the interaction of populations of *Panonychus ulmi* (Koch) (Acarina: Tetranychidae) and its predator, *Blepharidopterus angulatus* (Fall.) (Heteroptera: Miridae). J. Appl. Ecol. 2: 31–41.

1965b. The effect of sprays on the fauna of apple trees. II. Some aspects of the interaction between populations of *Blepharidopterus angulatus* (Fall.) (Heteroptera: Miridae) and its prey, *Panonychus ulmi* (Koch) (Acarina: Tetranychidae). J. Appl. Ecol. 2: 43–57.

1966a. The effect of temperature on development and hatching of the egg of *Blepharidopterus angulatus* (Fall.) (Heteroptera, Miridae). Bull. Entomol. Res. 57: 61–67.

1966b. The effect of sprays on the fauna of apple trees. IV. The recolonization of orchard plots by the predatory mirid *Blepharidopterus angulatus* and its effect on populations of *Panonychus ulmi*. J. Appl. Ecol. 3: 269–276.

Murdoch, W. W., J. Chesson, and P. L. Chesson. 1985. Biological control in theory and practice. Amer. Nat. 125: 344–366.

Murray, R. A., and M. G. Solomon. 1978. A rapid technique for analysing diets of invertebrate predators by electrophoresis. Ann. Appl. Biol. 90: 7–10.

Murtfeldt, M. E. 1894. Notes on the insects of Missouri for 1893. U.S. Dept. Agric. Div. Entomol. Bull. 32: 37–45.

Myers, J. G. 1927a. Natural enemies of the pear leaf-curling midge, *Perrisia pyri*, Bouché (Dipt., Cecidom.). Bull. Entomol. Res. 18: 129–138.

1927b. Ethological observations on some Pyrrhocoridae of Cuba. (Hemiptera-Heteroptera). Ann. Entomol. Soc. Amer. 20: 279–300.

Napompeth, B. 1973. Ecology and population dynamics of the corn planthopper, *Peregrinus maidis* (Ashmead) (Homoptera: Delphacidae), in Hawaii. Ph.D. dissertation, University of Hawaii, Manoa, Hawaii, U.S.A. 257 pp.

Naranjo, S. E., and R. L. Gibson. 1996. Phytophagy in predaceous Heteroptera: Effects on life history and population dynamics. Pp. 57–93 *in* O. Alomar and R. N. Wiedenmann [eds.], Zoophytophagous Heteroptera: Implications for Life History and Integrated Pest Management. Thomas Say Publ. Entomol.: Proceedings Entomological Society of America, Lanham, Maryland, U.S.A. 202 pp.

Neal, J. W., Jr., and R. H. Haldemann. 1992. Regulation of seasonal egg hatch by plant phenology in *Stethoconus japonicus* (Heteroptera: Miridae), a specialist predator of *Stephanitis pyrioides* (Heteroptera: Tingidae). Environ. Entomol. 21: 793–798.

Neal, J. W., Jr., R. H. Haldemann, and T. J. Henry. 1991. Biological control potential of a Japanese plant bug, *Stethoconus japonicus* (Heteroptera: Miridae), an adventive predator of azalea lace bug (Heteroptera: Tingidae). Ann. Entomol. Soc. Amer. 84: 287–293.

New, T. R. 1991. Insects as Predators. New South Wales University Press, Kensington, Australia. 178 pp.

Niemczyk, E. 1963. Heteroptera associated with apple orchards in the district of Nowy Sącz. Ekol. Pol. (A) 11: 295–300.

1966. Food ecology of *Psallus ambiguus* (Fall.) (Heteroptera: Miridae). P. 69 *in* I. Hodek [ed.], Ecology of Aphidophagous Insects. Proc. Symp. Liblice near Prague, Sept. 27–Oct. 1, 1965. Junk, The Hague, the Netherlands; Academia, Prague, Czechoslovakia. 360 pp.

1967. *Psallus ambiguus* (Fall.) (Heteroptera, Miridae) Część I: Morfologia i biologia. Pol. Pismo Entomol. 37: 797–842.

1968. *Psallus ambiguus* (Fall.) (Heteroptera, Miridae). Część II. Odżywianie się i rola w biocenozie sadów. Pol. Pismo Entomol. 38: 387–416.

1978. *Campylomma verbasci* Mey-Dur (Heteroptera, Miridae) as a predator of aphids and mites in apple orchards. Pol. Pismo Entomol. 48: 221–235.

Niemczyk, E., and M. Pruska. 1986. The occurrence of predators in different types of colonies of apple aphids. Pp. 303–310 *in* I. Hodek [ed.], Ecology of Aphidophaga. Proc. 2nd Symp. at Zvíkovské Podhradí, Sept. 2–8, 1984. Junk, Dordrecht, the Netherlands. 562 pp.

Novak, H., and R. Achtziger. 1995. Influence of heteropteran predators (Het., Anthocoridae, Miridae) on larval populations of hawthorn psyllids (Hom., Psyllidae). J. Appl. Entomol. 119: 479–486.

Nucifora, A., and C. Calabretta. 1986. Advances in integrated control of gerbera protected crops. Acta Hortic. (Wageningen) 176: 191–197.

Nuessly, G. S., and W. L. Sterling. 1994. Mortality of *Helicoverpa zea* (Lepidoptera: Noctuidae) eggs in cotton as a function of oviposition sites, predator species, and desiccation. Environ. Entomol. 23: 1189–1202.

O'Connor, B. A. 1952. The rice leaf hopper, *Sogata furcifera kolophon* Kirkaldy and "rice yellows." Fiji Agric. J. 23: 97–104.

Ooi, P. A. C. 1979. Flight activities of brown planthopper, whitebacked planthopper, and their predator *C. lividipennis* in Malaysia. Int. Rice Res. Newsl. 4(6): 12.

Ooi, P. A. C., and B. M. Shepard. 1994. Predators and parasitoids of rice insect pests. Pp. 585–612 *in* E. A. Heinrichs [ed.], Biology and Management of Rice Insects. Wiley, New York, New York, U.S.A. 779 pp.

Parrella, M. P., and J. A. ["F."] Bethke. 1982. Biological studies with *Cyrtopeltis modestus* (Hemiptera: Miridae): a facultative predator of *Liriomyza* spp. (Diptera: Agromyzidae). Pp. 180–185 *in* S. L. Poe [ed.], Proc. 3rd Annu. Ind. Conf. Leafminer, San Diego, California. Society of American Florists, Alexandria, Virginia, U.S.A.

Parrella, M. P., J. P. McCaffrey, and R. L. Horsburgh. 1981. Population trends of selected phytophagous arthropods and predators under different pesticide programs in Virginia apple orchards. J. Econ. Entomol. 74: 492–498.

Parrella, M. P., K. L. Robb, G. D. Christie, and J. A. Bethke. 1982. Control of *Liriomyza trifolii* with biological agents and insect growth regulators. California Agric. 36(11–12): 17–19.

Pemberton, C. E. 1948. History of the Entomology Department Experiment Station, H.S.P.A. 1904–1945. Hawaiian Plant. Rec. 52: 53–90.

Pemberton, C. E. [chairman]. 1954. Introduction of beneficial parasites and predators into Guam and the Trust Territory. Pp. 42–45 *in* Invertebrate Consultants Committee for the Pacific; Rep. 1949–1954. National Research Council Pacific Science Board, Washington, D.C., U.S.A. 56 pp.

Petacchi, R., and E. Rossi. 1991. Prime osservazioni su *Dicyphus (Dicyphus) errans* (Wolff) (Heteroptera Miridae) diffuso sul pomodoro in serre della Liguria. Boll. Zool. Agrar. Bachic. 23: 77–86.

Petherbridge, F. R., and M. A. Husain. 1918. A study of the capsid bugs found on apple trees. Ann. Appl. Biol. 4: 179–205.

Polhemus, D. A. 1994. An annotated checklist of the plant bugs of Colorado (Heteroptera: Miridae). Pan-Pac. Entomol. 70: 122–147.

Pophaly, D. J., T. Bhasker Rao, and M. B. Kalode. 1978. Biology & predation of the mirid bug, *Cyrtorhinus lividipennis* Reuter on plant and leafhoppers. Indian J. Plant Prot. 6: 7–14.

Quaglia, F., E. Rossi, R. Petacchi, and C. E. Taylor. 1993. Observations on an infestation by green peach aphids (Homoptera: Aphididae) on greenhouse tomatoes in Italy. J. Econ. Entomol. 86: 1019–1025.

Rapusas, H. R., D. G. Bottrell, and M. Coll. 1996. Intraspecific variation in chemical attraction of rice to insect predators. Biol. Control 6: 394–400.

Readshaw, J. L. 1971. An ecological approach to the control of mites in Australian orchards. J. Austral. Inst. Agric. Sci. 37: 226–230.

1975. The ecology of tetranychid mites in Australian orchards. J. Appl. Ecol. 12: 473–495.

Reissig, W. H., E. A. Heinrichs, and S. L. Valencia. 1982a. Insecticide-induced resurgence of the brown planthopper, *Nilaparvata lugens*, on rice varieties with different levels of resistance. Environ. Entomol. 11: 165–168.

1982b. Effects of insecticides on *Nilaparvata lugens* and its predators: Spiders, *Microvelia atrolineata*, and *Cyrtorhinus lividipennis*. Environ. Entomol. 11: 193–199.

Reyes, T. M., and B. P. Gabriel. 1975. The life history and consumption habits of *Cyrtorhinus lividipennis* Reuter (Hemiptera: Miridae). Philippine Entomol. 3: 79–88.

Reynolds, D. R., and M. R. Wilson. 1989. Aerial samples of macro-insects migrating at night over central India. J. Plant Prot. Trop. 6: 89–101.

Riedl, H. 1991. Beneficial arthropods for pear pest management. Pp. 101–118 *in* K. Williams [gen. ed.], New Directions in Tree Fruit Pest Management. Good Fruit Grower, Yakima, Washington, U.S.A. 214 pp.

Riley, J. R., D. R. Reynolds, and R. A. Farrow. 1987. The migration of *Nilaparvata lugens* (Stål) (Delphacidae) and other Hemiptera associated with rice during the dry season in the Philippines: a study using radar, visual observations, aerial netting and ground trapping. Bull. Entomol. Res. 77: 145–169.

Riley, J. R., D. R. Reynolds, S. Mukhopadhyay, M. R. Ghosh, and T. K. Sarkar. 1995. Long-distance migration of aphids and other small insects in northeast India. Eur. J. Entomol. 92: 639–653.

Riudavets, J., and C. Castañé. 1998. Identification and evaluation of native predators of *Frankliniella occidentalis* (Thysanoptera: Thripidae) in the Mediterranean. Environ. Entomol. 27: 86–93.

Riudavets, J., R. Gabarra, and C. Castañé. 1993. *Frankliniella occidentalis* predation by native natural enemies. Int. Organ. Biol. Control/West Palearctic Reg. Sect. Bull 16: 137–140.

Riudavets, J., C. Castañé, and R. Gabarra. 1995. Native predators of western flower thrips in horticultural crops. Pp. 255–258 *in* B. L. Parker, M. Skinner, and T. Lewis [eds.], Thrips Biology and Management. Plenum Press, New York, New York, U.S.A. 636 pp.

Room, P. M. 1979. Parasites and predators of *Heliothis* spp. (Lepidoptera: Noctuidae) in cotton in the Namoi Valley, New South Wales. J. Austral. Entomol. Soc. 18: 223–228.

Ruberson, J. R., and M. Coll. 1998. Research needs for the predaceous Heteroptera. Pp. 225–233 *in* M. Coll and J. R. Ruberson [eds.], Predatory Heteroptera: Their Ecology and Use in Biological Control. Thomas Say Publ. Entomol.: Proceedings Entomological Society of America, Lanham, Maryland, U.S.A. 233 pp.

Rutter, J. F., A. P. Mills, and L. J. Rosenberg. 1998. Weather associated with autumn and winter migrations of rice pests and other insects in south-eastern and eastern Asia. Bull. Entomol. Res. 88: 189–197.

Salamero, A., R. Gabarra, and R. Albajes. 1987. Observations on the predatory and phytophagous habits of *Dicyphus tamaninii* Wagner (Heteroptera; Miridae). Proc. Working Group, "Integrated Control in Glasshouses," EPRS/WPRS, Budapest, Hungary, 26–30 April 1987, Int. Organ. Biol. Control/West. Palearctic Reg. Sect. Bull. 10(2): 165–169.

Sanford, K. H. 1964a. Eggs and oviposition sites of some predaceous mirids on apple trees (Miridae: Hemiptera). Canad. Entomol. 96: 1185–1189.

1964b. Life history and control of *Atractotomus mali*, a new pest of apple in Nova Scotia (Miridae: Hemiptera). J. Econ. Entomol. 57: 921–925.

Sanford, K. H., and H. J. Herbert. 1966. The influence of spray programs on the fauna of apple orchards in Nova Scotia. XV. Chemical controls for winter moth, *Operophtera brumata* (L.), and their effects on phytophagous mite and predator populations. Canad. Entomol. 98: 991–999.

Sanford, K. H., and F. T. Lord. 1962. The influence of spray programs on the fauna of apple orchards in Nova Scotia. XIII. Effects of Perthane on predators. Canad. Entomol. 94: 928–934.

Saunders, E. 1892. The Hemiptera Heteroptera of the British Islands. A descriptive account of the families, genera, and species indigenous to Great Britain and Ireland, with notes as to localities, habitats, etc. L. Reeve, London, U.K. 350 pp.

Schelt, J. van., J. Klapwijk, M. Letard, and C. Aucouturier. 1996. The use of *Macrolophus caliginosus* as a whitefly predator in protected crops. Pp. 515–521 *in* D. Gerling and R. T. Mayer [eds.], *Bemisia*: 1995: Taxonomy, Biology, Damage, Control and Management. Intercept, Andover, U.K. 702 pp.

Schewket Bey, N. 1930. Zur Biologie der phytophagen Wanze *Dicyphus errans* Wolff (Capsidae). Z. Wiss. Insektenbiol. 25: 179–183.

Schuh, R. T. 1976. Pretarsal structure in the Miridae (Hemiptera) with a cladistic analysis of relationships within the family. Amer. Mus. Novit. No. 2601: 1–39.

1984. Revision of the Phylinae (Hemiptera, Miridae) of the Indo-Pacific. Bull. Amer. Mus. Nat. Hist. 177: 1–476.

Schuh, R. T., and J. A. Slater. 1995. True Bugs of the World (Hemiptera: Heteroptera): Classification and Natural History. Cornell University Press, Ithaca, New York, U.S.A. 336 pp.

Schuh, R. T., and P. Štys. 1991. Phylogenetic analysis of cimicomorphan family relationships (Heteroptera). J. New York Entomol. Soc. 99: 298–350.

Schumacher, F. 1917. Ueber Psociden-Feinde aus der Ordnung der Hemipteren. Z. Wiss. Insektenbiol. 13: 217–218.

Schwartz, M. D., and G. G. E. Scudder. 1998. Newly recognized Holarctic and introduced plant bugs in North America (Heteroptera: Miridae). Can. Entomol. 130: 267–283.

Scudder, G. G. E. 1997. True bugs (Heteroptera) of the Yukon. Pp. 241–336 *in* H. V. Danks and J. A. Downes [eds.], Insects of the Yukon. Monogr. Ser. 2. Biological Survey of Canada (Terrestrial Arthropods), Ottawa, Ontario, Canada. 1034 pp.

Shteynberg, D. M. 1962. The use of entomorphages [sic] to protect apple orchards in eastern Canada. Entomol. Obozr. 41(2): 300–305. [in Russian; English translation in Entomol. Rev. 41(2): 185–187].

Simmonds, F. J. 1969. Biological control of sugar cane pests: a general survey. Pp. 461–479 *in* J. R. Williams et al. [eds.], Pests of Sugar Cane. Elsevier, Amsterdam, the Netherlands. 568 pp.

Sivapragasam, A., and A. Asma. 1985. Development and reproduction of the mirid bug, *Cyrtorhinus lividipennis* (Heteroptera: Miridae) and its functional response to the brown planthopper. Appl. Entomol. Zool. 20: 373–379.

Slater, J. A., and R. M. Baranowski. 1978. How to Know the True Bugs (Hemiptera-Heteroptera). Brown, Dubuque, Iowa, U.S.A. 256 pp.

Snodgrass, G. L. 1991. *Deraecoris* [sic] *nebulosus* (Heteroptera: Miridae): little-known predator in cotton in the Mississippi Delta. Florida Entomol. 74: 340–344.

Solomon, M. G. 1975. The colonization of an apple orchard by predators of the fruit tree red spider mite. Ann. Appl. Biol. 80: 119–122.

1976. Natural enemies of the oystershell scale, *Quadraspidiotus pyri*. East Malling (U.K.) Res. Stn. Rep. 1975: 130.

1982. Phytophagous mites and their predators in apple orchards. Ann. Appl. Biol. 101: 201–203.

Song, Y. H., and K. L. Heong. 1997. Changes in searching responses with temperature of *Cyrtorhinus lividipennis* Reuter (Hemiptera: Miridae) on the eggs of the brown planthopper, *Nilaparvata lugens* (Stål) (Homoptera: Delphacidae). Res. Popul. Ecol. (Kyoto) 39: 201–206.

Southwood, T. R. E. 1956. The structure of the eggs of terrestrial Heteroptera and its relationship to the classification of the group. Trans. R. Entomol. Soc. London 108: 163–221.

1986. Plant surfaces and insects — an overview. Pp. 1–22 *in* B. Juniper and Sir R. [T. R. E.] Southwood [eds.], Insects and the Plant Surface. Arnold, London, U.K. 360 pp.

Southwood, T. R. E, and D. Leston. 1959. Land and water bugs of the British Isles. Warne, London, England. 436 pp.

Stapley, J. H. 1976. The brown planthopper and *Cyrtorhinus* spp. predators in the Solomon Islands. Rice Entomol. Newsl. 4: 17.

Stear, J. R. 1925. Three mirids predaceous on the rose leaf-hopper on apple. J. Econ. Entomol. 18: 633–636.

Steiner, H. M. 1938. Effects of orchard practices on natural enemies of the white apple leafhopper. J. Econ. Entomol. 31: 232–240.

Stephens, G. S. 1975. Transportation and culture of *Tytthus mundulus*. J. Econ. Entomol. 68: 753–754.

Sterling, W. L. 1982. Predaceous insects and spiders. Pp. 25–31 *in* G. T. Bohmfalk, R. E. Frisbie, W. L. Sterling, R. B. Metzer, and A. E. Knutson [eds.], Identification, Biology and Sampling of Cotton Insects. Texas Agric. Ext. Serv. B-933. 43 pp.

Stinner, B. R. 1975. Observations on predaceous Miridae. Proc. Pennsylvania Acad. Sci. 49: 101–102.

Stonedahl, G. M. 1988. Revision of the mirine genus *Phytocoris* Fallén (Heteroptera: Miridae) for western North America. Bull. Amer. Mus. Nat. Hist. 188: 1–257.

1990. Revision and cladistic analysis of the Holarctic genus *Atractotomus* Fieber (Heteroptera: Miridae: Phylinae). Bull. Amer. Mus. Nat. Hist. 198: 1–88.

Strawínski, K. 1964. Zoophagism of terrestrial Hemiptera-Heteroptera occurring in Poland. Ekol. Pol. (A) 12: 429–452.

Stultz, H. T. 1955. The influence of spray programs on the fauna of apple orchards in Nova Scotia. VIII. Natural enemies of the eye-spotted bud moth, *Spilonota ocellana* (D. & S.) (Lepidoptera: Olethreutidae). Canad. Entomol. 87: 79–85.

Susman, I. 1988. The cotton insects of Israel and aspects of the biology of *Deraeocoris pallens* Reuter (Heteroptera, Miridae). M.S. thesis, Tel Aviv University, Israel. 154 pp. [in Hebrew, English summary].

Sweetman, H. L. 1958. The Principles of Biological Control: Interrelation of Hosts and Pests and Utilization in Regulation of Animal and Plant Populations. Brown, Dubuque, Iowa, U.S.A. 560 pp.

Swezey, O. H. 1936. Biological control of the sugar cane leafhopper in Hawaii. Hawaiian Plant. Rec. 40: 57–101.
Takara, J., and T. Nishida. 1981. Eggs of the Oriental fruit fly for rearing the predaceous anthocorid, *Orius insidiosus* (Say). Proc. Hawaiian Entomol. Soc. 23: 441–445.
Tanada, Y., and F. G. Holdaway. 1954. Feeding habits of the tomato bug, *Cyrtopeltis (Engytatus) modestus* (Distant), with special reference to the feeding lesion on tomato. Hawaii Agric. Exp. Stn. Tech. Bull. 24: 1–40.
Tedders, W. L. 1995. Identity of spittlebug on pecan and life history of *Clastoptera achatina* (Homoptera: Cercopidae). J. Econ. Entomol. 88: 1641–1649.
Thistlewood, H. M. A., and R. F. Smith. 1996. Management of the mullein bug, *Campylomma verbasci* (Heteroptera: Miridae), in pome fruit orchards of Canada. Pp. 119–140 in O. Alomar and R. N. Wiedenmann [eds.], Zoophytophagous Heteroptera: Implications for Life History and Integrated Pest Management. Thomas Say Publ. Entomol.: Proceedings Entomological Society of America, Lanham, Maryland, U.S.A. 202 pp.
Thomas, W. A. 1945. *Cyrtopeltis varians* in some of the tobacco-growing areas of North Carolina. J. Econ. Entomol. 38: 498–499.
Timberlake, P. H. 1927. Biological control of insect pests in the Hawaiian Islands. Proc. Hawaiian Entomol. Soc. 6: 529–556.
Torreno, H. S. 1994. Predation behavior and efficiency of the bug, *Cyrtopeltis tenuis* (Hemiptera: Miridae), against the cutworm, *Spodoptera litura* (F.). Philippine Entomol. 9: 426–434.
Torreno, H. S., and E. D. Magallona. 1994. Biological relationship of the bug, *Cyrtopeltis tenuis* Reuter (Hemiptera: Miridae) with tobacco. Philippine Entomol. 9: 406–425.
Uhler, P. R. 1876. List of Hemiptera of the region west of the Mississippi River, including those collected during the Hayden explorations of 1873. Bull. U.S. Geol. Geogr. Surv. Terr. 1: 267–361.
Usinger, R. L. 1939. Distribution and host relationships of *Cyrtorhinus* (Hemiptera: Miridae). Proc. Hawaiian Entomol. Soc. 10: 271–273.
van de Baan, H. E., and B. A. Croft. 1990. Factors influencing insecticide resistance in *Psylla pyricola* (Homoptera: Psyllidae) and susceptibility in the predator *Deraeocoris brevis* (Heteroptera: Miridae). Environ. Entomol. 19: 1223–1228.
van den Berg, H., B. M. Shepard, J. A. Litsinger, and P. C. Pantua. 1988. Impact of predators and parasitoids on the eggs of *Rivula atimeta, Naranga aenescens* [Lepidoptera: Noctuidae] and *Hydrellia philippina* [Diptera: Ephydridae] in rice. J. Plant Prot. Trop. 5: 103–108.
van den Berg, H., J. A. Litsinger, B. M. Shepard, and P. C. Pantua. 1992. Acceptance of eggs of *Rivula atimeta, Naranga aenescens* [Lep.: Noctuidae] and *Hydrellia philippina* [Dipt.: Ephydridae] by insect predators on rice. Entomophaga 37: 21–28.
Verma, J. S. 1955. Biological studies to explain the failure of *Cyrtorhinus mundulus* (Breddin) as an egg-predator of *Peregrinus maidis* (Ashmead) in Hawaii. Proc. Hawaiian Entomol. Soc. 15: 623–634.
Vojdani, S., and A. Doftari. 1963. The alfalfa weevil, *Hypera postica* Gyll., a destructive beetle in Karaj. Tehran Univ. Publ. Dept. Plant Prot. Phytopharm., Karaj, Iran. 34 pp. [in Persian; English summary].
Walker, J. T. S., C. H. Wearing, and A. J. Hayes. 1989. *Panonychus ulmi* (Koch), European red mite (Acari: Tetranychidae). Pp. 217–221 in P. J. Cameron, R. L. Hill, J. Bain, and W. P. Thomas [eds.], A Review of Biological Control of Invertebrate Pests and Weeds in New Zealand 1874 to 1987. Tech. Commun. No. 10. CAB International Institute of Biological Control, Wallingford, Oxon, U.K. 424 pp.
Waterhouse, D. F., and K. R. Norris. 1989. Biological control: Pacific prospects — supplement 1. Australian Centre for International Agricultural Research, Canberra, Australia. 123 pp.
Westigard, P. H. 1973. The biology of and effect of pesticides on *Deraeocoris brevis piceatus* (Heteroptera: Miridae). Canad. Entomol. 105: 1105–1111.
Westigard, P. H., P. M. Lombard, and D. W. Berry. 1979. Integrated pest management of insects and mites attacking pears in southern Oregon. Oregon Agric. Exp. Stn. Bull. 634: 1–41.
Wheeler, A. G., Jr. 1976. Lygus bugs as facultative predators. Pp. 28–35 in D. R. Scott and L. E. O'Keeffe [eds.], Lygus Bug: Host Plant Interactions. University of Idaho Press, Moscow, Idaho, U.S.A. 38 pp.
1991. Plant bugs of *Quercus ilicifolia*: myriads of mirids (Heteroptera) in pitch pine-scrub oak barrens. J. New York Entomol. Soc. 99: 405–440.
2001. Biology of the Plant Bugs (Hemiptera: Miridae): Pests, Predators, Opportunists. Cornell University Press, Ithaca, New York, U.S.A. (in press).

Wheeler, A. G., Jr., and T. J. Henry. 1977. Miridae associated with Pennsylvania conifers 1. Species on arborvitae, false cypress, and juniper. Trans. Amer. Entomol. Soc. 103: 623–656.

1978a. Isometopinae (Hemiptera: Miridae) in Pennsylvania: biology and descriptions of fifth instars, with observations of predation on obscure scale. Ann. Entomol. Soc. Amer. 71: 607–614.

1978b. *Ceratocapsus modestus* (Hemiptera: Miridae), a predator of grape phylloxera: seasonal history and description of fifth instar. Melsheimer Entomol. Ser. No. 25: 6–10.

1992. A Synthesis of the Holarctic Miridae (Heteroptera): Distribution, Biology, and Origin, with Emphasis on North America. Thomas Say Found. Monogr. Vol. 25. Entomological Society of America, Lanham, Maryland, U.S.A. 282 pp.

Wheeler, A. G., Jr., J. T. Hayes, and J. L. Stephens. 1968. Insect predators of mummified pea aphids. Canad. Entomol. 100: 221–222.

Wheeler, A. G., Jr., B. R. Stinner, and T. J. Henry. 1975. Biology and nymphal stages of *Deraeocoris nebulosus* (Hemiptera: Miridae), a predator of arthropod pests on ornamentals. Ann. Entomol. Soc. Amer. 68: 1063–1068.

Whitman, D. W., M. S. Blum, and F. Slansky, Jr. 1994. Carnivory in phytophagous insects. Pp. 161–205 *in* T. N. Ananthakrishnan [ed.], Functional Dynamics of Phytophagous Insects. Science Publishers, Lebanon, New Hampshire, U.S.A. 304 pp.

Wiedenmann, R. N., and L. T. Wilson. 1996. Zoophytophagous Heteroptera: Summary and future research needs. Pp. 190–202 *in* O. Alomar and R. N. Wiedenmann [eds.], Zoophytophagous Heteroptera: Implications for Life History and Integrated Pest Management. Thomas Say Publ. Entomol.: Proceedings Entomological Society of America, Lanham, Maryland, U.S.A. 202 pp.

Williams, F. X. 1931. Handbook of the Insects and Other Invertebrates of Hawaiian Sugar Cane Fields. Experiment Station of the Hawaiian Sugar Planters' Association, Honolulu, Hawaii, U.S.A. 400 pp.

Wilson, J. W. 1948. A note on the predaceous habit of the mirid *Cyrtopeltis varians* (Dist). Florida Entomol. 21: 20.

Yasunaga, T., M. Takai, and Y. Nakatani. 1997. Species of the genus *Stethoconus* of Japan (Heteroptera, Miridae): predaceous deraeocorine plant bugs associated with lace bugs (Tingidae). Appl. Entomol. Zool. 32: 261–264.

Yeargan, K. V. 1998. Predatory Heteroptera in North American agroecosystems: An overview. Pp. 7–19 *in* M. Coll and J. R. Ruberson [eds.], Predatory Heteroptera: Their Ecology and Use in Biological Control. Thomas Say Publ. Entomol.: Proceedings Entomological Society of America, Lanham, Maryland, U.S.A. 233 pp.

Zimmerman, E. C. 1948a. Insects of Hawaii. Volume 3. Heteroptera. University of Hawaii Press, Honolulu, Hawaii, U.S.A. 255 pp.

1948b. Insects of Hawaii. Volume 4. Homoptera: Auchenorrhyncha. University of Hawaii Press, Honolulu, Hawaii, U.S.A. 268 pp.

CHAPTER **29**

Assassin Bugs
(Reduviidae excluding Triatominae)

Dunston P. Ambrose

1. INTRODUCTION

Reduviidae is the largest family of predaceous land Heteroptera, containing about 6250 species and subspecies in 913 genera and 25 subfamilies (Maldonado 1990). Reduviids are abundant, they occur worldwide, they are voracious predators (hence their name, "assassin bugs"), and most are general predators. Being larger than many other predaceous land bugs and encompassing in their development a greater range of size, reduviid predators consume not only more prey but also a wider array of prey (Schaefer 1988). Because they are polyphagous, reduviids may not be useful as predators on specific pests, but they are valuable predators in situations where a variety of insect pests occur. Moreover, they kill more prey than they need to satiate themselves. Thus, reduviids are important mortality factors and should be conserved and augmented for their utilization in biocontrol programs (Ambrose 1987, 1988, 1991, 1995, 1996b, 1999; Schaefer 1988; Schaefer and Ahmad 1987). Unfortunately, to date their potential for biocontrol of pests has been little studied.

Conservation and augmentation of reduviids can be achieved only if their systematics and ecology are studied thoroughly. One must know not only what the insect is, but also what its relatives and what its phylogenetic relationships are; such knowledge broadens and deepens the biological information and thereby makes it more useful (Schaefer 1988).

Despite the abundance of the world's reduviid fauna with its rich taxonomic, geographic, ecological, trophic, morphological, biological, and behavioral diversity, and despite its prey record and biocontrol potential (Ambrose 1996a,b, 1999), studies on Reduviidae are meager. This chapter reviews what is known of these bugs. The economic importance of hematophagous reduviids (Triatominae) is considered in **Chapter 18**.

The family contains more subfamilies than any other heteropteran family, and their composition and relationships remain unsettled (Wygodzinsky 1944, Villiers 1948, Carayon et al. 1956, China and Miller 1959, Usinger and Wygodzinsky 1964, Davis 1969). Hence, there is an absolute need for a complete comprehensive reassessment of the family's higher level classification and phylogenetic relationships.

2. BIOECOLOGY

2.1 Ecology

Analysis of the microhabitat of 347 Indian reduviids show that 85 species live exclusively under boulders, 31 on shrubs, 15 under the bark, and 11 in litter. Up to four reduviids may share a microhabitat. Of these reduviids, 36% dwell exclusively in tropical rain forests, 10% in semiarid zones, 8% in scrub jungles. Many species share two or more habitats.

Tropical rain forest reduviids are soft-cuticled, alate, devoid of warning coloration, and their instars are armored with straight as well as club-shaped hairs and often feign death. They lack tibial pads or fossula spongiosae and they pin and lift their prey by the straight or slightly curved rostrum. Eggs are laid in clusters and are exposed. Fecundity and hatchability are high. Eclosion and ecdysis occur usually in daytime. Precopulatory riding is common. On the other hand, scrub jungle and semiarid zone reduviids are hard-cuticled, pterygopolymorphic, possess warning coloration, and their instars are armored with setose hairs, spines, and tubercles. They possess both fore and midtibial pads and they chase and pounce upon their prey and pierce it with their curved or acutely curved rostrum. They withstand prolonged starvation. Eggs are laid singly, unexposed, in crevices or deep in the soil. Fecundity and hatchability are low. Eclosion and ecdysis occur usually at dusk and dawn. Nymphal camouflaging is common (see Ambrose 1999). Such information is not available for non-Indian reduviids.

Adaptive ecotypic as well as polymorphic diversities have been reported in reduviids (see Ambrose 1999).

Population dynamics studies indicate that reduviid populations are directly regulated by the prey populations (see Ambrose 1999).

2.2 Biology

Biologies of apiomerines, ectrichodiines, harpactorines, peiratines, reduviines, salyavatines, and stenopodainaes, covering information on preoviposition period, fecundity, incubation period, hatchability, stadial period, nymphal mortality, eclosion and ecdysis periodicities, adult longevity, sex ratio, nymphal morphology, nymphal food requirement, postembryonic development, life table, growth pattern, and parental care have been studied (see Table 1) (Readio 1924, 1927, 1931; Hoffmann 1934, 1935; Muller 1937; Cherian and Kylasam 1939; Balduf 1948; Balduf and Slater 1943; Pruthi 1947; Bose 1949; Miller 1953; Wallace 1953; Underhill 1954; Immel 1954, 1955; Stride 1956b; Gillet 1957; Subbiah and Mahadevan 1957; Vanderplank 1958; Odhiambo 1958, 1959; Joseph 1959; Edwards 1962, 1966; Parker 1965a,b; De Coursey 1963; Nyirra 1970; Veeresh and Puttarudriah 1970; Greene 1973; Swadener and Yonke 1973a,b,c, 1975; Louis 1973, 1974; Satpathy et al. 1975; Habib 1976; Guerra and Rodriguez 1976; Singh and Gangrade 1976; Sitaramaiah and Satyanarayana 1975; Livingstone and Ambrose 1978a,b; Gamarra 1981; Shepard et al. 1982; Ambrose 1983; Johnson 1983; Tawfik et al. 1983a,b; Bueno and Filho 1984; Sundararaju 1984; Ambrose and Livingstone 1979a,b, 1985a,b,c, 1986a,b, 1987a,b,c, 1989b,c, 1991; Ambrose et al. 1985; Putschkov 1986; Mbata et al. 1987; Vijayavathi 1987; Ambrose and Paniadima 1988; Lakkundi 1989; Kumaraswami 1991; Kumaraswami and Ambrose 1991, 1992, 1993; Sahayaraj 1991; Consoli and Filho 1992; Rukmani 1992; Sahayaraj and Ambrose 1992a,b, 1993; Scudder 1992; Kumar 1993; Vennison and Ambrose 1986a,b, 1988, 1989a,b, 1990a,b,c,d,e, 1991a,b, 1992a,b; DeFrietas 1994, 1995; Ambrose and Ambrose 1996; Das 1996; George et al. 1998a,b). Impact of female parental age, mating, prey, space, prey deprivation, decamouflaging, flooding, insecticides, soil moisture, and sensory inputs on development has been reported (see Ambrose 1999).

Courtship behavior of reduviids is much like that of other insects, although it has some characteristic features. The sequential acts of mating behavior can be categorized into arousal,

Table 1 Biological Characteristics of Reduviids

Subfamily	Fecundity (no)	Hatchability (%)	Stadial Period (days)	Adult Longevity		Sex Ratio	
				Male	Female	Male	Female
Apiomerinae	—	—	123.81	—	—	—	—
Ectrichodiinae	85.00	—	93.10	45.00	79.00	1.00	1.00
Harpactorinae	140.43 ± 139.31 (9.0–669.0)	81.87 ± 12.02 (30.0–100.0)	58.81 ± 19.51 (17.2–104.3)	67.78 (22.0–137.0)	70.89 (20.3–138.5)	1.00 (0.5–1.0)	0.99 (0.6–1.0)
Peiratinae	73.00 ± 32.70 (33.0–113.0)	24.40 ± 7.83 (22.8–59.7)	93.40 ± 12.26 (77.5–197.5)	56.40 (28–119)	72.30 (42.6–121.5)	1.00 (0.8–1.0)	0.98 (0.7–1.0)
Reduviinae	125.07 ± 17.5 (52–322)	46.17 ± 22.07 (29.7–84.8)	83.43 ± 71.8 (49.5–117.2)	96.7 (26–202)	111.8 (57.2–204)	0.92 (0.6–1.0)	1.00 (0.7–1.0)
Salyavatinae	87.20	75.60	173.50	8.60	12.60	1.00	0.70
Stenopodainae	49.20	77.00	71.80	52.30	36.50	0.70	1.00

approach, riding over (Harpactorinae), nuptial clasp, extension of genitalia and connection, copulation, and postcopulatory acts (see Ambrose 1999).

2.3 Predatory Behavior and Prey

Predatory behavior, prey records, and food requirements of reduviids have been reported worldwide by several authors. Predatory behaviors are mediated by sensory responses, and can be listed in sequence as arousal, approach, capturing, rostral probing, injection of toxic saliva and paralyzing, sucking, and postpredatory behavior (see Ambrose 1999).

Apiomerines and ectrichodiines sticky-trap prey by extending resin-coated forelegs in front of prey. Emesines wait and capture prey with long raptorial forelegs armed with spines and tubercles. Reduviines, salyavatines, and stenopodaines wait and grab their prey. Harpactorines pin and jab the prey with the long rostrum. Peiratines, reduviines, and ectrichodiines chase, pounce, and grab the prey with their well-developed tibial pads (see Ambrose 1999).

The impact of sensory inputs and structural adaptations, sex, nymphal camouflaging, prey deprivation, competition, space, and crowding on the predatory behavior has been studied in several reduviids (see Ambrose 1999).

Although reduviids feed on a wide variety of arthropods, they sometimes show both prey and stage preferences. Ectrichodiines feed exclusively on millipedes; peiratines prefer beetles and grasshoppers; harpactorines usually prey upon soft prey, like caterpillars, grubs, termites; reduviines prefer ants, bees, and termites; holoptilines, stenopodaines, salyavatines, saicines, and tribelocephalines prefer ants, termites, and blattids; and emesines prefer flies (Stride 1956a,b; Parker 1965a; Bass and Shepard 1974; Fuesini and Kumar 1975; Ables 1978; Kumaraswami 1991; Sahayaraj 1991; Kumaraswami and Ambrose 1992; Vennison and Ambrose 1992a; Sahayaraj 1994a, 1995; Sahayaraj and Ambrose 1994; Ambrose and Claver 1996; George and Seenivasagan 1998).

Stage preference studies of capture success and choice experiments showed this preference is greater in younger instars and gradually lessens as the reduviids grow (Joseph 1959; MacMahan 1982, 1983; Inoue 1985; Kumaraswami 1991; Ambrose and Sahayaraj 1993; Sahayaraj and Ambrose 1994; Ambrose 1995, 1996b; Sahayaraj and Sivakumar 1995; Das 1996).

2.4 Defensive and Offensive Behaviors

Assassin bugs have an array of defensive and offensive behaviors, accompanied by morphological adaptations. These behaviors threaten and aid escape from enemies and larger prey, and gain protection from cannibalism.

Camouflaging of nymphs is found in Cetherinae, Reduviinae, and Salyavatinae. The nymphs use both hind tarsomeres as shovels and actively throw and accumulate sand and debris. They cover the entire body except antennal flagellum, rostrum, the ventral surface of head, prosternum, and tibial pads with sand and debris. Fine, feltlike dorsal hairs and serrated cuticle hold the camouflaging materials. Hypodermal cells secrete a sticky material that hardens to become fine silvery filaments which intertwine with the camouflage material. Camouflaging efficiency (camouflage material carrying capacity) is greater in younger instars, gradually lessens as nymphs grow older, and disappears in adults (see Ambrose 1999).

Harpactorine instars often feign death. Harpactorine and ectrichodiine instars, by withdrawing their cephalic and thoracic appendages, also roll the body into a ball and lie motionless. Nodding of the head and rubbing of the rostrum against the transversely striated prosternal groove to produce a sound is better developed in reduviids with acutely curved rostrums (Peiratinae) than in reduviids with straight rostrums; these produce an obscure sound. Extension of rostrum, spitting of watery saliva, and stinging also occur. Emission of a volatile secretion with pungent odor from the dorsolateral abdominal scent gland or Brindley's gland is well pronounced in Harpactorinae and Reduviinae (see Ambrose 1999).

A combination of Batesian and Mullerian mimicry between lygaeids and similar-appearing reduviids provides protection from vertebrate predators. *Rhaphidosoma* species (Harpactorinae) and stenopodaines have elongate slender bodies that resemble closely the elongate slender grass and sledge stems and leaves among which they live (see Ambrose 1999).

3. BIOLOGICAL CONTROL

3.1 Pest Prey Record

More than 150 reduviid predators belonging to 53 genera and 7 subfamilies have been found preying upon a wide array of insect pests (Ashmead 1895; Morgan 1907;Webster 1907; Chittenden 1905, 1909, 1916; Maxwell-Lefroy and Howlett 1909; Morrill 1910; Garman 1916; Britton 1917; Horton 1918a,b; Watson 1918; Caffrey and Barber 1919; Pettey 1919; Austin 1922; Hutson 1922; Readio 1924, 1927; Severein 1924; Blatchley 1926; Mossop 1927; Pinto 1927; Romeo 1927; Ullyett 1930; Taylor 1932; Hayashi 1933; Hoffmann 1934, 1935; Cherian 1937; Musgrave 1938; Cherian and Kylasam 1939; Cherian and Brahmachari 1941; Beeson 1941; Corbett and Pagden 1941; Risbec 1941; Nei 1942; Clancy 1946; Bose 1949; Balduf 1950; Wallace 1953; Underhill 1954; Carayon et al. 1956; Stride 1956a,b; Werner and Butler 1957; Butani 1958; Vanderplank 1958; Joseph 1959; Evans 1962; Debach and Hagen 1964; Gifford 1964; Whitcomb and Bell 1964; Parker 1965a,b, 1969, 1971, 1972; Flower et al. 1967; Nyirra 1970; Mukerjee 1971; Eveleens et al. 1973; Nault et al. 1973; Swadener and Yonke 1973a,b,c; Debach 1974; Rao 1974; Fuseini and Kumar 1975; Misra 1975; Neuenschwander et al. 1975; Singh and Gangrade 1975; Sitaramaiah et al. 1975; Nayar et al. 1976; Sitaramaiah and Satyanarayana 1976; Ables 1978; Fadare 1978; Weaver 1978; Ponnamma et al. 1979; Ambrose 1980, 1987, 1988, 1995, 1996b, 1999; Haridass and Ananthakrishnan 1980; Richman et al. 1980; Abasa 1981; Rajagopal and Veeresh 1981; Rao et al. 1981, Thangavelu and Ananthasubramanian 1981; Mohamed et al. 1982; Nagarkatti 1982; Shepard et al. 1982; Hiremath and Thontadarya 1983; Gravana and Sterling 1983; Patil and Thontadarya 1983; Awadallah et al. 1984; Krishnaswamy et al. 1984; Rajagopal 1984; Sundararaju 1984; Martin and Brown 1984; Ren 1984; Henneberry and Clayton 1985; Singh 1985; Livingstone and Yacoob 1986; Pawar et al. 1986; Barrion et al. 1987; Ambrose and Livingstone 1987a, 1989a,b, 1991; Singh and Singh 1987; Singh et al. 1987; Ambrose and Paniadima 1988; David and Kumaraswami 1988; Muraleedharan et al. 1988; Vennison 1988; Ambrose and Vennison 1989; David and Natrajan 1989; Godfrey et al. 1989; Lakkundi 1989; Varma 1989; Ambrose and Sahayaraj 1990; Cohen 1990; Yano 1990; Ambrose and Livingstone 1991; Kumaraswami 1991; Sahayaraj 1991; James 1992, 1994a,b; Singh 1992; Vennison and Ambrose 1992a; Kumar 1993; Ruberson et al. 1994a,b; Zanuncio et al. 1994; Barrion and Litsinger 1994; James et al. 1994; Ambrose and Claver 1995a; Babu et al. 1995; Balu et al. 1995; Kuppusamy and Kannan 1995; Uthamasamy 1995; Mohanadas 1996). *Reduvius personatus* (L.) feeds on ticks (Morel 1974), which is interesting because this "masked bed bug hunter" may also feed on bed bugs (see discussion in Readio 1927). See Table 2.

3.2 Functional and Numerical Responses

Functional response of reduviids has been documented (Flower et al. 1967; Bass and Shepard 1974; Ables 1978; Awadallah et al. 1984; Ambrose and Kumaraswami 1990; Kumaraswami 1991; Sahayaraj 1991; Ambrose et al. 1994, 1996; Sahayaraj 1994b; Sahayaraj and Ambrose 1996b,c; Ambrose and Sahayaraj 1996; Kumar and Ambrose 1996; Ambrose and Claver 1995b, 1996, 1997; Das 1996).

Reduviids respond to an increasing prey population with a decreased attack ratio and searching time and an increased rate of discovery and number of prey attacked (Holling 1959). Age, sex,

Table 2 Field Cage Evaluation on the Biocontrol Efficacy of Five Reduviid Predators against Four Insect Pests

Reduviid	Pest	Crop	Duration (days)	Rate of Suppression (%)	Increase in Yield/Plant (mg)
R. kumarii	D. cingulatus	G. hirsutum	15	60.00	—
	D. cingulatus	G. hirsutum	15	50.10	—
	M. pustulata	G. hirsutum	21	69.20	225
	S. litura	G. hirustum	21	47.40	165
R. marginatus	D. cingulatus	G. hirsutum	15	55.00	—
	D. cingulatus	G. hirsutum	15	54.10	—
	M. pustulata	G. hirsutum	21	64.20	—
	S. litura	A. esculentus	11	61.40	—
	S. litura	G. hirsutum	21	59.30	—
C. brevipennis	D. cingulatus	G. hirsutum	14	59.45	—
	D. cingulatus	G. hirsutum	14	86.67	—
E. tibialis	D. cingulatus	G. hirsutum	14	56.76	—
	D. cingulatus	G. hirsutum	14	86.67	—
A. pedestris	H. armigera	A. esculentus	11	65.55	—

Pest: *Dysdercus cingulatus* Fabricius, *Helicoverpa armigera* Hübner, *Mylabris pustulata* (Thunberg), *Spodoptera litura* F.

Reduviid: *Rhynocoris kumarii* Ambrose and Livingstone, *R. marginatus* F. (Harpactorinae), *Catamiarus brevipennis* Serville, *Ectomocoris tibialis* Distant (Peiratinae), and *Acanthaspis pedestris* Stål (Reduviinae)

Crop: *Abelmoschus esculentus* Mennch. and *Gossypium hirsutum* L.

and size of predator and size of the prey determine functional response. Reduviids exhibit a positive numerical response by killing more prey in terms of available prey population per predator at a given time and by increasing their population through higher fecundity and survival (see Ambrose 1999).

3.3 Augmentation

Augmentation of reduviid predators was attempted by Edwards (1962), Van den Bosch and Telford (1964), Tawfik et al. (1983a,b), and Lakkundi and Parshad (1987). The impacts of ecophysiological conditions, such as levels of mating, competition, space, prey deprivation, crowding, prey type, female parental age, flooding, soil moisture, and decamouflaging of reduviine instars on oviposition pattern, hatchability, and postembryonic development of reduviids were studied to develop strategies to mass-rear the predators (see Ambrose 1999).

Five reduviid predators (Harpactorinae, Peiratinae, Reduviinae) were successfully mass-reared in the author's laboratory on head-crushed *Corcyra cephalonica* Stainton (to avoid the entangling of reduviids in the web of metamorphosing larvae) in large plastic troughs filled with sand, small stones, and leaves simulating their concealed microhabitat. Mass rearing increased the predatory value, conversion rate, and body weight; reduced the postembryonic development period; quickened oviposition; increased fecundity; and promoted a female-biased sex ratio. Nontarget reduviid predators exposed to chemical control in agroecosystems were found affected (see Ambrose 1999).

4. IMPORTANT SPECIES

The importance of biocontrol potential of three species of *Rhynocoris* (Harpactorinae), *R. fuscipes* (F.) *R. kumarii* Ambrose and Livingstone, and *R. marginatus* (F.); a peiratine reduviid, *Ectomocoris tibialis* (Distant); and a reduviine species, *Acanthaspis pedestris* Stål, has been realized against a wider array of insect pests of cotton, vegetables, castor, peanut, and cereals in India.

Detailed information on their distributions and life histories, biologies, augmentation, and biocontrol potential is available (see Ambrose 1999).

Although many reduviids were identified as preying upon a wider array of insect pests outside India, very few consistent investigations on their distributions, and life histories have been carried out. One such important group of neglected reduviid predators is *Phonoctonus* spp., preying upon species of *Dysdercus* and *Odontopus* (Pyrrhocoridae). Other promising groups of reduviid predators are the species of *Sinea* and *Zelus*, which may help regulate insect pests of cotton (see Ambrose 1999).

5. DAMAGE

Because reduviids are for the most part general predators, from time to time they may attack beneficial arthropods. A few reduviids appear to specialize on insects of some value.

Stingless bees of the genus *Trigona* are at times useful pollinators. Several apiomerine reduviids wait for these bees at flowers or at sites where resin is flowing (Weaver et al. 1975, Johnson 1983); the bees use the resin for nest construction. The apiomerines eat the bees, which may be attracted by a kairomone from the bugs (Weaver et al. 1975). It has been suggested that the bugs use the resin itself on their front legs to help seize their prey (reference in Poinar 1992), but the sticky material is probably produced by the bug itself (Swadener and Yonke 1975). Fossil "resin bugs" and stingless bees have been found together in Dominican amber from at least 25 million years ago (Poinar 1992), indicating that the relationship between bug and bee is an old one.

In India, *Acanthaspis siva* (Distant), a polymorphic reduviine reduviid predator, hunts for Indian honey bees, *Apis indica*, in their hives (Subbiah and Mahadevan 1957).

The roles of "generalist" and "specialist" reduviid predators (introduced or natural) against insect pests need to be studied. For instance, Whitcomb and Bell (1964) reported reduviids feeding on lady beetles in Arkansas cotton fields. Rosenheim and Wilhoit (1993), Rosenheim et al. (1993), and Cisenros and Rosenheim (1997) reported that the generalist reduviid *Zelus renardii* (instars and adults) prey upon the lacewing *Chrysoperla carnea* (Stephens) in cotton fields in the San Joaquin Valley (California, U.S.A.), a biological control agent against *Aphis gossypii*. According to them, *Z. renardii* imposes heavy mortality on *C. carnea* population and thereby renders natural and augmented populations of *C. carnea* ineffective as biological control agents. *Zelus renardii* decreased the survival of *C. carnea* from 47% (in its absence) to 0%. These authors attributed the decreased lacewing survival to predation by the reduviid predator rather than to competition for aphid prey. *Zelus renardii* was found not only to influence the prevalence of intraguild predation but also the intensity of the disruption of the aphid biological control. Moreover, none of the developmental stages of *Z. renardii* was found to be an effective control agent of the cotton aphid.

Thus, it is imperative to understand the interaction between reduviid predators, keeping in mind that some generalist predators may attack other specialist predators, with potentially negative effects on pest control. Proper assessment of the role of reduviid predators in regulation of insect pests in diverse crop systems (Dumar et al. 1962, Jones et al. 1983, Ambrose 1999), and the management of environment and habitat to increase predator population (Whitcomb and Bell 1964) need attention.

6. PROGNOSIS

The biocontrol potential of reduviids has been evaluated by Livingstone and Ambrose (1978a), Richman et al. (1980), Singh (1985), Singh and Gangrade (1975), Sahayaraj and Ambrose (1993), and Sahayaraj (1994b,1995). Pest suppression by reduviid predators was evaluated by comparing the rate of mortality of pests in field cages with predators and without predators, as well as in terms of reducing damage to plants. The pest-suppression potential of reduviids is determined by the type of predator, and the nature and age of the pests that reduviids are suppressing both at low predator and prey density suggest they have good searching and prey-regulating abilities. Field release and recovery studies indicate their synchronization with the environment.

7. CONCLUSION

Their worldwide distribution, abundance, diversity, larger pest prey record, amenability for mass culturing, ready synchronization, and freedom from hyperparasites and predators are the merits of reduviids as potential biological control agents. A holistic approach to their systematics and ecology with subsequent efforts on economical mass rearing, large-scale release, and biocontrol potential evaluation should be undertaken.

8. ACKNOWLEDGMENTS

I acknowledge the Department of Environment and Forests, Government of India for financial assistance (30/14/94-RE). I thank C. W. Schaefer and A. R. Panizzi for inviting me to write this chapter. C. W. Schaefer was a most gracious host when I visited him in 1997 and 1999 and reviewed this chapter. I dedicate this work to my beloved teacher, David Livingstone.

9. REFERENCES CITED

Abasa, R. O. 1981. *Harpactor tibialis* Stål (Hemiptera: Reduviidae), a predator of *Ascotis reciprocaria* Wlk. in Kenya coffee estates. Kenya J. Sci. Technol. B2: 53–55.
Ables, J. R. 1978. Feeding behaviour of an assassin bug, *Zelus renardii*. Ann. Entomol. Soc. Amer. 71: 476–478.
Ambrose, D. P. 1980. Bioecology, ecophysiology and ethology of Reduviids (Heteroptera) of the scrub jungles of Tamil Nadu; India. Ph.D. thesis, University of Madras, Madras, India.
1983. Bioecology of an alate assassin bug, *Acanthaspis quinquespinosa* (L) Fabr. (Heteroptera: Reduviidae). Pp.107–115 *in* S.C. Goel [eds.], Proc. Symp. Insect Ecol. Resource Manage., Muzzafarnagar, India. 296 pp.
1987. Assassin bugs of Tamil Nadu and their role in biological control (Insecta: Heteroptera: Reduviidae). Pp. 16–28 *in* K.J. Joseph and U. C. Abdurahiman [eds.], Advances in Biological Control Research in India. Dept. of Zool. University of Calicut, Calicut, India. 258 pp.
1988. Biological control of insect pests by augmenting assassin bugs (Insecta: Heteroptera: Reduviidae). Pp. 25–40 *in* K. S. P. Ananthasubramanian, P. Venkatesan, and S. Sivaraman [eds.], Bicovas II. Loyola College, Madras, India. 243 pp.
1991. Conservation and augmentation of predaceous bugs and their role in biological control. Pp. 225–257 *in* G.K. Veeresh, D. Rajagopal, and C. A. Viraktamath [eds.], Advances in Management and Conservation of Soil Fauna. Oxford & IBH Publ. Co. Pvt. Ltd., New Delhi, India. 925 pp.
1995. Reduviids as predators: their role in biologial control. Pp. 153–170 *in* T. N. Ananthakrishnan [ed.], Biological Control of Social Forests and Plantation Crop Insects. Oxford & IBH Publ. Co. Pvt. Ltd., New Delhi, India. 225 pp.
1996a. Biosystematics, distribution, diversity, population dynamics and biology of reduviids of Indian subcontinent — an overview. Pp. 93–102 *in* D. P. Ambose [ed.], Biological and Cultural Control of Insect Pests, an Indian Scenario. Adeline Publishers, Tirunelveli, India. 362 pp.
1996b. Assassin bugs (Insecta-Heteroptera-Reduviidae) in biocontrol: success and strategies, a review. Pp. 262–284 *in* D. P. Ambose [ed.], Biological and Cultural Control of Insect Pests, an Indian Scenario. Adeline Publishers, Tirunelveli, India. 362 pp.
1999. Assassin Bugs. Science Publishers, Inc., Enfield, New Hampshire, U.S.A. 337 pp.

Ambrose, A. D., and D. P. Ambrose. **1996.** Biological tools in the biosystematics of three subfamilies of Reduviidae (Insecta: Heteroptera). Pp. 40–48 *in* D. P. Ambrose [ed.], Biological and Cultural Control of Insect Pests, an Indian Scenario. Adeline Publishers, Tirunelveli, India. 362 pp.

Ambrose, D. P., and M. A. Claver. **1995a.** Food requirement of *Rhynocoris kumarii* Ambrose and Livingstone (Heteroptera: Reduviidae). J. Biol. Control 9: 47–50.

1995b. Functional response of *Rhynocoris fuscipes* (Heteroptera: Reduviidae) to *Riptortus clavatus* Thunb. (Heteroptera: Alydidae). J. Biol. Control 9: 74–77.

1996. Size preference and functional response of the reduviid predator *Rhynocoris marginatus* Fabricius (Heteroptera: Reduviidae) to its prey *Spodoptera litura* (Fabricius) (Lepidoptera: Noctuidae). J. Biol. Control 10: 29–37.

1997. Functional and numerical responses of the reduviid predator *Rhynocoris fuscipes* (Het., Reduviidae) to cotton leafworm *Spodoptera litura* F. (Lep., Noctuidae). J. Appl. Entomol. 121: 331–336.

1999. Suppression of cotton leafworm *Spodoptera litura* flower beetle *Mylabris pustulata* and red cotton bug *Dysdercus cingulatus* by *Rhynocoris marginatus* (Fabr.) (Het., Reduviidae) in cotton field cages. J. Appl. Entomol. 123: 225–229.

Ambrose, D. P., and N. S. Kumaraswami. **1990.** Functional response of the reduviid predator *Rhinocoris marginatus* Fabr. on the cotton stainer *Dysdercus cingulatus* Fabr. J. Biol. Control 4: 22–25.

Ambrose, D. P., and D. Livingstone. **1979a.** Population dynamics of three species of reduvilids from peninsular India. Bull. Entomol. 19: 201–203.

1979b. Impact of mating on the oviposition pattern and hatchability in *Acanthaspis pedestris* Stål (Reduviidae: Acanthaspidinae). Entomon 4: 269–275.

1985a. Mating behaviour and its impact on oviposition pattern, and hatchability in *Acanthaspis siva* Distant (Heteroptera: Reduviidae). Uttar Pradesh J. Zool. 5: 123–129.

1985b. Impact of mating on adult longevity, oviposition pattern, hatchability and incubation period in *Rhinocoris marginatus*. Environ. Ecol. 3: 99–102.

1985c. Development of *Coranus vitellinus* Distant (Hemiptera: Reduviidae: Harpactorinae). J. Soil Biol. Ecol. 5: 65–71.

1986a. Bioecology of *Rhinocoris fuscipes* Fabr. (Reduviidae) a potential predator on insect pests. Uttar Pradesh J. Zool. 6: 36–39.

1986b. Bioecology of *Sphedanolestes aterrimus* Distant (Heteroptera: Reduviidae). J. Bombay Nat. Hist. Soc. 83: 248–252.

1987a. Biology of a new harpactorine assassin bug *Rhinocoris kumarii* (Hemiptera: Reduviidae) in South India. J. Soil Biol. Ecol. 7: 48–58.

1987b. Biology of *Acanthaspis siva* Distant, a polymorphic assassin bug (Insecta, Heteroptera, Reduviidae). Mitt. Zool. Mus. Berlin 63: 321–330.

1987c. Mating behaviour and impact of mating on oviposition pattern and hatchability in *Rhinocoris kumarii*. Environ. Ecol. 5: 156–161.

1989a. Population dynamics of assassin bugs from peninsular India (Insecta-Heteroptera-Reduviidae). J. Bombay Nat. Hist. Soc. 86: 388–395.

1989b. Biology of the predaceous bug *Rhinocoris marginatus* Fabricius (Insecta-Heteroptera-Reduviidae). J. Bombay Nat. Hist. Soc. 86: 155–160.

1989c. Biosystematics of *Ectomocoris tibialis* Distant (Insecta-Heteroptera-Reduviidae) a predator of insect pests. Arq. Mus. Bocage 1: 281–291.

1991. Biology of *Neohaematorrhophus therassi* (Hemiptera: Reduviidae). Colemania: Insect Biosyst. Ecol. 1: 5–9.

Ambrose, D. P., and A. Paniadima. **1988.** Biology and behaviour of a harpactorine assassin bug *Sycanus pyrrhomelas* Walker (Hemiptera: Reduviidae) from South India. J. Soil Biol. Ecol. 8: 37–58.

Ambrose, D. P., and K. Sahayaraj. **1990.** Effect of space on the postembryonic development and predatory behaviour of *Ectomocoris tibialis* Distant (Insecta: Heteroptera: Reduviidae). Uttar Pradesh J. Zool. 10: 163–170.

1993. Predatory potential and stage preference of a reduviid predator *Allaeocranum signatum* Reuter on *Dysdercus cingulatus* Fabricius. J. Biol. Control 7: 12–14.

1996. Long term functional response of the reduviid predator *Acanthaspis pedestris* Stål (Heteroptera: Reduviidae) in relation to its prey *Pectinophora gossypiella* Saunders (Lepidoptera: Noctuidae) density. Hexapoda 8: 77–84.

Ambrose, D. P., and S. J. Vennison. 1989. Biology of a migratory assassin bug, *Polididus armatissimus* Stål (Insecta–Heteroptera–Reduviidae). Arq. Mus. Bocage 1: 271–280.

Ambrose, D. P., P. Ramakrishnan, and V. Kasinathan. 1985. Biology of *Catamiarus brevipennis* Serv. (Reduviidae) a predator of Bombay locust. Uttar Pradesh J. Zool. 5: 20–24.

Ambrose, D. P., P. J. E. George, and N. Kalidoss. 1994. Functional response of the reduviid predator *Acanthaspis siva* Distant (Heteroptera: Reduviidae) to *Camponotus compressus* Fabricius and *Ditopternis venusta* Walker. Environ. Ecol. 12: 877–879.

Ambrose, D. P., M. A. Claver, and P. Mariappan. 1996. Functional response of *Rhinocoris marginatus* Fabricius (Heteroptera: Reduviidae) to *Mylabris pustulata* Thunberg (Coleoptera: Meloidae). Fresenius Environ. Bull. 5: 85–89.

Ashmead, W. H. 1895. Notes on cotton insects found in Mississippi. Insect Life 7: 320.

Austin, G. D. 1922. A preliminary report on paddy fly investigation. Ceylon Dept. Agric. Bull. 59: 1–22.

Awadallah, K. T., M. F. S. Tawfik, and M. M. H. Abdellah. 1984. Suppression effect of the reduviid predator, *Alloeocranum biannulipes* (Montr. et Sign.) on populations of some stored product insect pests. Z. Angew. Entomol. 97: 249–253.

Babu, A., R. Seenivasagam, and C. Karuppasamy. 1995. Biological control resources in social forest stands. Pp. 7–24 *in* T. N. Ananthakrishnan [ed.], Biological Control of Social Forest and Plantation Crop Insects. Oxford & IBH Publishing Co. Pvt. Ltd., New Delhi, India. 225 pp.

Balduf, W. V. 1948. The weight of *Sinea diadema* (Fabr.). Ann. Entomol. Soc. Amer. 40: 588–597.

1950. Utilization of food by *Sinea diadema* (Fabr.). (Reduviidae: Hemiptera). Ann. Entomol. Soc. Amer. 43: 354–360.

Balduf, W. V., and J. Slater. 1943. Addition to the bionomics of *Sinea diadema* (Fabr.) (Reduviidae, Hemiptera). Proc. Entomol. Soc. Washington 45: 11–18.

Balu, A., S. R. M. Pillai, K. R. Sasidharan, B. Deeparaj, and B. Sunitha. 1995. Natural enemies of Babul (*Acacia nilotica* L.) (Wild. ex. Del. ssp. *indica* (Bth.)) defoliators *Selepa celtis* and *Tephrnia pulinda* (Insecta: Lepidoptera). Pp. 43–53 *in* T. N. Ananthakrishnan [ed.], Biological Control of Social Forest and Plantation Crop Insects. Oxford & IBH Publishing Co. Pvt. Ltd., New Delhi, India. 225 pp.

Barrion, A. T., and J. A. Litsinger. 1994. Taxonomy of rice insect pests and their arthropod parasites and predators. Pp. 13–359 *in* E.A. Heinrichs [ed.], Biology and Management of Rice Insects. Wiley Eastern Ltd., New Delhi, India. 779 pp.

Barrion, A. T., R. M. Anguda, and J. A. Litsinger. 1987. The natural enemies and chemical control of the Leucaena psyllid *Heteropsyila cubana* Crawford in the Philippines. Leucaena Res. Rep. 7: 45–49.

Bass, J. A., and M. Shepard. 1974. Predation by *Sycanus indagator* on larvae of *Galleria mellonella* and *Spodoptera frudgiperda*. Entomol. Exp. Appl. 17: 143–148.

Beeson, C. F. G. 1941. Ecology and Control of Forest Insect Pests of India and Neighbouring Countries. 1961 Vasane Press, Dheradun, India. 767pp.

Blatchley, W. S. 1926. Heteroptera or True Bugs of Eastern North America, with Especial Reference to the Faunas of Indiana and Florida. Nature Publ. Co., Indianapolis, Indiana, U.S.A. 1116 pp.

Bose, M. 1949. On the biology of *Coranus spiniscutis* Reuter, an assassin bug (Heteroptera: Reduviidae). Indian J. Entomol. 11: 203–208.

Britton, W. E. 1917. The fall webworm. Connecticut Bull. 203: 323.

Bueno, V. H. P., and E. Filho. 1984. *Montina confusa* (Stål, 1859) (Hemiptera: Reduviidae: Zelinae): I. Aspectos biologicos. Rev. Bras. Entomol. 28: 345–353.

Butani, D. K. 1958. Parasites and predators recorded on sugarcane pests in India. Indian J. Entomol. 20: 270–282.

Caffrey, D. T., and G. W. Barber. 1919. The grain bug *Zelus renardii*. Bull. U.S. Dept. Agric. 799: 35.

Carayon, J., R. L. Usinger, and P. Wygodzinsky. 1956. Notes on the higher classification of the Reduviidae with the description of a new tribe of the Phymatinae. Rev. Zool. Bot. Afr. 57: 256–281.

Cherian, M. C. 1937. Administration report of Government Entomologist, Coimbatore for 1936–37. Report Dept. Agric. Madras. Pp. 126–133.

Cherian, M. C., and K. Brahmachari. 1941. Notes on three predatory hemipteran from South India. Indian J. Entomol. 3: 115–118.

Cherian, M. C. , and M. S. Kylasam. 1939. On the biology and feeding habits of *Rhinocoris fuscipes* (Fabr.) (Heteroptera: Reduviidae). J. Bombay Nat. Hist. Soc. 61: 256–259.

China, W. E., and N. C. E. Miller. 1959. Check list and keys to the families and subfamilies of the Hemiptera: Heteroptera. Bull. British Mus. Nat. Hist. Entomol. 8: 1–45.

Chittenden, F. H. 1905. The corn root worm. U.S. Dept. Agric. Circ. 59: 5.

1909. The Colorado potato beetle. U.S. Dept. Agric. Bull. Entomol. 87: 11.

1916. The common cabbage worm. U.S. Dept. Agric. Farmers' Bull. 766: 9.

Cisenros, J. J., and A. J. Rosenheim. 1997. Ontogenetic change of prey preference in the generalist predator *Zelus renardii* and its influence on predator-predator interactions. Ecol. Entomol. 22: 339–407.

Clancy, D. W. 1946. Natural enemies of some Arizona cotton insects. J. Econ. Entomol. 39: 326–328.

Cohen, A. C. 1990. Feeding adaptations of some predaceous heteropterans. Ann. Entomol. Soc. Amer. 83: 1215–1223.

Consoli, F. L., and B. F. A. Filho. 1992. Ciclo biológico de *Montina confusa* (Stål, 1859) (Hemiptera: Reduviidae) alimentado com diferentes presas. Rev. Bras. Entomol. 36: 679–702.

Corbett, G. H., and H. T. Pagden. 1941. A review of some recent entomological investigations and observations. Malaya Agric. J. 29: 347–375.

Das, S. S. M. 1996. Biology and behaviour of chosen predatory hemipterans. Ph.D. thesis, Madurai Kamaraj University, Madurai, India.

David, B. V., and T. Kumaraswami. 1988. Elements of Economic Entomology. Popular Book Depot, Madras, India. 536 pp.

David, P. M. M., and S. Natrajan. 1989. Bug that destroys chilli fruits. The Hindu [Indian National Newspaper]. June 21, p. 24.

Davis, N. T. 1969. Contribution to the morphology and phylogeny of the Reduvioidea Part. IV. The harpactoroid complex. Ann. Entomol. Soc. Amer. 62: 72–94.

Debach, P. 1974. Biological Control by Natural Enemies. Cambridge University Press, Bentley House, London, U.K. 323 pp.

Debach, P., and K. S. Hagen. 1964. Manipulation of entomophagous species. Pp. 429–458 *in* P. DeBach [ed.], Biological Control of Insect Pests and Weeds. Chapman & Hall Ltd., London, U.K. 822 pp.

DeCoursey, R. M. 1963. The life history of *Fitchia aptera* Stål (Hemiptera-Heteroptera: Reduviidae). Bull. Brooklyn Entomol. Soc. 58: 151–156.

DeFrietas, S. 1994. Desenvolvimento post-embrionario e peso de adultos de *Montina confusa* Stål (Hemiptera: Reduviidae) criados sob diferentes regimes alimentares. An. Soc. Entomol. Brasil 23: 317–320.

1995. Capavidable de predação, sobrevivência e ciclo biologico do predator *Montina confusa* Stål (Heteroptera: Reduviidae) alimentado com lagartas da broca da cana-de-acucar, *Diatraea saccharalis* Fabr. (Lepidoptera: Pyralidae). An. Soc. Entomol. Brasil 24: 195–199.

Dumar, B. A., W. P. Boyser, and W. H. Whitcomb. 1962. Effect of time of day on surveys of predaceous insects in field crops. Florida Entomol. 45: 121–128.

Edwards, J. S. 1962. Observations on the development and predatory habits of two reduviids (Heteroptera), *Rhynococis carmelita* Stål and *Platrymeris rhadamanthus* Gerst. Proc. R. Entomol. Soc. London (A)37: 89–98.

1966. Observation on the life history and predatory behaviour of *Zelus exsanguis* (Heteroptera Reduviidae). Proc. R. Entomol. Soc. London (A): 41: 21–24

Evans, D. E. 1962. The food requirement of *Phonoctonus nigrofasciatus* Stål (Hemiptera: Reduviiidae). Entomol. Exp. Appl. 5: 33–39.

Eveleens, K. G., R. Van Den Bosch, and L. E. Ehler. 1973. Secondary outbreaks of beet armyworm by experimental insecticide applications in cotton in California. Environ. Entomol. 2: 497–503.

Fadare, T. A. 1978. Efficiency of *Phonoctonus* spp. (Hemiptera: Reduviidae) and regulators of populations of *Dysdercus* spp. (Hemiptera: Pyrrchocoridae). Nigerian J. Entomol. 1: 45–48.

Flower, H. G., L. Crestana, M. T. V. Camargode, J. J. Junior, M. M. L. Costa, N. B. Saes, D. Camargo, and J. C. A. Pinto. 1967. Functional response of the most prevalent potential predators to variation in the density of mole crickets (Orthoptera, Gryllotalpidae: *Scapteriscus borelli*). Naturalia 11/12: 47–52.

Fuesini, B. A., and R. Kumar. 1975. Biology and immature stage of cotton stainers (Heteroptera: Pyrrhocoridae) found in Ghana. Biol. J. Linn. Soc. 7: 83–111.

Gamarra, P. 1981. Desarrolla larvario de *Rhinocoris cuspidatus* Ribuat (Hemiptera: Reduviidae). Biol. Assoc. Exp. Entomol. 5: 117–127.

Garman, H. 1916. The locust borer. Kentucky Bull. 200: 121.

George, P. J. E., and R. Seenivasagan. 1998. Predatory efficiency of *Rhinocoris marginatus* (Fabricius) (Heteroptera: Reduviidae) on *Helicoverpa armigera* (Hubner) and *Spodoptera litura* (Fabricius). J. Biol. Control 12: 25–29.

George, P. J. E., R. Seenivasagan, and S. Kannan. 1998a. Biology and life table studies of *Acanthaspis pedestris* Stål (Heteroptera: Reduviidae) population on three lepidopteran insect pests. J. Biol. Control 12: 1–6.

1998b. Influence of prey species on the development and reproduction of *Acanthaspis siva* Distant (Heteroptera: Reduviidae). Entomon 23: 313–319.

Gifford, T. R. 1964. A brief review of sugar cane insects research in Florida 1960–1964. Proc. Soil Crop. Sci. Soc. Florida 24: 449–453.

Gillet, J. D. 1957. On the habits and life history of captive emesine bug (Hemiptera: Reduviidae). Proc. R. Entomol. Soc. London (A) 32: 193–195.

Godfrey, K. E., W. H. Whitcomb, and J. L. Stimae. 1989. Arthropod predators of velvetbean caterpillar, *Anticarsia gemmatalis* Hubner (Lepidoptera: Noctuidae) egg and larvae. Environ. Entomol. 18: 118–123.

Gravana, S., and W. L. Sterling. 1983. Natural predation of the cotton leaf worm (Lepidoptera : Noctuidae). J. Econ. Entomol. 76: 779–784.

Greene, G. L. 1973. Biological studies of a predator *Sycanus indagator* (Heteroptera: Reduviidae) I. Life history and feeding habits. Florida Entomol. 56: 120–258.

Guerra, S. L., and A. J. Rodriguez. 1976. Biologia otras observaciones del ciclo biologico de dos chinches asesinas *Zelus exsanguis* (Fab.) y *Sinea diadema* (Kol.) en la Comarca Lagunera. Invest. Agric. Noreste 2: 133–138.

Habib, M. E. M. 1976. Biological studies on *Zelus leucogrammaus* Perty. An. Soc. Entomol. Brasil 5: 120–129.

Haridass, E. T., and T. N. Ananthakrishnan. 1980. Models for the feeding behaviour of some reduviids from South India (Insecta: Heteroptera: Reduviidae). Proc. Indian Acad. Sci. Anim. Sci. 89: 387–402.

Hayashi, I. 1933. Ecological relatives between parasitic insects and their host. Bot. Zool. 1: 1289–1296 [in Japanese].

Henneberry, T. J., and T. E. Clayton. 1985. Consumption of pink bollworm (Lepidoptera: Gelechiidae) and tobacco budworm (Lepidoptera: Noctuidae) eggs by some predators commonly found in cotton fields. Environ. Entomol. 14: 416–419.

Hiremath, I. G., and T. S. Thontadarya. 1983. Natural enemies of sorghum earhead bug *Calocoris angustatus* Lethierry (Hemiptera: Miridae). Curr. Res. 2: 10–11.

Hoffmann, W. E. 1934. Biology of *Sycanus croceovittatus* Dohrn (Harpactorinae: Reduviidae). Lingnan Sci. J. 13: 503–505.

1935. The binomics and morphology of *Isyndus reticulatus* Stål (Hemiptera, Reduviidae). Lingnan Sci. J. 14: 145–153.

Holling, C. S. 1959. Some characteristics of simple type of predation and parasitism. Canad. Entomol. 91: 385–395.

Horton, J. R. 1918a. The citrus thrips. U.S. Dept. Agric. Bull. 616: 26.

1918b. Argentine ant in relation to citrus groves. U.S. Dept. Agric. Bull. 647: 30.

Hutson, J. C. 1922. Report of the entomologist. Ceylon Dept. Agric. 1921: 23–26.

Immel, R. 1954. Biologische Beobachtungen auf der Staubwanze (*Reduvius personatus* L.). Zool. Arts. (Leipzig) 152: 96–98.

1955. Zur Biologie and Physiologie von *Reduvius personatus* L. Morphol. Ökol. Tiere 44: 163–195.

Inoue, H. 1985. Group predatory behaviour by the assassin bug, *Agriosphodrus dohrni* Signoret (Hemiptera: Reduviidae). Res. Popul. Ecol. 27: 255–264.

James, D. G. 1992. Effect of temperature on development and survival of *Pristhesancus plagipennis* (Hemiptera: Reduviidae). Entomophaga 37: 259–264.

1994a. The development of suppression tactics for *Biprorulus bibax* (Hemiptera: Pentatomidae) as part of an integrated pest management programme in inland citrus of southeastern Australia. Bull. Entomol. Res. 84: 31–38.

1994b. Prey consumption by *Pristhesancus plagipennis* (Hemiptera: Reduviidae) during development. Austral. Entomol. Res. 21: 43–47.

James, D. G., C. J. Moore, and J. R. Aldrich. 1994. Identification, synthesis and bioactivity of a male-produced aggregation pheromone in assassin bug, *Pristehsancus plagipennis* (Hemiptera: Reduviidae). J. Chem. Ecol. 20: 3281–3295.

Johnson, L. K. 1983. *Apiomerus pictipes* (reduvie, chinche asesina, assassin bug). Pp. 684–687 *in* D. H. Janzen [ed.], Costa Rican Natural History. University of Chicago Press, Chicago, Illinois, U.S.A.

Jones, W. A., Jr., S. Y. Young, M. Shepard, and W. H. Whitcomb. 1983. Use of imported natural enemies against insect pests of soybean. Pp. 63–77 *in* H. N. Pitre [ed.] Natural Enemies of Arthropod Pests in Soybean. Southern Co-operative Ser. Bull. Florida 285.

Joseph, M. T. 1959. Biology binomics and economic importance of some reduviids collected from Delhi. Indian J. Entomol. 21: 46–58.

Krishnaswamy, N., O. P. Chowhan, and R. K. Das. 1984. Some common predators of rice pests in Assam, India. Indian Rice Res. Newsl. 9: 15–16.

Kumar, S. P. 1993. Biology and behaviour of chosen assassin bugs (Insecta: Heteroptera: Reduviidae). Ph.D thesis, Madurai Kamaraj University, Madurai, India.

Kumar, S. P., and D. P. Ambrose. 1996. Functional response of two reduviid predators *Rhinocoris longifrons* Stål and *Coranus obscurus* Kirby (Insecta: Heteroptera: Reduviidae) on *Odontotermus obesus* Rambur. Pp. 321–327 *in* D. P. Ambrose [ed.], Biological and Cultural Control of Insect Pests, an Indian Scenario. Adeline Publishers, Tirunelveli, India. 362 pp.

Kumaraswami, N. S. 1991. Bioecology and ethology of chosen predatory bugs and their potential in biological control. Ph.D. thesis, Madurai Kamaraj University, Madurai, India.

Kumaraswami, N. S., and D. P. Ambrose. 1991. Bioecology and ethology of *Edocla annulata* Distant (Hemiptera: Reduviidae: Acanthaspidinae) from South India. J. Soil Biol. Ecol. 11: 33–42.

1992. Biology and prey preference of *Sycanus versicolor* Dohrn (Hemiptera: Reduviidae). J. Biol. Control 6: 67–71.

1993. Biology of *Sphedanolestes pubinotum* Reuter (Heteroptera: Reduviidae) a potential predator of insect pests. Uttar Pradesh. J. Zool. 13: 11–16.

Kuppusamy, A., and S. Kannan. 1995. Biological control of bagworm (Lepidoptera: Psychidae) in social forest stands. Pp. 25–41 *in* T. N. Ananthakrishnan [ed.], Biological Control of Social Forests and Plantation Crop Insects. Oxford & IBH Publishing Co. Pvt. Ltd, New Delhi, India. 225 pp.

Lakkundi, N. S. 1989. Assessment of reduviids for their predation and possibilities of their utilization in biological control. Ph.D. thesis, Indian Agricultural Research Institute, New Delhi, India.

Lakkundi, N. H., and B. Parshad. 1987. A technique for mass multiplication of predator with sucking type of mouth parts with special reference to reduviids. J. Soil Boil. Ecol. 7: 65–69.

Livingstone, D., and D. P. Ambrose. 1978a. Feeding behaviour and predatory efficiency of some reduviids from the Palghat Gap, India. J. Madras Univ. B. 41: 1–25.

1978b. Bioecology, ecophysiology and ethology of the reduviids of the scrub jungles of the Palghat Gap, Part VII: bioecology of *Acanthaspis pedestris* Stål (Reduviidae: Acanthaspidinae) a micropterous, entomophagous species. J. Madras Univ. B 41: 97–118.

Livingstone, D., and M. H. S. Yacoob. 1986. Natural enemies and biologies of the egg parasitoids of Tingidae of southern India. Uttar Pradesh J. Zool. 1–12.

Louis, D. 1973. Life cycles and immature stages of Reduviidae (Hemiptera: Heteroptera) of the cocoa farms in Ghana. J. Entomol. Soc. Nigeria (Occas. Publ.) 13: 1–53.

1974. Biology of Reduviidae of cocoa forms in Ghana. Amer. Midl. Nat. 9: 68–89.

MacMahan, E. A. 1982. Bait and capture strategy of a termite eating assassin bug. Insectes Soc. 29: 346–351.

1983. Adaptations, feeding preference and biometric of a termite baiting assassin bug (Hemiptera: Reduviidae). Ann. Entomol. Soc. Amer. 76: 483–486.

Maldonado Capriles, J. M. 1990. Systematic Catalogue of the Reduviidae of the World (Insecta: Heteroptera). University of Puerto Rico, Mayaguez, Puerto Rico, U.S.A. 694 pp.

Martin, W. R. J. R., and J. M. Brown. 1984. The action of acephate in *Pseudoplusia includens* (Lepidoptera: Noctuidae) and *Pristhesancus papuensis* (Hemiptera: Reduviidae). Entomol. Exp. Appl. 35: 3–9.

Maxwell-Lefroy, H., and F. M. Howlett. 1909. Indian Insect Life. Thacker, Spink & Co., London, U.K.

Mbata, K. J., E. R. Hart, and R. E. Lewis. 1987. Reproductive behaviour in *Zelus renardii* Kolenati 1857 (Hemiptera, Reduviidae). Iowa State J. Res. 62: 261–265.

Miller, N. C. E. 1953. Notes on the biology of the Reduviidae of Southern Rhodesia. Trans. Zool. Soc. London 27: 541–562.

Misra, R. M. 1975. Notes on *Anthia sexguttata* Fabr. (Carabidae: Coleoptera) a new predator of *Pyrausta machaeralis* Walker and *Hyblea puera* Crammer. Indian For. 101: 604.

Mohamed, U. V. K., U. C. Abdurahiman, and O. K. Remadevi. 1982. Coconut caterpillar and its natural enemies. P. 23 *in* A study of the parasites and predators of *Nephantis serinopa* Myrick. Zoological Monograph No. 2. University of Calicut, Calicut, India. 162 pp.

Mohanadas. 1996. New records of some natural enemies of the teak detoliator *Hyblaea puera* Cramer (Lepidoptera: Hybiaeidae) from Kerala, India. Entomon 21: 251–254.

Morel, P. C. 1974. Les methods de lutte contre les tiques en fonction de leur biologie. Cah. Med. Vet. 43: 3–23.

Morgan, A. C. 1907. A predatory bug *Apiomerus spissipes* Say reported as an enemy of the cotton bollweevils. U.S. Dept. Agric. Bull. Entomol. 63: 1–54.

Morrill, A. W. 1910. Plant bugs injurious to cotton bolls. U.S. Dept. Agric. Bull. Entomol. 86: 110.

Mossop, M. C. 1927. Insect enemies of the eucalyptus snout beetle. Farming South Africa. 11: 430–431.

Mukerjee, A. B. 1971. Observations on the feeding habit of *Reduvius* sp. predators of *Corcyra cephalonica* (Stainton). Indian J. Entomol. 33: 230–231.

Muller, G. 1937. Zur Biologie von *Rhinocoris iracundus* Poda. Entomol. Zool. 58: 162–164.

Muraleedharan, N., R. Selvasundaran, and B. Radhakrishnan. 1988. Natural enemies of certain tea pests occurring in southern India. Insect Sci. Appl. 9: 647–654.

Musgrave, A. 1938. Notes on the assassin bugs. Austral. Mus. Mag. 354–355.

Nagarkatti, S. 1982. The utilization of biological control of *Heliothis management* in India. Pp. 159–167 *in* W. Reed and V. Kumble [eds.], Proc. Int. Workshop on *Heliothis* Management, ICRISAT, Patencheru, India. 418 pp.

Nault, L. R., L. J. Edward, and W. B. Styler. 1973. Aphid alarm pheromones, secretion and reception. Environ. Entomol. 2: 101–105.

Nayar, K. K., T. N. Ananthakrishnan, and B. V. David. 1976. General and Applied Entomology. Tata McGraw-Hill Publ. Co. Ltd., New Delhi, India. 589 pp.

Nei, R. K. 1942. Biological control of the codling moth in South Africa. J. Entomol. Soc. South. Africa 5: 118–137.

Neuenschwander, P., K. S. Gagen, and R. F. Smith. 1975. Predation on aphid in California's alfalfa fields. Hilgardia 43: 53–78.

Nyirra, Z. M. 1970. The biology and behaviour of *Rhinocoris albopunctatus* (Hemiptera: Reduviidae). Ann. Entomol. Soc. Amer. 63: 1224–1227.

Odhiambo, T. R. 1958. Some observations on the natural history of *Acanthaspis petax* Stål (Hemiptera: Reduviidae) living in termite mounds in Uganda. Proc. R. Entomol. Soc. London (A) 33: 167–175.

1959. An account of parental care in *Rhinocoris albospilus* Signoret (Hemiptera: Heteroptera: Reduviidae) with notes on its life history. Proc. R. Entomol. Soc. London. (A) 34: 175–185.

Parker, A. H. 1965a. The predatory behaviour and life history of *Pisilus tipuliformis* Fabr. (Hemiptera: Reduviidae). Entomol. Exp. Appl. 8: 1–12.

1965b. The maternal behaviour of *Pisilus tipuliformis* Fabr. (Hemiptera: Reduviidae). Entomol. Exp. Appl. 8: 13–19.

1969. The predatory and reproduction behaviour of *Rhinocoris bicolor* and *R. tropicus* (Hemiptera: Reduviidae). Entomol. Exp. Appl. 12: 107–117.

1971. The predatory and reproductive behaviour of *Vestula lineaticeps* (Sign.) (Hemiptera: Reduviidae). Bull. Entomol. Res. 61: 119–124.

1972. The predatory and sexual behaviour of *Phonoctonus fasciatus* and *P. subimpictus* Stål (Hemiptera: Reduviidae). Bull. Entomol. Res. 62: 139–150.

Patil, B. V., and T. S. Thontadarya. 1983. Natural enemy complex of the teak skeletonizer *Pyrausta mechaeralis* Walker (Lepidoptera: Pyralididae) in Karnataka. Entomon 8: 249–255.

Pawar, C. S., V. S. Bhatnagar, and D. R. Jadhav. 1986. *Heliothis* species and their natural enemies, with their potential in biological control. Proc. Indian Acad. Sci. Anim. Sci. 95: 695–703.

Pettey, F. W. 1919. Insect enemies of the codling moth in South Africa and their relation to its control. South Afric. J. Sci.16: 239–257.

Pinto, C. F. 1927. *Crithidia spinigeri* n. sp. parasita do apparelho digestivo de *Spiniger domesticus* (Hemiptero Reduviidae). Bol. Biol. 7: 86–87.

Poinar, G. O., Jr. 1992. Fossil evidence of resin utilization by insects. Biotropica 24: 466–468.

Ponnamma, K. N., C. Kurian, and K. M. A. Koya. 1979. Record of *Rhinocoris fuscipes* (Fabr.) (Heteroptera: Reduviidae) as a predator of *Myllocerus curicornis* (F.) (Coleoptera: Curculionidae) the ash weevil, pest of the coconut palm. Agric. Res. J. Kerala 17: 91–92.

Pruthi, H. S. 1947. Biology of the reduviid bug *Acanthaspis quinquespinosa* an enemy of white ants. Proc. 28th Indian Sci. Anim. Husb. 27: 117–122.

Putschkov, P. V. 1986. Local distribution and life cycles of assassin bugs (Heteroptera: Reduvidae) of the Ukraine USSR (habitat-overwintering-voltinism). Entomol. Rev. 65: 1–13.

Rajagopal, D. 1984. Observation on the natural enemies of *Odontotermes wallonensis* (Wasssmann) (Isoptera: Termitidae) in South India. J. Soil Biol. Ecol. 4: 102–107.

Rajagopal, D., and G. K. Veeresh. 1981. Termitophiles and termitariophiles of *Odontotermes walloensis* (Isoptera: Termitidae) in Karnataka, India. Colemania 1: 129–130.

Rao, R. S. N., S. V. V. Satyanarayana, and V. Soundarajan. 1981. Notes on new addition to the natural enemies of *Spodoptera litura* F. and *Myzus persicae* Sulz. on fine cored tobacco in Andhra Pradesh. Sci. Cult. 47: 98–99.

Rao, V. P. 1974. Biology and breeding techniques for parasites and predators of *Ostrina* sp. and *Heliothis* sp. CIBC Final Technical Report PL 480 Project, Bangalore, India. 86 pp.

Readio, P. A. 1924. Notes on the life history of a beneficial reduviid, *Sinea diadema* (Fabr.) Heteroptera. J. Econ. Entomol. 17: 80.

1927. Studies on the biology of the Reduviidae of America north of Mexico. Univ. Kansas Sci. Bull. 17: 5–291.

1931. Dormancy in *Reduvius personatus* (Linnaeus) Ann. Entomol. Soc. Amer. 24: 19–39.

Ren, S. Z. 1984. Studies on the genus *Coranus* Curtis from China (Heteroptera: Reduviidae). Entomotaxonomia 6: 279–284.

Richman, D. P., R. C. Hemanway, and W. H. Whitcomb. 1980. Field cage evaluation of predators of the soybean looper *Pseudoplusia includens* (Lepidoptera: Noctuidae). Environ. Entomol. 9: 315–317.

Risbec, J. 1941. Les insectes de l'arachide. Trav. Lab. Entomol. Sect. Sudon. Rub. Agron. 22 pp.

Romeo, A. 1927. Ossseruazional suakuni pentatomidie loreidi neidintorni di Rando zoo (Catania). Ann. Rep. 1st Sup. Agrar. Protici.(3)ii: 261–268.

Rosenheim, J. A., and L. R. Wilhoit. 1993. Predators that eat other predators disrupt cotton aphid control. California Agric. 47: 7–9.

Rosenheim, J. A., L. R. Wilhoit, and C. A. Armer. 1993. Influence of intraguild predation among generalist predators on the suppression of an herbivore population. Oecologia 96: 439–449.

Ruberson, J. R., G. A. Herzog, W. R. Lambert, and W. J. Lewis. 1994a. Management of the beet armyworm (Lepidoptera: Noctuidae) in cotton: role of natural enemies. Florida Entomol. 77: 440–453.

1994b. Management of the beet armyworm: integration of control approaches. Proc. 1994 Beltwide Cotton Prod. Conf., pp. 857–859.

Rukmani, J. 1992. Multidisciplinary tools in the biosystematics of Reduviidae (Insecta: Heteroptera: Reduviidae). Ph.D. thesis, Madurai Kamaraj University, Madurai, India.

Sahayaraj, K. 1991. Bioecology, ecophysiology and ethology of chosen predatory hemipterans and their potential in biological control (Insecta: Heteroptera: Reduviidae). Ph.D. thesis, Madurai Kamaraj University, Madurai, India.

1994a. Capturing success by reduviid predators *Rhinocoris kumarii* and *Rhinocoris marginatus* on different age groups of *Spodoptera litura*, a polyphagous pest (Heteroptera: Reduviidae). J. Ecobiol. 6: 221–224.

1994b. Biocontrol potential evaluation of the reduviid predator *Rhinocoris marginatus* Fabricius to the serious groundnut pest *Spodoptera litura* (Fabricius) by functional response study. Fresenius Environ. Bull. 3: 546–550.

1995. Bioefficacy and prey size suitability of *Rhinocoris marginatus* Fabricius to *Helicoverpa armigera* Hubner of groundnut (Insecta: Heteroptera: Reduviidae). Fresenius Environ. Bull. 4: 270–278.

Sahayaraj, K., and D. P. Ambrose. 1992a. Biology, redescription and predatory behaviour of the reduviid *Allaeocranum quadrisignatum* (Hemiptera: Reduviidae) from South India. J. Soil Biol. Ecol. 12: 120–133.

1992b. Biology and predatory potential of *Endochus umbrinus* Reuter (Heteroptera: Harpactorinae: Reduviidae) from South India. Bull. Entomol. 33: 42–55.

1993. Biology and predatory potential of *Coranus nodulosus* Ambrose and Sahayaraj on *Dysdercus cingulatus* Fabr. and *Oxycarenus hyalinipennis* Costa (Heteroptera: Reduviidae). Hexapoda 5: 17–24.

1994. Stage and host preference and functional response of a reduviid predator *Acanthaspis pedestris* Stål to four cotton pests. J. Biol. Control 8: 23–26.

1996a. Biocontrol potential of the reduviid predator *Neohaematorrhophus therasii* Ambrose and Livingstone (Heteroptera: Reduviidae). J. Adv. Zool. 17: 49–53.

1996b. Functional response of the reduviid predator *Neohamatorrhophus therasii* Ambrose and Livingstone to the cotton stainer *Dysdercus cingulatus* Fabricius. Pp. 328–331 *in* D. P. Ambrose [ed.], Biological and Cultural Control of Insect Pests, an Indian Scenario. Adeline Publishers, Tirunelveli, India. 362 pp.

1996c. Short term functional response and stage preference of the reduviid predator *Ectomocoris tibialis* Distant on the cotton stainer *Dysdercus cingulatus* Fabr. J. Appl. Zool. 81: 219–225.

Sahayaraj, K., and K. Sivakumar. 1995. Ground nut pest and pest stage preference of a reduviid predator *Rhinocoris kumarii* Ambrose and Livingstone (Heteroptera: Reduviidae). Fresenius Environ. Bull. 4: 263–269.

Satpathy, J. M., N. C. Patnaik, and A. P. Samalo. 1975. Observations on the biology and habits of *Sycanus affinis* Reuter (Hemiptera: Reduviidae) and its status as a predator. J. Bombay Nat. Hist. Soc. 72: 589–595.

Schaefer, C. W. 1988. Reduviidae (Hemiptera: Heteroptera) as agents of biological control. Pp. 27–33 *in* K. S. Ananthasubramanian, P. Venkatesan, and S. Sivaraman [eds.], Bicovas I, Loyola College, Madras, India. 226 pp.

Schaefer, C. W., and I. Ahmad. 1987. Parasities and predators of Pyrrhocoroidea (Hemiptera) and possible control of cotton stainers *Phonoctonus* spp. (Hemiptera: Reduviidae). Entomophaga 32: 269–275.

Scudder, G. G. E. 1992. The distribution and the life cycle of *Reduviius personatus* (L.) (Hemiptera: Reduviidae) in Canada. J. Entomol. Soc. British Columbia 89: 38–42.

Severein, H. H. P. 1924. Natural enemies of beet leaf hopper (*Eutettix tenella* Baker). J. Econ. Entomol. 17: 369–377.

Shepard, M., R. E. McWhorter, and E. W. King. 1982. Life history and illustrations of *Pristhesancus papuensis* (Hemiptera: Reduviidae). Canad. Entomol. 114: 1089–1092.

Singh, G. 1992. Management of oil palm pests and diseases in Malaysia in 2000. Pp. 195–212 *in* A. Aziz, S. A. Kadir, and H. S. Barlon [eds.], Pest Management and the Environment in 2000. CAB International, Wallingford, Oxon, U.K. 401 pp.

Singh, J., R. Arora, and A. S. Singh. 1987. First record of predators of cotton pests in the Punjab. J. Bombay Nat. Hist. Soc. 84: 456.

Singh, O. P. 1985. New record of *Rhinocoris fuscipes* Fabr. as a predator of *Dicladispa armigera* (Oliver). Agric. Sci. Dig. 5: 179–180.

Singh, O. P., and G. A. Gangrade. 1975. Parasites, predator and diseases of larvae of *Diacrisia obliqua* Walker (Lepidoptera: Arctiidae) on soybean. Curr. Sci. 44: 481–482.

1976. Biology of a reduviid predator, *Rhinocoris fuscipes* Fabricius on the larvae of *Diacrisia obliqua* Walker (Lepidoptera: Noctuidae). JNKVV Res. J. 10: 148–160.

Singh, O. P., and K. J. Singh. 1987. Record of *Rhinocoris fuscipes* Fabricius as a predator of green stink bug, *Nezara viridula* Linn. infesting soybean in India. J. Biol. Control 1: 143–146.

Sitarmaiah, S., and S. V. V. Satyanarayana. 1976. Biology of *Harpactor costalis* Stål (Heteroptera: Reduviidae) on tobacco caterpillar *Spodoptera litura* F. Tobacco Res. 2: 134–136.

Sitaramaiah, S., B. G. Joshi, G. R. Prasad, and S. V. V. Satyanarayana. 1975. *Harpactor costalis* Stål (Reduviidae: Heteroptera) new predator of the tobacco caterpillar (*Spodoptera litura* Fabr.). Sci. Cult. 41: 545–546.

Stride, G. O. 1956a. On the mimetic association between certain species of *Phonoctonus* (Hemiptera: Reduviidae) and the Pyrrhocoridae. J. Entomol. Soc. South. Africa 19: 12–27.

1956b. On the biology of certain West African species of *Phonoctonus* (Hemiptera: Reduviidae) mimetic predators of the Pyrrhocoridae. J. Entomol. Soc. South. Africa 19: 52–69.

Subbiah, M. S., and V. Mahadevan. 1957. The life history and biology of the reduviid *Acanthaspis siva*, a predator of the Indian honey bee *Apis indica*. Indian J. Vet. Sci. 27: 117–122.

Sundararaju, D. 1984. Cashew pests and their natural enemies in Goa. J. Plant. Crops 12: 38–46.

Swadener, S. O., and T. R. Yonke. 1973a. Immature stages and biology of *Apiomerus crassipes* (Hemiptera: Reduviidae). Ann. Entomol. Soc. Amer. 66: 188–196.

1973b. Immature stages and biology of *Zelus socius* (Hemiptera: Reduviidae). Canad. Entomol. 105: 231–238.

1973c. Immature stages and biology of *Sinea complexa* with notes on four additional reduviids (Hemiptera: Reduviidae). J. Kansas Entomol. Soc. 46: 124–136.

1975. Immature stage and biology of *Pselliopus cinctus* and *Pselliopus barberi* (Hemiptera: Reduviidae). J. Kansas Entomol. Soc. 48: 472–492.

Tawfik, M. F. S., K. T. Awadallah, and M. M. H. Abdullah. 1983a. Effect of prey on various stages of the predator, *Alloeocranum biannulipes* (Montr. et Sign.) (Hemiptera: Reduviidae). Bull. Soc. Entomol. Egypt 64: 251–258.

Tawfik, M. F. S., K. T. Awadallah, and N. A. Abozeid. 1983b. The biology of the reduviid *Allaeocranum biannulipes* (Montr.et.Sign.) a predator of stored-product insects. Bull. Soc. Entomol. Egypt 64: 231–237.

Taylor, J. S. 1932. Report on cotton insect and disease investigations. Part II. Notes on the American boll worm (*Heliothis obsoleta* Fabr.) on cotton, and on its parasite, *Microbracoss brevicornis* Western. Sci. Bull. Dept. Agric. South Africa 113: 18.

Thangavelu, K., and K. S. Ananthasubramanian. 1981. Natural enemies of *Rhypara chrominae* (Lygaeidae: Heteroptera) from Southern India. Proc. Indian Nat. Sci. Acad. B47: 632–636.

Ullyett, G. C. 1930. The life history, bionomics and control of cotton stainer (*Dysdercus* spp.) in South Africa. Sci. Bull Dept. Agric. South Africa 94: 3–9.

Underhill, R. A. 1954. Habits and life history of *Arilus cristatus* (Hemiptera: Reduviidae). Walla Walla College Publ. Dept. Biol. Sci. Biol. Stn. 11: 1–15.

Usinger, R. L., and P. Wygodzinsky. 1964. Description of a new species of *Mendanocoris* Miller, with notes on the systematic position of the genus (Reduviidae: Hemiptera: Insecta). Amer. Mus. Novit. 2204: 1–13.

Uthamasamy, S. 1995. Biological control of the introduced psyllid, *Heteropsylla cubana* Crawford (Homoptera: Psyllidae), infesting subabul *Leucaena leueocephala* (Lam.) de Wit — progress and prospects. Pp. 65–75 *in* T. N. Ananthakrishnan [ed.], Biological Control of Social Forest and Plantation Crop Insects. Oxford & IBH Publ. Co. Pvt. Ltd., New Delhi, India. 225 pp.

Van den Bosch, R., and A. D. Telford. 1964. Environmental modification and biological control. Pp. 8–44 *in* P. DeBach [ed.], Biological Control of Insect Pests and Weeds. Chapman & Hall Ltd., London, U.K. 822 pp.

Vanderplank, F. L. 1958. The assassin bug *Platymeris rhadamanthus* Gerst. (Hemiptera: Reduviidae) a useful predator of the rhinoceros beetles *Oryctes boas* F. and *Oryctes moneros* (Oliv). Rev. Entomol. Soc. South. Africa 1: 309–314.

Varma, R. V. 1989. New record of *Panthous bimaculatus* (Hemiptera: Reduviidae) as a predator of pests of *Alantha triphysa*. Entomon 14: 357–358.

Veeresh, G.K., and M. Puttarudriah. 1970. Studies on the morphology and biology of the assassin bug, *Isyndus heros* Fabricius (Hemiptera: Reduviidae). Mysore J. Agric. Sci. 4: 285–295.

Vennison, S. J. 1988. Bioecology and ethology of assassin bugs (Insecta: Heteroptera: Reduviidae). Ph.D. thesis, Madurai Kamaraj University, Madurai, India.

Vennison, S. J., and D. P. Ambrose. 1986a. Bioecology of dimorphic assassin bug, *Edocla slateri* Distant (Heteroptera: Reduviidae). Entomon 11: 255–258.

1986b. Impact of mating on oviposition pattern and hatchability in *Rhinocoris fuscipes* (Heteroptera: Reduviidae), potential predator of *Heliothis armigera*. J. Soil Biol. Ecol. 6: 57–61.

1988. Impact of space on stadial period, adult longevity, morphometry, oviposition, hatching and prey capturing in *Rhinocoris marginatus* Fabricius. (Insecta: Heteroptera: Reduviidae). J. Mitt. Zool. Mus. Berlin 64: 349–355.

1989a. Biology and predatory potential of a reduviid predator *Oncocephalus annulipes* Stål (Hemiptera: Reduviidae). J. Biol. Control 3: 24–27.

1989b. Impact of space on development, size, oviposition pattern and hatchability of *Acanthaspis pedestris* Stål (Heteroptera : Reduviidae). Uttar Pradesh J. Zool. 9: 65–72.

1990a. Biology and behaviour of *Sphedanolestes signatus* Distant (Insecta: Heteroptera: Reduviidae) a potential predator of *Helopeltis antonii* Signoret. Uttar Pradesh J. Zool.10: 30–43.

1990b. Biology of an assassin bug *Velitra sinensis* Walker (Insecta: Heteroptera: Reduviidae) from South India. Indian J. Entomol. 52: 310–319.

1990c. Biology and behaviour of the assassin bug *Ectomocoris vishnu* Distant (Hemiptera: Reduviidae: Piratinae) from South India. J. Soil Biol. Ecol. 10: 89–102.

1990d. Impact of mating on oviposition pattern and hatchability in *Acanthaspis quinquespinosa* Fabricius (Insecta: Heteroptera: Reduviidae). J. Adv. Zool. 11: 1–4.

1990e. Egg development in relation to soil moisture in two species of reduviids (Insecta: Heteroptera). J. Soil Biol. Ecol. 10: 116–118.

1991a. Biology and behaviour of *Euagoras plagiatus* Burmeister (Heteroptera: Reduviidae) from South India. J. Bombay Nat. Hist. Soc. 88: 222–228.

1991b. Biology and behaviour of *Ectomocoris xavierei* Vennison and Ambrose 1990. (Heteroptera: Reduviidae: Peiratinae). Trop. Zool. 4: 251–258.

1992a. Biology behaviour and biocontrol effeciency of a reduviid predator *Sycanus reclinatus* Dohrn (Heteroptera: Reduviidae) from southern India. J. Mitt. Zool. Mus. Berlin 68: 143–156.

1992b. Biology and behaviour of *Acanthaspis philomanmariae* Vennison and Ambrose (Acanthaspidinae) and *Coranus soosaii* Ambrose and Vennison (Harpactorinae) (Insecta: Heteroptera: Reduviidae) from South India. Hexapoda 4: 91–106.

Vijayavathi, B. S. 1987. Studies on the biology and ethology of *Acanthaspis pedestris* Stål (Reduviidae: Acanthaspidinae), an ant feeding assassin bug of the scrub jungles and semiarid zones of Western ghats, India. M. Phil. thesis, Bharathiar University, Coimbatore, India.

Villiers, A. 1948. Faune de l'Empire Francais. IX Hemipteres Reduviides de l'Afrique Noire. Office de la Recherche Scientifique Coloniale. Editions du Mus., Paris, France. 489 pp.

Wallace, H. R. 1953. Notes on the biology of *Coranus subapterus* DeGeer (Hemiptera: Reduviidae). Proc. R. Entomol. Soc. London (A) 28: 100–110.

Watson, T. R. 1918. Insects of citrus grove. Univ. Florida Agric. Exp. Stn. Bull. 144: 261.

Weaver, E. C., E. T. Clarke, and N. Weaver. 1975. Attractiveness of an assassin bug to stingless bees. J. Kansas Entomol. Soc. 48: 17–18.

Weaver, N. 1978. Chemical control of behavior-interspecific. Pp. 392–418 *in* M. Rockstein [ed.], Biochemistry of Insects, Academic Press, New York, New York, U.S.A.

Webster, F. M. 1907. The chinch bug. U.S. Dept. Agric. Bull. Entomol. 69: 95.

Werner, F. G., and G. D. Butler. 1957. The reduviids and nabids associated with Arizona crops. Arizona Agric. Exp. Stn. Tech. Bull. 133: 1–12.

Whitcomb, W. H., and K. Bell. 1964. Predaceous insects, spiders and mites and Arkansas cotton fields. Bull. Arkansas Agric. Exp. Stn. 690: 1–84.

Wygodzinsky, P. 1944. Notas sobre a biologia e o desenvolvimento do *Macrocephalus notatus* West. (Phymatidae: Reduvioidea: Hemiptera). Rev. Entomol. Rio de Janeiro 15: 139–143.

Yano, 1990. A host record of *Isyndus obscurus* (Dallas) (Hemiptera: Reduviidae). Jpn. J. Entomol. 58: 204.

Zanuncio, J. C., J. B. Alves, T. V. Zanuncio, and J. F. Garcia. 1994. Hemipteran predators of eucalyptus defoliator caterpillars. For. Ecol. Manage. 65: 65–73.

CHAPTER 30

Economic Importance of Predation by Big-Eyed Bugs (Geocoridae)

Merrill H. Sweet II

1. INTRODUCTION

Henry (1997) has divided the paraphyletic family Lygaeidae into smaller but monophyletic families; the subfamily Geocorinae becomes the family Geocoridae. There are 14 described genera; most species are in *Geocoris* and *Germalus*, and these include all the known economic species. The genus *Geocoris*, which is worldwide in distribution, included 124 described species when Readio and Sweet (1982) made their compilation. Most described species, 36 (27%), occur in the Palearctic zoogeographic region; 20% occur in the Oriental, 18% in the Nearctic, 15% in the Ethiopian, 10% in the Neotropical, 8% in the Australian, and the remaining 2% are island species. The systematics is far from settled. Readio and Sweet (1982) revised the *Geocoris* of eastern North America, which incidentally treated four of the five species most used in economic studies. Theirs is only key to the species east of the 100th meridian. Mead (1972) published a key to the Florida (U.S.A.) species. This does not mean that geocorids are not important in the rest of the world! It reflects instead the very strong research effort to use big-eyed bugs in integrated pest programs.

The family Geocoridae is easily recognized as the members have a characteristic general appearance. The insects are moderately small, ranging from 2.7 to 5 mm in length, oval in form, with conspicuously large stylate to semistylate eyes, hence their common name. The insects vary in color, which generally matches that of the background. Usually each geographic area has pale, mottled, striped, and black species. Moreover, the coloration within a species often varies considerably, sometimes showing seasonal variation (Readio and Sweet 1982). *Geocoris* species are therefore often very difficult to tell apart. In addition to highly variable color characters and patterns, there is the equally confounding problem of the structural uniformity of most species, which probably reflects the constraints of a form optimal for a predaceous lifestyle. Torre-Bueno (1946) wrote:

> It should be specifically pointed out that the closet naturalists of other times described species in this genus from a totally inadequate number of examples and without an acquaintance with the forms in the vast numbers in which they are found in the field; they reveled in the least nuances of color, in a group which is as variegated in this respect as the offspring of ranging tomcats. In consequence, their equally closet-bound successors, likewise with an insufficient number of specimens before them and seemingly with the same absence of the corrective of close observation in the field, have perpetuated these unstable color-forms as varieties, and keyed them out.

However, by fieldwork and mating tests and restudying the morphology, Readio and Sweet (1982) discovered that some of these varieties were good species, others were not. They noted that there were also many more undescribed species in western North America and Mexico. The author's fieldwork in South America indicates this to be true of the Americas in general. Clearly, revisional studies that are combined with biological fieldwork are badly needed to understand the species of *Geocoris*.

2. GENERAL STATEMENT OF ECONOMIC IMPORTANCE

The significance of geocorids in biological control is shown by several recent comprehensive reviews. These excellent works are not repeated here; rather, the reader is directed to Naranjo and Gibson (1996), to the entire symposium edited by Coll and Ruberson (1998), and to the section on Lygaeidae in Hagen et al. (1999). The influence of these papers on this review is here acknowledged. Why are geocorids so useful? First, geocorids are very generalized predators and will feed on eggs, larvae or nymphs, and adults of appropriate size. Several species of *Geocoris* are very abundant and widely distributed. They have frequently been observed in cultivated fields on legumes, cotton, sugar beets, and vegetables; in orchards on ornamentals; in residential lawns (often with chinch bugs); and in disturbed roadside areas. *Geocoris* species have been observed feeding on many species of Hemiptera, Coleoptera, Lepidoptera, small Diptera, Hymenoptera (including ants), Thysanoptera, Collembola, and Acarina (Tamaki and Weeks 1972b, Crocker and Whitcomb 1980, and Readio and Sweet 1982). Actually, the taxonomic identity is not so important as the size and behavior of the prey. Geocorids tend to prefer small prey, including eggs. All instars attack the same type of prey as the adults, only scaled for the size of the insect. Very large prey is avoided if active. With their big eyes the bugs locate prey visually, but extend their antennae to the prey item as they raise the proboscis to spear it. Often the bug walks about with the prey impaled on the labium. The insects fly readily to avoid danger and to seek new habitats or hunting grounds. Crocker and Whitcomb (1980) observed the bugs to fly with their prey on the beak, but the approach to a prey item is always made on foot. Unfortunately for mass-rearing efforts, the bugs are cannibalistic and readily consume smaller conspecifics, even adults (Readio and Sweet 1982).

Much of the recent literature on *Geocoris* concerns their feeding habits and demonstrates that *Geocoris* species have rather complex nutritional needs. Although *Geocoris* spp. are generalist predators, they evidently require plant food for optimal development and reproduction, which led many earlier workers to think that the insects were pests (York 1944; Sweet 1960; Tamaki and Weeks 1972; Naranjo and Stimac 1985, 1987; Bugg et al. 1987, 1991). King and Cook (1932) showed that although *G. punctipes* feeding caused some internal swelling in the tissues of cotton stems and leaf petioles similar to the reaction produced by the cotton fleahopper *Pseudotomoscelis seriatus* (Reuter) (Miridae), such feeding, in contrast to the mirid's feeding, caused no external swelling or apparent tissue damage. York (1944) found that *Geocoris* required both vegetable matter and prey for survival and suggested that plants were important only as a source of water. Sweet (1960) reported that *G. uliginosus*, *G. bullatus*, and *G. piceus* survived on sunflower seeds and water alone for as long as 3 to 4 months and would feed on seeds even in the presence of potential prey. He noted that seed and plant feeding allow these insects to persist when prey abundance is low, which helps to explain the high abundance of these predators in the field. By using radiolabeled inorganic ^{32}P, Ridgway and Jones (1968) verified that *G. pallens* fed on cotton. They concluded that because predators do use plants at least to obtain moisture, systemic insecticide treatment of the plants would decrease the populations of these predators. Stoner (1970) tested many plants and found *G. punctipes* could live over 20 days on certain seeds, about 9 days on cotton leaves, 12 days on dandelion pollen, and 3.5 days on water alone, but required arthropod prey for normal development and fecundity. Dunbar (1971) studied the feeding behavior of *G. bullatus* in Connecticut (U.S.A.) lawns and found that the predator was not injuring the grasses as might be thought, but was feeding upon other lawn insects such as the chinch bug *Blissus leucopterus* (Say).

Nevertheless, *Geocoris* species are definitely predaceous and therefore may in some cases hold promise as biological control agents (York 1944). Research by recent workers primarily involves biological studies of *Geocoris* to learn if these insects are important in controlling some of the pests on crops and, if so, to determine how *Geocoris* may be utilized to play an even more active role in an integrated pest management program. Nearly all the pertinent recent studies have been on North American species, the notable exception being the promising studies in India by Mukhopadhyay (1986, 1997; also Mukhopadhyay and Ghosh 1982). However, in the author's fieldwork in South America and Africa, all Geocoridae were found to be energetic predators. A key point of Readio and Sweet (1982) is that geocorids divide into two groups: one adapted to temporary environments and the other to permanent habitats, permanence being defined in terms of the life cycle of the insect. The species of temporary environments are usually macropterous and have high powers of dispersal, and those of permanent habitats are usually predominantly brachypterous and have low powers of dispersal. The latter group is essentially K-selected and the former is r-selected and very well adapted to exploit the largely early successional stages that are called agroecosystems. In such habitats the most abundant predators tend to be geocorids, nabids, and anthocorids (Irwin and Shepard 1980, Yeargan 1998).

Unlike the cimicomorphan families Anthocoridae, Nabidae, and Miridae, which lay their eggs into plants, *Geocoris* spp. like many other lygaeoids oviposit their elongate oval eggs singly, either horizontally on pubescent plant surfaces or in soil. In eastern North America eggs are often parasitized by the scelionid *Telenomus reynoldsi* Gordh and Coker and the trichogrammatid *Trichogramma pretiosum* Riley (Cave and Gaylor 1988a,b, 1989). The parasite is known from *G. punctipes, G. uliginosus, G. floridanus* Blatchley, and *G. pallens*. This parasitism by *Trichogramma* is unusual as it is normally a parasite of Lepidoptera eggs. Cave and Gaylor studied the effect of temperature and humidity on parasite development, the incidence of parasitism in the field, and the rate of fertilization and population growth of the egg parasites. Adult *Geocoris* in California (U.S.A.) (Clancy and Pierce 1966) and Arizona (U.S.A.) (Atim and Graham 1983) were parasitized by the tachinid *Hyalomya aldrichii* Townsend and the eggs by *T. reynoldsi* and *T.* near *opacus*.

One problem of predators is the degree to which they attack each other. Already mentioned is the strong cannibalistic tendency of geocorids. Guillebeau and All (1989, 1990) studied the interactions between the predaceous spider *Oxyopes salticus* and *Geocoris* spp.; *Geocoris* prefers *Heliothis* spp. eggs but feeds also on the spider. The big-eyed bug adults ate all the eggs or emerging spiderlings from unguarded egg sacs, but the adult *O. salticus* females fed on more active prey such as larger *Geocoris* nymphs and adults. The authors suggested that there would be an optimal density for the spider and bug that would maximize feeding on *Heliothis* eggs; too many of either bug or spider would be disadvantageous.

The influence of temperature on rate of development in *Geocoris* species is accelerated up to the upper lethal point, which is species specific, showing the importance of accurate species identification (Butler 1966; Champlain and Sholdt 1966; Dunbar and Bacon 1972a; Cohen 1982, 1983b). Tamaki and Weeks (1972) compared the rates of development with temperature of *G. bullatus, G. pallens,* and *G. punctipes* and found the overall regression lines to be similar.

A serious problem with biocontrol work with predators is their susceptibility to pesticides, including herbicides and defoliants as well as insecticides, which can reduce *Geocoris* populations. It is therefore important to use more closely targeted pesticides. Yokoyama et al. (1984) determined the acute toxicity of 22 pesticides, noting those materials that are particularly suitable for use in cotton but would harm biocontrol agents the least. For predatory heteropterans like geocorids systemic insecticides are particularly toxic. Ridgway et al. (1967) demonstrated that the application of systemic insecticides to cotton showed the importance of natural populations of *Geocoris, Nabis,* and *Orius* predators in "regulating" *Heliothis* spp. populations: heteropteran predators that fed on cotton were reduced in numbers by the systemic treatments, which had minimal impact on *Chrysopa* larvae and hymenopteran parasitoids.

3. THE IMPORTANT SPECIES

3.1 *Geocoris punctipes* (Say)

3.1.1 *Distribution, Life History, and Biology*

This species is the most studied of the Geocoridae and a book could be written on it alone. *Geocoris punctipes* is widely distributed across the southern two thirds of the United States into Mexico southward to Colombia, but the more southern records are less certain because there is a large complex of species related to *G. punctipes* in South America. Even within this complex *G. punctipes* is relatively large. The males average 3.7 mm long, the females 3.9 mm long. This larger size allows the insect to attack larger prey than can other *Geocoris* species. The insect is very pale; a distinctive line runs on to the vertex and another transversely crosses it from the lateral pits of the head. This character is shared with many South American species. The nymphs can be readily recognized from other pale geocorids by the pale bluish-gray appearance and the abdomen's reticulate coloration of numerous red lines. A dark groove extends onto the vertex of the head beyond the tylus (Readio and Sweet 1982). The species overwinters as an adult and is multivoltine with no evidence of having a diapause state; thus nymphs occur late in fall, although in Texas (U.S.A.) at the latitude of College Station in early spring only adults were found.

The characteristic habitat of this species is along roadsides and disturbed areas. It is abundant in agroecosystems associated with agricultural crops such as alfalfa, cotton, lettuce, peanuts, soybeans, and sugar beets (Readio and Sweet 1982). It appears to be the most studied geocorid species among the natural enemies of agricultural pests. It more readily than the other species climbs into vegetation in search of prey, so it is more often collected by sweeping and thus gives a more exaggerated abundance than the other geocorids. Throughout its wide range it commonly coexists in the same habitat with one or more of the following species: *G. pallens, G. lividipennis, G. bullatus, G. floridanus,* and *G. uliginosus.* Crocker and Whitcomb (1980) in Florida (U.S.A.) noted the tendency for *G. punctipes* to be more up in the vegetation than the more geophilous *G. uliginosus.* Crocker et al. (1975) studied the effect of sex, developmental stages, and temperature on predation on eggs of the soybean looper *Pseudoplusia includens* (Walker) by *G. punctipes.* They found that the larger bugs consumed more prey, the females more than males, and fed more at higher temperatures up to 30°C. Dunbar and Bacon (1972b) demonstrated that *G. punctipes* preferred insect eggs over live or dead insects, inactive prey over active prey, and suggested that this species in nature may be basically an egg predator. However, Champlain and Sholdt (1966) reared *G. punctipes* in the laboratory successfully on larvae of the beet armyworm *Spodoptera exigua* (Hübner) with green beans provided for moisture.

In the laboratory at 25°C, on a diet of beet armyworm larvae and green beans, Champlain and Sholdt (1966, 1967a,b) determined the developmental time at 25°C of *G. punctipes* eggs to be about 10 days, and a total nymphal development period of about 27 days. The stadia were first, 7.6; second, 4.8; third, 4.1; and fifth, 6.3 days. The adult longevity was 41.5 days for males, 67.7 days for females, but the maxima were 111 and 109 days, respectively. The preoviposition period was about 5 days and the mean fecundity was 178 eggs. Cohen (1989), using green beans (provided as a water source) and fresh *Heliothis virescens* eggs, found that adults of *G. punctipes* had a food absorption efficiency of 95.2% and growth efficiency of 52.9%. He also determined the ingestion efficiency and protein consumption of *G. punctipes* fed pea aphids. Cohen developed a method of encapsulating in wax an artificial diet that was fed successfully to *G. punctipes, G. pallens,* and several other predaceous insects (Cohen 1983). Cohen further simplified an artificial diet for *G. punctipes* by making a sausagelike paste of beef liver, ground beef, and fresh spinach with wheat germ enclosed in Parafilm®. *Geocoris punctipes* was reared on this diet for many generations (Cohen 1985a). Cohen in **Chapter 20** discusses how this bug feeds on prey, extracting food efficiently with long, flexible mouthparts.

3.1.2 Economic Benefit

As noted, research on the use of geocorids has focused on this species, hence the extensive efforts to develop methods of rearing the species and understanding its nutritional needs. However, as Slater and O'Donnell (1995) stated, "a flood of a papers has appeared (usually multiauthored) on potentially important predatory lygaeoids, especially species of *Geocoris* .... Despite this rather extensive literature one feels that it remains to be established whether these generalist predators are of major importance in controlling serious pests." This comment applies especially to *G. punctipes*, and time will tell how useful the research will be in effecting biological control.

There have been attempts to quantify the benefits of geocorid predation, some in laboratory settings, others in the field. The first field study using *G. punctipes* in biological control was by McGregor and McDononough (1917) on the control of the two-spotted mite *Tetranychus* (= *Paratetranychus*) *urticae* on cotton. They estimated that an adult *G. punctipes* can consume 80 mites a day and that a nymph growing to adult would consume 1600 mites. Watve and Clower (1976) observed *G. punctipes* feeding on the larvae and pupae of the bandedwing whitefly *Trialeurodes abutilonea* (Haldman). The number of soybean looper eggs consumed daily at four different temperatures was determined by Crocker et al. (1975). Lawrence and Watson (1979) and Chiravathanapong and Pitre (1980) showed that first instar bugs fed only on the eggs of *H. virescens* (F.), and the nymphs could feed only on the early larvae, but eggs were the most nutritious. Lingren et al. (1968a,b) released *G. punctipes* in field cages in cotton and showed that about 252,000 adults/acre were ineffective, but a release of 630,000/acre reduced the egg and larval populations of *H. virescens* by 88%. Waddill and Shepard (1974) showed by using field cages on soybeans that *G. punctipes* significantly reduced the Mexican bean beetle populations by preying on eggs and first instars of the beetles.

3.2 Geocoris bullatus (Say)

3.2.1 Distribution, Life History, and Biology

This gray species can be told from *G. punctipes* by its smaller size, the brachypterous condition of most adults, and the restriction of the longitudinal head groove to the tylar area alone. It is not easily distinguished from *G. pallens*. *Geocoris pallens* is light tan rather than grayish, and the dark punctures stand out in *G. bullatus,* whereas in *G. pallens* the punctures are smaller and light colored. *Geocoris pallens* is entirely macropterous and in *G. bullatus* the populations are usually about 70% brachypterous. In the nymph of *G. bullatus* the abdomen is a dark red and covered with greenish-blue spots. This is well illustrated in Dunbar (1971).

This is a common species of sandy habitats across the northern United States and Canada and is called the western big-eyed bug. Sweet (1960) discovered that *G. bullatus* readily feeds on sunflower seeds, even in the presence of prey, and survived for 3 to 4 months on a diet of sunflower seeds.

In the Yakima Valley of Washington (U.S.A.), Tamaki and Weeks (1972) carried out a 5-year study on this species, which coexisted in great numbers with *G. pallens*. Because the two species were so similar and taxonomic specialists differed in their interpretation of the specimens, Tamaki and Weeks ran mating tests and demonstrated that the species were reproductively isolated. They showed the species to have distinctly different life cycles: *G. pallens* overwintered as an adult and had at least two generations a year, whereas *G. bullatus* overwintered as an egg and as an adult. This author found *G. bullatus* also to lay diapause eggs in Connecticut (U.S.A.), but the adults did not survive the winter although they laid eggs until late in the autumn under cool conditions. These late-fall adults probably account for the reports of overwintering adults. However, the *G. bullatus* of Tamaki and Weeks might possibly represent still another species because Tamaki (1972) and Tamaki and Weeks (1972) make no mention of pterygopolymorphism in their Washington (U.S.A.) populations. The fecundity of *G. bullatus* was determined by Tamaki and Weeks (1972) to be 75 eggs per female on a mixed diet. This seems somewhat low. Fecundity became greatly reduced on suboptimal diets lacking plants, seeds, or prey. The bug did not lay eggs on a diet of seeds and beans or of pea aphids and beans unless sugar beet leaves were provided. Yet, on beet leaves alone the fecundity was zero. Tamaki and Weeks (1972) were able to increase the field populations of *G. bullatus* greatly by using supplemental crushed sunflower seeds.

The eggs of *G. bullatus* average 0.42 by 0.92 mm and have an average of 7.9 chorionic processes (range 6 to 10) on the anterior pole of the egg. The eggs are light pink, rather than tan as in *G. pallens*. The average lengths of the instars are first, 1.19 mm; second, 1.53; third, 1.93; fourth, 2.34; fifth, 3.24. The males averaged 3.66; the females, 4.07 mm. The egg hatching times at ambient temperatures varies from 22 days at 21°C to 5 days at 40°C.

3.2.2 Economic Benefit

Tamaki and Weeks (1972) found *G. bullatus* important in reducing the number of green peach aphids, *Myzus persicae* (Sulzer). It became most abundant (25/ft^2) in orchard floor vegetation, particularly under peach trees infested with *M. persicae*. *Geocoris bullatus* reduced the number of green peach aphids on fallen leaves returning to each tree by more than half; an average of more than 10,000 oviparae returned to the trees where predators were excluded from the orchard floor (Tamaki and Weeks 1972). They also found that in a mixture of caterpillar and aphid prey, *G. bullatus* attacked only a few caterpillars; thus the caterpillars interfered only slightly with predation on the aphids. In a greenhouse study, *G. bullatus* was released in cages with *M. persicae* infesting sugar beets and suppressed the aphids so there was only a ninefold increase in 28 days compared with a 2000-fold increase in cages without *Geocoris*. By providing chopped sunflower seeds as a supplemental food in sugar beet plots, the number of *Geocoris* eggs laid on sugar beet plants was doubled compared with controls (Tamaki and Weeks 1972). *Geocoris bullatus* also preys on *Lygus* spp., and Chow et al. (1983) determined the functional response of the predator to three different densities of *Lygus* nymphs at three different temperatures.

3.3 *Geocoris pallens* Stål

3.3.1 Distribution, Life History, and Biology

This species was described from California (U.S.A.) and probably is the species that Tamaki and Weeks (1972) worked with in Washington State. As discussed by Readio and Sweet (1982), the California specimens would not mate in the laboratory, so two species may be involved. However, this opens a Pandora's box because the western complex of *Geocoris* near *pallens* contains many new species and problematic populations, and Readio and Sweet recommended no taxonomic changes be made until the western populations of *Geocoris* have been reviewed. *Geocoris pallens* is smaller than *G. bullatus* and is entirely macropterous. The nymph is pale with a red abdomen spotted with transverse rows of numerous small white spots.

Geocoris pallens occurs throughout the western United States east to central Texas, mostly in open disturbed habitats and agricultural fields (Readio and Sweet 1982). It has a biology similar to that of *G. bullatus* but is more southern in its distribution.

Tamaki and Weeks (1972, 1973) studied this species in comparison with *G. bullatus* at Yakima Valley, Washington, in an apple-growing area. They found the egg to be tan, a little smaller than the eggs of *G. bullatus*, 0.37 × 0.88 mm, with a range of six to eight chorionic processes. The lengths of the instars are first, 1.08; second, 1.35; third, 1.64; fourth, 2.18; fifth, 2.84 mm. The lengths of the adults are male, 3.10; female, 3.50 mm, a little smaller than *G. bullatus*. The hatching time for the egg at 21°C was 22 days, at 40°C, 5.5 days in July. In June the hatching time at 21°C was 15 days, an interesting difference that was also seen in *G. bullatus* and may represent an acclimatization phenomenon between spring adults and summer adults. As with *G. bullatus*, the highest fecundities were on a mixed diet of sunflower seeds, insect prey, and sugar beet leaves. Yokoyama (1980) was able to rear *G. pallens* successfully in the laboratory on sunflower seeds, and eggs and nymphs of *Oncopeltus fasciatus* (Dallas). Yokoyama (1978) also found that *G. pallens* feeds on the extrafloral nectar of cotton and some population changes reflected the availability of nectar.

The insects readily dispersed by flying during the day when temperatures were above 24°C. Below 24°C, there was very little flight activity, even when disturbed. Working in the field in California, the present author found *G. pallens* to be very active, quickly flying when disturbed, and thus difficult to capture.

3.3.2 Economic Benefit

Tamaki and Weeks (1972) in their work on *G. bullatus* and *G. pallens* did not discriminate between the two species in their fieldwork on pest populations of *M. persicae*. They showed that the geocorids were the most abundant predators in the alfalfa interplant fields and maintained their populations even when the prey species had declined. *Geocoris pallens* like *G. bullatus* also suppressed *M. persicae* on sugar beets in experimental study plots (Tamaki and Weeks 1973).

In field studies in cotton, *G. pallens* was an effective predator of *Helicoverpa zea* larvae (van den Bosch et al. 1969), *Lygus* bugs (Leigh and Gonzalez 1976), cabbage looper (Ehler et al. 1973, Ehler 1977), pink bollworm (Irwin et al. 1974), and mites (Gonzalez and Wilson 1982).

Of converse significance, *G. pallens* in southern California (U.S.A.) (incidentally very close to the type-locality) was found predaceous on *Microlarinus* spp. eggs. *Microlarinus* weevils were imported for biological control of the noxious weed, puncture vine (*Tribulus terrestris* L.) (Goeden and Ricker 1967).

3.4 Other Species of Minor Economic Value

In North America, two other species are often present in agroecosystems and have received some research attention, but on a far lesser scale than the species listed. *Geocoris atricolor* Montandon is a black western North American species found in alfalfa and cotton fields. Dunbar and Bacon (1972a,b) reared the species in the laboratory on potato tuber moth larvae and string beans and determined under six different temperatures its developmental time for eggs and nymphs, and its fecundity.

Another black species found in eastern North America is *G. uliginosus* (Say), which is very common in disturbed habitats, especially in shaded areas where its dark coloration helps it to blend into the background. In Texas (U.S.A.) it is much less abundant in agroecosystems than *G. pallens* and *G. lividipennis*, and occurs in much more protected microhabitats. It has been listed as a control agent for chinch bugs as it is often found in lawns. It is geophilous and rarely climbs on plants, and is common in more closed field crops (Crocker and Whitcomb 1980, Readio and Sweet 1982). Braman et al. (1985) reared the insect at five different temperatures. It did not grow at 15°C. At 21°C the stadial lengths were egg, 17 days; first, 14; second 9.3; third, 8.6. fourth, 11; and fifth, 15.4 days. The males lived for 78 days, the females, 72 days.

Mukhopadhyay and Ghosh (1982) described two new species of *Geocoris*, *G. pseudoliteratus* and *G. bengalensis* from eastern India, and Mukhopadhyay (1985) gave an excellent account of the biology of these species, describing the immature forms and growth rates. If these species were to become economically important, this information would be very valuable. Kumar and Ananthakrishnan (1985) showed that *G. ochropterus* Fieber is a predator on thrips (*Caliothrips indicus*, *Ayyaria chaetophora*, and *Scirtothrips dorsalis*) which feed on peanuts (*Arachis hypogaea*). Mukhopadhyay et al. (1996) and Mukhopadhyay (1997) proposed that *G. ochropterus* be used more extensively to control insect pests and help avoid the excess use of insecticides.

4. PREDATORS OF MINOR HARM

4.1 *Germalus pacificus* Distant

The genus *Germalus* is relatively large and concentrated in the Indo-Malaya area; many species extend to larger islands of the South Pacific, in contrast with the more continental distribution of *Geocoris*. On Fiji the lantana bug *Teleonemia lantanae* Distant was imported to help control the noxious weed *Lantana camara* L., a plant originally imported from tropical America as an ornamental plant, but now a pest in eastern Australia and several islands, in particular, Fiji (Fyfe 1935, 1937). Unfortunately, *Germalus pacificus* has found this tingid to its liking, and is controlling a beneficial insect, to the detriment of the program. Already mentioned is the similar attack by *Geocoris pallens* on the eggs of a weevil imported to control the puncture vine in California (U.S.A.).

5. PROGNOSIS AND CONCLUSIONS

It takes no crystal ball to realize that with increasing populations and the demand for a better lifestyle, there will be great pressure to increase the food supply. However, at the same time there is need to avoid the use of insecticides and other poisonous chemicals. It is therefore imperative to develop a sustainable agriculture compatible with these worthy goals. But can it be done? Thus far, nearly all the research is being done in North America. Even in India where efforts to develop IPM are strong, the prevailing control methods are largely heavy use of insecticides, many of which are banned for use in the United States. Pollution knows no boundaries as was learned from the DDT disaster, so it is of paramount importance that the United States, with its truly enormous pest control system, reach out to help fellow inhabitants of this planet. Further, the IPM programs should be affordable for the so-called developing parts of the world.

Next, on a practical note, there is too much duplication of effort to publish paper after paper documenting the presence of the predator or parasite. What can be done to develop sustainable biological control that does not require expensive mass-rearing facilities? Can it be done?

6. REFERENCES CITED

Anderson, C. A., and K. V. Yeargan. 1998. Influence of soybean canopy closure on predator abundances and predation on *Helicoverpa zea* (Lepidoptera: Noctuidae) eggs. Environ. Entomol. 27: 1488–1495.

Atim, A. B., and H. M. Graham. 1983. Parasites of *Geocoris* spp. near Tucson, Arizona. Southwest. Entomol. 8: 210–215.

Bell, K. D., Jr., and W. H. Whitcomb. 1962. Efficiency of egg predators of the bollworm. Arkansas Farm Res. 11: 9.

Bosch, R. van den, T. F. Leigh, D. Gonzales, and R. E. Stinner. 1969. Cage studies on predators of bollworm in cotton. J. Econ. Entomol. 62: 1486–1489.

Braman, S. K., K. E. Godfrey, and K. V. Yeargan. 1985. Rates of development of a Kentucky population of *Geocoris uliginosus*. J. Agric. Entomol. 2: 185–191.

Bugg, R. L., F. L. Wackers, K. E. Bronson, J. D. Dutcher, and S. C. Phatak. 1991. Cool-season cover crops relay intercropped with cantaloupe; influence on a generalist predator, *Geocoris punctipes*. J. Econ. Entomol. 84: 408–416.

Bugg, R. L., L. E. Ehler, and L. T. Wilson. 1987. Effect of common knotweed (*Polygonum aviculare*) on abundance and efficiency of insect predators of crop pests. Hilgardia 55: 1–52.

Butler, G. D., Jr. 1966. Development of several predaceous Hemiptera in relation to temperature. J. Econ. Entomol. 59: 1306–1307.

Cave, R. D., and M. J. Gaylor. 1988a. Influence of temperature and humidity on development and survival of *Telenomus reynoldsi* parasitizing *Geocoris punctipes* eggs. Ann. Entomol. Soc. Amer. 81: 278–285.

1988b. Parasitism of *Geocoris* eggs by *Telenomus reynoldsi* and *Trichogramma pretiosum* in Alabama. Environ. Entomol. 17: 945–951.

1989. Longevity fertility and population growth statistics of *Telenomus reynoldsi*. Proc. Entomol. Soc. Washington 91: 588–593.

Cave, R. D., M. J. Gaylor, and J. T. Bradley. 1987. Host handling and recognition by *Telenomus reynoldsi*, an egg parasitoid of *Geocoris* spp. Ann. Entomol. Soc. Amer. 80: 217–223.

Champlain, R. A., and L. L. Sholdt. 1966. Rearing *Geocoris punctipes*, a *Lygus* bug predator, in the laboratory. J. Econ. Entomol. 59: 1301.

1967a. Life history of *Geocoris punctipes* in the laboratory. Ann. Entomol. Soc. Amer. 60: 881–883.

1967b. Temperature range for development of immature stages of *Geocoris punctipes*. Ann. Entomol. Soc. Amer. 60: 883–885.

Chiravathanapong, S. N., and H. N. Pitre. 1980. Effects of *Heliothis virescens* larval size on predation by *Geocoris punctipes*. Florida Entomol. 63: 146–151.

Chow, T., G. E. Long, and G. Tamaki. 1983. Effects of temperature and hunger on the functional response of *Geocoris bullatus* to *Lygus* spp. Environ. Entomol. 12: 1332–1338.

Clancy, C. A., and H. D. Pierce. 1966. Natural enemies of some *Lygus* bugs. J. Econ. Entomol. 59: 853–858.

Cohen, A. C. 1981. An artificial diet for *Geocoris punctipes*. Southwest. Entomol. 6: 109–113.

1982. Water and temperature relations of two hemipteran members of a predatory–prey complex. Environ. Entomol. 11: 715–719.

1983. Improved method of encapsulating artificial diet for rearing predators of harmful insects. J. Econ. Entomol. 76: 957–959.

1984. Food consumption, food utilization and metabolic rates of *Geocoris punctipes* fed *Heliothis virescens* eggs. Entomophaga 29: 361–368.

1985a. Simple method for rearing the insect predator *Geocoris punctipes* on a meat diet. J. Econ. Entomol. 78: 1173–1175.

1985b. Metabolic rates of two hemipteran members of a predator prey complex. Comp. Biochem. Physiol. (A) 81: 833–836.

1989. Ingestion efficiency and protein consumption by a heteropteran predator. Ann. Entomol. Soc. Amer. 82: 495–499.

1990. Feeding adaptations of some predaceous Hemiptera. Ann. Entomol. Soc. Amer. 83: 1215–1223.

Cohen, A. C., and D. N. Byrne. 1992. *Geocoris punctipes* as a predator of *Bemisia tabaci*: a laboratory evaluation. Entomol. Exp. Appl. 64: 195–202.

Cohen, A. C., and J. W. Debolt. 1983. Rearing *Geocoris punctipes* on insect eggs. Southwest. Entomol. 8: 61–64.

Cohen, A. C., and C. G. Jackson. 1989. Using rubidium to mark a predator, *Geocoris punctipes*. J. Entomol. Sci. 24: 57–61.

Cohen, A. C., and N. M. Urias. 1986. Meat-based artificial diets for *Geocoris punctipes*. Southwest. Entomol. 11: 171–176.

1988. Food utilization and egestion rates of the predator *Geocoris punctipes* fed artificial diets with rutin. J. Entomol. Sci. 23: 174–179.

Coll, M., and J. R. Ruberson [eds.] 1998. Predatory Heteroptera: Their Ecology and Use in Biological Control. Thomas Say Publ. Entomol.: Proceedings Entomological Society of America, Lanham, Maryland, U.S.A.

Crocker, R. L., and W. H. Whitcomb. 1980. Feeding niches of the big-eyed bugs *Geocoris bullatus*, *G. punctipes*, and *G. uliginosus*. Environ. Entomol. 9: 508–513.

Crocker, R. L., W. H. Whitcomb, and R. M. Ray. 1975. Effects of sex, developmental stage, and temperature on predation by *Geocoris punctipes*. Environ. Entomol. 4: 531–534.

Dinkins, R. L., J. R. Brazzel, and C. A. Wilson. 1970a. Seasonal incidence of major predaceous arthropods in Mississippi cotton fields. J. Econ. Entomol. 63: 814–817.

1970b. Species and relative abundance of *Chysopa, Geocoris,* and *Nabis* in Mississippi cotton fields. J. Econ. Entomol. 63: 660–661.
Dunbar, D. M. 1971. Big-eyed bugs in Connecticut lawns. Connecticut Agric. Exp. Stn. Circ. 244: 6 pp.
1972. Notes on the mating behavior of *Geocoris punctipes*. Ann. Entomol. Soc. Amer. 65: 764–765.
Dunbar, D. M., and O. G. Bacon. 1972a. Feeding, development, and reproduction of *Geocoris punctipes* on eight diets. Ann. Entomol. Soc. Amer. 65: 892–895.
1972b. Influence of temperature on development and reproduction of *Geocoris tricolor, G. pallens,* and *G. punctipes* from California. Environ. Entomol. 1: 596–599.
Ehler, I. E. 1977. Natural enemies of cabbage looper on cotton in the San Joaquin Valley. Hilgardia 45: 73–106.
Ehler, L. E., K. G. Eveleens, and R. Van den Bosch. 1973. An evaluation of some natural enemies of cabbage looper on cotton in California. Environ. Entomol. 26: 1009–1015.
Elsey, K. D. 1972. Predation of eggs of *Heliothis* spp. on tobacco. Environ. Entomol. 1: 433–438.
Ferguson, H. J., and R. M. McPherson. 1982. Effects of selected insecticides on soybean pest and predator species. Virginia J. Sci. 33: 72.
Fyfe, R. V. 1935. The lantana bug of Fiji. Agric. J. Fiji 8: 35–36.
1937. The lantana bug, *Teleonemia lantanae* Distant. J. Council Sci. Ind. Res. Australia 10: 181–186.
Godfrey, K. E., W. H. Whitcomb, and J. L. Stimac. 1989. Arthropod predators of velvetbean caterpillar, *Anticarsia gemmatalis* Hubner, eggs and larvae. Environ. Entomol. 18: 118–123.
Goeden, R. D., and D. W. Ricker. 1967. *Geocoris pallens* found to be predaceous on *Microlarinus* spp. introduced to California for the biological control of puncture vine, *Tribulus terrestris*. J. Econ. Entomol. 60: 725–729.
Gonzalez, D., and L. T. Wilson. 1982. A food web approach to economic thresholds a sequence of pests predaceous arthropods on California U.S.A. cotton. Entomophaga 27: 31–44.
Gordh, G., and R. A. Coker. 1973. A new species of *Telenomus* parasitic on *Geocoris* (Hymenoptera: Proctotrupoidea: Hemiptera: Lygaeidae) in California. Canad. Entomol. 105: 1407–1411.
Greene, G. L., W. H. Whitcomb, and R. Baker. 1974. Minimum rates of insecticide on soybeans. *Geocoris* and *Nabis* populations following treatment. Canad. Entomol. 105: 1407–1411.
Greene, R. G., and W. J. Sterling. 1988. Quantitative phosphorus-32 labeling method for analysis of predators of the cotton fleahopper. J. Econ. Entomol. 81: 1494–1498.
Guillebeau, L. P., and J. N. All. 1989. *Geocoris* spp. and the striped lynx spider cross predation and prey preferences. J. Econ. Entomol. 82: 1106–1110.
1990. Big-eyed bugs and the striped lynx spider: intraspecific and interspecific interference on predation of first instar corn earworm. J. Econ. Sci. 25: 30–33.
Hagen, K. S., N. J. Mills, G. Gordh, and J. A. McMurty. 1999. Terrestrial arthropod predators of insect and mite pests. Pp. 383–503 *in* T. S. Bellows and T. W. Fisher [eds.], Handbook of Biological Control: Principles and Applications. Academic Press, San Diego, California, U.S.A. 1046 pp.
Henry, T. J. 1997. Phylogenetic analysis of family groups within the infraorder Pentatomomorpha (Hemiptera: Heteroptera), with emphasis on the Lygaeoidea. Ann. Entomol. Soc. Amer. 90: 275–301.
Irwin, M. E., and M. Shepard. 1980. Sampling predaceous Hemiptera on soybean. Pp. 505–531 *in* M. Kogan and D. C. Herzog [eds.], Sampling Methods in Soybean Entomology. Springer-Verlag, New York, New York, U.S.A. 587 pp.
Irwin, M. E., R. W. Gill, and D. Gonzalez. 1974. Field cage studies of native egg predators of the pink bollworm in southern California cotton. J. Econ. Entomol. 67: 193–196.
King, W. V., and W. S. Cook. 1932. Feeding punctures of mirids and other plant-sucking insects and their effect on cotton. Tech. Bull. U.S. Dept. Agric. 296. 11 pp.
Kumar, N. S., and T. N. Ananthakrishnan. 1985. *Geocoris ochropterus* as a predator of some thrips. Proc. Indian Natl. Sci. Acad. B. 51: 185–193.
Lawrence, R. K., and T. F. Watson. 1979. Predator–prey relationship of *Geocoris punctipes* and *Heliothis virescens*. Environ. Entomol. 8: 245–248.
Leigh, T. F., and D. Gonzalez. 1976. Field cage evaluation of predators for control of *Lygus hesperus* on cotton. Environ. Entomol. 5: 948–952.
Leigh, T. F., J. H. Black, E. C. Jackson, and V. E. Burton. 1966. Insecticides and beneficial insects in cotton fields. California Agric. 20: 4–6.
Lingren, P. D., and R. L. Ridgway. 1967. Toxicity of five insecticides to several insect predators. J. Econ. Entomol. 60: 1639–1641.

Lingren, P. D., R. L. Ridgway, and S. L. Jones. 1968a. Consumption by several arthropods of eggs and larvae of two *Heliothis* species that attack cotton. Ann. Entomol. Soc. Amer. 61: 613–618.

Lingren, P. D., R. L. Ridgway, C. B. Cowan, Jr., J. W. Davis, and W. C. Watkins. 1968b. Biological control of bollworm and tobacco budworm by arthropod predators affected by insecticides. J. Econ. Entomol. 61: 1521–1525.

McGregor, E. A., and F. L. McDonough 1917. The red spider on cotton. U.S. Dept. Agric. Bull. 416. 72 pp.

McPherson, J. E., J. C. Smith, and W. A. Allen. 1982. Incidence of arthropod predators in different soybean cropping systems. Environ. Entomol. 11: 685–689.

Mead, F. W. 1972. Key to the species of bigeyed bugs. *Geocoris* spp. in Florida. Entomol. Circ. No. 121. Florida Dept. Agric. Consumer Serv. 2 pp.

Mukhopadhyay, A. 1986 (1985). Biological notes on two species of big-eyed bugs. (Insecta: Hemiptera: Lygaeidae: Geocorinae). J. Bombay Nat. Hist. Soc. 85: 298–310.

1997. Crop association of a geocorine predator (Insecta: Hemiptera) in India and its biocontrol potential. Proc. Zool. Soc. Calcutta 50: 12–18.

Mukhopadhyay, A., and L. K. Ghosh. 1982. Two new species of *Geocoris* Fallen (Heteroptera, Lygaeidae) with some notes on their food habits and habitats. Kontyu 50: 169–174.

Mukhopadhyay, A., D. C. Deb, S. Dey, and S. S. Singha. 1996. Potential of geocorid predators (Lygaeidae: Insecta) as biocontrol agents in India: an overview. IPM & Sustain. Agric. Entomol. Appr. 6: 115–118.

Naranjo, S. E., and R. L. Gibson. 1996. Phytophagy in predaceous Heteroptera: effects on life history and population dynamics. Pp. 57–93 *in* O. Alomar and R. N. Wiedenmann [eds.], Zoophytophagous Heteroptera: Implications for Life History and Integrated Pest Management. Thomas Say Publ. Entomol.: Proceedings Entomological Society of America, Lanham, Maryland, U.S.A. 202 pp.

Naranjo, S. E., and J. L. Stimac. 1985. Development, survival, and reproduction of *Geocoris punctipes* (Hemiptera: Lygaeidae). Effects of plant feeding on soybean and associated weeds. Environ. Entomol. 14: 523–530.

Oscar, G. J. de Lima, and T. F. Leigh. 1984. Effect of cotton genotypes on the western big eyed bug (Heteroptera: Miridae [sic]) *Geocoris pallens*. J. Econ. Entomol. 77: 898–902.

Pfannenstiel, R. S., and K. V. Yeargan. 1998. Association of predaceous Hemiptera with selected crops. Environ. Entomol. 27: 32–239.

Readio, J., and M. H. Sweet. 1982. A review of the Geocorinae of the United States east of the 100th meridian (Hemiptera: Lygaeidae). Misc. Publ. Entomol. Soc. Amer. 12: 1–91.

Richman, D. B., R. C. Hemenway, Jr., and W. H. Whitcomb. 1980. Field cage evaluation of predators of the soybean looper, *Pseudoplusia includens* (Lepidoptera: Noctuidae). Environ. Entomol. 9: 315–317.

Ridgway, R. L., P. D. Lingren, C. B. Cowan, Jr., and J. W. Davis. 1967. Populations of arthropod predators and *Heliothis* spp. after applications of systemic insecticides to cotton. J. Econ. Entomol. 60: 1012–1016.

Ridgway, R. L., and S. L. Jones. 1968. Plant feeding by *Geocoris pallens* and *Nabis americoferus*. Ann. Entomol. Soc. Amer. 61: 232–233.

Rodgers, D. J., and M. J. Sullivan. 1986. Nymphal performance of *Geocoris punctipes* (Hemiptera: Heteroptera: Lygaeidae) on attached and detached leaves of pest-resistant soybeans. J. Entomol. Sci. 22: 282–285.

Schuster, M. F., M. J. Lukefahr, and F. G. Maxwell. 1976. Impact of nectariless cotton on plant bugs and natural enemies. J. Econ. Entomol. 69: 400–402.

Slater, J. A., and J. E. O'Donnell. 1995. A Catalogue of the Lygacidac of the World (1960–1994). New York Entomological Society, New York, New York, U.S.A. 411 pp.

Stoner, A. 1970. Plant feeding by a predaceous insect. *Geocoris punctipes*. J. Econ. Entomol. 63: 1911–1915.

Sweet, M. H. 1960. The seed bugs: A contribution to the feeding habits of the Lygaeidae. Ann. Entomol. Soc. Amer. 53: 317–321.

Tamaki, G. 1972. The biology of *Geocoris bullatus* inhabiting orchard floors and its impact on *Myzus persicae* on peaches. Environ. Entomol. 1: 559–565.

Tamaki, G., and R. E. Weeks. 1972. Biology and ecology of two predators, *Geocoris pallens* Stål and *G. bullatus* (Say). U.S. Dept. Agric. Tech. Bull. 1446: 46 pp.

1973. The impact of predators on populations of green peach aphids on field-grown sugarbeets. Environ. Entomol. 2: 345–349.

Torre-Bueno, J. R. de la. 1946. A synopsis of the Hemiptera-Heteroptera of America north of Mexico. III. Family XI. Lygaeidae. Entomol. Amer. 26: 1–141.

Waddill, V. H., and B. M. Shepard. 1974. Potential of *Geocoris punctipes* and *Nabis* spp. as predators of *Epilachna varivestis.* Entomophaga 19: 421–426.

Watve, C. M., and D. F. Clower. 1976. Natural enemies of the banded wing whitefly in Louisiana. Environ. Entomol. 5: 1075–1078.

Wilson, L. T., and A. P. Gutierrez. 1980. Within plant distribution of predators on cotton, *Gossypium hirsutum:* comments on sampling and predator efficiencies. Hilgardia 48: 3–11.

Yeargan, K. V. 1998. Influence of soybean canopy closure on predator abundances and predation on *Helicoverpa zea* (Lepidoptera: Noctuidae) eggs. Environ. Entomol. 27: 1488–1495.

Yokoyama, V. Y. 1978. Relation of seasonal changes in extrafloral nectar and foliar protein and arthropod populations in cotton. Environ. Entomol. 7: 799–802.

1980. Method for rearing *Geocoris pallens* a predator in California cotton. Canad. Entomol. 112: 1–3.

Yokoyama, V. Y., and J. Pritchard. 1984. Effect of pesticides on mortality, fecundity and egg viability of *Geocoris pallens.* J. Econ. Entomol. 77: 876–879.

Yokoyama, V. Y., J. Pritchard, and R. V. Dowell. 1984. Laboratory toxicity of pesticides to *Geocoris pallens* a predator in California cotton. J. Econ. Entomol. 77: 10–15.

York, G. T. 1944. Food studies of *Geocoris* spp., predators of the beet leaf hopper. J. Econ. Entomol. 37: 25–29.

Zaitzeva, I. F. 1974. Predaceous species of heteropteran insects (Hemiptera-Heteroptera) of Georgia. Mater. Faune Gruzii 4: 73–88.

CHAPTER 31

Stilt Bugs (Berytidae)

Thomas J. Henry

1. INTRODUCTION

The family Berytidae is a small, morphologically diverse group of Heteroptera belonging to the infraorder Pentatomomorpha and superfamily Lygaeoidea (Henry 1997b). It is separated into three subfamilies (and six tribes), the Berytinae (Berytini and Berytinini), Gampsocorinae (Gampsocorini and Hoplinini), and Metacanthinae (Metacanthini and Metatropini) (Henry 1997a); 35 genera and 170 species are recognized for the world (Henry and Froeschner 1998, 2000).

Péricart (1984) treated seven genera and 31 species for the Mediterranean region and Henry (1997c) revised the New World fauna, including 13 genera and 52 species. Other useful taxonomic works include Gross' (1950) treatment of the Australian fauna, Kerzhner's (1964) keys to the former European U.S.S.R., a review of the Chinese berytids by Hsiao et al. (1977), Froeschner's (1981, 1985) synopsis of the fauna of Ecuador and the Galápagos Islands, Kanyukova's (1988) keys to the former Far-Eastern U.S.S.R., and Froeschner's (1999) synoptic catalog for Panama. Henry (1997a) provided a phylogenetic analysis of the family and revised and gave keys to the genera of the world, and Henry and Froeschner (1998, 2000) cataloged the family for the world.

The long legs and antennae of many species have earned berytids the common name stilt bugs. Stilt bugs range in size from the smallest, such as the New World species *Hoplinus scutellaris* Henry and *Pronotacantha stusaki* Henry, measuring only 2.30 to 2.50 mm long (Henry 1997c), to the largest, *Plyapomus longus* Štusák from St. Helena, up to 16 mm long (Štusák 1976). Most berytids have long, slender bodies, in addition to their elongate appendages, but many deviate from this general pattern, particularly some of the New World taxa (Henry 1997c). For example, species of the genus *Hoplinus* have spindle-shaped bodies, relatively short legs, and the head, pronotum, and often the veins of the hemelytra are armed with rows of stout spines. Members of the genus *Pronotacantha* are armed with three rows of long, erect spines on the pronotum, whereas species of *Parajalysus* have only three stout pronotal spines, one at the middle of the anterior lobe and one at each posterior angle of hind lobe. The Afrotropical *Dimophoberytus variabilis* Štusák (1965), a large, robust species also possessing three stout spines on the pronotum, is unique in having a row of stout spines on the pro- and mesofemora of males.

Many stilt bugs have greatly developed ostiolar processes that may extend upward and well above the hemelytra (Schaefer 1972, Henry 1997a). The ostiolar process in the metacanthine genus *Jalysus* ends in a sharp spine, whereas these processes in *Metacanthus* and *Pneustocerus* extend outward, then curve posteriorly before ending in a rounded apex. The ostiole in Berytini, however, is much more abbreviated and lacks the extended processes, having only flared scent channels;

species of the berytinine genus *Berytinus* have an even more extreme reduction and lack all signs of a channel. The scent channels of Gampsocorini are similar to those of the Berytini, but are lined with overlapping scalelike plates, and hoplinines lack apparent channels and often have reduced evaporative or mycoid areas, usually with a tubercle arising in the middle.

Henry (1997c) found that all Berytidae, except members of the genus *Berytinus*, have cleft or dentate claws. This character, considered a synapomorphy for the family, may have allowed stilt bugs to radiate to glandular-hairy or viscid host plants not available to most other Heteroptera. Southwood (1986) observed a similar claw structure in plant bugs of the tribe Dicyphini. He observed that *Dicyphus errans* (Wolff) actually is able to grip the trichomes of sticky or glandular plants, using the pretarsal cleft for support, and noted that the relatively long legs of this bug provide added leverage if it becomes entangled, allowing it to rotate the claws to escape. He also speculated that stilt bugs, with longer legs, would have even more leverage on sticky surfaces than *D. errans*.

Considerable attention also has been given to the chromosome structure in Berytidae (e.g., Southwood and Leston 1959, Ueshima 1979, Grozeva 1995). Most species in the three subfamilies Berytinae, Gampsocorinae, and Metacanthinae have a $2n = 14 + XY$ chromosome complement. Higher numbers, however, have been reported in the genera *Metatropis* and *Berytinus*. Nokkala and Grozeva (1997) restudied *M. rufescens* (Herrich-Schaeffer) and *B. minor* (Herrich-Schaeffer) and confirmed that these taxa have higher chromosome numbers than other Berytidae, observations that are consistent with Henry's (1997a) establishment of a new tribe for each of these two genera.

2. GENERAL STATEMENT OF ECONOMIC IMPORTANCE

Despite the small size of the family and the abundance of certain species, the habits and hosts of most are poorly known. Wheeler and Schaefer (1982) provided a world review of the host plants and noted that species of *Berytinus* tend to feed low on their favored fabaceous hosts, whereas gampsocorines, metacanthines, and most Berytini live primarily on glandular-hairy or viscid plants, particularly those in the families Cucurbitaceae, Geraniaceae, Lamiaceae, Malvaceae, Onagraceae, Scrophulariaceae, and Solanaceae. Some species of *Gampsocoris*, *Jalysus*, and *Metacanthus* also feed on certain pubescent grasses such as those found in the genera *Agrostis* and *Panicum* (Wheeler and Henry 1981, Henry 1997c). More recently, Henry and Froeschner (1998) provided an expanded list of nearly 300 hosts.

Compared with some other heteropteran families, such as the Miridae or Tingidae, relatively few stilt bugs have been implicated as pests and most accounts of them are somewhat anecdotal. In North America, only *J. wickhami* Van Duzee has been considered an important pest, primarily of commercial tomatoes, *Lycopersicon esculentum* Mill. Other species, such as *J. spinosus* (Say), have been observed feeding on gourds (Froeschner 1942), but little other injury has been documented. In Peru, Wille (1944) reported *Parajalysus spinosus* Distant causing serious damage to cacao, *Theobroma cacao* L., by piercing and sucking young shoots, twigs, and fruits in a fashion similar to that inflicted by the plant bug *Monalonion dissimulatum* Distant (Miridae: Bryocorinae); but later he considered it an important predator of *Heliothis virescens* (Wille 1951).

Perhaps even more important, berytids have strong predatory tendencies, making them potentially valuable in biological control. An increasing number of studies show that berytids previously thought to be entirely phytophagous supplement their diets with animal food, either as active predators (e.g., Hickman 1976) or as scavengers (e.g., Wheeler and Henry 1981). The preference of most stilt bugs for glandular plants seems to enhance their chances of scavenging prey entrapped in the sticky hairs of their various hosts (Henry 1997c). Inadvertent probing by stilt bugs, such as the Nearctic *Pronotacantha annulata* Uhler, may even occasionally inflict painful bites on humans (Henry 1997c).

The role stilt bugs play in agricultural ecosystems is more complex than once thought. Species once considered entirely phytophagous are now known to be strongly predatory or zoophagous.

Also, based on reports of several Neotropical species, a number of stilt bugs may play significant roles in plant pollination (Pruett 1996, Henry 1997c). Although little attention has been given to the majority of stilt bugs, their multiple feeding habits, particularly the predatory tendencies, warrant more attention to this small family of interesting and potentially important bugs.

3. THE MOST IMPORTANT SPECIES

3.1 *Jalysus wickhami* Van Duzee

3.1.1 *Distribution, Life History, and Biology*

Jalysus wickhami, recorded from every continental U.S. state (except Alaska, New Hampshire, and Vermont), ranges across southern Canada from British Columbia to Quebec, and south into Mexico as far as Hidalgo (Froeschner and Henry 1988, Scudder 1991, Henry 1997c, Henry and Froeschner 1998). Long confused with *J. spinosus* and thought to be confined to the western United States (McAtee 1919), *J. wickhami* was first reported in the eastern United States by Harris (1941). Slater and Baranowski (1978) provided the first accurate clarification of distribution and gave characters to separate *J. wickhami* from *J. spinosus*. Wheeler and Henry (1981) furnished more extensive details on distribution; illustrated the head, genital capsule, parameres, and nymphal characters; and clarified misidentifications reported in the literature, including important studies by Elsey (e.g., 1971, 1972a,b, 1973b) and Elsey and Stinner (1971). Unfortunately, Hagen et al. (1999), in a chapter reviewing terrestrial arthropod predators, used the name *J. spinosus* rather than the correct *J. wickhami* in their overview of Elsey and Stinner's work, even though they cited Froeschner and Henry's (1988) catalog that reflects the correct name. All the pre-1981 literature cited below refers to *J. wickhami*, as clarified by Wheeler and Henry (1981).

This species has the greatest host range of any known stilt bug. Wheeler and Henry (1981) listed 56 plants in 16 families as hosts of *J. wickhami*, most being characterized as glandular-hairy or viscid herbs, particularly those in the families Malvaceae, Onagraceae, Oxalidaceae, Scrophulariaceae, and Solanaceae. Common hosts in the western United States are species of *Eriogonum* (Polygonaceae) and *Sphaeralcea* (Malvaceae) (Beck and Allred 1966, Wheeler and Henry 1981). Other common hosts in the midwestern and eastern United States include *Gaura biennis* L. (Onagraceae) and *Verbascum thapsus* L. (Scrophulariaceae), as well as tobacco, *Nicotiana tabacum* L., and tomato (Wheeler and Henry 1981).

Readio (1923) described and illustrated the eggs, five instars, and adult and provided life-history notes for Kansas (U.S.A.). He observed that adults and nymphs remain on their hosts as late as early November and that adults overwinter in protected places, such as under leaves of common mullein, *V. thapsus*. Eggs are laid on the host plant either singly or in small clusters of two or more, and the life cycle is completed within 33 days from adult to adult. Phipps (1924) found that *J. wickhami* took 35 days to develop from egg to adult in Missouri (U.S.A.) and that three or four generations occur each year. Mating occurs tail to tail and can last from 30 to 180 min (Faust and Harrison 1968).

Although most early workers considered *J. wickhami* strictly phytophagous, its predatory tendencies now have been well documented (e.g., Lawson 1959; Elsey 1971, 1972a; Elsey and Stinner 1971). Gilmore (1938) gave the first report of *J. wickhami* preying on tobacco hornworm, *Manduca sexta* (L.), eggs and Kulash (1949) recorded it preying on tobacco aphids. Lawson (1959) noted that *J. wickhami* was a general predator that had been observed feeding on aphids, newly emerged larvae of *Apanteles* parasitoids, and on dead insects stuck to the hairs of tobacco leaves. *Jalysus wickhami* has long been considered the principal predator of the tobacco budworm, *Heliothis virescens* (F.), and tobacco hornworm eggs in North Carolina (Elsey and Lam 1978). Elsey and Stinner (1971) studied its biology in the laboratory and found that it developed from egg to adult in about 20 days at 25.5°C when provided hornworm eggs, in contrast to 35 days on only a diet of tomatoes (Phipps 1924). The duration of the stadia, excluding I, were 2 to 5 days for stadium II, 2 to 6 days for III, 2 to 5 days for IV, and 5 to 6 days for V. They also determined that *J. wickhami* requires animal food for optimal nymphal development, adult maintenance, and oogenesis. Jackson and Kester (1996) showed that *J. wickhami* provided with protein-rich food sources, such as eggs and prepupae of the braconid *Cotesia congregata* (Say) and eggs of the tobacco hornworm, lived nearly twice as long as those having access only to tobacco leaves. Although the kind of prey did not affect longevity, females were significantly more fecund when fed hornworm eggs than when given tobacco aphids (*Myzus nicotianae* Blackman). Kester and Jackson (1996) showed that *J. wickhami* does not always distinguish between pests and beneficial insects and that it could cause significant mortality to the hornworm parasitoid *C. congregata* under laboratory conditions, the first documented case of direct intraguild predation involving a predator and parasitoid. Despite its importance as a predator, *J. wickhami* has a limited feeding capacity and can prey only on eggs, dead insects, or slow-moving insects such as aphids (Elsey 1972b).

3.1.2   Damage

Jalysus wickhami has been considered a serious pest of tomatoes, resulting in the common name "tomato stilt bug" by the U.S. Department of Agriculture (Wheeler and Henry 1981). Somes (1914) observed adults and nymphs feeding on tomato fruit stems, which were thought to cause the flowers to darken and abort. Somes (1916) also reported that such feeding prevented many tomatoes from setting fruit, and in some tomato-growing districts of Missouri (U.S.A.) this insect caused serious losses. Phipps (1924) also regarded *J. wickhami* a serious tomato pest causing the "blossoms to wither up and drop off, preventing the plants from setting fruit." He also speculated that *J. wickhami* might vector certain tomato diseases such as "southern blight," but his suspicions have not been supported. Outbreaks of this species on tomato have been reported from Indiana, Kentucky, Maryland, Michigan, and Nebraska (U.S.A.) (Wheeler and Henry 1981). In 1995, Wheeler (personal communication) observed serious injury to both green and ripe tomatoes by large populations of *J. wickhami* in commercial polyhouses in Pennsylvania (U.S.A.). He characterized damage by both nymphs and adults as white blotches beneath the skin or halos on the fruit surface that limited the fruit's marketability.

3.1.3   Control

Because of its value as a predator on tobacco, few studies have been devoted to controlling *J. wickhami*, although increasing human health concerns over tobacco might well inspire renewed interest in the control of this beneficial insect. Elsey (1973a) recognized that various insecticides targeted at tobacco pests could adversely affect certain predators such as *J. wickhami*. He found that systemic insecticides applied to pretransplant soil lowered *J. wickhami* populations throughout the season, whereas the effects of foliar sprays varied from reducing populations briefly to having no effect at all. In a later study, Jackson and Lam (1989) evaluated the toxicity of 13 pesticides and found that the systemics aldicarb, carbofuran, and phenamiphos were particularly harmful to *J. wickhami*, but none of the others was toxic in field experiments.

3.2   *Metacanthus pulchellus* Dallas

3.2.1   Distribution, Life History, and Biology

This is a widespread species known from Australia, India, Indonesia, Japan, Korea, Malaya, New Guinea, the Philippines, and Sri Lanka (Henry and Froeschner 1998). It has a wide range of hosts, including plants in the genera *Cajanus, Datura, Dombeya, Hibiscus, Lagenaria, Lycopersicon,*

Nicotiana, Ononis, Passiflora, Paulownia, Solanum, and *Theobroma* (Wheeler and Schaefer 1982, Dhiman 1984b, Henry and Froeschner 1998).

Sharma (1983) and Dhiman (1992) provided accounts of the life history of *M. pulchellus*; Dhiman and Garg (1983) gave notes on overwintering habits and how temperature and rainfall affected populations at Saharanpur, India. Sharma (1983) noted a premating period of 4 to 7 days. Mating was observed both day and night. When allowed to mate several times, eggs laid per female ranged from 47 to 133 ($\bar{x} = 78.75$). Females allowed to mate only once laid from 4 to 13 eggs ($\bar{x} = 7.6$). The incubation period ranged from 3.2 to 4.8 days ($\bar{x} = 3.97$). There are five instars, the length of stadium I averaging 2.20 days; II, 2.25 days; III, 2.80 days; IV, 3.20 days; and V, 4.25 days. Under field conditions males lived on the average 29.55 days and females, 32.50 days. Eggs are deposited on various parts of the host plant, including the upper and lower surfaces of leaves, stems, flower buds, and immature fruits, but lower leaf surfaces are preferred. Eggs are cylindrical to oval, with an average length of 0.628 ± 0.21 mm and average width of 0.303 ± 0.013 mm.

3.2.2   Damage

Most of the more detailed studies pertaining to this species concern its impact as a pest of bottle gourd, *Lagenaria siceraria* (Molina) Standl., on which it causes stunting and yellow patches on the leaves (Sinha et al. 1981, Dhiman 1992). Singh and Patel (1973) observed *M. pulchellus* causing extensive damage to bottle gourd during spring and summer in Hissar, India. Dhiman (1984a) showed that *M. pulchellus* preferred feeding on bottle gourd over pumpkin (*Cucurbita maxima* Dusch.), sponge gourd (*Luffa cylindrica* L.), and bitter gourd (*Momordica charantia* L.). Dhiman (1984b) reported this species damaging the blossoms of ornamental *Dombeya* sp. (Sterculiaceae) from June to October in the Saharanpur District of India.

3.2.3   Control

Sinha et al. (1981) performed a series of field trials to evaluate the efficacy of certain insecticides against *M. pulchellus*. In these trials, ten insecticides were tested, including carbaryl (Sevin), endosulfan (Thiodan), malathion (Cythion), methylodemeton (Metasystox), dimethoate (Rogor), bendiocarb (Garvox), and fenitrothion (Accothion). In the second trial, only bendicarb and acephate (Orthene) were tried. Bendicarb consistently gave protection from bug infestations. Most other treatments were ineffective, but carbaryl and methylodemeton gave some control. Sharma (1983) also screened a number of insecticides and found that foliar sprays of cypermethrin, fenthion, malathion, methomyl, monocrotophos, or oxydemetonmethyl would effectively control stilt bugs on bottle gourd.

4. THE LESS IMPORTANT SPECIES

4.1   *Chinoneides tasmaniensis* (Gross)

Chinoneides tasmaniensis is known only from New South Wales and Tasmania, Australia (Henry 1997b). This species was originally described in *Neides* (Gross 1950) and later transferred to *Chinoneides* Štusák (Henry 1997a), a genus characterized by the weakly flared, slender ostiolar scent channel, micropterous hemelytra, widely set ocelli, and Y-shaped groove on the scutellum. Hickman (1976) reported fully macropterous individuals from Tasmania.

Hickman (1976) studied the life history in Tasmania on geranium (*Geranium* sp.: Geraniaceae) and pelargonium (*Pelargonium* sp.: Geraniaceae) and illustrated the egg and five instars. Eggs were deposited singly among the hairs on young stems and between veins on the undersides of leaves.

Incubation lasted 27 to 31 days in August and September but only 11 to 13 days in December. Adults and nymphs were observed throughout the year, but adults were most abundant in November. The bug is found in dry leaves in and around the bases of the host during the cold months. Mating occurred from late July to the end of January, but was most frequent during November. Tail-to-tail pairing lasted from 4 to 6 hours. The duration of stadium I was 8 to 14 days; II, 11 to 13 days; III, 5 to 16 days; IV, 5 to 38 days; and V, 9 to 22 days.

Nymphs and adults feed on both plant juices and on living aphids. One fifth instar fed on a young *Pelargonium* leaf for 54 min, leaving its abdomen noticeably distended. Hickman (1976), however, found that it was also necessary to provide *C. tasmaniensis* with aphids, as well as with fresh plant material, to be successful in rearing it to adulthood. The stilt bug fed not only on the aphid *Acyrthosiphon pelargonii* (Kaltenbach), but also on *Macrosiphum rosae* (L.) from rose and on aphids from other plants. In one observation, a late instar fed on a rose aphid for 114 min.

4.2   *Jalysus spinosus* (Say)

Jalysus spinosus is common over much of eastern North America, west to about the 100th meridian. Nearly all earlier literature referring to this species as a pest of tomato or as an important predator of tobacco pests pertains to the more polyphagous *J. wickhami* (Wheeler and Henry 1981). Little information is available on the life history of *J. spinosus*, but it is known to overwinter in the adult stage (Wheeler and Stimmel 1988). Wheeler and Henry (1981) illustrated the fifth instar and gave information on how to separate it from the closely related *J. wickhami*. Although this species breeds primarily on species of *Panicum* (Poaceae) and less commonly on enchanter's nightshade, *Circaea quadrisulcata* (Maxim.) Franch. and Sav. (Onagraceae) (Wheeler 1986), and *Tradescantia hirsuticaulis* Small (Commelinaceae) (Wheeler 1994), it has been recorded from other hosts. Wheeler and Henry (1981) observed it on inflorescences of *Solidago canadensis* L. (Asteraceae) and noted records from apple blossoms and flowers of *Euphorbia commutata* Engelm. (Euphorbiaceae). J. A. Slater (cited in Wheeler and Henry 1981) found nymphs and adults on ornamental gourds (*Lagenaria* sp.) in Iowa (U.S.A.), and Froeschner (1942) observed adults preying on aphids on gourds in Missouri (U.S.A.). Although primarily a monocot specialist, *J. spinosus* disperses to other hosts, particularly those in flower, and displays predatory tendencies typical of many members of the family, making it potentially important in agricultural and ornamental situations.

4.3   *Neoneides muticus* (Say)

Neoneides muticus (Say) is a widespread North American species known from all conterminous United States, across southern Canada, and into Mexico (Scudder 1991, Henry 1997c). Although placed in the European genus *Neides* Latreille for many years, Henry (1997a) gave generic status to *Neoneides* Štusák. This monotypic genus is characterized by the long, recurved spine on the frons, absence of a clypeal process, the punctate abdomen, tomentose pubescence, and shape of the male genital capsule.

Wheeler (1978) studied the seasonal history of *N. muticus* on common mullein, *Verbascum thapsus*, and illustrated the fifth instar. Overwintered adults became active on warm days in April at Ithaca, New York (U.S.A.), and mating and oviposition began by late April to early May. Eggs hatched in early May, but oviposition continued into June. In central Pennsylvania (U.S.A.) fourth and fifth instars were most common by June 10; the first adults appeared in late June to early July. The presence of early instars in mid-July seemed to indicate the beginning of a second generation. In New York, fifth instars were present into early October. Adults were active until subfreezing temperatures, at which time they sought shelter in mullein rosettes. Mating occurs tail to tail. Eggs are laid on the upper and lower surfaces of leaves. The duration of stadium I was 7 to 11 days; II, 6 to 9 days; III, 5 to 12 days; IV, 6 to 11 days; and V, 8 to 13 days.

This species has been recorded feeding on more than 20 plant species, including alfalfa (*Medicago sativa* L.), the hairy beardtongue *Penstemon hirsutus* (L.) Willd., moth mullein (*V. blattaria* L.), orange hawkweed (*Hieracium pratense* Tausch), hay-scented fern *Dennstaedtia punctilobula* (Michx.) Moore, and moss phlox (*Phlox subulata* L.) (Wheeler and Schaefer 1982, Wheeler 1997, Henry and Froeschner 1998). Although Wheeler (1978) reported *N. muticus* feeding on mullein leaves and flower buds, as well as taking water from leaf surfaces, he also observed it feeding on eggs and larvae of mullein thrips (*Haplothrips verbasci* Osborn). This is consistent with other observations, such as its feeding on the eggs of the pierid *Neophasia menapis* (Felder) (Hagen 1884), preying on the beet leafhopper *Circulifer tenellus* (Baker) (Knowlton and Harmston 1940), and serving as a minor predator of leafhoppers and other arthropods (Smith and Hagen 1956). Wheeler's (1978) mention that he was unable to rear newly hatched nymphs on hay-scented fern fronds suggests that prey are required for *N. muticus* to complete its life cycle, not too unlike what has been found for *Chinoneides tasmaniensis* (Hickman 1976).

4.4 *Parajalysus andinus* Horváth

Parajalysus andinus, a Neotropical species known from Bolivia, Brazil, Colombia, Peru, and Venezuela, is one of the most commonly collected species of *Parajalysus*, probably because of its abundance on the frequently studied and economically important host *Theobroma cacao* (Henry 1997c). Pruett (1996 and personal communication) observed large numbers *P. andinus* swarming about flowering cacao trees from September to December in Bolivia. He found significant amounts of cacao pollen on the bodies of both nymphs and adults and observed that peak populations of this species coincided with the peak bloom of cacao. Of 141 adults examined during one study, 78 had pollen on their bodies; of 187 nymphs observed, 89 had pollen. As a consequence, he considered *P. andinus* one of the most important pollinators of cacao, along with certain species of Cecidomyiidae, and speculated that cross-pollination by berytids contributed to increased yields in certain difficult-to-pollinate cultivars of cacao.

4.5 *Parajalysus spinosus* Distant

Parajalysus spinosus is widespread in the Neotropics, ranging from Mexico, south through Panama to Brazil, Colombia, Peru, and Venezuela (Henry 1997c). It has been taken on cacao and cassava, *Manihot tripartita* (Muell.) Arg. Wille (1944) reported damage similar to that caused by the mirid *Monalonion dissimulatus* Distant on cacao: "The affected parts are marked by pustules and black warty spots, which are especially noticeable on young green fruits, and they die and wilt."

In another paper, Wille (1951) considered *P. spinosus* the fourth most important predator of *Heliothis virescens* eggs on cotton in Peru. Such habits are in keeping with observations of *P. punctipes* Van Duzee probing cecidomyiid galls on *Machaerium aculeatum* Raddi (Fabaceae) in Brazil (Fernandes et al. 1987). Clearly, much more study is needed to determine fully the habits of *J. spinosus*. The lack of subsequent reports of *P. spinosus* damaging cacao suggests that Wille's (1944) observations were in error.

4.6 *Yemma exilis* Horváth

Yemma exilis, known only from Japan, has been implicated as a pest of soybeans, *Glycine max* (L.) Merr. (Kuwayama 1953, Kobayashi 1981). It also has been collected from kudzu (*Pueraria lobata*), paulownia (*Paulownia tomentosa* [Thumb.] Steud.), raspberry (*Rubus hirsutus*), and sesame (*Sesamum indicum* L.) (Tomokuni et al. 1993).

Kohno and Hirose (1997) have shown that *Y. exilis* may be an important predator. They found adults and nymphs preying on apterous viviparous females and nymphs of the cotton aphid *Aphis gossypii* Glover and adults and larvae of *Thrips palmi* Karny, as well as on lacewing eggs

(*Chrysoperla* sp.) on eggplant (*Solanum melongena* L.) in plastic greenhouses in Kurume, Japan. They conducted a series of laboratory feeding experiments with three replications (one with aphids and an eggplant leaf, one with thrips and an eggplant leaf, and one with eggplant only) to determine if *Y. exilis* is mainly predaceous. Their studies showed that *Y. exilis* survived significantly longer on an aphid/eggplant diet (up to 15 days) than it did on either thrips/eggplant (most to 7 days; only one to 15) or on eggplant alone (most only 6 days). As a result, they concluded that *Y. exilis* is an important predator, promising as a biological control agent of eggplant pests in Japan.

5. PROGNOSIS

Although several species can injure certain crops, the likelihood of additional stilt bugs becoming important as crop pests appears limited. What can be expected is an increased appreciation for the beneficial aspects of stilt bugs as more is learned about their biology and feeding behavior. As shown in this chapter, a growing number of berytids is being discovered to have predatory tendencies. Under certain conditions, stilt bugs may also come to play an even larger role in plant pollination. As more is learned about this small family of bugs, the more important they may become in biological control programs.

6. CONCLUSIONS

Only two stilt bugs are considered serious pests. A few others are credited with causing injury, but evidence supporting such claims is weak or anecdotal. On the other hand, a growing amount of evidence shows that stilt bugs can be important predators. The berytine *Chinoneides tasmaniensis* in Australia and the metacanthine *Y. exilis* in Japan well reflect the broad familial tendencies toward zoophagy. That even the Holarctic and mostly phytophagous *Berytinus minor* has been observed preying on aphids (Southwood and Leston 1959, Wheeler 1970) provides evidence that most or even nearly all stilt bugs exploit animal protein when given the opportunity.

That stilt bugs may be important plant pollinators is supported by Pruett's (1996) observations of *Parajalysus andinus* on cocoa in Bolivia. Observations of stilt bugs, such as *J. albidus* Štusák, visiting or swarming about the flowers of their hosts probably indicate at least a minor role in plant pollination (Henry 1997c).

Although two stilt bugs have attained some notoriety as pests, it is the predatory habits of berytids that make them potentially important in agriculture. Much more research is needed to better understand and evaluate the role of stilt bugs.

7. ACKNOWLEDGMENTS

I thank the editors, Carl W. Schaefer (Department of Ecology and Evolutionary Biology, University of Connecticut, Storrs, U.S.A.) and Antônio R. Panizzi (Centro Nacional de Pesquisa de Soja EMBRAPA, Londrina, PR, Brazil), for the invitation to write this chapter. I also thank Michele Touchet of the Systematic Entomology Laboratory (SEL), ARS, USDA, c/o National Museum of Natural History, Washington, D.C., U.S.A., and A. G. Wheeler, Jr. (Department of Entomology, Clemson University, Clemson, South Carolina, U.S.A.) for assistance in obtaining literature; and C. J. H. Pruett (Centro de Investigacion y Mejoramiento de la Caña de Azucar, Santa Cruz, Bolivia) for sharing his research notes on *P. andinus*. D. R. Smith (SEL), A. G. Wheeler, Jr., and R. A. Ochoa (SEL) kindly reviewed the manuscript.

8. REFERENCES CITED

Beck, D. E., and D. M. Allred. 1966. Tingidae, Neididae (Berytidae) and Pentatomidae of the Nevada test site. Great Basin Nat. 26: 9–16.
Dhiman, S. C. 1984a. Host specificity of *Metacanthus pulchellus* Dall. (Heteroptera: Berytidae). Uttar Pradesh J. Zool. 4(2): 205–207.
1984b. New record of the host plant of *Metacanthus pulchellus* Dall. and its seasonal occurrence. Geobios New Rep. 3: 65.
1992. Biology of *Metacanthus pulchellus* Dall. (Heteroptera: Berytidae) on bottle gourd. Bioecol. Control Insect Pests, pp. 50–59.
Dhiman, S. C., and G. D. Garg. 1983. Effect of some climatic factors on the occurrence of *Metacanthus pulchellus* Dall. (Heteroptera–Berytidae). Indian J. Ecol. 10: 90–93.
Elsey, K. D. 1971. Stilt bug predation on artificial infestation of tobacco hornworm eggs. J. Econ. Entomol. 64: 772–773.
1972a. Predation of eggs of *Heliothis* spp. on tobacco. Environ. Entomol. 1: 433–438.
1972b. Defenses of eggs of *Manduca sexta* against predation by *Jalysus spinosus*. Ann. Entomol. Soc. Amer. 65: 896–897.
1973a. *Jalysus spinosus*: effect of insecticide treatments on this predator of tobacco pests. Environ. Entomol. 2: 240–243.
1973b. *Jalysus spinosus*: spring biology and factors that influence occurrence of the predator on tobacco in North Carolina. Environ. Entomol. 2: 421–425.
Elsey, K. D., and J. J. Lam, Jr. 1978. *Jalysus spinosus*: instantaneous rate of population growth at different temperatures and factors influencing the success of storage. Ann. Entomol. Soc. Amer. 71: 322–324.
Elsey, K. D., and R. E. Stinner. 1971. Biology of *Jalysus spinosus*, an insect predator found on tobacco. Ann. Entomol. Soc. Amer. 64: 779–783.
Faust, R. M., and F. P. Harrison. 1968. The life history and habits of the stilt bug, *Jalysus spinosus*, in Maryland. J. Econ. Entomol. 61: 1110.
Fernandes, G. W., R. P. Martins, and E. T. Neto. 1987. Food web relationships involving *Anadiplosis* sp. galls (Diptera: Cecidomyiidae) on *Machaerium aculeatum* (Leguminosae). Rev. Bras. Bot. 10: 117–123.
Froeschner, R. C. 1942. Contributions to a synopsis of the Hemiptera of Missouri, part II. Coreidae, Aradidae, Neididae. Amer. Midl. Nat. 27: 591–609.
1981. Heteroptera or true bugs of Ecuador: a partial catalog. Smithsonian Contrib. Zool. 322: 1–147.
1985. A synopsis of the Heteroptera or true bugs of the Galapagos Islands. Smithsonian Contrib. Zool. 407: 1–84.
1999. True bugs (Heteroptera) of Panama: a synoptic catalog as a contribution to the study of Panamanian biodiversity. Mem. Amer. Entomol. Inst. 61: 1–393.
Froeschner, R. C., and T. J. Henry. 1988. Family Berytidae Fieber, 1851 (= Neididae Kirkaldy, 1902; Berytinidae Southwood and Leston, 1959). The stilt bugs. Pp. 56–60 *in* T. J. Henry and R. C. Froeschner [eds.], Catalog of the Heteroptera, or True Bugs, of Canada and the Continental United States. E. J. Brill, Leiden, the Netherlands. 958 pp.
Gilmore, J. U. 1938. Observations on the hornworms attacking tobacco in Tennessee and Kentucky. J. Econ. Entomol. 31: 706–712.
Gross, G. F. 1950. The stilt-bugs (Heteroptera-Neididae) of the Australian and New Zealand regions. Rec. S. Austral. Mus. 9(3): 313–326.
Grozeva, S. M. 1995. Karyotypes, male reproductive system and abdominal trichobothria of the Berytidae (Heteroptera) with phylogenetic considerations. Syst. Entomol. 20: 207–216.
Hagen, H. A. 1884. Enemies of *Pieris menapia*. Canad. Entomol. 16: 40.
Hagen, K. S., N. J. Mills, G. Gordh, and J. A. McMurtry. 1999. Terrestrial arthropod predators of insect and mite pests. Pp. 383–503 *in* T. S. Bellows and T. W. Fisher [eds.], Handbook of Biological Control: Principles and Applications of Biological Control. Academic Press, San Diego, California, U.S.A. 1046 pp.
Harris, H. M. 1941. Concerning Neididae, with new species and new records for North America. Bull. Brooklyn Entomol. Soc. 36: 105–109.

Henry, T. J. 1997a. Cladistic analysis and revision of the stilt bug genera of the world (Heteroptera: Berytidae). Contrib. Amer. Entomol. Inst. 30(1): 1–100.

1997b. Phylogenetic analysis of the family groups within the infraorder Pentatomomorpha (Hemiptera: Heteroptera), with emphasis on the Lygaeoidea. Ann. Entomol. Soc. Amer. 90: 257–301.

1997c. Monograph of the stilt bugs, or Berytidae (Heteroptera), of the Western Hemisphere. Mem. Entomol. Soc. Washington 19: 1–149.

Henry, T. J., and R. C. Froeschner. 1998. Catalog of the stilt bugs, or Berytidae, of the world (Insecta: Hemiptera: Heteroptera). Contrib. Amer. Entomol. Inst. 30(4): 1–72.

2000. Corrections and additions to the "Catalog of the stilt bugs, or Berytidae, of the world." Proc. Entomol. Soc. Washington 101: (in press).

Hickman, V. V. 1976. The biology of *Neides tasmaniensis* Gross (Hemiptera: Berytidae). J. Entomol. Soc. Aust. (N.S.W.) 9: 3–10.

Hsiao, T.-Y. et al. 1977. [Descriptions of new Berytidae by T.-Y. Hsiao]. A Handbook for the Determination of the Chinese Hemiptera-Heteroptera. Vol. 1. Science Press, Beijing. 330 pp. [in Chinese with English summary].

Jackson, D. M., and K. M. Kester. 1996. Effects of diet on longevity and fecundity of the spined stilt bug, *Jalysus wickhami*. Entomol. Exp. Appl. 80: 421–425.

Jackson, D. M., and J. J. Lam, Jr. 1989. *Jalysus wickhami* (Hemiptera: Berytidae): toxicity of pesticides applied to the soil or in the transplant water of flue-cured tobacco. J. Econ. Entomol. 82: 913–918.

Kanyukova, E. V. 1988. 26. Fam. Berytidae–stilt bugs. P. 882 *in* P. A. Ler [ed.], Key to the Insects of the Far-Eastern USSR. Acad. Sci., Far-Eastern Div., Leningrad, USSR Vol. 2. 972 pp. [in Russian].

Kerzhner, I. M. 1964. Order Hemiptera (Heteroptera) [Terrestrial families 13–27, pp. 684–845]. *In* G. Y. Bei-Bienko. Keys to the Insects of the European U.S.S.R. Vol. I. Apterygota, Palaeoptera, Hemimetabola. Zool. Inst., Acad. Sci. Leningrad, U.S.S.R. [Translated (1967) from Russian by Israel Program for Scientific Translations for the Smithsonian Institution and National Science Foundation, Washington, D.C.].

Kester, K. M., and D. M. Jackson. 1996. When good bugs go bad: intraguild predation by *Jalysus wickhami* on the parasitoid, *Cotesia congregata*. Entomol. Exp. Appl. 81: 271–276.

Knowlton, G. F., and F. C. Harmston. 1940. Utah insects. Hemiptera. Utah Agric. Exp. Stn. Mimeo Ser. 200 (Tech.). 6: 1–10.

Kobayashi, T. 1981. Misc. Publ. Tohoku Natl. Agric. Exp. Stn. 2: 1–39 [not seen, *fide* Kohno and Hirose 1997].

Kohno, K., and Y. Hirose. 1997. The stilt bug *Yemma exilis* (Heteroptera: Berytidae) as a predator of *Aphis gossypii* (Homoptera: Aphididae) and *Thrips palmi* (Thysanoptera: Thripidae) on eggplant. Appl. Entomol. Zool. 32(2): 406–409.

Kulash, E. M. 1949. The green peach aphid as a pest of tobacco. J. Econ. Entomol. 42: 677–680.

Kuwayama, S. 1953. Survey on the Fauna of Soybean Insect-Pests in Japan. Yokendo, Tokyo, Japan. 129 pp. [in Japanese; not seen, *fide* Kohno and Hirose 1997].

Lawson, F. R. 1959. The natural enemies of the hornworms on tobacco (Lepidoptera: Sphingidae). Ann. Entomol. Soc. Amer. 52: 741–755.

McAtee, W. L. 1919. Key to the Nearctic genera and species of Berytidae (Heteroptera). J. New York Entomol. Soc. 27: 79–92.

Nokkala, S., and S. M. Grozeva. 1997. Chromosomes in two stilt bug species *Metatropis rufescens* Herrich-Schaeffer ssp. *linneae* Wagner and *Berytinus minor* Herrich-Schaeffer (Berytidae, Heteroptera). Caryologia 50: 263–269.

Péricart, J. 1984. Hémiptères Berytidae Euro-Mediterranéens. Faune de France. France et Régions Limitrophes. Fédération Française des Société de Sciences Naturelles, Paris, France. 165 pp.

Phipps, C. R. 1924. A stilt-bug, *Jalysus spinosus* Say, destructive to the tomato. J. Econ. Entomol. 17: 390–393.

Pruett, C. J. H. 1996. Cacao, *Theobroma cacao* Linnaeus (Sterculiaceae): estudio sobre insectos polinizadores, Santa Cruz, Boliga, 1989–1991. IV Congreso Nacional de Biologia, Santa Cruz de Sierra, Bolivia. Facultad de Ciencias Agricolas de la Universidad Autonoma "Gabriel Rene Moreno," Santa Cruz, Bolivia. 36 pp.

Readio, P. A. 1923. The life history of *Jalysus spinosus* (Say) (Neididae, Heteroptera). Canad. Entomol. 55: 230–236.

Schaefer, C. W. 1972. Degree of metathoracic scent-gland development in the trichophorous Heteroptera (Hemiptera). Ann. Entomol. Soc. Amer. 65: 810–821.

Scudder, G. G. E. 1991. The stilt bugs (Heteroptera: Berytidae) of Canada. Canad. Entomol. 123: 425–438.
Sharma, M. L. 1983. Population build up and chemical control of the stilt bug, *Metacanthus pulchellus* Dallas, and the tobacco bug, *Nesidiocoris tenuis* (Reuter), and biology of *M. pulchellus* Dallas on bottle-gourd. Ms.Sc. thesis, College of Agriculture, Haryana Agricultural University, Hissar, India.
Singh, R., and H. K. Patel. 1973. Some studies on the damage behaviour and seasonal distribution of stilt bug (*Gampsocoris pulchellus* Dallas) of bottle gourd. Bhartiya Krishi Anusandhan Patrika 1(1): 15–18 [in Hindi].
Sinha, S. N., A. K. Chakrabarti, and U. Ramakrishnan. 1981. Field evaluation of some insecticides for the control of hemipteran bugs on bottlegourd. Indian J. Agric. Sci. 51: 906–910.
Slater, J. A., and R. M. Baranowski. 1978. How to Know the True Bugs (Hemiptera–Heteroptera). Wm. C. Brown Co., Dubuque, Iowa, U.S.A. 256 pp.
Smith, R. F, and K. S. Hagen. 1956. Enemies of spotted alfalfa aphid. California Agric. 10: 8–10.
Somes, M. P. 1914. A new insect pest on tomato. Pp. 16–17 *in* Entomologist's Report. Bienn. Rep. Missouri State Fruit Exp. Stn., Mountain Grove, Bull. 24: 4–19.
1916. Some insects of *Solanum carolinense* L., and their economic relations. J. Econ. Entomol. 9: 39–44.
Southwood, T. R. E. 1986. Plant surfaces and insects–an overview. Pp. 1–22 *in* B. Juniper and T. R. E. Southwood [eds.], Insect and the Plant Surface. E. Arnold Publ., London, U.K. 360 pp.
Southwood, T. R. E., and D. Leston. 1959. Land and Water Bugs of the British Isles. Frederick Warne, London, U.K. 436 pp.
Štusák, J. M. 1965. Berytidae (Heteroptera) of Congo (Léopoldville), Rwanda and Burundi. Acta Entomol. Mus. Natl. Pragae 36: 509–542.
1976. Family Berytidae. Pp. 410–427 *in* G. Schmitz [ed.], La Faune Terrestre de l'Ile de Sainte-Helene. Troisieme Partie. Ann. Mus. Royale L'Afr. Centr., Tervuren, Belgium Ser. 8, Sci. Zool. No. 215.
Tomokuni, M., T. Yasunaga, M. Takai, I. Yamashita, M. Kawamura, and T. Kawasawa. 1993. A field guide to Japanese bugs — terrestrial heteropterans. Zenkoku Noson Kyoiku Kyokai, Tokyo, Japan. 382 pp. [in Japanese].
Ueshima, N. 1979. Hemiptera. II. Heteroptera. Pp. 11–118 *in* Animal Cytogenetics. 3. Insecta. 6. Gerbrüder Borntraeger, Berlin, Germany.
Wheeler, A. G., Jr. 1970. *Berytinus minor* (Hemiptera: Berytidae) in North America. Canad. Entomol. 102: 876–886.
1978. *Neides muticus* (Hemiptera: Berytidae): life history and description of the fifth instar. Ann. Entomol. Soc. Amer. 71: 733–736.
1986. A new host association for the stilt bug *Jalysus spinosus* (Heteroptera: Berytidae). Entomol. News 97: 63–65.
1994. A new host for *Jalysus spinosus* (Heteroptera: Berytidae) and new host family (Commelinaceae) for stilt bugs. Entomol. News 105: 201–203.
1997. *Neoneides muticus* (Heteroptera: Berytidae): host plants and seasonality in mid-Appalachian shale barrens. Entomol. News 108: 175–178.
Wheeler, A. G., Jr., and T. J. Henry 1981. *Jalysus spinosus* and *J. wickhami*: taxonomic clarification, review of host plants and distribution, and keys to adults and 5th instars. Ann. Entomol. Soc. Amer. 74: 606–615.
Wheeler, A. G., Jr., and C. W. Schaefer. 1982. Review of stilt bug (Hemiptera: Berytidae) host plants. Ann. Entomol. Soc. Amer. 75: 498–506.
Wheeler, A. G., Jr., and J. F. Stimmel. 1988. Heteroptera overwintering in magnolia leaf litter in Pennsylvania. Entomol. News 99: 65–71.
Wille, J. E. 1944. Insect pests of cacao in Peru. Trop. Agric. 21: 143.
1951. Biological control of certain cotton insects and the application of new organic insecticides in Peru. J. Econ. Entomol. 44: 13–18.

CHAPTER 32

Predaceous Stinkbugs (Pentatomidae: Asopinae)

Patrick De Clercq

1. INTRODUCTION

Members of the subfamily Asopinae (common names: predatory stinkbugs or soldier bugs) are set apart from the other pentatomid subfamilies by their essentially predaceous feeding habits. Asopine first instars do not attack prey and only need moisture, mainly plant juices, to develop. Although for some predatory stinkbugs partial development on certain plant foods has been observed, nymphs from the second instar on require animal food to complete development. Further, nymphs and adults are often observed to take up plant juices or free water in addition to feeding on prey, suggesting that metabolic water or that from prey is insufficient for survival (Schumacher 1910, for more details see **Section 3.2**).

Unlike phytophagous pentatomids, predatory stinkbugs are characterized by having a crassate rostrum. The first segment is markedly thickened and free, only the base being embedded between the bucculae. This enables a fully forward extension of the rostrum and thus facilitates feeding on active prey. The bucculae form a rostral groove that does not reach the posterior margin of the head. The rostrum generally extends beyond the coxae of the middle legs, but never surpasses the base of the venter. The scutellum is mostly much shorter than the abdomen, but in some genera it is enlarged, covering most of the abdominal dorsum. The anterior tibiae have a short, acute spine on the lower surface (Schouteden 1907; Schumacher 1910; Miller 1956; DeCoursey and Allen 1968; Thomas 1992, 1994). McDonald (1966) stated that the genitalia of the Asopinae have no unique characters, although the asopine males are unique in combining the presence of genital plates (parandria) with a thecal shield.

The subfamily Asopinae is distributed throughout the world, but is most abundantly represented in the New World (Schumacher 1910). Its species-level taxonomy is largely unsettled. The only worldwide taxonomic revision is that of Schouteden (1907), who listed over 280 species in 55 genera and included keys to the genera. Thomas (1992) presented a revised taxonomic synopsis of the asopines of the Western Hemisphere and provided keys to all 110 species. Subsequently, Thomas (1994) published a taxonomic revision of the 187 Old World species, with keys to the genera and a review of the species, with nomenclatural changes. Altogether, Thomas (1992, 1994) recognizes 69 genera and 297 species of Asopinae, worldwide.

2. GENERAL STATEMENT OF ECONOMIC IMPORTANCE

Of the nearly 300 known species of asopines, only about 10% have been studied in more or less detail. Certain species, however, have received considerable attention throughout the world in regard to their potential to suppress agricultural pests.

Predatory stinkbugs are associated with a wide range of natural and agricultural habitats but many species appear to prefer shrubland and woods. They attack mainly slow-moving, soft-bodied insects, primarily larval forms of the Lepidoptera, Coleoptera, and Hymenoptera. However, they will also attack prey from other insect orders, or from other developmental stages (Schumacher 1910, McPherson 1982, Schaefer 1996). Froeschner (1988) pointed out that very few, if any, Asopinae are truly host specific. This author attributed the belief that certain asopines confine their feeding to a single insect species (e.g., Knight 1923) to disproportionate sampling. Nevertheless, whereas some asopine bugs are generalist predators attacking a wide array of prey in a diversity of habitats, others appear to be more closely associated with a limited number of insect species and occur in few habitats (Evans 1982b,d; Saint-Cyr and Cloutier 1996; Schaefer 1996). Schaefer (1996) hypothesized that there may be an evolutionary progression from drab asopines that feed rather generally to bright asopines preferring Chrysomelidae and, to a lesser extent, Coccinellidae.

The potential of predatory pentatomids to control a variety of foliage-feeding insect pests of orchards, forests, and field crops has been favorably assessed on a worldwide scale (e.g., Khloptseva 1991, Hough-Goldstein et al. 1993, Zanuncio et al. 1994a). To date, however, these bugs' use in agricultural practice is still negligible: the only asopine at present commercially available for biological control in North America and Europe is the spined soldier bug, *Podisus maculiventris* (Say). Within the context of integrated pest management, however, several predatory pentatomids are believed to have a future for augmentation programs to suppress economically important crop pests. In North America and Europe, two New World species of Asopinae, *Perillus bioculatus* (F.) and *P. maculiventris*, have received the most attention, for biocontrol of the Colorado potato beetle, *Leptinotarsa decemlineata* (Say), and of the Mexican bean beetle, *Epilachna varivestis* (Mulsant), and various caterpillar pests, respectively. *Podisus nigrispinus* (Dallas) has been the subject of multiple studies in South America and is particularly noted to be a promising control agent of leaf-feeding caterpillars in forests. *Eocanthecona furcellata* (Wolff) has received increasing attention in different regions of Southeast Asia and India for its potential to suppress outbreaks of lepidopterous and coleopterous defoliators.

3. THE IMPORTANT SPECIES

3.1 *Podisus maculiventris* (Say)

3.1.1 Distribution, Life History, and Biology

Compared with the other asopine bugs, the spined soldier bug, *P. maculiventris*, has received extensive attention, and several literature reviews of its biology and control potential are available (e.g., Mukerji and LeRoux 1965, McPherson 1982, Hough-Goldstein et al. 1993). This insect is the most common asopine species in North America; its natural distribution ranges from Mexico, the Bahamas, and parts of the West Indies into Canada. *Podisus maculiventris* is associated with a variety of natural habitats such as woods and shrubs, but it is also found in such varied agricultural ecosystems as orchards and several field crops (Stoner 1930, Hayslip et al. 1953, Mukerji and LeRoux 1965, Deitz et al. 1980, Richman and Mead 1980a, Evans 1982a).

Descriptions of the different life stages of the spined soldier bug have been provided by a number of workers. The egg stage was studied by Couturier (1938), Esselbaugh (1946), Warren and Wallis (1971), Lambdin and Lu (1984), and Javahery (1994). Descriptions and measurements of the five instars were provided by Coppel and Jones (1962), DeCoursey and Esselbaugh (1962), Richman and Whitcomb (1978), and Richman and Mead (1980a). Evans (1985) provided a simple key to nymphs of this and three other *Podisus* species common to North America. Keys to the adults of *P. maculiventris* and of other New World Asopinae are given by Hart (1919), Stoner (1920), Blatchley (1926), Torre-Bueno (1939), McPherson (1982), and more recently by Thomas (1992). Female adults are 12 to 14 mm long, and males average 11 mm (Richman and Whitcomb 1978; Richman and Mead 1980a; P. De Clercq, unpublished results). The genitalia are described by McDonald (1966).

Numerous studies have addressed different aspects of the life history of *P. maculiventris*; among these are Kirkland (1896, 1898), Whitmarsh (1916), Stoner (1920), Couturier (1938), Esselbaugh (1948), Coppel and Jones (1962), Mukerji and LeRoux (1965), Warren and Wallis (1971), Gusev et al. (1980), and De Clercq (1993). The development and eclosion of *P. maculiventris* eggs were described in detail by Couturier (1938) and Mukerji and LeRoux (1965). Newly emerged nymphs of *Podisus* are highly gregarious, whereas later instars become progressively more solitary with each molt (Evans and Root 1980). This early aggregative behavior may serve to protect against predators and unfavorable climatic conditions (Tostowaryk 1971, Lockwood and Story 1986, Aldrich 1988). First instars of Asopinae are not predaceous and take up only water or plant juices. Occasionally, they also feed on unhatched eggs (Prebble 1933, Mukerji and LeRoux 1965, Oetting and Yonke 1971, Warren and Wallis 1971, De Clercq and Degheele 1990a). The process of molting is described by Mukerji and LeRoux (1965). From the second stadium on, *Podisus* nymphs attack prey shortly after molting and feed up to 1 to 2 days before the next molt. Small nymphs tend to attack and feed collectively, particularly upon large prey; nevertheless, individual second instars have been observed to subdue prey much larger than themselves (e.g., Oetting and Yonke 1971, De Clercq and Degheele 1994). Larger nymphs and adults prefer to attack prey individually; late instars or adults that have captured a prey are, however, sometimes joined by other bugs to share in the meal (Tostowaryk 1971, De Clercq 1993). Under conditions of food shortage, nymphs and adults are highly cannibalistic (Mukerji and LeRoux 1965, Warren and Wallis 1971, Richman and Whitcomb 1978, De Clercq 1993). Adults usually move by crawling, but they are also noted to be good fliers. Although *Podisus* bugs are capable of flying at least several hundred meters nonstop, usually they make repeated short flights (Aldrich et al. 1984a, Aldrich 1986, Wiedenmann and O'Neil 1991b, De Clercq 1993). Feeding habits and hunting behavior are more extensively discussed below.

Several authors have reported on the bionomics of *P. maculiventris*, but the work of Couturier (1938) may be considered as a landmark study on this subject. Lower developmental thresholds for eggs and nymphs of the spined soldier bug are estimated to be 11 to 12°C (Couturier 1938, Sellke and von Winning 1939, Shagov and Shutova 1977, De Clercq and Degheele 1992b). Generally, eggs hatch after 5 to 7 days and development of the five instars takes about 3 weeks at constant temperatures around 23°C (Table 1). Goryshin et al. (1988b) found that under conditions of diurnal thermorhythms with different amplitudes, the duration of larval development was similar to that at corresponding constant temperatures. Data on immature development of *P. maculiventris* vary considerably among workers (Table 1). Given the wide distribution of the bug, differences in developmental rates may in part reflect the adaptation of the studied strains to the climatic conditions of their particular geographic origin. Development times of *P. maculiventris* immatures were generally shorter for Florida (U.S.A.) specimens (Richman and Whitcomb 1978, De Clercq and Degheele 1992b) than for those originating from New York (U.S.A.) (Couturier 1938) and Québec (Canada) (Mukerji and LeRoux 1965). Further, differences in survival and developmental rates of nymphs may also be explained by diet. For instance, longer developmental periods were observed when predator nymphs were fed on larvae of the Colorado potato beetle (Landis 1937, Couturier 1938, Pruszynski and Wegorek 1980, Drummond et al. 1984) or of the housefly *Musca domestica* L. (Golubeva et al. 1980). Besides prey species, developmental stage of the prey also influences predator development. In this context, Mukerji and LeRoux (1969a,b)

Table 1 Development of the Immature States of *P. maculiventris* Strains from Different Geographic Regions

Ref.	Origin	Temp. (°C)	Developmental Duration (days)	
			Egg Stage	Nymphal Stage
Couturier (1938)	New York, U.S.A.	23	6.5	27–30
		27.5	5	20–24
Mukerji and LeRoux (1965)	Québec, Canada	27	4–7	25–31
Warren and Wallis (1971)	Arkansas, U.S.A.	21	6–8	28.7
Richman and Whitcomb (1978)	Florida, U.S.A.	27	5	21.7
De Clercq and Deghelle (1992b)	Florida, U.S.A.	23	5.8	23.2
		27	5	18.5

found that growth rate of nymphs varied depending on the size of larval prey offered. Waddill and Shepard (1975a) reported that *P. maculiventris* failed to reach adulthood when fed eggs, first instars, or adults of the Mexican bean beetle, *Epilachna varivestis*; successful development was achieved when second to third instars or pupae were offered as prey. De Clercq and Degheele (1994) and De Clercq et al. (1998a) found that nymphal development was generally prolonged when eggs and first to second instars of the beet armyworm, *Spodoptera exigua* (Hübner), and the tomato looper, *Chrysodeixis chalcites* (Esper), were supplied as food; highest adult weights and shortest nymphal development were recorded when late instars and pupae were provided. Landis (1937), Orr and Boethel (1986), Stamp et al. (1991), Weiser and Stamp (1998), and other studies (reviewed in part by Coll 1998) demonstrated that immature development of *P. maculiventris* can also be affected by plant antibiosis on a tritrophic level. In these studies, survival and development were strongly influenced by food of the prey, indicating that toxins from food plants may render herbivorous insects toxic to their predators.

The sex ratio of *P. maculiventris* is reported to be equal under normal conditions of climate and food supply (Coppel and Jones 1962, Mukerji and LeRoux 1965, Warren and Wallis 1971). De Clercq and Degheele (1992b) and De Clercq (1993) noted a shift in sex ratio toward males at low (<20°C) and high (>30°C) temperatures; Mukerji and LeRoux (1965) observed a predominance of males when food was scarce. Adult males of *Podisus* and other asopine genera possess hypertrophied dorsal abdominal glands, whose secretions function as long-range attractant pheromones for adults and immatures of both sexes (Aldrich et al. 1984a,b,c, 1986, 1991; Aldrich 1985, 1988, 1995, 1998; Sant'Ana and Dickens 1999; Sant'Ana et al. 1999). Wind-tunnel bioassays on the effects of feeding history on the response of *P. maculiventris* to its synthetic pheromone suggested that predators may use the pheromone as a cue indicating the presence of prey, in addition to a mating cue (Shetty and Hough-Goldstein 1998). Further, calling *Podisus* males become vulnerable to a complex of scelionid (see also Girault 1907, Javahery 1968, Yeargan 1979, Izhevskii et al. 1980, Orr et al. 1986, Torres et al. 1996) and tachinid (see also Eger and Ables 1981, McPherson et al. 1982) parasitoids that exploit the pheromones as host-finding kairomones. The dorsal abdominal glands in females are much smaller than in males; their secretions are suggested to act as close-range aggregation pheromones (Aldrich et al. 1984c). Mating behavior of *Podisus* spp. has been investigated by Coppel and Jones (1962), Mukerji and LeRoux (1965), and Tostowaryk (1971). In the field, mating begins immediately after adults come out of hibernation. In the laboratory, adults first mate 3 to 4 days after emergence. Pairs remain *in copula* from a few hours up to 1 day. During this period, a mating pair will move about and attack prey; in such cases, the female drags the male. *Podisus* bugs copulate several times throughout their lifetime. Males of *Podisus* are polygynous, females polyandrous. Repeated matings, preferably with different males, appear to be necessary to ensure a constantly high egg fertility (Couturier 1938, Mukerji and LeRoux 1965, Tostowaryk 1971, Warren and Wallis 1971, De Clercq and Degheele 1990b). According to Legaspi et al. (1994), microscopic examination of the spermatheca is a good method for assessing the mating status of *P. maculiventris* females collected from the field. Unmated females oviposit as well, but none of the eggs hatches. In contrast to the findings of Wiedenmann and O'Neil (1990), who stated that unmated females of *P. maculiventris* can lay a full complement of eggs, Couturier (1938), Evans (1982a), Javahery (1986), Drummond et al. (1987), Legaspi and O'Neil (1993), and De Clercq and Degheele (1997) noted that unmated females of *Podisus* and other Asopinae produce fewer eggs than mated ones. Baker and Lambdin (1985) reported that fecundity and longevity of mated *P. maculiventris* females were not affected by the amount of time that females were paired with males.

Most females of *P. maculiventris* lay eggs throughout their entire lifetime and usually still contain eggs at the time of death. Eggs are laid in loose oval masses; mean numbers of eggs per mass recorded for *P. maculiventris* range from 15 to 30 (Couturier 1938, Esselbaugh 1946, Coppel and Jones 1962, Mukerji and LeRoux 1965, Warren and Wallis 1971, De Clercq 1993). Oviposition is reportedly initiated at about 15°C and increases with temperature up to about 30°C (Couturier 1938, Mukerji and LeRoux 1965, De Clercq 1993). Mean fecundity values reported for this bug vary considerably. De Clercq and Degheele (1992a, 1993c, 1997) found that laboratory-reared females of *P. maculiventris* produce 700 to 1000 eggs. Other authors, however, reported a total fecundity of only 200 to 300 eggs per female (Sellke and von Winning 1939, Mukerji and LeRoux 1965, Warren and Wallis 1971, Golubeva et al. 1980, Pruszynski and Wegorek 1980, Lambdin and Lu 1984, Baker and Lambdin 1985, Wiedenmann and O'Neil 1990, Legaspi and O'Neil 1993). Likewise, longevity data recorded for female bugs in the laboratory differ markedly. Sellke and von Winning (1939), Golubeva et al. (1980), Pruszynski and Wegorek (1980), Wiedenmann and O'Neil (1990), and De Clercq and Degheele (1992a, 1993c) reported a mean longevity of about 2 months, whereas Mukerji and LeRoux (1965), Warren and Wallis (1971), and De Clercq and Degheele (1997) mentioned longevities of 80 to 120 days. According to Baker and Lambdin (1985), however, females of *P. maculiventris* only live about 30 days under laboratory conditions. Males have been reported to live longer than females (Mukerji and LeRoux 1965, Warren and Wallis 1971). Differences in fecundity and longevity may be related to rearing methods and to the geographic origin of the studied strains; on the other hand, they may also reflect differences in health condition and adaptation to the laboratory environment. Furthermore, the observed differences in fecundity suggest that a search for biotypes with superior reproductive capacities may be rewarding for use in augmentative biocontrol (De Clercq et al. 1998c). Evans (1982a) demonstrated a positive relationship between body size and fecundity in *P. maculiventris* females that had been obtained by feeding nymphs varying amounts of prey. In contrast, Mohaghegh-Neyshabouri et al. (1996) and De Clercq et al. (1998c) could not establish a significant correlation between weight and size of females and their total fecundity in laboratory strains of *P. maculiventris* and *P. nigrispinus*. Mohaghegh et al. (1998a) also investigated effects of egg size and maternal age on developmental time and body weight of offspring in laboratory-reared *P. maculiventris*. They found that nymphs from small eggs and old females took longer to develop and that large eggs and young parents yielded heavier offspring.

In Canada and in central and northern parts of the United States, *P. maculiventris* is reported to have two to three generations in a year and to hibernate from October to April (Stoner 1920, Esselbaugh 1948, Mukerji and LeRoux 1965, Warren and Wallis 1971); in the warmer southern parts of the United States, this bug is active all year (Richman and Mead 1980a). In the European part of the former Soviet Union, where the predator was introduced for the biological control of *L. decemlineata*, two to three generations were also observed (Gusev et al. 1980).

In the field, Nearctic *Podisus* spp. are reported to hibernate usually as adults. Overwintering takes place in litter, in soil, under stones, under tree bark, etc. (Coppel and Jones 1962, Mukerji and LeRoux 1965, Tostowaryk 1971, Sazonova et al. 1976, Jones and Sullivan 1981, Pruszynski and Rosada 1987). There is some confusion concerning the ability of *P. maculiventris* to survive northern winters. Tadic (1975), for instance, assumed that the reason for failure in establishing the predator in the former Yugoslavia was its inability to overwinter. On the other hand, the insect has been reported to survive winter in the eastern United States, Canada, Poland, and in different parts of Russia and the Ukraine (Mukerji and LeRoux 1965, Sazonova et al. 1976, Gusev et al. 1980,

Vlasova et al. 1980, Pruszynski and Rosada 1987). Couturier (1938) and Vlasova et al. (1980), who studied the possibility of acclimatization of *P. maculiventris* under laboratory and field conditions, mentioned that adults could subsist during winters at temperatures of 0 to 10°C. They suggested that adults of the bug survive cold periods in a quiescent state and not in true diapause. Goryshin et al. (1988a,b) and Volkovich et al. (1991) reported that imaginal diapause can be induced by maintaining nymphal and adult stages at 20°C and daylengths of 11 to 12 hours; low prey availability and, to a lesser extent, low temperature shift the critical photoperiod toward longer daylengths. Chloridis et al. (1997) succeeded in inducing reproductive diapause in *P. maculiventris* by keeping nymphs and adults at 23°C and an 8-hour photoperiod and reported that nymphal development was faster under the diapause-inducing short photoperiod than under a 16-hour photoperiod. Warren and Bjegovic (1972) observed reproductive diapause in *P. placidus* (Uhler) at 27°C and a 12-hour photoperiod; under similar conditions, they could not, however, induce diapause in *P. maculiventris*. Likewise, De Clercq and Degheele (1993a) could not obtain diapause in laboratory cultures of *P. maculiventris* (originating from Florida, U.S.A.) and of the Neotropical species *P. nigrispinus* when maintained at 20°C and a 12-hour day length. These differences in ability to overwinter or to enter diapause may be related to the climatic conditions found at the geographic origins of the different strains studied; also, diapause may be lost through inadvertent selection during laboratory rearing. Saulich (1994) connected the fact that *P. maculiventris* has not become successfully acclimatized in different regions of Russia with its photoperiodic reactions. According to this worker, the low mean value for critical day length, the sensitivity to photoperiod from the third stadium on, and the long period for prediapause feeding, together restrict the bug's potential zone of acclimatation to the southernmost regions of Russia.

3.1.2 Food Preferences

Feeding mechanisms and adaptations in predatory stinkbugs are discussed in **Chapter 20**. Details on the feeding adaptations of *Podisus* can be found in Gallopin and Kitching (1972), Cohen (1990, 1995), Stamopoulos et al. (1993), and Muntyan and Yazlovetskii (1994).

In natural ecosystems, *P. maculiventris* is usually found in association with larval lepidopterans and coleopterans. Nevertheless, it reportedly feeds on more than 90 insect species from eight orders, including Ephemeroptera, Orthoptera, Heteroptera, Homoptera, Coleoptera, Lepidoptera, Diptera, and Hymenoptera (see McPherson 1980). The wide range of habitats occupied and of prey insects attacked by *P. maculiventris* suggests that this pentatomid is a fairly generalist predator. Nevertheless, several workers have noted important differences in developmental success and fecundity of the predator depending on the prey species offered (see above). A review of the literature suggests that *P. maculiventris* may have a particular preference for lepidopterous larvae. The availability of efficient procedures for stomach analysis (e.g., serological assays) may provide more information on the prey preferences of pentatomid predators in the field (see also Whalon and Parker 1978, Fichter and Stephen 1981, Greenstone and Morgan 1989, Greenstone and Trowell 1994).

Like other asopines, *P. maculiventris* has frequently been recorded feeding on plant juices (Olsen 1910, Stoner 1930, Prebble 1933, Couturier 1938, Morris 1963, Mukerji and LeRoux 1965, Stoner et al. 1974b, Ruberson et al. 1986, Wiedenmann and O'Neil 1990, De Clercq and Degheele 1992d, Valicente and O'Neil 1995, Weiser and Stamp 1998). Although plant feeding has been observed at all prey densities, it is of particular importance under conditions of low prey availability (see **Section 3.1.4**). Plant feeding primarily provides moisture, but it may also furnish additional nutrients to the predatory bugs at critical times. Nymphs have been reared to the third instar on certain plant materials; where adults were offered only free water or plant foods, they lived but failed to produce eggs (Prebble 1933, Moens 1965, Mukerji and LeRoux 1965, Wiedenmann and O'Neil 1990, De Clercq and Degheele 1992d). Ruberson et al. (1986) found that adding plant material to a diet of *L. decemlineata* larvae enhanced survival and shortened nymphal development and the preoviposition period in *P. maculiventris*. The same was true of *P. nigrispinus* feeding on

Eucalyptus seedlings (Zanuncio et al. 1993b) and of *Eocanthecona furcellata* feeding on leaves of *Mangifera indica* L. (Senrayan 1991). Weiser and Stamp (1998), however, noted no efforts or negative effects on development of *P. maculiventris* when plant materials were added to a diet of insect prey. In contrast to other predatory heteropterans (e.g., *Geocoris*, *Dicyphus*, *Macrolophus*), plant-feeding asopines have not been reported to injure crops.

3.1.3 Rearing

Several authors in North America and Europe have described rearing procedures for *Podisus*. Various types of cages have been used for group-rearing of nymphs and adults (e.g., Coppel and Jones 1962, Mukerji and LeRoux 1965, Warren and Wallis 1971, Danilkina 1986, De Clercq and Degheele 1993c, Zanuncio et al. 1994b). Most workers have emphasized the importance of providing sufficient hiding places and avoiding high population densities. Crowded conditions with few refugia are conducive to high mortality from cannibalism in both nymphal and adult cultures (Warren and Wallis 1971, De Clercq and Degheele 1993c, Zanuncio et al. 1993a). In addition, crowding in adult cultures may result in decreased egg production (Mukerji and LeRoux 1965).

Podisus bugs can be reared with relative ease on a variety of natural and unnatural hosts. Larvae of the greater wax moth, *Galleria mellonella* L., were frequently reported to be adequate food for breeding *Podisus* spp. (Coppel and Jones 1962, Mukerji and LeRoux 1965, Warren and Wallis 1971, Waddill and Shepard 1975a, Drummond et al. 1984, Goryshin et al. 1988a, De Clercq and Degheele 1993c, De Clercq et al. 1998b). Couturier (1938), Sellke and von Winning (1939), Pruszynski and Wegorek (1980), and De Clercq et al. (1988) provided larvae of another pyralid, the Mediterranean flour moth *Ephestia kuehniella* Zeller, as food. *Podius maculiventris* was also easily reared on caterpillars of several noctuids, including the beet armyworm *S. exigua* (Cohen 1990, De Clercq and Degheele 1993c), the fall armyworm *S. frugiperda* (J.E. Smith) (Yu 1987), the cabbage looper *Trichoplusia ni* (Hübner) (Biever and Chauvin 1992a), the soybean looper *Pseudoplusia includens* (Walker) (Orr et al. 1986), the tobacco budworm *Heliothis virescens* (F.) (Pfannenstiel et al. 1995), and the cotton bollworm *Helicoverpa zea* (Boddie) (Warren and Wallis 1971). All these lepidopterans can be reared on artificial diets. Caterpillars of other lepidopteran families were sometimes used as prey for rearing spined soldier bugs, e.g., the fall webworm *Hyphantria cunea* (Drury) (Kim et al. 1968, Warren and Wallis 1971, Sazonova et al. 1976) and the eastern tent caterpillar *Malacosoma americanum* (F.) (Warren and Wallis 1971). Also, frozen caterpillars of different species proved to be acceptable food (Warren and Wallis 1971, Warren and Bjegovic 1972, De Clercq et al. 1988). In a few studies, coleopterous larvae were supplied as food for *Podisus* cultures. Wiedenmann and O'Neil (1990), Stamp and Bowers (1991), Sant'Ana et al. (1997), and De Clercq et al. (1998b) fed the bugs on larvae of the mealworm *Tenebrio molitor* L. Landis (1937), Couturier (1938), Pruszynski and Wegorek (1980), Drummond et al. (1984), and Goryshin et al. (1988a) presented *P. maculiventris* with larvae of the Colorado potato beetle *L. decemlineata*, but development and fecundity of the predator indicated that the chrysomelid is a suboptimal host. Occasionally, laboratory cultures of *Podisus* were also fed on larvae of pine sawflies, *Neodiprion* spp. (Tostowaryk 1971, Warren and Wallis 1971). In the former Soviet Union, *P. maculiventris* has been successfully reared on larvae of the dipterans *Musca domestica* and *Calliphora erythrocephala* Meigen (Golubeva et al. 1980, Volkovich and Saulich 1992). Richman and Whitcomb (1978) concluded that the key element for successful rearing of *Podisus* may be provisioning the stinkbugs with more than one kind of prey. As Mackauer (1976) and van Lenteren (1991) pointed out, providing variation in rearing conditions (food, climate) may enhance fitness and minimize selection during laboratory propagation of these and other entomophages.

For many reasons, the availability of an adequate artificial diet is highly desirable for the mass production of entomophagous insects (Waage et al. 1985). This also applies to the rearing of predatory stinkbugs. Given the voracity of the asopine predators, extensive parallel cultures of prey insects are needed. An artificial diet that supports growth and reproduction could make the mass production of predatory stinkbugs less time-consuming and therefore more economical. Different semiartificial diets for *Podisus* bugs were developed by Khlistovskii et al. (1985), Lyashova et al. (1985), Saavedra et al. (1992a,b), and Sumenkova and Yazlovetskii (1992), but they all contained insects as components. Adidharma (1986) succeeded in rearing *P. nigrispinus* on completely defined liquid media, but nymphal survival was poor and adult weights were very low. Better results were obtained by De Clercq and Degheele (1992a, 1993b,c), De Clercq et al. (1998b), and Saavedra et al. (1995, 1997), who used oligidic diets based on bovine meat. Although development and reproduction on the meat diet were somewhat inferior to that on live prey, the nutritional value of the diet was sufficient to produce consecutive generations of *P. maculiventris* and *P. nigrispinus* in the laboratory. When the predators were returned to a diet of live prey after more than 15 generations on the artificial diet, viability and predatory performance were similar to that of bugs continuously reared on live prey (De Clercq and Degheele 1993b). Although the production of the meat diet is simple and the ingredients are relatively inexpensive, the processing of the diet is rather time-consuming. Hence, automation of the preparation technique may increase the practical value of this artificial diet for mass-rearing purposes.

For successful rearing of predatory stinkbugs, prey must be supplemented with a moisture source. Moisture can be provided as free water or via plant materials (e.g., bean pods) (Mukerji and LeRoux 1965, De Clercq and Degheele 1993c). Adding plant material to a diet of animal prey has been found to enhance survival and to shorten nymphal development of several laboratory-reared asopines (see **Section 3.2**).

Efficient methods for storage of entomophagous insects are needed to plan production. Eggs of *Podisus* spp. can tolerate short periods of cold exposure. Sellke and von Winning (1939) kept eggs in a refrigerator at 1 and 6°C for 7 and 14 days, respectively. Goryshin and Tuganova (1989) stored 1-day-old eggs of *P. maculiventris* for 7 days at 13 to 15°C and near 100% relative humidity without an apparent reduction in viability; older eggs were less suitable for storage. De Clercq and Degheele (1993a) reported that 1- to 4-day-old eggs of the Neotropical species *P. nigrispinus* (see **Section 3.2**) could be stored up to 6 days at 9°C and 90 to 100% relative humidity with no adverse effects on viability. Reducing temperature and especially humidity, and increasing the period of refrigeration, resulted in both studies in lower egg survival. Jenkins et al. (1998) stored 1- to 2-day-old eggs of *P. maculiventris* for 5 days at 5, 10, and 15°C without adverse effects on hatch. Adults of *Podisus* can be kept longer under cold conditions. For optimal storage of *P. maculiventris* during the winter, Couturier (1938) recommended maintaining adults at 5 to 7°C and high humidity. De Clercq and Degheele (1993a) found that adults of *P. nigrispinus* and *P. maculiventris* could be kept with good survival at 9°C and 75% relative humidity for at least 1 and 2 months, respectively; in some cases, however, poststorage longevity and reproduction were reduced. In both of these studies, adults were in a quiescent state during the cold period, rather than in diapause. Volkovich and Saulich (1992) recommended that adults of *P. maculiventris* be stored in induced diapause (see above). Diapause was reported to end after 1.5 months, but the bugs could then be stored at temperatures below 8°C for another 3 months (Volkovich 1995). According to the Russian workers, this two-step storage method is an important contribution to the establishment of an economic mass-breeding scheme for the spined soldier bug.

3.1.4 *Predator–Prey Relationships*

Predatory stinkbugs are reported to use visual, chemical, and tactile cues to locate and recognize their prey; vision is certainly the most important sense used by the bugs to locate prey. *Podisus maculiventris* and other asopine bugs react to moving prey at distances up to 10 cm, but their reactive distance to immobile prey is considerably less, detection often seeming to occur at antennal or rostral contact (Mukerji and LeRoux 1965, Tostowaryk 1972, Jermy 1980, Awan et al. 1989, De Clercq 1993, Usha Rani and Wakamura 1993, Heimpel and Hough-Goldstein 1994b). McLain (1979) observed terrestrial trail following with the use of antennae and rostrum in three species of predatory stinkbugs, including *P. maculiventris*. Evidence is accumulating that several asopines can also use airborne chemical cues for prey detection. Both *P. maculiventris* and *P. bioculatus* are sensitive to systemic volatiles produced by plants in response to prey feeding (Dickens 1999, Sant'Ana et al. 1999, Weissbecker et al. 1999). *Eocanthecona furcellata* is attracted to its lepidopterous prey based in part on a volatile component derived from the chlorophyll ingested by the prey (see **Section 3.4.4**). Pfannenstiel et al. (1995) demonstrated that *P. maculiventris* can use substrate-borne vibrations as cues for prey location.

In many cases, however, prey appear to be detected by chance. The lack of efficient mechanisms in predatory pentatomids to locate prey at greater distances is compensated by their intensive random searching. Searching behavior of *Podisus* has been documented by Morris (1963), Tostowaryk (1971), Gusev et al. (1980), Evans (1982b), O'Neil (1988a,b), Wiedenmann and O'Neil (1991b, 1992), and Heimpel and Hough-Goldstein (1994b). In experiments in the former Soviet Union, third instars of *P. maculiventris* released on a horizontal surface could locate larvae of *L. decemlineata* 10 cm away in about 2 min, and 20 cm away in 6 min. Fourth instars of *P. maculiventris* released in a potato field spread over a distance of 1.6 m in 1 day and 4 m in 2 days (Gusev et al. 1980). *Podisus maculiventris* nymphs disperse in soybeans more along than across rows (Waddill and Shepard 1975b); a similar dispersal pattern was recorded for nymphs of the two-spotted stinkbug, *Perillus bioculatus* (Cloutier and Bauduin 1995). Searching activity is determined by age of the predator. Tostowaryk (1971) reported that younger instars of *Podisus brevispinus* Phillips (see **Section 4.1**) search more actively than fifth instars and adults. Further, nymphs are more active at the beginning of each instar; after a few days of feeding, nymphs stop searching and spend most of their time resting until the next molt (Mukerji and LeRoux 1965, De Clercq and Degheele 1990a). As adults get older, their searching activity decreases (Morris 1963). Stamp and Wilkens (1993) pointed out that plant characteristics too are likely to affect searching efficiency of predatory stinkbugs, and Stamopoulos and Chloridis (1994) observed that the smooth leaf surface of cabbage plants suppressed predation rates of *P. maculiventris* on caterpillars of the large white butterfly, *Pieris brassicae* L.

O'Neil (1988a,b, 1997) and Wiedenmann and O'Neil (1991b, 1992) made an extensive study on the searching behavior of *Podisus maculiventris* under laboratory and field conditions using the Mexican bean beetle, *E. varivestis*, or the Colorado potato beetle, *L. decemlineata*, as prey. Attack rates of the bug in soybean or potato fields remained relatively constant even when the size of canopy increased, indicating that the predator searches more to compensate for changes in the size of canopy. They also observed that the predator searches a greater area and for longer at low prey density than at high density, and does not investigate areas in which it was previously successful. A negative exponential function was used to develop a model describing search area as a function of prey density. The authors argued that searching less area as prey density increases may keep attack rates low and consistent at low prey density, and searching a constant amount of area at high density may result in increasing attack rates at higher prey density. The predators were further observed to spend more time resting and plant-feeding than searching. The described search strategy, minimizing the area searched and maintaining a constant rate of attack, was assumed to result from balancing energetic costs and predation risks with the benefits gained from attacking prey, and to reflect budgeting of time between searching and other time-consuming activities (e.g., reproduction).

Recognition of prey is based primarily on antennal and rostral contact. Podisus and other asopine bugs are timid predators. After finding and orienting to prey, the bugs may spend from several minutes up to an hour stealthily approaching the prey. The stylets are inserted at any soft area of the prey body. During the struggle that follows, the only contact between predator and prey is by the labium and stylets. Several studies have suggested that predatory pentatomids inject a salivary toxin into the prey body by which they can quickly immobilize the prey. More detailed accounts on the process of attack by asopine bugs can be found in McDermott (1911), Baker (1927), Prebble (1933), Morris (1963), Mukerji and LeRoux (1965), Oetting and Yonke (1971), Tostowaryk (1971), Evans (1982b), Awan (1984), and De Clercq (1993).

Behavior of the prey considerably affects the ability of pentatomid predators to find and subdue them. Stamp (1992) and Stamp and Bowers (1991, 1993) found that noncryptic caterpillars like *Junonia* sp. are more susceptible to predation by *Podisus* bugs; the latter workers also reported that, unlike wasps, pentatomid predators caused little indirect, but substantial direct mortality of caterpillars. Different species of prey vary in their response to attack by the predators: some escape by dropping off the plant or by crawling away, whereas others exhibit defensive behavior such as thrashing, biting, or regurgitating; in some cases, prey insects injure or even kill attacking stinkbugs (Morris 1963; Iwao and Wellington 1970; Tostowaryk 1972; Hokyo and Kawauchi 1975; Marston et al. 1978; Evans 1982c; Awan 1984, 1985, 1990; De Clercq and Degheele 1990a; Usha Rani and Wakamura 1993; Saini 1994b; review by Stamp and Wilkens 1993). Based on laboratory findings with *Galleria mellonella* as prey, Mukerji and LeRoux (1969c) suggested that *P. maculiventris* will kill more small than large caterpillars in the field, due in part to the ability of large larvae to defend themselves. Predatory stinkbugs are reported to often locate gregarious caterpillars and sawfly larvae and then remain near the aggregation for hours to days, periodically attacking and feeding (Morris 1972, Tostowaryk 1972, Evans 1983). Although gregarious behavior of prey may increase the risk of being attacked, it has also been shown to reduce predation pressure by *Podisus* bugs (Tostowaryk 1972). Stamp et al. (1991, 1997), Bozer et al. (1996), Traugott and Stamp (1996), and Weiser and Stamp (1998) found that allelochemicals derived from the food plant may contribute to a prey's defense. Caterpillars fed allelochemicals present in tomato were easier for *P. maculiventris* to locate but often deterred predation by experienced predators. Further, allelochemicals had a greater negative impact on consumption and growth of the predator when prey were scarce. Berenbaum et al. (1992) demonstrated that because of its specific feeding adaptations, *P. maculiventris* is able to circumvent chemical defense behavior in Papilionidae. Also in this context, several authors have mentioned that *Podisus* bugs have a great dislike of hairy caterpillars (Whitmarsh 1916, Oetting and Yonke 1971, De Clercq et al. 1988).

Several studies have demonstrated the high voracity of *Podisus* bugs in the laboratory and have hence suggested their potential to reduce pest populations in the field. Predation rates of *P. maculiventris* have been measured on a variety of insect prey, including the Mexican bean beetle *E. varivestis* (Waddill and Shepard 1975a), the pine sawfly *Diprion similis* (Hartig) (Coppel and Jones 1962), the Colorado potato beetle *L. decemlineata* (Gusev et al. 1983, Hough-Goldstein and McPherson 1996), the cotton bollworm *Helicoverpa zea* and the tobacco budworm *Heliothis virescens* (Lopez et al. 1976), the beet armyworm *S. exigua* (De Clercq and Degheele 1994), and the tomato looper *Chrysodeixis chalcites* (De Clercq et al. 1998a). The predator displayed high predation rates against different developmental stages of the prey, including the immobile stages (i.e., eggs, pupae). Coppel and Jones (1962) found that individual nymphs were able to kill an average of 13.4 *D. similis* larvae during their development. Waddill and Shepard (1975a) reported that *P. maculiventris* nymphs consumed on average 160 eggs, 100 second instars, 15 fourth instars, or 18 pupae of the Mexican bean beetle; in contrast to the adults of another asopine, *Stiretrus anchorago* (F.) (see **Section 4.1**), adults of *P. maculiventris* could not use eggs of the coccinellid as food, but did consume approximately 3 second instars, 1.6 fourth instars, or 0.8 pupae of the prey per day. Gusev et al. (1983) noted that a single nymph of *P. maculiventris* could consume during its development an average of 293 eggs, 4.5 late instars, or 5 adults of the Colorado potato beetle, and an adult could consume during its life up to 4000 eggs or 100 late instars. De Clercq and Degheele (1994) found that individual nymphs of the predator were able to destroy approximately 1200 eggs, 150 second instars, 25 fifth instars, or 20 pupae of the beet armyworm during their development; a female adult could kill about 100 eggs, 20 second instars, 5 fifth instars, or 2.5 pupae of the same prey per day. Although the predator was highly destructive to eggs and small larvae of *Spodoptera exigua*, a clear preference was noted for caterpillars in the third to fourth stadia and older, depending on the size of the predator. Likewise, Lopez et al. (1976) noted that in comparison with three other insect predators, third instars and adults of *P. maculiventris* were more efficient predators of third-instar *Heliothis* caterpillars than of eggs or first-instar caterpillars.

Arnoldi et al. (1991) reported that a single adult of *P. maculiventris* could consume 1 to 2 adults of the mirid pests *Lygus lineolaris* (Palisot de Beauvois) and *Lygocoris communis* (Knight) per day; the predator consumed about twice as many late instars as adults of the latter mirid. Only a few studies have addressed the effects of environmental conditions on predation by predatory pentatomids. Waddill and Shepard (1975a) established a linear relationship between predation rate and temperature, and reported that predation by *P. maculiventris* on larvae of the Mexican bean beetle increased almost twice as rapidly with temperature as did that by *Stiretrus anchorago*. Effects of temperature on predation by other pentatomid predators were studied by Awan and Browning (1983), Awan (1988b), and Usha Rani (1992).

Functional responses of *Podisus* spp. have been the subject of several studies (Morris 1963; Mukerji and LeRoux 1969c; Tostowaryk 1972; Waddill and Shepard 1975a; O'Neil 1988a,b, 1989; O'Neil and Wiedenmann 1987; Wiedenmann and O'Neil 1991a, 1992; De Clercq et al. 1998a). Like many arthropod predators, *Podisus* bugs were usually found to exhibit a type II response, in which the number of prey attacked increases with prey density at a continually decreasing rate, leveling off to a plateau at the higher prey densities. Tostowaryk (1972) showed, however, that defensive reactions by the prey at higher densities may result in a domed response curve (i.e., a type IV response). Furthermore, O'Neil (1988a,b; 1989) and Wiedenmann and O'Neil (1991a, 1992) demonstrated the existence of a differential search strategy in *P. maculiventris* attacking the soybean pest *E. varivestis*. In these studies, attack rates in the field were consistently low at low densities, averaging 0.5 attacks per day, and increased linearly at high densities (see above). Also, these authors pointed out the importance of scale when studying predation. In laboratory experiments using petri dishes, female adults of the spined soldier bug were able to kill a maximum of 15 *E. varivestis* third instars per day, whereas in the field they only attacked about two prey larvae per day even at high prey densities. It was concluded that predation should be measured under conditions experienced by the predators in the field, i.e., by using realistic numbers of prey and a suitable searching arena.

O'Neil and Wiedenmann (1990), Legaspi and O'Neil (1993, 1994b), and Legaspi et al. (1996) emphasized that the key to understanding the contribution of predators to pest dynamics lies not only in the measurement of predation under field-realistic conditions, but also in the measurement of their life-history traits under prey densities experienced by the predators in the field. Life histories of predaceous stinkbugs confronted with low numbers of prey have received considerable attention. Developmental success of asopine nymphs depends upon the level of food supply. Nymphs that experience food shortage take longer to develop to adulthood and have lower survival; the bodies of the resulting adults will be smaller that those of their well-fed peers (Mukerji and LeRoux 1969a,b; Evans 1982d; Legaspi and O'Neil 1994a). Evans (1982a) demonstrated that body size of *Podisus* females is strongly correlated with oviposition rate, which led the author to suggest that small females maturing in late summer, with deteriorating food supplies, have only part of the reproductive capacity of females maturing in early summer. Evans (1982a) also stated that, unlike specialist asopine predators such as *Perillus circumcinctus* Stål, a generalist like *Podisus maculiventris* is characterized by relatively poor timing of oviposition, which results in most of its offspring suffering substantial food stress. O'Neil and Wiedenmann (1990), Wiedenmann and O'Neil (1990), De Clercq and Degheele (1992c), Legaspi and O'Neil (1993, 1994b), Legaspi et al. (1996), and Legaspi and Legaspi (1998) found evidence for an energy trade-off between survival and reproduction under conditions of food scarcity: females of *Podisus* spp. were able to survive under low prey input but reduced reproduction as a function of feeding interval. It was argued that an energy allocation scheme of survival first and reproduction second is an appropriate adaptation for arthropod predators, which are often faced with scarce or infrequent food supplies. Mukerji and LeRoux (1969a) and De Clercq and Degheele (1992c) reported increased longevities for adults confronted with low prey numbers. Mukerji and LeRoux (1969a,b), O'Neil and Wiedenmann (1990), De Clercq and Degheele (1992d), and Valicente and O'Neil (1995) further showed that adult stinkbugs provided only water or plant foods survived at least as long as well-fed predators;

adult bugs given only plant foods maintained body weight after an initial decline, whereas those given no food or water lost weight continually until death. By accessing moisture and perhaps some nutrients from plants, predatory pentatomids may sustain their field populations until prey becomes locally more abundant. Good survival of starved adult predators may also be related to suppressed metabolic rates: under laboratory conditions of low food availability, the predators spend most of their time resting and take up moisture from time to time.

Its polyphagous nature enables *P. maculiventris* to feed on alternative prey in case a preferred prey species or type becomes scarce. In this respect, Evans (1982d) and Arnoldi et al. (1991) noted that in time of scarcity of preferred prey, generalists like the spined soldier bug migrate to areas where other prey are more numerous. Ruberson et al. (1986) believed that plant feeding may serve to provide nutrients during migration. Several reports mention that *Podisus* bugs have a strong tendency to disperse as nymphs and adults (Warren and Wallis 1971, Waddill and Shepard 1975b, Ignoffo et al. 1977, Pruszynski and Wegorek 1980), but little is known of the importance of migration as an adaptive strategy for conditions of prey scarcity. Cannibalism is considered another adaptation to low prey densities, and Carayon (1961) went so far as to suggest that cannibalism is partly responsible for the limited efficacy of Asopinae in the field.

Although predatory stinkbugs have occasionally been reported to probe human skin and bite (Warren and Wallis 1971, De Clercq 1993), they do not aggressively attack persons (Froeschner 1988).

3.1.5   Control Potential

Some authors have questioned the value of *P. maculiventris* as a component of integrated pest management (IPM) programs. Based on the generalist nature of their feeding, Carayon (1961) and Evans (1982d) doubted the ability of most asopines, including *P. maculiventris*, to control a specific target pest. Further, generalist predators may interfere with the beneficial action exerted by other organisms. In this context, *P. maculiventris* has been observed to prey on larvae, pupae, and adults of ladybird beetles, including *Adalia bipunctata* (L.), *Harmonia axyridis* (Pallas), and *Hippodamia convergens* (Guérin-Méneville), both in the laboratory and the field (personal observations; Hough-Goldstein et al. 1996). Second, thermal conditions early in the season strongly influence the activity of the predator. During studies in an orchard of cherry and apple in New York State (U.S.A.), Evans (1982c) observed that activity of *P. maculiventris* at tents of *Malacosoma americanum* caterpillars was severely limited by cool spring temperatures, while the caterpillars continued to feed and grow at a normal rate; this enabled the prey larvae to escape predation while small and to grow sufficiently large to defend themselves effectively against the predator by the time the weather grew warmer. Based on the life history and search strategies shown by this bug (see **Section 3.1.4**), Wiedenmann et al. (1996) concluded that the contribution of *P. maculiventris* to pest management in field crops will be more to prevent or delay pest outbreaks than to suppress outbreaks once they have begun.

Nonetheless, many studies have favorably assessed the potential of *P. maculiventris* to suppress a variety of agricultural pests. In North America, several workers have evaluated the effects of predation by the spined soldier bug on the population dynamics of some major insect pests. Morrill (1906) thought the species to be of particular importance in the natural control of the eastern tent caterpillar *M. americanum*, the elm leaf beetle *Pyrrhalta luteola* (Müller), the cotton bollworm *Helicoverpa zea*, and the cotton leafworm *Alabama argillacea* Hübner. Coppel and Jones (1962) believed that the absence of insect parasites on larvae of the pine sawfly *D. similis* is compensated for by the presence of *Podisus* spp., including *P. maculiventris*. LeRoux (1960) and LeRoux et al. (1963) reported that adult *Podisus* bugs were largely responsible for the effective control of larval populations of the pistol case bearer *Coleophora serratella* (L.) and of tent caterpillars in Québec (Canada), apple orchards. The fall webworm *Hyphantria cunea*, a lepidopteran pest of broad-leaved forests, is native to North America but was inadvertently introduced to several regions in Europe and Asia. The spined soldier bug was introduced into Korea, Japan, and the former Yugoslavia to control this pest, with some promising first results. However, in most cases establishment of the predator was not attained because of its inability to overwinter (Kim et al. 1968, Morris 1972, Hokyo and Kawauchi 1975, Tadic 1975). Lopez et al. (1976) reported that releases of *P. maculiventris* on cotton in field cages at rates of 100,000 nymphs per acre resulted in a substantial reduction in the number of *Heliothis virescens* caterpillars; the stinkbug was a particularly efficient predator of larger caterpillars. Ables and McCommas (1982) obtained a 75% reduction of pest numbers within 48 hours of releasing adult bugs on cotton in greenhouses infested with third to fourth instars of the variegated cutworm *Peridroma saucia* (Hübner) at a predator–prey ratio of 1:3.5. Releasing fourth instars of the spined soldier bug in experimental greenhouses at a predator–prey ratio of 1:3 reduced damage to sweet pepper plants by caterpillars of the tomato looper *C. chalcites* by more than 60% (De Clercq et al. 1998a). In field-cage experiments in Greece, Stamopoulos and Chloridis (1994) reported that fifth instars and adults of *Podisus maculiventris* killed many third-instar caterpillars of the large white butterfly *Pieris brassicae* on cabbage; however, predation rates on third instars of the Colorado potato beetle on eggplant were about 50% greater than those recorded on *P. brassicae* caterpillars. Biever and Chauvin (1992a,b), Hough-Goldstein and McPherson (1996), and Aldrich and Cantelo (1999) found that releases of about five nymphs of *Podisus maculiventris* or *P. bioculatus* per plant could suppress high-density populations of the Colorado potato beetle and concluded that augmentative releases of both predatory stinkbugs could be an important component of IPM programs against *L. decemlineata* in the northeastern United States (see also **Section 3.3.4**). *Podisus maculiventris* was introduced from the United States into France in 1933 and, in 1974, also into the former Soviet Union and eastern Europe; since then, considerable work has been done to evaluate its value for the biological control of the Colorado potato beetle in these regions (Couturier 1938, Golubeva et al. 1980, Boiteau 1988, Pruszynski 1989). In the Black Sea coastal regions of Russia, the Ukraine, and Moldavia, seasonal releases of 40,000 to 100,000 second and third instars of the predator per hectare at a predator–prey ratio of 1:10 to 1:30 have resulted in a substantial reduction of both egg and larval populations of the beetle in early potatoes and eggplants (Izhevskii and Ziskind 1982, Boiteau 1988, Filipov et al. 1989, Khloptseva 1991, Novozhilov et al. 1991).

Since the end of the 1980s, *P. maculiventris* has been commercially available in North America for small-scale use. The predators are shipped as eggs or second instars. Recommended use is 50 to 100 nymphs for every 10 ft of row; additional releases may be necessary, because the bugs migrate rapidly away from the treated plots. Control for up to 85 days has been noted under optimal conditions (Thomson 1992). In the Netherlands, *P. maculiventris* has been commercialized since 1997 to control outbreaks of noctuid caterpillars in glasshouse vegetables and ornamentals. Release rates of 0.5 to 1 fourth instars/m^2 once or twice per growing season have yielded successful control in eggplants, sweet peppers, and some ornamental crops (*Strelitzia*, *Heliconia*) (unpublished data).

In most of the field releases mentioned above, nymphs were preferred to adults because the latter tend to fly quickly from the treated field plots. Therefore, Ignoffo et al. (1977) and Lambdin and Baker (1986) suggested de-winging *Podisus* adults prior to their release; de-winging does not affect viability of the bugs and increases predation in the field by retaining the predators on the plants. Given that releasing nymphs is highly labor intensive, Jenkins et al. (1998) suggested releasing pentatomid predators in the egg stage and to disperse the eggs in the crop by way of carrier gels.

As mentioned in **Section 3.1.1**, secretions from the dorsal abdominal glands of *P. maculiventris* males act as long-range attractants for adults of both sexes. Aldrich et al. (1984a) and Aldrich and Cantelo (1999) suggested the use of such pheromonal attractants to lure large numbers of spined soldier bugs to target areas in early spring, as the bugs emerge from overwintering. Pheromonal attractants for the spined soldier bug are commercially available for use in gardens and small orchards (Anonymous 1992). Thorpe et al. (1994) evaluated the effects of sticky barrier bands, augmentative releases of *P. maculiventris*, and the deployment of *P. maculiventris* pheromone on larval densities of the gypsy moth *Lymantria dispar* (L.) in the canopy of oak trees. Although the augmentation experiments were inconclusive, it was demonstrated that large numbers of immature *P. maculiventris* can be produced using a synthetic pheromone to harvest wild stinkbug adults. Sant'Ana et al. (1997)

showed that nymphs of *P. maculiventris* were attracted to the synthetic pheromone both in the laboratory and in the field, and concluded that the pheromone could be useful in manipulating field populations of immature predators. Aldrich and Cantelo (1999) investigated the potential of pheromone-mediated augmentation to control the Colorado potato beetle. They proposed the use of porous nursery cages retaining adult predators but allowing their offspring to escape. Dispersal of young nymphs into the crop is promoted by peripherally placed pheromone dispensers.

A key feature of IPM programs is the compatibility between the use of natural enemies and other phytoprotection measures. A growing number of studies have focused on the side effects of pesticides on the spined soldier bug and related species. Wilkinson et al. (1979) observed very high mortality of *P. maculiventris* nymphs and adults when exposed by tarsal contact to the organophosphate insecticides profenofos and sulprofos at field rates, but relatively low mortality for the synthetic pyrethroids fenvalerate and permethrin. McPherson et al. (1979) reported that the spined soldier bug was strongly affected by treatments with parathion-methyl under both laboratory and field conditions. Yu (1987, 1988, 1990) determined the toxicity of 13 insecticides applied topically to *P. maculiventris* females and compared the bugs' detoxification capacity to that of their lepidopteran prey. Yu also found that the predator was generally more susceptible to organophosphate and carbamate insecticides, but was more tolerant to pyrethroids, than was its prey. Zanuncio et al. (1993c) and Picanço et al. (1996) reported similar results for other asopine species. De Clercq et al. (1995a) investigated the susceptibility of *P. maculiventris* fifth instars to the insect growth regulators diflubenzuron and pyriproxyfen via direct contact, residual contact, and ingestion. Diflubenzuron was harmless to the predator by direct and residual contact but was highly toxic when ingested; pyriproxyfen caused severe deformities at ecdysis whatever the method of exposure. In a following study, De Clercq et al. (1995b) studied the transport of both insect growth regulators from prey (*Spodoptera exigua*) to predator and their retention inside the predator. De Cock et al. (1996) reported that *P. maculiventris* nymphs and adults suffered high mortality when exposed to the nitroguanidine insecticide imidacloprid and to the thiourea compound diafenthiuron in the laboratory. In their laboratory assays, Smagghe and Degheele (1995) observed no adverse effects of the nonsteroidal ecdysteroid agonists RH 5849 and RH 5992 (tebufenozide) on nymphs and adults of *Podisus* and suggested the use of these novel compounds for the integrated management of lepidopterous pests. Boyd and Boethel (1998a,b) noted that nymphs and adults of *P. maculiventris* experienced high mortality after exposure to foliage treated with the pyrrole chlorfenapyr. In their tests, both permethrin and the avermectin emamectin benzoate were less toxic than methyl parathion to adults, but all these compounds were equally toxic to nymphs. Based on laboratory experiments, Mohaghegh et al. (2000) concluded that the use of deltamethrin and of the microbial insecticide *Bacillus thuringiensis* Berliner ssp. *kurstaki* may be compatible with releases of *P. maculiventris*, but that its populations may be harmed when methomyl or teflubenzuron is applied.

In Russia, much effort has gone into developing an efficient IPM program against the Colorado potato beetle on eggplants. Gusev et al. (1983) stated that *B. thuringiensis* (Bitoxibacillin) at 0.5% was harmless to *P. maculiventris* and could be used successfully in conjunction with releases of the predator; a control scheme in which 0.4% trichlorfon was applied in May, 0.5% Bitoxibacillin was sprayed in July, and 13,000 nymphs of *P. maculiventris* were released in August to September resulted in a markedly greater yield than from untreated fields. Likewise, Izhevskii et al. (1988) and Novozhilov et al. (1991) recommended the use of chemical treatments against overwintered adults and first-generation larvae of *Leptinotarsa decemlineata* and mass releases of *Podisus* nymphs against second and third generations of the beetle.

Biever et al. (1982) and Abbas and Boucias (1984) investigated the feasibility of using *P. maculiventris* to distribute insect pathogens into crop systems. They established that the predator can disseminate nuclear polyhedrosis virus on plant surfaces while searching for prey; the virus had no adverse effects on the bug. Both nymphal and adult predators that had fed on infected prey excreted virulent polyhedral inclusion bodies. In addition, surface-contaminated stinkbugs could also transmit the virus. The release of virus-contaminated bugs resulted in a considerable suppression of field populations of such noctuid pests as the cabbage looper *Trichoplusia ni* and the velvetbean caterpillar *Anticarsia gemmatalis* Hübner. Oien and Ragsdale (1993) found that ingestion of *Nosema furnacalis* (Microsporidia) spores did not cause active infection in *P. maculiventris* and could thus be used in conjunction with the predator to suppress lepidopterous pests.

3.2 *Podisus nigrispinus* (Dallas) (= *Podisus connexivus* [Bergroth])

3.2.1 Distribution, Life History, and Biology

Podisus nigrispinus is one of the most common asopine species in the Neotropical area (Buckup 1960, Grazia et al. 1985). The insect is distributed from Costa Rica into Argentina (Thomas 1992) but has been studied most extensively in Brazil. This polyphagous predator attacks numerous pest insects in a variety of agricultural and sylvicultural ecosystems (Saini 1994a, Zanuncio et al. 1994a).

Because of the lack of a complete and accurate taxonomic revision before the excellent work of D. B. Thomas was published in 1992, there has been considerable confusion concerning the correct identity of several New World Asopinae. Thus, several papers on the biology and rearing of *P. nigrispinus* have been published under the name of the closely related species *P. sagitta* (F.) (De Clercq and Degheele 1995). Further, Thomas (1992) placed *P. connexivus* (Bergroth), subject of numerous studies in Brazil and Argentina, as a junior synonym of *P. nigrispinus*.

Detailed descriptions of immature and adult stages of *P. nigrispinus* were provided by Grazia et al. (1985), De Clercq and Degheele (1990a), and Saini (1994a). Female and male adults are 10 to 12 and 8.5 to 10 mm long, respectively (De Clercq and Degheele 1990a, Zanuncio et al. 1991a). The genitalia were described by Hildebrand (1987).

The life history of *P. nigrispinus* is in many aspects similar to that of the spined soldier bug, *P. maculiventris*. El-Refai and Degheele (1988) and De Clercq and Degheele (1992b) studied nymphal development of laboratory-reared specimens originating from Suriname at various constant temperatures, using *Galleria mellonella* larvae as prey. In the latter study, developmental thresholds for egg and nymphal stages were estimated to be 13.3 and 12.2°C. At 23°C, eggs hatched after about 6 days and nymphs took about 24 days to develop. Similar results were reported by Zanuncio et al. (1991a), Didonet et al. (1996a), Torres et al. (1998), and Saini (1994a), who studied the insect in Brazil and Argentina. Developmental rates of nymphs are further influenced by the type of food offered. Zanuncio et al. (1990, 1993d) reported shorter nymphal periods of *P. nigrispinus* nymphs fed on larvae of the silkworm *Bombyx mori* and the mealworm *Tenebrio molitor*, than on larvae of the housefly *Musca domestica*. Nymphs of *P. nigrispinus* showed lower growth rates when fed on larvae of the Colorado potato beetle *L. decemlineata* than on caterpillars of the wax moth *G. mellonella* or the beet armyworm *S. exigua* (De Clercq 1993). Mateeva (1994), however, noted shortest developmental durations for nymphs reared on larvae of *L. decemlineata* and on *Hyphantria cunea* caterpillars and longest for those fed on caterpillars of *Yponomeuta malinella* (Zeller). Developmental stage of the prey also affects predator development. De Clercq and Degheele (1994) and De Clercq et al. (1998a) observed that development was prolonged when eggs and early instar caterpillars of the noctuids *S. exigua* and *Chrysodeixis chalcites* were supplied as food. When presenting nymphs of *P. nigrispinus* with caterpillars of the cotton leafworm *Alabama argillacea*, Dos Santos et al. (1995, 1996) likewise observed that nymphal duration of the predator was inversely proportional to the size of the prey. Further, suitability of prey for predator development also depends on the defensive behavior of the prey (De Clercq et al. 1988, Saini 1994b). In addition, plant feeding affects nymphal development in *P. nigrispinus*. Zanuncio et al. (1993b) demonstrated that supplementing a diet consisting of *M. domestica* larvae with seedlings of *Eucalyptus urophylla* S.T. Blake yielded heavier adults in a significantly shorter period, which corroborates the findings of Ruberson et al. (1986) for *P. maculiventris*.

Sex ratios in *P. nigrispinus* are usually reported to be around 1:1.2 (males:females) (Zanuncio et al. 1991a, De Clercq and Degheele 1992b), although Saini (1994a) noted a slightly higher proportion of females (~1:1.7). Mating behavior was described by De Clercq and Degheele (1990a,b), Carvalho et al. (1994), and Torres et al. (1997). In the laboratory, females start producing eggs about 1 week after emergence. As for *P. maculiventris* (see **Section 3.1.1**), reported values of fecundity and longevity differ strongly. Whereas Zanuncio et al. (1991a,b) mentioned that this bug produces 100 to 200 eggs during a lifetime of about 1 month, De Clercq and Degheele (1990a,b; 1993c) recorded mean fecundities of 600 to 900 eggs per female and longevities of 2 to 3 months. Torres et al. (1998) reported intermediate fecundity and longevity values of about 400 eggs per female and 2 months, respectively, at temperatures fluctuating between 15 and 25°C. In their laboratory studies, De Clercq and Degheele (1990b) observed highest oviposition rates at 27°C and 75% relative humidity; extreme temperatures and humidities reduced the reproductive capacity of the predator. De Clercq and Degheele (1992c) further established that poorly fed *P. nigrispinus* females lived as least as long as their well-fed peers, but reduced their reproductive outputs as a function of food supply (see also **Section 3.1.4**). Effects of maternal age and egg weight on developmental time and body weight of offspring were studied by Mohaghegh et al. (1998b).

Little is known about the phenology of *P. nigrispinus* in the field. According to Saini (1994a), the insect has two to three generations in Buenos Aires, Argentina. Cold tolerance of the different life stages in the laboratory has been documented by De Clercq and Degheele (1993a) (see **Section 3.1.1**).

3.2.2 Food Preferences

Like *P. maculiventris*, *P. nigrispinus* is a markedly polyphagous predator that attacks mainly soft-bodied lepidopteran and coleopteran larvae. The predator has been reported to prey also on other developmental stages (De Clercq and Degheele 1994). In addition, it feeds on insects from other orders, including Diptera (Zanuncio et al. 1993b,d) and Heteroptera (Saini 1994a). The plant-feeding habit of *P. nigrispinus* has been documented by De Clercq and Degheele (1992d) and Zanuncio et al. (1993b) (see **Sections 3.1.2 and 3.1.4**).

3.2.3 Rearing

The rearing of *Podisus* bugs has been discussed extensively in **Section 3.1.3**. De Clercq et al. (1988) and De Clercq and Degheele (1993c) easily produced large numbers of *P. nigrispinus* on caterpillars of *G. mellonella* and *Spodoptera* spp. A meat-based artificial diet was sufficient to sustain continuous generations of this predatory bug (De Clercq and Degheele 1992a, 1993b,c; Saavedra et al. 1995, 1996, 1997). Much of the work done on the mass-breeding of *P. nigrispinus* and other asopines in Brazil has been reviewed by Zanuncio et al. (1994a); best results were obtained when using larvae of *B. mori*, *T. molitor*, and *M. domestica* as food.

3.2.4 Control Potential

Relatively few studies have focused on the predation characteristics of *P. nigrispinus*. Searching and attacking behavior was described by De Clercq (1993) and was generally similar to that of *P. maculiventris* (see **Section 3.1.4**). De Clercq and Degheele (1994) and De Clercq et al. (1998a) measured predation by *P. nigrispinus* and *P. maculiventris* on the beet armyworm *S. exigua* and the tomato looper *C. chalcites* in the laboratory. Both predators displayed high predation rates against different life stages of the noctuids, both active and immobile. There was a clear preference for third to fourth instars and older, depending on the size of the predator. Predation rates of *P. nigrispinus* were generally similar to those of the larger *P. maculiventris*. Nymphs of the former were, however, better able to use first instars of the beet armyworm. Male adults of *P. nigrispinus* had considerably lower attack rates than females, whereas attack rates in males and females of *P. maculiventris* were similar.

Saini (1994b), evaluating the influence of prey behavior on the predator efficiency of *P. nigrispinus* in the laboratory, found that the lepidopteran soybean pests *Rachiplusia nu* (Guenée) (Noctuidae) and *Colias lesbia* (F.) (Pieridae) were most commonly attacked because of their poor resistance to attack. Conversely, the noctuids *S. frugiperda* and *Helicoverpa zea* offered greater resistance and consequently were less preyed upon (see also **Section 3.1.4**).

Bellotti et al. (1992) stated that the effectiveness of such generalist predators as *P. nigrispinus* in the control of the cassava hornworm *Erinnyis ello* (L.) was limited by poor functional response during hornworm outbreaks. In Brazil, however, *Podisus* species are believed to be important natural enemies of caterpillar pests in cotton, coffee, and soybean (Gravena and Lara 1982). Furthermore, Zanuncio et al. (1994a) emphasized the potential role of *P. nigrispinus* and related species in the integrated management of a number of primary and secondary caterpillar pests of *Eucalyptus* in Brazil. Despite the lack of fundamental studies on its predation capacity and functional response, *P. nigrispinus* has been routinely mass-reared and released for several years to suppress lepidopterous defoliators in commercial *Eucalyptus* plantations (J. C. Zanuncio, personal communication).

Zanuncio et al. (1993c) pointed out that despite the action of asopine predators as natural control agents, caterpillar pests of *Eucalyptus* sometimes achieve high population densities requiring the use of selective insecticides. In this respect, these authors evaluated the impact of the pyrethroid deltamethrin in aerial applications against *Eucalyptus* caterpillars and their asopine predators. They reported that the treatments were effective against several economically important defoliator pests but had very low impact on populations of *Podisus* spp. Likewise, Picanço et al. (1996) reported that the asopines *P. nigrispinus* and *Supputius cincticeps* (Stål) (see **Section 4.1**) showed higher tolerance to deltamethrin than to fenthion, malathion, and cartap. In their laboratory trials on the selectivity of tebufenozide, Smagghe and Degheele (1995) observed no ill effects of this novel ecdysteroid agonist against different life stages of *P. nigrispinus* (see also **Section 3.1.5**). In contrast to reports on other asopines, Nascimento et al. (1998) found that development and reproduction of *P. nigrispinus* were negatively affected when it was continuously fed on caterpillars treated with *Bacillus thuringiensis* var. *kurstaki*. Because the pathogen could not be retrieved in the predator's hemolymph, adverse effects were attributed in part to lower nutritional quality of the contaminated prey. Torres et al. (2000) demonstrated that the predator was less susceptible to the microbial insecticide spinosad than its lepidopterous prey *Spodoptera frugiperda* and *Tuta absoluta* (Meyrick).

3.3 *Perillus bioculatus* (F.)

3.3.1 Distribution, Life History, and Biology

The two-spotted stinkbug or double-eyed soldier bug, *Perillus bioculatus*, is a predatory pentatomid native to North America. According to Knight (1923), this insect originated in the southern Rocky Mountains and apparently followed the migration of its primary prey, the Colorado potato beetle, *L. decemlineata*. Nowadays, *P. bioculatus* is distributed from Mexico into Canada (Franz and Szmidt 1960, McPherson 1982, Thomas 1992).

Knight (1923) was the first to provide a detailed description of the different life stages of *P. bioculatus*. Female adults are 10 to 12 mm long and males 8 to 10 mm. Knight (1923, 1952) demonstrated how temperatures and prey type influenced the red and black dorsal color pattern in adults of this bug. Details on egg morphology and development were presented by Knight (1923), Tamaki and Butt (1978), and Javahery (1994). A detailed description of the instars was given by DeCoursey and Esselbaugh (1962).

Bionomics of *P. bioculatus* were investigated by several workers in North America and Europe (e.g., Knight 1923; Trouvelot 1931; Landis 1937; Franz and Szmidt 1960; Moens 1963, 1965; Shagov 1967a,b,c, 1969; Tremblay 1967; Tremblay and Zouliamis 1968; Bjegovic 1971; Tamaki and Butt 1978). Eggs are laid in batches consisting of 10 to 25 eggs. At temperatures lower than 15°C, no embryonic development has been observed. At 20 to 25°C, eggs hatch within 5 to 8 days.

As in other asopines, first instars are highly gregarious. However, whereas the late instars of other stinkbugs are not gregarious or only slightly so, in *P. bioculatus* gregariousness is still apparent up to the fourth instar. Lachance and Cloutier (1997) found that temperature, interactions with conspecifics, stage and physiological age all affect dispersal of predator nymphs. Whereas early instars aggregate more than late instars, within a stadium physiologically more advanced nymphs usually form larger and more stable groups than less mature ones. Further, nymphs tend to cluster as temperature drops below 19°C. Usually, first instars do not feed and only require moisture, mainly in the form of plant juices, to become second instars. Tamaki and Butt (1978), however, reported occasional feeding of first instars on eggs of the Colorado potato beetle. Shortly after molting to the second instar, nymphs start to attack prey. Although Knight (1923) believed that *P. bioculatus* feeds almost exclusively on Colorado potato beetle, in the laboratory this predator was observed to feed on insects from different orders (Froeschner 1988, Heimpel 1991, Biever and Chauvin 1992a, Hough-Goldstein et al. 1996, Saint-Cyr and Cloutier 1996); in the field, however, it is usually found in association with coleopterous insects, especially Chrysomelidae (Knight 1923, Landis 1937, Strickland 1953, Franz 1957, Altieri and Whitcomb 1979, Logan et al. 1987, Eckberg and Cranshaw 1994). Specialization in *P. bioculatus* toward chrysomelid prey was confirmed by Saint-Cyr and Cloutier (1996), who demonstrated a genetically inheritable as well as maternally reinforcible affinity toward the Colorado potato beetle in the laboratory.

Development of the five instars takes on average 3 weeks at 20 to 25°C and depends mainly on temperature conditions and type of food. Shagov (1968) estimated lower developmental thresholds to be 14.6 to 16.5°C. Volkovich et al. (1990) found that thermorhythms accelerated nymphal growth over constant temperatures and estimated the developmental threshold for nymphs to be around 14°C. Developmental rates reported for nymphs fed on eggs are higher than those of nymphs fed on larval or adult prey. Landis (1937) found that feeding on eggs of *L. decemlineata* yielded higher survival and developmental rates than did feeding on prey larvae. Landis also reported that prey of the same species, but reared on different food plants, caused considerable variation in developmental rates and survival of predator nymphs (see also **Section 3.1.1**). Shorter developmental times and higher survival occurred when predator nymphs were offered a diet of *L. decemlineata* eggs, than when offered diets consisting of larvae or adults of the prey (Franz and Szmidt 1960). Moens (1963) reported a mean nymphal period of 26 days at 22 to 23°C and observed a significant decrease in survival and growth rate of predator nymphs when transferred from a diet consisting of larvae and adults of *L. decemlineata* to a diet consisting of caterpillars of the cabbage moth, *Mamestra brassicae* L. According to Tamaki and Butt (1978), nymphs take approximately 18 days to complete development at 24°C when fed on eggs of *L. decemlineata*, whereas they take approximately 21 days when fed on third and fourth instars of the beetle.

Sex ratios in laboratory-reared *P. bioculatus* are usually around 1:1 (Knight 1923, Franz and Szmidt 1960). At 20 to 25°C, females have a mean fecundity of 100 to 200 eggs, which is considerably lower than that reported for *Podisus maculiventris*. Feytaud (1938) stated that the lower threshold for oviposition is 29°C. In contrast, Volkovich et al. (1990) found that thermorhythms with a cold component below 14°C and where the absolute value of the amplitude exceeds 16 to 19°C result in optimal adult fecundity and survival. Tamaki and Butt (1978) constructed life tables for *Perillus bioculatus* at 22°C and calculated an intrinsic rate of increase (r_m) of 0.08 and a net replacement rate (R_0) of 46 females per female.

Males of *Perillus* do not have enlarged dorsal abdominal glands that release aggregation pheromones, like those of *Podisus*, but they do possess abdominal sternal glands. The function of their secretions is, however, still unclear. In the laboratory, starvation followed by engorgement of prey stimulated secretion of sternal glands in *Perillus bioculatus* males; also, secretion has been observed during courtship (Aldrich 1995, 1998).

Given the wide distribution of the two-spotted stinkbug throughout North America, the reported number of generations per year varies, depending on prevailing climatic conditions. Knight (1923) and Tamaki and Butt (1978) estimated that in the United States and Canada the predator has two to three generations per year. Also in different parts of Europe, where *P. bioculatus* was introduced to control Colorado potato beetle, two to three generations were observed (Moens 1963, Jasic 1975).

There is conflicting evidence on the overwintering ability of *P. bioculatus* in northern climates. Trouvelot (1931) noted that the stinkbug mainly occurs where winters are mild and where there is abundance of shelter for hibernation. Knight (1923) reported up to 95% overwintering mortality in New York State (U.S.A.). Likewise, Franz and Szmidt (1960), Le Berre and Portier (1963), and Jasic (1975) have reported high mortality of overwintered bugs in Germany, France, and Slovakia, respectively. However, Moens (1963) found that bugs that had continually been reared outdoors survived winter in Belgium well and showed greater hardiness and fecundity than those kept indoors. Tremblay and Zouliamis (1968) and Bjegovic (1971) also mentioned successful overwintering of the bug in some mountainous regions of southern Italy and in Yugoslavia, respectively. Shagov (1969) and Shutova et al. (1976), studying the overwintering of *P. bioculatus* in the Ukraine and southern parts of European Russia, found that it has a facultative diapause from September to December, whereas from January to March inhibition of metabolism is maintained by cold-induced torpor. Despite the low overwintering survival observed by other workers, Shutova et al. (1976) and Shagov and Chesnek (1978) have demonstrated that, particularly, diapausing adults possess considerable cold hardiness and can survive cold Russian winters. Jasic (1975) and Jermy (1980) concluded that overwintering success in *P. bioculatus* may be related to the ability to find sufficient food in the autumn to prepare for diapause. Food quality also influences photoperiodic response (Shagov 1977, Horton et al. 1998); the latter authors reported that females fed large *L. decemlineata* larvae showed a greater tendency to enter diapause than females fed a mix of eggs and small larvae. Details on the influence of thermoperiodic conditions on the photoperiodic induction of diapause can be found in Volkovich et al. (1990).

3.3.2 Rearing

Rearing procedures for *P. bioculatus* have been described in detail by Franz and Szmidt (1960) and Tremblay (1967). The former reported better results when the predator was offered *L. decemlineata* eggs than when reared on larvae or adults of the prey. They also reared nymphs successfully on frozen larvae of the Colorado potato beetle, but development was prolonged by more than 15 days and mortality was up to 80%. Biever and Chauvin (1992a) and De Clercq (unpublished data) found that nymphs and adults of *P. bioculatus* can be reared successfully on coddled larvae of the cabbage looper, *Trichoplusia ni* (Hübner), and of the cotton leafworm, *Spodoptera littoralis* (Boisduval), respectively; this method is less expensive than rearing on Colorado potato beetle, because the noctuids can be reared on artificial diet. Hough-Goldstein and McPherson (1996) reported successful mass-rearing of the predatory bug using a combination of larvae of the Mexican fruit fly, *Anastrepha ludens* (Loew), and of the European corn borer, *Ostrinia nubilalis* (Hübner).

3.3.3 Predator–Prey Relationships

During nymphal development, *P. bioculatus* can consume over 300 Colorado potato beetle eggs, 3 to 4 fourth instars, or 5 adult beetles; about 70% of this number is consumed by the last instar. Adults kill about 10 eggs, 1 adult beetle, and 0.5 to 1 fourth instars of *L. decemlineata* per day (Franz and Szmidt 1960, Moens 1963, Tamaki and Butt 1978). Feytaud (1938) stated that *P. bioculatus* prefers eggs and larvae of the Colorado potato beetle, but will eat adults; Le Berre and Portier (1963), however, observed a clear preference for eggs of the chrysomelid. Whereas mandibulate predators may eat only a portion of an egg mass, *P. bioculatus* has the habit of sucking out every egg in the mass; this behavior suggests an ability to suppress the first brood of the Colorado potato beetle if sufficiently high numbers of the predator are present early enough in the season (Hough-Goldstein et al. 1993). Hough-Goldstein and McPherson (1996) noted that nymphs of *P. bioculatus* did not discriminate by size of Colorado potato beetle larvae, whereas larger

P. maculiventris nymphs tended to select larger prey larvae. Cloutier (1997) reported that in *P. bioculatus* gregariousness enhances development of smaller nymphs by opportunistic feeding on prey killed by larger conspecific nymphs; this type of communal feeding allows small nymphs to have access to relatively large prey, which they would be unable to subdue themselves.

Although chemical cues were until recently not believed to be important for prey finding by asopine predators including *P. bioculatus* (see **Section 3.1.4**), Dickens (1999) and Weissbecker et al. (1999) found that the predator is responsive to plant volatiles released by potato plants in response to feeding by *L. decemlineata* larvae. Feytaud (1938) and Franz (1967) observed that *Perillus* bugs can see their prey from a few centimeters away. Gusev et al. (1980) stated that third instars of *P. bioculatus* can locate eggs of the Colorado potato beetle 10 and 20 cm away in 4 and 10 min, respectively. The Russian workers found Perillus nymphs to be less mobile than *Podisus* nymphs; the greater mobility of the latter was considered to be an adaptation to feeding on mobile prey (see also **Section 3.1.4**). Heimpel and Hough-Goldstein (1994b) argued that *Perillus* nymphs may detect their prey by short-range visual cues but that recognition of prey is primarily based on antennal contact. In both artificial arena and plant observations, these workers found nymphs of *P. bioculatus* to engage more intensively in area-restricted search than did nymphs of *Podisus maculiventris*. This different searching behavior was attributed to the less diverse array of prey attacked by *Perillus bioculatus* than by *Podisus maculiventris*. As a "specialist" of chrysomelid prey, *Perillus bioculatus* was believed to be better adapted to the aggregated distribution of eggs and young larvae of Chrysomelidae. Similar presumptions were made by Evans and Root (1980) for the predatory stinkbug *Apateticus bracteatus* (Fitch) (see **Section 4.1**), which appears to be a specialist on larval forms of Trirhabda spp. (Chrysomelidae) in goldenrod stands. In their study on the functional response of fifth instars of *P. bioculatus* attacking neonate larvae of *L. decemlineata*, Heimpel and Hough-Goldstein (1994a) found that neither type II nor type III functional response curves provided a satisfactory fit to the data. This was explained by the fact that area-restricted search following prey consumption may lead to additional successful feeding events at high densities of prey but not at low densities.

3.3.4 Control Potential

Although Trouvelot (1931) mentioned that the two-spotted stinkbug was occasionally responsible for significant reductions of *L. decemlineata* populations in North American potato fields, several other studies in North America have considered the potential of *P. bioculatus* for the natural control of the Colorado potato beetle rather limited. Knight (1923) noted that the stinkbug hardly ever appears in sufficiently large numbers during the first brood of the beetle in Minnesota (U.S.A.). According to Harcourt (1971), *L. decemlineata* populations in Canada are only little affected by predators such as *P. bioculatus*. The projections of Tamaki and Butt (1978) from life tables, predation rates, and defoliation curves also indicated little prospect for biological control of the Colorado potato beetle by the pentatomid in Washington State (U.S.A.). Likewise, Boiteau (1987, 1988) believed that pentatomid predators like *P. bioculatus* and *Podisus maculiventris* had little promise for natural control of *L. decemlineata* given the presumed lower predatory activity in Canadian potato fields. Based on the lower developmental thresholds of *L. decemlineata* and *Perillus bioculatus* (about 10 and 14°C, respectively) (Logan and Casagrande 1980, Volkovich et al. 1990), Giroux et al. (1995) also concluded that *P. bioculatus* is less effective at lower temperatures than other predators such as *Podisus maculiventris* and the ladybird beetle *Coleomegilla maculata* DeGeer.

Like *P. maculiventris*, *Perillus bioculatus* has been introduced into a number of European countries to control the Colorado potato beetle. Although extensive releases were made, no permanent establishment of these asopine predators has occurred (Franz and Szmidt 1960, Bjegovic 1971, Jermy 1980, Lipa 1985, Boiteau 1988, Pruszynski and Wegorek 1991). In his review of the European results with *P. bioculatus*, Jermy (1980) attributed failure to establish in part to the high tendency of released adults to disperse, which limits the probability of mating and survival of small inoculative populations. Further, Le Berre and Portier (1963) and Moens (1963) reported that activity and survival of *P. bioculatus* was severely affected by cold snaps in the spring. In addition, mass-rearing and release of the predator against the first generation of the Colorado potato beetle were considered more expensive than the use of insecticides (Kahlow 1963, Jermy 1980).

Nevertheless, results from several studies in both Europe and North America have indicated significant control of *L. decemlineata* by the two-spotted stinkbug in small plots and field cages. In their comprehensive review on the arthropod natural enemies of the Colorado potato beetle, Hough-Goldstein et al. (1993) summarized the conditions under which releases of *P. bioculatus* were successful: (1) if sufficient numbers were released; (2) if young nymphs (second to third instars) were used; and (3) if releases were timed so that only eggs and newly hatched larvae of the pest were present at the time of release. In an updated review on the use of predatory pentatomids against the Colorado potato beetle, Hough-Goldstein (1998) specified that the most critical period for the predators to suppress populations of the beetle is before and during the bloom period of potato.

In their report on the first results obtained with the bug in France, Le Berre and Portier (1963) mentioned that a release of about ten *Perillus* (late egg stage to second instars) per potato plant resulted in a 50% reduction of the egg population of *L. decemlineata*. Szmidt and Wegorek (1967) stated that even during heavy infestations by the beetle in Poland, releasing only 2.6 second to fourth instars of *P. bioculatus* per plant was sufficient to suppress the pest. According to Jermy (1966), 6.6 *Perillus* nymphs per plant at a density of 2.5 beetles per plant resulted in negligible damage by the pest in Hungary. Recent studies in the United States and Canada have also shown that *P. bioculatus* can be efficacious in field releases. Field-cage tests performed by Biever and Chauvin (1992a,b) established that releases of 5 to 10 third instars per plant of both *P. bioculatus* and *Podisus maculiventris* were able to reduce high-density populations of Colorado potato beetle by 50%; however, *Perillus bioculatus* provided greater protection of foliage than *Podisus maculiventris*, perhaps by killing the beetle at an earlier stage than did the latter. Subsequent field plot tests demonstrated that a release rate of three *Perillus bioculatus* nymphs per plant suppressed populations of the beetle by 62%; three releases of the predator provided greater foliage protection than a single release. Hough-Goldstein and Whalen (1993) reported that inundative releases of three second instars per plant reduced pest numbers to near zero. In contrast with the results of Biever and Chauvin (1992a), Hough-Goldstein and McPherson (1996) found that in small field plots *P. bioculatus* and *Podisus maculiventris* were equally effective at reducing populations of Colorado potato beetle larvae and protecting foliage from defoliation. To overcome drawbacks related to the transport and release of nymphs, Hough-Goldstein et al. (1996) investigated the possibility of releasing the pentatomid as eggs placed in screened plastic cups at the base of plants; this provided nearly as good control of Colorado potato beetle as when released as nymphs. When eggs were scattered in the foliage, only few survived, probably because of predation by other organisms. Jenkins et al. (1998) suggested the use of gels as carriers for eggs of *Perillus bioculatus* and *Podisus maculiventris*, but most of the tested gels had negative effects on egg hatch, particularly at the high viscosities needed to stick eggs to plant leaves. Cloutier and Bauduin (1995) demonstrated that augmentative releases of *Perillus bioculatus* can control first-generation eggs and larvae of *L. decemlineata* under Québec, Canada, short-season conditions, contradicting the concerns regarding the presumed ineffectiveness of the predator under cool climates. In their field experiments, these workers released second and third instars of the predator during spring oviposition of the beetle. They achieved 80% control of potential damage that might be expected from about 300 Colorado potato beetle eggs per plant by releasing three *P. bioculatus* nymphs per plant. This was obtained by releasing predators at 12 per plant on only 25% of plants, meaning that 75% of plot coverage was via active dispersal of the predator nymphs. In contrast to earlier reports indicating the preference of *Perillus* for eggs (see **Section 3.3.2**), high predation on first to third instars of the pest was observed in this study.

Several studies have assessed the sensitivity of *P. bioculatus* to commonly used pesticides. Knight (1923) observed that the pentatomid was much more sensitive to arsenicals than its prey, *L. decemlineata*, and stated that feeding on poisoned prey and sucking contaminated plant sap markedly decreased the predator population. In small field-cage experiments, all adult bugs exposed to field rates of endosulfan and carbaryl died, whereas calcium arsenate caused only 20% mortality (Franz and Szmidt 1960). These workers also found very low mortality with the fungicidal dithiocarbamates dithane and maneb, but copper- and tin-based fungicides had more severe effects. Wegorek and Pruszynski (1979) reported that chlorfenvinphos and phosalone were more toxic to the Colorado potato beetle than to *P. bioculatus*; carbaryl, propoxur, and a mixture of carbaryl and lindane were found to be more toxic to the predator, highest mortality being observed in the case of direct application. Topical application of neem seed extract to nymphs delayed molting and caused deformities after the molt in some insects, whereas rotenone and the synergist piperonyl butoxide caused significant mortality (Hough-Goldstein and Keil 1991). In the same study, cryolite, *Bacillus thuringiensis* var. *san diego*, horticultural oil, insecticidal soap, and the fungicides chlorothalonil, maneb, and metalaxyl were found relatively harmless to different life stages of the predator. Hough-Goldstein and Whalen (1993) reported that the systemic soil insecticides disulfoton, phorate, and aldicarb caused high mortality of *P. bioculatus* nymphs exposed to potato foliage 11 weeks after soil application, but ethoprop, carbofuran, and imidacloprid had no adverse effects; the foliar insecticides cyromazine, imidacloprid, and phosmet plus piperonyl butoxide had no effect on predator nymphs exposed to recently sprayed foliage. Carbofuran applications suppressed populations of *P. bioculatus* and other predators of the Colorado potato beetle in commercial potato fields in North Carolina (U.S.A.), but the predators recolonized the fields within 1 to 2 weeks following application (Hilbeck and Kennedy 1996).

Using an IPM approach to suppress the Colorado potato beetle with *B. thuringiensis* var. *tenebrionis*, Hough-Goldstein and Keil (1991) and Hough-Goldstein and Whalen (1993) found that *P. bioculatus* significantly added to the control level obtained with the bioinsecticide. Cloutier and Bauduin (1995) and Cloutier and Jean (1998) also established that spraying *B. thuringiensis* and releasing the stinkbug, both alone and in combination, produced significant control of the pest. In combined applications of *P. bioculatus* and the fungus *Beauveria bassiana* (Balsamo) Vuillemin, however, the predator appeared to provide no additional protection against the Colorado potato beetle than the fungus did alone (Poprawski et al. 1997). Finally, Overney et al. (1998) warned about the effects of protease inhibitors in transgenic crops on nontarget organisms: they demonstrated that oryzacystatins in transgenic potato plants not only affect digestive proteases of the Colorado potato beetle, but also those of its most important predator, *P. bioculatus*.

3.4 *Eocanthecona furcellata* (Wolff)

3.4.1 Distribution, Life History, and Biology

In recent years, *E. furcellata* has been studied increasingly as a control agent of agricultural insect pests in southeastern Asia and India. This Oriental pentatomid is documented as a predator of various, mostly lepidopteran, pests of agricultural crops and forest plantations in the Indian subcontinent (Vasantharaj and Basheer 1962, Ghorpade 1972, Kapoor et al. 1973, Ahmad et al. 1974, Rai 1978, Chandra 1979, Gope 1981, Prasad et al. 1983, Jakhmola 1983, Ahmad and Rana 1988, Senrayan 1991), Indonesia (Sipayung et al. 1992), Japan (Yasuda and Wakamura 1992), Taiwan (Chu and Chu 1975), Thailand (Burikam and Napompeth 1978, Kirtibutr 1987), and the Philippine Republic (Cahatian 1991, Tabasa 1991). *Eocanthecona furcellata* is reportedly distributed from the Okinawa region in Japan to Pakistan in the east and Indonesia in the south. However, Sipayung et al. (1992) noted the limited knowledge on the taxonomy of *Eocanthecona* and hypothesized that some bugs studied under the name of *E. furcellata* may in fact be different species. The predator was introduced into Florida (U.S.A.) as a potential biocontrol agent of the

Colorado potato beetle and tent caterpillar, but it apparently never became established there (Froeschner 1988).

Information on the morphology and life history of this predatory stinkbug is rather scattered. Male and female adults are 10 to 12 and 14 to 16 mm long, respectively. Sen et al. (1971) and Chu and Chu (1975) described the different life stages of *E. furcellata*. Ahmad and Rana (1988) provided a key to eight species of the genus (as *Canthecona*) from the Indo-Pakistan subcontinent, including *E. furcellata*. Sipayung et al. (1992) described the species of the *Eocanthecona–Cantheconidea* complex in Indonesia, including a detailed description of immatures and adults of *E. furcellata*. Thomas (1992, 1994) presented a key to the genus *Eocanthecona* and gave some details on the morphology of *E. furcellata*.

This species is adapted to warmer climates. At 25 to 30°C, eggs from bugs originating from India, Taiwan, the Philippine Republic, and Japan are reported to hatch in about 1 week (Sen et al. 1971, Chu and Chu 1975, Choudhary et al. 1989, Tabasa 1991, Yasuda and Wakamura 1992). Chu and Chu (1975) estimated a lower threshold for development of eggs, 16.3°C, but Usha Rani (1992) said that egg development is arrested at temperatures below 20°C. Constant temperatures of 35 to 40°C are also reported to be detrimental to eggs (Chu and Chu 1975, Usha Rani 1992). Literature data on development of immature *E. furcellata* are difficult to compare because of differences in temperatures and food used in the various studies. At constant temperatures of 25 to 30°C, nymphs take 15 to 25 days to develop to adulthood (Sen et al. 1971, Chu and Chu 1975, Zhu 1990, Tabasa 1991, Sipayung et al. 1992, Usha Rani 1992, Yasuda and Wakamura 1992, Kobayashi and Okada 1994, Ahmad et al. 1996). Lower developmental thresholds for nymphs are estimated to be around 15°C (Chu and Chu 1975). Usha Rani (1992) stated that temperatures of 20°C or below result in longer development and lower adult weights than does a temperature of 25°C. At or above 30°C, on the other hand, the bugs were more active and consumed more food, which yielded heavier adults; however, such high temperatures also caused an increase in cannibalistic behavior of the nymphs. In Madhya Pradesh, India, the total life cycle from egg to adult in an insectary took about 2 months during the cooler period from October to December (Singh et al., 1989).

Males and females are generally produced in equal numbers (Tabasa 1991, Kobayashi and Okada 1994). In the laboratory, mating takes place 3 to 7 days after emergence and the preoviposition period ranges from 4 to 13 days (Sen et al. 1971, Chu and Chu 1975, Tabasa 1991). The mating period is shortened at temperatures above 30°C and prolonged below 20°C (Usha Rani 1992). Oviposition is reportedly initiated at 20 to 25°C (Usha Rani 1992). Again, large differences exist in the literature concerning the level of reproduction of *E. furcellata* under laboratory conditions (see **Section 3.1.1**). Sen et al. (1971), Chu and Chu (1975), Choudhary et al. (1989), Zhu (1990), and Ahmad et al. (1996) found that females produce 50 to 300 eggs during a total life span of 15 to 30 days. This is in sharp contrast to the findings of Yasuda and Wakamura (1992), who mentioned a total fecundity of 500 to 600 eggs per female and an average longevity of 3 months.

There are only a few studies on the life cycle of *E. furcellata* in the field. Sen et al. (1971) estimated that the bug has about five generations per year in the Bihar region of India; they noted that the insect is active all year except from December to February and in April, when it was presumed to be hibernating. Based on degree-day accumulations, Chu and Chu (1975) also believed that *E. furcellata* has five to six generations in northern Taiwan. Prasad et al. (1983) and Choudhary et al. (1989) reported that this predator is active during the monsoon season in central India. Zhu (1990) mentioned that *E. furcellata* is not very resistant to low temperatures.

3.4.2 Food Preferences

This pentatomid is a polyphagous predator. As in other asopines, the predatory habit of *E. furcellata* is developed from the second instar on. It has been reported to feed on a variety of lepidopterous insects, including Noctuidae (Kapoor et al. 1973, Chandra 1979, Choudhary et al.

1989, Zhu 1990, Cahatian 1991, Tabasa 1991, Usha Rani 1992, Yasuda and Wakamura 1992, Dai and Yang 1993), Arctiidae (Singh et al. 1989, Senrayan and Ananthakrishnan 1991), Pyralidae (Jakhmola 1983, Zhu 1990), Hesperiidae (Zhu 1990), Pieridae (Chu and Chu 1975, Zhu 1990), Lasiocampidae (Zhu 1990), Limacodidae (Senrayan 1991, Sipayung et al. 1992), Saturnidae (Sen et al. 1971), and Thaumetopoeidae (Senrayan and Ananthakrishnan 1991). Further, it has been found to prey on the larvae of different chrysomelids (Kirtibutr 1987, Froeschner 1988).

Senrayan and Ananthakrishnan (1991) reported that the slug caterpillar *Parasa lepida* (Cramer) supported reproduction better than did *Eupterote mollifera* W. and *Pericallia ricini* F., both hairy caterpillars; further, feeding on young prey stages decreased egg output and longevity drastically (see also **Section 3.1.4**).

Plant feeding by *Eocanthesona furcellata* has been observed by several workers. Senrayan (1991) reported that nymphs managed to reach the third instar when provided only leaves of the mango tree *Mangifera indica*. When a diet of insect prey (*Parasa lepida*) was supplemented with *M. indica* leaves, developmental time was reduced and resulting adults were larger. It was concluded that plant feeding may help to sustain predator populations in times of prey scarcity. Tabasa (1991) further noted that damage inflicted to plants by sucking of plant sap is negligible (see also **Sections 3.1.2 and 3.1.4**).

3.4.3 Rearing

Rearing methods for *E. furcellata* have been described by Chu (1975), Sipayung et al. (1992), Yasuda and Wakamura (1992), and Usha Rani and Wakamura (1993), and are generally similar to those developed for other asopine bugs. Chu (1975) mentioned that handling ~40,000 eggs and breeding 4000 nymphs of this pentatomid required 8 hours, excluding prey-rearing operations. The predator is easily reared on several caterpillars, such as the noctuid *Spodoptera litura* (F.) (Yasuda and Wakamura 1992). Sipayung et al. (1992) and Yasuda and Wakamura (1992), however, found that development and reproduction of predators fed frozen caterpillars were better than or similar to that of bugs fed live prey; therefore, frozen caterpillars may be a suitable diet for the mass production of *E. furcellata*. Kobayashi and Okada (1994, 1995) reared nymphs of the predator on several alternative foods, including meat-based artificial diets (see **Section 3.1.3**).

3.4.4 Predator–Prey Relationships and Control Potential

Several workers have measured the predation capacity of *E. furcellata* on a variety of prey in the laboratory. Choudhary et al. (1989) reported that an individual predator consumed an average of 38 larvae of the noctuid *Rivula* sp. during the nymphal stage; adults consumed 112 caterpillars during their entire lifetime or about 6 larvae per day. Zhu (1990) found that adult *E. furcellata* consume one to three larvae of the pyralid *Corcyra cephalonica* (Stainton) per day. Usha Rani (1992) found that the prey capture efficiency of *E. furcellata* on *S. litura* third instars was low at temperatures below 25°C and increased with temperature up to 35°C; males consistently killed more prey than did females. This contrasts with the findings of Tabasa (1991), that *E. furcellata* females killed an average of 121 *S. litura* larvae during their lifetime, but males killed 107. Tabasa (1991) also found that the functional response curves to the density of *S. litura* larvae followed a Holling's type II response. Ahmad et al. (1996) reported that a single individual of *E. furcellata* consumes on average 160 second instars, 115 third instars, 90 fourth instars, and 86 fifth instars of the notodontid poplar pest *Pygaera* (= *Clostera*) *cupreata* (Butler).

Usha Rani and Wakamura (1993) investigated the searching behavior of *E. furcellata* toward larvae of *S. litura*. The stinkbug did not use size, shape, or movement of the prey as a primary cue for prey selection, although there was a preference for moving prey in a choice situation. As McLain (1979) found for *Podisus maculiventris*, *E. furcellata* was stimulated to search by the feces of the prey. Usha Rani et al. (1994) suggested that the chemotactic sensilla at the rostral tip are mainly responsible for prey detection and acceptance. Yasuda and Wakamura (1996) and Yasuda (1997; 1998a,b) observed that *E. furcellata* is attracted to odors from both intact and dead *S. litura* larvae, and to solvent extracts of the prey. Their data indicated the presence of at least two different chemical cues in prey location; one of these, (*E*)-phytol, is derived from the chlorophyll ingested by the prey (Yasuda 1998a,b). These findings suggest that searching by this and other predatory stinkbugs may not be as random as generally thought (see also **Section 3.1.4**). Usha Rani et al. (1994) noted that predation on larger caterpillars was limited by aggressive defense behavior of the prey. Like other asopines, *E. furcellata* appears to dislike hairy caterpillars (Senrayan and Ananthakrishnan 1991).

As early as the 1930s, an attempt was made to use *E. furcellata* to control insect pests in the tea and gambir plantations of Sumatra (de Jong 1931, Schneider 1940). Kirtibutr (1987) and Kumar et al. (1996) mentioned that *E. furcellata* plays an important role in the population dynamics of the tortoise beetle *Calopepla leayana* (Latreille), a chrysomelid pest of the yemane tree *Gmelina arborea* Roxb. in Thailand and India, respectively. These workers believed that the stinkbug is a promising biocontrol agent for this and other defoliator pests if the bugs can be mass-reared. Desmier de Chenon et al. (1989) and Sipayung et al. (1992) released mass-reared *E. furcellata* over several years to suppress caterpillar pests of oil palm in Indonesia, in particular the limacodids *Setothosea asigna* Van Eecke and *Setora nitens* Walker. Adult predators released in oil palm plantations dispersed quickly, established themselves, and reduced populations of *S. nitens* by about eight times. After treatment with viral suspensions, the predatory stinkbugs also helped the spread of the pathogens. The Indonesian workers concluded that it is preferable to release both adults and nymphs and believed that the release of *E. furcellata* is more effective when the population of caterpillars is relatively low, and the caterpillars are gregarious (as are limacodids). *Eocanthecona furcellata* has also been considered of importance in the management of several noctuid pests, such as *Anomis* spp. (Chu 1975), *Ophiusa janata* (L.) (Wiwat Suasa 1989), *Spodoptera litura* (Kapoor et al. 1973, Takai and Yasuoka 1993), *S. exigua* (Dai and Yang 1993), and *Helicoverpa armigera* (Hübner) (Cahatian 1991). During the early 1980s, several thousands of *E. furcellata* were released in Florida (U.S.A.) to combat the Colorado potato beetle *L. decemlineata* and the eastern tent caterpillar *Malacosoma americanum*, but the pentatomid probably never became established there (Froeschner 1988).

The high predatory activity of *E. furcellata* may not always be a beneficial trait. Sen et al. (1971) found this predator inflicted important losses on the outdoor-reared tasar silkworm *Antheraea mylitta* (Drury) in India. They estimated that an individual predator can destroy over 200 young silkworm larvae during its lifetime. Also, *E. furcellata* has been reported to attack larvae and pupae of the predatory coccinellid *Menochilus sexmaculatus* (F.) in the absence of lepidopteran prey (Srivastava et al. 1987).

3.5   *Picromerus bidens* L.

3.5.1   *Distribution, Life History, and Biology*

Since the beginning of the 20th century, several European studies have dealt with the biology and ecology of the Palearctic asopine *Picromerus bidens*, including those of Schumacher (1910), Butler (1923), Strawinski (1927), Gäbler (1937, 1938), Mayné and Breny (1947a,b, 1948a,b), Dupuis (1949), Leston (1955), Groves (1956), Javahery (1967, 1968), Hertzel (1982), and Ahmad and Önder (1990). The species was introduced into North America sometime before 1932, probably with nursery stock or other horticultural plants (Cooper 1967). Studies on it in Canada and the United States include Lattin and Donahue (1969), Oliveira and Juillet (1971), Kelton (1972), Larochelle and Larivière (1980), McPherson (1982), Larochelle (1984), and Javahery (1986, 1994). Larivière and Larochelle (1989) provided an extensive world review of the distribution and bionomics of *P. bidens*.

In the Old World, *P. bidens* is widely distributed in the western Palearctic and throughout Europe, from 64° north latitude southward to North Africa in the west and China in the east. Its known distribution in North America has been limited to Ontario, Québec, and the Maritime Provinces in Canada and to Maine, Massachusetts, New Hampshire, Vermont, New York, Rhode Island, and Pennsylvania in the United States (Larivière and Larochelle 1989, Wheeler 1999). Both in Europe and North America, the predator favors damp, shrubby areas and forests, where it is found on the vegetation within 2 m of the ground; it more rarely occurs in orchards and gardens (Schumacher 1910; Mayné and Breny 1940, 1948b; Dupuis 1949; Southwood and Leston 1959; Larivière and Larochelle 1989).

Adult females of *P. bidens* are 12 to 14 mm long, males 11 to 12 mm. Detailed descriptions of the adult were presented by Mayné and Breny (1948b), Dupuis (1949), and Kelton (1972). Dupuis (1949), Lodos and Önder (1983), and Ahmad and Önder (1990) provide keys to adults of several western Palearctic species of *Picromerus*. Eggs were described by Schumacher (1910), Butler (1923), Mayné and Breny (1948b), Cooper (1967), Southwood and Leston (1959), and Javahery (1994), and nymphs were described by Schumacher (1910), Gulde (1919), Butler (1923), Mayné and Breny (1948b), and Dupuis (1949).

The life cycle of *P. bidens* is regulated by the influence of ecological conditions on nymphal development and reproductive activity. In their review, Larivière and Larochelle (1989) summarized that in *P. bidens* two cycles are possible: a primary cycle, in which the species overwinters in the egg stage, and, less frequently, a secondary cycle, in which some adults overwinter. Egg development is characterized by an obligatory diapause. Javahery (1986, 1994) mentioned that eggs must be subjected to low temperatures (0 to 2°C) for at least 1 month to initiate embryonic development. In nature, eggs usually hatch in May, and 40 to 60 days are required for the five instars to complete development. Mayné and Breny (1948b) reported that nymphs fed *Ephestia kuehniella* larvae under laboratory conditions of 25 to 26°C and 80 to 95% relative humidity took 30 to 35 days to develop. In the field, the first adults appear in June to August (Strawinski 1927, Mayné and Breny 1948b, Dupuis 1949, Southwood and Leston 1959, Javahery 1986, Larivière and Larochelle 1989). About 15 to 30 days appear to be necessary for the reproductive organs to mature before mating (Mayné and Breny 1948b, Javahery 1986). Females mate several times throughout their lifetime (Javahery 1986); mating does not take place at temperatures below 15°C (Strawinsky 1927, Mayné and Breny 1948b). Schumacher (1910) and Mayné and Breny (1948b) observed total fecundities of 200 to 300 eggs; Javahery (1967, 1986) stated that the mean fecundity of *P. bidens* is 129 to 225 eggs and that fecundity of virgin females is only about half of that of mated ones (see also **Section 3.1.1**). Javahery (1986) reported that size of egg batches varies from only 2 to as many as 75 eggs. Also according to this author, fecundity and longevity decrease at 27 to 28°C, suggesting that *P. bidens* is a species of cooler climates. Observed adult longevities in the laboratory average 3 months (Javahery 1986), and have been estimated to be about 4 months in nature (Mayné and Breny 1948b). In the field, most adults have died by the end of October (Mayné and Breny 1948b, Javahery 1986). Some adults, however, overwinter; such adults are presumed to have not been reproductively active, leaving them sufficient energy to survive northern winters (Larivière and Larochelle 1989). In both Europe and North America, the species is generally recognized to be univoltine (Larivière and Larochelle 1989, Saulich and Musolin 1996).

Picromerus bidens travels little and disperses mainly by walking, but flight may be important for newly emerged adults to escape conditions of overpopulation in summer (Mayné and Breny 1948b, Javahery 1986, Larivière and Larochelle 1989).

3.5.2 Food Preferences

Picromerus bidens is associated with a wide diversity of plants and as a result it feeds on many insects. The predator attacks larvae and, to a lesser extent, pupae and adults of leaf-feeding insects, particularly of the Lepidoptera and Coleoptera. Further, it readily feeds on larvae of Tenthredinidae and other leaf-feeding Hymenoptera and has occasionally been reported to prey on insects of Dermaptera, Diptera, Heteroptera, Neuroptera, and Orthoptera (review by Larivière and Larochelle 1989). It also sucks plant juices to obtain water (Mayné and Breny 1948b, Javahery 1986). Cannibalism has been observed in periods of overpopulation or low food supply (Mayné and Breny 1948b; Javahery 1967, 1986). According to Javahery (1986) and Larivière and Larochelle (1989), *P. bidens* prefers small to medium-sized prey and is attracted to movement of the prey.

3.5.3 Rearing

Laboratory rearing of *P. bidens* has been described by Mayné and Breny (1948b) and Javahery (1986). Larvae of the pyralids *Galleria mellonella* and *E. kuehniella*, the noctuid *Mamestra brassicae*, the chrysomelids *L. decemlineata* and *Chrysomela* (*Melosoma*) *populi* L., and of the housefly *Musca domestica* have been successfully used for the laboratory production of this pentatomid. Moisture is provided from the first stadium on by plant materials or free water. Javahery (1986) mentioned that eggs of *P. bidens* can be stored at 2°C for about 6 months; during this period, sufficient humidity (~85% relative humidity) is necessary.

3.5.4 Control Potential

Unfortunately, few or no studies have attempted to quantify the predatory effectiveness of *P. bidens*. Nevertheless, many authors have emphasized its potential in reducing populations of insect pests in a variety of ecosystems. In Germany, Gäbler (1937, 1938) reported efficient control of larvae of the pine sawfly *Diprion pini* L. and emphasized the potential of *P. bidens* against caterpillars of *Lymantria monacha* L. Engel (1939) estimated that *P. bidens* and *Troilus luridus* (F.) (see **Section 4.2**) destroyed 38 to 70% of the population of the geometrid *Bupalus piniarius* (L.) in Germany. According to Clausen (1940), the use of *P. bidens* to control the bedbug *Cimex lectularius* L. was recommended as early as 1776; it was reported that a few predators confined in a heavily infested room completely exterminated the bedbugs within a few weeks. In Sweden, the predator had a significant impact on populations of the pine defoliator pest *Neodiprion sertifer* (Geoffroy) (Forsslund 1946). Pschorn-Walcher and Zinnert (1971) and Mallach (1974) reported high voracity against larvae of the sawflies *Pristiphora erichsonii* (Hartig) and *Microdiprion pallipes* (Fallén) in Germany. Asanova (1980) and Zeki and Toros (1990) mentioned *Picromerus bidens* as an important predator of several forestry pests in Kazakhstan and Turkey, respectively.

Mayné and Breny (1948b) evaluated the potential use of *P. bidens* for the control of the Colorado potato beetle in Belgium. These workers observed that second instars of the predator suffered high mortality upon introduction in potato fields and concluded that even under optimal conditions (i.e., in fields adjacent to the predator's natural habitat), *P. bidens* could not effectively protect the plants from attacks by *Leptinotarsa decemlineata*. Javahery (1986) and Larivière and Larochelle (1989) concluded that its wide range of prey species, slow dispersal, and relatively difficult mass propagation militate against the systematic use of *P. bidens* in biological control. Nevertheless, field experiments in Russia showed that the predator was effective in controlling economic populations of *L. decemlineata* in potato when applied at a rate of 100,000 eggs/ha (Volkov and Tkacheva 1997).

4. THE LESS IMPORTANT SPECIES

4.1 Nearctic and Neotropical Regions

Besides the well-documented species *Podisus maculiventris*, *P. nigrispinus*, and *Perillus bioculatus*, several other New World Asopinae have received some attention for their potential in biological control.

Podisus brevispinus Phillips (= *P. modestus* Uhler), *P. placidus* Uhler, and *P. serieventris* Uhler are the most common *Podisus* species in northeastern North America besides *P. maculiventris* (see also **Section 3.1**). Evans (1985) presented a key to the nymphs of these four species, and Coppel and Jones (1962) provided a key to eggs and adults and compared the developmental and reproductive biology of the four species. Keys to adults of these and other Nearctic *Podisus* species can also be found in Blatchley (1926). Prebble (1933) and Oetting and Yonke (1971) described and illustrated the nymphal stages of *P. serieventris* and *P. placidus*, respectively. Prebble (1933) and Tostowaryk (1971) detailed the biology of *P. serieventris* and *P. brevispinus*, respectively. Kirkland (1897, 1898) provided notes on the life history of *P. placidus*, and Tadic (1963) and Warren and Bjegovic (1972) studied the biology of the species in Arkansas (U.S.A.). Developmental and reproductive characteristics of *P. brevispinus*, *P. placidus*, and *P. serieventris* are like those of *P. maculiventris*. At 20 to 25°C, eggs hatch in about 1 week and nymphs take 25 to 30 days to develop. In the laboratory, females have been reported to lay 100 to 300 eggs. Although these three North American *Podisus* spp. are polyphagous, their feeding behavior and predatory effectiveness have been studied particularly in view of their potential to suppress outbreaks of pine sawflies (Diprionidae) (Coppel and Jones 1962; Tostowaryk 1971, 1972). Tostowaryk (1971) showed that three third-instar *P. brevispinus* devoured an entire colony of 57 first-instar *Neodiprion swainei* Middleton in 7 days; adults consumed about 0.5 fifth instars of the pest per day. Subsequently, Tostowaryk (1972) determined the functional response of *P. brevispinus* to densities of two colonial species of jack pine sawfly; he observed a domed response curve caused by the more effective defense reactions of the sawfly larvae at higher densities. Prompted by the promising laboratory studies made by Tadic (1963) in Arkansas, *P. placidus* and *P. maculiventris* were introduced into the former Yugoslavia in 1968 for the biological control of the fall webworm *Hyphantria cunea* and other defoliator pests; the pentatomids had no significant effects on caterpillar populations and were unable to survive the cold winters (Tadic 1975). *Podisus sagitta* (F.), a species closely related to *P. nigrispinus* (Thomas 1992, De Clercq and Degheele 1995), has been reported to be an effective predator of *Epilachna* larvae in Mexico (Plummer and Landis 1932).

Stiretrus anchorago (F.) is distributed from Canada to Panama and occurs in different color forms, most of which have received separate names at one time or another (Thomas 1992). Its biology was studied by Howard (1936), Waddill and Shepard (1974), Richman and Whitcomb (1978), Richman and Mead (1980b), and Oetting and Yonke (1971), who also provided descriptions of the immature stages. Several workers have reported that *S. anchorago* is not very promising for laboratory culturing. Although the insect goes through its life cycle rapidly (development from egg to adult takes 25 to 30 days at about 27°C), females are not prolific egg-layers (total fecundity has been reported to be 57 eggs at about 27°C). Further, this pentatomid appears to have a preference for larvae of the Mexican bean beetle *Epilachna varivestis*. Waddill and Shepard (1974) reported that females reared exclusively on larvae of the greater wax moth *Galleria mellonella* only produced infertile eggs. Likewise, Richman (1977) found that although *S. anchorago* was able to feed and develop on larvae of the curculionid *Hypera postica* (Gyllenhal) and of the noctuid *Trichoplusia ni*, none of the ensuing females oviposited. Waddill and Shepard (1975a) compared the predation by *P. maculiventris* and *S. anchorago* on Mexican bean beetle *E. varivestis* and velvetbean caterpillar *Anticarsia gemmatalis*. *Stiretrus anchorago* preferred the coccinellid, whereas *P. maculiventris* took approximately equal numbers of each prey species. Also, *P. maculiventris* accepted a wider range of *E. varivestis* stages than did *S. anchorago*. These data again suggest that *S. anchorago* is more restricted in its diet. Nevertheless, in their field-cage experiments on soybean, Richman et al. (1980) recorded high predation rates against the larvae of another noctuid soybean pest, *Pseudoplusia includens*. Cappaert et al. (1991) found *S. anchorago* to be the most abundant asopine feeding on the chrysomelid *L. decemlineata* in Mexico.

Alcaeorrhynchus grandis (Dallas) is the largest asopine occurring in the Western Hemisphere: length of males and females is 16 to 21 and 18 to 25 mm, respectively. It is an insect of warmer climates with a wide distribution from the southern United States into Argentina (Thomas 1992).

Descriptions and illustrations of the life stages of *A. grandis* were presented by Blatchley (1926), Richman and Mead (1978), Richman and Whitcomb (1978), Brailovsky and Mayorga (1994), and Zanuncio et al. (1994a). The bionomics of *A. grandis* specimens from Florida (U.S.A.) were studied by Richman and Whitcomb (1978). They reported that development from egg to adult at 27°C took about 2 months and that the bug was difficult to rear. Although little effort has been done to measure its predation capacity, *A. grandis* has been mentioned as an important predator of lepidopterous pests in various agricultural crops and forests (Araujo e Silva 1933, Whitcomb 1973, Zanuncio et al. 1994a, Malaguido and Panizzi 1998).

Euthyrhynchus floridanus (L.) occurs in a South American form, *E. macrocnemis* (Perty), and a North American form, *E. floridanus* (Thomas 1992). Oetting and Yonke (1975) provided detailed descriptions of the immature stages of *E. floridanus*; keys to and illustrations of the adult stage can be found in Blatchley (1926), Mead (1976), Thomas (1992), and Brailovsky and Mayorga (1994). The life history and bionomics of this species in the United States were investigated by Ables (1975), Mead (1976), Oetting and Yonke (1975), and Richman and Whitcomb (1978). *Euthyrhynchus floridanus* has the longest developmental periods reported for North American asopines. Eggs hatch after about 35 and 20 days, and nymphal development takes about 65 and 40 days at 24 and 27°C, respectively. Ables (1975) reported that females produce an average of 140 eggs during a life span of 45 days. According to Richman and Whitcomb (1978), this species is easily reared; Ables (1975), however, noted that only four to five generations can be reared per year because of the length of the bug's life cycle and because adult females will not lay eggs for at least 2 weeks after maturing. There are only few records on the field history and feeding behavior of *E. floridanus*. It is noted to be a nonspecific predator, attacking insects from different orders (Mead 1976, Logan et al. 1987).

Species belonging to the closely related genera *Apateticus* and *Apoecilus* are relatively large insects with a body length of 13 to 20 mm (Blatchley 1926, Torre-Bueno 1939, Thomas 1992). *Apoecilus bracteatus* (Fitch) and *A. cynicus* (Say) are two arboreal species occurring in the United States and Canada. *Apoecilus cynicus* has been reported to be a nonspecific predator of larvae from several insect orders (Whitmarsh 1916, Stoner 1920, Blatchley 1926, Jones and Coppel 1963). *Apoecilus bracteatus* appears to feed primarily on larvae of Chrysomelidae, including *Trirhabda* spp. (Evans and Root 1980, Eckberg and Cranshaw 1994) and *Galerucella* spp. (Diehl et al. 1997). Jones and Coppel (1963) described the immature stages of *A. cynicus* (as *Apateticus cynicus*), and reported that development from egg to adult at about 22°C takes about 45 days. Adults lived up to 6 months in the laboratory; however, females produced only 1 to 3 egg batches with an average of 57 eggs per batch. Both *Apoecilus cynicus* and *A. bracteatus* appear to be univoltine and to overwinter in the egg stage (Whitmarsh 1916, Stone 1939, Jones and Coppel 1963). *Apateticus lineolatus* (Herrich-Schaeffer) is the most common species of the genus in Mexico (Thomas 1992). In that country, it has been reported as a predator of *Epilachna varivestis* (Plummer and Landis 1932) and of *Leptinotarsa* spp. (Logan et al. 1987).

Oplomus dichrous (Herrich-Schaeffer) is another asopine common to Mexico and the southern United States (Thomas 1992). Its biology was studied by Plummer and Landis (1932) and Drummond et al. (1987). It is adapted to a higher range of temperatures. Its immature development is negligible between 12 and 15°C and has an optimum from 28 to 33°C. At 25°C, development from egg to adult averages 34 days and the preoviposition period is 18.5 days. Females are reported to lay 150 to 200 eggs during a reproductive period of about 25 days. Although *O. dichrous* feeds on a variety of insects, it has particularly been noted as a voracious predator of eggs and larvae of the Colorado potato beetle *L. decemlineata* (Drummond et al. 1987, Cappaert et al. 1991). Based on field observations, Logan et al. (1987) believed that *O. dichrous* nymphs may be important in regulating *L. decemlineata* populations in central Mexico. To evaluate the introduction of this species against the Colorado potato beetle in temperate regions of North America and Europe, Drummond et al. (1987) undertook studies of the predator in Rhode Island (U.S.A.). These indicated high levels of prey consumption, under both laboratory and field conditions. In field-cage studies,

95% control was attained with a release ratio of about 1 predator adult to 40 to 50 *L. decemlineata* eggs. However, poor synchrony with field populations of the pest and the unlikely overwintering ability in the northeastern United States indicated that *O. dichrous* may have little potential to control the Colorado potato beetle in cool climates. Moreover, laboratory rearing is complicated by diapause (Ruberson et al. 1998). Nevertheless, the authors concluded that the pentatomid may have some use in an inundative release program, although this would require releasing large numbers of (laboratory-reared) predators. Plummer and Landis (1932) observed *O. dichrous* as a predator of the Mexican bean beetle *E. varivestis*.

Tylospilus acutissimus Stål is distributed from the southern United States into Central America. Thomas (1992) mentioned that it is common in Arizona (U.S.A.) on mesquite trees, but it has been sampled from other plants, as well (Butler and Werner 1960). The biology of this predator was studied by Stoner et al. (1974a,b) (as *Podisus acutissimus*). Little is known on the preferred prey of *T. acutissimus*. In their laboratory cultures, Stoner et al. (1974a,b) fed the insect larvae and prepupae of the noctuid *Spodoptera exigua*.

Brontocoris tabidus (Signoret), erroneously referred to in part of the Brazilian literature as *Podisus nigrolimbatus* (Spinola), is one of the most common predatory stinkbugs in Brazil (Zanuncio et al. 1994a). Details on the morphology and biology of this bug have been presented by Gonçalves et al. (1990) and Barcelos et al. (1993, 1994). In the laboratory at 25°C, nymphs develop in about 3 weeks and females may lay over 300 eggs during a lifetime of up to 3 months. Although the predator is easily reared on various insect prey such as *Bombyx mori*, *Tenebrio molitor*, and *Musca domestica* (Barcelos et al. 1991, 1994; Zanuncio et al. 1993f, 1994a), Zanuncio et al. (1996) proposed the use of a meat-based artificial diet as an alternative food for its mass production. Zanuncio et al. (1994a) believed that *Brontocoris tabidus* is a promising biocontrol agent of lepidopterous defoliators in Brazilian *Eucalyptus* plantations. In contrast, Artola et al. (1982) who studied the closely related species *Brontocoris* (= *Podisus*) *nigrolimbatus* (Spinola) in Argentina and evaluated its potential to suppress the chrysomelid elm pest *Pyrrhalta luteola*, concluded that the predator is generally active only after damage has been caused. *Supputius cincticeps* (Stål) is another Neotropical asopine that has recently received attention for the integrated management of caterpillar pests in Brazilian forests, although it is not as common there as *Podisus nigrispinus* and *B. tabidus* (Zanuncio et al. 1994a). Its biology and laboratory rearing were investigated by Zanuncio et al. (1992; 1993e,g), Didonet et al. (1996a,b), and Silva et al. (1996).

4.2 Palearctic Region

Zicrona caerulea (L.) is a holarctic species with a wide distribution in North America and throughout Europe and Asia, extending up to China, Japan, and the Oriental region; it has also been found in North America (Ahmad et al. 1974; Lodos and Önder 1983; Thomas 1992, 1994). Lodos and Önder (1983) mentioned that *Z. caerulea* mainly occurs on herbaceous plants, whereas Mayné and Breny (1948b) and Augustin and Lévieux (1993) reported that the insect is also found in association with shrubs and trees; according to Schumacher (1911), however, *Z. caerulea* is more of a ground-dwelling predator with a dislike of flying. Descriptions of the immature and adult stages can be found in Schumacher (1911), Dupuis (1949), Kobayashi (1951), and Mayné (1965). This pentatomid has been reported to prey on larval forms of Coleoptera, Lepidoptera, Heteroptera, and Diptera (Schumacher 1911, Chen 1986, Barrion and Litsinger 1987). According to Schumacher (1911) and Mayné and Breny (1940), *Z. caerulea* is not a very powerful predator and attacks only small prey. In Europe and Asia, *Z. caerulea* is particularly noted as a predator of Chrysomelidae. Schumacher (1911) mentioned that a single adult can kill 12 larvae per day of the flea beetle *Altica ampelophaga* Guérin-Méneville, a pest of grapevine. Mineo and Iannazzo (1986) also observed *Z. caerulea* as a predator of *A. ampelophaga* in Italy, where the bug appears to have one generation per year. Likewise, Chen (1982) reported *Z. caerulea* to be an important natural enemy of larvae and adults of *Altica* spp. in China. According to the calculations of Balcells (1951), however,

Z. caerulea can reduce populations of *Altica* under optimal conditions by only one-seventh, because of its low reproductive potential. Nayek and Banerjee (1987) considered that the predatory activity of the asopine restricts the use of *A. cyanea* (Weber) for the biological control of the water primrose *Ludwigia adscendens* (L.), an economically important weed in many Oriental rice fields. The asopine also preyed upon larvae of the Colorado potato beetle in the former Yugoslavia (Bjegovic 1971). Zeki and Toros (1990) and Augustin and Lévieux (1993) stated that populations of poplar beetles (*Chrysomela* spp.) in Turkey and France, respectively, are heavily attacked by *Z. caerulea*.

Arma custos (F.), distributed throughout Europe and into China and Japan (Thomas 1994), is mainly found in arboreal habitats (Schumacher 1910, Dupuis 1949). The morphology of immature and adult stages was studied by Schumacher (1910) and Dupuis (1947, 1949). Schumacher (1910), Michalk (1938), and Dupuis (1949) presented some notes on its life history. Saulich and Volkovich (1994) and Volkovich and Saulich (1994) investigated the influence of photoperiod and of constant and alternating temperatures on the preimaginal stages of *A. custos* and observed that in Russia the insect has only one generation per year, not only in forest and steppe zones, but in more southern zones as well. In his attempts to overcome cultivation problems related to the univoltinism of *A. custos*, Saulich (1995) found that the bug has a facultative diapause controlled by a qualitative photoperiodic reaction of the long-day type; this made it possible to maintain an active laboratory culture independent of seasonal time. *Arma custos* is a nonspecific predator attacking larval forms of leaf-feeding Lepidoptera, Coleoptera (Chrysomelidae), and Hymenoptera (Tenthredinidae) (Schumacher 1910, Michalk 1938, Dupuis 1949, Zheng et al. 1992). Literature on the predatory potential of *A. custos* is scarce. Nonetheless, Saulich (1995) believed that the bug could be valuable in suppressing outbreaks of the Colorado potato beetle in Russia. Augustin and Lévieux (1993) mentioned *A. custos* as one of the main mortality factors of *Chrysomela* populations on poplars in France. In field studies in the Jilin province of China, Zheng et al. (1992) found that *A. custos* preys on 40 species of agricultural and forest pests, with a preference for larvae of the chrysomelid *Ambrostoma quadriimpressum* Motsch and of the limacodid *Cnidocampa flavescens* (Walker). When the bug was released into the forest at a predator–prey ratio of 1:5, population reductions of larvae and adults of *A. quadriimpressum*, young larvae of *C. flavescens*, and larvae of the notodontid *Clostera anachoreta* (F.) were 69.2, 39.9, 65.5, and 41.6%, respectively, 15 days after release.

Troilus luridus (F.) is another asopine attacking lepidopterous and coleopterous larvae in European and Asian forests (Dupuis 1949, Thomas 1994). Its biology was discussed by Schumacher (1911), Koehler (1948), Mayné and Breny (1940, 1948b), and Mayné (1965). According to Engel (1939), *T. luridus* has a significant impact on populations of the geometrid *Bupalus piniaria* L. in Europe.

4.3 Oriental, Australian, and Ethiopian Regions

There is only little information available on the Asopinae from these regions. *Andrallus spinidens* (F.) actually is cosmopolitan in distribution (Thomas 1992, 1994), but its biology and economic importance have been studied mostly in India and the Far East. A detailed description of *A. spinidens* can be found in Chopra and Sucheta (1986). Rajendra and Patel (1971), Pawar (1976), Manley (1982), and Singh and Singh (1989a) described the life history and biology of the species. In the laboratory, females are reported to produce 300 to 600 eggs and development from egg to adult takes about 3 weeks. In rice fields in Malaysia, Manley (1982) observed that large populations of *A. spinidens* were associated with outbreaks of the caterpillar pest *Melanitis leda* L. (Nymphalidae). Based on laboratory and field data, however, Manley (1982) believed that the predator may be of importance only under moderate to high densities of the pest. In India, *A. spinidens* was considered a promising natural enemy of *Rivula* sp. (Noctuidae), a pest of soybean. Singh and Singh (1989a) estimated that during its nymphal development the predator could consume about 40 *Rivula* larvae; a gravid female was estimated to kill an average of 4.7 caterpillars per day. In a further step toward the establishment of an IPM program against the pest, Singh and Singh

(1989b) evaluated the toxicity of 11 insecticides against eggs and adults of the predator. The results showed that phosalone applied at 0.035% caused the least egg and adult mortality, followed by malathion at 0.05%.

In their discussion of the different asopine bugs occurring in Indonesian oil palm plantations, Sipayung et al. (1992) mentioned that besides *Eocanthecona furcellata* (**Section 3.4**) and other species from the *Eocanthecona–Cantheconidea* complex, also *Montrouzieriellus falleni* (= *Platynopus melacanthus*)(Guérin-Méneville) is commonly observed attacking noctuid caterpillars. As early as 1930, Tothill et al. discussed the use of *M. falleni* in the control of the zygaenid coconut moth *Levuana iridescens* Bethune-Baker in Fiji; *M. falleni* nymphs averaged 5 *L. iridescens* larvae killed per day, and adults killed about 18 caterpillars per day.

In India, Singh et al. (1973) provided a note on the biology of *Amyotea malabarica* (F.). In the laboratory, this species showed a strong preference for the southern green stink bug *Nezara viridula* (L.) over other heteropteran or lepidopteran prey (Singh 1973). A single predator was reported to kill an average of 54 *N. viridula* individuals of different life stages from the second stadium to the end of adult life, i.e., about one prey item per day.

Ramsay (1963, 1964), Edwards and Suckling (1980), Awan (1983, 1984, 1985, 1987, 1988a,b, 1990), Awan and Browning (1983), and Awan et al. (1989) studied the biology and predation characteristics of *Oechalia schellembergi* (Guérin-Méneville) and *Cermatulus nasalis* (Westwood), two polyphagous asopines occurring in Australia and New Zealand (Thomas 1994, Larivière 1995). Their studies suggested that both species may have potential to reduce native populations of the eucalyptus tortoise beetle *Paropsis charybdis* Stål and of the noctuid *Helicoverpa punctigera* (Wallengren), which is a key pest of cotton, lucerne, and many other field crops. In laboratory experiments, high predation rates were recorded against different life stages of *H. punctigera* and *P. charybdis*. Temperatures below 20°C limited the predatory performance of the predators.

Descriptions and distributions of several African Asopinae can be found in Cachan (1952), Villiers (1952), Gillon (1972), Linnavuori (1982), and Thomas (1994). Unfortunately, very little is known about the biology and economic importance of the pentatomid predators from the African continent. Taylor (1965) mentioned that *Macrorhaphis acuta* Dallas and *Glypsus* spp. are important predators of larvae of the fruit-piercing moth *Achaea lienardi* Boisduval in South Africa. Abasa and Mathenge (1974) reported on the life history of *M. acuta* in the laboratory. Nyiira (1970), Couilloud (1989), and Poutouli (1992) mentioned that species of *Macrorhaphis*, *Afrius*, and *Glypsus* are important predators of various cotton pests, including *Diparopsis* spp., *Earias* spp., and *Heliothis* spp. Mugo (1992) suggested the use of *M. acuta* to control the giant looper *Boarmia selenaria* Schiffermiller (Geometridae) on coffee in Kenya. In South Africa, Urban and Eardley (1995) reported that *Macrorhaphis* is a key factor in the natural control of the tenthredinid sawfly *Nematus oligospilus* Foerster, a newly introduced defoliator pest of willow trees.

5. CONCLUSIONS

This literature review has demonstrated the potential value of several pentatomid predators for the management of a wide array of agricultural insect pests. Because only a small proportion of the species belonging to this subfamily have been studied to any extent, it is clear that an enormous potential remains to be investigated. Further studies are required to elucidate the basic biology and ecology of the Asopinae in different parts of the world. Especially needed are more quantitative studies to measure the role of predatory stinkbugs in regulating populations of economically important insect pests. Emphasis should be placed on the field evaluation of their effectiveness and climatic adaptedness.

Classical biological control involving asopine bugs has often failed because of their unadaptedness to the climatic conditions prevalent in the area of introduction. A further search for cold-adapted species may be warranted for the control of exotic pests in temperate zones. Alternatively, predatory stinkbugs may have potential for use in augmentative biocontrol in such varied agroecosystems as field crops, greenhouse crops, and forests. Clearly, the use of asopine predators in agricultural practice should be considered within the context of integrated pest control. One practical problem for augmentative releases of pentatomid predators will certainly be the artificial production of predators of high quality in sufficient numbers at reasonable cost. Future challenges here are the mechanization of rearing procedures (e.g., artificial diets) and the development of efficient storage techniques. Also, attention should be given to the development of suitable application systems for these predators. Another key factor is the compatibility between augmentation or conservation of the predators and the use of chemical and cultural control methods. Susceptibility of asopines toward commonly used and novel pesticides should be investigated in the laboratory as well as in the field.

6. ACKNOWLEDGMENTS

Thanks go to Carl W. Schaefer and Antônio R. Panizzi for providing many helpful suggestions that greatly improved the manuscript. I will always be grateful to the late Danny Degheele for his generous help and for having allowed me the freedom to pursue my fascination with predatory bugs. This chapter is affectionately dedicated to him.

7. REFERENCES CITED

Abasa, R. O., and W. M. Mathenge. 1974. Laboratory studies of the biology and food requirements of *Macrorhaphis acuta* (Hemiptera: Pentatomidae). Entomophaga 19: 213–218.

Abbas, M. S. T., and D. G. Boucias. 1984. Interaction between nuclear polyhedrosis virus-infected *Anticarsia gemmatalis* (Lepidoptera: Noctuidae) larvae and predator *Podisus maculiventris* (Say) (Hemiptera: Pentatomidae). Environ. Entomol. 13: 599–602.

Ables, J. R. 1975. Notes on the biology of the predaceous pentatomid *Euthyrhynchus floridanus* (L.). J. Georgia Entomol. Soc. 10: 353–356.

Ables, J. R., and D. W. McCommas. 1982. Efficacy of *Podisus maculiventris* as a predator of variegated cutworm on greenhouse cotton. J. Georgia Entomol. Soc. 17: 204–206.

Adidharma, D. 1986. The development and survival of *Podisus sagittus* (Hemiptera: Pentatomidae) on artificial diets. J. Austral. Entomol. Soc. 25: 15–16.

Ahmad, I., and F. Önder. 1990. A revision of the genus *Picromerus* Amyot et Serville (Hemiptera: Pentatomidae: Pentatominae: Asopini) from western Palaearctic with description of two new species from Turkey. Türk. Entomol. Derg. 14: 75–84.

Ahmad, I., and N. A. Rana. 1988. A revision of the genus *Canthecona* Amyot et Serville (Hemiptera: Pentatomidae: Asopini) from Indo-Pakistan subcontinent with description of two new species from Pakistan. Türk. Entomol. Derg. 12: 75–84.

Ahmad, I., Q. A. Abbasi, and A. A. Khan. 1974. Generic and supergeneric keys with reference to a checklist of pentatomid fauna of Pakistan (Heteroptera: Pentatomoidea) with notes on their distribution and food plants. Suppl. Entomol. Soc. Karachi 1: 1–103.

Ahmad, M., A. P. Singh, S. Sharma, R. K. Mishra, and M. D. J. Ahmad. 1996. Potential estimation of predatory bug, *Canthecona furcellata* Wolff (Hemiptera: Pentatomidae) against poplar defoliator, *Clostera cupreata* (Lepidoptera: Notodontidae). Ann. For. 4: 133–138.

Aldrich, J. R. 1985. Pheromone of a true bug (Hemiptera-Heteroptera): attractant for the predator, *Podisus maculiventris*, and kairomonal effects. Pp. 95–119 *in* T. E. Acree and D. M. Soderlund [eds.], Semiochemistry: Flavors and Pheromones. de Gruyter, Berlin, Germany.

1986. Seasonal variation of black pigmentation under the wings in a true bug (Hemiptera: Pentatomidae): a laboratory and field study. Proc. Entomol. Soc. Washington 88: 409–421.

1988. Chemical ecology of the Heteroptera. Annu. Rev. Entomol. 33: 211–238.

1995. Chemical communication in the true bugs and parasitoid exploitation. Pp. 318–363 *in* R. T. Cardé and W. J. Bell [eds.], Chemical Ecology of Insects 2. Chapman & Hall, New York, New York, U.S.A. 433 pp.

1998. Status of semiochemical research on predatory Heteroptera. Pp. 33–48 *in* M. Coll and J. R. Ruberson [eds.], Predatory Heteroptera: Their Ecology and Use in Biological Control. Thomas Say Publ. Entomol.: Proceedings Entomological Society of America, Lanham, Maryland, U.S.A. 233 pp.

Aldrich, J. R., and W. W. Cantelo. 1999. Suppression of Colorado potato beetle infestation by pheromone-mediated augmentation of the predatory spined soldier bug, *Podisus maculiventris* (Say) (Heteroptera: Pentatomidae). Agric. For. Entomol., in press.

Aldrich, J. R., J. P. Kochansky, and C. B. Abrams. 1984a. Attractant for a beneficial insect and its parasitoids: Pheromone of the predatory spined soldier bug, *Podisus maculiventris* (Hemiptera: Pentatomidae). Environ. Entomol. 13: 1031–1036.

Aldrich, J. R., J. P. Kochansky, W. R. Lusby, and J. D. Sexton. 1984b. Semiochemicals from a predaceous stink bug, *Podisus maculiventris* (Hemiptera: Pentatomidae). J. Washington Acad. Sci. 74: 39–46.

Aldrich, J. R., J. P. Kochansky, W. R. Lusby, and M. Borges. 1991. Pheromone blends of predaceous bugs (Heteroptera: Pentatomidae: *Podisus* spp.). Z. Naturforsch. 46: 264–269.

Aldrich, J. R., W. R. Lusby, J. P. Kochansky, and C. B. Abrams. 1984c. Volatile compounds from the predatory insect *Podisus maculiventris* (Hemiptera: Pentatomidae): male and female metathoracic scent gland and female dorsal abdominal gland secretions. J. Chem. Ecol. 10: 561–568.

Altieri, M. A., and W. H. Whitcomb. 1979. Predaceous arthropods associated with Mexican tea in North Florida. Florida Entomol. 62: 175–182.

Anonymous. 1992. Wanted: a few good soldier (bugs). Science 257: 1049.

Araujo e Silva, A. G., d'. 1933. Contribuição para o estudo da biologia de tres pentatomideos. O Campo 4: 23–25.

Arnoldi, D., R. K. Stewart, and G. Boivin. 1991. Field survey and laboratory evaluation of the predator complex of *Lygus lineolaris* and *Lygocoris communis* (Hemiptera: Miridae) in apple orchards. J. Econ. Entomol. 84: 830–836.

Artola, J. A., M. F. Garcia, and S. E. Dicindio. 1982. Bioecologia de *Podisus nigrolimbatus*, Spinola (Heteroptera, Pentatomidae), predator de *Pyrrhalta luteola* (Muller) (Coleoptera, Chrysomelidae). IDIA No. 401–404: 25–33.

Asanova, R. B. 1980. New and rare species of bugs for the fauna of northern Kazakhstan (Heteroptera). Trudy Instituta Zoologii, Akademiya Nauk Kazakhskoi SSR 39: 49–54 [in Russian].

Augustin, S., and J. Lévieux. 1993. Life history of the poplar beetle *Chrysomela tremulae* F. in the central region of France. Canad. Entomol. 125: 399–401.

Awan, M. S. 1983. A convenient recipe for rearing a predacious bug, *Oechalia schellenbergii* Guérin-Méneville (Hemiptera: Pentatomidae). Pakistan J. Zool. 15: 217–218.

1984. Foraging behaviour of the predacious pentatomid *Oechalia schellenbergii* (Guérin-Méneville) against its common prey, *Heliothis punctiger* Wallengren (Noctuidae, Lepidoptera). Z. Angew. Entomol. 98: 230–233.

1985. Anti-predator ploys of *Heliothis punctiger* (Lepidoptera: Noctuidae) caterpillars against the predator *Oechalia schellenbergii* (Hemiptera: Pentatomidae). Austral. J. Zool. 33: 885–890.

1987. Efficiency of capture of a predacious pentatomid against its common prey and relationship between temperature and predator's efficiency of capture. Pakistan J. Zool. 19: 291–300.

1988a. Development and mating behaviour of *Oechalia schellenbergii* (Guérin-Méneville) and *Cermatulus nasalis* (Westwood) (Hemiptera: Pentatomidae). J. Austral. Entomol. Soc. 27: 183–187.

1988b. Study of the interaction between temperatures and complexity of searching conditions and its influence on the voracity of a predacious pentatomid, *Oechalia schellenbergii* (Guérin-Méneville). Pakistan J. Zool. 20: 383–389.

1990. Predation by three hemipterans: *Troponabis nigrolineatus*, *Oechalia schellenbergii* and *Cermatulus nasalis*, on *Heliothis punctiger* larvae in two types of searching arenas. Entomophaga 35: 203–210.

Awan, M. S., and T. O. Browning. 1983. Effect of temperature on the functional response of a hemipteran predator, *Oechalia schellenbergii* Guérin-Méneville (Hemiptera: Pentatomidae). Pakistan J. Zool. 15: 17–21.

Awan, M. S., L. T. Wilson, and M. P. Hoffmann. 1989. Prey location by *Oechalia schellembergii*. Entomol. Exp. Appl. 51: 225–231.

Baker, A. D. 1927. Some remarks on the feeding process of the Pentatomidae (Hemiptera–Heteroptera). 19th Annu. Rep. Québec Soc. Prot. Pl. (1926–1927): 24–34.

Baker, A. M., and P. L. Lambdin. 1985. Fecundity, fertility, and longevity of mated and unmated spined soldier bug females. J. Agric. Entomol. 2: 378–382.
Balcells, E. 1951. Datos para el estudio del ciclo biologico de *Zicrona coerulea* L. Publ. Inst. Biol. Apl. 8: 127–150.
Barcelos, J. A. V., J. C. Zanuncio, E. C. do Nascimento, and T. V. Zanuncio. 1993. Caracterização dos estádios ninfais de *Podisus nigrolimbatus* (Spinola, 1852) (Hemiptera, Pentatomidae). Rev. Bras. Entomol. 37: 537–543.
Barcelos, J. A. V., J. C. Zanuncio, A. C. Oliveira, and E. C. do Nascimento. 1994. Performance em duas dietas e descrição dos adultos de *Brontocoris tabidus* (Signoret) (Heteroptera: Pentatomidae). An. Soc. Entomol. Brasil 23: 519–524.
Barcelos, J. A. V., J. C. Zanuncio, G. P. Santos, and F. P. Reis. 1991. Viabilidade da criação, em laboratório, de *Podisus nigrolimbatus* (Spinola, 1852) (Hemiptera: Pentatomidae) sobre duas dietas. Rev. Árvore 15: 316–322.
Barrion, A. T., and J. A. Litsinger. 1987. The bionomics, karyology and chemical control of the node-feeding black bug, *Scotinophara latiuscula* Breddin (Hemiptera: Pentatomidae) in the Philippines. J. Plant Prot. Trop. 4: 37–54.
Bellotti, A. C., B. V. Arias, and O. L. Guzman. 1992. Biological control of the cassava hornworm *Erinnyis ello* (Lepidoptera: Sphingidae). Florida Entomol. 75: 506–515.
Berenbaum, M. R., B. Moreno, and E. Green. 1992. Soldier bug predation on swallowtail caterpillars (Lepidoptera: Papilionidae): circumvention of defensive chemistry. J. Insect Behav. 5: 547–553.
Biever, K. D., and R. L. Chauvin. 1992a. Suppression of the Colorado potato beetle (Coleoptera: Chrysomelidae) with augmentative releases of predaceous stinkbugs (Hemiptera: Pentatomidae). J. Econ. Entomol. 85: 720–726.
1992b. Timing of infestation by the Colorado potato beetle (Coleoptera: Chrysomelidae) on the suppressive effect of field released stinkbugs (Hemiptera: Pentatomidae) in Washington. Environ. Entomol. 21: 1212–1219.
Biever, K. D., P. L. Andrews, and P. A. Andrews. 1982. Use of a predator, *Podisus maculiventris*, to distribute virus and initiate epizootics. J. Econ. Entomol. 75: 150–152.
Bjegovic, P. 1971. The natural enemies of the Colorado potato beetle, *Leptinotarsa decemlineata* (Say) and an attempt of its biological control in Yugoslavia. Zast. Bilja 21: 97–111 [in Serbo-Croatian].
Blatchley, W. S. 1926. Heteroptera or True Bugs of Eastern North America, with Especial Reference to the Faunas of Indiana and Florida. Nature Publ. Co., Indianapolis, Indiana, U.S.A. 1116 pp.
Boiteau, G. 1987. The significance of predators and cultural methods. Pp. 201–223 *in* G. Boiteau, R. P. Singh, and R. H. Parry [eds.], Potato Pest Management in Canada/Lutte contre les parasites de la pomme de terre au Canada, Proceedings of the Symposium "Improving Potato Pest Protection." Canada/New Brunswick Agreement on Food and Agriculture Development, Fredericton, New Brunswick, Canada.
1988. Control of the Colorado potato beetle, *Leptinotarsa decemlineata* (Say): learning from the Soviet experience. Bull. Entomol. Soc. Canada 20 (1): 9–14.
Boyd, M. L., and D. J. Boethel. 1998a. Residual toxicity of selected insecticides to heteropteran predaceous species (Heteroptera: Lygaeidae, Nabidae, Pentatomidae) on soybean. Environ. Entomol. 27: 154–160.
1998b. Susceptibility of predaceous hemipteran species to selected insecticides on soybean in Louisiana. J. Econ. Entomol. 91: 401–409.
Bozer, S. F., M. S. Traugott, and N. E. Stamp. 1996. Combined effects of allelochemical-fed and scarce prey on generalist insect predator *Podisus maculiventris*. Ecol. Entomol. 21: 328–334.
Brailovsky, H., and C. Mayorga. 1994. Hemiptera-Heteroptera de México XLV. La subfamilia Asopinae (Pentatomidae), en la estación de biología tropical "Los Tuxtlas," Veracruz, México. An. Inst. Biol. Univ. Nac. Autón. México, Ser. Zool. 65: 33–43.
Buckup, L. 1960. Pentatomídeos Neotropicais II. Contribuição para o conhecimento de Asopinae (Hemiptera: Pentatomidae) da América do Sul. Iheringia (Ser. Zool.) 15: 1–25.
Burikam, I., and B. Napompeth. 1978. Biological attributes of the predaceous pentatomid, *Canthecona furcellata* (Wolff) (Hemiptera: Pentatomidae). National Biological Control Research Center Thailand, Tech. Bull. 4 [not seen, *fide* Sipayung et al. 1992].
Butler, E. A. 1923. A Biology of the British Hemiptera–Heteroptera. H. F. and G. Witherby, London, U.K. 682 pp.

Butler, G. D., Jr., and F. G. Werner. 1960. Pentatomids associated with Arizona crops. Univ. Arizona Exp. Stn. Tech. Bull. 132.
Cachan, P. 1952. Les Pentatomidae de Madagascar (Hémiptères, Hétéroptères). Mém. Inst. Sci. Madagascar, Sér. E. Entomol. 1: 231–462.
Cahatian, P. O. 1991. Natural enemies of *Helicoverpa armigera* (Hübner) (Lepidoptera: Noctuidae) in the Philippines. M.S. thesis, Philippines University, Los Baños, Laguna, the Philippines.
Cappaert, D. L., F. A. Drummond, and P. A. Logan. 1991. Incidence of natural enemies of the Colorado potato beetle, *Leptinotarsa decemlineata* (Coleoptera: Chrysomelidae) on a native host in Mexico. Entomophaga 36: 369–378.
Carayon, J. 1961. Quelques remarques sur les Hémiptères-Hétéroptères: Leur importance comme insectes auxiliaires et les possibilités de leur utilisation dans la lutte biologique. Entomophaga 6: 133–141.
Carvalho, R. D. S., E. F. Vilela, M. Borges, and J. C. Zanuncio. 1994. Ritmo do comportamento de acasalamento e atividade sexual de *Podisus connexivus* Bergroth (Heteroptera: Pentatomidae: Asopinae). An. Soc. Entomol. Brasil 23: 197–202.
Chandra, D. 1979. A pentatomid, *Cantheconidea furcellata* Wolff, as a predator of *Spodoptera litura* Fabricius in Delhi. Bull. Entomol. 20: 158–159.
Chen, H. Q. 1982. A preliminary observation on *Altica* sp. Kunchong Zhishi 19(6): 21–23 [in Chinese].
Chen, Z. Y. 1986. Predatory stink bugs. Nat. Enemies of Insects 8: 207–208 [in Chinese].
Chloridis, A. S., D. S. Koveos, and D. C. Stamopoulos. 1997. Effect of photoperiod on the induction and maintenance of diapause and on development of the predatory bug *Podisus maculiventris* (Hem.: Pentatomidae). Entomophaga 42: 427–434.
Chopra, N. P., and Sucheta. 1986. Taxonomic studies on genus *Andrallus* Bergroth (Hemiptera: Pentatomidae: Asopinae). Bull. Entomol. 27: 37–40.
Choudhary, A. K., N. Khandwe, and O. P. Singh. 1989. Biology of *Cantheconidea furcellata* Wolff, a pentatomid predator of *Rivula* sp. J. Insect Sci. 2: 44–48.
Chu, Y. I. 1975. Rearing density of *Eocanthecona furcellata*, with special consideration to its mass production (Asopinae: Pentatomidae). Rostria 24: 135–140 [in Japanese].
Chu, Y. I., and C. M. Chu. 1975. Life history and the effect of temperature on the growth of *Eocanthecona furcellata* (Wolff). Plant Prot. Bull. Taiwan 17: 99–114 [in Chinese].
Clausen, C. P. 1940. Entomophagous Insects. McGraw-Hill, New York, New York, U.S.A. 688 pp.
Cloutier, C. 1997. Facilitated predation through interaction between life stages in the stinkbug predator *Perillus bioculatus* (Hemiptera, Pentatomidae). J. Insect Behav. 10: 581–598.
Cloutier, C., and F. Bauduin. 1995. Biological control of the Colorado potato beetle *Leptinotarsa decemlineata* (Coleoptera: Chrysomelidae) in Quebec by augmentative releases of the two-spotted stinkbug *Perillus bioculatus* (Hemiptera: Pentatomidae). Canad. Entomol. 127: 195–212.
Cloutier, C., and C. Jean. 1998. Synergism between natural enemies and biopesticides: a test case using the stinkbug *Perillus bioculatus* (Hemiptera: Pentatomidae) and *Bacillus thuringiensis* tenebrionis against Colorado potato beetle (Coleoptera: Chrysomelidae). J. Econ. Entomol. 91: 1096–1108.
Cohen, A. C. 1990. Feeding adaptations of some predaceous Hemiptera. Ann. Entomol. Soc. Amer. 83: 1215–1223.
1995. Extra-oral digestion in predaceous terrestrial Arthropoda. Annu. Rev. Entomol. 40: 85–103.
Coll, M. 1998. Living and feeding on plants in predatory Heteroptera. Pp. 89–129 *in* M. Coll and J. R. Ruberson [eds.], Predatory Heteroptera: Their Ecology and Use in Biological Control. Proceedings Thomas Say Publ. Entomol., Entomological Society of America, Lanham, Maryland, U.S.A. 233 pp.
Cooper, K. W. 1967. *Picromerus bidens* (Linn.), a beneficial, predatory European bug discovered in Vermont (Heteroptera: Pentatomidae). Entomol. News 78: 36–39.
Coppel, H. C., and P. A. Jones. 1962. Bionomics of *Podisus* spp. associated with the introduced pine sawfly, *Diprion similis* (Htg.) in Wisconsin. Wisconsin Acad. Sci. Arts Let. 51: 31–56.
Couilloud, R. 1989. Hétéroptères déprédateurs du cottonier en Afrique et à Madagascar (Pyrrhocoridae, Pentatomidae, Coreidae, Alydidae, Rhopalidae, Lygaeidae). Coton Fibres Trop. 44: 185–225.
Couturier, A. 1938. Contribution à l'étude biologique de *Podisus maculiventris* Say, prédateur américain du Doryphore. Ann. Epiphyt. Phytogén. 4: 95–165.
Dai, X. H., and Y. P. Yang. 1993. Bionomics of the beet armyworm *Spodoptera exigua* and its control. Plant Prot. 19(2): 20–21 [in Chinese].
Danilkina, E. N. 1986. A cage for rearing *Podisus*. Zashch. Rast. 10: 38 [in Russian].

De Clercq, P. 1993. Biology, ecology, rearing and predation potential of the predatory bugs *Podisus maculiventris* (Say) and *Podisus sagitta* (Fabricius) (Heteroptera: Pentatomidae) in the laboratory. Ph.D. thesis, University of Gent, Belgium.

De Clercq, P., and D. Degheele. 1990a. Description and life history of the predatory bug *Podisus sagitta* (Fab.) (Hemiptera: Pentatomidae). Canad. Entomol. 122: 1149–1156.

1990b. Effects of temperature and relative humidity on the reproduction of the predatory bug *Podisus sagitta* (Fab.) (Heteroptera: Pentatomidae). Meded. Fac. Landbouwwet. Rijksuniv. Gent 55: 439–443.

1992a. A meat-based diet for rearing the predatory stinkbugs *Podisus maculiventris* and *Podisus sagitta* (Het.: Pentatomidae). Entomophaga 37: 149–157.

1992b. Development and survival of *Podisus maculiventris* (Say) and *Podisus sagitta* (Fab.) (Heteroptera: Pentatomidae) at various constant temperatures. Canad. Entomol. 124: 125–133.

1992c. Influence of feeding interval on reproduction and longevity of *Podisus sagitta* (Het.: Pentatomidae). Entomophaga 37: 583–590.

1992d. Plant feeding by two species of predatory bugs of the genus *Podisus* (Heteroptera: Pentatomidae). Meded. Fac. Landbouwwet. Univ. Gent 57: 591–596.

1993a. Cold storage of the predatory bugs *Podisus maculiventris* (Say) and *Podisus sagitta* (Fabricius) (Heteroptera: Pentatomidae). Parasitica 49: 27–41.

1993b. Quality assessment of the predatory bugs *Podisus maculiventris* (Say) and *Podisus sagitta* (Fab.) (Heteroptera: Pentatomidae) after prolonged rearing on a meat-based artificial diet. Biocontrol Sci. Technol. 3: 133–139.

1993c. Quality of predatory bugs of the genus *Podisus* (Heteroptera: Pentatomidae) reared on natural and artificial diets. Pp. 129–142 *in* G. Nicoli, M. Benuzzi, and N. C. Leppla [eds.], Proceedings of the 7th Workshop of the IOBC Global Working Group "Quality Control of Mass Reared Arthropods," September 13–16, 1993, Rimini, Italy.

1994. Laboratory measurement of predation by *Podisus maculiventris* and *P. sagitta* (Hemiptera: Pentatomidae) on beet armyworm (Lepidoptera: Noctuidae). J. Econ. Entomol. 87: 76–83.

1995. *Podisus nigrispinus* (Dallas) and *Podisus sagitta* (Fabricius) (Heteroptera: Pentatomidae): correction of a misidentification. Canad. Entomol. 127: 265–266.

1997. Effects of mating status on body weight, oviposition, egg load, and predation in the predatory stinkbug *Podisus maculiventris* (Heteroptera: Pentatomidae). Ann. Entomol. Soc. Amer. 90: 121–127.

De Clercq, P., G. Keppens, G. Anthonis, and D. Degheele. 1988. Laboratory rearing of the predatory stinkbug *Podisus sagitta* (Fab.) (Heteroptera: Pentatomidae). Meded. Fac. Landbouwwet. Rijksuniv. Gent 53: 1213–1217.

De Clercq, P., A. De Cock, L. Tirry, E. Vinuela, and D. Degheele. 1995a. Toxicity of diflubenzuron and pyriproxyfen to the predatory bug *Podisus maculiventris*. Entomol. Exp. Appl. 74: 17–22.

De Clercq, P., E. Viñuela, G. Smagghe, and D. Degheele. 1995b. Transport and kinetics of diflubenzuron and pyriproxyfen in the beet armyworm, *Spodoptera exigua*, and its predator *Podisus maculiventris*. Entomol. Exp. Appl. 76: 189–194.

De Clercq, P., F. Merlevede, I. Mestdagh, K. Vandendurpel, J. Mohaghegh, and D. Degheele. 1998a. Predation on the tomato looper *Chrysodeixis chalcites* (Esper) (Lep., Noctuidae) by *Podisus maculiventris* (Say) and *Podisus nigrispinus* (Dallas) (Het., Pentatomidae). J. Appl. Entomol. 122: 93–98.

De Clercq, P., F. Merlevede, and L. Tirry. 1998b. Unnatural prey and artificial diets for rearing *Podisus maculiventris* (Heteroptera: Pentatomidae). Biol. Control 12: 137–142.

De Clercq, P., M. Vandewalle, and L. Tirry. 1998c. Impact of inbreeding on performance of the predator *Podisus maculiventris* (Heteroptera: Pentatomidae). BioControl 43: 299–310.

De Cock, A., P. De Clercq, L. Tirry, and D. Degheele. 1996. Toxicity of diafenthiuron and imidacloprid to the predatory bug *Podisus maculiventris* (Heteroptera: Pentatomidae). Environ. Entomol. 25: 476–480.

DeCoursey, R. M., and R. C. Allen. 1968. A generic key to the nymphs of the Pentatomidae of the eastern United States (Hemiptera: Heteroptera). Univ. Connecticut Occas. Pap. (Biol. Sci. Ser.) 1: 141–151.

DeCoursey, R. M., and C. O. Esselbaugh. 1962. Description of the nymphal stages of some North American Pentatomidae (Hemiptera-Heteroptera). Ann. Entomol. Soc. Amer. 55: 323–342.

Deitz, L. L., R. L. Rabb, J. W. Van Duyn, W. M. Brooks, J. R. Bradley, Jr., and R. E. Stinner. 1980. A guide to the identification and biology of soybean arthropods in North Carolina. North Carolina Agric. Exp. Stn. Tech. Bull. 238.

de Jong, J. K. 1931. Cantheconidea furcellata Wolff als natuurlijke vijand van rupsen in theetuinen. Archief Theecult. Ned. Indie 1: 17–24.

Desmier de Chenon, R., A. Sipayung, and P. Sudharto. 1989. The importance of natural enemies on leaf-eating caterpillars in oil palm plantations in Sumatra. Uses and possibilities. Porim International Palm Oil Development Conference, September 5–9, 1989 [not seen, *fide* **Sipayung** et al. 1992].

Dickens, J. C. 1999. Predator–prey interactions: olfactory adaptations of generalist and specialist predators. Agric. For. Entomol. 1: 47–54.

Didonet, J., J. C. Zanuncio, C. S. Sediyama, and M. C. Picanço. 1996a. Determinação das exigências térmicas de *Podisus nigrispinus* (Dallas, 1851) e de *Supputius cincticeps* Stal, 1860 (Heteroptera: Pentatomidae), em condições de laboratário. Rev. Bras. Entomol. 40: 61–63.

1996b. Influência da temperatura na reproducão e na longevidade de *Podisus nigrispinus* (Dallas) e de *Supputius cincticeps* Stal (Heteroptera, Pentatomidae). An. Soc. Entomol. Brasil 25: 117–123.

Diehl, J. K., N. J. Holliday, C. J. Lindgren, and R. E. Roughley. 1997. Insects associated with purple loosestrife, *Lythrum salicaria* L., in Southern Manitoba. Canad. Entomol. 129: 937–948.

Dos Santos, T. M., E. N. Silva, and F. S. Ramalho. 1995. Desenvolvimento ninfal de *Podisus connexivus* Bergroth (Hemiptera: Pentatomidae) alimentado com curuquerê-do-algodoeiro. Pesq. Agropec. Bras. 30: 163–167.

1996. Consumo alimentar e desenvolvimento de *Podisus nigrispinus* (Dallas) sobre *Alabama argillacea* (Huebner) em condições de laboratório. Pesq. Agropec. Bras. 31: 699–707.

Drummond, F. A., R. A. Casagrande, and E. Groden. 1987. Biology of *Oplomus dichrous* (Heteroptera: Pentatomidae) and its potential to control Colorado potato beetle (Coleoptera: Chrysomelidae). Environ. Entomol. 16: 633–638.

Drummond, F. A., R. L. James, R. A. Casagrande, and H. Faubert. 1984. Development and survival of *Podisus maculiventris* (Hemiptera: Pentatomidae), a predator of the Colorado potato beetle (Coleoptera: Chrysomelidae). Environ. Entomol. 13: 1283–1286.

Dupuis, C. 1947. Formes préimaginales d'Hémiptères Pentatomidae. I. Les nymphes des Asopinae: *Pinthaeus sanguinipes* F. et *Arma custos* F. Bull. Soc. Entomol. France 52: 54–57.

1949. Les Asopinae de la faune française (Hemiptera: Pentatomidae). Essai sommaire de synthèse morphologique, systématique et biologique. Rev. France Entomol. 16: 233–250.

Eckberg, T. B., and W. S. Cranshaw. 1994. Larval biology and control of the rabbitbrush beetle, *Trirhabda nitidicollis* Leconte (Coleoptera: Chrysomelidae). Southwest. Entomol. 19: 249–256.

Edwards, P. B., and D. M. Suckling. 1980. *Cermatulus nasalis* and *Oechalia schellembergii* (Hemiptera: Pentatomidae) as predators of eucalyptus tortoise beetle larvae, *Paropsis charybdis* (Coleoptera: Chrysomelidae), in New Zealand. New Zealand Entomol. 7: 158–164.

Eger, J. E., and J. R. Ables. 1981. Parasitism of Pentatomidae by Tachinidae in South Carolina and Texas. Southwest. Entomol. 6: 28–33.

El-Refai, S. A., and D. Degheele. 1988. Bio-ecology of *Podisus sagitta* (Fab.) (Heteroptera: Pentatomidae) as predator of lepidopterous larvae. Meded. Fac. Landbouwwet. Rijksuniv. Gent 53: 1219–1224.

Engel, H. 1939. Populationsdynamik des Kieferspanners in verschiedenen Biotopen. Verh. VII Intern. Kongr. Entomol., 1938, Berlin. 3: 1941–1949.

Esselbaugh, C. O. 1946. A study of the eggs of the Pentatomidae. Ann. Entomol. Soc. Amer. 39: 667–691.

1948. Notes on the bionomics of some midwestern Pentatomidae. Entomol. Amer. 28: 1–73.

Evans, E. W. 1982a. Consequences of body size for fecundity in the predatory stinkbug, *Podisus maculiventris* (Hemiptera: Pentatomidae). Ann. Entomol. Soc. Amer. 75: 418–420.

1982b. Feeding specialization in predatory insects: hunting and attack behavior of two stinkbug species (Hemiptera: Pentatomidae). Amer. Midl. Nat. 108: 96–104.

1982c. Influence of weather on predator/prey relations: stinkbugs and tent caterpillars. J. New York Entomol. Soc. 90: 241–246.

1982d. Timing of reproduction by predatory stinkbugs (Hemiptera: Heteroptera, Pentatomidae): patterns and consequences for a generalist and a specialist. Ecology 63: 147–158.

1983. Niche relations of predatory stinkbugs (*Podisus* spp., Pentatomidae) attacking tent caterpillars (*Malacosoma americanum*, Lasiocampidae). Amer. Midl. Nat. 109: 316–323.

1985. A key to the nymphs of four species of the genus *Podisus* (Hemiptera: Pentatomidae) of northeastern North America. Proc. Entomol. Soc. Washington 87: 94–97.

Evans, E. W., and R. B. Root. 1980. Group molting and other lifeways of a solitary hunter, *Apateticus bracteatus* (Hemiptera: Pentatomidae). Ann. Entomol. Soc. Amer. 73: 270–274.

Feytaud, J. 1938. Recherches sur le Doryphore. IV. L'acclimatation d'insectes entomophages américains ennemis du *Leptinotarsa decemlineata* Say. Ann. Epiphyt. Phytogén. 4: 27–93.

Fichter, B. L., and W. P. Stephen. 1981. Time related decay in prey antigens ingested by the predator *Podisus maculiventris* (Hemiptera, Pentatomidae) as detected by enzyme-linked immunosorbent assay. Oecologia 51: 404–407.

Filipov, N. A., A. F. Vorotyntseva, N. Ya. Prant, A. S. Stengach, A. V. Tuzlukov, and N. A. Cherna. 1989. The efficiency of *Podisus* against potato bugs when grown on different types of food. A biological method for controlling pests of vegetable crops. VO Agropromizdat, Moscow, 148–155 [in Russian].

Forsslund, K.-H. 1946. Något om röda tallstekelns (*Diprion sertifer* Geoffr.) skadegörelse. Medd. Skogsförsöksanst. 34: 365–390.

Franz, J. 1957. Beobachtungen über die natürliche Sterblichkeit des Kartoffelkäfers *Leptinotarsa decemlineata* (Say) in Kanada. Entomophaga 2: 197–212.

Franz, J. M. 1967. Beobachtungen über das Verhalten der Raubwanze *Perillus bioculatus* (Fabr.) (Pentatomidae) gegenüber ihrer Beute *Leptinotarsa decemlineata* (Say) (Chrysomelidae). Z. Pflanzenkr. Pflanzenpathol. Pflanzenschutz 73: 1–13.

Franz, J., and A. Szmidt. 1960. Beobachtungen beim Züchten von *Perillus bioculatus* (Fabr.) (Heteropt., Pentatomidae), einem aus Nordamerika importierten Räuber des Kartoffelkäfers. Entomophaga 5: 87–110.

Froeschner, R. C. 1988. Family Pentatomidae Leach, 1815. The stinkbugs. Pp. 544–597 *in* T. J. Henry and R. C. Froeschner [eds.], Catalog of the Heteroptera or True Bugs of Canada and the Continental United States. E. J. Brill, Leiden, the Netherlands. 958 pp.

Gäbler, H. 1937. *Picromerus bidens* L. als Feind der Lophyruslarven. Tharandter Forstl. Jahrb. 88(1): 51–58.

1938. Die Bedeutung einiger Wanzenarten als Feinde der Nonne. Z. Angew. Entomol. 25: 277–290.

Gallopin, G. C., and R. L. Kitching. 1972. Studies on the process of ingestion in the predatory bug *Podisus maculiventris* (Hemiptera: Pentatomidae). Canad. Entomol. 104: 231–237.

Ghorpade, K. D. 1972. Predaceous pentatomid bug *Cantheconidea furcellata* (Wolff) attacking *Latoia lepida* (Cramer) on mango near Bangalore. J. Bombay Nat. Hist. Soc. 72: 596–598.

Gillon, D. 1972. Les Hémiptères Pentatomides d'une savane préforestière de Côte d'Ivoire. Ann. Univ. Abidjan (Série E) 5: 265–371.

Girault, A. A. 1907. Hosts of egg-parasites in North and South America. Psyche 14: 27–39.

Giroux, S., R.-M. Duchesne, and D. Coderre. 1995. Predation of *Leptinotarsa decemlineata* (Coleoptera: Chrysomelidae) by *Coleomegilla maculata* (Coleoptera: Coccinellidae): comparative effectiveness of predator developmental stages and effect of temperature. Environ. Entomol. 24: 748–754.

Golubeva, N. N., L. A. Ziskind, S. S. Izhevskii, and L. A. Stradimova. 1980. Laboratory rearing of *Podisus*. Zashch. Rast. 1: 53–54 [in Russian].

Gonçalves, L., V. H. P. Bueno, and C. F. De Carvalho. 1990. Controle biológico em *Eucalyptus* spp.: 1. Etologia de ninfas e adultos de *Podisus nigrolimbatus* Spinola 1832 e *Podisus connexivus* Bergroth 1891 (Hemiptera: Pentatomidae: Asopinae). Inst. Pesqui. Estudos Florestais 43–44: 70–73.

Gope, B. 1981. A promising predator of Bihar hairy caterpillar and bunch caterpillar. Two Leaves Bud 28: 47–48.

Goryshin, N. I., and I. A. Tuganova. 1989. Optimization of short-term storage of eggs of the predatory bug *Podisus maculiventris* (Hemiptera, Pentatomidae). Zool. Zh. 68: 111–119 [in Russian].

Goryshin, N. I., A. Kh. Saulich, T. A. Volkovich, I. A. Borisenko, and N. P. Simonenko. 1988a. The influence of the food factor on the development and photoperiodic reaction of the predatory bug *Podisus maculiventris* (Hemiptera, Pentatomidae). Zool. Zh. 67: 1324–1332 [in Russian].

Goryshin, N. I., T. A. Volkovich, A. Kh. Saulich, M. Vagner, and I. A. Borisenko. 1988b. The role of temperature and photoperiod in the control of development and diapause of the predatory bug *Podisus maculiventris* (Hemiptera, Pentatomidae). Zool. Zh. 67: 1149–1161 [in Russian].

Gravena, S., and F. M. Lara. 1982. Controle integrado de pragas e receituário agronômico. Pp. 123–161 *in* F. Graziano Neto [ed.], Receituário Agronômico. Agroedições, São Paulo, Brazil.

Grazia, J., M. C. del Vecchio, and R. Hildebrand. 1985. Estudo das ninfas de Heterópteros predadores: I. *Podisus connexivus* Bergroth 1891 (Pentatomidae, Asopinae). An. Soc. Entomol. Brasil 14: 303–313.

Greenstone, M. H., and C. E. Morgan. 1989. Predation on *Heliothis zea* (Lepidoptera: Noctuidae): an instar-specific ELISA for stomach analysis. Ann. Entomol. Soc. Amer. 82: 45–49.
Greenstone, M. H., and S. C. Trowell. 1994. Arthropod predation: a simplified immunodot format for predator gut analysis. Ann. Entomol. Soc. Amer. 87: 214–217.
Groves, E. W. 1956. Gregarious behaviour in the larvae of *Picromerus bidens* (L.) (Hem., Pentatomidae). Entomol. Mon. Mag. 92: 65–66.
Gulde, J. 1919. Die Larvenstadien der Asopiden (Hemipt. Het.). Dtsch. Entomol. Z. (1919): 45–55.
Gusev, G. B., Yu. V. Zayats, N. V. Shmetser, and L. V. Perepelina. 1980. Details on the biology of Podisus [sic]. Zashch. Rast. 9: 38–39 [in Russian].
Gusev, G. V., Yu. V. Zayats, E. M. Topashchenko, and G. K. Rzhavina. 1983. Control of the Colorado beetle on aubergines. Zashch. Rast. 8: 34 [in Russian].
Harcourt, D. G. 1971. Population dynamics of *Leptinotarsa decemlineata* (Say) in eastern Ontario. III. Major population processes. Canad. Entomol. 103: 1049–1061.
Hart, C. A. 1919. The Pentatomoidea of Illinois with keys to the nearctic genera. Illinois Nat. Hist. Surv. Bull. 13: 157–223.
Hayslip, N. E., W. G. Genung, E. G. Kelsheimer, and J. W. Wilson. 1953. Insects attacking cabbage and other crucifers in Florida. Florida Agric. Exp. Stn. Bull. 543: 1–57.
Heimpel, G. E. 1991. Searching behavior and functional response of *Perillus bioculatus*, a predator of the Colorado potato beetle. M.S. thesis, University of Delaware, Newark, Delaware, U.S.A.
Heimpel, G. E., and J. A. Hough-Goldstein. 1994a. Components of the functional response of *Perillus bioculatus* (Hemiptera: Pentatomidae). Environ. Entomol. 23: 855–859.
1994b. Search tactics and response to cues by predatory stink bugs. Entomol. Exp. Appl. 73: 193–197.
Hertzel, G. 1982. Zur Phänologie und Fortpflanzungsbiologie einheimischer Pentatomiden-Arten (Heteroptera). Entomol. Nachr. Ber. 26: 69–72.
Hilbeck, A., and G. G. Kennedy. 1996. Predators feeding on the Colorado potato beetle in insecticide-free plots and insecticide-treated commercial potato fields in eastern North Carolina. Biol. Control 6: 273–282.
Hildebrand, R. 1987. The types of *Podisus* Herrich-Schaeffer, 1851, preserved in the M.N.H.N., Paris (Heteroptera, Pentatomidae, Asopinae). Rev. France Entomol. 9: 87–93.
Horton, D. R., T. Hinojosa, and S. R. Olson. 1998. Effects of photoperiod and prey type on diapause tendency and preoviposition period in *Perillus bioculatus* (Hemiptera: Pentatomidae). Canad. Entomol. 130: 315–320.
Hokyo, N., and S. Kawauchi. 1975. The effect of prey size and prey density on the functional response, survival, growth and development of a predatory pentatomid bug, *Podisus maculiventris* Say. Res. Popul. Ecol. 16: 207–218.
Hough-Goldstein, J. A. 1998. Use of predatory pentatomids in integrated management of the Colorado potato beetle (Coleoptera: Chrysomelidae). Pp. 209–223 *in* M. Coll and J. R. Ruberson [eds.], Predatory Heteroptera: Their Ecology and Use in Biological Control. Thomas Say Publ. Entomol.: Proceedings Entomological Society of America, Lanham, Maryland, U.S.A. 233 pp.
Hough-Goldstein, J., and C. B. Keil. 1991. Prospects for integrated control of the Colorado potato beetle (Coleoptera: Chrysomelidae) using *Perillus bioculatus* (Hemiptera: Pentatomidae) and various pesticides. J. Econ. Entomol. 84: 1645–1651.
Hough-Goldstein, J., and D. McPherson. 1996. Comparison of *Perillus bioculatus* and *Podisus maculiventris* (Hemiptera: Pentatomidae) as potential control agents of the Colorado potato beetle (Coleoptera: Chrysomelidae). J. Econ. Entomol. 89: 1116–1123.
Hough-Goldstein, J., and J. Whalen. 1993. Inundative release of predatory stink bugs for control of Colorado potato beetle. Biol. Control 3: 343–347.
Hough-Goldstein, J., J. Cox, and A. Armstrong. 1996a. *Podisus maculiventris* (Hemiptera: Pentatomidae) predation on ladybird beetles (Coleoptera: Coccinellidae). Florida Entomol. 79: 64–68.
Hough-Goldstein, J., J. A. Janis, and C. D. Ellers. 1996b. Release methods for *Perillus bioculatus* (F.), a predator of the Colorado potato beetle. Biol. Control 6: 114–122.
Hough-Goldstein, J. A., G. E. Heimpel, H. E. Bechmann, and C. E. Mason. 1993. Arthropod natural enemies of the Colorado potato beetle. Crop Prot. 12: 324–334.
Howard, N. F. 1936. Parasites and predators of the Mexican bean beetle in the United States. U.S. Dept. Agric. Circ. 418: 1–12.

Ignoffo, C. M., C. Garcia, W. A. Dickerson, G. T. Schmidt, and K. D. Biever. 1977. Imprisonment of entomophages to increase effectiveness: evaluation of a concept. J. Econ. Entomol. 70: 292–294.

Iwao, S., and W. G. Wellington. 1970. The influence of behavioral differences among tent-caterpillar larvae on predation by a pentatomid bug. Canad. J. Zool. 48: 896–898.

Izhevskii, S. S., N. N. Golubeva, and V. V. Buleza. 1980. Egg parasites of the family Scelionidae as possible parasites of the introduced predacious bug *Podisus maculiventris* (Hemiptera, Pentatomidae). Zool. Zh. 59: 73–78 [in Russian].

Izhevskii, S. S., and L. A. Ziskind. 1982. A descriptive model of the laboratory breeding of Podisus. Zashch. Rast. 6: 43 [in Russian].

Izhevskii, S. S., L. A. Ziskind, and V. L. Rybak. 1988. A complex of measures against the Colorado beetle. Zashch. Rast. 10: 45–46 [in Russian].

Jakhmola, S. S. 1983. Natural enemies of *Til* leafroller and capsuleborer, *Antigastra catalaunalis* (Dup.). Bull. Entomol. 24: 147–148.

Jasic, J. 1975. On the life cycle of *Perillus bioculatus* (Heteroptera, Pentatomidae) in Slovakia. Acta Entomol. Bohemoslov. 72: 383–390.

Javahery, M. 1967. The biology of some Pentatomoidea and their egg parasites. Ph.D. thesis, University of London, London, U.K.

1968. The egg parasite complex of British Pentatomoidea (Hemiptera): Taxonomy of Telenominae (Hymenoptera: Scelionidae). Trans. R. Entomol. Soc. London 120: 417–436.

1986. Biology and ecology of *Picromerus bidens* (Hemiptera: Pentatomidae) in southeastern Canada. Entomol. News 97: 87–98.

1994. Development of eggs in some true bugs (Hemiptera-Heteroptera). Part I. Pentatomoidea. Canad. Entomol. 126: 401–433.

Jenkins, D. J., J. Hough-Goldstein, and J. D. Pesek, Jr. 1998. Novel application of gels as potential carriers for beneficial insects. J. Econ. Entomol. 91: 419–427.

Jermy, T. 1966. Ergebnis des Grossversuches zur Einbürgerung von *Perillus bioculatus*. In Cir. No. 18 of the IOBC Working Group "Population Dynamics and Biological Control of the Colorado Beetle."

1980. The introduction of *Perillus bioculatus* into Europe to control the Colorado beetle. Bull. OEPP/EPPO Bull. 10: 475–479.

Jones, P. A., and H. C. Coppel. 1963. Immature stages and biology of *Apateticus cynicus* (Say) (Hemiptera: Pentatomidae). Canad. Entomol. 95: 770–779.

Jones, W. A., Jr., and M. J. Sullivan. 1981. Overwintering habitats, spring emergence patterns and winter mortality of some South Carolina Hemiptera. Environ. Entomol. 10: 409–414.

Kahlow, E. 1963. Ergebnisse und kritische Einschatzung einer Massenaufzucht von *Perillus bioculatus* (Fabr.). Beitr. Entomol. 13: 345–358.

Kapoor, K. N., J. P. Gujrati, and G. A. Gangrade. 1973. *Cantheconidea furcellata* Wolff as a predator of *Prodenia litura* Fabr. larvae. Indian J. Entomol. 35: 275.

Kelton, L. A. 1972. *Picromerus bidens* in Canada (Heteroptera: Pentatomidae). Canad. Entomol. 104: 1743–1744.

Khlistovskii, E. D., I. N. Oleshchenko, Zh. A. Shirinyan, and V. Ya. Ismailov. 1985. Artificial nutrient media for rearing larvae of predatory bugs of the family Pentatomidae. Zool. Zh. 64: 117–123 [in Russian].

Khloptseva, R. I. 1991. The use of entomophages in biological pest control in the USSR. Biocontrol News Inf. 12: 243–246.

Kim, C. W., J. I. Kim, and S. H. Kim. 1968. Biological control of fall webworm, *Hyphantria cunea* Drury in Korea. II. Studies on the natural enemies imported. Entomol. Res. Bull. 4: 37–56.

Kirkland, A. H. 1896. Predaceous Hemiptera-Heteroptera. Pp. 392–403 *in* E. H. Forbush and C. H. Fernald [eds.], The Gypsy Moth, *Porthetria dispar* (Linn.). Wright and Potter Printing Co., State Printers, Boston, Massachusetts, U.S.A.

1897. Notes on the life history and habits of certain predaceous Heteroptera. Pp. 51–59 *in* C. H. Fernald [ed.], Extermination of the Gypsy Moth. Rep. Massachusetts State Board Agric. Appendix.

1898. The species of *Podisus* occurring in the United States, pp. 112–138 *in* Rep. Secretary Massachusetts State Board Agric. on Gypsy Moths, Appendix.

Kirtibutr, N. 1987. Some insect pests of fast growing trees in Thailand. Pp. 125–137 *in* Proceedings of the Symposium on the Status of Forest Pests and Diseases in Southeast Asia, May 13–15, 1985. SEAMEO-BIOTROP, Regional Center for Tropical Biology, Bogor, Indonesia.

Knight, H. H. 1923. Studies on the life history and biology of *Perillus bioculatus* Fabricius, including observations on the nature of the color pattern. 19th Rep. State Entomol. Minnesota: 50–96.

1952. Review of the genus *Perillus* (Hemiptera: Pentatomidae) with description of a new species. Ann. Entomol. Soc. Amer. 45: 229–232.

Kobayashi, H., and T. Okada. 1994. Rearing of the predatory stink bug, *Eocanthecona furcellata* (Wolff) (Heteroptera: Pentatomidae), on alternative foods. Proc. Assoc. Plant Prot. Shikoku No. 29: 133–136 [in Japanese].

1995. Suitability of two kinds of meat-based artificial diets for the development of the predatory stinkbug, *Eocanthecona furcellata* (Wolff) (Heteroptera: Pentatomidae). Proc. Assoc. Plant Prot. Shikoku 30: 131–135 [in Japanese].

Kobayashi, T. 1951. The developmental stages of four species of the Japanese Pentatomidae (Hemiptera). Trans. Shikoku Entomol. Soc. 2: 7–16.

Koehler, W. 1948. *Troilus luridus* F. (Hem. Het.). Trav. C. R. Inst. Pol. Rech. For. (Ser. A) 51: 1–79 [in Polish].

Kumar, M., A. N. Shylesha, and N. S. A. Thakur. 1996. *Eocanthecona furcellata* (Wolff) (Heteroptera: Pentatomidae): a promising predator of *Craspedonta leayana* (Latr.) (Chrysomelidae: Coleoptera) on *Gmelina arborea* in Meghalaya. Insect Environ. 2: 56–57.

Lachance, S., and C. Cloutier. 1997. Factors affecting dispersal of *Perillus bioculatus* (Hemiptera: Pentatomidae), a predator of the Colorado potato beetle (Coleoptera: Chrysomelidae). Environ. Entomol. 26: 946–954.

Lambdin, P. L., and A. M. Baker. 1986. Evaluation of dewinged spined soldier bugs, *Podisus maculiventris* (Say), for longevity and suppression of the Mexican bean beetle, *Epilachna varivestis* Mulsant, on snapbeans. J. Entomol. Sci. 21: 263–266.

Lambdin, P. L., and G. Q. Lu. 1984. External morphology of eggs of the spined soldier bug, *Podisus maculiventris* (Say) (Hemiptera: Pentatomidae). Proc. Entomol. Soc. Washington 86: 374–377.

Landis, B. J. 1937. Insect hosts and nymphal development of *Podisus maculiventris* Say and *Perillus bioculatus* F. (Hemiptera: Pentatomidae). Ohio J. Sci. 37: 252–259.

Larivière, M.-C. 1995. Cydnidae, Acanthosomatidae, and Pentatomidae (Insecta: Heteroptera): systematics, geographical distribution, and bioecology. Fauna of New Zealand No. 35. Manaaki Whenua Press, Lincoln, Canterbury, New Zealand.

Larivière, M.-C., and A. Larochelle. 1989. *Picromerus bidens* (Heteroptera: Pentatomidae) in North America, with a world review of its distribution and bionomics. Entomol. News 100: 133–146.

Larochelle, A. 1984. Les punaises terrestres (Hémiptères: Géocorises) du Québec. Fabreries, Suppl. 3.

Larochelle, A., and M.-C. Larivière. 1980. *Picromerus bidens* L. (Heteroptera: Pentatomidae) en Amérique du Nord: répartition géographique, habitat et biologie. Bull. Invent. Ins. Québec 2(1): 10–18.

Lattin, J. D., and J. P. Donahue. 1969. The second record of *Picromerus bidens* (L.) in North America (Heteroptera: Pentatomidae: Asopinae). Proc. Entomol. Soc. Washington 71: 567–568.

Le Berre, J. R., and G. Portier. 1963. Utilisation d'un hétéroptère Pentatomidae *Perillus bioculatus* (Fabr.) dans la lutte contre le Doryphore *Leptinotarsa decemlineata* (Say): premiers résultats obtenus en France. Entomophaga 8: 183–190.

Legaspi, J. C., and B. C. Legaspi. 1998. Life-history trade-offs in insects, with emphasis on *Podisus maculiventris* (Heteroptera: Pentatomidae). Pp. 71–87 *in* M. Coll and J. R. Ruberson [eds.], Predatory Heteroptera: Their Ecology and Use in Biological Control. Thomas Say Publ. Entomol.: Proceedings Entomological Society of America, Lanham, Maryland, U.S.A. 233 pp.

Legaspi, J. C., and R. J. O'Neil. 1993. Life history of *Podisus maculiventris* given low numbers of *Epilachna varivestis* as prey. Environ. Entomol. 22: 1192–1200.

1994a. Developmental response of nymphs of *Podisus maculiventris* (Heteroptera: Pentatomidae) reared with low numbers of prey. Environ. Entomol. 23: 374–380.

1994b. Lipids and egg production of *Podisus maculiventris* (Heteroptera: Pentatomidae) under low rates of predation. Environ. Entomol. 23: 374–380.

Legaspi, J. C., R. J. O'Neil, and B. C. Legaspi, Jr. 1996. Trade-offs in body weights, egg loads, and fat reserves of field-collected *Podisus maculiventris* (Heteroptera: Pentatomidae). Environ. Entomol. 25: 155–164.

Legaspi, J. C., V. Russell, and B. C. Legaspi, Jr. 1994. Microscopic examination of the spermatheca as an indicator of mating success in *Podisus maculiventris* (Say). Southwest. Entomol. 19: 189–190.

LeRoux, E. J. 1960. Effects of "modified" and "commercial" spray programs on the fauna of apple orchards in Quebec. Ann. Entomol. Soc. Québec 6: 87–121.

LeRoux, E. J., R. O. Paradis, and M. Hudon. 1963. Major mortality factors in the population dynamics of the eye-spotted bud-moth, the pistol casebearer, the fruit-tree leaf roller, and the European corn borer in Quebec. Pp. 67–82 *in* E. J. LeRoux [ed.], Population Dynamics of Agricultural and Forest Insect Pests, Mem. Entomol. Soc. Can. 32.

Leston, D. 1955. The life-cycle of *Picromerus bidens* (L.) (Hem., Pentatomidae) in Britain. Entomol. Mon. Mag. 91: 109.

Linnavuori, R. E. 1982. Pentatomidae and Acanthosomatidae of Nigeria and the Ivory Coast, with remarks on species of the adjacent countries in West and Central Africa. Acta Zool. Fenn. 163: 1–176.

Lipa, J. J. 1985. Progress in biological control of the Colorado beetle (*Leptinotarsa decemlineata*) in Eastern Europe. Bull. OEPP/EPPO Bull. 15: 207–211.

Lockwood, J. A., and R. N. Story. 1986. Adaptive functions of nymphal aggregation in the southern green stink bug, *Nezara viridula* (L.) (Hemiptera: Pentatomidae). Environ. Entomol. 15: 739–749.

Lodos, N., and F. Önder. 1983. Contribution to the study on the Turkish Pentatomoidea (Heteroptera) VI. Asopinae (Amyot and Serville) 1843. Türk. Bit. Kor. Derg. 7: 221–230.

Logan, P. A., and R. A. Casagrande. 1980. Predicting Colorado potato beetle (*Leptinotarsa decemlineata* Say) density and potato yield loss. Environ. Entomol. 9: 659–663.

Logan, P. A., R. A. Casagrande, T. H. Hsiao, and F. A. Drummond. 1987. Collections of natural enemies of *Leptinotarsa decemlineata* (Col.: Chrysomelidae) in Mexico, 1980–1985. Entomophaga 32: 249–254.

Lopez, J. D., Jr., R. L. Ridgway, and R. E. Pinnell. 1976. Comparative efficacy of four insect predators of the bollworm and tobacco budworm. Environ. Entomol. 5: 1160–1164.

Lyashova, L. V., A. A. Zhemchuzhina, and E. P. Ovsyanko. 1985. Rearing medium for larvae of Podisus. Zashch. Rast. 1: 53.

Mackauer, M. 1976. Genetic problems in the production of biological control agents. Annu. Rev. Entomol. 21: 369–385.

Malaguido, A. B., and A. R. Panizzi. 1998. *Alcaeorrhynchus grandis* (Dallas): an eventual predator of *Chlosyne lacinia saundersii* Doubleday & Hewitson on sunflower in Northern Paraná state. An. Soc. Entomol. Brasil 27: 671–674.

Mallach, N. 1974. Zur Kenntnis der kleinen Kiefern-Buschhornblattwespe, *Diprion (Microdiprion) pallipes* (Fall.) (Hym.: Diprionidae). Teil 3. Populationsökologie. Z. Angew. Entomol. 75: 337–380.

Manley, G. V. 1982. Biology and life history of the rice field predator *Andrallus spinidens* F. (Hemiptera: Pentatomidae). Entomol. News 93: 19–24.

Marston, N. L., G. T. Schmidt, K. D. Biever, and W. A. Dickerson. 1978. Reaction of five species of soybean caterpillars to attack by the predator, *Podisus maculiventris*. Environ. Entomol. 7: 53–56.

Mateeva, A. 1994. Insect prey and nymphal development of the predatory bug *Podisus sagitta* (F.) (Hemiptera: Pentatomidae). Meded. Fac. Landbouwwet. Univ. Gent 59: 253–255.

Mayné, R. 1965. Les Hémiptères de la Réserve domaniale du Westhoek. Pentatomoidea. Service des Réserves Naturelles domaniales et de la Conservation de la Nature, Ministère de l'Agriculture, Belgique. Travaux 1: 1–47.

Mayné, R., and R. Breny. 1940. Prédateurs et parasites du doryphore. Bull. Inst. Agron. Stn. Rech. Gembloux 9: 61–80.

1947a. Les éclosions de *Picromerus bidens* L. dans la nature. Parasitica 3: 53–67.

1947b. Contribution à l'étude des circonstances climatiques influençant le pouvoir d'éclosion des oeufs de *Picromerus bidens* L. Parasitica 3: 133–141.

1948a. *Picromerus bidens* L.: la vie larvaire au premier âge. Parasitica 4: 1–20.

1948b. *Picromerus bidens* L.: Morphologie. Biologie. Détermination de sa valeur d'utilisation dans la lutte biologique contre le doryphore de la pomme de terre — la valeur économique antidoryphorique des Asopines indigènes belges. Parasitica 4: 189–224.

McDermott, F. A. 1911. The attack of a larval Hemiptera upon a caterpillar. Proc. Entomol. Soc. Washington 13: 90–91.

McDonald, F. J. D. 1966. The genitalia of North American Pentatomoidea (Hemiptera: Heteroptera). Quaest. Entomol. 2: 7–150.
McLain, D. K. 1979. Terrestrial trail-following by three species of predatory stink bugs. Florida Entomol. 62: 152–154.
McPherson, J. E. 1980. A list of the prey species of *Podisus maculiventris* (Hemiptera: Pentatomidae). Great Lakes Entomol. 13: 17–24.
1982. The Pentatomoidea (Hemiptera) of Northeastern North America. South Illinois University Press, Carbondale and Edwardsville, Illinois, U.S.A. 240 pp.
McPherson, R. M., J. B. Graves, and T. A. Allain. 1979. Dosage-mortality responses and field control of seven pentatomids, associated with soybean, exposed to methyl parathion. Environ. Entomol. 8: 1041–1043.
McPherson, R. M., J. R. Pitts, L. D. Newsom, J. B. Chapin, and D. C. Herzog. 1982. Incidence of tachinid parasitism of several stink bug (Heteroptera: Pentatomidae) species associated with soybean. J. Econ. Entomol. 75: 783–786.
Mead, F. W. 1976. A predatory stink bug, *Euthyrhynchus floridanus* (Linnaeus) (Hemiptera: Pentatomidae). Florida Dept. Agric. Consumer Serv., Entomol. Circ. 174.
Michalk, O. 1938. Die Wanzen (Hemiptera Heteroptera) der Leipziger Tieflandsbucht und der angrenzenden Gebiete. Sitzungsber. Naturforsch. Ges. Leipzig (63–64): 15–188.
Miller, N. C. E. 1956. The Biology of the Heteroptera. Leonard Hill, London, U.K. 162 pp.
Mineo, G., and M. Iannazzo. 1986. Brevi notizie sull' *Altica ampelophaga* Guérin (Col. Chrysomelidae). Redia 69: 543–553.
Moens, R. 1963. Essai d'acclimatation de *Perillus bioculatus* Fabr., prédateur du Doryphore. Meded. Landbouwhogesch. Gent 28: 792–810.
1965. Observations sur les exigences alimentaires de *Perillus bioculatus* Fabr. Meded. Landbouwhogesch. Gent 30: 1504–1515.
Mohaghegh-Neyshabouri, J., P. De Clercq, and D. Degheele. 1996. Influence of female body weight on reproduction in laboratory-reared *Podisus nigrispinus* and *Podisus maculiventris* (Heteroptera: Pentatomidae). Meded. Fac. Landbouwwet. Univ. Gent 61: 693–696.
Mohaghegh, J., P. De Clercq, and L. Tirry. 1998a. Effects of maternal age and egg weight on developmental time and body weight of offspring in *Podisus maculiventris* (Heteroptera: Pentatomidae). Ann. Entomol. Soc. Amer. 91: 315–322.
1998b. Maternal age and egg weight affect offspring performance in the predatory stink bug *Podisus nigrispinus*. BioControl 43: 163–174.
2000. Toxicity of selected insecticides to the spined soldier bug, *Podisus maculiventris* (Heteroptera: Pentatomidae). Biocontrol Sci. Technol., 10: 33–40.
Morrill, A. W. 1906. Some observations on the spined soldier bug (*Podisus maculiventris* Say). U.S. Div. Entomol. Bull. (n.s.) 60: 155–161.
Morris, R. F. 1963. The effect of predator age and prey defense on the functional response of *Podisus maculiventris* Say to the density of *Hyphantria cunea* Drury. Canad. Entomol. 95: 1009–1020.
1972. Predation by insects and spiders inhabiting colonial webs of *Hyphantria cunea*. Canad. Entomol. 104: 1197–1207.
Mugo, H. M. 1992. Possibility of using natural enemies to control giant looper (*Ascotis selenaria reciprocaria* Walk.) in Kenya. Kenya Coffee, Coffee Board of Kenya Monthly Bull. 57: 1453–1455.
Mukerji, M. K., and E. J. LeRoux. 1965. Laboratory rearing of a Quebec strain of the pentatomid predator, *Podisus maculiventris* (Say) (Hemiptera: Pentatomidae). Phytoprotection 46: 40–60.
1969a. A quantitative study of the food consumption and growth of *Podisus maculiventris* (Hemiptera: Pentatomidae). Canad. Entomol. 101: 387–403.
1969b. A study on energetics of *Podisus maculiventris* (Hemiptera: Pentatomidae). Canad. Entomol. 101: 449–460.
1969c. The effect of predator age on the functional response of *Podisus maculiventris* to the prey size of *Galleria mellonella*. Canad. Entomol. 101: 314–327.
Muntyan, E. M., and I. G. Yazlovetskii. 1994. Digestive carbohydrases in larval predaceous bugs *Podisus maculiventris* and *Perillus bioculatus*. Zh. Evol. Biokhim. Fiziol. 30: 161–167 [in Russian].

Nascimento, M. L., D. F. Capalbo, G. J. Moraes, E. A. De Nardo, A. H. N. Maia, and R. C. A. L. Oliveira. **1998.** Effect of a formulation of *Bacillus thuringiensis* Berliner var. *kurstaki* on *Podisus nigrispinus* Dallas (Heteroptera: Pentatomidae: Asopinae). J. Invertebr. Pathol. 72: 178–180.

Nayek, T. K., and T. C. Banerjee. **1987.** Life history and host specificity of *Altica cyanea* (Coleoptera: Chrysomelidae), a potential biological control agent for water primrose, *Ludwigia adscendens*. Entomophaga 32: 407–414.

Novozhilov, K. V., G. V. Gusev, I. N. Sazonova, A. G. Koval, S. N. Moralev, and F. I. Patrashku. **1991.** Prospect for a complex system of egg plant protection against Colorado beetle in Moldava. Pp. 78–89 *in* Ekologicheskie osnovy primeneniya insektoakaritsidov. VIZR, St. Petersburg, Russia [in Russian].

Nyiira, Z. M. **1970.** A note on the natural enemies of lepidopterous larvae in cotton bolls in Uganda. Ann. Entomol. Soc. Amer. 63: 1461–1462.

Oetting, R. D., and T. R. Yonke. **1971.** Immature stages and biology of *Podisus placidus* and *Stiretrus fimbriatus* (Hemiptera: Pentatomidae). Canad. Entomol. 103: 1505–1516.

1975. Immature stages and notes on the biology of *Euthyrhynchus floridanus* (L.) (Hemiptera: Pentatomidae). Ann. Entomol. Soc. Amer. 68: 659–662.

Oien, C. T., and D. W. Ragsdale. **1993.** Susceptibility of nontarget hosts to *Nosema furnacalis* (Microsporida: Nosematidae), a potential biological control agent of the European corn borer, *Ostrinia nubilalis* (Lepidoptera: Pyralidae). Biol. Control 3: 323–328.

Oliveira, D. de, and J. Juillet. **1971.** *Picromerus bidens* L. (Hémiptères: Pentatomides), nouveau prédateur de la mouche-à-scie du pin de pépinière, *Diprion frutetorum* Fab. Phytoprotection 52: 32–34.

Olsen, C. E. **1910.** Notes on breeding Hemiptera. J. New York Entomol. Soc. 70: 60–62.

O'Neil, R. J. **1988a.** A model of predation by *Podisus maculiventris* (Say) on Mexican bean beetle, *Epilachna varivestis* Mulsant, in soybeans. Canad. Entomol. 120: 601–608.

1988b. Predation by *Podisus maculiventris* (Say) on Mexican bean beetle, *Epilachna varivestis* Mulsant, in Indiana soybeans. Canad. Entomol. 120: 161–166.

1989. Comparison of laboratory and field measurements of the functional response of *Podisus maculiventris* (Heteroptera: Pentatomidae). J. Kansas Entomol. Soc. 62: 148–155.

1997. Functional response and search strategy of *Podisus maculiventris* (Heteroptera: Pentatomidae) attacking Colorado potato beetle (Coleoptera: Chrysomelidae). Environ. Entomol. 26: 1183–1190.

O'Neil, R. J., and R. N. Wiedenmann. **1987.** Adaptations of arthropod predators to agricultural systems. Florida Entomol. 70: 41–48.

1990. Body weight of *Podisus maculiventris* (Say) under various feeding regimens. Canad. Entomol. 122: 285–294.

Orr, D. B., and D. J. Boethel. **1986.** Influence of plant antibiosis through four trophic levels. Oecologia 70: 242–249.

Orr, D. B., J. S. Russin, and D. J. Boethel. **1986.** Reproductive biology and behavior of *Telenomus calvus* (Hymenoptera: Scelionidae), a phoretic egg parasitoid of *Podisus maculiventris* (Hemiptera: Pentatomidae). Canad. Entomol. 118: 1063–1072.

Overney, S., S. Yelle, and C. Cloutier. **1998.** Occurrence of digestive cysteine proteases in *Perillus bioculatus*, a natural predator of the Colorado potato beetle. Comp. Biochem. Physiol. Part B 120: 191–195.

Pawar, A. D. **1976.** *Andrallus spinidens* (Fabricius) (Asopinae: Pentatomidae: Hemiptera) as a predator of insect pests of rice in Himachal Pradesh, India. Rice Entomol. Newsl. 4: 23–24.

Pfannenstiel, R. S., R. E. Hunt, and K. V. Yeargan. **1995.** Orientation of a hemipteran predator to vibrations produced by feeding caterpillars. J. Insect Behav. 8: 1–9.

Picanço, M. C., R. N. C. Guedes, V. C. Batalha, and R. P. Campos. **1996.** Toxicity of insecticides to *Dione juno juno* (Lepidoptera: Heliconidae) and selectivity to two of its predaceous bugs. Trop. Sci. 36: 51–53.

Plummer, C. C., and B. J. Landis. **1932.** Records of some insects predacious on *Epilachna corrupta* Muls. in Mexico. Ann. Entomol. Soc. Amer. 25: 695–708.

Poprawski, T. J., R. I. Carruthers, J. Speese III, D. C. Vacek, and L. E. Wendel. **1997.** Early-season applications of the fungus *Beauveria bassiana* and introduction of the hemipteran predator *Perillus bioculatus* for control of Colorado potato beetle. Biol. Control 10: 48–57.

Poutouli, W. **1992.** Plantes-hôtes secondaires des hétéroptères recencés sur coton, maïs, niébé au Togo. Meded. Fac. Landbouwwet. Univ. Gent 57: 627–636.

Prasad, D., K. M. Singh, R. N. Singh, and D. N. Mehto. 1983. A new predator of new pest of jasmine in Delhi. Bull. Entomol. 24: 140–141.
Prebble, M. L. 1933. The biology of *Podisus serieventris* Uhler, in Cape Breton, Nova Scotia. Canad. J. Res. 9: 1–30.
Pruszynski, S. 1989. 100 years of introduction of useful insects in plant protection. Mater. Sesji Inst. Ochr. Rosl. 29: 102–112 [in Polish].
Pruszynski, S., and J. Rosada. 1987. Researches on the migration and hibernation of the predatory bug *Podisus maculiventris* Say using radiotracer ^{32}P. Prace Nauk. Inst. Ochr. Rosl. 28: 409–421 [in Polish].
Pruszynski, S., and W. Wegorek. 1980. Researches on biology and introduction of *Podisus maculiventris* (Say) - new for Poland predator of the Colorado potato beetle (*Leptinotarsa decemlineata* Say). Mat. Sesji Inst. Ochr. Rosl. 20: 127–136.
1991. Control of the Colorado beetle (*Leptinotarsa decemlineata*) in Poland. Bull. OEPP/EPPO Bull. 21: 11–16.
Pschorn-Walcher, H., and K. D. Zinnert. 1971. Investigations on the ecology and natural control of the larch sawfly (*Pristiphora erichsonii* Htg., Hym.: Tenthredinidae) in central Europe. Part II: Natural enemies: their biology and ecology, and their role as mortality factors in *P. erichsonii*. Tech. Bull. Commonw. Inst. Biol. Control 14: 1–50.
Rai, P. S. 1978. *Cantheconidea furcellata* (Wolff) (Pentatomidae: Heteroptera): a predator of leaf feeding caterpillars of rice. Curr. Sci. 47: 556–557.
Rajendra, M. K., and R. C. Patel. 1971. Studies on the life history of a predatory pentatomid bug, *Andrallus spinidens* (Fabr.). J. Bombay Nat. Hist. Soc. 68: 319–327.
Ramsey, G. W. 1963. Predaceous shield-bugs (Heteroptera: Pentatomidae) in New Zealand. New Zealand Entomol. 3(2): 3–6.
1964. Two species of shield bug that destroy caterpillars. New Zealand J. Agric. 109: 17.
Richman, D. B. 1977. Predation on the alfalfa weevil, *Hypera postica* (Gyllenhal), by *Stiretrus anchorago* (F.) (Hemiptera, Pentatomidae). Florida Entomol. 60: 192.
Richman, D. B., and F. W. Mead. 1978. Stages in the life cycle of a predatory stink bug, *Alcaeorrhynchus grandis* (Dallas) (Hemiptera: Pentatomidae). Florida Dept. Agric. Consumer Serv., Entomol. Circ. 192.
1980a. Stages in the life cycle of a predatory stink bug, *Podisus maculiventris* (Say) (Hemiptera: Pentatomidae). Florida Dept. Agric. Consumer Serv., Entomol. Circ. 216.
1980b. Stages in the life cycle of a predatory stink bug, *Stiretrus anchorago* (Fabricius) (Hemiptera: Pentatomidae). Florida Dept. Agric. Consumer Serv., Entomol. Circ. 210.
Richman, D. B., and W. H. Whitcomb. 1978. Comparative life cycles of four species of predatory stinkbugs (Hemiptera: Pentatomidae). Florida Entomol. 61: 113–119.
Richman, D. B., R. C. Hemenway, Jr., and W. H. Whitcomb. 1980. Field cage evaluation of predators of the soybean looper, *Pseudoplusia includens* (Lepidoptera: Noctuidae). Environ. Entomol. 9: 315–317.
Ruberson, J. R., T. J. Kring, and N. Elkassabany. 1998. Overwintering and the diapause syndrome of predatory Heteroptera. Pp. 49–69 *in* M. Coll and J. R. Ruberson [eds.], Predatory Heteroptera: Their Ecology and Use in Biological Control. Thomas Say Publ. Entomol.: Proceedings Entomological Society of America, Lanham, Maryland, U.S.A. 233 pp.
Ruberson, J. R., M. J. Tauber, and C. A. Tauber. 1986. Plant feeding by *Podisus maculiventris* (Heteroptera: Pentatomidae): effect on survival, development, and preoviposition period. Environ. Entomol. 15: 894–897.
Saavedra, J. L. D., J. C. Zanuncio, T. M. C. Della Lucia, and F. P. Reis. 1992a. Efeito da dieta artificial na fecundidade e fertilidade do predador *Podisus connexivus* Bergroth, 1891 (Hemiptera: Pentatomidae). An. Soc. Entomol. Brasil 21: 69–76.
Saavedra, J. L. D., J. C. Zanuncio, T. M. C. Della Lucia, and E. F. Vilela. 1992b. Dieta artificial par la crianza de *Podisus connexivus* (Hemiptera: Pentatomidae). Turrialba 42: 258–261.
Saavedra, J. L. D., J. C. Zanuncio, R. N. C. Guedes, and P. De Clercq. 1996. Continuous rearing of *Podisus nigrispinus* (Dallas) (Heteroptera: Pentatomidae) on an artificial diet. Meded. Fac. Landbouwwet. Univ. Gent 61: 767–772.
Saavedra, J. L. D., J. C. Zanuncio, E. F. Vilela, C. S. Sediyama, and P. De Clercq. 1995. Development of *Podisus nigrispinus* (Dallas) (Heteroptera: Pentatomidae) on meat-based artificial diets. Meded. Fac. Landbouwwet. Univ. Gent 60: 683–688.

Saavedra, J. L. D., J. C. Zanuncio, T. V. Zanuncio, and R. N. C. Guedes. 1997. Prey capture ability of *Podisus nigrispinus* (Dallas) (Het., Pentatomidae) reared for successive generations on a meridic diet. J. Appl. Entomol. 121: 327–330.

Saini, E. 1994a. Aspectos morfológicos y biológicos de *Podisus connexivus* Bergroth (Heteroptera: Pentatomidae). Rev. Soc. Entomol. Argentina 53: 35–42.

1994b. Preferencia alimentaria de *Podisus connexivus* Bergroth (Heteroptera-Pentatomidae) e influencia del comportamiento de lepidópteros plagas de la soja sobre la eficiencia del depredador. Rev. Inv. Agropec. 25: 151–157.

Saint-Cyr, J.-F., and C. Cloutier. 1996. Prey preference by the stinkbug *Perillus bioculatus*, a predator of the Colorado potato beetle. Biol. Control 7: 251–258.

Sant'Ana, J., and J. C. Dickens. 1998. Comparative electrophysiological studies of olfaction in predaceous bugs, *Podisus maculiventris* and *Podisus nigrispinus*. J. Chem. Ecol. 24: 965–984.

Sant'Ana, J., R. Bruni, A. A. Abdul-Baki, and J. R. Aldrich. 1997. Pheromone-induced movement of nymphs of the predator, *Podisus maculiventris* (Heteroptera: Pentatomidae). Biol. Control 10: 123–128.

Sant'Ana, J., R. F. P. da Silva, and J. C. Dickens. 1999. Olfactory reception of conspecific aggregation pheromone and plant odors by nymphs of the predator, *Podisus maculiventris*. J. Chem. Ecol. 25: 1813–1826.

Saulich, A. Kh. 1994. Role of abiotic factors in formation of the secondary ranges of adventive insect species. Entomol. Obozr. 73: 591–605 [in Russian].

1995. Natural predatory bug *Arma custos* as possible agent against *Leptinotarsa decemlineata*. Abstracts of the XIII International Plant Protection Congress, July, 2–7, 1995, The Hague, the Netherlands. Eur. J. Plant Pathol., abstr. 909.

Saulich, A. Kh., and D. L. Musolin. 1996. Univoltinism and its regulation in some temperate true bugs (Heteroptera). Eur. J. Entomol. 93: 507–518.

Saulich, A. Kh., and T. A. Volkovich. 1994. The thermal reactions of preimaginal stages in *Arma custos* (Hemiptera, Pentatomidae, Asopinae). Zool. Zh. 73: 43–53 [in Russian].

Sazonova, R. A., E. M. Shagov, and L. A. Stradimova. 1976. *Podisus* — a predator of the American white butterfly and the Colorado beetle. Zashch. Rast. 8: 52 [in Russian].

Schaefer, C. W. 1996. Bright bugs and bright beetles: asopine pentatomids (Hemiptera: Heteroptera) and their prey. Pp. 18–56 *in* O. Alomar and R. N. Wiedenmann [eds.], Zoophytophagous Heteroptera: Implications for Life History and IPM. Thomas Say Publ. Entomol.: Proceedings Entomological Society of America, Lanham, Maryland, U.S.A. 202 pp.

Schneider, F. 1940. Schadinsekten und ihre Bekämpfung in Ostindischen Gambirkulturen. Mitt. Schweiz. Entomol. Ges. 18: 77–207.

Schouteden, H. 1907. Heteroptera. Fam. Pentatomidae. Subfam. Asopinae (Amyoteinae). Pp. 1–82 *in* P. Wytsman [ed.], Genera Insectorum, fasc. 52. Verteneuil & Desmet, Bruxelles, Belgium.

Schumacher, F. 1910. Beiträge zur Kenntnis der Biologie der Asopiden. Z. Wiss. Insektenbiol. 6: 263–266, 376–383, 430–437.

1911. Beiträge zur Kenntnis der Biologie der Asopiden. Z. Wiss. Insektenbiol. 7: 40–47.

Sellke, K., and E. von Winning. 1939. Zuchtversuche mit der Raubwanze *Podisus maculiventris* Say (Pentatomidae; Asopinae) unter Berücksichtigung ihrer Eignung als natürlicher Feind des Kartoffelkäfers (*Leptinotarsa decemlineata* Say). Arb. Phys. Angew. Entomol. 6: 329–342.

Sen, S. K., M. S. Jolly, and T. R. Jammy. 1971. Biology and life cycle of *Canthecona furcellata* Wolff (Hem.: Pentatomiidae), predator of tasar silkworm *Antheraea mylitta* Drury. Indian J. Seric. 10: 53–56.

Senrayan, R. 1991. Plant feeding by *Eocanthecona furcellata* (Wolff) (Heteroptera: Asopinae) effect on development, survival and reproduction. Phytophaga (Madras) 3: 103–108.

Senrayan, R., and T. N. Ananthakrishnan. 1991. Influence of prey species and age of prey on the reproductive performance of a predatory stink bug (*Eocanthecona furcellata* (Wolff)) (Heteroptera: Asopinae). J. Biol. Control 5: 8–13.

Shagov, E. M. 1967a. On the photoperiodic reaction of *Perillus bioculatus* Fabr. (Heteroptera, Pentatomidae). Zool. Zh. 46: 948–950 [in Russian].

1967b. Preventing the onset of diapause in *Perillus bioculatus* Fabr. (Heteroptera, Pentatomidae) by exposure to artificial long-day conditions. Dokl. Akad. Nauk. SSSR (Biol.) 174: 965–968 [in Russian].

1967c. The effect of temperature and humidity on the embryonic development of *Perillus bioculatus* Fabr. (Heteroptera, Pentatomidae). Zool. Zh. 46: 1260–1262 [in Russian].

1968. The effect of temperature on the predacious bug *Perillus bioculatus* Fabr. (Heteroptera, Pentatomidae). Zool. Zh. 47: 563–570 [in Russian].

1969. Some characteristics of the physiology of *Perillus bioculatus* (Heteroptera, Pentatomidae) in the period of winter dormancy. Zool. Zh. 48: 827–835 [in Russian].

1977. Photoperiodic reaction of the predatory bug *Perillus* and its variation. Ekologiya 4: 96–99 [in Russian].

Shagov, E. M., and S. I. Chesnek. 1978. Cold hardiness in the predacious bug *Perillus bioculatus* (Heteroptera, Pentatomidae). Zool. Zh. 57: 398–406 [in Russian].

Shagov, E. M., and N. N. Shutova. 1977. The development of *Podisus* in conditions of constant temperature. Zashch. Rast. 4: 48–49 [in Russian].

Shetty, P. N., and J. A. Hough-Goldstein. 1998. Behavioral response of *Podisus maculiventris* (Hemiptera: Pentatomidae) to its synthetic pheromone. J. Entomol. Sci. 33: 72–81.

Shutova, N. N., E. M. Shagov, and Yu. V. Zayats. 1976. Possibilities of overwintering of the bug *Perillus*. Zashch. Rast. 11: 50–51 [in Russian].

Silva, E. N., T. M. Santos, and F. S. Ramalho. 1996. Desenvolvimento ninfal de *Supputius cincticeps* Stål (Hemiptera: Pentatomidae) alimentado com curuquerê-do-algodoeiro. An. Soc. Entomol. Brasil 25: 103–108.

Singh, K. J., and O. P. Singh. 1989a. Biology of a pentatomid predator, *Andrallus spinidens* (Fab.) on Rivula sp., a pest of soybean in Madhya Pradesh. J. Insect Sci. 2: 134–138.

1989b. Toxicity of some insecticides to *Andrallus spinidens* (Fabricius) (Pentatomidae: Hemiptera). J. Biol. Control 3: 67–68.

Singh, O. P., K. K. Nema, and S. N. Verma. 1989. Biology of *Cantheconidea furcellata* Wolff (Heteroptera: Pentatomidae), a predator on the larvae of *Spilosoma obliqua* Walker. Agric. Sci. Dig. Karnal 9: 79–80.

Singh, Z. 1973. Southern Green Stink Bug and Its Relationship to Soybeans. Metropolitan, Delhi, India.

Singh, Z., C. E. White, and W. H. Luckmann. 1973. Notes on *Amyotea malabarica*, a predator of *Nezara viridula* in India. J. Econ. Entomol. 66: 551–552.

Sipayung, A., R. Desmier de Chenon, and P. Sudharto. 1992. Study of the *Eocanthecona–Cantheconidea* (Hemiptera: Pentatomidae, Asopinae) predator complex in Indonesia. J. Plant Prot. Trop. 9: 85–103.

Smagghe, G., and D. Degheele. 1995. Selectivity of nonsteroidal ecdysteroid agonists RH 5849 and RH 5992 to nymphs and adults of predatory soldier bugs, *Podisus nigrispinus* and *P. maculiventris* (Hemiptera: Pentatomidae). J. Econ. Entomol. 88: 40–45.

Southwood, T. R. E., and D. Leston. 1959. Land and Water Bugs of the British Isles. Frederick Warne and Co., London, U.K. 436 pp.

Srivastava, A. S., R. R. Katiyar, K. D. Upadhyay, and S. V. Singh. 1987. *Canthecona furcellata* Wolff (Hemiptera: Pentatomidae) predating on *Menochilus sexmaculata* (F.) (Coleoptera: Coccinellidae). Indian J. Entomol. 49: 558.

Stamopoulos, D. C., and A. Chloridis. 1994. Predation rates, survivorship and development of *Podisus maculiventris* (Het.: Pentatomidae) on larvae of *Leptinotarsa decemlineata* (Col.: Chrysomelidae) and *Pieris brassicae* (Lep.: Pieridae), under field conditions. Entomophaga 39: 3–9.

Stamopoulos, D. C., G. Diamantidis, and A. Chloridis. 1993. Activités enzymatiques du tube digestif du prédateur *Podisus maculiventris* (Hem.: Pentatomidae). Entomophaga 38: 493–499.

Stamp, N. E. 1992. Relative susceptibility to predation of two species of caterpillar on plantain. Oecologia 92: 124–129.

Stamp, N. E., and M. D. Bowers. 1991. Indirect effect on survivorship of caterpillars due to presence of invertebrate predators. Oecologia 88: 325–330.

1993. Presence of predatory wasps and stinkbugs alters foraging behavior of cryptic and non-cryptic caterpillars on plantain (*Plantago lanceolata*). Oecologia 95: 376–384.

Stamp, N. E., and R. T. Wilkens. 1993. On the cryptic side of life: being unapparent to enemies and the consequences for foraging and growth of caterpillars. Pp. 283–330 *in* N. E. Stamp and T. M. Casey [eds.], Caterpillars: Ecological and Evolutionary Constraints on Foraging. Chapman & Hall, New York, New York, U.S.A. 587 pp.

Stamp, N. E., T. Erksine, and C. J. Paradise. 1991. Effects of rutin-fed caterpillars on an invertebrate predator depend on temperature. Oecologia 88: 289–295.

Stamp, N. E., Y. Yang, and T. L. Osier. 1997. Response of an insect predator to prey fed multiple allelochemicals under representative thermal regimes. Ecology 78: 203–214.

Stone, P. C. 1939. Notes on the predacious stink bug, *Apateticus cynicus* Say. Trans. Ill. State Acad. Sci. 32: 228.

Stoner, A., A. M. Metcalfe, and R. E. Weeks. 1974a. Development of *Podisus acutissimus* in relation to constant temperature. Ann. Entomol. Soc. Amer. 67: 718–719.

1974b. Plant feeding by a predaceous insect, *Podisus acutissimus*. Environ. Entomol. 3: 187–188.

Stoner, D. 1920. The Scutelleroidea of Iowa. Univ. Iowa Stud. Nat. Hist. 8(4): 1–140.

1930. Spined soldier-bug reared on celery leaf-tyer. Florida Entomol. 14: 21–22.

Strawinski, K. 1927. *Picromerus bidens* (L.) (Hem.-Heteroptera, Pentatomidae). Pol. Pismo Entomol. 6: 123–151 [in Polish].

Strickland, E. H. 1953. An annotated list of the Hemiptera (s.l.) of Alberta. Canad. Entomol. 85: 193–214.

Sumenkova, V. V., and I. G. Yazlovetskii. 1992. A simple artificial nutritional diet for the predacious bug *Podisus maculiventris* (Hemiptera: Pentatomidae). Zool. Zh. 71: 52–57 [in Russian].

Szmidt, A., and W. Wegorek. 1967. Populationsdynamische Wirkung von *Perillus bioculatus* (Fabr.) (Het.: Pentatomidae) auf den Kartoffelkäfer. Entomophaga 12: 403–408.

Tabasa, M. A. 1991. Life history and functional response of the predatory bug *Eocanthecona furcellata* (Wolff) (Hemiptera: Pentatomidae) in relation to its prey, *Spodoptera litura* (F.). M.S. thesis, Philippines University, Los Baños, Laguna, the Philippines.

Tadic, M. D. 1963. The possibility of introducing *Podisus placidus* Uhler, an American predator of *Hyphantria cunea*, into Yugoslavia. Arch. Poljopr. Nauke 16 (54): 40–50 [in Serbo-Croatian].

1975. Process of the adaptation of indigenous parasites and predators on the fall webworm (*Hyphantria cunea* Dr.) in Yugoslavia in 1963–1972. Zast. Bilja 26(133): 247–267 [in Serbo-Croatian].

Takai, M., and S. T. Yasuoka. 1993. Biological control of *Spodoptera litura* (Fabricius) in plastic greenhouse by the predatory stink bug, *Eocanthecona furcellata*. Proc. Assoc. Plant Prot. Shikoku 28: 103–108 [in Japanese].

Tamaki, G., and B. A. Butt. 1978. Impact of *Perillus bioculatus* on the Colorado potato beetle and plant damage. U.S. Dept. Agric. Tech. Bull. 1581.

Taylor, J. S. 1965. The fruit-piercing moth, *Achaea lienardi* Boisduval (Lepidoptera: Noctuidae), in the Eastern Cape Province. J. Entomol. Soc. South. Africa. 28: 50–56.

Thomas, D. B. 1992. Taxonomic synopsis of the asopine Pentatomidae (Heteroptera) of the Western Hemisphere. Thomas Say Foundation Monographs, Vol. 15. Entomological Society of America, Lanham, Maryland, U.S.A. 156 pp.

1994. Taxonomic synopsis of the Old World asopine genera (Heteroptera: Pentatomidae). Insecta Mundi 8: 145–212.

Thomson, W. T. 1992. A worldwide guide to beneficial animals (insects, mites, nematodes) used for pest control purposes. Thomson Publications, Fresno, California, U.S.A. 91 pp.

Thorpe, K. W., R. E. Webb, J. R. Aldrich, and K. M. Tatman. 1994. Effects of spined soldier bug (Hemiptera: Pentatomidae) augmentation and sticky barrier bands on gypsy moth (Lepidoptera: Lymantriidae) density in oak canopies. J. Entomol. Sci. 29: 339–346.

Torre–Bueno, J. R., de la. 1939. A synopsis of the Hemiptera–Heteroptera of America north of Mexico. Part I. Families Scutelleridae, Cydnidae, Pentatomidae, Aradidae, Dysodiidae and Temitaphididae. Entomol. Amer. 19: 207–304.

Torres, J. B., J. C. Zanuncio, M. C. Picanço, and A. C. de Oliveira. 1996. Parámetros poblacionales de tres parasitoides (Hymenoptera: Scelionidae, Encyrtidae) utilizando al depredador *Podisus nigrispinus* (Heteroptera: Pentatomidae) como hospedero. Rev. Biol. Trop. 44: 233–240.

Torres, J. B., J. C. Zanuncio, and M. C. de Oliveira. 1997. Mating frequency and its effect on female reproductive output in the stinkbug predator *Podisus nigrispinus* (Heteroptera: Pentatomidae). Meded. Fac. Landbouwwet. Univ. Gent 62: 491–498.

Torres, J. B., J. C. Zanuncio, and H. N. de Oliveira. 1998. Nymphal development and adult reproduction of the stinkbug predator *Podisus nigrispinus* (Het., Pentatomidae) under fluctuating temperatures. J. Appl. Entomol. 122: 509–514.

Torres, J. B., P. De Clercq, and R. Barros. 2000. Effect of spinosad on the predator *Podisus nigrispinus* and its lepidopterous prey. Meded. Fac. Landbouwwet. Univ. Gent 63: 211–218.

Tostowaryk, W. 1971. Life history and behavior of *Podisus modestus* (Hemiptera: Pentatomidae) in boreal forest in Quebec. Canad. Entomol. 103: 662–674.

1972. The effect of prey defense on the functional response of *Podisus modestus* (Hemiptera: Pentatomidae) to densities of the sawflies *Neodiprion swainei* and *N. pratti banksianae* (Hymenoptera: Neodiprionidae). Canad. Entomol. 104: 61–69.

Tothill, J. D., T. H. C. Taylor, and R. W. Paine. 1930. The Coconut Moth in Fiji. A history of its control by means of parasites. Imperial Institute of Entomology, London, U.K., 269 pp.

Traugott, M. S., and N. E. Stamp. 1996. Effects of chlorogenic acid- and tomatine-fed caterpillars on the behavior of an insect predator. J. Insect Behav. 9: 461–476.

Tremblay, E. 1967. Primi tentativi di introduzione in Italia (Campania) del *Perillus bioculatus* (Fabr.), nemico naturale della dorifora della patata (*Leptinotarsa decemlineata* Say). Ann. Fac. Sci. Agric. Univ. Napoli 1: 1–13.

Tremblay, E., and N. Zouliamis. 1968. Dati conclusivi sull'introduzione, sulla biologia ed impiego del *Perillus bioculatus* (Fabr.) (Heteroptera: Pentatomidae) nell'Italia meridionale. Boll. Lab. Entomol. Agric. Filipo Silvestri 26: 99–122.

Trouvelot, B. 1931. Recherches sur les parasites et prédateurs attaquant le doryphore en Amérique du Nord et envoi en France des premières colonies des espèces les plus actives. Ann. Epiphyt. 17: 408–445.

Urban, A. J., and C. D. Eardley. 1995. A recently introduced sawfly, *Nematus oligospilus* Foerster (Hymenoptera: Tenthredinidae), that defoliates willows in southern Africa. Afr. Entomol. 3: 23–27.

Usha Rani, P. 1992. Temperature-induced effects on predation and growth of *Eocanthecona furcellata* (Wolff) (Pentatomidae: Heteroptera). J. Biol. Control 6: 72–76.

Usha Rani, P., and S. Wakamura. 1993. Host acceptance behaviour of a predatory pentatomid, *Eocanthecona furcellata* (Wolff) (Heteroptera: Pentatomidae) towards larvae of *Spodoptera litura* (Lepidoptera: Noctuidae). Insect Sci. Appl. 14: 141–147.

Usha Rani, P., S. Wakamura, and K. Asoaka. 1994. Rostral tip appendages in carnivorous stink bug, *Eocanthecona furcellata* (Wolff) (Heteroptera: Pentatomidae). J. Entomol. Res. 18: 199–202.

Valicente, F. H., and R. J. O'Neil. 1995. Effects of host plants and feeding regimes on selected life history characteristics of *Podisus maculiventris* (Say) (Heteroptera: Pentatomidae). Biol. Control 5: 449–461.

van Lenteren, J. C. 1991. Quality control of natural enemies: hope or illusion? Pp. 1–14 *in* F. Bigler [ed.], Proceedings of the Fifth Workshop of the IOBC Global Working Group, Quality Control of Mass Reared Arthropods, March 25–28, 1991, Wageningen, the Netherlands.

Vasantharaj, D. B., and M. Basheer. 1962. Mass occurrence of the predatory stink bug, *Cantheconidea* (*Canthecona*) *furcellata* (Wolff) preying on *Amsacta albistriga* Walk. in South India. J. Bombay Nat. Hist. Soc. 58: 817–819.

Villiers, A. 1952. Hémiptères de l'Afrique Noire (Punaises et Cigales). Initiations africaines, IFAN, Dakar 9: 1–256.

Vlasova, V. A., L. A. Ziskind, and S. S. Izhevskii. 1980. The possibility of the acclimatisation of *Podisus*. Zashch. Rast. 4: 46–47 [in Russian].

Volkov, O. G., and L. B. Tkacheva. 1997. A natural enemy of the Colorado potato beetle — *Picromerus bidens*. Zashch. Karantin Rast. 3: 30 [in Russian].

Volkovich, T. A. 1995. Diapause of the predatory bug *Podisus maculiventris* as important part of the technological cycle in the mass rearing. Abstracts of the XIII International Plant Protection Congress, July 2–7, Den Haag, the Netherlands. Eur. J. Plant Pathol., abstr. 918.

Volkovich, T. A., and A. Kh. Saulich. 1992. New results in the mass-breeding technology of Podisus. Zashchita Rastenii No. 10: 47–48 [in Russian].

Volkovich, T. A., L. I. Kolechnichenko, and A. Kh. Saulich. 1990. The role of thermal rhythms in the development of *Perillus bioculatus* (Hemiptera, Pentatomidae). Zool. Zh. 69: 70–81 [in Russian].

1994. The predatory bug *Arma custos*: photoperiodic and temperature control of diapause and coloration. Zool. Zh. 73: 26–37 (in Russian).

Volkovich, T. A., A. Kh. Saulich, and N. I. Goryshin. 1991. Day-length sensitive stage and the accumulation of photoperiodic information in the predatory bug *Podisus maculiventris* Say (Heteroptera: Pentatomidae). Entomol. Obozr. 70: 14–22 [in Russian].

Waage, J. K., K. P. Carl, N. J. Mills, and D. J. Greathead. 1985. Rearing entomophagous insects. Pp. 45–66 *in* P. Singh and R. F. Moore [eds.], Handbook of Insect Rearing, Vol. I. Elsevier, Amsterdam, the Netherlands. 488 pp.

Waddill, V., and M. Shepard. 1974. Biology of a predaceous stinkbug, *Stiretrus anchorago* (Hemiptera: Pentatomidae). Florida Entomol. 57: 249–253.

1975a. A comparison of predation by the pentatomids, *Podisus maculiventris* (Say) and *Stiretrus anchorago* (F.), on the Mexican bean beetle, *Epilachna varivestis* Mulsant. Ann. Entomol. Soc. Amer. 68: 1023–1027.
1975b. Dispersal of *Podisus maculiventris* nymphs in soybeans. Environ. Entomol. 4: 233–234.
Warren, L. O., and P. Bjegovic. 1972. The influence of different photoperiods on the development of two predatory bugs, *Podisus placidus* Uhl. and *Podisus maculiventris* Say (Hemiptera: Pentatomidae). Zast. Bilja 23(117/118): 7–10 [in Serbo-Croatian].
Warren, L. O., and G. Wallis. 1971. Biology of the spined soldier bug, *Podisus maculiventris* (Hemiptera: Pentatomidae). J. Georgia Entomol. Soc. 6: 109–116.
Wegorek, W., and S. Pruszynski. 1979. The toxicity of preparations applied for the control of the Colorado potato beetle to the two-spotted shield bug (*Perillus bioculatus*) (Heteroptera, Pentatomidae). Prace Nauk. Inst. Ochr. Ros. 21(2): 119–127 [in Polish].
Weiser, L. A., and N. E. Stamp. 1998. Combined effects of allelochemicals, prey availability, and supplemental plant material on growth of a generalist predator. Entomol. Exp. Appl. 87: 181–189.
Weissbecker, B., J. J. A. van Loon, and M. Dicke. 1999. Electroantennogram responses of a predator, *Perillus bioculatus*, and its prey, *Leptinotarsa decemlineata*, to plant volatiles. J. Chem. Ecol. 25: 2313–2325.
Whalon, M. E., and B. L. Parker. 1978. Immunological identification of tarnished plant bug predators. Ann. Entomol. Soc. Amer. 71: 453–456.
Wheeler, A. G., Jr. 1999. Southern range extension of a palearctic stink bug, *Picromerus bidens* (Hemiptera: Pentatomidae), in North America. Entomol. News 110: 97–98.
Whitcomb, W. H. 1973. Natural populations of entomophagous arthropods and their effect on the agroecosystem. Pp. 150–169 *in* Proceedings of the Mississippi Symposium on Biological Control. University Press Mississippi State, Mississippi, U.S.A.
Whitmarsh, R. D. 1916. Life-history notes on *Apateticus cynicus* and *maculiventris*. J. Econ. Entomol. 9: 51–53.
Wiedenmann, R. N., and R. J. O'Neil. 1990. Effects of low rates of predation on selected life–history characteristics of *Podisus maculiventris* (Say) (Heteroptera: Pentatomidae). Canad. Entomol. 122: 271–283.
1991a. Laboratory measurement of the functional response of *Podisus maculiventris* (Say) (Heteroptera: Pentatomidae). Environ. Entomol. 20: 610–614.
1991b. Searching behavior and time budgets of the predator *Podisus maculiventris*. Entomol. Exp. Appl. 60: 83–93.
1992. Searching strategy of the predator *Podisus maculiventris* (Say) (Heteroptera: Pentatomidae). Environ. Entomol. 21: 1–9.
Wiedenmann, R. N., J. C. Legaspi, and R. J. O'Neil. 1996. Impact of prey density and facultative plant feeding on the life history of the predator *Podisus maculiventris* (Heteroptera: Pentatomidae). Pp. 94–118 *in* O. Alomar and R. N. Wiedenmann [eds.], Zoophytophagous Heteroptera: Implications for Life History and IPM. Thomas Say Publ. Entomol.: Proceedings Entomological Society of America, Lanham, Maryland, U.S.A. 202 pp.
Wilkinson, J. D., K. D. Biever, and C. M. Ignoffo. 1979. Synthetic pyrethroid and organophosphate insecticides against the parasitoid *Apanteles marginiventris* and the predators *Geocoris punctipes*, *Hippodamia convergens*, and *Podisus maculiventris*. J. Econ. Entomol. 72: 473–475.
Wiwat Suasa. 1989. Utilization of *Eocanthecona furcellata* (Wolff) (Hemiptera: Pentatomidae) for augmentative biological control of castor bean semilooper, *Ophiusa janata* (L.) (Lepidoptera: Noctuidae) in Thailand. Symposium on Biological Control of Pests in Tropical Agriculture Ecosystems, June 1–3, 1988. Biotrop Special Publication 36: 191–200.
Yasuda, T. 1997. Chemical cues from *Spodoptera litura* larvae elicit prey locating behavior by the predatory stink bug, *Eocanthecona furcellata*. Entomol. Exp. Appl. 82: 349–354.
1998a. Effects of (*E*)-phytol of several lepidopteran species in prey-locating behavior of a generalist predatory stink bug, *Eocanthecona furcellata* (Heteroptera: Pentatomidae). Entomol. Sci. 1: 159–164.
1998b. Role of chlorophyll content of prey diets in prey-locating behavior of a generalist predatory stink bug, *Eocanthecona furcellata*. Entomol. Exp. Appl. 86: 119–124.
Yasuda, T., and S. Wakamura. 1992. Rearing of the predatory stink bug, *Eocanthecona furcellata* (Wolff) (Heteroptera: Pentatomidae), on frozen larvae of *Spodoptera litura* (Fabricius) (Lepidoptera: Noctuidae). Appl. Entomol. Zool. 27: 303–305.

1996. Behavioral responses in prey location of the predatory stink bug, *Eocanthecona furcellata*, to chemical cues in the larvae of *Spodoptera litura*. Entomol. Exp. Appl. 81: 91–96.

Yeargan, K. V. 1979. Parasitism and predation of stink bug eggs in soybean and alfalfa fields. Environ. Entomol. 8: 715–719.

Yu, S. J. 1987. Biochemical defense capacity in the spined soldier bug (*Podisus maculiventris*) and its lepidopterous prey. Pestic. Biochem. Physiol. 28: 216–223.

1988. Selectivity of insecticides to the spined soldier bug (Heteroptera: Pentatomidae) and its lepidopterous prey. J. Econ. Entomol. 81: 119–122.

1990. Liquid chromatographic determination of permethrin esterase activity in six phytophagous and entomophagous insects. Pestic. Biochem. Physiol. 36: 237–241.

Zanuncio, J. C., J. B. Alves, J. E. M. Leite, N. R. da Silva, and R. C. Sartório. 1990. Desenvolvimento ninfal de *Podisus connexivus* Bergroth, 1891 (Hemiptera: Pentatomidae) alimentado com dois hospedeiros alternativos. Rev. Árvore 14: 164–174.

Zanuncio, J. C., E. C. do Nascimento, G. P. Santos, R. C. Sartório, and F. S. Araújo. 1991a. Aspectos biológicos do percevejo predador *Podisus connexivus* (Hemiptera: Pentatomidae). An. Soc. Entomol. Brasil 20: 243–249.

Zanuncio, J. C., M. F. de Freitas, J. B. Alves, and J. E. M. Leite. 1991b. Fecundidade de fêmeas de *Podisus connexivus* Bergroth, 1891 (Hemiptera: Pentatomidae) em diferentes tipos de hospedeiros. An. Soc. Entomol. Brasil 20: 369–378.

Zanuncio, T. V., J. C. Zanuncio, V. C. Batalha, and G. P. Santos. 1993f. Efeito da alimentação com lagartas de *Bombyx mori* e larvas de *Musca domestica* no desenvolvimento de *Podisus nigrolimbatus* (Hemiptera, Pentatomidae). Rev. Bras. Entomol. 37: 273–277.

Zanuncio, J. C., V. C. Batalha, T. V. Zanuncio, and G. P. Santos. 1993a. Influência da densidade ninfal na criação de *Podisus connexivus* Berg. (Hemiptera: Pentatomidae) alimentado com larvas de *Musca domestica* L. An. Soc. Entomol. Brasil 22: 449–453.

Zanuncio, J. C., A. T. Ferreira, T. V. Zanuncio, and J. F. Garcia. 1993b. Influence of feeding on *Eucalyptus urophylla* seedlings on the development of the predatory bug *Podisus connexivus* (Hemiptera: Pentatomidae). Meded. Fac. Landbouwwet. Univ. Gent 58: 469–475.

Zanuncio, J. C., R. N. C. Guedes, J. F. Garcia, and L. A. Rodrigues. 1993c. Impact of two formulations of deltamethrin in aerial application against *Eucalyptus* caterpillars and their predaceous bugs. Meded. Fac. Landbouwwet. Univ. Gent 58: 477–481.

Zanuncio, J. C., J. E. M. Leite, J. B. Alves, and G. P. Santos. 1993d. Duração do período ninfal e sobrevivência do predador *Podisus connexivus* Bergroth (Hemiptera, Pentatomidae), em três presas alternativas. Rev. Bras. Zool. 10: 327–332.

Zanuncio, T. V., L. A. Moreira, J. C. Zanuncio, and G. P. Santos. 1993e. Efeito da densidade ninfal na viabilidade e sobrevivência de *Supputius cincticeps* Stål, 1860 (Hemiptera, Pentatomidae) criado em laboratório com larvas de *Tenebrio molitor* (Coleoptera, Tenebrionidae). Rev. Bras. Entomol. 37: 483–487.

Zanuncio, T. V., J. C. Zanuncio, E. C. do Nascimento, and E. F. Vilela. 1993g. Descrição das ninfas do predador *Supputius cincticeps* StDl (Hemiptera: Pentatomidae). An. Soc. Entomol. Brasil 22: 221–229.

Zanuncio, T. V., J. C. Zanuncio, E. F. Vilela, and R. C. Sartório. 1992. Aspectos biológicos, da fase adulta, de *Supputius cincticeps* Stål, 1860 (Hemiptera: Pentatomidae), predador de lagartas desfolhadoras de eucalipto. IPEF, Piracicaba 45: 35–39.

Zanuncio, J. C., J. B. Alves, T. V. Zanuncio, and J. F. Garcia. 1994a. Hemipterous predators of eucalypt defoliator caterpillars. For. Ecol. Manage. 65: 65–73.

Zanuncio, J. C., J. E. M. Leite, G. P. Santos, and E. C. do Nascimento. 1994b. Novo metodologia para criação em laboratório de hemípteros predadores. Rev. Ceres 41(233): 88–93.

Zanuncio, J. C., J. L. D. Saavedra, H. N. Oliveira, D. Degheele, and P. De Clercq. 1996. Development of the predatory stinkbug *Brontocoris tabidus* (Signoret) (Heteroptera: Pentatomidae) on different proportions of an artificial diet and pupae of *Tenebrio molitor* L. (Coleoptera: Tenebrionidae). Biocontrol Sci. Technol. 6: 619–625.

Zeki, H., and S. Toros. 1990. Determination of natural enemies of *Chrysomela populi* L. and *Chrysomela tremulae* F. (Coleoptera, Chrysomelidae) harmful to poplars and the efficiency of their parasitoids in Central Anatolia region. Pp. 251–260 *in* Proceedings of the Second Turkish National Congress of Biological Control, Sept. 26–29, 1990, Turkey [in Turkish].

Zheng, Z. Y., Y. W. Chen, and Y. G. Wen. 1992. Experiments on the use of *Arma custos* (Fabricius) (Hem.: Pentatomidae) to control forest pests. Chin. J. Biol. Control 8: 155–156 [in Chinese].

Zhu, D. F. 1990. Studies on the biological characteristics of *Cantheconidea furcellata* (Hemiptera: Pentatomidae, Asopinae). Nat. Enemies Insects 12: 71–74 [in Chinese].

Insect Index

A

Abdastartus
   atrus, **105**
   sacchari, **105**
Acanthaspis
   pedestris, 700
   siva, 701
Acanthia, 98
Acanthocephala, 338
Acanthocerus clavipes, 384
Acanthochila armigera, **105**
Acanthocoris, **373**
   obscuricornis, 373
   scabrator, **373**
   sordidus, 373, **374**
Acanthomia, 350
   horrida, 350, 353
Acanthomya, 350
Acanthomyia, 350
Acaulona peruviana, 274
Achaea lienardi, 768
Acholla multispinosus, 555
Acissa
   uncatoides, 613
Acompocoris, 609, 611, **612**
   apinus, 612
   lepidus, 612
   pygmaeus, 612
Aconchus urbanus, **105**
Acrobasis nuxvorella, 669
Acroclisoides tectacorisi, 429
Acrosternum, **447**, 448
   acuata, 448
   armigera, 447
   bellum, 447
   hilare, 422
   impicticorne, 448
   marginatum, 448
Acyrthosiphon
   pelargonii, 730
   pisum, 673
Acysta perseae, **114**
Adalia bipunctata, 748
Adelges, 622
   piceae, 622

Adelphocoris
   lineolatus, 15, **52**, 53
   seticornis, 15
   suturalis, 15, 24
   triannulatus, 15
   variabilis, 15
Adrisa, **415**
Aedes, 578
   aegypti, 520, 603
Aelia, 20, 220, 422, 425, 448
   acuminata, 14, 18, 20, **448**
   americana, **448**
   furcula, **423**, 424, 425
   germari, **424**
   melanota, **424**
   obtusa, 424
   rostrata, 195, 424, **425**, 487
   sibirica, 14
   virgata, **449**
Aeschynteles maculatus, 227
Aethus
   indicus, 556
   laticollis orientalis, **409**
   nigrita, **411**
Aetorrhinus angulatus, 662
Afrius, 768
Afropiesma, 265
Agonoscelis rutilia, 14, **449**
Agramma atricapillum, **105**
Agrophopus, 310
Alabama argillacea, 675, 748, 751
Alcaeorrhynchus grandis, 764, 765
Alfalfa plant bug, 52, 53
Alfalfa weevil, 670, 673
Allotrobium, 112
Alnetoidea alneti, 663
Alophora, 192
   lepidofera, 192, 204
   nasalis, 282, 352, 354
Alophorella, 187
Altica, 766, 767
   ampelophaga, 766
   cyanea, 767
Alydus pilosulus, **327**
Amara, 161
Amarali, 146

Amblypelta, 14, 338, **339**, 340, 341, 370, 372, 379, 386
   ardleyi, 339
   bukharii, 339
   cocophaga, **342**, 371
   cocophaga cocophaga, 342
   cocophaga malaitensis, 342
   costalis, 339
   cristobalensis, 339
   danishi, 339
   lutescens, **339**, 341
   lutescens lutescens, 339–341
   lutescens papuensis, 339, 340
   madangana, 339
   manihotis, **341**, 342
   nitida, 340, **341**
   theobromae, **341**, 342
Ambrostoma quadriimpressum, 767
Ambrysus, 573
   hungerfordi, 573
   lunatus, 571–573
   mormon, 572
Amorbus, 338, **374**, 377
   alternatus, 375
   obscuricornis, **374**, 375, 377
Amyotea
   hamatus, 446
   malabarica, 768
Anagrus, 102
   takeyanus, 100, 102
Anaphes, 102
   tingitiphagus, 89
Anasa, 337, **343**
   andresii, **347**
   armigera, 343, **346**, 347
   guttifera, 343
   litigiosa, 343
   maculipes, 343
   repetita, 343
   scorbutica, 346, **347**
   tristis, 22, **343**, 346, 347
   uhleri, 343
Anastatus, 340, 341, 352, 363, 368, 370, 372, 378, 450
   axiagasti, 342
   biproruli, 429
   dasyni, 377
   gastropache, 378
   japonicus, 326, 360
   reduvii, 345, 489
   semiflavidus, 368
Anastrepha ludens, 755
Andrallus spinidens, 767
Andromeda lace bug, 671
Anisops, 592
Anomis, 761
Anopheles, 571
Anoplocnemis, 338, **347**
   curvipes, **347**, 348, 349, 350
   phasiana, 347, **349**
   testator, 350
   tristator, 347, **350**
Anoplolepis, 372

cristobalensis, 342
custodiens, 349, 372
longipes, 342, 372
Antestia bugs, 449
Antestia, 449
   cincticollis, 449
   trivialis, 449
   usambarica, 449
Antestiopsis, **449**
   clymeneis, 449
   crypta, 449
   facetoides, 449
   falsa, 449
   intricata, 449
   orbitalis bechuana, 449
   orbitalis ghesquierei, 449
Antheraea mylitta, 761
Anthocoris, 98, 554, 609–611, **612**, 613, 614, 625
   antevolens, 610, 612, **613**, 614
   austropicus, 612
   butleri, 610
   confusus, 612, **613**
   dimorphicus, 614
   gallarum-ulmi, 612
   kingi, 554
   limbatus, 614
   melanocerus, 612
   minki, 612
   musculus, 612, **613**, 614
   nemoralis, 610, 612, **613**, 614
   nemorum, 114, 554, 610, 612, 613, **614**
   pilosus, **615**
   sarothamni, 610, 612, 613
   sibiricus, **615**
   tomentosus, 612, **615**
Anticarsia gemmatalis, 648, 751, 764
Antilochus, 274
   coquebertii, 280, **290**
   russus, 280
Apanteles, 727
Apateticus, 765
   bracteatus, 756
   cynicus, 765
   lineolatus, 765
Aphanus, 207, 208
   sordidus, 208
Aphelocheirus, 572, 573
   aestivalis, 572, 573
   occidentalis, 573
Aphis
   gossyii, 620
   gossypii, 643, 660, 701, 731
   maidis, 615
   pomi, 668
   rumicis, 267
Apiomerus, 292, 347
   spissipes, 555
Apis indica, 701
Apoecilus, 765
   bracteatus, 765
   cynicus, 765

INSECT INDEX

Apple aphid(s), 668, 674, 676, 677
Apple brown bug, 676
Apple capsid, 57
Apple dimpling bug, 676
Apple mealybug, 670
Apple red bug(s), 57, 59
Apple sucker, 677
Appolodotus, 115
Aptus mirmicoides, 643
Aquarius, 601
    najas, 603
    paludum, 603
Aradus, 513
    antennalis, 514
    cinnamomeus, 20, **514**, 515, 516
    kormilevi, 514
    orientalis, 17
Araecerus fasciculatus, 44
Arhyssus
    hirtus, 310
    lateralis, 310
    parvicornis, **315**
Aridelus, 446
Arilus cristatus, 363, 555
Arma
    chinensis, 112
    custos, 14, 767
Arocatus, 224
    melanocephalus, 224
Arvelius albopunctatus, **425**, 426
Ash lace bug, 112
Asiarcha nigridorsis, 509
Asolcus, 483
Aspavia, 450
    armigera, **450**
Aspongopus, 505, 506
    brunneus, 505, 506
    fuscus, 505, 506
    janus, 505, 506, 509
    nepalensis, 506
    obscurus, 506
    remipes, 506
Atarsocoris brachiariae, 407, **408**, 409
Athaumastus
    flaviventris, 384
    haematicus, 384
Atractotomus, 676
    mali, **675**, 676
Atrademus, 173
Aufeius impressicollis, **315**
Aulacostermum nigrorubrum, 384
Australis, 358
Avocado lace bug, 114
Axiagastus cambelli, **450**
Ayyaria chaetophora, 719
Azalea lace bug, 98, 643, 671
Azalea plant bug, 555

B

Baclozygum
    brachypterum, 139, 140
    depressum, 140, **141**, 142
Bagrada cruciferarum, **450**
Balsam woolly aphid, 611, 612, 622
Bamboo chinch bug, 213
Banana lacewing bug, 102
Banana spotting bug(s), 339
Bandedwing whitefly, 717
Barnsleya, 146
Bathycles, 146
Bathycoelia thalassina, 21, 422, **450**
Bean bug, 325
Bean capsid, 52
Bean lace bug, 92
Bechuana, 449
Bed bug, 519, 521–524, 530, 763
Bee bug, 52
Beet armyworm(s), 608, 647, 673, 716, 740, 746, 751, 752
Beet leafhopper, 731
Belonochilus
    mexicanus, 224
    numenius, **223**, 224
Belostoma, 577, 578
    flumineum, 578
    micantulum, 579
Bemisia, 641
    argentifolii, 669
    tabaci, 660
Berry bug, 430
Berytinus, 726
    minor, 726, 732
Beskia
    aelops, 441
    cornuta, 440
Bicyrtes
    quadrifasciata, 489
    spinosa, 310
Biprorulus bibax, 422, **426**, 427–430, 445, 456
Birch catkin bug, 224
Black bean bug, 205
Black bug of sugarcane, 170, 176
Black cocoa ant, 44
Black corsair, 555
Black grass bugs, 643
Black kissing bug, 555
Black-kneed nirid, 662
Black squash mirid, 52
Blaptostethus, **615**
    pallescens, 615
    piceus pallescens, 615
Blepharidopterus
    angulatus, 554, **662**, 663, 675
    provancheri, **674**
Blissoxenos, 214
    eskii, 214
Blissus, 145, 147–150, 153–155, 158–160–164, 166, 172–175, 210, 212, 229, 230
    antillus, 163, **210**

arenarius, 148, 150
arenarius arenarius, 149
arenarius maritima, 148
brevisculus, 149
canadensis, **209**, 210
insularis, 148–150, 157, 158, **162**, 163, 164, 166, 210, 211
iowensis, 152, 153
hirtulus, 147
hirtus, 162
leucopterus, 144, 145, 148–154, 157, 158, 162–164, 166, 173, 187, 209, 212, 715
leucopterus hirtus, 148, 149, **156**, 157–161, 183
leucopterus leucopterus, **147**, 148–161, 210
occiduus, 150, 209, 210
planarius, 150
planus, 163
pulchellus, **210**
richardsoni, **210**
slateri, 163, 210
Boarmia selenaria, 768
Bochartia, 110
Bogosia rubens, 449
Bogosiella pomeroyi, 282
Boisea, 309, 315
coimbatorensis, 312
rubrolineata, 310, **311**, 312, 557
trivittata, **310**, 311, 312
Bolacidothrips graminis, 615
Boll shedder bug, 54
Bollworm, 675, 678
Bombyx mori, 751, 752, 766
Boxelder bug(s), 310, 311
Brachylybus
inflexus, 384
variegatus, 384
Brachyplatys testudonigra, 507
Bracon, 624
Brazilian chicken bug, 531
Brinjal lace bug, 105
Brinjal tingid, 104
Brontocoris
nigrolimbatus, 766
piniarius, 767
tabidus, 766
Bronze orange bug, 508, 509
Brown marmorated stink bug, 454
Brown plant hopper, 602, 604, 664, 665
Brown shield bug, 452
Brown smudge bug, 670
Brown stink bug, 434
Brown-winged green bug, 445
Brumus saturalis, 88
Bryocoropsis laticollis, 15, 19, 22
Bucculatrix thurberiella, 675
Buenoa, 592
Buffalo grass chinch bug, 150
Bupalus piniarius, 763
Buttonwood tingis, 90
Bycyrtes discisa, 432
Byrsinus varians, **409**

C

Cabbage looper(s), 647, 719, 743, 751, 755
Cabbage moth, 754
Cacao capsid bug, 22
Cacodmus, 519
Cacopsylla, 676
mali, 676
pyricola, 668
Cadamustus, 103
Cadmilos retiarius, **108**
Caenocoris nerii, **218**
Caledonia seed bug, 222
Calidea, **490**
dregii, 490
duodicimpunctata, 490
Caliothrips indicus, 719
Calliphara, 476
Calliphora erythrocephala, 743
Calocoris, 53
angustatus, **45**, 46
norvegicus, **53**
Calopepla leayana, 761
Camptischium clavipes, 384
Campylomma, 676
liebknechti, **676**
livida, 676
lividum, 676
nicolasi, **61**, **676**
verbasci, **61**, 62, 659, **676**, 677, 679
Campyloneura virgula, **666**, 667
Cantao, 476
parentum, 476
Canthecona, 759
Canthophorus melanopterus, **414**
Capsus ater, 15, 17, 18, 20
Carbula humerigera, 14
Cardiastethus, 610, 625
rugicollis, 617
Carlisis wahlbergi, 384
Carpocoris
fuscispinus, **451**
purpureipennis, 14
Carpophilus, 291
Carulaspis, 674
Cassava hornworm, 753
Cassava lace bug, 116
Catamiarus brevipennis, 700
Cavelerius, 167, 168, 170, 172, 230
excavatus, 167, 169, **170**, 172, 175, 176
illustris, 167
minor, 167
saccharivorus, **167**, 168, 169, 170, 171, 173
sweeti, 167, **170**, 171–173, 175, 177
tinctus, 171, 173
Centrodora darwini, 429
Cephalonomia, 209
Ceratocapsus, **674**, 675
dispersus, 675
mariliensis, 675
modestus, 675
punctulatus, 675

INSECT INDEX

Cercotmetus, **586**
Cermatulus nasalis, 768
Chaoborus, 571
Chauliops, 146, **205**, 206, 207
    bisontula, 205
    choprai, **205**, 206, 207
    fallax, 205, 206, 207
    horizontalis, 205
    lobatula, 205
    nigrescens, 205, 206, 207
    rutherfordi, 205, 206
Chelinidea, 338, 339, **375**, 376, 383
    canyona, 383
    tabulata, **376**, 383
    vittiger, **375**, 376, 383
Chilacis typhae, 14, 19
Chilmenes sexmaculatus, 88
Chilo agamemnon, 615
Chinch bug, 21, 715
Chinoneides, 729
    tasmaniensis, **729**, 730–732
Chiracanthium mildei, 91
Chlorochroa
    ligata, **451**
    sayi, 451
Chrysanthemum lace bug, 93
Chrysocoris, **492**
    purpureus, 492
    stockerus, 492
    stollii, 17, 19, 476, 492
Chrysodeixis chalcites, 740, 746, 751
Chrysomela, 767
    populi, 763
Chrysopa, 98, 113, 115, 617, 645, 715
    carnea, 641, 643
    plorabuna, 154
    rufilabris, 91
    septempunctata, 112
    sinica, 112
    vulgaris, 204
Chrysoperla, 732
    carnea, 701
Cimex, 519, 528
    abietis, 229
    columbarius, 521, **531**
    ferus, 639
    hemipterus, 520, 521, 523–525, **526**, 527, 532
    insuetus, **530**
    lectularius, 3, 519, 520, **521**, 522–527, 531, 532, 763
    militaris, 177
    pipistrelli, 554
    rotundatus, **526**
    sordidus, 207
Cimexopsis nyctalis, 554
Cinara, 612
Circulifer tenellus, 731
Citron bug, 363
Citrus mealybug, 669
Clavigralla, 348, **350**, 351, 352, 383, 386
    elongata, 350, **353**, 354
    gibbosa, 14, 353, **354**, 355

    horrida, 350
    scutellaris, 350, 354, 355
    shadabi, 350–352, **353**
    tomentosicollis, **350**, 351, 352, 353, 354
Cleptocoris turpis, 556
Clerada, 556
    apicicornis, 146
Cletomorpha
    fuscescens, 384
    hastata, 357
    lancigera, 384
    unifasciata, 384
Cletus, **355**
    bipunctatus, 355, **357**
    notatus, 355
    punctiger, 227, **355**, 356, 357
    rusticus, 14, 356, **357**
    signatus, 14, 18, 355
    trigonus, 355
Cloresmus, 338
    modestus, 384
    montanus, 384
Clostera, 760
    anachoreta, 767
Closterotomus, 53
    norwegicus, **53**
Clytiomyia helvola, 483
Cnidocampa flavescens, 767
Coccinella, 115
    munda, 154
    septempunctata, 88
    transversalis, 120
    undecimpunctata, 196
Cochlochila bullita, **88**
Cocoa capsids, 25
Cocoa mirids, 40, 41, 44
Coconut bug, 379
Coconut spathe bug, 450
Codling moth, 662, 663, 669, 670, 674, 676–678
Coleophora
    chalcites, 749
    laricella, 674
    serratella, 748
Coleostichus, 490
Coleotichus, 490, 493
    blackburniae, **490**
    costatus, 493
    sordidus, 490
Colias lesbia, 753
Collops
    quadrimaculatus, 154
    vittatus, 641
Colobathristes, 146
Colorado potato beetle, 642, 673, 738, 739, 743, 745, 746, 749–751, 753–759, 761, 763, 765–767
Common bed bug, 526, 527, 531
Common chinch bug, 147
Common green mirid, 55, 56
Compseuta ornatella, **105**
Conchuela, 451
Conocephalus longipennis, 325

Coon bugs, 200
Copium, 86
Coptosoma, 507
　cibraria, 507
　fuscomaculatum, 507
　nubila, 507
　ostensum, 507
　punctissimum, 507
　siamicum, 507
　variegatum, 507
Coranus, 221
　pallidus, 209
Corcyra cephalonica, 700, 760
Corecoris, 382
Coreus
　marginatus, 18
　marginatus orientalis, 14, 17
Coridius, 505, 506
　janus, 15, 19
Corizus, 310
　crassicornis, 310
　parumpunctatus, 14
Corn delphacid, 665, 666
Corn earworm, 619, 648, 675
Corsicus roseipennis, 639
Corythaica
　cyathicollis, **88**, 89, **116**
　monacha, **88**
　passiflorae, **88**
　planaris, **88**
Corythauma ayyari, **89**
Corythucha, 86, **90**, **91**, **93**, 121, 555
　arcuata, **91**, 93, 669
　ciliata, 87, **90**, 91, 116, 121, 555, 619
　cydoniae, 87, **91**, 92, 109, 555
　decens, **94**
　distincta, **116**
　gossypii, 85, **92**, 93
　marmorata, 85, **93**, 94
　morrilli, **94**, 555
　ulmi, 87
Cotesia congregata, 728
Cotton aphid, 660, 669, 731
Cotton bollworm, 743, 746, 748
Cotton fleahopper, 50, 659, 678, 714
Cotton harlequin bug, 491
Cotton lace bug, 92
Cotton leafperforator, 675
Cotton leafworm, 675, 678, 748, 751, 755
Cottonseed bug, 145, 203, 205
Courtesius, 273
Creontiades
　dilutus, 15, 18, 24, **53**, 54
　modestum, 16
　pallidus, **54**, 55
Crinocerus sanctus, **376**
Criocoris saliens, 675
Crocistethus waltli, **415**
Crophius disconotus, 200
Crusader-bug, 378

Cryphocricos, 572, 573
　hungerfordi, 572
　latus, 573
Cryptoblabes gnidiella, 615
Cryptomyzus ribis, 615
Cuban laurel thrips, 617, 618, 621
Cucumis sativus, 660
Culex, 579, 587
　modestus, 587
Cuspicona simplex, **451**
Cycloneda sanguinea, 89
Cydia pomenella, 614, 662
Cylindromyia
　brassicaria, 431, 483
　euchenor, 441
　fumipennis, 456
Cymoninus, 228
　basicornis, 228
　notabilis, **228**
Cymus
　aurescens, 227, 228
　basicornis, 227
　glandicolor, 14
Cyrtocoris gibbus, 505
Cyrtomenus, 409
　bergi, **410**, 411, 415
　ciliatus, **414**
　(*Cyrtomenus*) *bergi*, 410
　(*Cyrtomenus*) *ciliatus*, 409
　(*Cyrtomenus*) *crassus*, 409
　(*Cyrtomenus*) *mirabilis*, **409**, 410
　mirabilis, 409, 410
　(*Syllobus*) *emarginatus*, 409
Cyrtopeltis, 50, 51
　modestus, **50**, **667**
　tenuis, 21, **51**, **667**
Cyrtorhinus, 665
　fulvus, 662, **663**, 664
　lividipennis, 662, **664**, 665
Cyrtothirps omnivorous, 98
Cysteochila
　ablusa, **106**
　endeca, **106**

D

Dacus cucurbitae, 360
Daktulosphaira vitifoliae, 675
Dallacoris, 380
　pictus, **380**
Dark apple red bug, 59
Dasyneura pyri, 677, 678
Dasynus
　manihotis, **341**
　piperis, **376**, 384
Delicate apple capsid, 675
Dentisblissus venosus, 167
Deraeocoris, 555, 661, 679
　brevis, **668**, 669

lutescens, 658
nebulosus, 91, **669**
pallens, **669**, 670
punctulatus, 16
ruber, 658
rutilus, 657
signatus, **670**
Dersagrena flaviventris, 384
Diactor bilineatus, 385
Diaphnidia, 674
Diaphnocoris, 674
  pellucida, **674**
  provancheri, 674
Dichelops
  furcatus, **451**, 452
  melacanthus, **452**
  phoenix, 452
Diconocoris hewetti, **106**
Dicranocephalus agilis, 338
Dictyla, 117
  echii, **106**, 107, **117**
  monotropidia, **107**
  nassata, **107, 117**
Dictyonata strichnocera, **117**
Dictyotus caenosus, **452**
Dicyphus, 743
  errans, **667**, 726
  tamaninii, **659**, 660, 679
Dieuches, **207**, 208
  albostriatus, 208
  armatipes, 208
  armipes, 208
  annalatus, 208
  consimilis, 208
  humilis, 208
  patruelis, 208
  triangulus, 208
Dimophoberytus variabilis, 725
Dimorphopterus, 175, 176, 211
  blissioides, 211
  brachypterus, **211**
  cornutus novaeguieneae, **211**
  gibbus, 170, **175**, 176
  hessei, **211**
  japonicus, 169, 212, 214
  pallipes, 169, **212**
  similis, **210**
  spinolae, **211**
  zuluensis, 211
Dindymus
  flavipes, **291**
  versicolor, **290**, 291
Diolcus irroratus, **491**
Diparopsis, 768
Diplogomphus hewitti, **106**
Diplonychus, **578**
  rusticus, 578, 579
  rusticus (indicus), 577

Diprion
  pini, 763
  similis, 746, 748
Discocoris, 140, 142
  imperialis, 139
Dissolcus paraguayensis, 453
Distantiella theobroma, 16, 19, 22, **40**, 41, 44, 45
Distinct lace bug, 116
Dolichoderus thoracicus, 44
Dolycoris
  baccarum, 14, 18, 20, 22, 25, **430**, 431
  penicillatus, **452**
Double-eyed soldier bug, 753
Douglas-fir tussock moth, 674
Dufouriellus, **615**
  ater, 554, 615, 616
Dullinius
  conchatus, **107**
  unicolor, **107**
Dura plant bug, 179
Dusk cotton bug, 198
Dusky bug, 219
Dusky cottonseed bug, 145, 197
Dusky cotton stainer bug, 203
Dusky stink bug, 453
Dyakiella, 225
Dyctyotus caenosus, 455
Dysdercus, 203, 271–279, 281–288, 290, 291, 293, 294, 491, 701
  albidiventris, **289**
  andreae, 274
  cardinalis, 275, 283, **287**
  cingulatus, 15, 272, **276**, 277, **278**, 279, 280, 700
  delauneyi, 274
  fasciatus, 15, 19, 272, 273, 275, **280**, 281–283, 287, 294
  fasciatus fulvoniger, 288
  fulvoniger, 279, **289**
  howardi, 15, 274, **288**, **289**
  howardi minor, 274, 288
  intermedius, 272, 281, **283**
  koenigii, 15, 19, 272, 273, **276**, **278**, 279, 280
  maurus, **288**, 289
  melanoderes, **287**, 288
  mendesi, 274
  mimulus, 274, **289**
  mimus, **289**
  nigrofasciatus, 275, 283, **285**, 286
  obscuratus, 273, **289**
  pallidus, **288**
  peruvianus, **289**
  ruficollis, 274, 279, **288**
  sidae, 15, **286**, 287, 291
  singulatus, 15
  superstitiosus, 275, 279, **284**, 285, 286, 288, 557
  suturellus, **290**
  suturellus capitatus, 290
  transversalis, 279
  voelkeri, 279, **284**
  völkeri, **284**

E

Earias, 768
Eastern tent caterpillar, 743, 748, 761
Ectomocoris tibialis, 700
Ectophasia
  crassipennis, 483
  rubra, 483
Ectopiognatha major, 429
Edessa, 421
  meditabunda, **431**, 432
  rufomarginata, **452**, 453
Efferia, 447
Eggplant lace bug, 109
Elasmognathus hewetti, **106**
Elasmolomus, 207, 208
  pallens, **207**, 208, 209, 227
  sordidus, 15, 20, 207
Elasmopoda, 338
  valga, 385
Elasmostethus humeralis, 16
Elasmucha
  dorsalis, 16
  grisea, 224
  putoni, 16
  signoreti, 16
Elatophilus, 609, 611, **616**, 625
  hebraicus, 611, 616
  inimica, 616
  nigricornis, 611
Elm lace bug, 87
Elm leaf beetle, 748
Elomyia lateralis, 483
Empoasca fabae, 53, 673
Enarsiella, 507
Encarsia formosa, 661
Encyrtus anasae, 345
Endopsylla endogena, 98
Engytatus, 50
  modestus, **50**, 51, **667**
Eocanthecona, 758, 759
  furcellata, 738, 743, 745, **758**, 759–761, 768
  varivestis, 745–747
Ephestia kuehniella, 661, 743, 762, 763
Epilachna, 764
  varivestis, 738, 740, 764–766
Epineura helva, 285
Epiphyas postvittana, 676
Eremocoris, 229
  borealis, 229
  depressus, 229
  ferus, 229
Eremovescia, 555
Erinnyis ello, 753
Erythmelus, 91, 97
Eteonius sigillatus, **108**
Euander lacertosis, **228**
Eucallipterus tiliae, 663
Euclytia flava, 456
Eumecopus
  australasiae, 14, 20
  punctiventris, 14

Eumicrosoma
  benefica, 154, 161, 166
  blissae, 168, 169
Euoxycarenus, 199
Eupelmus reduvii, 345
Euphorus carcinus, 220
Eupocilia ambiguella, 643
Eupterote mollifera, 760
European alder leafminer, 662
European bug, 528, 529
European corn borer, 648, 755
European martin bug, 529
European tarnished plant bug, 57, 58
Eurydema, 422, 432
  pulchrum, **432**
  rugosum, 12, 14, 18, 20, 21, 27, **432**
Eurygaster, 220, 475, 476, **477**, 478, 481, **492**, 493, 494
  amerinda, 493
  austriaca, 477, 480, 484, **488**
  integriceps, 17, 19, 20, 22, 23, 195, 449, 476, **477**, 478–487, 493
  maura, 477, 480, **486**, 487, 493
  minidoka, 493
Eurystylus
  coelestialium, 16
  immaculatus, **55**
  marginatus, 55
  oldi, **55**
  rufocunealis, 55
Euschistus, 421, 434
  conspersus, **453**
  heros, **432**, 433, 434
  servus, **434**, 453
  servus euschistoides, 434
  servus servus, 434
  tristigmus, **453**
  variolarius, **453**
Eutettix tenellus, 267
Euthera tentatrix, 441, 453
Euthyrhynchus
  floridanus, 765
  macrocnemis, 765
Eutrichopodopsis nitens, 434, 439, 446
Excavatus, 167
Eyespotted bud moth, 613, 670
Eysarcoris
  aeneus, 14
  guttiger, 14
  lewisi, 14
  ventralis, 14

F

Fabrictilis, 357
  gonagra, **363**
Fall armyworm, 743
Fall webworm, 743, 748, 764
False chinch, 145
False chinch bug, 182, 184, 185, 186, 188, 219, 221
Fenusa dohrnii, 662

Filbert aphid, 610, 668
Fire ant(s), 165, 439
Florida potato plant bug, 382
Frankliniella occidentalis, 614, 621, 659
Franklinothrips, 110
    vespiformis, 114
Fromundus pygmaeus, **415**, 556
Fruit spotting bug, 341

G

Galeatus
    helianthi, **108**, 109
    involutus, **107**
    maculatur, **117**
    scrophicus, **108**, 109
Galerucella, 765
Galleria mellonella, 743, 746, 751, 752, 763, 764
Gampsocoris, 726
Gargaphia, 122
    arizonica, **117**
    lunulata, **109**
    opacula, **117**, 118
    sanchezi, **109**
    solani, 87, **109**
    tiliae, **110**
    torresi, **110**
Gastrodes, 229
    abietum, 229
    errugineus, 15
    grossipes, 229
Gelonus tasmanicus, **377**
Geoblissus, 147, 148, 212
    hirtulus, **212**
    rotundatus, 147
Geocoris, 95, 114, 161, 188, 205, 230, 556, 619, 640–642, 645, 646, 648, 679, 713–720, 743
    atricolor, 719
    bengalensis, 719
    bullatus, 161, 165, 188, 642, 714–716, **717**, 718, 719
    capricornutus, 204
    floridanus, 715, 716
    lividipennis, 716, 719
    lubra, 204
    ochropterus, 719
    pallens, 640, 641, 714–717, **718**, 719, 720
    pallens var. *decoratus*, 188
    piceus, 714
    pseudoliteratus, 719
    punctipes, 188, 556, 566, 641, 647, 648, 714, 715, **716**, 717
    uliginosus, 161, 165, 188, 714–716, 719
Geotomus, 556
    pygmaeus, **415**
Germalus, 713, 720
    pacificus, 120, **720**
Gerris, 601
    nitida, 602
    swakopenis, 603

Giant looper, 768
Giant red bug, 292
Globose scale, 677
Globulifera, 102
Gloomy scale, 672
Glypsus, 768
Grain stink bug, 173
Grape mealybug, 669
Grape phylloxera, 675
Graphosoma
    lineatum, 14
    rubrolineatum, 14
Graptopeltus
    albomaculatus, 227
    amurensis, 227
    angustatus, 15, 227
Graptostethus, 216
    manillensis, 217
    servus, 178, 179, **216**, 217
Gray bug, 219
Gray cluster bug, 189
Greater wax moth, 743, 764
Green cloverworm, 641
Greenhouse whitefly, 659–661, 667
Green leafhopper, 664
Green mirid, 53, 54
Green peach aphid, 667, 718
Gryon, 326, 340, 341, 352, 355, 370, 377, 378
    anasae, 345
    carinatifrons, 346, 364, 368
    chelinideae, 376
    clavigrallae, 355
    flavipes, 324
    gallardoi, 382
    gnidas, 352
    japonicum, 326, 357, 374
    largi, 292
    nixoni, 324
    pennsylvanicum, 346, 360, 361, 363, 368, 376
Gundhi bug, 323
Gymnoclytia
    immaculata, 441
    occidua, 456
Gymnosoma
    desertorum, 483
    rotundatum, 431, 446
Gynaikothrips
    ficorum, 617, 618
Gypsy moth, 749

H

Habrochila
    ghesquierei, **95**, 110
    laeta, **110**
    placida, 95, **110**, 111
Hadronotus, 345, 346, 377
    ajax, 346
    antestiae, 355
    carinatifrons, 346

gnidas, 352
homoeoceri, 341
Haedus vicarius, **95**, 96
Haematosiphon inodorus, **530**, 531, 554
Hairy chinch bug, 156–162
Halobates, 601
Halticus
    bractatus, **59**
    citri, **59**
Halyomorpha
    marmorea, **454**
    mista, **454**
Haplothrips verbasci, 731
Harlequin bug, 454, 455
Harmonia axyridis, 748
Harmostes
    reflexulus, 310
    serratus, 310
Harpactor segmentarius, 353
Hawthorn bug, 92
Hawthorn lace bug, 91, 92
Hegesidemus harbus, **112**
Helicoverpa
    armigera, 700, 761
    punctigera, 768
    zea, 675, 719, 743, 746, 748, 753
Heliothis, 715, 746, 768
    obsoleta, 619
    virescens, 564, 622, 673, 716, 717, 726, 727, 731, 743, 746, 749
Heliozeta helluo, 483
Helomyia lateralis, 483
Helopeltis, 41, 43, 44, **51**
    anacardii, 51
    antonii, **41**, 42, 43, 51
    bergrothi, 16, 19, 22
    bradyi, 41, 51
    clavifer, 16, 22, 51, 52
    corbisieri, 16
    schoutedeni, **42**, 43, 51
    theivora, **43**, 44, 51
    theobromae, **43**
Hemerobius, 120
Hermya confusa, 349
Hesperocimex sonorensis, **531**
Heterocordyllus malinus, 57, **59**, 60
Heterogaster urtica, 15
Heterotoma merioptera, 658
Hexacladia
    smithi, 434
Hierodula patellifera, 112
Hierodulam patellifea, 360
Himacerus mirmicoides, 91
Hippodamia convergens, 109, 641, 748
Hirtulus, 212
Holhymenia histrio, 385, 386
Hollyhock lace bug, 115
Homoeocerus, **377**
    ignotus, 377
    inornatus, 377
    laevilineus, 377

marginellus, 378
pallens, 377, **378**
prominulus, 378
puncticornis, 377
serrifer, 377
striicornis, 378
unipunctatus, 378
walkeri, 378
Hoplinus, 725
    scutellaris, 725
Hoplistoscelis, 647
    deceptivus, 640
Horehound bug, 449
Horridipamara nietneri, **227**
Horvathiolus, 143
Housefly, 739, 751, 763
Humped-back melon bug, 52
Hyaliodes
    harti, **670**
    vitripennis, 670
Hyalomyia, 192, 325
    aldrichii, 185, 188, 192, 715
    pusilla, 192
Hyalomyodes, 325, 434
Hyalopeplus pellucidus, 45, **55**
Hyalymenus tarsatus, **329**
Hydara tenuicornis, 385
Hylalomya, 185
Hypera postica, 670, 764
Hyphantria cunea, 643, 743, 748, 751, 764
Hyponygrus, 161
Hypselonotus, 386
    intermedius, 385
    lanceolatus var. *gentilis*, 385
    lineatus detersus, 385

I

Ilyocoris cimicoides, 571–573
Imported fire ants, 641
Indian honey bee, 701
Indian meal moth, 616
Invermay bug, 222
Iphicrates, 215
    angulatus, 215
    nigritus, 215
    papuensis, 215
    spinicaput, 214, 215
    weni, **214**
Irbisia, 45, 60
    solani, 555
Iridomyrmex myrmecodiae, 342
Ischnodemus, 167, 213
    brinki, 144, 175
    congoensis, 212
    diplanche, **212**
    fallicus, **212**
    fulvipes, **212**, 213
    noctulus, 213

perplexus, 213
Isyndus obscurus, 556
Ithamar hawaiiensis, 310

J

Jack pine sawfly, 764
Jadera, 309, 310
    haematoloma, 309, **312**, 313
    sanguinolenta, **313**
Jalysus, 725, 726
    albidus, 732
    spinosus, 726, 727, **730**, 731
    wickhami, 726, **727**, 728, 730
Junonia, 746

K

Kleidocerys
    geminates, 224
    resedae, **224**
    resedae germinatus, 224
    resedae resedae, 224

L

Labidura riparia, 165
Labops, 60
    hesperius, 23, **60**, 61, 643
    hirtus, 60
Laccocoris
    insularis, 573
    limigenus, 571–573
Laccotrephes, 583–586
    griseus, 584, 587
    grossus, **584**, 585
    kohlii, **584**
    maculatus, **585**
    ruber, 584
Lactistes
    minutus, **414**
    rastellus, 414
Ladybird beetles, 748
Lampromicra senator, **492**
Lantana bug, 720
Lantana lace bug, 119, 120
Larch casebearer, 674
Larch sawfly, 674
Large milkweed bug, 218
Large white butterfly, 745, 749
Largus, 272, 273, **291**, 292
    balteatus, 292
    californicus, 292
    cinctus, 292
    humilis, 292
    rufipennis, 292

    succinctus, 292
Lasiochilus, 610, 625
    pallidulus, 165
Lathromeromyia corythaumaii, 90
Lawn chinch bug, 162
Leaf-footed pine seedbug, 361
Leis axyridis, 112
Lepidosaphes ulmi, 677
Leptinotarsa, 765
    decemlineata, 673, 738, 741–743, 745, 746, 749–751, 753–758, 761, 763–766
Leptobyrsa, 101
    decora, **111**, **118**
    explanata, **100**
    rhododendri, 20
Leptocimex, 519
    boueti, **530**
Leptocoris, 309
    amicta, **315**
    augur, **312**, 315
    hexophthalma, **315**
    rubrolineata, **311**
    trivittata, 557
Leptocorisa, 321, 322, **328**, 329
    acuta, **322**, **323**, 328
    chinensis, 227, 322, 323
    discoidalis, 328
    maculiventris, **323**
    oratorius, 322, **323**, 324, 328
    palawanensis, 328
    solomonensis, 328
    varicornis, 14, 18, **322**
Leptodemus, 227, 556
    bicolor ventralis, 556
    irroratus, 227
    minutus, **226**, 227
Leptodictya tabida, **96**, 97
Leptoglossus, 24, 338, **357**, 358, 361, 363, 370, 386
    ashmeadi, 357
    australis, 291, **358**, 360, 363
    balteatus, 358
    bidentatus, **358**
    brevirostris, 357
    chilensis, 357
    cinctus, 357
    clypealis, 24, **360**, 361, 365
    concolor, 358, 368, 369
    conspersus, **361**
    corculus, 360, **361**, 362, 363, 364, 366, 489
    dilaticollis, 357
    fulvicornis, 357
    gonager, 363
    gonagra, 357, 358, **363**, 364, 368
    impictus, 357
    lonchoides, 357
    membranaceous, **358**
    occidentalis, 14, 24, 360, 361, **364**, 365, 366
    oppositus, **366**, 367
    phyllopus, 338, 340, 357, 364, 366, **367**, 368, 370
    stigma, 366, **368**, 369
    zonatus, 329, 339, 361, 366, **369**, 370, 489, 490

Leptopharsa
  gibbicarina, **111**, 121, 435
  heveae, **111**
  illudens, **115**
  manihotae, **116**
Leptoterna, 154, 187
Leptothorax curvispinosus ambiguus, 154
Leptoypha, 112
  minor, **112**
  mutica, **112**
Leptus, 108
Lethocerus, 578, **579**
  americanus, 579
  fakir, 578
  indicum, 579
Lettuce root aphid, 613
Leucopterus, 156, 212
Leucostoma
  acirostre, 310
  crassa, 181
  creticus, 181
  equestris, 181
  simplex, 192, 646
Levuana iridescens, 768
Light brown apple moth, 676
Lime aphids, 663
Limnocoris, 573
Limnogeton, 578
Limnogonus, 601
  cereiventris, 603
  fossarum, 602
  nitidus, 602, 603
Limnoporus, 601
Lincus, 422, 434, 435
  lobuliger, **434**, 435
  malevolus, **435**
  spurcus, 435
Linear bug, 225
Liorhyssus hyalinus, **314**, 315
Liothrips urichi, 618
Lipostemmata, 229
  humeralis, 229
  major, 229
  scutellatus, 229
Liriomyza trifolii, 667, 668
Little bean bug, 205, 206
Little milkweed bug, 218
Loania canadensis, 224
Lohita grandis, **292**
Long-tailed mealy bugs, 193
Loxa
  deducta, **454**
  flavicollis, **454**
Luridus, 453
Lybindus dichrous, 385
Lyctocoris, 611, **617**
  beneficus, 608, 617, 624
  campestris, 553, 617
  elongatus, 617
Lygaeus, 19, 143, 181, 216
  elegans, 177

  equestris, 15, **218**
  kalmii, 218
  militaris, 178
Lygidea mendax, 57, 59, 60
Lyglineolaris, 16
Lygocoris, 45, 56
  communis, 747
  communis novascotiensis, 555
  locorum, 16
  pabulinus, **55**, 56
  rugicollis, **56**, 57, 59
  viridanus, 55
Lygus, 11, 13, 21, 23, 25, 369, 555, 567, 642, 645, 658, 659, **672**, 673, 718, 719
  disponsi, 13, 16, 17, 19, 24, **57**
  elisus, 24
  gemellatus, 13, 16–20
  hesperus, 13, 16, 23, 24, 38, 39, **46**, 47, 49, 641, 642, 647, 673
  lineolaris, 13, 21, 23, 24, 38, **47**, 48, 49, 555, 673, 747
  perplexus, 57
  pratensis, 13, 16–20, 57
  pubulinus, 23
  punctatus, 13, 16–20
  rugulipennis, 13, 16–27, 38, **57**, 58, 63, 673
  saundersi, 16
  simonyi, 21, 58
  vosseleri, **58**
Lymantria
  dispar, 749
  monacha, 763

M

Macchiademus, 173, 174
  diplopterus, 145, **173**, 174
Machtima cruciger, 385
Macroceroea grandis, **292**
Macrocheraia
  grandis, 273, **292**, 293
  grandis sumatrana, 292
Macrolophus, 743
  caliginosus, **660**
  melanotoma, **660**, 661
  pygmaeus, 660
Macroparius, 220
Macropes, 167, 170, 171
  excavatus, 171, 177
  harringtonae, **214**
  hedini, 213
  maai, **214**
  obnubilus, **213**, 214, 230
  privus, 214
  punctatus, **214**
  spinimanus, 214
  subauratus, 214
  varipennis, 214
Macrorhaphis, 768
  acuta, 768

Macroscytus
  japonensis, 17
  transversus, **414**
Macrosiphum
  euphorbiae, 660
  rosae, 26, 730
Macrotracheliella, **617**
  laevis, 617
  nigra, 617
Malacocoris chlorizans, **675**
Malacosoma americanum, 743, 748, 761
Malayan black bug, 456
Malcus, 146, 206, 224,
  flavidipes, 225
  japonicus, **224**, 225
  scutellatus, 225
Mamestra brassicae, 643, 754, 763
Manduca sexta, 727
Martin bug, 528
Masked bed bug hunter, 525
Matsucoccus, 611, 616
  feytaudi, 616
  josephi, 616
  resinosae, 616
Mayetiola destructor, 485
Mcateella, 265
Mealworm, 743, 751
Mecidea, 556
Mediterranean flour moth, 661, 743
Megacoelum modestum, **53**
Megalotomus, 321, 324, 329
  quinquespinosus, **327**, 328
Megilla maculata, 109, 274
Melamphaus faber, **291**
Melanaethus
  parvus, 556
  subglaber, 556
Melanaspis
  obscura, 672
  tenebricosa, 672
Melanitis leda, 767
Melanocoris, 609, 611
Melanolestes, 556
  picipes, 555
Melanoplus, 192
Melon fly, 360
Melosoma, 763
Menochilus sexmaculatus, 761
Mesovelia, 601
Metancanthus, 725, 726
  elegans, 17
  pulchellus, **728**, 729
Metasalis populi, **112**
Metatropis, 726
  rufescens, 726
Metopoplax ditomoides, 227
Metrobates hesperius, 603
Mexican bean beetle, 647, 717, 738, 740, 745–747, 764, 766
Mexican chicken bug, 530, 554
Mexican fruit fly, 755

Mexican poultry bug, 554
Microdiprion pallipes, 763
Microlarinus, 719
Microphanurus, 483
Microporus nigrita, **411**
Microvelia, 601, 602
  douglasi atrolineata, 602
  pulchella, 602
Mictis, 338, **378**
  caja, 378
  longicornis, 378
  profana, 14, 17, 18, **378**, 379
  tenebrosa, 378
Miespa, 265
Militaris, 177
Miris
  dolabratus, 16, 20, 25, 154, 187
  ferus, 639
  vagans, 639
Mirperus jaculus, 350
Mogannia iwaskii, 170
Moissonia importunitas, 16, 21, **62**
Monalonion, **44**
  annulipes, 44
  bahiensi, 44
  bondari, 44
  dissimulatum, 44, 726, 731
  schaefferi, 44
Monanthia, **97, 107**
  ampliata, **121**
  bullita, 88
  cardui, 17
  globulifera, **88**
  nassata, **107**
  simplex, **118**
  tabida, **96**
Monosteira, 86
  edeia, **113**
  lobulifera, **113**
  unicostata, 85, 87, **97**, 98
  zizyphora, **113**
Monotropida, **107**
Montandoniola, 609, **618**
  moraguesi, 617, 618
Montrouzieriellus falleni, 768
Mormidea, 422
  notulifera, **436**
  quinqueluteum, **435**, 436
Mormonomyia, 192, 204
Morrill lace bug, 94
Mosquito bug, 44
Mozena obtusa, 339, 383, 385
Mulberry bug, 225
Mullein bug, 61, 659, 676, 679
Mullein thrips, 731
Murgantia histrionica, **454**
Musca domestica, 739, 743, 751, 752, 763, 766
Musgraveia sulciventris, 508, 509
Mygdonia, 338
Myiomma cixiiforme, 672
Mylabris pustulata, 700

Myodocha
  serripes, **228**
Myrmoplasta, 273
Myrmus miriformis, 310
Myzocallis coryli, 610, 668
Myzus
  nicotianae, 728
  persicae, 267, 620, 642, 667, 718, 719

N

Nabicula, 646
  limbata, 643
Nabis, 95, 619, 639–648, 715
  alternatus, 639–647
  americoferus, 639–642, 644–648
  brevis, 643
  capsiformis, 639, 640, 644–646, 648
  deceptivus, 645
  feroides, 645
  ferus, 161, 639, 643, 645
  flavomarginata, 643
  kalmii, 640, 645
  kinbergii, 556
  limbatus, 643
  pseudoferus, 91, 643, 645
  punctatus, 643
  punctipes, 639
  roseipennis, 639–648
  rufusculus, 639, 640, 643, 644, 646
  sinoferus, 643, 644
  sordidus, 645
  stenoferus, 644
Naphius apicalis, 208
Nardo
  cumaeus, 172
  phaeax, 172
Narnia, 338
Naucoris, 572, 573, 587
  maculatus, 572, 573
Neacoryphus, 143
Neides, 729, 730
Nematus oligospilus, 768
Neoblissus, 147, 148
  parasitaster, 147
Neobozius, 507
Neobrachelia, 453
Neochauliops, 205
  laciniata, 205
Neodiprion, 743
  sertifer, 763
  swainei, 764
Neogerris, 601
Neomegalotomus, 321, 324, 329
  pallescens, 324
  parvus, 322, **324**, 325, 329, 386
  rufipes, 324
Neoneides, 730
  muticus, **730**, 731

Neopamera, 228
Neophasia menapis, 731
Neorileya, 370, 434, 447, 448
  ashmeadi, 382
Nepa, 583, 587
  apiculata, 587
  cinerea, 583, 584
Nephotettix virescens, 664
Nesidiocoris tenuis, **51**, 52, **667**, 668
New World garden fleahopper, 58
Nezara
  antennata, 14, 454, **455**
  viridula, 14, 23, 291, 325, 328, 381, 421–423, 432–434, **436**, 437–439, 443, 444, 448–450, 452, 455, 509, 643, 768
Nidicola, 610
Niesthrea
  louisianica, 310, **313**
  sidae, **314**
Nilaparvata lugens, 169, 602, 664
Nipa [sic], 583
Northern false chinch bug, 182, 183
Notobitus, 338, 379
  meleagris, **379**
Notonecta, 587, 592
  glauca, 592, 593
  hoffmanni, 592
Notostira erratica, 16
Numicia viridis, 666
Nysius, 145, 146, 182–190, 192, 193, 194, 196, 204, 219, 220, 221, **222**, 223, 230, 556
  albidus, 219, **221**
  angustatus, 154, 182, 183
  binotatus, 219, **220**
  caledoniae, **222**
  californicus, 182
  ceylonicus, **221**
  clevelandensis, **189**, 190–193, 204
  coenosulus, 195, 222
  convexus, 193
  cymoides, 192, **219**, 220
  destructor, 186
  ericae, 154, 182–185, 187, 188, **219**
  expressus, 15
  graminicolus, **220**
  groenlandicus, 184, 186
  huttoni, 20, 23, 192, **193**, 194, 195, 196, 223
  inconspicuus, **221**
  jacobaea, 192
  kinbergi, 195, 222
  liliputanus, 193
  lineatus, 192
  minor, **221**
  minutus, 185, 186
  natalensis, **221**
  nemorivagus, 222
  niger, **182**, 183–185, 187, 188, 192, 219
  nigriscutellus, 222
  nysiiphagus, 221
  plebeius, **196**, 197, 227
  plebejus, **196**

raphanus, 182, 183, **185**, 186–188, 192, 193, 219
scutellatus, 183
senecionis, **220**
stali, **221**
strigosus, 185
tenellus, 182, 185, 186, 188
thymi, 219, **220**
turneri, 189, **222**
vinitor, 185, **189**, 190–193, 204, 222

O

Oak lace bug, 669
Obscure scale, 672, 679
Ochrimnus mimulus, 178
Ocimun tingid, 88
Odontopus, 701
   confusus, **291**
   nigricornis, 15, 272
Odontoscelis, 476
Oebalus, 422, 439, 440
   ornatus, 441
   poecilus, 435, 436, **439**, 440–442
   pugnax, 227, **441**
   ypsilongriseus, **441**, 442
Oechalia
   schellembergi, 14, 768
Oeciacus, 528
   hirundinis, **528**, 529
   vicarius, **528**, 529
Oecophylla, 372, 373
   longinoda, 371, 372, 373
   smaragdina, 44, 340, 342, 373
Oliva bug, 113
Omanocoris versicolor, 385
Oncocephatus pacificus, 556
Oncochila, 86
   simplex, 85, **118**
Oncopeltus, 143, 230
   famelicus, **218**
   fasciatus, 13, 15, 18, 19, 22, 143, 146, 181, 218, 719
Onespotted stink bug, 453
Onomaua lautus, 16
Onymocoris, 140, 142
Ooencyrtus, 326, 340, 346, 352, 370, 429, 434, 444
   albicrus, 372
   anasae, 345, 441
   erionotae, 324
   fasciatus, 447
   johnsoni, 376, 455
   kuvanae, 352
   leptoglossi, 363
   malayensis, 341, 377, 378
   nezarae, 326, 378, 445
   patriciae, 352, 354
   submetallicus, 444
   trinidadensis, 363
Ophiusa janata, 761
Oplobates, **618**
   woodwardi, 618

Oplomus dichrous, 765, 766
Orchrimmus mimulus, 228
Orgyia pseudotsugata, 674
Oriental chinch bug, 167, 168, 169, 173
Oriental green stink bug, 455
Orius, 98, 609, 610, **618**, 619–621, 625, 640, 641, 645, 648, 670, 715
   albidipennis, **618**
   australis, **619**
   indicus, **619**
   insidiosus, 91, 110, 554, 608, 609, 615, 618, **619**, 620, 641, 642, 645–647, 669
   laevigatus, **620**
   majusculus, 610, 621
   maxidentes, **620**
   minutus, 611, **620**
   niger, 107, 117, **620**
   pallidicornis, 608, 615, 618
   sauteri, **620**
   tantillus, **621**
   thripoborus, **621**
   thyestes, 621
   tristicolor, 609, 610, 615, 617, 618, **621**, 641
   vicinus, 610, **621**
Ornithocoris
   pallidus, 530, **531**
   toledoi, **531**
Orsillus, 229
   depressus, **223**
   maculatus, 223
Orthaea, 556
Orthocephalus funestus, 16, 19
Orthops sachalinus, 16
Oryzaephilus surinamensis, 624
Ostrinia
   nubilalis, 615, 643, 755
Oxycarenus, 145, **197**, 198–201, 203, 204, 226, 230
   albidipennis, 198
   amygdali, 203
   annulipes, 203, 205
   arctatus, 200, 202
   bicolor, 198
   bokalae, 198
   dudgeoni, 198, 199, 204
   exitiosus, 203, 205
   fieberi, 198, 199, 202, 204
   gossipinus, 198, 199, 202
   gossypii, 198
   hyalinipennis, 15, 19, 145, **197**, 198–205, 282
   laetus, 197, 198, 199, 202, 203, 205
   lavaterae, **226**
   luctuosos, 192, 197, 200, 202–205
   lugubris, 198
   maculatus, 199
   modestus, 199
   multiformis, 198
   pallens, 199
   rufiventris, 198, 202
Oxyrhachis tarandus, 201
Oystershell scale, 677

P

Pachybrachius
  biolobata, 228
  flavipes, 227
  lateralis, 227
  luridus, 227
  pacificus, 227
  rusticus, 227
  vinctus, 228
Pachycoris
  fabricii, **492**
  klugii, 314, **489**, 490, 492
Pachygrontha antennata, 15
Pachympallens, 207
Pagasa, 643
Painted bug, 450
Palomena angulosa, 12, 14, 21, **442**
Pamera, 556
Pangaeus
  aethiops, 412
  bilineatus, **412**
Panstrongylus megistus, 541
Paraclara magnifica, 349
Paradasynus, 339, **379**, 380
  rostratus, 339, 379, **380**
Parajalysus, 725, 731
  andinus, **731**, 732
  punctipes, 731
  spinosus, 726, **731**
Parallelaptera
  panis, 98
  polyphaga, 88
Paraphorantha peruviana, 274
Parapiesma, 265
Paraplesius unicolor, 14
Parasa lepida, 760
Parastrachia japonensis, 274
Paratriphleps, **622**
  laeviusculus, 608, 622
Paromius
  exguus, 227
  gracilis, 227
  jejunus, 227
  longulus, **227**, 228
  pallidus, 227
  piratoides, 227
Paropsis charybdis, 768
Pasira
  basiptera, 556
  lewisi, 555
Pasture land bug, 408, 409
Patapius spinosus, 616
Pea aphid(s), 644, 673, 716, 718
Pear midge, 677
Pear psylla, 613–616, 668, 676, 677
Pear psyllid, 642
Pecan nut casebearer, 669
Pectinophora gossypiella, 203, 673
Peirates, 556

Pelocoris
  binotulatus, 572
  femoratus, 571, 572
  poeyi, 572
Pemphigus, 613
  bursarius, 613
Pentatoma
  japonica, 14
  rufipes, 14
Pepper bug, 376
Pepper tingid, 106
Peregrinus maidis, 664, 666
Pericallia
  lepida, 760
  ricini, 760
Peridroma saucia, 749
Perillus, 754, 756, 757
  bidens, 525
  bioculatus, 738, 745, **753**, 754–758, 763
  circumcinctus, 747
Perkinsiella saccharicida, 665
Peruda brasiliana, **226**
Phaenacantha, 146, 225
  australiae, **225**
  bicolor, 225
  saccharicida, 225
  viridipennis, **225**
Phaenotropis cleopatra, **113**
Phasia
  crassipennis, 487
  oblonga, 483
  occidentalis, 187, 188
  subcoleopterata, 483
Phatnoma
  marmorata, **113**
  varians, **113**
Pheidole, 372, 413
  megacephala, 342, 349, 373
Phenacoccus aceris, 670
Philonthus varius, 161
Phonoctonus, 274, 275, 282, 285, 287, 701
  lutescens, 282
  subimpictus, 288
Phorantha, 187
  occidentis, 154, 155
Phthia, 338, 357, 386
  picta, 338, **380**, 381, 386
Physomerus grossipes, 385
Physopelta, 272, 273
  apterus, 272, 273
  gutta, 17
Phytocoris, 45, 658, 661, **673**, 674
  canadensis, 674
  conspurcatus, 674
  dimidiatus, 673
  erectus, 674
  intricatus, 673, 674
  populi, 673
  reuteri, 674
  tiliae, 673, 674

ulmi, 673, 674
varipes, 673
Phytophaga, 485
Phytoscaphus, 172
Picromerus, 762
  bidens, **761**, 762, 763
Pieris
  bioculatus, 749
  brassicae, 745, 749
Piesma, 265
  capitatum, 265, **268**
  cinereum, 265, **267**, 268
  maculatum, 265, **268**
  quadratum, 87, **265**, 266, 267, 268
Piezodorus, 421, 444
  guildinii, **443**, 444
  hybneri, **444**, 445
  lituratus, **455**
  rubrofasciatus, 444
Piezosternum, 507
Piezostethus afer, 554
Pigeon bug, 521, 531
Pilophorus
  perplexus, **677**
  walshii, 677
Pine bark bug, 514–516
Pine sawfly(ies), 743, 746, 748, 763, 764
Pineus, 611, 622, 623
  pini, 622, 623
Pink boll worm, 203, 640, 641, 673, 675, 719
Pirkimerus, 214
  davidi, 214
  esakii, 214
  japonicus, **214**
Pistol case bearer, 748
Planococcus citri, 669
Plantain lacewing bug, 102
Platyngomiriodes apiformis, **52**
Platynopus melacanthus, 768
Plautia, 422
  affinis, **445**, 446, **455**
  crossata stali, 14
  stali, **445**, 446, 454, 455
Plectrocnemia oblongipes, 385
Plerochila
  australis, **113**, 113
Plesiocoris, 56
  rugicollis, 23, **56**
Plodia interpunctella, 616
Plyapomus longus, 725
Pneustocerus, 725
Podisus, 739–750, 752–754, 756, 764, 766
  acutissimus, 766
  bioculatus, 745
  brevispinus, 745, 764
  connexivus, **751**
  maculiventris, 110, 325, 556, 648, **738**, 739–752, 754, 756, 757, 760, 763, 764
  modestus, 764

nigrispinus, 738, 741, 742, 744, **751**, 752, 753, 763, 764, 766
  nigrolimbatus, 766
  placidus, 742, 764
  sagitta, 751, 764
  serieventris, 764
Poecilocapsus lineatus, 45
Poecilocoris, 476
Poeciloscytus unifasciatus, 16
Polynema, 643
  boreum, 646
Polytes, 476
Pomponatius, 383
  typicus, 339
Potamocoris, 573
  binotulatus nigriculus, 573
  femoratus, 573
Potato aphid, 660
Potato leafhopper(s), 53, 619, 642, 647, 673
Potato mirid, 53
Priocnemicoris flaviceps, 385
Pristhesancus plagipennis, 429, 446, 509
Pristiphora erichsonii, 674, 763
Procheiloneuris, 490
Pronotacantha, 725
  annulata, 556, 726
  stusaki, 725
Proreus simulans, 211
Prosimulium, 571
Proxys, 456
  albopunctulatus, 456
  punctulatus, **455**
Psallus
  ambiguus, **677**, 678
  seriatus, **49**
Pselliopus cinctus, 154
Pseudacysta perseae, **114**
Pseudatomoscelis, 49
  seriatus, 16, 23–25, **49**, 50, 658, **678**, 714
Pseudococcus
  longispinus, 193
  maritimus, 669
Pseudolopya, 188
  taylori, 188
Pseudopachybrachius, 228
Pseudoplusia includens, 644, 648, 716, 743, 764
Pseudotheraptus, 339, **370**, 386
  devastans, 370, **372**, 373, 377
  wayi, 338, **370**, 371, 372, 373, 380
Psix glabiscrobis, 429
Psylla
  pyri, 642
  pyricola, 613
Pterochila horvathi, **113**
Pycnoderes quadrimaculatus, **52**
Pygaera cupreata, 760
Pyralis pictalis, 527
Pyroderces simplex, 615
Pyrrhalta luteola, 748, 766
Pyrrhocoris apterus, 15, 19, 271, 274, 275, 276, 281

Q

Quadraspidiotus
   juglansregiae, 672
   pyri, 677
Quince tingis, 91

R

Rachiplusia nu, 753
Radish bug, 185
Ragmus importunitus, 19, **62**
Ranatra, 583, 584, 586, 587
   chinensis, 584, **586**
   elongata, 584, **586**
   filiformis, 584
   fusca, **585**, 587
   linearis, 583, 584
   longipes, 587
   parvipes, **585**
   varipes, 584
Raphanus, 154
Raphidia, 515
Rasahus, 556
   biguttatus, 555
   thoracicus, 555
Raxa, 274
Red cotton bug, 276, 278, 279, 280
Red pumpkin bug, 505, 509
Red tree ant, 371
Redbanded shield bug, 444
Redbanded thrips, 671, 672
Redshouldered stink bug, 456
Reduviolus
   ferus, 639
   roseipennis, 639
Reduvius, 178
   personatus, 525, 699
Rhagovelia, 601
Rhaphidosoma, 699
Rheumatobates rileyi, 603
Rhinocapsus vanduzeei, 555
Rhodnius
   ecuatoriensis, 541
   pallescens, 541
   prolixus, 277, 381, 520, 541–544, 546
Rhododendron lace bug, 87, 100
Rhopalus maculatus, 14
Rhynocoris, 700
   albopilosus, 204
   albopunctatus, 274
   bicolor, 204, 353
   carmelita, 204
   fuscipes, 700
   iracundus, 91
   kumarii, 700
   lapidicola, 221
   lotarus, 204
   marginatus, 700

   tropicus, 113
Rhyparochromus, 15, 207
Rhypodes, 193, 194, 223
   anceps, 194, 223
Rice bug, 322
Rice planthopper(s), 662, 664
Rice stink bug, 227, 441
Rice weevil, 424
Riptortus, 321, 322, 325, 327, 328, 355
   attricornis, 329
   clavatus, **325**, 326–328, 330
   dentipes, **326**, 349, 350, 351
   linearis, 326, **327**
   pedestris, **328**
   serripes, **328**, 330
   tenuicornis, **326**
Rivula, 760, 767
Root black bug, 410
Royal palm bug, 140
Rutherglen bug, 189, 190

S

Saccharicoccus sacchari, 618
Saccharivorus, 167
Sahlbergella singularis, 16, 19, 22, 40, **44**, 45
Sawflies, 763
Scantius volucris, **291**
Scaptocoris
   carvalhoi, 407, **413**
   castanea, **406**, 407–410, 413
   divergens, **413**
   talpa, **413**
   terginus, 407
Scirtothrips dorsalis, 719
Scoloposcelis, 609, 611, **622**
   flavicornis, 622
   mississipensis, 622
Scotinophara coarctata, **456**
Scutellera, **491**
   fasciata, 491
   nobilis, 491
   perplexa, 491
Scutiphora pedicellata, 476
Sehirus luctuosus, **414**
Sejanus albisignatus, **678**
Selenothrips rubrocinctus, 672
Serinetha
   amicta, **315**
   augur, **312**
Setora nitens, 761
Setothosea asigna, 761
Shieldbacked pine seedbug, 488
Silkworm, 751
Silverleaf whitefly, 669
Sinea, 555, 701
   confusa, 641
   diadema, 107
Sirthenea flavipes, 556

Sitophilus oryzae, 424
Slaterobius quadristriatus, 229
Slug caterpillar, 760
Small bean bug, 205
Small green stink bug, 443
Small pox bug, 410
Small rice stink bug, 439, 440
Small squash bug, 52
Soapberry bug, 312
Sogatella furcifera, 664
Solenopsis, 165
    geminata, 165
    invicta, 439
    saevissima, 147
Solenostedium, 476
Solenosthethium liligerum, 476
Solierella, 188
    peckhami, 188
Solubea, 227, 439
    poecila, 439
Sorghum chinch-bug, 211
Sorghum earhead bug, 45
Sorghum earhead midge, 620
Southern chinch bug, 162, 164, 165, 166
Southern false chinch bug, 182
Southern green bug, 509
Southern green stink bug, 325, 328, 421, 436, 768
Southern pine beetle, 622
Soybean looper, 648, 716, 717, 743
Spanagonicus albofasciatus, 49, 658
Spartocera, **382**
    batatas, 382
    brevicornis, 382
    dentiventris, **382**
    diffusa, **382**
    fusca, **382**
    lativentris, 382
Sphaerocoris
    annulus, 492
    testudogrisea, **492**
Sphaerodema, 585
    urinator, 578
Sphaerolecanium prunastri, 677
Spilonota ocellana, 670
Spilostethus, 143, 146, 177, 178, 180, 181, 216, 230
    crudelis, 178
    furcula, 216
    furculus, 178, **216**
    hospes, 179, 180, **215**, 216, 217
    lemniscatus, 178
    longulus, **216**
    macilentus, 177, 178, 182
    militaris, 178, 181
    pandurus, 144, 146, **177**, 178–182, 184, 215, 216, 217
    pandurus asiaticus, 177
    pandurus elegans, 178
    pandurus pandurus, 177
    pandurus tetricus, 177
    rivularis, 178, **216**
    taeniatus, 178
Spined citrus bug, 422, 426, 456

Spined soldier bug(s), 556, 738, 739, 748–751
Spodoptera, 752
    exigua, 564, 642, 673, 716, 740, 746, 750–752, 761, 766
    frugiperda, 743, 753
    littoralis, 615, 755
    litura, 668, 700, 760, 761
    praefica, 642
Squash bug, 343, 346
Stalifera, 357
Stem rice stink bug, 447
Stenocoris, **328**
    filiformis, 328
    tipuloides, 328
Stenodema calcaratum, 16, 18, 24
Stenotus binotatus, 16, 20, 23
Stephanitis, 86, 87, 98, 101, 102, 105, 121
    blatchleyi, 101
    chinensis, **114**
    globulifera, **102**
    laudata, **114**
    mitrata, 101
    olyrae, 101
    parana, 101
    pyri, 87, 97, **98**, 99, 620
    pyrioides, 86, 87, **98**, 99, 100, 101, 102, 643, 671
    rhododendri, 85, 86, 87, **100**, 101, 102, 104
    takeyai, 87, 100, 101, **102**, 555, 558, 671
    typica, 85, 86, **102**, 103, 104
    typicus, **102**
Stethoconus, 95, 104, 111, 113, 121, 671, 679
    distanti, 95
    cyrtopeltis, 98
    frappai, 108
    frappini, 107
    japonicus, 100, 102, 114, **671**
    praefectus, 104, 671
    pyri, 98
Stibaropus
    formosanus, **407**
    indonesicus, **408**
    molginus, **408**, **414**
    pseudominor, **414**
    tabulatus, **413**
Stictopleurus punctatonervosus, 14
Stigmatonotum
    rufipes, 227
    sparsum, 227
Stiretrus anchorago, 746, 747, 764
Strawberry bug, 228
Stricticimex, 519
    antennatus, 554
    parvus, **530**
Subterranean chinch bug, 410
Sugarcane delphacid, 665, 666
Sunn bug, 195
Sunn pest, 220
Supputius cincticeps, 753, 766
Swallow bug, 529
Sweetpotato whitefly, 641, 660, 669, 673
Sycamore lace bug, 87, 90, 619
Sycamore seed bug, 224
Sycanus leucomesus, 225

T

Taeniothrips nigricornis, 619
Tanybryrsa cumbere, 85
Tarnished plant bug(s), 47, 48, 49, 63, 555, 644, 647
Taro delphacid, 663
Tarophagus colocasiae, 663
Tasar silkworm, 761
Tea mosquito bug, 43
Tectocoris, 491
    diophthalmus, 287, **491**
    lineola, 17, **491**
    purpureus, 476
Telenomus, 181, 192, 431, 434, 444, 446, 483, 643
    anasae, 345
    aradi, 515
    ashmeadi, 434
    chlorops, 483
    edessae, 432, 453
    mormideae, 434, 440, 442, 444, 448, 452
    opacus, 715
    ovivorus, 188
    pachycoris, 490
    podisi, 423, 434, 441, 453, 454, 456
    reynoldsi, 715
    schrottkyi, 453
    sokolovi, 455, 483
    triptus, 444, 445
    truncatus, 455, 483
Teleonemia, 86, 121
    australis, **113**
    elata, **118**, 119
    harleyi, **119**
    lantanae, **119**, 720
    nigrina, **114**
    prolixa, **119**
    scrupulosa, **119**, 120
    vanduzeei, **119**
Temnostethus, 609
Tenebrio molitor, 743, 751, 752, 766
Tent caterpillar, 759
Termatophylidea, **671**, 672
    maculata, 671, 672
    maculosa, 672
    opaca, 671, 672
    pilosa, 671
Tessaratoma javanica, 277, 280, 508
Tetraphleps, 609, 611, **622**
    abdulghani, 622
    aterrimus, 622
    bicuspis, 622
    raoi, 611, 622
Tetyra, 488
    bipunctata, **488**, 489
Thaicoris, 265
Thaumastocoris, 142
    petilus, 139
Theognis, 357
    occidentalis, **364**
Theraptus, 370
Theridion lunatum, 91

Theseus, 476
Thlastocoris laetus, 385
Thrips palmi, 620, 623, 731
Thyanta
    custator accerra, **456**
    pallidovirens, **456**
    perditor, **446**
Tibraca limbativentris, **447**
Tigava sesoris, **116**
Tingis, 86, 98
    ampliata, **121**
    beesoni, **114**, 115
    buddleiae, **115**
    cydoniae, **91**
Tobacco aphids, 728
Tobacco budworm, 640, 641, 644, 673, 675, 678, 727, 743, 746
Tobacco hornworm, 727, 728
Togo, 227
    hemipterus, **227**
    praetor, 227
    victor, 227
Tomato bug, 50, 51
Tomato girdler, 50
Tomato looper, 740, 746, 749, 752
Tomato stilt bug, 728
Tomato stink bug, 425, 426
Tominotus communis, 412, **414**
Tortoise beetle, 761, 768
Trepobates inermis, 603
Trialeurodes vaporariorum, 659
Triatoma
    barberi, 541
    brasiliensis, 541, 546
    dimidiata, 541, 546
    infestans, 541, 546
    maculata, 541
    pseudomaculata, 541
    rubrovaria, 541, 546
    sanguisuga, 539
    sordida, 541, 546
Tribolium castaneum, 624
Trichogramma, 108, 624, 715
    pretiosum, 715
Trichoplusia ni, 564, 743, 751, 755, 764
Trichopoda, 274, 325, 370
    giacomellii, 434, 439, 446, 509
    pennipes, 342, 345, 347, 363, 366, 368, 376, 382, 439, 454
    plumipes, 368
Trichopodopsis pennipes, 426
Trigona, 701
Triphleps, 618
Trirhabda, 765
Trissolcus, 374, 378, 483
    basalis, 434, 439, 444, 446–449, 452, 483, 487
    biproruli, 429
    caridei, 432
    euschisti, 434, 453
    flaviscapus, 429
    grandis, 424, 425, 455, 483, 487

latisulcus, 429
mitsukurii, 429
murgantiae, 455
nigribasalis, 483
oenone, 429, 446
oeneus, 429
ogyges, 429
painei, 450
plautiae, 445
rufiventris, 424, 425, 448, 483, 487, 488
scuticarinatus, 434, 444, 446–448
semistriatus, 455, 483, 487
simoni, 483, 487
urichi, 432
vassilievi, 483, 487
waloffae, 448, 483
Tristigmus, 453
Tritomegas
    bicolor, **415**
    sexmaculatus, **415**
Troilus luridus, 763, 767
Tropical bed bug, 526, 527
Tropidosteptes, 45
Tropidothorax leucopterus, 181
True apple red bug, 59
Tupiocoris notatus, 667
Tur pod bug, 354
Tuta absoluta, 753
Two-spotted corsair, 555
Two-spotted stinkbug, 745, 753, 754, 757
Tylospilus acutissimus, 766
Tynacantha marginata, 444
Typhlocyba pomaria, 662
Tytthus, 665
    mundulus, 664, **665**, 666

U

Urentius
    aegyptiacus, **104**
    echinus, **104**
    euonymus, **115**
    hystricellus, 86, **104**
    sentis, **104**
Urochela
    distincta, 506
    luteovaria, 506
    quadrinotata, 506
Urostylis
    annulicornis, 506
    yangi, 506

V

Variegated cutworm, 749
Vatiga
    illudens, **115**, 116

    manihotae, **116**
    viscosana, **116**
Velvetbean caterpillar, 641, 644, 751, 764
Veneza, 357
    stigma, **368**
Vinitor, 222
Vitellus, **456**
    antemna, 456
    rufolineus, 456
    warreni, 456

W

Walnut scale, 672
Wax moth, 751
Weaver ant, 44, 371
Wesmaelia pendula, 464
Western big-eyed bug, 717
Western boxelder bug, 311
Western chinch bug, 210
Western flower thrips, 619–621, 659–661, 672, 676
Western tarnished plant bug, 46, 47
Western yellow-striped armyworm, 642
Wheat bug, 22, 23, 193
Wheelbug, 363, 555
White apple leafhopper, 662
Whitebacked planthopper, 664
Winter cherry bug, 373, 374
Wollastoniella, 610, **623**
    parvicuneus, 623
    rotunda, 623
    testudo, 623
Wood rose bug, 217

X

Xenoencyrtus hemipterus, 375
Xenopyxis edessae, 432, 453
Xylastodorus, 139, 140
Xylocoris, 110, 609–611, 616, **623**, 625
    cursitans, 623
    flavipes, 616, **623**, 625
    galactinus, 554, 608, 616, 617, 623, **624**
    hiurai, 616
    luteolus, **140**
    nigromarginatus, 554
    sordidus, 616, 623, **624**
    vicarius, 165
Xyonysius, 182, 223
    californicus, **223**
    major, **223**

Y

Yemma exilis, **731**, 732
Yponomeuta malinella, 751

Z

Zelus, 555, 701
  longipes, 89
  nugax, 93, 116
  renardii, 567, 643, 701
Zicrona caerulea, 766, 767
Zulubius, 329
  acaciaphagus, **329**
  maculatus, 329

Plant Index

A

Abelmoschus, 179, 197
   esculentus, 275, 279, 284, 292, 700
   moschatus, 198
Abies, 229, 611, 612, 622
   fraseri, 362
   sibirica, 622
Abutilon, 105, 115, 198, 278, 314
   cabrae, 198
   indicum, 115, 198, 199, 202
   mauritianum, 198, 282, 287
   theophrasti, 313
Acacia, 142, 379, 415, 490, 493
   auriculiformis, 492
   cyclops, 329, 493
   iteaphylla, 20
   koa, 490
   longifolia, 613
   riparia, 119
   saligna, 493
   villosa, 341
Acantholimon, 481
Acanthophyllum, 481
Acanthospermum hispidum, 433, 437
Acer, 224, 411, 670
   macrophyllum, 312
   negundo, 310, 311
   saccharinum, 672
Acerola, 376
Achillea, 411
Acorn, 344
Actinocheita filicina, 369
Adansonia digitata, 281, 282, 284, 287
Adhatoda vasica, 215, 217
Aegle marmelos, 349
Aeluropus villosus, 212
Aerva, 221
   tomentosa, 221
African oil palm, 435
Agaretum conyzoides, 189
Ageratum
   conyzoides, 221
   houstonianum, 168
Agonis flexuosa, 140, 142

Agropyron, 60
   cristatum, 61, 209
   repens, 149
Agrostis, 157, 448, 480, 487, 726
   canina, 157
   hyemalis, 150
   palustris, 157
   stolonifera, 157
   tenuis, 157
Akh, 322, 492, 505
Akk, 179, 215
Albizia, 341, 378
   chinensis, 378
   lebbeck, 378
Albuca setosa, 216
Alcea, 198
Alder, 57, 199, 224, 662, 663, 667, 677
Alder tree, 224
Alfalfa, 21, 46, 47, 53, 54, 58, 59, 179, 184, 219, 221, 407, 431, 434, 442–444, 452, 453, 456, 619, 620, 640, 642–648, 670, 716, 719, 731
Allium sativum, 280
Allspice, 42
Almond(s), 97, 180, 312, 360, 361
Alnus, 57, 97, 199, 224, 662, 663
   longiceps, 224
   nitida, 207
   nitidula, 221
Alopecurus carolinianus, 150
Alpinia, 103, 213
Althara, 109
Althea
   officinalis, 198
   rosea, 198, 227
Amaranth, 48, 357
Amaranthus, 222, 267, 384
   bengalensis, 221
   cruentus, 48
   retroflexus, 48, 222
   spinosus, 222
   viridis, 221, 357
Ambrosia, 94, 289
   dumosa, 94
American beach grass, 149
American plane tree, 223
Ammophila breviligulata, 149
Amsinckia intermedia, 493

Amygdalus, 97, 98
Anacardium occidentale, 672
Anagallis arvensis, 194
Anarcadium occidentale, 39
Andromeda, 102
Andropogon, 155
    gerardii, 151
    scoparium, 149, 152
    virginicus, 151
Annatto, 341, 369
Annoma, 92
    diversifolia, 92
    muricata, 103
Annual bluegrass, 157
Annual ryegrass, 157
Anona tree, 92
Anthocephalus cadamba, 277
Antirrhinum majus, 114
Aperula, 102
Apple(s), 39, 42, 48, 54, 56, 57, 60–62, 98, 174, 190, 203, 291, 422, 453, 609, 611, 613–615, 619–621, 642, 662, 663, 669, 670, 674–679, 718, 730, 748
Apple-of-Peru, 378
Apricot, 174, 179, 190, 200, 203, 411
Arachis
    hypogaea, 207, 279, 412, 670, 719
    hypogea, 179
Arahar, 179
Aramina fiber, 284, 285
Aramina plant, 215
Aramina, 198
Arbor vitae, 223, 360
Arctostaphylos pungens, 360
Areca, 450
Arecanut, 454
Argemone mexicana, 108
Aristolochia bracteata, 280
Artemisia, 108, 219, 220, 481, 668
    absintium, 218
    annua, 219, 280
    campestris, 411
    heba-alba, 484
    maritima, 218, 220
Artichoke, 56, 430
Artocarpus integrifolia, 103
Artotheca calendula, 189
Arundinaria
    cobinii, 167, 215
    simoni, 213
Asclepias, 215, 218
    syriaca, 181
Ash, 311
Ash-leaf maple, 310
Asparagus, 53, 221, 346
Asparagus bean, 206
Asparagus officinalis, 53
Aster, 94
    ericoides, 186
Asters, 93, 94
Astragalus, 108, 423, 481, 484

Athaea, 89
    officinalis, 89
Atriplex, 266, 267
Atriplexes, 184
Aubergine, 104, 109, 180, 215
Australian bottle brush, 280
Avena fatua, 174
Averrhoa carambola, 368
Avocado(s), 43, 114, 203, 340, 341, 369–371, 377, 621
Axonopus compressus, 164, 211
Azadirachta indica, 42, 89, 280
Azalea(s), 86, 87, 98, 99, 100, 384, 555, 671

B

Baccharis, 178, 229
    halimifolia, 185
    neglecta, 185
    pilularis, 94
    sarothroides, 185
Bactris
    gasipaes, 226, 357
Bahiagrass, 163
Bajra, 198, 409
Bamboo, 167, 213–215, 225, 226, 230, 349, 379, 384
Bambusa, 214
    oldhamii, 379
Banana, 103, 340, 406, 407, 413
Banksia, 140, 142
Baobab, 281, 282, 284, 287
Barbados cherry, 376
Barleria, 110
Barleria cristata, 110
Barley, 53, 145, 151, 153, 156, 174, 209, 210, 423, 425, 440–442, 448, 477–479, 481–483, 485–487, 493
Barnyard grass, 96, 149
Basswood, 110
Bauhinia, 93, 106
    bakeana, 106
    purpurea, 106
    variagata, 106
Beach grasses, 150
Bean(s), 54, 56, 109, 189, 216, 326, 351, 353, 354, 359, 369, 376, 378, 379, 381, 406, 430, 442, 444, 445, 452, 456, 622, 647, 659, 718, 719
Beet(s), 186, 192, 194, 195, 220, 222, 414
Beetroots, 189
Beggar weed, 437
Bellis perennis, 108
Benniseed, 208
Bergvygie, 315
Bermuda grass, 149, 150, 163, 211
Beta, 266
    vulgaris, 47, 114, 266, 267
Betel nut palm, 450
Betula, 224, 516, 662
Bhindi, 179, 180, 199, 279, 286, 290, 292, 293
Bidens pilosa, 446
Big bluestem, 151

Bilberries, 414
Birch, 224, 662
Birdsfoot trefoil, 53
Bitter gourd, 359, 729
Bixa orellana, 369
Black beans, 406, 407, 413
Blackberries, 224
Black-eyed pea, 205
Black gram, 58, 179, 206
Black pepper, 42, 106
Blady grass, 190
Blue grama, 186
Bluegrass, 149, 150
Blueweed, 106
Bo tree, 208
Bolboschoenus, 105
Bombax
   buonopozense, 284
   ceiba, 292
   sessilis, 288
Bonavist, 206
Bottle gourd, 729
Boxelder tree(s), 310, 311
Bracatinga, 505
Brachiaria, 148, 323
   brisantha, 407
   decumbens, 407
   humidicola, 407
   iowensis, 149
Brachychiton, 198
Brassica(s), 196, 456
Brassica, 48, 88, 437
   juncea, 183, 450
   napa, 183, 195
   napobrassica, 195
   napus, 58, 195
   oleracea, 58, 195
Brazilian guava (araçá), 363, 368
Breynia oblongifolia, 492
Brinjal, 104, 105, 109, 179, 180, 349, 444
Broad bean, 179, 181
Broccoli, 184, 189, 454, 455
Bromous, 480, 487
Bromus, 448
   catharticus, 174
   diandrus, 174
Broom, 455, 610, 612, 613
Broom corn, 153
Broomsedge bluestem, 151
Brown knapweed, 220
Brussels sprout(s), 189, 454
Bryonopsis laciniosa, 358
Buchloë dactyloides, 150
Buckthorn, 422
Bucus, 610
Buddle
   asiatica, 115
   madagascarensis, 115
Buffalo gourd, 345
Buffalo grass, 150, 186
Butia, 140

frondosa, 507
Butterfly bush, 115
Butternut squash, 344
Buxus, 220

C

Cabbage, 27, 58, 184, 186, 189, 219, 359, 447, 454, 745, 749
Cacao, 22, 110, 339–342, 349, 357, 359, 370, 372, 377, 378, 384, 385, 492, 726, 731
Cajanus, 728
   cajan, 115, 179, 198, 279, 286, 287, 322
   indicus, 507
Calamagrostis epigeios, 211
Calandrinia caulescens, 194
Callistemon, 383
   lanceolatus, 280
Calluna, 220, 411
Calophyllum, 277
   inophyllum, 280
Calopogonium mucunoides, 206
Calotropis, 178, 180, 181, 182
   gigantea, 179, 181, 215, 217, 218
   procera, 179, 215, 217, 322, 492, 505
Camellia sinensis, 39
Camphor tree, 44, 114
Canada thistle, 116, 121
Canavalia, 110, 376
   ensiformis, 217, 348
Candillo, 218
Caneberry, 620
Canna, 382
Canna, 213
   edulis, 213
   flaccida, 213
   glauca, 213
   indica, 212
Canola, 48, 183, 184, 189
Cantaloupe, 52, 343, 346
Canthium, 217
Cape gooseberry, 215, 359, 374
Cape weed, 189
Capsella bursa-pastoris, 186, 194
Capsicum, 406
   annuum, 48, 374, 382
   frutescens, 374
Capsicums, 189
Carambola, 368
Cardamom, 103
Carduus 121
   lanceolatus, 116
Carex, 105
Carica papaya, 189, 192
Carolina foxtail, 150
Carpet grass, 164, 211
Carpetweed, 187
Carrot(s), 39, 53, 54, 189, 415

Carthamus, 108
   tinctorius, 47, 88, 121, 190
Carus, 105
Carya illinoensis, 669
Cashew, 39–43, 51, 340, 341, 368–373, 379, 380, 672
Cassava, 93, 115, 338–342, 348, 359, 373, 377, 378, 410, 411, 731
Cassia, 109
   occidentalis, 349
   siamea, 280
Cassinia leptophyla, 194
Castanea, 98, 506
Castor, 55, 700
Castor bean, 93, 208, 384, 437
Castor oil bean, 505
Catch fly, 194
Cattails, 224
Cattleya, 113
   schroederae, 113
Cauliflower, 189, 219, 450, 454
Cayenne pepper, 374
Ceanothus, 112
Ceiba pentandra, 40, 198, 282, 284, 286–288
Celosia argentea, 221
Cenchrus, 323
Centaurea, 199
   jacea, 220
   sibirica, 218
   splendens, 199
Centipede grass, 163
Cestrum nocturnum, 374
Chaenomeles, 98
Chamaecyparis lawsoniana, 223
Chamaesyce serpylliforia, 187
Chamomile, 220, 643
Chaparral broom, 94
Chayote, 52, 368
Chenopodium, 222, 266, 267
   album, 194
   quinoa, 186
Cherelles, 378
Cherry(ies), 189, 200, 311, 430, 445, 492, 748
Cherry tomatoes, 367
Chickpea, 327
Chickweed, 194
Chili, 444
Chilopsis linearis, 370
Chinese artichokes, 179
Chinese cabbage, 21, 222
Chinese date, 113
Chinese long bean, 349
Chinese potato, 88
Chinese soapnut, 179
Chinese tallow, 369
Chionanthus virginicus, 112
Chorista speciosa, 288
Christophine, 52
Chrozophora rottleri, 115
Chrysanthemum, 53, 93, 94, 108, 196, 197, 667
Chrysanthemum, 53, 108
   morifolium, 196, 667

Chrysothamnus nauseosus, 555
Chufa, 414
Cicer arietinum, 179
Cinchona, 43, 119
Cinnamomum, 102, 114
   camphora, 44, 103
Circaea quadrisulcata, 730
Cirsium, 121
   arvense, 116
   segetum, 199
Citrillus lanatus, 344
Citron, 364
Citron melons, 363
Citrus, 179, 189, 192, 219, 286, 290, 358, 359, 363, 364, 367, 369, 370, 372, 376, 378, 379, 422, 427–429, 449, 456, 508, 672
Citrus, 92, 118, 288, 292, 377, 450, 454
   aurantium, 118, 348
   australasica, 428
   paradisi, 348, 359
   reticulata, 359
   sinensis, 359
Clidemia hirta, 618
Clockflower, 217
Clover, 58, 59, 184, 219, 414, 436, 643
Cnicus, 116
Cocoa, 39–45, 51, 52, 113, 288, 326, 339, 341, 342, 373, 377, 378, 422, 450, 451, 506, 672, 731, 732
Coconut(s), 86, 104, 119, 338–342, 370–373, 379, 380, 422, 435, 768
Coconut palm, 103, 450
Cocos nucifera, 86, 103, 435, 450
Coeus parviflorus, 88
Coffea
   arabica, 449
   robusta, 449
Coffee, 95, 107, 110, 208, 313, 315, 359, 377, 406, 449, 610, 753, 768
Cola, 45, 198
   acuminata, 450
   nitida, 198
Coleus, 88
Collards, 454
Colocasia esculenta, 384
Colonial bent grass, 157
Colza, 183
Common bean(s), 322, 324, 325, 328, 329, 437, 443, 447
Common mullein, 61
Common nightshade, 209, 215
Common reed, 211
Common spruce, 229
Congo jute, 95, 96
Conifer(s), 229, 364, 611, 612, 622
Coniferous, 616
Conyza canadensis, 48
Corchorus
   capsularis, 217, 507
   olitorius, 217
Cordia, 105, 107
   alliodora, 107

Corn, 54, 96, 118, 145, 150, 151, 152, 154–159, 163, 164, 210, 313, 363, 364, 369, 370, 414, 434, 436, 438, 440, 441, 452, 453, 456, 608, 615, 617, 619, 620, 640, 642, 645, 647, 665, 666
Corn sperry, 189
Coronopus didymus, 194
Corylus, 555
  avellana, 179, 667
Corynephorus canescens, 411
Cosmos, 221
Cotoneaster, 91
Cotoneaster, 92, 98
  dammeri, 91, 92
  lactea, 92
Cotton, 21, 23, 25, 39, 43, 46–50, 54, 61, 104, 107, 110, 179, 180, 184, 186–190, 197–205, 215–218, 223, 226, 271–279, 281–294, 310, 312–315, 324, 347, 348, 359, 369, 376, 378, 381, 382, 384 385, 406–408, 412, 422, 431, 434, 440, 442, 443, 448–450, 453, 455, 490, 491, 610, 611, 618 619, 621, 622, 640–642, 645, 647, 669, 670, 673, 676, 678, 679, 700, 701, 714–717, 719, 731, 749, 753, 768
Cotton rose, 198–200
Cotyledon mollissima, 216
Country beans, 179
Country gooseberry, 491
Country mallow, 198, 199
Cowpea(s), 103, 205, 221, 275, 326–328, 348, 349, 351–355, 358, 359, 366, 367, 376, 379, 381, 384 385, 386, 450
Coyote brush, 94
Crab apple, 60, 95, 96
Crabgrass, 149, 157
Crassula, 189
Crataegus, 91, 97, 98, 667
  phaeopyrum, 92
Creeping bent grass, 157
Creeping red fescue, 157
Creeping thistle, 121
Crested wheat grass, 61, 209
Crookneck squash, 344
Crotalaria, 62, 324, 411, 437, 443
  juncea, 62, 217
Croton, 50
  discolor, 491
  linearis, 491
Crucifers, 21, 432, 454
Cubiu, 89
Cucumber(s), 52, 344–346, 359, 366, 369, 454, 614, 659, 660
Cucumis, 52
  cantalupensis, 52
  melo, 52, 359
  sativus, 52, 344, 359
Cucurbita, 52, 344
  ficifolia, 347
  foetidissima, 345
  maxima, 344, 346, 374, 381, 729
  melo, 344
  moschata, 343, 344, 346

  pepo, 344, 346, 347, 368, 381, 668
Cudweed, 185, 189, 190
Cupressus, 223
Curcuma longa, 103
Curnow's curse, 194
Currant, 56, 57
Custard apple, 340
Cyanchium vincetoxicum, 218
Cyanopsis tetragonoloba, 179
Cydonia, 97, 98
  oblonga, 91
Cynara scolymus, 56, 121
Cynodon, 148
  dactylon, 149, 163, 211
  plectostachys, 211
Cynoglossum, 106, 107
Cyperus esculentus, 414
Cyphomandra crassicaulis, 359
Cypress, 223

D

Dactylis, 448, 480, 487
  glomerata, 149
Dadap, 349
Dahlias, 385
Daisy, 108
Dalbergia sissoo, 377
Dandelion, 714
Dasylirion wheeleri, 360
Dates, 203
Datura, 178, 728
  mentel, 178
  stramonium, 178
Daucus carota, 39
Deccan hemp, 199, 278
Delonix regia, 284
Dennstaedtia punctilobula, 731
Descurainia sophia, 183
Desert lime, 426, 428
Desert willow, 370
Desmodium
  maurtianum, 206
  polycarpum, 280
  tortuosum, 437
Detanve, 218
Diervilsa japonica var. *sinica*, 506
Digitalis
  amandiana, 218
  chinensis, 218
Digitaria, 323
  decumbens, 148, 163
  sanguinalis, 149, 157
Dilatris viscosa, 175
Dinochloa utilis, 225
Diplanchne fusca, 212
Dolichos, 206
  biflorus, 206, 323
  lablab, 206, 353, 505, 507

Dombeya, 198, 199, 728, 729
  burgessiae, 199
  mastersii, 199
Douglas fir, 338, 364–366
Drosanthemum, 315
Dryanda, 140, 142
Durra, 211
Durra millet, 179
Dwarf bamboo, 213

E

Eastern white pine, 362, 366
Echinochloa, 323
  colonum, 148
  crusgalli, 96, 149, 323
Echinops, 94, 108
  sphaerocephalus, 218
Echium, 106, 107, 117
  plantagineum, 107, 117
  vulgare, 106, 107
Eggplant(s), 86, 88, 89, 104, 109, 180, 215, 279, 348, 349, 369, 374, 381, 384, 425, 431, 452, 505, 620, 623, 732, 749, 750
Egyptian clover, 179
Ehrharta
  erecta, 174
  longiflora, 174
  near *calycina*, 174
Elaegnus latifolia, 509
Elaeis guineensis, 93, 104, 111, 435
Elaeocoarpus, 142
Elettaria, 213
  cardamomum, 103, 203, 213
Eleusine
  coracana, 179, 182
  indica, 149, 210
Elm, 87, 224, 506, 675, 766
Emilia sonchifolia, 215, 222
Enchanter's nightshade, 730
Eragrostis, 158, 188
Eremochla ophiuroides, 163
Eremocitrus glauca, 426–428
Erigeron, 191, 220
  annus, 48, 191
  canadensis, 222
  linifolius, 196
Eriogonum, 727
Erythrina subumbrans, 349
Eucalypts, 359, 374, 377
Eucalyptus, 142, 174, 190, 338, 342, 374, 375, 406, 753, 766
  amaldulensis, 219
  deglupta, 342
  globulus, 141, 142
  occidentalis, 219
  pulchella, 141
  saligna, 525
  trachyphloia, 141
  urophylla, 750
  viminalis, 141
Eugenia
  jambolana, 179
  stipitata, 385
  uniflora, 292, 377
Euphorbia, 109, 118
  commutata, 730
  esula, 118
  geniculata, 323
  heterophylla, 433
  hirta, 215, 222
  pilulifera, 221
  pinea, 338
European filbert, 667, 675
Euterpe, 140
Evergreen azalea, 102

F

Fagus grandifolia, 613
False redtop, 151
Fathen, 194
Festuca, 157, 158, 448, 480, 487
  arundinacea, 157
  novae-zelandiae, 194
  rubra, 157
Ficus, 179, 224, 610, 617
  exasperata, 218
  religiosa, 208
Fig(s), 200, 203, 348, 357
Fig tree, 174
Finger lime, 428
Finger millet, 179
Fir, 229
Firethorn, 92
Firewheel, 188
Flame of the forest, 284
Flat-topped yate, 219
Flax, 53
Flaxseed, 189, 191
Fleabane, 48, 191, 196
Flixweed, 183, 184
Flossflower, 189
Flower of an hour, 198
Fragaria, 228
  ananassa, 39
  chisoensis, 196
Fraser fir, 362
Fraxinus
  excelsior, 621
  oregona, 112
  velutina, 112
French bean, 205–207
Fringetree, 112
Furze, 117

G

Gaillardia, 108
Gaillardia, 108
   pulchella, 188
Gambir, 761
Garbanzo, 179
Gardenia, 384
Gardenia, 106
   augusta, 217
   florida, 217
Garlic, 221
Gaura biennis, 727
Geranium, 667, 729
Geranium, 729
Gerbera, 668
Gerbera jamesonii, 668
Giló, 89
Gingelly, 179, 181, 182, 221
Ginger, 103
Gliricidia, 507
   sepium, 378
Globe artichoke, 121
Globemallows, 198
Gloriosa superba, 215, 217
Glory lily, 215, 217
Glycine, 414
   max, 109, 205, 313, 670, 731
Gmelina arborea, 114, 115, 761
Gnaphalium purpureum, 185, 189
Gogu, 199
Golden gram, 205
Goldenrain, 310, 312
Goldenrod, 93, 94
Gooseberry, 56, 57
Goosegrass, 149, 150, 210
Gorse, 117, 455
Gossypium, 109, 110, 179, 190, 197, 200, 215, 216, 313
   herbuceum, 292
   hirsutum, 39, 104, 218, 282, 287, 313, 700
   sturtianum, 200
Gourd(s), 52, 347, 726, 730
Grain sorghum, 47, 153, 189, 196, 220, 367, 369, 441
Gram, 179, 180, 353
Grape(s), 42, 54, 179, 188, 219, 311, 357, 379, 382, 415, 431, 445, 455, 670, 675
Grapefruit(s), 348, 357, 363, 364, 368, 427, 428
Grapevine(s), 184, 186, 187, 189, 192, 219, 220, 268, 349, 414, 610, 766
Green bean(s), 360, 367, 369, 422, 425, 621, 716
Green bristle grass, 149
Green gram, 180, 206, 284, 377, 379
Green pepper, 48
Grewia
   asiatica, 199
   subinaequalis, 198
Groundnut(s), 207, 208, 279, 349, 355, 359, 406
Guaco, 384
Guava, 39, 41–44, 55, 179, 290, 322, 339, 340, 348, 349, 358, 359, 363, 367–373, 376, 378–380, 440, 443, 445, 492

Guayule, 94, 186, 223
Guinea grass, 96
Gutel, 292

H

Hairy beardtongue, 731
Haldi, 103
Halimione, 266
Hardwood tree, 114, 670
Hawkweed, 194
Hawthorn, 91, 614, 667, 676, 677
Hay, 640
Hay-scented fern, 731
Hazel, 610
Hazelnuts, 179
Heather, 220
Hedychium, 103
Helianthus, 54, 94, 108
   annuus, 94, 108, 189, 190, 367, 670
Heliconia, 749
Hemarthria, 148
Hemp, 198, 199
Herba absinthi, 220
Heteranthelium piliferum, 480
Hevea brasiliensis, 111, 329, 377
Hibiscus, 491
Hibiscus, 93, 109, 110, 197, 198, 216, 226, 284, 292, 491, 728
   abelmoschus, 198, 199, 203
   cannabinus, 198, 275, 278, 284
   esculentus, 179, 197, 216, 286, 313
   fuscus, 198
   gossypinus, 198
   lancibracteatus, 198
   meeusei, 198
   micranthus, 288
   mutabilis, 198, 200
   rosa-sinensis, 198, 199, 384, 491
   sabdariffa, 179, 198, 284, 290
   surattensis, 198
   syriacus, 313
   tiliaceus, 198
   trionum, 198
Hieracium, 194
   pilosella, 117
   pratense, 731
Hollyhock, 115, 198, 199, 227, 279
Holoschoenus vulgaris, 105
Hops, 643
Hordeum
   murinum, 174, 424
   vulgare, 53
Horehound, 449
Horse gram, 206
Horse nettle, 109
Horseweed, 48
Hung Pin Tsoi, 215
Hyacinth bean, 351, 353, 355

Hydnocarpus
 anthelmintaca, 291
 wightiana, 291
Hyparrhenia involucrata, 176
Hyptis, 267

I

Impatiens, 56
Imperata arundinacea, 190
Indian corn, 153
Indian grass, 149
Indian long pepper, 44
Indian mallow, 199
Indian mustard, 183, 450
Indigo, 113
Indigofera, 113, 350, 443
 hirsuta, 443
 truxillensis, 443
Indocalamus migoi, 215
Intermediate wheat grass, 61
Inula viscosa, 660
Ipomoea, 110, 217, 374
 batatas, 52, 179, 217, 218, 382, 507
 caraica, 217
 purpurea, 217
 tuberosa, 217
Italian ryegrass, 157, 196

J

Jack bean, 217
Jack pine, 229, 362
Jamaica dogwood, 93
Jamon, 179, 181
Japanese andromeda, 102, 671
Japanese millet, 148
Japanese red pine, 362
Jasmine, 89, 113
Jasminum, 89, 113
 pubescens, 89
 sambac, 89
Jatropha
 curcas, 292, 314, 329, 347, 369, 385, 448, 489, 492
 gossypiifolia, 490
 podagrica, 490
Java plum, 179
Jimsonweed, 178
Job's tears, 170
Johnsongrass, 96
Jojoba, 219, 220
Jowar, 179
Juglans, 97, 98
Jujuba, 113
Juncus, 105
 actus, 105
 maritimus, 105
 subnodulosus, 105
Juniper, 223, 360, 616, 674
Juniperus, 223, 360, 674
 communis, 223
 excelsa, 223
Justicia orchiodes, 315
Jute, 217, 285

K

Kachnar, 106
Kadghi, 199
Kale, 184, 454
Kalmia
 latifolia, 99, 100
Kapok, 198, 199, 286, 288, 342
Karanj, 506
Kasturi blendi, 198
Kenaf, 198–200, 203–205, 220, 284, 288, 643
Kentucky bluegrass, 150, 157, 160, 161, 162, 168
Khata mitha, 491
Khejri, 385
Kidney, 205
Kidney bean(s), 325, 454, 507
Koa, 490
Kola, 450
Kosh trees, 207
Koshta, 217
Kudzu, 206, 731
Kudzu vine, 378
Kulthi, 323
Kurrijong, 198
Kusum tree, 312

L

Lablab, 206, 324, 505
Lac tree, 312
Lagenaria, 728, 730
 leucantha, 347
 siceraria, 729
Lalung, 190
Languas galeata, 103
Lantana, 119, 120
Lantana, 89, 118, 119, 312
 camara, 116, 118–120, 277, 720
 montevidensis, 119
Larch, 622
Larix, 514, 609, 611, 612
 decidua, 117
Lathyrus nuttallii, 116
Launea, 108
Laurel, 107
Lawn bent grasses, 157
Lawn grass, 149
Lawnson, 223
Lawson cypress, 223

Leafy spurge, 118
Leeks, 221
Lemon(s), 426–429
Lentils, 23, 444
Leonurus sibiricus, 437
Lepidium
    nitidium, 186
    sativum, 184
Lettuce, 118, 184, 186, 221, 716
Leucaena, 454
Leucaena leucocephala, 454
Liendre puerco, 148
Ligustrum, 98
    japonicum, 437
    lucidum, 437, 454
Lima bean, 48, 92, 327, 349, 505, 621
Lime(s), 349, 369, 378, 663
Linden, 226, 663
Lindera, 102
Linen flax, 194
Linseed, 189, 367, 444
Linum usitatissimum, 53
Litchi chinensis, 179
Little bluestem, 151, 152, 155, 158
Lobia, 180
Loblolly pine, 362, 489
Lodgepole pine, 365, 613, 622
Lolichi, 179
Lolium, 448, 480, 487
    multiflorum, 157, 174, 196
    perenne, 157, 161, 174
Lonchocarpus cyanescens, 350
London plane tree, 223
London rocket, 186, 188
Longleaf, 362
Loofah, 347, 359
Lopholaena coriifolia, 216
Loquat, 219
Lotus, 53
Lotus
    corniculatus, 53
    pedunculatus, 53
Lucerne, 24, 53, 179, 180, 189, 194, 195, 219, 406, 646
Ludwigia adscendens, 767
Luffa, 180, 358, 359
    acutangula, 179, 349, 359
    aegyptiaca, 347, 358, 359, 363
    cylindrica, 358, 363, 729
Lupin, 324, 436
Lupine, 411
Lupinus
    albus, 455
    angustifolium, 455
    arboreus, 292
Lychee(s), 340, 341, 369
Lycopersicon, 728
    esculentum, 40, 215, 381, 659, 726
Lycopersicum esculentum, 179
Lynoia, 102
Lyonia neziki, 99
Lysimachia vulgares, 218

M

Macadamia, 329, 339–341, 370, 438
Macassor oil tree, 312
Machaerium aculeatum, 731
Madar, 179, 215, 217, 218, 322, 492, 505
Magnolia, 357
Maize, 145, 148, 150, 153, 155, 168, 170, 174, 210, 225, 347, 349, 351, 354, 367, 369, 381, 406, 407, 410, 411, 615, 618, 620, 622
Malabar nut tree, 215, 217
Mallotus, 273
Malpighia puncifolia, 376
Malus, 60, 97, 98, 506
    domestica, 39, 662
    sylvestris, 95, 96
Malva, 198, 199
    rotundifolia, 198–200
    sylvestris, 226
Malvascus, 199
Malvastrum, 199
Mandarin(s), 426–428
Mandarin orange, 359
Mangifera, 92
    indica, 39, 179, 350, 374, 450, 760
Mango(es), 39, 43, 179, 322, 340, 348, 349, 359, 363, 368, 370, 371, 372, 374, 378, 385, 450, 760
Manihot, 109
    dulcis, 115
    esculenta, 93, 115, 350, 377
    tripartita, 731
    utilissima, 115
Manzanita, 360
Maple(s), 224, 311, 670
Marigold, 108
Marrow(s), 359, 363
Marrubium vulgare, 449
Marshmallow, 89, 198
Matricaria, 227
    chamomilla, 220
Meadow grass, 157
Medicago, 455
    sativa, 39, 179, 194, 219, 670, 731
Melaleuca, 142, 383
    quinquenervia, 339, 383
Melilotus, 455
Melochia corchorifolia, 199
Melon(s), 346, 366, 369, 381
Mentha, 88
Mesquite, 383, 385, 766
Mexican sycamore, 224
Microcitrus australasica, 508
Mikania micrantha, 119
Milkweeds, 182, 218
Millet, 148, 153, 167, 211, 359, 407
Millet ragi, 182
Millo, 148
Mimosa, 376
Mimosa
    pigra, 379
    scrabella, 505

Mimusops elengi, 280
Mint, 88
Miscanthus, 168
    sinensis, 168, 226
Mollugo, 221
    verticillata, 187
Momordica charantia, 359, 363, 729
Monanthia, 106
Monarda, 50
Moong, 206
Morinda, 107
    tinctoria, 107
Morning glory, 217, 224, 374
Morus, 225
    alba, 374
    bombycis, 225
Moss phlox, 731
Moth mullein, 731
Mothbean, 206
Mucuna urens, 218
Mulberry(ies), 146, 374, 445, 455
Mullein, 62, 109, 727, 730
Mung, 205
Mungbean, 326, 328, 349, 355
Musenosa, 107
Musenoso, 107
Mushkdana, 203
Muskmallow, 199
Muskmelon, 344, 345
Mustard(s), 184, 186, 437, 454, 455
Mysore cardamon plant, 213

N

Nassella tussock grass, 194
Native kumquat, 428
Nectarines, 357
Neem, 42, 439
Nepeta hindostana, 278
Nephelium litchi, 179
Nerium, 218
    indicum, 218
    oleander, 178, 179, 218
Nicandra physalodes, 378
Nicotiana, 105, 179, 729
    tabacum, 40, 216, 382, 667, 727
Nut(s), 180, 456, 659
Nutmeg, 322

O

Oak(s), 93, 506, 667, 670, 677, 679, 749
Oat(s), 17, 145, 153, 174, 210, 425, 436, 440–442
Ocimum, 88
    basilicum, 88
    kilmandscharicum, 88
    sanctum, 88

Oenothera, 50
Oil palm(s), 93, 104, 111, 119, 359, 435, 761, 768
Oilseed rape, 48
Okra, 179, 180, 197–200, 202, 203, 205, 216, 278, 279, 284, 286, 288, 292, 313, 425, 436, 447
Olea
    cuspidata, 108
    europaea, 108, 113
    ferruginea, 108
Oleander, 218
Oligosita itoi, 212
Olive(s), 113, 227, 414, 454
Onehunga weed, 194
Onion(s), 221, 410
Ononis, 729
Onosma, 107
Operculina, 217
Opismenus oppositus, 105
Opuntia, 339, 375, 383
    megacantha, 383
    polyacantha, 376
Orange(s), 227, 348, 358, 359, 363, 364, 368–370, 382, 384, 427, 428, 456
Orange hawkweed, 731
Orchard grass, 149
Oryza, 113
    sativa, 163, 210, 227, 507
Ouratea, 105
Ox-tongue, 186

P

Paederia foetida, 107
Palm, 140, 142, 371, 414, 434
Pangola grass, 148, 163
Panicum, 113, 210, 323, 726, 730
    bartowense, 164
    maximum, 96, 163
    mileaceum, 167, 211
    purpurascens, 148
    purpureum, 210
    repens, 163
    virgatum, 149
Papaya, 105, 192, 339, 340, 342, 431
Para grass, 148, 210
Parsnip, 54
Parthenium, 110
    argentatum, 94, 186, 223
Paspalum
    notatum, 163
    urvillei, 441
Passiflora, 88, 109, 385, 729
    edulis, 54, 218, 359
    edulis var. *flavicarpa*, 361
    foetida, 359
    quadrangularis, 359, 361, 385
Passionflower, 109, 349
Passion fruit, 54, 218, 341, 359, 361, 363, 370, 381
Pastinaca sativa, 54

PLANT INDEX

Pat, 217
Paullinia pinnata, 350
Paulownia, 731
Paulownia, 729, 731
   tomentosa, 731
Pavonia, 198
   hastata, 198
Pawpaw, 189, 340, 379
Pea(s), 206, 354, 359, 381, 406, 431, 443, 445, 447, 454
Peach(es), 39, 48, 53, 54, 174, 179, 190, 200, 203, 219, 220, 291, 292, 311, 341, 357, 359, 366, 367, 422, 445, 456, 509, 613, 718
Peach palm, 226
Peanut(s), 146, 179, 180, 207–209, 279, 367, 407, 410, 412, 414, 616, 642, 670, 700, 716, 719
Pear(s), 39, 56, 58, 61, 98, 174, 190, 219, 311, 422, 453, 609, 611, 613, 614, 668, 674, 677–679
Pearl millet, 153, 156, 179, 182, 198, 219
Pecan(s), 338, 341, 366, 367, 434, 438, 439, 647, 669
Peepal, 208
Pelargonium, 729, 730
   hortorum, 667, 729
Pennisetum
   americanum, 179
   glaucum, 179, 198
   thypoides, 182, 219, 284, 409
   typhoideum, 179
Penstemon hirsutus, 731
Pentaschistus thunbergii, 174
Pepper, 106, 369, 376, 377, 384, 412, 425
Pepper-grass, 184, 186
Perennial rye grasses, 157, 161, 162
Pergularia extensa, 218
Persddrolpeer, 199
Persea americana, 43, 114, 377
Persimmon(s), 203, 445
Pestalozzia, 111
Phaeomeria, 103
Phaseolus, 54, 109, 110, 206, 284, 353
   aureus, 205, 217, 348
   lunatus, 48, 505
   mungo, 179, 217
   vulgaris, 109, 206, 325, 326, 351, 353, 384, 385, 437, 647
Phleum pratense, 157, 160
Phlox subulata, 731
Phorandendron, 357
Phragmites, 211
   australis, 211
   communis, 211
   mauritanicus, 211
Phyllanthus
   distichus, 491
   emblica, 277
Phyllostachys
   pubescens, 214
   viridis, 211, 214, 379
Physalis
   alkekengi, 374
   minima, 216
   peruviana, 359, 374

Physic nut, 292, 347, 363, 369, 370, 448, 489, 490
Phytlephas, 140
Picea, 612, 622
   abies, 229
   excelsa, 229
   glauca, 365
   mugo, 365
Picris echioides, 186
Pieris, 100, 102
   andromedae, 102
   japonica, 102, 671
   ovalifolia, 99
Pigeon pea, 179, 198, 279, 286, 287, 322, 324, 327, 328, 349, 351, 353–355, 359, 376, 383, 448
Pigweed, 48, 189, 191
Pimenta dioica, 42
Pimentos, 406, 407
Pimiento, 381
Pin oak, 672
Pine(s), 21, 58, 223, 229, 338, 361, 364, 365, 488, 489, 506, 514–516, 611, 616, 617, 622, 763
Pineapple, 113, 383, 385
Pink dombeya, 199
Pinus, 223, 362, 365, 489, 516, 611, 616, 622
   attenuata, 365
   banksiana, 229, 362, 365
   cembroides, 365
   contorta, 365, 514
   densiflora, 362
   echinata, 362
   elliotti, 362
   glabra, 362
   insularis, 622
   monticola, 365
   mugo, 514
   nigra, 365, 514
   occidentalis, 622
   oocarpa, 622
   palustris, 229, 362
   ponderosa, 365, 669
   pungens, 229
   radiata, 622, 623
   resinosa, 362, 365
   rigida, 229, 362
   strobus, 362, 365, 514
   sylvestris, 21, 117, 229, 514, 612
   taeda, 229, 362, 489
   virginiana, 107, 362
Piper
   longum , 44
   nigrum, 42, 106, 376
Piscidia piscipula, 93
Pisonia, 105
Pista, 179
Pistachio(s), 24, 53, 179, 217, 312, 314, 360, 361, 365, 366
Pistacia vera, 53, 179, 217, 360
Pistia stratiotes, 229
Pitanga, 292
Pitch pine, 362
Plane tree, 620
Plantago, 108

Platanus
  acerfolia, 223
  mexicana, 224
  occidentalis, 87, 90, 223
  racemosa, 224
  wrightii, 224
Plectronia didymum, 217
Pleioblastus amarus, 214
Pluchea
  indica, 222
  odorata, 222
Plum(s), 190, 200, 311, 357, 359, 360, 445, 491
Plumbago, 277
Poa, 448, 480, 487
  annua, 157, 174
  caespitosa, 194
  colensoi, 194
  pratensis, 149, 157
  trivialis, 157
Polygonum
  aviculare, 189, 194
  punctatum, 440, 442
Polyscias guilfoylei, 277
Polytria amaura, 163, 210
Pomegranate(s), 174, 358, 359, 363, 367–370, 381, 445
Ponderosa pine, 365, 669
Pongamia planata, 506
Poplar(s), 48, 63, 492, 760, 767
Poplar trees, 97, 112
Populus trees, 23
Populus, 48, 97, 98, 613
  americanus, 112
  candicans, 112
  deltoides, 492
Port Orford cedar, 365
Port Orford cypress, 223
Portia tree, 199
Portulaca oleracea, 189, 220, 222, 357
Potato(es), 21, 23, 53, 54, 56, 58, 88, 89, 106, 109, 184, 186, 189, 192, 222, 357, 381, 382, 384, 406 407, 411, 414, 425, 426, 430, 431, 442, 444, 451, 452, 612, 620, 622, 719, 745, 749, 756, 758, 763
Pouzolzia guineensis, 218
Prairie cordgrass, 212
Prickly pear, 339, 375, 383
Prickly poppy, 108
Privet, 437, 454
Prosopis
  cineraria, 378
  glandulosa, 339, 383
  spicigera, 385
Protea, 199
Prune, 219
Prunus, 97, 98, 506
  armeniaca, 179
  persica, 39, 179, 509
Pseudarthria, 58
  hookeri, 58
Pseudotsuga, 611
  menziesii, 365
Psidium, 290
  araca, 368
  guajava, 39, 179, 368
Psophocarpus tetragonolobus, 206
Pueraria lobata, 206, 378, 731
Pulicaria, 108
Pulmonaria, 106
Pumpkin, 21, 25, 180, 343–345, 363, 368, 381, 383, 729
Puncture vine, 719, 720
Punica granatum, 359
Purslane, 357
Pyracantha, 91, 92
  coccinea, 92
  crenato-serrata, 92
Pyrethrum cinerariaefolium, 218
Pyrus, 91, 97, 98, 506
  communis, 39

Q

Quack grass, 149
Queensland arrowroot, 213
Queensland hemp, 198
Quercus, 93, 98, 411, 425, 506, 610, 667, 674
  palustris, 672
Quince, 91
Quinine, 43, 44, 119
Quinoa, 186

R

Rabbit brush, 555
Radish, 184–186, 188, 189, 437
Ragi, 179
Ragweed, 93
Raisin bush, 198
Ramossissima, 186
Ranunculus, 573
  callifornia, 493
Rape, 58, 183, 194, 195, 220, 455
Rapeseed, 189, 191, 442
Raphanus raphanistrum, 437
Raspberry(ies), 56, 57, 200, 414, 445, 670, 731
Red clover, 194, 195, 328
Red currant, 615
Red fescue, 157
Red gram, 179, 180, 217, 279
Red pepper, 374
Red pine, 362
Rhamnus, 422
Rheum caspicum, 218
Rhododendron, 87, 100, 101, 102
Rhododendron, 86, 99, 100, 101, 102, 761
  calendulaceum, 102
  maximum, 100
Rhus aromatica, 360
Ribes, 56, 98

Rice, 145, 146, 148, 153–156, 164, 170, 196, 197, 207, 210–213, 227, 228, 313, 314, 321–323, 326 328, 338, 355–357, 359, 381, 406, 407, 414, 422, 435, 436, 440–442, 446, 447, 450, 452, 456, 507, 602, 604, 621, 664, 665
Ricinus, 88, 109
    communis, 55, 93, 437, 505
Ridged gourd, 359
River red gum, 219
Robinia, 98
Ropoko, 212
Rosa, 98
Rose(s), 385, 730
Rose of China, 198, 199
Rose of Sharon, 313
Roselle, 179, 198, 203, 284, 290
Rosemary, 88
Rosmarinus officinalis, 88
Rough bluegrass, 157
Royal palm, 140
Roystonea, 140
    elata, 140
    regia, 140
Rubber, 119, 322, 339–341, 348, 349, 359, 377, 380
Rubber plant, 329
Rubus, 56, 670
    allegheniensis, 224
    hirsutus, 731
    idaeus, 445
Rudbeckia, 94
Rumex acestosella, 194
Russian-thistle, 184, 217
Rye, 145, 153, 411, 425, 421, 448

S

Saccharum, 167
    arundinaceum, 214, 226
    bengalense, 214
    munga, 214
    officinale, 179
    officinarum, 167, 210, 507
Safflower, 47, 88, 121, 189–191
Sage, 88
Sagebrush, 668
Salix, 57, 102, 612, 613, 615, 675
Salsola, 217, 266, 267
    kali var. *tenuifolia*, 217
Salt cedar tree, 186
Salvia, 109, 220
    officinalis, 88
Salvinia, 229
Sand spurrey, 194
Sandal tree, 507
Sannhemp, 217
Santalum album, 507
Sapindus mukorossi, 179
Sapium sebiferum, 329, 369
Sarothamnus scoparius, 117, 455, 610

Sativus, 39
Scarlet pimpernel, 194
Schizachyrium, 149, 152
    scoparium, 151, 155, 158
Schizocarpum reflexum, 369
Schleichera oleosa, 312
Scirpus, 267
Scots pine, 229, 514
Sechium edule, 52, 368
Sedge-root, 414
Sedum acre, 189
Semul, 292
Senecio, 108
Sesame, 179, 180, 181, 182, 208, 221, 446, 731
Sesame orientale, 208
Sesamum indicum, 179, 731
Sesbania, 443
    acuelata, 443
    bispinosa, 443
Setaria, 148
    glauca, 149
    lutescens, 149
    viridis, 149
Sheep sorrel, 194
Shepherd's purse, 186, 194
Shortleaf pine, 362
Shrubby basil, 88
Siberian fir, 622
Siberian larch, 612
Siberian motherwort, 437
Sicyos, 343
    angulatus, 343
Sida, 110, 198, 273, 274
    acuta, 198
    carpinifolia, 288, 289
    cordifolia, 115, 198
    mollissima, 198, 200
    rhombifolia, 198, 200, 202
Sidasthrum micranthum, 288
Silene gallica, 194
Silk cotton tree, 284
Silver maple, 672
Silver nightshade, 117
Simmondsia
    californica, 219
    chinensis, 219
Sinapsis alba, 52
Sinocalamus beecheyanus, 214
Sisymbrium irio, 186
Skaapbossie, 315
Slash pine, 362
Small-leaved linden, 226
Smooth gourd, 359
Snake gourd, 358, 359
Snapdragon, 114
Soapberry, 310, 312
Solanum, 50, 92, 104, 109, 216, 381, 729
    americanum, 363
    carolinense, 109
    elaeagnifolium, 117, 216
    gilo, 89

indicum, 104, 373
integrifolium, 104
mauritanum, 216
melongena, 86, 104, 179, 215, 374, 505, 732
nigrum, 209, 215, 374, 382
peruviana, 215
sessiliflorum, 89
sodomaeus, 215
torvum, 88, 215, 374
tuberosum, 54, 106
viarum, 116
xanthocarpum, 104
Solidago, 94
    altissima, 93
    canadensis, 200, 730
Soliva sessilis, 194
Sonchus oleraceus, 189, 222
Sorbus aria, 92
Sorghastrum nutans, 149
Sorghum, 39, 45, 46, 54, 55, 58, 145, 148, 151, 153–156, 168, 174, 180, 186, 197, 211, 212, 285, 313, 314, 323, 326, 351, 352, 367, 369, 370, 407, 410, 434, 440, 441, 446, 451, 453, 456, 619
Sorghum, 353
    bicolor, 39, 179, 215–217, 313
    halepense, 96
    vulgare, 215–217, 284
    vulgare var. *durra*, 179
    vulgare var. *sudanense*, 164
Sour orange, 118
Soushumber, 215
Southern pine, 363
Sowthistle, 189, 190
Soybean(s), 5, 155, 156, 205, 207, 285, 313, 321, 322, 324–330, 355, 356, 367, 369, 370, 386, 407, 422, 425, 431–440, 443–448, 450–452, 454–456, 507, 609, 619, 640–648, 670, 716, 717, 731, 745, 747, 753, 764, 767
Spaghetti squash, 344
Spanish cocklebur, 290
Spartina, 212
    pectinata, 212
Spartium scoparium, 117
Spelt, 153
Spergula arvensis, 189
Spergularia rubra, 194
Sphaeralcea, 198, 727
    miniata, 198
    umbellata, 198
Spinach, 412, 447, 716
Spirea, 224
Spirea vanhouttei, 224
Sponge gourd, 729
Sporobolus asper, 151
Sprout, 189
Spruce, 229
Spruce pine, 362
Squash, 22, 52, 343–347, 363, 366, 374, 381
St. Augustine grass, 149, 162, 163, 165, 166, 211
Stachys sieboldii, 179
Star bristle, 433, 437

Starfruit, 368
Stellaria media, 194
Stenotaphrum
    dimidiatum, 166
    secundatum, 149, 163, 166
Sterculia, 285, 291, 373
    africana, 283
    bequaerti, 198
    cinerea, 284
    diversifolia, 198
    foetida, 281, 288
    quinqueloba, 283
    rhynchocarpa, 281, 282, 287
    tragacantha, 198
Stone crop, 189
Strawberry(ies), 24, 47, 53, 57, 58, 146, 186, 189, 192, 194, 196, 208, 228, 229, 412, 415, 445, 452, 620
Strelitzia, 749
String bean, 385
Subterranean clover, 194
Suckling clover, 194
Sudan grass, 153, 163
Sugar beet(s), 21, 23, 24, 47, 53, 58, 114, 184, 188, 265–268, 453, 643, 714, 716, 718, 719
Sugarcane, 96, 105, 145, 146, 150, 164, 167–177, 179, 180, 210, 212, 217, 218, 225, 347, 406–408 410, 413, 414, 507, 615, 618, 665, 666
Sugar pine, 365
Sumac, 360
Summer squash, 346
Sunberry, 216
Sunflower(s), 54, 58, 93, 94, 108, 109, 180, 181, 189–191, 193, 198, 203–205, 217, 220, 221, 223, 290, 291, 360, 367, 369, 381, 384, 425, 430, 431, 433, 443, 452, 643, 670, 714, 717–719
Sunn-hemp, 21, 62
Sunnjute, 217
Surinam cherry, 292, 377
Swallowswort, 218
Swede, 195
Sweet basil, 88
Sweet corn, 453
Sweet orange, 359
Sweet peppers, 620, 749
Sweet potato, 52, 110, 179, 216–218, 340, 359, 374, 377, 381, 382, 385, 425, 507
Sweet sorghum, 145
Sweet william, 221
Swietenia macrophylla, 280
Switchgrass, 149
Sycamore, 87, 90, 91, 223, 224, 616
Sycamore tree, 555
Symphytum, 106

T

Tagetes, 108
Tall dropseed, 151
Tall fescue, 157

Tamarind, 106, 380
Tamarindus indicus, 106
Tamarix pentandra, 186
Tangerine(s), 359, 363, 368
Tapioca, 380
Taro, 384, 663
Tauhinu, 194
Tea, 39–43, 51, 114, 119, 322, 349, 623, 761
Teak, 111, 118–120
Tebeldi, 281, 284
Tectona grandis, 111, 118, 277
Tempate, 329
Tephrosia, 113
Thalia, 213
    dealbata, 213
    geniculata, 213
Thatching grass, 176
Theobroma, 729
    cacao, 39, 377, 672, 726, 731
Thespesia populnea, 199
Thinopyrum, 60
    intermedium, 61
Thistles, 116, 121
Three-leaf maple, 310
Thuja, 223
Thyme, 220
Thyme-leaved spurge, 187
Thymus, 220
Tickle grass, 150
Til, 179
Tilia, 110, 663
    cordata, 226
    parvifolia, 226
    vulgaris, 663
Timber tree, 377
Timothy, 160
Tinia, 98
Tobacco, 40, 51, 88, 179, 186, 189, 216, 221, 314, 363, 378, 381, 382, 407, 413, 414, 430, 431, 434, 436, 438, 448, 452, 640, 642, 645, 667, 668, 727, 728, 730
Tomato(es), 40, 48, 51, 52, 59, 88, 89, 109, 179, 180, 189, 192, 215, 324, 338, 347, 357–359, 361, 367, 369, 370, 374, 378–382, 406, 407, 425, 431, 438, 444, 447, 451, 453, 454, 456, 645, 646, 659–661, 666, 668, 679, 726–728, 730, 746
Tomato-fruited eggplant, 104
Tori, 179, 180
Torpedo grass, 163
Tradescantia hirsuticaulis, 730
Tree cactus, 383
Tree tomato, 359
Trewia nudiflora, 292, 293
Tribulus terrestris, 719
Trichodesma incanum, 117
Trichosanthes cucumerina, 358, 359
Trifolium, 58, 414
    alexandrium, 179
    dubium, 194
    medium, 455
    pratense, 194
    repens, 53, 194
Triplasis purpurea, 151
Triticum, 195, 409
    aestivum, 39, 277, 437
Triticus spelta, 617
Triumfetta, 110
    macrophylla, 198
    rhomboidea, 198, 218
Tropical soda apple, 116
Tsuga, 611
    canadensis, 365
Tulsi plant, 88
Turban squash, 381
Turfgrasses, 212
Turmeric, 103
Turnip(s), 184, 186, 188, 189, 195, 220, 367, 454
Twin cress, 194
Typha, 105, 224
    latifolia, 105

U

Ulex, 117
    europaeus, 117, 455
Ulmus, 87, 98, 224, 506, 675
Urena, 109
    lobata, 95, 96, 198, 215, 218, 275, 284, 285
Urid, 179, 206
Urochloa, 105
    mosambicensis, 211

V

Vaccinium, 98
Valtissius, 229
Vanda orchids, 222
Vasey grass, 441
Vegetable-sponge, 179
Velvetbean, 456
Velvet bent grass, 157
Velvetleaf, 310, 313, 314
Verbascum
    blattaria, 731
    thapsus, 61, 109, 727, 730
Verbascus, 121
Verbena, 114
Vernonia, 108, 215, 220
    cinerea, 178
Vernonia, 182
Vetiveria zizaniodes, 280
Vicia faba, 179
Vigna, 323
    acontifolia, 206
    aureus, 284
    mungo, 58, 206, 353
    radiata, 180, 206, 326, 353, 377
    sesquipedalis, 349

sinensis, 103, 206
　unguiculata, 110, 180, 206, 275, 284
Vines, 174, 220
Viper bugloss, 106
Virginia pine, 107, 362
Vitex trifolia, 115
Vitis, 267, 670, 675
　vinifera, 42

W

Wachendorfia paniculata, 175
Walnuts, 186
Water fern, 229
Water lettuce, 229
Watermelon, 344, 363
Water peppermint, 573
Water primrose, 767
Weed bitter, 118
Western plane tree, 224
Western white pine, 365
Wheat grasses, 60
Wheat, 22, 23, 39, 53, 58, 145, 151, 153, 155, 156, 174, 194–196, 208, 210, 221, 223, 277, 323, 407–409, 422–425, 436, 437, 440–442, 446–448, 451, 452, 456, 475, 477–488, 493, 494, 623, 640, 643, 716
Whin, 117
Whireweed, 189, 194
White cedar, 223
White cherry, 456
White clover, 53, 194, 195, 381
White mustard, 52
White spruce, 365
Wild cotton, 200
Wild cucumber, 343
Wild eggplant, 88
Wild melon, 358
Wild mustard, 220
Wild oats, 493
Wild olive, 108, 113
Willow, 57, 675, 677, 768

Winter cherry, 374
Winter squash, 346
Withenia somnifera, 278, 291
Wooden rose, 217
Wormwood, 108, 219, 220

X

Xanthium, 110
Xanthorrhoaea, 140

Y

Yam, 359
Yellow bristle grass, 149
Yellow foxtail, 149
Yellow popolo, 215
Yellow sage, 119, 120
Yellow straightneck squash, 344
Yemane, 761

Z

Zea, 110
　mays, 54, 148, 150, 152, 158, 210, 284, 313
Zingiber officinale, 103
Zizyphus, 217
　jujuba, 113, 217
　numularis, 491
Zonal geranium, 667
Zoysia, 163
　japonicum, 157
　matrella, 157
　tenufolia, 157
Zoysiagrass, 157, 163
Zucchini, 344, 668

DATE DUE

MAY 2 1 200?			
MAY 2 0 2001			
MAY 1 9 '0?			
MAY 1 7 20??			
MAY 2 7 2004			
SEP 2 6 2012			

WITHDRAWN

GAYLORD   PRINTED IN U.S.A.

SCI QL 521 .H48 2000

Heteroptera of economic importance